THE PRINCIPAL FUNCTIONAL GROUPS OF ORGANIC CHEMISTRY

	Example	*Acceptable Name(s) of Example*	*Characteristic Reaction Type*
Hydrocarbons			
Alkanes	CH_3CH_3	Ethane	Free-radical substitution of hydrogen by halogen
Alkenes	$H_2C{=}CH_2$	Ethene or ethylene	Electrophilic addition to double bond
Alkynes	$HC{\equiv}CH$	Ethyne or acetylene	Electrophilic addition to triple bond
Dienes	$H_2C{=}CHCH{=}CH_2$	1,3-Butadiene	Electrophilic addition to double bonds
Arenes		Benzene	Electrophilic aromatic substitution
Halogen-substituted derivatives of hydrocarbons			
Alkyl halides	CH_3CH_2Cl	Chloroethane or ·hloride	Nucleophilic substitution; elimination
Alkenyl halides	$H_2C{=}CH$…	…iene or …loride	Electrophilic addition to double bond; elimination
Aryl halides	C_6H_5Cl	…zene	Electrophilic aromatic substitution; nucleophilic aromatic substitution
Oxygen-containing organic compounds			
Alcohols	CH_3CH_2OH	Ethanol or ethyl alcohol	Dehydration; conversion to alkyl halides; esterification
Phenols	C_6H_5OH	Phenol	Electrophilic aromatic substitution
Ethers	$CH_3CH_2OCH_2CH_3$	Ethoxyethane or diethyl ether	Cleavage by hydrogen halides
Epoxides	$H_2C{-}CH_2$ bridged by O (three-membered ring)	Epoxyethane or ethylene oxide or oxirane	Nucleophilic ring opening
Aldehydes	$CH_3\overset{O}{\overset{\|}{C}}H$	Ethanal or acetaldehyde	Nucleophilic addition to carbonyl group
Ketones	$CH_3\overset{O}{\overset{\|}{C}}CH_3$	2-Propanone or acetone	Nucleophilic addition to carbonyl group
Carboxylic acids	$CH_3\overset{O}{\overset{\|}{C}}OH$	Ethanoic acid or acetic acid	Ionization of carboxyl; esterification

THE PRINCIPAL FUNCTIONAL GROUPS OF ORGANIC CHEMISTRY

	Example	*Acceptable Name(s) of Example*	*Characteristic Reaction Type*
Carboxylic acid derivatives			
Acyl halides	$CH_3\overset{O}{\overset{\|}{C}}Cl$	Ethanoyl chloride or acetyl chloride	Nucleophilic acyl substitution
Acid anhydrides	$CH_3\overset{O}{\overset{\|}{C}}O\overset{O}{\overset{\|}{C}}CH_3$	Ethanoic anhydride or acetic anhydride	Nucleophilic acyl substitution
Esters	$CH_3\overset{O}{\overset{\|}{C}}OCH_2CH_3$	Ethyl ethanoate or ethyl acetate	Nucleophilic acyl substitution
Amides	$CH_3\overset{O}{\overset{\|}{C}}NHCH_3$	*N*-Methylethanamide or *N*-methylacetamide	Nucleophilic acyl substitution
Nitrogen-containing organic compounds			
Amines	$CH_3CH_2NH_2$	Ethanamine or ethylamine	Nitrogen acts as a base or as a nucleophile
Nitriles	$CH_3C{\equiv}N$	Ethanenitrile or acetonitrile	Nucleophilic addition to carbon–nitrogen triple bond
Nitro compounds	$C_6H_5NO_2$	Nitrobenzene	Reduction of nitro group to amine
Sulfur-containing organic compounds			
Thiols	CH_3CH_2SH	Ethanethiol	Oxidation to a sulfenic, sulfinic, or sulfonic acid or to a disulfide
Sulfides	$CH_3CH_2SCH_2CH_3$	Diethyl sulfide	Alkylation to a sulfonium salt; oxidation to a sulfoxide or sulfone

Eighth Edition

Organic Chemistry

Francis A. Carey
University of Virginia

Robert M. Giuliano
Villanova University

Mc Graw Hill
Connect Learn Succeed™

The McGraw-Hill Companies

Connect Learn Succeed™

ORGANIC CHEMISTRY, EIGHTH EDITION
International Edition 2011

10 09 08 07 06 05 04 03 02 01
20 15 14 13 12 11
CTP MPM

When ordering this title, use ISBN 978-007-132397-0 or MHID 007-132397-X.

Printed in Singapore

www.mhhe.com

To our wives, Margot
Giuliano and Jill Carey,
and our children
Michael, Ellen, and
Christopher Giuliano
and Andrew, Robert,
and William Carey.

Frank Carey

Bob Giuliano

Welcome

It is a pleasure for me to welcome Robert Giuliano as coauthor of *Organic Chemistry*. Bob is Professor of Chemistry at Villanova University where he has served as departmental chair and regularly teaches introductory and graduate-level courses in organic chemistry. Professor Giuliano's research in carbohydrate synthesis has been recognized with his current appointments as a Regional Editor of the *Journal of Carbohydrate Chemistry* and as a member of the Editorial Board for *Current Topics in Medicinal Chemistry*.

I first met Bob shortly before he began graduate work in our department at the University of Virginia in the 1970s and was immediately impressed with his intelligence and sense of purpose. These same personal qualities characterized his efforts in shaping the present edition. More than that, his background in bioorganic topics provided a fresh and informed perspective on some of the most active areas of modern organic chemistry.

Although no later edition can be as satisfying to an author as the first, this eighth edition of *Organic Chemistry* is special because it signals the resumption of a professional collaboration begun decades ago.

Francis A. Carey

About the Authors

Francis A. Carey is a native of Philadelphia, educated at Drexel University (B.S. in chemistry), and at Penn State (Ph.D.). Following postdoctoral work at Harvard and military service, he served on the faculty of the University of Virginia from 1966 until retiring as Professor Emeritus in 2000.

In addition to this text, Professor Carey is coauthor (with Robert C. Atkins) of *Organic Chemistry: A Brief Course* and (with Richard J. Sundberg) of *Advanced Organic Chemistry*, a two-volume treatment designed for graduate students and advanced undergraduates.

Frank and his wife Jill, who is a teacher/director of a preschool and a church organist, are the parents of Andy, Bob, and Bill and the grandparents of Riyad and Ava.

Robert M. Giuliano was born in Altoona, Pennsylvania and attended Penn State (B.S. in chemistry) and the University of Virginia (Ph.D., under the direction of Francis Carey). Following postdoctoral studies with Bert Fraser-Reid at the University of Maryland, he joined the chemistry department faculty of Villanova University in 1982, where he is currently Professor. His research interests are in synthetic organic and carbohydrate chemistry, and in functionalized carbon nanomaterials.

Bob and his wife Margot, an elementary and preschool teacher he met while attending UVa, are the parents of Michael, Ellen, and Christopher.

Brief Contents

Contents

CHAPTER 4

Alcohols and Alkyl Halides 137

CHAPTER 5

Structure and Preparation of Alkenes: Elimination Reactions 184

CHAPTER 6

Addition Reactions of Alkenes 226

CHAPTER 7

Stereochemistry 278

CHAPTER 8

Nucleophilic Substitution 322

CHAPTER 9

Alkynes 359

CHAPTER 10

Conjugation in Alkadienes and Allylic Systems 388

CHAPTER 11

Arenes and Aromaticity 428

CHAPTER 12

Reactions of Arenes: Electrophilic and Nucleophilic Aromatic Substitution 478

CHAPTER 13

Spectroscopy 538

CHAPTER 14

Organometallic Compounds 606

CHAPTER 15

Alcohols, Diols, and Thiols 646

CHAPTER 16

Ethers, Epoxides, and Sulfides 686

CHAPTER 17

Aldehydes and Ketones: Nucleophilic Addition to the Carbonyl Group 724

CHAPTER 20

Enols and Enolates 866

CHAPTER 21

Amines 930

CHAPTER 22

Phenols 988

CHAPTER 26

Nucleosides, Nucleotides, and Nucleic Acids 1174

CHAPTER 27

Synthetic Polymers 1216

List of Important Features

Mechanisms

Tables

Boxed Essays

Descriptive Passage and Interpretive Problems

Preface

What Sets This Book Apart?

The central message of chemistry is that the properties of a substance come from its structure. What is less obvious, but very powerful, is the corollary. Someone with training in chemistry can look at the structure of a substance and tell you a lot about its properties. Organic chemistry has always been, and continues to be, the branch of chemistry that best connects structure with properties.

The goal of this text, as it has been through seven previous editions, is to provide students with the conceptual tools to understand and apply the relationship between the structures of organic compounds and their properties. Both the organization of the text and the presentation of individual topics were designed with this objective in mind.

A Functional Group Organization

The text is organized according to functional groups—structural units within a molecule that are most closely identified with characteristic properties. This organization offers two major advantages over alternative organizations based on mechanisms or reaction types.

1. The information content of individual chapters is more manageable when organized according to functional groups.
2. Patterns of reactivity are reinforced when a reaction used to prepare a particular functional group reappears as a characteristic reaction of a different functional group.

A Mechanistic Emphasis and Its Presentation

The text emphasizes mechanisms and encourages students to see similarities in mechanisms among different functional groups. Mechanisms are developed from observations; thus, reactions are normally presented first, followed by their mechanism.

In order to maintain consistency with what our students have already learned, this text presents multistep mechanisms in the same way as most general chemistry textbooks; that is, as a series of *elementary steps*. Additionally, we provide a brief comment about how each step contributes to the overall mechanism.

Section 1.12 "Curved Arrows and Chemical Reactions" introduces the student to the notational system employed in all of the mechanistic discussions in the text.

Numerous reaction mechanisms are accompanied by potential energy diagrams. Section 4.9 "Potential Energy Diagrams for Multistep Reactions: The S_N1 Mechanism" shows how the potential energy diagrams for three elementary steps are combined to give the diagram for the overall reaction.

Mechanism 4.3

Conversion of an Alcohol to an Alkyl Chloride with Thionyl Chloride

THE OVERALL REACTION:

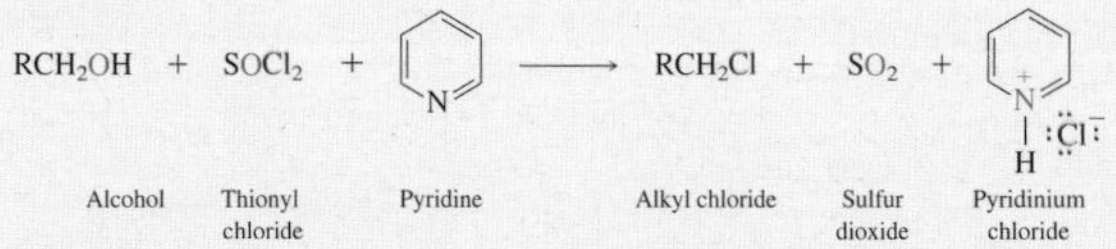

Step 1: The hydroxyl group oxygen attacks sulfur and displaces chloride to give a protonated chlorosulfite ester:

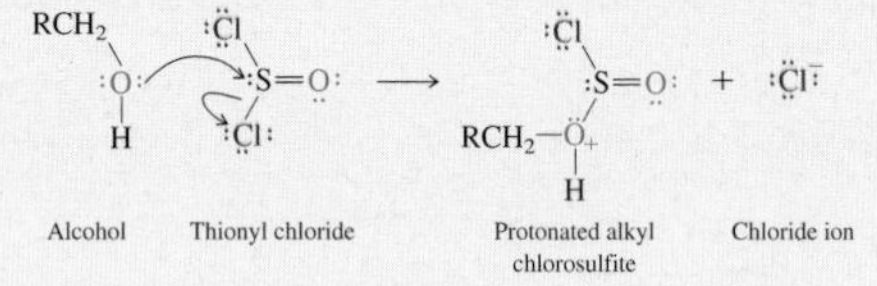

Step 2: Pyridine acts as a base to remove a proton and give a neutral alkyl chlorosulfite:

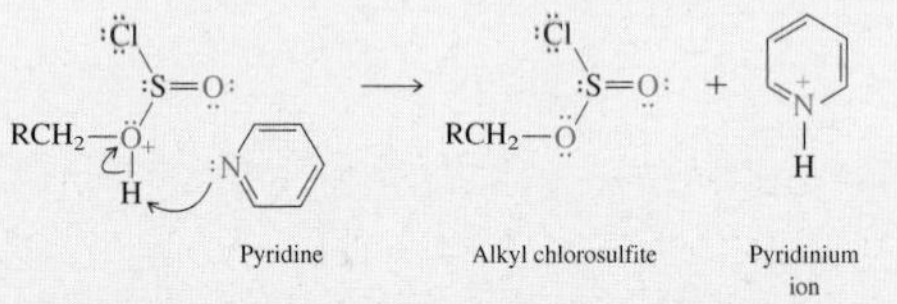

Step 3: Nucleophilic attack of chloride ion breaks the C—O bond to form the alkyl chloride:

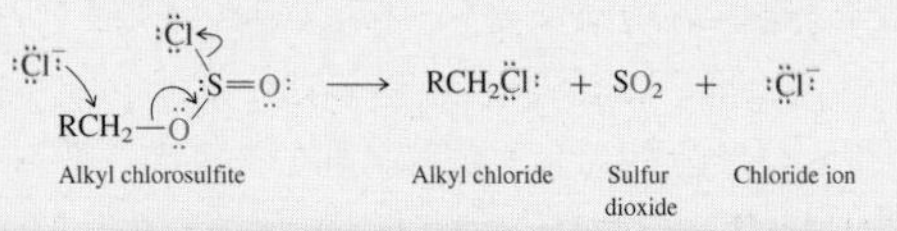

Enhanced Graphics

The teaching of organic chemistry has especially benefited as powerful modeling and graphics software have become routinely available. Computer-generated molecular models and electrostatic potential maps were integrated into the third edition of this text and their number has increased in succeeding editions. Also seeing increasing use are graphically correct representations of orbitals and the role of orbital interactions in chemical reactivity. The E2 mechanism of elimination, which involves a single elementary step, is supplemented by showing the orbital interactions that occur during that step.

Mechanism 5.4

E2 Elimination of an Alkyl Halide

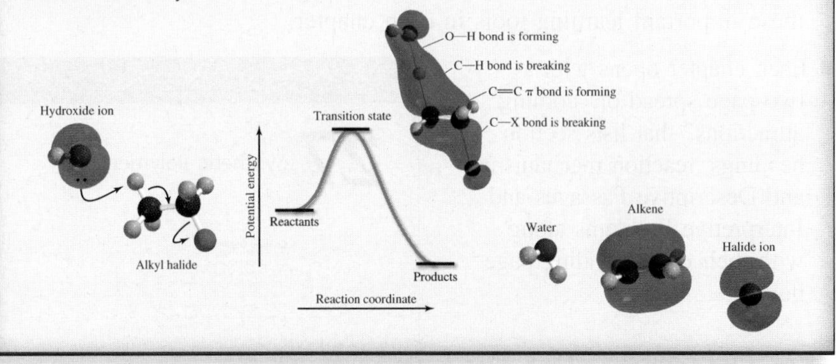

O—H bond is forming
C—H bond is breaking
C=C π bond is forming
C—X bond is breaking
Hydroxide ion
Transition state
Potential energy
Reactants
Alkyl halide
Products
Reaction coordinate
Water
Alkene
Halide ion

Generous and Effective Use of Tables

The relative reactivity of different compounds is relevant to both the theory and practice of organic chemistry. While it is helpful—even important—to know that one compound is more reactive than another, it is even better to know by how much. Our text provides more experimental information of this type than is customary. Chapter 8 "Nucleophilic Substitution," for example, contains several tables of *quantitative* relative rate data of which the following is but one example.

TABLE 8.2 Reactivity of Some Alkyl Bromides Toward Substitution by the S_N2 Mechanism*

Alkyl bromide	Structure	Class	Relative rate†
Methyl bromide	CH_3Br	Unsubstituted	221,000
Ethyl bromide	CH_3CH_2Br	Primary	1,350
Isopropyl bromide	$(CH_3)_2CHBr$	Secondary	1
tert-Butyl bromide	$(CH_3)_3CBr$	Tertiary	Too small to measure

*Substitution of bromide by lithium iodide in acetone.

†Ratio of second-order rate constant *k* for indicated alkyl bromide to *k* for isopropyl bromide at 25°C.

Annotated summary tables have been a staple of *Organic Chemistry* since the first edition. Some tables review reactions from earlier chapters, others the reactions or concepts of a current chapter. Still others walk the reader step-by-step through skill builders and concepts unique to organic chemistry. Well received by students and faculty alike, these summary tables remain one of the text's strengths.

Problems

Problem-solving strategies and skills are emphasized throughout. Understanding is progressively reinforced by problems that appear within topic sections. For many problems, sample solutions are given, including examples of handwritten solutions from the author.

Problem 8.16

A standard method for the synthesis of ethers is an S_N2 reaction between an alkoxide and an alkyl halide.

$$R—\ddot{O}:^- + R'—\ddot{X}: \longrightarrow R—\ddot{O}—R' + :\ddot{X}:^-$$

Show possible combinations of alkoxide and alkyl halide for the preparation of the following ethers. Which of these ethers can be prepared effectively by this method?

(a) $CH_3CH_2CHOCH(CH_3)_2$ (with CH_3 on the CH) (b) $CH_3CH_2CH_2OCH(CH_3)_2$

Sample Solution

(a) • There are two possible combinations of alkoxide and alkyl halide (shown as bromide).

$CH_3CH_2CH(CH_3)—\ddot{O}:^-$ + $BrCH(CH_3)_2$

$CH_3CH_2CH(CH_3)—Br$ + $^-:\ddot{O}—CH(CH_3)_2$

• In both cases the alkyl halide is secondary. Secondary alkyl halides undergo elimination (E2) rather than substitution (S_N2) with alkoxide bases.

Pedagogy

- A list of tables, mechanisms, boxes and Descriptive Passages and Interpretive Questions is included in the front matter (page xvii) as a quick reference to these important learning tools in each chapter.
- Each chapter opens with a two-page spread of "coming attractions" that lists section headings, reaction mechanisms and Descriptive Passages and Interpretive Problems along with their corresponding page numbers.

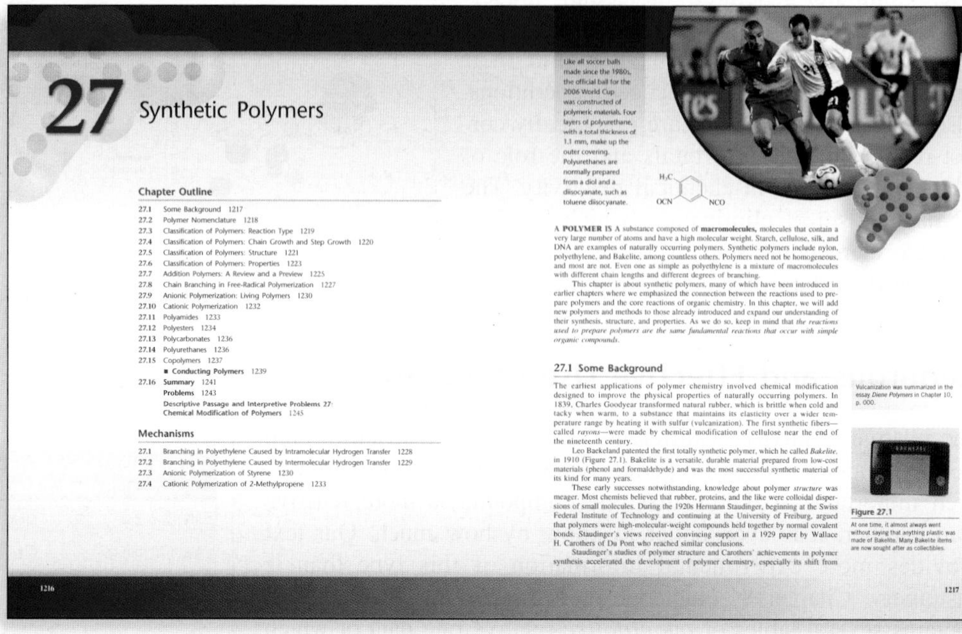

27 Synthetic Polymers

Chapter Outline

27.1 Some Background 1217
27.2 Polymer Nomenclature 1218
27.3 Classification of Polymers: Reaction Type 1219
27.4 Classification of Polymers: Chain Growth and Step Growth 1220
27.5 Classification of Polymers: Structure 1221
27.6 Classification of Polymers: Properties 1223
27.7 Addition Polymers: A Review and a Preview 1225
27.8 Chain Branching in Free-Radical Polymerization 1227
27.9 Anionic Polymerization: Living Polymers 1230
27.10 Cationic Polymerization 1232
27.11 Polyamides 1233
27.12 Polyesters 1234
27.13 Polycarbonates 1236
27.14 Polyurethanes 1236
27.15 Copolymers 1237
■ Conducting Polymers 1239
27.16 Summary 1241
Problems 1243
Descriptive Passage and Interpretive Problems 27: Chemical Modification of Polymers 1245

Mechanisms

27.1 Branching in Polyethylene Caused by Intramolecular Hydrogen Transfer 1228
27.2 Branching in Polyethylene Caused by Intermolecular Hydrogen Transfer 1229
27.3 Anionic Polymerization of Styrene 1230
27.4 Cationic Polymerization of 2-Methylpropene 1233

Like all soccer balls made since the 1980s, the official ball for the 2006 World Cup was constructed of polymeric materials. Four layers of polyurethane, with a total thickness of 1.1 mm, make up the outer covering. Polyurethanes are normally prepared from a diol and a diisocyanate, such as toluene diisocyanate.

A POLYMER IS A substance composed of **macromolecules,** molecules that contain a very large number of atoms and have a high molecular weight. Starch, cellulose, silk, and DNA are examples of naturally occurring polymers. Synthetic polymers include nylon, polyethylene, and Bakelite, among countless others. Polymers need not be homogeneous, and most are not. Even one as simple as polyethylene is a mixture of macromolecules with different chain lengths and different degrees of branching.

This chapter is about synthetic polymers, many of which have been introduced in earlier chapters where we emphasized the connection between the reactions used to prepare polymers and the core reactions of organic chemistry. In this chapter, we will add new polymers and methods to those already introduced and expand our understanding of their synthesis, structure, and properties. As we do so, keep in mind that *the reactions used to prepare polymers are the same fundamental reactions that occur with simple organic compounds.*

27.1 Some Background

The earliest applications of polymer chemistry involved chemical modification designed to improve the physical properties of naturally occurring polymers. In 1839, Charles Goodyear transformed natural rubber, which is brittle when cold and tacky when warm, to a substance that maintains its elasticity over a wider temperature range by heating it with sulfur (vulcanization). The first synthetic fibers—called *rayons*—were made by chemical modification of cellulose near the end of the nineteenth century.

Leo Baekeland patented the first totally synthetic polymer, which he called *Bakelite*, in 1910 (Figure 27.1). Bakelite is a versatile, durable material prepared from low-cost materials (phenol and formaldehyde) and was the most successful synthetic material of its kind for many years.

These early successes notwithstanding, knowledge about polymer *structure* was meager. Most chemists believed that rubber, proteins, and the like were colloidal dispersions of small molecules. During the 1920s Hermann Staudinger, beginning at the Swiss Federal Institute of Technology and continuing at the University of Freiburg, argued that polymers were high-molecular-weight compounds held together by normal covalent bonds. Staudinger's views received convincing support in a 1929 paper by Wallace H. Carothers of Du Pont who reached similar conclusions.

Staudinger's studies of polymer structure and Carothers' achievements in polymer synthesis accelerated the development of polymer chemistry, especially its shift from

Vulcanization was summarized in the essay *Diene Polymers* in Chapter 10, p. 000.

Figure 27.1
At one time, it almost always went without saying that anything plastic was made of Bakelite. Many Bakelite items are now sought after as collectibles.

1216 1217

- Summary tables allow the student easy access to a wealth of information in an easy-to-use format while reviewing information from previous chapters.
- End-of-Chapter Summaries highlight and consolidate all of the important concepts and reactions within a chapter.

14.17 SUMMARY

Section 14.1 Organometallic compounds contain a carbon–metal bond. They are named as alkyl (or aryl) derivatives of metals.

$CH_3CH_2CH_2CH_2Li$ — Butyllithium C_6H_5MgBr — Phenylmagnesium bromide

Section 14.2 Carbon is more electronegative than metals and carbon–metal bonds are polarized so that carbon bears a partial to complete negative charge and the metal bears a partial to complete positive charge.

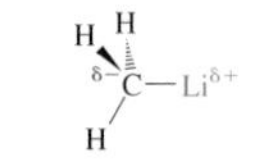

$HC\equiv\bar{C}: Na^+$

Methyllithium has a polar covalent carbon–lithium bond. Sodium acetylide has an ionic bond between carbon and sodium.

Section 14.3 See Table 14.3

Section 14.4 See Table 14.3

Section 14.5 Organolithium compounds and Grignard reagents are strong bases and react instantly with compounds that have —OH groups.

$$R{-}M + H{-}O{-}R' \longrightarrow R{-}H + M^+ \ ^-O{-}R'$$

These organometallic compounds cannot therefore be formed or used in solvents such as water and ethanol. The most commonly employed solvents are diethyl ether and tetrahydrofuran.

Section 14.6 See Tables 14.2 and 14.4

Section 14.7 See Table 14.4

Section 14.8 See Table 14.4

Section 14.9 When planning the synthesis of a compound using an organometallic reagent, or indeed any synthesis, the best approach is to reason backward from the product. This method is called **retrosynthetic analysis.** Retrosynthetic analysis of 1-methylcyclohexanol suggests it can be prepared by the reaction of methylmagnesium bromide and cyclohexanone.

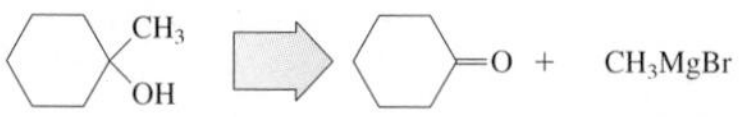

1-Methylcyclohexanol Cyclohexanone Methylmagnesium bromide

Audience

This electrostatic potential map is of the transition state for the reaction of hydroxide ion with chloromethane. The tetrahedral arrangement of bonds inverts like an umbrella in a storm during the reaction.

Organic Chemistry is designed to meet the needs of the "mainstream," two-semester undergraduate organic chemistry course. From the beginning and with each new edition, we have remained grounded in some fundamental notions. These include important issues concerning the intended audience. Is the topic appropriate for them with respect to their interests, aspirations, and experience? Just as important is the need to present an accurate picture of the present state of organic chemistry. How do we know what we know? What makes organic chemistry worth knowing? Where are we now? Where are we headed?

Even the art that opens each chapter has been designed with the audience in mind. The electrostatic potential maps combined with a graphic of a familiar object help connect the map to the chapter's content. Chapter 8, for example, opens by illustrating the umbrella-in-a-windstorm analogy used by virtually everyone who has ever taught nucleophilic substitution.

Descriptive Passage and Interpretive Problems

Many organic chemistry students later take standardized pre-professional examinations composed of problems derived from a descriptive passage, this text includes comparable passages and problems to familiarize students with this testing style.

Thus, *every* chapter concludes with a self-contained *Descriptive Passage and Interpretive Problems* unit that complements the chapter's content while emulating the "MCAT style." These 27 passages—listed on page xx—are accompanied by more than 100 total multiple-choice problems.

The passages focus on a wide range of topics—from structure, synthesis, mechanism, and natural products to using the Internet to calculate ^{13}C chemical shifts. They provide instructors with numerous opportunities to customize their own organic chemistry course while giving students practice in combining new information with what they have already learned.

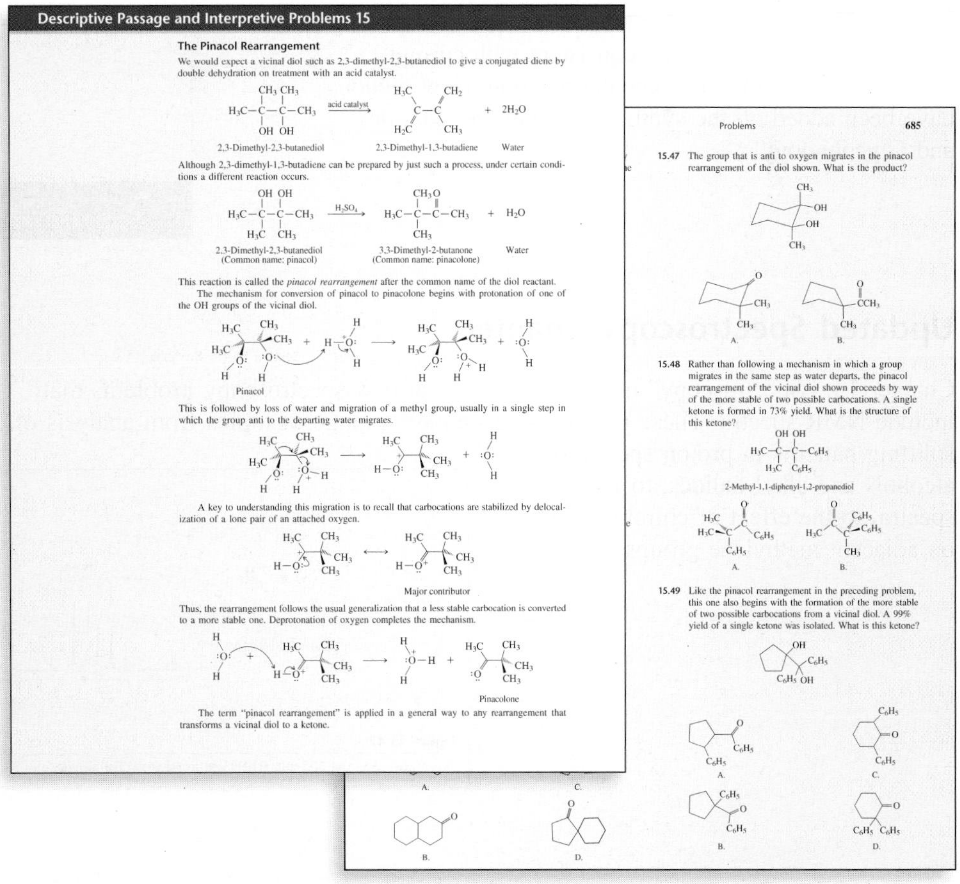
Descriptive Passage and Interpretive Problems 15
The Pinacol Rearrangement
Problems
685

What's New

Reorganization and Consolidation

By reorganizing certain related topics, the number of chapters in *Organic Chemistry* has been reduced from 29 in the 7th edition to 27 in the 8th. Thus, nucleophilic aromatic substitution has now joined electrophilic aromatic substitution in: Chapter 12 Reactions of Arenes: Electrophilic and Nucleophilic Aromatic Substitution while Chapter 20 Enols and Enolates combines the treatment of ester enolates with that of aldehyde and ketone enols and enolates.

Each of the two new chapters offers the advantage of a timely treatment of core material along with efficiency in its delivery. Nucleophilic aromatic substitution, for example, becomes a "first-semester" or "second-quarter" topic along with both nucleophilic aliphatic and electrophilic aromatic substitution.

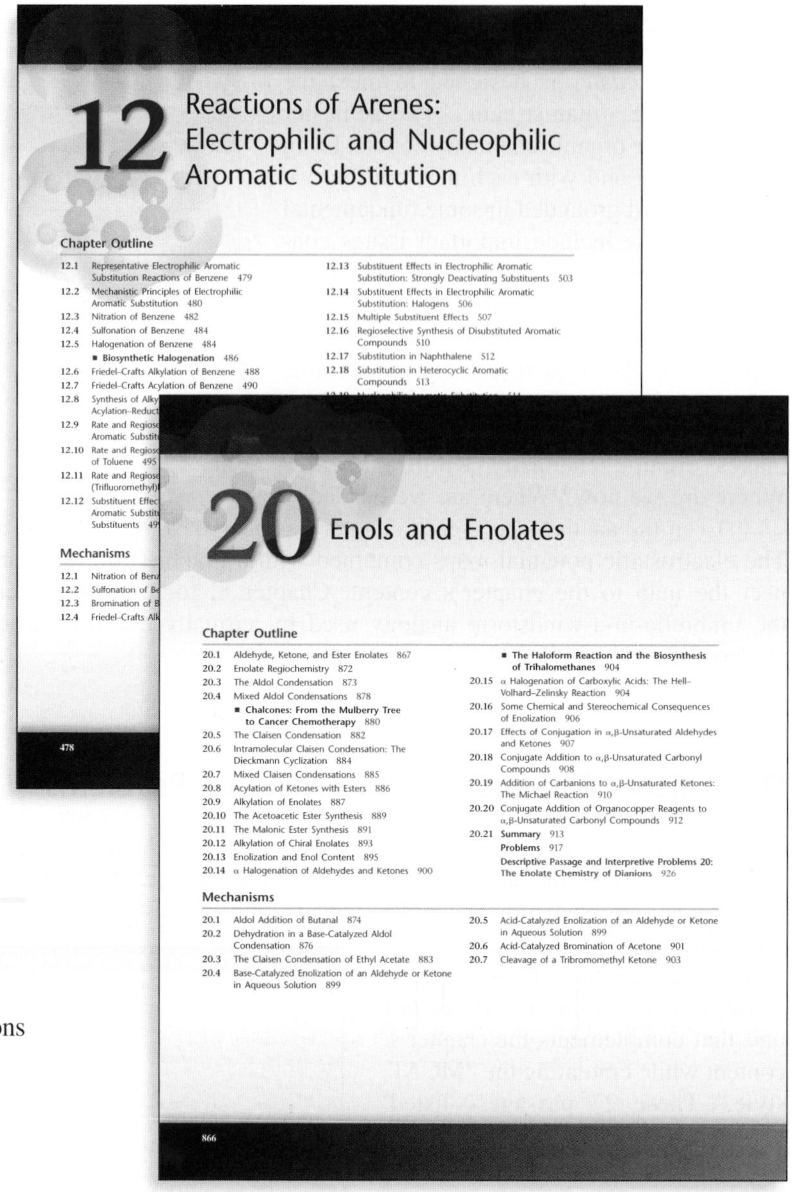

12 Reactions of Arenes: Electrophilic and Nucleophilic Aromatic Substitution

Chapter Outline

12.1 Representative Electrophilic Aromatic Substitution Reactions of Benzene 479
12.2 Mechanistic Principles of Electrophilic Aromatic Substitution 480
12.3 Nitration of Benzene 482
12.4 Sulfonation of Benzene 484
12.5 Halogenation of Benzene 484
■ **Biosynthetic Halogenation** 486
12.6 Friedel–Crafts Alkylation of Benzene 488
12.7 Friedel–Crafts Acylation of Benzene 490
12.8 Synthesis of Alky Acylation–Reduct
12.9 Rate and Regiose Aromatic Substit
12.10 Rate and Regiose of Toluene 495
12.11 Rate and Regiose (Trifluoromethyl)
12.12 Substituent Effec Aromatic Substit Substituents 49
12.13 Substituent Effects in Electrophilic Aromatic Substitution: Strongly Deactivating Substituents 503
12.14 Substituent Effects in Electrophilic Aromatic Substitution: Halogens 506
12.15 Multiple Substituent Effects 507
12.16 Regioselective Synthesis of Disubstituted Aromatic Compounds 510
12.17 Substitution in Naphthalene 512
12.18 Substitution in Heterocyclic Aromatic Compounds 513

Mechanisms

12.1 Nitration of Benz
12.2 Sulfonation of Be
12.3 Bromination of B
12.4 Friedel–Crafts Alk

478

20 Enols and Enolates

Chapter Outline

20.1 Aldehyde, Ketone, and Ester Enolates 867
20.2 Enolate Regiochemistry 872
20.3 The Aldol Condensation 873
20.4 Mixed Aldol Condensations 878
■ **Chalcones: From the Mulberry Tree to Cancer Chemotherapy** 880
20.5 The Claisen Condensation 882
20.6 Intramolecular Claisen Condensation: The Dieckmann Cyclization 884
20.7 Mixed Claisen Condensations 885
20.8 Acylation of Ketones with Esters 886
20.9 Alkylation of Enolates 887
20.10 The Acetoacetic Ester Synthesis 889
20.11 The Malonic Ester Synthesis 891
20.12 Alkylation of Chiral Enolates 893
20.13 Enolization and Enol Content 895
20.14 α Halogenation of Aldehydes and Ketones 900
■ **The Haloform Reaction and the Biosynthesis of Trihalomethanes** 904
20.15 α Halogenation of Carboxylic Acids: The Hell–Volhard–Zelinsky Reaction 904
20.16 Some Chemical and Stereochemical Consequences of Enolization 906
20.17 Effects of Conjugation in α,β-Unsaturated Aldehydes and Ketones 907
20.18 Conjugate Addition to α,β-Unsaturated Carbonyl Compounds 908
20.19 Addition of Carbanions to α,β-Unsaturated Ketones: The Michael Reaction 910
20.20 Conjugate Addition of Organocopper Reagents to α,β-Unsaturated Carbonyl Compounds 912
20.21 **Summary** 913
Problems 917
Descriptive Passage and Interpretive Problems 20: The Enolate Chemistry of Dianions 926

Mechanisms

20.1 Aldol Addition of Butanal 874
20.2 Dehydration in a Base-Catalyzed Aldol Condensation 876
20.3 The Claisen Condensation of Ethyl Acetate 883
20.4 Base-Catalyzed Enolization of an Aldehyde or Ketone in Aqueous Solution 899
20.5 Acid-Catalyzed Enolization of an Aldehyde or Ketone in Aqueous Solution 899
20.6 Acid-Catalyzed Bromination of Acetone 901
20.7 Cleavage of a Tribromomethyl Ketone 903

866

Updated Carbohydrates Chapter

Chapter 23: "Carbohydrates" has been significantly updated. The mechanism for the Fischer glycosidation has been revised to be more consistent with current understanding of this fundamental reaction. New sections have been added on the synthesis of oligosacccharides and Glycobiology.

Updated Spectroscopy Chapter

Chapter 13: "Spectroscopy" now includes several new spectroscopy problems that include NMR spectra. These problems encompass a range of topics from analysis of splitting patterns in proton spectra of alcohols and alkyl halides, to DEPT spectra, to the effect of chirality center on adjacent methylene groups.

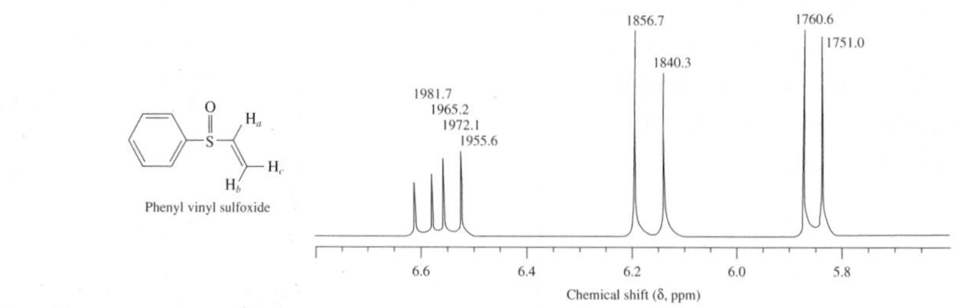

Figure 13.47

Vinyl proton region of the 300-MHz ^{1}H NMR spectrum of phenyl vinyl sulfoxide.

New Problems

Over one hundred seventy new problems have been added, many of which involve the synthesis of pharmaceuticals and natural products.

Bioorganic Emphasis

There is an increased emphasis on bioorganic chemistry, with new coverage of glycobiology, liposomes, G-coupled protein receptors, recombinant DNA technology, and other topics.

New Boxed Essays

Chapter 12: "Biosynthetic Halogenation" describes the enzymatic halogenation that occurs in the biosynthesis of the antibiotic pyrrolnitrin. A mechanism showing the electrophilic chlorinating species is included as well as a new problem that is based on an electrophilic halogenation with a reagent that has an oxygen-halogen bond.

Chapter 20: "Chalcones: From the Mulberry Tree to Cancer Chemotherapy" focuses on aromatase inhibitors in the treatment of breast cancer, and illustrates the use of a mixed aldol condensation in the synthesis of a naturally occurring chalcone.

Chapter 22: "James Bond, Oxidation Stress, and Antioxidant Phenols" shows how phenolic compounds such as vitamin E can protect cell membranes against damage caused by reactive oxygen species by terminating a free-radical chain reaction.

Chapter 27: "Conducting Polymers" describes organic light emitting diodes (OLEDs) which are conducting polymers that can be used in the production of displays used in cellular telephone and flat panel televisions.

Chalcones: From the Mulberry Tree to Cancer Chemotherapy

Excluding certain types of skin cancer, breast cancer is the most common cancer among women. The chance of developing invasive breast cancer during a woman's lifetime is approximately 1 in 8, and postmenopausal women are at greater risk for developing it because of changes in the production of the hormone estrogen. One approach to the treatment of breast cancer is through the use of drugs such as *tamoxifen* that restrict tumor growth by binding to an estrogen receptor in tumor cells.

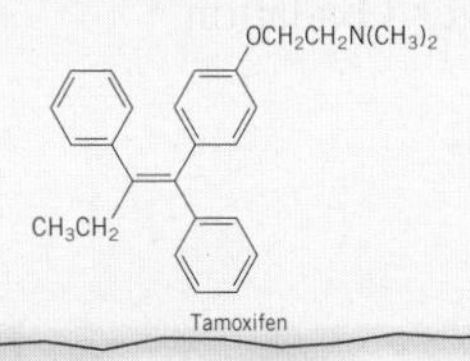

Tamoxifen

An alternative strategy is to reduce the body's production of estrogen by inhibiting the pathway for its biosynthesis. Aromatase inhibitors, which block the final step in the biosynthetic pathway to the estrogens estrone and estradiol, constitute one such class of anticancer drugs. For example, the enzyme aromatase catalyzes the conversion of androstenedione to estrone. By blocking the action of this enzyme, estrone production is inhibited, but the production of other necessary steroids is not affected.

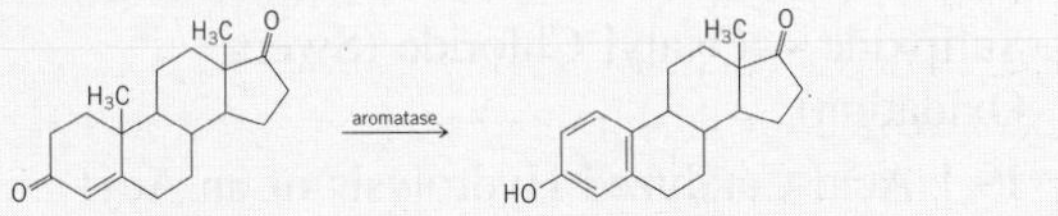

Androstenedione Estrone (an estrogen)

The potential of aromatase inhibitors in the treatment of breast cancer has spurred efforts to find more potent drugs. Bioassays of over 4000 plants led to the discovery of potent aromatase inhibitors from organic extracts of the paper mulberry tree, *Broussonetia papyrifera*. One of these compounds is a chalcone called morachalcone A. Chalcone and its derivatives are present in many plants, have a long history in folk medicine, and continue to be widely used as alternatives to conventional therapies.

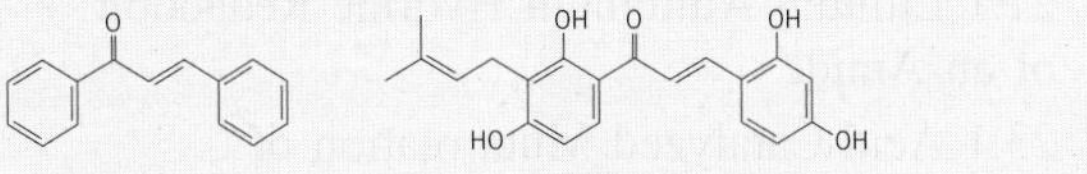

Chalcone Morachalcone A

In one synthesis of morachalcone A, an aldol condensation between an aldehyde and a ketone was used to prepare the compound shown. The remaining steps involved adding the five-carbon side chain and removing the three groups shown in red.

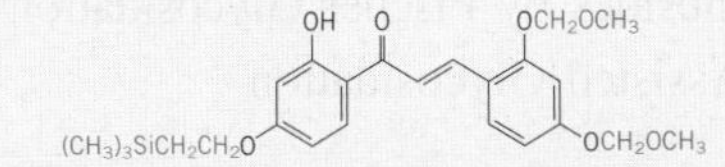

Problem 20.12

Two carbonyl compounds react in ethanol in the presence of KOH to yield the morachalcone precursor just shown. Write structural formulas for these two compounds.

The paper mulberry tree *Broussonetia papyrifera* is a source of a compound that holds promise as an anticancer drug.

New Sections

Every section was reviewed thoroughly and numerous changes were made on a continuing basis to ensure accuracy, relevance, and readability. Thus, many sections from previous editions, although substantially reworked, are not considered "new." The following lists those sections that are.

- Section 2.22: "Bonding in Water and Ammonia: Hybridization of Oxygen and Nitrogen"
- Section 4.13: "More on Activation Energy"
- Section 7.9: "The Chirality Axis"
- Section 20.12: "Alkylation of Chiral Enolates"
- Section 23.24: "Glycosides: Synthesis of Oligasaccharides"
- Section 23.25: "Glycobiology"
- Section 25.23: "G-Coupled Protein Receptors"
- Section 26.17: "Recombinant DNA Technology"

New Descriptive Passages

- Chapter 17 The Baeyer-Villiger Oxidation
- Chapter 20 Benzyne

New Mechanisms

- 4.3 Conversion of an Alcohol to an Alkyl Chloride with Thionyl Chloride
- 15.4 Oxidation of an Alcohol with Dimethyl Sulfoxide —Oxalyl Chloride (Swern Oxidation)
- 19.1 Acid-Catalyzed Hydrolysis of an Acyl Chloride via a Tetrahedral Intermediate
- 19.5 Reaction of an Ester with a Grignard Reagent
- 21.1 Lithium Aluminum Hydride Reduction of an Amide
- 23.1 Acid-Catalyzed Mutarotation of D-Glucopyranose
- 23.2 Preparation of Methyl D-Glucopyranosides by Fischer Glycosidation
- 23.3 Silver-Assisted Glycosidation

Mechanism 23.1

Acid-Catalyzed Mutarotation of D-Glucopyranose

THE OVERALL REACTION:

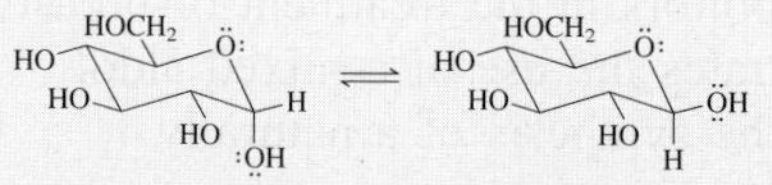

α-D-Glucopyranose β-D-Glucopyranose

Step 1: Protonation of the oxygen of the pyranose ring by the acid catalyst. In aqueous solution, the acid catalyst is the hydronium ion.

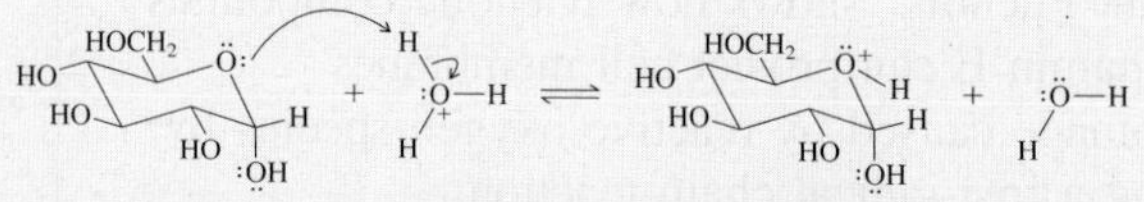

α-D-Glucopyranose Hydronium ion Conjugate acid of α-D-glucopyranose Water

Step 2: The pyranose ring opens by cleaving the bond between the anomeric carbon and the positively charged oxygen. This ring opening is facilitated by electron release from the OH group at the anomeric carbon and gives the conjugate acid of the open-chain form of D-glucose.

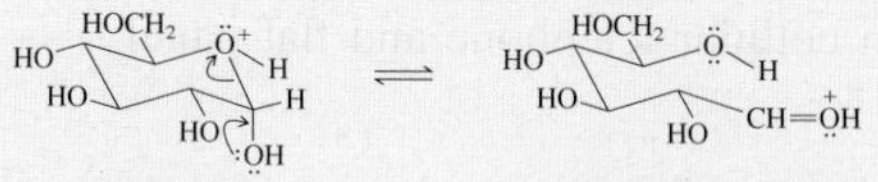

Conjugate acid of α-D-glucopyranose Conjugate acid of open-chain form of D-glucose

Step 3: The species formed in the preceding step cyclizes to give the conjugate acid of β-D-glucopyranose. This cyclization is analogous to the acid-catalyzed nucleophilic additions to aldehydes and ketones in Chapter 17.

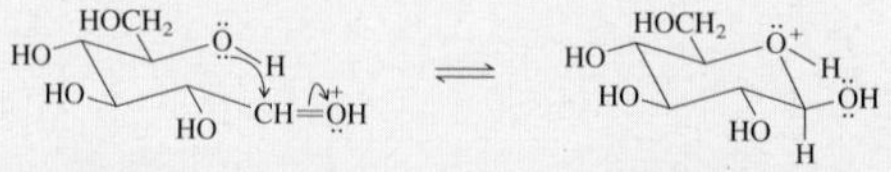

Conjugate acid of open-chain form of D-glucose Conjugate acid of β-D-glucopyranose

Step 4: The product of step 3 transfers a proton to water to regenerate the acid catalyst and yield the neutral form of the product.

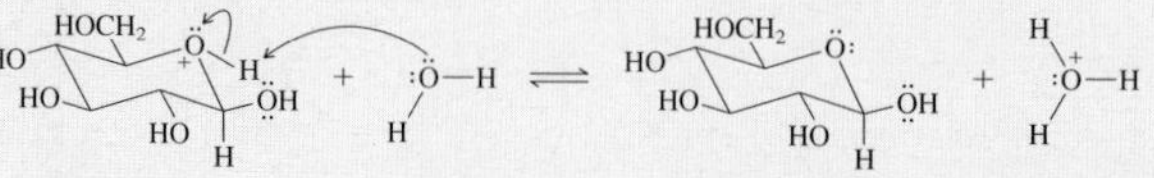

Conjugate acid of β-D-glucopyranose Water β-D-Glucopyranose Hydronium ion

Supplementary Materials

For the Instructor and Student:

McGraw-Hill Connect Chemistry

McGraw-Hill Connect Chemistry is a web-based assignment and assessment platform that gives students the means to better connect with their coursework, with their instructors, and with the important concepts that they will need to know for success now and in the future.

With Connect Chemistry, instructors can deliver assignments, quizzes and tests online. Many of the questions from the text are presented in an auto-gradable format and tied to the text's learning objectives. Instructors can edit existing questions and author entirely new problems. Track individual student performance—by question, assignment or in relation to the class overall—with detailed grade reports. Integrate grade reports easily with Learning Management Systems (LMS) such as WebCT and Blackboard. And much more.

By choosing Connect Chemistry, instructors are providing their students with a powerful tool for improving academic performance and truly mastering course material. Connect Chemistry allows students to practice important skills at their own pace and on their own schedule. Importantly, students' assessment results and instructors' feedback are all saved online—so students can continually review their progress and plot their course to success.

Like Connect Chemistry, Connect Chemistry *Plus* provides students with online assignments and assessments, plus 24/7 online access to an eBook—an online edition of the text—to aid them in successfully completing their work, wherever and whenever they choose.

For the Instructor

Presentation Center

Build instructional materials wherever, whenever, and however you want!

The Presentation Center is an online digital library containing assets such as photos, artwork, animations, PowerPoints, and other media types that can be used to create customized lectures, visually enhanced tests and quizzes, compelling course websites, or attractive printed support materials.

Access to your book, access to all books!

The Presentation Center library includes thousands of assets from many McGraw-Hill titles. This ever-growing resource gives instructors the power to utilize assets specific to an adopted textbook as well as content from all other books in the library.

Nothing could be easier!

Accessed from the instructor side of your textbook's website, Presentation Center's dynamic search engine allows you to explore by discipline, course, textbook chapter, asset type, or keyword. Simply browse, select, and download the files you need to build engaging course materials.

Course Management Software

With help from Blackboard or WebCT®, you can take complete control over your course content. These course cartridges also feature online testing and powerful student tracking. Contact your McGraw-Hill sales representative for more details.

Classroom Response Systems bring interactivity into the classroom or lecture hall. These wireless response systems, which are essentially remote that are easy to use and engage students, give the instructor and students immediate feedback from the entire class. Wireless response systems allow instructors to motivate student preparation, interactivity, and active learning. Nearly 600 questions covering the content of the *Organic Chemistry* text are available on the *Organic Chemistry* site for use with any classroom response system.

Test Bank an updated test bank with over 1300 questions is available with the 8th edition. The Test Bank is available as both word files and in a computerized test bank program, which utilizes testing software to quickly create customized exams. This user-friendly program allows instructors to sort questions by format; edit existing questions or add new ones; and scramble questions for multiple versions of the same test.

Solutions Manual This manual provides complete solutions to all end-of-chapter problems in the text. The Solutions Manual includes step-by-step solutions to each problem in the text as well as self-tests to assess student understanding.

Student Resources

Solutions Manual

The Solutions Manual provides step-by-step solutions guiding the student through the reasoning behind each problem in the text. There is also a self-test section at the end of each chapter which is designed to assess the student's mastery of the material.

Schaum's Outline of Organic Chemistry

This helpful study aid provides students with hundreds of solved and supplementary problems for the organic chemistry course.

Acknowledgments

It has been nearly five years since I received an email from Thomas Timp of McGraw-Hill inviting me to contribute ideas for a new edition of the Carey textbook. Having been a Carey student in graduate school, I was thrilled to have the opportunity to work with Frank again. Textbook writing was completely new to my experience, but I have been fortunate to have the support of many talented individuals. I have worked most closely with Jodi Rhomberg, Developmental Editor II, who made my entry into this project as smooth as possible and was extremely patient in dealing with a new author. The other individuals from McGraw-Hill that I wish to thank are Tami Hodge, Senior Sponsoring Editor, Ryan Blankenship, Publisher, Thomas Timp, Marketing Director, Tammy Ben, Editorial Assistant, Todd Turner, Chemistry Marketing Manager, and finally, Larissa Youshock, the Sales Representative for the region that includes Villanova. Revising a textbook has turned out to be a much larger undertaking than I had imagined. The encouragement that I have received from these individuals has helped sustain my efforts throughout the entire process.

I have benefitted from the support of the faculty, staff, and students of the Chemistry Department of Villanova University, in particular, my organic chemist colleagues, Joseph Bausch, Eduard Casillas, and Brian Ohta. The calculation of the electrostatic potential map of digitoxigenin that is part of the opener for chapter 24 was carried out by Brian Ohta. The synthesis of the aromatase inhibitor that is included in the new boxed essay on chalcones in chapter 20 is from research carried out by Ed Casillas and his students. Joe Bausch was the accuracy checker for this edition, and contributed many ideas during the revision. New mass spectra were recorded and formatted by Anthony Lagalante, who is also a faculty member in our department. New NMR and IR spectra were recorded and formatted by Walter Boyko, our NMR Laboratory Director. Other members of the Villanova community are also to be thanked for their support. I wish to mention particularly the staff of Villanova Graphics Services, and the staff of the Villanova Mailroom.

The other new member of the Carey team is Professor Neil Allison, University of Arkansas, who is the author of the Student Solutions Manual. I wish to thank Neil and his predecessor, Professor Robert Atkins, James Madison University, for their work on this excellent resource.

I remain indebted to my teachers and mentors in organic chemistry, to whom I am deeply grateful, especially Frank Carey, Bert Fraser-Reid, and the senior faculty in organic chemistry at Villanova who guided my development as a teacher and researcher. Hopefully, they would recognize the results of that guidance at many places in this edition. Frank's skill in solving problems in textbook writing was just as invaluable as his direction of my research when I was a student in his laboratory. Finally, I wish to thank my family, especially my wife Margot, and children Michael, Ellen, and Christopher. The end-of-chapter problem on native chemical ligation in chapter 25 is based on a discussion I had with Michael, who is a doctoral student in chemistry at the University of Wisconsin–Madison. I also wish to thank my parents, William and Edith, and my brother, William Jr., all of whom were with me at the time I began working on this project. I am grateful to all of you.

Robert Giuliano

Reviewers

Hundreds of teachers of organic chemistry have reviewed *Organic Chemistry* in its various editions. Those listed below are the most recent.

Christopher J. Abelt, *College of William and Mary*
Ruquia Ahmed-Schofield, *Xavier University of Louisiana*
Neil Allison, *University of Arkansas*
David Anderson, *University of Colorado at Colorado Springs*
Rodrigo B. Andrade, *Temple University*
Peter Andreana, *Wayne State University*
William F. Bailey, *University of Connecticut*
David C. Baker, *University of Tennessee*
Anamitro Banerjee, *University of North Dakota*
Vladimir Benin, *University of Dayton*
David Berkowitz, *University of Nebraska, Lincoln*
Chad Booth, *Texas State University*
Ned Bowden, *University of Iowa*
Amy L. Bradley, *Wilkes University*
Philip A. Brown, *North Carolina State University*
Scott Bur, *Gustavus Adolphus College*
Robert M. Carlson, *University of Minnesota Duluth*
Patrick E. Cassidy, *Ohio State University*
Christopher Callam, *Ohio State University*
Steven Castle, *Brigham Young University*
Ronald K. Castellano, *University of Florida*
Robert S. Coleman, *Ohio State University*
David M. Collard, *Georgia Institute of Technology*

Peter de Lijser, *California State University-Fullerton*
Andrew P. Dicks, *University of Toronto*
Matthew R. Dintzner, *DePaul University*
Ajit D. Dixit, *Wake Technical Community College*
Norma Dunlap, *Middle Tennessee State University*
Ihsan Erden, *San Francisco State University*
Steven Fleming, *Temple University*
Andreas H. Franz, *University of the Pacific*
Lee Friedman, *University of Maryland, College Park*
Charles M. Garner, *Baylor University*
Jacquelyn Gervay-Hague, *University of California–Davis*
Joseph J. Grabowski, *University of Pittsburgh*
Zhongwu Guo, *Wayne State University*
Kathleen M. Halligan, *York College of Pennsylvania*
Scott T. Handy, *Middle Tennessee State University*
Kenn Harding, *Texas A & M University*
John Haseltine, *Kennesaw State University*
F. J. Heldrich, *College of Charleston*
Paul Higgs, *University of Tennessee at Martin*
Edwin Hillinski, *Florida State University*
Michael T. Huggins, *University of West Florida*
Colleen S. Kaczmarek, *Connecticut College*
Bob Kane, *Baylor University*
Eric Kantorowski, *California Polytechnic State University*
Steven Kass, *University of Minnesota*
Robert Kerber, *Stony Brook University*
Rebecca M. Kissling, *Binghamton University*
Spencer Knapp, *Rutgers University*
Dalila G. Kovacs, *Grand Valley State University*
Michael Kurz, *University of Texas, San Antonio*
Kenneth Laali, *Kent State University*
Chunmei Li, *Stephen F. Austin State University*
Todd Lowary, *The University of Alberta*
Cecilia H. Marzabadi, *Seton Hall University*
Jeremy E. B. McCallum, *Loyola Marymount University*
Kevin P. C. Minbiole, *James Madison University*
Neil Miranda, *University of Illinois at Chicago*
Miguel O. Mitchell, *Salisbury University*
David R. Mootoo, *Hunter College/City University of New York*
Harry Morrison, *Purdue University*
Richard W. Morrison, *University of Georgia*
Stephen Nelsen, *University of Wisconsin-Madison*
Felix N. Ngassa, *Grand Valley State University*
Timo V. Ovaska, *Connecticut College*
Kimberly Pacheco, *University of Northern Colorado*
Cyril Parkanyi, *Florida Atlantic University*
Jon R. Parquette, *Ohio State University*
Michael W. Rathke, *Michigan State University*
Stanley Raucher, *University of Washington*
Ieva L. Reich, *University of Wisconsin–Madison*
Mike Rennekamp, *Columbus State Community College*
JamieLee I'olani Rizzo, *Pace University*
Suzanne Ruder, *Virginia Commonwealth University*
Raymond Sadeghi, *University of Texas, San Francisco*
Paul Sampson, *Kent State University*
Kay Sandberg, *North Carolina State University*
Joe C. Saunders, *Concord University*
James Schreck, *University of Northern Colorado*
Alexander J. Seed, *Kent State University*
Grigoriy Sereda, *University of South Dakota*
Irina P. Smoliakova, *University of North Dakota*
David Son, *Southern Methodist University*
David Spurgeon, *University of Arizona*
Chad Stearman, *Missouri State University*
Daniela Tapu, *Kennesaw State University*
Kevin Walker, *Michigan State University*
Milton J. Weider, *Metropolitan State College of Denver*
James K. Whitesell, *University of California, San Diego*
Ron Wikholm, *University of Connecticut*
Jane E. Wissinger, *University of Minnesota*
Laurie A. Witucki, *Grand Valley State University*
Evelyn A. Wolfe, *St. Francis College*
Catherine Woytowicz, *The George Washington University*
David G. J. Young, *East Tennessee State University*
Viktor V. Zhdankin, *University of Minnesota Duluth*

Organic Chemistry

1 Structure Determines Properties

Chapter Outline

Although function dictates form in the things we build, structure determines properties in molecules.

STRUCTURE* is the key to everything in chemistry. The properties of a substance depend on the atoms it contains and the way these atoms are connected. What is less obvious, but very powerful, is the idea that someone who is trained in chemistry can look at the structural formula of a substance and tell you a lot about its properties. This chapter begins your training toward understanding the relationship between structure and properties in organic compounds. It reviews some fundamental principles of the Lewis approach to molecular structure and bonding. By applying these principles, you will learn to recognize structural patterns that are more stable than others and develop skills in communicating structural information that will be used throughout your study of organic chemistry. A key relationship between structure and properties will be introduced by examining the fundamentals of acid–base chemistry from a structural perspective.

1.1 Atoms, Electrons, and Orbitals

Before discussing structure and bonding in *molecules,* let's first review some fundamentals of *atomic* structure. Each element is characterized by a unique **atomic number** Z, which is equal to the number of protons in its nucleus. A neutral atom has equal numbers of protons, which are positively charged, and electrons, which are negatively charged.

Electrons were believed to be particles from the time of their discovery in 1897 until 1924, when the French physicist Louis de Broglie suggested that they have wave-like properties as well. Two years later Erwin Schrödinger took the next step and calculated the energy of an electron in a hydrogen atom by using equations that treated the electron as if it were a wave. Instead of a single energy, Schrödinger obtained a series of them, each of which corresponded to a different mathematical description of the electron wave. These mathematical descriptions are called **wave functions** and are symbolized by the Greek letter ψ (psi).

According to the Heisenberg uncertainty principle, we can't tell exactly where an electron is, but we can tell where it is most likely to be. The probability of finding an electron at a particular spot relative to an atom's nucleus is given by the square of the wave function (ψ^2) at that point. Figure 1.1 illustrates the probability of finding an electron at various points in the lowest energy (most stable) state of a hydrogen atom. The darker the color in a region, the higher the probability. The probability of finding an electron at a particular point is greatest near the nucleus and decreases with increasing distance from the nucleus but never becomes zero. We commonly describe Figure 1.1 as an "electron cloud" to call attention to the spread-out nature of the electron probability.

*A glossary of the terms shown in boldface may be found immediately before the index at the back of the book.

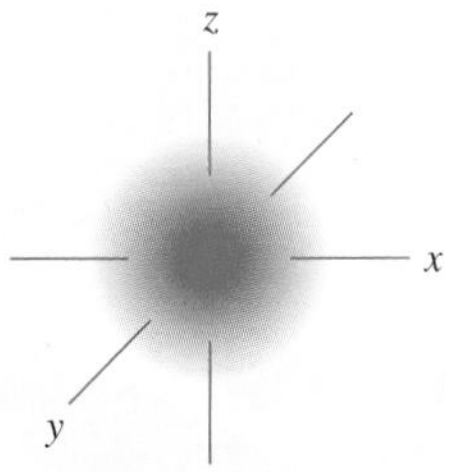

Figure 1.1

Probability distribution (ψ^2) for an electron in a 1*s* orbital.

Be careful, though. The "electron cloud" of a hydrogen atom, although drawn as a collection of many dots, represents only one electron.

Wave functions are also called **orbitals.** For convenience, chemists use the term "orbital" in several different ways. A drawing such as Figure 1.1 is often said to represent an orbital. We will see other kinds of drawings in this chapter, and use the word "orbital" to describe them too.

Orbitals are described by specifying their size, shape, and directional properties. Spherically symmetrical ones such as shown in Figure 1.1 are called *s orbitals.* The letter *s* is preceded by the **principal quantum number** n (n = 1, 2, 3, etc.), which specifies the **shell** and is related to the energy of the orbital. An electron in a 1*s* orbital is likely to be found closer to the nucleus, is lower in energy, and is more strongly held than an electron in a 2*s* orbital.

Instead of probability distributions, it is more common to represent orbitals by their **boundary surfaces,** as shown in Figure 1.2 for the 1*s* and 2*s* orbitals. The boundary surface encloses the region where the probability of finding an electron is high—on the order of 90–95%. Like the probability distribution plot from which it is derived, a picture of a boundary surface is usually described as a drawing of an orbital.

A hydrogen atom (Z = 1) has one electron; a helium atom (Z = 2) has two. The single electron of hydrogen occupies a 1*s* orbital, as do the two electrons of helium. We write their electron configurations as:

Hydrogen: $1s^1$ Helium: $1s^2$

In addition to being negatively charged, electrons possess the property of **spin.** The **spin quantum number** of an electron can have a value of either $+\frac{1}{2}$ or $-\frac{1}{2}$. According to the **Pauli exclusion principle,** two electrons may occupy the same orbital only when they have opposite, or "paired," spins. For this reason, no orbital can contain more than two electrons. Because two electrons fill the 1*s* orbital, the third electron in lithium (Z = 3) must occupy an orbital of higher energy. After 1*s*, the next higher energy orbital is 2*s*. The third electron in lithium therefore occupies the 2*s* orbital, and the electron configuration of lithium is

Lithium: $1s^22s^1$

A complete periodic table of the elements is presented on the inside back cover.

The **period** (or **row**) of the periodic table in which an element appears corresponds to the principal quantum number of the highest numbered occupied orbital (n = 1 in the case of hydrogen and helium). Hydrogen and helium are first-row elements; lithium (n = 2) is a second-row element.

With beryllium (Z = 4), the 2*s* level becomes filled and, beginning with boron (Z = 5), the next orbitals to be occupied are $2p_x$, $2p_y$, and $2p_z$. These three orbitals (Figure 1.3) are of equal energy and are characterized by boundary surfaces that are usually described as "dumbell-shaped." The axes of the three 2*p* orbitals are at right angles to one another. Each orbital consists of two "lobes," represented in Figure 1.3 by regions of different colors. Regions of a single orbital, in this case, each 2*p* orbital, may be separated by **nodal surfaces** where the wave function changes sign and the probability of finding an electron is zero.

Other methods are also used to contrast the regions of an orbital where the signs of the wave function are different. Some mark one lobe of a *p* orbital + and the other −. Others shade one lobe and leave the other blank. When this level of detail isn't necessary, no differentiation is made between the two lobes.

Figure 1.2

Boundary surfaces of a 1*s* orbital and a 2*s* orbital. The boundary surfaces enclose the volume where there is a 90–95% probability of finding an electron.

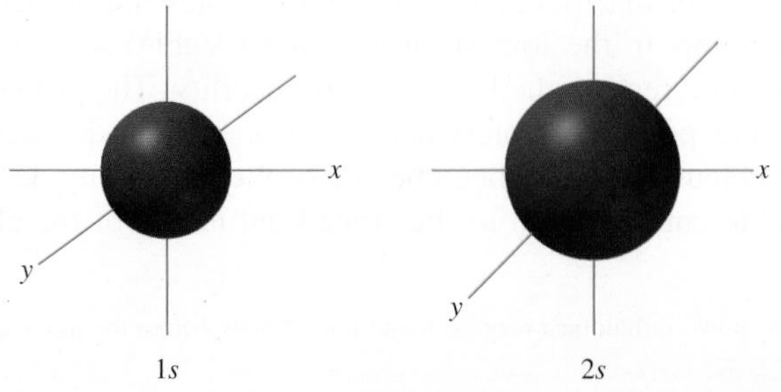

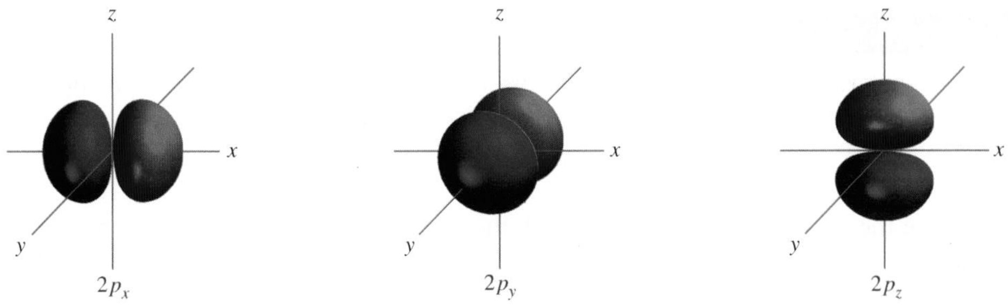

Figure 1.3

Boundary surfaces of the $2p$ orbitals. The wave function changes sign at the nucleus. The two halves of each orbital are indicated by different colors. The *yz*-plane is a nodal surface for the $2p_x$ orbital. The probability of finding a $2p_x$ electron in the *yz*-plane is zero. Analogously, the *xz*-plane is a nodal surface for the $2p_y$ orbital, and the *xy*-plane is a nodal surface for the $2p_z$ orbital.

The electron configurations of the first 12 elements, hydrogen through magnesium, are given in Table 1.1. In filling the $2p$ orbitals, notice that each is singly occupied before any one is doubly occupied. This general principle for orbitals of equal energy is known as **Hund's rule.** Of particular importance in Table 1.1 are *hydrogen, carbon, nitrogen,* and *oxygen.* Countless organic compounds contain nitrogen, oxygen, or both in addition to carbon, the essential element of organic chemistry. Most of them also contain hydrogen.

It is often convenient to speak of the **valence electrons** of an atom. These are the outermost electrons, the ones most likely to be involved in chemical bonding and reactions. For second-row elements these are the $2s$ and $2p$ electrons. Because four orbitals ($2s$, $2p_x$, $2p_y$, $2p_z$) are involved, the maximum number of electrons in the **valence shell** of any second-row element is 8. Neon, with all its $2s$ and $2p$ orbitals doubly occupied, has eight valence electrons and completes the second row of the periodic table. For **main-group elements,** the number of valence electrons is equal to its group number in the periodic table.

TABLE 1.1 Electron Configurations of the First Twelve Elements of the Periodic Table

Element	Atomic number Z	Number of electrons in indicated orbital					
		$1s$	$2s$	$2p_x$	$2p_y$	$2p_z$	$3s$
Hydrogen	1	1					
Helium	2	2					
Lithium	3	2	1				
Beryllium	4	2	2				
Boron	5	2	2	1			
Carbon	6	2	2	1	1		
Nitrogen	7	2	2	1	1	1	
Oxygen	8	2	2	2	1	1	
Fluorine	9	2	2	2	2	1	
Neon	10	2	2	2	2	2	
Sodium	11	2	2	2	2	2	1
Magnesium	12	2	2	2	2	2	2

Detailed solutions to all of the problems are found in the *Solutions Manual* along with a brief discussion and advice on how to do problems of the same type.

Problem 1.1

How many electrons does carbon have? How many are valence electrons? What third-row element has the same number of valence electrons as carbon?

Once the $2s$ and $2p$ orbitals are filled, the next level is the $3s$, followed by the $3p_x$, $3p_y$, and $3p_z$ orbitals. Electrons in these orbitals are farther from the nucleus than those in the $2s$ and $2p$ orbitals and are of higher energy.

In-chapter problems that contain multiple parts are accompanied by a sample solution to part (a).

Problem 1.2

Referring to the periodic table as needed, write electron configurations for all the elements in the third period.

Sample Solution The third period begins with sodium and ends with argon. The atomic number Z of sodium is 11, and so a sodium atom has 11 electrons. The maximum number of electrons in the $1s$, $2s$, and $2p$ orbitals is ten, and so the eleventh electron of sodium occupies a $3s$ orbital. The electron configuration of sodium is $1s^2 2s^2 2p_x^2 2p_y^2 2p_z^2 3s^1$.

Neon, in the second period, and argon, in the third, have eight electrons in their valence shell; they are said to have a complete **octet** of electrons. Helium, neon, and argon belong to the class of elements known as **noble gases** or **rare gases.** The noble gases are characterized by an extremely stable "closed-shell" electron configuration and are very unreactive.

Structure determines properties and the properties of atoms depend on atomic structure. All of an element's protons are in its nucleus, but the element's electrons are distributed among orbitals of various energy and distance from the nucleus. More than anything else, we look at its electron configuration when we wish to understand how an element behaves. The next section illustrates this with a brief review of ionic bonding.

1.2 Ionic Bonds

Atoms combine with one another to give **compounds** having properties different from the atoms they contain. The attractive force between atoms in a compound is a **chemical bond.** One type of chemical bond, called an **ionic bond,** is the force of attraction between oppositely charged species **(ions)** (Figure 1.4). Positively charged ions are referred to as **cations;** negatively charged ions are **anions.**

Whether an element is the source of the cation or anion in an ionic bond depends on several factors, for which the periodic table can serve as a guide. In forming ionic compounds, elements at the left of the periodic table typically lose electrons, giving a cation that has the same electron configuration as the preceding noble gas. Loss of an electron from sodium, for example, yields Na^+, which has the same electron configuration as neon.

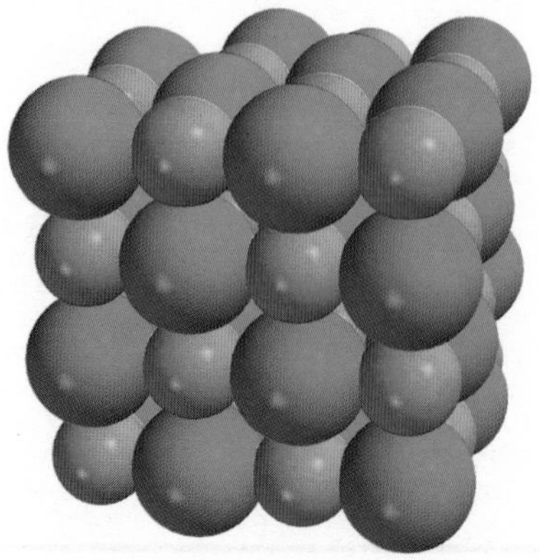

Figure 1.4

An ionic bond is the force of attraction between oppositely charged ions. Each Na^+ ion in the crystal lattice of solid NaCl is involved in ionic bonding to each of six surrounding Cl^- ions and vice versa. The smaller balls are Na^+ and the larger balls are Cl^-.

$$\underset{\substack{\text{Sodium atom}\\ 1s^22s^22p^63s^1}}{Na(g)} \longrightarrow \underset{\substack{\text{Sodium ion}\\ 1s^22s^22p^6}}{Na^+(g)} + \underset{\text{Electron}}{e^-}$$

[The symbol (g) indicates that the species is present in the gas phase.]

Problem 1.3

Species that have the same number of electrons are described as *isoelectronic.* What +2 ion is isoelectronic with Na^+? What −2 ion?

A large amount of energy, called the **ionization energy,** must be transferred to any atom to dislodge an electron. The ionization energy of sodium, for example, is 496 kJ/mol (119 kcal/mol). Processes that absorb energy are said to be **endothermic.** Compared with other elements, sodium and its relatives in group 1A have relatively low ionization energies. In general, ionization energy increases across a row in the periodic table.

The SI (*Système International d'Unites*) unit of energy is the *joule* (J). An older unit is the *calorie* (cal). Many chemists still express energy changes in units of kilocalories per mole (1 kcal/mol = 4.184 kJ/mol).

Elements at the right of the periodic table tend to gain electrons to reach the electron configuration of the next higher noble gas. Adding an electron to chlorine, for example, gives the anion Cl^-, which has the same closed-shell electron configuration as the noble gas argon.

$$\underset{\substack{\text{Chlorine atom}\\ 1s^22s^22p^63s^23p^5}}{Cl(g)} + \underset{\text{Electron}}{e^-} \longrightarrow \underset{\substack{\text{Chloride ion}\\ 1s^22s^22p^63s^23p^6}}{Cl^-(g)}$$

Problem 1.4

What −2 ion is isoelectronic with Cl^-?

Energy is released when a chlorine atom captures an electron. Energy-releasing reactions are described as **exothermic,** and the energy change for an exothermic process has a negative sign. The energy change for addition of an electron to an atom is referred to as its **electron affinity** and is –349 kJ/mol (–83.4 kcal/mol) for chlorine.

Problem 1.5

Which of the following ions possess a noble gas electron configuration? Which ions are isoelectronic?

(a) K^+
(b) He^+
(c) H^-
(d) O^-
(e) F^-
(f) Ca^{2+}

Sample Solution (a) Potassium has atomic number 19, and so a potassium atom has 19 electrons. The ion K^+, therefore, has 18 electrons, the same as the noble gas argon. The electron configurations of both K^+ and Ar are $1s^22s^22p^63s^23p^6$. K^+ and Ar are isoelectronic.

Transfer of an electron from a sodium atom to a chlorine atom yields a sodium cation and a chloride anion, both of which have a noble gas electron configuration:

$$\underset{\text{Sodium atom}}{Na(g)} + \underset{\text{Chlorine atom}}{Cl(g)} \longrightarrow \underset{\text{Sodium chloride}}{Na^+Cl^-(g)}$$

Were we to simply add the ionization energy of sodium (496 kJ/mol) and the electron affinity of chlorine (−349 kJ/mol), we would conclude that the overall process is endothermic by +147 kJ/mol. The energy liberated by adding an electron to chlorine is insufficient to override the energy required to remove an electron from sodium. This analysis, however, fails to consider the force of attraction between the oppositely charged ions Na^+ and Cl^-, which exceeds 500 kJ/mol and is more than sufficient to make the overall process exothermic. Attractive forces between oppositely charged particles are termed **electrostatic,** or **coulombic, attractions** and are what we mean by an **ionic bond** between two atoms.

Ionic bonding was proposed by the German physicist Walther Kossel in 1916, in order to explain the ability of substances such as molten sodium chloride to conduct an electric current. He was the son of Albrecht Kossel, winner of the 1910 Nobel Prize in physiology or medicine for early studies in nucleic acids.

Problem 1.6

What is the electron configuration of C^+? Of C^-? Does either one of these ions have a noble gas (closed-shell) electron configuration?

Ionic bonds are very common in *inorganic* compounds, but rare in *organic* ones. The ionization energy of carbon is too large and the electron affinity too small for carbon to realistically form a C^{4+} or C^{4-} ion. What kinds of bonds, then, link carbon to other elements in millions of organic compounds? Instead of losing or gaining electrons, carbon *shares* electrons with other elements (including other carbon atoms) to give what are called covalent bonds.

1.3 Covalent Bonds, Lewis Structures, and the Octet Rule

Gilbert Newton Lewis (born Weymouth, Massachusetts, 1875; died Berkeley, California, 1946) has been called the greatest American chemist.

The **covalent,** or **shared electron pair,** model of chemical bonding was first suggested by G. N. Lewis of the University of California in 1916. Lewis proposed that a *sharing* of two electrons by two hydrogen atoms permits each one to have a stable closed-shell electron configuration analogous to helium.

H· ·H → H:H

Two hydrogen atoms, each with a single electron — Hydrogen molecule: covalent bonding by way of a shared electron pair

Structural formulas of this type in which electrons are represented as dots are called **Lewis structures.** It is customary to represent a shared electron-pair bond by a dash (—). Thus, H:H becomes H—H.

The amount of energy required to dissociate a hydrogen molecule H_2 to two separate hydrogen atoms is its **bond dissociation enthalpy.** For H_2 it is quite large, amounting to +435 kJ/mol (+104 kcal/mol). The main contributor to the strength of the covalent bond in H_2 is the increased binding force exerted on its two electrons. Each electron in H_2 "feels" the attractive force of two nuclei, rather than one as it would in an isolated hydrogen atom.

Only the electrons in an atom's valence shell are involved in covalent bonding. Fluorine, for example, has nine electrons, but only seven are in its valence shell. Pairing a valence electron of one fluorine atom with one of a second fluorine gives a fluorine molecule (F_2) in which each fluorine has eight valence electrons and an electron configuration equivalent to that of the noble gas neon. Shared electrons count toward satisfying the octet of both atoms.

Two fluorine atoms, each with seven electrons in its valence shell — Fluorine molecule: covalent bonding by way of a shared electron pair

The six valence electrons of each fluorine that are not involved in bonding comprise three **unshared pairs.**

Unshared pairs are also called *lone pairs.*

The Lewis model limits second-row elements (Li, Be, B, C, N, O, F, Ne) to a total of eight electrons (shared plus unshared) in their valence shells. Hydrogen is limited to two. Most of the elements that we'll encounter in this text obey the **octet rule:** *In forming compounds they gain, lose, or share electrons to achieve a stable electron configuration characterized by eight valence electrons.* When the octet rule is satisfied for carbon, nitrogen, oxygen, and fluorine, each has an electron configuration analogous to the noble gas neon. The Lewis structures of methane (CH_4), ammonia (NH_3), water (H_2O), and hydrogen fluoride (HF) given in Table 1.2 illustrate the octet rule.

TABLE 1.2 Lewis Formulas of Methane, Ammonia, Water, and Hydrogen Fluoride

Compound	Atom	Number of valence electrons in atom	Atom and sufficient number of hydrogen atoms to complete octet	Lewis formula
Methane	Carbon	4	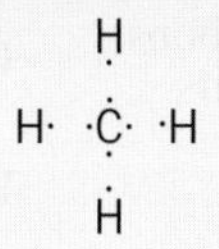	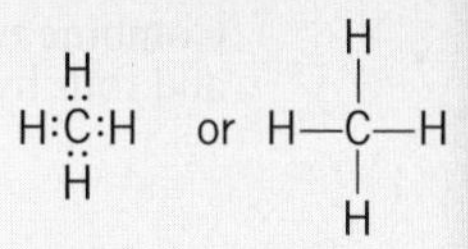
Ammonia	Nitrogen	5	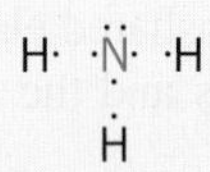	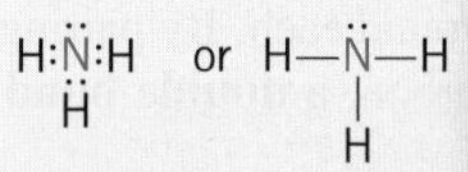
Water	Oxygen	6		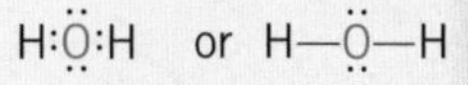
Hydrogen fluoride	Fluorine	7	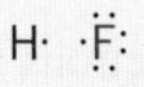	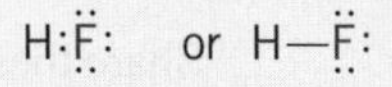

With four valence electrons, carbon normally forms four covalent bonds as shown in Table 1.2 for CH_4. In addition to C—H bonds, most organic compounds contain covalent C—C bonds. Ethane (C_2H_6) is an example.

Combine two carbons and six hydrogens 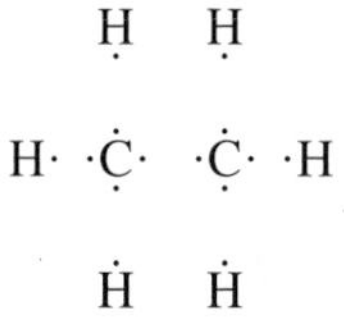 to write a Lewis structure for ethane 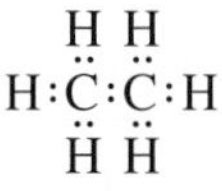 or 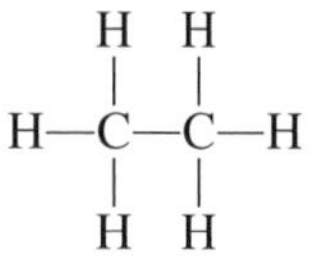

Problem 1.7

Write Lewis structures, including unshared pairs, for each of the following. Carbon has four bonds in each compound.

(a) Propane (C_3H_8)
(b) Methanol (CH_4O)
(c) Methyl fluoride (CH_3F)
(d) Ethyl fluoride (C_2H_5F)

Sample Solution (a) The Lewis structure of propane is analogous to that of ethane but the chain is three carbons long instead of two.

Combine three carbons and eight hydrogens 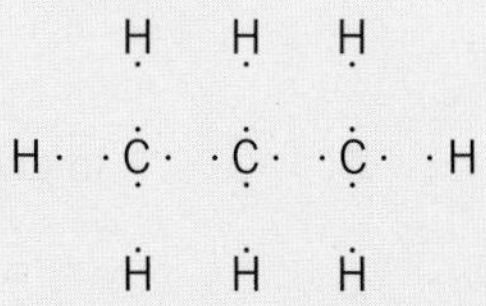

to write a Lewis structure for propane 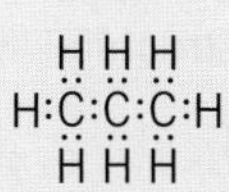 or 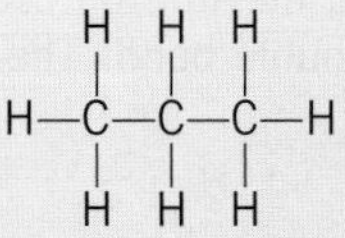

The ten covalent bonds in the Lewis structure shown account for 20 valence electrons, which is the same as that calculated from the molecular formula (C_3H_8). The eight hydrogens of C_3H_8 contribute 1 electron each and the three carbons 4 each, for a total of 20 (8 from the hydrogens and 12 from the carbons). Therefore, all the valence electrons are in covalent bonds; propane has no unshared pairs.

1.4 Double Bonds and Triple Bonds

Lewis's concept of shared electron pair bonds allows for four-electron double bonds and six-electron triple bonds. Ethylene (C_2H_4) has 12 valence electrons, which can be distributed as follows:

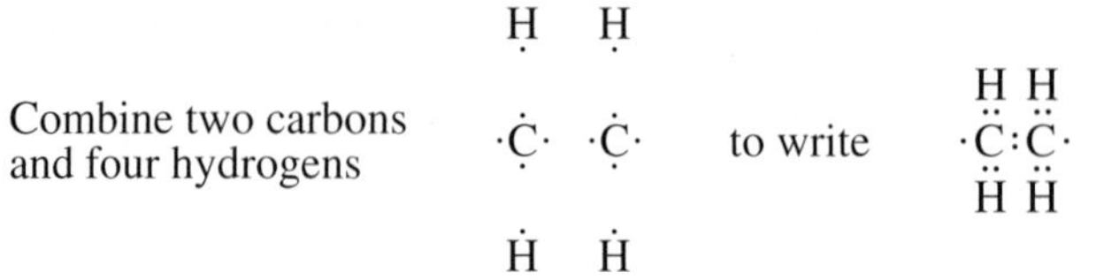
Combine two carbons and four hydrogens
to write

The structural formula produced has a single bond between the carbons and seven electrons around each. By pairing the unshared electron of one carbon with its counterpart of the other carbon, a **double bond** results and the octet rule is satisfied for both carbons.

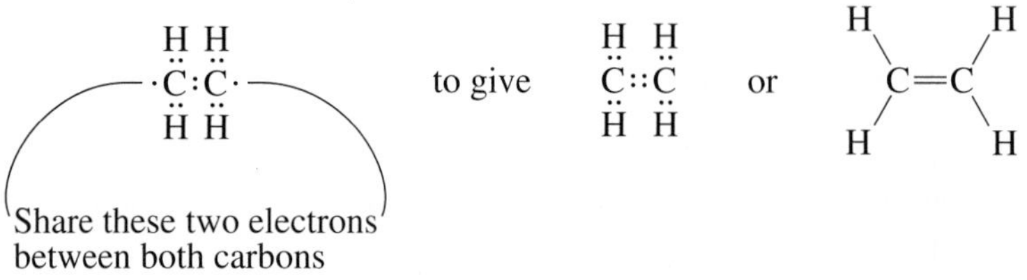
Share these two electrons between both carbons
to give
or

Likewise, the ten valence electrons of acetylene (C_2H_2) can be arranged in a structural formula that satisfies the octet rule when six of them are shared in a **triple bond** between the carbons.

H:C:::C:H or H—C≡C—H

Carbon dioxide (CO_2) has two carbon–oxygen double bonds, thus satisfying the octet rule for both carbon and oxygen.

:Ö::C::Ö: or :Ö=C=Ö:

Problem 1.8

All of the hydrogens are bonded to carbon in both of the following. Write a Lewis structure that satisfies the octet rule for each.

(a) Formaldehyde (CH_2O) (b) Hydrogen cyanide (HCN)

Sample Solution (a) Formaldehyde has 12 valence electrons; 4 from carbon, 2 from two hydrogens, and 6 from oxygen. Connect carbon to oxygen and both hydrogens by covalent bonds.

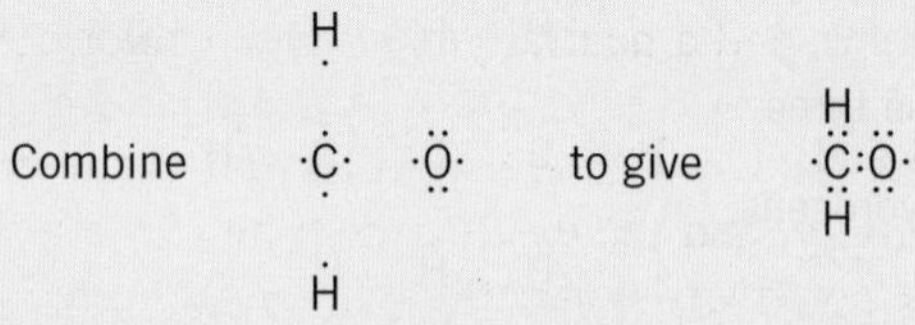
Combine
to give

Pair the unpaired electron on carbon with the unpaired electron on oxygen to give a carbon–oxygen double bond. The resulting structural formula satisfies the octet rule.

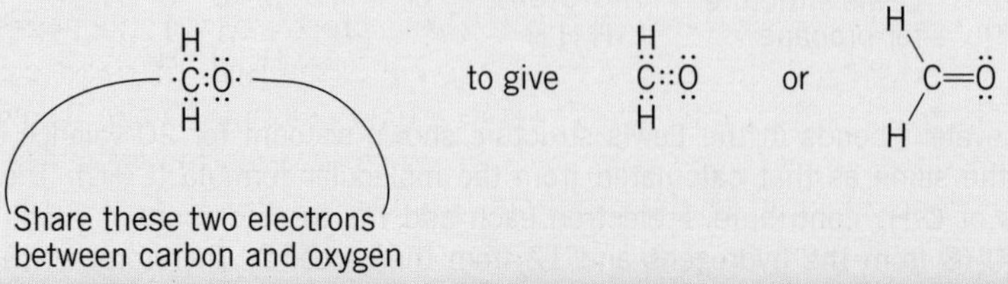
Share these two electrons between carbon and oxygen
to give
or

1.5 Polar Covalent Bonds, Electronegativity, and Bond Dipoles

Electrons in covalent bonds are not necessarily shared equally by the two atoms that they connect. If one atom has a greater tendency to attract electrons toward itself than the other, the electron distribution is *polarized,* and the bond is described as **polar covalent.** The tendency of an atom to attract the electrons in a covalent bond toward itself defines its **electronegativity.** An electronegative element attracts electrons; an electropositive one donates them.

Hydrogen fluoride, for example, has a polar covalent bond. Fluorine is more electronegative than hydrogen and pulls the electrons in the H—F bond toward itself, giving fluorine a partial negative charge and hydrogen a partial positive charge. Two ways of representing the polarization in HF are:

$^{\delta+}$H—F$^{\delta-}$	$\overset{+\!\!\longrightarrow}{\text{H—F}}$
(The symbols δ+ and δ− indicate partial positive and partial negative charge, respectively)	(The symbol $+\!\!\longrightarrow$ represents the direction of polarization of electrons in the H—F bond)

A third way of illustrating the electron polarization in HF is graphically, by way of an **electrostatic potential map,** which uses the colors of the rainbow to show the charge distribution. Blue through red tracks regions of greater positive charge to greater negative charge. (For more details, see the boxed essay *Electrostatic Potential Maps* in this section.)

Positively charged region of molecule
Negatively charged region of molecule

Contrast the electrostatic potential map of HF with those of H_2 and F_2.

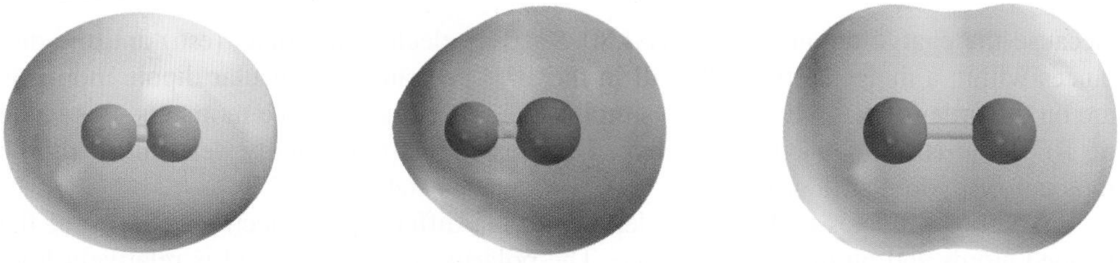

H—H $^{\delta+}$H—F$^{\delta-}$ F—F

The covalent bond in H_2 joins two hydrogen atoms. Because the bonded atoms are identical, so are their electronegativities. There is no polarization of the electron distribution, the H—H bond is nonpolar, and a neutral yellow-green color dominates the electrostatic potential map. Likewise, the F—F bond in F_2 is nonpolar and its electrostatic potential map resembles that of H_2. The covalent bond in HF, on the other hand, unites two atoms of different electronegativity, and the electron distribution is very polarized. Blue is the dominant color near the positively polarized hydrogen, and red the dominant color near the negatively polarized fluorine.

The most commonly used electronegativity scale was devised by Linus Pauling. Table 1.3 keys Pauling's electronegativity values to the periodic table.

Electronegativity *increases* from left to right across a row in the periodic table. Of the second-row elements, the most electronegative is fluorine, the least electronegative is lithium. Electronegativity *decreases* going down a column. Of the halogens, fluorine is the most electronegative, then chlorine, then bromine, then iodine. Indeed, fluorine is the most electronegative of all the elements; oxygen is second.

In general, the greater the electronegativity difference between two elements, the more polar the bond between them.

Linus Pauling (1901–1994) was born in Portland, Oregon, and was educated at Oregon State University and at the California Institute of Technology, where he earned a Ph.D. in chemistry in 1925. In addition to research in bonding theory, Pauling studied the structure of proteins and was awarded the Nobel Prize in Chemistry for that work in 1954. Pauling won a second Nobel Prize (the Peace Prize) in 1962 for his efforts to limit the testing of nuclear weapons. He was one of only four scientists to have won two Nobel Prizes. The first double winner was a woman. Can you name her?

TABLE 1.3 Selected Values from the Pauling Electronegativity Scale

	Group number						
Period	**1A**	**2A**	**3A**	**4A**	**5A**	**6A**	**7A**
1	H 2.1						
2	Li 1.0	Be 1.5	B 2.0	C 2.5	N 3.0	O 3.5	F 4.0
3	Na 0.9	Mg 1.2	Al 1.5	Si 1.8	P 2.1	S 2.5	Cl 3.0
4	K 0.8	Ca 1.0					Br 2.8
5							I 2.5

Problem 1.9

In which of the compounds CH_4, NH_3, H_2O, SiH_4, or H_2S is δ+ for hydrogen the greatest? In which one does hydrogen bear a partial negative charge?

Table 1.4 compares the polarity of various bond types according to their **bond dipole moments.** A dipole exists whenever opposite charges are separated from each other, and a **dipole moment μ** is the product of the amount of the charge *e* multiplied by the distance *d* between the centers of charge.

$$\mu = e \times d$$

Because the charge on an electron is 4.80×10^{-10} electrostatic units (esu) and the distances within a molecule typically fall in the 10^{-8} cm range, molecular dipole moments are on the order of 10^{-18} esu·cm. To simplify the reporting of dipole moments, this value of 10^{-18} esu·cm is defined as a **debye, D.** Thus the experimentally determined dipole moment of hydrogen fluoride, 1.7×10^{-18} esu·cm is stated as 1.7 D.

The debye unit is named in honor of Peter Debye, a Dutch scientist who did important work in many areas of chemistry and physics and was awarded the Nobel Prize in Chemistry in 1936.

The bond dipoles in Table 1.4 depend on the difference in electronegativity of the bonded atoms and on the bond distance. The polarity of a C—H bond is relatively low; substantially less than a C—O bond, for example. Don't lose sight of an even more important difference between a C—H bond and a C—O bond, and that is the *direction*

TABLE 1.4 Selected Bond Dipole Moments

Bond*	Dipole moment, D	Bond*	Dipole moment, D
H—F	1.7	C—F	1.4
H—Cl	1.1	C—O	0.7
H—Br	0.8	C—N	0.4
H—I	0.4	C=O	2.4
H—C	0.3	C=N	1.4
H—N	1.3	C≡N	3.6
H—O	1.5		

*The direction of the dipole moment is toward the more electronegative atom. In the listed examples hydrogen and carbon are the positive ends of the dipoles. Carbon is the negative end of the dipole associated with the C—H bond.

Electrostatic Potential Maps

All of the material in this text, and most of chemistry generally, can be understood on the basis of what physicists call the *electromagnetic force.* Its major principle is that opposite charges attract and like charges repel. As you learn organic chemistry, a good way to start to connect structure to properties such as chemical reactivity is to find the positive part of one molecule and the negative part of another. Most of the time, these will be the reactive sites.

Imagine that you bring a positive charge toward a molecule. The interaction between that positive charge and some point in the molecule will be attractive if the point is negatively charged, repulsive if it is positively charged, and the strength of the interaction will depend on the magnitude of the charge. Computational methods make it possible to calculate and map these interactions. It is convenient to display this map using the colors of the rainbow from red to blue. Red is the negative (electron-rich) end and blue is the positive (electron-poor) end.

The electrostatic potential map of hydrogen fluoride (HF) was shown in the preceding section and is repeated here. Compare it with the electrostatic potential map of lithium hydride (LiH).

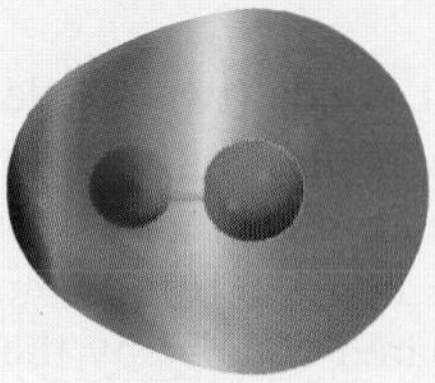

H—F

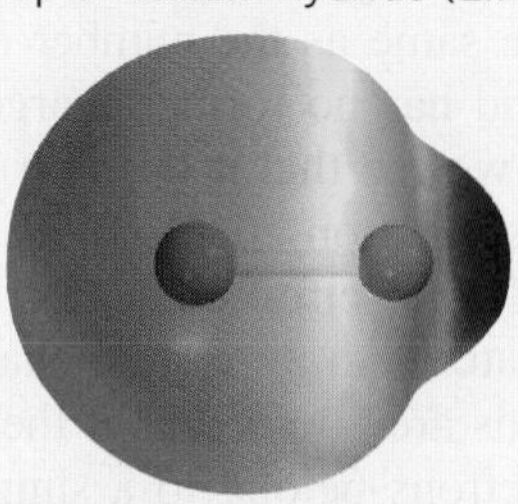

H—Li

The H—F bond is polarized so that hydrogen is partially positive (blue) and fluorine partially negative (red). Because hydrogen is more electronegative than lithium, the H—Li bond is polarized in the opposite sense, making hydrogen partially negative (red) and lithium partially positive (blue).

We will use electrostatic potential maps often to illustrate charge distribution in both organic and inorganic molecules. However, we need to offer one cautionary note. Electrostatic potential mapping within a single molecule is fine, but we need to be careful when comparing maps of different molecules. The reason for this is that the entire red-to-blue palette is used to map the electrostatic potential regardless of whether the charge difference is large or small. This is apparent in the H—F and H—Li electrostatic potential maps just shown. If, as shown in the following map, we use the same range for H—F that was used for H—Li we see that H is green instead of blue and the red of F is less intense.

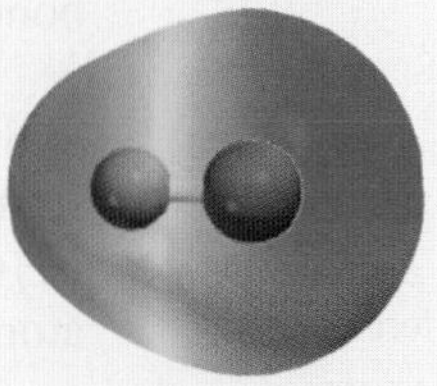

H—F

Thus, electrostatic potential maps can give an exaggerated picture of the charge distribution when the entire palette of colors is used. In most cases, that won't matter to us inasmuch as we are mostly concerned with the distribution within a single molecule. When we want to compare trends in a series of molecules, we'll use a common scale and will point that out. For example, the electrostatic potentials of H_2, F_2, and HF that were compared on page 11 were mapped using the same color scale.

of the dipole moment. In a C—H bond the electrons are drawn away from H, toward C. In a C—O bond, electrons are drawn from C toward O. As we'll see in later chapters, the kinds of reactions that a substance undergoes can often be related to the size and direction of key bond dipoles.

Problem 1.10

(a) Indicate the direction of the dipole for the following bonds using the symbol

$+\!\!\!\longrightarrow$

H—O H—N C—O C=O C—N C=N C≡N

(b) Indicate the direction of each dipole using $\delta+$ and $\delta-$.

1.6 Formal Charge

Lewis structures frequently contain atoms that bear a positive or negative charge. If the molecule as a whole is neutral, the sum of its positive charges must equal the sum of its negative charges. An example is nitric acid, HNO_3.

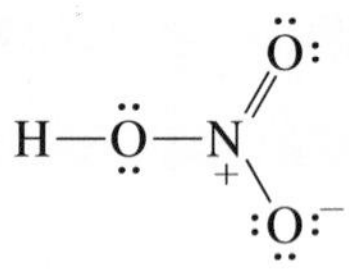

As written, the Lewis formula for nitric acid depicts different bonding patterns for its three oxygens. One oxygen is doubly bonded to nitrogen, another is singly bonded to both nitrogen and hydrogen, and the third has a single bond to nitrogen and a negative charge. Nitrogen is positively charged. The positive and negative charges are called **formal charges,** and the Lewis structure of nitric acid would be incomplete were they to be omitted.

The number of valence electrons in an atom of a main-group element such as nitrogen is equal to its group number. In the case of nitrogen this is five.

We calculate formal charges by counting the number of electrons "owned" by each atom in a Lewis structure and comparing this *electron count* with that of the neutral atom. Figure 1.5 illustrates how electrons are counted for each atom in nitric acid. Counting electrons for the purpose of calculating the formal charge differs from counting electrons to see if the octet rule is satisfied. A second-row element has a filled valence shell if the sum of all the electrons, shared and unshared, is eight. Electrons that connect two atoms by a covalent bond count toward filling the valence shell of both atoms. When calculating the formal charge, however, only half the number of electrons in covalent bonds can be considered to be "owned" by an atom.

It will always be true that a covalently bonded hydrogen has no formal charge (formal charge = 0).

To illustrate, let's start with the hydrogen of nitric acid. As shown in Figure 1.5, hydrogen is associated with only two electrons: those in its covalent bond to oxygen. It shares those two electrons with oxygen, and so the electron count of each hydrogen is $\frac{1}{2}(2) = 1$. Because this is the same as the number of electrons in a neutral hydrogen atom, the hydrogen in nitric acid has no formal charge.

It will always be true that a nitrogen with four covalent bonds has a formal charge of +1. (A nitrogen with four covalent bonds cannot have unshared pairs, because of the octet rule.)

Moving now to nitrogen, we see that it has four covalent bonds (two single bonds + one double bond), and so its electron count is $\frac{1}{2}(8) = 4$. A neutral nitrogen has five electrons in its valence shell. The electron count for nitrogen in nitric acid is one less than that of a neutral nitrogen atom, so its formal charge is +1.

It will always be true that an oxygen with two covalent bonds and two unshared pairs has no formal charge.

Electrons in covalent bonds are counted as if they are shared equally by the atoms they connect, but unshared electrons belong to a single atom. Thus, the oxygen that is doubly bonded to nitrogen has an electron count of six (four electrons as two unshared pairs + two electrons from the double bond). Because this is the same as a neutral oxygen atom, its formal charge is 0. Similarly, the OH oxygen has two bonds plus two unshared electron pairs, giving it an electron count of six and no formal charge.

It will always be true that an oxygen with one covalent bond and three unshared pairs has a formal charge of −1.

The green oxygen in Figure 1.5 owns three unshared pairs (six electrons) and shares two electrons with nitrogen to give it an electron count of seven. This is one more than the number of electrons in the valence shell of an oxygen atom, and so its formal charge is −1.

The method described for calculating formal charge has been one of reasoning through a series of logical steps. It can be reduced to the following equation:

$$\text{Formal charge} = \text{Group number in periodic table} - \text{Electron count}$$

where

$$\text{Electron count} = \tfrac{1}{2}(\text{Number of shared electrons}) + \text{Number of unshared electrons}$$

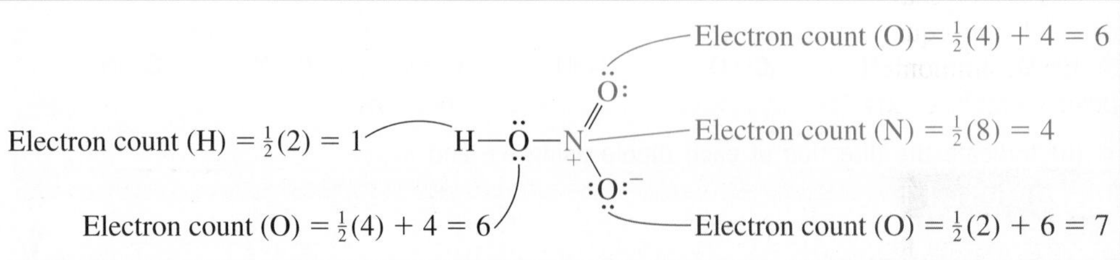
Electron count (O) = ½(4) + 4 = 6
Electron count (H) = ½(2) = 1
Electron count (N) = ½(8) = 4
Electron count (O) = ½(4) + 4 = 6
Electron count (O) = ½(2) + 6 = 7

Figure 1.5

Counting electrons in nitric acid. The electron count of each atom is equal to half the number of electrons it shares in covalent bonds plus the number of electrons in its own unshared pairs.

Problem 1.11

Like nitric acid, each of the following inorganic compounds will be frequently encountered in this text. Calculate the formal charge on each of the atoms in the Lewis structures given.

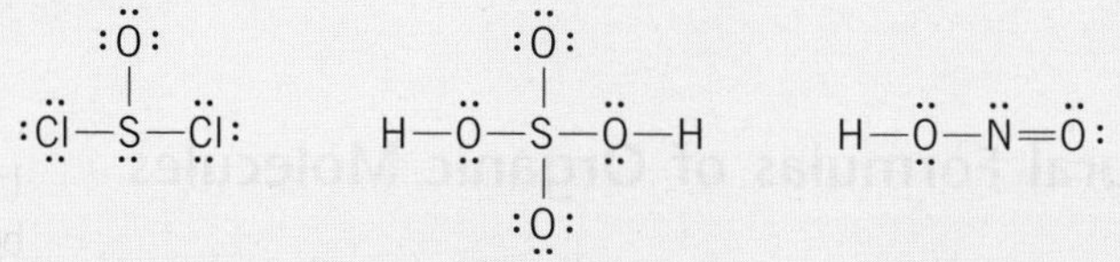

(a) Thionyl chloride (b) Sulfuric acid (c) Nitrous acid

Sample Solution (a) The formal charge is the difference between the number of valence electrons in the neutral atom and the electron count in the Lewis structure. (The number of valence electrons is the same as the group number in the periodic table for the main-group elements.)

	Valence electrons of neutral atom	Electron count	Formal charge
Sulfur:	6	$\frac{1}{2}(6) + 2 = 5$	+1
Oxygen:	6	$\frac{1}{2}(2) + 6 = 7$	−1
Chlorine:	7	$\frac{1}{2}(2) + 6 = 7$	0

The formal charges are shown in the Lewis structure of thionyl chloride as 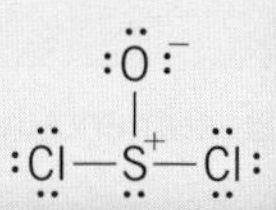

So far we've only considered neutral molecules—those in which the sums of the positive and negative formal charges were equal. With ions, of course, these sums will not be equal. Ammonium cation and borohydride anion, for example, are ions with net charges of +1 and −1, respectively. Nitrogen has a formal charge of +1 in ammonium ion, and boron has a formal charge of −1 in borohydride. None of the hydrogens in the Lewis structures shown for these ions bears a formal charge.

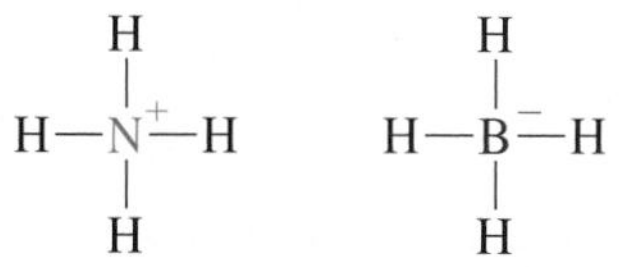

Ammonium ion Borohydride ion

Problem 1.12

Verify that the formal charges on nitrogen in ammonium ion and boron in borohydride ion are as shown.

Formal charges are based on Lewis structures in which electrons are considered to be shared equally between covalently bonded atoms. Actually, polarization of N—H bonds in ammonium ion and of B—H bonds in borohydride leads to some transfer of positive and negative charge, respectively, to the hydrogens.

Problem 1.13

Calculate the formal charge on each nitrogen in the following Lewis structure (azide ion) and the net charge on the species.

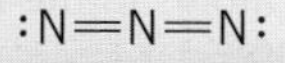

Determining formal charges on individual atoms of Lewis structures is an important element in good "electron bookkeeping." So much of organic chemistry can be made more understandable by keeping track of electrons that it is worth taking some time at the beginning to become proficient at the seemingly simple task of counting them.

1.7 Structural Formulas of Organic Molecules

Most organic compounds are more complicated than the examples we've seen so far and require a more systematic approach to writing structural formulas for them. The approach outlined in Table 1.5 begins (step 1) with the **molecular formula** that tells us which atoms and how many of each are present in the compound. From the molecular formula we calculate the number of valence electrons (step 2).

TABLE 1.5 A Systematic Approach to Writing Lewis Structures

Step	Illustration
1. The molecular formula is determined experimentally.	Ethanol and dimethyl ether both have the molecular formula C_2H_6O.
2. Based on the molecular formula, count the number of valence electrons.	In C_2H_6O, each hydrogen contributes 1 valence electron, each carbon contributes 4, and oxygen contributes 6 for a total of 20.
3. Given the connectivity, connect bonded atoms by a shared electron pair bond (:) represented by a dash (—).	Oxygen and the two carbons are connected in the order CCO in ethanol and COC in dimethyl ether. The connectivity and the fact that carbon normally has four bonds in neutral molecules allow us to place the hydrogens of ethanol and dimethyl ether. 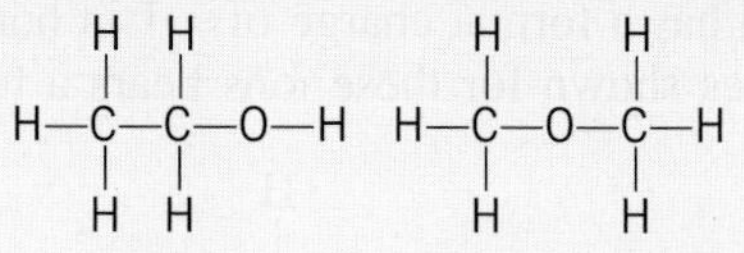 Ethanol Dimethyl ether
4. Count the number of electrons in the bonds (twice the number of bonds), and subtract this from the total number of valence electrons to give the number of electrons that remain to be added.	The structural formulas in step 3 contain eight bonds, accounting for 16 electrons. Because C_2H_6O contains 20 valence electrons, 4 more are needed.
5. Add electrons in pairs so that as many atoms as possible have eight electrons. It is usually best to begin with the most electronegative atom. (Hydrogen is limited to two electrons.) *Under no circumstances can a second-row element such as C, N, or O have more than eight valence electrons.*	Both carbons already have complete octets in the structures illustrated in step 3. The remaining four electrons are added to each oxygen as two unshared pairs to complete its octet. The Lewis structures are: 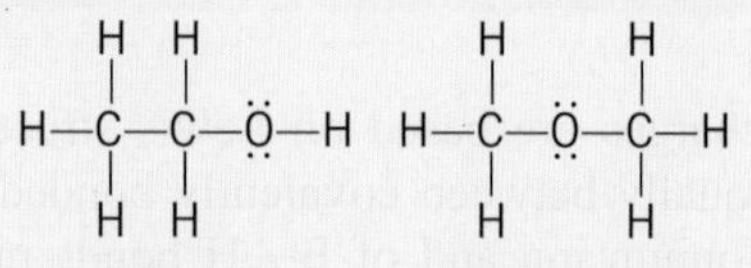 Ethanol Dimethyl ether
6. If one or more atoms (excluding hydrogens) has fewer than eight electrons, use an unshared pair from an adjacent atom to form a double or triple bond to complete the octet. Use one double bond for each deficiency of two electrons to complete the octet for each atom.	All the carbon and oxygen atoms in the structural formulas of ethanol and dimethyl ether have complete octets. No double bonds are needed.
7. Calculate formal charges.	None of the atoms in the Lewis structures shown in step 5 bears a formal charge.

In step 3 we set out a partial structure that shows the order in which the atoms are connected. This is called the **connectivity** of the molecule and is almost always determined by experiment. Most of the time carbon has four bonds, nitrogen has three, and oxygen two. It frequently happens in organic chemistry that two or more different compounds have the same molecular formula, but different connectivities. Ethanol and dimethyl ether—the examples shown in the table—are different compounds with different properties, yet have the same molecular formula (C_2H_6O). Ethanol is a liquid with a boiling point of 78°C. Dimethyl ether is a gas at room temperature; its boiling point is −24°C.

Different compounds that have the same molecular formula are classified as **isomers.** Isomers can be either **constitutional isomers** (differ in connectivity) or **stereoisomers** (differ in arrangement of atoms in space). Constitutional isomers are also sometimes called **structural isomers.** Ethanol and dimethyl ether are constitutional isomers of each other. Stereoisomers will be introduced in Section 3.11.

The framework of covalent bonds revealed by the connectivity information accounts for 16 of the 20 valence electrons in C_2H_6O (step 4). The remaining four valence electrons are assigned to each oxygen as two unshared pairs in step 5 to complete the Lewis structures of ethanol and dimethyl ether.

The suffix *-mer* in the word "isomer" is derived from the Greek word *meros,* meaning "part," "share," or "portion." The prefix *iso-* is also from Greek (*isos,* meaning "the same"). Thus isomers are different molecules that have the same parts (elemental composition).

Problem 1.14

Write structural formulas for all the constitutional isomers that have the given molecular formula.

(a) C_2H_7N (b) C_3H_7Cl (c) C_3H_8O

Sample Solution (a) The molecular formula C_2H_7N requires 20 valence electrons. Two carbons contribute a total of eight, nitrogen contributes five, and seven hydrogens contribute a total of seven. Nitrogen and two carbons can be connected in the order CNN or CNC. Assuming four bonds to each carbon and three to nitrogen, we write these connectivities as:

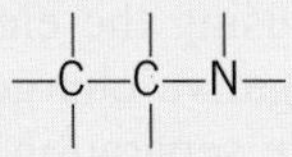

and

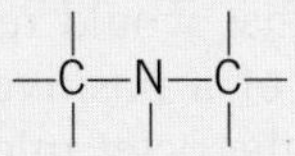

Place a hydrogen on each of the seven available bonds of each framework.

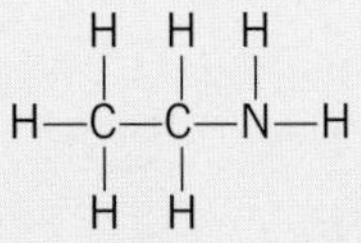

and

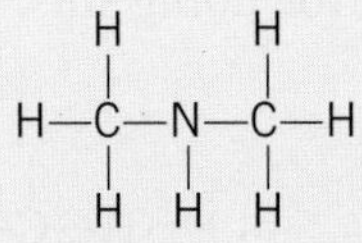

The nine bonds in each structural formula account for 18 electrons. Add an unshared pair to each nitrogen to complete its octet and give a total of 20 valence electrons as required by the molecular formula.

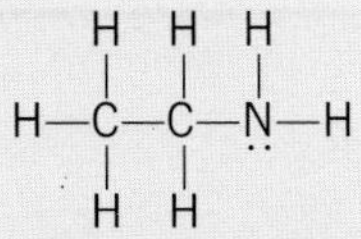

and

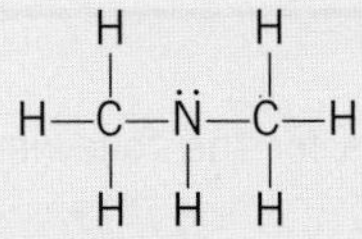

These two are constitutional isomers.

Now let's consider a slightly more complex molecule, methyl nitrite, in which we have to include multiple bonds when writing the Lewis structure (step 6). Methyl nitrite has the molecular formula CH_3NO_2. All the hydrogens are attached to the carbon, and the order of atom connections is CONO. First count the number of valence electrons.

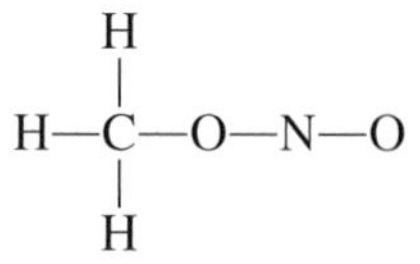

Each hydrogen contributes 1 valence electron, carbon 4, nitrogen 5, and each oxygen 6, for a total of 24. The partial structure shown contains 6 bonds equivalent to 12 electrons, so an additional 12 electrons must be added. Add these 12 electrons in pairs to oxygen and nitrogen. Both oxygens will end up with eight electrons, but nitrogen because it is less electronegative, will have only six.

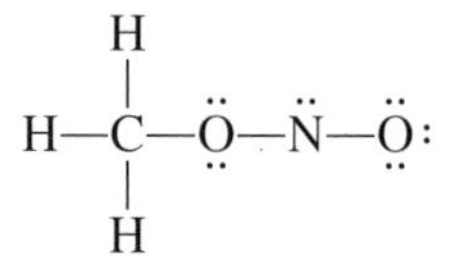

All atoms have complete octets except nitrogen, which has a deficiency of two electrons. Use an electron pair from the terminal oxygen to form a double bond to nitrogen to complete its octet. The resulting structure is the best (most stable) for methyl nitrite. All atoms (except hydrogen) have eight electrons in their valence shell.

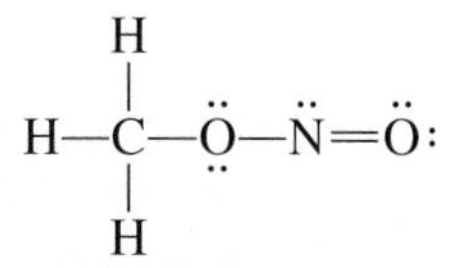

You may wonder why the electron pair from the terminal oxygen and not the oxygen in the middle was used to form the double bond to nitrogen in the Lewis structure. The structure that would result from using the electron pair on the middle oxygen has a separation of a positive and a negative charge. Although the resulting Lewis structure satisfies the octet rule, its charge separation makes it less stable than the other. When two or more Lewis structures satisfy the octet rule, we need to know how to choose the one that best represents the true structure. The following section describes a procedure for doing this.

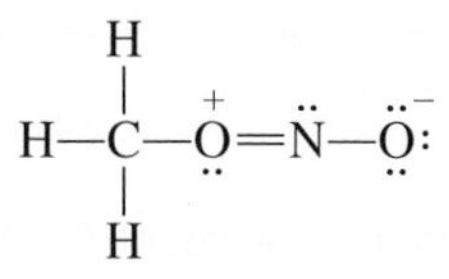

Problem 1.15

Write Lewis structures for the following compounds.

(a) 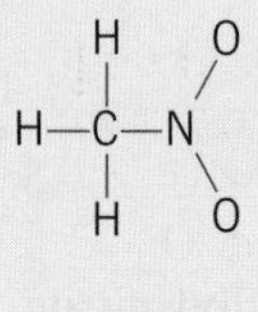

CH_3NO_2

(b) 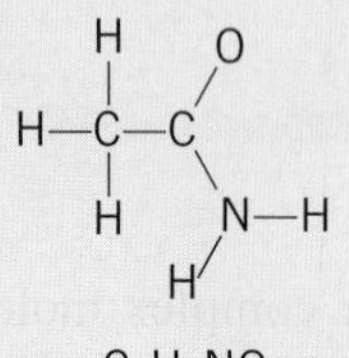

C_2H_5NO

(c) 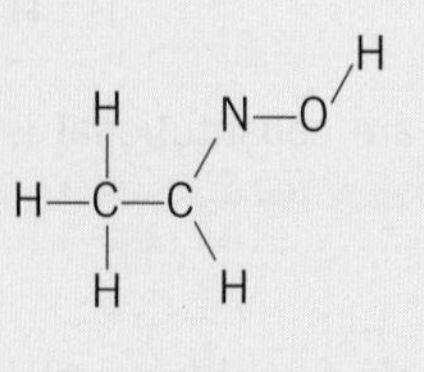

C_2H_5NO

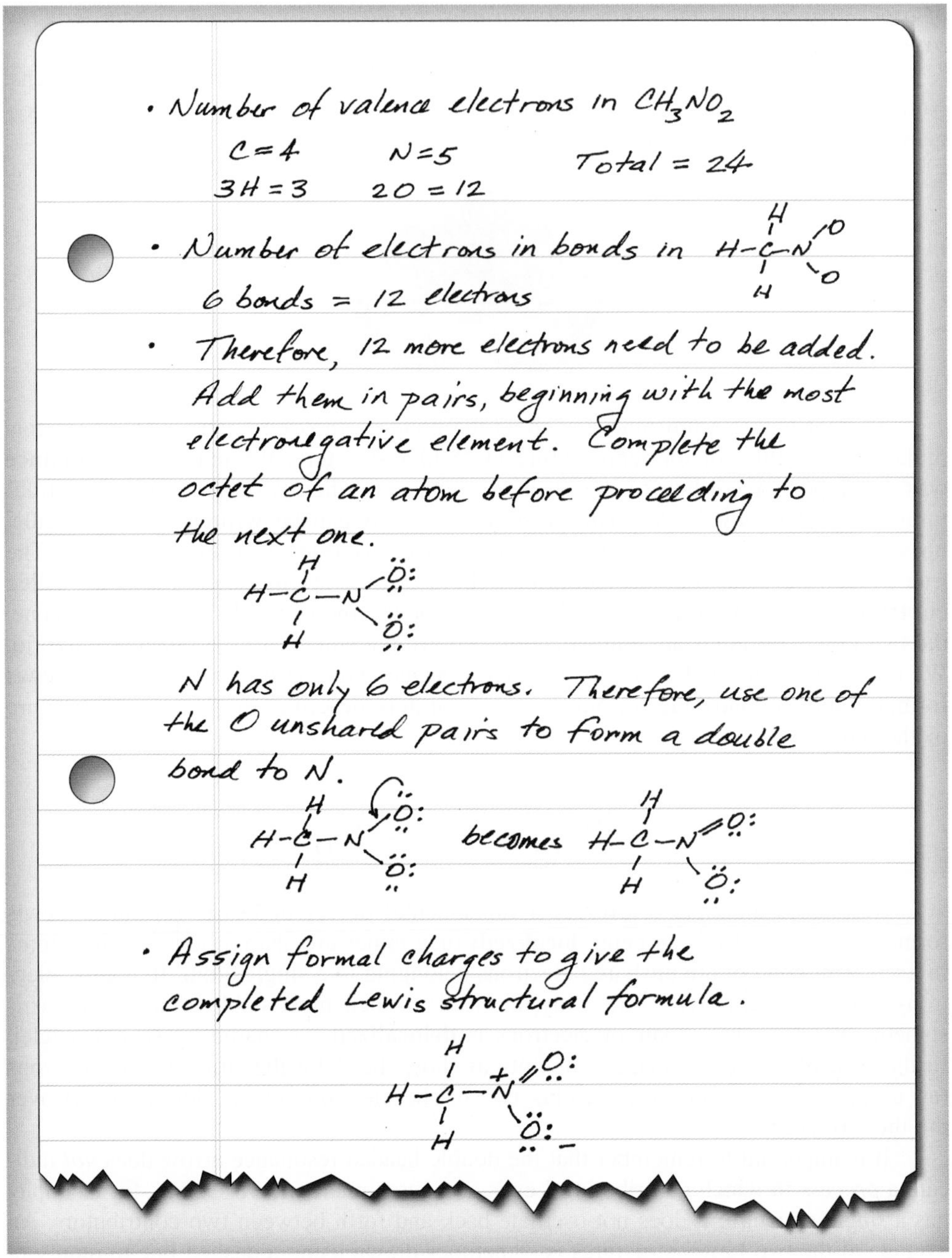
• Number of valence electrons in CH_3NO_2
C = 4   N = 5   Total = 24
3H = 3   2O = 12
• Number of electrons in bonds in
6 bonds = 12 electrons
• Therefore, 12 more electrons need to be added. Add them in pairs, beginning with the most electronegative element. Complete the octet of an atom before proceeding to the next one.
N has only 6 electrons. Therefore, use one of the O unshared pairs to form a double bond to N.
becomes
• Assign formal charges to give the completed Lewis structural formula.

1.8 Resonance

Sometimes more than one Lewis formula can be written for a molecule, especially if the molecule contains a double or triple bond. A simple example is ozone (O_3), for which we can write the Lewis structure:

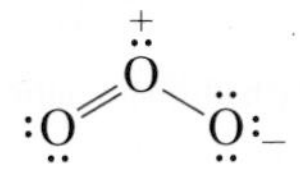

Ozone occurs naturally in large quantities in the upper atmosphere where it screens the surface of the Earth from much of the sun's ultraviolet rays.

The Lewis formula for ozone, however, is inconsistent with the experimentally determined structure. On the basis of the Lewis formula, we would expect ozone to have two different O—O bond lengths, one of them similar to the O—O single bond distance of 147 pm in hydrogen peroxide (HO—OH) and the other similar to the 121 pm O—O double bond distance in O_2. In fact, both bond distances are the same (128 pm)—somewhat shorter than a single bond, somewhat longer than a double bond. *The structure*

We will express bond distances in picometers (pm), which is an SI unit (1 pm = 10^{-12} m). To convert pm to angstrom units (1 Å = 10^{-10} m), divide by 100.

of ozone requires that the central oxygen must be identically bonded to both terminal oxygens.

An electrostatic potential map shows the equivalence of the two terminal oxygens. Notice, too, that the central oxygen is blue (positively charged) and both terminal oxygens are red (negatively charged).

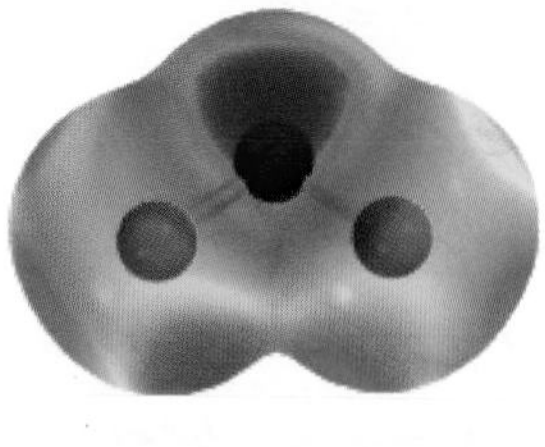

To deal with circumstances such as the bonding in ozone, yet retain Lewis formulas as a useful tool for representing molecular structure, the notion of **resonance** was developed. According to the resonance concept, when two or more Lewis structures that *differ only in the distribution of electrons* can be written for a molecule, no single Lewis structure is sufficient to describe its true electron distribution. The true structure is said to be a **resonance hybrid** of the various Lewis formulas, called **contributing structures,** that can be written for the molecule. In the case of ozone, the two Lewis formulas are equivalent and contribute equally to the resonance hybrid. We use a double-headed arrow to signify resonance and read it to mean that the Lewis formulas shown contribute to, but do not separately describe, the electronic structure of the molecule.

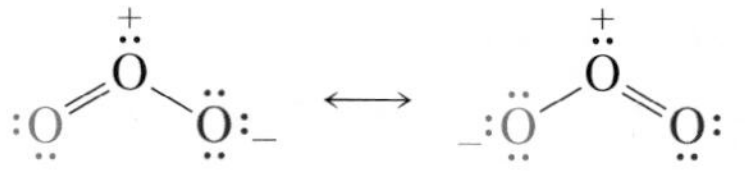

Resonance attempts to correct a fundamental defect in Lewis formulas. Lewis formulas show electrons as being **localized;** they either are shared between two atoms in a covalent bond or are unshared electrons belonging to a single atom. In reality, electrons distribute themselves in the way that leads to their most stable arrangement. This sometimes means that a pair of electrons is **delocalized,** or shared by several nuclei. In the case of ozone, resonance attempts to show the delocalization of four electrons (an unshared pair of one oxygen plus two of the electrons in the double bond) over the three oxygens.

It is important to remember that the double-headed resonance arrow does *not* indicate a *process* in which contributing Lewis structures interconvert. Ozone, for example, has a *single* structure; it does not oscillate back and forth between two contributors. An average of the two Lewis structures is sometimes drawn using a dashed line to represent a "partial" bond. In the dashed-line notation the central oxygen is linked to the other two by bonds that are halfway between a single bond and a double bond, and the terminal oxygens each bear one half of a unit negative charge. The structure below represents the resonance hybrid for ozone.

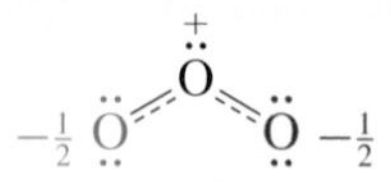

Dashed-line notation

Writing the various Lewis formulas that contribute to a resonance hybrid can be made easier by using **curved arrows** to keep track of delocalized electrons. We can convert one Lewis structure of ozone to another by moving electron pairs as shown:

The main use of curved arrows is to show electron flow in chemical reactions and will be described in Section 1.12.

Move electron pairs as shown by curved arrows 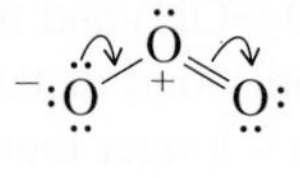 to transform one Lewis formula to another 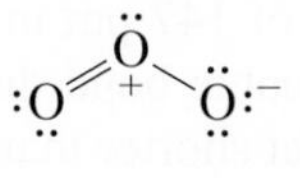

Curved arrows show the origin and destination of a pair of electrons. In the case of ozone, one arrow begins at an unshared pair and becomes the second half of a double bond. The other begins at a double bond and becomes an unshared pair of the other oxygen.

Problem 1.16

All of the bonds in the carbonate ion (CO_3^{2-}) are between C and O. Write Lewis structures for the major resonance contributors, and use curved arrows to show their relationship. Apply the resonance concept to explain why all of the C—O bond distances in carbonate are equal.

In most cases, the various resonance structures of a molecule are not equivalent and do not contribute equally to the resonance hybrid. The electron distribution in the molecule resembles that of its major contributor more closely than any of its alternative resonance structures. Therefore, it is important that we develop some generalizations concerning the factors that make one resonance form more important (more stable) than another. Table 1.6 outlines the structural features that alert us to situations when resonance needs to be considered and lists criteria for evaluating the relative importance of the contributing structures.

TABLE 1.6 Introduction to the Rules of Resonance

Rule	Illustration
I. *When can resonance be considered?*	
1. The connectivity must be the same in all contributing structures; only the electron positions may vary among the various contributing structures.	The Lewis formulas A and B are *not* resonance forms of the same compound. They are *isomers* (different compounds with the same molecular formula). 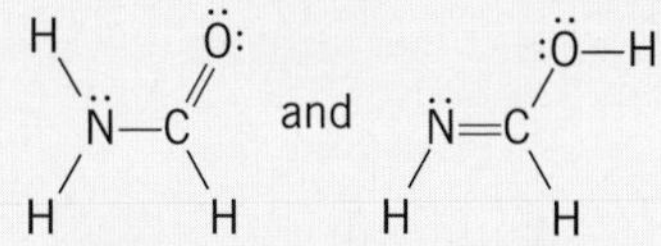 A B The Lewis formulas A, C, and D are resonance forms of a single compound. 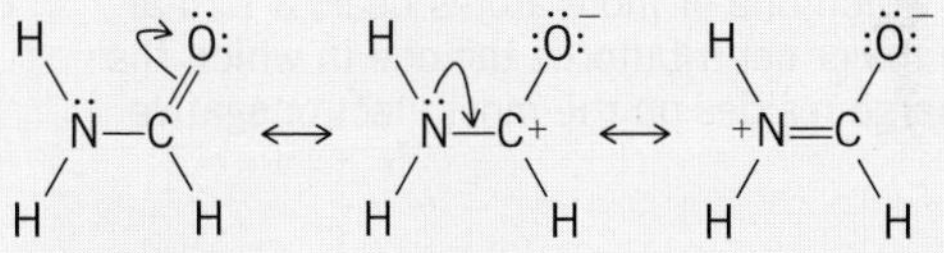 A C D
2. Each contributing structure must have the same number of electrons and the same *net* charge. The formal charges of individual atoms may vary among the various Lewis structures.	Structures A, C, and D (preceding example) all have 18 valence electrons and a net charge of 0, even though they differ in respect to formal charges on individual atoms. Structure E has 20 valence electrons and a net charge of −2. It is not a resonance structure of A, C, or D. 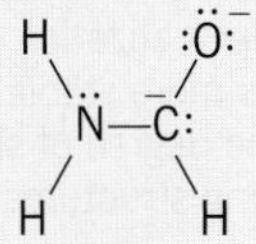 E
3. Each contributing structure must have the same number of *unpaired* electrons.	Structural formula F has the same atomic positions and the same number of electrons as A, C, and D, but is not a resonance form of any of them. F has two unpaired electrons; all the electrons in A, C, and D are paired. 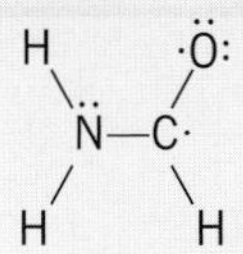 F

Continued

TABLE 1.6 Introduction to the Rules of Resonance (*Continued*)

Rule	Illustration
4. Contributing structures in which the octet rule is exceeded for second-row elements make no contribution. (The octet rule may be exceeded for elements beyond the second row.)	Lewis structures G and H are resonance contributors to the structure of nitromethane. Structural formula I is not a permissible Lewis structure because it has ten electrons around nitrogen. 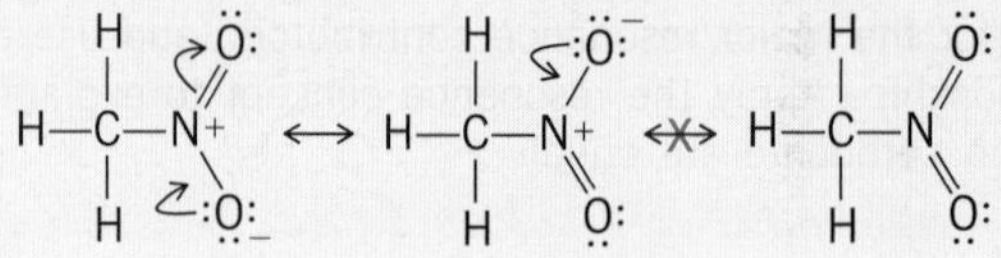 G H I
II. *Which resonance form contributes more?*	
5. As long as the octet rule is not exceeded for second-row elements, the contributing structure with the greater number of covalent bonds contributes more to the resonance hybrid. Maximizing the number of bonds and satisfying the octet rule normally go hand in hand. This rule is more important than rules 6 and 7.	Of the two Lewis structures for formaldehyde, the major contributor J has one more bond than the minor contributor K. 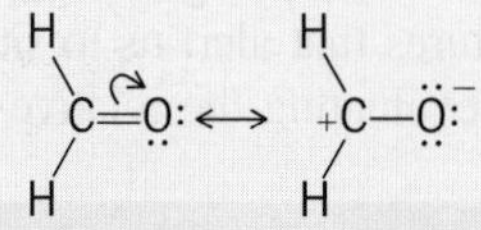 J (Major contributor) K (Minor contributor)
6. When two or more structures satisfy the octet rule, the major contributor is the one with the smallest separation of oppositely charged atoms.	The two structures L and M for methyl nitrite have the same number of bonds, but L is the major contributor because it lacks the separation of positive and negative charge that characterizes M. 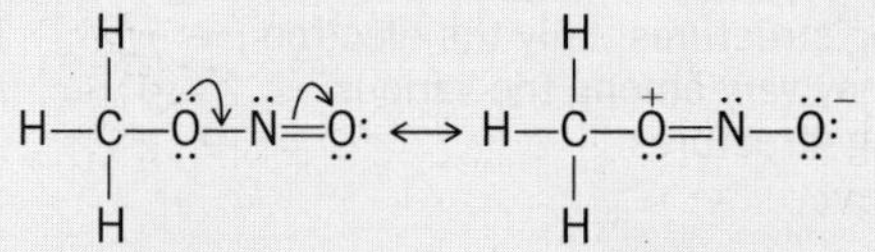 L (Major contributor) M (Minor contributor)
7. Among structural formulas that satisfy the octet rule and in which one or more atoms bears a formal charge, the major contributor is the one in which the negative charge resides on the most electronegative atom.	The major contributing structure for cyanate ion is N because the negative charge is on its oxygen. 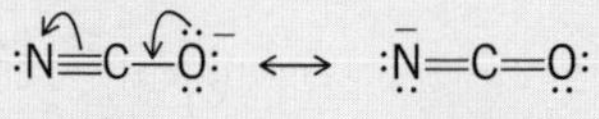 N (Major contributor) O (Minor contributor) In O the negative charge is on nitrogen. Oxygen is more electronegative than nitrogen and can better support a negative charge.
III. *What is the effect of resonance?*	
8. Electron delocalization stabilizes a molecule. Resonance is a way of showing electron delocalization. Therefore, the true electron distribution is more stable than any of the contributing structures. The degree of stabilization is greatest when the contributing structures are of equal stability.	Structures P, Q, and R for carbonate ion are equivalent and contribute equally to the electron distribution. The true structure of carbonate ion is a hybrid of P, Q, and R and is more stable than any of them. 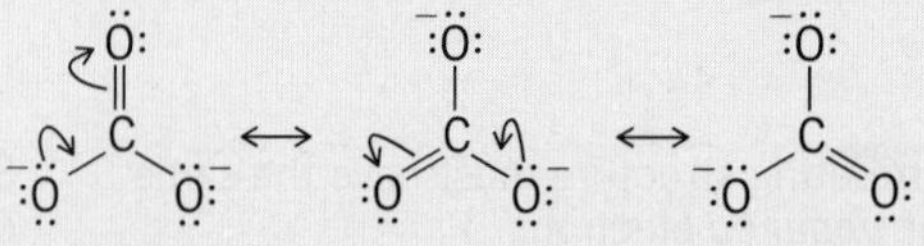 P Q R

Problem 1.17

Write the resonance structure obtained by moving electrons as indicated by the curved arrows. Compare the stabilities of the two Lewis structures according to the guidelines in Table 1.6. Are the two Lewis structures equally stable, or is one more stable than the other? Why?

Sample Solution (a) The curved arrow shows how we move an unshared electron pair assigned to oxygen so that it becomes shared by carbon and oxygen. This converts a single bond to a double bond and leads to a formal charge of +1 on oxygen.

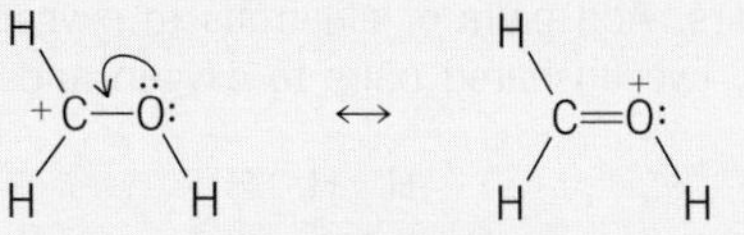

The Lewis structure on the right is more stable because it has one more covalent bond than the original structure. Carbon did not have an octet of electrons in the original structure, but the octet rule is satisfied for both carbon and oxygen in the new Lewis structure.

It is good chemical practice to represent molecules by their most stable Lewis structure. However, the ability to write alternative resonance forms and to assess their relative contributions can provide insight into both molecular structure and chemical behavior.

1.9 Writing Organic Structures

Organic chemists use a number of shortcuts to speed the writing of structural formulas. Sometimes we leave out unshared electron pairs, but only when we are sure of our ability to count electrons to know when they are present and when they are not. In **condensed formulas** we leave out some, many, or all of the covalent bonds and use subscripts to indicate the number of identical groups attached to a particular atom. These successive levels of simplification are illustrated for diethyl ether, written first as

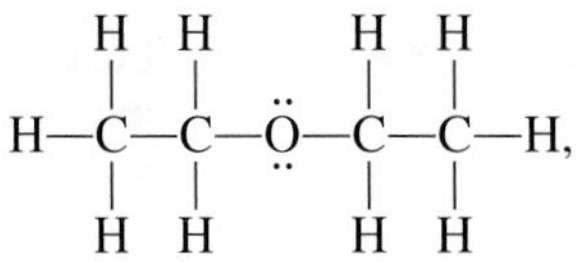 then $CH_3CH_2OCH_2CH_3$, then $(CH_3CH_2)_2O$.

Problem 1.18

Expand the following condensed formulas so as to show all the bonds and unshared electron pairs.

(a) $HOCH_2CH_2NH_2$
(b) $(CH_3)_3CH$
(c) $ClCH_2CH_2Cl$
(d) CH_3CHCl_2
(e) $CH_3NHCH_2CH_3$
(f) $(CH_3)_2CHCH{=}O$

Continued

Sample Solution (a) The molecule contains two carbon atoms, which are bonded to each other. Both carbons bear two hydrogens. One carbon bears the group HO—; the other is attached to $—NH_2$.

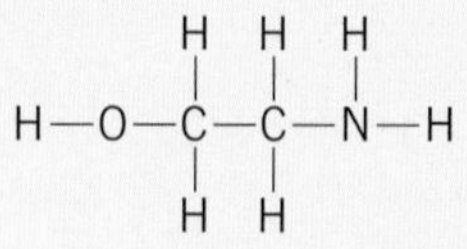

When showing connectivity, it is not necessary to concern yourself with the spatial orientation of the atoms. There are many other correct ways to represent the constitution shown. What is important is to show the connectivity OCCN (or its equivalent NCCO) and to have the correct number of hydrogens on each atom.

To locate unshared electron pairs, first count the total number of valence electrons brought to the molecule by its component atoms. Each hydrogen contributes 1, each carbon 4, nitrogen 5, and oxygen 6, for a total of 26. Ten bonds are revealed by the connectivity, accounting for 20 electrons; therefore 6 electrons must be contained in unshared pairs. Add pairs of electrons to oxygen and nitrogen so that their octets are complete, two unshared pairs to oxygen and one to nitrogen.

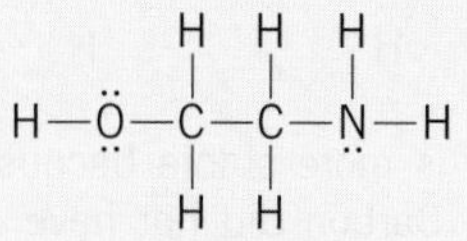

As you practice, you will begin to remember patterns of electron distribution. A neutral oxygen with two bonds has two unshared electron pairs. A neutral nitrogen with three bonds has one unshared pair.

With practice, drawing structural formulas for organic molecules soon becomes routine and can be simplified even more. For example, a chain of carbon atoms can be represented by drawing all of the C—C bonds while omitting individual carbon atoms. The resulting structural formulas are further simplified by stripping away the hydrogens.

$CH_3CH_2CH_2CH_3$ becomes 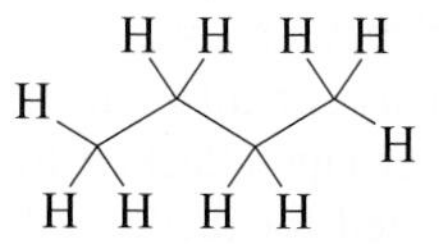 simplified to 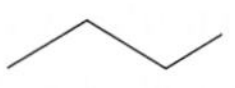

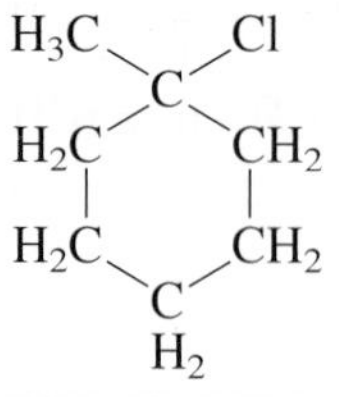 simplifies to 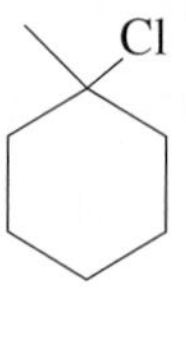

In the skeletal representations on the right, called a **bond-line formula** or **carbon skeletal diagram,** we assume that there is a carbon atom at every vertex and at the end of a line. Hydrogens attached to a vertex or to the end of a line are left out. When atoms that are neither carbon nor hydrogen, termed **heteroatoms,** are present, hydrogens attached to them are shown, as illustrated in the following examples. The second example is the amino acid isoleucine, which has the molecular formula $C_6H_{13}NO_2$.

$CH_3CH_2CH_2CH_2OH$ simplifies to 

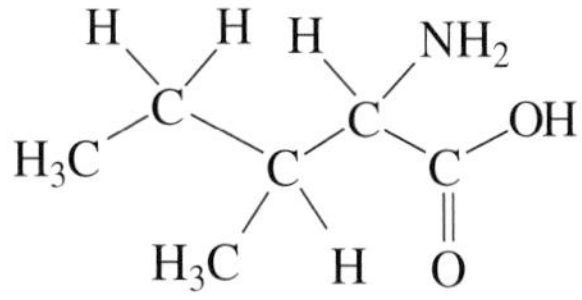

simplifies to

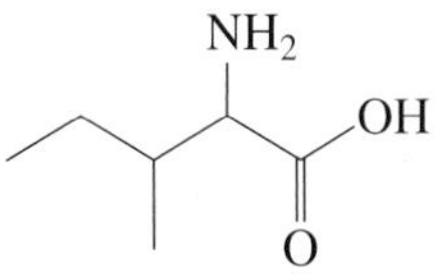

The structural language of organic chemistry has been developed so that complex molecules can be described in a clear, yet economical way. A molecule as complex as cholesterol can be drawn rapidly in a bond-line formula, while drawing even a condensed formula would require a prohibitive amount of time.

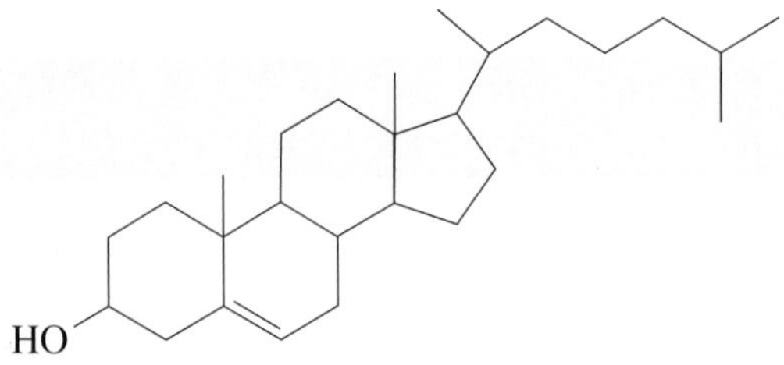

Cholesterol

Problem 1.19

Expand the following bond-line formulas to show all the atoms including carbon and hydrogen.

(a) 

(c) 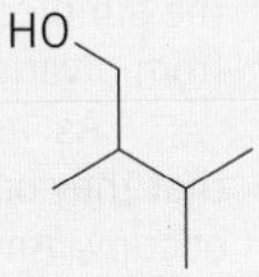

(b) 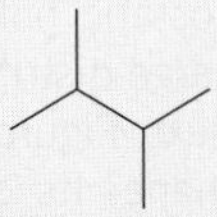

(d) 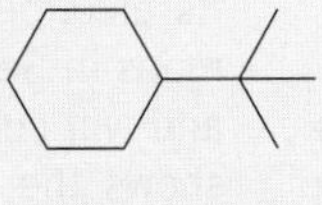

Sample Solution (a) There is a carbon at each bend in the chain and at the ends of the chain. Each of the ten carbon atoms bears the appropriate number of hydrogens to give it four bonds.

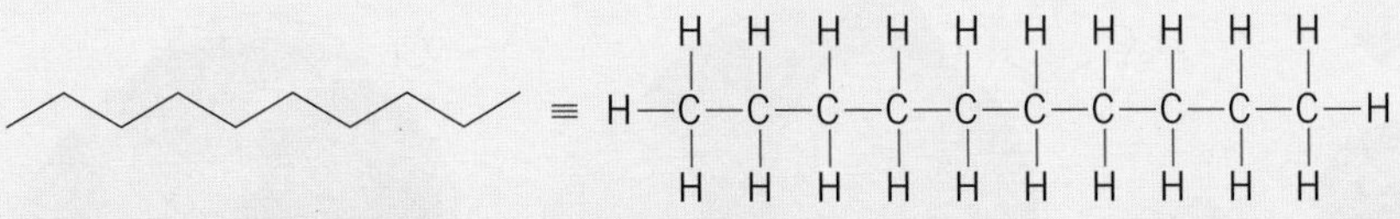

Alternatively, the structure could be written as $CH_3CH_2CH_2CH_2CH_2CH_2CH_2CH_2CH_2CH_3$ or in condensed form as $CH_3(CH_2)_8CH_3$.

Problem 1.20

(a) What is the molecular formula for each of the molecules in Problem 1.19?
(b) What is the molecular formula for cholesterol?

1.10 The Shapes of Some Simple Molecules

So far we have emphasized structure in terms of "electron bookkeeping." We now turn our attention to molecular geometry and will see how we can begin to connect the three-dimensional shape of a molecule to its Lewis formula. Table 1.7 lists some simple compounds illustrating the geometries that will be seen most often in our study of organic chemistry.

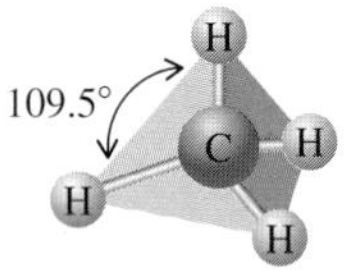

Tetrahedral methane

Methane (CH_4) is a tetrahedral molecule; its four hydrogens occupy the corners of a tetrahedron with carbon at its center. Several types of molecular models of methane are shown in Figure 1.6, and Table 1.7 recalls their tetrahedral geometry by way of a ball-and-spoke model. Table 1.7 also shows a common method of representing three-dimensionality through the use of different bond styles. A solid wedge (▬) stands for a

Molecular Modeling

As early as the nineteenth century many chemists built scale models to better understand molecular structure. We can gain a clearer idea about the features that affect structure and reactivity when we examine the three-dimensional shape of a molecule. Several types of molecular models are shown for methane in Figure 1.6. Probably the most familiar are ball-and-stick models (Figure 1.6*b*), which direct approximately equal attention to the atoms and the bonds that connect them. Framework models (Figure 1.6*a*) and space-filling models (Figure 1.6*c*) represent opposite extremes. Framework models emphasize the pattern of bonds of a molecule while ignoring the sizes of the atoms. Space-filling models emphasize the volume occupied by individual atoms at the cost of a clear depiction of the bonds; they are most useful in cases in which one wishes to examine the overall molecular shape and to assess how closely two nonbonded atoms approach each other.

The earliest ball-and-spoke models were exactly that: wooden balls in which holes were drilled to accommodate dowels that connected the atoms. Plastic versions, including relatively inexpensive student sets, became available in the 1960s and proved to be a valuable learning aid. Precisely scaled stainless steel framework and plastic space-filling models, although relatively expensive, were once standard equipment in most research laboratories.

Computer graphics-based representations are rapidly replacing classical molecular models. Indeed, the term "molecular modeling" as now used in organic chemistry implies computer generation of models. The methane models shown in Figure 1.6 were all drawn on a personal computer using software that possesses the feature of displaying and printing the same molecule in framework, ball-and-spoke, and space-filling formats. In addition to permitting models to be constructed rapidly, even the simplest software allows the model to be turned and viewed from a variety of perspectives.

As useful as molecular models are, they are limited in that they only show the location of the atoms and the space they occupy. Another important dimension to molecular structure is its electron distribution. We introduced electrostatic potential maps in Section 1.5 as a way of illustrating charge distribution and will continue to use them throughout the text. Figure 1.6*d* shows the electrostatic potential map of methane. Its overall shape is similar to the volume occupied by the space-filling model. The most electron-rich regions are closer to carbon and the most electron-poor ones are closer to the hydrogens.

(*a*) (*b*) (*c*) (*d*)

Figure 1.6

Molecular models of methane (CH_4). (*a*) Framework (tube) models show the bonds connecting the atoms, but not the atoms themselves. (*b*) Ball-and-stick (ball-and-spoke) models show the atoms as balls and the bonds as rods. (*c*) Space-filling models portray overall molecular size; the radius of each sphere approximates the van der Waals radius of the atom. (*d*) An electrostatic potential map of methane.

TABLE 1.7 VSEPR and Molecular Geometry

Compound	Structural formula	Repulsive electron pairs	Arrangement of electron pairs	Molecular shape	Molecular model
Methane (CH_4)	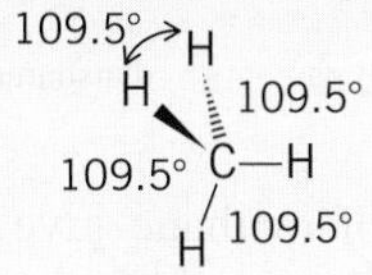109.5° H H 109.5° 109.5° C—H H 109.5°	Carbon has four bonded pairs	Tetrahedral	**Tetrahedral**	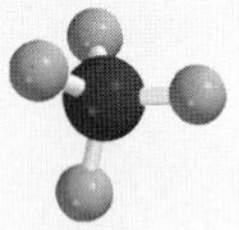
Water (H_2O)	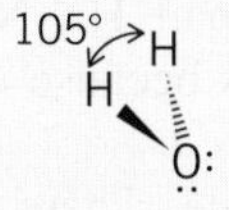105° H H O:	Oxygen has two bonded pairs + two unshared pairs	Tetrahedral	**Bent**	
Ammonia (NH_3)	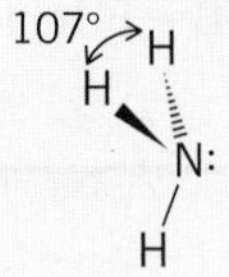107° H H N: H	Nitrogen has three bonded pairs + one unshared pair	Tetrahedral	**Trigonal pyramidal**	
Boron trifluoride (BF_3)	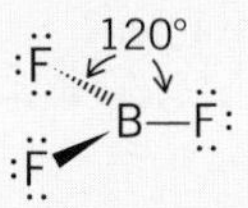120° :F B—F: :F	Boron has three bonded pairs	Trigonal planar	**Trigonal planar**	
Formaldehyde (H_2CO)	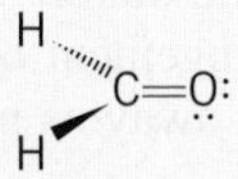H C=O: H	Carbon has two bonded pairs + one double bond, which is counted as one bonded pair	Trigonal planar	**Trigonal planar**	
Carbon dioxide (CO_2)	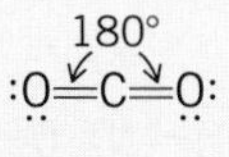180° :O=C=O:	Carbon has two double bonds, which are counted as two bonded pairs	Linear	**Linear**	

bond that projects toward you, a dashed wedge (⋯) for one that points away from you, and a simple line (—) for a bond that lies in the plane of the paper.

The tetrahedral geometry of methane is often explained with the **valence shell electron-pair repulsion (VSEPR) model.** The VSEPR model rests on the idea that an electron pair, either a bonded pair or an unshared pair, associated with a particular atom will be as far away from the atom's other electron pairs as possible. Thus, a tetrahedral geometry permits the four bonds of methane to be maximally separated and is characterized by H—C—H angles of 109.5°, a value referred to as the **tetrahedral angle.**

Although reservations have been expressed concerning VSEPR as an *explanation* for molecular geometries, it remains a useful *tool* for predicting the shapes of organic compounds.

Water, ammonia, and methane share the common feature of an approximately tetrahedral arrangement of four electron pairs. Because we describe the shape of a molecule according to the positions of its atoms only rather than by the orientation of its electron pairs, water is said to be *bent,* and ammonia is *trigonal pyramidal.*

The H—O—H angle in water (105°) and the H—N—H angles in ammonia (107°) are slightly smaller than the tetrahedral angle. These bond-angle contractions are easily accommodated by VSEPR by reasoning that bonded pairs take up less space than unshared pairs. A bonded pair feels the attractive force of two nuclei and is held more tightly than an unshared pair localized on a single atom. Thus, repulsive forces increase in the order:

Increasing force of repulsion between electron pairs

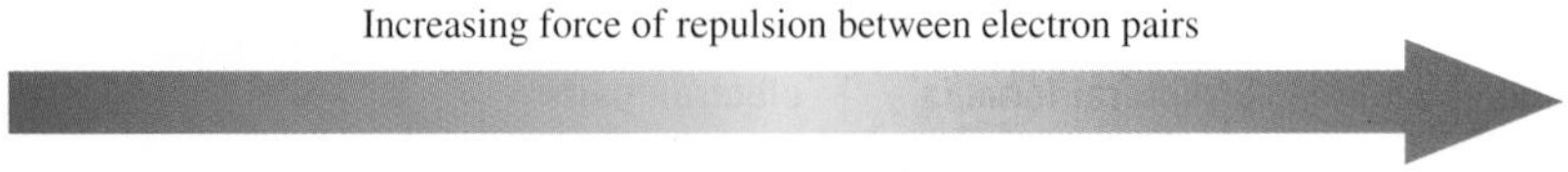

Bonded pair-bonded pair < Unshared pair-bonded pair < Unshared pair-unshared pair
Least repulsive — Most repulsive

Repulsions among the four bonded pairs of methane give the normal tetrahedral angle of 109.5°. Repulsions among the unshared pair of nitrogen in ammonia and the three bonded pairs cause the bonded pair-bonded pair H—N—H angles to be smaller than 109.5°. In water, a larger repulsive force exists because of two unshared pairs, and the H—O—H angle is compressed further to 105°.

Boron trifluoride is a *trigonal planar* molecule. There are six electrons, two for each B—F bond, associated with the valence shell of boron. These three bonded pairs are farthest apart when they are coplanar, with F—B—F bond angles of 120°.

Problem 1.21

The salt sodium borohydride, $NaBH_4$, has an ionic bond between Na^+ and the anion BH_4^-. What are the H—B—H angles in borohydride anion?

Multiple bonds are treated as a single unit in the VSEPR model. Formaldehyde is a trigonal planar molecule in which the electrons of the double bond and those of the two single bonds are maximally separated. A linear arrangement of atoms in carbon dioxide allows the electrons in one double bond to be as far away as possible from the electrons in the other double bond.

Problem 1.22

Specify the shape of the following:

(a) H—C≡N: (Hydrogen cyanide)
(b) H_4N^+ (Ammonium ion)
(c) $:\overset{-}{\ddot{N}}=\overset{+}{N}=\overset{-}{\ddot{N}}:$ (Azide ion)
(d) CO_3^{2-} (Carbonate ion)

Sample Solution (a) The structure shown accounts for all the electrons in hydrogen cyanide. No unshared electron pairs are associated with carbon, and so the structure is determined by maximizing the separation between its single bond to hydrogen and the triple bond to nitrogen. Hydrogen cyanide is a *linear* molecule.

1.11 Molecular Dipole Moments

We can combine our knowledge of molecular geometry with a feel for the polarity of chemical bonds to predict whether a molecule has a dipole moment or not. The **molecular dipole moment** is the resultant of all of the individual bond dipole moments of a substance. Some molecules, such as carbon dioxide, have polar bonds, but lack a dipole moment because their geometry causes the individual C=O bond dipoles to cancel.

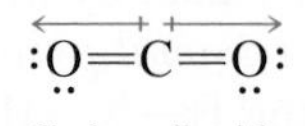

Dipole moment = 0 D

Carbon dioxide

Carbon tetrachloride, with four polar C—Cl bonds and a tetrahedral shape, has no net dipole moment, because the result of the four bond dipoles, as shown in Figure 1.7, is

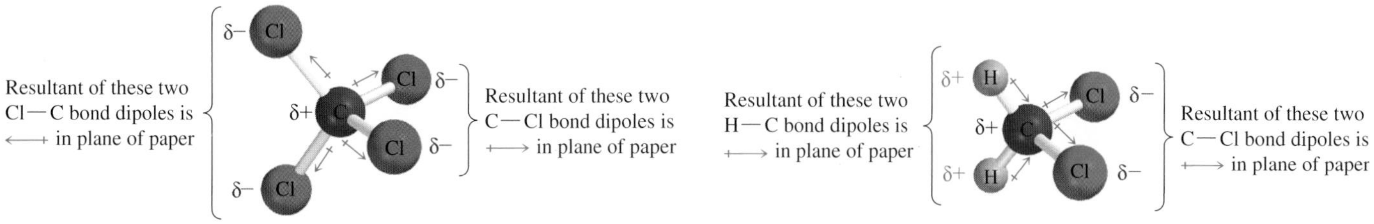

(*a*) There is a mutual cancellation of individual bond dipoles in carbon tetrachloride. It has no dipole moment.

(*b*) The H—C bond dipoles reinforce the C—Cl bond moment in dichloromethane. The molecule has a dipole moment of 1.62 D.

Figure 1.7

Contribution of individual bond dipole moments to the molecular dipole moments of (*a*) carbon tetrachloride (CCl_4) and (*b*) dichloromethane (CH_2Cl_2).

zero. Dichloromethane, on the other hand, has a dipole moment of 1.62 D. The C—H bond dipoles reinforce the C—Cl bond dipoles.

Problem 1.23

Which of the following compounds would you expect to have a dipole moment? If the molecule has a dipole moment, specify its direction.

(a) BF_3 (c) CH_4 (e) CH_2O

(b) H_2O (d) CH_3Cl (f) HCN

Sample Solution (a) Boron trifluoride is planar with 120° bond angles. Although each boron–fluorine bond is polar, their combined effects cancel and the molecule has no dipole moment.

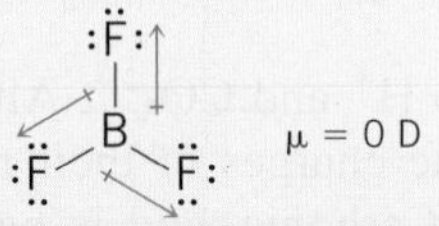

The opening paragraph of this chapter emphasized that the connection between structure and properties is central to understanding organic chemistry. We have just seen one such connection. From the Lewis structure of a molecule, we can use electronegativity to tell us about the polarity of bonds and combine that with VSEPR to predict whether the molecule has a dipole moment. In the next several sections we'll see a connection between structure and *chemical reactivity* as we review acids and bases.

1.12 Curved Arrows and Chemical Reactions

In Section 1.8 we introduced curved arrows as a tool for systematically converting one resonance contributor to another. Their more common use is to track electron flow in chemical reactions. Organic chemistry involves a vast number of reactions—far too many to memorize! By learning to track electron flow, you will be able to see trends in reactivity and to predict the outcome of almost any given reaction.

There are two kinds of curved arrows. A double-barbed arrow (⤻) shows the movement of a *pair* of electrons, either a bonded pair or a lone pair. A single-barbed, or fishhook, arrow (⤻) shows the movement of *one* electron. For now, we'll concern ourselves only with reactions that involve electron pairs and focus on double-barbed arrows.

We'll start with some simple examples—reactions involving only one electron pair. Suppose the molecule A—B dissociates to cation A^+ and anion B^-. A chemical equation for this ionization could be written as:

$$AB \longrightarrow A^+ + B^-$$

Alternatively, we could write:

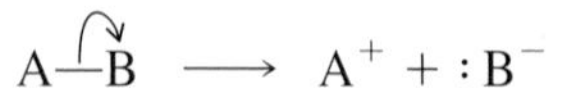

The reaction is the same but the second equation provides more information by including the bond that is broken during ionization and showing the flow of electrons. The curved arrow begins where the electrons are originally—in the bond—and points to atom B as their destination where they become an unshared pair of the anion B^-.

To illustrate, the ionization of carbonic acid corresponds to that just described for A—B. A neutral molecule (H_2CO_3) dissociates to a cation (H^+) and an anion (HCO_3^-).

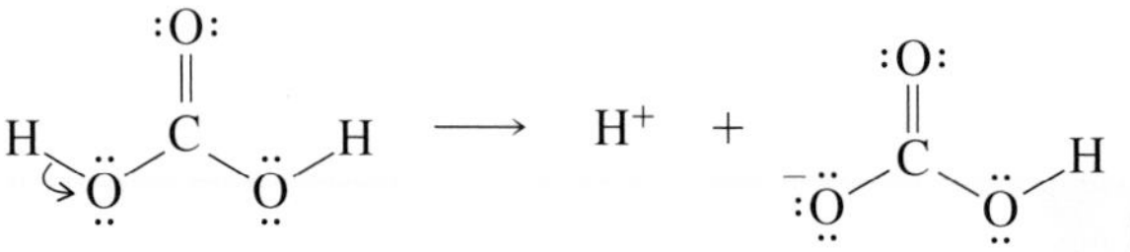

Carbonic acid | Proton | Hydrogen carbonate ion

More generally, the reactant need not always be a neutral molecule. It can be an ion such as the hydrogen carbonate ion produced in the preceding reaction.

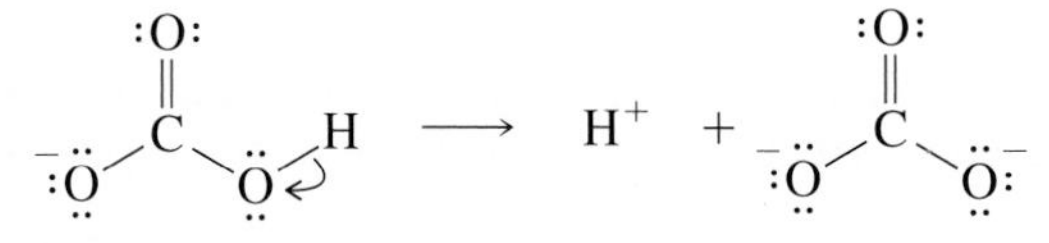

Hydrogen carbonate ion | Proton | Carbonate ion

In this case, HCO_3^- dissociates to H^+ and CO_3^{2-}. Although the dissociations of H_2CO_3 and HCO_3^- differ in respect to the charges of their reactants and products, the overall net charge is conserved in both. Carbonic acid is neutral and dissociates to a +1 ion and a −1 ion. The net charge is 0 for both reactant and products. Hydrogen carbonate has a charge of −1 and dissociates to a +1 ion and a −2 ion. The net charge is −1 on both sides of the equation. Charge, as well as mass, is conserved in all chemical reactions.

Problem 1.24

Using the curved arrow to guide your reasoning, show the products of the following dissociations. Include formal charges and unshared electron pairs. Check your answers to ensure that charge is conserved. [The reaction in (a) will reappear in an important way in Section 4.8; the reaction in (b) reappears in Section 8.6.]

Sample Solution (a) The curved arrow tells us that the C—O bond breaks and the pair of electrons in that bond becomes an unshared electron pair of oxygen.

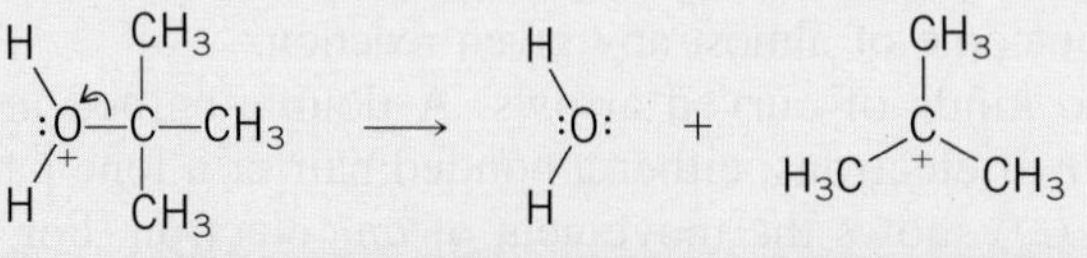

Water is one product of the reaction. The organic species produced is a cation. Its central carbon has six electrons in its valence shell and a formal charge of +1. Charge is conserved in the reaction. The net charge on both the left and right side of the equation is +1.

The reverse of a dissociation is a combination, such as the formation of a covalent bond between a cation A^+ and an anion B^-.

$$A^+ + :B^- \longrightarrow A\text{—}B$$

Here the curved arrow begins at the unshared pair of B^- and points to A^+. *Electrons flow from sites of higher electron density to lower.* The unshared electron pair of B^- becomes the shared electron pair in the A—B bond. Charge is conserved.

Problem 1.25

Write equations, including curved arrows, describing the reverse reactions of Problem 1.24.

Sample Solution (a) First write the equation for the reverse process. Next, use a curved arrow to show that the electron pair in the C—O bond in the product originates as an unshared electron pair of oxygen in water.

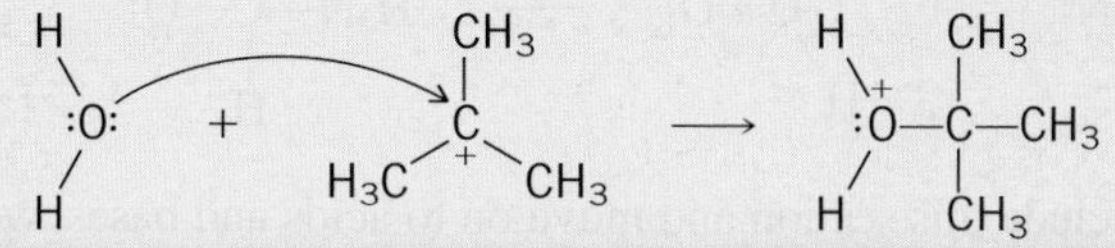

Many reactions combine bond making with bond breaking and require more than one curved arrow.

$$^-A: + B\text{—}C \longrightarrow A\text{—}B + :C^-$$

An example is a reaction that will be discussed in detail in Section 8.3.

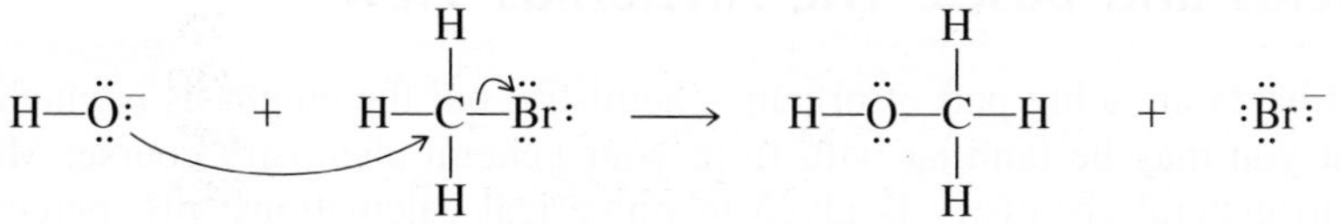

An unshared electron pair of a negatively charged oxygen becomes a shared electron pair in a C—O bond. Again, notice that electrons flow from electron-rich to electron-poor sites. Hydroxide ion is negatively charged and, therefore, electron-rich while the carbon of H_3CBr is partially positive because of the polarization of the C—Br bond (Section 1.5).

A very common process involves transfer of a proton from one atom to another. An example is the reaction that occurs when hydrogen bromide gas dissolves in water.

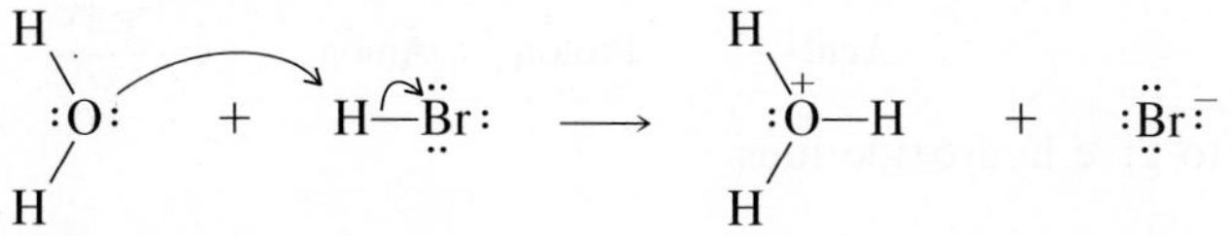

Numerous other proton-transfer reactions will appear in the remainder of this chapter.

Problem 1.26

When a solution of NaSH in water is acidified, H_2S is produced. Complete the following equation and track the flow of electrons via curved arrows. Is the net charge the same on both sides of the equation?

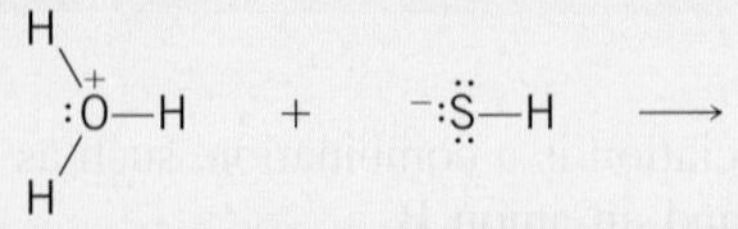

Curved-arrow notation is also applied to reactions in which double and triple bonds are made or broken. Only one component (one electron pair) of the double or triple bond is involved. Examples include:

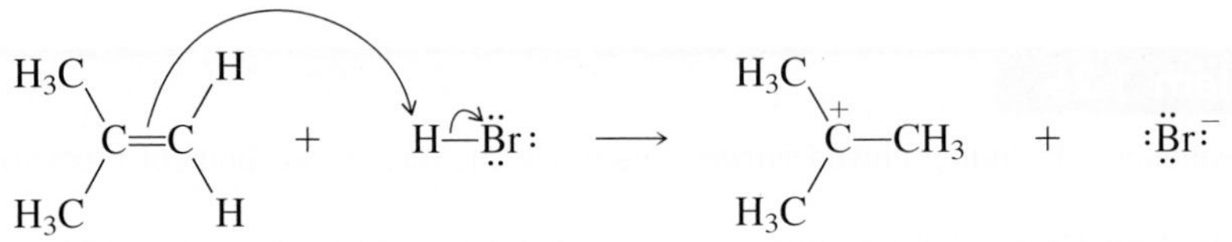

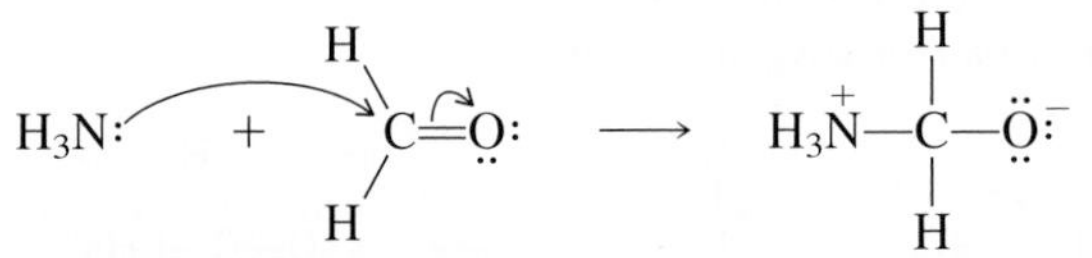

Before we conclude this section and move on to acids and bases, we should emphasize an important point.

- *Resist the temptation to use curved arrows to show the movement of atoms.* Curved arrows always show *electron* flow.

Although our eyes are drawn to the atoms when we look at a chemical equation, following the electrons provides a clearer understanding of how reactants become products.

1.13 Acids and Bases: The Arrhenius View

Acids and bases are a big part of organic chemistry, but the emphasis is much different from what you may be familiar with from your general chemistry course. Most of the attention in general chemistry is given to numerical calculations: pH, percent ionization, buffer problems, and so on. Some of this returns in organic chemistry, but mostly we are concerned with the roles that acids and bases play as reactants, products, and catalysts in chemical reactions. We'll start by reviewing some general ideas about acids and bases.

According to the theory proposed by Svante Arrhenius, a Swedish chemist and winner of the 1903 Nobel Prize in Chemistry, an **acid** is a substance that ionizes to give protons when dissolved in water.

$$\underset{\text{Acid}}{H{-}A} \rightleftharpoons \underset{\text{Proton}}{H^+} + \underset{\text{Anion}}{:A^-}$$

A **base** ionizes to give hydroxide ions.

$$\underset{\text{Base}}{M{-}\ddot{\underset{..}{O}}{-}H} \rightleftharpoons \underset{\text{Cation}}{M^+} + \underset{\text{Hydroxide ion}}{{}^{-}:\ddot{\underset{..}{O}}{-}H}$$

Acids differ in the degree to which they ionize. Those that ionize completely are called *strong acids;* those that do not are *weak acids.* Likewise, *strong bases* ionize completely; *weak bases* do not.

The strength of a weak acid is measured by its **acidity constant,** which is the equilibrium constant K_a for its ionization in aqueous solution.

$$K_a = \frac{[H^+][:A^-]}{[HA]}$$

A convenient way to express the strength of an acid is by its **pK_a**, defined as:

$$pK_a = -\log_{10}K_a$$

Thus, acetic acid with $K_a = 1.8 \times 10^{-5}$ has a pK_a of 4.7. The advantage of pK_a over K_a is that it avoids exponentials. You are probably more familiar with K_a, but most organic chemists and biochemists use pK_a. It is a good idea to be comfortable with both systems, so you should practice converting K_a to pK_a and vice versa.

Problem 1.27

Salicylic acid, the starting material for the preparation of aspirin, has a K_a of 1.06×10^{-3}. What is its pK_a?

Problem 1.28

Hydrogen cyanide (HCN) has a pK_a of 9.1. What is its K_a?

1.14 Acids and Bases: The Brønsted–Lowry View

A more general theory of acids and bases was devised independently by Johannes Brønsted (Denmark) and Thomas M. Lowry (England) in 1923. In the Brønsted–Lowry approach, an acid is a **proton donor,** and a base is a **proton acceptor.** The reaction that occurs between an acid and a base is *proton transfer.*

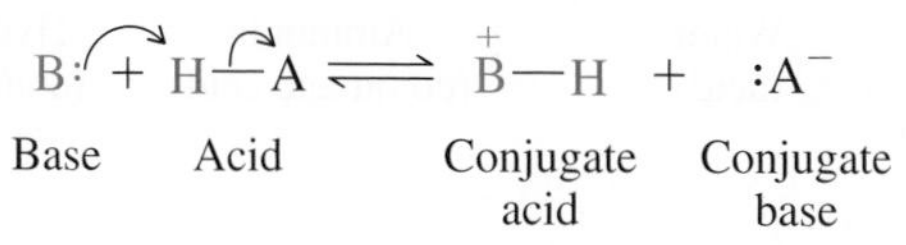
B: + H—A ⇌ B—H + :A⁻
Base
Acid
Conjugate acid
Conjugate base

In the equation shown, the base uses an unshared pair of electrons to remove a proton from an acid. The base is converted to its **conjugate acid,** and the acid is converted to its **conjugate base.** A base and its conjugate acid always differ by a single proton. Likewise, an acid and its conjugate base always differ by a single proton.

In the Brønsted–Lowry view, an acid doesn't dissociate in water; it transfers a proton to water. Water acts as a base.

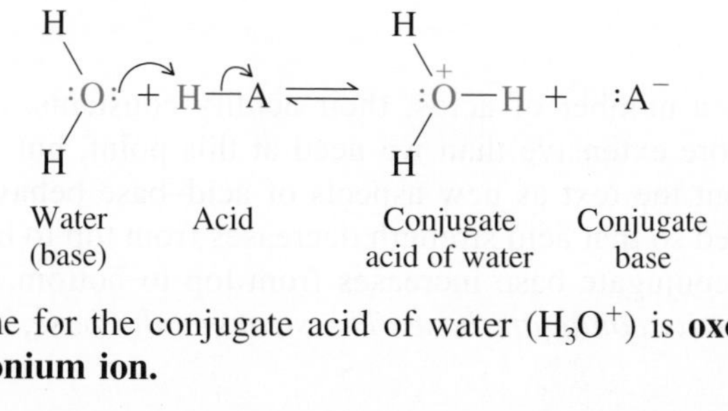
Water (base)
Acid
Conjugate acid of water
Conjugate base

The systematic name for the conjugate acid of water (H_3O^+) is **oxonium ion.** Its common name is **hydronium ion.**

Problem 1.29

Write an equation for proton transfer from hydrogen chloride (HCl) to

(a) Ammonia ($:NH_3$)
(b) Trimethylamine [$(CH_3)_3N:$]

Identify the acid, base, conjugate acid, and conjugate base and use curved arrows to track electron movement.

Sample Solution We are told that a proton is transferred from HCl to $:NH_3$. Therefore, HCl is the Brønsted acid and $:NH_3$ is the Brønsted base.

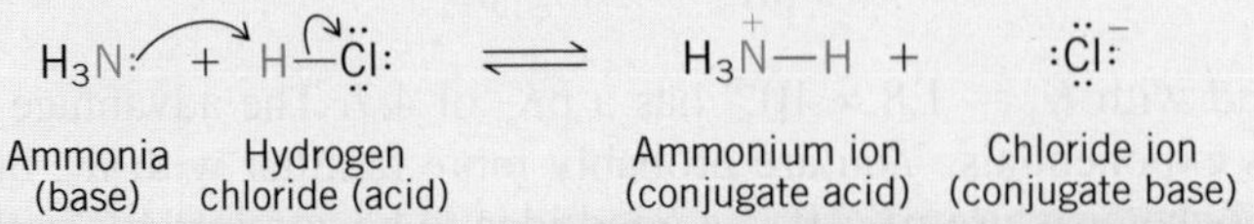

The acidity constant K_a has the same form in Brønsted–Lowry as in the Arrhenius approach, but is expressed in the concentration of H_3O^+ rather than H^+. The concentration terms $[H_3O^+]$ and $[H^+]$ are considered equivalent quantities in equilibrium constant expressions.

$$K_a = \frac{[H_3O^+][:A^-]}{[HA]}$$

Even though water is a reactant (a Brønsted base), its concentration does not appear in the expression for K_a because it is the solvent. The convention for equilibrium constant expressions is to omit concentration terms for pure solids, liquids, and solvents.

Water can also be a Brønsted acid, donating a proton to a base. Sodium amide ($NaNH_2$), for example, is a source of the strongly basic amide ion, which reacts with water to give ammonia.

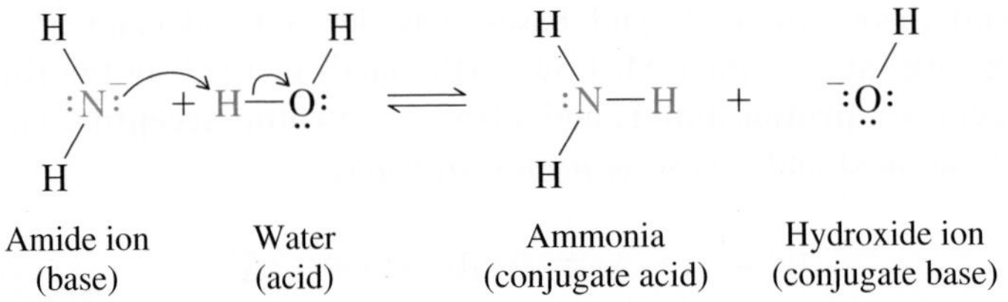

Problem 1.30

Potassium hydride (KH) is a source of the strongly basic hydride ion ($:H^-$).

Using curved arrows to track electron movement, write an equation for the reaction of hydride ion with water. What is the conjugate acid of hydride ion?

Table 1.8 lists a number of acids, their acidity constants, and their conjugate bases. The list is more extensive than we need at this point, but we will return to it repeatedly throughout the text as new aspects of acid–base behavior are introduced. The table is organized so that acid strength decreases from top to bottom. Conversely, the strength of the conjugate base increases from top to bottom. Thus, *the stronger the acid, the weaker its conjugate base. The stronger the base, the weaker its conjugate acid.*

TABLE 1.8 Acidity Constants (pK_a) of Acids

Acid	pK_a	Formula	Conjugate base	Discussed in section
Hydrogen iodide	−10.4	HI	I^-	1.16
Hydrogen bromide	−5.8	HBr	Br^-	1.16
Sulfuric acid	−4.8	$HOSO_2OH$	$HOSO_2O^-$	1.17
Hydrogen chloride	−3.9	HCl	Cl^-	1.16
Ethyloxonium ion	−2.4	$CH_3CH_2\overset{+}{O}H_2$	CH_3CH_2OH	4.8
Hydronium ion*	−1.7	H_3O^+	H_2O	1.17
Nitric acid	−1.4	$HONO_2$	$^-ONO_2$	1.16
Hydrogen sulfate ion	2.0	$HOSO_2O^-$	$^-OSO_2O^-$	1.17
Hydrogen fluoride	3.1	HF	F^-	1.16
Anilinium ion	4.6	$C_6H_5\overset{+}{N}H_3$	$C_6H_5NH_2$	21.4
Acetic acid	4.7	$CH_3\overset{O}{\overset{\|}{C}}OH$	$CH_3\overset{O}{\overset{\|}{C}}O^-$	1.16; 19.4
Pyridinium ion	5.2	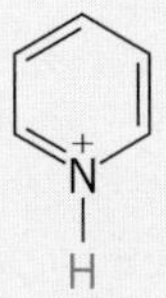	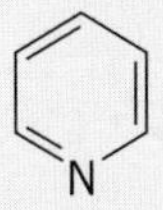	1.15; 21.4
Carbonic acid	6.4	H_2CO_3	HCO_3^-	18.9
Hydrogen sulfide	7.0	H_2S	HS^-	8.11; 15.12
2,4-Pentanedione	9	$CH_3\overset{O}{\overset{\|}{C}}CH_2\overset{O}{\overset{\|}{C}}CH_3$	$CH_3\overset{O}{\overset{\|}{C}}\underset{\cdot\cdot}{\overset{-}{C}}H\overset{O}{\overset{\|}{C}}CH_3$	20.1
Hydrogen cyanide	9.1	HCN	CN^-	
Ammonium ion	9.3	NH_4^+	NH_3	1.15; 21.4
Glycine	9.6	$H_3\overset{+}{N}CH_2\overset{O}{\overset{\|}{C}}O^-$	$H_2NCH_2\overset{O}{\overset{\|}{C}}O^-$	25.3
Phenol	10	C_6H_5OH	$C_6H_5O^-$	1.17; 22.4
Hydrogen carbonate ion	10.2	HCO_3^-	CO_3^{2-}	18.9
Methanethiol	10.7	CH_3SH	CH_3S^-	15.12
Dimethylammonium ion	10.7	$(CH_3)_2\overset{+}{N}H_2$	$(CH_3)_2NH$	21.4
Methylammonium ion	10.7	$CH_3\overset{+}{N}H_3$	CH_3NH_2	21.4
Ethyl acetoacetate	11	$CH_3\overset{O}{\overset{\|}{C}}CH_2\overset{O}{\overset{\|}{C}}OCH_2CH_3$	$CH_3\overset{O}{\overset{\|}{C}}\underset{\cdot\cdot}{\overset{-}{C}}H\overset{O}{\overset{\|}{C}}OCH_2CH_3$	20.1
Piperidinium ion	11.2	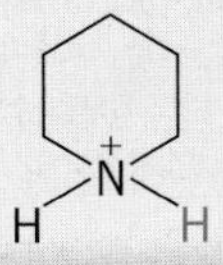	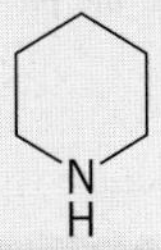	21.4

Continued

*For acid–base reactions in which water is the solvent, the pK_a of H_3O^+ is zero and the pK_a of H_2O is 14.

TABLE 1.8 Acidity Constants (pK_a) of Acids (*Continued*)

Acid	pK_a	Formula	Conjugate base	Discussed in section
Diethyl malonate	13	$CH_3CH_2O\overset{O}{C}CH_2\overset{O}{C}OCH_2CH_3$	$CH_3CH_2O\overset{O}{C}\ddot{\bar{C}}H\overset{O}{C}OCH_2CH_3$	20.1
Methanol	15.2	CH_3OH	CH_3O^-	1.16
2-Methylpropanal	15.5	$(CH_3)_2CH\overset{O}{C}H$	$(CH_3)_2\ddot{\bar{C}}\overset{O}{C}H$	20.1
Water*	15.7	H_2O	HO^-	1.16
Ethanol	16	CH_3CH_2OH	$CH_3CH_2O^-$	1.16
Cyclopentadiene	16	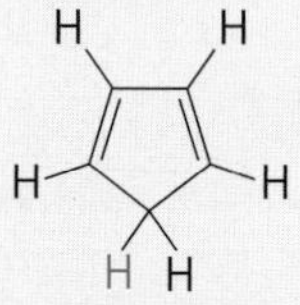	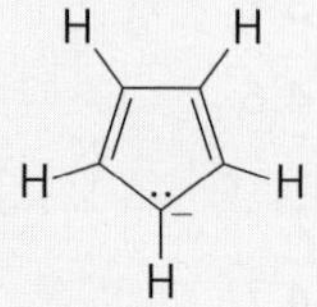	11.21
Isopropyl alcohol	17	$(CH_3)_2CHOH$	$(CH_3)_2CHO^-$	1.16
tert-Butyl alcohol	18	$(CH_3)_3COH$	$(CH_3)_3CO^-$	1.16
Acetone	19	$CH_3\overset{O}{C}CH_3$	$CH_3\overset{O}{C}\ddot{\bar{C}}H_2$	20.1
Ethyl acetate	25.6	$CH_3\overset{O}{C}OCH_2CH_3$	$H_2\ddot{\bar{C}}\overset{O}{C}OCH_2CH_3$	20.1
Acetylene	26	$HC{\equiv}CH$	$HC{\equiv}CH{:}^-$	9.5
Hydrogen	35	H_2	H^-	20.8
Ammonia	36	NH_3	H_2N^-	1.16
Diisopropylamine	36	$[(CH_3)_2CH]_2NH$	$[(CH_3)_2CH]_2N^-$	20.1
Benzene	43	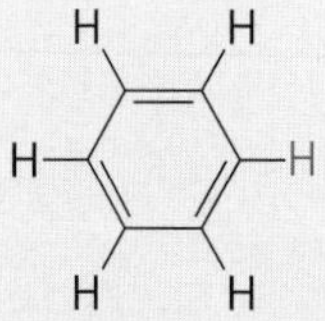	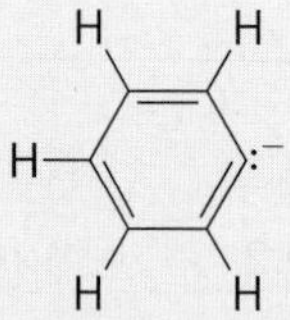	14.5
Ethylene	45	$H_2C{=}CH_2$	$H_2C{=}\ddot{\bar{C}}H$	9.4; 9.5
Methane	60	CH_4	$\ddot{\bar{C}}H_3$	1.16; 14.5
Ethane	62	CH_3CH_3	$CH_3\ddot{\bar{C}}H_2$	14.5

Web collections of pK_a data include those of H. Reich (University of Wisconsin) at *http://www.chem.wisc.edu/areas/reich/pkatable/kacont.htm* and D. Ripin and D. A. Evans (Harvard) at *http://www2.lsdiv.harvard.edu/labs/evans/pdf/evans_pKa_table.pdf.*

*For acid-base reactions in which water is the solvent, the pK_a of H_3O^+ is zero and the pK_a of H_2O is 14.

1.15 What Happened to pK_b?

The Brønsted–Lowry approach involving conjugate relationships between acids and bases makes a separate **basicity constant K_b** unnecessary. Rather than having separate tables listing K_a for acids and K_b for bases, the usual practice is to give only K_a or pK_a as was done in Table 1.8. Assessing relative basicities requires only that we remember that the weaker the acid, the stronger the conjugate base and find the appropriate acid–base pair in the table.

Suppose, for example, we wished to compare the basicities of ammonia and pyridine.

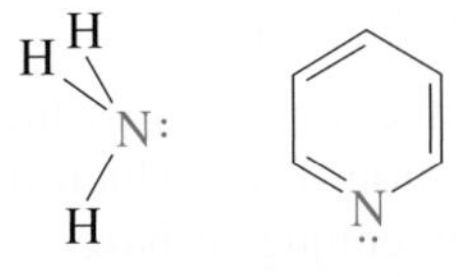

The stronger base is derived from the weaker conjugate acid. Therefore, add a proton to ammonia to give its conjugate acid (ammonium ion) and a proton to pyridine to give its conjugate acid (pyridinium ion), then look up the pK_a values for each.

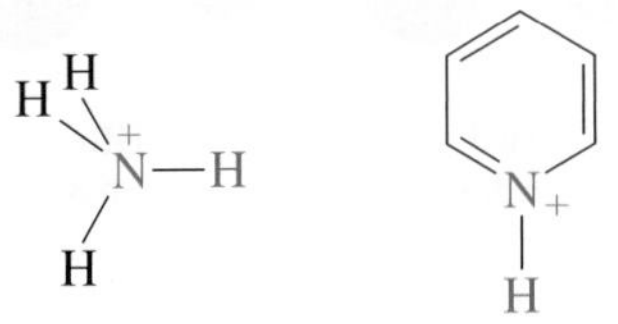

	Ammonium ion	Pyridinium ion
pK_a	9.3	5.2
	weaker acid	stronger acid

Ammonium ion is a weaker acid than pyridinium ion; therefore, ammonia is a stronger base than pyridine.

The conjugate bases listed in Table 1.8 that are anions are commonly encountered as their sodium or potassium salts. Thus, sodium methoxide ($NaOCH_3$), for example, is a source of methoxide ion (CH_3O^-), which is the conjugate base of methanol.

Problem 1.31

Which is the stronger base in each of the following pairs? (*Note:* This information will prove useful when you get to Chapter 9.)

(a) Sodium ethoxide ($NaOCH_2CH_3$) or sodium amide ($NaNH_2$)
(b) Sodium acetylide ($NaC{\equiv}CH$) or sodium amide ($NaNH_2$)
(c) Sodium acetylide ($NaC{\equiv}CH$) or sodium ethoxide ($NaOCH_2CH_3$)

Sample Solution (a) $NaOCH_2CH_3$ contains the ions Na^+ and $CH_3CH_2O^-$. $NaNH_2$ contains the ions Na^+ and H_2N^-. $CH_3CH_2O^-$ is the conjugate base of ethanol; H_2N^- is the conjugate base of ammonia.

Base	$CH_3CH_2O^-$	H_2N^-
Conjugate acid	CH_3CH_2OH	NH_3
pK_a of conjugate acid	16	36

The conjugate acid of $CH_3CH_2O^-$ is stronger than the conjugate acid of H_2N^-. Therefore, H_2N^- is a stronger base than $CH_3CH_2O^-$.

If pK_b is ever needed, it is easily calculated from the pK_a of its conjugate acid with the equation:

$$pK_b = 14 - pK_a \text{ of conjugate acid}$$

1.16 How Structure Affects Acid Strength

The acids in Table 1.8 span a range of more than 70 pK_a units (10^{70} in K_a). In this section we'll introduce some generalizations that will permit us to connect molecular structure with acidity, at least insofar as trends in related compounds are concerned. The main ways in which structure affects acidity depend on:

1. The strength of the bond to the atom from which the proton is lost
2. The electronegativity of the atom from which the proton is lost
3. Electron delocalization in the conjugate base

Bond Strength. The effect of bond strength is easy to see by comparing the acidities of the hydrogen halides.

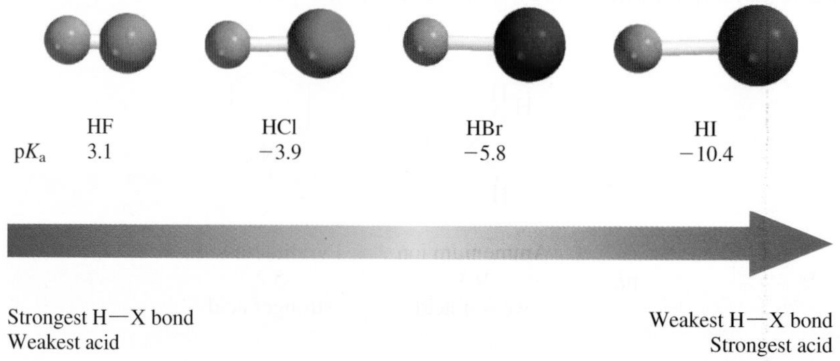
HF
HCl
HBr
HI
pKa
3.1
−3.9
−5.8
−10.4
Strongest H—X bond
Weakest acid
Weakest H—X bond
Strongest acid

In general, bond strength decreases going down a group in the periodic table. As the halogen X becomes larger, the H—X bond becomes longer and weaker and acid strength increases. This is the dominant factor in the series HCl, HBr, HI and also contributes to the relative weakness of HF.

With HF, a second factor concerns the high charge-to-size ratio of F^-. Other things being equal, processes that give ions in which the electric charge is constrained to a small volume are less favorable than processes in which the charge is more spread out. The strong H—F bond and the high charge-to-size ratio of F^- combine to make HF the weakest acid of the hydrogen halides.

Because of the conjugate relationship between acidity and basicity, the strongest acid (HI) has the weakest conjugate base (I^-), and the weakest acid (HF) has the strongest conjugate base (F^-).

Problem 1.32

Which is the stronger acid, H_2O or H_2S? Which is the stronger base, HO^- or HS^-? Check your predictions against the data in Table 1.8.

Electronegativity. The effect of electronegativity on acidity is evident in the following series involving bonds between hydrogen and the second-row elements C, N, O, and F.

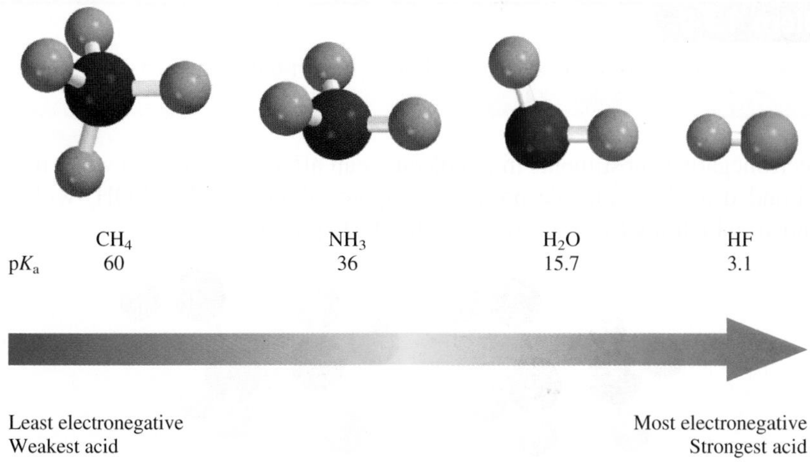
CH4
NH3
H2O
HF
pKa
60
36
15.7
3.1
Least electronegative
Weakest acid
Most electronegative
Strongest acid

As the atom (A) to which H is bonded becomes more electronegative, the polarization $^{\delta+}H{-}A^{\delta-}$ becomes more pronounced and H is more easily transferred as H^+. An alternative approach to the same conclusion is based on the equation for proton transfer, especially with regard to the flow of electrons as shown by curved arrows.

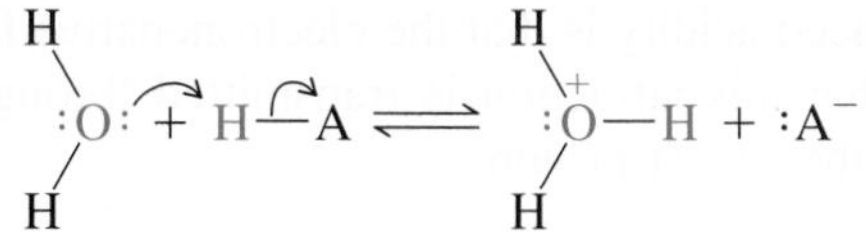

Here we see that when the H—A bond breaks, both electrons in the bond are retained by A. The more electronegative atom A is, the easier it becomes for the electrons to flow in its direction.

Bond strength is more important than electronegativity when comparing elements in the same group of the periodic table as the pK_a's for the hydrogen halides show. Fluorine is the most electronegative and iodine the least electronegative of the halogens, but HF is the weakest acid while HI is the strongest. Electronegativity is the more important factor when comparing elements in the same row of the periodic table.

Problem 1.33

Try to do this problem without consulting Table 1.8.

(a) Which is the stronger acid: $(CH_3)_3\overset{+}{N}H$ or $(CH_3)_2\overset{+}{\underset{..}{O}}H$?

(b) Which is the stronger base: $(CH_3)_3N\colon$ or $(CH_3)_2\ddot{O}\colon$?

Sample Solution (a) The ionizable proton is bonded to N in $(CH_3)_3\overset{+}{N}H$ and to O in $(CH_3)_2\overset{+}{\underset{..}{O}}H$.

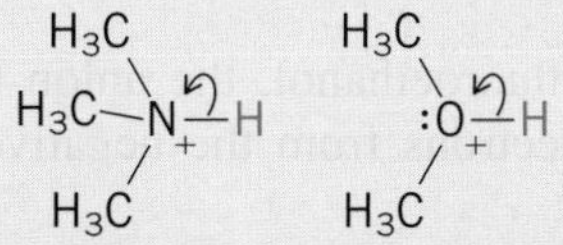

Nitrogen and oxygen are in the same row of the periodic table, so their relative electronegativities are the determining factor. Oxygen is more electronegative than nitrogen; therefore $(CH_3)_2\overset{+}{\underset{..}{O}}H$ is a stronger acid than $(CH_3)_3\overset{+}{N}H$.

In many acids the acidic proton is bonded to oxygen. Such compounds can be considered as derivatives of water. Among organic compounds, the ones most closely related to water are alcohols. Most alcohols are somewhat weaker acids than water; methanol is slightly stronger.

	HO—H	CH_3O—H	CH_3CH_2O—H	$(CH_3)_2CHO$—H	$(CH_3)_3CO$—H
	Water	Methanol	Ethanol	Isopropyl alcohol	*tert*-Butyl alcohol
pK_a	15.7	15.2	16	17	18

Problem 1.34

Which is a stronger base, ethoxide ($CH_3CH_2\ddot{O}:^-$) or *tert*-butoxide [$(CH_3)_3C\ddot{O}:^-$]?

A *substituent* is an atom or group other than hydrogen in a molecule.

Electronegative substituents in a molecule can affect acidity even when they are not directly bonded to the ionizable proton. Compare ethanol (CH_3CH_2OH) with a related compound in which a CF_3 group replaces the CH_3 group.

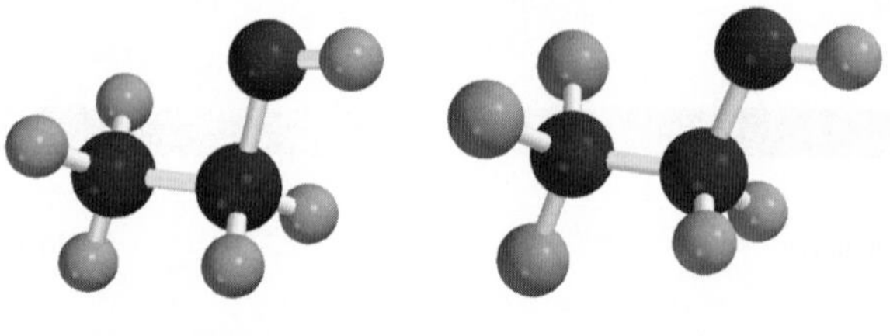

	$CH_3CH_2O—H$	$CF_3CH_2O—H$
	Ethanol	2,2,2-Trifluoroethanol
pK_a	16	11.3

We see that the substitution of C—H bonds by C—F increases the acidity of the O—H proton by 4.7 pK_a units, which corresponds to a difference of $10^{4.7}$ in K_a. The simplest explanation for this enhanced acidity is that the electronegative fluorines attract electrons toward themselves and that this attraction is transmitted through the bonds, increasing the positive character of the O—H proton.

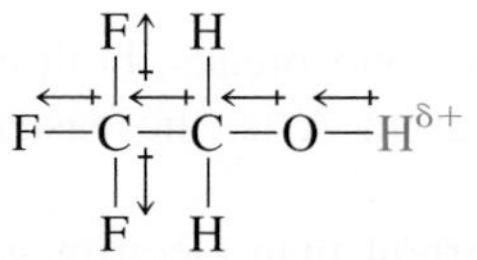

The greater positive character, hence the increased acidity, of the O—H proton of 2,2,2-trifluoroethanol can be seen in the electrostatic potential maps displayed in Figure 1.8.

We can also explain the greater acidity of CF_3CH_2OH relative to CH_3CH_2OH by referring to the equations for their ionization.

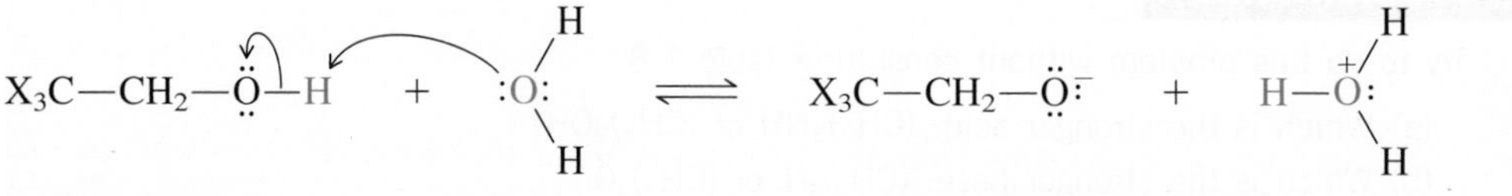

X = H: Ethanol
X = F: 2,2,2-Trifluoroethanol

X = H: Conjugate base of ethanol
X = F: Conjugate base of 2,2,2-trifluoroethanol

The conjugate base of 2,2,2-trifluoroethanol, the anion $CF_3CH_2O^-$, is stabilized by its three fluorines, which attract electrons from the negatively charged oxygen, dispersing

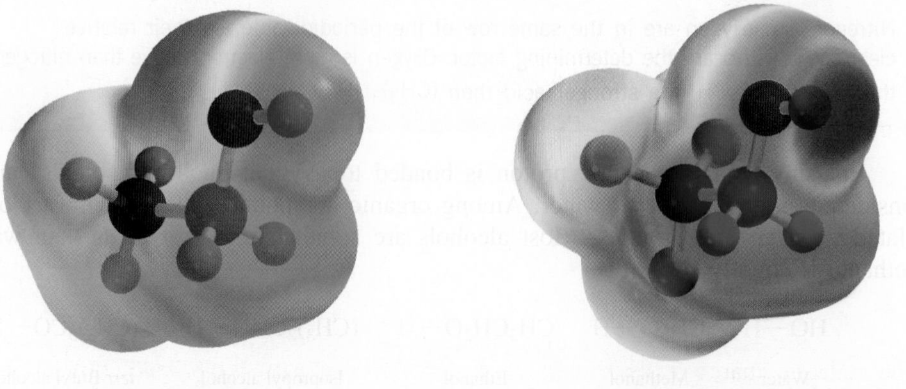

Ethanol (CH_3CH_2OH) 2,2,2-Trifluoroethanol (CF_3CH_2OH)

Figure 1.8

Electrostatic potential maps of ethanol and 2,2,2-trifluoroethanol. As indicated by the more blue, less green color in the region near the OH proton in 2,2,2-trifluoroethanol, this proton bears a greater degree of positive charge and is more acidic than the OH proton in ethanol. The color scale is the same in both maps.

the negative charge. Because of this stabilization, the equilibrium for ionization of CF_3CH_2OH lies farther to the right than that of CH_3CH_2OH.

Structural effects that are transmitted through bonds are called **inductive effects.** A substituent *induces* a polarization in the bonds between it and some remote site.

The same kind of inductive effects that make CF_3CH_2OH a stronger acid than CH_3CH_2OH makes the trifluoro derivative of acetic acid more than 4 pK_a units stronger than acetic acid.

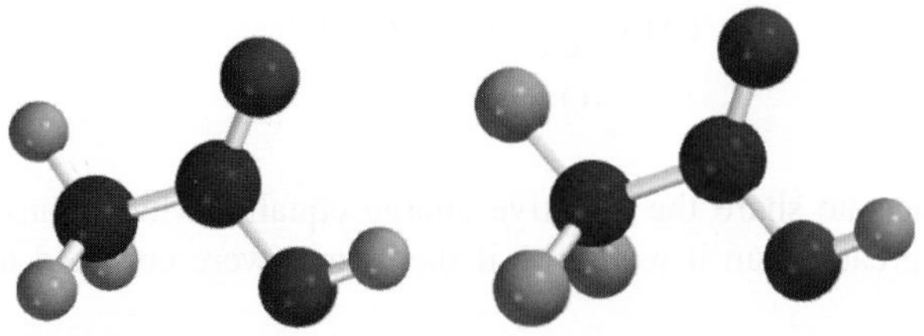

	$CH_3\overset{O}{\overset{\|}{C}}O{-}H$	$CF_3\overset{O}{\overset{\|}{C}}O{-}H$
	Acetic acid	Trifluoroacetic acid
pK_a	4.7	0.50

Problem 1.35

Hypochlorous and hypobromous acid (HOCl and HOBr) are weak acids. Write chemical equations for the ionization of each in water and predict which one is the stronger acid.

Inductive effects depend on the electronegativity of the substituent and the number of bonds between it and the affected site. As the number of bonds between the two units increases, the inductive effect decreases.

Electron Delocalization in the Conjugate Base. With a pK_a of −1.4, nitric acid is almost completely ionized in water. If we look at the Lewis structure of nitric acid in light of what we have said about inductive effects, we can see why. The N atom in nitric acid is not only electronegative in its own right, but bears a formal charge of +1, which enhances its ability to attract electrons away from the —OH group.

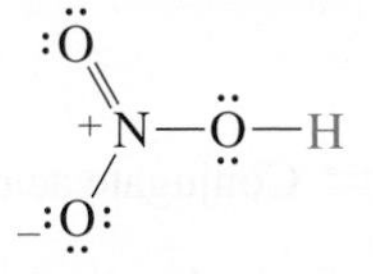

But inductive effects are only part of the story. When nitric acid transfers its proton to water, nitrate ion is produced.

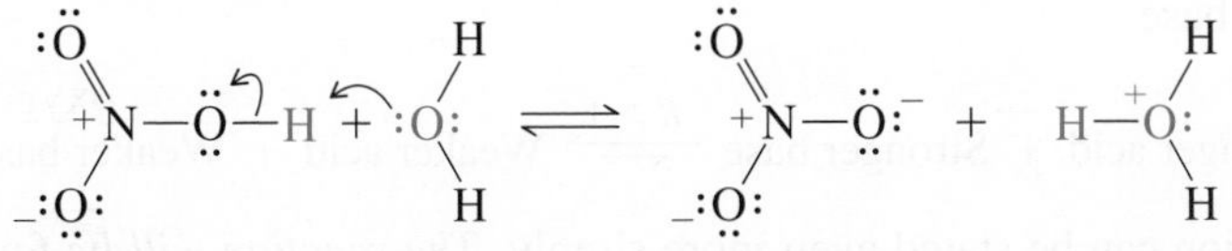

Nitric acid (acid) Water (base) Nitrate ion (conjugate base) Hydronium ion (conjugate acid)

Nitrate ion is stabilized by electron delocalization, which we can represent in terms of resonance between three equivalent contributing structures:

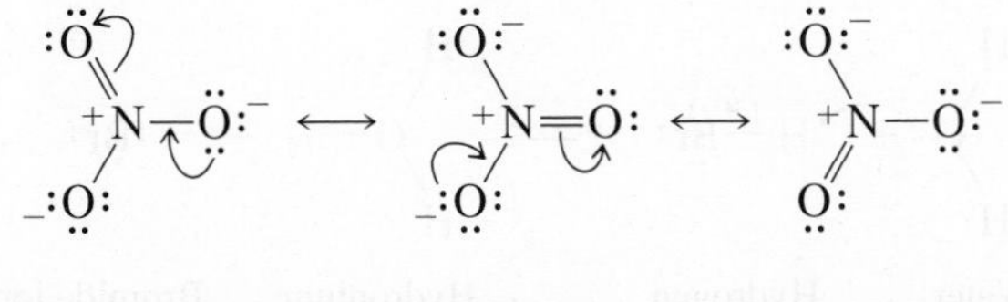

The negative charge is shared equally by all three oxygens. Stabilization of nitrate ion by electron delocalization increases the equilibrium constant for its formation.

Problem 1.36

What is the average formal charge on each oxygen in nitrate ion?

A similar electron delocalization stabilizes acetate ion and related species.

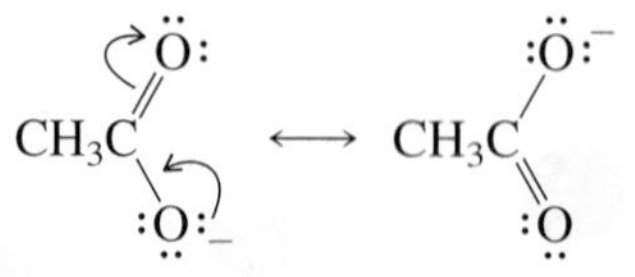

Both oxygens of acetate share the negative charge equally, which translates into a K_a for acetic acid that is greater than it would be if the charge were confined to a single oxygen.

Problem 1.37

Show by writing appropriate resonance structures that the two compounds shown form the same conjugate base on ionization. Which atom in the conjugate base, O or S, bears the greater share of negative charge?

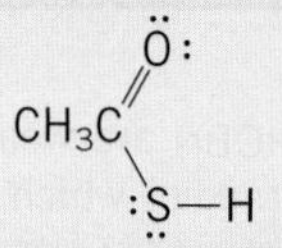

and

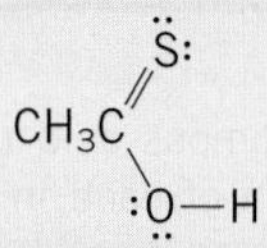

Organic chemistry involves a good bit of reasoning by analogy and looking for trends. At the beginning of this section we listed three ways that structure can affect acidity. The last two—electronegativity of the atom from which the proton is lost, and electron delocalization in the conjugate base—are both related to the stability of the conjugate base. A useful trend emerges: *factors that stabilize the conjugate base increase the acidity of the parent acid.*

1.17 Acid–Base Equilibria

In any proton-transfer reaction:

$$\text{Acid} + \text{Base} \rightleftharpoons \text{Conjugate acid} + \text{Conjugate base}$$

we are concerned with the question of whether the position of equilibrium lies to the side of products or reactants. There is an easy way to determine this. The reaction proceeds in the direction that converts the stronger acid and the stronger base to the weaker acid and the weaker base.

$$\text{Stronger acid} + \text{Stronger base} \xrightleftharpoons{K > 1} \text{Weaker acid} + \text{Weaker base}$$

This generalization can be stated even more simply. *The reaction will be favorable when the stronger acid is on the left and the weaker acid is on the right.* The equilibrium lies to the side of the acid that holds the proton more tightly.

Consider first the case of adding a strong acid such as HBr to water. The equation for the Brønsted acid–base reaction that occurs between them is:

For acid–base reactions in which water is the solvent, the pK_a of H_3O^+ = 0. See Table 1.8.

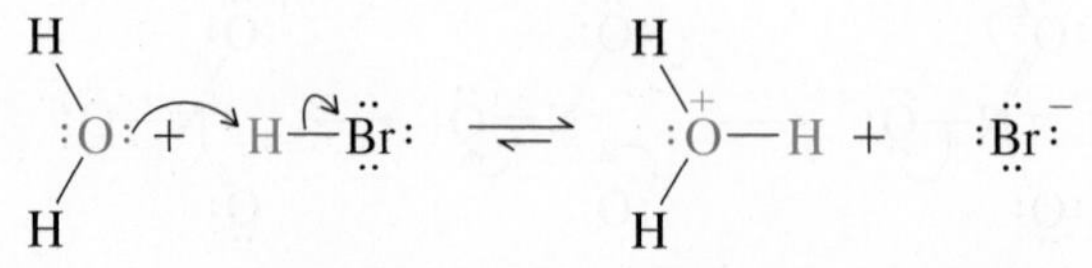

Water	Hydrogen bromide pK_a = −5.8 stronger acid	Hydronium ion pK_a = 0 weaker acid	Bromide ion

We identify the acid on the left and the acid on the right and compare their pK_a's to decide which is stronger. (Remember, the more negative the pK_a, the stronger the acid.) The acid on the left is HBr, which has a pK_a of −5.8. The acid on the right is H_3O^+, which has a pK_a of 0. The stronger acid (HBr) is on the left and the weaker acid (H_3O^+) is on the right, so the position of equilibrium lies to the right. The equilibrium constant K_{eq} for an acid–base reaction is given by the ratio of the K_a of the reactant acid to the K_a of the product acid.

$$K_{eq} = \frac{K_a \text{ of reactant acid}}{K_a \text{ of product acid}}$$

Since $10^{-pK_a} = K_a$, we rewrite the expression as:

$$K_{eq} = \frac{10^{-pK_a} \text{ of reactant acid}}{10^{-pK_a} \text{ of product acid}}$$

and substitute the pK_a values of HBr and H_3O^+ to calculate K_{eq}.

$$K_{eq} = \frac{10^{5.8}}{10^{0}} = 10^{5.8}$$

For all practical purposes, the equilibrium constant is so large that we consider HBr to be completely ionized in water.

Compare the reaction of HBr with water to that of acetic acid with water.

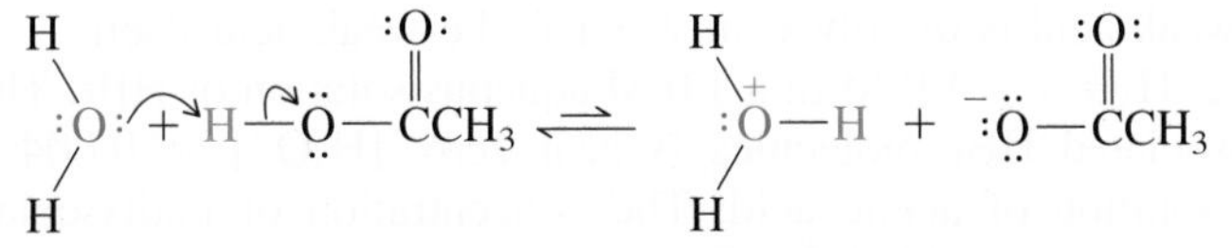

Water	Acetic acid	Hydronium ion	Acetate ion
	pK_a = 4.7 weaker acid	pK_a = 0 stronger acid	

Here, the weaker acid (acetic acid) is on the left and the stronger acid (hydronium ion) is on the right. The equilibrium constant $K_{eq} = 10^{-4.7}$, and the position of equilibrium lies far to the left.

Problem 1.38

What is the equilibrium constant for the following acid–base reactions?

(a) ammonia and acetic acid

(b) fluoride ion and acetic acid

(c) ethanol and hydrobromic acid

Sample Solution (a) Always start with an equation for an acid–base reaction. Ammonia is a Brønsted base and accepts a proton from the —OH group of acetic acid. Ammonia is converted to its conjugate acid, and acetic acid to its conjugate base.

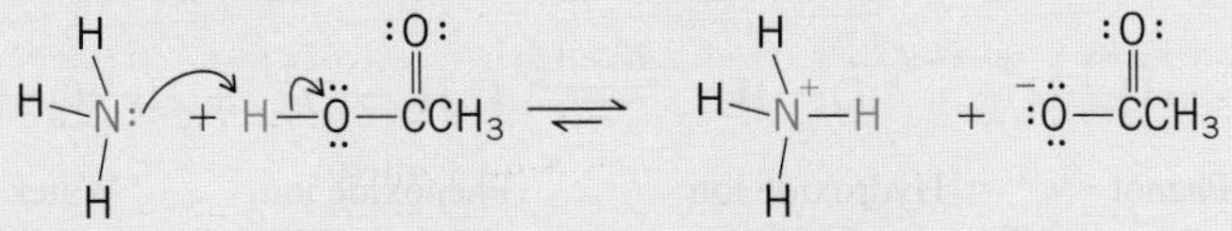

Ammonia	Acetic acid	Ammonium ion	Acetate ion
	pK_a = 4.7 stronger acid	pK_a = 9.3 weaker acid	

From their respective pK_a's, we see that acetic acid is a much stronger acid than ammonium ion. Therefore, the equilibrium lies to the right. The equilibrium constant for the process is

$$K_{eq} = \frac{10^{-pK_a} \text{ of acetic acid (reactant)}}{10^{-pK_a} \text{ of ammonium ion (product)}} = \frac{10^{-4.7}}{10^{-9.3}} = 10^{4.6}$$

Continued

An unexpected fact emerges by working through this exercise. We see that although acetic acid is a weak acid and ammonia is a weak base, the acid–base reaction between them is virtually complete.

Two important points come from using relative pK_a's to analyze acid–base equilibria:

1. They permit clear-cut distinctions between strong and weak acids and bases. *A strong acid is one that is stronger than H_3O^+.* Conversely, a weak acid is one that is weaker than H_3O^+.
   Example: The pK_a's for the first and second ionizations of sulfuric acid are −4.8 and 2.0, respectively. Sulfuric acid ($HOSO_2OH$) is a strong acid; hydrogen sulfate ion ($HOSO_2O^-$) is a weak acid. A *strong base is one that is stronger than HO^-.*
   Example: A common misconception is that the conjugate base of a weak acid is strong. This is sometimes, but not always, true. It is true, for example, for ammonia, which is a very weak acid (pK_a 36). Its conjugate base amide ion (H_2N^-) is a much stronger base than HO^-. It is not true, however, for acetic acid; both acetic acid and its conjugate base acetate ion are weak. The conjugate base of a weak acid will be strong only when the acid is a weaker acid than water.
2. The strongest acid present in significant amounts at equilibrium after a strong acid is dissolved in water is H_3O^+. The strongest acid present in significant amounts when a weak acid is dissolved in water is the weak acid itself.
   Example: $[H_3O^+]$ = 1.0 M in a 1.0 M aqueous solution of HBr. The concentration of undissociated HBr molecules is near zero. $[H_3O^+]$ = 0.004 M in a 1.0 M aqueous solution of acetic acid. The concentration of undissociated acetic acid molecules is near 1.0 M. Likewise, HO^- is the strongest base that can be present in significant quantities in aqueous solution.

Problem 1.39

Rank the following in order of decreasing concentration in a solution prepared by dissolving 1.0 mol of sulfuric acid in enough water to give 1.0 L of solution. (It is not necessary to do any calculations.)

H_2SO_4, HSO_4^-, SO_4^{2-}, H_3O^+

Analyzing acid–base reactions according to the Brønsted–Lowry picture provides yet another benefit. Table 1.8, which lists acids according to their strength in descending order along with their conjugate bases, can be used to predict the direction of proton transfer. Acid–base reactions in which a proton is transferred from an acid to a base that lies below it in the table have favorable equilibrium constants. Proton transfers from an acid to a base that lies above it in the table are unfavorable. Thus the equilibrium constant for proton transfer from phenol to hydroxide ion is greater than 1, but that for proton transfer from phenol to hydrogen carbonate ion is less than 1.

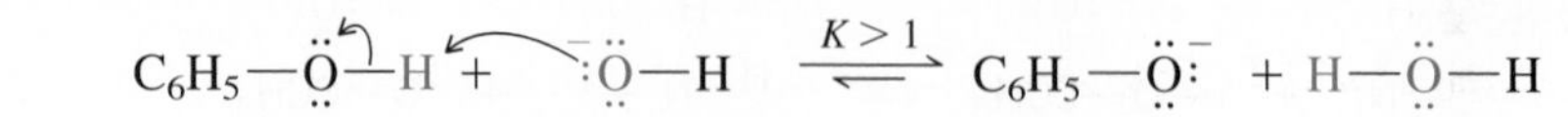

Phenol Hydroxide ion Phenoxide ion Water

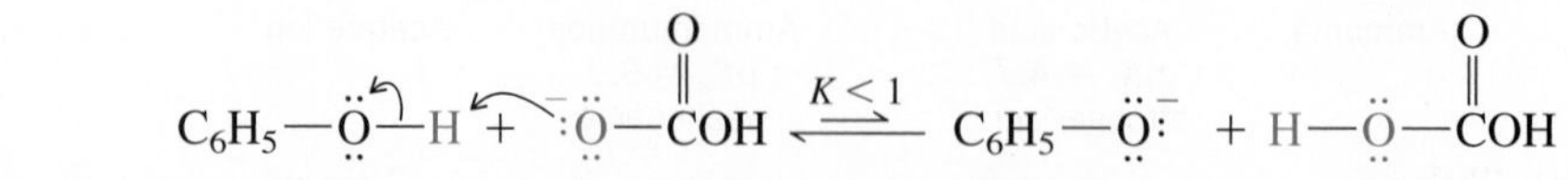

Phenol Hydrogen carbonate ion Phenoxide ion Carbonic acid

Hydroxide ion lies below phenol in Table 1.8; hydrogen carbonate ion lies above phenol. The practical consequence of the reactions shown is that NaOH is a strong enough base to convert phenol to phenoxide ion, but $NaHCO_3$ is not.

Problem 1.40

Verify that the position of equilibrium for the reaction between phenol and hydroxide ion lies to the right by comparing the pK_a of the acid on the left to the acid on the right. Which acid is stronger? Do the same for the reaction of phenol with hydrogen carbonate ion.

1.18 Lewis Acids and Lewis Bases

The same G. N. Lewis who gave us electron-dot formulas also suggested a way to classify acids and bases that is more general than the Brønsted–Lowry approach. Where Brønsted and Lowry viewed acids and bases as donors and acceptors of protons (positively charged), Lewis took the opposite view and focused on electron pairs (negatively charged). According to Lewis *an acid is an electron-pair acceptor, and a base is an electron-pair donor.*

If we apply Lewis's definitions narrowly, we can write an equation for the reaction between a Lewis acid and a Lewis base as:

$$A^{+} + :B^{-} \rightleftharpoons A{-}B$$

Lewis acid   Lewis base

An unshared pair of electrons from the Lewis base is used to form a covalent bond between the Lewis acid and the Lewis base. The Lewis acid and the Lewis base are shown as ions in the equation, but they need not be. If both are neutral molecules, the corresponding equation becomes:

$$A + :B \rightleftharpoons \overset{-}{A}{-}\overset{+}{B}$$

Lewis acid   Lewis base

We can illustrate this latter case by the reaction:

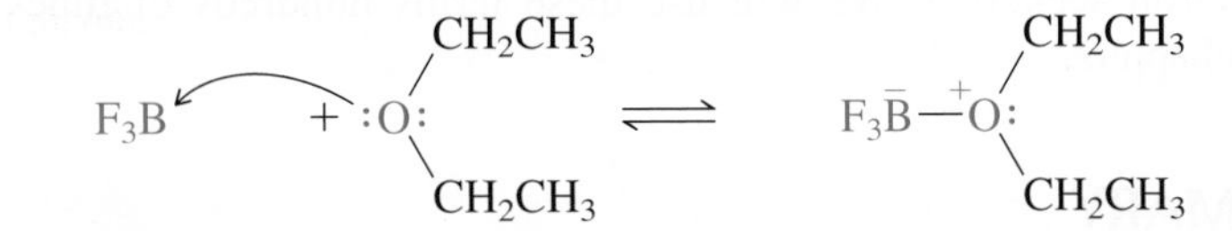

Boron trifluoride (Lewis acid)   Diethyl ether (Lewis base)   "Boron trifluoride etherate" (Lewis acid/Lewis base complex)

Verify that the formal charges on boron and oxygen in "boron trifluoride etherate" are correct.

The product of this reaction, a **Lewis acid/Lewis base complex** called informally "boron trifluoride etherate," may look unusual but it is a stable species with properties different from those of the reactants. Its boiling point (126°C), for example, is much higher than that of boron trifluoride—a gas with a boiling point of −100°C—and diethyl ether, a liquid that boils at 34°C.

Problem 1.41

Write an equation for the Lewis acid/Lewis base reaction between boron trifluoride and dimethyl sulfide [$(CH_3)_2S$]. Use curved arrows to track the flow of electrons and show formal charges if present.

The Lewis acid/Lewis base idea also includes certain **substitution** reactions in which one atom or group replaces another.

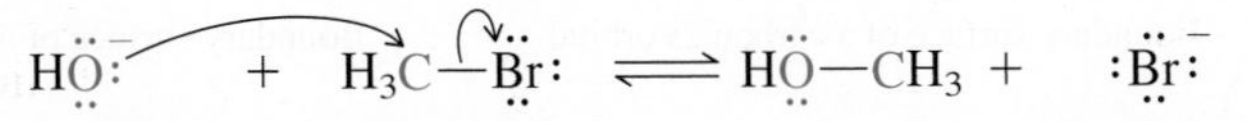

Hydroxide ion (Lewis base)   Bromomethane (Lewis acid)   Methanol   Bromide ion

The carbon atom in bromomethane can accept an electron pair if its covalent bond with bromine breaks with both electrons in that bond becoming an unshared pair of bromide ion. Thus, bromomethane acts as a Lewis acid in this reaction.

Notice the similarity of the preceding reaction to one that is more familiar to us.

Hydroxide ion (Lewis base) + Hydrogen bromide (Lewis acid) ⇌ Water + Bromide ion

Clearly, the two reactions are analogous and demonstrate that the reaction between hydroxide ion and hydrogen bromide is simultaneously a Brønsted–Lowry acid–base reaction and a Lewis acid/Lewis base reaction. *Brønsted–Lowry acid–base reactions constitute a subcategory of Lewis acid/Lewis base reactions.*

Many important biochemical reactions involve Lewis acid/Lewis base chemistry. Carbon dioxide is rapidly converted to hydrogen carbonate ion in the presence of the enzyme *carbonic anhydrase*.

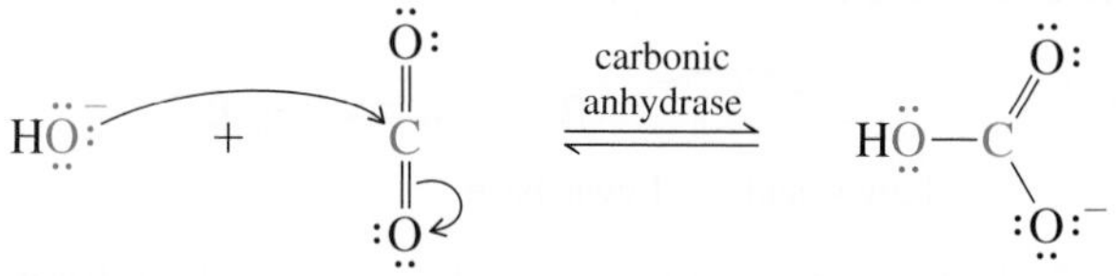

Hydroxide ion (Lewis base) + Carbon dioxide (Lewis acid) ⇌ Hydrogen carbonate ion

Recall that the carbon atom of carbon dioxide bears a partial positive charge because of the electron-attracting power of its attached oxygens. When hydroxide ion (the Lewis base) bonds to this positively polarized carbon, a pair of electrons in the carbon–oxygen double bond leaves carbon to become an unshared pair of oxygen.

Examine the table of contents. What chapters include terms related to *nucleophile* or *electrophile* in their title?

Lewis bases use an unshared pair to form a bond to some other atom and are also referred to as **nucleophiles** ("nucleus seekers"). Conversely, Lewis acids are **electrophiles** ("electron seekers"). We will use these terms hundreds of times throughout the remaining chapters.

1.19 SUMMARY

This chapter sets the stage for all of the others by reminding us that the relationship between structure and properties is what chemistry is all about. It begins with a review of Lewis structures, moves to a discussion of the Arrhenius, Brønsted–Lowry, and Lewis pictures of acids and bases, and concludes with the effects of structure on acidity and basicity.

Section 1.1 A review of some fundamental knowledge about atoms and electrons leads to a discussion of **wave functions, orbitals,** and the **electron configurations** of atoms. Neutral atoms have as many electrons as the number of protons in the nucleus. These electrons occupy orbitals in order of increasing energy, with no more than two electrons in any one orbital. The most frequently encountered atomic orbitals in this text are *s* orbitals (spherically symmetrical) and *p* orbitals ("dumbbell"-shaped).

Boundary surface of a carbon 2*s* orbital

Boundary surface of a carbon 2*p* orbital

Section 1.2 An **ionic bond** is the force of electrostatic attraction between two oppositely charged ions. Atoms at the upper right of the periodic table, especially fluorine and oxygen, tend to gain electrons to form anions. Elements toward the left of the periodic table, especially metals such as sodium, tend to lose electrons to form cations. Ionic bonds in which carbon is the cation or anion are rare.

Section 1.3 The most common kind of bonding involving carbon is **covalent bonding.** A covalent bond is the sharing of a pair of electrons between two atoms. **Lewis structures** are written on the basis of the **octet rule,** which limits second-row elements to no more than eight electrons in their valence shells. In most of its compounds, carbon has four bonds.

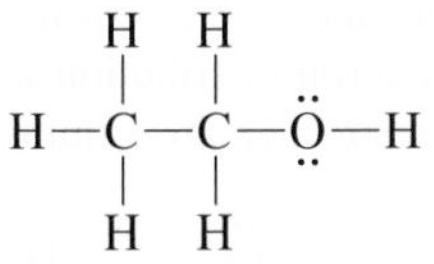

Each carbon has four bonds in ethyl alcohol; oxygen and each carbon are surrounded by eight electrons.

Section 1.4 Many organic compounds have **double** or **triple bonds** to carbon. Four electrons are involved in a double bond; six in a triple bond.

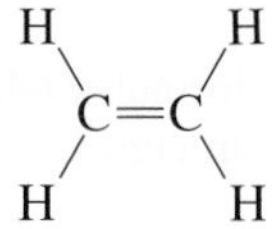

Ethylene has a carbon–carbon double bond.

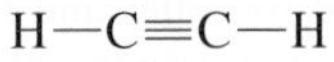

Acetylene has a carbon–carbon triple bond.

Section 1.5 When two atoms that differ in **electronegativity** are covalently bonded, the electrons in the bond are drawn toward the more electronegative element.

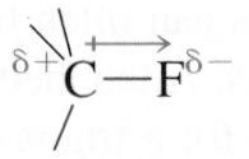

The electrons in a carbon–fluorine bond are drawn away from carbon, toward fluorine.

Section 1.6 Counting electrons and assessing charge distribution in molecules is essential to understanding how structure affects properties. A particular atom in a Lewis structure may be neutral, positively charged, or negatively charged. The **formal charge** of an atom in the Lewis structure of a molecule can be calculated by comparing its electron count with that of the neutral atom itself.

$$\text{Formal charge} = \text{Group number in periodic table} - \text{Electron count}$$

where

$$\text{Electron count} = \tfrac{1}{2}(\text{Number of shared electrons}) + \text{Number of unshared electrons}$$

Section 1.7 Table 1.5 in this section sets forth the procedure to be followed in writing Lewis structures for organic molecules. It begins with experimentally determined information: the **molecular formula** and the **connectivity** (order in which the atoms are connected).

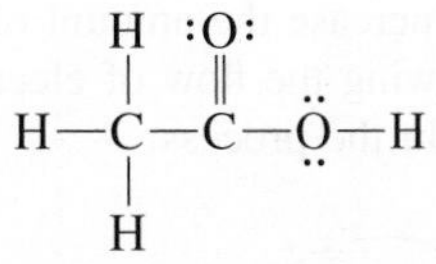

The Lewis structure of acetic acid

Different compounds that have the same molecular formula are called **isomers.** If they are different because their atoms are connected in a different order, they are called **constitutional isomers.**

Formamide (*left*) and formaldoxime (*right*) are constitutional isomers; both have the same molecular formula (CH_3NO), but the atoms are connected in a different order.

Section 1.8 Many molecules can be represented by two or more Lewis structures that differ only in the placement of electrons. In such cases the electrons are delocalized, and the real electron distribution is a hybrid of the **contributing structures.** The rules for resonance are summarized in Table 1.6.

Two Lewis structures (resonance contributors) of formamide; the atoms are connected in the same order, but the arrangement of the electrons is different.

Section 1.9 Condensed formulas and bond-line formulas (skeletal diagrams) are used to economize the drawing of organic structures.

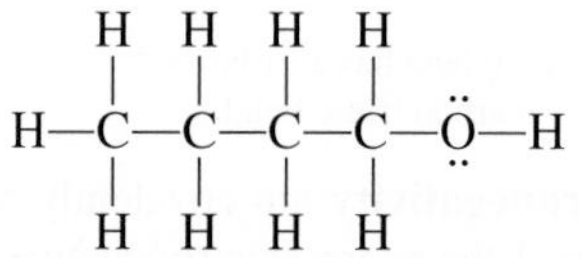

$CH_3CH_2CH_2CH_2OH$
or
$CH_3(CH_2)_2CH_2OH$
condensed formula

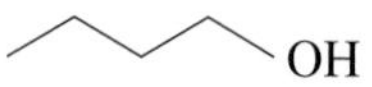

bond-line formula

Section 1.10 The shapes of molecules can often be predicted on the basis of **valence shell electron-pair repulsions.** A tetrahedral arrangement gives the maximum separation of four electron pairs (*left*); a trigonal planar arrangement is best for three electron pairs (*center*), and a linear arrangement for two electron pairs (*right*).

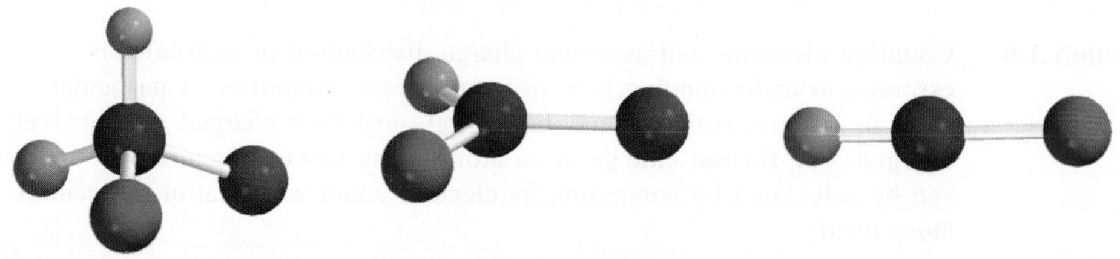

Section 1.11 Knowing the shape of a molecule and the polarity of its various bonds allows the presence or absence of a **molecular dipole moment** and its direction to be predicted.

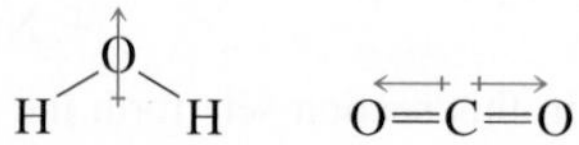

Both water and carbon dioxide have polar bonds, but water has a dipole moment while carbon dioxide does not.

Section 1.12 Curved arrows increase the amount of information provided by a chemical equation by showing the flow of electrons associated with bond making and bond breaking. In the process:

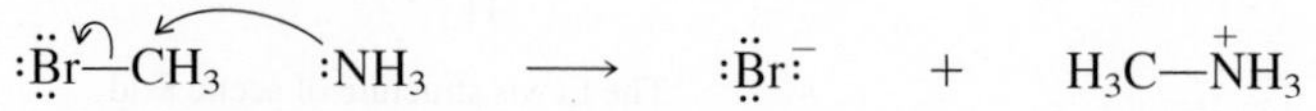

an electron pair of nitrogen becomes the pair of electrons in a C—N bond. The C—Br bond breaks, with the pair of electrons in that bond becoming an unshared pair of bromide ion.

Section 1.13 According to the Arrhenius definitions, an acid ionizes in water to produce protons (H^+) and a base produces hydroxide ions (HO^-). The strength of an acid is given by its equilibrium constant K_a for ionization in aqueous solution:

$$K_a = \frac{[H^+][:A^-]}{[HA]}$$

or more conveniently by its pK_a:

$$pK_a = -\log_{10} K_a$$

Section 1.14 According to the Brønsted–Lowry definitions, an acid is a proton donor and a base is a proton acceptor.

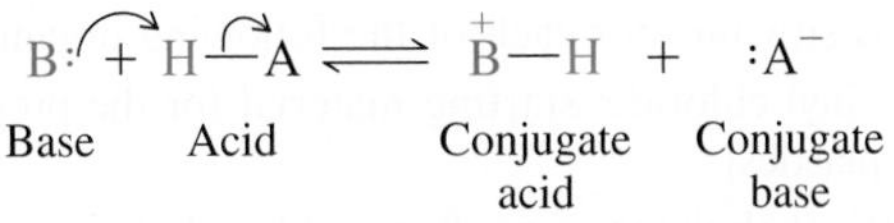
B:
H—A
B—H
:A⁻
Base
Acid
Conjugate acid
Conjugate base

The Brønsted–Lowry approach to acids and bases is more generally useful than the Arrhenius approach.

Section 1.15 **Basicity constants** are not necessary in the Brønsted–Lowry approach. Basicity is measured according to the pK_a of the conjugate acid. The weaker the conjugate acid, the stronger the base.

Section 1.16 The strength of an acid depends on the atom to which the proton is bonded. The two main factors are the strength of the H—X bond and the electronegativity of X. Bond strength is more important for atoms in the same group of the periodic table; electronegativity is more important for atoms in the same row. Electronegative atoms elsewhere in the molecule can increase the acidity by **inductive effects.**

Electron **delocalization** in the conjugate base, usually expressed via resonance between Lewis structures, increases acidity by stabilizing the conjugate base.

Section 1.17 The position of equilibrium in an acid–base reaction lies to the side of the weaker acid.

$$\text{Stronger acid} + \text{Stronger base} \underset{}{\overset{K > 1}{\rightleftharpoons}} \text{Weaker acid} + \text{Weaker base}$$

This is a very useful relationship. You should practice writing equations according to the Brønsted–Lowry definitions of acids and bases and familiarize yourself with Table 1.8, which gives the pK_a's of various Brønsted acids.

Section 1.18 The Lewis definitions of acids and bases provide for a more general view of acid–base reactions than either the Arrhenius or Brønsted–Lowry picture. A **Lewis acid** is an electron-pair acceptor. A **Lewis base** is an electron-pair donor. The Lewis approach incorporates the Brønsted–Lowry approach as a subcategory in which the atom that accepts the electron pair in the Lewis acid is a hydrogen.

PROBLEMS

1.42 Each of the following species will be encountered at some point in this text. They all have the same number of electrons binding the same number of atoms and the same arrangement of bonds; they are *isoelectronic.* Specify which atoms, if any, bear a formal charge in the Lewis structure given and the net charge for each species.

(a) :N≡N:
(b) :C≡N:
(c) :C≡C:
(d) :N≡O:
(e) :C≡O:

1.43 All the following compounds are characterized by ionic bonding between a group 1 metal cation and a tetrahedral anion. Write an appropriate Lewis structure for each anion, remembering to specify formal charges where they exist.

(a) $NaBF_4$ (c) K_2SO_4

(b) $LiAlH_4$ (d) Na_3PO_4

1.44 The connectivity of carbon oxysulfide is OCS.

(a) Write a Lewis structure for carbon oxysulfide that satisfies the octet rule.

(b) What is the molecular geometry according to VSEPR?

(c) Does carbon oxysulfide have a dipole moment? If so, what is its direction?

1.45 *Peroxides* are compounds that contain an O—O bond. Write Lewis formulas for two isomeric peroxides having the molecular formula $C_2H_6O_2$. Include all unshared electron pairs.

1.46 Write a Lewis structure for each of the following organic molecules:

(a) C_2H_3Cl [vinyl chloride: starting material for the preparation of poly(vinyl chloride), or PVC, plastics]

(b) $C_2HBrClF_3$ (halothane: a nonflammable inhalation anesthetic; all three fluorines are bonded to the same carbon)

(c) $C_2Cl_2F_4$ (Freon 114: formerly used as a refrigerant and as an aerosol propellant; each carbon bears one chlorine)

1.47 Write a Lewis formula for the CH_3NO isomer characterized by the structural unit indicated. None of the atoms in the Lewis structure should have a formal charge.

(a) $C-N=O$ (c) $O-C=N$

(b) $C=N-O$ (d) $O=C-N$

1.48 Consider Lewis formulas A, B, and C:

$$\underset{A}{H_2\ddot{C}-N\equiv N:} \qquad \underset{B}{H_2C=N=\ddot{N}:} \qquad \underset{C}{H_2C-\ddot{N}=\ddot{N}:}$$

(a) Are A, B, and C constitutional isomers, or are they resonance contributors?

(b) Which have a negatively charged carbon?

(c) Which have a positively charged carbon?

(d) Which have a positively charged nitrogen?

(e) Which have a negatively charged nitrogen?

(f) What is the net charge on each?

(g) Which is a more stable structure, A or B? Why?

(h) Which is a more stable structure, B or C? Why?

(i) What is the CNN geometry in each according to VSEPR?

1.49 Complete the following Lewis structures. Indicate any unshared pairs of electrons and formal charges. The formulas represent neutral molecules.

(a) 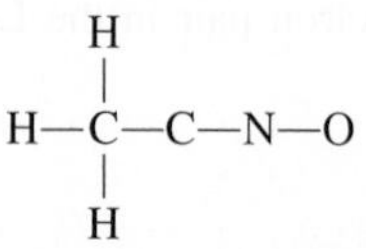

(b) 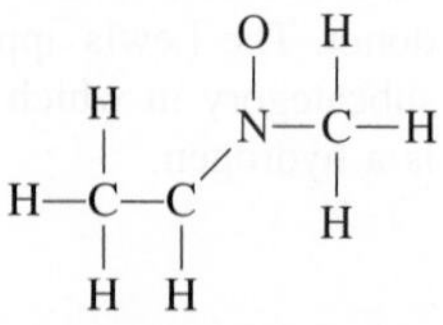

1.50 In each of the following pairs, determine whether the two represent resonance contributors of a single species or depict different substances. If two structures are not resonance contributors, explain why.

(a) $:\ddot{N}-N\equiv N:$ and $:\dot{N}=N=\dot{N}:$

(b) $:\ddot{N}-N\equiv N:$ and $:\ddot{N}-N=\ddot{N}:$

(c) $:\ddot{N}-N\equiv N:$ and $:\ddot{N}-\ddot{N}-\ddot{N}:$

1.51 (a) Which one of the following is *not* a permissible contributing structure? Why?

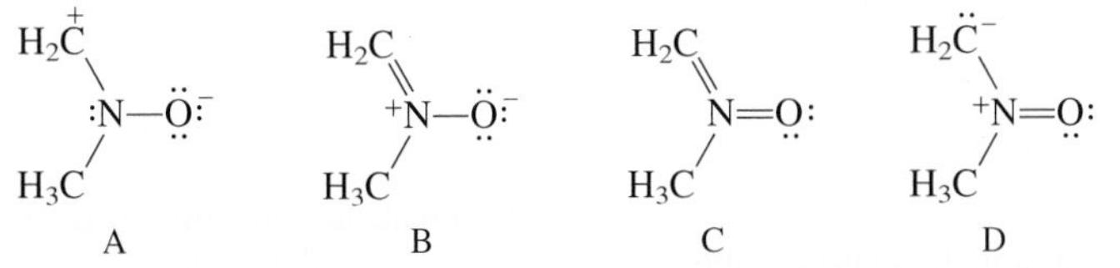

(b) Rank the three remaining structures in order of their contribution to the resonance hybrid. Explain your reasoning.

(c) Using curved arrows, show the electron movement that connects the three resonance contributors.

1.52 Write a more stable contributing structure for each of the following. Use curved arrows to show how to transform the original Lewis formula to the new one. Be sure to specify formal charges, if any.

(a) $H_3C—\ddot{N}=\overset{+}{N}:$

(b) 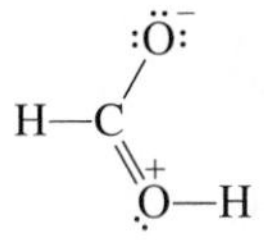

(c) $H_2\overset{+}{C}—\overset{-}{\underset{\cdot\cdot}{C}}H_2$

(d) $H_2\overset{+}{C}—CH=CH—\overset{-}{\underset{\cdot\cdot}{C}}H_2$

(e) $H_2\overset{+}{C}—CH=CH—\ddot{\underset{\cdot\cdot}{O}}:^-$

(f) 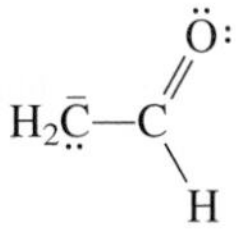

(g) $H—\overset{+}{C}=\ddot{O}:$

(h) $H_2\overset{+}{C}—\ddot{\underset{\cdot\cdot}{O}}H$

(i) 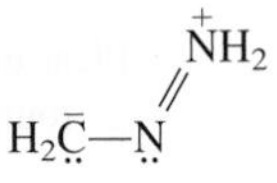

1.53 *Dimethyl sulfoxide* (DMSO) is a byproduct of paper making and has a number of uses, especially as a solvent. It is a neutral molecule having the connectivity $(CH_3)_2SO$.

(a) Write a Lewis structure of DMSO that obeys the octet rule. Show all unshared pairs and any formal charges.

(b) The octet rule may be exceeded for elements beyond the second period of the periodic table. Write a Lewis formula for DMSO with ten valence electrons around sulfur.

1.54 Write structural formulas for all the constitutionally isomeric compounds having the given molecular formula.

(a) C_4H_{10} (c) $C_2H_4Cl_2$ (e) C_3H_9N

(b) C_5H_{12} (d) C_4H_9Br

1.55 Write structural formulas for all the constitutional isomers of

(a) C_3H_8 (b) C_3H_6 (c) C_3H_4

1.56 Write structural formulas for all the constitutional isomers of molecular formula C_3H_6O that contain

(a) Only single bonds (b) One double bond

1.57 For each of the following molecules that contain polar covalent bonds, indicate the positive and negative ends of the dipole, using the symbol $\mapsto$. Refer to Table 1.3 as needed.

(a) HCl (c) HI (e) HOCl

(b) ICl (d) H_2O

1.58 The compounds FCl and ICl have dipole moments μ that are similar in magnitude (0.9 and 0.7 D, respectively) but opposite in direction. In one compound, chlorine is the positive end of the dipole; in the other it is the negative end. Specify the direction of the dipole moment in each compound, and explain your reasoning.

1.59 Which compound in each of the following pairs would you expect to have the greater dipole moment μ? Why?

(a) NaCl or HCl (e) $CHCl_3$ or CCl_3F

(b) HF or HCl (f) CH_3NH_2 or CH_3OH

(c) HF or BF_3 (g) CH_3NH_2 or CH_3NO_2

(d) $(CH_3)_3CH$ or $(CH_3)_3CCl$

1.60 Expand the following structural representations so as to more clearly show all the atoms and any unshared electron pairs.

(a) 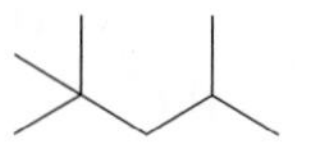

A component of high-octane gasoline

(b) 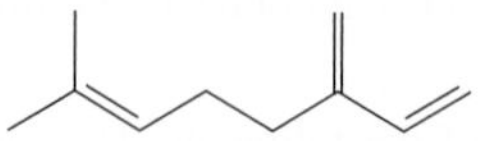

Occurs in bay and verbena oil

(c) 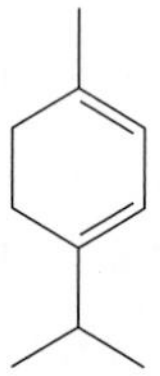

Pleasant-smelling substance found in marjoram oil

(d) 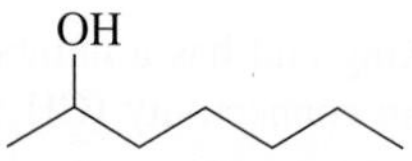

Present in oil of cloves

(e) 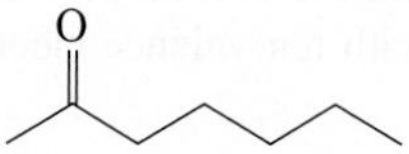

Found in Roquefort cheese

(f) 

Benzene: parent compound of a large family of organic substances

(g) 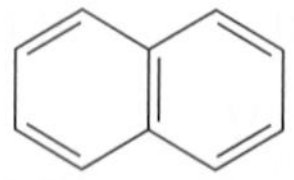

Naphthalene: sometimes used as a moth repellent

(h) 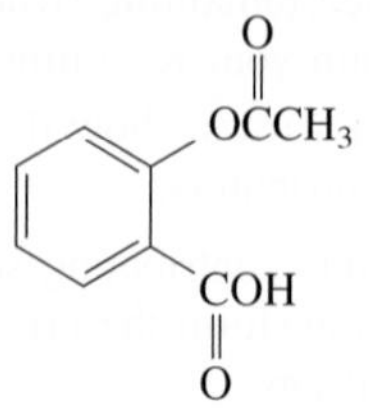

Aspirin

(i) 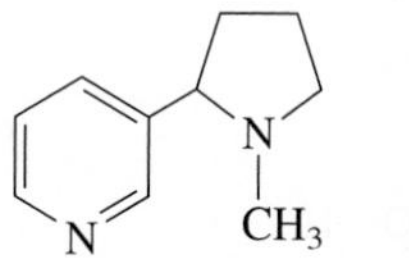

Nicotine: a toxic substance present in tobacco

(j) 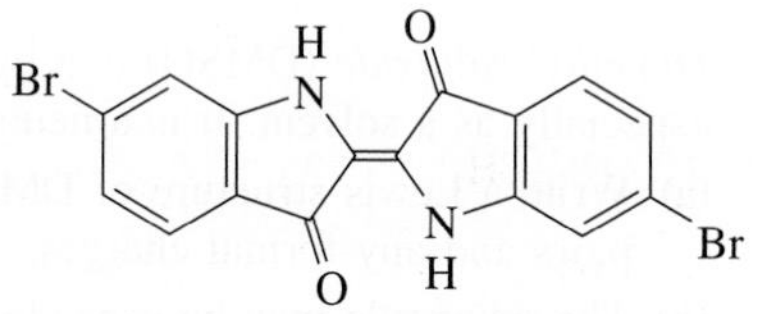

Tyrian purple: a purple dye extracted from a species of Mediterranean sea snail

1.61 Molecular formulas of organic compounds are customarily presented in the fashion $C_2H_5BrO_2$. The number of carbon and hydrogen atoms are presented first, followed by the other atoms in alphabetical order. Give the molecular formulas corresponding to each of the compounds in the preceding problem. Are any of them isomers?

1.62 The structure of montelukast, an antiasthma drug, is shown here.

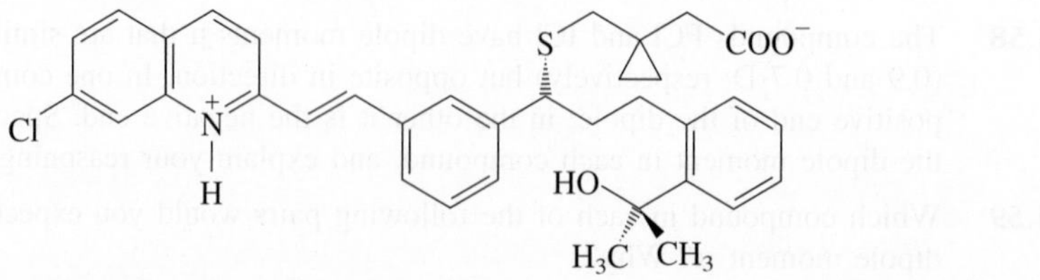

(a) What is the molecular formula of montelukast?

(b) Use Table 1.8 to identify the most acidic and most basic sites in the molecule. (Although you won't find an exact match in structure, make a prediction based on analogy with similar groups in simpler molecules.)

(c) Write the structure of the product formed by treating montelukast with one equivalent of sodium hydroxide.

(d) Write the structure of the product formed by treating montelukast with one equivalent of HCl.

1.63 (a) One acid has a pK_a of 2, the other has a pK_a of 8. What is the ratio of their K_a's?

(b) Two acids differ by 10,000 in their K_a's. If the pK_a of the weaker acid is 5, what is the pK_a of the stronger acid?

1.64 Calculate K_a for each of the following acids, given its pK_a. Rank the compounds in order of decreasing acidity.

(a) Aspirin: pK_a = 3.48

(b) Vitamin C (ascorbic acid): pK_a = 4.17

(c) Formic acid (present in sting of ants): pK_a = 3.75

(d) Oxalic acid (poisonous substance found in certain berries): pK_a = 1.19

1.65 Rank the following in order of decreasing acidity. Although none of these specific structures appear in Table 1.8, you can use analogous structures in the table to guide your reasoning.

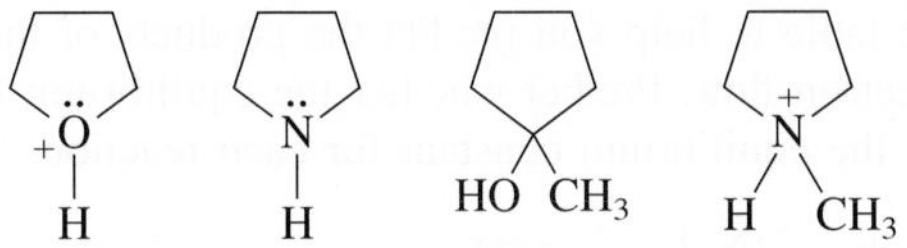

1.66 Rank the following in order of decreasing basicity. As in the preceding problem, Table 1.8 should prove helpful.

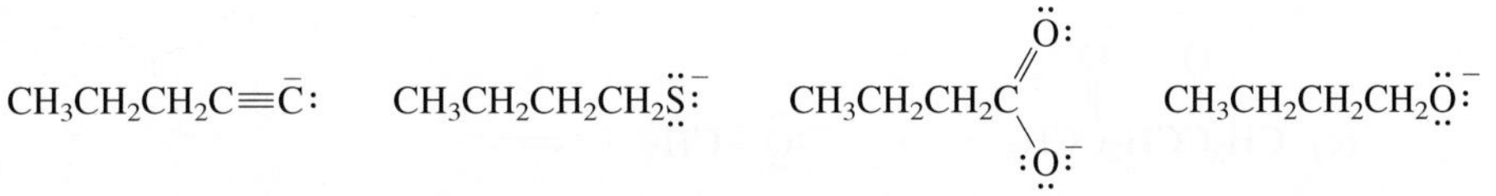

1.67 Consider 1.0 M aqueous solutions of each of the following. Which solution is more basic?

(a) Sodium cyanide (NaCN) or sodium fluoride (NaF)

(b) Sodium carbonate (Na_2CO_3) or sodium acetate ($CH_3\overset{O}{\overset{\|}{C}}ONa$)

(c) Sodium sulfate (Na_2SO_4) or sodium methanethiolate ($NaSCH_3$)

1.68 (a) Which is the stronger acid: $(CH_3)_3NH^+$ or $(CH_3)_3PH^+$?

(b) Which is the stronger base: $(CH_3)_3N$: or $(CH_3)_3P$:?

1.69 Write an equation for the Brønsted–Lowry acid–base reaction that occurs when each of the following acids reacts with water. Show all unshared electron pairs and formal charges, and use curved arrows to track electron movement.

(a) H—C≡N:

(b) 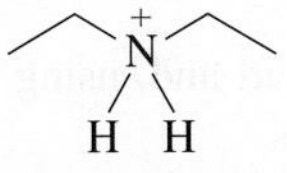

(c) 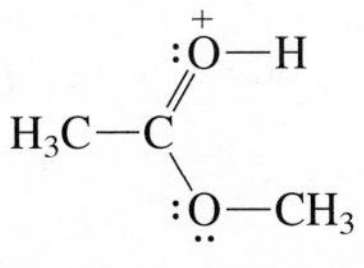

1.70 Write an equation for the Brønsted–Lowry acid–base reaction that occurs when each of the following bases reacts with water. Show all unshared electron pairs and formal charges, and use curved arrows to track electron movement.

(a) $H_3C—C≡\bar{C}$:

(b) 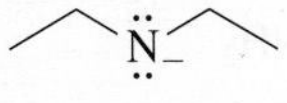

(c) 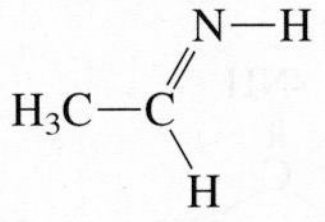

1.71 All of the substances shown in the following acid–base reactions are found in Table 1.8, and the equilibrium lies to the right in each case. Following the curved arrows, complete each equation to show the products formed. Identify the acid, base, conjugate acid, and conjugate base. Calculate the equilibrium constant for each reaction.

(a) 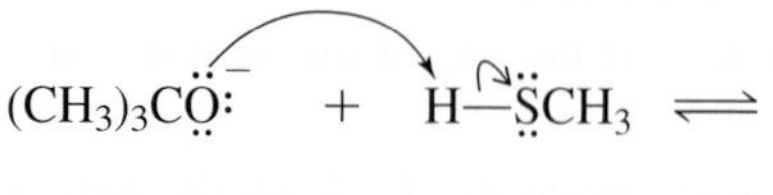

(b) 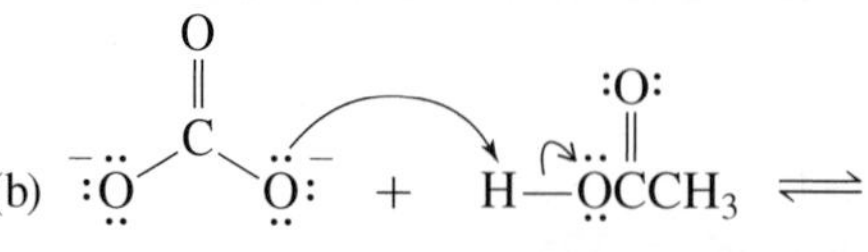

(c) 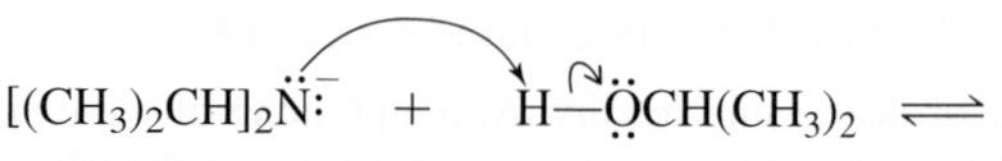

1.72 Each of the following acid–base reactions involves substances found in Table 1.8. Use the pK_a data in the table to help you predict the products of the reactions. Use curved arrows to show electron flow. Predict whether the equilibrium lies to the left or to the right and calculate the equilibrium constant for each reaction.

(a) $HC\equiv CH \quad + \quad {}^{-}\!:\ddot{N}H_2 \quad \rightleftharpoons$

(b) $HC\equiv CH \quad + \quad {}^{-}\!:\ddot{\underset{..}{O}}-CH_3 \quad \rightleftharpoons$

(c) $CH_3\overset{\overset{O}{\|}}{C}CH_2\overset{\overset{O}{\|}}{C}CH_3 \quad + \quad {}^{-}\!:\ddot{\underset{..}{O}}-CH_3 \quad \rightleftharpoons$

(d) $CH_3\overset{\overset{O}{\|}}{C}OCH_3 \quad + \quad {}^{-}\!:\ddot{\underset{..}{O}}-CH_3 \quad \rightleftharpoons$

(e) $CH_3\overset{\overset{O}{\|}}{C}OCH_3 \quad + \quad {}^{-}\!:\ddot{N}[CH(CH_3)_2]_2 \quad \rightleftharpoons$

1.73 With a pK_a of 1.2, squaric acid is unusually acidic for a compound containing only C, H, and O.

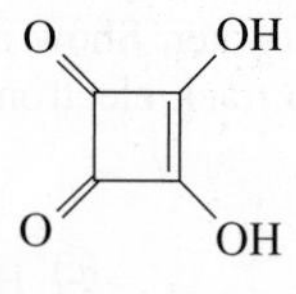

Squaric acid

Write a Lewis dot structure for the conjugate base of squaric acid and, using curved arrows, show how the negative charge is shared by two oxygens.

1.74 Of two possible structures A and B for the conjugate acid of guanidine, the more stable is the one that is better stabilized by electron delocalization. Which one is it? Write resonance structures showing this electron delocalization.

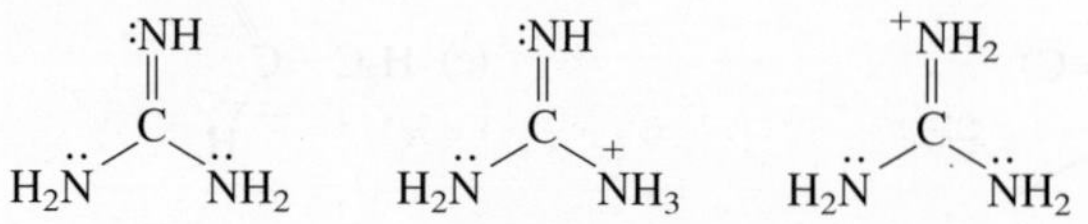

Guanidine A B

Descriptive Passage and Interpretive Problems 1

Amide Lewis Structures

Lewis dot formulas are the major means by which structural information is communicated in organic chemistry. These structural formulas show the atoms, bonds, location of unshared pairs, and formal charges.

Two or more Lewis structures, differing only in the placement of electrons, can often be written for a single compound. In such cases the separate Lewis structures are said to be in *resonance,* and the true electron distribution is a *hybrid* of the electron distributions of the *contributing structures.*

The amide function is an important structural unit in peptides and proteins. Formamide, represented by the Lewis structure shown, is the simplest amide. It is a planar molecule with a dipole moment of 3.7 D. Lewis structures I–IV represent species that bear some relationship to the Lewis structure for formamide.

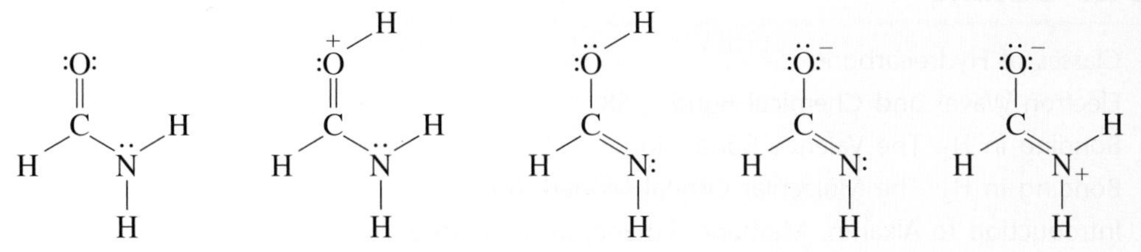

Formamide I II III IV

1.75 Which Lewis formula is a resonance contributor to the structure of formamide?

A. I C. III
B. II D. IV

1.76 Which Lewis formula is a constitutional isomer of formamide?

A. I C. III
B. II D. IV

1.77 Which Lewis formula corresponds to the conjugate acid of formamide?

A. I C. III
B. II D. IV

1.78 Which Lewis formula corresponds to the conjugate base of formamide?

A. I C. III
B. II D. IV

1.79 According to VSEPR, which Lewis formula has a pyramidal arrangement of bonds to nitrogen?

A. I C. III
B. II D. IV

2 Alkanes and Cycloalkanes: Introduction to Hydrocarbons

Chapter Outline

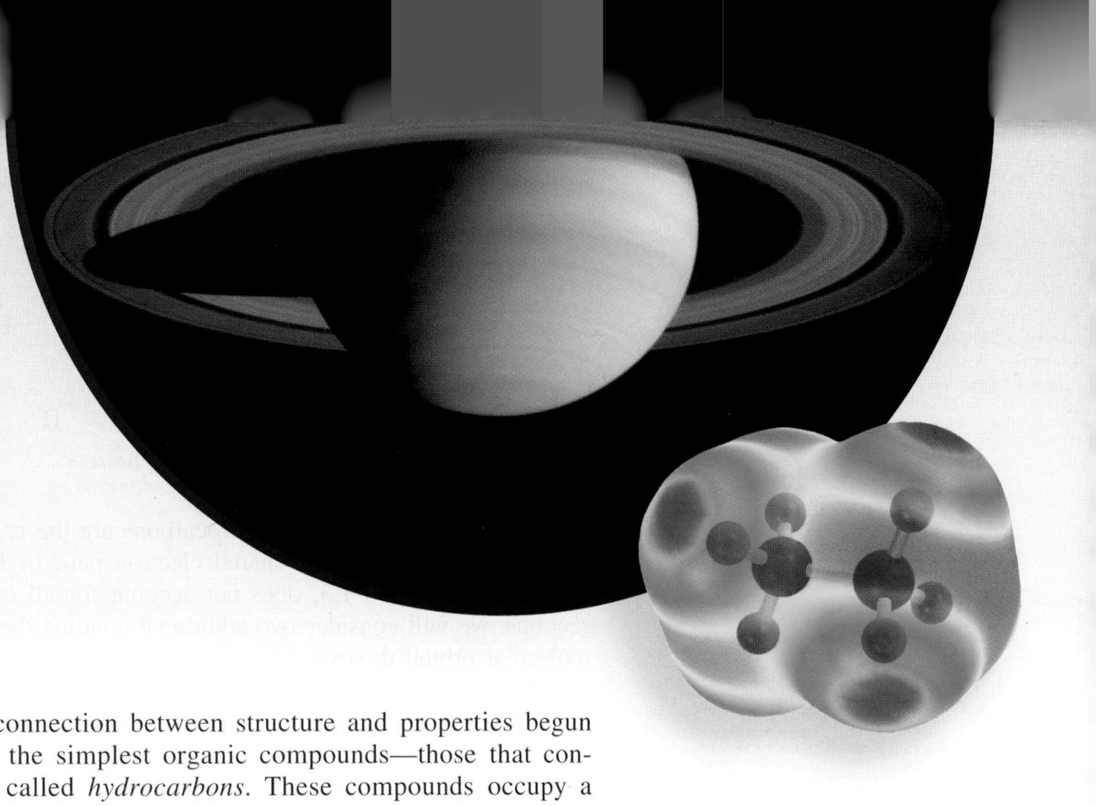

Saturn's moon Titan is the only planetary satellite in the solar system with an atmosphere. Ethane is present in Titan's atmosphere along with the two major components nitrogen and methane.

THIS CHAPTER continues the connection between structure and properties begun in Chapter 1. In it we focus on the simplest organic compounds—those that contain only carbon and hydrogen, called *hydrocarbons*. These compounds occupy a key position in the organic chemical landscape. Their framework of carbon–carbon bonds provides the scaffolding on which more reactive groups, called *functional groups,* are attached. We'll have more to say about functional groups beginning in Chapter 4; for now, we'll explore aspects of structure and bonding in hydrocarbons, especially alkanes.

We'll expand our picture of bonding by introducing two approaches that grew out of the idea that electrons can be described as waves: the *valence bond* and *molecular orbital* models. In particular, one aspect of the valence bond model, called *orbital hybridization,* will be emphasized.

A major portion of this chapter deals with how we name organic compounds. The system used throughout the world is based on a set of rules for naming hydrocarbons, then extending these rules to encompass other families of organic compounds.

2.1 Classes of Hydrocarbons

Hydrocarbons are divided into two main classes: aliphatic and aromatic. This classification dates from the nineteenth century, when organic chemistry was devoted almost entirely to the study of materials from natural sources, and terms were coined that reflected a substance's origin. Two sources were fats and oils, and the word *aliphatic* was derived from the Greek word *aleiphar* meaning "fat." Aromatic hydrocarbons, irrespective of their own odor, were typically obtained by chemical treatment of pleasant-smelling plant extracts.

Aliphatic hydrocarbons include three major groups: *alkanes, alkenes,* and *alkynes.* **Alkanes** are hydrocarbons in which all the bonds are single bonds, **alkenes** contain at least one carbon–carbon double bond, and **alkynes** contain at least one carbon–carbon triple bond. Examples of the three classes of aliphatic hydrocarbons are the two-carbon compounds *ethane, ethylene,* and *acetylene.*

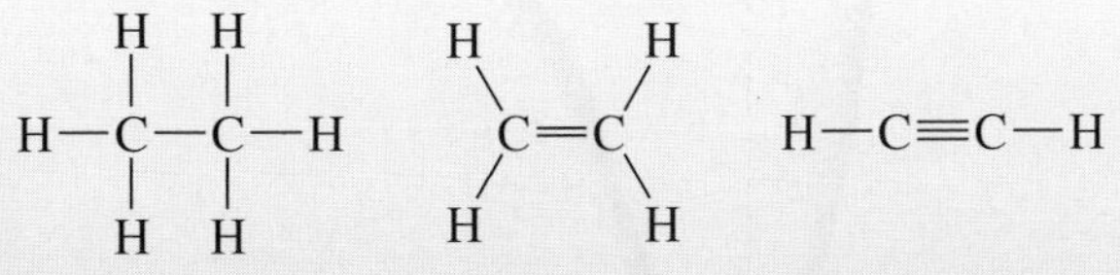

Ethane (alkane) — Ethylene (alkene) — Acetylene (alkyne)

Another name for aromatic hydrocarbons is **arenes.** The most important aromatic hydrocarbon is *benzene.*

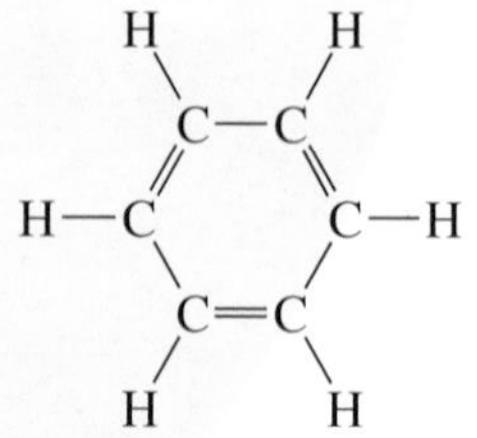

Benzene
(arene)

Different properties in these hydrocarbons are the result of the different types of bonding involving carbon. The shared electron pair, or Lewis model of chemical bonding described in Section 1.3, does not account for all of the differences. In the following sections, we will consider two additional bonding theories; the valence bond model and molecular orbital theory.

2.2 Electron Waves and Chemical Bonds

De Broglie's and Schrödinger's contributions to our present understanding of electrons were described in Section 1.1.

G.N. Lewis proposed his shared electron-pair model of bonding in 1916, almost a decade before Louis de Broglie's theory of wave–particle duality. De Broglie's radically different view of an electron, and Erwin Schrödinger's success in using wave equations to calculate the energy of an electron in a hydrogen *atom,* encouraged the belief that bonding in *molecules* could be explained on the basis of interactions between electron waves. This thinking produced two widely used theories of chemical bonding; one is called the *valence bond model,* the other the *molecular orbital model.*

All of the forces in chemistry, except for nuclear chemistry, are electrical. Opposite charges attract; like charges repel. This simple fact can take you a long way.

Before we describe these theories in the context of organic molecules, let's first think about bonding between two hydrogen atoms in the most fundamental terms. We'll begin with two hydrogen atoms that are far apart and see what happens as the distance between them decreases. The forces involved are electron–electron (−−) repulsions, nucleus–nucleus (++) repulsions, and electron–nucleus (−+) attractions. All of these forces *increase* as the distance between the two hydrogens *decreases.* Because the electrons are so mobile, however, they can choreograph their motions so as to minimize their mutual repulsion while maximizing their attractive forces with the protons. Thus, as shown in Figure 2.1, a net, albeit weak, attractive force exists between the two hydrogens even when the atoms are far apart. This interaction becomes stronger as the two atoms approach each other—the electron of each hydrogen increasingly feels the attractive force

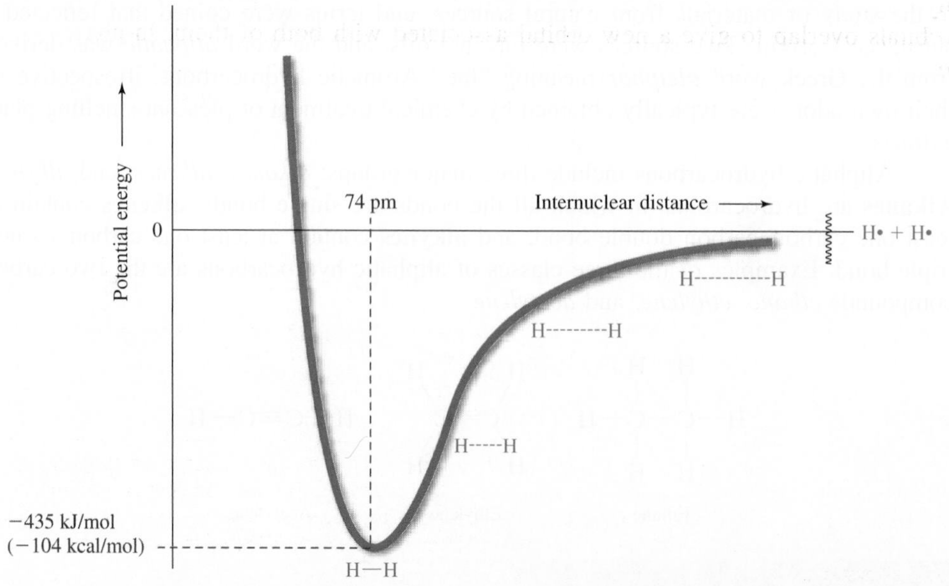

Figure 2.1

Plot of potential energy versus distance for two hydrogen atoms. At long distances, there is a weak attractive force. As the distance decreases, the potential energy decreases, and the system becomes more stable because each electron now "feels" the attractive force of two protons rather than one. The lowest energy state corresponds to a separation of 74 pm, which is the normal bond distance in H_2. At shorter distances, nucleus–nucleus and electron–electron repulsions are greater than electron–nucleus attractions, and the system becomes less stable.

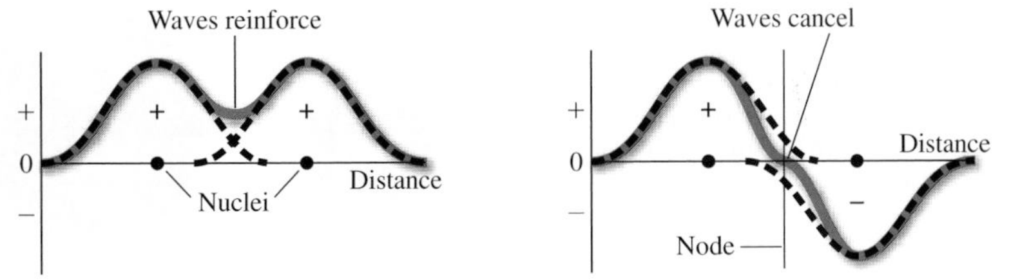

(*a*) Amplitudes of wave functions added

(*b*) Amplitudes of wave functions subtracted

Figure 2.2

Interference between waves. (*a*) Constructive interference occurs when two waves combine in phase with each other. The amplitude of the resulting wave at each point is the sum of the amplitudes of the original waves. (*b*) Destructive interference decreases the amplitude when two waves are out of phase with each other.

of two protons rather than one, the total energy decreases, and the system becomes more stable. A potential energy minimum is reached when the separation between the nuclei reaches 74 pm, which corresponds to the H—H bond length in H_2. At distances shorter than this, the nucleus–nucleus and electron–electron repulsions dominate, and the system becomes less stable.

Valence bond and molecular orbital theory both incorporate the wave description of an atom's electrons into this picture of H_2, but in somewhat different ways. Both assume that electron waves behave like more familiar waves, such as sound and light waves. One important property of waves is called interference in physics. *Constructive interference* occurs when two waves combine so as to reinforce each other (in phase); *destructive interference* occurs when they oppose each other (out of phase) (Figure 2.2).

Recall from Section 1.1 that electron waves in atoms are characterized by their wave function, which is the same as an orbital. For an electron in the most stable state of a hydrogen atom, for example, this state is defined by the 1*s* wave function and is often called the 1*s* orbital. The *valence bond* model bases the connection between two atoms on the overlap between half-filled orbitals of the two atoms. The *molecular orbital* model assembles a set of molecular orbitals by combining the atomic orbitals of *all* of the atoms in the molecule.

For a molecule as simple as H_2, valence bond and molecular orbital theory produce very similar pictures. The next two sections describe these two approaches.

2.3 Bonding in H_2: The Valence Bond Model

The characteristic feature of **valence bond theory** is that it pictures a covalent bond between two atoms in terms of an in-phase overlap of a half-filled orbital of one atom with a half-filled orbital of the other, illustrated for the case of H_2 in Figure 2.3. Two hydrogen atoms, each containing an electron in a 1*s* orbital, combine so that their orbitals overlap to give a new orbital associated with both of them. In-phase orbital overlap (constructive interference) increases the probability of finding an electron in the region between the two nuclei where it feels the attractive force of both of them.

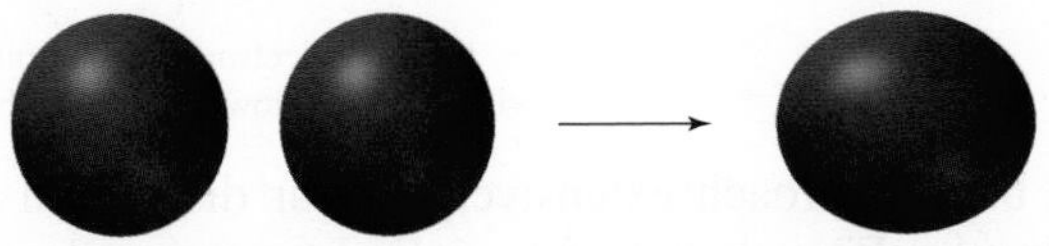

1*s* orbitals of two hydrogen atoms. Each orbital contains one electron.

In-phase overlap of two 1*s* orbitals gives new orbital encompassing both hydrogen atoms. This orbital contains two electrons.

Figure 2.3

Valence bond picture of bonding in H_2. Overlap of half-filled 1*s* orbitals of two hydrogen atoms gives a new orbital that encompasses both atoms and contains both electrons. The electron density (electron probability) is highest in the region between the two atoms. When the wave functions are of the same sign, constructive interference increases the probability of finding an electron in the region where the two orbitals overlap.

Figure 2.4

Valence bond picture of bonding in H_2 as illustrated by electrostatic potential maps. The $1s$ orbitals of two hydrogen atoms overlap to give an orbital that contains both electrons of an H_2 molecule.

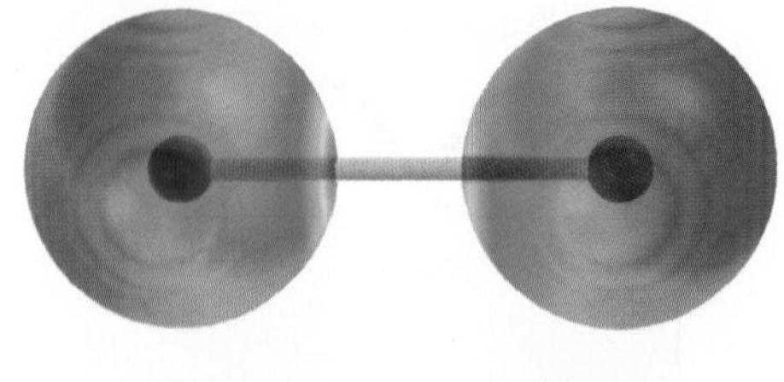

(*a*) The $1s$ orbitals of two separated hydrogen atoms, sufficiently far apart so that essentially no interaction takes place between them. Each electron is associated with only a single proton.

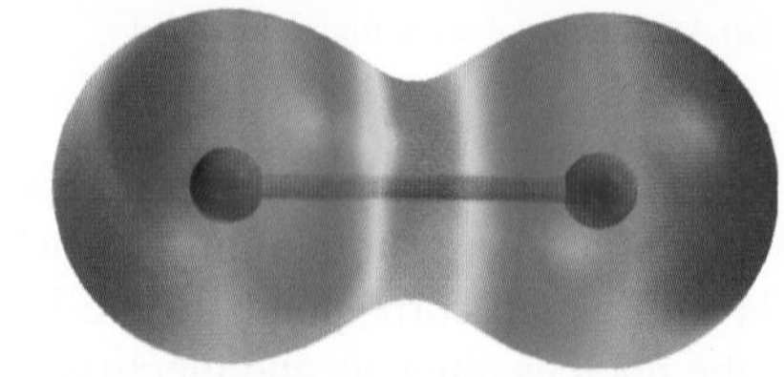

(*b*) As the hydrogen atoms approach each other, their $1s$ orbitals begin to overlap and each electron begins to feel the attractive force of both protons.

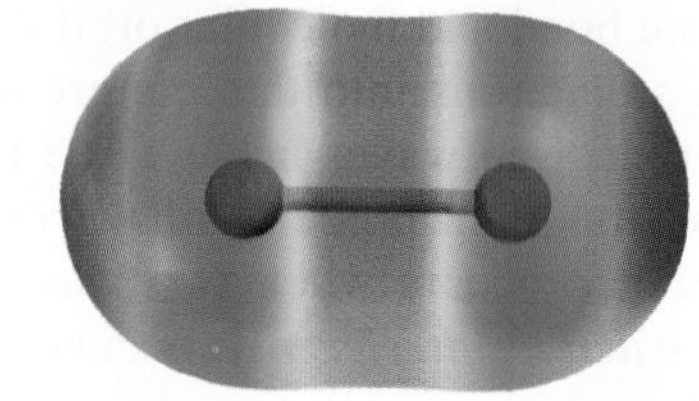

(*c*) The hydrogen atoms are close enough so that appreciable overlap of the two $1s$ orbitals occurs. The concentration of electron density in the region between the two protons is more readily apparent.

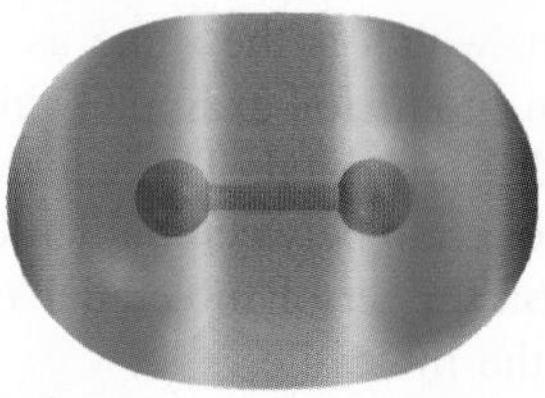

(*d*) A molecule of H_2. The center-to-center distance between the hydrogen atoms is 74 pm. The two individual $1s$ orbitals have been replaced by a new orbital that encompasses both hydrogens and contains both electrons. The electron density is greatest in the region between the two hydrogens.

Figure 2.4 uses electrostatic potential maps to show this build-up of electron density in the region between two hydrogen atoms as they approach each other closely enough for their orbitals to overlap.

A bond in which the orbitals overlap along a line connecting the atoms (the *internuclear axis*) is called a **sigma (σ) bond.** The electron distribution in a σ bond is cylindrically symmetrical; were we to slice through a σ bond perpendicular to the internuclear axis, its cross section would appear as a circle. Another way to see the shape of the electron distribution is to view the molecule end-on.

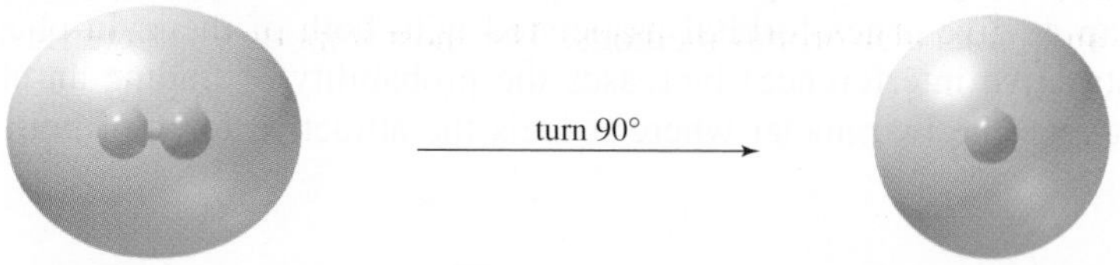

Orbitals overlap along a line connecting the two atoms

Circular electron distribution when viewing down the H—H bond

We will use the valence bond approach extensively in our discussion of organic molecules and expand on it shortly. First though, let's introduce the molecular orbital method to see how it uses the $1s$ orbitals of two hydrogen atoms to generate the orbitals of an H_2 molecule.

2.4 Bonding in H_2: The Molecular Orbital Model

The molecular orbital theory of chemical bonding rests on the notion that, as electrons in atoms occupy *atomic orbitals,* electrons in molecules occupy *molecular orbitals.* Just as our first task in writing the electron configuration of an atom is to identify the atomic

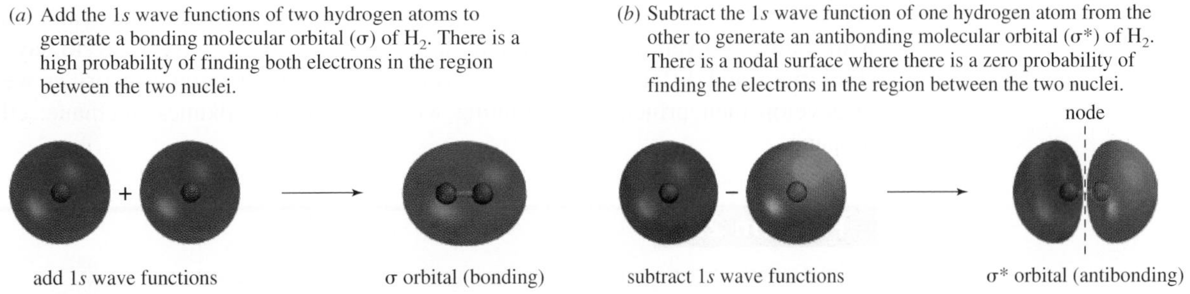
(a) Add the 1s wave functions of two hydrogen atoms to generate a bonding molecular orbital (σ) of H_2. There is a high probability of finding both electrons in the region between the two nuclei.
(b) Subtract the 1s wave function of one hydrogen atom from the other to generate an antibonding molecular orbital (σ*) of H_2. There is a nodal surface where there is a zero probability of finding the electrons in the region between the two nuclei.
node
+
−
add 1s wave functions
σ orbital (bonding)
subtract 1s wave functions
σ* orbital (antibonding)

Figure 2.5

Generation of σ and σ* molecular orbitals of H_2 by combining 1s orbitals of two hydrogen atoms.

orbitals that are available to it, so too must we first describe the orbitals available to a molecule. In the molecular orbital method this is done by representing molecular orbitals as combinations of atomic orbitals, the *linear combination of atomic orbitals-molecular orbital* (LCAO-MO) method.

Two molecular orbitals (MOs) of H_2 are generated by combining the 1*s* atomic orbitals (AOs) of two hydrogen atoms. In one combination, the two wave functions are added; in the other they are subtracted. The two new orbitals that are produced are portrayed in Figure 2.5. The additive combination generates a **bonding orbital;** the subtractive combination generates an **antibonding orbital.** Both the bonding and antibonding orbitals have σ symmetry, meaning that they are symmetrical with respect to the internuclear axis. The two are differentiated by calling the bonding orbital σ and the antibonding orbital σ* ("sigma star"). The bonding orbital is characterized by a region of high electron probability between the two atoms, whereas the antibonding orbital has a nodal surface between them.

A molecular orbital diagram for H_2 is shown in Figure 2.6. The customary format shows the starting AOs at the left and right sides and the MOs in the middle. It must always be true that *the number of MOs is the same as the number of AOs that combine to produce them.* Thus, when the 1*s* AOs of two hydrogen atoms combine, two MOs result. The bonding MO (σ) is lower in energy and the antibonding MO (σ*) higher in energy than either of the original 1*s* orbitals.

When assigning electrons to MOs, the same rules apply as for writing electron configurations of atoms. Electrons fill the MOs in order of increasing orbital energy, and the maximum number of electrons in any orbital is two. Both electrons of H_2 occupy the bonding orbital, have opposite spins, and both are held more strongly than they would be in separated hydrogen atoms. There are no electrons in the antibonding orbital.

For a molecule as simple as H_2, it is hard to see much difference between the valence bond and molecular orbital methods. The most important differences appear in molecules with more than two atoms. In those cases, the valence bond method continues to view a molecule as a collection of bonds between connected atoms. The molecular

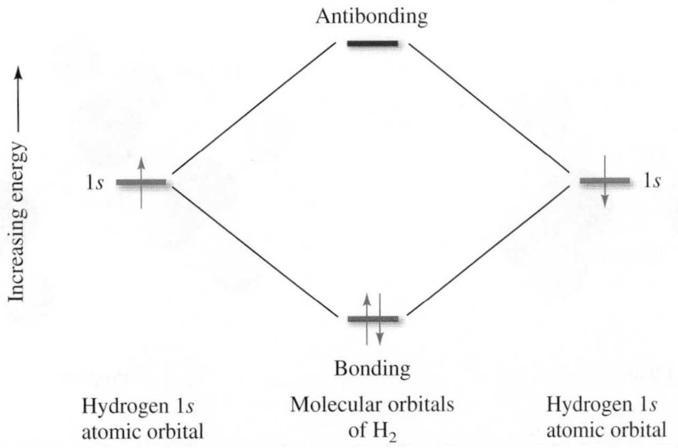
Antibonding
Increasing energy
1s
1s
Bonding
Hydrogen 1s atomic orbital
Molecular orbitals of H_2
Hydrogen 1s atomic orbital

Figure 2.6

Two molecular orbitals (MOs) are generated by combining two hydrogen 1*s* atomic orbitals (AOs). The bonding MO is lower in energy than either of the AOs that combine to produce it. The antibonding MO is of higher energy than either AO. Each arrow indicates one electron, and the electron spins are opposite in sign. Both electrons of H_2 occupy the bonding MO.

orbital method, however, leads to a picture in which the same electron can be associated with many, or even all, of the atoms in a molecule. We'll have more to say about the similarities and differences in valence bond and molecular orbital theory as we continue to develop their principles, beginning with the simplest alkanes: methane, ethane, and propane.

Problem 2.1

(a) Construct a diagram similar to Figure 2.6 for diatomic helium.
(b) Why is helium monatomic instead of diatomic?

2.5 Introduction to Alkanes: Methane, Ethane, and Propane

Alkanes have the general molecular formula C_nH_{2n+2}. The simplest one, methane (CH_4), is also the most abundant. Large amounts are present in our atmosphere, in the ground, and in the oceans. Methane has been found on Mars, Jupiter, Saturn, Uranus, Neptune, and Pluto, and even on Halley's Comet. About 2–8% of the atmosphere of Titan, Saturn's largest moon, is methane. When it rains on Titan, it rains methane.

Ethane (C_2H_6: CH_3CH_3) and propane (C_3H_8: $CH_3CH_2CH_3$) are second and third, respectively, to methane in many ways. Ethane is the alkane next to methane in structural simplicity, followed by propane. Ethane ($\approx$ 10%) is the second and propane ($\approx$ 5%) the third most abundant component of natural gas, which is $\approx$ 75% methane. Natural gas is colorless and nearly odorless, as are methane, ethane, and propane. The characteristic odor of the natural gas we use for heating our homes and cooking comes from trace amounts of unpleasant-smelling sulfur-containing compounds, called thiols, that are deliberately added to it to warn us of potentially dangerous leaks.

Methane is the lowest boiling alkane, followed by ethane, then propane.

	CH_4	CH_3CH_3	$CH_3CH_2CH_3$
	Methane	Ethane	Propane
Boiling point:	−160°C	−89°C	−42°C

Boiling points cited in this text are at 1 atm (760 mm Hg) unless otherwise stated.

It is generally true that as the number of carbon atoms increases, so does the boiling point. All the alkanes with four carbons or fewer are gases at room temperature and atmospheric pressure. With the highest boiling point of the three, propane is the easiest one to liquefy. We are all familiar with "propane tanks." These are steel containers in which a propane-rich mixture of hydrocarbons called *liquefied petroleum gas* (LPG) is maintained in a liquid state under high pressure as a convenient clean-burning fuel.

The structural features of methane, ethane, and propane are summarized in Figure 2.7. All of the carbon atoms have four bonds, all of the bonds are single bonds, and the bond angles are close to tetrahedral. In the next section we'll see how to adapt the valence bond model to accommodate the observed structures.

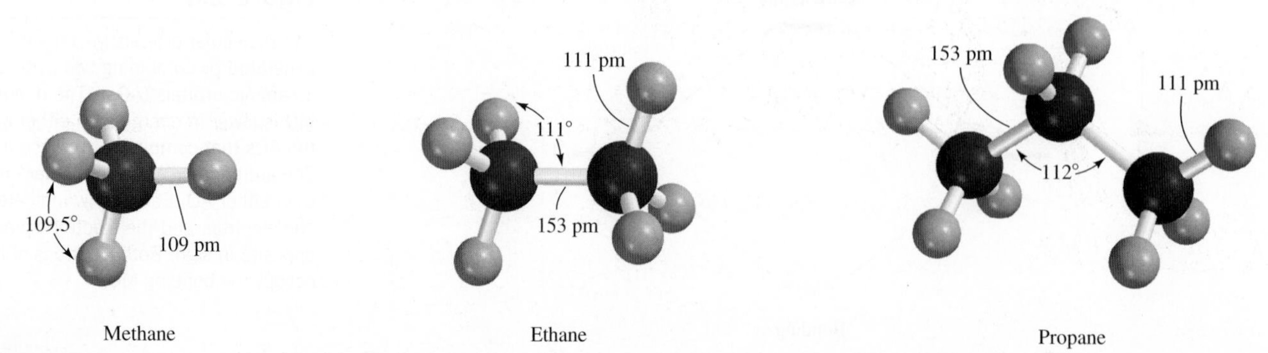

Figure 2.7

Structures of methane, ethane, and propane showing bond distances and bond angles.

Methane and the Biosphere

One of the things that environmental scientists do is to keep track of important elements in the biosphere—in what form do these elements normally occur, to what are they transformed, and how are they returned to their normal state? Careful studies have given clear, although complicated, pictures of the "nitrogen cycle," the "sulfur cycle," and the "phosphorus cycle," for example. The "carbon cycle" begins and ends with atmospheric carbon dioxide. It can be represented in an abbreviated form as:

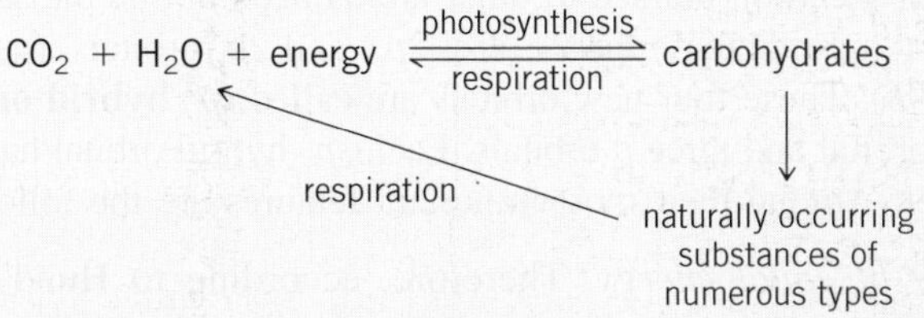
CO2 + H2O + energy
photosynthesis
respiration
carbohydrates
respiration
naturally occurring substances of numerous types

Methane is one of literally millions of compounds in the carbon cycle, but one of the most abundant. It is formed when carbon-containing compounds decompose in the absence of air (*anaerobic* conditions). The organisms that bring this about are called *methanoarchaea*. Cells can be divided into three types: *archaea, bacteria,* and *eukarya*. Methanoarchaea are one kind of archaea and may rank among the oldest living things on Earth. They can convert a number of carbon-containing compounds, including carbon dioxide and acetic acid, to methane.

Virtually anywhere water contacts organic matter in the absence of air is a suitable place for methanoarchaea to thrive—at the bottom of ponds, bogs, and rice fields, for example. *Marsh gas* (*swamp gas*) is mostly methane. Methanoarchaea live inside termites and grass-eating animals. One source quotes 20 L/day as the methane output of a large cow.

The scale on which methanoarchaea churn out methane, estimated to be 10^{11}–10^{12} lb/year, is enormous. About 10% of this amount makes its way into the atmosphere, but most of the rest simply ends up completing the carbon cycle. It exits the anaerobic environment where it was formed and enters the aerobic world where it is eventually converted to carbon dioxide by a variety of processes.

When we consider sources of methane we have to add "old" methane, methane that was formed millions of years ago but became trapped beneath the Earth's surface, to the "new" methane just described. *Firedamp*, an explosion hazard to miners, occurs in layers of coal and is mostly methane. Petroleum deposits, formed by microbial decomposition of plant material under anaerobic conditions, are always accompanied by pockets of natural gas, which is mostly methane.

An interesting thing happens when methane from biological processes leaks from sites under the deep-ocean floor. If the pressure is high enough (50 atm) and the water cold enough (4°C), the methane doesn't simply bubble to the surface. Individual methane molecules become trapped inside clusters of 6–18 water molecules forming *methane clathrates* or *methane hydrates* (Figure 2.8). Aggregates of these hydrates stay at the bottom of the ocean in what looks like a lump of dirty ice, ice that burns. Far from being mere curiosities, methane hydrates are potential sources of energy on a scale greater than that of all known oil reserves combined. The extraction of methane from hydrates was demonstrated on a small scale in 2002. Estimates suggest some modest contribution of methane from hydrates to the global energy supply by 2020.

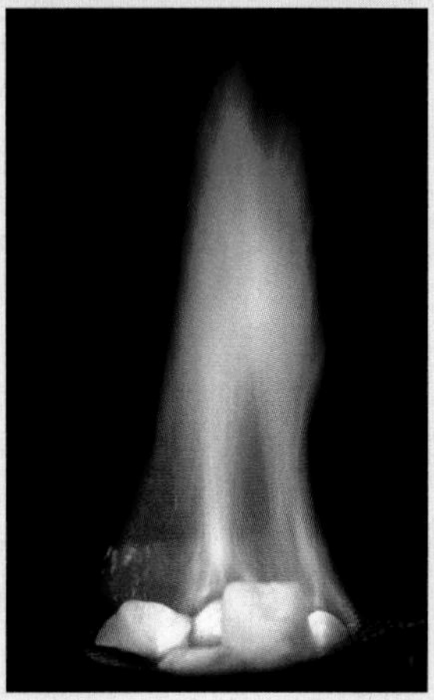

Figure 2.8

Methane burning as it is released from a clathrate.

Not all of the attention that methane hydrates receive is as a potential source of energy though. Environmental scientists are looking into the possibility that methane hydrate dissociation was responsible for a major global warming event that occurred 55 million years ago, lasted 40,000 years, and raised the temperature of the Earth some 5°C. They speculate that a modest warming of the oceans encouraged the dissociation of hydrates, releasing methane—a potent greenhouse gas—into the atmosphere. The resulting greenhouse effect raised the temperature of the Earth, causing more methane to be released from the oceans into the atmosphere, which, in turn, increased the greenhouse warming. Eventually a new, warmer equilibrium state was reached.

Turning to the ocean itself, biologists suspect that the methane in hydrates is a key nutrient for bacteria and other inhabitants of deep-sea ecosystems, including those that lie beneath the ocean floor.

2.6 sp^3 Hybridization and Bonding in Methane

Before we describe the bonding in methane, it is worth emphasizing that bonding theories attempt to describe a molecule on the basis of its component atoms; bonding theories do not attempt to explain *how* bonds form. The world's methane does *not* come from the reaction of carbon atoms with hydrogen atoms; it comes from biological processes. The boxed essay *Methane and the Biosphere* tells you more about the origins of methane and other organic compounds.

We *begin* with the experimentally determined three-dimensional structure of a molecule, *then* propose bonding models that are consistent with the structure. We do not claim that the observed structure is a result of the bonding model. Indeed, there may be two or more equally satisfactory models. Structures are facts; bonding models are theories that we use to try to understand the facts.

A vexing puzzle in the early days of valence bond theory concerned the fact that methane is CH_4 and that the four bonds to carbon are directed toward the corners of a tetrahedron. Valence bond theory is based on the in-phase overlap of half-filled orbitals of the connected atoms. But with an electron configuration of $1s^22s^2\ 2p_x^{\ 1}2p_y^{\ 1}$ carbon has only two half-filled orbitals (Figure 2.9*a*). How, then, can it have four bonds?

In the 1930s Linus Pauling offered an ingenious solution to this puzzle. He suggested that the electron configuration of a carbon bonded to other atoms need not be the same as a free carbon atom. By mixing ("hybridizing") the $2s$, $2p_x$, $2p_y$, and $2p_z$ orbitals, four new orbitals are obtained (Figure 2.9*b*). These four new orbitals are called ***sp*³ hybrid orbitals** because they come from one *s* orbital and three *p* orbitals. Each sp^3 hybrid orbital has 25% *s* character and 75% *p* character. Among their most important features are the following:

1. *All four* sp^3 *orbitals are of equal energy.* Therefore, according to Hund's rule (Section 1.1) the four valence electrons of carbon are distributed equally among them, making four half-filled orbitals available for bonding.
2. *The axes of the* sp^3 *orbitals point toward the corners of a tetrahedron.* Therefore, sp^3 hybridization of carbon is consistent with the tetrahedral structure of methane. Each C—H bond is a σ bond in which a half-filled 1*s* orbital of hydrogen overlaps with a half-filled sp^3 orbital of carbon along a line drawn between them (Figure 2.10).
3. σ *Bonds involving* sp^3 *hybrid orbitals of carbon are stronger than those involving unhybridized* 2s *or* 2p *orbitals.* Each sp^3 hybrid orbital has two lobes of unequal size, making the electron density greater on one side of the nucleus than the other. In a C—H σ bond, it is the larger lobe of a carbon sp^3 orbital that overlaps with a hydrogen 1*s* orbital. This concentrates the electron density in the region between the two atoms.

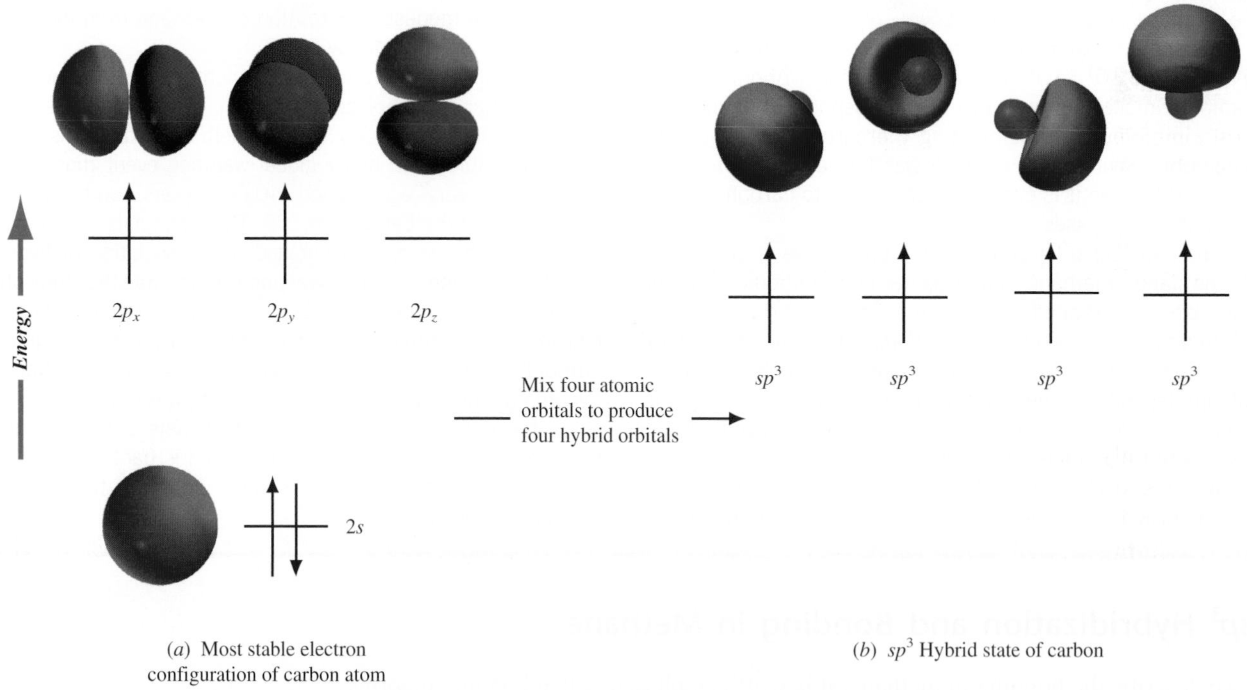

Figure 2.9

sp^3 Hybridization. (*a*) Electron configuration of carbon in its most stable state. (*b*) Mixing the *s* orbital with the three *p* orbitals generates four sp^3 hybrid orbitals. The four sp^3 hybrid orbitals are of equal energy; therefore, the four valence electrons are distributed evenly among them. The axes of the four sp^3 orbitals are directed toward the corners of a tetrahedron.

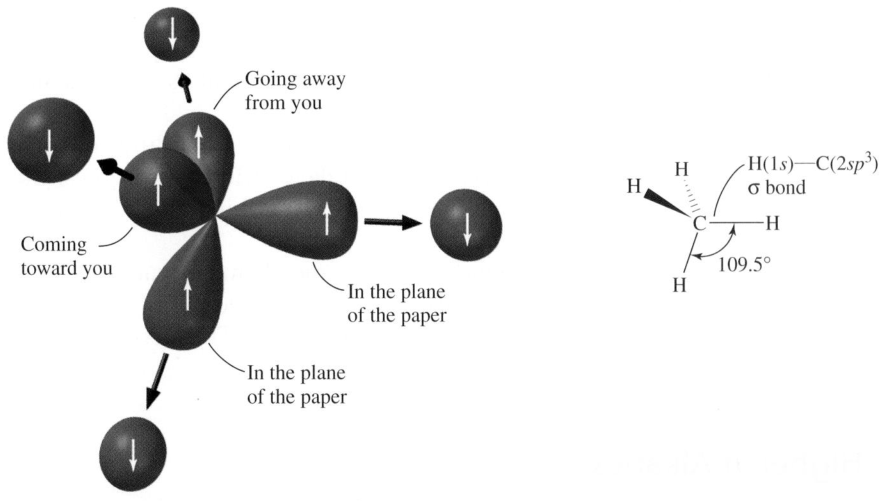

Figure 2.10

Each half-filled sp^3 orbital overlaps with a half-filled hydrogen $1s$ orbital along a line between them giving a tetrahedral arrangement of four σ bonds. Only the major lobe of each sp^3 orbital is shown. Each orbital contains a smaller back lobe, which has been omitted for clarity.

The orbital hybridization model accounts for carbon having four bonds rather than two, the bonds are stronger than they would be in the absence of hybridization, and they are arranged in a tetrahedral fashion around carbon.

2.7 Bonding in Ethane

The orbital hybridization model of covalent bonding is readily extended to carbon–carbon bonds. As Figure 2.11 illustrates, ethane is described in terms of a carbon–carbon σ bond joining two CH_3 **(methyl)** groups. Each methyl group consists of an sp^3-hybridized carbon attached to three hydrogens by sp^3–$1s$ σ bonds. Overlap of the remaining half-filled sp^3 orbital of one carbon with that of the other generates a σ bond between them. Here is a third kind of σ bond, one that has as its basis the overlap of two half-filled sp^3-hybridized orbitals. *In general, you can expect that carbon will be* sp^3*-hybridized when it is directly bonded to four atoms.*

Problem 2.2

Describe the bonding in propane according to the orbital hybridization model.

We will return to the orbital hybridization model to discuss bonding in other aliphatic hydrocarbons—alkenes and alkynes—later in the chapter. At this point, however, we'll explore alkanes as a class in more detail.

2.8 Isomeric Alkanes: The Butanes

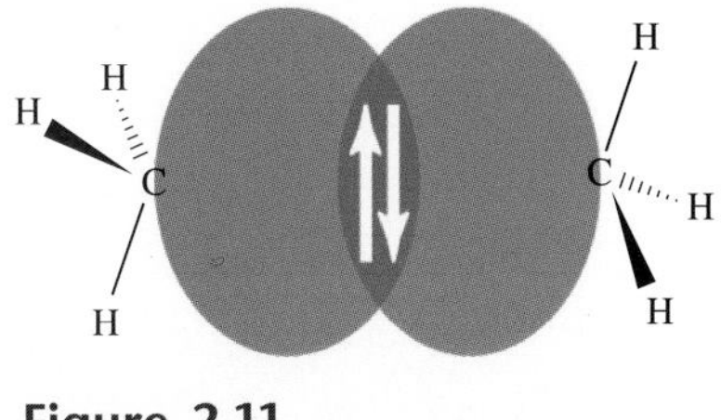

Figure 2.11

The C—C σ bond in ethane, pictured as an overlap of a half-filled sp^3 orbital of one carbon with a half-filled sp^3 orbital of the other. Each C—H σ bond represents the overlap of a half-filled sp^3 orbital of carbon with a half-filled $1s$ orbital of hydrogen.

Methane is the only alkane of molecular formula CH_4, ethane the only one that is C_2H_6, and propane the only one that is C_3H_8. Beginning with C_4H_{10}, however, constitutional isomers (Section 1.7) are possible; two alkanes have this particular molecular formula. In one, called ***n*-butane,** four carbons are joined in a continuous chain. The *n* in *n*-butane stands for "normal" and means that the carbon chain is unbranched. The second isomer has a branched carbon chain and is called **isobutane.**

$CH_3CH_2CH_2CH_3$ $\qquad$ CH_3CHCH_3 (with CH_3 on the central carbon) or $(CH_3)_3CH$

	n-Butane	Isobutane
Boiling point:	−0.4°C	−10.2°C
Melting point:	−139°C	−160.9°C

As just noted (Section 2.7), CH_3 is called a *methyl* group. In addition to having methyl groups at both ends, *n*-butane contains two CH_2, or **methylene** groups. Isobutane contains three methyl groups bonded to a CH unit. The CH unit is called a **methine** group.

"Butane" lighters contain about 5% *n*-butane and 95% isobutane in a sealed container. The pressure produced by the two compounds (about 3 atm) is enough to keep them in the liquid state until opening a small valve emits a fine stream of the vaporized mixture across a spark, which ignites it.

n-Butane and isobutane have the same molecular formula but differ in connectivity. They are *constitutional isomers* of each other and have different properties. Both are gases at room temperature, but *n*-butane boils almost 10°C higher than isobutane and has a melting point that is over 20°C higher.

Bonding in *n*-butane and isobutane continues the theme begun with methane, ethane, and propane. All of the carbon atoms are sp^3-hybridized, all of the bonds are σ bonds, and the bond angles at carbon are close to tetrahedral. This generalization holds for all alkanes regardless of the number of carbons they have.

2.9 Higher *n*-Alkanes

n-Alkanes are alkanes that have an unbranched carbon chain. *n*-Pentane and *n*-hexane are *n*-alkanes possessing five and six carbon atoms, respectively.

$CH_3CH_2CH_2CH_2CH_3$ — *n*-Pentane

$CH_3CH_2CH_2CH_2CH_2CH_3$ — *n*-Hexane

These condensed formulas can be abbreviated even more by indicating within parentheses the number of methylene groups in the chain. Thus, *n*-pentane may be written as $CH_3(CH_2)_3CH_3$ and *n*-hexane as $CH_3(CH_2)_4CH_3$. This shortcut is especially convenient with longer-chain alkanes. The laboratory synthesis of the "ultralong" alkane $CH_3(CH_2)_{388}CH_3$ was achieved in 1985; imagine trying to write a structural formula for this compound in anything other than an abbreviated way!

Problem 2.3

An *n*-alkane of molecular formula $C_{28}H_{58}$ has been isolated from a certain fossil plant. Write a condensed structural formula for this alkane.

n-Alkanes have the general formula $CH_3(CH_2)_xCH_3$ and constitute a **homologous series** of compounds. A homologous series is one in which successive members differ by a —CH_2— group.

Unbranched alkanes are sometimes referred to as "straight-chain alkanes," but, as we'll see in Chapter 3, their chains are not straight but instead tend to adopt the "zigzag" shape as portrayed in the bond-line formulas.

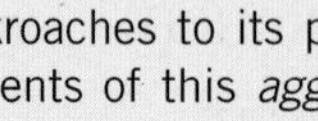

Bond-line formula of *n*-pentane

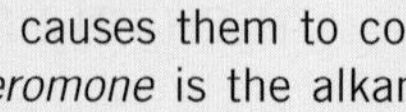

Bond-line formula of *n*-hexane

Problem 2.4

Much of the communication between insects involves chemical messengers called *pheromones.* A species of cockroach secretes a substance from its mandibular glands that alerts other cockroaches to its presence and causes them to congregate. One of the principal components of this *aggregation pheromone* is the alkane shown. Give the molecular formula of this substance, and represent it by a condensed formula.

2.10 The C_5H_{12} Isomers

Three isomeric alkanes have the molecular formula C_5H_{12}. The unbranched isomer is *n*-pentane. The isomer with a single methyl branch is called *isopentane*. The third isomer has a three-carbon chain with two methyl branches. It is called *neopentane*.

n-Pentane: $CH_3CH_2CH_2CH_2CH_3$ or $CH_3(CH_2)_3CH_3$ or 

Isopentane: $CH_3CHCH_2CH_3$ (with CH_3 on C-2) or $(CH_3)_2CHCH_2CH_3$ or 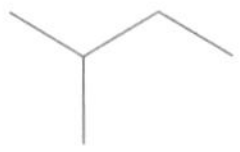

Neopentane: CH_3CCH_3 (with CH_3 above and below C-2) or $(CH_3)_4C$ or 

Table 2.1 presents the number of possible alkane isomers as a function of the number of carbon atoms they contain. As the table shows, the number of isomers increases enormously with the number of carbon atoms and raises two important questions:

1. How can we tell when we have written all the possible isomers corresponding to a particular molecular formula?
2. How can we name alkanes so that each one has a unique name?

The answer to the first question is that you cannot easily calculate the number of isomers. The data in Table 2.1 were determined by a mathematician who concluded that no simple expression can calculate the number of isomers. The best way to ensure that you have written all the isomers of a particular molecular formula is to work systematically, beginning with the unbranched chain and then shortening it while adding branches one by one. It is essential that you be able to recognize when two different-looking structural formulas are actually the same molecule written in different ways. The key point is the *connectivity* of the carbon chain. For example,

TABLE 2.1 The Number of Constitutionally Isomeric Alkanes of Particular Molecular Formulas

Molecular formula	Number of constitutional isomers
CH_4	1
C_2H_6	1
C_3H_8	1
C_4H_{10}	2
C_5H_{12}	3
C_6H_{14}	5
C_7H_{16}	9
C_8H_{18}	18
C_9H_{20}	35
$C_{10}H_{22}$	75
$C_{15}H_{32}$	4,347
$C_{20}H_{42}$	366,319
$C_{40}H_{82}$	62,491,178,805,831

the following structural formulas do *not* represent different compounds; they are just a portion of the many ways we could write a structural formula for isopentane. Each one has a continuous chain of four carbons with a methyl branch located one carbon from the end of the chain.

All of these C_5H_{12} structures represent the same compound.

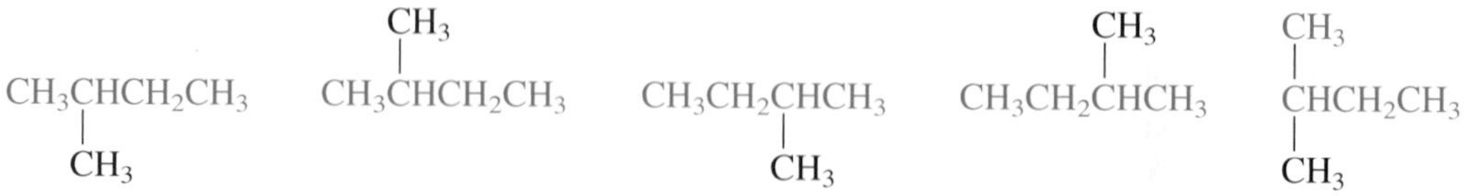

Problem 2.5

Write condensed and bond-line formulas for the five isomeric C_6H_{14} alkanes.

- The unbranched isomer is

$CH_3CH_2CH_2CH_2CH_2CH_3$ = 

- Shortening the chain by one carbon and adding a CH_3 branch gives two more isomers.

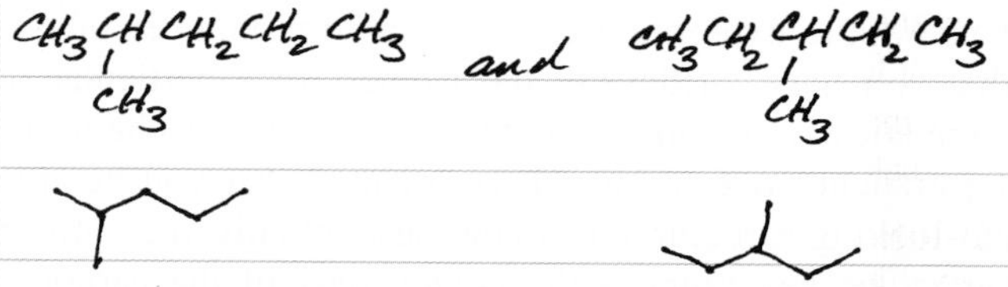

- A four-carbon chain can have two CH_3 branches on the same carbon or on adjacent carbons.

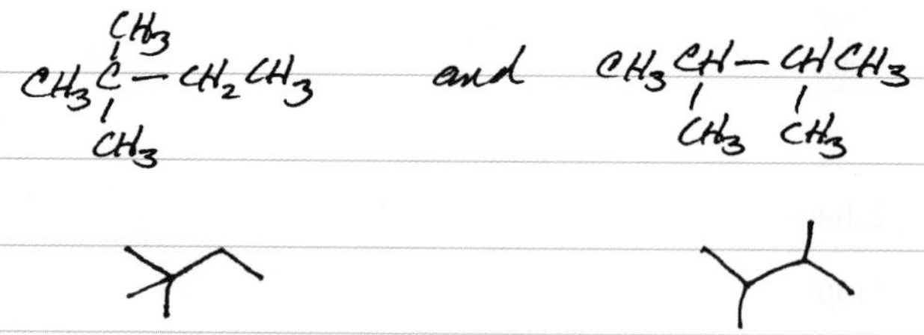

The answer to the second question—how to provide a name that is unique to a particular structure—is presented in the following section. It is worth noting, however, that being able to name compounds in a *systematic* way is a great help in deciding whether two structural formulas represent isomers or are the same compound written in two different ways. By following a precise set of rules, you will always get the same systematic name for a compound, regardless of how it is written. Conversely, two different compounds will always have different names.

2.11 IUPAC Nomenclature of Unbranched Alkanes

We have just seen that the three C_5H_{12} isomers all incorporate "pentane" in their names and are differentiated by the prefixes "*n*-", "iso", and "neo." Extending this approach to alkanes beyond C_5H_{12} fails because we run out of descriptive prefixes before all the isomers have unique names. As difficult as it would be to invent different names for the 18 constitutional isomers of C_8H_{18}, for example, it would be even harder to remember which structure corresponded to which name. For this and other reasons, organic chemists have developed systematic ways to name compounds based on their structure. The most widely used approach is called the **IUPAC rules;** *IUPAC* stands for the International Union of Pure and Applied Chemistry. (See the boxed essay, *What's in a Name? Organic Nomenclature.*)

Alkane names form the foundation of the IUPAC system; more complicated compounds are viewed as being derived from alkanes. The IUPAC names assigned to unbranched alkanes are shown in Table 2.2. Methane, ethane, propane, and butane are retained for CH_4, CH_3CH_3, $CH_3CH_2CH_3$, and $CH_3CH_2CH_2CH_3$, respectively. Thereafter, the number of carbon atoms in the chain is specified by a Greek prefix preceding the suffix *-ane,* which identifies the compound as a member of the alkane family. Notice that the prefix *n*- is not part of the IUPAC system. The IUPAC name for $CH_3CH_2CH_2CH_3$ is butane, not *n*-butane.

Problem 2.6

Refer to Table 2.2 as needed to answer the following questions:

(a) Beeswax (Figure 2.12) contains 8–9% hentriacontane. Write a condensed structural formula for hentriacontane.

(b) Octacosane has been found to be present in a certain fossil plant. Write a condensed structural formula for octacosane.

(c) What is the IUPAC name of the alkane described in Problem 2.4 as a component of the cockroach aggregation pheromone?

Sample Solution (a) Note in Table 2.2 that hentriacontane has 31 carbon atoms. All the alkanes in Table 2.2 have unbranched carbon chains. Hentriacontane has the condensed structural formula $CH_3(CH_2)_{29}CH_3$.

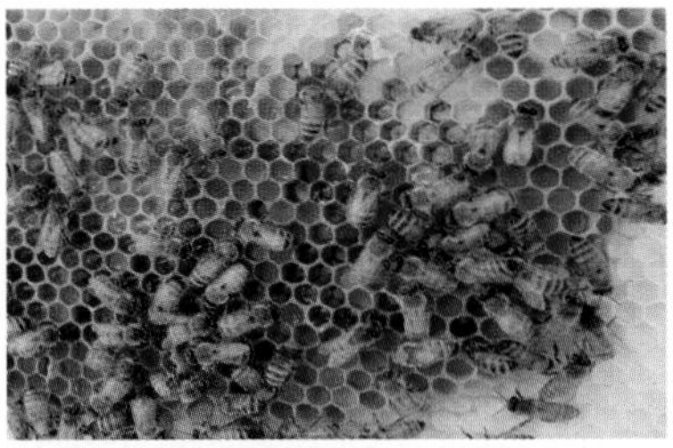

Figure 2.12

Worker bees build the hive with wax secreted from their abdominal glands.

TABLE 2.2 IUPAC Names of Unbranched Alkanes

Number of carbon atoms	Name	Number of carbon atoms	Name	Number of carbon atoms	Name
1	Methane	11	Undecane	21	Henicosane
2	Ethane	12	Dodecane	22	Docosane
3	Propane	13	Tridecane	23	Tricosane
4	Butane	14	Tetradecane	24	Tetracosane
5	Pentane	15	Pentadecane	30	Triacontane
6	Hexane	16	Hexadecane	31	Hentriacontane
7	Heptane	17	Heptadecane	32	Dotriacontane
8	Octane	18	Octadecane	40	Tetracontane
9	Nonane	19	Nonadecane	50	Pentacontane
10	Decane	20	Icosane*	100	Hectane

*Spelled "eicosane" prior to 1979 version of IUPAC rules.

What's in a Name? Organic Nomenclature

Systematic Names and Common Names *Systematic names are derived according to a prescribed set of rules, common names are not.*

Many compounds are better known by **common names** than by their **systematic names.**

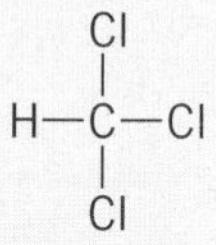

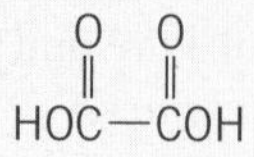

Common name:	Chloroform	Oxalic acid
Systematic name:	Trichloromethane	Ethanedioic acid

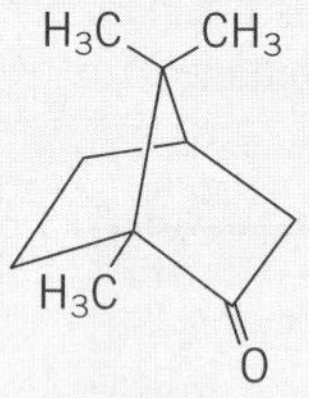

Common name: Camphor

Systematic name: 1,7,7-Trimethylbicyclo[2.2.1]heptan-2-one

Common names, despite their familiarity in certain cases, suffer serious limitations compared with systematic ones. The number of known compounds (millions!) already exceeds our capacity to give each one a unique common name, and most common names are difficult to connect directly to a structural formula. A systematic approach *based on structure* not only conveys structural information, but also generates a unique name for each structural variation.

Evolution of the IUPAC Rules *A single compound can have several acceptable systematic names but no two compounds can have the same name.*

As early as 1787 with the French publication of *Méthode de nomenclature chimique,* chemists suggested guidelines for naming compounds according to chemical composition. Their proposals were more suited to inorganic compounds than organic ones, and it was not until the 1830s that comparable changes appeared in organic chemistry. Later (1892), a group of prominent chemists met in Geneva, Switzerland, where they formulated the principles on which our present system of organic nomenclature is based.

During the twentieth century, what we now know as the *International Union of Pure and Applied Chemistry* (IUPAC) carried out major revisions and extensions of organic nomenclature culminating in the **IUPAC Rules.** The widely used 1979 and 1993 versions of these rules were joined in 2004 by an ambitious set of recommendations.* Despite their sweeping scope, the 2004 recommendations do not render the 1979 and 1993 IUPAC rules obsolete. Rather, the three documents together comprise a collection of systems that give the chemist wide latitude in choosing which of several acceptable ways to name a particular compound.

General and Preferred IUPAC Names One of the goals of the 2004 IUPAC recommendations was to establish the concept of the "preferred IUPAC name" **(PIN)** suitable for use in formal communications (patents, international commerce, occupational health and safety, environmental regulation, etc.). The PIN is distinguished from other acceptable IUPAC names by referring to the latter as "general" IUPAC names. In the following example, the PIN is the same as the 1993 IUPAC name and differs from the 1979 name only in the placement of the number "2."

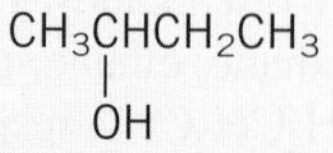

1979	2-Butanol
1993	Butan-2-ol
2004	Butan-2-ol

Because the 1979, 1993, and 2004 IUPAC names tend to be similar, it is not difficult to convert a name to a structure once the fundamental IUPAC principles are learned.

There is no single standard for assigning a compound's PIN. The IUPAC exercises its judgment as to which of the various permissible names for a particular compound can best serve the needs of the chemical community.

Nomenclature in This Text Our practice will be to name compounds in the manner of most active chemists and to use nomenclature as a tool to advance our understanding of organic chemistry. Because the 2004 recommendations are unfamiliar, "general" IUPAC conventions will be emphasized and PINs will not be designated. Similarly, when the 1979 IUPAC names are more widely used than those of the 1993 sequel, the 1979 names are emphasized. Exercises illustrating the relationship among the 1979, 1993, and 2004 IUPAC names are included where appropriate.

Other Nomenclatures Chemical Abstracts Service, a division of the American Chemical Society, surveys all the world's leading scientific journals and publishes brief abstracts of their chemistry papers. *Chemical Abstracts* nomenclature has evolved in a direction geared to computerized literature searches and, although once similar to IUPAC, it is now much different. In general, it is easier to make the mental connection between a structure and its IUPAC name than its *Chemical Abstracts* name.

The **generic name** of a drug is not derived from systematic nomenclature. The group responsible for most generic names in the United States is the U.S. Adopted Names (USAN) Council, a private organization founded by the American Medical Association, the American Pharmacists Association, and the U.S. Pharmacopeial Convention.

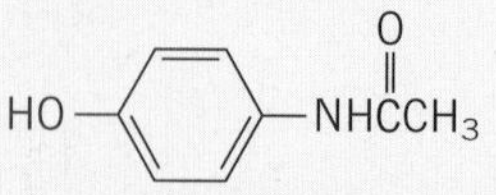

Tylenol

USAN	Acetaminophen
INN	Paracetamol
IUPAC	*N*-(4-Hydroxyphenyl)acetamide

The USAN name is recognized as the official name by the U.S. Food and Drug Administration. International Proprietary Names (INN) are generic names as designated by the World Health Organization.

*The 1979 and 1993 IUPAC rules may be accessed at http://www.acdlabs.com/iupac/nomenclature. The link for the 2004 recommendations is www.iupac.org/reports/provisional/abstract04/favre_310305.html.

In Problem 2.5 you were asked to write structural formulas for the five isomeric alkanes of molecular formula C_6H_{14}. In the next section you will see how the IUPAC rules generate a unique name for each isomer.

2.12 Applying the IUPAC Rules: The Names of the C_6H_{14} Isomers

We can present and illustrate the most important of the IUPAC rules for alkane nomenclature by naming the five C_6H_{14} isomers. By definition (see Table 2.2), the unbranched C_6H_{14} isomer is hexane.

$$CH_3CH_2CH_2CH_2CH_2CH_3$$

IUPAC name: hexane

The IUPAC rules name branched alkanes as *substituted derivatives* of the unbranched alkanes listed in Table 2.2. Consider the C_6H_{14} isomer represented by the structure

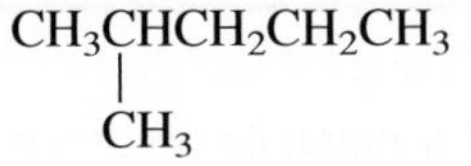

Step 1
Pick out the *longest continuous carbon chain,* and find the IUPAC name in Table 2.2 that corresponds to the unbranched alkane having that number of carbons. This is the parent alkane from which the IUPAC name is to be derived.

In this case, the longest continuous chain has *five* carbon atoms; the compound is named as a derivative of pentane. The key word here is *continuous.* It does not matter whether the carbon skeleton is drawn in an extended straight-chain form or in one with many bends and turns. All that matters is the number of carbons linked together in an uninterrupted sequence.

Step 2
Identify the substituent groups attached to the parent chain.

The parent pentane chain bears a methyl (CH_3) group as a substituent.

Step 3
Number the longest continuous chain in the direction that gives the lowest number to the substituent at the first point of branching.

The numbering scheme

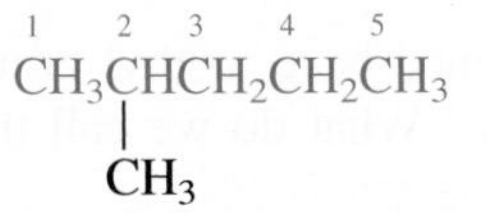

is equivalent to

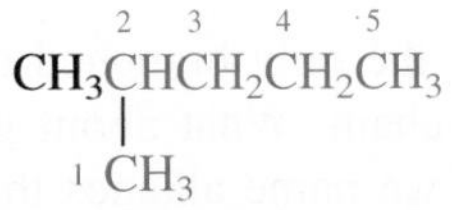

Both schemes count five carbon atoms in their longest continuous chain and bear a methyl group as a substituent at the second carbon. An alternative numbering sequence that begins at the other end of the chain is incorrect:

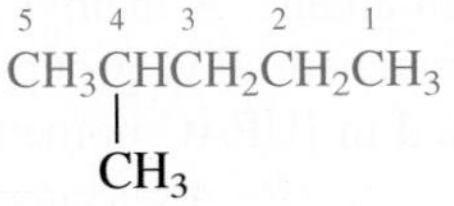

(methyl group attached to C-4)

Step 4
Write the name of the compound. The parent alkane is the last part of the name and is preceded by the names of the substituents and their numerical locations **(locants).** Hyphens separate the locants from the words.

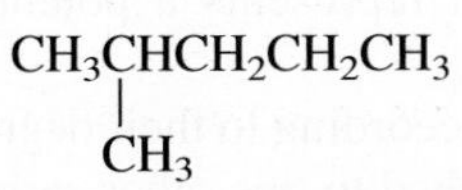

IUPAC name: 2-methylpentane

The same sequence of four steps gives the IUPAC name for the isomer that has its methyl group attached to the middle carbon of the five-carbon chain.

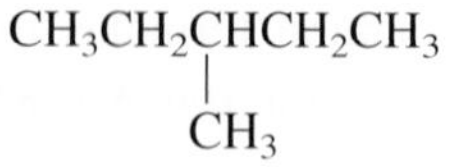

IUPAC name: 3-methylpentane

Both remaining C_6H_{14} isomers have two methyl groups as substituents on a four-carbon chain. Thus the parent chain is butane. When the same substituent appears more than once, use the multiplying prefixes *di-*, *tri-*, *tetra-*, and so on. A separate locant is used for each substituent, and the locants are separated from each other by commas and from the words by hyphens.

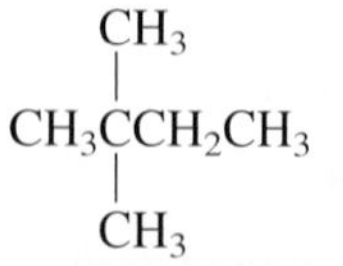

IUPAC name: 2,2-dimethylbutane

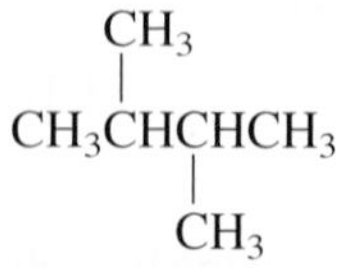

IUPAC name: 2,3-dimethylbutane

Problem 2.7

Phytane is the common name of a naturally occurring alkane produced by the alga *Spirogyra* and is a constituent of petroleum. The IUPAC name for phytane is 2,6,10,14-tetramethylhexadecane. Write a structural formula for phytane.

Problem 2.8

Derive the IUPAC names for

(a) The isomers of C_4H_{10}

(b) The isomers of C_5H_{12}

(c) $(CH_3)_3CCH_2CH(CH_3)_2$

(d) $(CH_3)_3CC(CH_3)_3$

Sample Solution (a) There are two C_4H_{10} isomers. Butane (see Table 2.2) is the IUPAC name for the isomer that has an unbranched carbon chain. The other isomer has three carbons in its longest continuous chain with a methyl branch at the central carbon; its IUPAC name is 2-methylpropane.

$CH_3CH_2CH_2CH_3$

IUPAC name: butane

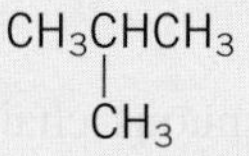

IUPAC name: 2-methylpropane

So far, the only branched alkanes that we've named have methyl groups attached to the main chain. What about groups other than CH_3? What do we call these groups, and how do we name alkanes that contain them?

2.13 Alkyl Groups

The method of naming alkyl groups described in this section is permitted by the 1979, 1993, and 2004 versions of the IUPAC rules.

An **alkyl group** lacks one of the hydrogens of an alkane. A methyl group ($—CH_3$) is an alkyl group derived from methane (CH_4). Unbranched alkyl groups in which the point of attachment is at the end of the chain are named in IUPAC nomenclature by replacing the *-ane* endings of Table 2.2 by *-yl.*

$CH_3CH_2—$	$CH_3(CH_2)_5CH_2—$	$CH_3(CH_2)_{16}CH_2—$
Ethyl group	**Heptyl** group	**Octadecyl** group

The dash at the end of the chain represents a potential point of attachment for some other atom or group.

Carbon atoms are classified according to their degree of substitution by other carbons. A **primary** carbon is *directly* attached to one other carbon. Similarly, a **secondary** carbon is directly attached to two other carbons, a **tertiary** carbon to three, and a **quaternary** carbon to four. Alkyl groups are designated as primary, secondary, or tertiary according to the degree of substitution of the carbon at the potential point of attachment.

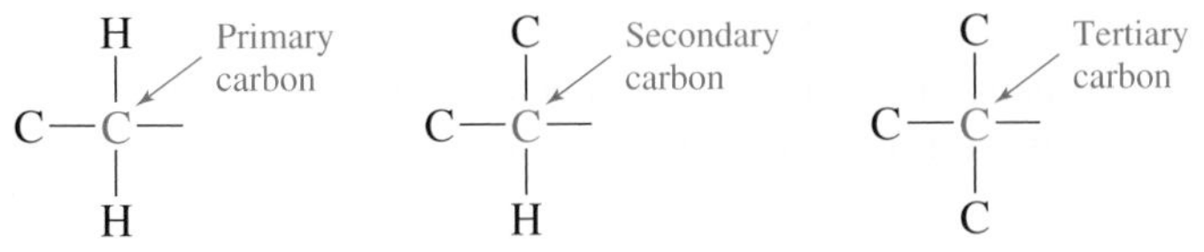

Primary alkyl group Secondary alkyl group Tertiary alkyl group

Ethyl (CH_3CH_2—), heptyl [$CH_3(CH_2)_5CH_2$—], and octadecyl [$CH_3(CH_2)_{16}CH_2$—] are examples of primary alkyl groups.

Branched alkyl groups are named by using the longest continuous chain *that begins at the point of attachment* as the base name. Thus, the systematic names of the two C_3H_7 alkyl groups are propyl and 1-methylethyl. Both are better known by their common names, *n*-propyl and isopropyl, respectively.

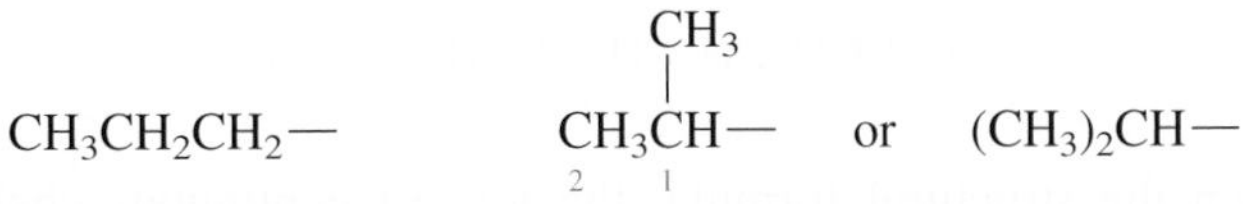

Propyl group (common name: ***n*-propyl**) **1-Methylethyl** group (common name: **isopropyl**)

An isopropyl group is a *secondary* alkyl group. Its point of attachment is to a secondary carbon atom, one that is directly bonded to two other carbons.

The C_4H_9 alkyl groups may be derived either from the unbranched carbon skeleton of butane or from the branched carbon skeleton of isobutane. Those derived from butane are the butyl (*n*-butyl) group and the 1-methylpropyl (*sec*-butyl) group.

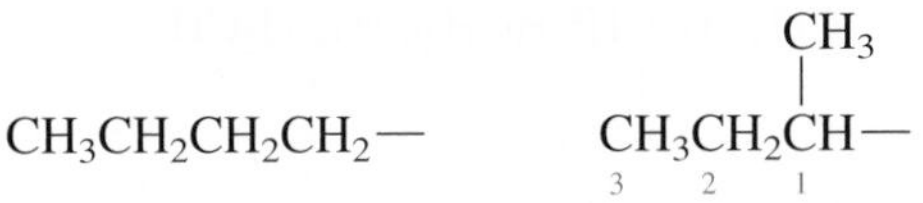

Butyl group (common name: ***n*-butyl**) **1-Methylpropyl** group (common name: ***sec*-butyl**)

Those derived from isobutane are the 2-methylpropyl (isobutyl) group and the 1,1-dimethylethyl (*tert*-butyl) group. Isobutyl is a primary alkyl group because its potential point of attachment is to a primary carbon. *tert*-Butyl is a tertiary alkyl group because its potential point of attachment is to a tertiary carbon.

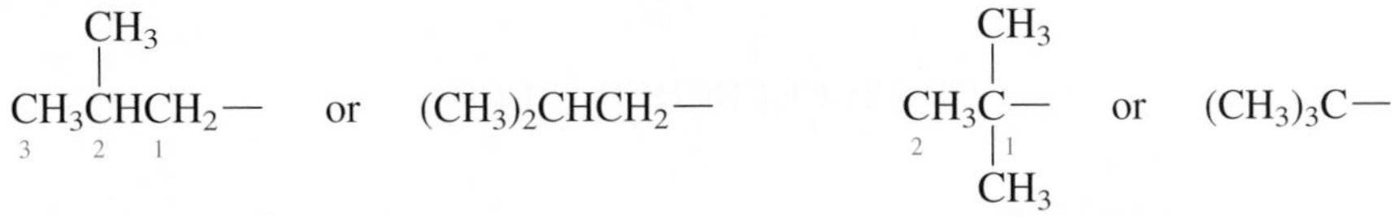

2-Methylpropyl group (common name: **isobutyl**) **1,1-Dimethylethyl** group (common name: ***tert*-butyl**)

Problem 2.9

Give the structures and IUPAC names of all the C_5H_{11} alkyl groups, and identify them as primary, secondary, or tertiary, as appropriate.

Sample Solution Consider the alkyl group having the same carbon skeleton as $(CH_3)_4C$. All the hydrogens are equivalent; replacing any one of them by a potential point of attachment is the same as replacing any of the others.

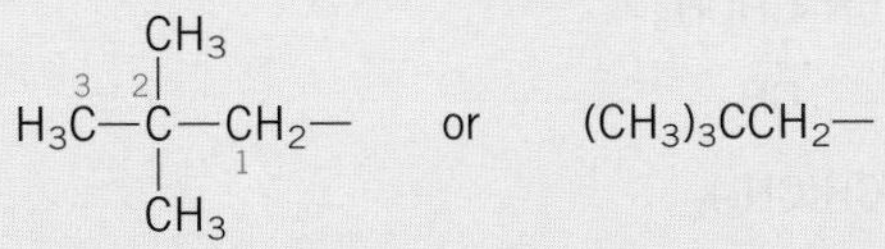

Numbering always begins at the point of attachment and continues through the longest continuous chain. In this case the chain is three carbons and there are two methyl groups at C-2. The IUPAC name of this alkyl group is *2,2-dimethylpropyl.* (The common name for this group is *neopentyl.*) It is a *primary* alkyl group because the carbon that bears the potential point of attachment (C-1) is itself directly bonded to one other carbon.

The names and structures of the most frequently encountered alkyl groups are given on the inside back cover.

In addition to methyl and ethyl groups, *n*-propyl, isopropyl, *n*-butyl, *sec*-butyl, isobutyl, *tert*-butyl, and neopentyl groups will appear often throughout this text. You should be able to recognize these groups on sight and to give their structures when needed.

2.14 IUPAC Names of Highly Branched Alkanes

By combining the basic principles of IUPAC notation with the names of the various alkyl groups, we can develop systematic names for highly branched alkanes. We'll start with the following alkane, name it, then increase its complexity by successively adding methyl groups at various positions.

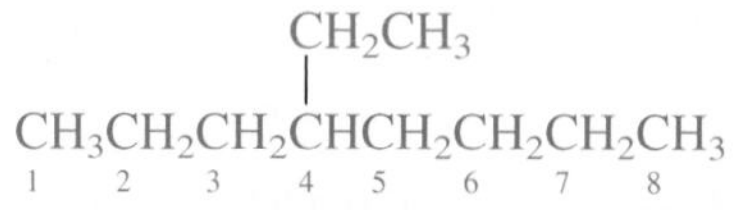

As numbered on the structural formula, the longest continuous chain contains eight carbons, and so the compound is named as a derivative of octane. Numbering begins at the end nearest the branch, and so the ethyl substituent is located at C-4, and the name of the alkane is *4-ethyloctane*.

What happens to the IUPAC name when a methyl group replaces one of the hydrogens at C-3?

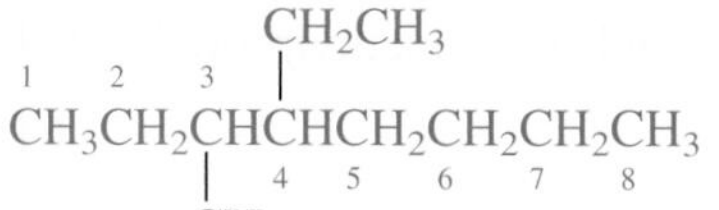

The compound becomes an octane derivative that bears a C-3 methyl group and a C-4 ethyl group. *When two or more different substituents are present, they are listed in alphabetical order in the name*. The IUPAC name for this compound is *4-ethyl-3-methyloctane*.

Replicating prefixes such as *di-*, *tri-*, and *tetra-* (see Section 2.12) are used as needed but are ignored when alphabetizing. Adding a second methyl group to the original structure, at C-5, for example, converts it to *4-ethyl-3,5-dimethyloctane*.

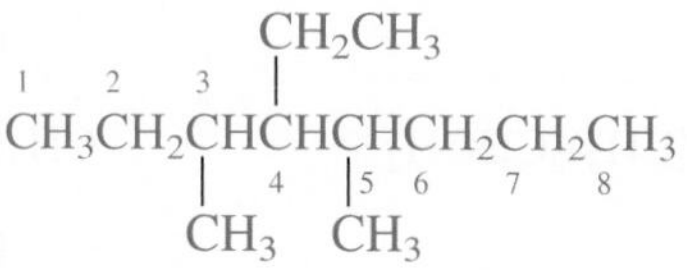

Italicized prefixes such as *sec-* and *tert-* are ignored when alphabetizing except when they are compared with each other. *tert*-Butyl precedes isobutyl, and *sec*-butyl precedes *tert*-butyl.

Problem 2.10

Give an acceptable IUPAC name for each of the following alkanes:

(a) 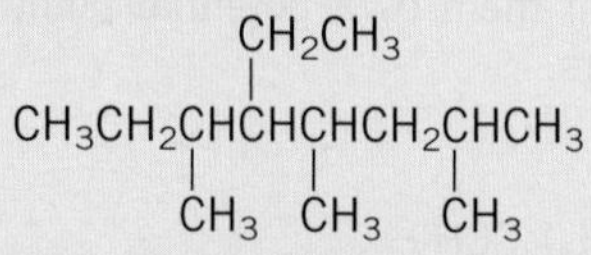

(b) $(CH_3CH_2)_2CHCH_2CH(CH_3)_2$

(c) 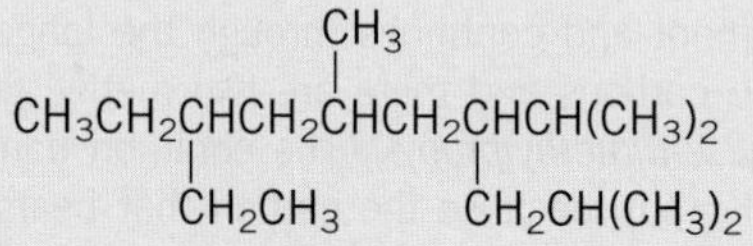

Continued

Sample Solution (a) This problem extends the preceding discussion by adding a third methyl group to 4-ethyl-3,5-dimethyloctane, the compound just described. It is, therefore, an *ethyltrimethyloctane*. Notice, however, that the numbering sequence needs to be changed in order to adhere to the rule of numbering from the end of the chain nearest the first branch. When numbered properly, this compound has a methyl group at C-2 as its first-appearing substituent.

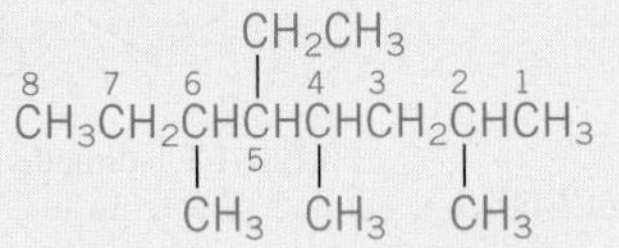

5-Ethyl-2,4,6-trimethyloctane

An additional feature of IUPAC nomenclature that concerns the direction of numbering is the "first point of difference" rule. Consider the two directions in which the following alkane may be numbered:

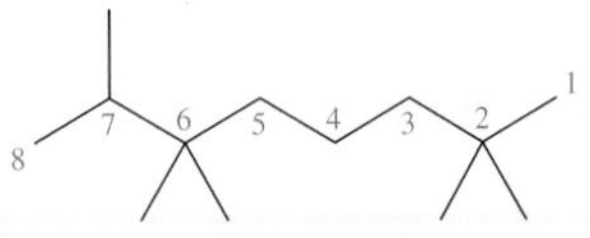

2,2,6,6,7-Pentamethyloctane (correct)

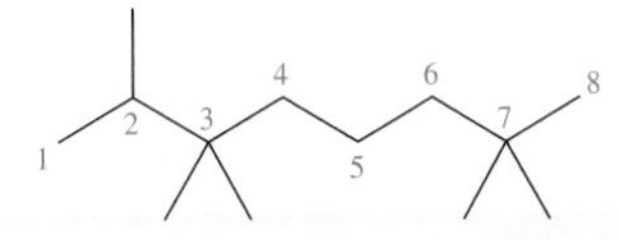

2,3,3,7,7-Pentamethyloctane (incorrect!)

When deciding on the proper direction, a point of difference occurs when one order gives a lower locant than another. Thus, although 2 is the first locant in both numbering schemes, the tie is broken at the second locant, and the rule favors 2,2,6,6,7, which has 2 as its second locant, whereas 3 is the second locant in 2,3,3,7,7. Notice that locants are *not* added together, but examined one by one.

Finally, when equal locants are generated from two different numbering directions, choose the direction that gives the lower number to the substituent that appears first in the name. (Remember, substituents are listed alphabetically.)

The IUPAC nomenclature system is inherently logical and incorporates healthy elements of common sense into its rules. Granted, some long, funny-looking, hard-to-pronounce names are generated. Once one knows the code (rules of grammar) though, it becomes a simple matter to convert those long names to unique structural formulas.

Tabular summaries of the IUPAC rules for alkane and alkyl group nomenclature appear in Tables 2.5 and 2.6 on pages 93–94.

2.15 Cycloalkane Nomenclature

Cycloalkanes are alkanes that contain a ring of three or more carbons. They are frequently encountered in organic chemistry and are characterized by the molecular formula C_nH_{2n}. Some examples include:

Cycloalkanes are one class of *alicyclic* (*aliphatic cyclic*) hydrocarbons.

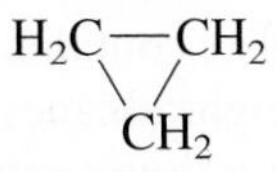

usually represented as

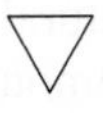

Cyclopropane

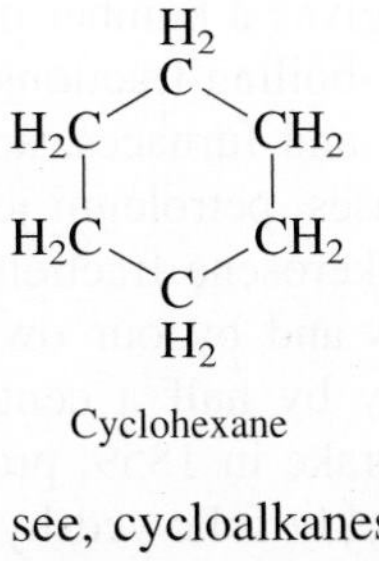

usually represented as

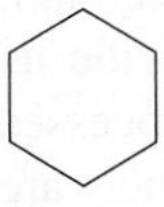

Cyclohexane

As you can see, cycloalkanes are named, under the IUPAC system, by adding the prefix *cyclo-* to the name of the unbranched alkane with the same number of carbons as the ring. Substituents are identified in the usual way. Their positions are specified by numbering the carbon atoms of the ring in the direction that gives the lowest number to the substituents at the first point of difference.

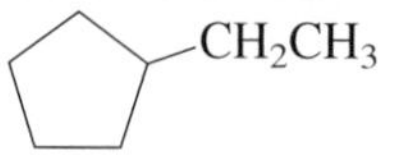
CH2CH3

Ethylcyclopentane

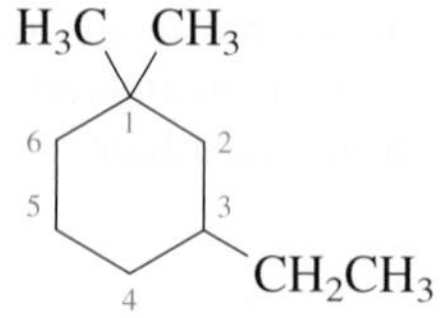
H3C CH3
1
2
3
4
5
6
CH2CH3

3-Ethyl-1,1-dimethylcyclohexane
(not 1-ethyl-3,3-dimethylcyclohexane, because first point of difference rule requires 1,1,3 substitution pattern rather than 1,3,3)

When the ring contains fewer carbon atoms than an alkyl group attached to it, the compound is named as an alkane, and the ring is treated as a *cycloalkyl* substituent:

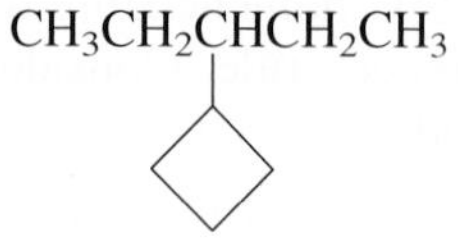
CH3CH2CHCH2CH3

3-Cyclobutylpentane

Problem 2.11

Name each of the following compounds:

(a) 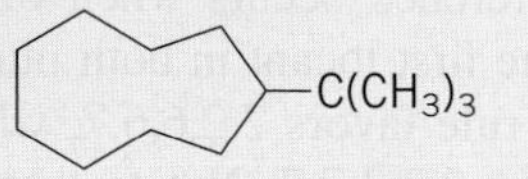
C(CH3)3

(c) 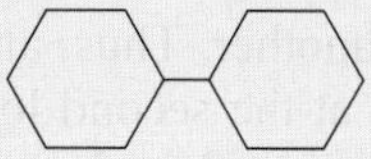

(b) 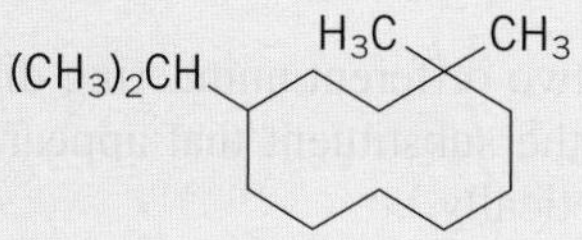
H3C
CH3
(CH3)2CH

Sample Solution (a) The molecule has a *tert*-butyl group bonded to a nine-membered cycloalkane. It is *tert*-butylcyclononane. Alternatively, the *tert*-butyl group could be named systematically as a 1,1-dimethylethyl group, and the compound would then be named (1,1-dimethylethyl)cyclononane. Parentheses are used when necessary to avoid ambiguity. In this case the parentheses alert the reader that the locants 1,1 refer to substituents on the alkyl group and not to ring positions.

2.16 Sources of Alkanes and Cycloalkanes

The word *petroleum* is derived from the Latin words for "rock" *(petra)* and "oil" *(oleum)*.

As noted earlier, natural gas is mostly methane but also contains ethane and propane, along with smaller amounts of other low-molecular-weight alkanes. Natural gas is often found associated with petroleum deposits. Petroleum is a liquid mixture containing hundreds of substances, including approximately 150 hydrocarbons, roughly half of which are alkanes or cycloalkanes. Distillation of crude oil gives a number of fractions, which by custom have the names given in Figure 2.13. High-boiling fractions such as kerosene and gas oil find wide use as fuels for diesel engines and furnaces, and the nonvolatile residue can be processed to give lubricating oil, greases, petroleum jelly, paraffin wax, and asphalt. Jet fuels are obtained from the naphtha–kerosene fractions.

Although both are closely linked in our minds and by our own experience, the petroleum industry predated the automobile industry by half a century. The first oil well, drilled in Titusville, Pennsylvania, by Edwin Drake in 1859, provided "rock oil," as it was then called, on a large scale. This was quickly followed by the development of a process to "refine" it so as to produce kerosene. As a fuel for oil lamps, kerosene

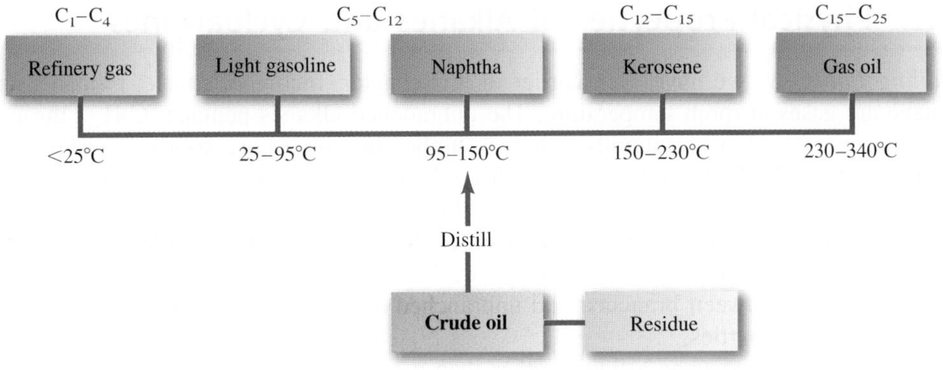

Figure 2.13

Distillation of crude oil yields a series of volatile fractions having the names indicated, along with a nonvolatile residue. The number of carbon atoms that characterize the hydrocarbons in each fraction is approximate.

burned with a bright, clean flame and soon replaced the vastly more expensive whale oil then in use (Figure 2.14). Other oil fields were discovered, and uses for other petroleum products were found—illuminating city streets with gas lights, heating homes with oil, and powering locomotives. There were oil refineries long before there were automobiles. By the time the first Model T rolled off Henry Ford's assembly line in 1908, John D. Rockefeller's Standard Oil holdings had already made him one of the half-dozen wealthiest people in the world.

Figure 2.14

The earliest major use for petroleum was as a fuel for oil lamps.

Modern petroleum **refining** involves more than distillation, however, and includes two major additional operations:

1. **Cracking.** The more volatile, lower-molecular-weight hydrocarbons are useful as automotive fuels and as a source of petrochemicals. Cracking increases the proportion of these hydrocarbons at the expense of higher-molecular-weight ones by processes that involve the cleavage of carbon–carbon bonds induced by heat (*thermal cracking*) or with the aid of certain catalysts (*catalytic cracking*).
2. **Reforming.** The physical properties of the crude oil fractions known as *light gasoline* and *naphtha* (see Figure 2.13) are appropriate for use as a motor fuel, but their ignition characteristics in high-compression automobile engines are poor and give rise to preignition, or "knocking." Reforming converts the hydrocarbons in petroleum to aromatic hydrocarbons and highly branched alkanes, both of which show less tendency for knocking than unbranched alkanes and cycloalkanes.

The tendency of a gasoline to cause "knocking" in an engine is given by its octane number. The lower the octane number, the greater the tendency. The two standards are heptane (assigned a value of 0) and "isooctane" (2,2,4-trimethylpentane, which is assigned a value of 100). The octane number of a gasoline is equal to the percentage of isooctane in a mixture of isooctane and heptane that has the same tendency to cause knocking as that sample of gasoline.

Petroleum is not the only place where alkanes occur naturally. Solid *n*-alkanes, especially those with relatively long chains, have a waxy constituency and coat the outer surface of many living things where they help prevent the loss of water. Pentacosane [$CH_3(CH_2)_{23}CH_3$] is present in the waxy outer layer of most insects. Hentriacontane [$CH_3(CH_2)_{29}CH_3$] is a component of beeswax (see Problem 2.6) as well as the wax that coats the leaves of tobacco, peach trees, pea plants, and numerous others. The C_{23}, C_{25}, C_{27}, C_{29}, and C_{31} *n*-alkanes have been identified in the surface coating of the eggs of honeybee queens.

Cyclopentane and cyclohexane are present in petroleum, but as a rule, unsubstituted cycloalkanes are rarely found in natural sources. Compounds that contain rings of various types, however, are quite abundant.

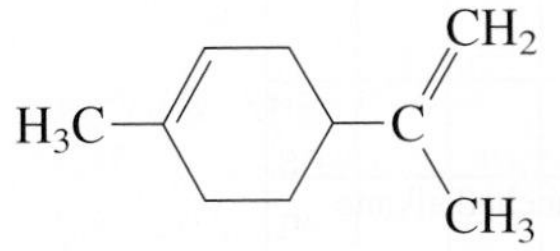

Limonene
(present in lemons and oranges)

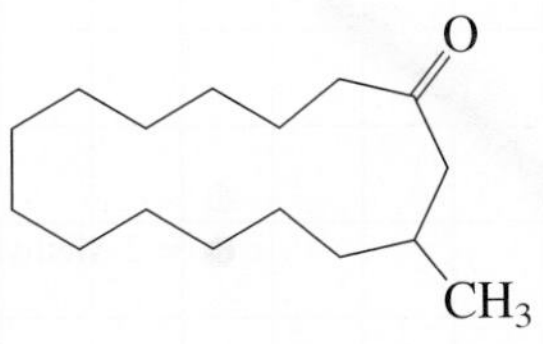

Muscone
(responsible for odor of musk; used in perfumery)

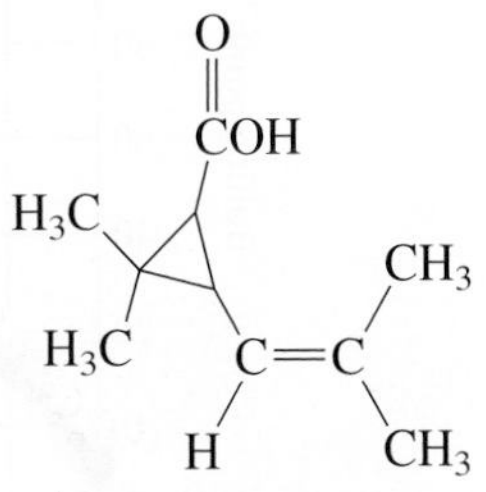

Chrysanthemic acid
(obtained from chrysanthemum flowers)

2.17 Physical Properties of Alkanes and Cycloalkanes

Boiling Point. As we have seen earlier in this chapter, methane, ethane, propane, and butane are gases at room temperature. The unbranched alkanes pentane (C_5H_{12}) through heptadecane ($C_{17}H_{36}$) are liquids, whereas higher homologs are solids. As shown in Figure 2.15, the boiling points of unbranched alkanes increase with the number of carbon atoms. Figure 2.15 also shows that the boiling points for 2-methyl-branched alkanes are lower than those of the unbranched isomer. By exploring at the molecular level the reasons for the increase in boiling point with the number of carbons and the difference in boiling point between branched and unbranched alkanes, we can continue to connect structure with properties.

A substance exists as a liquid rather than a gas because attractive forces between molecules **(intermolecular attractive forces)** are greater in the liquid than in the gas phase. Attractive forces between neutral species (atoms or molecules, but not ions) are referred to as **van der Waals forces** and may be of three types:

1. dipole–dipole (including hydrogen bonding)
2. dipole/induced-dipole
3. induced-dipole/induced-dipole

Induced-dipole/induced-dipole attractive forces are often called *London forces,* or *dispersion forces.*

These forces are electrical in nature, and in order to vaporize a substance, enough energy must be added to overcome them. Most alkanes have no measurable dipole moment, and therefore the only van der Waals force to be considered is the **induced-dipole/induced-dipole attractive force.**

It might seem that two nearby molecules A and B of the same nonpolar substance would be unaffected by each other.

In fact, the electric fields of both A and B are dynamic and fluctuate in a complementary way that results in a temporary dipole moment and a weak attraction between them.

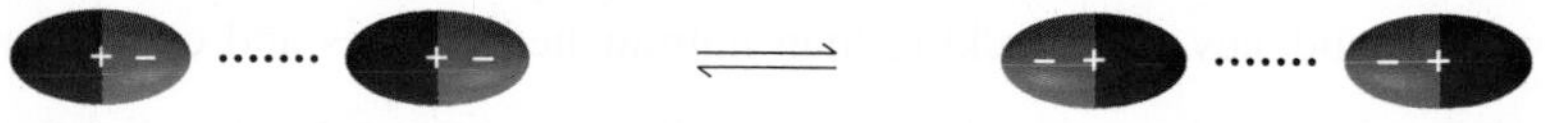

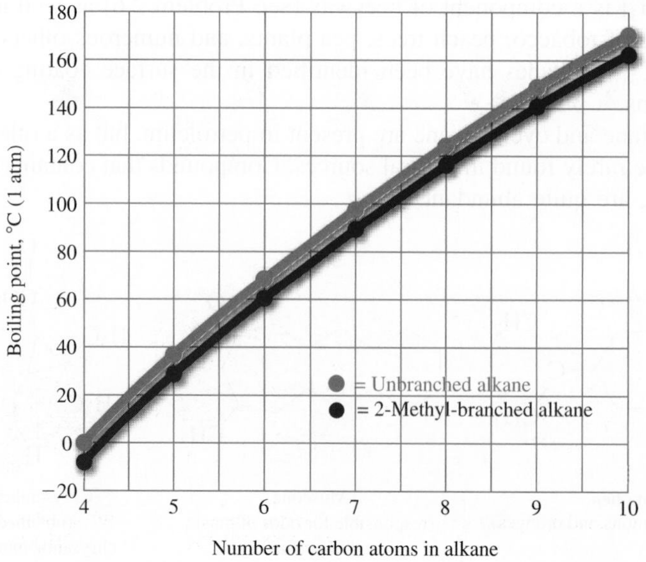

Figure 2.15

Boiling points of unbranched alkanes and their 2-methyl-branched isomers. (Temperatures in this text are expressed in degrees Celsius, °C. The SI unit of temperature is the kelvin, K. To convert degrees Celsius to kelvins add 273.15.)

Extended assemblies of induced-dipole/induced-dipole attractions can accumulate to give substantial intermolecular attractive forces. An alkane with a higher molecular weight has more atoms and electrons and, therefore, more opportunities for intermolecular attractions and a higher boiling point than one with a lower molecular weight.

As noted earlier in this section, branched alkanes have lower boiling points than their unbranched isomers. Isomers have, of course, the same number of atoms and electrons, but a molecule of a branched alkane has a smaller surface area than an unbranched one. The extended shape of an unbranched alkane permits more points of contact for intermolecular associations. Compare the boiling points of pentane and its isomers:

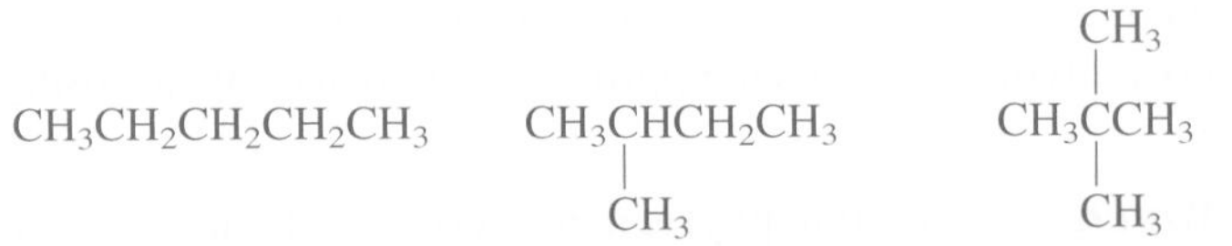
CH3CH2CH2CH2CH3
CH3CHCH2CH3
CH3
CH3
CH3CCH3
CH3

Pentane (bp 36°C) | 2-Methylbutane (bp 28°C) | 2,2-Dimethylpropane (bp 9°C)

The shapes of these isomers are clearly evident in the space-filling models depicted in Figure 2.16. Pentane has the most extended structure and the largest surface area available for "sticking" to other molecules by way of induced-dipole/induced-dipole attractive forces; it has the highest boiling point. 2,2-Dimethylpropane has the most compact, most spherical structure, engages in the fewest induced-dipole/induced-dipole attractions, and has the lowest boiling point.

Induced-dipole/induced-dipole attractions are very weak forces individually, but a typical organic substance can participate in so many of them that they are collectively the most important of all the contributors to intermolecular attraction in the liquid state. They are the only forces of attraction possible between nonpolar molecules such as alkanes.

Problem 2.12

Match the boiling points with the appropriate alkanes.
Alkanes: octane, 2-methylheptane, 2,2,3,3-tetramethylbutane, nonane
Boiling points (°C, 1 atm): 106, 116, 126, 151

Cyclopentane has a higher boiling point (49.3°C) than pentane (36°C), indicating greater forces of association in the cyclic alkane than in the alkane.

(*a*) Pentane: $CH_3CH_2CH_2CH_2CH_3$ (*b*) 2-Methylbutane: $(CH_3)_2CHCH_2CH_3$ (*c*) 2,2-Dimethylpropane: $(CH_3)_4C$

Figure 2.16

Tube (*top*) and space-filling (*bottom*) models of (*a*) pentane, (*b*) 2-methylbutane, and (*c*) 2,2-dimethylpropane. The most branched isomer, 2,2-dimethylpropane, has the most compact, most spherical three-dimensional shape.

Melting Point. Solid alkanes are soft, generally low-melting materials. The forces responsible for holding the crystal together are the same induced-dipole/induced-dipole interactions that operate between molecules in the liquid, but the degree of organization is greater in the solid phase. By measuring the distances between the atoms of one molecule and its neighbor in the crystal, it is possible to specify a distance of closest approach characteristic of an atom called its **van der Waals radius.** In space-filling molecular models, such as those of pentane, 2-methylbutane, and 2,2-dimethylpropane shown in Figure 2.16, the radius of each sphere corresponds to the van der Waals radius of the atom it represents. The van der Waals radius for hydrogen is 120 pm. When two alkane molecules are brought together so that a hydrogen of one molecule is within 240 pm of a hydrogen of the other, the balance between electron–nucleus attractions versus electron–electron and nucleus–nucleus repulsions is most favorable. Closer approach is resisted by a strong increase in repulsive forces.

Solubility in Water. A familiar physical property of alkanes is contained in the adage "oil and water don't mix." Alkanes—indeed all hydrocarbons—are virtually insoluble in water. In order for a hydrocarbon to dissolve in water, the framework of hydrogen bonds between water molecules would become more ordered in the region around each molecule of the dissolved hydrocarbon. This increase in order, which corresponds to a decrease in entropy, signals a process that can be favorable only if it is reasonably exothermic. Such is not the case here. The hydrogen bonding among water molecules is too strong to be disrupted by nonpolar hydrocarbons. Being insoluble, and with densities in the 0.6–0.8 g/mL range, alkanes float on the surface of water. The exclusion of nonpolar molecules, such as alkanes, from water is called the **hydrophobic effect.** We will encounter it again at several points later in the text.

2.18 Chemical Properties: Combustion of Alkanes

Alkanes are so unreactive that George A. Olah of the University of Southern California was awarded the 1994 Nobel Prize in Chemistry in part for developing novel substances that do react with alkanes.

An older name for alkanes is **paraffin hydrocarbons.** *Paraffin* is derived from the Latin words *parum affinis* ("with little affinity") and testifies to the low level of reactivity of alkanes.

Table 1.8 shows that hydrocarbons are extremely weak acids. Among the classes of hydrocarbons, acetylene is a stronger acid than methane, ethane, ethylene, or benzene, but even its K_a is 10^{10} smaller than that of water.

	$HC{\equiv}CH$	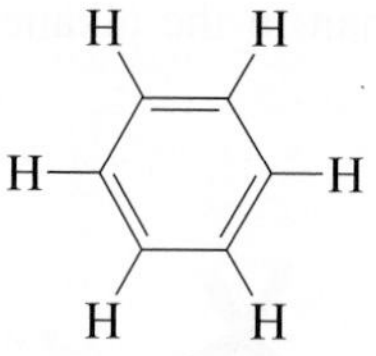	$H_2C{=}CH_2$	CH_4	CH_3CH_3
pK_a:	26	43	45	60	62

Although essentially inert in acid–base reactions, alkanes do participate in oxidation–reduction reactions as the compound that undergoes oxidation. Burning in air **(combustion)** is the best known and most important example. Combustion of hydrocarbons is exothermic and gives carbon dioxide and water as the products.

$$\underset{\text{Methane}}{CH_4} + \underset{\text{Oxygen}}{2O_2} \longrightarrow \underset{\text{Carbon dioxide}}{CO_2} + \underset{\text{Water}}{2H_2O} \qquad \Delta H^\circ = -890\text{ kJ }(-212.8\text{ kcal})$$

$$\underset{\text{2-Methylbutane}}{(CH_3)_2CHCH_2CH_3} + \underset{\text{Oxygen}}{8O_2} \longrightarrow \underset{\text{Carbon dioxide}}{5CO_2} + \underset{\text{Water}}{6H_2O} \qquad \Delta H^\circ = -3529\text{ kJ }(-843.4\text{ kcal})$$

Problem 2.13

Write a balanced chemical equation for the combustion of cyclohexane.

The heat released on combustion of a substance is called its **heat of combustion.** The heat of combustion is equal to $-\Delta H°$ for the reaction written in the direction shown. By convention

$$\Delta H° = H°_{\text{products}} - H°_{\text{reactants}}$$

where $H°$ is the heat content, or **enthalpy,** of a compound in its standard state, that is, the gas, pure liquid, or crystalline solid at a pressure of 1 atm. In an exothermic process the enthalpy of the products is less than that of the starting materials, and $\Delta H°$ is a negative number.

Table 2.3 lists the heats of combustion of several alkanes. Unbranched alkanes have slightly higher heats of combustion than their 2-methyl-branched isomers, but the most important factor is the number of carbons. The unbranched alkanes and the 2-methyl-branched alkanes constitute two separate *homologous series* (see Section 2.9) in which there is a regular increase of about 653 kJ/mol (156 kcal/mol) in the heat of combustion for each additional CH_2 group.

Problem 2.14

Using the data in Table 2.3, estimate the heat of combustion of

(a) 2-Methylnonane (in kcal/mol) (b) Icosane (in kJ/mol)

Sample Solution (a) The last entry for the group of 2-methylalkanes in the table is 2-methylheptane. Its heat of combustion is 1306 kcal/mol. Because 2-methylnonane has two more methylene groups than 2-methylheptane, its heat of combustion is 2 × 156 kcal/mol higher.

Heat of combustion of 2-methylnonane = 1306 + 2(156) = 1618 kcal/mol

TABLE 2.3 Heats of Combustion ($-\Delta H°$) of Representative Alkanes

		$-\Delta H°$	
Compound	**Formula**	**kJ/mol**	**kcal/mol**
Unbranched alkanes			
Hexane	$CH_3(CH_2)_4CH_3$	4,163	995.0
Heptane	$CH_3(CH_2)_5CH_3$	4,817	1,151.3
Octane	$CH_3(CH_2)_6CH_3$	5,471	1,307.5
Nonane	$CH_3(CH_2)_7CH_3$	6,125	1,463.9
Decane	$CH_3(CH_2)_8CH_3$	6,778	1,620.1
Undecane	$CH_3(CH_2)_9CH_3$	7,431	1,776.1
Dodecane	$CH_3(CH_2)_{10}CH_3$	8,086	1,932.7
Hexadecane	$CH_3(CH_2)_{14}CH_3$	10,701	2,557.6
2-Methyl-branched alkanes			
2-Methylpentane	$(CH_3)_2CHCH_2CH_2CH_3$	4,157	993.6
2-Methylhexane	$(CH_3)_2CH(CH_2)_3CH_3$	4,812	1,150.0
2-Methylheptane	$(CH_3)_2CH(CH_2)_4CH_3$	5,466	1,306.3

Thermochemistry

Thermochemistry is the study of the heat changes that accompany chemical processes. It has a long history dating back to the work of the French chemist Antoine Laurent Lavoisier in the late eighteenth century. Thermochemistry provides quantitative information that complements the qualitative description of a chemical reaction and can help us understand why some reactions occur and others do not. It is of obvious importance when assessing the relative value of various materials as fuels, when comparing the stability of isomers, or when determining the practicality of a particular reaction. In the field of bioenergetics, thermochemical information is applied to the task of sorting out how living systems use chemical reactions to store and use the energy that originates in the sun.

By allowing compounds to react in a calorimeter, it is possible to measure the heat evolved in an exothermic reaction or the heat absorbed in an endothermic one. Thousands of reactions have been studied to produce a rich library of thermochemical data. These data take the form of heats of reaction and correspond to the value of the standard enthalpy change $\Delta H°$ for a particular reaction of a particular substance.

In this section you have seen how heats of combustion can be used to determine relative stabilities of isomeric alkanes. In later sections we shall expand our scope to include the experimentally determined heats of certain other reactions, such as *bond dissociation enthalpies* (Section 4.19) and *heats of hydrogenation* (Section 6.2), to see how $\Delta H°$ values from various sources can aid our understanding of structure and reactivity.

The **standard heat of formation** ($\Delta H_f°$), is the enthalpy change for formation of one mole of a compound directly from its elements, and is one type of heat of reaction. In cases such as the formation of CO_2 or H_2O from the combustion of carbon or hydrogen, respectively, the heat of formation of a substance can be measured directly. In most other cases, heats of formation are not measured experimentally but are calculated from the measured heats of other reactions. Consider, for example, the heat of formation of methane. The reaction that defines the formation of methane from its elements,

$$\underset{\text{Carbon}}{\text{C (graphite)}} + \underset{\text{Hydrogen}}{2\text{H}_2(g)} \longrightarrow \underset{\text{Methane}}{\text{CH}_4(g)}$$

can be expressed as the sum of three reactions:

$$\begin{array}{llr} \text{(1) C (graphite)} + \text{O}_2(g) & \longrightarrow \text{CO}_2(g) & \Delta H° = -393 \text{ kJ} \\ \text{(2) } 2\text{H}_2(g) + \text{O}_2(g) & \longrightarrow 2\text{H}_2\text{O}(l) & \Delta H° = -572 \text{ kJ} \\ \text{(3) CO}_2(g) + 2\text{H}_2\text{O}(l) & \longrightarrow \text{CH}_4(g) + 2\text{O}_2(g) & \Delta H° = +890 \text{ kJ} \\ \hline \text{C (graphite)} + 2\text{H}_2(g) & \longrightarrow \text{CH}_4(g) & \Delta H° = -75 \text{ kJ} \end{array}$$

Equations (1) and (2) are the heats of formation of one mole of carbon dioxide and two moles of water, respectively. Equation (3) is the reverse of the combustion of methane, and so the heat of reaction is equal to the heat of combustion but opposite in sign. The sum of equations (1)–(3) is the enthalpy change for formation of one mole of methane from its elements. Thus, $\Delta H_f° = -75$ kJ/mol.

The heats of formation of most organic compounds are derived from heats of reaction by arithmetic manipulations similar to that shown. Chemists find a table of $\Delta H_f°$ values to be convenient because it replaces many separate tables of $\Delta H°$ values for individual reaction types and permits $\Delta H°$ to be calculated for any reaction, real or imaginary, for which the heats of formation of reactants and products are available. It is more appropriate for our purposes, however, to connect thermochemical data to chemical processes as directly as possible, and therefore we will cite heats of particular reactions, such as heats of combustion and heats of hydrogenation, rather than heats of formation.

Heats of combustion can be used to measure the relative stability of isomeric hydrocarbons. They tell us not only which isomer is more stable than another, but by how much. Consider a group of C_8H_{18} alkanes:

$$\underset{\text{Octane}}{CH_3(CH_2)_6CH_3} \qquad \underset{\text{2-Methylheptane}}{(CH_3)_2CHCH_2CH_2CH_2CH_2CH_3}$$

$$\underset{\text{2,2-Dimethylhexane}}{(CH_3)_3CCH_2CH_2CH_2CH_3} \qquad \underset{\text{2,2,3,3-Tetramethylbutane}}{(CH_3)_3CC(CH_3)_3}$$

Figure 2.17 compares the heats of combustion of these C_8H_{18} isomers on a *potential energy diagram.* **Potential energy** is comparable with enthalpy; it is the energy a molecule has exclusive of its kinetic energy. A molecule with more potential energy is less stable than an isomer with less potential energy. These C_8H_{18} isomers all undergo combustion to the same final state according to the equation:

$$C_8H_{18} + \tfrac{25}{2}O_2 \longrightarrow 8CO_2 + 9H_2O$$

therefore, the differences in their heats of combustion translate directly to differences in their potential energies. *When comparing isomers, the one with the lowest potential energy (in this case, the smallest heat of combustion) is the most stable.* Among the C_8H_{18} alkanes, the most highly branched isomer, 2,2,3,3-tetramethylbutane, is the most stable,

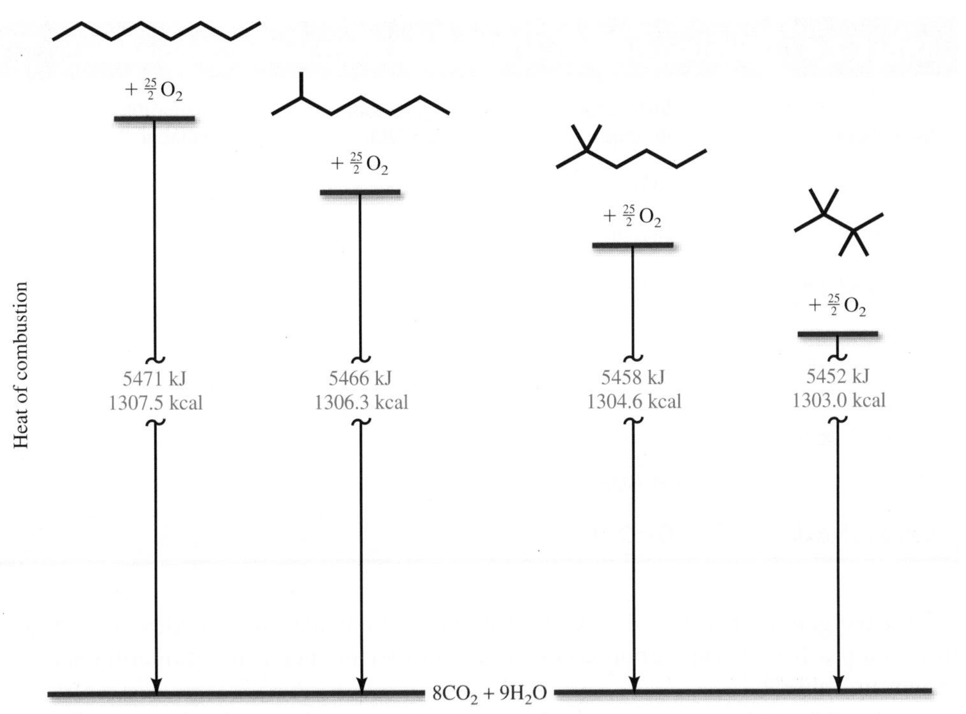

Figure 2.17

Energy diagram comparing heats of combustion of isomeric C_8H_{18} alkanes.

and the unbranched isomer octane is the least stable. It is generally true for alkanes that a more branched isomer is more stable than a less branched one.

The small differences in stability between branched and unbranched alkanes result from an interplay between attractive and repulsive forces *within* a molecule **(intramolecular forces).** These forces are nucleus–nucleus repulsions, electron–electron repulsions, and nucleus–electron attractions, the same set of fundamental forces we met when talking about chemical bonding (Section 2.2) and van der Waals forces between molecules (Section 2.17). When the energy associated with these interactions is calculated for all of the nuclei and electrons within a molecule, it is found that the attractive forces increase more than the repulsive forces as the structure becomes more compact. Sometimes, though, two atoms in a molecule are held too closely together. We'll explore the consequences of that in Chapter 3.

Problem 2.15

Without consulting Table 2.3, arrange the following compounds in order of decreasing heat of combustion: pentane, 2-methylbutane, 2,2-dimethylpropane, hexane.

2.19 Oxidation–Reduction in Organic Chemistry

As we have just seen, the reaction of alkanes with oxygen to give carbon dioxide and water is called *combustion.* A more fundamental classification of reaction types places it in the **oxidation–reduction** category. To understand why, let's review some principles of oxidation–reduction, beginning with the **oxidation number** (also known as **oxidation state**).

There are a variety of methods for calculating oxidation numbers. In compounds that contain a single carbon, such as methane (CH_4) and carbon dioxide (CO_2), the oxidation number of carbon can be calculated from the molecular formula. For neutral molecules the algebraic sum of all the oxidation numbers must equal zero. Assuming, as is customary, that the oxidation state of hydrogen is +1, the oxidation state of carbon in CH_4 then is calculated to be −4. Similarly, assuming an oxidation state of

TABLE 2.4 Oxidation Number of Carbon in One-Carbon Compounds

Compound	Structural formula	Molecular formula	Oxidation number
Methane	CH_4	CH_4	−4
Methanol	CH_3OH	CH_4O	−2
Formaldehyde	$H_2C{=}O$	CH_2O	0
Formic acid	$HC(=O)OH$	CH_2O_2	+2
Carbonic acid	$HOC(=O)OH$	H_2CO_3	+4
Carbon dioxide	$O{=}C{=}O$	CO_2	+4

-2 for oxygen, carbon is $+4$ in CO_2. This kind of calculation provides an easy way to develop a list of one-carbon compounds in order of increasing oxidation state, as shown in Table 2.4.

The carbon in methane has the lowest oxidation number (-4) of any of the compounds in Table 2.4. Methane contains carbon in its most *reduced* form. Carbon dioxide and carbonic acid have the highest oxidation numbers ($+4$) for carbon, corresponding to its most *oxidized* state. When methane or any alkane undergoes combustion to form carbon dioxide, carbon is oxidized and oxygen is reduced.

A useful generalization from Table 2.4 is the following:

Oxidation of carbon corresponds to an increase in the number of bonds between carbon and oxygen or to a decrease in the number of carbon–hydrogen bonds. Conversely, reduction corresponds to an increase in the number of carbon–hydrogen bonds or to a decrease in the number of carbon–oxygen bonds.

From Table 2.4 it can be seen that each successive increase in oxidation state increases the number of bonds between carbon and oxygen and decreases the number of carbon–hydrogen bonds. Methane has four C—H bonds and no C—O bonds; carbon dioxide has four C—O bonds and no C—H bonds.

Among the various classes of hydrocarbons, alkanes contain carbon in its most reduced state, and alkynes contain carbon in its most oxidized state.

Increasing oxidation state of carbon
(decreasing hydrogen content)

CH_3CH_3	$H_2C{=}CH_2$	$HC{\equiv}CH$
Ethane (6 C—H bonds)	Ethylene (4 C—H bonds)	Acetylene (2 C—H bonds)

We can extend the generalization by recognizing that the pattern of oxidation states is not limited to increasing oxygen or hydrogen content. Any element *more electronegative* than carbon will have the same effect on oxidation number as oxygen. Thus, the oxidation numbers of carbon in CH_3Cl and CH_3OH are the same (–2). The reaction of chlorine with methane (to be discussed in Section 4.16) involves *oxidation* at carbon.

$$\underset{\text{Methane}}{CH_4} + \underset{\text{Chlorine}}{Cl_2} \longrightarrow \underset{\text{Chloromethane}}{CH_3Cl} + \underset{\text{Hydrogen chloride}}{HCl}$$

Any element *less electronegative* than carbon will have the same effect on oxidation number as hydrogen. Thus, the oxidation numbers of carbon in CH_3Li and CH_4 are the same (−4), and the reaction of CH_3Cl with lithium (to be discussed in Section 14.3) involves *reduction* at carbon.

CH_3Cl	+ 2 Li	⟶	CH_3Li	LiCl
Chloromethane	Lithium		Methyllithium	Lithium chloride

The oxidation number of carbon decreases from −2 in CH_3Cl to −4 in CH_3Li.

The generalization illustrated by the preceding examples can be expressed in terms broad enough to cover these reactions and many others, as follows: *Oxidation of carbon occurs when a bond between a carbon and an atom that is less electronegative than carbon is replaced by a bond to an atom that is more electronegative than carbon. The reverse process is reduction.*

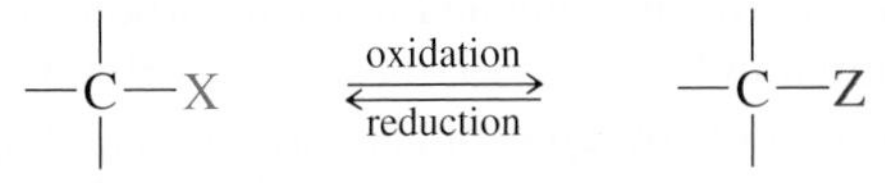

X is less electronegative than carbon — Z is more electronegative than carbon

Problem 2.16

Both of the following reactions will be encountered in Chapter 4. One is oxidation–reduction, the other is not. Which is which?

$$(CH_3)_3COH + HCl \longrightarrow (CH_3)_3CCl + H_2O$$
$$(CH_3)_3CH + Br_2 \longrightarrow (CH_3)_3CBr + HBr$$

Many, indeed most organic compounds contain carbon in more than one oxidation state. Consider ethanol (CH_3CH_2OH), for example. One carbon is connected to three hydrogens; the other carbon to two hydrogens and one oxygen. Although we could calculate the actual oxidation numbers, we rarely need to in organic chemistry. Most of the time we are only concerned with whether a particular reaction is an oxidation or a reduction. The ability to recognize when oxidation or reduction occurs is of value when deciding on the kind of reactant with which an organic molecule must be treated to convert it into a desired product.

Problem 2.17

Which of the following reactions requires an oxidizing agent, a reducing agent, or neither?

(a) CH_3CH_2OH ⟶ 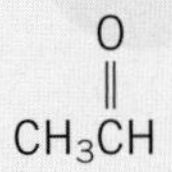

(b) CH_3CH_2Br ⟶ CH_3CH_2Li

(c) $H_2C{=}CH_2$ ⟶ CH_3CH_2OH

(d) $H_2C{=}CH_2$ ⟶ 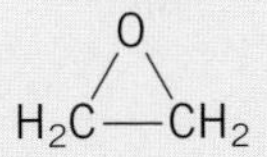

Sample Solution The CH_3 carbon is unchanged in the reaction; however, the carbon of CH_2OH now has two bonds to oxygen. Therefore, the reaction requires an oxidizing agent.

2.20 *sp*² Hybridization and Bonding in Ethylene

We conclude this introduction to hydrocarbons by describing the orbital hybridization model of bonding in ethylene and acetylene, parents of the alkene and alkyne families, respectively.

Ethylene is planar with bond angles close to 120° (Figure 2.18); therefore, some hybridization state other than *sp*³ is required. The hybridization scheme is determined by the number of atoms to which carbon is directly attached. In *sp*³ hybridization, four

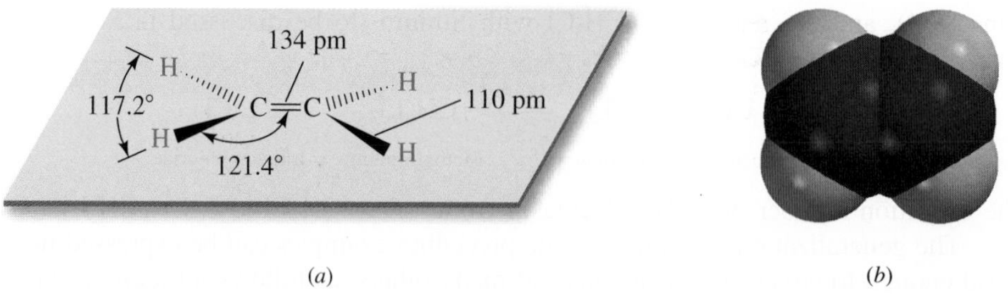

Figure 2.18

(*a*) All the atoms of ethylene lie in the same plane, the bond angles are close to 120°, and the carbon–carbon bond distance is significantly shorter than that of ethane. (*b*) A space-filling model of ethylene.

atoms are attached to carbon by σ bonds, and so four equivalent sp^3 hybrid orbitals are required. In ethylene, three atoms are attached to each carbon, so three equivalent hybrid orbitals are needed. As shown in Figure 2.19, these three orbitals are generated by mixing the carbon 2*s* orbital with *two* of the 2*p* orbitals and are called **sp^2 hybrid orbitals.** One of the 2*p* orbitals is left unhybridized. The three sp^2 orbitals are of equal energy; each has one-third *s* character and two-thirds *p* character. Their axes are coplanar, and each has a shape much like that of an sp^3 orbital. The three sp^2 orbitals and the unhybridized *p* orbital each contain one electron.

Each carbon of ethylene uses two of its sp^2 hybrid orbitals to form σ bonds to two hydrogen atoms, as illustrated in the first part of Figure 2.20. The remaining sp^2 orbitals, one on each carbon, overlap along the internuclear axis to give a σ bond connecting the two carbons.

Each carbon atom still has, at this point, an unhybridized 2*p* orbital available for bonding. These two half-filled 2*p* orbitals have their axes perpendicular to the framework of σ bonds of the molecule and overlap in a side-by-side manner to give a **pi (π) bond.** The carbon–carbon double bond of ethylene is viewed as a combination of a σ bond plus a π bond. The additional increment of bonding makes a carbon–carbon double bond both stronger and shorter than a carbon–carbon single bond.

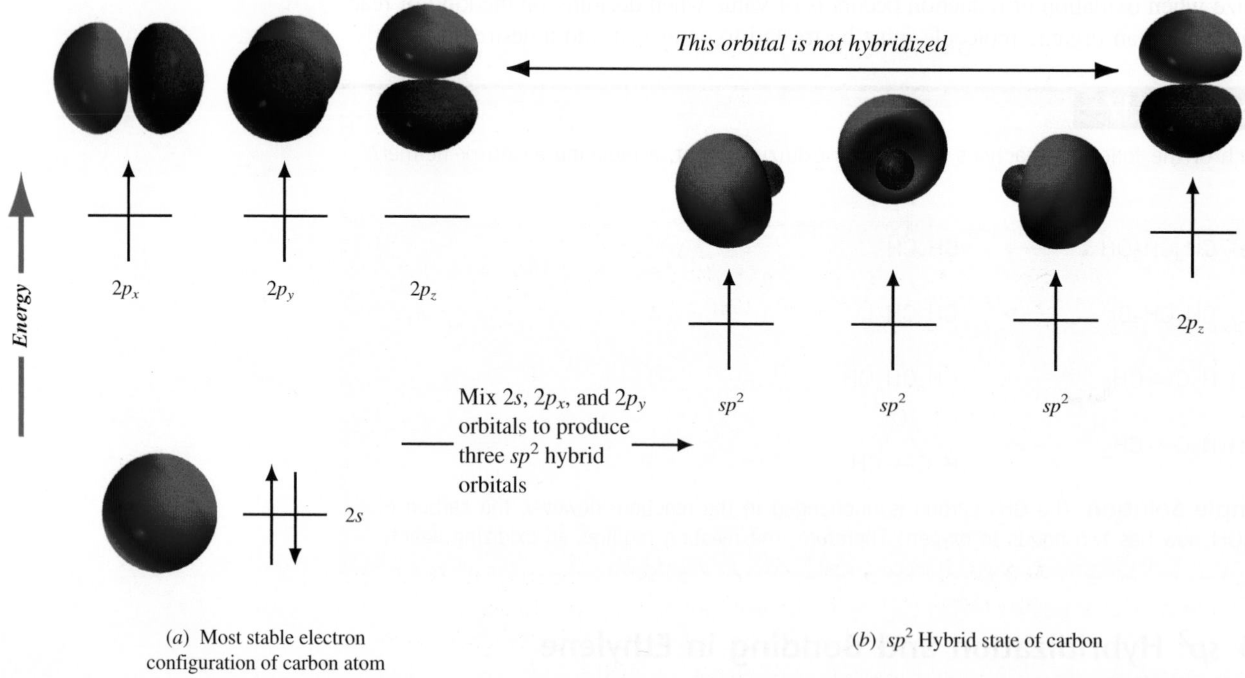

Figure 2.19

sp^2 Hybridization. (*a*) Electron configuration of carbon in its most stable state. (*b*) Mixing the *s* orbital with two of the three *p* orbitals generates three sp^2 hybrid orbitals and leaves one of the 2*p* orbitals untouched. The axes of the three sp^2 orbitals lie in the same plane and make angles of 120° with one another.

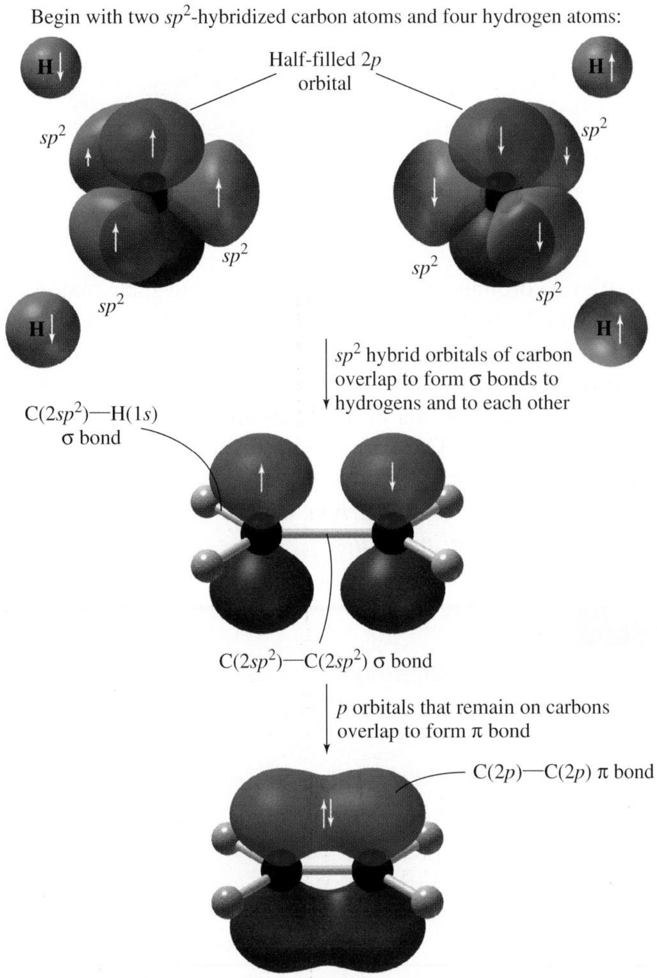

Figure 2.20

The carbon–carbon double bond in ethylene has a σ component and a π component. The σ component arises from overlap of sp^2-hybridized orbitals along the internuclear axis. The π component results from a side-by-side overlap of $2p$ orbitals.

Electrons in a π bond are called **π electrons.** The probability of finding a π electron is highest in the region above and below the plane of the molecule. The plane of the molecule corresponds to a nodal plane, where the probability of finding a π electron is zero.

In general, you can expect that carbon will be sp^2*-hybridized when it is directly bonded to three atoms in a neutral molecule.*

Problem 2.18

Identify the orbital overlaps involved in the indicated bond in the compound shown (propene). Is this a π bond or a σ bond?

$$H_2C{=}CH\underset{\uparrow}{—}CH_3$$

2.21 *sp* Hybridization and Bonding in Acetylene

One more hybridization scheme is important in organic chemistry. It is called ***sp* hybridization** and applies when carbon is directly bonded to two atoms, as in acetylene. The structure of acetylene is shown in Figure 2.21 along with its bond distances and bond angles. Its most prominent feature is its linear geometry.

Because each carbon in acetylene is bonded to two other atoms, the orbital hybridization model requires each carbon to have two equivalent orbitals available for σ bonds as outlined in Figure 2.22. According to this model the carbon $2s$ orbital and one of its $2p$ orbitals combine to generate two *sp* hybrid orbitals, each of which has 50% *s* character and 50%

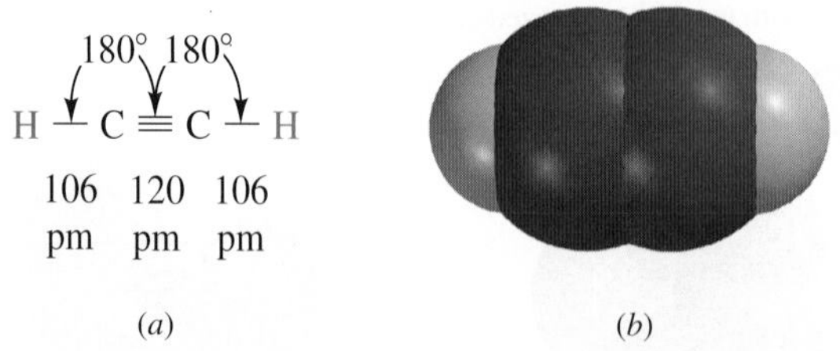

Figure 2.21

Acetylene is a linear molecule as indicated in (*a*) the structural formula and (*b*) a space-filling model.

p character. These two *sp* orbitals share a common axis, but their major lobes are oriented at an angle of 180° to each other. Two of the original 2*p* orbitals remain unhybridized.

As portrayed in Figure 2.23, the two carbons of acetylene are connected to each other by a 2*sp*–2*sp* σ bond, and each is attached to a hydrogen substituent by a 2*sp*–1*s* σ bond. The unhybridized 2*p* orbitals on one carbon overlap with their counterparts on the other to form two π bonds. The carbon–carbon triple bond in acetylene is viewed as a multiple bond of the σ + π + π type.

In general, you can expect that carbon will be sp-*hybridized when it is directly bonded to two atoms in a neutral molecule.*

Problem 2.19

The hydrocarbon shown, called *vinylacetylene,* is used in the synthesis of neoprene, a synthetic rubber. Identify the orbital overlaps involved in the indicated bond. How many σ bonds are there in vinylacetylene? How many π bonds?

$$H_2C{=}CH{-}C{\equiv}CH$$
(↑ indicated bond: CH—C)

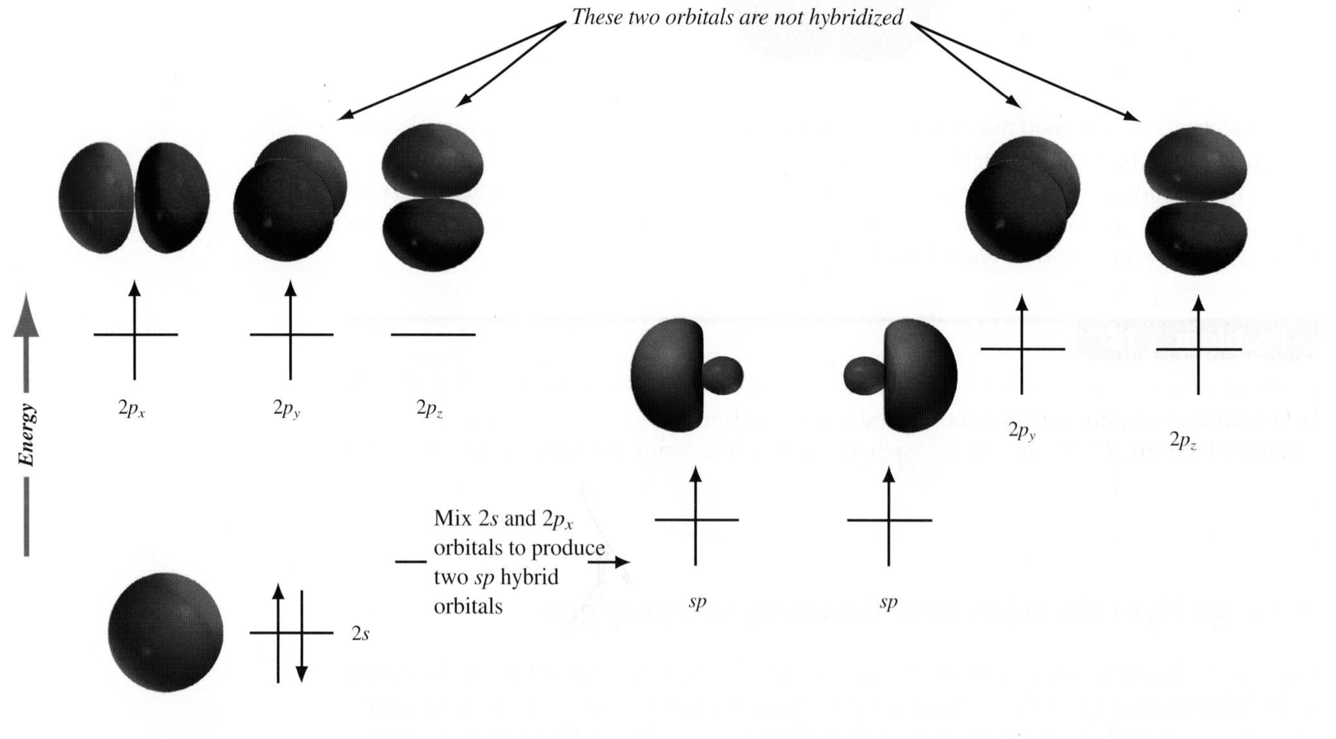

(*a*) Most stable electron configuration of carbon atom

(*b*) *sp* Hybrid state of carbon

Figure 2.22

sp Hybridization. (*a*) Electron configuration of carbon in its most stable state. (*b*) Mixing the *s* orbital with one of the three *p* orbitals generates two *sp* hybrid orbitals and leaves two of the 2*p* orbitals untouched. The axes of the two *sp* orbitals make an angle of 180° with each other.

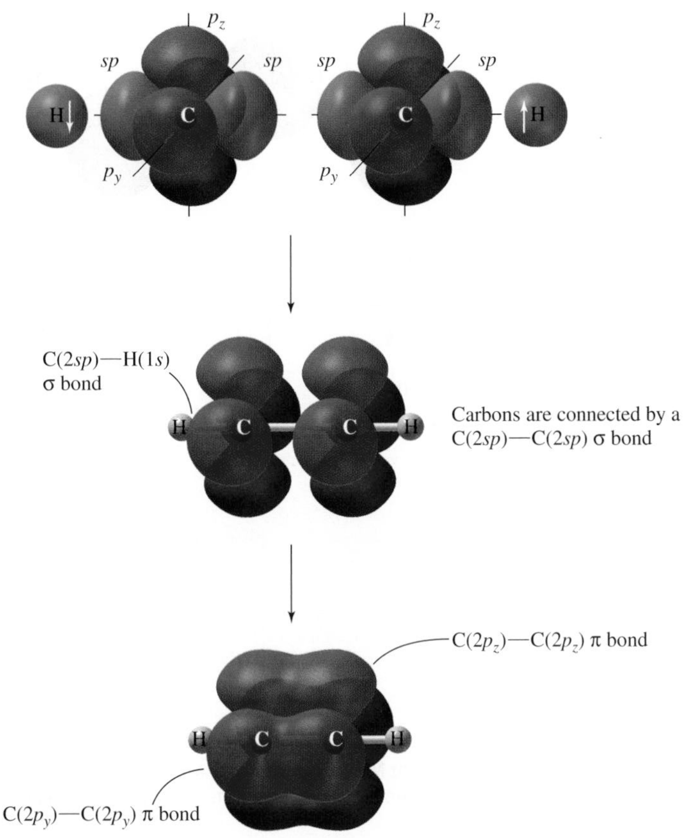

Figure 2.23

Bonding in acetylene based on *sp* hybridization of carbon. The carbon–carbon triple bond is viewed as consisting of one σ bond and two π bonds.

2.22 Bonding in Water and Ammonia: Hybridization of Oxygen and Nitrogen

The valence bond model and the accompanying idea of orbital hybridization have applications beyond those of bonding in hydrocarbons. A very simple extension to inorganic chemistry describes the bonding to the nitrogen of ammonia and the oxygen of water.

Like the carbon of methane, nitrogen is surrounded by four electron pairs in ammonia—three bonded pairs and an unshared pair. Oxygen in water has two bonded pairs and two unshared pairs. As we saw when discussing the tetrahedral geometry of CH_4, the pyramidal geometry of NH_3, and the bent geometry of H_2O in Section 1.10, all three are consistent with a tetrahedral arrangement of electron pairs. By associating a tetrahedral arrangement of electron pairs with sp^3 orbital hybridization, we see that CH_4, NH_3, and H_2O share a common valence bond description, differing only in the number of their valence electrons. C, N, and O are each sp^3-hybridized, with σ bonds connecting them to four, three, and two hydrogens, respectively. The unshared pair of nitrogen in NH_3 occupies an sp^3-hybridized orbital. The two unshared pairs of oxygen in H_2O occupy two sp^3-hybridized orbitals.

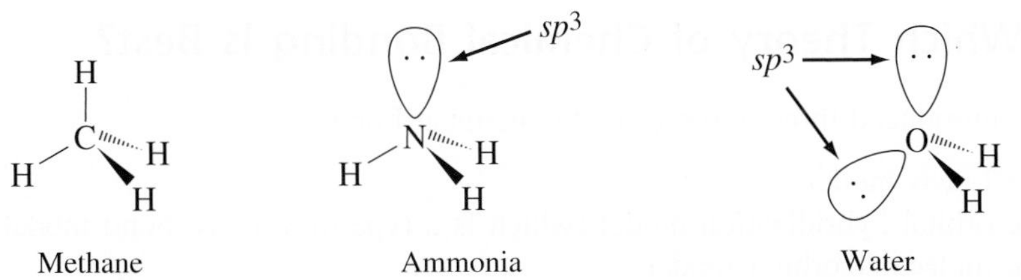

As noted in Section 1.10, the H—N—H angle in ammonia (107°), and the H—O—H angle in water (105°) are slightly smaller than the tetrahedral angle of 109.5° because of the larger volume required by unshared pairs.

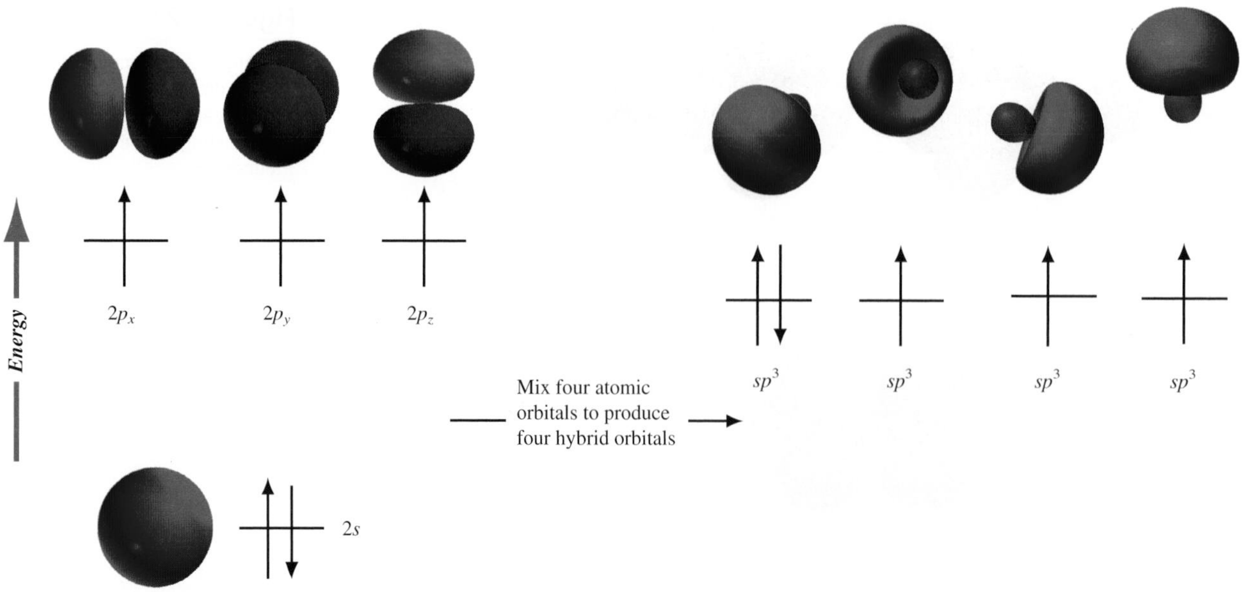

(*a*) Most stable electron configuration of nitrogen atom

(*b*) sp^3 Hybrid state of nitrogen

Figure 2.24

sp^3 Hybridization of nitrogen: (*a*) Electron configuration of nitrogen in its most stable state. (*b*) Mixing the *s* orbital with the three *p* orbitals generates four sp^3 hybrid orbitals. The four sp^3 hybrid orbitals are of equal energy. One sp^3 orbital contains a pair of electrons; the other three each contain a single electron.

Figure 2.24 illustrates sp^3 hybridization of nitrogen in ammonia. Except for the fact that nitrogen has one more valence electron than carbon, the diagram is identical to that shown for methane in Figure 2.9.

Problem 2.20

Construct an orbital diagram to show the hybrid orbitals of sp^3-hybridized oxygen in water.

Problem 2.21

Two contributing structures A and B that satisfy the octet rule can be written for formamide:

A B

(a) What is the hybridization of carbon and nitrogen in each?
(b) Formamide is planar or nearly so. Which structural formula better fits this fact?

2.23 Which Theory of Chemical Bonding Is Best?

We have introduced three approaches to chemical bonding:

1. The Lewis model
2. The orbital hybridization model (which is a type of valence bond model)
3. The molecular orbital model

Which one should you learn?

Generally speaking, the three models offer complementary information. Organic chemists use all three, emphasizing whichever one best suits a particular feature of structure or reactivity. Until recently, the Lewis and orbital hybridization models were used far more than the molecular orbital model. But that is changing.

The Lewis rules are relatively straightforward, easiest to master, and the most familiar. You will find that your ability to write Lewis formulas increases rapidly with experience. *Get as much practice as you can early in the course. Success in organic chemistry depends on writing correct Lewis structures.*

Orbital hybridization descriptions, because they too are based on the shared electron-pair bond, enhance the information content of Lewis formulas by distinguishing among various types of atoms, electrons, and bonds. As you become more familiar with a variety of structural types, you will find that the term *sp^3-hybridized carbon* triggers associations in your mind that are different from those of some other term, such as *sp^2-hybridized carbon,* for example.

Molecular orbital theory can provide insights into structure and reactivity that the Lewis and orbital hybridization models can't. It is the least intuitive of the three methods, however, and requires the most training, background, and chemical knowledge to apply. We have *discussed* molecular orbital theory so far only in the context of the bonding in H_2. We have *used* the results of molecular orbital theory, however, several times without acknowledging it until now. Electrostatic potential maps are obtained by molecular orbital calculations. Four molecular orbital calculations provided the drawings that we used in Figure 2.4 to illustrate how electron density builds up between the atoms in the valence bond (!) treatment of H_2. Molecular orbital theory is well suited to quantitative applications and is becoming increasingly available for routine use via software that runs on personal computers. You will see the results of molecular orbital theory often in this text, but the theory itself will be developed only at an introductory level.

2.24 SUMMARY

Section 2.1 The classes of hydrocarbons are **alkanes, alkenes, alkynes,** and **arenes.** Alkanes are hydrocarbons in which all of the bonds are *single* bonds and are characterized by the molecular formula C_nH_{2n+2}.

Section 2.2 Two theories of bonding, valence bond and molecular orbital theory, are based on the wave nature of an electron. Constructive interference between the electron wave of one atom and that of another gives a region between the two atoms in which the probability of sharing an electron is high—a bond.

Section 2.3 In valence bond theory a covalent bond is described in terms of in-phase overlap of a half-filled orbital of one atom with a half-filled orbital of another. When applied to bonding in H_2, the orbitals involved are the 1*s* orbitals of two hydrogen atoms and the bond is a σ bond.

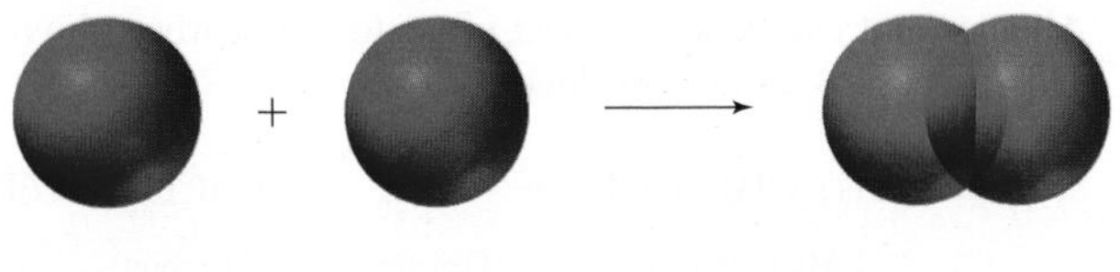

Section 2.4 In molecular orbital theory, the molecular orbitals (MOs) are approximated by combining the atomic orbitals (AOs) of all of the atoms in a molecule. The number of MOs must equal the number of AOs that are combined.

Section 2.5 The first three alkanes are methane (CH_4), ethane (CH_3CH_3), and propane ($CH_3CH_2CH_3$).

Section 2.6 Bonding in methane is most often described by an orbital hybridization model, which is a modified form of valence bond theory. Four equivalent sp^3 hybrid orbitals of carbon are generated by mixing the 2*s*, $2p_x$, $2p_y$, and $2p_z$ orbitals. In-phase overlap of each half-filled sp^3 hybrid orbital with a half-filled hydrogen 1*s* orbital gives a σ bond.

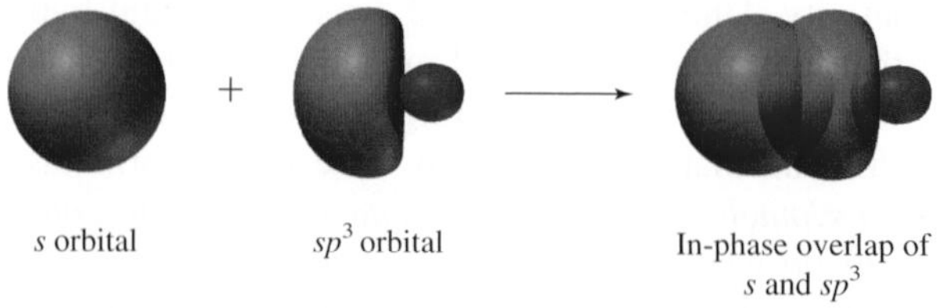
s orbital
sp3 orbital
In-phase overlap of s and sp3

Section 2.7 The carbon–carbon bond in ethane is a σ bond viewed as an overlap of a half-filled sp^3 orbital of one carbon with a half-filled sp^3 orbital of another.

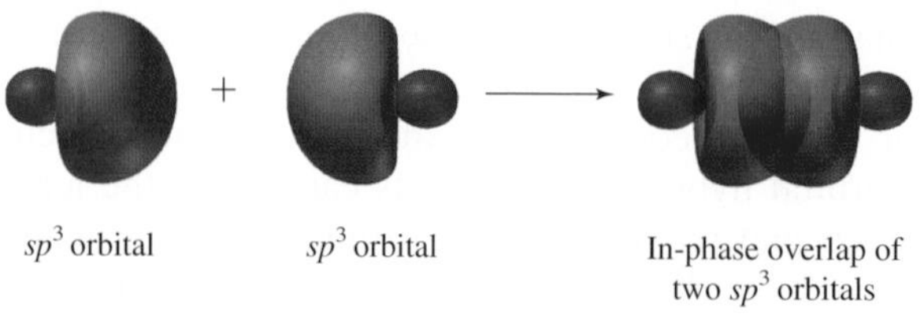
sp3 orbital
sp3 orbital
In-phase overlap of two sp3 orbitals

Section 2.8 Two constitutionally isomeric alkanes have the molecular formula C_4H_{10}. One has an unbranched chain ($CH_3CH_2CH_2CH_3$) and is called ***n*-butane;** the other has a branched chain [$(CH_3)_3CH$] and is called **isobutane.** Both *n*-butane and isobutane are **common names.**

Section 2.9 Unbranched alkanes of the type $CH_3(CH_2)_xCH_3$ are often referred to as *n*-alkanes, and are said to belong to a **homologous series.**

Section 2.10 There are three constitutional isomers of C_5H_{12}: *n*-pentane ($CH_3CH_2CH_2CH_2CH_3$), isopentane [$(CH_3)_2CHCH_2CH_3$], and neopentane [$(CH_3)_4C$].

Sections 2.11–2.15 A single alkane may have several different names; a name may be a common name, or it may be a *systematic name* developed by a well-defined set of rules. The most widely used system is **IUPAC** nomenclature. Table 2.5 summarizes the rules for alkanes and cycloalkanes. Table 2.6 gives the rules for naming alkyl groups.

Section 2.16 Natural gas is an abundant source of methane, ethane, and propane. Petroleum is a liquid mixture of many hydrocarbons, including alkanes. Alkanes also occur naturally in the waxy coating of leaves and fruits.

Section 2.17 Alkanes and cycloalkanes are nonpolar and insoluble in water. The forces of attraction between alkane molecules are **induced-dipole/induced-dipole** attractive forces. The boiling points of alkanes increase as the number of carbon atoms increases. Branched alkanes have lower boiling points than their unbranched isomers. There is a limit to how closely two atoms can approach each other, which is given by the sum of their **van der Waals radii.**

Section 2.18 Alkanes and cycloalkanes burn in air to give carbon dioxide, water, and heat. This process is called **combustion.**

$$\underset{\text{2-Methylbutane}}{(CH_3)_2CHCH_2CH_3} + \underset{\text{Oxygen}}{8O_2} \longrightarrow \underset{\text{Carbon dioxide}}{5CO_2} + \underset{\text{Water}}{6H_2O}$$

$$\Delta H^\circ = -3529 \text{ kJ } (-843.4 \text{ kcal})$$

The heat evolved on burning an alkane increases with the number of carbon atoms. The relative stability of isomers may be determined by comparing their respective **heats of combustion.** The more stable of two isomers has the lower heat of combustion.

Section 2.19 Combustion of alkanes is an example of **oxidation–reduction.** Although it is possible to calculate oxidation numbers of carbon in organic molecules, it is more convenient to regard oxidation of an organic substance as an increase in its oxygen content or a decrease in its hydrogen content.

TABLE 2.5 Summary of IUPAC Nomenclature of Alkanes and Cycloalkanes

Rule	Example
A. Alkanes **1.** Find the longest continuous chain of carbon atoms, and assign a basis name to the compound corresponding to the IUPAC name of the unbranched alkane having the same number of carbons.	The longest continuous chain in the alkane shown is six carbons. 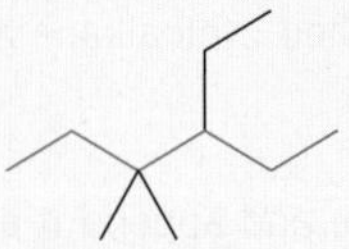 This alkane is named as a derivative of hexane.
2. List the substituents attached to the longest continuous chain in alphabetical order. Use the prefixes *di-*, *tri-*, *tetra-*, and so on, when the same substituent appears more than once. Ignore these prefixes when alphabetizing.	The alkane bears two methyl groups and an ethyl group. It is an *ethyldimethylhexane.* 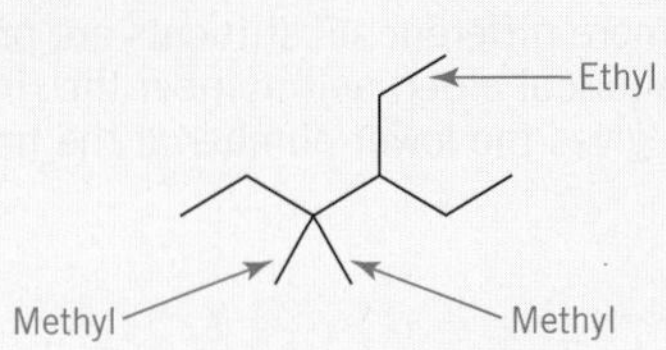
3. Number from the end of the chain in the direction that gives the lower locant to a substituent at the first point of difference.	When numbering from left to right, the substituents appear at carbons 3, 3, and 4. When numbering from right to left the locants are 3, 4, and 4; therefore, number from left to right. 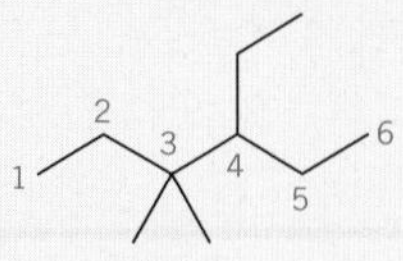 Correct 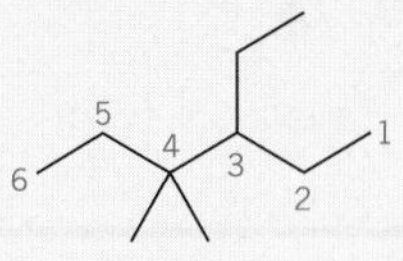 Incorrect The correct name is *4-ethyl-3,3-dimethylhexane.*
4. When two different numbering schemes give equivalent sets of locants, choose the direction that gives the lower locant to the group that appears first in the name.	In the following example, the substituents are located at carbons 3 and 4 regardless of the direction in which the chain is numbered. 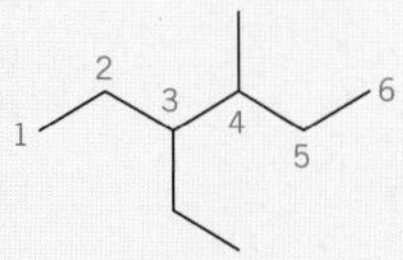 Correct 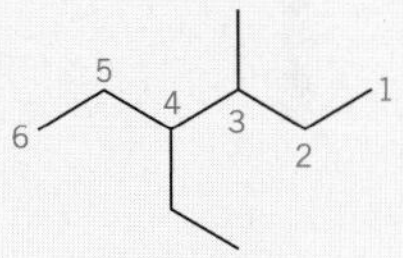 Incorrect Ethyl precedes methyl in the name; therefore *3-ethyl-4-methylhexane* is correct.
5. When two chains are of equal length, choose the one with the greater number of substituents as the parent. (Although this requires naming more substituents, the substituents have simpler names.)	Two different chains contain five carbons in the alkane: 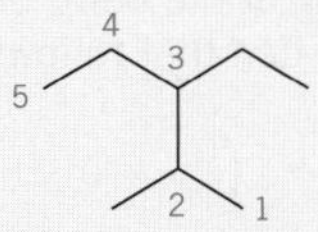 Correct 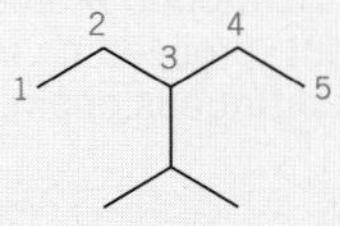 Incorrect The correct name is *3-ethyl-2-methylpentane* (disubstituted chain), rather than 3-isopropylpentane (monosubstituted chain). *Continued*

TABLE 2.5 Summary of IUPAC Nomenclature of Alkanes and Cycloalkanes (*Continued*)

Rule	Example
B. Cycloalkanes 1. Count the number of carbons in the ring, and assign a basis name to the cycloalkane corresponding to the IUPAC name of the unbranched cycloalkane having the same number of carbons.	The compound shown contains five carbons in its ring. 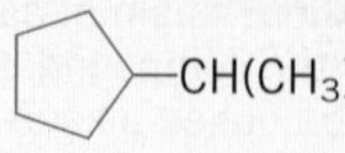 It is named as a derivative of *cyclopentane.*
2. Name the alkyl group, and append it as a prefix to the cycloalkane. No locant is needed if the compound is a monosubstituted cycloalkane. It is understood that the alkyl group is attached to C-1.	The previous compound is *isopropylcyclopentane.* Alternatively, the alkyl group can be named according to the rules summarized in Table 2.7, whereupon the name becomes (*1-methylethyl*)*cyclopentane.* Parentheses are used to set off the name of the alkyl group as needed to avoid ambiguity.
3. When two or more different substituents are present, list them in alphabetical order, and number the ring in the direction that gives the lower number at the first point of difference.	The compound shown is *1,1-diethyl-4-hexylcyclooctane.* 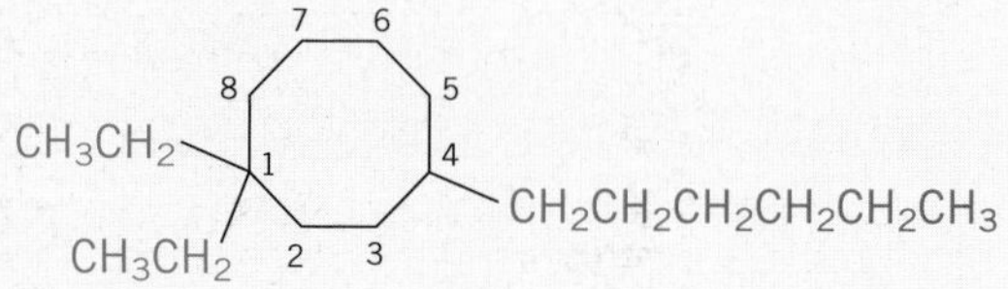
4. Name the compound as a cycloalkyl-substituted alkane if the substituent has more carbons than the ring.	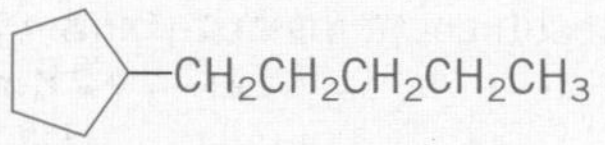 is *pentylcyclopentane* 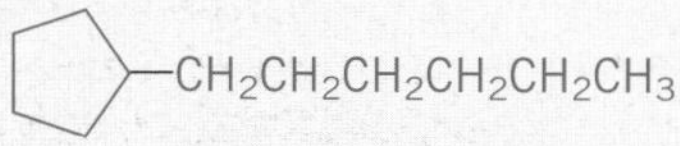 is *1-cyclopentylhexane*

Section 2.20 Carbon is ***sp*²-hybridized** in ethylene, and the double bond has a σ component and a π component. The sp^2 hybridization state is derived by mixing the 2*s* and two of the three 2*p* orbitals. Three equivalent sp^2 orbitals result, and their axes are coplanar. Overlap of a half-filled sp^2 orbital of one carbon with a half-filled sp^2 orbital of another gives a σ bond between the two carbons. Each carbon still has one unhybridized *p* orbital available

TABLE 2.6 Summary of IUPAC Nomenclature of Alkyl Groups

Rule	Example
1. Number the carbon atoms beginning at the point of attachment, proceeding in the direction that follows the longest continuous chain.	The longest continuous chain that begins at the point of attachment in the group shown contains six carbons. $\mathrm{CH_3CH_2CH_2\overset{1\vert}{C}\overset{2}{C}H_2\overset{3}{C}H\overset{4}{C}H_2\overset{5}{C}H_2\overset{6}{C}H_3}$ with CH_3 groups on C-1 and C-3
2. Assign a basis name according to the number of carbons in the corresponding unbranched alkane. Drop the ending *-ane* and replace it with *-yl.*	The alkyl group shown in step 1 is named as a substituted *hexyl* group.
3. List the substituents attached to the basis group in alphabetical order using replicating prefixes when necessary.	The alkyl group in step 1 is a *dimethylpropylhexyl* group.
4. Locate the substituents according to the numbering of the main chain described in step 1.	The alkyl group is a *1,3-dimethyl-1-propylhexyl* group.

for bonding, and "side-by-side" overlap of half-filled p orbitals of adjacent carbons gives a π bond between them.

The π bond in ethylene is generated by overlap of half-filled p orbitals of adjacent carbons.

Section 2.21 Carbon is ***sp*-hybridized** in acetylene, and the triple bond is of the $\sigma + \pi + \pi$ type. The 2s orbital and one of the 2p orbitals combine to give two equivalent sp orbitals that have their axes in a straight line. A σ bond between the two carbons is supplemented by two π bonds formed by overlap of the remaining half-filled p orbitals.

The triple bond of acetylene has a σ bond component and two π bonds; the two π bonds are shown here and are perpendicular to each other.

Section 2.22 The hybridization model for carbon can also be extended to oxygen and nitrogen. Oxygen and nitrogen are sp^3-hybridized in water and ammonia, respectively.

Section 2.23 Lewis structures, orbital hybridization, and molecular orbital descriptions of bonding are all used in organic chemistry. Lewis structures are used the most, MO descriptions the least. All will be used in this text.

PROBLEMS

2.22 The general molecular formula for alkanes is C_nH_{2n+2}. What is the general molecular formula for:

(a) Cycloalkanes
(b) Alkenes
(c) Alkynes
(d) Cyclic hydrocarbons that contain one double bond

2.23 A certain hydrocarbon has a molecular formula of C_5H_8. Which of the following is *not* a structural possibility for this hydrocarbon?

(a) It is a cycloalkane.
(b) It contains one ring and one double bond.
(c) It contains two double bonds and no rings.
(d) It is an alkyne.

2.24 Which of the hydrocarbons in each of the following groups are isomers?

(a)   $(CH_3)_3CH$ 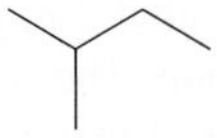

(b) 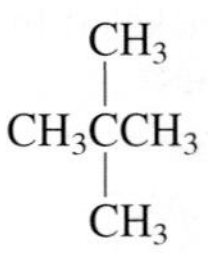  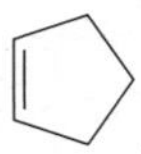 

(c) 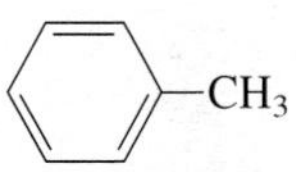 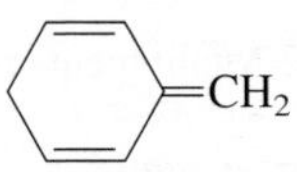  $HC{\equiv}CCH_2CH_2C{\equiv}CCH_3$

2.25 Write structural formulas and give the IUPAC names for the nine alkanes that have the molecular formula C_7H_{16}.

2.26 From among the 18 constitutional isomers of C_8H_{18}, write structural formulas, and give the IUPAC names for those that are named as derivatives of

(a) Heptane (b) Hexane (c) Pentane (d) Butane

2.27 *Pristane* is an alkane that is present to the extent of about 14% in shark liver oil. Its IUPAC name is 2,6,10,14-tetramethylpentadecane. Write its structural formula.

2.28 All the parts of this problem refer to the alkane having the carbon skeleton shown.

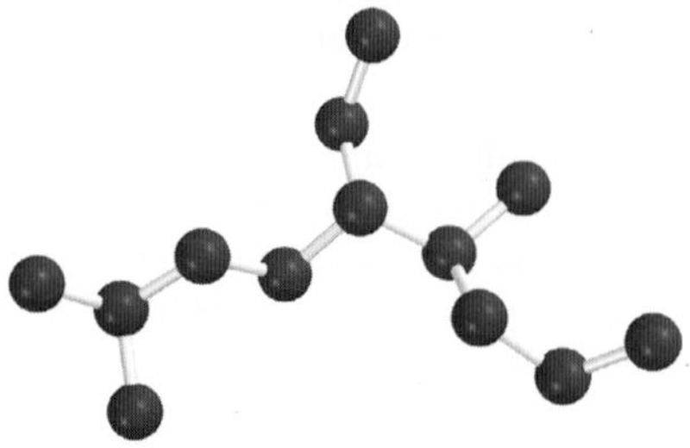

(a) What is the molecular formula of this alkane?

(b) What is its IUPAC name?

(c) How many methyl groups are present in this alkane? Methylene groups? Methine groups?

(d) How many carbon atoms are primary? Secondary? Tertiary? Quaternary?

2.29 Give the IUPAC name for each of the following compounds:

(a) $CH_3(CH_2)_{25}CH_3$

(b) $(CH_3)_2CHCH_2(CH_2)_{14}CH_3$

(c) $(CH_3CH_2)_3CCH(CH_2CH_3)_2$

(d) 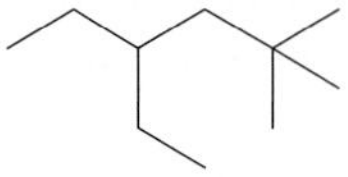

(e) 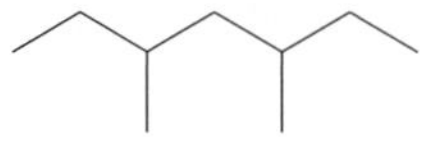

(f) 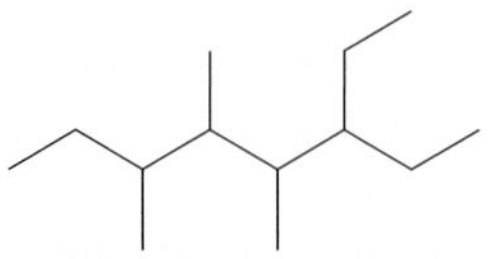

2.30 Write a structural formula for each of the following compounds:

(a) 6-Isopropyl-2,3-dimethylnonane

(b) 4-*tert*-Butyl-3-methylheptane

(c) 4-Isobutyl-1,1-dimethylcyclohexane

(d) *sec*-Butylcycloheptane

(e) Cyclobutylcyclopentane

2.31 Using the method outlined in Section 2.13, give an IUPAC name for each of the following alkyl groups, and classify each one as primary, secondary, or tertiary:

(a) $CH_3(CH_2)_{10}CH_2—$

(b) $—CH_2CH_2CHCH_2CH_2CH_3$ (with CH_2CH_3 on the third carbon)

(c) $—C(CH_2CH_3)_3$

(d) 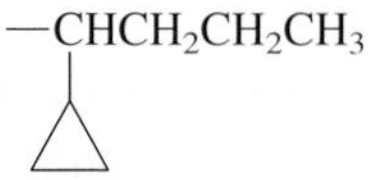

(e) 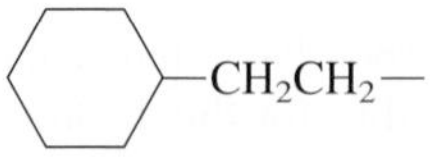

(f) 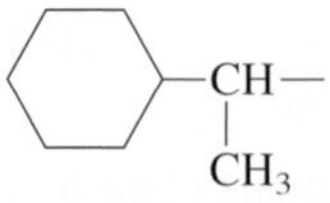

2.32 The 2004 IUPAC names for alkyl groups are derived from the alkane having the same carbon chain as the alkyl group. The *-e* ending of that alkane is replaced by *-yl*, and the chain is numbered from the end that gives the carbon at the point of attachment its lower number. This number immediately precedes the *-yl* ending and is bracketed by hyphens.

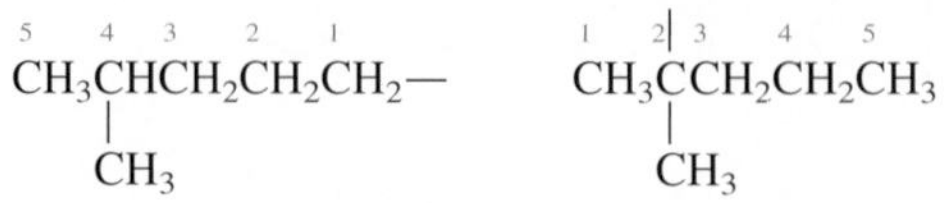

4-Methylpentan-1-yl 2-Methylpentan-2-yl

Name the C_4H_9 alkyl groups according to this system.

2.33 Write the structural formula of a compound of molecular formula $C_4H_8Cl_2$ in which

(a) All the carbons belong to methylene groups

(b) None of the carbons belong to methylene groups

2.34 Female tiger moths signify their presence to male moths by giving off a sex attractant (pheromone). The sex attractant has been isolated and found to be a 2-methyl-branched alkane having a molecular weight of 254. What is this material?

2.35 Write a balanced chemical equation for the combustion of each of the following compounds:

(a) Decane (c) Methylcyclononane

(b) Cyclodecane (d) Cyclopentylcyclopentane

2.36 The heats of combustion of methane and butane are 890 kJ/mol (212.8 kcal/mol) and 2876 kJ/mol (687.4 kcal/mol), respectively. When used as a fuel, would methane or butane generate more heat for the same mass of gas? Which would generate more heat for the same volume of gas?

2.37 In each of the following groups of compounds, identify the one with the largest heat of combustion and the one with the smallest. (Try to do this problem without consulting Table 2.3.)

(a) Hexane, heptane, octane

(b) 2-Methylpropane, pentane, 2-methylbutane

(c) 2-Methylbutane, 2-methylpentane, 2,2-dimethylpropane

(d) Pentane, 3-methylpentane, 3,3-dimethylpentane

(e) Ethylcyclopentane, ethylcyclohexane, ethylcycloheptane

2.38 (a) Given $\Delta H°$ for the reaction

$$H_2(g) + \tfrac{1}{2}O_2(g) \longrightarrow H_2O(l) \qquad \Delta H° = -286 \text{ kJ}$$

along with the information that the heat of combustion of ethane is 1560 kJ/mol and that of ethylene is 1410 kJ/mol, calculate $\Delta H°$ for the hydrogenation of ethylene:

$$H_2C{=}CH_2(g) + H_2(g) \longrightarrow CH_3CH_3(g)$$

(b) If the heat of combustion of acetylene is 1300 kJ/mol, what is the value of $\Delta H°$ for its hydrogenation to ethylene? To ethane?

(c) What is the value of $\Delta H°$ for the hypothetical reaction

$$2H_2C{=}CH_2(g) \longrightarrow CH_3CH_3(g) + HC{\equiv}CH(g)$$

2.39 We have seen in this chapter that, among isomeric alkanes, the unbranched isomer is the least stable and has the highest boiling point; the most branched isomer is the most stable and has the lowest boiling point. Does this mean that one alkane boils lower than another *because* it is more stable? Explain.

2.40 Higher octane gasoline typically contains a greater proportion of branched alkanes relative to unbranched ones. Are branched alkanes better fuels because they give off more energy on combustion? Explain.

2.41 The reaction shown is important in the industrial preparation of dichlorodimethylsilane for eventual conversion to silicone polymers.

$$2CH_3Cl + Si \longrightarrow (CH_3)_2SiCl_2$$

(a) Is carbon oxidized, reduced, or neither in this reaction?

(b) On the basis of the molecular model of $(CH_3)_2SiCl_2$, deduce the hybridization state of silicon in this compound. What is the principal quantum number *n* of the silicon *s* and *p* orbitals that are hybridized?

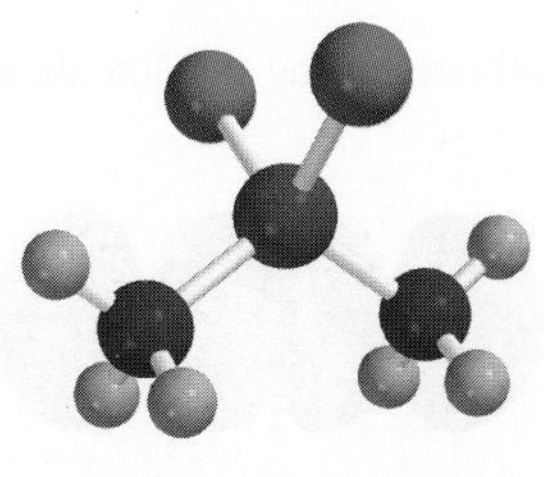

2.42 Alkanes spontaneously burst into flame in the presence of elemental fluorine. The reaction that takes place between pentane and F_2 gives CF_4 and HF as the only products.

(a) Write a balanced equation for this reaction.

(b) Is carbon oxidized, reduced, or does it undergo no change in oxidation state in this reaction?

2.43 What is the hybridization of each carbon in $CH_3CH{=}CHC{\equiv}CH$? What are the CCC bond angles?

2.44 What is the hybridization of oxygen or nitrogen in each of the following?

(a) $(CH_3)_2C{=}\ddot{\underset{..}{O}}$

(b) $CH_3CH_2C{\equiv}N{:}$

(c) $H_2C{=}\ddot{N}\ddot{N}H_2$

(d) $CH_3CH_2\ddot{\underset{..}{O}}CH_2CH_3$

(e) $(CH_3CH_2)_2\ddot{N}H$

2.45 Construct a figure analogous to Figures 2.9, 2.19, or 2.22 to show the hybridization of nitrogen in $CH_3C{\equiv}N{:}$ and $H_2C{=}\underset{..}{N}H$.

2.46 Which atoms in the following reaction undergo changes in their oxidation state? Which atom is oxidized? Which one is reduced?

$$2CH_3CH_2OH + 2Na \longrightarrow 2CH_3CH_2ONa + H_2$$

2.47 Compound A undergoes the following reactions:

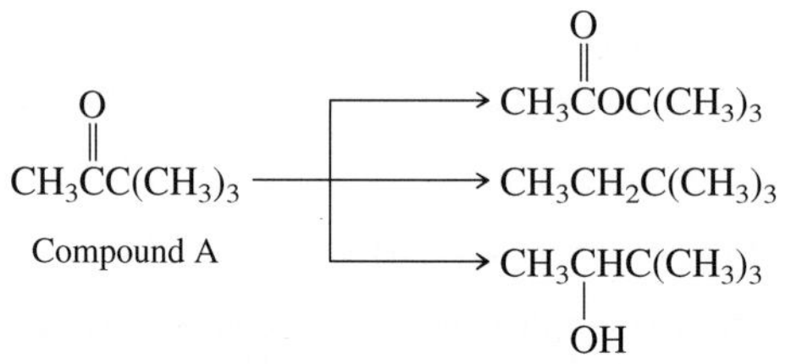
O
CH3CC(CH3)3
Compound A
O
CH3COC(CH3)3
CH3CH2C(CH3)3
CH3CHC(CH3)3
OH

(a) Which of the reactions shown require(s) an oxidizing agent?

(b) Which of the reactions shown require(s) a reducing agent?

2.48 Each of the following reactions will be encountered at some point in this text. Classify each one according to whether the organic substrate is oxidized or reduced in the process.

(a) $CH_3C{\equiv}CH + 2Na + 2NH_3 \longrightarrow CH_3CH{=}CH_2 + 2NaNH_2$

(b)
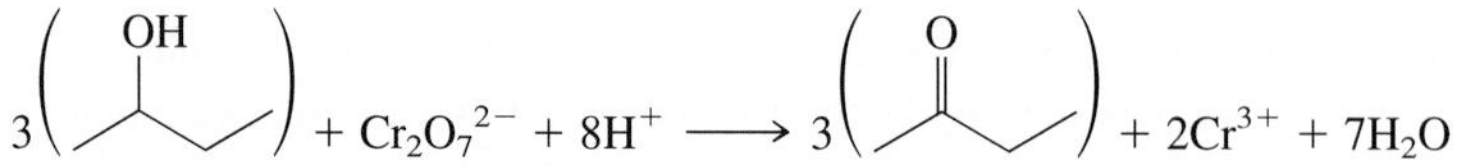
OH
3
+ Cr2O7^2− + 8H+
3
O
+ 2Cr^3+ + 7H2O

(c) $HOCH_2CH_2OH + HIO_4 \longrightarrow 2H_2C{=}O + HIO_3 + H_2O$

(d)
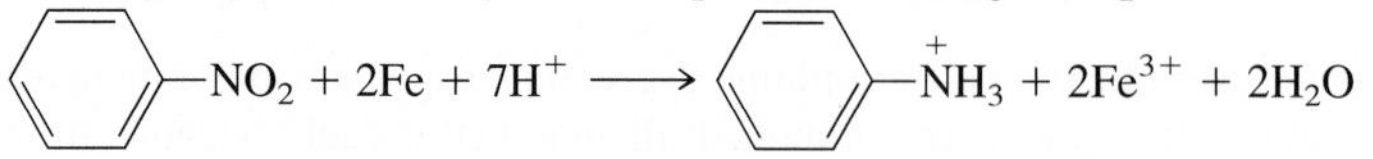
−NO2 + 2Fe + 7H+
+
−NH3 + 2Fe^3+ + 2H2O

2.49 Of the overlaps between an *s* and a *p* orbital as shown in the illustration, one is bonding, one is antibonding, and the third is nonbonding (neither bonding nor antibonding). Which orbital overlap corresponds to which interaction? Why?

2.50 Does the overlap of two *p* orbitals in the fashion shown correspond to a σ bond or to a π bond? Explain.

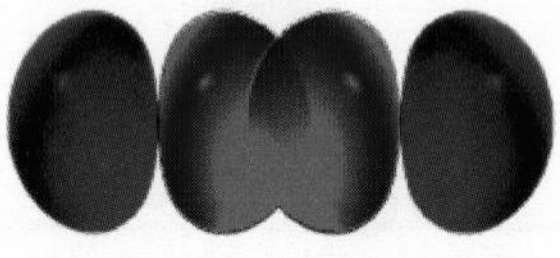

Descriptive Passage and Interpretive Problems 2

Some Biochemical Reactions of Alkanes

Alkanes occur naturally in places other than petroleum deposits. In insects, for example. The waxy alkanes dispersed in its cuticle help protect an insect from dehydration. Some insects use volatile alkanes to defend themselves or communicate with others of the same species. Alkanes even serve as starting materials that the insect converts to other biologically important substances.

The major biosynthetic pathway leading to alkanes is by enzyme-catalyzed decarboxylation (loss of CO_2) of fatty acids, compounds of the type $CH_3(CH_2)_nCO_2H$ in which *n* is an even number and the chain has 14 or more carbons.

$$CH_3(CH_2)_nCO_2H \longrightarrow CH_3(CH_2)_{n-1}CH_3 + CO_2$$

Biochemical conversion of alkanes to other substances normally begins with oxidation.

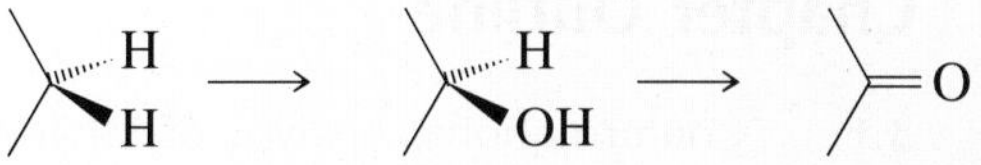

In addition to alkanes, the oxidation of drugs and other substances occurs mainly in the liver and is catalyzed by the enzyme cytochrome P-450. Molecular oxygen and nicotinamide adenine dinucleotide (NAD) are also required.

Reactions of this type also occur in insects but are not limited to them. Oxidation by microorganisms has been extensively studied and is often selective for certain kinds of C—H bonds. The fungus *Pseudomonas oleovorans,* for example, oxidizes the CH_3 groups at the end of the carbon chain of 4-methyloctane faster than the CH_3 branch and faster than the CH_2 and CH units within the chain.

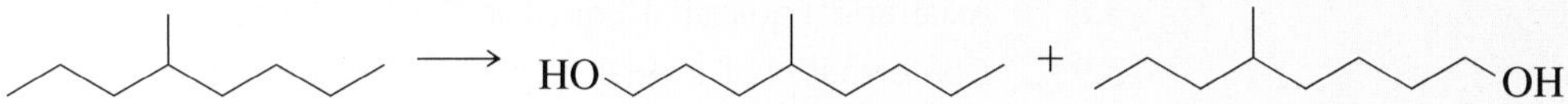

2.51 Tridecane [$CH_3(CH_2)_{11}CH_3$] is a major component of the repellent which the stink bug *Piezodorus guildinii* releases from its scent glands when attacked. What fatty acid gives tridecane on decarboxylation?

A. $CH_3(CH_2)_{10}CO_2H$
B. $CH_3(CH_2)_{11}CO_2H$
C. $CH_3(CH_2)_{12}CO_2H$
D. $CH_3(CH_2)_{13}CO_2H$

2.52 Assuming a selectivity analogous to that observed in the microbiological oxidation of 4-methyloctane by *Pseudomonas oleovorans,* which of the following is expected to give two constitutionally isomeric alcohols on oxidation?

A. Heptane
B. 3-Methylheptane
C. 4-Methylheptane
D. 4,4-Dimethylheptane

2.53 Female German cockroaches convert the alkane shown to a substance that attracts males.

$$CH_3CH_2\underset{\displaystyle CH_3}{\underset{|}{C}}H(CH_2)_7\underset{\displaystyle CH_3}{\underset{|}{C}}H(CH_2)_{16}CH_2CH_3$$

Oxidation at C-2 of the alkane gives the sex attractant, which has a molecular formula $C_{31}H_{62}O$ and the same carbon skeleton as the alkane. What is the structure of the sex attractant?

A. $$CH_3\overset{\displaystyle OH}{\overset{|}{C}}H\underset{\displaystyle CH_3}{\underset{|}{C}}H(CH_2)_7\underset{\displaystyle CH_3}{\underset{|}{C}}H(CH_2)_{16}CH_2CH_3$$

B. $$CH_3CH_2\underset{\displaystyle CH_3}{\underset{|}{C}}H(CH_2)_7\underset{\displaystyle CH_3}{\underset{|}{C}}H(CH_2)_{16}\overset{\displaystyle OH}{\overset{|}{C}}HCH_3$$

C. $$CH_3\overset{\displaystyle O}{\overset{\|}{C}}\underset{\displaystyle CH_3}{\underset{|}{C}}H(CH_2)_7\underset{\displaystyle CH_3}{\underset{|}{C}}H(CH_2)_{16}CH_2CH_3$$

D. $$CH_3CH_2\underset{\displaystyle CH_3}{\underset{|}{C}}H(CH_2)_7\underset{\displaystyle CH_3}{\underset{|}{C}}H(CH_2)_{16}\overset{\displaystyle O}{\overset{\|}{C}}CH_3$$

2.54 Biological oxidation of the hydrocarbon adamantane by the fungus *Absidia glauca* gives a mixture of two alcohols. Classify the carbon in adamantane that is oxidized in forming the major product.

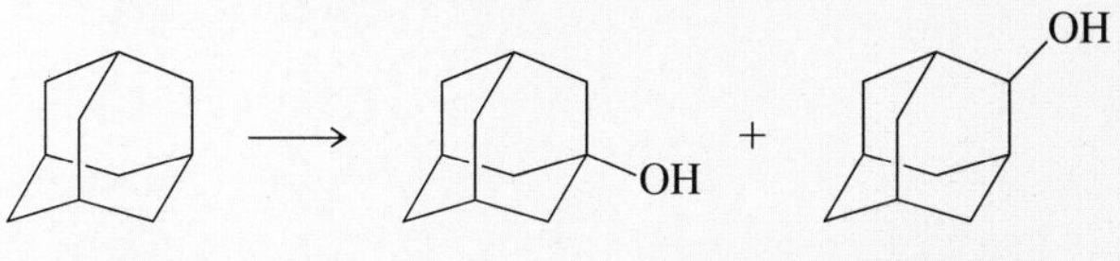

Adamantane Major Minor

A. Primary
B. Secondary
C. Tertiary
D. Quaternary

3 Alkanes and Cycloalkanes: Conformations and cis–trans Stereoisomers

Chapter Outline

The macroscopic structure of diamond owes its many facets to the diamond-cutter's art. Microscopically, it is an assembly of tetrahedral carbon atoms bonded together in a way related to the much simpler cyclohexane molecule.

HYDROGEN PEROXIDE is formed in the cells of plants and animals but is toxic to them. Consequently, living systems have developed mechanisms to rid themselves of hydrogen peroxide, usually by enzyme-catalyzed reduction to water. An understanding of how reactions take place, be they reactions in living systems or reactions in test tubes, begins with a thorough knowledge of the structure of the reactants, products, and catalysts. Even a simple molecule such as hydrogen peroxide (four atoms!) may be structurally more complicated than you think. Suppose we wanted to write the structural formula for H_2O_2 in enough detail to show the positions of the atoms relative to one another. We could write two different planar geometries A and B that differ by a 180° rotation about the O—O bond. We could also write an infinite number of nonplanar structures, of which C is but one example, that differ from one another by tiny increments of rotation about the O—O bond.

Structures A, B, and C represent different **conformations** of hydrogen peroxide. *Conformations are different spatial arrangements of a molecule that are generated by rotation about single bonds.* Although we can't tell from simply looking at these structures, we now know from experimental studies that C is the most stable conformation.

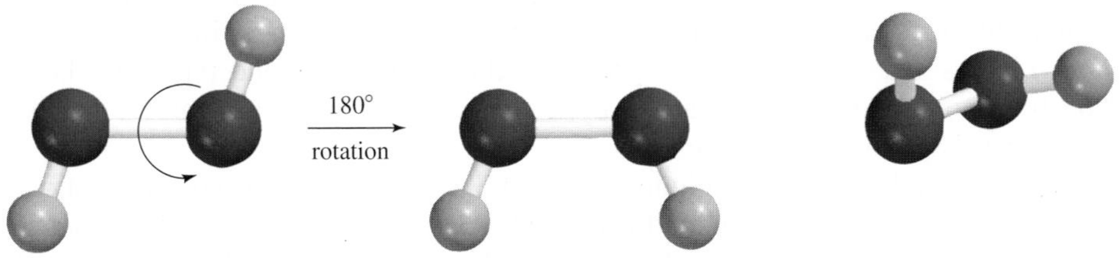

A B C

Conformational analysis is the study of how conformational factors affect the structure of a molecule and its properties. In this chapter we'll examine the conformations of various alkanes and cycloalkanes, focusing most of our attention on three of them: *ethane, butane,* and *cyclohexane.* You will see that even simple organic molecules can exist in many conformations. Conformational analysis will help us to visualize organic molecules in 3D and to better understand their structure and properties.

3.1 Conformational Analysis of Ethane

Ethane is the simplest hydrocarbon that can have distinct conformations. Two, the **staggered conformation** and the **eclipsed conformation,** deserve special mention and are illustrated with molecular models in Figure 3.1.

In the staggered conformation, each C—H bond of one carbon bisects an H—C—H angle of the other carbon.

In the eclipsed conformation, each C—H bond of one carbon is aligned with a C—H bond of the other carbon.

Different conformations of the same compound are sometimes called *rotamers* (short for rotational isomers).

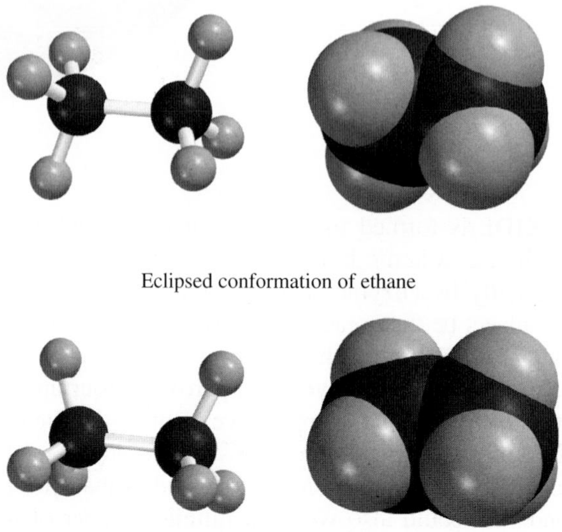

Figure 3.1

The staggered and eclipsed conformations of ethane shown as ball-and-spoke models (*left*) and as space-filling models (*right*).

The staggered and eclipsed conformations interconvert by rotation around the C—C bond, and do so very rapidly. We'll see just how rapidly later in this section.

Among the various ways in which the staggered and eclipsed forms are portrayed, wedge-and-dash, sawhorse, and Newman projection drawings are especially useful. These are shown for the staggered conformation of ethane in Figure 3.2 and for the eclipsed conformation in Figure 3.3.

We used *wedge-and-dash* drawings in earlier chapters, and so Figures 3.2*a* and 3.3*a* are familiar to us. A *sawhorse* drawing (Figures 3.2*b* and 3.3*b*) shows the conformation of a molecule without having to resort to different styles of bonds. In a *Newman projection* (Figures 3.2*c* and 3.3*c*), we sight down the C—C bond, and represent the front carbon by a point and the back carbon by a circle. Each carbon has three other bonds that are placed symmetrically around it.

Newman projections were devised by Professor Melvin S. Newman of Ohio State University.

The structural feature that Figures 3.2 and 3.3 illustrate is the spatial relationship between bonds on adjacent carbons. Each H—C—C—H unit in ethane is characterized by a *torsion angle* or *dihedral angle,* which is the angle between the H—C—C plane and the C—C—H plane. The torsion angle is easily seen in a Newman projection of ethane as the angle between C—H bonds of adjacent carbons.

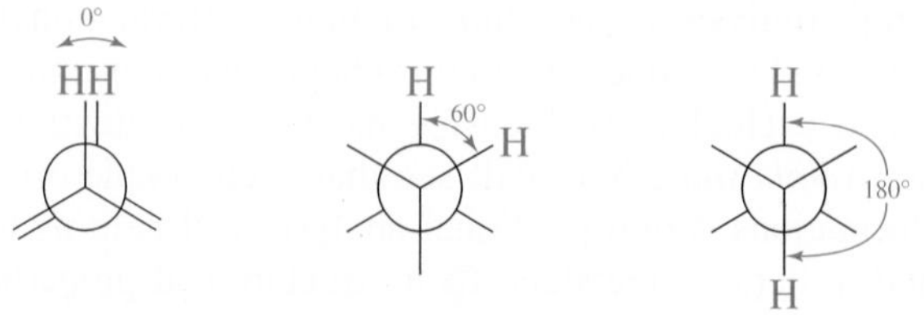

Eclipsed bonds are characterized by a torsion angle of 0°. When the torsion angle is approximately 60°, we say that the spatial relationship is **gauche;** and when it is 180°

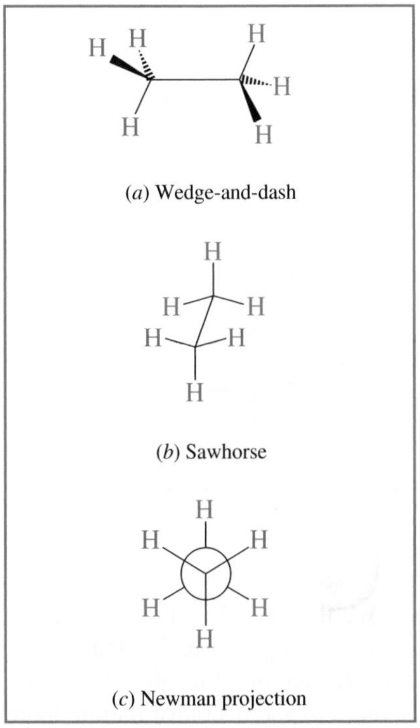
(a) Wedge-and-dash
(b) Sawhorse
(c) Newman projection

Figure 3.2

Some commonly used drawings of the staggered conformation of ethane.

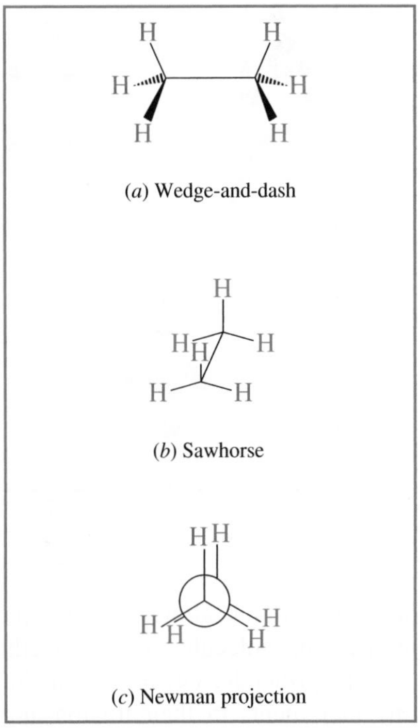
(a) Wedge-and-dash
(b) Sawhorse
(c) Newman projection

Figure 3.3

Some commonly used drawings of the eclipsed conformation of ethane.

Sawhorses are beams with four legs that are used in pairs to support a plank for sawing, or in other uses as a support structure or road marker.

we say that it is **anti.** Staggered conformations have only gauche or anti relationships between bonds on adjacent atoms.

Problem 3.1

Identify the alkanes corresponding to each of the drawings shown.

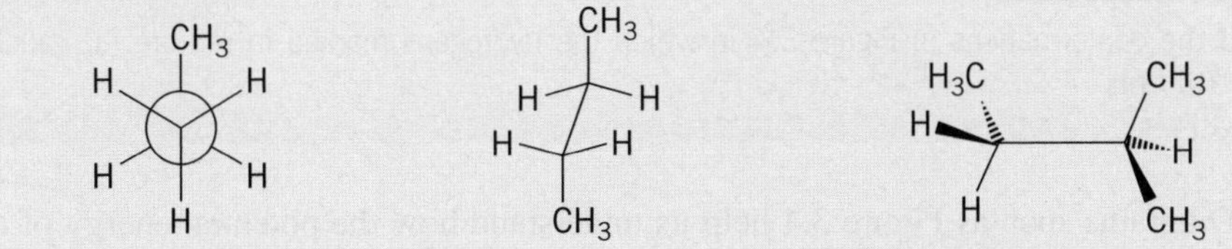

(a) Newman projection (b) Sawhorse (c) Wedge-and-dash

Sample Solution (a) The Newman projection of this alkane resembles that of ethane, except one of the hydrogens has been replaced by a methyl group. The drawing is a Newman projection of propane, $CH_3CH_2CH_3$.

Of the two conformations of ethane, the staggered is 12 kJ/mol (2.9 kcal/mol) more stable than the eclipsed. The staggered conformation is the most stable conformation, the eclipsed is the least stable conformation. Two main explanations have been offered for the difference in stability between the two conformations. One explanation holds that repulsions between bonds on adjacent atoms *destabilize* the eclipsed conformation. The other suggests that better electron delocalization *stabilizes* the staggered conformation. The latter of these two explanations is now believed to be the correct one.

Conformations in which the torsion angles between adjacent bonds are other than 60° are said to have **torsional strain.** Eclipsed bonds produce the most torsional strain; staggered bonds none. Because three pairs of eclipsed bonds are responsible for 12 kJ/mol (2.9 kcal/mol) of torsional strain in ethane, it is reasonable to assign an "energy cost"

Figure 3.4

Potential energy diagram for rotation about the carbon–carbon bond in ethane. Two of the hydrogens are shown in red and four in green so as to indicate more clearly the bond rotation.

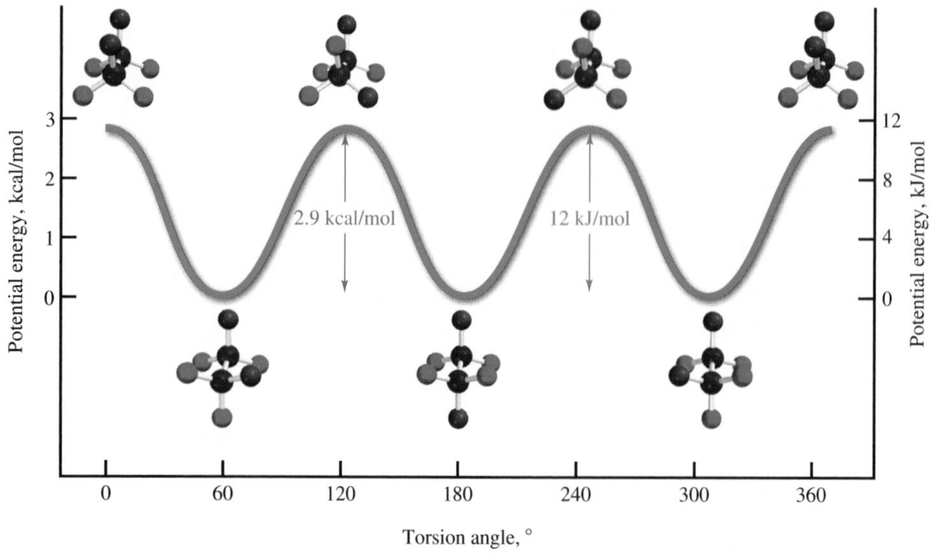

Steric is derived from the Greek word *stereos* for "solid" and refers to the three-dimensional or spatial aspects of chemistry.

of 4 kJ/mol (1 kcal/mol) to each pair. In this chapter we'll learn of additional sources of strain in molecules, which together with torsional strain comprise **steric strain.**

In principle, ethane has an infinite number of conformations that differ by only tiny increments in their torsion angles. Not only is the staggered conformation more stable than the eclipsed, it is the most stable of all of the conformations; the eclipsed is the least stable. Figure 3.4 shows how the potential energy of ethane changes for a 360° rotation about the carbon–carbon bond. Three equivalent eclipsed conformations and three equivalent staggered conformations occur during the 360° rotation; the eclipsed conformations appear at the highest points on the curve (*potential energy maxima*), the staggered ones at the lowest (*potential energy minima*).

At any instant, almost all of the molecules are in staggered conformations; hardly any are in eclipsed conformations.

Problem 3.2

Find the conformations in Figure 3.4 in which the hydrogens marked in red are (a) gauche and (b) anti.

Diagrams such as Figure 3.4 help us understand how the potential energy of a system changes during a process. The process can be a simple one such as the one described here—rotation around a carbon–carbon bond. Or it might be more complicated—a chemical reaction, for example. We will see applications of potential energy diagrams to a variety of processes throughout the text.

Let's focus our attention on a portion of Figure 3.4. The region that lies between a torsion angle of 60° and 180° tracks the conversion of one staggered conformation of ethane to the next one. Both staggered conformations are equivalent and equal in energy, but for one staggered conformation to get to the next, it must first pass through an eclipsed conformation and needs to gain 12 kJ/mol (2.9 kcal/mol) of energy to reach it. This amount of energy is the **activation energy (E_{act})** for the process. Molecules must become energized in order to undergo a chemical reaction or, as in this case, to undergo rotation around a carbon–carbon bond. Kinetic (thermal) energy is absorbed by a molecule from collisions with other molecules and is transformed into potential energy. When the potential energy exceeds E_{act}, the unstable arrangement of atoms that exists at that instant can relax to a more stable structure, giving off its excess potential energy in collisions with other molecules or with the walls of a container. The point of maximum potential energy encountered by the reactants as they proceed to products is called the **transition state.** The eclipsed conformation is the transition state for the conversion of one staggered conformation of ethane to another.

The structure that exists at the transition state is sometimes referred to as the *transition structure* or the *activated complex.*

Rotation around carbon–carbon bonds is one of the fastest processes in chemistry. Among the ways that we can describe the rate of a process is by its *half-life,* which is

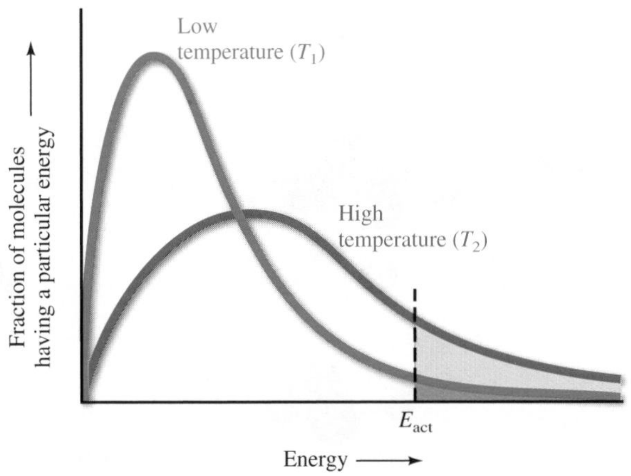

Figure 3.5

Distribution of energies. The number of molecules with energy greater than E_{act} at temperature T_1 is shown as the darker green-shaded area. At some higher temperature T_2, the curve is flatter, and more molecules have energies in excess of E_{act}.

the length of time it takes for one half of the molecules to have reacted. It takes less than 10^{-6} s for half of the molecules in a sample of ethane to have gone from one staggered conformation to another at 25°C.

As with all chemical processes, the rate of rotation about the carbon–carbon bond increases with temperature. The reason for this is apparent from Figure 3.5, where it can be seen that most of the molecules in a sample have energies that are clustered around some average value; some have less energy, a few have more. Only molecules with a potential energy greater than E_{act}, however, are able to go over the transition state and proceed on. The number of these molecules is given by the shaded areas under the curve in Figure 3.5. The energy distribution curve flattens out at higher temperatures, and a greater proportion of molecules have energies in excess of E_{act} at T_2 (higher) than at T_1 (lower). The effect of temperature is quite pronounced; an increase of only 10°C produces a two- to threefold increase in the rate of a typical chemical process.

3.2 Conformational Analysis of Butane

The next alkane that we will examine is butane. In particular, we consider conformations related by rotation about the bond between the middle two carbons (CH_3CH_2—CH_2CH_3). Unlike ethane, in which the staggered conformations are equivalent, two different staggered conformations occur in butane, shown in Figure 3.6. The methyl groups are gauche to each other in one, anti in the other. Both conformations are staggered, so are free of torsional strain, but two of the methyl hydrogens of the gauche conformation lie within 210 pm of each other. This distance is less than the sum of their van der Waals radii (240 pm), and there is a repulsive force between them. The destabilization of a molecule that results when two of its atoms are too close to each other is called **van der Waals strain,** or **steric hindrance,** and contributes to the total steric strain. In the case of butane, van der Waals strain makes the gauche conformation approximately 3.3 kJ/mol (0.8 kcal/mol) less stable than the anti.

Figure 3.7 illustrates the potential energy relationships among the various conformations of butane around the central carbon–carbon bond. The staggered conformations are more stable than the eclipsed. At any instant, almost all the molecules exist in staggered conformations, and more are present in the anti conformation than in the gauche. The point of maximum potential energy lies some 25 kJ/mol (6.1 kcal/mol) above the anti conformation. The total strain in this structure is approximately equally divided between the torsional strain associated with three pairs of eclipsed bonds (12 kJ/mol; 2.9 kcal/mol) and the van der Waals strain between the eclipsed methyl groups.

Problem 3.3

Sketch a potential energy diagram for rotation around a carbon–carbon bond in propane. Identify each potential energy maximum and minimum with a structural formula that shows the conformation of propane at that point. Does your diagram more closely resemble that of ethane or of butane? Would you expect the activation energy for bond rotation in propane to be more than or less than that of ethane? Of butane?

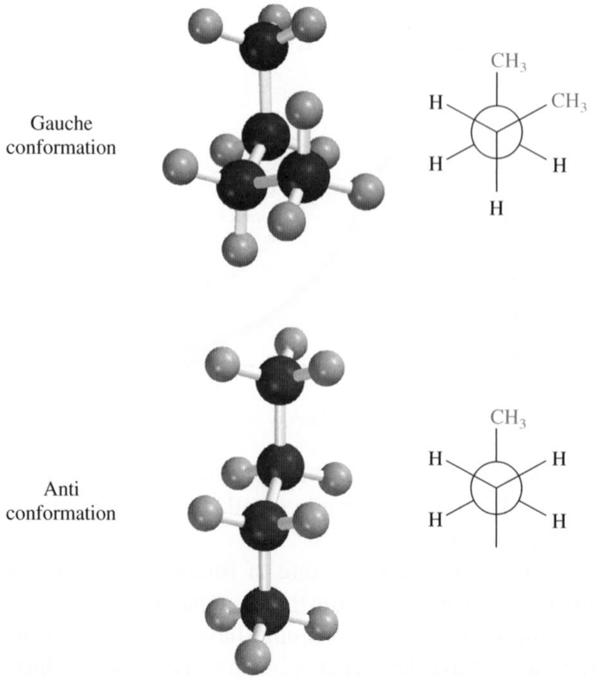

Figure 3.6

The gauche and anti conformations of butane shown as ball-and-spoke models (*left*) and as Newman projections (*right*). The gauche conformation is less stable than the anti because of the van der Waals strain between the methyl groups.

Problem 3.4

Acetylcholine is a neurotransmitter in the central nervous system in humans. Sighting down the C-1 to C-2 bond, complete the Newman projection formulas for the anti and gauche conformations of acetylcholine.

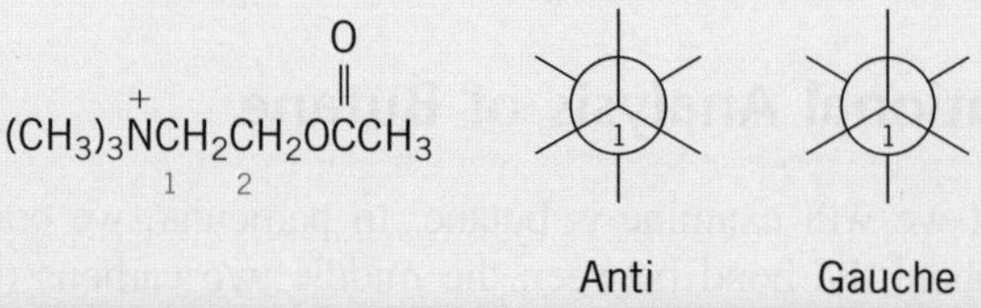

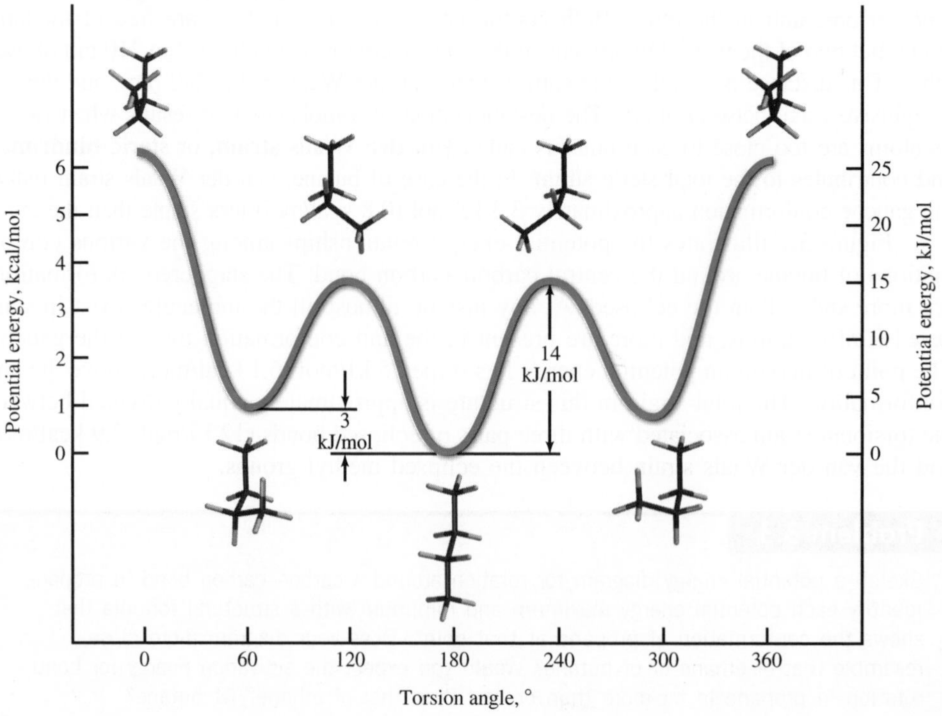

Figure 3.7

Potential energy diagram for rotation around the central carbon–carbon bond in butane.

Molecular Mechanics Applied to Alkanes and Cycloalkanes

Of the numerous applications of computer technology to chemistry, one that has been enthusiastically embraced by organic chemists examines molecular structure from a perspective similar to that gained by manipulating molecular models but with an additional quantitative dimension. *Molecular mechanics* is a computational method that allows us to assess the stability of a molecule by comparing selected features of its structure with those of ideal "unstrained" standards. Molecular mechanics makes no attempt to explain why the van der Waals radius of hydrogen is 120 pm, why the bond angles in methane are 109.5°, why the C—C bond distance in ethane is 153 pm, or why the staggered conformation of ethane is 12 kJ/mol more stable than the eclipsed, but instead uses these and other experimental observations as benchmarks to which the corresponding features of other substances are compared.

If we assume that there are certain "ideal" values for bond angles, bond distances, and so on, it follows that deviations from these ideal values will destabilize a particular structure and increase its potential energy. This increase in potential energy is referred to as the **strain energy** of the structure. Other terms for this increase include *steric energy* and *steric strain.* Arithmetically, the total strain energy (E_s) of an alkane or cycloalkane can be considered as

$$E_s = E_{\text{bond stretching}} + E_{\text{angle bending}} + E_{\text{torsional}} + E_{\text{van der Waals}}$$

where

$E_{\text{bond stretching}}$ is the strain that results when C—C and C—H bond distances are distorted from their ideal values of 153 pm and 111 pm, respectively.

$E_{\text{angle bending}}$ is the strain that results from the expansion or contraction of bond angles from the normal values of 109.5° for sp^3-hybridized carbon.

$E_{\text{torsional}}$ is the strain that results from deviation of torsion angles from their stable staggered relationship.

$E_{\text{van der Waals}}$ is the strain that results from "nonbonded interactions."

Nonbonded interactions are the forces between atoms that aren't bonded to one another; they may be either attractive or repulsive. It often happens that the shape of a molecule may cause two atoms to be close in space even though they are separated from each other by many bonds. Induced-dipole/induced-dipole interactions make van der Waals forces in alkanes weakly attractive at most distances, but when two atoms are closer to each other than the sum of their van der Waals radii, nuclear–nuclear and electron–electron repulsive forces between them dominate the $E_{\text{van der Waals}}$ term. The resulting destabilization is called van der Waals strain.

At its most basic level, separating the total strain of a structure into its components is a qualitative exercise. For example, a computer-drawn model of the eclipsed conformation of butane using ideal bond angles and bond distances (Figure 3.8) reveals that two pairs of hydrogens are separated by a distance of only 175 pm, a value considerably smaller than the sum of their van der Waals radii (2 × 120 pm = 240 pm). Thus, this conformation is destabilized not only by the torsional strain associated with its eclipsed bonds, but also by van der Waals strain.

At a higher level, molecular mechanics is applied quantitatively to strain energy calculations. Each component of strain is separately described by a mathematical expression developed and refined so that it gives solutions that match experimental observations for reference molecules. These empirically derived and tested expressions are then used to calculate the most stable structure of a substance. The various structural features are interdependent; van der Waals strain, for example, might be decreased at the expense of introducing some angle strain, torsional strain, or both. The computer program searches for the combination of bond angles, distances, torsion angles, and nonbonded interactions that gives the molecule the lowest total strain. This procedure is called *strain energy minimization* and is based on the commonsense notion that the most stable structure is the one that has the least strain.

The first widely used molecular mechanics program was developed by Professor N. L. Allinger of the University of Georgia and was known in its various versions as *MM2, MM3,* and so on. Molecular mechanics has been refined to the extent that many structural features can be calculated more easily and more accurately than they can be measured experimentally.

Many college and university chemistry laboratory courses now include experiments in which students model molecular structures and properties on personal computers. Typically, the software used combines molecular mechanics and molecular orbital methods. A preliminary model is created on the screen based on a structural drawing, a tentative geometry calculated by molecular mechanics, then refined by molecular orbital methods to produce an "optimized" structure.

Figure 3.8

Ball-and-spoke and space-filling models of methyl–methyl eclipsed conformation of butane.

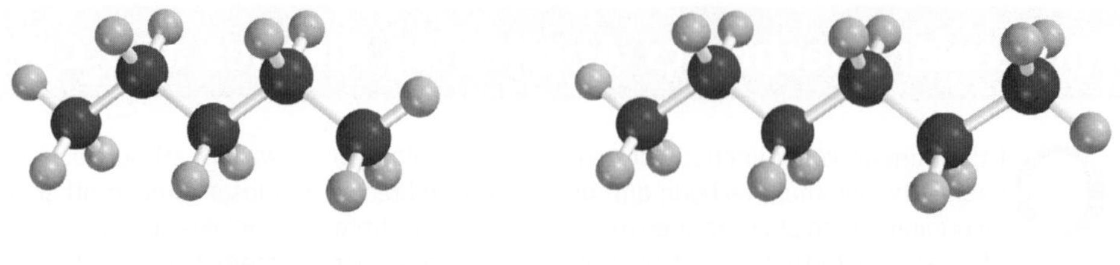

Figure 3.9

Ball-and-spoke models of pentane and hexane in their all-anti (zigzag) conformations.

3.3 Conformations of Higher Alkanes

Higher alkanes having unbranched carbon chains are, like butane, most stable in their all-anti conformations. The energy difference between gauche and anti conformations is similar to that of butane, and appreciable quantities of the gauche conformation are present in liquid alkanes at 25°C. In depicting the conformations of higher alkanes it is often more helpful to look at them from the side rather than end-on as in a Newman projection. Viewed from this perspective, the most stable conformations of pentane and hexane have their carbon "backbones" arranged in a zigzag fashion, as shown in Figure 3.9. All the bonds are staggered, and the chains are characterized by anti arrangements of C—C—C—C units.

3.4 The Shapes of Cycloalkanes: Planar or Nonplanar?

During the nineteenth century it was widely believed—incorrectly, as we'll soon see—that cycloalkane rings are planar. A leading advocate of this view was the German chemist Adolf von Baeyer. He noted that compounds containing rings other than those based on cyclopentane and cyclohexane were rarely encountered naturally and were difficult to synthesize. Baeyer connected both observations with cycloalkane stability, which he suggested was related to how closely the internal angles of planar rings match the tetrahedral value of 109.5°. For example, the 60° bond angle of cyclopropane and the 90° bond angles of a planar cyclobutane ring are much smaller than the tetrahedral angle of 109.5°. Baeyer suggested that three- and four-membered rings suffer from what we now call angle strain. **Angle strain** is the strain a molecule has because one or more of its bond angles deviate from the ideal value; in the case of alkanes the ideal value is 109.5°.

Although better known now for his incorrect theory that cycloalkanes were planar, Baeyer was responsible for notable advances in the chemistry of organic dyes such as indigo and was awarded the 1905 Nobel Prize in Chemistry for his work in that area.

According to Baeyer, cyclopentane should be the most stable of all the cycloalkanes because the ring angles of a planar pentagon, 108°, are closer to the tetrahedral angle than those of any other cycloalkane. A prediction of the *Baeyer strain theory* is that the cycloalkanes beyond cyclopentane should become increasingly strained and correspondingly less stable. The angles of a regular hexagon are 120°, and the angles of larger polygons deviate more and more from the ideal tetrahedral angle.

Problems with the Baeyer strain theory become apparent when we use heats of combustion (Table 3.1) to probe the relative energies of cycloalkanes. The most important column in the table is the heat of combustion per methylene (CH_2) group. Because all of the cycloalkanes have molecular formulas of the type C_nH_{2n}, dividing the heat of combustion by n allows direct comparison of ring size and potential energy. Cyclopropane has the highest heat of combustion per methylene group, which is consistent with the idea that its potential energy is raised by angle strain. Cyclobutane has less angle strain at each of its carbon atoms and a lower heat of combustion per methylene group. Cyclopentane, as expected, has a lower value still. Notice, however, that contrary to the prediction of the Baeyer strain theory, cyclohexane has a *smaller* heat of combustion per methylene group than cyclopentane. If angle strain were greater in cyclohexane than in cyclopentane, the opposite would have been observed.

Furthermore, the heats of combustion per methylene group of the very large rings are all about the same and similar to that of cyclohexane. Rather than rising because of increasing angle strain in large rings, the heat of combustion per methylene group remains constant at approximately 653 kJ/mol (156 kcal/mol), the value cited in Section 2.18 as the difference between successive members of a homologous series of alkanes. We conclude,

TABLE 3.1 Heats of Combustion ($-\Delta H°$) of Cycloalkanes

Cycloalkane	Number of CH_2 groups	Heat of combustion kJ/mol	Heat of combustion kcal/mol	Heat of combustion per CH_2 group kJ/mol	Heat of combustion per CH_2 group kcal/mol
Cyclopropane	3	2,091	499.8	697	166.6
Cyclobutane	4	2,721	650.3	681	162.7
Cyclopentane	5	3,291	786.6	658	157.3
Cyclohexane	6	3,920	936.8	653	156.0
Cycloheptane	7	4,599	1099.2	657	157.0
Cyclooctane	8	5,267	1258.8	658	157.3
Cyclononane	9	5,933	1418.0	659	157.5
Cyclodecane	10	6,587	1574.3	659	157.5
Cycloundecane	11	7,237	1729.8	658	157.3
Cyclododecane	12	7,845	1875.1	654	156.3
Cyclotetradecane	14	9,139	2184.2	653	156.0
Cyclohexadecane	16	10,466	2501.4	654	156.3

therefore, that the bond angles of large cycloalkanes are not much different from the bond angles of alkanes themselves. The prediction of the Baeyer strain theory that angle strain increases steadily with ring size is contradicted by experimental fact.

The Baeyer strain theory is useful to us in identifying angle strain as a destabilizing effect. Its fundamental flaw is its assumption that the rings of cycloalkanes are planar. *With the exception of cyclopropane, cycloalkanes are nonplanar.* Sections 3.5–3.13 describe the shapes of cycloalkanes. We'll begin with cyclopropane.

3.5 Small Rings: Cyclopropane and Cyclobutane

Conformational analysis is far simpler in cyclopropane than in any other cycloalkane. Cyclopropane's three carbon atoms are, of geometric necessity, coplanar, and rotation about its carbon–carbon bonds is impossible. You saw in Section 3.4 how angle strain in cyclopropane leads to an abnormally large heat of combustion. Let's now look at cyclopropane in more detail to see how our orbital hybridization bonding model may be adapted to molecules of unusual geometry.

Strong sp^3–sp^3 σ bonds are not possible for cyclopropane, because the 60° bond angles of the ring do not permit the orbitals to be properly aligned for effective overlap (Figure 3.10). The less effective overlap that does occur leads to what chemists refer to as "bent" bonds. The electron density in the carbon–carbon bonds of cyclopropane

In keeping with the "bent-bond" description of Figure 3.10, the carbon–carbon bond distance in cyclopropane (151 pm) is slightly shorter than that of ethane (153 pm) and cyclohexane (154 pm).

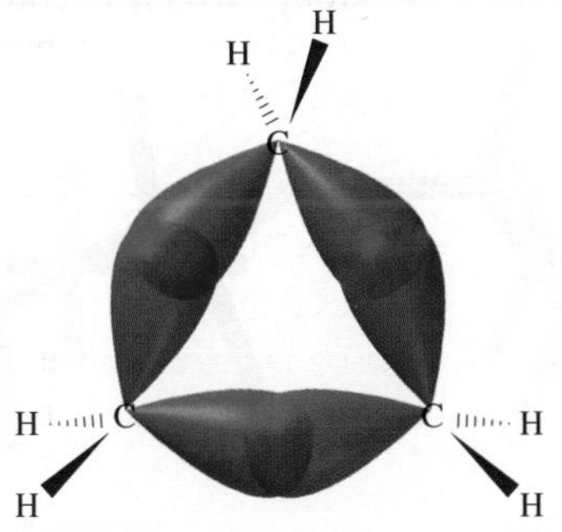

(*a*)

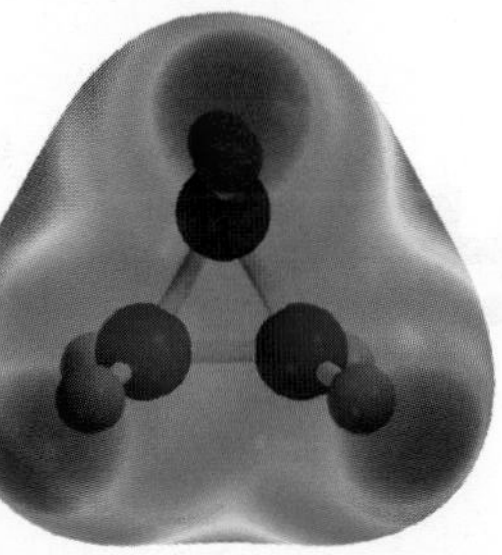

(*b*)

Figure 3.10

"Bent bonds" in cyclopropane. (*a*) The orbitals involved in carbon–carbon bond formation overlap in a region that is displaced from the internuclear axis. (*b*) The three areas of greatest negative electrostatic potential (red) correspond to those predicted by the bent-bond description.

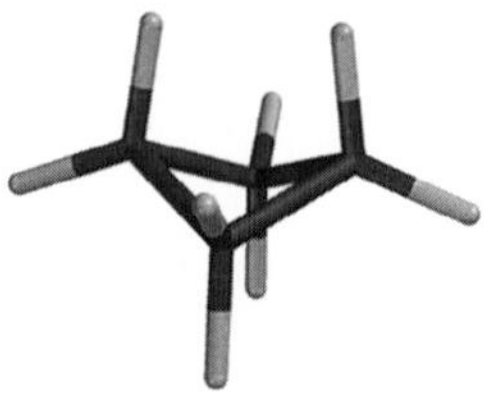

Figure 3.11

Nonplanar ("puckered") conformation of cyclobutane. The nonplanar conformation reduces the eclipsing of bonds on adjacent carbons that characterizes the planar conformation.

does not lie along the internuclear axis but is distributed along an arc between the two carbon atoms. The ring bonds of cyclopropane are weaker than other carbon–carbon σ bonds.

In addition to angle strain, cyclopropane is destabilized by torsional strain. Each C—H bond of cyclopropane is eclipsed with two others.

All adjacent pairs
of bonds are eclipsed

Cyclobutane has less angle strain than cyclopropane and can reduce the torsional strain that goes with a planar geometry by adopting the nonplanar "puckered" conformation shown in Figure 3.11, in which hydrogen atoms are twisted away from one another. A fully staggered arrangement in cyclobutane is not possible, but eclipsing interactions are decreased in the puckered form.

Problem 3.5

The heats of combustion of ethylcyclopropane and methylcyclobutane have been measured as 3352 and 3384 kJ/mol (801.2 and 808.8 kcal/mol). Assign the correct heat of combustion to each isomer.

3.6 Cyclopentane

Neighboring C—H bonds are eclipsed in any planar cycloalkane. Thus all planar conformations are destabilized by torsional strain.

Angle strain in the planar conformation of cyclopentane is relatively small because the 108° angles of a regular pentagon are not much different from the normal 109.5° bond angles of sp^3-hybridized carbon. The torsional strain, however, is substantial, because five bonds are eclipsed on the top face of the ring, and another set of five are eclipsed on the bottom face (Figure 3.12*a*). Some, but not all, of this torsional strain is relieved in nonplanar conformations. Two nonplanar conformations of cyclopentane, the **envelope** (Figure 3.12*b*) and the **half-chair** (Figure 3.12*c*), are of similar energy.

In the envelope conformation four of the carbon atoms are coplanar. The fifth carbon is out of the plane of the other four. There are three coplanar carbons in the half-chair conformation, with one carbon atom displaced above that plane and another below it. In both the envelope and the half-chair conformations, in-plane and out-of-plane carbons exchange positions rapidly. Equilibration between conformations of cyclopentane is very fast and occurs at rates similar to that of rotation about the carbon–carbon bond of ethane.

Figure 3.12

The (*a*) planar, (*b*) envelope, and (*c*) half-chair conformations of cyclopentane.

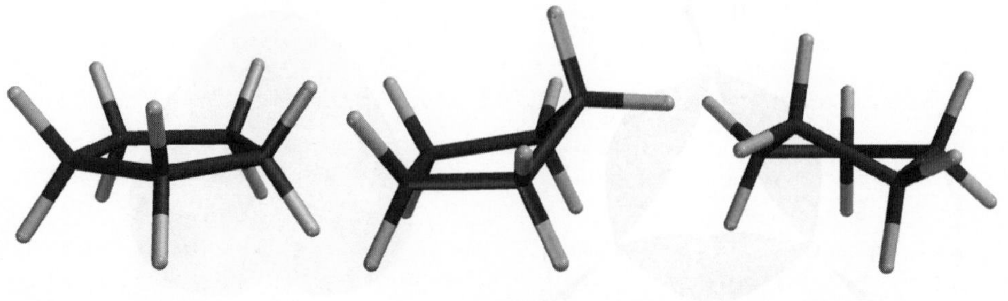
(*a*) Planar
(*b*) Envelope
(*c*) Half-chair

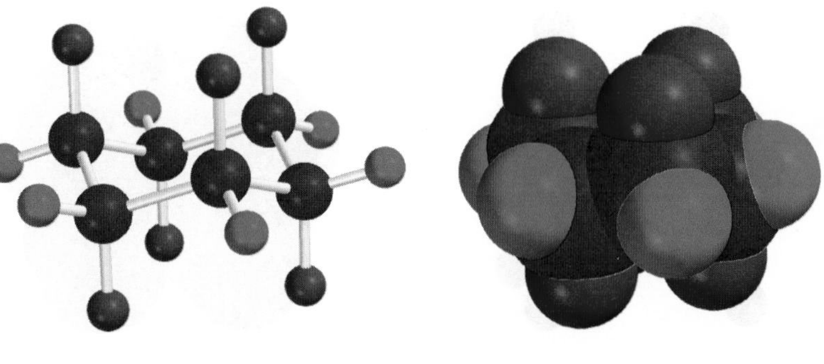

(*a*) (*b*)

Figure 3.13

(*a*) A ball-and-spoke model and (*b*) a space-filling model of the chair conformation of cyclohexane.

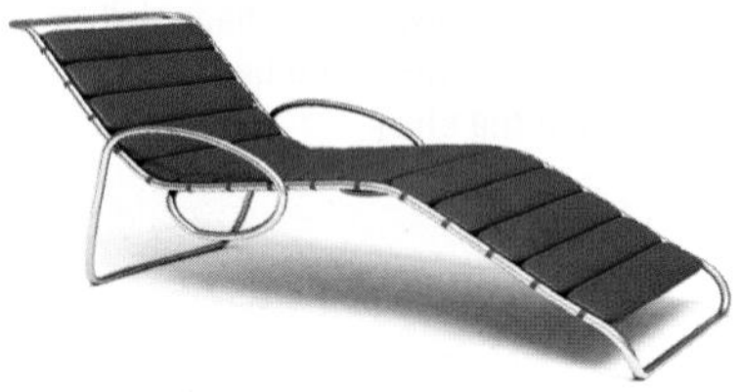

Chair cyclohexane bears some resemblance to a chaise lounge.

3.7 Conformations of Cyclohexane

Experimental evidence indicating that six-membered rings are nonplanar began to accumulate in the 1920s. Eventually, Odd Hassel of the University of Oslo established that the most stable conformation of cyclohexane has the shape shown in Figure 3.13. This is called the **chair** conformation. With C—C—C bond angles of 111°, the chair conformation is nearly free of angle strain. All its bonds are staggered, making it free of torsional strain as well. The staggered arrangement of bonds in the chair conformation of cyclohexane is apparent in a Newman-style projection.

Hassel shared the 1969 Nobel Prize in Chemistry with Sir Derek Barton of Imperial College (London). Barton demonstrated how Hassel's structural results could be extended to an analysis of conformational effects on chemical reactivity.

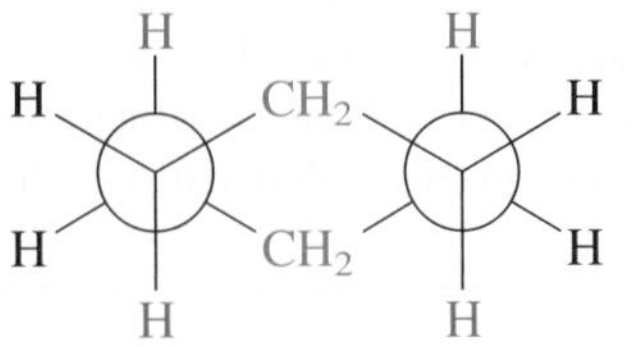

Staggered arrangement of bonds in chair conformation of cyclohexane

The cyclohexane chair is best viewed from the side-on perspective, which is useful for describing its conformational properties. You may draw chair cyclohexane in this pespective using different techniques, but your final drawing must have the following features. Bonds that are across the ring from each other are parallel, as indicated for the pairs of red, green, and blue bonds in the drawing shown on the left. Notice also that the bonds shown in red are drawn with longer lines to show the side-on perspective. In reality, all of the C—C bonds of cyclohexane are of the same length. Bonds are slanted as indicated. Although not planar, the cyclohexane ring should be level with respect to carbons 2 and 4 and carbons 1 and 5. The side-on perspective of cyclohexane is sometimes depicted with wedge bonds for C-1 to C-2 and C-4 to C-5 and a bold line for C-2 to C-3.

A second, but much less stable, nonplanar conformation called the **boat** is shown in Figure 3.14. Like the chair, the boat conformation has bond angles that are approximately tetrahedral and is relatively free of angle strain. It is, however, destabilized by the torsional strain associated with eclipsed bonds on four of its carbons. The close approach of the two "flagpole" hydrogens shown in Figure 3.14 contributes a small amount of van der Waals strain as well. Both sources of strain are reduced by rotation about the carbon–carbon bond to give the slightly more stable **twist boat,** or **skew boat,** conformation (Figure 3.15).

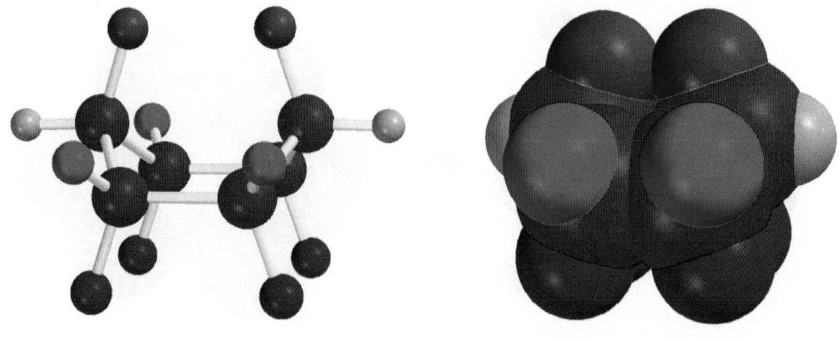

(*a*) (*b*)

Figure 3.14

(*a*) A ball-and-spoke model and (*b*) a space-filling model of the boat conformation of cyclohexane. Torsional strain from eclipsed bonds and van der Waals strain involving the "flagpole" hydrogens (red) make the boat less stable than the chair.

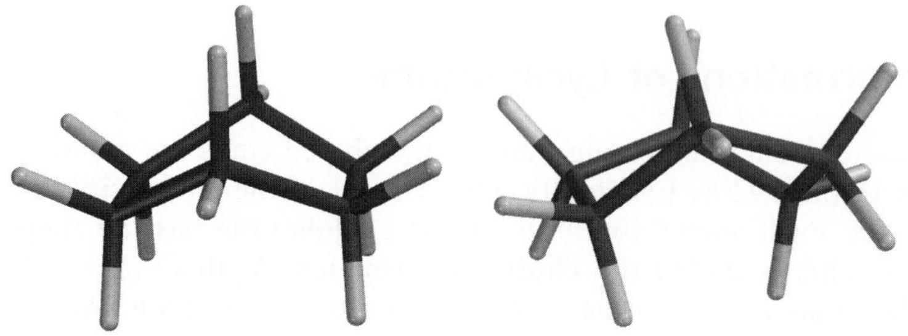

(*a*) (*b*)

Figure 3.15

(*a*) The boat and (*b*) skew boat conformations of cyclohexane. Some of the torsional strain in the boat is relieved by rotation about C—C bonds in going to the skew boat. This motion also causes the flagpole hydrogens to move away from one another, reducing the van der Waals strain between them.

The various conformations of cyclohexane are in rapid equilibrium with one another, but at any moment almost all of the molecules exist in the chair conformation. Less than five molecules per 100,000 are present in the skew boat conformation at 25°C. Thus, the discussion of cyclohexane conformational analysis that follows focuses exclusively on the chair conformation.

3.8 Axial and Equatorial Bonds in Cyclohexane

One of the most important findings to come from conformational studies of cyclohexane is that its 12 hydrogen atoms can be divided into two groups, as shown in Figure 3.16. Six of the hydrogens, called **axial** hydrogens, have their bonds parallel to a vertical axis that passes through the ring's center. These axial bonds alternately are directed up and down on adjacent carbons. The second set of six hydrogens, called **equatorial** hydrogens, are located approximately along the equator of the molecule. Notice that the four bonds to each carbon are arranged tetrahedrally, consistent with an sp^3 hybridization of carbon.

The conformational features of six-membered rings are fundamental to organic chemistry, so it is essential that you have a clear understanding of the directional properties of axial and equatorial bonds and be able to represent them accurately. Figure 3.17 offers some guidance on the drawing of chair cyclohexane rings.

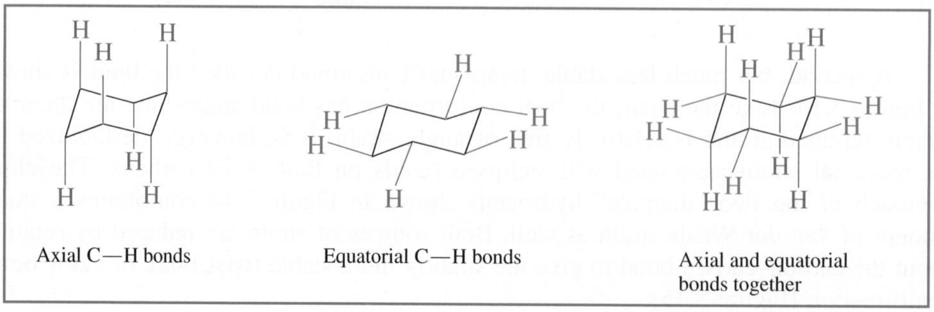
Axial C—H bonds
Equatorial C—H bonds
Axial and equatorial bonds together

Figure 3.16

Axial and equatorial bonds in cyclohexane.

(1) Begin with the chair conformation of cyclohexane.

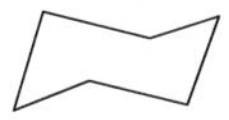

(2) Draw the axial bonds before the equatorial ones, alternating their direction on adjacent atoms. Always start by placing an axial bond "up" on the uppermost carbon or "down" on the lowest carbon.

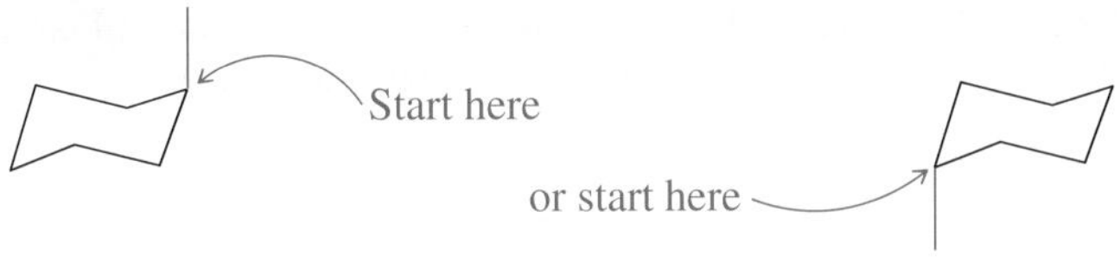
Start here
or start here

Then alternate to give

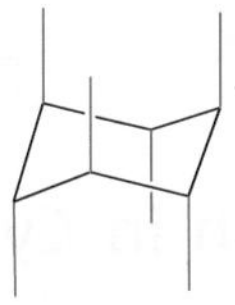

in which all the axial bonds are parallel to one another

(3) Place the equatorial bonds so as to approximate a tetrahedral arrangement of the bonds to each carbon. The equatorial bond of each carbon should be parallel to the ring bonds of its two nearest neighbor carbons.

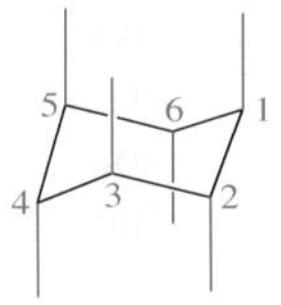
5
6
1
4
3
2

Place equatorial bond at C-1 so that it is parallel to the bonds between C-2 and C-3 and between C-5 and C-6.

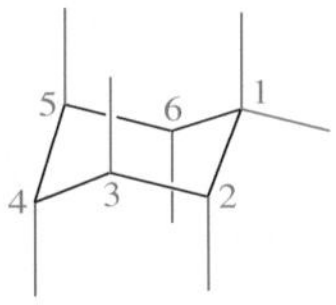
5
6
1
4
3
2

Following this pattern gives the complete set of equatorial bonds.

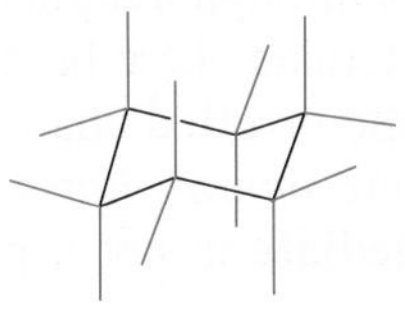

(4) Practice drawing cyclohexane chairs oriented in either direction.

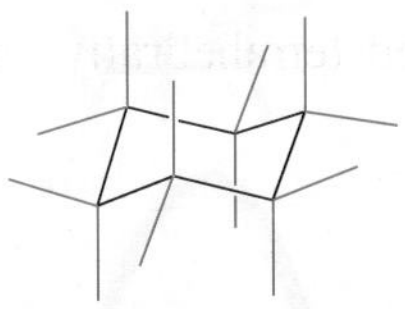

and

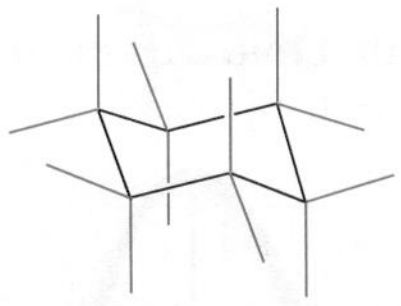

Figure 3.17

A guide to representing the orientations of the bonds in the chair conformation of cyclohexane.

It is no accident that sections of our chair cyclohexane drawings resemble sawhorse projections of staggered conformations of alkanes. The same spatial relationships seen in alkanes carry over to substituents on a six-membered ring. In the structure

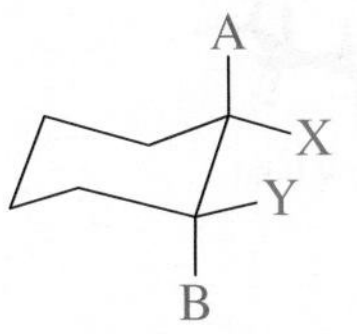
A
X
Y
B

(The substituted carbons have the spatial arrangement shown)

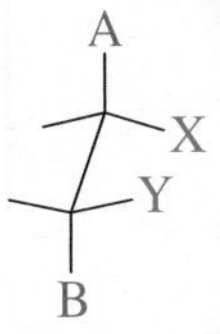
A
X
Y
B

substituents A and B are anti to each other, and the other relationships—A and Y, X and Y, and X and B—are gauche.

Problem 3.6

Given the following partial structure, add a substituent X to C-1 so that it satisfies the indicated stereochemical requirement.

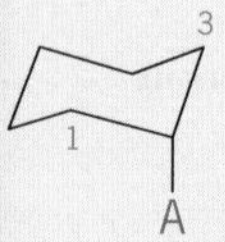

(a) Anti to A (c) Anti to C-3
(b) Gauche to A (d) Gauche to C-3

Sample Solution (a) In order to be anti to A, substituent X must be axial. The blue lines in the drawing show the A—C—C—X torsion angle to be 180°.

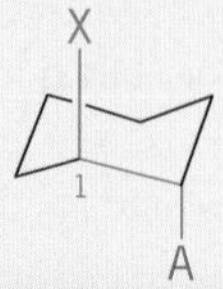

3.9 Conformational Inversion in Cyclohexane

We have seen that alkanes are not locked into a single conformation. Rotation around the central carbon–carbon bond in butane occurs rapidly, interconverting anti and gauche conformations. Cyclohexane, too, is conformationally mobile. Through a process known as **ring inversion,** or **chair–chair interconversion,** one chair conformation is converted to another chair.

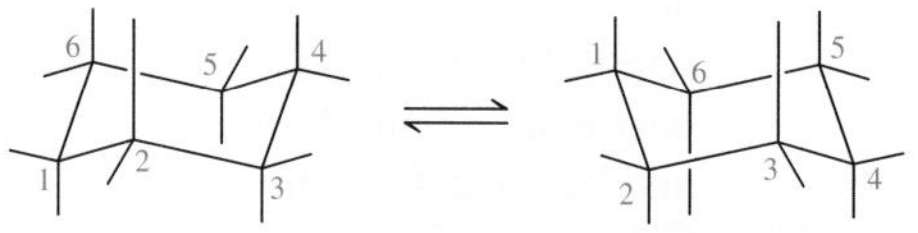

A potential energy diagram for chair–chair interconversion in cyclohexane is shown in Figure 3.18. In the first step, the chair conformation is converted to a skew boat. In this step, cyclohexane passes through a higher-energy half-chair conformation. The skew boat is converted to an alternate skew boat, via the boat conformation. The second skew boat then proceeds to the inverted chair via another half-chair conformation. The skew boat conformations are *intermediates* in the process of ring inversion. Unlike a transition state, an **intermediate** is not a potential energy maximum but is

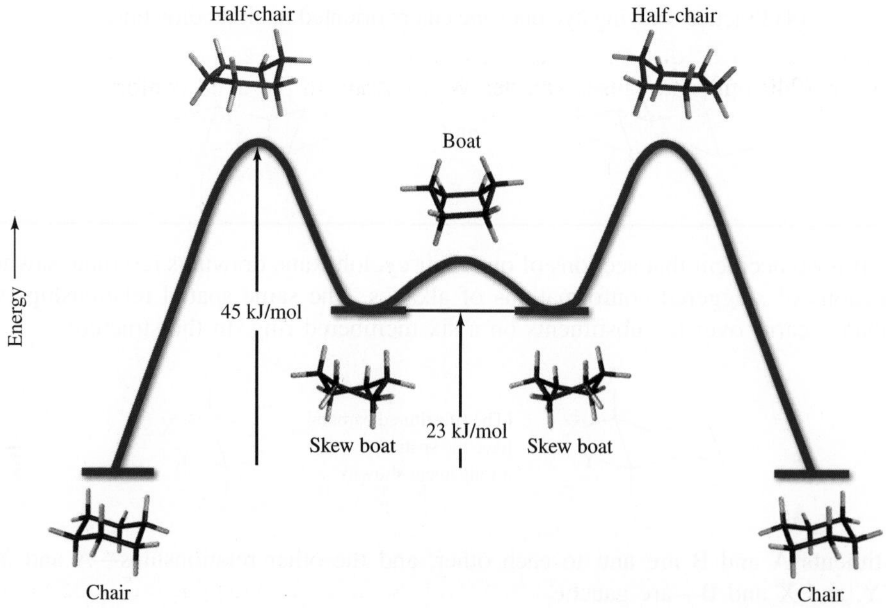

Figure 3.18

Energy diagram for ring inversion in cyclohexane. The energy of activation is the difference in energy between the chair and half-chair conformations. The skew boat conformations are intermediates. The boat and half-chair conformations are transition states.

a local minimum on the potential energy profile. The half-chair conformations are highest in energy because they have the most eclipsing interactions. The difference in energy between the chair and half-chair conformations is the activation energy for the chair–chair interconversion, which is 45 kJ/mol (10.8 kcal/mol). It is a very rapid process with a half-life of 10^{-5} s at 25°C.

The most important result of ring inversion is that any substituent that is axial in the original chair conformation becomes equatorial in the ring-inverted form and vice versa.

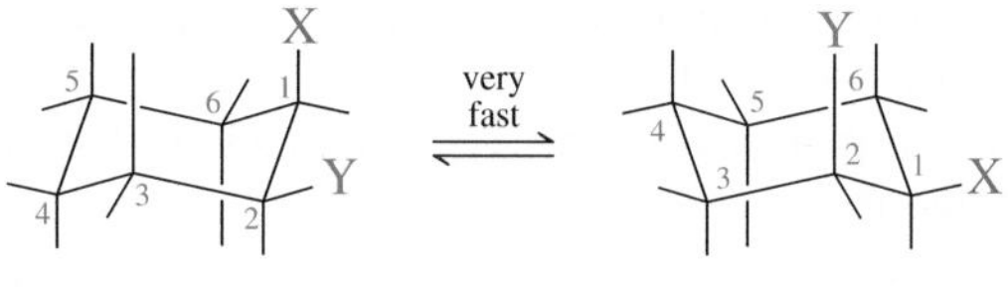

The consequences of this point are developed for a number of monosubstituted cyclohexane derivatives in the following section, beginning with methylcyclohexane.

3.10 Conformational Analysis of Monosubstituted Cyclohexanes

Ring inversion in methylcyclohexane differs from that of cyclohexane in that the two chair conformations are not equivalent. In one chair the methyl group is axial; in the other it is equatorial. At room temperature approximately 95% of the molecules of methylcyclohexane are in the chair conformation that has an equatorial methyl group, whereas only 5% of the molecules have an axial methyl group.

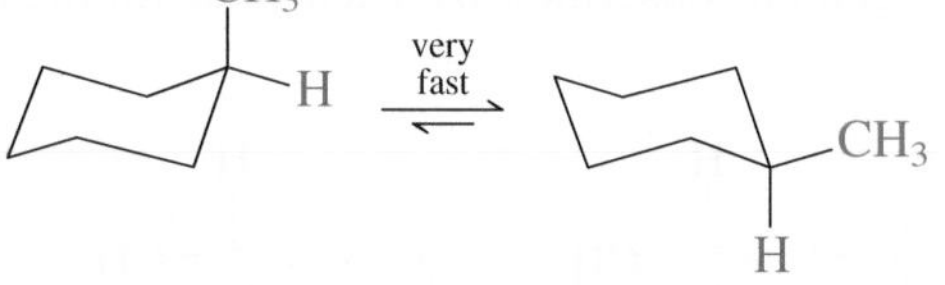

When two conformations of a molecule are in equilibrium with each other, the one with the lower free energy predominates. Why is equatorial methylcyclohexane more stable than axial methylcyclohexane?

See the box entitled *Enthalpy, Free Energy, and Equilibrium Constant* accompanying this section for a discussion of these relationships.

A methyl group is less crowded when it is equatorial than when it is axial. One of the hydrogens of an axial methyl group is within 190–200 pm of the axial hydrogens at C-3 and C-5. This distance is less than the sum of the van der Waals radii of two hydrogens (240 pm) and causes van der Waals strain in the axial conformation. When the methyl group is equatorial, it experiences no significant crowding.

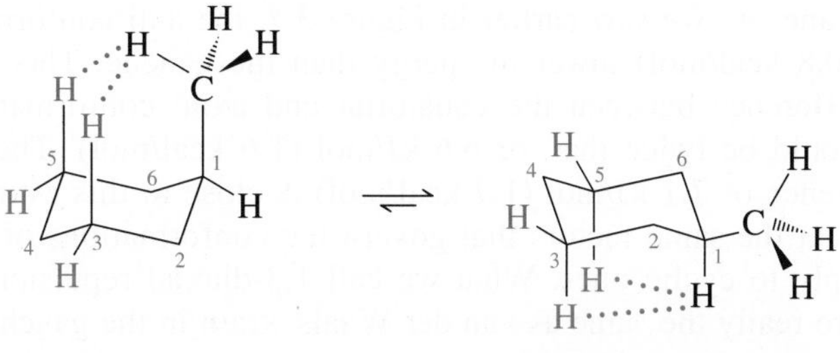

Van der Waals strain between hydrogen of axial CH_3 and axial hydrogens at C-3 and C-5

Negligible van der Waals strain between hydrogen at C-1 and axial hydrogens at C-3 and C-5

The greater stability of an equatorial methyl group, compared with an axial one, is another example of a *steric effect* (Section 3.2). An axial substituent is said to be crowded because of **1,3-diaxial repulsions** between itself and the other two axial substituents located on the same side of the ring.

Problem 3.7

The following questions relate to a cyclohexane ring depicted in the chair conformation shown.

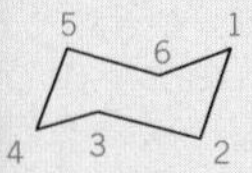

(a) Is a methyl group at C-6 that is "down" axial or equatorial?

(b) Is a methyl group that is "up" at C-1 more or less stable than a methyl group that is up at C-4?

(c) Place a methyl group at C-3 in its most stable orientation. Is it up or down?

Sample Solution (a) First indicate the directional properties of the bonds to the ring carbons. A substituent is down if it is below the other substituent on the same carbon atom. A methyl group that is down at C-6 is therefore axial.

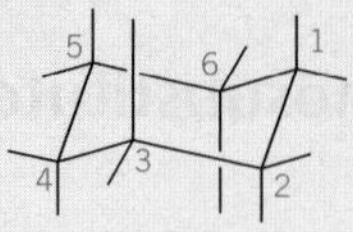

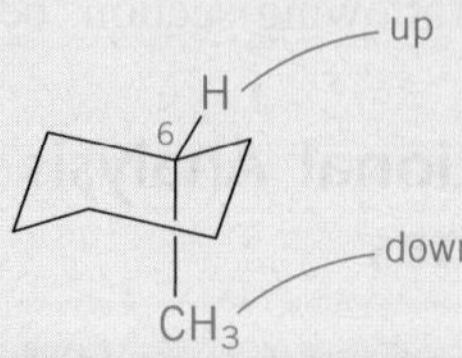

We can relate the conformational preference for an equatorial methyl group in methylcyclohexane to the conformation of butane. The red bonds in the following structural formulas trace paths through four carbons, beginning at an equatorial methyl group. The zigzag arrangement described by each path mimics the anti conformation of butane.

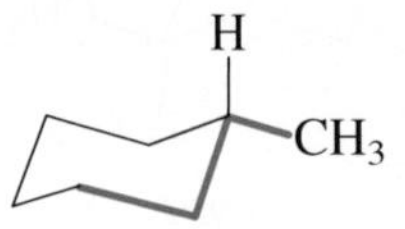

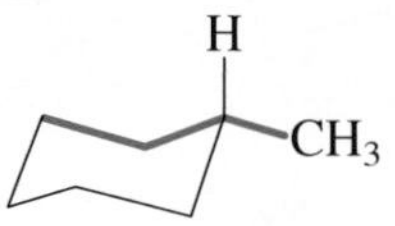

When the methyl group is axial, each path mimics the gauche conformation of butane.

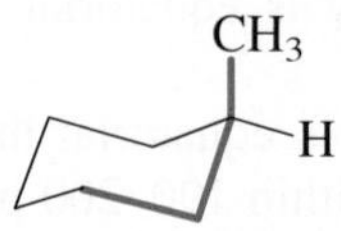

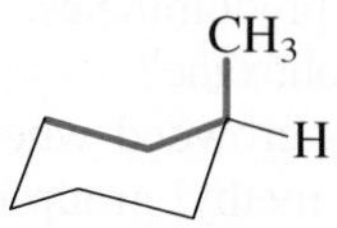

The preference for an equatorial methyl group in methylcyclohexane is therefore analogous to the preference for the anti conformation in butane. Two gauche butane-like structural units are present in axial methylcyclohexane that are absent in equatorial methylcyclohexane. As we saw earlier in Figure 3.7, the anti conformation of butane is 3.3 kJ/mol (0.8 kcal/mol) lower in energy than the gauche. Therefore, the calculated energy difference between the equatorial and axial conformations of methylcyclohexane should be twice that, or 6.6 kJ/mol (1.6 kcal/mol). The experimentally measured difference of 7.1 kJ/mol (1.7 kcal/mol) is close to this estimate. This gives us confidence that the same factors that govern the conformations of noncyclic compounds also apply to cyclic ones. What we call 1,3-diaxial repulsions in substituted cyclohexanes are really the same as van der Waals strain in the gauche conformations of alkanes.

Other substituted cyclohexanes are similar to methylcyclohexane. Two chair conformations exist in rapid equilibrium, and the one in which the substituent is equatorial is more stable. The relative amounts of the two conformations depend on the effective size of the substituent. The size of a substituent, in the context of cyclohexane conformations, is related to the degree of branching at the atom connected to the ring. A single atom, such as a halogen, does not take up much space, and its preference for an equatorial orientation is less than that of a methyl group.

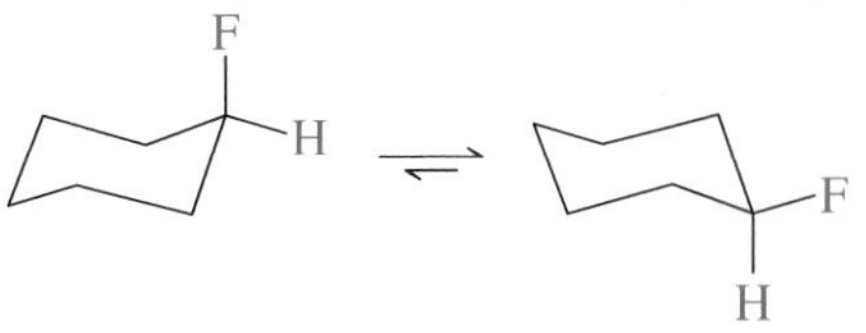

40% 60%

The halogens F, Cl, Br, and I do not differ much in their preference for the equatorial position. As the atomic radius increases in the order F < Cl < Br < I, so does the carbon–halogen bond distance, and the two effects tend to cancel.

A branched alkyl group such as isopropyl exhibits a greater preference for the equatorial orientation than does methyl.

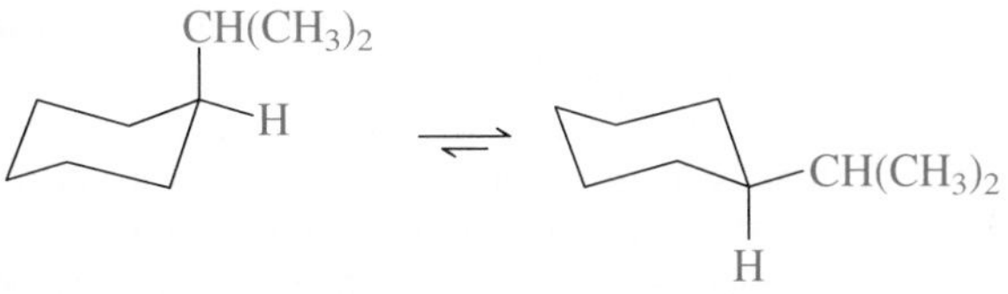

3% 97%

A *tert*-butyl group is so large that *tert*-butylcyclohexane exists almost entirely in the conformation in which the *tert*-butyl group is equatorial. The amount of axial *tert*-butylcyclohexane present is too small to measure.

Highly branched groups such as *tert*-butyl are commonly described as "bulky."

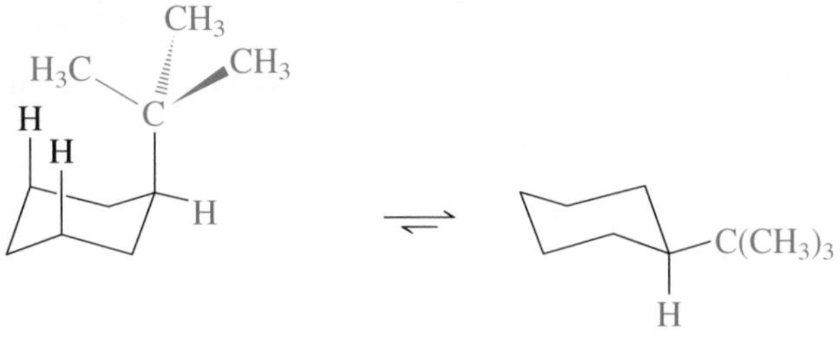

Less than 0.01%
(Serious 1,3-diaxial repulsions involving *tert*-butyl group)

Greater than 99.99%
(Decreased van der Waals strain)

Problem 3.8

Draw the most stable conformation of 1-*tert*-butyl-1-methylcyclohexane.

- Draw chair cyclohexane 
- $(CH_3)_3C$ and CH_3 are on same carbon

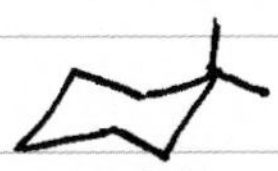

- $(CH_3)_3C$ is equatorial because it is bulkier than CH_3

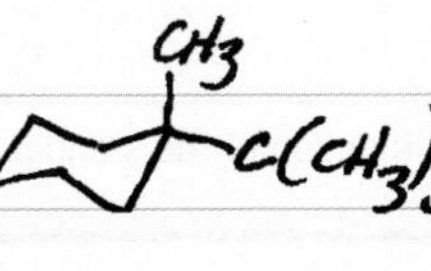

Enthalpy, Free Energy, and Equilibrium Constant

One of the fundamental equations of thermodynamics concerns systems at equilibrium and relates the equilibrium constant K to the difference in **standard free energy** ($\Delta G°$) between the products and the reactants.

$$\Delta G° = G°_{products} - G°_{reactants} = -RT \ln K$$

where T is the absolute temperature in kelvins and the constant R equals 8.314 J/mol · K (1.99 cal/mol · K).

For the equilibrium between the axial and equatorial conformations of a monosubstituted cyclohexane,

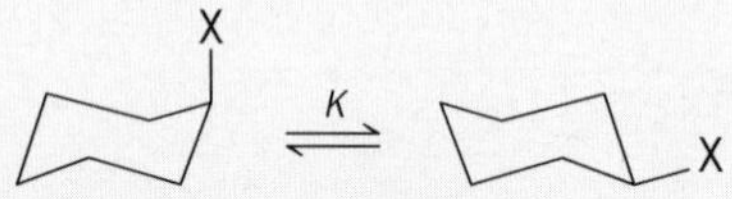

the equilibrium constant is given by the expression

$$K = \frac{[\text{products}]}{[\text{reactants}]}$$

Inserting the appropriate values for R, T (298 K), and K gives the values of $\Delta G°$ listed in the following table for the various substituents discussed in Section 3.10.

The relationship between $\Delta G°$ and K is plotted in Figure 3.19. A larger value of K is associated with a more negative $\Delta G°$.

Free energy and enthalpy are related by the expression

$$\Delta G° = \Delta H° - T\Delta S°$$

where $\Delta S°$ is the difference in *entropy* between the products and reactants. A positive $\Delta S°$ is accompanied by an increase in the disorder of a system. A positive $\Delta S°$ leads to a $\Delta G°$ that is more negative than $\Delta H°$ and a larger K than expected on the basis of enthalpy considerations alone. Conversely, a negative $\Delta S°$ gives a smaller K than expected. In the case of conformational equilibration between the chair forms of a substituted cyclohexane, $\Delta S°$ is close to zero and $\Delta G°$ and $\Delta H°$ are approximately equal.

				$\Delta G°_{298K}$	
Substituent X	**Percent axial**	**Percent equatorial**	***K***	**kJ/mol**	**kcal/mol**
—F	40	60	1.5	−1.0	−0.24
$—CH_3$	5	95	19	−7.3	−1.7
$—CH(CH_3)_2$	3	97	32.3	−8.6	−2.1
$—C(CH_3)_3$	<0.01	>99.99	>9999	−22.8	−5.5

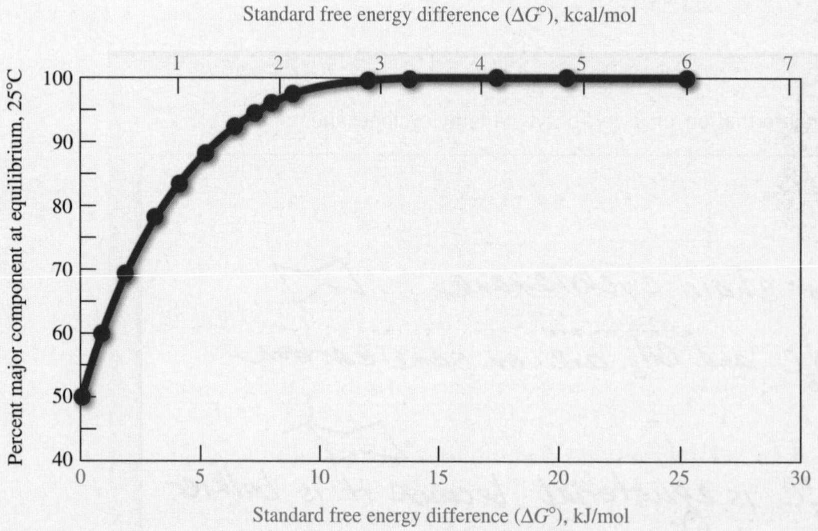

Figure 3.19

Distribution of two products at equilibrium at 25°C as a function of the standard free energy difference ($\Delta G°$) between them.

3.11 Disubstituted Cycloalkanes: cis–trans Stereoisomers

When a cycloalkane bears two substituents on different carbons—methyl groups, for example—these substituents may be on the same or on opposite sides of the ring. When substituents are on the same side, we say they are ***cis*** to each other; if they are on opposite sides, they are ***trans*** to each other. Both terms come from the Latin, in which *cis* means "on this side" and *trans* means "across."

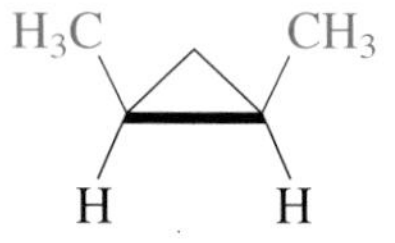
H_3C
CH_3
H
H

cis-1,2-Dimethylcyclopropane

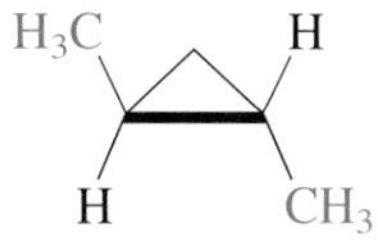
H_3C
H
H
CH_3

trans-1,2-Dimethylcyclopropane

Problem 3.9

Exclusive of compounds with double bonds, four hydrocarbons are *constitutional* isomers of *cis*- and *trans*-1,2-dimethylcyclopropane. Identify these compounds.

The cis and trans forms of 1,2-dimethylcyclopropane are stereoisomers. **Stereoisomers** are isomers that have their atoms bonded in the same order—that is, they have the same constitution, but they differ in the arrangement of atoms in space. Stereoisomers of the cis–trans type are sometimes referred to as *geometric isomers.* You learned in Section 2.18 that constitutional isomers could differ in stability. What about stereoisomers?

The prefix *stereo-* is derived from the Greek word *stereos,* meaning "solid." *Stereochemistry* is the term applied to the three-dimensional aspects of molecular structure and reactivity.

We can measure the energy difference between *cis*- and *trans*-1,2-dimethylcyclopropane by comparing their heats of combustion. As illustrated in Figure 3.20, the two compounds are isomers, and so the difference in their heats of combustion is a direct measure of the difference in their energies. Because the heat of combustion of *trans*-1,2-dimethylcyclopropane is 5 kJ/mol (1.2 kcal/mol) less than that of its cis stereoisomer, it follows that *trans*-1,2-dimethylcyclopropane is 5 kJ/mol (1.2 kcal/mol) more stable than *cis*-1,2-dimethylcyclopropane.

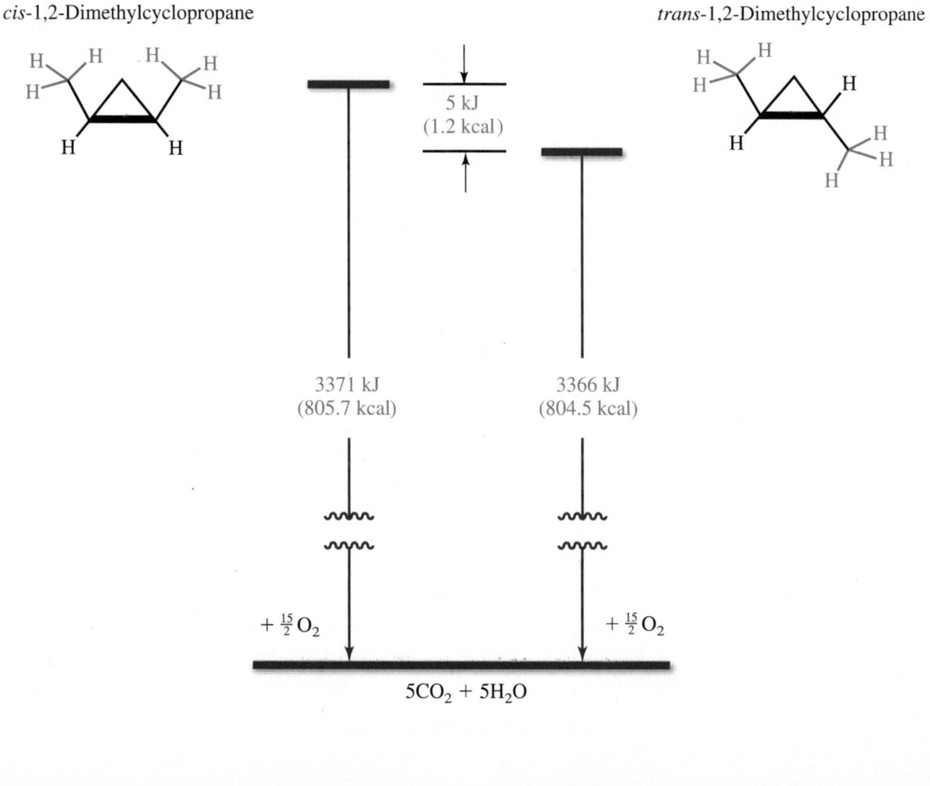
cis-1,2-Dimethylcyclopropane
trans-1,2-Dimethylcyclopropane
5 kJ
(1.2 kcal)
3371 kJ
(805.7 kcal)
3366 kJ
(804.5 kcal)
$+ \frac{15}{2} O_2$
$+ \frac{15}{2} O_2$
$5CO_2 + 5H_2O$

Figure 3.20

The enthalpy difference between *cis*- and *trans*-1,2-dimethylcyclopropane can be determined from their heats of combustion. Van der Waals strain between methyl groups on the same side of the ring makes the cis stereoisomer less stable than the trans.

In this case, the relationship between stability and stereochemistry is easily explained on the basis of van der Waals strain. The methyl groups on the same side of the ring in *cis*-1,2-dimethylcyclopropane crowd each other and increase the potential energy of this stereoisomer. Steric hindrance between methyl groups is absent in *trans*-1,2-dimethylcyclopropane.

Problem 3.10

Chrysanthemic acid, from the chrysanthemum flower, is a naturally occurring insecticide, with the structure indicated here. Draw the structures of the cis and trans stereoisomers of chrysanthemic acid.

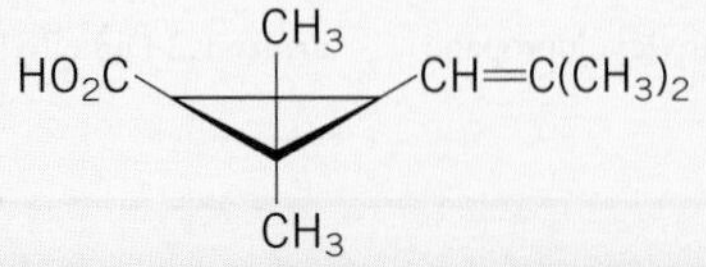

Disubstituted cyclopropanes exemplify one of the simplest cases involving stability differences between stereoisomers. A three-membered ring has no conformational mobility, so cannot reduce the van der Waals strain between cis substituents on adjacent carbons without introducing other strain. The situation is different in disubstituted derivatives of cyclohexane.

3.12 Conformational Analysis of Disubstituted Cyclohexanes

We'll begin with *cis*- and *trans*-1,4-dimethylcyclohexane as represented by wedge-and-dash structural formulas.

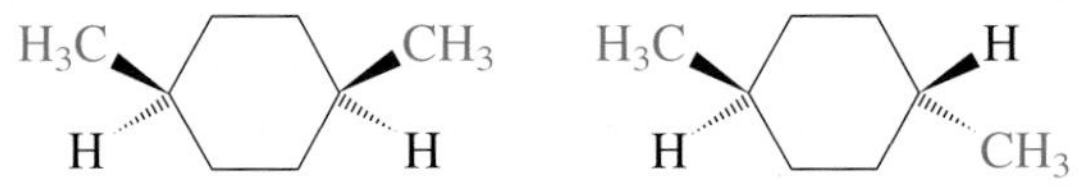

cis-1,4-Dimethylcyclohexane *trans*-1,4-Dimethylcyclohexane

Wedge-and-dash drawings fail to show conformation, and it's important to remember that the rings of *cis*- and *trans*-1,2-dimethylcyclohexane exist in a chair conformation. This fact must be taken into consideration when evaluating the relative stabilities of the stereoisomers.

Their heats of combustion (Table 3.2) reveal that *trans*-1,4-dimethylcyclohexane is 7 kJ/mol (1.7 kcal/mol) more stable than the cis stereoisomer. It is unrealistic to believe that van der Waals strain between cis substituents is responsible, because the methyl groups are too far away from each other. To understand why *trans*-1,4-dimethylcyclohexane is more stable than *cis*-1,4-dimethylcyclohexane, we need to examine each stereoisomer in its most stable conformation.

cis-1,4-Dimethylcyclohexane can adopt either of two equivalent chair conformations, *each having one axial methyl group and one equatorial methyl group.* The two are in rapid equilibrium with each other by ring interconversion. The equatorial methyl group becomes axial, and the axial methyl group becomes equatorial.

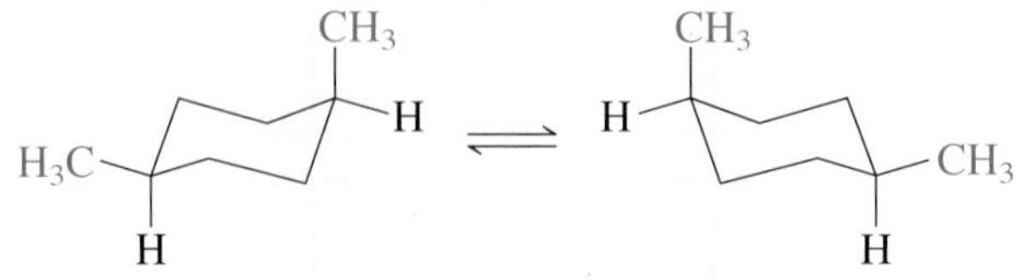

(One methyl group is axial, the other equatorial) (One methyl group is axial, the other equatorial)

(Both methyl groups are up)

cis-1,4-Dimethylcyclohexane

TABLE 3.2 Heats of Combustion of Isomeric Dimethylcyclohexanes

Compound	Orientation of methyl groups in most stable conformation	Heat of combustion		Difference in heat of combustion		More stable stereoisomer
		kJ/mol	kcal/mol	kJ/mol	kcal/mol	
cis-1,2-Dimethylcyclohexane	Axial–equatorial	5223	1248.3	6	1.5	trans
trans-1,2-Dimethylcyclohexane	Diequatorial	5217	1246.8			
cis-1,3-Dimethylcyclohexane	Diequatorial	5212	1245.7	7	1.7	cis
trans-1,3-Dimethylcyclohexane	Axial–equatorial	5219	1247.4			
cis-1,4-Dimethylcyclohexane	Axial–equatorial	5219	1247.4	7	1.7	trans
trans-1,4-Dimethylcyclohexane	Diequatorial	5212	1245.7			

The methyl groups are described as cis because both are up relative to the hydrogen present at each carbon. If both methyl groups were down, they would still be cis to each other. Notice that ring inversion does not alter the cis relationship between the methyl groups. Nor does it alter their up-versus-down quality; substituents that are up in one conformation remain up in the ring inverted form.

The most stable conformation of trans-1,4-dimethylcyclohexane has both methyl groups in equatorial orientations. The two chair conformations of *trans*-1,4-dimethylcyclohexane are not equivalent. One has two equatorial methyl groups; the other, two axial methyl groups.

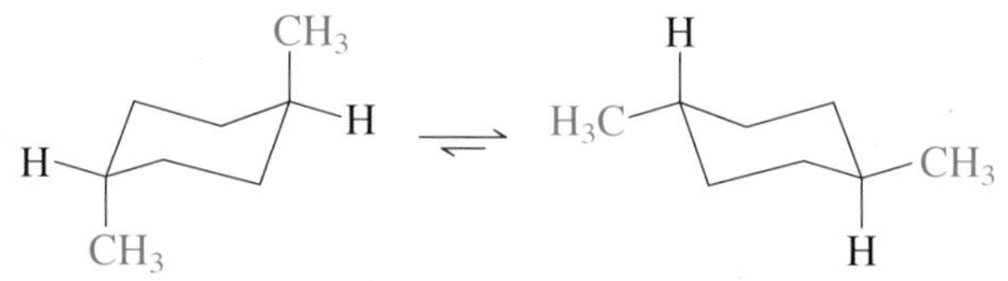

(Both methyl groups are axial: less stable chair conformation) (Both methyl groups are equatorial: more stable chair conformation)
(One methyl group is up, the other down)
trans-1,4-Dimethylcyclohexane

The more stable chair—the one with both methyl groups equatorial—is adopted by most of the *trans*-1,4-dimethylcyclohexane molecules.

trans-1,4-Dimethylcyclohexane is more stable than *cis*-1,4-dimethylcyclohexane because both of the methyl groups are equatorial in its most stable conformation. One methyl group must be axial in the cis stereoisomer. Remember, it is a general rule that any substituent is more stable in an equatorial orientation than in an axial one. It is worth pointing out that the 7 kJ/mol (1.7 kcal/mol) energy difference between *cis*- and *trans*-1,4-dimethylcyclohexane is the same as the energy difference between the axial and equatorial conformations of methylcyclohexane. There is a simple reason for this: in both instances the less stable structure has one axial methyl group, and the 7 kJ/mol (1.7 kcal/mol) energy difference can be considered the "energy cost" of having a methyl group in an axial rather than an equatorial orientation.

Like the 1,4-dimethyl derivatives, *trans*-1,2-dimethylcyclohexane has a lower heat of combustion (see Table 3.2) and is more stable than *cis*-1,2-dimethylcyclohexane. The cis stereoisomer has two chair conformations of equal energy, each containing one axial and one equatorial methyl group.

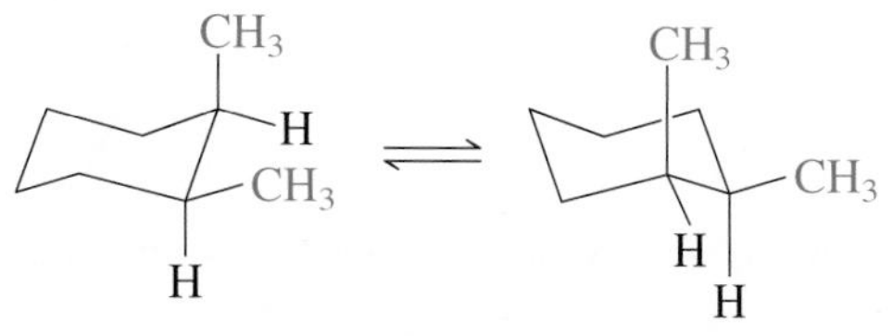

cis-1,2-Dimethylcyclohexane

Both methyl groups are equatorial in the most stable conformation of *trans*-1,2-dimethylcyclohexane.

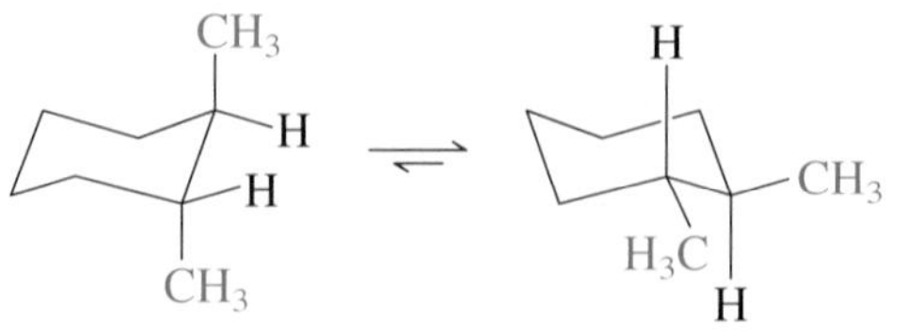

(Both methyl groups are axial: less stable chair conformation)

(Both methyl groups are equatorial: more stable chair conformation)

trans-1,2-Dimethylcyclohexane

As in the 1,4-dimethylcyclohexanes, the 6 kJ/mol (1.5 kcal/mol) energy difference between the more stable (trans) and the less stable (cis) stereoisomer is attributed to the strain associated with the presence of an axial methyl group in the cis stereoisomer.

Probably the most interesting observation in Table 3.2 concerns the 1,3-dimethylcyclohexanes. Unlike the 1,2- and 1,4-dimethylcyclohexanes, in which the trans stereoisomer is more stable than the cis, we find that *cis*-1,3-dimethylcyclohexane is 7 kJ/mol (1.7 kcal/mol) more stable than *trans*-1,3-dimethylcyclohexane. Why?

The most stable conformation of *cis*-1,3-dimethylcyclohexane has both methyl groups equatorial.

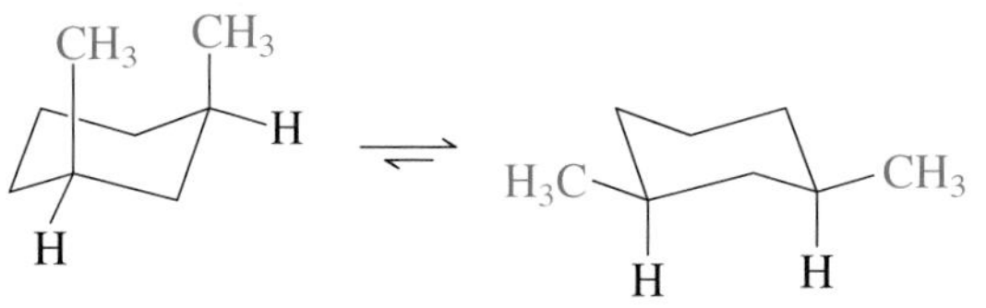

(Both methyl groups are axial: less stable chair conformation)

(Both methyl groups are equatorial: more stable chair conformation)

cis-1,3-Dimethylcyclohexane

The two chair conformations of *trans*-1,3-dimethylcyclohexane are equivalent to each other. Both contain one axial and one equatorial methyl group.

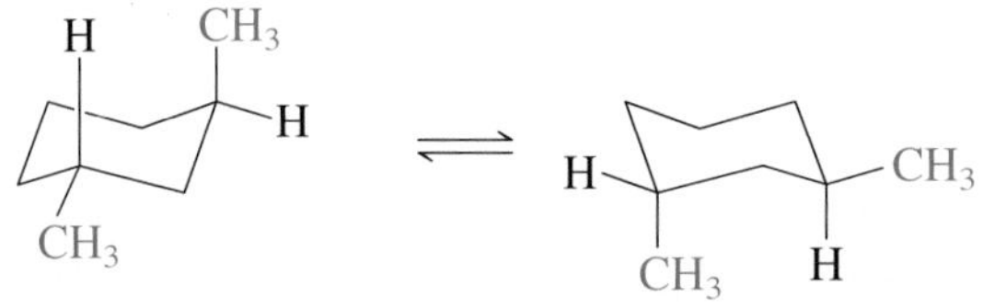

(One methyl group is axial, the other equatorial)

(One methyl group is axial, the other equatorial)

trans-1,3-Dimethylcyclohexane

Thus the trans stereoisomer, with one axial methyl group, is less stable than *cis*-1,3-dimethylcyclohexane where both methyl groups are equatorial.

Problem 3.11

Based on what you know about disubstituted cyclohexanes, which of the following two stereoisomeric 1,3,5-trimethylcyclohexanes would you expect to be more stable?

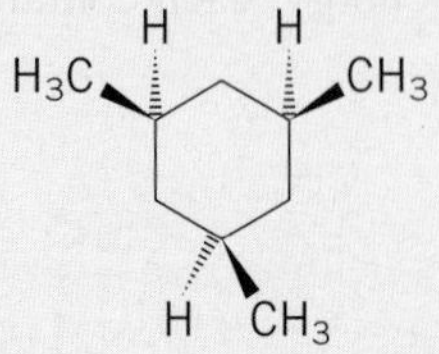

cis-1,3,5-Trimethylcyclohexane

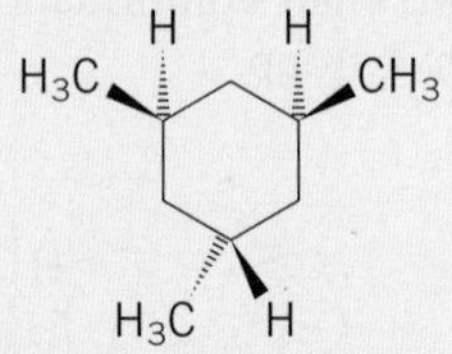

trans-1,3,5-Trimethylcyclohexane

If a disubstituted cyclohexane has two different substituents, then the most stable conformation is the chair that has the larger substituent in an equatorial orientation. This is most apparent when one of the substituents is a bulky group such as *tert*-butyl. Thus, the most stable conformation of *cis*-1-*tert*-butyl-2-methylcyclohexane has an equatorial *tert*-butyl group and an axial methyl group.

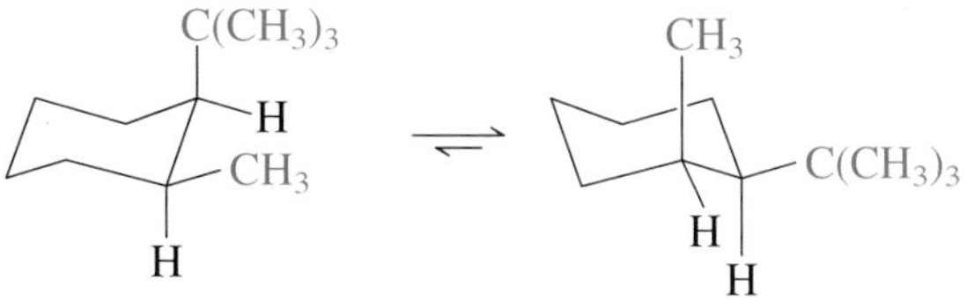

(Less stable conformation: larger group is axial) (More stable conformation: larger group is equatorial)

cis-1-*tert*-Butyl-2-methylcyclohexane

Problem 3.12

Write structural formulas for the most stable conformation of each of the following compounds:

(a) *trans*-1-*tert*-Butyl-3-methylcyclohexane
(b) *cis*-1-*tert*-Butyl-3-methylcyclohexane
(c) *trans*-1-*tert*-Butyl-4-methylcyclohexane
(d) *cis*-1-*tert*-Butyl-4-methylcyclohexane

Sample Solution

(a) • tert-Butyl is the largest substituent, so will be equatorial in the most stable conformation.

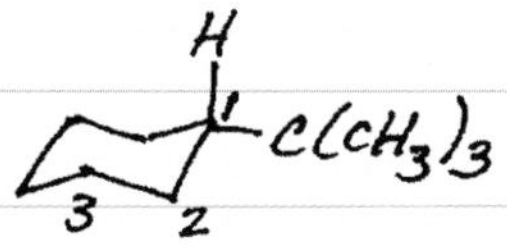

• Place bonds at C-3 in proper orientations and attach a CH_3 group so that it is trans to $-C(CH_3)_3$.

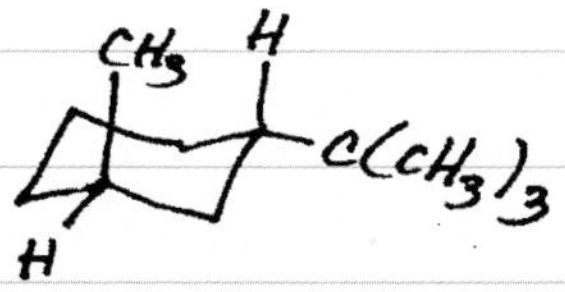

Cyclohexane rings that bear *tert*-butyl substituents are examples of conformationally biased molecules. A *tert*-butyl group has such a pronounced preference for the equatorial orientation that it will strongly bias the equilibrium to favor such conformations. This does not mean that ring inversion does not occur, however. Ring inversion does occur, but at any instant only a tiny fraction of the molecules exist in conformations having axial *tert*-butyl groups. It is not strictly correct to say that *tert*-butylcyclohexane and its derivatives are "locked" into a single conformation; conformations related by ring inversion are in rapid equilibrium with one another, but the distribution between them strongly favors those in which the *tert*-butyl group is equatorial.

3.13 Medium and Large Rings

In 1978, a German–Swiss team of organic chemists reported the synthesis of a cycloalkane with 96 carbons in its ring (cyclo-$C_{96}H_{192}$).

Beginning with cycloheptane, which has four conformations of similar energy, conformational analysis of cycloalkanes becomes more complicated. The same fundamental principles apply to medium and large rings as apply to smaller ones—but there are more atoms and more bonds to consider and more conformational possibilities.

3.14 Polycyclic Ring Systems

Polycyclic compounds are those that contain more than one ring. The IUPAC classifies polycyclic structures according to the minimum number of bond cleavages required to generate a noncyclic structure. The structure is *bicyclic* if two bond disconnections yield an open-chain structure, *tricyclic* if three, *tetracyclic* if four, and so on. Adamantane, a naturally occurring hydrocarbon found in petroleum, for example, is tricyclic because three bond cleavages are needed before an open-chain structure results.

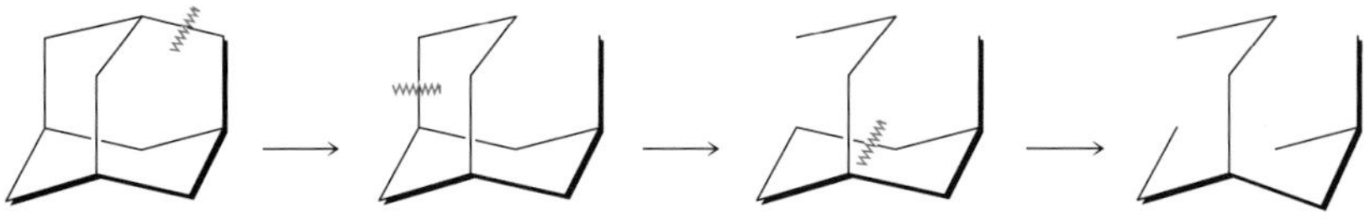

Adamantane

The correct number of rings may be determined by different sets of disconnections, and the final open-chain structure need not be the same for different sets. All that matters is finding the minimum number of disconnections.

Problem 3.13

Geohopanoids, compounds based on the carbon skeleton shown for *hopane*, have been found in every geological sediment that has been examined for their presence, and rank among the most abundant natural products on Earth. It is fairly easy to see that hopane contains five rings. Verify that hopane is pentacyclic by applying the bond-disconnection rule.

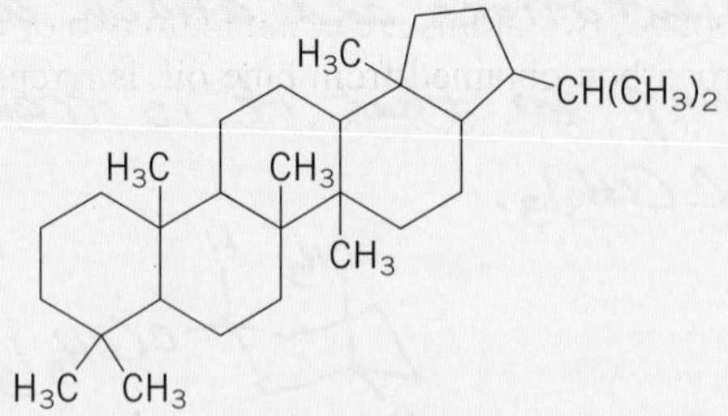
H3C
CH(CH3)2
H3C
CH3
CH3
H3C
CH3

Hopane ($C_{30}H_{52}$)

In addition to classifying polycyclic compounds according to the number of rings they contain, we also classify them with respect to the way in which the rings are joined. In a **spiro** compound, one atom is common to two rings.

The simplest spiro alkane is *spiro[2.2]pentane,* a molecular model of which illustrates an interesting structural feature of spiro compounds. The two rings lie at right angles to each other.

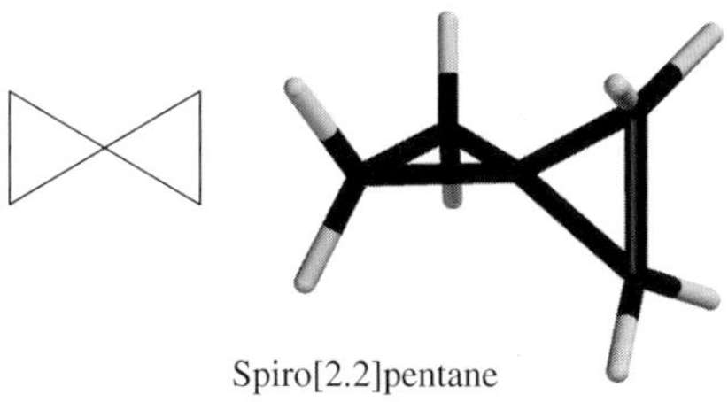

Spiro[2.2]pentane

The IUPAC names of spiro alkanes take the form *spiro[number.number]alkane.* The *alkane* suffix is simply the name of the unbranched alkane having the same number of carbons as those in the two rings. The numbers inside the brackets are, in ascending order, the number of carbons unique to each ring. Thus, eight carbons make up the two rings of spiro[3.4]octane; the spiro carbon is bridged by three carbons of one ring and four carbons of the other.

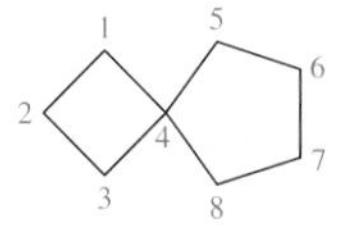
1
5
6
2
4
7
3
8

Spiro[3.4]octane

When substituents are present, numbering begins in the smaller ring adjacent to the spiro carbon and proceeds consecutively around the smaller ring away from the spiro carbon, through it, then around the larger ring. As with alkanes, the direction is chosen so as to give the lower locant at the first point of difference, substituents are listed in alphabetical order, and the locants and substituents appear first in the name.

Problem 3.14

Vetiver, a soothing oil popular in aromatherapy (Figure 3.21), contains β-vetivone, which can be viewed as a derivative of compound A. What is the IUPAC name of A?

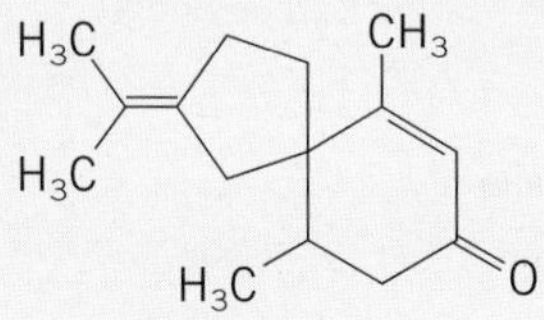
H3C
CH3
H3C
H3C
O

β–Vetivone

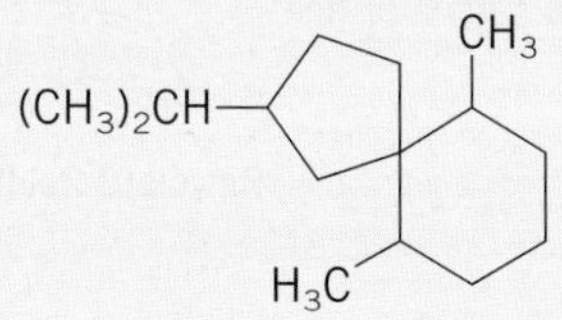
CH3
(CH3)2CH
H3C

Compound A

Figure 3.21

Vetiver grass is the source of vetiver oil.

In a **bridged** compound, two atoms are common to two or more rings. *Camphene,* a naturally occurring hydrocarbon obtained from pine oil, is a representative bridged bicyclic hydrocarbon. It is convenient to regard camphene as a six-membered ring (indicated by the blue bonds in the following structure) in which the two compounds designated by asterisks (*) are bridged by a CH_2 group. The two designated carbons are known as *bridgehead* carbons.

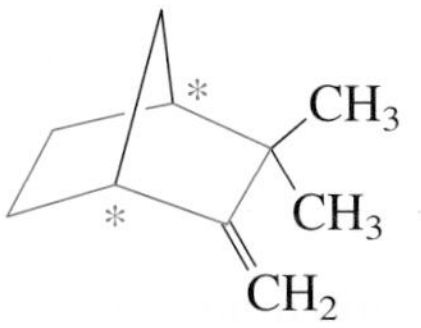
*
CH3
*
CH3
CH2

Camphene

Problem 3.15

Use the bond-cleavage criterion to verify that camphene is bicyclic.

Bridged bicyclic alkanes are named in the manner: *bicyclo[number.number.number] alkane.* As illustrated for bicyclo[3.2.1]octane, the parent alkane is the one with the same number of carbons as the total in the bicyclic skeleton.

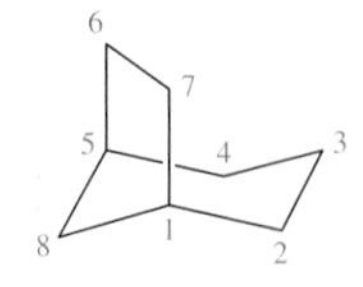
6
7
5
4
3
8
1
2

Bicyclo[3.2.1]octane

The bracketed numbers identify the number of carbons in the three bridges in descending order. Numbering begins at a bridgehead position and proceeds consecutively in the direction of the largest bridge and continues through the next largest. The atoms in the smallest bridge are numbered last.

Problem 3.16

Write structural formulas for each of the following bicyclic hydrocarbons:

(a) Bicyclo[2.2.1]heptane
(b) 1,7,7-Trimethylbicyclo[2.2.1]heptane
(c) Bicyclo[3.1.1]heptane

Sample Solution (a) The bicyclo[2.2.1]heptane ring system is one of the most frequently encountered bicyclic structural types. It contains seven carbon atoms, as indicated by the suffix *-heptane.* The bridging groups contain two, two, and one carbon, respectively.

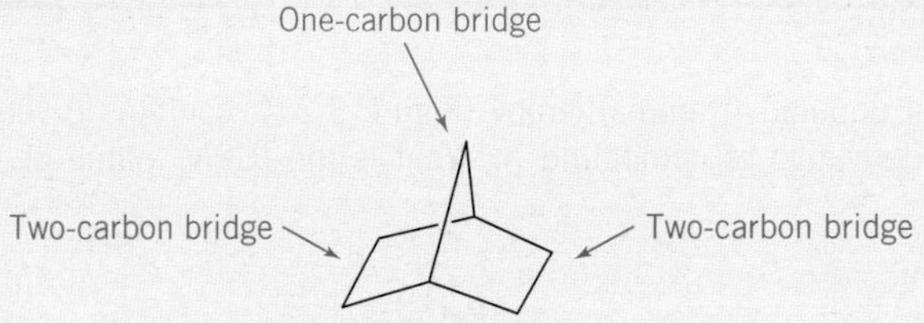
One-carbon bridge
Two-carbon bridge
Two-carbon bridge

Bicyclo[2.2.1]heptane

Many compounds contain rings that share a common side. Such compounds are normally referred to as *fused-ring* compounds, but for classification and naming purposes they are placed in the "bridged" category. The bridge in these cases is the common side and is given a value of zero atoms. The two stereoisomeric bicyclo[4.4.0]decanes, called *cis-* and *trans*-decalin, are important examples.

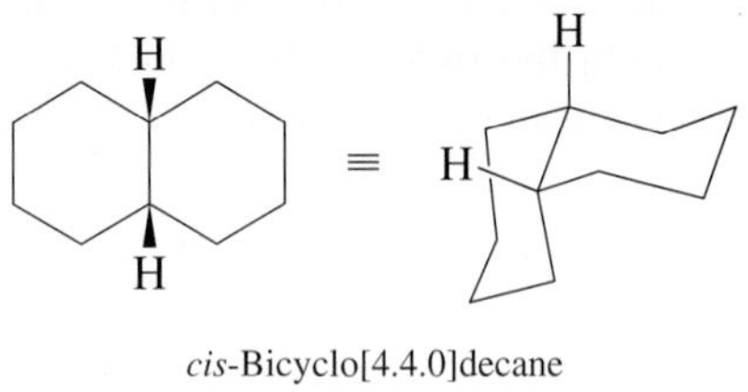
H
H
H
H

cis-Bicyclo[4.4.0]decane
(*cis*-decalin)

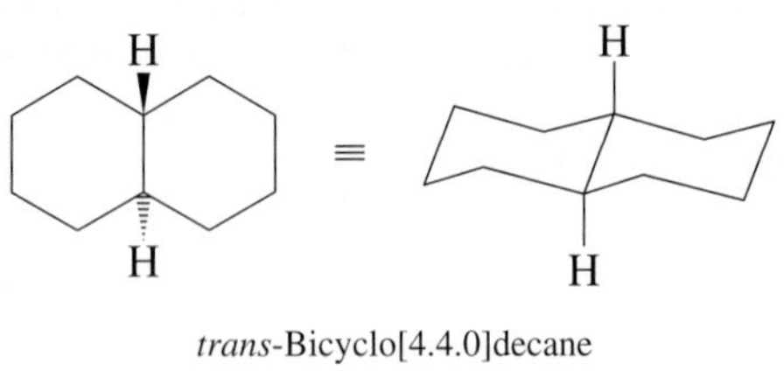
H
H
H
H

trans-Bicyclo[4.4.0]decane
(*trans*-decalin)

The hydrogen atoms at the ring junctions are on the same side in *cis*-decalin and on opposite sides in *trans*-decalin. Both rings adopt the chair conformation in each stereoisomer.

Decalin ring systems appear as structural units in a large number of naturally occurring substances, particularly the steroids. Cholic acid, for example, a steroid present in bile that promotes digestion, incorporates *cis*-decalin and *trans*-decalin units into a rather complex *tetracyclic* structure.

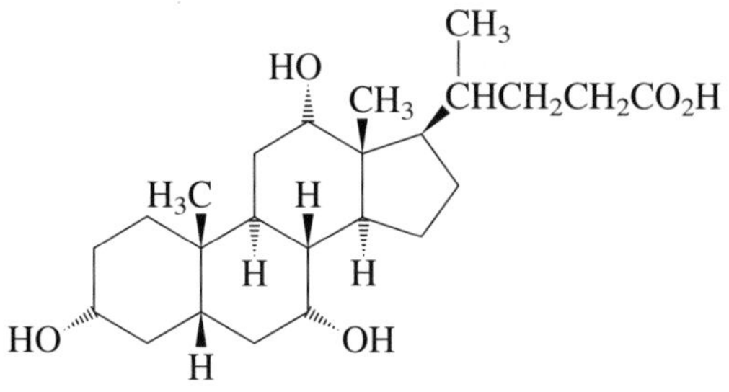

Cholic acid

Problem 3.17

Geosmin is a natural product that smells like dirt. It is produced by several microorganisms and can be obtained from beet extracts. Complete the following decalin ring skeleton, placing the substituents of geosmin in their proper orientations.

3.15 Heterocyclic Compounds

Not all cyclic compounds are hydrocarbons. Many substances include an atom other than carbon, called a *heteroatom* (Section 1.9), as part of a ring. A ring that contains at least one heteroatom is called a *heterocycle,* and a substance based on a heterocyclic ring is a **heterocyclic compound.** Each of the following heterocyclic ring systems will be encountered in this text:

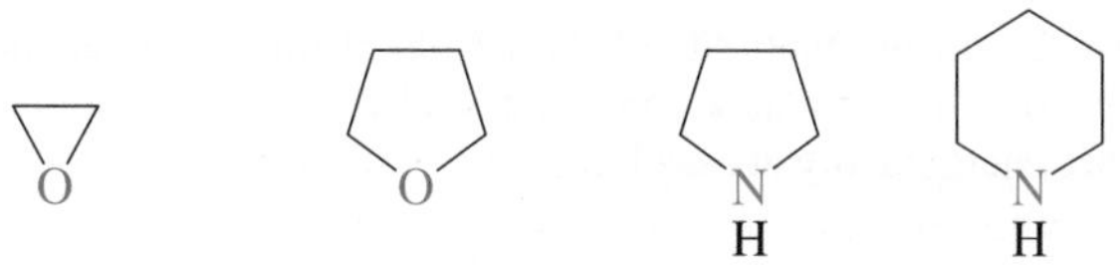

Ethylene oxide Tetrahydrofuran Pyrrolidine Piperidine

The names cited are common names, which have been in widespread use for a long time and are acceptable in IUPAC nomenclature. We will introduce the systematic nomenclature of these ring systems as needed in later chapters.

The shapes of heterocyclic rings are very much like those of their all-carbon analogs. Thus, six-membered heterocycles such as piperidine exist in a chair conformation analogous to cyclohexane.

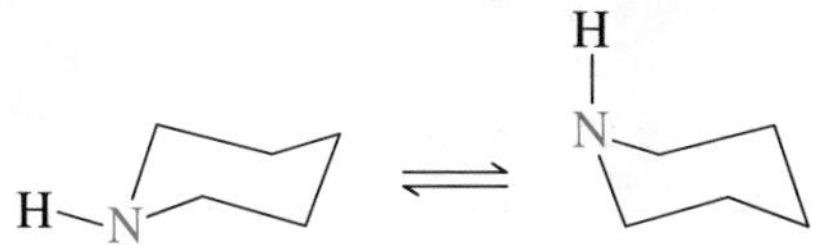

The hydrogen attached to nitrogen can be either axial or equatorial, and both chair conformations are approximately equal in stability.

Problem 3.18

Draw what you would expect to be the most stable conformation of the piperidine derivative in which the hydrogen bonded to nitrogen has been replaced by methyl.

Sulfur-containing heterocycles are also common. Compounds in which sulfur is the heteroatom in three-, four-, five-, and six-membered rings, as well as larger rings, are all well known. Two interesting heterocyclic compounds that contain sulfur–sulfur bonds are *lipoic acid* and *lenthionine.*

Cyclic structures also exist in inorganic chemistry. The most stable form of elemental sulfur is an 8-membered ring of sulfur atoms.

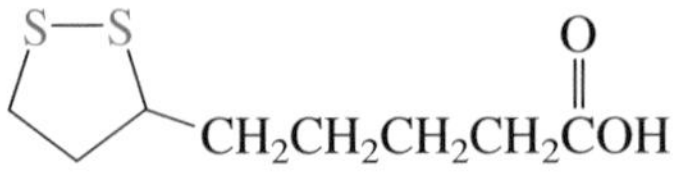

Lipoic acid: a growth factor required by a variety of different organisms

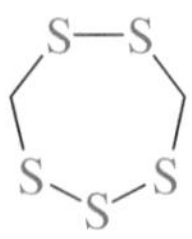

Lenthionine: contributes to the odor of shiitake mushrooms

Many heterocyclic systems contain double bonds and are related to arenes. The most important representatives of this class are introduced in Sections 11.22 and 11.23.

3.16 SUMMARY

In this chapter we explored the three-dimensional shapes of alkanes and cycloalkanes. The most important point to be taken from the chapter is that a molecule adopts the shape that minimizes its total strain. The sources of strain in alkanes and cycloalkanes are:

1. ***Bond length distortion:*** destabilization of a molecule that results when one or more of its bond distances are different from the normal values
2. ***Angle strain:*** destabilization that results from distortion of bond angles from their normal values
3. ***Torsional strain:*** destabilization that results when bonds on adjacent atoms are not staggered
4. ***Van der Waals strain:*** destabilization that results when atoms or groups on nonadjacent atoms are too close to one another

The various spatial arrangements available to a molecule by rotation about single bonds are called **conformations,** and **conformational analysis** is the study of the differences in stability and properties of the individual conformations. Rotation around carbon–carbon single bonds is normally very fast, occurring hundreds of thousands of times per second at room temperature. Molecules are rarely frozen into a single conformation but engage in rapid equilibration among the conformations that are energetically accessible.

Section 3.1 The most stable conformation of ethane is the **staggered** conformation. It is approximately 12 kJ/mol (3 kcal/mol) more stable than the **eclipsed,** which is the least stable conformation.

Staggered conformation of ethane (most stable conformation)

Eclipsed conformation of ethane (least stable conformation)

The difference in energy between the staggered and eclipsed forms is due almost entirely to the torsional strain in the eclipsed conformation. At any instant, almost all the molecules of ethane reside in the staggered conformation.

Section 3.2 The two staggered conformations of butane are not equivalent. The **anti** conformation is more stable than the **gauche.**

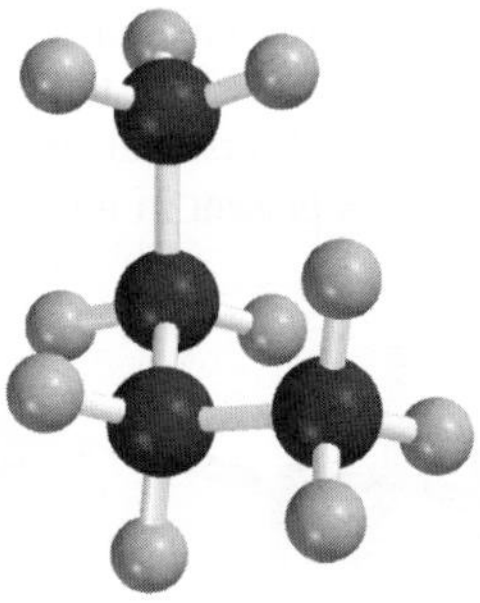

Anti conformation of butane Gauche conformation of butane

Neither conformation suffers torsional strain, because each has a staggered arrangement of bonds. The gauche conformation is less stable because of van der Waals strain involving the methyl groups.

Section 3.3 Higher alkanes adopt a zigzag conformation of the carbon chain in which all the bonds are staggered.

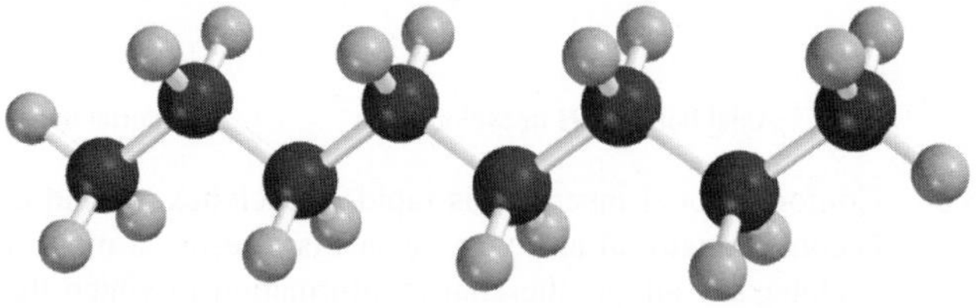

Octane

Section 3.4 At one time all cycloalkanes were believed to be planar. It was expected that cyclopentane would be the least strained cycloalkane because the angles of a regular pentagon (108°) are closest to the tetrahedral angle of 109.5°. Heats of combustion established that this is not so. With the exception of cyclopropane, the rings of all cycloalkanes are nonplanar.

Section 3.5 Cyclopropane is planar and destabilized by angle strain and torsional strain. Cyclobutane is nonplanar and less strained than cyclopropane.

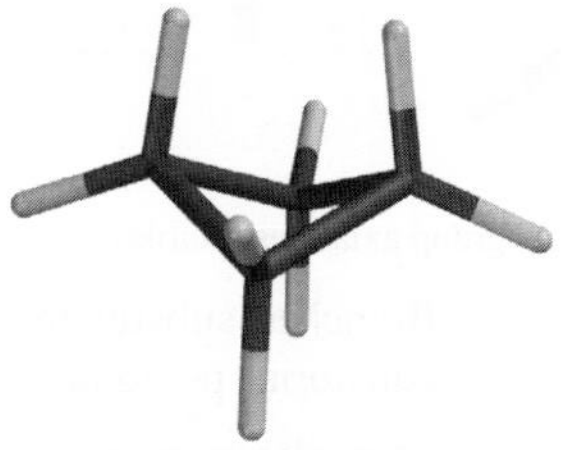

Cyclopropane Cyclobutane

Section 3.6 Cyclopentane has two nonplanar conformations that are of similar stability: the **envelope** and the **half-chair.**

Envelope conformation of cyclopentane Half-chair conformation of cyclopentane

Section 3.7 Three conformations of cyclohexane have approximately tetrahedral angles at carbon: the chair, the boat, and the skew boat. The chair is by far the most stable; it is free of torsional strain, but the boat and skew boat are not. When a cyclohexane ring is present in a compound, it almost always adopts a chair conformation.

Chair

Skew boat

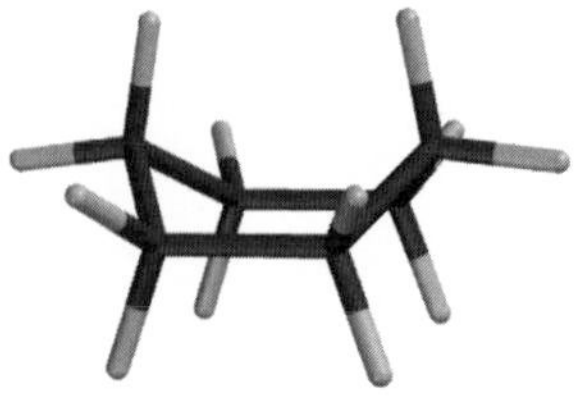

Boat

Section 3.8 The C—H bonds in the chair conformation of cyclohexane are not all equivalent but are divided into two sets of six each, called **axial** and **equatorial.**

Axial bonds to H in cyclohexane

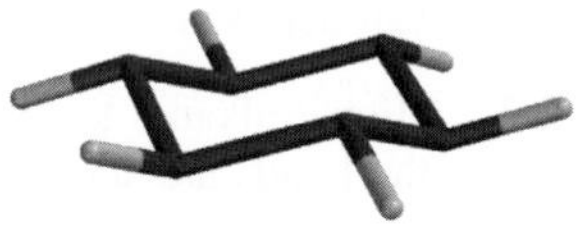

Equatorial bonds to H in cyclohexane

Section 3.9 Conformational inversion is rapid in cyclohexane and causes all axial bonds to become equatorial and vice versa. As a result, a monosubstituted derivative of cyclohexane adopts the chair conformation in which the substituent is equatorial (see next section). *No bonds are made or broken in this process.*

Section 3.10 A substituent is less crowded and more stable when it is equatorial than when it is axial on a cyclohexane ring. Ring inversion of a monosubstituted cyclohexane allows the substituent to become equatorial.

Methyl group axial (less stable) Methyl group equatorial (more stable)

Branched substituents, especially *tert*-butyl, have an increased preference for the equatorial position.

Sections 3.11–3.12 **Stereoisomers** are isomers that have the same constitution but differ in the arrangement of atoms in space. *Cis*- and *trans*-1,3-dimethylcyclohexane are stereoisomers. The cis isomer is more stable than the trans.

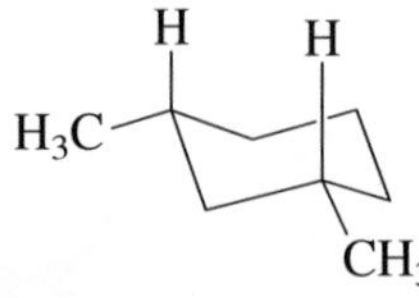

Most stable conformation of *cis*-1,3-dimethylcyclohexane (no axial methyl groups)

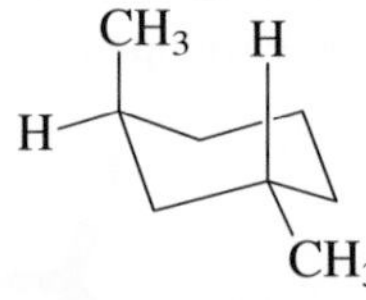

Most stable conformation of *trans*-1,3-dimethylcyclohexane (one axial methyl group)

Section 3.13 Higher cycloalkanes have angles at carbon that are close to tetrahedral and are sufficiently flexible to adopt conformations that reduce their torsional strain. They tend to be populated by several different conformations of similar stability.

Section 3.14 Cyclic hydrocarbons can contain more than one ring. **Spiro** hydrocarbons are characterized by the presence of a single carbon that is common to two rings. Bicyclic alkanes contain two rings that share two or more atoms.

Section 3.15 Substances that contain one or more atoms other than carbon as part of a ring are called **heterocyclic** compounds. Rings in which the heteroatom is oxygen, nitrogen, or sulfur rank as both the most common and the most important.

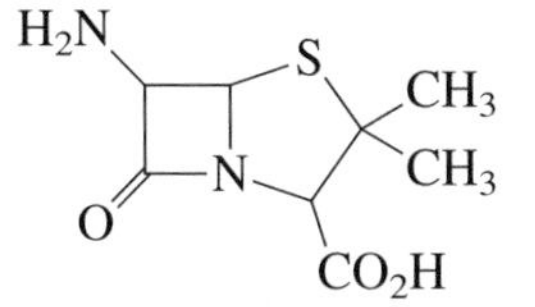

6-Aminopenicillanic acid
(bicyclic and heterocyclic)

PROBLEMS

3.19 Give the IUPAC names of each of the following alkanes.

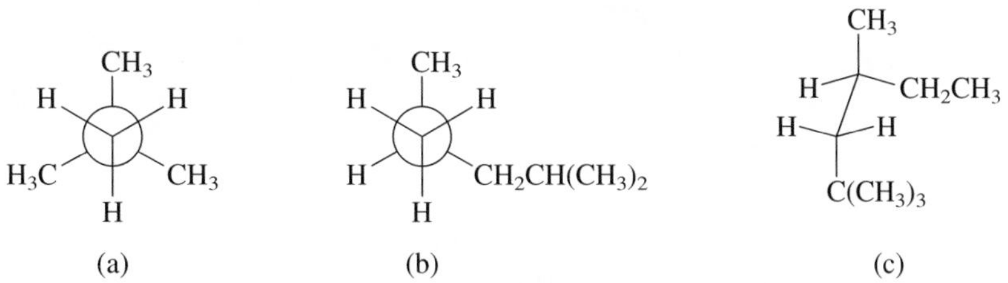

3.20 Sight down the C-2—C-3 bond, and draw Newman projection formulas for the
(a) Most stable conformation of 2,2-dimethylbutane
(b) Two most stable conformations of 2-methylbutane
(c) Two most stable conformations of 2,3-dimethylbutane

3.21 One of the staggered conformations of 2-methylbutane in Problem 3.20b is more stable than the other. Which one is more stable? Why?

3.22 Sketch an approximate potential energy diagram for rotation about the carbon–carbon bond in 2,2-dimethylpropane similar to that shown in Figures 3.4 and 3.7. Does the form of the potential energy curve of 2,2-dimethylpropane more closely resemble that of ethane or that of butane?

3.23 Repeat Problem 3.22 for the case of 2-methylbutane.

3.24 Identify all atoms that are (a) anti and (b) gauche to bromine in the conformation shown for $CH_3CH_2CH_2Br$.

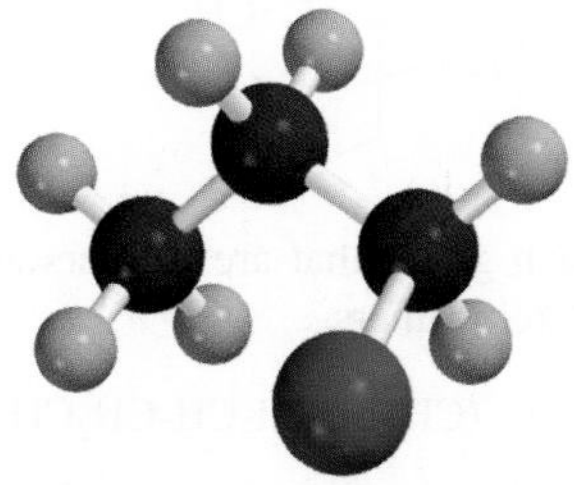

3.25 Even though the methyl group occupies an equatorial site, the conformation shown is not the most stable one for methylcyclohexane. Explain.

3.26 Which do you expect to be the more stable conformation of *cis*-1,3-dimethylcyclobutane, A or B? Why?

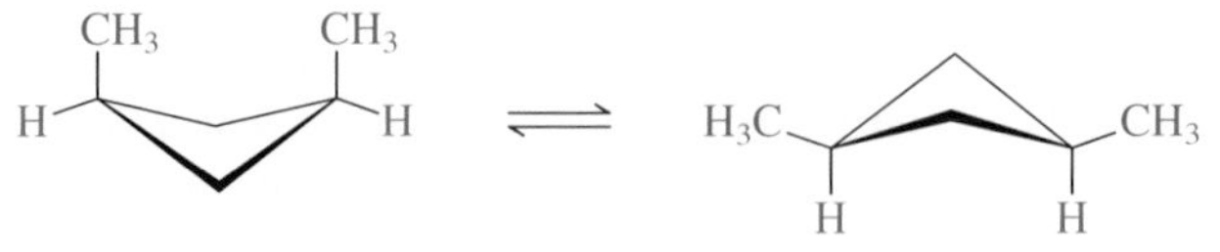

A B

3.27 Determine whether the two structures in each of the following pairs represent *constitutional isomers,* different *conformations* of the same compound, or *stereoisomers* that cannot be interconverted by rotation about single bonds.

(a) 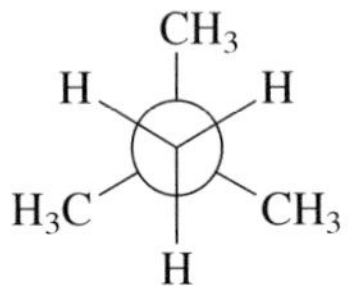

and 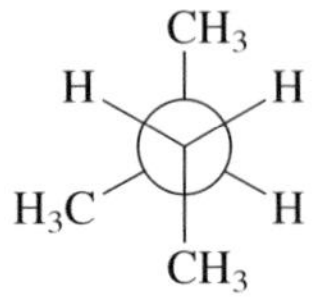

(b) 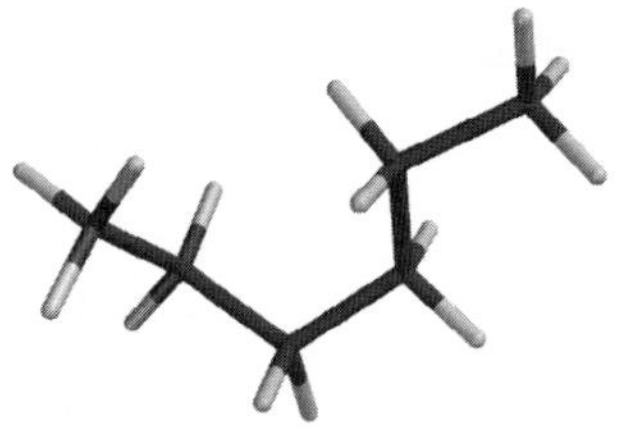 and 

(c) 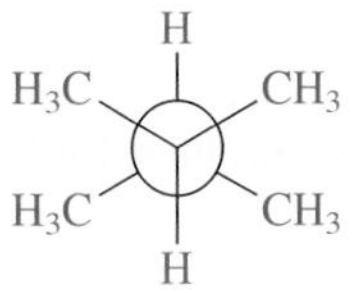

and 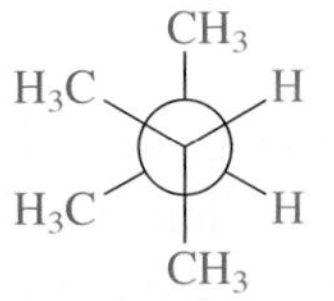

(d) *cis*-1,2-Dimethylcyclopentane and *trans*-1,3-dimethylcyclopentane

(e) 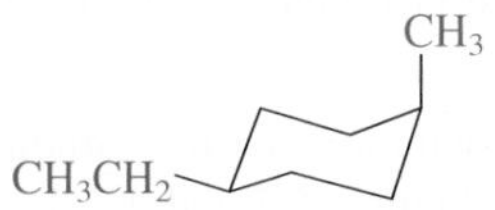

and 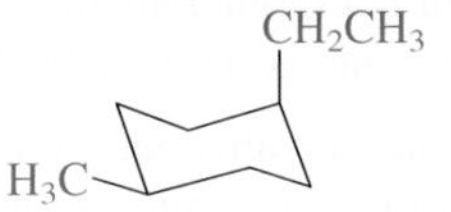

(f) 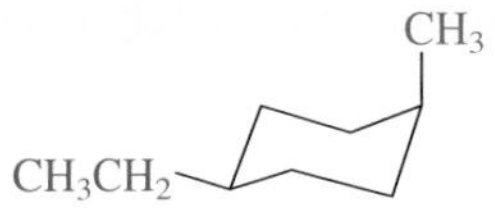

and 

(g) 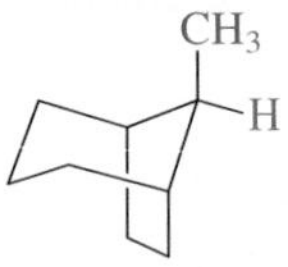

and 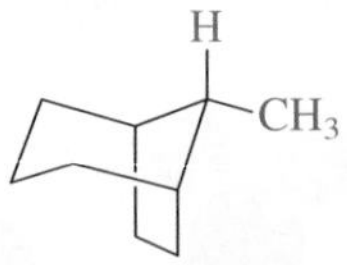

3.28 Select the compounds in each group that are isomers and specify whether they are constitutional isomers or stereoisomers.

(a) $(CH_3)_3CCH_2CH_2CH_3$ $(CH_3)_3CCH_2CH_2CH_2CH_3$ $(CH_3)_2CHCH(CH_3)CH_2CH_3$

(b) 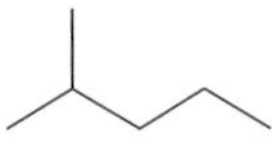 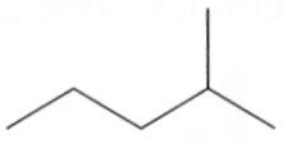 

(c)  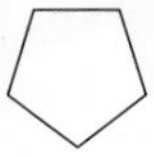 

(d) 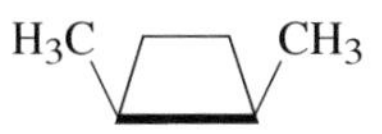 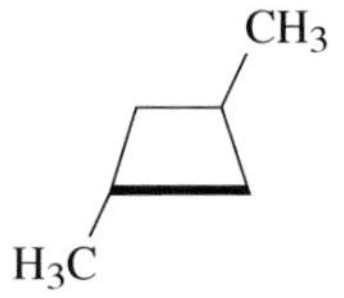 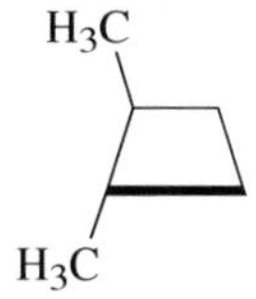

(e) 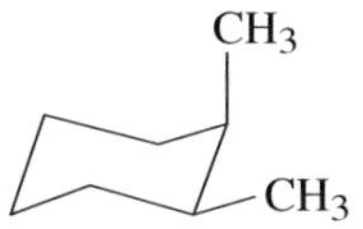 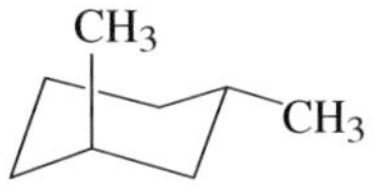 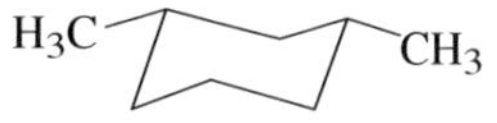

3.29 Excluding compounds that contain methyl or ethyl groups, write structural formulas for all the bicyclic isomers of (a) C_5H_8 and (b) C_6H_{10}.

3.30 In each of the following groups of compounds, identify the one with the largest heat of combustion and the one with the smallest. In which cases can a comparison of heats of combustion be used to assess relative stability?

(a) Cyclopropane, cyclobutane, cyclopentane

(b) *cis*-1,2-Dimethylcyclopentane, methylcyclohexane, 1,1,2,2-tetramethylcyclopropane

(c)  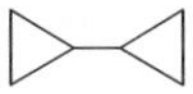 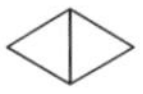

(d) 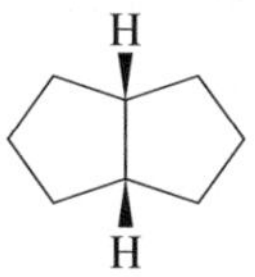 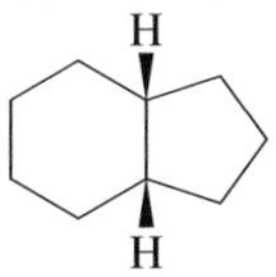 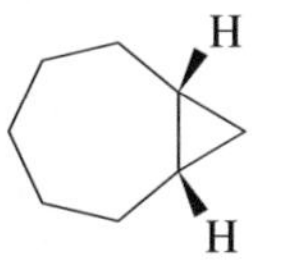

3.31 Ambroxol is a drug used to treat bronchopulmonary disease.

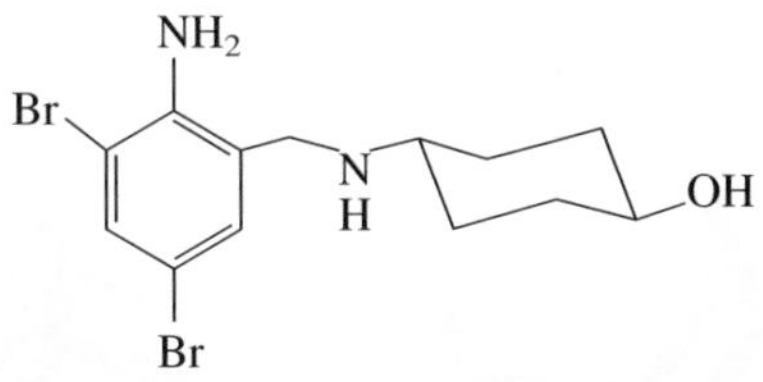

Draw the structure of ambroxol in the alternative chair conformation. Which of the two conformations is more stable?

3.32 Write a structural formula for the most stable conformation of each of the following compounds:

(a) 2,2,5,5-Tetramethylhexane (Newman projection of conformation about C-3—C-4 bond)

(b) 2,2,5,5-Tetramethylhexane (zigzag conformation of entire molecule)

(c) *cis*-1-Isopropyl-3-methylcyclohexane

(d) *trans*-1-Isopropyl-3-methylcyclohexane

(e) *cis*-1-*tert*-Butyl-4-ethylcyclohexane

(f) *cis*-1,1,3,4-Tetramethylcyclohexane

(g) 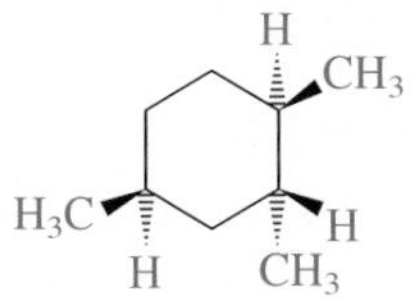

3.33 Identify the more stable stereoisomer in each of the following pairs, and give the reason for your choice:

(a) *cis*- or *trans*-1-Isopropyl-2-methylcyclohexane

(b) *cis*- or *trans*-1-Isopropyl-3-methylcyclohexane

(c) *cis*- or *trans*-1-Isopropyl-4-methylcyclohexane

(d) 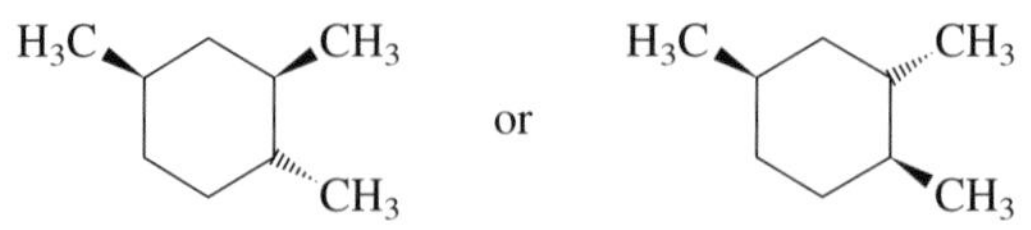

(e) 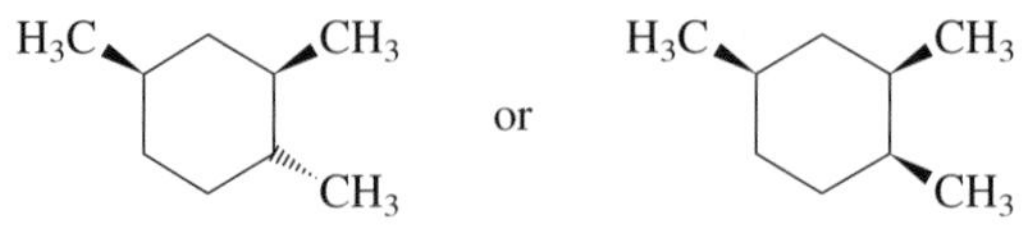

(f) 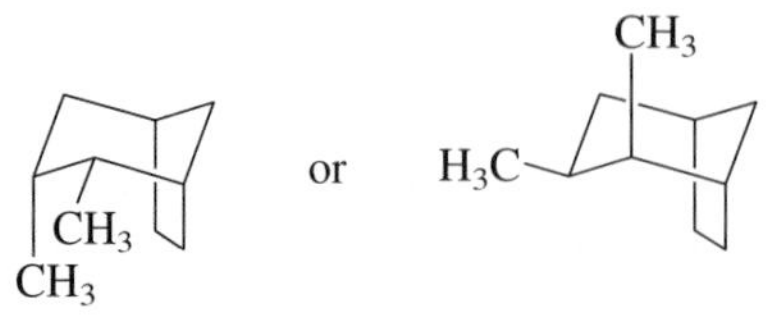

3.34 One stereoisomer of 1,1,3,5-tetramethylcyclohexane is 15 kJ/mol (3.7 kcal/mol) less stable than the other. Indicate which isomer is the less stable, and identify the reason for its decreased stability.

3.35 The heats of combustion of the more and less stable stereoisomers of the 1,2-, 1,3-, and 1,4-dimethylcyclohexanes are given here. The values are higher for the 1,2-dimethylcyclohexanes than for the 1,3- and 1,4-isomers. Suggest an explanation.

Dimethylcyclohexane	1,2	1,3	1,4
Heats of combustion (kJ/mol):			
More stable stereoisomer	5217	5212	5212
Less stable stereoisomer	5223	5219	5219

3.36 One of the following two stereoisomers is 20 kJ/mol (4.9 kcal/mol) less stable than the other. Indicate which isomer is the less stable, and identify the reason for its decreased stability.

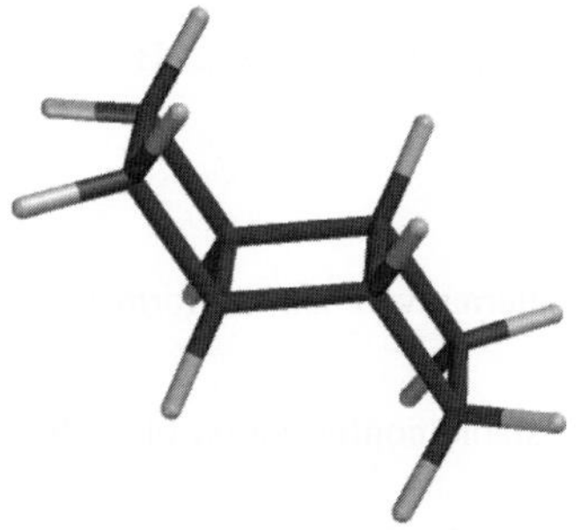

A

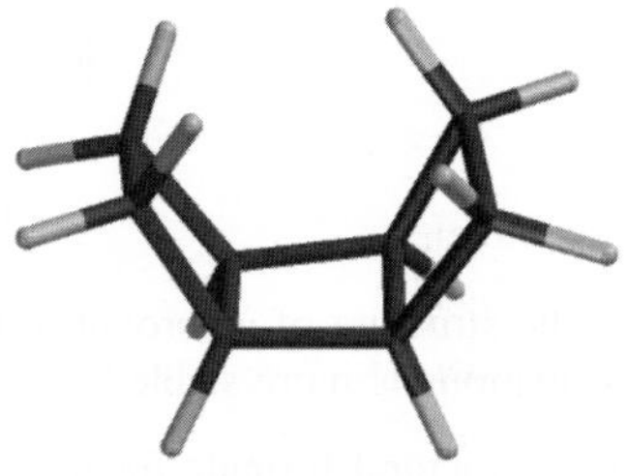

B

3.37 Cubane (C_8H_8) is the common name of a polycyclic hydrocarbon that was first synthesized in the early 1960s. As its name implies, its structure is that of a cube. How many rings are present in cubane?

Cubane

3.38 Biological oxidation of hydrocarbons is a commonly observed process.

(a) To what class of hydrocarbons does the reactant in the following equation belong? What is its IUPAC name?

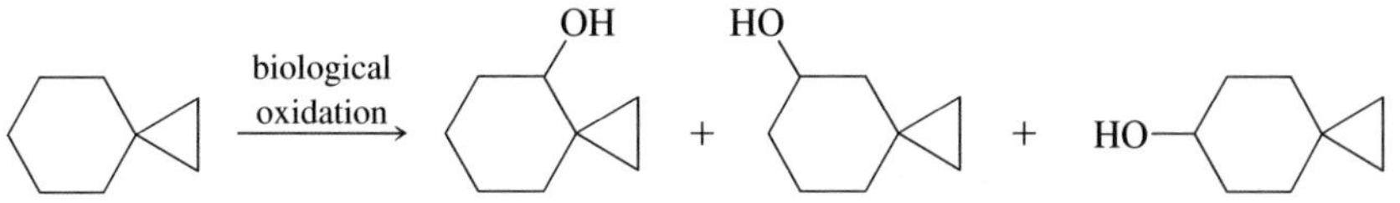

(b) Identify by IUPAC locant the carbon that is oxidized in the formation of each product.

(c) How are alcohols A, B, and C related? Are they constitutional isomers or stereoisomers?

3.39 The following are representations of two forms of glucose. The six-membered ring is known to exist in a chair conformation in each form. Draw clear representations of the most stable conformation of each. Are they two different conformations of the same molecule, or are they stereoisomers that cannot be interconverted by rotation about single bonds? Which substituents (if any) occupy axial sites?

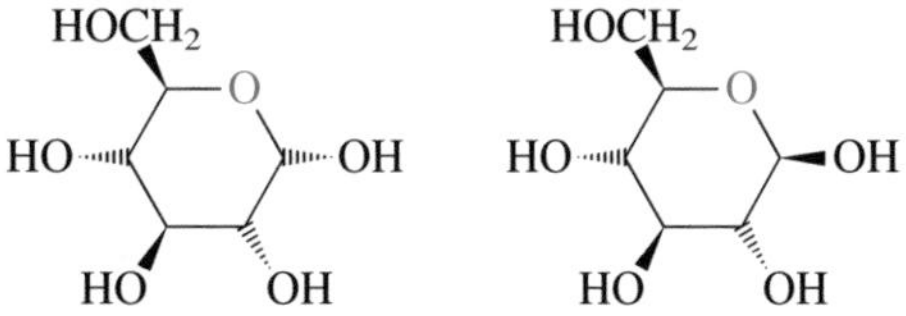

3.40 A typical steroid skeleton is shown along with the numbering scheme used for this class of compounds. Specify in each case whether the designated substituent is axial or equatorial.

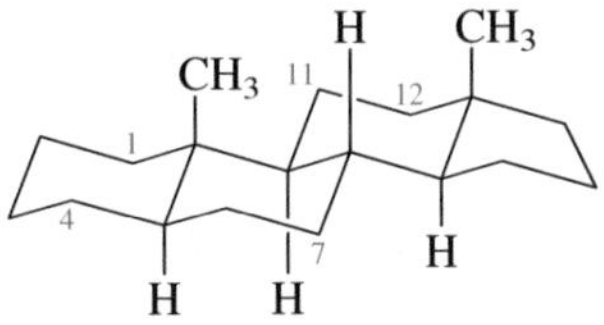

(a) Substituent at C-1 cis to the methyl groups

(b) Substituent at C-4 cis to the methyl groups

(c) Substituent at C-7 trans to the methyl groups

(d) Substituent at C-11 trans to the methyl groups

(e) Substituent at C-12 cis to the methyl groups

3.41 Repeat Problem 3.40 for the stereoisomeric steroid skeleton having a cis ring fusion between the first two rings.

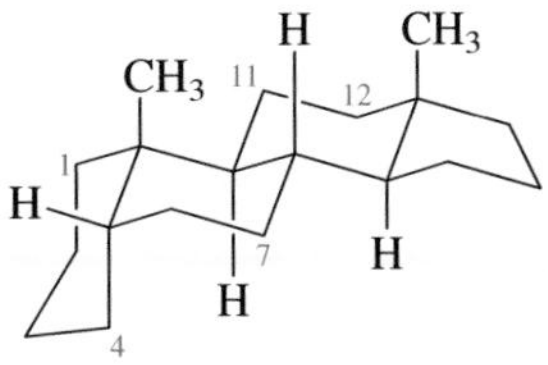

3.42 (a) Write Newman projections for the gauche and anti conformations of 1,2-dichloroethane ($ClCH_2CH_2Cl$).

(b) The measured dipole moment of $ClCH_2CH_2Cl$ is 1.12 D. Which one of the following statements about 1,2-dichloroethane is false?

(1) It may exist entirely in the anti conformation.

(2) It may exist entirely in the gauche conformation.

(3) It may exist as a mixture of anti and gauche conformations.

Descriptive Passage and Interpretive Problems 3

Cyclic Forms of Carbohydrates

Five- and six-membered ring structures are common in carbohydrates and are often in equilibrium with each other. The five-membered ring structures are called furanose forms; the six-membered ring structures are pyranose forms. D-Ribose, especially in its β-furanose form, is a familiar carbohydrate.

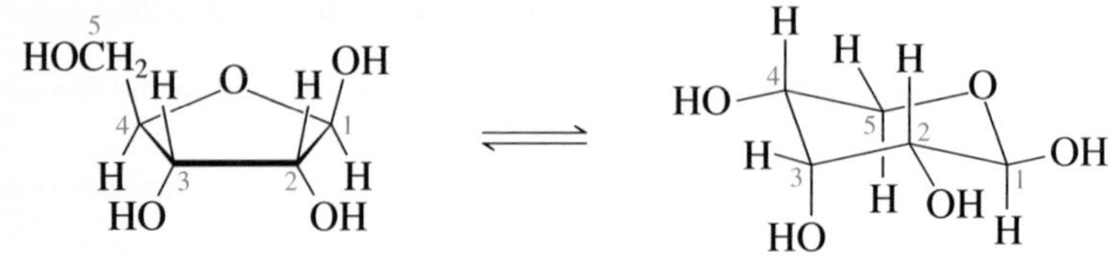

D-Ribose: β–Furanose form β–Pyranose form

3.43 The β–furanose and β–pyranose forms of D-ribose are:

A. Conformational isomers C. Resonance forms
B. Constitutional isomers D. Stereoisomers

3.44 What is the orientation of the OH groups at C-2 and C-3 in the β–pyranose form of D-ribose?

A. Both are axial.
B. Both are equatorial.
C. C-2 is axial; C-3 is equatorial.
D. C-2 is equatorial; C-3 is axial.

3.45 The OH groups at C-2 and C-3 in the β–pyranose form of D-ribose are:

A. cis and gauche C. trans and gauche
B. cis and anti D. trans and anti

3.46 All of the OH groups of the β–pyranose form of D-xylose are equatorial. Which of the following is the β–furanose form of D-xylose?

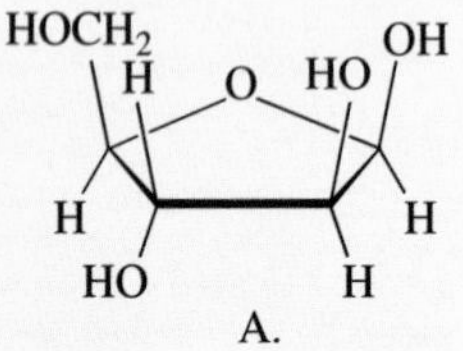

A.

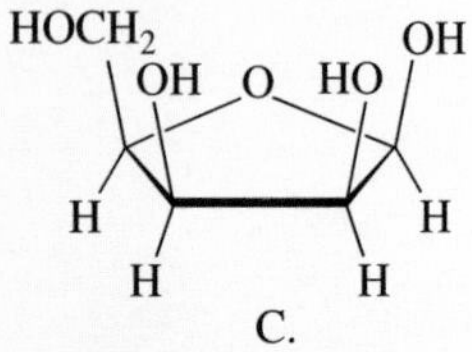

C.

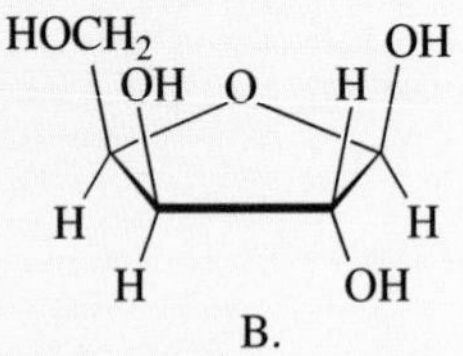

B.

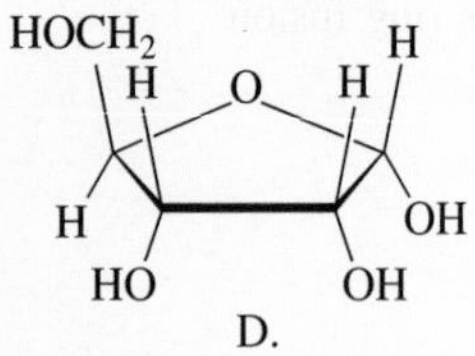

D.

3.47 The carbohydrate shown here is a component of a drug used in veterinary medicine. Which is its most stable pyranose conformation?

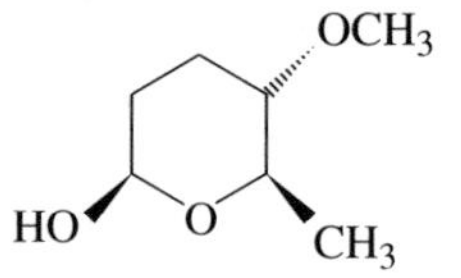

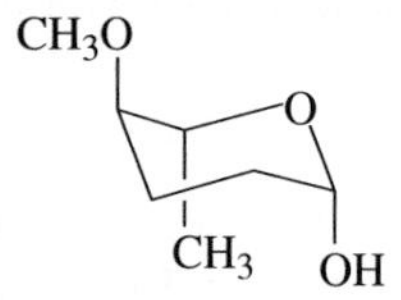

A.

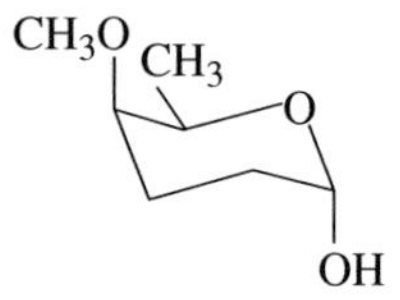

C.

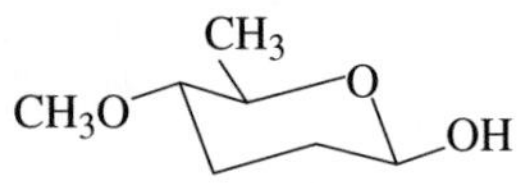

B.

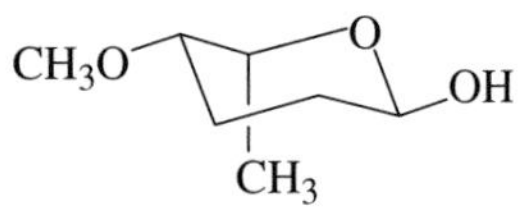

D.

Alcohols and Alkyl Halides

4

Chapter Outline

Mechanisms

As a motion picture tells a story, so too a mechanism tells us how a chemical reaction takes place. If we could slow down and look at a reaction "frame-by-frame," as we can with a film reel, we would observe intermediates and transition states that appear during it.

OUR FIRST THREE CHAPTERS established some fundamental principles concerning the *structure* of organic molecules and introduced the connection between structure and *reactivity* with a review of acid–base reactions. In this chapter we explore structure and reactivity in more detail by developing two concepts: *functional groups* and *reaction mechanisms.* A **functional group** is the atom or group in a molecule most responsible for the reaction the compound undergoes under a prescribed set of conditions. *How* the structure of the reactant is transformed to that of the product is what we mean by the reaction **mechanism.**

Organic compounds are grouped into families according to the functional groups they contain. Two of the most important families are **alcohols** and **alkyl halides.** Alcohols and alkyl halides are especially useful because they are versatile starting materials for preparing numerous other families. *Indeed, alcohols or alkyl halides—often both—will appear in virtually all of the remaining chapters of this text.*

The major portion of the present chapter concerns the conversion of alcohols to alkyl halides by reaction with hydrogen halides:

$$\underset{\text{Alcohol}}{\text{R—OH}} + \underset{\text{Hydrogen halide}}{\text{H—X}} \longrightarrow \underset{\text{Alkyl halide}}{\text{R—X}} + \underset{\text{Water}}{\text{H—OH}}$$

It is convenient in equations such as this to represent generic alcohols and alkyl halides as ROH and RX, respectively, where **"R"** stands for an alkyl group. In addition to convenience, this notation lets us focus more clearly on the functional group transformation; the OH functional group of an alcohol is replaced by a halogen, usually chlorine (X = Cl) or bromine (X = Br).

While developing the connections between structure, reaction, and mechanism, we will also extend the fundamentals of IUPAC nomenclature to functional group families, beginning with alcohols and alkyl halides.

4.1 Functional Groups

The families of hydrocarbons—*alkanes, alkenes, alkynes,* and *arenes*—were introduced in Section 2.1. The double bond is a functional group in an alkene, the triple bond a functional group in an alkyne, and the benzene ring itself is a functional group in an arene. Alkanes (RH) are not considered to have a functional group, although as we'll see later in this chapter, reactions that replace a hydrogen atom can take place. In general though, hydrogen atoms of alkanes are relatively unreactive and any other group attached to the hydrocarbon framework will be the functional group.

Table 4.1 lists the major families of organic compounds covered in this text and their functional groups.

TABLE 4.1 Functional Groups in Some Important Classes of Organic Compounds

Class	Generalized abbreviation*	Representative example	Name of example†
Alcohol	ROH	CH_3CH_2OH	Ethanol
Alkyl halide	RCl	CH_3CH_2Cl	Chloroethane
Amine‡	RNH_2	$CH_3CH_2NH_2$	Ethanamine
Epoxide	$\underset{\diagdown O \diagup}{R_2C—CR_2}$	$\underset{\diagdown O \diagup}{H_2C—CH_2}$	Oxirane
Ether	ROR	$CH_3CH_2OCH_2CH_3$	Diethyl ether
Nitrile	RC≡N	$CH_3CH_2C{\equiv}N$	Propanenitrile
Nitroalkane	RNO_2	$CH_3CH_2NO_2$	Nitroethane
Sulfide	RSR	CH_3SCH_3	Dimethyl sulfide
Thiol	RSH	CH_3CH_2SH	Ethanethiol
Aldehyde	$R\overset{O}{\overset{\|}{C}}H$	$CH_3\overset{O}{\overset{\|}{C}}H$	Ethanal
Ketone	$R\overset{O}{\overset{\|}{C}}R$	$CH_3\overset{O}{\overset{\|}{C}}CH_2CH_3$	2-Butanone
Carboxylic acid	$R\overset{O}{\overset{\|}{C}}OH$	$CH_3\overset{O}{\overset{\|}{C}}OH$	Ethanoic acid
Carboxylic acid derivatives			
Acyl halide	$R\overset{O}{\overset{\|}{C}}X$	$CH_3\overset{O}{\overset{\|}{C}}Cl$	Ethanoyl chloride
Acid anhydride	$R\overset{O}{\overset{\|}{C}}O\overset{O}{\overset{\|}{C}}R$	$CH_3\overset{O}{\overset{\|}{C}}O\overset{O}{\overset{\|}{C}}CH_3$	Ethanoic anhydride
Ester	$R\overset{O}{\overset{\|}{C}}OR$	$CH_3\overset{O}{\overset{\|}{C}}OCH_2CH_3$	Ethyl ethanoate
Amide	$R\overset{O}{\overset{\|}{C}}NR_2$	$CH_3\overset{O}{\overset{\|}{C}}NH_2$	Ethanamide

*When more than one R group is present, the groups may be the same or different.

†Most compounds have more than one acceptable name.

‡The example given is a *primary* amine (RNH_2). *Secondary* amines have the general structure R_2NH; *tertiary* amines are R_3N.

Problem 4.1

(a) Write a structural formula for a sulfide having the molecular formula C_3H_8S. (b) What two thiols have the molecular formula C_3H_8S?

Sample Solution (a) According to Table 4.1, sulfides have the general formula RSR and the Rs may be the same or different. The only possible connectivity for a sulfide with three carbons is C—S—C—C. Therefore, the sulfide is $CH_3SCH_2CH_3$.

Carbonyl group chemistry is discussed in a block of four chapters (Chapters 17–20).

We have already touched on some of these functional-group families in our discussion of acids and bases. We have seen that alcohols resemble water in their acidity and that carboxylic acids, although weak acids, are stronger acids than alcohols. Carboxylic acids belong to one of the most important groups of organic compounds—those that contain carbonyl groups (C=O). They and other carbonyl-containing compounds rank among the most abundant and biologically significant classes of naturally occurring substances. In this chapter we focus our attention on two classes of organic compounds listed in Table 4.1: alkyl halides and alcohols.

Problem 4.2

Many compounds contain more than one functional group. Prostaglandin E_1, a hormone that regulates the relaxation of smooth muscles, contains two different kinds of carbonyl groups. Classify each one (aldehyde, ketone, carboxylic acid, ester, amide, acyl chloride, or acid anhydride). Identify the most acidic proton in prostaglandin E_1 and use Table 1.8 to estimate its pK_a.

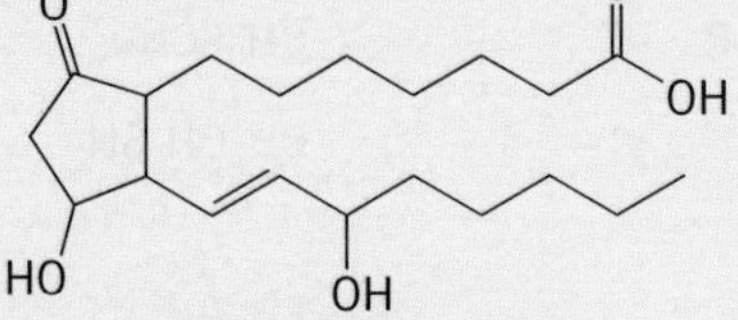

Prostaglandin E_1

4.2 IUPAC Nomenclature of Alkyl Halides

The IUPAC rules permit certain common alkyl group names to be used. These include *n*-propyl, isopropyl, *n*-butyl, *sec*-butyl, isobutyl, *tert*-butyl, and neopentyl (Section 2.13).

The IUPAC rules permit alkyl halides to be named in two different ways, called *functional class* nomenclature and *substitutive* nomenclature. In **functional class nomenclature** the alkyl group and the halide *(fluoride, chloride, bromide,* or *iodide)* are designated as separate words. The alkyl group is named on the basis of its longest continuous chain beginning at the carbon to which the halogen is attached.

CH_3F — Methyl fluoride

$CH_3CH_2CH_2CH_2CH_2Cl$ — Pentyl chloride

$CH_3CH_2\overset{1}{C}H\overset{2}{C}H_2\overset{3}{C}H_2\overset{4}{C}H_3$ (Br on C-1) — 1-Ethylbutyl bromide

Cyclohexyl iodide

Substitutive nomenclature of alkyl halides treats the halogen as a *halo (fluoro-, chloro-, bromo-,* or *iodo-) substituent* on an alkane chain. The carbon chain is numbered in the direction that gives the substituted carbon the lower number.

$\overset{5}{C}H_3\overset{4}{C}H_2\overset{3}{C}H_2\overset{2}{C}H_2\overset{1}{C}H_2F$ — 1-Fluoropentane

$\overset{1}{C}H_3\overset{2}{C}H\overset{3}{C}H_2\overset{4}{C}H_2\overset{5}{C}H_3$ (Br on C-2) — 2-Bromopentane

$\overset{1}{C}H_3\overset{2}{C}H_2\overset{3}{C}H\overset{4}{C}H_2\overset{5}{C}H_3$ (I on C-3) — 3-Iodopentane

When the carbon chain bears both a halogen and an alkyl substituent, the two are considered of equal rank, and the chain is numbered so as to give the lower number to the substituent nearer the end of the chain.

$\overset{1}{C}H_3\overset{2}{C}H\overset{3}{C}H_2\overset{4}{C}H_2\overset{5}{C}H\overset{6}{C}H_2\overset{7}{C}H_3$ (CH_3 on C-2, Cl on C-5) — 5-Chloro-2-methylheptane

$\overset{1}{C}H_3\overset{2}{C}H\overset{3}{C}H_2\overset{4}{C}H_2\overset{5}{C}H\overset{6}{C}H_2\overset{7}{C}H_3$ (Cl on C-2, CH_3 on C-5) — 2-Chloro-5-methylheptane

> **Problem 4.3**
>
> Write structural formulas and give the functional class and substitutive names of all the isomeric alkyl chlorides that have the molecular formula C_4H_9Cl.

Substitutive names are preferred, but functional class names are sometimes more convenient or more familiar and are frequently encountered in organic chemistry.

Functional class names are part of the IUPAC system; they are not "common names."

4.3 IUPAC Nomenclature of Alcohols

Functional class names of alcohols are derived by naming the alkyl group that bears the hydroxyl substituent (—OH) and then adding *alcohol* as a separate word. The chain is always numbered beginning at the carbon to which the hydroxyl group is attached.

Substitutive names of alcohols are developed by identifying the longest continuous chain that bears the hydroxyl group and replacing the *-e* ending of the corresponding alkane by the suffix *-ol*. The position of the hydroxyl group is indicated by number, choosing the sequence that assigns the lower locant to the carbon that bears the hydroxyl group.

The 2004 IUPAC recommendations alter the substitutive names of alcohols by bracketing the numerical locant for the substituted carbon with hyphens and placing it immediately before the *-ol* ending.

	CH_3CH_2OH	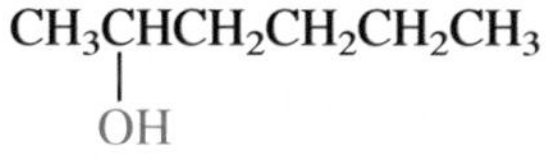	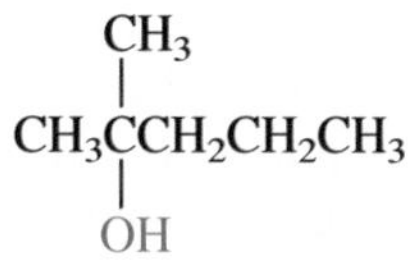
Functional class name:	Ethyl alcohol	1-Methylpentyl alcohol	1,1-Dimethylbutyl alcohol
Substitutive name:	Ethanol	2-Hexanol	2-Methyl-2-pentanol
2004 name:	Ethanol	Hexan-2-ol	2-Methylpentan-2-ol

Several alcohols are commonplace substances, well known by common names that reflect their origin (wood alcohol, grain alcohol) or use (rubbing alcohol). Wood alcohol is *methanol* (methyl alcohol, CH_3OH), grain alcohol is *ethanol* (ethyl alcohol, CH_3CH_2OH), and rubbing alcohol is *2-propanol* [isopropyl alcohol, $(CH_3)_2CHOH$].

Hydroxyl groups take precedence over ("outrank") alkyl groups and halogens in determining the direction in which a carbon chain is numbered. The OH group is assumed to be attached to C-1 of a cyclic alcohol, and need be numbered only when using the 2004 IUPAC rules.

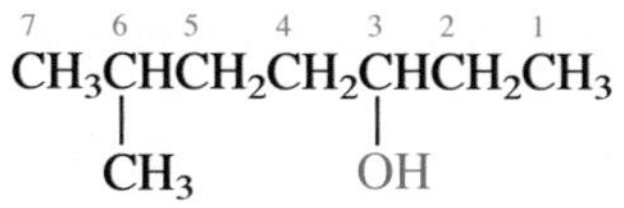	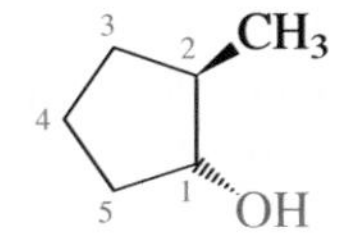	3 2 1 $FCH_2CH_2CH_2OH$
6-Methyl-3-heptanol	*trans*-2-Methylcyclopentanol	3-Fluoro-1-propanol
6-Methylheptan-3-ol	*trans*-2-Methylcyclopentan-1-ol	3-Fluoropropan-1-ol

> **Problem 4.4**
>
> Write structural formulas, and give the functional class and substitutive names of all the isomeric alcohols that have the molecular formula $C_4H_{10}O$.

4.4 Classes of Alcohols and Alkyl Halides

Alcohols and alkyl halides are classified as primary, secondary, or tertiary according to the degree of substitution of the carbon that bears the functional group (Section 2.13). Thus, *primary alcohols* and *primary alkyl halides* are compounds of the type RCH_2G (where G is the functional group), *secondary alcohols* and *secondary alkyl halides* are compounds of the type R_2CHG, and *tertiary alcohols* and *tertiary alkyl halides* are compounds of the type R_3CG.

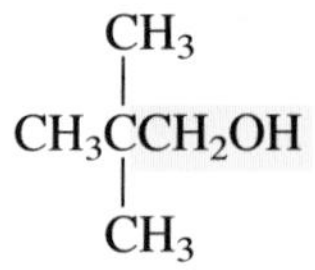

2,2-Dimethyl-1-propanol
(a primary alcohol)

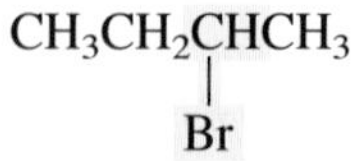

2-Bromobutane
(a secondary alkyl halide)

1-Methylcyclohexanol
(a tertiary alcohol)

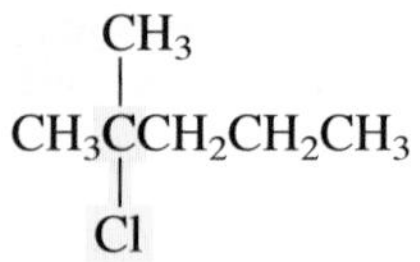

2-Chloro-2-methylpentane
(a tertiary alkyl halide)

Problem 4.5

Classify the isomeric $C_4H_{10}O$ alcohols as being primary, secondary, or tertiary.

Many of the properties of alcohols and alkyl halides are affected by whether their functional groups are attached to primary, secondary, or tertiary carbons. We will see a number of cases in which a functional group attached to a primary carbon is more reactive than one attached to a secondary or tertiary carbon, as well as other cases in which the reverse is true.

4.5 Bonding in Alcohols and Alkyl Halides

The carbon that bears the functional group is sp^3-hybridized in alcohols and alkyl halides. Figure 4.1 illustrates bonding in methanol. The bond angles at carbon are approximately tetrahedral, as is the C—O—H angle. A similar orbital hybridization model applies to alkyl halides, with the halogen connected to sp^3-hybridized carbon by a σ bond. Carbon–halogen bond distances in alkyl halides increase in the order C—F (140 pm) < C—Cl (179 pm) < C—Br (197 pm) < C—I (216 pm).

Carbon–oxygen and carbon–halogen bonds are polar covalent bonds, and carbon bears a partial positive charge in alcohols ($^{\delta+}$C—O$^{\delta-}$) and in alkyl halides ($^{\delta+}$C—X$^{\delta-}$). Alcohols and alkyl halides are polar molecules. The dipole moments of methanol and chloromethane are very similar to each other and to water.

Water
(μ = 1.8 D)

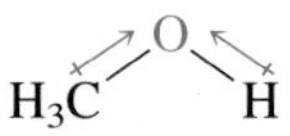

Methanol
(μ = 1.7 D)

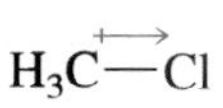

Chloromethane
(μ = 1.9 D)

Problem 4.6

Bromine is less electronegative than chlorine, yet methyl bromide and methyl chloride have very similar dipole moments. Why?

Figure 4.2 maps the electrostatic potential in methanol and chloromethane. Both are similar in that the sites of highest negative potential (red) are near the electronegative atoms: oxygen and chlorine. The polarization of the bonds to oxygen and chlorine, as well as their unshared electron pairs, contribute to the concentration of negative charge on these atoms.

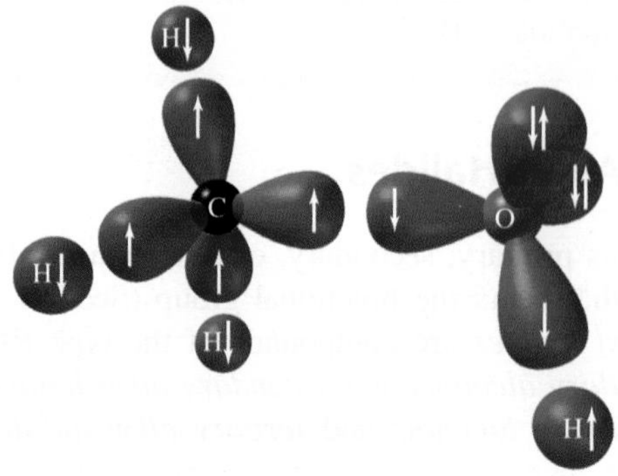

(a)

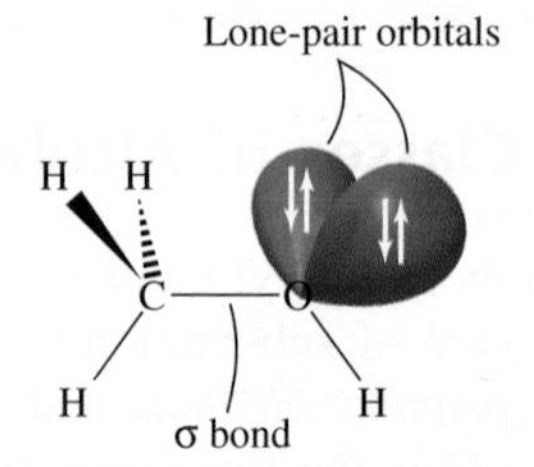

C—O—H angle = 108.5°
C—O bond distance = 142 pm

(b)

Figure 4.1

Orbital hybridization model of bonding in methanol. (*a*) The orbitals used in bonding are the 1*s* orbital of hydrogen and sp^3-hybridized orbitals of carbon and oxygen. (*b*) The bond angles at carbon and oxygen are close to tetrahedral, and the carbon–oxygen σ bond is about 10 pm shorter than a carbon–carbon single bond.

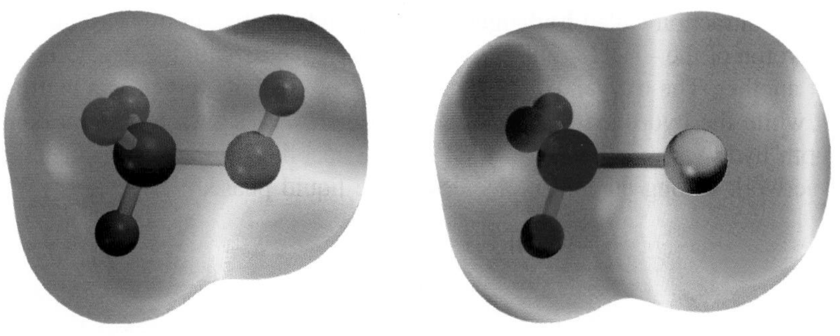

Methanol (CH_3OH) Chloromethane (CH_3Cl)

Figure 4.2

Electrostatic potential maps of methanol and chloromethane. The electrostatic potential is most negative near oxygen in methanol and near chlorine in chloromethane. The most positive region is near the O—H proton in methanol and near the methyl group in chloromethane.

Relatively simple notions of attractive forces between opposite charges are sufficient to account for many of the properties of chemical substances. You will find it helpful to keep the polarity of carbon–oxygen and carbon–halogen bonds in mind as we develop the properties of alcohols and alkyl halides in later sections.

4.6 Physical Properties of Alcohols and Alkyl Halides: Intermolecular Forces

Boiling Point. When describing the effect of alkane structure on boiling point in Section 2.17, we pointed out that van der Waals attractive forces between neutral molecules are of three types.

1. Induced-dipole/induced-dipole forces (dispersion forces; London forces)
2. Dipole/induced-dipole forces
3. Dipole–dipole forces

Induced-dipole/induced-dipole forces are the only intermolecular attractive forces available to nonpolar molecules such as alkanes and are important in polar molecules as well. In addition, polar molecules also engage in dipole–dipole and dipole/induced-dipole attractions. The **dipole–dipole attractive force** is easiest to visualize and is illustrated in Figure 4.3. Two molecules of a polar substance experience a mutual attraction between the positively polarized region of one molecule and the negatively polarized region of the other. As its name implies, the **dipole/induced-dipole force** combines features of both the induced-dipole/induced-dipole and dipole–dipole attractive forces. A polar region of one molecule alters the electron distribution in a nonpolar region of another in a direction that produces an attractive force between them.

Figure 4.3

A dipole–dipole attractive force. Two molecules of a polar substance associate so that the positively polarized region of one and the negatively polarized region of the other attract each other.

We can gain a sense of the relative importance of these intermolecular forces by considering three compounds similar in size and shape: the alkane propane, the alkyl halide fluoroethane, and the alcohol ethanol. Both of the polar compounds, ethanol and fluoroethane, have higher boiling points than the nonpolar one, propane. We attribute this to a combination of dipole/induced-dipole and dipole–dipole attractive forces that are present in the liquid states of ethanol and fluoroethane, but absent in propane.

	$CH_3CH_2CH_3$	CH_3CH_2F	CH_3CH_2OH
	Propane (μ = 0 D)	Fluoroethane (μ = 1.9 D)	Ethanol (μ = 1.7 D)
Boiling point:	−42°C	−32°C	78°C

The most striking difference, however, is that despite the similarity in their dipole moments, ethanol has a much higher boiling point than fluoroethane. This suggests that the attractive forces in ethanol are unusually strong. They are an example of a special type of dipole–dipole attraction called **hydrogen bonding** and involve, in this case, the positively polarized proton of the —OH group of one ethanol molecule with the negatively polarized oxygen of another. The oxygen of the —OH group of alcohols serves as a hydrogen bond *acceptor,* while the hydrogen attached to the oxygen serves as a hydrogen bond *donor.* Having both hydrogen bond acceptor and donor capability in the same molecule creates a strong network among ethanol molecules in the liquid phase.

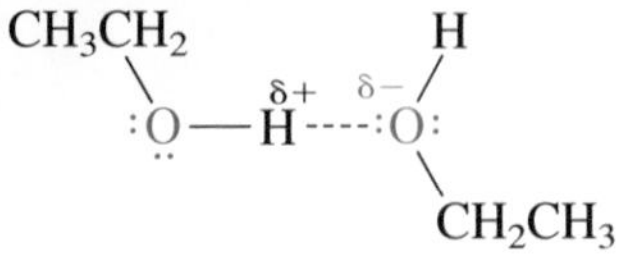

CH_3CH_2
H
δ+ δ−
:O—H----:O:
CH_2CH_3

Figure 4.4 shows the association of two ethanol molecules to form a hydrogen-bonded complex. The proton in the hydrogen bond (O—H---O) is not shared equally between the two oxygens, but is closer to and more strongly bonded to one oxygen than the other. Typical hydrogen bond strengths are on the order of 20 kJ/mol (about 5 kcal/mol), making them some 15–20 times weaker than most covalent bonds. Extended networks of hydrogen bonds are broken when individual ethanol molecules escape from the liquid to the vapor phase, but the covalent bonds remain intact.

Hydrogen bonds between —OH groups are stronger than those between —NH groups, as a comparison of the boiling points of water (H_2O, 100°C) and ammonia (NH_3, −33°C) demonstrates.

Among organic compounds, hydrogen bonding involves only OH or NH protons, as in:

O—H---O   O—H---N   N—H---O   N—H---N

The hydrogen must be bonded to a strongly electronegative element in order for the bond to be polar enough to support hydrogen bonding. Therefore, C—H groups do not participate in hydrogen bonds.

Problem 4.7

The constitutional isomer of ethanol, dimethyl ether (CH_3OCH_3), is a gas at room temperature. Suggest an explanation for this observation.

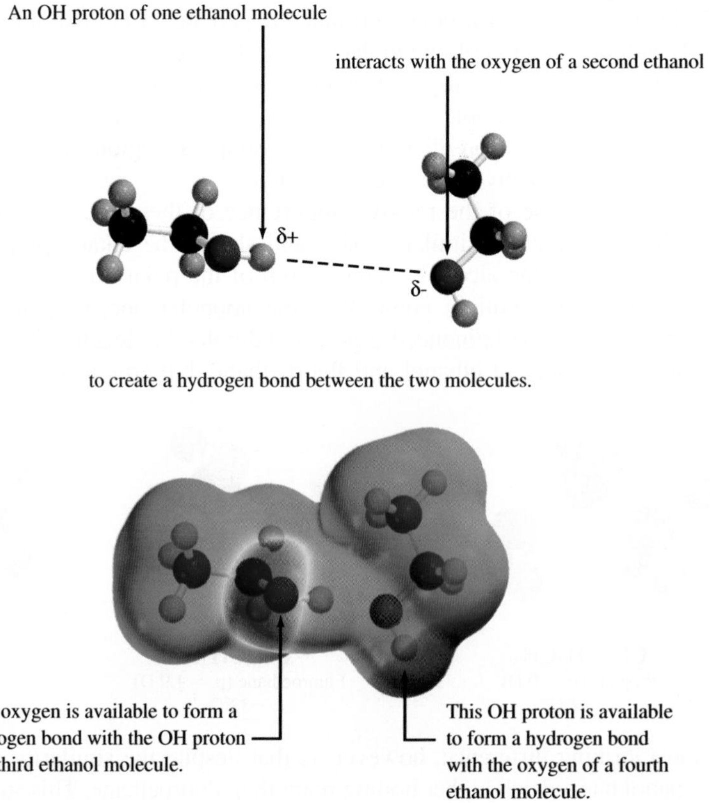

An OH proton of one ethanol molecule
interacts with the oxygen of a second ethanol
δ+
δ-
to create a hydrogen bond between the two molecules.
This oxygen is available to form a hydrogen bond with the OH proton of a third ethanol molecule.
This OH proton is available to form a hydrogen bond with the oxygen of a fourth ethanol molecule.

Figure 4.4

Hydrogen bonding in ethanol involves the oxygen of one molecule and the proton of the —OH group of another. A network of hydrogen-bonded complexes composed of many molecules characterizes the liquid phase of ethanol.

TABLE 4.2 Boiling Point of Some Alkyl Halides and Alcohols

Name of alkyl group	Formula	Substituent X and boiling point, °C (1 atm)				
		X = F	X = Cl	X = Br	X = I	X = OH
Methyl	CH_3X	−78	−24	3	42	65
Ethyl	CH_3CH_2X	−32	12	38	72	78
Propyl	$CH_3CH_2CH_2X$	−3	47	71	103	97
Pentyl	$CH_3(CH_2)_3CH_2X$	65	108	129	157	138
Hexyl	$CH_3(CH_2)_4CH_2X$	92	134	155	180	157

More than other dipole–dipole attractions, intermolecular hydrogen bonds are strong enough to impose a relatively high degree of structural order on systems in which they occur. We'll see, in Chapters 25 and 26, that the three-dimensional structures adopted by proteins and nucleic acids, the organic chemicals of life, are strongly influenced by hydrogen bonds.

Table 4.2 lists the boiling points of some representative alkyl halides and alcohols. When comparing the boiling points of related compounds as a function of the *alkyl group,* we find that the boiling point increases with the number of carbon atoms, as it does with alkanes.

The importance of hydrogen bonding in alcohols is evident in the last column of the table where it can be seen that the boiling points of alcohols are consistently higher than the corresponding alkyl fluoride, chloride, or bromide.

Among alkyl halides, the boiling point increases with increasing size of the halogen; alkyl fluorides have the lowest boiling points, alkyl iodides the highest. Dispersion forces are mainly responsible. Induced-dipole/induced-dipole attractions are favored when the electron cloud around an atom is easily distorted. This property of an atom is its **polarizability** and is more pronounced when the electrons are farther from the nucleus (iodine) than when they are closer (fluorine). Thus, induced-dipole/induced-dipole attractions are strongest in alkyl iodides, weakest in alkyl fluorides, and the boiling points of alkyl halides reflect this.

The boiling points of the chlorinated derivatives of methane increase with the number of chlorine atoms because the induced-dipole/induced-dipole attractive forces increase with each replacement of hydrogen by chlorine.

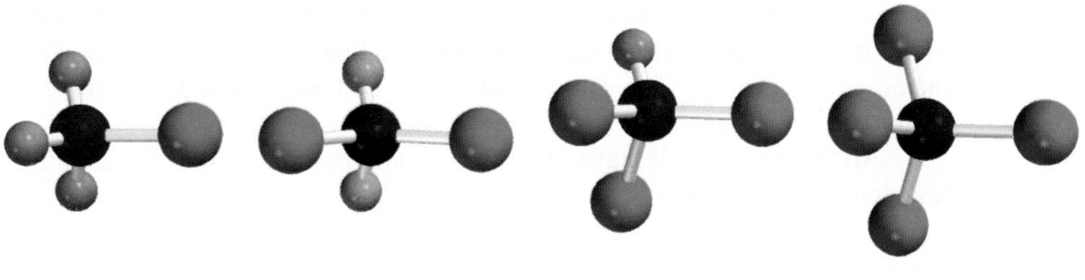

	CH_3Cl	CH_2Cl_2	$CHCl_3$	CCl_4
	Chloromethane	Dichloromethane	Trichloromethane	Tetrachloromethane
Boiling point:	−24°C	40°C	61°C	77°C

Fluorine is unique among the halogens in that increasing the number of fluorines does not lead to higher and higher boiling points.

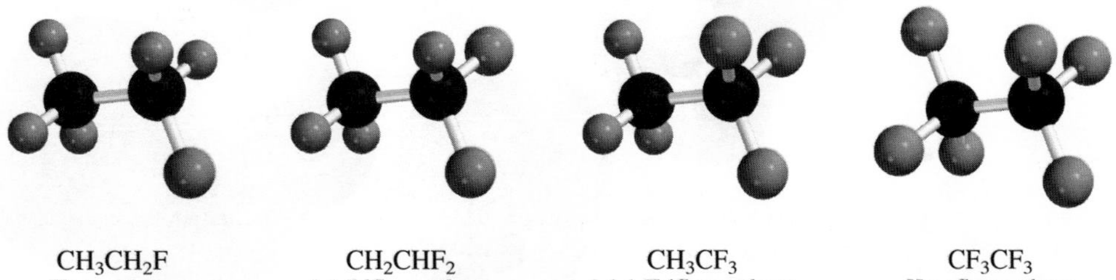

	CH_3CH_2F	CH_3CHF_2	CH_3CF_3	CF_3CF_3
	Fluoroethane	1,1-Difluoroethane	1,1,1-Trifluoroethane	Hexafluoroethane
Boiling point:	−32°C	−25°C	−47°C	−78°C

These boiling points illustrate why we should do away with the notion that boiling points always increase with increasing molecular weight.

Thus, although the difluoride CH_3CHF_2 boils at a higher temperature than CH_3CH_2F, the trifluoride CH_3CF_3 boils at a lower temperature than either of them. Even more striking is the observation that the hexafluoride CF_3CF_3 is the lowest boiling of any of the fluorinated derivatives of ethane. The boiling point of CF_3CF_3 is, in fact, only 11°C higher than that of ethane itself. The reason for this behavior has to do with the very low polarizability of fluorine and a decrease in induced-dipole/induced-dipole forces that accompanies the incorporation of fluorine substituents into a molecule. Their weak intermolecular attractive forces give fluorinated hydrocarbons certain desirable physical properties such as that found in the "no stick" *Teflon* coating of frying pans. Teflon is a *polymer* (Section 6.21 and Chapter 27) made up of long chains of —CF_2CF_2—units.

Solubility in Water. Alkyl halides and alcohols differ markedly from one another in their solubility in water. All alkyl halides are insoluble in water, but low-molecular-weight alcohols (methyl, ethyl, *n*-propyl, and isopropyl) are soluble in water in all proportions. Their ability to participate in intermolecular hydrogen bonding not only affects the boiling points of alcohols, but also enhances their water solubility. Hydrogen-bonded networks of the type shown in Figure 4.5, in which alcohol and water molecules associate with one another, replace the alcohol–alcohol and water–water hydrogen-bonded networks present in the pure substances.

Higher alcohols become more "hydrocarbon-like" and less water-soluble. 1-Octanol, for example, dissolves to the extent of only 1 mL in 2000 mL of water. As the alkyl chain gets longer, the hydrophobic effect (Section 2.17) becomes more important, to the point that it, more than hydrogen bonding, governs the solubility of alcohols.

Density. Alkyl fluorides and chlorides are less dense, and alkyl bromides and iodides more dense, than water.

	$CH_3(CH_2)_6CH_2F$	$CH_3(CH_2)_6CH_2Cl$	$CH_3(CH_2)_6CH_2Br$	$CH_3(CH_2)_6CH_2I$
Density (20°C):	0.80 g/mL	0.89 g/mL	1.12 g/mL	1.34 g/mL

Because alkyl halides are insoluble in water, a mixture of an alkyl halide and water separates into two layers. When the alkyl halide is a fluoride or chloride, it is the upper layer and water is the lower. The situation is reversed when the alkyl halide is a bromide or an iodide. In these cases the alkyl halide is the lower layer. Polyhalogenation increases the density. The compounds CH_2Cl_2, $CHCl_3$, and CCl_4, for example, are all more dense than water.

All liquid alcohols have densities of approximately 0.8 g/mL and are, therefore, less dense than water.

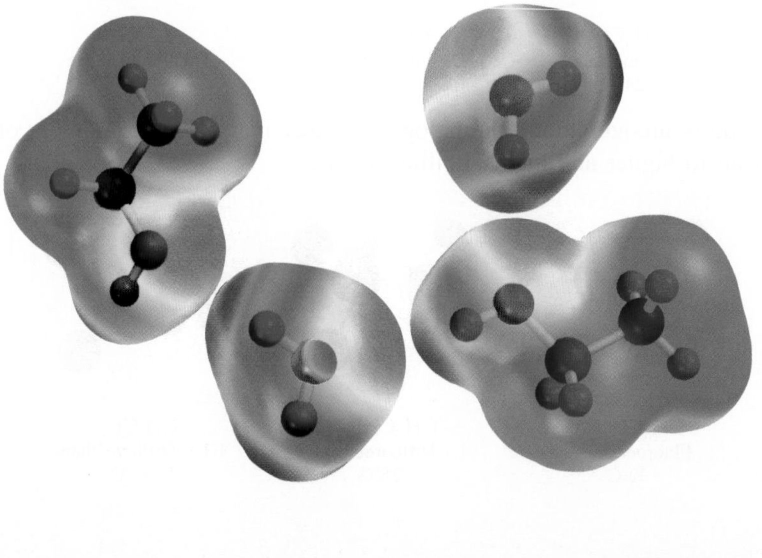

Figure 4.5

Hydrogen bonding between molecules of ethanol and water.

4.7 Preparation of Alkyl Halides from Alcohols and Hydrogen Halides

Much of what organic chemists do is directed toward practical goals. Chemists in the pharmaceutical industry synthesize new compounds as potential drugs for the treatment of disease. Agricultural chemicals designed to increase crop yields include organic compounds used for weed control, insecticides, and fungicides. Among the "building block" molecules used as starting materials to prepare new substances, alcohols and alkyl halides are especially valuable.

The reactions to be described in the remainder of this chapter use either an alkane or an alcohol as the starting material for preparing an alkyl halide. By knowing how to prepare alkyl halides, we can better appreciate the material in later chapters, where alkyl halides figure prominently in key functional group transformations. Just as important, the preparation of alkyl halides will serve as our focal point as we examine the principles of reaction mechanisms. We'll begin with the preparation of alkyl halides from alcohols by reaction with hydrogen halides.

$$\underset{\text{Alcohol}}{R{-}OH} + \underset{\text{Hydrogen halide}}{H{-}X} \longrightarrow \underset{\text{Alkyl halide}}{R{-}X} + \underset{\text{Water}}{H{-}OH}$$

The order of reactivity of the hydrogen halides parallels their acidity: HI > HBr > HCl >> HF. Hydrogen iodide is used infrequently, however, and the reaction of alcohols with hydrogen fluoride is not a useful method for the preparation of alkyl fluorides.

Among the various classes of alcohols, tertiary alcohols are observed to be the most reactive and primary alcohols the least reactive.

Increasing reactivity of alcohols
toward hydrogen halides

RCH_2OH	<	R_2CHOH	<	R_3COH
Primary Least reactive		Secondary		Tertiary Most reactive

Tertiary alcohols are converted to alkyl chlorides in high yield within minutes on reaction with hydrogen chloride at room temperature and below.

$$\underset{\substack{\text{2-Methyl-2-propanol}\\(\textit{tert}\text{-butyl alcohol})}}{(CH_3)_3COH} + \underset{\text{Hydrogen chloride}}{HCl} \xrightarrow{25°C} \underset{\substack{\text{2-Chloro-2-methylpropane}\\(\textit{tert}\text{-butyl chloride) (78–88\%)}}}{(CH_3)_3CCl} + \underset{\text{Water}}{H_2O}$$

The efficiency of a synthetic transformation is normally expressed as a *percent yield*, or percentage of the theoretical yield. *Theoretical yield* is the amount of product that could be formed if the reaction proceeded to completion and did not lead to any products other than those given in the equation.

Secondary and primary alcohols do not react with HCl at rates fast enough to make the preparation of the corresponding alkyl chlorides a method of practical value. Therefore, the more reactive hydrogen halide HBr is used; even then, elevated temperatures are required to increase the rate of reaction.

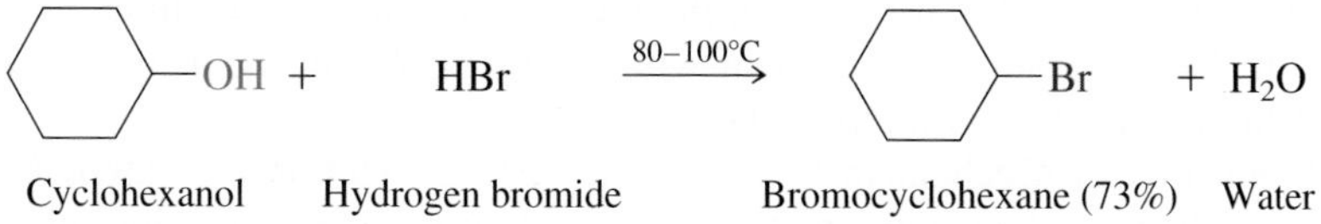

Cyclohexanol Hydrogen bromide Bromocyclohexane (73%) Water

$$\underset{\text{1-Heptanol}}{CH_3(CH_2)_5CH_2OH} + \underset{\substack{\text{Hydrogen}\\\text{bromide}}}{HBr} \xrightarrow{120°C} \underset{\substack{\text{1-Bromoheptane}\\(87–90\%)}}{CH_3(CH_2)_5CH_2Br} + \underset{\text{Water}}{H_2O}$$

The same kind of transformation may be carried out by heating an alcohol with sodium bromide and sulfuric acid.

$$\underset{\substack{\text{1-Butanol}\\(n\text{-butyl alcohol})}}{CH_3CH_2CH_2CH_2OH} \xrightarrow[\text{heat}]{NaBr,\ H_2SO_4} \underset{\substack{\text{1-Bromobutane (70–83\%)}\\(n\text{-butyl bromide})}}{CH_3CH_2CH_2CH_2Br}$$

We'll often write chemical equations in the abbreviated form just shown, in which reagents, especially inorganic ones, are not included in the body of the equation but instead are indicated over the arrow. Inorganic products—in this case, water—are usually omitted.

Problem 4.8

Write chemical equations for the reaction that takes place between each of the following pairs of reactants:

(a) 2-Butanol and hydrogen bromide
(b) 3-Ethyl-3-pentanol and hydrogen chloride
(c) 1-Tetradecanol and hydrogen bromide

Sample Solution (a) An alcohol and a hydrogen halide react to form an alkyl halide and water. In this case 2-bromobutane was isolated in 73% yield.

CH3CHCH2CH3 +
OH
HBr
CH3CHCH2CH3 + H2O
Br

2-Butanol Hydrogen bromide 2-Bromobutane Water

4.8 Mechanism of the Reaction of Alcohols with Hydrogen Halides: Hammond's Postulate

The reaction of an alcohol with a hydrogen halide is a **substitution.** A halogen, usually chlorine or bromine, replaces a hydroxyl group as a substituent on carbon. Calling the reaction a substitution tells us the relationship between the organic reactant and product but does not reveal the mechanism. The **mechanism** is the step-by-step pathway of bond cleavage and bond formation that leads from reactants to products. In developing a mechanistic picture for a particular reaction, we combine some basic principles of chemical reactivity with experimental observations to deduce the most likely sequence of steps.

Consider the reaction of *tert*-butyl alcohol with hydrogen chloride:

$$(CH_3)_3COH + HCl \longrightarrow (CH_3)_3CCl + H_2O$$

tert-Butyl alcohol, Hydrogen chloride, *tert*-Butyl chloride, Water

The generally accepted mechanism for this reaction is presented as a series of three equations in Mechanism 4.1. We say "generally accepted" because a reaction mechanism can never be proven correct. A mechanism is our best present assessment of how a reaction proceeds and must account for all experimental observations. If new experimental data appear that conflict with the mechanism, the mechanism must be modified to accommodate them. If the new data are consistent with the proposed mechanism, our confidence grows that the mechanism is likely to be correct.

Each equation in Mechanism 4.1 represents a single **elementary step.** An elementary step is one that involves only one transition state. A particular reaction might proceed by way of a single elementary step, in which it is described as a **concerted reaction,** or by a series of elementary steps as in Mechanism 4.1. To be valid a proposed mechanism must meet a number of criteria, one of which is that the sum of the equations for the elementary steps must correspond to the equation for the overall reaction. Before we examine each step in detail, you should verify that the process in Mechanism 4.1 satisfies this requirement.

Step 1: Proton Transfer

We saw in Chapter 1, especially in Table 1.8, that alcohols resemble water in respect to their Brønsted acidity (ability to donate a proton *from oxygen*). They also resemble

Mechanism 4.1

Formation of *tert*-Butyl Chloride from *tert*-Butyl Alcohol and Hydrogen Chloride

THE OVERALL REACTION:

$$(CH_3)_3COH \quad + \quad HCl \quad \longrightarrow \quad (CH_3)_3CCl \quad + \quad HOH$$

tert-Butyl alcohol — Hydrogen chloride — *tert*-Butyl chloride — Water

Step 1: Protonation of *tert*-butyl alcohol to give an alkyloxonium ion:

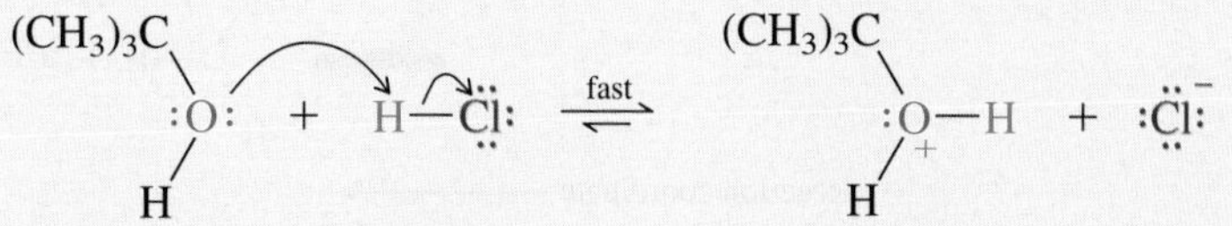

tert-Butyl alcohol — Hydrogen chloride — *tert*-Butyloxonium ion — Chloride ion

Step 2: Dissociation of *tert*-butyloxonium ion to give a carbocation:

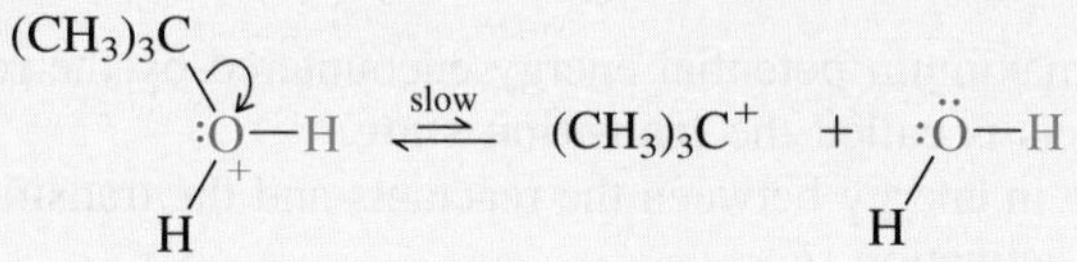

tert-Butyloxonium ion — *tert*-Butyl cation — Water

Step 3: Capture of *tert*-butyl cation by chloride ion:

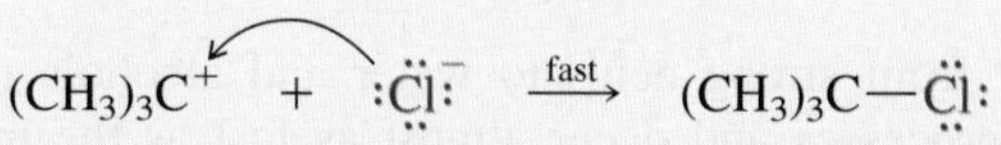

tert-Butyl cation — Chloride ion — *tert*-Butyl chloride

water in their Brønsted basicity (ability to accept a proton *on oxygen*). Just as proton transfer to a water molecule gives oxonium ion (hydronium ion, H_3O^+), proton transfer to an alcohol gives an **alkyloxonium ion** ($ROH_2{}^+$).

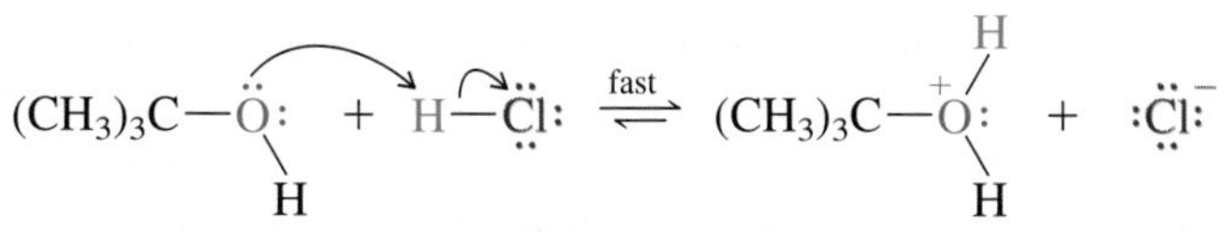

tert-Butyl alcohol — Hydrogen chloride — *tert*-Butyloxonium ion — Chloride ion

Recall from Section 1.12 that curved arrows are used to indicate the *movement of electrons* in chemical reactions.

Furthermore, a strong acid such as HCl that ionizes completely when dissolved in water, also ionizes completely when dissolved in an alcohol. Many important reactions of alcohols involve strong acids either as reactants or as catalysts. In all these reactions the first step is formation of an alkyloxonium ion by proton transfer from the acid to the alcohol.

The **molecularity** of an elementary step is given by the number of species that undergo a chemical change in that step. Transfer of a proton from hydrogen chloride to *tert*-butyl alcohol is **bimolecular** because two molecules [HCl and $(CH_3)_3COH$] undergo chemical change.

The *tert*-butyloxonium ion formed in step 1 is an **intermediate.** It was not one of the initial reactants, nor is it formed as one of the final products. Rather it is formed in one elementary step, consumed in another, and lies on the pathway from reactants to products.

Potential energy diagrams are especially useful when applied to reaction mechanisms. A **potential energy diagram** for proton transfer from hydrogen chloride to *tert*-butyl alcohol is shown in Figure 4.6. The potential energy of the system is plotted against the "reaction

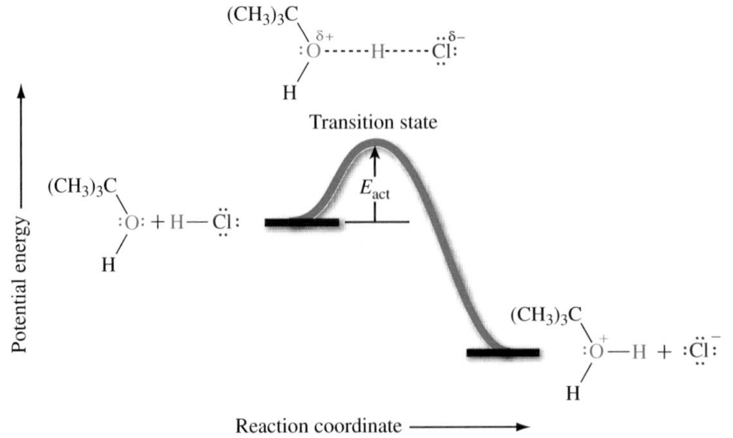

Figure 4.6

Potential energy diagram for proton transfer from hydrogen chloride to *tert*-butyl alcohol (step 1 of Mechanism 4.1).

coordinate," which is a measure of the degree to which the reacting molecules have progressed on their way to products. These aspects of the diagram are worth noting:

The 1967 Nobel Prize in Chemistry was shared by Manfred Eigen, a German chemist who developed novel methods for measuring the rates of very fast reactions such as proton transfers.

- The point of maximum potential energy encountered by the reactants as they proceed to products is called the **transition state.**
- The difference in energy between the reactants and the transition state is known as the energy of activation, $E_{act.}$
- Because this is an elementary step, it involves a single transition state.
- The step is known to be exothermic, so the products are placed lower in energy than the reactants. It is exothermic because HCl is a stronger acid than the alkyloxonium ion.
- Proton transfers from strong acids to water and alcohols rank among the most rapid chemical processes and occur almost as fast as the molecules collide with one another. Thus the height of the energy barrier, the E_{act} for proton transfer, must be quite low.

The concerted nature of proton transfer contributes to its rapid rate. The energy cost of breaking the H—Cl bond is partially offset by the energy released in forming the new bond between the transferred proton and the oxygen of the alcohol. Thus, the activation energy is far less than it would be for a hypothetical two-step process in which the H—Cl bond breaks first, followed by bond formation between H^+ and the alcohol.

The species present at the transition state is not a stable structure and cannot be isolated or examined directly. In general, the bonds in transition states are partially rather than fully formed. Its structure is assumed to be one in which the proton being transferred is partially bonded to both chlorine and oxygen simultaneously, although not necessarily to the same extent.

Dashed lines in transition-state structures represent *partial* bonds, that is, bonds in the process of being made or broken.

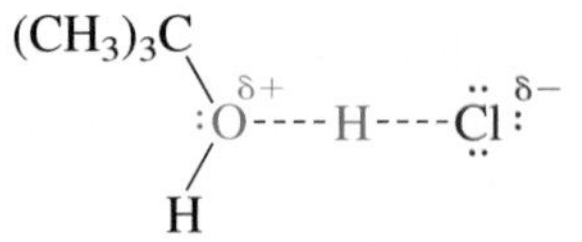

Inferring the structure of a transition state on the basis of the reactants and products of the elementary step in which it is involved is a time-honored practice in organic chemistry. Speaking specifically of transition states, George S. Hammond suggested that *if two states are similar in energy, they are similar in structure.* This rationale is known as **Hammond's postulate.** One of its corollaries is that the structure of a transition state more closely resembles the immediately preceding or following state to which it is closer in energy. In the case of the exothermic proton transfer in Figure 4.6, the transition state is closer in energy to the reactants and so resembles them more closely than it does the products of this step. We often call this an "early" transition state. The next step of this mechanism will provide us with an example of a "late" transition state.

Hammond made his proposal in 1955 while at Iowa State University. He later did pioneering work in organic photochemistry at CalTech.

Step 2: Carbocation Formation

In the second step of the process described in Mechanism 4.1, the alkyloxonium ion dissociates to a molecule of water and a **carbocation,** an ion that contains a positively charged carbon.

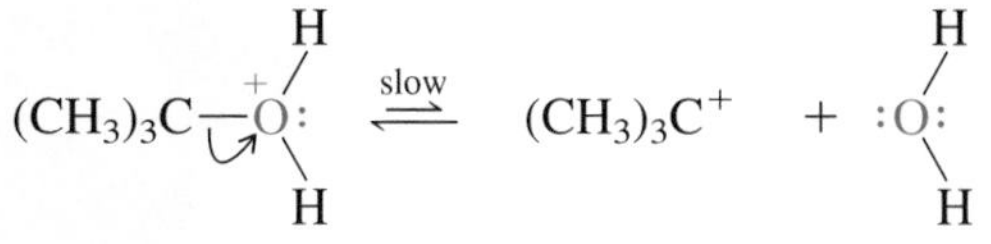

tert-Butyloxonium ion   *tert*-Butyl cation   Water

One way to name carbocations in the IUPAC system is to add the word "cation" to the name of the alkyl group.

Only one species, *tert*-butyloxonium ion, undergoes a chemical change in this step. Therefore, the step is **unimolecular.**

Like *tert*-butyloxonium ion, *tert*-butyl cation is an intermediate along the reaction pathway. It is, however, a relatively unstable species and its formation by dissociation of the alkyloxonium ion is endothermic. Step 2 is the slowest step in the mechanism and has the highest activation energy. Figure 4.7 shows a potential energy diagram for this step.

- Because this step is endothermic, the products of it are placed higher in energy than the reactants.
- The transition state is closer in energy to the carbocation (*tert*-butyl cation), so, according to Hammond's postulate, its structure more closely resembles the carbocation than it resembles *tert*-butyloxonium ion. The transition state has considerable "carbocation character," meaning that a significant degree of positive charge has developed at carbon.

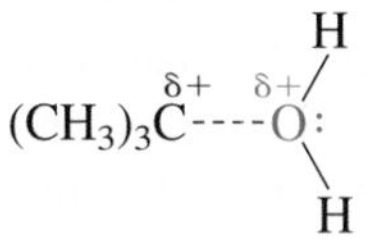

There is ample evidence from a variety of sources that carbocations are intermediates in some chemical reactions, but they are almost always too unstable to isolate. The simplest reason for the instability of carbocations is that the positively charged carbon has only six electrons in its valence shell—the octet rule is not satisfied for the positively charged carbon.

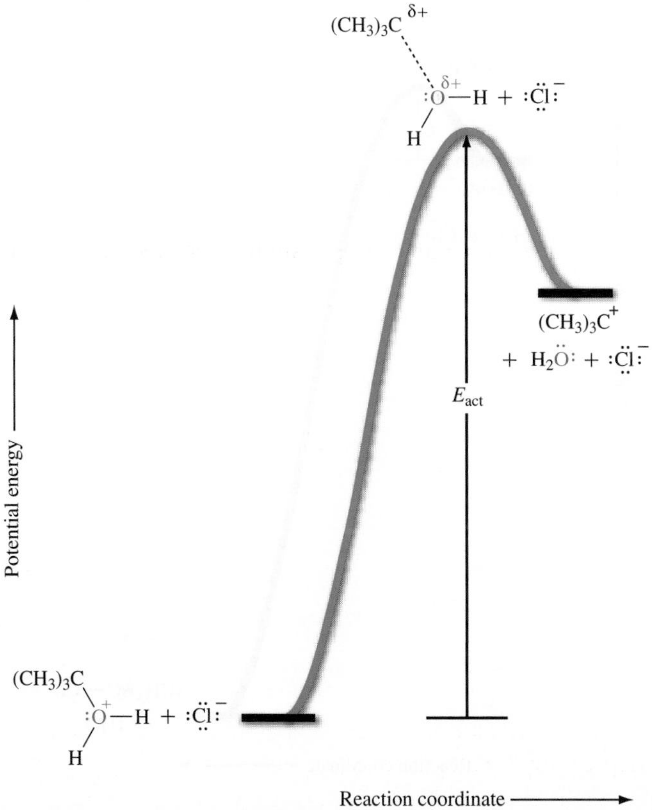

Figure 4.7

Potential energy diagram for dissociation of *tert*-butyloxonium ion to *tert*-butyl cation (step 2 of Mechanism 4.1).

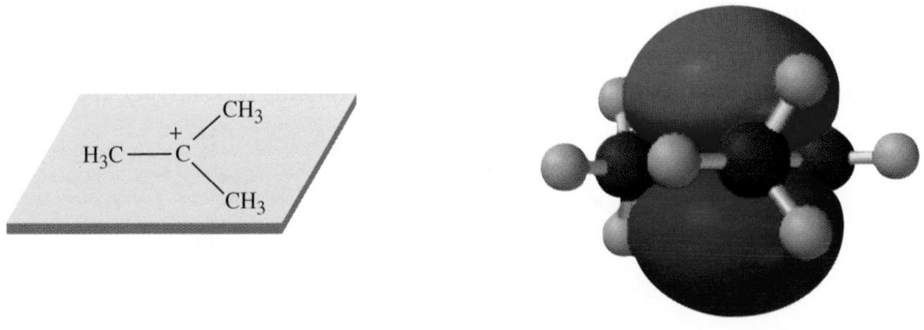

(*a*) (*b*)

Figure 4.8

tert-Butyl cation. (*a*) The positively charged carbon is sp^2-hybridized. Each methyl group is attached to the positively charged carbon by a σ bond, and these three bonds lie in the same plane. (*b*) The sp^2-hybridized carbon has an empty 2*p* orbital, the axis of which is perpendicular to the plane of the carbon atoms.

The properties of *tert*-butyl cation can be understood by focusing on its structure, which is shown in Figure 4.8. With only six valence electrons, which are distributed among three coplanar σ bonds, the positively charged carbon is sp^2-hybridized. The unhybridized 2*p* orbital that remains on the positively charged carbon contains no electrons; its axis is perpendicular to the plane of the bonds connecting that carbon to the three methyl groups.

The positive charge on carbon and the vacant *p* orbital combine to make carbocations strongly **electrophilic** ("electron-loving" or "electron-seeking"). Electrophiles are Lewis acids (Section 1.18). They are electron-pair acceptors and react with Lewis bases (electron-pair donors). Step 3, which follows and completes the mechanism, is a Lewis acid/Lewis base reaction. We'll return to carbocations and describe them in more detail in Section 4.10.

Step 3: Reaction of tert-Butyl Cation with Chloride Ion

The Lewis bases that react with electrophiles are called **nucleophiles** ("nucleus seekers"). They have an unshared electron pair that they can use in covalent bond formation. The nucleophile in step 3 of Mechanism 4.1 is chloride ion.

tert-Butyl cation (electrophile) | Chloride ion (nucleophile) | *tert*-Butyl chloride

Step 3 is bimolecular because two species, the carbocation and chloride ion, react together. Figure 4.9 shows a potential energy diagram for this step.

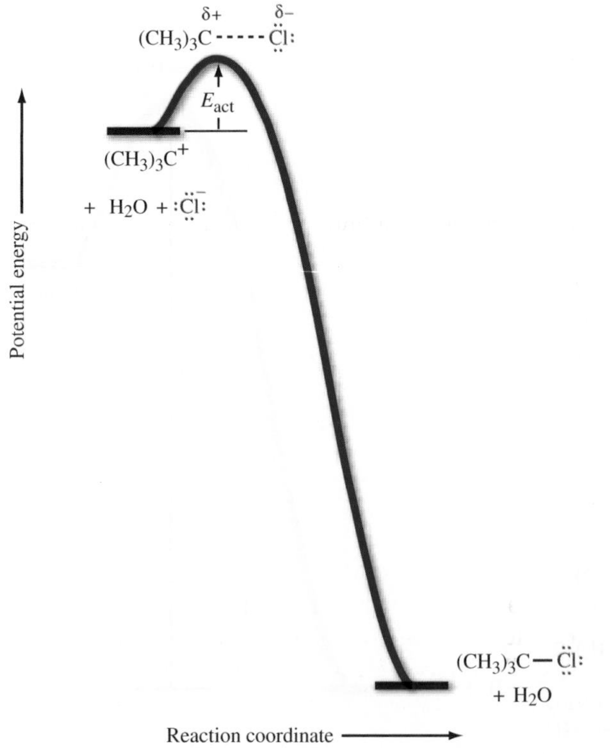

Figure 4.9

Potential energy diagram for reaction of *tert*-butyl cation with chloride anion (step 3 of Mechanism 4.1).

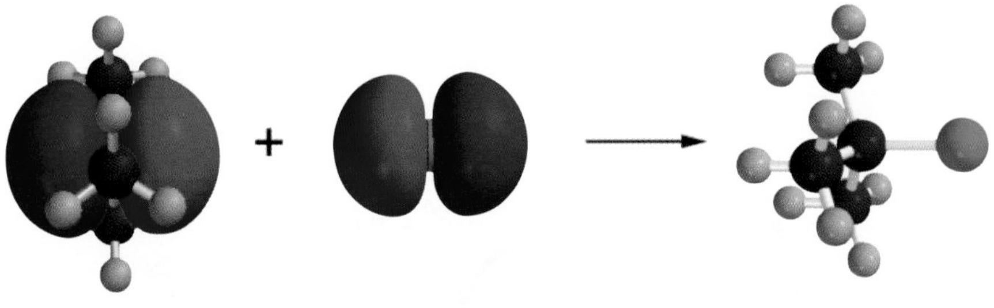

Figure 4.10

Combination of *tert*-butyl cation and chloride anion to give *tert*-butyl chloride. In-phase overlap between a vacant *p* orbital of $(CH_3)_3C^+$ and a filled *p* orbital of Cl^- gives a C—Cl σ bond.

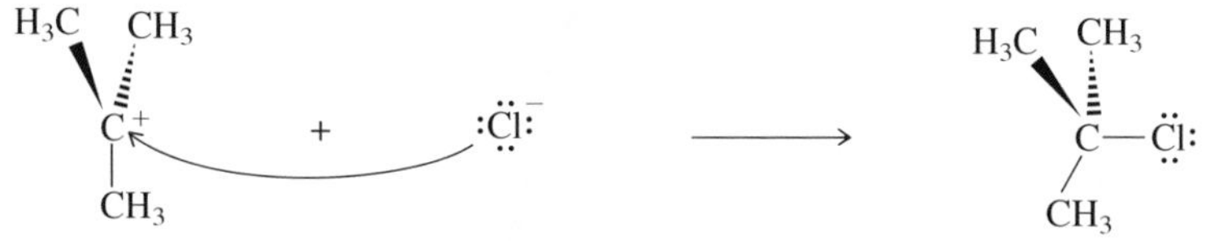

- The step is exothermic; it leads from the carbocation intermediate to the stable isolated products of the reaction.
- The activation energy for this step is small, and bond formation between a positive ion and a negative ion occurs rapidly.
- The transition state for this step involves partial bond formation between *tert*-butyl cation and chloride ion.

$$(CH_3)_3\overset{\delta+}{C}\text{----}\overset{\delta-}{\ddot{Cl}}:$$

As shown in Figure 4.10, the crucial electronic interaction is between an unshared electron pair of Cl^- and the vacant 2*p* orbital of the positively charged carbon of $(CH_3)_3C^+$.

4.9 Potential Energy Diagrams for Multistep Reactions: The S_N1 Mechanism

We've just seen how the reaction of *tert*-butyl alcohol with hydrogen chloride, written as a series of elementary steps in Mechanism 4.1, can be supplemented with potential energy diagrams (Figures 4.6, 4.7, and 4.9). We'll complete the energy picture by combining the three separate diagrams into one that covers the entire process. This composite diagram (Figure 4.11) has three peaks and two valleys. The peaks correspond to transition states, one for each of the three elementary steps. The valleys correspond to the reactive intermediates—*tert*-butyloxonium ion and *tert*-butyl cation—species formed in one step and consumed in another.

With the potential energies shown on a common scale, we see that the transition state for formation of $(CH_3)_3C^+$ is the highest energy point on the diagram. A reaction can proceed no faster than its slowest step, which is referred to as the **rate-determining step.** In the reaction of *tert*-butyl alcohol with hydrogen chloride, formation of the carbocation by dissociation of the alkyloxonium ion is rate-determining.

Substitution reactions, of which the reaction of alcohols with hydrogen halides is but one example, will be discussed in more detail in Chapter 8. There, we will make extensive use of a notation originally introduced by Sir Christopher Ingold. Ingold proposed the symbol, S_N, to stand for *substitution nucleophilic,* to be followed by the number *1* or *2* according to whether the rate-determining step is unimolecular or bimolecular. The reaction of *tert*-butyl alcohol with hydrogen chloride, for example, is said to follow an **S_N1 mechanism** because its slow step (dissociation of *tert*-butyloxonium ion) is unimolecular. Only the alkyloxonium ion undergoes a chemical change in this step.

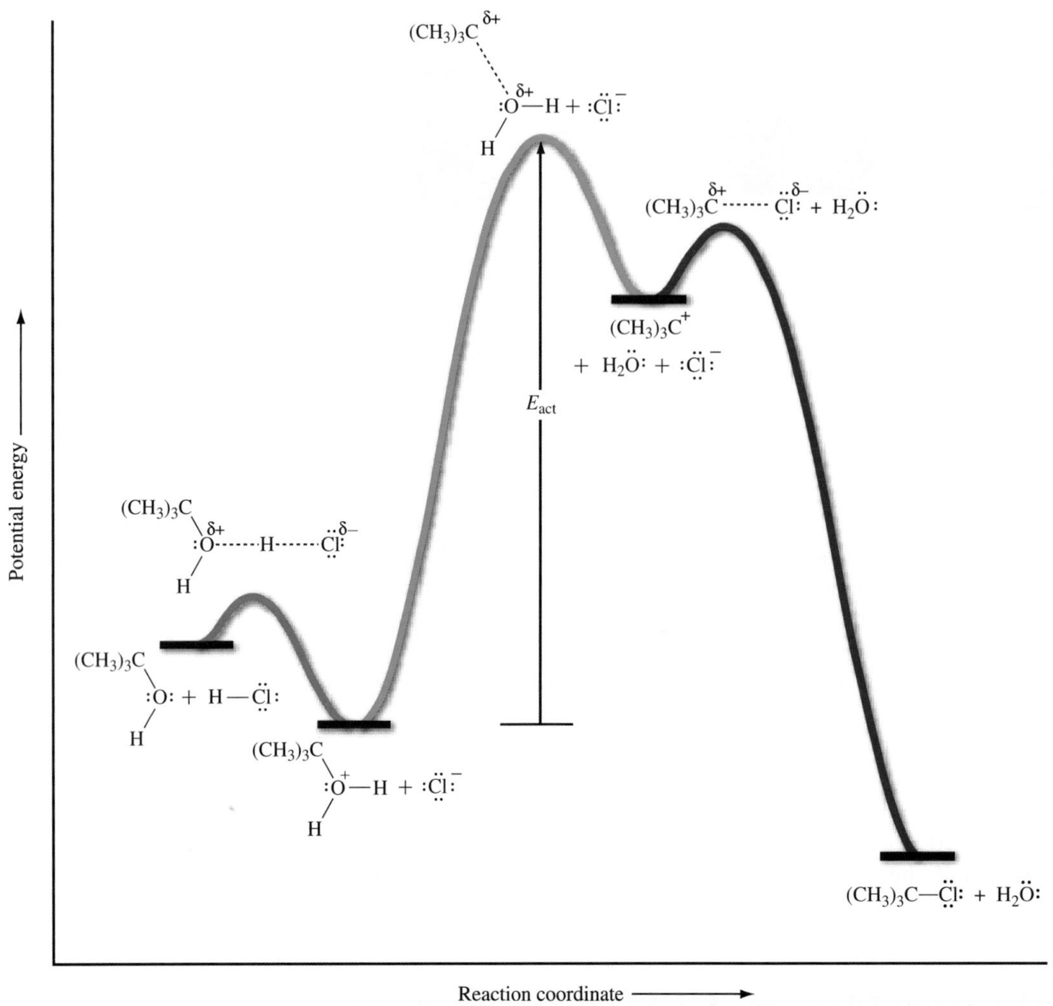
Potential energy
Reaction coordinate
E_{act}

Figure 4.11

Potential energy diagram for the reaction of *tert*-butyl alcohol and hydrogen chloride according to the S_N1 mechanism (Mechanism 4.1).

Problem 4.9

Assume the reaction

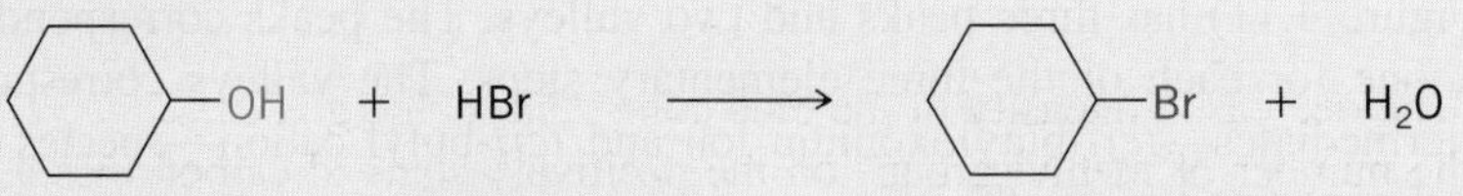
OH
HBr
Br
H_2O
Cyclohexanol
Hydrogen bromide
Bromocyclohexane
Water

follows an S_N1 mechanism, and write a chemical equation for the rate-determining step. Use curved arrows to show the flow of electrons.

4.10 Structure, Bonding, and Stability of Carbocations

As we have just seen, the rate-determining step in the reaction of *tert*-butyl alcohol with hydrogen chloride is formation of the carbocation $(CH_3)_3C^+$. Convincing evidence from a variety of sources tells us that carbocations can exist, but are relatively unstable. When carbocations are involved in chemical reactions, it is as reactive intermediates, formed slowly in one step and consumed rapidly in the next one.

Numerous other studies have shown that *alkyl groups directly attached to the positively charged carbon stabilize a carbocation.* Figure 4.12 illustrates this generalization

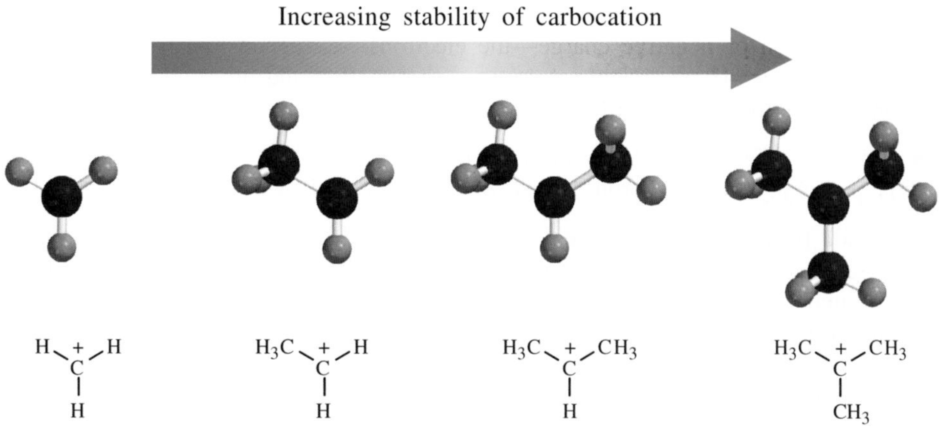

Methyl cation — Least stable | Ethyl cation (primary) | Isopropyl cation (secondary) | *tert*-Butyl cation (tertiary) — Most stable

Figure 4.12

The order of carbocation stability is methyl < primary < secondary < tertiary. Alkyl groups that are directly attached to the positively charged carbon stabilize carbocations.

for CH_3^+, $CH_3CH_2^+$, $(CH_3)_2CH^+$, and $(CH_3)_3C^+$. Among this group, CH_3^+ is the least stable and $(CH_3)_3C^+$ the most stable.

Carbocations are classified according to the degree of substitution at the positively charged carbon. The positive charge is on a primary carbon in $CH_3CH_2^+$, a secondary carbon in $(CH_3)_2CH^+$, and a tertiary carbon in $(CH_3)_3C^+$. Ethyl cation is a primary carbocation, isopropyl cation a secondary carbocation, and *tert*-butyl cation a tertiary carbocation.

As carbocations go, CH_3^+ is particularly unstable, and its existence as an intermediate in chemical reactions has never been demonstrated. Primary carbocations, although more stable than CH_3^+, are still too unstable to be involved as intermediates in chemical reactions. The threshold of stability is reached with secondary carbocations. Many reactions, including the reaction of secondary alcohols with hydrogen halides, are believed to involve secondary carbocations. The evidence in support of tertiary carbocation intermediates is stronger yet.

Problem 4.10

Of the isomeric $C_5H_{11}^+$ carbocations, which one is the most stable?

Alkyl groups stabilize carbocations by releasing electron density to the positively charged carbon, thereby dispersing the positive charge. Figure 4.13 illustrates this charge dispersal by comparing the electrostatic potential maps of CH_3^+, $CH_3CH_2^+$, $(CH_3)_2CH^+$, and $(CH_3)_3C^+$. The decreased intensity of the blue color reflects the greater dispersal of positive charge as the number of methyl groups on the positively charged carbon increases.

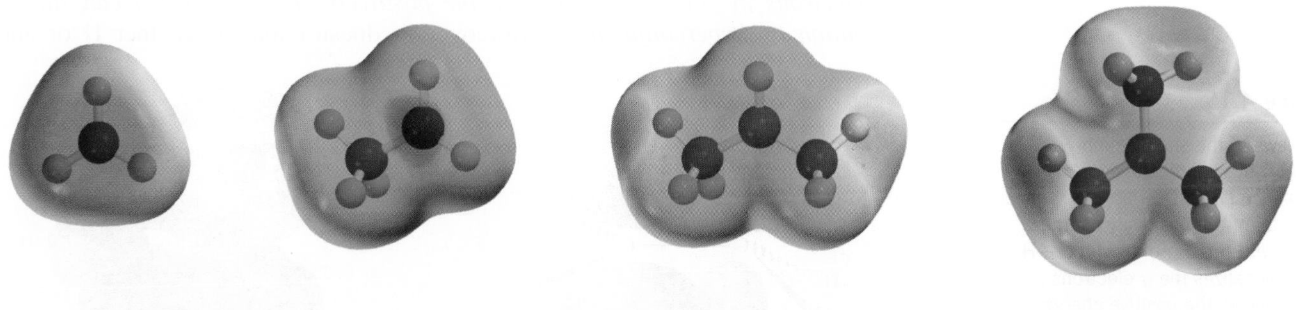

Methyl cation (CH_3^+) | Ethyl cation ($CH_3CH_2^+$) | Isopropyl cation [$(CH_3)_2CH^+$] | *tert*-Butyl cation [$(CH_3)_3C^+$]

Figure 4.13

Electrostatic potential maps of carbocations. The positive charge (blue) is most concentrated in CH_3^+ and most spread out in $(CH_3)_3C^+$. (The electrostatic potentials were mapped on the same scale to allow direct comparison.)

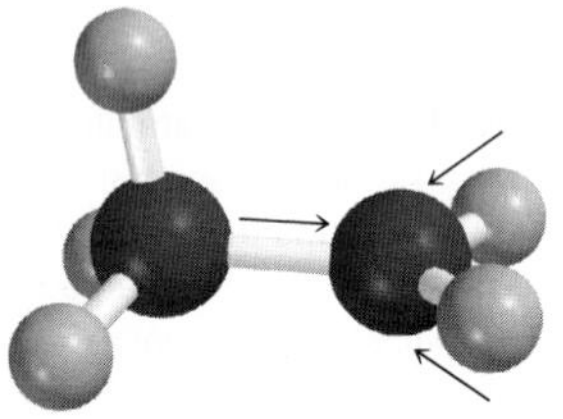

Figure 4.14

The charge in ethyl cation is stabilized by polarization of the electron distribution in the σ bonds to the positively charged carbon atom. Alkyl groups release electrons better than hydrogen.

Dispersal of positive charge goes hand in hand with delocalization of electrons. The redistribution of negative charge—the electrons—is responsible for spreading out the positive charge. There are two main ways that methyl and other alkyl groups act as electron sources to stabilize carbocations:

- Inductive effect (by polarization of σ bonds)
- Hyperconjugation (electron delocalization via orbital overlap)

Recall from Section 1.16 that an inductive effect is an electron-donating-or-withdrawing effect of a substituent that is transmitted by the polarization of σ bonds. As illustrated for $CH_3CH_2^+$ in Figure 4.14, the positively charged carbon draws the electrons in its σ bonds toward itself and away from the atoms attached to it. Electrons in a C—C bond are more polarizable than those in a C—H bond, so replacing hydrogens by alkyl groups reduces the net charge on the positively charged carbon. Alkyl groups are electron-releasing substituents with respect to their inductive effect. The more alkyl groups that are directly attached to the positively charged carbon, the more stable the carbocation.

Problem 4.11

Which would you expect to be more stable: $(CH_3)_3C^+$ or $(CF_3)_3C^+$? Why?

Hyperconjugation refers to the delocalization of electrons in a σ bond through a system of overlapping orbitals. Again consider $CH_3CH_2^+$, this time directing your attention to the electrons in the C—H bonds of the CH_3 group. Figure 4.15*a* illustrates how an orbital associated with the CH_3 group of $CH_3CH_2^+$ can overlap with the vacant *p* orbital of the positively charged carbon to give an extended orbital that encompasses them both. This allows the electrons in the C—H bonds of the CH_3 group to be shared by both carbons and disperses the positive charge.

The valence-bond approach to hyperconjugation expressed in Figure 4.15*a* finds a parallel in the molecular orbital picture of $CH_3CH_2^+$. One of the filled bonding MOs of $CH_3CH_2^+$ (Figure 4.15*b*) is essentially a combination of the *p* orbital of the positively charged carbon and orbitals associated with the CH_3 group. The pair of electrons in this orbital are shared by the CH_3 group and the positively charged carbon.

When applying hyperconjugation to carbocations more complicated than $CH_3CH_2^+$, it is helpful to keep track of the various bonds. Begin with the positively charged carbon and label the three bonds originating from it with the Greek letter α. Proceed down the chain, labeling the bonds extending from the next carbon β, those from the next carbon γ, and so on.

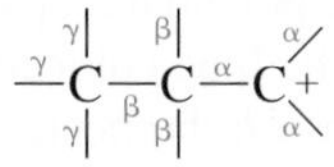

Only electrons in bonds that are β to the positively charged carbon can stabilize a carbocation by hyperconjugation. Moreover, it doesn't matter whether H or another

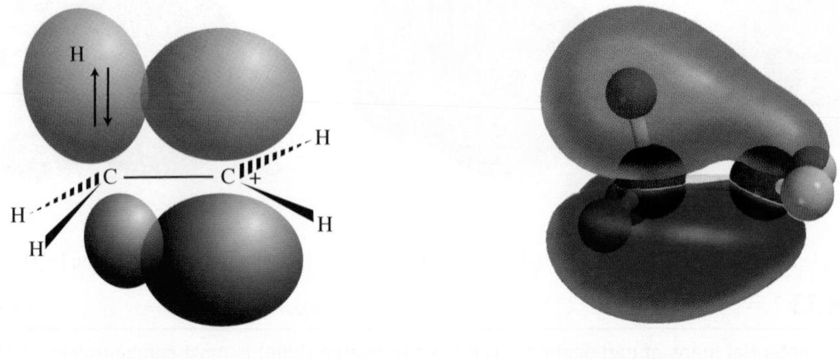

(*a*) Valence bond (*b*) Molecular orbital

Figure 4.15

Two views of the stabilization of $CH_3CH_2^+$ by hyperconjugation. (*a*) *Valence bond:* Overlap of the vacant 2*p* orbital of the positively charged carbon with the σ orbital of a C—H bond delocalizes the σ electrons and disperses the positive charge. (*b*) *Molecular orbital:* One of the molecular orbitals of $CH_3CH_2^+$ encompasses both the CH_3 group and the positively charged carbon; it is a bonding MO and contains two electrons.

carbon is at the far end of the β bond; stabilization by hyperconjugation will still operate. The key point is that electrons in bonds that are β to the positively charged carbon are more stabilizing than electrons in an α^{+}C—H bond. Thus, successive replacement of first one, then two, then three hydrogens of CH_3^+ by alkyl groups increases the opportunities for hyperconjugation, which is consistent with the observed order of carbocation stability: $CH_3^+ < CH_3CH_2^+ < (CH_3)_2CH^+ < (CH_3)_3C^+$.

Problem 4.12

For the general case of R = any alkyl group, how many bonded pairs of electrons are involved in stabilizing R_3C^+ by hyperconjugation? How many in R_2CH^+? In RCH_2^+?

To summarize, the most important factor to consider in assessing carbocation stability is the degree of substitution at the positively charged carbon.

CH_3^+	< RCH_2^+	< R_2CH^+	< R_3C^+
Methyl Least stable	Primary	Secondary	Tertiary Most stable

We will see numerous reactions that involve carbocation intermediates as we proceed through the text, so it is important to understand how their structure determines their stability.

4.11 Effect of Alcohol Structure on Reaction Rate

For a proposed reaction mechanism to be valid, the sum of its elementary steps must equal the equation for the overall reaction and the mechanism must be consistent with all experimental observations. The S_N1 process set forth in Mechanism 4.1 satisfies the first criterion. What about the second?

One important experimental fact is that the rate of reaction of alcohols with hydrogen halides increases in the order primary < secondary < tertiary. This reactivity order parallels the carbocation stability order and is readily accommodated by the mechanism we have outlined.

The rate-determining step in the S_N1 mechanism is dissociation of the alkyloxonium ion to the carbocation.

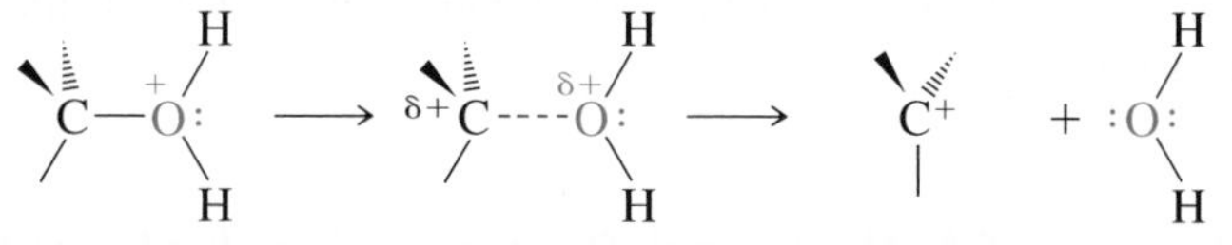

Alkyloxonium ion — Transition state — Carbocation — Water

The rate of this step is proportional to the concentration of the alkyloxonium ion:

Rate = k 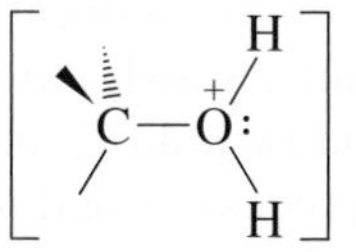

where k is a constant of proportionality called the *rate constant*. The value of k is related to the activation energy for alkyloxonium ion dissociation and is different for different alkyloxonium ions. A low activation energy implies a large value of k and a rapid rate of alkyloxonium ion dissociation. Conversely, a large activation energy is characterized by a small k for dissociation and a slow rate.

The rate of any chemical reaction increases with increasing temperature. Thus the value of k for a reaction is not constant, but increases as the temperature increases.

The transition state is closer in energy to the carbocation and, according to Hammond's postulate, more closely resembles it than the alkyloxonium ion. Thus, structural features that stabilize carbocations stabilize transition states leading to them. It follows,

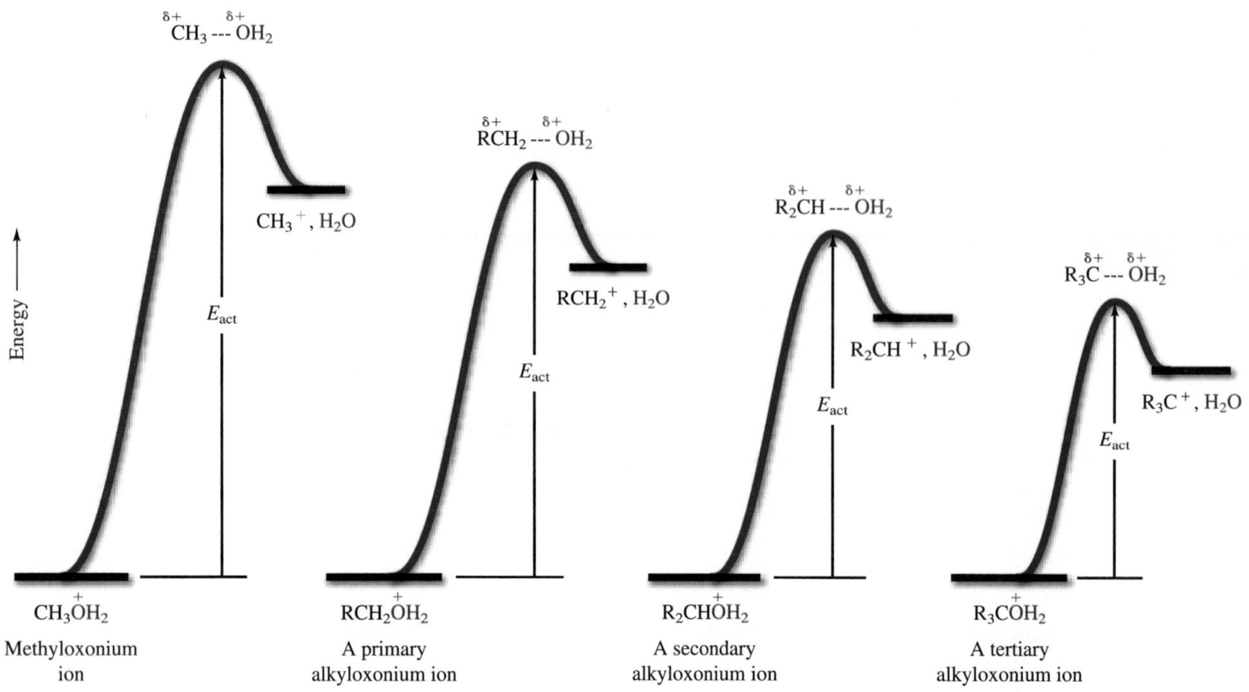

Figure 4.16

Energies of activation for formation of carbocations from alkyloxonium ions of methyl, primary, secondary, and tertiary alcohols.

therefore, that alkyloxonium ions derived from tertiary alcohols have a lower energy of activation for dissociation and are converted to their corresponding carbocations faster than those derived from secondary and primary alcohols. Simply put: *more stable carbocations are formed faster than less stable ones.* Figure 4.16 expresses this principle via a potential energy diagram.

The S_N1 mechanism is generally accepted to be correct for the reaction of tertiary and secondary alcohols with hydrogen halides. It is almost certainly *not* correct for methyl alcohol and primary alcohols because methyl and primary carbocations are believed to be much too unstable, and the activation energies for their formation much too high, for them to be reasonably involved. The next section describes how methyl and primary alcohols are converted to their corresponding halides by a mechanism related to, but different from, S_N1.

4.12 Reaction of Methyl and Primary Alcohols with Hydrogen Halides: The S_N2 Mechanism

Unlike tertiary and secondary carbocations, methyl and primary carbocations are too high in energy to be intermediates in chemical reactions. However, methyl and primary alcohols are converted, albeit rather slowly, to alkyl halides on treatment with hydrogen halides. Therefore, they must follow a different mechanism, one that avoids carbocation intermediates. This alternative process is outlined in Mechanism 4.2 for the reaction of 1-heptanol with hydrogen bromide.

The first step of this new mechanism is exactly the same as that seen earlier for the reaction of *tert*-butyl alcohol with hydrogen chloride—formation of an alkyloxonium ion by proton transfer from the hydrogen halide to the alcohol. Like the earlier example, this is a rapid, reversible Brønsted acid–base reaction.

The major difference between the two mechanisms is the second step. The second step in the reaction of *tert*-butyl alcohol with hydrogen chloride is the unimolecular dissociation of *tert*-butyloxonium ion to *tert*-butyl cation and water. Heptyloxonium ion, however, instead of dissociating to an unstable primary carbocation, reacts differently.

Mechanism 4.2

Formation of 1-Bromoheptane from 1-Heptanol and Hydrogen Bromide

THE OVERALL REACTION:

$$CH_3(CH_2)_5CH_2OH + HBr \longrightarrow CH_3(CH_2)_5CH_2Br + H_2O$$

1-Heptanol | Hydrogen bromide | 1-Bromoheptane | Water

Step 1: Protonation of 1-heptanol to give the corresponding alkyloxonium ion:

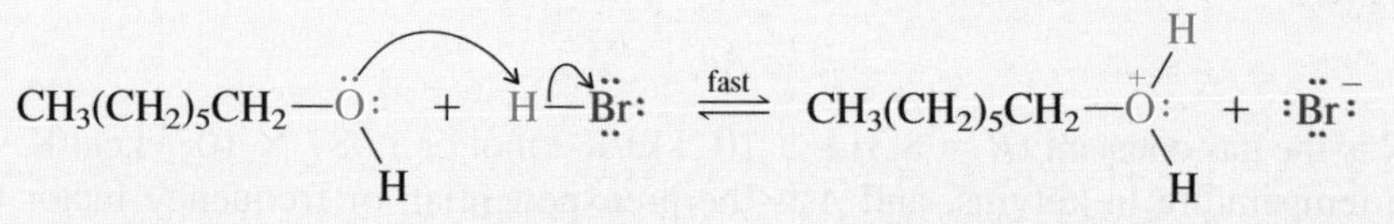

1-Heptanol | Hydrogen bromide | Heptyloxonium ion | Bromide ion

Step 2: Nucleophilic attack on the alkyloxonium ion by bromide ion:

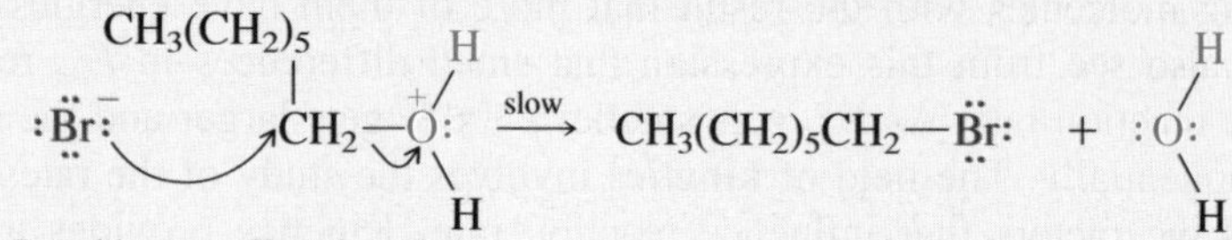

Bromide ion | Heptyloxonium ion | 1-Bromoheptane | Water

It is attacked by bromide ion, which acts as a nucleophile. We can represent step 2 and its transition state as:

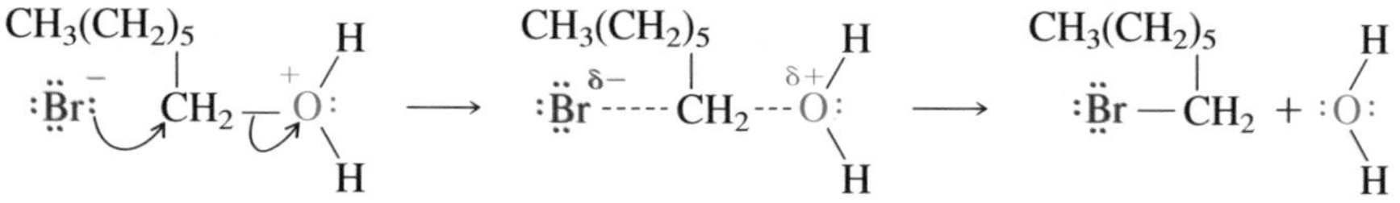

Transition state

Bromide ion forms a bond to the primary carbon by "pushing off" a water molecule. This step is bimolecular because it involves both bromide and heptyloxonium ion. Step 2 is slower than the proton transfer in step 1, so it is rate-determining. Using Ingold's terminology, we classify nucleophilic substitutions that have a bimolecular rate-determining step by the mechanistic symbol **S_N2.**

Problem 4.13

Sketch a potential energy diagram for the reaction of 1-heptanol with hydrogen bromide, paying careful attention to the positioning and structures of the intermediates and transition states.

Problem 4.14

1-Butanol and 2-butanol are converted to their corresponding bromides on being heated with hydrogen bromide. Write a suitable mechanism for each reaction, and assign each the appropriate symbol (S_N1 or S_N2).

It is important to note that although methyl and primary alcohols react with hydrogen halides by a mechanism that involves fewer steps than the corresponding reactions of secondary and tertiary alcohols, fewer steps do not translate to faster reaction rates. Remember, the observed order of reactivity of alcohols with hydrogen halides is tertiary > secondary > primary. Reaction rate is governed by the activation energy of the slowest step, regardless of how many steps there are.

4.13 More on Activation Energy

In Section 4.7, you saw that the reaction of 1-heptanol with hydrogen bromide requires elevated temperature to proceed at a synthetically acceptable rate.

$$CH_3(CH_2)_5CH_2OH + HBr \xrightarrow{120°C} CH_3(CH_2)_5CH_2Br + H_2O$$

(87–90%)

A quantitative relationship between the energy of activation (E_{act}), the rate constant (k) and temperature (T) is expressed by the Arrhenius equation:

$$k = Ae^{-E_{act}/RT}$$

where R is the gas constant ($R = 8.314 \times 10^{-3}$ kJ/K · mol or 1.987×10^{-3} kcal/K · mol), T is the temperature in kelvins, and A is the preexponential, or frequency factor that is related to the collision frequency and geometry. Because temperature appears in the denominator of the negative exponent, the rate constant k increases with increasing temperature. In physical terms, raising the temperature increases the average kinetic energy of the reacting molecules with the result that more of them have energies greater than E_{act}. We can also see from this expression that small differences in E_{act} result in large differences in reaction rate. As E_{act} gets smaller $e^{-E_{act}/RT}$ gets larger and the rate constant increases exponentially. The field of **kinetics** involves the study of the rates of chemical reactions and the factors that influence reaction rate. Kinetics provides a quantitative understanding for many of the structure–reactivity relationships that we encounter in organic chemistry.

E_{act} in the Arrhenius equation is only an approximate measure of the energy barrier for a chemical reaction as it takes into account only the enthalpy of activation. The true activation barrier, known as $\Delta G^{\ddagger}$, accounts for both the enthalpy and the entropy of activation.

Problem 4.15

A certain reaction has an E_{act} of 75.3 kJ/mol (18.0 kcal/mol) and a frequency factor of 5.00×10^7 L · mol^{-1} · sec^{-1}. Use the Arrhenius equation to calculate the rate constant for this reaction at (a) 40°C and (b) 50°C.

4.14 Other Methods for Converting Alcohols to Alkyl Halides

Alkyl halides are such useful starting materials for preparing other functional group types that chemists have developed several different methods for converting alcohols to alkyl halides. Two methods, based on the inorganic reagents *thionyl chloride* and *phosphorus tribromide,* bear special mention.

Thionyl chloride reacts with alcohols to give alkyl chlorides. The reaction is typically carried out in the presence of pyridine, which acts both as solvent and a base that reacts with the HCl that is produced.

$$ROH + SOCl_2 \longrightarrow RCl + SO_2 + HCl$$

Alcohol; Thionyl chloride; Alkyl chloride; Sulfur dioxide; Hydrogen chloride

Because tertiary alcohols are so readily converted to chlorides with hydrogen chloride, thionyl chloride is used mainly to prepare primary and secondary alkyl chlorides.

$$(CH_3CH_2)_2CHCH_2OH \xrightarrow[\text{pyridine}]{SOCl_2} (CH_3CH_2)_2CHCH_2Cl$$

2-Ethyl-1-butanol → 1-Chloro-2-ethylbutane (82%)

Mechanism 4.3 shows the conversion of a primary alcohol to the alkyl chloride using thionyl chloride. The reaction involves the formation of an alkyl chlorosulfite that is attacked by the nucleophilic chloride ion.

Mechanism 4.3

Conversion of an Alcohol to an Alkyl Chloride with Thionyl Chloride

THE OVERALL REACTION:

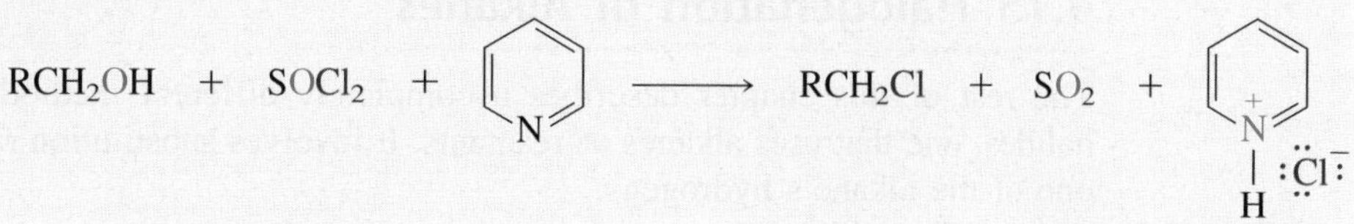

Alcohol | Thionyl chloride | Pyridine | Alkyl chloride | Sulfur dioxide | Pyridinium chloride

Step 1: The hydroxyl group oxygen attacks sulfur and displaces chloride to give a protonated chlorosulfite ester:

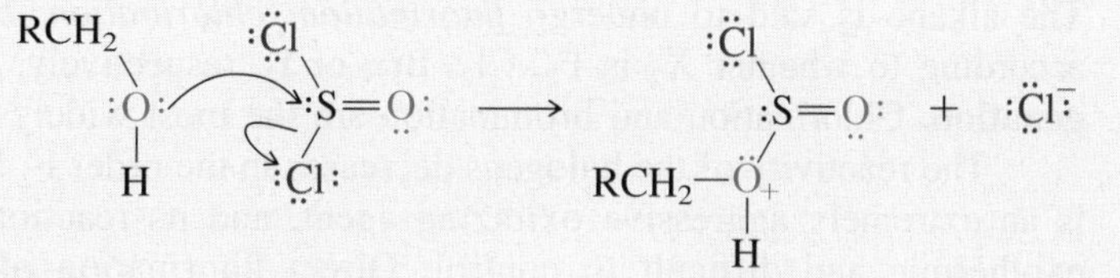

Alcohol | Thionyl chloride | Protonated alkyl chlorosulfite | Chloride ion

Step 2: Pyridine acts as a base to remove a proton and give a neutral alkyl chlorosulfite:

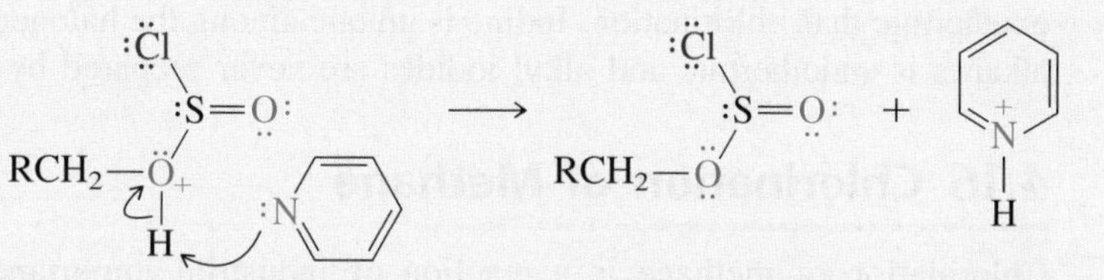

Pyridine | Alkyl chlorosulfite | Pyridinium ion

Step 3: Nucleophilic attack of chloride ion breaks the C—O bond to form the alkyl chloride:

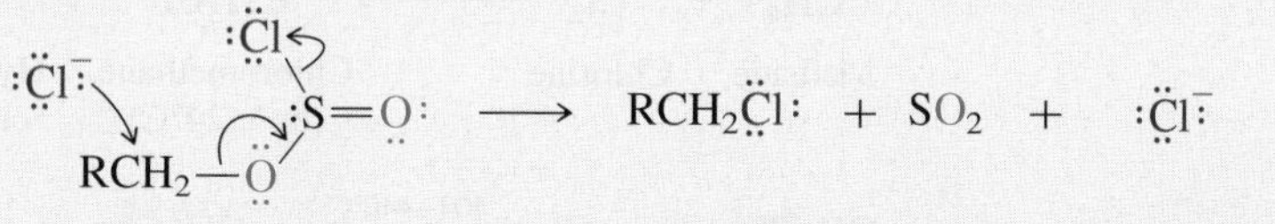

Alkyl chlorosulfite | Alkyl chloride | Sulfur dioxide | Chloride ion

Phosphorus tribromide reacts with alcohols to give alkyl bromides and phosphorous acid.

$$3ROH + PBr_3 \longrightarrow 3RBr + H_3PO_3$$

Alcohol | Phosphorus tribromide | Alkyl bromide | Phosphorous acid

Phosphorous acid is water-soluble and may be removed by washing the alkyl halide with water or with dilute aqueous base.

$$(CH_3)_2CHCH_2OH \xrightarrow{PBr_3} (CH_3)_2CHCH_2Br$$

Isobutyl alcohol | Isobutyl bromide (55–60%)

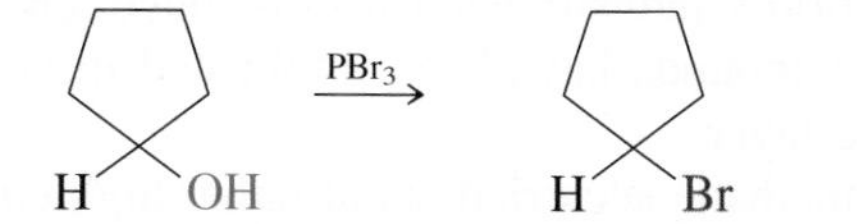

Cyclopentanol | Cyclopentyl bromide (78–84%)

The mechanism for the reaction of phosphorus tribromide with alcohols is similar to that for the reaction of alcohols with thionyl chloride. The alcohol reacts with PBr_3 giving an intermediate of the type $ROPBr_2$ which reacts with the nucleophilic bromide ion.

4.15 Halogenation of Alkanes

The rest of this chapter describes a completely different method for preparing alkyl halides, one that uses alkanes as reactants. It involves substitution of a halogen atom for one of the alkane's hydrogens.

$$\underset{\text{Alkane}}{R{-}H} + \underset{\text{Halogen}}{X_2} \longrightarrow \underset{\text{Alkyl halide}}{R{-}X} + \underset{\text{Hydrogen halide}}{H{-}X}$$

The alkane is said to undergo *fluorination, chlorination, bromination,* or *iodination* according to whether X_2 is F_2, Cl_2, Br_2, or I_2, respectively. The general term is **halogenation.** Chlorination and bromination are the most widely used.

Volume II of *Organic Reactions,* an annual series that reviews reactions of interest to organic chemists, contains the statement "Most organic compounds burn or explode when brought in contact with fluorine."

The reactivity of the halogens decreases in the order $F_2 > Cl_2 > Br_2 > I_2$. Fluorine is an extremely aggressive oxidizing agent, and its reaction with alkanes is strongly exothermic and difficult to control. Direct fluorination of alkanes requires special equipment and techniques, is not a reaction of general applicability, and will not be discussed further.

Chlorination of alkanes is less exothermic than fluorination, and bromination less exothermic than chlorination. Iodine is unique among the halogens in that its reaction with alkanes is endothermic and alkyl iodides are never prepared by iodination of alkanes.

4.16 Chlorination of Methane

Chlorination of methane is a reaction of industrial importance and is carried out in the gas phase to give a mixture of chloromethane (CH_3Cl), dichloromethane (CH_2Cl_2), trichloromethane ($CHCl_3$), and tetrachloromethane (CCl_4).

Chlorination of methane provides approximately one third of the annual U.S. production of chloromethane. The reaction of methanol with hydrogen chloride is the major synthetic method for the preparation of chloromethane.

$$\underset{\text{Methane}}{CH_4} + \underset{\text{Chlorine}}{Cl_2} \xrightarrow{400-440°C} \underset{\substack{\text{Chloromethane}\\(\text{bp } -24°C)}}{CH_3Cl} + \underset{\substack{\text{Hydrogen}\\\text{chloride}}}{HCl}$$

$$\underset{\text{Chloromethane}}{CH_3Cl} + \underset{\text{Chlorine}}{Cl_2} \xrightarrow{400-440°C} \underset{\substack{\text{Dichloromethane}\\(\text{bp } 40°C)}}{CH_2Cl_2} + \underset{\substack{\text{Hydrogen}\\\text{chloride}}}{HCl}$$

$$\underset{\text{Dichloromethane}}{CH_2Cl_2} + \underset{\text{Chlorine}}{Cl_2} \xrightarrow{400-440°C} \underset{\substack{\text{Trichloromethane}\\(\text{bp } 61°C)}}{CHCl_3} + \underset{\substack{\text{Hydrogen}\\\text{chloride}}}{HCl}$$

$$\underset{\text{Trichloromethane}}{CHCl_3} + \underset{\text{Chlorine}}{Cl_2} \xrightarrow{400-440°C} \underset{\substack{\text{Tetrachloromethane}\\(\text{bp } 77°C)}}{CCl_4} + \underset{\substack{\text{Hydrogen}\\\text{chloride}}}{HCl}$$

Dichloromethane, trichloromethane, and tetrachloromethane are widely known by their common names methylene chloride, chloroform, and carbon tetrachloride, respectively.

One of the chief uses of chloromethane is as a starting material from which silicone polymers are made. Dichloromethane is widely used as a paint stripper. Trichloromethane was once used as an inhalation anesthetic, but its toxicity caused it to be replaced by safer materials many years ago. Tetrachloromethane is the starting material for the preparation of several chlorofluorocarbons (CFCs), at one time widely used as refrigerant gases. Most of the world's industrialized nations have agreed to phase out all uses of CFCs because these compounds have been implicated in atmospheric processes that degrade the Earth's ozone layer.

The chlorination of methane is carried out at rather high temperatures (400–440°C), even though each substitution in the series is exothermic. The high temperature provides the energy to initiate the reaction. The term *initiation step* has a specific meaning in organic chemistry, one that is related to the mechanism of the reaction. This mechanism, to be presented in Section 4.18, is fundamentally different from the mechanism by which alcohols react with hydrogen halides. Alcohols are converted to alkyl halides in reactions involving ionic (or "polar") intermediates—alkyloxonium ions and carbocations. The intermediates in the chlorination of methane and other alkanes are quite different; they are neutral ("nonpolar") species called *free radicals.*

4.17 Structure and Stability of Free Radicals

Free radicals are species that contain unpaired electrons. The octet rule notwithstanding, not all compounds have all of their electrons paired. Oxygen (O_2) is the most familiar example of a compound with unpaired electrons; it has two of them. Compounds that have an odd number of electrons, such as nitrogen dioxide (NO_2), must have at least one unpaired electron.

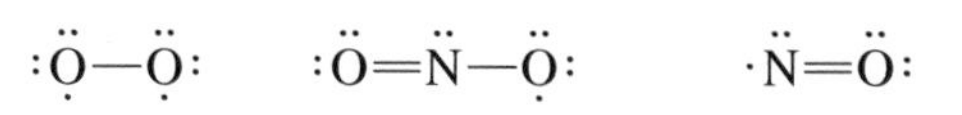

Oxygen Nitrogen dioxide Nitrogen monoxide

For more on the role of NO in physiology, see the boxed essay *Oh NO! It's Inorganic!* in Chapter 25.

Nitrogen monoxide ("nitric oxide") is another stable free radical. Although known for hundreds of years, NO has only recently been discovered to be an extremely important biochemical messenger and moderator of so many biological processes that it might be better to ask "Which ones is it not involved in?"

The free radicals that we usually see in carbon chemistry are much less stable than these. Simple alkyl radicals, for example, require special procedures for their isolation and study. We will encounter them here only as reactive intermediates, formed in one step of a reaction mechanism and consumed in the next. Alkyl radicals are classified as primary, secondary, or tertiary according to the number of carbon atoms directly attached to the carbon that bears the unpaired electron.

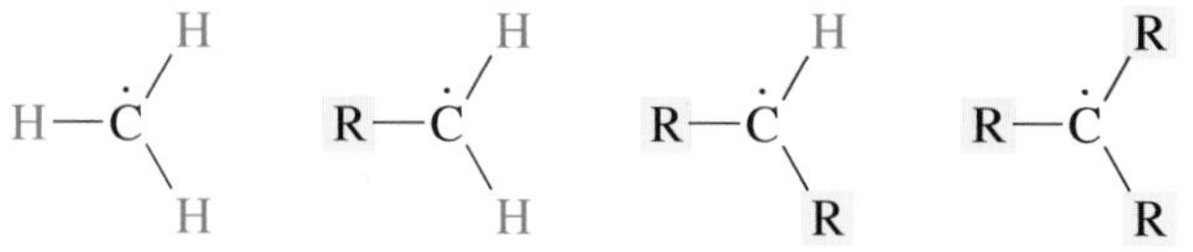

Methyl radical Primary radical Secondary radical Tertiary radical

An alkyl radical is neutral and has one more electron than the corresponding carbocation. Thus, bonding in methyl radical may be approximated by simply adding an electron to the vacant $2p$ orbital of sp^2-hybridized carbon in methyl cation (Figure 4.17*a*).

(*a*)
Planar CH_3
Carbon is sp^2-hybridized (120° bond angles). Unpaired electron is in $2p$ orbital.

(*b*)
Pyramidal CH_3
Carbon is sp^3-hybridized (109.5° bond angles). Unpaired electron is in sp^3-hybridized orbital.

Figure 4.17

Bonding in methyl radical. Model (*a*) is more consistent with experimental observations.

Figure 4.18

Spin density (yellow) in methyl and ethyl radical. (*a*) The unpaired electron in methyl radical is localized in a *p* orbital of sp^2-hybridized carbon. (*b*) The unpaired electron in ethyl radical is shared by the sp^2-hybridized carbon and by the hydrogens of the CH_3 group.

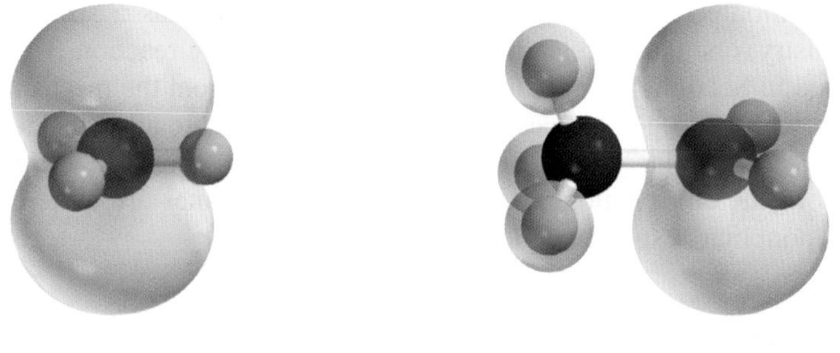

(*a*) $\cdot CH_3$ (*b*) $H_3C—\dot{C}H_2$

Alternatively, we could assume that carbon is sp^3-hybridized and place the unpaired electron in an sp^3 orbital (Figure 4.17*b*).

Of the two extremes, experimental studies indicate that the planar sp^2 model describes the bonding in alkyl radicals better than the pyramidal sp^3 model. Methyl radical is planar, and more highly substituted radicals such as *tert*-butyl radical are flattened pyramids closer in shape to that expected for sp^2-hybridized carbon than for sp^3.

Free radicals, like carbocations, have an unfilled 2*p* orbital and are stabilized by substituents, such as alkyl groups, that can donate electrons by hyperconjugation. We can illustrate the stabilization of free radicals by hyperconjugation by comparing the spin density in methyl and ethyl radical. **Spin density** is a measure of unpaired electron density at a particular point in a molecule—it tells us where the unpaired electron is most likely to be. As Figure 4.18 shows, the spin density in methyl radical is localized on the sp^2-hybridized carbon, but is shared by the sp^2-hybridized carbon and the methyl hydrogens in ethyl radical. More highly substituted radicals are more stable than less highly substituted ones, and the order of free-radical stability parallels that of carbocations.

Increasing stability of free radical

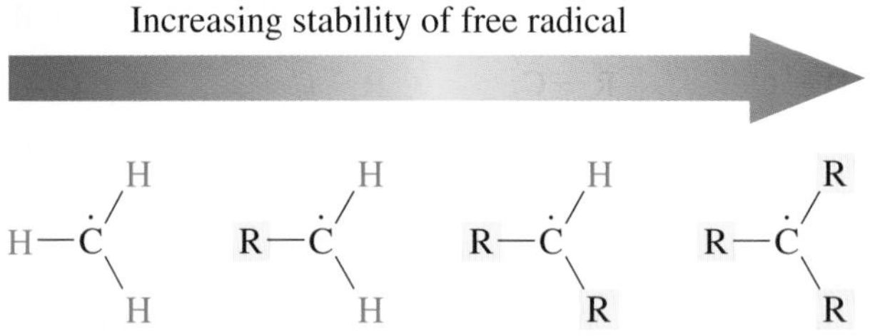

Methyl radical (least stable) | Primary radical | Secondary radical | Tertiary radical (most stable)

Problem 4.16

Write a structural formula for the most stable of the free radicals that have the formula C_5H_{11}.

Some of the evidence indicating that alkyl substituents stabilize free radicals comes from bond enthalpies. The strength of a bond is measured by the energy required to break it. A covalent bond can be broken in two ways. In a **homolytic cleavage** a bond between two atoms is broken so that each of them retains one of the electrons in the bond.

A curved arrow shown as a single-barbed fishhook signifies the movement of one electron. "Normal" curved arrows track the movement of a *pair* of electrons.

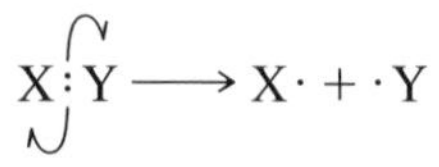

Homolytic bond cleavage

TABLE 4.3 Some Bond Dissociation Enthalpies*

Bond	Bond dissociation enthalpy (D)		Bond	Bond dissociation enthalpy (D)	
	kJ/mol	kcal/mol		kJ/mol	kcal/mol
Diatomic molecules					
H—H	436	104	H—F	571	136
F—F	159	38	H—Cl	432	103
Cl—Cl	243	58	H—Br	366	87.5
Br—Br	193	46	H—I	298	71
I—I	151	36			
Alkanes					
CH_3—H	439	105	CH_3—CH_3	375	90
CH_3CH_2—H	421	100.5	CH_3CH_2— CH_3	369	88
$CH_3CH_2CH_2$—H	423	101			
$(CH_3)_2CH$—H	413	99			
$(CH_3)_2CHCH_2$—H	422	101	$(CH_3)_2CH$—CH_3	370	88
$(CH_3)_3C$—H	400	95	$(CH_3)_3C$—CH_3	362	86
Alkyl halides					
CH_3—F	459	110	$(CH_3)_2CH$—Cl	355	85
CH_3—Cl	351	84	$(CH_3)_2CH$—Br	297	72
CH_3—Br	292	70			
CH_3—I	238	57			
CH_3CH_2—Cl	350	83	$(CH_3)_3C$—Cl	349	83
$CH_3CH_2CH_2$—Cl	354	85	$(CH_3)_3C$—Br	292	69

*Bond dissociation enthalpies refer to the bond indicated in each structural formula and were calculated from standard enthalpy of formation values as recorded in the NIST Standard Reference Database Number 69, http://webbook.nist.gov/chemistry/.

In contrast, in a **heterolytic cleavage** one fragment retains both electrons.

$$X{:}Y \longrightarrow X^{+} + {:}Y^{-}$$

Heterolytic bond cleavage

We assess the relative stability of alkyl radicals by measuring the enthalpy change ($\Delta H°$) for the homolytic cleavage of a C—H bond in an alkane:

$$R{-}H \longrightarrow R\cdot + \cdot H$$

The more stable the radical, the lower the energy required to generate it by homolytic cleavage of a C—H bond.

The energy required for homolytic bond cleavage is called the **bond dissociation enthalpy (*D*).** A list of some bond dissociation enthalpies is given in Table 4.3.

As the table indicates, C—H bond dissociation enthalpies in alkanes are approximately 400–440 kJ/mol (95–105 kcal/mol). Cleaving the H—CH_3 bond in methane gives methyl radical and requires 439 kJ/mol (105 kcal/mol). The dissociation enthalpy of the H—CH_2CH_3 bond in ethane, which gives a primary radical, is somewhat less (421 kJ/mol, or 100.5 kcal/mol) and is consistent with the notion that ethyl radical (primary) is more stable than methyl.

The dissociation enthalpy of the terminal C—H bond in propane is almost the same as that of ethane. The resulting free radical is primary ($R\dot{C}H_2$) in both cases.

$$CH_3CH_2CH_2{-}H \longrightarrow CH_3CH_2\dot{C}H_2 + H\cdot \qquad \Delta H^\circ = +423\ \text{kJ}\ (101\ \text{kcal})$$

Propane; *n*-Propyl radical (primary); Hydrogen atom

Note, however, that Table 4.3 includes two entries for propane. The second entry corresponds to the cleavage of a bond to one of the hydrogens of the methylene (CH_2) group. It requires slightly less energy to break a C—H bond in the methylene group than in the methyl group.

$$CH_3\underset{\displaystyle H}{\underset{|}{C}}HCH_3 \longrightarrow CH_3\dot{C}HCH_3 + H\cdot \qquad \Delta H^\circ = +413\ \text{kJ}\ (99\ \text{kcal})$$

Propane; Isopropyl radical (secondary); Hydrogen atom

Because the starting material (propane) and one of the products (H·) are the same in both processes, the difference in bond dissociation enthalpies is equal to the energy difference between an *n*-propyl radical (primary) and an isopropyl radical (secondary). As depicted in Figure 4.19, the secondary radical is 10 kJ/mol (2 kcal/mol) more stable than the primary radical.

Similarly, by comparing the bond dissociation enthalpies of the two different types of C—H bonds in 2-methylpropane, we see that a tertiary radical is 22 kJ/mol (6 kcal/mol) more stable than a primary radical.

$$CH_3\underset{\displaystyle CH_3}{\underset{|}{C}}HCH_2{-}H \longrightarrow CH_3\underset{\displaystyle CH_3}{\underset{|}{C}}H\dot{C}H_2 + H\cdot \qquad \Delta H^\circ = +422\ \text{kJ}\ (101\ \text{kcal})$$

2-Methylpropane; Isobutyl radical (primary); Hydrogen atom

$$CH_3\underset{\displaystyle CH_3}{\underset{|}{\overset{\displaystyle H}{\overset{|}{C}}}}CH_3 \longrightarrow CH_3\underset{\displaystyle CH_3}{\underset{|}{\dot{C}}}CH_3 + H\cdot \qquad \Delta H^\circ = +400\ \text{kJ}\ (95\ \text{kcal})$$

2-Methylpropane; *tert*-Butyl radical (tertiary); Hydrogen atom

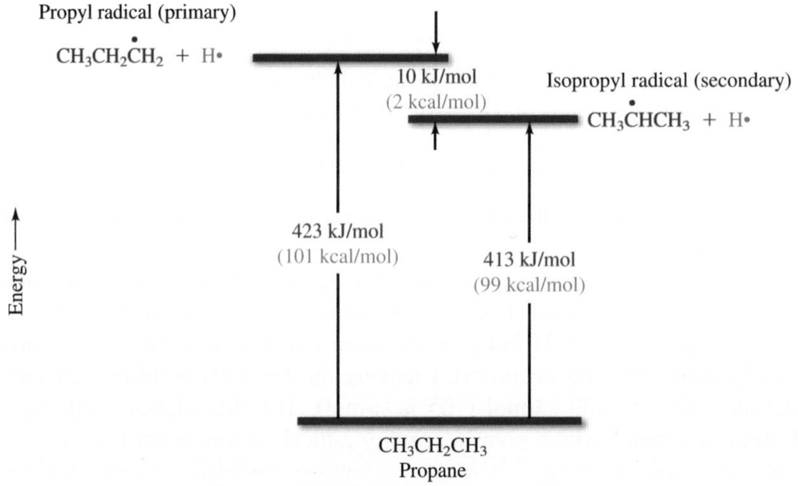

Figure 4.19

The bond dissociation enthalpies of methylene and methyl C—H bonds in propane reveal a difference in stabilities between two isomeric free radicals. The secondary radical is more stable than the primary.

From Bond Enthalpies to Heats of Reaction

You have seen that measurements of heats of reaction, such as heats of combustion, can provide quantitative information concerning the relative stability of constitutional isomers (Section 2.18) and stereoisomers (Section 3.11). The box in Section 2.18 described how heats of reaction can be manipulated arithmetically to generate heats of formation (ΔH_f°) for many molecules. The following material shows how two different sources of thermochemical information, heats of formation and bond dissociation enthalpies (see Table 4.3), can reveal whether a particular reaction is exothermic or endothermic and by how much.

Consider the chlorination of methane to chloromethane. The heats of formation of the reactants and products appear beneath the equation. These heats of formation for the chemical compounds are taken from published tabulations; the heat of formation of chlorine is zero, as it is for all elements.

$$\underset{-74.8}{CH_4} + \underset{0}{Cl_2} \longrightarrow \underset{-83.7}{CH_3Cl} + \underset{-92.3}{HCl}$$

ΔH_f° (kJ/mol): −74.8, 0, −83.7, −92.3

The overall heat of reaction is given by

$$\Delta H^\circ = \sum(\text{heats of formation of products}) - \sum(\text{heats of formation of reactants})$$

$$\Delta H^\circ = (-83.7 \text{ kJ} - 92.3 \text{ kJ}) - (-74.8 \text{ kJ}) = -101.2 \text{ kJ}$$

Thus, the chlorination of methane is calculated to be an exothermic reaction on the basis of heat of formation data.

The same conclusion is reached using bond dissociation enthalpies. The following equation shows the bond dissociation enthalpies of the reactants and products taken from Table 4.3:

$$CH_4 + Cl_2 \longrightarrow CH_3Cl + HCl$$

BDE (kJ/mol): 439, 243, 351, 432

Because stronger bonds are formed at the expense of weaker ones, the reaction is exothermic and

$$\Delta H^\circ = \sum(\text{BDE of bonds broken}) - \sum(\text{BDE of bonds formed})$$

$$\Delta H^\circ = (439 \text{ kJ} + 243 \text{ kJ}) - (351 \text{ kJ} + 432 \text{ kJ}) = -101 \text{ kJ}$$

This value is in good agreement with that obtained from heat of formation data.

Compare chlorination of methane with iodination. The relevant bond dissociation enthalpies are given in the equation.

$$CH_4 + I_2 \longrightarrow CH_3I + HI$$

BDE (kJ/mol): 439, 151, 238, 298

$$\Delta H^\circ = \sum(\text{BDE of bonds broken}) - \sum(\text{BDE of bonds formed})$$

$$\Delta H^\circ = (439 \text{ kJ} + 151 \text{ kJ}) - (238 \text{ kJ} + 298 \text{ kJ}) = +54 \text{ kJ}$$

A positive value for ΔH° signifies an **endothermic** reaction. The reactants are more stable than the products, and so iodination of alkanes is not a feasible reaction. You would not want to attempt the preparation of iodomethane by iodination of methane.

A similar analysis for fluorination of methane gives $\Delta H^\circ = -432$ kJ for its heat of reaction. Fluorination of methane is about four times as exothermic as chlorination. A reaction this exothermic, if it also occurs at a rapid rate, can proceed with explosive violence.

Bromination of methane is exothermic, but less so than chlorination. The value calculated from bond dissociation enthalpies is $\Delta H^\circ = -26$ kJ. Although bromination of methane is energetically favorable, economic considerations cause most of the methyl bromide prepared commercially to be made from methanol by reaction with hydrogen bromide.

Problem 4.17

Carbon–carbon bond dissociation enthalpies have been measured for many alkanes. Without referring to Table 4.3, identify the alkane in each of the following pairs that has the lower carbon–carbon bond dissociation enthalpy, and explain the reason for your choice.

(a) Ethane or propane
(b) Propane or 2-methylpropane
(c) 2-Methylpropane or 2,2-dimethylpropane

Sample Solution (a) First write the equations that describe homolytic carbon–carbon bond cleavage in each alkane.

$$\underset{\text{Ethane}}{CH_3{-}CH_3} \longrightarrow \underset{\text{Two methyl radicals}}{\cdot CH_3 + \cdot CH_3}$$

$$\underset{\text{Propane}}{CH_3CH_2{-}CH_3} \longrightarrow \underset{\text{Ethyl radical}}{CH_3\dot{C}H_2} + \underset{\text{Methyl radical}}{\cdot CH_3}$$

Cleavage of the carbon–carbon bond in ethane yields two methyl radicals, whereas propane yields an ethyl radical and one methyl radical. Ethyl radical is more stable than methyl, and so less energy is required to break the carbon–carbon bond in propane than in ethane. The measured carbon–carbon bond dissociation enthalpy in ethane is 375 kJ/mol (90 kcal/mol), and that in propane is 369 kJ/mol (88 kcal/mol).

Like carbocations, most free radicals are exceedingly reactive species—too reactive to be isolated but capable of being formed as transient intermediates in chemical reactions. Methyl radical, as we shall see in the following section, is an intermediate in the chlorination of methane.

4.18 Mechanism of Methane Chlorination

The generally accepted process for the chlorination of methane is presented in Mechanism 4.4. As we noted earlier (Section 4.16), the reaction is normally carried out in the gas phase at high temperature. The reaction itself is strongly exothermic, but energy must be put into the system to get it going. This energy goes into breaking the weakest bond in the system, which, as we see from the bond dissociation enthalpy data in Table 4.3, is the Cl—Cl bond with a bond dissociation enthalpy of 243 kJ/mol (58 kcal/mol). The step in which Cl—Cl bond homolysis occurs is called the **initiation step.**

The bond dissociation enthalpy of the other reactant, methane, is much higher. It is 439 kJ/mol (105 kcal/mol).

Each chlorine atom formed in the initiation step has seven valence electrons and is very reactive. Once formed, a chlorine atom abstracts a hydrogen atom from methane as shown in step 2 in Mechanism 4.4. Hydrogen chloride, one of the isolated products from the overall reaction, is formed in this step. A methyl radical is also formed, which then reacts with a molecule of Cl_2 in step 3 giving chloromethane, the other product of the overall reaction, along with a chlorine atom. The chlorine atom then cycles back to step 2, and the process repeats. Steps 2 and 3 are called the **propagation steps** of the

Mechanism 4.4

Free-Radical Chlorination of Methane

(*a*) Initiation

Step 1: Dissociation of a chlorine molecule into two chlorine atoms:

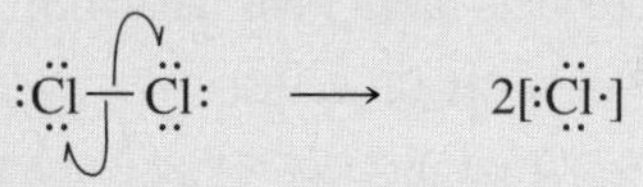

Chlorine molecule Two chlorine atoms

(*b*) Chain propagation

Step 2: Hydrogen atom abstraction from methane by a chlorine atom:

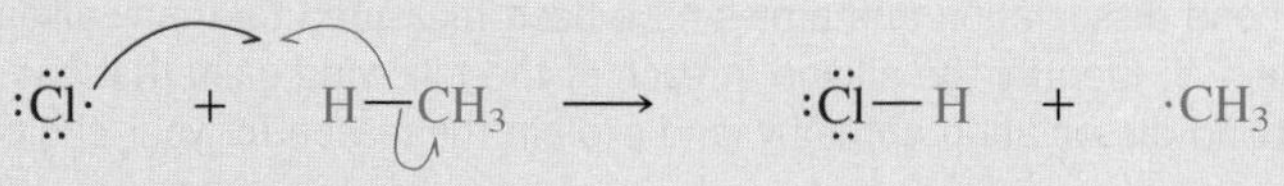

Chlorine atom Methane Hydrogen chloride Methyl radical

Step 3: Reaction of methyl radical with molecular chlorine:

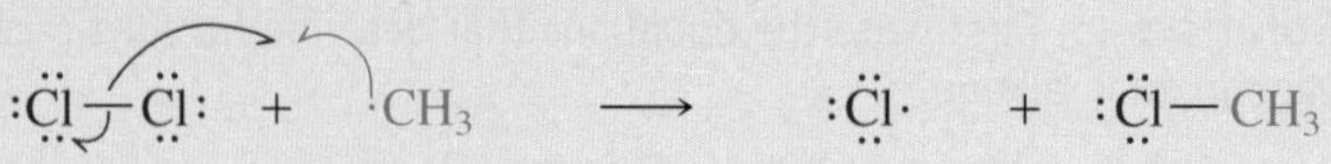

Chlorine molecule Methyl radical Chlorine atom Chloromethane

Steps 2 and 3 then repeat many times.

(*c*) Sum of steps 2 and 3

$$CH_4 + Cl_2 \longrightarrow CH_3Cl + HCl$$

Methane Chlorine Chloromethane Hydrogen chloride

reaction and, when added together, give the overall equation for the reaction. Because one initiation step can result in a great many propagation cycles, the overall process is called a free-radical **chain reaction.**

Problem 4.18

Write equations for the initiation and propagation steps for the formation of dichloromethane by free-radical chlorination of chloromethane.

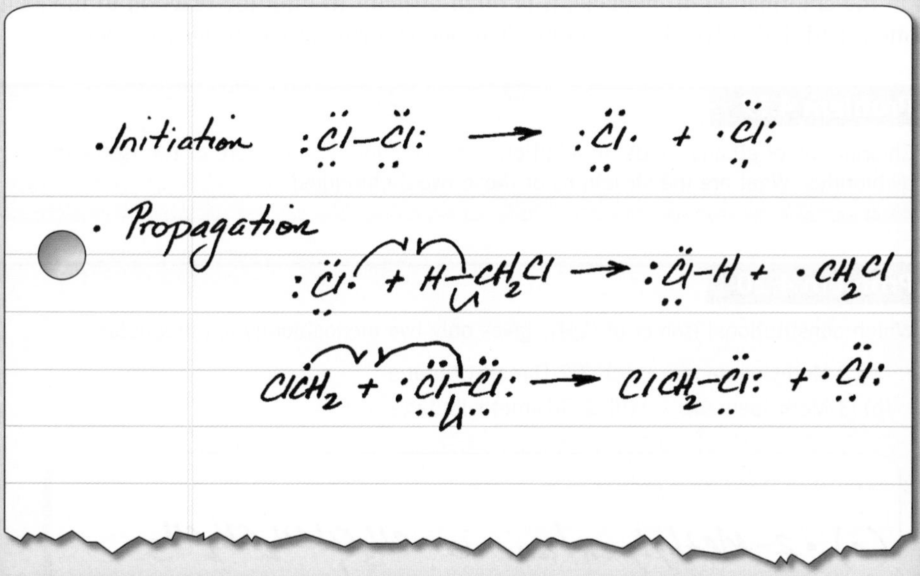

In practice, side reactions intervene to reduce the efficiency of the propagation steps. The chain sequence is interrupted whenever two odd-electron species combine to give an even-electron product. Reactions of this type are called **chain-terminating steps.** Some commonly observed chain-terminating steps in the chlorination of methane are shown in the following equations.

Combination of a methyl radical with a chlorine atom:

Methyl radical Chlorine atom Chloromethane

Combination of two methyl radicals:

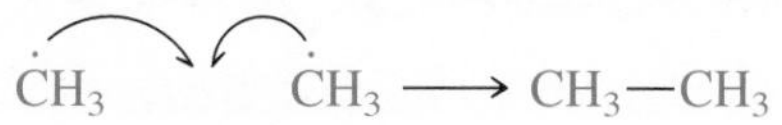

Two methyl radicals Ethane

Combination of two chlorine atoms:

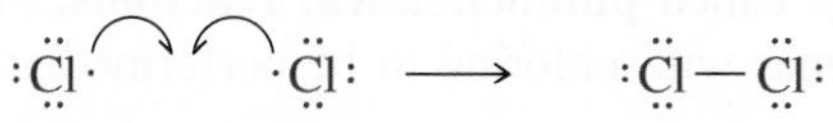

Two chlorine atoms Chlorine molecule

Termination steps are, in general, less likely to occur than the propagation steps. Each of the termination steps requires two free radicals to encounter each other in a medium that contains far greater quantities of other materials (methane and chlorine molecules) with which they can react. Although some chloromethane undoubtedly arises via direct combination of methyl radicals with chlorine atoms, most of it is formed by the propagation sequence shown in Mechanism 4.4.

4.19 Halogenation of Higher Alkanes

Like the chlorination of methane, chlorination of ethane is carried out on an industrial scale as a high-temperature gas-phase reaction.

$$\underset{\text{Ethane}}{CH_3CH_3} + \underset{\text{Chlorine}}{Cl_2} \xrightarrow{420°C} \underset{\substack{\text{Chloroethane (78\%)} \\ \text{(ethyl chloride)}}}{CH_3CH_2Cl} + \underset{\text{Hydrogen chloride}}{HCl}$$

As in the chlorination of methane, it is often difficult to limit the reaction to monochlorination, and derivatives having more than one chlorine atom are also formed.

Problem 4.19

Chlorination of ethane yields, in addition to ethyl chloride, a mixture of two isomeric dichlorides. What are the structures of these two dichlorides?

Problem 4.20

Which constitutional isomer of C_6H_{14} gives only two monochlorination products?

(a) 2-Methylpentane (c) 2,2-Dimethylbutane
(b) 3-Methylpentane (d) 2,3-Dimethylbutane

(a) • 2-Methylpentane is $CH_3\underset{\displaystyle CH_3}{\underset{|}{C}H}CH_2CH_2CH_3$

• 5-monochloro substitution products are possible

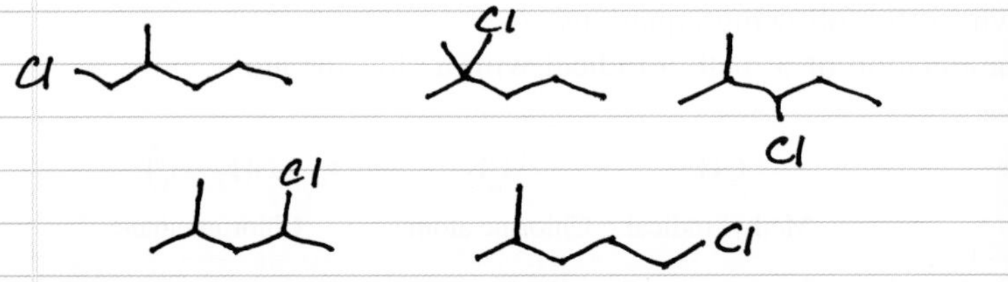

In the laboratory it is more convenient to use light, either visible or ultraviolet, as the source of energy to initiate the reaction. Reactions that occur when light energy is absorbed by a molecule are called **photochemical reactions.** Photochemical techniques permit the reaction of alkanes with chlorine to be performed at room temperature.

Photochemical energy is indicated by writing "light" or "$h\nu$" above or below the arrow. The symbol $h\nu$ is equal to the energy of a light photon and will be discussed in more detail in Section 13.1.

Cyclobutane Chlorine Chlorocyclobutane (73%) (cyclobutyl chloride) Hydrogen chloride

Methane, ethane, and cyclobutane share the common feature that each one can give only a *single* monochloro derivative. All the hydrogens of cyclobutane, for example, are equivalent, and substitution of any one gives the same product as substitution of any other. Chlorination of alkanes in which the hydrogens are not all equivalent is more complicated in that a mixture of every possible monochloro derivative is formed, as the chlorination of butane illustrates:

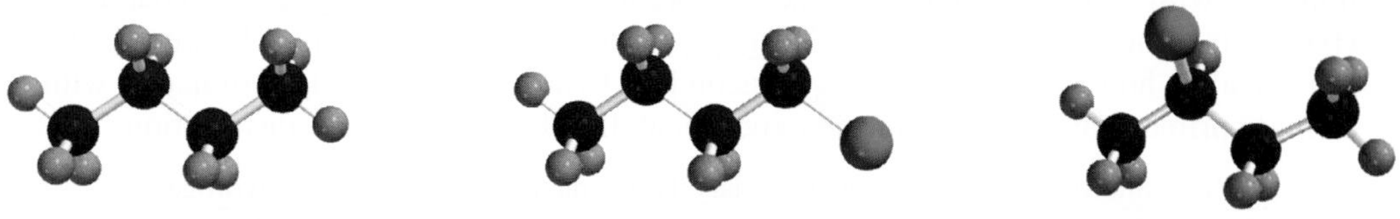

$$\underset{\text{Butane}}{CH_3CH_2CH_2CH_3} \xrightarrow[h\nu,\ 35°C]{Cl_2} \underset{\substack{\text{1-Chlorobutane (28\%)} \\ (n\text{-butyl chloride})}}{CH_3CH_2CH_2CH_2Cl} + \underset{\substack{\text{2-Chlorobutane (72\%)} \\ (sec\text{-butyl chloride})}}{CH_3\underset{\displaystyle Cl}{\underset{|}{C}}HCH_2CH_3}$$

The percentages cited in the preceding equation reflect the composition of the monochloride fraction of the product mixture rather than the isolated yield of each component.

These two products arise because in one of the propagation steps a chlorine atom may abstract a hydrogen atom from either a methyl or a methylene group of butane.

$$\underset{\text{Butane}}{CH_3CH_2CH_2CH_2—H} + \cdot\ddot{\underset{..}{Cl}}: \longrightarrow \underset{n\text{-Butyl radical}}{CH_3CH_2CH_2\dot{C}H_2} + H—\ddot{\underset{..}{Cl}}:$$

$$\underset{\text{Butane}}{CH_3\underset{\displaystyle H}{\underset{|}{C}}HCH_2CH_3} + \cdot\ddot{\underset{..}{Cl}}: \longrightarrow \underset{sec\text{-Butyl radical}}{CH_3\dot{C}HCH_2CH_3} + H—\ddot{\underset{..}{Cl}}:$$

The resulting free radicals react with chlorine to give the corresponding alkyl chlorides. Butyl radical gives only 1-chlorobutane; *sec*-butyl radical gives only 2-chlorobutane.

$$\underset{n\text{-Butyl radical}}{CH_3CH_2CH_2\dot{C}H_2} + :\ddot{\underset{..}{Cl}}—\ddot{\underset{..}{Cl}}: \longrightarrow \underset{\substack{\text{1-Chlorobutane} \\ (n\text{-butyl chloride})}}{CH_3CH_2CH_2CH_2\ddot{\underset{..}{Cl}}:} + \cdot\ddot{\underset{..}{Cl}}:$$

$$\underset{sec\text{-Butyl radical}}{CH_3\dot{C}HCH_2CH_3} + :\ddot{\underset{..}{Cl}}—\ddot{\underset{..}{Cl}}: \longrightarrow \underset{\substack{\text{2-Chlorobutane} \\ (sec\text{-butyl chloride})}}{CH_3\underset{:\ddot{\underset{..}{Cl}}:}{\underset{|}{C}}HCH_2CH_3} + \cdot\ddot{\underset{..}{Cl}}:$$

If every collision of a chlorine atom with a butane molecule resulted in hydrogen abstraction, the *n*-butyl/*sec*-butyl radical ratio and, therefore, the 1-chloro/2-chlorobutane ratio, would be given by the relative numbers of hydrogens in the two equivalent methyl groups of $CH_3CH_2CH_2CH_3$ (six) compared with those in the two equivalent methylene groups (four). The product distribution expected on this basis would be 60% 1-chlorobutane and 40% 2-chlorobutane. The *experimentally observed* product distribution, however, is 28% 1-chlorobutane and 72% 2-chlorobutane. *sec*-Butyl radical is therefore formed in greater amounts, and *n*-butyl radical in lesser amounts, than expected.

This behavior stems from the greater stability of secondary compared with primary free radicals. The transition state for the step in which a chlorine atom abstracts a hydrogen from carbon has free-radical character at carbon.

$$\underset{\substack{\text{Transition state for abstraction} \\ \text{of a primary hydrogen}}}{\overset{\delta\cdot}{Cl}---H---\overset{\delta\cdot}{C}H_2CH_2CH_2CH_3} \qquad \underset{\substack{\text{Transition state for abstraction} \\ \text{of a secondary hydrogen}}}{\overset{\delta\cdot}{Cl}---H---\overset{\overset{\displaystyle CH_3}{|}}{\overset{\delta\cdot}{C}}HCH_2CH_3}$$

A secondary hydrogen is abstracted faster than a primary hydrogen because the transition state with secondary radical character is more stable than the one with primary radical character. The same factors that stabilize a secondary radical stabilize a transition state with secondary radical character more than one with primary radical character. Hydrogen atom abstraction from a CH_2 group occurs faster than from a CH_3 group. We can calculate how much faster a *single* secondary hydrogen is abstracted compared with a *single* primary hydrogen from the experimentally observed product distribution.

$$\frac{72\%\ \text{2-chlorobutane}}{28\%\ \text{1-chlorobutane}} = \frac{\text{rate of secondary H abstraction} \times 4 \text{ secondary hydrogens}}{\text{rate of primary H abstraction} \times 6 \text{ primary hydrogens}}$$

$$\frac{\text{Rate of secondary H abstraction}}{\text{Rate of primary H abstraction}} = \frac{72}{28} \times \frac{6}{4} = \frac{3.9}{1}$$

A single secondary hydrogen in butane is abstracted by a chlorine atom 3.9 times faster than a single primary hydrogen.

Problem 4.21

Assuming the relative rate of secondary to primary hydrogen atom abstraction to be the same in the chlorination of propane as it is in that of butane, calculate the relative amounts of propyl chloride and isopropyl chloride obtained in the free-radical chlorination of propane.

A similar study of the chlorination of 2-methylpropane established that a tertiary hydrogen is removed 5.2 times faster than each primary hydrogen.

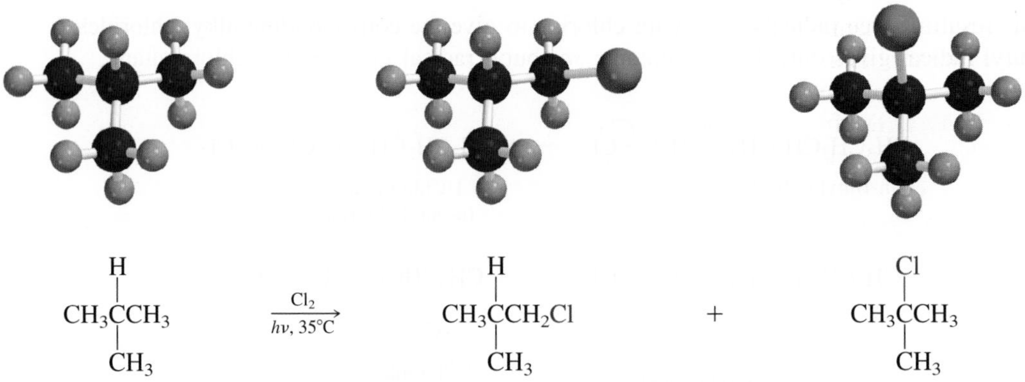

2-Methylpropane → 1-Chloro-2-methylpropane (63%) (isobutyl chloride) + 2-Chloro-2-methylpropane (37%) (*tert*-butyl chloride)

In summary, the chlorination of alkanes is not very selective. The various kinds of hydrogens present in a molecule (tertiary, secondary, and primary) differ by only a factor of 5 in the relative rate at which each reacts with a chlorine atom.

	R_3CH >	R_2CH_2 >	RCH_3
	(tertiary)	(secondary)	(primary)
Relative rate (chlorination)	5.2	3.9	1

Bromine reacts with alkanes by a free-radical chain mechanism analogous to that of chlorine. There is an important difference between chlorination and bromination, however. Bromination is highly selective for substitution of *tertiary hydrogens.* The spread in reactivity among primary, secondary, and tertiary hydrogens is greater than 10^3.

	R_3CH >	R_2CH_2 >	RCH_3
	(tertiary)	(secondary)	(primary)
Relative rate (bromination)	1640	82	1

In practice, this means that when an alkane contains primary, secondary, and tertiary hydrogens, it is usually only the tertiary hydrogen that is replaced by bromine.

$$\underset{\text{2-Methylpentane}}{CH_3\underset{\substack{|\\CH_3}}{\overset{\substack{H\\|}}{C}}CH_2CH_2CH_3} + \underset{\text{Bromine}}{Br_2} \xrightarrow[60°C]{h\nu} \underset{\substack{\text{2-Bromo-2-methylpentane}\\\text{(76\% isolated yield)}}}{CH_3\underset{\substack{|\\CH_3}}{\overset{\substack{Br\\|}}{C}}CH_2CH_2CH_3} + \underset{\substack{\text{Hydrogen}\\\text{bromide}}}{HBr}$$

The percentage cited in this reaction is the isolated yield of purified product. Isomeric bromides constitute only a tiny fraction of the product.

We can understand why bromination is more selective than chlorination by using bond dissociation enthalpies (Table 4.3) to calculate the energy changes for the propagation step in which each halogen atom abstracts a hydrogen from ethane.

$$\underset{\text{Ethane}}{CH_3CH_2\text{—}H} + \underset{\text{Chlorine atom}}{\cdot\ddot{\underset{..}{Cl}}:} \longrightarrow \underset{\text{Ethyl radical}}{CH_3\dot{C}H_2} + \underset{\text{Hydrogen chloride}}{H\text{—}\ddot{\underset{..}{Cl}}:} \quad \Delta H° = -11\text{ kJ }(-2.5\text{ kcal})$$

$$\underset{\text{Ethane}}{CH_3CH_2\text{—}H} + \underset{\text{Bromine atom}}{\cdot\ddot{\underset{..}{Br}}:} \longrightarrow \underset{\text{Ethyl radical}}{CH_3\dot{C}H_2} + \underset{\text{Hydrogen bromide}}{H\text{—}\ddot{\underset{..}{Br}}:} \quad \Delta H° = +54\text{ kJ }(+13\text{ kcal})$$

The alkyl radical-forming step is *exothermic for chlorination, endothermic for bromination.* Applying Hammond's postulate to these elementary steps, we conclude that alkyl radical character is more highly developed in the transition state for abstraction of hydrogen by a bromine atom than by a chlorine atom. Thus, bromination is more sensitive to the stability of the free-radical intermediate than chlorination and more selective.

The greater selectivity of bromination can be illustrated by comparing the reaction coordinate diagrams for the formation of primary and tertiary free radicals by hydrogen atom abstraction from 2-methylpropane with bromine versus chlorine radicals (Figure 4.20).

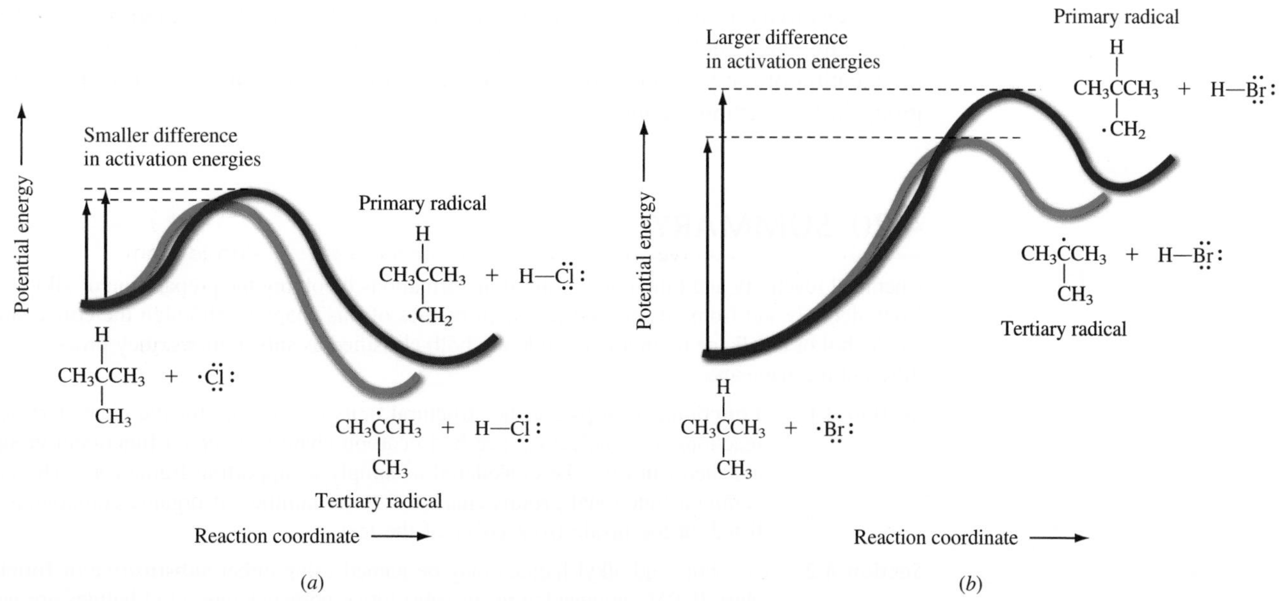

Figure 4.20

These reaction coordinate diagrams for the formation of primary and tertiary alkyl radicals by halogen atom abstraction from 2-methylpropane illustrate a larger difference in the activation energies for the reaction with a bromine atom (b) than with a chlorine atom (a). This difference is consistent with the higher selectivity of bromination.

Abstraction of hydrogen by a chlorine atom is exothermic for the formation of either the primary or tertiary radical. The transition states are early and more reactant-like, and the difference in E_{act} (ΔE_{act}) is small. Conversely, abstraction of hydrogen by a bromine atom, illustrated in Figure 4.20*b,* is endothermic in both cases. The transition states are more product-like and possess more radical character; therefore, the difference in radical stability is more strongly expressed, and ΔE_{act} is larger. The larger ΔE_{act} is associated with greater product selectivity, since the tertiary bromide is obtained from the tertiary free radical.

Problem 4.22

Give the structure of the principal organic product formed by free-radical bromination of each of the following:

(a) Methylcyclopentane
(b) 2,2,4-Trimethylpentane
(c) 1-Isopropyl-1-methylcyclopentane

Sample Solution (a) Write the structure of the starting hydrocarbon, and identify any tertiary hydrogens that are present. The only tertiary hydrogen in methylcyclopentane is the one attached to C-1. This is the one replaced by bromine.

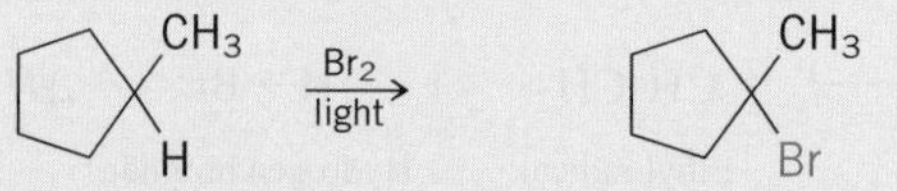

Methylcyclopentane 1-Bromo-1-methylcyclopentane

This difference in selectivity between chlorination and bromination of alkanes needs to be kept in mind when one wishes to prepare an alkyl halide from an alkane:

1. Because chlorination of an alkane yields every possible monochloride, it is used only when all the hydrogens in an alkane are equivalent.
2. Bromination of alkanes is mainly used to prepare tertiary alkyl bromides.

Selectivity is not an issue in the conversion of alcohols to alkyl halides. Except for certain cases such as those involving carbocation rearrangement (see Problem 8.41), the location of the halogen substituent in the product corresponds to that of the hydroxyl group in the starting alcohol.

4.20 SUMMARY

Chemical reactivity and functional group transformations involving the preparation of alkyl halides from alcohols and from alkanes are the main themes of this chapter. Although the conversions of an alcohol or an alkane to an alkyl halide are both classified as substitutions, they proceed by very different mechanisms.

Section 4.1 **Functional groups** are the structural units responsible for the characteristic reactions of a molecule. The hydrocarbon chain to which a functional group is attached can often be considered as simply a supporting framework. The most common functional groups characterize the families of organic compounds listed on the inside front cover of the text.

Section 4.2 Alcohols and alkyl halides may be named using either **substitutive** or **functional class** IUPAC nomenclature. In substitutive nomenclature alkyl halides are named as halogen derivatives of alkanes. The parent is the longest continuous chain that bears the halogen substituent, and in the absence of other substituents the chain is numbered from the direction that gives the lower number to the carbon that bears the halogen. The functional class names of alkyl halides begin with the name of the alkyl group and end with the halide as a separate word.

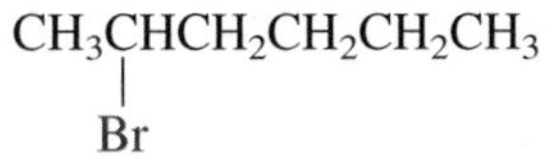

Substitutive name: 2-Bromohexane
Functional class name: 1-Methylpentyl bromide

Section 4.3 The substitutive names of alcohols are derived by replacing the *-e* ending of an alkane with *-ol.* The longest chain containing the OH group becomes the basis for the name. Functional class names of alcohols begin with the name of the alkyl group and end in the word *alcohol.*

Substitutive name: 2-Hexanol or Hexan-2-ol
Functional class name: 1-Methylpentyl alcohol

Section 4.4 Alcohols (X = OH) and alkyl halides (X = F, Cl, Br, or I) are classified as primary, secondary, or tertiary according to the degree of substitution at the carbon that bears the functional group.

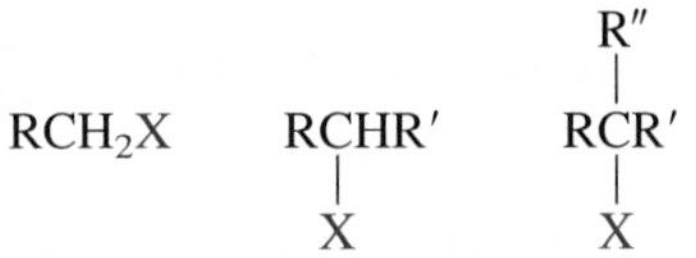

Primary Secondary Tertiary

Section 4.5 The halogens (especially fluorine and chlorine) and oxygen are more electronegative than carbon, and the carbon–halogen bond in alkyl halides and the carbon–oxygen bond in alcohols are polar. Carbon is the positive end of the dipole and halogen or oxygen the negative end.

Section 4.6 Dipole/induced-dipole and dipole–dipole attractive forces make alcohols higher boiling than alkanes of similar molecular size. The attractive force between —OH groups is called **hydrogen bonding.**

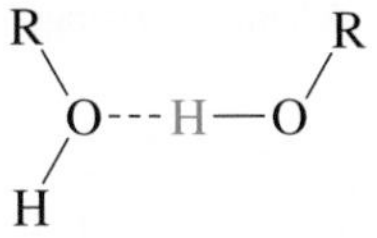

Hydrogen bonding between the hydroxyl group of an alcohol and water makes the water-solubility of alcohols greater than that of hydrocarbons. Low-molecular-weight alcohols [CH_3OH, CH_3CH_2OH, $CH_3CH_2CH_2OH$, and $(CH_3)_2CHOH$] are soluble in water in all proportions. Alkyl halides are insoluble in water.

Section 4.7 See Table 4.4

Section 4.8 Secondary and tertiary alcohols react with hydrogen halides by a mechanism that involves formation of a carbocation intermediate in the rate-determining step.

$$\textbf{1.}\quad \underset{\text{Alcohol}}{ROH} + \underset{\text{Hydrogen halide}}{HX} \underset{}{\overset{\text{fast}}{\rightleftharpoons}} \underset{\text{Alkyloxonium ion}}{R\overset{+}{O}H_2} + \underset{\text{Halide anion}}{X^-}$$

$$\textbf{2.}\quad \underset{\text{Alkyloxonium ion}}{R\overset{+}{O}H_2} \xrightarrow{\text{slow}} \underset{\text{Carbocation}}{R^+} + \underset{\text{Water}}{H_2O}$$

$$\textbf{3.}\quad \underset{\text{Carbocation}}{R^+} + \underset{\text{Halide ion}}{X^-} \xrightarrow{\text{fast}} \underset{\text{Alkyl halide}}{RX}$$

Section 4.9 The potential energy diagrams for separate elementary steps can be merged into a diagram for the overall process. The diagram for the reaction of a secondary or tertiary alcohol with a hydrogen halide is characterized by two intermediates and three transition states. The reaction is classified as a *unimolecular nucleophilic substitution,* abbreviated as S_N1.

TABLE 4.4 Conversions of Alcohols and Alkanes to Alkyl Halides

Reaction (section) and comments	General equation and specific example(s)
Reactions of alcohols with hydrogen halides (Section 4.7) Alcohols react with hydrogen halides to yield alkyl halides. The reaction is useful as a synthesis of alkyl halides. The reactivity of hydrogen halides decreases in the order HI > HBr > HCl > HF. Alcohol reactivity decreases in the order tertiary > secondary > primary.	$ROH + HX \longrightarrow RX + H_2O$ Alcohol, Hydrogen halide, Alkyl halide, Water 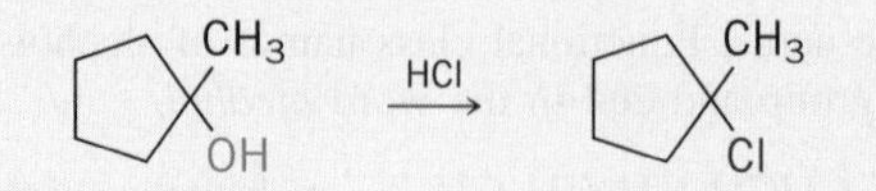 1-Methylcyclopentanol → 1-Chloro-1-methylcyclopentane (96%)
Reaction of alcohols with thionyl chloride (Section 4.14) Thionyl chloride converts alcohols to alkyl chlorides.	$ROH + SOCl_2 \longrightarrow RCl + SO_2 + HCl$ Alcohol, Thionyl chloride, Alkyl chloride, Sulfur dioxide, Hydrogen chloride $CH_3CH_2CH_2CH_2CH_2OH \xrightarrow[\text{pyridine}]{SOCl_2} CH_3CH_2CH_2CH_2CH_2Cl$ 1-Pentanol → 1-Chloropentane (80%)
Reaction of alcohols with phosphorus tribromide (Section 4.14) As an alternative to converting alcohols to alkyl bromides with hydrogen bromide, the inorganic reagent phosphorus tribromide is sometimes used.	$3ROH + PBr_3 \longrightarrow 3RBr + H_3PO_3$ Alcohol, Phosphorus tribromide, Alkyl bromide, Phosphorous acid $CH_3CH(OH)CH_2CH_2CH_3 \xrightarrow{PBr_3} CH_3CH(Br)CH_2CH_2CH_3$ 2-Pentanol → 2-Bromopentane (67%)
Free-radical halogenation of alkanes (Sections 4.15 through 4.19) Alkanes react with halogens by substitution of a halogen for a hydrogen on the alkane. The reactivity of the halogens decreases in the order $F_2 > Cl_2 > Br_2 > I_2$. The ease of replacing a hydrogen decreases in the order tertiary > secondary > primary > methyl. Chlorination is not very selective and so is used only when all the hydrogens of the alkane are equivalent. Bromination is highly selective, replacing tertiary hydrogens much more readily than secondary or primary ones.	$RH + X_2 \longrightarrow RX + HX$ Alkane, Halogen, Alkyl halide, Hydrogen halide 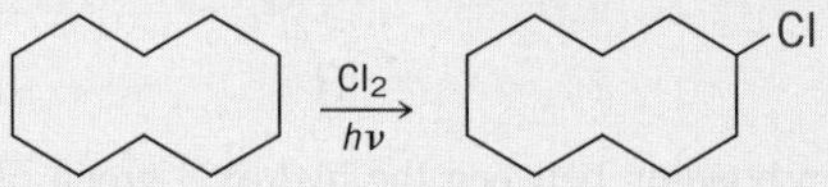 Cyclodecane → Cyclodecyl chloride (64%) $(CH_3)_2CHC(CH_3)_3 \xrightarrow[h\nu]{Br_2} (CH_3)_2C(Br)C(CH_3)_3$ 2,2,3-Trimethylbutane → 2-Bromo-2,3,3-trimethylbutane (80%)

Section 4.10 Carbocations contain a positively charged carbon with only three atoms or groups attached to it. This carbon is sp^2-hybridized and has a vacant $2p$ orbital.

Carbocations are stabilized by alkyl substituents attached directly to the positively charged carbon. Alkyl groups are *electron-releasing* substituents. Stability increases in the order:

$$\text{(least stable)} \quad CH_3^+ < RCH_2^+ < R_2CH^+ < R_3C^+ \quad \text{(most stable)}$$

Carbocations are strong **electrophiles** (Lewis acids) and react with **nucleophiles** (Lewis bases).

Section 4.11 The rate at which alcohols are converted to alkyl halides depends on the rate of carbocation formation: tertiary alcohols are most reactive; primary alcohols are least reactive.

Section 4.12 Primary alcohols and methanol do not react with hydrogen halides by way of carbocation intermediates. The nucleophilic species (Br^- for example) attacks the alkyloxonium ion and displaces a water molecule from carbon in a bimolecular step. This step is rate-determining, and the mechanism is S_N2.

Section 4.13 A quantitative relationship between the energy of activation (E_{act}), the rate constant (k) and temperature (T) is expressed by the Arrhenius equation:

$$k = Ae^{-E_{act}/RT}$$

An increase in temperature will increase the value of the rate constant (k). Small differences in activation energy lead to large differences in reaction rate.

Section 4.14 See Table 4.4

Section 4.15 See Table 4.4

Section 4.16 Methane reacts with Cl_2 to give chloromethane, dichloromethane, trichloromethane, and tetrachloromethane.

Section 4.17 Chlorination of methane, and halogenation of alkanes generally, proceed by way of **free-radical** intermediates. Alkyl radicals are neutral and have an unpaired electron on carbon.

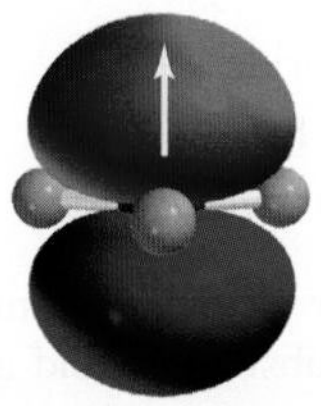

Like carbocations, free radicals are stabilized by alkyl substituents. The order of free-radical stability parallels that of carbocation stability.

Section 4.18 The elementary steps (1) through (3) describe a free-radical chain mechanism for the reaction of an alkane with a halogen.

1. (initiation step)

$$X_2 \longrightarrow 2X\cdot$$

Halogen molecule → Two halogen atoms

2. (propagation step)

$$RH + X\cdot \longrightarrow R\cdot + HX$$

Alkane + Halogen atom → Alkyl radical + Hydrogen halide

3. (propagation step)

$$R\cdot + X_2 \longrightarrow RX + \cdot X$$

Alkyl radical + Halogen molecule → Alkyl halide + Halogen atom

Section 4.19 See Table 4.4

PROBLEMS

4.23 Write structural formulas for each of the following alcohols and alkyl halides:

(a) Cyclobutanol (c) 3-Heptanol

(b) *sec*-Butyl alcohol (d) *trans*-2-Chlorocyclopentanol

(e) 2,6-Dichloro-4-methyl-4-octanol
(f) *trans*-4-*tert*-Butylcyclohexanol
(g) 1-Cyclopropylethanol
(h) 2-Cyclopropylethanol

4.24 Name each of the following compounds according to substitutive IUPAC nomenclature:

(a) $(CH_3)_2CHCH_2CH_2CH_2Br$
(b) $(CH_3)_2CHCH_2CH_2CH_2OH$
(c) Cl_3CCH_2Br
(d) $Cl_2CHCHBr$
|
Cl
(e) CF_3CH_2OH
(f)
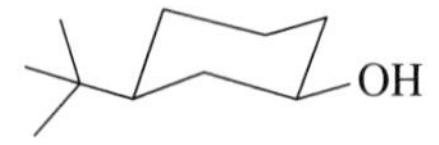

(g)
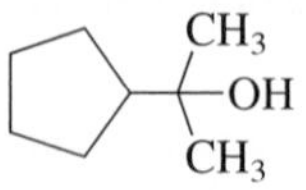

(h)
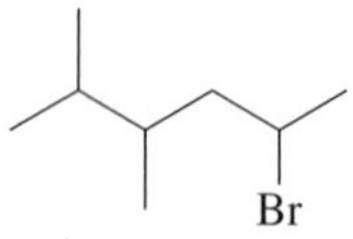

(i)
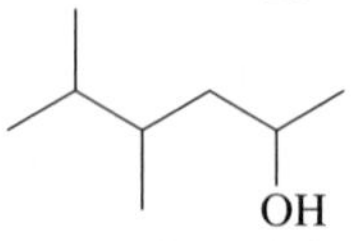

4.25 Handbooks are notorious for listing compounds according to their common names. One gives the name "*sec*-isoamyl alcohol" for a compound that could be called 1,2-dimethylpropyl alcohol according to the IUPAC functional class rules. The best name for this compound is the substitutive IUPAC name. What is it?

4.26 Each of the following is a functional class name developed according to the 1993 and 2004 IUPAC recommendations. Alkyl group names of this type are derived by naming the longest continuous chain that includes the point of attachment, numbering in the direction so as to give the substituted carbon the lower number. The *-e* ending of the corresponding alkane is replaced by *-yl,* which is preceded by the number corresponding to the substituted carbon bracketed by hyphens. Write a structural formula for each alkyl halide.

(a) 6-Methylheptan-3-yl chloride
(b) 2,2-Dimethylpentan-3-yl bromide
(c) 3,3-Dimethylcyclopentan-1-yl alcohol

4.27 Write structural formulas for all the constitutionally isomeric alcohols of molecular formula $C_5H_{12}O$. Assign a substitutive and a functional class name to each one, and specify whether it is a primary, secondary, or tertiary alcohol.

4.28 A hydroxyl group is a somewhat "smaller" substituent on a six-membered ring than is a methyl group. That is, the preference of a hydroxyl group for the equatorial orientation is less pronounced than that of a methyl group. Given this information, write structural formulas for all the isomeric methylcyclohexanols, showing each one in its most stable conformation. Give the substitutive IUPAC name for each isomer.

4.29 By assuming that the heat of combustion of the cis isomer was larger than the trans, structural assignments were made many years ago for the stereoisomeric 2-, 3-, and 4-methylcyclohexanols. This assumption is valid for two of the stereoisomeric pairs but is incorrect for the other. For which pair of stereoisomers is the assumption incorrect? Why?

4.30 (a) *Menthol,* used to flavor various foods and tobacco, is the most stable stereoisomer of 2-isopropyl-5-methylcyclohexanol. Draw its most stable conformation. Is the hydroxyl group cis or trans to the isopropyl group? To the methyl group?

(b) *Neomenthol* is a stereoisomer of menthol. That is, it has the same constitution but differs in the arrangement of its atoms in space. Neomenthol is the second most stable stereoisomer of 2-isopropyl-5-methylcyclohexanol; it is less stable than menthol but more stable than any other stereoisomer. Write the structure of neomenthol in its most stable conformation.

4.31 *Epichlorohydrin* is the common name of an industrial chemical used as a component in epoxy cement. The molecular formula of epichlorohydrin is C_3H_5ClO. Epichlorohydrin has an epoxide functional group; it does not have a methyl group. Write a structural formula for epichlorohydrin.

4.32 (a) Complete the structure of the pain-relieving drug *ibuprofen* on the basis of the fact that ibuprofen is a carboxylic acid that has the molecular formula $C_{13}H_{18}O_2$, X is an isobutyl group, and Y is a methyl group.

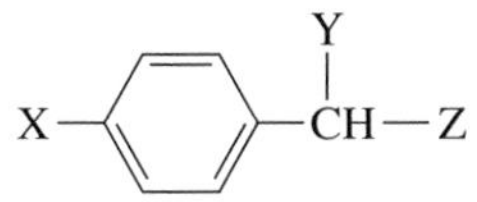

(b) *Mandelonitrile* may be obtained from peach flowers. Derive its structure from the template in part (a) given that X is hydrogen, Y is the functional group that characterizes alcohols, and Z characterizes nitriles.

4.33 *Isoamyl acetate* is the common name of the substance most responsible for the characteristic odor of bananas. Write a structural formula for isoamyl acetate, given the information that it is an ester in which the carbonyl group bears a methyl substituent and there is a 3-methylbutyl group attached to one of the oxygens.

4.34 *n-Butyl mercaptan* is the common name of a foul-smelling substance obtained from skunk fluid. It is a thiol of the type RX, where R is an *n*-butyl group and X is the functional group that characterizes a thiol. Write a structural formula for this substance.

4.35 Some of the most important organic compounds in biochemistry are the α-*amino acids,* represented by the general formula shown.

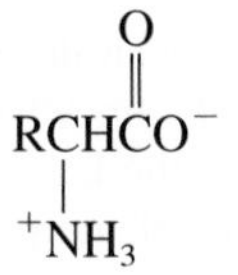

Write structural formulas for the following α-amino acids.

(a) Alanine (R = methyl)
(b) Valine (R = isopropyl)
(c) Leucine (R = isobutyl)
(d) Isoleucine (R = *sec*-butyl)
(e) Serine (R = XCH_2, where X is the functional group that characterizes alcohols)
(f) Cysteine (R = XCH_2, where X is the functional group that characterizes thiols)
(g) Aspartic acid (R = XCH_2, where X is the functional group that characterizes carboxylic acids)

4.36 The compound *zoapatanol* was isolated from the leaves of a Mexican plant. Classify each oxygen in zoapatanol according to the functional group to which it belongs. If an oxygen is part of an alcohol, classify the alcohol as primary, secondary, or tertiary.

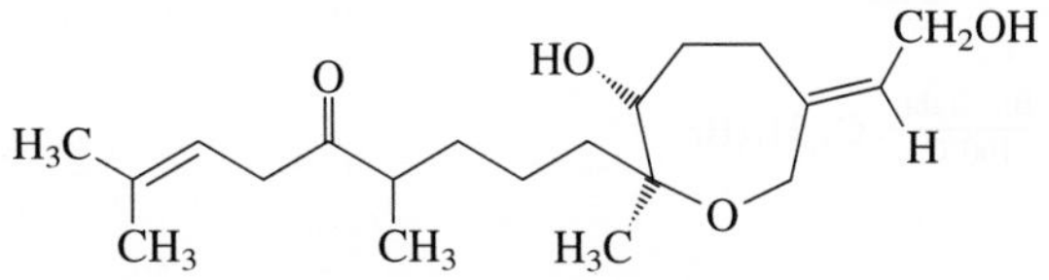

4.37 Consult Table 4.1 and classify each nitrogen-containing functional group in the anesthetic *lidocaine* according to whether it is an amide, or a primary, secondary, or tertiary amine.

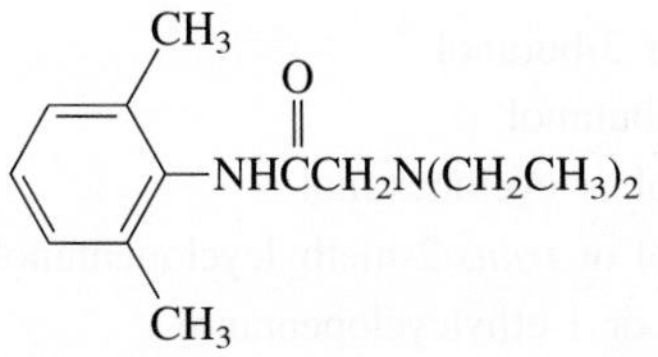

4.38 *Uscharidin* is the common name of a poisonous natural product having the structure shown. Locate all of the following in uscharidin:

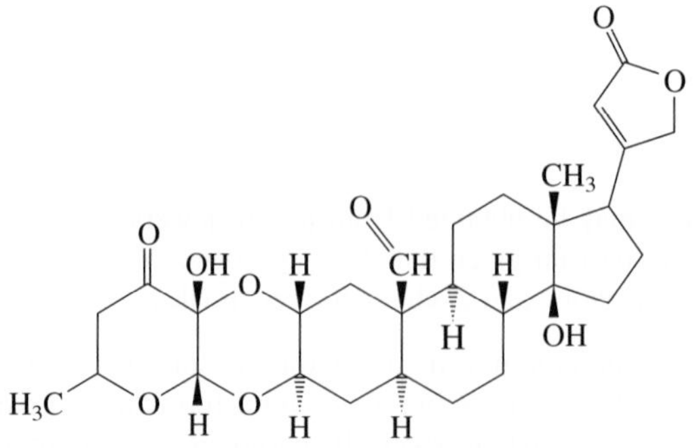

(a) Alcohol, aldehyde, ketone, and ester functional groups
(b) Methylene groups
(c) Primary carbons

4.39 Write a chemical equation for the reaction of 1-butanol with each of the following:
(a) Sodium amide ($NaNH_2$)
(b) Hydrogen bromide, heat
(c) Sodium bromide, sulfuric acid, heat
(d) Phosphorus tribromide
(e) Thionyl chloride

4.40 Each of the following reactions has been described in the chemical literature and involves an organic starting material somewhat more complex than those we have encountered so far. Nevertheless, on the basis of the topics covered in this chapter, you should be able to write the structure of the principal organic product of each reaction.

(a) 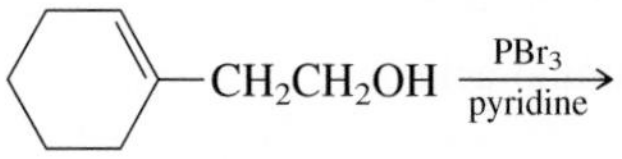

(b) 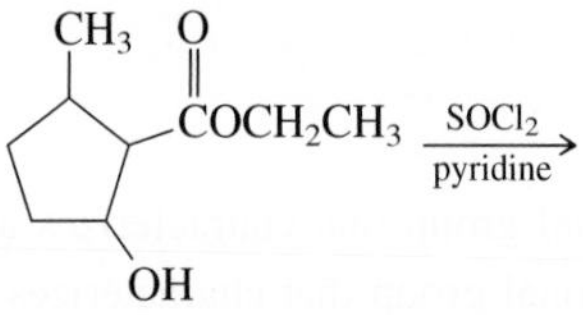

(c) 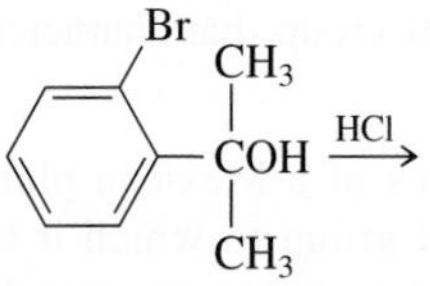

(d) 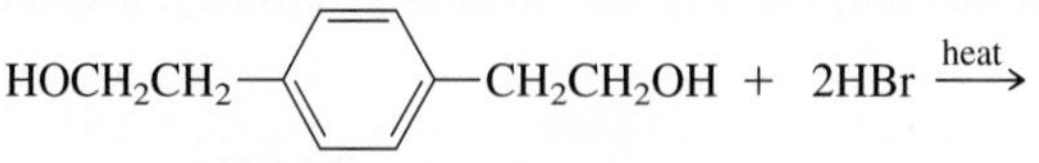

(e) 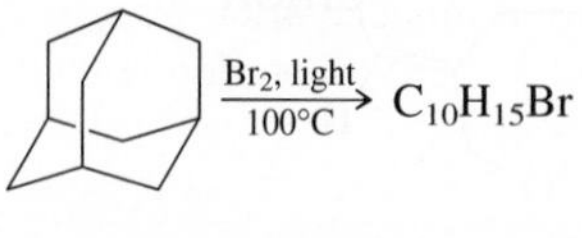

4.41 Select the compound in each of the following pairs that will be converted to the corresponding alkyl bromide more rapidly on being treated with hydrogen bromide. Explain the reason for your choice.
(a) 1-Butanol or 2-butanol
(b) 2-Methyl-1-butanol or 2-butanol
(c) 2-Methyl-2-butanol or 2-butanol
(d) 2-Methylbutane or 2-butanol
(e) 1-Methylcyclopentanol or cyclohexanol
(f) 1-Methylcyclopentanol or *trans*-2-methylcyclopentanol
(g) 1-Cyclopentylethanol or 1-ethylcyclopentanol

4.42 Assuming that the rate-determining step in the reaction of 2-methyl-2-butanol with hydrogen chloride to give 2-chloro-2-methylbutane is unimolecular, write an equation for this step. Use curved arrows to show the flow of electrons.

4.43 Assuming that the rate-determining step in the reaction of 1-hexanol with hydrogen bromide to give 1-bromohexane is an attack by a nucleophile on an alkyloxonium ion, write an equation for this step. Use curved arrows to show the flow of electrons.

4.44 Although useful in agriculture as a soil fumigant, methyl bromide is an ozone-depleting chemical, and its production is being phased out. The industrial preparation of methyl bromide is from methanol, by reaction with hydrogen bromide. Write a mechanism for this reaction and classify it as S_N1 or S_N2.

4.45 Two stereoisomers of 1-bromo-4-methylcyclohexane are formed when *trans*-4-methylcyclohexanol reacts with hydrogen bromide. Write structural formulas of:

(a) *trans*-4-Methylcyclohexanol

(b) The carbocation intermediate in this reaction

(c) The two stereoisomers of 1-bromo-4-methylcyclohexane

4.46 (a) Use the bond dissociation enthalpy data in Table 4.3 to calculate $\Delta H°$ for the propagation step

$$CH_4 + \cdot\ddot{\underset{\cdot\cdot}{Br}}: \longrightarrow \cdot CH_3 + H—\ddot{\underset{\cdot\cdot}{Br}}:$$

(b) The activation energy for this step is 76 kJ/mol (18.3 kcal/mol). Sketch a potential energy diagram for this step, labeling reactants, products, and transition state.

(c) Does the structure of the transition state more closely resemble reactants or products? Why?

4.47 The bond dissociation enthalpies of *n*-propyl and isopropyl chloride are the same within experimental error (Table 4.3). However, it is incorrect to conclude that the data indicate equal stabilities of *n*-propyl and isopropyl radical. Why? Why are the bond dissociation enthalpies of propane a better indicator of the free-radical stabilities?

4.48 Under controlled conditions at −78°C it is possible to fluorinate 2,2-dimethylpropane to yield $(CF_3)_4C$. Write a balanced chemical equation for this reaction.

4.49 In a search for fluorocarbons having anesthetic properties, 1,2-dichloro-1,1-difluoropropane was subjected to photochemical chlorination. Two isomeric products were obtained, one of which was identified as 1,2,3-trichloro-1,1-difluoropropane. What is the structure of the second compound?

4.50 In both the following exercises, assume that all the methylene groups in the alkane are equally reactive as sites of free-radical chlorination.

(a) Photochemical chlorination of heptane gave a mixture of monochlorides containing 15% 1-chloroheptane. What other monochlorides are present? Estimate the percentage of each of these additional $C_7H_{15}Cl$ isomers in the monochloride fraction.

(b) Photochemical chlorination of dodecane gave a monochloride fraction containing 19% 2-chlorododecane. Estimate the percentage of 1-chlorododecane present in that fraction.

4.51 Photochemical chlorination of 2,2,4-trimethylpentane gives four isomeric monochlorides.

(a) Write structural formulas for these four isomers.

(b) The two primary chlorides make up 65% of the monochloride fraction. Assuming that all the primary hydrogens in 2,2,4-trimethylpentane are equally reactive, estimate the percentage of each of the two primary chlorides in the product mixture.

4.52 Photochemical chlorination of pentane gave a mixture of three isomeric monochlorides. The principal monochloride constituted 46% of the total, and the remaining 54% was approximately a 1:1 mixture of the other two isomers. Write structural formulas for the three monochloride isomers and specify which one was formed in greatest amount. (Recall that a secondary hydrogen is abstracted three times faster by a chlorine atom than a primary hydrogen.)

4.53 Cyclopropyl chloride has been prepared by the free-radical chlorination of cyclopropane. Write a stepwise mechanism for this reaction.

4.54 Deuterium oxide (D_2O) is water in which the protons (^1H) have been replaced by their heavier isotope deuterium (^2H). It is readily available and is used in a variety of mechanistic studies in organic chemistry and biochemistry. When D_2O is added to an alcohol (ROH), deuterium replaces the proton of the hydroxyl group.

$$ROH + D_2O \rightleftharpoons ROD + DOH$$

The reaction takes place extremely rapidly, and if D_2O is present in excess, all the alcohol is converted to ROD. This hydrogen–deuterium exchange can be catalyzed by either acids or bases. If D_3O^+ is the catalyst in acid solution and DO^- the catalyst in base, write reasonable reaction mechanisms for the conversion of ROH to ROD under conditions of (a) acid catalysis and (b) base catalysis.

Descriptive Passage and Interpretive Problems 4

More About Potential Energy Diagrams

Chapter 5 describes *elimination* reactions and their mechanisms. In one example, heating *tert*-butyl bromide in ethanol gives the alkene 2-methylpropene by a two-step mechanism:

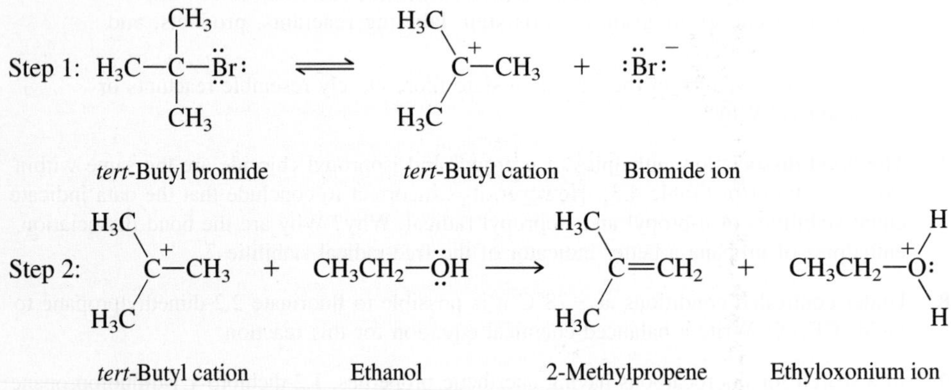

Step 1:
tert-Butyl bromide
tert-Butyl cation
Bromide ion
Step 2:
tert-Butyl cation
Ethanol
2-Methylpropene
Ethyloxonium ion

A potential energy diagram for the reaction provides additional information to complement the mechanism expressed in the equations for the two elementary steps.

The energy relationships in the diagram are not only useful in their own right, but also aid in understanding the structural changes occurring at the transition state. Hammond's postulate tells us that if two states occur consecutively, the closer they are in energy, the more similar they are in structure.

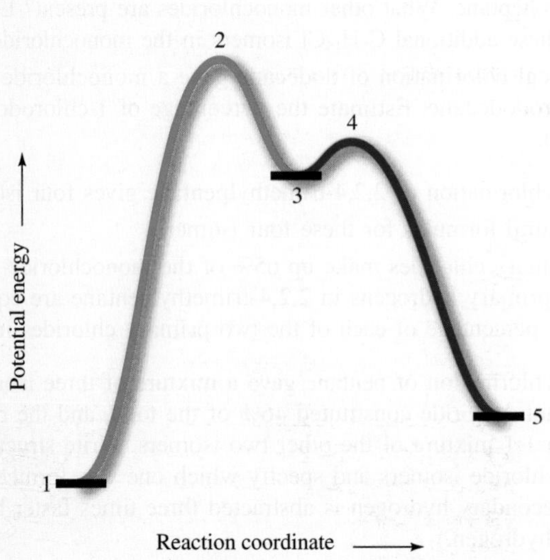

1
2
3
4
5
Potential energy
Reaction coordinate

4.55 Which equation corresponds to the overall reaction for which steps 1 and 2 describe the mechanism?

A. 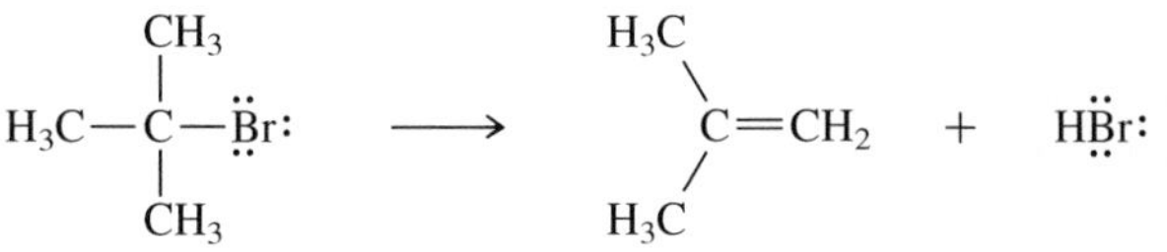

B. 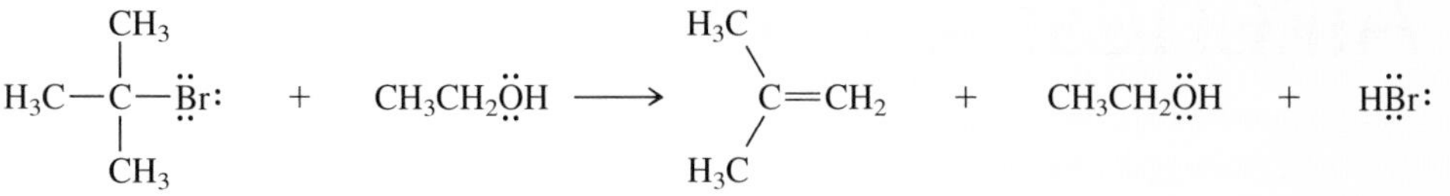

C. 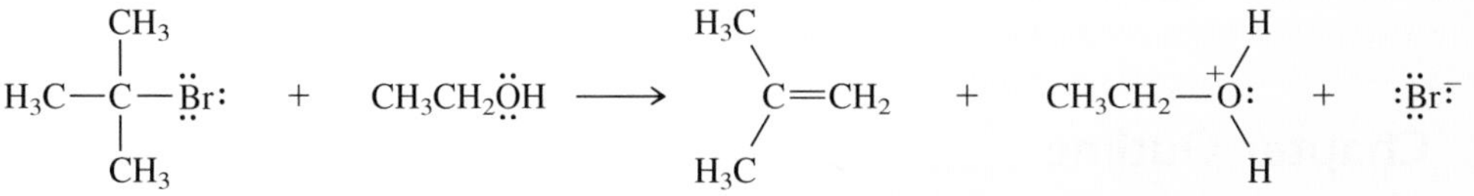

D. 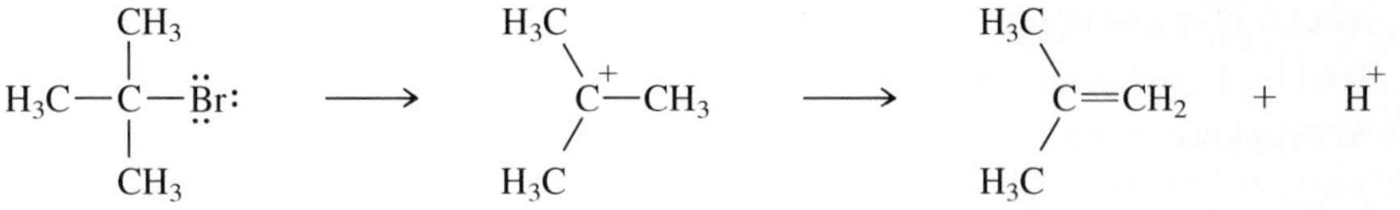

4.56 Ethanol is:

A. a catalyst

B. a reactive intermediate

C. a Brønsted acid

D. a Brønsted base

4.57 According to the potential energy diagram, the overall reaction is:

A. endothermic

B. exothermic

4.58 Classify the elementary steps in the mechanism according to their molecularity.

A. Step 1 is unimolecular; step 2 is bimolecular.

B. Step 1 is bimolecular; step 2 is unimolecular.

C. Both steps are unimolecular.

D. Both steps are bimolecular.

4.59 Classify states 2–4 in the potential energy diagram.

A. 2, 3, and 4 are transition states

B. 2, 3, and 4 are reactive intermediates

C. 2 and 4 are transition states; 3 is a reactive intermediate

D. 2 and 4 are reactive intermediates; 3 is a transition state

4.60 According to the diagram, the activation energy of the slow step is given by the energy difference between states

A. 1 and 2

B. 2 and 3

C. 3 and 4

D. 1 and 5

4.61 What best describes the species at the rate-determining transition state?

A. 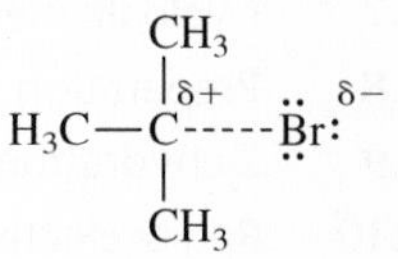

B. 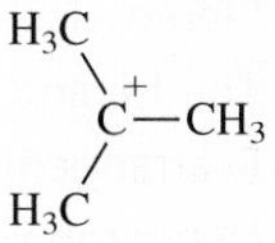

C. 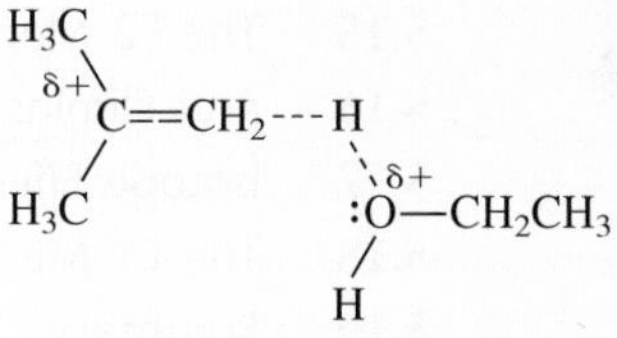

D. 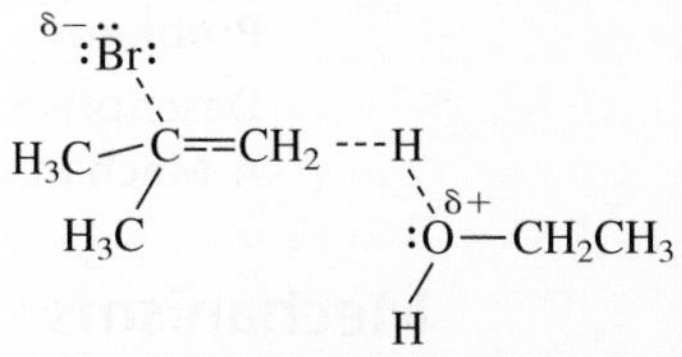

4.62 By applying Hammond's postulate to the potential energy diagram for this reaction, we can say that:

A. the structure of 2 is more carbocation-like than 4

B. the structure of 2 is less carbocation-like than 4

C. the structure of 2 resembles 1 more than it resembles 3

D. the structure of 4 resembles 5 more than it resembles 3

5 Structure and Preparation of Alkenes: Elimination Reactions

Chapter Outline

Mechanisms

Ethylene is a naturally occurring hormone, present in all plants. It stimulates these tomatoes to ripen.

ALKENES are hydrocarbons that contain a carbon–carbon double bond. A carbon–carbon double bond is both an important structural unit and an important functional group in organic chemistry. The shape of an organic molecule is influenced by its presence, and the double bond is the site of most of the chemical reactions that alkenes undergo.

This chapter is the first of two dealing with alkenes; it describes their structure, bonding, and preparation. Chapter 6 examines their chemical reactions.

5.1 Alkene Nomenclature

We give alkenes IUPAC names by replacing the *-ane* ending of the corresponding alkane with *-ene*. The two simplest alkenes are ethene and propene. Both are also well known by their common names *ethylene* and *propylene*.

$H_2C{=}CH_2$	$CH_3CH{=}CH_2$
IUPAC name: **ethene**	IUPAC name: **propene**
Common name: ethylene	Common name: propylene

The longest continuous chain that includes the double bond forms the base name of the alkene, and the chain is numbered in the direction that gives the doubly bonded carbons their lower numbers. The locant (or numerical position) of only one of the doubly bonded carbons is specified in the name; it is understood that the other doubly bonded carbon must follow in sequence. The locant may precede the parent chain (1979 IUPAC rules) or the *-ene* suffix (1993 and 2004 recommendations).

$\overset{1}{H_2C}{=}\overset{2}{C}H\overset{3}{C}H_2\overset{4}{C}H_3$	$\overset{6}{C}H_3\overset{5}{C}H_2\overset{4}{C}H_2\overset{3}{C}H{=}\overset{2}{C}H\overset{1}{C}H_3$
1-Butene	2-Hexene
or	or
But-1-ene	Hex-2-ene

Carbon–carbon double bonds take precedence over alkyl groups and halogens in determining the main carbon chain and the direction in which it is numbered.

$\overset{4}{C}H_3\overset{3}{C}H(CH_3)\overset{2}{C}H{=}\overset{1}{C}H_2$	$Br\overset{6}{C}H_2\overset{5}{C}H_2\overset{4}{C}H_2\overset{3}{C}H(\overset{2}{C}H{=}\overset{1}{C}H_2)CH_2CH_2CH_3$
3-Methyl-1-butene	6-Bromo-3-propyl-1-hexene
or	or
3-Methylbut-1-ene	6-Bromo-3-propylhex-1-ene

Hydroxyl groups, however, outrank the double bond, and a chain that contains both an —OH group and a double bond is numbered in the direction that gives the carbon attached to the —OH group the lower number. Compounds that contain both a double bond and a hydroxyl group combine the suffixes *-en* + *-ol* to signify that both functional groups are present.

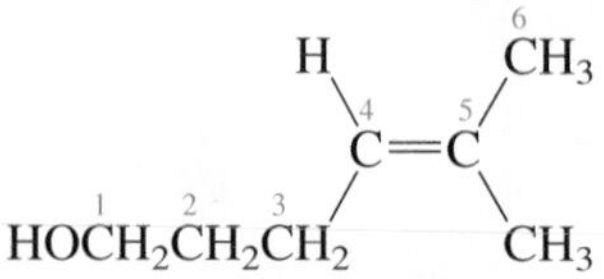

5-Methyl-4-hexen-1-ol
or
5-Methylhex-4-en-1-ol

Problem 5.1

Name each of the following using IUPAC nomenclature:

(a) $(CH_3)_2C{=}C(CH_3)_2$
(b) $(CH_3)_3CCH{=}CH_2$
(c) $(CH_3)_2C{=}CHCH_2CH_2CH_3$
(d) $H_2C{=}CHCH_2CHCH_3$ (with Cl on C-4)
(e) $H_2C{=}CHCH_2CHCH_3$ (with OH on C-4)

Sample Solution (a) The longest continuous chain in this alkene contains four carbon atoms. The double bond is between C-2 and C-3, and so it is named as a derivative of 2-butene.

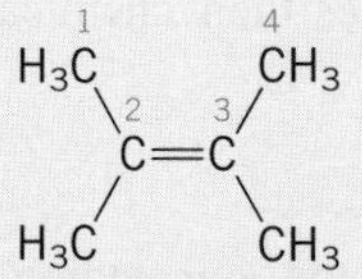

2,3-Dimethyl-2-butene
or
2,3-Dimethylbut-2-ene

The two methyl groups are substituents attached to C-2 and C-3 of the main chain.

The common names of certain frequently encountered *alkyl* groups, such as isopropyl and *tert*-butyl, are acceptable in the IUPAC system. Three *alkenyl* groups—**vinyl, allyl,** and **isopropenyl**—are treated the same way.

Vinyl chloride is an industrial chemical produced in large amounts (10^{10} lb/year in the United States) and is used in the preparation of poly(vinyl chloride). Poly(vinyl chloride), often called simply *vinyl*, has many applications, including siding for houses, wall coverings, and PVC piping.

$H_2C{=}CH—$ as in $H_2C{=}CHCl$ or 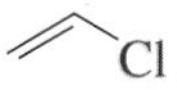

Vinyl — Vinyl chloride

$H_2C{=}CHCH_2—$ as in $H_2C{=}CHCH_2OH$ or 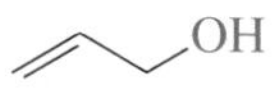

Allyl — Allyl alcohol

$H_2C{=}C(CH_3)—$ as in $H_2C{=}CCl(CH_3)$ or 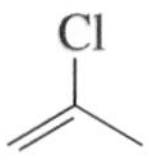

Isopropenyl — Isopropenyl chloride

When a CH_2 group is doubly bonded to a ring, the prefix *methylene* is added to the name of the ring.

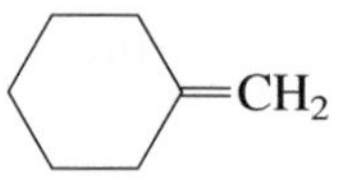

Methylenecyclohexane

Cycloalkenes and their derivatives are named by adapting cycloalkane terminology to the principles of alkene nomenclature.

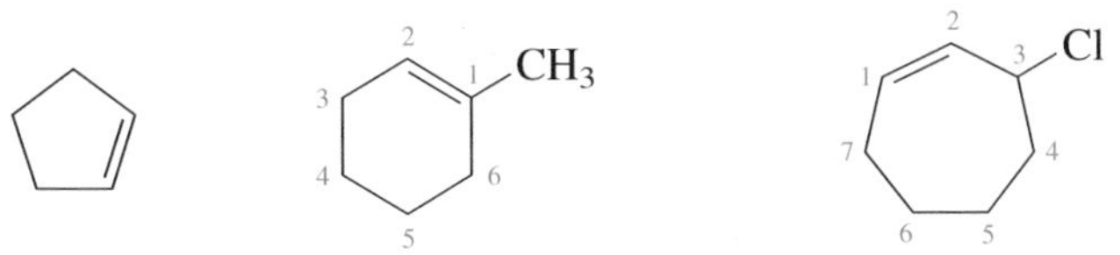

Cyclopentene 1-Methylcyclohexene 3-Chlorocycloheptene (not 1-chloro-2-cycloheptene)

No locants are needed in the absence of substituents; it is understood that the double bond connects C-1 and C-2. Substituted cycloalkenes are numbered beginning with the double bond, proceeding through it, and continuing in sequence around the ring. The direction is chosen so as to give the lower of two possible numbers to the substituent.

Problem 5.2

Write structural formulas and give the IUPAC names of all the monochloro-substituted derivatives of cyclopentene.

5.2 Structure and Bonding in Alkenes

The structure of ethylene and the orbital hybridization model for its double bond were presented in Section 2.20 and are briefly reviewed in Figure 5.1. Ethylene is planar, each carbon is sp^2-hybridized, and the double bond is considered to have a σ component and a π component. The σ component arises from overlap of sp^2 hybrid orbitals along a line connecting the two carbons, the π component via a "side-by-side" overlap of two *p* orbitals. Regions of high electron density, attributed to the π electrons, appear above and below the plane of the molecule and are clearly evident in the electrostatic potential map. Most of the reactions of ethylene and other alkenes involve these electrons.

On the basis of their bond-dissociation enthalpies, the C=C bond in ethylene is stronger than the C—C single bond in ethane, but it is not twice as strong.

$$H_2C{=}CH_2 \longrightarrow 2\,\dot{\underset{\cdot}{C}}H_2 \quad \Delta H^\circ = +730 \text{ kJ } (172 \text{ kcal})$$

Ethylene Methylene

$$H_3C{-}CH_3 \longrightarrow 2\,\dot{C}H_3 \quad \Delta H^\circ = +375 \text{ kJ } (90 \text{ kcal})$$

Ethane Methyl

While it is not possible to apportion the C=C bond energy of ethylene between its σ and π components, the data suggest that the π bond is weaker than the σ bond.

Problem 5.3

What assumptions would you have to make in order to calculate the π-bond strength in ethylene from the bond-dissociation data?

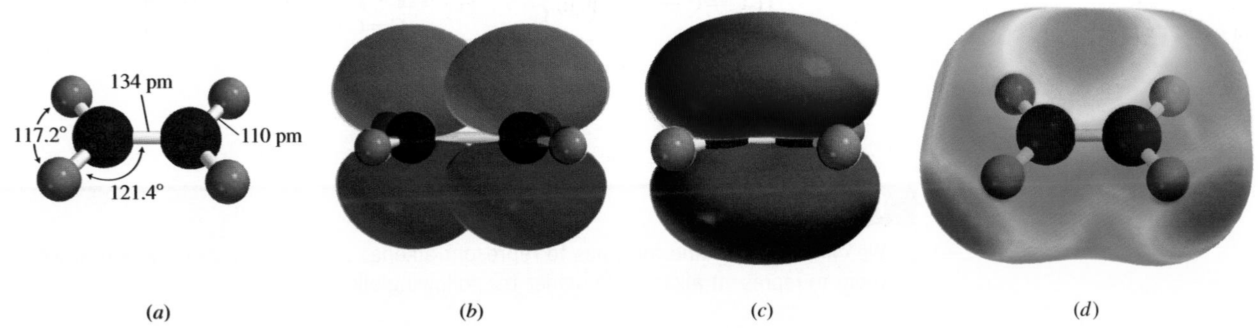

Figure 5.1

(*a*) The planar framework of σ bonds in ethylene showing bond distances and angles. (*b*) and (*c*) The half-filled *p* orbitals of two sp^2-hybridized carbons overlap to produce a π bond. (*d*) The electrostatic potential map shows a region of high negative potential due to the π electrons above and below the plane of the atoms.

Ethylene

Ethylene was known to chemists in the eighteenth century and isolated in pure form in 1795. An early name for ethylene was *gaz oléfiant* (French for "oil-forming gas"), to describe the fact that an oily liquid product is formed when two gases—ethylene and chlorine—react with each other.

$$\underset{\substack{\text{Ethylene}\\(\text{bp: }-104^\circ\text{C})}}{H_2C{=}CH_2} + \underset{\substack{\text{Chlorine}\\(\text{bp: }-34^\circ\text{C})}}{Cl_2} \longrightarrow \underset{\substack{\text{1,2-Dichloroethane}\\(\text{bp: }83^\circ\text{C})}}{ClCH_2CH_2Cl}$$

The term *gaz oléfiant* was the forerunner of the general term *olefin,* formerly used as the name of the class of compounds we now call *alkenes.*

Ethylene occurs naturally in small amounts as a plant hormone. It is formed in a complex series of steps from a compound containing a cyclopropane ring:

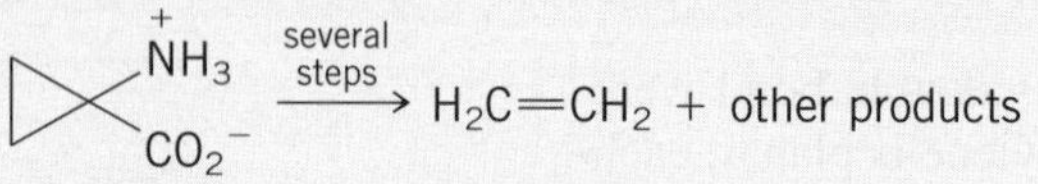

1-Amino-cyclopropane-carboxylic acid — Ethylene

Even minute amounts of ethylene can stimulate the ripening of fruits for example, and the rate of ripening increases with the concentration of ethylene. This property is used to advantage in the marketing of bananas. Bananas are picked green in the tropics, kept green by being stored with adequate ventilation to limit the amount of ethylene present, and then induced to ripen at their destination by passing ethylene over the fruit.

Ethylene is the cornerstone of the world's mammoth petrochemical industry and is produced in vast quantities. In a typical year the amount of ethylene produced in the United States (5×10^{10} lb) exceeds the combined weight of all of its people. In one process, ethane from natural gas is heated to bring about its dissociation into ethylene and hydrogen:

$$\underset{\text{Ethane}}{CH_3CH_3} \xrightarrow{750^\circ\text{C}} \underset{\text{Ethylene}}{H_2C{=}CH_2} + \underset{\text{Hydrogen}}{H_2}$$

This **dehydrogenation** is simultaneously both a source of ethylene and one of the methods by which hydrogen is prepared on an industrial scale. Most of this hydrogen is subsequently used to reduce nitrogen to ammonia for the preparation of fertilizer.

Similarly, dehydrogenation of propane gives propene:

$$\underset{\text{Propane}}{CH_3CH_2CH_3} \xrightarrow{750^\circ\text{C}} \underset{\text{Propene}}{CH_3CH{=}CH_2} + \underset{\text{Hydrogen}}{H_2}$$

Propene is the second most important petrochemical and is produced on a scale about half that of ethylene.

Almost any hydrocarbon can serve as a starting material for production of ethylene and propene. Cracking of petroleum (Section 2.16) gives ethylene and propene by processes involving cleavage of carbon–carbon bonds of higher molecular weight hydrocarbons.

The major uses of ethylene and propene are as starting materials for the preparation of polyethylene and polypropylene plastics, fibers, and films. These and other applications will be described in Chapter 6.

There are two different types of carbon–carbon bonds in propene, $CH_3CH{=}CH_2$. The double bond is of the $\sigma + \pi$ type, and the bond to the methyl group is a σ bond formed by sp^3–sp^2 overlap.

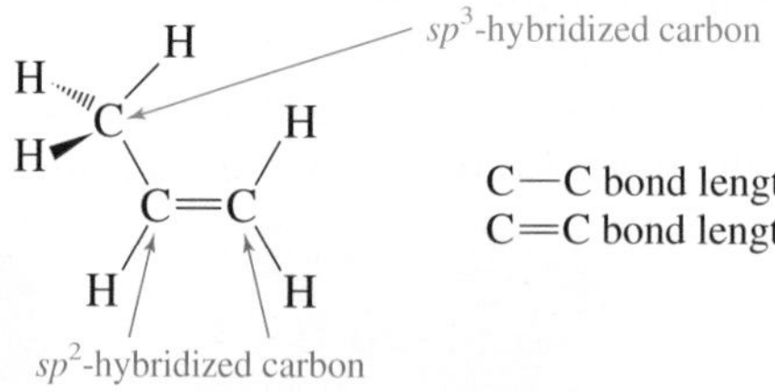

C—C bond length = 150 pm
C=C bond length = 134 pm

Problem 5.4

We can use bond-line formulas to represent alkenes in much the same way that we use them to represent alkanes. Consider the following alkene:

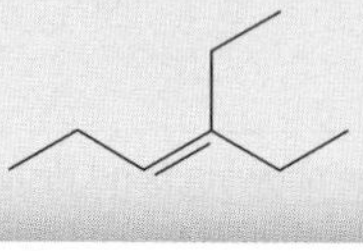

(a) What is the molecular formula of this alkene?
(b) What is its IUPAC name?
(c) How many carbon atoms are sp^2-hybridized in this alkene? How many are sp^3-hybridized?
(d) How many σ bonds are of the sp^2–sp^3 type? How many are of the sp^3–sp^3 type?

Sample Solution (a) Recall when writing bond-line formulas for hydrocarbons that a carbon occurs at each end and at each bend in a carbon chain. The appropriate number of hydrogens are attached so that each carbon has four bonds. Thus the compound shown is

$$CH_3CH_2CH{=}C(CH_2CH_3)_2$$

The general molecular formula for an alkene is C_nH_{2n}. Ethylene is C_2H_4; propene is C_3H_6. Counting the carbons and hydrogens of the compound shown (C_8H_{16}) reveals that it, too, corresponds to C_nH_{2n}.

5.3 Isomerism in Alkenes

Although ethylene is the only two-carbon alkene, and propene the only three-carbon alkene, there are *four* isomeric alkenes of molecular formula C_4H_8:

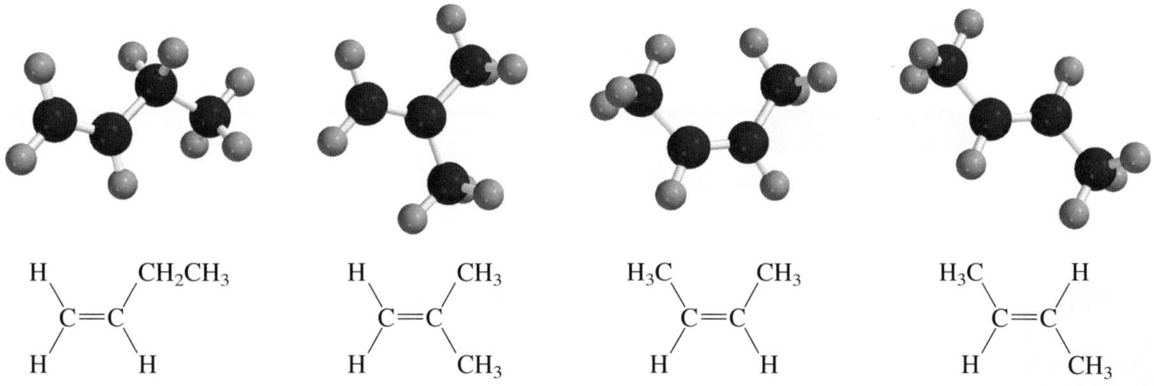

1-Butene | 2-Methylpropene | *cis*-2-Butene | *trans*-2-Butene

1-Butene has an unbranched carbon chain with a double bond between C-1 and C-2. It is a constitutional isomer of the other three. Similarly, 2-methylpropene, with a branched carbon chain, is a constitutional isomer of the other three.

The pair of isomers designated *cis*- and *trans*-2-butene have the same constitution; both have an unbranched carbon chain with a double bond connecting C-2 and C-3. They differ from each other in that the cis isomer has both of its methyl groups on the same side of the double bond, but the methyl groups in the trans isomer are on opposite sides of the double bond. Recall from Section 3.11 that isomers that have the same constitution but differ in the arrangement of their atoms in space are classified as *stereoisomers*. *cis*-2-Butene and *trans*-2-butene are stereoisomers, and the terms *cis* and *trans* specify the *configuration* of the double bond.

Stereoisomeric alkenes are sometimes referred to as *geometric isomers*.

Cis–trans stereoisomerism in alkenes is not possible when one of the doubly bonded carbons bears two identical substituents. Thus, neither 1-butene nor 2-methylpropene can have stereoisomers.

1-Butene (no stereoisomers possible) | 2-Methylpropene (no stereoisomers possible)

Problem 5.5

How many alkenes have the molecular formula C_5H_{10}? Write their structures and give their IUPAC names. Specify the configuration of stereoisomers as cis or trans as appropriate.

In principle, *cis*-2-butene and *trans*-2-butene may be interconverted by rotation about the C-2—C-3 double bond. However, unlike rotation about single bonds, which is quite fast, rotation about double bonds is restricted. Interconversion of the cis and trans isomers of 2-butene has an activation energy which is 10–15 times greater than that for rotation about the single bond of an alkane and does *not* occur under normal circumstances.

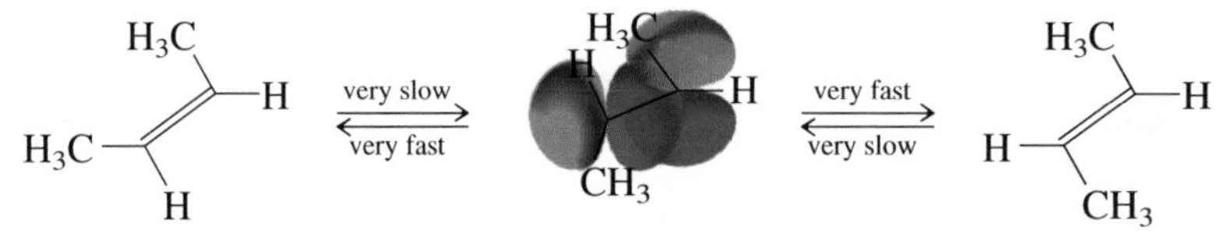

cis-2-Butene (stable) — transition state for rotation about C=C (unstable: *p* orbitals perpendicular to each other; poor geometry for π bonding) — *trans*-2-Butene (stable)

π-Bonding in *cis*- and *trans*-2-butene is strong because of the favorable parallel alignment of the *p* orbitals at C-2 and C-3. Interconverting the two stereoisomers, however, requires these *p* orbitals to be at right angles to each other, decreases their overlap, and weakens the π component of the double bond.

Problem 5.6

Are *cis*-2-hexene and *trans*-3-hexene stereoisomers? Explain.

Oleic acid is prepared from olive oil.

5.4 Naming Stereoisomeric Alkenes by the *E–Z* Notational System

When the groups on either end of a double bond are the same or are structurally similar to each other, it is a simple matter to describe the configuration of the double bond as cis or trans. Oleic acid, for example, has a cis double bond. Cinnamaldehyde has a trans double bond.

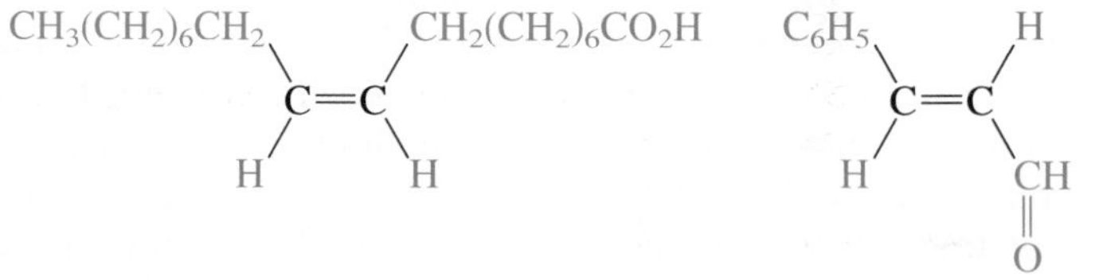

Oleic acid — Cinnamaldehyde

Problem 5.7

Female houseflies attract males by sending a chemical signal known as a *pheromone.* The substance emitted by the female housefly that attracts the male has been identified as *cis*-9-tricosene, $C_{23}H_{46}$. Write a structural formula, including stereochemistry, for this compound.

Cinnamaldehyde gives cinnamon its flavor.

The terms *cis* and *trans* are ambiguous, however, when it is not obvious which substituent on one carbon is "similar" or "analogous" to a reference substituent on the other. A completely unambiguous system for specifying double-bond stereochemistry has been adopted by the IUPAC based on an *atomic number* criterion for ranking substituents on the doubly bonded carbons. When atoms of higher atomic number are on the *same* side of the double bond, we say that the double bond has the ***Z*** configuration, where *Z* stands for the German word ***zusammen,*** meaning "together." When atoms of higher atomic number are on *opposite* sides of the double bond, the configuration is ***E,*** standing for the German word ***entgegen,*** meaning "opposite."

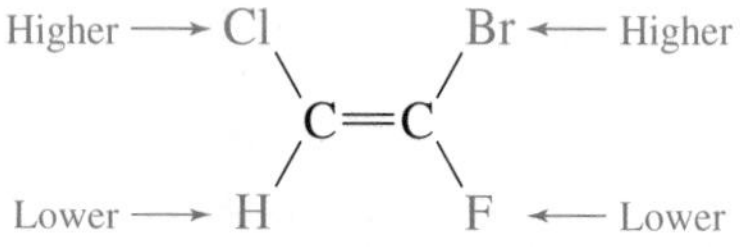

Z configuration
Higher ranked substituents (Cl and Br) are on same side of double bond

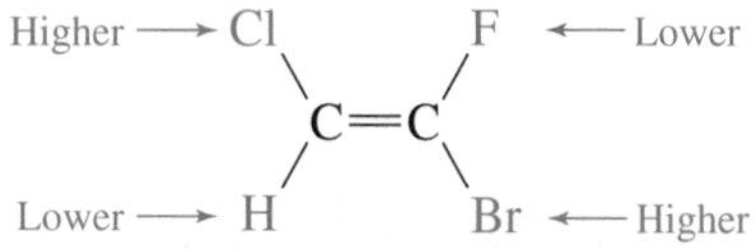

E configuration
Higher ranked substituents (Cl and Br) are on opposite sides of double bond

The groups on the double bonds of most alkenes are, of course, often more complicated than in this example. The rules for ranking substituents, especially alkyl groups, are described in Table 5.1.

The priority rules in Table 5.1 were developed by R. S. Cahn and Sir Christopher Ingold (England) and Vladimir Prelog (Switzerland) in the context of a different aspect of organic stereochemistry; they will appear again in Chapter 7.

Problem 5.8

Determine the configuration of each of the following alkenes as *Z* or *E* as appropriate:

(a) 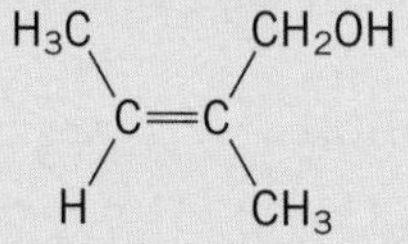

(c) 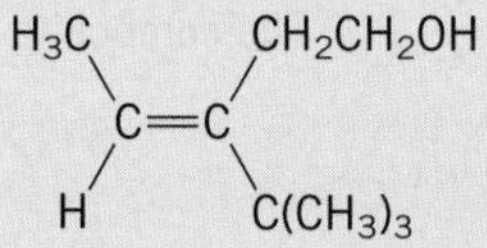

(b) 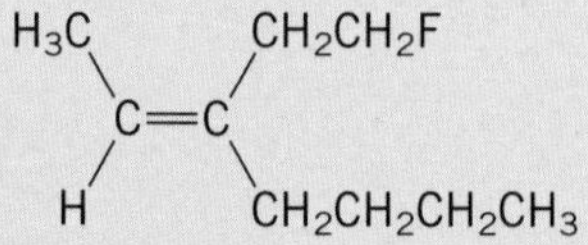

(d) 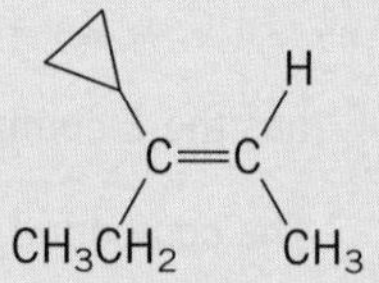

Sample Solution (a) One of the doubly bonded carbons bears a methyl group and a hydrogen. According to the rules of Table 5.1, methyl outranks hydrogen. The other carbon atom of the double bond bears a methyl and a —CH_2OH group. The —CH_2OH group is of higher priority than methyl.

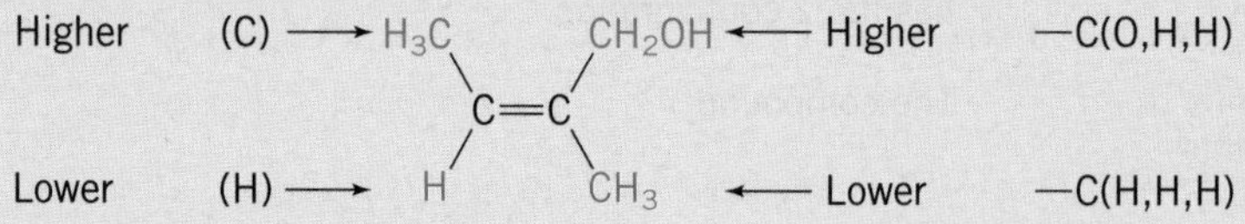

Higher ranked groups are on the same side of the double bond; the configuration is *Z*.

A table on the inside back cover (right page) lists some of the more frequently encountered atoms and groups in order of increasing precedence. You should not attempt to memorize this table, but should be able to derive the relative placement of one group versus another.

Problem 5.9

Name the first three compounds in Table 5.1.

Sample Solution

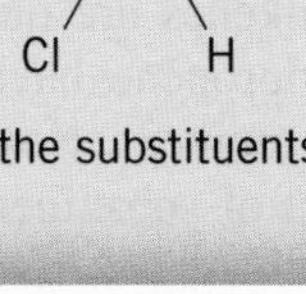

Number the chain as shown and list the substituents alphabetically. The compound is (*Z*)-1-bromo-1-chloropropene.

TABLE 5.1 Cahn–Ingold–Prelog Priority Rules

Rule	Example
1. Higher atomic number takes precedence over lower. Bromine (atomic number 35) outranks chlorine (atomic number 17). Methyl (C, atomic number 6) outranks hydrogen (atomic number 1).	The compound 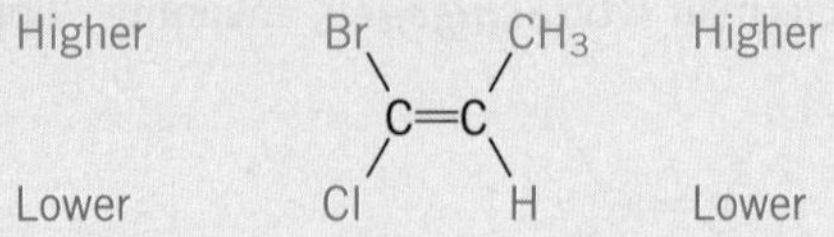Higher Br CH_3 Higher C=C Lower Cl H Lower has the *Z* configuration. Higher ranked atoms (Br and C of CH_3) are on the same side of the double bond.
2. When two atoms directly attached to the same carbon of the double bond are identical, compare the atoms attached to these two on the basis of their atomic numbers. Precedence is determined at the first point of difference: Ethyl [—C(C,H,H)] outranks methyl [—C(H,H,H)] Similarly, *tert*-butyl outranks isopropyl, and isopropyl outranks ethyl: $—C(CH_3)_3 > —CH(CH_3)_2 > —CH_2CH_3$ —C(C,C,C) > —C(C,C,H) > —C(C,H,H)	The compound 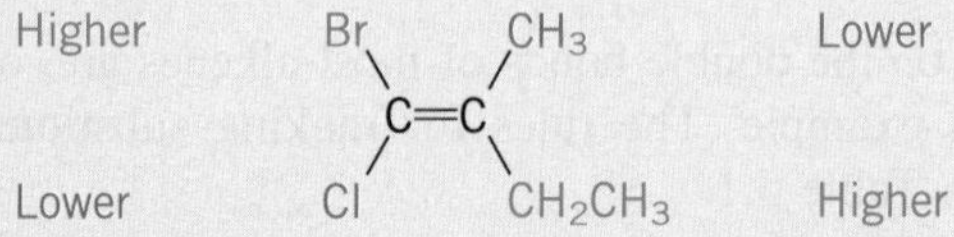Higher Br CH_3 Lower C=C Lower Cl CH_2CH_3 Higher has the *E* configuration.
3. Work outward from the point of attachment, comparing all the atoms attached to a particular atom before proceeding further along the chain: $—CH(CH_3)_2$ [—C(C,C,H)] outranks $—CH_2CH_2OH$ [—C(C,H,H)]	The compound 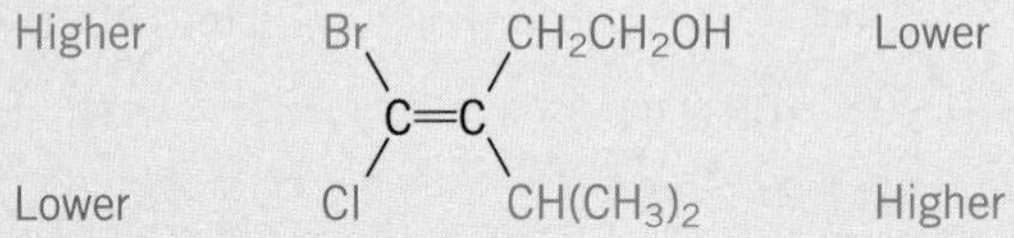Higher Br CH_2CH_2OH Lower C=C Lower Cl $CH(CH_3)_2$ Higher has the *E* configuration.
4. When working outward from the point of attachment, always evaluate substituent atoms one by one, never as a group. Because oxygen has a higher atomic number than carbon, $—CH_2OH$ [—C(O,H,H)] outranks $—C(CH_3)_3$ [—C(C,C,C)]	The compound 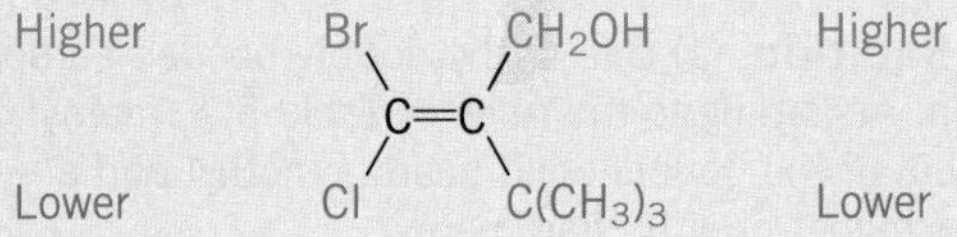Higher Br CH_2OH Higher C=C Lower Cl $C(CH_3)_3$ Lower has the *Z* configuration.
5. An atom that is multiply bonded to another atom is considered to be replicated as a substituent on that atom: 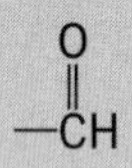O —CH is treated as if it were —C(O,O,H) The group —CH=O [—C(O,O,H)] outranks $—CH_2OH$ [—C(O,H,H)]	The compound 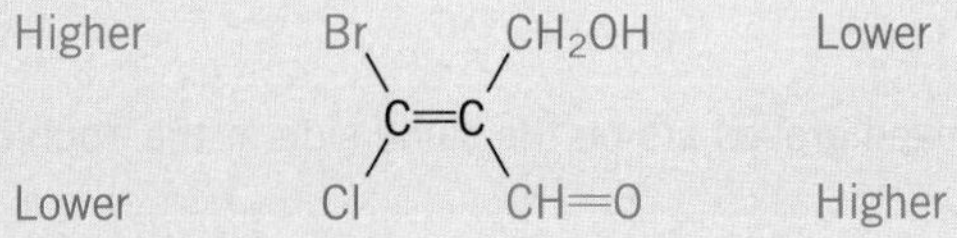Higher Br CH_2OH Lower C=C Lower Cl CH=O Higher has the *E* configuration.

5.5 Physical Properties of Alkenes

Alkenes resemble alkanes in most of their physical properties. The lower molecular weight alkenes through C_4H_8 are gases at room temperature and atmospheric pressure.

The dipole moments of most alkenes are quite small. Among the C_4H_8 isomers, 1-butene, *cis*-2-butene, and 2-methylpropene have dipole moments in the 0.3–0.5 D range; *trans*-2-butene has no dipole moment. Nevertheless, we can learn some things about alkenes by looking at the effect of substituents on dipole moments.

Experimental measurements of dipole moments give size, but not direction. We normally deduce the overall direction by examining the individual bond dipoles. With alkenes the basic question concerns the alkyl groups attached to C=C. *Does an alkyl group donate electrons to or withdraw electrons from a double bond?* This question can be approached by comparing the effect of an alkyl group, methyl for example, with other substituents.

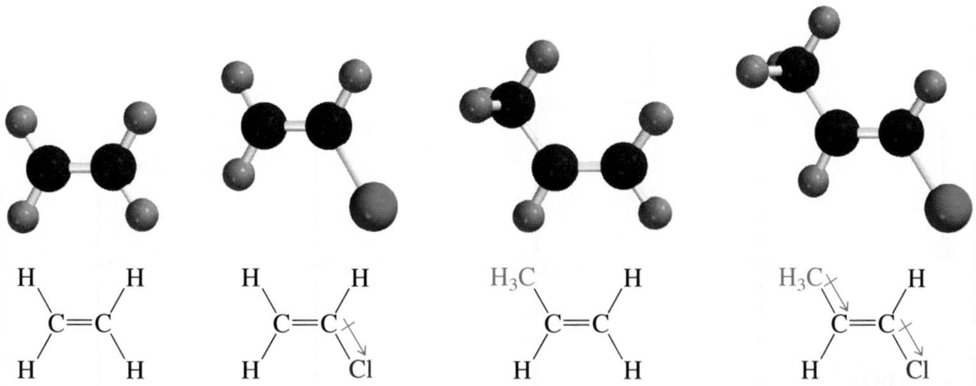

Ethylene	Vinyl chloride	Propene	*trans*-1-Chloropropene
μ = 0 D	μ = 1.4 D	μ = 0.3 D	μ = 1.7 D

Ethylene, of course, has no dipole moment. Replacing one of its hydrogens by chlorine gives vinyl chloride, which has a dipole moment of 1.4 D. The effect is much smaller when one of the hydrogens is replaced by methyl; propene has a dipole moment of only 0.3 D. Now place CH_3 and Cl trans to each other on the double bond. If methyl releases electrons better than H, then the dipole moment of *trans*-CH_3CH=CHCl should be larger than that of H_2C=CHCl, because the effects of CH_3 and Cl reinforce each other. If methyl is electron attracting, the opposite should occur, and the dipole moment of *trans*-CH_3CH=CHCl will be smaller than 1.4 D. In fact, the dipole moment of *trans*-CH_3CH=CHCl is larger than that of H_2C=CHCl, indicating that a methyl group is an electron-donating substituent on the double bond.

A methyl group releases electrons to a double bond in much the same way that it releases electrons to the positively charged carbon of a carbocation—by an inductive effect and by hyperconjugation (Figure 5.2). Other alkyl groups behave similarly and, as we go along, we'll see several ways in which the electron-releasing effects of alkyl substituents influence the properties of alkenes. The first is described in the following section.

Problem 5.10

Arrange the following in order of increasing dipole moment.

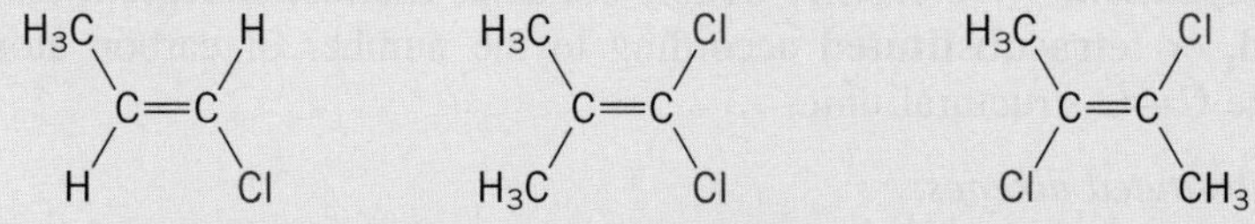

(*E*)-1-Chloropropene 1,1-Dichloro-2-methylpropene (*E*)-2,3-Dichloro-2-butene

sp^2-hybridized carbons of an alkene are more electronegative than sp^3-hybridized carbon and are stabilized by electron-donating substituents.

Methyl group is a better electron-donating substituent than hydrogen.

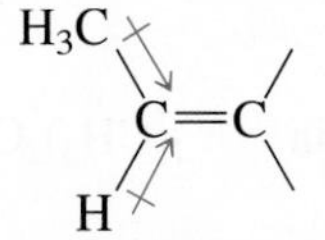

Figure 5.2

Alkyl groups donate electrons to sp^2-hybridized carbons of an alkene.

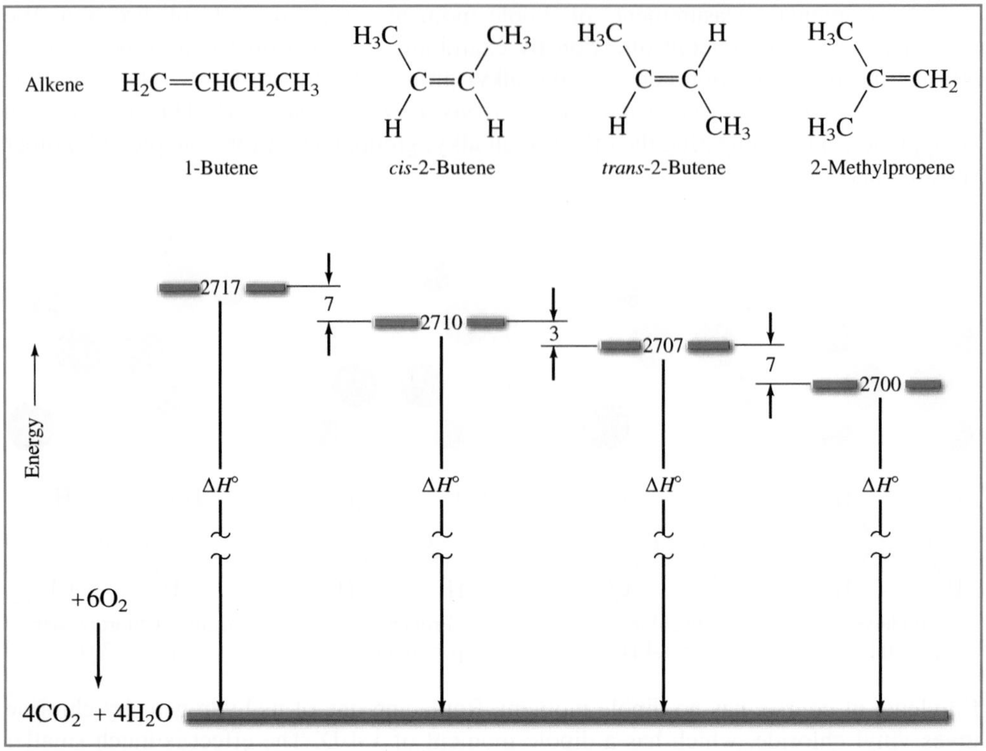

Figure 5.3

Heats of combustion of C_4H_8 alkene isomers. All energies are in kilojoules. (An energy difference of 3 kJ is equivalent to 0.7 kcal, 7 kJ is equivalent to 1.7 kcal.)

5.6 Relative Stabilities of Alkenes

Earlier (Sections 2.18, 3.11) we saw how to use heats of combustion to compare the stabilities of isomeric alkanes. We can do the same thing with isomeric alkenes. Consider the heats of combustion of the four isomeric alkenes of molecular formula C_4H_8. All undergo combustion according to the equation

$$C_4H_8 + 6O_2 \rightarrow 4CO_2 + 4H_2O$$

When the heats of combustion of the isomers are plotted on a common scale as in Figure 5.3, we see that the isomer of highest energy (the least stable one) is 1-butene, $H_2C{=}CHCH_2CH_3$. The isomer of lowest energy (most stable) is 2-methylpropene $(CH_3)_2C{=}CH_2$.

Analogous data for a host of alkenes tell us that the most important factors governing alkene stability are:

1. *Degree of substitution* (alkyl substituents stabilize a double bond)
2. *Van der Waals strain* (destabilizing when alkyl groups are cis to each other)

Degree of substitution. We classify double bonds as **monosubstituted, disubstituted, trisubstituted,** or **tetrasubstituted** according to the number of carbon atoms *directly* attached to the C=C structural unit.

Monosubstituted alkenes:

$RCH{=}CH_2$ as in $CH_3CH_2CH{=}CH_2$ (1-butene)

Disubstituted alkenes:
(R and R′ may be the same or different)

$RCH{=}CHR'$ as in $CH_3CH{=}CHCH_3$ (*cis-* or *trans*-2-butene)

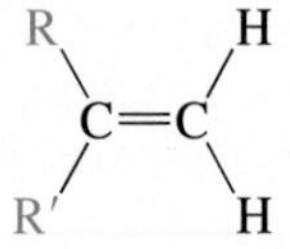

as in $(CH_3)_2C{=}CH_2$ (2-methylpropene)

Trisubstituted alkenes:
(R, R′, and R″ may be the same or different)

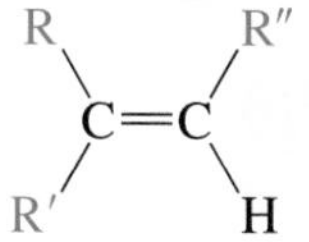

as in (CH$_3$)$_2$C=CHCH$_2$CH$_3$ (2-methyl-2-pentene)

Tetrasubstituted alkenes:
(R, R′, R″, and R‴ may be the same or different)

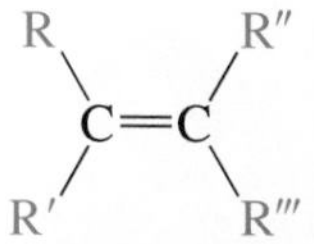

as in

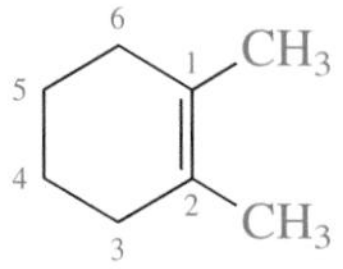

(1,2-dimethylcyclohexene)

Note that carbons 3 and 6 of the ring count as substituents on the double bond.

Problem 5.11

Write structural formulas and give the IUPAC names for all the alkenes of molecular formula C_6H_{12} that contain a trisubstituted double bond. (Don't forget to include stereoisomers.)

From the heats of combustion of the C_4H_8 alkenes in Figure 5.3 we see that each of the disubstituted alkenes

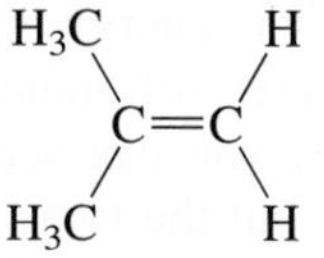

2-Methylpropene

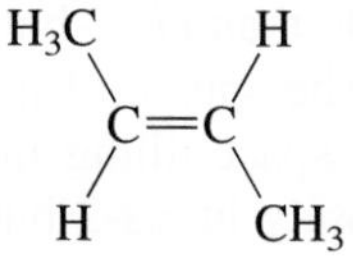

trans-2-Butene

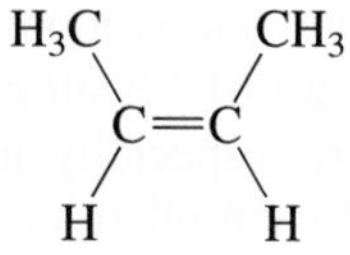

cis-2-Butene

is more stable than the monosubstituted alkene

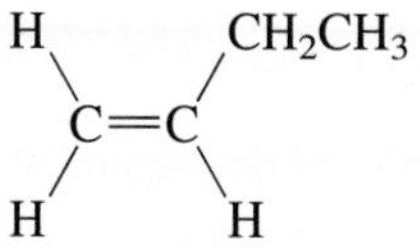

1-Butene

In general, alkenes with more highly substituted double bonds are more stable than isomers with less substituted double bonds.

Branching in 2-methylpropene accounts for its greater stability relative to *cis*- and *trans*-2-butene, which are not branched.

Problem 5.12

Give the structure of the most stable C_6H_{12} alkene.

Like the sp^2-hybridized carbons of carbocations and free radicals, the sp^2-hybridized carbons of double bonds are electron attracting, and alkenes are stabilized by substituents that release electrons to these carbons. As we saw in the preceding section, alkyl groups are better electron-releasing substituents than hydrogen and are, therefore, better able to stabilize an alkene.

Figure 5.4

Ball-and-spoke and space-filling models of *cis*- and *trans*-2-butene. The space-filling model shows the serious van der Waals strain between two of the hydrogens in *cis*-2-butene. The molecule adjusts by expanding those bond angles that increase the separation between the crowded atoms. The combination of angle strain and van der Waals strain makes *cis*-2-butene less stable than trans-2-butene.

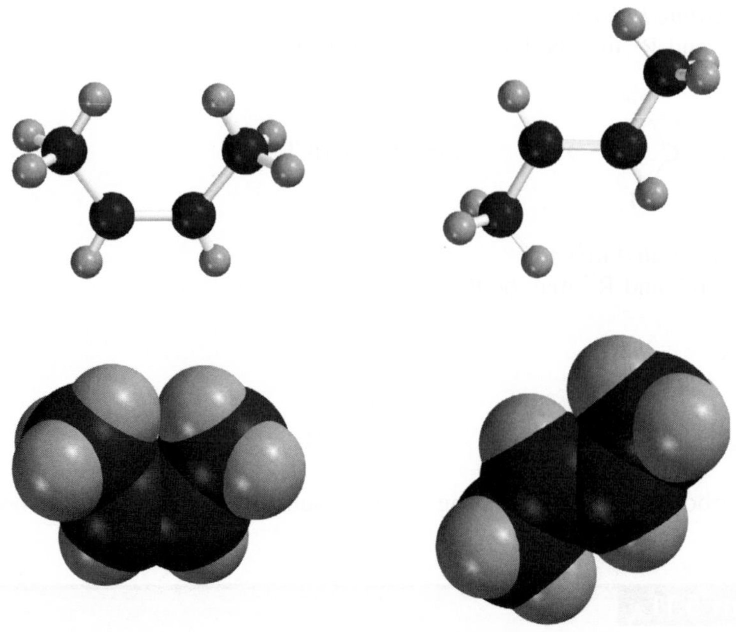

cis-2 Butene   *trans*-2 Butene

An effect that results when two or more atoms or groups interact so as to alter the electron distribution in a system is called an **electronic effect.** The greater stability of more highly substituted alkenes is an example of an electronic effect.

van der Waals strain. *Alkenes are more stable when large substituents are trans to each other than when they are cis.* As we saw in Figure 5.3, *trans*-2-butene has a lower heat of combustion and is more stable than *cis*-2-butene. The energy difference between the two is 3 kJ/mol (0.7 kcal/mol). The source of this energy difference is illustrated in Figure 5.4, where, especially in the space-filling models, you can see that the methyl groups approach each other very closely in *cis*-2-butene, but the trans isomer is free of strain. An effect that results when two or more atoms are close enough in space that a repulsion occurs between them is one type of **steric effect.** The greater stability of trans alkenes compared with their cis counterparts is an example of a steric effect.

A similar steric effect was seen in Section 3.11, where van der Waals strain between methyl groups on the same side of the ring made *cis*-1,2-dimethylcyclopropane less stable than its trans stereoisomer.

Problem 5.13

Arrange the following alkenes in order of decreasing stability: 1-pentene; (*E*)-2-pentene; (*Z*)-2-pentene; 2-methyl-2-butene.

The difference in stability between stereoisomeric alkenes is even more pronounced with larger alkyl groups on the double bond. A particularly striking example compares *cis*- and *trans*-2,2,5,5-tetramethyl-3-hexene, in which the heat of combustion of the cis stereoisomer is 44 kJ/mol (10.5 kcal/mol) higher than that of the trans. The cis isomer is destabilized by the large van der Waals strain between the bulky *tert*-butyl groups on the same side of the double bond.

The common names of these alkenes are *cis*- and *trans*-di-*tert*-butylethylene. In cases such as this the common names are somewhat more convenient than the IUPAC names because they are more readily associated with molecular structure.

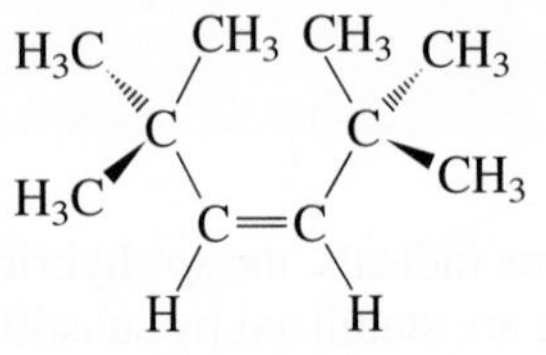

cis-2,2,5,5-Tetramethyl-3-hexene
(less stable)

Energy difference =
44 kJ/mol
(10.5 kcal/mol)

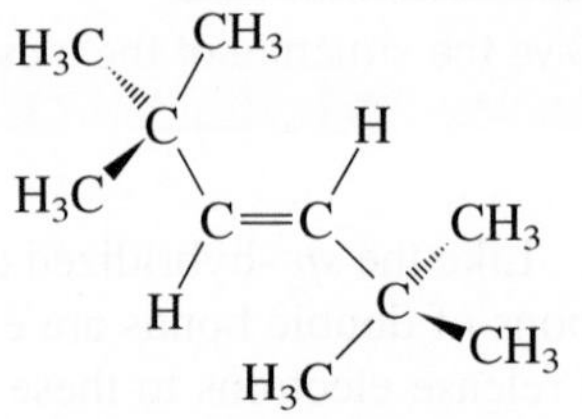

trans-2,2,5,5-Tetramethyl-3-hexene
(more stable)

Problem 5.14

Despite numerous attempts, the alkene 3,4-di-*tert*-butyl-2,2,5,5-tetramethyl-3-hexene has never been synthesized. Can you explain why?

5.7 Cycloalkenes

Double bonds are accommodated by rings of all sizes. The smallest cycloalkene, cyclopropene, was first synthesized in 1922. A cyclopropene ring is present in sterculic acid, a substance derived from one of the components of the oil present in the seeds of a tree *(Sterculia foelida)* that grows in the Philippines and Indonesia.

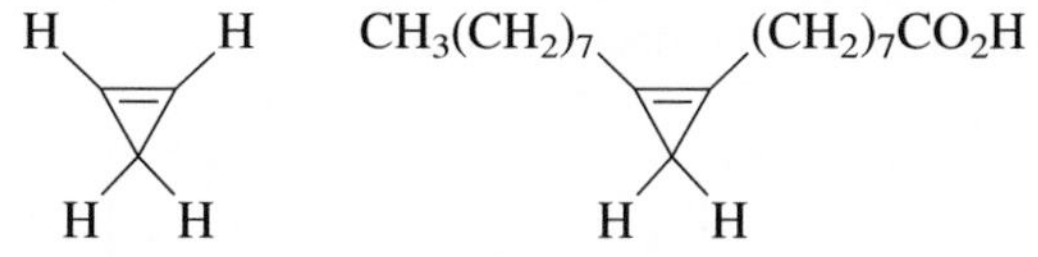

Cyclopropene Sterculic acid

As we saw in Section 3.5, cyclopropane is destabilized by angle strain because its 60° bond angles are much smaller than the normal 109.5° angles associated with sp^3-hybridized carbon. Cyclopropene is even more strained because of the distortion of the bond angles at its doubly bonded carbons from their normal sp^2-hybridization value of 120°. Cyclobutene has, of course, less angle strain than cyclopropene, and the angle strain in cyclopentene, cyclohexene, and higher cycloalkenes is negligible.

The presence of the double bond in cycloalkenes affects the conformation of the ring. The conformation of cyclohexene is a half-chair, with carbons 1, 2, 3, and 6 in the same plane, and carbons 4 and 5 above and below the plane. Conversion to the alternative half-chair occurs readily, with an energy barrier of 22.2 kJ/mol (5.3 kcal/mol), which is about one half that required for chair-to-chair interconversion in cyclohexane. Substituents at carbons 3 and 6 are tilted from their usual axial and equatorial orientations in cyclohexane and are referred to as *pseudoaxial* and *pseudoequatorial.*

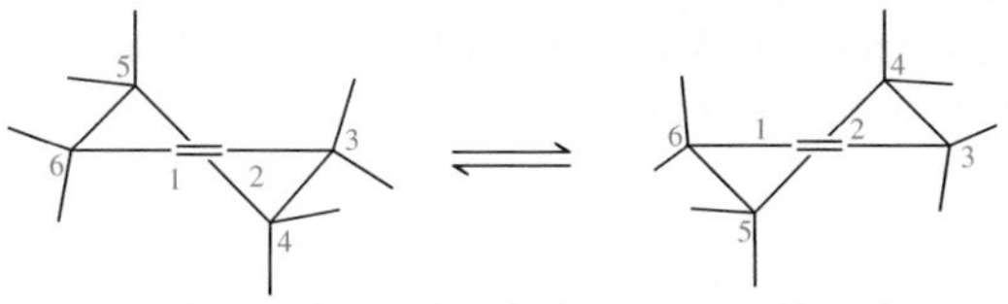

So far we have represented cycloalkenes by structural formulas in which the double bonds are of the cis configuration. If the ring is large enough, however, a trans stereoisomer is also possible. The smallest trans cycloalkene that is stable enough to be isolated and stored in a normal way is *trans*-cyclooctene.

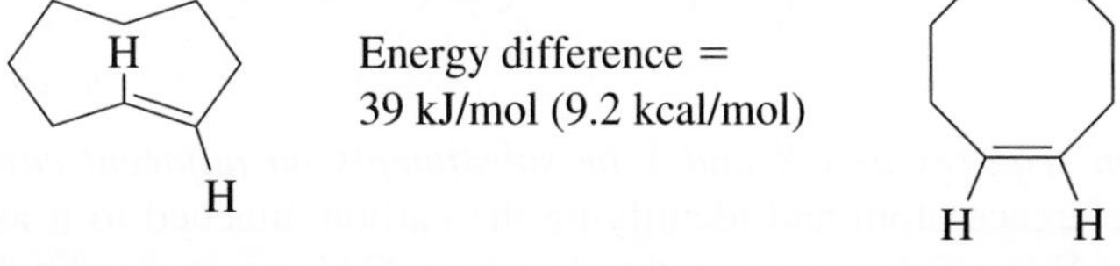

(*E*)-Cyclooctene (*trans*-cyclooctene) Less stable

(*Z*)-Cyclooctene (*cis*-cyclooctene) More stable

trans-Cycloheptene has been prepared and studied at low temperature (–90°C) but is too reactive to be isolated and stored at room temperature. Evidence has also been presented for the fleeting existence of the even more strained *trans*-cyclohexene as a reactive intermediate in certain reactions.

Problem 5.15

Place a double bond in the carbon skeleton shown so as to represent

(a) (*Z*)-1-Methylcyclodecene
(b) (*E*)-1-Methylcyclodecene
(c) (*Z*)-3-Methylcyclodecene
(d) (*E*)-3-Methylcyclodecene
(e) (*Z*)-5-Methylcyclodecene
(f) (*E*)-5-Methylcyclodecene

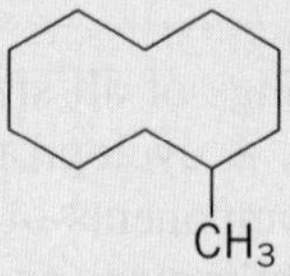

Sample Solution (a) and (b) Because the methyl group must be at C-1, there are only two possible places to put the double bond:

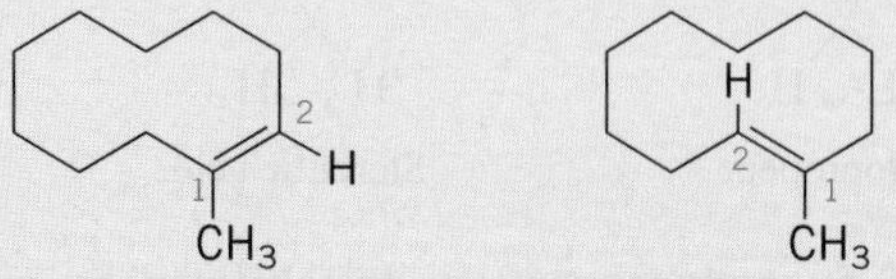

(*Z*)-1-Methylcyclodecene (*E*)-1-Methylcyclodecene

In the *Z* stereoisomer the two lower priority substituents—the methyl group and the hydrogen—are on the same side of the double bond. In the *E* stereoisomer these substituents are on opposite sides of the double bond. The ring carbons are the higher ranking substituents at each end of the double bond.

Because larger rings have more carbons with which to span the ends of a double bond, the strain associated with a trans cycloalkene decreases with increasing ring size. The strain eventually disappears when a 12-membered ring is reached and *cis*- and *trans*-cyclododecene are of approximately equal stability. When the rings are larger than 12-membered, trans cycloalkenes are more stable than cis. In these cases, the ring is large enough and flexible enough that it is energetically similar to a noncyclic alkene. As in noncyclic cis alkenes, van der Waals strain between carbons on the same side of the double bond destabilizes a cis cycloalkene.

5.8 Preparation of Alkenes: Elimination Reactions

The rest of this chapter describes how alkenes are prepared by elimination; that is, reactions of the type:

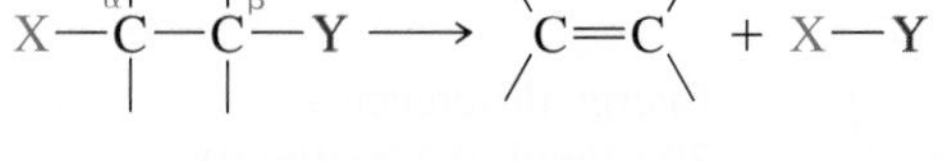

Alkene formation requires that X and Y be substituents on adjacent carbon atoms. By making X the reference atom and identifying the carbon attached to it as the α carbon, we see that atom Y is a substituent on the β carbon. Carbons succeedingly more remote from the reference atom are designated γ, δ, and so on. Only β elimination reactions will be discussed in this chapter. [**Beta (β) elimination reactions** are also known as *1,2 eliminations.*]

You are already familiar with one type of β elimination, having seen in Section 5.2 that ethylene and propene are prepared on an industrial scale by the high-temperature *dehydrogenation* of ethane and propane. Both reactions involve β elimination of H_2.

$$\underset{\text{Ethane}}{CH_3CH_3} \xrightarrow{750°C} \underset{\text{Ethylene}}{H_2C{=}CH_2} + \underset{\text{Hydrogen}}{H_2}$$

$$\underset{\text{Propane}}{CH_3CH_2CH_3} \xrightarrow{750°C} \underset{\text{Propene}}{CH_3CH{=}CH_2} + \underset{\text{Hydrogen}}{H_2}$$

Many reactions classified as dehydrogenations occur within the cells of living systems at 25°C. H_2 is not one of the products, however. Instead, the hydrogens are lost in separate steps of an enzyme-catalyzed process. The enzyme indicated in the reaction:

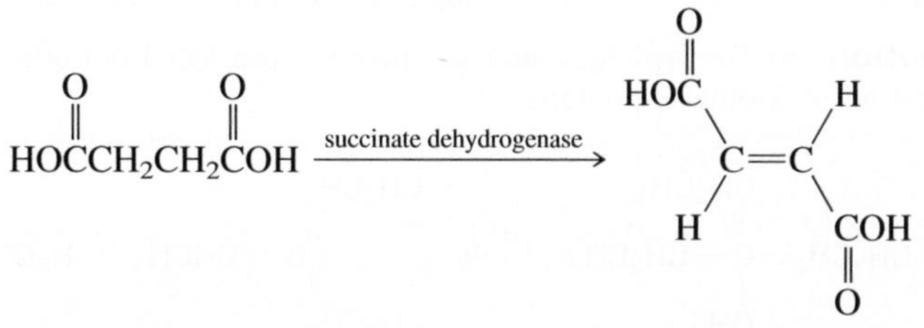

Succinic acid Fumaric acid

is a special kind, known as a *flavoprotein.*

Dehydrogenation of alkanes is not a practical *laboratory* synthesis for the vast majority of alkenes. The principal methods by which alkenes are prepared in the laboratory are two other β eliminations: the *dehydration* of alcohols and the *dehydrohalogenation* of alkyl halides. A discussion of these two methods makes up the remainder of this chapter.

A quote from a biochemistry text is instructive here. "This is not an easy reaction in organic chemistry. It is, however, a very important type of reaction in metabolic chemistry and is an integral step in the oxidation of carbohydrates, fats, and several amino acids." G. L. Zubay, *Biochemistry,* 4th ed., William C. Brown Publishers, 1996, p. 333.

5.9 Dehydration of Alcohols

In the **dehydration** of alcohols, the H and OH are lost from adjacent carbons. An acid catalyst is necessary.

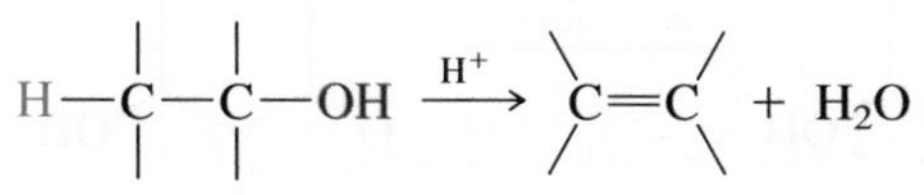

Alcohol Alkene Water

Before dehydrogenation of ethane became the dominant method, ethylene was prepared by heating ethyl alcohol with sulfuric acid.

$$\underset{\text{Ethyl alcohol}}{CH_3CH_2OH} \xrightarrow[160°C]{H_2SO_4} \underset{\text{Ethylene}}{H_2C{=}CH_2} + \underset{\text{Water}}{H_2O}$$

Other alcohols behave similarly. Secondary alcohols undergo elimination at lower temperatures than primary alcohols.

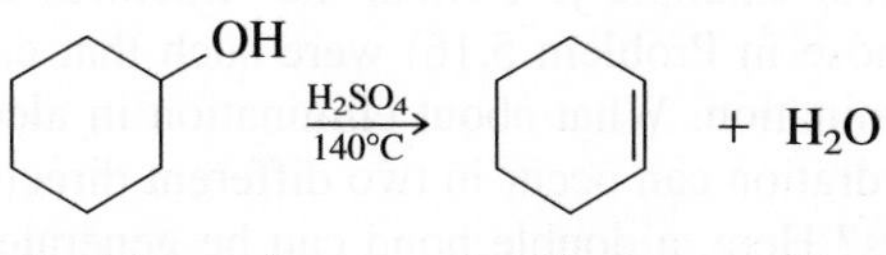

Cyclohexanol Cyclohexene (79–87%) Water

Tertiary alcohols dehydrate at lower temperatures than secondary alcohols.

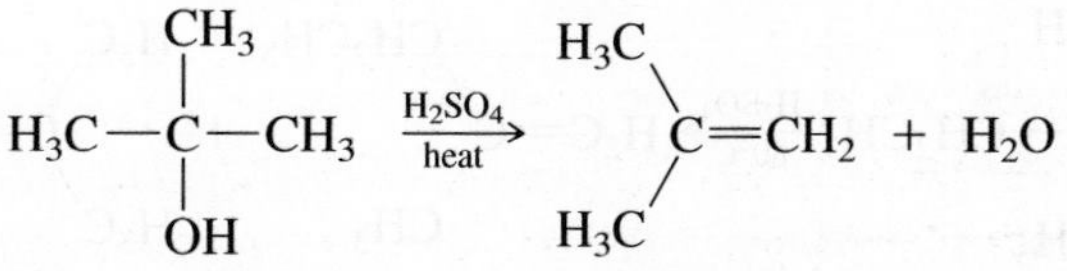

2-Methyl-2-propanol 2-Methylpropene (82%) Water

HSO_4^- and H_3PO_4 are very similar in acid strength. Both are much weaker than H_2SO_4, which is a strong acid.

Reaction conditions, such as the acid used and the temperature, are chosen to maximize the formation of alkene by elimination. Sulfuric acid (H_2SO_4) and phosphoric acid (H_3PO_4) are the acids most frequently used in alcohol dehydrations. Potassium hydrogen sulfate ($KHSO_4$) is also often used.

Problem 5.16

Identify the alkene obtained on dehydration of each of the following alcohols:

(a) 3-Ethyl-3-pentanol
(b) 1-Propanol
(c) 2-Propanol
(d) 2,3,3-Trimethyl-2-butanol

Sample Solution (a) The hydrogen and the hydroxyl are lost from adjacent carbons in the dehydration of 3-ethyl-3-pentanol.

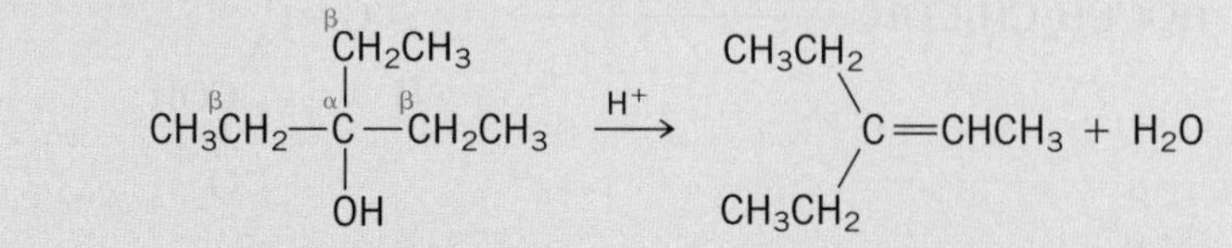

3-Ethyl-3-pentanol — 3-Ethyl-2-pentene — Water

The hydroxyl group is lost from a carbon that bears three equivalent ethyl substituents. Beta elimination can occur in any one of three equivalent directions to give the same alkene, 3-ethyl-2-pentene.

Some biochemical processes involve alcohol dehydration as a key step. An example is the conversion of a compound called 3-dehydroquinic acid to 3-dehydroshikimic acid.

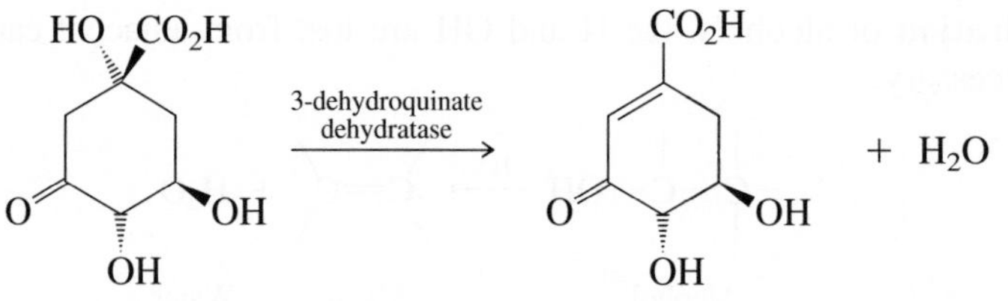

3-Dehydroquinic acid — 3-Dehydroshikimic acid — Water

This reaction is catalyzed by an enzyme called a *dehydratase* and is one step along the pathway by which plants convert glucose to certain amino acids.

5.10 Regioselectivity in Alcohol Dehydration: The Zaitsev Rule

Except for the biochemical example just cited, the structures of all of the alcohols in Section 5.9 (including those in Problem 5.16) were such that each one could give only a single alkene by β elimination. What about elimination in alcohols such as 2-methyl-2-butanol, in which dehydration can occur in two different directions to give alkenes that are constitutional isomers? Here, a double bond can be generated between C-1 and C-2 or between C-2 and C-3. Both processes occur but not nearly to the same extent. Under the usual reaction conditions 2-methyl-2-butene is the major product, and 2-methyl-1-butene the minor one.

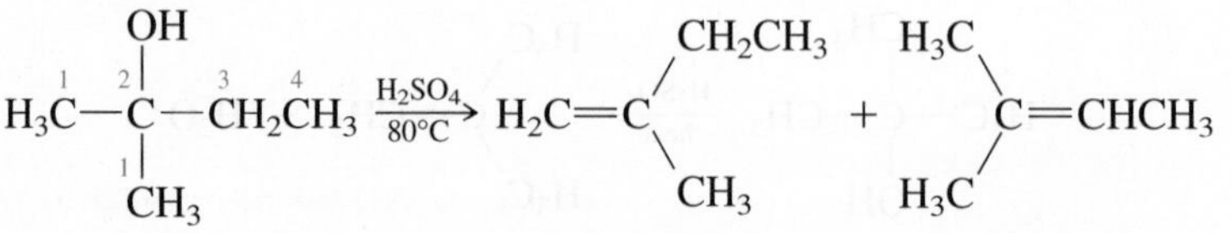

2-Methyl-2-butanol — 2-Methyl-1-butene (10%) — 2-Methyl-2-butene (90%)

Dehydration of this alcohol is selective in respect to its *direction*. Elimination occurs in the direction that leads to the double bond between C-2 and C-3 more than between C-2 and C-1. Reactions that can proceed in more than one direction, but in which one direction is preferred, are said to be **regioselective.**

As a second example, consider the regioselective dehydration of 2-methylcyclohexanol to yield a mixture of 1-methylcyclohexene (major) and 3-methylcyclohexene (minor).

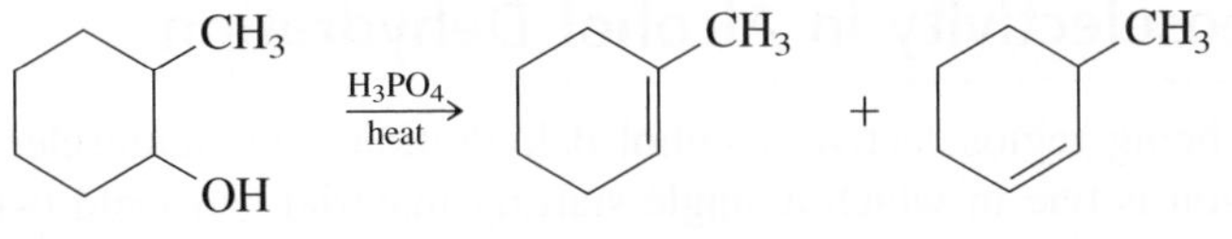

2-Methylcyclohexanol 1-Methylcyclohexene (84%) 3-Methylcyclohexene (16%)

Although Russian, Zaitsev published most of his work in German scientific journals, where his name was transliterated as *Saytzeff.* The spelling used here *(Zaitsev)* corresponds to the currently preferred style.

In 1875, Alexander M. Zaitsev of the University of Kazan (Russia) set forth a generalization describing the regioselectivity of β eliminations. **Zaitsev's rule** summarizes the results of numerous experiments in which alkene mixtures were produced by β elimination. In its original form, Zaitsev's rule stated that *the alkene formed in greatest amount is the one that corresponds to removal of the hydrogen from the β carbon having the fewest hydrogens.*

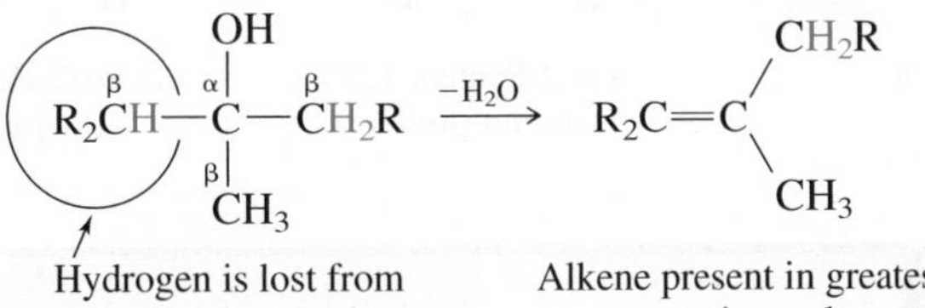

Hydrogen is lost from β carbon having the fewest attached hydrogens

Alkene present in greatest amount in product

Zaitsev's rule as applied to the acid-catalyzed dehydration of alcohols is now more often expressed in a different way: *β elimination reactions of alcohols yield the most highly substituted alkene as the major product.* Because, as was discussed in Section 5.6, the most highly substituted alkene is also normally the most stable one, Zaitsev's rule is sometimes expressed as a preference for *predominant formation of the most stable alkene that could arise by β elimination.*

Problem 5.17

Each of the following alcohols has been subjected to acid-catalyzed dehydration and yields a mixture of two isomeric alkenes. Identify the two alkenes in each case, and predict which one is the major product on the basis of the Zaitsev rule.

(a) $(CH_3)_2CCH(CH_3)_2$ (with OH on the C of $(CH_3)_2C$)

(b) 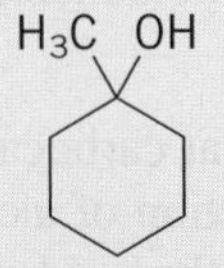

(c) 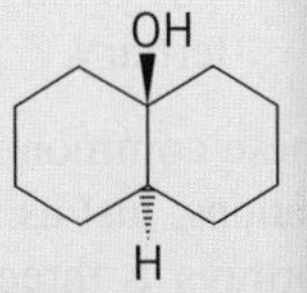

Sample Solution (a) Dehydration of 2,3-dimethyl-2-butanol can lead to either 2,3-dimethyl-1-butene by removal of a C-1 hydrogen or to 2,3-dimethyl-2-butene by removal of a C-3 hydrogen.

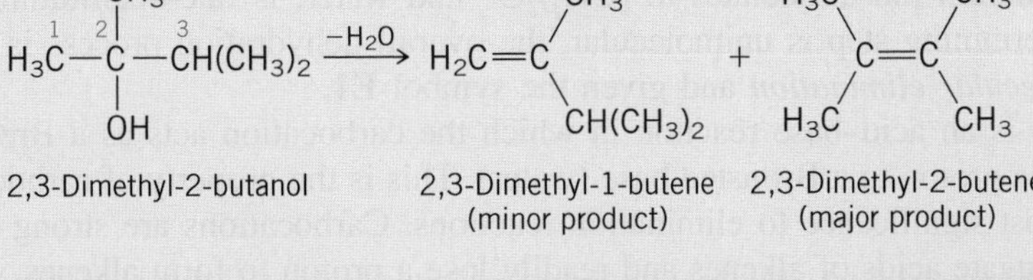

2,3-Dimethyl-2-butanol 2,3-Dimethyl-1-butene (minor product) 2,3-Dimethyl-2-butene (major product)

Continued

The major product is 2,3-dimethyl-2-butene. It has a tetrasubstituted double bond and is more stable than 2,3-dimethyl-1-butene, which has a disubstituted double bond. The major alkene arises by loss of a hydrogen from the β carbon that has fewer attached hydrogens (C-3) rather than from the β carbon that has the greater number of hydrogens (C-1).

5.11 Stereoselectivity in Alcohol Dehydration

In addition to being regioselective, alcohol dehydrations are stereoselective. A **stereoselective** reaction is one in which a single starting material can yield two or more stereoisomeric products, but gives one of them in greater amounts than any other. Alcohol dehydrations tend to produce the more stable stereoisomer of an alkene. Dehydration of 3-pentanol, for example, yields a mixture of *trans*-2-pentene and *cis*-2-pentene in which the more stable trans stereoisomer predominates.

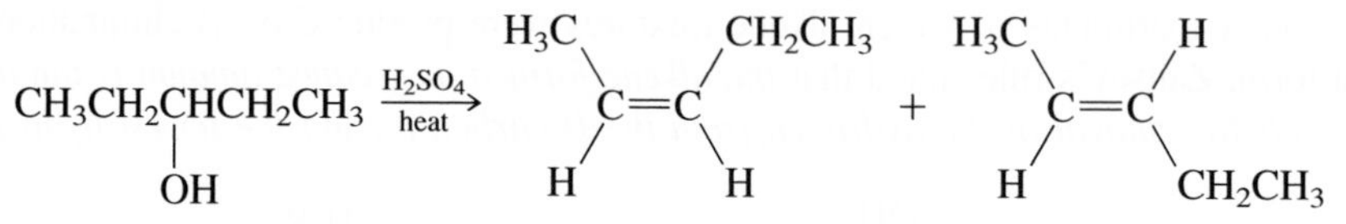

$CH_3CH_2CHCH_2CH_3$
OH
H_2SO_4
heat
H_3C
CH_2CH_3
C=C
H
H
+
H_3C
H
C=C
H
CH_2CH_3

3-Pentanol — *cis*-2-Pentene (25%) (minor product) — *trans*-2-Pentene (75%) (major product)

Problem 5.18

What three alkenes are formed in the acid-catalyzed dehydration of 2-pentanol?

The biological dehydrogenation of succinic acid described in Section 5.8 is 100% stereoselective. Only fumaric acid, which has a trans double bond, is formed. High levels of stereoselectivity are characteristic of enzyme-catalyzed reactions.

5.12 The E1 and E2 Mechanisms of Alcohol Dehydration

The dehydration of alcohols resembles the reaction of alcohols with hydrogen halides (Section 4.7) in two important ways.

1. Both reactions are promoted by acids.
2. The relative reactivity of alcohols increases in the order primary < secondary < tertiary.

These common features suggest that carbocations are key intermediates in alcohol dehydrations, just as they are in the reaction of alcohols with hydrogen halides. Mechanism 5.1 portrays a three-step process for the acid-catalyzed dehydration of *tert*-butyl alcohol. Steps 1 and 2 describe the generation of *tert*-butyl cation by a process similar to that which led to its formation as an intermediate in the reaction of *tert*-butyl alcohol with hydrogen chloride.

Like the reaction of *tert*-butyl alcohol with hydrogen chloride, step 2 in which *tert*-butyloxonium ion dissociates to $(CH_3)_3C^+$ and water, is rate-determining. Because the rate-determining step is unimolecular, the overall dehydration process is referred to as a *unimolecular elimination* and given the symbol **E1.**

Step 3 is an acid–base reaction in which the carbocation acts as a Brønsted acid, transferring a proton to a Brønsted base (water). This is the property of carbocations that is of the most significance to elimination reactions. Carbocations are strong acids; they are the conjugate acids of alkenes and readily lose a proton to form alkenes. Even weak bases such as water are sufficiently basic to abstract a proton from a carbocation.

Step 3 in Mechanism 5.1 shows water as the base that abstracts a proton from the carbocation. Other Brønsted bases present in the reaction mixture that can function in the same way include *tert*-butyl alcohol and hydrogen sulfate ion.

Mechanism 5.1

The E1 Mechanism for Acid-Catalyzed Dehydration of *tert*-Butyl Alcohol

THE OVERALL REACTION:

$$(CH_3)_3COH \xrightarrow[\text{heat}]{H_2SO_4} (CH_3)_2C{=}CH_2 + H_2O$$

tert-Butyl alcohol — 2-Methylpropene — Water

Step 1: Protonation of *tert*-butyl alcohol:

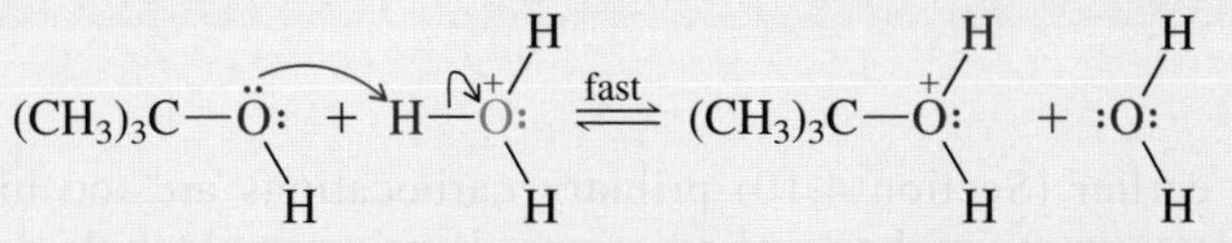

tert-Butyl alcohol — Hydronium ion — *tert*-Butyloxonium ion — Water

Step 2: Dissociation of *tert*-butyloxonium ion:

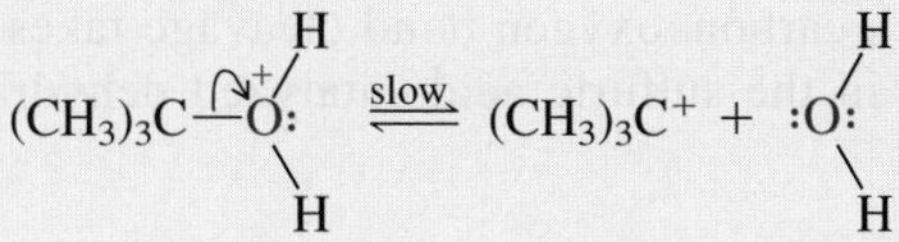

tert-Butyloxonium ion — *tert*-Butyl cation — Water

Step 3: Deprotonation of *tert*-butyl cation:

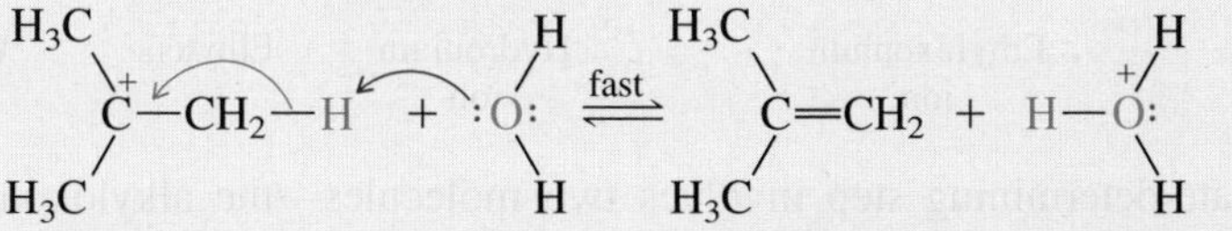

tert-Butyl cation — Water — 2-Methylpropene — Hydronium ion

Problem 5.19

Write a structural formula for the carbocation intermediate formed in the dehydration of each of the alcohols in Problem 5.17 (Section 5.10). Using curved arrows, show how each carbocation is deprotonated by water to give a mixture of alkenes.

Sample Solution (a) The carbon that bears the hydroxyl group in the starting alcohol is the one that becomes positively charged in the carbocation.

$$(CH_3)_2\underset{|\atop OH}{C}CH(CH_3)_2 \xrightarrow[-H_2O]{H^+} (CH_3)_2\overset{+}{C}CH(CH_3)_2$$

Water may remove a proton from either C-1 or C-3 of this carbocation. Loss of a proton from C-1 yields the minor product 2,3-dimethyl-1-butene. (This alkene has a disubstituted double bond.)

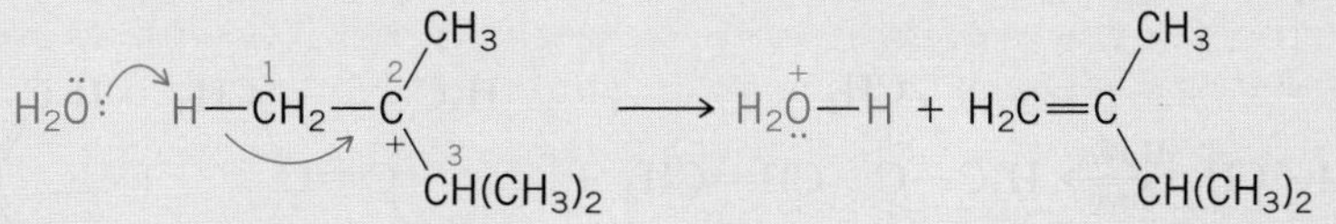

2,3-Dimethyl-1-butene

Continued

Loss of a proton from C-3 yields the major product 2,3-dimethyl-2-butene. (This alkene has a tetrasubstituted double bond.)

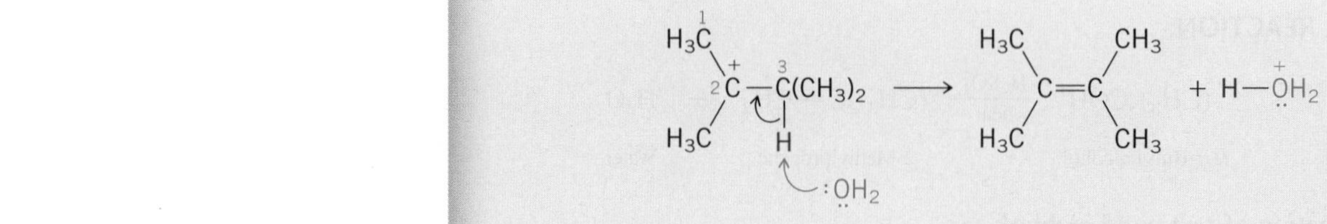

2,3-Dimethyl-2-butene

As noted earlier (Section 4.10) primary carbocations are too high in energy to be intermediates in most chemical reactions. If primary alcohols don't form primary carbocations, then how do they undergo elimination? A modification of our general mechanism for alcohol dehydration offers a reasonable explanation. For primary alcohols it is believed that a proton is lost from the alkyloxonium ion in the same step in which carbon–oxygen bond cleavage takes place. For example, the rate-determining step in the sulfuric acid-catalyzed dehydration of ethanol may be represented as:

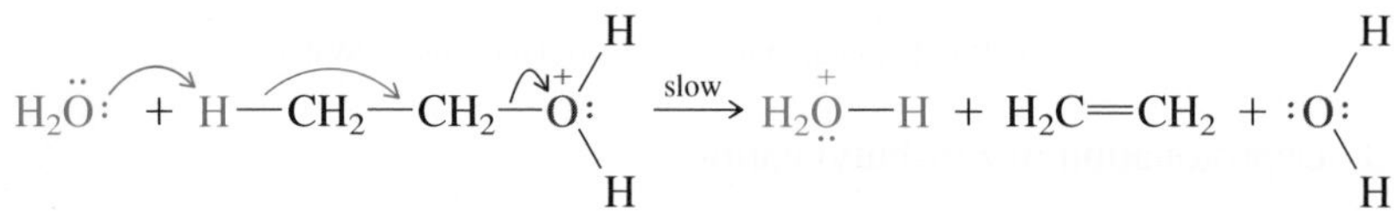

Water — Ethyloxonium ion — Hydronium ion — Ethylene — Water

Because the rate-determining step involves two molecules—the alkyloxonium ion and water—the overall reaction is classified as a *bimolecular elimination* and given the symbol **E2.**

Like tertiary alcohols, secondary alcohols normally undergo dehydration by way of carbocation intermediates.

In Chapter 4 you learned that carbocations could be captured by halide anions to give alkyl halides. In the present chapter, a second type of carbocation reaction has been introduced—a carbocation can lose a proton to form an alkene. In the next section a third aspect of carbocation behavior will be described, the *rearrangement* of one carbocation to another.

5.13 Rearrangements in Alcohol Dehydration

Some alcohols undergo dehydration to yield alkenes having carbon skeletons different from the starting alcohols. Not only has elimination taken place, but the arrangement of atoms in the alkene is different from that in the alcohol. A **rearrangement** has occurred. An example of an alcohol dehydration that is accompanied by rearrangement is the case of 3,3-dimethyl-2-butanol. This is one of many such experiments carried out by F. C. Whitmore and his students at Pennsylvania State University in the 1930s as part of a general study of rearrangement reactions.

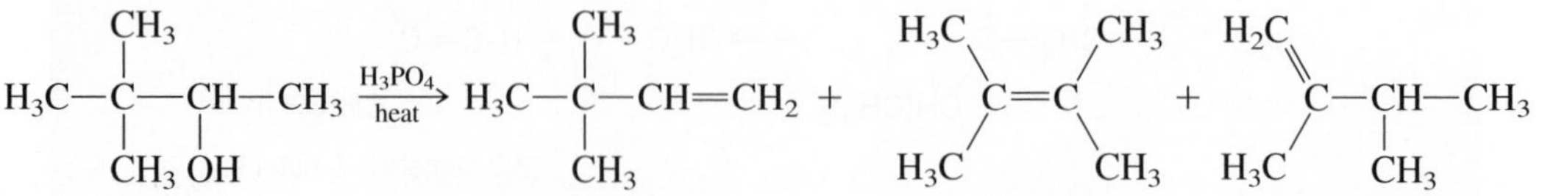

3,3-Dimethyl-2-butanol — 3,3-Dimethyl-1-butene (3%) — 2,3-Dimethyl-2-butene (64%) — 2,3-Dimethyl-1-butene (33%)

Mechanism 5.2

Carbocation Rearrangement in Dehydration of 3,3-Dimethyl-2-butanol

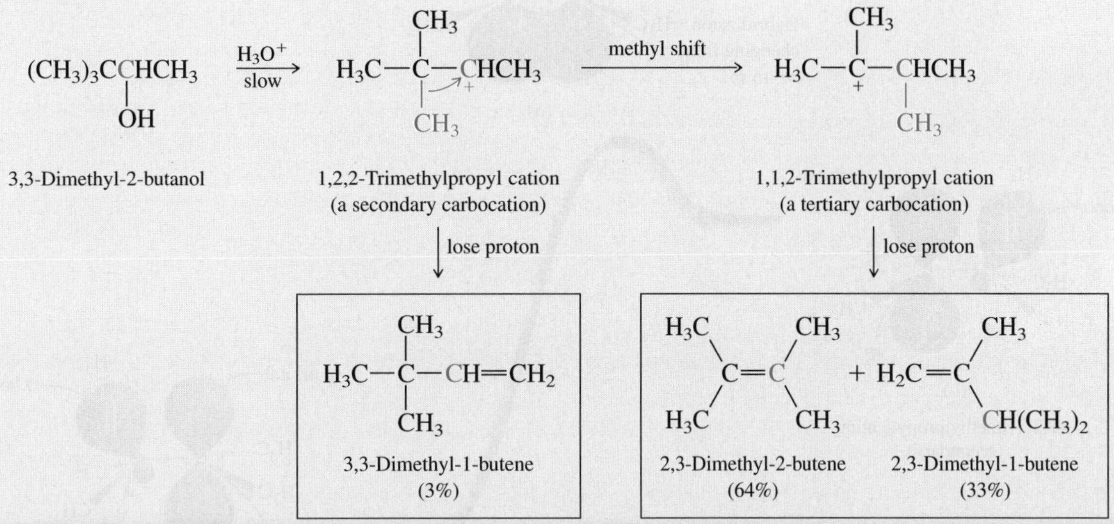

A mixture of three alkenes was obtained in 80% yield, having the composition shown. The alkene having the same carbon skeleton as the starting alcohol, 3,3-dimethyl-1-butene, constituted only 3% of the alkene mixture. The two alkenes present in greatest amount, 2,3-dimethyl-2-butene and 2,3-dimethyl-1-butene, both have carbon skeletons different from that of the starting alcohol.

Whitmore proposed that the carbon skeleton rearrangement occurred in a separate step following carbocation formation. Once the alcohol was converted to the corresponding carbocation, that carbocation could either lose a proton to give an alkene having the same carbon skeleton or rearrange to a different carbocation, as shown in Mechanism 5.2. The rearranged alkenes arise by loss of a proton from the rearranged carbocation.

Why do carbocations rearrange? The answer is straightforward once we recall that tertiary carbocations are more stable than secondary carbocations (Section 4.10). Thus, rearrangement of a secondary to a tertiary carbocation is energetically favorable. As shown in Mechanism 5.2, the carbocation that is formed first in the dehydration of 3,3-dimethyl-2-butanol is secondary; the rearranged carbocation is tertiary. Rearrangement occurs, and almost all of the alkene products come from the tertiary carbocation.

How do carbocations rearrange? To understand this we need to examine the structural change that takes place at the transition state. Referring to the initial (secondary) carbocation intermediate in Mechanism 5.2, rearrangement occurs when a methyl group shifts from C-2 of the carbocation to the positively charged carbon. The methyl group migrates with the pair of electrons that made up its original σ bond to C-2. In the curved arrow notation for this methyl migration, the arrow shows the movement of both the methyl group and the electrons in the σ bond.

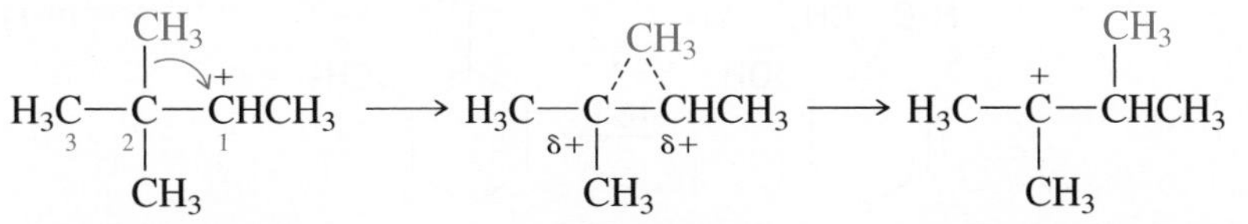

1,2,2-Trimethylpropyl cation (secondary, less stable) — Transition state for methyl migration (dashed lines indicate partial bonds) — 1,1,2-Trimethylpropyl cation (tertiary, more stable)

At the transition state for rearrangement, the methyl group is partially bonded both to its point of origin and to the carbon that will be its destination.

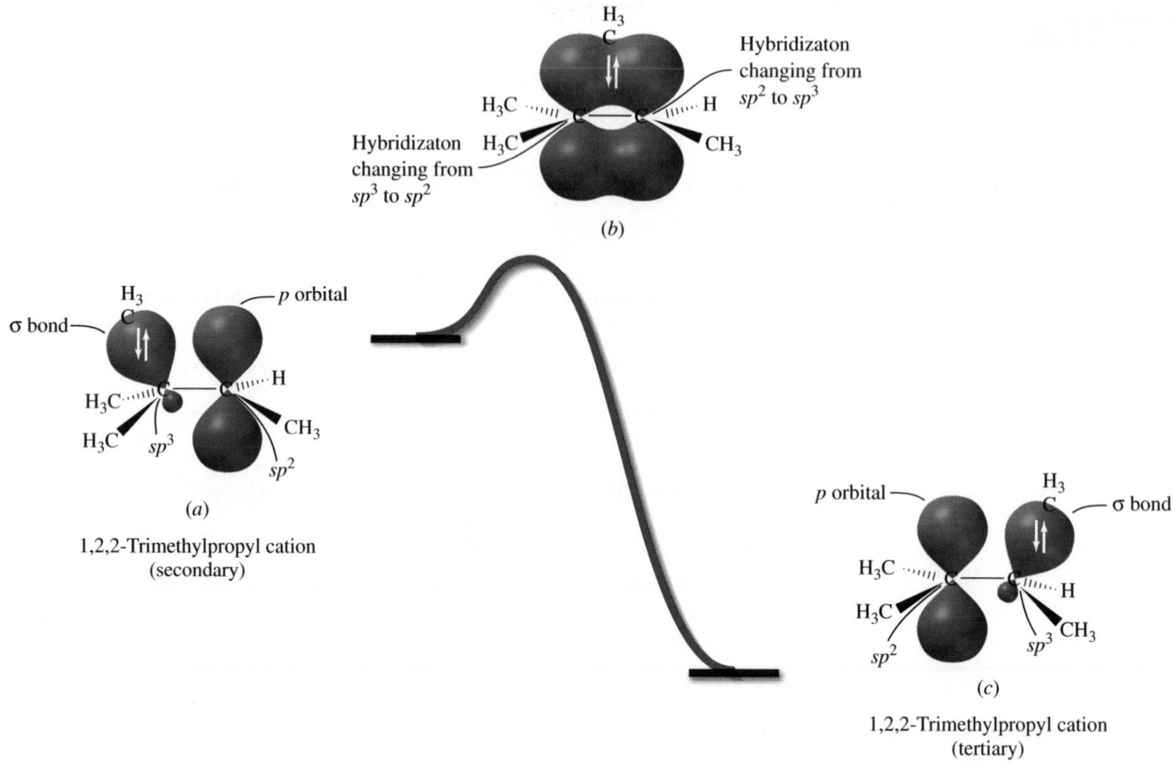

Figure 5.5

Methyl migration in 1,2,2-trimethylpropyl cation. Structure (*a*) is the initial secondary carbocation; structure (*b*) is the transition state for methyl migration, and structure (*c*) is the final tertiary carbocation.

This rearrangement is shown in orbital terms in Figure 5.5. The relevant orbitals of the secondary carbocation are shown in structure (*a*), those of the transition state for rearrangement in (*b*), and those of the tertiary carbocation in (*c*). Delocalization of the *electrons* of the C—CH_3 σ bond into the vacant *p* orbital of the positively charged carbon by hyperconjugation is present in both (*a*) and (*c*), requires no activation energy, and stabilizes each carbocation. Migration of the *atoms* of the methyl group, however, occurs only when sufficient energy is absorbed by (*a*) to achieve the transition state (*b*). The activation energy is modest, and carbocation rearrangements are normally quite fast.

Once a carbocation is formed, anything that happens afterward occurs rapidly.

Problem 5.20

The alkene mixture obtained on dehydration of 2,2-dimethylcyclohexanol contains appreciable amounts of 1,2-dimethylcyclohexene. Give a mechanistic explanation for the formation of this product.

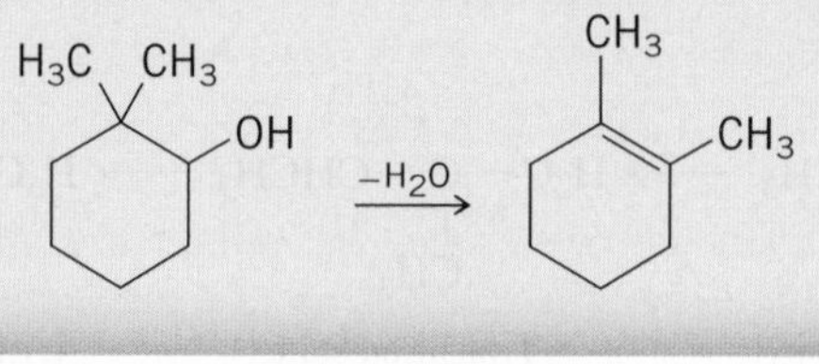

Alkyl groups other than methyl can also migrate to a positively charged carbon.

Many carbocation rearrangements involve migration of a hydrogen. These are called **hydride shifts.** The same requirements apply to hydride shifts as to alkyl group migrations; they proceed in the direction that leads to a more stable carbocation; the

Mechanism 5.3

Hydride Shift in Dehydration of 1-Butanol

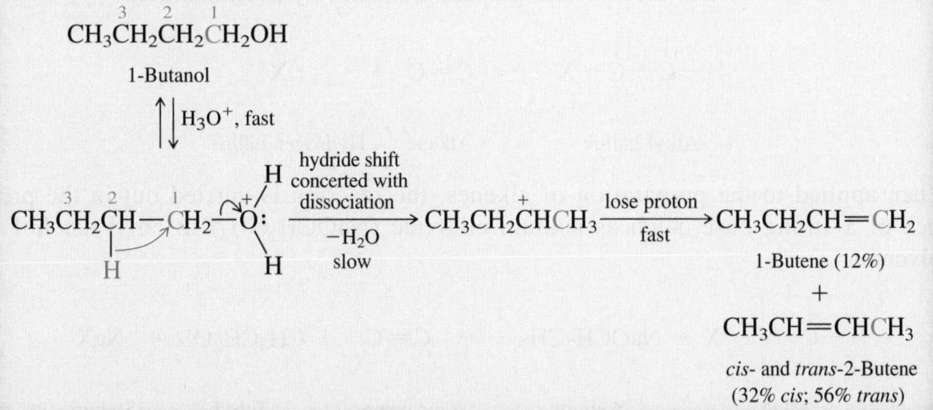
CH3CH2CH2CH2OH
1-Butanol
H3O+, fast
hydride shift concerted with dissociation
−H2O
slow
lose proton
fast
1-Butene (12%)
CH3CH=CHCH3
cis- and trans-2-Butene
(32% cis; 56% trans)

origin and destination of the migrating hydrogen are adjacent carbons, one of which must be positively charged; and the hydrogen migrates with a pair of electrons.

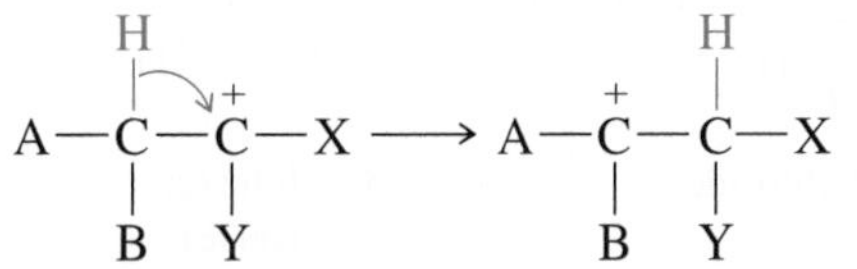

Hydride shift

Hydride shifts often occur during the dehydration of primary alcohols. Thus, although 1-butene would be expected to be the only alkene formed on dehydration of 1-butanol, it is in fact accompanied by a mixture of *cis*- and *trans*-2-butene.

$$CH_3CH_2CH_2CH_2OH \xrightarrow[140-170^\circ C]{H_2SO_4} CH_3CH_2CH{=}CH_2 + CH_3CH{=}CHCH_3$$

1-Butene (12%) — Mixture of *cis*-2-butene (32%) and *trans*-2-butene (56%)

Mechanism 5.3 shows how the butyloxonium ion formed on protonation of 1-butanol can give a secondary carbocation by a hydride shift from C-2 to C-1 in the step in which a molecule of water is lost. Deprotonation of this secondary carbocation leads to 1-butene, *cis*-2-butene, and *trans*-2-butene.

Problem 5.21

Using curved arrows, show how the butyloxonium ion in Mechanism 5.3 can react with water by an E2 mechanism to form 1-butene. (*Hint:* See page 204 for an example of electron flow in the E2 mechanism.)

The ratio of isomeric alkenes obtained from a primary alcohol such as 1-butanol depends on the relative rates of dehydration by Mechanism 5.3 and the E2 process in Problem 5.21. However, the strongly acidic reaction conditions also promote alkene equilibration, which increases the proportion of 2-butenes at the expense of 1-butene. The mechanism of this equilibration is based on principles that we will consider in Chapter 6.

This concludes discussion of our second functional group transformation involving *alcohols:* the first was the conversion of alcohols to alkyl halides (Chapter 4), and the second the conversion of alcohols to alkenes. In the remaining sections of the chapter the conversion of *alkyl halides* to alkenes by dehydrohalogenation is described.

5.14 Dehydrohalogenation of Alkyl Halides

Dehydrohalogenation is the loss of a hydrogen and a halogen from an alkyl halide. It is one of the most useful methods for preparing alkenes by β elimination.

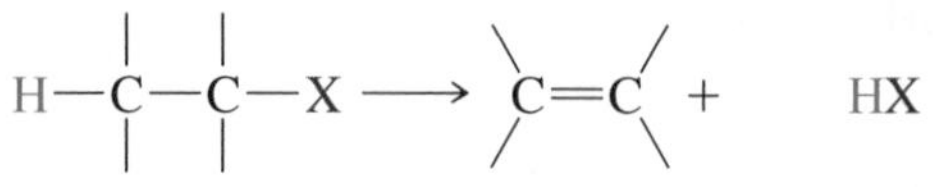

Alkyl halide Alkene Hydrogen halide

Sodium ethoxide is prepared by the reaction of sodium metal with ethanol.

When applied to the preparation of alkenes, the reaction is carried out in the presence of a strong base, such as sodium ethoxide ($NaOCH_2CH_3$) in ethyl alcohol as solvent.

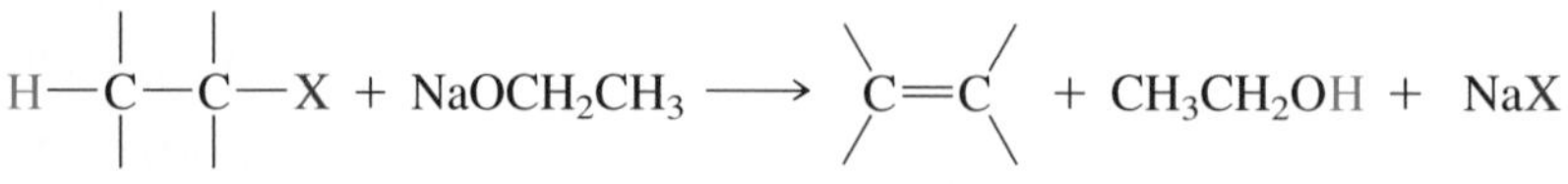

Alkyl halide Sodium ethoxide Alkene Ethyl alcohol Sodium halide

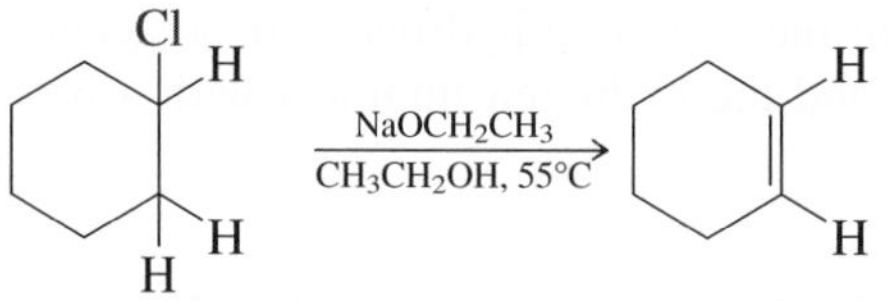

Cyclohexyl chloride Cyclohexene (100%)

Dimethyl sulfoxide has the structure $(CH_3)_2\overset{+}{\ddot{S}}—\ddot{O}:^-$ and is commonly referred to as DMSO. It is a relatively inexpensive solvent, obtained as a byproduct in paper manufacture.

Similarly, sodium methoxide ($NaOCH_3$) is a suitable base and is used in methyl alcohol. Potassium hydroxide in ethyl alcohol is another base–solvent combination often employed in the dehydrohalogenation of alkyl halides. Potassium *tert*-butoxide [$KOC(CH_3)_3$] is the preferred base when the alkyl halide is primary; it is used in either *tert*-butyl alcohol or dimethyl sulfoxide as solvent.

$$CH_3(CH_2)_{15}CH_2CH_2Cl \xrightarrow[\text{DMSO, 25°C}]{KOC(CH_3)_3} CH_3(CH_2)_{15}CH{=}CH_2$$

1-Chlorooctadecane 1-Octadecene (86%)

The regioselectivity of dehydrohalogenation of alkyl halides follows the Zaitsev rule; β elimination predominates in the direction that leads to the more highly substituted alkene.

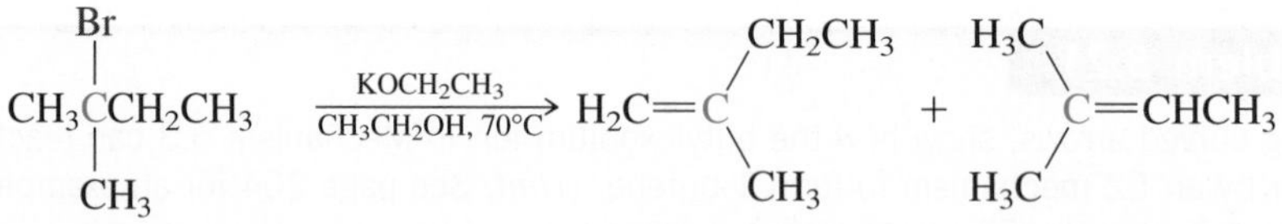

2-Bromo-2-methylbutane 2-Methyl-1-butene (29%) 2-Methyl-2-butene (71%)

Problem 5.22

Write the structures of all the alkenes that can be formed by dehydrohalogenation of each of the following alkyl halides. Apply the Zaitsev rule to predict the alkene formed in greatest amount in each case.

(a) 2-Bromo-2,3-dimethylbutane
(b) *tert*-Butyl chloride
(c) 3-Bromo-3-ethylpentane
(d) 2-Bromo-3-methylbutane
(e) 1-Bromo-3-methylbutane
(f) 1-Iodo-1-methylcyclohexane

Sample Solution (a) First analyze the structure of 2-bromo-2,3-dimethylbutane with respect to the number of possible β elimination pathways.

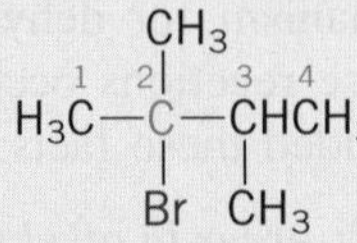

Bromine must be lost from C-2; hydrogen may be lost from C-1 or from C-3

The two possible alkenes are

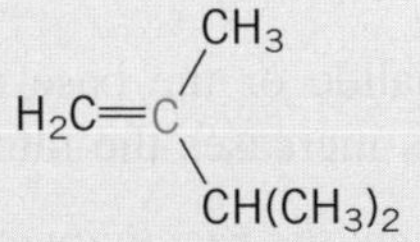

2,3-Dimethyl-1-butene (minor product)

and

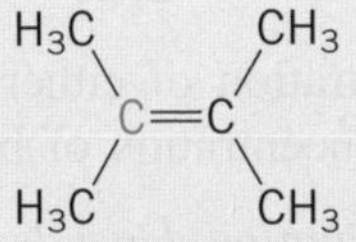

2,3-Dimethyl-2-butene (major product)

The major product, predicted on the basis of Zaitsev's rule, is 2,3-dimethyl-2-butene. It has a tetrasubstituted double bond. The minor alkene has a disubstituted double bond.

In addition to being regioselective, dehydrohalogenation of alkyl halides is stereoselective and favors formation of the more stable stereoisomer. Usually, as in the case of 5-bromononane, the trans (or *E*) alkene is formed in greater amounts than its cis (or *Z*) stereoisomer.

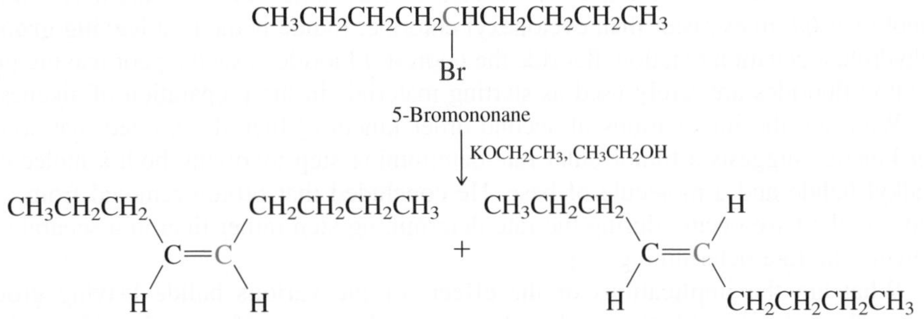
5-Bromononane
$KOCH_2CH_3$, CH_3CH_2OH

cis-4-Nonene (23%) + *trans*-4-Nonene (77%)

Problem 5.23

Write structural formulas for all the alkenes that can be formed in the reaction of 2-bromobutane with potassium ethoxide ($KOCH_2CH_3$).

Dehydrohalogenation of cycloalkyl halides leads exclusively to cis cycloalkenes when the ring has fewer than ten carbons. As the ring becomes larger, it can accommodate either a cis or a trans double bond, and large-ring cycloalkyl halides give mixtures of cis and trans cycloalkenes.

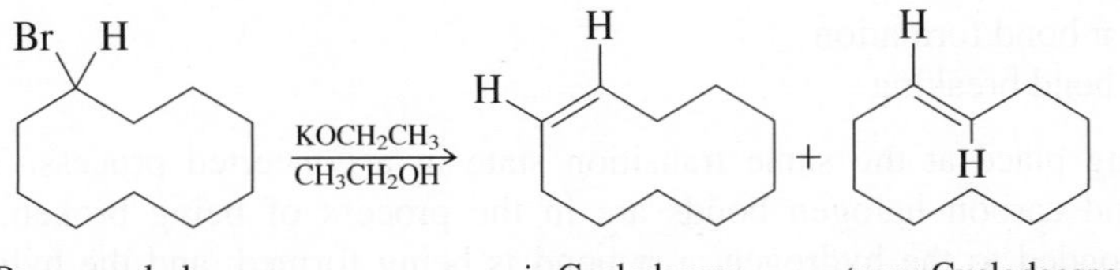
$KOCH_2CH_3$
CH_3CH_2OH

Bromocyclodecane → *cis*-Cyclodecene [(*Z*)-cyclodecene] (85%) + *trans*-Cyclodecene [(*E*)-cyclodecene] (15%)

5.15 The E2 Mechanism of Dehydrohalogenation of Alkyl Halides

In the 1920s, Sir Christopher Ingold proposed a mechanism for dehydrohalogenation that is still accepted as the best description of how these reactions occur. Some of the information on which Ingold based his mechanism included these facts:

1. *The reaction exhibits second-order kinetics; it is first-order in alkyl halide and first-order in base.*

$$\text{Rate} = k[\text{alkyl halide}][\text{base}]$$

Doubling the concentration of either the alkyl halide or the base doubles the reaction rate. Doubling the concentration of both reactants increases the rate by a factor of 4.

2. *The rate of elimination depends on the halogen, the reactivity of alkyl halides increasing with decreasing strength of the carbon–halogen bond.*

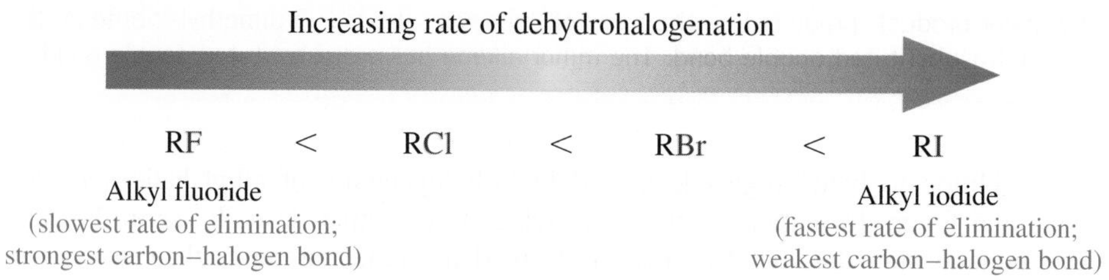
Increasing rate of dehydrohalogenation
RF < RCl < RBr < RI
Alkyl fluoride
(slowest rate of elimination; strongest carbon–halogen bond)
Alkyl iodide
(fastest rate of elimination; weakest carbon–halogen bond)

Cyclohexyl bromide, for example, is converted to cyclohexene by sodium ethoxide in ethanol over 60 times faster than cyclohexyl chloride. Iodide is the best **leaving group** in a dehydrohalogenation reaction, fluoride the poorest. Fluoride is such a poor leaving group that alkyl fluorides are rarely used as starting materials in the preparation of alkenes.

What are the implications of second-order kinetics? Ingold reasoned that second-order kinetics suggests a bimolecular rate-determining step involving both a molecule of the alkyl halide and a molecule of base. He concluded that proton removal from the β carbon by the base occurs during the rate-determining step rather than in a separate step following the rate-determining step.

What are the implications of the effects of the various halide leaving groups? Because the halogen with the weakest bond to carbon reacts fastest, Ingold concluded that the carbon–halogen bond breaks in the rate-determining step. The weaker the carbon–halogen bond, the easier it breaks.

On the basis of these observations, Ingold proposed a one-step bimolecular E2 mechanism for dehydrohalogenation.

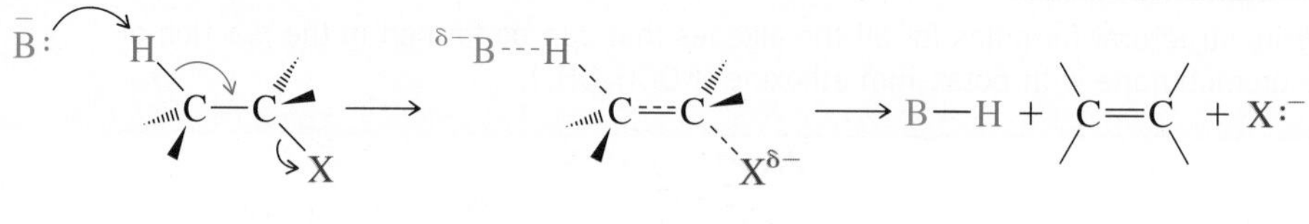

Transition state for bimolecular elimination

In the E2 mechanism the four key elements

1. B—H bond making
2. C—H bond breaking
3. C=C π bond formation
4. C—X bond breaking

are all taking place at the same transition state in a concerted process. The carbon–hydrogen and carbon–halogen bonds are in the process of being broken, the base is becoming bonded to the hydrogen, a π bond is being formed, and the hybridization of carbon is changing from sp^3 to sp^2. An energy diagram for the E2 mechanism is shown in Mechanism 5.4.

Mechanism 5.4

E2 Elimination of an Alkyl Halide

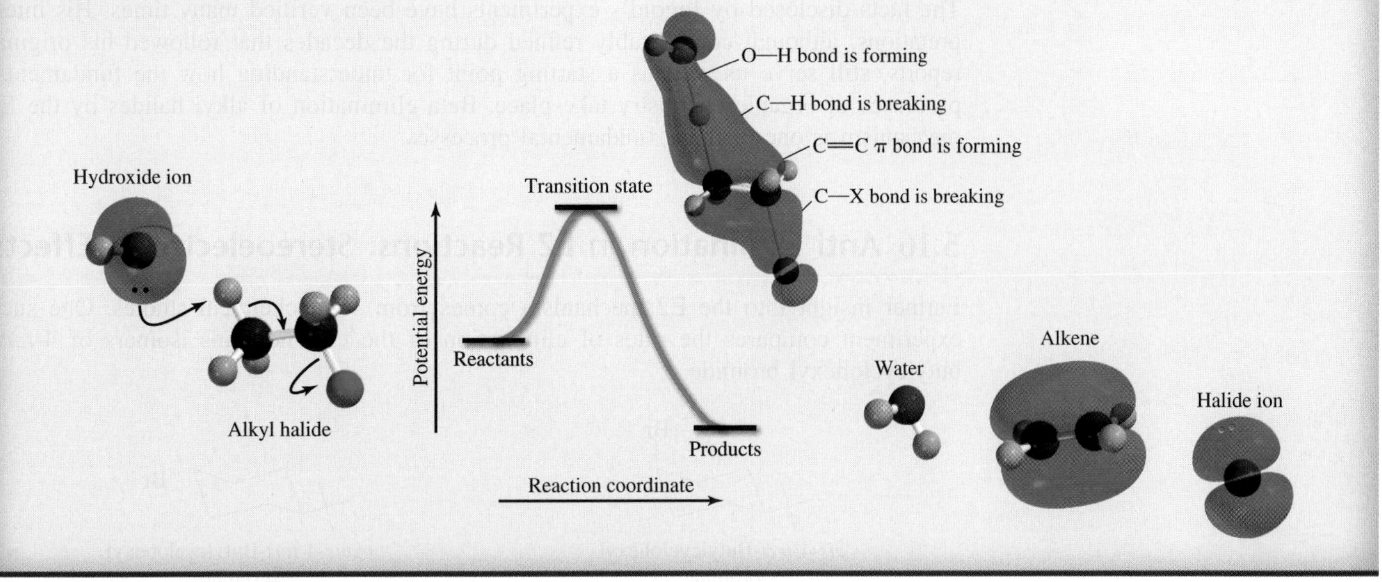

The E2 mechanism is followed whenever an alkyl halide—be it primary, secondary, or tertiary—undergoes elimination in the presence of a strong base.

Problem 5.24

(a) Use curved arrows to track electron movement in the dehydrohalogenation of *tert*-butyl chloride by sodium methoxide by the E2 mechanism.

(b) Do the same to show the formation of both elimination products from 2-bromo-2-methylbutane (page 208).

Sample Solution

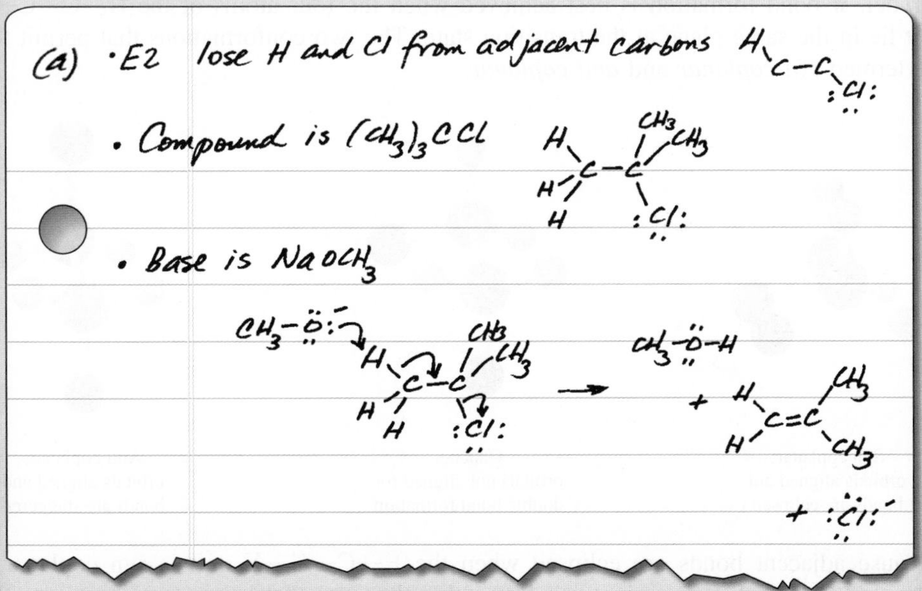

The regioselectivity of elimination is accommodated in the E2 mechanism by noting that a partial double bond develops at the transition state. Because alkyl groups stabilize double bonds, they also stabilize a partially formed π bond in the transition state. The more stable alkene therefore requires a lower energy of activation for its formation and predominates in the product mixture because it is formed faster than a less stable one.

Ingold was a pioneer in applying quantitative measurements of reaction rates to the understanding of organic reaction mechanisms. Many of the reactions to be described in this text were studied by him and his students during the period of about 1920 to 1950. The facts disclosed by Ingold's experiments have been verified many times. His interpretations, although considerably refined during the decades that followed his original reports, still serve us well as a starting point for understanding how the fundamental processes of organic chemistry take place. Beta elimination of alkyl halides by the E2 mechanism is one of those fundamental processes.

5.16 Anti Elimination in E2 Reactions: Stereoelectronic Effects

Further insight into the E2 mechanism comes from stereochemical studies. One such experiment compares the rates of elimination of the cis and trans isomers of 4-*tert*-butylcyclohexyl bromide.

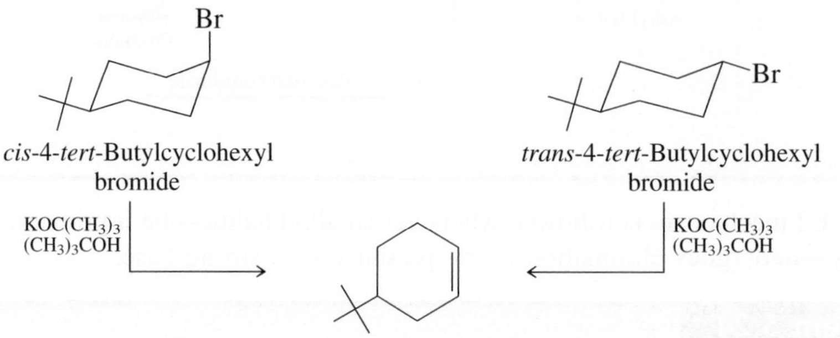

4-*tert*-Butylcyclohexene

Although both stereoisomers yield 4-*tert*-butylcyclohexene as the only alkene, they do so at quite different rates. The cis isomer reacts over 500 times faster than the trans.

The difference in reaction rate results from different degrees of π bond development in the E2 transition state. Since π overlap of *p* orbitals requires their axes to be parallel, π bond formation is best achieved when the four atoms of the H—C—C—X unit lie in the same plane at the transition state. The two conformations that permit this are termed *syn coplanar* and *anti coplanar.*

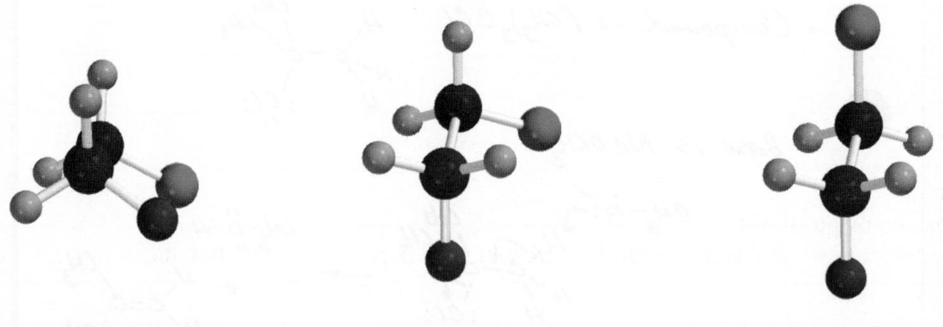

Syn coplanar; orbitals aligned but bonds are eclipsed — Gauche; orbitals not aligned for double bond formation — Anti coplanar; orbitals aligned and bonds are staggered

Because adjacent bonds are eclipsed when the H—C—C—X unit is syn coplanar, a transition state with this geometry is less stable than one that has an anti coplanar relationship between the proton and the leaving group.

Bromine is axial and anti coplanar to two axial hydrogens in the most stable conformation of *cis*-4-*tert*-butylcyclohexyl bromide and has the proper geometry for ready E2 elimination. The transition state is reached with little increase in strain, and elimination occurs readily.

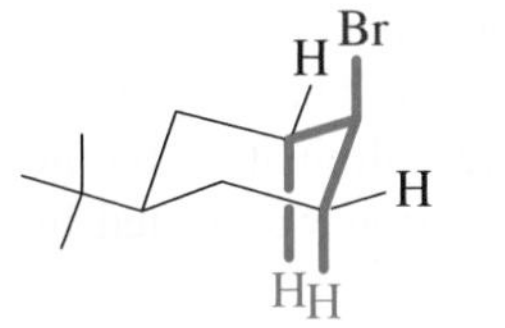

cis-4-*tert*-Butylcyclohexyl bromide
(faster E2 rate:
H and Br are anti coplanar)

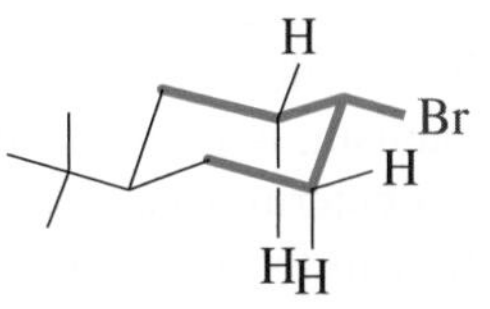

trans-4-*tert*-Butylcyclohexyl bromide
(slower E2 rate:
no H atoms anti to Br)

In its most stable conformation, the trans stereoisomer has no β hydrogens anti to Br; all four are gauche. Strain increases significantly in going to the E2 transition state, and the rate of elimination is slower than for the cis stereoisomer.

Problem 5.25

Use curved arrows to show the bonding changes in the reaction of *cis*-4-*tert*-butylcyclohexyl bromide with potassium *tert*-butoxide [$KOC(CH_3)_3$]. Be sure your drawing correctly represents the spatial relationship between the leaving group and the proton that is lost.

Effects that arise because one spatial arrangement of electrons (or orbitals or bonds) is more stable than another are called **stereoelectronic effects.** *There is a stereoelectronic preference for the anti coplanar arrangement of proton and leaving group in E2 reactions.* Although coplanarity of the *p* orbitals is the best geometry for the E2 process, modest deviations from it can be tolerated.

Stereoelectronic effects are also important in the dehydrohalogenation of acyclic alkyl halides by an E2 pathway. Again, the most favorable arrangement for the hydrogen and the halide being lost is anti coplanar. In the formation of 2-methyl-2-butene from 2-bromo-2-methylbutane shown on page 208, the elimination of HBr occurs readily from the conformation on the left but not from the one on the right.

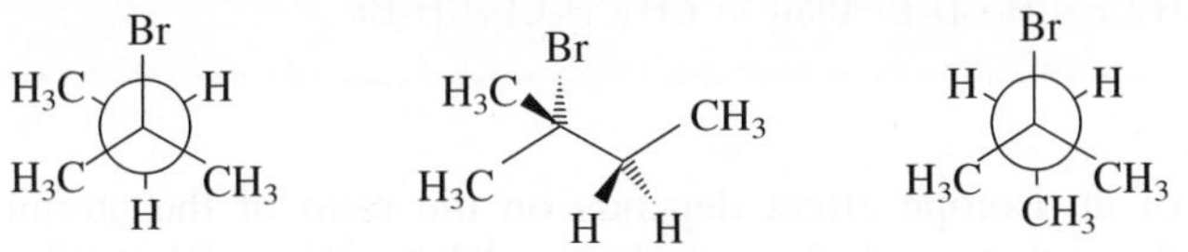

2-Bromo-2-methylbutane

5.17 Isotope Effects and the E2 Mechanism

The E2 mechanism as outlined in the preceding two sections receives support from studies of the dehydrohalogenation of alkyl halides that contain deuterium (D = ^2H) instead of protium (^1H) at the β carbon. The fundamental *kinds* of reactions a substance undergoes are the same regardless of which isotope is present, but the reaction *rates* can be different.

A C—D bond is ≈ 12 kJ/mol stronger than a C—H bond, making the activation energy for breaking a C—D bond slightly greater than that of an analogous C—H bond. Consequently, the rate constant *k* for an elementary step in which a C—D bond breaks is smaller than for a C—H bond. This difference in rate is expressed as a ratio of the respective rate constants (k_H/k_D) and is a type of **kinetic isotope effect.** Because it compares ^2H to ^1H, it is also referred to as a **deuterium isotope effect.**

Typical deuterium isotope effects for reactions in which C—H bond breaking is rate-determining lie in the range k_H/k_D = 3–8. If the C—H bond breaks after the rate-determining step, the overall reaction rate is affected only slightly and k_H/k_D = 1–2. *Thus, measuring the deuterium isotope effect can tell us if a C—H bond breaks in the rate-determining step.*

According to the E2 mechanism for dehydrohalogenation, a base removes a proton from the β carbon in the same step as the halide is lost. This step, indeed it is the only step in the mechanism, is rate-determining. Therefore, elimination by the E2 mechanism should exhibit a deuterium isotope effect. This prediction was tested by comparing the rate of elimination in the reaction:

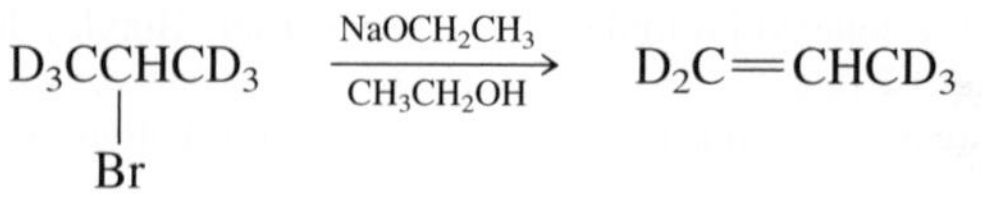

with that of $(CH_3)_2CHBr$. The measured value was $k_H/k_D = 6.7$, consistent with the idea that the β hydrogen is removed by the base in the rate-determining step, not after it.

Problem 5.26

Choose the compound in the following pairs that undergoes E2 elimination at the faster rate.

(a) $CH_3CH_2CH_2CD_2Br$ or $CH_3CH_2CD_2CH_2Br$

(b) 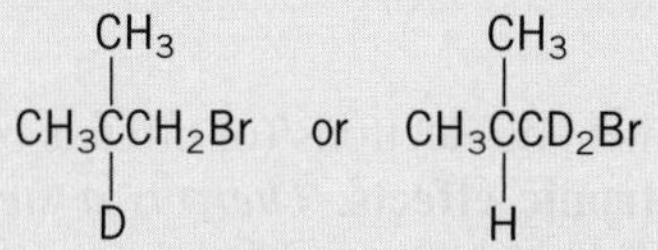

(c) 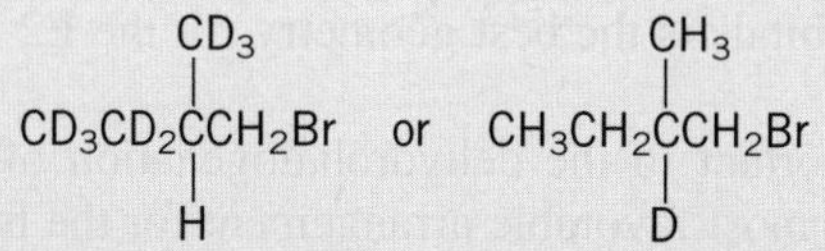

Sample Solution (a) A double bond is formed between C-1 and C-2 when either of the two compounds undergoes elimination. Bromine is lost from C-1, and H (or D) is lost from C-2. A C—H bond breaks faster than a C—D bond; therefore E2 elimination is faster in $CH_3CH_2CH_2CD_2Br$ than in $CH_3CH_2CD_2CH_2Br$.

The size of an isotope effect depends on the ratio of the atomic masses of the isotopes; thus, those that result from replacing ^1H by ^2H or ^3H (tritium) are easiest to measure. This, plus the additional facts that most organic compounds contain hydrogen and many reactions involve breaking C—H bonds, have made rate studies involving hydrogen isotopes much more common than those of other elements.

In later chapters we'll see several additional examples of reactions in which deuterium isotope effects were measured in order to test proposed mechanisms.

5.18 The E1 Mechanism of Dehydrohalogenation of Alkyl Halides

The E2 mechanism is a concerted process in which the carbon–hydrogen and carbon–halogen bonds both break in the same elementary step. What if these bonds break in separate steps?

One possibility is the two-step process of Mechanism 5.5, in which the carbon–halogen bond breaks first to give a carbocation intermediate, followed by deprotonation of the carbocation in a second step.

The alkyl halide, in this case 2-bromo-2-methylbutane, ionizes to a carbocation and a halide anion by a heterolytic cleavage of the carbon–halogen bond. Like the dissociation of an alkyloxonium ion to a carbocation, this step is rate-determining. Because the rate-determining step is unimolecular—it involves only the alkyl halide and not the base—it is an E1 mechanism.

Mechanism 5.5

The E1 Mechanism for Dehydrohalogenation of 2-Bromo-2-methylbutane in Ethanol

THE OVERALL REACTION:

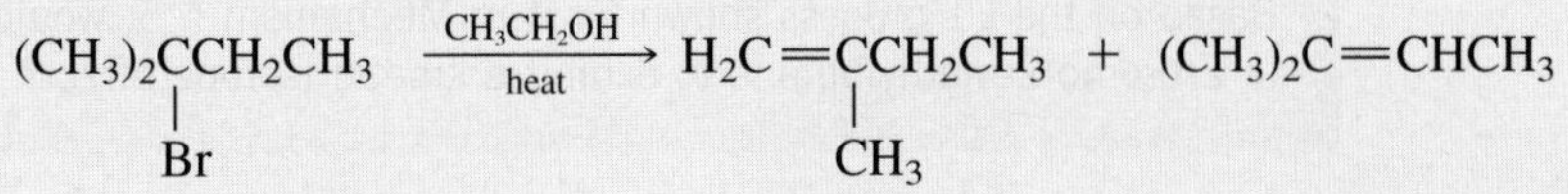

2-Bromo-2-methylbutane — 2-Methyl-1-butene (25%) — 2-Methyl-2-butene (75%)

THE MECHANISM:

Step 1: Alkyl halide dissociates by heterolytic cleavage of carbon–halogen bond. (Ionization step)

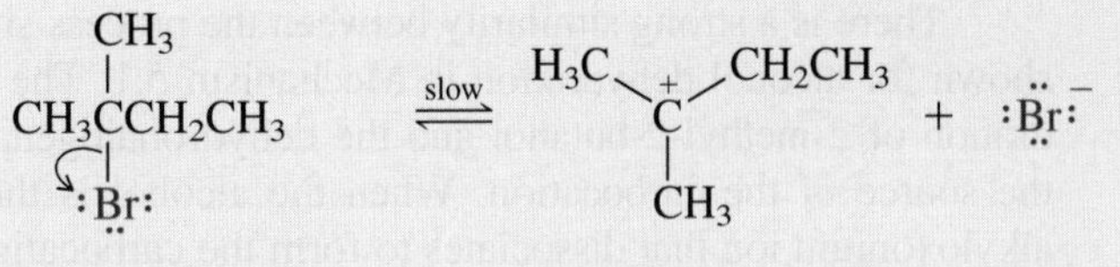

2-Bromo-2-methylbutane — 1,1-Dimethylpropyl cation — Bromide ion

Step 2: Ethanol acts as a base to remove a proton from the carbocation to give the alkene products. (Deprotonation step)

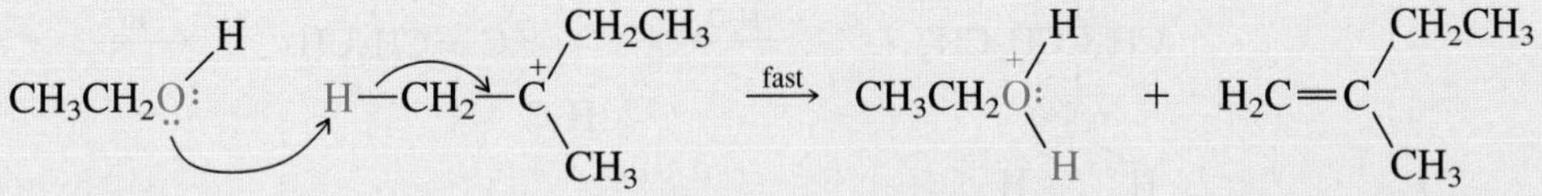

Ethanol — 1,1-Dimethylpropyl cation — Ethyloxonium ion — 2-Methyl-1-butene

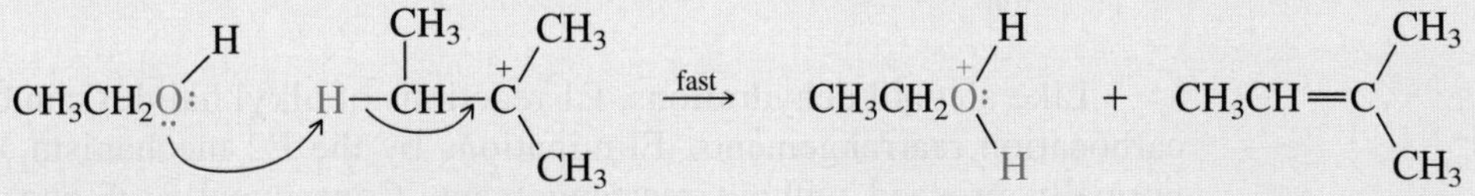

Ethanol — 1,1-Dimethylpropyl cation — Ethyloxonium ion — 2-Methyl-2-butene

Typically, elimination by the E1 mechanism is observed only for tertiary and some secondary alkyl halides, and then only when the base is weak or in low concentration. Unlike eliminations that follow an E2 pathway and exhibit second-order kinetic behavior:

$$\text{Rate} = k[\text{alkyl halide}][\text{base}]$$

those that follow an E1 mechanism obey a first-order rate law.

$$\text{Rate} = k[\text{alkyl halide}]$$

The reactivity order parallels the ease of carbocation formation.

Increasing rate of elimination by the E1 mechanism

RCH_2X < R_2CHX < R_3CX

Primary alkyl halide: slowest rate of E1 elimination

Tertiary alkyl halide: fastest rate of E1 elimination

Because the carbon–halogen bond breaks in the slow step, the rate of the reaction depends on the leaving group. Alkyl iodides have the weakest carbon–halogen bond and are the most reactive; alkyl fluorides have the strongest carbon–halogen bond and are the least reactive.

Problem 5.27

Based on the E1 process shown for it in Mechanism 5.5, would you expect elimination in 2-bromo-2-methylbutane to exhibit a kinetic isotope effect?

The best examples of E1 eliminations are those carried out in the absence of added base. In the example cited in Mechanism 5.5, the base that abstracts the proton from the carbocation intermediate is a very weak one; it is a molecule of the solvent, ethyl alcohol. At even modest concentrations of strong base, elimination by the E2 mechanism is much faster than E1 elimination.

There is a strong similarity between the process shown in Mechanism 5.5 and the one shown for alcohol dehydration in Mechanism 5.1. The main difference between the dehydration of 2-methyl-2-butanol and the dehydrohalogenation of 2-bromo-2-methylbutane is the source of the carbocation. When the alcohol is the substrate, it is the corresponding alkyloxonium ion that dissociates to form the carbocation. The alkyl halide ionizes directly to the carbocation.

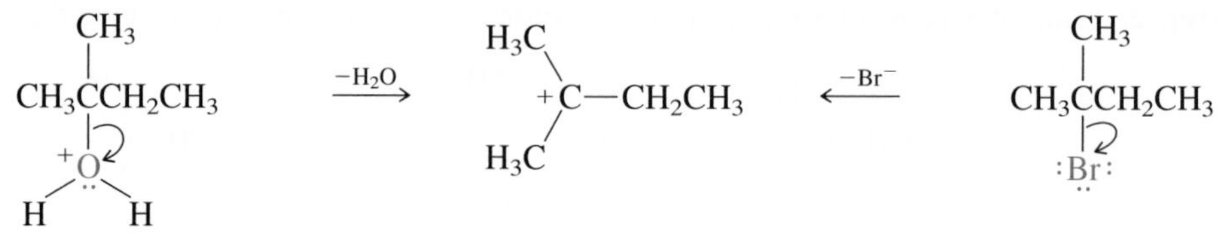
CH3
CH3CCH2CH3
O
H H
−H2O
H3C
+C—CH2CH3
H3C
−Br−
CH3
CH3CCH2CH3
:Br:

Alkyloxonium ion Carbocation Alkyl halide

Like alcohol dehydrations, E1 reactions of alkyl halides can be accompanied by carbocation rearrangements. Eliminations by the E2 mechanism, on the other hand, normally proceed without rearrangement. Consequently, if one wishes to prepare an alkene from an alkyl halide, conditions favorable to E2 elimination should be chosen. In practice this simply means carrying out the reaction in the presence of a strong base.

5.19 SUMMARY

Section 5.1 Alkenes and cycloalkenes contain carbon–carbon double bonds. According to **IUPAC nomenclature,** alkenes are named by substituting *-ene* for the *-ane* suffix of the alkane that has the same number of carbon atoms as the longest continuous chain that includes the double bond. The chain is numbered in the direction that gives the lower number to the first-appearing carbon of the double bond. The double bond takes precedence over alkyl groups and halogens in dictating the direction of numbering, but is outranked by a hydroxyl group.

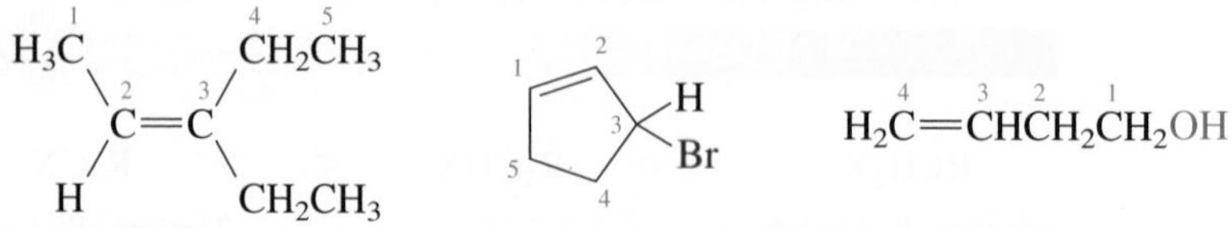
H3C
CH2CH3
C=C
H
CH2CH3
H
Br
H2C=CHCH2CH2OH

3-Ethyl-2-pentene 3-Bromocyclopentene 3-Buten-1-ol

Section 5.2 Bonding in alkenes is described according to an sp^2 orbital hybridization model. The double bond unites two sp^2-hybridized carbon atoms and is made of a σ component and a π component. The σ bond arises by overlap of an sp^2 hybrid orbital on each carbon. The π bond is weaker than the σ bond and results from a side-by-side overlap of half-filled *p* orbitals.

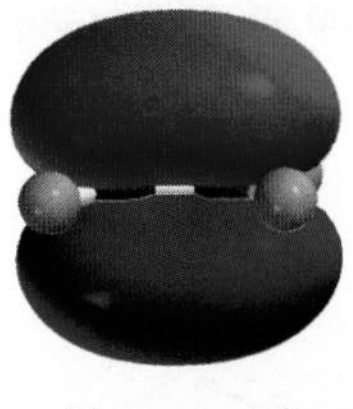

Sections 5.3–5.4 Isomeric alkenes may be either **constitutional isomers** or **stereoisomers.** There is a sizable barrier to rotation about a carbon–carbon double bond, which corresponds to the energy required to break the π component of the double bond. Stereoisomeric alkenes do not interconvert under normal conditions. Their configurations are described according to two notational systems. One system adds the prefix *cis-* to the name of the alkene when similar substituents are on the same side of the double bond and the prefix *trans-* when they are on opposite sides. The other ranks substituents according to a system of rules based on atomic number. The prefix *Z* is used for alkenes that have higher ranked substituents on the same side of the double bond; the prefix *E* is used when higher ranked substituents are on opposite sides.

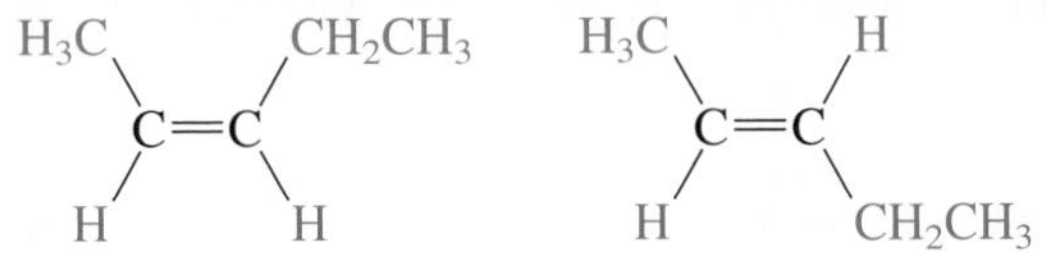

cis-2-Pentene [(*Z*)-2-pentene] *trans*-2-Pentene [(*E*)-2-pentene]

Section 5.5 Alkenes are nonpolar. Alkyl substituents donate electrons to an sp^2-hybridized carbon to which they are attached slightly better than hydrogen does.

Section 5.6 Electron release from alkyl substituents stabilizes a double bond. In general, the order of alkene stability is:

1. Tetrasubstituted alkenes ($R_2C{=}CR_2$) are the most stable.
2. Trisubstituted alkenes ($R_2C{=}CHR$) are next.
3. Among disubstituted alkenes, *trans*-RCH═CHR is normally more stable than *cis*-RCH═CHR. Exceptions are cycloalkenes, cis cycloalkenes being more stable than trans when the ring contains fewer than 12 carbons.
4. Monosubstituted alkenes ($RCH{=}CH_2$) have a more stabilized double bond than ethylene (unsubstituted) but are less stable than disubstituted alkenes.

The greater stability of more highly substituted double bonds is an example of an **electronic effect.** The decreased stability that results from van der Waals strain between cis substituents is an example of a **steric effect.**

Section 5.7 Cycloalkenes that have trans double bonds in rings smaller than 12 members are less stable than their cis stereoisomers. *trans*-Cyclooctene can be isolated and stored at room temperature, but *trans*-cycloheptene is not stable above −30°C.

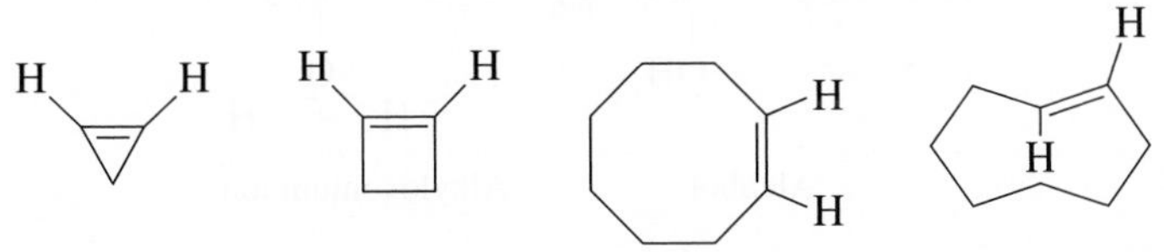

Cyclopropene Cyclobutene *cis*-Cyclooctene *trans*-Cyclooctene

Cyclohexene adopts a half-chair conformation.

Section 5.8 Alkenes are prepared by **β elimination** of alcohols and alkyl halides. These reactions are summarized with examples in Table 5.2. In both cases, β elimination

TABLE 5.2 Preparation of Alkenes by Elimination Reactions of Alcohols and Alkyl Halides

Reaction (section) and comments	General equation and specific example
Dehydration of alcohols (Sections 5.9–5.13) Dehydration requires an acid catalyst; the order of reactivity of alcohols is tertiary > secondary > primary. Elimination is regioselective and proceeds in the direction that produces the most highly substituted double bond. When stereoisomeric alkenes are possible, the more stable one is formed in greater amounts. An E1 (elimination unimolecular) mechanism via a carbocation intermediate is followed with secondary and tertiary alcohols. Primary alcohols react by an E2 (elimination bimolecular) mechanism. Sometimes elimination is accompanied by rearrangement.	$R_2CH\underset{OH}{\underset{\mid}{C}}R_2 \xrightarrow{H^+} R_2C{=}CR_2 + H_2O$ Alcohol Alkene Water 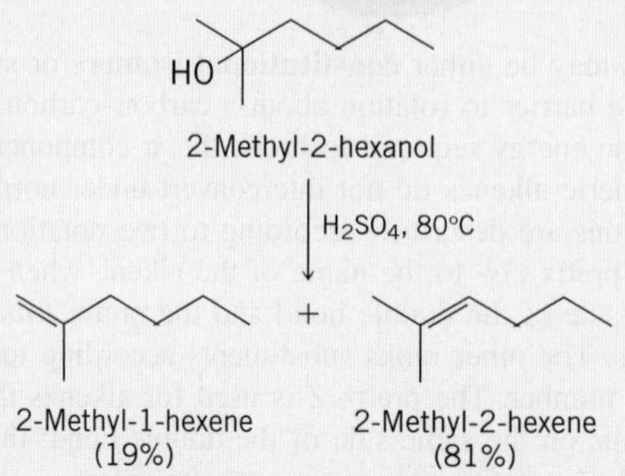HO 2-Methyl-2-hexanol H_2SO_4, 80°C + 2-Methyl-1-hexene (19%) 2-Methyl-2-hexene (81%)
Dehydrohalogenation of alkyl halides (Sections 5.14–5.16) Strong bases cause a proton and a halide to be lost from adjacent carbons of an alkyl halide to yield an alkene. Regioselectivity is in accord with the Zaitsev rule. The order of halide reactivity is I > Br > Cl > F. A concerted E2 reaction pathway is followed, carbocations are not involved, and rearrangements do not occur. An anti coplanar arrangement of the proton being removed and the halide being lost characterizes the transition state.	$R_2CH\underset{X}{\underset{\mid}{C}}R'_2 + :B^- \longrightarrow R_2C{=}CR'_2 + H{-}B + X^-$ Alkyl halide Base Alkene Conjugate acid of base Halide 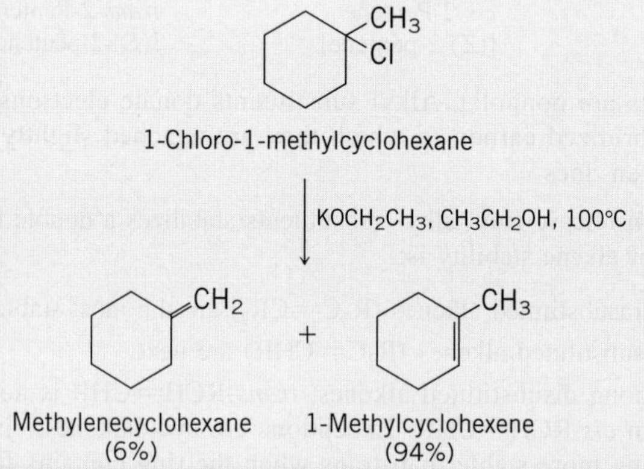CH_3 Cl 1-Chloro-1-methylcyclohexane $KOCH_2CH_3$, CH_3CH_2OH, 100°C CH_2 + CH_3 Methylenecyclohexane (6%) 1-Methylcyclohexene (94%)

proceeds in the direction that yields the more highly substituted double bond **(Zaitsev's rule).**

Sections 5.9–5.11 See Table 5.2.

Section 5.12 Secondary and tertiary alcohols undergo **dehydration** by an E1 mechanism involving carbocation intermediates.

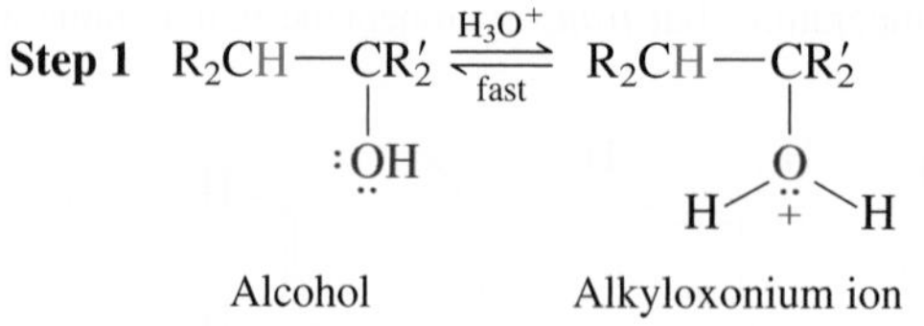
Step 1
$R_2CH{-}CR'_2$
$:\ddot{O}H$
H_3O^+
fast
$R_2CH{-}CR'_2$
O
H + H
Alcohol
Alkyloxonium ion

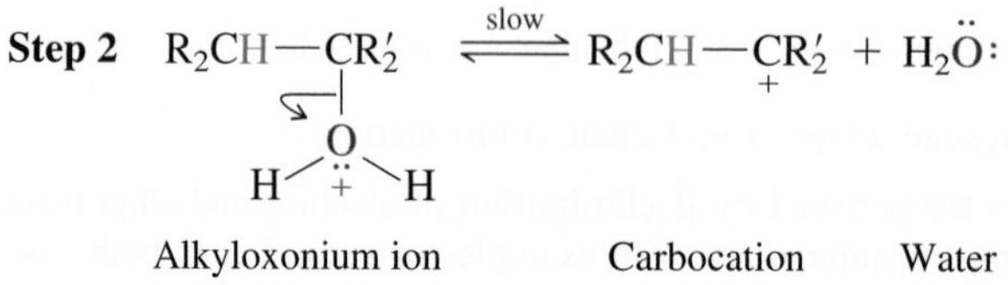
Step 2
$R_2CH{-}CR'_2$
O
H + H
slow
$R_2CH{-}\overset{}{\underset{+}{C}}R'_2 + H_2\ddot{O}:$
Alkyloxonium ion
Carbocation
Water

Step 3

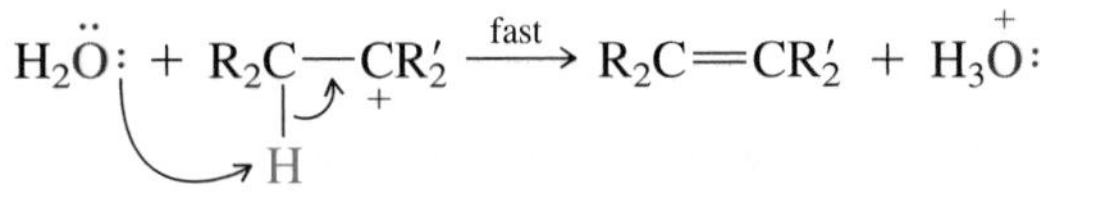

Water Carbocation Alkene Hydronium ion

Primary alcohols do not dehydrate as readily as secondary or tertiary alcohols, and their dehydration does not involve a primary carbocation. A proton is lost from the β carbon in the same step in which carbon–oxygen bond cleavage occurs. The mechanism is E2.

Section 5.13 Alkene synthesis via alcohol dehydration is sometimes accompanied by carbocation **rearrangement.** A less stable carbocation can rearrange to a more stable one by an alkyl group migration or by a hydride shift, opening the possibility for alkene formation from two different carbocations.

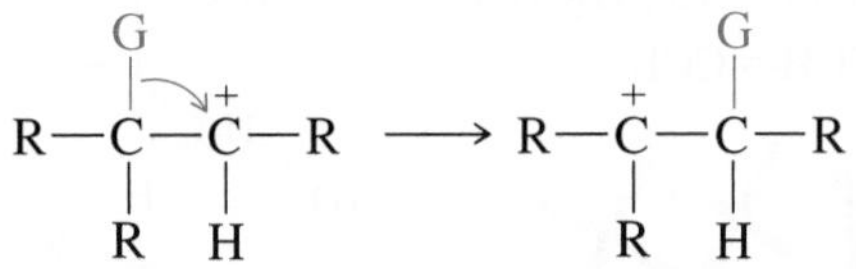

Secondary carbocation Tertiary carbocation

(G is a migrating group; it may be either a hydrogen or an alkyl group)

Section 5.14 See Table 5.2.

Section 5.15 **Dehydrohalogenation** of alkyl halides by alkoxide bases is not complicated by rearrangements, because carbocations are not intermediates. The mechanism is E2. It is a concerted process in which the base abstracts a proton from the β carbon while the bond between the halogen and the α carbon undergoes heterolytic cleavage.

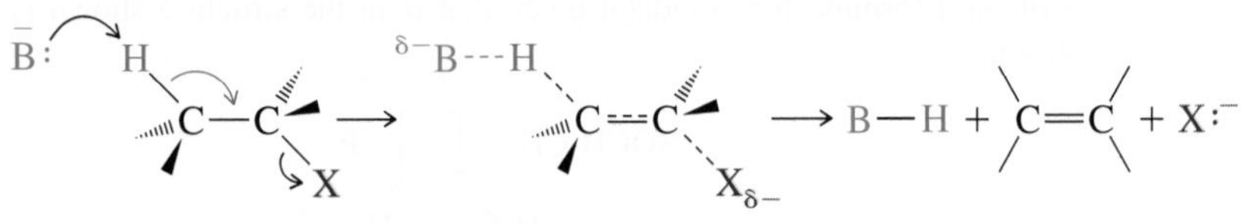

Transition state

Section 5.16 The preceding equation shows the proton H and the halogen X in the *anti coplanar* relationship that is required for elimination by the E2 mechanism.

Section 5.17 A β C—D bond is broken more slowly in the E2 dehydrohalogenation of alkyl halides than a β C—H bond. The ratio of the rate constants k_H/k_D is a measure of the **deuterium isotope effect** and has a value in the range 3–8 when a carbon–hydrogen bond breaks in the rate-determining step of a reaction.

Section 5.18 In the absence of a strong base, alkyl halides eliminate by an E1 mechanism. Rate-determining ionization of the alkyl halide to a carbocation is followed by deprotonation of the carbocation.

Step 1

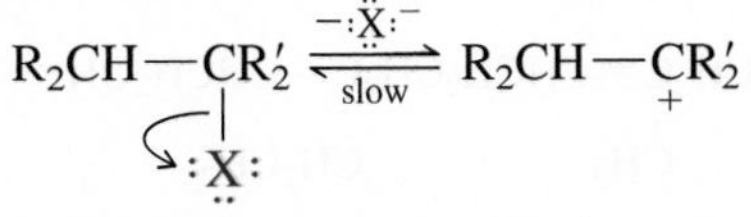

Alkyl halide Carbocation

Step 2

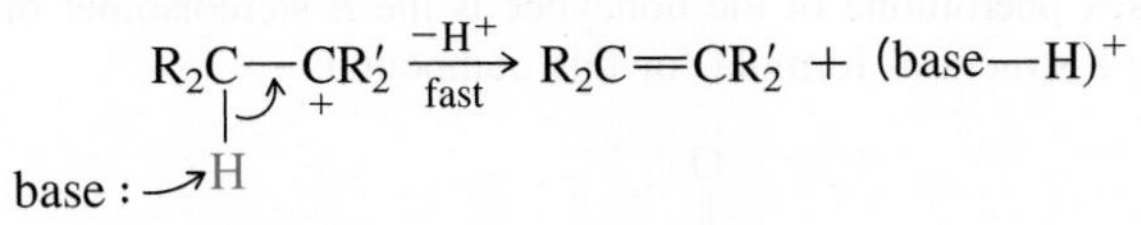

Carbocation Alkene

PROBLEMS

5.28 Write structural formulas for each of the following:

(a) 1-Heptene
(b) 3-Ethyl-2-pentene
(c) *cis*-3-Octene
(d) *trans*-1,4-Dichloro-2-butene
(e) (*Z*)-3-Methyl-2-hexene
(f) (*E*)-3-Chloro-2-hexene
(g) 1-Bromo-3-methylcyclohexene
(h) 1-Bromo-6-methylcyclohexene
(i) 4-Methyl-4-penten-2-ol
(j) Vinylcycloheptane
(k) 1,1-Diallylcyclopropane
(l) *trans*-1-Isopropenyl-3-methylcyclohexane

5.29 Write a structural formula and give two acceptable IUPAC names for each alkene of molecular formula C_7H_{14} that has a *tetrasubstituted* double bond.

5.30 Give an IUPAC name for each of the following compounds:

(a) $(CH_3CH_2)_2C{=}CHCH_3$
(b) $(CH_3CH_2)_2C{=}C(CH_2CH_3)_2$
(c) $(CH_3)_3CCH{=}CCl_2$

(d) 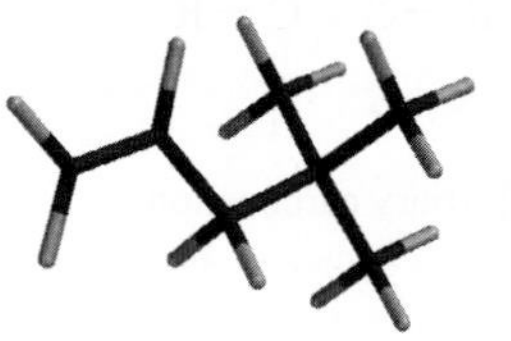

(e) 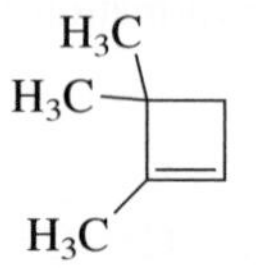
H₃C
H₃C
H₃C

(f) 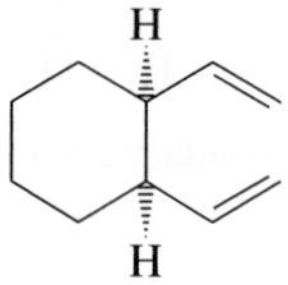
H
H

(g) 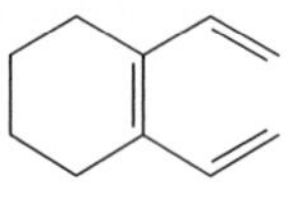

5.31 (a) A hydrocarbon isolated from fish oil and from plankton was identified as 2,6,10,14-tetramethyl-2-pentadecene. Write its structure.

(b) Alkyl isothiocyanates are compounds of the type RN=C=S. Write a structural formula for *allyl isothiocyanate,* a pungent-smelling compound isolated from mustard.

(c) Grandisol is one component of the sex attractant of the boll weevil. Write a structural formula for grandisol given that R in the structure shown is an isopropenyl group.

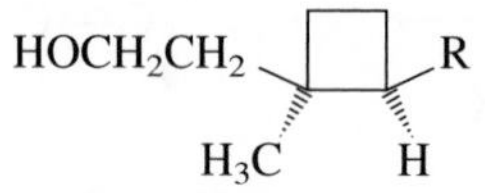
HOCH₂CH₂
R
H₃C
H

5.32 (a) The sex attractant of the Mediterranean fruit fly is (*E*)-6-nonen-l-ol. Write a structural formula for this compound, showing the stereochemistry of the double bond.

(b) Geraniol is a naturally occurring substance present in the fragrant oil of many plants. It has a pleasing, rose-like odor. Geraniol is the *E* isomer of

$$(CH_3)_2C{=}CHCH_2CH_2\underset{\displaystyle |\atop\displaystyle CH_3}{C}{=}CHCH_2OH$$

Write a structural formula for geraniol, showing its stereochemistry.

(c) Nerol is a naturally occurring substance that is a stereoisomer of geraniol. Write its structure.

(d) The sex attractant of the codling moth is the 2*Z*,6*E* stereoisomer of

$$CH_3CH_2CH_2\underset{\displaystyle |\atop\displaystyle CH_3}{C}{=}CHCH_2CH_2\underset{\displaystyle |\atop\displaystyle CH_2CH_3}{C}{=}CHCH_2OH$$

Write the structure of this substance in a way that clearly shows its stereochemistry.

(e) The sex pheromone of the honeybee is the *E* stereoisomer of the compound shown. Write a structural formula for this compound.

$$CH_3\overset{\displaystyle O\atop\displaystyle \|}{C}(CH_2)_4CH_2CH{=}CHCO_2H$$

(f) A growth hormone from the cecropia moth has the structure shown. Express the stereochemistry of the double bonds according to the *E–Z* system.

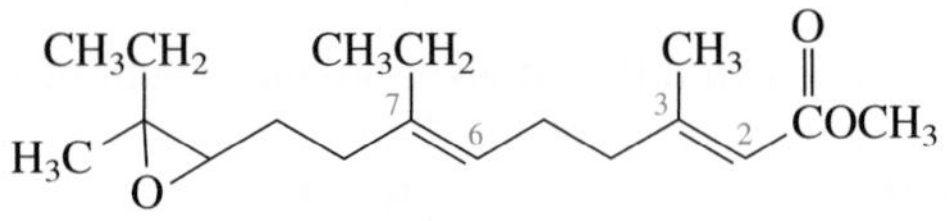

5.33 Match each alkene with the appropriate heat of combustion:
Heats of combustion (kJ/mol): 5293; 4658; 4650; 4638; 4632
Heats of combustion (kcal/mol): 1264.9; 1113.4; 1111.4; 1108.6; 1107.1
(a) 1-Heptene
(b) 2,4-Dimethyl-1-pentene
(c) 2,4-Dimethyl-2-pentene
(d) (*Z*)-4,4-Dimethyl-2-pentene
(e) 2,4,4-Trimethyl-2-pentene

5.34 Choose the more stable alkene in each of the following pairs. Explain your reasoning.
(a) 1-Methylcyclohexene or 3-methylcyclohexene
(b) Isopropenylcyclopentane or allylcyclopentane

(c) 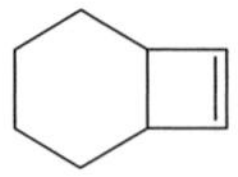 or 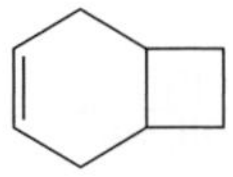

Bicyclo[4.2.0]oct-7-ene   Bicyclo[4.2.0]oct-3-ene

(d) (*Z*)-Cyclononene or (*E*)-cyclononene
(e) (*Z*)-Cyclooctadecene or (*E*)-cyclooctadecene

5.35 (a) Suggest an explanation for the fact that 1-methylcyclopropene is some 42 kJ/mol (10 kcal/mol) less stable than methylenecyclopropane.

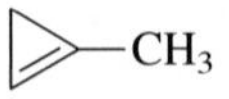

is less stable than
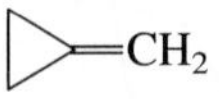

1-Methylcyclopropene   Methylenecyclopropane

(b) On the basis of your answer to part (a), compare the expected stability of 3-methylcyclopropene with that of 1-methylcyclopropene and that of methylenecyclopropane.

5.36 How many alkenes would you expect to be formed from each of the following alkyl bromides under conditions of E2 elimination? Identify the alkenes in each case.
(a) 1-Bromohexane
(b) 2-Bromohexane
(c) 3-Bromohexane
(d) 2-Bromo-2-methylpentane
(e) 2-Bromo-3-methylpentane
(f) 3-Bromo-2-methylpentane
(g) 3-Bromo-3-methylpentane
(h) 3-Bromo-2,2-dimethylbutane

5.37 Write structural formulas for all the alkene products that could reasonably be formed from each of the following compounds under the indicated reaction conditions. Where more than one alkene is produced, specify the one that is the major product.
(a) 1-Bromo-3,3-dimethylbutane (potassium *tert*-butoxide, *tert*-butyl alcohol, 100°C)
(b) 1-Methylcyclopentyl chloride (sodium ethoxide, ethanol, 70°C)
(c) 3-Methyl-3-pentanol (sulfuric acid, 80°C)
(d) 2,3-Dimethyl-2-butanol (phosphoric acid, 120°C)
(e) 3-Iodo-2,4-dimethylpentane (sodium ethoxide, ethanol, 70°C)
(f) 2,4-Dimethyl-3-pentanol (sulfuric acid, 120°C)

5.38 Choose the compound of molecular formula $C_7H_{13}Br$ that gives each alkene shown as the *exclusive* product of E2 elimination.

(a) 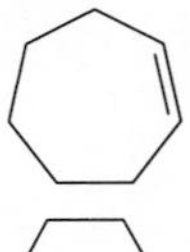

(c) 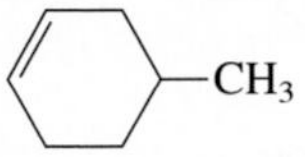

(b) 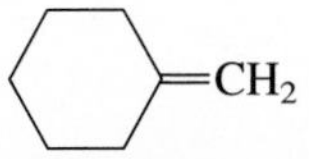

(d) 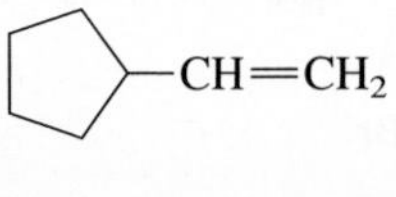

(e) 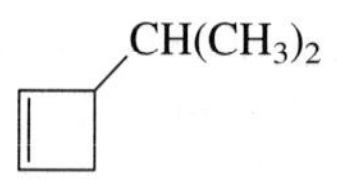

(f) 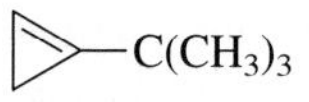

5.39 Give the structures of two different alkyl bromides both of which yield the indicated alkene as the *exclusive* product of E2 elimination.

(a) $CH_3CH{=}CH_2$

(b) $(CH_3)_2C{=}CH_2$

(c) $BrCH{=}CBr_2$

(d) 

5.40 Predict the major organic product of each of the following reactions. In spite of the structural complexity of some of the starting materials, the functional group transformations are all of the type described in this chapter.

(a) 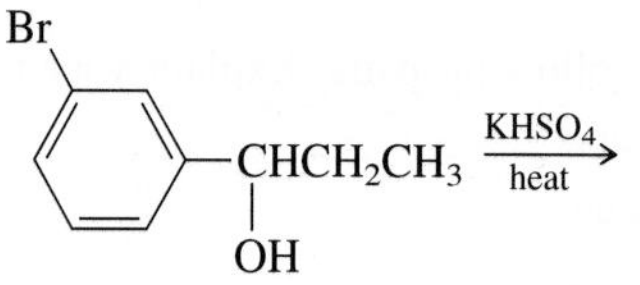

(b) $ICH_2CH(OCH_2CH_3)_2 \xrightarrow[(CH_3)_3COH,\ heat]{KOC(CH_3)_3}$

(c) 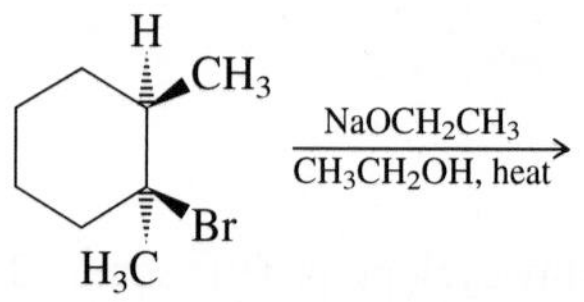

(d) 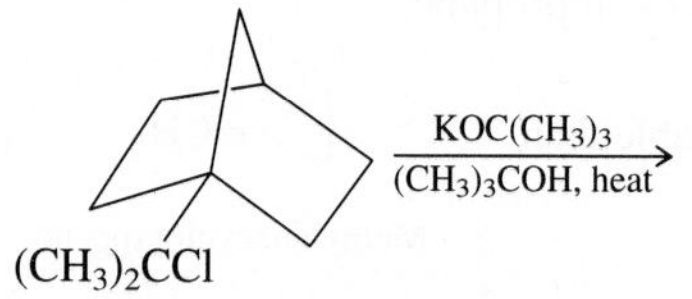

(e) 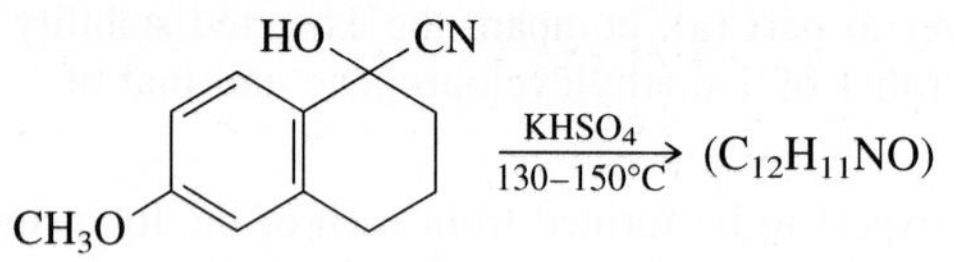

(f) $HOC(CH_2CO_2H)_2 \xrightarrow[140\text{–}145°C]{H_2SO_4} (C_6H_6O_6)$ (with CO_2H on the central carbon)

Citric acid

(g) 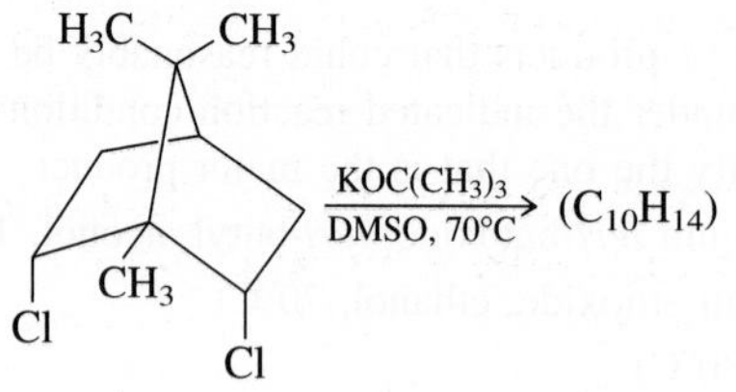

(h) 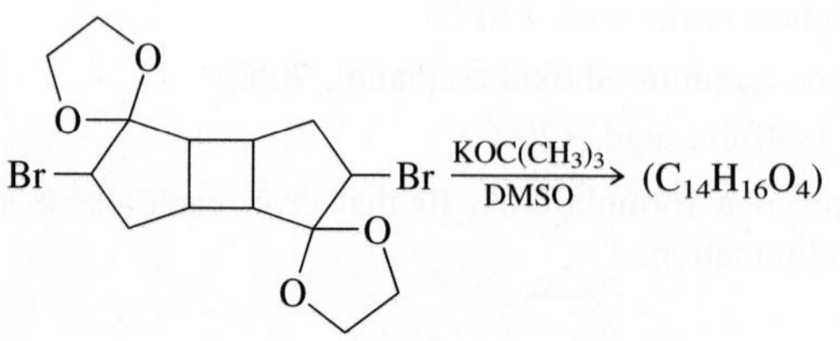

(i) 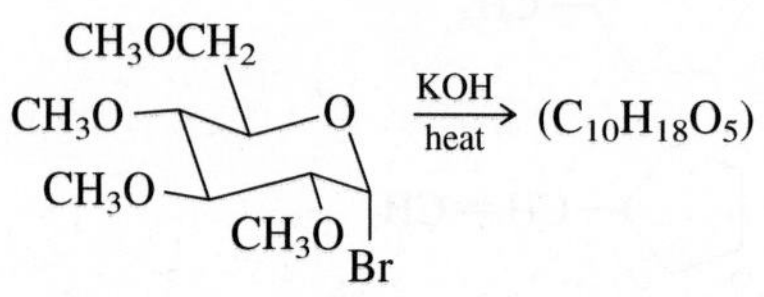

(j) 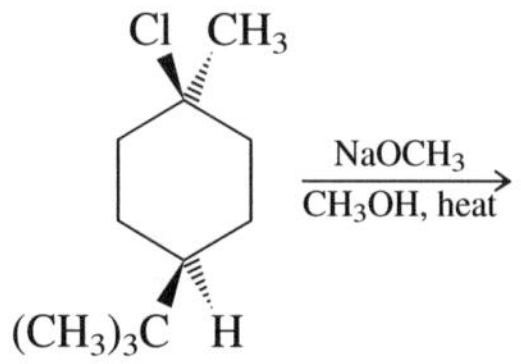

5.41 The rate of the reaction

$$(CH_3)_3\,CCl + NaSCH_2CH_3 \rightarrow (CH_3)_2C{=}CH_2 + CH_3CH_2SH + NaCl$$

is first-order in $(CH_3)_3CCl$ and first-order in $NaSCH_2CH_3$. Give the symbol (E1 or E2) for the most reasonable mechanism, and use curved arrows to show the flow of electrons.

5.42 Menthyl chloride and neomenthyl chloride have the structures shown. One of these stereoisomers undergoes elimination on treatment with sodium ethoxide in ethanol much more readily than the other. Which reacts faster, menthyl chloride or neomenthyl chloride? Why?

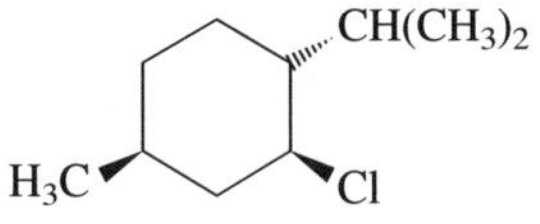

Menthyl chloride

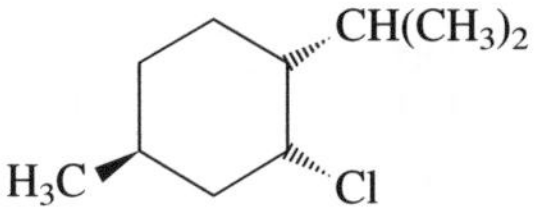

Neomenthyl chloride

5.43 The stereoselectivity of elimination of 5-bromononane on treatment with potassium ethoxide was described in Section 5.14. Draw Newman projections of 5-bromononane showing the conformations that lead to *cis*-4-nonene and *trans*-4-nonene, respectively. Identify the proton that is lost in each case, and suggest a mechanistic explanation for the observed stereoselectivity.

5.44 You have available 2,2-dimethylcyclopentanol (**A**) and 2-bromo-1,1-dimethylcyclopentane (**B**) and wish to prepare 3,3-dimethylcyclopentene (**C**). Which would you choose as the more suitable reactant, **A** or **B**, and with what would you treat it?

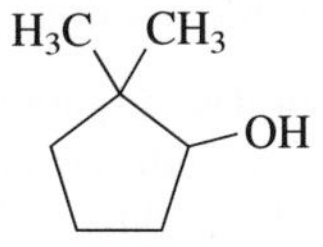

A

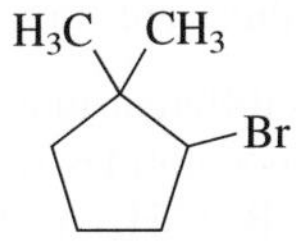

B

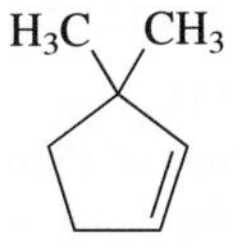

C

5.45 In the acid-catalyzed dehydration of 2-methyl-1-propanol, what carbocation would be formed if a hydride shift accompanied cleavage of the carbon–oxygen bond in the alkyloxonium ion? What ion would be formed as a result of a methyl shift? Which pathway do you think will predominate, a hydride shift or a methyl shift?

5.46 Each of the following carbocations has the potential to rearrange to a more stable one. Write the structure of the rearranged carbocation, and use curved arrows to show how it is formed.

(a) $CH_3CH_2\overset{+}{C}H_2$

(b) $(CH_3)_2CH\overset{+}{C}HCH_3$

(c) $(CH_3)_3C\overset{+}{C}HCH_3$

(d) $(CH_3CH_2)_3C\overset{+}{C}H_2$

(e) 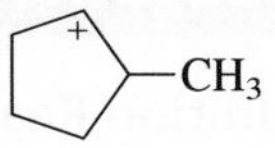

5.47 Write a sequence of steps depicting the mechanisms of each of the following reactions. Use curved arrows to show electron flow.

(a) 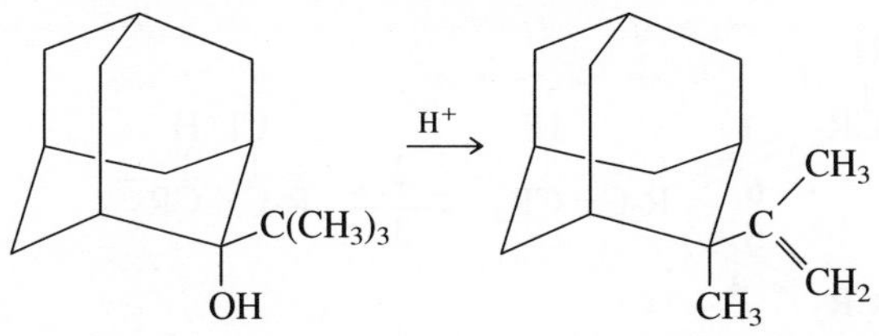

(b)

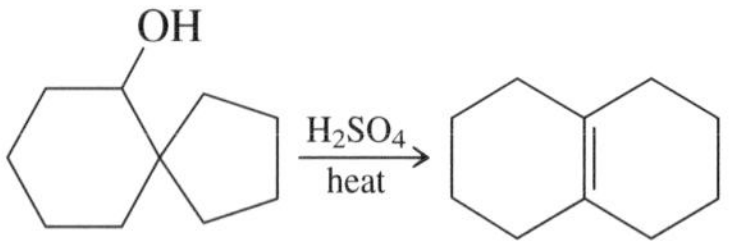

(c)

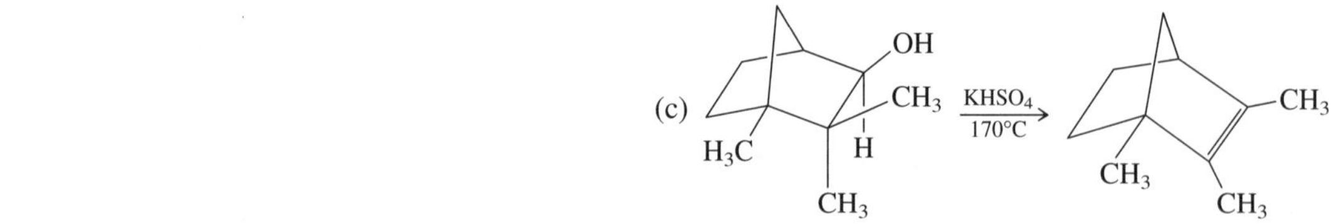

5.48 In Problem 5.20 (Section 5.13) we saw that acid-catalyzed dehydration of 2,2-dimethylcyclohexanol afforded 1,2-dimethylcyclohexene. To explain this product we must write a mechanism for the reaction in which a methyl shift transforms a secondary carbocation to a tertiary one. Another product of the dehydration of 2,2-dimethylcyclohexanol is isopropylidenecyclopentane. Write a mechanism to rationalize its formation, using curved arrows to show the flow of electrons.

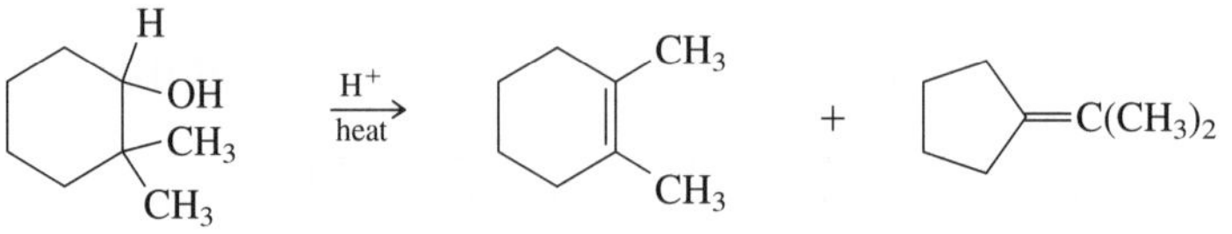

2,2-Dimethylcyclohexanol 1,2-Dimethylcyclohexene Isopropylidenecyclopentane

5.49 Acid-catalyzed dehydration of 2,2-dimethyl-1-hexanol gave a number of isomeric alkenes including 2-methyl-2-heptene as shown in the following equation.

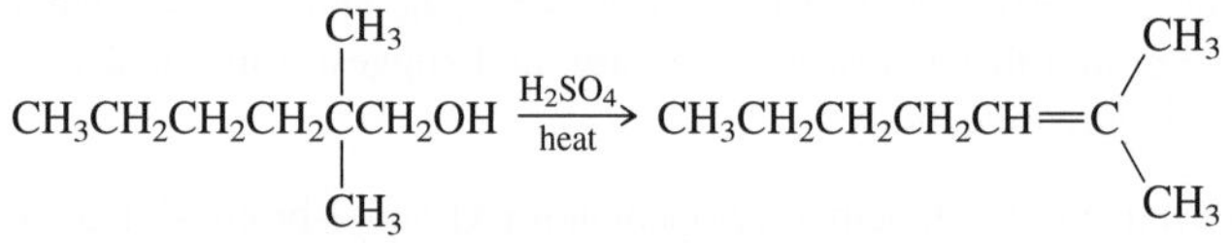

(a) Write a stepwise mechanism for the formation of 2-methyl-2-heptene, using curved arrows to show the flow of electrons.

(b) What other alkenes do you think are formed in this reaction?

5.50 Compound A (C_4H_{10}) gives two different monochlorides on photochemical chlorination. Treatment of either of these monochlorides with potassium *tert*-butoxide in dimethyl sulfoxide gives the same alkene B (C_4H_8) as the only product. What are the structures of compound A, the two monochlorides, and alkene B?

5.51 Compound A (C_6H_{14}) gives three different monochlorides on photochemical chlorination. One of these monochlorides is inert to E2 elimination. The other two monochlorides yield the same alkene B (C_6H_{12}) on being heated with potassium *tert*-butoxide in *tert*-butyl alcohol. Identify compound A, the three monochlorides, and alkene B.

Descriptive Passage and Interpretive Problems 5

A Mechanistic Preview of Addition Reactions

The following flow chart connects three of the reactions we have discussed that involve carbocation intermediates. *Each arrow may represent more than one elementary step in a mechanism.*

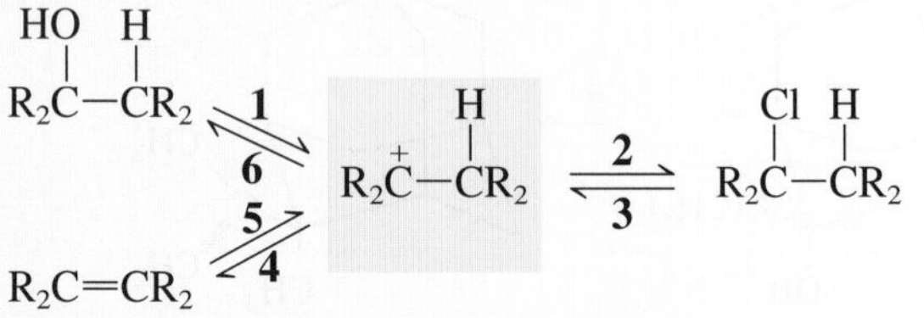

Arrows **1** and **2** summarize the conversion of alcohols to alkyl halides, **3** and **4** the dehydrohalogenation of an alkyl halide to an alkene by the E1 mechanism, and **1** and **4** the formation of an alkene by dehydration of an alcohol.

The reaction indicated by arrow **5** constitutes a major focus of the next chapter. There we will explore reactions that give overall *addition* to the double bond by way of carbocation intermediates. One such process converts alkenes to alkyl halides (**5** + **2**), another converts alkenes to alcohols (**5** + **6**).

5.52 Based on the S_N1 mechanism for the reaction of tertiary alcohols with HCl as summarized in arrows **1** and **2,** which arrow(s) represent(s) more than one elementary step?

A. Arrow **1**
B. Arrow **2**
C. Both **1** and **2**
D. Neither **1** nor **2**

5.53 Based on the E1 mechanism for the acid-catalyzed dehydration of a tertiary alcohol as summarized in arrows **1** and **4,** which arrow(s) represent(s) more than one elementary step?

A. Arrow **1**
B. Arrow **4**
C. Both **1** and **4**
D. Neither **1** nor **4**

5.54 Based on the E1 mechanism for the conversion of a tertiary alkyl chloride to an alkene as summarized in arrows **3** and **4,** which arrow(s) represent(s) more than one elementary step?

A. Arrow **3**
B. Arrow **4**
C. Both **3** and **4**
D. Neither **3** nor **4**

5.55 Based on the E1 mechanism for the conversion of a tertiary alkyl chloride to an alkene as summarized in arrows **3** and **4,** which arrow(s) correspond(s) to exothermic processes?

A. Arrow **3**
B. Arrow **4**
C. Both **3** and **4**
D. Neither **3** nor **4**

5.56 What term best describes the relationship between an alkene and a carbocation?

A. Isomers
B. Resonance contributors
C. Alkene is conjugate acid of carbocation
D. Alkene is conjugate base of carbocation

5.57 The overall equation for the addition of HCl to alkenes is:

$$R_2C{=}CR_2 \;+\; HCl \;\longrightarrow\; R_2C(Cl){-}CR_2(H)$$

If the transition state for proton transfer from HCl to the alkene (arrow **5**) resembles a carbocation and this step is rate-determining, what should be the effect of alkene structure on the rate of the overall reaction?

	Fastest rate		Slowest rate
A.	$H_2C{=}CH_2$	$CH_3CH{=}CHCH_3$	$(CH_3)_2C{=}C(CH_3)_2$
B.	$CH_3CH{=}CHCH_3$	$(CH_3)_2C{=}C(CH_3)_2$	$H_2C{=}CH_2$
C.	$CH_3CH{=}CHCH_3$	$H_2C{=}CH_2$	$(CH_3)_2C{=}C(CH_3)_2$
D.	$(CH_3)_2C{=}C(CH_3)_2$	$CH_3CH{=}CHCH_3$	$H_2C{=}CH_2$

5.58 For the addition of HCl to alkenes according to the general equation given in the preceding problem, assume the mechanism involves rate-determining formation of the more stable carbocation (arrow **5**) and predict the alkyl chloride formed by reaction of HCl with $(CH_3)_2C{=}CH_2$.

A. $(CH_3)_2CHCH_2Cl$
B. $(CH_3)_3CCl$

5.59 *Zaitsev's rule* was presented in this chapter. In the next chapter we will introduce *Markovnikov's rule*, which is related to arrow **5.** To which arrow does *Zaitsev's rule* most closely relate from a mechanistic perspective?

A. **1**
B. **2**
C. **3**
D. **4**

6 Addition Reactions of Alkenes

Chapter Outline

Mechanisms

Bromine with two bonds and a positive charge? In a three-membered ring? In 1937, two Columbia chemists proposed just such a species as an intermediate in the reaction of ethylene with bromine. They were right.

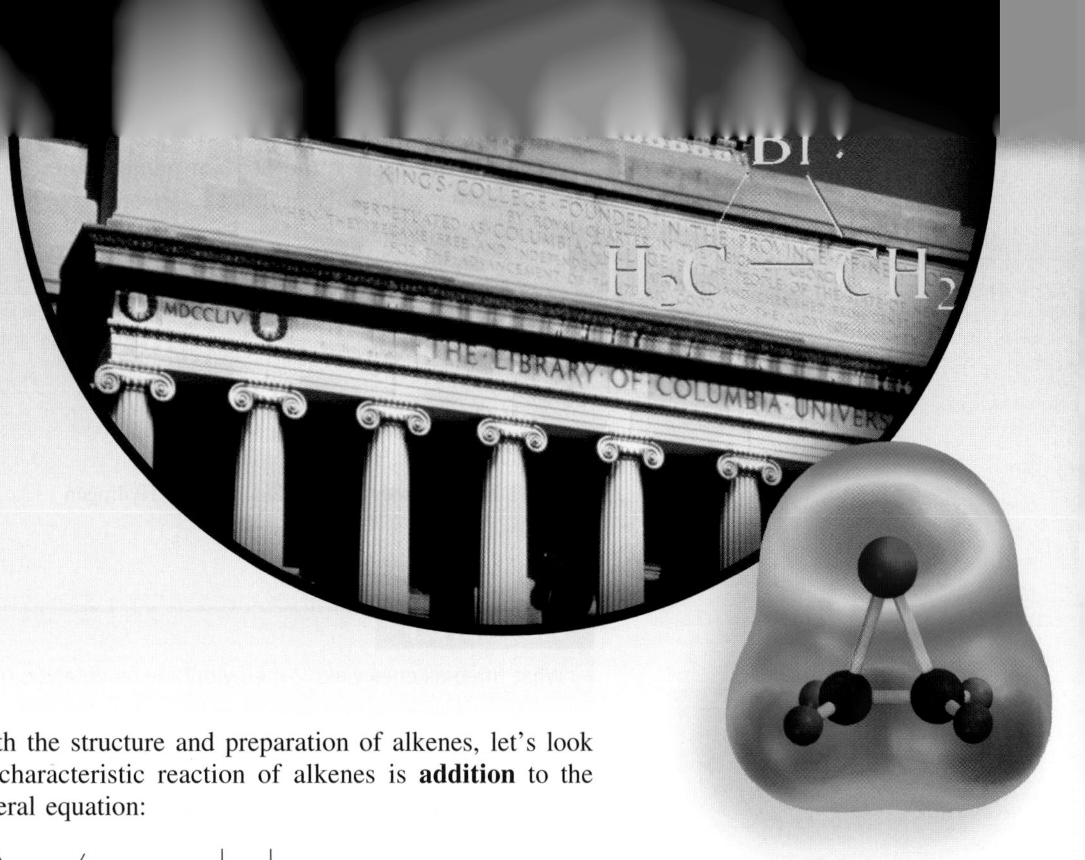

NOW THAT WE'RE familiar with the structure and preparation of alkenes, let's look at their chemical reactions. The characteristic reaction of alkenes is **addition** to the double bond according to the general equation:

$$\mathrm{A{-}B} + \mathrm{>C{=}C<} \longrightarrow \mathrm{A{-}\overset{|}{\underset{|}{C}}{-}\overset{|}{\underset{|}{C}}{-}B}$$

The range of compounds represented as A—B in this equation offers a wealth of opportunity for converting alkenes to a number of other structural types.

Alkenes are commonly described as **unsaturated hydrocarbons** because they have the capacity to react with substances that add to them. Alkanes, on the other hand, are **saturated hydrocarbons** and are incapable of undergoing addition reactions.

6.1 Hydrogenation of Alkenes

The relationship between reactants and products in addition reactions can be illustrated by the *hydrogenation* of alkenes to yield alkanes. **Hydrogenation** is the addition of H_2 to a multiple bond. An example is the reaction of hydrogen with ethylene to form ethane.

$$\underset{\text{Ethylene}}{H_2C\overset{\pi}{=}CH_2} + \underset{\text{Hydrogen}}{H\overset{\sigma}{-}H} \xrightarrow{\text{Pt, Pd, Ni, or Rh}} \underset{\text{Ethane}}{H\overset{\sigma}{-}CH_2{-}CH_2\overset{\sigma}{-}H} \qquad \Delta H^\circ = -136\ \text{kJ}\ (-32.6\ \text{kcal})$$

The bonds in the product are stronger than the bonds in the reactants; two C—H σ bonds of an alkane are formed at the expense of the H—H σ bond and the π component of the alkene's double bond. The reaction is exothermic and is characterized by a negative sign for ΔH°. Indeed, *hydrogenation of all alkenes is exothermic.* The heat given off is called the **heat of hydrogenation** and cited without a sign. In other words, heat of hydrogenation $= -\Delta H^\circ$.

The uncatalyzed addition of hydrogen to an alkene, although exothermic, is very slow. The rate of hydrogenation increases dramatically, however, in the presence of certain finely divided metal catalysts. *Platinum* is the hydrogenation catalyst most often used; *palladium, nickel,* and *rhodium* are also effective. Metal-catalyzed addition of

The French chemist Paul Sabatier received the 1912 Nobel Prize in Chemistry for his discovery that finely divided nickel is an effective hydrogenation catalyst.

hydrogen is normally rapid at room temperature, and the alkane is produced in high yield, usually as the only product.

$$(CH_3)_2C{=}CHCH_3 + H_2 \xrightarrow{Pt} (CH_3)_2CHCH_2CH_3$$

2-Methyl-2-butene | Hydrogen | 2-Methylbutane (100%)

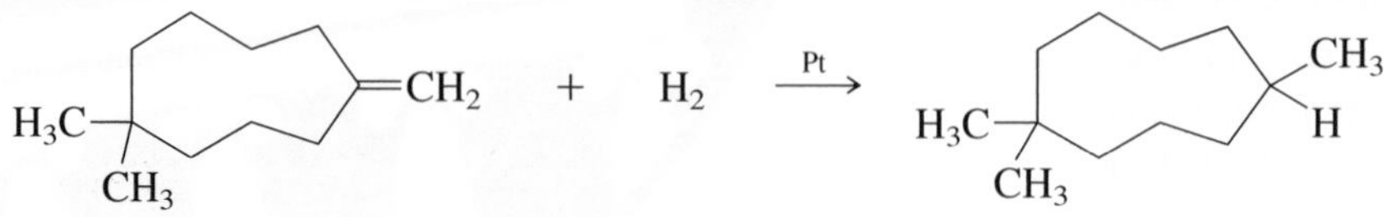

5,5-Dimethyl(methylene)cyclononane | Hydrogen | 1,1,5-Trimethylcyclononane (73%)

Problem 6.1

What three alkenes yield 2-methylbutane on catalytic hydrogenation?

The solvent used in catalytic hydrogenation is chosen for its ability to dissolve the alkene and is typically ethanol, hexane, or acetic acid. The metal catalysts are insoluble in these solvents (or, indeed, in any solvent). Two phases, the solution and the metal, are present, and the reaction takes place at the interface between them. Reactions involving a substance in one phase with a different substance in a second phase are called **heterogeneous reactions.**

Catalytic hydrogenation of an alkene is believed to proceed by the series of steps shown in Mechanism 6.1. As already noted, addition of hydrogen to the alkene is very slow in the absence of a metal catalyst, meaning that any uncatalyzed mechanism must have a very high activation energy. The metal catalyst accelerates the rate of hydrogenation by providing an alternative pathway that involves a sequence of several low activation energy steps.

6.2 Heats of Hydrogenation

Remember that a catalyst affects the rate of a reaction but not the energy relationships between reactants and products. Thus, the heat of hydrogenation of a particular alkene is the same irrespective of what catalyst is used.

In much the same way as heats of combustion, heats of hydrogenation are used to compare the relative stabilities of alkenes. Both methods measure the differences in the energy of *isomers* by converting them to a product or products common to all. Catalytic hydrogenation of 1-butene, *cis*-2-butene, or *trans*-2-butene yields the same product—butane. As Figure 6.1 shows, the measured heats of hydrogenation reveal that *trans*-2-butene is 4 kJ/mol (1.0 kcal/mol) lower in energy than *cis*-2-butene and that *cis*-2-butene is 7 kJ/mol (1.7 kcal/mol) lower in energy than 1-butene.

Heats of hydrogenation can be used to *estimate* the stability of double bonds as structural units, even in alkenes that are not isomers. Table 6.1 lists the heats of hydrogenation for a representative collection of alkenes.

The pattern of alkene stability determined from heats of hydrogenation parallels exactly the pattern deduced from heats of combustion.

Decreasing heat of hydrogenation and increasing stability of the double bond

$H_2C{=}CH_2$	$RCH{=}CH_2$	$RCH{=}CHR$	$R_2C{=}CHR$	$R_2C{=}CR_2$
Ethylene	Monosubstituted	Disubstituted	Trisubstituted	Tetrasubstituted

Mechanism 6.1

Hydrogenation of Alkenes

Step 1: Hydrogen molecules react with metal atoms at the catalyst surface. The relatively strong hydrogen–hydrogen σ bond is broken and replaced by two weak metal–hydrogen bonds.

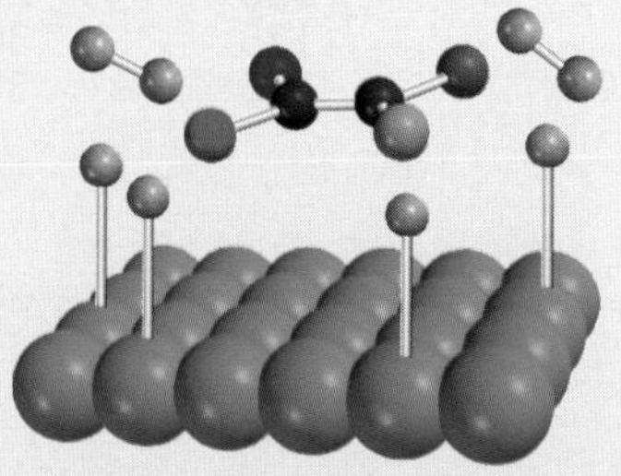

Step 2: The alkene reacts with the metal catalyst. The π component of the double bond between the two carbons is replaced by two relatively weak carbon–metal σ bonds.

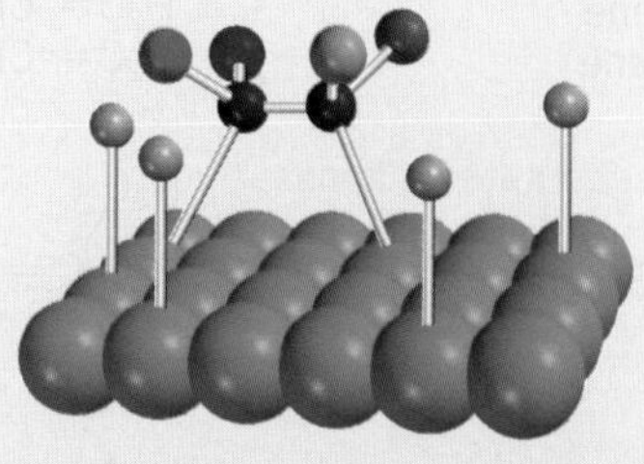

Step 3: A hydrogen atom is transferred from the catalyst surface to one of the carbons of the double bond.

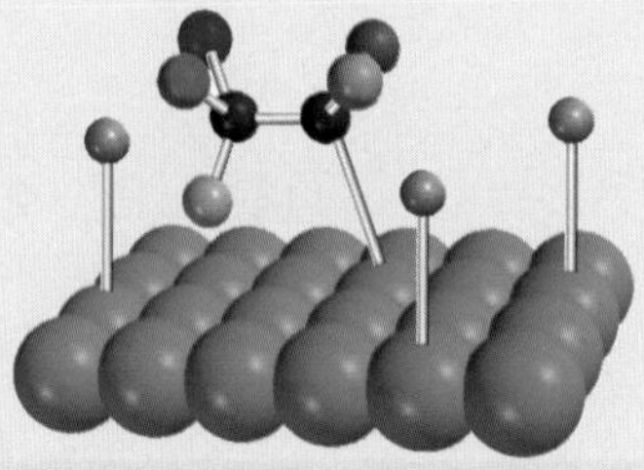

Step 4: The second hydrogen atom is transferred, forming the alkane. The sites on the catalyst surface at which the reaction occurred are free to accept additional hydrogen and alkene molecules.

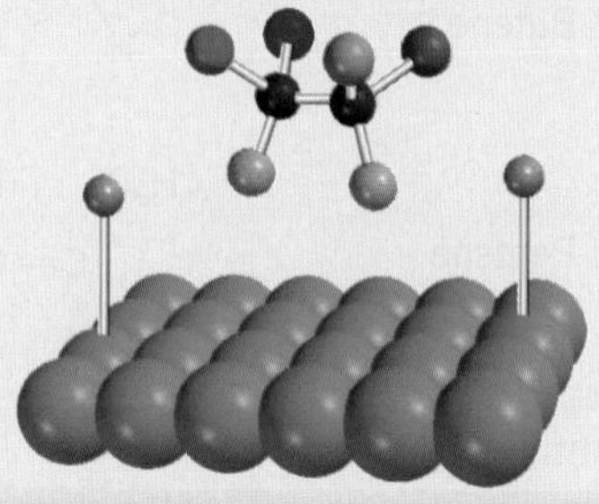

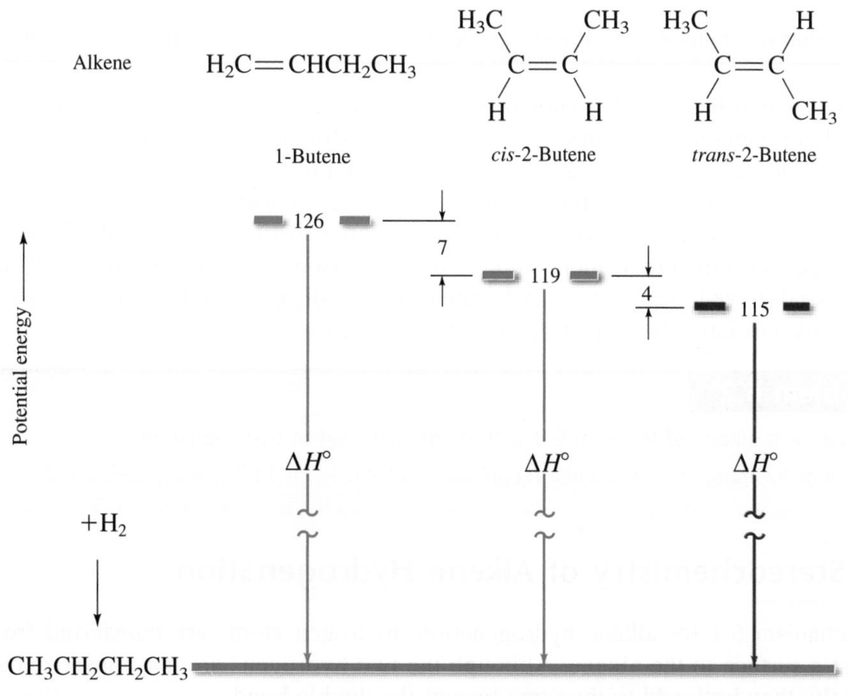

Figure 6.1

Heats of hydrogenation of butene isomers. All energies are in kilojoules per mole.

TABLE 6.1 Heats of Hydrogenation of Some Alkenes

Alkene	Structure	Heat of hydrogenation kJ/mol	kcal/mol
Ethylene	$H_2C{=}CH_2$	136	32.6
Monosubstituted alkenes			
Propene	$H_2C{=}CHCH_3$	125	29.9
1-Butene	$H_2C{=}CHCH_2CH_3$	126	30.1
1-Hexene	$H_2C{=}CHCH_2CH_2CH_2CH_3$	126	30.2
Cis-disubstituted alkenes			
cis-2-Butene	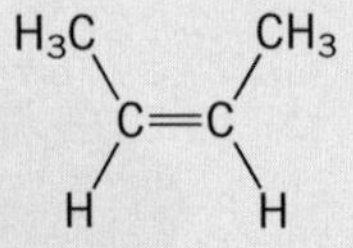	119	28.4
cis-2-Pentene	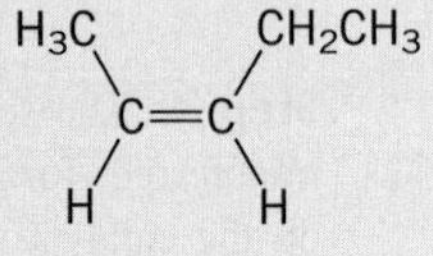	117	28.1
Trans-disubstituted alkenes			
trans-2-Butene	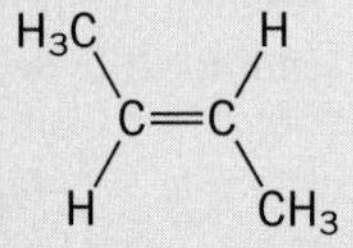	115	27.4
trans-2-Pentene	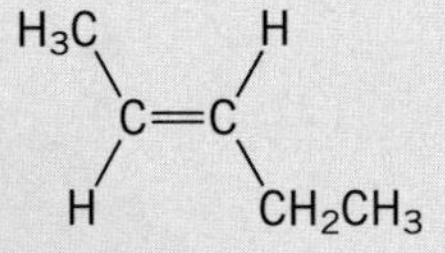	114	27.2
Trisubstituted alkenes			
2-Methyl-2-pentene	$(CH_3)_2C{=}CHCH_2CH_3$	112	26.7
Tetrasubstituted alkenes			
2,3-Dimethyl-2-butene	$(CH_3)_2C{=}C(CH_3)_2$	110	26.4

Ethylene, which has no alkyl substituents to stabilize its double bond, has the highest heat of hydrogenation. Alkenes that are similar in structure to one another have similar heats of hydrogenation. For example, the heats of hydrogenation of the monosubstituted (terminal) alkenes propene, 1-butene, and 1-hexene are almost identical. Cis-disubstituted alkenes have lower heats of hydrogenation than monosubstituted alkenes but higher heats of hydrogenation than their more stable trans stereoisomers. Alkenes with trisubstituted double bonds have lower heats of hydrogenation than disubstituted alkenes, and tetrasubstituted alkenes have the lowest heats of hydrogenation.

Problem 6.2

Match each alkene of Problem 6.1 with its correct heat of hydrogenation.
Heats of hydrogenation in kJ/mol (kcal/mol): 112 (26.7); 118 (28.2); 126 (30.2)

6.3 Stereochemistry of Alkene Hydrogenation

In Mechanism 6.1 for alkene hydrogenation, hydrogen atoms are transferred from the catalyst's surface to the alkene. Although the two hydrogens are not transferred simultaneously, they both add to the same face of the double bond.

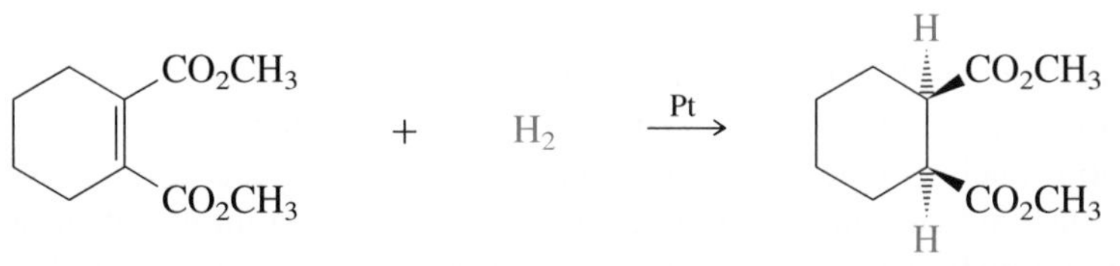

Dimethyl cyclohexene-1,2-dicarboxylate

Dimethyl cyclohexane-*cis*-1,2-dicarboxylate (100%)

The term **syn addition** describes the stereochemistry of reactions such as hydrogenation in which two atoms or groups add to the *same face* of a double bond. When atoms or groups add to *opposite faces* of the double bond, the process is called **anti addition.**

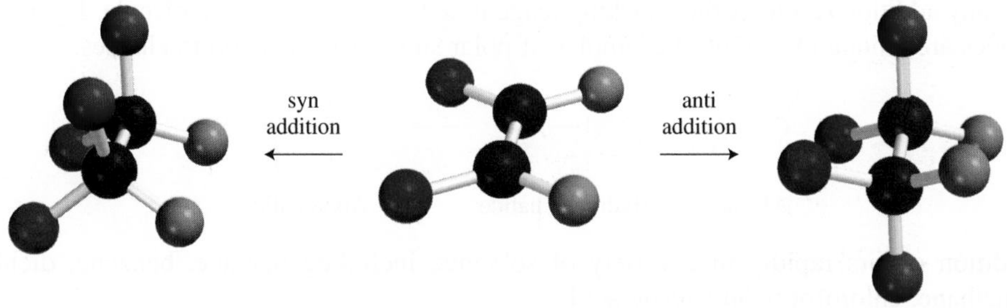

Another aspect of alkene hydrogenation is its *stereoselectivity*. Recall that a stereoselective reaction is one in which a single starting material can give two or more stereoisomeric products but yields one of them in greater amounts than the other (or even to the exclusion of the other). The catalytic hydrogenation of α-pinene (a constituent of turpentine) is an example of a stereoselective reaction. Syn addition of hydrogen can in principle lead to either *cis*-pinane or *trans*-pinane, depending on which face of the double bond accepts the hydrogen atoms (shown in red in the equation).

Stereoselectivity was defined and introduced in connection with the formation of stereoisomeric alkenes in elimination reactions (Section 5.11).

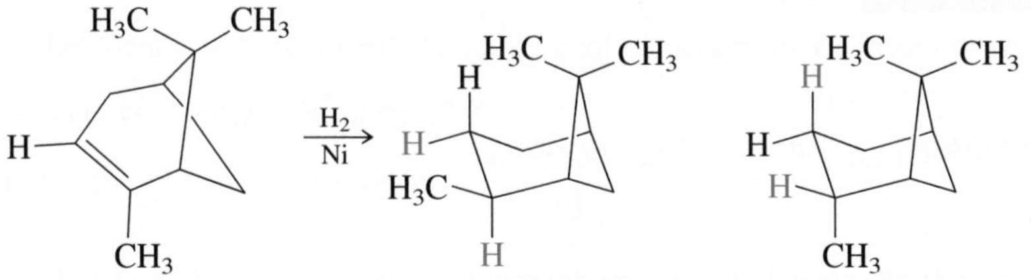

α-Pinene

cis-Pinane (only product)

trans-Pinane (not formed)

cis-Pinane and *trans*-pinane are common names that denote the relationship between the pair of methyl groups on the bridge and the third methyl group.

Hydrogenation of α-pinene is 100% stereoselective and gives only *cis*-pinane. No *trans*-pinane is formed.

The stereoselectivity of this reaction depends on how the alkene approaches the catalyst surface. As the molecular model in Figure 6.2 shows, one of the methyl groups

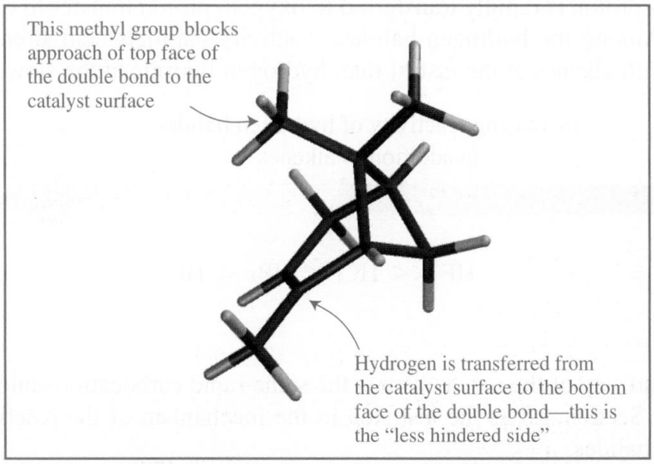

Figure 6.2

The methyl group that lies over the double bond of α-pinene shields one face of it, preventing a close approach to the surface of the catalyst. Hydrogenation of α-pinene occurs preferentially from the bottom face of the double bond.

on the bridge carbon lies directly over the double bond and blocks that face from easy access to the catalyst. The bottom face of the double bond is more exposed, and both hydrogens are transferred from the catalyst surface to that face.

Reactions such as catalytic hydrogenation that take place at the "less hindered" side of a reactant are common in organic chemistry and are examples of steric effects on *reactivity*. Previously we saw steric effects on *structure* and *stability* in the case of cis and trans stereoisomers of substituted cycloalkanes (Sections 3.11 and 3.12) and alkenes (Sections 5.6 and 5.7).

6.4 Electrophilic Addition of Hydrogen Halides to Alkenes

In many addition reactions the attacking reagent, unlike H_2, is a polar molecule. Hydrogen halides are among the simplest examples of polar substances that add to alkenes.

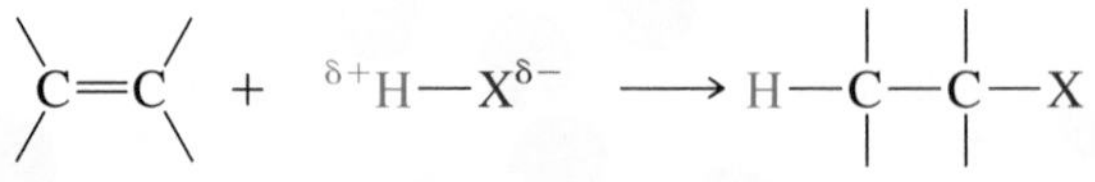

+
δ+H—Xδ−
H—C—C—X

Alkene Hydrogen halide Alkyl halide

Addition occurs rapidly in a variety of solvents, including pentane, benzene, dichloromethane, chloroform, and acetic acid.

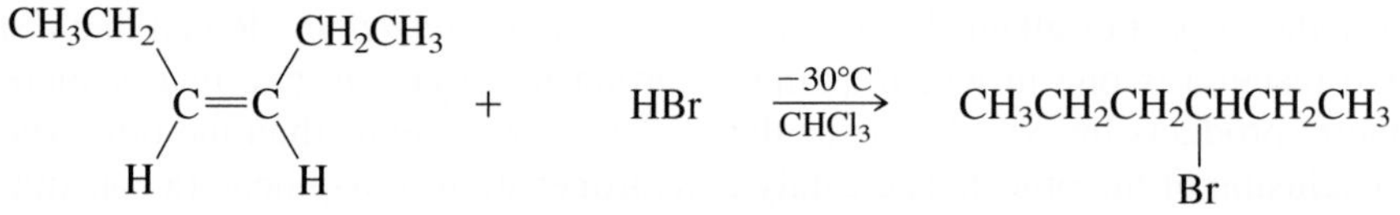

CH3CH2
CH2CH3
C=C
H
H
+
HBr
−30°C
CHCl3
CH3CH2CH2CHCH2CH3
Br

cis-3-Hexene Hydrogen bromide 3-Bromohexane (76%)

Problem 6.3

The heats of reaction were measured for addition of HBr to *cis*- and *trans*-2-butene.

$$CH_3CH{=}CHCH_3 + HBr \longrightarrow CH_3CH_2\underset{\displaystyle Br}{\underset{|}{C}}HCH_3$$

cis-2-butene: $\Delta H° = -77$ kJ (-18.4 kcal)

trans-2-butene: $\Delta H° = -72$ kJ (-17.3 kcal)

Use these data to calculate the energy difference between *cis*- and *trans*-2-butene. How does this energy difference compare to that based on heats of hydrogenation (Table 6.1) and heats of combustion (Figure 5.3)?

Mechanism 6.2 shows the two-step sequence for addition of hydrogen halides to alkenes. The first step is an acid–base reaction in which the hydrogen halide donates a proton to the alkene, forming a carbocation. Unlike other acid–base reactions that we have seen in which a proton is rapidly transferred to oxygen, proton transfer to carbon is almost always slow. Among the hydrogen halides, reactivity parallels acid strength. Hydrogen iodide reacts with alkenes at the fastest rate, hydrogen fluoride at the slowest.

Increasing reactivity of hydrogen halides in addition to alkenes

$$HF \ll HCl < HBr < HI$$

Slowest rate of addition; weakest acid — Fastest rate of addition; strongest acid

The second step of the mechanism is the same rapid carbocation–anion combination that we saw in Section 4.8 as the last step in the mechanism of the reaction of alcohols with hydrogen halides.

Mechanism 6.2

Electrophilic Addition of a Hydrogen Halide to an Alkene

THE OVERALL REACTION:

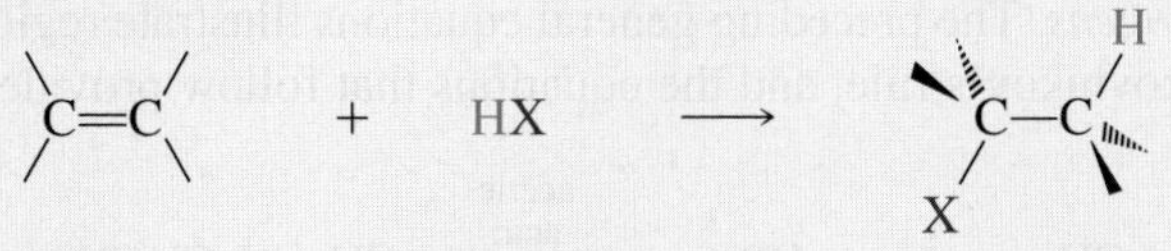

Alkene Hydrogen halide Alkyl halide

THE MECHANISM:

Step 1: Protonation of the carbon–carbon double bond by the hydrogen halide:

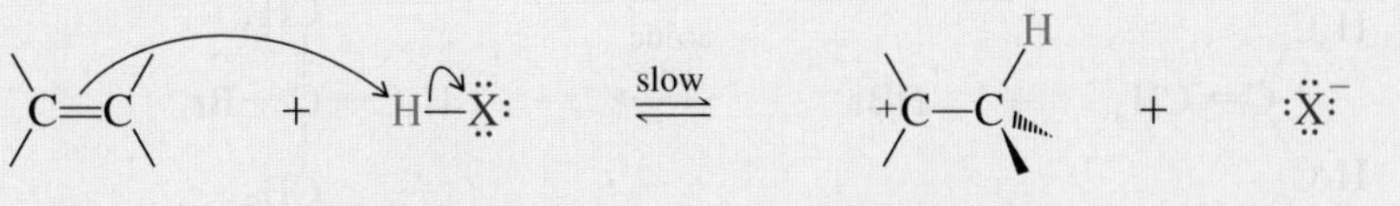

Alkene (base) Hydrogen halide (acid) Carbocation (conjugate acid) Halide ion (conjugate base)

Step 2: Carbocation–anion combination:

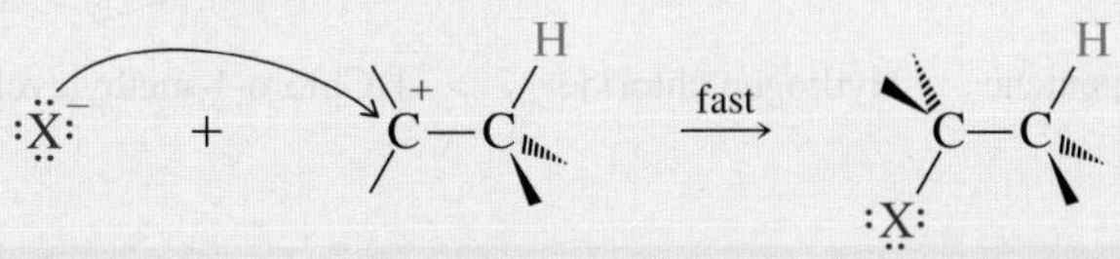

Halide ion Carbocation Alkyl halide

This general mechanism is called **electrophilic addition.** It is triggered by the acid acting as an electrophile toward the π electrons of the double bond. Figure 6.3 shows the complementary distribution of charge in an alkene and a hydrogen halide. The proton of the hydrogen halide is positively polarized (electrophilic) and the region of highest negative character in the alkene is where the π electrons are—above and below the plane of the bonds to the sp^2-hybridized carbons.

The characteristic chemical *property* of a C=C *structural* unit is susceptibility to attack by electrophiles. Electrons flow from the π component of the double bond toward the electrophile and ultimately become a shared-electron pair in a covalent bond. We'll see numerous other examples of electrophilic addition to alkenes in this chapter. First, however, we need to extend our discussion of hydrogen halide addition to alkenes of various types.

Figure 6.3

Electrostatic potential maps of HCl and ethylene. When the two react, the interaction is between the electron-rich site (red) of ethylene and electron-poor region (blue) of HCl. The electron-rich region of ethylene is associated with the π electrons of the double bond, and H is the electron-poor atom of HCl.

6.5 Regioselectivity of Hydrogen Halide Addition: Markovnikov's Rule

In principle, a hydrogen halide can add to an unsymmetrical alkene (an alkene in which the two carbons of the double bond are not equivalently substituted) in either of two directions. In practice, addition is so highly regioselective as to be considered regiospecific.

$$RCH{=}CH_2 + H{-}X \longrightarrow \underset{X}{RCH}{-}\underset{H}{CH_2} \quad \text{rather than} \quad \underset{H}{RCH}{-}\underset{X}{CH_2}$$

$$R_2C{=}CH_2 + H{-}X \longrightarrow \underset{X}{R_2C}{-}\underset{H}{CH_2} \quad \text{rather than} \quad \underset{H}{R_2C}{-}\underset{X}{CH_2}$$

$$R_2C{=}CHR + H{-}X \longrightarrow \underset{X}{R_2C}{-}\underset{H}{CHR} \quad \text{rather than} \quad \underset{H}{R_2C}{-}\underset{X}{CHR}$$

Recall from section 5.10 that a regioselective reaction is one that can produce two (or more) constitutional isomers from a single reactant, but gives one in greater amounts than the other. A regiospecific reaction is one that is 100% regioselective.

In 1870, Vladimir Markovnikov, a colleague of Alexander Zaitsev at the University of Kazan, noticed a pattern in the hydrogen halide addition to alkenes and organized his observations into a simple statement. **Markovnikov's rule** states that *when an unsymmetrically substituted alkene reacts with a hydrogen halide, the hydrogen adds to the carbon that has the greater number of hydrogens, and the halogen adds to the carbon having fewer hydrogens.* The preceding general equations illustrate regioselective addition according to Markovnikov's rule, and the equations that follow provide some examples.

$CH_3CH_2CH{=}CH_2$ + HBr
acetic acid
$CH_3CH_2CHCH_3$
Br

1-Butene | Hydrogen bromide | 2-Bromobutane (80%)

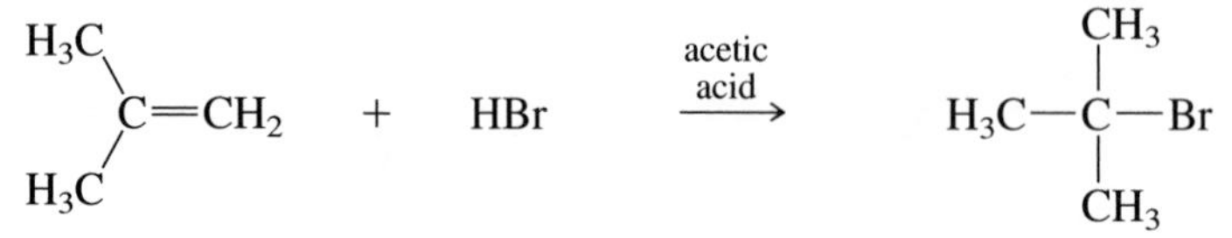

H_3C
$C{=}CH_2$
H_3C
+ HBr
acetic acid
CH_3
$H_3C{-}C{-}Br$
CH_3

2-Methylpropene | Hydrogen bromide | 2-Bromo-2-methylpropane (90%)

CH_3
+ HCl
0°C
CH_3
Cl

1-Methylcyclopentene | Hydrogen chloride | 1-Chloro-1-methylcyclopentane (100%)

Problem 6.4

Write the structure of the major organic product formed in the reaction of hydrogen chloride with each of the following:

(a) 2-Methyl-2-butene (c) 2-Methyl-1-butene

(b) *cis*-2-Butene (d) 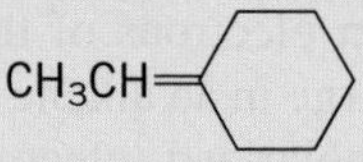

$CH_3CH{=}$

Sample Solution (a) Hydrogen chloride adds to the double bond of 2-methyl-2-butene in accordance with Markovnikov's rule. The proton adds to the carbon that has one attached hydrogen, chlorine to the carbon that has none.

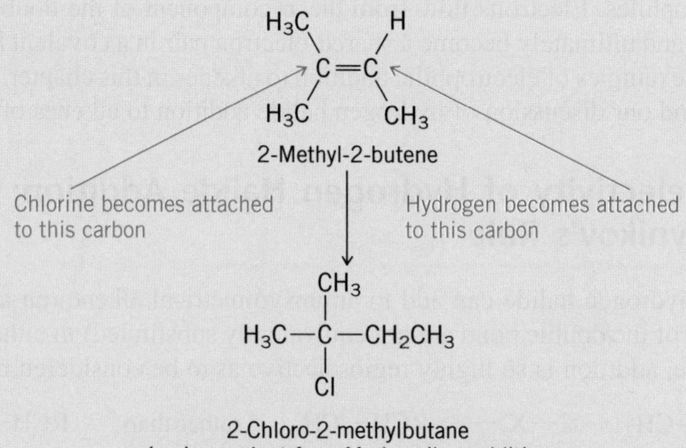

H_3C
H
$C{=}C$
H_3C
CH_3
2-Methyl-2-butene
Chlorine becomes attached to this carbon
Hydrogen becomes attached to this carbon
CH_3
$H_3C{-}C{-}CH_2CH_3$
Cl

2-Chloro-2-methylbutane
(major product from Markovnikov addition of hydrogen chloride to 2-methyl-2-butene)

Markovnikov's rule, like Zaitsev's, organizes experimental observations in a form suitable for predicting the major product of a reaction. The reasons why will appear when we examine the mechanism of electrophilic addition in more detail.

6.6 Mechanistic Basis for Markovnikov's Rule

Let's compare the carbocation intermediates for addition of a hydrogen halide (HX) to an unsymmetrical alkene of the type $RCH{=}CH_2$ (a) according to Markovnikov's rule and (b) opposite to Markovnikov's rule.

(a) *Addition according to Markovnikov's rule:*

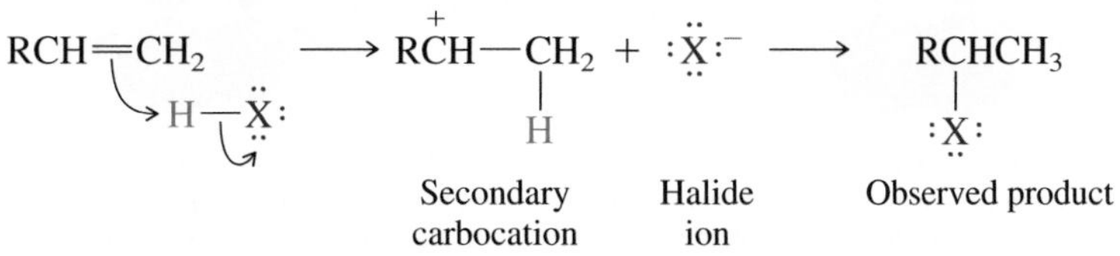

(b) *Addition opposite to Markovnikov's rule:*

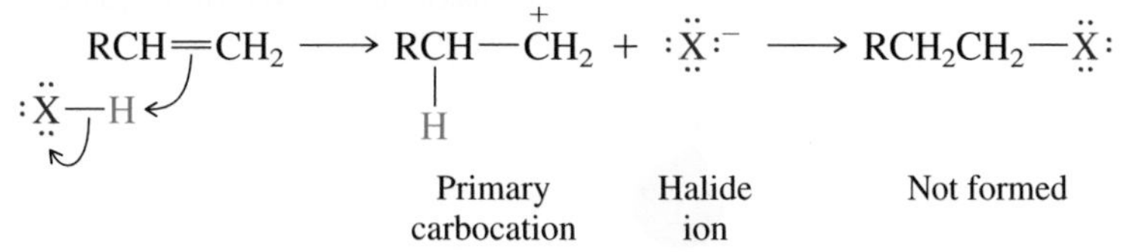

According to Hammond's postulate, the transition state for protonation of the double bond has much of the character of a carbocation, and the activation energy for formation of the more stable carbocation (secondary) is less than that for formation of the less stable (primary) one. Figure 6.4 illustrates these two competing modes of addition. Both carbocations are rapidly captured by X^- to give an alkyl halide, with the major product derived from the carbocation that is formed faster. The energy difference between a primary carbocation and a secondary carbocation is so great and their rates of formation are so different that essentially all the product is derived from the secondary carbocation.

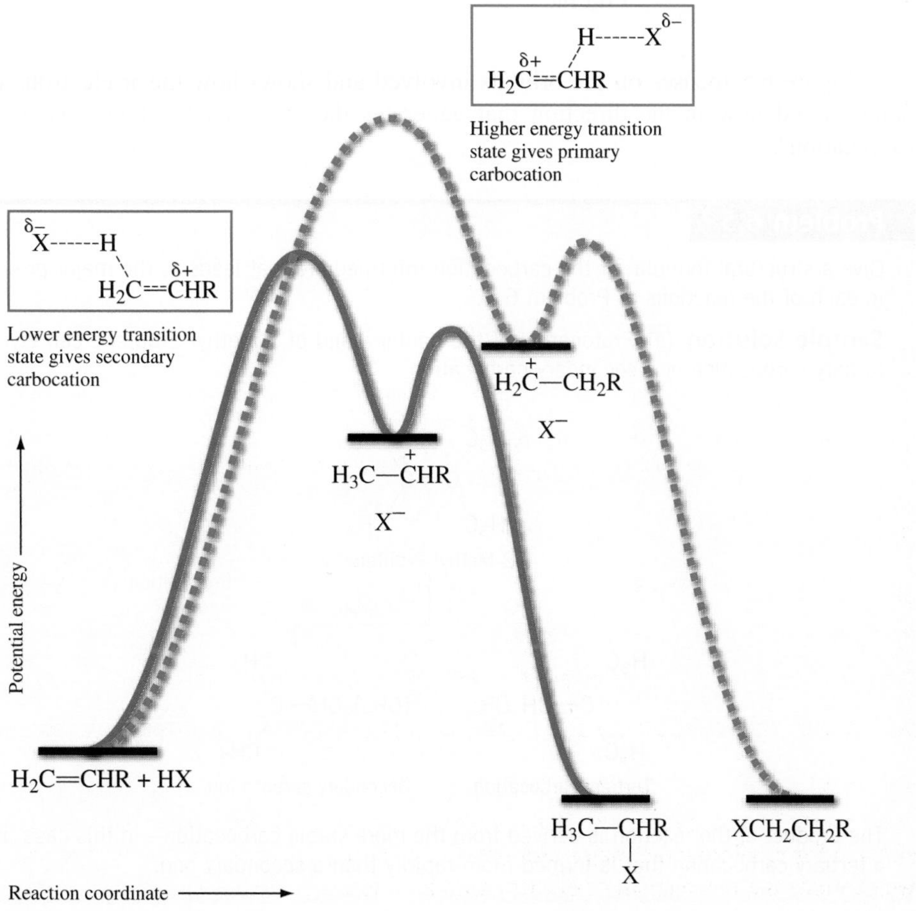

Figure 6.4

Energy diagrams comparing addition of a hydrogen halide HX with an alkene $H_2C{=}CHR$ according to Markovnikov's rule (solid red) and opposite to Markovnikov's rule (dashed blue). The energy of activation is less and the reaction is faster for the reaction that proceeds through the more stable secondary carbocation.

Figure 6.5

Electron flow and orbital interactions in the transfer of a proton from a hydrogen halide to an alkene of the type $H_2C{=}CHR$.

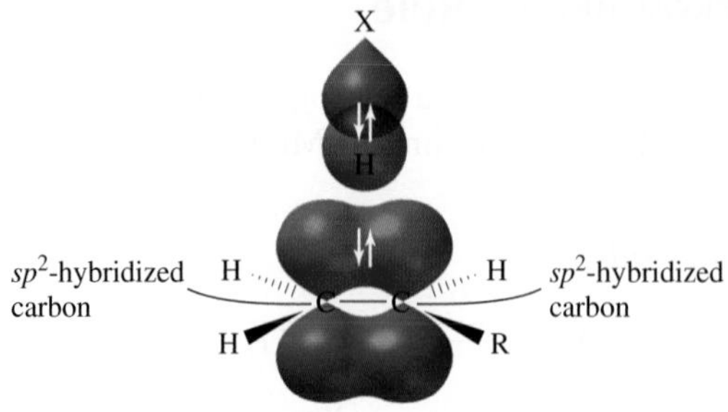
X
H
sp²-hybridized carbon
H
H
sp²-hybridized carbon
H
R

(*a*) The hydrogen halide (HX) and the alkene ($H_2C{=}CHR$) approach each other. The electrophile is the hydrogen halide, and the site of electrophilic attack is the orbital containing the π electrons of the double bond.

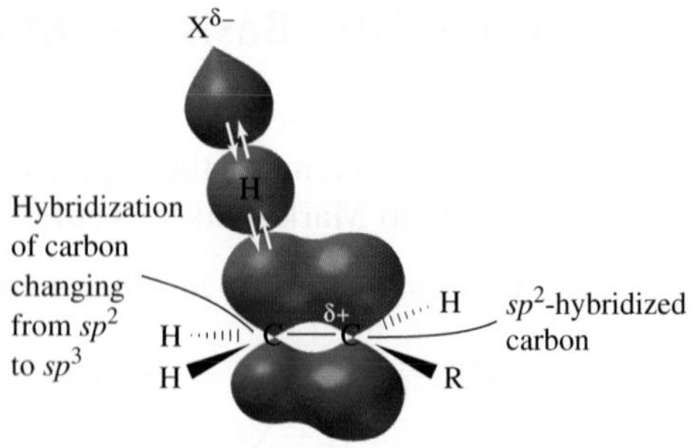
Xδ−
H
Hybridization of carbon changing from sp² to sp³
δ+
H
H
H
R
sp²-hybridized carbon

(*b*) Electrons flow from the π orbital of the alkene to the hydrogen halide. The π electrons flow in the direction that generates a partial positive charge on the carbon atom that bears the electron-releasing alkyl group (R). The hydrogen–halogen bond is partially broken and a C—H σ bond is partially formed at the transition state.

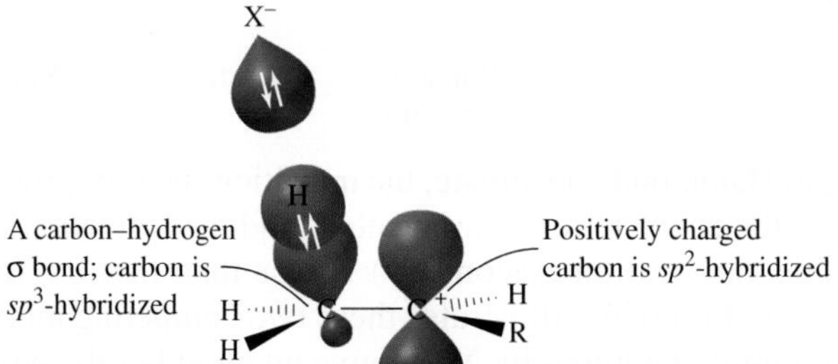
X⁻
H
A carbon–hydrogen σ bond; carbon is sp³-hybridized
H
H
Positively charged carbon is sp²-hybridized
H
R

(*c*) Loss of the halide ion (X^-) from the hydrogen halide and C—H σ bond formation complete the formation of the more stable carbocation intermediate $CH_3\overset{+}{C}HR$.

Figure 6.5 focuses on the orbitals involved and shows how the π electrons of the double bond flow in the direction that generates the more stable of the two possible carbocations.

Problem 6.5

Give a structural formula for the carbocation intermediate that leads to the major product in each of the reactions of Problem 6.4.

Sample Solution (a) Protonation of the double bond of 2-methyl-2-butene can give a tertiary carbocation or a secondary carbocation.

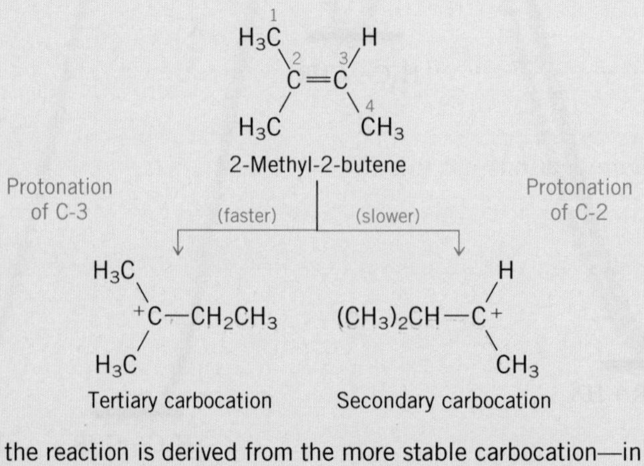
2-Methyl-2-butene
Protonation of C-3
(faster)
(slower)
Protonation of C-2
Tertiary carbocation
Secondary carbocation

The product of the reaction is derived from the more stable carbocation—in this case, it is a tertiary carbocation that is formed more rapidly than a secondary one.

Rules, Laws, Theories, and the Scientific Method

As we have just seen, Markovnikov's rule can be expressed in two ways:

1. When a hydrogen halide adds to an alkene, hydrogen adds to the carbon of the alkene that has the greater number of hydrogens attached to it, and the halogen to the carbon that has the fewer hydrogens.
2. When a hydrogen halide adds to an alkene, protonation of the double bond occurs in the direction that gives the more stable carbocation.

The first of these statements is close to the way Vladimir Markovnikov expressed it in 1870; the second is the way we usually phrase it now. These two statements differ in an important way—a way that is related to the **scientific method.**

Adherence to the scientific method is what defines science. The scientific method has four major elements: observation, law, theory, and hypothesis.

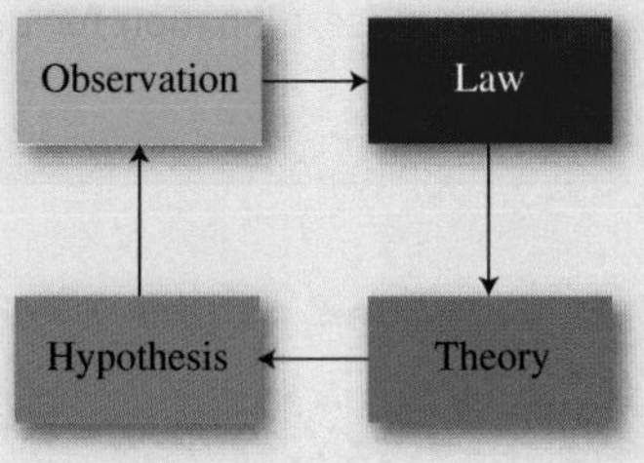
Observation
Law
Hypothesis
Theory

Most *observations* in chemistry come from experiments. If we do enough experiments we may see a pattern running through our observations. A *law* is a mathematical (the law of gravity) or verbal (the law of diminishing returns) description of that pattern. Establishing a law can lead to the framing of a *rule* that lets us predict the results of future experiments. This is what the 1870 version of Markovnikov's rule is: a statement based on experimental observations that has predictive value.

A *theory* is our best present interpretation of why things happen the way they do. The modern version of Markovnikov's rule, which is based on mechanistic reasoning and carbocation stability, recasts the rule in terms of theoretical ideas. Mechanisms, and explanations grounded in them, belong to the theory part of the scientific method.

It is worth remembering that a theory can never be proven correct. It can only be proven incorrect, incomplete, or inadequate. Thus, theories are always being tested and refined. As important as anything else in the scientific method is the *testable hypothesis.* Once a theory is proposed, experiments are designed to test its validity. If the results are consistent with the theory, our belief in its soundness is strengthened. If the results conflict with it, the theory is flawed and must be modified. Section 6.7 describes some observations that support the theory that carbocations are intermediates in the addition of hydrogen halides to alkenes.

In general, alkyl substituents increase the reactivity of a double bond toward electrophilic addition. Alkyl groups are electron-releasing, and the more *electron-rich* a double bond, the better it can share its π electrons with an electrophile. Along with the observed regioselectivity of addition, this supports the idea that carbocation formation, rather than carbocation capture, is rate-determining.

6.7 Carbocation Rearrangements in Hydrogen Halide Addition to Alkenes

Our belief that carbocations are intermediates in the addition of hydrogen halides to alkenes is strengthened by the fact that rearrangements sometimes occur. For example, the reaction of hydrogen chloride with 3-methyl-1-butene is expected to produce 2-chloro-3-methylbutane. Instead, a mixture of 2-chloro-3-methylbutane and 2-chloro-2-methylbutane results.

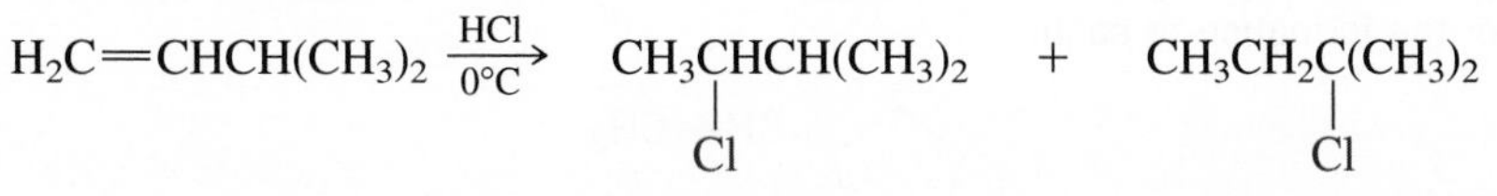
$H_2C{=}CHCH(CH_3)_2$
HCl
0°C
$CH_3CHCH(CH_3)_2$
Cl
+
$CH_3CH_2C(CH_3)_2$
Cl

3-Methyl-1-butene | 2-Chloro-3-methylbutane (40%) | 2-Chloro-2-methylbutane (60%)

Addition begins in the usual way, by protonation of the double bond to give, in this case, a secondary carbocation.

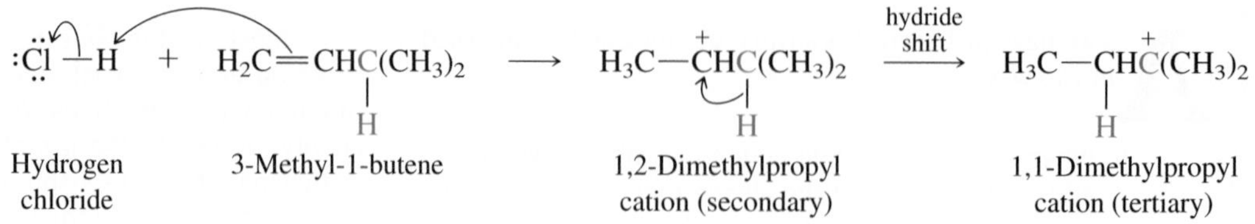

This carbocation can be captured by chloride to give 2-chloro-3-methylbutane (40%) or it can rearrange by way of a hydride shift to give a tertiary carbocation. The tertiary carbocation reacts with chloride ion to give 2-chloro-2-methylbutane (60%). The similar yields of the two alkyl chloride products indicate that the rate of attack by chloride on the secondary carbocation and the rate of rearrangement must be very similar.

Problem 6.6

(a) Addition of hydrogen chloride to 3,3-dimethyl-1-butene gives a mixture of two isomeric chlorides in approximately equal amounts. Suggest reasonable structures for these two compounds, and offer a mechanistic explanation for their formation.

Sample Solution

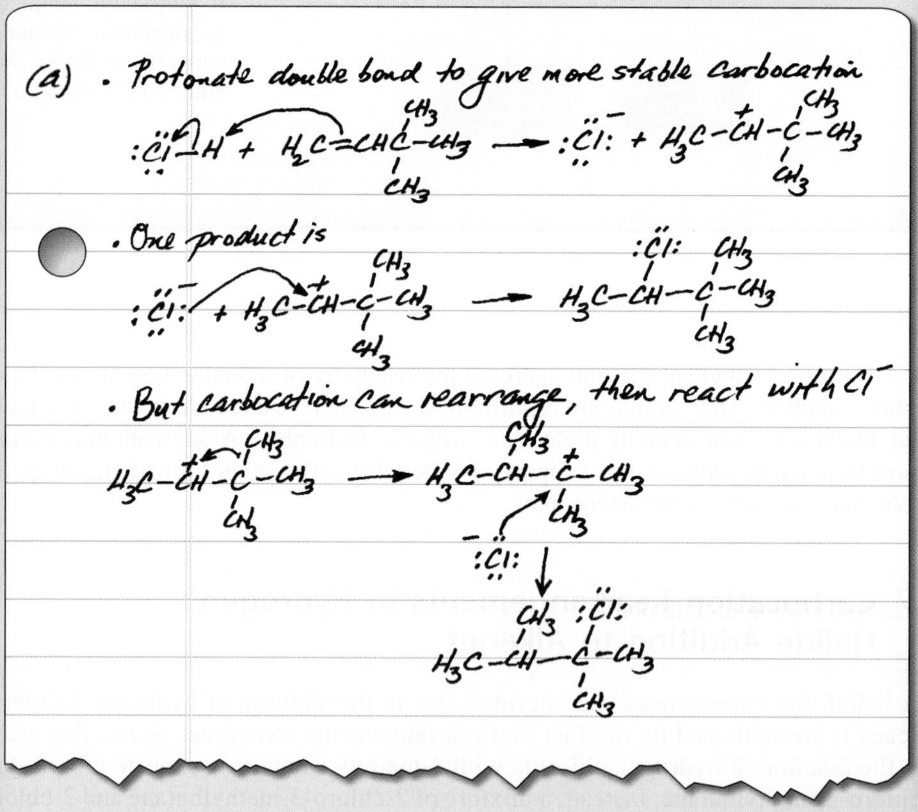

(b) Vinylcyclopentane gives a mixture of a secondary and a tertiary alkyl chloride when treated with hydrogen chloride. Suggest reasonable structures and a mechanism for the formation of each.

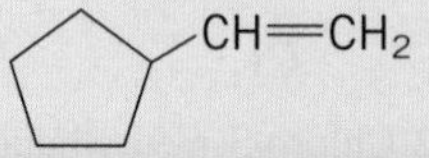

6.8 Addition of Sulfuric Acid to Alkenes

Acids other than hydrogen halides also add to the carbon–carbon bond of alkenes. Concentrated sulfuric acid, for example, reacts with certain alkenes to form alkyl hydrogen sulfates.

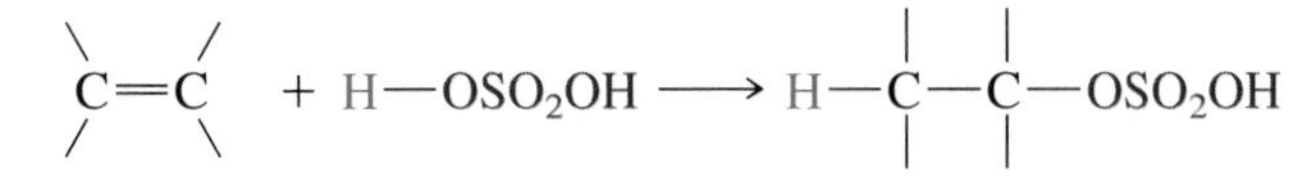

C=C
+ H—OSO₂OH
H—C—C—OSO₂OH

Alkene Sulfuric acid Alkyl hydrogen sulfate

Notice in the following example that a proton adds to the carbon that has the greater number of hydrogens, and the hydrogen sulfate anion ($^{-}OSO_2OH$) adds to the carbon that has the fewer hydrogens.

$$CH_3CH{=}CH_2 + HOSO_2OH \longrightarrow CH_3\underset{\displaystyle OSO_2OH}{\underset{|}{C}}HCH_3$$

Propene Sulfuric acid Isopropyl hydrogen sulfate

Markovnikov's rule is obeyed because the mechanism of sulfuric acid addition to alkenes is analogous to that described earlier for the electrophilic addition of hydrogen halides.

Alkyl hydrogen sulfates can be converted to alcohols by heating them with water. This is called **hydrolysis,** because a bond is cleaved by reaction with water. It is the oxygen–sulfur bond that is broken when an alkyl hydrogen sulfate undergoes hydrolysis.

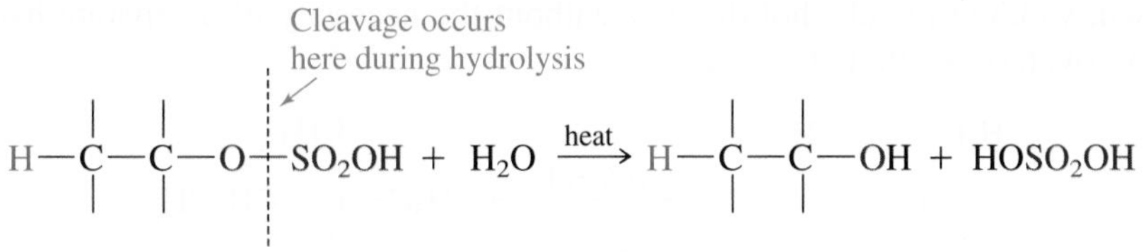

Cleavage occurs
here during hydrolysis
H—C—C—O—SO₂OH + H₂O
heat
H—C—C—OH + HOSO₂OH

Alkyl hydrogen sulfate Water Alcohol Sulfuric acid

The combination of sulfuric acid addition to propene, followed by hydrolysis of the resulting isopropyl hydrogen sulfate, is the major method by which over 10^9 lb of isopropyl alcohol is prepared each year in the United States.

$$CH_3CH{=}CH_2 \xrightarrow{H_2SO_4} CH_3\underset{\displaystyle OSO_2OH}{\underset{|}{C}}HCH_3 \xrightarrow[\text{heat}]{H_2O} CH_3\underset{\displaystyle OH}{\underset{|}{C}}HCH_3$$

Propene Isopropyl hydrogen sulfate Isopropyl alcohol

We say that propene has undergone **hydration.** Overall, H and OH have added across the carbon–carbon double bond.

Problem 6.7

Applying the same hydration method to the mixture of 1-butene, *cis*-2-butene, and *trans*-2-butene obtained during petroleum refining provides 2-butanol on an industrial scale.

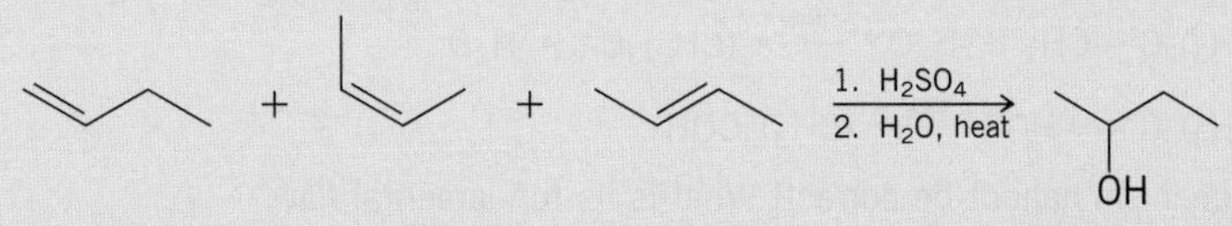

1. H₂SO₄
2. H₂O, heat
OH

Mixture of 1-butene, *cis*-2-butene, and *trans*-2-butene 2-Butanol

Why do each of the alkenes give the same alkyl hydrogen sulfate in reaction 1? What is its structure?

In synthetic transformations involving more than one step it is convenient simply to list all the reagents with a single arrow. Individual synthetic steps are indicated by number. Numbering the individual steps is essential so as to avoid the implication that everything is added to the reaction mixture at the same time.

Hydration of alkenes by this method is limited to monosubstituted alkenes and disubstituted alkenes of the type RCH=CHR.

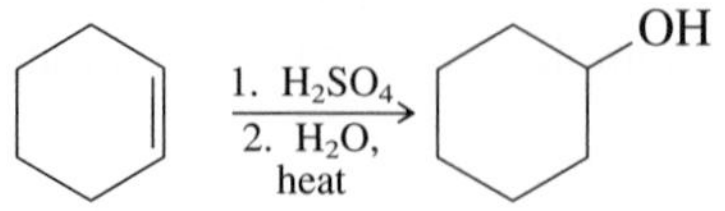

Cyclohexene — Cyclohexanol (75%)

Disubstituted alkenes of the type $R_2C{=}CH_2$, along with trisubstituted and tetrasubstituted alkenes, do not form alkyl hydrogen sulfates under these conditions but instead react in a more complicated way with concentrated sulfuric acid (to be discussed in Section 6.21).

6.9 Acid-Catalyzed Hydration of Alkenes

Another method for the hydration of alkenes is by reaction with water under conditions of acid catalysis.

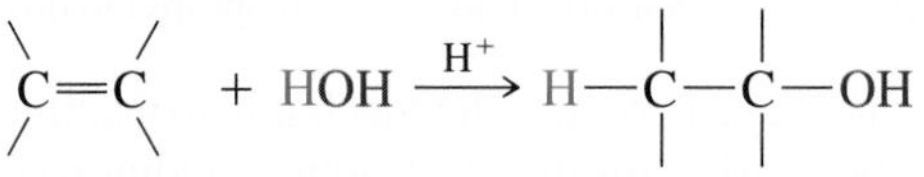

Alkene — Water — Alcohol

Unlike the addition of concentrated sulfuric acid to form alkyl hydrogen sulfates, this reaction is carried out in a *dilute acid* medium. A 50% water/sulfuric acid solution is often used, yielding the alcohol directly without the necessity of a separate hydrolysis step. Markovnikov's rule is followed:

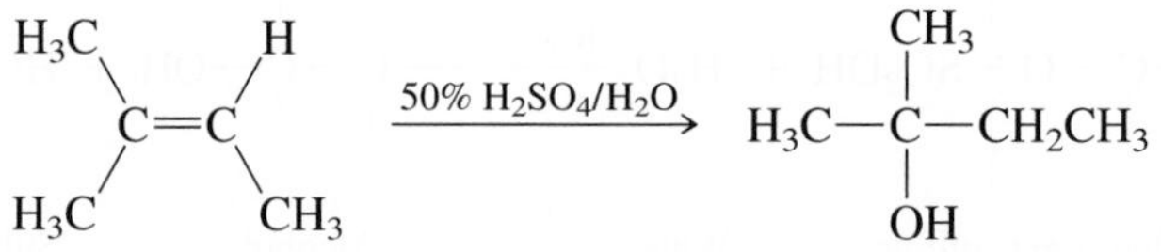

2-Methyl-2-butene — 2-Methyl-2-butanol (90%)

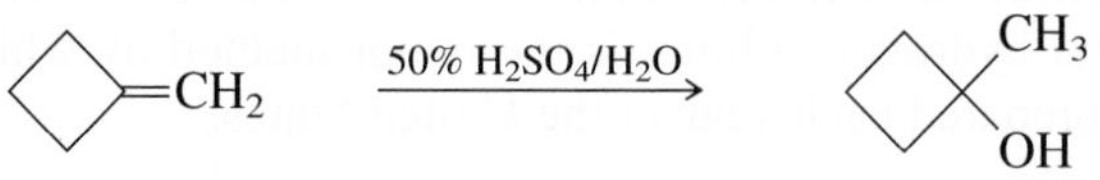

Methylenecyclobutane — 1-Methylcyclobutanol (80%)

Mechanism 6.3 extends the general principles of electrophilic addition to acid-catalyzed hydration. In the first step of the mechanism, proton transfer to 2-methylpropene forms the *tert*-butyl cation. This is followed in step 2 by reaction of the carbocation with a molecule of water acting as a nucleophile. The alkyloxonium ion formed in this step is simply the conjugate acid of *tert*-butyl alcohol. Deprotonation of the alkyloxonium ion in step 3 yields the alcohol and regenerates the acid catalyst.

Problem 6.8

Instead of the three-step process of Mechanism 6.3, the following two-step mechanism might be considered:

1. $(CH_3)_2C{=}CH_2 + H_3O^+ \xrightarrow{\text{slow}} (CH_3)_3C^+ + H_2O$
2. $(CH_3)_3C^+ + HO^- \xrightarrow{\text{fast}} (CH_3)_3COH$

This mechanism cannot be correct! What is its fundamental flaw?

The notion that carbocation formation is rate-determining follows from our previous experience and by observing how the reaction rate is affected by the structure of the alkene. Table 6.2 gives some data showing that alkenes that yield relatively stable carbocations react faster than those that yield less stable carbocations. Protonation of ethylene, the least

Mechanism 6.3

Acid-Catalyzed Hydration of 2-Methylpropene

THE OVERALL REACTION:

$$\underset{\text{2-Methylpropene}}{(CH_3)_2C{=}CH_2} + \underset{\text{Water}}{H_2O} \xrightarrow{H_3O^+} \underset{\textit{tert}\text{-Butyl alcohol}}{(CH_3)_3COH}$$

THE MECHANISM:

Step 1: Protonation of the carbon–carbon double bond in the direction that leads to more stable carbocation:

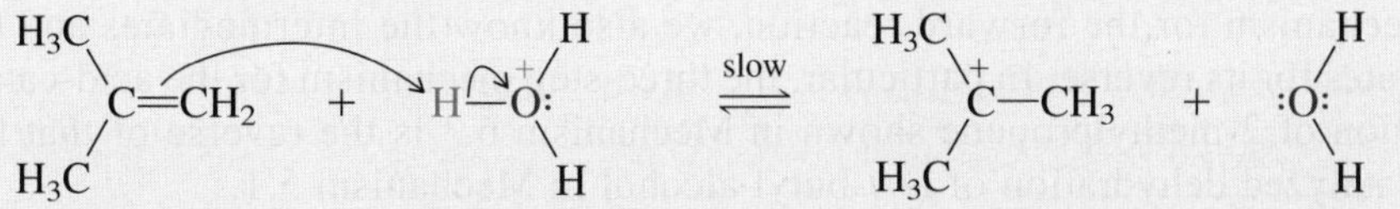

2-Methylpropene | Hydronium ion | *tert*-Butyl cation | Water

Step 2: Water acts as a nucleophile to capture *tert*-butyl cation:

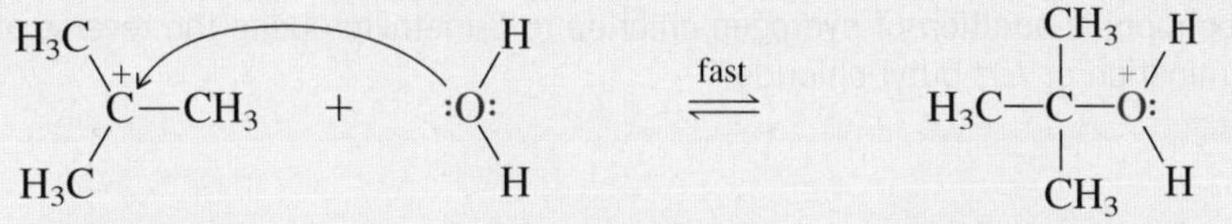

tert-Butyl cation | Water | *tert*-Butyloxonium ion

Step 3: Deprotonation of *tert*-butyloxonium ion. Water acts as a Brønsted base:

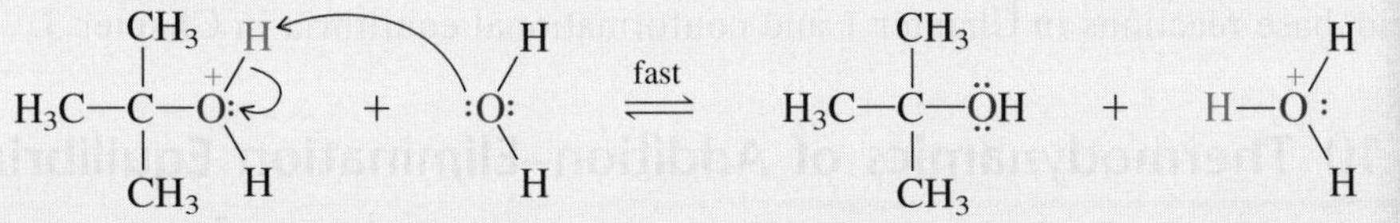

tert-Butyloxonium ion | Water | *tert*-Butyl alcohol | Hydronium ion

reactive alkene in the table, yields a primary carbocation; protonation of 2-methylpropene, the most reactive in the table, yields a tertiary carbocation. As we have seen on other occasions, the more stable the carbocation, the faster is its rate of formation.

Problem 6.9

The rates of hydration of the two alkenes shown differ by a factor of over 7000 at 25°C. Which isomer is the more reactive? Why?

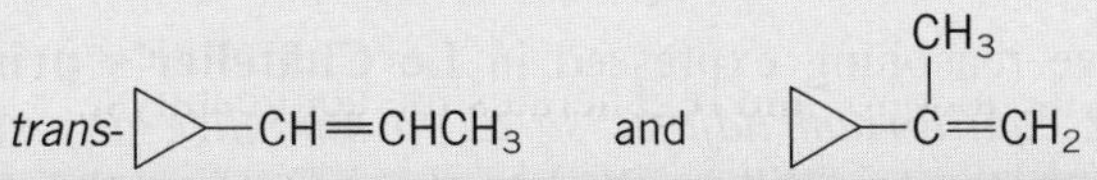

TABLE 6.2 Relative Rates of Acid-Catalyzed Hydration of Some Representative Alkenes

Alkene	Structural formula	Relative rate of acid-catalyzed hydration*
Ethylene	$H_2C{=}CH_2$	1.0
Propene	$CH_3CH{=}CH_2$	1.6×10^6
2-Methylpropene	$(CH_3)_2C{=}CH_2$	2.5×10^{11}

*In water, 25°C.

You may have noticed that the acid-catalyzed hydration of an alkene and the acid-catalyzed dehydration of an alcohol are the reverse of each other. For example:

$$(CH_3)_2C{=}CH_2 + H_2O \overset{H^+}{\rightleftharpoons} (CH_3)_3COH$$

2-Methylpropene   Water   *tert*-Butyl alcohol

An important principle, called **microscopic reversibility,** connects the mechanisms of the forward and reverse reactions. It states that *in any equilibrium, the sequence of intermediates and transition states encountered as reactants proceed to products in one direction must also be encountered, and in precisely the reverse order, in the opposite direction.* Just as the reaction is reversible with respect to reactants and products, so each tiny increment of progress along the mechanistic pathway is reversible. Once we know the mechanism for the forward reaction, we also know the intermediates and transition states for its reverse. In particular, the three-step mechanism for the acid-catalyzed hydration of 2-methylpropene shown in Mechanism 6.3 is the reverse of that for the acid-catalyzed dehydration of *tert*-butyl alcohol in Mechanism 5.1.

It would be a good idea to verify the statement in the last sentence of this paragraph by revisiting Mechanisms 5.1 (p. 203) and 6.3.

Problem 6.10

Is the electrophilic addition of hydrogen chloride to 2-methylpropene the reverse of the E1 or E2 elimination of *tert*-butyl chloride?

Reaction mechanisms help us understand the "how" of reversible reactions, but not the "how much." To gain an appreciation for the factors that influence equilibria in addition reactions we need to expand on some ideas introduced when we discussed acid–base reactions in Chapter 1 and conformational equilibria in Chapter 3.

6.10 Thermodynamics of Addition–Elimination Equilibria

We have seen that both the forward and reverse reactions represented by the hydration–dehydration equilibrium are useful synthetic methods.

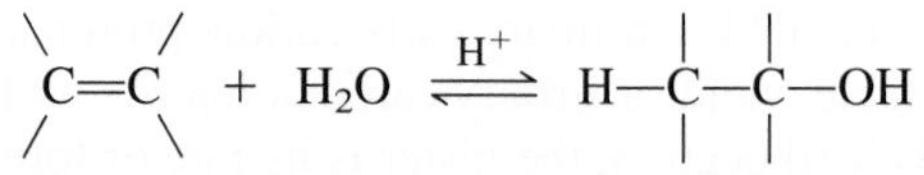

Alkene   Water   Alcohol

We can prepare alcohols from alkenes, and alkenes from alcohols, but how do we control the position of equilibrium so as to maximize the yield of the compound we want?

The qualitative reasoning expressed in **Le Châtelier's principle** is a helpful guide: *a system at equilibrium adjusts so as to minimize any stress applied to it.* For hydration–dehydration equilibria, the key stress factor is the water concentration. Adding water to a hydration–dehydration equilibrium mixture causes the system to respond by consuming water. More alkene is converted to alcohol, and the position of equilibrium shifts to the right. When we prepare an alcohol from an alkene, we use a reaction medium in which the molar concentration of water is high—dilute sulfuric acid, for example.

On the other hand, alkene formation is favored when the concentration of water is kept low. The system responds to the absence of water by causing more alcohol molecules to dehydrate, forming more alkene. The amount of water in the reaction mixture is kept low by using concentrated acids as catalysts. Distilling the reaction mixture is an effective way of removing water as it is formed, causing the equilibrium to shift to the left. If the alkene is low-boiling, it too can be removed by distillation. This offers the additional benefit of protecting the alkene from acid-catalyzed isomerization after it is formed.

Problem 6.11

We studied the forward phase of the reaction

$$(CH_3)_3COH + HCl \rightleftharpoons (CH_3)_3CCl + H_2O$$

in Section 4.8 and will study its reverse in Section 8.6. Which would provide a more complete conversion of one mole of *tert*-butyl alcohol to *tert*-butyl chloride, a concentrated or a dilute solution containing 1 mol of HCl in water? Explain.

Le Châtelier's principle helps us predict qualitatively how an equilibrium will respond to changes in experimental conditions. For a quantitative understanding, we need to examine reactions from a thermodynamic point of view.

At constant temperature and pressure, the direction in which a reaction proceeds—that is, the direction in which it is **spontaneous**—is the one that leads to a decrease in **free energy (G)**

Free energy is also called "Gibbs free energy." The official term is **Gibbs energy,** in honor of the nineteenth century American physicist J. Willard Gibbs.

$$\Delta G = G_{\text{products}} - G_{\text{reactants}} \qquad \text{spontaneous when } \Delta G < 0$$

The free energy of the reactants and products depends on what they are and how much of each is present. The sign of G is always positive, but ΔG can be positive or negative. If only the reactants are present at the beginning, $G_{\text{reactants}}$ has some value but G_{products} is zero; therefore, ΔG is negative and the reaction is spontaneous in the direction written. As the reaction proceeds, $G_{\text{reactants}}$ decreases while G_{products} increases until both are equal and $\Delta G = 0$. At this point the system is at equilibrium. Both the forward and reverse reactions continue to take place, but at equal rates.

Because reactions are carried out under a variety of conditions, it is convenient to define a *standard state* for substances and experimental conditions. The standard state is the form (solid, liquid, or gas) assumed by the pure substance at 1 atm pressure. For substances in aqueous solution, the standard-state concentration is 1 M. Standard-state values are designated by a superscript ° following the thermodynamic symbol as in $\Delta G°$.

For a reversible reaction

$$a\text{A} + b\text{B} \rightleftharpoons c\text{C} + d\text{D}$$

the relationship between ΔG and $\Delta G°$ is

$$\Delta G = \Delta G° + RT\ln\frac{[\text{C}]^c[\text{D}]^d}{[\text{A}]^a[\text{B}]^b}$$

where $R = 8.314$ J/(mol·K) or 1.99 cal/(mol·K) and T is the kelvin temperature. At equilibrium $\Delta G = 0$, and $\frac{[\text{C}]^c[\text{D}]^d}{[\text{A}]^a[\text{B}]^b}$ becomes the equilibrium constant K. Substituting these values in the preceding equation and rearranging, we get

$$\Delta G° = -RT\ln K$$

Reactions for which the sign of $\Delta G°$ is negative are described as **exergonic;** those for which $\Delta G°$ is positive are **endergonic.** Exergonic reactions have an equilibrium constant greater than 1; endergonic reactions have equilibrium constants less than 1.

Free energy has both an enthalpy (H) and an entropy (S) component.

$$G = H - TS$$

At constant temperature, $\Delta G° = \Delta H° - T\Delta S°$

For the hydration of 2-methylpropene, the standard-state thermodynamic values are given beside the equation.

$$(CH_3)_2C{=}CH_2(g) + H_2O(\ell) \rightleftharpoons (CH_3)_3COH(\ell)$$

$\Delta G° = -5.4$ kJ (-1.3 kcal)	Exergonic
$\Delta H° = -52.7$ kJ (-12.6 kcal)	Exothermic
$\Delta S° = -0.16$ kJ/K (-0.038 kcal/K)	Entropy decreases

The negative sign for $\Delta G°$ tells us the reaction is exergonic. From the relationship

$$\Delta G° = -RT\ln K$$

we can calculate the equilibrium constant at 25°C as $K = 9$.

Problem 6.12

You can calculate the equilibrium constant for the dehydration of $(CH_3)_3COH$ (the reverse of the preceding reaction) by reversing the sign of $\Delta G°$ in the expression $\Delta G° = -RT\ln K$, but there is an easier way. Do you know what it is? What is K for the dehydration of $(CH_3)_3COH$?

The $\Delta H°$ term is dominated by bond strength. A negative sign for $\Delta H°$ almost always means that bonding is stronger in the products than in the reactants. Stronger bonding reduces the free energy of the products and contributes to a more negative $\Delta G°$. Such is the normal case for addition reactions. Hydrogenation, hydration, and hydrogen halide additions to alkenes, for example, are all characterized by negative values for $\Delta H°$.

The $\Delta S°$ term is a measure of the increase or decrease in the order of a system. A more ordered system has less entropy and is less probable than a disordered one. The main factors that influence $\Delta S°$ in a chemical reaction are the number of moles of material on each side of the balanced equation and their physical state. The liquid phase of a substance has more entropy (less order) than the solid, and the gas phase has much more entropy than the liquid. Entropy increases when more molecules are formed at the expense of fewer ones, as for example in elimination reactions. Conversely, addition reactions convert more molecules to fewer ones and are characterized by a negative sign for $\Delta S°$.

The negative signs for both $\Delta H°$ and $\Delta S°$ in typical addition reactions of alkenes cause the competition between addition and elimination to be strongly temperature-dependent. Addition is favored at low temperatures, elimination at high temperatures. The economically important hydrogenation–dehydrogenation equilibrium that connects ethylene and ethane illustrates this.

$$\underset{\text{Ethylene}}{H_2C{=}CH_2(g)} + \underset{\text{Hydrogen}}{H_2(g)} \rightleftharpoons \underset{\text{Ethane}}{CH_3CH_3(g)}$$

Hydrogenation of ethylene converts two gas molecules on the left to one gas molecule on the right, leading to a decrease in entropy. The hydrogenation is sufficiently exothermic and $\Delta H°$ sufficiently negative, however, that the equilibrium lies far to the right over a relatively wide temperature range.

Very high temperatures—typically in excess of 750°C—reverse the equilibrium. At these temperatures, the $-T\Delta S°$ term in

$$\Delta G° = \Delta H° - T\Delta S°$$

becomes so positive that it eventually overwhelms $\Delta H°$ in magnitude, and the equilibrium shifts to the left. In spite of the fact that *dehydrogenation* is very endothermic, billions of pounds of ethylene are produced each year by this process.

Problem 6.13

Does the presence or absence of a catalyst such as finely divided platinum, palladium, or nickel affect the equilibrium constant for the ethylene–ethane conversion?

Problem 6.14

The gas phase reaction of ethanol with hydrogen bromide can occur either by elimination or substitution.

$$CH_3CH_2OH(g) \overset{HBr}{\rightleftharpoons} H_2C{=}CH_2(g) + H_2O(g)$$

$$CH_3CH_2OH(g) + HBr(g) \rightleftharpoons CH_3CH_2Br(g) + H_2O(g)$$

Which product, ethylene or ethyl bromide, will increase relative to the other as the temperature is raised? Why?

6.11 Hydroboration–Oxidation of Alkenes

Acid-catalyzed hydration converts alkenes to alcohols according to Markovnikov's rule. Frequently, however, one needs an alcohol having a structure that corresponds to hydration of an alkene with a regioselectivity opposite to that of Markovnikov's rule. The conversion of 1-decene to 1-decanol is an example of such a transformation.

$$CH_3(CH_2)_7CH{=}CH_2 \longrightarrow CH_3(CH_2)_7CH_2CH_2OH$$

1-Decene 1-Decanol

The synthetic method used to accomplish this is an indirect one known as **hydroboration–oxidation.** It was developed by Professor Herbert C. Brown and his coworkers at Purdue University as part of a broad program designed to apply boron-containing reagents to organic chemical synthesis. The number of applications is so large (hydroboration–oxidation is just one of them) and the work so novel that Brown was a corecipient of the 1979 Nobel Prize in Chemistry.

Hydroboration is a reaction in which a boron hydride, a compound of the type R_2BH, adds to a carbon–carbon π bond. A carbon–hydrogen bond and a carbon–boron bond result.

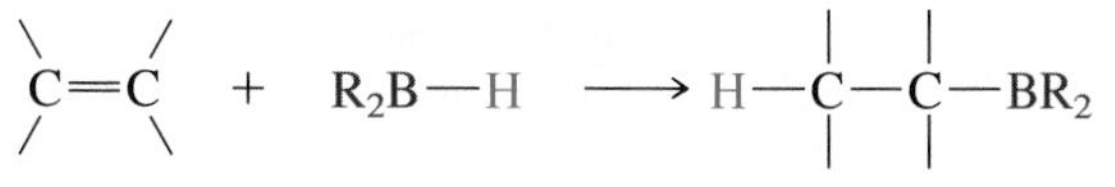

C=C
+
R₂B—H
H—C—C—BR₂

Alkene Boron hydride Organoborane

Following hydroboration, the organoborane is oxidized by treatment with hydrogen peroxide in aqueous base. This is the *oxidation* stage of the sequence; hydrogen peroxide is the oxidizing agent, and the organoborane is converted to an alcohol.

With sodium hydroxide as the base, boron of the alkylborane is converted to the water-soluble and easily removed sodium salt of boric acid.

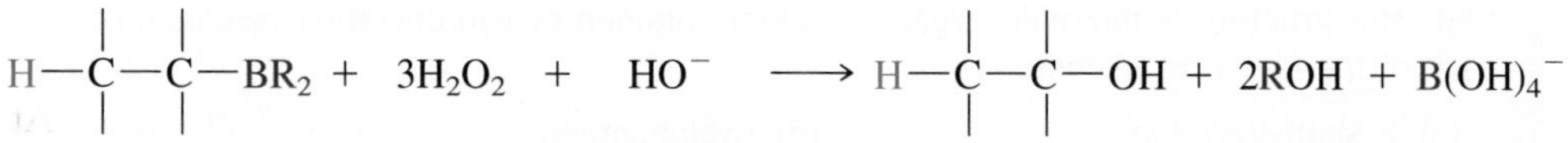

H—C—C—BR₂ + 3H₂O₂ + HO⁻
H—C—C—OH + 2ROH + B(OH)₄⁻

Organoborane Hydrogen peroxide Hydroxide ion Alcohol Alcohol Borate ion

Hydroboration–oxidation leads to the overall hydration of an alkene. Notice, however, that water is not a reactant. The hydrogen that becomes bonded to carbon comes from the organoborane, and the hydroxyl group from hydrogen peroxide.

With this as introduction, let us now look at the individual steps in more detail for the case of hydroboration–oxidation of 1-decene. A boron hydride that is often used is *diborane* (B_2H_6). Diborane adds to 1-decene to give tridecylborane according to the balanced equation:

$$6CH_3(CH_2)_7CH{=}CH_2 + B_2H_6 \xrightarrow{\text{diglyme}} 2[CH_3(CH_2)_7CH_2CH_2]_3B$$

1-Decene Diborane Tridecylborane

Diglyme, shown above the arrow in the equation, is the solvent in this example. Diglyme is an acronym for *di*ethylene *gly*col di*m*ethyl *e*ther, and its structure is $CH_3OCH_2CH_2OCH_2CH_2OCH_3$.

There is a pronounced tendency for boron to become bonded to the less substituted carbon of the double bond. Thus, the hydrogen atoms of diborane add to C-2 of 1-decene, and boron to C-1. This is believed to be mainly a steric effect, but the regioselectivity of addition does correspond to Markovnikov's rule in the sense that hydrogen is the negatively polarized atom in a B—H bond and boron the positively polarized one.

Oxidation of tridecylborane gives 1-decanol. The net result is the conversion of an alkene to an alcohol with a regioselectivity opposite to that of acid-catalyzed hydration.

$$[CH_3(CH_2)_7CH_2CH_2]_3B \xrightarrow[NaOH]{H_2O_2} CH_3(CH_2)_7CH_2CH_2OH$$

Tridecylborane 1-Decanol

It is customary to combine the two stages, hydroboration and oxidation, in a single equation with the operations numbered sequentially above and below the arrow.

$$CH_3(CH_2)_7CH{=}CH_2 \xrightarrow[2.\ H_2O_2,\ HO^-]{1.\ B_2H_6,\ \text{diglyme}} CH_3(CH_2)_7CH_2CH_2OH$$

1-Decene 1-Decanol (93%)

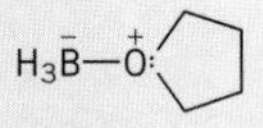

Borane–tetrahydrofuran complex

A more convenient hydroborating agent is the borane–tetrahydrofuran complex ($H_3B{\cdot}THF$). It is very reactive, adding to alkenes within minutes at 0°C, and is used in tetrahydrofuran as the solvent.

$$(CH_3)_2C{=}CHCH_3 \xrightarrow[2.\ H_2O_2,\ HO^-]{1.\ H_3B{\cdot}THF} (CH_3)_2CH\underset{\displaystyle OH}{\underset{|}{C}}HCH_3$$

2-Methyl-2-butene 3-Methyl-2-butanol (98%)

Carbocation intermediates are not involved in hydroboration–oxidation. Hydration of double bonds takes place without rearrangement, even in alkenes as highly branched as the following:

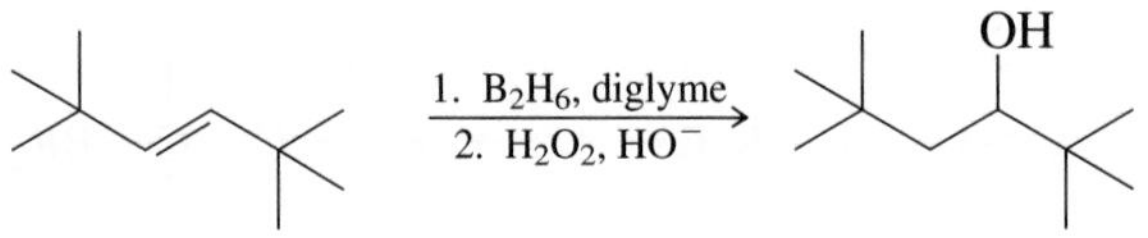

(*E*)-2,2,5,5-Tetramethyl-3-hexene 2,2,5,5-Tetramethyl-3-hexanol (82%)

Problem 6.15

Write the structure of the major organic product obtained by hydroboration–oxidation of each of the following alkenes:

(a) 2-Methylpropene
(b) *cis*-2-Butene
(c) 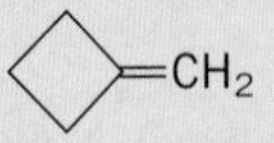

(d) Cyclopentene
(e) 3-Ethyl-2-pentene
(f) 3-Ethyl-1-pentene

Sample Solution (a) In hydroboration–oxidation H and OH are introduced with a regioselectivity opposite to that of Markovnikov's rule. In the case of 2-methylpropene, this leads to 2-methyl-1-propanol as the product.

$$(CH_3)_2C{=}CH_2 \xrightarrow[2.\ \text{oxidation}]{1.\ \text{hydroboration}} (CH_3)_2CH{-}CH_2OH$$

2-Methylpropene 2-Methyl-1-propanol

Hydrogen becomes bonded to the carbon that has the fewer hydrogens, hydroxyl to the carbon that has the greater number of hydrogens.

6.12 Stereochemistry of Hydroboration–Oxidation

A second aspect of hydroboration–oxidation concerns its stereochemistry. As illustrated for the case of 1-methylcyclopentene, H and OH add to the same face of the double bond.

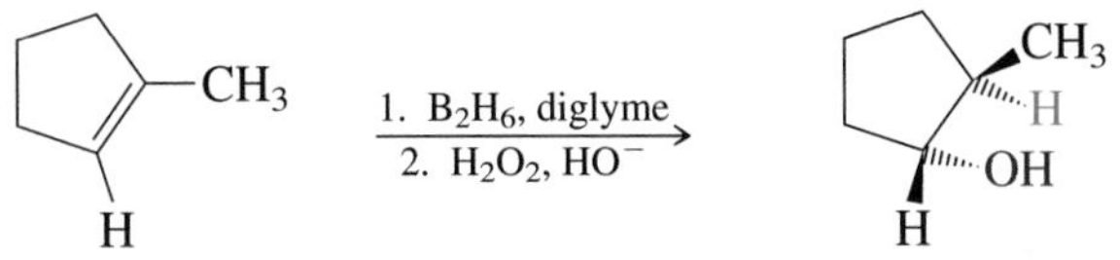

1-Methylcyclopentene

trans-2-Methylcyclopentanol (only product, 86% yield)

Overall, the reaction leads to syn addition of H and OH to the double bond. This fact has an important bearing on the mechanism of the process.

Problem 6.16

Hydroboration–oxidation of α-pinene (page 231), like catalytic hydrogenation, is stereoselective. Addition takes place at the less hindered face of the double bond, and a single alcohol is produced in high yield (89%). Suggest a reasonable structure for this alcohol.

6.13 Mechanism of Hydroboration–Oxidation

The regioselectivity and syn stereochemistry of hydroboration–oxidation, coupled with a knowledge of the chemical properties of alkenes and boranes, contribute to our understanding of the reaction mechanism.

In order to simplify our mechanistic analysis of hydroboration, we will consider the formation of the monoalkylborane that results from the addition of borane (BH_3) to 1-methylcyclopentene. Borane is electrophilic; it has a vacant $2p$ orbital available to accept a pair of electrons. The source of this electron pair is the π bond of an alkene. It is believed, as shown in Mechanism 6.4, that the first step produces an unstable intermediate called a *π complex.* In this π complex boron and the two carbon atoms of the double bond are joined by a *three-center, two-electron bond,* by which we mean that three atoms share two electrons. Three-center, two-electron bonds are frequently encountered in boron chemistry. The π complex is formed by a transfer of electron density from the π orbital of the alkene to the $2p$ orbital of boron. This leaves each carbon of the complex with a small positive charge, while boron is slightly negative. The negative character of boron in this intermediate makes it easy for one of its hydrogens to migrate with a pair of electrons (a hydride shift) from boron to carbon. The transition state for this process is shown in step 2(*a*) of Mechanism 6.4; completion of the migration in step 2(*b*) yields the alkylborane. According to this mechanism, the carbon–boron bond and the carbon–hydrogen bond are formed on the same face of the alkene, consistent with the observed syn addition.

Borane (BH_3) does not exist as such at room temperature and atmospheric pressure. Two molecules of BH_3 combine to give diborane (B_2H_6), which is the more stable form.

The regioselectivity of addition is consistent with the electron distribution in the complex. Hydrogen is transferred with a pair of electrons to the carbon atom that can best support a positive charge, namely, the one that bears the methyl group.

Steric effects may be an even more important factor in controlling the regioselectivity of addition. Boron, with its attached substituents, is much larger than a hydrogen atom and becomes bonded to the less crowded carbon of the double bond; hydrogen becomes bonded to the more crowded carbon.

The electrophilic character of boron is again evident when we consider the oxidation of organoboranes (Mechanism 6.5). The conjugate base of hydrogen peroxide is

Mechanism 6.4

Hydroboration of 1-Methylcyclopentene

Step 1: A molecule of borane (BH_3) attacks the alkene. Electrons flow from the π orbital of the alkene to the $2p$ orbital of boron. A π complex is formed.

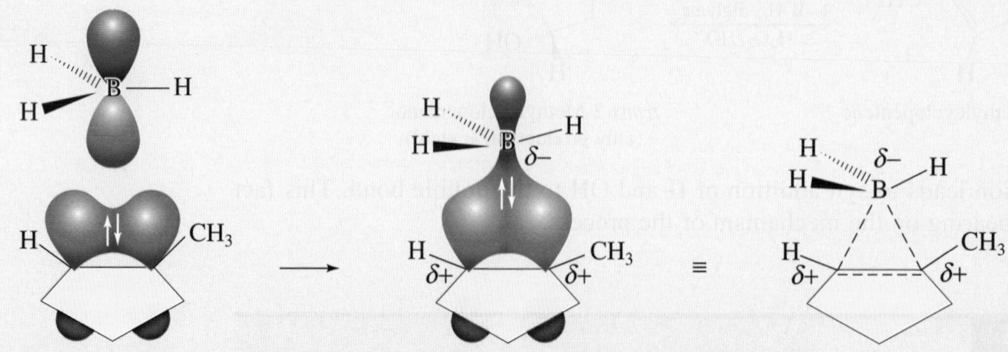

Alternative representations of π-complex intermediate

Step 2: The π complex rearranges to an organoborane. Hydrogen migrates from boron to carbon, carrying with it the two electrons in its bond to boron.

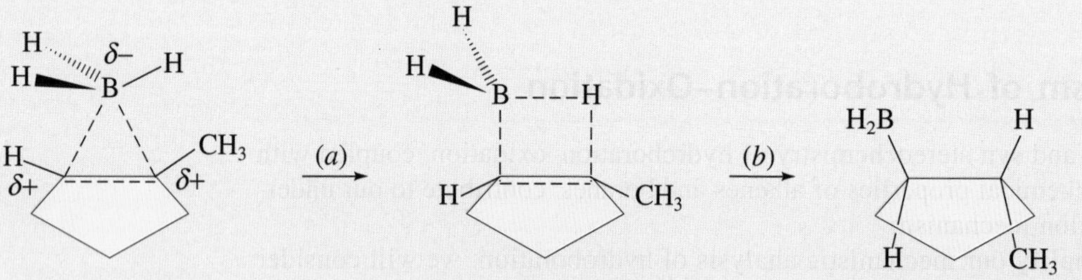

Transition state for hydride migration in π-complex intermediate

Product of addition of borane (BH_3) to 1-methylcyclopentene

formed in an acid–base reaction in step 1 and attacks boron in step 2. The empty $2p$ orbital of boron makes it electrophilic and permits nucleophilic reagents such as HOO^- to add to it.

The combination of a negative charge on boron and the weak oxygen–oxygen bond causes an alkyl group to migrate from boron to oxygen in step 3, which is the step in which the critical carbon–oxygen bond is formed. What is especially significant about this alkyl group migration is that the stereochemical orientation of the new carbon–oxygen bond is the same as that of the original carbon–boron bond. This is crucial to the overall syn stereochemistry of the hydroboration–oxidation sequence. Migration of the alkyl group from boron to oxygen is said to have occurred with **retention of configuration** at carbon. The alkoxyborane intermediate formed in step 3 undergoes subsequent base-promoted oxygen–boron bond cleavage in step 4 to give the alcohol product.

The mechanistic complexity of hydroboration–oxidation stands in contrast to the simplicity with which these reactions are carried out experimentally. Both the hydroboration and oxidation steps are extremely rapid reactions and are performed at room temperature with conventional laboratory equipment. Ease of operation, along with the fact that hydroboration–oxidation leads to syn hydration of alkenes with a regioselectivity opposite to Markovnikov's rule, makes this procedure one of great value to the synthetic chemist.

Mechanism 6.5

Oxidation of an Organoborane

Step 1: Hydrogen peroxide is converted to its anion in basic solution:

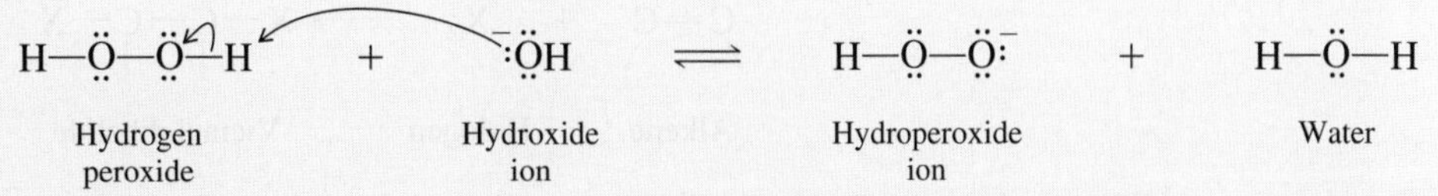

Step 2: Anion of hydrogen peroxide acts as a nucleophile, attacking boron and forming an oxygen–boron bond:

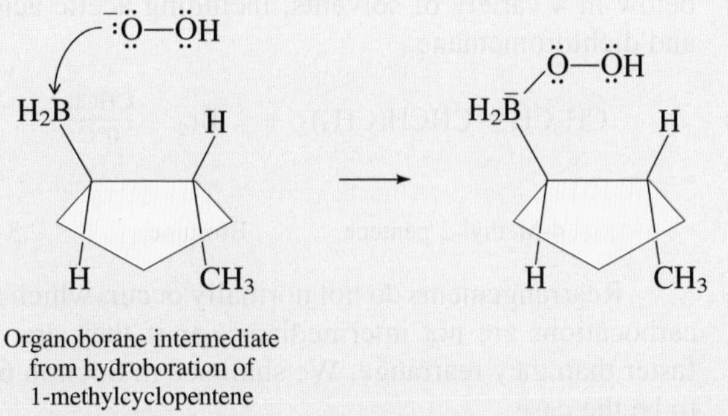

Step 3: Carbon migrates from boron to oxygen, displacing hydroxide ion. Carbon migrates with the pair of electrons in the carbon–boron bond; these become the electrons in the carbon–oxygen bond.

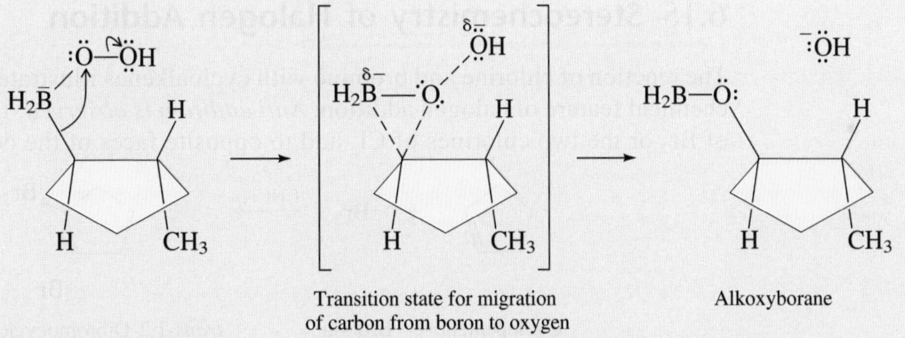

Step 4: Hydrolysis cleaves the boron–oxygen bond, yielding the alcohol:

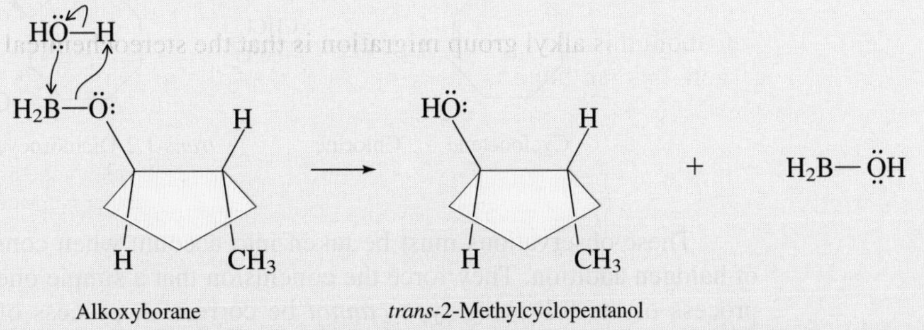

6.14 Addition of Halogens to Alkenes

In contrast to the free-radical substitution observed when halogens react with *alkanes,* halogens normally react with *alkenes* by electrophilic addition.

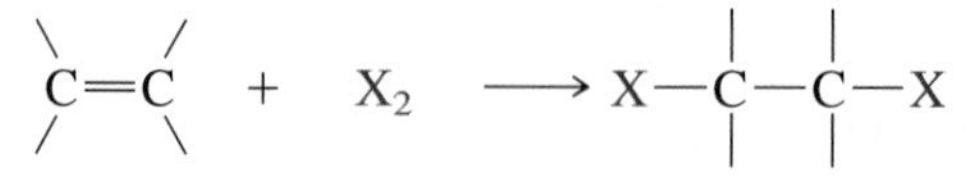

Alkene Halogen Vicinal dihalide

> Like the word *vicinity, vicinal* comes from the Latin *vicinalis,* which means "neighboring."

The products of these reactions are called **vicinal** dihalides. Two substituents, in this case the halogens, are vicinal if they are attached to adjacent carbons. The halogen is either chlorine (Cl_2) or bromine (Br_2), and addition takes place rapidly at room temperature and below in a variety of solvents, including acetic acid, carbon tetrachloride, chloroform, and dichloromethane.

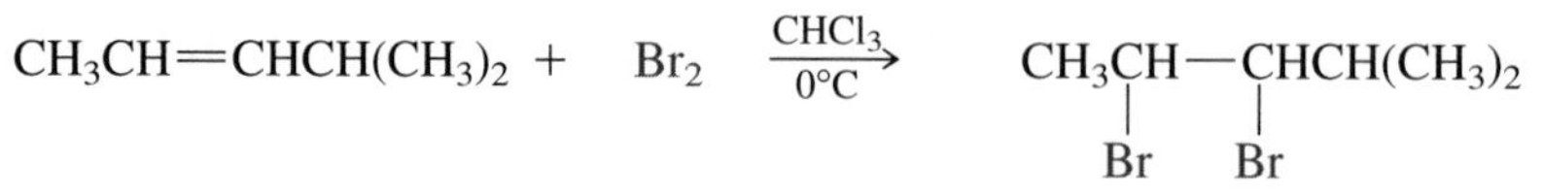

4-Methyl-2-pentene Bromine 2,3-Dibromo-4-methylpentane (100%)

Rearrangements do not normally occur, which can mean either of two things. Either carbocations are not intermediates, or if they are, they are captured by a nucleophile faster than they rearrange. We shall see in Section 6.16 that the first of these is believed to be the case.

Fluorine addition to alkenes is a violent reaction, difficult to control, and accompanied by substitution of hydrogens by fluorine. Vicinal diiodides, on the other hand, tend to lose I_2 and revert to alkenes, making them an infrequently encountered class of compounds.

6.15 Stereochemistry of Halogen Addition

The reaction of chlorine and bromine with cycloalkenes illustrates an important stereochemical feature of halogen addition. *Anti addition is observed;* the two bromine atoms of Br_2 or the two chlorines of Cl_2 add to opposite faces of the double bond.

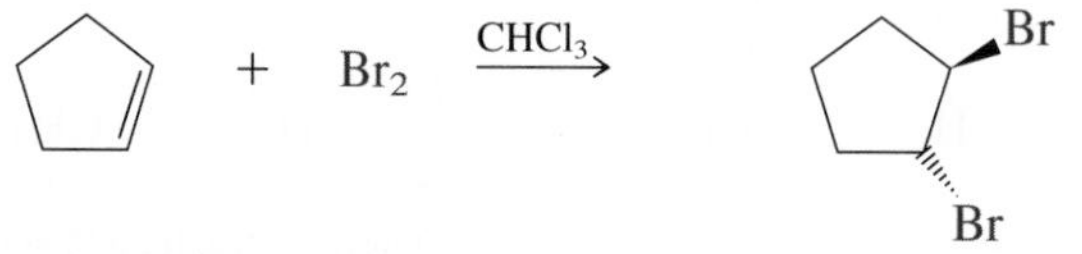

Cyclopentene Bromine *trans*-1,2-Dibromocyclopentane (80% yield; none of the cis isomer is formed)

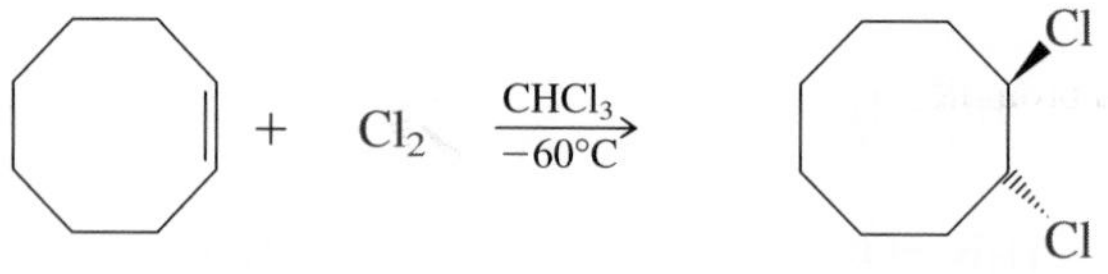

Cyclooctene Chlorine *trans*-1,2-Dichlorocyclooctane (73% yield; none of the cis isomer is formed)

These observations must be taken into account when considering the mechanism of halogen addition. They force the conclusion that a simple one-step "bond-switching" process of the following type *cannot* be correct. A process of this type requires syn addition; it is *not* consistent with the anti addition that we actually see.

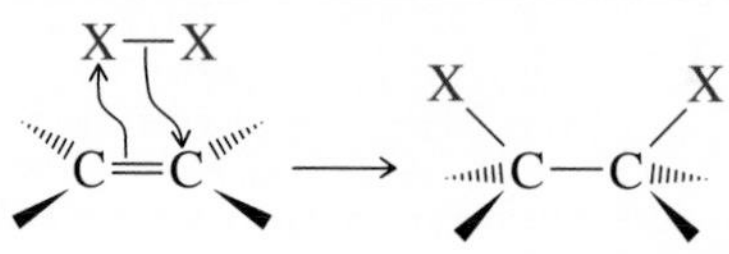

Problem 6.17

The mass 82 isotope of bromine (^{82}Br) is radioactive and is used as a tracer to identify the origin and destination of individual atoms in chemical reactions and biological transformations. A sample of 1,1,2-tribromocyclohexane was prepared by adding $^{82}Br—^{82}Br$ to ordinary (nonradioactive) 1-bromocyclohexene. How many of the bromine atoms in the 1,1,2-tribromocyclohexane produced are radioactive? Which ones are they?

6.16 Mechanism of Halogen Addition to Alkenes: Halonium Ions

Many of the features of the generally accepted mechanism for the addition of halogens to alkenes can be introduced by referring to the reaction of ethylene with bromine:

$$H_2C{=}CH_2 + Br_2 \longrightarrow BrCH_2CH_2Br$$

Ethylene Bromine 1,2-Dibromoethane

Until it was banned in the United States in 1984, 1,2-dibromoethane (ethylene dibromide, or EDB) was produced on a large scale for use as a pesticide and soil fumigant.

Electrons flow from the π system of ethylene to Br_2, causing the weak bromine–bromine bond to break. By analogy to the customary mechanisms for electrophilic addition, we might represent this as the formation of a carbocation in a bimolecular elementary step.

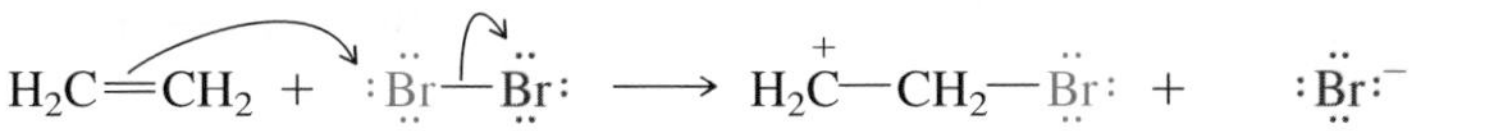
H₂C=CH₂ + :Br—Br: ⟶ H₂C—CH₂—Br: + :Br:⁻

Ethylene (nucleophile) Bromine (electrophile) 2-Bromoethyl cation (carbocation) Bromide ion (leaving group)

Such a carbocation, however, is not formed but is bypassed in favor of a cyclic **bromonium ion,** a more stable structure in which the positive charge resides on bromine, not carbon.

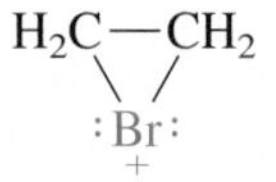
H₂C—CH₂
:Br:
+

Ethylenebromonium ion

The chief reason why ethylenebromonium ion, in spite of its strained three-membered ring, is more stable than 2-bromoethyl cation is that both carbons and bromine have octets of electrons, whereas one carbon has only six electrons in the carbocation.

Mechanism 6.6 for electrophilic addition of Br_2 to ethylene is characterized by the direct formation of a cyclic bromonium ion as its first elementary step via the transition state:

Transition state for bromonium ion formation from an alkene and bromine

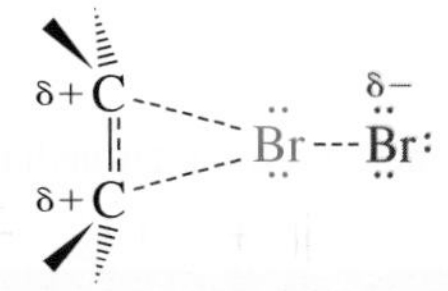
δ+C
δ+C
Br---Br:
δ−

Step 2 is the conversion of the bromonium ion to 1,2-dibromoethane by reaction with bromide ion (Br^-).

Table 6.3 shows that the effect of substituents on the rate of addition of bromine to alkenes is substantial and consistent with a rate-determining step in which electrons flow from the alkene to the halogen. Alkyl groups on the carbon–carbon double bond release electrons, stabilize the transition state for bromonium ion formation, and increase the reaction rate.

Problem 6.18

Arrange the compounds 2-methyl-1-butene, 2-methyl-2-butene, and 3-methyl-1-butene in order of decreasing reactivity toward bromine.

Mechanism 6.6

Electrophilic Addition of Bromine to Ethylene

THE OVERALL REACTION:

$$H_2C{=}CH_2 + Br_2 \longrightarrow BrCH_2CH_2Br$$

Ethylene | Bromine | 1,2-Dibromoethane

THE MECHANISM:

Step 1: Reaction of ethylene and bromine to form a bromonium ion intermediate:

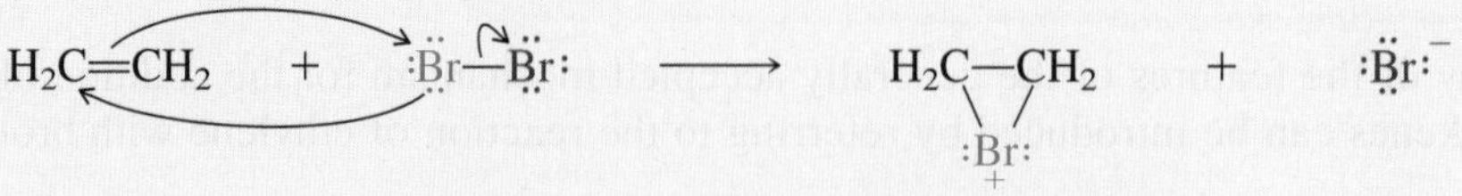

Ethylene | Bromine | Ethylenebromonium ion | Bromide ion

Step 2: Nucleophilic attack of bromide anion on the bromonium ion:

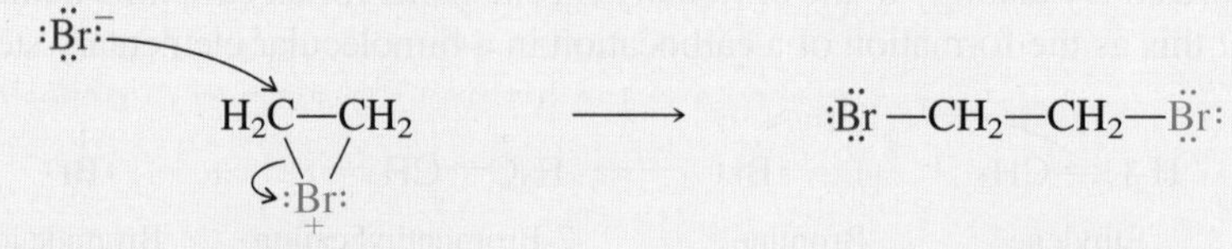

Bromide ion | Ethylenebromonium ion | 1,2-Dibromoethane

Step 2 of Mechanism 6.6 is a nucleophilic attack by Br^- at one of the carbons of the cyclic bromonium ion. For reasons that will be explained in Chapter 8, reactions of this type normally take place via a transition state in which the nucleophile approaches carbon from the side opposite the bond that is to be broken. Recalling that the vicinal dibromide formed from cyclopentene is exclusively the trans stereoisomer, we see that attack by Br^- from the side opposite the C—Br bond of the bromonium ion intermediate can give only *trans*-1,2-dibromocyclopentane in accordance with the experimental observations.

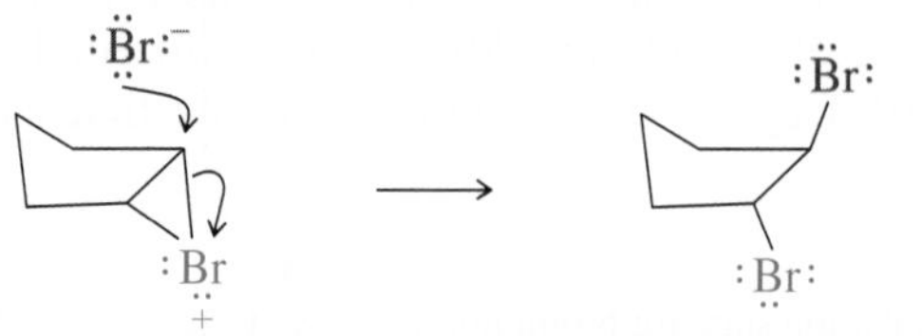

Bromonium ion intermediate | *trans*-1,2-Dibromocyclopentane

TABLE 6.3 Relative Rates of Reaction of Some Representative Alkenes with Bromine

Alkene	Structural formula	Relative rate of reaction with bromine*
Ethylene	$H_2C{=}CH_2$	1.0
Propene	$CH_3CH{=}CH_2$	61
2-Methylpropene	$(CH_3)_2C{=}CH_2$	5,400
2,3-Dimethyl-2-butene	$(CH_3)_2C{=}C(CH_3)_2$	920,000

*In methanol, 25°C.

The idea that a cyclic bromonium ion was an intermediate was a novel concept when it was first proposed. Much additional evidence, including the isolation of a stable cyclic bromonium ion, has been obtained since then to support it. Similarly, cyclic **chloronium ions** are believed to be involved in the addition of chlorine to alkenes. In the next section we shall see how cyclic chloronium and bromonium ions **(halonium ions)** are intermediates in a second reaction involving alkenes and halogens.

6.17 Conversion of Alkenes to Vicinal Halohydrins

In *aqueous* solution chlorine and bromine react with alkenes to form **vicinal halohydrins,** compounds that have a halogen and a hydroxyl group on adjacent carbons.

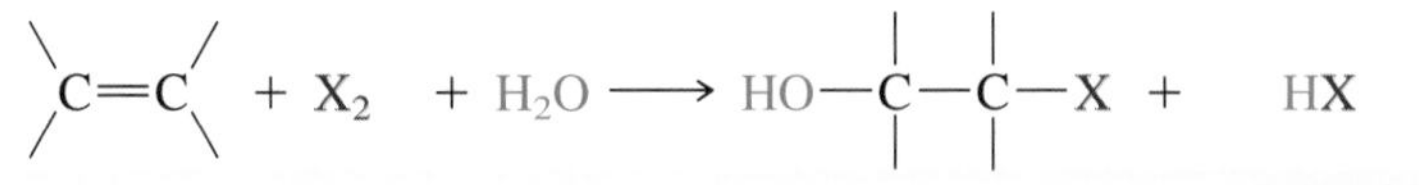

Alkene Halogen Water Halohydrin Hydrogen halide

$$H_2C{=}CH_2 + Br_2 \xrightarrow{H_2O} HOCH_2CH_2Br$$

Ethylene Bromine 2-Bromoethanol (70%)

Anti addition occurs. The halogen and the hydroxyl group add to opposite faces of the double bond.

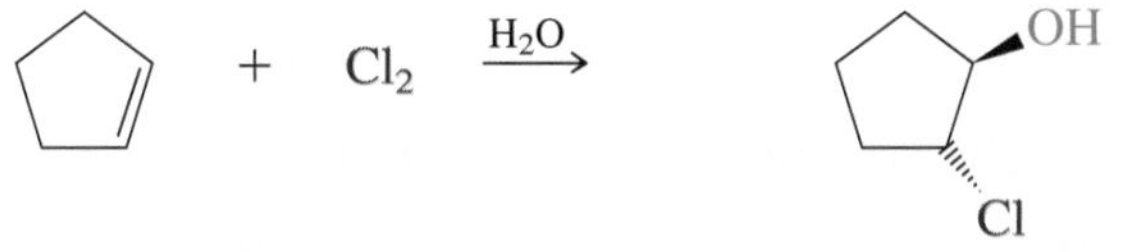

Cyclopentene Chlorine *trans*-2-Chlorocyclopentanol (52–56% yield; cis isomer not formed)

Halohydrin formation, as depicted in Mechanism 6.7, is mechanistically related to halogen addition to alkenes. A halonium ion intermediate is formed, which is attacked by water in aqueous solution.

The regioselectivity of addition is established when water attacks one of the carbons of the halonium ion. In the following example, the structure of the product tells us that water attacks the more highly substituted carbon.

$$(CH_3)_2C{=}CH_2 \xrightarrow[H_2O]{Br_2} (CH_3)_2C(OH){-}CH_2Br$$

2-Methylpropene 1-Bromo-2-methyl-2-propanol (77%)

Mechanism 6.7

Formation of a Bromohydrin

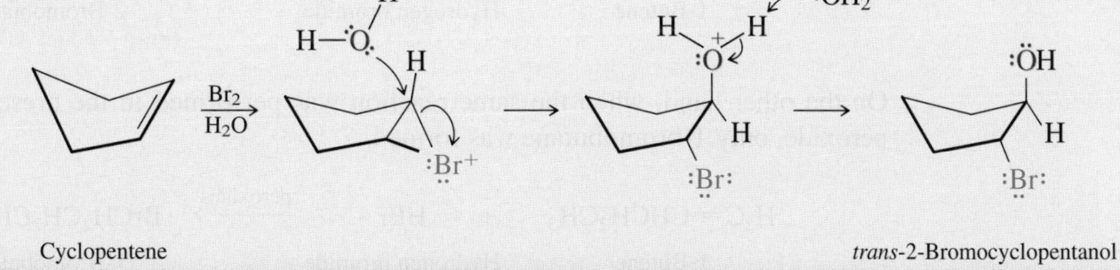

Cyclopentene *trans*-2-Bromocyclopentanol

The transition state for attack on the bromonium ion by water has some of the character of a carbocation. We know that more substituted carbocations are more stable than less substituted ones; therefore, when the bromonium ion ring opens, it does so by breaking the bond between bromine and the more substituted carbon.

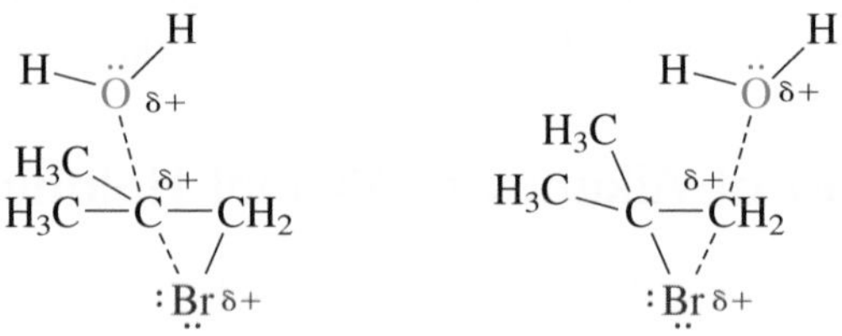

More stable transition state; has some of the character of a tertiary carbocation

Less stable transition state; has some of the character of a primary carbocation

Problem 6.19

Give the structure of the product formed when each of the following alkenes reacts with bromine in water:

(a) 2-Methyl-1-butene
(b) 2-Methyl-2-butene
(c) 3-Methyl-1-butene
(d) 1-Methylcyclopentene

Sample Solution (a) The hydroxyl group becomes bonded to the more substituted carbon of the double bond, and bromine bonds to the less substituted one.

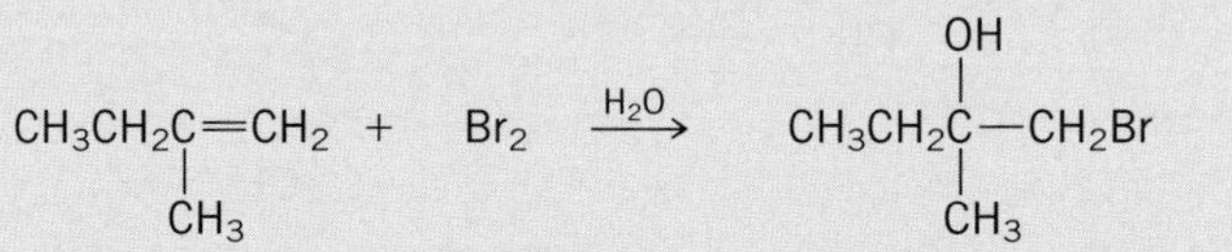

2-Methyl-1-butene Bromine 1-Bromo-2-methyl-2-butanol

6.18 Free-Radical Addition of Hydrogen Bromide to Alkenes

For a long time the regioselectivity of addition of hydrogen bromide to alkenes was unpredictable. Sometimes addition occurred according to Markovnikov's rule, but at other times, seemingly under the same conditions, it occurred opposite to Markovnikov's rule. In 1929, Morris S. Kharasch and his students at the University of Chicago began a systematic investigation of this puzzle. After hundreds of experiments, Kharasch concluded that addition occurred opposite to Markovnikov's rule when peroxides, that is, organic compounds of the type ROOR, were present in the reaction mixture. He and his colleagues found, for example, that carefully purified 1-butene reacted with hydrogen bromide to give only 2-bromobutane—the product expected on the basis of Markovnikov's rule.

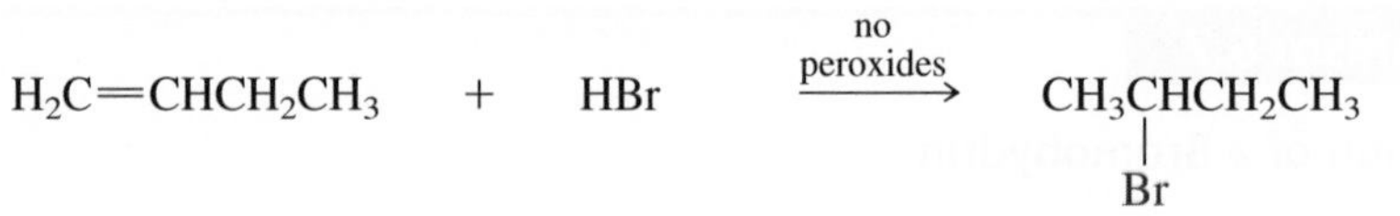

1-Butene Hydrogen bromide 2-Bromobutane (only product; 90% yield)

On the other hand, when the same reaction was performed in the presence of an added peroxide, only 1-bromobutane was formed.

$$H_2C{=}CHCH_2CH_3 + HBr \xrightarrow{\text{peroxides}} BrCH_2CH_2CH_2CH_3$$

1-Butene Hydrogen bromide 1-Bromobutane (only product; 95% yield)

Kharasch called this the **peroxide effect** and demonstrated that it could occur even if peroxides were not deliberately added to the reaction mixture. Unless alkenes are protected from atmospheric oxygen, they become contaminated with small amounts of alkyl hydroperoxides, compounds of the type ROOH. These alkyl hydroperoxides act in the same way as deliberately added peroxides, promoting addition in the direction opposite to that predicted by Markovnikov's rule.

Problem 6.20

Kharasch's earliest studies in this area were carried out in collaboration with graduate student Frank R. Mayo. Mayo performed over 400 experiments in which allyl bromide (3-bromo-1-propene) was treated with hydrogen bromide under a variety of conditions, and determined the distribution of the "normal" and "abnormal" products formed during the reaction. What two products were formed? Which is the product of addition in accordance with Markovnikov's rule? Which one corresponds to addition opposite to the rule?

Kharasch proposed that hydrogen bromide can add to alkenes by two different mechanisms, both of which are regiospecific. The first mechanism is electrophilic addition and follows Markovnikov's rule.

The second mechanism is the one followed when addition occurs opposite to Markovnikov's rule. Unlike electrophilic addition via a carbocation intermediate, this alternative mechanism is a chain reaction involving free-radical intermediates. It is presented in Mechanism 6.8.

Peroxides are *initiators;* they are not incorporated into the product but act as a source of radicals necessary to get the chain reaction started. The oxygen–oxygen bond of a peroxide is relatively weak, and the free-radical addition of hydrogen bromide to alkenes begins when a peroxide molecule's O—O bond breaks homolytically, giving two alkoxy radicals. This is depicted in step 1 of Mechanism 6.8. A bromine atom is generated in step 2 when one of these alkoxy radicals abstracts a hydrogen atom from hydrogen bromide. Once a bromine atom becomes available, the propagation phase of the chain reaction begins. In one step of the propagation phase (step 3), a bromine atom adds to the alkene in the direction that produces the more stable alkyl radical. Remember, the order of free-radical stability is tertiary > secondary > primary (see Section 4.17).

Addition of a bromine atom to C-1 gives a secondary alkyl radical.

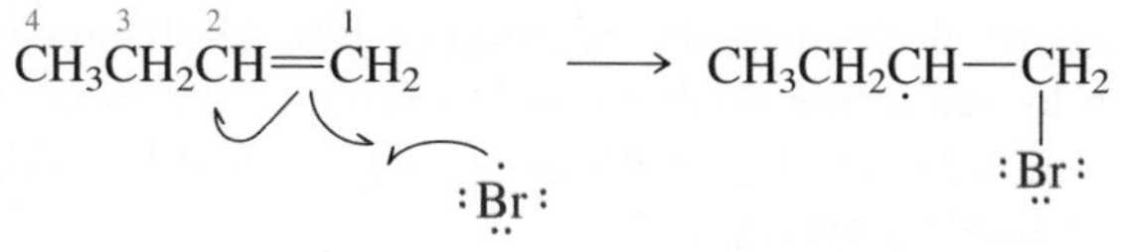
4 3 2 1
$CH_3CH_2CH{=}CH_2$
$:\dot{Br}:$
$CH_3CH_2\dot{C}H{-}CH_2$
$:Br:$

Secondary alkyl radical

Addition of a bromine atom to C-2 gives a primary alkyl radical.

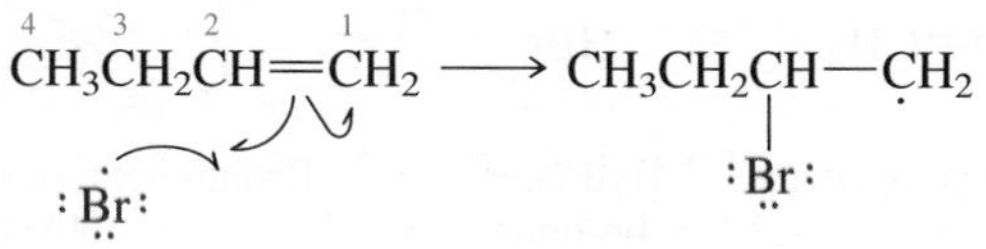
4 3 2 1
$CH_3CH_2CH{=}CH_2$
$:\dot{Br}:$
$CH_3CH_2CH{-}\dot{C}H_2$
$:Br:$

Primary alkyl radical

A secondary alkyl radical is more stable than a primary radical and is formed faster. Bromine adds to C-1 of 1-butene faster than it adds to C-2. Once the bromine atom has added to the double bond, the regioselectivity of addition is set. The alkyl radical then abstracts a hydrogen atom from hydrogen bromide to give the alkyl bromide product as shown in step 4 of Mechanism 6.8. Steps 3 and 4 propagate the chain, making 1-bromobutane the major product.

Mechanism 6.8

Free-Radical Addition of Hydrogen Bromide to 1-Butene

THE OVERALL REACTION:

$$\underset{\text{1-Butene}}{CH_3CH_2CH{=}CH_2} + \underset{\text{Hydrogen bromide}}{HBr} \xrightarrow[\text{light or heat}]{\text{ROOR}} \underset{\text{1-Bromobutane}}{CH_3CH_2CH_2CH_2Br}$$

THE MECHANISM:

(*a*) Initiation

Step 1: Homolytic Dissociation of a peroxide into two alkoxy radicals:

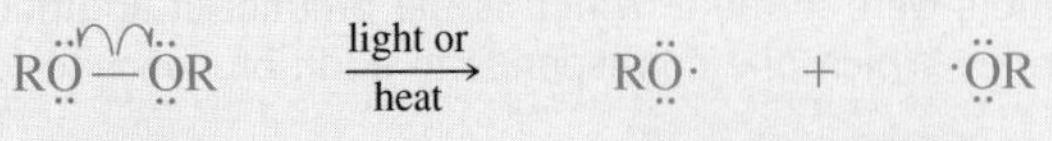

Peroxide   Two alkoxy radicals

Step 2: Hydrogen atom abstraction from hydrogen bromide by an alkoxy radical:

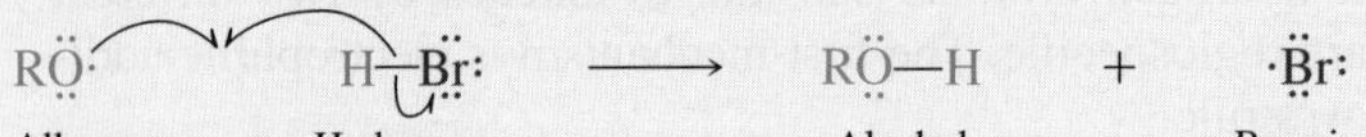

Alkoxy radical   Hydrogen bromide   Alcohol   Bromine atom

(*b*) Chain propagation

Step 3: Addition of a bromine atom to the alkene:

1-Butene   Bromine atom   1-(Bromomethyl)propyl radical

Step 4: Abstraction of a hydrogen atom from hydrogen bromide by the free radical formed in step 3:

1-(Bromomethyl)propyl radical   Hydrogen bromide   1-Bromobutane   Bromine atom

The regioselectivity of electrophilic addition of HBr to alkenes is controlled by the tendency of a *proton* to add to the double bond to produce the more stable *carbocation.* Under free-radical conditions the regioselectivity is governed by addition of a *bromine atom* to give the more stable *alkyl radical.*

Free-radical addition of hydrogen bromide to the double bond can also be initiated photochemically, either with or without added peroxides.

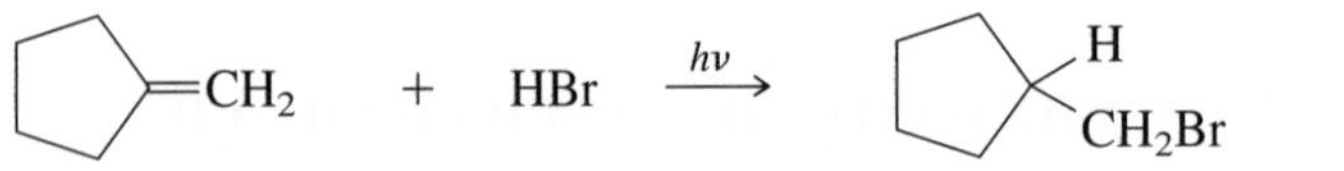

Methylenecyclopentane   Hydrogen bromide   (Bromomethyl)cyclopentane (60%)

Among the hydrogen halides, only hydrogen bromide reacts with alkenes by both electrophilic and free-radical addition mechanisms. Hydrogen iodide and hydrogen chloride always add to alkenes by electrophilic addition and follow Markovnikov's rule. Hydrogen bromide normally reacts by electrophilic addition, but if peroxides are present or if the reaction is initiated photochemically, the free-radical mechanism is followed.

Problem 6.21

Give the major organic product formed when hydrogen bromide reacts with each of the alkenes in Problem 6.4 in the absence of peroxides and in their presence.

Sample Solution (a) The addition of hydrogen bromide in the absence of peroxides exhibits a regioselectivity just like that of hydrogen chloride addition; Markovnikov's rule is followed.

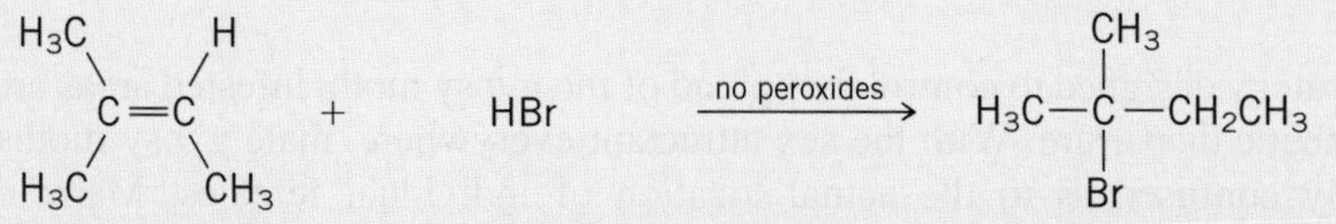
H3C
H
C=C
H3C
CH3
+
HBr
no peroxides
CH3
H3C—C—CH2CH3
Br

2-Methyl-2-butene   Hydrogen bromide   2-Bromo-2-methylbutane

Under free-radical conditions in the presence of peroxides, addition takes place with a regioselectivity opposite to that of Markovnikov's rule.

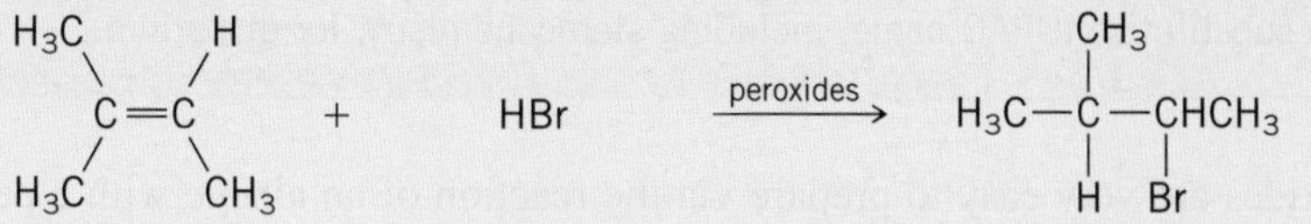
H3C
H
C=C
H3C
CH3
+
HBr
peroxides
CH3
H3C—C—CHCH3
H
Br

2-Methyl-2-butene   Hydrogen bromide   2-Bromo-3-methylbutane

Although the possibility of having two different reaction paths available to an alkene and hydrogen bromide may seem like a complication, it can be an advantage in organic synthesis. From a single alkene one may prepare either of two different alkyl bromides, with control of regioselectivity, simply by choosing reaction conditions that favor electrophilic addition or free-radical addition of hydrogen bromide.

6.19 Epoxidation of Alkenes

You have seen that cyclic halonium ion intermediates are formed when sources of electrophilic *halogen* attack a double bond. Likewise, three-membered oxygen-containing rings are formed by the reaction of alkenes with sources of electrophilic *oxygen.*

Three-membered rings that contain oxygen are called **epoxides.** At one time, epoxides were named as oxides of alkenes. Ethylene oxide and propylene oxide, for example, are the common names of two industrially important epoxides.

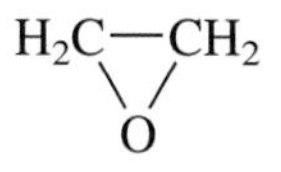
H2C—CH2
O

Ethylene oxide

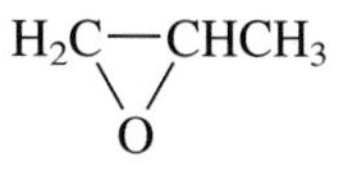
H2C—CHCH3
O

Propylene oxide

Substitutive IUPAC nomenclature names epoxides as *epoxy* derivatives of alkanes. According to this system, ethylene oxide becomes epoxyethane, and propylene oxide becomes 1,2-epoxypropane. The *epoxy-* prefix is listed in alphabetical order like other substituents.

A second method for naming epoxides in the IUPAC system is described in Section 16.1.

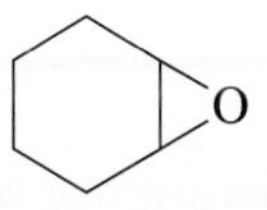
O

1,2-Epoxycyclohexane

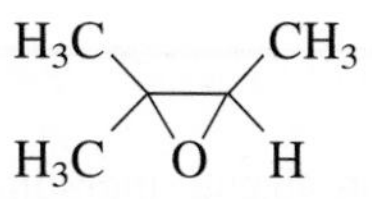
H3C
CH3
H3C
O
H

2,3-Epoxy-2-methylbutane

Functional group transformations of epoxides rank among the fundamental reactions of organic chemistry, and epoxides are commonplace natural products. The female gypsy moth, for example, attracts the male by emitting an epoxide known as *disparlure.*

Gypsy moths were accidentally introduced into United States forests around 1869 in Medford, Massachusetts. They have become persistent pests throughout the Northeast and Middle Atlantic states, defoliating millions of acres of woodlands.

On detecting the presence of this pheromone, the male follows the scent to its origin and mates with the female.

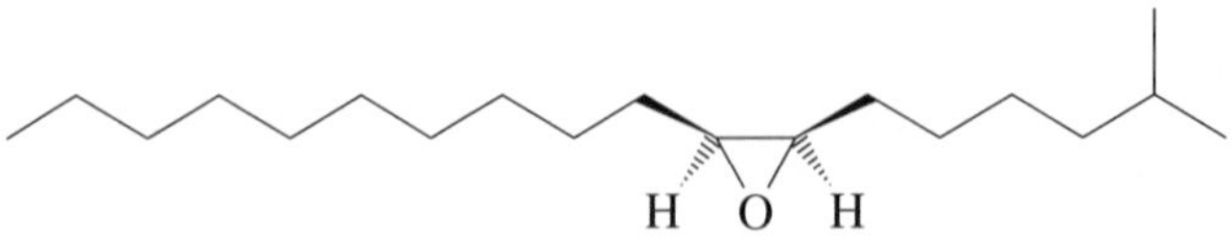

Disparlure

In one strategy designed to control the spread of the gypsy moth, infested areas are sprayed with synthetic disparlure. With the sex attractant everywhere, male gypsy moths become hopelessly confused as to the actual location of individual females. Many otherwise fertile female gypsy moths then live out their lives without producing hungry gypsy moth caterpillars.

Problem 6.22

Give the substitutive IUPAC name, including stereochemistry, for disparlure.

Epoxides are very easy to prepare via the reaction of an alkene with a peroxy acid. This process is known as **epoxidation.**

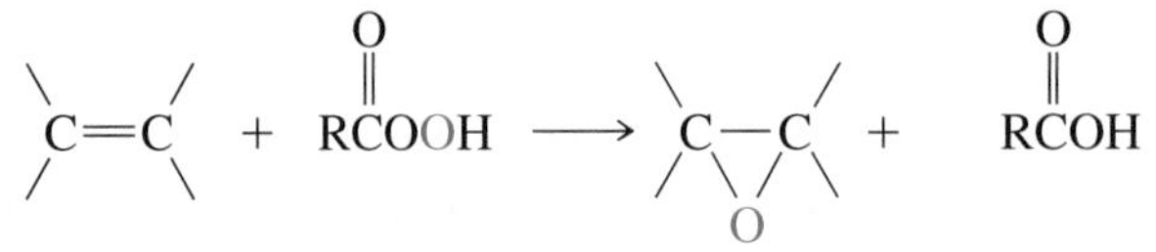

Alkene   Peroxy acid   Epoxide   Carboxylic acid

A commonly used peroxy acid is peroxyacetic acid (CH_3CO_2OH). Peroxyacetic acid is normally used in acetic acid as the solvent, but epoxidation reactions tolerate a variety of solvents and are often carried out in dichloromethane or chloroform.

$$H_2C{=}CH(CH_2)_9CH_3 + CH_3\overset{O}{\overset{\|}{C}}OOH \longrightarrow H_2C\overset{O}{\frown}CH(CH_2)_9CH_3 + CH_3\overset{O}{\overset{\|}{C}}OH$$

1-Dodecene   Peroxyacetic acid   1,2-Epoxydodecane (52%)   Acetic acid

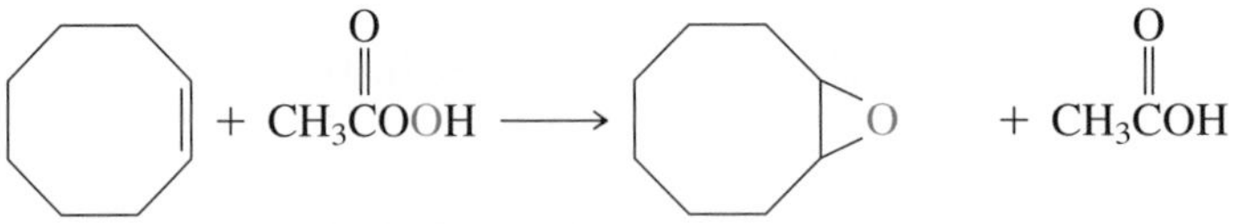

Cyclooctene   Peroxyacetic acid   1,2-Epoxycyclooctane (86%)   Acetic acid

Epoxidation of alkenes with peroxy acids is a syn addition to the double bond. Substituents that are cis to each other in the alkene remain cis in the epoxide; substituents that are trans in the alkene remain trans in the epoxide.

Problem 6.23

Give the structure of the alkene, including stereochemistry, that you would choose as the starting material in a preparation of synthetic disparlure.

As shown in Table 6.4, electron-releasing alkyl groups on the double bond increase the rate of epoxidation. This suggests that the peroxy acid acts as an electrophile toward the alkene.

TABLE 6.4 Relative Rates of Epoxidation of Some Representative Alkenes with Peroxyacetic Acid

Alkene	Structural formula	Relative rate of epoxidation*
Ethylene	$H_2C{=}CH_2$	1.0
Propene	$CH_3CH{=}CH_2$	22
2-Methylpropene	$(CH_3)_2C{=}CH_2$	484
2-Methyl-2-butene	$(CH_3)_2C{=}CHCH_3$	6,526

*In acetic acid, 26°C.

Alkene epoxidation is believed to be concerted, occurring by way of a single bimolecular elementary step, as shown in Mechanism 6.9.

6.20 Ozonolysis of Alkenes

Ozone (O_3) is the triatomic form of oxygen. It is a neutral but polar molecule that can be represented as a hybrid of its two most stable Lewis structures.

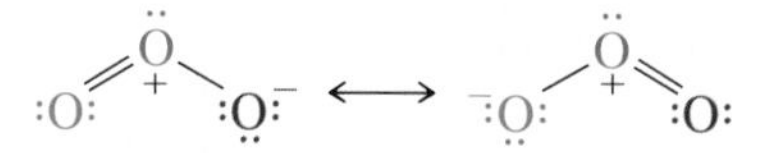

Ozone is a powerful electrophile and undergoes a remarkable reaction with alkenes in which both the σ and π components of the carbon–carbon double bond are cleaved to give a product referred to as an **ozonide.**

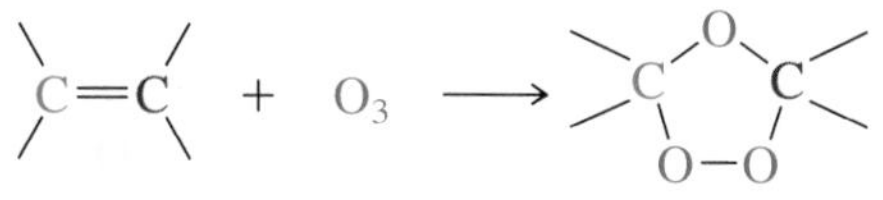
Alkene
Ozone
Ozonide

Ozonides undergo hydrolysis in water, giving carbonyl compounds.

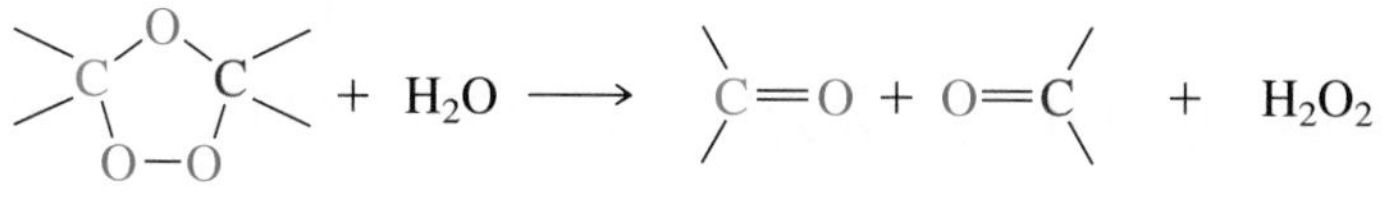
Ozonide
Water
Two carbonyl compounds
Hydrogen peroxide

Mechanism 6.9

Epoxidation of an Alkene

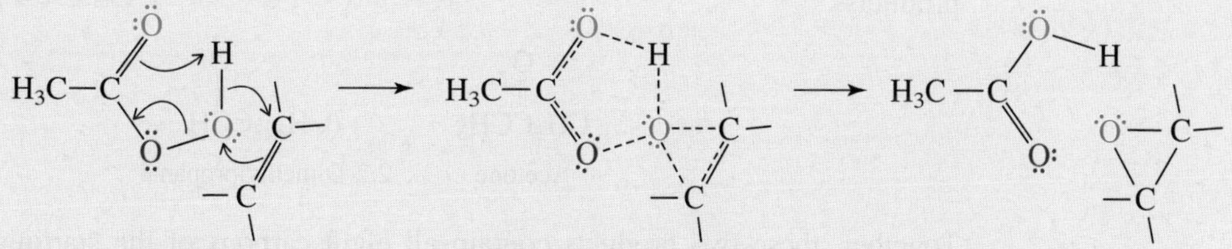
Peroxyacetic acid and alkene
Transition state for oxygen transfer from the OH group of the peroxy acid to the alkene
Acetic acid and epoxide

Two aldehydes, two ketones, or one aldehyde and one ketone may be formed. Aldehydes have at least one hydrogen on the carbonyl group; ketones have two carbon substituents—alkyl groups, for example—on the carbonyl. Carboxylic acids have a hydroxyl substituent attached to the carbonyl group.

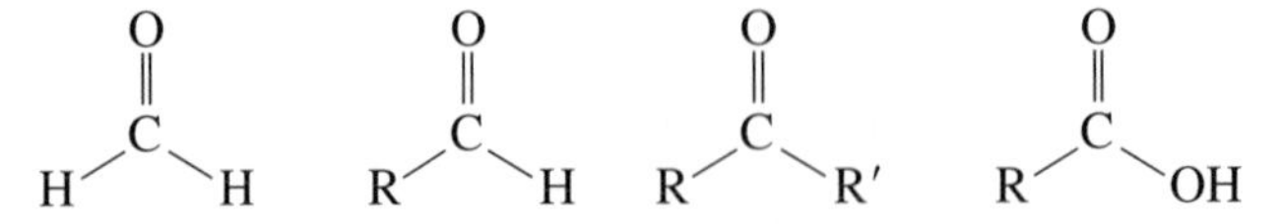

Formaldehyde Aldehyde Ketone Carboxylic acid

Aldehydes are easily oxidized to carboxylic acids under conditions of ozonide hydrolysis. When one wishes to isolate the aldehyde itself, a reducing agent such as zinc is included during the hydrolysis step. Zinc reduces the ozonide and reacts with any oxidants present (excess ozone and hydrogen peroxide) to prevent them from oxidizing any aldehyde formed. An alternative, more modern technique follows ozone treatment of the alkene in methanol with reduction by dimethyl sulfide (CH_3SCH_3).

The two-stage reaction sequence is called **ozonolysis** and is represented by the general equation

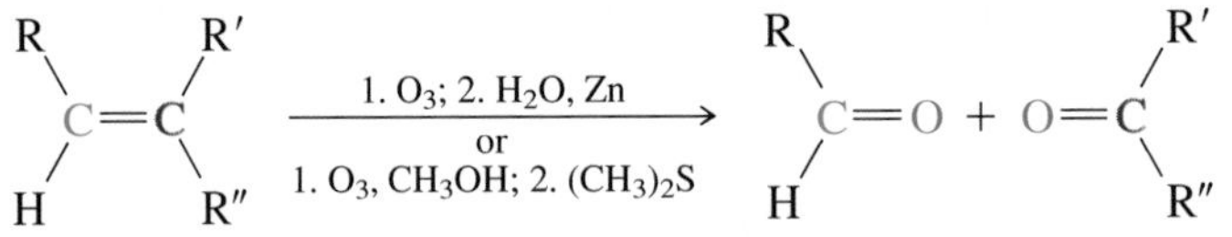

Alkene Aldehyde Ketone

Each carbon of the double bond becomes the carbon of a carbonyl group.

Ozonolysis has both synthetic and analytical applications in organic chemistry. In synthesis, ozonolysis of alkenes provides a method for the preparation of aldehydes and ketones.

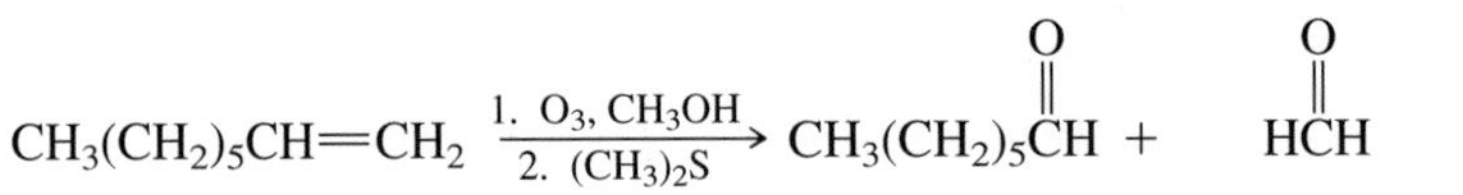

1-Octene Heptanal (75%) Formaldehyde

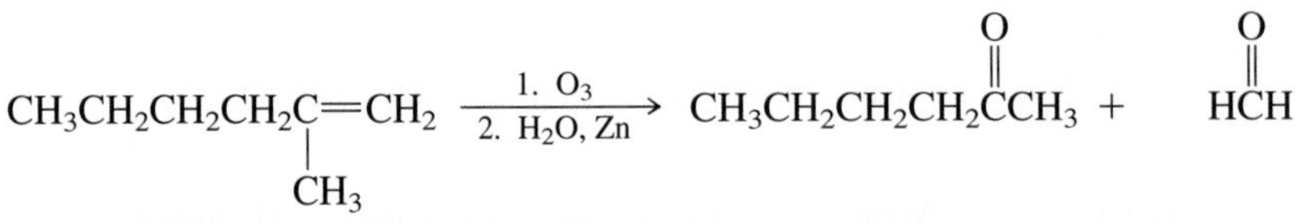

2-Methyl-1-hexene 2-Hexanone (60%) Formaldehyde

When the objective is analytical, the products of ozonolysis are isolated and identified, thereby allowing the structure of the alkene to be deduced. In one such example, an alkene having the molecular formula C_8H_{16} was obtained from a chemical reaction and was then subjected to ozonolysis, giving acetone and 2,2-dimethylpropanal as the products.

Acetone 2,2-Dimethylpropanal

Together, these two products contain all eight carbons of the starting alkene. The two carbonyl carbons correspond to those that were doubly bonded in the original alkene. One of the doubly bonded carbons therefore bears two methyl substituents; the other bears a hydrogen and a *tert*-butyl group. The alkene is identified as 2,4,4-trimethyl-2-pentene, $(CH_3)_2C{=}CHC(CH_3)_3$, as shown in Figure 6.6.

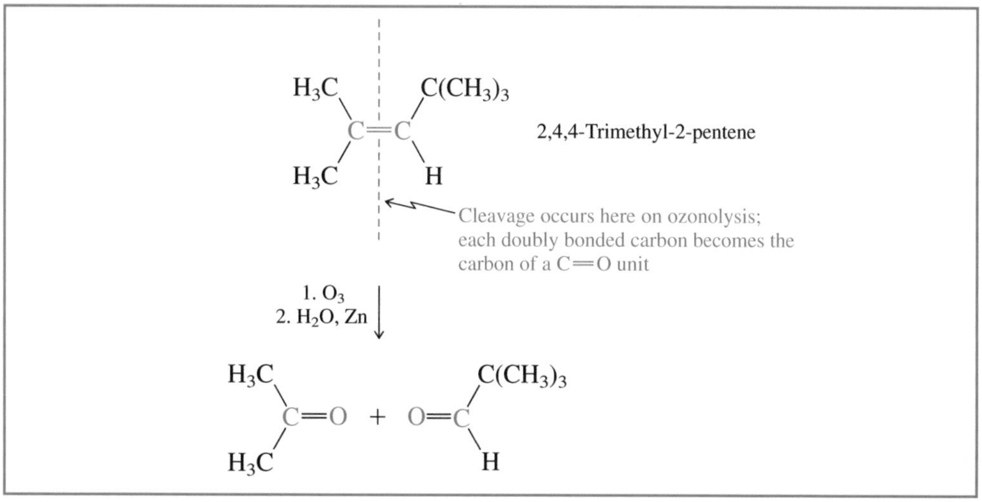
2,4,4-Trimethyl-2-pentene
Cleavage occurs here on ozonolysis; each doubly bonded carbon becomes the carbon of a C=O unit
1. O3
2. H2O, Zn

Figure 6.6

Ozonolysis of 2,4,4-trimethyl-2-pentene. On cleavage, each of the doubly bonded carbons becomes the carbon of a carbonyl (C=O) group.

Problem 6.24

The same reaction that gave 2,4,4-trimethyl-2-pentene also yielded an isomeric alkene. This second alkene produced formaldehyde and 4,4-dimethyl-2-pentanone on ozonolysis. Identify this alkene.

$$CH_3\overset{\overset{\displaystyle O}{\|}}{C}CH_2C(CH_3)_3$$

4,4-Dimethyl-2-pentanone

6.21 Reactions of Alkenes with Alkenes: Polymerization

Chapter 27 is entirely devoted to polymers.

Although 2-methylpropene undergoes acid-catalyzed hydration in *dilute* sulfuric acid to form *tert*-butyl alcohol (Section 6.9), a different reaction occurs in more concentrated solutions of sulfuric acid. Rather than form the expected alkyl hydrogen sulfate (Section 6.8), 2-methylpropene is converted to a mixture of two isomeric C_8H_{16} alkenes.

$$2(CH_3)_2C{=}CH_2 \xrightarrow{65\%\ H_2SO_4} H_2C{=}\underset{\underset{\displaystyle CH_3}{|}}{C}CH_2C(CH_3)_3 + (CH_3)_2C{=}CHC(CH_3)_3$$

2-Methylpropene — 2,4,4-Trimethyl-1-pentene — 2,4,4-Trimethyl-2-pentene

With molecular formulas corresponding to twice that of the starting alkene, the products of this reaction are referred to as **dimers** of 2-methylpropene, which is, in turn, called the **monomer.** The suffix *-mer* is derived from the Greek *meros,* meaning "part." Three monomeric units produce a *trimer,* four a *tetramer,* and so on. A high-molecular-weight material comprising a large number of monomer subunits is called a **polymer.**

Problem 6.25

The two dimers of 2-methylpropene shown in the equation can be converted to 2,2,4-trimethylpentane (known by its common name *isooctane*) for use as a gasoline additive. Can you suggest a method for this conversion?

The two dimers of $(CH_3)_2C{=}CH_2$ are formed by the process shown in Mechanism 6.10. In step 1 protonation of the double bond generates a small amount of *tert*-butyl

Mechanism 6.10

Acid-Catalyzed Dimerization of 2-Methylpropene

THE MECHANISM:

Step 1: Protonation of the carbon–carbon double bond in the direction that leads to more stable carbocation:

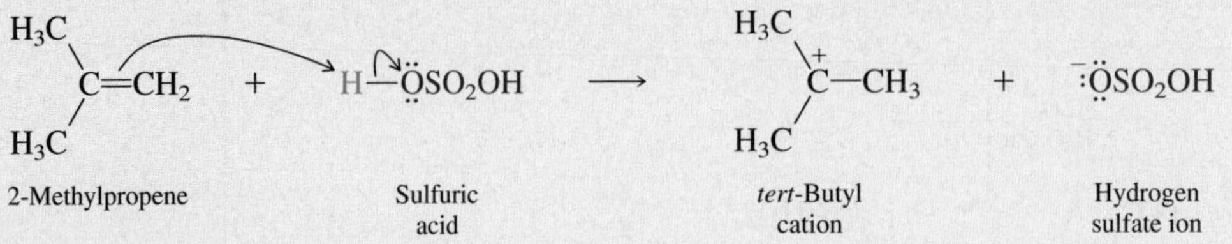
2-Methylpropene
Sulfuric acid
tert-Butyl cation
Hydrogen sulfate ion

Step 2: The carbocation acts as an electrophile toward the alkene. A carbon–carbon bond is formed, resulting in a new carbocation—one that has eight carbons:

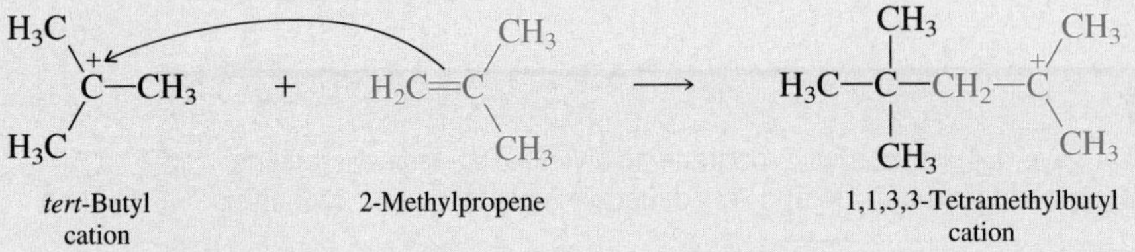
tert-Butyl cation
2-Methylpropene
1,1,3,3-Tetramethylbutyl cation

Step 3: Loss of a proton from this carbocation can produce either 2,4,4-trimethyl-1-pentene or 2,4,4-trimethyl-2-pentene:

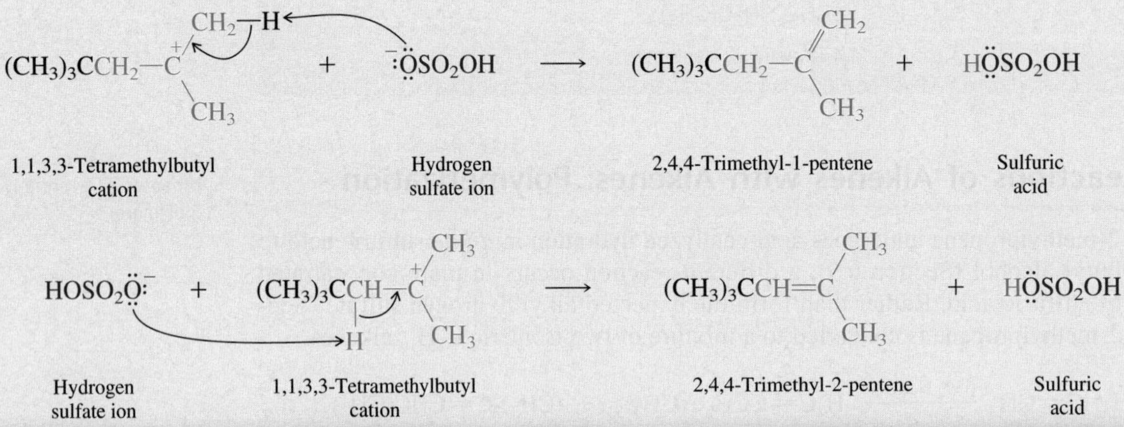
1,1,3,3-Tetramethylbutyl cation
Hydrogen sulfate ion
2,4,4-Trimethyl-1-pentene
Sulfuric acid
Hydrogen sulfate ion
1,1,3,3-Tetramethylbutyl cation
2,4,4-Trimethyl-2-pentene
Sulfuric acid

cation in equilibrium with the alkene. The carbocation is an electrophile and attacks a second molecule of 2-methylpropene in step 2, forming a new carbon–carbon bond and generating a C_8 carbocation. This new carbocation loses a proton in step 3 to form a mixture of 2,4,4-trimethyl-1-pentene and 2,4,4-trimethyl-2-pentene.

Dimerization in concentrated sulfuric acid occurs mainly with those alkenes that form tertiary carbocations. In some cases reaction conditions can be developed that favor the formation of higher molecular weight polymers. Because these reactions proceed by way of carbocation intermediates, the process is referred to as **cationic polymerization.**

We made special mention in Section 5.1 of the enormous volume of ethylene and propene production in the petrochemical industry. The boxed essay "Ethylene and Propene: The Most Important Industrial Organic Chemicals" summarizes the principal uses of these alkenes. Most of the ethylene is converted to **polyethylene,** a high-molecular-weight polymer of ethylene. Polyethylene cannot be prepared by cationic polymerization, but is the simplest example of a polymer that is produced on a large scale by **free-radical polymerization.**

Ethylene and Propene: The Most Important Industrial Organic Chemicals

Having examined the properties of alkenes and introduced the elements of polymers and polymerization, let's now look at some commercial applications of ethylene and propene.

Ethylene We discussed ethylene production in an earlier boxed essay (Section 5.1), where it was pointed out that the output of the U.S. petrochemical industry exceeds 5×10^{10} lb/year. Approximately 90% of this material is used for the preparation of four compounds (polyethylene, ethylene oxide, vinyl chloride, and styrene), with polymerization to polyethylene accounting for half the total. Both vinyl chloride and styrene are polymerized to give poly(vinyl chloride) and polystyrene, respectively (see Table 6.5). Ethylene oxide is a starting material for the preparation of ethylene glycol for use as an antifreeze in automobile radiators and in the production of polyester fibers.

Propene The major use of propene is in the production of polypropylene. Two other propene-derived organic chemicals, acrylonitrile and propylene oxide, are also starting materials for polymer synthesis. Acrylonitrile is used to make acrylic fibers (see Table 6.5), and propylene oxide is one component in the preparation of *polyurethane* polymers. Cumene itself has no direct uses but rather serves as the starting material in a process that yields two valuable industrial chemicals: acetone and phenol.

We have not indicated the reagents employed in the reactions by which ethylene and propene are converted to the compounds shown. Because of patent requirements, different companies often use different processes. Although the processes may be different, they share the common characteristic of being extremely efficient. The industrial chemist faces the challenge of producing valuable materials, at low cost. Success in the industrial environment requires both an understanding of chemistry and an appreciation of the economics associated with alternative procedures.

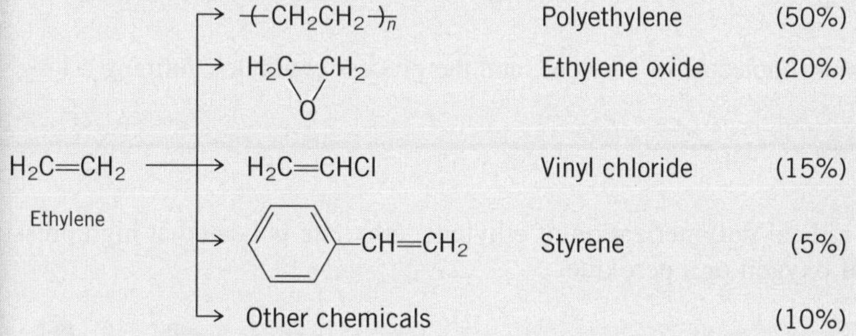

$H_2C{=}CH_2$
Ethylene
$+CH_2CH_2+_n$ Polyethylene (50%)
$H_2C—CH_2$ O Ethylene oxide (20%)
$H_2C{=}CHCl$ Vinyl chloride (15%)
—CH=CH$_2$ Styrene (5%)
Other chemicals (10%)

Among the "other chemicals" prepared from ethylene are ethanol and acetaldehyde:

CH_3CH_2OH
O
CH_3CH

Ethanol (industrial solvent; used in preparation of ethyl acetate; unleaded gasoline additive)

Acetaldehyde (used in preparation of acetic acid)

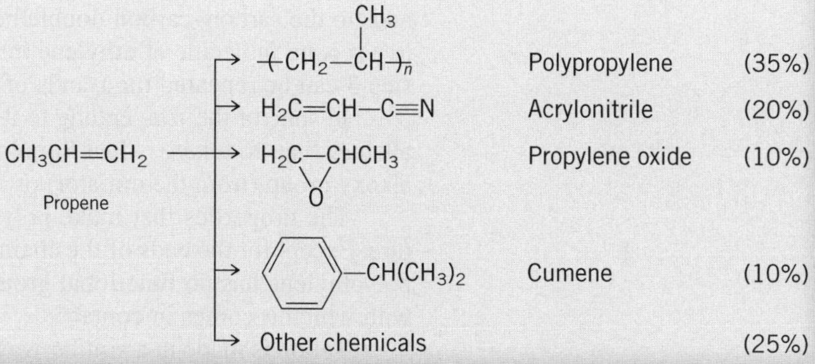

$CH_3CH{=}CH_2$
Propene
CH_3
$+CH_2—CH+_n$ Polypropylene (35%)
$H_2C{=}CH—C{\equiv}N$ Acrylonitrile (20%)
$H_2C—CHCH_3$ O Propylene oxide (10%)
—CH(CH$_3$)$_2$ Cumene (10%)
Other chemicals (25%)

Mechanism 6.11

Free-Radical Polymerization of Ethylene

Step 1: Homolytic dissociation of a peroxide produces alkoxy radicals that serve as free-radical initiators:

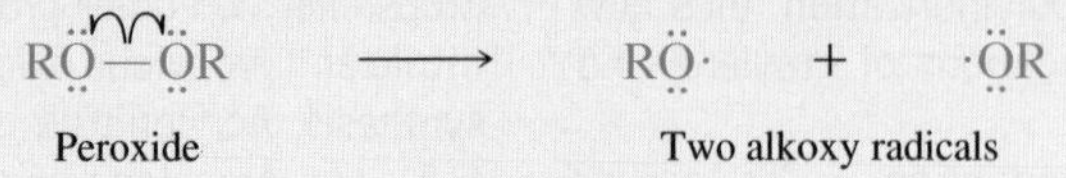
RÖ—ÖR
RÖ· + ·ÖR
Peroxide
Two alkoxy radicals

Step 2: An alkoxy radical adds to the carbon-carbon double bond:

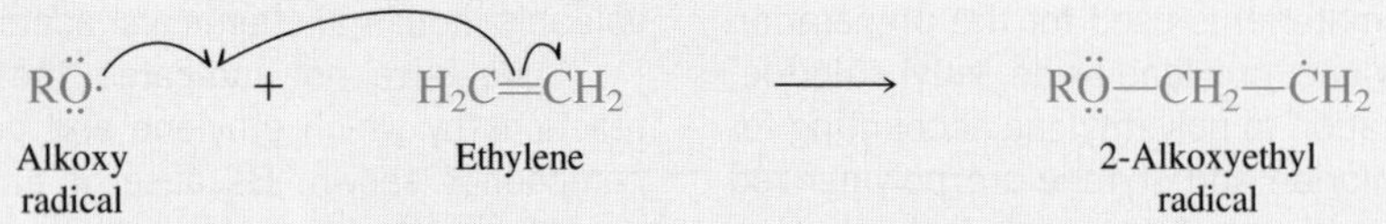
RÖ· + H₂C=CH₂
RÖ—CH₂—ĊH₂
Alkoxy radical
Ethylene
2-Alkoxyethyl radical

Step 3: The radical produced in step 2 adds to a second molecule of ethylene:

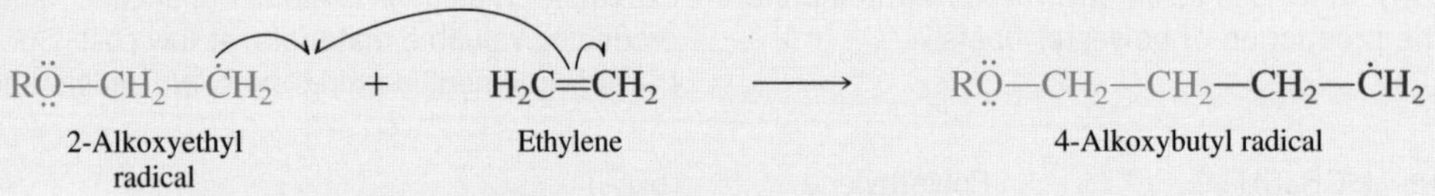
RÖ—CH₂—ĊH₂ + H₂C=CH₂
RÖ—CH₂—CH₂—CH₂—ĊH₂
2-Alkoxyethyl radical
Ethylene
4-Alkoxybutyl radical

The radical formed in step 3 then adds to a third molecule of ethylene, and the process continues, forming a long chain of methylene groups.

In the free-radical polymerization of ethylene, ethylene is heated at high pressure in the presence of oxygen or a peroxide.

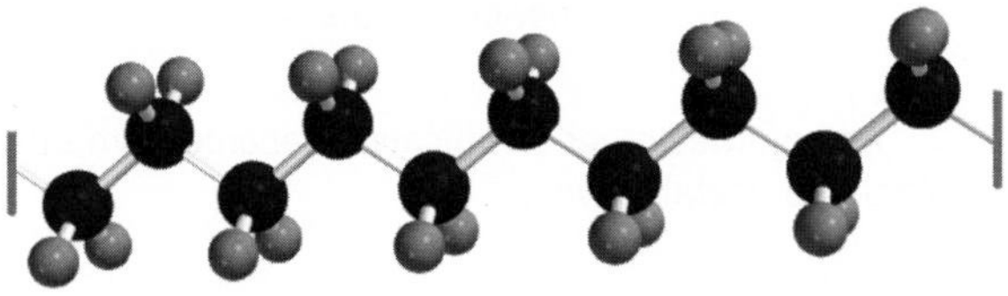

$$nH_2C{=}CH_2 \xrightarrow[\text{O}_2\text{ or peroxides}]{200^\circ\text{C, 2000 atm}} -CH_2-CH_2-(CH_2-CH_2)_{n-2}-CH_2-CH_2-$$

Ethylene — Polyethylene

In this reaction n can have a value of thousands.

Mechanism 6.11 shows the steps in the free-radical polymerization of ethylene. Dissociation of a peroxide initiates the process in step 1. The resulting peroxy radical adds to the carbon–carbon double bond in step 2, giving a new radical, which then adds to a second molecule of ethylene in step 3. The carbon–carbon bond-forming process in step 3 can be repeated thousands of times to give long carbon chains.

In spite of the *-ene* ending to its name, polyethylene is much more closely related to alk*anes* than to alk*enes*. It is simply a long chain of CH_2 groups bearing at its ends an alkoxy group (from the initiator) or a carbon–carbon double bond.

The properties that make polyethylene so useful come from its alkane-like structure. Except for the ends of the chain, which make up only a tiny portion of the molecule, polyethylene has no functional groups so is almost completely inert to most substances with which it comes in contact.

Teflon is made in a similar way by free-radical polymerization of tetrafluoroethene. Carbon–fluorine bonds are quite strong (slightly stronger than C—H bonds), and like polyethylene, Teflon is a very stable, inert material. We are all familiar with the most characteristic property of Teflon, its "nonstick" surface. This can be understood by comparing Teflon and polyethylene. The high electronegativity of fluorine makes C—F bonds less polarizable than C—H bonds, causing the dispersion forces in Teflon to be less than

TABLE 6.5 Some Compounds with Carbon–Carbon Double Bonds Used to Prepare Polymers

A. Alkenes of the type $H_2C{=}CH{-}X$ used to form polymers of the type $\left(CH_2{-}\underset{X}{\underset{|}{CH}}\right)_n$

Compound	Structure	–X in polymer	Application
Ethylene	$H_2C{=}CH_2$	$-H$	Polyethylene films as packaging material; "plastic" squeeze bottles are molded from high-density polyethylene.
Propene	$H_2C{=}CH{-}CH_3$	$-CH_3$	Polypropylene fibers for use in carpets and automobile tires; consumer items (luggage, appliances, etc.); packaging material.
Styrene	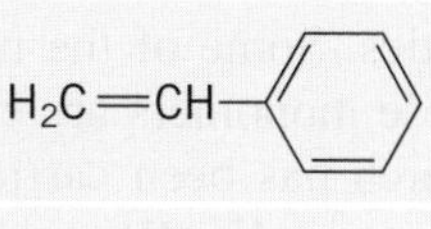	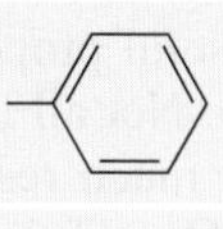	Polystyrene packaging, housewares, luggage, radio and television cabinets.
Vinyl chloride	$H_2C{=}CH{-}Cl$	$-Cl$	Poly(vinyl chloride) (PVC) has replaced leather in many of its applications; PVC tubes and pipes are often used in place of copper.
Acrylonitrile	$H_2C{=}CH{-}C{\equiv}N$	$-C{\equiv}N$	Wool substitute in sweaters, blankets, etc.

B. Alkenes of the type $H_2C{=}CX_2$ used to form polymers of the type $\left(CH_2{-}CX_2\right)_n$

Compound	Structure	X in polymer	Application
1,1-Dichloroethene (vinylidene chloride)	$H_2C{=}CCl_2$	Cl	Saran used as air- and watertight packaging film.
2-Methylpropene	$H_2C{=}C(CH_3)_2$	CH_3	Polyisobutylene is component of "butyl rubber," one of earliest synthetic rubber substitutes.

C. Others

Compound	Structure	Polymer	Application
Tetrafluoroethene	$F_2C{=}CF_2$	$\left(CF_2{-}CF_2\right)_n$ (Teflon)	Nonstick coating for cooking utensils; bearings, gaskets, and fittings.
Methyl methacrylate	$H_2C{=}\underset{CH_3}{\underset{\mid}{C}}CO_2CH_3$	$\left(CH_2{-}\overset{CO_2CH_3}{\overset{\mid}{\underset{CH_3}{\underset{\mid}{C}}}}\right)_n$	When cast in sheets, is transparent; used as glass substitute (Lucite, Plexiglas).
2-Methyl-1,3-butadiene	$H_2C{=}\underset{CH_3}{\underset{\mid}{C}}CH{=}CH_2$	$\left(CH_2\underset{CH_3}{\underset{\mid}{C}}{=}CH{-}CH_2\right)_n$ (Polyisoprene)	Synthetic rubber.

Source: R. C. Atkins and F. A. Carey, *Organic Chemistry: A Brief Course,* 3rd ed. McGraw-Hill, New York, 2002, p. 237.

those in polyethylene. Thus, the surface of Teflon is even less "sticky" than the already slick surface of polyethylene.

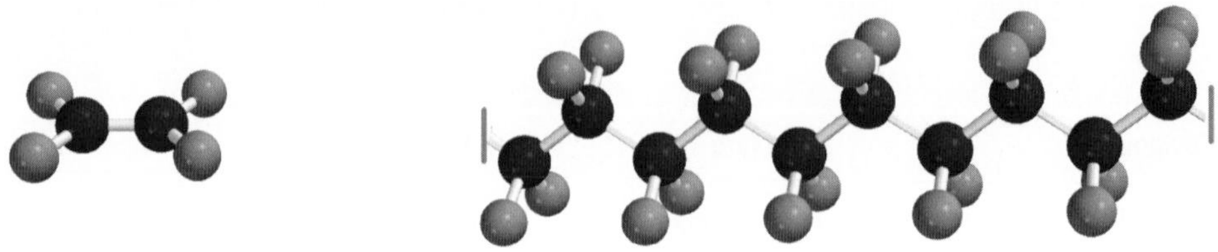

$$nF_2C{=}CF_2 \xrightarrow[\text{peroxides}]{80°C,\ 40\text{–}100\ \text{atm}} {-}CF_2{-}CF_2{-}(CF_2{-}CF_2)_{n-2}{-}CF_2{-}CF_2{-}$$

Tetrafluoroethene | Teflon

A large number of compounds with carbon–carbon double bonds have been polymerized to yield materials having useful properties. Some of the more important or familiar of these are listed in Table 6.5. Not all these monomers are effectively polymerized under free-radical conditions, and much research has been carried out to develop alternative polymerization techniques. One of these, **coordination polymerization,** employs novel transition-metal catalysts. Polyethylene produced by coordination polymerization has a higher density than that produced by free-radical polymerization and somewhat different—in many applications, more desirable—properties. Coordination polymerization of ethylene was developed independently by Karl Ziegler in Germany and applied to propene by Giulio Natta in Italy. They shared the Nobel Prize in Chemistry in 1963 for this work. Coordination polymerization gives a form of **polypropylene** suitable for plastics and fibers. When propene is polymerized under free-radical conditions, the polypropylene has physical properties (such as a low melting point) that make it useless for most applications.

Coordination polymerization is described in more detail in Sections 7.16, 14.16, and 27.7.

6.22 SUMMARY

Alkenes are **unsaturated hydrocarbons** and react with substances that add to the double bond.

Section 6.1 See Table 6.6.

Section 6.2 Hydrogenation of alkenes is exothermic. Heats of hydrogenation can be measured and used to assess the stability of various types of double bonds. The information parallels that obtained from heats of combustion.

Section 6.3 Hydrogenation of alkenes is a syn addition.

Sections 6.4–6.7 See Table 6.6. Hydrogen halide addition to alkenes proceeds by electrophilic attack of the reagent on the π electrons of the double bond. Carbocations are intermediates. Addition to unsymmetrical alkenes is regioselective.

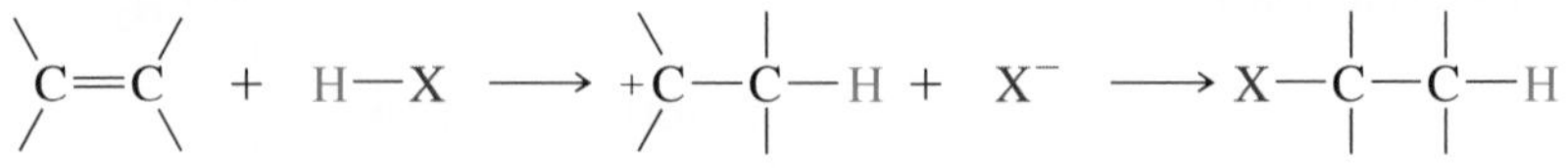

Alkene | Hydrogen halide | Carbocation | Halide ion | Alkyl halide

Protonation of the double bond occurs in the direction that gives the more stable of two possible carbocations.

Sections 6.8–6.9 See Table 6.6

Section 6.10 Addition and elimination reactions are often reversible, and proceed spontaneously in the direction in which the free energy G decreases. The reaction is at equilibrium when $\Delta G = 0$. Free energy is related to enthalpy (H) and entropy (S) by the equations

$$G = H - TS \qquad \text{and} \qquad \Delta G = \Delta H - T\Delta S$$

TABLE 6.6 Addition Reactions of Alkenes

Reaction (section) and comments	General equation and specific example
Catalytic hydrogenation (Sections 6.1–6.3) Alkenes react with hydrogen in the presence of a platinum, palladium, rhodium, or nickel catalyst to form the corresponding alkane.	$R_2C{=}CR_2 + H_2 \xrightarrow{\text{Pt, Pd, Rh, or Ni}} R_2CHCHR_2$ Alkene Hydrogen Alkane 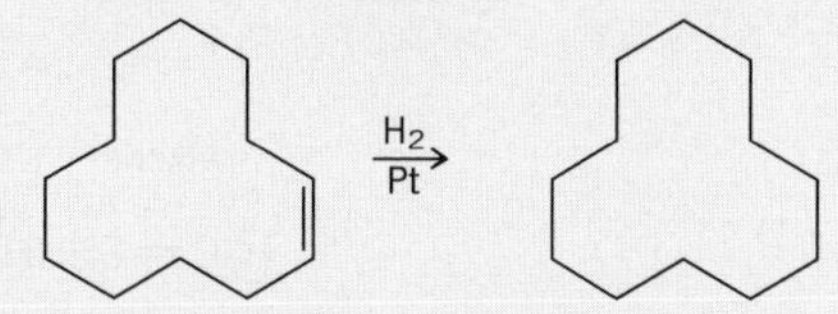 *cis*-Cyclododecene Cyclododecane (100%)
Addition of hydrogen halides (Sections 6.4–6.7) A proton and a halogen add to the double bond of an alkene to yield an alkyl halide. Addition proceeds in accordance with Markovnikov's rule; hydrogen adds to the carbon that has the greater number of hydrogens, halide to the carbon that has the fewer hydrogens.	$RCH{=}CR'_2 + HX \longrightarrow RCH_2{-}CR'_2(X)$ Alkene Hydrogen halide Alkyl halide 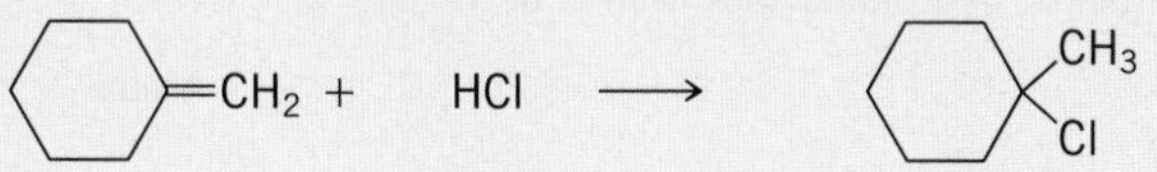 Methylenecyclohexane Hydrogen chloride 1-Chloro-1-methylcyclohexane (75–80%)
Addition of sulfuric acid (Section 6.8) Alkenes react with sulfuric acid to form alkyl hydrogen sulfates. A proton and a hydrogen sulfate ion add to the double bond in accordance with Markovnikov's rule. Alkenes that yield tertiary carbocations on protonation tend to polymerize in concentrated sulfuric acid (Section 6.21).	$RCH{=}CR'_2 + HOSO_2OH \longrightarrow RCH_2{-}CR'_2(OSO_2OH)$ Alkene Sulfuric acid Alkyl hydrogen sulfate $H_2C{=}CHCH_2CH_3 + HOSO_2OH \longrightarrow H_3C{-}CH(OSO_2OH)CH_2CH_3$ 1-Butene Sulfuric acid *sec*-Butyl hydrogen sulfate
Acid-catalyzed hydration (Section 6.9) Addition of water to the double bond of an alkene takes place in aqueous acid. Addition occurs according to Markovnikov's rule. A carbocation is an intermediate and is captured by a molecule of water acting as a nucleophile.	$RCH{=}CR'_2 + H_2O \xrightarrow{H^+} RCH_2CR'_2(OH)$ Alkene Water Alcohol $H_2C{=}C(CH_3)_2 \xrightarrow{50\%\ H_2SO_4/H_2O} (CH_3)_3COH$ 2-Methylpropene *tert*-Butyl alcohol (55–58%)
Hydroboration-oxidation (Sections 6.11–6.13) This two-step sequence achieves hydration of alkenes in a stereospecific syn manner, with a regioselectivity opposite to Markovnikov's rule. An organoborane is formed by electrophilic addition of diborane to an alkene. Oxidation of the organoborane intermediate with hydrogen peroxide completes the process. Rearrangements do not occur.	$RCH{=}CR'_2 \xrightarrow[\text{2. } H_2O_2,\ HO^-]{\text{1. } B_2H_6,\ \text{diglyme}} RCH(OH)CHR'_2$ Alkene Alcohol $(CH_3)_2CHCH_2CH{=}CH_2 \xrightarrow[\text{2. } H_2O_2,\ HO^-]{\text{1. } H_3B\cdot THF} (CH_3)_2CHCH_2CH_2CH_2OH$ 4-Methyl-1-pentene 4-Methyl-1-pentanol (80%)

Continued

TABLE 6.6 Addition Reactions of Alkenes (*Continued*)

Reaction (section) and comments	General equation and specific example
Addition of halogens (Sections 6.14–6.16) Bromine and chlorine add to alkenes to form vicinal dihalides. A cyclic halonium ion is an intermediate. Anti addition is observed.	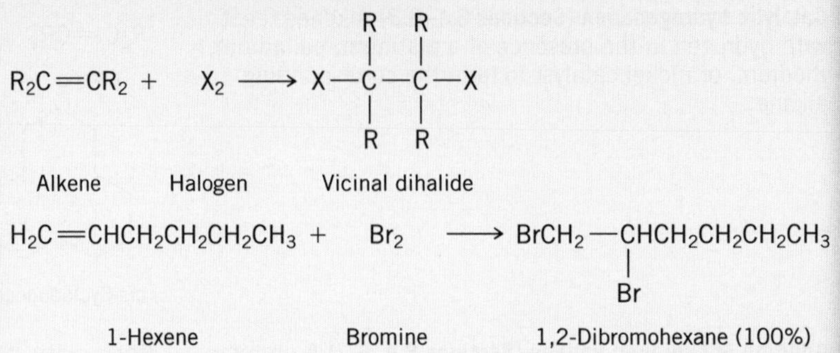Alkene Halogen Vicinal dihalide 1-Hexene Bromine 1,2-Dibromohexane (100%)
Halohydrin formation (Section 6.17) When treated with bromine or chlorine in aqueous solution, alkenes are converted to vicinal halohydrins. A halonium ion is an intermediate. The halogen adds to the carbon that has the greater number of hydrogens. Addition is anti.	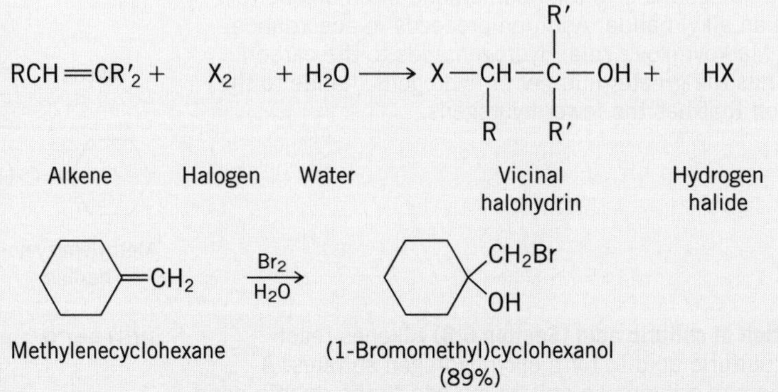Alkene Halogen Water Vicinal halohydrin Hydrogen halide Methylenecyclohexane (1-Bromomethyl)cyclohexanol (89%)
Epoxidation (Section 6.19) Peroxy acids transfer oxygen to the double bond of alkenes to yield epoxides. The reaction is a stereospecific syn addition.	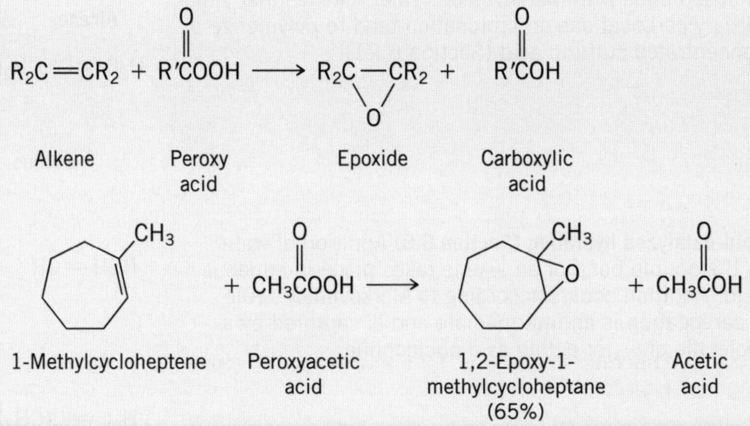Alkene Peroxy acid Epoxide Carboxylic acid 1-Methylcycloheptene Peroxyacetic acid 1,2-Epoxy-1-methylcycloheptane (65%) Acetic acid

The standard free energy change $\Delta G°$ is related to the equilibrium constant K by the equation

$$\Delta G° = -RT\ln K$$

Sections 6.11–6.17, 6.19 See Table 6.6

Section 6.18 Hydrogen bromide is unique among the hydrogen halides in that it can add to alkenes either by electrophilic or free-radical addition. Under photochemical conditions or in the presence of peroxides, free-radical addition is observed, and HBr adds to the double bond with a regioselectivity opposite to that of Markovnikov's rule.

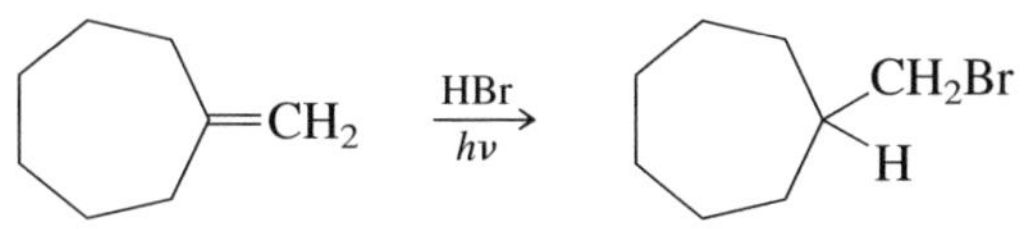

Methylenecycloheptane (Bromomethyl)cycloheptane (61%)

Section 6.20 Alkenes are cleaved to carbonyl compounds by **ozonolysis.** This reaction is useful both for synthesis (preparation of aldehydes, ketones, or carboxylic acids) and analysis. When applied to analysis, the carbonyl compounds are isolated and identified, allowing the substituents attached to the double bond to be deduced.

$$\underset{\text{3-Ethyl-2-pentene}}{CH_3CH{=}C(CH_2CH_3)_2} \xrightarrow[\text{2. Zn, } H_2O]{\text{1. } O_3} \underset{\text{Acetaldehyde}}{CH_3\overset{\overset{O}{\|}}{C}H} + \underset{\text{3-Pentanone}}{CH_3CH_2\overset{\overset{O}{\|}}{C}CH_2CH_3}$$

Section 6.21 In their **polymerization,** many individual alkene molecules combine to give a high-molecular-weight product. Among the methods for alkene polymerization, *cationic polymerization, coordination polymerization,* and *free-radical polymerization* are the most important. An example of cationic polymerization is

$$\underset{\text{2-Methylpropene}}{2n(CH_3)_2C{=}CH_2} \xrightarrow{H^+} \underset{\text{Polyisobutylene}}{\left(\underset{CH_3}{\overset{CH_3}{\underset{|}{\overset{|}{C}}}}-CH_2-\underset{CH_3}{\overset{CH_3}{\underset{|}{\overset{|}{C}}}}-CH_2 \right)_n}$$

PROBLEMS

6.26 Write the structure of the major organic product formed in the reaction of 1-pentene with each of the following:

(a) Hydrogen chloride
(b) Hydrogen bromide
(c) Hydrogen bromide in the presence of peroxides
(d) Hydrogen iodide
(e) Dilute sulfuric acid
(f) Diborane in diglyme, followed by basic hydrogen peroxide
(g) Bromine in carbon tetrachloride
(h) Bromine in water
(i) Peroxyacetic acid
(j) Ozone
(k) Product of part (j) treated with zinc and water
(l) Product of part (j) treated with dimethyl sulfide $(CH_3)_2S$.

6.27 Repeat Problem 6.26 for 2-methyl-2-butene.

6.28 Repeat Problem 6.26 for 1-methylcyclohexene.

6.29 Match the following alkenes with the appropriate heats of hydrogenation:

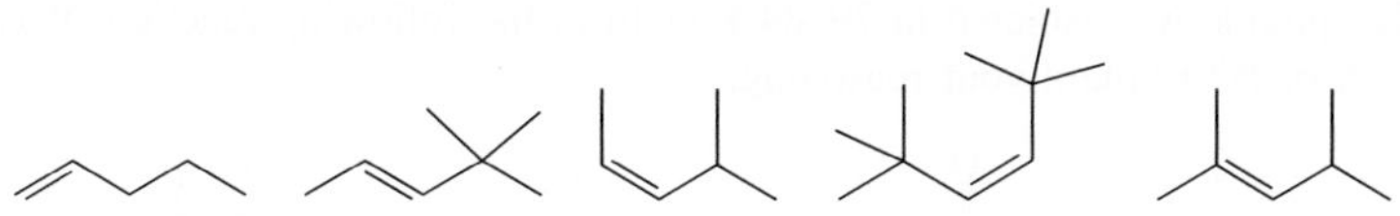

A B C D E

Heats of hydrogenation in kJ/mol (kcal/mol): 151(36.2); 122(29.3); 114(27.3); 111(26.5); 105(25.1).

6.30 (a) How many alkenes yield 2,2,3,4,4-pentamethylpentane on catalytic hydrogenation?
(b) How many yield 2,3-dimethylbutane?
(c) How many yield methylcyclobutane?

6.31 Two alkenes undergo hydrogenation to yield a mixture of *cis*- and *trans*-1,4-dimethylcyclohexane. Which two are these? A third, however, gives only *cis*-1,4-dimethylcyclohexane. What compound is this?

6.32 Specify reagents suitable for converting 3-ethyl-2-pentene to each of the following:
(a) 2,3-Dibromo-3-ethylpentane
(b) 3-Chloro-3-ethylpentane
(c) 2-Bromo-3-ethylpentane
(d) 3-Ethyl-3-pentanol
(e) 3-Ethyl-2-pentanol
(f) 2,3-Epoxy-3-ethylpentane
(g) 3-Ethylpentane

6.33 (a) Which primary alcohol of molecular formula $C_5H_{12}O$ cannot be prepared from an alkene by hydroboration–oxidation? Why?
(b) Write equations describing the preparation of three isomeric primary alcohols of molecular formula $C_5H_{12}O$ from alkenes.
(c) Write equations describing the preparation of the tertiary alcohol of molecular formula $C_5H_{12}O$ by acid-catalyzed hydration of two different alkenes.

6.34 All the following reactions have been reported in the chemical literature. Give the structure of the principal organic product in each case.

(a) $CH_3CH_2CH{=}CHCH_2CH_3 + HBr \xrightarrow{\text{no peroxides}}$

(b) $(CH_3)_2CHCH_2CH_2CH_2CH{=}CH_2 \xrightarrow[\text{peroxides}]{HBr}$

(c) 2-*tert*-Butyl-3,3-dimethyl-1-butene $\xrightarrow[\text{2. } H_2O_2,\ HO^-]{\text{1. } B_2H_6}$

(d) 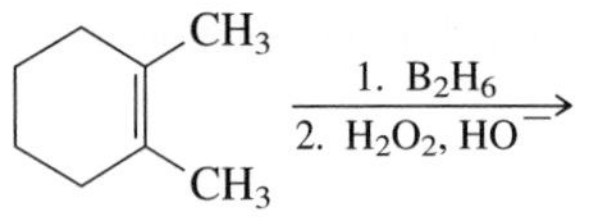

(e) $H_2C{=}C(CH_3)CH_2CH_2CH_3 + Br_2 \xrightarrow{CHCl_3}$

(f) $(CH_3)_2C{=}CHCH_3 + Br_2 \xrightarrow{H_2O}$

(g) 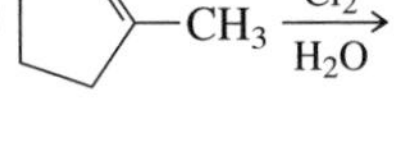

(h) $(CH_3)_2C{=}C(CH_3)_2 + CH_3\overset{O}{\overset{\|}{C}}OOH \longrightarrow$

(i) 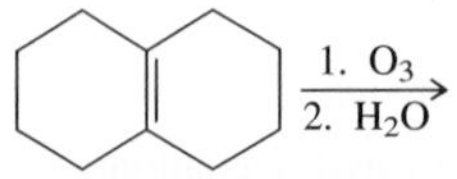

6.35 A single epoxide was isolated in 79–84% yield in the following reaction. Was this epoxide A or B? Explain your reasoning.

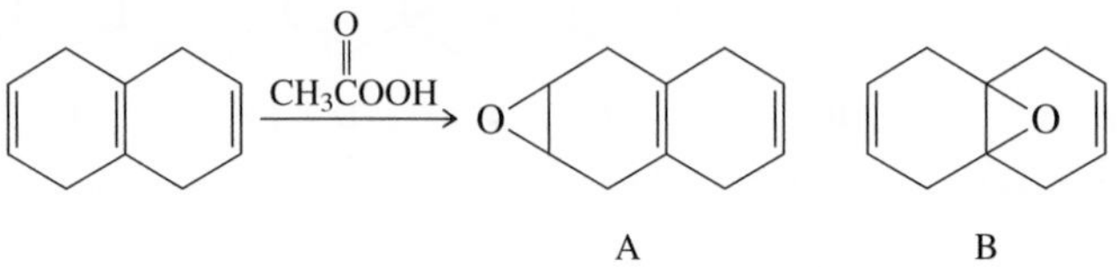

A B

6.36 Suggest a sequence of reactions suitable for preparing each of the following compounds from the indicated starting material. You may use any necessary organic or inorganic reagents.

(a) 1-Propanol from 2-propanol

(b) 1-Bromopropane from 2-bromopropane

(c) 1,2-Dibromopropane from 2-bromopropane

(d) 1-Bromo-2-propanol from 2-propanol

(e) 1-Bromo-2-methyl-2-propanol from *tert*-butyl bromide

(f) 1,2-Epoxypropane from 2-propanol

(g) *tert*-Butyl alcohol from isobutyl alcohol

(h) *tert*-Butyl iodide from isobutyl iodide

(i) *trans*-2-Chlorocyclohexanol from cyclohexyl chloride

(j) Cyclopentyl iodide from cyclopentane

(k) *trans*-1,2-Dichlorocyclopentane from cyclopentane

(l) $H\overset{O}{\overset{\|}{C}}CH_2CH_2CH_2\overset{O}{\overset{\|}{C}}H$ from cyclopentanol

6.37 Suggest reasonable mechanisms for each of the following reactions. Use curved arrows to show electron flow.

(a) 
+ HBr
ROOR
Br

(b) 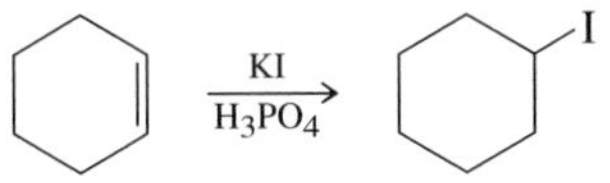
KI
H3PO4
I

(c) 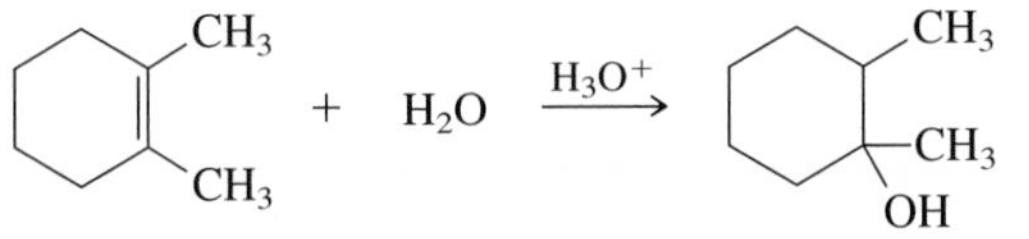
CH3
CH3
+ H2O
H3O+
CH3
CH3
OH
(mixture of stereoisomers)

(d) 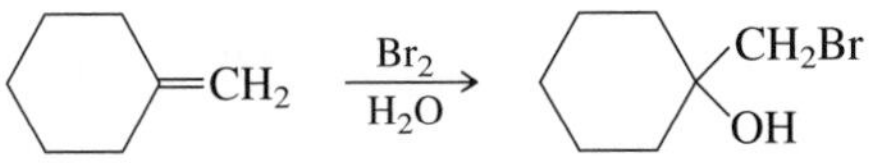
CH2
Br2
H2O
CH2Br
OH

6.38 Two different compounds having the molecular formula $C_8H_{15}Br$ are formed when 1,6-dimethylcyclohexene reacts with hydrogen bromide in the dark and in the absence of peroxides. The same two compounds are formed from 1,2-dimethylcyclohexene. What are these two compounds?

6.39 On catalytic hydrogenation over a rhodium catalyst, the compound shown gave a mixture containing *cis*-1-*tert*-butyl-4-methylcyclohexane (88%) and *trans*-1-*tert*-butyl-4-methylcyclohexane (12%). With this stereochemical result in mind, consider the reactions in (a) and (b).

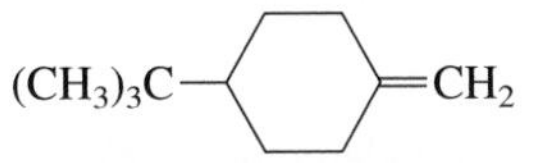
(CH3)3C
CH2

4-*tert*-Butyl(methylene)cyclohexane

(a) What two products are formed in the epoxidation of 4-*tert*-butyl(methylene)cyclohexane? Which one do you think will predominate?

(b) What two products are formed in the hydroboration–oxidation of 4-*tert*-butyl(methylene)-cyclohexane? Which one do you think will predominate?

6.40 Compound A undergoes catalytic hydrogenation much faster than does compound B. Why?

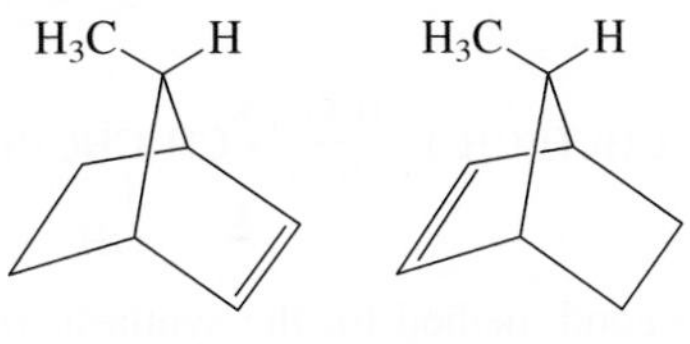
H3C
H
H3C
H

A B

6.41 Catalytic hydrogenation of 1,4-dimethylcyclopentene yields a mixture of two products. Identify them. One of them is formed in much greater amounts than the other (observed ratio = 10:1). Which one is the major product?

6.42 What two products can be formed by syn addition of hydrogen to 2,3-dimethylbicyclo[2.2.1]-2-heptene?

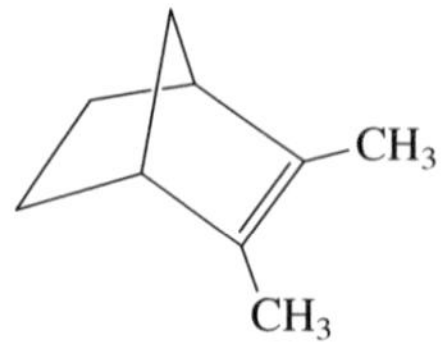

2,3-Dimethylbicyclo[2.2.1]-2-heptene

6.43 Hydrogenation of 3-carene is, in principle, capable of yielding two stereoisomeric products. Write their structures. Only one of them was actually obtained on catalytic hydrogenation over platinum. Which one do you think is formed?

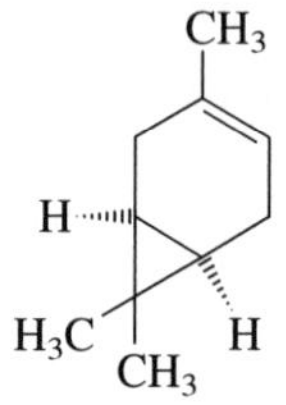

3-Carene

6.44 Complete the following table by adding + and − signs to the $\Delta H°$ and $\Delta S°$ columns so as to correspond to the effect of temperature on a reversible reaction.

Reaction is	Sign of $\Delta H°$	Sign of $\Delta S°$
(a) Exergonic at all temperatures		
(b) Exergonic at low temperature; endergonic at high temperature		
(c) Endergonic at all temperatures		
(d) Endergonic at low temperature; exergonic at high temperature		

6.45 The iodination of ethylene at 25°C is characterized by the thermodynamic values shown.

$$H_2C{=}CH_2(g) + I_2(g) \rightleftharpoons ICH_2CH_2I(g) \qquad \Delta H° = -48\ \text{kJ};\ \Delta S° = -0.13\ \text{kJ/K}$$

(a) Calculate $\Delta G°$ and K at 25°C.

(b) Is the reaction exergonic or endergonic at 25°C?

(c) What happens to K as the temperature is raised?

6.46 In a widely used industrial process, the mixture of ethylene and propene that is obtained by dehydrogenation of natural gas is passed into concentrated sulfuric acid. Water is added, and the solution is heated to hydrolyze the alkyl hydrogen sulfate. The product is almost exclusively a single alcohol. Is this alcohol ethanol, 1-propanol, or 2-propanol? Why is this particular one formed almost exclusively?

6.47 On the basis of the mechanism of acid-catalyzed hydration, can you suggest a reason why the reaction

$$H_2C{=}CHCH(CH_3)_2 \xrightarrow[H_2O]{H_2SO_4} CH_3\underset{\displaystyle OH}{\underset{|}{C}}HCH(CH_3)_2$$

would probably *not* be a good method for the synthesis of 3-methyl-2-butanol?

6.48 As a method for the preparation of alkenes, a weakness in the acid-catalyzed dehydration of alcohols is that the initially formed alkene (or mixture of alkenes) sometimes isomerizes under the conditions of its formation. Write a stepwise mechanism showing how 2-methyl-1-butene might isomerize to 2-methyl-2-butene in the presence of sulfuric acid.

6.49 The reaction of thiocyanogen (N≡CS—SC≡N) with *cis*-cyclooctene proceeds by anti addition.

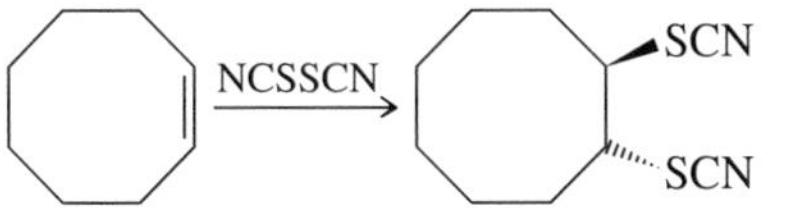
NCSSCN
SCN
SCN

A bridged *sulfonium ion* is presumed to be an intermediate. Write a stepwise mechanism for this reaction.

6.50 On the basis of the mechanism of cationic polymerization, predict the alkenes of molecular formula $C_{12}H_{24}$ that can most reasonably be formed when 2-methylpropene [$(CH_3)_2C{=}CH_2$] is treated with sulfuric acid.

6.51 On being heated with a solution of sodium ethoxide in ethanol, compound A ($C_7H_{15}Br$) yielded a mixture of two alkenes B and C, each having the molecular formula C_7H_{14}. Catalytic hydrogenation of the major isomer B or the minor isomer C gave only 3-ethylpentane. Suggest structures for compounds A, B, and C consistent with these observations.

6.52 Compound A ($C_7H_{15}Br$) is not a primary alkyl bromide. It yields a single alkene (compound B) on being heated with sodium ethoxide in ethanol. Hydrogenation of compound B yields 2,4-dimethylpentane. Identify compounds A and B.

6.53 Compounds A and B are isomers of molecular formula $C_9H_{19}Br$. Both yield the same alkene C as the exclusive product of elimination on being treated with potassium *tert*-butoxide in dimethyl sulfoxide. Hydrogenation of alkene C gives 2,3,3,4-tetramethylpentane. What are the structures of compounds A and B and alkene C?

6.54 Alcohol A ($C_{10}H_{18}O$) is converted to a mixture of alkenes B and C on being heated with potassium hydrogen sulfate ($KHSO_4$). Catalytic hydrogenation of B and C yields the same product. Assuming that dehydration of alcohol A proceeds without rearrangement, deduce the structures of alcohol A and alkene C.

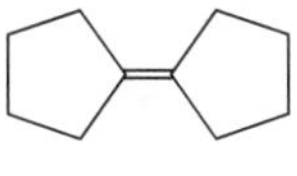

Compound B

6.55 A mixture of three alkenes (A, B, and C) was obtained by dehydration of 1,2-dimethylcyclohexanol. The composition of the mixture was A (3%), B (31%), and C (66%). Catalytic hydrogenation of A, B, or C gave 1,2-dimethylcyclohexane. The three alkenes can be equilibrated by heating with sulfuric acid to give a mixture containing A (0%), B (15%), and C (85%). Identify A, B, and C.

6.56 Reaction of 3,3-dimethyl-1-butene with hydrogen iodide yields two compounds A and B, each having the molecular formula $C_6H_{13}I$, in the ratio A:B = 90:10. Compound A, on being heated with potassium hydroxide in *n*-propyl alcohol, gives only 3,3-dimethyl-1-butene. Compound B undergoes elimination under these conditions to give 2,3-dimethyl-2-butene as the major product. Suggest structures for compounds A and B, and write a reasonable mechanism for the formation of each.

6.57 Dehydration of 2,2,3,4,4-pentamethyl-3-pentanol gave two alkenes A and B. Ozonolysis of the lower boiling alkene A gave formaldehyde ($H_2C{=}O$) and 2,2,4,4-tetramethyl-3-pentanone. Ozonolysis of B gave formaldehyde and 3,3,4,4-tetramethyl-2-pentanone. Identify A and B, and suggest an explanation for the formation of B in the dehydration reaction.

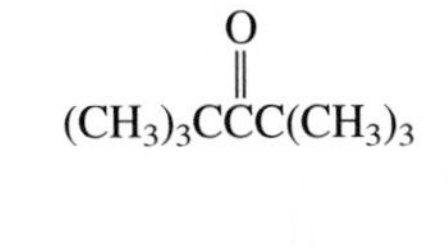
O
$(CH_3)_3CCC(CH_3)_3$

2,2,4,4-Tetramethyl-3-pentanone

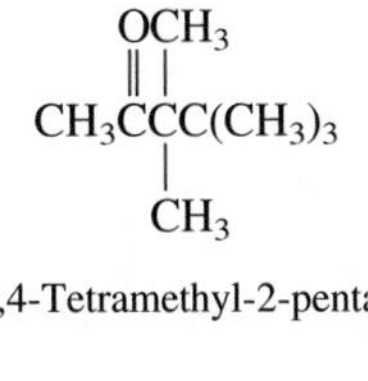
O CH₃
$CH_3CCC(CH_3)_3$
CH_3

3,3,4,4-Tetramethyl-2-pentanone

6.58 Compound A ($C_7H_{13}Br$) is a tertiary bromide. On treatment with sodium ethoxide in ethanol, A is converted into B (C_7H_{12}). Ozonolysis of B gives C as the only product. Deduce the structures of A and B. What is the symbol for the reaction mechanism by which A is converted to B under the reaction conditions?

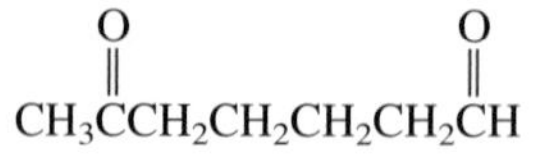

Compound C

6.59 East Indian sandalwood oil contains a hydrocarbon given the name *santene* (C_9H_{14}). Ozonolysis of santene gives compound A. What is the structure of santene?

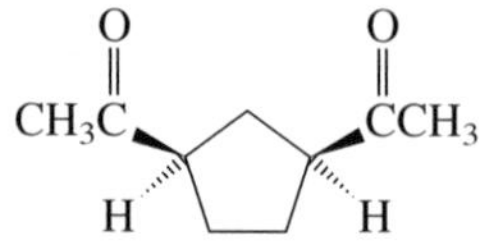

Compound A

6.60 *Sabinene* and Δ^3-*carene* are isomeric natural products with the molecular formula $C_{10}H_{16}$. (a) Ozonolysis of sabinene followed by hydrolysis in the presence of zinc gives compound A. What is the structure of sabinene? What other compound is formed on ozonolysis? (b) Ozonolysis of Δ^3-carene gives compound B. What is the structure of Δ^3-carene?

Compound A

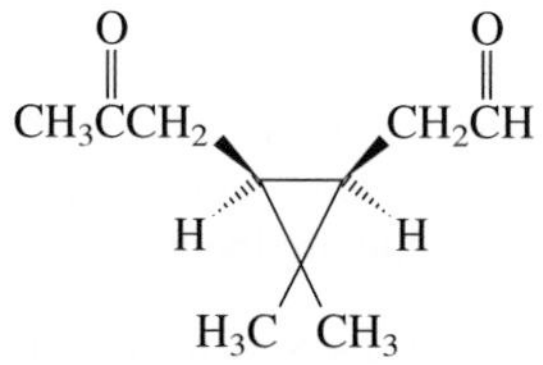

Compound B

6.61 The sex attractant by which the female housefly attracts the male has the molecular formula $C_{23}H_{46}$. Catalytic hydrogenation yields an alkane of molecular formula $C_{23}H_{48}$. Ozonolysis yields

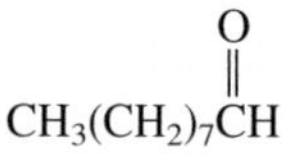 and 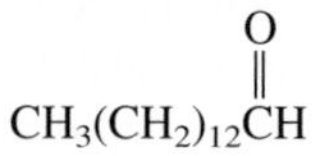

What is the structure of the housefly sex attractant?

6.62 A certain compound of molecular formula $C_{19}H_{38}$ was isolated from fish oil and from plankton. On hydrogenation it gave 2,6,10,14-tetramethylpentadecane. Ozonolysis gave $(CH_3)_2C{=}O$ and a 16-carbon aldehyde. What is the structure of the natural product? What is the structure of the aldehyde?

6.63 The sex attractant of the female arctiid moth contains, among other components, a compound of molecular formula $C_{21}H_{40}$ that yields

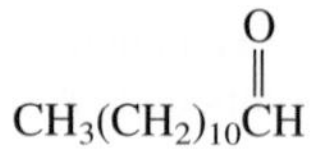 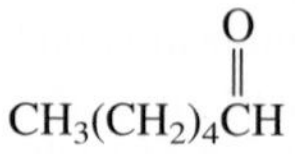 and 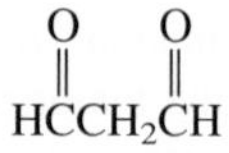

on ozonolysis. What is the constitution of this material?

6.64 The following reaction was performed as part of a research program sponsored by the National Institutes of Health to develop therapeutic agents for the treatment of cocaine addiction. Using what you have seen about the reactions of halogens with alkenes, propose a mechanism for this process.

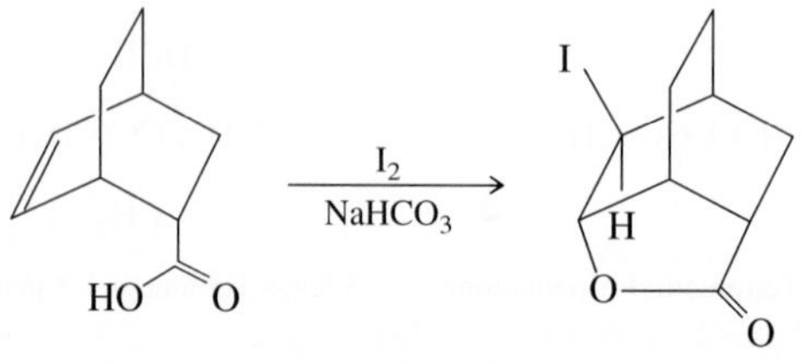

Descriptive Passage and Interpretive Problems 6

Oxymercuration

Concerns about mercury's toxicity have led to decreased use of mercury-based reagents in synthetic organic chemistry. Alternatives exist for many of the transformations formerly carried out with mercury compounds while carrying much less risk. The chemistry of several of the reactions, however, is sufficiently interesting to examine here.

Among the synthetically useful reactions of Hg(II) salts with organic compounds, the most familiar is a two-stage procedure for alkene hydration called **oxymercuration–demercuration.** Its application in the conversion of 3,3-dimethyl-1-butene to 3,3-dimethyl-2-butanol illustrates the procedure.

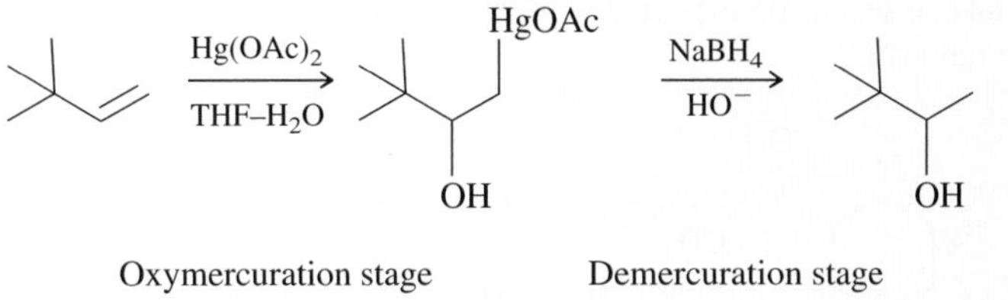

The reaction is performed in two operations, the first of which is oxymercuration. In this stage the alkene is treated with mercury(II) acetate [$Hg(O_2CCH_3)_2$, abbreviated as $Hg(OAc)_2$]. Mercury(II) acetate is a source of the electrophile $^+$HgOAc, which bonds to C-1 of the alkene. The oxygen of water, one of the components in the $THF–H_2O$ solvent mixture, bonds to C-2. The demercuration operation uses sodium borohydride ($NaBH_4$, a reducing agent) to convert C—Hg to C—H.

From the overall reaction, we see that oxymercuration–demercuration

1. accomplishes hydration of the double bond in accordance with Markovnikov's rule, and
2. carbocation rearrangements do not occur.

Additional information from stereochemical studies with other alkenes has established that

3. anti addition of HgOAc and OH characterizes the oxymercuration stage, and
4. the replacement of HgOAc by H in the demercuration stage is not stereospecific.

The structure of the intermediate in oxymercuration has received much attention and can be approached by considering what is likely to happen when the electrophile $^+$HgOAc reacts with the double bond of an alkene.

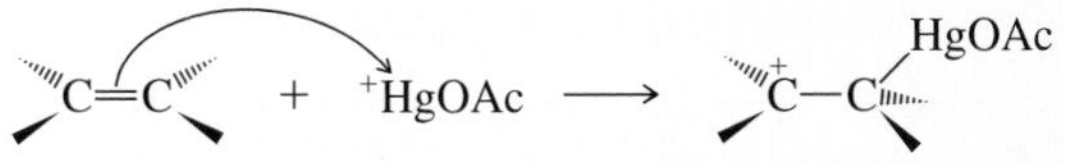

Recall from Section 4.10 that electrons in bonds that are β to a positively charged carbon stabilize a carbocation by hyperconjugation.

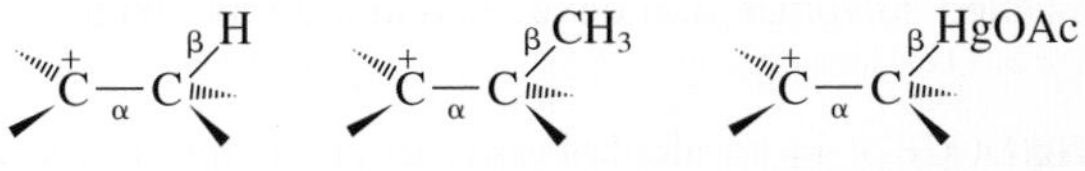

The electrons in a C—Hg σ bond are more loosely held than C—H or C—C electrons, making stabilization by hyperconjugation more effective for β-C—Hg than for β-C—H or β-C—C. Hyperconjugative stabilization of the intermediate in oxymercuration is normally shown using dashed lines to represent partial bonds. The intermediate is referred to as a "bridged" *mercurinium ion.*

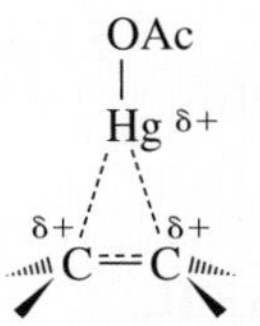

The problems that follow explore various synthetic aspects of oxymercuration–demercuration. Experimental procedures sometimes vary depending on the particular transformation. The source of the electrophile may be a mercury(II) salt other than $Hg(OAc)_2$, the nucleophile may be other than H_2O, and the reaction may be intramolecular rather than intermolecular.

6.65 Oxymercuration of methylcyclopentene gives which of the following products?

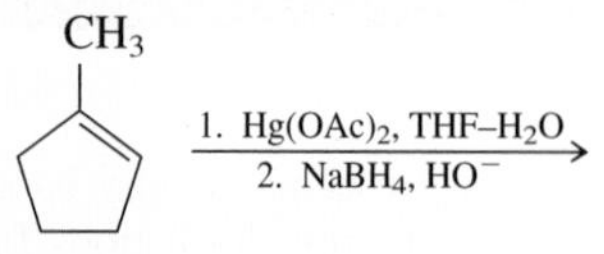

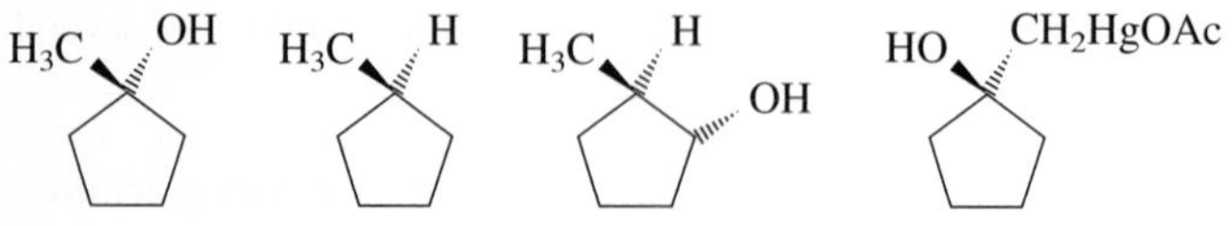

A. B. C. D.

6.66 Which alkene would be expected to give the following alcohol by oxymercuration–demercuration?

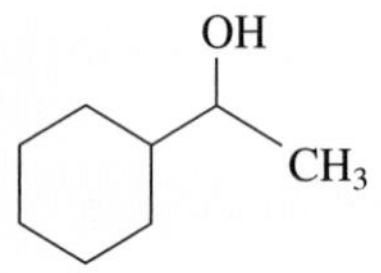

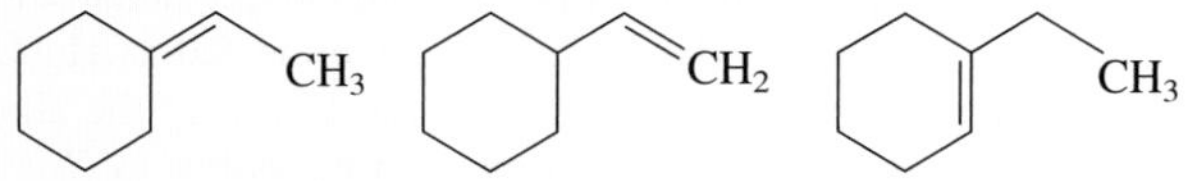

A. B. C.

6.67 Given that 2-methyl-1-pentene undergoes oxymercuration–demercuration approximately 35 times faster than 2-methyl-2-pentene, predict the major product from oxymercuration–demercuration of limonene.

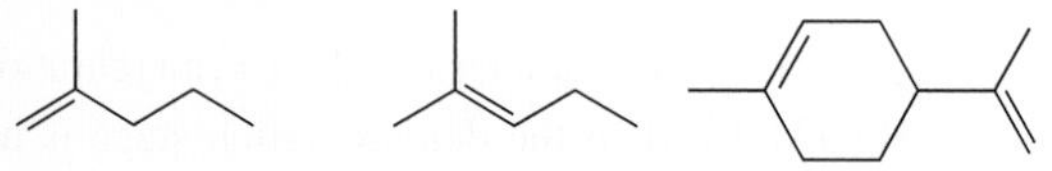

2-Methyl-1-pentene 2-Methyl-2-pentene Limonene

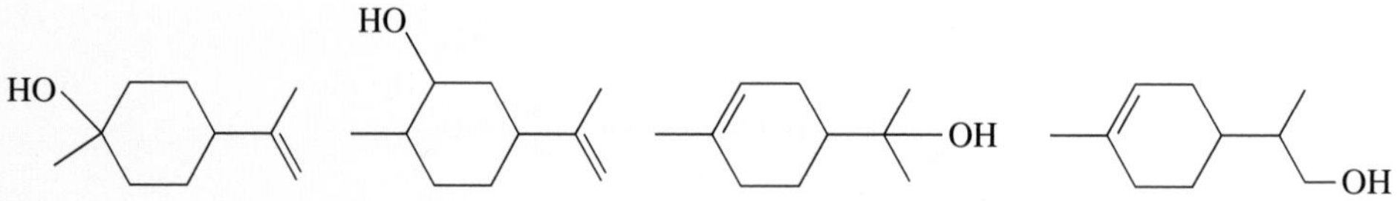

A. B. C. D.

6.68 In a procedure called solvomercuration–demercuration an alkene is treated with $Hg(OAc)_2$ or $Hg(O\overset{\overset{O}{\|}}{C}CF_3)_2$ in an alcohol solvent rather than in the THF–H_2O mixture used in oxymercuration. The oxygen of the alcohol solvent reacts with the mercurinium ion during solvomercuration. What is the product of the following solvomercuration–demercuration?

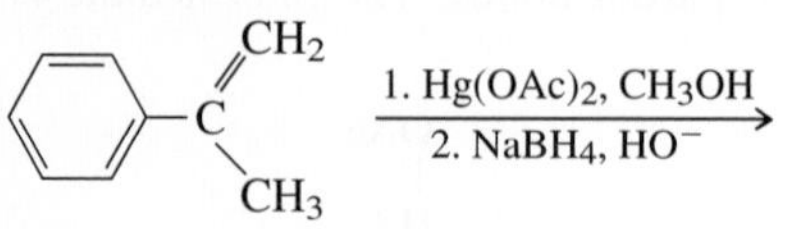

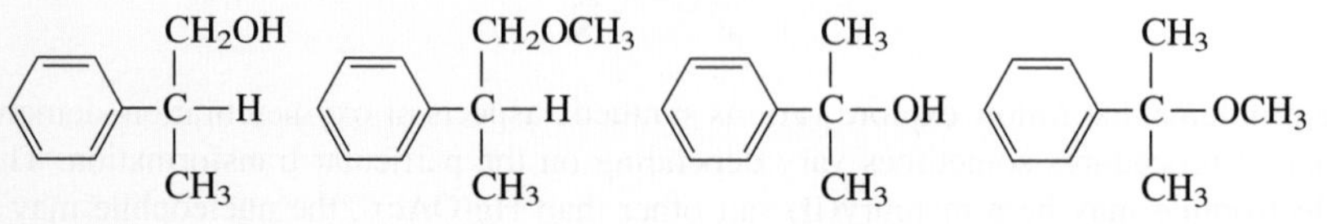

A. B. C. D.

6.69 From among the same product choice as Problem 6.68, which one is the major product of the following reaction?

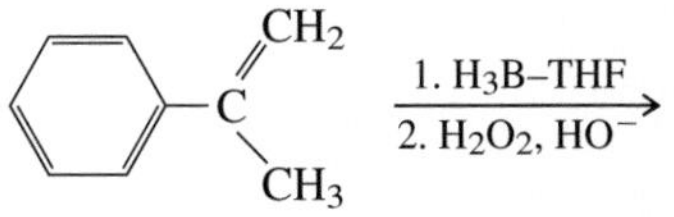
CH2
C
CH3
1. H3B–THF
2. H2O2, HO−

6.70 Oxymercuration–demercuration of allyl alcohol gives 1,2-propanediol.

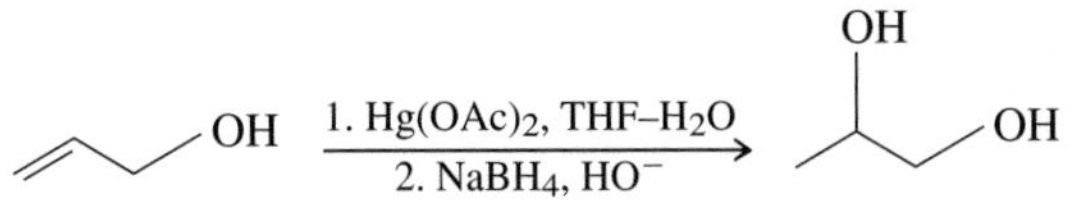
OH
1. Hg(OAc)2, THF–H2O
2. NaBH4, HO−
OH
OH

Under the same conditions, however, 4-penten-1-ol yields a compound having the molecular formula $C_5H_{10}O$.

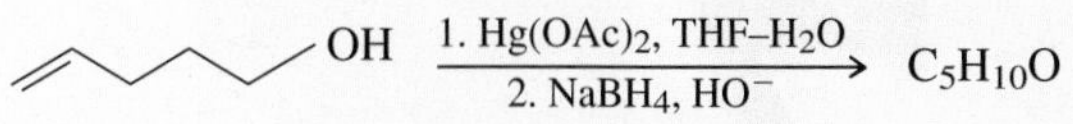
OH
1. Hg(OAc)2, THF–H2O
2. NaBH4, HO−
C5H10O

What is the most reasonable structure for the product of this reaction?

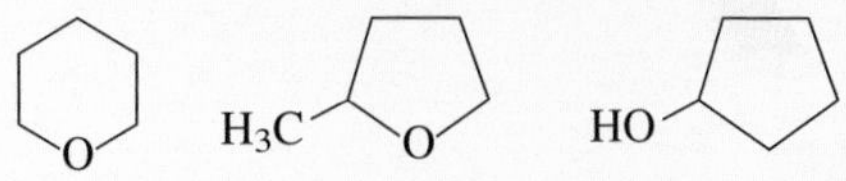
O
H3C
O
HO

A. B. C.

7 Stereochemistry

Chapter Outline

Bromochlorofluoromethane molecules come in right- and left-handed versions.

STEREOCHEMISTRY is chemistry in three dimensions. Its foundations were laid by Jacobus van't Hoff* and Joseph Achille Le Bel in 1874. Van't Hoff and Le Bel independently proposed that the four bonds to carbon were directed toward the corners of a tetrahedron. One consequence of a tetrahedral arrangement of bonds to carbon is that two compounds may be different because the arrangement of their atoms in space is different. Isomers that have the same constitution but differ in the spatial arrangement of their atoms are called **stereoisomers.** We have already had considerable experience with certain types of stereoisomers—those involving cis and trans substitution in alkenes and in cycloalkanes.

Our major objectives in this chapter are to develop a feeling for molecules as three-dimensional objects and to become familiar with stereochemical principles, terms, and notation. A full understanding of organic and biological chemistry requires an awareness of the spatial requirements for interactions between molecules; this chapter provides the basis for that understanding.

7.1 Molecular Chirality: Enantiomers

Everything has a mirror image, but not all things are superimposable on their mirror images. Mirror-image superimposability characterizes many objects we use every day. Cups and saucers, forks and spoons, chairs and beds are all identical with their mirror images. Many other objects though—and this is the more interesting case—are not. Your left hand and your right hand, for example, are mirror images of each other but can't be made to coincide point for point, palm to palm, knuckle to knuckle, in three dimensions. In 1894, William Thomson (Lord Kelvin) coined a word for this property. He defined an object as **chiral** if it is not superimposable on its mirror image. Applying Thomson's term to chemistry, we say that a *molecule is chiral if its two mirror-image forms are not superimposable in three dimensions.* The word *chiral* is derived from the Greek word *cheir,* meaning "hand," and it is entirely appropriate to speak of the "handedness" of molecules. The opposite of chiral is **achiral.** A molecule that *is* superimposable on its mirror image is achiral.

*Van't Hoff was the recipient of the first Nobel Prize in Chemistry in 1901 for his work in chemical dynamics and osmotic pressure—two topics far removed from stereochemistry.

In organic chemistry, chirality most often occurs in molecules that contain a carbon that is attached to four different groups. An example is bromochlorofluoromethane (BrClFCH).

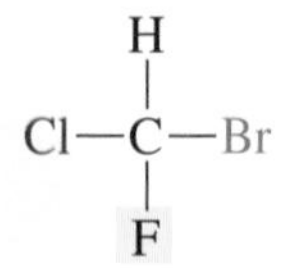

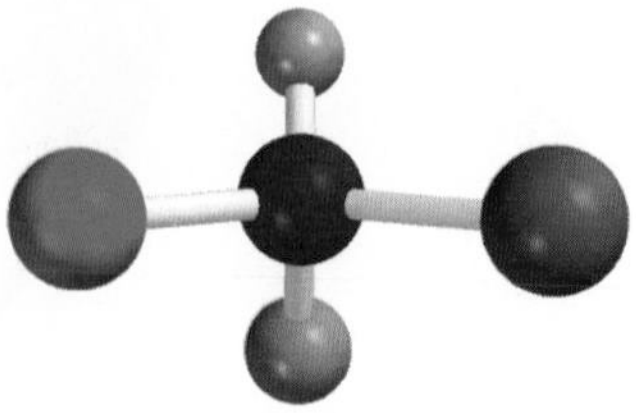

Bromochlorofluoromethane

As shown in Figure 7.1, the two mirror images of bromochlorofluoromethane cannot be superimposed on each other. *Because the two mirror images of bromochlorofluoromethane are not superimposable, BrClFCH is chiral.*

Figure 7.1

A molecule with four different groups attached to a single carbon is chiral. Its two mirror-image forms are not superimposable.

(*a*) Structures A and B are mirror-image representations of bromochlorofluoromethane (BrClFCH).

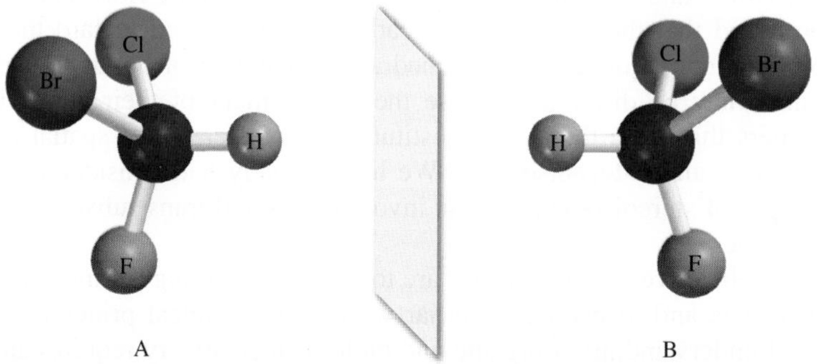

(*b*) To test for superimposability, reorient B by turning it 180°.

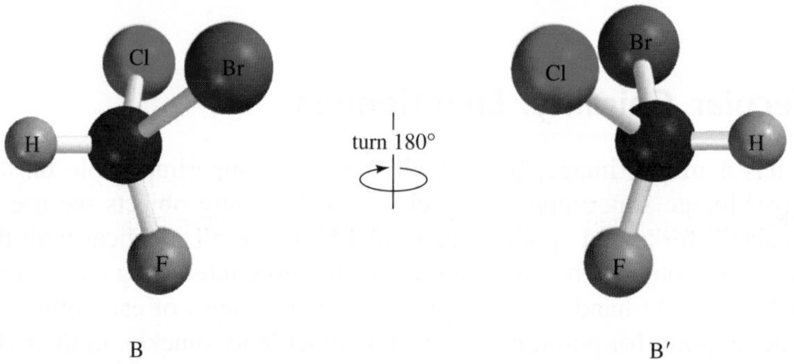

(*c*) Compare A and B′. The two do not match. A and B′ cannot be superimposed on each other. Bromochlorofluoromethane is therefore a chiral molecule. The two mirror-image forms are called enantiomers.

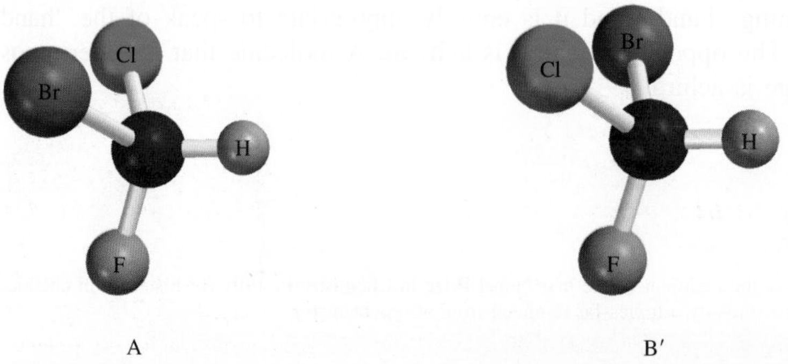

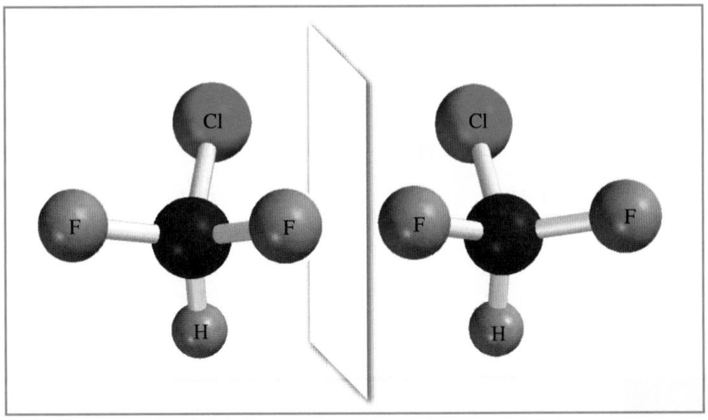

Figure 7.2

Mirror-image forms of chlorodifluoromethane are superimposable on each other. Chlorodifluoromethane is achiral.

The mirror images of bromochlorofluoromethane have the same constitution. That is, the atoms are connected in the same order. But they differ in the arrangement of their atoms in space; they are **stereoisomers.** Stereoisomers that are related as an object and its nonsuperimposable mirror image are classified as **enantiomers.** The word *enantiomer* describes a particular relationship between two objects. One cannot look at a single molecule in isolation and ask if it is an enantiomer any more than one can look at an individual human being and ask, "Is that person a cousin?" Furthermore, just as an object has one, and only one, mirror image, a chiral molecule can have one, and only one, enantiomer.

Bromochlorofluoromethane is a known compound, and samples selectively enriched in each enantiomer have been described in the chemical literature. In 1989 two chemists at Polytechnic University (Brooklyn, New York) described a method for the preparation of BrClFCH that is predominantly one enantiomer.

Notice in Figure 7.1*c,* where the two enantiomers of bromochlorofluoromethane are similarly oriented, that the difference between them corresponds to an interchange of the positions of bromine and chlorine. It will generally be true for species of the type C(*w,x,y,z*), where *w, x, y,* and *z* are different atoms or groups, that an exchange of two of them converts a structure to its enantiomer, but an exchange of three returns the original structure, albeit in a different orientation.

Consider next a molecule such as chlorodifluoromethane (ClF_2CH), in which two of the atoms attached to carbon are the same. Figure 7.2 shows two molecular models of ClF_2CH drawn so as to be mirror images. As is evident from these drawings, it is a simple matter to merge the two models so that all the atoms match. *Because mirror-image representations of chlorodifluoromethane are superimposable on each other, ClF_2CH is achiral.*

The surest test for chirality is a careful examination of mirror-image forms for superimposability. Working with models provides the best practice in dealing with molecules as three-dimensional objects and is strongly recommended.

7.2 The Chirality Center

As we've just seen, molecules of the general type

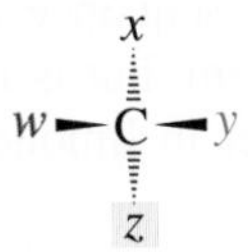

are chiral when *w, x, y,* and *z* are different. In 1996, the IUPAC recommended that a tetrahedral carbon atom that bears four different atoms or groups be called a **chirality center,** which is the term that we will use. Several earlier terms, including *asymmetric center, asymmetric carbon, chiral center, stereogenic center,* and *stereocenter,* are still widely used.

The 1996 IUPAC recommendations for stereochemical terms can be viewed at www.chem.qmw.ac.uk/iupac/stereo

Noting the presence of one (but not more than one) chirality center is a simple, rapid way to determine if a molecule is chiral. For example, C-2 is a chirality center in 2-butanol; it bears H, OH, CH_3, and CH_3CH_2 as its four different groups. By way of contrast, none of the carbon atoms bear four different groups in the achiral alcohol 2-propanol.

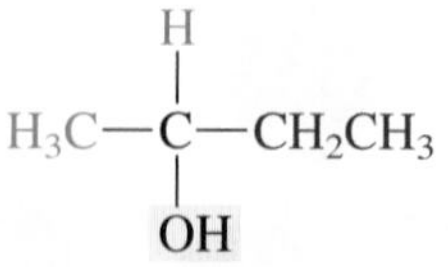

2-Butanol
Chiral; four different groups at C-2

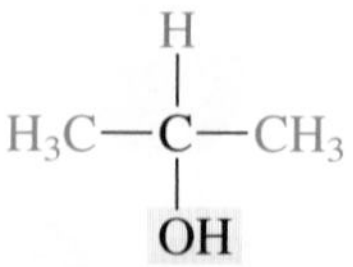

2-Propanol
Achiral; two of the groups at C-2 are the same

Problem 7.1

Examine the following for chirality centers:

(a) 2-Bromopentane
(b) 3-Bromopentane
(c) 1-Bromo-2-methylbutane
(d) 2-Bromo-2-methylbutane

Sample Solution A carbon with four different groups attached to it is a chirality center. (a) In 2-bromopentane, C-2 satisfies this requirement. (b) None of the carbons in 3-bromopentane has four different substituents, and so none of its atoms is a chirality center.

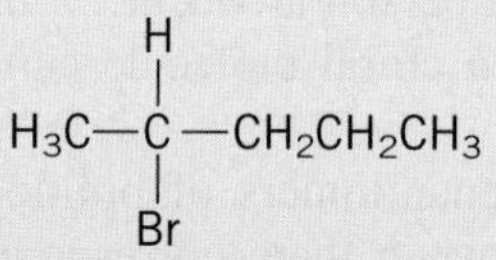

2-Bromopentane

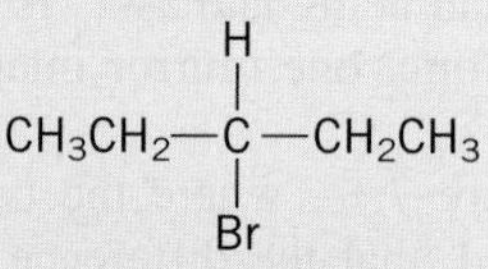

3-Bromopentane

Molecules with chirality centers are very common, both as naturally occurring substances and as the products of chemical synthesis. (Carbons that are part of a double bond or a triple bond can't be chirality centers.)

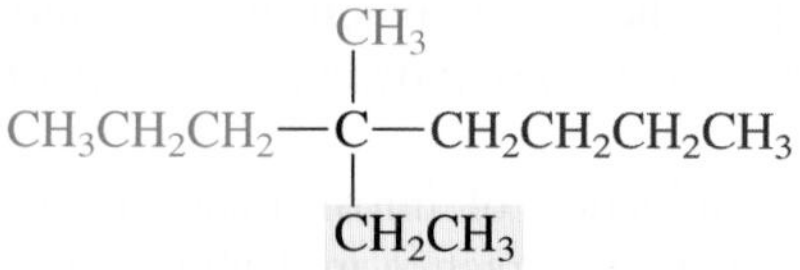

4-Ethyl-4-methyloctane
(a chiral alkane)

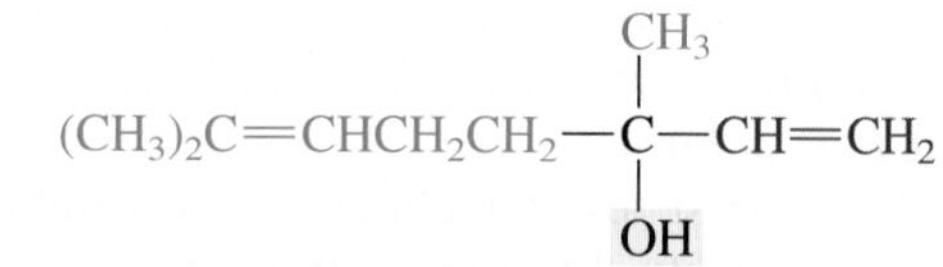

Linalool
(a pleasant-smelling oil obtained from orange flowers)

A carbon atom in a ring can be a chirality center if it bears two different groups and the path traced around the ring from that carbon in one direction is different from that traced in the other. The carbon atom that bears the methyl group in 1,2-epoxypropane, for example, is a chirality center. The sequence of groups is O—CH_2 as one proceeds clockwise around the ring from that atom, but is H_2C—O in the counterclockwise direction. Similarly, C-4 is a chirality center in limonene.

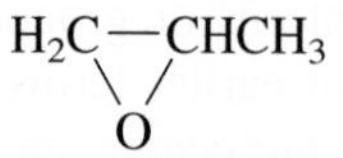

1-2-Epoxypropane
(product of epoxidation of propene)

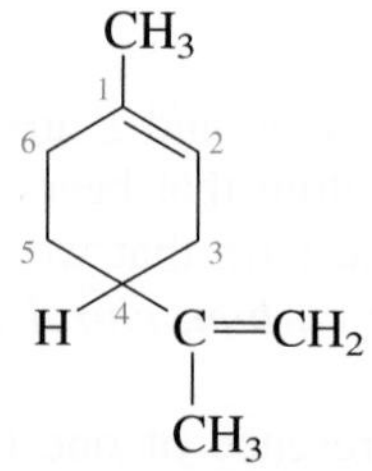

Limonene
(a constituent of lemon oil)

Problem 7.2

Identify the chirality centers, if any, in

(a) 2-Cyclopentenol and 3-cyclopentenol
(b) 1,1,2-Trimethylcyclobutane and 1,1,3-trimethylcyclobutane

Sample Solution (a) The hydroxyl-bearing carbon in 2-cyclopentenol is a chirality center. There is no chirality center in 3-cyclopentenol, because the sequence of atoms $1 \rightarrow 2 \rightarrow 3 \rightarrow 4 \rightarrow 5$ is equivalent regardless of whether one proceeds clockwise or counterclockwise.

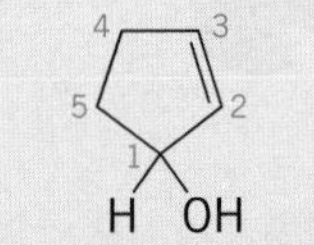

2-Cyclopentenol

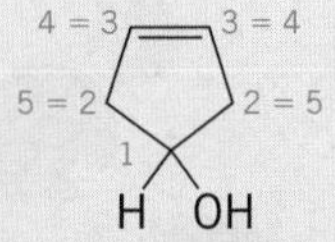

3-Cyclopentenol
(does not have a chirality center)

Even isotopes qualify as different substituents at a chirality center. The stereochemistry of biological oxidation of a derivative of ethane that is chiral because of deuterium ($D = {}^2H$) and tritium ($T = {}^3H$) atoms at carbon, has been studied and shown to proceed as follows:

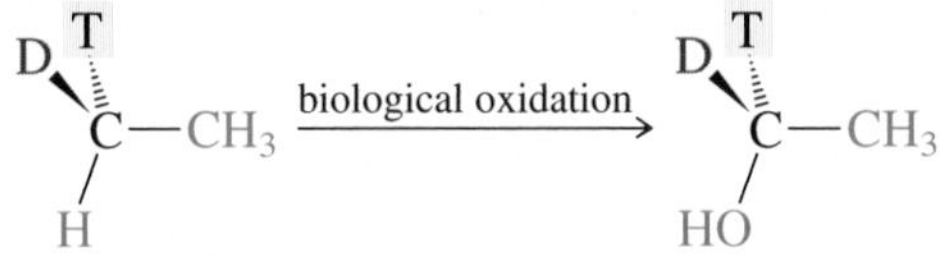

The stereochemical relationship between the reactant and the product, revealed by the isotopic labeling, shows that oxygen becomes bonded to carbon on the same side from which H is lost. As you will see in this and the chapters to come, determining the three-dimensional aspects of a chemical or biochemical transformation can be a subtle, yet powerful, tool for increasing our understanding of how these reactions occur.

One final, very important point: *Everything we have said in this section concerns molecules that have one and only one chirality center; molecules with more than one chirality center may or may not be chiral.* Molecules that have more than one chirality center will be discussed in Sections 7.11 through 7.14.

7.3 Symmetry in Achiral Structures

Certain structural features can sometimes help us determine by inspection whether a molecule is chiral or achiral. For example, a molecule that has a *plane of symmetry* or a *center of symmetry* is superimposable on its mirror image and is achiral.

A **plane of symmetry** bisects a molecule so that one half of the molecule is the mirror image of the other half. The achiral molecule chlorodifluoromethane, for example, has the plane of symmetry shown in Figure 7.3.

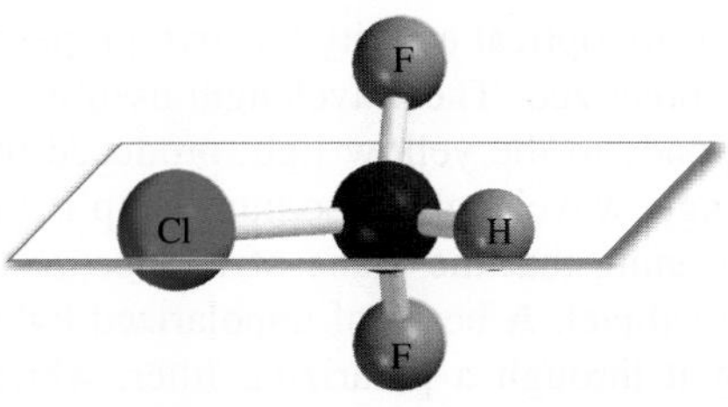

Figure 7.3

A plane of symmetry defined by the atoms H—C—Cl divides chlorodifluoromethane into two mirror-image halves. Note that the Cl and H atoms lie within the plane and reflect upon themselves.

Problem 7.3

Locate any planes of symmetry in each of the following compounds. Which of the compounds are chiral? Which are achiral?

(a) (*E*)-1,2-Dichloroethene
(b) (*Z*)-1,2-Dichloroethene
(c) *cis*-1,2-Dichlorocyclopropane
(d) *trans*-1,2-Dichlorocyclopropane

Sample Solution (a) (*E*)-1,2-Dichloroethene is planar. The molecular plane is a plane of symmetry. Identifying a plane of symmetry tells us the molecule is achiral.

Among the compounds given in Problem 7.3, only (*E*)-1,2-dichloroethene has a center of symmetry.

A point in the center of a molecule is a **center of symmetry** if any line drawn from it to some element of the structure will, when extended an equal distance in the opposite direction, encounter an identical element. *Trans*-1,3-cyclobutanediol has a plane of symmetry as well as a center of symmetry. The center of symmetry is the center of the molecule. A line starting at one of the hydroxyl groups and drawn through the center of the molecule encounters the equidistant hydroxyl group on the opposite side. Mirror images A and B are superimposable, and *trans*-1,3-cyclobutanediol is achiral.

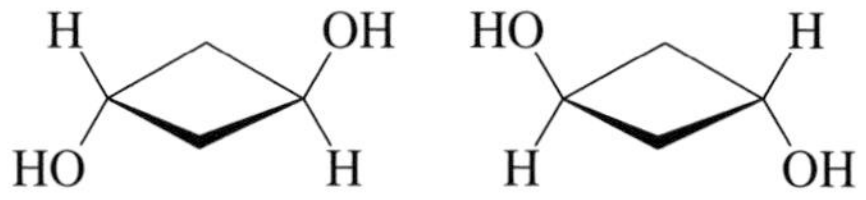
H
OH
HO
H
HO
H
H
OH

A B

Problem 7.4

(a) Where is the plane of symmetry in *trans*-1,3-cyclobutanediol?
(b) Does *cis*-1,3-cyclobutanediol possess a center of symmetry? A plane of symmetry? Is it chiral or achiral?

Planes of symmetry are easier to identify and more common than centers of symmetry. Because either one is sufficient to make a molecule achiral, it makes sense to look first for a plane of symmetry. A molecule lacking a plane or center of symmetry is *likely* to be chiral, but the superimposability test must be applied to be certain.

7.4 Optical Activity

The phenomenon of optical activity was discovered by the French physicist Jean-Baptiste Biot in 1815.

The experimental facts that led van't Hoff and Le Bel to propose that molecules having the same constitution could differ in the arrangement of their atoms in space concerned the physical property of optical activity. **Optical activity** is the ability of a chiral substance to rotate the plane of plane-polarized light and is measured using an instrument called a **polarimeter** (Figure 7.4).

The light used to measure optical activity has two properties: it consists of a single wavelength and it is plane-polarized. The wavelength used most often is 589 nm (called the *D line*), which corresponds to the yellow light produced by a sodium lamp. Except for giving off light of a single wavelength, a sodium lamp is like any other lamp in that its light is unpolarized, meaning that the plane of its electric field vector can have any orientation along the line of travel. A beam of unpolarized light is transformed to plane-polarized light by passing it through a polarizing filter, which removes all the waves

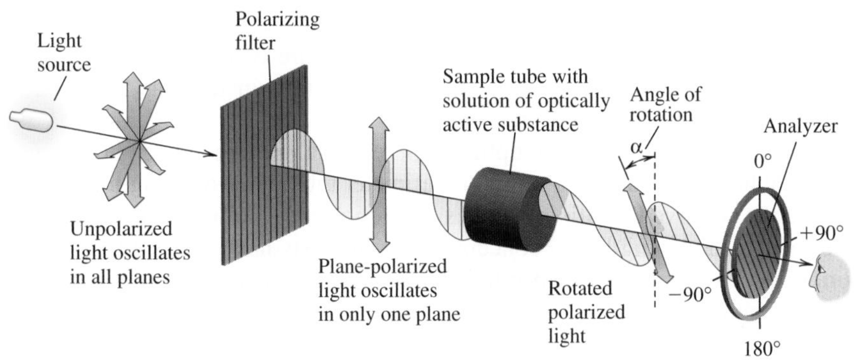

Figure 7.4

The sodium lamp emits light moving in all planes. When the light passes through the first polarizing filter, only one plane emerges. The plane-polarized beam enters the sample compartment, which contains a solution enriched in one of the enantiomers of a chiral substance. The plane rotates as it passes through the solution. A second polarizing filter (called the analyzer) is attached to a movable ring calibrated in degrees that is used to measure the angle of rotation α. (Adapted from M. Silberberg, *Chemistry,* 5th ed., McGraw-Hill Higher Education, New York, 2009, p. 640.)

except those that have their electric field vector in the same plane. This plane-polarized light now passes through the sample tube containing the substance to be examined, either in the liquid phase or as a solution in a suitable solvent (usually water, ethanol, or chloroform). The sample is "optically active" if it rotates the plane of polarized light. The direction and magnitude of rotation are measured using a second polarizing filter (the "analyzer") and cited as α, the observed rotation.

To be optically active, the sample must contain a chiral substance and one enantiomer must be present in excess of the other. A substance that does not rotate the plane of polarized light is said to be optically inactive. *All achiral substances are optically inactive.*

What causes optical rotation? The plane of polarization of a light wave undergoes a minute rotation when it encounters a chiral molecule. Enantiomeric forms of a chiral molecule cause a rotation of the plane of polarization in exactly equal amounts but in opposite directions. A solution containing equal quantities of enantiomers therefore exhibits no net rotation because all the tiny increments of clockwise rotation produced by molecules of one "handedness" are canceled by an equal number of increments of counterclockwise rotation produced by molecules of the opposite handedness.

Mixtures containing equal quantities of enantiomers are called **racemic mixtures.** *Racemic mixtures are optically inactive.* Conversely, when one enantiomer is present in excess, a net rotation of the plane of polarization is observed. At the limit, where all the molecules are of the same handedness, we say the substance is **optically pure.** Optical purity, or percent **enantiomeric excess,** is defined as:

Optical purity = percent enantiomeric excess
= percent of major enantiomer − percent of minor enantiomer

Thus, a material that is 50% optically pure contains 75% of one enantiomer and 25% of the other.

Problem 7.5

A sample of the chiral molecule limonene is 95% optically pure. What percentage of each enantiomer is present?

Rotation of the plane of polarized light in the clockwise sense is taken as positive (+), and rotation in the counterclockwise sense is taken as a negative (−) rotation. Older terms for positive and negative rotations were *dextrorotatory* and *levorotatory,* from the Latin prefixes *dextro-* ("to the right") and *levo-* ("to the left"), respectively. At one time, the symbols *d* and *l* were used to distinguish between enantiomeric forms of a substance. Thus the dextrorotatory enantiomer of 2-butanol was called *d*-2-butanol, and the levorotatory form *l*-2-butanol; a racemic mixture of the two was referred to as *dl*-2-butanol. Current custom favors using algebraic signs instead, as in (+)-2-butanol, (−)-2-butanol, and (±)-2-butanol, respectively.

The observed rotation α of an optically pure substance depends on how many molecules the light beam encounters. A filled polarimeter tube twice the length of another produces twice the observed rotation, as does a solution twice as concentrated. To account for the effects of path length and concentration, chemists have defined the term **specific rotation,** given the symbol [α]. Specific rotation is calculated from the observed rotation according to the expression

$$[\alpha] = \frac{100\,\alpha}{cl}$$

If concentration is expressed as grams per milliliter of solution instead of grams per 100 mL, an equivalent expression is

$$[\alpha] = \frac{\alpha}{cl}$$

where *c* is the concentration of the sample in grams per 100 mL of solution, and *l* is the length of the polarimeter tube in decimeters. (One decimeter is 10 cm.)

Specific rotation is a physical property of a substance, just as melting point, boiling point, density, and solubility are. For example, the lactic acid obtained from milk is exclusively a single enantiomer. We cite its specific rotation in the form $[\alpha]_D^{25} = +3.8°$. The temperature in degrees Celsius and the wavelength of light at which the measurement was made are indicated as superscripts and subscripts, respectively. Optical purity can also be calculated from specific rotation:

$$\text{Optical purity} = \frac{\text{specific rotation of sample}}{\text{specific rotation of pure enantiomer}} \times 100$$

Problem 7.6

Cholesterol, when isolated from natural sources, is obtained as a single enantiomer. The observed rotation α of a 0.3-g sample of cholesterol in 15 mL of chloroform solution contained in a 10-cm polarimeter tube is −0.78°. Calculate the specific rotation of cholesterol.

Problem 7.7

A sample of synthetic cholesterol was prepared consisting entirely of (+)-cholesterol. This synthetic (+)-cholesterol was mixed with some natural (−)-cholesterol. The mixture had a specific rotation $[\alpha]_D^{20}$ of −13°. What fraction of the mixture was (+)-cholesterol? (*Note:* You need to use the solution to Problem 7.6 for the specific rotation of (−)-cholesterol.)

It is convenient to distinguish between enantiomers by prefixing the sign of rotation to the name of the substance. For example, we refer to one of the enantiomers of 2-butanol as (+)-2-butanol and the other as (−)-2-butanol. Optically pure (+)-2-butanol has a specific rotation $[\alpha]_D^{27}$ of +13.5°; optically pure (−)-2-butanol has an exactly opposite specific rotation $[\alpha]_D^{27}$ of −13.5°.

7.5 Absolute and Relative Configuration

The exact three-dimensional spatial arrangement of substituents at a chirality center is its **absolute configuration.** Neither the sign nor the magnitude of rotation by itself can tell us the absolute configuration of a substance. Thus, one of the following structures is (+)-2-butanol and the other is (−)-2-butanol, but without additional information we can't tell which is which.

In several places throughout the chapter we will use red and blue frames to call attention to structures that are enantiomeric.

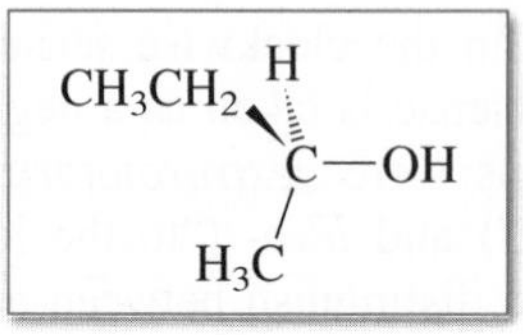
CH3CH2
H
C—OH
H3C

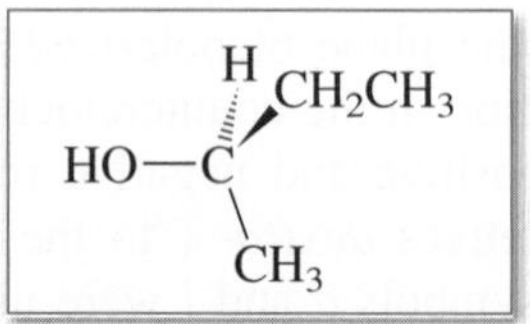
H
CH2CH3
HO—C
CH3

Although no absolute configuration was known for any substance until the mid-twentieth century, organic chemists had experimentally determined the configurations of thousands of compounds relative to one another (their **relative configurations**) through chemical interconversion. To illustrate, consider (+)-3-buten-2-ol. Hydrogenation of this compound yields (+)-2-butanol.

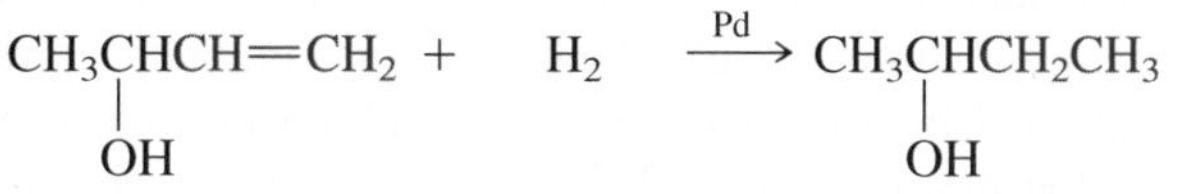

3-Buten-2-ol $[\alpha]_D^{27}$ +33.2° | Hydrogen | 2-Butanol $[\alpha]_D^{27}$ +13.5°

Because hydrogenation of the double bond does not involve any of the bonds to the chirality center, the spatial arrangement of substituents in (+)-3-buten-2-ol must be the same as that of the substituents in (+)-2-butanol. The fact that these two compounds have the same sign of rotation when they have the same relative configuration is established by the hydrogenation experiment; it could not have been predicted in advance of the experiment.

Compounds that have the same relative configuration can have optical rotations of opposite sign. For example, treatment of (−)-2-methyl-1-butanol with hydrogen bromide converts it to (+)-1-bromo-2-methylbutane.

2-Methyl-1-butanol $[\alpha]_D^{25}$ −5.8° | Hydrogen bromide | 1-Bromo-2-methylbutane $[\alpha]_D^{25}$ +4.0° | Water

This reaction does not involve any of the bonds to the chirality center, and so both the starting alcohol (−) and the product bromide (+) have the same relative configuration.

An elaborate network connecting signs of rotation and relative configurations was developed that included the most important compounds of organic and biological chemistry. When, in 1951, the absolute configuration of a salt of (+)-tartaric acid was determined, the absolute configurations of all the compounds whose configurations had been related to (+)-tartaric acid stood revealed as well. Thus, returning to the pair of 2-butanol enantiomers that began this section, their absolute configurations are now known to be as shown.

The salt of tartaric acid analyzed by X-ray crystallography was the sodium rubidium salt of (+)-tartaric acid. X-ray crystallography of biomolecules is described in the boxed essays in Sections 14.13 and 26.8.

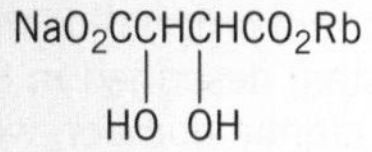

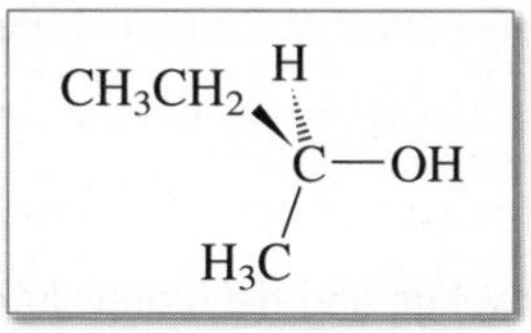

(+)-2-Butanol

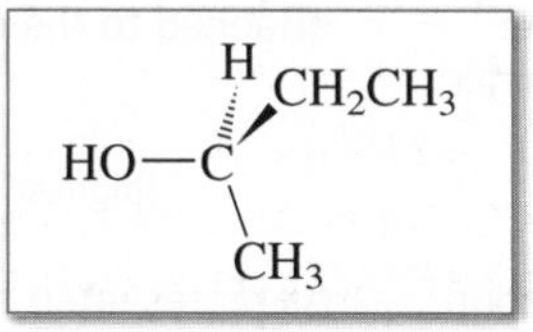

(−)-2-Butanol

Problem 7.8

Does the molecular model shown represent (+)-2-butanol or (−)-2-butanol?

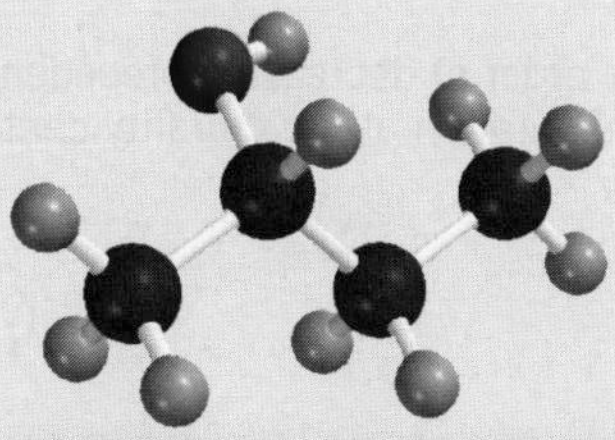

7.6 The Cahn–Ingold–Prelog *R–S* Notational System

Just as it makes sense to have a nomenclature system by which we can specify the constitution of a molecule in words rather than pictures, so too is it helpful to have one that lets us describe stereochemistry. We have already had some experience with this idea when we distinguished between *E* and *Z* stereoisomers of alkenes.

In the *E–Z* system, substituents are ranked by atomic number according to a set of rules devised by R. S. Cahn, Sir Christopher Ingold, and Vladimir Prelog (Section 5.4). Actually, Cahn, Ingold, and Prelog first developed their ranking system to deal with the problem of the absolute configuration at a chirality center, and this is the system's major application. Table 7.1 shows how the **Cahn–Ingold–Prelog system,** called the **sequence rules,** is used to specify the absolute configuration at the chirality center in (+)-2-butanol.

As outlined in Table 7.1, (+)-2-butanol has the *S* configuration. Its mirror image is (−)-2-butanol, which has the *R* configuration.

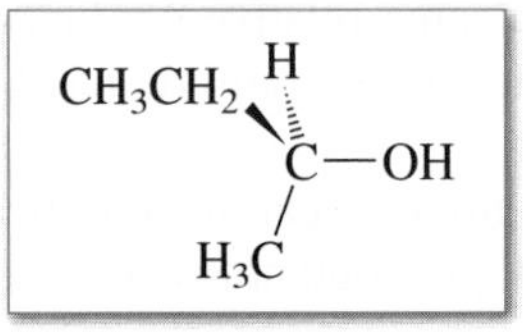

CH_3CH_2
H
C—OH
H_3C

(*S*)-2-Butanol

and

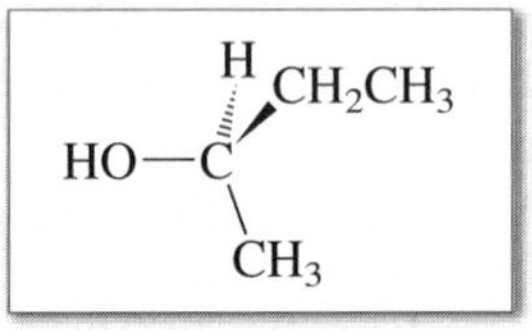

H
CH_2CH_3
HO—C
CH_3

(*R*)-2-Butanol

Often, the *R* or *S* configuration and the sign of rotation are incorporated into the name of the compound, as in (*R*)-(−)-2-butanol and (*S*)-(+)-2-butanol.

TABLE 7.1 Absolute Configuration According to the Cahn–Ingold–Prelog Notational System

Step number	Example
Given that the absolute configuration of (+)-2-butanol is 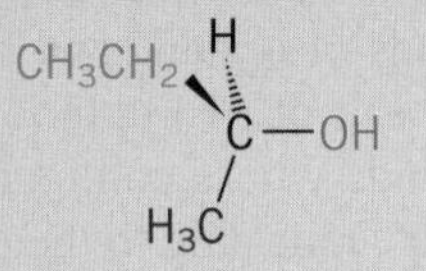 (+)-2-Butanol	
1. Identify the substituents at the chirality center, and rank them in order of decreasing precedence according to the system described in Section 5.4. Precedence is determined by atomic number, working outward from the point of attachment at the chirality center.	In order of decreasing precedence, the four substituents attached to the chirality center of 2-butanol are HO— > CH_3CH_2— > CH_3— > H— (highest) ... (lowest)
2. Orient the molecule so that the lowest ranked atom or group points away from you.	As represented in the wedge-and-dash drawing at the top of this table, the molecule is already appropriately oriented. Hydrogen is the lowest ranked atom attached to the chirality center and points away from us.
3. Draw the three highest ranked substituents as they appear to you when the molecule is oriented so that the lowest ranked group points away from you.	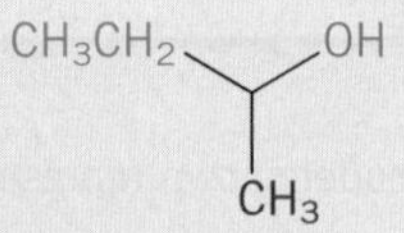
4. If the order of decreasing precedence of the three highest ranked substituents appears in a clockwise sense, the absolute configuration is *R* (Latin *rectus,* "right," "correct"). If the order of decreasing precedence is counterclockwise, the absolute configuration is *S* (Latin *sinister,* "left").	The order of decreasing precedence is *counterclockwise.* The configuration at the chirality center is *S.* 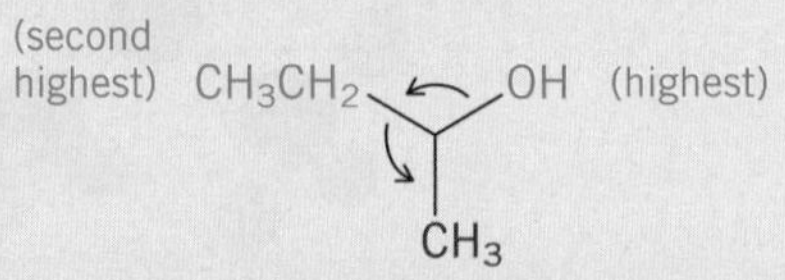 (third highest)

Problem 7.9

Assign absolute configurations as *R* or *S* to each of the following compounds:

(a) 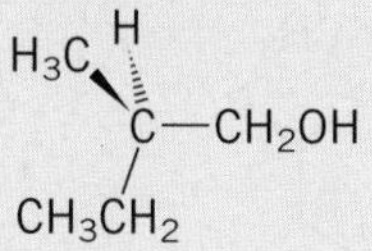
H₃C H
C—CH₂OH
CH₃CH₂
(+)-2-Methyl-1-butanol

(c) 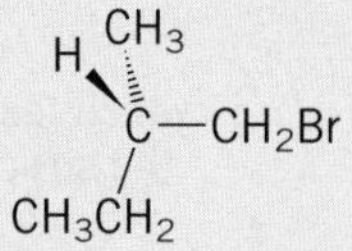
H CH₃
C—CH₂Br
CH₃CH₂
(+)-1-Bromo-2-methylbutane

(b) 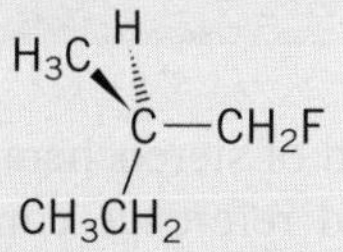
H₃C H
C—CH₂F
CH₃CH₂
(+)-1-Fluoro-2-methylbutane

(d) 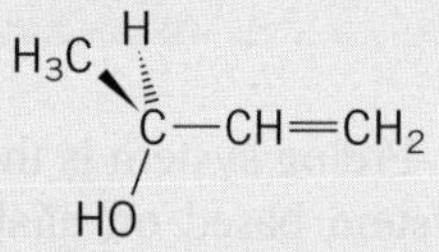
H₃C H
C—CH=CH₂
HO
(+)-3-Buten-2-ol

Sample Solution (a) The highest ranking substituent at the chirality center of 2-methyl-1-butanol is CH_2OH; the lowest is H. Of the remaining two, ethyl outranks methyl.

Order of precedence: $CH_2OH > CH_3CH_2 > CH_3 > H$

The lowest ranking group (hydrogen) points away from us in the drawing. The three highest ranking groups trace a clockwise path from $CH_2OH \rightarrow CH_3CH_2 \rightarrow CH_3$.

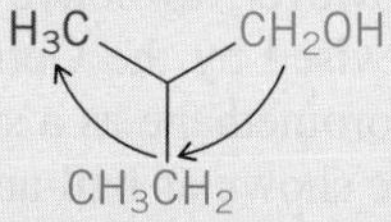
H₃C
CH₂OH
CH₃CH₂

This compound therefore has the *R* configuration. It is (*R*)-(+)-2-methyl-1-butanol.

Compounds in which a chirality center is part of a ring are handled in an analogous fashion. To determine, for example, whether the configuration of (+)-4-methylcyclohexene is *R* or *S*, treat the right- and left-hand paths around the ring as if they were independent groups.

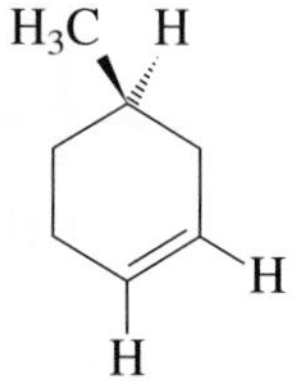
H₃C H
H
H
(+)-4-Methylcyclohexene

is treated as

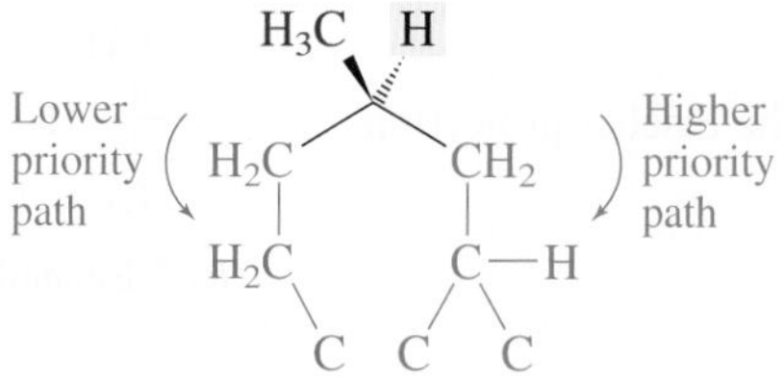
H₃C H
Lower priority path
H₂C
CH₂
Higher priority path
H₂C
C—H
C C C

With the lowest ranked group (hydrogen) directed away from us, we see that the order of decreasing sequence rule precedence is *clockwise*. The absolute configuration is *R*.

Problem 7.10

Draw three-dimensional representations of

(a) The *R* enantiomer of

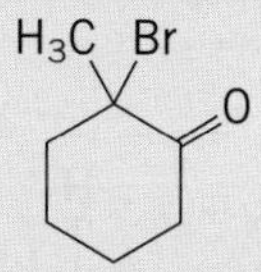
H₃C Br
O

(b) The *S* enantiomer of

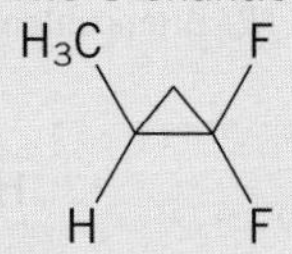
H₃C F
H F

Sample Solution (a) The chirality center is the one that bears the bromine. In order of decreasing precedence, the substituents attached to the chirality center are

$$\text{Br} > \overset{\text{O}}{\overset{\|}{\text{C}}} > \text{—CH}_2\text{C} > \text{CH}_3$$

Continued

When the lowest ranked substituent (the methyl group) is away from us, the order of decreasing precedence of the remaining groups must appear in a clockwise sense in the *R* enantiomer.

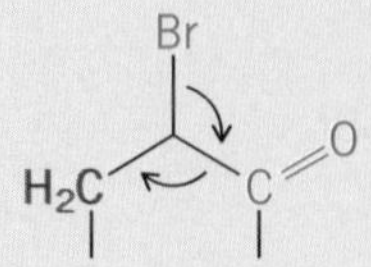

which leads to the structure

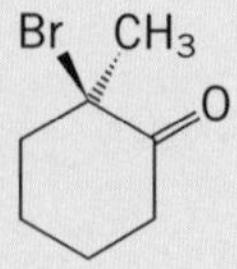

(*R*)-2-Bromo-2-methylcyclohexanone

The Cahn–Ingold–Prelog system is the standard method of stereochemical notation. It replaced an older system based on analogies to specified reference compounds that used the prefixes D and L, a system that is still used for carbohydrates and amino acids. We will use D and L notation when we get to Chapters 23–26, but won't need it until then.

7.7 Fischer Projections

Stereochemistry deals with the three-dimensional arrangement of a molecule's atoms, and we have attempted to show stereochemistry with wedge-and-dash drawings and computer-generated models. It is possible, however, to convey stereochemical information in an abbreviated form using a method devised by the German chemist Emil Fischer.

Fischer was the foremost organic chemist of the late nineteenth century. He won the 1902 Nobel Prize in Chemistry for his pioneering work in carbohydrate and protein chemistry.

Let's return to bromochlorofluoromethane as a simple example of a chiral molecule. The two enantiomers of BrClFCH are shown as ball-and-spoke models, as wedge-and-dash drawings, and as **Fischer projections** in Figure 7.5. Fischer projections are always generated the same way: the molecule is oriented so that the vertical bonds at the chirality center are directed away from you and the horizontal bonds point toward you. A projection of the bonds onto the page is a cross. The chirality center lies at the center of the cross but is not explicitly shown.

It is customary to orient the molecule so that the carbon chain is vertical with the lowest numbered carbon at the top as shown for the Fischer projection of (*R*)-2-butanol.

The Fischer projection

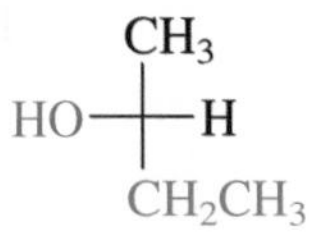

(*R*)-2-Butanol

corresponds to

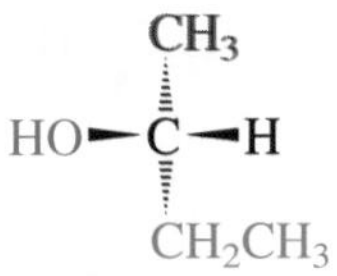

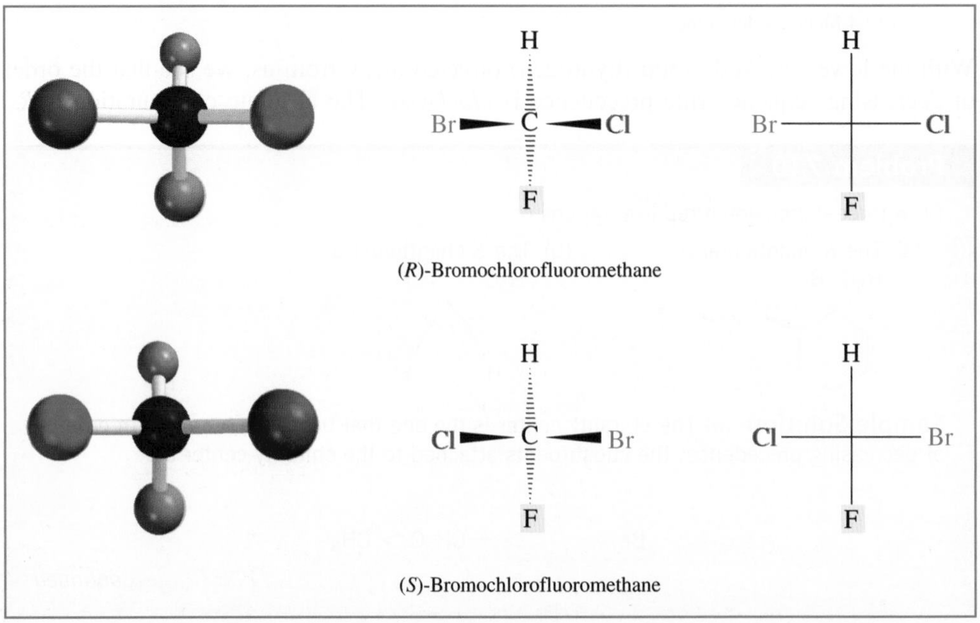

Figure 7.5

Ball-and-spoke models (*left*), wedge-and-dash drawings (*center*), and Fischer projections (*right*) of the *R* and *S* enantiomers of bromochlorofluoromethane.

To verify that the Fischer projection has the *R* configuration at its chirality center, rotate the three-dimensional representation so that the lowest-ranked atom (H) points away from you. Be careful to maintain the proper stereochemical relationships during the operation.

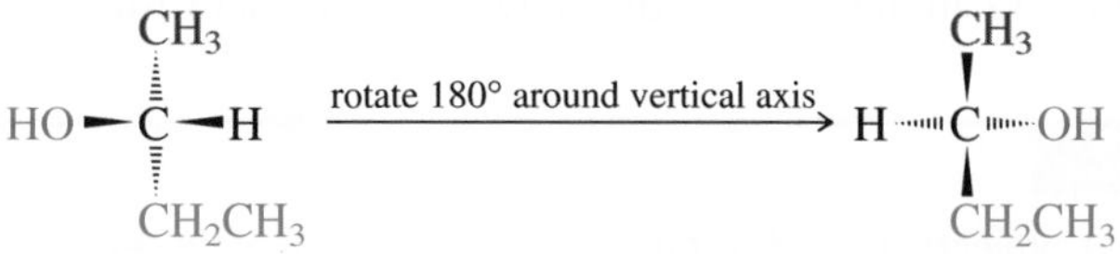

rotate 180° around vertical axis

With H pointing away from us, we can see that the order of decreasing precedence OH > CH_2CH_3 > CH_3 traces a clockwise path, verifying the configuration as *R*.

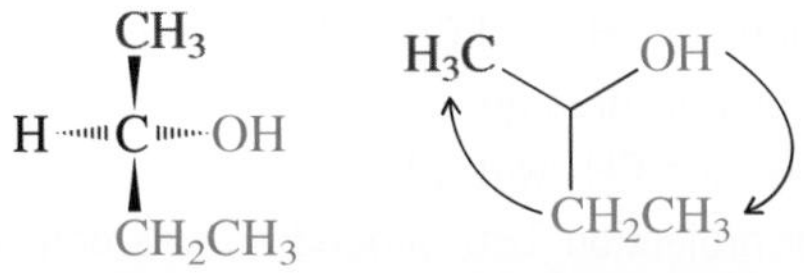

Problem 7.11

What is the absolute configuration (*R* or *S*) of the compounds represented by the Fischer projections shown here?

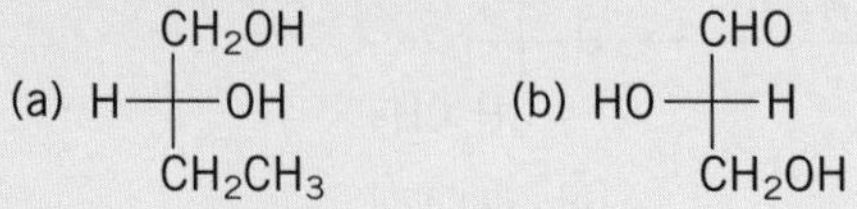

Sample Solution

(a) • The Fischer projection given in the problem stands for:

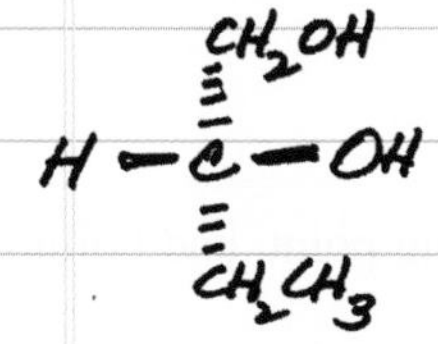

precedence is

OH > CH_2OH > CH_2CH_3 > H

• Orient molecule so that H points away

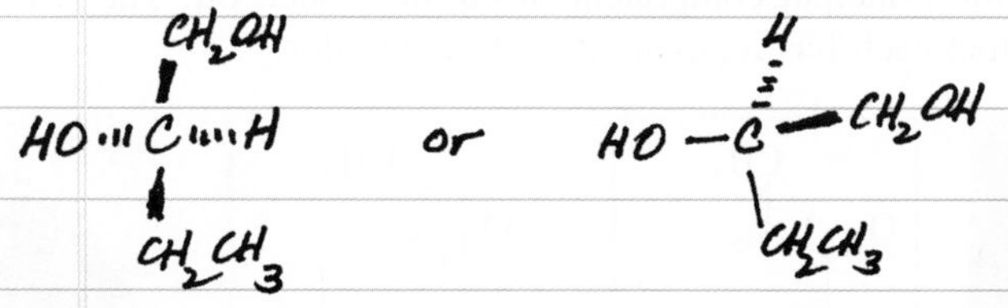

• order of decreasing precedence is

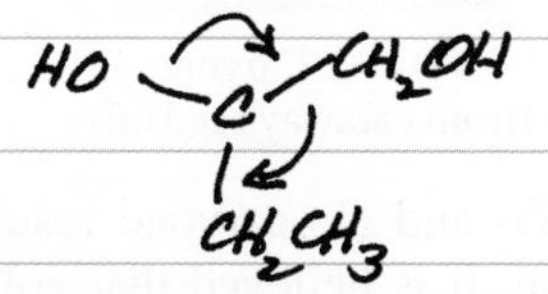

clockwise therefore <u>R</u>

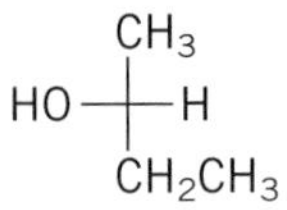

As you work with Fischer projections, you may notice that some routine structural changes lead to predictable outcomes—outcomes that may reduce the number of manipulations you need to do to solve stereochemistry problems. Instead of listing these shortcuts, Problem 7.12 invites you to discover some of them for yourself.

Problem 7.12

Using the Fischer projection of (*R*)-2-butanol given in the margin, explain how each of the following affects the configuration of the chirality center.

(a) Switching the positions of H and OH.
(b) Switching the positions of CH_3 and CH_2CH_3.
(c) Switching the positions of three groups.
(d) Switching H with OH, and CH_3 with CH_2CH_3.
(e) Rotating the Fischer projection 180° around an axis perpendicular to the page.

Sample Solution (a) Exchanging the positions of H and OH in the Fischer projection of (*R*)-2-butanol converts it to the mirror-image Fischer projection. The configuration of the chirality center goes from *R* to *S*.

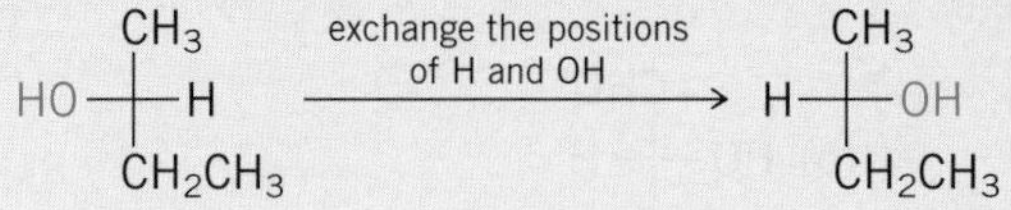

(*R*)-2-butanol (*S*)-2-butanol

Switching the positions of two groups in a Fischer projection reverses the configuration of the chirality center.

We mentioned in Section 7.6 that the D,L system of stereochemical notation, while outdated for most purposes, is still widely used for carbohydrates and amino acids. Likewise, Fischer projections find their major application in these same two families of compounds.

7.8 Properties of Enantiomers

The usual physical properties such as density, melting point, and boiling point are identical for both enantiomers of a chiral compound.

Enantiomers can have striking differences, however, in properties that depend on the arrangement of atoms in space. Take, for example, the enantiomeric forms of carvone. (*R*)-(−)-Carvone is the principal component of spearmint oil. Its enantiomer, (*S*)-(+)-carvone, is the principal component of caraway seed oil. The two enantiomers do not smell the same; each has its own characteristic odor.

Spearmint leaves

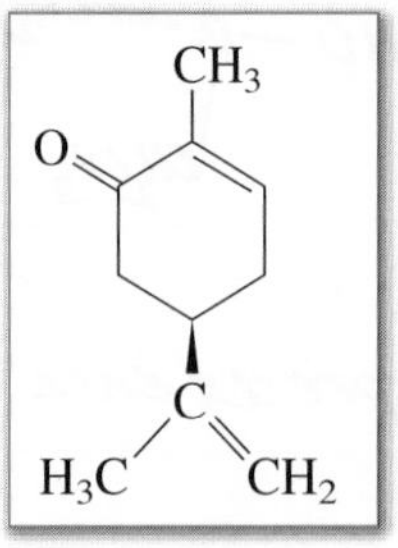

(*R*)-(−)-Carvone
(from spearmint oil)

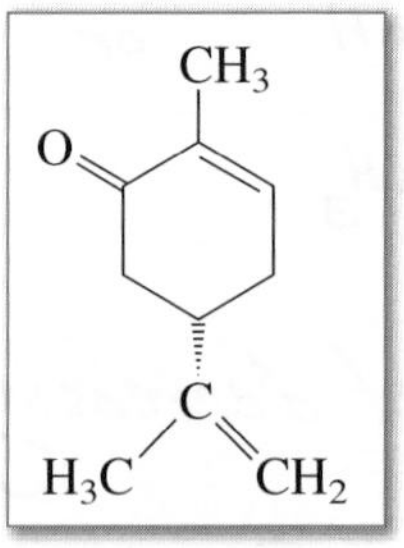

(*S*)-(+)-Carvone
(from caraway seed oil)

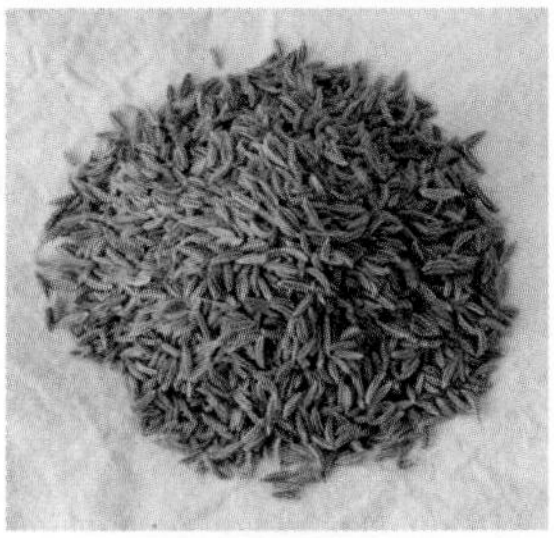

Caraway seeds

The difference in odor between (*R*)- and (*S*)-carvone results from their different behavior toward receptor sites in the nose. It is believed that volatile molecules occupy only those odor receptors that have the proper shape to accommodate them. Because the receptor sites are themselves chiral, one enantiomer may fit one kind of receptor while the other enantiomer fits a different kind. An analogy that can be drawn is to hands and gloves. Your left hand and your right hand are enantiomers. You can place your left hand into a left glove but not into a right one. The receptor (the glove) can accommodate one enantiomer of a chiral object (your hand) but not the other.

The term *chiral recognition* refers to a process in which some chiral receptor or reagent interacts selectively with one of the enantiomers of a chiral molecule. Very high levels of chiral recognition are common in biological processes. (−)-Nicotine, for example, is much more toxic than (+)-nicotine, and (+)-adrenaline is more active than (−)-adrenaline in constricting blood vessels. (−)-Thyroxine, an amino acid of the thyroid gland that speeds up metabolism, is one of the most widely used of all prescription drugs—about 10 million people in the United States take (−)-thyroxine on a daily basis. Its enantiomer, (+)-thyroxine has none of the metabolism-regulating effects, but was formerly given to heart patients to lower their cholesterol levels.

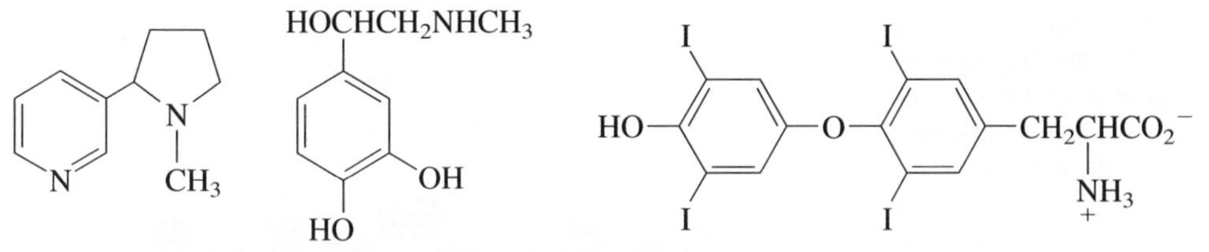

Nicotine Adrenaline Thyroxine

(Can you find the chirality center in each of these?)

7.9 The Chirality Axis

We have, so far, restricted our discussion of chiral molecules to those that contain a chirality center. Although these are the most common, they are not the only kinds of chiral molecules. A second group consists of molecules that contain a **chirality axis**—an axis about which a set of atoms or groups is arranged so that the spatial arrangement is not superimposable on its mirror image. We can think of two enantiomers characterized by a chirality axis as being analogous to a left-handed screw and a right-handed screw.

Among molecules with a chirality axis, substituted derivatives of biaryls have received much attention. **Biaryls** are compounds in which two aromatic rings are joined by a single bond: biphenyl and 1,1′-binaphthyl, for example.

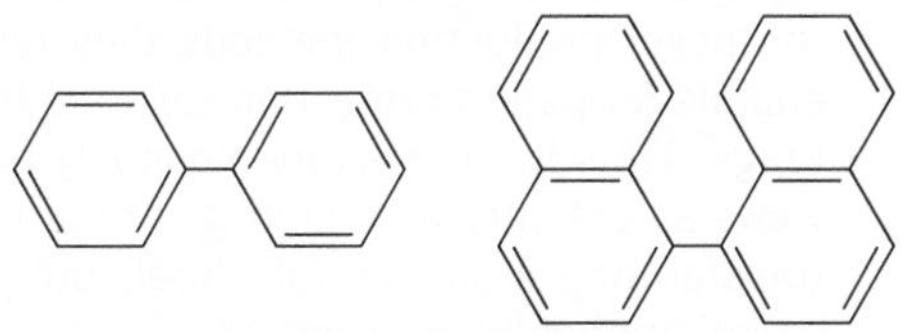
Biphenyl 1,1′-Binaphthyl

Although the individual rings in biphenyl and 1,1′-binaphthyl are flat, the molecules themselves are not. Rotation about the single bond connecting the two rings in biphenyl reduces the steric strain between nearby hydrogens of one ring (red) and those of the other (green). This rotation makes the "twisted" conformation more stable than one in which all of the atoms lie in the same plane.

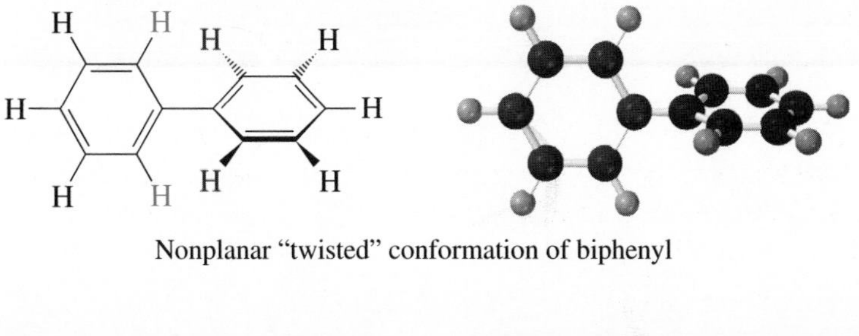

Nonplanar "twisted" conformation of biphenyl

The experimentally measured angle between the two rings of biphenyl in the gas phase is 44°.

Chiral Drugs

A recent estimate places the number of prescription and over-the-counter drugs marketed throughout the world at about 2000. Approximately one third of these are either naturally occurring substances themselves or are prepared by chemical modification of natural products. Most of the drugs derived from natural sources are chiral and are almost always obtained as a single enantiomer rather than as a racemic mixture. Not so with the over 500 chiral substances represented among the more than 1300 drugs that are the products of synthetic organic chemistry. Until recently, such substances were, with few exceptions, prepared, sold, and administered as racemic mixtures even though the desired therapeutic activity resided in only one of the enantiomers. Spurred by a number of factors ranging from safety and efficacy to synthetic methodology and economics, this practice is undergoing rapid change as more and more chiral synthetic drugs become available in enantiomerically pure form.

Because of the high degree of chiral recognition inherent in most biological processes (Section 7.8), it is unlikely that both enantiomers of a chiral drug will exhibit the same level, or even the same kind, of effect. At one extreme, one enantiomer has the desired effect, and the other exhibits no biological activity at all. In this case, which is relatively rare, the racemic form is simply a drug that is 50% pure and contains 50% "inert ingredients." Real cases are more complicated. For example, the *S* enantiomer is responsible for the pain-relieving properties of ibuprofen, normally sold as a racemic mixture. The 50% of racemic ibuprofen that is the *R* enantiomer is not completely wasted, however, because enzyme-catalyzed reactions in our body convert much of it to active (*S*)-ibuprofen.

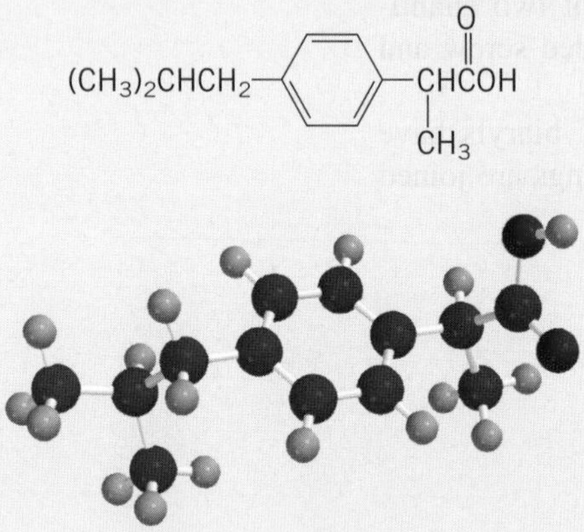
(CH3)2CHCH2
O
CHCOH
CH3

Ibuprofen

A much more serious drawback to using chiral drugs as racemic mixtures is illustrated by thalidomide, briefly employed as a sedative and antinausea drug in Europe during the period 1959–1962. The desired properties are those of (*R*)-thalidomide. (*S*)-Thalidomide, however, has a very different spectrum of biological activity and was shown to be responsible for over 2000 cases of serious birth defects in children born to women who took it while pregnant.

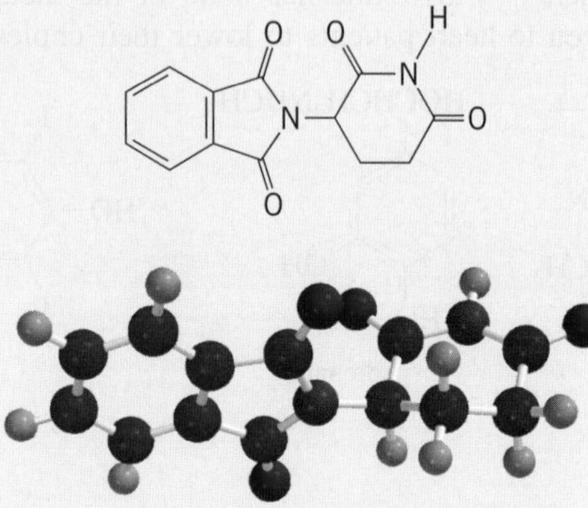
O
O
H
N
N
O
O

Thalidomide

Basic research aimed at controlling the stereochemistry of chemical reactions has led to novel methods for the synthesis of chiral molecules in enantiomerically pure form. Aspects of this work were recognized with the award of the 2001 Nobel Prize in Chemistry to William S. Knowles (Monsanto), Ryoji Noyori (Nagoya University), and K. Barry Sharpless (Scripps Research Institute). Most major pharmaceutical companies are examining their existing drugs to see which are the best candidates for synthesis as single enantiomers and, when preparing a new drug, design its synthesis so as to provide only the desired enantiomer. One incentive to developing enantiomerically pure versions of existing drugs is that the novel production methods they require may make them eligible for patent protection separate from that of the original drugs. Thus the temporary monopoly position that patent law views as essential to fostering innovation can be extended by transforming a successful chiral, but racemic, drug into an enantiomerically pure version.

Problem 7.13

Find the chirality center in the molecular model of thalidomide shown above and identify its configuration as *R* or *S*.

Rotation about the bond joining the two rings is very fast in biphenyl, about the same as in ethane, but is slowed when the carbons adjacent to the ones joining the two rings bear groups other than hydrogen.

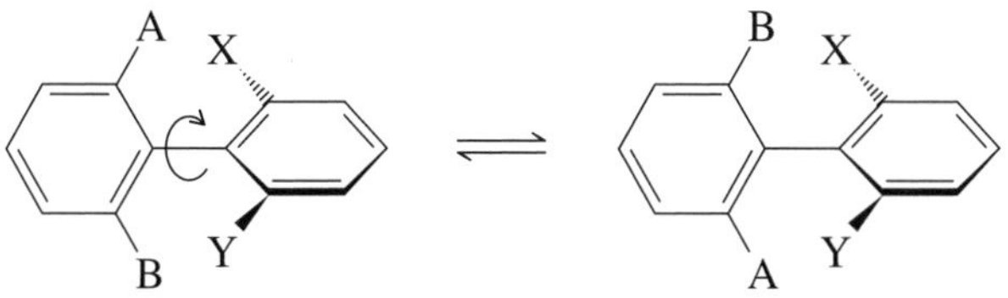

If the substituents are large enough, the steric strain that accompanies their moving past each other during rotation about the single bond can decrease the rate of equilibration so much that it becomes possible to isolate the two conformations under normal laboratory conditions.

When A ≠ B, and X ≠ Y, the two conformations are nonsuperimposable mirror images of each other; that is, they are enantiomers. The bond connecting the two rings lies along a chirality axis.

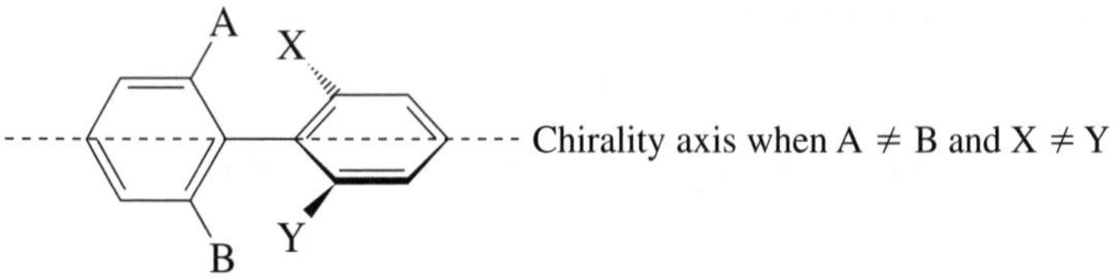

The first compound demonstrated to be chiral because of restricted rotation about a single bond was 6,6′-dinitrobiphenyl-2,2′-dicarboxylic acid in 1922.

(+)-6,6′-Dinitrobiphenyl-2,2′-dicarboxylic acid $[\alpha]_D^{29}$ +127° (methanol)

(−)-6,6′-Dinitrobiphenyl-2,2′-dicarboxylic acid $[\alpha]_D^{29}$ −127° (methanol)

Problem 7.14

The 3,3′-5,5′ isomer of the compound just shown has a chirality axis, but its separation into isolable enantiomers would be extremely difficult. Why?

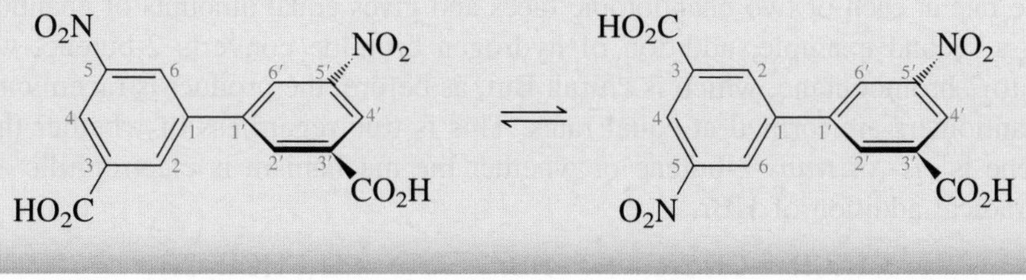

Structures such as chiral biaryls, which are related by rotation around a single bond yet are capable of independent existence, are sometimes called **atropisomers,** from the Greek *a* meaning not, and *tropos* meaning turn. They represent a subcategory of conformers.

Derivatives of 1,1′-binaphthyl exhibit atropisomerism, due to hindered rotation about the single bond that connects the two naphthalene rings. A commercially important application of chiral binaphthyls is based on a substituted derivative known as BINAP, a

Chemists don't agree on the minimum energy barrier for bond rotation that allows isolation of enantiomeric atropisomers at room temperature, but it is on the order of 100 kJ/mol (24 kcal/mol). Recall that the activation energy for rotation about C—C single bonds in alkanes is about 12 kJ/mol (3 kcal/mol).

component of a hydrogenation catalyst. In this catalyst, ruthenium is bound by the two phosphorus atoms present on the groups attached to the naphthalene rings.

BINAP is an abbreviation for 2,2′-bis(diphenylphosphino)-1,1′-binaphthyl.

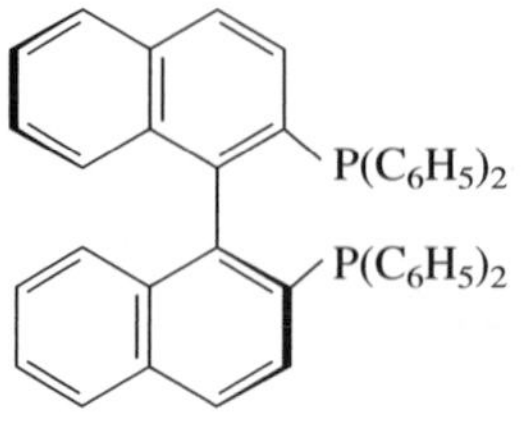
$P(C_6H_5)_2$
$P(C_6H_5)_2$

(*S*)-(−)-BINAP

We will explore the use of the ruthenium BINAP catalysts in the enantioselective synthesis of chiral drugs in Chapter 14.

7.10 Reactions That Create a Chirality Center

Many of the reactions we've already encountered can yield a chiral product from an achiral starting material. Epoxidation of propene, for example, creates a chirality center by adding oxygen to the double bond.

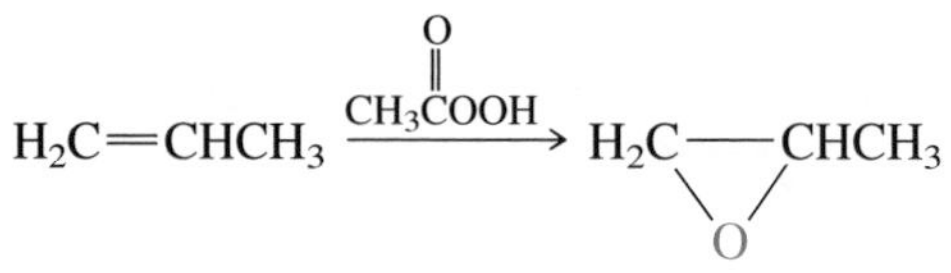
$H_2C{=}CHCH_3$
CH_3COOH (with C=O)
H_2C—$CHCH_3$
O

Propene (achiral) → 1,2-Epoxypropane (chiral)

In this, as in other reactions in which achiral reactants yield chiral products, the product is formed as a *racemic mixture* and is *optically inactive.* Remember, for a substance to be optically active, not only must it be chiral but one enantiomer must be present in excess of the other.

It is a general principle that *optically active products cannot be formed from an optically inactive starting material unless at least one optically active reactant or catalyst is present.* This principle holds irrespective of whether the addition is syn or anti, concerted or stepwise. No matter how many steps are involved in a reaction, if the reactants are achiral, formation of one enantiomer is just as likely as the other, and a racemic mixture results.

Figure 7.6 shows why equal amounts of (*R*)- and (*S*)-1,2-epoxypropane are formed in the epoxidation of propene. Transfer of oxygen to the top face of the double bond gives (*R*)-1,2-epoxypropane; oxygen transfer to the bottom face gives (*S*). The two faces are termed **prochiral** because addition to either face converts an achiral reactant to a chiral product. They are further classified as **enantiotopic** because the product from reaction at one face is the enantiomer of the product from reaction at the other. An achiral reagent reacts at the same rate at each of two enantiotopic faces and gives equal amounts of enantiomers.

The Descriptive Passage and Interpretive Problems at the end of this chapter explore prochirality in more detail.

In a second example, addition of hydrogen bromide converts 2-butene, which is achiral, to 2-bromobutane, which is chiral. But, as before, the product is racemic because both enantiomers are formed at equal rates. This is true regardless of whether the starting alkene is *cis*- or *trans*-2-butene or whether the mechanism is electrophilic addition or free-radical addition of HBr.

$$CH_3CH{=}CHCH_3 \;+\; HBr \longrightarrow CH_3\underset{\displaystyle Br}{\underset{|}{C}}HCH_2CH_3$$

cis- or *trans*-2-Butene (achiral) + Hydrogen bromide (achiral) → 2-Bromobutane (chiral, but racemic)

Whatever happens at one enantiotopic face of the double bond of *cis*- or *trans*-2-butene happens at the same rate at the other, resulting in a 1:1 mixture of (*R*)- and (*S*)-2-bromobutane.

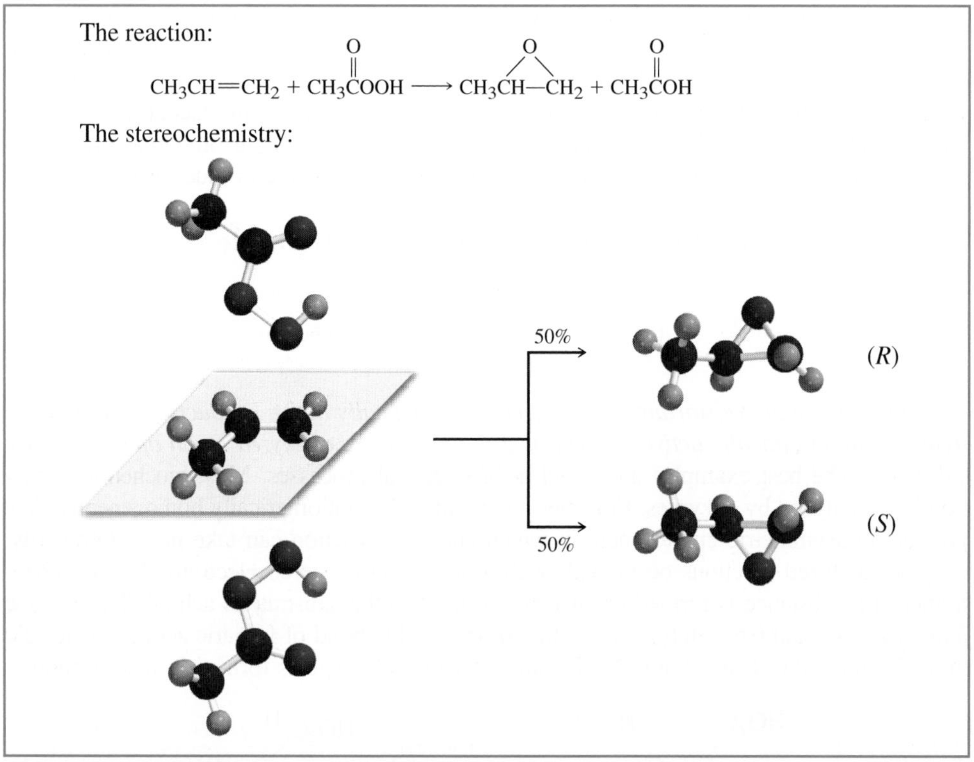
The reaction:
CH3CH=CH2 + CH3COOH ⟶ CH3CH—CH2 + CH3COH
The stereochemistry:
50%
50%
(R)
(S)

Figure 7.6

Epoxidation of propene produces equal amounts of (*R*)- and (*S*)-1,2-epoxypropane.

Problem 7.15

What two stereoisomeric alkanes are formed in the catalytic hydrogenation of (*E*)-3-methyl-2-hexene? What are the relative amounts of each?

Addition to double bonds is not the only kind of reaction that converts an achiral molecule to a chiral one. Other possibilities include substitution reactions such as the formation of 2-chlorobutane by free-radical chlorination of butane. Here again, the product is chiral, but racemic.

$$CH_3CH_2CH_2CH_3 + Cl_2 \xrightarrow{\text{heat}} CH_3\underset{\displaystyle Cl}{\underset{|}{C}}HCH_2CH_3 + HCl$$

Butane	Chlorine	2-Chlorobutane	Hydrogen chloride
(achiral)	(achiral)	(chiral, but racemic)	

1-Chlorobutane is also formed in this reaction.

C-2 of butane is a **prochirality center;** replacing one of its attached hydrogens by chlorine converts it to a chirality center in 2-chlorobutane. The hydrogens at C-2 of butane are enantiotopic; replacing one of them by some different atom or group gives the enantiomer of the structure obtained by replacing the other. Enantiotopic hydrogens are equally reactive toward an achiral reagent, and a racemic product results.

Problem 7.16

Citric acid has three CO_2H groups. Which, if any, of them are enantiotopic?

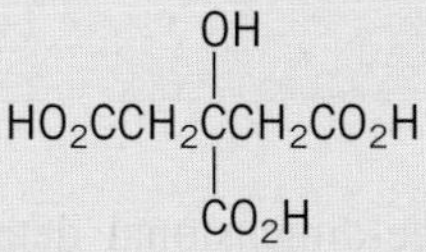
OH
HO2CCH2CCH2CO2H
CO2H

Citric acid

When a reactant is chiral but optically inactive because it is *racemic,* any products derived from its reactions with optically inactive reagents will be *optically inactive.* For example, 2-butanol is chiral and may be converted with hydrogen bromide to 2-bromobutane, which is also chiral. If racemic 2-butanol is used, each enantiomer will react at the same rate with the achiral reagent. Whatever happens to (*R*)-(−)-2-butanol is mirrored in a corresponding reaction of (*S*)-(+)-2-butanol, and a racemic, optically inactive product results.

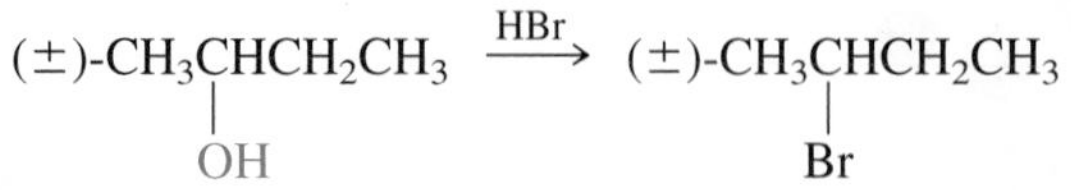
(±)-CH3CHCH2CH3
OH
HBr
(±)-CH3CHCH2CH3
Br

2-Butanol (chiral but racemic) — 2-Bromobutane (chiral but racemic)

Optically inactive starting materials can give optically active products only if they are treated with an optically active reagent or if the reaction is catalyzed by an optically active substance. The best examples are found in biochemical processes. Most biochemical reactions are catalyzed by enzymes. Enzymes are chiral and enantiomerically homogeneous; they provide an asymmetric environment in which chemical reaction can take place. Ordinarily, enzyme-catalyzed reactions occur with such a high level of stereoselectivity that one enantiomer of a substance is formed exclusively even when the substrate is achiral. The enzyme *fumarase,* for example, catalyzes hydration of the double bond of fumaric acid to malic acid in apples and other fruits. Only the *S* enantiomer of malic acid is formed in this reaction.

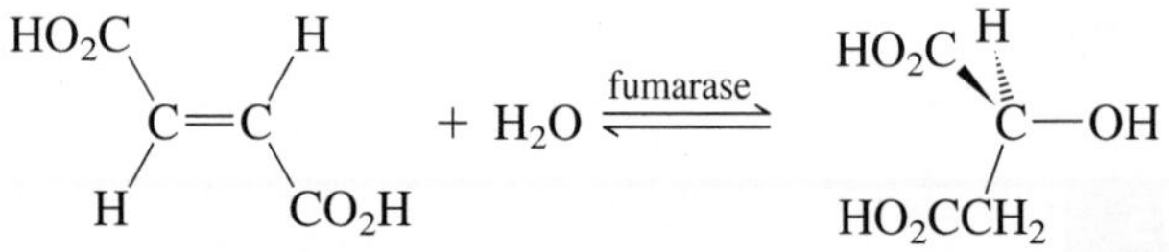
HO2C
H
H
CO2H
+ H2O
fumarase
HO2C
H
C—OH
HO2CCH2

Fumaric acid — (*S*)-(−)-Malic acid

The reaction is reversible, and its stereochemical requirements are so pronounced that neither the cis isomer of fumaric acid (maleic acid) nor the *R* enantiomer of malic acid can serve as a substrate for the fumarase-catalyzed hydration–dehydration equilibrium.

The stereospecific formation of the *S* enantiomer of malic acid from fumaric acid also occurs in the tricarboxylic acid (TCA) cycle. The (*S*)-(−)-malic acid that is produced is converted to oxaloacetic acid by the enzyme *malate dehydrogenase.*

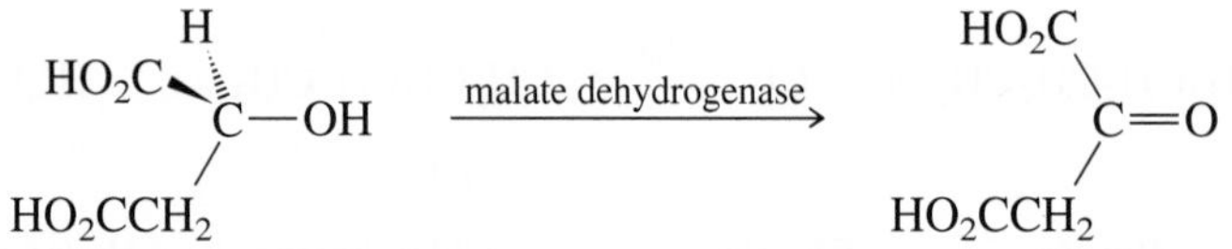
H
HO2C
C—OH
HO2CCH2
malate dehydrogenase
HO2C
C=O
HO2CCH2

(*S*)-(−)-Malic acid — Oxaloacetic acid

Problem 7.17

Biological reduction of pyruvic acid, catalyzed by the enzyme *lactate dehydrogenase,* gives (+)-lactic acid, represented by the Fischer projection shown. What is the configuration of (+)-lactic acid according to the Cahn–Ingold–Prelog *R–S* notational system?

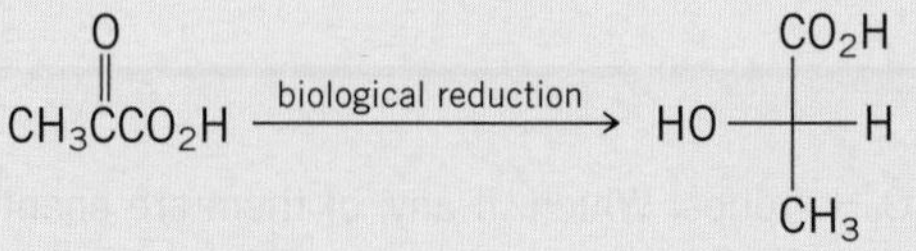
O
CH3CCO2H
biological reduction
CO2H
HO
H
CH3

Pyruvic acid — (+)-Lactic acid

We'll continue with the three-dimensional details of chemical reactions later in this chapter. First though, we need to develop some additional stereochemical principles concerning structures with more than one chirality center.

7.11 Chiral Molecules with Two Chirality Centers

When a molecule contains two chirality centers, as does 2,3-dihydroxybutanoic acid, how many stereoisomers are possible?

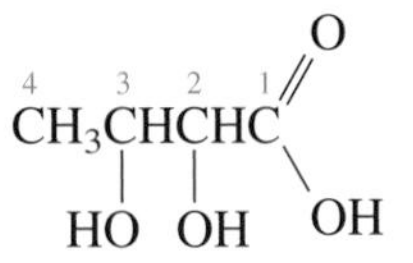

2,3-Dihydroxybutanoic acid

We can use straightforward reasoning to come up with the answer. The absolute configuration at C-2 may be *R* or *S*. Likewise, C-3 may have either the *R* or the *S* configuration. The four possible combinations of these two chirality centers are

(2*R*,3*R*)	(stereoisomer I)	(2*S*,3*S*)	(stereoisomer II)
(2*R*,3*S*)	(stereoisomer III)	(2*S*,3*R*)	(stereoisomer IV)

Figure 7.7 presents structural formulas for these four stereoisomers. Stereoisomers I and II are enantiomers of each other; the enantiomer of (*R*,*R*) is (*S*,*S*). Likewise stereoisomers III and IV are enantiomers of each other, the enantiomer of (*R*,*S*) being (*S*,*R*).

Stereoisomer I is not a mirror image of III or IV, so it is not an enantiomer of either one. Stereoisomers that are not related as an object and its mirror image are called diastereomers; ***diastereomers** are stereoisomers that are not mirror images.* Thus, stereoisomer I is a diastereomer of III and a diastereomer of IV. Similarly, II is a diastereomer of III and IV.

To convert a molecule with two chirality centers to its enantiomer, the configuration at *both* centers must be changed. Reversing the configuration at only one chirality center converts it to a diastereomeric structure.

Enantiomers must have equal and opposite specific rotations. Diastereomers can have different rotations, with respect to both sign and magnitude. Thus, as Figure 7.7 shows,

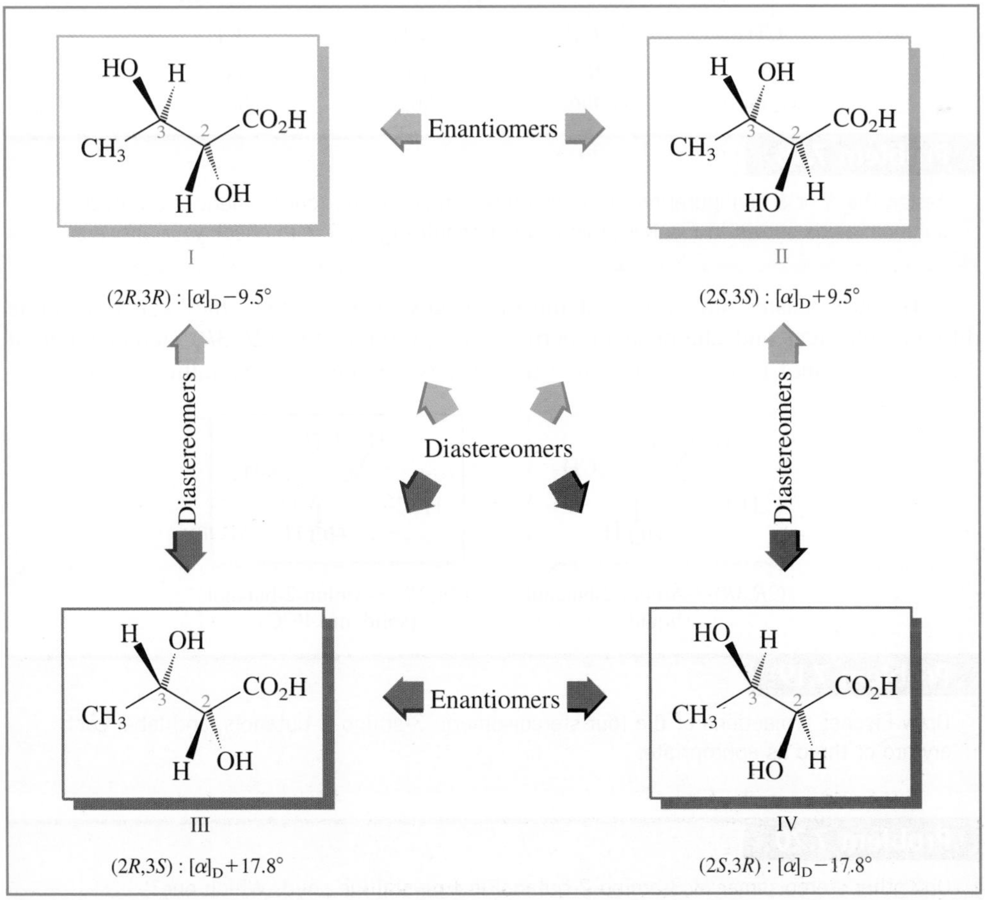

Figure 7.7

Stereoisomeric 2,3-dihydroxybutanoic acids. Stereoisomers I and II are enantiomers. Stereoisomers III and IV are enantiomers. All other relationships are diastereomeric (see text).

Figure 7.8

Representations of (2*R*,3*R*)-dihydroxybutanoic acid. (*a*) The staggered conformation is the most stable, but is not properly arranged to show stereochemistry as a Fischer projection. (*b*) Rotation about the C-2–C-3 bond gives the eclipsed conformation, and projection of the eclipsed conformation onto the page gives (*c*) a correct Fischer projection.

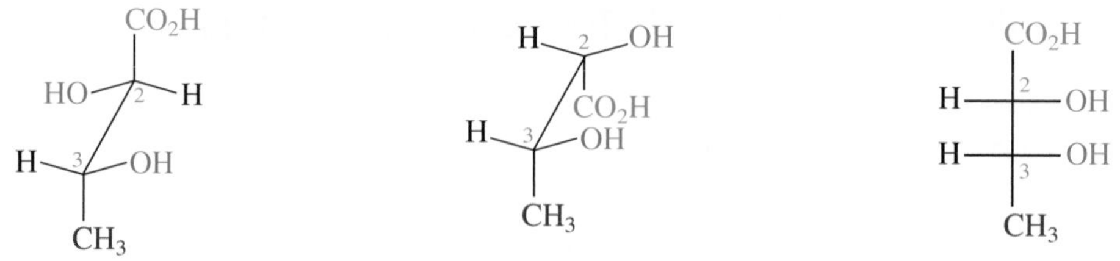

(*a*) (*b*) (*c*)

the (2*R*,3*R*) and (2*S*,3*S*) enantiomers (I and II) have specific rotations that are equal in magnitude but opposite in sign. The (2*R*,3*S*) and (2*S*,3*R*) enantiomers (III and IV) likewise have specific rotations that are equal to each other but opposite in sign. The magnitudes of rotation of I and II are different, however, from those of their diastereomers III and IV.

In writing Fischer projections of molecules with two chirality centers, the molecule is arranged in an *eclipsed* conformation for projection onto the page, as shown in Figure 7.8. Again, horizontal lines in the projection represent bonds coming toward you; vertical lines represent bonds pointing away.

Organic chemists use an informal nomenclature system based on Fischer projections to distinguish between diastereomers. When the carbon chain is vertical and like substituents are on the same side of the Fischer projection, the molecule is described as the **erythro** diastereomer. When like substituents are on opposite sides of the Fischer projection, the molecule is described as the **threo** diastereomer. Thus, as seen in the Fischer projections of the stereoisomeric 2,3-dihydroxybutanoic acids, compounds I and II are erythro stereoisomers and III and IV are threo.

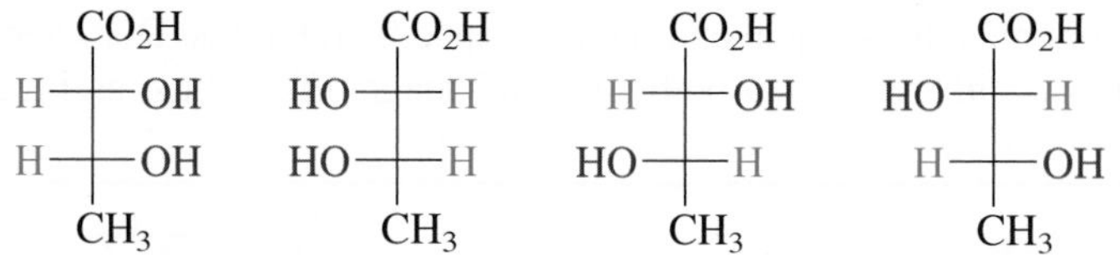

I erythro | II erythro | III threo | IV threo

Problem 7.18

Assign the *R* or *S* configuration to the chirality centers in the four isomeric 2,3-dihydroxybutanoic acids shown in Fischer projections. Consult Figure 7.7 to check your answers.

Because diastereomers are not mirror images of each other, they can have quite different physical and chemical properties. For example, the (2*R*,3*R*) stereoisomer of 3-amino-2-butanol is a liquid, but the (2*R*,3*S*) diastereomer is a crystalline solid.

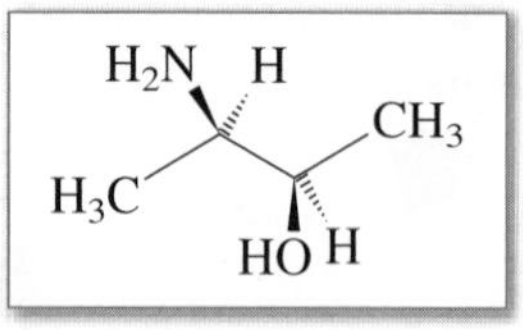

(2*R*,3*R*)-3-Amino-2-butanol (liquid)

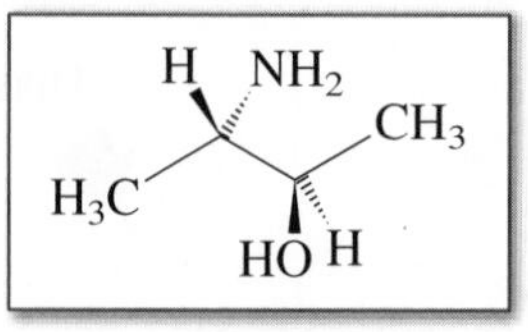

(2*R*,3*S*)-3-Amino-2-butanol (solid, mp 49°C)

Problem 7.19

Draw Fischer projections of the four stereoisomeric 3-amino-2-butanols, and label each erythro or threo as appropriate.

Problem 7.20

One other stereoisomer of 3-amino-2-butanol is a crystalline solid. Which one?

The situation is the same when the two chirality centers are present in a ring. There are four stereoisomeric 1-bromo-2-chlorocyclopropanes: a pair of enantiomers in which the halogens are trans and a pair in which they are cis. The cis compounds are diastereomers of the trans.

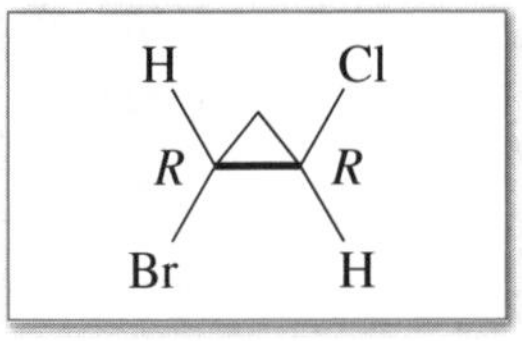

Enantiomers

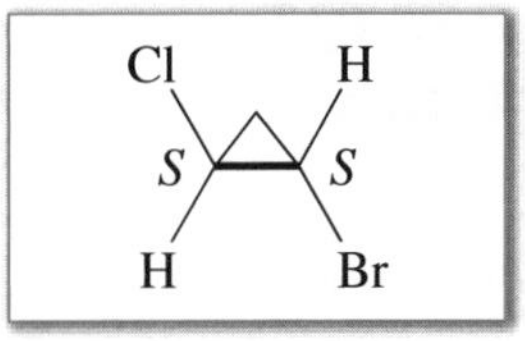

(1*R*,2*R*)-1-Bromo-2-chlorocyclopropane (1*S*,2*S*)-1-Bromo-2-chlorocyclopropane

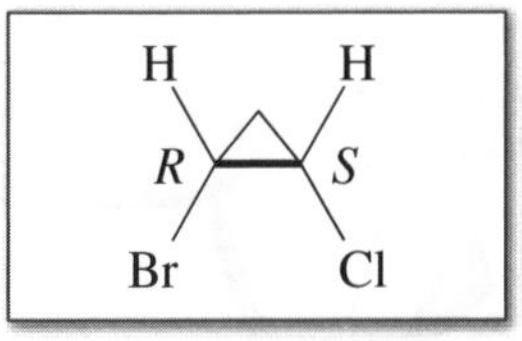

Enantiomers

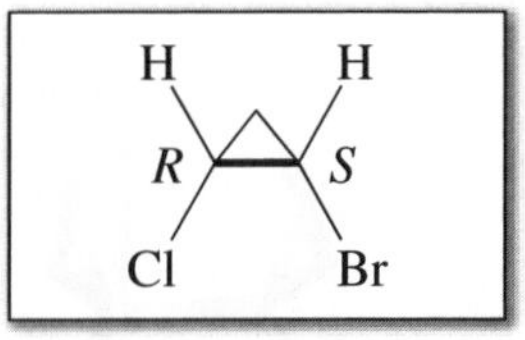

(1*R*,2*S*)-1-Bromo-2-chlorocyclopropane (1*S*,2*R*)-1-Bromo-2-chlorocyclopropane

A good thing to remember is that the cis and trans isomers of a particular compound are diastereomers of each other.

In Section 7.5, the term "relative configuration" was used to describe the stereochemical relationship between a single chirality center in one molecule to a chirality center in a different molecule. Relative configuration is also used to describe the way multiple chirality centers within the same molecule are related. The two erythro stereoisomers of 2,3-dihydroxybutanoic acid possess the same relative configuration. The relationship of one chirality center to the other is the same in both, but different from that in the threo stereoisomer.

Problem 7.21

Which stereoisomers of 1-bromo-2-chlorocyclopropane possess the same relative configuration?

7.12 Achiral Molecules with Two Chirality Centers

Now think about a molecule, such as 2,3-butanediol, which has two chirality centers that are equivalently substituted.

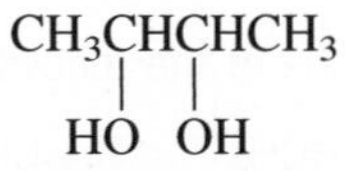

2,3-Butanediol

Only *three,* not four, stereoisomeric 2,3-butanediols are possible. These three are shown in Figure 7.9. The (2*R*,3*R*) and (2*S*,3*S*) forms are enantiomers and have equal and opposite optical rotations. A third combination of chirality centers, (2*R*,3*S*), however, gives an *achiral* structure that is superimposable on its (2*S*,3*R*) mirror image. Because it is achiral, this third stereoisomer is *optically inactive.* We call achiral molecules that have chirality centers **meso forms.** The meso form in Figure 7.9 is known as *meso*-2,3-butanediol.

One way to demonstrate that *meso*-2,3-butanediol is achiral is to recognize that its eclipsed conformation has a plane of symmetry that passes through and is perpendicular to the C-2—C-3 bond, as illustrated in Figure 7.10*a.* The anti conformation is achiral as well. As Figure 7.10*b* shows, this conformation is characterized by a center of symmetry at the midpoint of the C-2—C-3 bond.

Figure 7.9

Stereoisomeric 2,3-butanediols shown in their eclipsed conformations for convenience. Stereoisomers (*a*) and (*b*) are enantiomers. Structure (*c*) is a diastereomer of (*a*) and (*b*), and is achiral. It is called *meso*-2,3-butanediol.

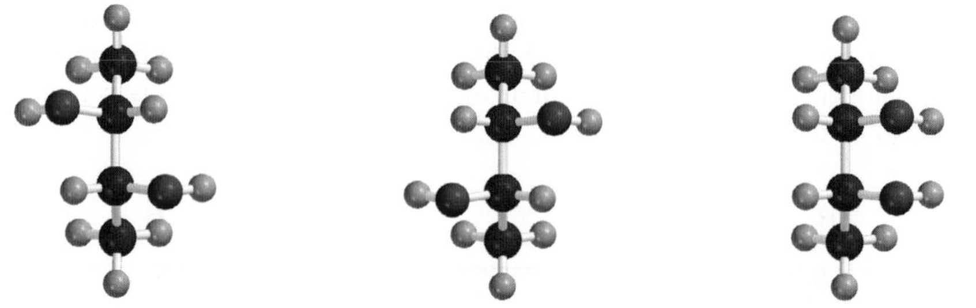

(2*R*,3*R*)-2,3-Butanediol (*a*) | (2*S*,3*S*)-2,3-Butanediol (*b*) | *meso*-2,3-Butanediol (*c*)

Figure 7.10

(*a*) The eclipsed conformation of *meso*-2,3-butanediol has a plane of symmetry. (*b*) The anti conformation of *meso*-2,3-butanediol has a center of symmetry.

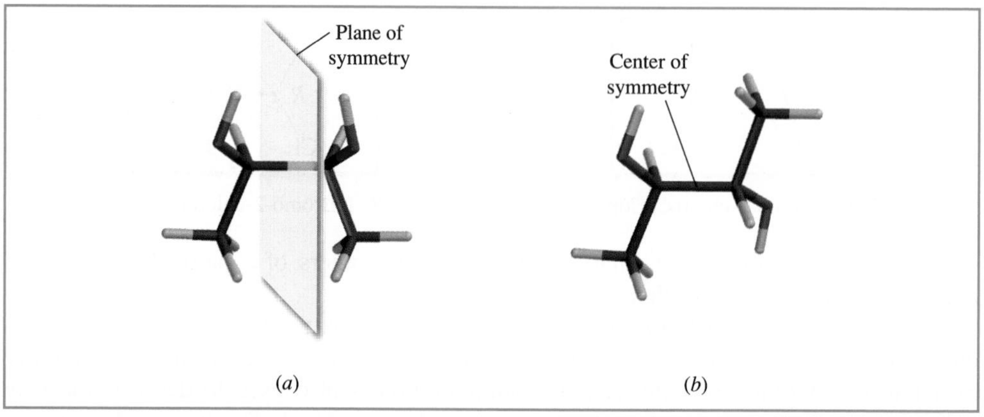

Fischer projection formulas can help us identify meso forms. Of the three stereoisomeric 2,3-butanediols, notice that only in the meso stereoisomer does a dashed line through the center of the Fischer projection divide the molecule into two mirror-image halves.

In the same way that a Fischer formula is a projection of the eclipsed conformation onto the page, the line drawn through its center is a projection of the plane of symmetry that is present in the eclipsed conformation of *meso*-2,3-butanediol.

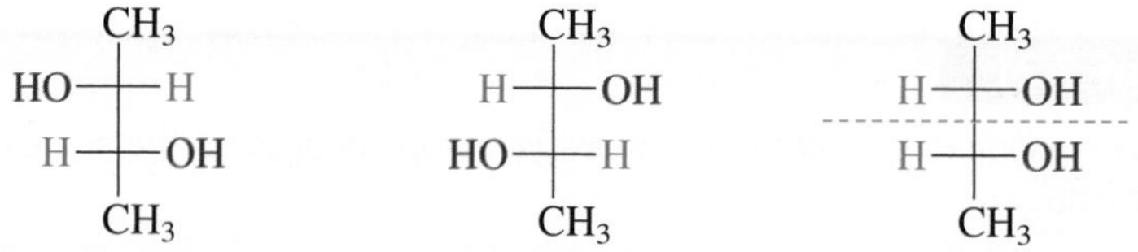

(2*R*,3*R*)-2,3-Butanediol | (2*S*,3*S*)-2,3-Butanediol | *meso*-2,3-Butanediol

When using Fischer projections for this purpose, however, be sure to remember what three-dimensional objects they stand for. One should not, for example, test for superimposition of the two chiral stereoisomers by a procedure that involves moving any part of a Fischer projection out of the plane of the paper in any step.

Problem 7.22

A meso stereoisomer is possible for one of the following compounds. Which one?

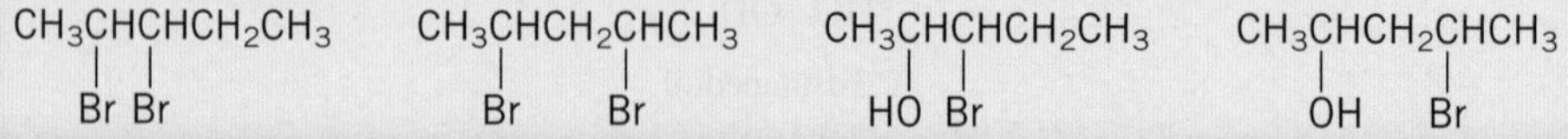

Turning to cyclic compounds, we see that there are three, not four, stereoisomeric 1,2-dibromocyclopropanes. Of these, two are enantiomeric *trans*-1,2-dibromocyclopropanes. The cis diastereomer is a meso form; it has a plane of symmetry.

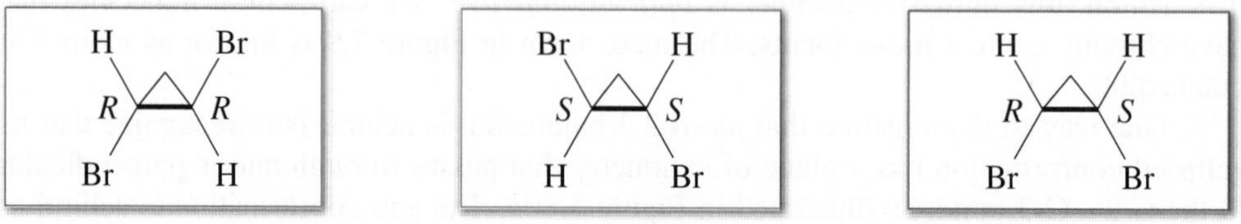

(1*R*,2*R*)-1,2-Dibromocyclopropane | (1*S*,2*S*)-1,2-Dibromocyclopropane | *cis*-1,2-Dibromocyclopropane

Problem 7.23

One of the stereoisomers of 1,3-dimethylcyclohexane is a meso form. Which one?

7.13 Molecules with Multiple Chirality Centers

Many naturally occurring compounds contain several chirality centers. By an analysis similar to that described for the case of two chirality centers, it can be shown that the maximum number of stereoisomers for a particular constitution is 2^n, where n is equal to the number of chirality centers.

Problem 7.24

Using *R* and *S* descriptors, write all the possible combinations for a molecule with three chirality centers.

When two or more of a molecule's chirality centers are equivalently substituted, meso forms are possible, and the number of stereoisomers is then less than 2^n. Thus, 2^n represents the *maximum* number of stereoisomers for a molecule containing n chirality centers.

The best examples of substances with multiple chirality centers are the *carbohydrates*. One class of carbohydrates, called *aldohexoses,* has the constitution

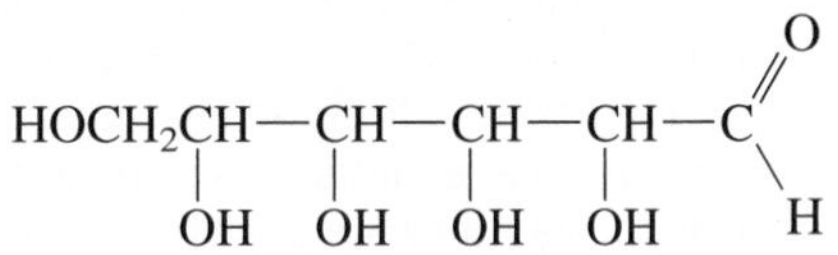

An aldohexose

Because there are four chirality centers and no possibility of meso forms, there are 2^4, or 16, stereoisomeric aldohexoses. All 16 are known, having been isolated either as natural products or as the products of chemical synthesis.

Problem 7.25

A second category of six-carbon carbohydrates, called *ketohexoses,* has the constitution shown. How many stereoisomeric 2-ketohexoses are possible?

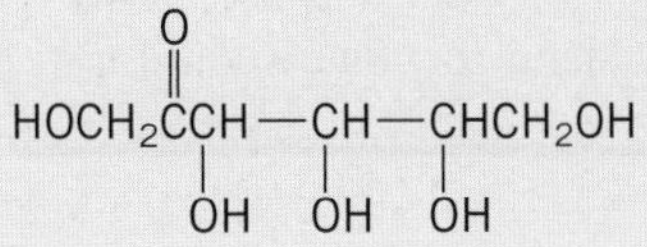

A 2-ketohexose

Steroids are another class of natural products with multiple chirality centers. One such compound is *cholic acid,* which can be obtained from bile. Its structural formula is given in Figure 7.11. Cholic acid has 11 chirality centers, and so a total (including cholic acid) of 2^{11}, or 2048, stereoisomers have this constitution. Of these 2048 stereoisomers, how many are diastereomers of cholic acid? Remember! Diastereomers are stereoisomers that are not enantiomers, and any object can have only one mirror image. Therefore, of the 2048 stereoisomers, one is cholic acid, one is its enantiomer, and the other 2046 are diastereomers of cholic acid. Only a small fraction of these compounds are known, and (+)-cholic acid is the only one ever isolated from natural sources.

Eleven chirality centers may seem like a lot, but it is nowhere close to a world record. It is a modest number when compared with the more than 100 chirality centers typical for most small proteins and the billions of chirality centers present in human DNA.

A molecule that contains both chirality centers and double bonds has additional opportunities for stereoisomerism. For example, the configuration of the chirality center

Chirality of Disubstituted Cyclohexanes

Disubstituted cyclohexanes present us with a challenging exercise in stereochemistry. Consider the seven possible dichlorocyclohexanes: 1,1-; *cis-* and *trans*-1,2-; *cis-* and *trans*-1,3-; and *cis-* and *trans*-1,4-. Which are chiral? Which are achiral?

Four isomers—the ones that are achiral because they have a plane of symmetry—are relatively easy to identify:

Achiral Dichlorocyclohexanes

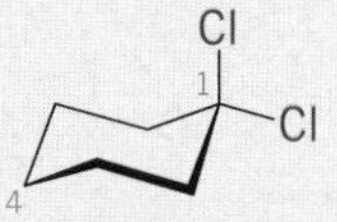

1,1
(plane of symmetry through C-1 and C-4)

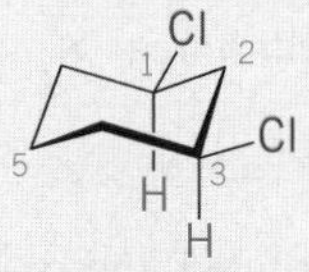

cis-1,3
(plane of symmetry through C-2 and C-5)

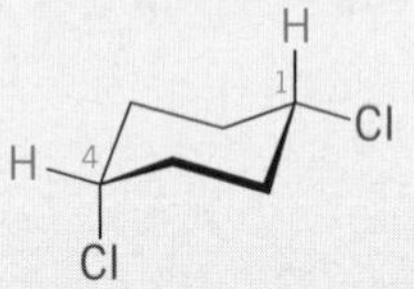

cis-1,4
(plane of symmetry through C-1 and C-4)

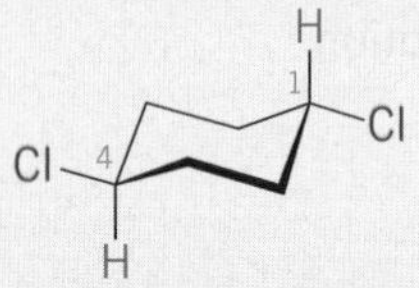

trans-1,4
(plane of symmetry through C-1 and C-4)

The remaining three isomers are chiral:

Chiral Dichlorocyclohexanes

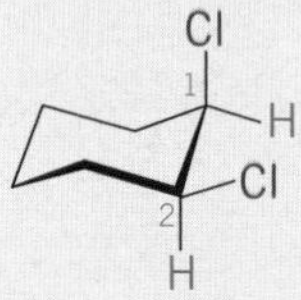

cis-1,2

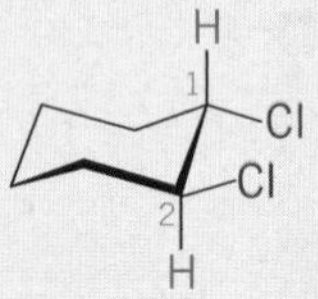

trans-1,2

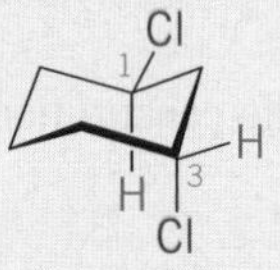

trans-1,3

Among all the isomers, *cis*-1,2-dichlorocyclohexane is unique in that the ring-inverting process typical of cyclohexane derivatives converts it to its enantiomer.

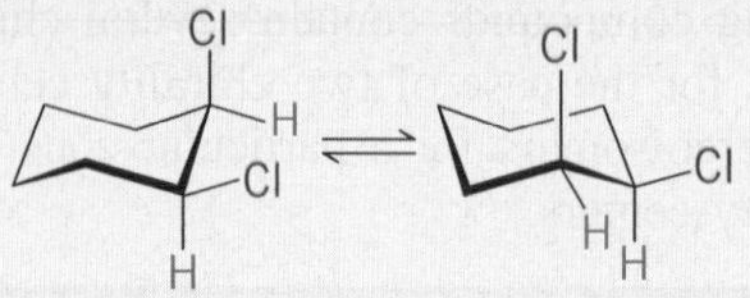

A A′

which is equivalent to

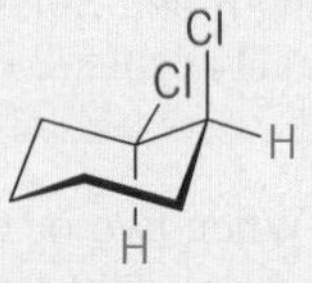

A′

Structures A and A′ are nonsuperimposable mirror images of each other. Thus although *cis*-1,2-dichlorocyclohexane is chiral, it is optically inactive when chair–chair interconversion occurs. Such interconversion is rapid at room temperature and converts optically active A to a racemic mixture of A and A′. Because A and A′ are enantiomers interconvertible by a conformational change, they are sometimes referred to as **conformational enantiomers.**

The same kind of spontaneous racemization occurs for any *cis*-1,2 disubstituted cyclohexane in which both substituents are the same. Because such compounds are chiral, it is incorrect to speak of them as meso compounds, which are achiral molecules that have chirality centers. Rapid chair–chair interconversion, however, converts them to a 1:1 mixture of enantiomers, and this mixture is optically inactive.

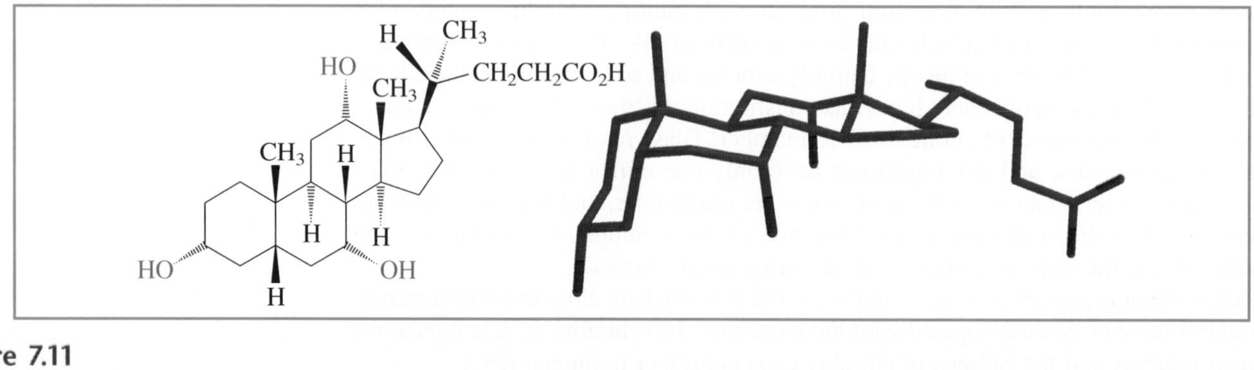

Figure 7.11

Cholic acid. Its 11 chirality centers are those carbons at which stereochemistry is indicated in the structural drawing at the left. The molecular model at the right more clearly shows the overall shape of the molecule.

in 3-penten-2-ol may be either *R* or *S*, and the double bond may be either *E* or *Z*. Therefore 3-penten-2-ol has four stereoisomers even though it has only one chirality center.

n in 2^n includes double bonds capable of stereochemical variation (*E*, *Z*) as well as chirality centers.

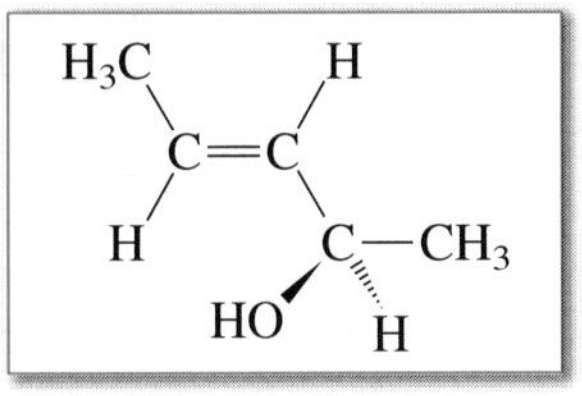

(2*R*,3*E*)-3-Penten-2-ol

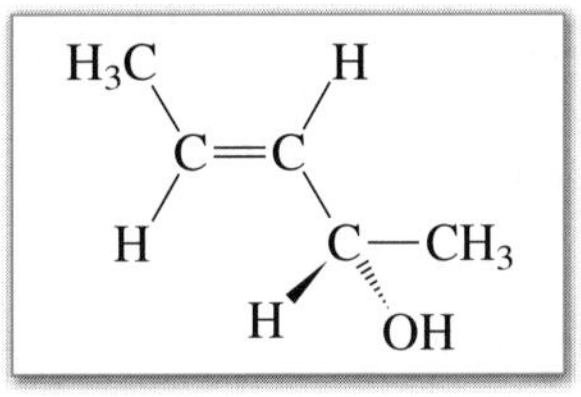

(2*S*,3*E*)-3-Penten-2-ol

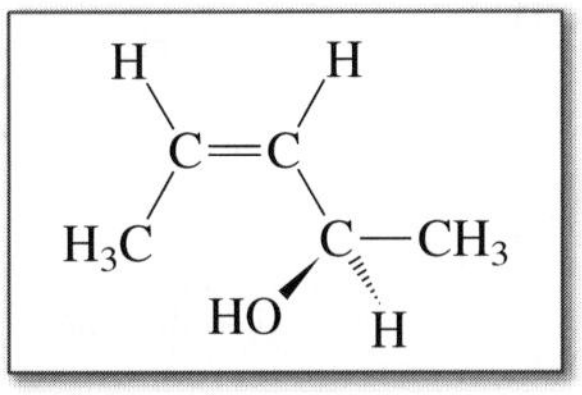

(2*R*,3*Z*)-3-Penten-2-ol

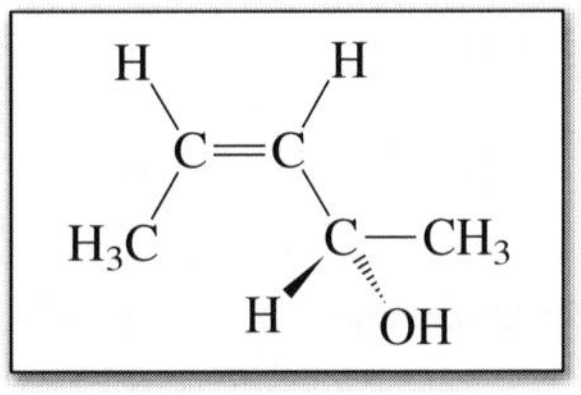

(2*S*,3*Z*)-3-Penten-2-ol

The relationship of the (2*R*,3*E*) stereoisomer to the others is that it is the enantiomer of (2*S*,3*E*)-3-penten-2-ol and is a diastereomer of the (2*R*,3*Z*) and (2*S*,3*Z*) isomers.

7.14 Reactions That Produce Diastereomers

Once we grasp the idea of stereoisomerism in molecules with two or more chirality centers, we can explore further details of addition reactions of alkenes.

When bromine adds to (*Z*)- or (*E*)-2-butene, the product 2,3-dibromobutane contains two equivalently substituted chirality centers:

$$\underset{(Z)\text{- or }(E)\text{-2-butene}}{CH_3CH{=}CHCH_3} \xrightarrow{Br_2} \underset{\text{2,3-Dibromobutane}}{CH_3\underset{\displaystyle Br}{\underset{|}{C}}H\underset{\displaystyle Br}{\underset{|}{C}}HCH_3}$$

Three stereoisomers are possible: a pair of enantiomers and a meso form.

Two factors combine to determine which stereoisomers are actually formed in the reaction.

1. The (*E*)- or (*Z*)-configuration of the starting alkene
2. The anti stereochemistry of addition (Section 6.15)

Figure 7.12 shows the stereochemical differences associated with anti addition of bromine to (*E*)- and (*Z*)-2-butene, respectively. The trans alkene (*E*)-2-butene yields only *meso*-2,3-dibromobutane, but the cis alkene (*Z*)-2-butene gives a racemic mixture of (2*R*,3*R*)- and (2*S*,3*S*)-2,3-dibromobutane.

Bromine addition to alkenes is a **stereospecific reaction,** a reaction in which stereoisomeric starting materials yield products that are stereoisomers of each other. In this case the starting materials, in separate reactions, are the *E* and *Z* stereoisomers of 2-butene. The chiral dibromides formed from (*Z*)-2-butene are stereoisomers (diastereomers) of the meso dibromide from (*E*)-2-butene.

Notice too that, consistent with the principle developed in Section 7.10, optically inactive starting materials (achiral alkenes and bromine) yield optically inactive products (a racemic mixture or a meso structure) in these reactions.

Problem 7.26

Epoxidation of alkenes is a stereospecific syn addition. Which stereoisomer of 2-butene reacts with peroxyacetic acid to give *meso*-2,3-epoxybutane? Which one gives a racemic mixture of (2*R*,3*R*)- and (2*S*,3*S*)-2,3-epoxybutane?

Figure 7.12

Addition of Br_2 to (*E*)- and (*Z*)-2-butene is stereospecific. Stereoisomeric products are formed from stereoisomeric reactants.

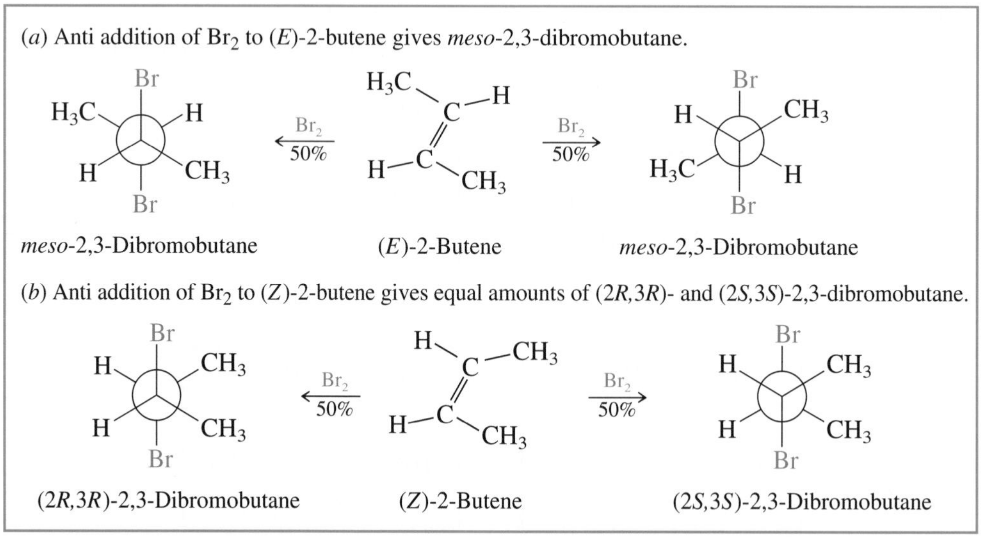

A reaction that introduces a second chirality center into a starting material that already has one need not produce equal quantities of two possible diastereomers. Consider catalytic hydrogenation of 2-methyl(methylene)cyclohexane. As you might expect, both *cis*- and *trans*-1,2-dimethylcyclohexane are formed.

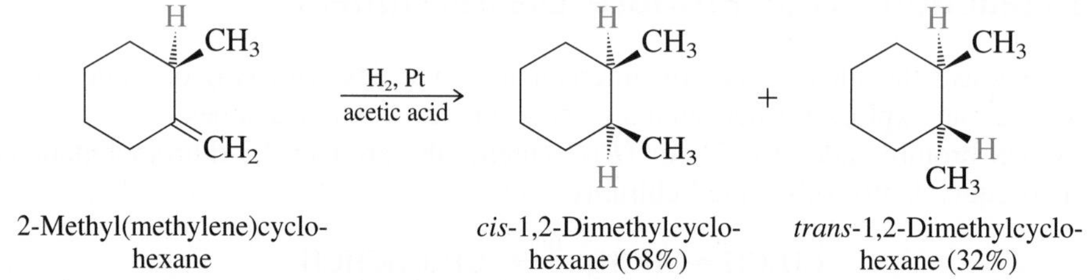

The relative amounts of the two products, however, are not equal; more *cis*-1,2-dimethylcyclohexane is formed than *trans*-. The reason for this is that it is the less hindered face of the double bond that approaches the catalyst surface and is the face to which hydrogen is transferred. Hydrogenation of 2-methyl(methylene)cyclohexane occurs preferentially at the side of the double bond opposite that of the methyl group and leads to a faster rate of formation of the cis stereoisomer of the product.

Problem 7.27

Could the fact that hydrogenation of 2-methyl(methylene)cyclohexane gives more *cis*-1,2-dimethylcyclohexane than *trans*- be explained on the basis of the relative stabilities of the two stereoisomeric products?

The two faces of the double bond in 2-methyl(methylene)cyclohexane are prochiral. They are not, however, enantiotopic as in the alkenes we discussed in Section 7.10. In those earlier examples, when addition to the double bond created a new chirality center, attack at one face gave one enantiomer; attack at the other gave the other enantiomer. In the case of 2-methyl(methylene)cyclohexane, which already has one chirality center, attack at opposite faces of the double bond gives two products that are diastereomers of each other. Prochiral faces of this type are called **diastereotopic.**

The hydrogenation of 2-methyl(methylene)cyclohexane is an example of a *stereoselective reaction,* meaning one in which stereoisomeric products are formed in unequal amounts from a single starting material.

A common misconception is that a stereospecific reaction is simply one that is 100% stereoselective. The two terms are not synonymous, however. A stereospecific reaction is one which, when carried out with stereoisomeric starting materials, gives a product from one reactant that is a stereoisomer of the product from the other. A stereoselective reaction is one in which a single starting material gives a predominance of a single stereoisomer when two or more are possible. *Stereospecific* is more closely connected with features of the reaction than with the reactant. Thus terms such as syn *addition* and anti *elimination* describe the stereospecificity of reactions. *Stereoselective* is more closely connected with structural effects in the reactant as expressed in terms such as *addition to the less hindered side.* For example, syn addition describes stereospecificity in the catalytic hydrogenation of alkenes, whereas the preference for addition to the less hindered face of the double bond describes stereoselectivity.

Note that the terms *regioselective* and *regiospecific*, however, are defined in terms of each other. A regiospecific reaction is one that is 100% regioselective.

7.15 Resolution of Enantiomers

The separation of a racemic mixture into its enantiomeric components is termed **resolution.** The first resolution, that of tartaric acid, was carried out by Louis Pasteur in 1848. Tartaric acid is a byproduct of wine making and is almost always found as its dextrorotatory 2*R*,3*R* stereoisomer, shown here in a perspective drawing and in a Fischer projection.

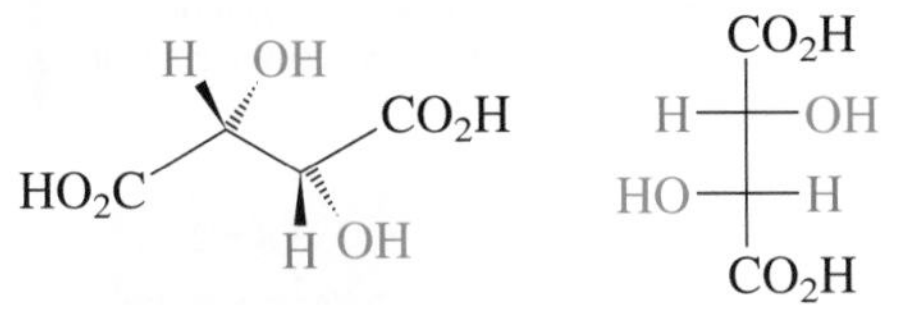
H
OH
CO_2H
HO_2C
H
OH
CO_2H
H
OH
HO
H
CO_2H

(2*R*,3*R*)-Tartaric acid (mp 170°C, $[\alpha]_D$ +12°)

Problem 7.28

There are two other stereoisomeric tartaric acids. Write their Fischer projections, and specify the configuration at their chirality centers.

Occasionally, an optically inactive sample of tartaric acid was obtained. Pasteur noticed that the sodium ammonium salt of optically inactive tartaric acid was a mixture of two mirror-image crystal forms. With microscope and tweezers, Pasteur carefully separated the two. He found that one kind of crystal (in aqueous solution) was dextrorotatory, whereas the mirror-image crystals rotated the plane of polarized light an equal amount but were levorotatory.

Although Pasteur was unable to provide a structural explanation—that had to wait for van't Hoff and Le Bel a quarter of a century later—he correctly deduced that the enantiomeric quality of the crystals was the result of enantiomeric molecules. The rare form of tartaric acid was optically inactive because it contained equal amounts of (+)-tartaric acid and (−)-tartaric acid. It had earlier been called *racemic acid* (from Latin *racemus,* meaning "a bunch of grapes"), a name that subsequently gave rise to our present term for an equal mixture of enantiomers.

Problem 7.29

Could the unusual, optically inactive form of tartaric acid studied by Pasteur have been *meso*-tartaric acid?

Pasteur's technique of separating enantiomers not only is laborious but requires that the crystals of the enantiomers be distinguishable. This happens very rarely. Consequently, alternative and more general approaches for resolving enantiomers have been developed. Most are based on a strategy of temporarily converting the enantiomers of a racemic mixture to diastereomeric derivatives, separating these diastereomers, then regenerating the enantiomeric starting materials.

Figure 7.13 illustrates this strategy. Say we have a mixture of enantiomers, which, for simplicity, we label as C(+) and C(−). Assume that C(+) and C(−) bear some functional group that can combine with a reagent P to yield adducts C(+)-P and C(−)-P. Now, if reagent P is chiral, and if only a single enantiomer of P, say, P(+), is added to a racemic mixture of C(+) and C(−), as shown in the first step of Figure 7.13, then the

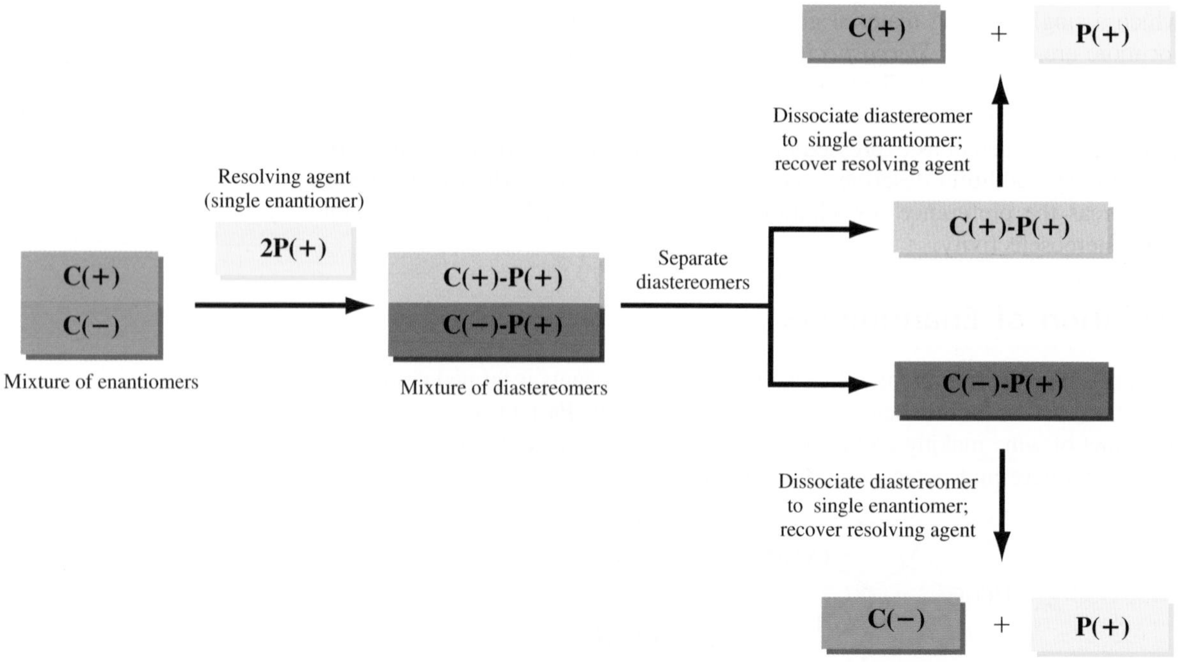

Figure 7.13

The general procedure for resolving a chiral substance into its enantiomers. Reaction with a single enantiomer of a chiral resolving agent P(+) converts the racemic mixture of enantiomers C(+) and C(−) to a mixture of diastereomers C(+)-P(+) and C(−)-P(+). The mixture of diastereomers is separated—by fractional crystallization, for example. A chemical reaction is then carried out to convert diastereomer C(+)-P(+) to C(+) and the resolving agent P(+). Likewise, diastereomer C(−)-P(+) is converted to C(−) and P(+). C(+) has been separated from C(−), and the resolving agent P(+) can be recovered for further use.

Most resolving agents are isolated as single enantiomers from natural sources. *S*-(−)-Malic acid is obtained from apples.

products of the reaction are C(+)-P(+) and C(−)-P(+). These products are not mirror images; they are diastereomers. Diastereomers can have different physical properties, which can serve as a means of separating them. The mixture of diastereomers is separated, usually by recrystallization from a suitable solvent. In the last step, an appropriate chemical transformation liberates the enantiomers and restores the resolving agent.

Whenever possible, the chemical reactions involved in the formation of diastereomers and their conversion to separate enantiomers are simple acid–base reactions. For example, naturally occurring (*S*)-(−)-malic acid is often used to resolve amines. One such amine that has been resolved in this way is 1-phenylethylamine. Amines are bases, and malic acid is an acid. Proton transfer from (*S*)-(−)-malic acid to a racemic mixture of (*R*)- and (*S*)-1-phenylethylamine gives a mixture of diastereomeric salts.

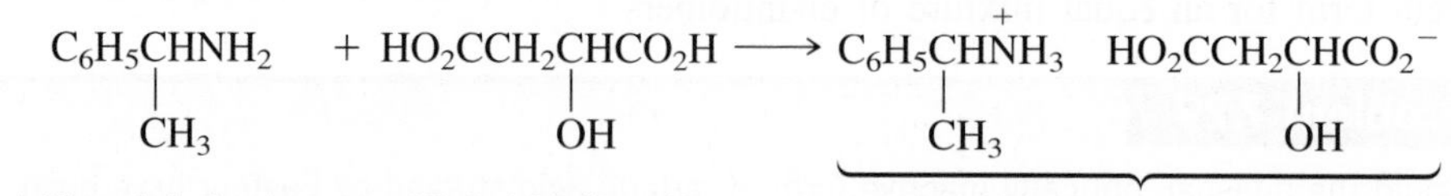

The diastereomeric salts are separated and the individual enantiomers of the amine liberated by treatment with a base:

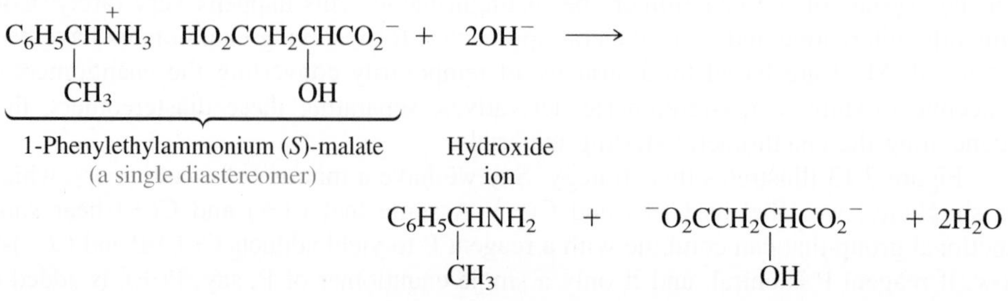

Problem 7.30

In the resolution of 1-phenylethylamine using (*S*)-(−)-malic acid, the compound obtained by recrystallization of the mixture of diastereomeric salts is (*R*)-1-phenylethylammonium (*S*)-malate. The other component of the mixture is more soluble and remains in solution. What is the configuration of the more soluble salt?

This method is widely used for the resolution of chiral amines and carboxylic acids. Analogous methods based on the formation and separation of diastereomers have been developed for other functional groups; the precise approach depends on the kind of chemical reactivity associated with the functional groups present in the molecule.

As the experimental tools for biochemical transformations have become more powerful and procedures for carrying out these transformations in the laboratory more routine, the application of biochemical processes to mainstream organic chemical tasks including the production of enantiomerically pure chiral molecules has grown.

Another approach, called **kinetic resolution,** depends on the different rates of reaction of two enantiomers with a chiral reagent. A very effective form of kinetic resolution uses enzymes as chiral catalysts to selectively bring about the reaction of one enantiomer in a racemic mixture **(enzymatic resolution).** *Lipases,* or *esterases,* enzymes that catalyze ester hydrolysis, are often used. In a typical procedure, one enantiomer of the acetate ester of a racemic alcohol undergoes hydrolysis and the other is left unchanged when hydrolyzed in the presence of an esterase from hog liver.

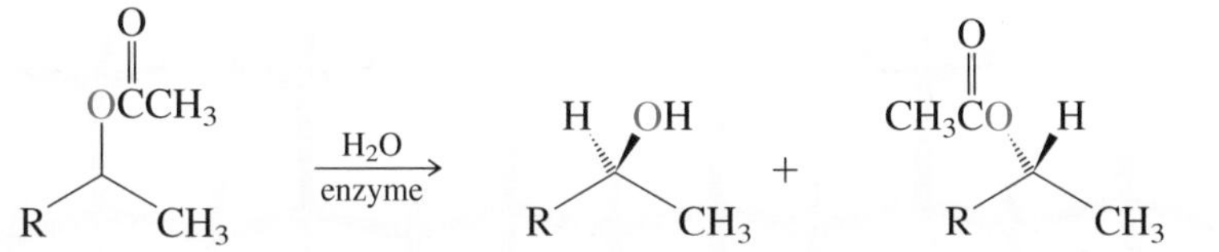

Acetate ester of racemic alcohol (*S*)-Alcohol Acetate ester of (*R*)-alcohol

High yields of the enantiomerically pure alcohol and enantiomerically pure ester are regularly achieved. The growing interest in chiral drugs (see the boxed essay on this topic, p. 294) has stimulated the development of large-scale enzymatic resolution as a commercial process.

7.16 Stereoregular Polymers

Before the development of the Ziegler–Natta catalyst systems (Section 6.21), polymerization of propene was not a reaction of much value. The reason for this has a stereochemical basis. Consider a section of *polypropylene:*

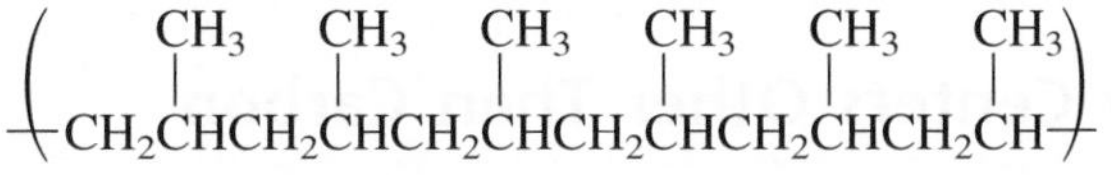

Representing the polymer chain in an extended zigzag conformation, as shown in Figure 7.14, reveals several distinct structural possibilities differing with respect to the relative configurations of the carbons that bear the methyl groups.

One structure, represented in Figure 7.14*a,* has all the methyl groups oriented in the same direction with respect to the polymer chain. This stereochemical arrangement is called **isotactic.** Another form called **syndiotactic,** shown in Figure 7.14*b,* has its methyl groups alternating front and back along the chain. Both isotactic and syndiotactic polypropylene are known as **stereoregular polymers** because each is characterized by a precise stereochemistry at the carbon atom that bears the methyl group. A third possibility, shown in Figure 7.14*c,* is described as **atactic.** Atactic polypropylene has a random orientation of its methyl groups; it is not a stereoregular polymer.

Polypropylene chains associate with one another because of attractive van der Waals forces. The extent of this association is relatively large for isotactic and syndiotactic polymers, because the stereoregularity of the polymer chains permits efficient packing. Atactic polypropylene, on the other hand, does not associate as strongly. It has a lower density and lower melting point than the stereoregular forms. The physical properties of stereoregular polypropylene are more useful for most purposes than those of atactic polypropylene.

Most polypropylene products are made from isotactic polypropylene.

Figure 7.14

Polymers of propene. The main chain is shown in a zigzag conformation. Every other carbon bears a methyl substituent and is a chirality center. (*a*) All the methyl groups are on the same side of the carbon chain in isotactic polypropylene. (*b*) Methyl groups alternate from one side to the other in syndiotactic polypropylene. (*c*) The spatial orientation of the methyl groups is random in atactic polypropylene.

(*a*) Isotactic polypropylene

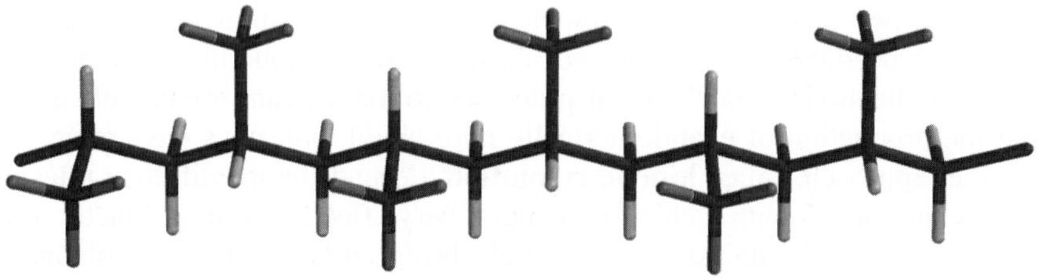

(*b*) Syndiotactic polypropylene

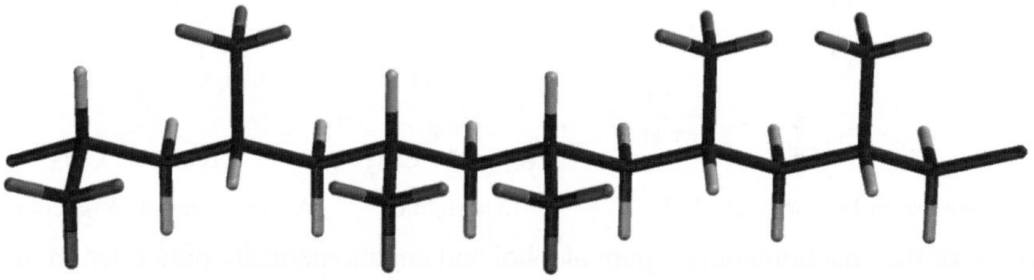

(*c*) Atactic polypropylene

When propene is polymerized under free-radical conditions, the polypropylene that results is atactic. Catalysts of the Ziegler–Natta type, however, permit the preparation of either isotactic or syndiotactic polypropylene. We see here an example of how proper choice of experimental conditions can affect the stereochemical course of a chemical reaction to the extent that entirely new materials with unique properties result.

7.17 Chirality Centers Other Than Carbon

Our discussion to this point has been limited to molecules in which the chirality center is carbon. Atoms other than carbon may also be chirality centers. Silicon, like carbon, has a tetrahedral arrangement of bonds when it bears four substituents. A large number of organosilicon compounds in which silicon bears four different groups have been resolved into their enantiomers.

Trigonal pyramidal molecules are chiral if the central atom bears three different groups. If one is to resolve substances of this type, however, the pyramidal inversion that interconverts enantiomers must be slow at room temperature. Pyramidal inversion at nitrogen is so fast that attempts to resolve chiral amines fail because of their rapid racemization.

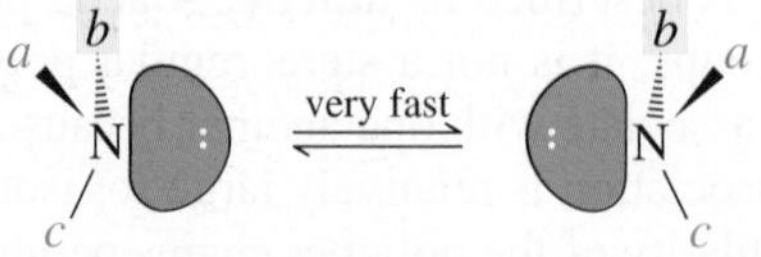

Phosphorus is in the same group of the periodic table as nitrogen, and tricoordinate phosphorus compounds (phosphines), like amines, are trigonal pyramidal. Phosphines, however, undergo pyramidal inversion much more slowly than amines, and a number of optically active phosphines have been prepared.

Tricoordinate sulfur compounds are chiral when sulfur bears three different groups. The rate of pyramidal inversion at sulfur is rather slow. The most common compounds in which sulfur is a chirality center are sulfoxides such as:

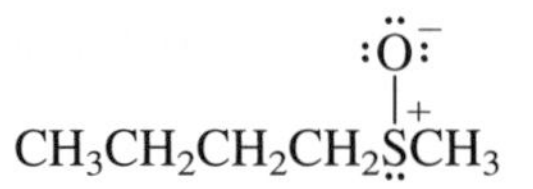

Butyl methyl sulfoxide

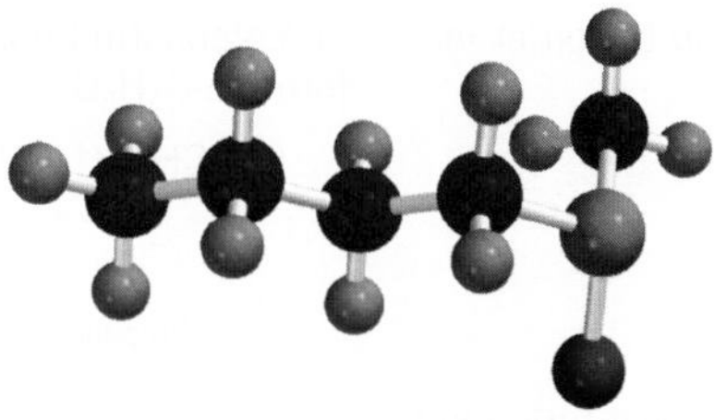

(*S*)-(+)-Butyl methyl sulfoxide

The absolute configuration at sulfur is specified by the Cahn–Ingold–Prelog method with the provision that the unshared electron pair is considered to be the lowest ranking substituent. Modafinil, a drug used to treat sleep disorders, is a chiral sulfoxide, and is dispensed as a racemic mixture.

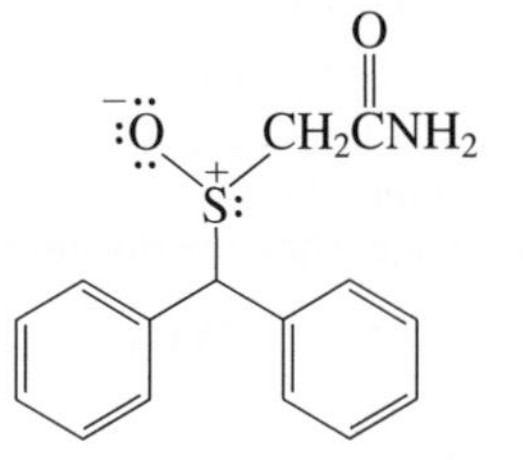

7.18 SUMMARY

Chemistry in three dimensions is known as **stereochemistry.** At its most fundamental level, stereochemistry deals with molecular structure; at another level, it is concerned with chemical reactivity. Table 7.2 summarizes some basic definitions relating to molecular structure and stereochemistry.

Section 7.1 A molecule is **chiral** if it cannot be superimposed on its mirror image. *Nonsuperimposable mirror images* are **enantiomers** of one another. Molecules in which mirror images are superimposable are achiral.

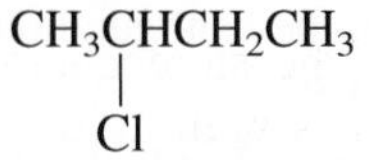

2-Chlorobutane
(chiral)

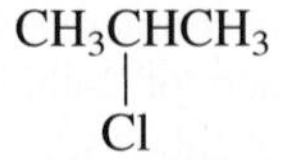

2-Chloropropane
(achiral)

Section 7.2 The most common kind of chiral molecule contains a carbon atom that bears four different atoms or groups. Such an atom is called a **chirality center.** Table 7.2 shows the enantiomers of 2-chlorobutane. C-2 is a chirality center in 2-chlorobutane.

Section 7.3 A molecule that has a plane of symmetry or a center of symmetry is achiral. *cis*-4-Methylcyclohexanol (Table 7.2) has a plane of symmetry that bisects the molecule into two mirror-image halves and is achiral. The same can be said for *trans*-4-methylcyclohexanol.

Section 7.4 **Optical activity,** or the degree to which a substance rotates the plane of polarized light, is a physical property used to characterize chiral substances. Enantiomers have equal and opposite optical rotations. To be optically active a substance must be chiral, and one enantiomer must be present in excess of the other. A **racemic mixture** is optically inactive and contains equal quantities of enantiomers.

Section 7.5 **Relative configuration** compares the arrangement of atoms in space to some reference. The prefix *cis* in *cis*-4-methylcyclohexanol, for example, describes relative configuration by referencing the orientation of the CH_3 group to the OH. **Absolute configuration** is an exact description of the arrangement of atoms in space.

TABLE 7.2 Classification of Isomers

Definition	Example
Isomers are different compounds that have the same molecular formula. They may be either constitutional isomers or stereoisomers.	
1. *Constitutional isomers* are isomers that differ in the order in which their atoms are connected.	Three constitutionally isomeric compounds have the molecular formula C_3H_8O: $CH_3CH_2CH_2OH$ (1-Propanol) $CH_3CH(OH)CH_3$ (2-Propanol) $CH_3CH_2OCH_3$ (Ethyl methyl ether)
2. *Stereoisomers* are isomers that have the same constitution but differ in the arrangement of their atoms in space.	
(a) *Enantiomers* are stereoisomers that are related as an object and its nonsuperimposable mirror image.	The two enantiomeric forms of 2-chlorobutane are  (*R*)-(−)-2-Chlorobutane and (*S*)-(+)-2-Chlorobutane
(b) *Diastereomers* are stereoisomers that are not mirror images.	The cis and trans isomers of 4-methylcyclohexanol are stereoisomers, but they are not related as an object and its mirror image; they are diastereomers.  *cis*-4-Methylcyclohexanol *trans*-4-Methylcyclohexanol

Section 7.6 Absolute configuration in chiral molecules is best specified using the prefixes *R* and *S* of the Cahn–Ingold–Prelog notational system. Substituents at a chirality center are ranked in order of decreasing precedence. If the three highest ranked substituents trace a clockwise path (highest→second highest→third highest) when the lowest ranked substituent is held away from you, the configuration is *R*. If the path is counterclockwise, the configuration is *S*. Table 7.2 shows the *R* and *S* enantiomers of 2-chlorobutane.

Section 7.7 A **Fischer projection** shows how a molecule would look if its bonds were projected onto a flat surface. Horizontal lines represent bonds pointing toward you; vertical lines represent bonds pointing away from you. The projection is normally drawn so that the carbon chain is vertical, with the lowest numbered carbon at the top.

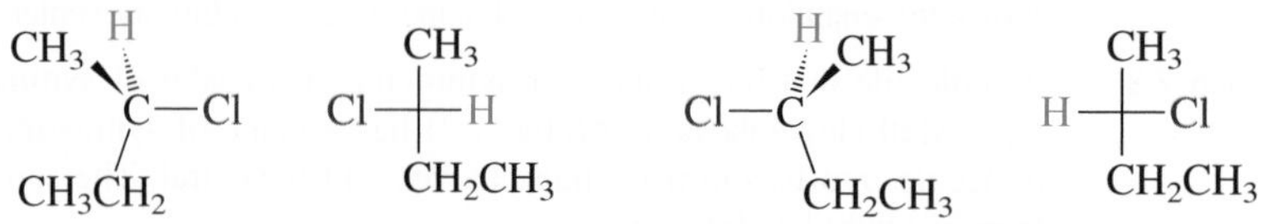

(*R*)-2-Chlorobutane   (*S*)-2-Chlorobutane

Section 7.8 Both enantiomers of the same substance are identical in most of their physical properties. The most prominent differences are biological ones, such as taste and odor, in which the substance interacts with a chiral receptor site. Enantiomers also have important consequences in medicine, in which the two enantiomeric forms of a drug can have much different effects on a patient.

Section 7.9 Molecules without chirality centers can be chiral. Biphenyls that are substituted can exhibit an **axis of chirality.** When A ≠ B, and X ≠ Y, the two conformations are nonsuperimposable mirror images of each other; that is, they are enantiomers. The bond connecting the two rings lies along a chirality axis.

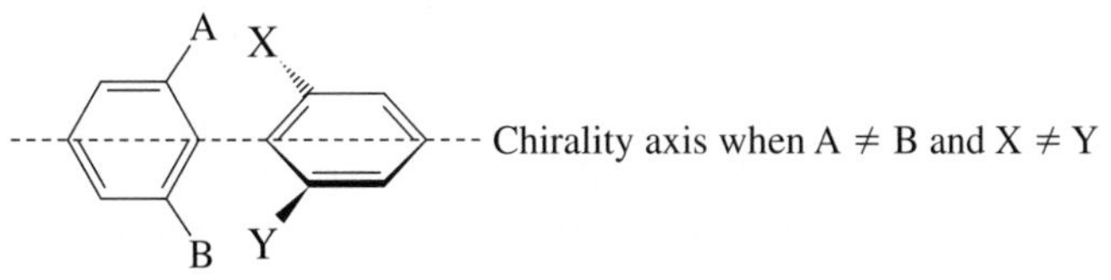
A
X
B
Y
Chirality axis when A ≠ B and X ≠ Y

Section 7.10 A chemical reaction can convert an achiral substance to a chiral one. If the product contains a single chirality center, it is formed as a racemic mixture. Optically active products can be formed from optically inactive starting materials only if some optically active agent is present. The best examples are biological processes in which enzymes catalyze the formation of only a single enantiomer.

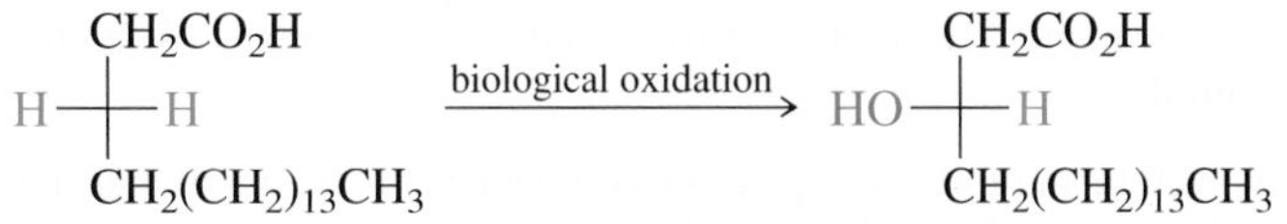
CH_2CO_2H
H
H
$CH_2(CH_2)_{13}CH_3$
biological oxidation
CH_2CO_2H
HO
H
$CH_2(CH_2)_{13}CH_3$

Stearic acid — (*S*)-3-Hydroxystearic acid

Section 7.11 When a molecule has two chirality centers and these two chirality centers are not equivalent, four stereoisomers are possible.

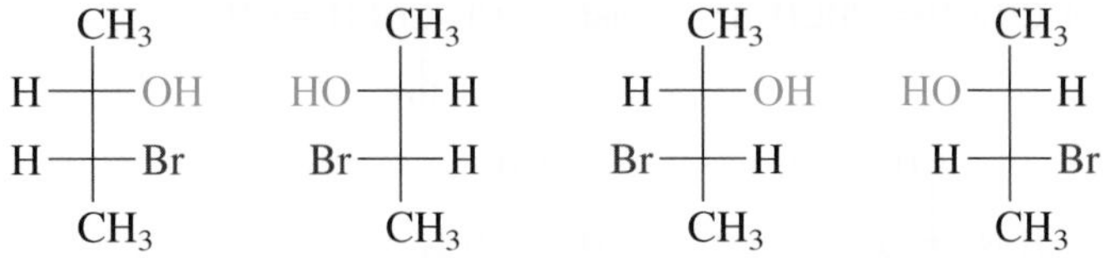
CH_3
H
OH
H
Br
CH_3
CH_3
HO
H
Br
H
CH_3
CH_3
H
OH
Br
H
CH_3
CH_3
HO
H
H
Br
CH_3

Enantiomers of *erythro*-3-bromo-2-butanol — Enantiomers of *threo*-3-bromo-2-butanol

Stereoisomers that are not mirror images are classified as **diastereomers.** Each enantiomer of *erythro*-3-bromo-2-butanol is a diastereomer of each enantiomer of *threo*-3-bromo-2-butanol.

Section 7.12 Achiral molecules that contain chirality centers are called **meso forms.** Meso forms typically contain (but are not limited to) two equivalently substituted chirality centers. They are optically inactive.

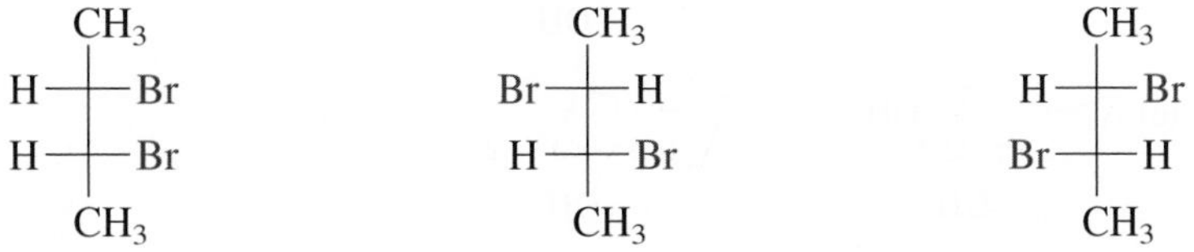
CH_3
H
Br
H
Br
CH_3
CH_3
Br
H
H
Br
CH_3
CH_3
H
Br
Br
H
CH_3

meso-2,3-Dibromobutane — (2*R*,3*R*)-2,3-Dibromobutane — (2*S*,3*S*)-2,3-Dibromobutane

Section 7.13 For a particular constitution, the maximum number of stereoisomers is 2^n, where n is the number of structural units capable of stereochemical variation—usually this is the number of chirality centers, but can include *E* and *Z* double bonds as well. The number of stereoisomers is reduced to less than 2^n when there are meso forms.

Section 7.14 Addition reactions of alkenes may generate one (Section 7.10) or two (Section 7.14) chirality centers. When two chirality centers are produced, their relative stereochemistry depends on the configuration (*E* or *Z*) of the alkene and whether the addition is syn or anti.

Section 7.15 **Resolution** is the separation of a racemic mixture into its enantiomers. It is normally carried out by converting the mixture of enantiomers to a mixture of diastereomers, separating the diastereomers, then regenerating the enantiomers.

Section 7.16 Certain polymers such as polypropylene contain chirality centers, and the relative configurations of these centers affect the physical properties of the polymers. Like substituents appear on the same side of a zigzag carbon chain in an **isotactic** polymer, alternate along the chain in a **syndiotactic** polymer, and appear in a random manner in an **atactic** polymer. Isotactic and syndiotactic polymers are referred to as **stereoregular** polymers.

Section 7.17 Atoms other than carbon can be chirality centers. Examples include those based on tetracoordinate silicon and tricoordinate sulfur as the chirality center. In principle, tricoordinate nitrogen can be a chirality center in compounds of the type N(*x, y, z*), where *x, y,* and *z* are different, but inversion of the nitrogen pyramid is so fast that racemization occurs virtually instantly at room temperature.

PROBLEMS

7.31 Which of the isomeric alcohols having the molecular formula $C_5H_{12}O$ are chiral? Which are achiral?

7.32 Write structural formulas for all the compounds that are trichloro derivatives of cyclopropane. (Don't forget to include stereoisomers.) Which are chiral? Which are achiral?

7.33 In each of the following pairs of compounds one is chiral and the other is achiral. Identify each compound as chiral or achiral, as appropriate.

(a) $ClCH_2CHCH_2OH$ (OH on the central CH) and $HOCH_2CHCH_2OH$ (Cl on the central CH)

(b) $CH_3CH{=}CHCH_2Br$ and $CH_3CHCH{=}CH_2$ (Br on the CH)

(c) 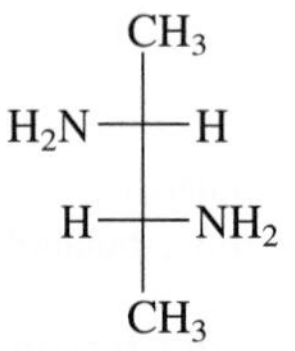 and 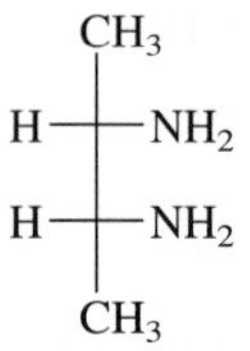

(d) 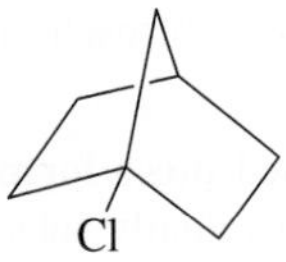 and 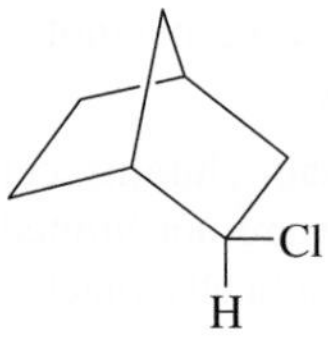

(e) 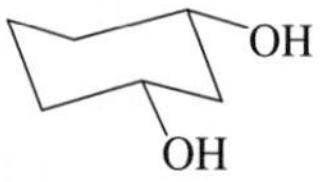 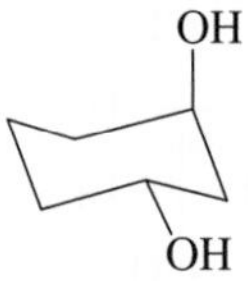

(f) 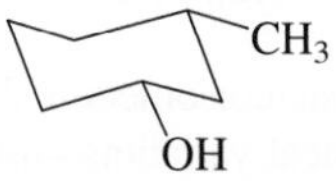 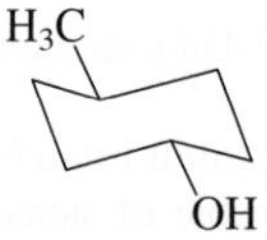

7.34 Compare 2,3-pentanediol and 2,4-pentanediol with respect to the number of stereoisomers possible for each constitution. Which stereoisomers are chiral? Which are achiral?

7.35 In 1996, it was determined that the absolute configuration of (−)-bromochlorofluoromethane is *R*. Which of the following is (are) (−)-BrClFCH?

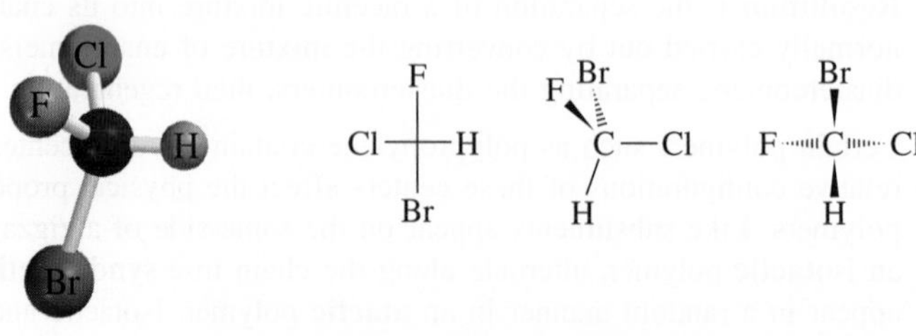

7.36 Specify the configuration of the chirality center as *R* or *S* in each of the following.

(a) (−)-2-Octanol

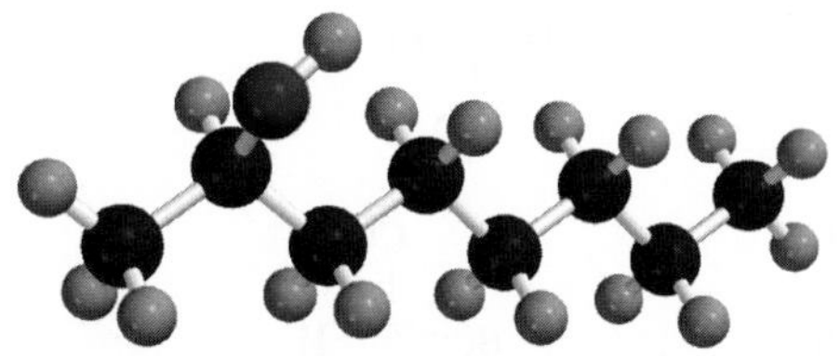

(b) Monosodium L-glutamate (only this stereoisomer is of any value as a flavor-enhancing agent)

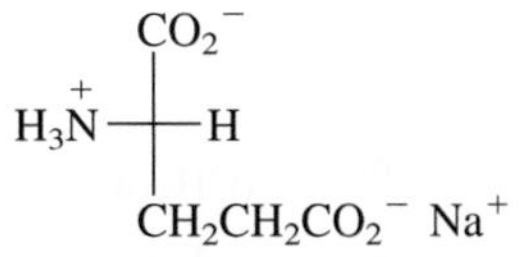

7.37 A subrule of the Cahn–Ingold–Prelog system specifies that higher mass number takes precedence over lower when distinguishing between isotopes.

(a) Determine the absolute configurations of the reactant and product in the biological oxidation of isotopically labeled ethane described in Section 7.2.

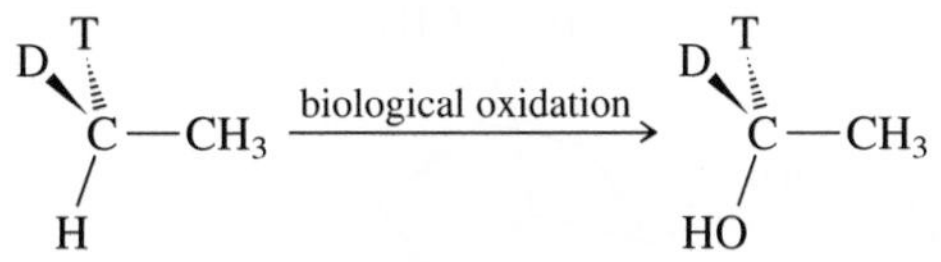

(b) Because OH becomes bonded to carbon at the same side from which H is lost, the oxidation proceeds with retention of configuration (Section 6.13). Compare this fact with the *R* and *S* configurations you determined in part (a) and reconcile any *apparent* conflicts.

7.38 Identify the relationship in each of the following pairs. Do the drawings represent constitutional isomers or stereoisomers, or are they just different ways of drawing the same compound? If they are stereoisomers, are they enantiomers or diastereomers?

(a) 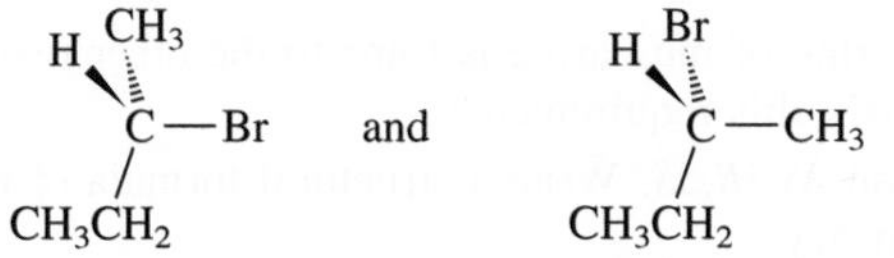

(b) 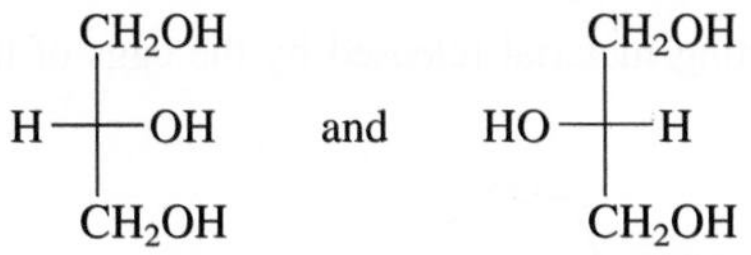

(c) 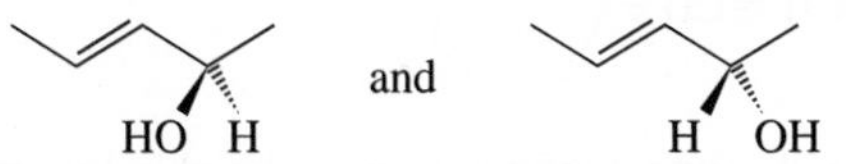

(d) 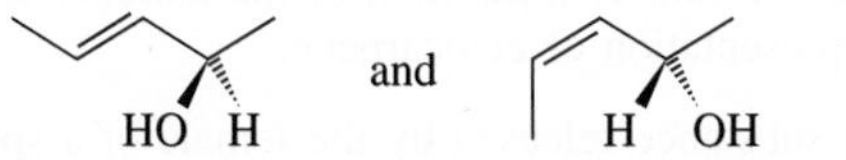

(e) 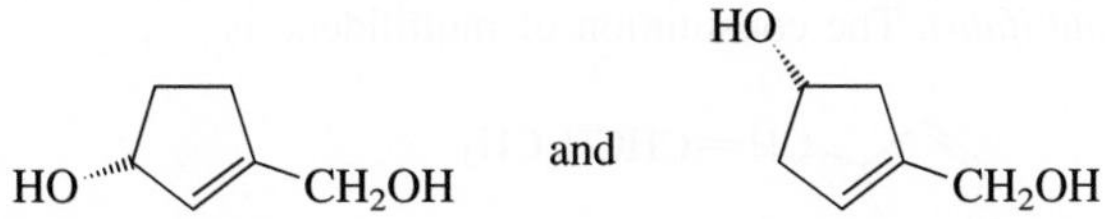

(f) 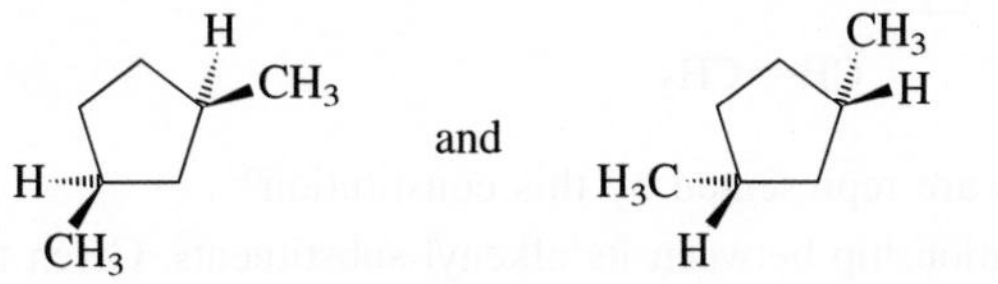

(g) 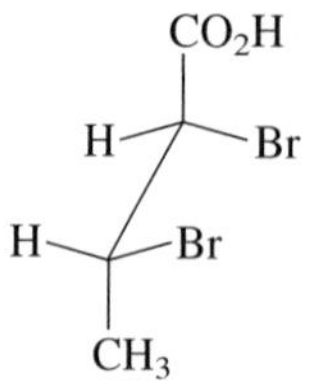
CO2H
H
Br
H
Br
CH3
and
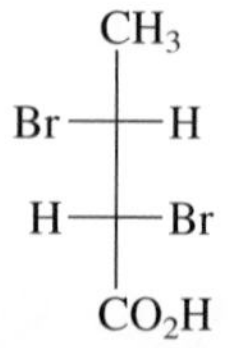
CH3
Br
H
H
Br
CO2H

(h) 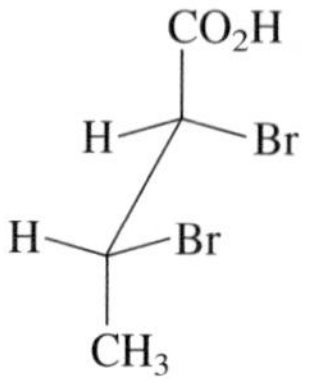
CO2H
H
Br
H
Br
CH3
and
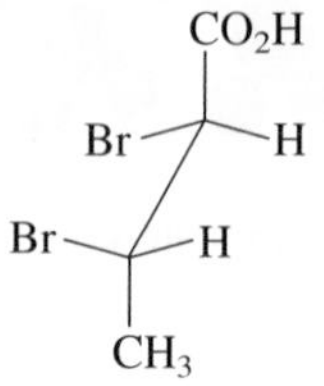
CO2H
Br
H
Br
H
CH3

(i) 
(CH3)3C
I
and
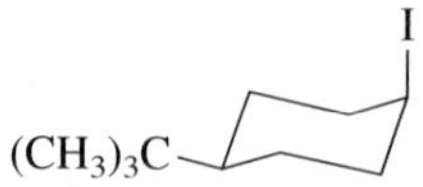
(CH3)3C
I

(j) 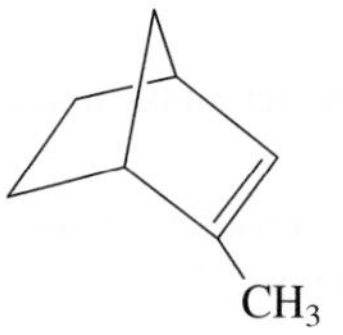
CH3
and
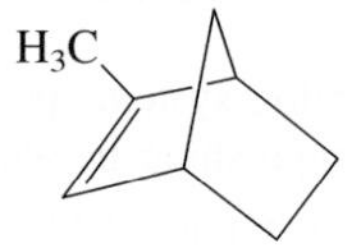
H3C

(k) 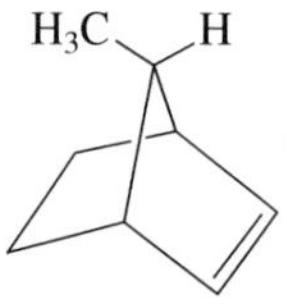
H3C
H
and
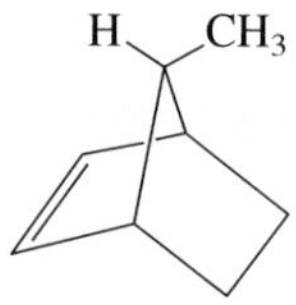
H
CH3

7.39 *Muscarine* is a poisonous substance present in the mushroom *Amanita muscaria.* Its structure is represented by the constitution shown here.

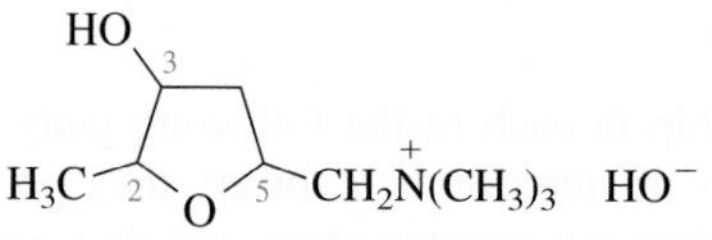
HO
3
H3C
2
O
5
CH2N(CH3)3 HO−

(a) Including muscarine, how many stereoisomers have this constitution?

(b) One of the substituents on the ring of muscarine is trans to the other two. How many of the stereoisomers satisfy this requirement?

(c) Muscarine has the configuration 2*S*,3*R*,5*S*. Write a structural formula of muscarine showing its correct stereochemistry.

7.40 *Ectocarpene* is a volatile, sperm cell-attracting material released by the eggs of the seaweed *Ectocarpus siliculosus.* Its constitution is

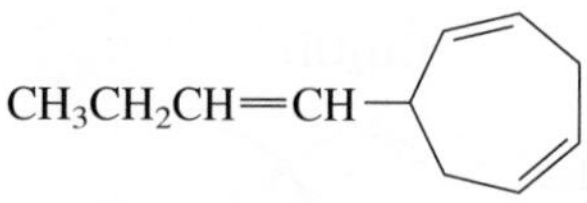
CH3CH2CH=CH

All the double bonds are cis, and the absolute configuration of the chirality center is *S.* Write a stereochemically accurate representation of ectocarpene.

7.41 *Multifidene* is a sperm cell-attracting substance released by the female of a species of brown algae *(Cutleria multifida).* The constitution of multifidene is

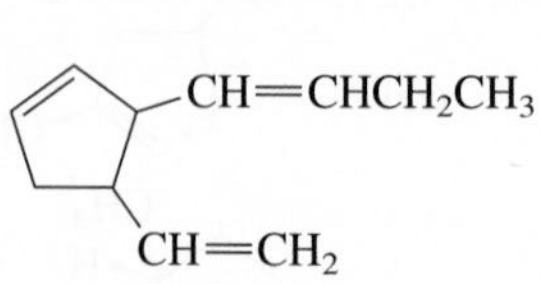
CH=CHCH2CH3
CH=CH2

(a) How many stereoisomers are represented by this constitution?

(b) Multifidene has a cis relationship between its alkenyl substituents. Given this information, how many stereoisomers are possible?

(c) The butenyl side chain has the *Z* configuration of its double bond. On the basis of all the data, how many stereoisomers are possible?

(d) Draw stereochemically accurate representations of all the stereoisomers that satisfy the structural requirements of multifidene.

(e) How are these stereoisomeric multifidenes related (enantiomers or diastereomers)?

7.42 Sphingosine is a component of membrane lipids, including those found in nerve and muscle cells. How many stereoisomers are possible?

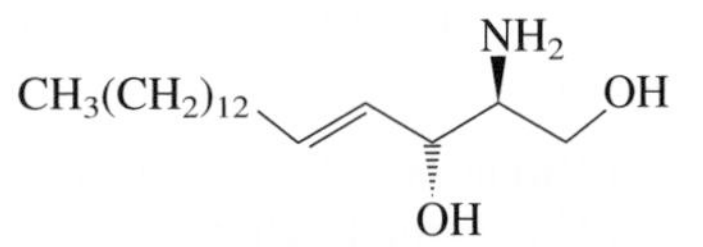
NH2
CH3(CH2)12
OH
OH

7.43 In Problem 4.30 you were asked to draw the preferred conformation of menthol on the basis of the information that menthol is the most stable stereoisomer of 2-isopropyl-5-methylcyclohexanol. We can now completely describe (−)-menthol structurally by noting that it has the *R* configuration at the hydroxyl-substituted carbon.

(a) Draw the preferred conformation of (−)-menthol.

(b) (+)-Isomenthol has the same constitution as (−)-menthol. The configurations at C-1 and C-2 of (+)-isomenthol are the opposite of the corresponding chirality centers of (−)-menthol. Write the preferred conformation of (+)-isomenthol.

7.44 A certain natural product having $[\alpha]_D$ + 40.3° was isolated. Two structures have been independently proposed for this compound. Which one do you think is more likely to be correct? Why?

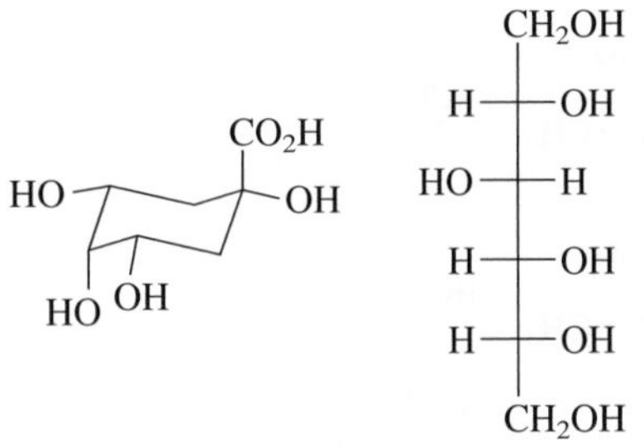
CO2H
HO
OH
HO OH
CH2OH
H
OH
HO
H
H
OH
H
OH
CH2OH

7.45 One of the principal substances obtained from archaea (one of the oldest forms of life on Earth) is derived from a 40-carbon diol. Given the fact that this diol is optically active, is it compound A or is it compound B?

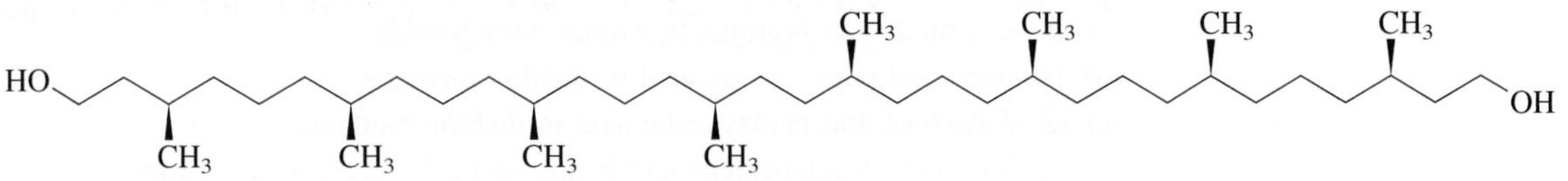
HO
CH3
CH3
CH3
CH3
CH3
CH3
CH3
CH3
OH

Compound A

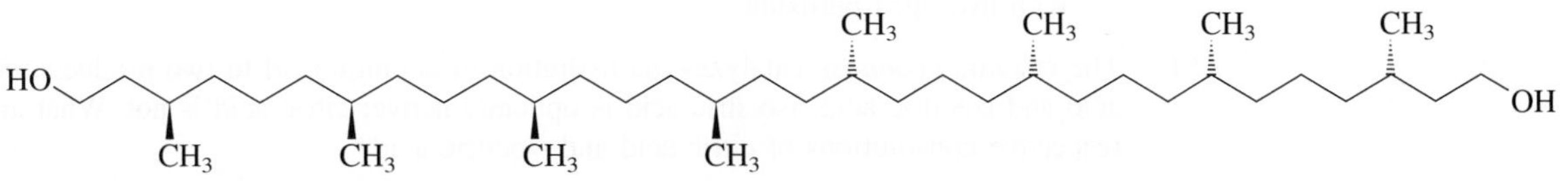
HO
CH3
CH3
CH3
CH3
CH3
CH3
CH3
CH3
OH

Compound B

7.46 (a) An aqueous solution containing 10 g of optically pure fructose was diluted to 500 mL with water and placed in a polarimeter tube 20 cm long. The measured rotation was −5.20°. Calculate the specific rotation of fructose.

(b) If this solution were mixed with 500 mL of a solution containing 5 g of racemic fructose, what would be the specific rotation of the resulting fructose mixture? What would be its optical purity?

7.47 The compounds shown here are widely used in medicine. Ampicillin is an antibiotic, and simvastatin is a cholesterol-lowering drug. Locate the chirality centers in each.

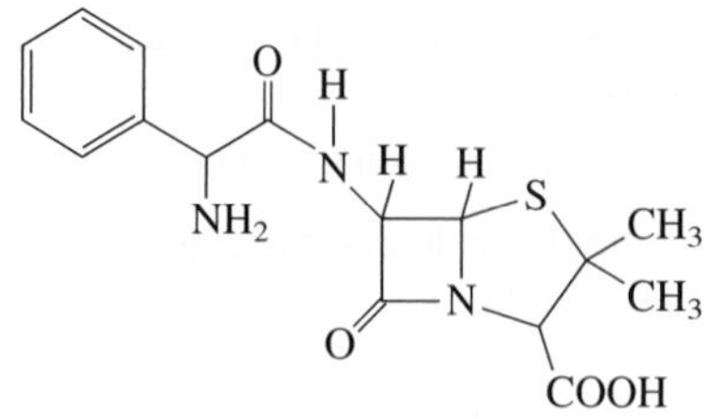

Ampicillin

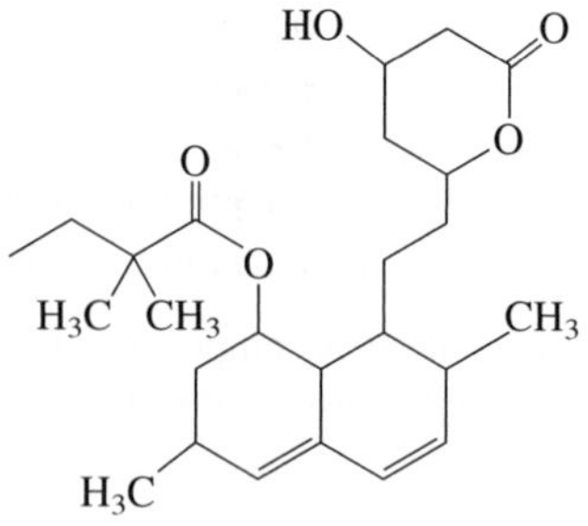

Simvastatin (Zocor)

7.48 Droxidopa is used to treat Parkinson's disease. Droxidopa has the *S* configuration at C-2 and the *R* configuration at C-3. Write Fischer projections for (a) droxidopa and (b) the enantiomer of droxidopa. Orient the Fischer projections so that the COOH group is at the top and the aryl group is at the bottom.

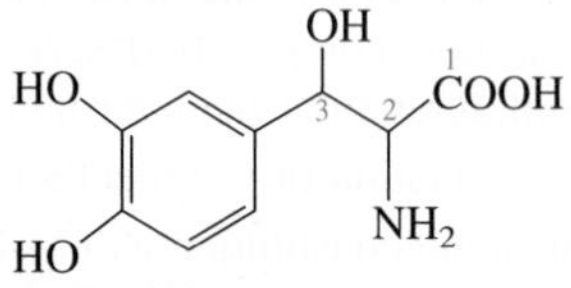

7.49 Each of the following reactions gives a mixture of two stereoisomers. Write their structures. Are they enantiomers or diastereomers? Are they chiral or achiral? Are they formed in equal amounts?

(a) $BrCH_2CH_2Br + Cl_2 \xrightarrow{\text{light}} C_2H_3Br_2Cl + HCl$

(b) 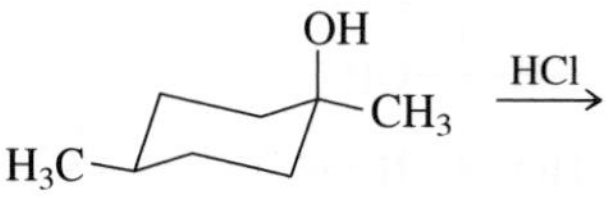

(c) 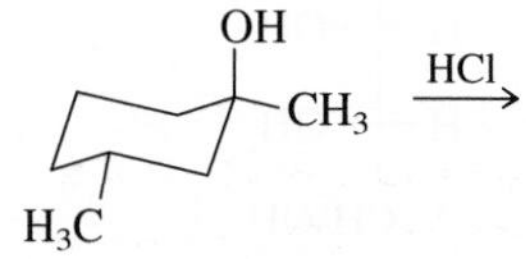

7.50 Write the organic products of each of the following reactions. If two stereoisomers are formed, show both. Label all chirality centers *R* or *S* as appropriate.

(a) 1-Butene and hydrogen iodide

(b) (*E*)-2-Pentene and bromine in carbon tetrachloride

(c) (*Z*)-2-Pentene and bromine in carbon tetrachloride

(d) 1-Butene and peroxyacetic acid in dichloromethane

(e) (*Z*)-2-Pentene and peroxyacetic acid in dichloromethane

(f) 1,5,5-Trimethylcyclopentene and hydrogen in the presence of platinum

(g) 1,5,5-Trimethylcyclopentene and diborane in tetrahydrofuran followed by oxidation with hydrogen peroxide

7.51 The enzyme *aconitase* catalyzes the hydration of aconitic acid to two products: citric acid and isocitric acid. Isocitric acid is optically active; citric acid is not. What are the respective constitutions of citric acid and isocitric acid?

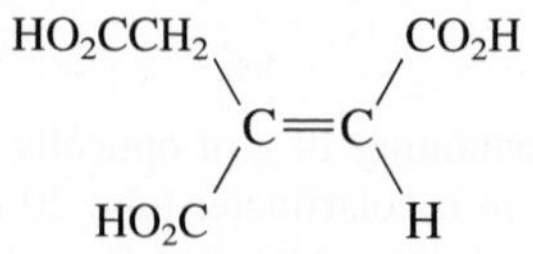

Aconitic acid

7.52 Consider the ozonolysis of *trans*-4,5-dimethylcyclohexene having the configuration shown.

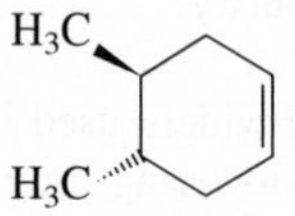

Structures A, B, and C are three stereoisomeric forms of the reaction product.

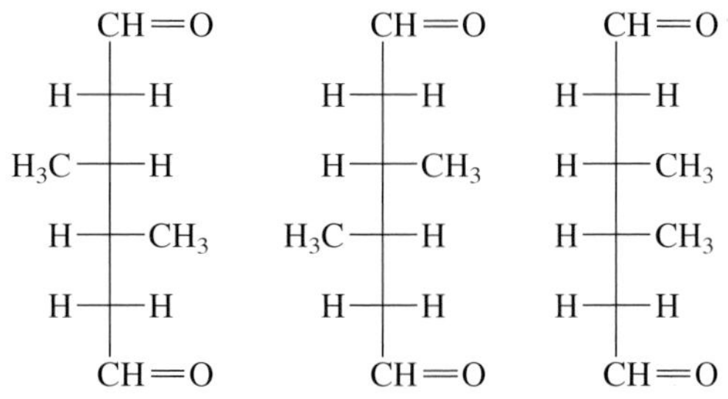

A B C

(a) Which, if any, of the compounds A, B, and C are chiral?

(b) What product is formed in the reaction?

(c) What product would be formed if the methyl groups were cis to each other in the starting alkene?

7.53 (a) On being heated with potassium ethoxide in ethanol (70°C), the deuterium-labeled alkyl bromide shown gave a mixture of 1-butene, *cis*-2-butene, and *trans*-2-butene. On the basis of your knowledge of the E2 mechanism, predict which alkene(s), if any, contained deuterium.

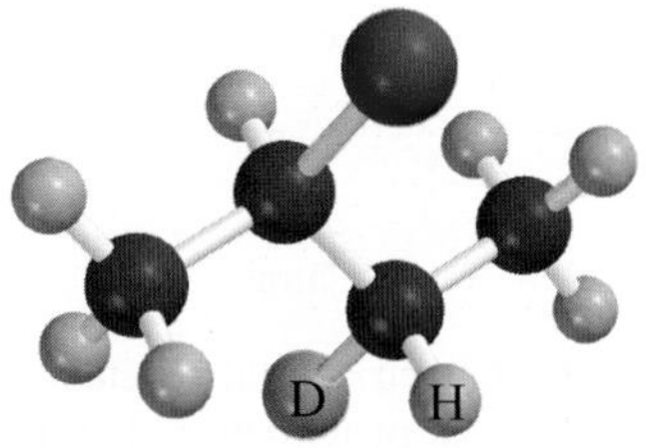

(b) The bromide shown in part (a) is the erythro diastereomer. How would the deuterium content of the alkenes formed by dehydrohalogenation of the threo diastereomer differ from those produced in part (a)?

7.54 A compound (C_6H_{10}) contains a five-membered ring. When Br_2 adds to it, two diastereomeric dibromides are formed. Suggest reasonable structures for the compound and the two dibromides.

7.55 When optically pure 2,3-dimethyl-2-pentanol was subjected to dehydration, a mixture of two alkenes was obtained. Hydrogenation of this alkene mixture gave 2,3-dimethylpentane, which was 50% optically pure. What were the two alkenes formed in the elimination reaction, and what were the relative amounts of each?

7.56 When (*R*)-3-buten-2-ol is treated with a peroxy acid, two stereoisomeric epoxides are formed in a 60:40 ratio. The minor stereoisomer has the structure shown.

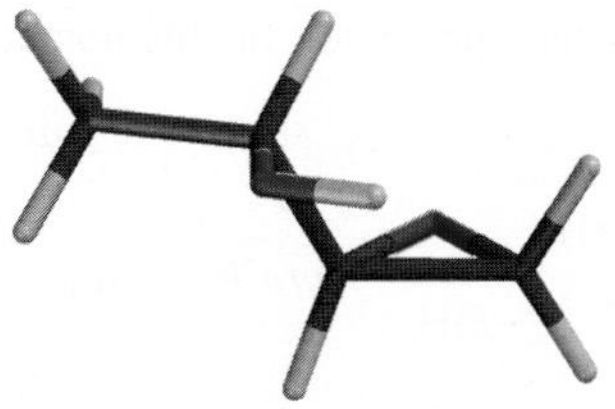

(a) Write the structure of the major stereoisomer.

(b) What is the relationship between the two epoxides? Are they enantiomers or diastereomers?

(c) What four stereoisomeric products are formed when racemic 3-buten-2-ol is epoxidized under the same conditions? How much of each stereoisomer is formed?

7.57 Among compounds (a)–(d), identify those that have a chirality axis.

	A	B	X	Y
(a)	$(CH_3)_3C$—	H—	H—	H—
(b)	$(CH_3)_3C$—	$(CH_3)_3C$—	H—	H—
(c)	$(CH_3)_3C$—	H—	$(CH_3)_3C$—	H—
(d)	$(CH_3)_3C$—	$(CH_3)_3C$—	$(CH_3)_3C$—	$(CH_3)_3C$—

Descriptive Passage and Interpretive Problems 7

Prochirality

Consider two chemical changes: one occurring at a tetrahedral sp^3 carbon C(*x,x,y,z*), the other at a trigonal sp^2 carbon C(*x,y,z*), where *x*, *y*, and *z* are different atoms or groups attached to C. Each reactant is achiral; both are converted to the chiral product C(*w,x,y,z*). In the first case *w* replaces one of the *x* atoms or groups, in the other *w* adds to the trigonal carbon.

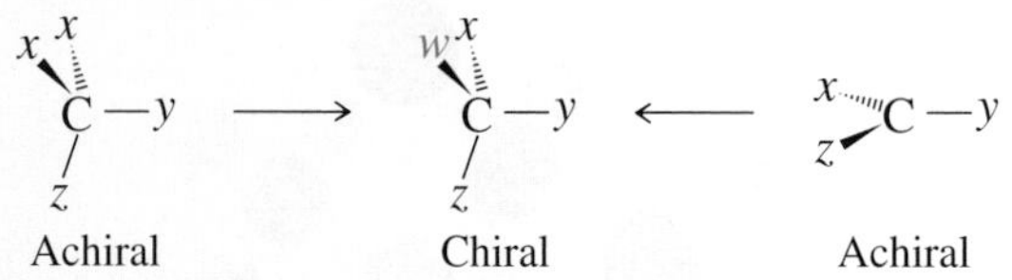

Both transformations convert C in each achiral reactant to a chirality center in the product. The two achiral reactants are classified as **prochiral.** C is a **prochirality center** in C(*x,x,y,z*) and has two **prochiral faces** in C(*x,y,z*).

In achiral molecules with tetrahedral prochirality centers, substitution of one of the two *x* groups by *w* gives the enantiomer of the product that results from substitution of the other. The two *x* groups occupy mirror-images sites and are **enantiotopic.**

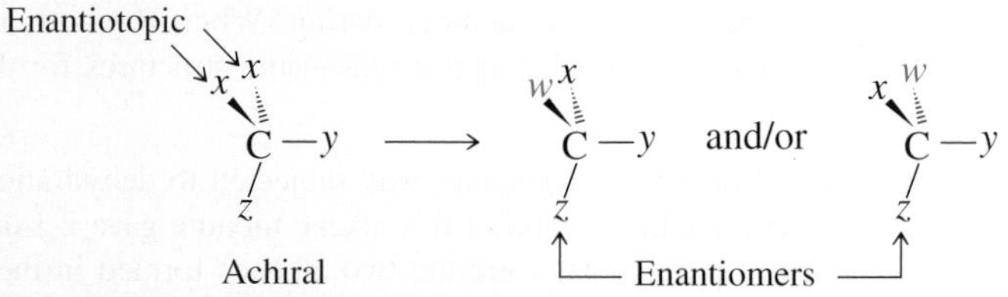

Enantiotopic groups are designated as *pro-R* or *pro-S* by a modification of Cahn–Ingold–Prelog notation. One is assigned a higher priority than the other without disturbing the priorities of the remaining groups, and the *R,S* configuration of the resulting chirality center is determined in the usual way. If it is *R,* the group assigned the higher rank is *pro-R*. If *S,* this group is *pro-S*. Ethanol and citric acid illustrate the application of this notation to two prochiral molecules.

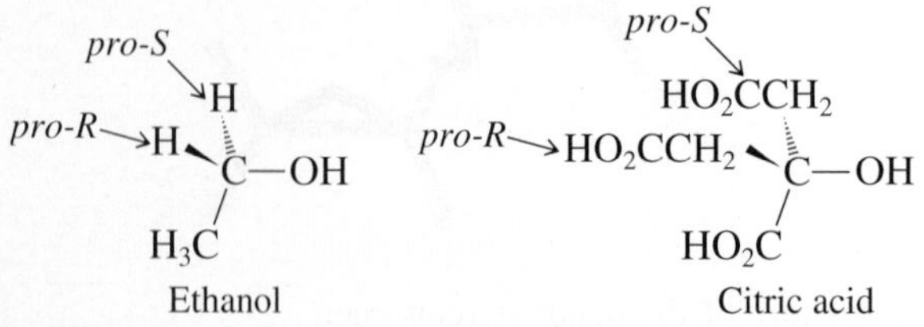

Citric acid played a major role in the development of the concept of prochirality. Its two CH_2CO_2H chains groups behave differently in a key step of the Krebs cycle, so differently that some wondered whether citric acid itself were really involved. Alexander Ogston (Oxford) provided the answer in 1948 when he pointed out that the two CH_2CO_2H groups are differentiated when citric acid interacts with the chiral environment of an enzyme.

The two prochiral faces of a trigonal atom C(*x*,*y*,*z*) are enantiotopic and designated *Re* and *Si* according to whether *x*, *y*, and *z* trace a clockwise (*Re*) or counterclockwise (*Si*) path in order of decreasing Cahn–Ingold–Prelog precedence. An acetaldehyde molecule that lies in the plane of the paper, for example, presents either the *Re* or *Si* face according to how it is oriented.

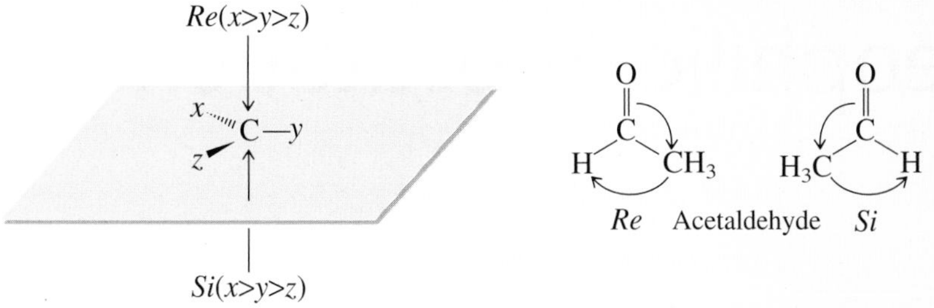

The stereochemical aspects of many enzyme-catalyzed reactions have been determined. The enzyme *alcohol dehydrogenase* catalyzes the oxidation of ethanol to acetaldehyde by removing the *pro-R* hydrogen (abbreviated as H_R). When the same enzyme catalyzes the reduction of acetaldehyde to ethanol, hydrogen is transferred to the *Re* face.

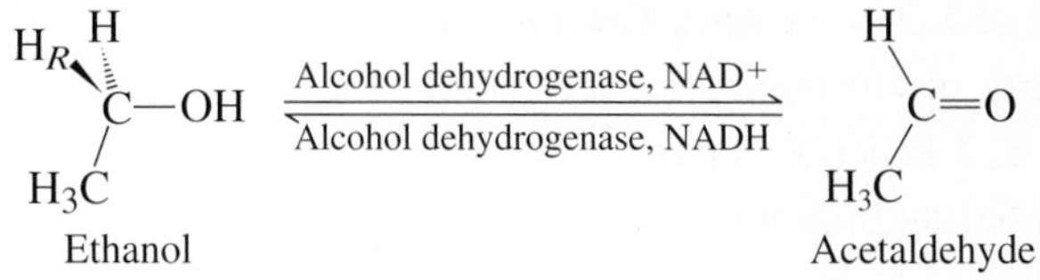

7.58 Which molecule is prochiral?
A. Ethane C. Butane
B. Propane D. Cyclopropane

7.59 How many of the carbons in 2-methylpentane [$(CH_3)_2CHCH_2CH_2CH_3$] are prochirality centers?
A. One C. Three
B. Two D. Four

7.60 What are the *pro-R* and *pro-S* designations for the enantiotopic hydrogens in 1-propanol?

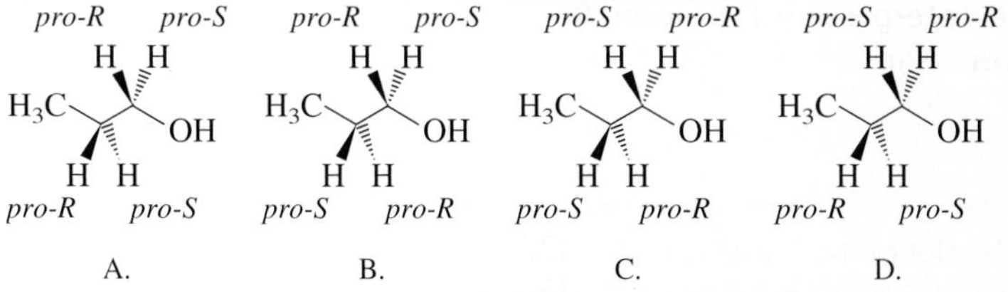

7.61 The enzyme fumarase catalyzes the addition of water to the double bond of fumaric acid.

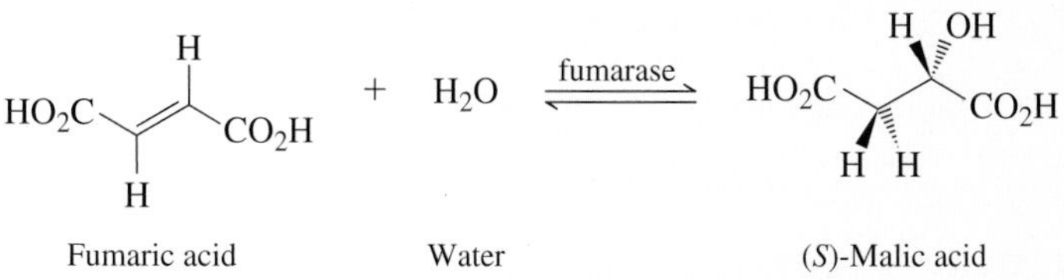

The —OH group and the *pro-R* hydrogen of the CH_2 group of (*S*)-(−)malic acid come from water. What stereochemical pathway describes the addition of water to the double bond?
A. syn Addition B. anti Addition

7.62 To which prochiral face of the double bond of fumaric acid does the —OH group add to in the fumarase-catalyzed hydration of fumaric acid described in the preceding problem?
A. *Re* B. *Si*

7.63 A method for the stereoselective synthesis of chiral epoxides gave the product shown in high enantiomeric excess. To which faces of the doubly bonded carbons is oxygen transferred?

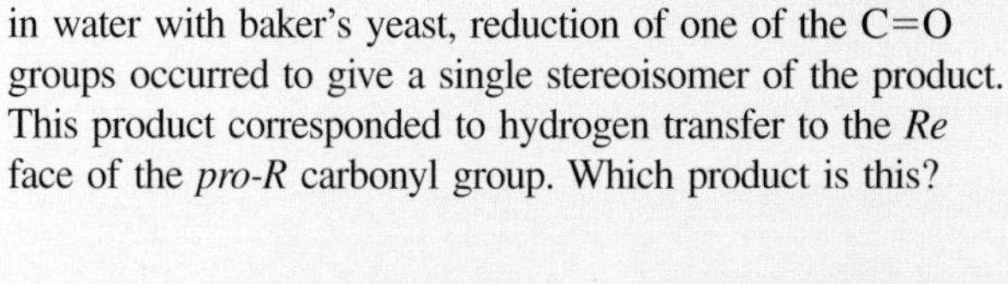

A. *Re Re* C. *Si Si*
B. *Re Si* D. *Si Re*

7.64 When the achiral dione shown (below left) was incubated in water with baker's yeast, reduction of one of the C=O groups occurred to give a single stereoisomer of the product. This product corresponded to hydrogen transfer to the *Re* face of the *pro-R* carbonyl group. Which product is this?

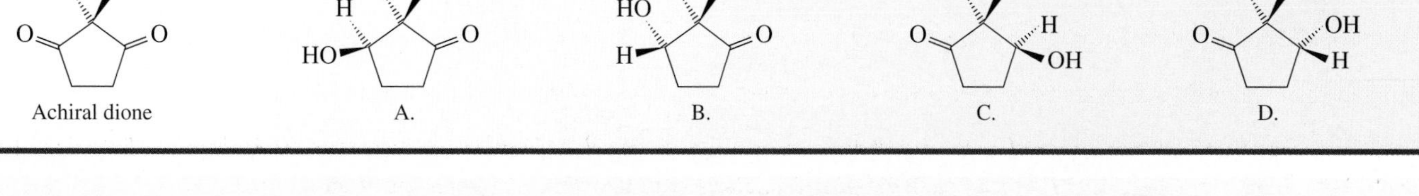

8 Nucleophilic Substitution

Chapter Outline

Mechanisms

This electrostatic potential map is of the transition state for the reaction of hydroxide ion with chloromethane. The tetrahedral arrangement of bonds inverts like an umbrella in a storm during the reaction.

WHEN WE DISCUSSED elimination reactions in Chapter 5, we learned that a Lewis base can react with an alkyl halide to form an alkene. In the present chapter, you will find that the same kinds of reactants can also undergo a different reaction, one in which the Lewis base acts as a **nucleophile** to substitute for the halogen substituent on carbon.

$$\underset{\text{Alkyl halide}}{R{-}\ddot{\underset{\cdot\cdot}{X}}\!:} + \underset{\text{Lewis base}}{Y\!:^-} \longrightarrow \underset{\text{Product of nucleophilic substitution}}{R{-}Y} + \underset{\text{Halide anion}}{:\!\ddot{\underset{\cdot\cdot}{X}}\!:^-}$$

We first encountered nucleophilic substitution in Chapter 4, in the reaction of alcohols with hydrogen halides to form alkyl halides. Now we'll see how alkyl halides can themselves be converted to other classes of organic compounds by nucleophilic substitution.

This chapter has a mechanistic emphasis designed to achieve a practical result. By understanding the mechanisms by which alkyl halides undergo nucleophilic substitution, we can choose experimental conditions best suited to carrying out a particular functional group transformation. The difference between a successful reaction that leads cleanly to a desired product and one that fails is often a subtle one. Mechanistic analysis helps us to appreciate these subtleties and use them to our advantage.

8.1 Functional Group Transformation by Nucleophilic Substitution

Nucleophilic substitution reactions of alkyl halides are related to elimination reactions in that the halogen acts as a leaving group on carbon and is lost as an anion. The carbon–halogen bond of the alkyl halide is broken **heterolytically:** the two electrons in that bond are lost with the leaving group.

The carbon–halogen bond in an alkyl halide is polar

$$\overset{\delta+}{R}{-}\overset{\delta-}{X} \qquad X = I, Br, Cl, F$$

and is cleaved on attack by a nucleophile so that the two electrons in the bond are retained by the halogen

$$^-Y\!: \;\; R{-}\ddot{\underset{\cdot\cdot}{X}}\!: \longrightarrow R{-}Y + :\!\ddot{\underset{\cdot\cdot}{X}}\!:^-$$

The most frequently encountered nucleophiles are anions, which are used as their lithium, sodium, or potassium salts. If we use M to represent lithium, sodium, or potassium, some representative nucleophilic reagents are

MOR	(a metal *alkoxide,* a source of the nucleophilic anion $R\ddot{\underset{..}{O}}{:}^-$)
$MO\overset{O}{\overset{\|\|}{C}}R$	(a metal *carboxylate,* a source of the nucleophilic anion $R\overset{:O:}{\overset{\|\|}{C}}-\ddot{\underset{..}{O}}{:}^-$)
MSH	(a metal *hydrogen sulfide,* a source of the nucleophilic anion $H\ddot{\underset{..}{S}}{:}^-$)
MCN	(a metal *cyanide,* a source of the nucleophilic anion ${}^-{:}C\equiv N{:}$)
MN_3	(a metal *azide,* a source of the nucleophilic anion ${:}\overset{-}{\underset{..}{N}}=\overset{+}{N}=\overset{-}{\underset{..}{N}}{:}$)

Table 8.1 illustrates an application of each of these to a functional group transformation. The anionic portion of the salt substitutes for the halogen of an alkyl halide. The metal cation portion becomes a lithium, sodium, or potassium halide.

$$M^+ \; {}^-Y{:} \; + \; R-\ddot{\underset{..}{X}}{:} \longrightarrow R-Y \; + \; M^+ \; {:}\ddot{\underset{..}{X}}{:}^-$$

Nucleophilic reagent — Alkyl halide — Product of nucleophilic substitution — Metal halide

Notice that all the examples in Table 8.1 involve alkyl halides, that is, compounds in which the halogen is attached to an sp^3-hybridized carbon. Alkenyl halides and aryl halides, compounds in which the halogen is attached to sp^2-hybridized carbons, are essentially unreactive under these conditions, and the principles to be developed in this chapter do not apply to them.

Alkenyl halides are also referred to as *vinylic halides.*

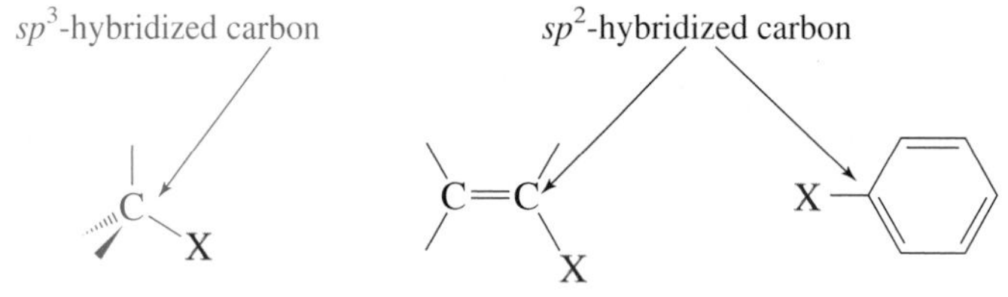

Alkyl halide — Alkenyl halide — Aryl halide

To ensure that reaction occurs in homogeneous solution, solvents are chosen that dissolve both the alkyl halide and the ionic salt. Alkyl halides are soluble in organic solvents, but the salts often are not. Inorganic salts are soluble in water, but alkyl halides are not. Mixed solvents such as ethanol–water mixtures that can dissolve enough of both the substrate and the nucleophile to give fairly concentrated solutions are frequently used.

TABLE 8.1 Representative Functional Group Transformations by Nucleophilic Substitution Reactions of Alkyl Halides

Nucleophile and comments	General equation and specific example
Alkoxide ion ($R\ddot{\underset{..}{O}}{:}^-$) The oxygen atom of a metal alkoxide is nucleophilic and replaces the halogen of an alkyl halide. The product is an *ether.*	$R\ddot{\underset{..}{O}}{:}^- + R'-\ddot{\underset{..}{X}}{:} \longrightarrow R\ddot{\underset{..}{O}}R' + {:}\ddot{\underset{..}{X}}{:}^-$ Alkoxide ion + Alkyl halide → Ether + Halide ion $(CH_3)_2CHCH_2ONa + CH_3CH_2Br \xrightarrow{\text{isobutyl alcohol}} (CH_3)_2CHCH_2OCH_2CH_3 + NaBr$ Sodium isobutoxide + Ethyl bromide → Ethyl isobutyl ether (66%) + Sodium bromide

Continued

TABLE 8.1 Representative Functional Group Transformations by Nucleophilic Substitution Reactions of Alkyl Halides (*Continued*)

Nucleophile and comments	General equation and specific example
Carboxylate ion ($RC(=O)\ddot{O}{:}^-$) An *ester* is formed when the negatively charged oxygen of a carboxylate replaces the halogen of an alkyl halide.	$RC(=O)\ddot{O}{:}^- + R{-}\ddot{X}{:} \longrightarrow R'C(=O)OR' + {:}\ddot{X}{:}^-$ Carboxylate ion, Alkyl halide, Ester, Halide ion $KOC(=O)(CH_2)_{16}CH_3 + CH_3CH_2I \xrightarrow[\text{water}]{\text{acetone}} CH_3CH_2OC(=O)(CH_2)_{16}CH_3 + KI$ Potassium octadecanoate, Ethyl iodide, Ethyl octadecanoate (95%), Potassium iodide
Hydrogen sulfide ion ($H\ddot{S}{:}^-$) Using hydrogen sulfide as a nucleophile permits the conversion of alkyl halides to compounds of the type RSH. These compounds are the sulfur analogs of alcohols and are known as *thiols*.	$H\ddot{S}{:}^- + R'{-}\ddot{X}{:} \longrightarrow R'\ddot{S}H + {:}\ddot{X}{:}^-$ Hydrogen sulfide ion, Alkyl halide, Thiol, Halide ion $KSH + CH_3CH(Br)(CH_2)_6CH_3 \xrightarrow[\text{water}]{\text{ethanol}} CH_3CH(SH)(CH_2)_6CH_3 + KBr$ Potassium hydrogen sulfide, 2-Bromononane, 2-Nonanethiol (74%), Potassium bromide
Cyanide ion (${:}\bar{C}{\equiv}N{:}$) The negatively charged carbon atom of cyanide ion is usually the site of its nucleophilic character. Use of cyanide ion as a nucleophile permits the extension of a carbon chain by carbon–carbon bond formation. The product is an *alkyl cyanide*, or *nitrile*.	${:}N{\equiv}\bar{C}{:} + R'{-}\ddot{X}{:} \longrightarrow R'C{\equiv}N{:} + {:}\ddot{X}{:}^-$ Cyanide ion, Alkyl halide, Alkyl cyanide, Halide ion 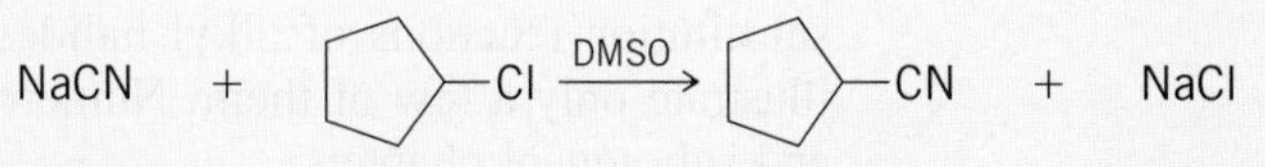 Sodium cyanide, Cyclopentyl chloride, Cyclopentyl cyanide (70%), Sodium chloride
Azide ion (${:}\bar{N}{=}\overset{+}{N}{=}\bar{N}{:}$) Sodium azide is a reagent used for carbon–nitrogen bond formation. The product is an *alkyl azide*.	${:}\bar{N}{=}\overset{+}{N}{=}\bar{N}{:} + R'{-}\ddot{X}{:} \longrightarrow R'N{=}\overset{+}{N}{=}\bar{N}{:} + {:}\ddot{X}{:}^-$ Azide ion, Alkyl halide, Alkyl azide, Halide ion $NaN_3 + CH_3(CH_2)_4I \xrightarrow{\text{1-propanol–water}} CH_3(CH_2)_4N_3 + NaI$ Sodium azide, Pentyl iodide, Pentyl azide (52%), Sodium iodide
Iodide ion (${:}\ddot{I}{:}^-$) Alkyl chlorides and bromides are converted to *alkyl iodides* by treatment with sodium iodide in acetone. NaI is soluble in acetone, but NaCl and NaBr are insoluble and crystallize from the reaction mixture, making the reaction irreversible.	${:}\ddot{I}{:}^- + R'{-}\ddot{X}{:} \xrightarrow{\text{acetone}} R'{-}\ddot{I}{:} + {:}\ddot{X}{:}^-$ Iodide ion, Alkyl chloride or bromide, Alkyl iodide, Chloride or bromide ion $CH_3CH(Br)CH_3 + NaI \xrightarrow{\text{acetone}} CH_3CH(I)CH_3 + NaBr\ (\text{solid})$ 2-Bromopropane, Sodium iodide, 2-Iodopropane (63%), Sodium bromide

The use of DMSO as a solvent in *elimination* reactions was mentioned earlier, in Section 5.14.

Many salts, as well as most alkyl halides, possess significant solubility in dimethyl sulfoxide (DMSO) or *N, N*-dimethylformamide (DMF), which makes them good solvents for carrying out nucleophilic substitution reactions (Section 8.10).

Problem 8.1

Write a structural formula for the principal organic product formed in the reaction of methyl bromide with each of the following compounds:

(a) NaOH (sodium hydroxide)
(b) $KOCH_2CH_3$ (potassium ethoxide)
(c) 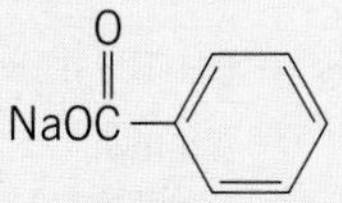 (sodium benzoate)
(d) LiN_3 (lithium azide)
(e) KCN (potassium cyanide)
(f) NaSH (sodium hydrogen sulfide)
(g) NaI (sodium iodide)

Sample Solution (a) The nucleophile in sodium hydroxide is the negatively charged hydroxide ion. The reaction that occurs is nucleophilic substitution of bromide by hydroxide. The product is methyl alcohol.

$$\mathrm{H\ddot{O}:^-} + \mathrm{H_3C{-}\ddot{Br}:} \longrightarrow \mathrm{H_3C{-}\ddot{O}H} + \mathrm{:\ddot{Br}:^-}$$

Hydroxide ion (nucleophile)	Methyl bromide (substrate)	Methyl alcohol (product)	Bromide ion (leaving group)

With Table 8.1 as background, you can begin to see how useful alkyl halides are in synthetic organic chemistry. Alkyl halides may be prepared from alcohols by nucleophilic substitution, from alkanes by free-radical halogenation, and from alkenes by addition of hydrogen halides. They then become available as starting materials for the preparation of other functionally substituted organic compounds by replacement of the halide leaving group with a nucleophile. The range of compounds that can be prepared by nucleophilic substitution reactions of alkyl halides is quite large; the examples shown in Table 8.1 illustrate only a few of them. Numerous other examples will be added to the list in this and subsequent chapters.

8.2 Relative Reactivity of Halide Leaving Groups

Among alkyl halides, alkyl iodides undergo nucleophilic substitution at the fastest rate, alkyl fluorides the slowest.

Increasing rate of substitution by nucleophiles

RF << RCl < RBr < RI

Least reactive — Most reactive

The order of alkyl halide reactivity in nucleophilic substitutions is the same as their order in eliminations. Iodine has the weakest bond to carbon, and iodide is the best leaving group. Alkyl iodides are several times more reactive than alkyl bromides and from 50 to 100 times more reactive than alkyl chlorides. Fluorine has the strongest bond to carbon, and fluoride is the poorest leaving group. Alkyl fluorides are rarely used as substrates in nucleophilic substitution because they are several thousand times less reactive than alkyl chlorides.

Problem 8.2

A single organic product was obtained when 1-bromo-3-chloropropane was allowed to react with one molar equivalent of sodium cyanide in aqueous ethanol. What was this product?

Leaving-group ability is also related to basicity. A strongly basic anion is usually a poorer leaving group than a weakly basic one. Fluoride is the most basic and the poorest leaving group among the halide anions, iodide the least basic and the best leaving group.

The relationship between leaving-group ability and basicity is explored in more detail in Section 8.12

8.3 The S_N2 Mechanism of Nucleophilic Substitution

The mechanisms by which nucleophilic substitution takes place have been the subject of much study. Extensive research by Sir Christopher Ingold and Edward D. Hughes and their associates at University College, London, during the 1930s emphasized kinetic and stereochemical measurements to probe the mechanisms of these reactions.

Kinetics. Kinetic studies measure the speed of a reaction, especially with respect to how the concentration of reactants (and catalysts, if any) affect the reaction rate. Having already seen that the rate of nucleophilic substitution depends on the leaving group (I > Br > Cl >> F), we know that the carbon–halogen bond must break in the slow step of the reaction. Consequently, we expect that the reaction rate will depend on the concentration of the alkyl halide. This is confirmed by kinetic studies of the reaction

$$\underset{\text{Methyl bromide}}{CH_3Br} + \underset{\text{Hydroxide ion}}{HO^-} \longrightarrow \underset{\text{Methyl alcohol}}{CH_3OH} + \underset{\text{Bromide ion}}{Br^-}$$

which follows the rate law:

$$\text{Rate} = k[CH_3Br][HO^-]$$

The reaction rate is directly proportional to the concentration of both methyl bromide and hydroxide ion. It is first order in each reactant, or *second order* overall. The most reasonable conclusion is that both hydroxide ion and methyl bromide react together in a *bimolecular* elementary step and that this step is rate-determining.

The mechanism proposed by Hughes and Ingold, called by them **substitution nucleophilic bimolecular** (S_N2) is shown as an equation in Mechanism 8.1 and as a potential energy diagram in Figure 8.1.

The S_N2 mechanism was introduced earlier in Section 4.12.

Mechanism 8.1

The S_N2 Mechanism of Nucleophilic Substitution

THE OVERALL REACTION:

$$\underset{\text{Methyl bromide}}{CH_3Br} + \underset{\text{Hydroxide ion}}{HO^-} \longrightarrow \underset{\text{Methyl alcohol}}{CH_3OH} + \underset{\text{Bromide ion}}{Br^-}$$

THE MECHANISM: The reaction proceeds in a single step. Hydroxide ion acts as a nucleophile. While the C—Br bond is breaking, the C—O bond is forming.

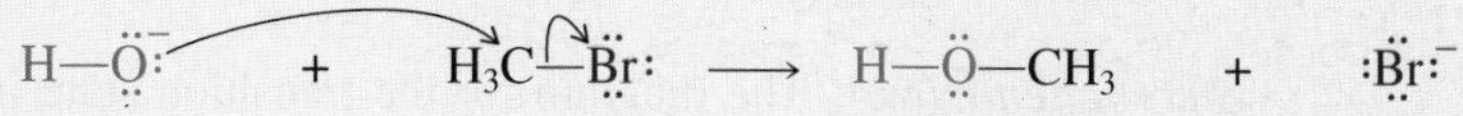

Hydroxide ion Methyl bromide Methyl alcohol Bromide ion

THE TRANSITION STATE: Hydroxide ion attacks carbon from the side opposite the C—Br bond.

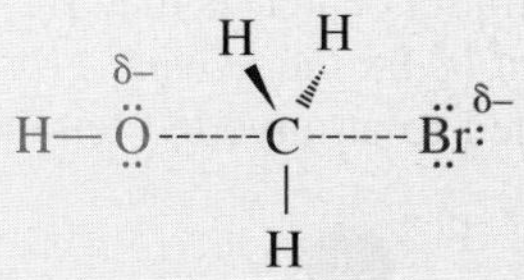

Carbon is partially bonded to both hydroxide and bromide. The arrangement of bonds undergoes tetrahedral inversion from

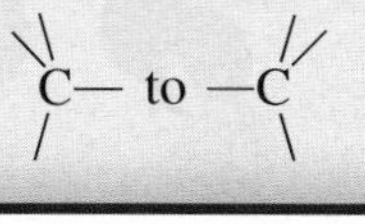

as the reaction progresses.

Figure 8.1

Potential energy diagram for the reaction of methyl bromide with hydroxide ion by the S_N2 mechanism.

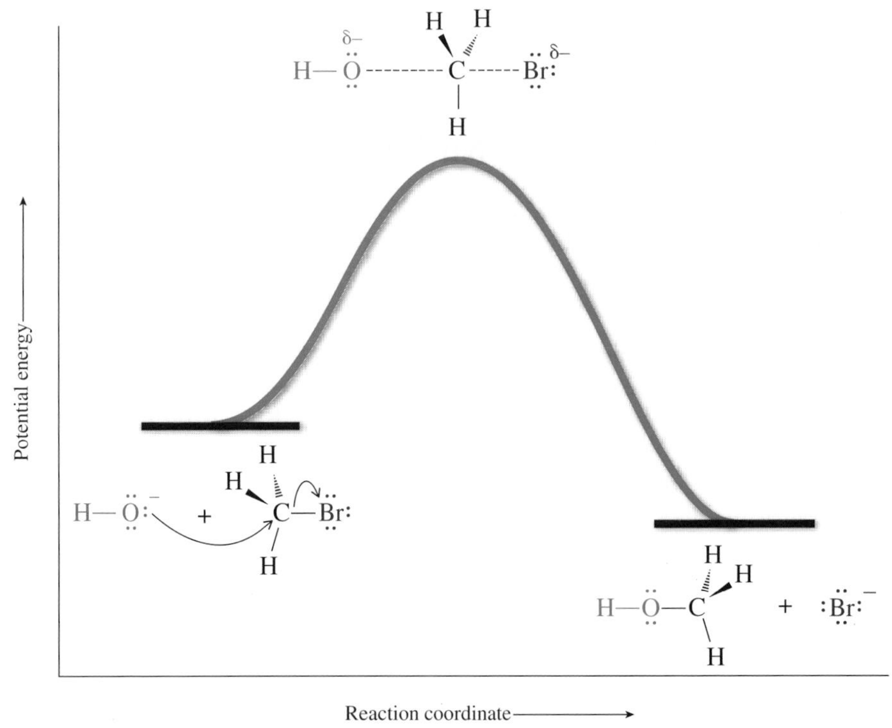

It is a one-step concerted process in which both the alkyl halide and the nucleophile are involved at the transition state. Cleavage of the bond between carbon and the leaving group is assisted by formation of a bond between carbon and the nucleophile. In effect, the nucleophile "pushes off" the leaving group from its point of attachment to carbon. Carbon is partially bonded to both the incoming nucleophile and the departing halide at the transition state. Progress is made toward the transition state as the nucleophile begins to share a pair of its electrons with carbon and the halide ion leaves, taking with it the pair of electrons in its bond to carbon.

Problem 8.3

Is the two-step sequence depicted in the following equations consistent with the second-order kinetic behavior observed for the hydrolysis of methyl bromide?

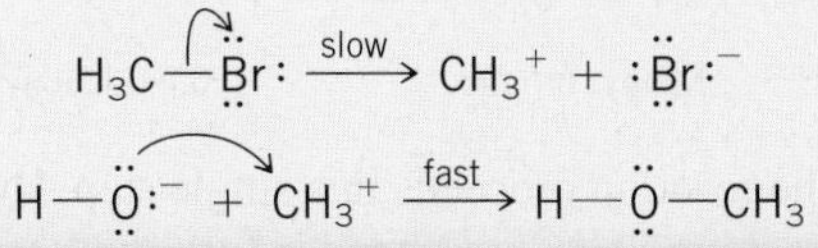

Stereochemistry. The diagram for the transition state in Mechanism 8.1 and Figure 8.1 for the reaction of methyl bromide with hydroxide anticipates a key stereochemical feature of the S_N2 mechanism. *The nucleophile attacks carbon from the side opposite the bond to the leaving group.* Another way of expressing the same point, especially when substitution occurs at a chirality center, is that *S_N2 reactions proceed with* ***inversion of configuration*** *at the carbon that bears the leaving group.* The tetrahedral arrangement of bonds in the reactant is converted to an inverted tetrahedral arrangement in the product.

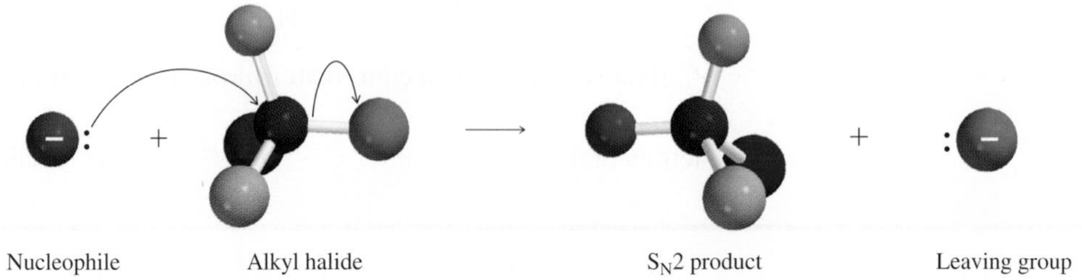

This stereochemical fact comes from studies of nucleophilic substitutions of optically active alkyl halides. In one such experiment, Hughes and Ingold determined that the reaction of optically active 2-bromooctane with hydroxide ion gave 2-octanol, having the opposite configuration at the chirality center.

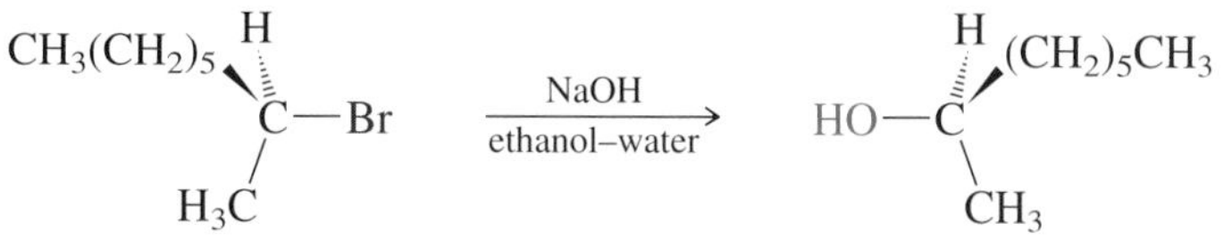
NaOH
ethanol–water

(*S*)-(+)-2-Bromooctane (*R*)-(−)-2-Octanol

Although the alkyl halide and alcohol given in this example have opposite configurations when they have opposite signs of rotation, it cannot be assumed that this will be true for all alkyl halide/alcohol pairs.

Nucleophilic substitution had occurred with inversion of configuration, consistent with the following transition state:

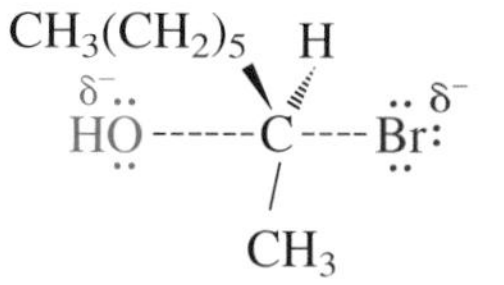

Problem 8.4

The Fischer projection for (+)-2-bromooctane is shown. Write the Fischer projection of the (−)-2-octanol formed from it by nucleophilic substitution with inversion of configuration.

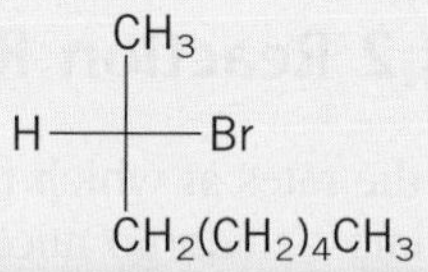

Problem 8.5

Would you expect the 2-octanol formed by S_N2 hydrolysis of (−)-2-bromooctane to be optically active? If so, what will be its absolute configuration and sign of rotation? What about the 2-octanol formed by hydrolysis of racemic 2-bromooctane?

Countless experiments have confirmed that substitution by the S_N2 mechanism is stereospecific and suggests that there exists a *stereoelectronic* requirement for the nucleophile to approach carbon from the side opposite the bond to the leaving group. The results of molecular orbital calculations help us understand why.

The first example of a stereoelectronic effect in this text concerned anti elimination in E2 reactions of alkyl halides (Section 5.16).

When a nucleophile such as hydroxide ion reacts with methyl bromide, electrons flow from the highest occupied molecular orbital (HOMO) of HO^- to the lowest unoccupied molecular orbital (LUMO) of CH_3Br. Directing our attention to the LUMO of CH_3Br, we find three main regions where the HOMO of the nucleophile can overlap with the LUMO. One of these—the blue region shown at the right—can be ignored because it is associated only with Br, and nucleophilic attack from that direction does not produce a C—O bond.

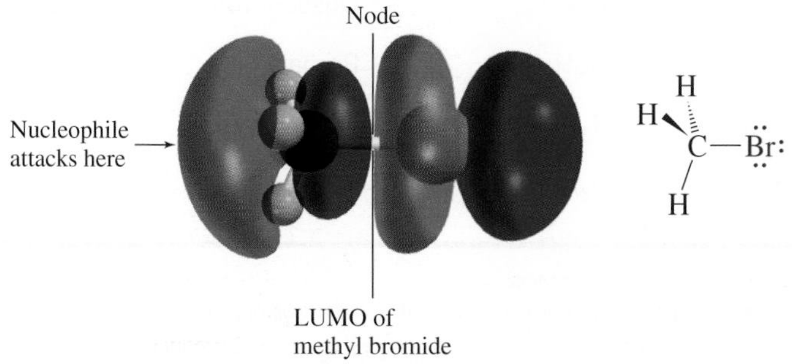
Node
Nucleophile attacks here
LUMO of methyl bromide

The region between carbon and bromine contains a nodal surface; therefore, no net bonding results from its overlap with the HOMO of HO^-. The remaining possibility, *which is also the one that coincides with experimental observation,* is overlap of the HOMO of HO^- with the LUMO of CH_3Br in the region opposite the C—Br bond. It involves a major region of the LUMO, avoids a node, and gives a C—O bond with inversion of configuration at carbon.

The S_N2 mechanism is believed to describe most substitutions in which simple primary and secondary alkyl halides react with negatively charged nucleophiles. All the examples that introduced nucleophilic substitution in Table 8.1 proceed by the S_N2 mechanism (or a mechanism very much like S_N2—remember, mechanisms can never be established with certainty but represent only our best present explanations of experimental observations).

Problem 8.6

Sketch the structure of the S_N2 transition state for the following reaction taken from Table 8.1. Na^+ is a spectator ion and can be omitted from the transition state.

$$(CH_3)_2CHBr + NaI \xrightarrow{\text{acetone}} (CH_3)_2CHI + NaBr$$

We saw in Section 8.2 that the rate of nucleophilic substitution depends strongly on the leaving group—alkyl iodides are the most reactive, alkyl fluorides the least. In the next section, we'll see that the structure of the alkyl group can have an even larger effect.

8.4 Steric Effects and S_N2 Reaction Rates

There are very large differences in the rates at which the various kinds of alkyl halides—methyl, primary, secondary, or tertiary—undergo nucleophilic substitution. As Table 8.2 shows for the reaction:

$$\underset{\text{Alkyl bromide}}{RBr} + \underset{\text{Lithium iodide}}{LiI} \xrightarrow{\text{acetone}} \underset{\text{Alkyl iodide}}{RI} + \underset{\text{Lithium bromide}}{LiBr}$$

the rates of nucleophilic substitution of a series of alkyl bromides differ by a factor of over 10^6 when comparing the most reactive member of the group (methyl bromide) and the least reactive member (*tert*-butyl bromide).

The large rate difference between methyl, ethyl, isopropyl, and *tert*-butyl bromides reflects the **steric hindrance** each offers to nucleophilic attack. The nucleophile must approach the alkyl halide from the side opposite the bond to the leaving group, and, as illustrated in Figure 8.2, this approach is hindered by alkyl substituents on the carbon that is being attacked. The three hydrogens of methyl bromide offer little resistance to approach of the nucleophile, and a rapid reaction occurs. Replacing one of the hydrogens by a methyl group somewhat shields the carbon from attack by the nucleophile

TABLE 8.2 Reactivity of Some Alkyl Bromides Toward Substitution by the S_N2 Mechanism*

Alkyl bromide	Structure	Class	Relative rate†
Methyl bromide	CH_3Br	Unsubstituted	221,000
Ethyl bromide	CH_3CH_2Br	Primary	1,350
Isopropyl bromide	$(CH_3)_2CHBr$	Secondary	1
tert-Butyl bromide	$(CH_3)_3CBr$	Tertiary	Too small to measure

*Substitution of bromide by lithium iodide in acetone.

†Ratio of second-order rate constant *k* for indicated alkyl bromide to *k* for isopropyl bromide at 25°C.

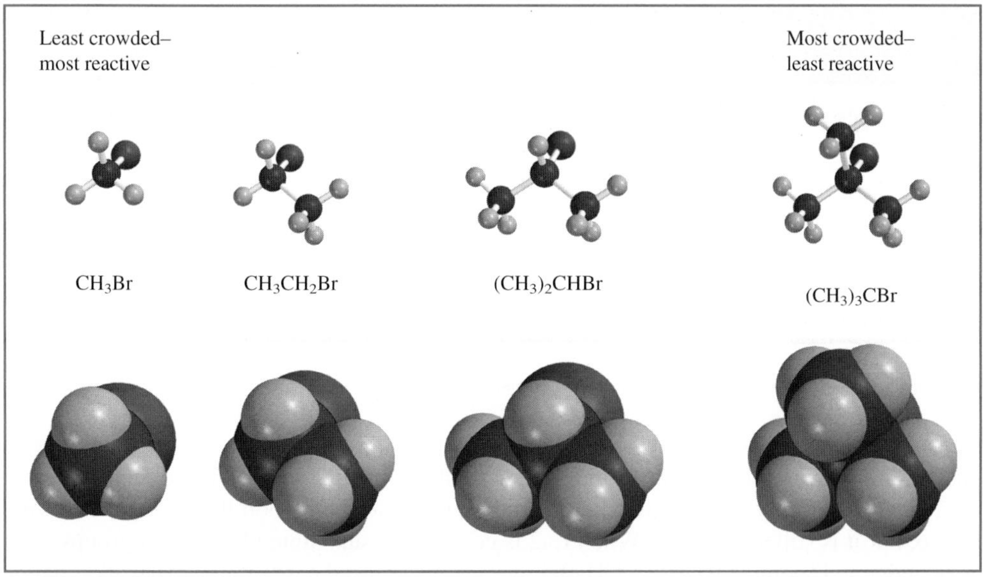
Least crowded–
most reactive
Most crowded–
least reactive
CH3Br
CH3CH2Br
(CH3)2CHBr
(CH3)3CBr

Figure 8.2

Ball-and-spoke (top) and space-filling (bottom) models of alkyl bromides, showing how substituents shield the carbon atom that bears the leaving group from attack by a nucleophile. The nucleophile must attack from the side opposite the bond to the leaving group.

and causes ethyl bromide to be less reactive than methyl bromide. Replacing all three hydrogens by methyl groups almost completely blocks approach to the tertiary carbon of $(CH_3)_3CBr$ and shuts down bimolecular nucleophilic substitution.

In general, S_N2 reactions of alkyl halides show the following dependence of rate on structure:

Increasing rate of substitution by the S_N2 mechanism →

R_3CX	<	R_2CHX	<	RCH_2X	<	CH_3X
Tertiary Least reactive, most crowded		Secondary		Primary		Methyl Most reactive, least crowded

Problem 8.7

Identify the compound in each of the following pairs that reacts with sodium iodide in acetone at the faster rate:

(a) 1-Chlorohexane or cyclohexyl chloride
(b) 1-Bromopentane or 3-bromopentane
(c) 2-Chloropentane or 2-fluoropentane
(d) 2-Bromo-2-methylhexane or 2-bromo-5-methylhexane
(e) 2-Bromopropane or 1-bromodecane

Sample Solution (a) Compare the structures of the two chlorides. 1-Chlorohexane is a primary alkyl chloride; cyclohexyl chloride is secondary. Primary alkyl halides are less crowded at the site of substitution than secondary ones and react faster in substitution by the S_N2 mechanism. 1-Chlorohexane is more reactive.

$CH_3CH_2CH_2CH_2CH_2CH_2Cl$

1-Chlorohexane
(primary, more reactive)

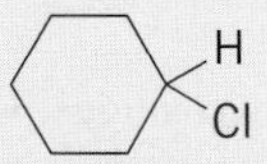
H
Cl

Cyclohexyl chloride
(secondary, less reactive)

Alkyl groups at the carbon atom *adjacent* to the point of nucleophilic attack also decrease the rate of the S_N2 reaction. Compare the rates of nucleophilic substitution in the series of primary alkyl bromides shown in Table 8.3. Taking ethyl bromide as the standard

TABLE 8.3 Effect of Chain Branching on Reactivity of Primary Alkyl Bromides Toward Substitution Under S_N2 Conditions*

Alkyl bromide	Structure	Relative rate†
Ethyl bromide	CH_3CH_2Br	1.0
Propyl bromide	$CH_3CH_2CH_2Br$	0.8
Isobutyl bromide	$(CH_3)_2CHCH_2Br$	0.036
Neopentyl bromide	$(CH_3)_3CCH_2Br$	0.00002

*Substitution of bromide by lithium iodide in acetone.

†Ratio of second-order rate constant *k* for indicated alkyl bromide to *k* for ethyl bromide at 25°C.

and successively replacing its C-2 hydrogens by methyl groups, we see that each additional methyl group decreases the rate of displacement of bromide by iodide. The effect is slightly smaller than for alkyl groups that are attached directly to the carbon that bears the leaving group, but it is still substantial. When C-2 is completely substituted by methyl groups, as it is in neopentyl bromide [$(CH_3)_3CCH_2Br$], we see the unusual case of a primary alkyl halide that is practically inert to substitution by the S_N2 mechanism because of steric hindrance.

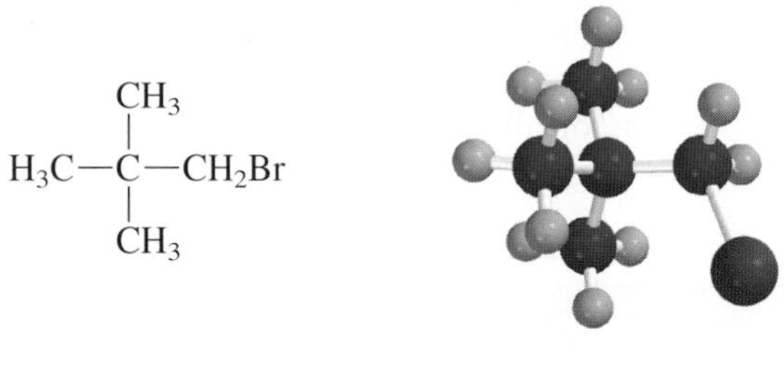

Neopentyl bromide
(1-Bromo-2,2-dimethylpropane)

8.5 Nucleophiles and Nucleophilicity

The Lewis base that acts as the nucleophile often is, but need not always be, an anion. Neutral Lewis bases such as amines (R_3N:), phosphines (R_3P:), and sulfides ($R_2\ddot{S}$:) can also serve as nucleophiles.

Dimethyl sulfide Methyl iodide Trimethylsulfonium iodide

Other common examples of substitutions involving neutral nucleophiles include **solvolysis** reactions—substitutions in which the nucleophile is the solvent in which the reaction is carried out. Solvolysis in water *(hydrolysis)* converts an alkyl halide to an alcohol.

$$RX + 2H_2O \longrightarrow ROH + H_3O^+ + X^-$$

Alkyl halide Water Alcohol Hydronium ion Halide ion

The reaction occurs in two stages. Only the first stage involves nucleophilic substitution. It is the rate-determining step.

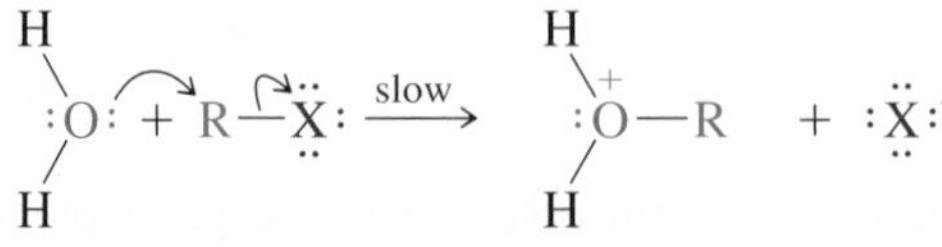

Water Alkyl halide Alkyloxonium ion Halide ion

The second stage is a Brønsted–Lowry acid–base reaction and is fast.

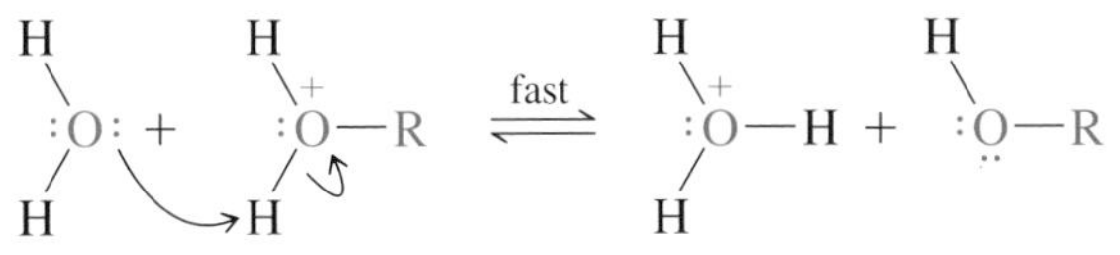

Water | Alkyloxonium ion | Hydronium ion | Alcohol

Analogous reactions take place in other solvents that, like water, contain an —OH group. Solvolysis in methanol (*methanolysis*) gives a methyl ether.

$$RX + 2CH_3OH \longrightarrow ROCH_3 + CH_3\overset{+}{O}H_2 + X^-$$

Alkyl halide | Methanol | Alkyl methyl ether | Methyloxonium ion | Halide ion

Problem 8.8

Adapt the preceding mechanism for the hydrolysis of RX so that it describes the methanolysis of ethyl bromide.

Because attack by the nucleophile is the rate-determining step of the S_N2 mechanism, the rate of substitution can, and does, vary from nucleophile to nucleophile. Just as some alkyl halides are more reactive than others, some nucleophiles are more reactive than others. Nucleophilic strength, or **nucleophilicity,** is a measure of how fast a Lewis base displaces a leaving group from a suitable substrate. By measuring the rate at which various Lewis bases react with methyl iodide in methanol, a list of their nucleophilicities relative to methanol as the standard nucleophile has been compiled. It is presented in Table 8.4.

As long as the nucleophilic atom is the same, the more basic the nucleophile, the more reactive it is. An alkoxide ion (RO^-) is more basic and more nucleophilic than a carboxylate ion (RCO_2^-).

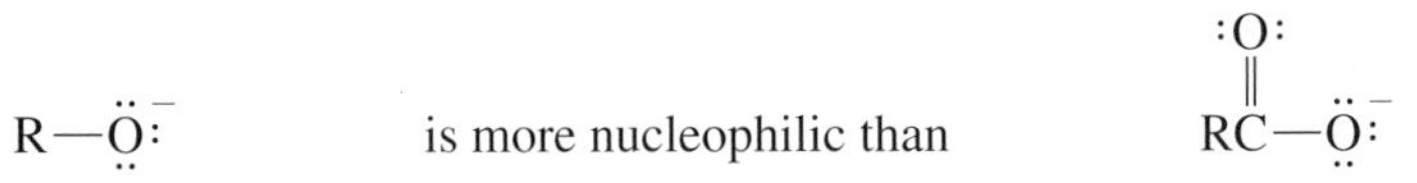

Stronger base
Conjugate acid is ROH:
$pK_a = 16$

Weaker base
Conjugate acid is RCO_2H:
$pK_a = 5$

The connection between basicity and nucleophilicity holds when comparing atoms in the *same row* of the periodic table. Thus, HO^- is more basic and more nucleophilic than F^-, and H_3N is more basic and more nucleophilic than H_2O. *It does not hold when proceeding down a column in the periodic table.* For example, I^- is the least basic of the halide ions but is the most nucleophilic. F^- is the most basic halide ion but the least nucleophilic.

TABLE 8.4 Nucleophilicity of Some Common Nucleophiles

Reactivity class	Nucleophile	Relative reactivity*
Very good nucleophiles	I^-, HS^-, RS^-	$>10^5$
Good nucleophiles	Br^-, HO^-, RO^-, CN^-, N_3^-	10^4
Fair nucleophiles	NH_3, Cl^-, F^-, RCO_2^-	10^3
Weak nucleophiles	H_2O, ROH	1
Very weak nucleophiles	RCO_2H	10^{-2}

*Relative reactivity is k(nucleophile)/k(methanol) for typical S_N2 reactions and is approximate. Data pertain to methanol as the solvent.

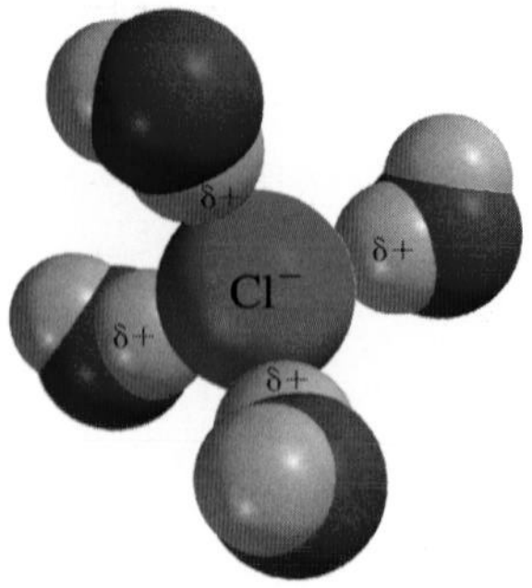

Figure 8.3

Solvation of a chloride ion by water.

Neutral Lewis bases such as water, alcohols, and carboxylic acids are much weaker nucleophiles than their conjugate bases. When comparing species that have the same nucleophilic atom, a negatively charged nucleophile is more reactive than a neutral one.

$R-\ddot{O}:^-$ Alkoxide ion	is more nucleophilic than	$R-\ddot{O}-H$ Alcohol
$RC(=O)-\ddot{O}:^-$ Carboxylate ion	is more nucleophilic than	$R-C(=O)-\ddot{O}-H$ Carboxylic acid

The factor that seems most responsible for the inverse relationship between basicity and nucleophilicity among the halide ions is the degree to which they are *solvated* by ion–dipole forces of the type illustrated in Figure 8.3. Smaller anions, because of their high charge-to-size ratio, are more strongly solvated than larger ones. In order to act as a nucleophile, the halide must shed some of the solvent molecules that surround it. Among the halide anions, ion–dipole forces are strongest for F^- and weakest for I^-. Thus, the nucleophilicity of F^- is suppressed more than that of Cl^-, Cl^- more than Br^-, and Br^- more than I^-. Similarly, HO^- is smaller, more solvated, and less nucleophilic than HS^-. The importance of solvation in reducing the nucleophilicity of small anions more than larger ones can be seen in the fact that, when measured in the gas phase where solvation forces don't exist, the order of halide nucleophilicity reverses and tracks basicity: $F^- > Cl^- > Br^- > I^-$.

8.6 The S_N1 Mechanism of Nucleophilic Substitution

Having just learned that tertiary alkyl halides are practically inert to substitution by the S_N2 mechanism because of steric hindrance, we might wonder whether they undergo nucleophilic substitution at all. We'll see in this section that they do, but by a mechanism different from S_N2.

Hughes and Ingold observed that the hydrolysis of *tert*-butyl bromide, which occurs readily, is characterized by a *first-order* rate law:

$$\underset{\textit{tert}\text{-Butyl bromide}}{(CH_3)_3CBr} + \underset{\text{Water}}{2H_2O} \longrightarrow \underset{\textit{tert}\text{-Butyl alcohol}}{(CH_3)_3COH} + \underset{\text{Hydronium ion}}{H_3O^+} + \underset{\text{Bromide ion}}{Br^-}$$

$$\text{Rate} = k[(CH_3)_3CBr]$$

They found that the rate of hydrolysis depends only on the concentration of *tert*-butyl bromide. Adding the stronger nucleophile hydroxide ion, moreover, causes no change in the rate of substitution, nor does this rate depend on the concentration of hydroxide. Just as second-order kinetics was interpreted as indicating a bimolecular rate-determining step, first-order kinetics was interpreted as evidence for a *unimolecular* rate-determining step—a step that involves only the alkyl halide.

The S_N1 mechanism was introduced earlier in Section 4.9.

The proposed process is outlined in Mechanism 8.2 (page 336) and is called S_N1, standing for **substitution nucleophilic unimolecular.** The first step, a unimolecular dissociation of the alkyl halide to form a carbocation as the key intermediate, is rate-determining. An energy diagram for the process is shown in Figure 8.5 (page 336).

Problem 8.9

Suggest a structure for the product of nucleophilic substitution obtained on solvolysis of *tert*-butyl bromide in methanol, and outline a reasonable mechanism for its formation.

Enzyme-Catalyzed Nucleophilic Substitutions of Alkyl Halides

Nucleophilic substitution is one of a variety of mechanisms by which living systems detoxify halogenated organic compounds introduced into the environment. Enzymes that catalyze these reactions are known as *haloalkane dehalogenases.* The hydrolysis of 1,2-dichloroethane to 2-chloroethanol, for example, is a biological nucleophilic substitution catalyzed by the dehalogenase shown in Figure 8.4.

$$\underset{\text{1,2-Dichloroethane}}{ClCH_2CH_2Cl} + \underset{\text{Water}}{2H_2O} \xrightarrow[\text{enzyme}]{\text{dehalogenase}} \underset{\text{2-Chloroethanol}}{ClCH_2CH_2OH} + \underset{\text{Hydronium ion}}{H_3O^+} + \underset{\text{Chloride ion}}{Cl^-}$$

This haloalkane dehalogenase is believed to act by using one of its side-chain carboxylates to displace chloride by an S_N2 mechanism. (Recall the reaction of carboxylate ions with alkyl halides from Table 8.1.)

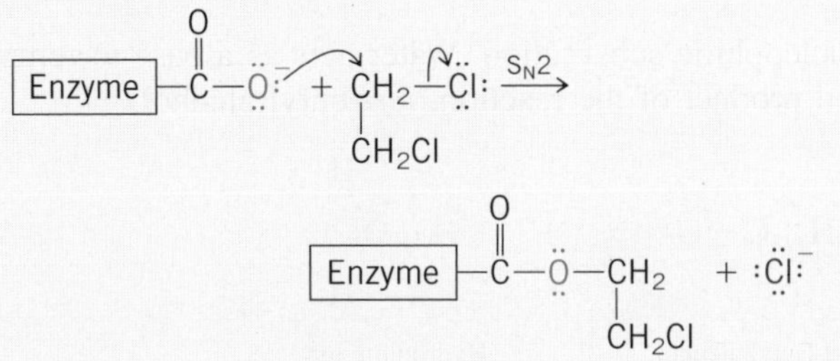

The product of nucleophilic substitution then reacts with water, restoring the enzyme to its original state and giving the observed products of the reaction.

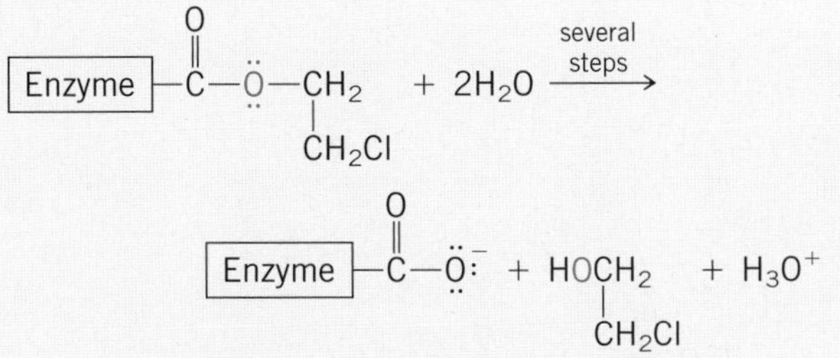

This stage of the reaction proceeds by a mechanism that will be discussed in Chapter 19. Both stages are faster than the reaction of 1,2-dichloroethane with water in the absence of the enzyme.

Enzyme-catalyzed hydrolysis of racemic 2-chloropropanoic acid is a key step in the large-scale preparation (2000 tons per year!) of (*S*)-2-chloropropanoic acid used for the preparation of agricultural chemicals.

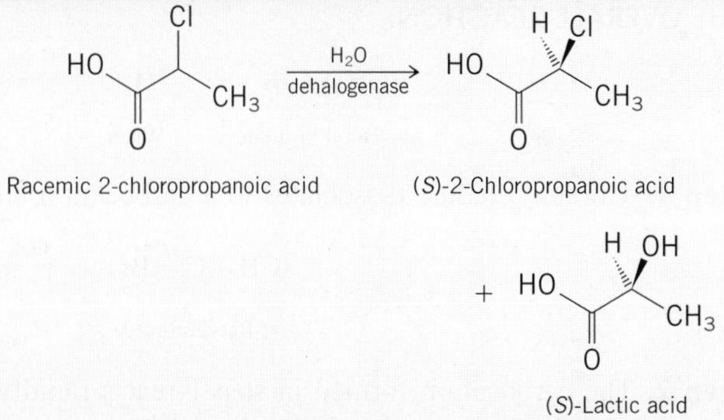

In this enzymatic resolution (Section 7.15), the dehalogenase enzyme catalyzes the hydrolysis of the *R*-enantiomer of 2-chloropropanoic acid to (*S*)-lactic acid. The desired (*S*)-2-chloropropanoic acid is unaffected and recovered in a nearly enantiomerically pure state.

Some of the most common biological S_N2 reactions involve attack at methyl groups, especially a methyl group of *S*-adenosylmethionine. Examples of these will be given in Chapter 16.

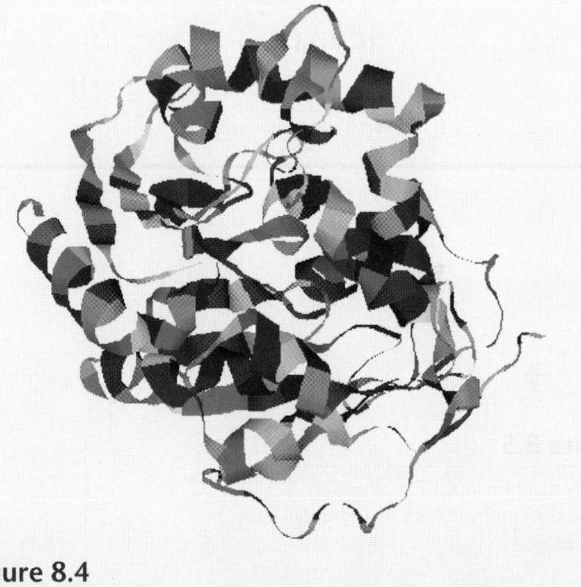

Figure 8.4

A ribbon diagram of the dehalogenase enzyme that catalyzes the hydrolysis of 1,2-dichloroethane. The progression of amino acids along the chain is indicated by a color change. The nucleophilic carboxylate group is near the center of the diagram.

The S_N1 mechanism is an *ionization* mechanism. The nucleophile does not participate until after the rate-determining step has taken place. Thus, the effects of nucleophile and alkyl halide structure are expected to be different from those observed for reactions proceeding by the S_N2 pathway. How the structure of the alkyl halide affects the rate of S_N1 reactions is the topic of the next section.

Mechanism 8.2

The S_N1 Mechanism of Nucleophilic Substitution

THE OVERALL REACTION:

$$(CH_3)_3CBr + 2H_2O \longrightarrow (CH_3)_3COH + H_3O^+ + Br^-$$

tert-Butyl bromide Water *tert*-Butyl alcohol Hydronium ion Bromide ion

Step 1: The alkyl halide dissociates to a carbocation and a halide ion.

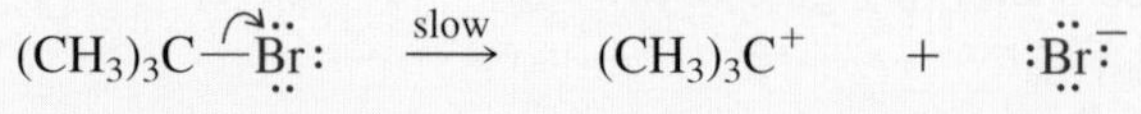

tert-Butyl bromide *tert*-Butyl cation Bromide ion

Step 2: The carbocation formed in step 1 reacts rapidly with a water molecule. Water is a nucleophile. This step completes the nucleophilic substitution stage of the mechanism and yields an alkyloxonium ion.

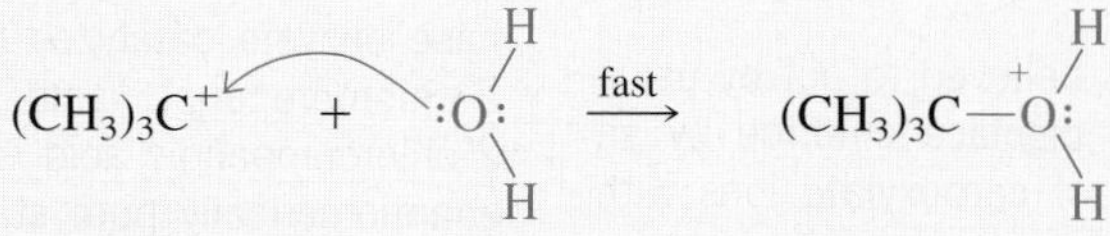

tert-Butyl cation Water *tert*-Butyloxonium ion

Step 3: This step is a fast acid-base reaction that follows the nucleophilic substitution. Water acts as a base to remove a proton from the alkyloxonium ion to give the observed product of the reaction, *tert*-butyl alcohol.

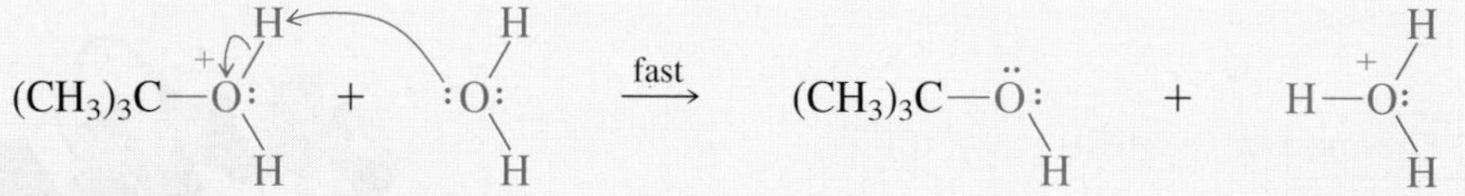

tert-Butyloxonium ion Water *tert*-Butyl alcohol Hydronium ion

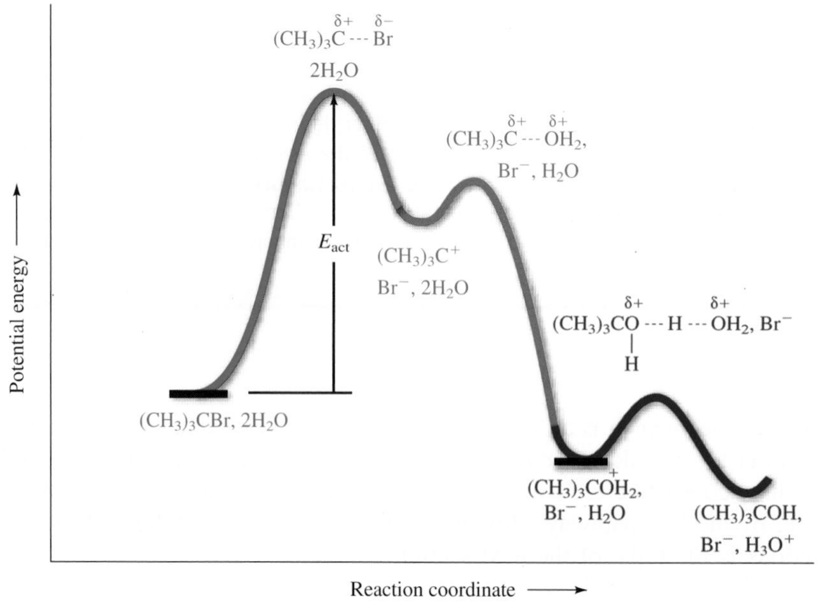

Figure 8.5

Energy diagram illustrating the S_N1 mechanism for hydrolysis of *tert*-butyl bromide.

TABLE 8.5 Reactivity of Some Alkyl Bromides Toward Substitution by the S_N1 Mechanism*

Alkyl bromide	Structure	Class	Relative rate
Methyl bromide	CH_3Br	Unsubstituted	0.6
Ethyl bromide	CH_3CH_2Br	Primary	1.0
Isopropyl bromide	$(CH_3)_2CHBr$	Secondary	26
tert-Butyl bromide	$(CH_3)_3CBr$	Tertiary	~100,000,000

*Solvolysis in aqueous formic acid.

8.7 Carbocation Stability and S_N1 Reaction Rates

In order to compare S_N1 substitution rates in a range of alkyl halides, experimental conditions are chosen in which competing substitution by the S_N2 mechanism is very slow. One such set of conditions is solvolysis in aqueous formic acid (HCO_2H):

$$\underset{\text{Alkyl halide}}{RX} + \underset{\text{Water}}{2H_2O} \xrightarrow{\text{formic acid}} \underset{\text{Alcohol}}{ROH} + \underset{\text{Hydronium ion}}{H_3O^+} + \underset{\text{Halide ion}}{X^-}$$

Neither formic acid nor water is very nucleophilic, and so S_N2 substitution is suppressed. The relative rates of hydrolysis of a group of alkyl bromides under these conditions are presented in Table 8.5.

The relative reactivity of alkyl halides in S_N1 reactions is exactly the opposite of S_N2:

S_N1 reactivity: methyl < primary < secondary < tertiary

S_N2 reactivity: tertiary < secondary < primary < methyl

Clearly, the steric crowding that influences S_N2 reaction rates plays no role in S_N1 reactions. The order of alkyl halide reactivity in S_N1 reactions is the same as the order of carbocation stability: the more stable the carbocation, the more reactive the alkyl halide.

Increasing rate of substitution by the S_N1 mechanism

CH_3X	<	RCH_2X	<	R_2CHX	<	R_3CX
Methyl Forms least stable carbocation		Primary		Secondary		Tertiary Forms most stable carbocation

We have seen this situation before in the reaction of alcohols with hydrogen halides (Section 4.11), in the acid-catalyzed dehydration of alcohols (Section 5.12), and in the conversion of alkyl halides to alkenes by the E1 mechanism (Section 5.18). As in these other reactions, the stabilization of the carbocation intermediate by alkyl substituents is the decisive factor. The more stable the carbocation, the faster it is formed.

Problem 8.10

Identify the compound in each of the following pairs that reacts at the faster rate in an S_N1 reaction:

(a) Isopropyl bromide or isobutyl bromide

(b) Cyclopentyl iodide or 1-methylcyclopentyl iodide

(c) Cyclopentyl bromide or 1-bromo-2,2-dimethylpropane

(d) *tert*-Butyl chloride or *tert*-butyl iodide

Continued

Sample Solution (a) Isopropyl bromide, $(CH_3)_2CHBr$, is a secondary alkyl halide, whereas isobutyl bromide, $(CH_3)_2CHCH_2Br$, is primary. Because the rate-determining step in an S_N1 reaction is carbocation formation and secondary carbocations are more stable than primary ones, isopropyl bromide is more reactive than isobutyl bromide in nucleophilic substitution by the S_N1 mechanism.

Methyl and primary carbocations are so high in energy that their intermediacy in nucleophilic substitutions is unlikely. When ethyl bromide undergoes hydrolysis in aqueous formic acid, substitution probably takes place by an S_N2 process in which water is the nucleophile (Section 8.5). *In general, methyl and primary alkyl halides never react by the S_N1 mechanism; tertiary alkyl halides never react by S_N2.*

8.8 Stereochemistry of S_N1 Reactions

Although S_N2 reactions are stereospecific and proceed with inversion of configuration at carbon, the situation is not as clear-cut for S_N1. When the leaving group departs from a chirality center of an optically active halide, the positively charged carbon that results is sp^2-hybridized and cannot be a chirality center. The three bonds to that carbon define a plane of symmetry.

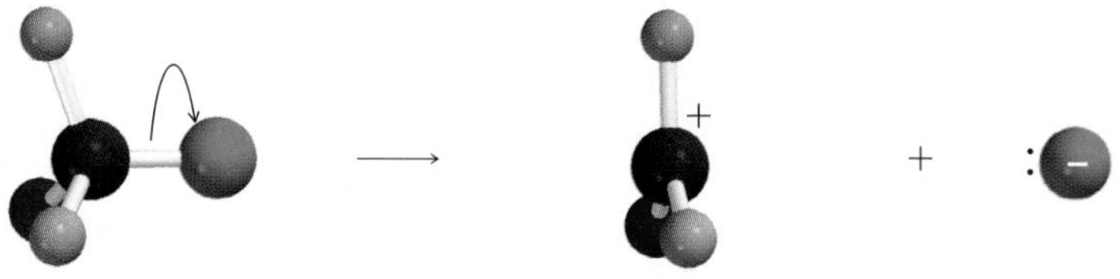

Alkyl halide is chiral Carbocation is achiral

If a nucleophile can approach each face of the carbocation equally well, substitution by the S_N1 mechanism should give a 1:1 mixture of enantiomers irrespective of whether the starting alkyl halide is *R, S,* or racemic. S_N1 reactions should give racemic products from optically active starting materials.

But they rarely do. For example, although the hydrolysis of optically active 2-bromooctane follows a first-order rate law, the resulting 2-octanol is only 34% racemized. Inversion of configuration is the major pathway.

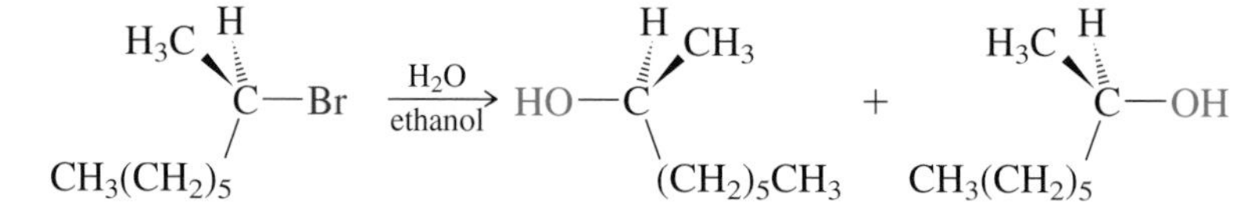

(*R*)-(−)-2-Bromooctane (*S*)-(+)-2-Octanol (*R*)-(−)-2-Octanol

66% net inversion corresponds to 83% *S*, 17% *R*

Partial but not complete loss of optical activity in S_N1 reactions is explained as shown in Figure 8.6. The key feature of this mechanism is that when the carbocation is formed, it is not completely free of the leaving group. Although ionization is complete, the leaving group has not yet diffused very far away from the carbon to which it was attached and partially blocks approach of the nucleophile from that direction. Nucleophilic attack on this species, called an *ion pair,* occurs faster from the side opposite the leaving group. Depending on how closely associated the two ions are, the product is formed with predominant to complete inversion of configuration. Once the leaving group has diffused away, however, both faces of the carbocation are equally accessible to nucleophiles and equal quantities of enantiomeric products result.

The stereochemistry of S_N1 substitution depends on the relative rates of competing processes—attack by the nucleophile on the ion pair versus separation of the ions. Consequently, the observed stereochemistry varies considerably according to the alkyl

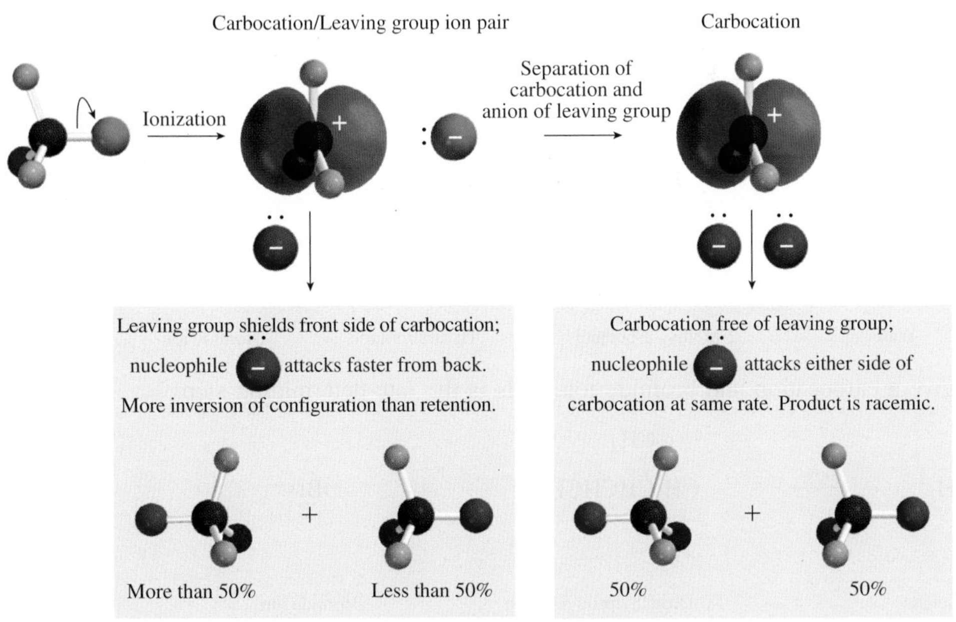
Carbocation/Leaving group ion pair
Carbocation
Ionization
Separation of carbocation and anion of leaving group
Leaving group shields front side of carbocation; nucleophile attacks faster from back. More inversion of configuration than retention.
Carbocation free of leaving group; nucleophile attacks either side of carbocation at same rate. Product is racemic.
More than 50%
Less than 50%
50%
50%

Figure 8.6

S_N1 stereochemistry. The carbocation formed by ionization of an alkyl halide is shielded on its "front" side by the leaving group. The nucleophile attacks this carbocation-halide ion pair faster from the less shielded "back" side and the product is formed with net inversion of configuration. In a process that competes with nucleophilic attack on the ion pair, the leaving group diffuses away from the carbocation. The nucleophile attacks the carbocation at the same rate from either side to give equal amounts of enantiomers.

halide, nucleophile, and experimental conditions. Some, such as the one we just discussed (hydrolysis of 2-bromooctane), give predominant, but incomplete, inversion of configuration. Others give products that are almost entirely racemic.

Problem 8.11

What two stereoisomeric substitution products would you expect to isolate from the hydrolysis of *cis*-1,4-dimethylcyclohexyl bromide? From hydrolysis of *trans*-1,4-dimethylcyclohexyl bromide?

8.9 Carbocation Rearrangements in S_N1 Reactions

Additional evidence for carbocation intermediates in certain nucleophilic substitutions comes from observing rearrangements of the kind normally associated with such species. For example, hydrolysis of the secondary alkyl bromide 2-bromo-3-methylbutane yields the rearranged tertiary alcohol 2-methyl-2-butanol as the only substitution product.

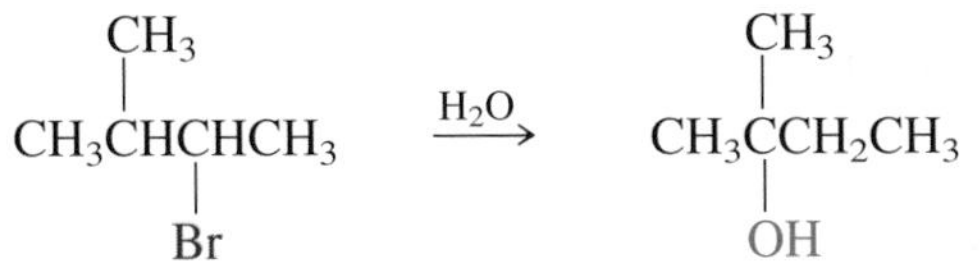
CH3
CH3CHCHCH3
Br
H2O
CH3
CH3CCH2CH3
OH

2-Bromo-3-methylbutane 2-Methyl-2-butanol (93%)

Mechanism 8.3 for this reaction assumes rate-determining ionization of the alkyl halide (step 1), followed by a hydride shift that converts a secondary carbocation to a more stable tertiary one (step 2). The tertiary carbocation then reacts with water to yield the observed product (steps 3 and 4).

Problem 8.12

Why does the carbocation intermediate in the hydrolysis of 2-bromo-3-methylbutane rearrange by way of a hydride shift rather than a methyl shift?

Rearrangements, when they do occur, are taken as evidence for carbocation intermediates and point to the S_N1 mechanism as the reaction pathway. Rearrangements are never observed in S_N2 reactions of alkyl halides.

Mechanism 8.3

Carbocation Rearrangement in the S_N1 Hydrolysis of 2-Bromo-3-methylbutane

THE OVERALL REACTION:

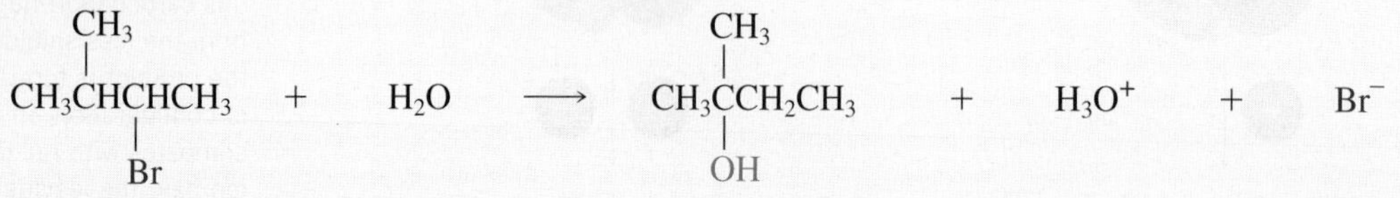

2-Bromo-3-methylbutane   Water   2-Methyl-2-butanol   Hydronium ion   Bromide ion

Step 1: The alkyl halide ionizes to give a carbocation and bromide ion. This is the rate-determining step.

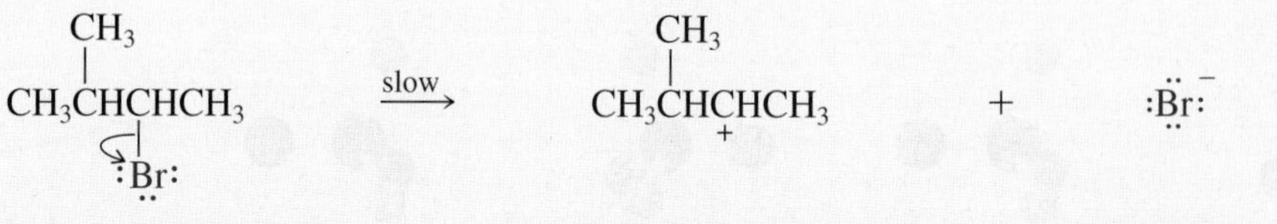

2-Bromo-3-methylbutane   1,2-Dimethylpropyl cation   Bromide ion

Step 2: The carbocation formed in step 1 is secondary; it rearranges by a hydride shift to form a more stable tertiary carbocation.

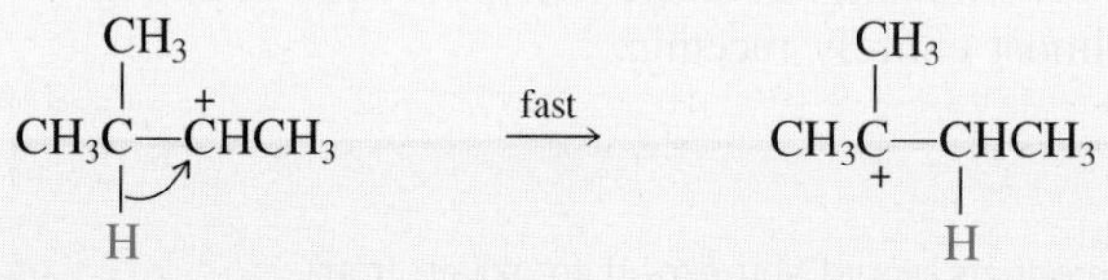

1,2-Dimethylpropyl cation   1,1-Dimethylpropyl cation

Step 3: The tertiary carbocation is attacked by water acting as a nucleophile.

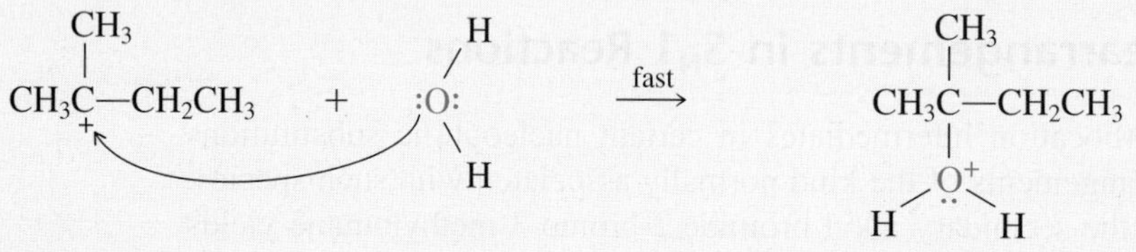

1,1-Dimethylpropyl cation   Water   1,1-Dimethylpropyloxonium ion

Step 4: Proton transfer from the alkyloxonium ion to water completes the process.

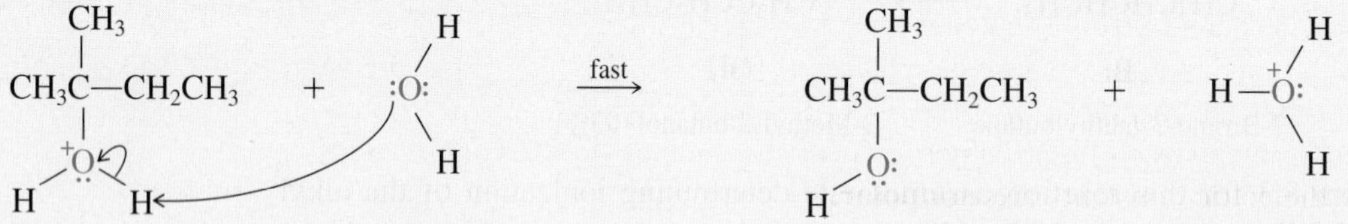

1,1-Dimethylpropyloxonium ion   Water   2-Methyl-2-butanol   Hydronium ion

8.10 Effect of Solvent on the Rate of Nucleophilic Substitution

The major effect of the solvent is on the *rate* of nucleophilic substitution, not on what the products are. Thus we need to consider two related questions:

1. What properties of the *solvent* influence the rate most?
2. How does the rate-determining step of the *mechanism* respond to the properties of the solvent?

We begin by looking at the solvents commonly employed in nucleophilic substitutions, then proceed to examine how these properties affect the S_N1 and S_N2 mechanisms. Because these mechanisms are so different from each other, we discuss each one separately.

Classes of Solvents. Table 8.6 lists a number of solvents in which nucleophilic substitutions are carried out and classifies them according to two criteria: whether they are *protic* or *aprotic,* and *polar* or *nonpolar.*

TABLE 8.6 Properties of Some Solvents Used in Nucleophilic Substitution

Solvent	Structural formula	Protic or Aprotic	Dielectric constant ε*	Polarity
				Most polar
Water	H_2O	Protic	78	↑
Formic acid	$H\overset{\overset{O}{\|}}{C}OH$	Protic	58	
Dimethyl sulfoxide	$(CH_3)_2\overset{+}{S}-O^-$	Aprotic	49	
Acetonitrile	$CH_3C\equiv N$	Aprotic	37	
N,N-Dimethylformamide	$(CH_3)_2N\overset{\overset{O}{\|}}{C}H$	Aprotic	37	
Methanol	CH_3OH	Protic	33	
Acetic acid	$CH_3\overset{\overset{O}{\|}}{C}OH$	Protic	6	
				Least polar

*Dielectric constants are approximate and temperature-dependent.

The kinds of solvents classified as **protic** are the same as those most capable of hydrogen-bonding interactions. Most have —OH groups, as do the examples in Table 8.6 (water, formic acid, methanol, and acetic acid). The **aprotic** solvents in the table (dimethyl sulfoxide, *N,N*-dimethylformamide, and acetonitrile) lack —OH groups and have all of their hydrogens bonded to carbon.

Unlike protic and aprotic, which constitute an "either-or" pair, polar and nonpolar belong to a continuous gradation with no sharply defined boundary separating them.

The polarity of a solvent is related to its **dielectric constant (ε),** which is a measure of the ability of a material to moderate the force of attraction between oppositely charged particles. The standard dielectric is a vacuum, assigned a value ε of exactly 1, to which the polarities of other materials are then compared. The higher the dielectric constant ε, the better the medium is able to support separated positively and negatively charged species. Solvents with high dielectric constants are classified as **polar** solvents; those with low dielectric constants are **nonpolar.**

Problem 8.13

Diethyl ether ($CH_3CH_2OCH_2CH_3$) has a dielectric constant of 4. What best describes its solvent properties: polar protic, nonpolar protic, polar aprotic, or nonpolar aprotic?

Solvent Effects on the Rate of Substitution by the S_N2 Mechanism. Polar solvents are required in typical bimolecular substitutions because ionic substances, such as the sodium and potassium salts cited earlier in Table 8.1, are not sufficiently soluble in nonpolar solvents to give a high enough concentration of the nucleophile to allow the reaction to occur at a rapid rate. Other than the requirement that the solvent be polar enough to dissolve ionic compounds, however, the effect of solvent polarity on the rate of S_N2 reactions is small. What is more important is whether the polar solvent is protic or aprotic. Protic solvents such as water, formic acid, methanol, and acetic acid all have —OH groups that allow them to form hydrogen bonds to anionic nucleophiles.

$$\underset{\text{Solvent}}{R\overset{\delta^-}{O}-H^{\delta^+}} + \underset{\text{Nucleophile}}{:Y^-} \longrightarrow \underset{\text{Hydrogen-bonded complex}}{R\overset{\delta^-}{O}-H\text{---}Y^{\delta^-}}$$

This clustering of *protic* solvent molecules *(solvation)* around it suppresses the nucleophilicity of the anion and retards the rate of bimolecular substitution.

Aprotic solvents, on the other hand, lack —OH groups and do not solvate anions very strongly, leaving the anions much more able to express their nucleophilic character. Table 8.7 compares the second-order rate constants k for S_N2 substitution of 1-bromobutane by azide ion (a good nucleophile) in several polar aprotic solvents with the corresponding k's for the much slower reactions in polar protic solvents.

$$\underset{\text{1-Bromobutane}}{CH_3CH_2CH_2CH_2Br} + \underset{\text{Azide ion}}{N_3^-} \longrightarrow \underset{\text{1-Azidobutane}}{CH_3CH_2CH_2CH_2N_3} + \underset{\text{Bromide ion}}{Br^-}$$

TABLE 8.7 Relative Rate of S_N2 Displacement of 1-Bromobutane by Azide in Various Solvents*

Solvent	Structural formula	Dielectric constant ε	Type of solvent	Relative rate
Methanol	CH_3OH	33	Polar protic	1
Water	H_2O	78	Polar protic	7
Dimethyl sulfoxide	$(CH_3)_2\overset{+}{S}-O^-$	49	Polar aprotic	1300
N,*N*-Dimethylformamide	$(CH_3)_2NCH=O$	37	Polar aprotic	2800
Acetonitrile	$CH_3C\equiv N$	37	Polar aprotic	5000

*Ratio of second-order rate constant for substitution in indicated solvent to that for substitution in methanol at 25°C.

Problem 8.14

Unlike protic solvents, which solvate anions, aprotic solvents form complexes with cations better than with anions. Use a dashed line to show the interaction between dimethyl sulfoxide $[(CH_3)_2\overset{+}{\ddot{S}}-\ddot{\underset{..}{O}}:^-]$ with a cation, using sodium azide (NaN_3) as the source of the cation.

The large rate enhancements observed for bimolecular nucleophilic substitutions in polar aprotic solvents offer advantages in synthesis. One example is the preparation of alkyl cyanides (nitriles) by the reaction of sodium cyanide with alkyl halides:

$$\underset{\text{Hexyl halide}}{CH_3(CH_2)_4CH_2X} + \underset{\text{Sodium cyanide}}{NaCN} \longrightarrow \underset{\text{Hexyl cyanide}}{CH_3(CH_2)_4CH_2CN} + \underset{\text{Sodium halide}}{NaX}$$

When the reaction was carried out in aqueous methanol as the solvent, hexyl bromide was converted to hexyl cyanide in 71% yield. Although this is perfectly acceptable for a synthetic reaction, it required heating for a period of over *20 hours*. Changing the solvent to dimethyl sulfoxide increased the reaction rate to the extent that the less reactive (and less expensive) substrate hexyl chloride could be used and the reaction was complete (91% yield) in only *20 minutes!*

The *rate* at which reactions occur can be important in the laboratory, and understanding how solvents affect rate is of practical value. As we proceed through the text, however, and see how nucleophilic substitution is applied to a variety of functional group transformations, be aware that the nature of both the substrate and the nucleophile, more than anything else, determines what *product* is formed.

Solvent Effects on the Rate of Substitution by the S_N1 Mechanism. Table 8.8 gives the relative rate of solvolysis of *tert*-butyl chloride in several protic solvents listed in order of increasing dielectric constant. As the table illustrates, the rate of solvolysis of *tert*-butyl chloride (which is equal to its rate of ionization) increases dramatically as the solvent becomes more polar.

TABLE 8.8 Relative Rate of S_N1 Solvolysis of *tert*-Butyl Chloride as a Function of Solvent Polarity*

Solvent	Dielectric constant ε	Relative rate
Acetic acid	6	1
Methanol	33	4
Formic acid	58	5,000
Water	78	150,000

*Ratio of first-order rate constant for solvolysis in indicated solvent to that for solvolysis in acetic acid at 25°C.

According to the S_N1 mechanism, a molecule of an alkyl halide ionizes to a positively charged carbocation and a negatively charged halide ion in the rate-determining step (see Figure 8.5). As the alkyl halide approaches the transition state for this step, a partial positive charge develops on the carbon and a partial negative charge on the halogen. The effects of a nonpolar and a polar solvent on the energy of the transition state are contrasted in Figure 8.7. Polar and nonpolar solvents are similar in their interaction with the starting alkyl halide, but differ markedly in how they stabilize the transition state. A solvent with a low dielectric constant has little effect on the energy of the transition state, whereas one with a high dielectric constant stabilizes

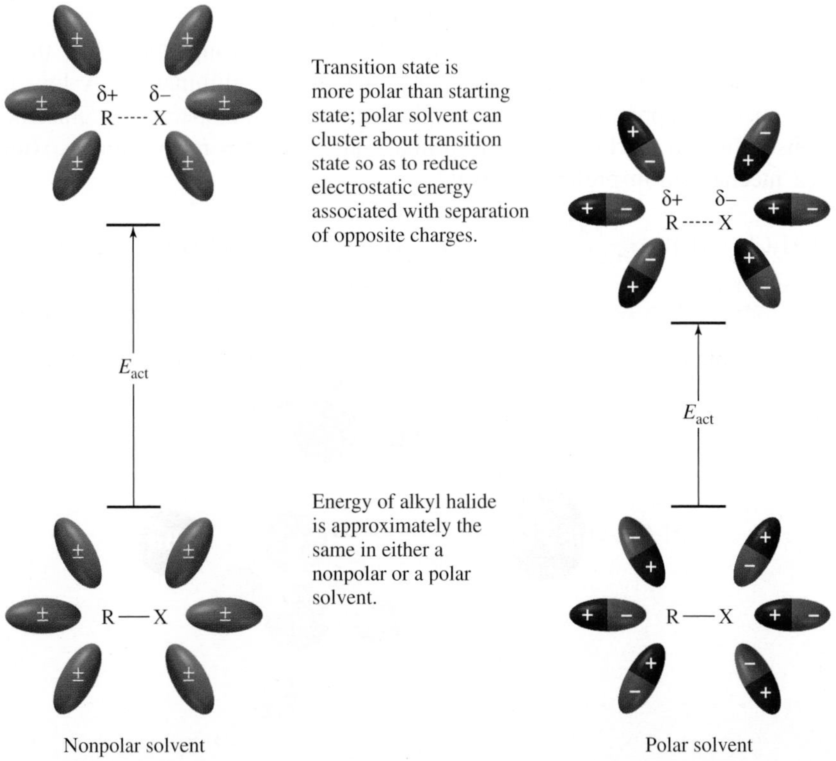
δ+ δ−
R----- X
Transition state is more polar than starting state; polar solvent can cluster about transition state so as to reduce electrostatic energy associated with separation of opposite charges.
E_{act}
Energy of alkyl halide is approximately the same in either a nonpolar or a polar solvent.
R—X
Nonpolar solvent
Polar solvent

Figure 8.7

A polar solvent stabilizes the transition state of an S_N1 reaction and increases its rate.

the charge-separated transition state, lowers the activation energy, and increases the rate of the reaction.

If the polar solvent is a protic one, stabilization of the transition state is even more pronounced because of the hydrogen bonding that develops as the leaving group becomes negatively charged.

8.11 Substitution and Elimination as Competing Reactions

We have seen that an alkyl halide and a Lewis base can react together in either a substitution or an elimination reaction.

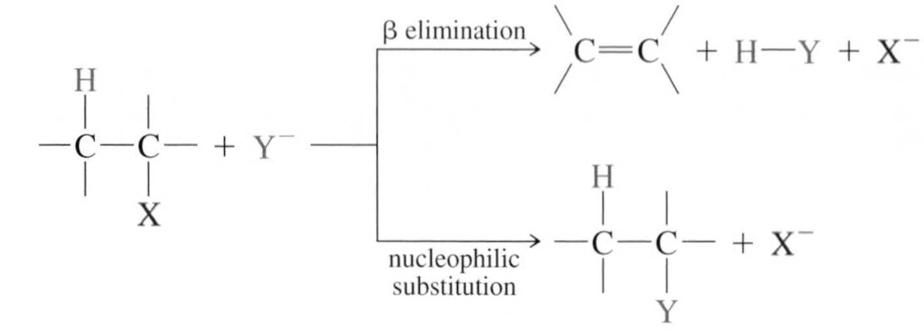

Substitution can take place by the S_N1 or the S_N2 mechanism, elimination by E1 or E2.

How can we predict whether substitution or elimination will be the principal reaction observed with a particular combination of reactants? The two most important factors are the *structure of the alkyl halide* and the *basicity of the anion.* It is useful to approach the question from the premise that the characteristic reaction of alkyl halides with Lewis bases is *elimination,* and that substitution predominates only under certain special circumstances. In a typical reaction, a secondary alkyl halide such as isopropyl bromide reacts with a Lewis base such as sodium ethoxide mainly by elimination:

$$\underset{\displaystyle Br}{CH_3CHCH_3} \xrightarrow[CH_3CH_2OH,\ 55^\circ C]{NaOCH_2CH_3} CH_3CH{=}CH_2 + \underset{\displaystyle OCH_2CH_3}{CH_3CHCH_3}$$

Isopropyl bromide — Propene (87%) — Ethyl isopropyl ether (13%)

Figure 8.8 illustrates the close relationship between the E2 and S_N2 pathways for this case, and the results cited in the preceding equation clearly show that E2 is faster than S_N2 when a secondary alkyl halide reacts with a strong base.

As crowding at the carbon that bears the leaving group decreases, the rate of nucleophilic substitution becomes faster than the rate of elimination. A low level of steric hindrance to approach of the nucleophile is one of the special circumstances that permit substitution to predominate, and primary alkyl halides react with alkoxide bases by an S_N2 mechanism in preference to E2:

$$CH_3CH_2CH_2Br \xrightarrow[CH_3CH_2OH,\ 55^\circ C]{NaOCH_2CH_3} CH_3CH{=}CH_2 + CH_3CH_2CH_2OCH_2CH_3$$

Propyl bromide — Propene (9%) — Ethyl propyl ether (91%)

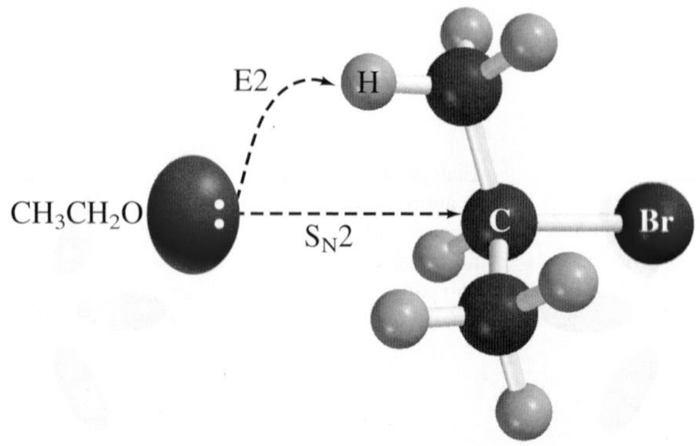

Figure 8.8

When a Lewis base reacts with an alkyl halide, either substitution or elimination can occur. Substitution (S_N2) occurs when the Lewis base acts as a nucleophile and attacks carbon to displace bromide. Elimination (E2) occurs when the Lewis base abstracts a proton from the β carbon. The alkyl halide shown is isopropyl bromide, and elimination (E2) predominates over substitution with alkoxide bases.

If, however, the base itself is a crowded one, such as potassium *tert*-butoxide, even primary alkyl halides undergo elimination rather than substitution:

$$CH_3(CH_2)_{15}CH_2CH_2Br \xrightarrow[(CH_3)_3COH,\ 40°C]{KOC(CH_3)_3} CH_3(CH_2)_{15}CH{=}CH_2 + CH_3(CH_2)_{15}CH_2CH_2OC(CH_3)_3$$

1-Bromooctadecane — 1-Octadecene (87%) — *tert*-Butyl octadecyl ether (13%)

A second factor that can tip the balance in favor of substitution is weak basicity of the nucleophile. Nucleophiles that are less basic than hydroxide react with both primary and secondary alkyl halides to give the product of nucleophilic substitution in high yield. To illustrate, cyanide ion is much less basic than hydroxide and reacts with 2-chlorooctane to give the corresponding alkyl cyanide as the major product.

Cyanide is a weaker base than hydroxide because its conjugate acid HCN (pK_a 9.1) is a stronger acid than water (pK_a 15.7).

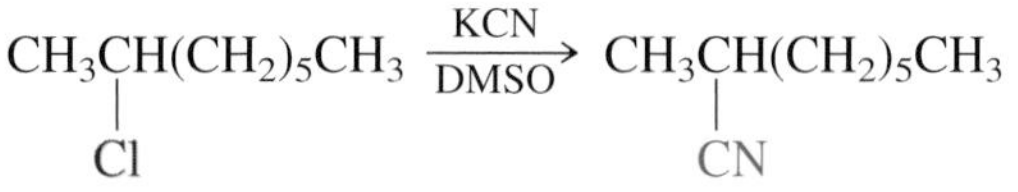
CH3CH(CH2)5CH3
Cl
KCN
DMSO
CH3CH(CH2)5CH3
CN

2-Chlorooctane — 2-Cyanooctane (70%)

Azide ion $:\!\overset{-}{N}{=}\overset{+}{N}{=}\overset{-}{N}\!:$ is an even weaker base than cyanide. It is a good nucleophile and reacts with secondary alkyl halides mainly by substitution:

The conjugate acid of azide ion is called *hydrazoic acid* (HN_3). It has a pK_a of 4.6, and so is similar to acetic acid in its acidity.

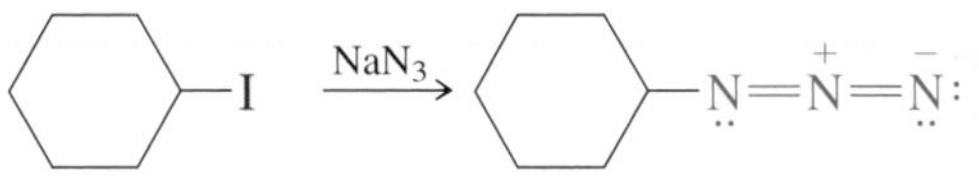
I
NaN3
N=N=N:

Cyclohexyl iodide — Cyclohexyl azide (75%)

Hydrogen sulfide ion HS^-, and anions of the type RS^-, are substantially less basic than hydroxide ion and react with both primary and secondary alkyl halides to give mainly substitution products.

Hydrogen sulfide (pK_a 7.0) is a stronger acid than water (pK_a 15.7). Therefore HS^- is a much weaker base than HO^-.

Tertiary alkyl halides are so sterically hindered to nucleophilic attack that the presence of any anionic Lewis base favors elimination. Usually substitution predominates over elimination in tertiary alkyl halides only when anionic Lewis bases are absent. In the solvolysis of the tertiary bromide 2-bromo-2-methylbutane, for example, the ratio of substitution to elimination is 64:36 in pure ethanol but falls to 1:99 in the presence of 2 M sodium ethoxide.

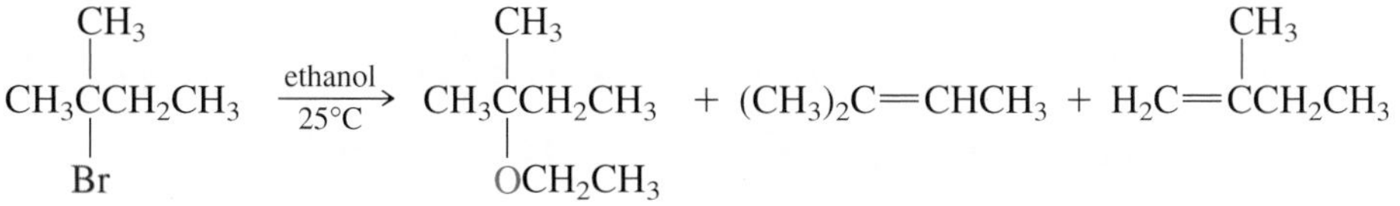
CH3
CH3CCH2CH3
Br
ethanol
25°C
CH3
CH3CCH2CH3
OCH2CH3
+ (CH3)2C=CHCH3 +
CH3
H2C=CCH2CH3

2-Bromo-2-methylbutane — 2-Ethoxy-2-methylbutane (Major product in absence of sodium ethoxide) — 2-Methyl-2-butene — 2-Methyl-1-butene (Alkene mixture is major product in presence of sodium ethoxide)

The substitution product in this case is formed by an S_N1 mechanism both in the presence and absence of sodium ethoxide. The alkenes are formed by an E1 mechanism in the absence of sodium ethoxide and by a combination of E2 (major) and E1 (minor) in its presence.

Problem 8.15

Predict the major organic product of each of the following reactions:

(a) Cyclohexyl bromide and potassium ethoxide
(b) Ethyl bromide and potassium cyclohexanolate
(c) *sec*-Butyl bromide solvolysis in methanol
(d) *sec*-Butyl bromide solvolysis in methanol containing 2 M sodium methoxide

Continued

Sample Solution (a) Cyclohexyl bromide is a secondary halide and reacts with alkoxide bases by elimination rather than substitution. The major organic products are cyclohexene and ethanol.

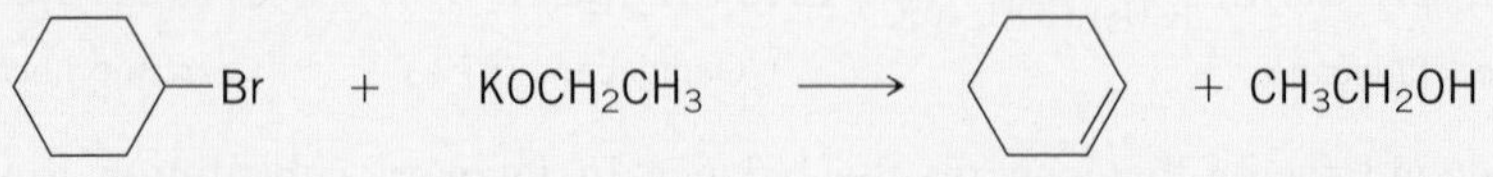
Br
+ KOCH₂CH₃
+ CH₃CH₂OH

Cyclohexyl bromide | Potassium ethoxide | Cyclohexene | Ethanol

Regardless of the alkyl halide, raising the temperature increases the rate of both substitution and elimination. The rate of elimination, however, usually increases faster than substitution, so that at higher temperatures the proportion of elimination products increases at the expense of substitution products.

As a practical matter, elimination can always be made to occur quantitatively. Strong bases, especially bulky ones such as *tert*-butoxide ion, react even with primary alkyl halides by an E2 process at elevated temperatures. The more difficult task is to find conditions that promote substitution. In general, the best approach is to choose conditions that favor the S_N2 mechanism—an unhindered substrate, a good nucleophile that is not strongly basic, and the lowest practical temperature consistent with reasonable reaction rates.

Problem 8.16

A standard method for the synthesis of ethers is an S_N2 reaction between an alkoxide and an alkyl halide.

$$R{-}\ddot{\underset{..}{O}}{:}^- \; + \; R'{-}\ddot{\underset{..}{X}}{:} \longrightarrow R{-}\ddot{\underset{..}{O}}{-}R' \; + \; {:}\ddot{\underset{..}{X}}{:}^-$$

Show possible combinations of alkoxide and alkyl halide for the preparation of the following ethers. Which of these ethers can be prepared effectively by this method?

(a) $CH_3CH_2CH(CH_3)OCH(CH_3)_2$ (b) $CH_3CH_2CH_2OCH(CH_3)_2$

Sample Solution

(a) • There are two possible combinations of alkoxide and alkyl halide (shown as bromide).

$CH_3CH_2CH(CH_3){-}\ddot{O}{:}^-$ + $BrCH(CH_3)_2$

$CH_3CH_2CH(CH_3){-}Br$ + ${}^-{:}\ddot{O}{-}CH(CH_3)_2$

• In both cases the alkyl halide is secondary. Secondary alkyl halides undergo elimination (E2) rather than substitution (S_N2) with alkoxide bases.

Functional group transformations that rely on substitution by the S_N1 mechanism are not as generally applicable as those of the S_N2 type. Hindered substrates are prone to elimination, and rearrangement is possible when carbocation intermediates are involved. Only in cases in which elimination is impossible are S_N1 reactions used for functional group transformations.

8.12 Nucleophilic Substitution of Alkyl Sulfonates

A few other classes of organic compounds undergo nucleophilic substitution reactions analogous to those of alkyl halides; the most important of these are sulfonates.

Sulfonic acids such as methanesulfonic acid and *p*-toluenesulfonic acid are strong acids, comparable in acidity with sulfuric acid.

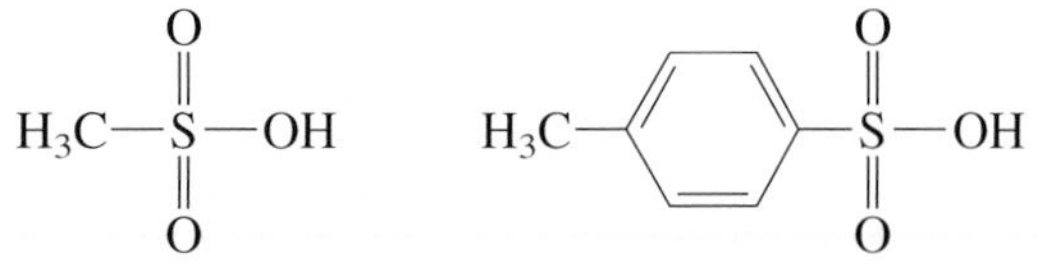

Methanesulfonic acid
p-Toluenesulfonic acid

Alkyl sulfonates are derivatives of sulfonic acids in which the proton of the —OH group is replaced by an alkyl group. They are prepared by treating an alcohol with the appropriate sulfonyl chloride, usually in the presence of the weak base pyridine.

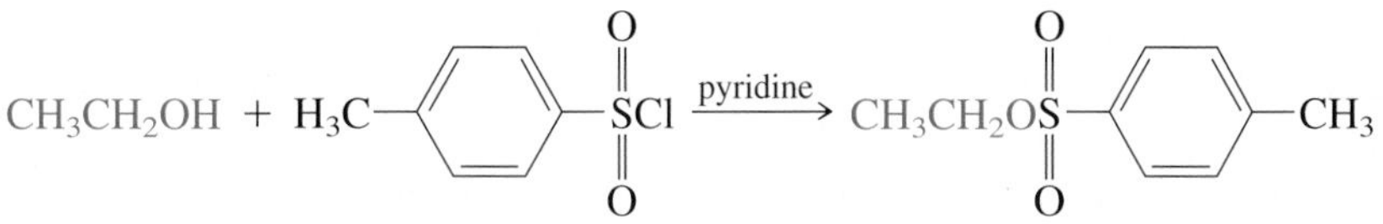

pyridine
Ethanol
p-Toluenesulfonyl chloride
Ethyl *p*-toluenesulfonate (72%)

Alkyl sulfonates resemble alkyl halides in their ability to undergo elimination and nucleophilic substitution.

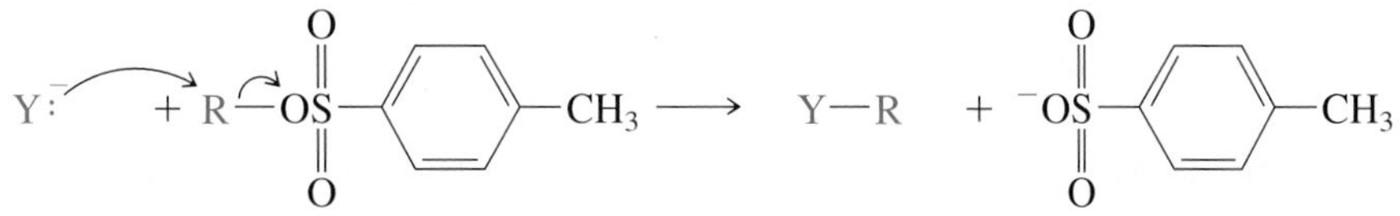

Nucleophile
p-Toluenesulfonate ester
Product of nucleophilic substitution
p-Toluenesulfonate anion

The sulfonates used most frequently are the *p*-toluenesulfonates. They are commonly known as *tosylates* and abbreviated as ROTs.

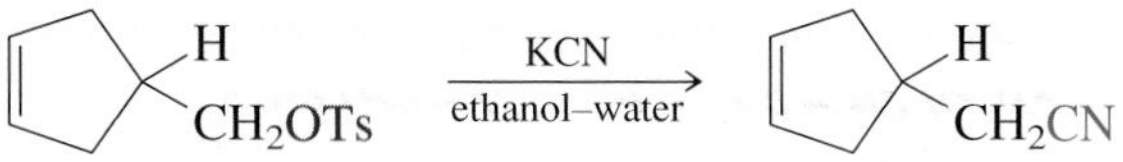

KCN
ethanol–water
(3-Cyclopentenyl)methyl *p*-toluenesulfonate
4-(Cyanomethyl)cyclopentene (86%)

p-Toluenesulfonate (TsO^-) is a very good leaving group. As Table 8.9 reveals, alkyl *p*-toluenesulfonates undergo nucleophilic substitution at rates that are even faster than those of alkyl iodides. The sulfonate group of a tosylate possesses a stronger electron-attracting ability than an iodide. Since the leaving group must accept the electrons of the bond that is broken in a nucleophilic substitution, sulfonates make excellent substrates for S_N2 reactions. A correlation of leaving-group abilities with carbon–halogen bond strengths was noted earlier, in Section 8.2. Note also the correlation with the basicity of the leaving group. Iodide is the weakest base among the halide anions and is the best leaving group, fluoride the strongest base and the poorest leaving group. A similar

TABLE 8.9 Approximate Relative Leaving-Group Abilities*

Leaving group	Relative rate	Conjugate acid of leaving group	pK_a of conjugate acid
F^-	10^{-5}	HF	3.1
Cl^-	10^{0}	HCl	−3.9
Br^-	10^{1}	HBr	−5.8
I^-	10^{2}	HI	−10.4
H_2O	10^{1}	H_3O^+	−1.7
$CH_3SO_2O^-$	10^{4}	CH_3SO_2OH	−2.6
TsO^-	10^{5}	TsOH	−2.8
$CF_3SO_2O^-$	10^{8}	CF_3SO_2OH	−6.0

*Values are approximate and vary according to substrate.

Trifluoromethanesulfonates are called *triflates.*

correlation with basicity is seen among oxygen-containing leaving groups. The weaker the base, the better the leaving group. Trifluoromethanesulfonic acid (CF_3SO_2OH) is a much stronger acid than *p*-toluenesulfonic acid, and therefore trifluoromethanesulfonate is a much weaker base than *p*-toluenesulfonate and a much better leaving group.

Notice too that strongly basic leaving groups are absent from Table 8.9. In general, any species that has pK_a greater than about 2 for its conjugate acid cannot be a leaving group in a nucleophilic substitution. Thus, hydroxide (HO^-) is far too strong a base to be displaced from an alcohol (ROH), and alcohols do not undergo nucleophilic substitution. In strongly acidic media, alcohols are protonated to give alkyloxonium ions, and these do undergo nucleophilic substitution, because the leaving group is a weakly basic water molecule. S_N2 reactions are most favorable when the more basic nucleophile displaces a less basic leaving group.

Because halides are poorer leaving groups than *p*-toluenesulfonate, alkyl *p*-toluenesulfonates can be converted to alkyl halides by S_N2 reactions involving chloride, bromide, or iodide as the nucleophile.

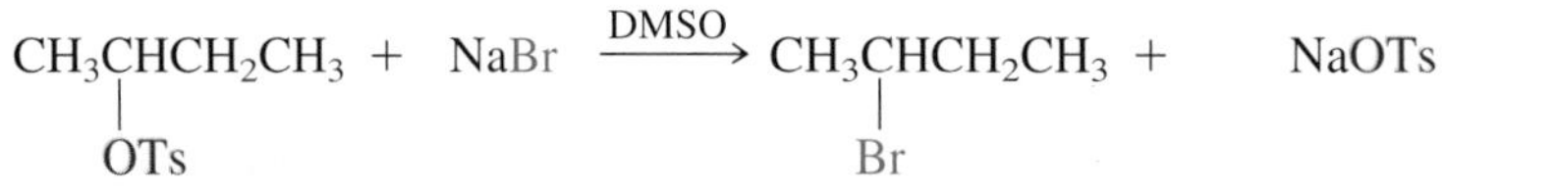

sec-Butyl *p*-toluenesulfonate + Sodium bromide → *sec*-Butyl bromide (82%) + Sodium *p*-toluenesulfonate

Problem 8.17

Write a chemical equation showing the preparation of octadecyl *p*-toluenesulfonate.

Problem 8.18

Write equations showing the reaction of octadecyl *p*-toluenesulfonate with each of the following reagents:

(a) Potassium acetate ($KOC(=O)CH_3$)
(b) Potassium iodide (KI)
(c) Potassium cyanide (KCN)
(d) Potassium hydrogen sulfide (KSH)
(e) Sodium butanethiolate ($NaSCH_2CH_2CH_2CH_3$)

Sample Solution All these reactions of octadecyl *p*-toluenesulfonate have been reported in the chemical literature, and all proceed in synthetically useful yield. You should begin by identifying the nucleophile in each of the parts to this problem. The nucleophile replaces the *p*-toluenesulfonate leaving group in an S_N2 reaction. In part (a) the nucleophile is acetate ion, and the product of nucleophilic substitution is octadecyl acetate.

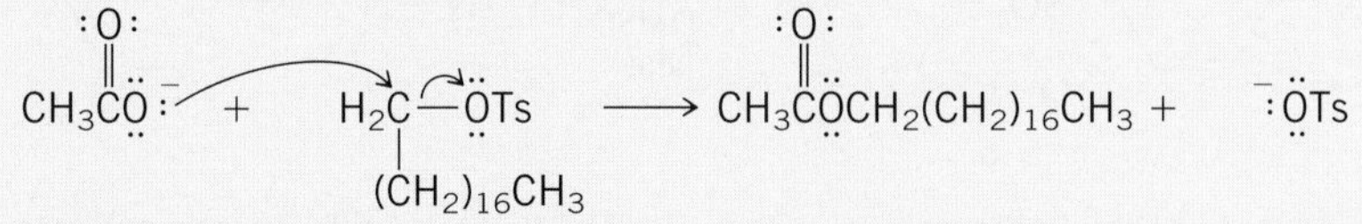

Acetate ion   Octadecyl tosylate   Octadecyl acetate   Tosylate ion

An advantage that sulfonates have over alkyl halides is that their preparation from alcohols does not involve any of the bonds to carbon. The alcohol oxygen becomes the oxygen that connects the alkyl group to the sulfonyl group. Thus, the configuration of a sulfonate is exactly the same as that of the alcohol from which it was prepared. If we wish to study the stereochemistry of nucleophilic substitution in an optically active substrate, for example, we know that a tosylate will have the same configuration and the same optical purity as the alcohol from which it was prepared.

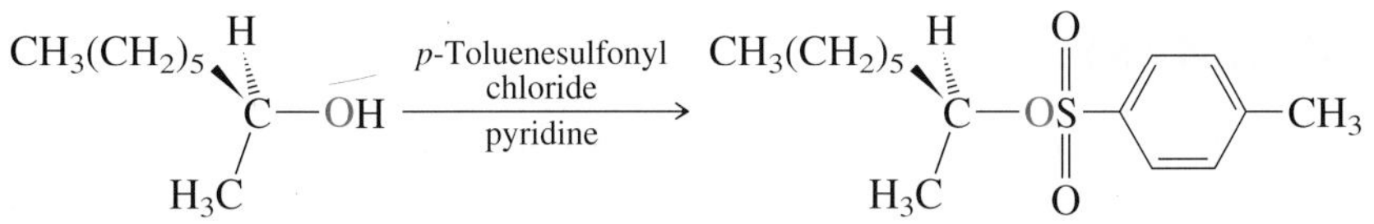

(*S*)-(+)-2-Octanol
$[\alpha]_D^{25}$ +9.9°
(optically pure)

(*S*)-(+)-1-Methylheptyl *p*-toluenesulfonate
$[\alpha]_D^{25}$ +7.9°
(optically pure)

The same cannot be said about reactions with alkyl halides as substrates. The conversion of optically active 2-octanol to the corresponding halide *does* involve a bond to the chirality center, and so the optical purity and absolute configuration of the alkyl halide need to be independently established.

The mechanisms by which sulfonates undergo nucleophilic substitution are the same as those of alkyl halides. Inversion of configuration is observed in S_N2 reactions of alkyl sulfonates and predominant inversion accompanied by racemization in S_N1 processes.

Problem 8.19

The hydrolysis of sulfonates of 2-octanol is stereospecific and proceeds with complete inversion of configuration. Write a structural formula that shows the stereochemistry of the 2-octanol formed by hydrolysis of an optically pure sample of (*S*)-(+)-1-methylheptyl *p*-toluenesulfonate, identify the product as *R* or *S*, and deduce its specific rotation.

Sulfonates are subject to the same limitations as alkyl halides. Competition from elimination needs to be considered when planning a functional group transformation that requires an anionic nucleophile, because tosylates undergo elimination reactions, just as alkyl halides do. For example, 1-methylpentyl *p*-toluenesulfonate undergoes elimination in the presence of sodium methoxide in methanol to give a mixture of 1-hexene and (*E*)/(*Z*)-2-hexenes.

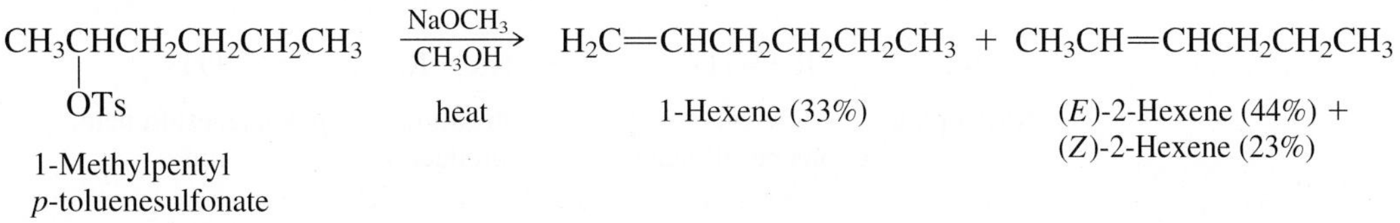

1-Methylpentyl *p*-toluenesulfonate   1-Hexene (33%)   (*E*)-2-Hexene (44%) + (*Z*)-2-Hexene (23%)

Because the leaving group is attached to a secondary carbon and methoxide is strongly basic, elimination predominates. However, note that the weakly basic azide ion reacts with the secondary *p*-toluenesulfonate shown by substitution (See Problem 8.43).

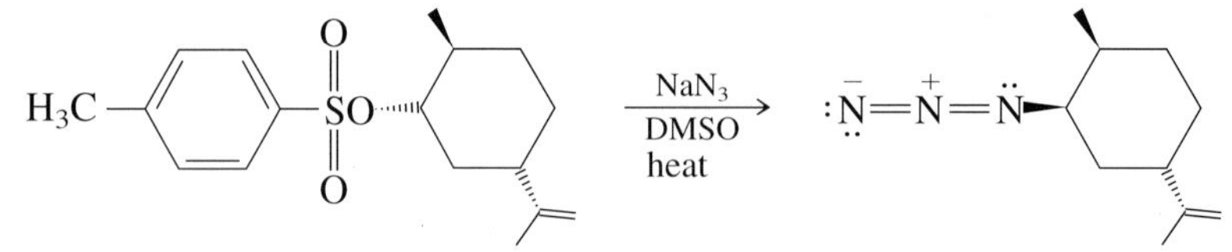

(+)-Dihydrocarveol-*p*-toluenesulfonate

(1*S*,2*R*,4*S*)-2-Azido-4-isopropylidene-1-methylcyclohexane (76%)

Alcohols are sometimes viewed as central intermediates in the synthesis of complex molecules, because they are available from other materials and they can be converted to other types of compounds. The ability to convert alcohols to sulfonates for use in nucleophilic substitutions opens up a wide range of possibilities for the synthesis of valuable organic compounds. We'll return to the use of alcohols in synthesis in Chapters 14 and 15.

8.13 SUMMARY

Section 8.1 Nucleophilic substitution is one of the main methods for functional group transformations. Examples of synthetically useful nucleophilic substitutions were given in Table 8.1. It is a good idea to return to that table and review its entries now that the details of nucleophilic substitution have been covered.

Sections 8.2–8.10 These sections show how a variety of experimental observations led to the proposal of the S_N1 and the S_N2 mechanisms for nucleophilic substitution. Summary Table 8.10 integrates the material in these sections.

Section 8.11 When nucleophilic substitution is used for synthesis, the competition between substitution and elimination must favor substitution. However, *the normal reaction of a secondary alkyl halide with a base as strong or stronger than hydroxide is elimination (E2).* Substitution by the S_N2 mechanism predominates only when the base is weaker than hydroxide or the alkyl halide is primary. Elimination predominates when tertiary alkyl halides react with any anion.

Section 8.12 Nucleophilic substitution can occur with leaving groups other than halide. Alkyl *p*-toluenesulfonates *(tosylates),* which are prepared from alcohols by reaction with *p*-toluenesulfonyl chloride, are often used.

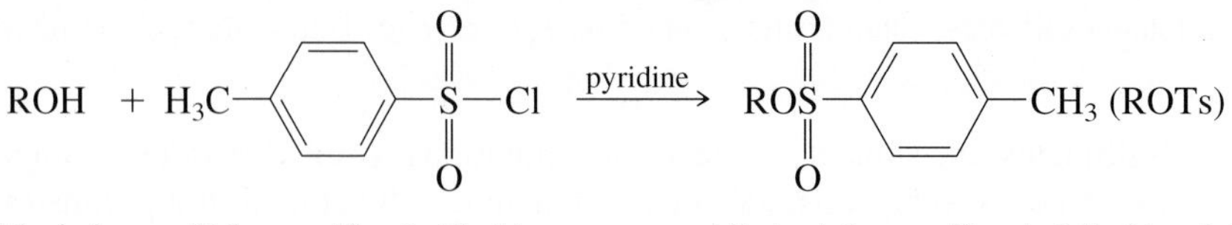

Alcohol *p*-Toluenesulfonyl chloride Alkyl *p*-toluenesulfonate (alkyl tosylate)

In its ability to act as a leaving group, *p*-toluenesulfonate is even more reactive than iodide.

Nucleophile Alkyl *p*-toluenesulfonate Substitution product *p*-Toluenesulfonate ion

TABLE 8.10 Comparison of S_N1 and S_N2 Mechanisms of Nucleophilic Substitution in Alkyl Halides

	S_N1	S_N2
Characteristics of mechanism	Two elementary steps: Step 1: $R{-}\ddot{X}: \rightleftharpoons R^+ + :\ddot{X}:^-$ Step 2: $R^+ + :Nu^- \longrightarrow R{-}Nu$ Ionization of alkyl halide (step 1) is rate-determining. (Section 8.6)	Single step: $^-Nu: \; R{-}\ddot{X}: \longrightarrow Nu{-}R + :\ddot{X}:^-$ Nucleophile displaces leaving group; bonding to the incoming nucleophile accompanies cleavage of the bond to the leaving group. (Section 8.3)
Rate-determining transition state	$^{\delta+}R{---}\ddot{X}:^{\delta-}$ (Section 8.6)	$^{\delta-}Nu{---}R{---}\ddot{X}:^{\delta-}$ (Section 8.3)
Molecularity	Unimolecular (Section 8.6)	Bimolecular (Section 8.3)
Kinetics and rate law	First order: Rate = k[alkyl halide] (Section 8.6)	Second order: Rate = k[alkyl halide][nucleophile] (Section 8.3)
Relative reactivity of halide leaving groups	RI > RBr > RCl >> RF (Section 8.2)	RI > RBr > RCl >> RF (Section 8.2)
Effect of structure on rate	$R_3CX > R_2CHX > RCH_2X > CH_3X$ Rate is governed by stability of carbocation that is formed in ionization step. Tertiary alkyl halides can react only by the S_N1 mechanism; they never react by the S_N2 mechanism. (Section 8.7)	$CH_3X > RCH_2X > R_2CHX > R_3CX$ Rate is governed by steric effects (crowding in transition state). Methyl and primary alkyl halides can react only by the S_N2 mechanism; they never react by the S_N1 mechanism. (Section 8.4)
Effect of nucleophile on rate	Rate of substitution is independent of both concentration and nature of nucleophile. Nucleophile does not participate until after rate-determining step. (Section 8.6)	Rate depends on both nature of nucleophile and its concentration. (Sections 8.3 and 8.5)
Effect of solvent on rate	Rate increases with increasing polarity of solvent as measured by its dielectric constant ε. (Section 8.10)	Polar aprotic solvents give fastest rates of substitution; solvation of $Nu:^-$ is minimal and nucleophilicity is greatest. (Section 8.10)
Stereochemistry	Not stereospecific: racemization accompanies inversion when leaving group is located at a chirality center. (Section 8.8)	Stereospecific: 100% inversion of configuration at reaction site. Nucleophile attacks carbon from side opposite bond to leaving group. (Section 8.3)
Potential for rearrangements	Carbocation intermediate capable of rearrangement. (Section 8.9)	No carbocation intermediate; no rearrangement.

PROBLEMS

8.20 Write the structure of the principal organic product to be expected from the reaction of 1-bromopropane with each of the following:

(a) Sodium iodide in acetone

(b) Sodium acetate ($CH_3\overset{O}{\overset{\|}{C}}ONa$) in acetic acid

(c) Sodium ethoxide in ethanol

(d) Sodium cyanide in dimethyl sulfoxide

(e) Sodium azide in aqueous ethanol

(f) Sodium hydrogen sulfide in ethanol

(g) Sodium methanethiolate ($NaSCH_3$) in ethanol

8.21 All the reactions of 1-bromopropane in the preceding problem give the product of nucleophilic substitution in high yield. High yields of substitution products are also obtained in all but one of the analogous reactions using 2-bromopropane as the substrate. In one case, however, 2-bromopropane is converted to propene, especially when the reaction is carried out at elevated temperature (about 55°C). Which reactant is most effective in converting 2-bromopropane to propene?

8.22 Each of the following nucleophilic substitution reactions has been reported in the chemical literature. Many of them involve reactants that are somewhat more complex than those we have dealt with to this point. Nevertheless, you should be able to predict the product by analogy to what you know about nucleophilic substitution in simple systems.

(a) $BrCH_2\overset{\overset{O}{\|}}{C}OCH_2CH_3 \xrightarrow[\text{acetone}]{\text{NaI}}$

(b) 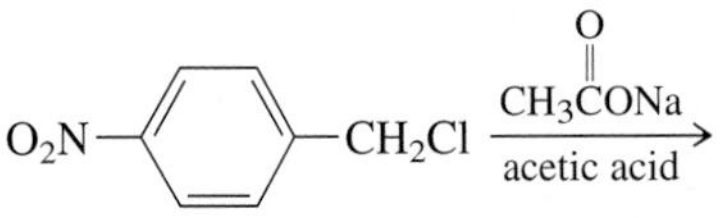

(c) $CH_3CH_2OCH_2CH_2Br \xrightarrow[\text{ethanol–water}]{\text{NaCN}}$

(d) 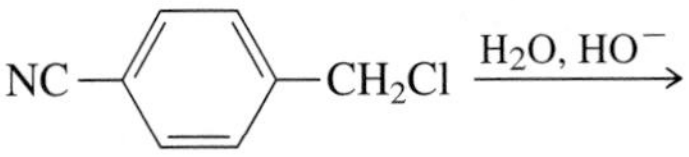

(e) $ClCH_2\overset{\overset{O}{\|}}{C}OC(CH_3)_3 \xrightarrow[\text{acetone–water}]{\text{NaN}_3}$

(f) 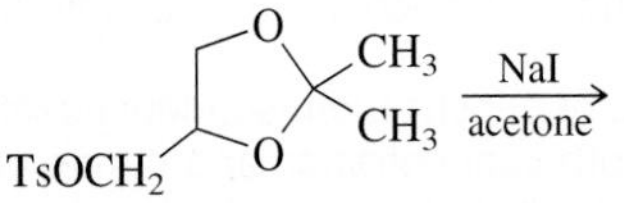

(g) 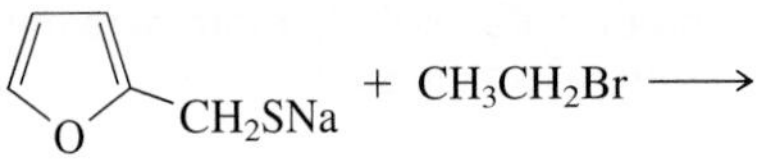

(h) 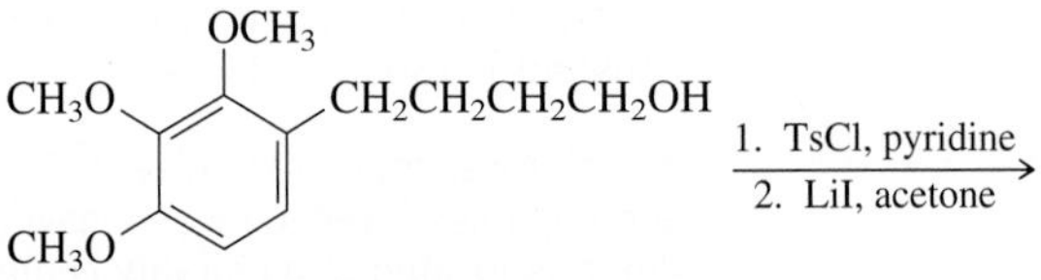

8.23 Each of the reactions shown involves nucleophilic substitution. The product of reaction (a) is an isomer of the product of reaction (b). What kind of isomer? By what mechanism does nucleophilic substitution occur? Write the structural formula of the product of each reaction.

(a) 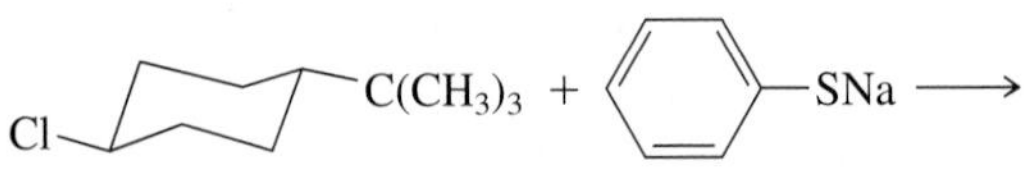

(b) 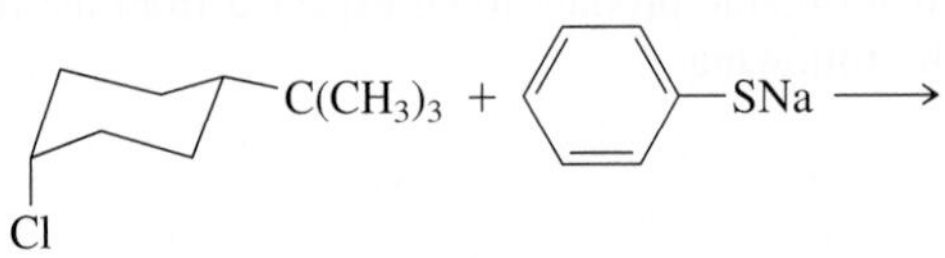

8.24 There is an overall 29-fold difference in reactivity of 1-chlorohexane, 2-chlorohexane, and 3-chlorohexane toward potassium iodide in acetone.

(a) Which one is the most reactive? Why?

(b) Two of the isomers differ by only a factor of 2 in reactivity. Which two are these? Which one is the more reactive? Why?

8.25 In each of the following indicate which reaction will occur faster. Explain your reasoning.

(a) $CH_3CH_2CH_2CH_2Br$ or $CH_3CH_2CH_2CH_2I$ with sodium cyanide in dimethyl sulfoxide

(b) 1-Chloro-2-methylbutane or 1-chloropentane with sodium iodide in acetone

(c) Hexyl chloride or cyclohexyl chloride with sodium azide in aqueous ethanol

(d) Solvolysis of 1-bromo-2,2-dimethylpropane or *tert*-butyl bromide in ethanol

(e) Solvolysis of isobutyl bromide or *sec*-butyl bromide in aqueous formic acid

(f) Reaction of 1-chlorobutane with sodium acetate in acetic acid or with sodium methoxide in methanol

(g) Reaction of 1-chlorobutane with sodium azide or sodium *p*-toluenesulfonate in aqueous ethanol

8.26 Photochemical chlorination of $(CH_3)_3CCH_2C(CH_3)_3$ gave a mixture of two monochlorides in a 4:1 ratio. The structures of these two products were assigned on the basis of their S_N1 hydrolysis rates in aqueous ethanol. The major product (compound A) underwent hydrolysis much more slowly than the minor one (compound B). Deduce the structures of compounds A and B.

8.27 The compound KSCN is a source of *thiocyanate* ion.

(a) Write the two most stable Lewis structures for thiocyanate ion and identify the atom in each that bears a formal charge of −1.

(b) Two constitutionally isomeric products of molecular formula C_5H_9NS were isolated in a combined yield of 87% in the reaction shown. (DMF stands for *N,N*-dimethylformamide, a polar aprotic solvent.) Suggest reasonable structures for these two compounds.

$$CH_3CH_2CH_2CH_2Br \xrightarrow[\text{DMF}]{\text{KSCN}}$$

8.28 Sodium nitrite ($NaNO_2$) reacted with 2-iodooctane to give a mixture of two constitutionally isomeric compounds of molecular formula $C_8H_{17}NO_2$ in a combined yield of 88%. Suggest reasonable structures for these two isomers.

8.29 Reaction of ethyl iodide with triethylamine [$(CH_3CH_2)_3N$:] yields a crystalline compound $C_8H_{20}NI$ in high yield. This compound is soluble in polar solvents such as water but insoluble in nonpolar ones such as diethyl ether. It does not melt below about 200°C. Suggest a reasonable structure for this product.

8.30 Write an equation, clearly showing the stereochemistry of the starting material and the product, for the reaction of (*S*)-1-bromo-2-methylbutane with sodium iodide in acetone. What is the configuration (*R* or *S*) of the product?

8.31 Identify the product in each of the following reactions:

(a) $$ClCH_2CH_2\underset{\underset{\displaystyle Cl}{|}}{C}HCH_2CH_3 \xrightarrow[\text{acetone}]{\text{NaI (1 mol)}} C_5H_{10}ClI$$

(b) $BrCH_2CH_2Br + NaSCH_2CH_2SNa \rightarrow C_4H_8S_2$

(c) $ClCH_2CH_2CH_2CH_2Cl + Na_2S \rightarrow C_4H_8S$

8.32 Give the mechanistic symbols (S_N1, S_N2, E1, E2) that are most consistent with each of the following statements:

(a) Methyl halides react with sodium ethoxide in ethanol only by this mechanism.

(b) Unhindered primary halides react with sodium ethoxide in ethanol mainly by this mechanism.

(c) When cyclohexyl bromide is treated with sodium ethoxide in ethanol, the major product is formed by this mechanism.

(d) The substitution product obtained by solvolysis of *tert*-butyl bromide in ethanol arises by this mechanism.

(e) In ethanol that contains sodium ethoxide, *tert*-butyl bromide reacts mainly by this mechanism.

(f) These reaction mechanisms represent concerted processes.

(g) Reactions proceeding by these mechanisms are stereospecific.

(h) These reaction mechanisms involve carbocation intermediates.

(i) These reaction mechanisms are the ones most likely to have been involved when the products are found to have a different carbon skeleton from the substrate.

(j) Alkyl iodides react faster than alkyl bromides in reactions that proceed by these mechanisms.

8.33 Outline an efficient synthesis of each of the following compounds from the indicated starting material and any necessary organic or inorganic reagents:

(a) Cyclopentyl cyanide from cyclopentane

(b) Cyclopentyl cyanide from cyclopentene

(c) Cyclopentyl cyanide from cyclopentanol

(d) $NCCH_2CH_2CN$ from ethyl alcohol

(e) Isobutyl iodide from isobutyl chloride

(f) Isobutyl iodide from *tert*-butyl chloride

(g) Isopropyl azide from isopropyl alcohol

(h) Isopropyl azide from 1-propanol

(i) (*S*)-*sec*-Butyl azide from (*R*)-*sec*-butyl alcohol

(j) (*S*)-$CH_3CH_2CHCH_3$ (with SH on the CH) from (*R*)-*sec*-butyl alcohol
  SH

8.34 Select the combination of alkyl bromide and potassium alkoxide that would be the most effective in the syntheses of the following ethers:

(a) $CH_3OC(CH_3)_3$

(b) 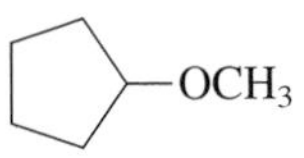
OCH3

(c) $(CH_3)_3CCH_2OCH_2CH_3$

8.35 (*Note to the student:* This problem previews an important aspect of Chapter 9 and is well worth attempting in order to get a head start on the material presented there.)

Alkynes of the type RC≡CH may be prepared by nucleophilic substitution reactions in which one of the starting materials is sodium acetylide (Na^+ :$\bar{C}$≡CH).

(a) Devise a method for the preparation of $CH_3CH_2C{\equiv}CH$ from sodium acetylide and any necessary organic or inorganic reagents.

(b) Given the information that the pK_a for acetylene (HC≡CH) is 26, comment on the scope of this preparative procedure with respect to R in RC≡CH. Could you prepare $(CH_3)_2CHC{\equiv}CH$ or $(CH_3)_3CC{\equiv}CH$ in good yield by this method?

8.36 Give the structures, including stereochemistry, of compounds A and B in the following sequence of reactions:

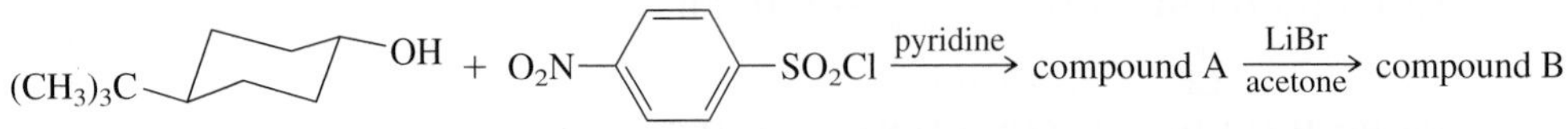

(CH3)3C
OH
+ O2N
SO2Cl
pyridine
compound A
LiBr
acetone
compound B

8.37 (a) Suggest a reasonable series of synthetic transformations for converting *trans*-2-methylcyclopentanol to *cis*-2-methylcyclopentyl acetate.

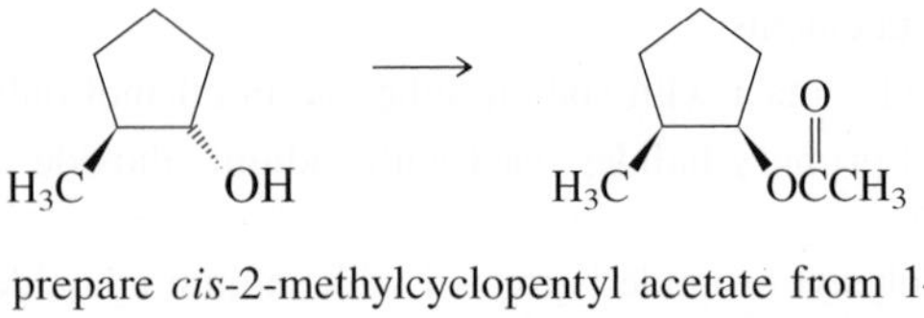

H3C
OH
H3C
O
OCCH3

(b) How could you prepare *cis*-2-methylcyclopentyl acetate from 1-methylcyclopentanol?

8.38 Optically pure (*S*)-(+)-2-butanol was converted to its methanesulfonate according to the reaction shown.

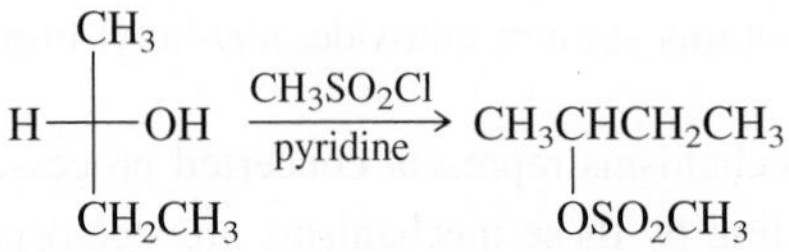

CH3
H
OH
CH2CH3
CH3SO2Cl
pyridine
CH3CHCH2CH3
OSO2CH3

(a) Write the Fischer projection of the *sec*-butyl methanesulfonate formed in this reaction.

(b) The *sec*-butyl methanesulfonate in part (a) was treated with $NaSCH_2CH_3$ to give a product having an optical rotation α_D of $-25°$. Write the Fischer projection of this product. By what mechanism is it formed? What is its absolute configuration (*R* or *S*)?

(c) When treated with PBr_3, optically pure (*S*)-(+)-2-butanol gave 2-bromobutane having an optical rotation $\alpha_D = -38°$. This bromide was then allowed to react with $NaSCH_2CH_3$ to give a product having an optical rotation α_D of $+23°$. Write the Fischer projection for (−)-2-bromobutane and specify its configuration as *R* or *S*. Does the reaction of 2-butanol with PBr_3 proceed with predominant inversion or retention of configuration?

(d) What is the optical rotation of optically pure 2-bromobutane?

8.39 The methanesulfonate ester shown here was used as a building block in the synthesis of an HIV protease inhibitor. The sulfonate ester is made from an alcohol. What alcohol? Show an equation for the preparation of the sulfonate ester from the alcohol and the other reagents that are needed.

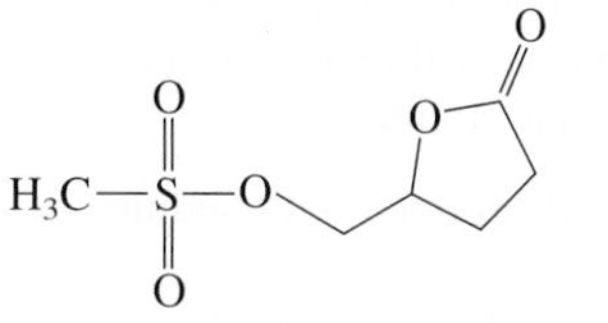

8.40 The ratio of elimination to substitution is exactly the same (26% elimination) for 2-bromo-2-methylbutane and 2-iodo-2-methylbutane in 80% ethanol/20% water at 25°C.

(a) By what mechanism does substitution most likely occur in these compounds under these conditions?

(b) By what mechanism does elimination most likely occur in these compounds under these conditions?

(c) Which substrate undergoes substitution faster?

(d) Which substrate undergoes elimination faster?

(e) What two substitution products are formed from each substrate?

(f) What two elimination products are formed from each substrate?

(g) Why do you suppose the ratio of elimination to substitution is the same for the two substrates?

8.41 The reaction of 2,2-dimethyl-1-propanol with HBr is very slow and gives 2-bromo-2-methylbutane as the major product.

$$CH_3C(CH_3)_2CH_2OH \xrightarrow[65°C]{HBr} CH_3C(CH_3)(Br)CH_2CH_3$$

Give a mechanistic explanation for these observations.

8.42 Solvolysis of 2-bromo-2-methylbutane in acetic acid containing potassium acetate gave three products. Identify them.

8.43 In contrast to the displacement by azide shown in Section 8.12, the sulfonate shown here gives significant amounts of elimination product. Suggest an explanation for this difference in reactivity. (*Hint:* First draw the reactant in its chair form, and do the same for the sulfonate in Section 8.12.)

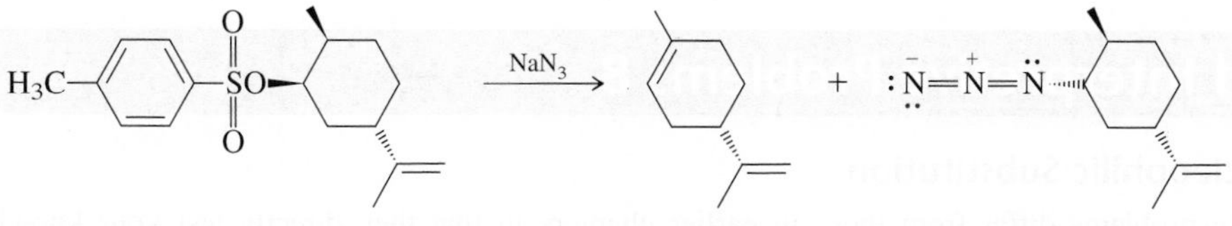

46% 37%

8.44 If the temperature is not kept below 25°C during the reaction of primary alcohols with *p*-toluenesulfonyl chloride in pyridine, it is sometimes observed that the isolated product is not the desired alkyl *p*-toluenesulfonate but is instead the corresponding alkyl chloride. Suggest a mechanistic explanation for this observation.

8.45 In a classic experiment, Edward Hughes (a colleague of Ingold's at University College, London) studied the rate of racemization of 2-iodooctane by sodium iodide in acetone and compared it with the rate of incorporation of radioactive iodine into 2-iodooctane.

$$RI + [I^*]^- \longrightarrow RI^* + I^-$$

(I^* = radioactive iodine)

How will the rate of racemization compare with the rate of incorporation of radioactivity if

(a) Each act of exchange proceeds stereospecifically with retention of configuration?

(b) Each act of exchange proceeds stereospecifically with inversion of configuration?

(c) Each act of exchange proceeds in a stereorandom manner, in which retention and inversion of configuration are equally likely?

8.46 Based on what we know about nucleophiles and leaving groups, we suspect that the reaction of (*R*)-2-chlorobutane with sodium iodide in acetone would not be useful as a synthesis of (*S*)-2-iodobutane. Explain.

8.47 The reaction of cyclopentyl bromide with sodium cyanide to give cyclopentyl cyanide

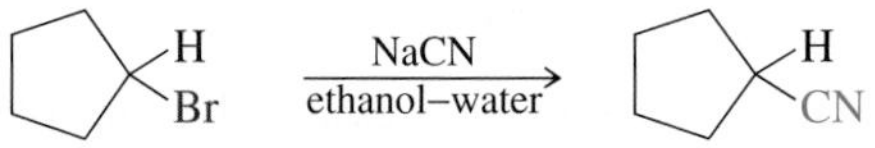

Cyclopentyl bromide Cyclopentyl cyanide

proceeds faster if a small amount of sodium iodide is added to the reaction mixture. Can you suggest a reasonable mechanism to explain the catalytic function of sodium iodide?

8.48 A compound known as *matsuone* is the sex pheromone of the pine bast scale, and it was synthesized in several steps from the alcohol citronellol. A key intermediate in the synthesis was the nitrile (*S*)-4,8-dimethyl-7-nonenenitrile. Show how this nitrile could be prepared from citronellol in two steps.

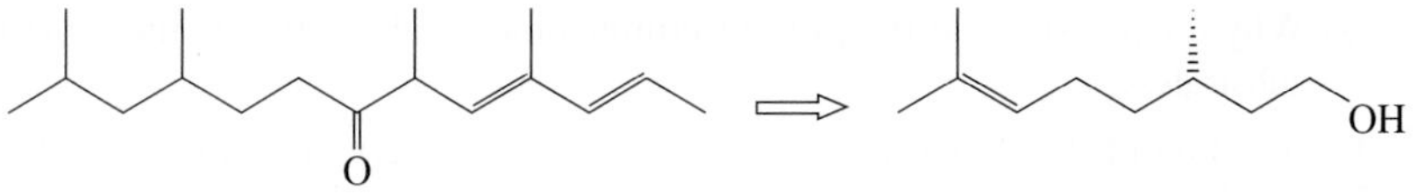

Matsuone Citronellol

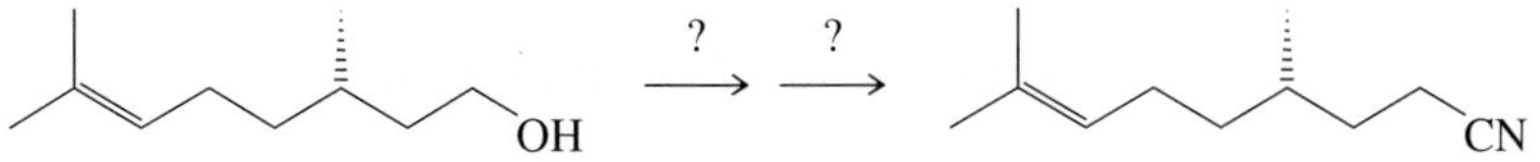

Citronellol (4*S*)-4,8-Dimethyl-7-nonenenitrile

(The arrow ⟹ signifies that the compound shown on the left (the target) can be made from the one on the right. A step shown in this reverse way is referred to as a *retrosynthetic* step. The concept of retrosynthesis is developed further in Chapter 14.)

Descriptive Passage and Interpretive Problems 8

Nucleophilic Substitution

These problems differ from those in earlier chapters in that they directly test your knowledge of core material rather than using a descriptive passage to extend the material or introduce new ideas. The number of factors that contribute to nucleophilic substitution can be daunting. The really major ones, though, are few and readily applied to specific reactions by using the S_N1 and S_N2 mechanisms to guide your analysis.

8.49 Which compound undergoes substitution by the S_N1 mechanism at the fastest rate?

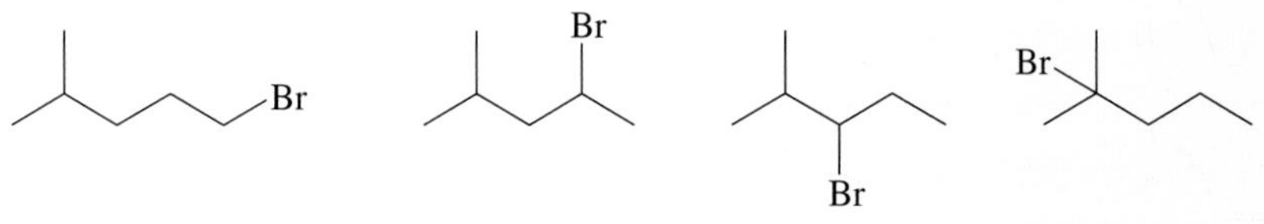

A. B. C. D.

8.50 Which compound undergoes substitution by the S_N2 mechanism at the fastest rate?

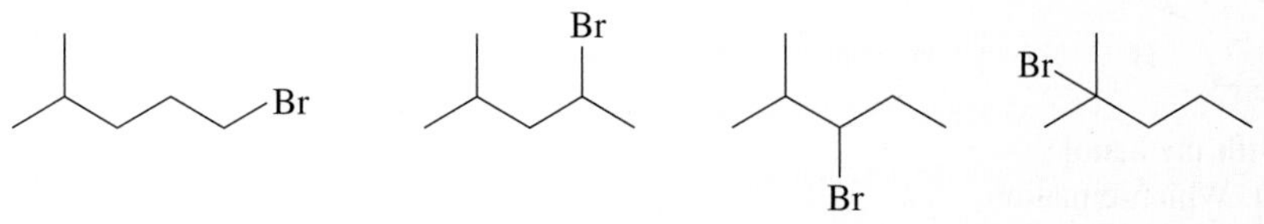

A. B. C. D.

8.51 Which reaction takes place at the fastest rate?

A. $CH_3CH_2CH_2CH_2Cl$ + NaSH $\xrightarrow{\text{Ethanol, 25°C}}$ $CH_3CH_2CH_2CH_2SH$ + NaCl

B. $CH_3CH_2CH_2CH_2Cl$ + NaSH $\xrightarrow{\text{Ethanol, 25°C}}$ $CH_3CH_2CH{=}CH_2$ + NaCl + H_2S

C. $CH_3CH_2CH_2CH_2Br$ + NaSH $\xrightarrow{\text{Ethanol, 25°C}}$ $CH_3CH_2CH_2CH_2SH$ + NaBr

D. $CH_3CH_2CH_2CH_2Br$ + NaSH $\xrightarrow{\text{Ethanol, 25°C}}$ $CH_3CH_2CH{=}CH_2$ + NaBr + H_2S

8.52 Identify the mechanism most responsible for the major product in the following reaction.

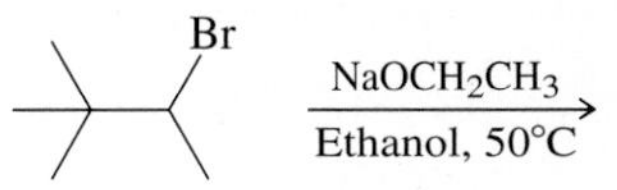

A. S_N1 C. E1
B. S_N2 D. E2

8.53 What is the major product of the reaction shown?

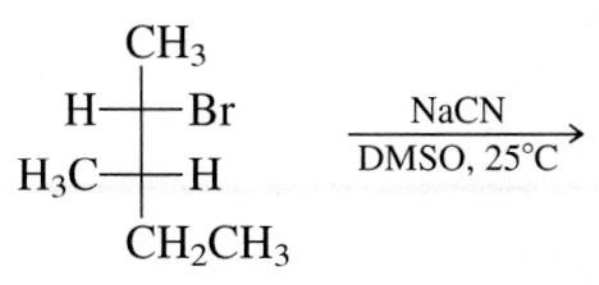

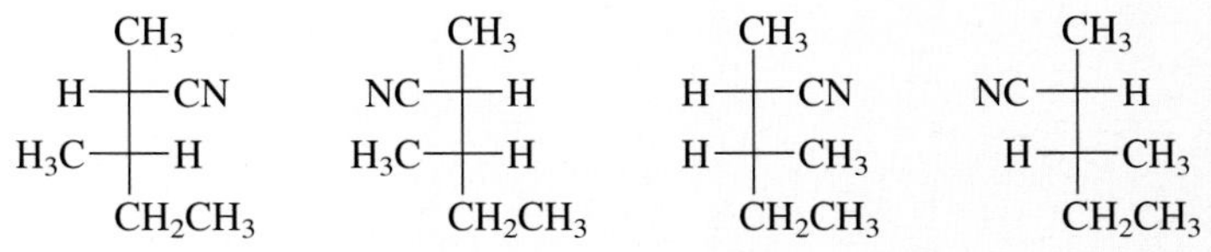

A. B. C. D.

8.54 What are reactant X and product Y in the following sequence of reactions?

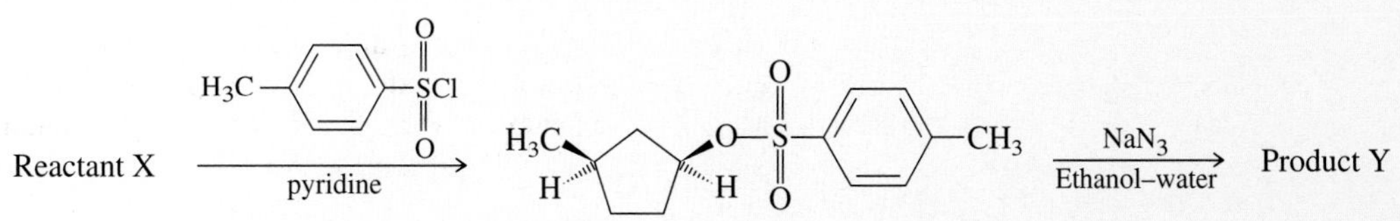

Reactant X | Product Y

A. 

B. 

C. 

D. 

8.55 Trimethyloxonium tetrafluoroborate reacts with methanol (CH_3OH) to give dimethyl ether (CH_3OCH_3). Which equation, including the curved arrows, best represents the rate-determining step in the mechanism?

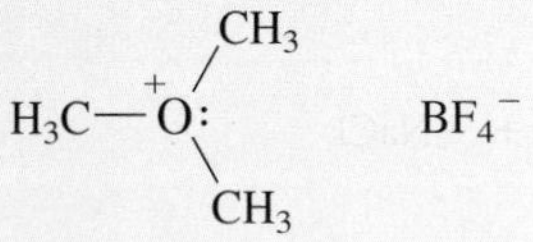

Trimethyloxonium tetrafluoroborate

A. 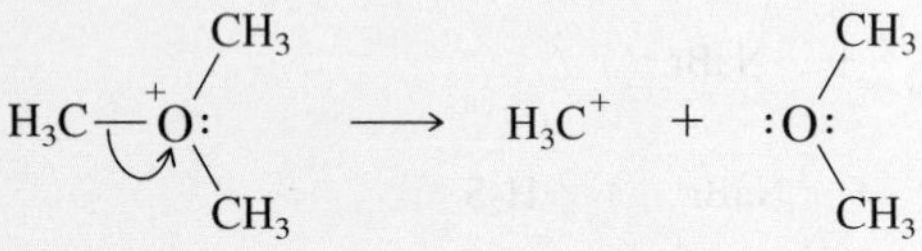

B. 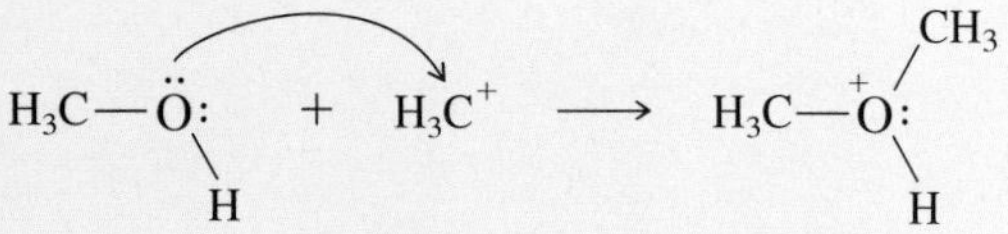

C. 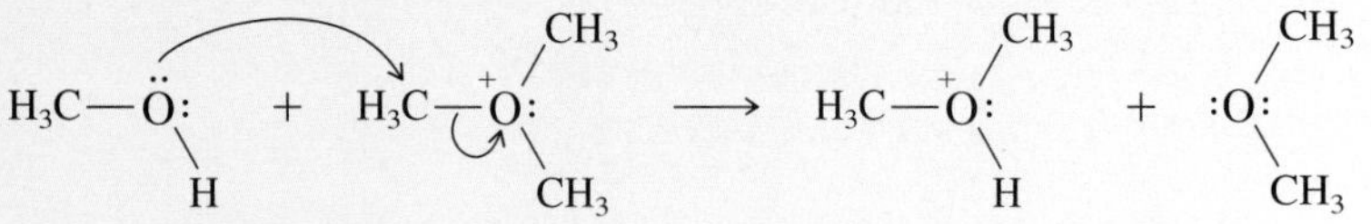

D. 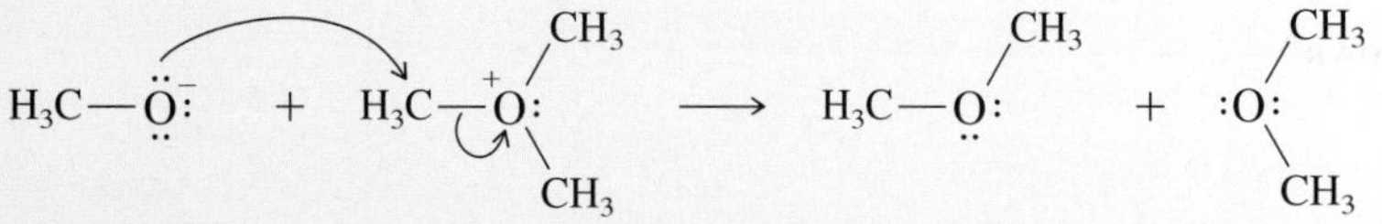

Alkynes

9

Chapter Outline

Mechanisms

With a temperature of 3300°C, the flame of an oxyacetylene torch exceeds that obtained from any other combustible gas mixture.

HYDROCARBONS that contain a carbon–carbon triple bond are called **alkynes.** Noncyclic alkynes have the molecular formula C_nH_{2n-2}. *Acetylene* (HC≡CH) is the simplest alkyne. We call compounds that have their triple bond at the end of a carbon chain (RC≡CH) *monosubstituted,* or **terminal, alkynes.** Disubstituted alkynes (RC≡CR′) have *internal* triple bonds. You will see in this chapter that a carbon–carbon triple bond is a functional group, reacting with many of the same reagents that react with the double bonds of alkenes.

The most distinctive aspect of the chemistry of acetylene and terminal alkynes is their acidity. As a class, compounds of the type RC≡CH are the most acidic of all hydrocarbons. The structural reasons for this property, as well as the ways in which it is used to advantage in chemical synthesis, are important elements of this chapter.

9.1 Sources of Alkynes

Acetylene was discovered in 1836 by Edmund Davy and characterized by the French chemist P. E. M. Berthelot in 1862. It did not command much attention until its large-scale preparation from calcium carbide near the end of the nineteenth century stimulated interest in industrial applications. In the first stage of that synthesis, limestone and coke, a material rich in elemental carbon obtained from coal, are heated in an electric furnace to form calcium carbide.

This reaction was accidentally discovered in 1892 by the Canadian inventor Thomas L. Willson while looking for a method to make aluminum.

$$\underset{\substack{\text{Calcium oxide}\\\text{(from limestone)}}}{CaO} + \underset{\substack{\text{Carbon}\\\text{(from coke)}}}{3C} \xrightarrow{1800\text{–}2100°C} \underset{\text{Calcium carbide}}{CaC_2} + \underset{\text{Carbon monoxide}}{CO}$$

Calcium carbide is the calcium salt of the doubly negative carbide ion ($:\bar{C}{\equiv}\bar{C}:$). Carbide ion is strongly basic and reacts with water to form acetylene:

$$\underset{\text{Calcium carbide}}{Ca^{2+}\left[:C{\equiv}C:\right]^{2-}} + \underset{\text{Water}}{2H_2O} \longrightarrow \underset{\text{Calcium hydroxide}}{Ca(OH)_2} + \underset{\text{Acetylene}}{HC{\equiv}CH}$$

Problem 9.1

Use curved arrows to show how calcium carbide reacts with water to give acetylene.

Beginning in the middle of the twentieth century, alternative methods of acetylene production became practical. One of these is the dehydrogenation of ethylene.

$$\underset{\text{Ethylene}}{H_2C{=}CH_2} \xrightleftharpoons{\text{heat}} \underset{\text{Acetylene}}{HC{\equiv}CH} + \underset{\text{Hydrogen}}{H_2}$$

The reaction is endothermic, and the equilibrium favors ethylene at low temperatures but shifts to favor acetylene above 1150°C. Indeed, at very high temperatures most hydrocarbons, even methane, are converted to acetylene. Acetylene has value not only by itself but is also the starting material from which higher alkynes are prepared.

More than 1000 natural products contain carbon–carbon triple bonds. Many, such as stearolic acid and tariric acid are **fatty acids**—carboxylic acids with unbranched chains of 12–20 carbon atoms—or are derived from them.

$$\underset{\text{Stearolic acid}}{CH_3(CH_2)_7C{\equiv}C(CH_2)_7\overset{\overset{\displaystyle O}{\|}}{C}OH} \qquad \underset{\text{Tariric acid}}{CH_3(CH_2)_{10}C{\equiv}C(CH_2)_4\overset{\overset{\displaystyle O}{\|}}{C}OH}$$

The main biosynthetic route to acetylenic fatty acids in plants appears to be by enzyme-catalyzed oxidation of analogous compounds with carbon–carbon double bonds. The enzymes responsible *(acetylenases)* belong to a class called *desaturases.*

$$\underset{\text{Linoleic acid}}{(Z,Z)\text{-}CH_3(CH_2)_4CH{=}CHCH_2CH{=}CH(CH_2)_7\overset{\overset{\displaystyle O}{\|}}{C}OH} \longrightarrow \underset{\text{Crepenynic acid}}{(Z)\text{-}CH_3(CH_2)_4C{\equiv}CCH_2CH{=}CH(CH_2)_7\overset{\overset{\displaystyle O}{\|}}{C}OH}$$

Cultures of the bacterium *Micromonospora chersina* produce dynemicin A, a novel substance containing a double bond and two triple bonds in a ten-membered ring (an *enediyne*). Dynemicin A has attracted interest because of its unusual structure as well as its interesting biological activity. It has the ability to cleave DNA by a novel mechanism, which may lead to the development of anticancer drugs that are based on the enediyne structure.

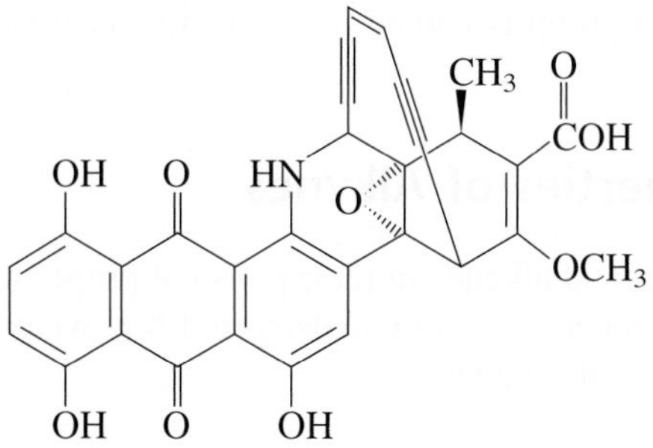

Dynemicin A

The poison dart frogs of Central and South America store toxic substances such as the acetylenic alkaloid histrionicotoxin within their bodies to deter attacks by other animals. Histrionicotoxin disrupts neuromuscular transmission in animals through its interaction with receptors for the neurotransmitter acetylcholine.

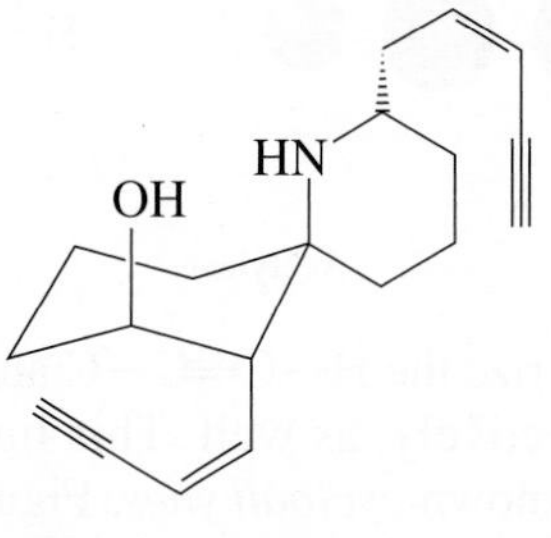

Histrionicotoxin

Acetylcholine has the structure:

$$CH_3\overset{\overset{\displaystyle O}{\|}}{C}OCH_2CH_2\overset{+}{N}(CH_3)_3$$

The biosynthesis of acetylcholine is described in Section 24.1. See also: Problem 19.55.

Rather than waiting to be attacked before launching a defense, poison dart frogs advertise themselves with bright colors. Potential predators recognize the threat these frogs present and avoid them. Frogs do not biosynthesize their alkaloids directly, but obtain them or their precursors from ants that they feed upon. The ants probably get their alkaloids from plants.

Diacetylene (HC≡C—C≡CH) has been identified as a component of the hydrocarbon-rich atmospheres of Uranus, Neptune, and Pluto. It is also present in the atmospheres of Titan and Triton, satellites of Saturn and Neptune, respectively.

9.2 Nomenclature

In naming alkynes the usual IUPAC rules for hydrocarbons are followed, and the suffix *-ane* is replaced by *-yne*. Both acetylene and ethyne are acceptable IUPAC names for HC≡CH. The position of the triple bond along the chain is specified by number in a manner analogous to alkene nomenclature.

$HC{\equiv}CCH_3$	$HC{\equiv}CCH_2CH_3$	$CH_3C{\equiv}CCH_3$	$(CH_3)_3CC{\equiv}CCH_3$
Propyne or Prop-1-yne	1-Butyne or But-1-yne	2-Butyne or But-2-yne	4,4-Dimethyl-2-pentyne or 4,4-Dimethylpent-2-yne

Problem 9.2

Write structural formulas and give the IUPAC names for all the alkynes of molecular formula C_5H_8.

If a compound contains both a double bond and a triple bond, the chain is numbered so as to give the first multiple bond the lowest number, irrespective of whether it is a double bond or a triple bond. Ties are broken in favor of the double bond. An *en* suffix for the double bond precedes *yne* and is separated from it by the *yne* locant. Thus, the compound vinylacetylene $H_2C{=}CH{-}C{\equiv}CH$ is named but-1-en-3-yne according to the latest IUPAC rules.

Vinylacetylene is a high-volume industrial chemical used in the preparation of neoprene.

When the —C≡CH group is named as a substituent, it is designated as an *ethynyl* group.

9.3 Physical Properties of Alkynes

Alkynes resemble alkanes and alkenes in their physical properties. They share with these other hydrocarbons the properties of low density and low water-solubility. Their boiling points are similar to those of alkanes.

9.4 Structure and Bonding in Alkynes: *sp* Hybridization

Acetylene is linear, with a carbon–carbon bond distance of 120 pm and carbon–hydrogen bond distances of 106 pm.

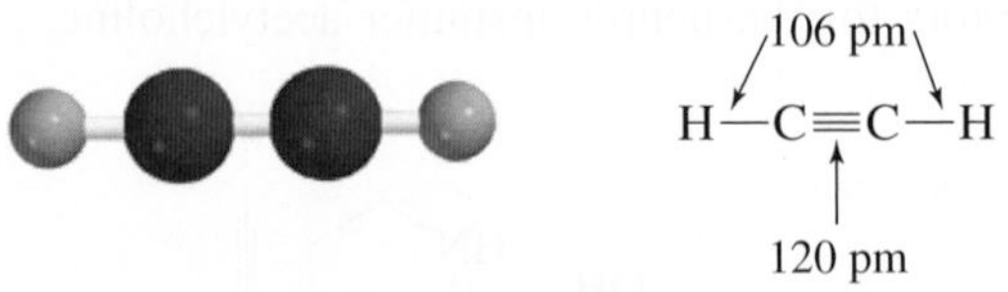

Acetylene

Linear geometries characterize the H—C≡C—C and C—C≡C—C units of terminal and internal triple bonds, respectively, as well. This linear geometry is responsible for the relatively small number of known *cycloalkynes.* Figure 9.1 shows a molecular model of cyclononyne in which the bending of the C—C≡C—C unit is clearly evident.

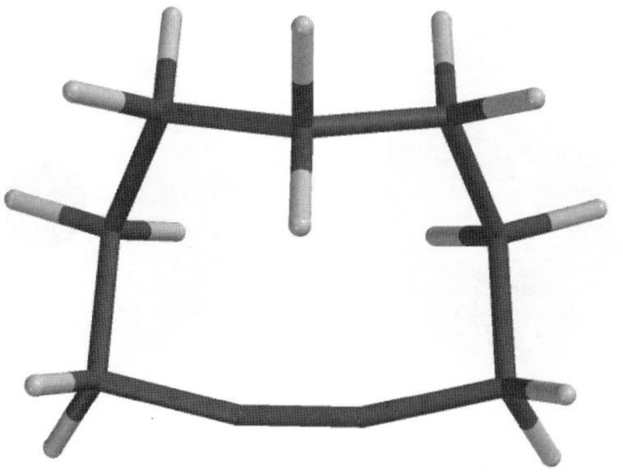

Figure 9.1

Molecular model of cyclononyne showing bending of the bond angles associated with the triply bonded carbons. This model closely matches the structure determined experimentally. Notice how the staggering of bonds on adjacent atoms governs the overall shape of the ring.

Angle strain destabilizes cycloalkynes to the extent that cyclononyne is the smallest one that is stable enough to be stored for long periods. The next smaller one, cyclooctyne, has been isolated, but is relatively reactive and polymerizes on standing.

An *sp* hybridization model for the carbon–carbon triple bond was developed in Section 2.21 and is reviewed for acetylene in Figure 9.2. Figure 9.3 compares the electrostatic potential maps of ethylene and acetylene and shows how the two π bonds in acetylene cause a band of high electron density to encircle the molecule.

Table 9.1 compares some structural features of alkanes, alkenes, and alkynes. As we progress through the series in the order ethane $\rightarrow$ ethylene $\rightarrow$ acetylene:

1. The geometry at carbon changes from tetrahedral $\rightarrow$ trigonal planar $\rightarrow$ linear.
2. The C—C and C—H bonds become shorter and stronger.
3. The acidity of the C—H bonds increases.

All of these trends can be accommodated by the orbital hybridization model. The bond angles are characteristic for the sp^3, sp^2, and *sp* hybridization states of carbon and don't require additional comment. The bond distances, bond strengths, and acidities are related to the *s* character in the orbitals used for bonding. *s* Character is the fraction of the hybrid orbital contributed by an *s* orbital. Thus, an sp^3 orbital has one quarter *s* character and three quarters *p*, an sp^2 orbital has one third *s* and two thirds *p*, and an *sp* orbital one half *s* and one half *p*. We then use this information to analyze how various qualities of the hybrid orbital reflect those of its *s* and *p* contributors.

Take C—H bond distance and bond strength, for example. Recalling that an electron in a 2*s* orbital is, on average, closer to the nucleus and more strongly held than an electron in a 2*p* orbital, it follows that an electron in an orbital with more *s* character will be closer to the nucleus and more strongly held than an electron in an orbital with less *s* character. Thus, when an *sp* orbital of carbon overlaps with a hydrogen 1*s* orbital to give a C—H σ bond, the electrons are held more strongly and the bond is stronger and shorter than one between hydrogen and sp^2-hybridized carbon. Similar reasoning holds for the shorter C—C bond distance of acetylene compared with ethylene, although here the additional π bond in acetylene is also a factor.

The hybridization model for bonding in acetylene is depicted in Figure 2.23.

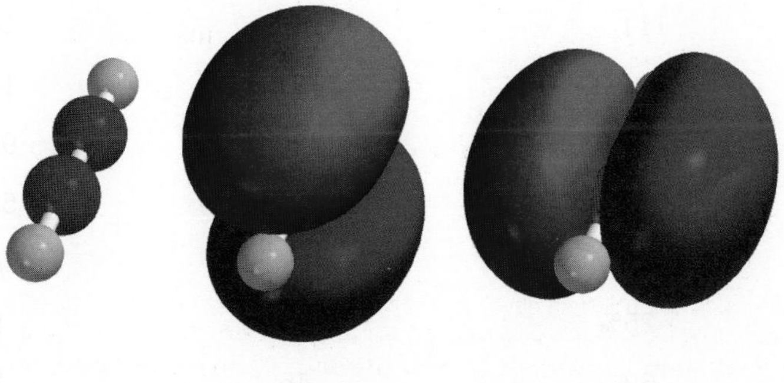

(*a*) (*b*) (*c*)

Figure 9.2

The carbon atoms of acetylene are connected by a $\sigma + \pi + \pi$ triple bond. (*a*) Both carbon atoms are *sp*-hybridized, and each is bonded to a hydrogen by a σ bond. The two π bonds are perpendicular to each other and are shown separately in (*b*) and (*c*).

Figure 9.3

Electrostatic potential maps of ethylene and acetylene. The region of highest negative charge (red) is associated with the π bonds and lies between the two carbons in both. This electron-rich region is above and below the plane of the molecule in ethylene. Because acetylene has two π bonds, a band of high electron density encircles the molecule.

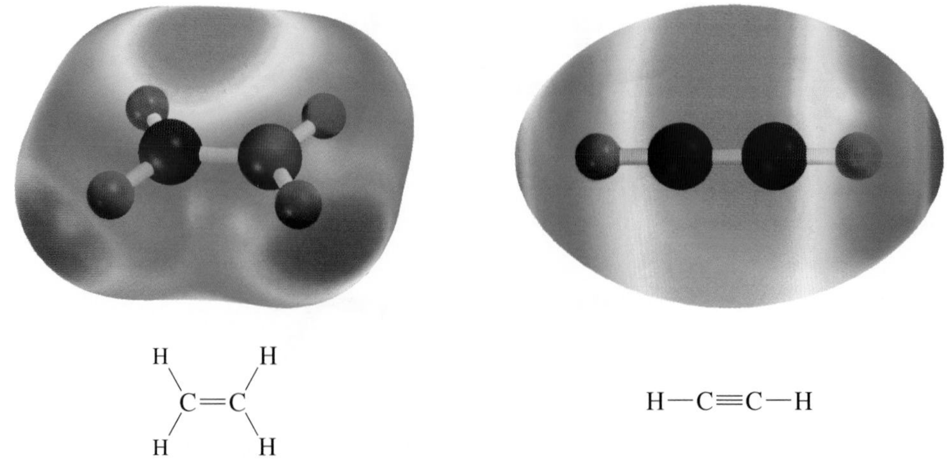

Ethylene Acetylene

The pattern is repeated in higher alkynes as shown when comparing propyne and propene. The bonds to the *sp*-hybridized carbons of propyne are shorter than the corresponding bonds to the sp^2-hybridized carbons of propene.

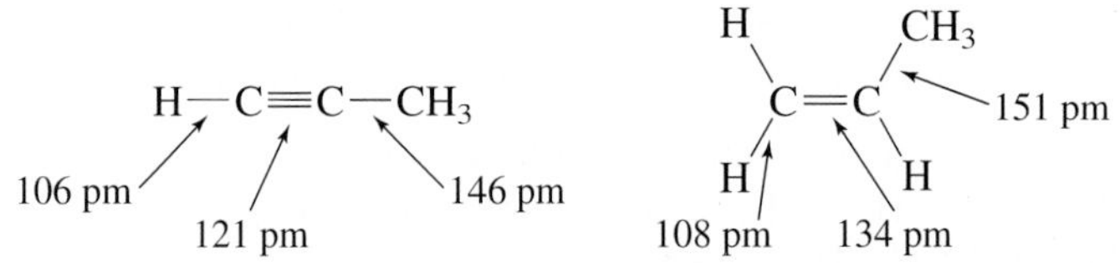
106 pm
121 pm
146 pm
151 pm
108 pm
134 pm

Propyne Propene

A good way to think about the effect of the *s* character is to associate it with electronegativity. As its *s* character increases, so does a carbon's electronegativity (the electrons in the bond involving that orbital are closer to carbon). The hydrogens in C—H bonds behave as if they are attached to an increasingly more electronegative carbon in the series ethane → ethylene → acetylene.

TABLE 9.1 Structural Features of Ethane, Ethylene, and Acetylene

Feature	Ethane	Ethylene	Acetylene
Systematic name	Ethane	Ethene	Ethyne
Molecular formula	C_2H_6	C_2H_4	C_2H_2
Structural formula	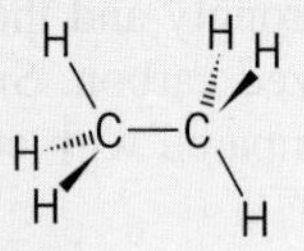	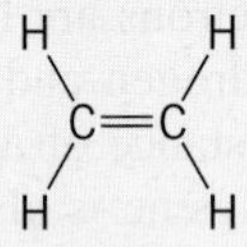	H—C≡C—H
C—C bond distance, pm	153	134	120
C—H bond distance, pm	111	110	106
H—C—C bond angles	111.0°	121.4°	180°
C—C bond dissociation enthalpy kJ/mol (kcal/mol)	375 (90)	720 (172)	961 (230)
C—H bond dissociation enthalpy, kJ/mol (kcal/mol)	421 (100.5)	464 (111)	547 (131)
Hybridization of carbon	sp^3	sp^2	sp
s character in C—H bonds	25%	33%	50%
Approximate pK_a	62	45	26

Problem 9.3

How do bond distances and bond strengths change with electronegativity in the series NH_3, H_2O, and HF?

The property that most separates acetylene from ethane and ethylene is its acidity. It, too, can be explained on the basis of the greater electronegativity of *sp*-hybridized carbon compared with sp^3 and sp^2.

9.5 Acidity of Acetylene and Terminal Alkynes

The C—H bonds of hydrocarbons show little tendency to ionize, and alkanes, alkenes, and alkynes are all very weak acids. The acid-dissociation constant K_a for methane, for example, is too small to be measured directly but is estimated to be about 10^{-60} (pK_a 60).

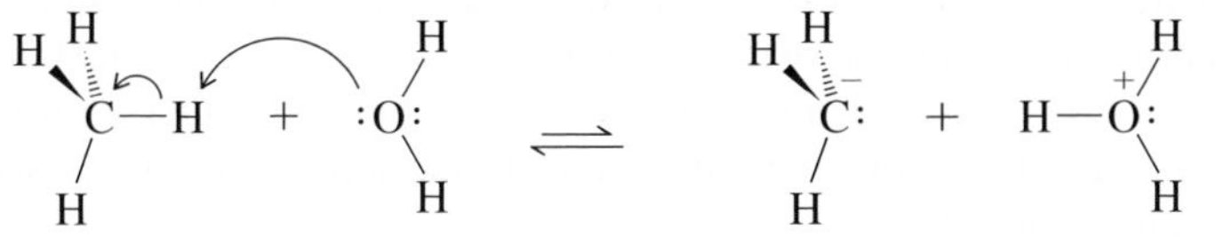

Methane (pK_a = 60) | Water | Methanide ion (a *carbanion*) | Hydronium ion (pK_a = −1.7)

The conjugate base of a hydrocarbon is called a **carbanion.** It is an anion in which the negative charge is borne by carbon. Because it is derived from a very weak acid, a carbanion such as $^{-}$:CH_3 is an exceptionally strong base.

Using the relationship from the preceding section that the electronegativity of carbon increases with its *s* character ($sp^3 < sp^2 < sp$), the order of hydrocarbon acidity is seen to increase with increasing *s* character of carbon.

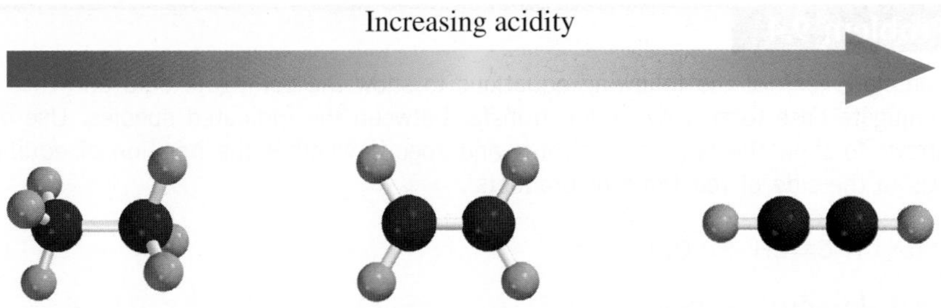

CH_3CH_3 < $H_2C{=}CH_2$ < HC≡CH

pK_a: 62 (weakest acid) | 45 | 26 (strongest acid)

Ionization of acetylene gives acetylide ion in which the unshared electron pair occupies an orbital with 50% *s* character.

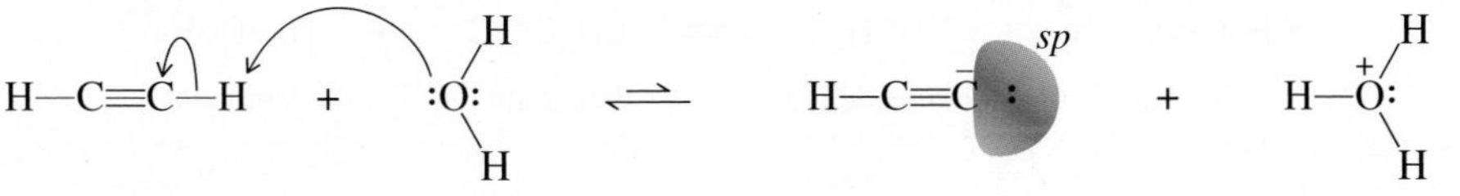

Acetylene (pK_a = 26) | Water | Acetylide ion | Hydronium ion (pK_a = −1.7)

In the corresponding ionizations of ethylene and ethane, the unshared pair occupies an orbital with 33% (sp^2) and 25% (sp^3) *s* character, respectively.

Terminal alkynes (RC≡CH) resemble acetylene in acidity.

$$(CH_3)_3CC{\equiv}CH \qquad pK_a = 25.5$$

3,3-Dimethyl-1-butyne

Although acetylene and terminal alkynes are far stronger acids than other hydrocarbons, we must remember that they are, nevertheless, very weak acids—much weaker than water and alcohols, for example. Hydroxide ion is too weak a base to convert acetylene to its anion in meaningful amounts. The position of the equilibrium described by the following equation lies overwhelmingly to the left:

$$H{-}C{\equiv}C{-}H + {}^{-}\!:\!\ddot{O}H \rightleftharpoons H{-}C{\equiv}C\!:^{-} + H{-}\ddot{O}H$$

Acetylene	Hydroxide ion	Acetylide ion	Water
(weaker acid)	(weaker base)	(stronger base)	(stronger acid)
$pK_a = 26$			$pK_a = 15.7$

Because acetylene is a far weaker acid than water and alcohols, these substances are not suitable solvents for reactions involving acetylide ions. Acetylide is instantly converted to acetylene by proton transfer from compounds that contain —OH groups.

Amide ion is a much stronger base than acetylide ion and converts acetylene to its conjugate base quantitatively.

$$H{-}C{\equiv}C{-}H + {}^{-}\!:\!\ddot{N}H_2 \rightleftharpoons H{-}C{\equiv}C\!:^{-} + H{-}\ddot{N}H_2$$

Acetylene	Amide ion	Acetylide ion	Ammonia
(stronger acid)	(stronger base)	(weaker base)	(weaker acid)
$pK_a = 26$			$pK_a = 36$

Solutions of sodium acetylide (HC≡CNa) may be prepared by adding *sodium amide* ($NaNH_2$) to acetylene in liquid ammonia as the solvent. Terminal alkynes react similarly to give species of the type RC≡CNa.

Problem 9.4

Complete each of the following equations to show the conjugate acid and the conjugate base formed by proton transfer between the indicated species. Use curved arrows to show the flow of electrons, and specify whether the position of equilibrium lies to the side of reactants or products.

(a) $CH_3C{\equiv}CH + {}^{-}\!:\!\ddot{O}CH_3 \rightleftharpoons$

(b) $HC{\equiv}CH + H_2\ddot{C}^{-}CH_3 \rightleftharpoons$

(c) $H_2C{=}CH_2 + {}^{-}\!:\!\ddot{N}H_2 \rightleftharpoons$

(d) $CH_3C{\equiv}CCH_2OH + {}^{-}\!:\!\ddot{N}H_2 \rightleftharpoons$

Sample Solution (a) The equation representing the acid–base reaction between propyne and methoxide ion is:

$$CH_3C{\equiv}C{-}H + {}^{-}\!:\!\ddot{O}CH_3 \rightleftharpoons CH_3C{\equiv}C\!:^{-} + H{-}\ddot{O}CH_3$$

Propyne	Methoxide ion	Propynide ion	Methanol
(weaker acid)	(weaker base)	(stronger base)	(stronger acid)

Alcohols are stronger acids than acetylene, and so the position of equilibrium lies to the left. Methoxide ion is not a strong enough base to remove a proton from acetylene.

Anions of acetylene and terminal alkynes are nucleophilic and react with methyl and primary alkyl halides to form carbon–carbon bonds by nucleophilic substitution. Some useful applications of this reaction will be discussed in the following section.

9.6 Preparation of Alkynes by Alkylation of Acetylene and Terminal Alkynes

Organic synthesis makes use of two major reaction types:

1. Carbon–carbon bond-forming reactions
2. Functional group transformations

Both strategies are applied to the preparation of alkynes. In this section we shall see how to prepare alkynes by carbon–carbon bond-forming reactions. By attaching alkyl groups to acetylene, more complex alkynes can be prepared.

$$\underset{\text{Acetylene}}{H{-}C{\equiv}C{-}H} \longrightarrow \underset{\text{Monosubstituted or terminal alkyne}}{R{-}C{\equiv}C{-}H} \longrightarrow \underset{\text{Disubstituted derivative of acetylene}}{R{-}C{\equiv}C{-}R'}$$

Reactions that attach alkyl groups to molecular fragments are called **alkylation** reactions. One way in which alkynes are prepared is by alkylation of acetylene.

Alkylation of acetylene involves a sequence of two separate operations. In the first one, acetylene is converted to its conjugate base by treatment with sodium amide.

$$\underset{\text{Acetylene}}{HC{\equiv}CH} + \underset{\text{Sodium amide}}{NaNH_2} \longrightarrow \underset{\text{Sodium acetylide}}{HC{\equiv}CNa} + \underset{\text{Ammonia}}{NH_3}$$

Next, an alkyl halide (the *alkylating agent*) is added to the solution of sodium acetylide. Acetylide ion acts as a nucleophile, displacing halide from carbon and forming a new carbon–carbon bond. Substitution occurs by an S_N2 mechanism.

$$\underset{\text{Sodium acetylide}}{HC{\equiv}CNa} + \underset{\text{Alkyl halide}}{RX} \longrightarrow \underset{\text{Alkyne}}{HC{\equiv}CR} + \underset{\text{Sodium halide}}{NaX} \quad \text{via} \quad HC{\equiv}C{:}^{-}\ R{-}X$$

The synthetic sequence is normally carried out in liquid ammonia, diethyl ether, or tetrahydrofuran as the solvent.

$$\underset{\text{Sodium acetylide}}{HC{\equiv}CNa} + \underset{\text{1-Bromobutane}}{CH_3CH_2CH_2CH_2Br} \xrightarrow{NH_3} \underset{\text{1-Hexyne (70–77\%)}}{CH_3CH_2CH_2CH_2C{\equiv}CH}$$

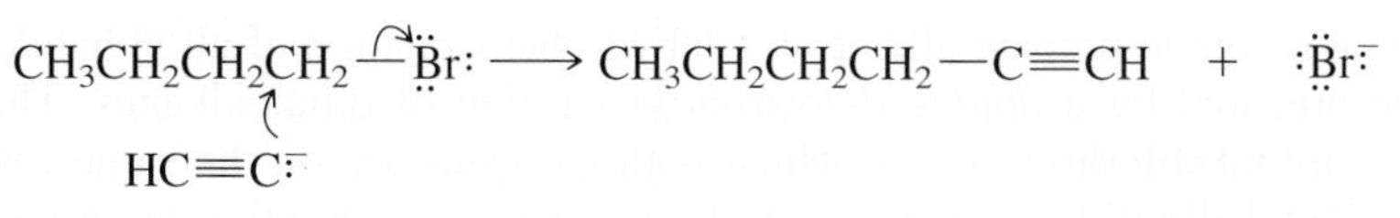

An analogous sequence starting with terminal alkynes (RC≡CH) yields alkynes of the type RC≡CR′.

$$\underset{\text{4-Methyl-1-pentyne}}{(CH_3)_2CHCH_2C{\equiv}CH} \xrightarrow[NH_3]{NaNH_2} (CH_3)_2CHCH_2C{\equiv}CNa \xrightarrow{CH_3Br} \underset{\text{5-Methyl-2-hexyne (81\%)}}{(CH_3)_2CHCH_2C{\equiv}CCH_3}$$

Dialkylation of acetylene can be achieved by carrying out the sequence twice.

$$\underset{\text{Acetylene}}{HC{\equiv}CH} \xrightarrow[\text{2. } CH_3CH_2Br]{\text{1. } NaNH_2,\ NH_3} \underset{\text{1-Butyne}}{HC{\equiv}CCH_2CH_3} \xrightarrow[\text{2. } CH_3Br]{\text{1. } NaNH_2,\ NH_3} \underset{\text{2-Pentyne (81\%)}}{CH_3C{\equiv}CCH_2CH_3}$$

As in other nucleophilic substitution reactions, alkyl *p*-toluenesulfonates may be used in place of alkyl halides.

Problem 9.5

Outline efficient syntheses of each of the following alkynes from acetylene and any necessary organic or inorganic reagents:

(a) 1-Heptyne (b) 2-Heptyne (c) 3-Heptyne

Sample Solution (a) An examination of the structural formula of 1-heptyne reveals it to have a pentyl group attached to an acetylene unit. Alkylation of acetylene, by way of its anion, with a pentyl halide is a suitable synthetic route to 1-heptyne.

$$\underset{\text{Acetylene}}{HC{\equiv}CH} \xrightarrow[NH_3]{NaNH_2} \underset{\text{Sodium acetylide}}{HC{\equiv}CNa} \xrightarrow{CH_3CH_2CH_2CH_2CH_2Br} \underset{\text{1-Heptyne}}{HC{\equiv}CCH_2CH_2CH_2CH_2CH_3}$$

The major limitation to this reaction is that synthetically acceptable yields are obtained only with methyl halides and primary alkyl halides. Acetylide anions are very basic, much more basic than hydroxide, for example, and react with secondary and tertiary alkyl halides by elimination.

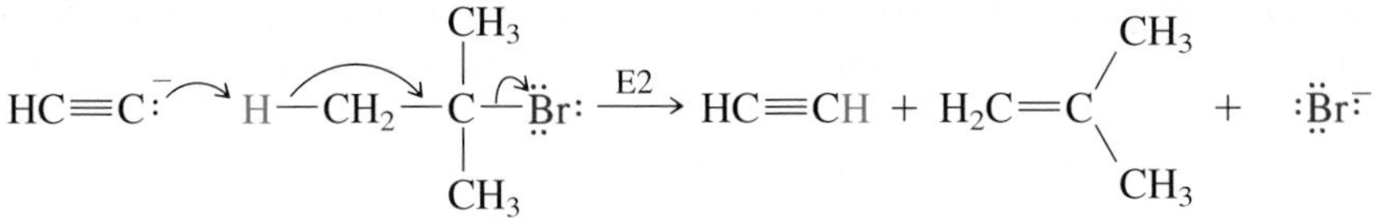

Acetylide *tert*-Butyl bromide Acetylene 2-Methylpropene Bromide

The desired S_N2 substitution pathway is observed only with methyl and primary alkyl halides.

Problem 9.6

Which of the alkynes of molecular formula C_5H_8 can be prepared in good yield by alkylation or dialkylation of acetylene? Explain why the preparation of the other C_5H_8 isomers would not be practical.

A second strategy for alkyne synthesis, involving functional group transformation reactions, is described in the following section.

9.7 Preparation of Alkynes by Elimination Reactions

Just as it is possible to prepare alkenes by dehydrohalogenation of alkyl halides, so may alkynes be prepared by a *double dehydrohalogenation* of dihaloalkanes. The dihalide may be a **geminal dihalide,** one in which both halogens are on the same carbon, or it may be a **vicinal dihalide,** one in which the halogens are on adjacent carbons.

Double dehydrohalogenation of a geminal dihalide

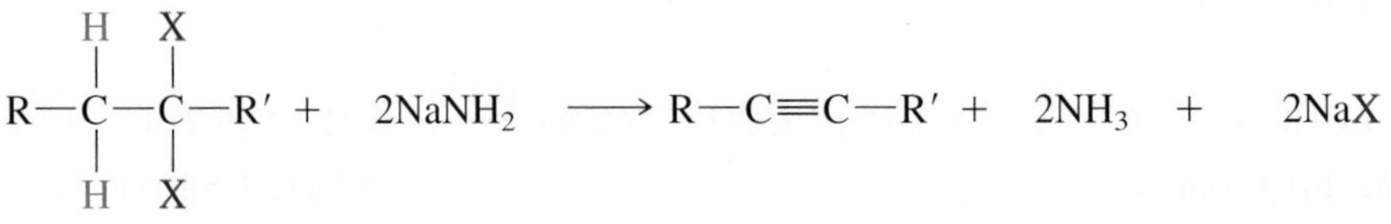

Geminal dihalide Sodium amide Alkyne Ammonia Sodium halide

Double dehydrohalogenation of a vicinal dihalide

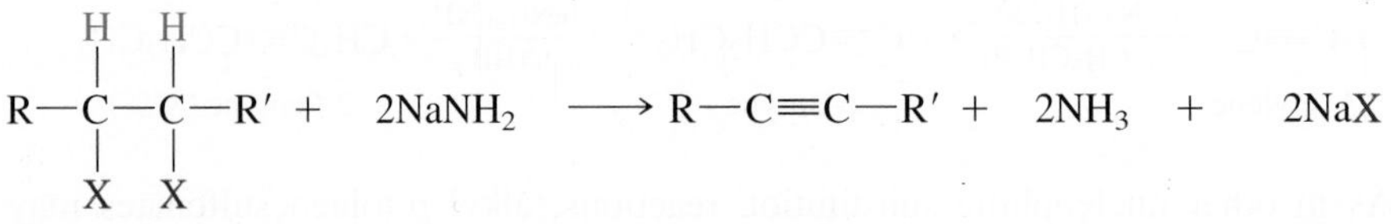

Vicinal dihalide Sodium amide Alkyne Ammonia Sodium halide

The most frequent applications of these procedures lie in the preparation of terminal alkynes. Because the terminal alkyne product is acidic enough to transfer a proton to amide anion, one equivalent of base in addition to the two equivalents required for double dehydrohalogenation is needed. Adding water or acid after the reaction is complete converts the sodium salt to the corresponding alkyne.

Double dehydrohalogenation of a geminal dihalide

$$(CH_3)_3CCH_2CHCl_2 \xrightarrow[NH_3]{3NaNH_2} (CH_3)_3CC{\equiv}CNa \xrightarrow{H_2O} (CH_3)_3CC{\equiv}CH$$

1,1-Dichloro-3,3-dimethylbutane — Sodium salt of alkyne product (not isolated) — 3,3-Dimethyl-1-butyne (56–60%)

Double dehydrohalogenation of a vicinal dihalide

$$CH_3(CH_2)_7\underset{\displaystyle Br}{\underset{|}{C}}HCH_2Br \xrightarrow[NH_3]{3NaNH_2} CH_3(CH_2)_7C{\equiv}CNa \xrightarrow{H_2O} CH_3(CH_2)_7C{\equiv}CH$$

1,2-Dibromodecane — Sodium salt of alkyne product (not isolated) — 1-Decyne (54%)

Double dehydrohalogenation to form terminal alkynes may also be carried out by heating geminal and vicinal dihalides with potassium *tert*-butoxide in dimethyl sulfoxide.

Problem 9.7

Give the structures of three isomeric dibromides that could be used as starting materials for the preparation of 3,3-dimethyl-1-butyne.

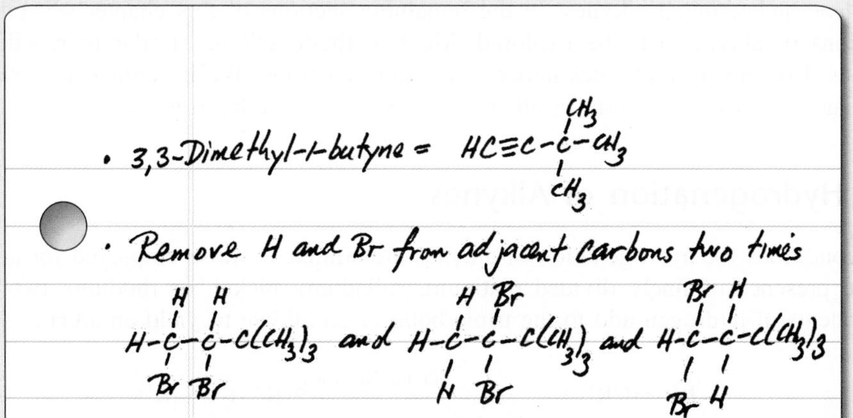

Because vicinal dihalides are prepared by addition of chlorine or bromine to alkenes (Section 6.14), alkenes, especially terminal alkenes, can serve as starting materials for the preparation of alkynes as shown in the following example:

$$(CH_3)_2CHCH{=}CH_2 \xrightarrow{Br_2} (CH_3)_2CH\underset{\displaystyle Br}{\underset{|}{C}}HCH_2Br \xrightarrow[2.\ H_2O]{1.\ NaNH_2,\ NH_3} (CH_3)_2CHC{\equiv}CH$$

3-Methyl-1-butene — 1,2-Dibromo-3-methylbutane — 3-Methyl-1-butyne (52%)

Problem 9.8

Show, by writing an appropriate series of equations, how you could prepare propyne from each of the following compounds as starting materials. You may use any necessary organic or inorganic reagents.

(a) 2-Propanol
(b) 1-Propanol
(c) Isopropyl bromide
(d) 1,1-Dichloroethane
(e) Ethyl alcohol

Sample Solution (a) Because we know that we can convert propene to propyne by the sequence of reactions

$$\underset{\text{Propene}}{CH_3CH{=}CH_2} \xrightarrow{Br_2} \underset{\text{1,2-Dibromopropane}}{CH_3\underset{\displaystyle Br}{\underset{|}{C}}HCH_2Br} \xrightarrow[\text{2. } H_2O]{\text{1. } NaNH_2,\ NH_3} \underset{\text{Propyne}}{CH_3C{\equiv}CH}$$

all that remains to completely describe the synthesis is to show the preparation of propene from 2-propanol. Acid-catalyzed dehydration is suitable.

$$\underset{\text{2-Propanol}}{(CH_3)_2CHOH} \xrightarrow[\text{heat}]{H_2SO_4} \underset{\text{Propene}}{CH_3CH{=}CH_2}$$

9.8 Reactions of Alkynes

We have already discussed one important chemical property of alkynes, the acidity of acetylene and terminal alkynes. In the remaining sections of this chapter several other reactions of alkynes will be explored. Most of them will be similar to reactions of alkenes. Like alkenes, alkynes undergo addition reactions. We'll begin with a reaction familiar to us from our study of alkenes, namely, catalytic hydrogenation.

9.9 Hydrogenation of Alkynes

The conditions for hydrogenation of alkynes are similar to those employed for alkenes. In the presence of finely divided platinum, palladium, nickel, or rhodium, two molar equivalents of hydrogen add to the triple bond of an alkyne to yield an alkane.

$$\underset{\text{Alkyne}}{RC{\equiv}CR'} + \underset{\text{Hydrogen}}{2H_2} \xrightarrow{\text{Pt, Pd, Ni, or Rh}} \underset{\text{Alkane}}{RCH_2CH_2R'}$$

$$\underset{\text{4-Methyl-1-hexyne}}{CH_3CH_2\underset{\displaystyle CH_3}{\underset{|}{C}}HCH_2C{\equiv}CH} + \underset{\text{Hydrogen}}{2H_2} \xrightarrow{Ni} \underset{\text{3-Methylhexane (77\%)}}{CH_3CH_2\underset{\displaystyle CH_3}{\underset{|}{C}}HCH_2CH_2CH_3}$$

Problem 9.9

Write a series of equations showing how you could prepare octane from acetylene and any necessary organic and inorganic reagents.

The heat of hydrogenation of an alkyne is greater than twice the heat of hydrogenation of an alkene. When two moles of hydrogen add to an alkyne, addition of the first mole (triple bond → double bond) is more exothermic than the second (double bond → single bond).

Substituents affect the heats of hydrogenation of alkynes in the same way they affect alkenes. Compare the heats of hydrogenation of 1-butyne and 2-butyne, both of which give butane on taking up two moles of H_2.

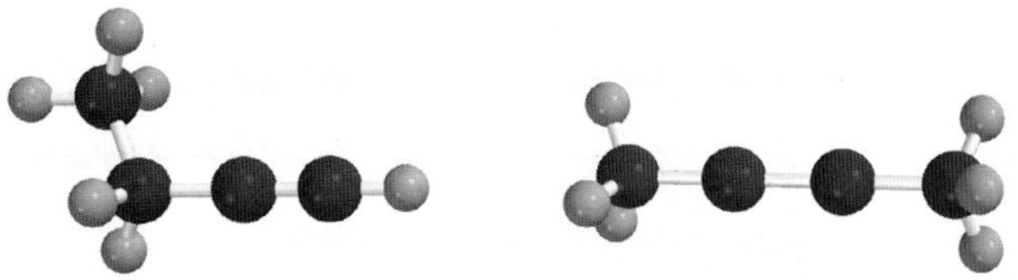

	$CH_3CH_2C{\equiv}CH$ 1-Butyne	$CH_3C{\equiv}CCH_3$ 2-Butyne
Heat of hydrogenation:	292 kJ/mol (69.9 kcal/mol)	275 kJ/mol (65.6 kcal/mol)

The internal triple bond of 2-butyne is stabilized relative to the terminal triple bond of 1-butyne. Alkyl groups release electrons to *sp*-hybridized carbon, stabilizing the alkyne and decreasing the heat of hydrogenation.

Like the hydrogenation of alkenes, hydrogenation of alkynes is a syn addition; cis alkenes are intermediates in the hydrogenation of alkynes to alkanes.

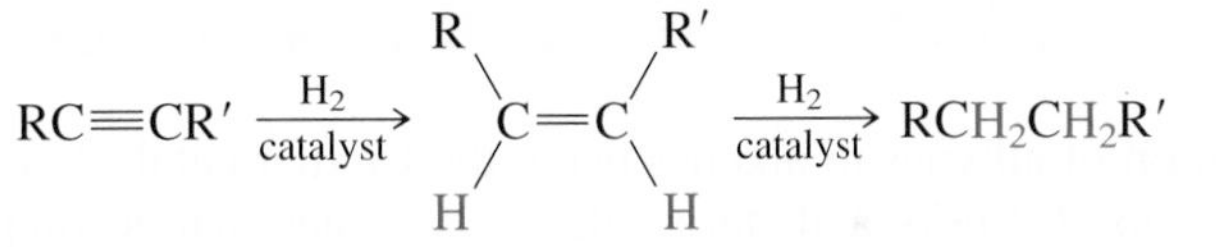

Alkyne cis Alkene Alkane

The fact that cis alkenes are intermediates in the hydrogenation of alkynes suggests that partial hydrogenation of an alkyne would provide a method for preparing:

1. Alkenes from alkynes, and
2. cis Alkenes free of their trans stereoisomers

Both objectives are met with special hydrogenation catalysts. The most frequently used one is the **Lindlar catalyst,** a palladium on calcium carbonate combination to which lead acetate and quinoline have been added. Lead acetate and quinoline partially deactivate ("poison") the catalyst, making it a poor catalyst for alkene hydrogenation while retaining its ability to catalyze the addition of H_2 to the triple bond.

The structure of quinoline is shown on page 463. In subsequent equations, we will simply use the term *Lindlar Pd* to stand for all of the components of the Lindlar catalyst.

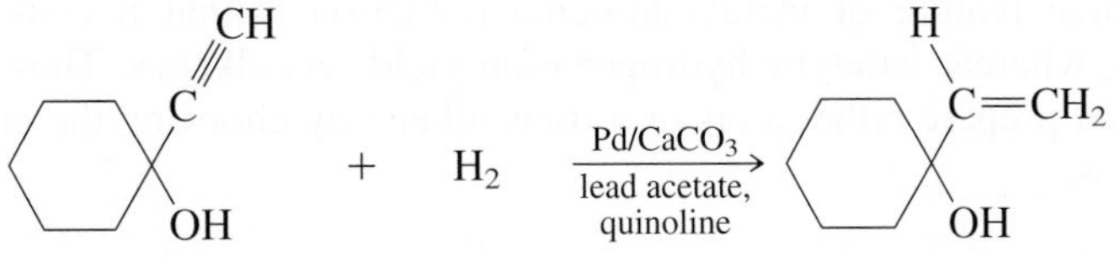

1-Ethynylcyclohexanol Hydrogen 1-Vinylcyclohexanol (90–95%)

Hydrogenation of alkynes with internal triple bonds gives cis alkenes.

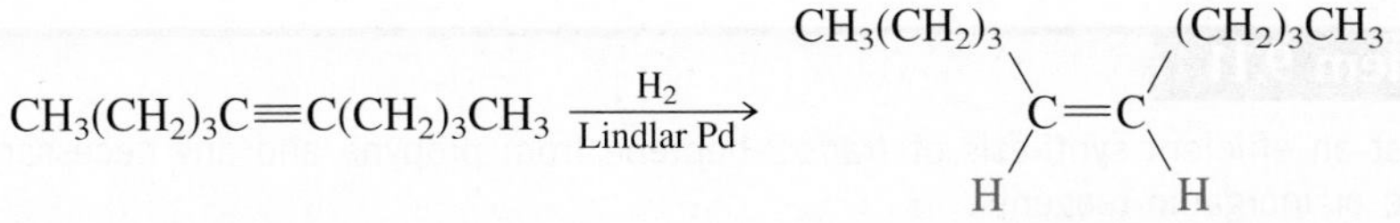

5-Decyne *cis*-5-Decene (87%)

Problem 9.10

Write a series of equations showing how to prepare *cis*-5-decene from acetylene and 1-bromobutane as the source of all its carbons, using any necessary organic or inorganic reagents. (*Hint:* You may find it helpful to review Section 9.6.)

- Can prepare cis-5-decene from 5-decyne by hydrogenation over Lindlar catalyst as per preceding equation.
- Therefore, need to prepare 5-decyne from acetylene.

$$HC{\equiv}CH \xrightarrow[NH_3]{NaNH_2} HC{\equiv}CNa \xrightarrow{CH_3(CH_2)_3Br} HC{\equiv}C(CH_2)_3CH_3$$

$$HC{\equiv}C(CH_2)_3CH_3 \xrightarrow{NaNH_2,\ NH_3} NaC{\equiv}C(CH_2)_3CH_3$$

$$NaC{\equiv}C(CH_2)_3CH_3 \xrightarrow{CH_3(CH_2)_3Br} CH_3(CH_2)_3C{\equiv}C(CH_2)_3CH_3$$

Hydrogenation of alkynes to alkenes using the Lindlar catalyst is attractive because it sidesteps the regioselectivity and stereoselectivity issues that accompany the dehydration of alcohols and dehydrohalogenation of alkyl halides. In terms of regioselectivity, the position of the double bond is never in doubt—it appears in the carbon chain at exactly the same place where the triple bond was. In terms of stereoselectivity, only the cis alkene forms. Recall that dehydration and dehydrohalogenation normally give a cis–trans mixture in which the cis isomer is the minor product.

In the following section, we'll see another method for converting alkynes to alkenes. The reaction conditions are very different from those of Lindlar hydrogenation. So is the stereochemistry.

9.10 Metal–Ammonia Reduction of Alkynes

A useful alternative to catalytic partial hydrogenation for converting alkynes to alkenes is reduction by a Group 1 metal (lithium, sodium, or potassium) in liquid ammonia. The unique feature of metal–ammonia reduction is that it converts alkynes to trans alkenes, whereas catalytic hydrogenation yields cis alkenes. Thus, from the same alkyne one can prepare either a cis or a trans alkene by choosing the appropriate reaction conditions.

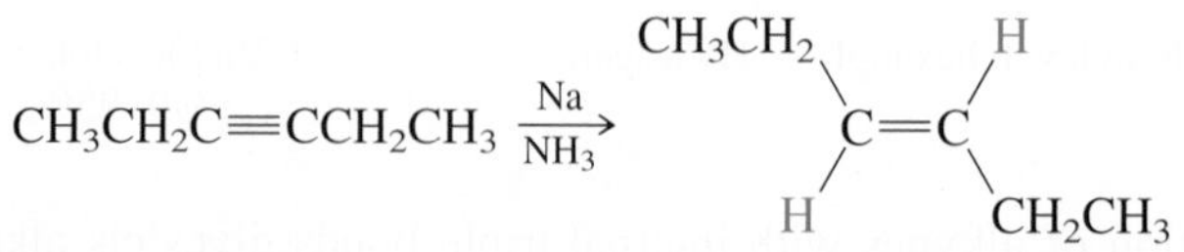
CH3CH2C≡CCH2CH3
Na
NH3
CH3CH2
H
C=C
H
CH2CH3

3-Hexyne → *trans*-3-Hexene (82%)

Problem 9.11

Suggest an efficient synthesis of *trans*-2-heptene from propyne and any necessary organic or inorganic reagents.

Mechanism 9.1

Sodium–Ammonia Reduction of an Alkyne

THE OVERALL REACTION:

$$\underset{\text{Alkyne}}{RC{\equiv}CR'} + \underset{\text{Sodium}}{2Na} + \underset{\text{Ammonia}}{2NH_3} \longrightarrow \underset{\text{Trans alkene}}{RCH{=}CHR'} + \underset{\text{Sodium amide}}{2NaNH_2}$$

Step 1: Electron transfer from sodium to the alkyne. The product is an anion radical.

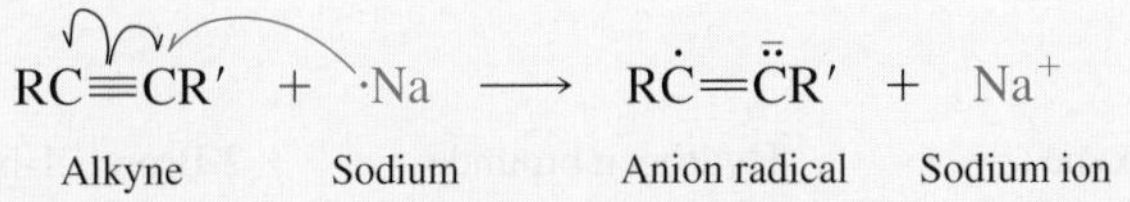
RC≡CR′ + ·Na ⟶ RĊ=C̈R′ + Na+
Alkyne
Sodium
Anion radical
Sodium ion

Step 2: The anion radical is a strong base and abstracts a proton from ammonia.

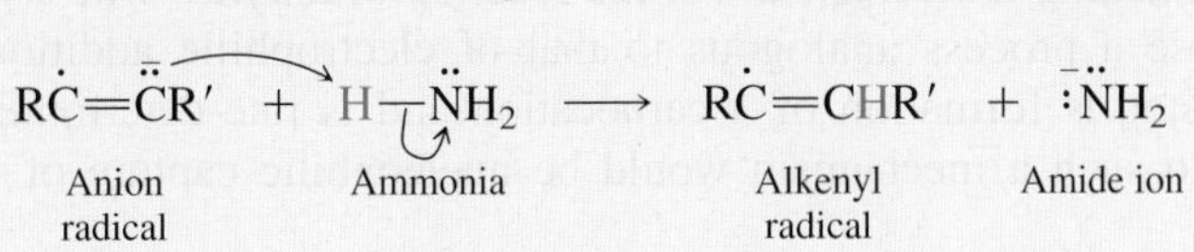
RĊ=C̈R′ + H—N̈H2 ⟶ RĊ=CHR′ + :N̈H2
Anion radical
Ammonia
Alkenyl radical
Amide ion

Step 3: Electron transfer to the alkenyl radical.

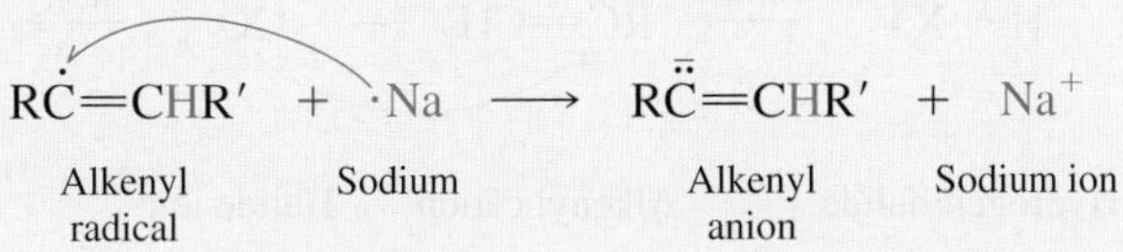
RĊ=CHR′ + ·Na ⟶ RC̈=CHR′ + Na+
Alkenyl radical
Sodium
Alkenyl anion
Sodium ion

Step 4: Proton transfer from ammonia converts the alkenyl anion to an alkene.

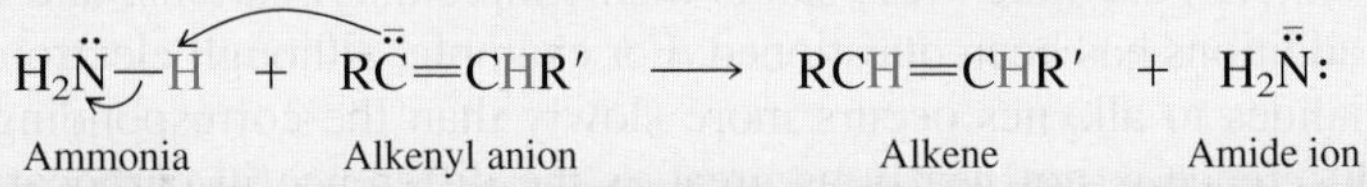
H2N̈—H + RC̈=CHR′ ⟶ RCH=CHR′ + H2N̈:
Ammonia
Alkenyl anion
Alkene
Amide ion

The stereochemistry of metal–ammonia reduction of alkynes differs from that of catalytic hydrogenation because the mechanisms of the two reactions are different. The mechanism of hydrogenation of alkynes is similar to that of catalytic hydrogenation of alkenes (Sections 6.1–6.3). Metal–ammonia reduction of alkynes is outlined in Mechanism 9.1.

The mechanism includes two single-electron transfers (steps 1 and 3) and two proton transfers (steps 2 and 4). Experimental evidence indicates that step 2 is rate-determining, and that the (*E*)-and (*Z*)-alkenyl radicals formed in this step interconvert rapidly.

(*Z*)-Alkenyl radical (less stable)

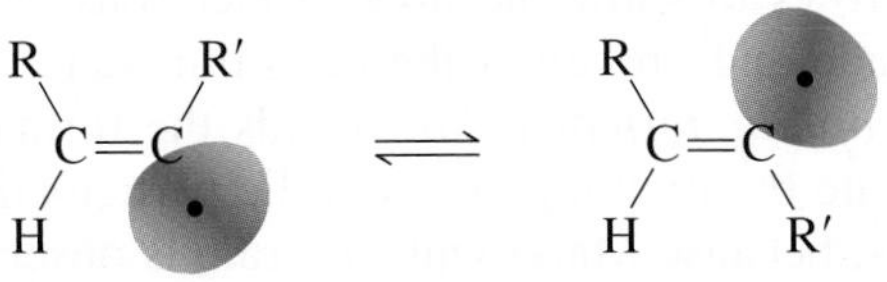
R
R′
C=C
H
R
C=C
H
R′

(*E*)-Alkenyl radical (more stable)

Reduction of these alkenyl radicals (step 3) gives a mixture of the (*E*)- and (*Z*)-alkenyl anions in which the more stable *E* stereoisomer predominates. Unlike the corresponding alkenyl radicals, the (*E*)- and (*Z*)-alkenyl anions are configurationally stable under the reaction conditions and yield an *E*/*Z* ratio of alkenes in step 4 that reflects the *E*/*Z* ratio of the alkenyl anions formed in step 3.

9.11 Addition of Hydrogen Halides to Alkynes

Alkynes react with many of the same electrophilic reagents that add to the carbon–carbon double bond of alkenes. Hydrogen halides, for example, add to alkynes to form alkenyl halides.

Alkyne Hydrogen halide Alkenyl halide

The regioselectivity of addition follows Markovnikov's rule. A proton adds to the carbon that has the greater number of hydrogens, and halide adds to the carbon with the fewer hydrogens.

1-Hexyne Hydrogen bromide 2-Bromo-1-hexene (60%)

When formulating a mechanism for the reaction of alkynes with hydrogen halides, we could propose a process analogous to that of electrophilic addition to alkenes in which the first step is formation of a carbocation and is rate-determining. The second step according to such a mechanism would be nucleophilic capture of the carbocation by a halide ion.

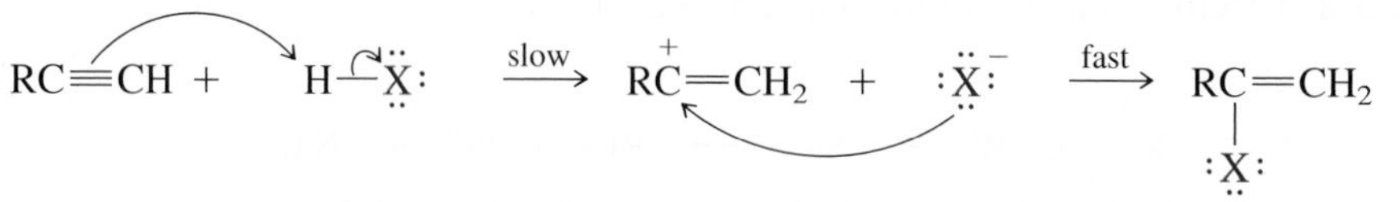

Alkyne Hydrogen halide Alkenyl cation Halide ion Alkenyl halide

Evidence from a variety of sources, however, indicates that alkenyl cations (also called *vinylic cations*) are much less stable than simple alkyl cations, and their involvement in these additions has been questioned. For example, although electrophilic addition of hydrogen halides to alkynes occurs more slowly than the corresponding additions to alkenes, the difference is not nearly as great as the difference in carbocation stabilities would suggest.

Furthermore, kinetic studies reveal that electrophilic addition of hydrogen halides to alkynes follows a rate law that is third-order overall and second-order in hydrogen halide.

$$\text{Rate} = k[\text{alkyne}][\text{HX}]^2$$

This third-order rate dependence suggests a transition state involving two molecules of the hydrogen halide and one of the alkyne. Figure 9.4 depicts a one-step termolecular process using curved arrows to show the flow of electrons, and dashed lines to indicate the bonds being made and broken at the transition state. This mechanism, called *Ad_E3* for *addition-electrophilic-termolecular,* avoids the formation of a very unstable alkenyl cation intermediate by invoking nucleophilic participation by the halogen at an early stage. Nevertheless, because Markovnikov's rule is observed, it seems likely that some degree of positive character develops at carbon and controls the regioselectivity of addition.

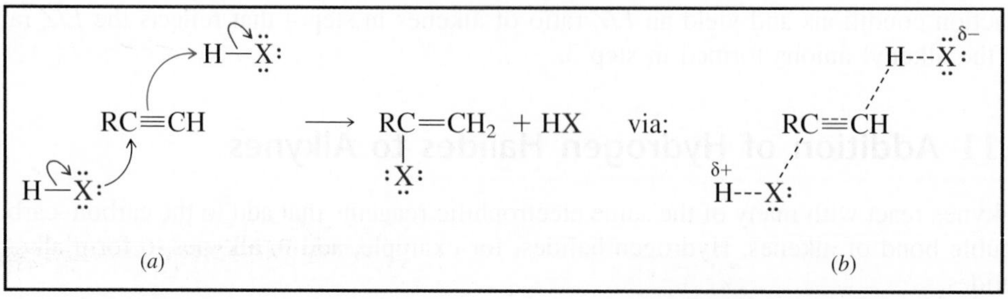

Figure 9.4

(*a*) Curved arrow notation, and (*b*) transition-state for electrophilic addition of a hydrogen halide HX to an alkyne by the Ad_E3 mechanism.

In the presence of excess hydrogen halide, geminal dihalides are formed by sequential addition of two molecules of hydrogen halide to the carbon–carbon triple bond.

$$\underset{\text{Alkyne}}{RC{\equiv}CR'} + \xrightarrow{HX} \underset{\text{Alkenyl halide}}{RCH{=}\underset{X}{\underset{|}{C}}R'} \xrightarrow{HX} \underset{\text{Geminal dihalide}}{RCH_2\overset{X}{\overset{|}{\underset{X}{\underset{|}{C}}}}R'}$$

The second mole of hydrogen halide adds to the initially formed alkenyl halide in accordance with Markovnikov's rule. Overall, both protons become bonded to the same carbon and both halogens to the adjacent carbon.

$$\underset{\text{3-Hexyne}}{CH_3CH_2C{\equiv}CCH_2CH_3} + \underset{\text{Hydrogen fluoride}}{2HF} \longrightarrow \underset{\text{3,3-Difluorohexane (76\%)}}{CH_3CH_2CH_2\overset{F}{\overset{|}{\underset{F}{\underset{|}{C}}}}CH_2CH_3}$$

Problem 9.12

Design a synthesis of 1,1-dichloroethane from each of the following. Write a series of equations, showing reactants and products, as illustrated in the Sample Solution.

(a) Ethylene (b) Vinyl chloride ($H_2C{=}CHCl$) (c) 1,1-Dibromoethane

Sample Solution (a) Reasoning backward, we recognize 1,1-dichloroethane as the product of addition of two molecules of hydrogen chloride to acetylene. Thus, the synthesis requires converting ethylene to acetylene as a key feature. As described in Section 9.7, this may be accomplished by conversion of ethylene to a vicinal dihalide, followed by double dehydrohalogenation. A suitable synthesis based on this analysis is as shown:

$$\underset{\text{Ethylene}}{H_2C{=}CH_2} \xrightarrow{Br_2} \underset{\text{1,2-Dibromoethane}}{BrCH_2CH_2Br} \xrightarrow[2.\ H_2O]{1.\ NaNH_2} \underset{\text{Acetylene}}{HC{\equiv}CH} \xrightarrow{2HCl} \underset{\text{1,1-Dichloroethane}}{CH_3CHCl_2}$$

Hydrogen bromide (but not hydrogen chloride or hydrogen iodide) adds to alkynes by a free-radical mechanism when peroxides are present in the reaction mixture. As in the free-radical addition of hydrogen bromide to alkenes (Section 6.18), a regioselectivity opposite to Markovnikov's rule is observed.

$$\underset{\text{1-Hexyne}}{CH_3CH_2CH_2CH_2C{\equiv}CH} + \underset{\text{Hydrogen bromide}}{HBr} \xrightarrow{\text{peroxides}} \underset{\text{1-Bromo-1-hexene (79\%)}}{CH_3CH_2CH_2CH_2CH{=}CHBr}$$

9.12 Hydration of Alkynes

By analogy to the hydration of alkenes, hydration of an alkyne is expected to yield an alcohol. The alcohol, however, would be a special kind, one in which the hydroxyl group is a substituent on a carbon–carbon double bond. This type of alcohol is called an **enol** (the double bond suffix *-ene* plus the alcohol suffix *-ol*). An important property of enols is their rapid isomerization to aldehydes or ketones under the conditions of their formation.

$$\underset{\text{Alkyne}}{RC{\equiv}CR'} + \underset{\text{Water}}{H_2O} \xrightarrow{\text{slow}} \underset{\substack{\text{Enol}\\\text{(not isolated)}}}{RCH{=}\overset{OH}{\overset{|}{C}}R'} \xrightarrow{\text{fast}} \underset{\substack{R' = \text{H; aldehyde}\\R' = \text{alkyl; ketone}}}{RCH_2\overset{O}{\overset{\|}{C}}R'}$$

Mechanism 9.2

Conversion of an Enol to a Ketone

THE OVERALL REACTION:

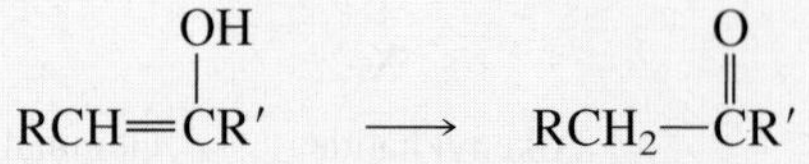

Enol — Ketone (aldehyde if R′=H)

Step 1: In aqueous acid, the first step is proton transfer to the carbon-carbon double bond.

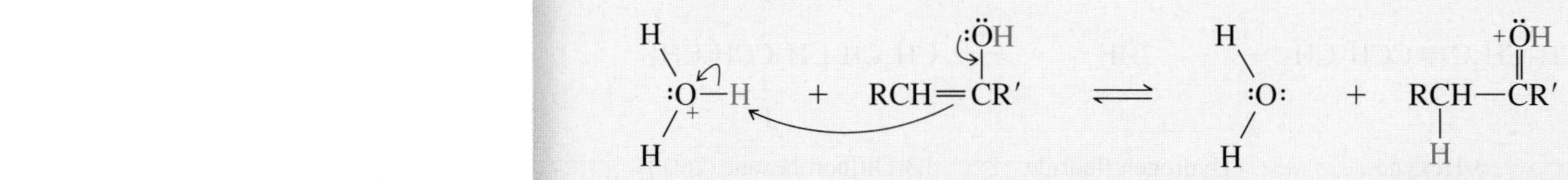

Hydronium ion — Enol — Water — Conjugate acid of ketone

Step 2: The conjugate acid of the ketone transfers a proton from oxygen to a water molecule, yielding a ketone.

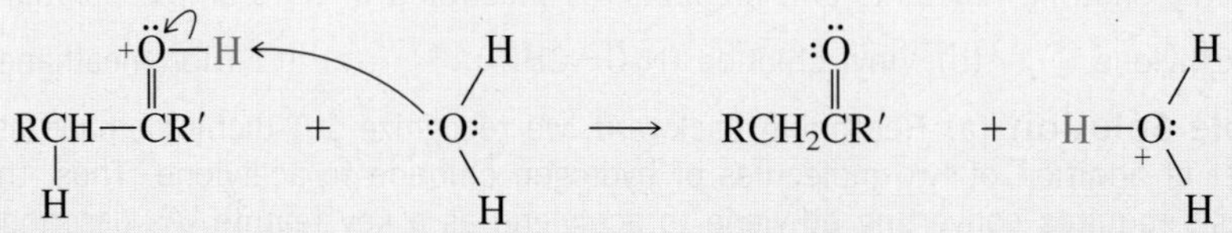

Conjugate acid of ketone — Water — Ketone — Hydronium ion

The aldehyde or ketone is called the **keto** form, and the keto ⇆ enol equilibration is referred to as *keto–enol isomerism* or *keto–enol tautomerism.* **Tautomers** are constitutional isomers that equilibrate by migration of an atom or group, and their equilibration is called **tautomerism.** Keto–enol isomerism involves the sequence of proton transfers shown in Mechanism 9.2.

The first step, protonation of the double bond of the enol, is analogous to the protonation of the double bond of an alkene. It takes place more readily, however, because the carbocation formed in this step is stabilized by resonance involving delocalization of a lone pair of oxygen.

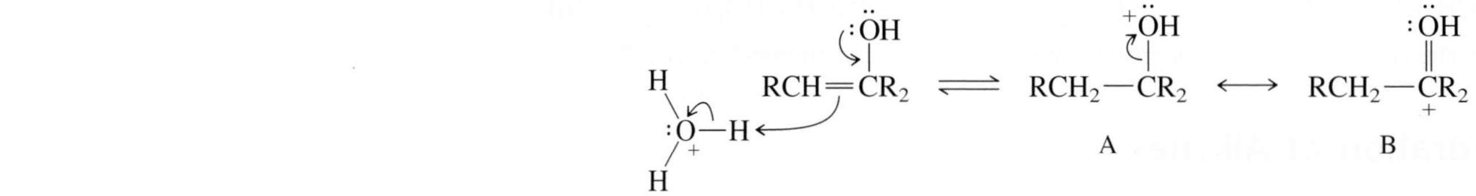

A — B

Of the two resonance forms A and B, B has only six electrons around its positively charged carbon. A satisfies the octet rule for both carbon and oxygen. It is more stable than B and more stable than a carbocation formed by protonation of a typical alkene.

Problem 9.13

Give the structure of the enol formed by hydration of 2-butyne, and write a series of equations showing its conversion to its corresponding ketone isomer.

In general, ketones are more stable than their enol precursors and are the products actually isolated when alkynes undergo acid-catalyzed hydration. The standard method for alkyne hydration employs aqueous sulfuric acid as the reaction medium and mercury(II) sulfate or mercury(II) oxide as a catalyst.

$$\underset{\text{4-Octyne}}{CH_3CH_2CH_2C{\equiv}CCH_2CH_2CH_3} + H_2O \xrightarrow{H^+,\ Hg^{2+}} \underset{\text{4-Octanone (89\%)}}{CH_3CH_2CH_2CH_2\overset{\overset{\displaystyle O}{\|}}{C}CH_2CH_2CH_3}$$

Hydration of alkynes follows Markovnikov's rule; terminal alkynes yield methyl-substituted ketones.

$$\underset{\text{1-Octyne}}{HC{\equiv}CCH_2CH_2CH_2CH_2CH_2CH_3} + H_2O \xrightarrow[HgSO_4]{H_2SO_4} \underset{\text{2-Octanone (91\%)}}{CH_3\overset{\overset{\displaystyle O}{\|}}{C}CH_2CH_2CH_2CH_2CH_2CH_3}$$

Problem 9.14

Show by a series of equations how you could prepare 2-octanone from acetylene and any necessary organic or inorganic reagents. How could you prepare 4-octanone?

Because of the regioselectivity of alkyne hydration, acetylene is the only alkyne structurally capable of yielding an aldehyde under these conditions.

$$\underset{\text{Acetylene}}{HC{\equiv}CH} + \underset{\text{Water}}{H_2O} \longrightarrow \underset{\substack{\text{Vinyl alcohol}\\ \text{(not isolated)}}}{H_2C{=}CHOH} \longrightarrow \underset{\text{Acetaldehyde}}{CH_3\overset{\overset{\displaystyle O}{\|}}{C}H}$$

At one time acetaldehyde was prepared on an industrial scale by this method. Modern methods involve direct oxidation of ethylene and are more economical.

The industrial synthesis of acetaldehyde from ethylene is shown on page 667.

9.13 Addition of Halogens to Alkynes

Alkynes react with chlorine and bromine to yield tetrahaloalkanes. Two molecules of the halogen add to the triple bond.

$$\underset{\text{Alkyne}}{RC{\equiv}CR'} + \underset{\substack{\text{Halogen}\\ \text{(chlorine or}\\ \text{bromine)}}}{2X_2} \longrightarrow \underset{\text{Tetrahaloalkane}}{R\overset{\overset{\displaystyle X}{|}}{\underset{\underset{\displaystyle X}{|}}{C}}{-}\overset{\overset{\displaystyle X}{|}}{\underset{\underset{\displaystyle X}{|}}{C}}R'}$$

$$\underset{\text{Propyne}}{CH_3C{\equiv}CH} + \underset{\text{Chlorine}}{2Cl_2} \longrightarrow \underset{\text{1,1,2,2-Tetrachloropropane (63\%)}}{CH_3\overset{\overset{\displaystyle Cl}{|}}{\underset{\underset{\displaystyle Cl}{|}}{C}}CHCl_2}$$

Some Things That Can Be Made from Acetylene . . . But Aren't

Acetylene had several uses around the time of World War I, primarily because it burned with a hot, luminous flame. The oxyacetylene torch and automobile and bicycle headlamps made by the Prest-O-Lite Company are representative of this period.

In an attempt to find a route to acetylene other than from calcium carbide, Prest-O-Lite sponsored research carried out by George O. Curme at Pittsburgh's Mellon Institute. Curme's research, which was directed toward converting the gases produced during petroleum refining to acetylene, led to methods better suited for making ethylene than acetylene. Viewed from our present perspective, Curme's petroleum-based route to ethylene ranks as a major discovery. It wasn't at the time though, because ethylene had virtually no uses before the 1920s. Curme's second great contribution was the research he carried out to see what useful products he could make from ethylene. The first was ethylene glycol, which became Prestone antifreeze. Others followed, and now ethylene is clearly the most important industrial organic chemical—perhaps the most important of all industrial chemicals.

What about acetylene? Based on the reactions described in this chapter we can write the following equations, all of which lead to useful compounds.

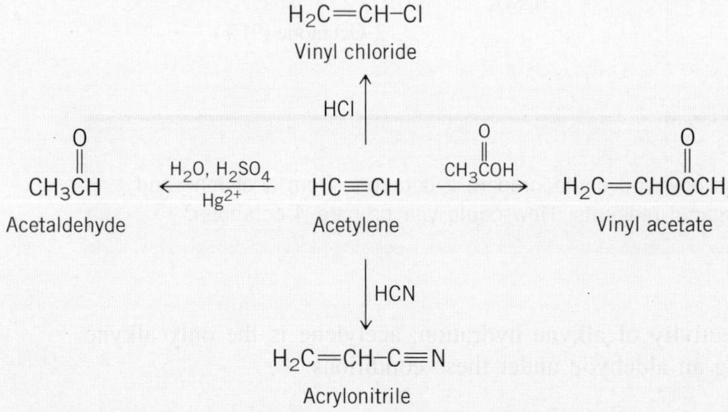
H2C=CH–Cl
Vinyl chloride
HCl
CH3CH
Acetaldehyde
H2O, H2SO4
Hg2+
HC≡CH
Acetylene
CH3COH
H2C=CHOCCH3
Vinyl acetate
HCN
H2C=CH–C≡N
Acrylonitrile

In fact, very little of each of these products is made from acetylene. Ethylene is the starting material for the preparation of vinyl chloride, vinyl acetate, and acetaldehyde. Propene is the starting material for acrylonitrile.

Economics dictate the choice of alkene in each case. Acetylene, because of the high energy cost of preparing it, is much more expensive than ethylene and propene. At present, acetylene is used as a starting material only in those few countries where local coal versus petroleum prices favor it. Ethylene comes from petroleum, acetylene can be made from coal. In time, as petroleum becomes increasingly expensive, acetylene-based syntheses may become competitive with ethylene-based ones.

A dihaloalkene is an intermediate and is the isolated product when the alkyne and the halogen are present in equimolar amounts. The stereochemistry of addition is anti.

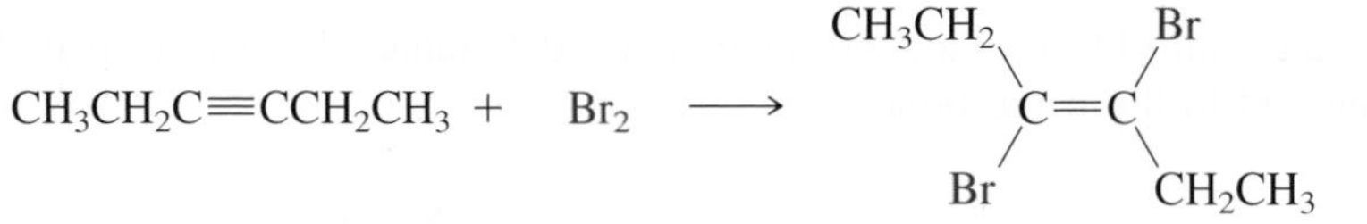

3-Hexyne — Bromine — (*E*)-3,4-Dibromo-3-hexene (90%)

9.14 Ozonolysis of Alkynes

Carboxylic acids are produced when alkynes are subjected to ozonolysis.

$$RC{\equiv}CR' \xrightarrow[2.\ H_2O]{1.\ O_3} R\overset{O}{\overset{\|}{C}}OH + HO\overset{O}{\overset{\|}{C}}R'$$

Recall that when carbonic acid is formed as a reaction product, it dissociates to carbon dioxide and water.

$$CH_3CH_2CH_2CH_2C{\equiv}CH \xrightarrow[2.\ H_2O]{1.\ O_3} CH_3CH_2CH_2CH_2CO_2H + HO\overset{O}{\overset{\|}{C}}OH$$

1-Hexyne — Pentanoic acid (51%) — Carbonic acid

Ozonolysis is sometimes used as a tool in structure determination. By identifying the carboxylic acids produced, we can deduce the structure of the alkyne. As with many other chemical methods of structure determination, however, it has been superseded by spectroscopic methods.

Problem 9.15

A certain hydrocarbon had the molecular formula $C_{16}H_{26}$ and contained two triple bonds. Ozonolysis gave $CH_3(CH_2)_4CO_2H$ and $HO_2CCH_2CH_2CO_2H$ as the only products. Suggest a reasonable structure for this hydrocarbon.

9.15 SUMMARY

Section 9.1 **Alkynes** are hydrocarbons that contain a carbon–carbon *triple bond.* Simple alkynes having no other functional groups or rings have the general formula C_nH_{2n-2}. Acetylene is the simplest alkyne.

Section 9.2 Alkynes are named in much the same way as alkenes, using the suffix *-yne* instead of *-ene.*

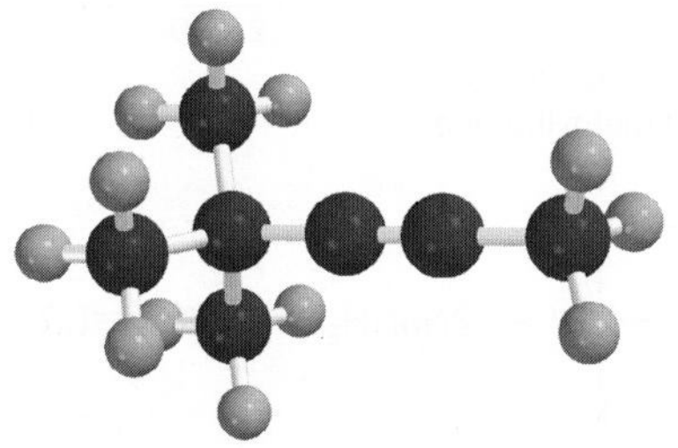

4,4-Dimethyl-2-pentyne

Section 9.3 The physical properties (boiling point, solubility in water, dipole moment) of alkynes resemble those of alkanes and alkenes.

Section 9.4 Acetylene is linear and alkynes have a linear geometry of their X—C≡C—Y units. The carbon–carbon triple bond in alkynes is composed of a σ and two π components.

The triply bonded carbons are *sp*-hybridized. The σ component of the triple bond contains two electrons in an orbital generated by the overlap of *sp*-hybridized orbitals on adjacent carbons. Each of these carbons also has two 2*p* orbitals, which overlap in pairs so as to give two π orbitals, each of which contains two electrons.

TABLE 9.2 Preparation of Alkynes

Reaction (section) and comments	General equation and specific example
Alkylation of acetylene and terminal alkynes (Section 9.6) The acidity of acetylene and terminal alkynes permits them to be converted to their conjugate bases on treatment with sodium amide. These anions are good nucleophiles and react with methyl and primary alkyl halides to form carbon–carbon bonds. Secondary and tertiary alkyl halides cannot be used, because they yield only elimination products under these conditions.	$RC{\equiv}CH + NaNH_2 \longrightarrow RC{\equiv}CNa + NH_3$ Alkyne, Sodium amide, Sodium alkynide, Ammonia $RC{\equiv}CNa + R'CH_2X \longrightarrow RC{\equiv}CCH_2R' + NaX$ Sodium alkynide, Primary alkyl halide, Alkyne, Sodium halide $(CH_3)_3CC{\equiv}CH \xrightarrow[2.\ CH_3I]{1.\ NaNH_2,\ NH_3} (CH_3)_3CC{\equiv}CCH_3$ 3,3-Dimethyl-1-butyne → 4,4-Dimethyl-2-pentyne (96%)
Double dehydrohalogenation of geminal dihalides (Section 9.7) An E2 elimination reaction of a geminal dihalide yields an alkenyl halide. If a strong enough base is used, sodium amide, for example, a second elimination step follows the first and the alkenyl halide is converted to an alkyne.	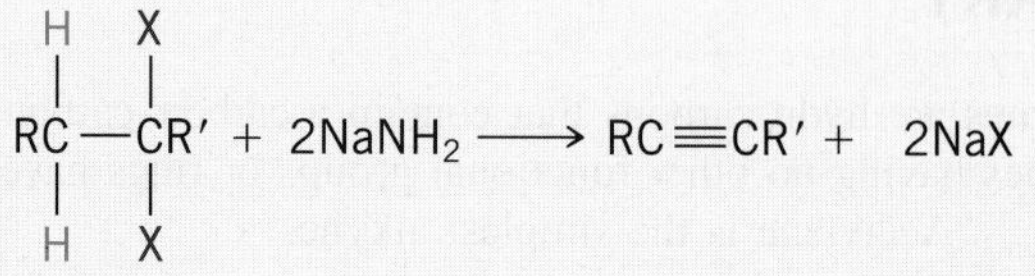 Geminal dihalide, Sodium amide, Alkyne, Sodium halide $(CH_3)_3CCH_2CHCl_2 \xrightarrow[2.\ H_2O]{1.\ 3NaNH_2,\ NH_3} (CH_3)_3CC{\equiv}CH$ 1,1-Dichloro-3,3-dimethylbutane → 3,3-Dimethyl-1-butyne (56–60%)
Double dehydrohalogenation of vicinal dihalides (Section 9.7) Dihalides in which the halogens are on adjacent carbons undergo two elimination processes analogous to those of geminal dihalides.	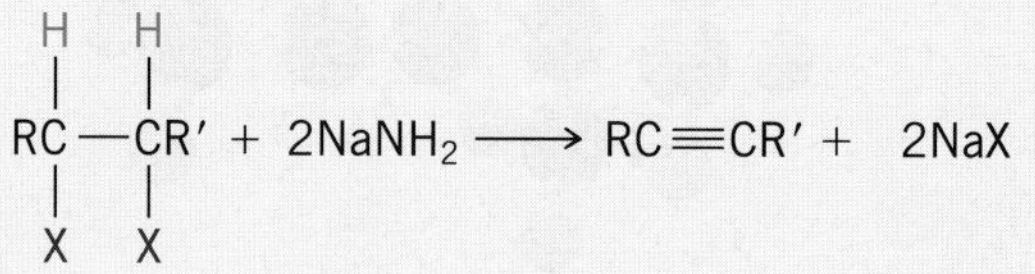 Vicinal dihalide, Sodium amide, Alkyne, Sodium halide 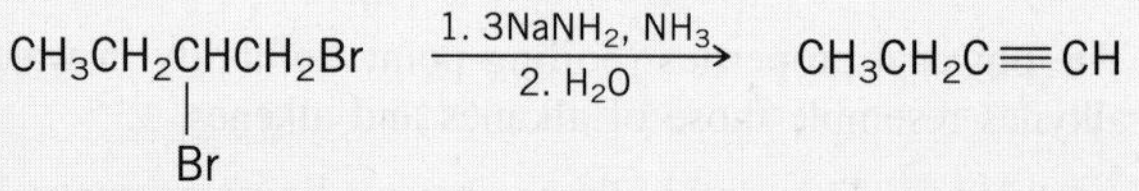 1,2-Dibromobutane → 1-Butyne (78–85%)

Section 9.5 Acetylene and terminal alkynes are more *acidic* than other hydrocarbons. They have pK_a's of approximately 26, compared with about 45 for alkenes and about 60 for alkanes. Sodium amide is a strong enough base to remove a proton from acetylene or a terminal alkyne, but sodium hydroxide is not.

$$CH_3CH_2C{\equiv}CH + NaNH_2 \longrightarrow CH_3CH_2C{\equiv}CNa + NH_3$$

1-Butyne, Sodium amide, Sodium 1-butynide, Ammonia

Sections 9.6–9.7 Table 9.2 summarizes the methods for preparing alkynes.

Section 9.8 Like alkenes, alkynes undergo addition reactions.

TABLE 9.3 Conversion of Alkynes to Alkenes and Alkanes

Reaction (section) and comments	General equation and specific example
Hydrogenation of alkynes to alkanes (Section 9.9) Alkynes are completely hydrogenated, yielding alkanes, in the presence of the customary metal hydrogenation catalysts.	$RC{\equiv}CR' + 2H_2 \xrightarrow{\text{metal catalyst}} RCH_2CH_2R'$ Alkyne Hydrogen Alkane 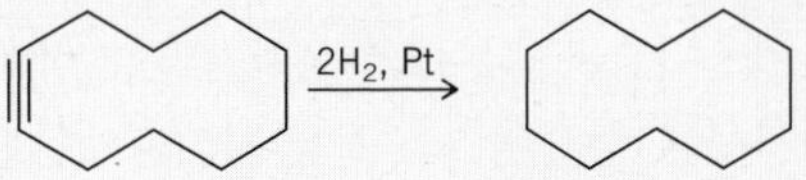 Cyclodecyne Cyclodecane (71%)
Hydrogenation of alkynes to alkenes (Section 9.9) Hydrogenation of alkynes may be halted at the alkene stage by using special catalysts. Lindlar palladium is the metal catalyst employed most often. Hydrogenation occurs with syn stereochemistry and yields a cis alkene.	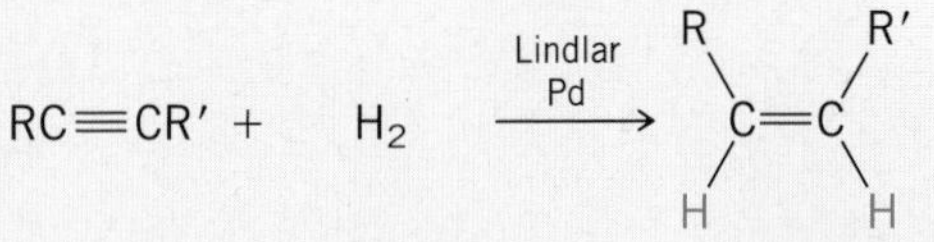 Alkyne Hydrogen Cis alkene 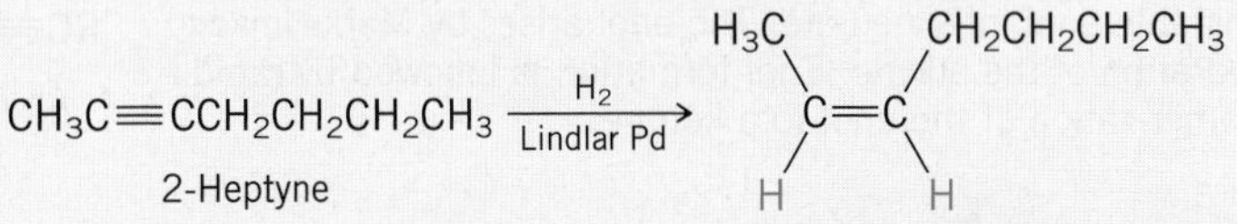 2-Heptyne *cis*-2-Heptene (59%)
Metal–ammonia reduction (Section 9.10) Group 1 metals—sodium is the one usually employed—in liquid ammonia as the solvent convert alkynes to trans alkenes. The reaction proceeds by a four-step sequence in which electron-transfer and proton-transfer steps alternate.	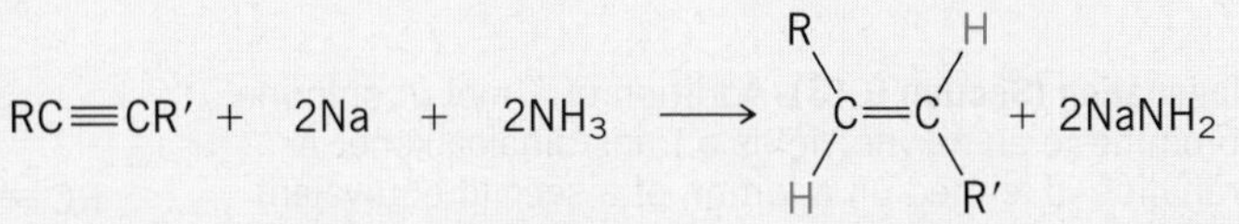 Alkyne Sodium Ammonia Trans alkene Sodium amide 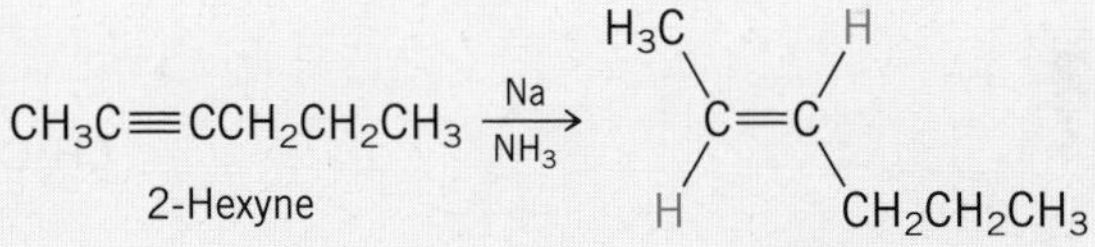 2-Hexyne *trans*-2-Hexene (69%)

Sections 9.9–9.10 Table 9.3 summarizes reactions that reduce alkynes to alkenes and alkanes.

Sections 9.11–9.13 Table 9.4 summarizes electrophilic addition to alkynes.

Section 9.14 Carbon–carbon triple bonds can be cleaved by ozonolysis. The cleavage products are carboxylic acids.

$$CH_3CH_2CH_2C{\equiv}CCH_3 \xrightarrow[2.\ H_2O]{1.\ O_3} CH_3CH_2CH_2\overset{\overset{\displaystyle O}{\|}}{C}OH + HO\overset{\overset{\displaystyle O}{\|}}{C}CH_3$$

2-Hexyne Butanoic acid Acetic acid

TABLE 9.4 Electrophilic Addition to Alkynes

Reaction (section) and comments	General equation and specific example
Addition of hydrogen halides (Section 9.11) Hydrogen halides add to alkynes in accordance with Markovnikov's rule to give alkenyl halides. In the presence of 2 mol of hydrogen halide, a second addition occurs to give a geminal dihalide.	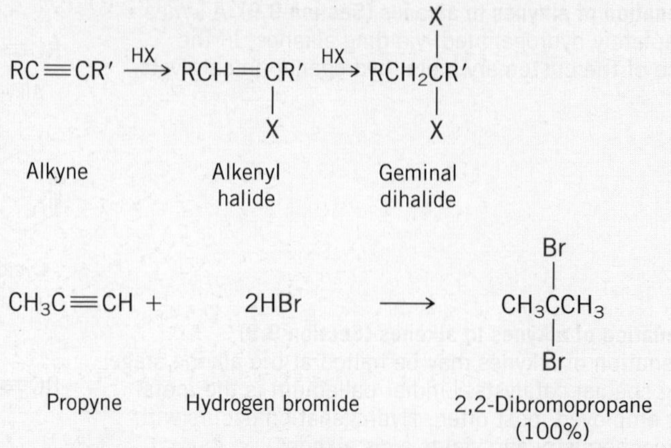
Acid-catalyzed hydration (Section 9.12) Water adds to the triple bond of alkynes to yield ketones by way of an unstable enol intermediate. The enol arises by Markovnikov hydration of the alkyne. Enol formation is followed by rapid isomerization of the enol to a ketone.	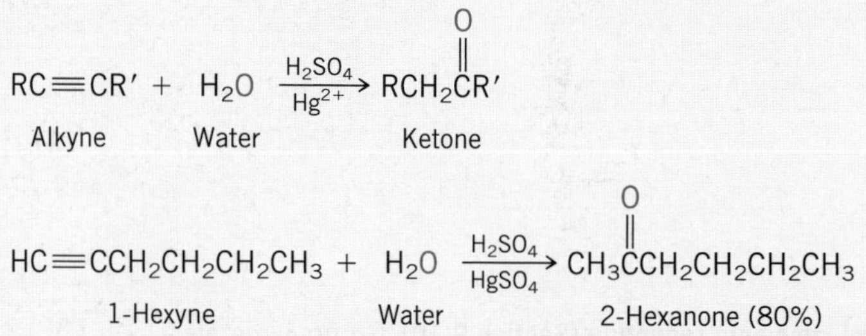
Halogenation (Section 9.13) Addition of 1 mol of chlorine or bromine to an alkyne yields a trans dihaloalkene. A tetrahalide is formed on addition of a second equivalent of the halogen.	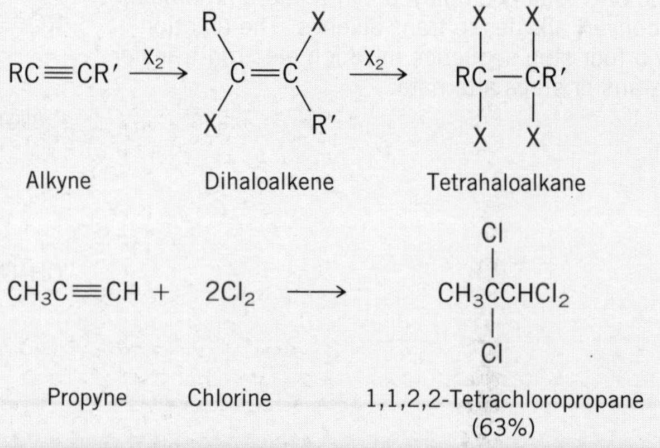

PROBLEMS

9.16 Write structural formulas and give acceptable IUPAC names for all the alkynes of molecular formula C_6H_{10}.

9.17 Provide the IUPAC name for each of the following alkynes:

(a) $CH_3CH_2CH_2C{\equiv}CH$

(b) $CH_3CH_2C{\equiv}CCH_3$

(c) 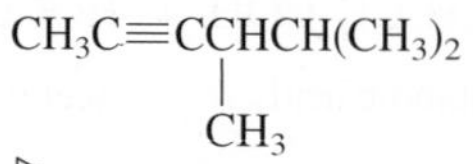

(d) 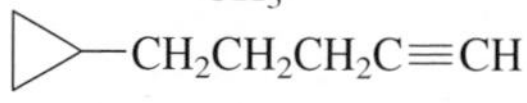

(e) 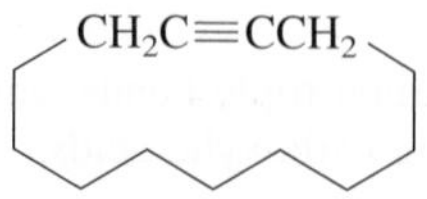

(f) 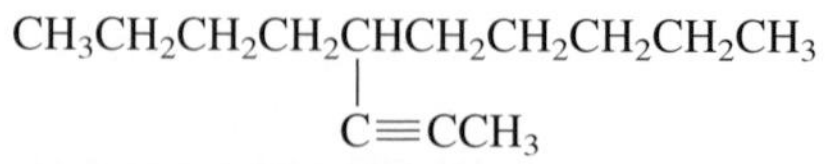

(g) $(CH_3)_3CC{\equiv}CC(CH_3)_3$

9.18 Write a structural formula for each of the following:

(a) 1-Octyne
(b) 2-Octyne
(c) 3-Octyne
(d) 4-Octyne
(e) 2,5-Dimethyl-3-hexyne
(f) 4-Ethyl-1-hexyne
(g) Ethynylcyclohexane
(h) 3-Ethyl-3-methyl-1-pentyne

9.19 All the compounds in Problem 9.18 are isomers except one. Which one?

9.20 Write structural formulas for all the alkynes of molecular formula C_8H_{14} that yield 3-ethylhexane on catalytic hydrogenation.

9.21 An unknown acetylenic amino acid obtained from the seed of a tropical fruit has the molecular formula $C_7H_{11}NO_2$. On catalytic hydrogenation over platinum this amino acid yielded homoleucine (an amino acid of known structure shown here) as the only product. What is the structure of the unknown amino acid?

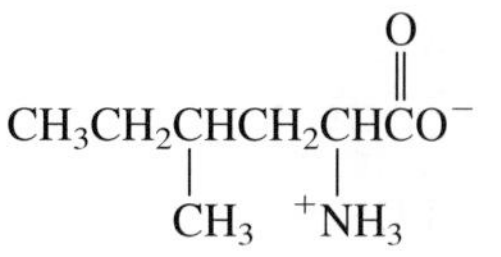

Homoleucine

9.22 Show by writing appropriate chemical equations how each of the following compounds could be converted to 1-hexyne:

(a) 1,1-Dichlorohexane
(b) 1-Hexene
(c) Acetylene
(d) 1-Iodohexane

9.23 Show by writing appropriate chemical equations how each of the following compounds could be converted to 3-hexyne:

(a) 1-Butene (b) 1,1-Dichlorobutane (c) Acetylene

9.24 When 1,2-dibromodecane was treated with potassium hydroxide in aqueous ethanol, it yielded a mixture of three isomeric compounds of molecular formula $C_{10}H_{19}Br$. Each of these compounds was converted to 1-decyne on reaction with sodium amide in dimethyl sulfoxide. Identify these three compounds.

9.25 Write the structure of the major organic product isolated from the reaction of 1-hexyne with

(a) Hydrogen (2 mol), platinum
(b) Hydrogen (1 mol), Lindlar palladium
(c) Lithium in liquid ammonia
(d) Sodium amide in liquid ammonia
(e) Product in part (d) treated with 1-bromobutane
(f) Product in part (d) treated with *tert*-butyl bromide
(g) Hydrogen chloride (1 mol)
(h) Hydrogen chloride (2 mol)
(i) Chlorine (1 mol)
(j) Chlorine (2 mol)
(k) Aqueous sulfuric acid, mercury(II) sulfate
(l) Ozone followed by hydrolysis

9.26 Write the structure of the major organic product isolated from the reaction of 3-hexyne with

(a) Hydrogen (2 mol), platinum
(b) Hydrogen (1 mol), Lindlar palladium
(c) Lithium in liquid ammonia
(d) Hydrogen chloride (1 mol)
(e) Hydrogen chloride (2 mol)
(f) Chlorine (1 mol)
(g) Chlorine (2 mol)
(h) Aqueous sulfuric acid, mercury(II) sulfate
(i) Ozone followed by hydrolysis

9.27 When 2-heptyne was treated with aqueous sulfuric acid containing mercury(II) sulfate, two products, each having the molecular formula $C_7H_{14}O$, were obtained in approximately equal amounts. What are these two compounds?

9.28 The alkane formed by hydrogenation of (*S*)-4-methyl-1-hexyne is optically active, but the one formed by hydrogenation of (*S*)-3-methyl-1-pentyne is not. Explain. Would you expect the products of hydrogenation of these two compounds in the presence of Lindlar palladium to be optically active?

9.29 All the following reactions have been described in the chemical literature and proceed in good yield. In some cases the reactants are more complicated than those we have so far encountered. Nevertheless, on the basis of what you have already learned, you should be able to predict the principal product in each case.

(a) $NaC{\equiv}CH + ClCH_2CH_2CH_2CH_2CH_2CH_2I \rightarrow$

(b) 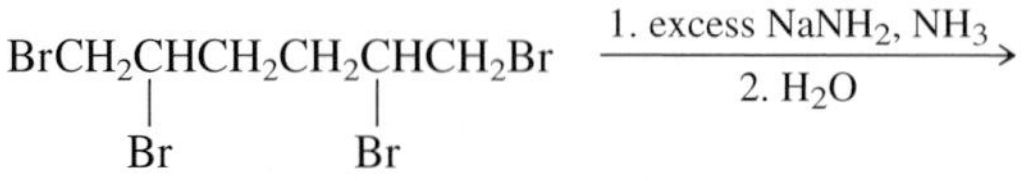

(c) 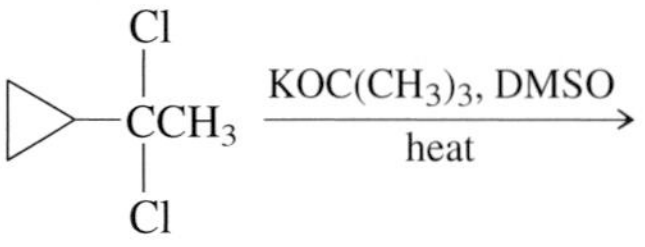

(d) 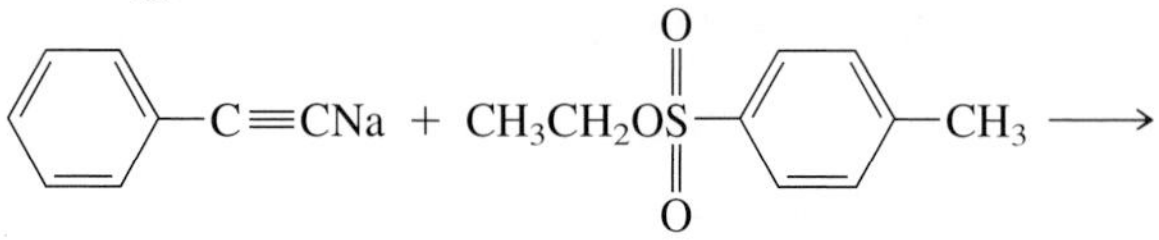

(e) Cyclodecyne $\xrightarrow[\text{2. } H_2O]{\text{1. } O_3}$

(f) 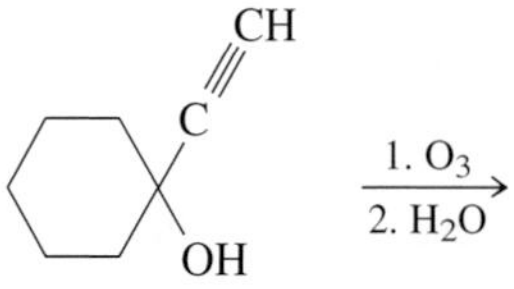

(g) 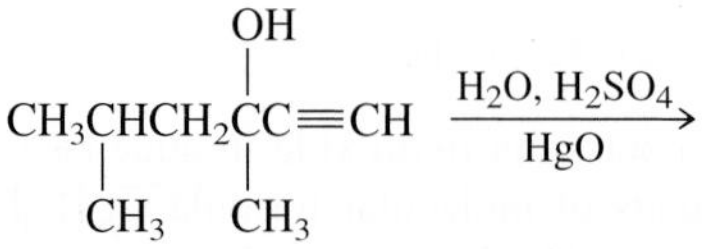

(h) (*Z*)-$CH_3CH_2CH_2CH_2CH{=}CHCH_2(CH_2)_7C{\equiv}CCH_2CH_2OH \xrightarrow[\text{2. } H_2O]{\text{1. Na, } NH_3}$

(i) 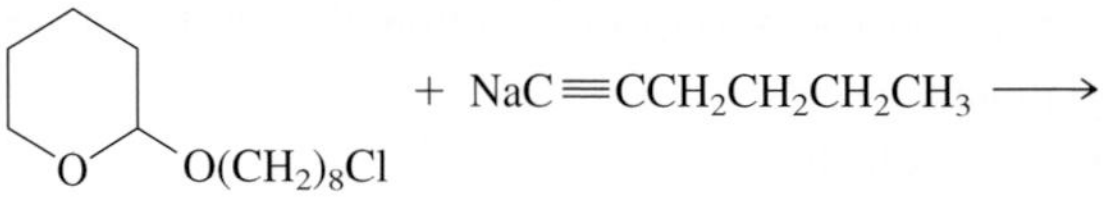

(j) Product of part (i) $\xrightarrow[\text{Lindlar Pd}]{H_2}$

9.30 (a) Oleic acid and stearic acid are naturally occurring compounds, which can be isolated from various fats and oils. In the laboratory, each can be prepared by hydrogenation of a compound known as *stearolic acid,* which has the formula $CH_3(CH_2)_7C{\equiv}C(CH_2)_7CO_2H$. Oleic acid is obtained by hydrogenation of stearolic acid over Lindlar palladium; stearic acid is obtained by hydrogenation over platinum. What are the structures of oleic acid and stearic acid?

(b) Sodium–ammonia reduction of stearolic acid yields a compound known as *elaidic acid.* What is the structure of elaidic acid?

9.31 The ketone 2-heptanone has been identified as contributing to the odor of a number of dairy products, including condensed milk and cheddar cheese. Describe a synthesis of 2-heptanone from acetylene and any necessary organic or inorganic reagents.

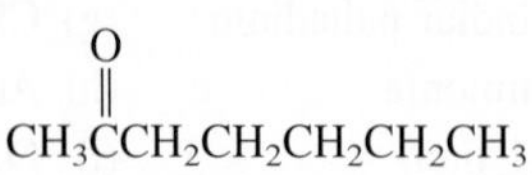

2-Heptanone

9.32 Alkynes undergo hydroboration to give alkenylboranes, which can be oxidized to give carbonyl compounds with hydrogen peroxide. The net result of the two-step sequence is hydration, which gives aldehydes from terminal alkynes.

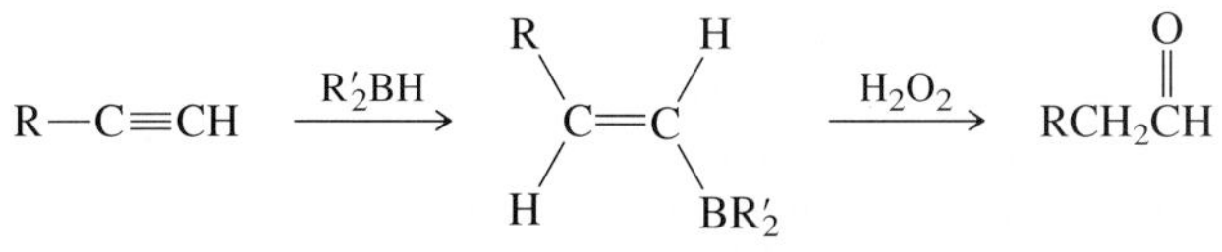

Alkenylborane

The oxidation step involves an enol intermediate. Using Mechanism 9.2 as a guide, write the structure of the enol that is formed in the conversion of 1-hexyne to hexanal.

$$\underset{\text{1-Hexyne}}{CH_3CH_2CH_2CH_2C{\equiv}CH} \xrightarrow[\text{2. } H_2O_2,\ NaOH]{\text{1. } R'_2BH} \underset{\text{Hexanal}}{CH_3CH_2CH_2CH_2CH_2\overset{O}{\overset{\|}{C}}H}$$

9.33 (*Z*)-9-Tricosene [(*Z*)-$CH_3(CH_2)_7CH{=}CH(CH_2)_{12}CH_3$] is the sex pheromone of the female housefly. Synthetic (*Z*)-9-tricosene is used as bait to lure male flies to traps that contain insecticide. Using acetylene and alcohols of your choice as starting materials, along with any necessary inorganic reagents, show how you could prepare (*Z*)-9-tricosene.

9.34 Show by writing a suitable series of equations how you could prepare each of the following compounds from the designated starting materials and any necessary organic or inorganic reagents:

(a) 2,2-Dibromopropane from 1,1-dibromopropane
(b) 2,2-Dibromopropane from 1,2-dibromopropane
(c) 1,1,2,2-Tetrachloropropane from 1,2-dichloropropane
(d) 2,2-Diiodobutane from acetylene and ethyl bromide
(e) 1-Hexene from 1-butene and acetylene
(f) Decane from 1-butene and acetylene
(g) Cyclopentadecyne from cyclopentadecene
(h) 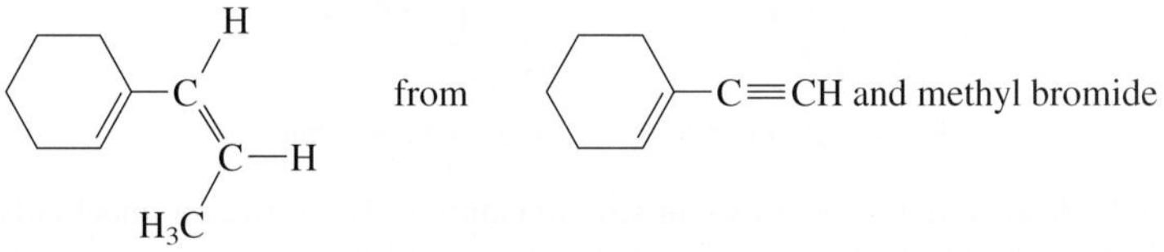

(i) *meso*-2,3-Dibromobutane from 2-butyne

9.35 Assume that you need to prepare 4-methyl-2-pentyne and discover that the only alkynes on hand are acetylene and propyne. You also have available methyl iodide, isopropyl bromide, and 1,1-dichloro-3-methylbutane. Which of these compounds would you choose in order to perform your synthesis, and how would you carry it out?

9.36 Diphenylacetylene can be synthesized by the double dehydrohalogenation of 1,2-dibromo-1,2-diphenylethene. The sequence starting from (*E*)-1,2-diphenylethene (stilbene) consists of bromination to give the dibromide, followed by dehydrohalogenation to give a vinylic bromide, then a second dehydrohalogenation to give diphenylacetylene.

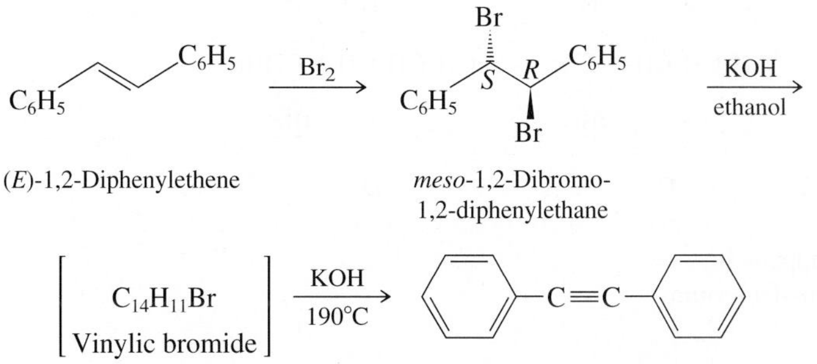

Diphenylacetylene

(a) What is the structure, including stereochemistry, of the vinylic bromide?
(b) If the sequence starts with (*Z*)-1,2-dibromo-1,2-diphenylethene, what is (are) the structure(s) of the intermediate dibromide(s)? What is the structure of the vinylic bromide?

9.37 Compound A has the molecular formula $C_{14}H_{25}Br$ and was obtained by reaction of sodium acetylide with 1,12-dibromododecane. On treatment of compound A with sodium amide, it was converted to compound B ($C_{14}H_{24}$). Ozonolysis of compound B gave the diacid $HO_2C(CH_2)_{12}CO_2H$. Catalytic hydrogenation of compound B over Lindlar palladium gave compound C ($C_{14}H_{26}$), and hydrogenation over platinum gave compound D ($C_{14}H_{28}$). Sodium–ammonia reduction of compound B gave compound E ($C_{14}H_{26}$). Both C and E yielded $O{=}CH(CH_2)_{12}CH{=}O$ on ozonolysis. Assign structures to compounds A through E so as to be consistent with the observed transformations.

Descriptive Passage and Interpretive Problems 9

Thinking Mechanistically About Alkynes

The preparation and properties of alkynes extend some topics explored in earlier chapters:

- Alkynes can be prepared by elimination reactions related to the E2 dehydrohalogenation of alkyl halides used to prepare alkenes.
- Alkynes can be prepared by S_N2 reactions in which a nucleophile of the type $RC{\equiv}C{:}^-$ reacts with a primary alkyl halide.
- Alkynes undergo addition reactions, especially electrophilic addition, with many of the same compounds that add to alkenes.

The greater *s* character of *sp* hybrid orbitals compared with sp^3 and sp^2 gives alkynes certain properties beyond those seen in alkanes and alkenes. It is convenient to think of *sp*-hybridized carbon as more electronegative than its sp^2 or sp^3 counterparts.

- The $\equiv$C—H unit of an alkyne is more acidic than a C—H unit of an alkene or alkane, allowing acetylene and terminal alkynes to be converted to their conjugate bases $\equiv C{:}^-$ by $NaNH_2$.
- Unlike alkenes, alkynes are reduced by metals, especially Li, Na, and K.
- Unlike alkenes, alkynes can undergo nucleophilic as well as electrophilic addition.

E^+ is an electrophile $Nu{:}^-$ is a nucleophile

Problems 9.38–9.42 emphasize mechanistic reasoning. By thinking mechanistically you reduce the need to memorize facts while increasing your ability to analyze and understand new material. Nucleophilic addition to alkynes, for example, is not covered in this chapter but is the focus of problem 9.42, which can be solved by thinking mechanistically.

9.38 Which of the following best describes what happens in the first step in the mechanism of the reaction shown?

$$CH_3CH_2CH_2CHBr_2 + 3NaNH_2 \xrightarrow{NH_3} CH_3CH_2C{\equiv}CNa + 2NaBr + 3NH_3$$

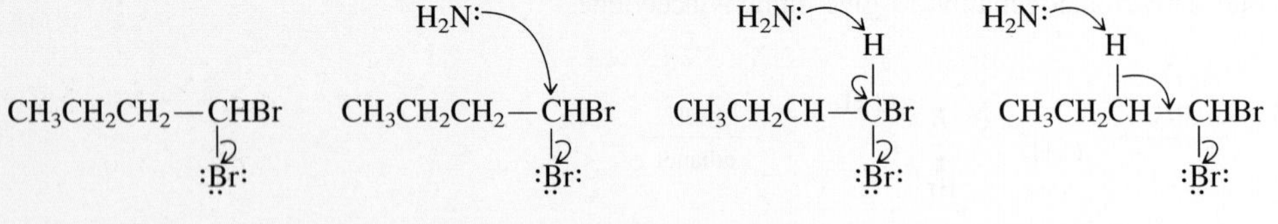

A. B. C. D.

9.39 Which of the following best describes what happens in the first step in the mechanism of the hydrogen–deuterium exchange reaction shown?

$$CH_3C{\equiv}CD \xrightarrow[NH_3]{NaNH_2} CH_3C{\equiv}CH$$

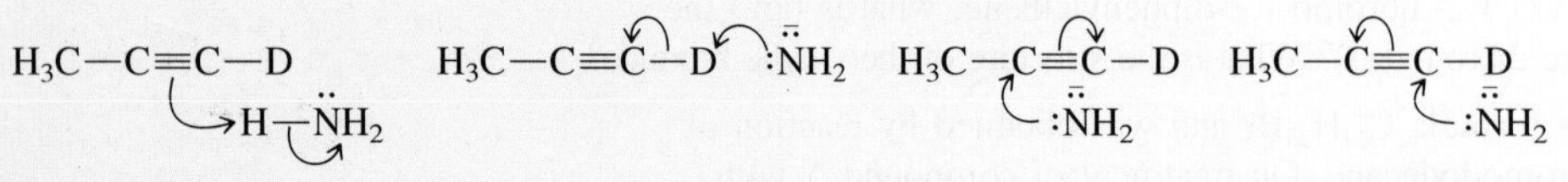

A. B. C. D.

9.40 Electrophilic addition of fluorosulfonic acid (FSO_2OH) to propyne proceeds by way of a very unstable vinyl cation intermediate. What is the most reasonable structure, including geometry, of this intermediate? (*Hint:* Use VSEPR to deduce the geometry.)

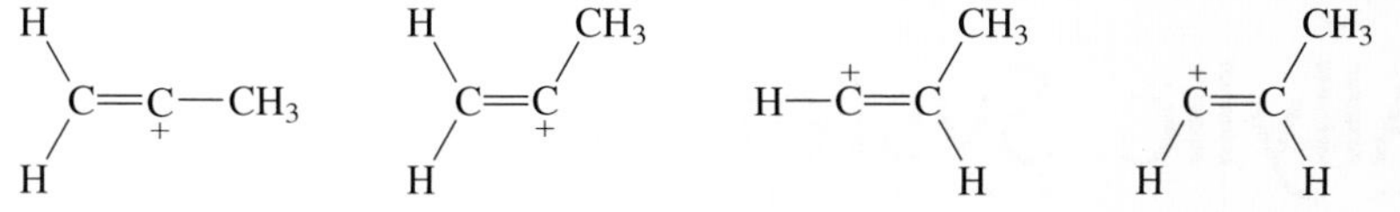

A. B. C. D.

9.41 Rates of Br_2 addition were measured for a series of alkynes, giving the data shown.

Alkyne	Relative rate
$HC\equiv CH$	1.0
$HC\equiv CCH_3$	13.4
$H_3CC\equiv CCH_3$	120
$H_3C-C(CH_3)_2-C\equiv CCH_3$	558

Assuming that Br_2 addition to alkynes proceeds through rate-determining formation of a cyclic bromonium ion, what generalizations can you make about the structure of the rate-determining transition state?

Positive charge development at carbons in original triple bond	More important effect of substituents on triple bond
A. One carbon only	Electron donation
B. One carbon only	Steric hindrance
C. Both carbons	Electron donation
D. Both carbons	Steric hindrance

9.42 Nucleophilic addition can occur with alkynes that bear strong electron-attracting substituents such as CF_3 on the triple bond. Predict the product of nucleophilic addition of CH_3OD to 3,3,3-trifluoropropyne. The stereochemistry of addition is anti, and the first step in the mechanism is bond formation between CH_3O^- and one of the carbons of the triple bond.

$$HC\equiv CCF_3 \xrightarrow[CH_3OD]{NaOCH_3}$$

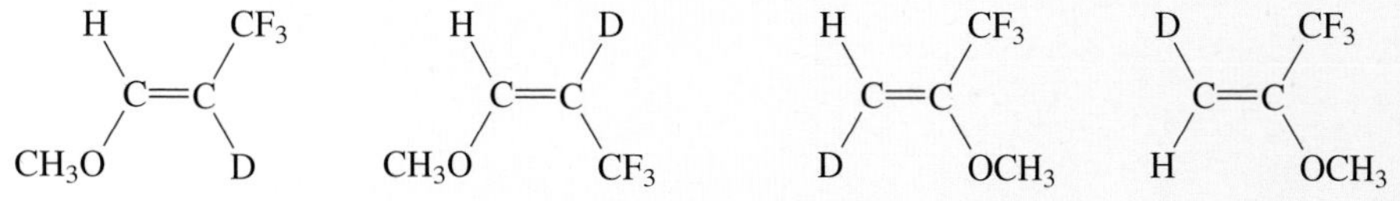

A. B. C. D.

10 Conjugation in Alkadienes and Allylic Systems

Chapter Outline

Mechanisms

A locomotive can pull a heavy load but has no space to carry freight. Boxcars can carry freight but lack pulling power. A train can do what its component parts can't do separately. Conjugation (joining together) of chemical structural units alters their properties too.

NOT ALL THE PROPERTIES of alkenes are revealed by focusing exclusively on the functional group behavior of the double bond. A double bond can affect the properties of a second functional unit to which it is directly attached. It can be a substituent, for example, on a positively charged carbon in an **allylic carbocation,** on a carbon that bears an unpaired electron in an **allylic free radical,** on a negatively charged carbon in an **allylic anion,** or it can be a substituent on a second double bond in a **conjugated diene.**

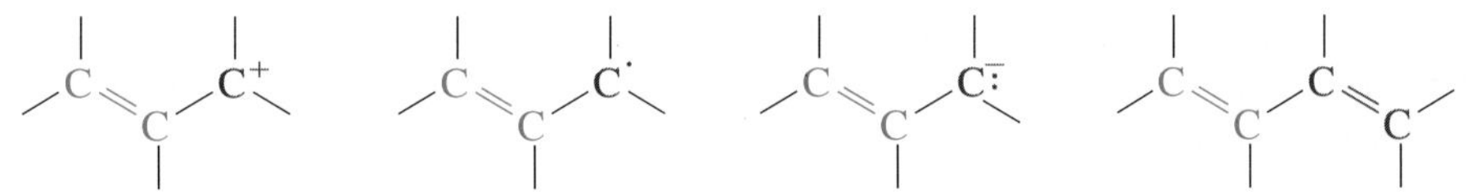

Allylic carbocation   Allylic free radical   Allylic anion   Conjugated diene

Conjugare is a Latin verb meaning "to link or yoke together," and allylic carbocations, allylic free radicals, allylic anions, and conjugated dienes are all examples of **conjugated systems.** In this chapter we'll see how conjugation permits two functional units within a molecule to display a kind of reactivity that is qualitatively different from that of either unit alone.

Allyl is derived from the botanical name for garlic (*Allium sativum*). It was found in 1892 that the major component obtained by distilling garlic oil is $H_2C{=}CHCH_2SSCH_2CH{=}CH_2$, and the word *allyl* was coined for the $H_2C{=}CHCH_2—$ group on the basis of this origin.

10.1 The Allyl Group

The group $H_2C{=}CHCH_2—$ is known as *allyl,* which is both a common name and a permissible IUPAC name. It is most often encountered in functionally substituted derivatives, and compounds containing this group are much better known by their functional class IUPAC names than by their substitutive ones:

$H_2C{=}CHCH_2OH$	$H_2C{=}CHCH_2Cl$	$H_2C{=}CHCH_2Br$
Allyl alcohol (2-propen-1-ol)	Allyl chloride (3-chloro-1-propene)	Allyl bromide (3-bromo-1-propene)

The sp^3-hybridized carbon of a C=C—C unit is classified as an **allylic** carbon and atoms or groups attached to it are *allylic substituents.* The two sp^2-hybridized carbons are **vinylic** and atoms or groups attached to them are *vinylic substituents.*

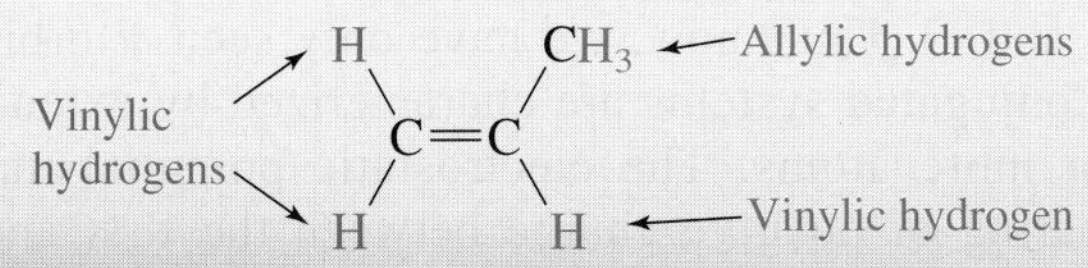
Allylic hydrogens
Vinylic hydrogens
Vinylic hydrogen

Allylic is also a general term for molecules that have a functional group at an allylic position. Thus, the following compounds represent an *allylic alcohol* and an *allylic chloride,* respectively.

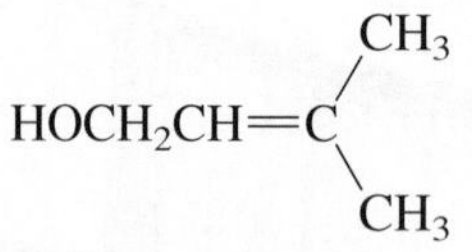

3-Methyl-2-buten-1-ol
(an allylic alcohol)

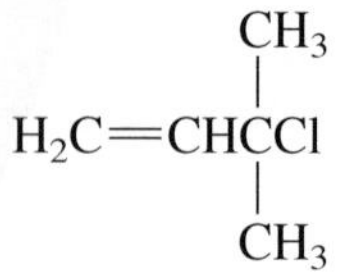

3-Chloro-3-methyl-1-butene
(an allylic chloride)

10.2 Allylic Carbocations

Allylic carbocations are carbocations in which an allylic carbon bears the positive charge. Allyl cation is the simplest allylic carbocation.

Allyl cation

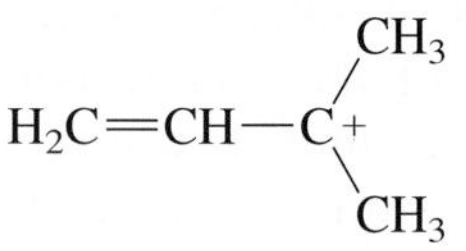

1,1-Dimethylallyl cation

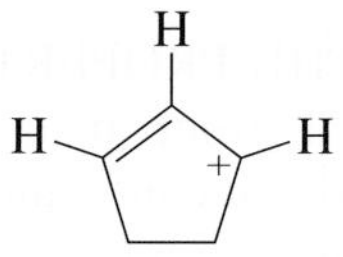

2-Cyclopentenyl cation

Allylic carbocations are more stable than simple alkyl cations because the C=C group acts as an electron-donating substituent to the positively charged carbon. We can represent this electron donation in resonance terms as:

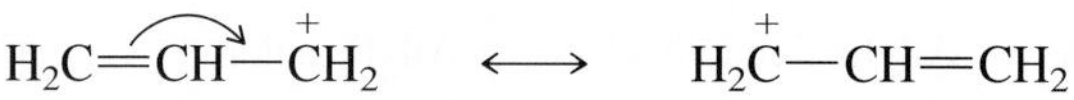

The double bond is *conjugated* to the positively charged carbon. Instead of being localized on a single carbon, the positive charge is shared by the carbons at each end of the three-carbon allyl unit. Likewise, the electrons in the π bond are delocalized over three carbons instead of two. The two resonance forms of allyl cation are equivalent, and the positive charge is shared equally by the carbons at each end.

Another way to show charge dispersal and π-electron delocalization in allylic systems is by a dashed-line representation in the resonance hybrid:

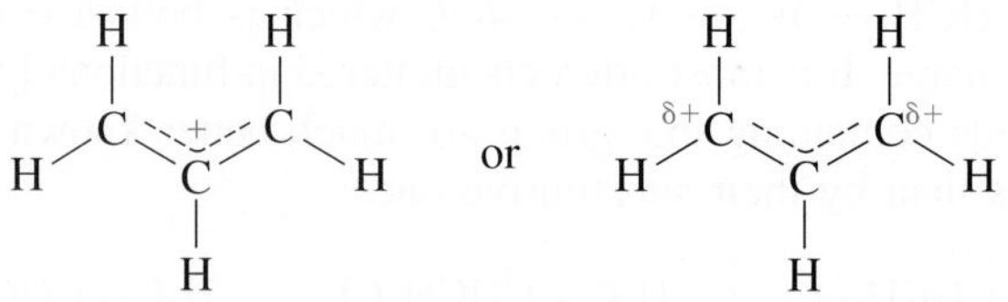

It is important to recognize that the center carbon does not bear a positive charge and the + sign above the middle of the dashed line is meant only to signify that the allylic unit as a group is positively charged.

Figure 10.1 gives an orbital overlap view of bonding in allyl cation. The planar structure of $H_2C{=}CHCH_2^+$ (Figure 10.1*a*) provides a framework that allows for continuous overlap of the 2*p* orbitals of the three adjacent sp^2-hybridized carbons (Figure 10.1*b* and *c*). Until now, we have only seen π orbitals involving two adjacent carbons. Conjugated systems are characterized by extended π orbitals that encompass three or more atoms. The electrostatic potential map (Figure 10.1*d*) shows the equal sharing of positive charge between the first and third carbons of $H_2C{=}CHCH_2^+$.

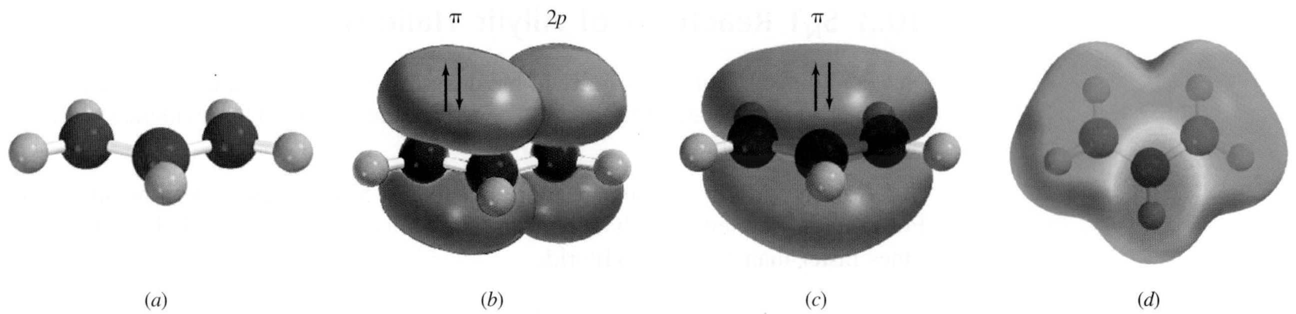
π
2p
π
(a)
(b)
(c)
(d)

Figure 10.1

Bonding in allyl cation. (*a*) All of the atoms of $H_2C{=}CHCH_2^+$ lie in the same plane. Each carbon is sp^2-hybridized. (*b*) The alignment of the π component of the double bond and the vacant *p* orbital permits overlap between them. (*c*) A π orbital encompasses all three carbons of $H_2C{=}CHCH_2^+$. The two electrons in this orbital are delocalized over three carbons. (*d*) An electrostatic potential map shows the positive charge to be shared equally by the two end carbons.

In contrast to the symmetrical charge distribution in the parent allyl cation, unequal charge distributions can result when substituents are present. Consider the methyl-substituted allylic ion represented by the following two resonance structures.

Major contributor (positive charge on secondary carbon) — Contributes less (positive charge on primary carbon)

The major contributor is the one that has its positive charge on a secondary carbon and more closely depicts the charge distribution in the delocalized ion. The electrostatic potential map of the delocalized ion in Figure 10.2 shows greater positive charge in the region surrounding this carbon. The regiochemical results of unequal charge distribution in substituted allylic carbocations will be seen in Sections 10.3 and 10.13.

Problem 10.1

Write a second resonance structure for each of the following carbocations. Is the charge shared equally by both allylic carbons? If not, which one bears more of the charge?

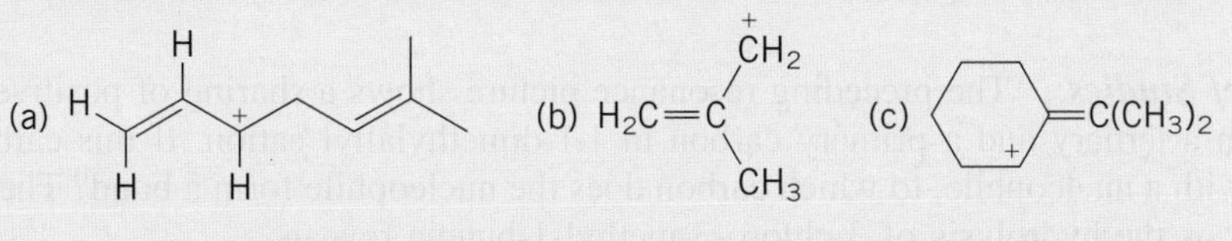
(a)
(b) H2C=C
CH2
CH3
(c)
=C(CH3)2

Sample Solution (a) First, identify the allylic unit by picking out the $C{=}C{-}C^+$ sequence. Of the two double bonds in this structure, only the one at the left is part of $C{=}C{-}C^+$. The double bond at the right is separated from the positively charged carbon by a CH_2 group, so is not conjugated to it. Move electrons in pairs from the double bond toward the positively charged carbon to generate a second resonance structure.

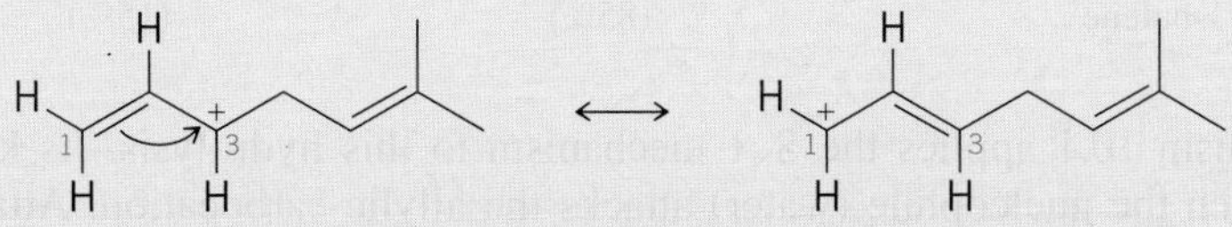

The two contributing structures are not equivalent; therefore, the positive charge is not shared equally between C-1 and C-3. C-1 is a primary carbon, C-3 is secondary. More of the positive charge resides on C-3 than on C-1. The original structure (left) contributes more to the resonance hybrid than the other (right).

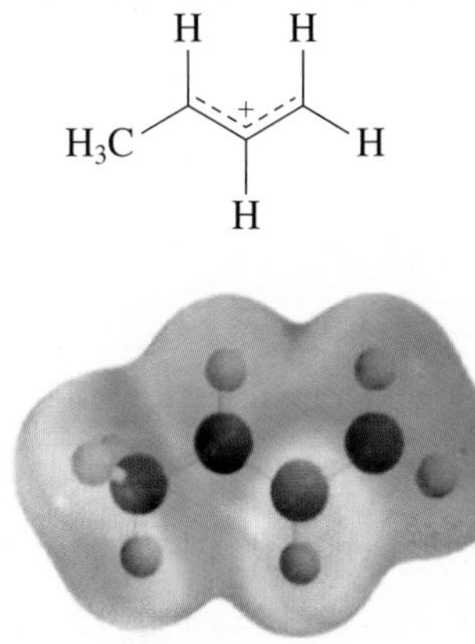

Figure 10.2

The electrostatic potential map (bottom) shows the unequal charge distribution in a methyl-substituted allyl cation. The region surrounding the more positive allylic carbon is more blue; that surrounding the less positive one is more green.

10.3 S_N1 Reactions of Allylic Halides

Nucleophilic substitutions of allylic halides provide much of the experimental evidence on which our understanding of allylic carbocations rests. This evidence includes both rate and product studies.

S_N1 Rates: Allylic halides are more reactive in S_N1 processes than nonallylic halides. For example, the tertiary allylic chloride shown undergoes solvolysis in ethanol over 100 times faster than *tert*-butyl chloride.

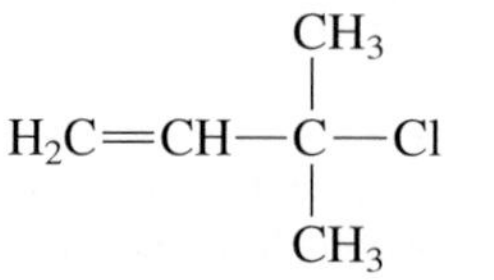

3-Chloro-3-methyl-1-butene
More reactive: k(rel) 123

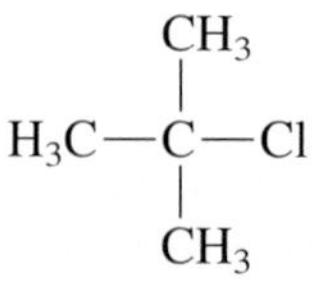

tert-Butyl chloride
Less reactive: k(rel) 1.0

Both chlorides react by an S_N1 mechanism, and their relative rates reflect their activation energies for carbocation formation. Because the allylic chloride is more reactive, we reason that it ionizes more rapidly because it forms a more stable carbocation.

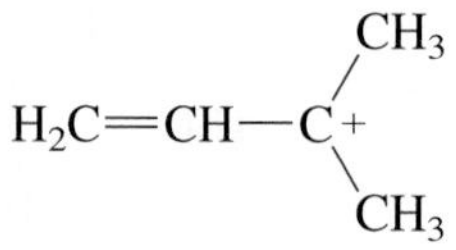

1,1-Dimethylallyl cation
(more stable)

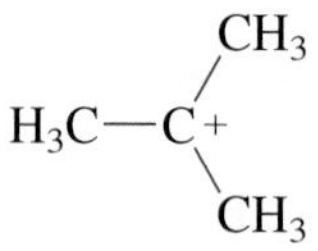

tert-Butyl cation
(less stable)

The greater stability of 1,1-dimethylallyl cation tells us that a vinyl group ($H_2C{=}CH{-}$) is a better electron-releasing substituent than methyl ($H_3C{-}$). This greater stability can be attributed to allylic resonance.

A rule of thumb is that a C═C substituent stabilizes a carbocation about as well as two methyl groups. Although allyl cation ($H_2C{=}CHCH_2^+$) is a primary carbocation, it is about as stable as a typical secondary carbocation such as isopropyl cation, $(CH_3)_2CH^+$.

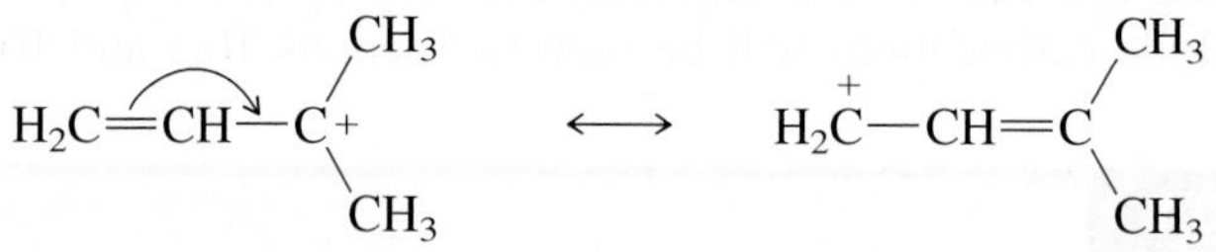

positive charge on tertiary carbon; major contributor

Product Studies: The preceding resonance picture shows a sharing of positive charge between a tertiary and a primary carbon in 1,1-dimethylallyl cation. If this carbocation reacts with a nucleophile, to which carbon does the nucleophile form a bond? The answer is *both,* as the hydrolysis of 3-chloro-3-methyl-1-butene reveals.

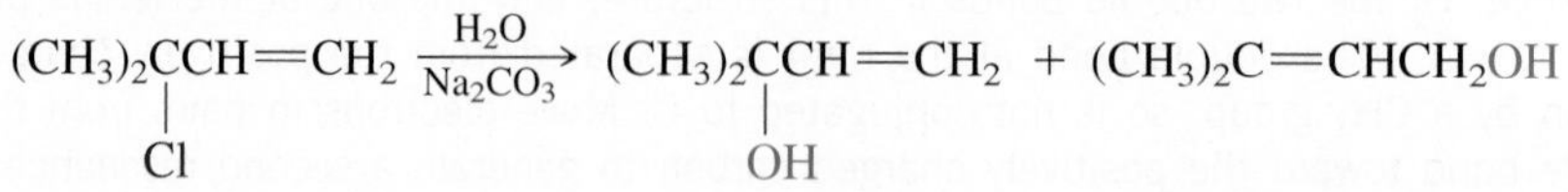

3-Chloro-3-methyl-1-butene → 2-Methyl-3-buten-2-ol (85%) + 3-Methyl-2-buten-1-ol (15%)

Mechanism 10.1 applies the S_N1 mechanism to this hydrolysis. Its key feature is step 2 in which the nucleophile (water) attacks the allylic carbocation. Attack occurs at both allylic carbons, but at different rates. The major product, a tertiary alcohol, results from attack by water at the tertiary carbon. The minor product, a primary alcohol, comes from attack by water at the primary carbon.

The oxygen of water bonds to the carbon that bears more of the positive charge; therefore, the tertiary alcohol is the major product.

Mechanism 10.1

Hydrolysis of an Allylic Halide (3-Chloro-3-methyl-1-butene)

THE OVERALL REACTION:

$$(CH_3)_2C(Cl)CH{=}CH_2 + H_2O \longrightarrow (CH_3)_2C(OH)CH{=}CH_2 + (CH_3)_2C{=}CHCH_2OH$$

3-Chloro-3-methyl-1-butene | Water | 2-Methyl-3-buten-2-ol | 3-Methyl-2-buten-1-ol

Step 1: The alkyl halide ionizes to give a carbocation. This step is rate-determining.

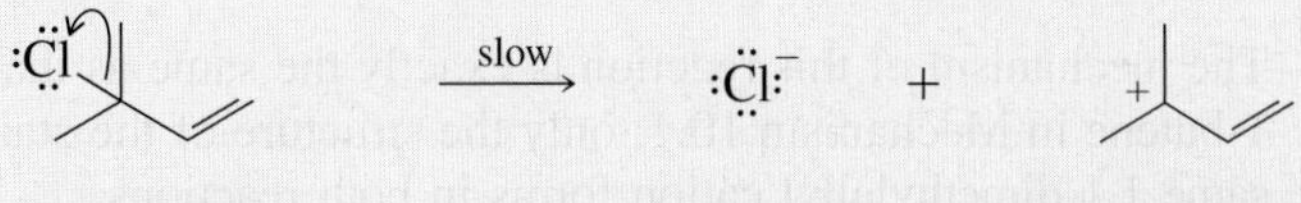

3-Chloro-3-methyl-1-butene | Chloride ion | 1,1-Dimethylallyl cation

The positive charge in the 1,1-dimethylallyl cation is shared between two allylic carbons.

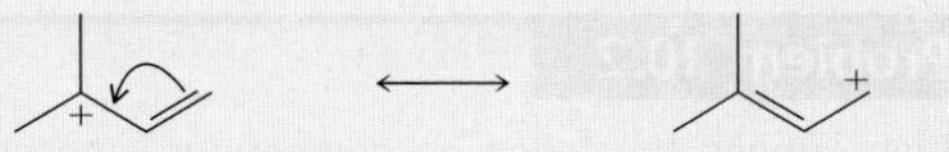

Major contributor | Minor contributor

Step 2: The carbocation (shown in its most stable resonance form) reacts with water. Water acts as a nucleophile; its oxygen can bond to either the tertiary carbon (a) or the primary carbon (b).

(a)

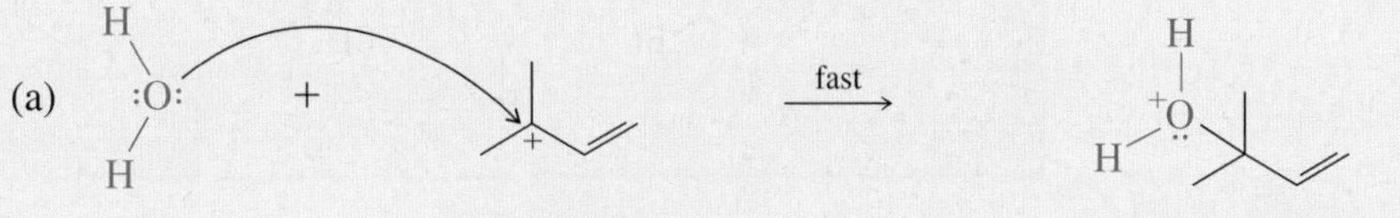

Water | 1,1-Dimethylallyl cation (major contributor) | 1,1-Dimethylallyloxonium ion

(b)

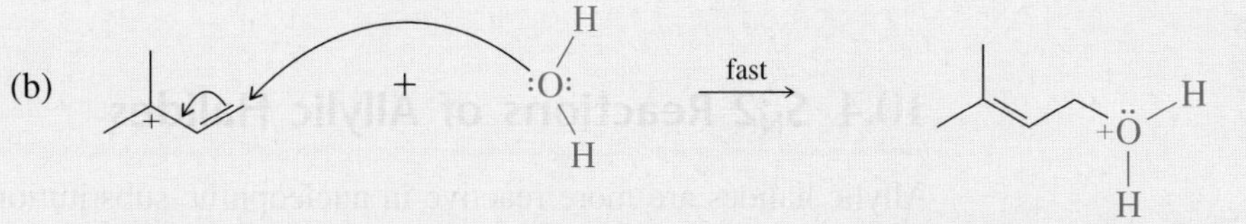

1,1-Dimethylallyl cation (major contributor) | Water | 3,3-Dimethylallyloxonium ion

Step 3: The alkyloxonium ions formed in step 2 are converted to the corresponding alcohols by proton transfer. Water is the proton acceptor.

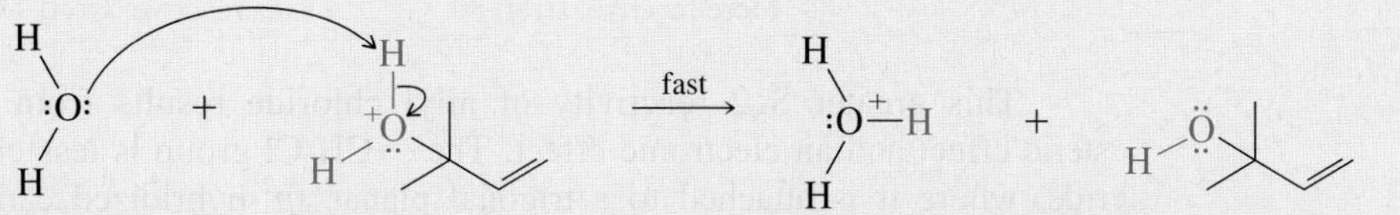

Water | 1,1-Dimethylallyloxonium ion | Hydronium ion | 2-Methyl-3-buten-2-ol

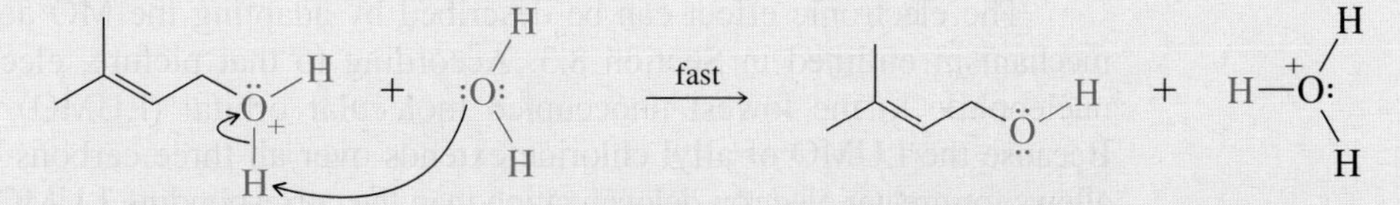

3,3-Dimethylallyloxonium ion | Water | 3-Methyl-2-buten-1-ol | Hydronium ion

Be sure you understand that we are not dealing with an equilibrium between two isomeric carbocations. *There is only one carbocation.* It has a delocalized structure, so is not adequately represented by a single Lewis formula.

The same carbocation intermediate, and the same two alcohols, are formed in the hydrolysis of 1-chloro-3-methyl-2-butene:

$$(CH_3)_2C{=}CHCH_2Cl \xrightarrow[Na_2CO_3]{H_2O} (CH_3)_2\underset{\displaystyle OH}{\underset{|}{C}}CH{=}CH_2 + (CH_3)_2C{=}CHCH_2OH$$

1-Chloro-3-methyl-2-butene	2-Methyl-3-buten-2-ol (85%)	3-Methyl-2-buten-1-ol (15%)

The mechanism of this reaction is exactly the same as that shown for 3-chloro-3-methyl-1-butene in Mechanism 10.1; only the structure of the starting material has changed. The same 1,1-dimethylallyl cation forms in both reactions.

Reactions of allylic systems that yield products in which double-bond migration has occurred are said to have proceeded with **allylic rearrangement.**

Problem 10.2

From among the following compounds, choose the two that yield the same carbocation on ionization.

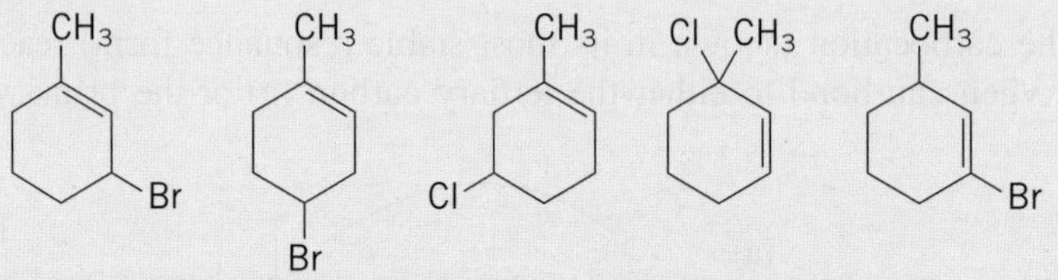

Later in this chapter we'll see how allylic carbocations are involved in electrophilic addition to dienes and how the principles developed in this section apply there as well.

10.4 S_N2 Reactions of Allylic Halides

Allylic halides are more reactive in nucleophilic substitutions than nonallylic ones, even in reactions that don't involve carbocations. In a typical S_N2 reaction, allyl chloride reacts with potassium iodide in acetone 80 times faster than 1-chloropropane.

$H_2C{=}CHCH_2{-}Cl$	$CH_3CH_2CH_2{-}Cl$
Allyl chloride	1-Chloropropane
More reactive: *k*(rel) 80	Less reactive: *k*(rel) 1.0

This greater S_N2 reactivity of allyl chloride results from a combination of a steric effect and an electronic effect. The $-CH_2Cl$ group is less crowded in allyl chloride, where it is attached to a trigonal planar sp^2-hybridized carbon of $H_2C{=}CH-$, than in 1-chloropropane, where it is attached a tetrahedral sp^3-hybridized carbon of CH_3CH_2-.

The electronic effect can be described by adapting the MO description of the S_N2 mechanism outlined in Section 8.3. According to that picture, electrons flow from the nucleophile to the lowest unoccupied molecular orbital (LUMO) of the alkyl halide. Because the LUMO of allyl chloride extends over all three carbons of the allyl group, it allows for greater electron delocalization than the corresponding LUMO of 1-chloropropane. A lower energy LUMO translates to a lower activation energy and a faster rate of nucleophilic substitution.

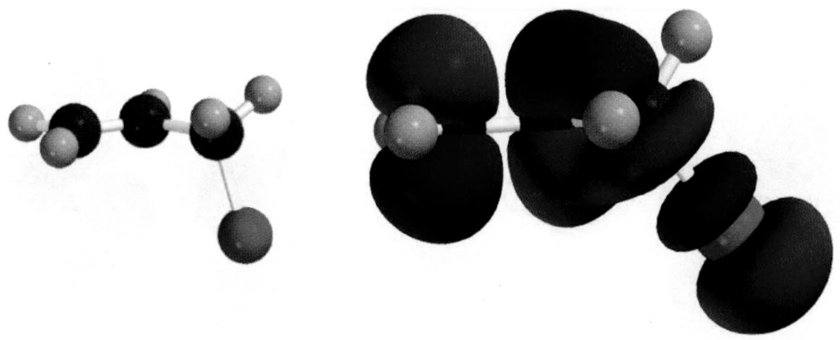

Allyl chloride LUMO of allyl chloride

Typical S_N2 displacements occur when primary allylic halides react with good nucleophiles.

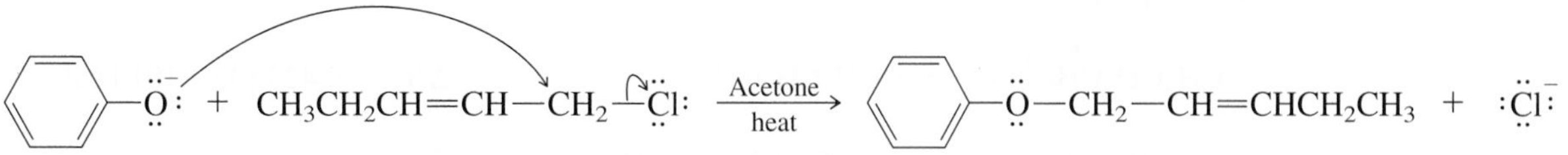

Phenoxide ion 1-Chloro-2-pentene 2-Pentenyl phenyl ether (57%) Chloride ion

With secondary and tertiary allylic halides or under solvolysis conditions, S_N1 reactions can compete with S_N2, and a mixture of direct displacement and allylic rearrangement products results.

Problem 10.3

3-Chloro-1-pentene was treated with phenoxide ion as shown for 1-chloro-2-pentene in the preceding equation. Two $C_{11}H_{14}O$ constitutional isomers were obtained. Suggest a reasonable structure for each isomer.

Phenoxide ion is about 6 p*K* units less basic than ethoxide. Unlike ethoxide, phenoxide reacts with secondary alkyl halides by substitution, not elimination.

10.5 Allylic Free Radicals

Just as allyl cation is stabilized by electron delocalization, so is allyl radical:

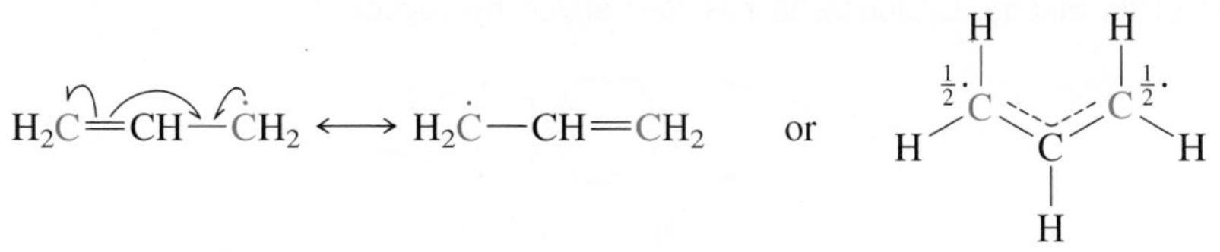

Allyl radical

Allyl radical is a conjugated system in which three electrons are delocalized over three carbons. The resonance structures indicate that the unpaired electron has an equal probability of being found at C-1 or C-3. C-2 shares none of the unpaired electron.

Figure 10.3 shows electron delocalization in allyl radical by mapping its spin density (see Section 4.17). In Figure 10.3*a* we see that, except for a small amount of spin density on the hydrogen at C-2, C-1 and C-3 share equally all of the spin density of allyl radical. Looking at allyl radical from a different direction in Figure 10.3*b* shows that the region of space where the spin density is greatest corresponds to the 2*p* orbitals of C-1 and C-3 that are part of the allylic π electron system.

Stabilization of allylic radicals by delocalization of the unpaired electron causes reactions that generate them to proceed more readily than those that give simple alkyl

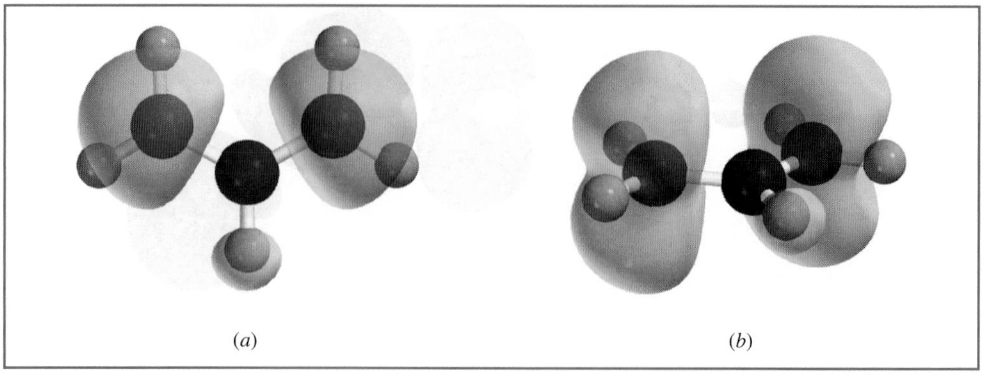

Figure 10.3

(*a*) The spin density (yellow) in allyl radical is equally divided between the two allylic carbons. There is a much smaller spin density at the C-2 hydrogen. (*b*) The odd electron is in an orbital that is part of the allylic π system.

radicals. Compare, for example, the bond dissociation enthalpies of the primary C—H bonds of propane and propene:

$$\underset{\text{Propane}}{CH_3CH_2CH_2\text{—}H} \longrightarrow \underset{\text{Propyl radical}}{CH_3CH_2\dot{C}H_2} + \underset{\text{Hydrogen atom}}{\cdot H} \qquad \Delta H^\circ = +423\text{ kJ }(+101\text{ kcal})$$

$$\underset{\text{Propene}}{H_2C\text{=}CHCH_2\text{—}H} \longrightarrow \underset{\text{Allyl radical}}{H_2C\text{=}CH\dot{C}H_2} + \underset{\text{Hydrogen atom}}{\cdot H} \qquad \Delta H^\circ = +368\text{ kJ }(+88\text{ kcal})$$

Breaking a bond to a primary hydrogen in propene requires less energy, by 55 kJ/mol (13 kcal/mol), than in propane. The free radical produced from propene is allylic and stabilized by electron delocalization; the one from propane is not.

Problem 10.4

Identify the allylic hydrogens in

(a) Cyclohexene
(b) 1-Methylcyclohexene
(c) 2,3,3-Trimethyl-1-butene
(d) 1-Octene

Sample Solution (a) Allylic hydrogens are bonded to an allylic carbon. An allylic carbon is an sp^3-hybridized carbon that is attached directly to an sp^2-hybridized carbon of an alkene. Cyclohexene has four allylic hydrogens.

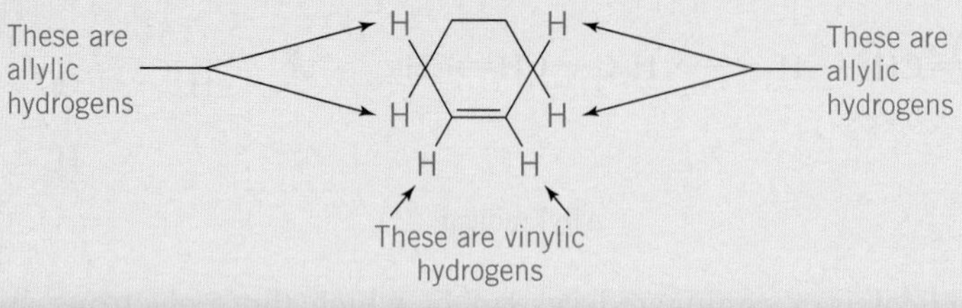

10.6 Allylic Halogenation

Of the reactions that involve carbon radicals, the most familiar are the chlorination and bromination of alkanes (Sections 4.15 through 4.19):

$$\underset{\text{Alkane}}{RH} + \underset{\text{Halogen}}{X_2} \xrightarrow[\text{light}]{\text{heat or}} \underset{\text{Alkyl halide}}{RX} + \underset{\text{Hydrogen halide}}{HX}$$

Although alkenes typically react with chlorine and bromine by *addition* to the double bond at room temperature and below (Section 6.14), *substitution* becomes competitive at higher temperatures, especially when the concentration of the halogen is low. When substitution does occur, it is highly selective for the allylic position. This forms the basis of an industrial preparation of allyl chloride:

$$\underset{\text{Propene}}{H_2C{=}CHCH_3} + \underset{\text{Chlorine}}{Cl_2} \xrightarrow{500°C} \underset{\substack{\text{Allyl chloride}\\(80\text{–}85\%)}}{H_2C{=}CHCH_2Cl} + \underset{\text{Hydrogen chloride}}{HCl}$$

Halogenation proceeds by the free-radical chain mechanism shown in Mechanism 10.2.

Allylic brominations are normally carried out using one of a number of specialized reagents developed for that purpose. *N*-Bromosuccinimide (NBS) is the most used. An alkene is dissolved in carbon tetrachloride, *N*-bromosuccinimide is added, and the reaction mixture is heated, illuminated with a sunlamp, or both. Small amounts of peroxides are sometimes added as free-radical initiators. The products are an allylic halide and succinimide.

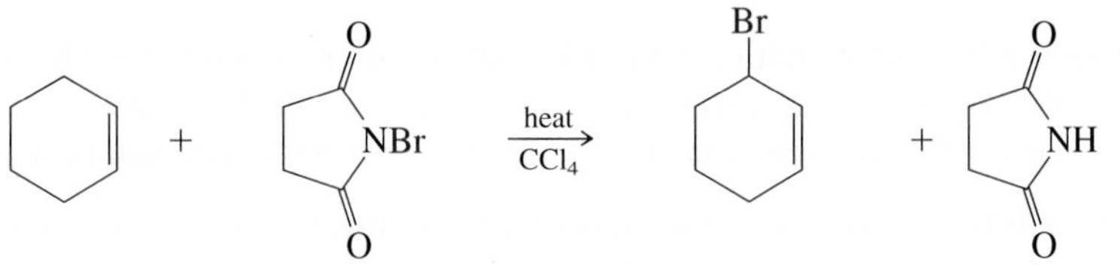

Cyclohexene + *N*-Bromosuccinimide (NBS) → 3-Bromocyclohexene (82–87%) + Succinimide

N-Bromosuccinimide will be seen again as a reagent for selective bromination in Section 11.11.

N-Bromosuccinimide provides a low concentration of molecular bromine, which reacts with alkenes by a mechanism analogous to that of other free-radical halogenations. The

Mechanism 10.2

Allylic Chlorination of Propene

THE OVERALL REACTION:

$$\underset{\text{Propene}}{H_2C{=}CHCH_3} + \underset{\text{Chlorine}}{Cl_2} \xrightarrow{500°C} \underset{\text{Allyl chloride}}{H_2C{=}CHCH_2Cl} + \underset{\text{Hydrogen chloride}}{HCl}$$

Initiation step: A chlorine molecule dissociates to two atoms.

$$\underset{\text{Chlorine}}{:\ddot{Cl}{-}\ddot{Cl}:} \longrightarrow \underset{\text{Chlorine atoms}}{:\ddot{Cl}\cdot + \cdot\ddot{Cl}:}$$

Propagation steps: In the first propagation step a chlorine atom abstracts a hydrogen atom from the allylic carbon of propene forming allyl radical.

$$\underset{\text{Propene}}{H_2C{=}CHCH_2{-}H} + \underset{\text{Chlorine atom}}{\cdot\ddot{Cl}:} \longrightarrow \underset{\text{Allyl radical}}{H_2C{=}CH\dot{C}H_2} + \underset{\text{Hydrogen chloride}}{H{-}\ddot{Cl}:}$$

The allyl radical formed in the first propagation step reacts with Cl_2 to form allyl chloride.

$$\underset{\text{Allyl radical}}{H_2C{=}CH\dot{C}H_2} + \underset{\text{Chlorine}}{:\ddot{Cl}{-}\ddot{Cl}:} \longrightarrow \underset{\text{Allyl chloride}}{H_2C{=}CHCH_2{-}\ddot{Cl}:} + \underset{\text{Chlorine atom}}{\cdot\ddot{Cl}:}$$

The chlorine atom generated in this propagation step then abstracts a hydrogen atom from another molecule of propene and the two propagation steps repeat over and over again.

reaction that produces bromine from *N*-bromosuccinimide involves the hydrogen bromide formed during free-radical bromination.

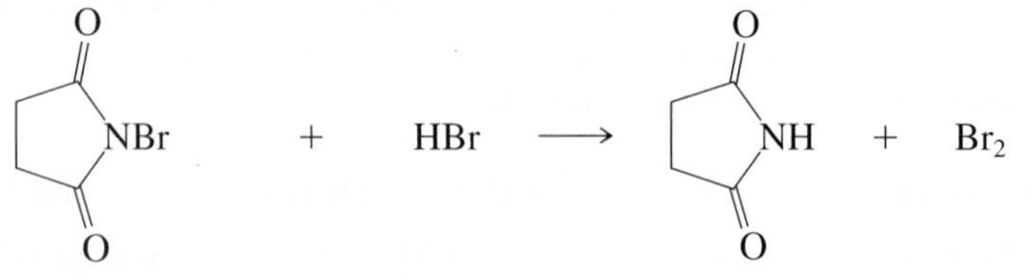

N-Bromosuccinimide + Hydrogen bromide ⟶ Succinimide + Bromine

Problem 10.5

Assume that *N*-bromosuccinimide serves as a source of Br_2, and write equations for the propagation steps in the formation of 3-bromocyclohexene by allylic bromination of cyclohexene.

Although allylic bromination and chlorination offer methods for attaching a reactive functional group to a hydrocarbon framework, we need to be aware of two important limitations. For allylic halogenation to be effective in a particular synthesis:

1. All the allylic hydrogens in the starting alkene must be equivalent, and
2. Both resonance forms of the allylic radical must be equivalent.

In the two examples cited so far, the chlorination of propene and the bromination of cyclohexene, both requirements are met.

All the allylic hydrogens of propene are equivalent. $H_2C{=}CH{-}CH_3$

The two resonance forms of allyl radical are equivalent. $H_2C{=}CH{-}\dot{C}H_2 \longleftrightarrow H_2\dot{C}{-}CH{=}CH_2$

All the allylic hydrogens of cyclohexene are equivalent.

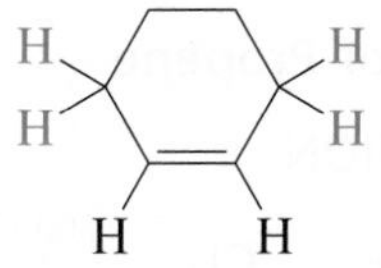

The two resonance forms of 2-cyclohexenyl radical are equivalent.

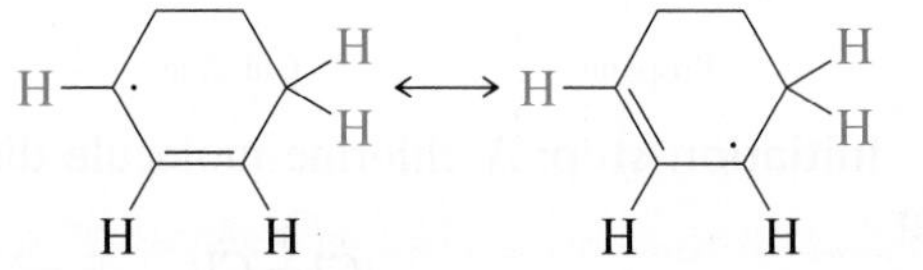

Unless both criteria are met, mixtures of constitutionally isomeric allylic halides result. The resonance forms of the allylic radical intermediate in the bromination of 1-octene, for example, are not equivalent and give both 3-bromo-1-octene and 1-bromo-2-octene, the latter as a mixture of cis and trans isomers.

$$H_2C{=}CHCH_2(CH_2)_4CH_3 \xrightarrow[\text{heat, } CCl_4]{\text{NBS, benzoyl peroxide}} H_2C{=}CHCH(Br)(CH_2)_4CH_3 + BrCH_2CH{=}CH(CH_2)_4CH_3$$

1-Octene → 3-Bromo-1-octene (17%) + 1-Bromo-2-octene (83%) (*cis*+*trans*)

via: $H_2C{=}CH{-}\dot{C}H(CH_2)_4CH_3 \longleftrightarrow H_2\dot{C}{-}CH{=}CH(CH_2)_4CH_3$

Problem 10.6

Evaluate 2,3,3-trimethyl-1-butene as a candidate for free-radical bromination. How many allylic bromides would you expect to result from its treatment with *N*-bromosuccinimide?

10.7 Allylic Anions

Allyl anion is the parent of the third allylic species we will examine. Like allyl cation and allyl radical, allyl anion is planar and stabilized by electron delocalization. In this case, the unshared pair on the negatively charged carbon plus the two π electrons of the double bond are shared by the three carbons of the allyl unit.

Allyl anion

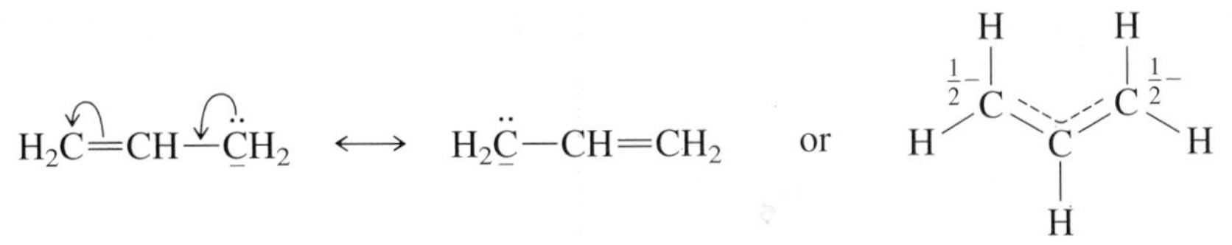

Allyl anion is the conjugate base of propene; thus, the extent to which electron delocalization stabilizes allyl anion can be assessed by comparing the pK_a of propene to the alkane propane. The allylic hydrogens of propene are much more acidic than the hydrogens of propane, none of which are allylic.

$$H_2C{=}CH{-}CH_3 \quad pK_a \approx 43 \qquad CH_3CH_2{-}CH_3 \quad pK_a \approx 62$$

Propene   Propane

Therefore, allyl anion $H_2C{=}CH{-}\ddot{C}H_2$ is a weaker base and holds its unshared electron pair more strongly than propyl anion $CH_3CH_2\ddot{C}H_2$.

The electrostatic potential maps in Figure 10.4 illustrate the contrast between the dispersed negative charge in $H_2C{=}CH{-}\ddot{C}H_2$ and the localized charge in $CH_3CH_2\ddot{C}H_2$.

In addition to the resonance stabilization of allyl anion, part of the enhanced acidity of propene can be assigned to an inductive effect. The CH_3 group in $H_2C{=}CH{-}CH_3$ is attached to an sp^2-hybridized carbon, which is more electron-attracting and acid-strengthening than its sp^3-hybridized counterpart in $CH_3CH_2{-}CH_3$.

Problem 10.7

The pK_a of $CH_3CH{=}CHCH{=}CH_2$ has been estimated to be about 35. Give the structure of its conjugate base and use resonance to show the sharing of negative charge among the various carbons.

Allylic anions are almost always associated with some metal cation partner and the combination belongs to a class called *organometallic compounds.* Organometallic compounds have a large number of synthetic applications to be explored in Chapter 14.

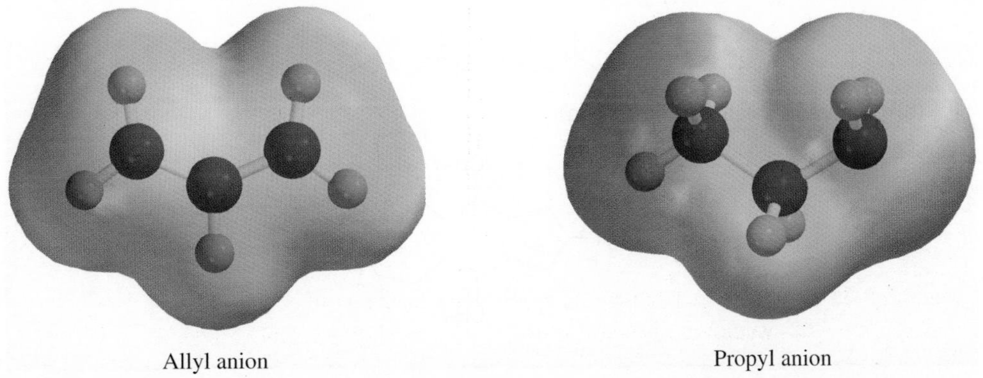

Figure 10.4

Electrostatic potential maps for allyl and propyl anion. The charge is dispersed in allyl and shared equally by C-1 and C-3. The charge is localized at C-1 in propyl. The color scale is the same for both maps.

10.8 Classes of Dienes

Allylic carbocations, radicals, and anions are conjugated systems involved as reactive intermediates in chemical reactions. The next type of conjugated system that we will examine, conjugated dienes, consists of stable molecules.

A hydrocarbon that contains two double bonds is called an **alkadiene,** and the relationship between the double bonds is described as *isolated, conjugated,* or *cumulated.* **Isolated diene** units are those in which two carbon–carbon double bond units are separated from each other by one or more sp^3-hybridized carbon atoms. 1,4-Pentadiene and 1,5-cyclooctadiene have isolated double bonds:

$$H_2C{=}CHCH_2CH{=}CH_2$$

1,4-Pentadiene
(Penta-1,4-diene)

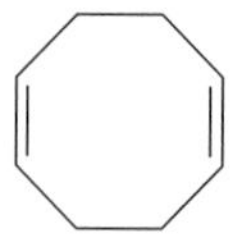

1,5-Cyclooctadiene
(Cycloocta-1,5-diene)

Conjugated dienes are those in which two carbon–carbon double bond units are connected to each other by a single bond. 1,3-Pentadiene and 1,3-cyclooctadiene contain conjugated double bonds:

$$H_2C{=}CH{-}CH{=}CHCH_3$$

1,3-Pentadiene
(Penta-1,3-diene)

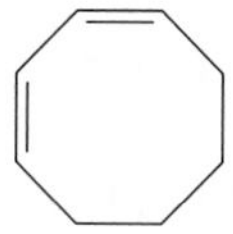

1,3-Cyclooctadiene
(Cycloocta-1,3-diene)

Problem 10.8

Are the double bonds in 1,4-cyclooctadiene isolated or conjugated?

Allene is an acceptable IUPAC name for 1,2-propadiene.

Cumulated dienes are those in which one carbon atom is common to two carbon–carbon double bonds. The simplest cumulated diene is 1,2-propadiene, also called **allene,** and compounds of this class are generally referred to as *allenes.*

$$H_2C{=}C{=}CH_2$$

1,2-Propadiene
(Propa-1,2-diene)

Problem 10.9

Many naturally occurring substances contain several carbon–carbon double bonds: some isolated, some conjugated, and some cumulated. Identify the types of carbon–carbon double bonds found in each of the following substances:

(a) β-Springene (a scent substance from the dorsal gland of springboks)

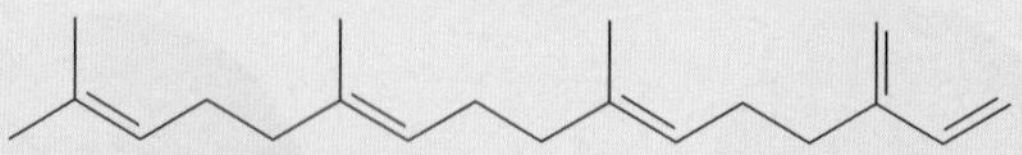

(b) Cembrene (occurs in pine resin)

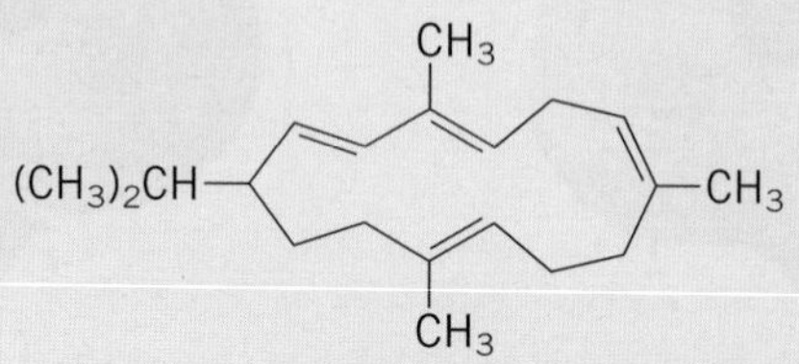

(c) The sex attractant of the male dried-bean beetle

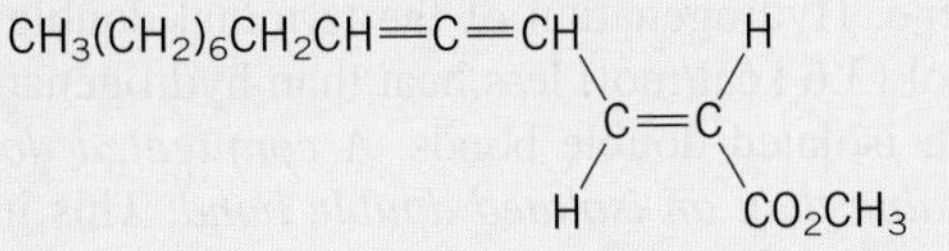

Sample Solution (a) β-Springene has three isolated double bonds and a pair of conjugated double bonds:

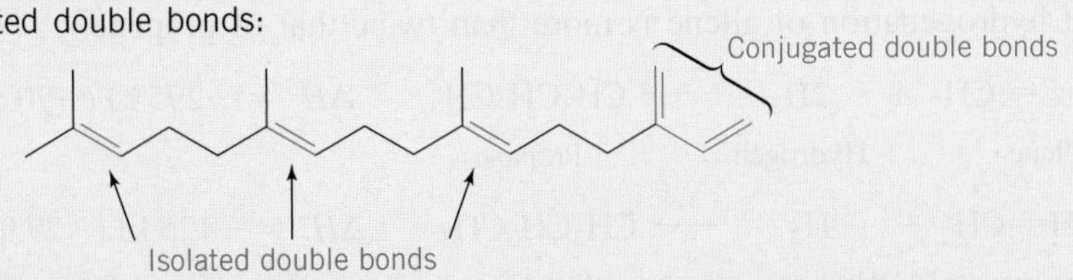

Isolated double bonds are separated from other double bonds by at least one sp^3-hybridized carbon. Conjugated double bonds are joined by a single bond.

Alkadienes are named according to the IUPAC rules by replacing the *-ane* ending of an alkane with *-adiene* and locating the position of each double bond by number. Compounds with three carbon–carbon double bonds are called *alkatrienes* and named accordingly, those with four double bonds are *alkatetraenes,* and so on.

10.9 Relative Stabilities of Dienes

Which is the most stable arrangement of double bonds in an alkadiene—isolated, conjugated, or cumulated?

$H_2C{=}CH{-}CH_2{-}CH{=}CH_2$	$H_2C{=}CH{-}CH{=}CH{-}CH_3$	$H_3C{-}CH{=}C{=}CH{-}CH_3$
1,4-Pentadiene (isolated)	1,3-Pentadiene (conjugated)	2,3-Pentadiene (cumulated)

As we saw in Chapter 6, the stabilities of alkenes may be assessed by comparing their heats of hydrogenation. Figure 10.5 shows the heats of hydrogenation of an isolated diene (1,4-pentadiene) and a conjugated diene (1,3-pentadiene), along with the alkenes 1-pentene and (*E*)-2-pentene. The figure shows that an isolated pair of double bonds behaves much like two independent alkene units. The measured heat of hydrogenation of the two double bonds in 1,4-pentadiene is 252 kJ/mol (60.2 kcal/mol), exactly twice the heat of hydrogenation of 1-pentene. Furthermore, the heat evolved on hydrogenation of each double bond must be 126 kJ/mol (30.1 kcal/mol) because 1-pentene is an intermediate in the hydrogenation of 1,4-pentadiene to pentane.

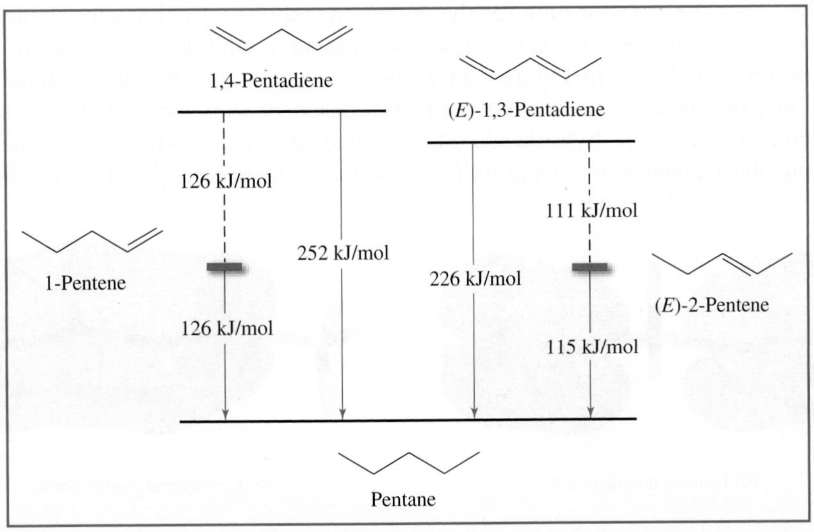

Figure 10.5

Heats of hydrogenation are used to assess the stabilities of isolated versus conjugated double bonds. Comparing the measured heats of hydrogenation (solid lines) of the four compounds shown gives the values shown by the dashed lines for the heats of hydrogenation of the terminal double bond of 1,4-pentadiene and (*E*)-1,3-pentadiene. A conjugated double bond is approximately 15 kJ/mol more stable than an isolated double bond.

By the same reasoning, hydrogenation of the terminal double bond in the conjugated diene (*E*)-1,3-pentadiene releases only 111 kJ/mol (26.5 kcal/mol) when it is hydrogenated to (*E*)-2-pentene. Hydrogenation of the terminal double bond in the conjugated diene evolves 15 kJ/mol (3.6 kcal/mol) less heat than hydrogenation of a terminal double bond in the diene with isolated double bonds. *A conjugated double bond is 15 kJ/mol (3.6 kcal/mol) more stable than an isolated double bond.* This increased stability due to conjugation is the **delocalization energy, resonance energy,** or **conjugation energy.**

The cumulated double bonds of an allenic system are of relatively high energy. The heat of hydrogenation of allene is more than twice that of propene.

$$\underset{\text{Allene}}{H_2C{=}C{=}CH_2} + \underset{\text{Hydrogen}}{2H_2} \longrightarrow \underset{\text{Propane}}{CH_3CH_2CH_3} \qquad \Delta H° = -295\ \text{kJ}\ (-70.5\ \text{kcal})$$

$$\underset{\text{Propene}}{CH_3CH{=}CH_2} + \underset{\text{Hydrogen}}{H_2} \longrightarrow \underset{\text{Propane}}{CH_3CH_2CH_3} \qquad \Delta H° = -125\ \text{kJ}\ (-29.9\ \text{kcal})$$

Problem 10.10

Another way in which energies of isomers may be compared is by their heats of combustion. Match the heat of combustion with the appropriate diene.

Dienes: 1,2-Pentadiene, (*E*)-1,3-pentadiene, 1,4-pentadiene

Heats of combustion: 3186 kJ/mol, 3217 kJ/mol, 3251 kJ/mol

Thus, the order of alkadiene stability decreases in the order: conjugated diene (most stable) → isolated diene → cumulated diene (least stable). To understand this ranking, we need to look at structure and bonding in alkadienes in more detail.

10.10 Bonding in Conjugated Dienes

At 146 pm the C-2—C-3 distance in 1,3-butadiene is relatively short for a carbon–carbon single bond. This is most reasonably seen as a hybridization effect. In ethane both carbons are sp^3-hybridized and are separated by a distance of 153 pm. The carbon–carbon single bond in propene unites sp^3- and sp^2-hybridized carbons and is shorter than that of ethane. Both C-2 and C-3 are sp^2-hybridized in 1,3-butadiene, and a decrease in bond distance between them reflects the tendency of carbon to attract electrons more strongly as its *s* character increases.

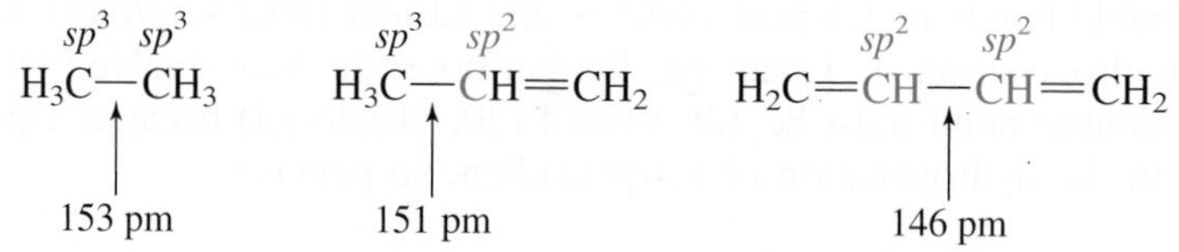
sp3 sp3
H3C—CH3
153 pm
sp3 sp2
H3C—CH=CH2
151 pm
sp2 sp2
H2C=CH—CH=CH2
146 pm

The factor most responsible for the increased stability of conjugated double bonds is the greater delocalization of their π electrons compared with the π electrons of isolated double bonds. As shown in Figure 10.6*a*, the π electrons of an isolated diene system occupy, in pairs, two noninteracting π orbitals. Each of these π orbitals encompasses two carbon atoms. An sp^3-hybridized carbon isolates the two π orbitals from each other, preventing the exchange of electrons between them. In a conjugated diene, however,

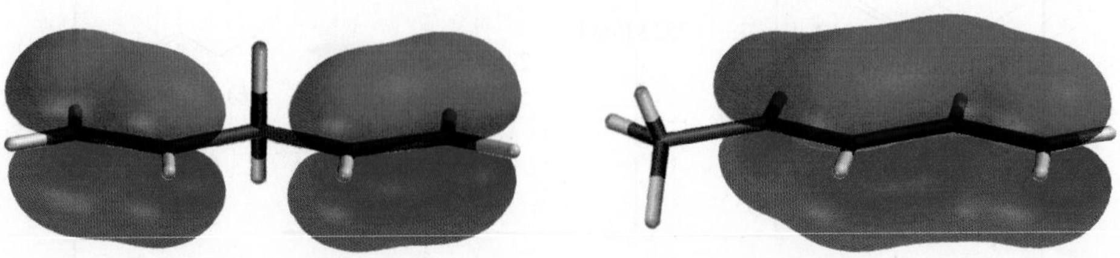

(*a*) Isolated double bonds (*b*) Conjugated double bonds

Figure 10.6

(*a*) Isolated double bonds are separated from one another by one or more sp^3-hybridized carbons and cannot overlap to give an extended π orbital. (*b*) In a conjugated diene, overlap of two π orbitals gives an extended π system encompassing four carbon atoms.

mutual overlap of the two π orbitals, represented in Figure 10.6*b*, gives an orbital system in which each π electron is delocalized over four carbon atoms. Delocalization of electrons lowers their energy and gives a more stable molecule.

Additional evidence for electron delocalization in 1,3-butadiene can be obtained by considering its conformations. Overlap of the two π electron systems is optimal when the four carbon atoms are coplanar. Two conformations allow this coplanarity: they are called the *s*-cis and *s*-trans conformations.

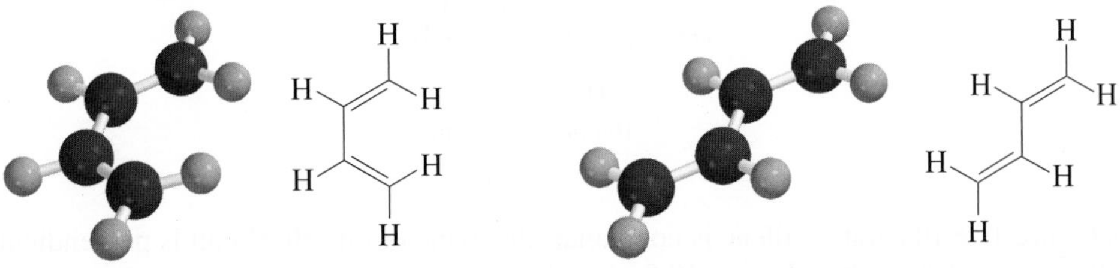

s-Cis conformation of 1,3-butadiene *s*-Trans conformation of 1,3-butadiene

The letter *s* in *s*-cis and *s*-trans refers to conformations around the C—C single bond in the diene. The *s*-trans conformation of 1,3-butadiene is 12 kJ/mol (2.8 kcal/mol) more stable than the *s*-cis, which is destabilized by van der Waals strain between the hydrogens at C-1 and C-4.

The *s*-cis and *s*-trans conformations of 1,3-butadiene interconvert by rotation around the C-2—C-3 bond, as illustrated in Figure 10.7. The conformation at the midpoint of this rotation, the *perpendicular conformation,* has its 2*p* orbitals in a geometry that prevents extended conjugation. It has localized double bonds. This is an example of a stereoelectronic effect (Section 5.16); electron delocalization in a conjugated system is most effective when the interacting orbitals are suitably aligned. In the case of conjugated dienes, the most favorable alignment occurs when the axes of the *p* orbitals are parallel. The main contributor to the energy of activation for rotation about the single bond in 1,3-butadiene is the decrease in electron delocalization that accompanies conversion of the *s*-cis or *s*-trans conformation to the perpendicular conformation.

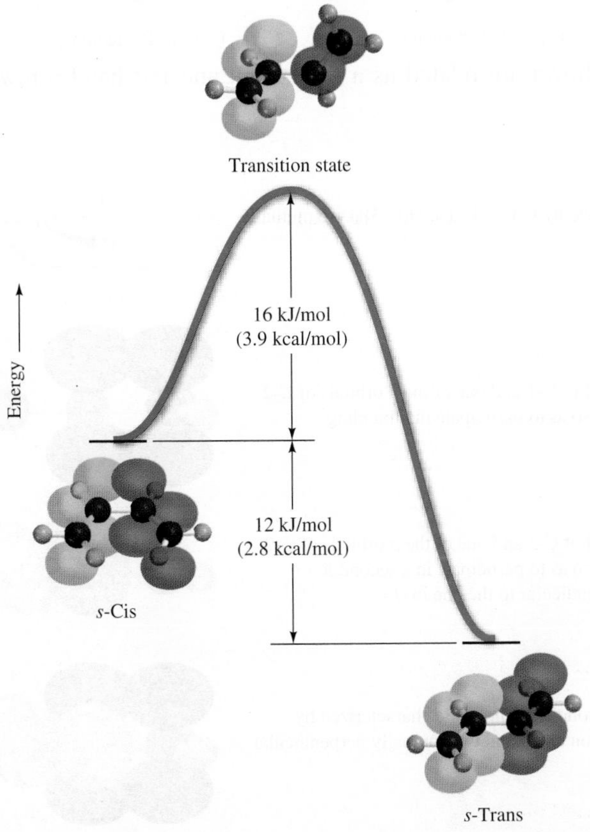
Transition state
16 kJ/mol
(3.9 kcal/mol)
Energy
12 kJ/mol
(2.8 kcal/mol)
s-Cis
s-Trans

Figure 10.7

Conformations and electron delocalization in 1,3-butadiene. The *s*-cis and the *s*-trans conformations permit the 2*p* orbitals to be aligned parallel to one another for maximum π electron delocalization. The *s*-trans conformation is more stable than the *s*-cis. Stabilization resulting from π electron delocalization decreases in going to the transition state for rotation about the C-2—C-3 single bond. The green and yellow colors are meant to differentiate the orbitals and do not indicate their phases.

10.11 Bonding in Allenes

The three carbons of allene lie in a straight line, with relatively short carbon–carbon bond distances of 131 pm. The middle carbon, because it has two π bonds, is *sp*-hybridized. The end carbons of allene are sp^2-hybridized.

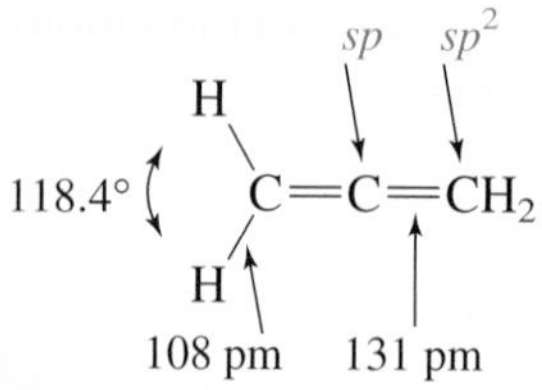
sp
sp²
H
118.4°
C=C=CH₂
H
108 pm
131 pm

Allene

As Figure 10.8 illustrates, allene is nonplanar; the plane of one HCH unit is perpendicular to the plane of the other. Figure 10.8 also shows the reason for this unusual geometry. The 2*p* orbital of each of the terminal carbons overlaps with a different 2*p* orbital of the central carbon. Because the 2*p* orbitals of the central carbon are perpendicular to each other, the perpendicular nature of the two HCH units follows naturally.

The nonplanarity of allenes has an interesting stereochemical consequence. 1,3-Disubstituted allenes are chiral; they are not superimposable on their mirror images. Even an allene as simple as 2,3-pentadiene ($CH_3CH{=}C{=}CHCH_3$) has been obtained as separate enantiomers.

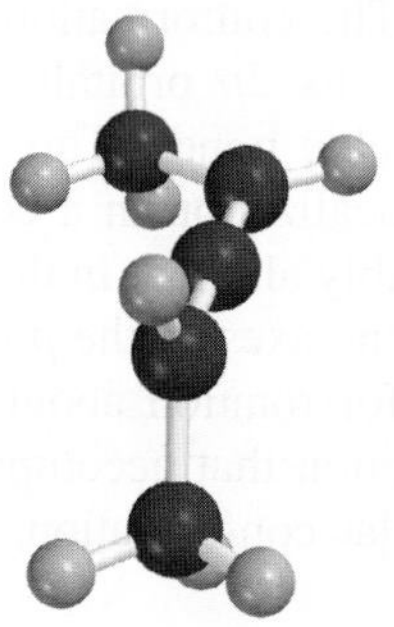

(+)-2,3-Pentadiene

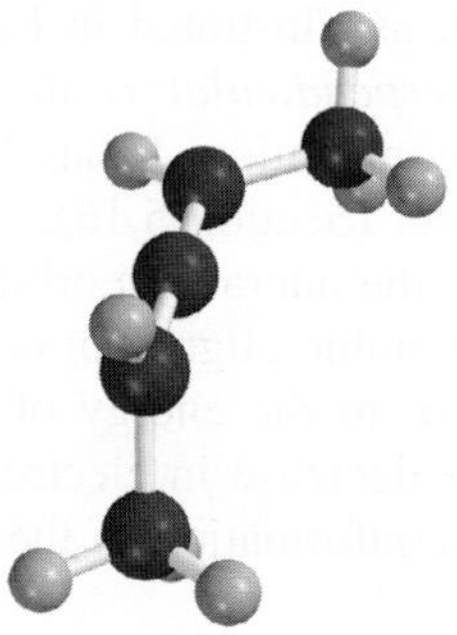

(−)-2,3-Pentadiene

The enantiomers shown are related as a right-hand and left-hand screw, respectively.

Figure 10.8

Bonding and geometry in 1,2-propadiene (allene). The green and yellow colors are meant to differentiate the orbitals and do not indicate their phases.

(*a*) Planes defined by H(C-1)H and H(C-3)H are mutually perpendicular.

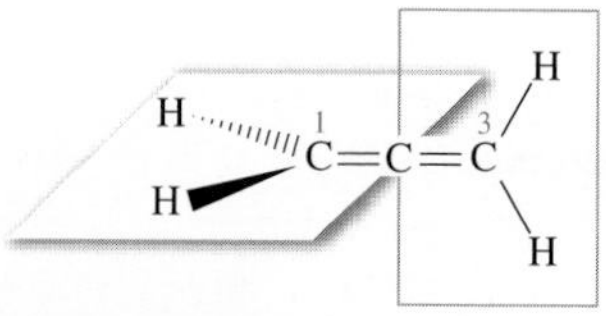
H
1
3
C=C=C
H
H
H

(*b*) The *p* orbital of C-1 and one of the *p* orbitals of C-2 can overlap so as to participate in π bonding.

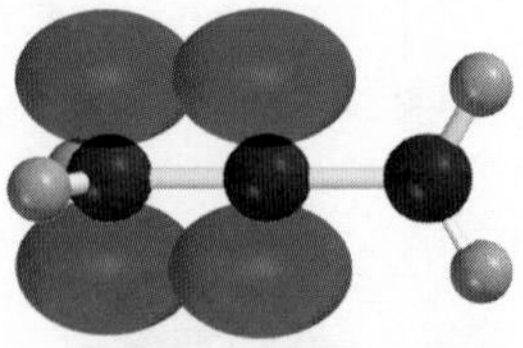

(*c*) The *p* orbital of C-3 and one of the *p* orbitals of C-2 can overlap so as to participate in a second π orbital perpendicular to the one in (*b*).

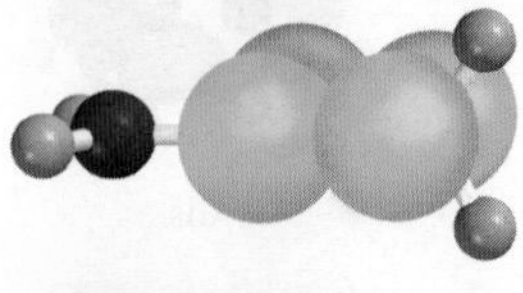

(*d*) Allene is a nonplanar molecule characterized by a linear carbon chain and two mutually perpendicular π bonds.

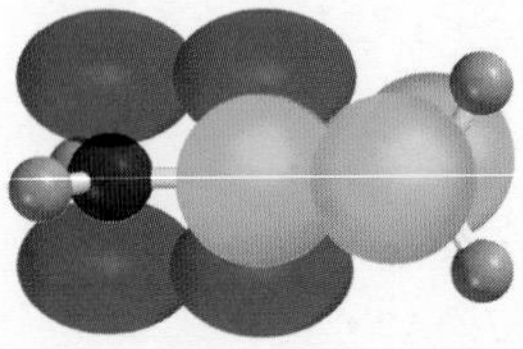

Chiral allenes are another example of molecules that are chiral, but that do not contain a chirality center. Like the chiral biaryl derivatives that were described in Section 7.9, chiral allenes contain an axis of chirality. The axis of chirality in 2,3-pentadiene is a line passing through the three carbons of the allene unit (carbons 2, 3, and 4).

The Cahn–Ingold–Prelog *R–S* notation has been extended to include molecules with a chirality axis. See the article by Mak in the November 2004 issue of the *Journal of Chemical Education* for a brief discussion of assigning *R* or *S* to chiral molecules that do not contain a chirality center.

Problem 10.11

Is 2-methyl-2,3-pentadiene chiral? What about 2-chloro-2,3-pentadiene?

10.12 Preparation of Dienes

The conjugated diene 1,3-butadiene is used in the manufacture of synthetic rubber for automobile tires and is prepared on an industrial scale in vast quantities. Production in the United States is currently 5×10^9 lb/year. One industrial process is similar to that used for the preparation of ethylene: In the presence of a suitable catalyst, butane undergoes thermal dehydrogenation to yield 1,3-butadiene.

$$CH_3CH_2CH_2CH_3 \xrightarrow[\text{chromia–alumina}]{590\text{–}675°C} H_2C{=}CHCH{=}CH_2 + 2H_2$$

The use of 1,3-butadiene in the preparation of synthetic rubber is discussed in the boxed essay *Diene Polymers* that appears on page 406.

Laboratory syntheses of conjugated dienes involve elimination reactions of unsaturated alcohols and alkyl halides. In the two examples that follow, the conjugated diene is produced in high yield even though an isolated diene is also possible.

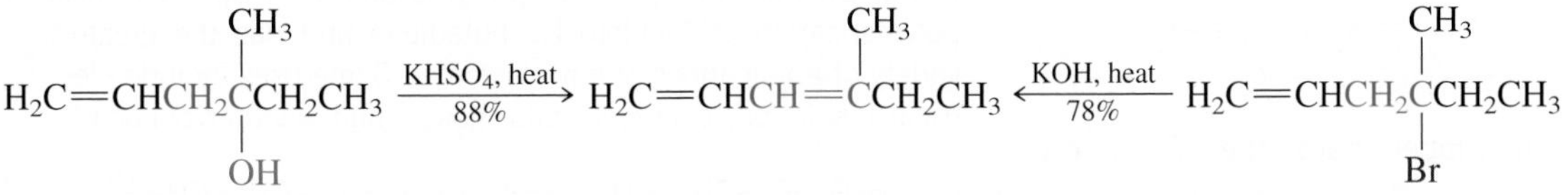
CH3
H2C=CHCH2CCH2CH3
OH
KHSO4, heat
88%
CH3
H2C=CHCH=CCH2CH3
KOH, heat
78%
CH3
H2C=CHCH2CCH2CH3
Br

3-Methyl-5-hexen-3-ol 4-Methyl-1,3-hexadiene 4-Bromo-4-methyl-1-hexene

As we saw in Chapter 5, dehydrations and dehydrohalogenations are typically regioselective in the direction that leads to the most stable double bond. Conjugated dienes are more stable than isolated dienes and are formed faster via a lower energy transition state.

Problem 10.12

What dienes containing isolated double bonds are capable of being formed, but are not observed, in the two preceding equations describing elimination in 3-methyl-5-hexen-3-ol and 4-bromo-4-methyl-1-hexene?

Dienes with isolated double bonds can be formed when the structure of the alkyl halide doesn't permit the formation of a conjugated diene.

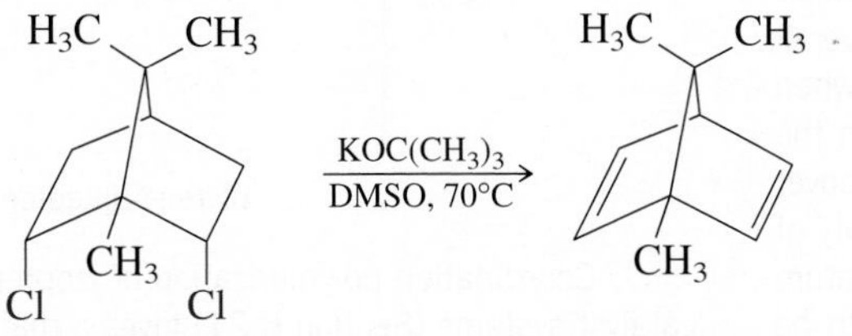
H3C
CH3
CH3
Cl
Cl
KOC(CH3)3
DMSO, 70°C
H3C
CH3
CH3

2,6-Dichlorocamphane Bornadiene (83%)

We will not discuss the preparation of allenes. They are prepared less readily than isolated or conjugated dienes and require special methods.

Diene Polymers

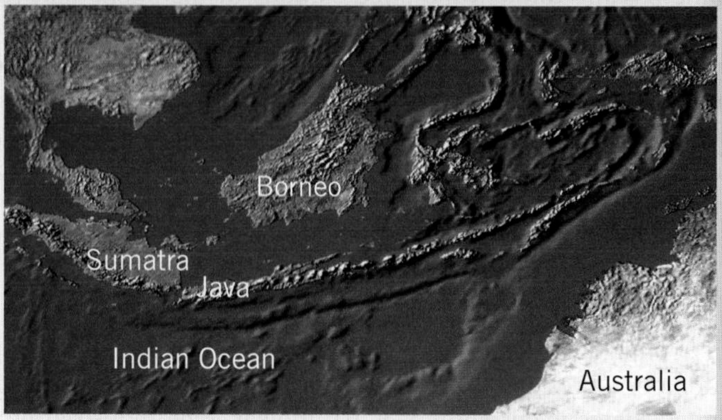
Borneo
Sumatra
Java
Indian Ocean
Australia

We begin with two trees, both cultivated on plantations in Southeast Asia. One, *Hevea brasiliensis,* is a source of natural rubber and was imported from Brazil in the nineteenth century. The other, *Isonandra gutta,* is native to Sumatra, Java, and Borneo and gives a latex from which gutta-percha is obtained.

Some 500 years ago during Columbus's second voyage to what are now the Americas, he and his crew saw children playing with balls made from the latex of trees that grew there. Later, Joseph Priestley called this material "rubber" to describe its ability to erase pencil marks by rubbing, and in 1823 Charles Macintosh demonstrated how rubber could be used to make waterproof coats and shoes. Shortly thereafter Michael Faraday determined an empirical formula of C_5H_8 for rubber. It was eventually determined that rubber is a polymer of 2-methyl-1,3-butadiene.

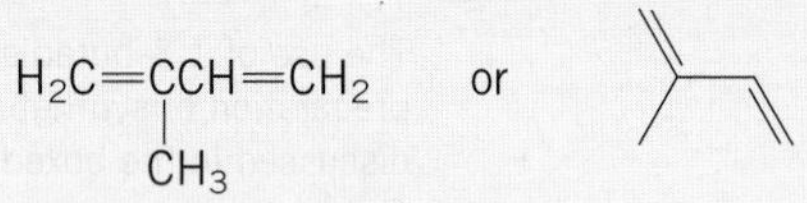
H2C=CCH=CH2
CH3
or

2-Methyl-1,3-butadiene (common name: *isoprene*)

The structure of rubber corresponds to 1,4 addition of several thousand isoprene units to one another:

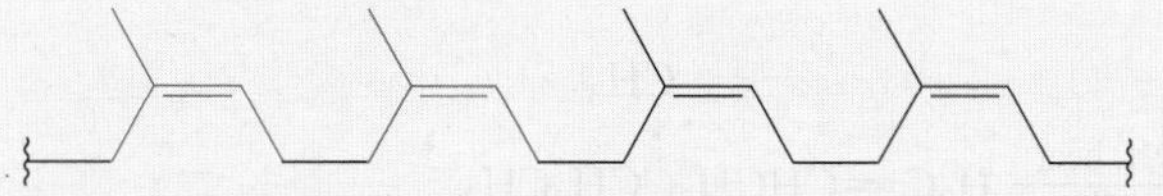

All the double bonds in rubber have the *Z* (or cis) configuration.

Gutta-percha is a different polymer of isoprene. Its chains are shorter than those of natural rubber and have *E* (or trans) double bonds.

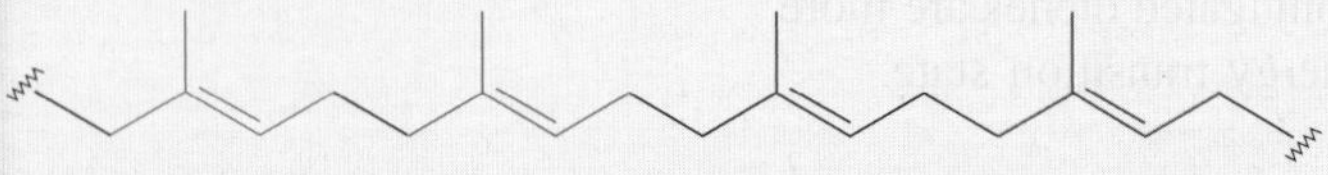

Gutta-percha is flexible when heated, but is harder and more durable than rubber at room temperature. It was, at one time, the material of choice for golf ball covers. Gutta-percha's main claim to fame though lies out of sight on the floors of the world's oceans. The first global communication network—the telegraph—relied on insulated copper wire to connect senders and receivers. Gutta-percha proved so superior to natural rubber in resisting deterioration, especially underwater, that it coated the thousands of miles of insulated telegraph cable that connected most of the countries of the world by the close of the nineteenth century.

In natural rubber the attractive forces between neighboring polymer chains are relatively weak, and there is little overall structural order. The chains slide easily past one another when stretched and return, in time, to their disordered state when the distorting force is removed. The ability of a substance to recover its original shape after distortion is its *elasticity.* The elasticity of natural rubber is satisfactory only within a limited temperature range; it is too rigid when cold and too sticky when warm to be very useful. Rubber's elasticity is improved by *vulcanization,* a process discovered by Charles Goodyear in 1839. When natural rubber is heated with sulfur, a chemical reaction occurs in which neighboring polyisoprene chains become connected through covalent bonds to sulfur. Although these sulfur "bridges" permit only limited movement of one chain with respect to another, their presence ensures that the rubber will snap back to its original shape once the distorting force is removed.

As the demand for rubber increased, so did the chemical industry's efforts to prepare a synthetic substitute. One of the first **elastomers** (a synthetic polymer that possesses elasticity) to find a commercial niche was *neoprene,* discovered by chemists at Du Pont in 1931. Neoprene is produced by free-radical polymerization of 2-chloro-1,3-butadiene and has the greatest variety of applications of any elastomer. Some uses include electrical insulation, conveyer belts, hoses, and weather balloons.

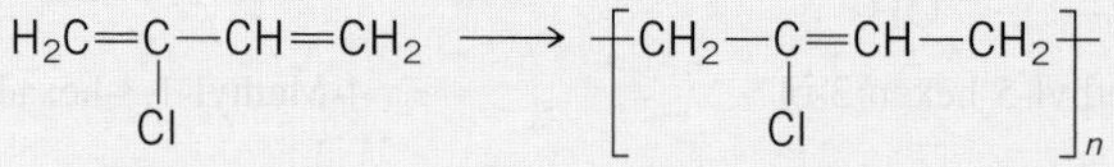
H2C=C—CH=CH2
Cl
CH2—C=CH—CH2
Cl
n

2-Chloro-1,3-butadiene Neoprene

The elastomer produced in greatest amount is *styrene-butadiene rubber* (SBR). Annually, just under 10^9 lb of SBR is produced in the United States, and almost all of it is used in automobile tires. As its name suggests, SBR is prepared from styrene and 1,3-butadiene. It is an example of a **copolymer,** a polymer assembled from two or more different monomers. Free-radical polymerization of a mixture of styrene and 1,3-butadiene gives SBR.

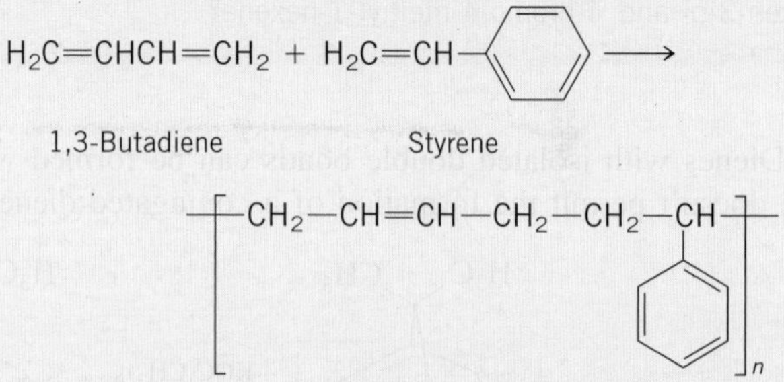
H2C=CHCH=CH2 + H2C=CH
1,3-Butadiene
Styrene
CH2—CH=CH—CH2—CH2—CH
n

Styrene-butadiene rubber

Coordination polymerization of isoprene using Ziegler–Natta catalyst systems (Section 6.21) gives a material similar in properties to natural rubber, as does polymerization of 1,3-butadiene. Poly(1,3-butadiene) is produced in about two thirds the quantity of SBR each year. It, too, finds its principal use in tires.

10.13 Addition of Hydrogen Halides to Conjugated Dienes

Our discussion of chemical reactions of alkadienes will be limited to those of conjugated dienes. The reactions of isolated dienes are essentially the same as those of individual alkenes. The reactions of allenes are—like their preparation—so specialized that their treatment is better suited to an advanced course in organic chemistry.

Electrophilic addition is the characteristic reaction of alkenes, and conjugated dienes undergo addition with the same electrophiles that react with alkenes, and by similar mechanisms. Hydrogen chloride, for example, adds to the diene unit of 1,3-cyclopentadiene to give 3-chlorocyclopentene. Mechanism 10.3 is analogous to the electrophilic addition of HCl to alkenes.

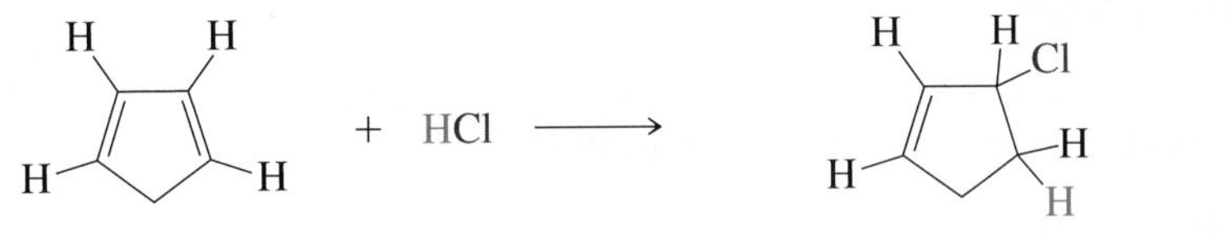

1,3-Cyclopentadiene 3-Chlorocyclopentene (70–90%)

As with alkenes, the regioselectivity of electrophilic addition to conjugated dienes is governed by the stability of the resulting carbocation. Protonation of a conjugated diene always occurs at the end of the diene unit because an allylic carbocation results.

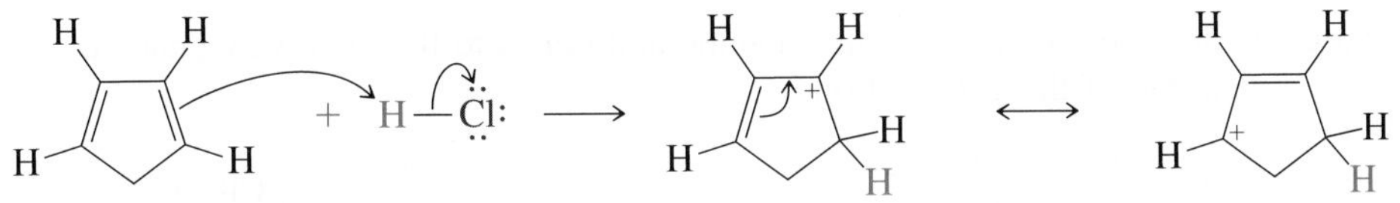

1,3-Cyclopentadiene Resonance forms of 2-cyclopentenyl cation

Problem 10.13

Carbons 1 and 4 of 1,3-cyclopentadiene are equivalent and give the same carbocation on protonation. Likewise, carbons 2 and 3 are equivalent. Write the structure of the carbocation formed by protonation of C-2 or C-3 to verify that it is not allylic and therefore not as stable as the one formed by protonation of C-1 or C-4.

Both resonance forms of the allylic carbocation from 1,3-cyclopentadiene are equivalent and so attack by chloride at either of the carbons that share the positive charge gives the same product, 3-chlorocyclopentene.

Such is not the case with 1,3-butadiene. Protonation of the diene is still regiospecific for the end carbon, but the two resonance forms of the resulting allylic carbocation are not equivalent.

$$H_2C{=}CH{-}CH{=}CH_2 \xrightarrow{H{-}Br:} H_2C{=}CH{-}\overset{+}{C}H{-}CH_3 \longleftrightarrow H_2\overset{+}{C}{-}CH{=}CH{-}CH_3$$

1,3-Butadiene Resonance forms of 1-methylallyl cation

Consequently, a mixture of two regioisomeric allylic bromides is formed when HBr adds to 1,3-butadiene.

$$H_2C{=}CH{-}CH{=}CH_2 \xrightarrow[-80°C]{HBr} H_2C{=}CH{-}\underset{Br}{\underset{|}{CH}}{-}CH_3 + BrCH_2{-}CH{=}CH{-}CH_3$$

1,3-Butadiene 3-Bromo-1-butene (81%) 1-Bromo-2-butene (19%)

Both products are formed from the same allylic carbocation. The major product corresponds to addition of a proton to C-1 of 1,3-butadiene and bromine to C-2. This mode of addition is called **1,2 addition** (also called **direct addition**). The minor product has

Mechanism 10.3

Addition of Hydrogen Chloride to 1,3-Cyclopentadiene

THE OVERALL REACTION:

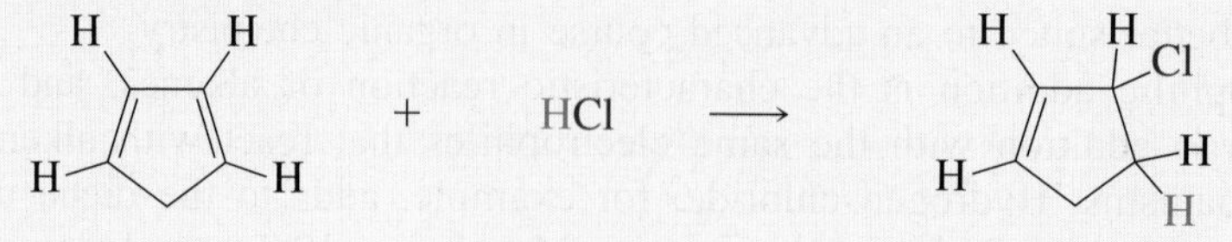

1,3-Cyclopentadiene Hydrogen chloride 3-Chlorocyclopentene

THE MECHANISM:

Step 1: A proton is transferred from HCl to a carbon at the end of the diene system to give an allylic carbocation.

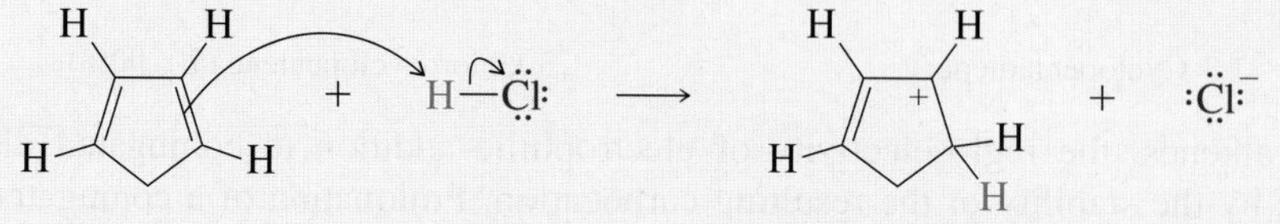

1,3-Cyclopentadiene Hydrogen chloride 2-Cyclopentenyl cation Chloride ion

Step 2: Chloride ion acts as a nucleophile and bonds to the positively charged carbon of the carbocation.

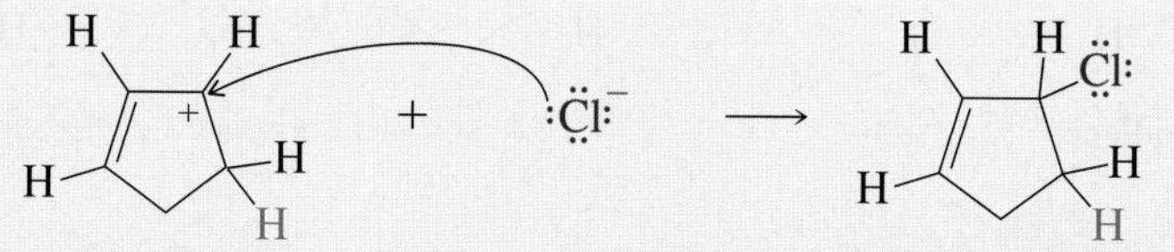

2-Cyclopentenyl cation Chloride ion 3-Chlorocyclopentene

its proton and bromide at C-1 and C-4, respectively, and is formed by **1,4 addition** (also called **conjugate addition**).

At −80°C the product from 1,2 addition predominates because it is formed faster than the 1,4-addition product. The product distribution is governed by **kinetic control.**

At room temperature, a much different product ratio is observed. Under these conditions the 1,4-addition product predominates.

$$H_2C{=}CH{-}CH{=}CH_2 \xrightarrow[\text{room temperature}]{HBr} H_2C{=}CH{-}\underset{\substack{|\\ Br}}{CH}{-}CH_3 + BrCH_2{-}CH{=}CH{-}CH_3$$

1,3-Butadiene 3-Bromo-1-butene (44%) (1,2 addition) 1-Bromo-2-butene (56%) (1,4 addition)

To understand why temperature affects the product composition, an important fact must be added. *The 1,2- and 1,4-addition products interconvert at elevated temperature in the presence of hydrogen bromide.*

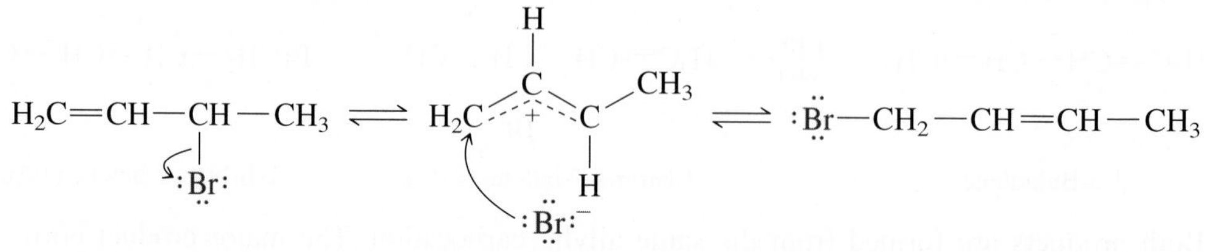

1,2-addition product
formed faster

1,4-addition product
more stable

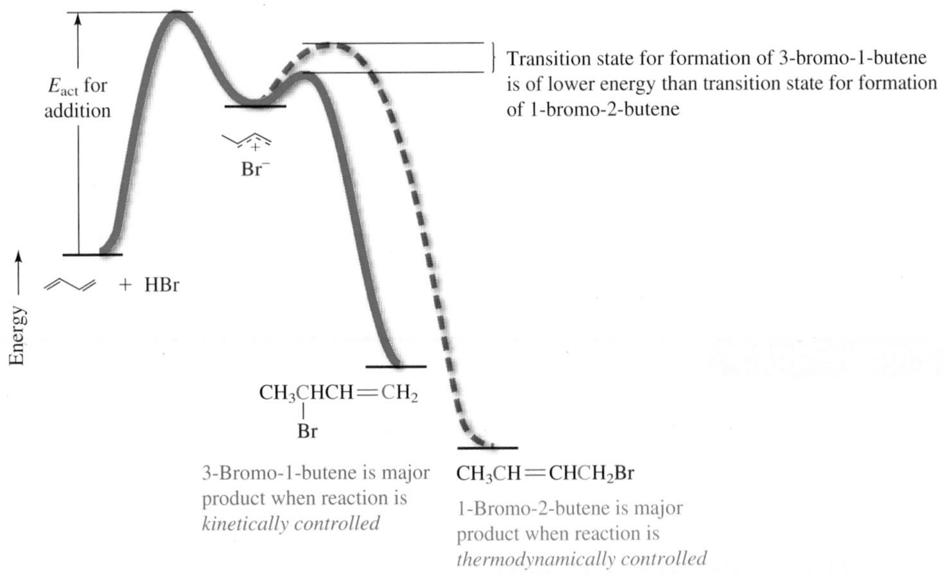

Figure 10.9

Energy diagram showing relationship of kinetic control to thermodynamic control in addition of hydrogen bromide to 1,3-butadiene.

At 45°C, for example, interconversion is rapid and gives an equilibrium mixture containing 85% of the 1,4-addition product and 15% of the 1,2 product. This demonstrates that the 1,4 product is more stable, presumably because it has a disubstituted double bond, whereas the double bond in the 1,2 product is monosubstituted.

When addition occurs under conditions in which the products can equilibrate, the composition of the reaction mixture no longer reflects the relative rates of formation of the products but tends to reflect their *relative stabilities.* Reactions of this type are governed by **thermodynamic control.**

The energy diagram of Figure 10.9 illustrates kinetic and thermodynamic control in the addition of hydrogen bromide to 1,3-butadiene. At low temperature, addition takes place irreversibly. Isomerization is slow because insufficient thermal energy is available to permit the products to surmount the energy barrier for ionization. At higher temperatures isomerization is possible, and the more stable product predominates.

Before leaving this section, we should point out that the numbers in the terms 1,2 and 1,4 addition refer to carbons within the C=C—C=C structural unit wherever it may be in the molecule and not to the IUPAC numbering. For example, 1,2 and 1,4 addition to 2,4-hexadiene would involve the carbons shown.

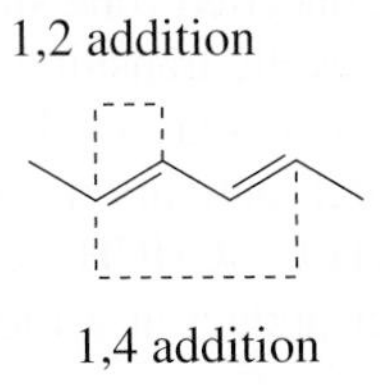

Problem 10.14

Write structural formulas for the products of 1,2 and 1,4 addition of hydrogen chloride to 2,4-hexadiene.

10.14 Halogen Addition to Dienes

Mixtures of 1,2- and 1,4-addition products are obtained when 1,3-butadiene reacts with chlorine or bromine.

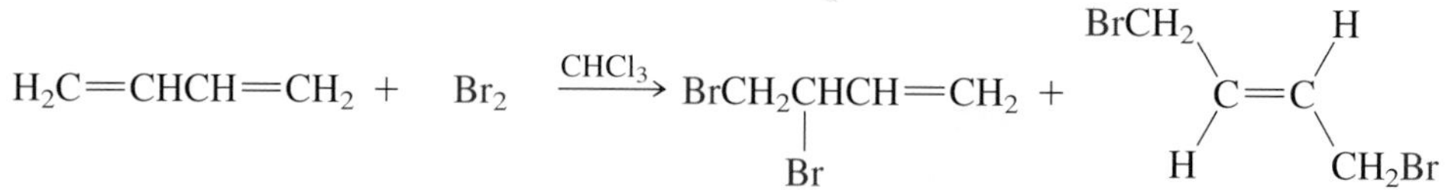
H₂C=CHCH=CH₂ + Br₂
CHCl₃
BrCH₂CHCH=CH₂ +
Br
BrCH₂
H
C=C
H
CH₂Br

1,3-Butadiene — Bromine — 3,4-Dibromo-1-butene (37%) — (*E*)-1,4-Dibromo-2-butene (63%)

The tendency for 1,4 addition is pronounced, and *E* double bonds are generated almost exclusively.

Problem 10.15

Exclusive of stereoisomers, how many products are possible in the electrophilic addition of 1 mol of bromine to 2-methyl-1,3-butadiene?

10.15 The Diels–Alder Reaction

We've already mentioned the value of carbon–carbon bond-forming reactions in organic synthesis. Imagine how useful it would be to have a reaction in which *two* carbon–carbon bonds are formed in a single operation simply by combining two compounds without having to add acids, bases, or other catalysts. For developing such a reaction, Otto Diels and Kurt Alder of the University of Kiel (Germany) shared the 1950 Nobel Prize in Chemistry. The **Diels–Alder reaction** is the conjugate addition of an alkene to a diene.

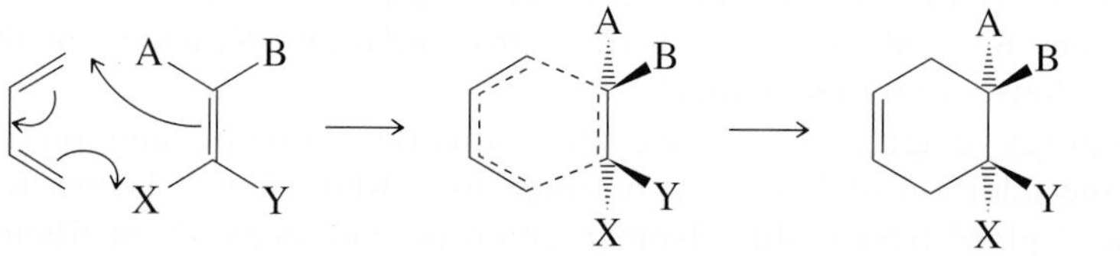
A
B
X
Y
A
B
Y
X
A
B
Y
X

Diene + dienophile — Transition state for cycloaddition — Diels–Alder adduct

The alkene that adds to the diene is called the **dienophile** ("diene seeker"). The reaction is classified as a **cycloaddition,** and the product contains a cyclohexene ring.

The reaction occurs in a single step, without an intermediate, by a mechanism in which six atoms undergo bonding changes in the same transition state by way of cyclic reorganization of their π electrons. Concerted (one-step) reactions such as the Diels–Alder cycloaddition that proceed through a cyclic transition state are called **pericyclic reactions.**

The simplest of all Diels–Alder reactions, cycloaddition of ethylene to 1,3-butadiene, does not proceed readily. It has a high activation energy and a low reaction rate. However, the cycloaddition of acrolein ($H_2C{=}CHCH{=}O$) to 1,3-butadiene occurs readily to give a high yield of the Diels–Alder adduct at a modest temperature.

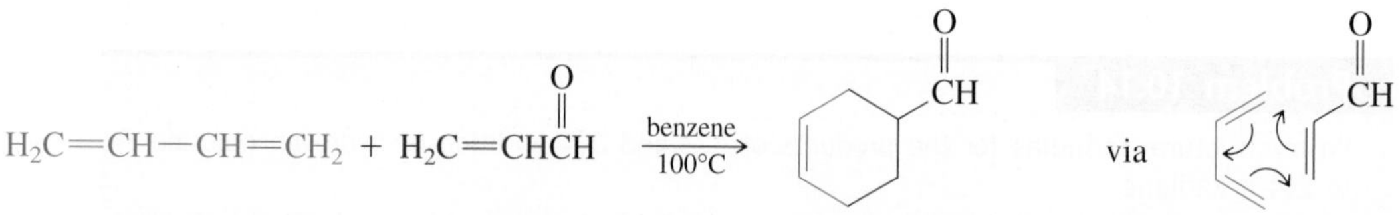
H₂C=CH—CH=CH₂ + H₂C=CHCH
O
benzene
100°C
O
CH
via
O
CH

1,3-Butadiene — Acrolein — Cyclohexene-4-carboxaldehyde (100%)

1. The most reactive dienophiles are typically those that contain a substituent such as a carbonyl ($>C{=}O$) or cyano ($-C{\equiv}N$) group attached directly to the double bond.
2. The diene, in this case 1,3-butadiene, must be able to adopt the *s*-cis conformation in order for the Diels–Alder reaction to occur.

Substituents such as carbonyl and cyano are electron-withdrawing and activate the dienophile toward cycloaddition. The electron-attracting properties of these groups arise from bond dipoles, which place a partial positive charge on the carbonyl or cyano carbon (see Section 1.5). Acrolein has one electron-withdrawing group attached to its dienophilic double bond. Diethyl fumarate and maleic anhydride, the most reactive and synthetically useful dienophiles you will encounter in this section, both have two.

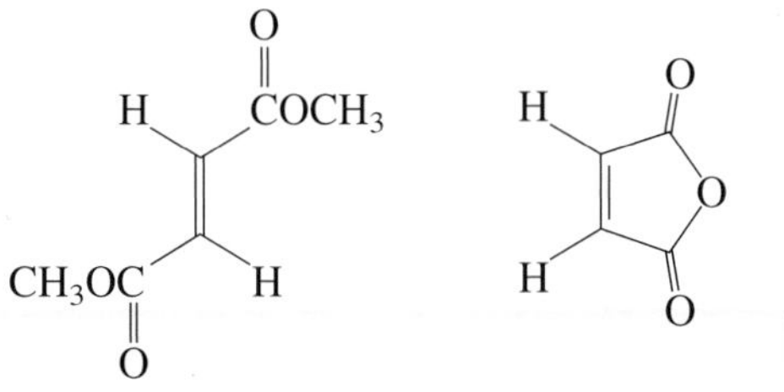

Dimethyl fumarate   Maleic anhydride

In order to achieve the necessary geometry in the Diels–Alder transition state, the diene must be able to adopt the *s*-cis conformation. In Section 10.10, we saw that the *s*-cis conformation of 1,3-butadiene is 12 kJ/mol (2.8 kcal/mol) less stable than the *s*-trans form. This is a relatively small energy difference, so 1,3-butadiene is reactive in the Diels–Alder reaction. Dienes that cannot readily adopt the *s*-cis conformation are less reactive. For example, 4-methyl-1,3-pentadiene is a thousand times less reactive in the Diels–Alder reaction than *trans*-1,3-pentadiene because its *s*-cis conformation is destabilized by the steric effect imposed by the additional methyl group.

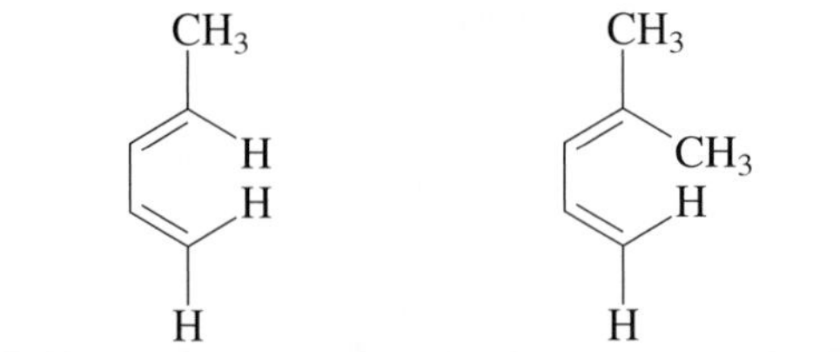

trans-1,3-Pentadiene   4-Methyl-1,3-pentadiene

Problem 10.16

In each pair, indicate the dienophile that you would expect to be more reactive in a Diels–Alder reaction.

(a) $CH_3\overset{\overset{O}{\|}}{C}CH{=}CH\overset{\overset{O}{\|}}{C}OCH_3$ or $HOCH_2CH{=}CHCH_2OH$

(b) 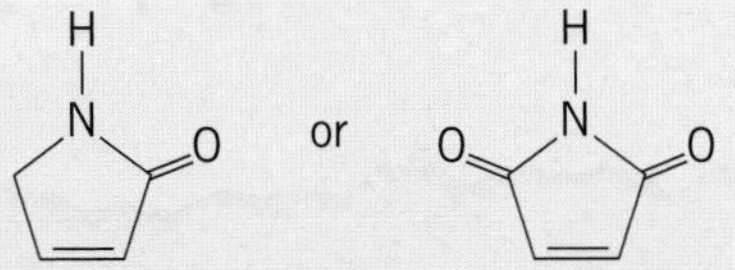

(c) $CH_3SCH{=}CH_2$ or $CH_3\underset{\underset{O}{\|}}{\overset{\overset{O}{\|}}{S}}CH{=}CH_2$

Sample Solution (a) The first dienophile has two carbonyl groups attached to the double bond, which make it more reactive in the Diels–Alder reaction. In the second dienophile, these two carbons are CH_2OH carbons, which are not activating.

Problem 10.17

2,3-Di-*tert*-butyl-1,3-butadiene is extremely unreactive in Diels–Alder reactions. Explain.

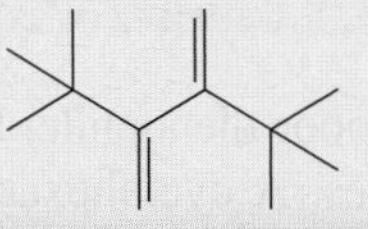

The product of a Diels–Alder reaction always contains one more ring than the reactants. Maleic anhydride contains one ring, so the product of its addition to 2-methyl-1,3-butadiene contains two.

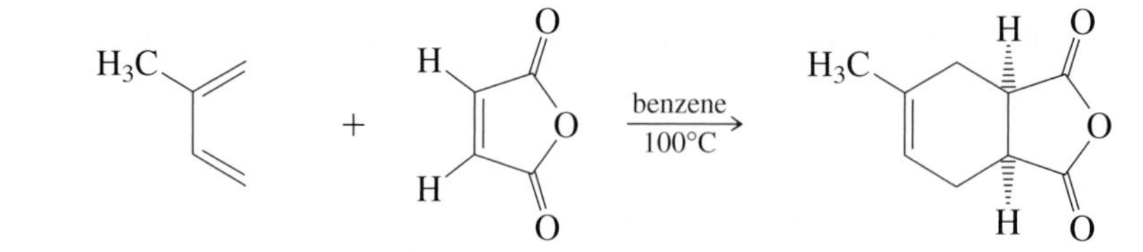

2-Methyl-1,3-butadiene   Maleic anhydride   1-Methylcyclohexene-4,5-dicarboxylic anhydride (100%)

Problem 10.18

Other dicarbonyl compounds such as quinones (Section 22.14) are reactive dienophiles.

(a) Benzoquinone reacts with 2-chloro-1,3-butadiene to give a single product, $C_{10}H_9ClO_2$, in 95% yield. Write a structural formula for this product.

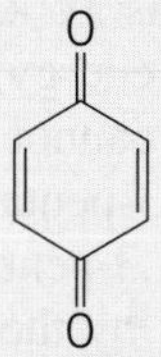

Benzoquinone

Sample Solution

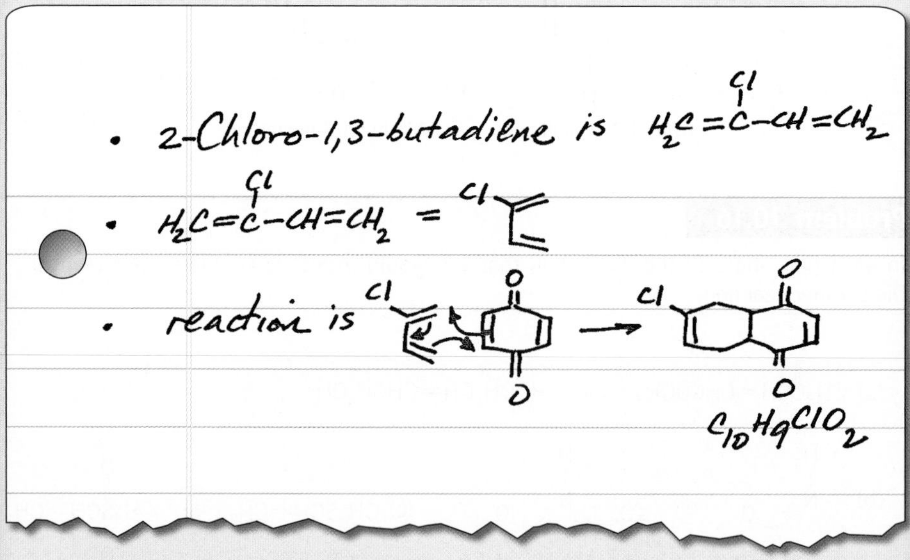

(b) Cyanobenzoquinone undergoes a Diels–Alder reaction with 1,3-butadiene to give a single cycloadduct in 84% yield. What is its structure?

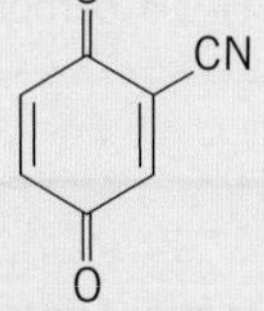

Cyanobenzoquinone

Acetylene, like ethylene, is a poor dienophile, but alkynes that bear C═O or C≡N substituents react readily with dienes. A cyclohexadiene derivative is the product.

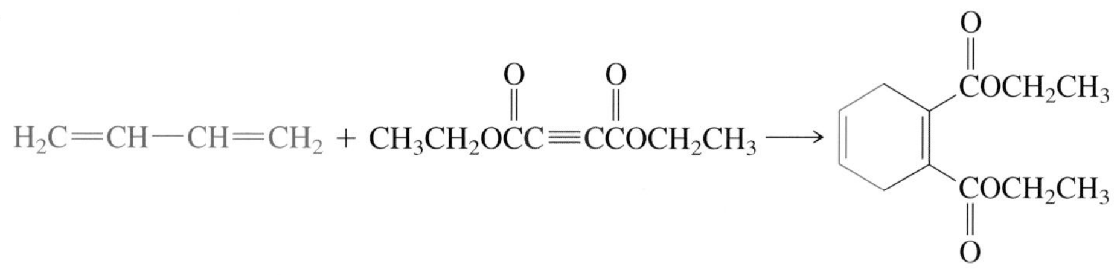

1,3-Butadiene   Diethyl acetylenedicarboxylate   Diethyl 1,4-cyclohexadiene-1,2-dicarboxylate (98%)

The Diels–Alder reaction is stereospecific. Substituents that are cis in the dienophile remain cis in the product; substituents that are trans in the dienophile remain trans in the product.

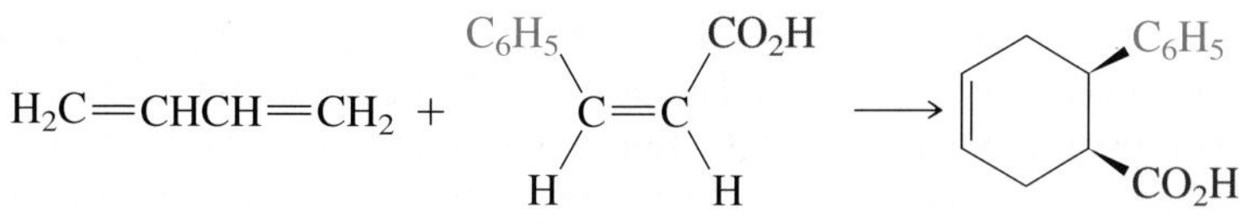

1,3-Butadiene   *cis*-Cinnamic acid   Only product

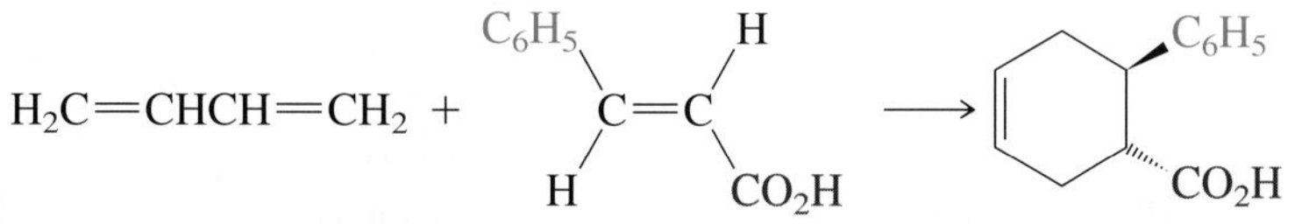

1,3-Butadiene   *trans*-Cinnamic acid   Only product

Recall from Section 7.14 that a stereospecific reaction is one in which each stereoisomer of a particular starting material yields a different stereoisomeric form of the reaction product. In the examples shown, the product from Diels-Alder cycloaddition of 1-3-butadiene to *cis*-cinnamic acid is a stereoisomer of the product from *trans*-cinnamic acid. Each product, although chiral, is formed as a racemic mixture.

Problem 10.19

What combination of diene and dienophile would you choose in order to prepare each of the following compounds?

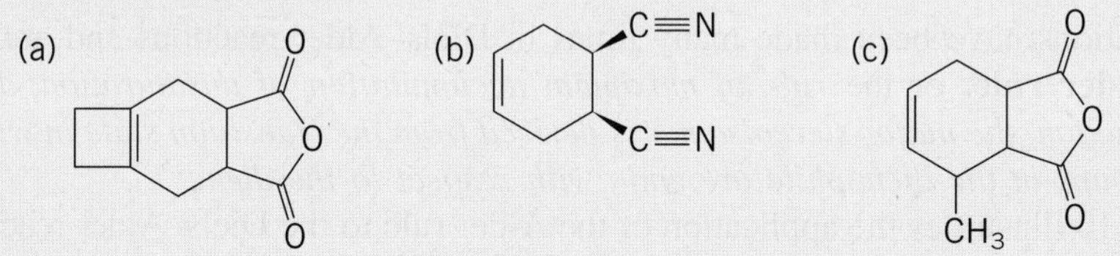

Sample Solution (a) Using curved arrows, we represent a Diels–Alder reaction as

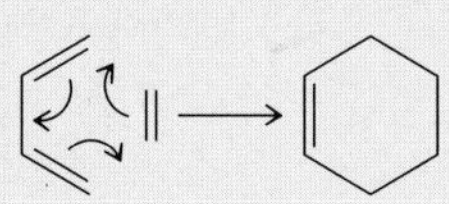

To deduce the identity of the diene and dienophile that lead to a particular Diels–Alder adduct, we use curved arrows in the reverse fashion to "undo" the cyclohexene derivative. Start with the π component of the double bond in the six-membered ring, and move electrons in pairs.

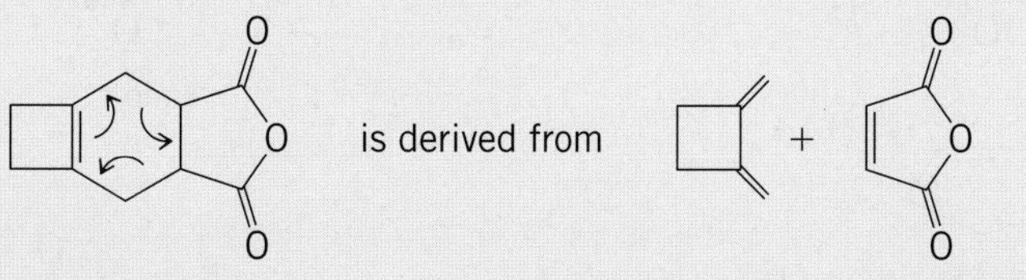

Diels–Alder adduct   Diene   Dienophile

Cyclic dienes yield bridged bicyclic Diels–Alder adducts. Since they are constrained to the *s*-cis conformation, cyclic dienes such as 1,3-cyclopentadiene are highly reactive in the Diels–Alder reaction.

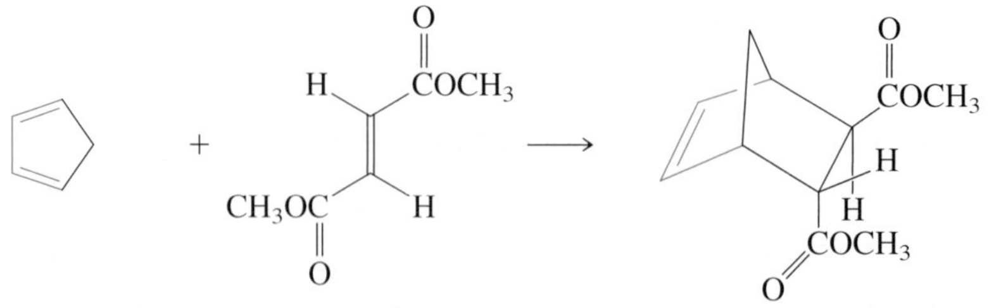

1,3-Cyclopentadiene — Dimethyl fumarate — Dimethyl bicyclo[2.2.1]hept-2-ene-*trans*-5,6-dicarboxylate

Again, the reaction is stereospecific; the trans relationship of the two ester groups in the dienophile is retained in the product. One ester group lies on the same face of the bicyclic system as the CH_2 bridge, the other lies on the opposite face. The two faces are called exo ("outside") and endo ("inside"), respectively. If dimethyl maleate, which is the cis stereoisomer of dimethyl fumarate, is used, the cis relationship of the two ester functions is retained, but the product is a mixture of diastereomers. Both ester groups are endo in the major product, exo in the minor one. In addition to being stereospecific, *the reaction is stereoselective.*

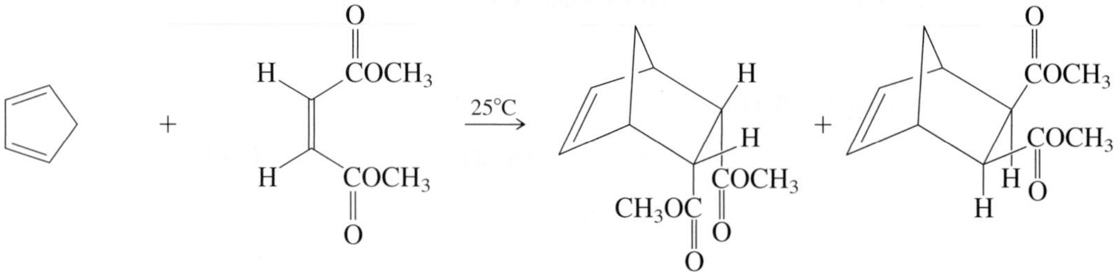

endo (67% yield) — exo (23% yield)

1,3-Cyclopentadiene — Dimethyl maleate — Dimethyl bicyclo[2.2.1]hept-2-ene-*cis*-5,6-dicarboxylate

Similar observations have been made many times in Diels–Alder reactions and are the basis of the **Alder rule,** or the *rule of maximum accumulation of unsaturation: In a Diels–Alder reaction, the major stereoisomer is derived from the transition state in which unsaturated groups in the dienophile are endo with respect to the diene.*

Figure 10.10 illustrates the application of the Alder rule to the Diels–Alder reaction just described. In the arrangement that leads to the endo diastereomer, the diene unit lies closer to the carbonyl groups of the dienophile than in the arrangement that leads to the

Figure 10.10

Diels–Alder transition states for the reaction of 1,3-cyclopentadiene with dimethyl maleate.

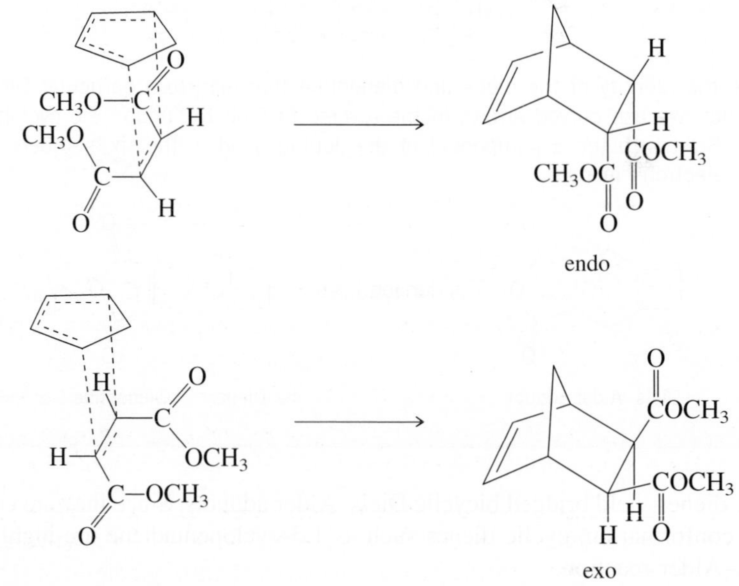

exo product. Transition states leading to the endo and exo cycloadducts, respectively, show that in the transition state leading to the endo diastereomer, the developing double bond lies closer to the carbonyl groups in the dienophile than it does in the transition state leading to the exo product.

Problem 10.20

The Alder rule is not perfect. It correctly predicts the major product from the Diels–Alder reaction of 1,3-cyclopentadiene with methyl acrylate, but is incorrect in the case of methyl methacrylate. On the basis of these facts, give the structure of the major product isolated in each case.

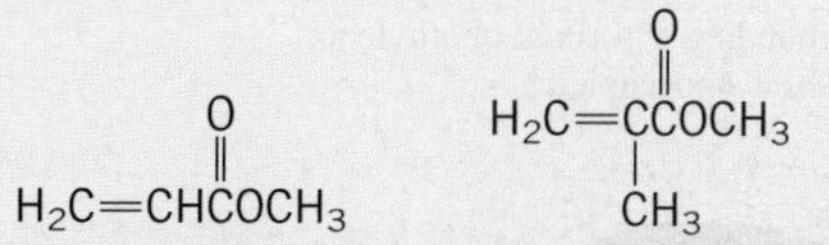

Methyl acrylate Methyl methacrylate

The mechanism of the Diels–Alder reaction is best understood on the basis of a molecular orbital approach. To understand this approach we need to take a more detailed look at the π orbitals of alkenes and dienes.

10.16 The π Molecular Orbitals of Ethylene and 1,3-Butadiene

The valence bond approach has served us well to this point as a tool to probe structure and reactivity in organic chemistry. An appreciation for the delocalization of π electrons through a system of overlapping *p* orbitals has given us insights into conjugated systems that are richer in detail than those obtained by examining Lewis formulas. An even deeper understanding can be gained by applying qualitative molecular orbital theory to these π electron systems. We shall see that useful information can be gained by directing attention to what are called the **frontier orbitals** of molecules. The frontier orbitals are the *highest occupied molecular orbital* (the **HOMO**) and the *lowest unoccupied molecular orbital* (the **LUMO**). When electrons are transferred *from* a molecule, it is the electrons in the HOMO that are involved, because they are the ones most weakly held. When electrons are transferred *to* a molecule, they go into the LUMO, because that is the lowest energy vacant orbital.

Ethylene. Let's begin by examining the π molecular orbitals of ethylene. Recall from Section 2.4 that the number of molecular orbitals is equal to the number of atomic orbitals that combine to form them. We saw that the 1*s* orbitals of two hydrogen atoms overlap to give both a bonding (σ) and an antibonding (σ*) orbital. The same principle applies to π orbitals. As Figure 10.11 illustrates for the case of ethylene, the 2*p* orbitals of adjacent carbons overlap to give both a bonding (π) and an antibonding (π*) orbital. Notice that the σ electrons are not explicitly considered in Figure 10.11. These electrons are strongly held, and the collection of σ bonds can be thought of as an inert framework that supports the valence electrons of the π orbital.

Both the π and π* molecular orbitals of ethylene are *antisymmetric* with respect to the plane of the molecule. The bonding π orbital has no nodes other than this plane, whereas the antibonding π* orbital has a nodal plane between the two carbons. The more nodes an orbital has, the higher is its energy.

As is true for all orbitals, a π orbital may contain a maximum of two electrons. Ethylene has two π electrons, and these occupy the bonding π molecular orbital, which is the HOMO. The antibonding π* molecular orbital is vacant, and is the LUMO.

Problem 10.21

Which molecular orbital of ethylene (π or π*) is the most important one to look at in a reaction in which ethylene is attacked by an electrophile?

Figure 10.11

The bonding (π) and antibonding (π^*) molecular orbitals of ethylene. The plane of the molecule is a nodal surface in both orbitals; the antibonding orbital has an additional nodal surface perpendicular to the plane of the molecule.

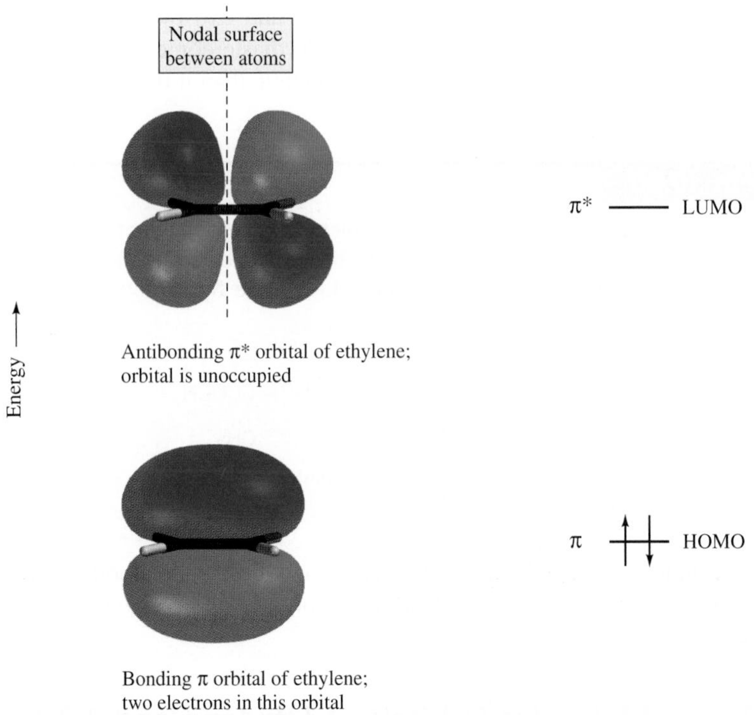
Nodal surface between atoms
Energy
π* LUMO
Antibonding π* orbital of ethylene; orbital is unoccupied
π HOMO
Bonding π orbital of ethylene; two electrons in this orbital

1,3-Butadiene. The π molecular orbitals of 1,3-butadiene are shown in Figure 10.12. The four sp^2-hybridized carbons contribute four $2p$ atomic orbitals, and their overlap leads to four π molecular orbitals. Two are bonding (π_1 and π_2) and two are antibonding

Figure 10.12

The π molecular orbitals of 1,3-butadiene.

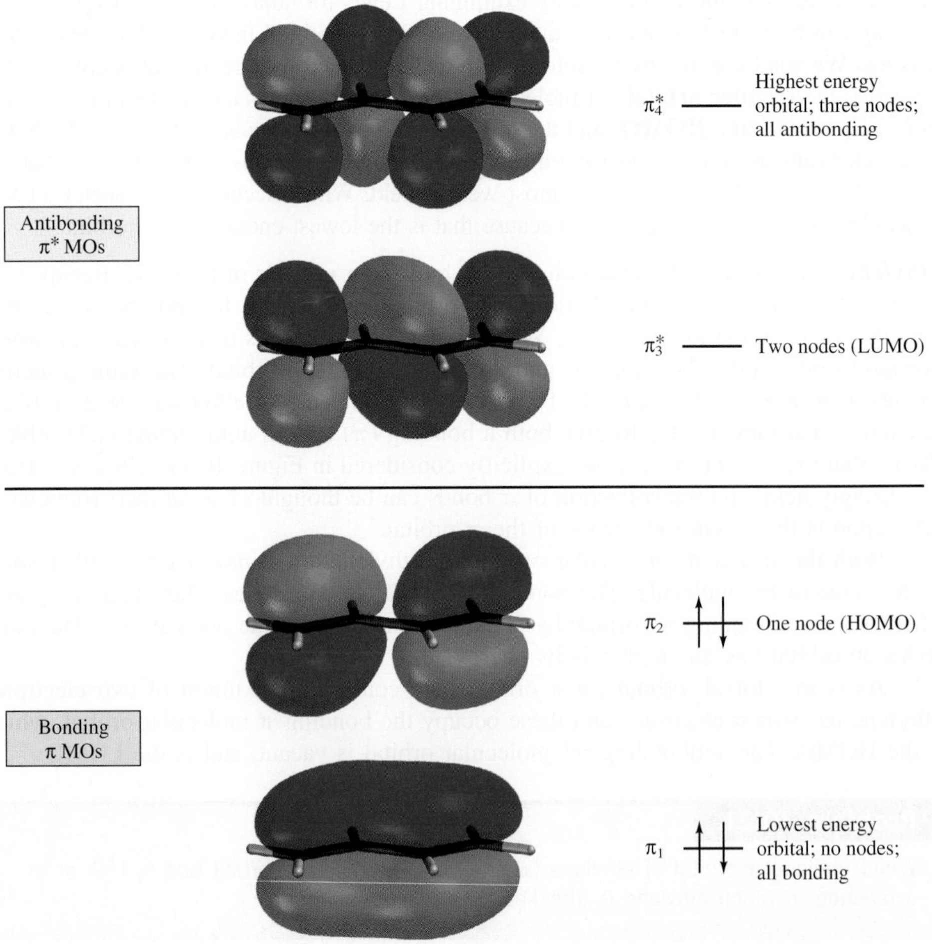
Antibonding π* MOs
π*4 Highest energy orbital; three nodes; all antibonding
π*3 Two nodes (LUMO)
Bonding π MOs
π2 One node (HOMO)
π1 Lowest energy orbital; no nodes; all bonding

(π_3^* and π_4^*). Each π molecular orbital encompasses all four carbons of the diene. There are four π electrons, and these are distributed in pairs between the two orbitals of lowest energy (π_1 and π_2). Both bonding orbitals are occupied; π_2 is the HOMO. Both antibonding orbitals are vacant; π_3* is the LUMO.

10.17 A π Molecular Orbital Analysis of the Diels–Alder Reaction

Let us now examine the Diels–Alder cycloaddition from a molecular orbital perspective. Chemical experience, such as the observation that substituents that increase the reactivity of a dienophile tend to be those that attract electrons, suggests that electrons flow from the diene to the dienophile during the reaction. Thus, the orbitals to be considered are the HOMO of the diene and the LUMO of the dienophile. As shown in Mechanism 10.4 for the case of ethylene and 1,3-butadiene, the symmetry properties of the HOMO of the diene and the LUMO of the dienophile permit bond formation between the ends of the diene system and the two carbons of the dienophile double bond because the necessary orbitals overlap in phase with each other. Cycloaddition of a diene and an alkene is said to be a **symmetry-allowed** reaction.

Contrast the Diels–Alder reaction with a reaction that looks superficially similar, the cycloaddition of two ethylene molecules to give cyclobutane.

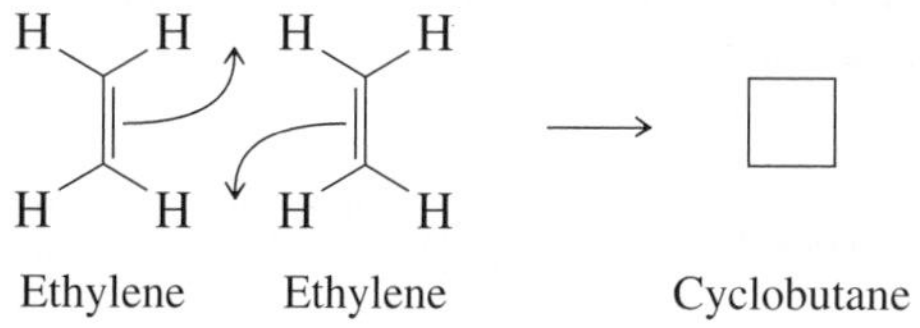
Ethylene
Ethylene
Cyclobutane

Reactions of this type are rare and seem to proceed in a stepwise fashion rather than by way of a concerted mechanism involving a single transition state.

Figure 10.13 shows the interaction between the HOMO of one ethylene molecule and the LUMO of another. In particular, notice that two of the carbons that are to become σ-bonded to each other in the product experience an antibonding interaction during the cycloaddition process. This raises the activation energy for cycloaddition and leads the reaction to be classified as a **symmetry-forbidden** reaction. Reaction, were it to occur, would take place slowly and by a mechanism in which the two new σ bonds are formed in separate steps rather than by way of a concerted process involving a single transition state.

Mechanism 10.4

Orbital Interactions in the Diels–Alder Reaction

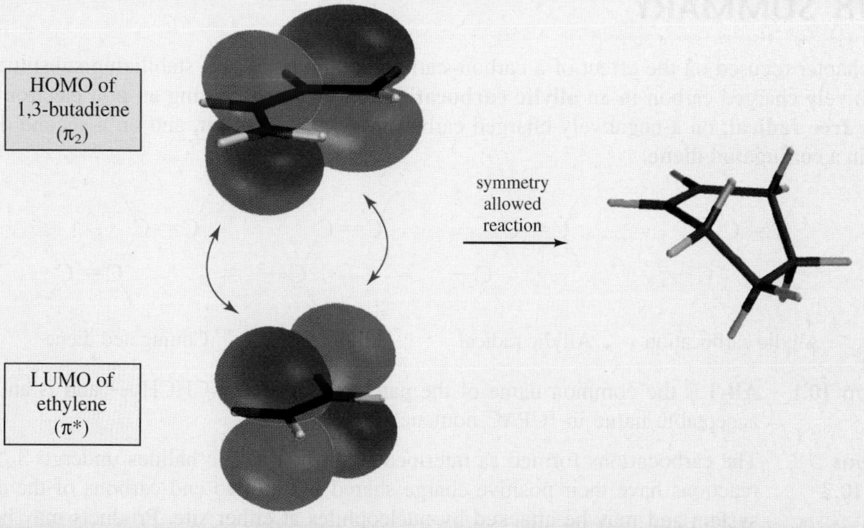
HOMO of
1,3-butadiene
(π_2)
symmetry
allowed
reaction
LUMO of
ethylene
(π^*)

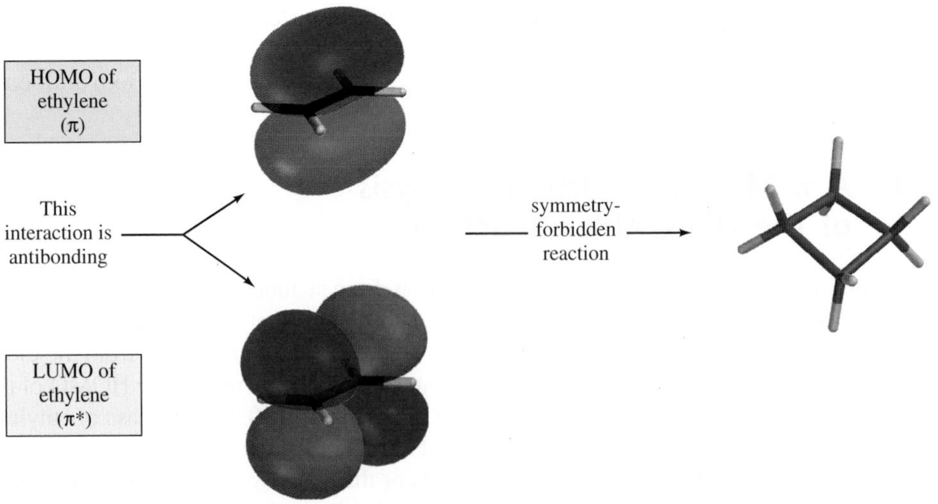
HOMO of
ethylene
(π)
This
interaction is
antibonding
symmetry-
forbidden
reaction
LUMO of
ethylene
(π*)

Figure 10.13

The HOMO of one ethylene molecule and the LUMO of another do not have the proper symmetry to permit two σ bonds to be formed in the transition state for concerted cycloaddition.

Problem 10.22

Use frontier orbital analysis to decide whether the dimerization of 1,3-butadiene shown here is symmetry-allowed or -forbidden.

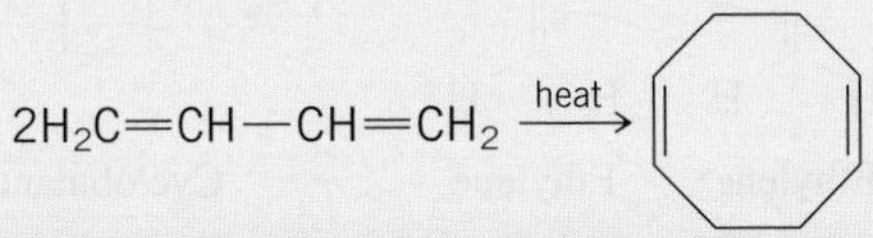
2H2C=CH—CH=CH2
heat

Frontier orbital analysis is a powerful theory that aids our understanding of a great number of organic reactions. Its early development is attributed to Professor Kenichi Fukui of Kyoto University, Japan. The application of frontier orbital methods to Diels–Alder reactions represents one part of what organic chemists refer to as the *Woodward–Hoffmann rules,* a beautifully simple analysis of organic reactions by Professor R. B. Woodward of Harvard University and Professor Roald Hoffmann of Cornell University. Professors Fukui and Hoffmann were corecipients of the 1981 Nobel Prize in Chemistry for their work. Woodward's death in 1979 prevented his being considered for a share of the 1981 prize with Fukui and Hoffmann. Woodward had earlier won a Nobel Prize (1965) for his achievements in organic synthesis.

10.18 SUMMARY

This chapter focused on the effect of a carbon–carbon double bond as a stabilizing substituent on a positively charged carbon in an **allylic carbocation,** on a carbon bearing an odd electron in an **allylic free radical,** on a negatively charged carbon in an **allylic anion,** and on a second double bond in a **conjugated diene.**

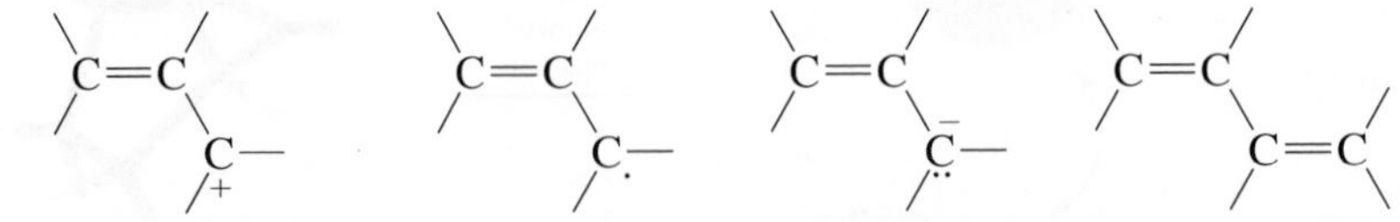

Allylic carbocation Allylic radical Allylic anion Conjugated diene

Section 10.1 **Allyl** is the common name of the parent group $H_2C{=}CHCH_2$— and is an acceptable name in IUPAC nomenclature.

Sections 10.2–10.3 The carbocations formed as intermediates when allylic halides undergo S_N1 reactions have their positive charge shared by the two end carbons of the allylic system and may be attacked by nucleophiles at either site. Products may be formed with the same pattern of bonds as the starting allylic halide or with *allylic rearrangement.*

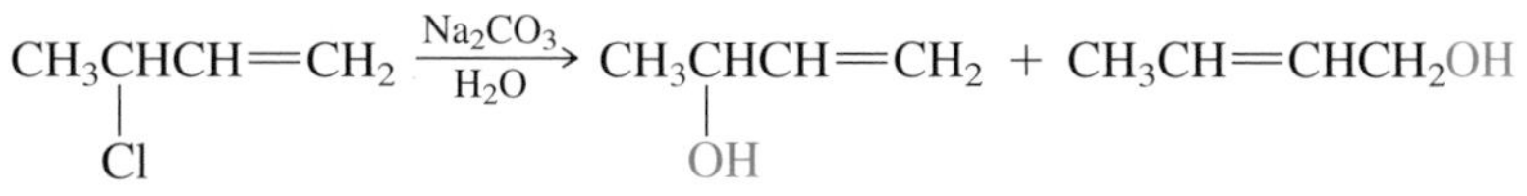

3-Chloro-1-butene 3-Buten-2-ol (65%) 2-Buten-1-ol (35%)

via: $CH_3\overset{+}{C}H—CH=CH_2 \longleftrightarrow CH_3CH=CH—\overset{+}{C}H_2$

Section 10.4 Allylic halides react faster than comparable alkyl halides under conditions that favor the S_N2 mechanism.

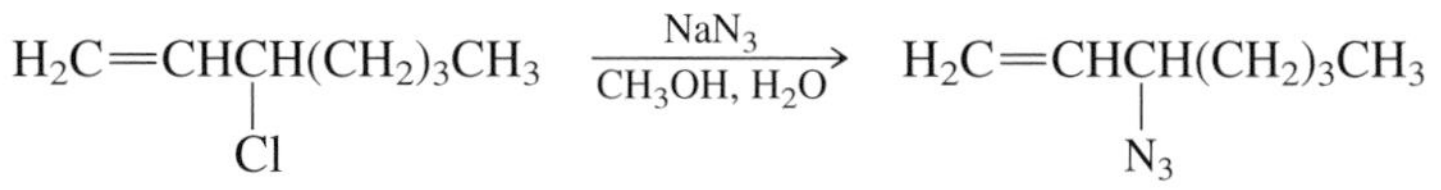

3-Chloro-1-heptene 3-Azido-1-heptene (78%)

Sections 10.5–10.6 Alkenes react with *N*-bromosuccinimide (NBS) to give allylic bromides. NBS serves as a source of Br_2, and substitution occurs by a free-radical mechanism. The reaction is used for synthetic purposes only when the two resonance forms of the allylic radical are equivalent. Otherwise a mixture of isomeric allylic bromides is produced.

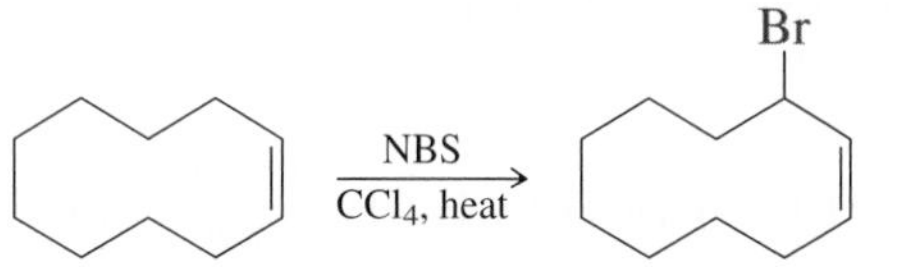

Cyclodecene 3-Bromocyclodecene (56%)

via:

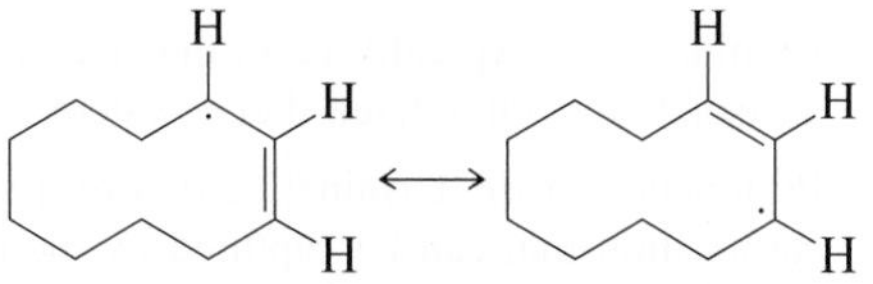

Section 10.7 Allylic anions are stabilized by electron delocalization. The negative charge is shared by the carbons at each end of the allyl unit.

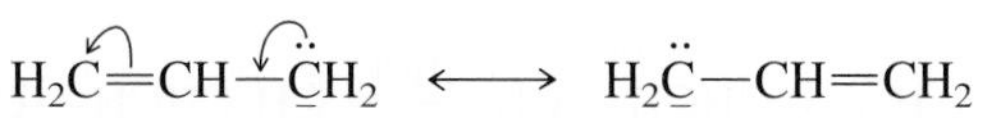

An allylic hydrogen is more acidic ($pK_a \approx 43$) than a hydrogen in an alkane ($pK_a \approx 62$).

Section 10.8 Dienes are classified as having **isolated, conjugated,** or **cumulated** double bonds.

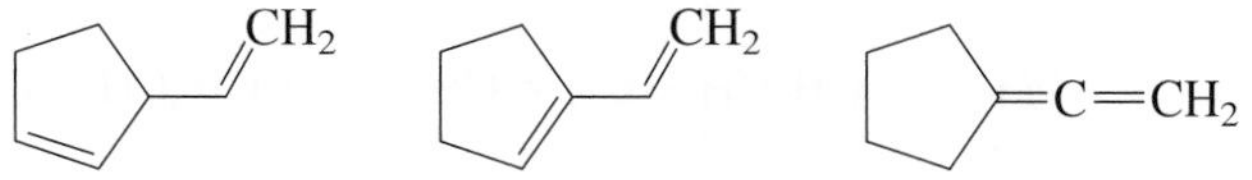

Isolated Conjugated Cumulated

Section 10.9 Conjugated dienes are more stable than isolated dienes, and cumulated dienes are the least stable of all.

Section 10.10 Conjugated dienes are stabilized by electron delocalization to the extent of 12–16 kJ/mol (3–4 kcal/mol). Overlap of the *p* orbitals of four adjacent sp^2-hybridized carbons in a conjugated diene gives an extended π system through which the electrons are delocalized.

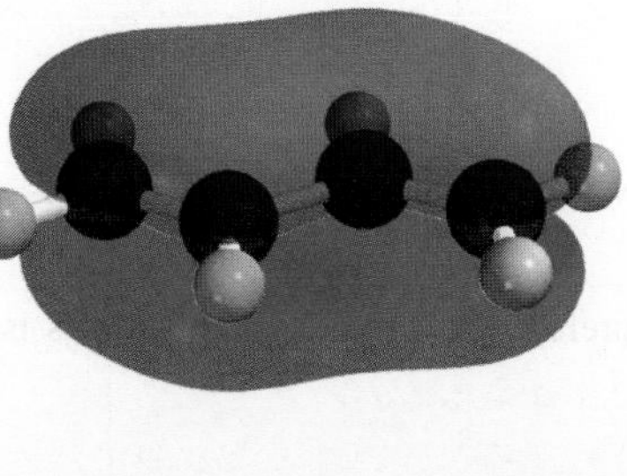

The two most stable conformations of conjugated dienes are the *s*-cis and *s*-trans. The *s*-trans conformation is normally more stable than the *s*-cis. Both conformations are planar, which allows the *p* orbitals to overlap to give an extended π system.

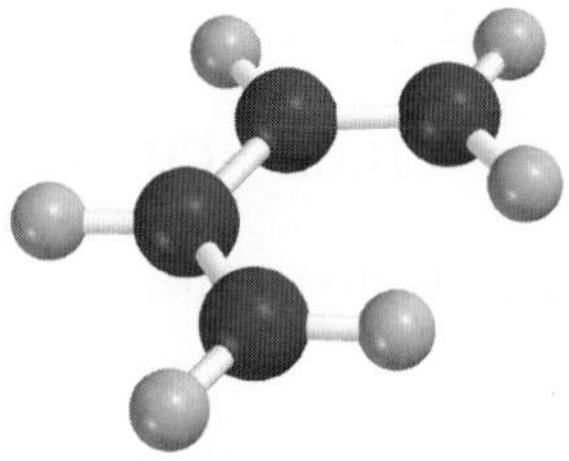

s-cis

s-trans

Section 10.11 1,2-Propadiene ($H_2C{=}C{=}CH_2$), also called **allene,** is the simplest cumulated diene. The two π bonds in an allene share an *sp*-hybridized carbon and are at right angles to each other. Certain allenes such as 2,3-pentadiene ($CH_3CH{=}C{=}CHCH_3$) possess a *chirality axis* and are chiral.

Section 10.12 1,3-Butadiene is an industrial chemical and is prepared by dehydrogenation of butane. Elimination reactions such as dehydration and dehydrohalogenation are common routes to alkadienes.

$$H_2C{=}CHCH_2\underset{\displaystyle OH}{\overset{\displaystyle CH_3}{C}}CH_2CH_3 \xrightarrow[\text{heat}]{KHSO_4} H_2C{=}CHCH{=}\overset{\displaystyle CH_3}{C}CH_2CH_3$$

3-Methyl-5-hexen-3-ol → 4-Methyl-1,3-hexadiene (88%)

Elimination is typically regioselective and gives a conjugated diene rather than an isolated or cumulated diene system of double bonds.

Section 10.13 Protonation at the terminal carbon of a conjugated diene system gives an allylic carbocation that can be captured by the halide nucleophile at either of the two sites that share the positive charge. Nucleophilic attack at the carbon adjacent to the one that is protonated gives the product of *1,2 addition*. Capture at the other site gives the product of *1,4 addition*.

$$H_2C{=}CHCH{=}CH_2 \xrightarrow{HCl} CH_3\underset{\displaystyle Cl}{CH}CH{=}CH_2 + CH_3CH{=}CHCH_2Cl$$

1,3-Butadiene; 3-Chloro-1-butene (78%); 1-Chloro-2-butene (22%)

via: $CH_3\overset{+}{C}H{-}CH{=}CH_2 \longleftrightarrow CH_3CH{=}CH{-}\overset{+}{C}H_2$

Section 10.14 1,4 Addition predominates when Cl_2 and Br_2 add to conjugated dienes.

Section 10.15 Conjugate addition of an alkene (the *dienophile*) to a conjugated diene gives a cyclohexene derivative in a process called the *Diels–Alder reaction.* It is concerted and stereospecific; substituents that are cis to each other on the dienophile remain cis in the product.

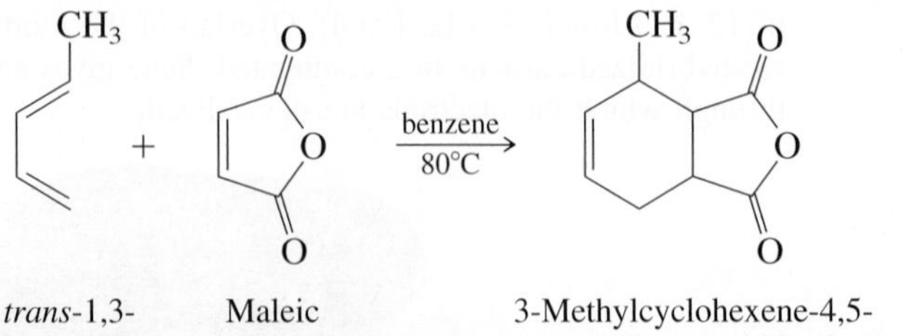

trans-1,3-Pentadiene; Maleic anhydride; 3-Methylcyclohexene-4,5-dicarboxylic anhydride (81%)

The preference for endo cycloadducts is known as the Alder rule.

Sections 10.16–10.17 The Diels–Alder reaction is believed to proceed in a single step. A deeper level of understanding of the bonding changes in the transition state can be obtained by examining the nodal properties of the highest occupied molecular orbital (HOMO) of the diene and the lowest unoccupied molecular orbital (LUMO) of the dienophile.

PROBLEMS

10.23 Write structural formulas for each of the following:

(a) 3,4-Octadiene
(b) (*E,E*)-3,5-Octadiene
(c) (*Z,Z*)-1,3-Cyclooctadiene
(d) (*Z,Z*)-1,4-Cyclooctadiene
(e) (*E,E*)-1,5-Cyclooctadiene
(f) (2*E*,4*Z*,6*E*)-2,4,6-Octatriene
(g) 5-Allyl-1,3-cyclopentadiene
(h) *trans*-1,2-Divinylcyclopropane
(i) 2,4-Dimethyl-1,3-pentadiene

10.24 Give an acceptable IUPAC name for each of the following compounds:

(a) $H_2C{=}CH(CH_2)_5CH{=}CH_2$

(b) 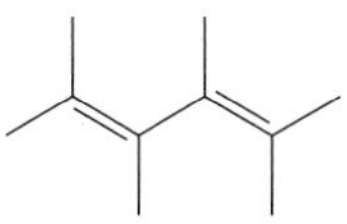

(c) $(H_2C{=}CH)_3CH$

(d) 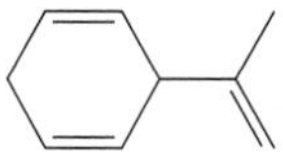

(e) 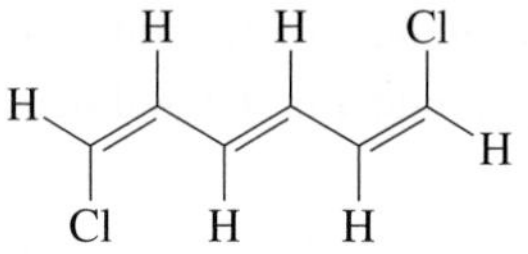

(f) $H_2C{=}C{=}CHCH{=}CHCH_3$

(g) 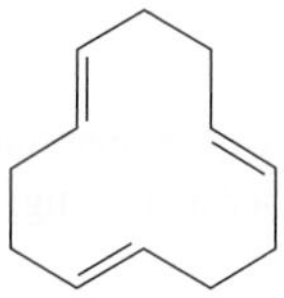

(h) 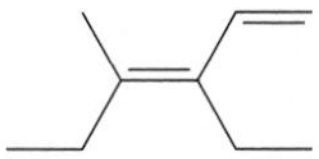

10.25 (a) What compound of molecular formula C_6H_{10} gives 2,3-dimethylbutane on catalytic hydrogenation over platinum?

(b) What two compounds of molecular formula $C_{11}H_{20}$ give 2,2,6,6-tetramethylheptane on catalytic hydrogenation over platinum?

10.26 Write structural formulas for all the

(a) Conjugated dienes (b) Isolated dienes (c) Cumulated dienes

that give 2,4-dimethylpentane on catalytic hydrogenation.

10.27 A certain species of grasshopper secretes an allenic substance of molecular formula $C_{13}H_{20}O_3$ that acts as an ant repellent. The carbon skeleton and location of various substituents in this substance are indicated in the partial structure shown. Complete the structure, adding double bonds where appropriate.

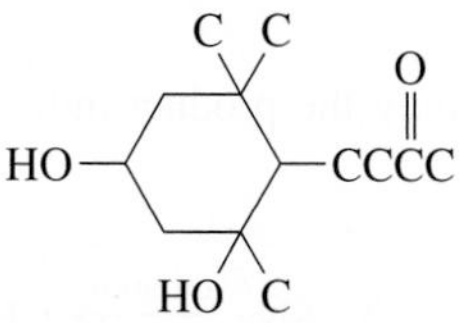

10.28 Show how to prepare each of the following compounds from propene and any necessary organic or inorganic reagents:

(a) Allyl bromide
(b) 1,2-Dibromopropane
(c) 1,3-Dibromopropane
(d) 1-Bromo-2-chloropropane
(e) 1,2,3-Tribromopropane
(f) Allyl alcohol
(g) Pent-1-en-4-yne ($H_2C{=}CHCH_2C{\equiv}CH$)
(h) 1,4-Pentadiene

10.29 Show, by writing a suitable sequence of chemical equations, how to prepare each of the following compounds from cyclopentene and any necessary organic or inorganic reagents:

(a) 2-Cyclopentenol
(b) 3-Iodocyclopentene
(c) 3-Cyanocyclopentene
(d) 1,3-Cyclopentadiene
(e)

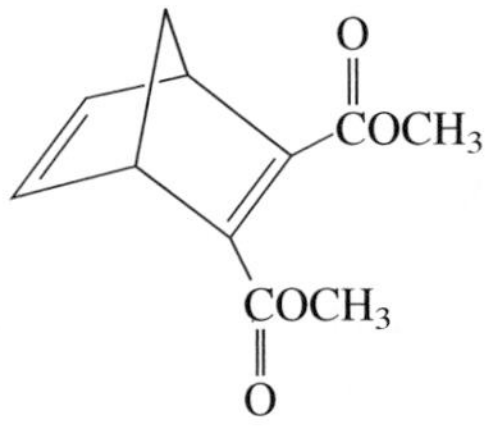
O
COCH3
COCH3
O

10.30 Give the structure, exclusive of stereochemistry, of the principal organic product formed on reaction of 2,3-dimethyl-1,3-butadiene with each of the following:

(a) 2 mol H_2, platinum catalyst
(b) 1 mol HCl (product of 1,2-addition)
(c) 1 mol HCl (product of 1,4-addition)
(d) 1 mol Br_2 (product of 1,2-addition)
(e) 1 mol Br_2 (product of 1,4-addition)
(f) 2 mol Br_2
(g)

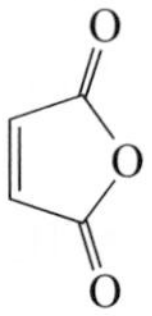
O
O
O

10.31 Repeat the previous problem for the reactions of 1,3-cyclohexadiene.

10.32 Which of the following two dienes can undergo the Diels–Alder reaction? Explain.

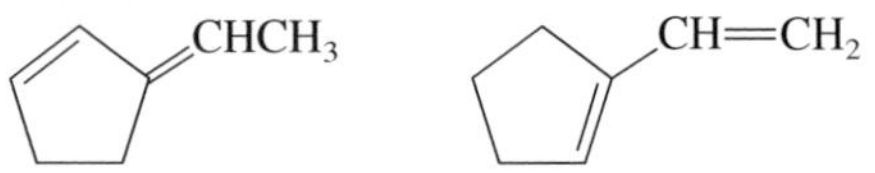
CHCH3
CH=CH2

10.33 Two constitutional isomers of molecular formula $C_8H_{12}O$ are formed in the following reaction. Ignoring stereochemistry, suggest reasonable structures for these Diels–Alder adducts.

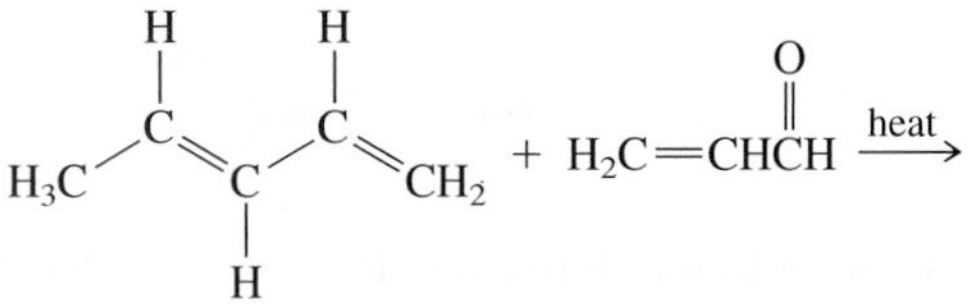
H
H
H3C
C
C
C
CH2
H
+ H2C=CHCH
O
heat

10.34 Allene can be converted to a trimer (compound A) of molecular formula C_9H_{12}. Compound A reacts with dimethyl acetylenedicarboxylate to give compound B. Deduce the structure of compound A.

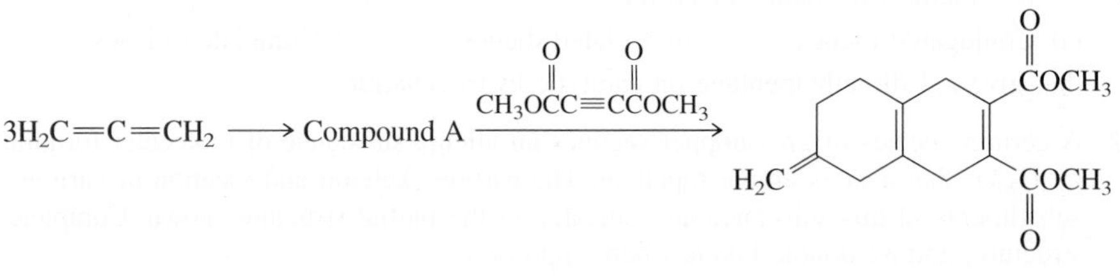
3H2C=C=CH2
Compound A
CH3OCC≡CCOCH3
O O
H2C
COCH3
COCH3
O
O

Compound B

10.35 The following reaction gives only the product indicated. By what mechanism does this reaction most likely occur?

CH3CH=CHCH2Cl +
SNa
ethanol
CH3CH=CHCH2S

10.36 Suggest reasonable explanations for each of the following observations:

(a) The first-order rate constant for the solvolysis of $(CH_3)_2C{=}CHCH_2Cl$ in ethanol is over 6000 times greater than that of allyl chloride (25°C).

(b) After a solution of 3-buten-2-ol in aqueous sulfuric acid had been allowed to stand for 1 week, it was found to contain both 3-buten-2-ol and 2-buten-1-ol.

(c) Treatment of $CH_3CH{=}CHCH_2OH$ with hydrogen bromide gave a mixture of 1-bromo-2-butene and 3-bromo-1-butene.

(d) Treatment of 3-buten-2-ol with hydrogen bromide gave the same mixture of bromides as in part (c).

(e) The major product in parts (c) and (d) was 1-bromo-2-butene.

10.37 What is the 1,2-addition product of the reaction shown?

$$H_2C{=}CH\underset{\displaystyle CH_3}{\underset{|}{C}}{=}CH_2 \xrightarrow{HCl}$$

10.38 2-Chloro-1,3-butadiene (chloroprene) is the monomer from which the elastomer *neoprene* is prepared. 2-Chloro-1,3-butadiene is the thermodynamically controlled product formed by addition of hydrogen chloride to vinylacetylene ($H_2C{=}CHC{\equiv}CH$). The principal product under conditions of kinetic control is the allenic chloride 4-chloro-1,2-butadiene. Suggest a mechanism to account for the formation of each product.

10.39 Which of the following are chiral?

(a) 2-Methyl-2,3-hexadiene

(b) 4-Methyl-2,3-hexadiene

(c) 2,4-Dimethyl-2,3-pentadiene

10.40 (a) Describe the molecular geometry expected for 1,2,3-butatriene ($H_2C{=}C{=}C{=}CH_2$).

(b) Two stereoisomers are expected for 2,3,4-hexatriene ($CH_3CH{=}C{=}C{=}CHCH_3$). What should be the relationship between these two stereoisomers?

10.41 Suggest reagents suitable for carrying out each step in the following synthetic sequence:

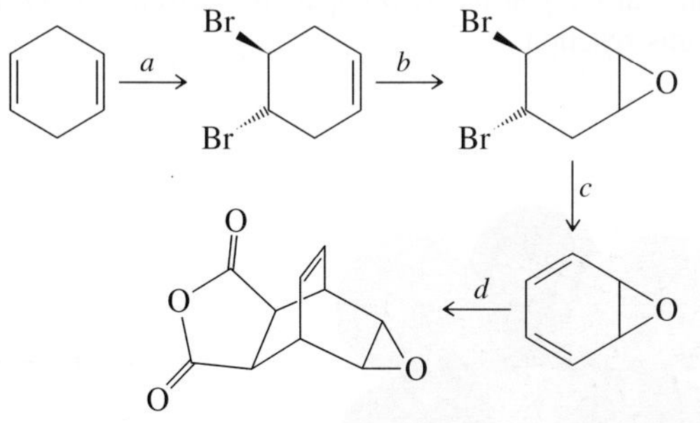

10.42 A very large number of Diels–Alder reactions are recorded in the chemical literature, many of which involve relatively complicated dienes, dienophiles, or both. On the basis of your knowledge of Diels–Alder reactions, predict the constitution of the Diels–Alder adduct that you would expect to be formed from the following combinations of dienes and dienophiles:

(a) 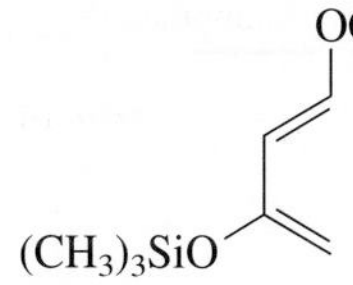

$+\ CH_3O_2CC{\equiv}CCO_2CH_3$

(b) 

$+\ CH_3O_2CC{\equiv}CCO_2CH_3$

(c) 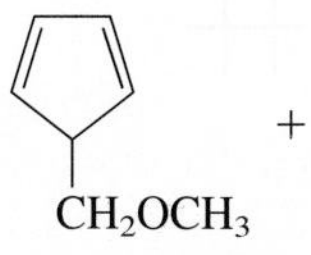

$+\ H_2C{=}CHNO_2$

10.43 Bromination of 1,5-cyclooctadiene with *N*-bromosuccinimide (NBS) gives a mixture of two constitutional isomers of $C_8H_{11}Br$. Suggest reasonable structures for these two isomers.

1,5-Cyclooctadiene

10.44 Refer to the molecular orbital diagrams of allyl cation (Figure 10.14) and those presented earlier in this chapter for ethylene and 1,3-butadiene (Figures 10.11 and 10.12) to decide which of the following cycloaddition reactions are allowed and which are forbidden according to the Woodward–Hoffmann rules.

(a) 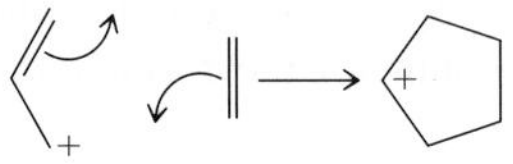

(b) 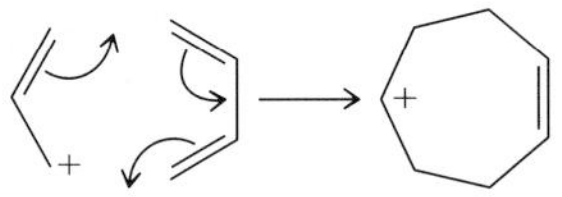

10.45 Alkenes slowly undergo a reaction in air called *autoxidation* in which allylic hydroperoxides are formed.

Cyclohexene Oxygen 3-Hydroperoxycyclohexene

Keeping in mind that oxygen has two unpaired electrons ($\cdot\ddot{\underset{\cdot\cdot}{O}}:\ddot{\underset{\cdot\cdot}{O}}\cdot$), suggest a reasonable mechanism for this reaction.

Figure 10.14

The π molecular orbitals of allyl cation. The allyl cation has two π electrons and they are in the orbital marked π_1.

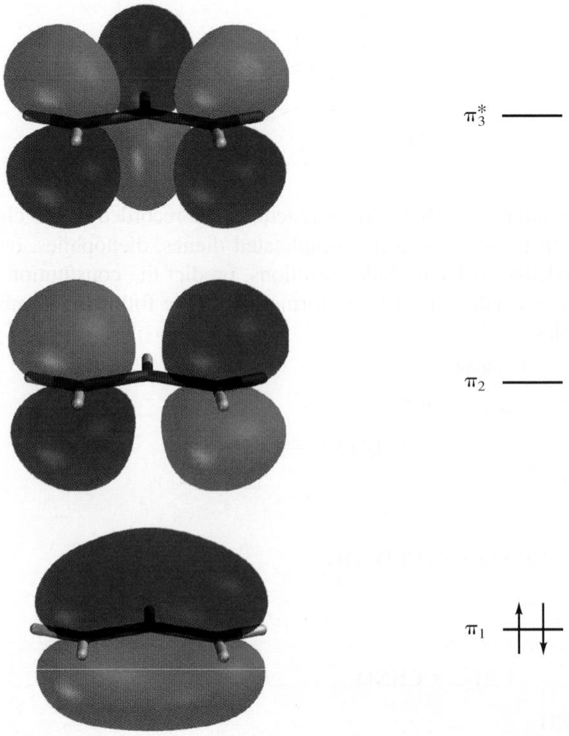

10.46 The antifungal agent terbenifine is synthesized from *N*-methyl-(1-naphthylmethyl)amine, which reacts with compound A to make compound B. Compound B is converted to terbenifine by several steps. What is the structure of compound A?

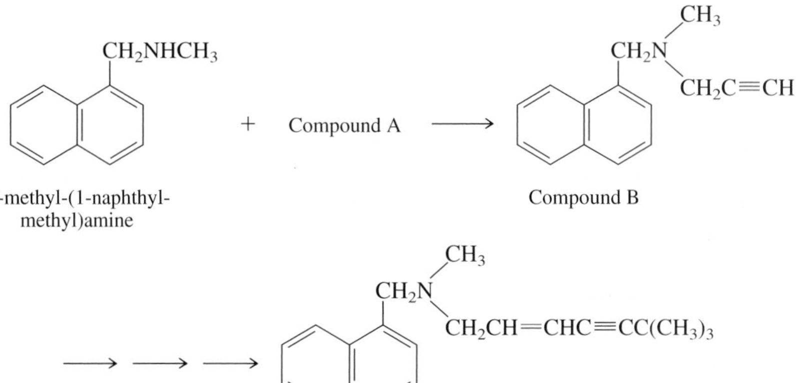

N-methyl-(1-naphthyl-methyl)amine

Compound B

Terbenifine

10.47 Compound C is a key intermediate in a chemical synthesis of paclitaxel, a drug used to treat breast, ovarian, and lung cancer. It is prepared by a reaction between compounds A and B. What is the structure of compound B?

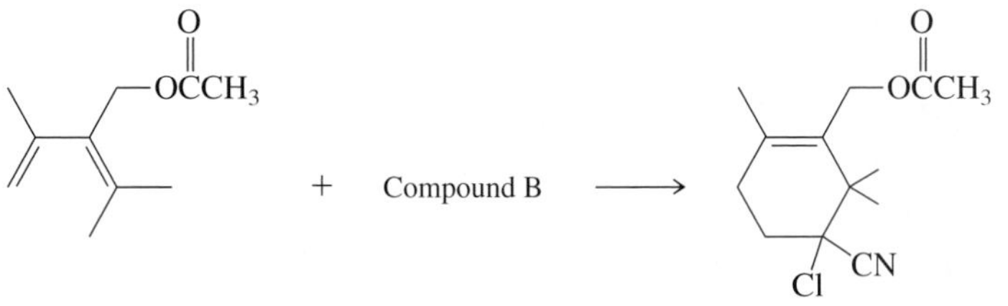

Compound A

Compound C

Descriptive Passage and Interpretive Problems 10

Intramolecular and Retro Diels–Alder Reactions

Not only is the Diels–Alder reaction useful in its own right, but variations on the general theme of cycloaddition have enhanced its versatility as a synthetic tool. In a customary Diels–Alder cycloaddition, the diene and the dienophile are functional groups in separate molecules. The reaction is **intermolecular** (between two molecules).

Intermolecular Diels–Alder reaction:

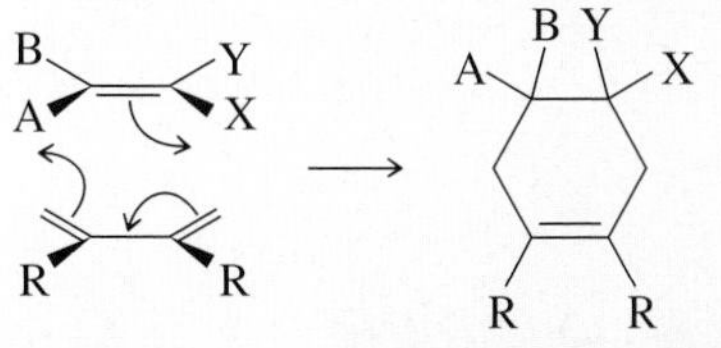

Cycloaddition can also be **intramolecular** (within a single molecule) when a conjugated diene and an isolated C=C double bond are both present in the same molecule.

Intramolecular Diels–Alder reaction:

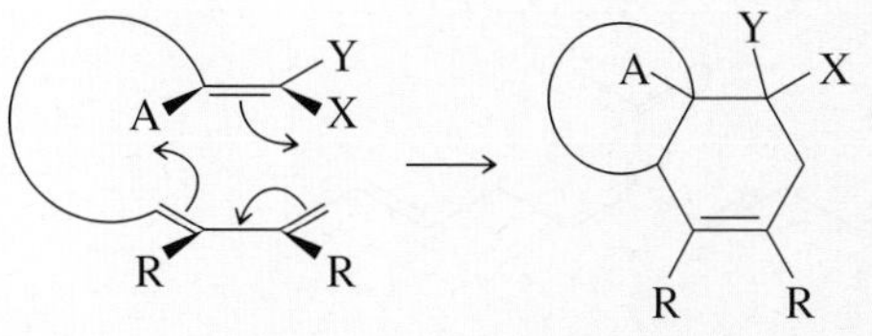

An intramolecular Diels–Alder reaction generates *two* new rings in a single operation. One of the new rings is, as in the intermolecular reaction, a cyclohexene. The other new ring is typically five- or six-membered.

Diels–Alder reactions are reversible; cyclohexenes can dissociate to a diene and an alkene by a **retro Diels–Alder reaction.**

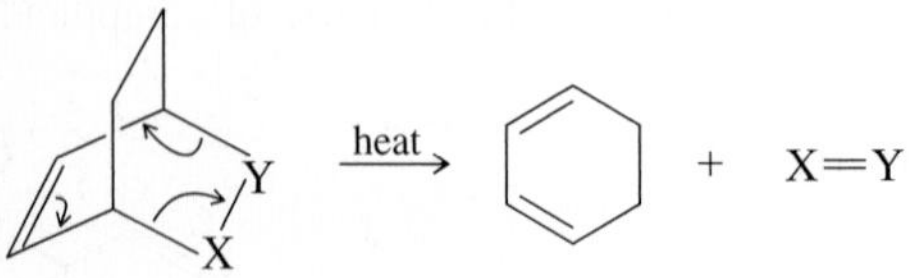

When applied to synthesis, the product X=Y of this *cycloelimination* is typically ethylene or carbon dioxide or can contain a triple bond as in acetylene or N_2. The bicyclic reactant can itself be prepared by a Diels–Alder cycloaddition or by some indirect method.

10.48 The compound shown undergoes an intramolecular Diels–Alder reaction at room temperature. What is the structure of the product? (No need to show stereochemistry.)

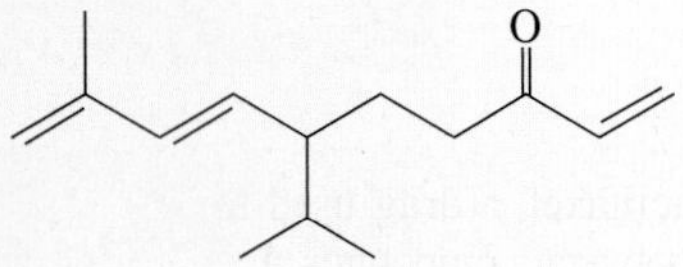

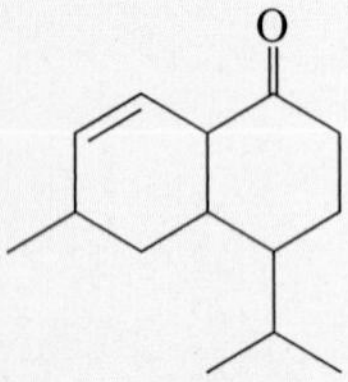

A.

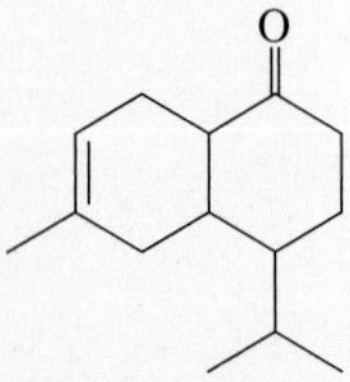

B.

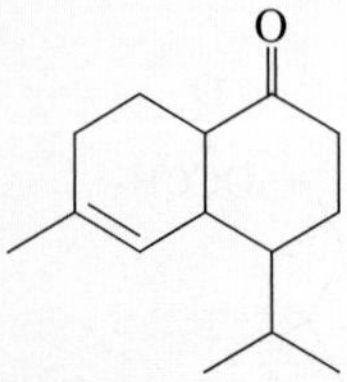

C.

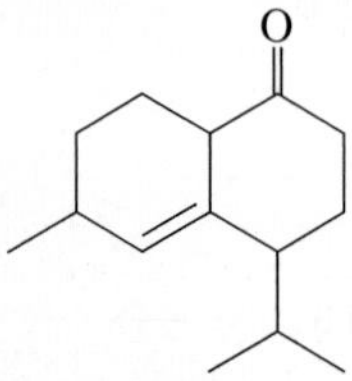

D.

10.49 What is the structure of the intramolecular Diels–Alder product of the compound shown? (No need to show stereochemistry.)

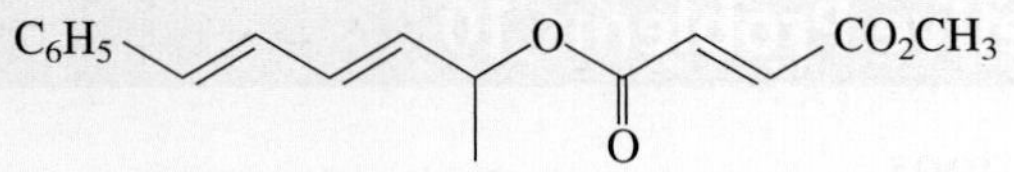

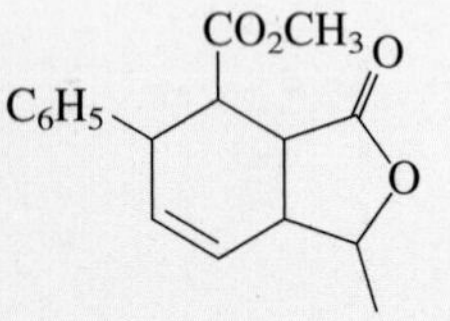

A.

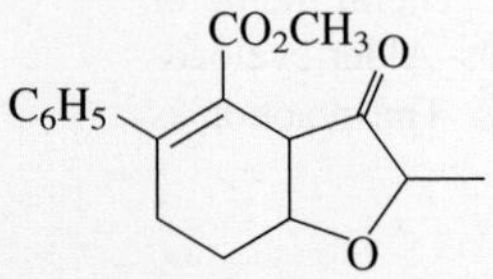

B.

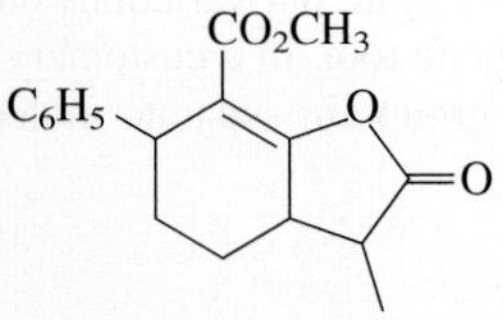

C.

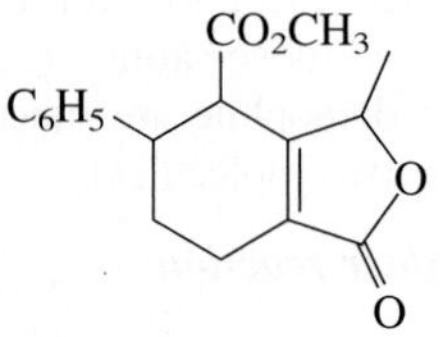

D.

10.50 What compound would you use to prepare the compound shown by an intramolecular Diels–Alder reaction?

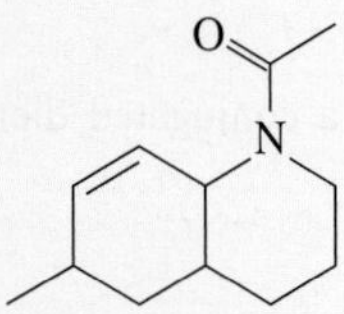

A.

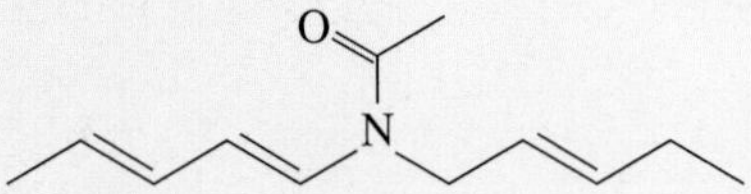

C.

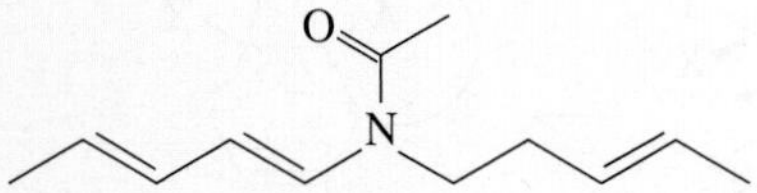

B.

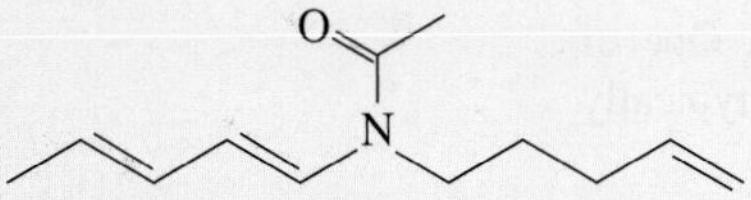

D.

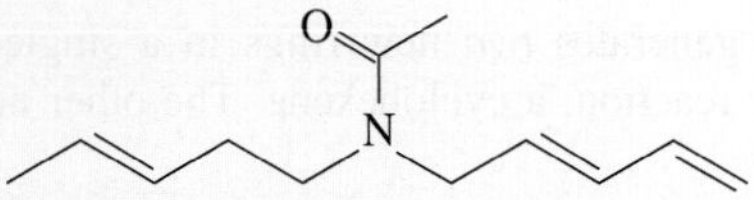

10.51 The customary laboratory source of 1,3-cyclopentadiene is a compound called "dicyclopentadiene" ($C_{10}H_{12}$). Dicyclopentadiene is the Diels–Alder cycloaddition product of two molecules of 1,3-cyclopentadiene. One molecule acts as diene, the other as a dienophile. Heating dicyclopentadiene causes it to undergo a retro Diels–Alder reaction to give 1,3-cyclopentadiene.

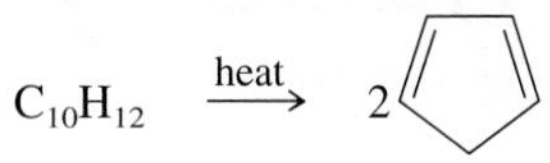
C10H12
heat
2

What is the structure of dicyclopentadiene?

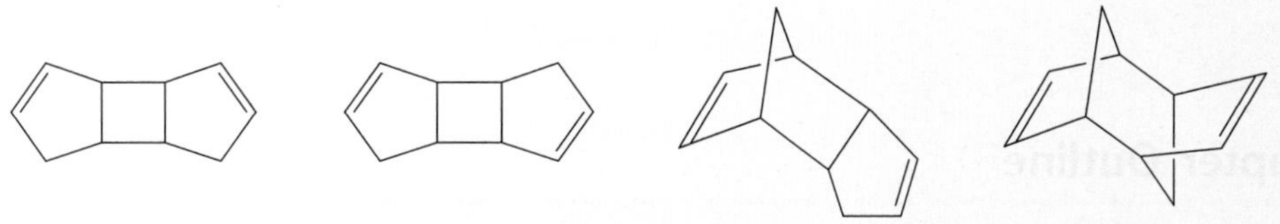

A. B. C. D.

10.52 Compound X is formed by way of a cycloaddition followed by a cycloelimination. What is its structure?

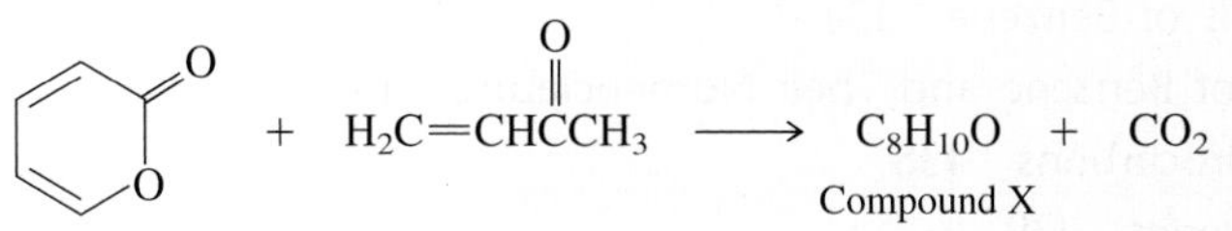
O
O
+ H2C=CHCCH3
O
C8H10O + CO2
Compound X

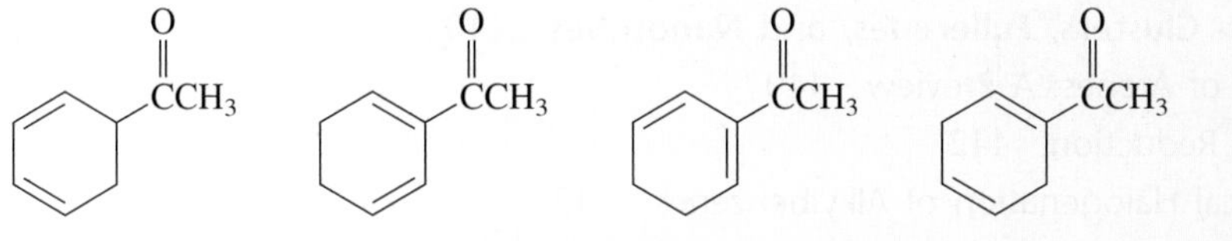
O
CCH3
O
CCH3
O
CCH3
O
CCH3

A. B. C. D.

10.53 What is compound X?

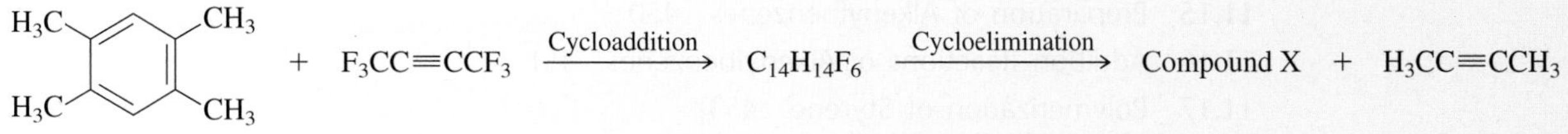
H3C
CH3
H3C
CH3
+ F3CC≡CCF3
Cycloaddition
C14H14F6
Cycloelimination
Compound X + H3CC≡CCH3

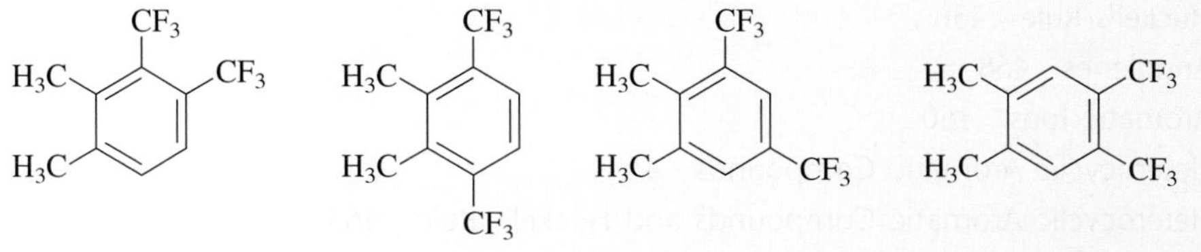
CF3
H3C
CF3
H3C
CF3
H3C
H3C
CF3
CF3
H3C
H3C
CF3
H3C
CF3
H3C
CF3

A. B. C. D.

11 Arenes and Aromaticity

Chapter Outline

Mechanisms

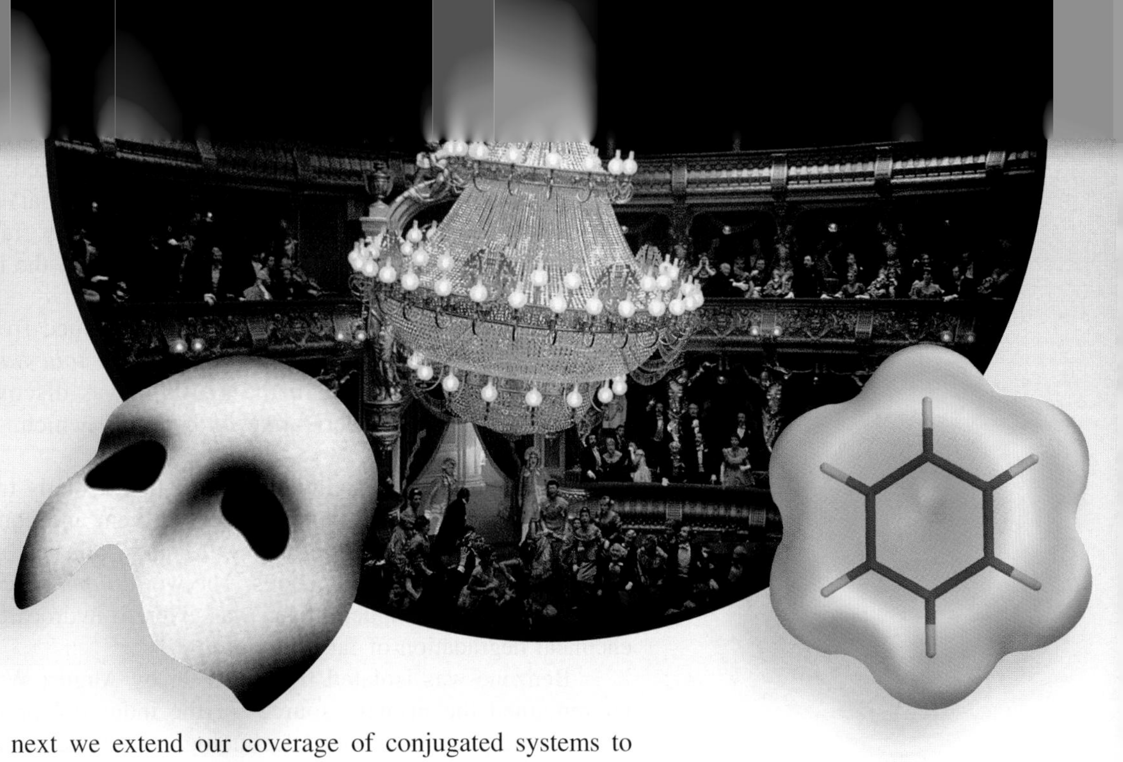

Illuminating gas was used for lighting in nineteenth-century Europe, including the infamous chandelier in the Opéra de Paris, the setting for the *Phantom of the Opera.* Methane, ethylene, and hydrogen are the main components of illuminating gas, but other hydrocarbons are present in small amounts. One of these is benzene.

IN THIS CHAPTER and the next we extend our coverage of conjugated systems to include **arenes.** Arenes are hydrocarbons based on the benzene ring as a structural unit. Benzene, toluene, and naphthalene, for example, are arenes.

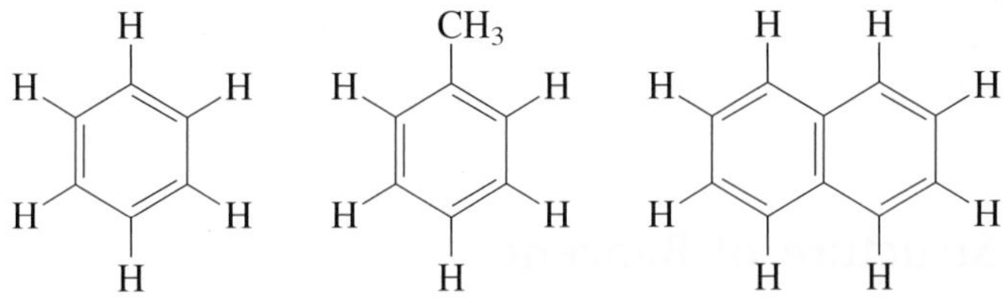

Benzene Toluene Naphthalene

One factor that makes conjugation in arenes special is its cyclic nature. A conjugated system that closes on itself can have properties that are much different from those of open-chain polyenes. Arenes are also referred to as aromatic hydrocarbons. Used in this sense, the word *aromatic* has nothing to do with odor but means instead that arenes are much more stable than we expect them to be based on their formulation as conjugated trienes. Our goal in this chapter is to develop an appreciation for the concept of **aromaticity**—to see what properties of benzene and its derivatives reflect its special stability and to explore the reasons for it. This chapter also examines the effect of a benzene ring as a substituent. The chapter following this one describes reactions that involve the ring itself.

Let's begin by tracing the history of benzene, its origin, and its structure. Many of the terms we use, including *aromaticity* itself, are of historical origin. We'll begin with the discovery of benzene.

11.1 Benzene

In 1825, Michael Faraday isolated a new hydrocarbon from illuminating gas, which he called "bicarburet of hydrogen." Nine years later Eilhardt Mitscherlich of the University of Berlin prepared the same substance by heating benzoic acid with lime and found it to be a hydrocarbon having the empirical formula C_nH_n.

Faraday is better known in chemistry for his laws of electrolysis and in physics for proposing the relationship between electric and magnetic fields and for demonstrating the principle of electromagnetic induction.

$$C_6H_5CO_2H + CaO \xrightarrow{\text{heat}} C_6H_6 + CaCO_3$$

Benzoic acid Calcium oxide Benzene Calcium carbonate

Eventually, because of its relationship to benzoic acid, this hydrocarbon came to be named *benzin,* then later *benzene,* the name by which it is known today.

Benzoic acid had been known for several hundred years by the time of Mitscherlich's experiment. Many trees exude resinous materials called *balsams* when cuts are made in their bark. Some of these balsams are very fragrant, which once made them highly prized articles of commerce, especially when the trees that produced them could be found only in exotic, faraway lands.

Compounds related to benzene were obtained from similar plant extracts. For example, a pleasant-smelling resin known as *tolu balsam* was obtained from the South American tolu tree. In the 1840s it was discovered that distillation of tolu balsam gave a methyl derivative of benzene, which, not surprisingly, came to be named *toluene*.

Although benzene and toluene are not particularly fragrant compounds themselves, their origins in aromatic plant extracts led them and compounds related to them to be classified as *aromatic hydrocarbons*. Alkanes, alkenes, and alkynes belong to another class, the **aliphatic hydrocarbons.** The word *aliphatic* comes from the Greek *aleiphar* (meaning "oil" or "unguent") and was given to hydrocarbons that were obtained by the chemical degradation of fats.

Benzene was isolated from coal tar by August W. von Hofmann in 1845. Coal tar remained the primary source for the industrial production of benzene for many years, until petroleum-based technologies became competitive about 1950. Current production is about 6 million tons per year in the United States. A substantial portion of this benzene is converted to styrene for use in the preparation of polystyrene plastics and films.

Toluene is also an important organic chemical. Like benzene, its early industrial production was from coal tar, but most of it now comes from petroleum.

11.2 The Structure of Benzene

The chemical properties of benzene were puzzling to those trying to understand its structure in the 1860s. The molecular formula of benzene is C_6H_6 and indicates that, like alkenes and alkynes, benzene is unsaturated and should undergo addition reactions. Under conditions in which bromine, for example, adds to alkenes and alkynes, however, benzene proved to be inert. Benzene does react with Br_2 in the presence of iron(III) bromide as a catalyst, but even then addition is not observed. Substitution occurs instead!

$$\underset{\text{Benzene}}{C_6H_6} + \underset{\text{Bromine}}{Br_2} \xrightarrow{CCl_4} \text{no observable reaction}$$

$$\underset{\text{Benzene}}{C_6H_6} + \underset{\text{Bromine}}{Br_2} \xrightarrow{FeBr_3} \underset{\text{Bromobenzene}}{C_6H_5Br} + \underset{\text{Hydrogen bromide}}{HBr}$$

Furthermore, only one monobromination product of benzene was ever obtained, which suggests that all the hydrogen atoms of benzene are equivalent.

Chemists came to regard the six carbon atoms of benzene as a fundamental structural unit. Reactions could be carried out that altered its substituents, but the integrity of the benzene unit remained intact. Benzene must have something "special" that makes it inert to many of the reagents that add to alkenes and alkynes.

In 1866, only a few years after publishing his ideas concerning what we now recognize as the structural theory of organic chemistry, August Kekulé applied it to the structure of benzene. He based his reasoning on three premises:

1. Benzene is C_6H_6.
2. All the hydrogens of benzene are equivalent.
3. The structural theory requires that there be four bonds to each carbon.

Kekulé advanced the venturesome notion that the six carbon atoms of benzene were joined together in a ring. Four bonds to each carbon could be accommodated by a system of alternating single and double bonds with one hydrogen on each carbon.

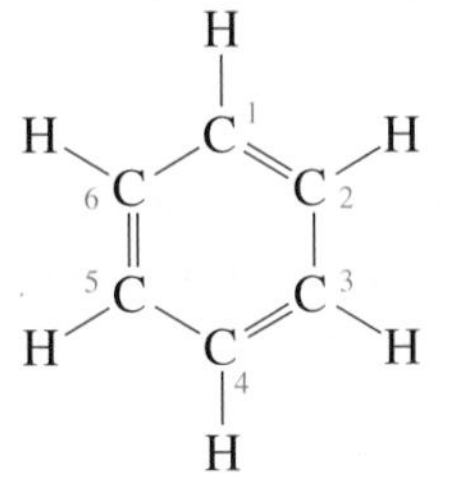

In 1861, Johann Josef Loschmidt, who was later to become a professor at the University of Vienna, privately published a book containing a structural formula for benzene similar to the one Kekulé would propose five years later. Loschmidt's book reached few readers, and his ideas were not well known.

A flaw in the **Kekulé structure** for benzene was soon discovered. Kekulé's structure requires that 1,2- and 1,6-disubstitution create different compounds (isomers).

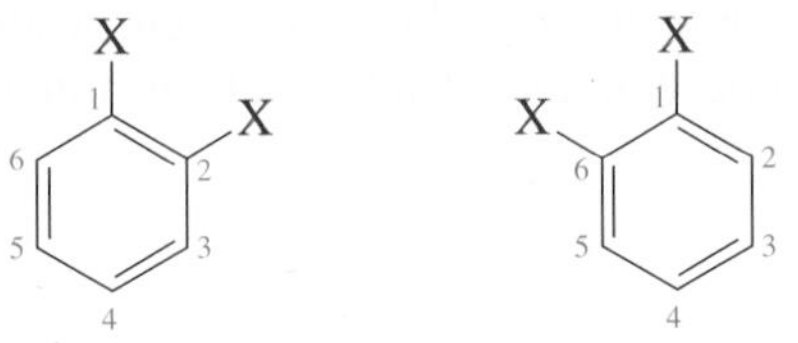

1,2-Disubstituted derivative of benzene

1,6-Disubstituted derivative of benzene

The two substituted carbons are connected by a double bond in one structure but by a single bond in the other. Because no such cases of isomerism in benzene derivatives were known, and none could be found, Kekulé suggested that two isomeric cyclohexatrienes could exist but interconverted too rapidly to be separated.

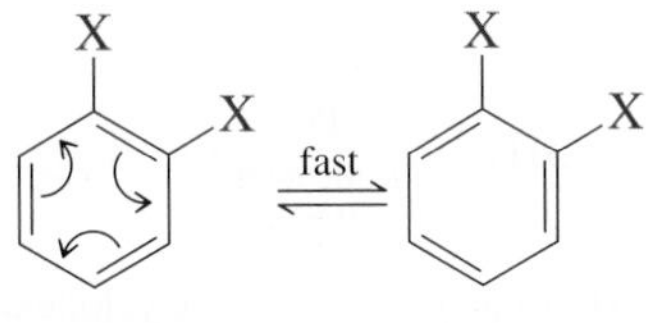

We now know that benzene is not cyclohexatriene, nor is it a pair of rapidly equilibrating isomers. Benzene is planar and its carbon skeleton has the shape of a regular hexagon. This is no evidence that it has alternating single and double bonds. As shown in Figure 11.1, all of the carbon–carbon bonds are the same length (140 pm) and the 120° bond angles correspond to perfect sp^2 hybridization. Interestingly, the 140-pm bond distances in benzene are exactly midway between the typical sp^2–sp^2 single-bond distance of 146 pm and the sp^2–sp^2 double-bond distance of 134 pm.

The two Kekulé structures for benzene have the same arrangement of atoms, but differ in the placement of electrons. Thus they are resonance forms, and neither one by itself correctly describes the bonding in the actual molecule. As a hybrid of the two Kekulé structures, benzene is often represented by a hexagon containing an inscribed circle.

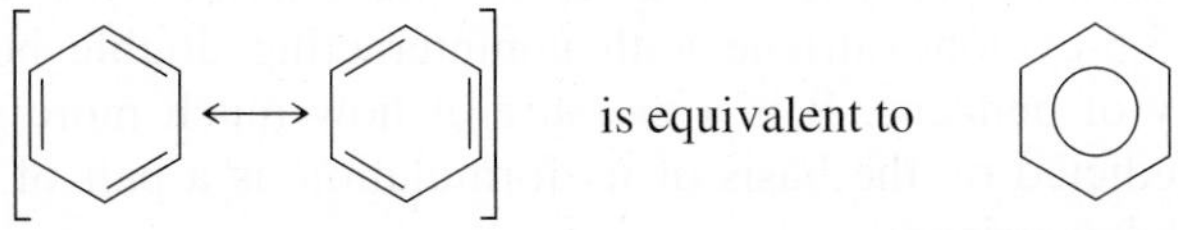

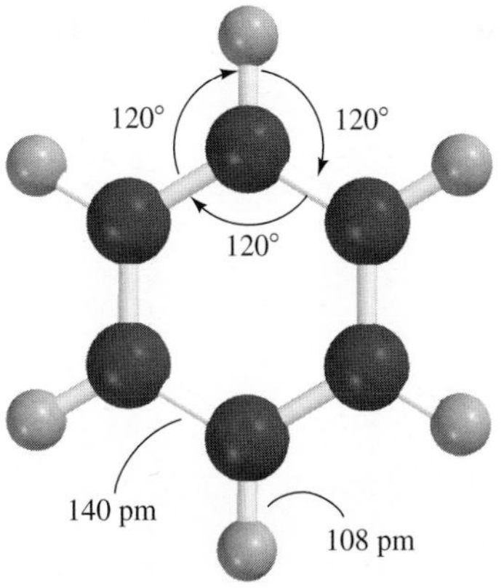

Figure 11.1

Bond distances and bond angles of benzene.

The circle-in-a-hexagon symbol was first suggested by the British chemist Sir Robert Robinson to represent what he called the "aromatic sextet"—the six delocalized π electrons of the three double bonds. Robinson's symbol is a convenient time-saving shorthand device, but Kekulé-type formulas are better for counting and keeping track of electrons, especially in chemical reactions.

Robinson won the 1947 Nobel Prize in Chemistry for his studies of natural products. He may also have been the first to use curved arrows to track electron movement.

Problem 11.1

Write structural formulas for toluene ($C_6H_5CH_3$) and for benzoic acid ($C_6H_5CO_2H$) (a) as resonance hybrids of two Kekulé forms and (b) with the Robinson symbol.

Because the carbons that are singly bonded in one resonance form are doubly bonded in the other, the resonance description is consistent with the observed carbon–carbon bond distances in benzene. These distances not only are all identical but also are intermediate between typical single-bond and double-bond lengths.

We have come to associate electron delocalization with increased stability. On that basis alone, benzene ought to be stabilized. It differs from other conjugated systems that we have seen, however, in that its π electrons are delocalized over a *cyclic conjugated* system. Both Kekulé structures of benzene are of equal energy, and one of the principles of resonance theory is that stabilization is greatest when the contributing structures are of similar energy. Cyclic conjugation in benzene, then, leads to a greater stabilization than is observed in noncyclic conjugated trienes. How much greater can be estimated from heats of hydrogenation.

11.3 The Stability of Benzene

Hydrogenation of benzene and other arenes is more difficult than hydrogenation of alkenes and alkynes. Two of the more active catalysts are rhodium and platinum, and it is possible to hydrogenate arenes in the presence of these catalysts at room temperature and modest pressure. Benzene consumes three molar equivalents of hydrogen to give cyclohexane.

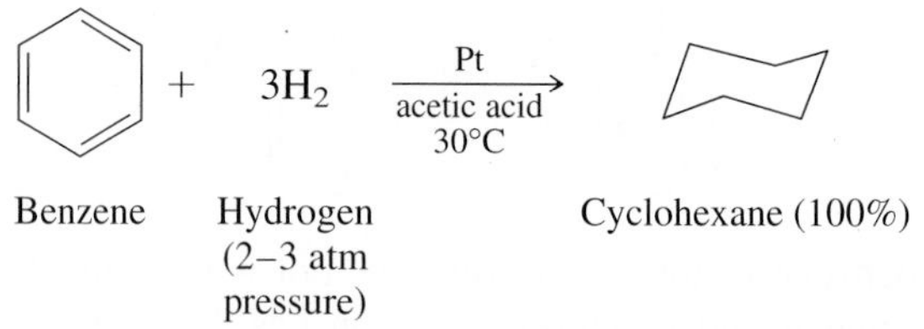
+
3H2
Pt
acetic acid
30°C
Benzene
Hydrogen
(2–3 atm
pressure)
Cyclohexane (100%)

Nickel catalysts, although less expensive than rhodium and platinum, are also less active. Hydrogenation of arenes in the presence of nickel requires high temperatures (100–200°C) and pressures (100 atm).

The measured heat of hydrogenation of benzene to cyclohexane is, of course, the same regardless of the catalyst and is 208 kJ/mol (49.8 kcal/mol). To put this value into perspective, compare it with the heats of hydrogenation of cyclohexene and 1,3-cyclohexadiene, as shown in Figure 11.2. The most striking feature of Figure 11.2 is that the heat of hydrogenation of benzene, with three "double bonds," is less than the heat of hydrogenation of the two double bonds of 1,3-cyclohexadiene. The heat of hydrogenation of benzene is 152 kJ/mol (36 kcal/mol) *less* than expected for a hypothetical 1,3,5-cyclohexatriene with noninteracting double bonds. This is the **resonance energy** of benzene. It is a measure of how much more stable benzene is than would be predicted on the basis of its formulation as a pair of rapidly interconverting 1,3,5-cyclohexatrienes.

We reach a similar conclusion when comparing benzene with the open-chain conjugated triene (*Z*)-1,3,5-hexatriene. Here we compare two real molecules, both conjugated trienes, but one is cyclic and the other is not. The heat of hydrogenation of (*Z*)-1,3,5-hexatriene is 337 kJ/mol (80.5 kcal/mol), a value which is 129 kJ/mol (30.7 kcal/mol) greater than that of benzene.

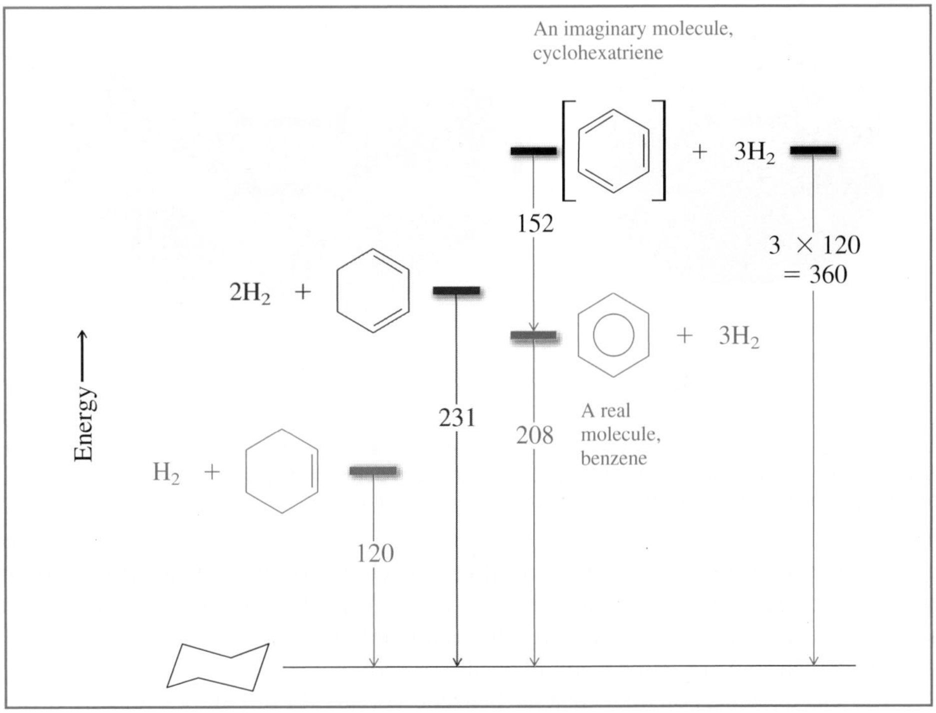

Figure 11.2

Heats of hydrogenation of cyclohexene, 1,3-cyclohexadiene, a hypothetical 1,3,5-cyclohexatriene, and benzene. All heats of hydrogenation are in kilojoules per mole.

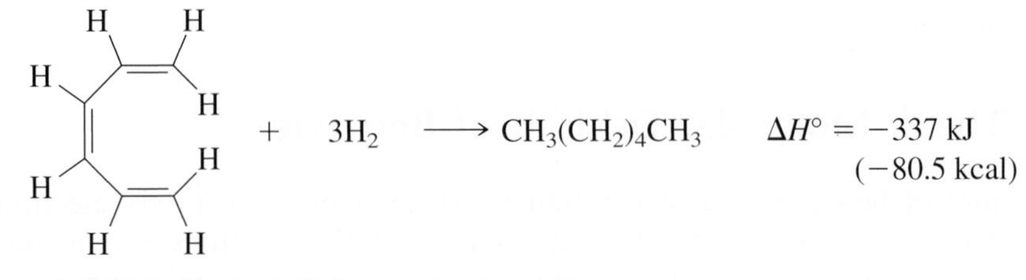

The precise value of the resonance energy of benzene depends, as comparisons with 1,3,5-cyclohexatriene and (*Z*)-1,3,5-hexatriene illustrate, on the compound chosen as the reference. What is important is that the resonance energy of benzene is quite large, six to ten times that of a conjugated triene. It is this very large increment of resonance energy that places benzene and related compounds in a separate category that we call *aromatic*.

Problem 11.2

The heats of hydrogenation of cycloheptene and 1,3,5-cycloheptatriene are 110 kJ/mol (26.3 kcal/mol) and 305 kJ/mol (73.0 kcal/mol), respectively. In both cases cycloheptane is the product. What is the resonance energy of 1,3,5-cycloheptatriene? How does it compare with the resonance energy of benzene?

11.4 An Orbital Hybridization View of Bonding in Benzene

The structural facts that benzene is planar, all of the bond angles are 120°, and each carbon is bonded to three other atoms, suggest sp^2 hybridization for carbon and the framework of σ bonds shown in Figure 11.3*a*.

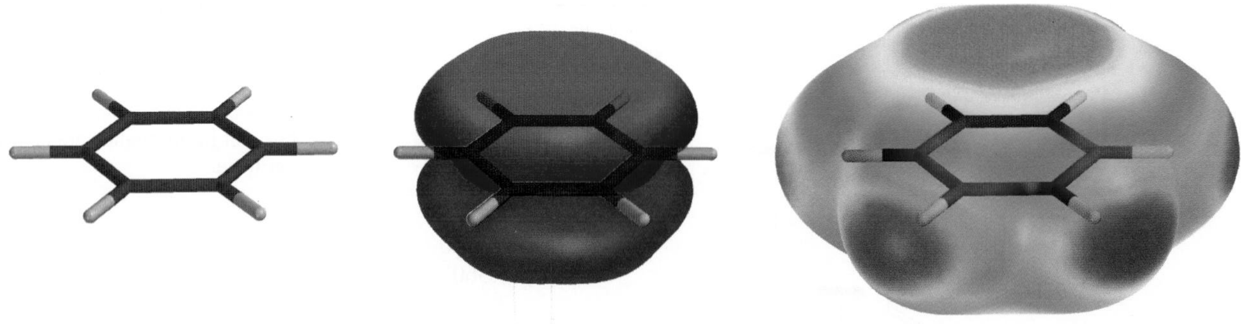

(a) (b) (c)

Figure 11.3

(*a*) The framework of bonds shown in the tube model of benzene are σ bonds. (*b*) Each carbon is sp^2-hybridized and has a $2p$ orbital perpendicular to the σ framework. Overlap of the $2p$ orbitals generates a π system encompassing the entire ring. (*c*) Electrostatic potential map of benzene. The red area in the center corresponds to the region above and below the plane of the ring where the π electrons are concentrated.

In addition to its three sp^2 hybrid orbitals, each carbon has a half-filled $2p$ orbital that can participate in π bonding. Figure 11.3*b* shows the continuous π system that encompasses all of the carbons that result from overlap of these $2p$ orbitals. The six π electrons of benzene are delocalized over all six carbons.

The electrostatic potential map of benzene (Figure 11.3*c*) shows regions of high electron density above and below the plane of the ring, which is where we expect the most loosely held electrons (the π electrons) to be. In Chapter 12 we will see how this region of high electron density is responsible for the characteristic chemical reactivity of benzene and its relatives.

11.5 The π Molecular Orbitals of Benzene

The picture of benzene as a planar framework of σ bonds with six electrons in a delocalized π orbital is a useful, but superficial, one. Six electrons cannot simultaneously occupy any one orbital, be it an atomic orbital or a molecular orbital. We can fix this with the more accurate molecular orbital picture shown in Figure 11.4. We learned in Section 2.4 that when atomic orbitals (AOs) combine to give molecular orbitals (MOs), the final number of MOs must equal the original number of AOs. Thus, the six $2p$ AOs of six sp^2-hybridized carbons combine to give six π MOs of benzene.

The orbitals in Figure 11.4 are arranged in order of increasing energy. Three orbitals are bonding; three are antibonding. Each of the three bonding MOs contains two electrons, accounting for the six π electrons of benzene. There are no electrons in the antibonding MOs. Benzene is said to have a **closed-shell** π-electron configuration.

Figure 11.4 also shows the orbital overlaps and nodal properties of the benzene MOs. Recall that a wave function changes sign on passing through a nodal plane and is zero at a node (Section 1.1). All of the orbital interactions in the lowest energy orbital π_1 are bonding; therefore, π_1 has no nodes. The other two bonding orbitals π_2 and π_3 each have one nodal plane. The first two antibonding orbitals π_4^* and π_5^* each have two nodal planes. The highest energy orbital π_6^* has three nodal planes. All adjacent *p* orbitals are out of phase with one another in π_6^*—all of the interactions are antibonding.

The pattern of orbital energies is different for benzene than it would be if the six π electrons were confined to three noninteracting double bonds. The delocalization provided by cyclic conjugation in benzene causes its π electrons to be held more strongly than they would be in the absence of cyclic conjugation. Stronger binding of its π electrons is the factor most responsible for the special stability—the aromaticity—of benzene.

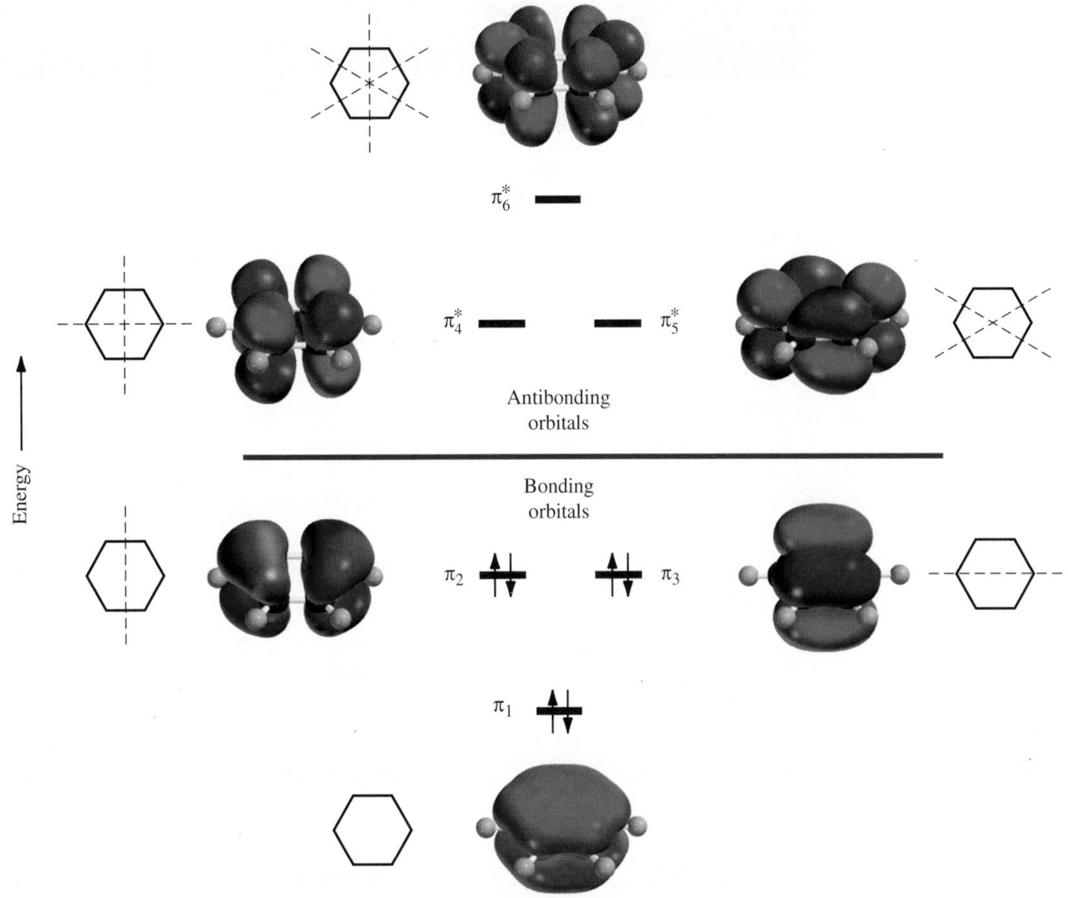

Figure 11.4

The π molecular orbitals of benzene arranged in order of increasing energy and showing nodal surfaces. The six π electrons of benzene occupy the three lowest energy orbitals, all of which are bonding.

Later in this chapter we'll explore the criteria for aromaticity in more detail to see how they apply to cyclic polyenes of different ring sizes. The next several sections introduce us to the chemistry of compounds that contain a benzene ring as a structural unit. We'll start with how we name them.

11.6 Substituted Derivatives of Benzene and Their Nomenclature

All compounds that contain a benzene ring are aromatic, and substituted derivatives of benzene make up the largest class of aromatic compounds. Many such compounds are named by attaching the name of the substituent as a prefix to *benzene.*

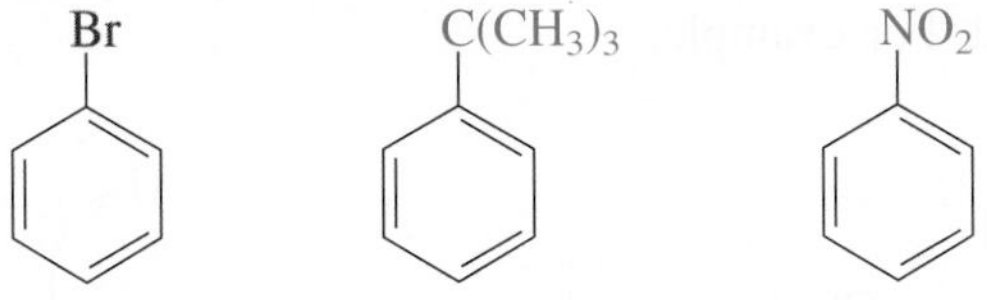

Bromobenzene *tert*-Butylbenzene Nitrobenzene

Many simple monosubstituted derivatives of benzene have common names of long standing that have been retained in the IUPAC system. Table 11.1 lists some of the most important ones.

Dimethyl derivatives of benzene are called *xylenes.* There are three xylene isomers, the *ortho* (*o*)-, *meta* (*m*)-, and *para* (*p*)- substituted derivatives.

TABLE 11.1 Names of Some Frequently Encountered Derivatives of Benzene

Structure	Systematic name	Common name*
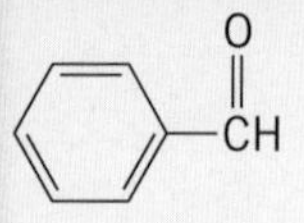	Benzenecarbaldehyde	Benzaldehyde
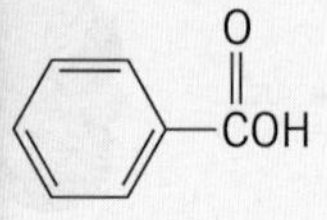	Benzenecarboxylic acid	Benzoic acid
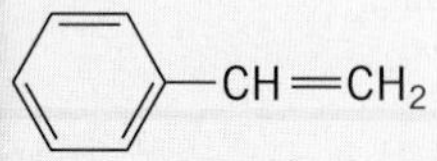	Vinylbenzene	Styrene
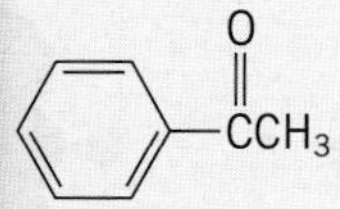	Methyl phenyl ketone	Acetophenone
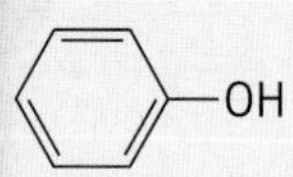	Benzenol	Phenol
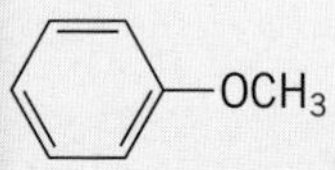	Methoxybenzene	Anisole
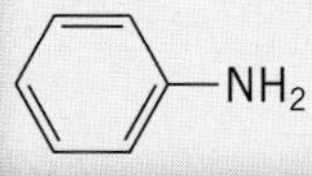	Benzenamine	Aniline

*These common names are acceptable in IUPAC nomenclature and are the names that will be used in this text.

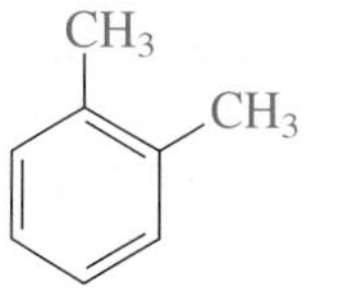

o-Xylene
(1,2-dimethylbenzene)

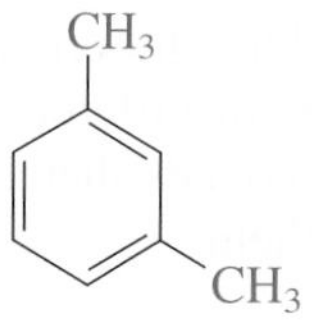

m-Xylene
(1,3-dimethylbenzene)

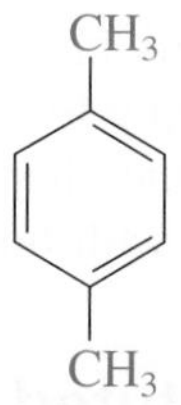

p-Xylene
(1,4-dimethylbenzene)

The prefix **ortho** signifies a 1,2-disubstituted benzene ring, **meta** signifies 1,3 disubstitution, and **para** signifies 1,4 disubstitution. The prefixes *o, m,* and *p* can be used when a substance is named as a benzene derivative or when a specific base name (such as acetophenone) is used. For example,

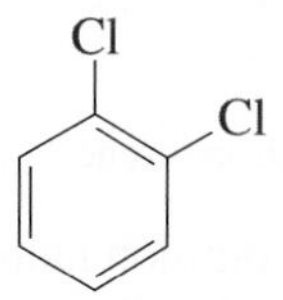

o-Dichlorobenzene
(1,2-dichlorobenzene)

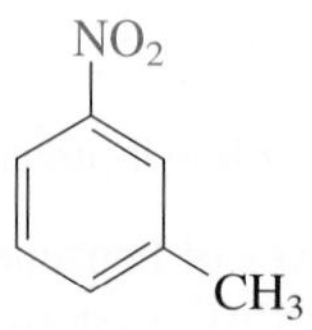

m-Nitrotoluene
(3-nitrotoluene)

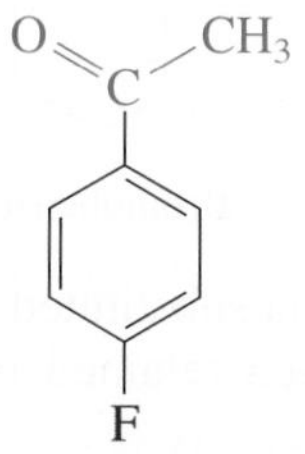

p-Fluoroacetophenone
(4-fluoroacetophenone)

Problem 11.3

Write a structural formula for each of the following compounds:

(a) *o*-Ethylanisole
(b) *m*-Chlorostyrene
(c) *p*-Nitroaniline

Sample Solution (a) The parent compound in *o*-ethylanisole is anisole. Anisole, as shown in Table 11.1, has a methoxy (CH_3O—) substituent on the benzene ring. The ethyl group in *o*-ethylanisole is attached to the carbon adjacent to the one that bears the methoxy substituent.

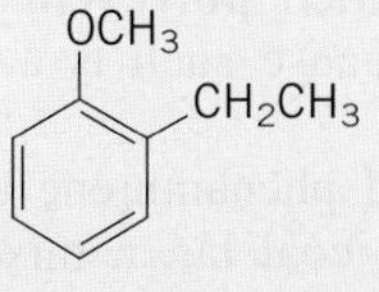

o-Ethylanisole

The *o, m,* and *p* prefixes are *not* used when three or more substituents are present on benzene; numerical locants must be used instead.

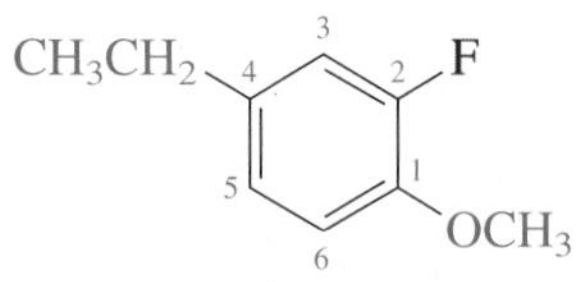

4-Ethyl-2-fluoroanisole

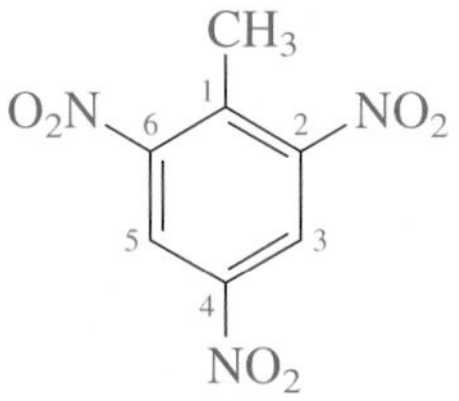

2,4,6-Trinitrotoluene

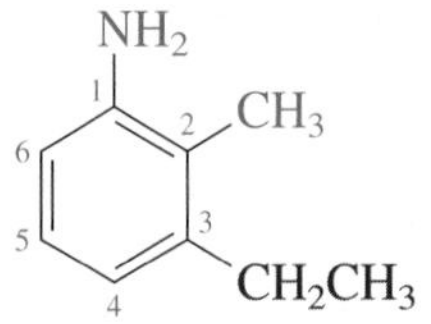

3-Ethyl-2-methylaniline

In these examples the base name of the benzene derivative determines the carbon at which numbering begins: anisole has its methoxy group at C-1, toluene its methyl group at C-1, and aniline its amino group at C-1. The direction of numbering is chosen to give the next substituted position the lowest number irrespective of what substituent it bears. *The order of appearance of substituents in the name is alphabetical.* When no simple base name other than benzene is appropriate, positions are numbered so as to give the lowest locant at the first point of difference. Thus, each of the following examples is named as a 1,2,4-trisubstituted derivative of benzene rather than as a 1,3,4-derivative:

The "first point of difference" rule was introduced in Section 2.14.

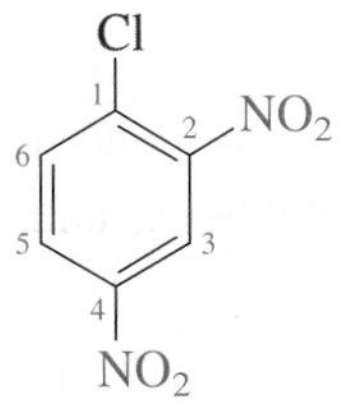

1-Chloro-2,4-dinitrobenzene

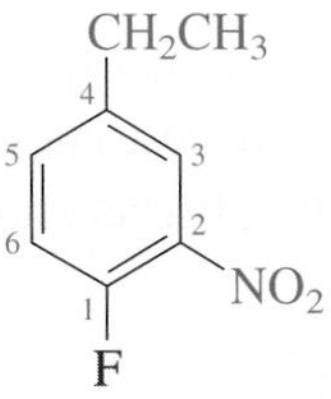

4-Ethyl-1-fluoro-2-nitrobenzene

When the benzene ring is named as a substituent, the word **phenyl** stands for C_6H_5—. Similarly, an arene named as a substituent is called an *aryl* group. A **benzyl group** is $C_6H_5CH_2$—.

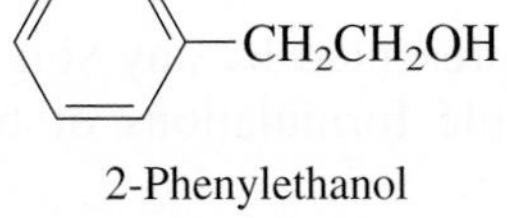

2-Phenylethanol

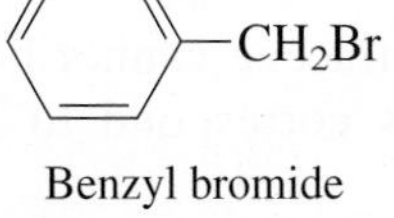

Benzyl bromide

Biphenyl is the accepted IUPAC name for the compound in which two benzene rings are connected by a single bond.

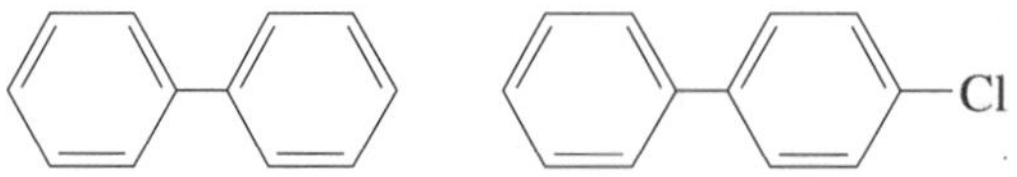

Biphenyl p-Chlorobiphenyl

11.7 Polycyclic Aromatic Hydrocarbons

Members of a class of arenes called **polycyclic aromatic hydrocarbons** possess substantial resonance energies because each is a collection of benzene rings fused together.

Naphthalene is a white crystalline solid melting at 80°C that sublimes readily. It has a characteristic odor and was formerly used as a moth repellent.

Naphthalene, anthracene, and phenanthrene are the three simplest members of this class. They are all present in coal tar, a mixture of organic substances formed when coal is converted to coke by heating at high temperatures (about 1000°C) in the absence of air. Naphthalene is bicyclic (has two rings), and its two benzene rings share a common side. Anthracene and phenanthrene are both tricyclic aromatic hydrocarbons. Anthracene has three rings fused in a "linear" fashion; an "angular" fusion characterizes phenanthrene. The structural formulas of naphthalene, anthracene, and phenanthrene are shown along with the numbering system used to name their substituted derivatives:

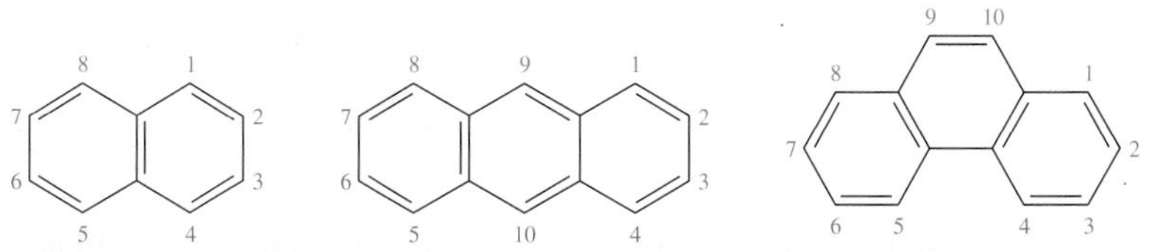

Arene:	Naphthalene	Anthracene	Phenanthrene
Resonance energy:	255 kJ/mol (61 kcal/mol)	347 kJ/mol (83 kcal/mol)	381 kJ/mol (91 kcal/mol)

Problem 11.4

How many monochloro derivatives of anthracene are possible? Write their structural formulas and give their IUPAC names.

In general, the most stable resonance structure for a polycyclic aromatic hydrocarbon is the one with the greatest number of rings that correspond to Kekulé formulations of benzene. Naphthalene provides a fairly typical example:

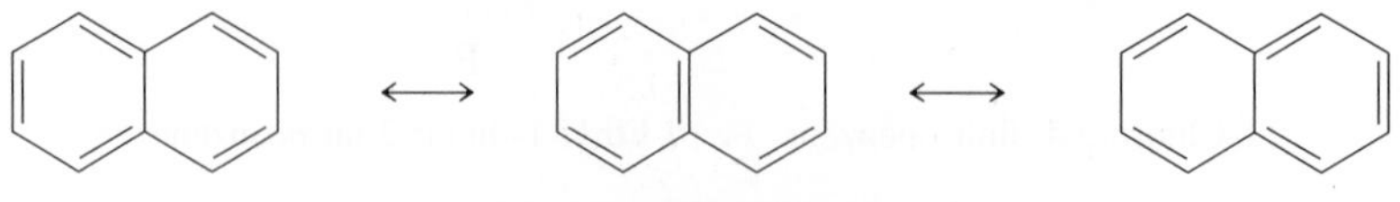

Only left ring corresponds to Kekulé benzene.

Most stable resonance form; major contributor
Both rings correspond to Kekulé benzene.

Only right ring corresponds to Kekulé benzene.

Notice that anthracene cannot be represented by any single Lewis structure in which all three rings correspond to Kekulé formulations of benzene, but phenanthrene can.

Problem 11.5

Chrysene is an aromatic hydrocarbon found in coal tar. Convert the molecular model to a Lewis structure in which all of the rings correspond to Kekulé formulas of benzene.

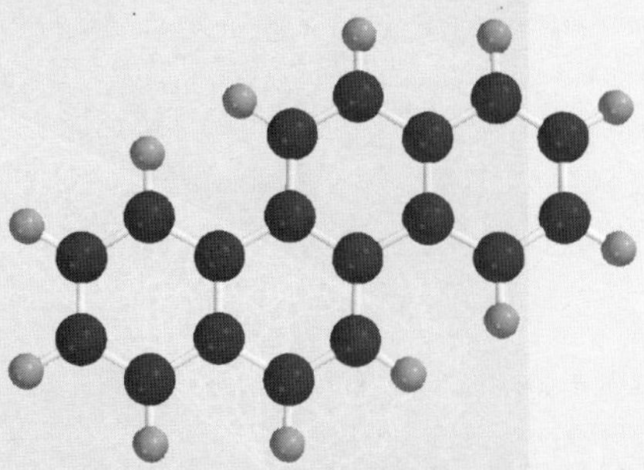

Chrysene

Problem 11.6

Bromine adds to the central ring of anthracene to give a 1,4-addition product. Write the structure of the product that would be formed if addition took place on one of the outer rings. By writing resonance structures for the product shown here and the one formed by addition to the outer ring, can you suggest why addition to the central ring is preferred?

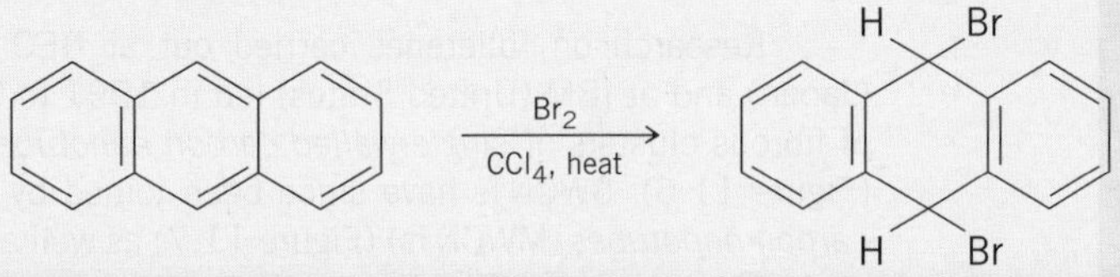

A large number of polycyclic aromatic hydrocarbons are known. Many have been synthesized in the laboratory, and several of the others are products of combustion. Benzo[a]pyrene, for example, is present in tobacco smoke, contaminates food cooked on barbecue grills, and collects in the soot of chimneys. Benzo[a]pyrene is a **carcinogen** (a cancer-causing substance). It is converted in the liver to an epoxy diol that can induce mutations leading to the uncontrolled growth of certain cells.

In 1775, the British surgeon Sir Percivall Pott suggested that scrotal cancer in chimney sweeps was caused by soot. This was the first proposal that cancer could be caused by chemicals present in the workplace.

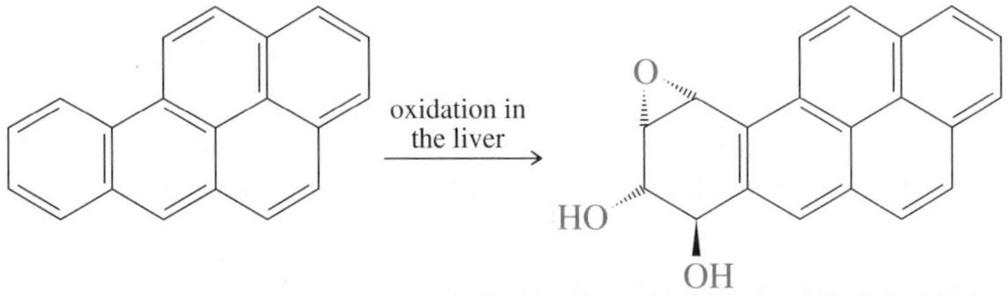

Benzo[a]pyrene

7,8-Dihydroxy-9,10-epoxy-7,8,9,10-tetrahydrobenzo[a]pyrene

11.8 Physical Properties of Arenes

In general, arenes resemble other hydrocarbons in their physical properties. They are nonpolar, insoluble in water, and less dense than water. In the absence of polar substituents, intermolecular forces are weak and limited to van der Waals attractions of the induced-dipole/induced-dipole type.

At one time, benzene was widely used as a solvent. This use virtually disappeared when statistical studies revealed an increased incidence of leukemia among workers exposed to atmospheric levels of benzene as low as 1 ppm. Toluene has replaced benzene as an

Carbon Clusters, Fullerenes, and Nanotubes

In general, the term *nanoscale* applies to dimensions on the order of 1–100 nanometers (1 nm = 10^{-9} m), and one goal of *nanotechnology* is to develop useful nanoscale devices (*nanodevices*). Because typical covalent bonds range from 0.1–0.2 nm, chemical structures hold promise as candidates on which to base nanodevices. Among them, much recent attention has been given to carbon-containing materials and even elemental carbon itself.

Until 1985, chemists recognized two elementary forms (*allotropes*) of carbon: diamond and graphite. Then, Professors Harold W. Kroto (University of Sussex), Robert F. Curl, and Richard E. Smalley (both of Rice University) reported that laser-induced evaporation of graphite gave a species with a molecular formula of C_{60} and proposed the spherical cluster of carbon atoms now called **buckminsterfullerene** (Figure 11.5) for it. Other closed carbon clusters, some larger than C_{60} and some smaller, were also formed in the experiment. These forms of carbon are now known as *fullerenes,* and Kroto, Smalley, and Curl were awarded the 1996 Nobel Prize in Chemistry for discovering them.

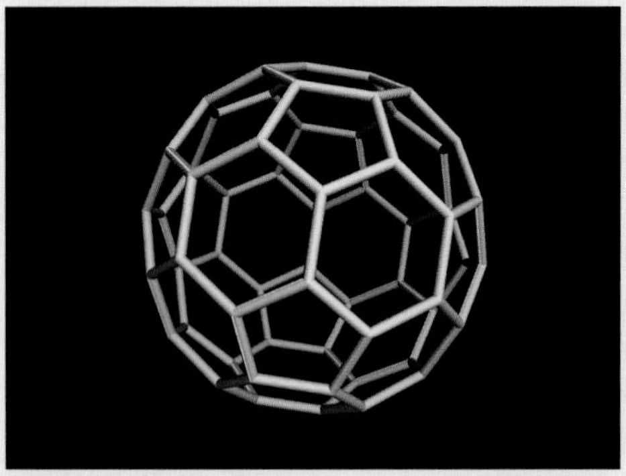

Figure 11.5

Buckminsterfullerene (C_{60}). All of the carbon atoms are equivalent and are sp^2-hybridized; each one simultaneously belongs to one five-membered ring and two benzene-like six-membered rings.

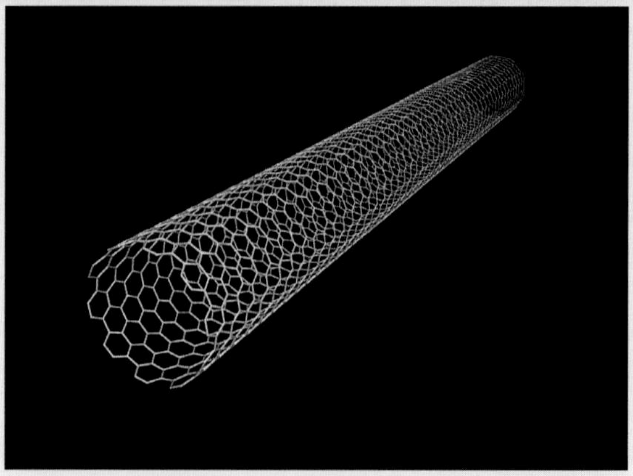

Figure 11.6

A single-walled carbon nanotube (SWCNT). SWCNTs can be regarded as a two-dimensional graphite sheet rolled into a cylinder. Credit: The single-walled carbon nanotube in Figure 11.6 is from the Spartan model provided by Dr. Warren J. Hehre of Wavefunction, Inc.

Research on fullerenes carried out at NEC Corporation (Japan) and at IBM (United States) led in 1991 to the isolation of fibrous clusters of *single-walled carbon nanotubes* (SWCNTs) (Figure 11.6). SWCNTs have since been joined by *multiwalled carbon nanotubes* (MWCNTs) (Figure 11.7) as well as nanotubes containing elements other than carbon.

CNTs themselves are of interest because of their electrical and mechanical properties, and functionally modified ones are being examined in applications ranging from medical diagnosis and therapy to photovoltaic systems. The methods used to add functionality to a CNT include among others: (1) covalent attachment of a reactive group to the CNT via a chemical reaction, and (2) noncovalent coating of the outer surface of the CNT with a substance that itself bears a functional substituent.

Continued

inexpensive organic solvent because it has similar solvent properties but has not been determined to be carcinogenic in the cell systems and at the dose levels that benzene is.

11.9 Reactions of Arenes: A Preview

We'll examine the chemical properties of aromatic compounds from two different perspectives:

1. *One mode of chemical reactivity involves the ring itself as a functional group and includes*
   (a) Reduction and oxidation
   (b) Electrophilic aromatic substitution

Reduction of arenes by catalytic hydrogenation was described in Section 11.3. A different method using Group 1 metals as reducing agents, which gives 1,4-cyclohexadiene derivatives, will be presented in Section 11.10. Oxidations of aromatic compounds are discussed in Chapter 22. **Electrophilic aromatic substitution** is the most important reaction type exhibited by benzene and its derivatives and constitutes the entire subject matter of Chapter 12.

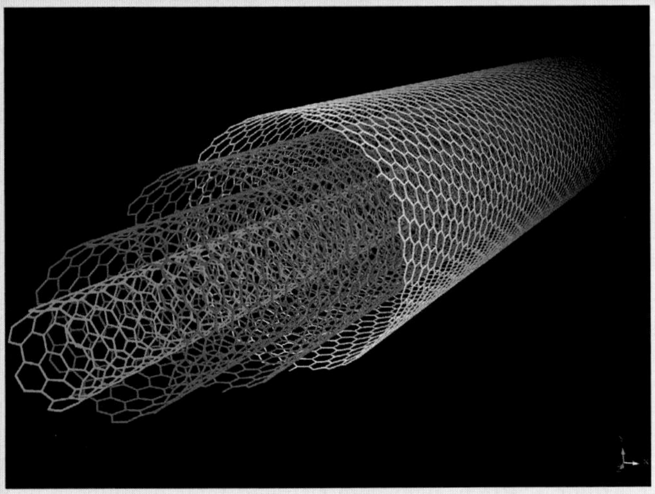

Figure 11.7

A multiwalled carbon nanotube (MWCNT).

The cover of this edition illustrates a carbon nanotube that is functionalized with DNA. The cover of our seventh edition featured the covalently modified SWCNT shown in Figure 11.8. There, an iron-containing organic molecule called *ferrocene* (Section 14.13) was tethered to the CNT to give a nanodevice capable of converting sunlight to electricity. On absorption of visible light, the ferrocene-containing side chain transfers an electron to the nanotube. The resulting species is analogous to a battery in that it has a positively charged component (the ferrocene unit) separated from a negatively charged one (the CNT). Like the graphite rod in a flashlight battery, the CNT is the cathode.

An approach to noncovalent functionalization takes advantage of attractive van der Waals forces to coat the outer surface of a nanotube with a substance that bears the functionality. This approach was successfully demonstrated using a synthetic polymer designed to mimic the properties of mucin. (Mucin is the main substance responsible for the lubricating properties of mucus and is important in processes that involve cell surfaces.) When the CNTs were coated with a synthetic "mucin" that had been modified by attaching specific structural units known to bind to the surfaces of particular cells, the combination bound to the test cells and was not toxic to them. The latter observation is significant because CNTs themselves are toxic. Binding of functionalized CNTs to specific cells is important because it provides, among other things, a method for targeting only those cells for drug delivery.

The unique properties of CNTs, in particular their high surface area and conductivity, have already attracted the attention of researchers in several areas of biology and chemistry. The ability to attach molecules with specific functions to the surfaces of CNTs opens up many possibilities for the applications of these novel materials as nanoscale devices.

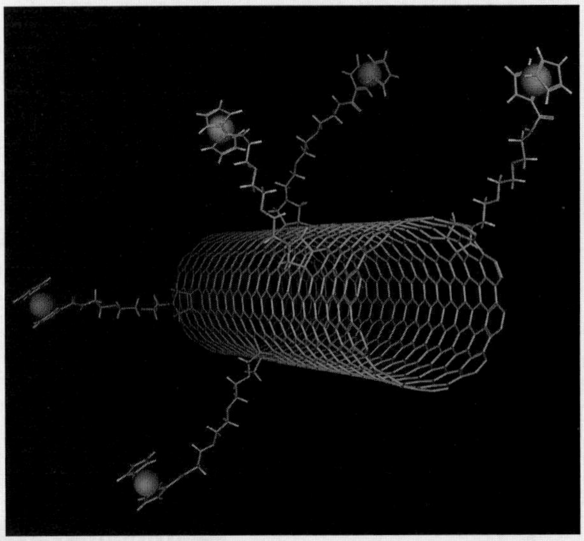

Figure 11.8

A functionalized SWCNT bearing ferrocene-containing side chains capable of absorbing visible light and converting it to electrical energy.

2. *The second family of reactions are those in which the aryl group acts as a substituent and affects the reactivity of a functional unit to which it is attached.*

A carbon atom that is directly attached to a benzene ring is called a **benzylic carbon** (analogous to the allylic carbon of C=C—**C**). A phenyl group (C_6H_5—) is an even better conjugating substituent than a vinyl group (H_2C=CH—), and benzylic carbocations and radicals are more highly stabilized than their allylic counterparts. The double bond of an alkenylbenzene is stabilized to about the same extent as that of a conjugated diene.

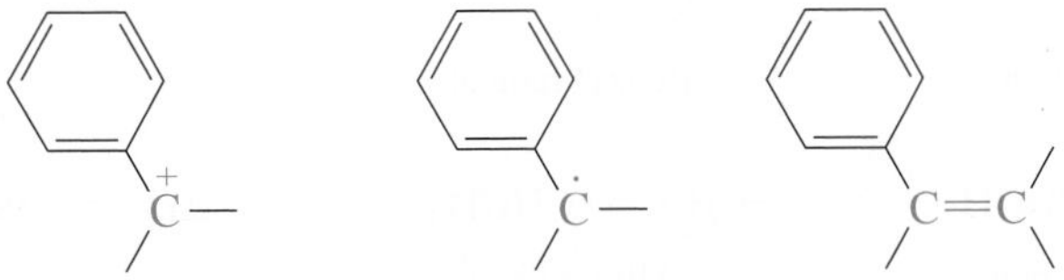

Benzylic carbocation Benzylic radical Alkenylbenzene

Reactions involving benzylic cations, benzylic radicals, and alkenylbenzenes will be discussed in Sections 11.11 through 11.17.

11.10 The Birch Reduction

We saw in Section 9.10 that the combination of a Group 1 metal and liquid ammonia is a powerful reducing system capable of reducing alkynes to trans alkenes. In the presence of an alcohol, this same combination reduces arenes to *nonconjugated dienes.* Thus, treatment of benzene with sodium and methanol or ethanol in liquid ammonia converts it to 1,4-cyclohexadiene.

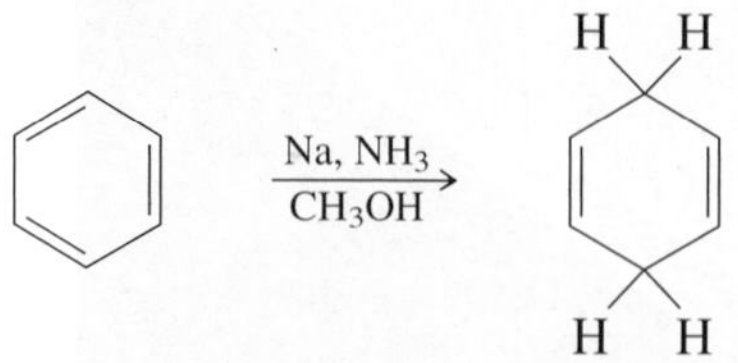

Benzene 1,4-Cyclohexadiene (80%)

Metal–ammonia–alcohol reductions of aromatic rings are known as **Birch reductions,** after the Australian chemist Arthur J. Birch, who demonstrated their usefulness beginning in the 1940s.

The mechanism by which the Birch reduction of benzene takes place (Mechanism 11.1) is analogous to the mechanism for the metal–ammonia reduction of alkynes. It involves a sequence of four steps in which steps 1 and 3 are single-electron transfers from the metal and steps 2 and 4 are proton transfers from the alcohol.

The Birch reduction not only provides a method to prepare dienes from arenes, which cannot be accomplished by catalytic hydrogenation, but also gives a nonconjugated diene system rather than the more stable conjugated one.

Alkyl-substituted arenes give 1,4-cyclohexadienes in which the alkyl group is a substituent on the double bond.

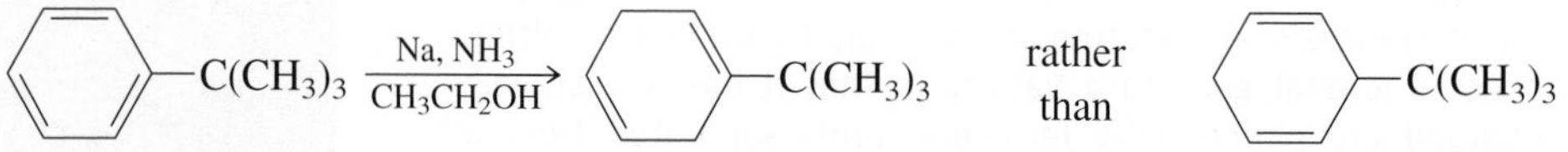

tert-Butylbenzene 1-*tert*-Butyl-1,4-cyclohexadiene (86%) 3-*tert*-Butyl-1,4-cyclohexadiene

Problem 11.7

A single organic product was isolated after Birch reduction of *p*-xylene. Suggest a reasonable structure for this substance.

Substituents other than alkyl groups may also be present on the aromatic ring, but their reduction is beyond the scope of the present discussion.

11.11 Free-Radical Halogenation of Alkylbenzenes

The benzylic position in alkylbenzenes is analogous to the allylic position in alkenes. Thus a benzylic C—H bond, like an allylic one, is weaker than a C—H bond of an alkane, as the bond dissociation enthalpies of toluene, propene, and 2-methylpropane attest:

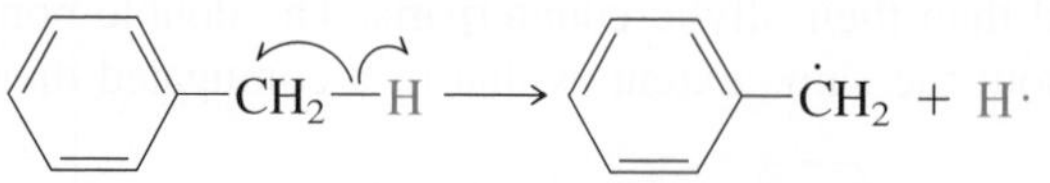

$\Delta H° = 356$ kJ (85 kcal)

Toluene Benzyl radical

$$H_2C{=}CHCH_2{-}H \longrightarrow H_2C{=}CH\dot{C}H_2 + H\cdot \qquad \Delta H° = 368 \text{ kJ (88 kcal)}$$

Propene Allyl radical

$$(CH_3)_3C{-}H \longrightarrow (CH_3)_3C\cdot + H\cdot \qquad \Delta H° = 380 \text{ kJ (91 kcal)}$$

2-Methylpropane *tert*-Butyl radical

Mechanism 11.1

The Birch Reduction

THE OVERALL REACTION:

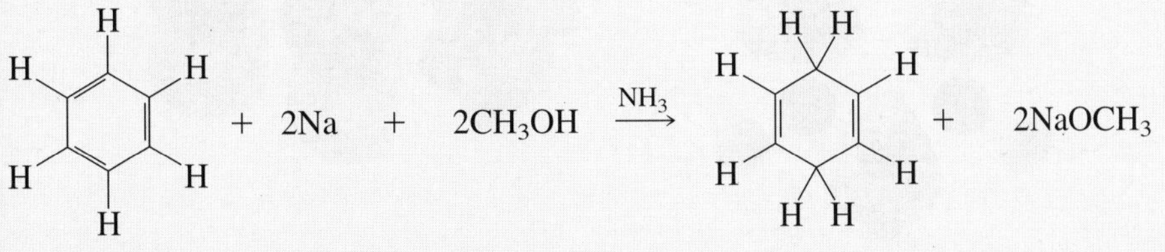

Benzene | Sodium | Methanol | 1,4-Cyclohexadiene | Sodium methoxide

THE MECHANISM:

Step 1: An electron is transferred from sodium (the reducing agent) to the π system of the aromatic ring. The product is an anion radical.

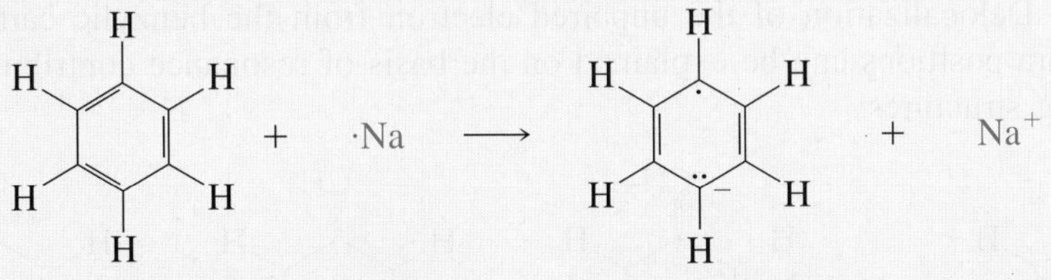

Benzene | Sodium | Benzene anion radical | Sodium ion

Step 2: The anion radical is a strong base and abstracts a proton from methanol.

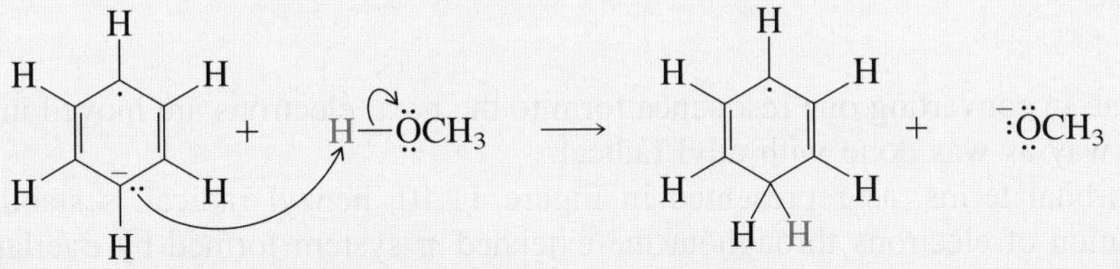

Benzene anion radical | Methanol | Cyclohexadienyl radical | Methoxide ion

Step 3: The cyclohexadienyl radical produced in step 2 is converted to an anion by electron transfer from sodium.

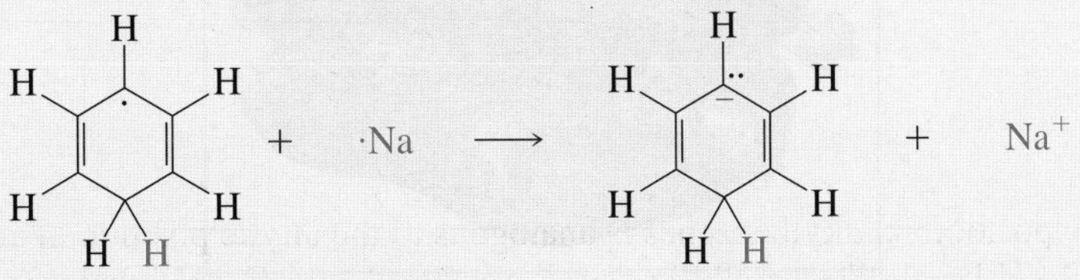

Cyclohexadienyl radical | Sodium | Cyclohexadienyl anion | Sodium ion

Step 4: Proton transfer from methanol to the anion gives 1,4-cyclohexadiene.

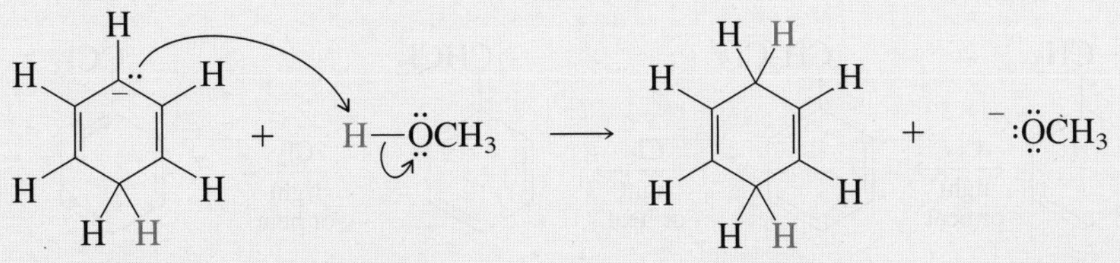

Cyclohexadienyl anion | Methanol | 1,4-Cyclohexadiene | Methoxide ion

Figure 11.9

Two views of the spin density in benzyl radical. The unpaired electron is shared mainly by the benzylic carbon and the ortho and para carbons of the ring.

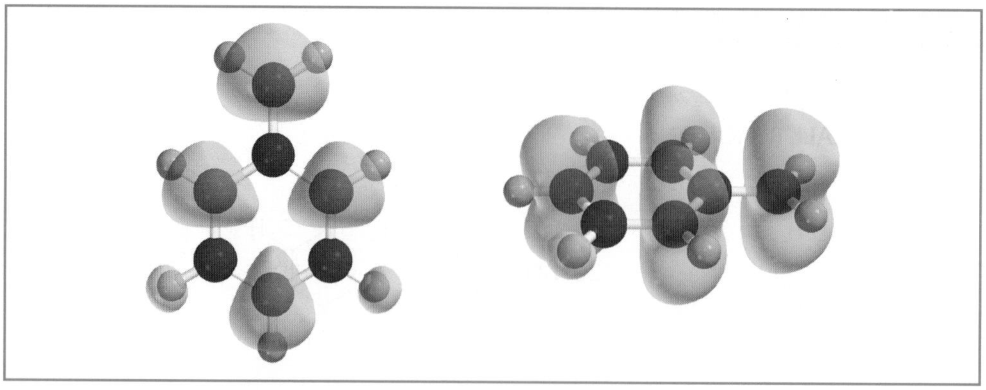

We attributed the decreased bond dissociation enthalpy in propene to stabilization of allyl radical by electron delocalization. Similarly, electron delocalization stabilizes benzyl radical and weakens the benzylic C—H bond.

The unpaired electron in benzyl radical is shared by the benzylic carbon and by the ring carbons that are ortho and para to it as shown by the spin density surface in Figure 11.9. Delocalization of the unpaired electron from the benzylic carbon to the ortho and para positions can be explained on the basis of resonance contributions from the following structures:

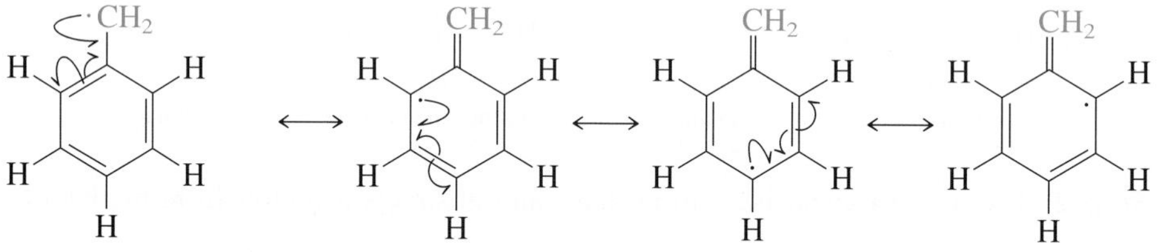

Major contributor

Notice that, in converting one resonance form to the next, electrons are moved in exactly the same way as was done with allyl radical.

In orbital terms, as represented in Figure 11.10, benzyl radical is stabilized by delocalization of electrons throughout the extended π system formed by overlap of the *p* orbital of the benzylic carbon with the π system of the ring.

Figure 11.10

The lowest energy π molecular orbital of benzyl radical shows the interaction of the 2*p* orbital of the benzylic carbon with the π system of the aromatic ring.

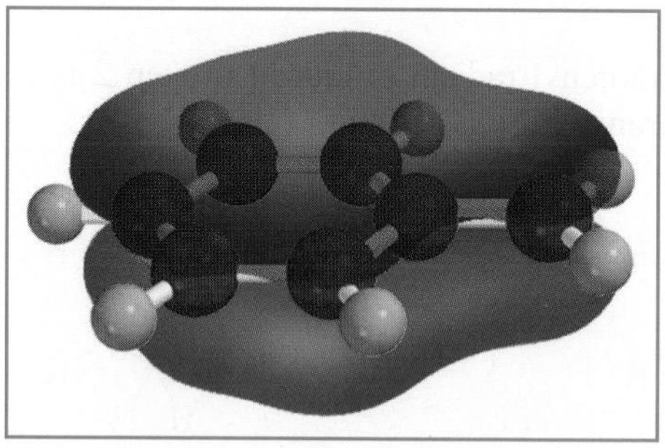

The comparative ease with which a benzylic hydrogen is abstracted leads to high selectivity in free-radical halogenations of alkylbenzenes. Thus, chlorination of toluene takes place exclusively at the benzylic carbon and is an industrial process for the preparation of the compounds shown.

The common names of (dichloromethyl)benzene and (trichloromethyl)benzene are benzal chloride and benzotrichloride, respectively.

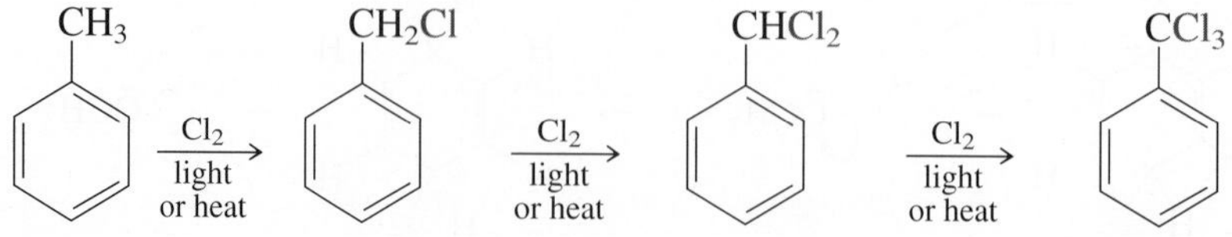

Toluene   Benzyl chloride   (Dichloromethyl)-benzene   (Trichloromethyl)-benzene

The propagation steps in the formation of benzyl chloride involve benzyl radical as an intermediate.

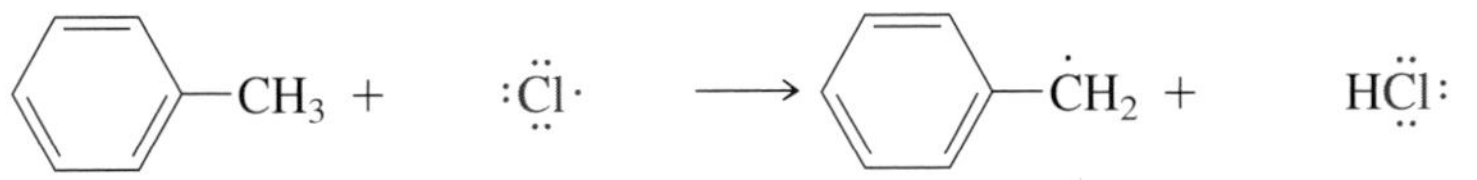

Toluene — Chlorine atom — Benzyl radical — Hydrogen chloride

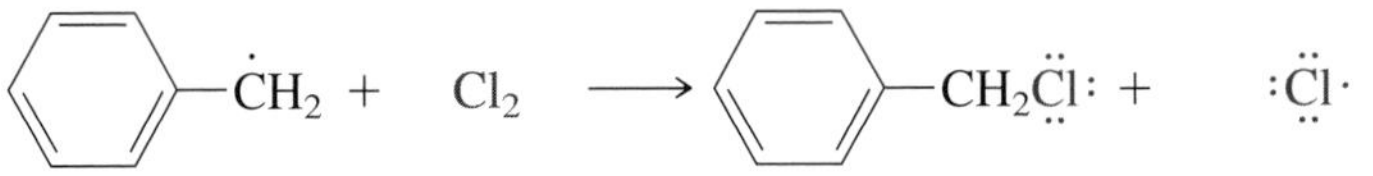

Benzyl radical — Chlorine — Benzyl chloride — Chlorine atom

(Dichloromethyl)benzene and (trichloromethyl)benzene arise by further side-chain chlorination of benzyl chloride.

Problem 11.8

The unpaired electron in benzyl radical is shared by the benzylic carbon plus the ortho and para carbons of the ring. Yet, chlorine becomes attached only to the benzylic carbon. Can you think of a reason why? (*Hint:* Write a structural formula for the compound formed by attachment of chlorine to one of the ring carbons.)

Benzylic bromination is a more commonly used laboratory procedure than chlorination and is typically carried out under conditions of photochemical initiation.

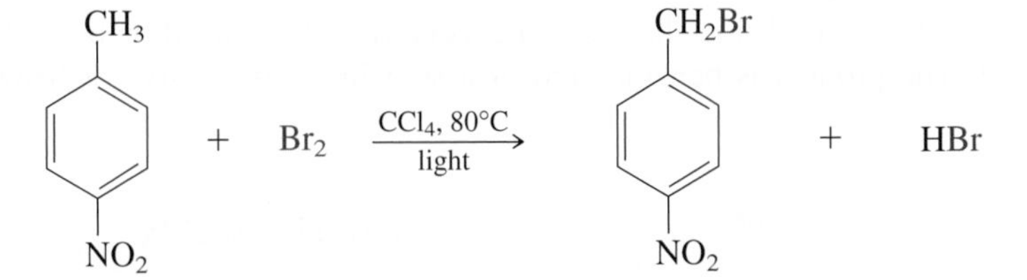

p-Nitrotoluene — Bromine — *p*-Nitrobenzyl bromide (71%) — Hydrogen bromide

As we saw when discussing allylic bromination in Section 10.6, *N*-bromosuccinimide (NBS) is a convenient free-radical brominating agent. Benzylic brominations with NBS are normally performed in carbon tetrachloride as the solvent in the presence of peroxides, which are added as initiators. As the example illustrates, free-radical bromination is selective for substitution of benzylic hydrogens.

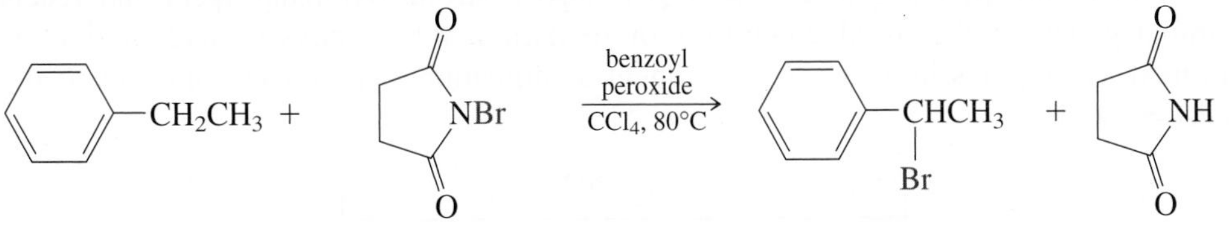

Ethylbenzene — *N*-Bromosuccinimide (NBS) — 1-Bromo-1-phenylethane (87%) — Succinimide

Benzoyl peroxide is a commonly used free-radical initiator. It has the formula

$$C_6H_5\overset{O}{\overset{\|}{C}}OO\overset{O}{\overset{\|}{C}}C_6H_5$$

Problem 11.9

The reaction of *N*-bromosuccinimide with the following compounds has been reported in the chemical literature. Each compound yields a single product in 95% yield. Identify the product formed from each starting material.

(a) *p-tert*-Butyltoluene

(b) 4-Methyl-3-nitroanisole

Continued

Sample Solution (a) The only benzylic hydrogens in *p-tert*-butyltoluene are those of the methyl group that is attached directly to the ring. Substitution occurs there to give *p-tert*-butylbenzyl bromide.

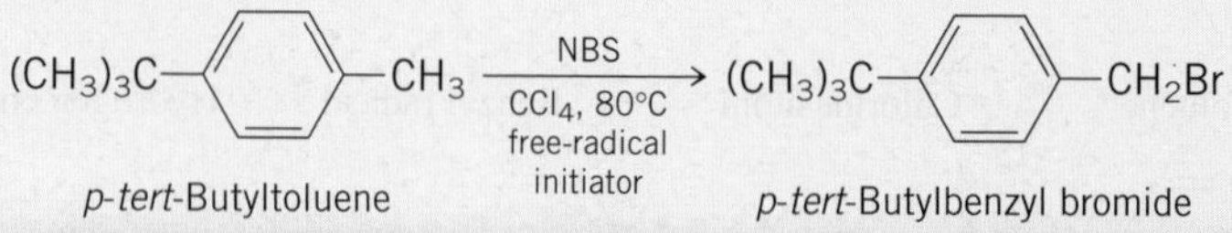
(CH3)3C
CH3
NBS
CCl4, 80°C
free-radical
initiator
(CH3)3C
CH2Br
p-tert-Butyltoluene
p-tert-Butylbenzyl bromide

11.12 Oxidation of Alkylbenzenes

A striking example of the activating effect that a benzene ring has on reactions that take place at benzylic positions may be found in the reactions of alkylbenzenes with oxidizing agents. Chromic acid (H_2CrO_4), for example, prepared by adding sulfuric acid to aqueous sodium dichromate, is a strong oxidizing agent but does not react either with benzene or with alkanes.

$$RCH_2CH_2R' \xrightarrow[H_2O,\ H_2SO_4,\ heat]{Na_2Cr_2O_7} \text{no reaction}$$

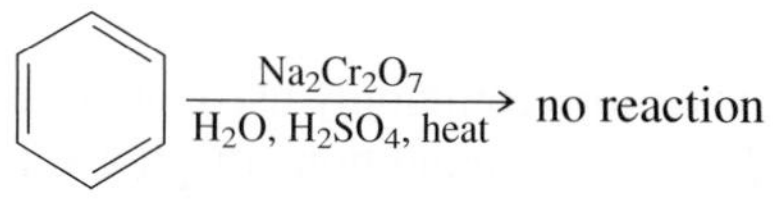
Na2Cr2O7
H2O, H2SO4, heat
no reaction

On the other hand, an alkyl side chain on a benzene ring is oxidized on being heated with chromic acid. The product is benzoic acid or a substituted derivative of benzoic acid.

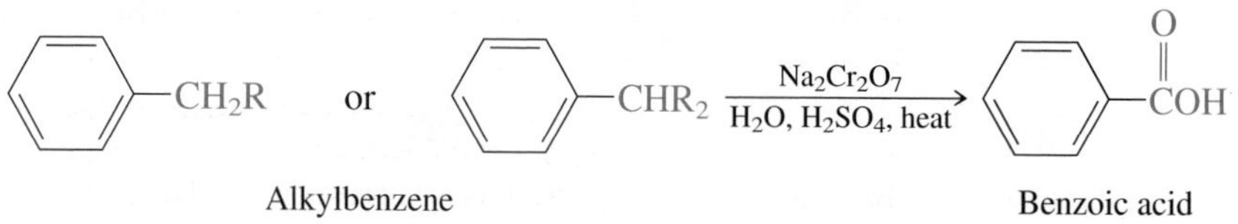
CH2R
or
CHR2
Na2Cr2O7
H2O, H2SO4, heat
O
COH
Alkylbenzene
Benzoic acid

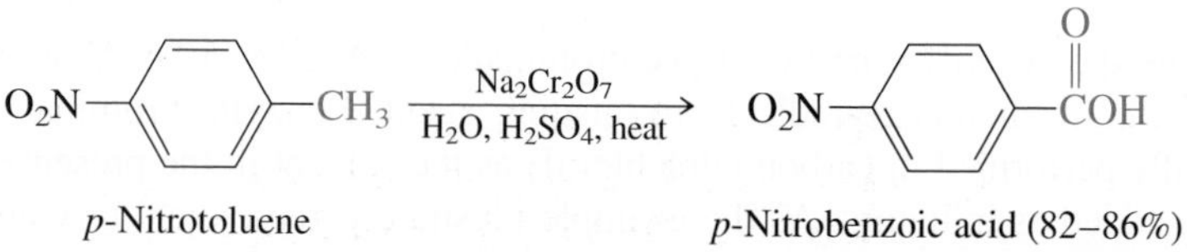
O2N
CH3
Na2Cr2O7
H2O, H2SO4, heat
O2N
O
COH
p-Nitrotoluene
p-Nitrobenzoic acid (82–86%)

Potassium permanganate ($KMnO_4$) is also a strong oxidizing agent and reacts similarly. Under the usual conditions of oxidation, the carboxylic acid product is formed as its potassium salt. A subsequent acidification step converts the salt to the desired acid.

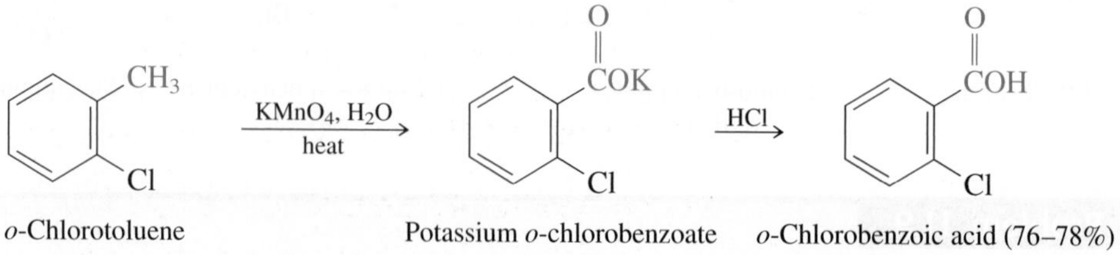
CH3
Cl
KMnO4, H2O
heat
O
COK
Cl
HCl
O
COH
Cl
o-Chlorotoluene
Potassium o-chlorobenzoate
o-Chlorobenzoic acid (76–78%)

When two alkyl groups are present on the ring, both are oxidized.

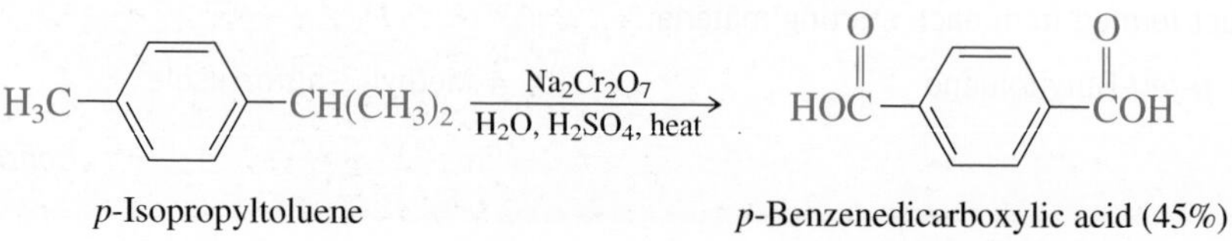
H3C
CH(CH3)2
Na2Cr2O7
H2O, H2SO4, heat
O
HOC
O
COH
p-Isopropyltoluene
p-Benzenedicarboxylic acid (45%)

Note that alkyl groups, regardless of their chain length, are converted to carboxyl groups (—CO_2H) attached directly to the ring. An exception is a substituent of the type —CR_3. Because it lacks benzylic hydrogens, such a group is not susceptible to oxidation under these conditions.

Problem 11.10

Chromic acid oxidation of 4-*tert*-butyl-1,2-dimethylbenzene yielded a single compound having the molecular formula $C_{12}H_{14}O_4$. What was this compound?

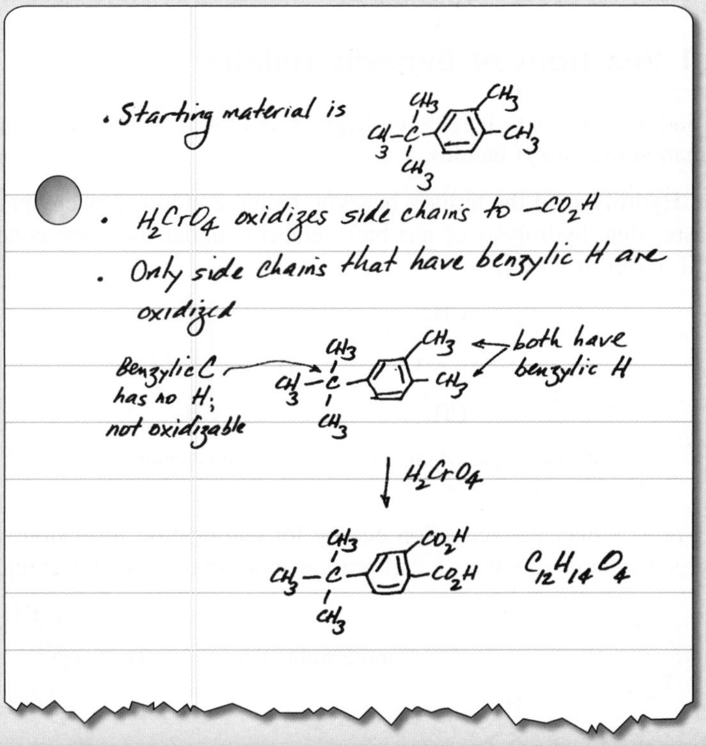

Problem 11.11

What product is expected from chromic acid oxidation of 2,3-dihydroindene?

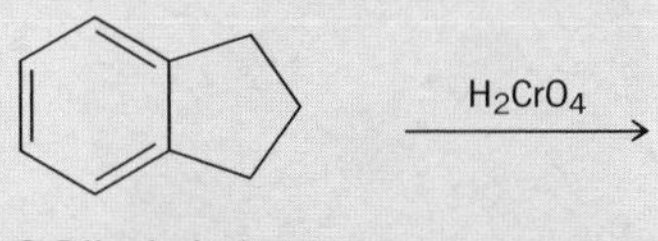

2,3-Dihydroindene

Side-chain oxidation of alkylbenzenes is important in certain metabolic processes. One way in which the body rids itself of foreign substances is by oxidation in the liver to compounds that are more easily excreted in the urine. Toluene, for example, is oxidized to benzoic acid and is eliminated rather readily.

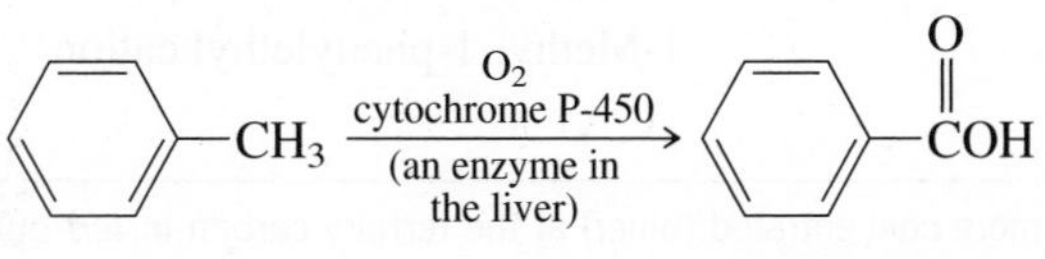

Toluene Benzoic acid

Benzene, with no alkyl side chain and no benzylic hydrogens, undergoes a different reaction under these conditions. Oxidation of the ring occurs to convert benzene to its epoxide.

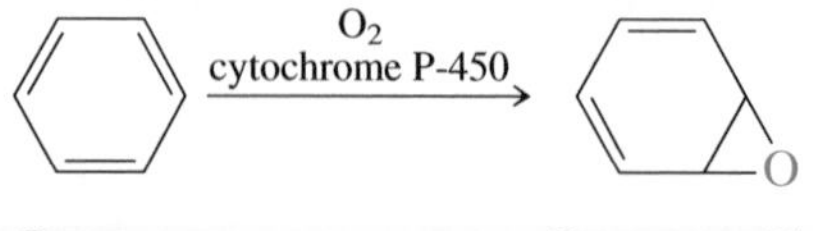

Benzene Benzene oxide

Benzene oxide and compounds derived from it are carcinogenic and can react with DNA to induce mutations. This difference in the site of biological oxidation—ring versus side-chain—seems to be responsible for the fact that benzene is carcinogenic but toluene is not.

11.13 S_N1 Reactions of Benzylic Halides

Like allylic halides, benzylic halides undergo nucleophilic substitution, both S_N1 and S_N2, faster than simple alkyl halides.

S_N1 Rates. Hydrolysis of the tertiary benzylic halide 2-chloro-2-phenylpropane occurs 620 times faster than hydrolysis of *tert*-butyl chloride under the same conditions (90% acetone—10% water at 25°C).

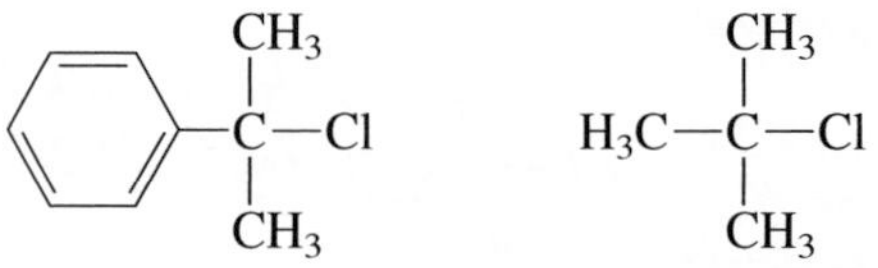

2-Chloro-2-phenylpropane — More reactive: *k*(rel) 620

tert-Butyl chloride — Less reactive: *k*(rel) 1.0

Because S_N1 rates reflect the activation energy for carbocation formation, we conclude that a phenyl substituent stabilizes a carbocation more than a methyl group.

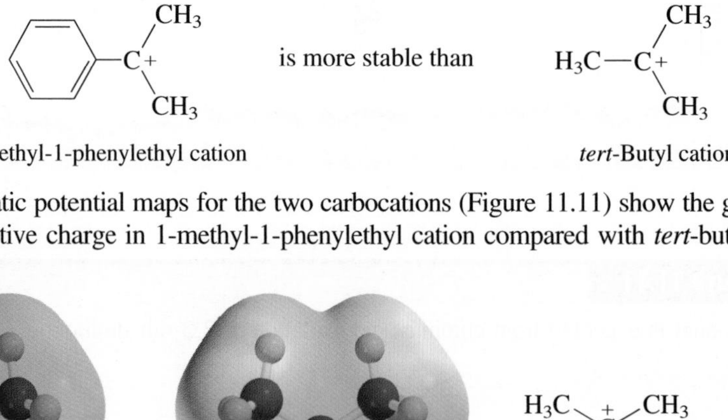

1-Methyl-1-phenylethyl cation *tert*-Butyl cation

The electrostatic potential maps for the two carbocations (Figure 11.11) show the greater dispersal of positive charge in 1-methyl-1-phenylethyl cation compared with *tert*-butyl cation.

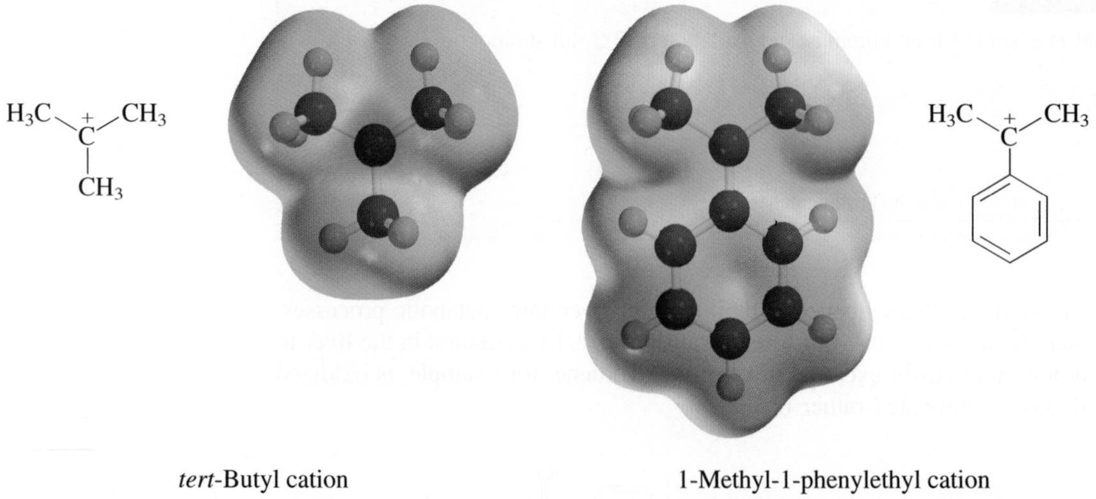

Figure 11.11

The respective electrostatic potential maps show the positive charge is more concentrated (bluer) at the tertiary carbon in *tert*-butyl cation than at the benzylic tertiary carbon in 1-methyl-1-phenylethyl cation. The color range is the same for both models.

Resonance structures indicate the positive charge is shared by the benzylic carbon and the ring carbons ortho and para to it.

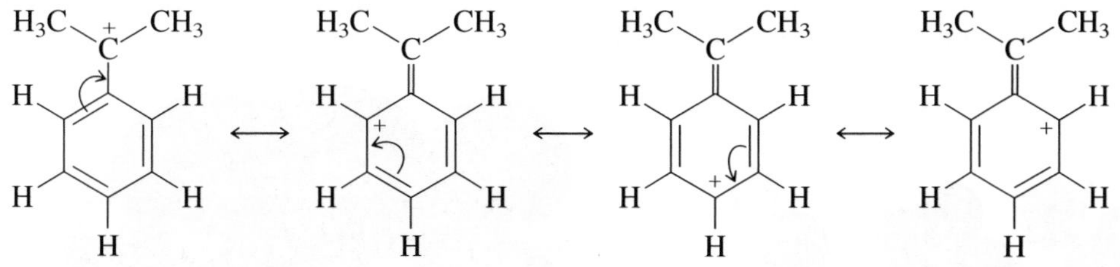

Major contributor

The major contributor is the Lewis structure that retains the aromaticity of the benzene ring.

S_N1 Products. Unlike the case with allylic carbocations, however, dispersal of the positive charge does not result in nucleophilic attack at more than one carbon. There is no "benzylic rearrangement" analogous to allylic rearrangement (Section 10.3), because the aromatic stabilization would be lost if the nucleophile became bonded to one of the ring carbons. Thus, when conditions are chosen that favor S_N1 substitution over E2 elimination (solvolysis, weakly basic nucleophile), benzylic halides give a single substitution product in high yield.

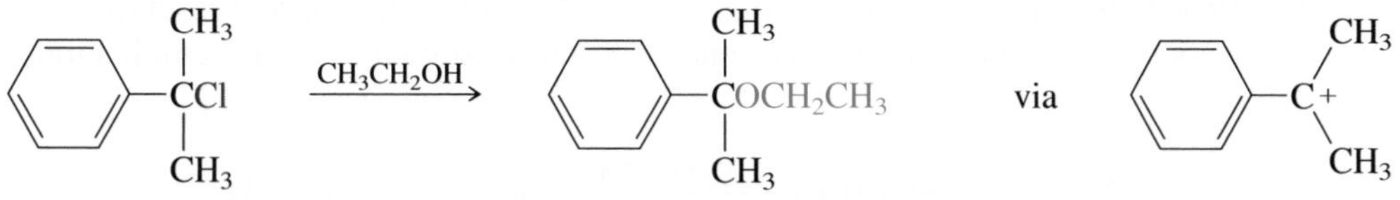

2-Chloro-2-phenylpropane 2-Ethoxy-2-phenylpropane (87%)

Additional phenyl substituents stabilize carbocations even more. Triphenylmethyl cation is particularly stable. Its perchlorate salt is ionic and stable enough to be isolated and stored indefinitely.

The triphenylmethyl group is often referred to as a *trityl* group.

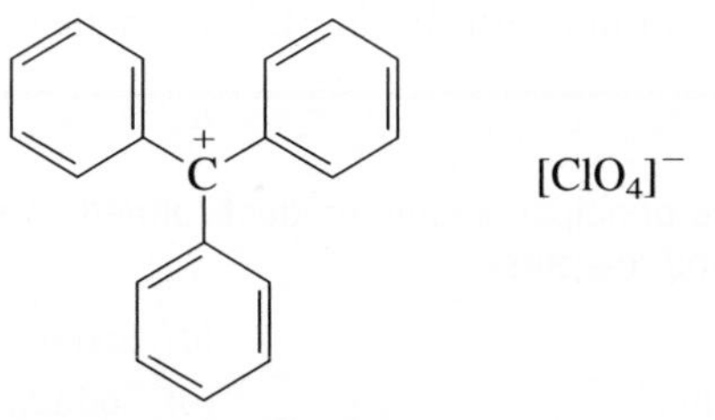

Triphenylmethyl perchlorate

Problem 11.12

Write two different resonance forms for triphenylmethyl cation. One structure should show the positive charge at an ortho position, the other at a para position.

11.14 S_N2 Reactions of Benzylic Halides

S_N2 Rates. Benzyl chloride undergoes S_N2 substitution with potassium iodide in acetone 197 times faster than 1-chloropropane; it is also more than twice as reactive as allyl chloride.

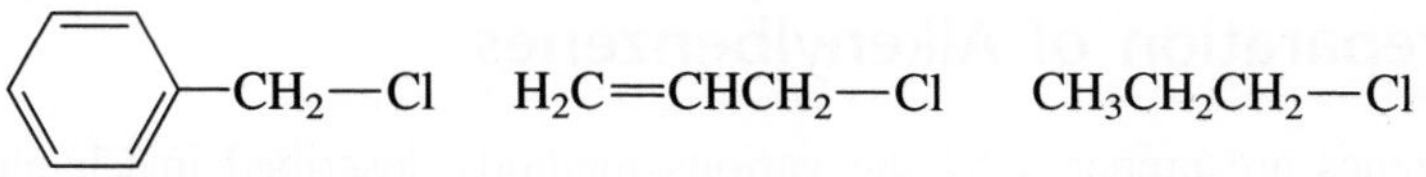

Benzyl chloride Allyl chloride 1-Chloropropane

Most reactive: *k*(rel) 197 *k*(rel) 80 Least reactive: *k*(rel) 1.0

S_N2 substitution at a benzylic carbon is analogous to substitution at an allylic carbon (Section 10.4) in that π delocalization facilitates the flow of electrons from the nucleophile to the LUMO of the halide. In a benzylic halide, the LUMO encompasses both the CH_2Cl unit and the π system of the aromatic ring.

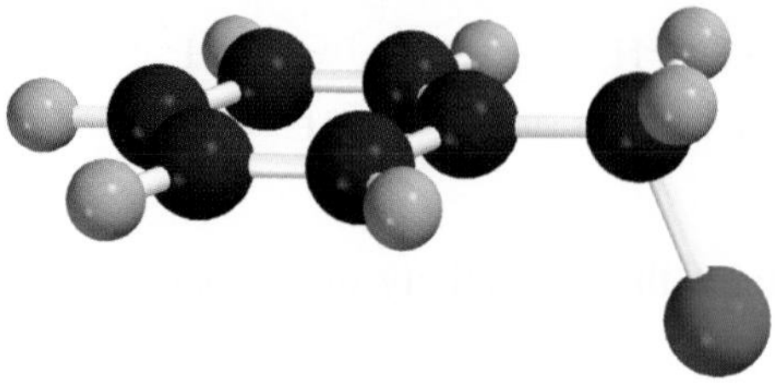

Benzyl chloride

LUMO of benzyl chloride

As electrons flow into the LUMO, they feel the attractive force of the benzylic carbon and the ring carbons as well, which lowers the energy of the transition state and increases the reaction rate.

S_N2 Products. Primary benzylic halides are ideal substrates for S_N2 reactions. In addition to being very reactive, they are unable to undergo competing E2 elimination.

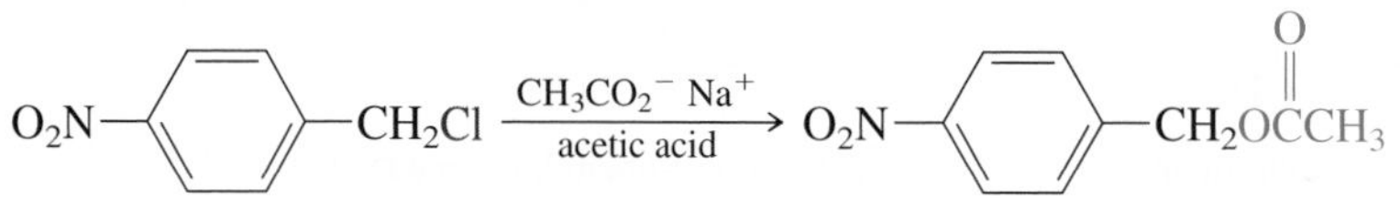
O₂N
CH₂Cl
CH₃CO₂⁻ Na⁺
acetic acid
O₂N
O
CH₂OCCH₃

p-Nitrobenzyl chloride *p*-Nitrobenzyl acetate (78–82%)

Benzylic halides that are secondary resemble secondary alkyl halides in that they undergo substitution only when the nucleophile is weakly basic. If the nucleophile is a strong base such as sodium ethoxide, elimination by the E2 mechanism is faster than substitution.

Problem 11.13

Give the structure of the principal organic product formed on reaction of benzyl bromide with each of the following reagents:

(a) Sodium ethoxide
(b) Potassium *tert*-butoxide
(c) Sodium azide
(d) Sodium hydrogen sulfide
(e) Sodium iodide (in acetone)

Sample Solution (a) Benzyl bromide is a primary bromide and undergoes S_N2 reactions readily. It has no hydrogens β to the leaving group and so cannot undergo elimination. Ethoxide ion acts as a nucleophile, displacing bromide and forming benzyl ethyl ether.

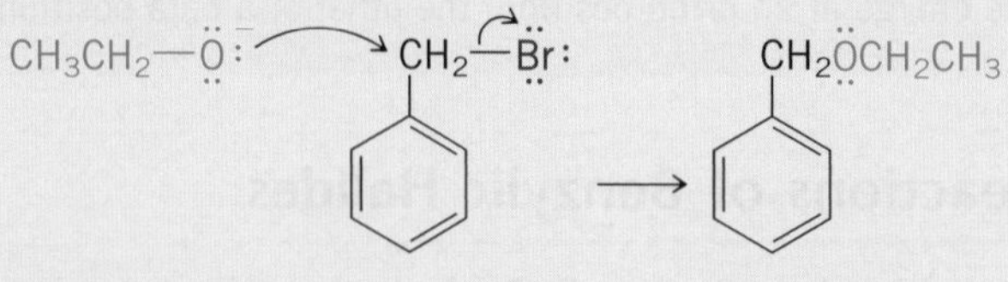
CH₃CH₂—O:
CH₂—Br:
CH₂OCH₂CH₃

Ethoxide ion Benzyl bromide Benzyl ethyl ether

11.15 Preparation of Alkenylbenzenes

Alkenylbenzenes are prepared by the various methods described in Chapter 5 for the preparation of alkenes: *dehydrogenation, dehydration,* and *dehydrohalogenation.*

Dehydrogenation of alkylbenzenes is not a convenient laboratory method but is used industrially to convert ethylbenzene to styrene.

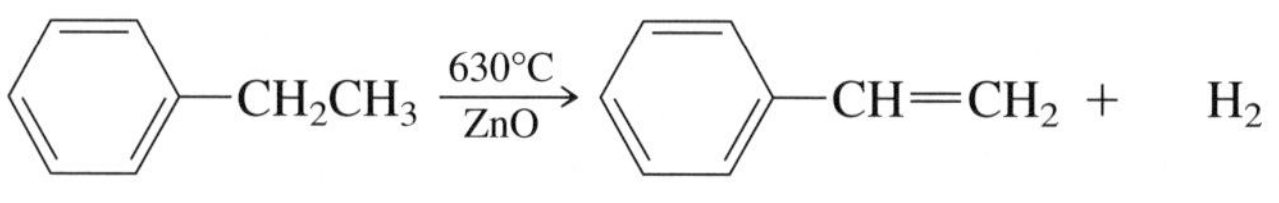

Ethylbenzene Styrene Hydrogen

Practically all of the 1.3×10^{10} lb of ethylbenzene produced annually in the United States is converted to styrene.

Acid-catalyzed dehydration of benzylic alcohols is a useful route to alkenylbenzenes. So too is dehydrohalogenation under E2 conditions.

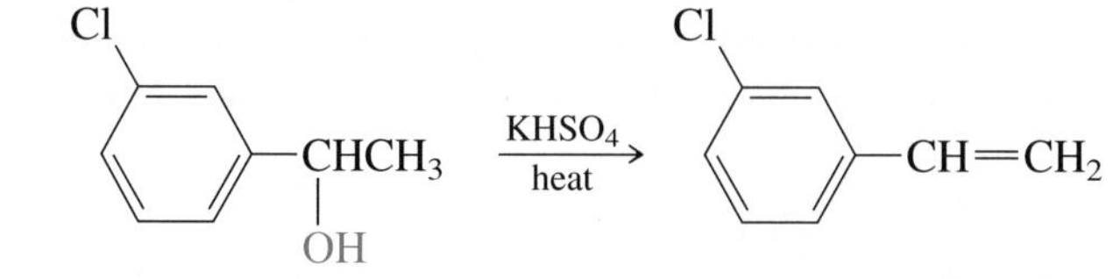

1-(*m*-Chlorophenyl)ethanol *m*-Chlorostyrene (80–82%)

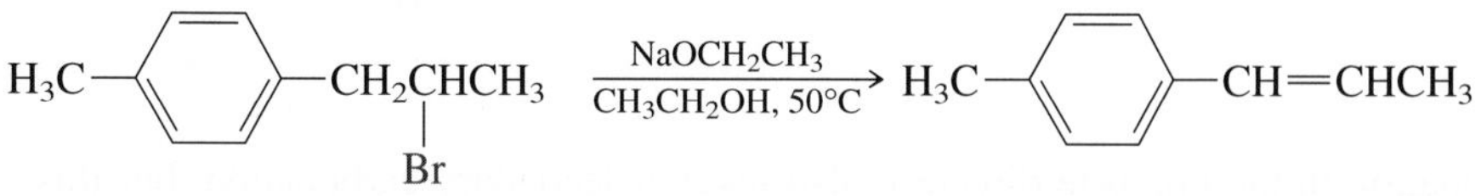

2-Bromo-1-(*p*-methylphenyl)propane 1-(*p*-Methylphenyl)propene (99%)

11.16 Addition Reactions of Alkenylbenzenes

Most of the reactions of alkenes that were discussed in Chapter 6 find a parallel in the reactions of alkenylbenzenes.

Hydrogenation of the side-chain double bond of an alkenylbenzene is much easier than hydrogenation of the aromatic ring and can be achieved with high selectivity, leaving the ring unaffected.

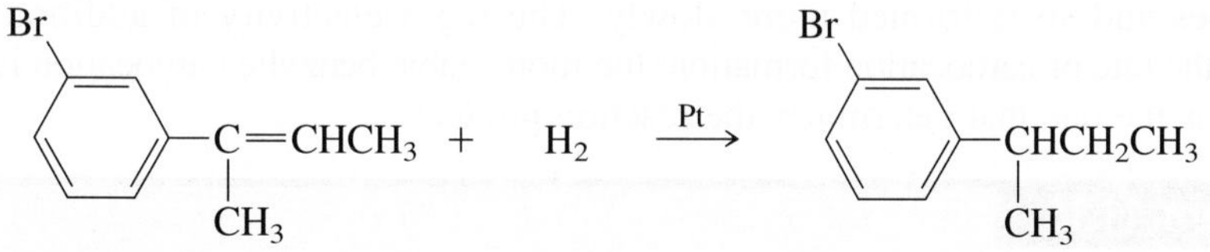

2-(*m*-Bromophenyl)-2-butene Hydrogen 2-(*m*-Bromophenyl)butane (92%)

Problem 11.14

Both 1,2-dihydronaphthalene and 1,4-dihydronaphthalene may be selectively hydrogenated to 1,2,3,4-tetrahydronaphthalene.

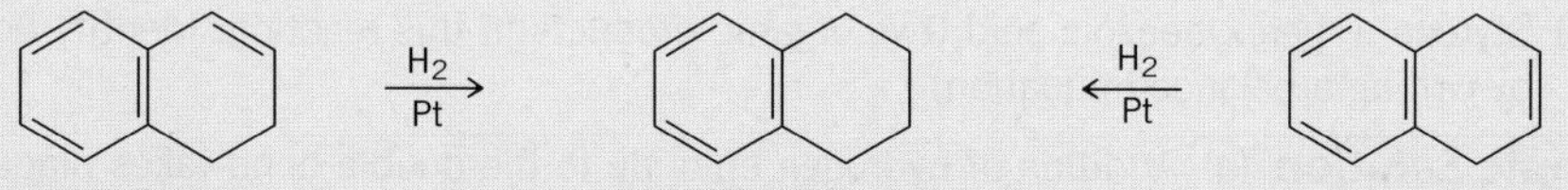

1,2-Dihydronaphthalene 1,2,3,4-Tetrahydronaphthalene 1,4-Dihydronaphthalene

One of these isomers has a heat of hydrogenation of 101 kJ/mol (24.1 kcal/mol), and the heat of hydrogenation of the other is 113 kJ/mol (27.1 kcal/mol). Match the heat of hydrogenation with the appropriate dihydronaphthalene.

The double bond in the alkenyl side chain undergoes addition reactions that are typical of alkenes when treated with electrophilic reagents.

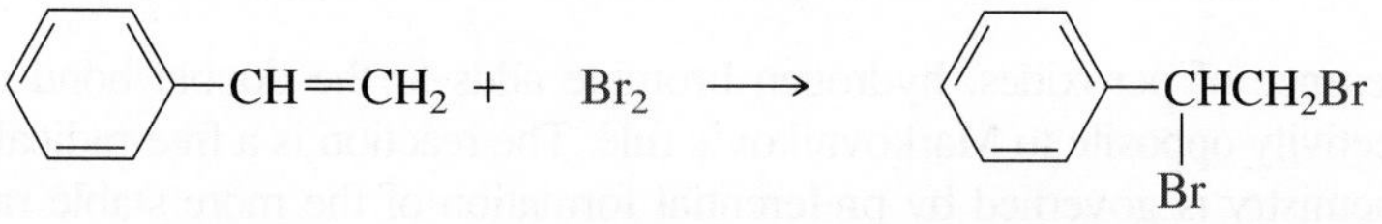

Styrene Bromine 1,2-Dibromo-1-phenylethane (82%)

The regioselectivity of electrophilic addition is governed by the ability of an aromatic ring to stabilize an adjacent carbocation. This is clearly seen in the addition of hydrogen chloride to indene. Only a single chloride is formed.

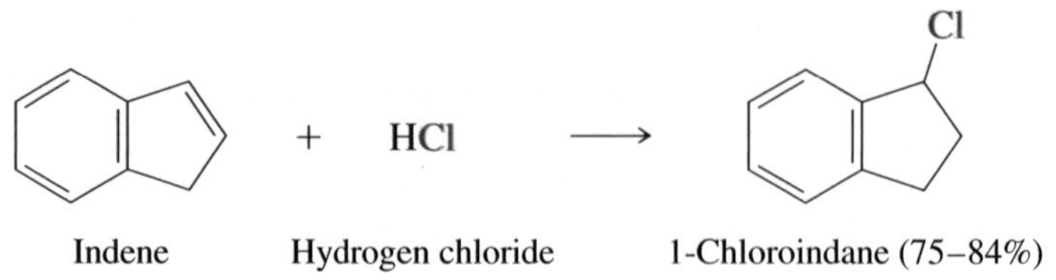

Only the benzylic chloride is formed because protonation of the double bond occurs in the direction that gives a carbocation that is both secondary and benzylic.

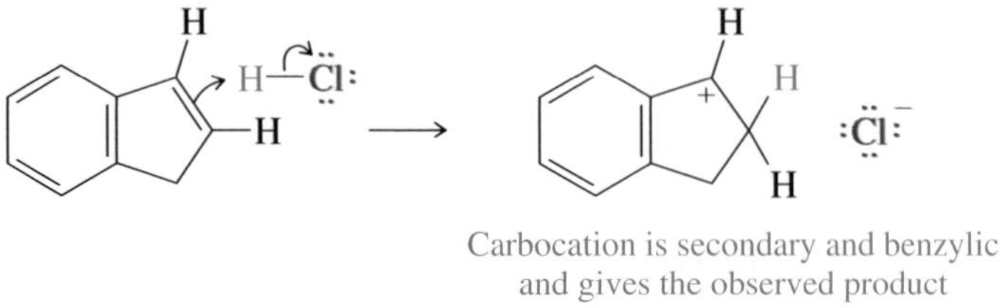

Protonation in the opposite direction also gives a secondary carbocation, but this carbocation is not benzylic.

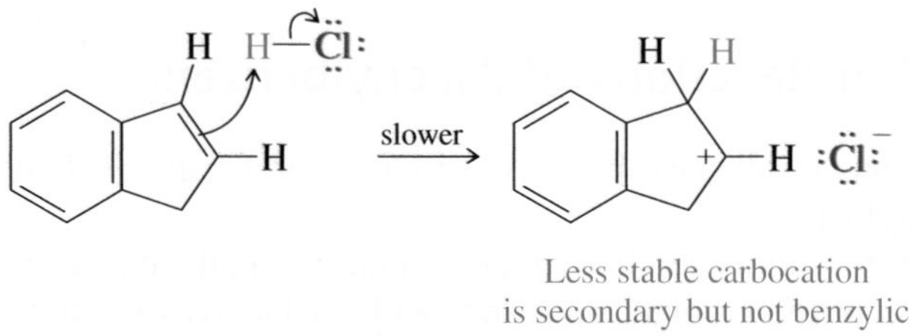

This carbocation does not receive the extra increment of stabilization that its benzylic isomer does and so is formed more slowly. The regioselectivity of addition is controlled by the rate of carbocation formation; the more stable benzylic carbocation is formed faster and is the one that determines the reaction product.

Problem 11.15

Each of the following reactions has been reported in the chemical literature and gives a single organic product in high yield. Write the structure of the product for each reaction.

(a) 2-Phenylpropene + hydrogen chloride

(b) 2-Phenylpropene treated with diborane in tetrahydrofuran followed by oxidation with basic hydrogen peroxide

(c) Styrene + bromine in aqueous solution

(d) Styrene + peroxybenzoic acid (two organic products in this reaction; identify both by writing a balanced equation)

Sample Solution (a) Addition of hydrogen chloride to the double bond takes place by way of a tertiary benzylic carbocation.

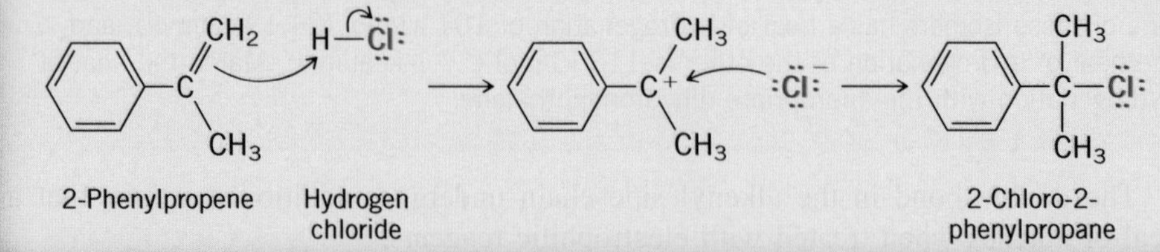

In the presence of peroxides, hydrogen bromide adds to the double bond of styrene with a regioselectivity opposite to Markovnikov's rule. The reaction is a free-radical addition, and the regiochemistry is governed by preferential formation of the more stable radical.

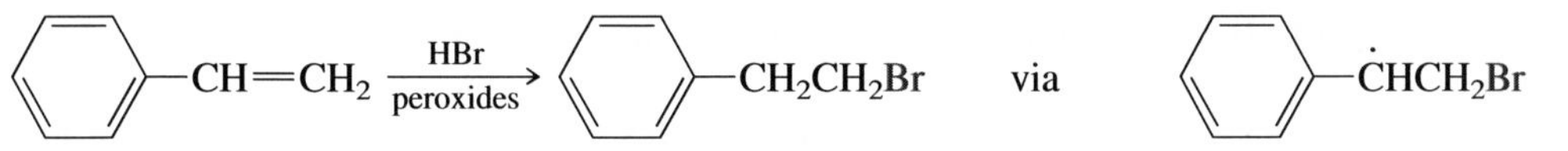

Styrene   1-Bromo-2-phenylethane (major product)   2-Bromo-1-phenylethyl radical (secondary; benzylic)

11.17 Polymerization of Styrene

The annual production of styrene in the United States is approximately 1.1 × 10^{10} lb, with about 65% of this output used to prepare polystyrene plastics and films. Styrofoam coffee cups are made from polystyrene. Polystyrene can also be produced in a form that is very strong and impact-resistant and is used widely in luggage, television and radio cabinets, and furniture.

As described in the boxed essay *Diene Polymers* in Chapter 10, most synthetic rubber is a copolymer of styrene and 1,3-butadiene.

Polymerization of styrene can be carried out under free-radical (Mechanism 11.2), cationic, anionic, or Ziegler–Natta conditions (see Section 14.16).

Mechanism 11.2

Free-Radical Polymerization of Styrene

Step 1: Polymerization of styrene usually employs a peroxide as an initiator. The peroxide dissociates on heating to produce two alkoxy radicals.

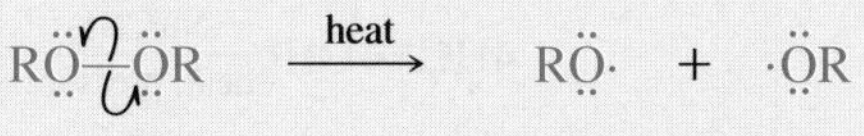

Peroxide   Two alkoxy radicals

Step 2: The free radical produced in step 1 adds to the double bond of styrene. Addition occurs in the direction that produces a benzylic radical.

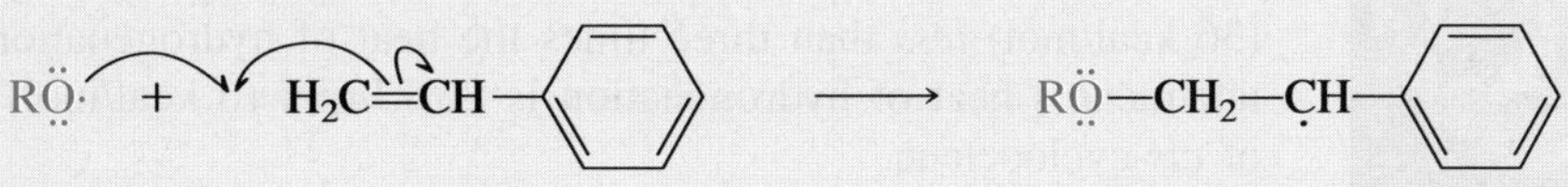

Alkoxy radical   Styrene   A benzylic radical

Step 3: The benzylic radical produced in step 2 adds to a molecule of styrene. Again addition occurs in the direction that produces a benzylic radical.

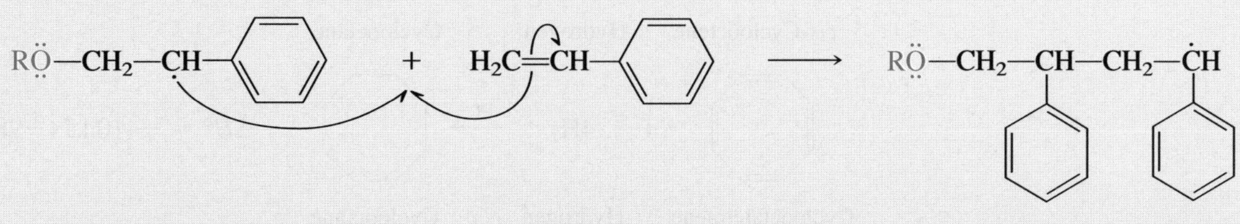

Benzylic radical from step 2   Styrene   Chain-extended benzylic radical

Step 4: The radical produced in step 3 reacts with another styrene molecule, and the process repeats over and over to produce a long-chain polymer having phenyl substituents at every other carbon in the chain.

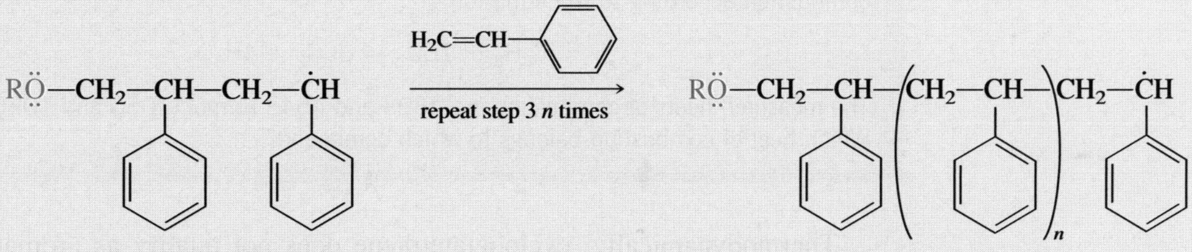

Benzylic radical from step 3   Growing polystyrene chain

11.18 Cyclobutadiene and Cyclooctatetraene

During our discussion of benzene and its derivatives, it may have occurred to you that cyclobutadiene and cyclooctatetraene might be stabilized by cyclic π electron delocalization in a manner analogous to that of benzene.

Cyclobutadiene Cyclooctatetraene

The same thought occurred to early chemists. However, the complete absence of naturally occurring compounds based on cyclobutadiene and cyclooctatetraene contrasted starkly with the abundance of compounds containing a benzene unit. Attempts to synthesize cyclobutadiene and cyclooctatetraene met with failure and reinforced the growing conviction that these compounds would prove to be quite unlike benzene if, in fact, they could be isolated at all.

Willstätter's most important work, for which he won the 1915 Nobel Prize in Chemistry, was directed toward determining the structure of chlorophyll.

The first breakthrough came in 1911 when Richard Willstätter prepared cyclooctatetraene by a lengthy degradation of *pseudopelletierine,* a natural product obtained from the bark of the pomegranate tree. Today, cyclooctatetraene is prepared from acetylene in a reaction catalyzed by nickel cyanide.

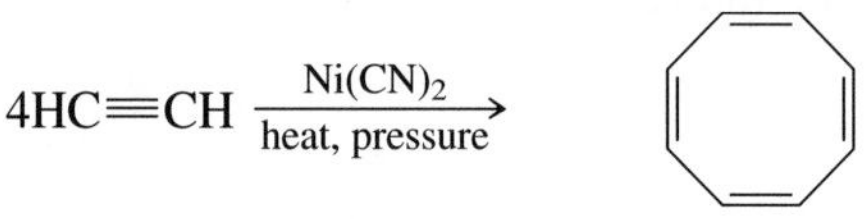

Acetylene Cyclooctatetraene (70%)

Pomegranate trees are best known for their large, juicy, seed-laden fruit.

Cyclooctatetraene is relatively stable, but lacks the "special stability" of benzene. Unlike benzene, which we saw has a heat of hydrogenation that is 152 kJ/mol (36 kcal/mol) *less* than three times the heat of hydrogenation of cyclohexene, cyclooctatetraene's heat of hydrogenation is 26 kJ/mol (6 kcal/mol) *more* than four times that of *cis*-cyclooctene.

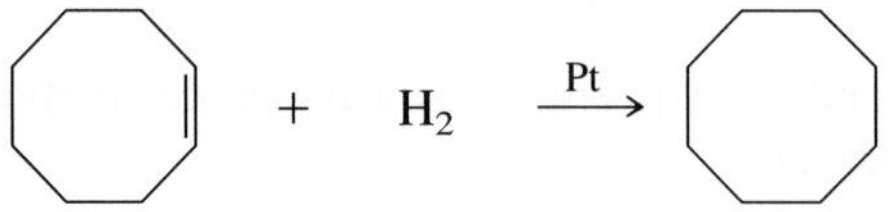

$\Delta H° = -96$ kJ (-23 kcal)

cis-Cyclooctene Hydrogen Cyclooctane

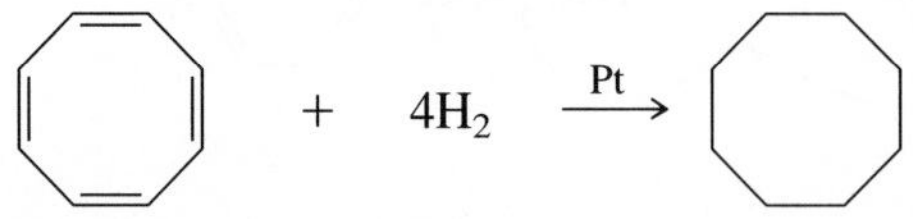

$\Delta H° = -410$ kJ (-98 kcal)

Cyclooctatetraene Hydrogen Cyclooctane

Problem 11.16

Both cyclooctatetraene and styrene have the molecular formula C_8H_8 and undergo combustion according to the equation

$$C_8H_8 + 10O_2 \rightarrow 8CO_2 + 4H_2O$$

The measured heats of combustion are 4393 and 4543 kJ/mol (1050 and 1086 kcal/mol). Which heat of combustion belongs to which compound?

Thermodynamically, cyclooctatetraene does not qualify as aromatic. Nor does its structure offer any possibility of the π electron delocalization responsible for aromaticity. As shown in Figure 11.12, cyclooctatetraene is *nonplanar* with four short

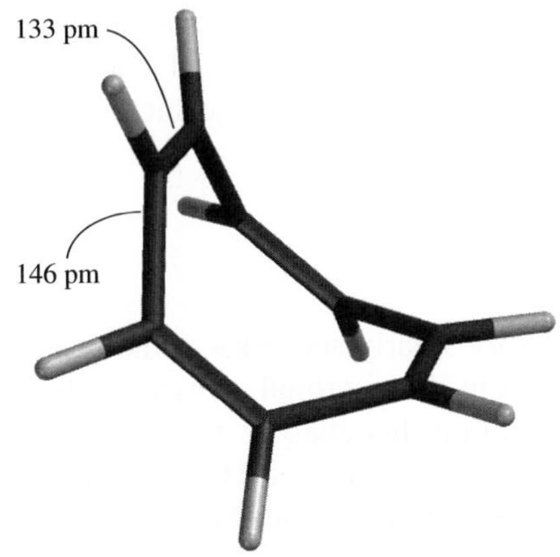

Figure 11.12

Molecular geometry of cyclooctatetraene. The ring is not planar, and the bond distances alternate between short double bonds and long single bonds.

and four long carbon–carbon bond distances. Cyclooctatetraene is satisfactorily represented by a single Lewis structure having alternating single and double bonds in a tub-shaped eight-membered ring. Experimental studies and theoretical calculations indicate that the structure of cyclooctatetraene shown in Figure 11.12 is about 75 kJ/mol (18 kcal/mol) more stable than the planar delocalized alternative. Cyclooctatetraene is not aromatic.

What about cyclobutadiene?

Cyclobutadiene escaped chemical characterization for more than 100 years. Despite numerous attempts, all synthetic efforts met with failure. It became apparent not only that cyclobutadiene was not aromatic but that it was exceedingly unstable. Beginning in the 1950s, a variety of novel techniques succeeded in generating cyclobutadiene as a transient, reactive intermediate.

Problem 11.17

One of the chemical properties that makes cyclobutadiene difficult to isolate is that it reacts readily with itself to give a dimer:

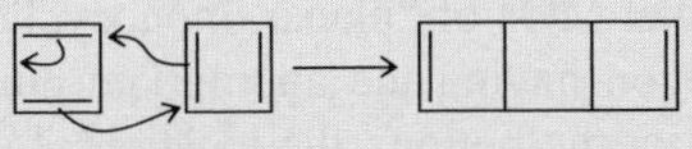

What reaction of dienes does this resemble?

High-level molecular orbital calculations of cyclobutadiene itself and experimentally measured bond distances of a stable, highly substituted derivative both reveal a pattern of alternating short and long bonds characteristic of a rectangular, rather than square, geometry.

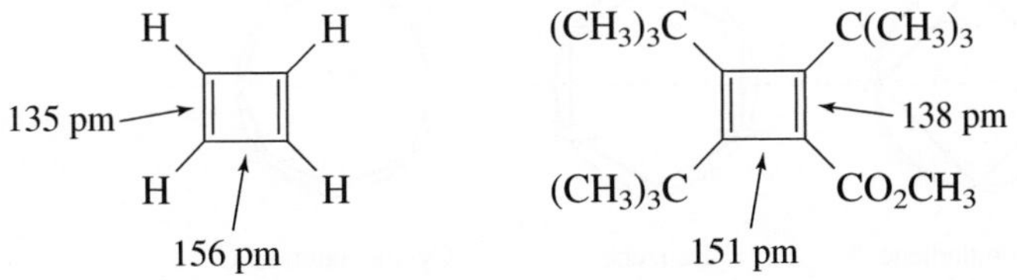

Cyclobutadiene　　Sterically hindered cyclobutadiene derivative

Experimental measurements place delocalized cyclobutadiene approximately 150 kJ/mol (36 kcal/mol) higher in energy than a structure with noninteracting double bonds.

Thus, both square cyclobutadiene and planar cyclooctatetraene are *antiaromatic.* **Antiaromatic** molecules are *destabilized by delocalization of their π electrons.* Consequently, both cyclobutadiene and cyclooctatetraene adopt structures that minimize the delocalization of these electrons.

Cyclic conjugation, although necessary for aromaticity, is not sufficient for it. Some other factor or factors must contribute to the special stability of benzene and compounds based on the benzene ring. To understand these factors, let's return to the molecular orbital description of benzene.

11.19 Hückel's Rule

Hückel was a German physical chemist. Before his theoretical studies of aromaticity, Hückel collaborated with Peter Debye in developing what remains the most widely accepted theory of electrolyte solutions.

One of molecular orbital theory's early successes came in 1931 when Erich Hückel discovered an interesting pattern in the π orbital energy levels of benzene, cyclobutadiene, and cyclooctatetraene. By limiting his analysis to monocyclic conjugated polyenes and restricting the structures to planar geometries, Hückel found that whether a hydrocarbon of this type was aromatic depended on its number of π electrons. He set forth what we now call **Hückel's rule:**

> *Among planar, monocyclic, fully conjugated polyenes, only those possessing* $(4n + 2)$ π *electrons, where* n *is a whole number, will have special stability; that is, be aromatic.*

Thus for this group of hydrocarbons, those with $(4n + 2) = 2, 6, 10, 14, \ldots$ π electrons will be aromatic. These values correspond to $(4n + 2)$ when $n = 0, 1, 2, 3 \ldots$

Hückel proposed his theory before ideas of antiaromaticity emerged. We can amplify his generalization by noting that among the hydrocarbons covered by Hückel's rule, those with $(4n)$ π electrons not only are not aromatic, they are antiaromatic.

Benzene, cyclobutadiene, and cyclooctatetraene provide clear examples of Hückel's rule. Benzene, with six π electrons is a $(4n + 2)$ system and is predicted to be aromatic by the rule. Square cyclobutadiene and planar cyclooctatetraene are $4n$ systems with four and eight π electrons, respectively, and are antiaromatic.

The $(4n + 2)$ π electron standard follows from the pattern of orbital energies in monocyclic, completely conjugated polyenes. The π energy levels were shown for benzene earlier in Figure 11.4 and are repeated in Figure 11.13*b*. Figure 11.13*a* and 11.13*c* show the π energy levels for square cyclobutadiene and planar cyclooctatetraene, respectively.

The circle mnemonic was devised by Arthur A. Frost, a theoretical chemist at Northwestern University.

The energy diagrams in Figure 11.13 illustrate a simple method, called the **Frost circle** for setting out the Hückel MOs of "planar, monocyclic, completely conjugated polyenes." By inscribing a polygon having the appropriate number of sides within a circle so that one of its vertices lies at the bottom, the location of each of the polygon's corners defines a π electron energy level. Their vertical separation is proportional to the energy difference between the MOs. A horizontal line drawn through the center of the circle separates the bonding and antibonding MOs; an orbital that lies directly on the line is nonbonding.

For qualitative purposes, the circle itself isn't even necessary. We could locate the Hückel MOs by simply working with the polygons themselves. The circle is needed only

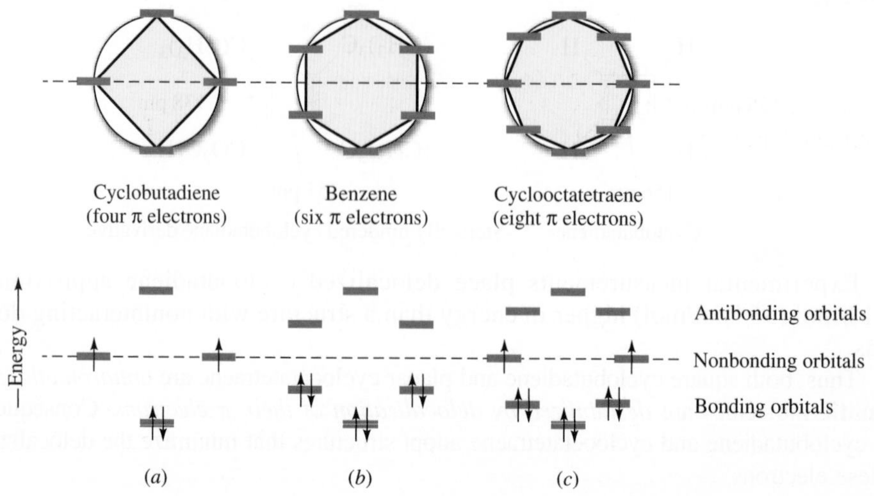

Figure 11.13

Frost's circle and the π molecular orbitals of (*a*) square cyclobutadiene, (*b*) benzene, and (*c*) planar cyclooctatetraene.

when Frost's method is used quantitatively. In those cases the radius of the circle has a prescribed value, allowing each MO to be assigned a specific energy.

The pattern of orbital energies in Figure 11.13 provides a convincing explanation for why benzene is aromatic while square cyclobutadiene and planar cyclooctatetraene are not. We start by counting π electrons; cyclobutadiene has four, benzene six, and cyclooctatetraene has eight. These π electrons are assigned to MOs in accordance with the usual rules—lowest energy orbitals first, a maximum of two electrons per orbital, and when two orbitals are of equal energy, each gets one electron before either orbital gets two (Hund's rule).

Benzene As seen earlier in Figure 11.4 (Section 11.5), the six π electrons of benzene are distributed in pairs among its three bonding π MOs, giving a closed-shell electron configuration. All the bonding orbitals are filled, and all the electron spins are paired.

Cyclobutadiene Square cyclobutadiene has one bonding π MO, two equal-energy nonbonding π MOs and one antibonding π^* MO. After the bonding MO is filled, the remaining two electrons are assigned to different nonbonding MOs in accordance with Hund's rule. This results in a species with two unpaired electrons—a diradical. In a square geometry, cyclobutadiene lacks a closed-shell electron configuration. It is not stabilized and, with two unpaired electrons, should be very reactive.

Cyclooctatetraene Six of the eight π electrons of planar cyclooctatetraene occupy three bonding orbitals. The remaining two π electrons occupy, one each, the two equal-energy nonbonding orbitals. Planar cyclooctatetraene should, like square cyclobutadiene, be a diradical.

An important conclusion we draw from the qualitative MO diagrams is that the geometry required for maximum π electron delocalization, a planar ring with *p* orbitals aligned and equal C—C bond distances, gives relatively unstable electron configurations for square cyclobutadiene and planar cyclooctatetraene. Both escape to alternative geometries that have electron configurations which, although not aromatic, at least have all their electron spins paired. For cyclobutadiene the stable geometry is rectangular; for cyclooctatetraene it is tub-shaped.

Benzene's structure allows effective π electron conjugation and gives a closed-shell electron configuration. To understand why it also conveys special stability, we need to go one step further and compare the Hückel π MOs of benzene to those of a hypothetical "cyclohexatriene" with alternating single and double bonds. Without going into quantitative detail, we'll simply note that the occupied orbitals of a structure in which the π electrons are restricted to three noninteracting double bonds are of higher energy (less stable) than the occupied Hückel MOs of benzene.

Before looking at other applications of Hückel's rule, it is worth pointing out that its opening phrase: "Among planar, monocyclic, fully conjugated polyenes" does *not* mean that *only* "planar, monocyclic, fully conjugated polyenes" can be aromatic. It merely limits the rule to compounds of this type. There are thousands of aromatic compounds that are not monocyclic—naphthalene and related polycyclic aromatic hydrocarbons (Section 11.7), for example. All compounds based on benzene rings are aromatic. Cyclic conjugation *is* a requirement for aromaticity, however, and in those cases the conjugated system must contain $(4n + 2)$ π electrons. Cyclic conjugated systems with $4n$ π electrons are antiaromatic.

Problem 11.18

Give an explanation for each of the following observations:

(a) Compound A has six π electrons but is not aromatic.
(b) Compound B has six π electrons but is not aromatic.
(c) Compound C has 12 π electrons and is aromatic.

Continued

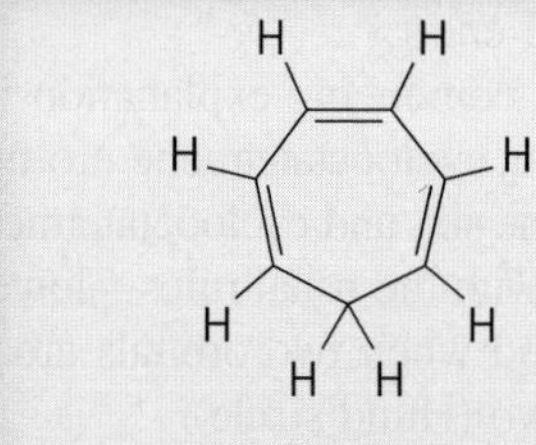

Compound A

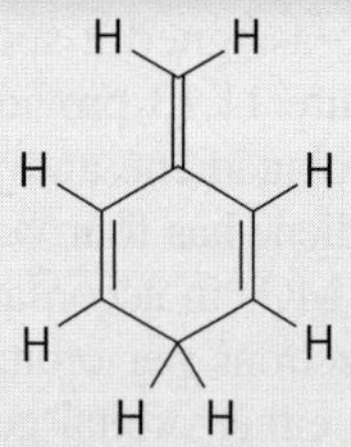

Compound B

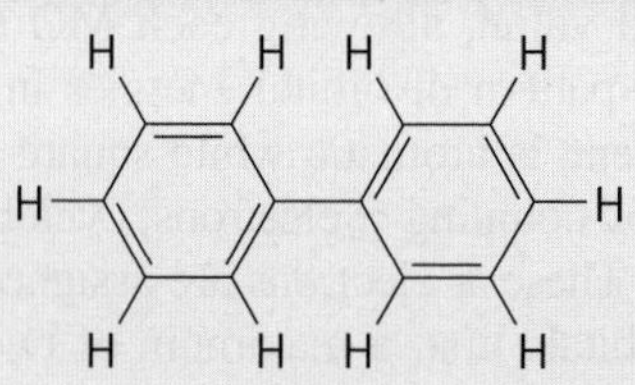

Compound C

Sample Solution (a) Cycloheptatriene (compound A) is not aromatic because, although it does contain six π electrons, its conjugated system of three double bonds does not close on itself—it lacks cyclic conjugation. The CH_2 group prevents cyclic delocalization of the π electrons.

In the next section we'll explore Hückel's rule for values of *n* greater than 1 to see how it can be extended beyond cyclobutadiene, benzene, and cyclooctatetraene.

11.20 Annulenes

The general term **annulene** refers to completely conjugated monocyclic hydrocarbons with more than six carbons. Cyclobutadiene and benzene retain their names, but higher members of the group are named **[*x*]annulene,** where *x* is the number of carbons in the ring. Thus, cyclooctatetraene becomes [8]annulene, cyclodecapentaene becomes [10]annulene and so on.

Problem 11.19

Use Frost's circle to construct orbital energy diagrams for (a) [10]annulene and (b) [12]annulene. Is either aromatic according to Hückel's rule?

Sample Solution (a) [10]Annulene is a ten-membered ring with five conjugated double bonds. Drawing a polygon with ten sides with its vertex pointing downward within a circle gives the orbital template. Place the orbitals at the positions where each vertex contacts the circle. The ten π electrons of [10]annulene satisfy the $(4n + 2)$ rule for $n = 2$ and occupy the five bonding orbitals in pairs. [10]Annulene is aromatic according to Hückel's rule.

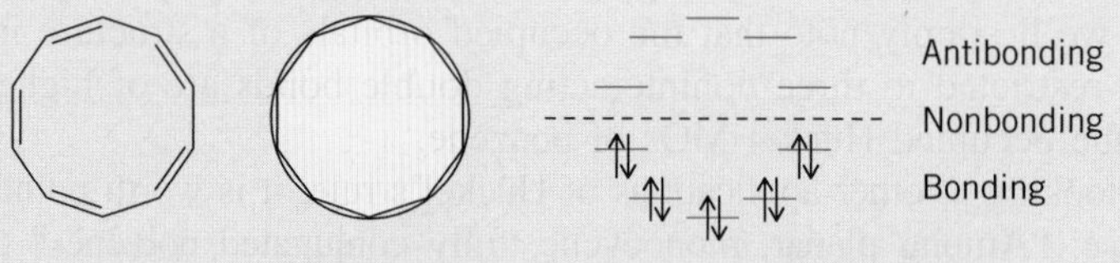
Antibonding
Nonbonding
Bonding

[10] Annulene Frost's circle π-Electron configuration

The size of each angle of a regular polygon is given by the expression

$$180° \times \frac{(\text{number of sides}) - 2}{(\text{number of sides})}$$

The prospect of observing aromatic character in conjugated polyenes having 10, 14, 18, and so on π electrons spurred efforts toward the synthesis of higher annulenes. A problem immediately arises in the case of the all-cis isomer of [10]annulene, the structure of which is shown in the preceding problem. Geometry requires a ten-sided regular polygon to have 144° bond angles; sp^2 hybridization at carbon requires 120° bond angles. Therefore, aromatic stabilization due to conjugation in all-*cis*-[10]annulene is opposed by the destabilizing effect of 24° of angle strain at each of its carbon atoms. All-*cis*-[10]annulene has been prepared. It is not very stable and is highly reactive.

A second isomer of [10]annulene (the cis, trans, cis, cis, trans stereoisomer) can have bond angles close to 120° but is destabilized by a close contact between two hydrogens directed toward the interior of the ring. To minimize the van der Waals strain between these hydrogens, the ring adopts a nonplanar geometry, which limits its ability to be stabilized by π electron delocalization. It, too, has been prepared and is not very stable. Similarly, the next higher $(4n + 2)$ system, [14]annulene, is also somewhat destabilized by van der Waals strain and is nonplanar.

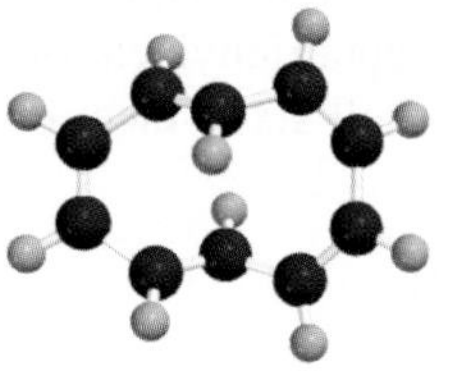
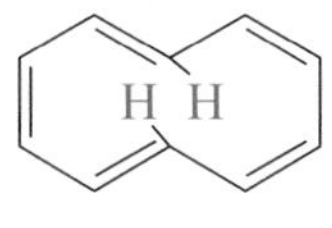

cis,trans,cis,cis,trans-[10]Annulene

Planar geometry required for aromaticity destabilized by van der Waals repulsions between indicated hydrogens

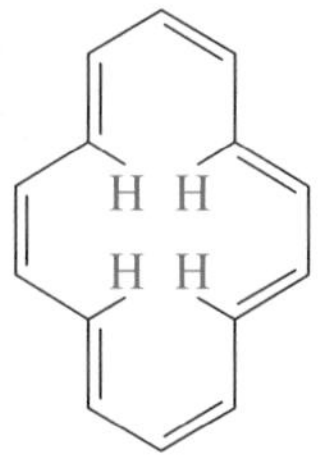

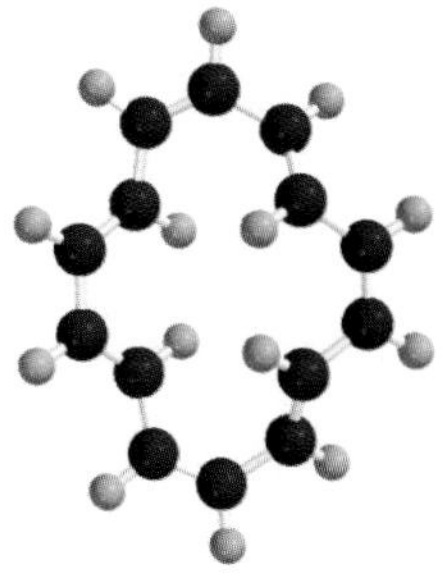

[14]Annulene

When the ring contains 18 carbon atoms, it is large enough to be planar while still allowing its interior hydrogens to be far enough apart so as to not interfere with one another. The [18]annulene shown is planar, or nearly so, and has all its carbon–carbon bond distances in the range 137–143 pm, very much like those of benzene. Its resonance energy is estimated to be about 418 kJ/mol (100 kcal/mol). Although its structure and resonance energy attest to the validity of Hückel's rule, which predicts "special stability" for [18]annulene, its chemical reactivity does not. [18]Annulene behaves more like a polyene than like benzene in that it is hydrogenated readily, undergoes addition rather than substitution with bromine, and forms a Diels–Alder adduct with maleic anhydride.

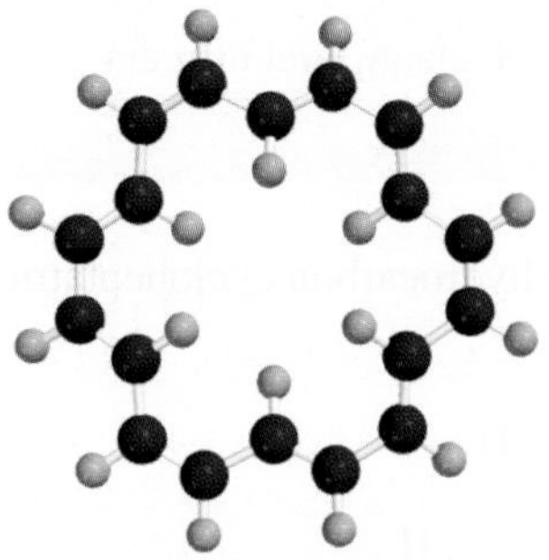
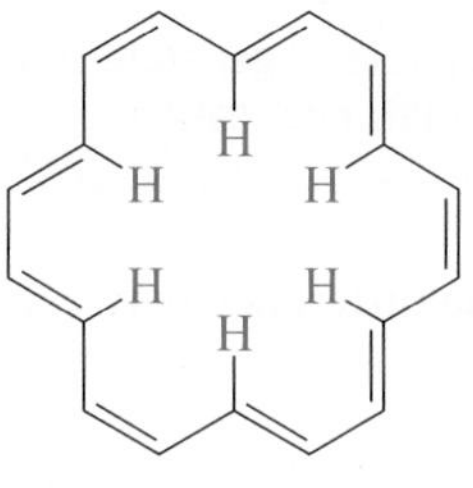

[18]Annulene

No serious repulsions among six interior hydrogens; molecule is planar and aromatic.

As noted earlier, planar annulenes with $4n$ π electrons are antiaromatic. A member of this group, [16]annulene, has been prepared. It is nonplanar and shows a pattern of alternating short (average 134 pm) and long (average 146 pm) bonds typical of a nonaromatic cyclic polyene.

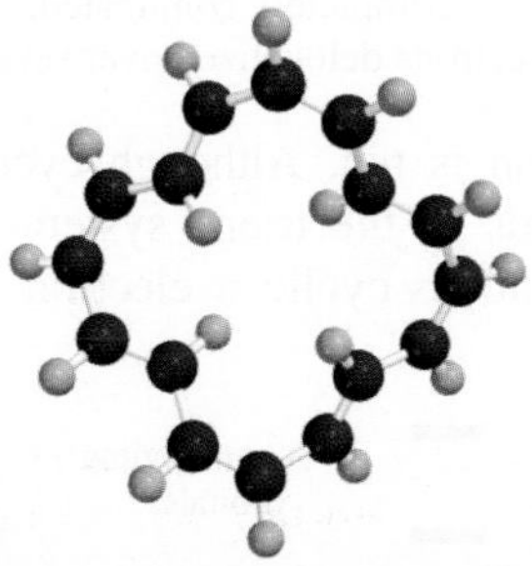
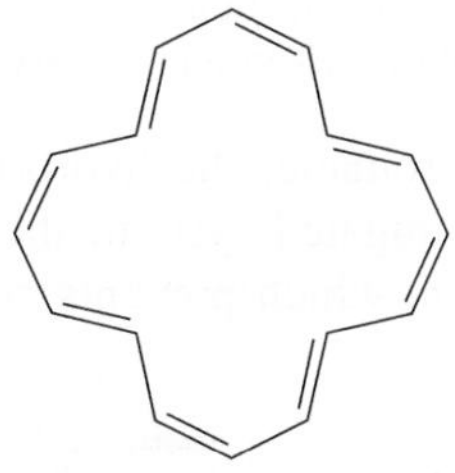

[16]Annulene

Problem 11.20

What does a comparison of the heats of combustion of benzene (3265 kJ/mol; 781 kcal/mol), cyclooctatetraene (4543 kJ/mol; 1086 kcal/mol), [16]annulene (9121 kJ/mol; 2182 kcal/mol), and [18]annulene (9806 kJ/mol; 2346 kcal/mol) reveal?

Most of the synthetic work directed toward the higher annulenes was carried out by Franz Sondheimer and his students, first at Israel's Weizmann Institute and later at the University of London. Sondheimer's research systematically explored the chemistry of these hydrocarbons and provided experimental verification of Hückel's rule.

11.21 Aromatic Ions

Hückel realized that his molecular orbital analysis of conjugated systems could be extended beyond neutral hydrocarbons. He pointed out that cycloheptatrienyl cation, also called *tropylium ion,* contained a completely conjugated closed-shell six-π electron system analogous to that of benzene.

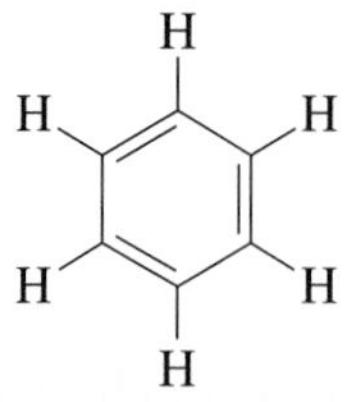

Benzene:
completely conjugated,
six π electrons delocalized over six carbons

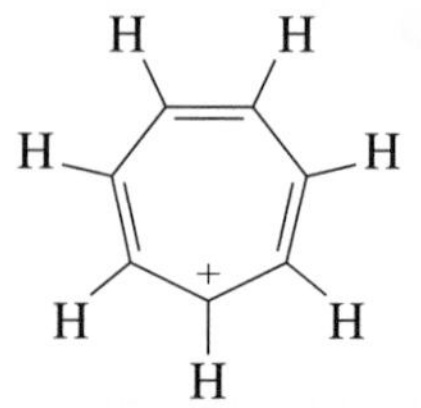

Cycloheptatrienyl cation:
completely conjugated,
six π electrons delocalized over seven carbons

Figure 11.14 shows a molecular orbital diagram for cycloheptatrienyl cation. There are seven π MOs, three of which are bonding and contain the six π electrons of the cation. Cycloheptatrienyl cation is a Hückel ($4n + 2$) system and is an aromatic ion.

Problem 11.21

Show how you could adapt Frost's circle to generate the orbital energy level diagram shown in Figure 11.14 for cycloheptatrienyl cation.

It is important to recognize the difference between the hydrocarbon cycloheptatriene and cycloheptatrienyl cation.

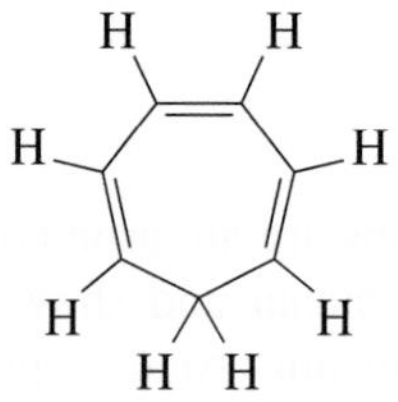

Cycloheptatriene:
lacks cyclic conjugation,
interrupted by CH_2 group

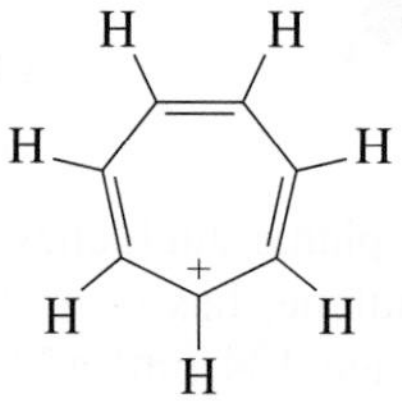

Cycloheptatrienyl cation:
completely conjugated,
six π electrons delocalized over seven carbons

The carbocation is aromatic; the hydrocarbon is not. Although cycloheptatriene has six π electrons in a conjugated system, the ends of the triene system are separated by an sp^3-hybridized carbon, which prevents continuous cyclic π electron delocalization.

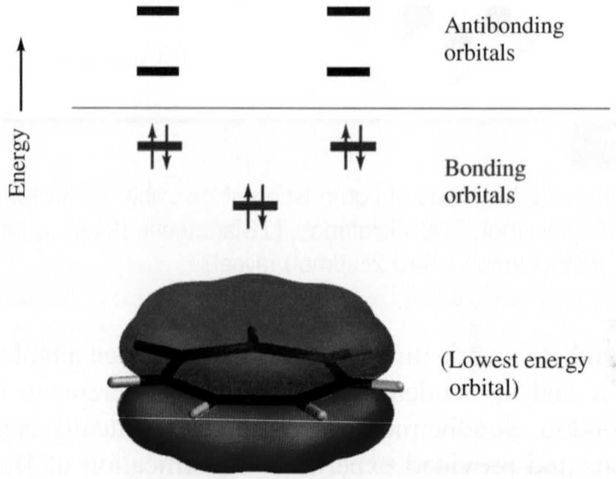

Figure 11.14

The π molecular orbitals of cycloheptatrienyl cation.

Problem 11.22

Cycloheptatrienyl radical ($C_7H_7\cdot$) contains a cyclic, completely conjugated system of π electrons. Is it aromatic? Is it antiaromatic? Explain.

When we say cycloheptatriene is not aromatic but cycloheptatrienyl cation is, we are not comparing the stability of the two to each other. Cycloheptatriene is a stable hydrocarbon but does not possess the *special stability* required to be called *aromatic.* Cycloheptatrienyl cation, although aromatic, is still a carbocation and reasonably reactive toward nucleophiles. Its special stability does not imply a rock-like passivity, but rather a much greater ease of formation than expected on the basis of the Lewis structure drawn for it. A number of observations indicate that cycloheptatrienyl cation is far more stable than most other carbocations. To emphasize its aromatic nature, chemists often write the structure of cycloheptatrienyl cation in the Robinson circle-in-a-ring style.

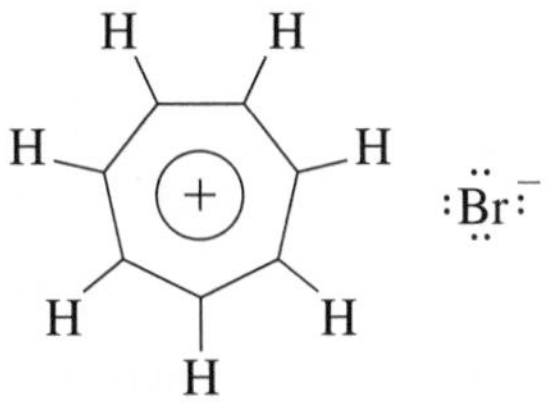

Tropylium bromide

Tropylium bromide was first prepared, but not recognized as such, in 1891. The work was repeated in 1954, and the ionic properties of tropylium bromide were demonstrated. The ionic properties of tropylium bromide are apparent in its unusually high melting point (203°C), its solubility in water, and its complete lack of solubility in diethyl ether.

Problem 11.23

Write resonance structures for tropylium cation sufficient to show the delocalization of the positive charge over all seven carbons.

The five-membered cyclopentadienyl system contrasts with cycloheptatrienyl. Here, the cation has four π electrons, is antiaromatic, very unstable, and very difficult to generate. Cyclopentadienyl anion, however, has six π electrons delocalized over five carbons and is aromatic.

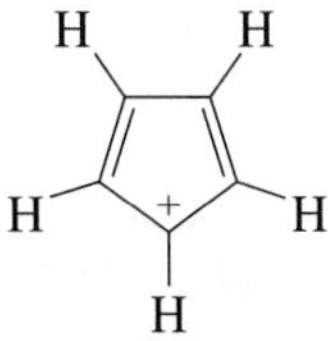

Cyclopentadienyl cation

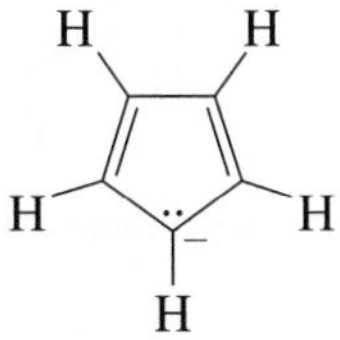

Cyclopentadienyl anion

is equivalent to

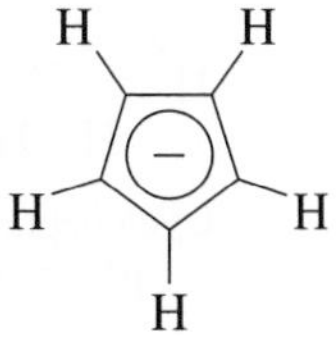

Figure 11.15 shows the Hückel MOs of cyclopentadienyl anion. Like benzene and cycloheptatrienyl cation, cyclopentadienyl anion has six π electrons and a closed-shell electron configuration.

Problem 11.24

Show how you could adapt Frost's circle to generate the orbital energy level diagram shown in Figure 11.15 for cyclopentadienyl anion.

Figure 11.15

The π molecular orbitals of cyclopentadienyl anion.

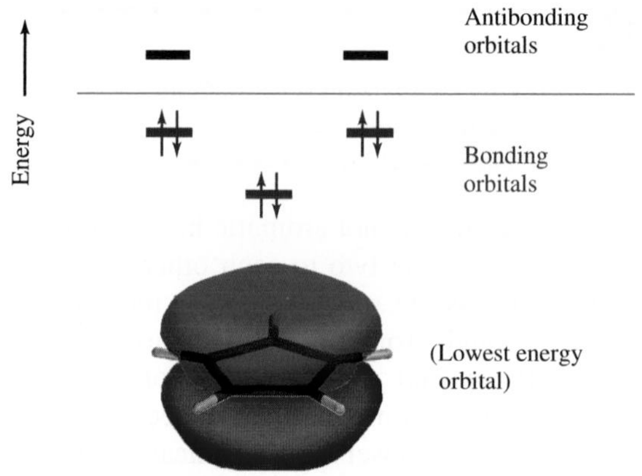

The acidity of cyclopentadiene provides convincing evidence for the special stability of cyclopentadienyl anion.

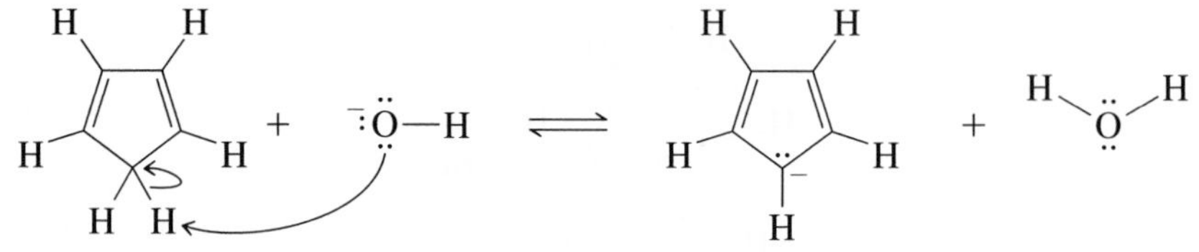

With a pK_a of 16, cyclopentadiene is only a slightly weaker acid than water (pK_a = 15.7). It is much more acidic than other hydrocarbons—its K_a for ionization is 10^{10} times greater than acetylene, for example—because its conjugate base is aromatic and stabilized by electron delocalization.

Problem 11.25

Write resonance structures for cyclopentadienyl anion sufficient to show the delocalization of the negative charge over all five carbons.

There is a striking difference in the acidity of cyclopentadiene compared with cycloheptatriene. Cycloheptatriene has a pK_a of 36, which makes it 10^{20} times weaker in acid strength than cyclopentadiene.

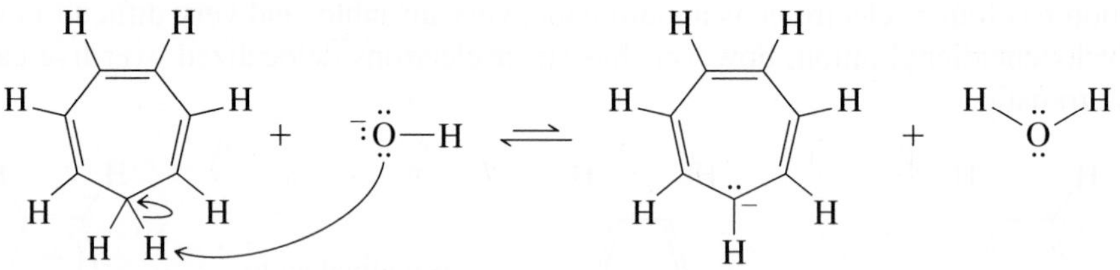

Even though resonance tells us that the negative charge in cycloheptatrienyl anion can be shared by all seven of its carbons, this delocalization offers little in the way of stabilization. Indeed with eight π electrons, cycloheptatrienyl anion is antiaromatic and relatively unstable.

Hückel's rule is now taken to apply to planar, monocyclic, completely conjugated systems generally, not just to neutral hydrocarbons.

A planar, monocyclic, continuous system of p *orbitals possesses aromatic stability when it contains (4*n *+ 2)* π *electrons.*

Other aromatic ions include cyclopropenyl cation (two π electrons) and cyclooctatetraene dianion (ten π electrons).

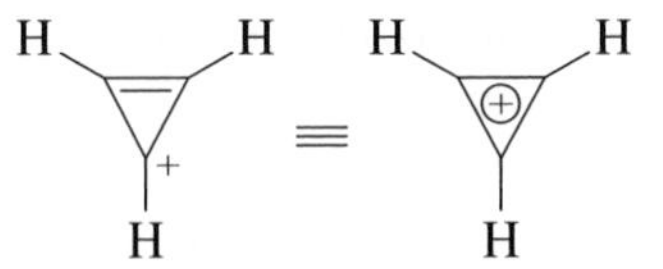

Cyclopropenyl
cation

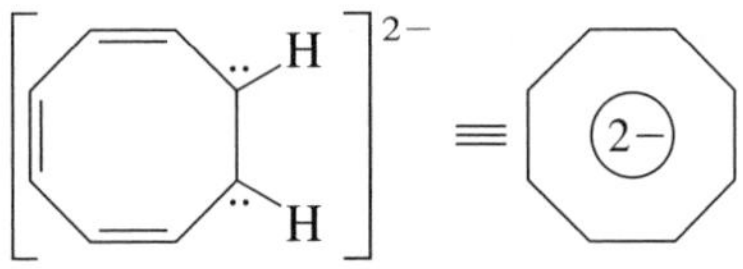

Cyclooctatetraene
dianion

Here, we've taken liberties with the Robinson symbol. Instead of restricting it to a sextet of electrons, organic chemists use it as an all-purpose symbol for cyclic electron delocalization.

Problem 11.26

Is either of the following ions aromatic? Is either antiaromatic?

(a)

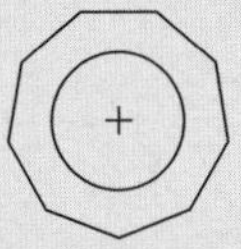

Cyclononatetraenyl
cation

(b)

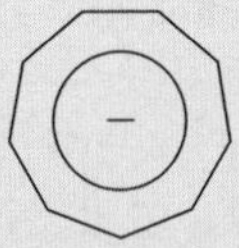

Cyclononatetraenide
anion

Sample Solution (a) The crucial point is the number of π electrons in a cyclic conjugated system. If there are $(4n + 2)$ π electrons, the ion is aromatic. Electron counting is easiest if we write the ion as a single Lewis structure and remember that each double bond contributes two π electrons, a negatively charged carbon contributes two, and a positively charged carbon contributes none.

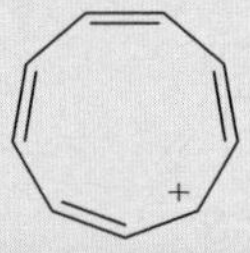

Cyclononatetraenyl cation has eight π electrons; it is *not aromatic.* It is antiaromatic ($4n = 8$; $n = 2$) if planar.

11.22 Heterocyclic Aromatic Compounds

Cyclic compounds that contain at least one atom other than carbon within their ring are called **heterocyclic compounds,** and those that possess aromatic stability are called **heterocyclic aromatic compounds.** Some representative heterocyclic aromatic compounds are *pyridine, pyrrole, furan,* and *thiophene.* The structures and the IUPAC numbering system used in naming their derivatives are shown. In their stability and chemical behavior, all these compounds resemble benzene more than they resemble alkenes.

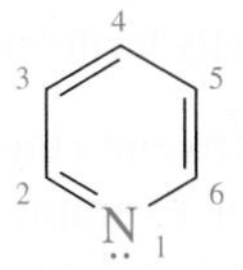

Pyridine

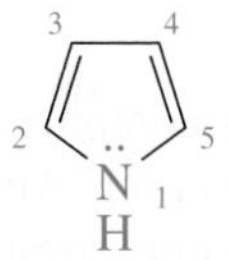

Pyrrole

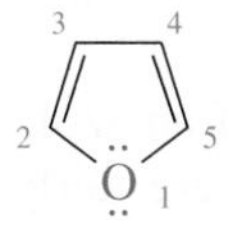

Furan

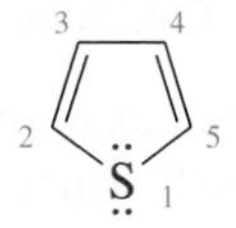

Thiophene

Pyridine, pyrrole, and thiophene, like benzene, are present in coal tar. Furan is prepared from a substance called *furfural* obtained from corncobs.

Heterocyclic aromatic compounds can be polycyclic as well. A benzene ring and a pyridine ring, for example, can share a common side in two different ways. One way gives a compound called *quinoline;* the other gives *isoquinoline.*

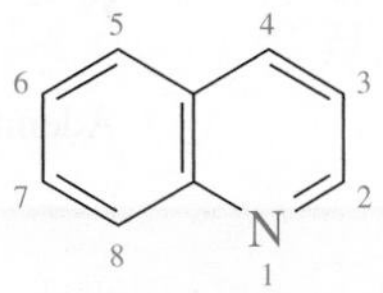

Quinoline

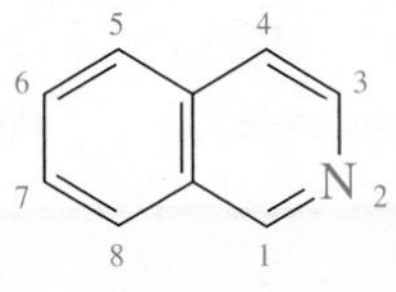

Isoquinoline

Analogous compounds derived by fusion of a benzene ring to a pyrrole, furan, or thiophene nucleus are called *indole, benzofuran,* and *benzothiophene.*

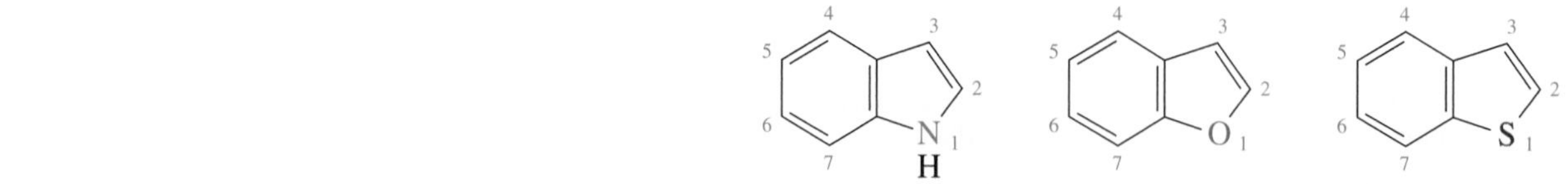

Indole Benzofuran Benzothiophene

Problem 11.27

Unlike quinoline and isoquinoline, which are of comparable stability, the compounds indole and isoindole are quite different from each other. Which one is more stable? Explain the reason for your choice.

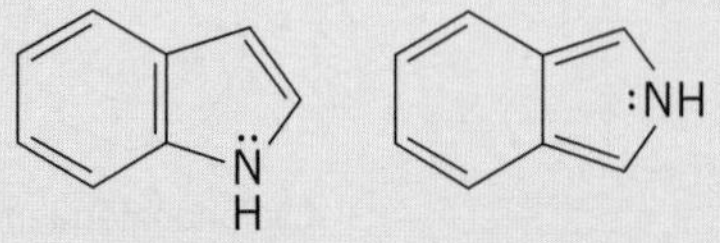

Indole Isoindole

A large group of heterocyclic aromatic compounds are related to pyrrole by replacement of one of the ring carbons β to nitrogen by a second heteroatom. Compounds of this type are called *azoles.*

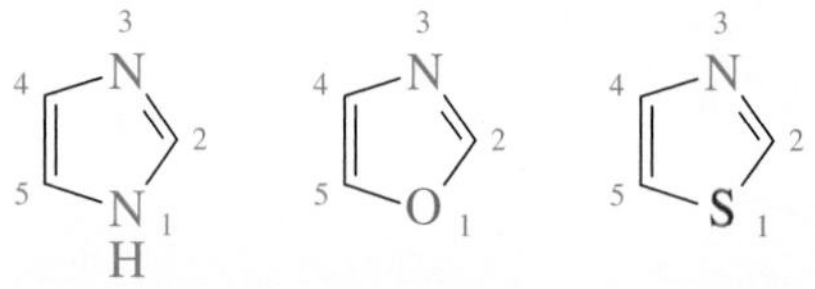

Imidazole Oxazole Thiazole

A widely prescribed drug for the treatment of gastric ulcers with the generic name *cimetidine* is a synthetic imidazole derivative. *Firefly luciferin* is a thiazole derivative that is the naturally occurring light-emitting substance present in fireflies.

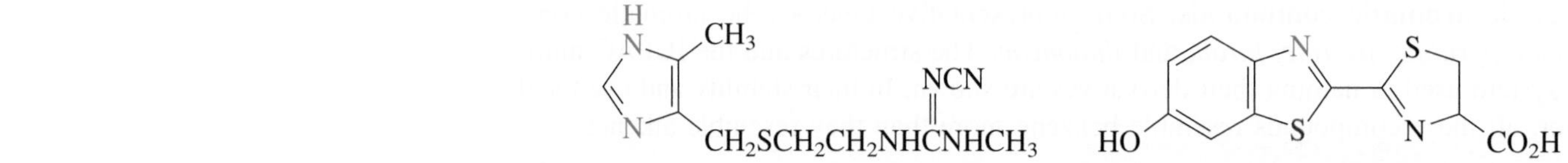

Cimetidine Firefly luciferin

Firefly luciferin is an example of an azole that contains a benzene ring fused to the five-membered ring. Such structures are fairly common. Another example is *benzimidazole,* present as a structural unit in vitamin B_{12}. Some compounds related to benzimidazole include *purine* and its amino-substituted derivative *adenine,* one of the so-called heterocyclic bases found in DNA and RNA (Chapter 26).

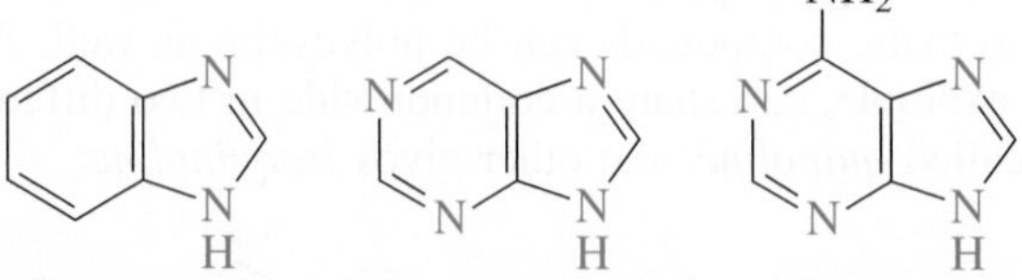

Benzimidazole Purine Adenine

Problem 11.28

Can you deduce the structural formulas of *benzoxazole* and *benzothiazole?*

11.23 Heterocyclic Aromatic Compounds and Hückel's Rule

Hückel's rule can be extended to heterocyclic aromatic compounds. A heteroatom such as oxygen or nitrogen can contribute either zero or two of its unshared electrons as needed to the π system so as to satisfy the $(4n + 2)$ π electron requirement.

The unshared pair in pyridine, for example, is not needed to satisfy the six π electron requirement for aromaticity, so is associated entirely with nitrogen and is not delocalized into the aromatic π system.

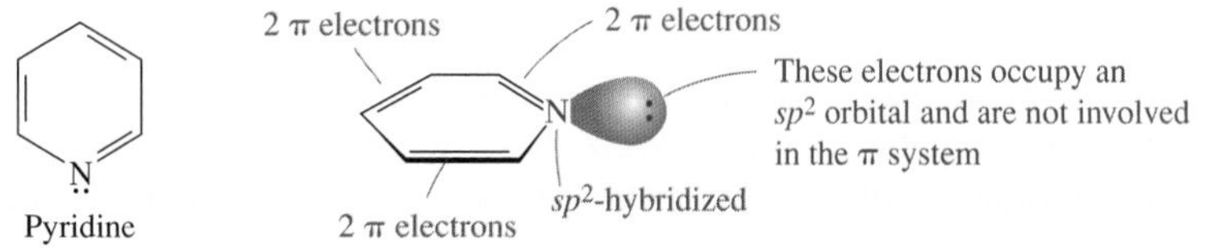

The unshared pair in the Lewis structure for pyrrole, on the other hand, must be added to the four π electrons of the two double bonds in order to meet the six π electron requirement.

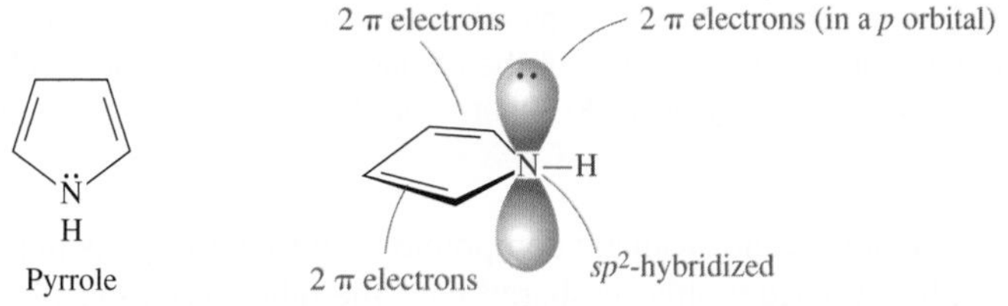

In both pyridine and pyrrole the unshared electron pair occupies that orbital which provides the most stable structure. It is a different orbital in each case. In pyridine it is an sp^2-hybridized orbital localized on nitrogen. In pyrrole it is a p orbital of nitrogen that overlaps with the p orbitals of the ring carbons to give a delocalized π system.

The electrostatic potential maps in Figure 11.16 show how pyridine and pyrrole differ with respect to their charge distribution. The unshared electron pair in pyridine gives rise to a region of high electron density (red) near nitrogen. A similar concentration of charge is absent in pyrrole because the corresponding electrons are delocalized among the five ring atoms.

Problem 11.29

Write two different resonance forms for pyrrole in which nitrogen has a formal charge of +1. Are comparable resonance forms possible for pyridine?

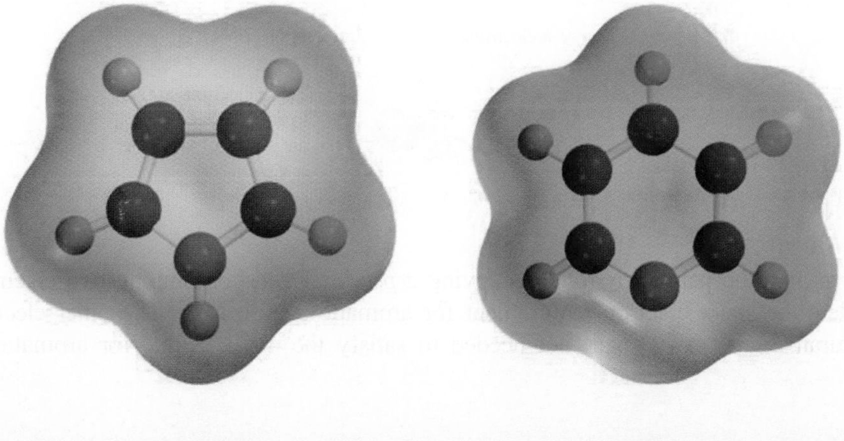

Figure 11.16

Electrostatic potential maps of pyridine and pyrrole. The color range is the same for both. In pyridine the unshared electron pair is responsible for the concentration of electron density (*red*) near nitrogen. In pyrrole the corresponding electron pair is delocalized into the π system of the ring.

The difference in bonding in pyridine and pyrrole is reflected in their properties. Although both are weak bases, pyridine is 10^7–10^9 times more basic than pyrrole. When pyridine acts as a Brønsted base, protonation of nitrogen converts an unshared pair (N:) to a bonded pair (N—H) while leaving the aromatic π system intact.

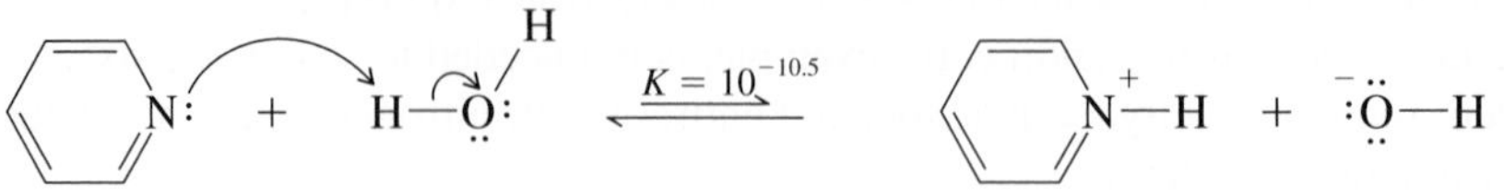
K = 10^−10.5

Pyridine | Water, Weaker acid (pK_a = 15.7) | Pyridinium ion, Stronger acid (pK_a = 5.2) | Hydroxide ion

With pyrrole, however, the pair of electrons shown as an unshared pair in its Lewis formula is actually part of the aromatic π system. Were these two electrons to be involved in covalent bonding to a proton, all of the stabilization associated with aromaticity would be lost.

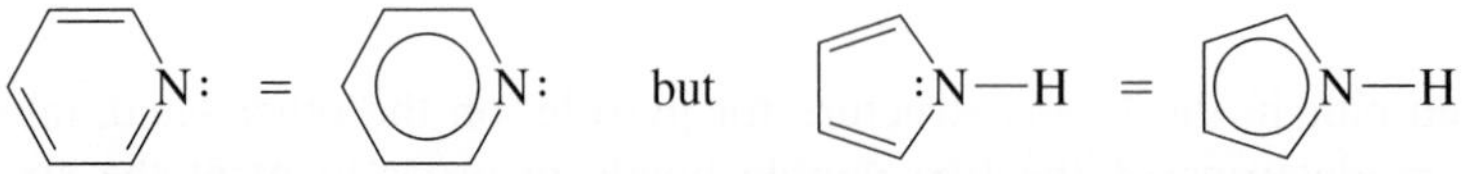
but

Problem 11.30

Estimate the pK_a of the conjugate acid of pyrrole given that pyrrole is about 10^7–10^9 times less basic than pyridine and that the pK_a of the conjugate acid of pyridine is 5.2. Is the conjugate acid of pyridine strong or weak? What about the conjugate acid of pyrrole?

Imidazole is a heterocyclic aromatic compound with two nitrogens in a five-membered ring. One nitrogen has a pyridine-like unshared pair; the other has a pyrrole-like pair that is incorporated into the aromatic π system. Imidazole is somewhat more basic than pyridine. When imidazole acts as a Brønsted base, protonation of its pyridine-like nitrogen permits aromaticity to be retained by leaving the pyrrole-like nitrogen untouched.

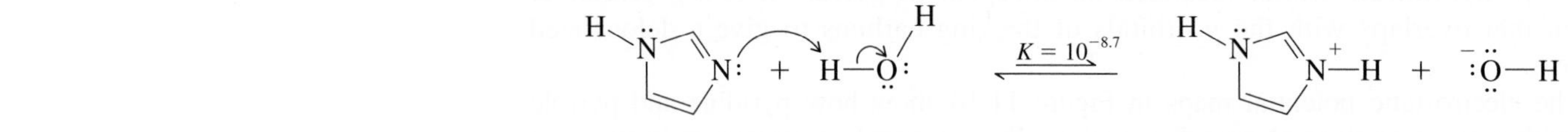
K = 10^−8.7

Imidazole | Water, Weaker acid (pK_a = 15.7) | Imidazolium ion, Stronger acid (pK_a = 7) | Hydroxide ion

Problem 11.31

Refer to the structure of imidazolium ion in the preceding equation and write a second resonance contributor that obeys the octet rule and has its positive charge on the other nitrogen. Use curved arrows to show how you reorganized the electrons.

Turning to oxygen as a heteroatom, the question of two unshared pairs on the same atom arises.

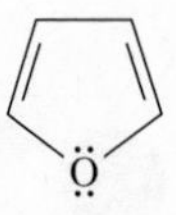

Furan

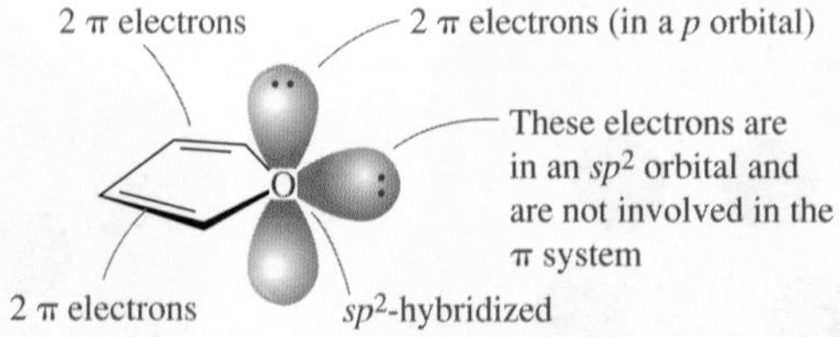
2 π electrons
2 π electrons (in a p orbital)
These electrons are in an sp² orbital and are not involved in the π system
2 π electrons
sp²-hybridized

One pair is like the pair in pyrrole, occupying a p orbital and contributing two electrons to complete the six-π electron requirement for aromatic stabilization. The other electron pair in furan is an "extra" pair, not needed to satisfy the $4n + 2$ rule for aromaticity, and occupies an sp^2-hybridized orbital like the unshared pair in pyridine. The degree of aromaticity among heterocyclic aromatic compounds varies and is different from that found in benzene. Furan and pyrrole have less resonance stabilization than pyridine and all three have less than benzene. Furan is reactive as a diene in Diels–Alder reactions (see Problem 10.42(b)).

11.24 SUMMARY

Section 11.1 Benzene is the parent of a class of compounds called **arenes,** which are aromatic hydrocarbons.

Section 11.2 An important property of aromatic hydrocarbons is that they are much more stable and less reactive than other unsaturated compounds. Benzene, for example, does not react with many of the reagents that react rapidly with alkenes. When reaction does take place, substitution rather than addition is observed. The Kekulé formulas for benzene seem inconsistent with its low reactivity and with the fact that all of the C—C bonds in benzene are the same length (140 pm).

One explanation for the structure and stability of benzene and other arenes is based on resonance, according to which benzene is regarded as a hybrid of the two Kekulé structures.

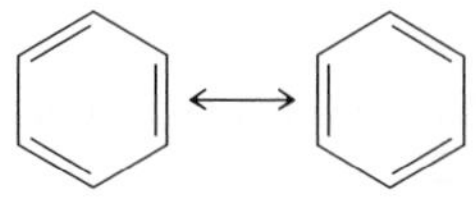

Section 11.3 The extent to which benzene is more stable than either of the Kekulé structures is its **resonance energy,** which is estimated to be 152 kJ/mol (36 kcal/mol) from heats of hydrogenation data.

Section 11.4 According to the orbital hybridization model, benzene has six π electrons, which are shared by all six sp^2-hybridized carbons. Regions of high π electron density are located above and below the plane of the ring.

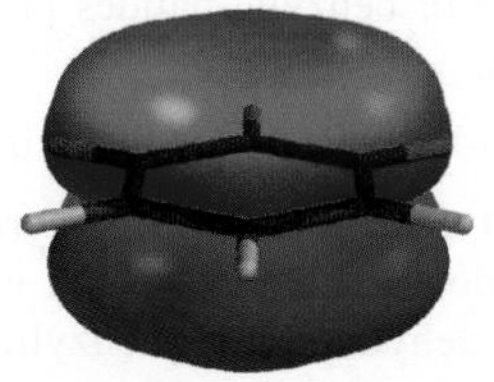

Section 11.5 A molecular orbital description of benzene has three π orbitals that are bonding and three that are antibonding. Each of the bonding orbitals is fully occupied (two electrons each), and the antibonding orbitals are vacant.

Section 11.6 Many aromatic compounds are simply substituted derivatives of benzene and are named accordingly. Many others have names based on some other parent aromatic compound.

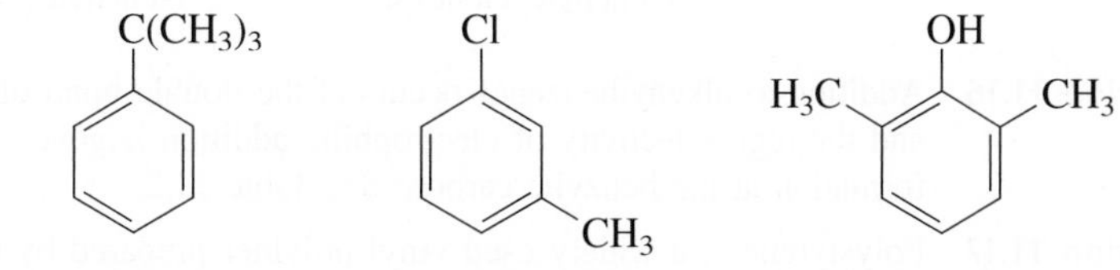

tert-Butylbenzene   *m*-Chlorotoluene   2,6-Dimethylphenol

Section 11.7 **Polycyclic aromatic hydrocarbons,** of which anthracene is an example, contain two or more benzene rings fused together.

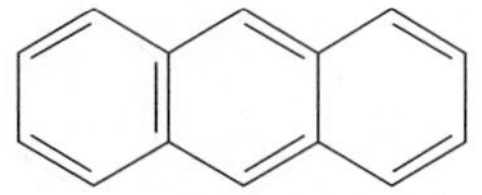

Anthracene

Section 11.8 The physical properties of arenes resemble those of other hydrocarbons.

Section 11.9 Chemical reactions of arenes can take place on the ring itself, or on a side chain. Reactions that take place on the side chain are strongly influenced by the stability of benzylic radicals and benzylic carbocations.

Benzylic radical Benzylic carbocation

Section 11.10 An example of a reaction in which the ring itself reacts is the **Birch reduction.** The ring of an arene is reduced to a nonconjugated diene by treatment with a Group 1 metal (usually sodium) in liquid ammonia in the presence of an alcohol.

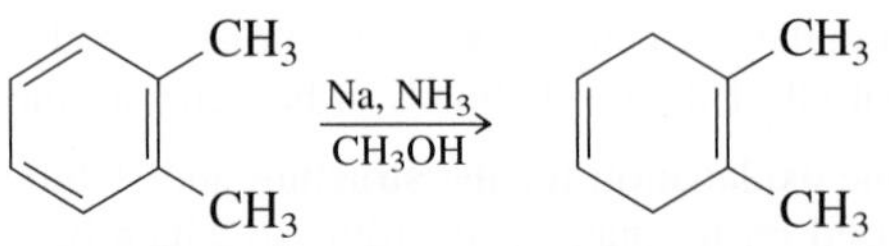

o-Xylene 1,2-Dimethyl-1,4-cyclohexadiene (92%)

Sections 11.11–11.12 Free-radical halogenation and oxidation involve reactions at the benzylic carbon. See Table 11.2.

Section 11.13 Benzylic carbocations are intermediates in S_N1 reactions of benzylic halides and are stabilized by electron delocalization.

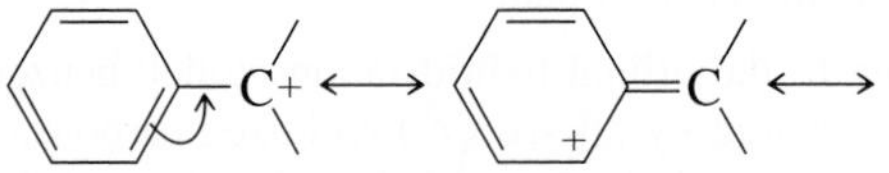

S_N1 reactions of benzylic halides proceed at rates faster than those of simple alkyl halides.

Section 11.14 Benzylic halides react faster than simple alkyl halides in S_N2 reactions.

Section 11.15 The simplest alkenylbenzene is styrene ($C_6H_5CH{=}CH_2$). An aryl group stabilizes a double bond to which it is attached. Alkenylbenzenes are usually prepared by dehydration of benzylic alcohols or dehydrohalogenation of benzylic halides.

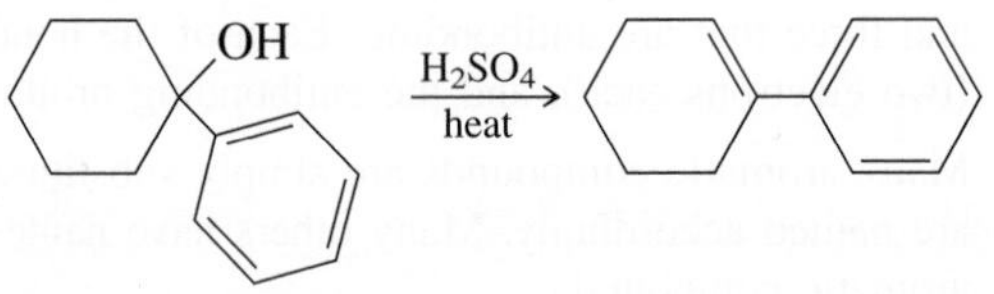

1-Phenylcyclohexanol 1-Phenylcyclohexene

Section 11.16 Addition to alkenylbenzenes occurs at the double bond of the alkenyl substituent, and the regioselectivity of electrophilic addition is governed by carbocation formation at the benzylic carbon. See Table 11.2.

Section 11.17 Polystyrene is a widely used vinyl polymer prepared by the free-radical polymerization of styrene.

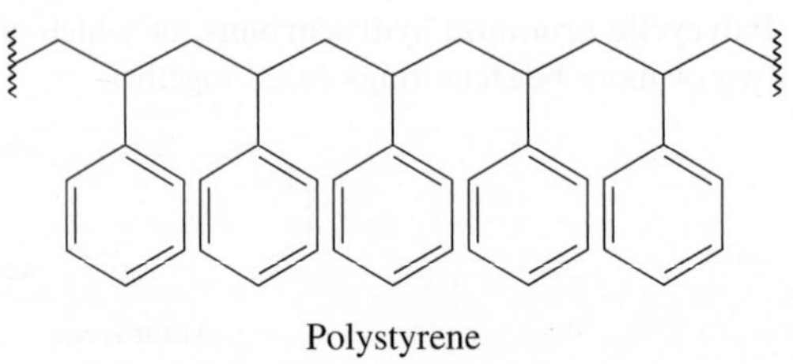

Polystyrene

TABLE 11.2 Reactions Involving Alkyl and Alkenyl Side Chains in Arenes and Arene Derivatives

Reaction (section) and comments	General equation and specific example
Halogenation (Section 11.11) Free-radical halogenation of alkylbenzenes is highly selective for substitution at the benzylic position. In the example shown, elemental bromine was used. Alternatively, *N*-bromosuccinimide (NBS) is a convenient reagent for benzylic bromination.	$ArCHR_2 \xrightarrow[\text{benzoyl peroxide, } CCl_4\text{, 80°C}]{\text{NBS}} ArCR_2(Br)$ Arene → 1-Arylalkyl bromide 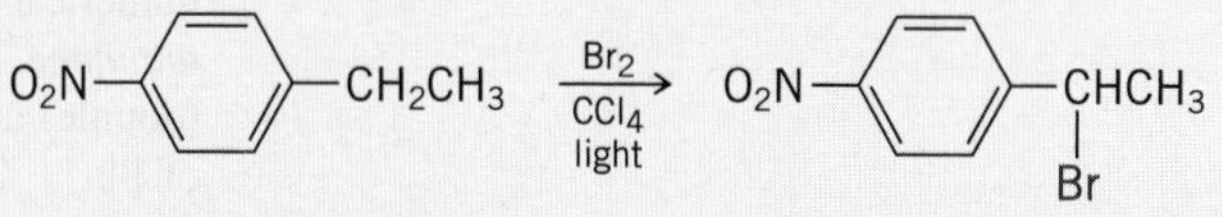 *p*-Ethylnitrobenzene → 1-(*p*-Nitrophenyl)ethyl bromide (77%)
Oxidation (Section 11.12) Oxidation of alkylbenzenes occurs at the benzylic position of the alkyl group and gives a benzoic acid derivative. Oxidizing agents include sodium or potassium dichromate in aqueous sulfuric acid. Potassium permanganate ($KMnO_4$) is also effective.	$ArCHR_2 \xrightarrow{\text{oxidize}} ArCO_2H$ Arene → Arenecarboxylic acid 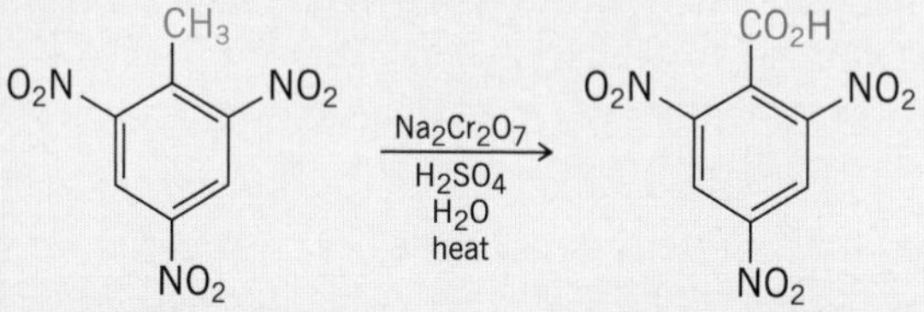 2,4,6-Trinitrotoluene → 2,4,6-Trinitrobenzoic acid (57–69%)
Hydrogenation (Section 11.16) Hydrogenation of aromatic rings is somewhat slower than hydrogenation of alkenes, and it is a simple matter to reduce the double bond of an unsaturated side chain in an arene while leaving the ring intact.	$ArCH{=}CR_2 + H_2 \xrightarrow{\text{Pt}} ArCH_2CHR_2$ Alkenylarene + Hydrogen → Alkylarene 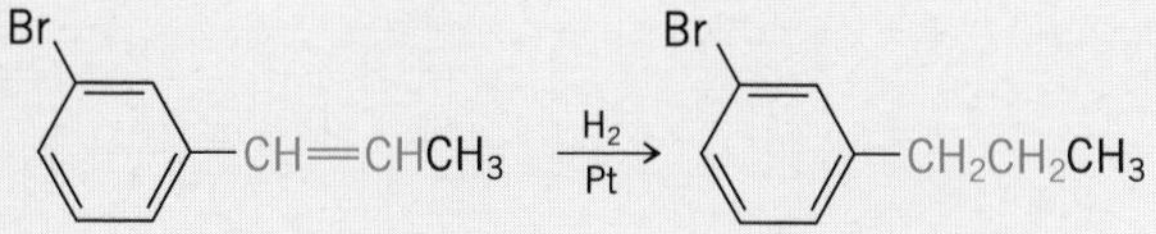 1-(*m*-Bromophenyl)propene → *m*-Bromopropylbenzene (85%)
Electrophilic addition (Section 11.16) An aryl group stabilizes a benzylic carbocation and controls the regioselectivity of addition to a double bond involving the benzylic carbon. Markovnikov's rule is obeyed.	$ArCH{=}CH_2 \xrightarrow{^{\delta+}E{-}Y^{\delta-}} ArCH(Y){-}CH_2E$ Alkenylarene → Product of electrophilic addition 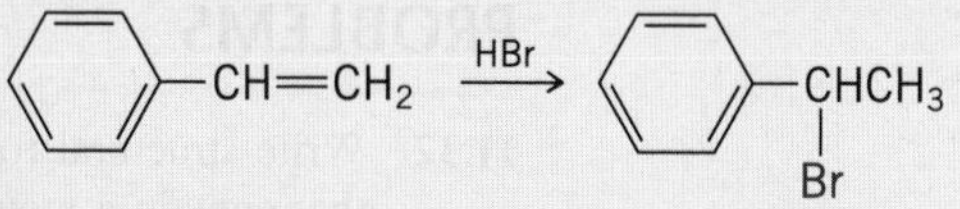 Styrene → 1-Phenylethyl bromide (85%)

Section 11.18 Although cyclic conjugation is a necessary requirement for aromaticity, this alone is not sufficient. If it were, cyclobutadiene and cyclooctatetraene would be aromatic. They are not.

Cyclobutadiene (not aromatic) Benzene (aromatic) Cyclooctatetraene (not aromatic)

Section 11.19 An additional requirement for aromaticity is that the number of π electrons in conjugated, planar, monocyclic species must be equal to $4n + 2$, where n is an integer. This is called **Hückel's rule.** Benzene, with six π electrons, satisfies Hückel's rule for $n = 1$. Square cyclobutadiene (four π electrons) and planar cyclooctatetraene (eight π electrons) do not. Both are examples of systems with $4n$ π electrons and are **antiaromatic.**

Section 11.20 **Annulenes** are monocyclic, completely conjugated polyenes synthesized for the purpose of testing Hückel's rule. They are named by using a bracketed numerical prefix to indicate the number of carbons, followed by the word *annulene.* [4*n*]- Annulenes are characterized by rings with alternating short (double) and long (single) bonds and are *antiaromatic*. The expected aromaticity of [$4n + 2$]-annulenes is diminished by angle and van der Waals strain unless the ring contains 18 or more carbons.

Section 11.21 Species with six π electrons that possess "special stability" include certain ions, such as *cyclopentadienide* anion and *cycloheptatrienyl* cation.

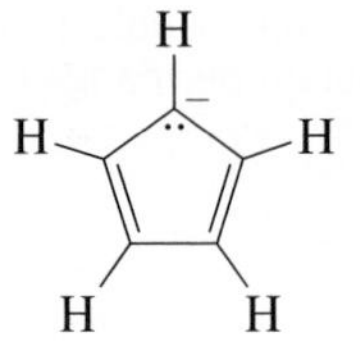

Cyclopentadienide anion (six π electrons)

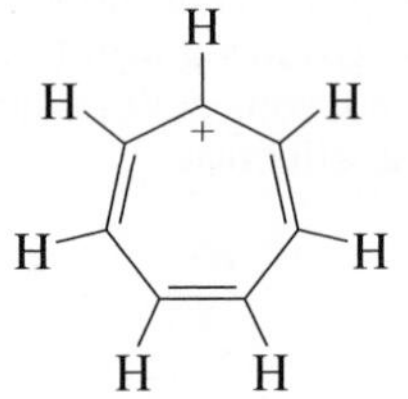

Cycloheptatrienyl cation (six π electrons)

Section 11.22 **Heterocyclic aromatic compounds** are compounds that contain at least one atom other than carbon within an aromatic ring.

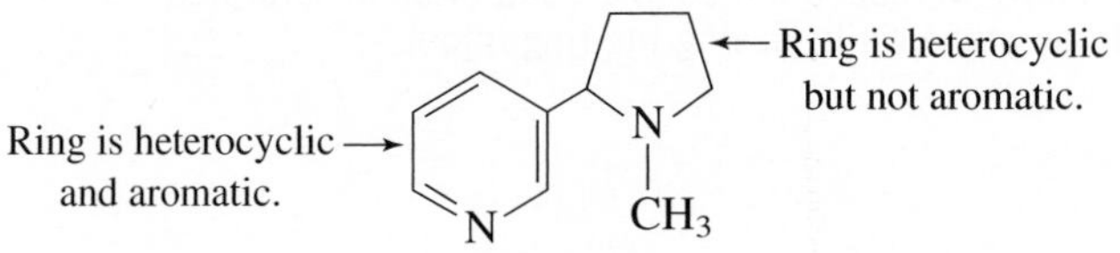

Nicotine

Section 11.23 Hückel's rule can be extended to heterocyclic aromatic compounds. Unshared electron pairs of the heteroatom may be used as π electrons as necessary to satisfy the $4n + 2$ rule.

PROBLEMS

11.32 Write structural formulas and give the IUPAC names for all the isomers of $C_6H_5C_4H_9$ that contain a monosubstituted benzene ring.

11.33 Write a structural formula corresponding to each of the following:

(a) Allylbenzene
(b) (*E*)-1-Phenyl-1-butene
(c) (*Z*)-2-Phenyl-2-butene
(d) (*R*)-1-Phenylethanol
(e) *o*-Chlorobenzyl alcohol
(f) *p*-Chlorophenol
(g) 2-Nitrobenzenecarboxylic acid
(h) *p*-Diisopropylbenzene
(i) 2,4,6-Tribromoaniline
(j) *m*-Nitroacetophenone
(k) 4-Bromo-3-ethylstyrene

11.34 Using numerical locants and the names in Table 11.1 as a guide, give an acceptable IUPAC name for each of the following compounds:

(a) Estragole (principal component of wormwood oil)

(b) Diosphenol (used in veterinary medicine to control parasites in animals)

(c) *m*-Xylidine (used in synthesis of lidocaine, a local anesthetic)

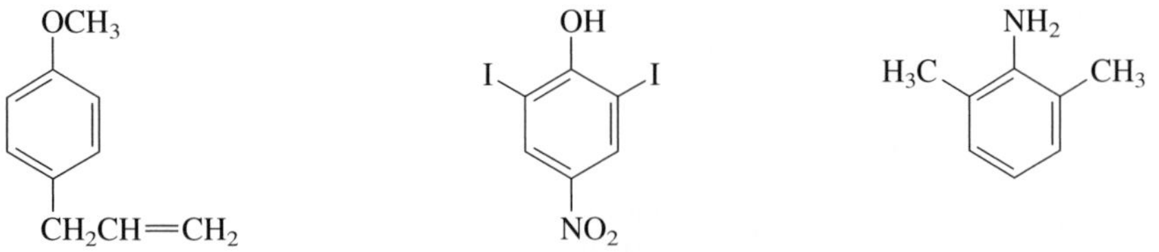

11.35 Write structural formulas and give acceptable names for all the isomeric

(a) Nitrotoluenes

(b) Dichlorobenzoic acids

(c) Tribromophenols

(d) Tetrafluorobenzenes

(e) Naphthalenecarboxylic acids

11.36 Each of the following may be represented by at least one alternative resonance structure in which all the six-membered rings correspond to Kekulé forms of benzene. Write such a resonance form for each.

(a) 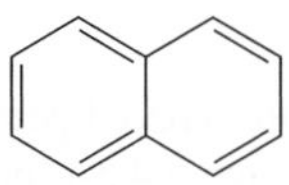

(b) 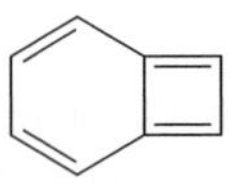

(c) 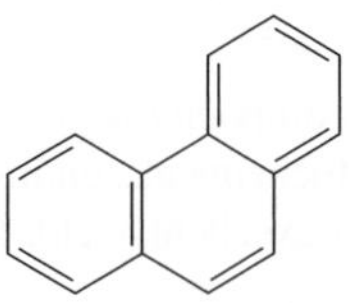

11.37 Anthracene undergoes a Diels–Alder reaction with maleic anhydride to give a cycloadduct with the formula $C_{18}H_{12}O_3$. What is its structure?

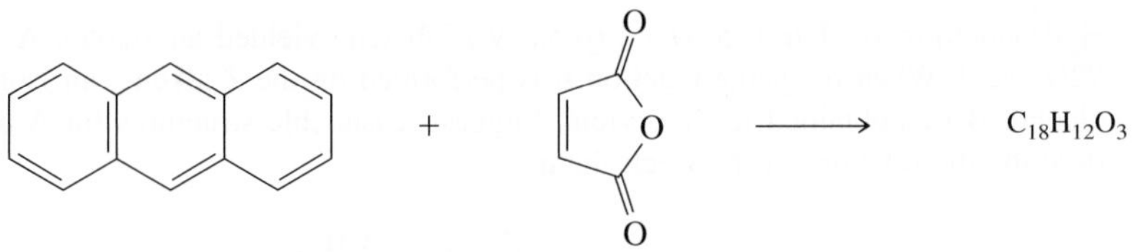

11.38 Give the structure of the expected product from the reaction of isopropylbenzene with

(a) Hydrogen (3 mol), Pt

(b) Sodium and ethanol in liquid ammonia

(c) Sodium dichromate, water, sulfuric acid, heat

(d) *N*-Bromosuccinimide in CCl_4, heat, benzoyl peroxide

(e) The product of part (d) treated with sodium ethoxide in ethanol

The common name of isopropylbenzene is *cumene*.

11.39 Each of the following reactions has been described in the chemical literature and gives a single organic product in good yield. Identify the product of each reaction.

(a) 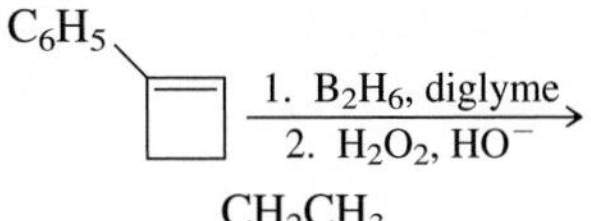

(b) 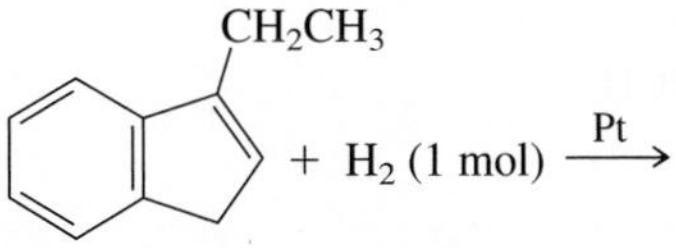

(c) 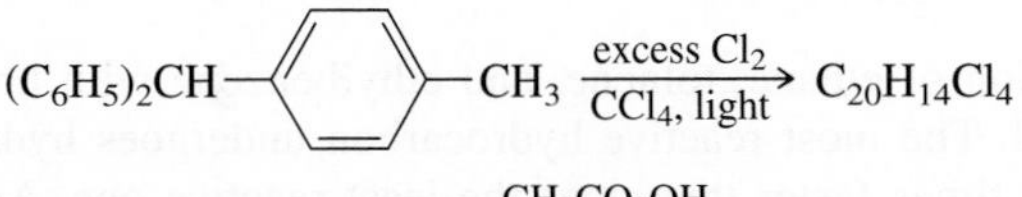

(d) $(E)\text{-}C_6H_5CH{=}CHC_6H_5 \xrightarrow[\text{acetic acid}]{CH_3CO_2OH}$

(e) 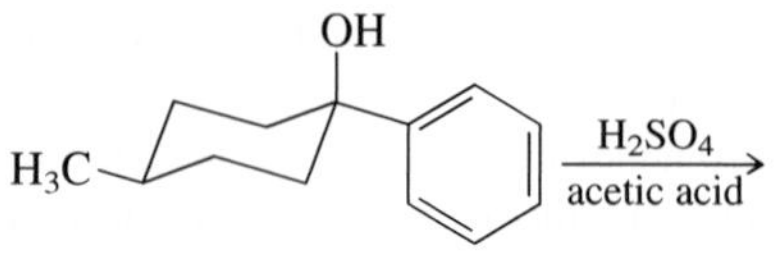
OH
H_3C
H_2SO_4
acetic acid

(f) 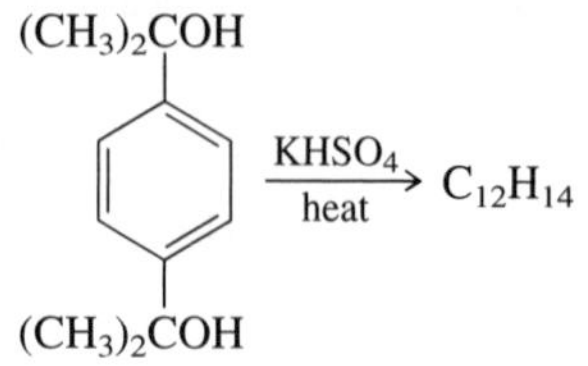
$(CH_3)_2COH$
$KHSO_4$
heat
$C_{12}H_{14}$
$(CH_3)_2COH$

(g) 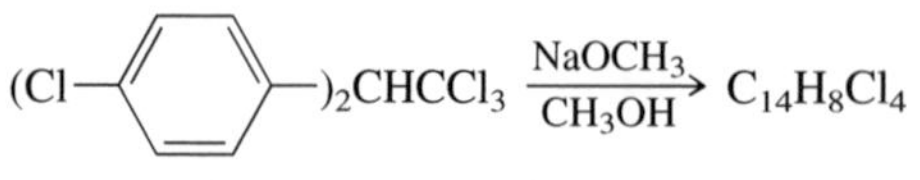
(Cl
$)_2CHCCl_3$
$NaOCH_3$
CH_3OH
$C_{14}H_8Cl_4$

(DDT)

(h) 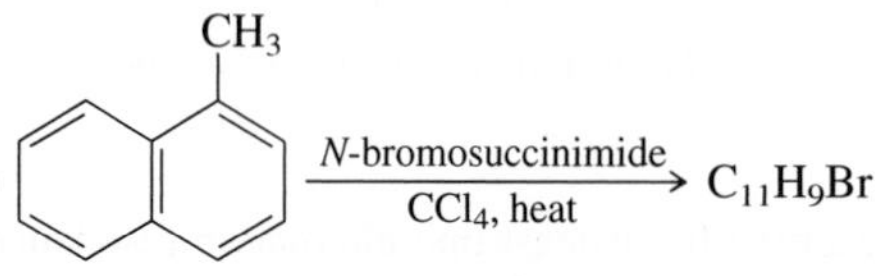
CH_3
N-bromosuccinimide
CCl_4, heat
$C_{11}H_9Br$

(i) 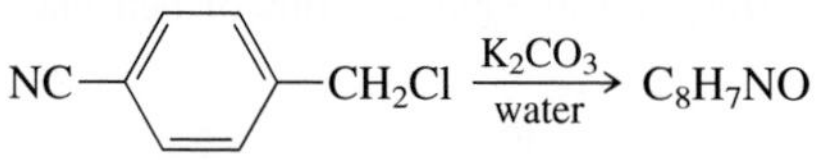
NC
CH_2Cl
K_2CO_3
water
C_8H_7NO

11.40 A certain compound A, when treated with *N*-bromosuccinimide and benzoyl peroxide under photochemical conditions in refluxing carbon tetrachloride, gave 3,4,5-tribromobenzyl bromide in excellent yield. Deduce the structure of compound A.

11.41 A compound was obtained from a natural product and had the molecular formula $C_{14}H_{20}O_3$. It contained three methoxy ($—OCH_3$) groups and a $—CH_2CH{=}C(CH_3)_2$ substituent. Oxidation with either chromic acid or potassium permanganate gave 2,3,5-trimethoxybenzoic acid. What is the structure of the compound?

11.42 Hydroboration–oxidation of (*E*)-2-(*p*-anisyl)-2-butene yielded an alcohol A, mp 60°C, in 72% yield. When the same reaction was performed on the *Z* alkene, an isomeric liquid alcohol B was obtained in 77% yield. Suggest reasonable structures for A and B, and describe the relationship between them.

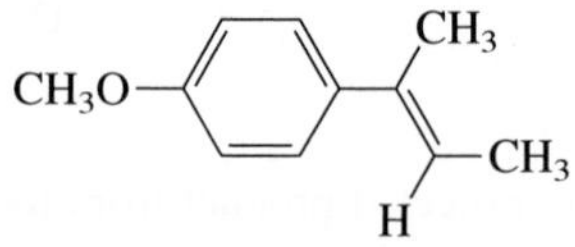
CH_3O
CH_3
CH_3
H

(*E*)-2-(*p*-Anisyl)-2-butene

11.43 Suggest reagents suitable for carrying out each of the following conversions. In most cases more than one synthetic operation will be necessary.

(a) $C_6H_5CH_2CH_3 \longrightarrow C_6H_5CH(Br)CH_3$

(b) $C_6H_5CH(Br)CH_3 \longrightarrow C_6H_5CH(Br)CH_2Br$

(c) $C_6H_5CH{=}CH_2 \longrightarrow C_6H_5C{\equiv}CH$

(d) $C_6H_5C{\equiv}CH \longrightarrow C_6H_5CH_2CH_2CH_2CH_3$

(e) $C_6H_5CH_2CH_2OH \longrightarrow C_6H_5CH_2CH_2C{\equiv}CH$

(f) $C_6H_5CH_2CH_2Br \longrightarrow C_6H_5CH(OH)CH_2Br$

11.44 The relative rates of reaction of ethane, toluene, and ethylbenzene with bromine atoms have been measured. The most reactive hydrocarbon undergoes hydrogen atom abstraction a million times faster than does the least reactive one. Arrange these hydrocarbons in order of decreasing reactivity.

11.45 Write the principal resonance structures of *o*-methylbenzyl cation and *m*-methylbenzyl cation. Which one has a tertiary carbocation as a contributing resonance form?

11.46 Suggest an explanation for the observed order of S_N1 reactivity of the following compounds.

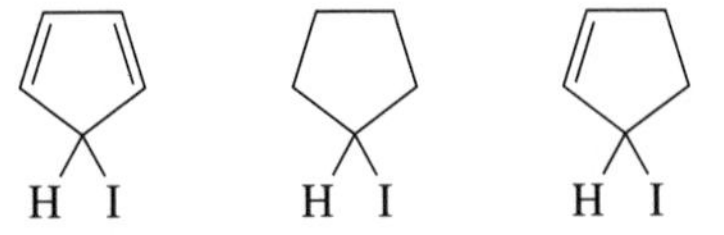

Least reactive — Most reactive

11.47 A standard method for preparing sodium cyclopentadienide (C_5H_5Na) is by the reaction of cyclopentadiene with a solution of $NaNH_2$ in liquid ammonia. Write a net ionic equation for this reaction, identify the acid and the base, and use curved arrows to track the flow of electrons.

11.48 The same anion is formed by loss of the most acidic proton from 1-methyl-1,3-cyclopentadiene as from 5-methyl-1,3-cyclopentadiene. Explain.

11.49 Cyclooctatetraene has two different tetramethyl derivatives with methyl groups on four adjacent carbon atoms. They are both completely conjugated and are not stereoisomers. Write their structures.

11.50 Evaluate each of the following processes applied to cyclooctatetraene, and decide whether the species formed is aromatic or not.

(a) Addition of one more π electron, to give $C_8H_8^-$

(b) Addition of two more π electrons, to give $C_8H_8^{2-}$

(c) Removal of one π electron, to give $C_8H_8^+$

(d) Removal of two π electrons, to give $C_8H_8^{2+}$

11.51 Evaluate each of the following processes applied to cyclononatetraene, and decide whether the species formed is aromatic or not:

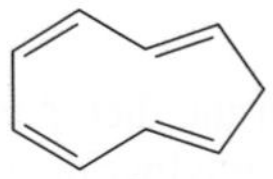

Cyclononatetraene

(a) Addition of one more π electron, to give $C_9H_{10}^-$

(b) Addition of two more π electrons, to give $C_9H_{10}^{2-}$

(c) Loss of H^+ from the sp^3-hybridized carbon

(d) Loss of H^+ from one of the sp^2-hybridized carbons

11.52 (a) Figure 11.17 is an electrostatic potential map of *calicene,* so named because its shape resembles a chalice (*calix* is the Latin word for "cup"). Both the electrostatic potential map and its calculated dipole moment (μ = 4.3 D) indicate that calicene is an unusually polar hydrocarbon. Which of the dipolar resonance forms, A or B, better corresponds to the electron distribution in the molecule? Why is this resonance form more important than the other?

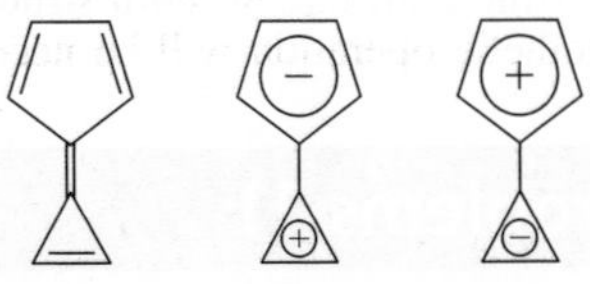

Calicene A B

(b) Which one of the following should be stabilized by resonance to a greater extent? (*Hint:* Consider the reasonableness of dipolar resonance forms.)

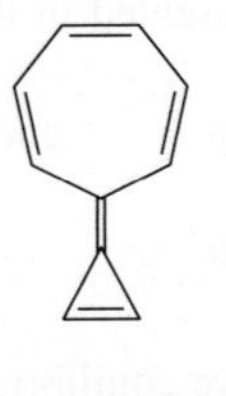

C

or

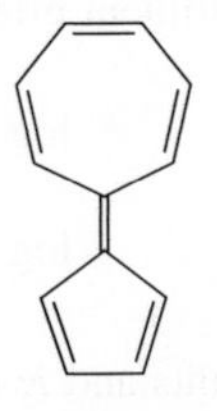

D

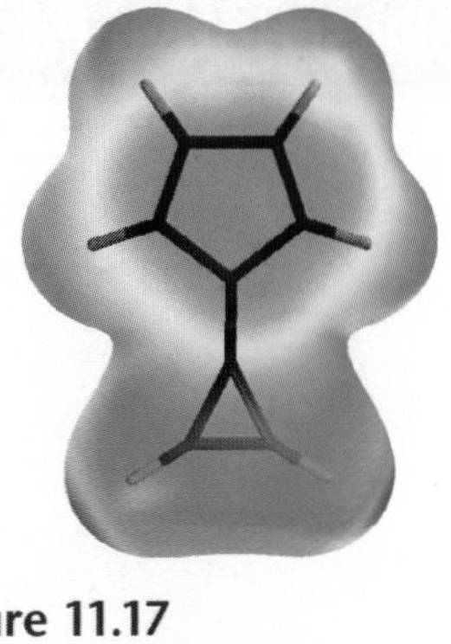

Figure 11.17

Electrostatic potential map of calicene (Problem 11.52).

11.53 Like calicene (Problem 11.52), the hydrocarbon azulene is planar and its electron distribution can be represented by a resonance contributor in which both rings satisfy Hückel's rule. Which is the major contributor, A or B?

Azulene

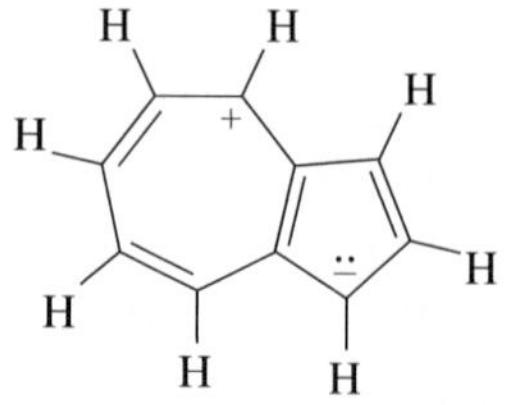

A

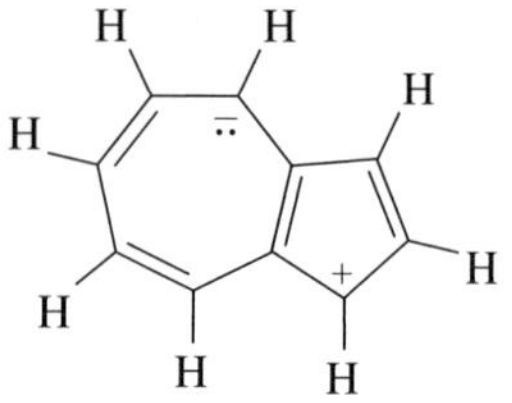

B

11.54 Classify each of the following molecules as aromatic or not, according to Hückel's rule. Are any antiaromatic?

(a) 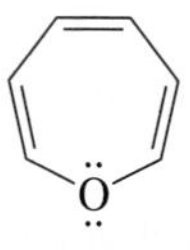

(c) 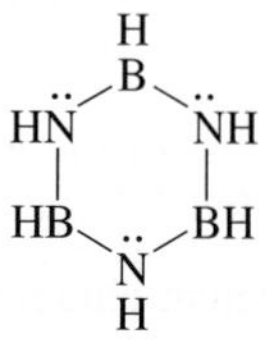

(b) 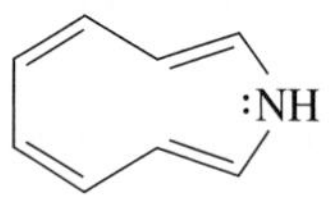

(d) 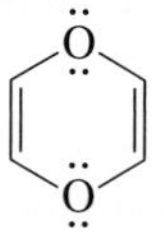

11.55 Furan is less stabilized by resonance than benzene and undergoes a 1,4 addition of bromine to give an unstable dibromide $C_4H_4Br_2O$. What is the structure of this compound?

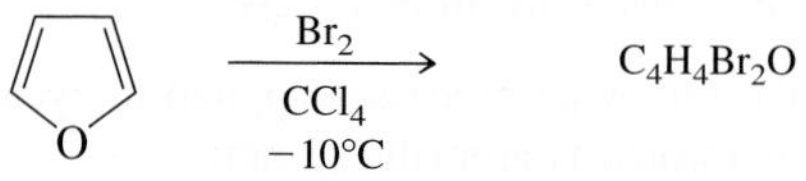

Furan

11.56 Pellagra is a disease caused by a deficiency of *niacin* ($C_6H_5NO_2$) in the diet. Niacin can be synthesized in the laboratory by the side-chain oxidation of 3-methylpyridine with chromic acid or potassium permanganate. Suggest a reasonable structure for niacin.

11.57 *Nitroxoline* is the generic name by which 5-nitro-8-hydroxyquinoline is sold as an antibacterial drug. Write its structural formula.

11.58 *Acridine* is a heterocyclic aromatic compound obtained from coal tar that is used in the synthesis of dyes. The molecular formula of acridine is $C_{13}H_9N$, and its ring system is analogous to that of anthracene except that one CH group has been replaced by N. The two most stable resonance structures of acridine are equivalent to each other, and both contain a pyridine-like structural unit. Write a structural formula for acridine.

Descriptive Passage and Interpretive Problems 11

The Hammett Equation

We have seen numerous examples of substituent effects on rates and equilibria of organic reactions and have developed a *qualitative* feel for various groups as electron-donating or electron-withdrawing. Beginning in the 1930s, Lewis P. Hammett of Columbia University developed a *quantitative* treatment of substituent effects represented in the equations:

$$\log \frac{k}{k_0} = \sigma\rho \qquad \text{and}$$

$$\log \frac{K}{K_0} = \sigma\rho$$

where k and k_0 are rate constants and K and K_0 are equilibrium constants. σ and ρ are experimentally determined constants characteristic of a substituent (σ) and a reaction (ρ).

TABLE 11.3 Substituent Constants (σ)

Substituent X	σ_{meta}	σ_{para}
CH_3	−0.06	−0.14
CF_3	0.46	0.53
CN	0.62	0.70
CH_3O	0.10	−0.12
Cl	0.37	0.24
F	0.34	0.15
NO_2	0.71	0.81

Source: O. Exner, *Correlation Analysis in Chemistry,* N.B. Chapman and J. Shorter, eds. Plenum Press, New York, 1978, Chapter 10.

k_0 and K_0 are, respectively, the rate and equilibrium constants for the unsubstituted parent compound; that is, the substituent is H. The standard substituent H is assigned a σ value of 0. The standard reaction, assigned a value of $\rho = 1.0$, is the ionization of substituted benzoic acids.

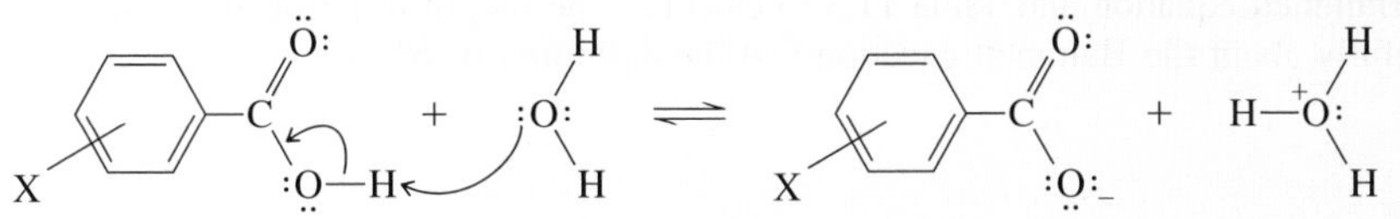

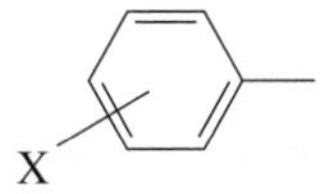

is a meta- or para-substituted aryl group

Defining ρ as 1.0 and inserting the measured K_a values for benzoic acid and its substituted derivative in the Hammett equation gives the value of σ for the substituent. When X is electron-withdrawing, the acid is stronger than benzoic acid and the sign of σ is +. Conversely, σ has a − sign for electron-releasing, acid-weakening groups. For individual substituents, σ differs according to whether it is meta or para to the reaction site. Table 11.3 gives σ values for the atoms and groups needed for Problems 11.59–11.65.

11.59 ρ for the hydrolysis of a series of tertiary benzylic chlorides is −4.5.

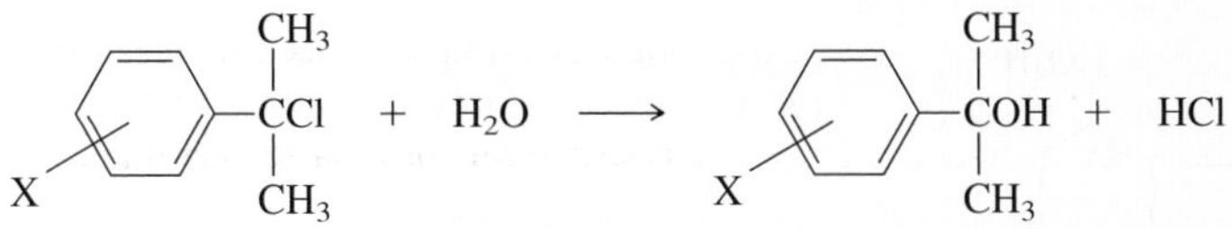

Which compound undergoes this reaction at the fastest rate?

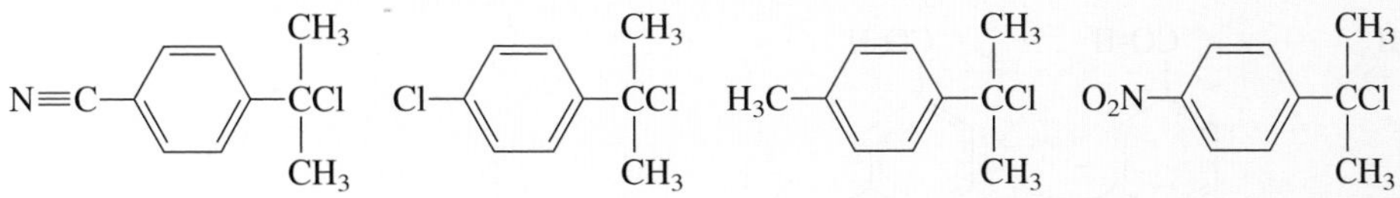

A. B. C. D.

11.60 ρ for the reaction of a series of benzylic chlorides with potassium iodide in acetone is +0.8.

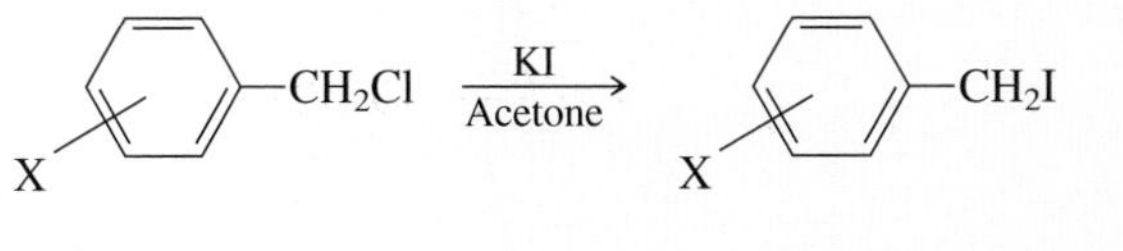

Which compound undergoes this reaction at the fastest rate?

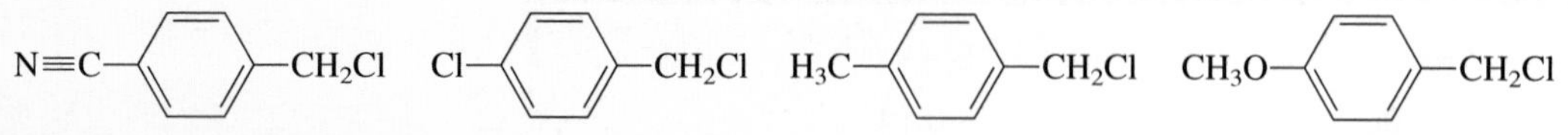

A. B. C. D.

11.61 ρ for the E2 elimination of a series of 2-arylethyl bromides with sodium ethoxide in ethanol is +2.1.

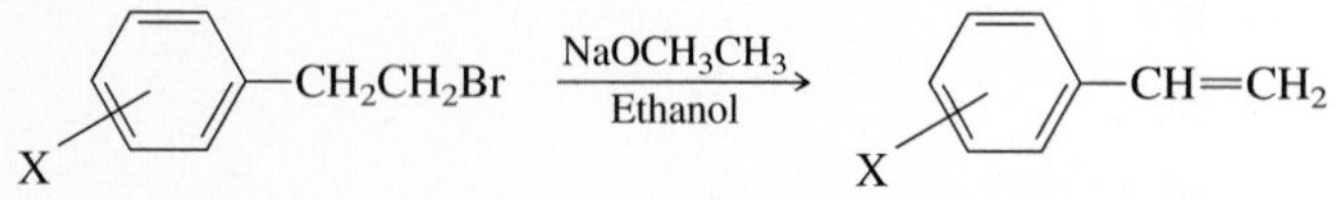

Which compound undergoes this reaction at the fastest rate?

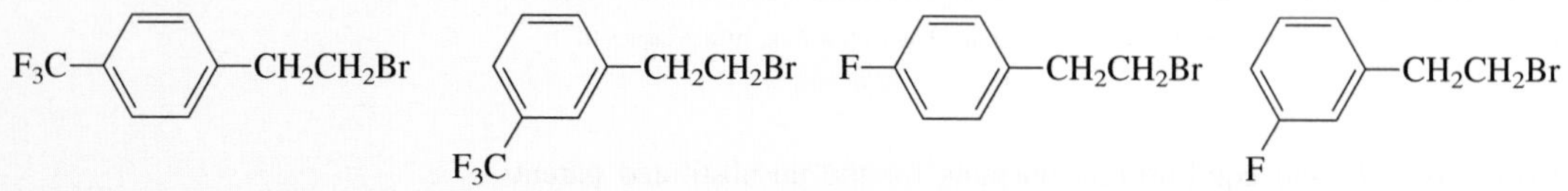

A. B. C. D.

11.62 The pK_a of benzoic acid is 4.2. Use the Hammett equation and Table 11.3 to calculate the pK_a of *m*-nitrobenzoic acid. (*Hint:* A calculator isn't required if you think carefully about the Hammett equation and the definition of pK_a.)

A. 3.0

B. 3.5

C. 4.9

D. 6.0

11.63 Use the table of σ values to rank the following substituted benzoic acids in order of *increasing* acidity. (Lowest pK_a = strongest acid)

Weakest acid **Strongest acid**

A. 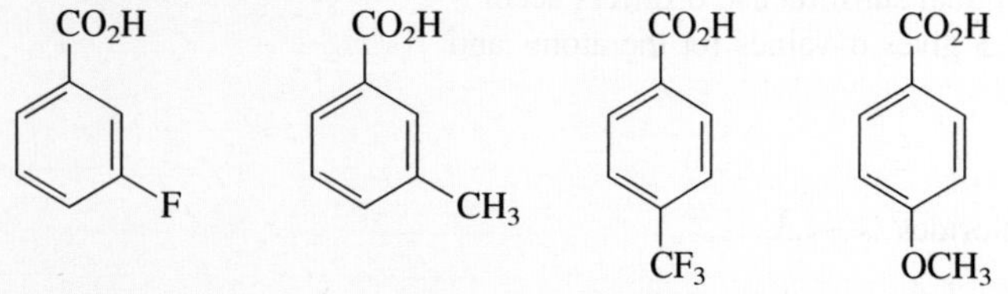

B. 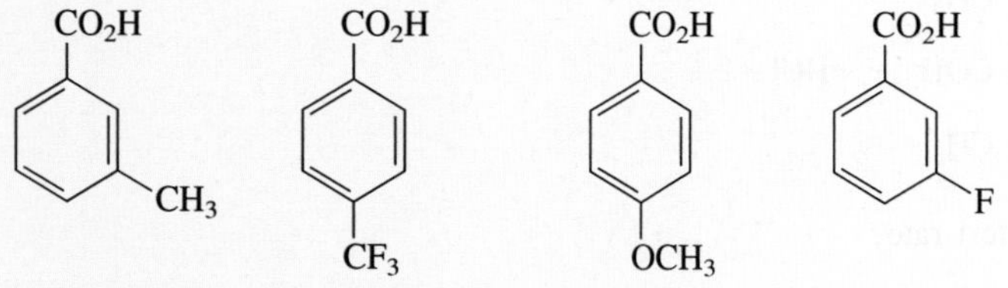

C. 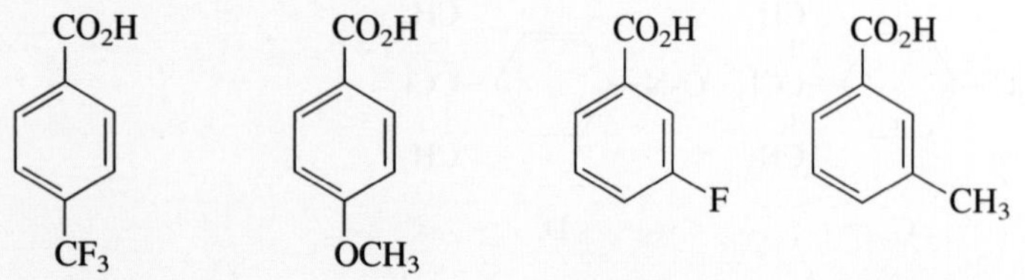

D. 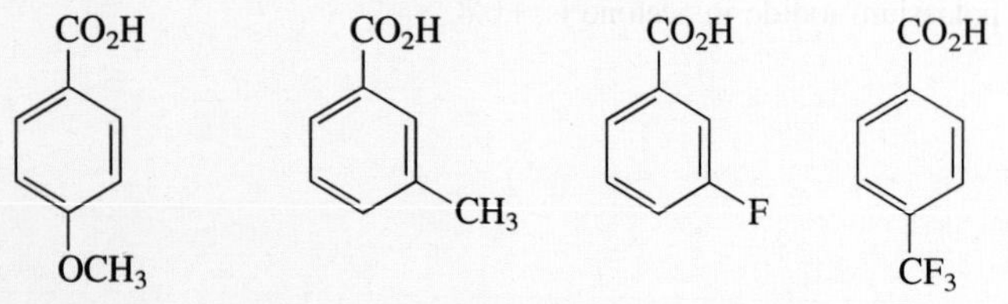

11.64 Nucleophilic substitution of the vinylic chloride shown follows an unusual two-step mechanism. The nucleophile adds to the double bond in the first step; chloride ion is expelled in the second.

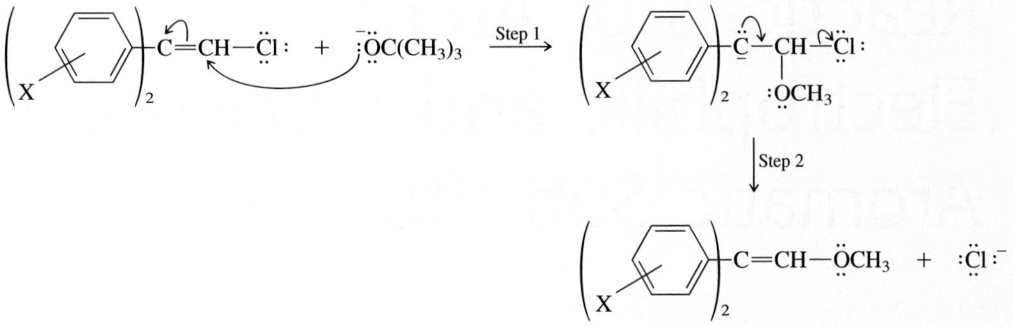

The measured value of ρ for the overall reaction is reported to be at least +4.5. Which is the most reasonable choice for the rate-determining step based on this information?

A. Step 1

B. Step 2

11.65 Transition states and ρ values are shown for E2 elimination in two series of compounds that differ in their leaving group. Based on their ρ values, in which series of reactants is there a greater degree of C—H bond breaking at the transition state?

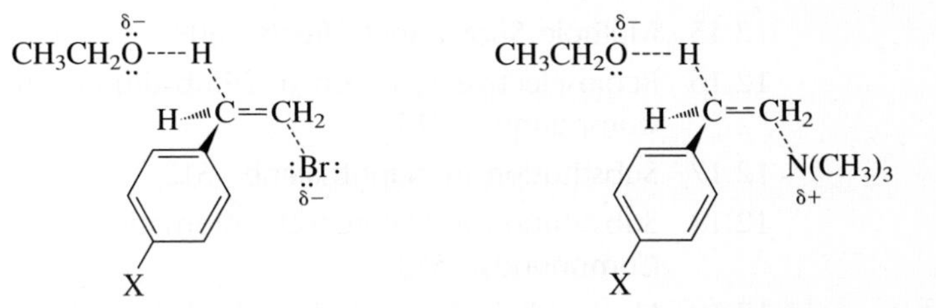

Reactant is $ArCH_2CH_2Br$; $\rho = +2.1$ Reactant is $ArCH_2CH_2\overset{+}{N}(CH_3)_3$; $\rho = +3.8$

A. $ArCH_2CH_2Br$

B. $ArCH_2CH_2\overset{+}{N}(CH_3)_3$

12 Reactions of Arenes: Electrophilic and Nucleophilic Aromatic Substitution

Chapter Outline

Mechanisms

The blackboard shows the flow of electrons in the reaction of benzene with nitronium ion. The electrostatic potential maps show the complementarity between the π-electron system of benzene and the nitrogen of nitronium ion.

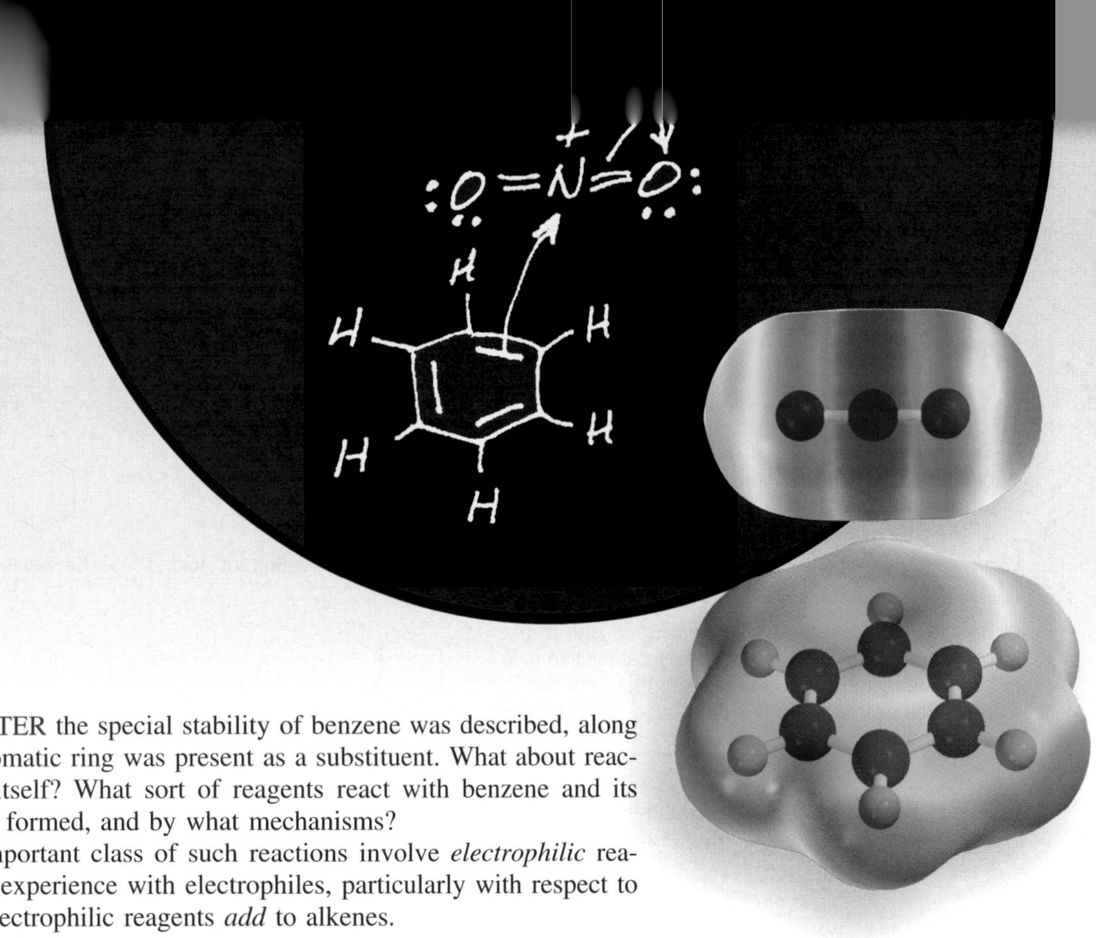

IN THE PRECEDING CHAPTER the special stability of benzene was described, along with reactions in which an aromatic ring was present as a substituent. What about reactions that occur on the ring itself? What sort of reagents react with benzene and its derivatives, what products are formed, and by what mechanisms?

The largest and most important class of such reactions involve *electrophilic* reagents. We already have some experience with electrophiles, particularly with respect to their reaction with alkenes. Electrophilic reagents *add* to alkenes.

$$\mathrm{C{=}C} + \overset{\delta+}{\mathrm{E}}{-}\overset{\delta-}{\mathrm{Y}} \longrightarrow \mathrm{E{-}C{-}C{-}Y}$$

Alkene | Electrophilic reagent | Product of electrophilic addition

A different reaction occurs with arenes. *Substitution is observed instead of addition.* The electrophilic portion of the reagent replaces one of the hydrogens on the ring:

$$\mathrm{Ar{-}H} + \overset{\delta+}{\mathrm{E}}{-}\overset{\delta-}{\mathrm{Y}} \longrightarrow \mathrm{Ar{-}E} + \mathrm{H{-}Y}$$

Arene | Electrophilic reagent | Product of electrophilic aromatic substitution

This reaction is known as **electrophilic aromatic substitution.** It is one of the fundamental processes of organic chemistry and the major concern of this chapter.

What about nucleophilic substitution in aryl halides?

$$\mathrm{Ar{-}\ddot{\underset{..}{X}}:} + :\mathrm{Nu^-} \longrightarrow \mathrm{Ar{-}Nu} + :\ddot{\underset{..}{\mathrm{X}}}:^-$$

Aryl halide | Nucleophile | Product of nucleophilic aromatic substitution | Halide ion

In Chapter 8, we noted that aryl halides are normally much less reactive toward nucleophilic substitution than alkyl halides. In the present chapter we'll see examples of novel, useful, and mechanistically interesting **nucleophilic aromatic substitutions** and explore the structural features responsible for these reactions.

12.1 Representative Electrophilic Aromatic Substitution Reactions of Benzene

The scope of electrophilic aromatic substitution is quite large; both the aromatic compound and the electrophilic reagent are capable of wide variation. Indeed, it is this breadth of scope that makes electrophilic aromatic substitution so important. Electrophilic aromatic

TABLE 12.1 Representative Electrophilic Aromatic Substitution Reactions of Benzene

Reaction and comments	Equation
1. **Nitration** Warming benzene with a mixture of nitric acid and sulfuric acid gives nitrobenzene. A nitro group (—NO_2) replaces one of the ring hydrogens.	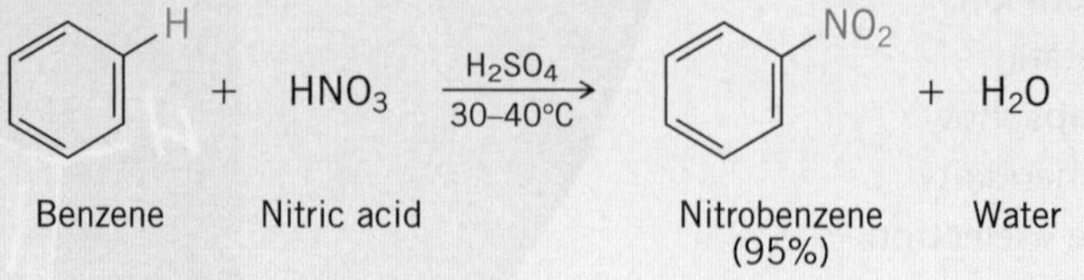 Benzene + HNO_3 (Nitric acid) $\xrightarrow[30-40°C]{H_2SO_4}$ Nitrobenzene (95%) + H_2O (Water)
2. **Sulfonation** Treatment of benzene with hot concentrated sulfuric acid gives benzenesulfonic acid. A sulfonic acid group (—SO_2OH) replaces one of the ring hydrogens.	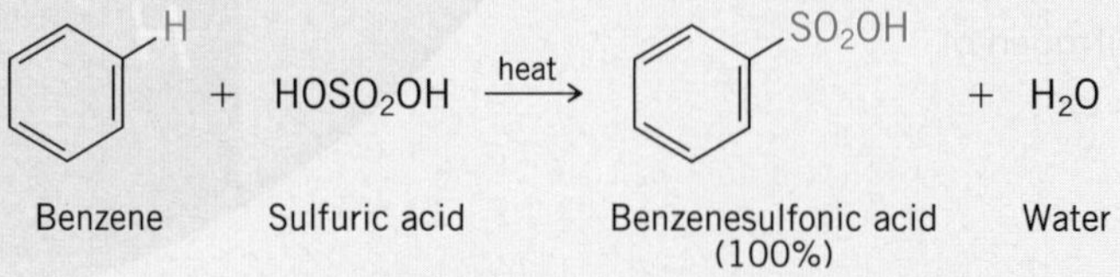 Benzene + $HOSO_2OH$ (Sulfuric acid) $\xrightarrow{heat}$ Benzenesulfonic acid (100%) + H_2O (Water)
3. **Halogenation** Bromine reacts with benzene in the presence of iron(III) bromide as a catalyst to give bromobenzene. Chlorine reacts similarly in the presence of iron(III) chloride to give chlorobenzene.	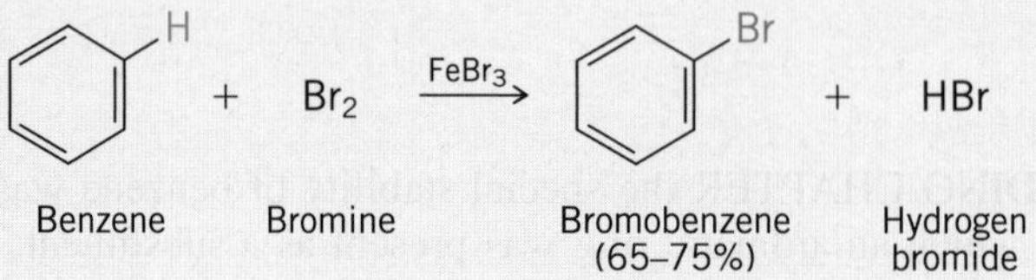 Benzene + Br_2 (Bromine) $\xrightarrow{FeBr_3}$ Bromobenzene (65–75%) + HBr (Hydrogen bromide)
4. **Friedel–Crafts alkylation** Alkyl halides react with benzene in the presence of aluminum chloride to yield alkylbenzenes.	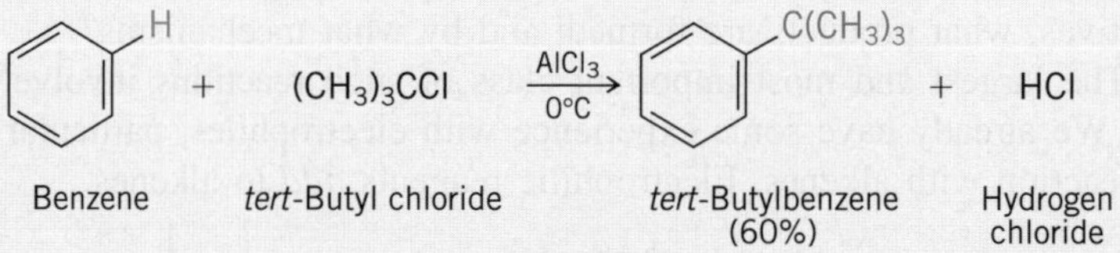 Benzene + $(CH_3)_3CCl$ (*tert*-Butyl chloride) $\xrightarrow[0°C]{AlCl_3}$ *tert*-Butylbenzene (60%) + HCl (Hydrogen chloride)
5. **Friedel–Crafts acylation** An analogous reaction occurs when acyl halides react with benzene in the presence of aluminum chloride. The products are aryl ketones.	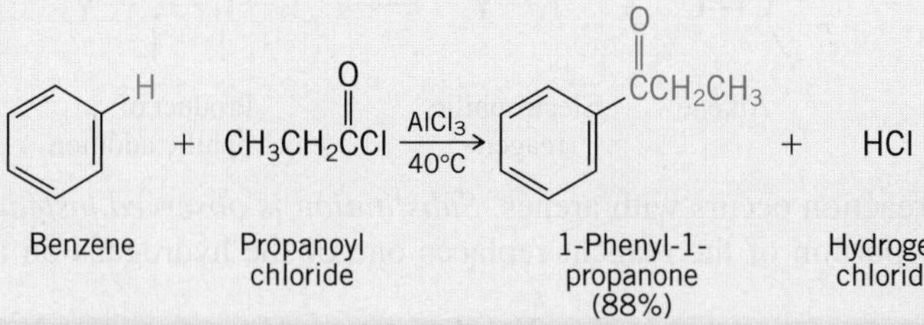 Benzene + CH_3CH_2CCl (Propanoyl chloride) $\xrightarrow[40°C]{AlCl_3}$ 1-Phenyl-1-propanone (88%) + HCl (Hydrogen chloride)

substitution is the method by which substituted derivatives of benzene are prepared. We can gain a feeling for these reactions by examining a few typical examples in which benzene is the substrate. These examples are listed in Table 12.1, and each will be discussed in more detail in Sections 12.3 through 12.7. First, however, let us look at the general mechanism of electrophilic aromatic substitution.

12.2 Mechanistic Principles of Electrophilic Aromatic Substitution

Recall from Chapter 6 the general mechanism for electrophilic addition to alkenes:

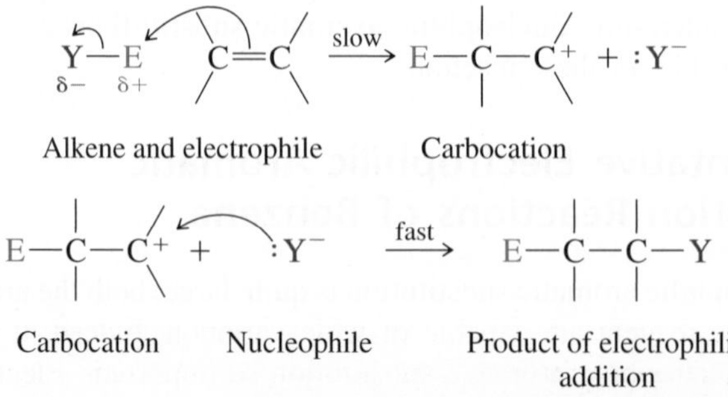

The first step is rate-determining. In it a carbocation forms when the pair of π electrons of the alkene is used to form a bond with the electrophile. The carbocation then undergoes rapid capture by some Lewis base present in the medium.

The first step in the reaction of electrophilic reagents with benzene is similar. An electrophile accepts an electron pair from the π system of benzene to form a carbocation:

Recall that an electron-pair acceptor is a Lewis acid. Electrophiles are Lewis acids.

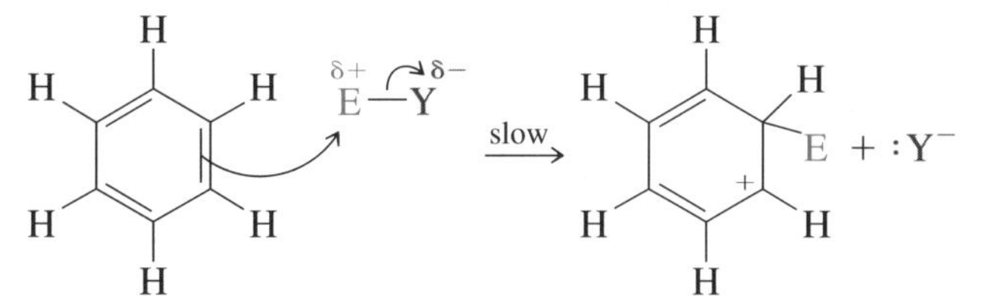

Benzene and electrophile Carbocation

The carbocation formed in this step is an **arenium ion** or **cyclohexadienyl cation,** also known as a **σ-complex.** It is an allylic carbocation and is stabilized by the electron delocalization represented by resonance among the contributing structures:

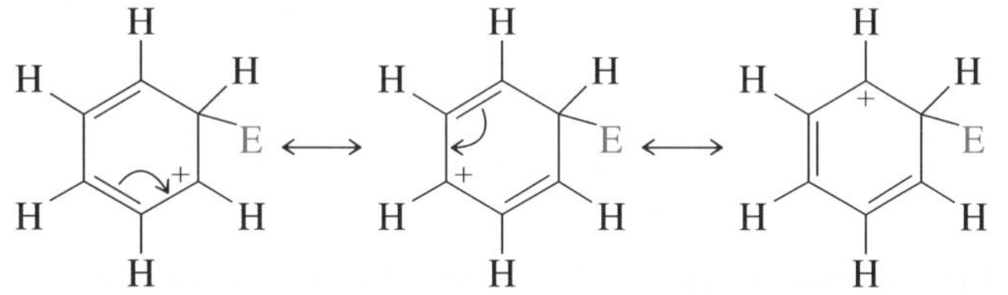
Resonance structures of a cyclohexadienyl cation

Most of the resonance stabilization of benzene is lost when it is converted to the cyclohexadienyl cation intermediate. In spite of being allylic, a cyclohexadienyl cation is *not* aromatic and possesses only a fraction of the resonance stabilization of benzene.

Once formed, the cyclohexadienyl cation rapidly loses a proton, restoring the aromaticity of the ring and giving the product of electrophilic aromatic substitution.

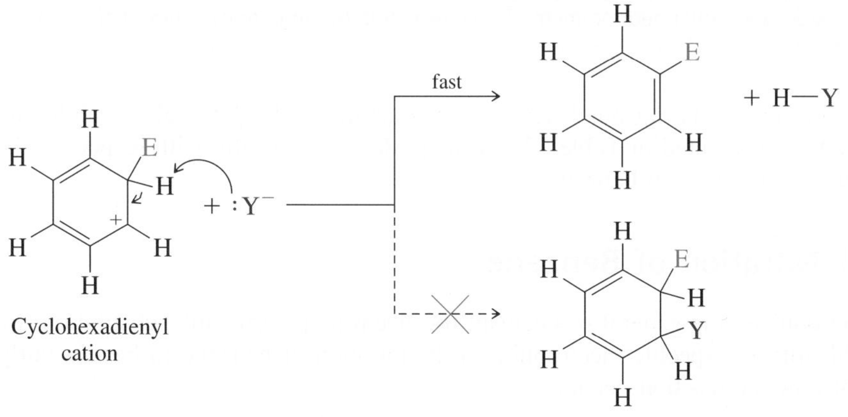

If the Lewis base (:Y$^-$) had acted as a nucleophile and bonded to carbon, the product would have been a nonaromatic cyclohexadiene derivative. Addition and substitution products arise by alternative reaction paths of a cyclohexadienyl cation. Substitution occurs preferentially because there is a substantial driving force favoring rearomatization.

Figure 12.1 is a potential energy diagram describing the general mechanism of electrophilic aromatic substitution. For electrophilic aromatic substitution reactions to

Figure 12.1

Potential energy diagram for electrophilic aromatic substitution.

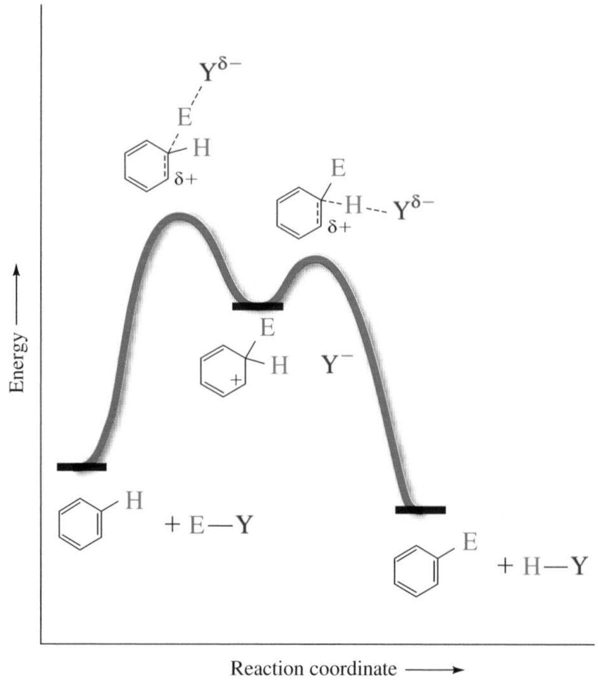

overcome the high activation energy that characterizes the first step, the electrophile must be a reactive one. Many of the electrophilic reagents that react rapidly with alkenes do not react at all with benzene. Peroxy acids and diborane, for example, fall into this category. Others, such as bromine, react with benzene only in the presence of catalysts that increase their electrophilicity. The low level of reactivity of benzene toward electrophiles stems from the loss of aromaticity that accompanies transfer of a pair of its six π electrons to an electrophile.

Problem 12.1

Based on Hammond's postulate, does the structure of the transition state for formation of the carbocation intermediate more closely resemble benzene or cyclohexadienyl cation?

With this as background, let us now examine each of the electrophilic aromatic substitutions presented in Table 12.1 in more detail, especially with respect to the electrophile that reacts with benzene.

12.3 Nitration of Benzene

Having outlined the general mechanism for electrophilic aromatic substitution, we need only identify the specific electrophile in the nitration of benzene to have a fairly clear idea of how the reaction occurs.

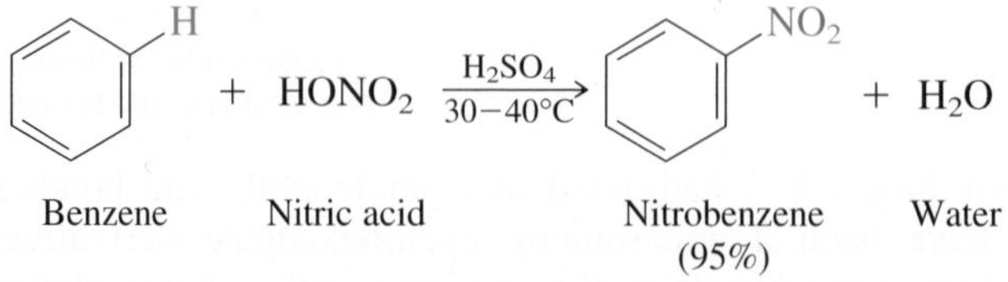

The role of nitronium ion in the nitration of benzene was demonstrated by Sir Christopher Ingold—the same person who suggested the S_N1 and S_N2 mechanisms of nucleophilic substitution and who collaborated with Cahn and Prelog on the *R* and *S* notational system.

The electrophile (E^+) in this reaction is *nitronium ion* ($:\ddot{O}=\overset{+}{N}=\ddot{O}:$). The charge distribution in nitronium ion is evident both in its Lewis structure and in the electrostatic potential map on page 479, which also shows the complementary relationship between the electron-poor region near nitrogen of NO_2^+ and the electron-rich region associated

Mechanism 12.1

Nitration of Benzene

Step 1: Reaction of nitronium cation with the π system of the aromatic ring. (The molecular model depicts the cyclohexadienyl cation intermediate.)

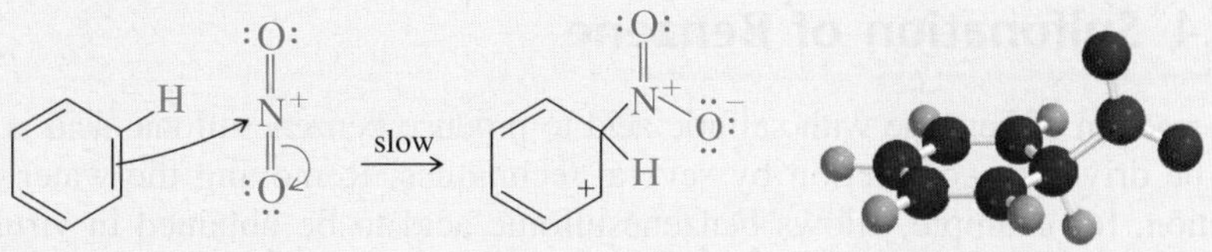

Benzene and nitronium ion — Cyclohexadienyl cation intermediate

Step 2: Loss of a proton from the cyclohexadienyl cation.

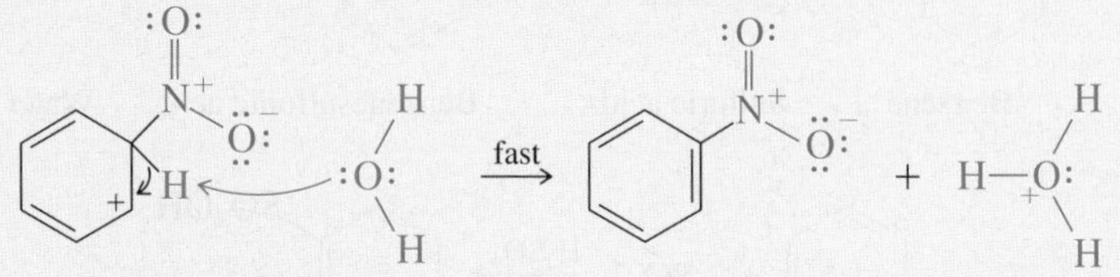

Cyclohexadienyl cation intermediate — Water — Nitrobenzene — Hydronium ion

with the π electrons of benzene. Nitronium ion is generated by the reaction of nitric acid with sulfuric acid, resulting in the protonation of nitric acid and loss of water:

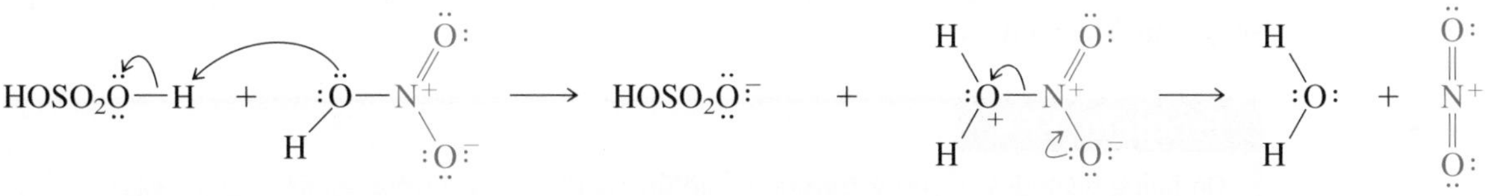

Sulfuric acid — Nitric acid — Hydrogen sulfate ion — Protonated nitric acid — Water — Nitronium ion

Mechanism 12.1 adapts the general mechanism of electrophilic aromatic substitution to the nitration of benzene. The first step is rate-determining; in it benzene reacts with nitronium ion to give the cyclohexadienyl cation intermediate. In the second step, the aromaticity of the ring is restored by loss of a proton from the cyclohexadienyl cation.

One way we know that step 1 is rate-determining is that nitration of benzene does not exhibit a deuterium isotope effect (see Section 5.17). Loss of deuterium ($D={}^2H$) during nitration of C_6H_5D occurs at the same rate as loss of a single ^1H, which tells us that the C—D bond must break *after* the rate-determining step, not during it.

Nitration by electrophilic aromatic substitution is not limited to benzene alone, but is a general reaction of compounds that contain a benzene ring. It would be a good idea to write out the answers to the following two problems to ensure that you understand the relationship of starting materials to products and the mechanism of aromatic nitration before continuing to the next section.

Problem 12.2

Nitration of 1,4-dimethylbenzene (*p*-xylene) gives a single product having the molecular formula $C_8H_9NO_2$ in high yield. What is this product?

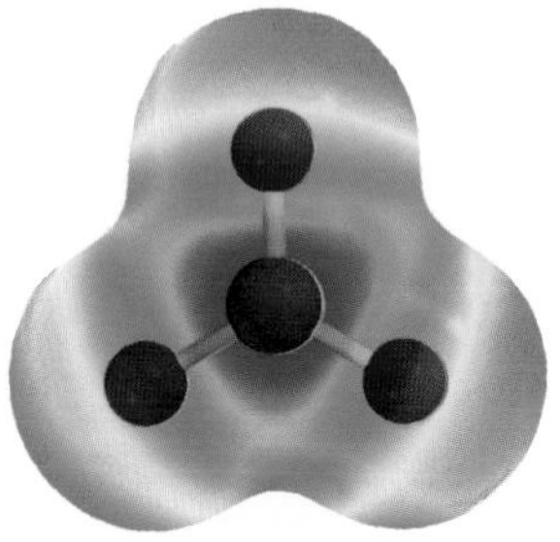

Figure 12.2

Electrostatic potential map of sulfur trioxide. The region of greatest positive charge surrounds sulfur.

Problem 12.3

Using :Ö=N̈⁺=Ö: as the electrophile, write a reasonable mechanism for the reaction given in Problem 12.2. Use curved arrows to show the flow of electrons.

12.4 Sulfonation of Benzene

The reaction of benzene with sulfuric acid to produce benzenesulfonic acid is reversible and can be driven to completion by several techniques. Removing the water formed in the reaction, for example, allows benzenesulfonic acid to be obtained in virtually quantitative yield. When a solution of sulfur trioxide in sulfuric acid is used as the sulfonating agent, the rate of sulfonation is much faster and the equilibrium is displaced entirely to the side of products, according to the equation

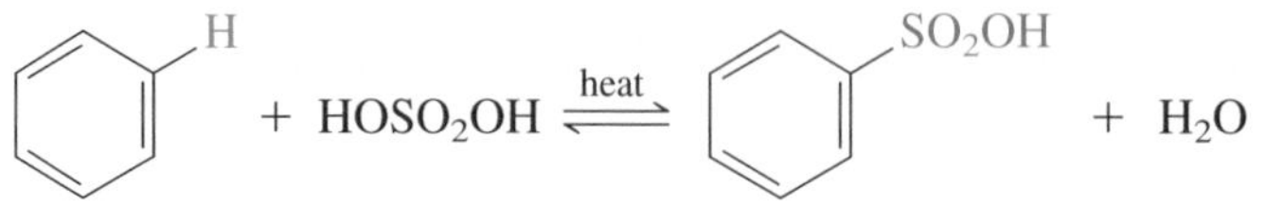
H
+ $HOSO_2OH$
heat
SO_2OH
+ H_2O

Benzene Sulfuric acid Benzenesulfonic acid Water

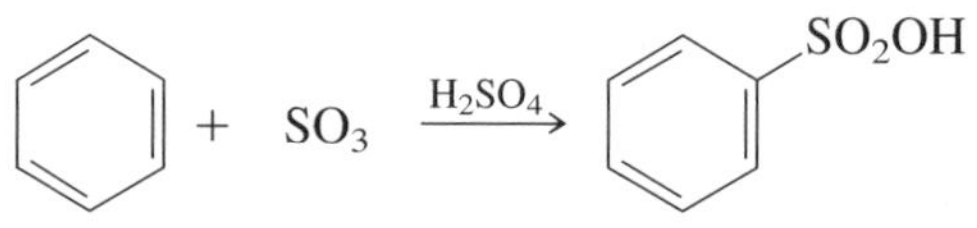
+ SO_3
H_2SO_4
SO_2OH

Benzene Sulfur trioxide Benzenesulfonic acid

Among the variety of electrophilic species present in concentrated sulfuric acid, sulfur trioxide (Figure 12.2) is probably the actual electrophile in aromatic sulfonation. We can represent the sulfonation of benzene by sulfur trioxide by the sequence of steps shown in Mechanism 12.2.

Problem 12.4

On being heated with sulfur trioxide in sulfuric acid, 1,2,4,5-tetramethylbenzene was converted to a product of molecular formula $C_{10}H_{14}O_3S$ in 94% yield. Suggest a reasonable structure for this product.

12.5 Halogenation of Benzene

According to the usual procedure for preparing bromobenzene, bromine is added to benzene in the presence of metallic iron (customarily a few carpet tacks) and the reaction mixture is heated.

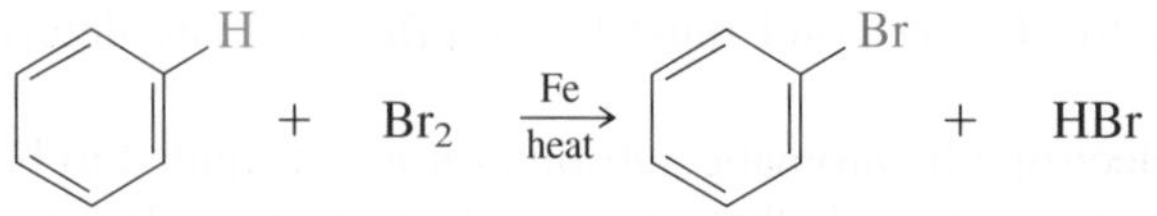
H
+ Br_2
Fe
heat
Br
+ HBr

Benzene Bromine Bromobenzene (65–75%) Hydrogen bromide

Bromine, although it adds rapidly to alkenes, is too weak an electrophile to react at an appreciable rate with benzene. A catalyst that increases the electrophilic properties of bromine must be present. Somehow carpet tacks can do this. How?

The active catalyst is not iron itself but iron(III) bromide, formed by reaction of iron and bromine.

Iron(III) bromide ($FeBr_3$) is also called *ferric bromide.*

$$2Fe + 3Br_2 \longrightarrow 2FeBr_3$$

Iron Bromine Iron(III) bromide

Mechanism 12.2

Sulfonation of Benzene

Step 1: Sulfur trioxide reacts with benzene in the rate-determining step. (The molecular model depicts the cyclohexadienyl cation intermediate.)

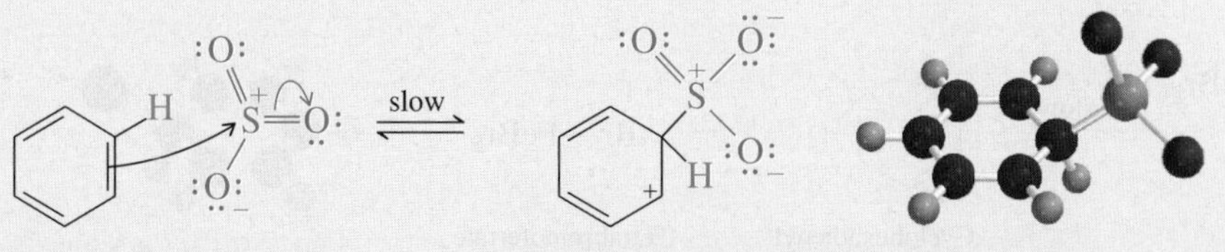

Benzene and sulfur trioxide — Cyclohexadienyl cation intermediate

Step 2: A proton is abstracted from the sp^3-hybridized carbon of the intermediate to restore the aromaticity of the ring. The species shown that abstracts the proton is a hydrogen sulfate ion formed by ionization of sulfuric acid.

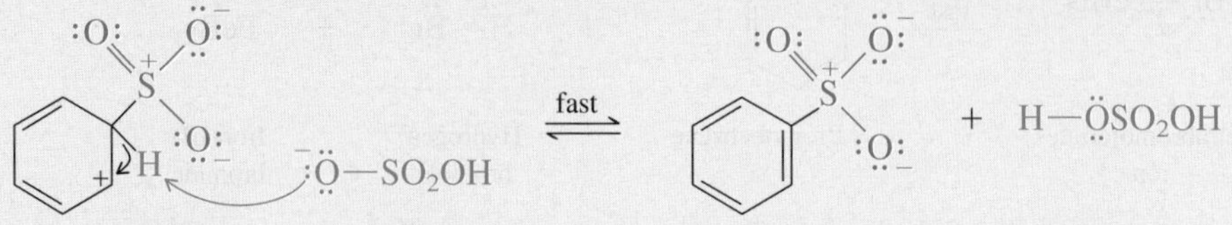

Cyclohexadienyl cation intermediate — Hydrogen sulfate ion — Benzenesulfonate ion — Sulfuric acid

Step 3: A rapid proton transfer from the oxygen of sulfuric acid to the oxygen of benzenesulfonate completes the process.

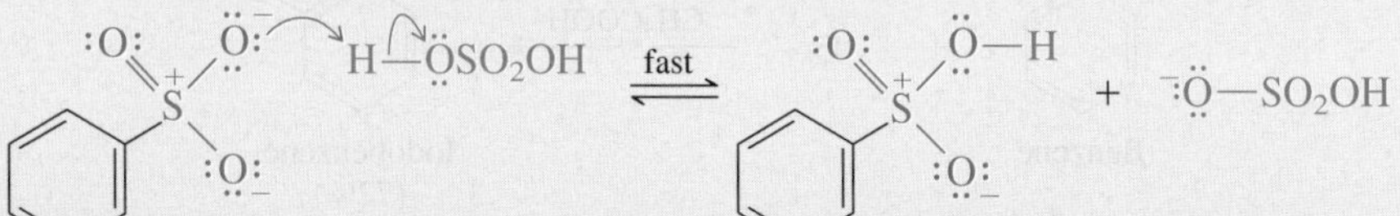

Benzenesulfonate ion — Sulfuric acid — Benzenesulfonic acid — Hydrogen sulfate ion

Iron(III) bromide, a weak Lewis acid, combines with bromine to form a Lewis acid/Lewis base complex.

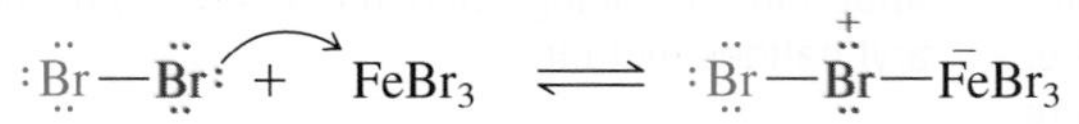

Lewis base — Lewis acid — Lewis acid/Lewis base complex

Complexation of bromine with iron(III) bromide makes bromine more electrophilic, and it reacts with benzene to give a cyclohexadienyl intermediate as shown in step 1 of Mechanism 12.3. In step 2, as in nitration and sulfonation, loss of a proton from the cyclohexadienyl cation is rapid and gives the product of electrophilic aromatic substitution.

Only small quantities of iron(III) bromide are required. It is a catalyst for the bromination and, as Mechanism 12.3 indicates, is regenerated in the course of the reaction. We'll see later in this chapter that some aromatic substrates are much more reactive than benzene and react rapidly with bromine even in the absence of a catalyst.

Chlorination is carried out in a manner similar to bromination and follows a similar mechanism to give aryl chlorides. Fluorination and iodination of arenes are rarely performed. Fluorine is so reactive that its reaction with benzene is difficult to control. Iodination is very slow and has an unfavorable equilibrium constant. However, iodine, in the presence of a powerful oxidizing agent can be used for electrophilic aromatic iodination. In the following example, the oxidant peroxyacetic acid reacts with iodine to

Mechanism 12.3

Bromination of Benzene

Step 1: The bromine–iron(III) bromide complex is the active electrophile that reacts with benzene. Two of the π electrons of benzene are used to form a bond to bromine and give a cyclohexadienyl cation intermediate. (The molecular model depicts the cyclohexadienyl cation intermediate.)

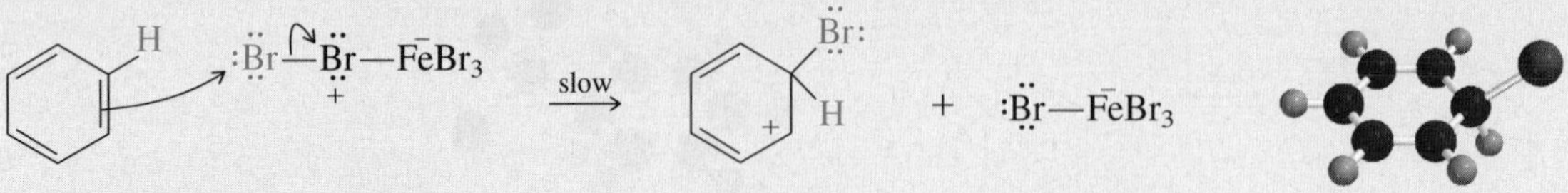

Benzene and bromine–iron(III) bromide complex — Cyclohexadienyl cation intermediate — Tetrabromoferrate ion

Step 2: Loss of a proton from the cyclohexadienyl cation yields bromobenzene.

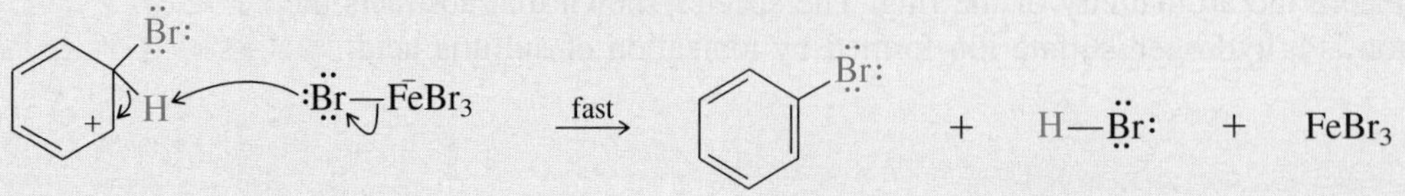

Cyclohexadienyl cation intermediate — Tetrabromoferrate ion — Bromobenzene — Hydrogen bromide — Iron(III) bromide

generate an electrophilic iodinating species that reacts with benzene to give iodobenzene (see Problem 12.5, in the Boxed Essay).

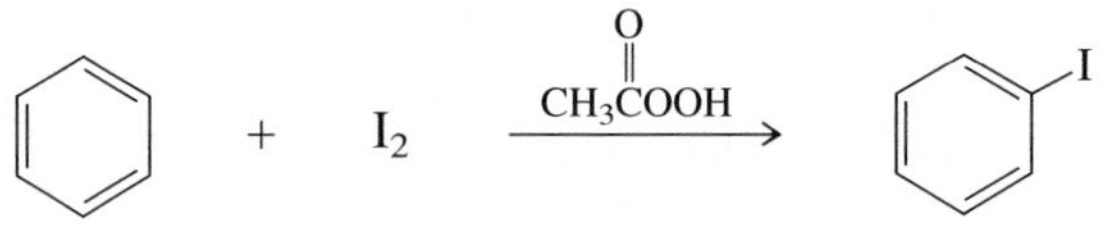

Benzene — Iodobenzene (72%)

The formation of a reactive halogenating species through oxidation occurs in biosynthetic halogenation, described in the Boxed Essay. Syntheses of aryl fluorides and aryl iodides

Biosynthetic Halogenation

Over 4000 natural products contain halogens. Most common are the halogens chlorine, bromine, and iodine, and many of the compounds that contain them are important in biology and medicine. Some naturally occurring aryl halides include:

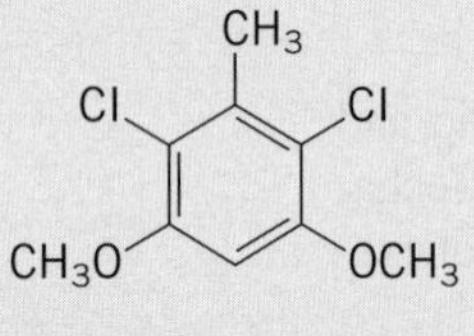

2,6-Dichloro-3,5-dimethoxytoluene: an antifungal compound isolated from lily plants

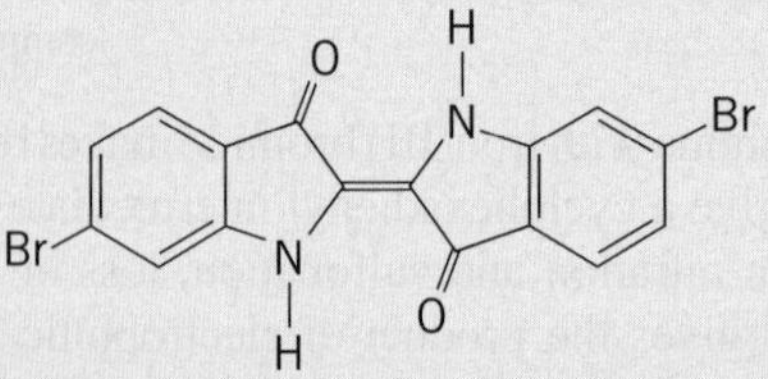

Dibromoindigo: principal constituent of a dye known as Tyrian purple and prized by ancient cultures, isolated from a species of Mediterranean sea snail

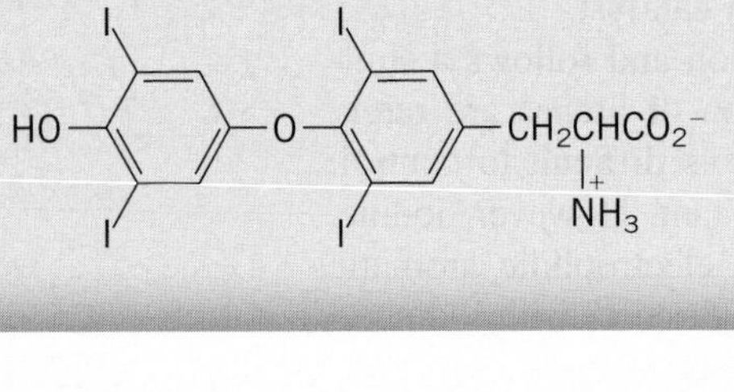

Thyroxine: a hormone of the thyroid gland; the (*S*)-enantiomer is a widely used drug prescribed to increase metabolic rate

Continued

The presence of the halogen in these and in other halogenated natural products has a strong effect on their properties, yet chemists have long been puzzled by the ways in which a halogen is introduced in the biosynthesis of such compounds. What are the biological halogenating agents, what enzymes catalyze the halogenation, and how do these enzymes affect the mechanism? Recent studies have unlocked the answers to some of these questions. Biosynthetic halogenation can occur through multiple pathways, but many halogenase enzymes use electrophilic halogenating species that are produced by oxidation of halide ions. The relatively low abundance of fluorinated natural products (only a dozen or so are known) may be the result of the higher energy required for oxidation of fluoride ion relative to the other halogens.

An example of an electrophilic aromatic halogenation occurs in the biosynthesis of the antifungal antibiotic pyrrolnitrin. Having just considered the halogenation of aromatic compounds by electrophilic halogen that is generated through the use of a Lewis acid catalyst such as $FeCl_3$, let's take a look at the halogenation step in pyrrolnitrin biosynthesis. Pyrrolnitrin is derived biosynthetically from the amino acid tryptophan. The benzene ring of tryptophan is halogenated by a halogenase enzyme known as PrnA. The electrophilic chlorinating species is thought to be hypochlorous acid (HOCl), which is generated in a separate step. Although hypochlorous acid is not electrophilic enough to chlorinate tryptophan by itself, it can be activated through hydrogen bonding with a nearby amino group found in the catalytic site of the enzyme. Loss of a proton to rearomatize the ring occurs from the σ complex with participation of a carboxyl group of another amino acid, to give the product 7-chlorotryptophan.

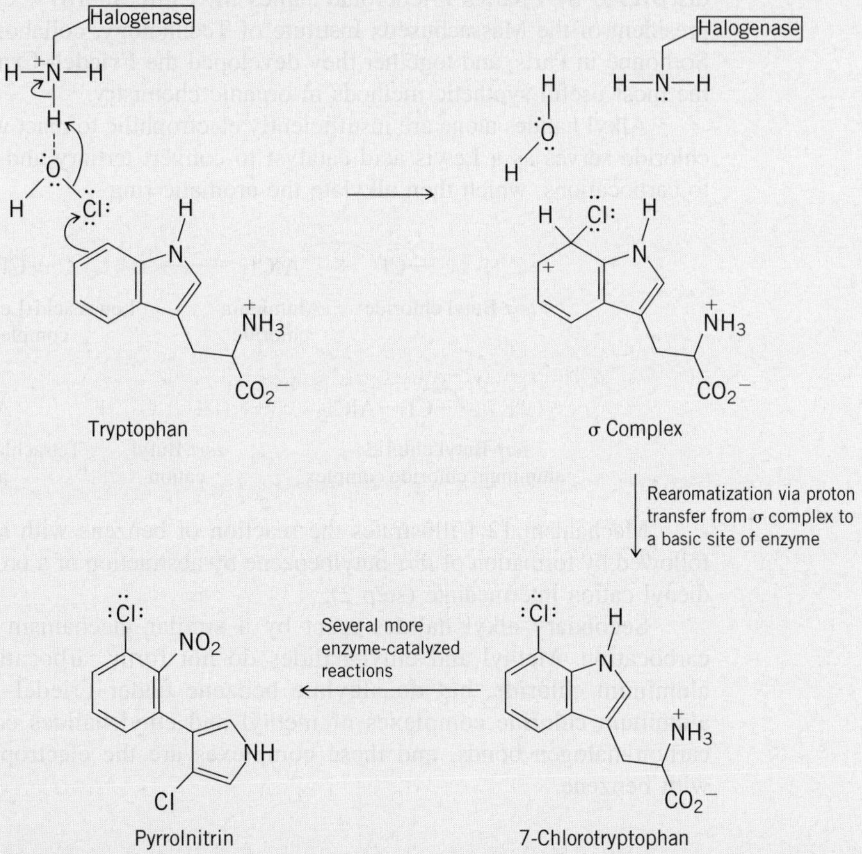

The remaining steps in the biosynthesis convert 7-chlorotryptophan to pyrrolnitrin. One of these steps is a second halogenation, this time on the pyrrole ring.

Problem 12.5

The halogenating species in the conversion of benzene to iodobenzene with iodine and peroxyacetic acid, shown in Section 12.5, is thought to be acetyl hypoiodite. Write a mechanism for the electrophilic iodination of benzene with this species.

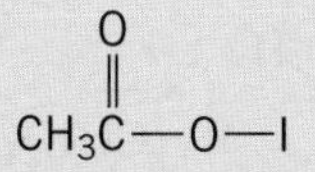

Acetyl hypoiodite

are normally carried out by way of functional group transformations of arylamines; these reactions will be described in Chapter 21.

12.6 Friedel–Crafts Alkylation of Benzene

Alkyl halides react with benzene in the presence of aluminum chloride to yield alkylbenzenes.

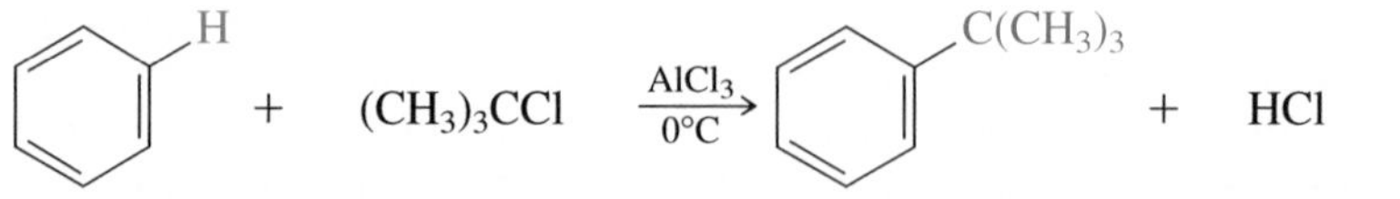

Benzene | *tert*-Butyl chloride | *tert*-Butylbenzene (60%) | Hydrogen chloride

Alkylation of benzene with alkyl halides in the presence of aluminum chloride was discovered by Charles Friedel and James M. Crafts in 1877. Crafts, who later became president of the Massachusetts Institute of Technology, collaborated with Friedel at the Sorbonne in Paris, and together they developed the **Friedel–Crafts reaction** into one of the most useful synthetic methods in organic chemistry.

Alkyl halides alone are insufficiently electrophilic to react with benzene. Aluminum chloride serves as a Lewis acid catalyst to convert tertiary and secondary alkyl halides to carbocations, which then alkylate the aromatic ring.

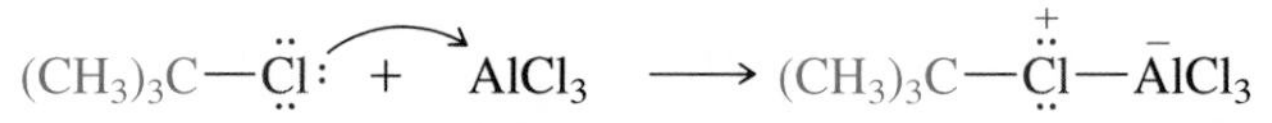

tert-Butyl chloride | Aluminum chloride | Lewis acid/Lewis base complex

tert-Butyl chloride–aluminum chloride complex | *tert*-Butyl cation | Tetrachloroaluminate anion

Mechanism 12.4 illustrates the reaction of benzene with *tert*-butyl cation (step 1) followed by formation of *tert*-butylbenzene by abstraction of a proton from the cyclohexadienyl cation intermediate (step 2).

Secondary alkyl halides react by a similar mechanism involving a secondary carbocation. Methyl and ethyl halides do not form carbocations when treated with aluminum chloride, but do alkylate benzene under Friedel–Crafts conditions. The aluminum chloride complexes of methyl and ethyl halides contain highly polarized carbon–halogen bonds, and these complexes are the electrophilic species that react with benzene.

$$H_3C-\overset{+}{X}-\overset{-}{Al}X_3 \qquad CH_3CH_2-\overset{+}{X}-\overset{-}{Al}X_3$$

Methyl halide/aluminum halide complex | Ethyl halide/aluminum halide complex

Other limitations to Friedel–Crafts reactions will be encountered in this chapter and are summarized in Table 12.4 (page 523).

One drawback to Friedel–Crafts alkylation is that rearrangements can occur, especially when primary alkyl halides are used. For example, Friedel–Crafts alkylation of benzene with isobutyl chloride (a primary alkyl halide) yields only *tert*-butylbenzene.

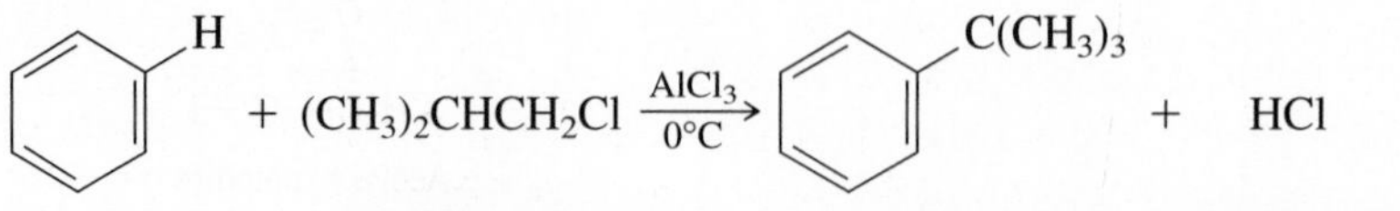

Benzene | Isobutyl chloride | *tert*-Butylbenzene (66%) | Hydrogen chloride

Mechanism 12.4

Friedel–Crafts Alkylation

Step 1: Once generated by the reaction of *tert*-butyl chloride and aluminum chloride, *tert*-butyl cation is attacked by the π electrons of benzene, and a carbon-carbon bond is formed. (The molecular model depicts the cyclohexadienyl cation intermediate.)

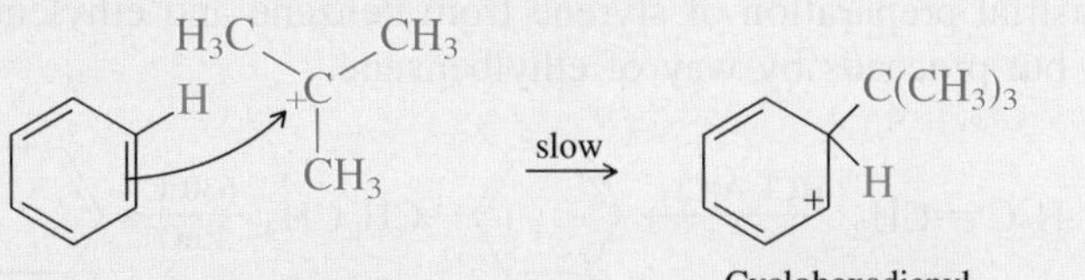

slow

Benzene and *tert*-butyl cation — Cyclohexadienyl cation intermediate

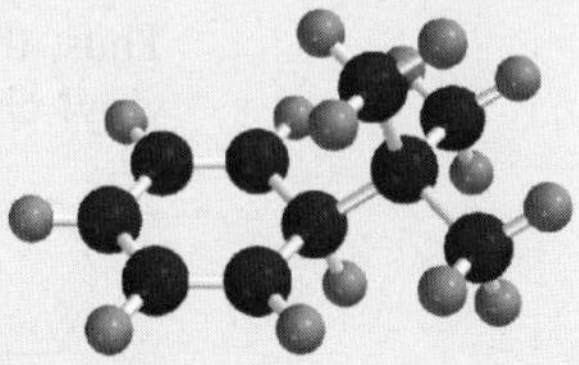

Step 2: Loss of a proton from the cyclohexadienyl cation intermediate yields *tert*-butylbenzene.

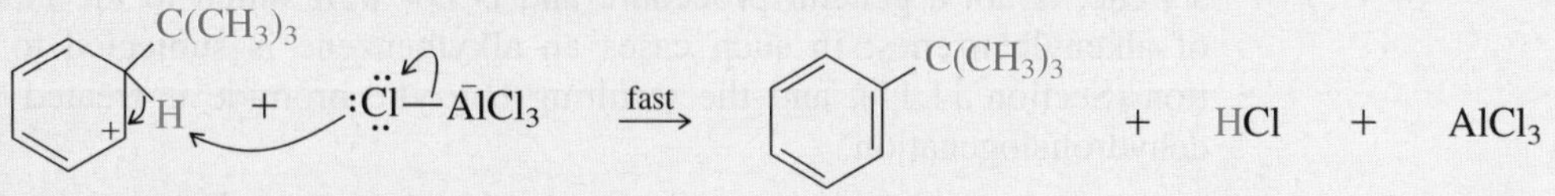

fast
HCl
AlCl3

Cyclohexadienyl cation intermediate — Tetrachloroaluminate ion — *tert*-Butylbenzene — Hydrogen chloride — Aluminum chloride

Here, the electrophile is *tert*-butyl cation formed by a hydride migration that accompanies ionization of the carbon–chlorine bond.

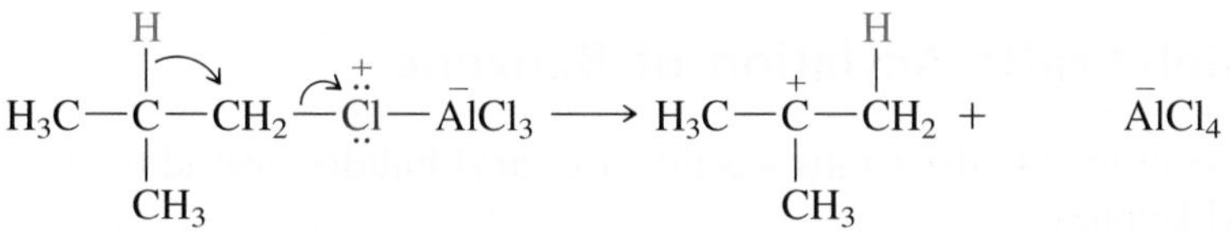

Isobutyl chloride/aluminum chloride complex — *tert*-Butyl cation — Tetrachloroaluminate ion

We saw rearrangements involving hydride shifts earlier in Sections 5.13 and 6.7.

Problem 12.6

In an attempt to prepare propylbenzene, a chemist alkylated benzene with 1-chloropropane and aluminum chloride. However, two isomeric hydrocarbons were obtained in a ratio of 2:1, the desired propylbenzene being the minor component. What do you think was the major product? How did it arise?

Because electrophilic aromatic substitution is simply another reaction available to a carbocation, other carbocation precursors can be used in place of alkyl halides. For example, alkenes, which are converted to carbocations by protonation, can be used to alkylate benzene.

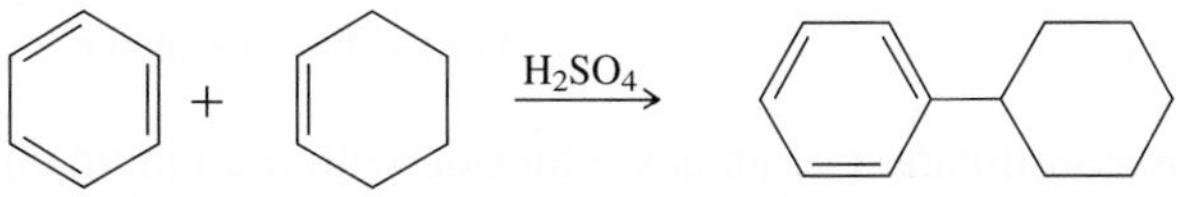

H2SO4

Benzene — Cyclohexene — Cyclohexylbenzene (65–68%)

Problem 12.7

Write a reasonable mechanism for the formation of cyclohexylbenzene from the reaction of benzene, cyclohexene, and sulfuric acid.

Problem 12.8

tert-Butylbenzene can be prepared by alkylation of benzene using an alkene or an alcohol as the carbocation source. What alkene? What alcohol?

Alkenyl halides such as vinyl chloride ($H_2C{=}CHCl$) do *not* form carbocations on treatment with aluminum chloride and so cannot be used in Friedel–Crafts reactions. Thus, the industrial preparation of styrene from benzene and ethylene does not involve vinyl chloride but proceeds by way of ethylbenzene.

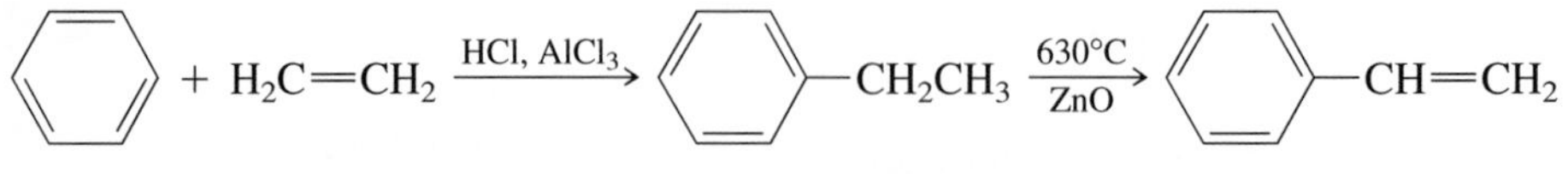

Benzene Ethylene Ethylbenzene Styrene

Dehydrogenation of alkylbenzenes, although useful in the industrial preparation of styrene, is not a general procedure and is not well suited to the laboratory preparation of alkenylbenzenes. In such cases an alkylbenzene is subjected to benzylic bromination (Section 11.11), and the resulting benzylic bromide is treated with base to effect dehydrohalogenation.

Problem 12.9

Outline a synthesis of 1-phenylcyclohexene from benzene and cyclohexene.

12.7 Friedel–Crafts Acylation of Benzene

An acyl group has the general formula

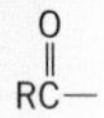

Another version of the Friedel–Crafts reaction uses **acyl halides** instead of alkyl halides and yields aryl ketones.

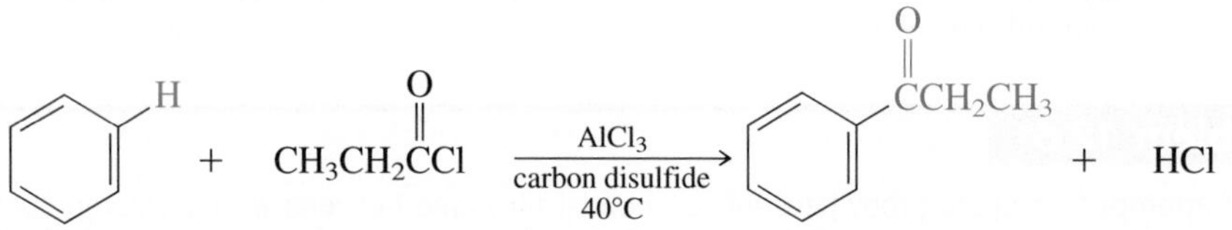

Benzene Propanoyl chloride 1-Phenyl-1-propanone (88%) Hydrogen chloride

The electrophile in a Friedel–Crafts **acylation** is an **acyl cation** (also referred to as an **acylium ion**). Acyl cations are stabilized by electron delocalization. The acyl cation derived from propanoyl chloride is represented by the following two forms. Note that the triple-bonded contributor is an octet-satisfied structure.

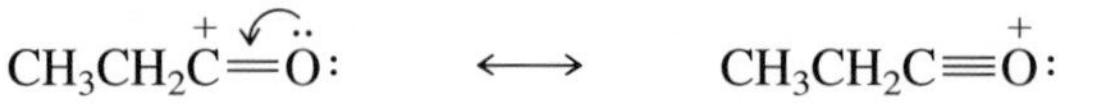

Most stable contributor

Acyl cations form by coordination of an acyl chloride with aluminum chloride, followed by cleavage of the carbon–chlorine bond.

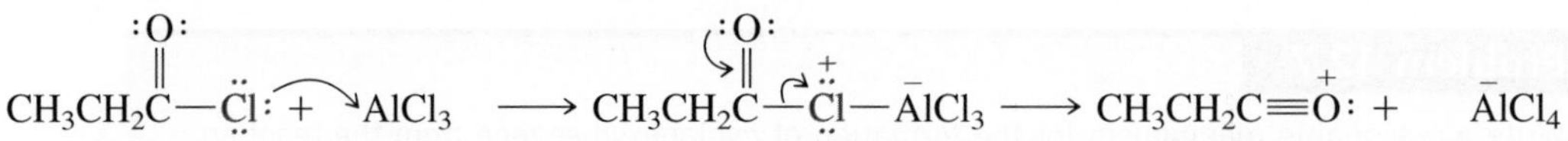

Propanoyl chloride Aluminum chloride Lewis acid/Lewis base complex Propanoyl cation Tetrachloroaluminate ion

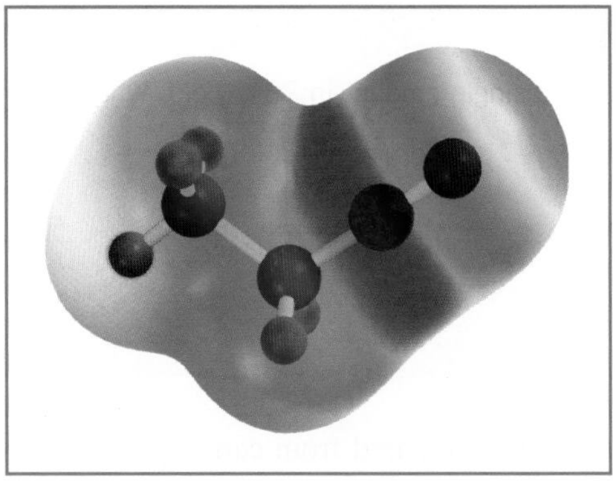

Figure 12.3

Electrostatic potential map of propanoyl cation [($CH_3CH_2C{\equiv}\overset{+}{O}$:)]. The region of greatest positive charge is associated with the carbon of the C=O group.

The electrophilic site of an acyl cation is its acyl carbon. An electrostatic potential map of the acyl cation from propanoyl chloride (Figure 12.3) illustrates the concentration of positive charge at the acyl carbon, as shown by the blue color. The reaction between this cation and benzene is analogous to that of other electrophiles (Mechanism 12.5).

An important difference between Friedel–Crafts alkylations and acylations is that acyl cations *do not rearrange.* The acyl group is transferred to the benzene ring unchanged. An acyl cation is so strongly stabilized by resonance that it is more stable than any ion that could conceivably arise from it by a hydride or alkyl group shift.

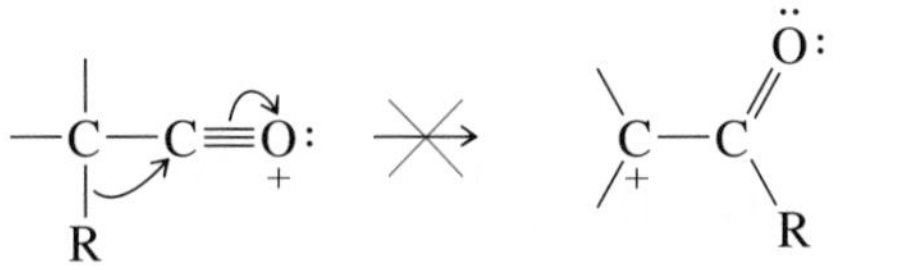

More stable cation; all atoms have octets of electrons

Less stable cation; six electrons at carbon

Mechanism 12.5

Friedel–Crafts Acylation

Step 1: The acyl cation reacts with benzene. A pair of π electrons of benzene is used to form a covalent bond to the carbon of the acyl cation. (The molecular model depicts the cyclohexadienyl cation intermediate.)

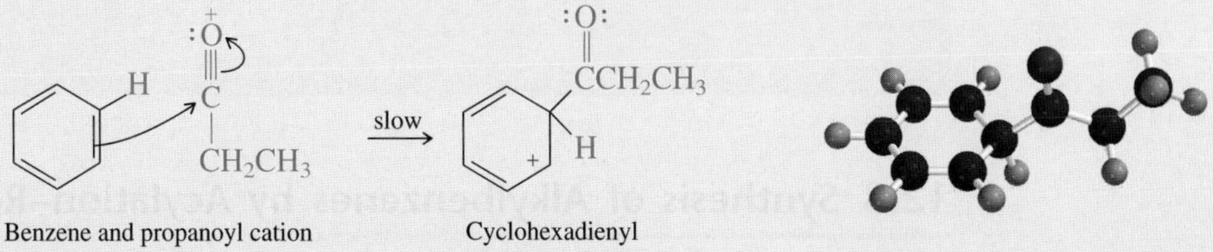

Benzene and propanoyl cation

Cyclohexadienyl cation intermediate

Step 2: Aromaticity of the ring is restored when it loses a proton to give the aryl ketone.

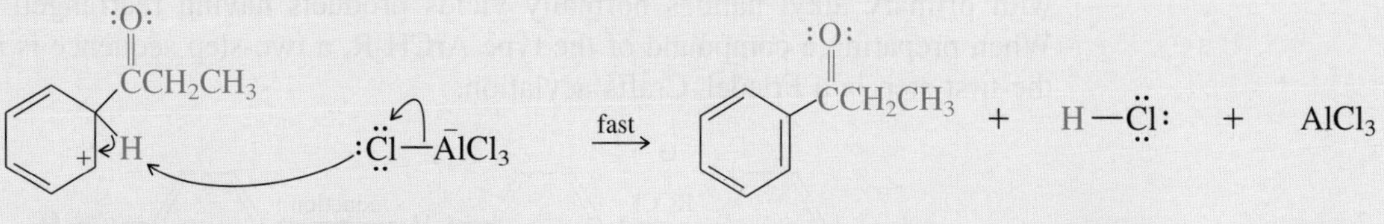

Cyclohexadienyl cation intermediate

Tetrachloroaluminate ion

1-Phenyl-1-propanone

Hydrogen chloride

Aluminum chloride

Problem 12.10

The reaction shown gives a single product in 88% yield. What is that product?

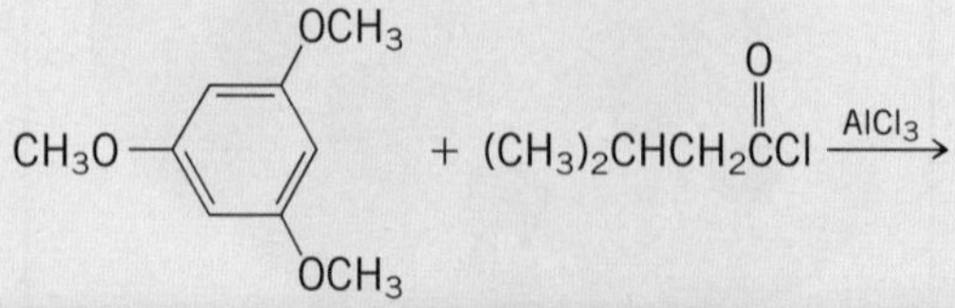

Acyl chlorides are readily prepared from carboxylic acids by reaction with thionyl chloride.

$$\underset{\text{Carboxylic acid}}{RCOH} + \underset{\text{Thionyl chloride}}{SOCl_2} \longrightarrow \underset{\text{Acyl chloride}}{RCCl} + \underset{\text{Sulfur dioxide}}{SO_2} + \underset{\text{Hydrogen chloride}}{HCl}$$

Carboxylic acid anhydrides, compounds of the type RCOCR, are also sources of acyl cations and, in the presence of aluminum chloride, acylate benzene. One acyl unit of an acid anhydride becomes attached to the benzene ring, and the other becomes part of a carboxylic acid.

Acetophenone is one of the commonly encountered benzene derivatives listed in Table 11.1.

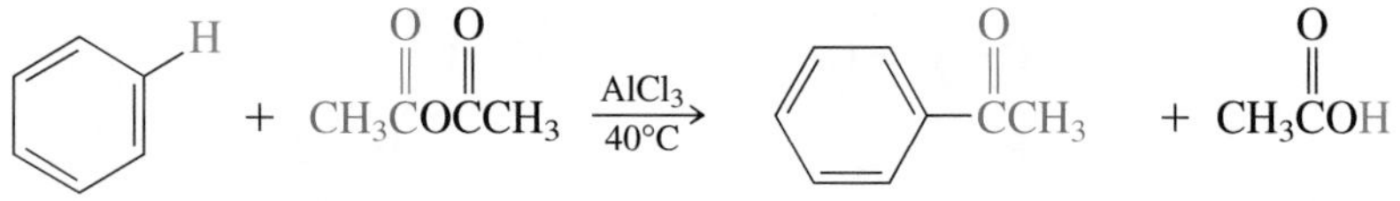

Benzene Acetic anhydride Acetophenone (76–83%) Acetic acid

Problem 12.11

Succinic anhydride, the structure of which is shown, is a cyclic anhydride often used in Friedel–Crafts acylations. Give the structure of the product obtained when benzene is acylated with succinic anhydride in the presence of aluminum chloride.

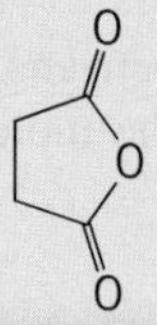

12.8 Synthesis of Alkylbenzenes by Acylation–Reduction

Because acylation of an aromatic ring can be accomplished without rearrangement, it is frequently used as the first step in a procedure for the *alkylation* of aromatic compounds by *acylation–reduction.* As we saw in Section 12.6, Friedel–Crafts alkylation of benzene with primary alkyl halides normally yields products having rearranged alkyl groups. When preparing a compound of the type $ArCH_2R$, a two-step sequence is used in which the first step is a Friedel–Crafts acylation.

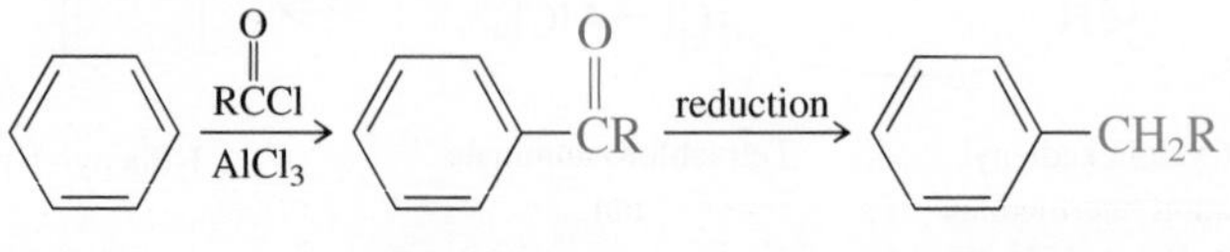

Benzene Aryl ketone Alkylbenzene

The second step is a reduction of the carbonyl group (C=O) to a methylene group (CH_2).

The most commonly used method for reducing an aryl ketone to an alkylbenzene employs a zinc–mercury amalgam in concentrated hydrochloric acid and is called the **Clemmensen reduction.** Zinc is the reducing agent.

The synthesis of butylbenzene illustrates the acylation–reduction sequence.

An amalgam is a mixture or alloy of mercury with another metal. For many years silver amalgams were used in dental fillings.

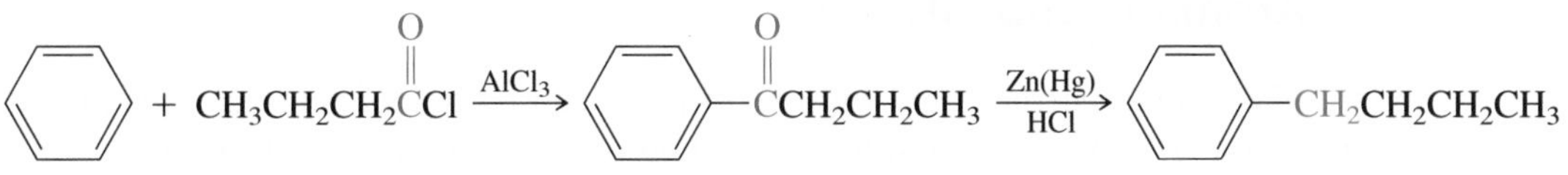

Benzene   Butanoyl chloride   1-Phenyl-1-butanone (86%)   Butylbenzene (73%)

Direct alkylation of benzene using 1-chlorobutane and aluminum chloride would yield *sec*-butylbenzene by rearrangement and so could not be used.

Problem 12.12

Using benzene and any necessary organic or inorganic reagents, suggest efficient syntheses of

(a) Isobutylbenzene, $C_6H_5CH_2CH(CH_3)_2$

(b) (2,2-Dimethylpropyl)benzene, $C_6H_5CH_2C(CH_3)_3$

Sample Solution (a) Friedel–Crafts alkylation of benzene with isobutyl chloride is not suitable, because it yields *tert*-butylbenzene by rearrangement.

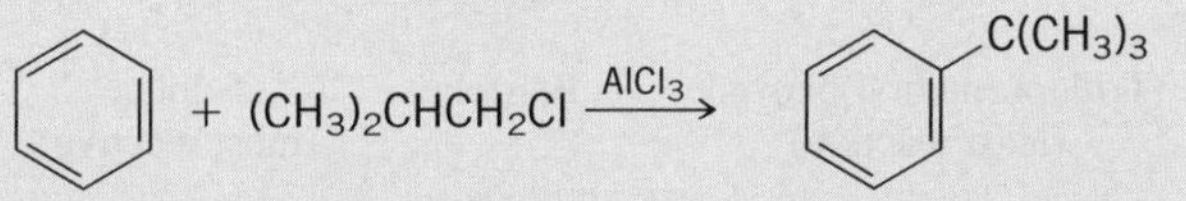

Benzene   Isobutyl chloride   *tert*-Butylbenzene (66%)

The two-step acylation–reduction sequence is required. Acylation of benzene puts the side chain on the ring with the correct carbon skeleton. Clemmensen reduction converts the carbonyl group to a methylene group.

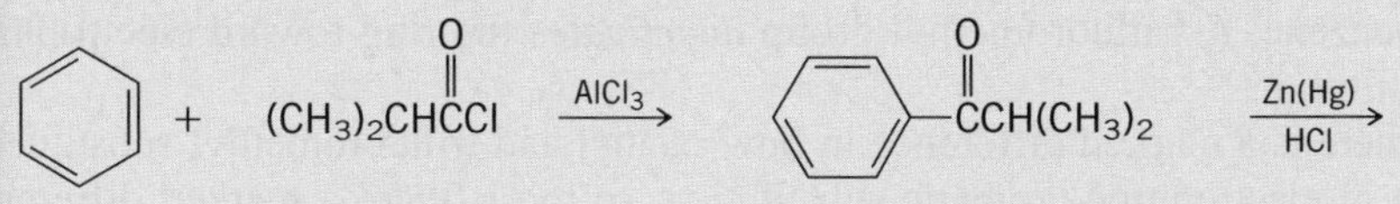

Benzene   2-Methylpropanoyl chloride   2-Methyl-1-phenyl-1-propanone (84%)

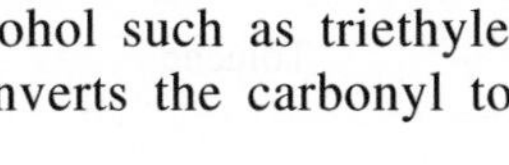

Isobutylbenzene (80%)

Another way to reduce aldehyde and ketone carbonyl groups is by **Wolff–Kishner reduction.** Heating an aldehyde or a ketone with hydrazine (H_2NNH_2) and sodium or potassium hydroxide in a high-boiling alcohol such as triethylene glycol ($HOCH_2CH_2OCH_2CH_2OCH_2CH_2OH$, bp 287°C) converts the carbonyl to a CH_2 group.

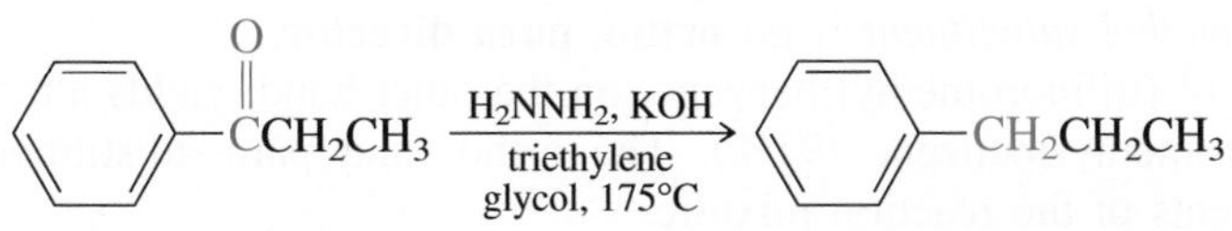

1-Phenyl-1-propanone   Propylbenzene (82%)

Both the Clemmensen and the Wolff–Kishner reductions convert an aldehyde or ketone carbonyl to a methylene group. Neither will reduce the carbonyl group of a carboxylic acid, nor are carbon–carbon double or triple bonds affected by these methods.

12.9 Rate and Regioselectivity in Electrophilic Aromatic Substitution

So far we've been concerned only with electrophilic substitution of benzene. Two important questions arise when we turn to substitution on rings that already bear at least one substituent:

1. What is the effect of a substituent on the *rate* of electrophilic aromatic substitution?
2. What is the effect of a substituent on the *regioselectivity* of electrophilic aromatic substitution?

To illustrate substituent effects on rate, consider the nitration of benzene, toluene, and (trifluoromethyl)benzene.

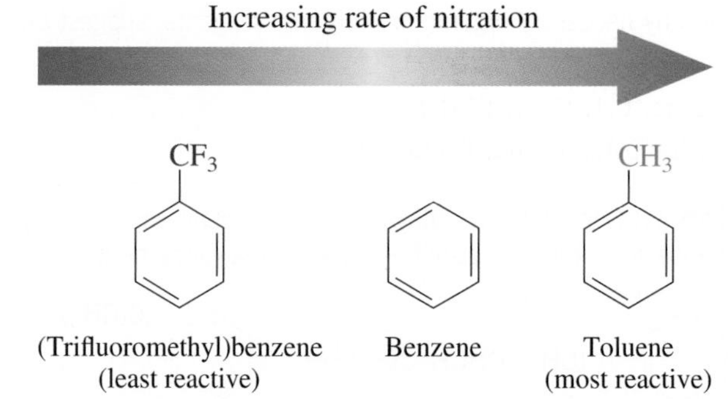
Increasing rate of nitration
CF_3
CH_3
(Trifluoromethyl)benzene
(least reactive)
Benzene
Toluene
(most reactive)

The range of nitration rates among these three compounds is quite large; it covers a spread of approximately 1-millionfold. Toluene undergoes nitration some 20–25 times faster than benzene. Because toluene is more reactive than benzene, we say that a methyl group *activates* the ring toward electrophilic aromatic substitution. (Trifluoromethyl)benzene, on the other hand, undergoes nitration about 40,000 times more slowly than benzene. A trifluoromethyl group *deactivates* the ring toward electrophilic aromatic substitution.

Just as there is a marked difference in how methyl and trifluoromethyl substituents affect the rate of electrophilic aromatic substitution, so too is there a marked difference in how they affect its regioselectivity.

Three products are possible from nitration of toluene: *o*-nitrotoluene, *m*-nitrotoluene, and *p*-nitrotoluene. All are formed, but not in equal amounts. Together, the ortho- and para-substituted isomers make up 97% of the product mixture; the meta only 3%.

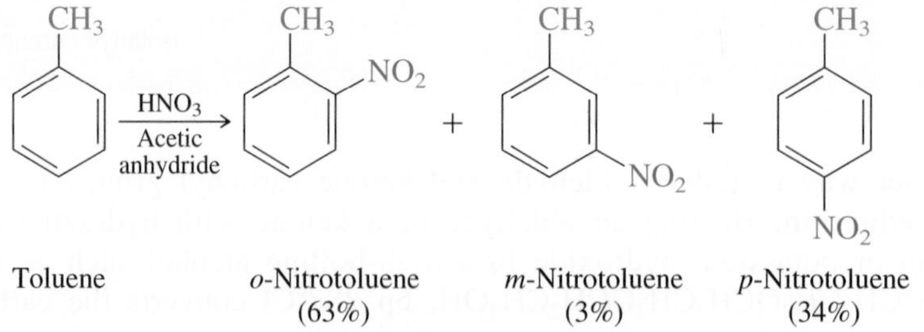
CH_3
HNO_3
Acetic anhydride
NO_2
+
Toluene
o-Nitrotoluene (63%)
m-Nitrotoluene (3%)
p-Nitrotoluene (34%)

Because substitution in toluene occurs primarily at positions ortho and para to methyl, we say that *a methyl substituent is an* **ortho, para director.**

Nitration of (trifluoromethyl)benzene, on the other hand, yields almost exclusively *m*-nitro(trifluoromethyl)benzene (91%). The ortho- and para-substituted isomers are minor components of the reaction mixture.

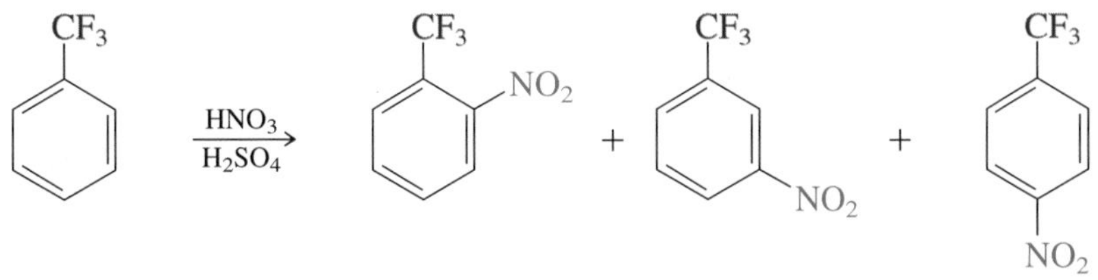

(Trifluoromethyl)benzene — *o*-Nitro(trifluoromethyl)benzene (6%) — *m*-Nitro(trifluoromethyl)benzene (91%) — *p*-Nitro(trifluoromethyl)benzene (3%)

Because substitution in (trifluoromethyl)benzene occurs primarily at positions meta to the substituent, *a trifluoromethyl group is a* **meta director.**

The regioselectivity of substitution, like the rate, is strongly affected by the substituent. In the following several sections we will examine the relationship between the structure of the substituent and its effect on rate and regioselectivity of electrophilic aromatic substitution.

12.10 Rate and Regioselectivity in the Nitration of Toluene

Why is there such a marked difference between methyl and trifluoromethyl substituents in their influence on electrophilic aromatic substitution? Methyl is **activating** and ortho, para-directing; trifluoromethyl is **deactivating** and meta-directing. The first point to remember is that the regioselectivity of substitution is set once the cyclohexadienyl cation intermediate is formed. If we can explain why

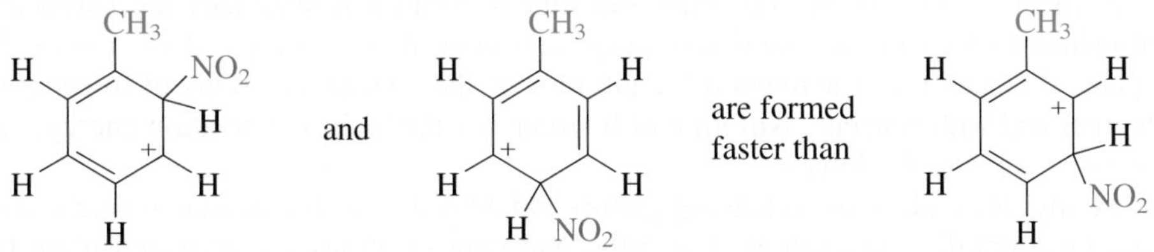

in the rate-determining step, we will understand the reasons for the regioselectivity. A principle we have used before serves us well here: *a more stable carbocation is formed faster than a less stable one.* The most likely reason for the directing effect of a CH_3 group must be that the carbocations that give *o*- and *p*-nitrotoluene are more stable than the one that gives *m*-nitrotoluene.

One way to assess the relative stabilities of these carbocations is to examine electron delocalization in them using a resonance description. The cyclohexadienyl cations leading to *o*- and *p*-nitrotoluene have tertiary carbocation character. Each has a contributing structure in which the positive charge resides on the carbon that bears the methyl group.

Ortho nitration

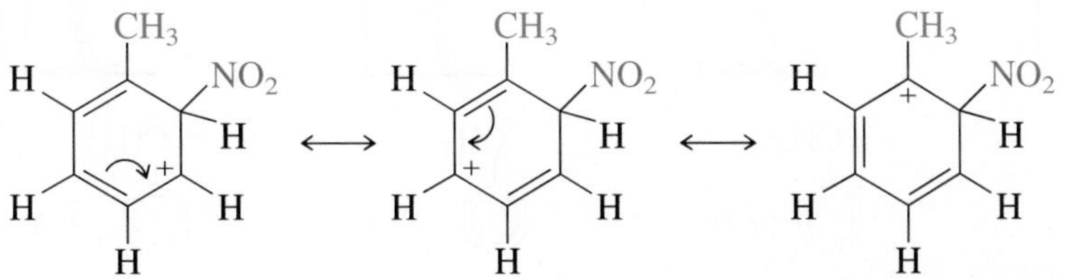

This resonance contributor is a tertiary carbocation

Para nitration

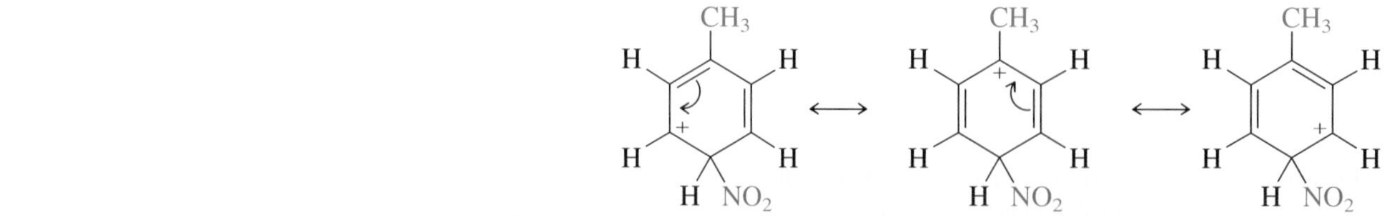

The three contributing resonance forms of the intermediate leading to meta substitution are all secondary carbocations.

Meta nitration

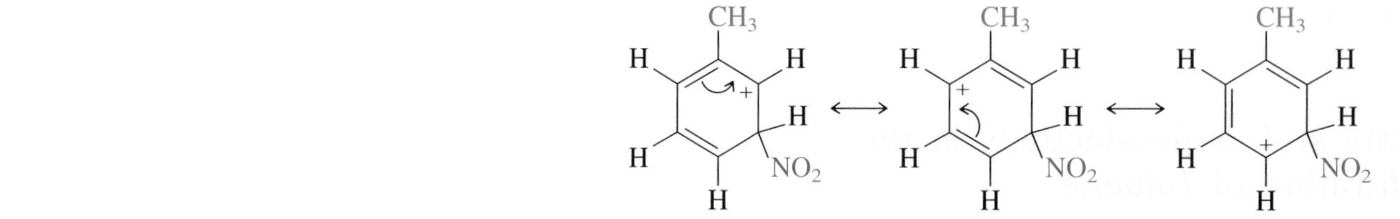

A methyl group is ortho, para-directing because the carbocations leading to *o*- and *p*-nitrotoluene are more stable and formed faster than the one leading to *m*-nitrotoluene. The greater stability of the carbocations for ortho and para substitution comes from their tertiary carbocation character. All of the contributing carbocation structures for meta substitution are secondary.

A methyl group is an activating substituent because it stabilizes the carbocation intermediate formed in the rate-determining step more than hydrogen does. Figure 12.4 compares the energies of activation for nitration at the various positions of toluene with each other and with benzene. Nitration of benzene has the highest activation energy, para-nitration of toluene the lowest.

Methyl is an **electron-releasing group** and activates *all* the available ring carbons toward electrophilic substitution. The ortho and para positions are activated more than

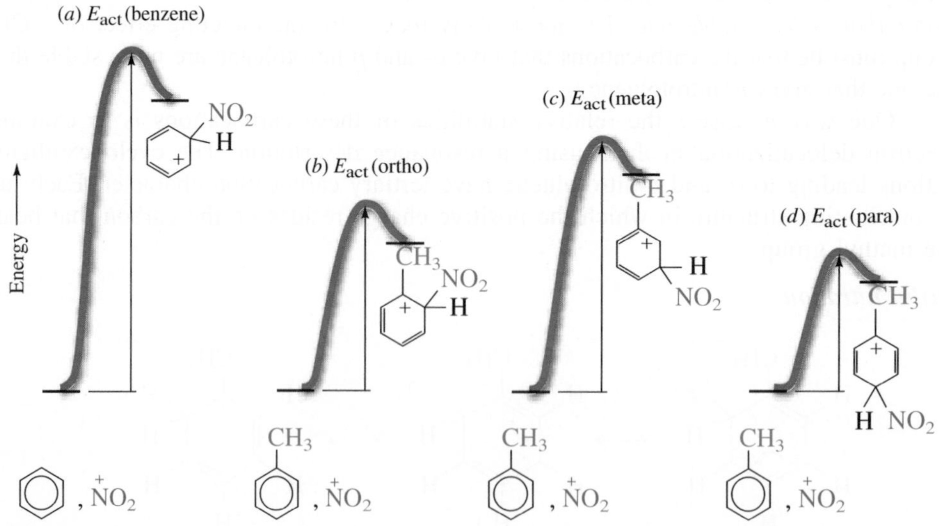

Figure 12.4

Comparative energy diagrams for reaction of nitronium ion with (*a*) benzene and at the (*b*) ortho, (*c*) meta, and (*d*) para positions of toluene. E_{act} (benzene) > E_{act} (meta) > E_{act} (ortho) > E_{act} (para).

meta. At 25°C, the relative rates of nitration at the various positions of toluene compared with a single carbon of benzene are:

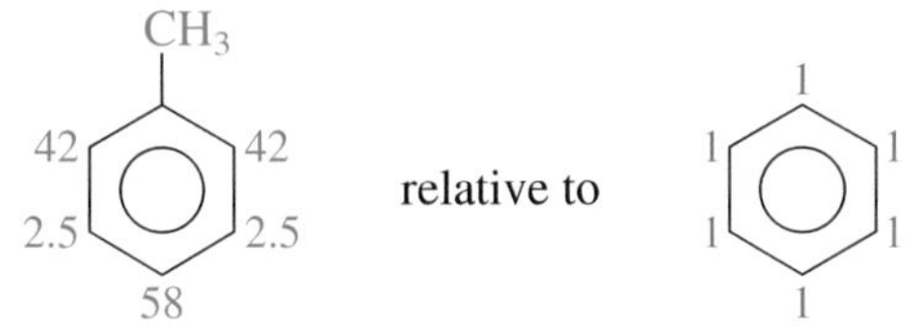

These relative rate data per position are experimentally determined and are known as **partial rate factors.** They offer a convenient way to express substituent effects in electrophilic aromatic substitutions.

The major influence of the methyl group is its *electronic* effect on carbocation stability. To a small extent, the methyl group sterically hinders the approach of the electrophile to the ortho positions, making substitution slightly slower at a single ortho carbon than at the para carbon. However, para substitution is at a statistical disadvantage because there are two equivalent ortho positions but only one para position.

Problem 12.13

The partial rate factors for nitration of *tert*-butylbenzene are as shown.

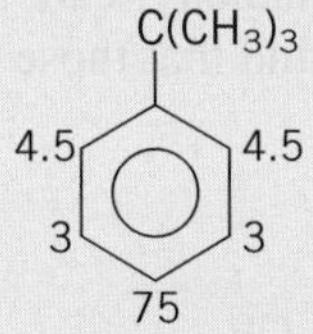

(a) How reactive is *tert*-butylbenzene toward nitration compared with benzene?

(b) How reactive is *tert*-butylbenzene toward nitration compared with toluene?

(c) Predict the distribution among the various mononitration products of *tert*-butylbenzene.

Sample Solution (a) Benzene has six equivalent sites at which nitration can occur. Summing the individual relative rates of nitration at each position in *tert*-butylbenzene compared with benzene, we obtain

$$\frac{tert\text{-Butylbenzene}}{\text{Benzene}} = \frac{2(4.5) + 2(3) + 75}{6(1)} = \frac{90}{6} = 15$$

tert-Butylbenzene undergoes nitration 15 times faster than benzene.

All alkyl groups, not just methyl, are electron-releasing, activating substituents and ortho, para directors. This is because any alkyl group, be it methyl, ethyl, isopropyl, *tert*-butyl, or any other, stabilizes a carbocation site to which it is directly attached. When R = alkyl,

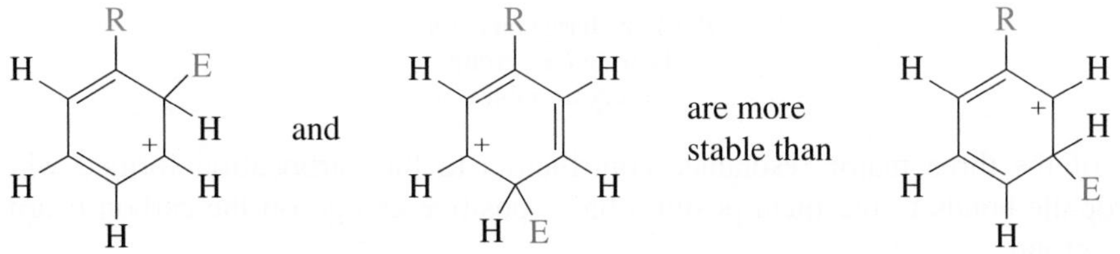

where E^+ is any electrophile. All three structures are more stable for R = alkyl than for R = H and are formed faster.

12.11 Rate and Regioselectivity in the Nitration of (Trifluoromethyl)benzene

Turning now to electrophilic aromatic substitution in (trifluoromethyl)benzene, we consider the electronic properties of a trifluoromethyl group. Because of their high

Recall from Section 1.16 that effects that are transmitted by the polarization of σ bonds are called *inductive effects.*

electronegativity the three fluorine atoms polarize the electron distribution in their σ bonds to carbon, so that carbon bears a partial positive charge.

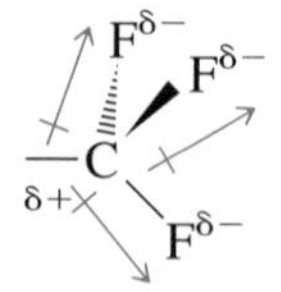
Fδ−
Fδ−
C
δ+
Fδ−

Unlike a methyl group, which is slightly electron-releasing, trifluoromethyl is a powerful **electron-withdrawing group.** Consequently, CF_3 *destabilizes* a carbocation site to which it is attached.

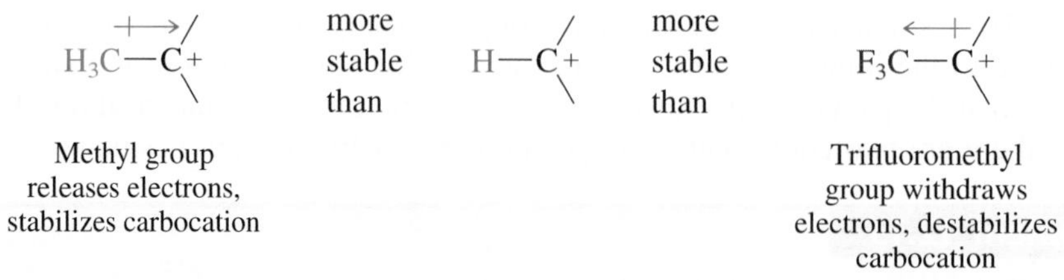
H₃C—C+
more stable than
H—C+
more stable than
F₃C—C+
Methyl group releases electrons, stabilizes carbocation
Trifluoromethyl group withdraws electrons, destabilizes carbocation

When we examine the cyclohexadienyl cation intermediates involved in the nitration of (trifluoromethyl)benzene, we find that those leading to ortho and para substitution are strongly *destabilized.*

Ortho nitration

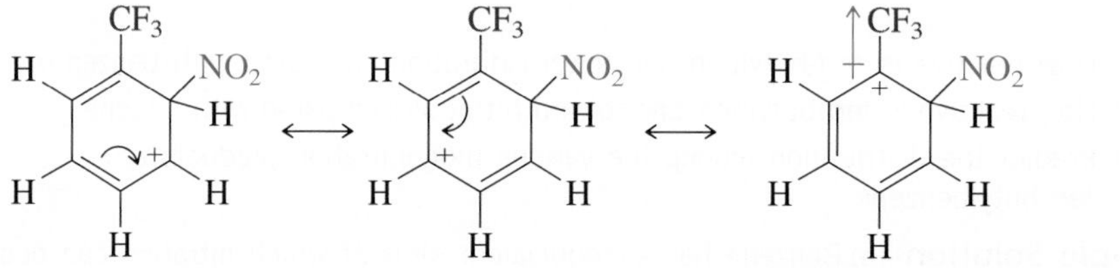
CF₃
NO₂
H

Positive charge on carbon bearing CF_3 group (very unstable)

Para nitration

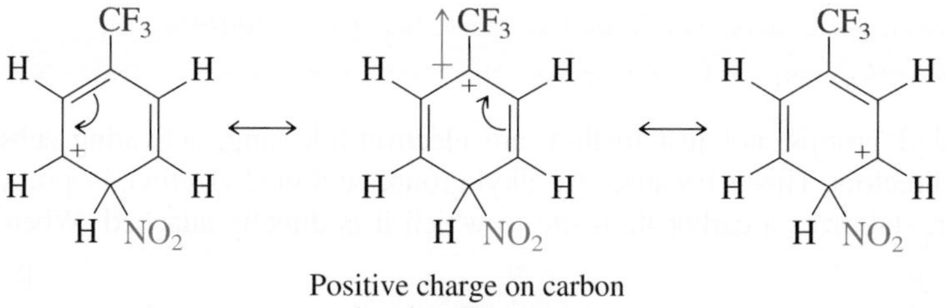
CF₃
H
H NO₂

Positive charge on carbon bearing CF_3 group (very unstable)

None of the three major resonance contributors to the carbocation formed when the electrophile bonds to the meta position has a positive charge on the carbon bearing the —CF_3 group.

Meta nitration

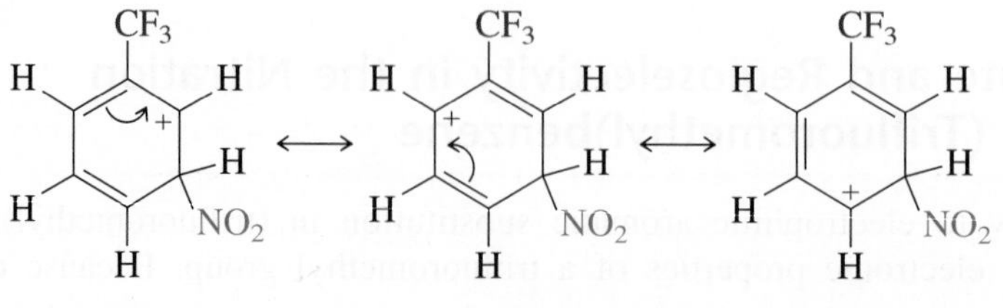
CF₃
H
NO₂

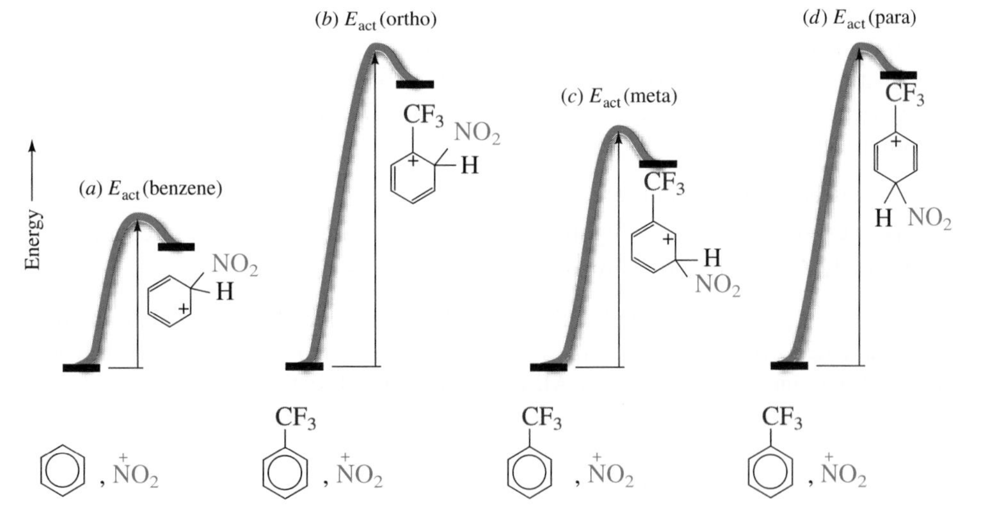

Figure 12.5

Comparative energy diagrams for nitronium ion attachment to (*a*) benzene and at the (*b*) ortho, (*c*) meta, and (*d*) para positions of (trifluoromethyl)-benzene. E_{act} (ortho) > E_{act} (para) > E_{act} (meta) > E_{act} (benzene).

Bonding of NO_2^+ to the meta position gives a more stable intermediate than bonding at either the ortho or the para position, and so meta substitution predominates. Even the carbocation intermediate corresponding to meta nitration, however, is very unstable and is formed with difficulty. The trifluoromethyl group is only one bond farther removed from the positive charge here than it is in the ortho and para intermediates and so still exerts a significant, although somewhat diminished, destabilizing inductive effect.

All the ring positions of (trifluoromethyl)benzene are deactivated compared with benzene. The meta position is simply deactivated *less* than the ortho and para positions. The partial rate factors for nitration of (trifluoromethyl)benzene are

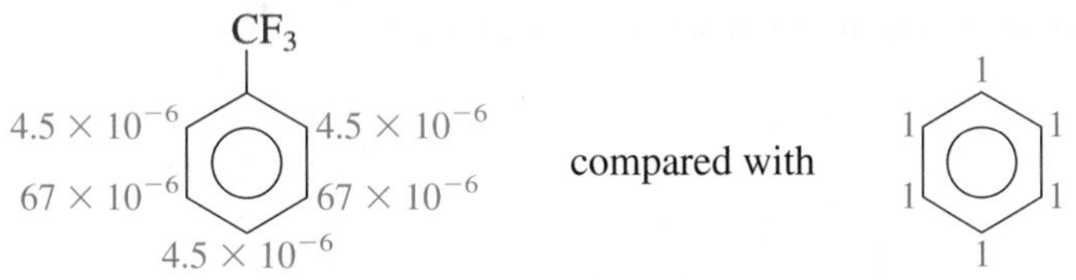

Figure 12.5 compares the energy profile for nitration of benzene with those for the ortho, meta, and para positions of (trifluoromethyl)benzene. The presence of the electron-withdrawing trifluoromethyl group raises the activation energy at all the ring positions, but the increase is least for the meta position.

Problem 12.14

The compounds benzyl chloride ($C_6H_5CH_2Cl$), (dichloromethyl)benzene ($C_6H_5CHCl_2$), and (trichloromethyl)benzene ($C_6H_5CCl_3$) all undergo nitration more slowly than benzene. The proportion of *m*-nitro-substituted product is 4% in one, 34% in another, and 64% in another. Classify the substituents —CH_2Cl, —$CHCl_2$, and —CCl_3 according to each one's effect on rate and regioselectivity in electrophilic aromatic substitution.

12.12 Substituent Effects in Electrophilic Aromatic Substitution: Activating Substituents

Our analysis of substituent effects has so far centered on two groups: methyl and trifluoromethyl. We have seen that a methyl substituent is electron-releasing, activating, and ortho, para-directing. A trifluoromethyl group is strongly electron-withdrawing, deactivating, and meta-directing. What about other substituents?

Table 12.2 summarizes orientation and rate effects in electrophilic aromatic substitution for some frequently encountered substituents. It is arranged in order of decreasing activating power: the most strongly *activating* substituents are at the top, the most strongly *deactivating* substituents are at the bottom. The main features of the table can be summarized as follows:

1. All activating substituents are ortho, para directors.
2. Halogen substituents are slightly deactivating but are ortho, para-directing.
3. Strongly deactivating substituents are meta directors.

Some of the most powerful activating substituents are those in which an oxygen atom is attached directly to the ring. These substituents include the hydroxyl group as well as alkoxy and acyloxy groups. All are ortho, para directors.

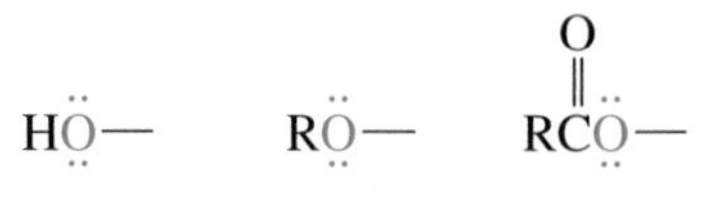

Hydroxyl Alkoxy Acyloxy

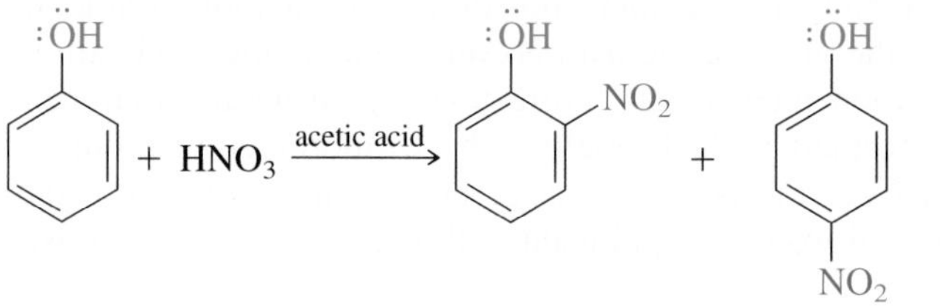

Phenol o-Nitrophenol (44%) p-Nitrophenol (56%)

Phenol and *anisole* are among the commonly encountered benzene derivatives listed in Table 11.1. Electrophilic aromatic substitution in phenol is discussed in more detail in Section 22.8.

Hydroxyl, alkoxy, and acyloxy groups activate the ring to such an extent that bromination occurs rapidly even in the absence of a catalyst.

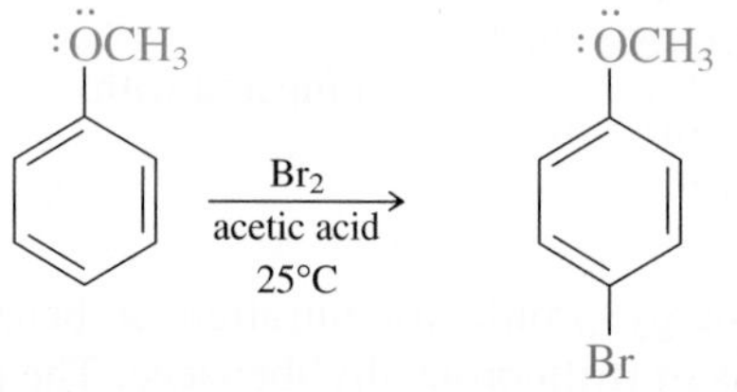

Anisole *p*-Bromoanisole (90%)

The *inductive* effect of hydroxyl and alkoxy groups, because of the electronegativity of oxygen, is to withdraw electrons and would seem to require that such substituents be deactivating. This electron-withdrawing inductive effect, however, is overcome by a much larger electron-releasing *resonance* effect involving the unshared electron pairs of oxygen. Bonding of the electrophile at positions ortho and para to a substituent of the type —ÖR gives a cation stabilized by delocalization of an unshared electron pair of oxygen into the π system of the ring.

Ortho attachment of E^+

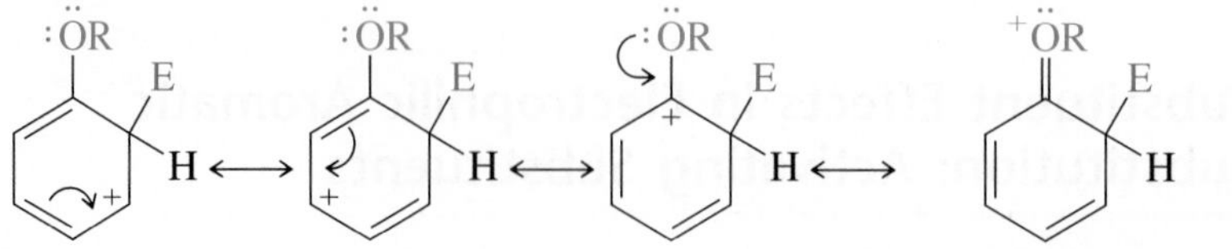

Most stable resonance contributor; oxygen and all carbons have octets of electrons

TABLE 12.2 Classification of Substituents in Electrophilic Aromatic Substitution Reactions

Effect on rate	Substituent		Effect on orientation
Very strongly activating	$-\ddot{\mathrm{N}}\mathrm{H}_2$	(amino)	Ortho, para-directing
	$-\ddot{\mathrm{N}}\mathrm{HR}$	(alkylamino)	
	$-\ddot{\mathrm{N}}\mathrm{R}_2$	(dialkylamino)	
	$-\ddot{\mathrm{O}}\mathrm{H}$	(hydroxyl)	
Strongly activating	$-\ddot{\mathrm{N}}\mathrm{H}\overset{\mathrm{O}}{\overset{\|}{\mathrm{C}}}\mathrm{R}$	(acylamino)	Ortho, para-directing
	$-\ddot{\mathrm{O}}\mathrm{R}$	(alkoxy)	
	$-\ddot{\mathrm{O}}\overset{\mathrm{O}}{\overset{\|}{\mathrm{C}}}\mathrm{R}$	(acyloxy)	
Activating	—R	(alkyl)	Ortho, para-directing
	—Ar	(aryl)	
	$-\mathrm{CH{=}CR_2}$	(alkenyl)	
Standard of comparison	—H	(hydrogen)	
Deactivating	—X	(halogen)	Ortho, para-directing
	(X = F, Cl, Br, I)		
	$-\mathrm{CH_2X}$	(halomethyl)	
Strongly deactivating	$-\overset{\mathrm{O}}{\overset{\|}{\mathrm{C}}}\mathrm{H}$	(formyl)	Meta-directing
	$-\overset{\mathrm{O}}{\overset{\|}{\mathrm{C}}}\mathrm{R}$	(acyl)	
	$-\overset{\mathrm{O}}{\overset{\|}{\mathrm{C}}}\mathrm{OH}$	(carboxylic acid)	
	$-\overset{\mathrm{O}}{\overset{\|}{\mathrm{C}}}\mathrm{OR}$	(ester)	
	$-\overset{\mathrm{O}}{\overset{\|}{\mathrm{C}}}\mathrm{Cl}$	(acyl chloride)	
	$-\mathrm{C{\equiv}N}$	(cyano)	
	$-\overset{:\ddot{\mathrm{O}}}{\overset{\|}{\underset{\underset{:\ddot{\mathrm{O}}:^{-}}{\vert}}{\mathrm{S}^{+}}}}-\ddot{\mathrm{O}}\mathrm{H}$	(sulfonic acid)	
Very strongly deactivating	$-\mathrm{CF_3}$	(trifluoromethyl)	Meta-directing
	$-\mathrm{NO_2}$	(nitro)	

Para attachment of E^+

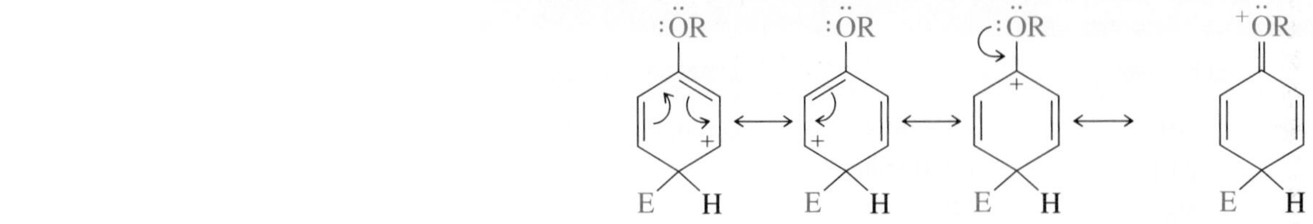

Most stable resonance contributor; oxygen and all carbons have octets of electrons

Oxygen-stabilized carbocations of this type are far more stable than tertiary carbocations. They are best represented by structures in which the positive charge is on oxygen because all the atoms then have octets of electrons. Their stability permits them to be formed rapidly, resulting in rates of electrophilic aromatic substitution that are much faster than that of benzene.

Meta attachment of E^+

The lone pair on oxygen cannot be directly involved in carbocation stabilization when the electrophile bonds to the meta carbon.

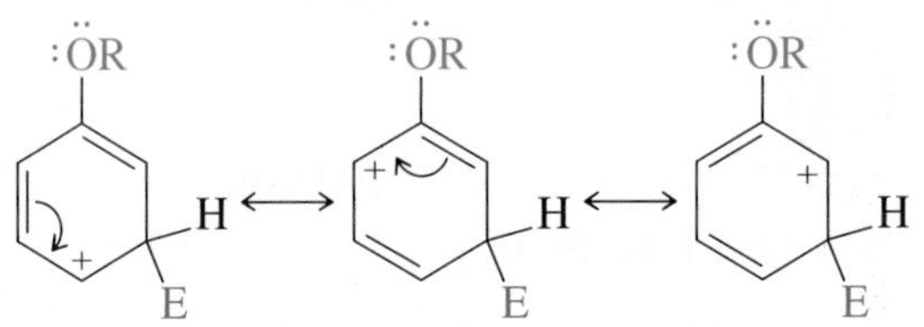

Oxygen lone pair cannot be used to stabilize positive charge in any of these structures; all have six electrons around positively charged carbon.

The greater stability of the carbocation intermediates arising from bonding of the electrophile to the ortho and para carbons compared with those at the carbon meta to oxygen explains the ortho, para-directing property of —$\ddot{O}$H, —$\ddot{O}$R, and —$\ddot{O}$C(O)R groups.

Nitrogen-containing substituents related to the amino group are even better electron-releasing groups and more strongly activating than the corresponding oxygen-containing substituents.

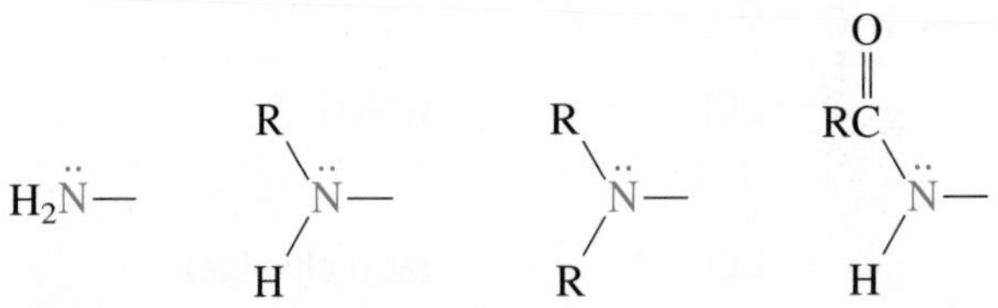

Amino Alkylamino Dialkylamino Acylamino

The nitrogen atom in each of these groups bears an electron pair that, like the unshared pairs of oxygen, stabilizes a carbocation to which it is attached. Nitrogen is less electronegative than oxygen, so is a better electron pair donor and stabilizes the cyclohexadienyl cation intermediates in electrophilic aromatic substitution to an even greater degree.

Aniline and its derivatives are so reactive in electrophilic aromatic substitution that special strategies are usually necessary to carry out these reactions effectively. This topic is discussed in Section 21.14.

Problem 12.15

Write structural formulas for the cyclohexadienyl cations formed from aniline ($C_6H_5NH_2$) during

(a) Ortho bromination (four resonance structures)
(b) Meta bromination (three resonance structures)
(c) Para bromination (four resonance structures)

Sample Solution (a) There are the customary three resonance contributors for the cyclohexadienyl cation plus a contributor (the most stable one) derived by delocalization of the nitrogen lone pair into the ring.

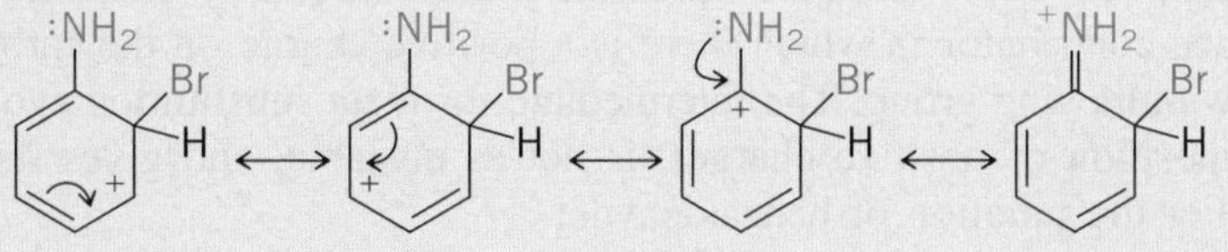

Most stable contributing structure

Alkyl groups are, as we saw when we discussed the nitration of toluene in Section 12.10, activating and ortho, para-directing substituents. Aryl and alkenyl substituents resemble alkyl groups in this respect; they too are activating and ortho, para-directing.

Problem 12.16

Treatment of biphenyl (see Section 11.6 to remind yourself of its structure) with a mixture of nitric acid and sulfuric acid gave two principal products both having the molecular formula $C_{12}H_9NO_2$. What are these two products?

The next group of substituents in Table 12.2 that we'll discuss are the ones near the bottom of the table, those that are meta-directing and strongly deactivating.

12.13 Substituent Effects in Electrophilic Aromatic Substitution: Strongly Deactivating Substituents

As Table 12.2 indicates, a number of substituents are *meta-directing and strongly deactivating*. We have already discussed one of these, the trifluoromethyl group. Several others have a carbonyl group attached directly to the aromatic ring.

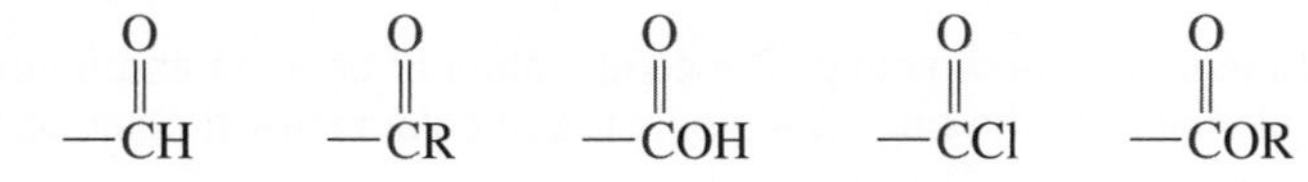

Aldehyde Ketone Carboxylic acid Acyl chloride Ester

The behavior of aromatic aldehydes is typical. Nitration of benzaldehyde takes place several thousand times more slowly than that of benzene and yields *m*-nitrobenzaldehyde as the major product.

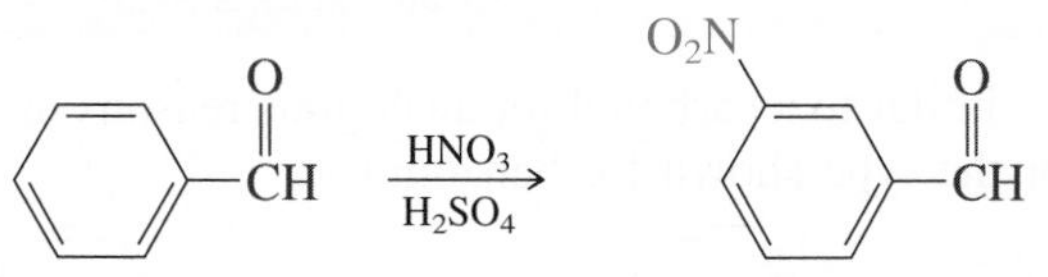

Benzaldehyde *m*-Nitrobenzaldehyde (75–84%)

To understand the effect of a carbonyl group attached directly to the ring, consider its polarization. The electrons in the carbon–oxygen double bond are drawn toward oxygen and away from carbon, leaving the carbon attached to the ring with a partial positive charge. Using benzaldehyde as an example,

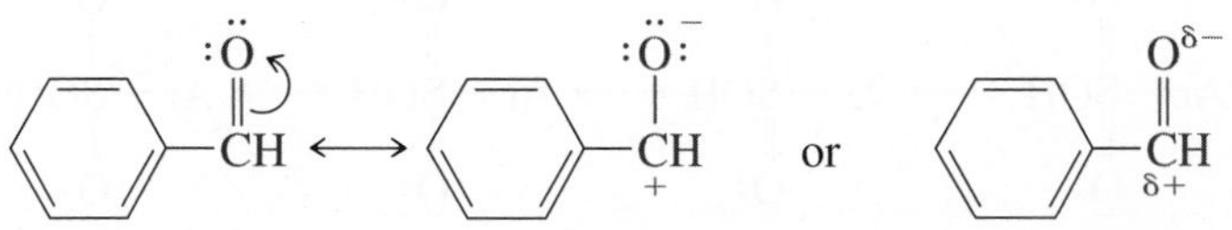

Because the carbon atom attached to the ring is positively polarized, a carbonyl group is *strongly electron-withdrawing* and behaves in much the same way as a trifluoromethyl group to destabilize all the cyclohexadienyl cation intermediates in electrophilic aromatic substitution. Reaction at any ring position in benzaldehyde is slower than in benzene. The intermediates for ortho and para substitution are particularly unstable because each has a resonance contributor in which there is a positive charge on the carbon that bears the electron-withdrawing group. The intermediate for meta substitution avoids this unfavorable juxtaposition of positive charges, is not as unstable, and gives rise to most of the product. For the nitration of benzaldehyde:

Ortho nitration

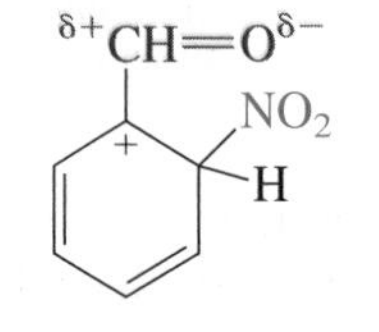

Unstable because of adjacent positively polarized atoms

Meta nitration

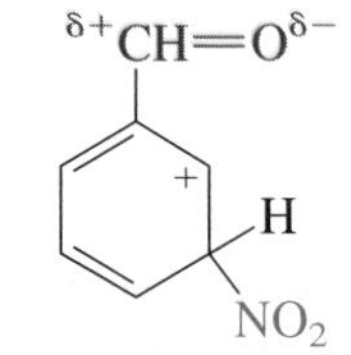

Positively polarized atoms not adjacent; most stable intermediate

Para nitration

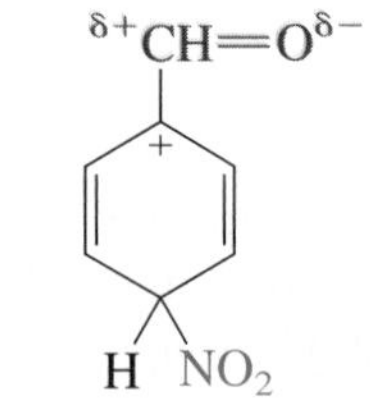

Unstable because of adjacent positively polarized atoms

Problem 12.17

Each of the following reactions has been reported in the chemical literature, and the major organic product has been isolated in good yield. Write a structural formula for the product of each reaction.

(a) Treatment of benzoyl chloride ($C_6H_5\overset{O}{\overset{\|}{C}}Cl$) with chlorine and iron(III) chloride

(b) Treatment of methyl benzoate ($C_6H_5\overset{O}{\overset{\|}{C}}OCH_3$) with nitric acid and sulfuric acid

(c) Nitration of 1-phenyl-1-propanone ($C_6H_5\overset{O}{\overset{\|}{C}}CH_2CH_3$)

Sample Solution (a) Benzoyl chloride has a carbonyl group attached directly to the ring. A $-\overset{O}{\overset{\|}{C}}Cl$ substituent is meta-directing. The combination of chlorine and iron(III) chloride introduces a chlorine onto the ring. The product is *m*-chlorobenzoyl chloride.

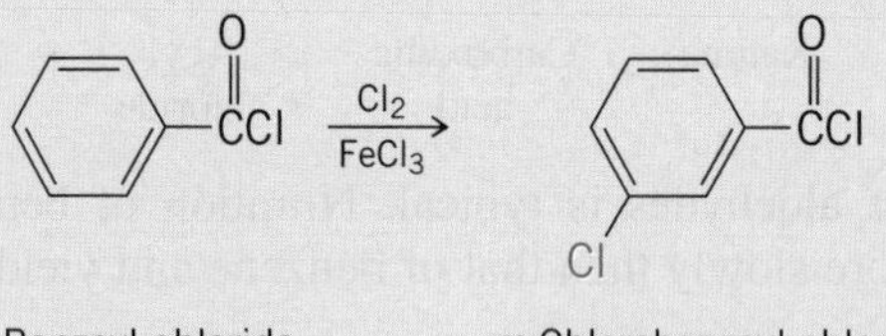

Benzoyl chloride → *m*-Chlorobenzoyl chloride (isolated in 62% yield)

A cyano group is similar to a carbonyl for analogous reasons involving contributing resonance structures of the type shown for benzonitrile.

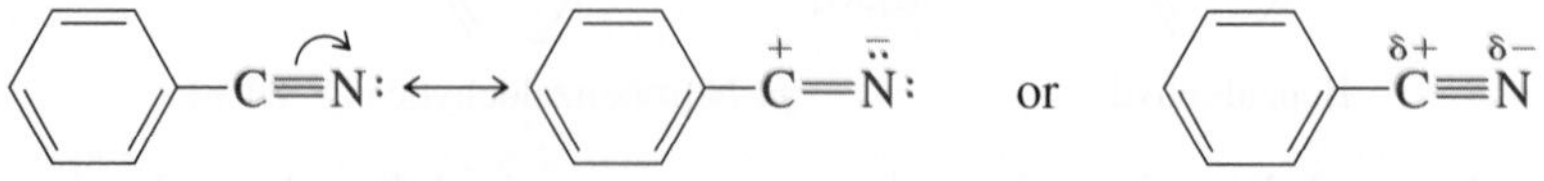

Cyano groups are electron-withdrawing, deactivating, and meta-directing.

Sulfonic acid groups are electron-withdrawing because sulfur has a formal positive charge in several of the contributing forms of $-SO_3H$.

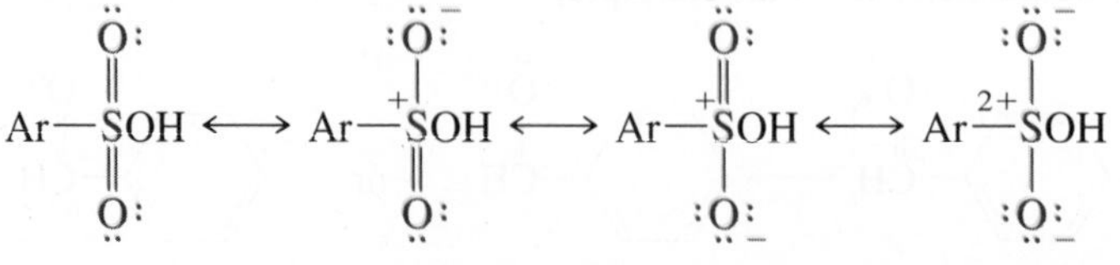

When benzene undergoes disulfonation, *m*-benzenedisulfonic acid is formed. The first sulfonic acid group to go on directs the second one meta to itself.

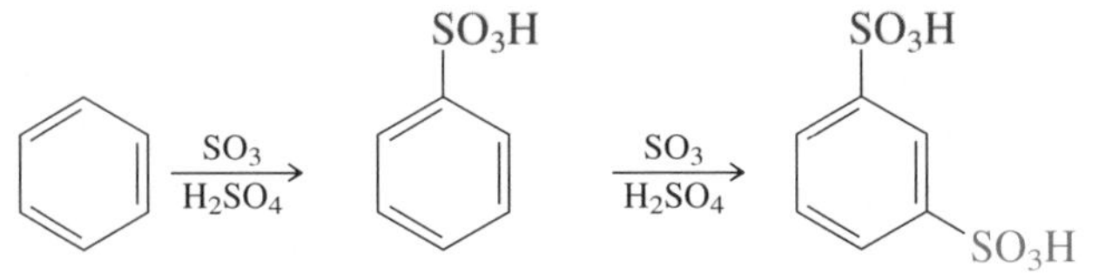

Benzene | Benzenesulfonic acid | *m*-Benzenedisulfonic acid (90%)

The nitrogen atom of a nitro group bears a full positive charge in its two most stable contributing structures.

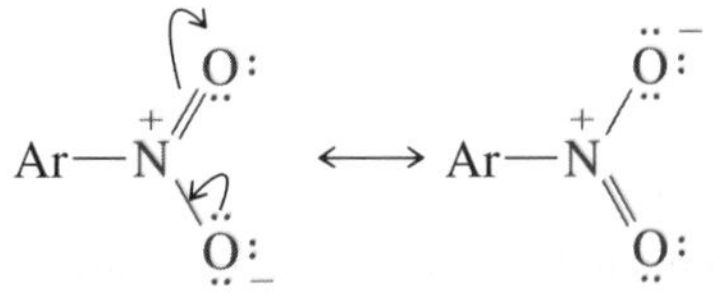

This makes the nitro group a powerful electron-withdrawing, deactivating substituent and a meta director.

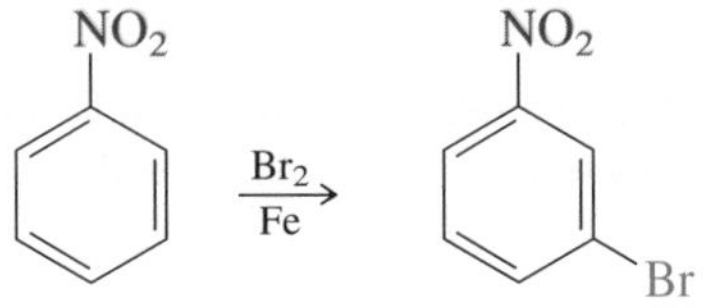

Nitrobenzene | *m*-Bromonitrobenzene (60–75%)

Problem 12.18

Would you expect the substituent $-\overset{+}{N}(CH_3)_3$ to more closely resemble $-\ddot{N}(CH_3)_2$ or $-NO_2$ in its effect on rate and regioselectivity in electrophilic aromatic substitution? Why?

$-\ddot{N}(CH_3)_2$: unshared pair on N; strongly activating; ortho, para director

$-NO_2$: positive charge on N; strongly deactivating; meta director

$-\overset{+}{N}(CH_3)_3$: positive charge on N

Therefore, resembles $-NO_2$ more than $-N(CH_3)_2$

Therefore, predict strongly deactivating, meta director

12.14 Substituent Effects in Electrophilic Aromatic Substitution: Halogens

Returning to Table 12.2, notice that *halogen substituents direct an incoming electrophile to the ortho and para positions but deactivate the ring toward substitution.* Nitration of chlorobenzene is a typical example; its rate is some 30 times slower than the corresponding nitration of benzene, and the major products are *o*-chloronitrobenzene and *p*-chloronitrobenzene.

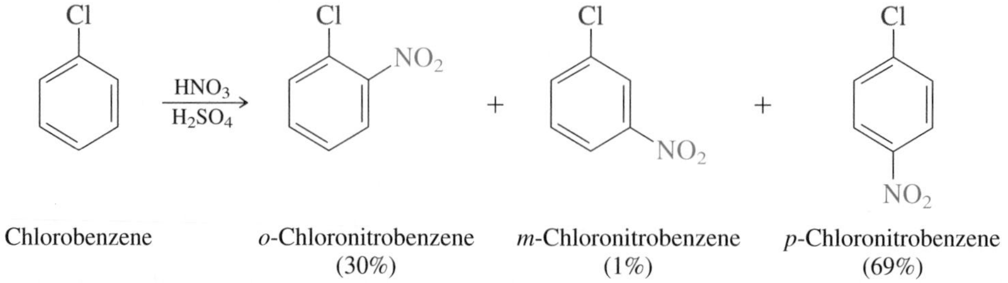

Problem 12.19

Reaction of chlorobenzene with *p*-chlorobenzyl chloride and aluminum chloride gave a mixture of two products in good yield (76%). What were these two products?

Rate and product studies of electrophilic aromatic substitution in halobenzenes reveal a fairly consistent pattern of reactivity. The partial rate factors for chlorination show that, with one exception, all the ring positions of fluoro-, chloro-, and bromobenzene are deactivated. The exception is the para position of fluorobenzene, which is slightly more reactive than a single position of benzene.

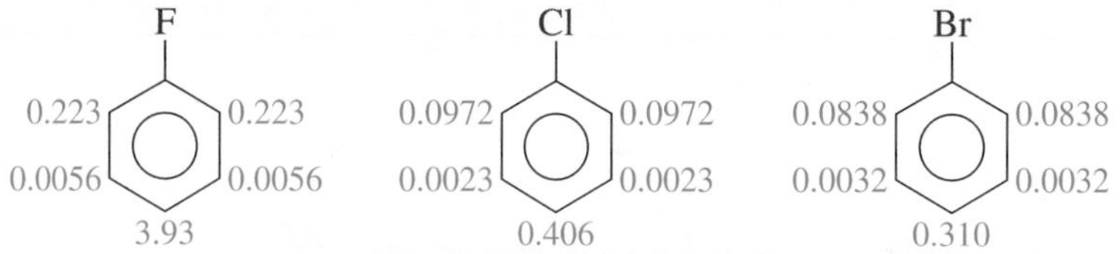

The range of reactivity is not large. Benzene undergoes chlorination only about 1.4 times faster than the most reactive of the group (fluorobenzene) and 14 times faster than the least reactive (bromobenzene). In each halobenzene the para position is the most reactive, followed by ortho.

Because we have come to associate activating substituents with ortho, para-directing effects and deactivating substituents with meta, the properties of halogen substituents appear on initial inspection to be unusual. The seeming inconsistency between regioselectivity and rate can be understood by analyzing the inductive and resonance effects of a halogen substituent.

Through its inductive effect, a halogen (X) withdraws electrons from the ring by polarization of the σ framework. The effect is greatest for fluorine, least for iodine.

Inductive effect of halogen attracts electrons from ring

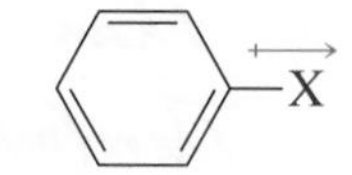

This polarization, in turn, causes the ring carbons to bind the π electrons more tightly, decreases their "availability" to an approaching electrophile, raises the activation energy for electrophilic aromatic substitution, and decreases the reaction rate. Figure 12.6 illustrates this effect by comparing the electrostatic potential maps of fluorobenzene and benzene.

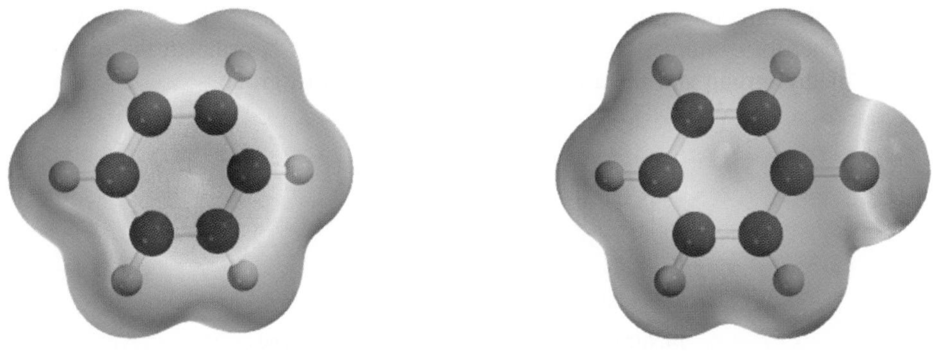

Benzene Fluorobenzene

Figure 12.6

Electrostatic potential maps of benzene and fluorobenzene. The high electronegativity of fluorine causes the π electrons of fluorobenzene to be more strongly held than those of benzene. This difference is reflected in the more pronounced red color associated with the π electrons of benzene. The color scale is the same for both models.

Like $-\ddot{O}H$ and $-\ddot{N}H_2$ groups, however, halogen substituents possess unshared electron pairs that can be donated to a positively charged carbon. This electron donation into the π system stabilizes the intermediates for ortho and para substitution.

Ortho attachment of E^+ ***Para attachment of E^+***

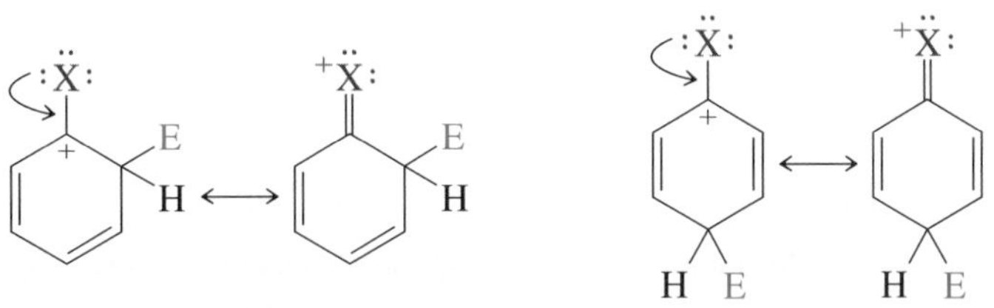

Comparable stabilization of the intermediate for meta substitution is not possible. Thus, resonance involving their lone pairs causes halogens to be ortho, para-directing substituents.

The resonance effect is greatest for fluorine and much smaller for the other halogens. For resonance stabilization to be effective, the lone-pair *p* orbital of the substituent must overlap with the π system of the ring. The 2*p* orbital of fluorine is well suited for such overlap, but the 3*p* orbital of chlorine is not because of its more diffuse character and the longer C—Cl bond distance. The situation is even worse for Br and I.

By stabilizing the cyclohexadienyl cation intermediate, lone-pair donation from fluorine counteracts the inductive effect to the extent that the rate of electrophilic aromatic substitution in fluorobenzene is, in most cases, only slightly less than that of benzene. With the other halogens, lone-pair donation is sufficient to make them ortho, para directors, but is less than that of fluorine.

12.15 Multiple Substituent Effects

When a benzene ring bears two or more substituents, both its reactivity and the site of further substitution can usually be predicted from the cumulative effects of its substituents.

In the simplest cases all the available sites are equivalent, and substitution at any one of them gives the same product.

Problems 12.2, 12.4, and 12.10 offer additional examples of reactions in which only a single product of electrophilic aromatic substitution is possible.

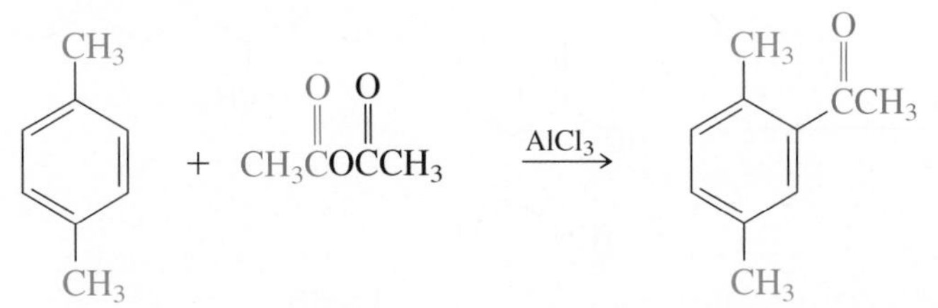

1,4-Dimethylbenzene (*p*-xylene) 2,5-Dimethylacetophenone (99%)

Often the directing effects of substituents reinforce each other. Bromination of *p*–nitrotoluene, for example, takes place at the position that is ortho to the ortho, para-directing methyl group and meta to the meta-directing nitro group.

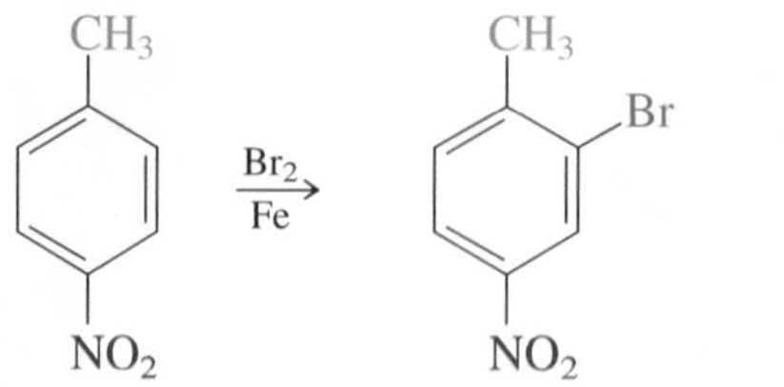

p-Nitrotoluene — 2-Bromo-4-nitrotoluene (86–90%)

In almost all cases, including most of those in which the directing effects of individual substituents oppose each other, *it is the more activating substituent that controls the regioselectivity of electrophilic aromatic substitution.* Thus, bromination occurs ortho to the *N*-methylamino group in 4-chloro-*N*-methylaniline because this group is a very powerful activating substituent while the chlorine is weakly deactivating.

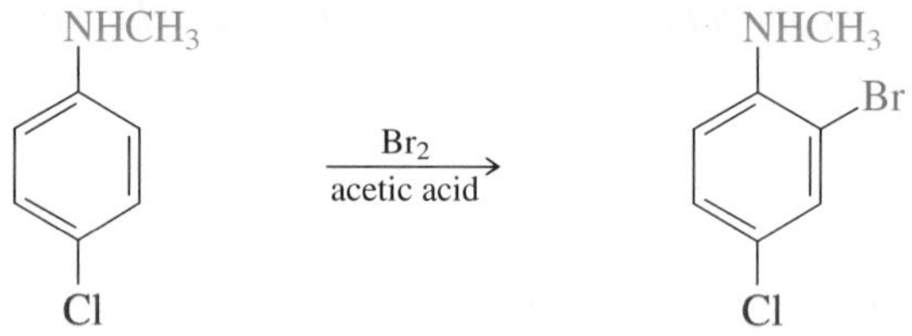

4-Chloro-*N*-methylaniline — 2-Bromo-4-chloro-*N*-methylaniline (87%)

Problem 12.20

The reactant in the preceding equation (4-chloro-*N*-methylaniline) is so reactive toward electrophilic aromatic substitution that no catalyst is necessary to bring about its bromination. Write a reasonable mechanism for the preceding reaction based on Br_2 as the electrophile.

Problem 12.21

Compound A was used as an intermediate in the synthesis of the drug labetelol, a type of drug known as a "β-blocker" that is used for the treatment of hypertension. What is the structure of compound A?

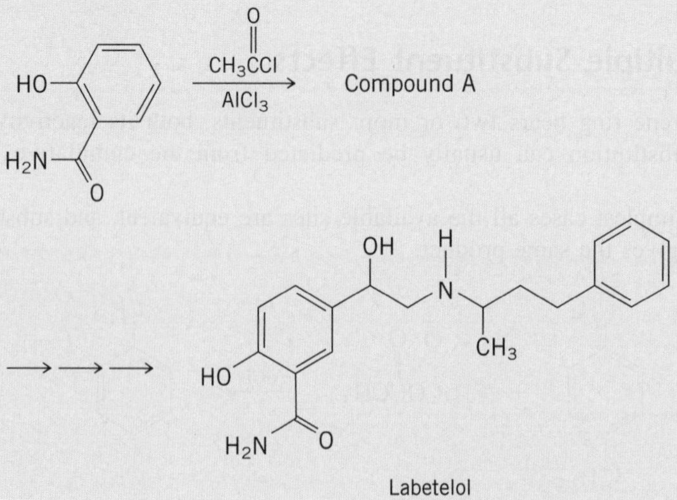

Labetelol
(a β-blocker)

When two positions are comparably activated by alkyl groups, substitution usually occurs at the less hindered site. Nitration of *p-tert*-butyltoluene takes place at positions ortho to the methyl group in preference to those ortho to the larger *tert*-butyl group. This is an example of a *steric effect.*

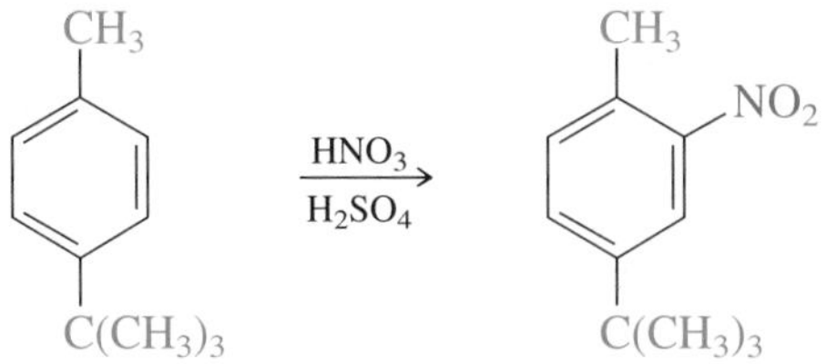

p-tert-Butyltoluene — 4-*tert*-Butyl-2-nitrotoluene (88%)

Nitration of *m*-xylene is directed ortho to one methyl group and para to the other.

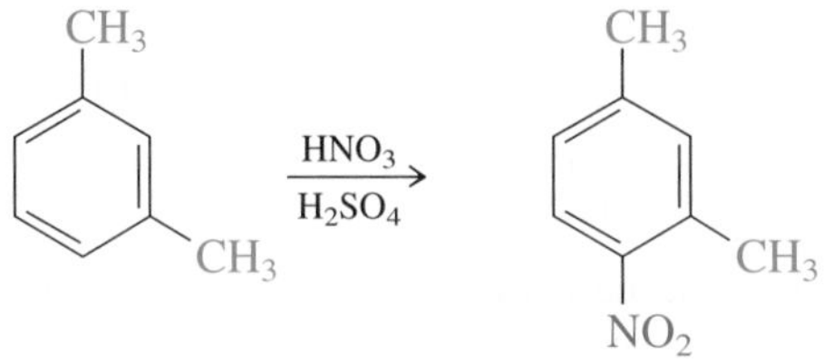

m-Xylene — 2,4-Dimethyl-1-nitrobenzene (98%)

The ortho position between the two methyl groups is less reactive because it is more sterically hindered.

Problem 12.22

Write the structure of the principal organic product obtained on nitration of each of the following:

(a) *p*-Methylbenzoic acid
(b) *m*-Dichlorobenzene
(c) *m*-Dinitrobenzene
(d) *p*-Methoxyacetophenone
(e) *p*-Methylanisole
(f) 2,6-Dibromoanisole

Sample Solution (a) Of the two substituents in *p*-methylbenzoic acid, the methyl group is more activating and so controls the regioselectivity of electrophilic aromatic substitution. The position para to the ortho, para-directing methyl group already bears a substituent (the carboxyl group), and so substitution occurs ortho to the methyl group. This position is meta to the *m*-directing carboxyl group, and the orienting properties of the two substituents reinforce each other. The product is 4-methyl-3-nitrobenzoic acid.

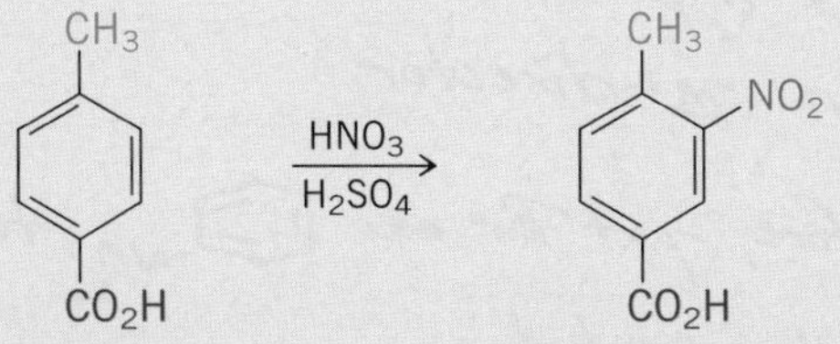

p-Methylbenzoic acid — 4-Methyl-3-nitrobenzoic acid

An exception to the rule that regioselectivity is controlled by the most activating substituent occurs when the directing effects of alkyl groups and halogen substituents oppose each other. Alkyl groups and halogen substituents are weakly activating and weakly deactivating, respectively, and the difference between them is too small to allow a simple generalization.

Problem 12.52 illustrates how partial rate factor data may be applied to such cases.

12.16 Regioselective Synthesis of Disubstituted Aromatic Compounds

Because the regioselectivity of electrophilic aromatic substitution is controlled by the directing effects of substituents already present, the preparation of disubstituted aromatic compounds requires that careful thought be given to the order of introduction of the two groups.

Compare the independent preparations of *m*-bromoacetophenone and *p*-bromoacetophenone from benzene. Both syntheses require a Friedel–Crafts acylation and a bromination, but the major product is determined by the *order* in which the two reactions are carried out. When the meta-directing acetyl group is introduced first, the final product is *m*-bromoacetophenone.

Aluminum chloride is a stronger Lewis acid than iron(III) bromide and has been used as a catalyst in electrophilic bromination when, as in the example shown, the aromatic ring bears a strongly deactivating substituent.

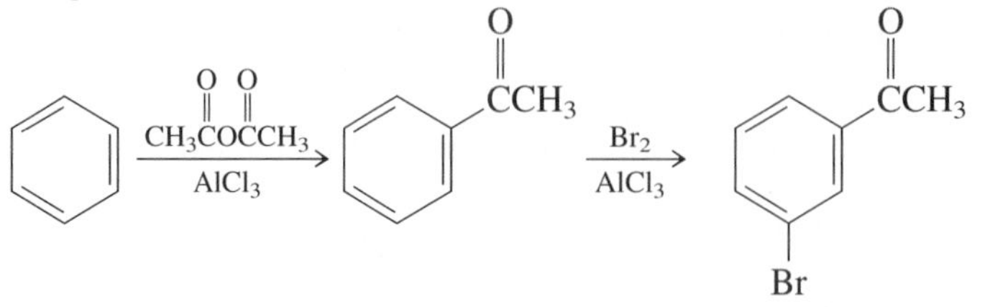

Benzene | Acetophenone (76–83%) | *m*-Bromoacetophenone (59%)

When the ortho, para-directing bromine is introduced first, the major product is *p*-bromoacetophenone (along with some of its ortho isomer, from which it is separated by distillation).

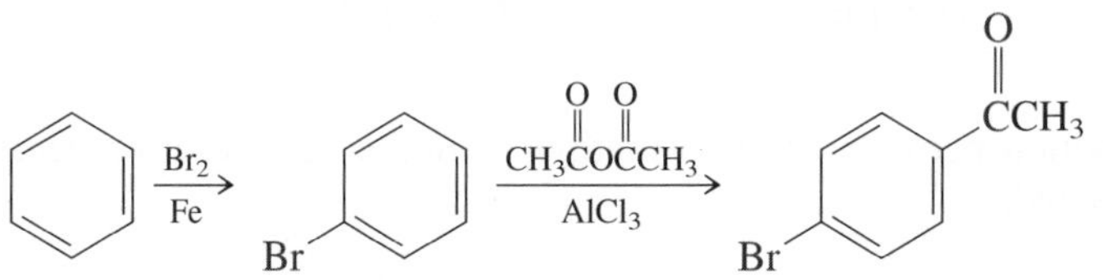

Benzene | Bromobenzene (65–75%) | *p*-Bromoacetophenone (69–79%)

Problem 12.23

Write chemical equations showing how you could prepare *m*-bromonitrobenzene as the principal organic product, starting with benzene and using any necessary organic or inorganic reagents. How could you prepare *p*-bromonitrobenzene?

Sample Solution

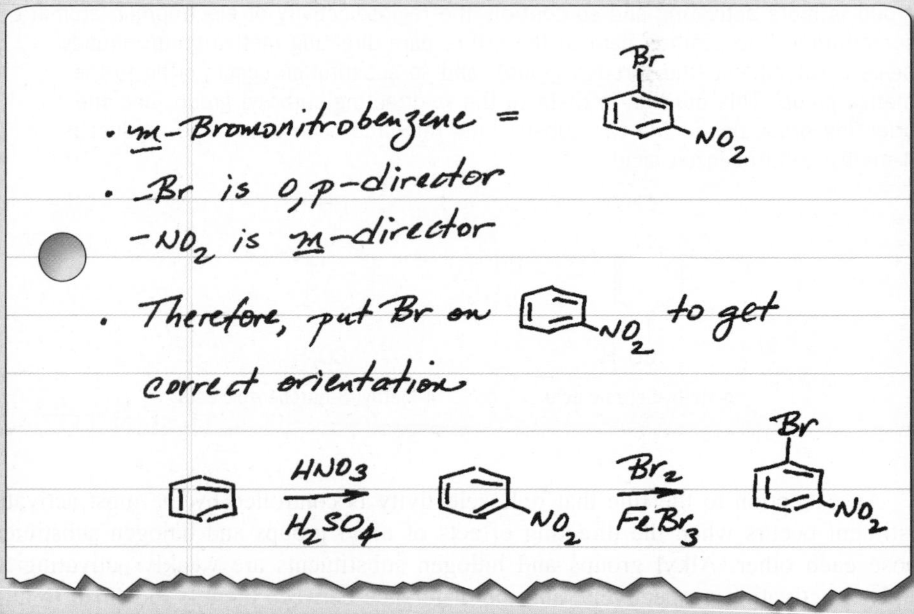

A less obvious example of a situation in which the success of a synthesis depends on the order of introduction of substituents is illustrated by the preparation of *m*-nitroacetophenone. Here, even though both substituents are meta-directing, the only practical synthesis is the one in which Friedel–Crafts acylation is carried out first.

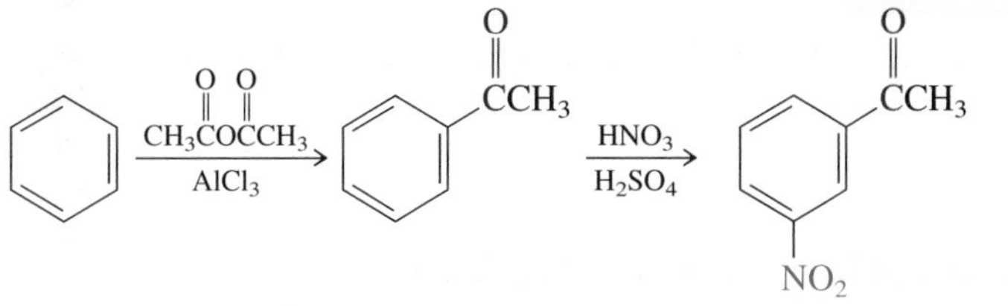

Benzene — Acetophenone (76–83%) — *m*-Nitroacetophenone (55%)

When the reverse order of steps is attempted, it is observed that the Friedel–Crafts acylation of nitrobenzene fails.

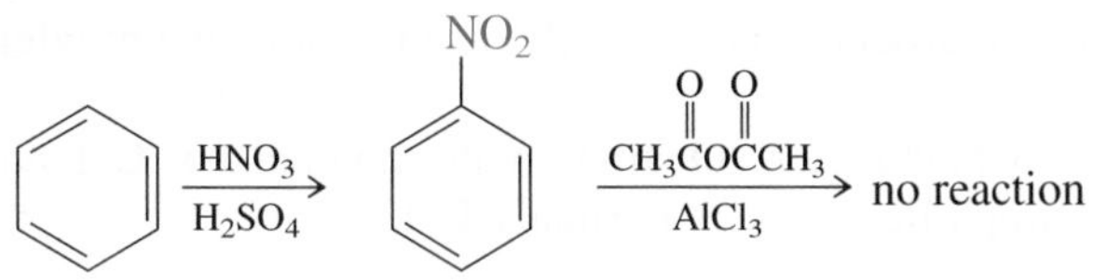

Benzene — Nitrobenzene (95%)

Neither Friedel–Crafts acylation nor alkylation can be carried out on nitrobenzene. The presence of a strongly deactivating substituent such as a nitro group on an aromatic ring so depresses its reactivity that Friedel–Crafts reactions do not take place. Nitrobenzene is so unreactive that it is sometimes used as a solvent in Friedel–Crafts reactions. The practical limit for Friedel–Crafts alkylation and acylation is effectively a monohalobenzene. *An aromatic ring more deactivated than a monohalobenzene cannot be alkylated or acylated under Friedel–Crafts conditions.*

Sometimes the orientation of two substituents in an aromatic compound precludes its straightforward synthesis. *m*-Chloroethylbenzene, for example, has two ortho, para-directing groups in a meta relationship and so can't be prepared either from chlorobenzene or ethylbenzene. In cases such as this we couple electrophilic aromatic substitution with functional group manipulation to produce the desired compound.

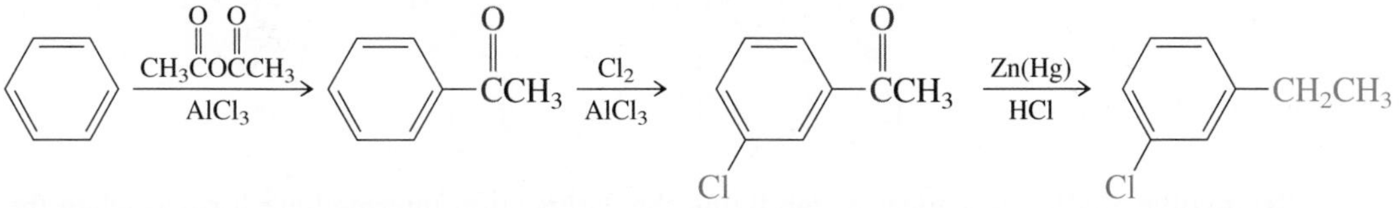

Benzene — Acetophenone — *m*-Chloroacetophenone — *m*-Chloroethylbenzene

The key here is to recognize that an ethyl substituent can be introduced by Friedel–Crafts acylation followed by a Clemmensen or Wolff–Kishner reduction step later in the synthesis. Introducing the chlorine prior to reduction places it meta to the acyl group, giving the correct substitution pattern.

A related problem concerns the synthesis of *p*-nitrobenzoic acid. Here, two meta-directing substituents are para to each other. This compound has been prepared from toluene according to the procedure shown:

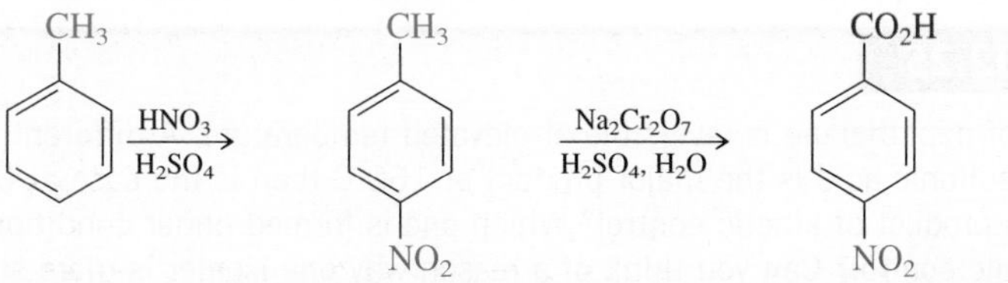

p-Nitrotoluene (separate from ortho isomer) — *p*-Nitrobenzoic acid (82–86%)

Because it may be oxidized to a carboxyl group (Section 11.12), a methyl group can be used to introduce the nitro substituent in the proper position.

Problem 12.24

Suggest an efficient synthesis of *m*-nitrobenzoic acid from toluene.

12.17 Substitution in Naphthalene

Polycyclic aromatic hydrocarbons undergo electrophilic aromatic substitution when treated with the same reagents that react with benzene. In general, polycyclic aromatic hydrocarbons are more reactive than benzene. Most lack the symmetry of benzene, however, and mixtures of products may be formed even on monosubstitution. Among polycyclic aromatic hydrocarbons, we will discuss only naphthalene, and that only briefly.

Two sites are available for substitution in naphthalene: C-1 and C-2. The more reactive site of electrophilic attack is normally C-1.

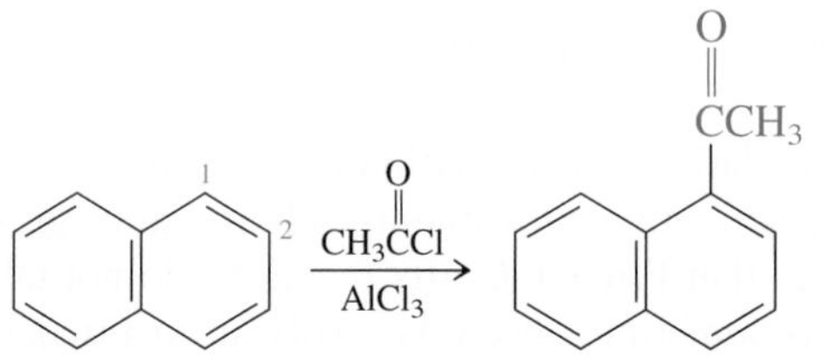

Naphthalene 1-Acetylnaphthalene (90%)

C-1 is more reactive because the intermediate formed when the electrophile bonds there is a relatively stable carbocation. A benzene-type pattern of bonds is retained in one ring, and the positive charge is delocalized by allylic resonance.

Attachment of E^+ to C-1

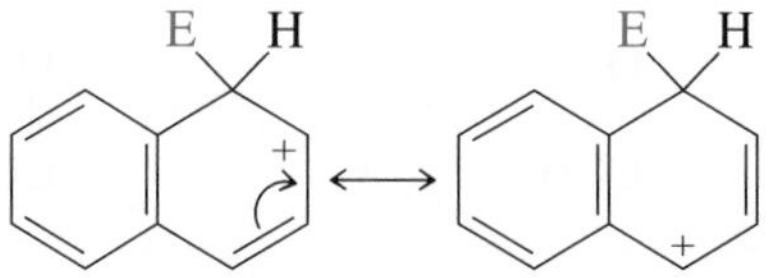

To involve allylic resonance in stabilizing the carbocation intermediate formed when the electrophile bonds to C-2, the benzene-like character of the other ring is sacrificed.

Attachment of E^+ to C-2

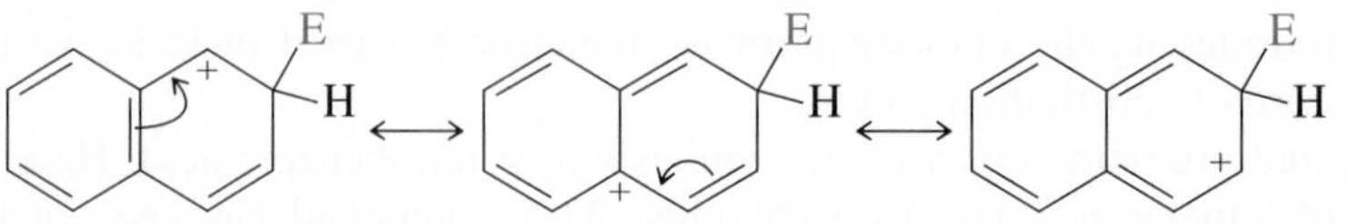

Problem 12.25

Sulfonation of naphthalene is reversible at elevated temperature. A different isomer of naphthalenesulfonic acid is the major product at 160°C than is the case at 0°C. Which isomer is the product of kinetic control? Which one is formed under conditions of thermodynamic control? Can you think of a reason why one isomer is more stable than the other?

12.18 Substitution in Heterocyclic Aromatic Compounds

Their great variety of structural types causes heterocyclic aromatic compounds to range from exceedingly reactive to practically inert toward electrophilic aromatic substitution.

Pyridine lies near one extreme in being far less reactive than benzene toward substitution by electrophilic reagents. In this respect it resembles strongly deactivated aromatic compounds such as nitrobenzene. It is incapable of being acylated or alkylated under Friedel–Crafts conditions, but can be sulfonated at high temperature. Electrophilic substitution in pyridine, when it does occur, takes place at C-3.

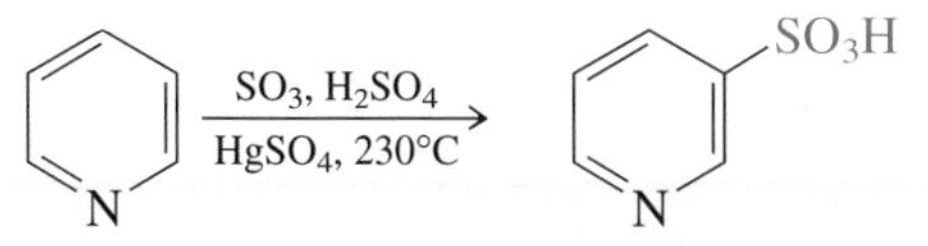

Pyridine → Pyridine-3-sulfonic acid (71%)

One reason for the low reactivity of pyridine is that nitrogen is more electronegative than carbon, which causes the π electrons of pyridine to be held more tightly and raises the activation energy for bonding to an electrophile. Another is that the nitrogen of pyridine is protonated in sulfuric acid and the resulting pyridinium ion is even more deactivated than pyridine itself.

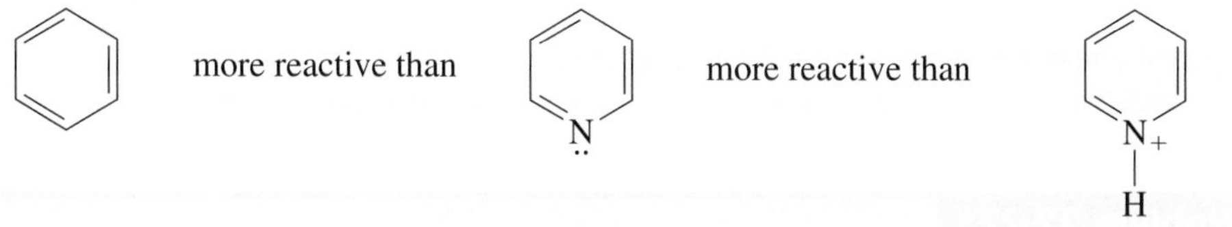

Benzene Pyridine Pyridinium ion

Lewis acid catalysts such as aluminum chloride and iron(III) halides also bond to nitrogen to strongly deactivate the ring toward Friedel–Crafts reactions and halogenation.

Pyrrole, furan, and thiophene, on the other hand, have electron-rich aromatic rings and are extremely reactive toward electrophilic aromatic substitution—more like phenol and aniline than benzene. Like benzene they have six π electrons, but these π electrons are delocalized over *five* atoms, not six, and are not held as strongly as those of benzene. Even when the ring atom is as electronegative as oxygen, substitution takes place readily.

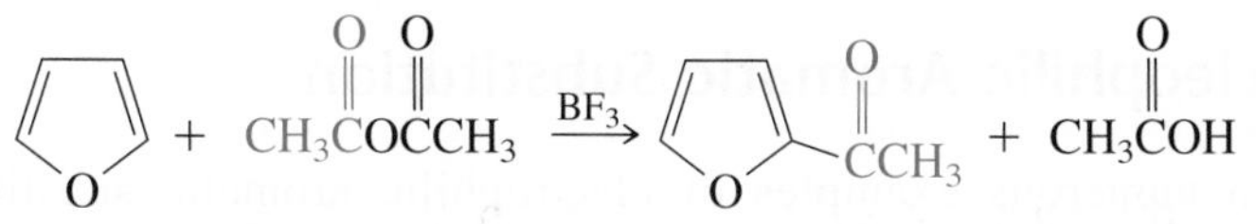

Furan Acetic anhydride 2-Acetylfuran (75–92%) Acetic acid

The regioselectivity of substitution in furan is explained using a resonance description. When the electrophile bonds to C-2, the positive charge is shared by three atoms: C-3, C-5, and O.

Attachment of E^+ to C-2

Carbocation *more stable;* positive charge shared by C-3, C-5, and O.

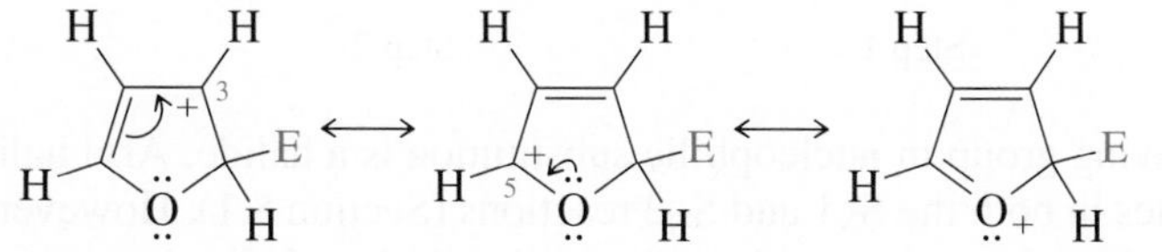

When the electrophile bonds to C-3, the positive charge is shared by only two atoms, C-2 and O, and the carbocation intermediate is less stable and formed more slowly.

Attachment of E^+ to C-3

Carbocation *less stable;* positive charge shared by C-2 and O.

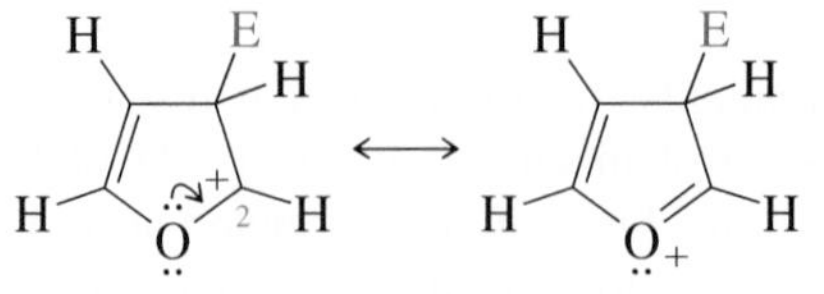

The regioselectivity of substitution in pyrrole and thiophene is like that of furan and for similar reasons.

Problem 12.26

Under acid-catalyzed conditions, the C-2 hydrogen of *N*-methylpyrrole is replaced by deuterium faster than the one at C-3 according to the equation:

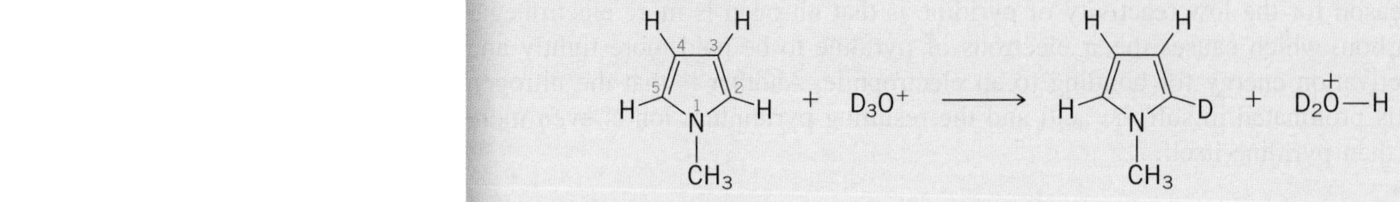

Suggest a reasonable mechanism for this reaction.

Problem 12.27

When benzene is prepared from coal tar, it is contaminated with thiophene, from which it cannot be separated by distillation because of very similar boiling points. Shaking a mixture of benzene and thiophene with sulfuric acid causes sulfonation of the thiophene ring but leaves benzene untouched. The sulfonation product of thiophene dissolves in the sulfuric acid layer, from which the benzene layer is separated; the benzene layer is then washed with water and distilled. Give the structure of the sulfonation product of thiophene.

12.19 Nucleophilic Aromatic Substitution

We have seen numerous examples of electrophilic aromatic substitution in this chapter. What about *nucleophilic* aromatic substitution? Under what circumstances might it be possible for a nucleophile to add to an aromatic ring to give a cyclohexadienyl anion that could rearomatize by expelling some negatively charged leaving group (LG)?

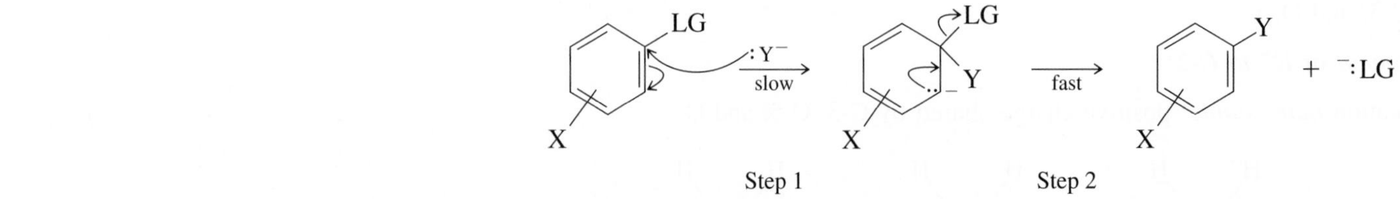

A common leaving group in nucleophilic substitution is a halide. Aryl halides are unreactive as substrates in both the S_N1 and S_N2 reactions (Section 8.1). However, as we will see in the following sections, the pathway for nucleophilic substitution that is suggested by the preceding equation is accessible to aryl halides, especially those in which substituent X on the aromatic ring is an electron-withdrawing group.

12.20 Nucleophilic Substitution in Nitro-Substituted Aryl Halides

Chlorobenzene is inert to aqueous sodium hydroxide at room temperature. Reaction temperatures of over 300°C are required for nucleophilic substitution of chlorobenzene to proceed at a reasonable rate.

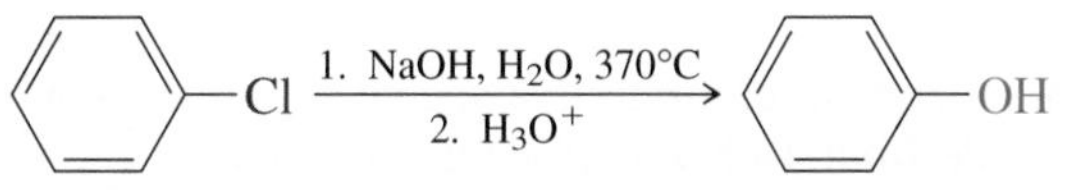
Cl
1. NaOH, H_2O, 370°C
2. H_3O^+
OH

Chlorobenzene   Phenol (97%)

One group of aryl halides that do undergo nucleophilic substitution readily consists of those that bear a nitro group ortho or para to the halogen. *p*-Chloronitrobenzene reacts with sodium methoxide at 85°C to give *p*-nitroanisole. The position of the nitro group on the ring is important.

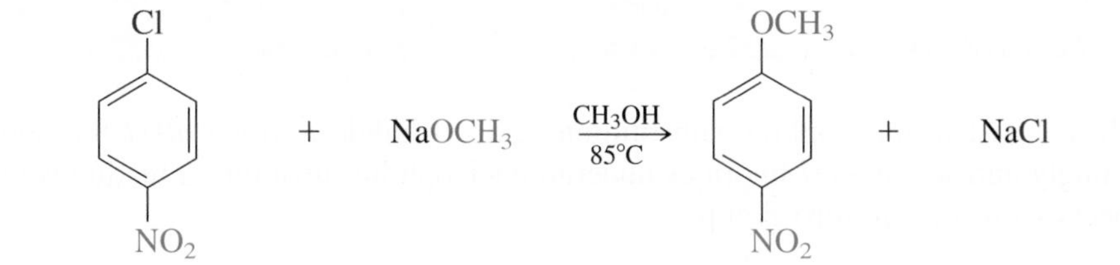
Cl
NO_2
+
$NaOCH_3$
CH_3OH
85°C
OCH_3
NO_2
+
NaCl

p-Chloronitrobenzene   Sodium methoxide   *p*-Nitroanisole (92%)   Sodium chloride

An *ortho*-nitro group exerts a comparable rate-enhancing effect. *m*-Chloronitrobenzene, although much more reactive than chlorobenzene itself, is thousands of times less reactive than either *o*- or *p*-chloronitrobenzene.

The effect of *o*- and *p*-nitro substituents is cumulative, as the following rate data demonstrate:

Increasing rate of reaction with sodium methoxide in methanol (50°C)

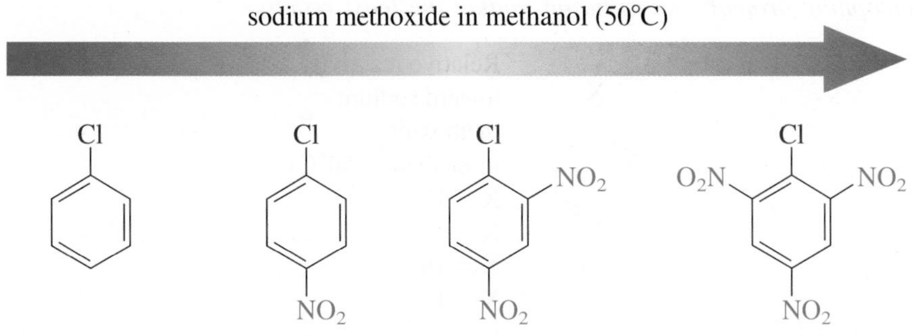
Cl
Cl
NO_2
Cl
NO_2
NO_2
O_2N
Cl
NO_2
NO_2

Chlorobenzene	1-Chloro-4-nitrobenzene	1-Chloro-2,4-dinitrobenzene	2-Chloro-1,3,5-trinitrobenzene
Relative rate: 1.0	7×10^{10}	2.4×10^{15}	(too fast to measure)

Aryl halides that bear *o*- or *p*-nitro groups (or both) are reactive enough to undergo nucleophilic substitution even with neutral nucleophiles such as ammonia.

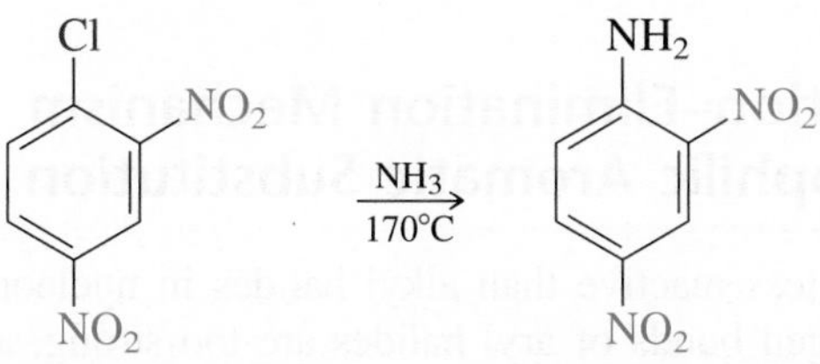
Cl
NO_2
NO_2
NH_3
170°C
NH_2
NO_2
NO_2

1-Chloro-2,4-dinitrobenzene   2,4-Dinitroaniline (68–76%)

Problem 12.28

Write the structure of the expected product from the reaction of 1-chloro-2,4-dinitrobenzene with each of the following reagents:

(a) CH_3CH_2ONa

(b) $C_6H_5CH_2SNa$

(c) CH_3NH_2

Sample Solution (a) Sodium ethoxide is a source of the nucleophile $CH_3CH_2O^-$, which displaces chloride from 1-chloro-2,4-dinitrobenzene.

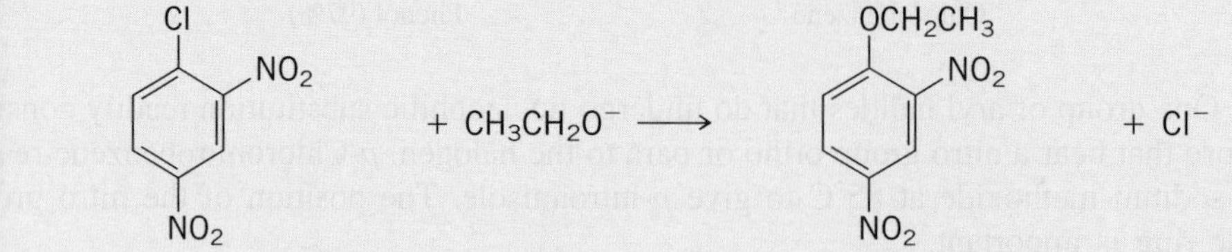
Cl
NO2
NO2
+ CH3CH2O−
OCH2CH3
NO2
NO2
+ Cl−

1-Chloro-2,4-dinitrobenzene Ethoxide anion 1-Ethoxy-2,4-dinitrobenzene

In contrast to nucleophilic substitution in alkyl halides, where *alkyl* fluorides are exceedingly unreactive, *aryl* fluorides undergo nucleophilic substitution readily when the ring bears an *o*- or a *p*-nitro group.

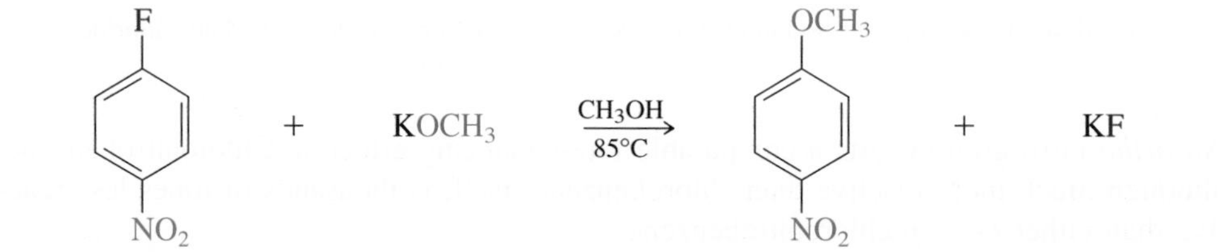
F
NO2
+
KOCH3
CH3OH
85°C
OCH3
NO2
+
KF

p-Fluoronitrobenzene Potassium methoxide *p*-Nitroanisole (93%) Potassium fluoride

Indeed, the order of leaving-group reactivity in nucleophilic aromatic substitution is the opposite of that seen in aliphatic substitution. *Fluoride is the most reactive leaving group in nucleophilic aromatic substitution, iodide the least reactive.*

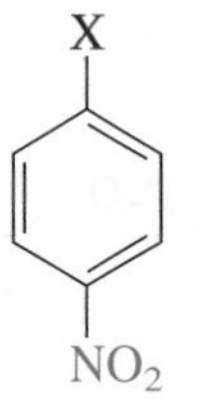
X
NO2

Relative reactivity toward sodium methoxide in methanol (50°C):

X = F	312
X = Cl	1.0
X = Br	0.8
X = I	0.4

Kinetic studies of these reactions reveal that they follow a second-order rate law:

$$\text{Rate} = k[\text{Aryl halide}][\text{Nucleophile}]$$

which suggests a bimolecular rate-determining step. In this case, then, we look for a mechanism in which both the aryl halide and the nucleophile are involved in the slowest step. Such a mechanism is described in the following section.

12.21 The Addition–Elimination Mechanism of Nucleophilic Aromatic Substitution

Aryl halides are much less reactive than alkyl halides in nucleophilic substitution reactions. The carbon–halogen bonds of aryl halides are too strong, and aryl cations are too high in energy, to permit aryl halides to ionize readily in S_N1-type processes. Furthermore, as Figure 12.7 shows, the optimal transition-state geometry required for S_N2 processes

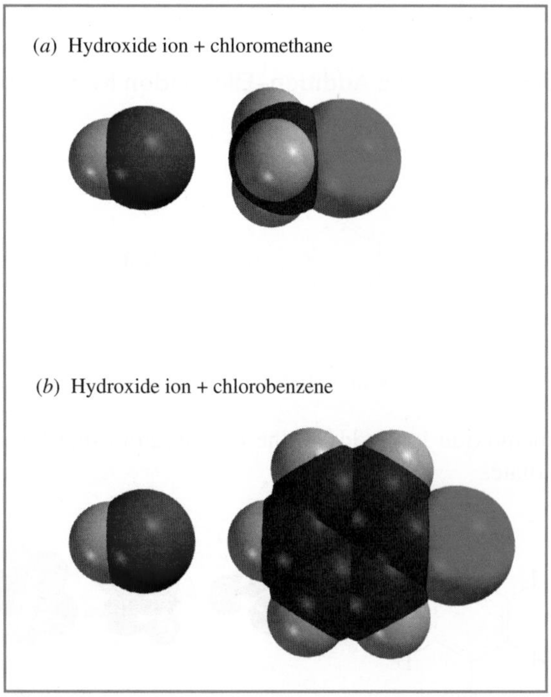
(a) Hydroxide ion + chloromethane
(b) Hydroxide ion + chlorobenzene

Figure 12.7

Nucleophilic substitution, with inversion of configuration, is blocked by the benzene ring of an aryl halide. (*a*) *Alkyl halide:* The new bond is formed by attack of the nucleophile at carbon from the side opposite the bond to the leaving group. Inversion of configuration is observed. (*b*) *Aryl halide:* The aromatic ring blocks the approach of the nucleophile to carbon at the side opposite the bond to the leaving group. Inversion of configuration is impossible.

cannot be achieved. Nucleophilic attack from the side opposite the carbon–halogen bond is blocked by the aromatic ring.

The generally accepted mechanism for nucleophilic aromatic substitution in nitro-substituted aryl halides, illustrated for the reaction of *p*-fluoronitrobenzene with sodium methoxide, is outlined in Mechanism 12.6. It is a two-step **addition–elimination mechanism,** in which addition of the nucleophile to the aryl halide is followed by elimination of the halide leaving group. The mechanism is consistent with the following experimental observations:

This mechanism is sometimes called S_NAr (*substitution-nucleophilic-aromatic*).

1. *Kinetics:* Consistent with the observed second-order rate law, the rate-determining step (step 1) involves both the aryl halide and the nucleophile.
2. *Rate-enhancing effect of the nitro group:* The nucleophilic addition step is rate-determining because the aromatic character of the ring must be sacrificed to form the cyclohexadienyl anion intermediate. Only when the anionic intermediate is stabilized by the presence of a strong electron-withdrawing substituent ortho or para to the leaving group will the activation energy for its formation be low enough to provide a reasonable reaction rate. We can illustrate the stabilization that a *p*-nitro group provides by examining the resonance structures for the cyclohexadienyl anion formed from methoxide and *p*-fluoronitrobenzene:

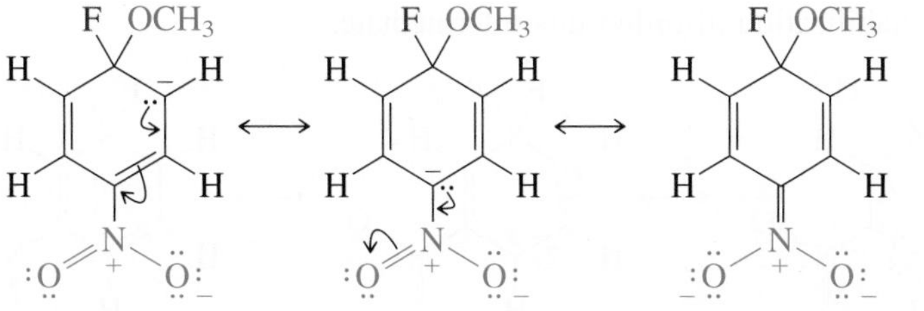
F OCH3
H
H
H
H
N
O
O

Most stable contributing structure; negative charge is on oxygen

Problem 12.29

Write the most stable contributing structure for the cyclohexadienyl anion formed by reaction of methoxide ion with *o*-fluoronitrobenzene.

Mechanism 12.6

Nucleophilic Aromatic Substitution in *p*-Fluoronitrobenzene by the Addition–Elimination Mechanism

THE OVERALL REACTION:

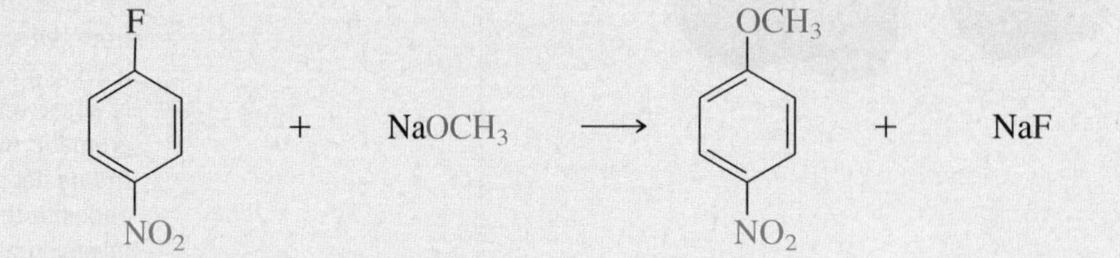

p-Fluoronitrobenzene Sodium methoxide *p*-Nitroanisole Sodium fluoride

Step 1: Addition stage. The nucleophile, in this case methoxide ion, adds to the carbon atom that bears the leaving group to give a cyclohexadienyl anion intermediate.

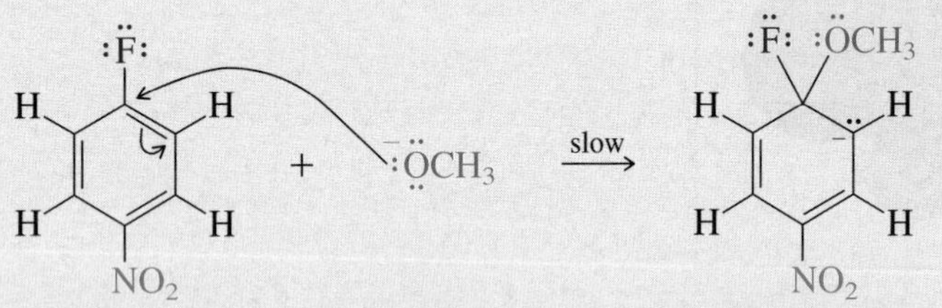

p-Fluoronitrobenzene Methoxide ion Cyclohexadienyl anion intermediate

Step 2: Elimination stage. Loss of halide from the cyclohexadienyl intermediate restores the aromaticity of the ring and gives the product of nucleophilic aromatic substitution.

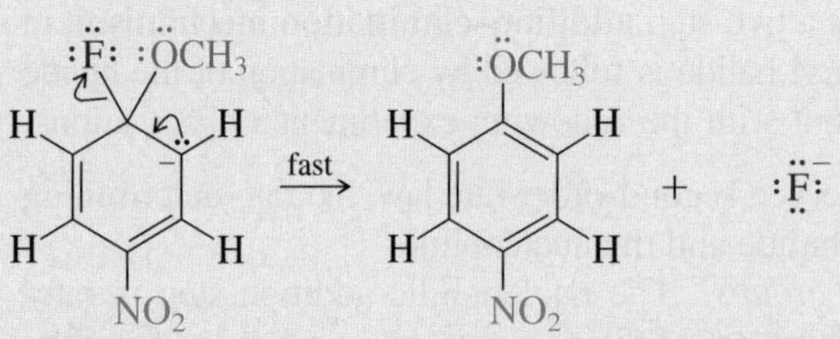

Cyclohexadienyl anion intermediate *p*-Nitroanisole Fluoride ion

m-Fluoronitrobenzene reacts with sodium methoxide 10^5 times more slowly than its ortho and para isomers. According to the resonance description, direct conjugation of the negatively charged carbon with the nitro group is not possible in the cyclohexadienyl anion intermediate from *m*-fluoronitrobenzene, and the decreased reaction rate reflects the decreased stabilization afforded this intermediate.

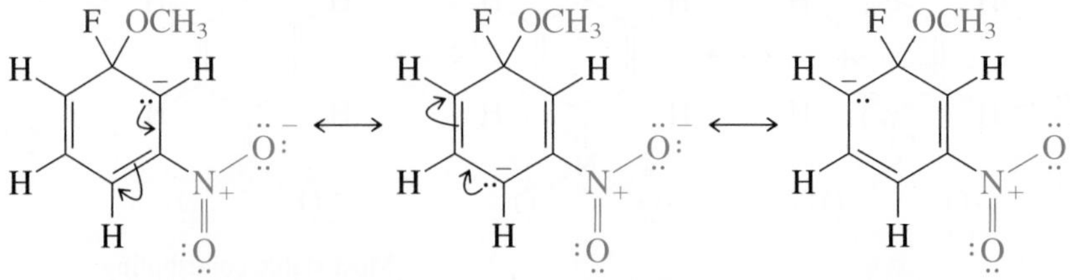

(Negative charge is restricted to carbon in all resonance contributors)

Problem 12.30

Reaction of 1,2,3-tribromo-5-nitrobenzene with sodium ethoxide in ethanol gave a single product, $C_8H_7Br_2NO_3$, in quantitative yield. Suggest a reasonable structure for this compound.

3. *Leaving-group effects:* Because aryl fluorides have the strongest carbon–halogen bond and react fastest, the rate-determining step cannot involve carbon–halogen bond cleavage. According to Mechanism 12.6 the carbon–halogen bond breaks in the rapid elimination step that follows the rate-determining addition step. The unusually high reactivity of aryl fluorides arises because fluorine is the most electronegative of the halogens, and its greater ability to attract electrons increases the rate of formation of the cyclohexadienyl anion intermediate in the first step of the mechanism.

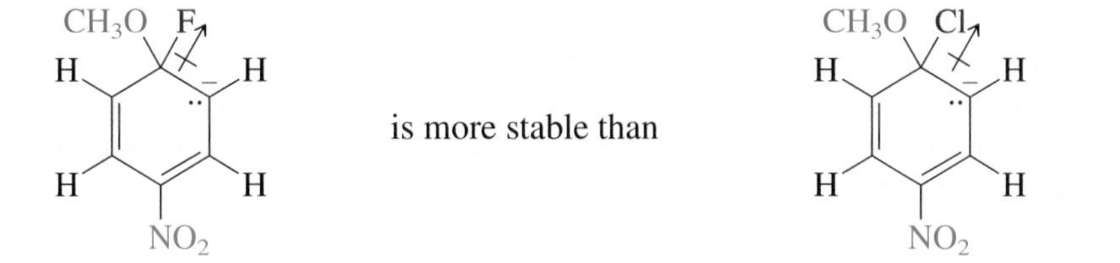

Fluorine stabilizes cyclohexadienyl anion by withdrawing electrons.

Chlorine is less electronegative than fluorine and does not stabilize cyclohexadienyl anion to as great an extent.

Aryl fluorides with two nitro groups are very reactive toward **nucleophilic aromatic substitution.** The reaction of 1-fluoro-2,4-dinitrobenzene, known as Sanger's reagent, with the amino acid phenylalanine occurs at room temperature. This reaction forms the basis of a method used in the analysis of proteins that is described in Section 25.11.

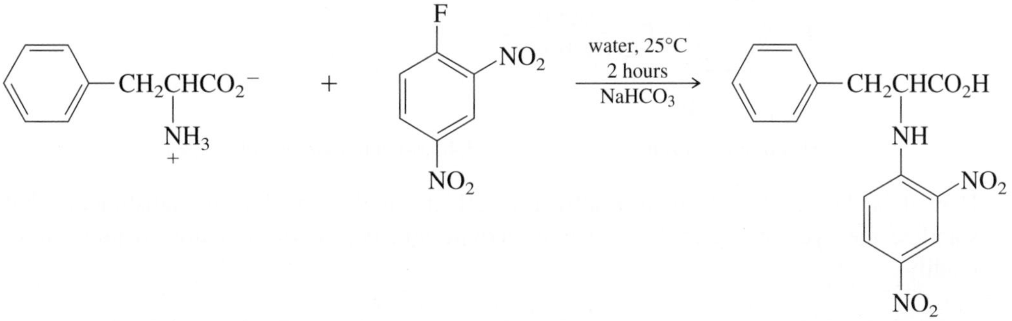

Phenylalanine

1-Fluoro-2,4-dinitrobenzene

N-(2,4-Dinitrophenyl)-phenylalanine (67%)

Before leaving this mechanistic discussion, we should mention that the addition–elimination mechanism for nucleophilic aromatic substitution illustrates a principle worth remembering. The words *activating* and *deactivating* as applied to substituent effects in organic chemistry are without meaning when they stand alone. When we say that a group is activating or deactivating, we need to specify the reaction type that is being considered. A nitro group is a strongly *deactivating* substituent in *electrophilic* aromatic substitution, where it markedly destabilizes the key cyclohexadienyl cation intermediate:

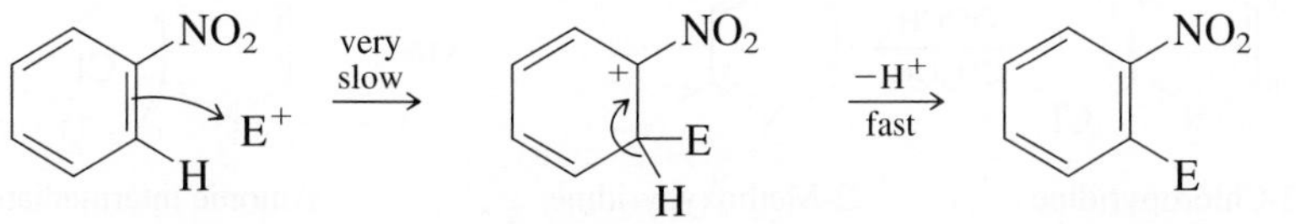

Nitrobenzene and an electrophile

Cyclohexadienyl cation intermediate; nitro group is destabilizing

Product of electrophilic aromatic substitution

A nitro group is a strongly *activating* substituent in *nucleophilic* aromatic substitution, where it stabilizes the key cyclohexadienyl anion intermediate:

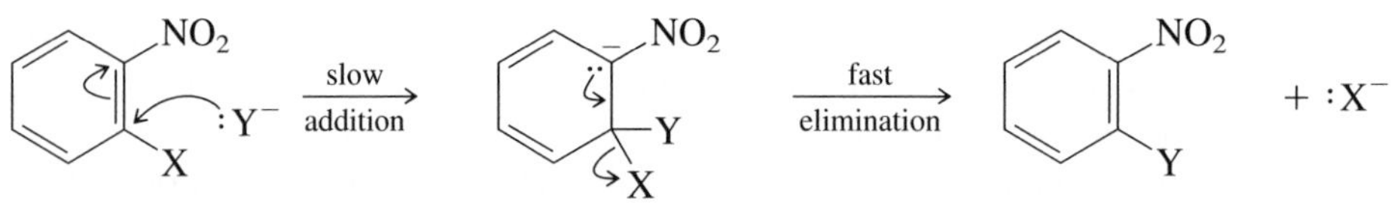

o-Halonitrobenzene (X = F, Cl, Br, or I) and a nucleophile

Cyclohexadienyl anion intermediate; nitro group is stabilizing

Product of nucleophilic aromatic substitution

A nitro group behaves the same way in both reactions: it attracts electrons. Reaction is retarded when electrons flow from the aromatic ring to the attacking species (electrophilic aromatic substitution). Reaction is facilitated when electrons flow from the attacking species to the aromatic ring (nucleophilic aromatic substitution). By being aware of the connection between reactivity and substituent effects, you will sharpen your appreciation of how chemical reactions occur.

12.22 Related Nucleophilic Aromatic Substitutions

The most common types of aryl halides in nucleophilic aromatic substitutions are those that bear *o*- or *p*-nitro substituents. Among other classes of reactive aryl halides, a few merit special consideration. One class includes highly fluorinated aromatic compounds such as hexafluorobenzene, which undergoes substitution of one of its fluorines on reaction with nucleophiles such as sodium methoxide.

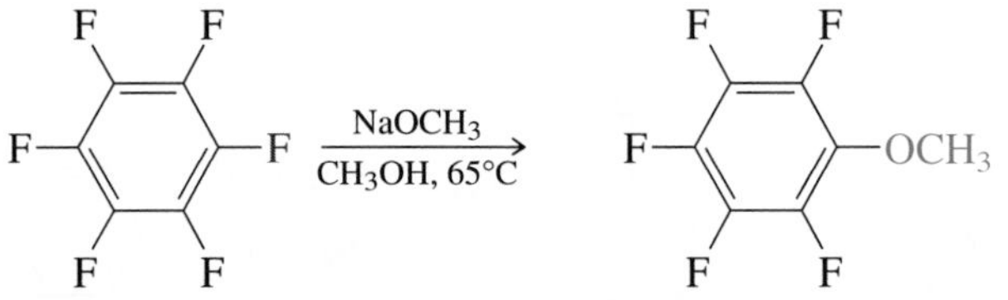

Hexafluorobenzene

2,3,4,5,6-Pentafluoroanisole (72%)

Here it is the combined electron-attracting effects of the six fluorine substituents that stabilize the cyclohexadienyl anion intermediate and permit the reaction to proceed so readily.

Problem 12.31

Write equations describing the addition–elimination mechanism for the reaction of hexafluorobenzene with sodium methoxide, clearly showing the structure of the rate-determining intermediate.

Halides derived from certain heterocyclic aromatic compounds are often quite reactive toward nucleophiles. 2-Chloropyridine, for example, reacts with sodium methoxide some 230 million times faster than chlorobenzene at 50°C.

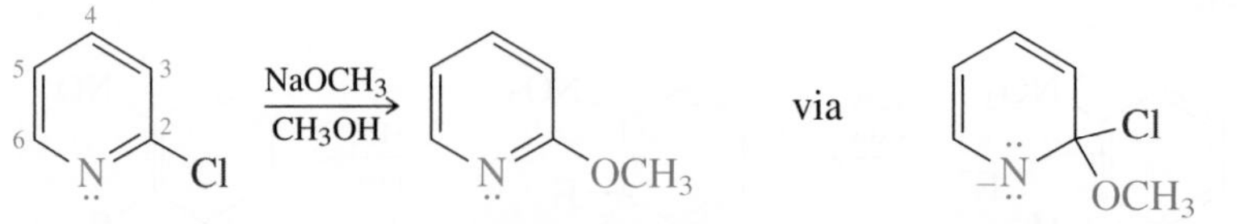

2-Chloropyridine

2-Methoxypyridine

Anionic intermediate

Again, rapid reaction is attributed to the stability of the intermediate formed in the addition step. In contrast to chlorobenzene, where the negative charge of the intermediate must be borne by carbon, the anionic intermediate in the case of 2-chloropyridine has its negative charge on nitrogen. Because nitrogen is more electronegative than carbon, the intermediate is more stable and is formed faster than the one from chlorobenzene.

Problem 12.32

Offer an explanation for the observation that 4-chloropyridine is more reactive toward nucleophiles than 3-chloropyridine.

The reactivity of 2-chloropyridines and analogous compounds can be enhanced by the presence of strongly electron-withdrawing groups. In the following example, the chlorine leaving group is activated toward nucleophilic aromatic substitution by both of the ring nitrogens and is ortho to the electron-withdrawing cyano group as well. Substitution by ammonia takes place at 0°C to give a 96% yield of product.

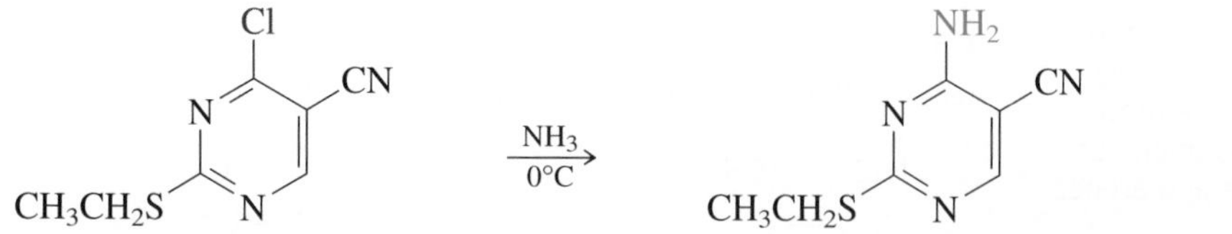

4-Chloro-5-cyano-2-(thioethyl)pyrimidine 4-Amino-5-cyano-2-(thioethyl)pyrimidine (96%)

Problem 12.33

Write contributing resonance structures to show how the negative charge in the intermediate in the preceding reaction is shared by three ring atoms and the nitrogen of the cyano group. Here is one of the contributing structures to get you started.

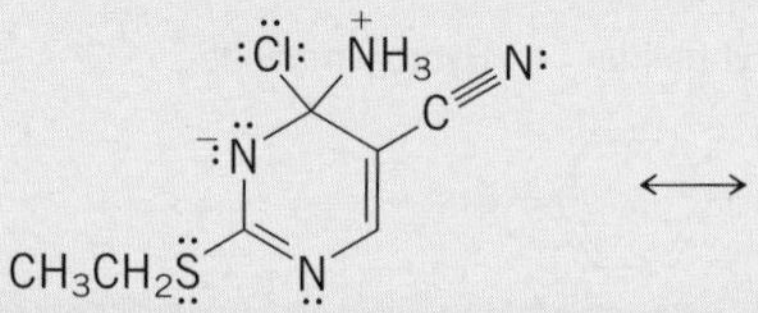

Very strong bases can bring about nucleophilic aromatic substitution by a mechanism other than the one we have been discussing. The intermediate in this other mechanism, outlined in the Descriptive Passage at the end of this chapter, may surprise you.

12.23 SUMMARY

Section 12.1 On reaction with electrophilic reagents, compounds that contain a benzene ring undergo **electrophilic aromatic substitution.** Table 12.1 in Section 12.1 and Table 12.3 in this summary give examples.

Section 12.2 The mechanism of electrophilic aromatic substitution involves two stages: bonding of the electrophile by the π electrons of the ring (slow, rate-determining), followed by rapid loss of a proton to restore the aromaticity of the ring.

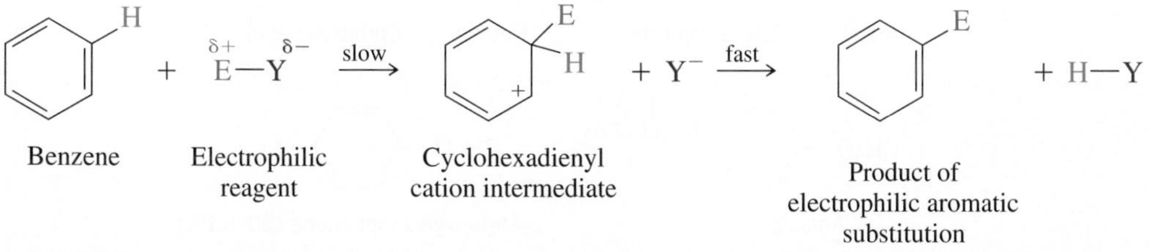

Sections 12.3–12.5 See Table 12.3.

Sections 12.6–12.7 See Tables 12.3 and 12.4.

TABLE 12.3 Representative Electrophilic Aromatic Substitution Reactions

Reaction (section) and comments	General equation and specific example
Nitration (Section 12.3) The active electrophile in the nitration of benzene and its derivatives is nitronium cation (:Ö=N⁺=Ö:). It is generated by reaction of nitric acid and sulfuric acid. Very reactive arenes—those that bear strongly activating substituents—undergo nitration in nitric acid alone.	$ArH + HNO_3 \xrightarrow{H_2SO_4} ArNO_2 + H_2O$ Arene Nitric acid Nitroarene Water 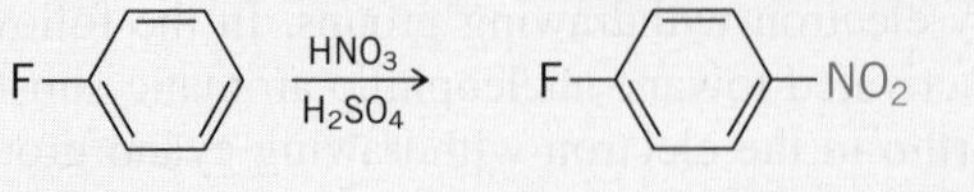 Fluorobenzene *p*-Fluoronitrobenzene (80%)
Sulfonation (Section 12.4) Sulfonic acids are formed when aromatic compounds are treated with sources of sulfur trioxide. These sources can be concentrated sulfuric acid (for very reactive arenes) or solutions of sulfur trioxide in sulfuric acid (for benzene and arenes less reactive than benzene).	$ArH + SO_3 \longrightarrow ArSO_3H$ Arene Sulfur trioxide Arenesulfonic acid 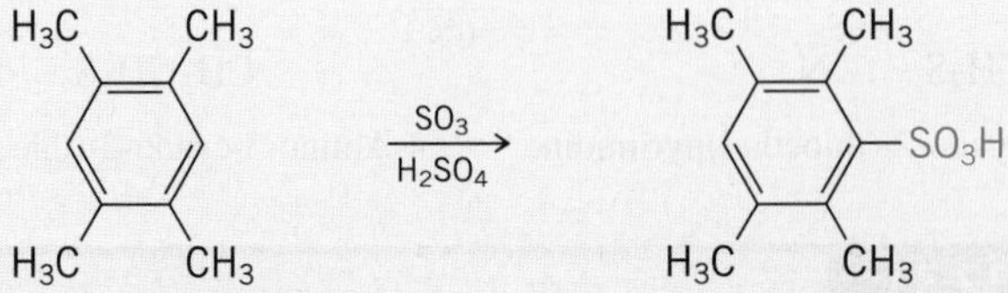 1,2,4,5-Tetramethylbenzene 2,3,5,6-Tetramethylbenzenesulfonic acid (94%)
Halogenation (Section 12.5) Chlorination and bromination of arenes are carried out by treatment with the appropriate halogen in the presence of a Lewis acid catalyst. Very reactive arenes undergo halogenation in the absence of a catalyst.	$ArH + X_2 \xrightarrow{FeX_3} ArX + HX$ Arene Halogen Aryl halide Hydrogen halide 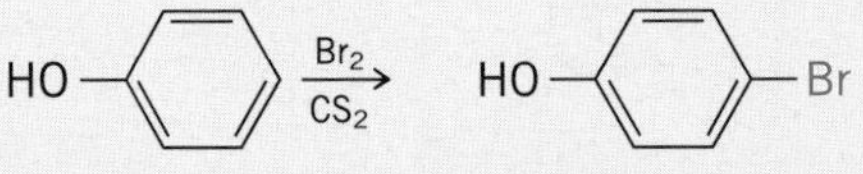 Phenol *p*-Bromophenol (80–84%)
Friedel–Crafts alkylation (Section 12.6) Carbocations, usually generated from an alkyl halide and aluminum chloride, alkylate the aromatic ring. The arene must be at least as reactive as a halobenzene. Rearrangements can occur, especially with primary alkyl halides.	$ArH + RX \xrightarrow{AlCl_3} ArR + HX$ Arene Alkyl halide Alkylarene Hydrogen halide 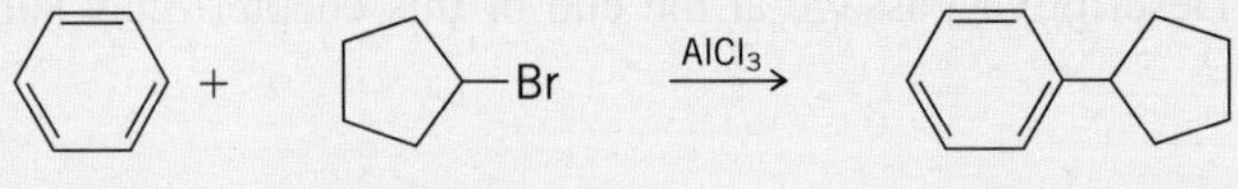 Benzene Cyclopentyl bromide Cyclopentylbenzene (54%)
Friedel–Crafts acylation (Section 12.7) Acyl cations (acylium ions) generated by treating an acyl chloride or acid anhydride with aluminum chloride acylate aromatic rings to yield ketones. The arene must be at least as reactive as a halobenzene. Acyl cations are relatively stable, and do not rearrange.	$ArH + R\overset{O}{\overset{\|}{C}}Cl \xrightarrow{AlCl_3} Ar\overset{O}{\overset{\|}{C}}R + HCl$ Arene Acyl chloride Ketone Hydrogen chloride $ArH + R\overset{O}{\overset{\|}{C}}O\overset{O}{\overset{\|}{C}}R \xrightarrow{AlCl_3} Ar\overset{O}{\overset{\|}{C}}R + R\overset{O}{\overset{\|}{C}}OH$ Arene Acid anhydride Ketone Carboxylic acid 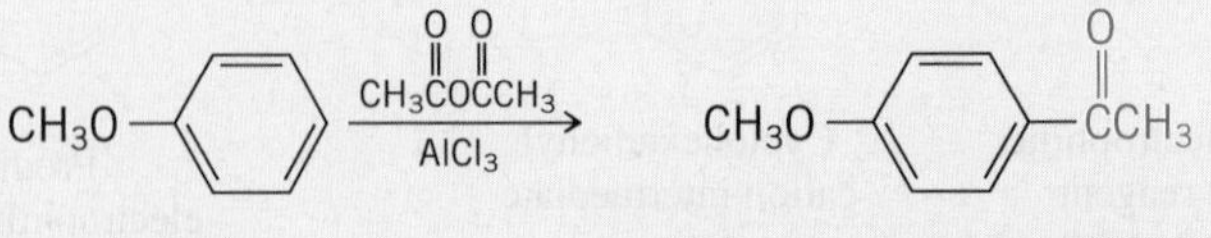 Anisole *p*-Methoxyacetophenone (90–94%)

TABLE 12.4 Limitations on Friedel–Crafts Reactions

1. The organic halide that reacts with the arene must be an alkyl halide (Section 12.6) or an acyl halide (Section 12.7).	*These will react with benzene under Friedel–Crafts conditions:* 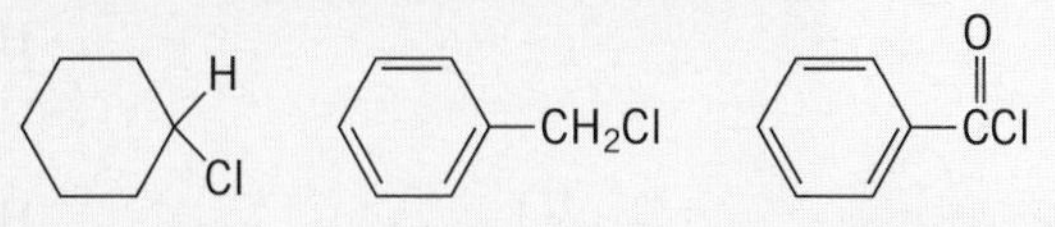 Alkyl halide, Benzylic halide, Acyl halide
Vinylic halides and aryl halides do not form carbocations under conditions of the Friedel–Crafts reaction and so cannot be used in place of an alkyl halide or an acyl halide.	*These will not react with benzene under Friedel–Crafts conditions:* 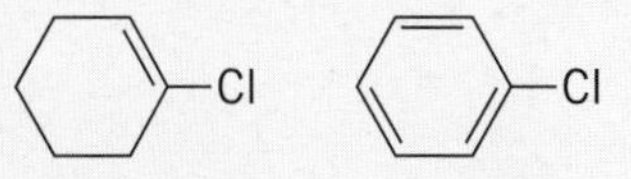 Vinylic halide, Aryl halide
2. Rearrangement of alkyl groups can occur (Section 12.6).	Rearrangement is especially prevalent with primary alkyl halides of the type RCH_2CH_2X and R_2CHCH_2X. Aluminum chloride induces ionization with rearrangement to give a more stable carbocation. Benzylic halides and acyl halides do not rearrange.
3. Strongly deactivated aromatic rings do not undergo Friedel–Crafts alkylation or acylation (Section 12.16). Friedel–Crafts alkylations and acylations fail when applied to compounds of the following type, where EWG is a strongly electron-withdrawing group: 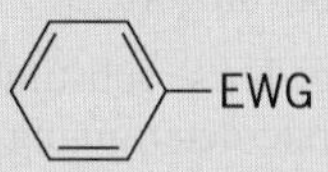	*EWG:* $-CF_3$, $-NO_2$, $-SO_3H$, $-C{\equiv}N$, $-\overset{O}{\overset{\Vert}{C}}H$, $-\overset{O}{\overset{\Vert}{C}}R$, $-\overset{O}{\overset{\Vert}{C}}OH$, $-\overset{O}{\overset{\Vert}{C}}OR$, $-\overset{O}{\overset{\Vert}{C}}Cl$
4. It is sometimes difficult to limit Friedel–Crafts alkylation to monoalkylation. Only monoacylation occurs during Friedel–Crafts acylation.	The first *alkyl* group that goes on makes the ring more reactive toward further substitution because alkyl groups are activating substituents. Monoacylation is possible because the first *acyl* group to go on is strongly electron-withdrawing and deactivates the ring toward further substitution.

Section 12.8 Friedel–Crafts acylation, followed by Clemmensen or Wolff–Kishner reduction is a standard sequence used to introduce a primary alkyl group onto an aromatic ring.

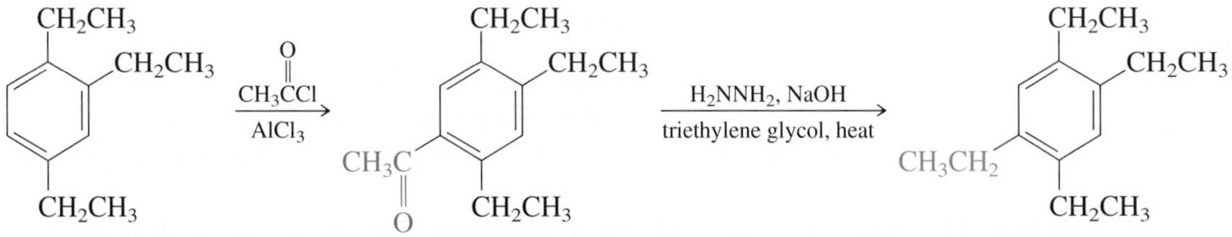

1,2,4-Triethylbenzene — 2,4,5-Triethylacetophenone (80%) — 1,2,4,5-Tetraethylbenzene (73%)

Section 12.9 Substituents on an aromatic ring can influence both the *rate* and *regioselectivity* of electrophilic aromatic substitution. Substituents are classified as *activating* or *deactivating* according to whether they cause electrophilic aromatic substitution to occur more rapidly or less rapidly than benzene. With respect to regioselectivity, substituents are either *ortho, para-directing* or *meta-directing.* A methyl group is activating and ortho, para-directing. A trifluoromethyl group is deactivating and meta-directing.

Sections 12.10–12.14 How substituents control rate and regioselectivity in electrophilic aromatic substitution results from their effect on carbocation stability. An electron-releasing substituent stabilizes the cyclohexadienyl cation intermediates leading to ortho and para substitution more than meta.

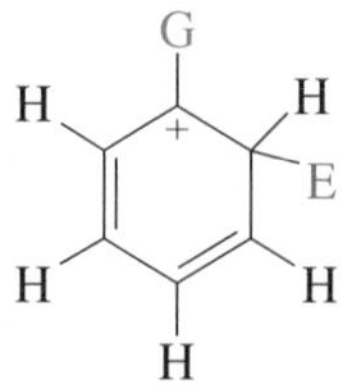

Stabilized when G is electron-releasing

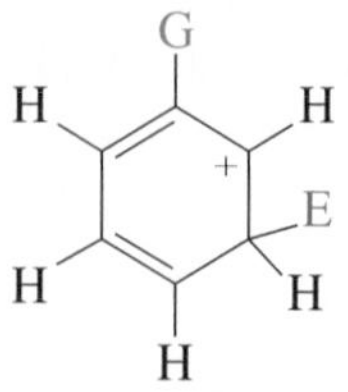

Less stabilized when G is electron-releasing

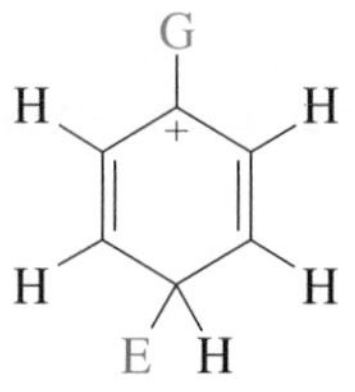

Stabilized when G is electron-releasing

Conversely, an electron-withdrawing substituent destabilizes the cyclohexadienyl cations leading to ortho and para substitution more than meta. Thus, meta substitution predominates.

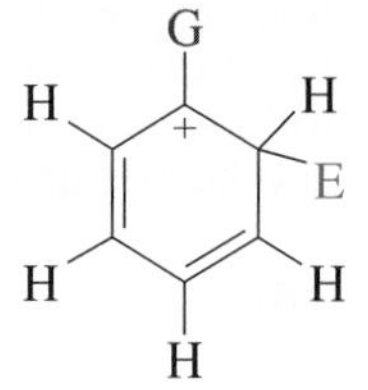

Destabilized when G is electron-withdrawing

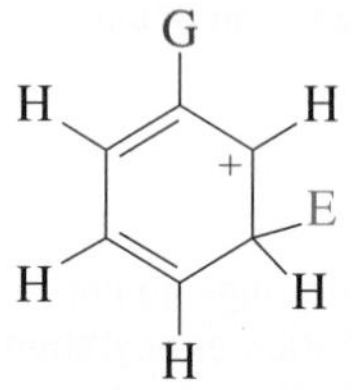

Less destabilized when G is electron-withdrawing

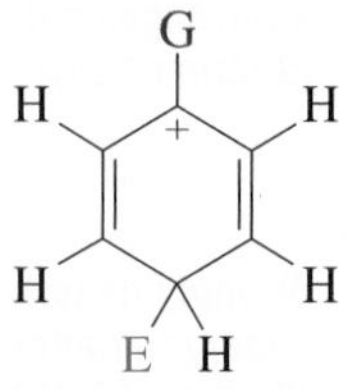

Destabilized when G is electron-withdrawing

Substituents can be arranged into three major categories:

1. **Activating and ortho, para-directing:** These substituents stabilize the cyclohexadienyl cation formed in the rate-determining step. They include $-\ddot{N}R_2$, $-\ddot{O}R$, $-R$, $-Ar$, and related species. The most strongly activating members of this group are bonded to the ring by a nitrogen or oxygen atom that bears an unshared pair of electrons.
2. **Deactivating and ortho, para-directing:** The halogens are the most prominent members of this class. They withdraw electron density from all the ring positions by an inductive effect, making halobenzenes less reactive than benzene. Lone-pair electron donation stabilizes the cyclohexadienyl cation intermediates for ortho and para substitution more than those for meta substitution.
3. **Deactivating and meta-directing:** These substituents are strongly electron-withdrawing and destabilize carbocations. They include

$$-CF_3,\ -\overset{\overset{\displaystyle O}{\|}}{C}R,\ -C{\equiv}N,\ -NO_2$$

and related species. All the ring positions are deactivated, but because the *meta* positions are deactivated less than the ortho and para, meta substitution is favored.

Section 12.15 When two or more substituents are present on a ring, the regioselectivity of electrophilic aromatic substitution is generally controlled by the directing effect of the more powerful *activating* substituent.

Section 12.16 The order in which substituents are introduced onto a benzene ring needs to be considered in order to prepare the desired isomer in a multistep synthesis.

Section 12.17 Polycyclic aromatic hydrocarbons undergo the same kind of electrophilic aromatic substitution reactions as benzene.

Section 12.18 Heterocyclic aromatic compounds may be more reactive or less reactive than benzene. Pyridine is much less reactive than benzene, but pyrrole, furan, and thiophene are more reactive. Pyridine undergoes substitution at the carbon-3 position, whereas pyrrole, furan, and thiophene give mainly carbon-2-substituted products.

Section 12.19 Aryl halides are less reactive than alkyl halides in reactions in which C—X bond breaking is rate-determining, especially in nucleophilic substitution reactions.

When an aromatic ring is substituted with an electron-withdrawing group such as a nitro group ($X=NO_2$), nucleophilic aromatic substitution can occur by a process that involves the formation of a cyclohexadienyl anion.

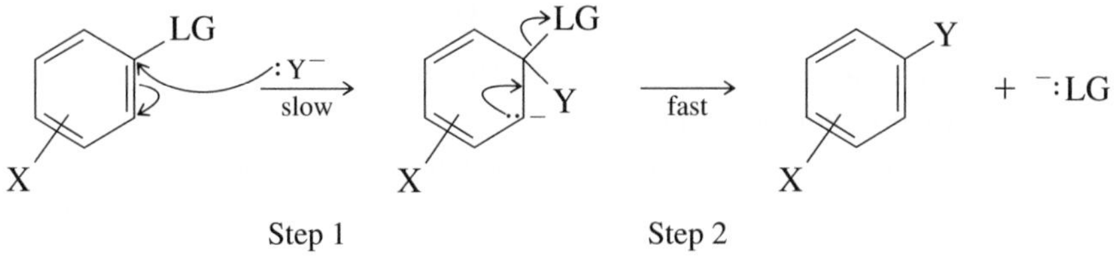

Section 12.20 Nucleophilic substitution in ArX is facilitated by the presence of a strong electron-withdrawing group, such as NO_2, ortho or para to the halogen.

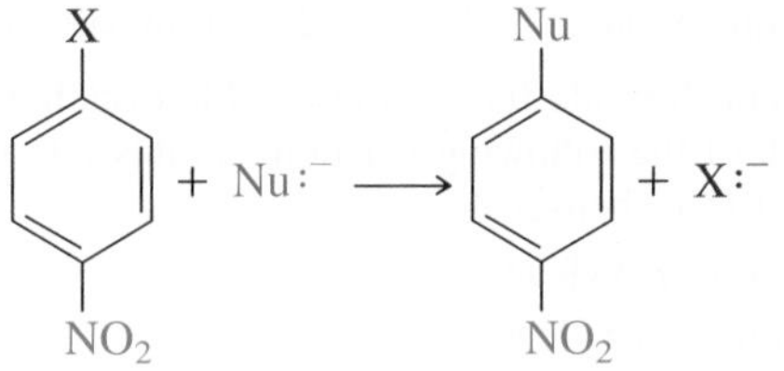

In reactions of this type, fluoride is the best leaving group of the halogens and iodide the poorest.

Section 12.21 Nucleophilic aromatic substitutions of the type just shown follow an **addition–elimination mechanism.**

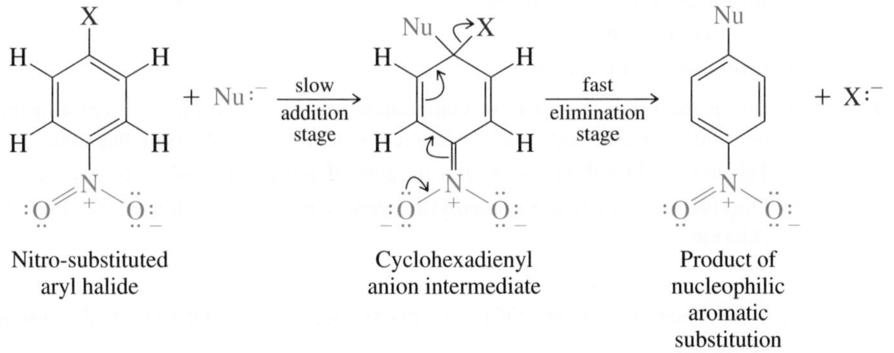

The rate-determining intermediate is a cyclohexadienyl anion and is stabilized by electron-withdrawing substituents.

Section 12.22 Other aryl halides that give stabilized anions can undergo nucleophilic aromatic substitution by the addition–elimination mechanism. Two examples are hexafluorobenzene and 2-chloropyridine.

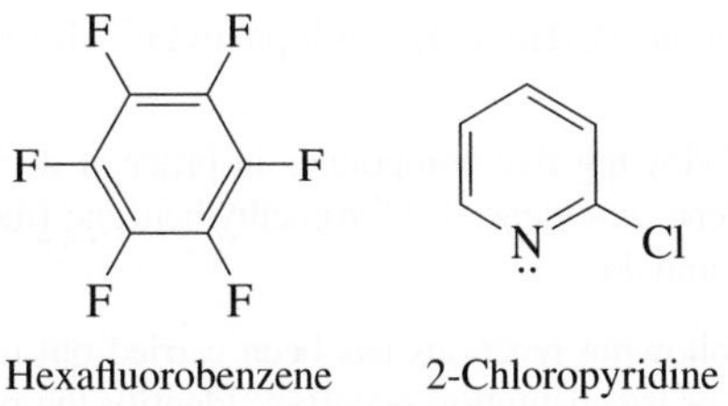

PROBLEMS

12.34 Give reagents suitable for carrying out each of the following reactions, and write the major organic products. If an ortho, para mixture is expected, show both. If the meta isomer is the expected major product, write only that isomer.

(a) Nitration of benzene

(b) Nitration of the product of part (a)

(c) Bromination of toluene
(d) Bromination of (trifluoromethyl)benzene
(e) Sulfonation of anisole
(f) Sulfonation of acetanilide ($C_6H_5NH\overset{\overset{O}{\|}}{C}CH_3$)
(g) Chlorination of bromobenzene
(h) Friedel–Crafts alkylation of anisole with benzyl chloride
(i) Friedel–Crafts acylation of benzene with benzoyl chloride ($C_6H_5\overset{\overset{O}{\|}}{C}Cl$)
(j) Nitration of the product from part (i)
(k) Clemmensen reduction of the product from part (i)
(l) Wolff–Kishner reduction of the product from part (i)

12.35 Write a structural formula for the most stable cyclohexadienyl cation intermediate formed in each of the following reactions. Is this intermediate more or less stable than the one formed from benzene?
(a) Bromination of *p*-xylene
(b) Chlorination of *m*-xylene
(c) Nitration of acetophenone
(d) Friedel–Crafts acylation of anisole with acetyl chloride ($CH_3\overset{\overset{O}{\|}}{C}Cl$)
(e) Nitration of isopropylbenzene
(f) Bromination of nitrobenzene
(g) Sulfonation of furan
(h) Bromination of pyridine

12.36 In each of the following pairs of compounds choose which one will react faster with the indicated reagent, and write a chemical equation for the faster reaction:
(a) Toluene or chlorobenzene with a mixture of nitric acid and sulfuric acid
(b) Fluorobenzene or (trifluoromethyl)benzene with benzyl chloride and aluminum chloride
(c) Methyl benzoate ($C_6H_5\overset{\overset{O}{\|}}{C}OCH_3$) or phenyl acetate ($C_6H_5O\overset{\overset{O}{\|}}{C}CH_3$) with bromine in acetic acid
(d) Acetanilide ($C_6H_5NH\overset{\overset{O}{\|}}{C}CH_3$) or nitrobenzene with sulfur trioxide in sulfuric acid
(e) *p*-Dimethylbenzene (*p*-xylene) or *p*-di-*tert*-butylbenzene with acetyl chloride and aluminum chloride
(f) Benzophenone ($C_6H_5\overset{\overset{O}{\|}}{C}C_6H_5$) or biphenyl ($C_6H_5$—$C_6H_5$) with chlorine and iron(III) chloride

12.37 Arrange the following five compounds in order of decreasing rate of bromination: benzene, toluene, *o*-xylene, *m*-xylene, 1,3,5-trimethylbenzene (the relative rates are 2×10^7, 5×10^4, 5×10^2, 60, and 1).

12.38 Each of the following reactions has been carried out under conditions such that disubstitution or trisubstitution occurred. Identify the principal organic product in each case.
(a) Nitration of *p*-chlorobenzoic acid (dinitration)
(b) Bromination of aniline (tribromination)
(c) Bromination of *o*-aminoacetophenone (dibromination)
(d) Nitration of benzoic acid (dinitration)
(e) Bromination of *p*-nitrophenol (dibromination)
(f) Reaction of biphenyl with *tert*-butyl chloride and iron(III) chloride (dialkylation)
(g) Sulfonation of phenol (disulfonation)

12.39 Write equations showing how to prepare each of the following from benzene or toluene and any necessary organic or inorganic reagents. If an ortho, para mixture is formed in any step of your synthesis, assume that you can separate the two isomers.

(a) Isopropylbenzene
(b) *p*-Isopropylbenzenesulfonic acid
(c) 2-Bromo-2-phenylpropane
(d) 4-*tert*-Butyl-2-nitrotoluene
(e) *m*-Chloroacetophenone
(f) *p*-Chloroacetophenone
(g) 3-Bromo-4-methylacetophenone
(h) 2-Bromo-4-ethyltoluene
(i) 1-Bromo-3-nitrobenzene
(j) 1-Bromo-2,4-dinitrobenzene
(k) 3-Bromo-5-nitrobenzoic acid
(l) 2-Bromo-4-nitrobenzoic acid
(m) Diphenylmethane
(n) 1-Phenyloctane
(o) 1-Phenyl-1-octene
(p) 1-Phenyl-1-octyne
(q) 1,4-Di-*tert*-butyl-1,4-cyclohexadiene

12.40 Write equations showing how you could prepare each of the following from anisole and any necessary organic or inorganic reagents. If an ortho, para mixture is formed in any step of your synthesis, assume that you can separate the two isomers.

(a) *p*-Methoxybenzenesulfonic acid
(b) 2-Bromo-4-nitroanisole
(c) 4-Bromo-2-nitroanisole
(d) *p*-Methoxystyrene

12.41 How many products are capable of being formed from toluene in each of the following reactions?

(a) Mononitration (HNO_3, H_2SO_4, 40°C).
(b) Dinitration (HNO_3, H_2SO_4, 80°C).
(c) Trinitration (HNO_3, H_2SO_4, 110°C). The explosive TNT (trinitrotoluene) is the major product obtained on trinitration of toluene. Which trinitrotoluene isomer is TNT?

12.42 The most stable resonance contributor for the acylium ion is one in which the positive charge is on oxygen rather than carbon; however, the addition of an acylium ion to benzene in the Friedel–Crafts reaction takes place at the carbon. Consider the structure of the cyclohexadienyl cation that would result if the addition of acylium ion took place on oxygen, and explain why this pathway is unfavorable.

$$CH_3CH_2\overset{+}{C}{=}\ddot{O}: \quad \longleftrightarrow \quad CH_3CH_2C{\equiv}\overset{+}{O}:$$

12.43 Friedel–Crafts acylation of the individual isomers of xylene with acetyl chloride and aluminum chloride yields a single product, different for each xylene isomer, in high yield in each case. Write the structures of the products of acetylation of *o*-, *m*-, and *p*-xylene.

12.44 Reaction of benzanilide ($C_6H_5NH\overset{O}{\overset{\|}{C}}C_6H_5$) with chlorine in acetic acid yields a mixture of two monochloro derivatives formed by electrophilic aromatic substitution. Suggest reasonable structures for these two isomers.

12.45 Each of the following reactions has been reported in the chemical literature and gives a predominance of a single product in synthetically acceptable yield. Write the structure of the product. Only monosubstitution is involved in each case, unless otherwise indicated.

(a) 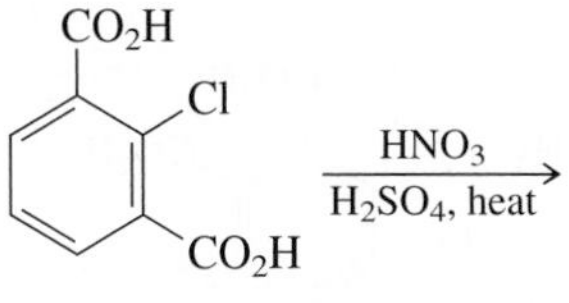

(d) 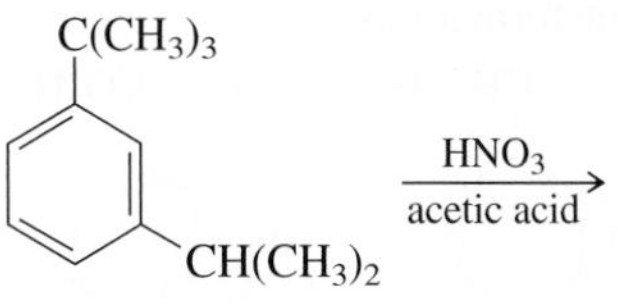

(b) 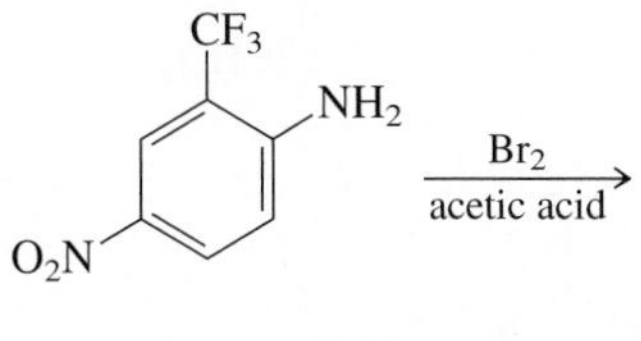

(e) 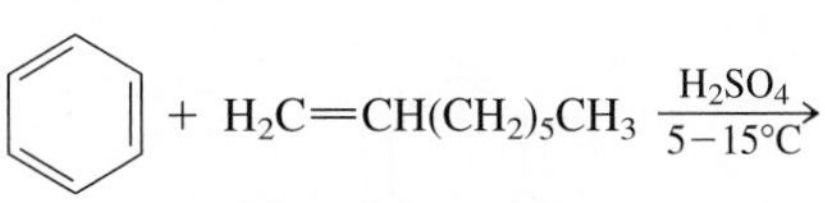

(c) 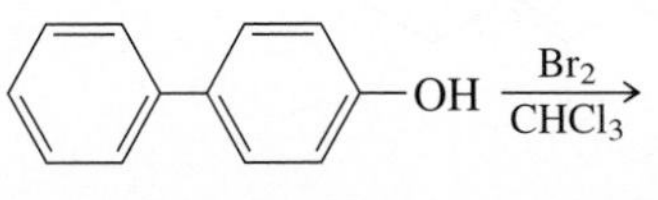

(f) 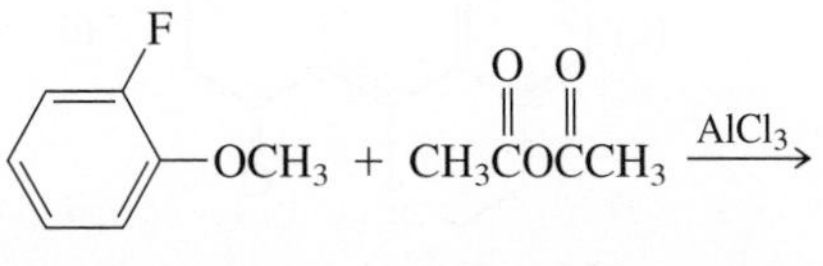

(g) 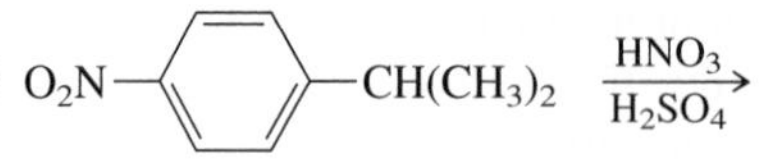

(k) 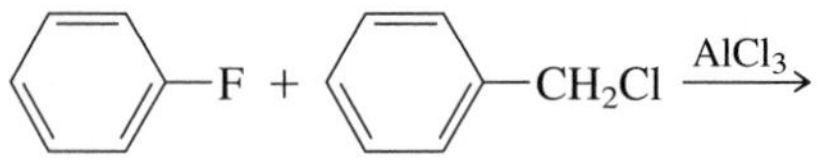

(h) 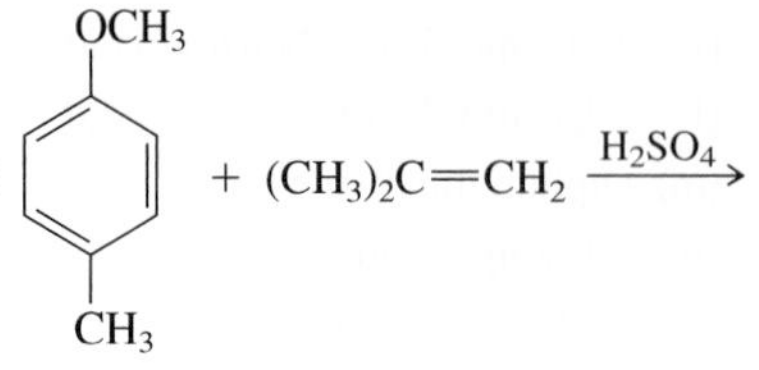

(l) 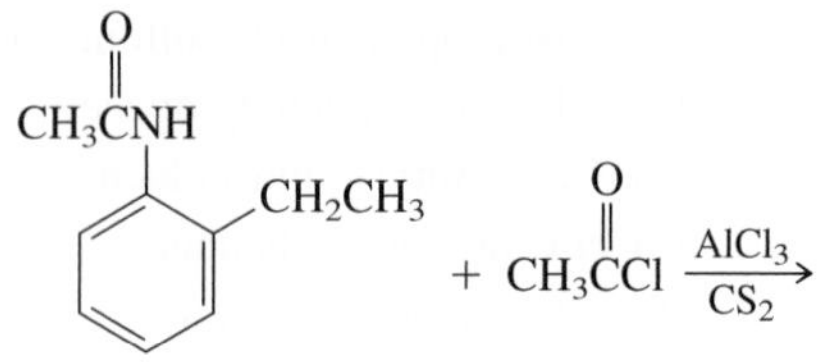

(i) 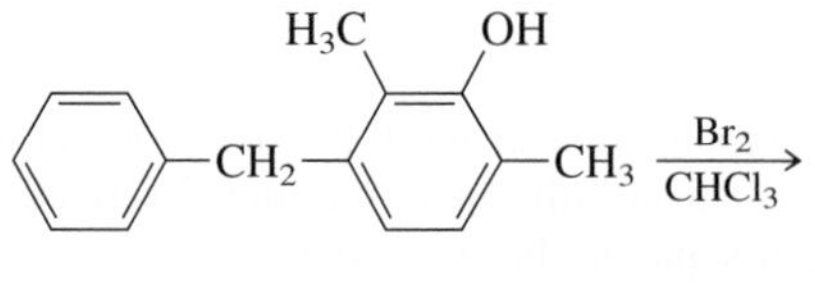

(m) 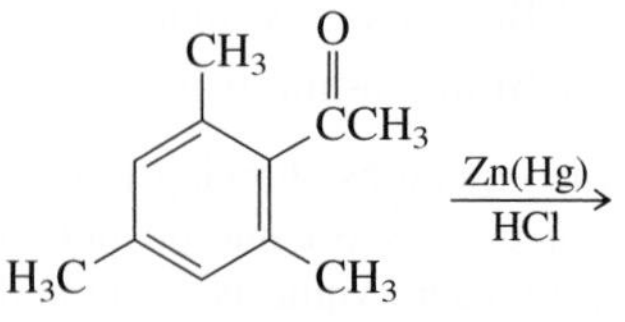

(j) 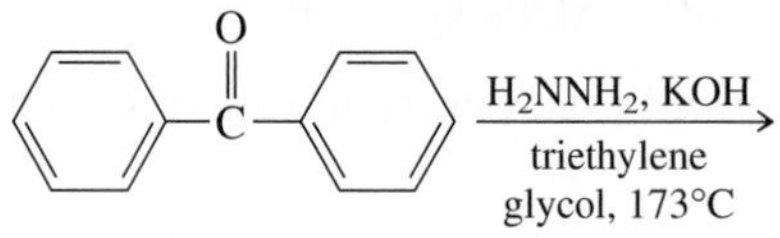

(n) 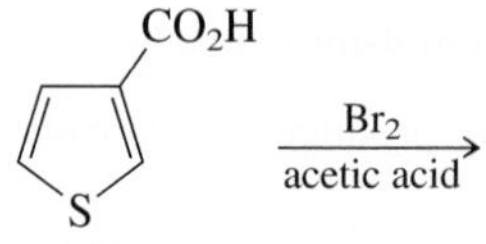

12.46 What combination of acyl chloride or acid anhydride and arene would you choose to prepare each of the following compounds by Friedel–Crafts acylation?

(a) $C_6H_5\overset{O}{\overset{\|}{C}}CH_2C_6H_5$

(b) 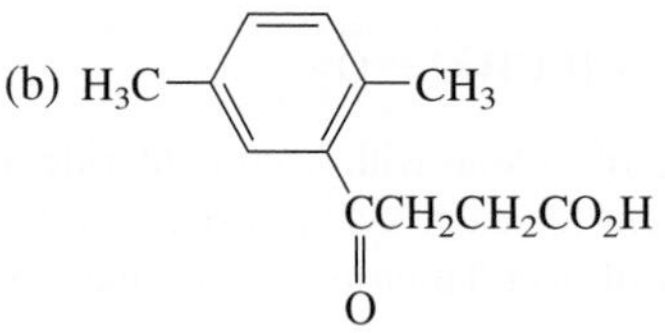

(c) 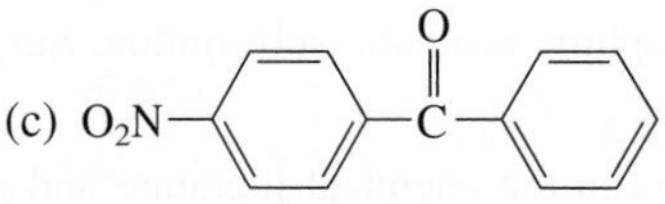

(d) 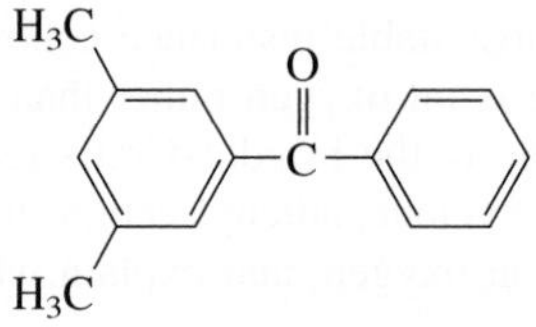

(e) 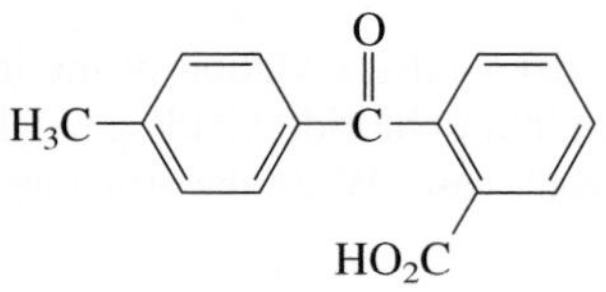

12.47 Suggest a suitable series of reactions for carrying out each of the following synthetic transformations:

(a) 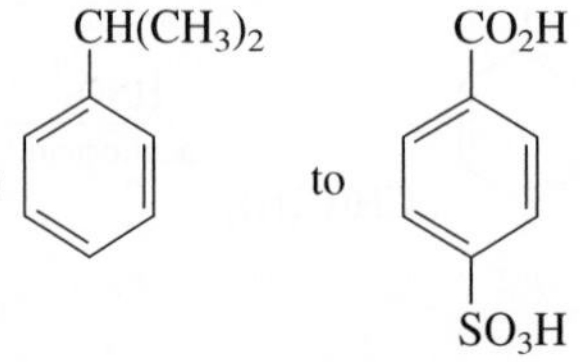

(b) 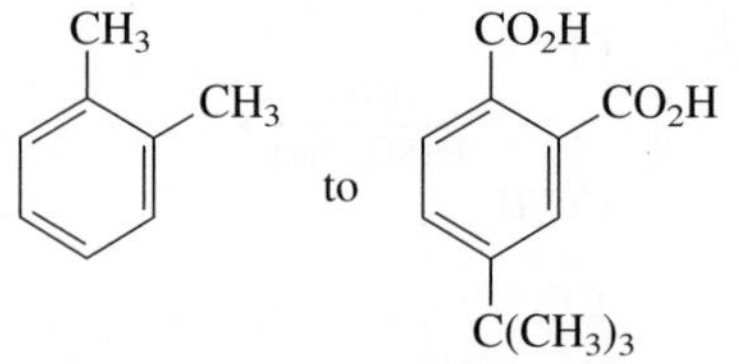

(c) 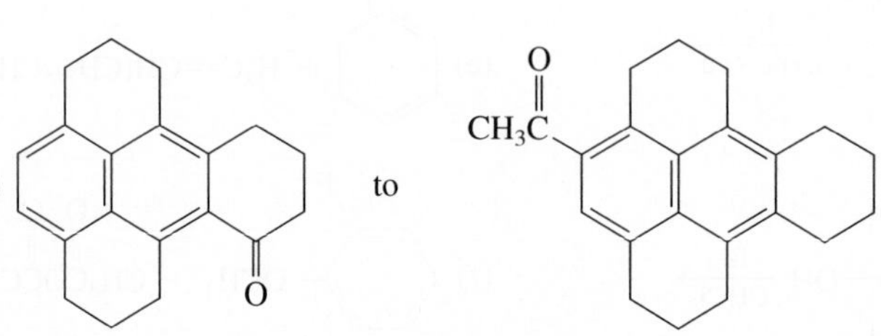

(d) 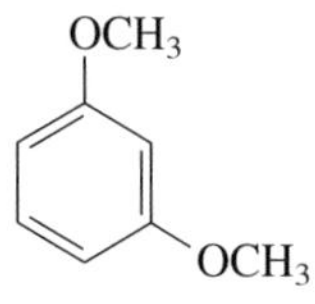 to 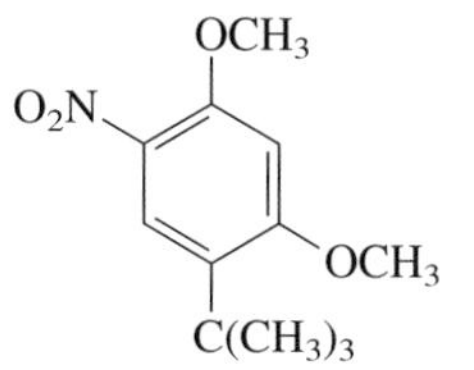

12.48 A standard synthetic sequence for building a six-membered cyclic ketone onto an existing aromatic ring is shown in outline as follows. Specify the reagents necessary for each step.

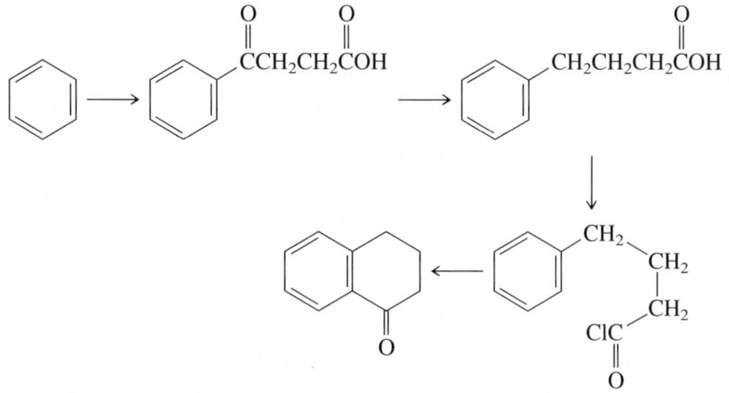

12.49 Each of the compounds indicated undergoes an intramolecular Friedel–Crafts acylation reaction to yield a cyclic ketone. Write the structure of the expected product in each case.

(a) 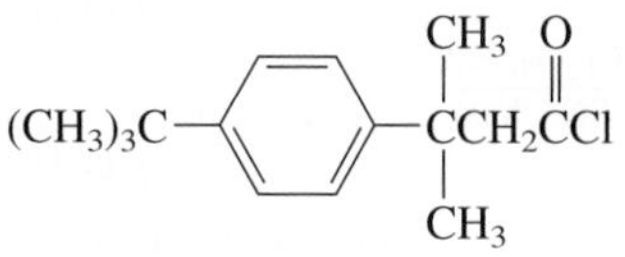

(c) 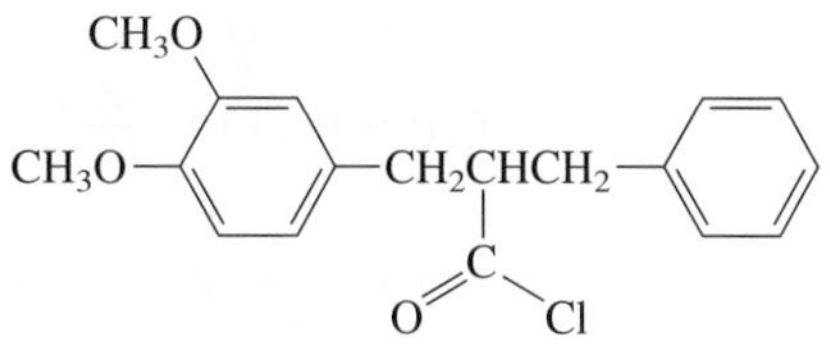

(b) 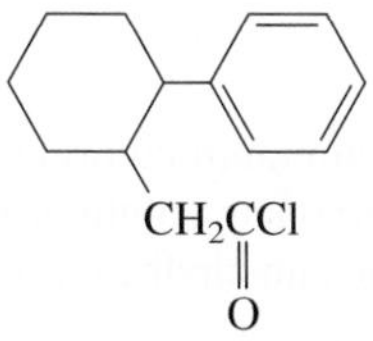

12.50 Of the groups shown, which is the most likely candidate for substituent X based on the partial rate factors for chlorination?

$—CF_3$ $—C(CH_3)_3$ $—Br$ $—SO_3H$ $—CH{=}O$

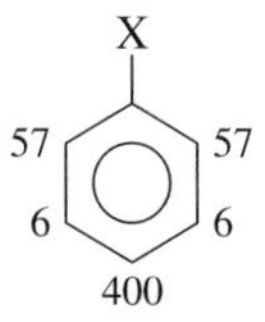

12.51 The partial rate factors for chlorination of biphenyl are as shown.

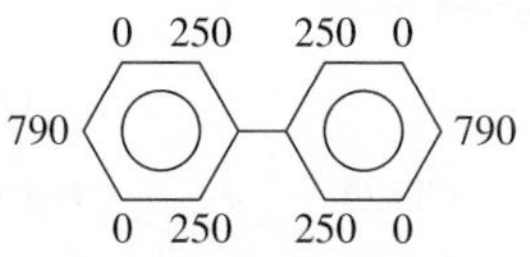

(a) What is the relative rate of chlorination of biphenyl compared with benzene?

(b) If, in a particular chlorination reaction, 10 g of *o*-chlorobiphenyl was formed, how much *p*-chlorobiphenyl would you expect to find?

12.52 Partial rate factors may be used to estimate product distributions in disubstituted benzene derivatives. The reactivity of a particular position in *o*-bromotoluene, for example, is given by the product of the partial rate factors for the corresponding position in toluene and bromobenzene. On the basis of the partial rate factor data given here for Friedel–Crafts acylation, predict the major product of the reaction of *o*-bromotoluene with acetyl chloride and aluminum chloride.

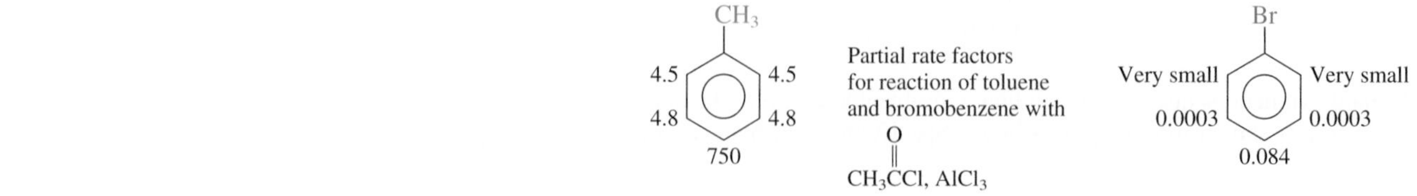

12.53 When 2-isopropyl-1,3,5-trimethylbenzene is heated with aluminum chloride (trace of HCl present) at 50°C, the major material present after 4 h is 1-isopropyl-2,4,5-trimethylbenzene. Suggest a reasonable mechanism for this isomerization.

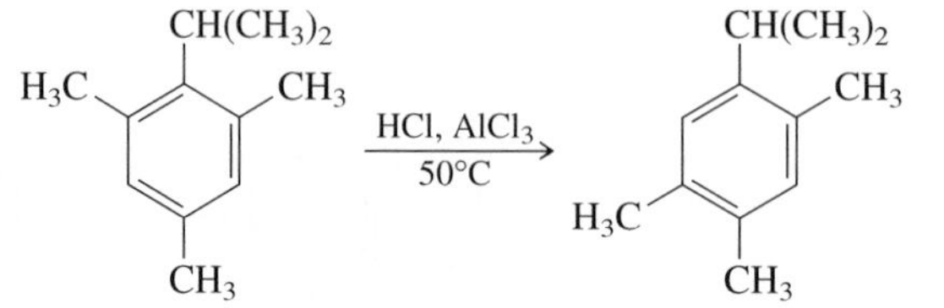

12.54 When a dilute solution of 6-phenylhexanoyl chloride in carbon disulfide was slowly added (over a period of eight days!) to a suspension of aluminum chloride in the same solvent, it yielded a product A ($C_{12}H_{14}O$) in 67% yield. Oxidation of A gave benzene-1,2-dicarboxylic acid.

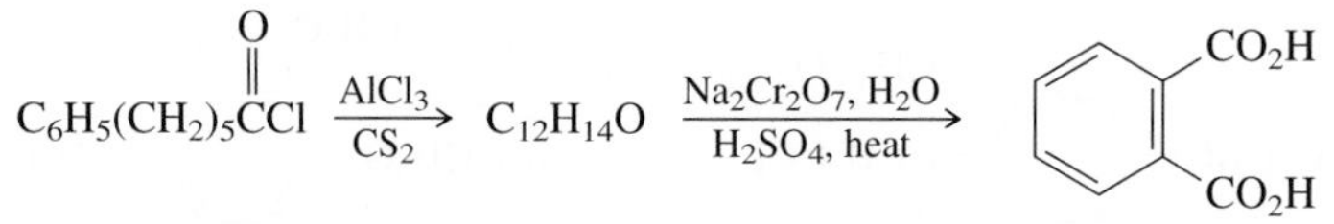

6-Phenylhexanoyl chloride — Compound A — Benzene-1,2-dicarboxylic acid

Formulate a reasonable structure for compound A.

12.55 Reaction of hexamethylbenzene with methyl chloride and aluminum chloride gave a salt A, which, on being treated with aqueous sodium bicarbonate solution, yielded compound B. Suggest a mechanism for the conversion of hexamethylbenzene to B by correctly inferring the structure of A.

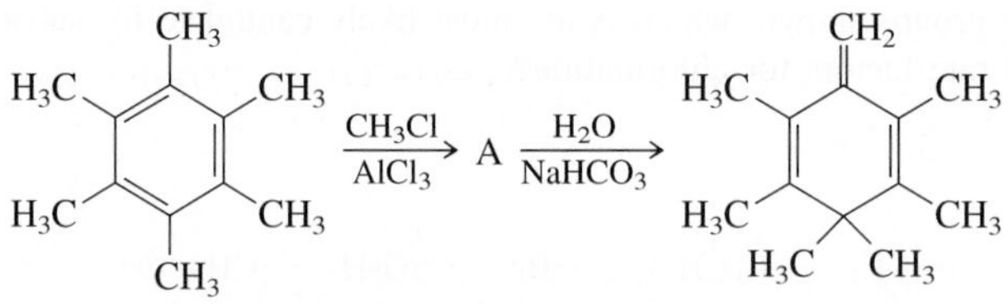

Hexamethylbenzene — Compound B

12.56 The synthesis of compound C was achieved by using compounds A and B as the source of all carbon atoms. Suggest a synthetic sequence involving no more than three steps by which A and B may be converted to C.

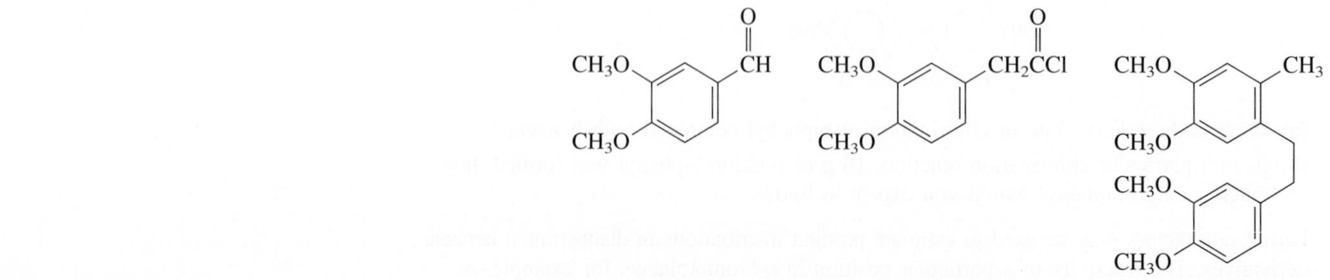

Compound A — Compound B — Compound C

12.57 When styrene is refluxed with aqueous sulfuric acid, two "styrene dimers" are formed as the major products. One of these styrene dimers is 1,3-diphenyl-1-butene; the other is 1-methyl-3-phenylindan. Suggest a reasonable mechanism for the formation of each of these compounds.

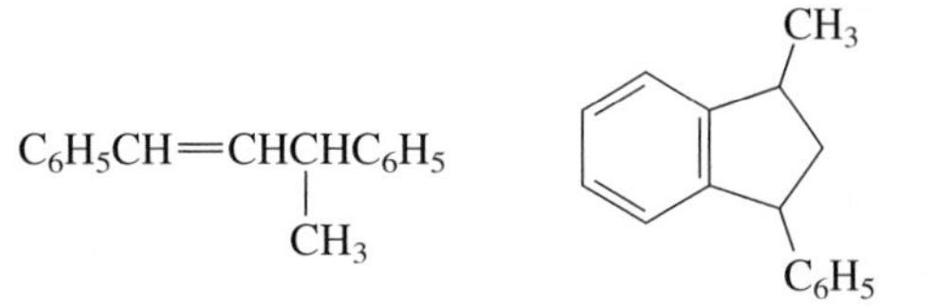

1,3-Diphenyl-1-butene   1-Methyl-3-phenylindan

12.58 Treatment of the alcohol whose structure is shown here with sulfuric acid gave as the major organic product a tricyclic hydrocarbon of molecular formula $C_{16}H_{16}$. Suggest a reasonable structure for this hydrocarbon.

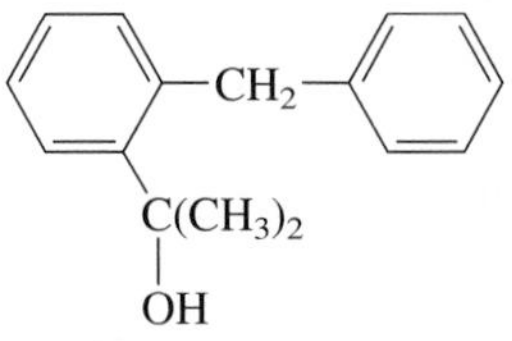

12.59 Rank the following aryl halides in increasing order of reactivity toward sodium methoxide in methanol (4 = most reactive; 1 = least reactive).

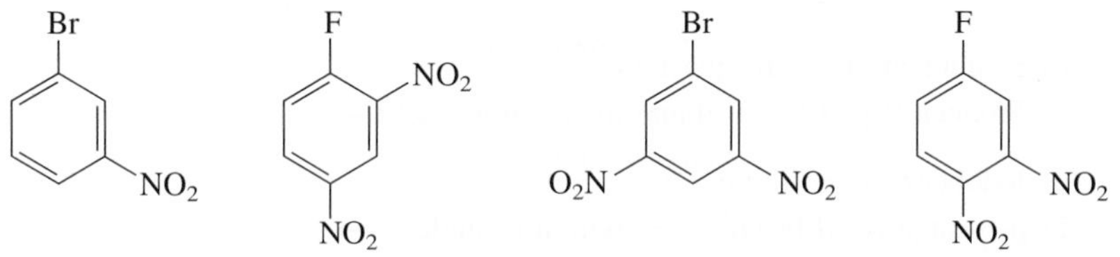

12.60 Halogens are not the only atoms or groups that can be displaced by nucleophilic aromatic substitution as the following reaction shows.

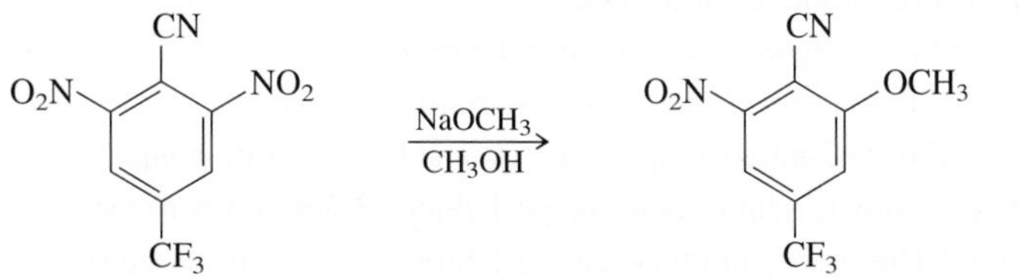

What is the leaving group in this reaction? (Formal charges have been intentionally omitted.)

:Ö=N=Ö: :Ö=N̈—Ö: :Ö=N̈—Ö:

(a)   (b)   (c)

12.61 Which is the best synthesis of 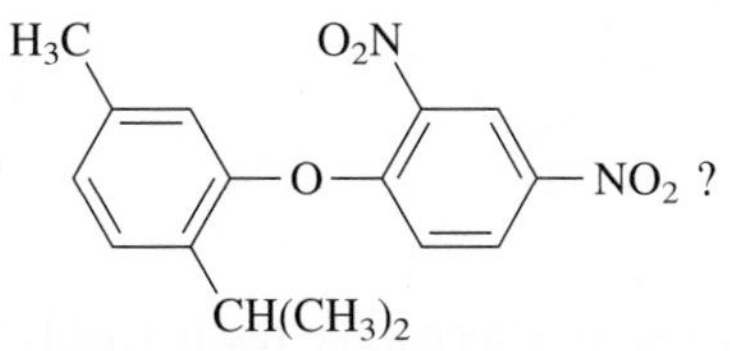 ?

(a) 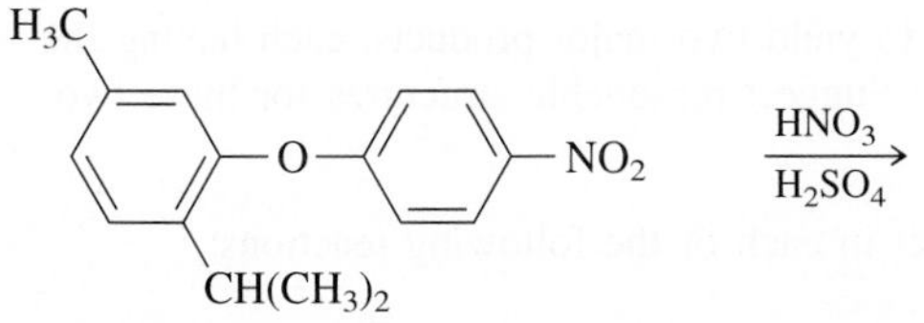

(b) 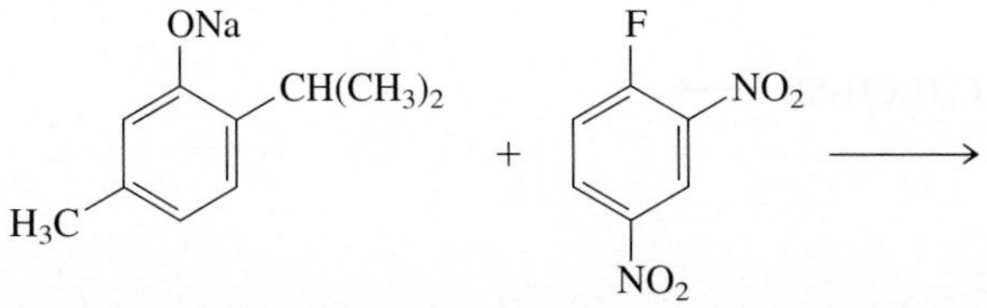

(c) 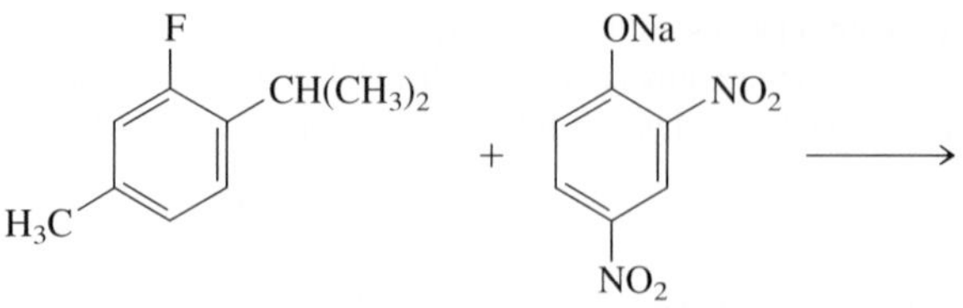

12.62 Write a structural formula for each of the following:

(a) *m*-Chlorotoluene
(b) 2,6-Dibromoanisole
(c) *p*-Fluorostyrene
(d) 4,4′-Diiodobiphenyl
(e) 2-Bromo-1-chloro-4-nitrobenzene
(f) 1-Chloro-1-phenylethane
(g) *p*-Bromobenzyl chloride
(h) 2-Chloronaphthalene
(i) 1,8-Dichloronaphthalene
(j) 9-Fluorophenanthrene

12.63 Identify the major organic product of each of the following reactions. If two regioisomers are formed in appreciable amounts, show them both.

(a) Chlorobenzene + acetyl chloride $\xrightarrow{AlCl_3}$
(b) Bromobenzene + magnesium $\xrightarrow{\text{diethyl ether}}$
(c) Product of part (b) + dilute hydrochloric acid ⟶
(d) Iodobenzene + lithium $\xrightarrow{\text{diethyl ether}}$
(e) *p*-Bromobenzyl bromide + sodium cyanide ⟶
(f) Hexafluorobenzene + sodium hydrogen sulfide ⟶

12.64 Choose the compound in each of the following pairs that reacts faster with sodium methoxide in methanol at 50°C:

(a) Chlorobenzene or *o*-chloronitrobenzene
(b) *o*-Chloronitrobenzene or *m*-chloronitrobenzene
(c) 4-Chloro-3-nitroacetophenone or 4-chloro-3-nitrotoluene
(d) 2-Fluoro-1,3-dinitrobenzene or 1-fluoro-3,5-dinitrobenzene
(e) 1,4-Dibromo-2-nitrobenzene or 1-bromo-2,4-dinitrobenzene

12.65 In each of the following reactions, the amine piperidine reacts with an aryl halide. Give the structure of the expected product.

(a) 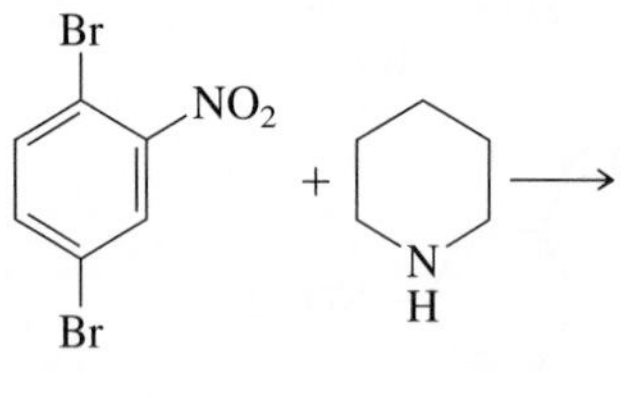

(b) 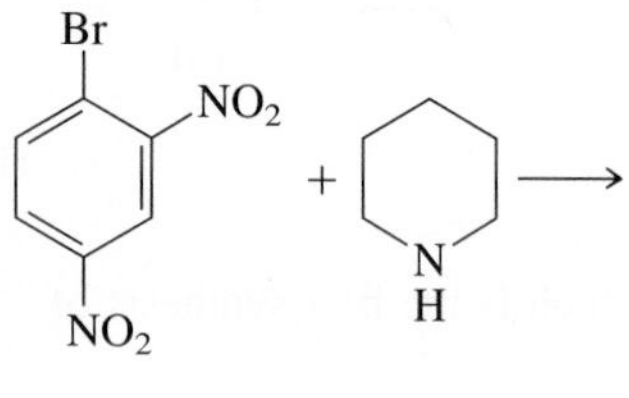

12.66 1,2,3,4,5-Pentafluoro-6-nitrobenzene reacts readily with sodium methoxide in methanol at room temperature to yield two major products, each having the molecular formula $C_7H_3F_4NO_3$. Suggest reasonable structures for these two compounds.

12.67 Predict the major organic product in each of the following reactions:

(a) 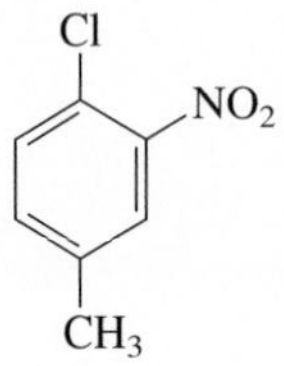 + $C_6H_5CH_2SK$ ⟶

(b) 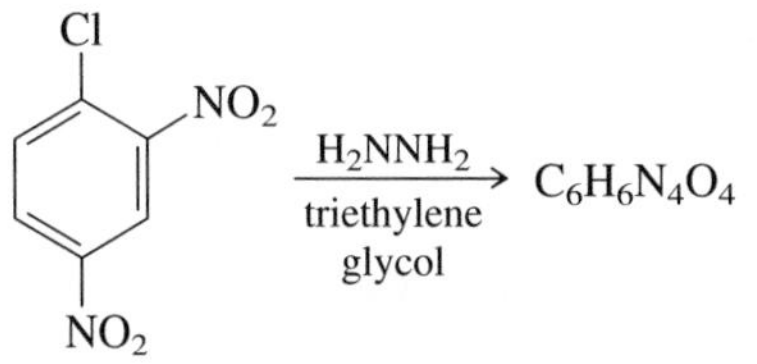
Cl
NO2
H2NNH2
triethylene glycol
C6H6N4O4
NO2

(c) 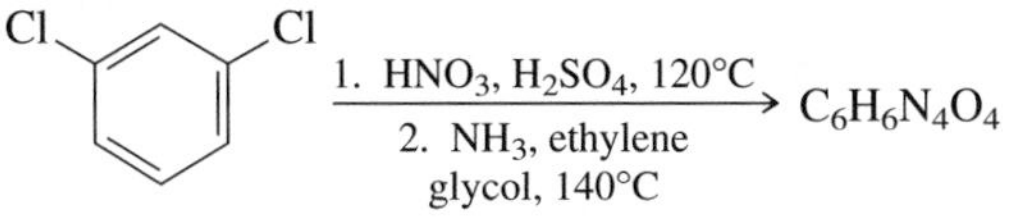
Cl
Cl
1. HNO3, H2SO4, 120°C
2. NH3, ethylene glycol, 140°C
C6H6N4O4

(d) 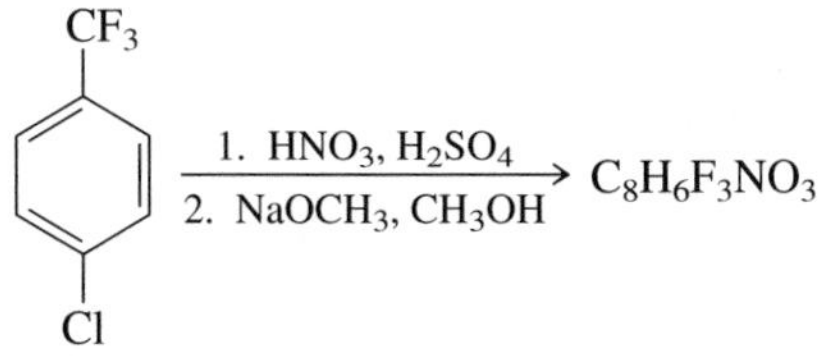
CF3
1. HNO3, H2SO4
2. NaOCH3, CH3OH
C8H6F3NO3
Cl

(e) 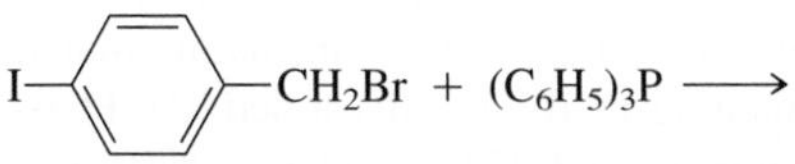
I
CH2Br + (C6H5)3P

(f) 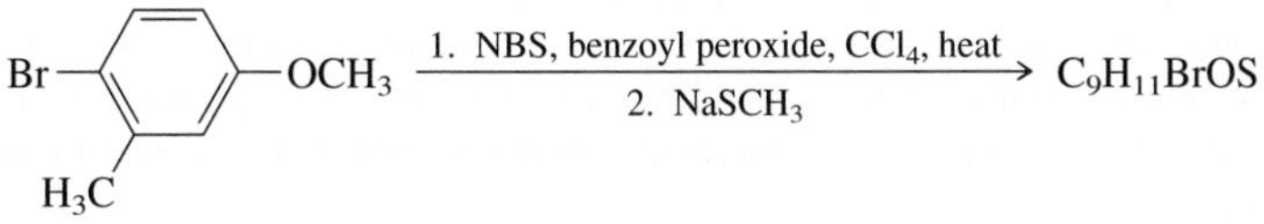
Br
OCH3
1. NBS, benzoyl peroxide, CCl4, heat
2. NaSCH3
C9H11BrOS
H3C

12.68 The herbicide *trifluralin* is prepared by the following sequence of reactions. Identify compound A and deduce the structure of trifluralin.

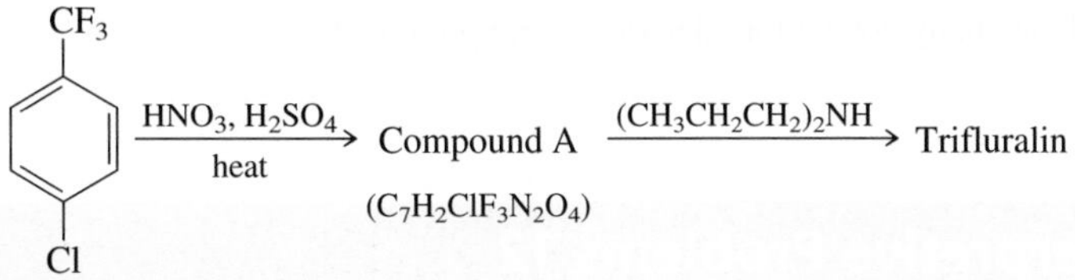
CF3
HNO3, H2SO4
heat
Compound A
(C7H2ClF3N2O4)
(CH3CH2CH2)2NH
Trifluralin
Cl

12.69 An article in the October 1998 issue of the *Journal of Chemical Education* (p. 1266) describes the following reaction.

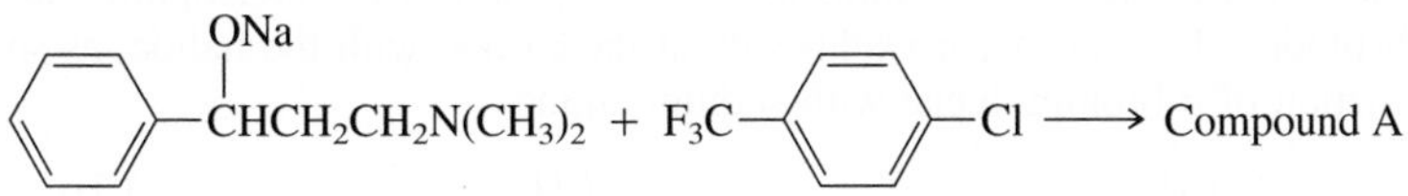
ONa
CHCH2CH2N(CH3)2 + F3C
Cl
Compound A

Fluoxetine hydrochloride *(Prozac)* is a widely prescribed antidepressant drug introduced by Eli Lilly & Co. in 1986. It differs from Compound A in having an —$NHCH_3$ group in place of —$N(CH_3)_2$. What is the structure of Prozac?

12.70 Nitro-substituted aromatic compounds that do not bear halide leaving groups react with nucleophiles according to the equation

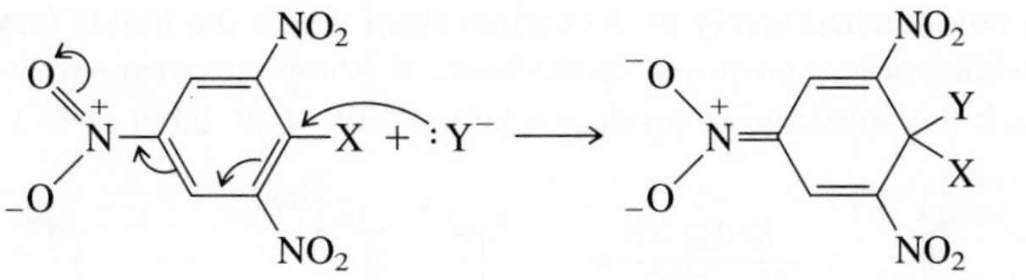
NO2
X + :Y⁻
NO2
NO2
Y
X
NO2

The product of this reaction, as its sodium salt, is called a *Meisenheimer complex* after the German chemist Jacob Meisenheimer, who reported on their formation and reactions in 1902. A Meisenheimer complex corresponds to the product of the nucleophilic addition stage in the addition–elimination mechanism for nucleophilic aromatic substitution.

(a) Give the structure of the Meisenheimer complex formed by addition of sodium ethoxide to 2,4,6-trinitroanisole.

(b) What other combination of reactants yields the same Meisenheimer complex as that of part (a)?

12.71 A careful study of the reaction of 2,4,6-trinitroanisole with sodium methoxide revealed that two different Meisenheimer complexes were present. Suggest reasonable structures for these two complexes.

12.72 Suggest a reasonable mechanism for the following reaction:

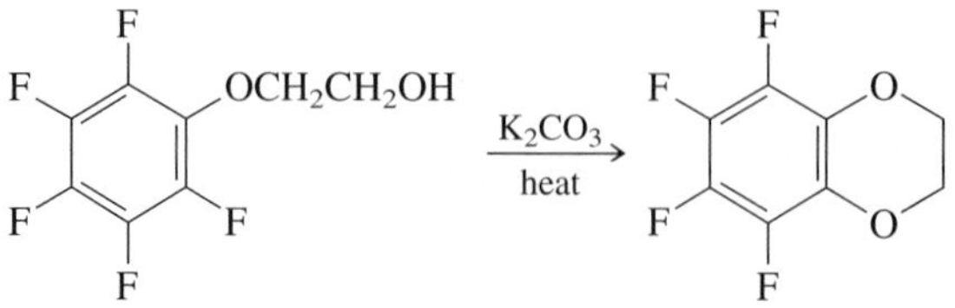
F
F
OCH2CH2OH
F
F
F
K2CO3
heat
F
F
O
O
F
F

12.73 Mixtures of chlorinated derivatives of biphenyl, called *polychlorinated biphenyls,* or *PCBs,* were once prepared industrially on a large scale as insulating materials in electrical equipment. As equipment containing PCBs was discarded, the PCBs entered the environment at a rate that reached an estimated 25,000 lb/year. PCBs are very stable and accumulate in the fatty tissue of fish, birds, and mammals. They have been shown to be *teratogenic,* meaning that they induce mutations in the offspring of affected individuals. Some countries have banned the use of PCBs. A large number of chlorinated biphenyls are possible, and the commercially produced material is a mixture of many compounds.

(a) How many monochloro derivatives of biphenyl are possible?

(b) How many dichloro derivatives are possible?

(c) How many octachloro derivatives are possible?

(d) How many nonachloro derivatives are possible?

Descriptive Passage and Interpretive Problems 12

Benzyne

Very strong bases such as sodium or potassium amide react readily with aryl halides, even those without electron-withdrawing substituents, to give products of nucleophilic substitution by the base. Substitution does not occur exclusively at the carbon with the halide, as shown for the following reaction of *o*-bromotoluene with sodium amide.

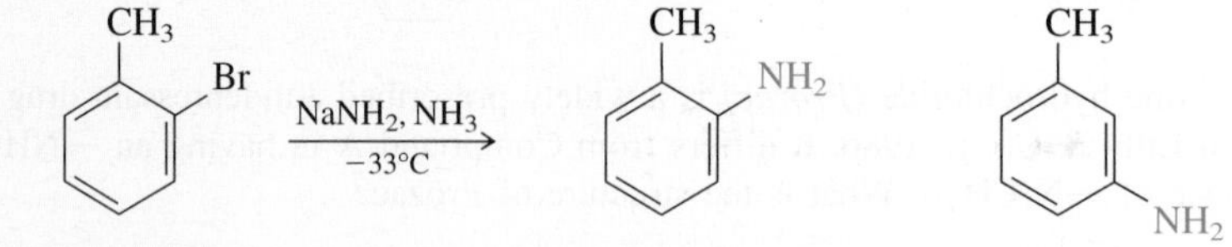
CH3
Br
NaNH2, NH3
−33°C
CH3
NH2
CH3
NH2
o-Bromotoluene
o-Methylaniline
m-Methylaniline

This experiment is inconsistent with substitution by an addition–elimination mechanism, because the nucleophile is not attached solely to the carbon from which the halide leaving group departed. An alternative mechanism was proposed on the basis of isotope experiments with ^{14}C-labeled chlorobenzene, in which the substitution product retained half of its label at C-1 and half at C-2.

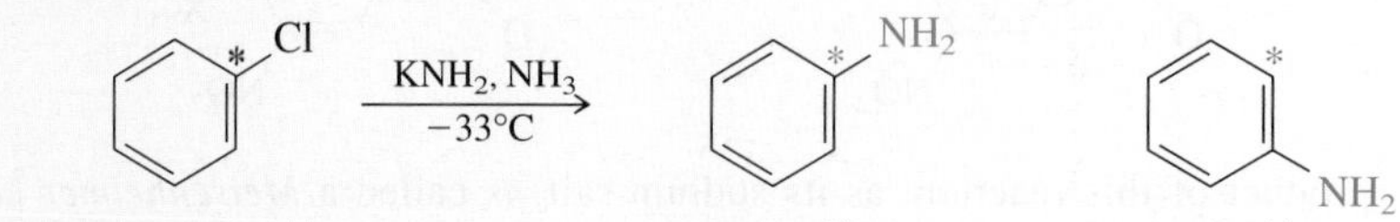
Cl
*
KNH2, NH3
−33°C
NH2
*
*
NH2
Chlorobenzene-1-14C
Aniline-1-14C
(48%)
Aniline-2-14C
(52%)

On the basis of the labeling experiment, an alternative mechanism was proposed for the substitution reaction of aryl halides with strong base-the elimination-addition mechanism. In the first step, the elimination stage, amide anion removes a proton from the carbon on the ring adjacent to the one with the halogen. The product is an unstable intermediate known as *benzyne*.

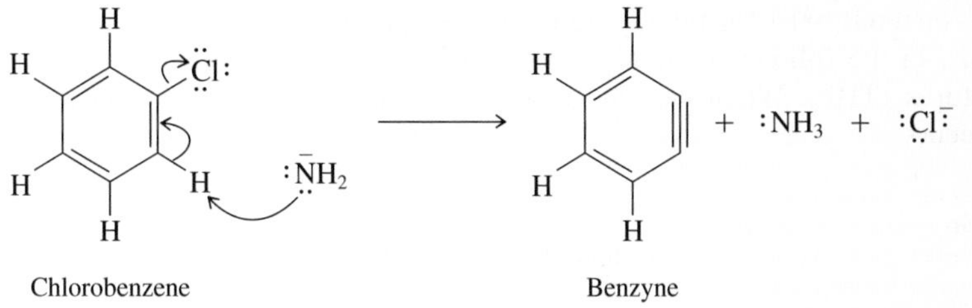

In the second step, amide anion now acts as a nucleophile, adding to one of the carbons of the triple bond. This addition step gives an aryl anion.

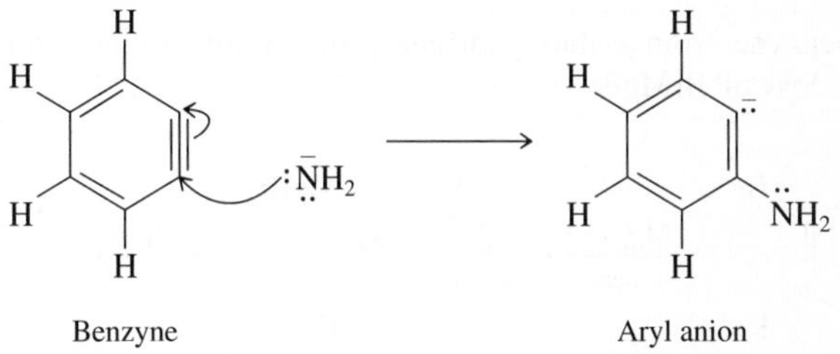

In the final step, the aryl anion abstracts a proton from ammonia to give aniline.

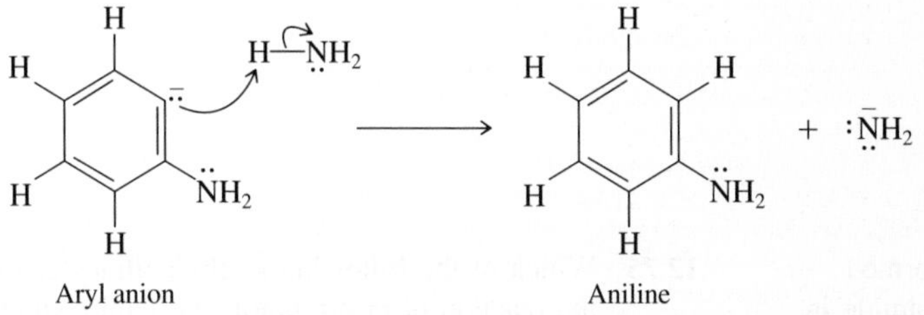

The triple bond in benzyne is different from the usual triple bond of an alkyne. In benzyne, one of the π components of the triple bond results from *p–p* overlap and is part of the delocalized π system of the aromatic ring. The second π component, which results from overlapping adjacent sp^2-hybridized orbitals, lies in the plane of the ring and is not part of the aromatic π system, as shown in Figure 12.8 and the electrostatic potential map.

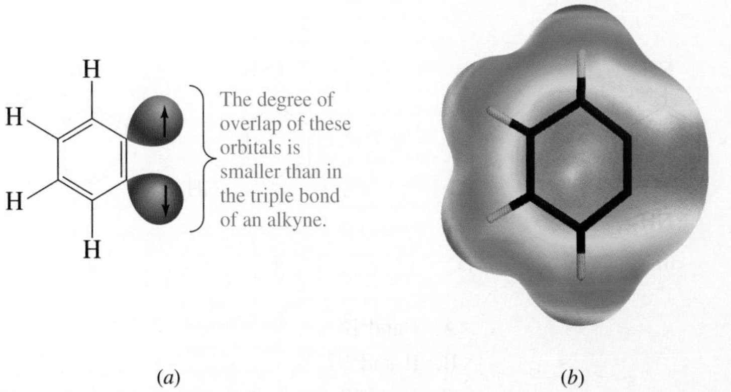

Figure 12.8

(*a*) The sp^2 orbitals in the plane of the ring in benzyne are not properly aligned for good overlap, and π bonding is weak. (*b*) The electrostatic potential map shows a region of high electron density associated with the "triple bond."

The intermediacy of benzyne in the elimination–addition mechanism for aryl halides accounts for the regioselectivity observed in the substitution reactions of labeled chlorobenzene and *o*-bromotoluene because both can give only a single aryne intermediate. Attack at either of the aryne carbons gives rise to the products.

The triple bond in benzyne is strained and is a dienophile in Diels–Alder reactions. Alternative methods exist for the generation of benzyne in cycloadditions and other synthetic applications. In the following example, *o*-bromofluorobenzene is treated with magnesium in tetrahydrofuran (THF). When carried out in the presence of cyclohexadiene, a Diels–Alder reaction occurs.

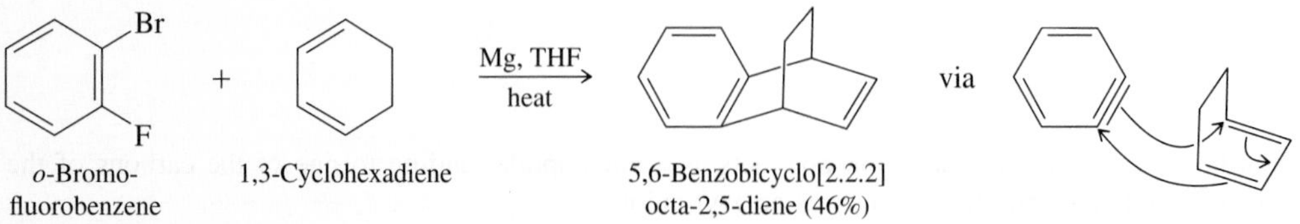

o-Bromo-fluorobenzene — 1,3-Cyclohexadiene — 5,6-Benzobicyclo[2.2.2]octa-2,5-diene (46%)

The formation of benzyne from *o*-fluorobenzene proceeds by initial formation of a Grignard reagent, followed by loss of BrMgF.

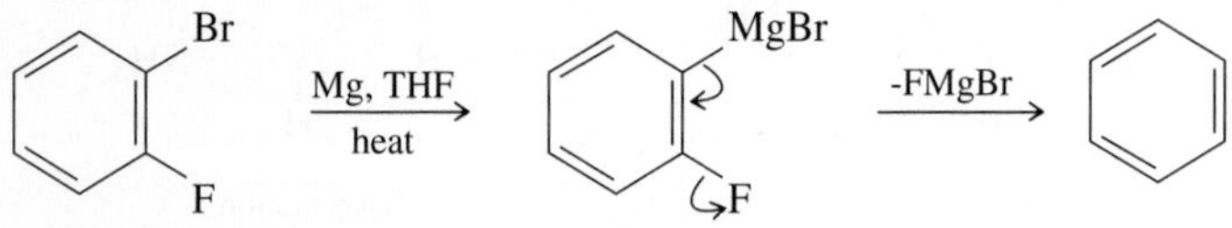

o-Bromo-fluorobenzene — *o*-Fluorophenylmagnesium bromide — Benzyne

12.74 Which of the following methylanilines can be formed by the reaction of *p*-bromotoluene with sodium amide in ammonia at −33°C?

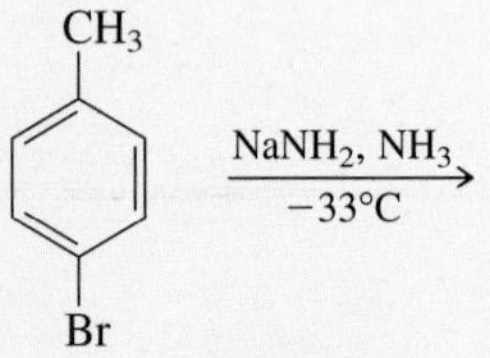

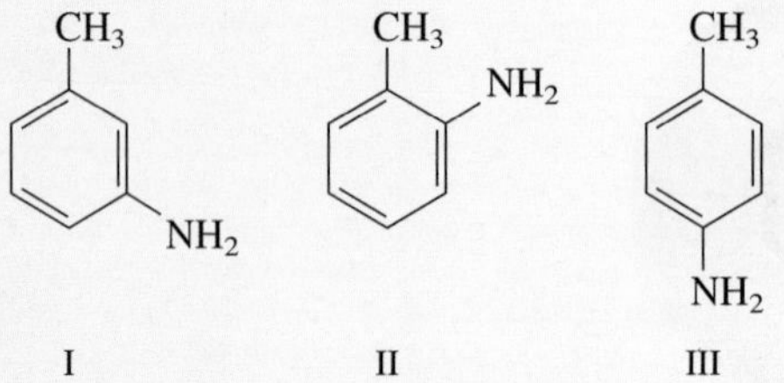

I II III

A. I and II
B. II and III
C. I and III
D. I, II, and III

12.75 Which of the following methylanilines can be formed by the reaction of *m*-bromotoluene with sodium amide in ammonia at −33°C?

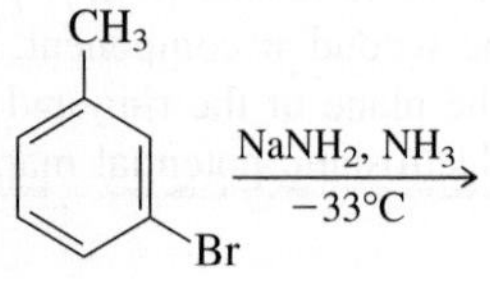

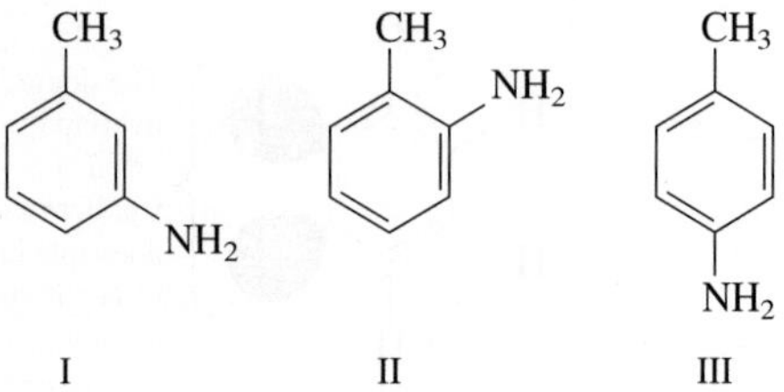

I II III

A. I and II
B. II and III
C. I and III
D. I, II, and III

12.76 Which one of the following isomers of bromodimethylbenzene *cannot* undergo nucleophilic aromatic substitution by treatment with sodium amide in liquid ammonia?

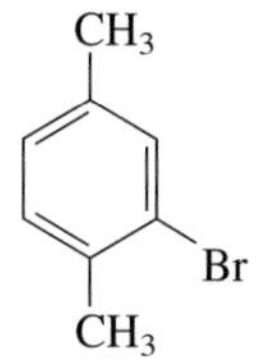

A.

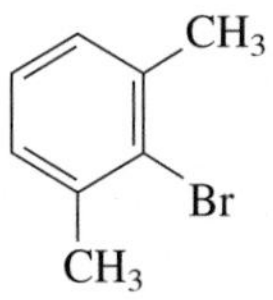

B.

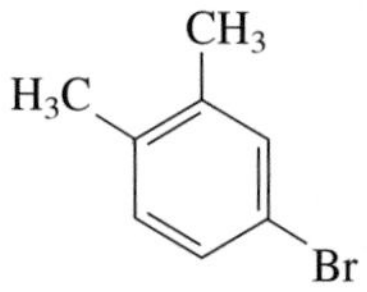

C.

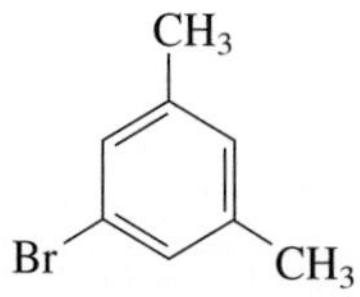

D.

12.77 Two isomeric phenols are obtained in comparable amounts on hydrolysis of *p*-iodotoluene with 1 M sodium hydroxide at 300°C. What are the structures of each?

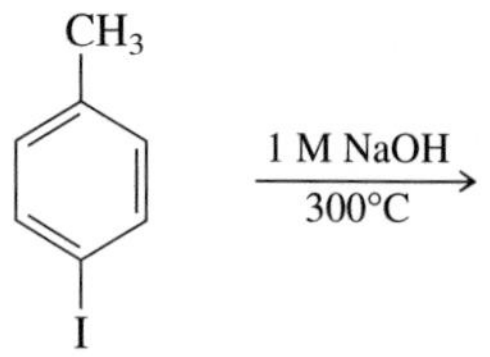

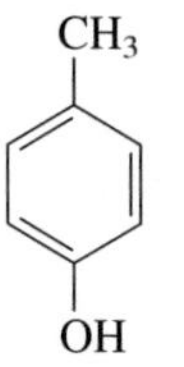

I

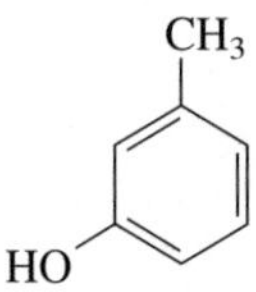

II

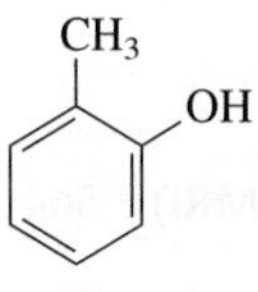

III

IV

A. I and II

B. I and III

C. II and III

D. III and IV

12.78 What is the structure of the cycloaddition product formed when benzyne is generated in the presence of furan?

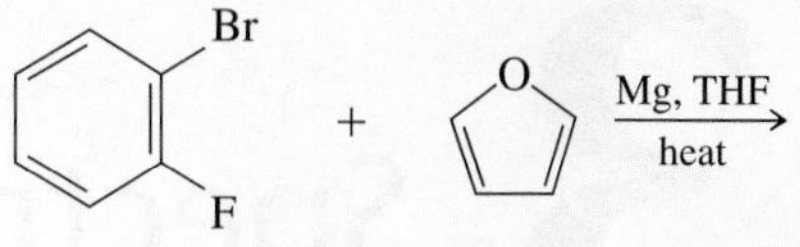

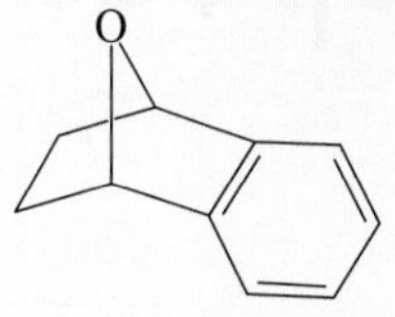

A.

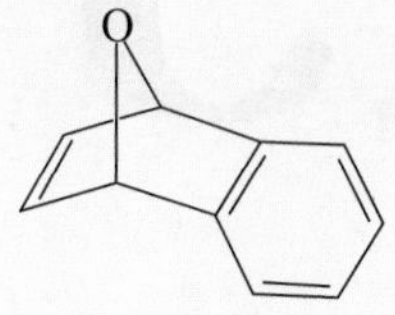

B.

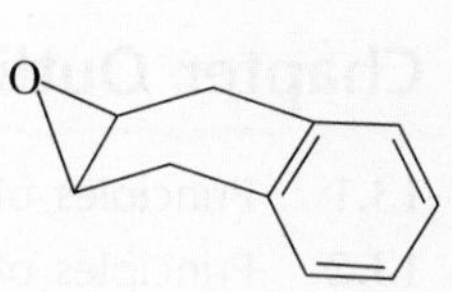

C.

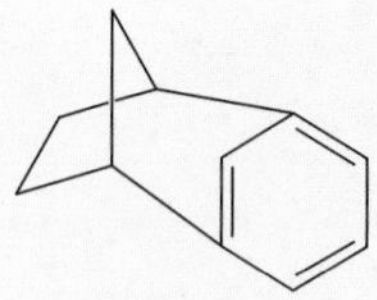

D.

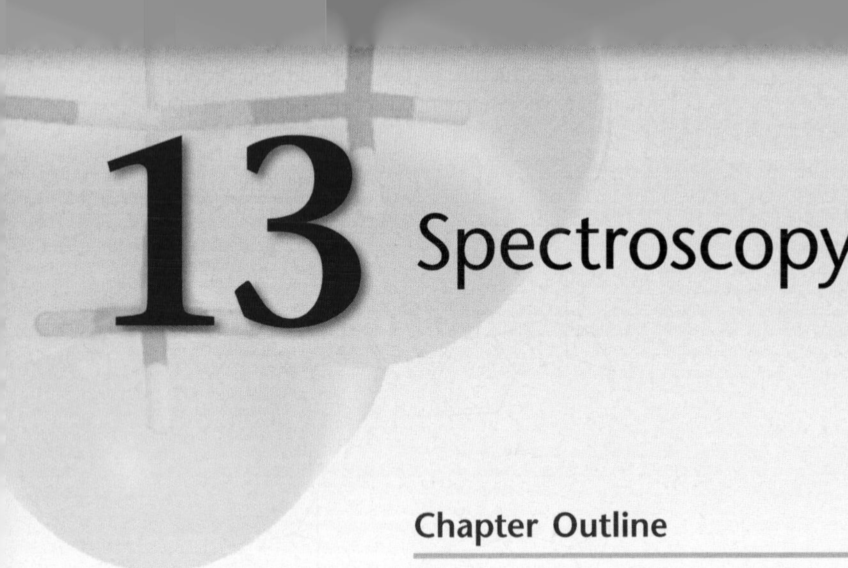

13 Spectroscopy

Chapter Outline

Many organosilicon compounds such as tetramethylsilane $[(CH_3)_4Si]$ are made from SiO_2, which occurs naturally in many forms, including quartz. The hydrogens and carbons of tetramethylsilane are the references to which other hydrogens and carbons are compared in nuclear magnetic resonance spectroscopy.

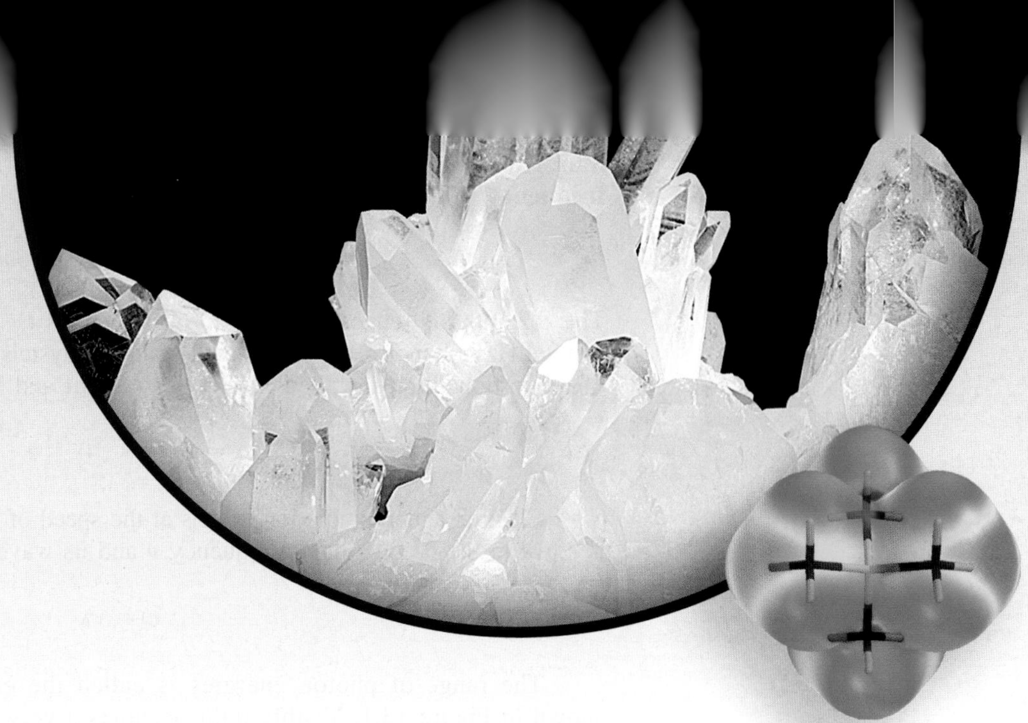

UNTIL THE SECOND HALF of the twentieth century, the structure of a substance—a newly discovered natural product, for example—was determined using information obtained from chemical reactions. This information included the identification of functional groups by chemical tests, along with the results of experiments in which the substance was broken down into smaller, more readily identifiable fragments. Typical of this approach is the demonstration of the presence of a double bond in an alkene by catalytic hydrogenation and determination of its location by ozonolysis. After considering all the available chemical evidence, the chemist proposed a candidate structure (or structures) consistent with the observations. Proof of structure was provided either by converting the substance to some already known compound or by an independent synthesis.

Qualitative tests and chemical degradation have given way to instrumental methods of structure determination. The main methods and the structural clues they provide are:

- **Nuclear magnetic resonance (NMR) spectroscopy,** which tells us about the carbon skeleton and the environments of the hydrogens attached to it.
- **Infrared (IR) spectroscopy,** which reveals the presence or signals the absence of key functional groups.
- **Ultraviolet-visible (UV-VIS) spectroscopy,** which probes the electron distribution, especially in molecules that have conjugated π electron systems.
- **Mass spectrometry (MS),** which gives the molecular weight and formula, both of the molecule itself and various structural units within it.

As diverse as these techniques are, all of them are based on the absorption of energy by a molecule, and all measure how a molecule responds to that absorption. In describing these techniques our emphasis will be on their application to structure determination. We'll start with a brief discussion of electromagnetic radiation, which is the source of the energy that a molecule absorbs in NMR, IR, and UV-VIS spectroscopy. Mass spectrometry is unique in that, instead of electromagnetic radiation, its energy source is a stream of charged particles such as electrons.

13.1 Principles of Molecular Spectroscopy: Electromagnetic Radiation

Electromagnetic radiation, of which visible light is but one example, has the properties of both particles and waves. The particles are called **photons,** and each possesses an amount of energy referred to as a **quantum.** In 1900, the German physicist

"Modern" physics dates from Planck's proposal that energy is quantized, which set the stage for the development of quantum mechanics. Planck received the 1918 Nobel Prize in Physics.

Max Planck proposed that the energy of a photon (E) is directly proportional to its **frequency** (ν).

$$E = h\nu$$

The SI units of frequency are reciprocal seconds (s^{-1}), given the name *hertz* and the symbol Hz in honor of the nineteenth-century physicist Heinrich R. Hertz. The constant of proportionality h is called **Planck's constant** and has the value

$$h = 6.63 \times 10^{-34}\ \text{J} \cdot \text{s}$$

Electromagnetic radiation travels at the speed of light ($c = 3.0 \times 10^8$ m/s), which is equal to the product of its frequency ν and its wavelength λ:

$$c = \nu\lambda$$

1 nm = 10^{-9} m

The range of photon energies is called the *electromagnetic spectrum* and is shown in Figure 13.1. Visible light occupies a very small region of the electromagnetic spectrum. It is characterized by wavelengths of 400 nm (violet) to 800 nm (red). When examining Figure 13.1 be sure to keep the following two relationships in mind:

1. *Frequency is inversely proportional to wavelength;* the greater the frequency, the shorter the wavelength.
2. *Energy is directly proportional to frequency;* electromagnetic radiation of higher frequency possesses more energy than radiation of lower frequency.

Gamma rays and X-rays are streams of very high energy photons. Radio waves are of relatively low energy. Ultraviolet radiation is of higher energy than the violet end of visible light. Infrared radiation is of lower energy than the red end of visible light. When a molecule is exposed to electromagnetic radiation, it may absorb a photon, increasing its energy by an amount equal to the energy of the photon. Molecules are highly selective with respect to the frequencies they absorb. Only photons of certain specific frequencies are absorbed by a molecule. The particular photon energies absorbed by a molecule depend on molecular structure and are measured with instruments called **spectrometers.** The data obtained are very sensitive indicators of molecular structure.

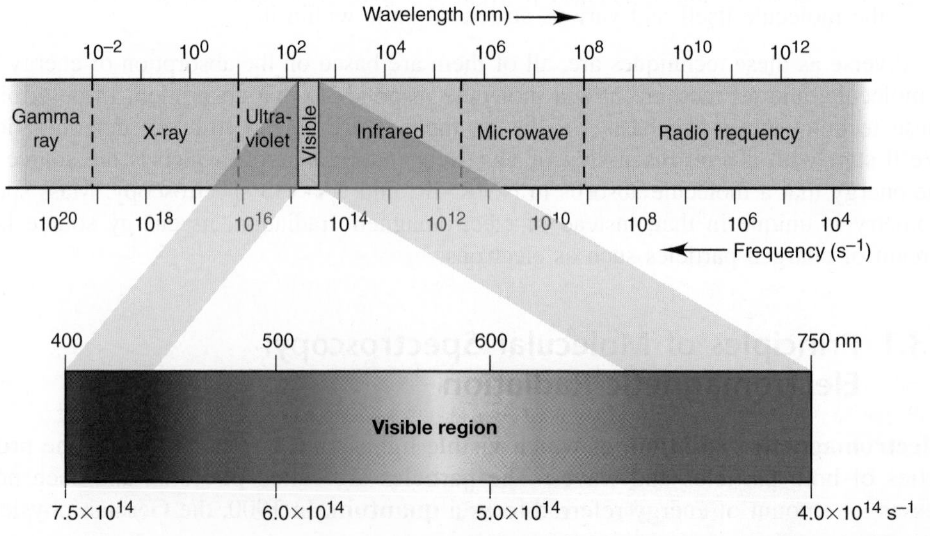

Figure 13.1

The electromagnetic spectrum. (Reprinted, with permission, from M. Silberberg, *Chemistry,* 5th ed., McGraw-Hill Higher Education, 2009, p. 271.)

13.2 Principles of Molecular Spectroscopy: Quantized Energy States

What determines whether electromagnetic radiation is absorbed by a molecule? The most important requirement is that the energy of the photon must equal the energy difference between two states, such as two nuclear spin states (NMR), two vibrational states (IR), or two electronic states (UV-VIS). In physics, the term for this is *resonance*—the transfer of energy between two objects that occurs when their frequencies are matched. In molecular spectroscopy, we are concerned with the transfer of energy from a photon to a molecule. Consider, for example, two energy states of a molecule designated E_1 and E_2 in Figure 13.2. The energy difference between them is $E_2 - E_1$, or ΔE. Unlike kinetic energy, which is continuous, meaning that all values of kinetic energy are available to a molecule, only certain energies are possible for electronic, vibrational, and nuclear spin states. These energy states are said to be **quantized.** More of the molecules exist in the lower energy state E_1 than in the higher energy state E_2. Excitation of a molecule from a lower state to a higher one requires the addition of an increment of energy equal to ΔE. Thus, when electromagnetic radiation strikes a molecule, only the frequency with energy equal to ΔE is absorbed. All other frequencies are transmitted.

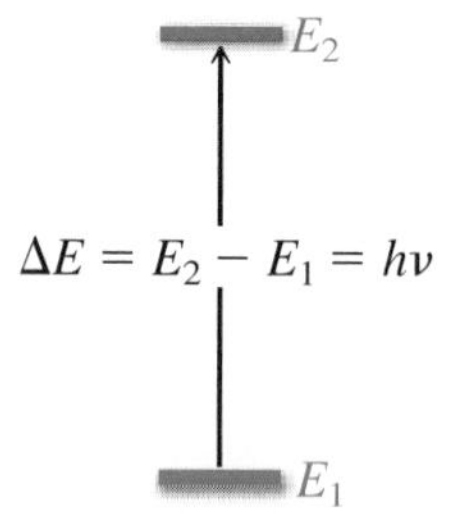

Figure 13.2

Two energy states of a molecule. Absorption of energy equal to $E_2 - E_1$ excites a molecule from its lower energy state to the next higher state.

Spectrometers are designed to measure the absorption of electromagnetic radiation by a sample. Basically, a spectrometer consists of a source of radiation, a compartment containing the sample through which the radiation passes, and a detector. The frequency of radiation is continuously varied, and its intensity at the detector is compared with that at the source. When the frequency is reached at which the sample absorbs radiation, the detector senses a decrease in intensity. The relation between frequency and absorption is plotted as a **spectrum,** which consists of a series of peaks at characteristic frequencies. Its interpretation can furnish structural information. Each type of spectroscopy is developed independently of the others, and so the data format is different for each one. An NMR spectrum looks different from an IR spectrum, and both look different from a UV-VIS spectrum.

With this as background, we will now discuss spectroscopic techniques individually. NMR, IR, and UV-VIS spectroscopy provide complementary information, and all are useful. Among them, NMR provides the information that is most directly related to molecular structure and is the one we'll examine first.

13.3 Introduction to ^{1}H NMR Spectroscopy

Nuclear magnetic resonance spectroscopy depends on the absorption of energy when the nucleus of an atom is excited from its lowest energy spin state to the next higher one. The nuclei of several elements can be studied by NMR. The two elements that are the most common in organic molecules (carbon and hydrogen) have isotopes (^{1}H and ^{13}C) capable of giving NMR spectra that are rich in structural information. A proton nuclear magnetic resonance (^{1}H NMR) spectrum tells us about the environments of the various hydrogens in a molecule; a carbon-13 nuclear magnetic resonance (^{13}C NMR) spectrum does the same for the carbon atoms. Separately and together ^{1}H and ^{13}C NMR take us a long way toward determining a substance's molecular structure. We'll develop most of the general principles of NMR by discussing ^{1}H NMR, then extend them to ^{13}C NMR. The ^{13}C NMR discussion is shorter, not because it is less important than ^{1}H NMR, but because many of the same principles apply to both techniques.

Nuclear magnetic resonance of protons was first detected in 1946 by Edward Purcell (Harvard) and by Felix Bloch (Stanford). Purcell and Bloch shared the 1952 Nobel Prize in Physics.

Like an electron, a proton has two spin states with quantum numbers of $+\frac{1}{2}$ and $-\frac{1}{2}$. There is no difference in energy between these two nuclear spin states; a proton is just as likely to have a spin of $+\frac{1}{2}$ as $-\frac{1}{2}$. Absorption of electromagnetic radiation can only occur when the two spin states have different energies. A way to make them different is to place the sample in a magnetic field. A spinning proton behaves like a tiny bar magnet and has a magnetic moment associated with it (Figure 13.3). In the presence of an external magnetic field B_0, the spin state in which the magnetic moment of the nucleus is aligned with B_0 is lower in energy than the one in which it opposes B_0.

Figure 13.3

(*a*) In the absence of an external magnetic field, the nuclear spins of the protons are randomly oriented. (*b*) In the presence of an external magnetic field B_0, the nuclear spins are oriented so that the resulting nuclear magnetic moments are aligned either parallel or antiparallel to B_0. The lower energy orientation is the one parallel to B_0, and more nuclei have this orientation.

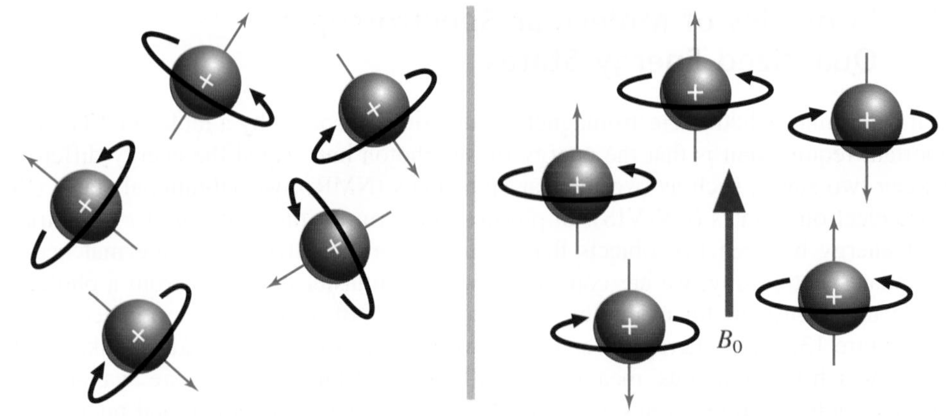

(*a*) No external magnetic field (*b*) Apply external magnetic field B_0

The SI unit for magnetic field strength is the tesla (T), named after Nikola Tesla, a contemporary of Thomas Edison and who, like Edison, was an inventor of electrical devices.

As shown in Figure 13.4, the energy difference between the two states is directly proportional to the strength of the applied field. Net absorption of electromagnetic radiation requires that the lower state be more highly populated than the higher one, and quite strong magnetic fields are required to achieve the separation necessary to give a detectable signal. A magnetic field of 4.7 T, which is about 100,000 times stronger than Earth's magnetic field, separates the two spin states of a proton by only 8×10^{-5} kJ/mol (1.9×10^{-5} kcal/mol). From Planck's equation $\Delta E = h\nu$, this energy gap corresponds to radiation having a frequency of 2×10^{8} Hz (200 MHz), which lies in the radio-frequency (rf) region of the electromagnetic spectrum (see Figure 13.1).

Frequency of electromagnetic radiation (s^{-1} or Hz) $\xrightarrow{\text{is proportional to}}$ Energy difference between nuclear spin states (kJ/mol or kcal/mol) $\xrightarrow{\text{is proportional to}}$ Magnetic field (T)

Problem 13.1

Most of the NMR spectra in this text were recorded on a spectrometer having a field strength of 4.7 T (200 MHz for ^1H). The first generation of widely used NMR spectrometers were 60-MHz instruments. What was the magnetic field strength of these earlier spectrometers? What is the field strength of the 920-MHz instruments now commercially available?

The response of an atom to the strength of the external magnetic field is different for different elements, and for different isotopes of the same element. The resonance frequencies of most nuclei are sufficiently different that an NMR experiment is sensitive

Figure 13.4

An external magnetic field causes the two nuclear spin states to have different energies. The difference in energy ΔE is proportional to the strength of the applied field.

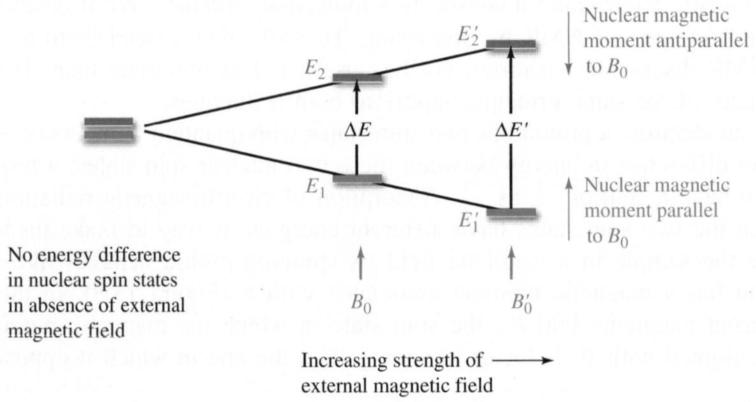

only to a particular isotope of a single element. The frequency for ^1H is 200 MHz at 4.7 T, but that of ^{13}C is 50.4 MHz. Thus, when recording the NMR spectrum of an organic compound, we see signals only for ^1H or ^{13}C, but not both; ^1H and ^{13}C NMR spectra are recorded in separate experiments with different instrument settings.

Problem 13.2

What will be the ^{13}C frequency of an NMR spectrometer that operates at 100 MHz for protons?

The essential features of an NMR spectrometer consist of a powerful magnet to align the nuclear spins, a radiofrequency (rf) transmitter as a source of energy to excite a nucleus from its lowest energy state to the next higher one, and a way to monitor the absorption of rf radiation and display the spectrum.

NMR spectra are acquired using *pulsed Fourier-transform* nuclear magnetic resonance (FT-NMR) spectrometers (Figure 13.5). The sample is placed in a magnetic field and irradiated with a short, intense burst of rf radiation (the *pulse*), which excites *all* of the protons in the molecule at the same time. The magnetic field associated with the new orientation of nuclear spins induces an electrical signal in the receiver that decreases as the nuclei return to their original orientation. The resulting *free-induction decay* (FID) is a composite of the decay patterns of all of the protons in the molecule. The FID pattern is stored in a computer and converted into a spectrum by a mathematical process known as a *Fourier transform.* The pulse-relaxation sequence takes only about a second. The signal-to-noise ratio is enhanced by repeating the sequence many times, then averaging the data. Noise is random and averaging causes it to vanish; signals always appear at the same frequency and accumulate. All of the operations—the interval between pulses, collecting, storing, and averaging the data and converting it to a spectrum by a Fourier transform—are under computer control, which makes the actual recording of an FT-NMR spectrum a routine operation.

Richard R. Ernst of the Swiss Federal Institute of Technology won the 1991 Nobel Prize in Chemistry for devising pulse-relaxation NMR techniques.

13.4 Nuclear Shielding and ^1H Chemical Shifts

Our discussion so far has concerned ^1H nuclei in general without regard for the environments of individual protons in a molecule. Protons in a molecule are connected to other atoms—carbon, oxygen, nitrogen, and so on—by covalent bonds. The electrons in these bonds, indeed all the electrons in a molecule, affect the magnetic environment of the protons. Alone, a proton would feel the full strength of the external field, but a proton in an organic molecule responds to both the external field plus any local fields within the molecule. An external magnetic field affects the motion of the electrons in a molecule, inducing local fields characterized by lines of force that circulate in the *opposite* direction from the applied field (Figure 13.6). Thus, the net field felt by a proton in a molecule will always be less than the applied field, and the proton is said to be **shielded.** All of the protons of a molecule are shielded from the applied field by the electrons, but some are less shielded than others. The term *deshielded* is often used to describe this decreased shielding of one proton relative to another.

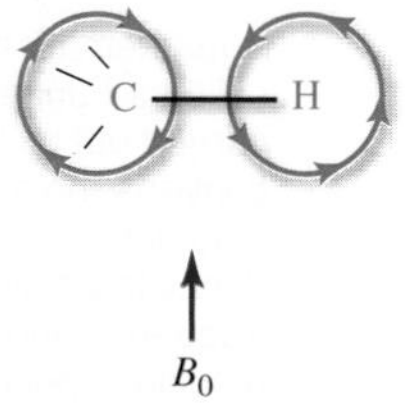

Figure 13.6

The induced magnetic field of the electrons in the carbon–hydrogen bond opposes the external magnetic field. The resulting magnetic field experienced by the proton and the carbon is slightly less than B_0.

The more shielded a proton is, the greater must be the strength of the applied field in order to achieve resonance and produce a signal. A more shielded proton absorbs rf radiation at higher field strength **(upfield)** compared with one at lower field strength **(downfield).** Different protons give signals at different field strengths. *The dependence of the resonance position of a nucleus that results from its molecular environment is called its* **chemical shift.** This is where the real power of NMR lies. The chemical shifts of various protons in a molecule can be different and are characteristic of particular structural features.

Figure 13.7 shows the ^1H NMR spectrum of chloroform ($CHCl_3$) to illustrate how the terminology just developed applies to a real spectrum.

Instead of measuring chemical shifts in absolute terms, we measure them with respect to a standard—*tetramethylsilane* $(CH_3)_4Si$, abbreviated *TMS.* The protons of TMS

1. Dissolve sample in deuterated chloroform ($CDCl_3$) and place in NMR tube.
2. Insert NMR tube into vertical cavity (bore) of the magnet.
3. Bore of magnet contains a probe that acts as a transmitter of radiofrequency (RF) pulses and receiver of signals from the sample. The transmitter is housed in a console along with other electronic equipment.
4. A short (5 μs), intense RF pulse is sent from the RF transmitter in the console to the probe. Absorption of RF energy tips the magnetic vector of the nuclei in the sample.
5. The magnetic field associated with the new orientation of the nuclei returns (relaxes) to the original state. Nuclei relax rapidly but at different rates that depend on their chemical environment. As the magnetic field changes, it generates an electrical impulse that is transmitted from the probe to a receiver in the console as a "free induction decay."
6. The pulse–relax sequence is repeated many times and the free-induction decay data stored in a computer in the console.
7. A mathematical operation called a Fourier transform carried out by the computer converts the amplitude-versus-time data of the free-induction decay to amplitude versus frequency and displays the resulting spectrum on the screen or prints it.

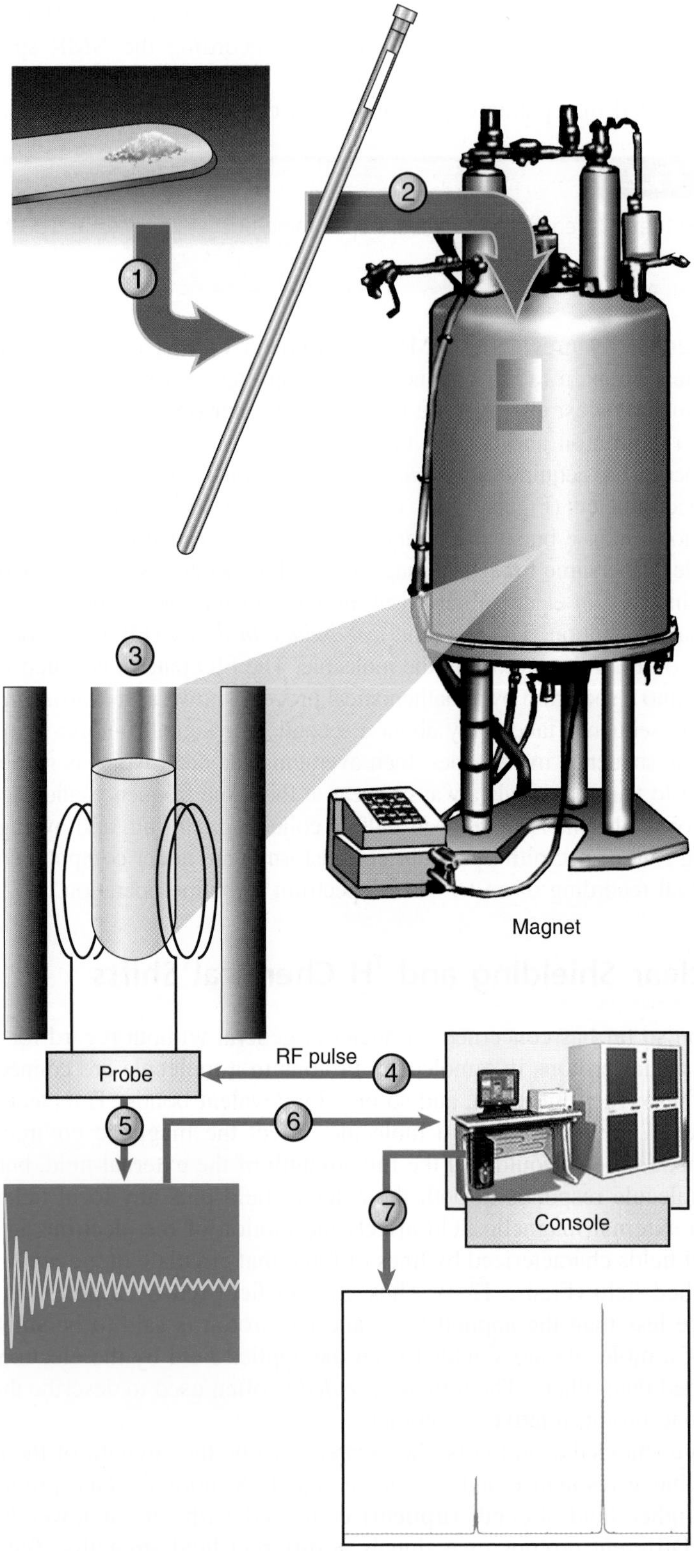

Figure 13.5

How an NMR spectrum is acquired using a pulse-Fourier transform (FT) NMR spectrometer.

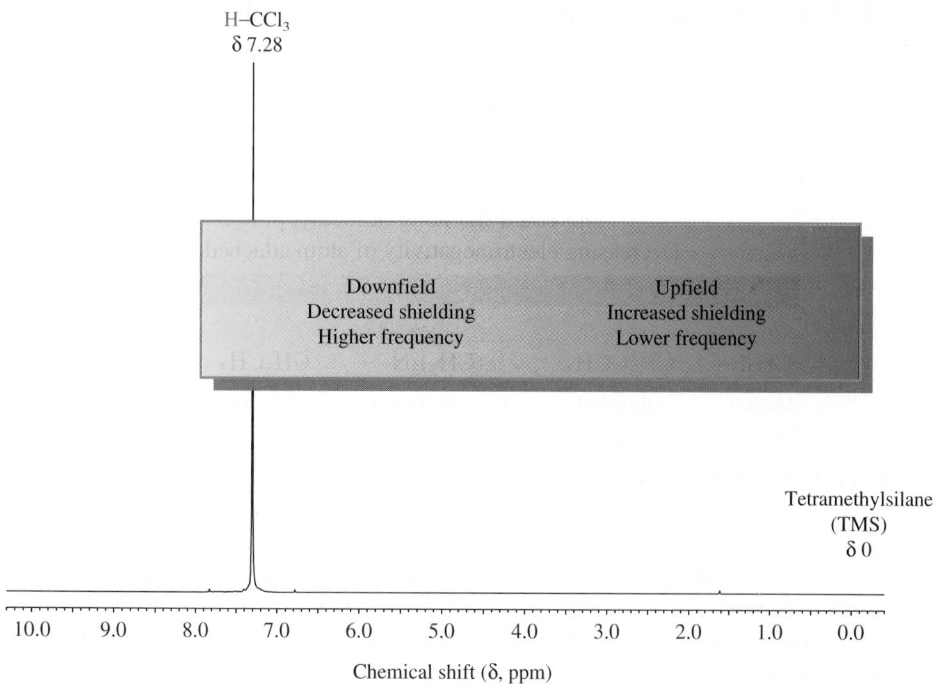

H–CCl3
δ 7.28
Downfield
Decreased shielding
Higher frequency
Upfield
Increased shielding
Lower frequency
Tetramethylsilane
(TMS)
δ 0
10.0 9.0 8.0 7.0 6.0 5.0 4.0 3.0 2.0 1.0 0.0
Chemical shift (δ, ppm)

Figure 13.7

The 200-MHz ^{1}H NMR spectrum of chloroform ($CHCl_3$). Chemical shifts are measured along the *x*-axis in parts per million (ppm) from tetramethylsilane as the reference, which is assigned a value of zero.

are more shielded than those of most organic compounds, so all of the signals in a sample ordinarily appear at lower field than those of the TMS reference. When measured using a 100-MHz instrument, the signal for the proton in chloroform ($CHCl_3$), for example, appears 728 Hz downfield from the TMS signal. But because frequency is proportional to magnetic field strength, the same signal would appear 1456 Hz downfield from TMS on a 200-MHz instrument. We simplify the reporting of chemical shifts by converting them to parts per million (ppm) downfield from TMS, which is assigned a value of 0. The TMS need not actually be present in the sample, nor even appear in the spectrum in order to serve as a reference. When chemical shifts are reported this way, they are identified by the symbol δ and are independent of the magnetic field strength.

$$\text{Chemical shift } (\delta) = \frac{\text{position of signal} - \text{position of TMS peak}}{\text{spectrometer frequency}} \times 10^6$$

Thus, the chemical shift for the proton in chloroform is:

$$\delta = \frac{1456 \text{ Hz} - 0 \text{ Hz}}{200 \times 10^6 \text{ Hz}} \times 10^6 = 7.28$$

Problem 13.3

The ^{1}H NMR signal for bromoform ($CHBr_3$) appears at 2065 Hz when recorded on a 300-MHz NMR spectrometer. (a) What is the chemical shift of this proton? (b) If the spectrum was recorded on a 400-MHz instrument, what would be the chemical shift of the $CHBr_3$ proton? (c) How many hertz downfield from TMS is the signal when recorded on a 400-MHz instrument?

NMR spectra are usually run in solution and, although chloroform is a good solvent for most organic compounds, it's rarely used because its own signal at δ 7.28 would be so intense that it would obscure signals in the sample. Because the magnetic properties of deuterium ($D = {}^2H$) are different from those of ^{1}H, $CDCl_3$ gives no signals at all in a ^{1}H NMR spectrum and is used instead. Indeed, $CDCl_3$ is the most commonly used solvent in ^{1}H NMR spectroscopy. Likewise, D_2O is used instead of H_2O for water-soluble substances such as carbohydrates.

13.5 Effects of Molecular Structure on ^{1}H Chemical Shifts

Nuclear magnetic resonance spectroscopy is such a powerful tool for structure determination because *protons in different environments experience different degrees of shielding and have different chemical shifts.* In compounds of the type CH_3X, for example, the shielding of the methyl protons increases as X becomes less electronegative.

Increased shielding of methyl protons
Decreasing electronegativity of atom attached to CH_3

CH_3F	CH_3OCH_3	$(CH_3)_3N$	CH_3CH_3	$(CH_3)_4Si$
Methyl fluoride	Dimethyl ether	Trimethylamine	Ethane	Tetramethylsilane
Chemical shift of methyl protons (δ):				
4.3	3.2	2.2	0.9	0.0

Problem 13.3 in the preceding section was based on the chemical-shift difference between the proton in $CHCl_3$ and the proton in $CHBr_3$ and its relation to shielding.

Inasmuch as the shielding is due to the electrons, it isn't surprising to find that the chemical shift depends on the degree to which X draws electrons away from the methyl group. A similar trend is seen in the methyl halides, in which the protons in CH_3F are the least shielded (δ 4.3) and those of CH_3I (δ 2.2) are the most.

The decreased shielding caused by electronegative substituents is primarily an inductive effect and, like other inductive effects, falls off rapidly as the number of bonds between the substituent and the proton increases. Compare the chemical shifts of the protons in propane and 1-nitropropane.

δ 0.9 δ 1.3 δ 0.9
$H_3C—CH_2—CH_3$
Propane

δ 4.3 δ 2.0 δ 1.0
$O_2N—CH_2—CH_2—CH_3$
1-Nitropropane

The strongly electron-withdrawing nitro group deshields the protons on C-1 by 3.4 ppm (δ 4.3 − 0.9). The effect is smaller on the protons at C-2 (0.7 ppm), and almost completely absent at C-3.

The deshielding effects of electronegative substituents are cumulative, as the chemical shifts for various chlorinated derivatives of methane indicate.

	CH_3Cl	CH_2Cl_2	$CHCl_3$
	Chloromethane	Dichloromethane	Trichloromethane
Chemical shift (δ):	3.1	5.3	7.3

Problem 13.4

Identify the most shielded and least shielded protons in

(a) 2-Bromobutane
(b) 1,1,2-Trichloropropane
(c) Tetrahydrofuran: 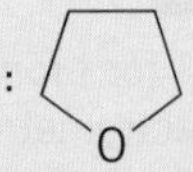

O

Sample Solution (a) Bromine is electronegative and will have its greatest electron-withdrawing effect on protons that are separated from it by the fewest bonds. Therefore, the proton at C-2 will be the least shielded, and those at C-4 the most shielded.

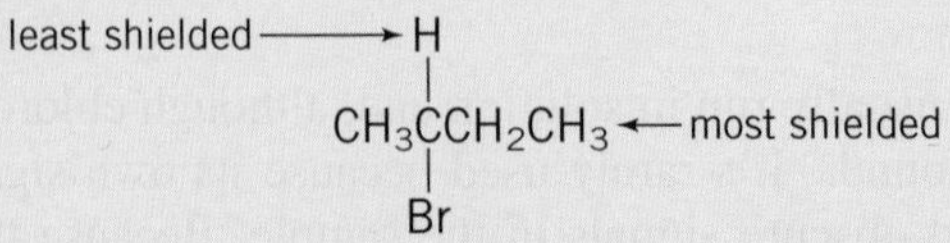

least shielded
H
CH3CCH2CH3
most shielded
Br

The observed chemical shifts are δ 4.1 for the proton at C-2 and δ 1.1 for the protons at C-4. The protons at C-1 and C-3 appear in the range δ 1.7–2.0.

Table 13.1 collects chemical-shift information for protons of various types. The major portion of the table concerns protons bonded to carbon. Within each type, methyl

TABLE 13.1 Approximate Chemical Shifts of Representative Protons

Compound class or type of proton		Chemical shift (δ), ppm*
Protons bonded to carbon		
Alkane	RCH_3, R_2CH_2, R_3CH	0.9–1.8
Allylic	H—C—C=C	1.5–2.6
Terminal alkyne	H—C≡C	1.8–3.1
C—H adjacent to C=O	H—C—C=O	2.0–2.5
C—H adjacent to C≡N	H—C—C≡N	2.1–2.3
Benzylic	H—C—Ar	2.3–2.8
Amine	H—C—NR_2	2.2–2.9
Alkyl chloride	H—C—Cl	3.1–4.1
Alkyl bromide	H—C—Br	2.7–4.1
Alcohol or ether	H—C—O	3.3–3.7
Vinylic	H—C=C	4.5–6.5
Aryl	H—Ar	6.5–8.5
Aldehyde	R(H)C=O	9–10
Protons bonded to nitrogen or oxygen		
Amine	H—NR_2	1–3†
Alcohol	H—OR	0.5–5†
Phenol	H—OAr	6–8†
Carboxylic acid	H—OC(=O)R	10–13†

*Approximate values relative to tetramethylsilane; other groups within the molecule can cause a proton signal to appear outside of the range cited.

†The chemical shifts of O—H and N—H protons are temperature- and concentration-dependent.

(CH_3) protons are more shielded than methylene (CH_2), and methylene protons are more shielded than methine (CH). The differences, however, are small.

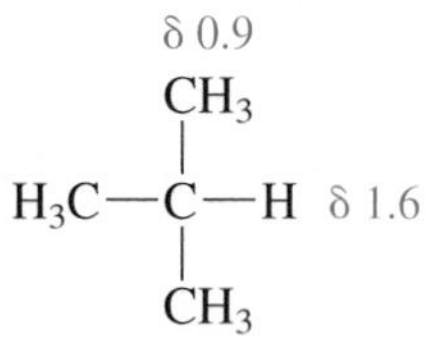

2-Methylpropane

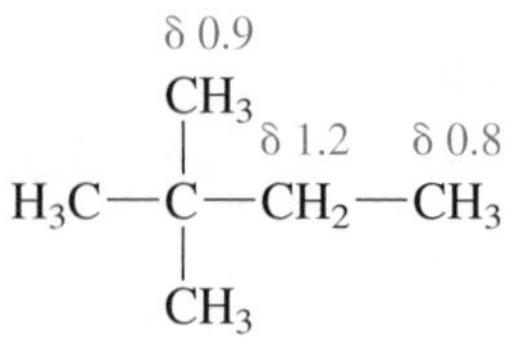

2,2-Dimethylbutane

Given that the chemical shift of methane is δ 0.2, we attribute the decreased shielding of the protons of RCH_3, R_2CH_2, and R_3CH to the number of carbons attached to primary, secondary, and tertiary carbons, respectively. Carbon is more electronegative than hydrogen, so replacing the hydrogens of CH_4 by one, then two, then three carbons decreases the shielding of the remaining protons.

Likewise, the generalization that sp^2-hybridized carbon is more electronegative than sp^3-hybridized carbon is consistent with the decreased shielding of allylic and benzylic protons.

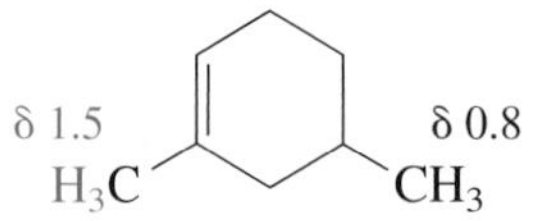

1,5-Dimethylcyclohexene

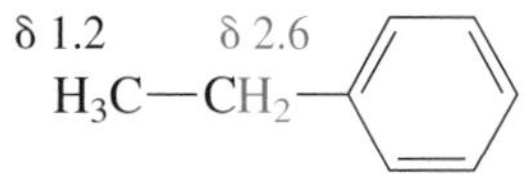

Ethylbenzene

Hydrogens that are directly attached to double bonds (vinylic protons) or to aromatic rings (aryl protons) are especially deshielded.

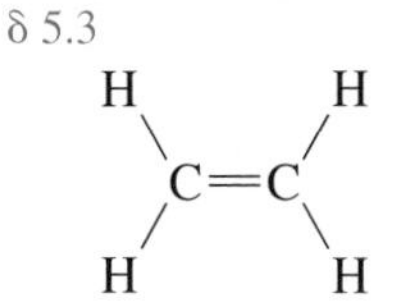

Ethylene

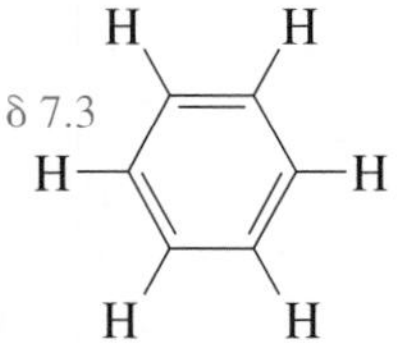

Benzene

The main contributor to the deshielding of vinylic and aryl protons is the induced magnetic field associated with π electrons. We saw earlier in Section 13.4 that the local field resulting from electrons in a C—H σ bond opposes the applied field and shields a molecule's protons. The hydrogens of ethylene and benzene, however, lie in a region of the molecule where the induced magnetic field of the π electrons reinforces the applied field, *deshielding* the protons (Figure 13.8). In the case of benzene, this is described as a **ring current** effect that originates in the circulating π electrons. It has interesting consequences, some of which are described in the boxed essay *Ring Currents—Aromatic and Antiaromatic* on page 551.

The induced field of C=C and aryl groups contributes to the deshielding of allylic and benzylic hydrogens.

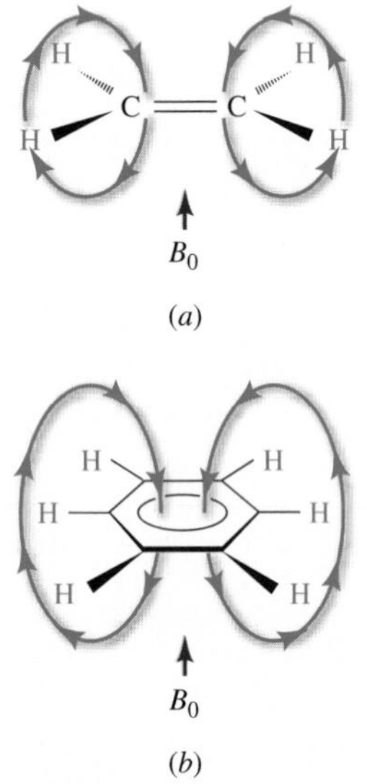

Figure 13.8

The induced magnetic field of the π electrons of (*a*) ethylene and (*b*) benzene reinforces the applied field in the regions near vinyl and aryl protons and deshields them.

Problem 13.5

(a) Assign the chemical shifts δ 1.6, δ 2.2, and δ 4.8 to the appropriate protons of methylenecyclopentane

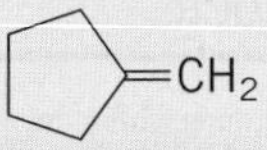

Sample Solution

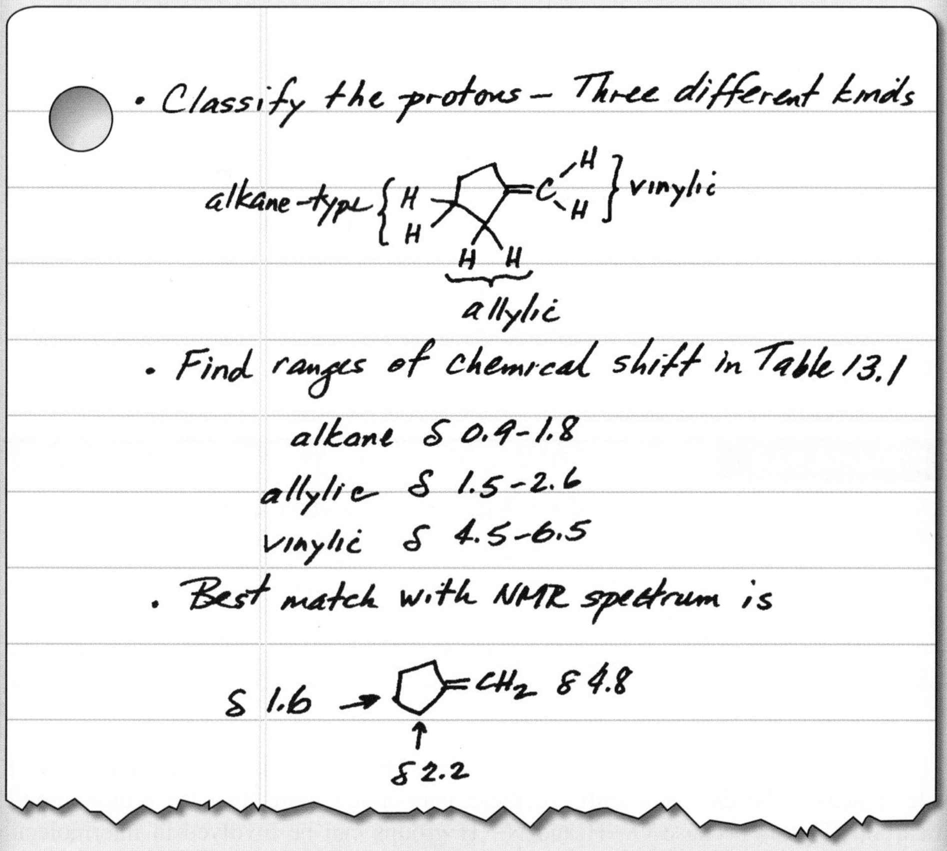
• Classify the protons – Three different kinds
alkane-type
vinylic
allylic
• Find ranges of chemical shift in Table 13.1
alkane δ 0.9–1.8
allylic δ 1.5–2.6
vinylic δ 4.5–6.5
• Best match with NMR spectrum is
δ 1.6
=CH₂ δ 4.8
δ 2.2

(b) Do the same for the chemical shifts δ 2.0, 5.1, and 7.2 of benzyl acetate.

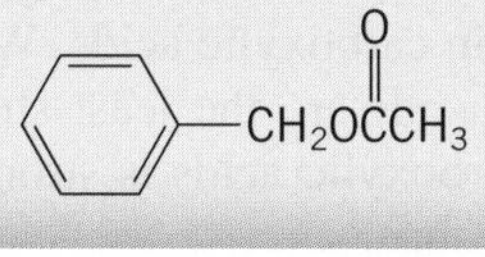
CH₂OCCH₃
O

Acetylenic hydrogens are unusual in that they are more shielded than we would expect for protons bonded to *sp*-hybridized carbon. This is because the π electrons circulate around the triple bond, not along it (Figure 13.9*a*). Therefore, the induced magnetic field is parallel to the long axis of the triple bond and shields the acetylenic proton (Figure 13.9*b*). Acetylenic protons typically have chemical shifts in the range δ 1.8–3.1.

δ 1.9
$H—C \equiv C—CH_2CH_2CH_2CH_3$
1-Hexyne

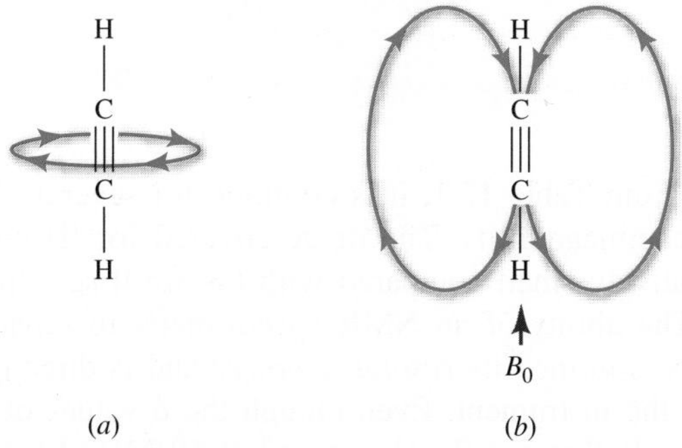
H
C
C
H
B_0
(a)
(b)

Figure 13.9

(*a*) The π electrons of acetylene circulate in a region surrounding the long axis of the molecule. (*b*) The induced magnetic field associated with the π electrons opposes the applied field and shields the protons.

The induced field of a carbonyl group (C═O) deshields protons in much the same way that C═C does, and its oxygen makes it even more electron withdrawing.

Thus, protons attached to C=O in aldehydes are the least shielded of any protons bonded to carbon. They have chemical shifts in the range δ 9–10.

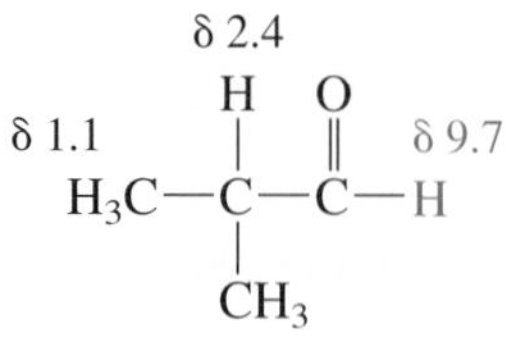

2-Methylpropanal

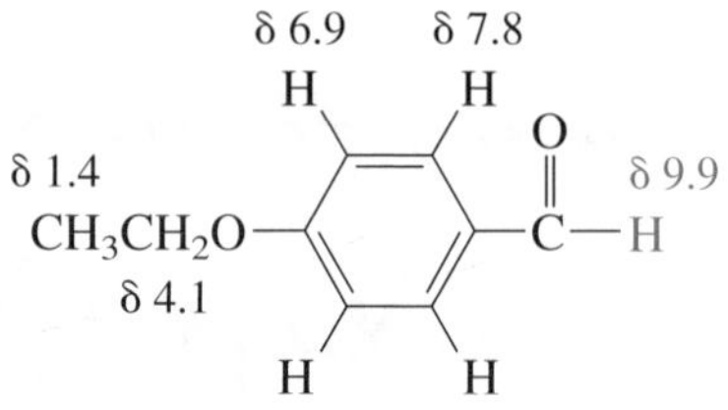

p-Ethoxybenzaldehyde

Protons on carbons adjacent to a carbonyl group are deshielded slightly more than allylic hydrogens.

Problem 13.6

Assign the chemical shifts δ 1.1, δ 1.7, δ 2.0, and δ 2.3 to the appropriate protons of 2-pentanone.

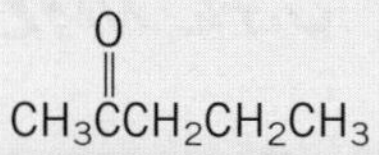

The second portion of Table 13.1 deals with O—H and N—H protons. As the table indicates, the chemical shifts of these vary much more than for protons bonded to carbon. This is because O—H and N—H groups can be involved in intermolecular hydrogen bonding, the extent of which depends on molecular structure, temperature, concentration, and solvent. Generally, an increase in hydrogen bonding decreases the shielding. This is especially evident in carboxylic acids. With δ values in the 10–12 ppm range, O—H protons of carboxylic acids are the least shielded of all of the protons in Table 13.1. Hydrogen bonding in carboxylic acids is stronger than in most other classes of compounds that contain O—H groups.

Problem 13.7

Assign the chemical shifts δ 1.6, δ 4.0, δ 7.5, δ 8.2, and δ 12.0 to the appropriate protons of 2-(*p*-nitrophenyl)propanoic acid.

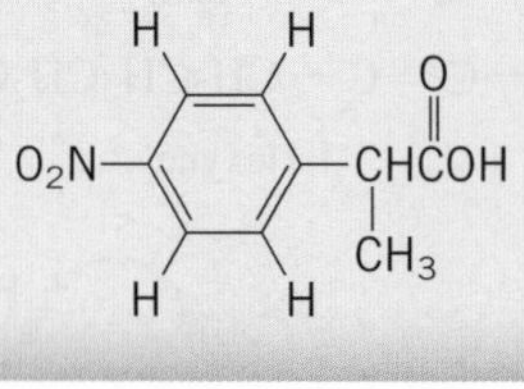

As you can see from Table 13.1, it is common for several different kinds of protons to have similar chemical shifts. The range covered for ^1H chemical shifts is only 12 ppm, which is relatively small compared with (as we'll see) the 200-ppm range for ^{13}C chemical shifts. The ability of an NMR spectrometer to separate signals that have similar chemical shifts is termed its *resolving power* and is directly related to the magnetic field strength of the instrument. Even though the δ values of their chemical shifts don't change, two signals that are closely spaced at 60 MHz become well separated at 300 MHz.

Ring Currents: Aromatic and Antiaromatic

We saw in Chapter 12 that aromaticity reveals itself in various ways. Qualitatively, aromatic compounds are more stable and less reactive than alkenes. Quantitatively, their heats of hydrogenation are smaller than expected. Theory, especially Hückel's rule, furnishes a structural basis for aromaticity. Now let's examine some novel features of the NMR spectra of aromatic compounds.

We have mentioned that the protons in benzene appear at relatively low field because of deshielding by the magnetic field associated with the circulating π electrons. The amount of deshielding is sufficiently large—on the order of 2 ppm more than the corresponding effect in alkenes—that its presence is generally accepted as evidence for aromaticity. We speak of this deshielding as resulting from an *aromatic ring current.*

Something interesting happens when we go beyond benzene to apply the aromatic ring current test to annulenes.

[18]Annulene satisfies the Hückel ($4n + 2$) π electron rule for aromaticity, and many of its properties indicate aromaticity (Section 11.20). As shown in Figure 13.10*a*, [18]annulene contains two different kinds of protons; 12 lie on the ring's periphery ("outside"), and 6 reside near the middle of the molecule ("inside"). The 2:1 ratio of outside/inside protons makes it easy to assign the signals in the ^{1}H NMR spectrum. The outside protons have a chemical shift δ of 9.3 ppm, which makes them even less shielded than those of benzene. The six inside protons, on the other hand, have a *negative* chemical shift (δ −3.0), meaning that the signal for these protons appears at *higher field* (to the right) of the TMS peak. *The inside protons of [18]annulene are more than 12 ppm more shielded than the outside protons.*

As shown in Figure 13.10*a*, both the shielding of the inside protons and the deshielding of the outside ones result from the same aromatic ring current. When the molecule is placed in an external magnetic field B_0, its circulating π electrons produce their own magnetic field. This induced field opposes the applied field B_0 in the center of the molecule, shielding the inside protons. Because the induced magnetic field closes on itself, the outside protons lie in a region where the induced field reinforces B_0. The aromatic ring current in [18]annulene shields the 6 inside protons and deshields the 12 outside ones.

Exactly the opposite happens in [16]annulene (Figure 13.10*b*). Now it is the outside protons (δ 5.3) that are more shielded. The inside protons (δ 10.6) are less shielded than the outside ones and less shielded than the protons of both benzene and [18]annulene. This reversal of the shielding and deshielding regions in going from [18] to [16]annulene can only mean that the directions of their induced magnetic fields are reversed. Thus [16]annulene, which has $4n$ π electrons and is antiaromatic, not only lacks an aromatic ring current, its π electrons produce exactly the opposite effect when placed in a magnetic field.

Score one for Hückel.

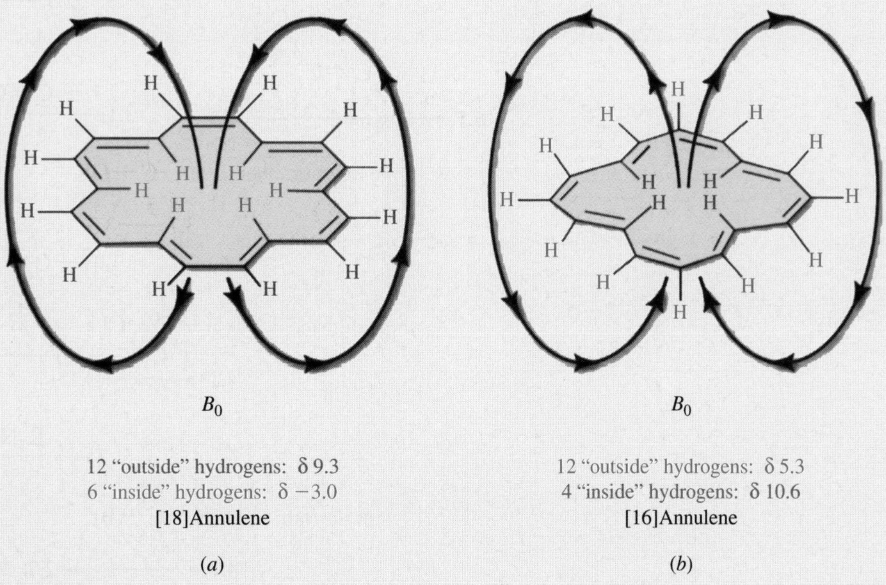

B_0
12 "outside" hydrogens: δ 9.3
6 "inside" hydrogens: δ −3.0
[18]Annulene
(*a*)
B_0
12 "outside" hydrogens: δ 5.3
4 "inside" hydrogens: δ 10.6
[16]Annulene
(*b*)

Figure 13.10

More shielded (red) and less shielded (blue) protons in (*a*) [18]annulene and (*b*) [16]annulene. The induced magnetic field associated with the aromatic ring current in [18]annulene shields the inside protons and deshields the outside protons. The opposite occurs in [16]annulene, which is antiaromatic.

13.6 Interpreting ¹H NMR Spectra

A good place to start when interpreting an NMR spectrum is to take a look at the "¹H spectral window" to get an idea of what structural units might be present from the chemical shifts that are observed. Figure 13.11 presents the same chemical-shift values as Table 13.1, graphically showing the relative positions of protons that are parts of different structural units.

An NMR spectrum also provides other useful information, including:

1. *The number of signals,* which tells us how many different kinds of protons there are.
2. *The intensity of the signals* as measured by the area under each peak, which tells us the relative ratios of the different kinds of protons.
3. *The multiplicity, or splitting, of each signal,* which tells us how many protons are vicinal to the one giving the signal.

Protons that have different chemical shifts are said to be **chemical-shift-nonequivalent** (or **chemically nonequivalent**). A separate NMR signal is given for each

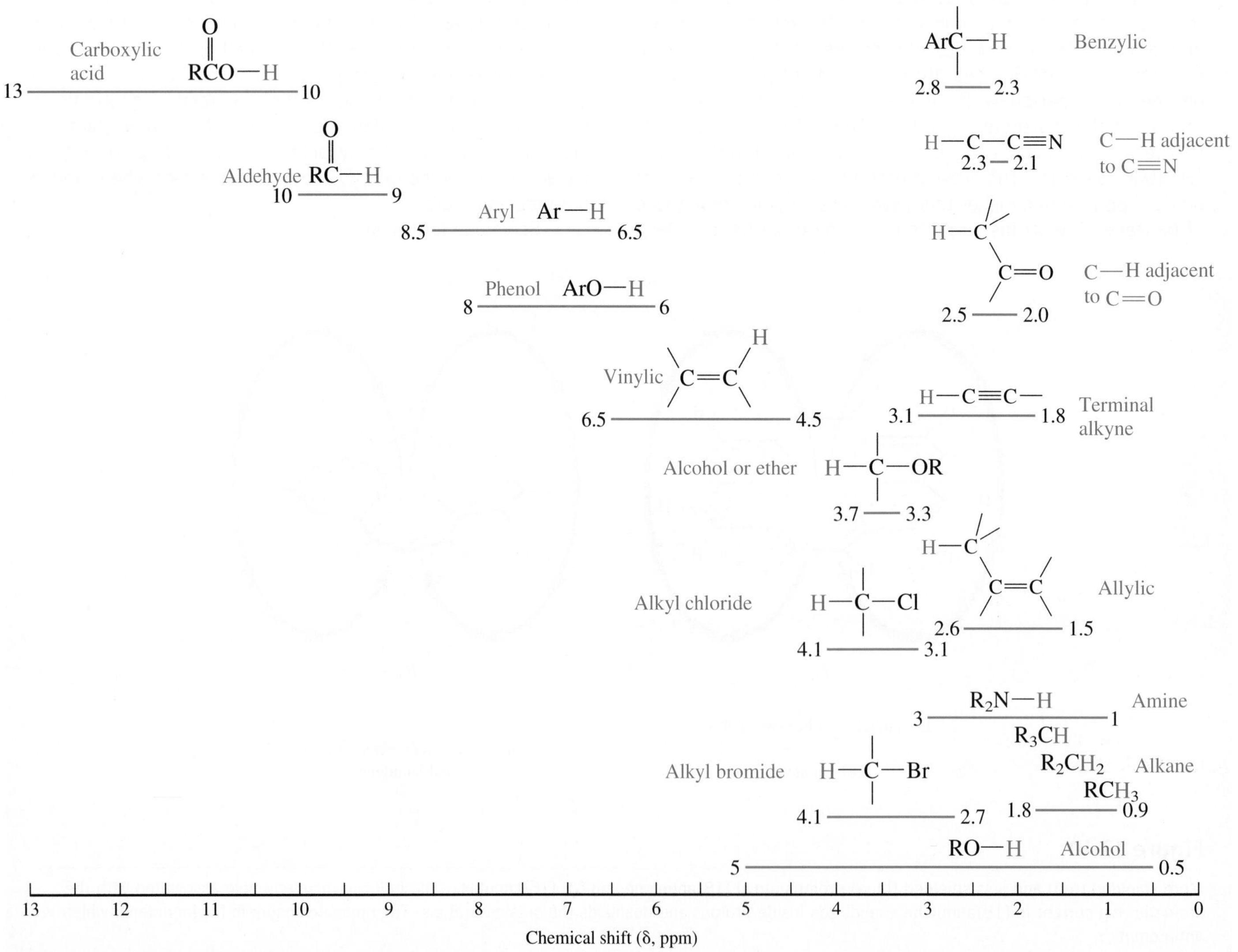

Figure 13.11

Approximate chemical-shift ranges for protons of various structural types in parts per million (δ) from tetramethylsilane. Protons in specific compounds may appear outside of the cited range depending on the shielding or deshielding effect of substituents. The chemical shifts of O—H and N—H protons depend on the conditions (solvent, temperature, concentration) under which the spectrum is recorded.

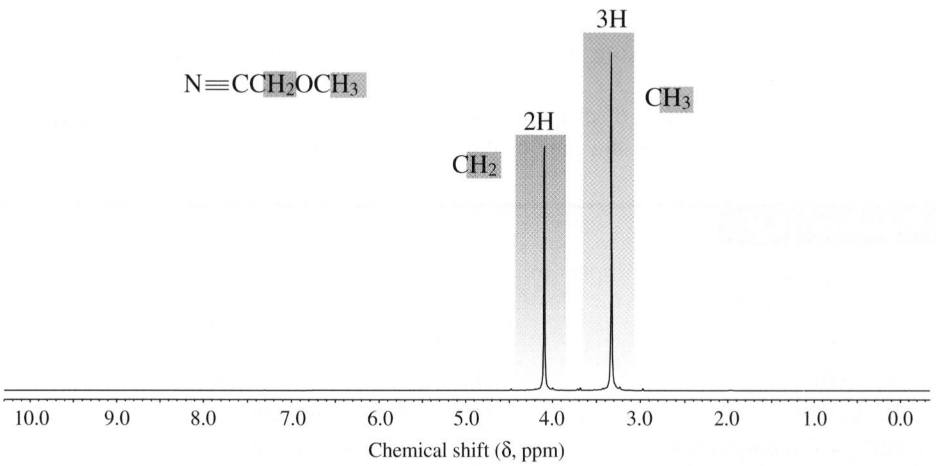

Figure 13.12

The 200-MHz ^1H NMR spectrum of methoxyacetonitrile (CH_3OCH_2CN).

chemical-shift-nonequivalent proton in a substance. Figure 13.12 shows the 200-MHz ^1H NMR spectrum of methoxyacetonitrile (CH_3OCH_2CN), a molecule with protons in two different environments. The three protons in the CH_3O group constitute one set, the two protons in the OCH_2CN group the other, and both can be assigned on the basis of their chemical shifts. The protons in the OCH_2CN group are connected to a carbon that bears two electronegative substituents (O and C≡N), are less shielded, and have a larger chemical shift (δ 4.1) than those of the CH_3O group (δ 3.3), which are attached to a carbon that bears only one electronegative atom (O).

Another way to assign the peaks is by comparing their intensities. The three equivalent protons of the CH_3O group give rise to a more intense peak than the two equivalent protons of the OCH_2CN group. This is clear by comparing the heights of the peaks in the spectrum in this case, but in general it is better to compare peak areas. This is done electronically at the time the NMR spectrum is recorded, and the **integrated areas** are displayed on the computer screen or printed out. Peak areas are proportional to the number of equivalent protons responsible for that signal.

It is important to remember that integration of peak areas gives relative, not absolute, proton counts. Thus, a 3:2 ratio of areas can, as in the case of CH_3OCH_2CN, correspond to a 3:2 ratio of protons. But in some other compound a 3:2 ratio of areas might correspond to a 6:4 or 9:6 ratio of protons.

Problem 13.8

The 200-MHz ^1H NMR spectrum of 1,4-dimethylbenzene looks exactly like that of CH_3OCH_2CN except the chemical shifts of the two peaks are δ 2.2 and δ 7.0. Assign the peaks to the appropriate protons of 1,4-dimethylbenzene.

Protons in equivalent environments have the same chemical shift. Often it is an easy matter to decide, simply by inspection, whether protons are equivalent or not. In more difficult cases, mentally replacing a proton in a molecule by a "test group" can help. We'll illustrate the procedure for a simple case—the protons of propane. To see if they have the same chemical shift, replace one of the methyl protons at C-1 by chlorine, then do the same thing for a proton at C-3. Both replacements give the same molecule, 1-chloropropane. Therefore the methyl protons at C-1 are equivalent to those at C-3.

$CH_3CH_2CH_3$	$ClCH_2CH_2CH_3$	$CH_3CH_2CH_2Cl$
Propane	1-Chloropropane	1-Chloropropane

If the two structures produced by mental replacement of two different hydrogens in a molecule by a test group are the same, the hydrogens are chemically equivalent. Thus, the six methyl protons of propane are all chemically equivalent to one another and have the same chemical shift.

Replacement of either one of the methylene protons of propane generates 2-chloropropane. Both methylene protons are equivalent. Neither of them is equivalent to any of the methyl protons.

The ^1H NMR spectrum of propane contains two signals: one for the six equivalent methyl protons, the other for the pair of equivalent methylene protons.

Problem 13.9

How many signals would you expect to find in the ^1H NMR spectrum of each of the following compounds?

(a) 1-Bromobutane
(b) 1-Butanol
(c) Butane
(d) 1,4-Dibromobutane
(e) 2,2-Dibromobutane
(f) 2,2,3,3-Tetrabromobutane
(g) 1,1,4-Tribromobutane
(h) 1,1,1-Tribromobutane

Sample Solution (a) To test for chemical-shift equivalence, replace the protons at C-1, C-2, C-3, and C-4 of 1-bromobutane by some test group such as chlorine. Four constitutional isomers result:

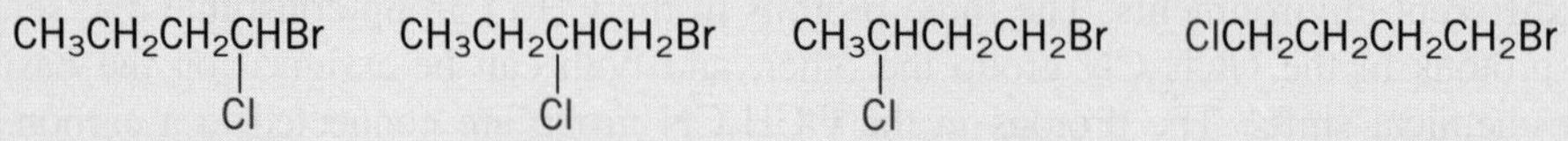
$CH_3CH_2CH_2CHBr$ (Cl)
$CH_3CH_2CHCH_2Br$ (Cl)
$CH_3CHCH_2CH_2Br$ (Cl)
$ClCH_2CH_2CH_2CH_2Br$

1-Bromo-1-chlorobutane | 1-Bromo-2-chlorobutane | 1-Bromo-3-chlorobutane | 1-Bromo-4-chlorobutane

Thus, separate signals will be seen for the protons at C-1, C-2, C-3, and C-4. Barring any accidental overlap, we expect to find four signals in the NMR spectrum of 1-bromobutane.

Chemical-shift nonequivalence can occur when two environments are stereochemically different. The two vinyl protons of 2-bromopropene have different chemical shifts.

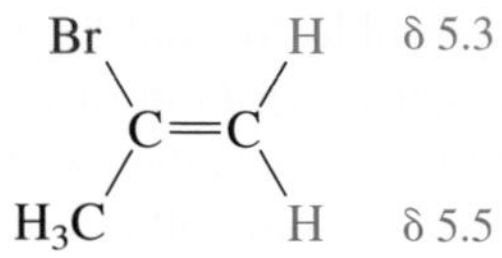
Br
H δ 5.3
C=C
H_3C
H δ 5.5

2-Bromopropene

One of the vinyl protons is cis to bromine; the other trans. Replacing one of the vinyl protons by some test group, say, chlorine, gives the *Z* isomer of 2-bromo-1-chloropropene; replacing the other gives the *E* stereoisomer. The *E* and *Z* forms of 2-bromo-1-chloropropene are diastereomers. Protons that yield diastereomers on being replaced by some test group are *diastereotopic* (Section 7.14) and can have different chemical shifts. Because their environments are similar, however, the chemical shift difference is usually small, and it sometimes happens that two diastereotopic protons accidentally have the same chemical shift. Recording the spectrum on a higher field NMR spectrometer is often helpful in resolving signals with similar chemical shifts.

Problem 13.10

How many signals would you expect to find in the ^1H NMR spectrum of each of the following compounds?

(a) Vinyl bromide
(b) 1,1-Dibromoethene
(c) *cis*-1,2-Dibromoethene
(d) *trans*-1,2-Dibromoethene
(e) Allyl bromide
(f) 2-Methyl-2-butene

Sample Solution (a) Each proton of vinyl bromide is unique and has a chemical shift different from the other two. The least shielded proton is attached to the carbon that bears the bromine. The pair of protons at C-2 are diastereotopic with respect to each other; one is cis to bromine and the other is trans to bromine. There are three proton signals in the NMR spectrum of vinyl bromide. Their observed chemical shifts are as indicated.

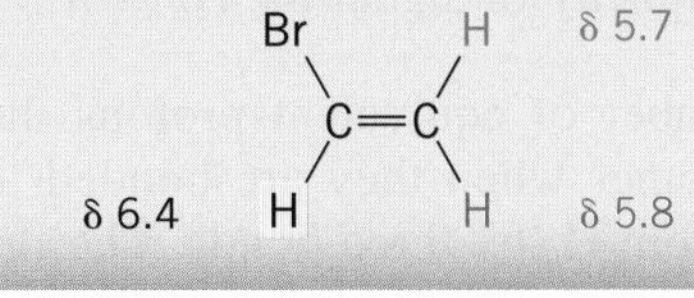

When enantiomers are generated by replacing first one proton and then another by a test group, the pair of protons are *enantiotopic* (Section 7.10). The methylene protons at C-2 of 1-propanol, for example, are enantiotopic.

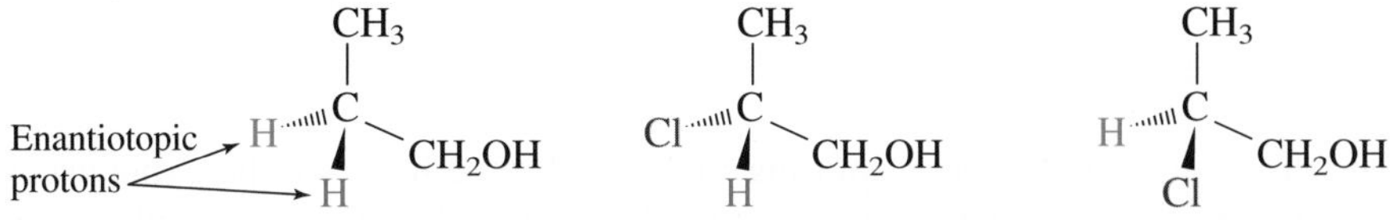

1-Propanol (*R*)-2-Chloro-1-propanol (*S*)-2-Chloro-1-propanol

Replacing one of these protons by chlorine as a test group gives (*R*)-2-chloro-1-propanol; replacing the other gives (*S*)-2-chloro-1-propanol. Enantiotopic protons have the same chemical shift, regardless of the field strength of the NMR spectrometer.

Enantiotopic protons can have different chemical shifts in an optically pure chiral solvent. Because the customary solvent ($CDCl_3$) used in NMR measurements is achiral, this phenomenon is not observed in routine work.

At the beginning of this section we noted that an NMR spectrum provides structural information based on chemical shift, the number of peaks, their relative areas, and the multiplicity, or splitting, of the peaks. We have discussed the first three of these features of ^{1}H NMR spectroscopy. Let's now turn our attention to peak splitting to see what kind of information it offers.

13.7 Spin–Spin Splitting in ^{1}H NMR Spectroscopy

The ^{1}H NMR spectrum of CH_3OCH_2CN (see Figure 13.12) displayed in the preceding section is relatively simple because both signals are singlets; that is, each one consists of a single peak. It is quite common though to see a signal for a particular proton appear not as a singlet, but as a collection of peaks. The signal may be split into two peaks (a doublet), three peaks (a triplet), four peaks (a quartet), or even more. Figure 13.13 shows the ^{1}H NMR spectrum of 1,1-dichloroethane (CH_3CHCl_2), which is characterized by a doublet centered at δ 2.1 for the methyl protons and a quartet at δ 5.9 for the methine proton.

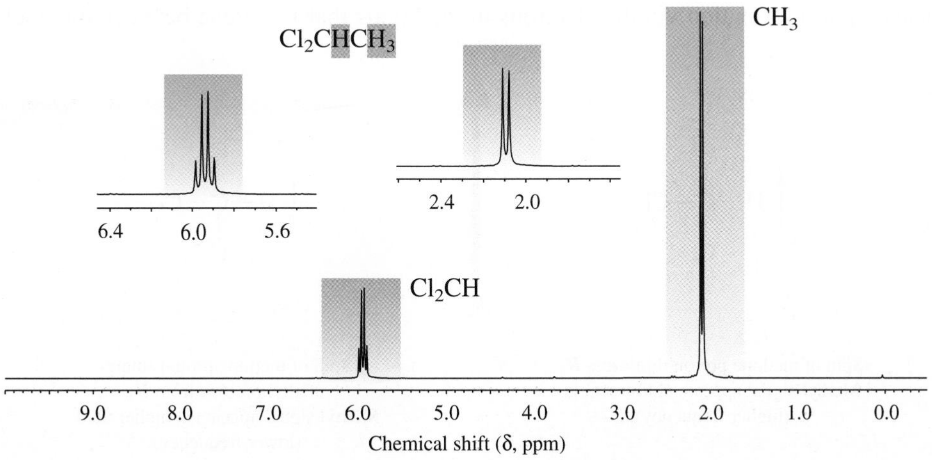

Figure 13.13

The 200-MHz ^{1}H NMR spectrum of 1,1-dichloroethane (Cl_2CHCH_3), showing the methine proton as a quartet and the methyl protons as a doublet. The peak multiplicities are seen more clearly in the scale-expanded insets.

More complicated splitting patterns conform to an extension of the "$n + 1$" rule and will be discussed in Section 13.11.

The number of peaks into which the signal for a particular proton is split is called its **multiplicity.** For simple cases the rule that allows us to predict splitting in ^{1}H NMR spectroscopy is

$$\text{Multiplicity of signal for } H_a = n + 1$$

where n is equal to the number of equivalent protons that are vicinal to H_a. Two protons are vicinal to each other when they are bonded to adjacent atoms. Protons vicinal to H_a are separated from H_a by three bonds. The three methyl protons of 1,1-dichloroethane are equivalent and vicinal to the methine proton and split its signal into a quartet. The single methine proton, in turn, splits the methyl protons' signal into a doublet.

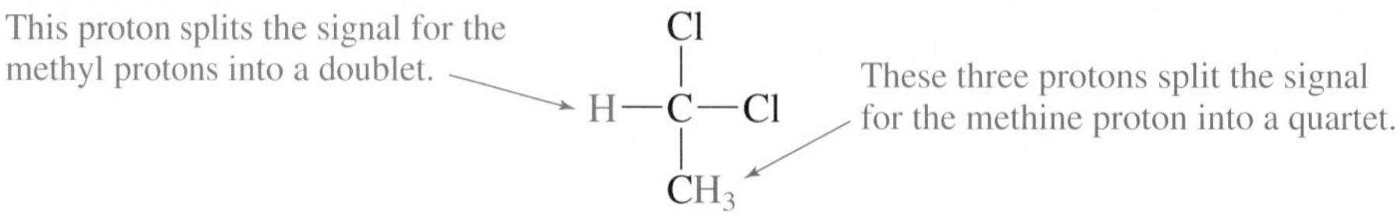
This proton splits the signal for the methyl protons into a doublet.
These three protons split the signal for the methine proton into a quartet.

The physical basis for peak splitting in 1,1-dichloroethane can be explained with the aid of Figure 13.14, which examines how the chemical shift of the methyl protons is affected by the spin of the methine proton. There are two magnetic environments for the methyl protons: one in which the magnetic moment of the methine proton is parallel to the applied field, and the other in which it is antiparallel to it. When the magnetic moment of the methine proton is parallel to the applied field, it reinforces it. This decreases the shielding of the methyl protons and causes their signal to appear at slightly lower field strength (higher frequency). Conversely, when the magnetic moment of the methine proton is antiparallel to the applied field, it opposes it and increases the shielding of the methyl protons. Instead of a single peak for the methyl protons, there are two of approximately equal intensity: one at slightly higher field than the "true" chemical shift, the other at slightly lower field.

Turning now to the methine proton, its signal is split by the methyl protons into a quartet. The same kind of analysis applies here and is outlined in Figure 13.15. The methine proton "sees" eight different combinations of nuclear spins for the methyl protons. In one combination, the magnetic moments of all three methyl protons reinforce the applied field. At the other extreme, the magnetic moments of all three methyl protons oppose the applied field. There are three combinations in which the magnetic moments of two methyl protons reinforce the applied field, whereas one opposes it. Finally, there are three combinations in which the magnetic moments of two methyl protons oppose the applied field and one reinforces it. These eight possible combinations give rise to four distinct peaks for the methine proton, with a ratio of intensities of 1:3:3:1.

We describe the observed splitting of NMR signals as **spin–spin splitting** and the physical basis for it as **spin–spin coupling.** It has its origin in the communication of nuclear spin information via the electrons in the bonds that intervene between the nuclei.

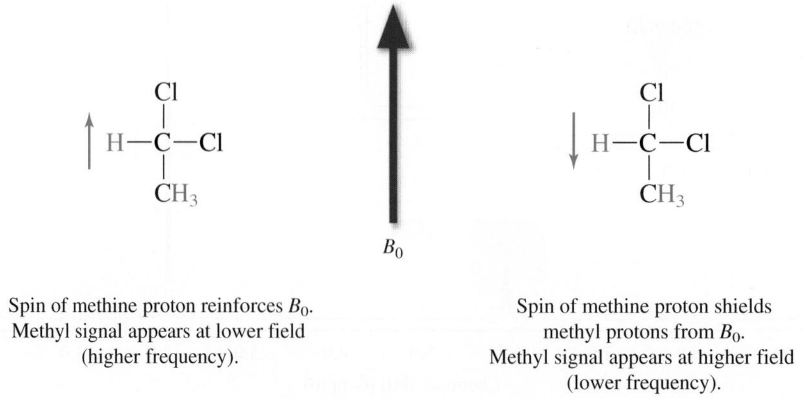
B_0
Spin of methine proton reinforces B_0. Methyl signal appears at lower field (higher frequency).
Spin of methine proton shields methyl protons from B_0. Methyl signal appears at higher field (lower frequency).

Figure 13.14

The magnetic moments (blue arrows) of the two possible spin states of the methine proton affect the chemical shift of the methyl protons in 1,1-dichloroethane. When the magnetic moment is parallel to the external field B_0 (green arrow), it adds to the external field and a smaller B_0 is needed for resonance. When it is antiparallel to the external field, it subtracts from it and shields the methyl protons.

There are eight possible combinations of the nuclear spins of the three methyl protons in CH_3CHCl_2.

These eight combinations cause the signal of the $CHCl_2$ proton to be split into a quartet, in which the intensities of the peaks are in the ratio 1:3:3:1.

Figure 13.15

The methyl protons of 1,1-dichloroethane split the signal of the methine proton into a quartet.

Its effect is greatest when the number of bonds is small. Vicinal protons are separated by three bonds, and coupling between vicinal protons, as in 1,1-dichloroethane, is called **three-bond coupling,** or **vicinal coupling.** Four-bond couplings are weaker and not normally observable.

A very important characteristic of spin–spin splitting is that protons that have the same chemical shift do not split each other's signal. Ethane, for example, shows only a single sharp peak in its NMR spectrum. Even though there is a vicinal relationship between the protons of one methyl group and those of the other, they do not split each other's signal because they are equivalent.

Problem 13.11

Describe the appearance of the ^1H NMR spectrum of each of the following compounds. How many signals would you expect to find, and into how many peaks will each signal be split?

(a) 1,2-Dichloroethane
(b) 1,1,1-Trichloroethane
(c) 1,1,2-Trichloroethane
(d) 1,2,2-Trichloropropane
(e) 1,1,1,2-Tetrachloropropane

Sample Solution (a) All the protons of 1,2-dichloroethane ($ClCH_2CH_2Cl$) are chemically equivalent and have the same chemical shift. Protons that have the same chemical shift do not split each other's signal, and so the NMR spectrum of 1,2-dichloroethane consists of a single sharp peak.

Coupling of nuclear spins requires that the nuclei split each other's signal equally. The separation between the two halves of the methyl doublet in 1,1-dichloroethane is equal to the separation between any two adjacent peaks of the methine quartet. The extent to which two nuclei are coupled is given by the **coupling constant *J*** and in simple cases is equal to the separation between adjacent lines of the signal of a particular proton. The three-bond coupling constant $^3J_{ab}$ in 1,1-dichloroethane has a value of 7 Hz. *The size of the coupling constant is independent of the field strength;* the separation between adjacent peaks in 1,1-dichloroethane is 7 Hz, irrespective of whether the spectrum is recorded at 200 MHz or 500 MHz.

13.8 Splitting Patterns: The Ethyl Group

One of the most characteristic patterns of peaks is that of the ethyl group, represented in the NMR spectrum of ethyl bromide in Figure 13.16. In compounds of the type CH_3CH_2X, especially where X is an electronegative atom or group, such as bromine in

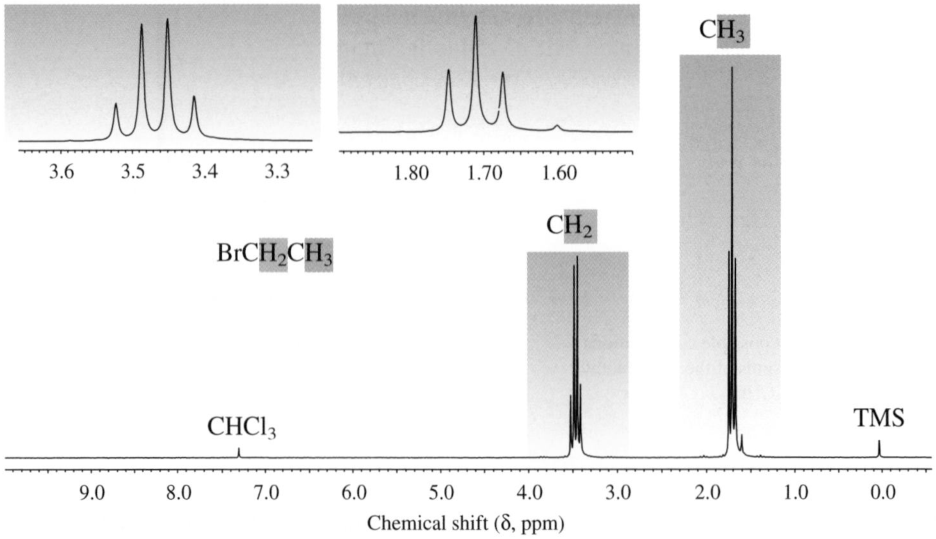
3.6
3.5
3.4
3.3
1.80
1.70
1.60
$BrCH_2CH_3$
CH_2
CH_3
$CHCl_3$
TMS
9.0 8.0 7.0 6.0 5.0 4.0 3.0 2.0 1.0 0.0
Chemical shift (δ, ppm)

Figure 13.16

The 200-MHz ^1H NMR spectrum of ethyl bromide ($BrCH_2CH_3$), showing the characteristic triplet–quartet pattern of an ethyl group. The small peak at δ 1.6 is an impurity.

ethyl bromide, the ethyl group appears as a *triplet–quartet pattern.* The signal for the methylene protons is split into a quartet by coupling with the three methyl protons. The signal for the methyl protons is a triplet because of vicinal coupling to the two protons of the adjacent methylene group.

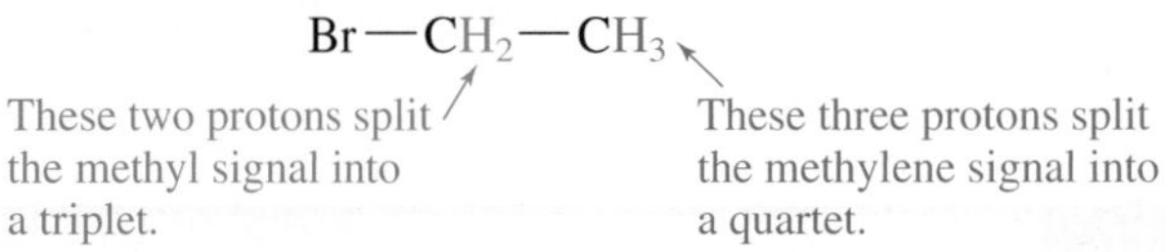
Br—CH_2—CH_3
These two protons split the methyl signal into a triplet.
These three protons split the methylene signal into a quartet.

We have discussed in the preceding section why methyl groups split the signals due to vicinal protons into a quartet. Splitting by a methylene group gives a triplet corresponding to the spin combinations shown in Figure 13.17 for ethyl bromide. The relative intensities of the peaks of this triplet are 1:2:1.

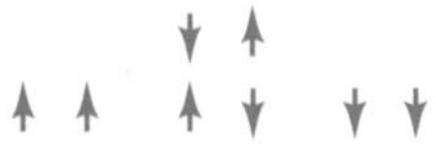

There are four possible combinations of the nuclear spins of the two methylene protons in CH_3CH_2Br.

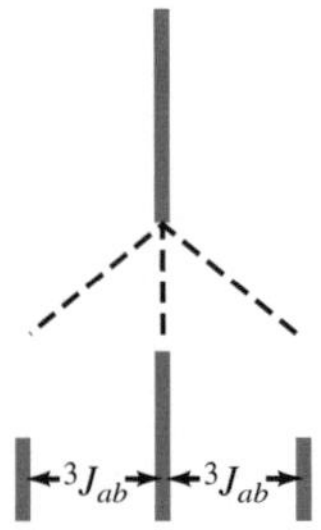
$^3J_{ab}$
$^3J_{ab}$

These four combinations cause the signal of the CH_3 protons to be split into a triplet, in which the intensities of the peaks are in the ratio 1:2:1.

Figure 13.17

The methylene protons of ethyl bromide split the signal of the methyl protons into a triplet.

Problem 13.12

Describe the appearance of the ^1H NMR spectrum of each of the following compounds. How many signals would you expect to find, and into how many peaks will each signal be split?

(a) $ClCH_2OCH_2CH_3$
(b) $CH_3CH_2OCH_3$
(c) $CH_3CH_2OCH_2CH_3$
(d) *p*-Diethylbenzene
(e) $ClCH_2CH_2OCH_2CH_3$

Sample Solution (a) Along with the triplet–quartet pattern of the ethyl group, the NMR spectrum of this compound will contain a singlet for the two protons of the chloromethyl group.

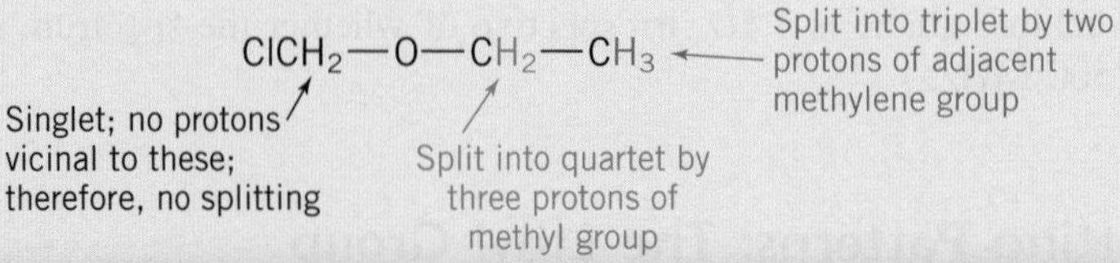
$ClCH_2$—O—CH_2—CH_3
Split into triplet by two protons of adjacent methylene group
Singlet; no protons vicinal to these; therefore, no splitting
Split into quartet by three protons of methyl group

Table 13.2 summarizes the splitting patterns and peak intensities expected for coupling to various numbers of protons.

TABLE 13.2 Splitting Patterns of Common Multiplets

Number of protons to which nucleus is equally coupled	Appearance of multiplet	Intensities of lines in multiplet
1	Doublet	1:1
2	Triplet	1:2:1
3	Quartet	1:3:3:1
4	Quintet	1:4:6:4:1
5	Sextet	1:5:10:10:5:1
6	Septet	1:6:15:20:15:6:1

The intensities correspond to the coefficients of a binomial expansion (Pascal's triangle).

13.9 Splitting Patterns: The Isopropyl Group

The NMR spectrum of isopropyl chloride (Figure 13.18) illustrates the appearance of an isopropyl group. The signal for the six equivalent methyl protons at δ 1.5 is split into a doublet by the proton of the H—C—Cl unit. In turn, the H—C—Cl proton signal at δ 4.2 is split into a septet by the six methyl protons. A *doublet–septet* pattern is characteristic of an isopropyl group.

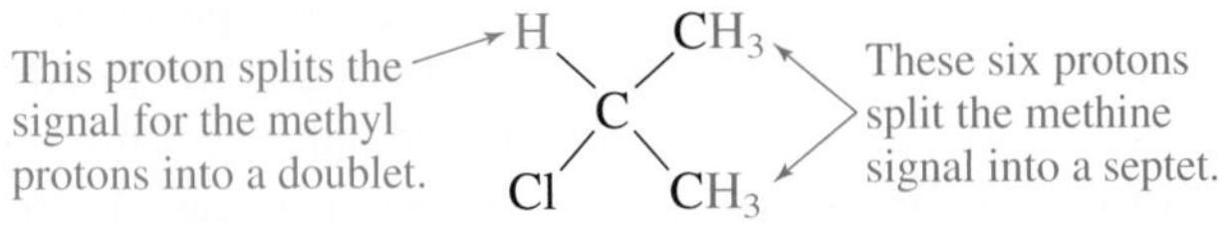

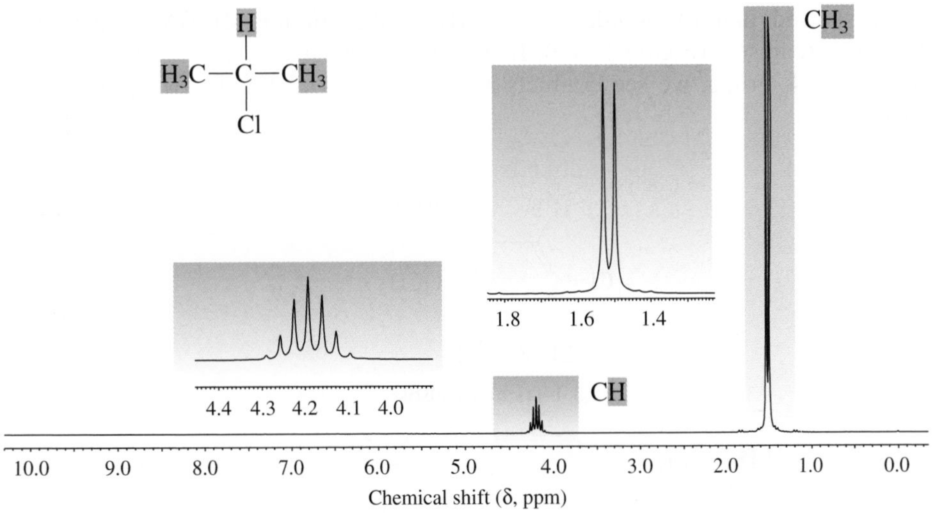

Figure 13.18

The 200-MHz ^1H NMR spectrum of isopropyl chloride, showing the doublet–septet pattern of an isopropyl group.

13.10 Splitting Patterns: Pairs of Doublets

We often see splitting patterns in which the intensities of the individual peaks do not match those given in Table 13.2, but are distorted in that the signals for coupled protons "lean" toward each other. This leaning is a general phenomenon, but is most easily illustrated for the case of two nonequivalent vicinal protons as shown in Figure 13.19.

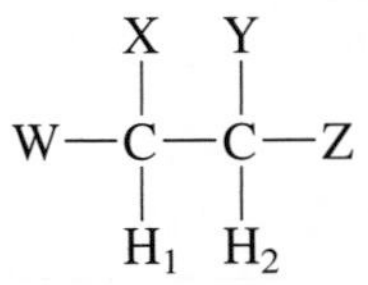

The appearance of the splitting pattern of protons 1 and 2 depends on their coupling constant J and the chemical shift difference $\Delta\nu$ between them. When the ratio $\Delta\nu/J$ is

Figure 13.19

The appearance of the splitting pattern of two coupled protons depends on their coupling constant *J* and the chemical shift difference $\Delta\nu$ between them. As the ratio $\Delta\nu/J$ decreases, the doublets become increasingly distorted. When the two protons have the same chemical shift, no splitting is observed.

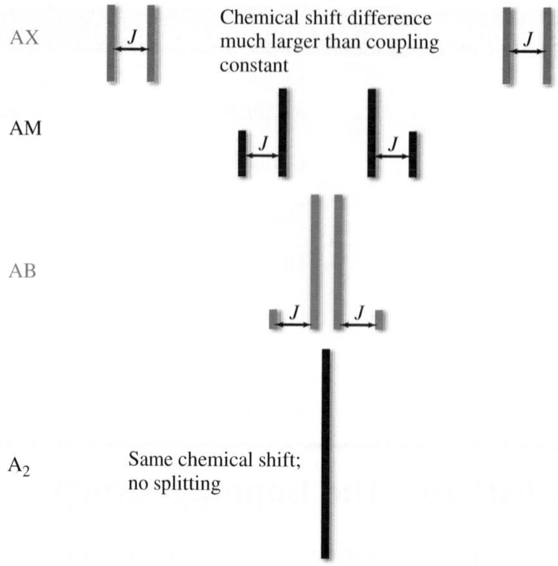

large, two symmetrical 1:1 doublets are observed. We refer to this as the "AX" case, using two letters that are remote in the alphabet to stand for signals well removed from each other on the spectrum. Keeping the coupling constant the same while reducing $\Delta\nu$ leads to a steady decrease in the intensity of the outer two peaks with a simultaneous increase in the inner two as we progress from AX through AM to AB. At the extreme (A_2), the two protons have the same chemical shift, the outermost lines have disappeared, and no splitting is observed. Because of its appearance, it is easy to misinterpret an AB or AM pattern as a quartet, rather than the pair of skewed doublets it really is.

A skewed pair of doublets is clearly visible in the ^{1}H NMR spectrum of 2,3,4-trichloroanisole (Figure 13.20). In addition to the singlet at δ 3.9 for the protons of the —OCH_3 group, we see doublets at δ 6.8 and δ 7.3 for the two protons of the aromatic ring.

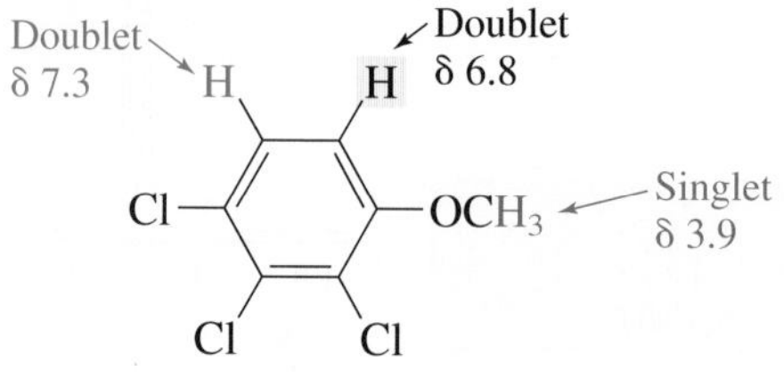

2,3,4-Trichloroanisole

Figure 13.20

The 200-MHz ^{1}H NMR spectrum of 2,3,4-trichloroanisole, showing the splitting of the ring protons into a pair of doublets that "lean" toward each other.

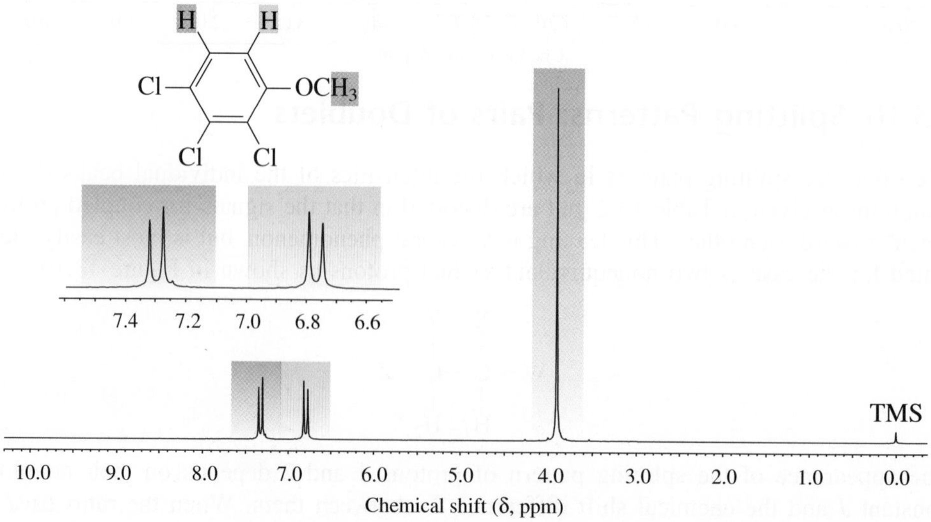

A pair of doublets frequently occurs with *geminal* protons (protons bonded to the same carbon). Geminal protons are separated by two bonds, and geminal coupling is referred to as *two-bond coupling* (^{2}J) in the same way that vicinal coupling is referred to as *three-bond coupling* (^{3}J). An example of geminal coupling is provided by the compound 1-chloro-1-cyanoethene, in which the two hydrogens appear as a pair of doublets. The splitting in each doublet is 2 Hz.

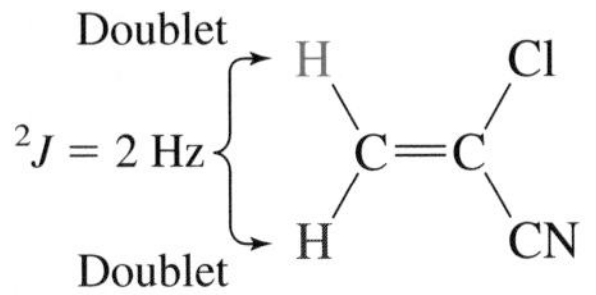

1-Chloro-1-cyanoethene

The protons in 1-chloro-1-cyanoethene are *diastereotopic* (Section 13.6). They are nonequivalent and have different chemical shifts. Remember, splitting can only occur between protons that have different chemical shifts.

Splitting due to geminal coupling is seen only in CH_2 groups and only when the two protons have different chemical shifts. All three protons of a methyl (CH_3) group are equivalent and cannot split one another's signal, and, of course, there are no protons geminal to a single methine (CH) proton.

13.11 Complex Splitting Patterns

All the cases we've discussed so far have involved splitting of a proton signal by coupling to other protons that were equivalent to one another. Indeed, we have stated the splitting rule in terms of the multiplicity of a signal as being equal to $n + 1$, where n is equal to the number of equivalent protons to which the proton that gives the signal is coupled. What if all the vicinal protons are *not* equivalent?

Figure 13.21*a* shows the signal for the proton marked Ar**CH**$_a$=CH_2 in *m*-nitrostyrene, which appears as a set of four peaks in the range δ 6.7–6.9. These four peaks are in fact a "doublet of doublets." The proton in question is *unequally coupled* to the two protons at the end of the vinyl side chain. The size of the vicinal coupling constant between protons trans to each other on a double bond is normally larger than that between cis protons. In this case the trans coupling constant is 16 Hz and the cis coupling constant is 12 Hz. Thus, as shown in Figure 13.21*b*, the signal for H_a is split into a doublet with a spacing of 16 Hz by one vicinal proton, and each line of this doublet is then split into another doublet with a spacing of 12 Hz.

You will find it revealing to construct a splitting diagram similar to that of Figure 13.21 for the case in which the cis and trans **H**—C=C—**H** coupling constants are equal. Under those circumstances the four-line pattern simplifies to a triplet, as it should for a proton equally coupled to two vicinal protons.

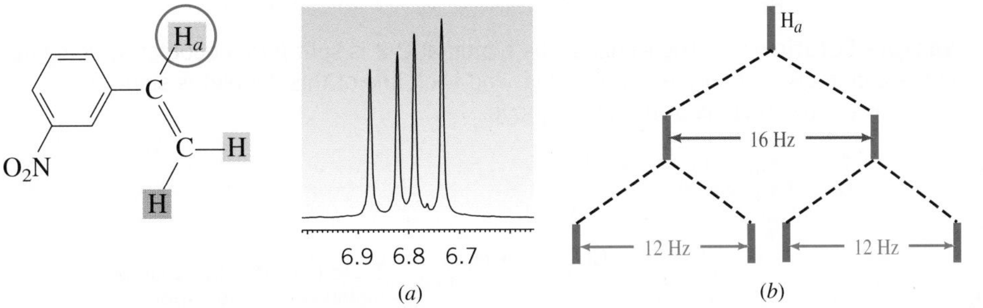

Figure 13.21

Splitting of a signal into a doublet of doublets by unequal coupling to two vicinal protons. (*a*) Appearance of the signal for the proton marked H_a in *m*-nitrostyrene as a set of four peaks. (*b*) Origin of these four peaks through successive splitting of the signal for H_a.

The "n + 1 rule" should be amended to read: *When a proton* H_a *is coupled to* H_b, H_c, H_d, *etc., and* $J_{ab} \neq J_{ac}, \neq J_{ad}$, *etc., the original signal for* H_a is split into $n + 1$ *peaks by n* H_b *protons, each of these lines is further split into n* + 1 *peaks by n* H_c *protons, and each of these into n* + 1 *lines by n* H_d *protons, and so on.* Bear in mind that because of overlapping peaks, the number of lines actually observed can be less than that expected on the basis of the splitting rule.

Diastereotopic hydrogens can complicate the NMR spectra of chiral molecules, especially those in which there is a CH_2 group adjacent to a chirality center. Consider H_a and H_b in (*S*)-1,2-diphenyl-2-propanol, which has a chirality center at C-2 (Figure 13.22). Replacement of H_a with a test group, for example, deuterium, gives a compound that is diastereomeric to the one that is generated by replacement of H_b. Recall that when diastereomers are produced by replacement of protons with test groups, the protons are diastereotopic and may have different chemical shifts (see Section 13.6). In

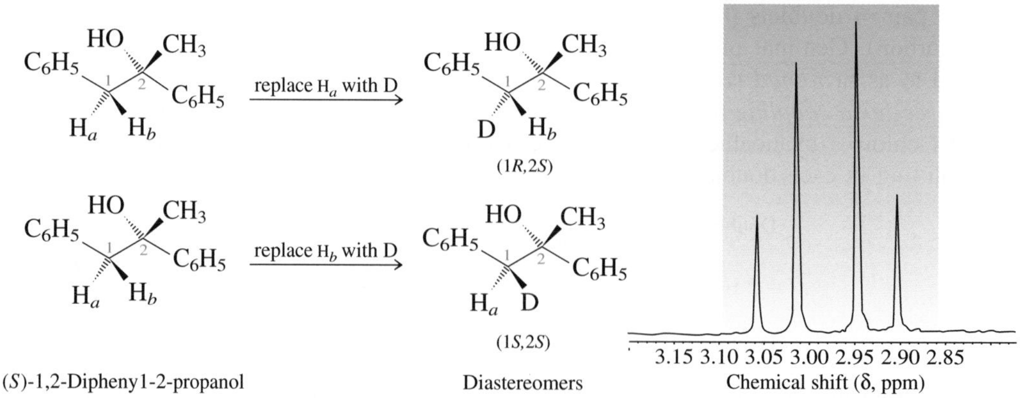

Figure 13.22

The methylene protons H_a and H_b of (*S*)-1,2-diphenyl-2-propanol are diastereotopic and appear as a pair of doublets in the 300-MHz ^1H NMR spectrum.

1,2-diphenyl-2-propanol, H_a and H_b are diastereotopic, have different chemical shifts, and appear as a pair of doublets. The pair of doublets appears as an AM splitting pattern (see Figure 13.19) because the chemical-shift differences between H_a and H_b are not much larger than the coupling constant.

The enantiomeric (*R*)-1,2-diphenyl-2-propanol, as well as a racemic mixture containing both *R* and *S* forms would produce NMR spectra identical to the one shown in Figure 13.22. As noted earlier, diastereotopic hydrogens can have very small, or even coincidental chemical shifts. In such cases, splitting would not normally be observed.

Problem 13.13

Describe the splitting pattern expected for the proton at

(a) C-2 in (*Z*)-1,3-dichloropropene

(b) C-2 in 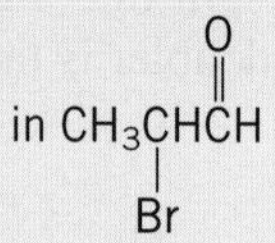

Sample Solution (a) The signal of the proton at C-2 is split into a doublet by coupling to the proton cis to it on the double bond, and each line of this doublet is split into a triplet by the two protons of the CH_2Cl group.

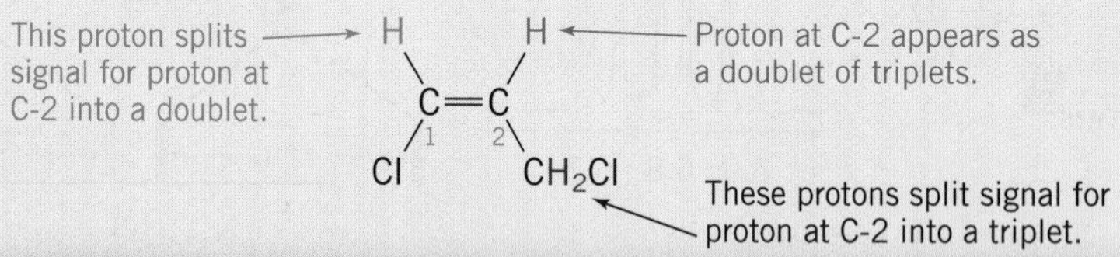

Problem 13.14

A portion of the ^1H NMR spectrum of the amino acid phenylalanine is shown in Figure 13.23. Why are eight lines observed for H_a and H_b?

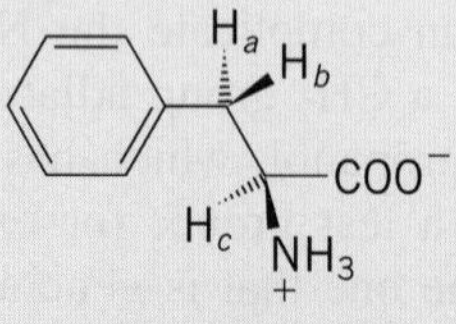

Phenylalanine

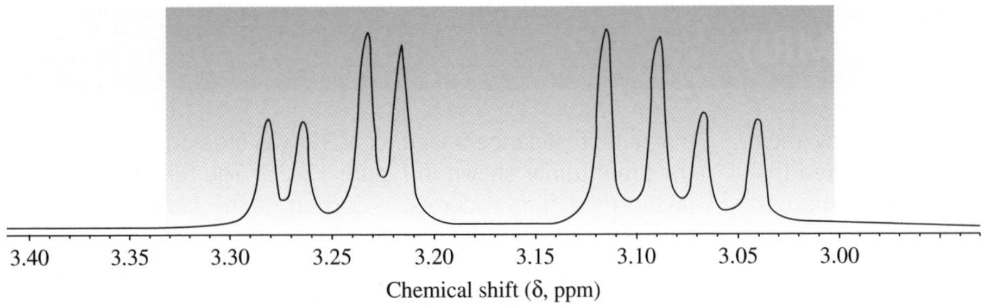

Figure 13.23

A portion of the 300-MHz ^{1}H NMR spectrum of phenylalanine for Problem 13.14 showing H_a and H_b.

13.12 ^{1}H NMR Spectra of Alcohols

The —OH proton of a primary alcohol RCH_2OH is vicinal to two protons, and its signal would be expected to be split into a triplet. Under certain conditions signal splitting of alcohol protons is observed, but usually it is not. Figure 13.24 presents the NMR spectrum of benzyl alcohol, showing the methylene and hydroxyl protons as singlets at δ 4.7 and 2.5, respectively. (The aromatic protons also appear as a singlet, but that is because they all accidentally have the same chemical shift and so cannot split each other.)

The reason that splitting of the hydroxyl proton of an alcohol is not observed is that it is involved in rapid exchange reactions with other alcohol molecules. Transfer of a proton from an oxygen of one alcohol molecule to the oxygen of another is quite fast and effectively *decouples* it from other protons in the molecule. Factors that slow down this exchange of OH protons, such as diluting the solution, lowering the temperature, or increasing the crowding around the OH group, can cause splitting of hydroxyl resonances.

Problem 13.15

Splitting of hydroxyl protons can be observed when the solvent in which the spectrum is recorded is dimethyl sulfoxide (DMSO) because hydrogen bonding to the oxygen of $(CH_3)_2\overset{+}{\ddot{S}}—\ddot{O}\!:^-$ slows down the rate of proton exchange between —OH groups. Explain how you could use this fact to distinguish among primary, secondary, and tertiary alcohols.

The chemical shift of the hydroxyl proton is variable, with a range of δ 0.5–5, depending on the solvent, the temperature at which the spectrum is recorded, and the concentration of the solution. The alcohol proton shifts to lower field in more concentrated solutions.

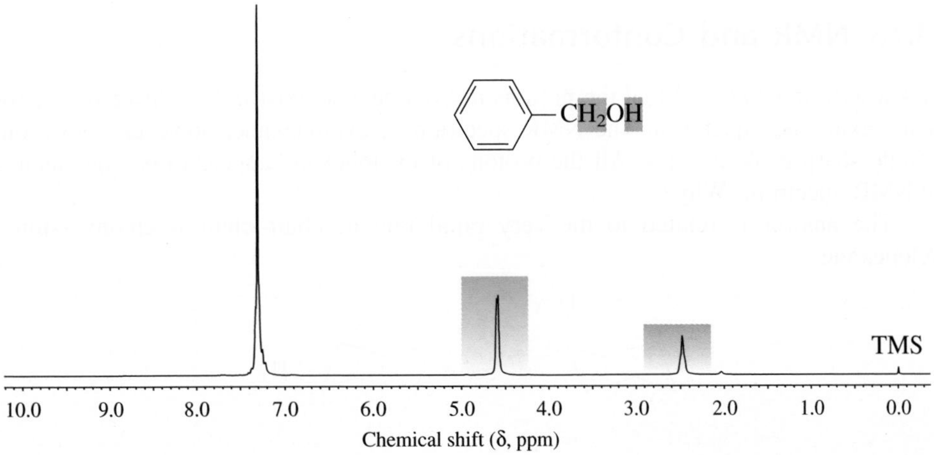

Figure 13.24

The 200-MHz ^{1}H NMR spectrum of benzyl alcohol. The hydroxyl proton and the methylene protons are vicinal but do not clearly split each other because of the rapid intermolecular exchange of hydroxyl protons.

Magnetic Resonance Imaging (MRI)

It isn't often that someone goes to the emergency room because of a headache, and when the staff discovered that the man who did was due in court for sentencing the next day, some of them felt that there might not be anything wrong with him at all. There was.

The man's behavior toward the staff soon convinced them that he should be admitted, kept overnight, and seen by a neurologist the next day. After a preliminary examination, a magnetic resonance image, or MRI, was ordered which revealed the brain tumor shown in Figure 13.25. The tumor was located in the right frontal cortex, a portion of the brain known to be involved in controlling impulsive behavior.

The man had behaved normally until middle age; then his personality underwent significant changes, involving certain impulsive behaviors and criminal tendencies. These, as well as other behaviors, had not responded to drugs or counseling. Even

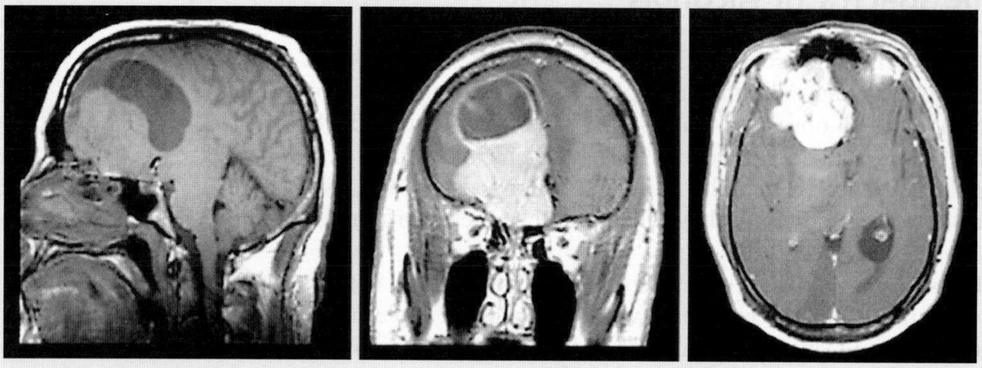

(*a*) (*b*) (*c*)

Figure 13.25

MRI scans of a brain tumor in the prefrontal cortex. The tumor is in the right hemisphere. The contrast-enhanced views (*b* and *c*) distinguish between the tumor (bright white) and a large accompanying cyst (gray oval with white outline in *b*). The tumor itself begins just beyond the nose (black shadow at top of *c*). (*a*) Sagittal view (patient facing left, head in profile). (*b*) Coronal view (view from the top of the head). (*c*) Axial view (diagonal section from the top of the head). (Images used, with permission, from J.M. Burns, R.H. Swerdlow: Right orbitofrontal tumor with pedophilia symptom and constructional apraxia sign.) *Archives of Neurology,* vol. 60:437–440; 2003. © Copyright, American Medical Association.

Continued

An easy way to verify that a particular signal belongs to a hydroxyl proton is to add D_2O. The hydroxyl proton is replaced by deuterium according to the equation:

$$RCH_2OH + D_2O \rightleftharpoons RCH_2OD + DOH$$

Deuterium does not give a signal under the conditions of ^1H NMR spectroscopy. Thus, replacement of a hydroxyl proton by deuterium leads to the disappearance of the OH peak of the alcohol. Protons bonded to nitrogen and sulfur also undergo exchange with D_2O. Those bound to carbon normally do not, which makes this a useful technique for assigning the proton resonances of OH, NH, and SH groups.

13.13 NMR and Conformations

We know from Chapter 3 that the protons in cyclohexane exist in two different environments: axial and equatorial. The NMR spectrum of cyclohexane, however, shows only a single sharp peak at δ 1.4. All the protons of cyclohexane appear to be equivalent in the NMR spectrum. Why?

The answer is related to the very rapid rate of chair–chair interconversion in cyclohexane.

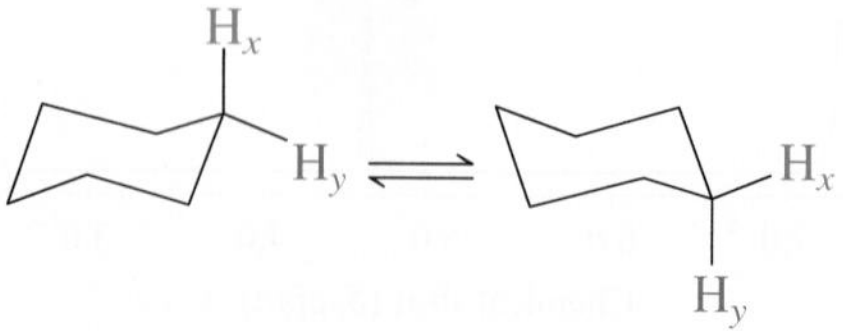

though he had earned a master's degree, the man performed poorly on some simple mental tests and was unable to sketch the face of a clock or write a legible, coherent sentence.

Once the tumor was found, it was surgically removed. The man's ability to curb his impulses was restored, his mental, graphical, and writing skills improved to the normal range, and he successfully completed a rehabilitation program. About a year later though, the headaches and some of the earlier behaviors returned. When a new MRI showed that the tumor had regrown, it was removed and again the symptoms disappeared.

At a turning point in this man's life, an MRI made all the difference. MRI is NMR. The word *nuclear* is absent from the name to avoid confusion with nuclear medicine, which involves radioactive isotopes. MRI is noninvasive, requires no imaging or contrast agents, and is less damaging than X-rays. In the time since the first MRI of a living creature—a clam—was successfully obtained in the early 1970s, MRI has become a standard diagnostic tool. Two of its early developers, Paul Lauterbur (University of Illinois) and Peter Mansfield (University of Nottingham) were recognized with the 2003 Nobel Prize in Physiology or Medicine.

An MRI scanner is an NMR machine large enough to accommodate a human being, has a powerful magnet, operates in the pulse-FT mode, and detects protons—usually the protons in water and, to a lesser extent, lipids. The principles are the same as those of conventional FT-NMR spectroscopy but, because the goal is different, the way the data are collected and analyzed differs too. Some key features of MRI include:

1. A selective pulse is used in order to excite protons in a particular slice of the object to be imaged.
2. Unlike conventional NMR, the magnetic field in MRI is not uniform. A linear gradient is applied in addition to the static field so that the field strength varies as a function of position in the object but is precisely known. Because the frequency of a proton signal is directly proportional to the strength of the applied magnetic field, the measured resonance frequency is linearly related to the position in the magnetic field gradient.
3. Computer software carries out the essential task of reconstructing the 2D or 3D image from the NMR signals. The data are generally presented as a series of slices through the imaged object. Three different views of a tumor are shown in Figure 13.25 with different slice orientations.
4. The intensity of the signal—its relative lightness or darkness in the image—depends on the concentration and spin relaxation times of the various protons. Spin relaxation time is the time it takes for the perturbed magnetization associated with a proton to return to its equilibrium value. The relaxation time is quite sensitive to the environment and is different for water in blood and various tissues.

New applications of nuclear magnetic resonance in biomedical science continue to appear. Functional MRI (fMRI) is an offshoot of MRI. Unlike MRI, which is used for diagnosis in a clinical setting, fMRI is a research tool that detects regions of the brain that are actively responding to stimuli. Increased brain activity is accompanied by an increase in blood flow to the region involved. This alters the ratio of oxygenated hemoglobin to its nonoxygenated counterpart. Because the two hemoglobins have different magnetic properties, the nuclear spin relaxation times of the protons in water are affected and can be studied by MRI. In the short time since its development, fMRI has been used successfully to study memory and cognition in relation to brain activity.

NMR is too slow to "see" the individual conformations of cyclohexane, but sees instead the *average* environment of the protons. Because chair–chair interconversion in cyclohexane converts each axial proton to an equatorial one and vice versa, the average environments of all the protons are the same. A single peak is observed that has a chemical shift midway between the true chemical shifts of the axial and the equatorial protons.

The rate of interconversion can be slowed down by lowering the temperature. At temperatures of about −100°C, separate signals are seen for the axial and equatorial protons of cyclohexane.

13.14 ^{13}C NMR Spectroscopy

We pointed out in Section 13.3 that both ^{1}H and ^{13}C are nuclei that can provide useful structural information when studied by NMR. Although a ^{1}H NMR spectrum helps us infer much about the carbon skeleton of a molecule, a ^{13}C NMR spectrum has the obvious advantage of probing the carbon skeleton directly. ^{13}C NMR spectroscopy is analogous to ^{1}H NMR in that the number of signals informs us about the number of different kinds of carbons, and their chemical shifts are related to particular chemical environments.

However, unlike ^{1}H, which is the most abundant of the hydrogen isotopes (99.985%), only 1.1% of the carbon atoms in a sample are ^{13}C. Moreover, the intensity of the signal produced by ^{13}C nuclei is far weaker than the signal produced by the same number of ^{1}H nuclei. In order for ^{13}C NMR to be a useful technique in structure determination, a

Figure 13.26

(*a*) The 200-MHz ^{1}H NMR spectrum and (*b*) the ^{13}C NMR spectrum of 1-chloropentane.

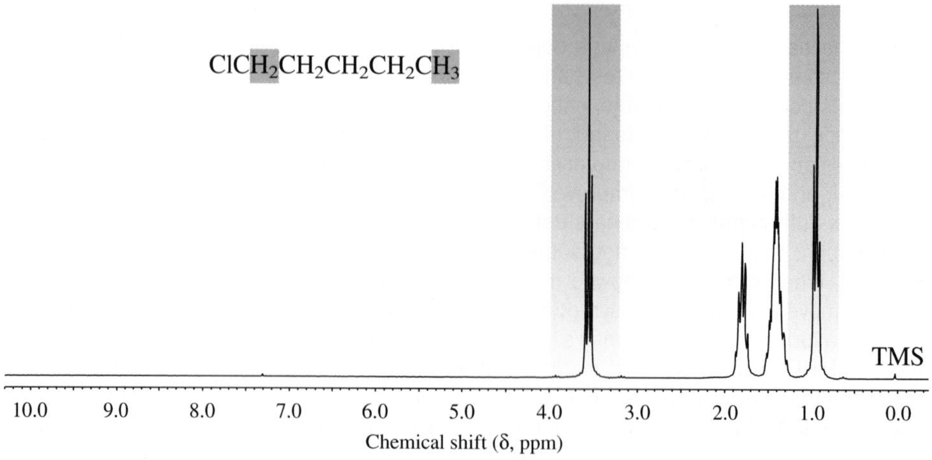

(*a*)

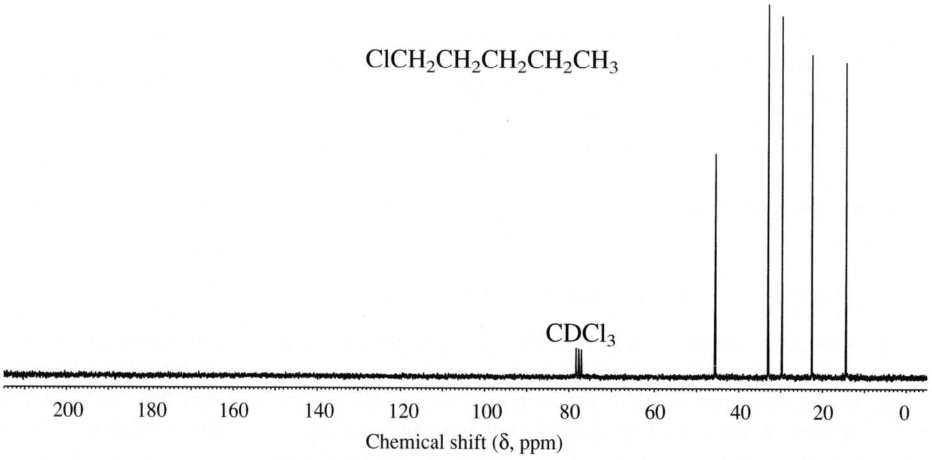

(*b*)

vast increase in the signal-to-noise ratio is required. Pulsed FT-NMR provides for this, and its development was the critical breakthrough that led to ^{13}C NMR becoming the routine tool that it is today.

To orient ourselves in the information that ^{13}C NMR provides, let's compare the ^{1}H and ^{13}C NMR spectra of 1-chloropentane (Figures 13.26*a* and 13.26*b,* respectively). The ^{1}H NMR spectrum shows reasonably well-defined triplets for the protons of the CH_3 and CH_2Cl groups (δ 0.9 and 3.55, respectively). The signals for the six CH_2 protons at C-2, C-3, and C-4 of $CH_3CH_2CH_2CH_2CH_2Cl$, however, appear as two unresolved multiplets at δ 1.4 and 1.8.

The ^{13}C NMR spectrum, on the other hand, is very simple: *a separate, distinct peak is observed for each carbon.*

Notice, too, how well-separated these ^{13}C signals are: they cover a range of over 30 ppm, compared with less than 3 ppm for the proton signals of the same compound. In general, the window for proton signals in organic molecules is about 12 ppm; ^{13}C chemical shifts span a range of over 200 ppm. The greater spread of ^{13}C chemical shifts makes it easier to interpret the spectra.

Problem 13.16

How many signals would you expect to see in the ^{13}C NMR spectrum of each of the following compounds?

(a) Propylbenzene
(b) Isopropylbenzene
(c) 1,2,3-Trimethylbenzene
(d) 1,2,4-Trimethylbenzene
(e) 1,3,5-Trimethylbenzene

Sample Solution (a) The two ring carbons that are ortho to the propyl substituent are equivalent and so must have the same chemical shift. Similarly, the two ring carbons that are meta to the propyl group are equivalent to each other. The carbon atom para to the substituent is unique, as is the carbon that bears the substituent. Thus, there will be four signals for the ring carbons, designated *w, x, y,* and *z* in the structural formula. These four signals for the ring carbons added to those for the three nonequivalent carbons of the propyl group yield a total of seven signals.

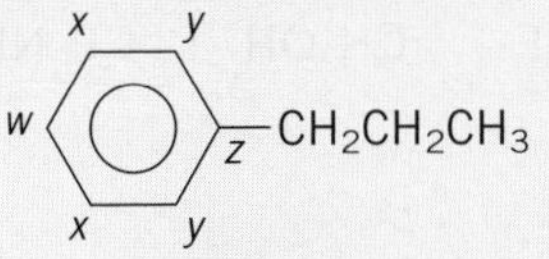

Propylbenzene

13.15 ^{13}C Chemical Shifts

Just as chemical shifts in ^{1}H NMR are measured relative to the *protons* of tetramethylsilane, chemical shifts in ^{13}C NMR are measured relative to the *carbons* of tetramethylsilane. Table 13.3 lists typical chemical-shift ranges for some representative types of carbon atoms.

In general, the factors that most affect ^{13}C chemical shifts are

1. The electronegativity of the groups attached to carbon
2. The hybridization of carbon

Electronegativity Effects. Electronegative substituents affect ^{13}C chemical shifts in the same way as they affect ^{1}H chemical shifts, by withdrawing electrons. For ^{1}H NMR, recall that because carbon is more electronegative than hydrogen, the protons in methane (CH_4) are more shielded than primary hydrogens (RCH_3), primary hydrogens are more shielded than secondary (R_2CH_2), and secondary more shielded than tertiary (R_3CH). The same holds true for carbons in ^{13}C NMR, but the effects can be 10–20 times greater.

TABLE 13.3 Chemical Shifts of Representative Carbons

Type of carbon	Chemical shift (δ) ppm*	Type of carbon	Chemical shift (δ) ppm*
Hydrocarbons		**Functionally substituted carbons**	
RCH_3	0–35	RCH_2Br	20–40
R_2CH_2	15–40	RCH_2Cl	25–50
R_3CH	25–50	RCH_2NH_2	35–50
R_4C	30–40	RCH_2OH and RCH_2OR	50–65
$RC{\equiv}CR$	65–90	$RC{\equiv}N$	110–125
$R_2C{=}CR_2$	100–150	$R\overset{O}{\overset{\|}{C}}OH$ and $R\overset{O}{\overset{\|}{C}}OR$	160–185
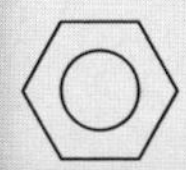	110–175	$R\overset{O}{\overset{\|}{C}}H$ and $R\overset{O}{\overset{\|}{C}}R$	190–220

*Approximate values relative to tetramethylsilane.

	$(CH_3)_4C$	$(CH_3)_3CH$	$CH_3CH_2CH_3$	CH_3CH_3	CH_4
Classification:	Quaternary	Tertiary	Secondary	Primary	
Chemical shift (δ), ppm:					
H		1.7	1.3	0.9	0.2
C	28	25	16	8	−2

Likewise, for functionally substituted methyl groups:

	CH_3F	CH_3OH	CH_3NH_2	CH_4
Chemical shift (δ), ppm:				
H	4.3	3.4	2.5	0.2
C	75	50	27	−2

Figure 13.26 compared the appearance of the ^1H and ^{13}C NMR spectra of 1-chloropentane and drew attention to the fact each carbon gave a separate peak, well separated from the others. Let's now take a closer look at the ^{13}C NMR spectrum of 1-chloropentane with respect to assigning these peaks to individual carbons.

	Cl	—CH_2—	CH_2—	CH_2—	CH_2—	CH_3
^{13}C chemical shift (δ), ppm:		45	33	29	22	14

The most obvious feature of these ^{13}C chemical shifts is that the closer the carbon is to the electronegative chlorine, the more deshielded it is. Peak assignments will not always be this easy, but the correspondence with electronegativity is so pronounced that *spectrum simulators* are available that allow reliable prediction of ^{13}C chemical shifts from structural formulas. These simulators are based on arithmetic formulas that combine experimentally derived chemical shift increments for the various structural units within a molecule.

The Descriptive Passage and Interpretive Problems at the end of this chapter illustrate the spectrum simulator approach to calculating ^{13}C chemical shifts.

Problem 13.17

The ^{13}C NMR spectrum of 1-bromo-3-chloropropane contains peaks at δ 30, δ 35, and δ 43. Assign these signals to the appropriate carbons.

Hybridization Effects. Here again, the effects are similar to those seen in ^1H NMR. As illustrated by 4-phenyl-1-butene, sp^3-hybridized carbons are more shielded than sp^2-hybridized ones.

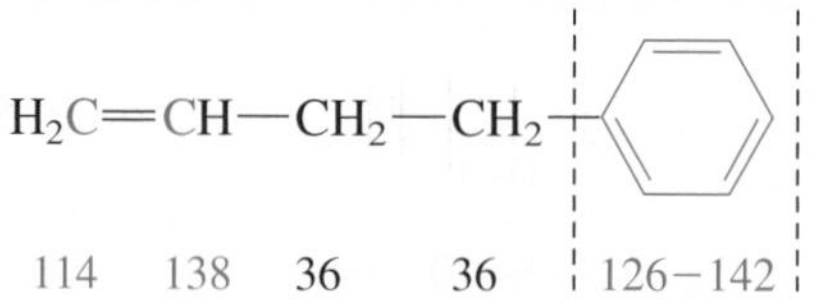

^{13}C chemical shift (δ), ppm: 114 138 36 36 126−142

Of the sp^2-hybridized carbons, C-1 is the most shielded because it is bonded to only one other carbon. The least shielded carbon is the ring carbon to which the side chain is attached. It is the only sp^2-hybridized carbon connected to three other carbons.

Problem 13.18

Consider carbons *x*, *y*, and *z* in *p*-methylanisole. One has a chemical shift of δ 20, another has δ 55, and the third δ 157. Match the chemical shifts with the appropriate carbons.

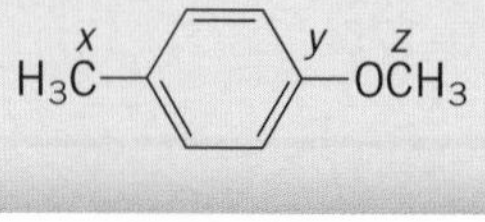

Acetylenes are anomalous in ^{13}C, as in ^{1}H NMR. *sp*-Hybridized carbons are less shielded than sp^3-hybridized ones, but more shielded than sp^2-hybridized ones.

$$H—C\equiv C—CH_2—CH_2—CH_3$$

^{13}C chemical shift (δ), ppm: 68 84 22 20 13

Electronegativity and hybridization effects combine to make the carbon of a carbonyl group especially deshielded. Normally, the carbon of C=O is the least shielded one in a ^{13}C NMR spectrum.

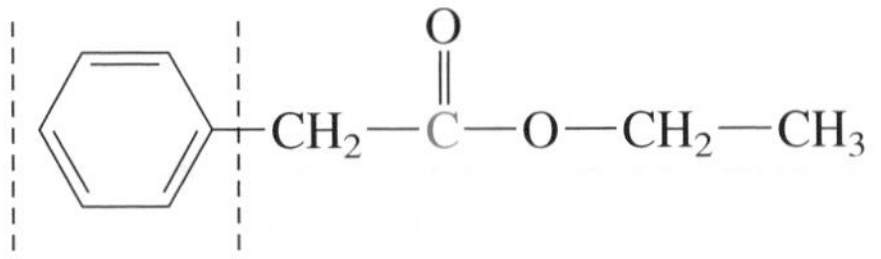

^{13}C chemical shift (δ), ppm: 127–134 41 171 61 14

Problem 13.19

Which would you expect to be more shielded, the carbonyl carbon of an aldehyde or a ketone? Why?

We will have more to say about ^{13}C chemical shifts in later chapters when various families of compounds, especially those that contain carbonyl groups, are discussed in more detail.

13.16 ^{13}C NMR and Peak Intensities

Two features that are fundamental to ^{1}H NMR spectroscopy—integrated areas and splitting patterns—are not very important in ^{13}C NMR.

Although it is a simple matter to integrate ^{13}C signals, it is rarely done because the observed ratios can be more misleading than helpful. The pulsed FT technique that is standard for ^{13}C NMR has the side effect of distorting the signal intensities, especially for carbons that lack attached hydrogens. Examine Figure 13.27, which shows the ^{13}C NMR spectrum of 3-methylphenol (*m*-cresol). Notice that, contrary to what we might expect for a compound with seven peaks for seven different carbons, the intensities of these peaks are not nearly the same. The two least intense signals, those at δ 140 and δ 157, correspond to carbons that lack attached hydrogens.

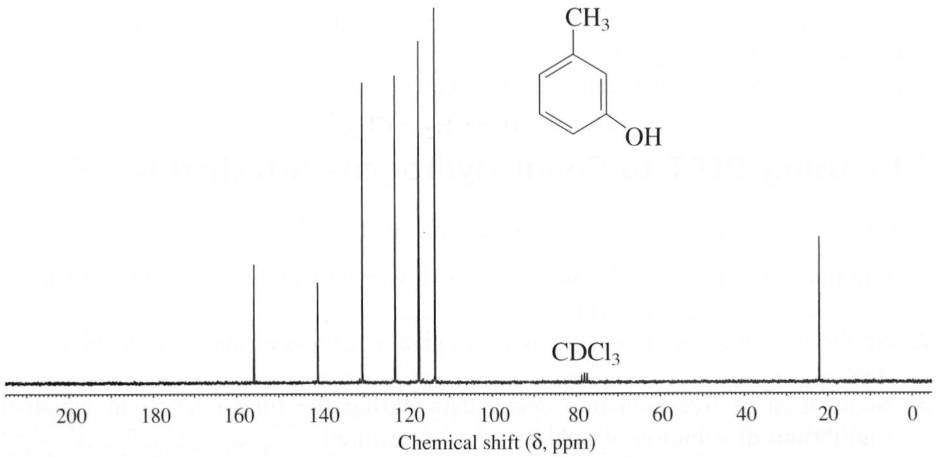

Figure 13.27

The ^{13}C NMR spectrum of *m*-cresol. Each of the seven carbons of *m*-cresol gives a separate peak. Integrating the spectrum would not provide useful information because the intensities of the peaks are so different, even though each one corresponds to a single carbon.

Problem 13.20

To which of the compounds of Problem 13.16 does the ^{13}C NMR spectrum of Figure 13.28 belong?

Figure 13.28

The ^{13}C NMR spectrum of the unknown compound of Problem 13.20.

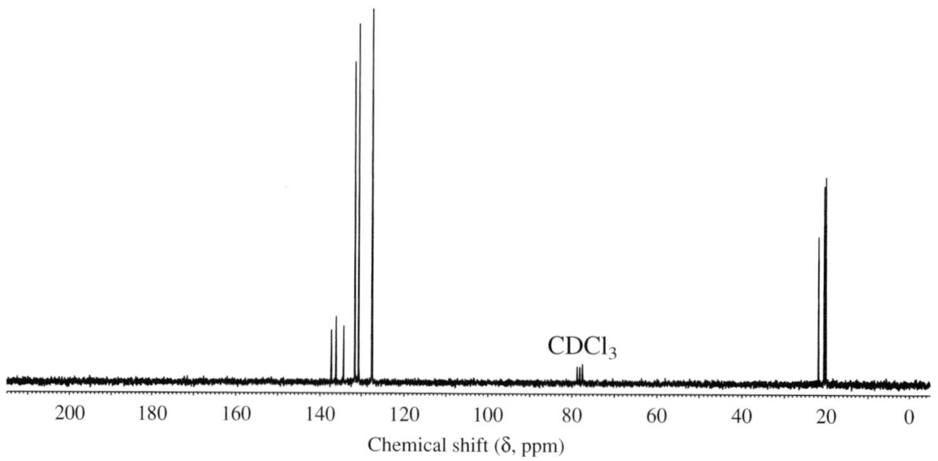

13.17 ^{13}C—^{1}H Coupling

You may have noticed another characteristic of ^{13}C NMR spectra—all of the peaks are singlets. With a spin of $\pm\frac{1}{2}$, a ^{13}C nucleus is subject to the same splitting rules that apply to ^{1}H, and we might expect to see splittings due to ^{13}C—^{13}C and ^{13}C—^{1}H couplings. We don't. Why?

The lack of splitting due to ^{13}C—^{13}C coupling is easy to understand. ^{13}C NMR spectra are measured on samples that contain ^{13}C at the "natural abundance" level. Only 1% of all the carbons in the sample are ^{13}C, and the probability that any molecule contains more than one ^{13}C atom is quite small.

Splitting due to ^{13}C—^{1}H coupling is absent for a different reason, one that has to do with the way the spectrum is run. Because a ^{13}C signal can be split not only by the protons to which it is directly attached, but also by protons separated from it by two, three, or even more bonds, the number of splittings might be so large as to make the spectrum too complicated to interpret. Thus, the spectrum is measured under conditions, called **broadband decoupling,** that suppress such splitting.

What we gain from broadband decoupling in terms of a simple-looking spectrum comes at the expense of some useful information. For example, being able to see splitting corresponding to one-bond ^{13}C—^{1}H coupling would immediately tell us the number of hydrogens directly attached to each carbon. The signal for a carbon with no attached hydrogens (a *quaternary* carbon) would be a singlet, the hydrogen of a CH group would split the carbon signal into a doublet, and the signals for the carbons of a CH_2 and a CH_3 group would appear as a triplet and a quartet, respectively. Although it is possible, with a technique called *off-resonance decoupling,* to observe such one-bond couplings, identifying a signal as belonging to a quaternary carbon or to the carbon of a CH, CH_2, or CH_3 group is normally done by a method called DEPT, which is described in the next section.

13.18 Using DEPT to Count Hydrogens Attached to ^{13}C

In general, a simple pulse FT-NMR experiment involves the following stages:

1. Equilibration of the nuclei between the lower and higher spin states under the influence of a magnetic field
2. Application of a radiofrequency pulse to give an excess of nuclei in the higher spin state
3. Acquisition of free-induction decay data during the time interval in which the equilibrium distribution of nuclear spins is restored
4. Mathematical manipulation (Fourier transform) of the data to plot a spectrum

The pulse sequence (stages 2–3) can be repeated hundreds of times to enhance the signal-to-noise ratio. The duration of time for stage 2 is on the order of milliseconds, and that for stage 3 is about 1 second.

Major advances in NMR have been made by using a second rf transmitter to irradiate the sample at some point during the sequence. There are several such techniques,

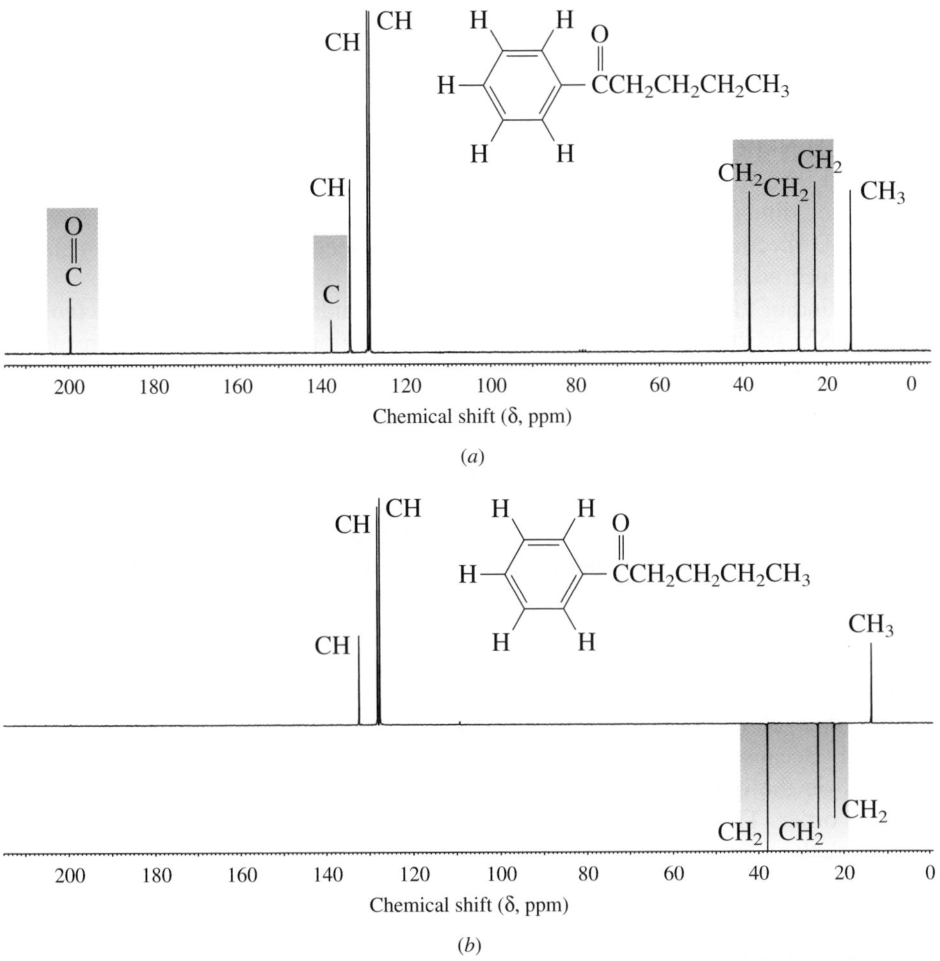

Figure 13.29

^{13}C NMR spectra of 1-phenyl-1-pentanone. (*a*) Normal spectrum. (*b*) DEPT spectrum recorded using a pulse sequence in which CH_3 and CH carbons appear as positive peaks, CH_2 carbons as negative peaks, and carbons without any attached hydrogens are nulled.

of which we'll describe just one, called **distortionless enhancement of polarization transfer,** abbreviated as **DEPT.**

In the DEPT routine, a second transmitter excites ^{1}H, which affects the appearance of the ^{13}C spectrum. A typical DEPT experiment is illustrated for the case of 1-phenyl-1-pentanone in Figure 13.29. In addition to the normal spectrum shown in Figure 13.29*a,* four more spectra are run using prescribed pulse sequences. In one (Figure 13.29*b*), the signals for carbons of CH_3 and CH groups appear normally, whereas those for CH_2 groups are inverted and those for C without any attached hydrogens are nulled. In the others (not shown) different pulse sequences produce combinations of normal, nulled, and inverted peaks that allow assignments to be made to the various types of carbons with confidence.

Problem 13.21

DEPT spectra for a compound with the formula $C_6H_{12}O$ are shown in Figure 13.30. Assign a structure. (More than one answer is possible.)

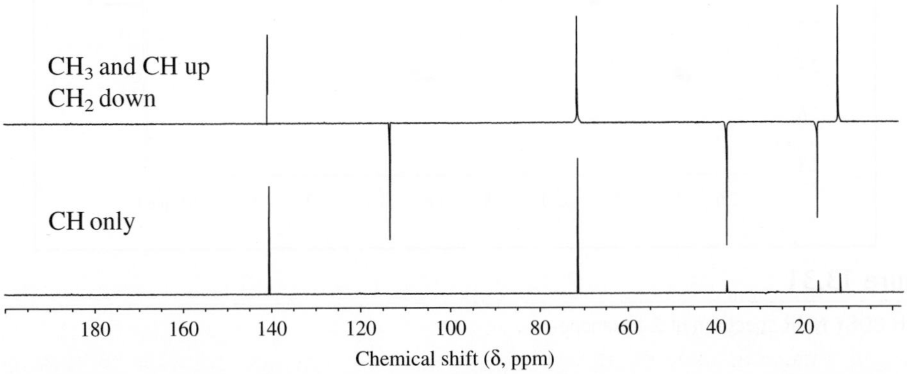

Figure 13.30

DEPT spectra for Problem 13.21

13.19 2D NMR: COSY and HETCOR

The more information you can extract from an NMR spectrum, the better your chances at arriving at a unique structure. Like spin–spin splitting, which complicates the appearance of an ^{1}H NMR spectrum but provides additional information, 2D NMR looks more complicated than it is while making structure determination easier.

The key dimension in NMR is the frequency axis. All of the spectra we have seen so far are 1D spectra because they have only one frequency axis. In 2D NMR a standard pulse sequence adds a second frequency axis.

One kind of 2D NMR is called **COSY,** which stands for **correlated spectroscopy.** With a COSY spectrum you can determine by inspection which signals correspond to spin-coupled protons. Identifying coupling relationships is a valuable aid to establishing a molecule's *connectivity*.

Figure 13.31 is the COSY spectrum of 2-hexanone. Both the *x*- and *y*-axes are frequency axes expressed as chemical shifts. Displaying the 1D ^{1}H NMR spectrum of 2-hexanone along the *x*- and *y*-axes makes it easier to interpret the 2D information, which is the collection of contoured objects contained within the axes. To orient ourselves, first note that many of the contours lie along the diagonal that runs from the lower left to the upper right. This diagonal bisects the 2D NMR into two mirror-image halves. The off-diagonal contours are called *cross peaks* and contain the connectivity information we need.

Each cross peak has *x* and *y* coordinates. One coordinate corresponds to the chemical shift of a proton, the other to the chemical shift of a proton to which it is coupled. Because the diagonal splits the 2D spectrum in half, each cross peak is duplicated on the other side of the other diagonal with the same coordinates, except in reverse order. This redundancy means that we really need to examine only half of the cross peaks.

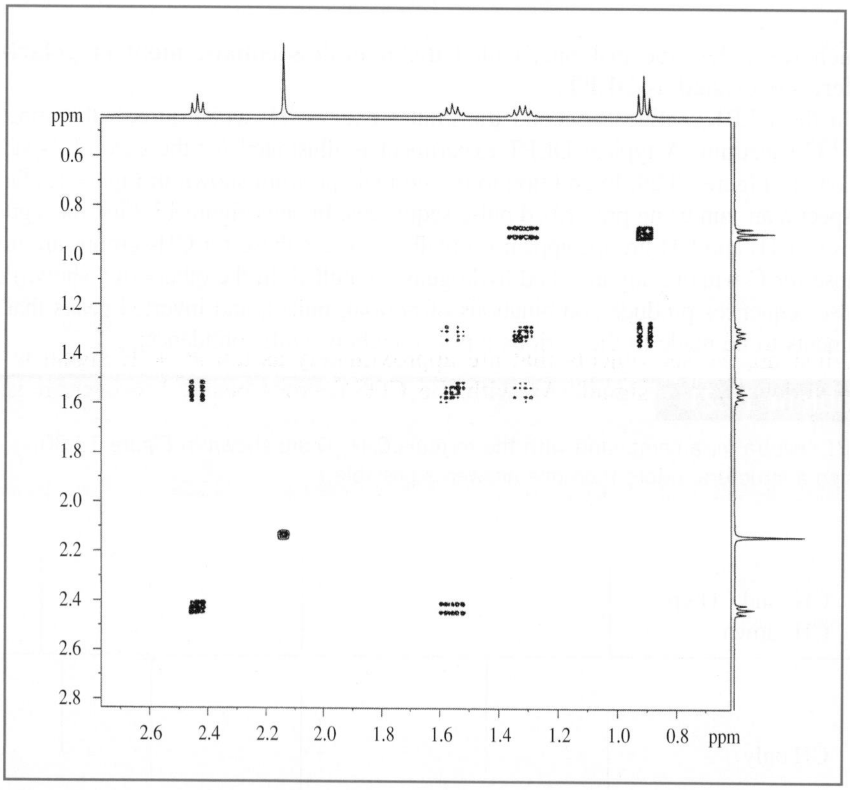

Figure 13.31

^{1}H-^{1}H COSY NMR spectrum of 2-hexanone.

To illustrate, start with the lowest field signal (δ 2.4) of 2-hexanone. We assign this signal, a triplet, to the protons at C-3 on the basis of its chemical shift and the splitting evident in the 1D spectrum.

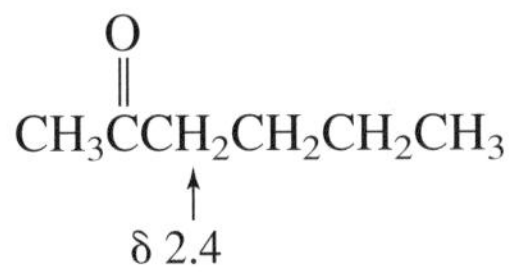

We look for cross peaks with the same *x* coordinate by drawing a vertical line from δ 2.4, finding a cross peak with a *y* coordinate of δ 1.6. *This means that the protons responsible for the signal at δ 2.4 are coupled to the ones at δ 1.6.* Therefore, the chemical shift of the C-4 protons is δ 1.6.

Now work from these C-4 protons. Drawing a vertical line from δ 1.6 on the *x*-axis finds two cross peaks. One cross peak simply confirms the coupling to the protons at C-3. The other has a *y* coordinate of δ 1.3 and, therefore, must correspond to the protons at C-5.

A vertical line drawn from δ 1.3 intersects the cross peaks at both δ 1.6 and δ 0.9. The former confirms the coupling of C-5 to C-4; the latter corresponds to the C-5 to C-6 coupling and identifies the signal at δ 0.9 as belonging to the protons at C-6.

Finally, a vertical line drawn from δ 2.1 intersects no cross peaks. The singlet at δ 2.1, as expected, is due to the protons at C-1, which are not coupled to any of the other protons in the molecule.

The complete connectivity and assignment of ^{1}H chemical shifts is

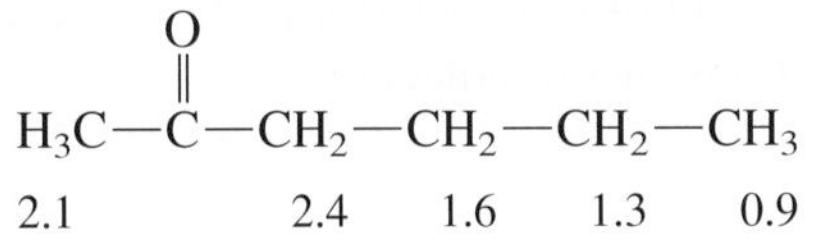

Although the 1D ^{1}H spectrum of 2-hexanone is simple enough to be interpreted directly, you can see that COSY offers one more tool we can call on in more complicated cases.

A second 2D NMR method called **HETCOR (heteronuclear chemical shift correlation)** is a type of COSY in which the two frequency axes are the chemical shifts for different nuclei, usually ^{1}H and ^{13}C. With HETCOR it is possible to relate a peak in a ^{13}C spectrum to the ^{1}H signal of the protons attached to that carbon. As we did with COSY, we'll use 2-hexanone to illustrate the technique.

The HETCOR spectrum of 2-hexanone is shown in Figure 13.32. It is considerably simpler than a COSY spectrum, lacking diagonal peaks and contoured cross peaks. Instead, we see objects that are approximately as tall as a ^{1}H signal is wide, and as wide as a ^{13}C signal. As with the COSY cross peaks, however, it is their coordinates that matter, not their size or shape. Interpreting the spectrum is straightforward. The ^{13}C peak at δ 30 correlates with the ^{1}H singlet at δ 2.1, which because of its multiplicity and chemical shift corresponds to the protons at C-1. Therefore, this ^{13}C peak can be assigned to C-1 of 2-hexanone. Repeating this procedure for the other carbons gives:

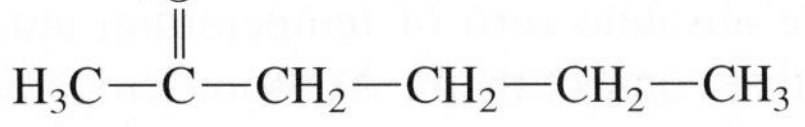

	H_3C	CH_2	CH_2	CH_2	CH_3
^{1}H chemical shift (δ), ppm:	2.1	2.4	1.6	1.3	0.9
^{13}C chemical shift (δ), ppm:	30	43	26	22	14

The chemical shift of the carbonyl carbon (δ 209) is not included because it has no attached hydrogens.

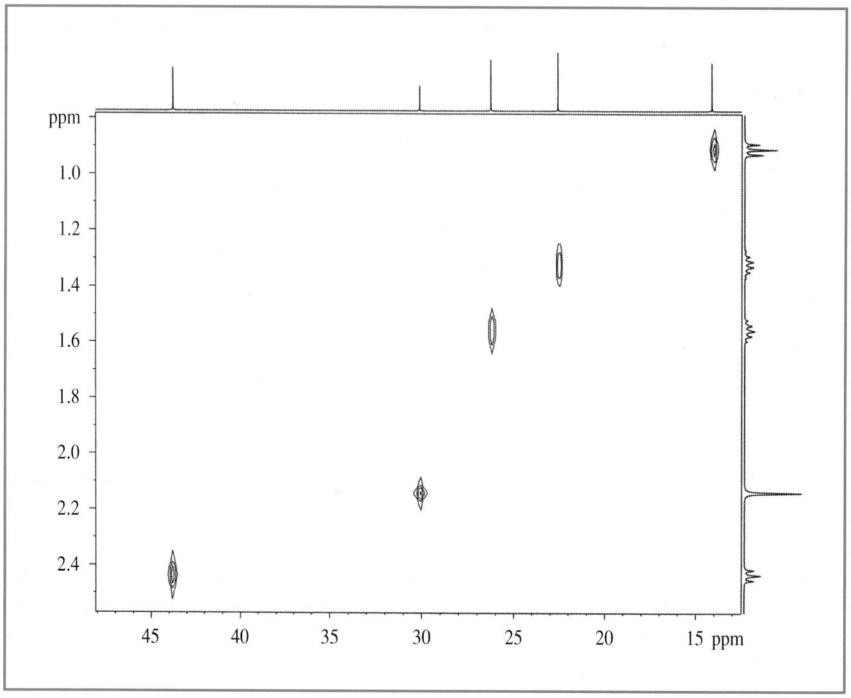

Figure 13.32

^{1}H-^{13}C HETCOR NMR spectrum of 2-hexanone.

A number of 2D NMR techniques are available for a variety of purposes. They are especially valuable when attempting to determine the structure of complicated natural products and the conformations of biomolecules.

13.20 Introduction to Infrared Spectroscopy

IR's earliest recognition came during World War II when it provided a key clue to the unusual β-lactam structure of the "miracle drug" penicillin.

Before the advent of NMR spectroscopy, infrared (IR) spectroscopy was the instrumental method most often applied to organic structure determination. Although NMR, in general, tells us more about the structure of an unknown compound, IR remains an important tool because of its usefulness in identifying the presence of certain *functional groups* within a molecule. Structural units, including functional groups, vibrate in characteristic ways and it is this sensitivity to *group vibrations* that is the basis of IR spectroscopy.

Among the ways a molecule responds to the absorption of energy is by vibrational motions such as the stretching and contracting of bonds and the opening and closing (bending) of bond angles. Vibrational motion and its energy are quantized. Only certain vibrational energy states are allowed.

We can visualize molecular vibrations by thinking of atoms and bonds as balls and springs.

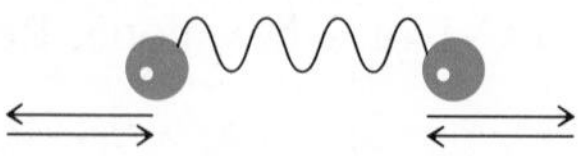

Zero-point energy is the term given to the energy of a molecule at absolute zero.

Even at the absolute zero of temperature, atoms in a molecule vibrate with respect to the bonds that connect them. At room temperature, the molecules are distributed among various vibrational energy states. *Frequency* is a property of the vibration and is related to the difference between vibrational energy states by $\Delta E = h\nu$ (see Section 13.1). Promoting a molecule from a lower to a higher vibrational energy state increases the *amplitude* of the vibration.

For a sense of the variety of vibrational modes available to a molecule, consider a CH_2 group. Stretching and contracting the pair of C—H bonds can occur in two different ways.

Spectra by the Thousands

The best way to get good at interpreting spectra is by experience. Look at as many spectra and do as many spectroscopy problems as you can.

Among Web sites that offer spectroscopic problems, two stand out (Figure 13.33). One, called *WebSpectra,* was developed by Professor Craig A. Merlic (UCLA):

www.chem.ucla.edu/~webspectra

The other is the *Organic Structure Elucidation* workbook, created by Professor Bradley D. Smith (Notre Dame):

www.nd.edu/~smithgrp/structure/workbook.html

WebSpectra includes 75 problems. All the problems display the ^{1}H and ^{13}C spectra, several with DEPT or COSY enhancements. A number include IR spectra. *Organic Structure Elucidation* contains 64 problems, all with ^{1}H and ^{13}C NMR, IR, and mass spectra. The exercises in both *WebSpectra* and *Organic Structure Elucidation* are graded according to difficulty. Give them a try.

Vast numbers of NMR, IR, and mass spectra are freely accessible via the *Spectral Data Base System* (SDBS) maintained by the Japanese National Institute of Advanced Industrial Science and Technology at:

http://riodb01.ibase.aist.go.jp/sdbs/cgi-bin/cre_index.cgi?lang=eng

The SDBS contains 14,800 ^{1}H NMR, 13,100 ^{13}C NMR, 51,600 IR, and 24,200 mass spectra. Not only does the SDBS contain more spectra than anyone could possibly browse through, it incorporates some very useful search features. If you want spectra for a particular compound, entering the name of the compound calls up links to its spectra, which can then be displayed. If you don't know what the compound is, but know one or more of the following:

- Molecular formula
- ^{1}H or ^{13}C chemical shift of one or more peaks
- Mass number of mass spectra fragments

entering the values singly or in combination returns the names of the best matches in the database. You can then compare the spectra of these candidate compounds with the spectra of the sample to identify it.

As extensive as the SDBS is, don't be disappointed if the exact compound you are looking for is not there. There are, after all, millions of organic compounds. However, much of structure determination (and organic chemistry in general) is based on analogy. Finding the spectrum of a related compound can be almost as helpful as finding the one you really want.

These Web resources, in conjunction with the figures and problems in your text, afford a wealth of opportunities to gain practice and experience in modern techniques of structure determination.

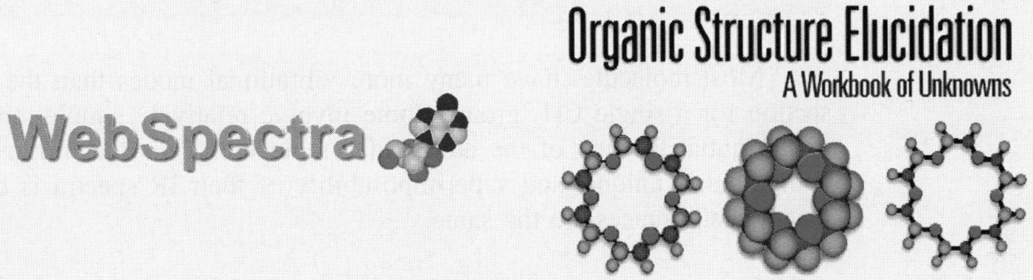

Figure 13.33

These two welcome screens open the door to almost 150 spectroscopy problems. The screens are used with permission of Professors Craig A. Merlic *(WebSpectra)* and Bradley D. Smith *(Organic Structure Elucidation).* See the text for the respective URLs.

In the *symmetric* stretch, both C—H bonds stretch at the same time and contract at the same time. In the *antisymmetric* stretch, one C—H bond stretches while the other contracts.

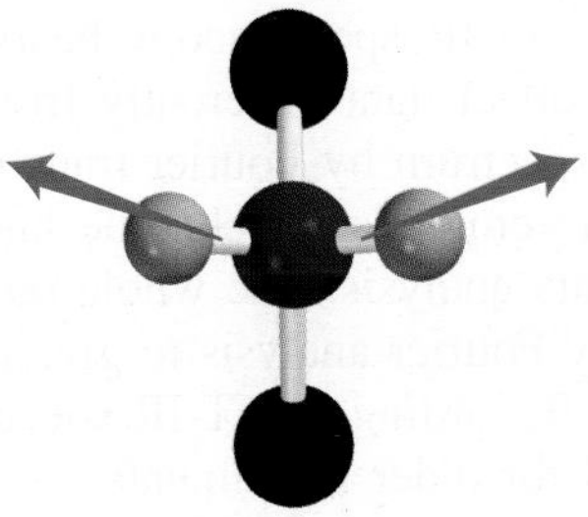
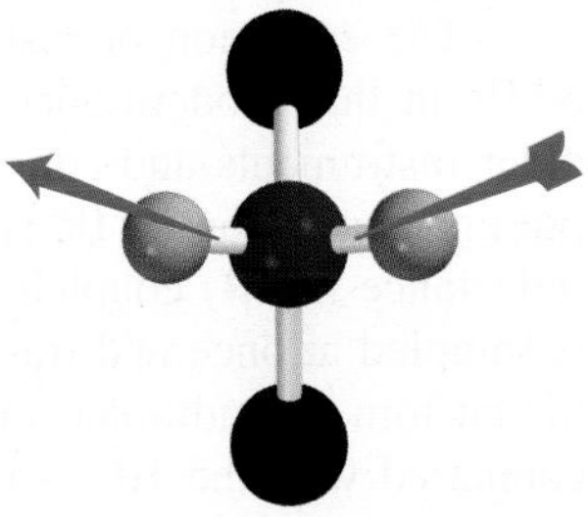

Stretching vibrations: Symmetric Antisymmetric

In addition to stretching vibrations, a CH_2 group can bend, and each bending mode has its own set of energy states.

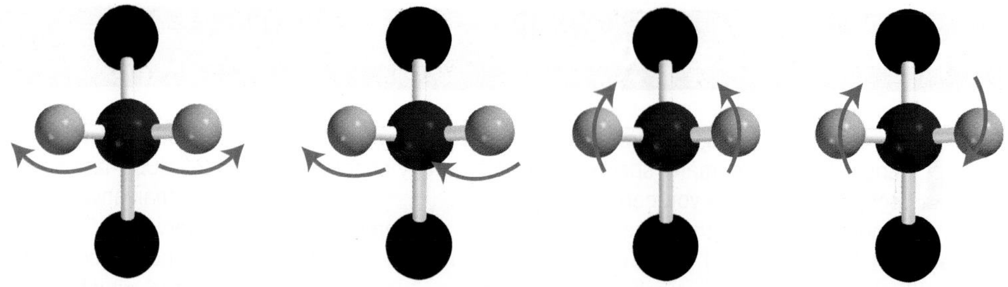

Bending vibrations: Scissoring Rocking Wagging Twisting

The energy difference between adjacent vibrational states is tens of thousands of times larger than what we saw for nuclear spin states in NMR.

A molecule absorbs that portion of electromagnetic radiation having a frequency that matches the energy difference between two vibrational energy levels. This radiation lies in the infrared region of the electromagnetic spectrum (see Figure 13.1). The wavelength λ of the infrared region that is the most useful for structure determination is 2.5–16 μm, where 1 μm = 10^{-6} m. Instead of wavelengths or SI units of frequency (s^{-1}), IR spectroscopy uses **wavenumbers,** which are equal to λ^{-1} and expressed in units of reciprocal centimeters (cm^{-1}). Thus, the region 2.5–16 μm corresponds to 4000–625 cm^{-1}. Wavenumbers are directly proportional to energy; 4000 cm^{-1} is the high-energy end of the scale for IR spectra, and 625 cm^{-1} is the low-energy end.

Problem 13.22

Vibrational frequencies are sensitive to isotopic replacement. The O—H stretching frequency is near 3600 cm^{-1}, but that of O—D is about 2630 cm^{-1}. Which are closer in energy, two adjacent O—H or two adjacent O—D vibrational states?

Most molecules have many more vibrational modes than the ones shown in this section for a single CH_2 group. Some involve relatively simple structural units, others a substantial fraction of the atoms in a molecule. Thus the infrared spectrum of each compound is unique, and superimposability of their IR spectra is convincing evidence that two substances are the same.

13.21 Infrared Spectra

IR spectra can be obtained regardless of the physical state of a sample—solid, liquid, gas, or dissolved in some solvent. If the sample is a liquid, a drop or two is placed between two sodium chloride disks, through which the IR beam is passed. Solids may be dissolved in a suitable solvent such as carbon tetrachloride or chloroform. More commonly, a solid sample is mixed with potassium bromide and the mixture pressed into a thin wafer, which is placed in the path of the IR beam. Newer instruments require little or no sample preparation.

All IR spectra in this text were recorded without solvent using an ATR instrument.

The evolution of instrumentation in IR spectroscopy bears a similarity to that of NMR in that modern spectrometers collect data differently from the methods used by older instruments and convert it to a spectrum by Fourier transform (FT) methods. The present generation of IR spectrometers employs a technique known as attenuated total reflectance (ATR) coupled with FT data analysis. The whole range of vibrational states is sampled at once and transformed by Fourier analysis to give a spectrum formatted in the custom of traditional instruments. Recording an FT-IR spectrum takes about 1 min, compared with the 10–15 min needed for older instruments.

Figure 13.34 orients us with respect to where we can expect to find IR absorptions for various structural units. Peaks in the range of 4000–1600 cm^{-1} are usually emphasized because this is the region in which the vibrations characteristic of particular functional groups are found. We'll look at some of these functional groups in more

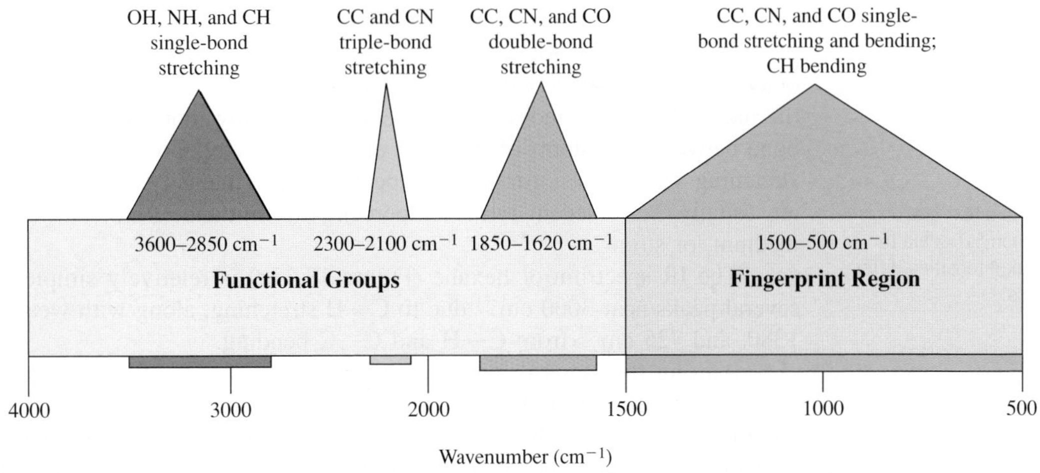

Figure 13.34

Structural units are commonly found in specific regions of the infrared spectrum.

detail in Section 13.22. The region 1500–500 cm^{-1} is known as the **fingerprint region;** it is here that the pattern of peaks varies most from compound to compound.

An IR spectrum usually contains more peaks than we can assign, or even need to assign. We gain information by associating selected absorptions with particular structural units and functional groups, as well as noting what structural units can be excluded from consideration because a key peak that characterizes it is absent from the spectrum.

Figure 13.35*a–d* shows the IR spectra of four hydrocarbons: hexane, 1-hexene, benzene, and hexylbenzene. Each spectrum consists of a series of absorption peaks of

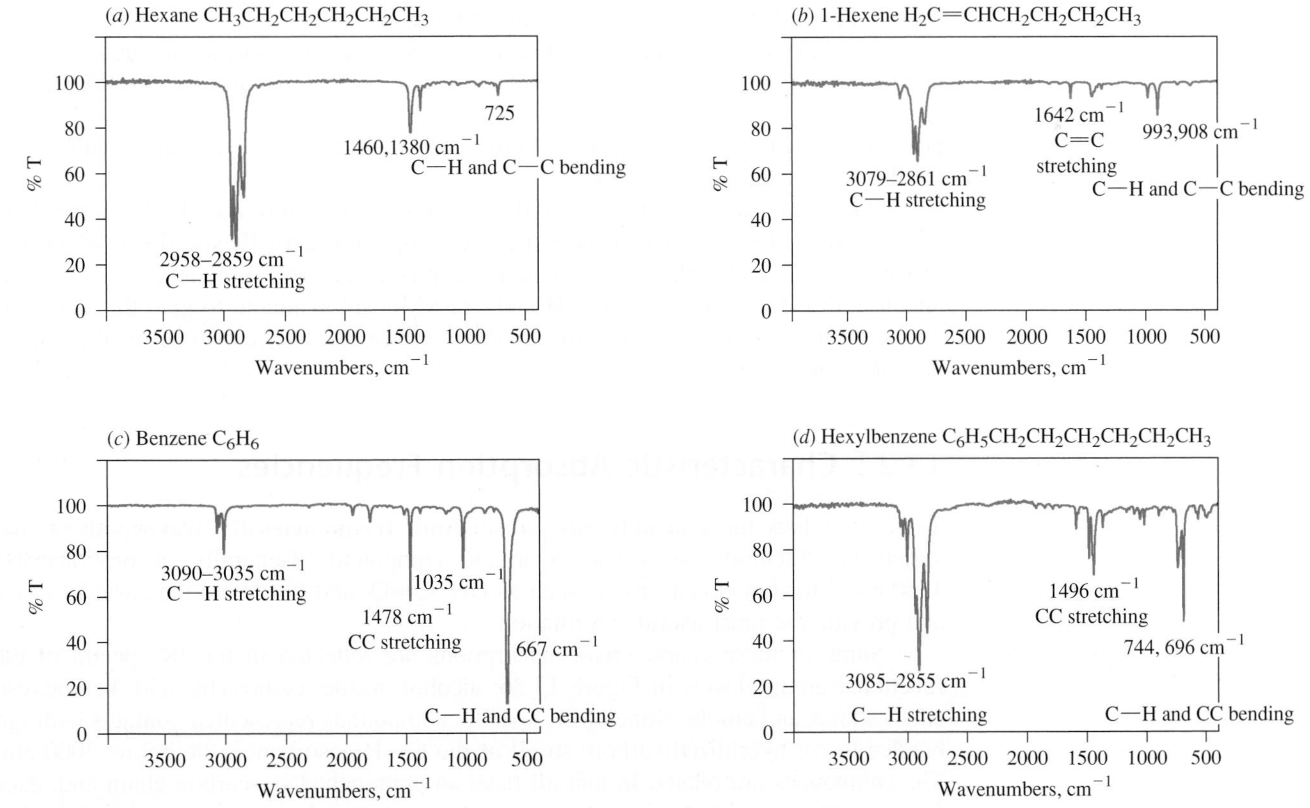

Figure 13.35

IR spectra of the hydrocarbons: (*a*) hexane, (*b*) 1-hexene, (*c*) benzene, and (*d*) hexylbenzene.

varying shape and intensity. Unlike NMR, in which intensities are related to the number of nuclei responsible for each signal, some IR vibrations give more intense peaks than others. To give an observable peak in the infrared, a vibration must produce a change in the molecular dipole moment, and peaks are usually more intense when they involve a bond between two atoms of different electronegativity. Consequently, C—C single-bond stretching vibrations normally give peaks of low intensity. The intensities of IR peaks are usually expressed in terms of percent transmittance (%T) and described as weak, medium, or strong.

All of the spectra in this text are displayed on a common %T scale to better show how peak intensities differ among various groups.

The IR spectrum of hexane (Figure 13.35*a*) is relatively simple, characterized by several peaks near 3000 cm^{-1} due to C—H stretching, along with weaker peaks at 1460, 1380, and 725 cm^{-1} from C—H and C—C bending.

Among the several ways in which the spectrum of the alkene 1-hexene (Figure 13.35*b*) differs from hexane, the most useful from the perspective of structure determination is found in the C—H stretching region. Although all the peaks for C—H stretching in *hexane* appear below 3000 cm^{-1}, *1-hexene* exhibits a peak at 3079 cm^{-1}. Peaks for C—H stretching above 3000 cm^{-1} are characteristic of hydrogens bonded to sp^2-hybridized carbon. The IR spectrum of 1-hexene also displays a weak peak at 1642 cm^{-1} corresponding to its C=C stretching vibration. The peaks at 993 and 908 cm^{-1} in the spectrum of 1-hexene, absent in the spectrum of hexane, are bending vibrations of the H_2C=C group.

Problem 13.23

Ethylene lacks a peak in its IR spectrum for C=C stretching. Why?

Benzene (Figure 13.35*c*) has *only* sp^2-hybridized carbons, and *all* of its peaks for C—H stretching lie above 3000 cm^{-1}. CC stretching gives a weak peak at 1478 cm^{-1}. The most intense peak in benzene (667 cm^{-1}) results from a vibration in which one of the C—H bonds bends out of the plane of the ring.

The hexylbenzene spectrum (Figure 13.35*d*) bears similarities to those of hexane and benzene. Peaks for C—H stretching are found both above and below 3000 cm^{-1} for sp^2 and sp^3 C—H stretching, respectively. The benzene ring is represented in the weak peak at 1496 cm^{-1}. The three peaks between 750 and 690 cm^{-1} include bending modes for the hexyl chain and the ring.

Rarely can the structure of a hydrocarbon ever be determined by IR alone. Figure 13.35 alerts us to the fact that most organic compounds give IR spectra in which many of the peaks are due to the carbon skeleton and its attached hydrogens. Chemists pay less attention to these peaks now that ^1H and ^{13}C NMR are available to gain the same information. What IR does best—identifying the presence or absence of functional groups—is described in the following section.

13.22 Characteristic Absorption Frequencies

Table 13.4 lists the **characteristic absorption frequencies** (in wavenumbers) for a variety of structural units found in organic compounds. Generally, absorptions above 1500 cm^{-1} for functional groups such as OH, C=O, and C≡N are the easiest to assign and provide the most useful information.

Some of these characteristic absorptions are reflected in the IR spectra of eight functional-group classes in Figure 13.36: alcohol, nitrile, carboxylic acid, ketone, ester, ether, amine, and amide. None of the specific compounds represented contains hydrogens bonded to sp^2-hybridized carbon, so all of the C—H absorbances lie below 3000 cm^{-1}. The compounds are related in that all have an unbranched six-carbon chain and, except for the peaks associated with the functional group, their spectra are similar, though not identical.

TABLE 13.4 Infrared Absorption Frequencies of Some Common Structural Units

Structural unit	Frequency, cm^{-1}	Structural unit	Frequency, cm^{-1}
	Stretching vibrations		
Single bonds		**Double bonds**	
—O—H (alcohols)	3200–3600	$\backslash C{=}C /$	1620–1680
—O—H (carboxylic acids)	2500–3600		
$\backslash N—H$	3350–3500	$\backslash C{=}O$	
sp C—H	3310–3320	Aldehydes and ketones	1710–1750
sp^2 C—H	3000–3100	Carboxylic acids	1700–1725
sp^3 C—H	2850–2950	Acid anhydrides	1800–1850 and 1740–1790
sp^2 C—O	1200	Acyl halides	1770–1815
sp^3 C—O	1025–1200	Esters	1730–1750
		Amides	1680–1700
		Triple bonds	
		—C≡C—	2100–2200
		—C≡N	2240–2280
	Bending vibrations of diagnostic value		
Alkenes:		***Substituted derivatives of benzene:***	
$RCH{=}CH_2$	910, 990	Monosubstituted	730–770 and 690–710
$R_2C{=}CH_2$	890	Ortho-disubstituted	735–770
cis-RCH=CHR′	665–730	Meta-disubstituted	750–810 and 680–730
trans-RCH=CHR′	960–980	Para-disubstituted	790–840
$R_2C{=}CHR'$	790–840		

Problem 13.24

Which of the following is the most likely structure of the compound characterized by the IR spectrum shown in Figure 13.37?

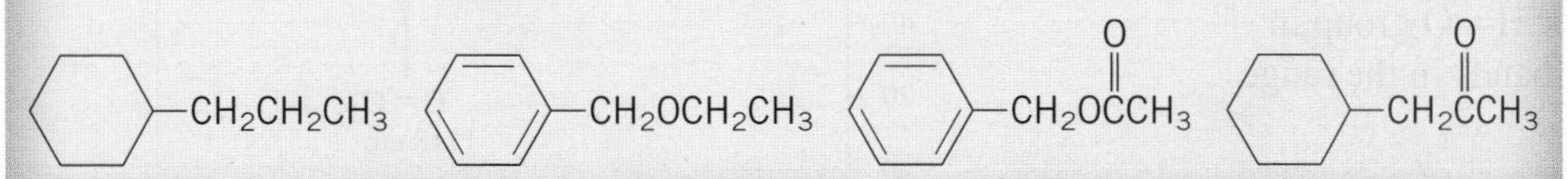

In later chapters, when families of compounds are discussed in detail, the IR frequencies associated with each type of functional group will be revisited.

(*a*) **Alcohols:** A broad peak at 3200–3400 cm^{-1} is characteristic of hydrogen-bonded OH groups. In dilute solution, hydrogen bonding is less, and a sharp second peak for "free" OH groups appears near 3600 cm^{-1}.

The peak at 1070 cm^{-1} lies in the range given in Table 13.4 (1025–1200 cm^{-1}) for C—O stretching and can be assigned to it.

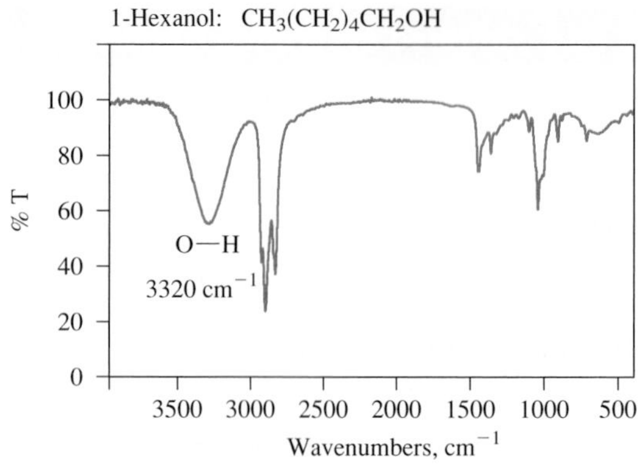

(*b*) **Nitriles:** The C≡N triple bond absorption is easily identifiable in the IR spectrum of a nitrile as a sharp peak of medium intensity at 2240–2280 cm^{-1}.

Very few other groups absorb in this region, the most notable being C≡C triple bonds (2100–2200 cm^{-1}).

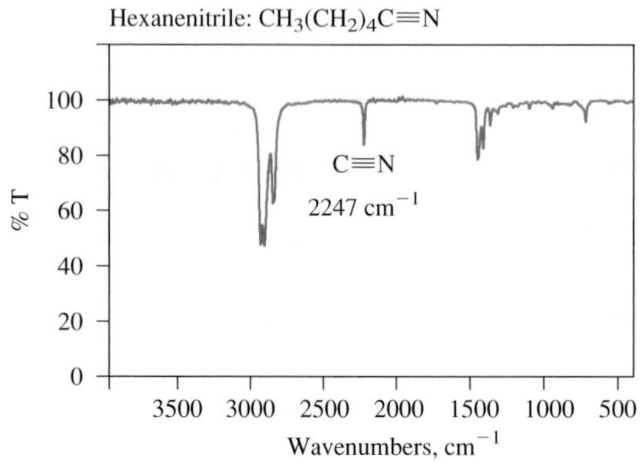

(*c*) **Carboxylic acids:** Carboxylic acids have two characteristic absorptions: a broad peak for O—H stretching in the range 2500–3600 cm^{-1} and a strong peak for C═O stretching at 1700–1725 cm^{-1}.

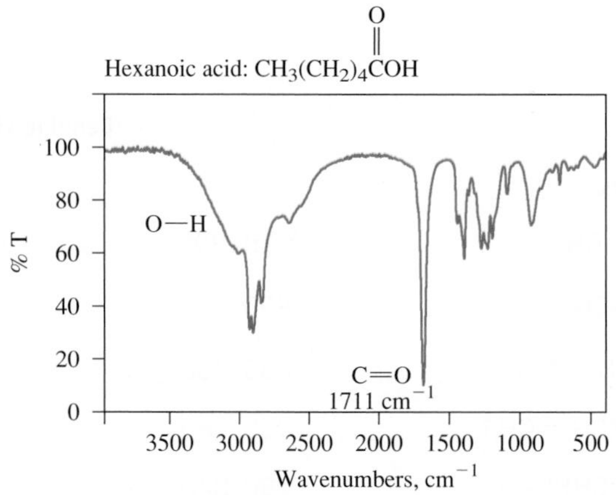

(*d*) **Aldehydes and ketones:** As in other carbonyl-containing compounds, the C═O stretching vibration gives the strongest peak in the IR spectra of aldehydes and ketones.

The C═O stretching frequencies of aldehydes are similar to those of ketones.

The C—H stretch of the CH═O group in aldehydes appears as a pair of bands in the range 2700–2900 cm^{-1}.

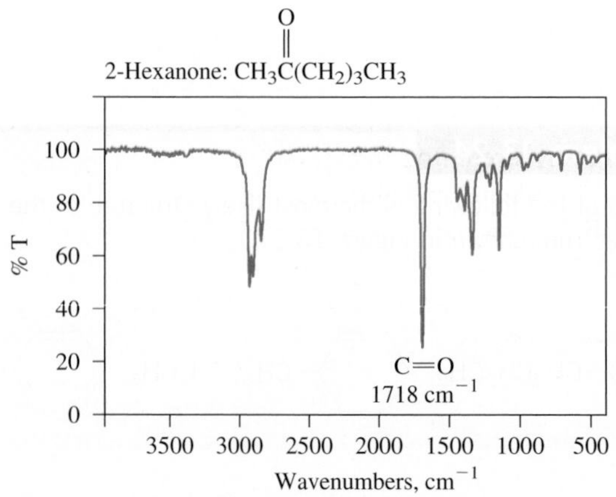

Figure 13.36

IR spectra of (*a*) 1-hexanol, (*b*) hexanenitrile, (*c*) hexanoic acid, (*d*) 2-hexanone, (*e*) methyl hexanoate, (*f*) dihexyl ether, (*g*) hexylamine, and (*h*) hexanamide.

(Continued)

(*e*) **Esters:** In addition to a strong C═O absorption (1730–1750 cm^{-1}), esters exhibit peaks for symmetric and antisymmetric C—O—C stretching at 1050–1300 cm^{-1}.

Methyl hexanoate: $CH_3(CH_2)_4COCH_3$ (C═O)

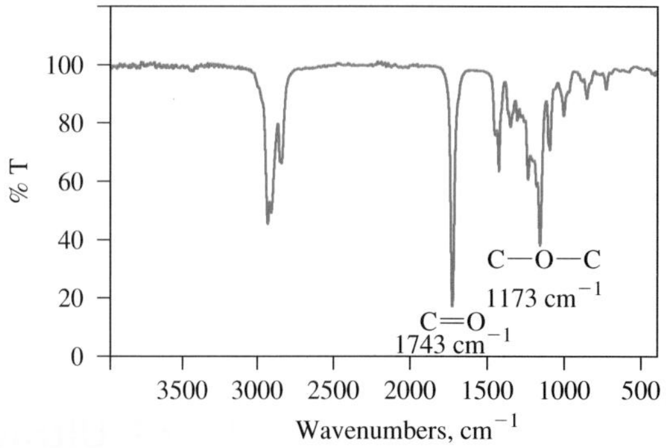

(*f*) **Ethers:** Peaks for C—O—C stretching in ethers appear in the range 1070–1150 cm^{-1}. Ethers of the type ROR′ where R and R′ are different have two peaks in this region.

Dihexyl ether: $[CH_3(CH_2)_4CH_2]_2O$

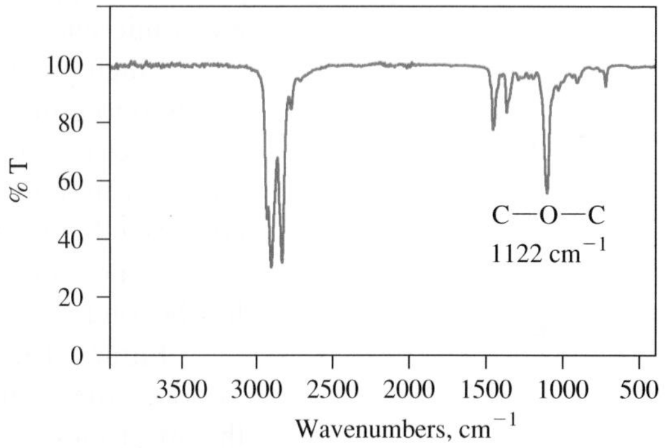

(*g*) **Amines:** Primary amines (RNH_2) have two peaks for the NH_2 group in the 3300–3500 cm^{-1} region, one for symmetric and the other for antisymmetric N—H stretching. Secondary amines (RNHR′) have only one peak (3310–3350 cm^{-1}).

An NH bending peak at 650–900 cm^{-1} occurs in both RH_2 and RNHR′. Primary amines also have an NH bending absorption at 1580–1650 cm^{-1}.

C—N stretching peaks are found at 1020–1250 cm^{-1}.

Hexylamine: $CH_3(CH_2)_4CH_2NH_2$

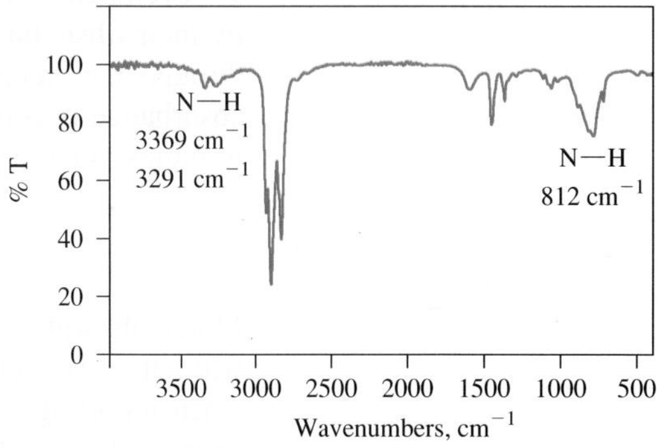

(*h*) **Amides:** Amides of the type $RC(O)NH_2$ have peaks for both symmetric and antisymmetric N—H stretching in the 3400–3150 cm^{-1} region.

The C═O absorption for amides appears at slightly lower frequency (1650–1700 cm^{-1}) than for ketones. Amides have a peak for NH_2 bending at a slightly lower frequency (1600–1650 cm^{-1}) than C═O.

Hexanamide: $CH_3(CH_2)_4CNH_2$ (C═O)

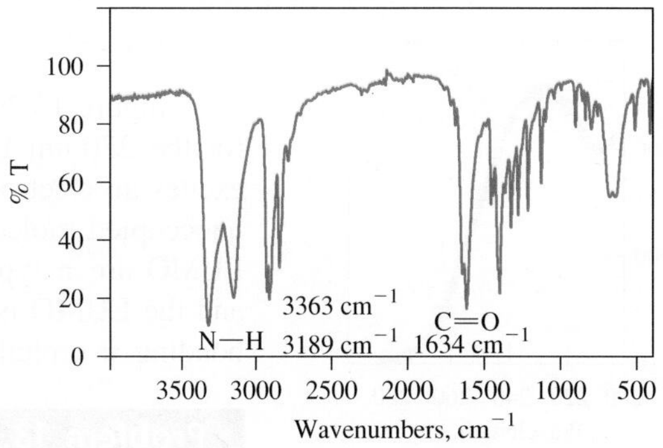

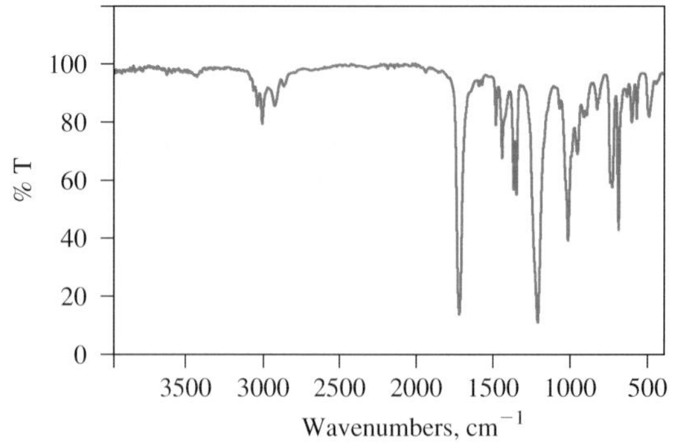

Figure 13.37

The IR spectrum of the unknown compound in Problem 13.24.

13.23 Ultraviolet-Visible (UV-VIS) Spectroscopy

The main application of UV-VIS spectroscopy, which depends on transitions between electronic energy levels, is in identifying conjugated π electron systems.

Much greater energies separate electronic states than vibrational states. The energy required to promote an electron from one electronic state to the next lies in the visible and ultraviolet range of the electromagnetic spectrum (see Figure 13.1). We usually identify radiation in the UV-VIS range by its wavelength in nanometers. Thus, the visible region corresponds to 400–800 nm. Red light is the low-energy (long wavelength) end of the visible spectrum, violet light the high-energy (short wavelength) end. Ultraviolet light lies beyond the visible spectrum with wavelengths in the 200–400-nm range.

Figure 13.38 shows the UV spectrum of the conjugated diene *cis,trans*-1,3-cyclooctadiene, measured in ethanol as the solvent. As is typical of most UV spectra, the absorption is rather broad and is often spoken of as a "band" rather than a "peak." The wavelength at an absorption maximum is referred to as the λ_{max} of the band. For 1,3-cyclooctadiene, λ_{max} is 230 nm. In addition to λ_{max}, UV-VIS bands are characterized by their **absorbance** (A), which is a measure of how much of the radiation that passes through the sample is absorbed. To correct for concentration and path length effects, absorbance is converted to **molar absorptivity** (ϵ) by dividing it by the concentration c in moles per liter and the path length l in centimeters.

$$\epsilon = \frac{A}{c \cdot l}$$

Molar absorptivity, when measured at λ_{max}, is cited as ϵ_{max}. It is normally expressed without units. Both λ_{max} and ϵ_{max} are affected by the solvent, which is therefore included when reporting UV-VIS spectroscopic data. Thus, you might find a literature reference expressed in the form

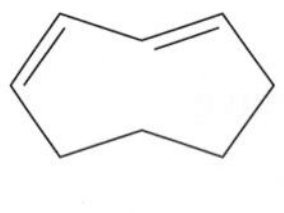

cis,trans-1,3-Cyclooctadiene

$\lambda_{max}^{ethanol}$ 230 nm

$\epsilon_{max}^{ethanol}$ 2630

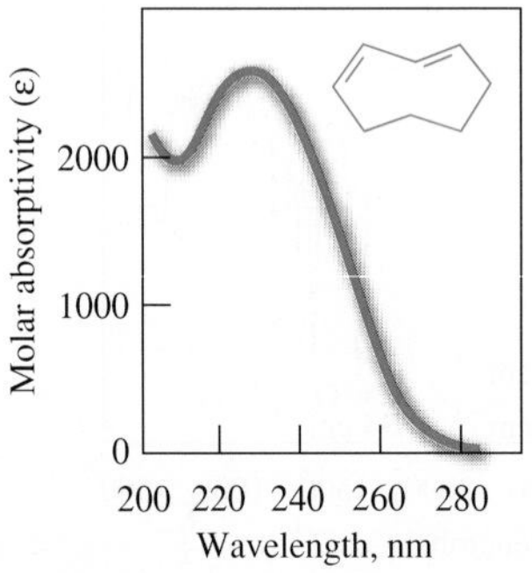

Figure 13.38

The UV spectrum of *cis,trans*-1,3-cyclooctadiene.

Figure 13.39 illustrates the transition between electronic energy states responsible for the 230-nm UV band of *cis,trans*-1,3-cyclooctadiene. Absorption of UV radiation excites an electron from the highest occupied molecular orbital (HOMO) to the lowest unoccupied molecular orbital (LUMO). In alkenes and polyenes, both the HOMO and LUMO are π type orbitals (rather than σ); the HOMO is the highest energy π orbital and the LUMO is the lowest energy π^* orbital. Exciting one of the π electrons from a bonding π orbital to an antibonding π^* orbital is referred to as a $\pi \rightarrow \pi^*$ transition.

Problem 13.25

λ_{max} for the $\pi \rightarrow \pi^*$ transition in ethylene is 170 nm. Is the HOMO–LUMO energy difference in ethylene greater than or less than that of *cis,trans*-1,3-cyclooctadiene (230 nm)?

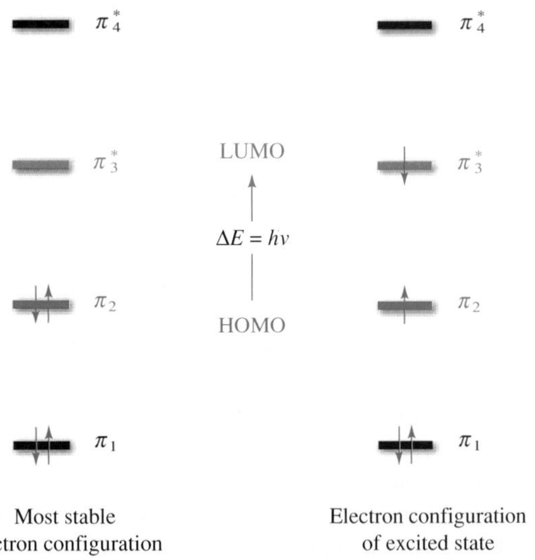
π4*
π3*
π2
π1
LUMO
ΔE = hν
HOMO
π4*
π3*
π2
π1
Most stable electron configuration
Electron configuration of excited state

Figure 13.39

The $\pi \rightarrow \pi^*$ transition in *cis,trans*-1,3-cyclooctadiene involves excitation of an electron from the highest occupied molecular orbital (HOMO) to the lowest unoccupied molecular orbital (LUMO).

The HOMO–LUMO energy gap and, consequently, λ_{max} for the $\pi \rightarrow \pi^*$ transition varies with the substituents on the double bonds. The data in Table 13.5 illustrate two substituent effects: adding methyl substituents to the double bond, and extending conjugation. Both cause λ_{max} to shift to longer wavelengths, but the effect of conjugation is the larger of the two. Based on data collected for many dienes it has been found that each methyl substituent on the double bonds causes a shift to longer wavelengths of about 5 nm, whereas extending the conjugation causes a shift of about 36 nm for each additional double bond.

Problem 13.26

Which one of the C_5H_8 isomers shown has its λ_{max} at the longest wavelength?

A striking example of the effect of conjugation on light absorption occurs in *lycopene,* one of the pigments in ripe tomatoes. Lycopene has a conjugated system of 11 double bonds and absorbs *visible light.* It has several UV-VIS bands, each characterized by a separate λ_{max}. Its longest wavelength absorption is at 505 nm. Note the inverse

TABLE 13.5 Absorption Maxima of Some Representative Alkenes and Polyenes*

Compound	Structure	λ_{max} (nm)
Ethylene	$H_2C{=}CH_2$	170
2-Methylpropene	$H_2C{=}C(CH_3)_2$	188
1, 3-Butadiene	$H_2C{=}CHCH{=}CH_2$	217
4-Methyl-1,3-pentadiene	$H_2C{=}CHCH{=}C(CH_3)_2$	234
2,5-Dimethyl-2,4-hexadiene	$(CH_3)_2C{=}CHCH{=}C(CH_3)_2$	241
(2*E*,4*E*,6*E*)-2,4,6-Octatriene	$CH_3CH{=}CHCH{=}CHCH{=}CHCH_3$	263
(2*E*,4*E*,6*E*,8*E*)-2,4,6,8-Decatetraene	$CH_3CH{=}CH(CH{=}CH)_2CH{=}CHCH_3$	299
(2*E*,4*E*,6*E*,8*E*,10*E*)-2,4,6,8,10-Dodecapentaene	$CH_3CH{=}CH(CH{=}CH)_3CH{=}CHCH_3$	326

*The value of λ_{max} refers to the longest wavelength $\pi \rightarrow \pi^*$ transition.

relationship between the color of a compound and the wavelength of light absorbed. Lycopene absorbs light in the blue region of the visible spectrum, yet appears red. The red color of lycopene is produced by the light that is not absorbed.

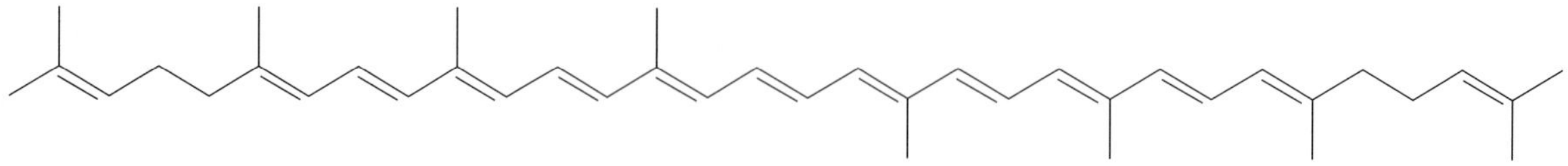

Lycopene

Many organic compounds such as lycopene are colored because their HOMO–LUMO energy gap is small enough that λ_{max} appears in the visible range of the spectrum. All that is required for a compound to be colored, however, is that it possess some absorption in the visible range. It often happens that a compound will have its λ_{max} in the UV region but that the peak is broad and extends into the visible. Absorption of the blue-to-violet components of visible light occurs, and the compound appears yellow.

Don't confuse the *n* in $n \rightarrow \pi^*$ with the *n* of Hückel's rule.

An important enzyme in biological electron transport called *cytochrome P450* gets its name from its UV absorption. The "P" stands for "pigment" because it is colored, and the "450" corresponds to the 450-nm absorption of one of its derivatives.

A second type of absorption that is important in UV-VIS examination of organic compounds is the $n \rightarrow \pi^*$ transition of the carbonyl (C=O) group. One of the electrons in a lone-pair orbital of oxygen is excited to an antibonding orbital of the carbonyl group. The *n* in $n \rightarrow \pi^*$ identifies the electron as one of the nonbonded electrons of oxygen. This transition gives rise to relatively weak absorption peaks ($\epsilon_{max} < 100$) in the region 270–300 nm. The structural unit associated with an electronic transition in UV-VIS spectroscopy is called a **chromophore.** UV-visible spectroscopy has applications in biochemistry, where chromophores such as the heterocyclic bases found in nucleic acids (see Section 11.22) and coenzymes involved in biochemical reactions (Section 15.10) can be studied.

13.24 Mass Spectrometry

Mass spectrometry differs from the other instrumental methods discussed in this chapter in a fundamental way. It does not depend on the absorption of electromagnetic radiation but rather examines ions produced from a molecule in the gas phase. Several techniques have been developed for ionization in mass spectrometry. In one method, the molecule is bombarded with high-energy electrons. If an electron having an energy of about 10 electronvolts (10 eV = 230.5 kcal/mol) collides with an organic molecule, the energy transferred as a result of that collision is sufficient to dislodge one of the molecule's electrons.

$$\underset{\text{Molecule}}{A{:}B} + \underset{\text{Electron}}{e^-} \longrightarrow \underset{\text{Cation radical}}{A\overset{+}{\cdot}B} + \underset{\text{Two electrons}}{2e^-}$$

We say the molecule AB has been ionized by **electron impact.** The species that results, called the **molecular ion,** is positively charged and has an odd number of electrons—it is a **cation radical.** The molecular ion has the same mass (less the negligible mass of a single electron) as the molecule from which it is formed.

Although energies of about 10 eV are required, energies of about 70 eV are used. Electrons this energetic not only cause ionization of a molecule but also impart a large amount of energy to the molecular ion, enough energy to break chemical bonds. The molecular ion dissipates this excess energy by dissociating into smaller fragments. Dissociation of a cation radical produces a neutral fragment and a positively charged fragment.

$$\underset{\text{Cation radical}}{A\overset{+}{\cdot}B} \longrightarrow \underset{\text{Cation}}{A^+} + \underset{\text{Radical}}{B\cdot}$$

Ionization and fragmentation produce a mixture of particles, some neutral and some positively charged. To understand what follows, we need to examine the design of

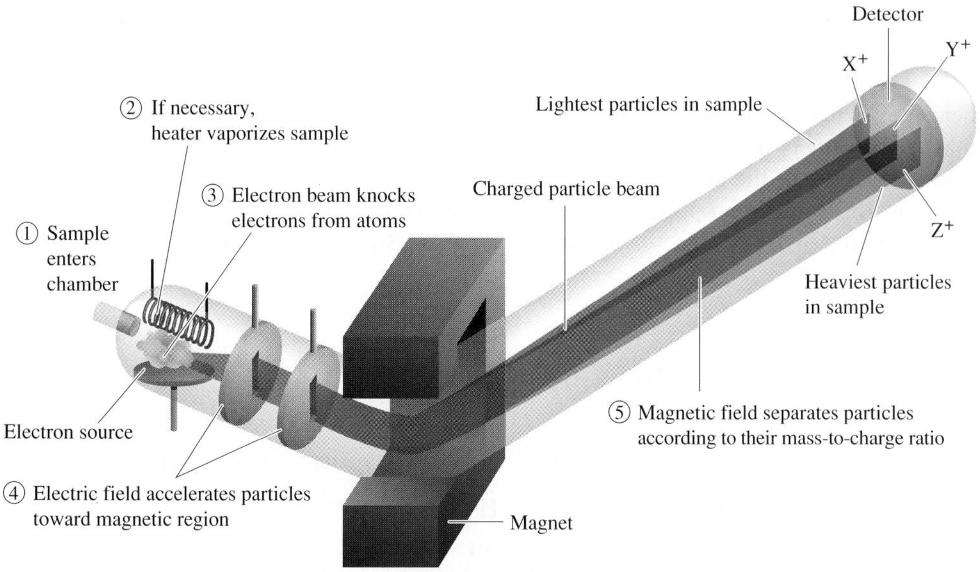

Figure 13.40

Diagram of a mass spectrometer. Only positive ions are detected. The cation X^+ has the lowest mass-to-charge ratio and its path is deflected most by the magnet. The cation Z^+ has the highest mass-to-charge ratio and its path is deflected least. (Adapted, with permission, from M. Silberberg, *Chemistry*, McGraw-Hill Higher Education, 2009, p. 55.)

an electron-impact mass spectrometer, shown in Figure 13.40. The sample is bombarded with 70-eV electrons, and the resulting positively charged ions (the molecular ion as well as fragment ions) are directed into an analyzer tube surrounded by a magnet. This magnet deflects the ions from their original trajectory, causing them to adopt a circular path, the radius of which depends on their mass-to-charge ratio *(m/z)*. Ions of small *m/z* are deflected more than those of larger *m/z*. By varying either the magnetic field strength or the degree to which the ions are accelerated on entering the analyzer, ions of a particular *m/z* can be selectively focused through a narrow slit onto a detector, where they are counted. Scanning all *m/z* values gives the distribution of positive ions, called a **mass spectrum,** characteristic of a particular compound.

Most mass spectrometers are interfaced with computerized data-handling systems capable of displaying the mass spectrum according to a number of different formats. Bar graphs on which relative intensity is plotted versus *m/z* are the most common. Figure 13.41 shows the mass spectrum of benzene in bar graph form.

The mass spectrum of benzene is relatively simple and illustrates some of the information that mass spectrometry provides. The most intense peak in the mass spectrum is called the **base peak** and is assigned a relative intensity of 100. Ion abundances are proportional to peak intensities and are reported as intensities relative to the base peak. The base peak in the mass spectrum of benzene corresponds to the molecular ion (M^+) at m/z = 78.

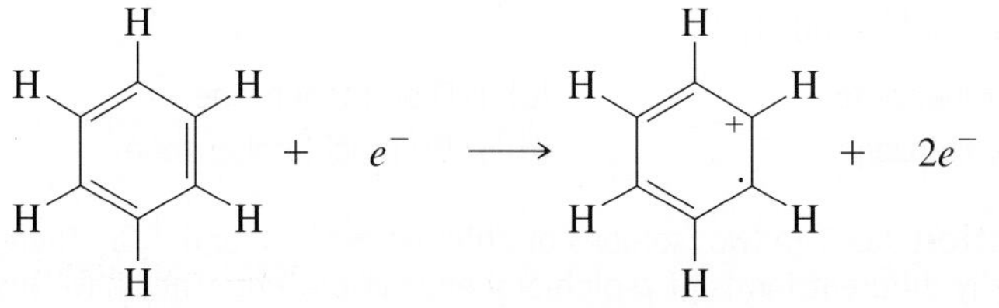

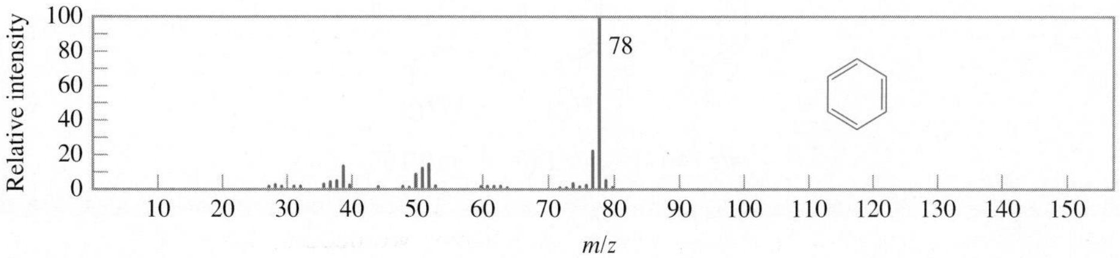

Figure 13.41

The mass spectrum of benzene. The peak at m/z = 78 corresponds to the C_6H_6 molecular ion.

Benzene does not undergo extensive fragmentation; none of the fragment ions in its mass spectrum are as abundant as the molecular ion.

There is a small peak one mass unit higher than M^+ in the mass spectrum of benzene. What is the origin of this peak? What we see in Figure 13.41 as a single mass spectrum is actually a superposition of the spectra of three isotopically distinct benzenes. Most of the benzene molecules contain only ^{12}C and ^{1}H and have a molecular mass of 78. Smaller proportions of benzene molecules contain ^{13}C in place of one of the ^{12}C atoms or ^{2}H in place of one of the protons. Both these species have a molecular mass of 79.

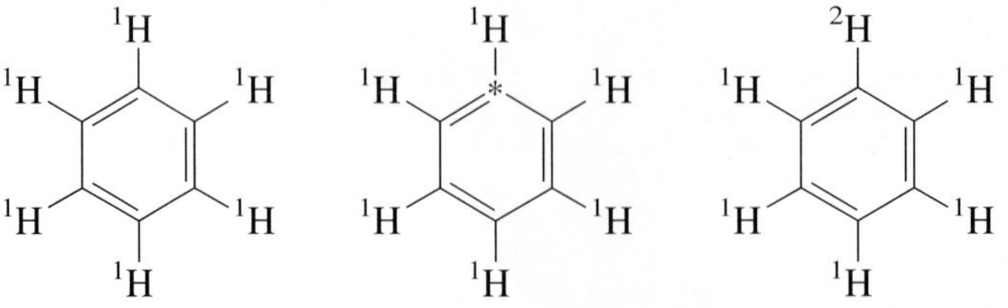

93.4% (all carbons are ^{12}C) Gives M^+ 78 | 6.5% (* = ^{13}C) Gives M^+ 79 | 0.1% (all carbons are ^{12}C) Gives M^+ 79

Not only the molecular ion peak but all the peaks in the mass spectrum of benzene are accompanied by a smaller peak one mass unit higher. Indeed, because all organic compounds contain carbon and most contain hydrogen, similar **isotopic clusters** will appear in the mass spectra of all organic compounds.

Isotopic clusters are especially apparent when atoms such as bromine and chlorine are present in an organic compound. The natural ratios of isotopes in these elements are

$$\frac{^{35}\mathrm{Cl}}{^{37}\mathrm{Cl}} = \frac{100}{32.7} \qquad \frac{^{79}\mathrm{Br}}{^{81}\mathrm{Br}} = \frac{100}{97.5}$$

Figure 13.42 presents the mass spectrum of chlorobenzene. There are two prominent molecular ion peaks, one at *m/z* 112 for $C_6H_5{}^{35}Cl$ and the other at *m/z* 114 for $C_6H_5{}^{37}Cl$. The peak at *m/z* 112 is three times as intense as the one at *m/z* 114.

Problem 13.27

Knowing what to look for with respect to isotopic clusters can aid in interpreting mass spectra. How many peaks would you expect to see for the molecular ion in each of the following compounds? At what *m/z* values would these peaks appear? (Disregard the small peaks due to ^{13}C and ^{2}H.)

(a) *p*-Dichlorobenzene
(b) *o*-Dichlorobenzene
(c) *p*-Dibromobenzene
(d) *p*-Bromochlorobenzene

Sample Solution (a) The two isotopes of chlorine are ^{35}Cl and ^{37}Cl. There will be three isotopically different forms of *p*-dichlorobenzene present. They have the structures shown as follows. Each one will give an M^+ peak at a different value of *m/z*.

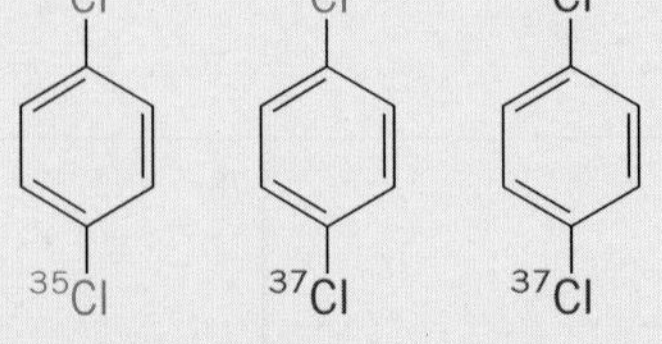

m/z 146 | *m/z* 148 | *m/z* 150

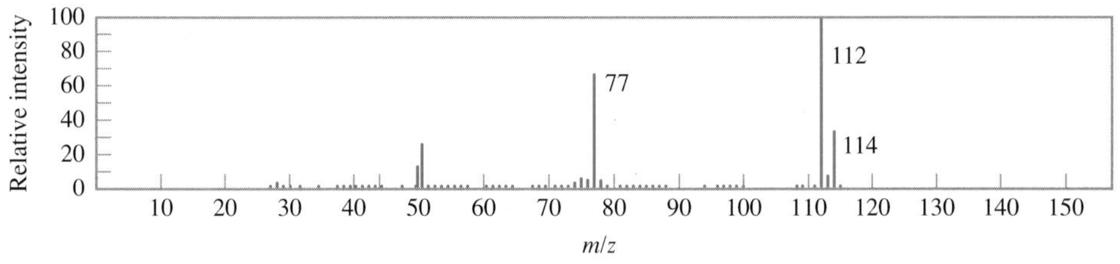

Figure 13.42

The mass spectrum of chlorobenzene.

Unlike the case of benzene, in which ionization involves loss of a π electron from the ring, electron-impact-induced ionization of chlorobenzene involves loss of an electron from an unshared pair of chlorine. The molecular ion then fragments by carbon–chlorine bond cleavage.

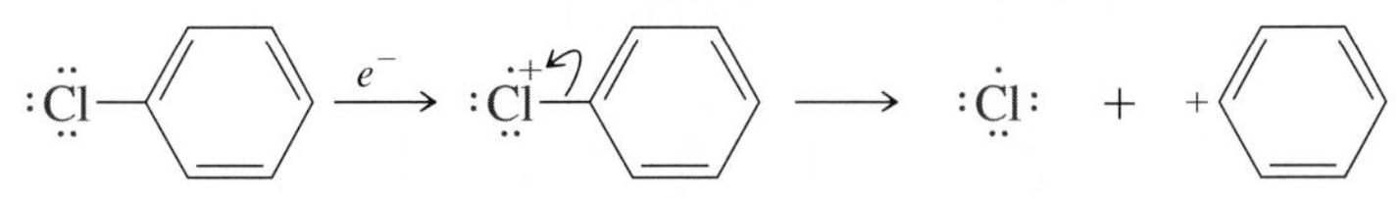

Chlorobenzene | Molecular ion of chlorobenzene | Chlorine atom | Phenyl cation *m/z* 77

The peak at *m/z* 77 in the mass spectrum of chlorobenzene in Figure 13.42 is attributed to this fragmentation. Because there is no peak of significant intensity two atomic mass units higher, we know that the cation responsible for the peak at *m/z* 77 cannot contain chlorine.

Some classes of compounds are so prone to fragmentation that the molecular ion peak is very weak. The base peak in most unbranched alkanes, for example, is *m/z* 43, which is followed by peaks of decreasing intensity at *m/z* values of 57, 71, 85, and so on. These peaks correspond to cleavage of each possible carbon–carbon bond in the molecule. This pattern is evident in the mass spectrum of decane, depicted in Figure 13.43. The points of cleavage are indicated in the following diagram:

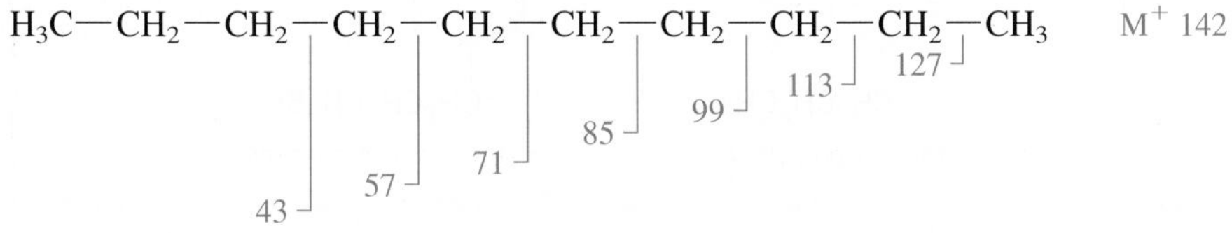

Many fragmentations in mass spectrometry proceed so as to form a stable carbocation, and the principles that we have developed regarding carbocation stability apply. Alkylbenzenes of the type $C_6H_5CH_2R$ undergo cleavage of the bond to the benzylic carbon to give *m/z* 91 as the base peak. The mass spectrum in Figure 13.44 and the following fragmentation diagram illustrate this for propylbenzene.

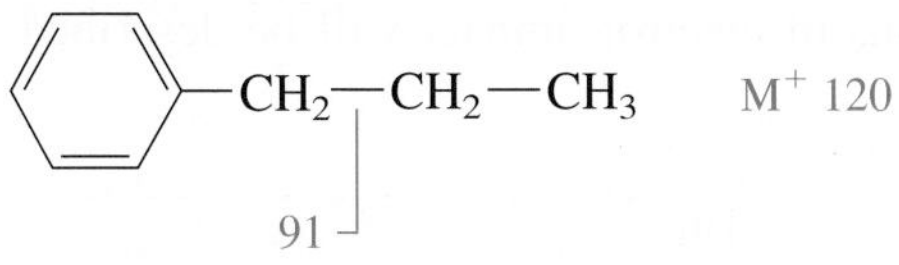

Although this cleavage is probably driven by the stability of benzyl cation, evidence has been obtained suggesting that tropylium cation, formed by rearrangement of benzyl cation, is actually the species responsible for the peak.

The structure of tropylium cation is given in Section 11.21.

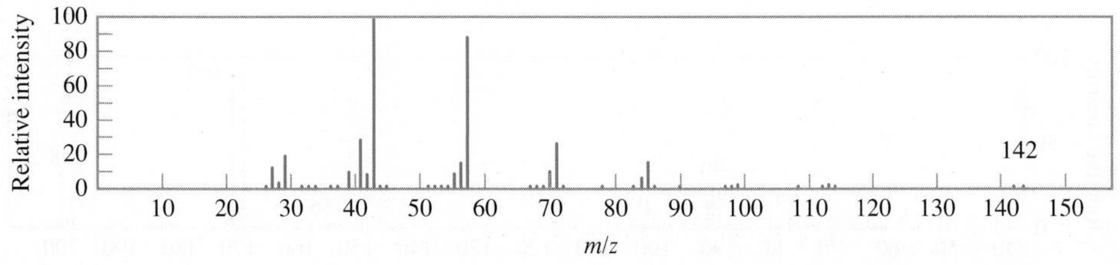

Figure 13.43

The mass spectrum of decane. The peak for the molecular ion is extremely small. The most prominent peaks arise by fragmentation.

Figure 13.44

The mass spectrum of propylbenzene. The most intense peak is $C_7H_7^+$.

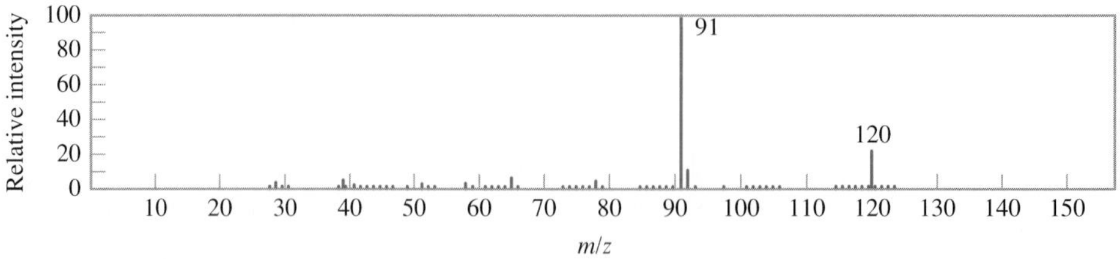

Problem 13.28

The base peak appears at *m/z* 105 for one of the following compounds and at *m/z* 119 for the other two. Match the compounds with the appropriate *m/z* values for their base peaks.

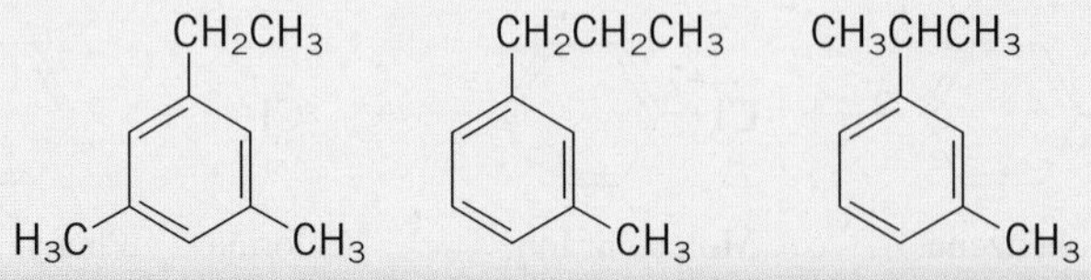

Problem 13.29

Mass spectra for 1-bromo-4-propylbenzene and (3-bromopropyl)benzene are shown in Figure 13.45. Match each spectrum to the appropriate compound. Write a structure for the ion that corresponds to the base peak in each spectrum.

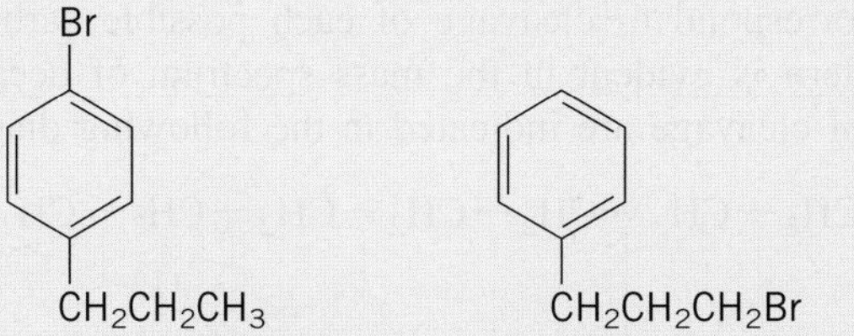

1-Bromo-4-propylbenzene (3-Bromopropyl)benzene

An alternate method of ionization is described in the boxed essay "Peptide Mapping and MALDI Mass Spectrometry" in Chapter 25.

Understanding how molecules fragment upon electron impact permits a mass spectrum to be analyzed in sufficient detail to deduce the structure of an unknown compound. Thousands of compounds of known structure have been examined by mass spectrometry, and the fragmentation patterns that characterize different classes are well documented. As various groups are covered in subsequent chapters, aspects of their fragmentation behavior under conditions of electron impact will be described.

Figure 13.45

Mass spectra of 1-bromo-4-propylbenzene and (3-bromopropyl)benzene.

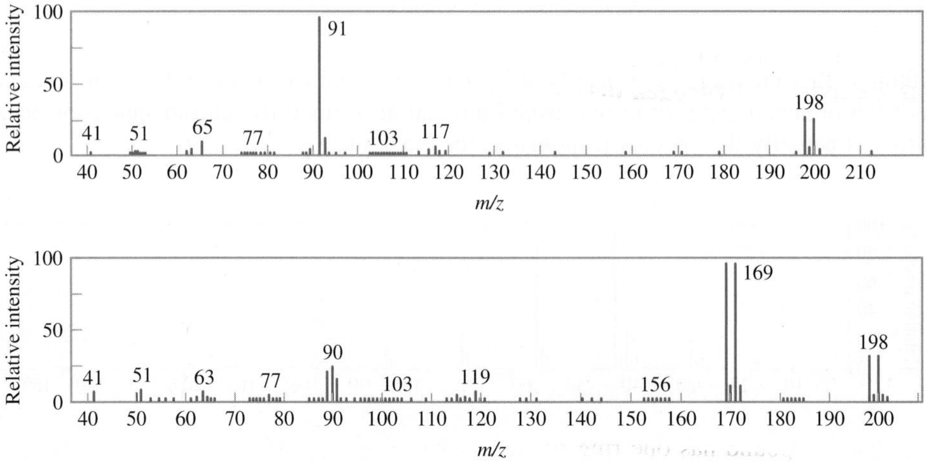

13.25 Molecular Formula as a Clue to Structure

As we have just seen, interpreting the fragmentation patterns in a mass spectrum in terms of a molecule's structural units makes mass spectrometry much more than just a tool for determining molecular weights. Nevertheless, even the molecular weight can provide more information than you might think.

A relatively simple example is the **nitrogen rule.** A molecule with an odd number of nitrogens has an odd molecular weight; a molecule with only C, H, and O or with an even number of nitrogens has an even molecular weight.

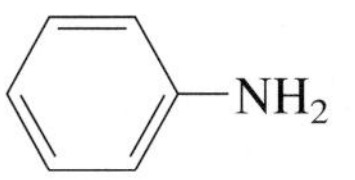

Aniline (C_6H_7N)

p-Nitroaniline ($C_6H_6N_2O_2$)

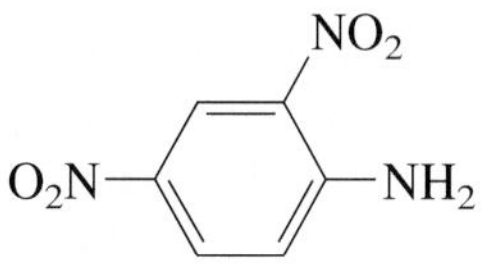

2,4-Dinitroaniline ($C_6H_5N_3O_4$)

Molecular weight: 93 138 183

A second example concerns different compounds that have the same molecular weight, but different molecular formulas, such as heptane and cyclopropyl acetate.

$CH_3(CH_2)_5CH_3$

Heptane (C_7H_{16})

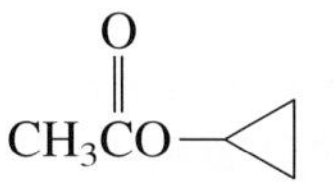

Cyclopropyl acetate ($C_5H_8O_2$)

Because we normally round off molecular weights to whole numbers, both have a molecular weight of 100 and both have a peak for their molecular ion at *m/z* 100 in a typical mass spectrum. Recall, however, that mass spectra contain isotopic clusters that differ according to the isotopes present in each ion. Using the exact values for the major isotopes of C, H, and O, we calculate *exact masses* of *m/z* of 100.1253 and 100.0524 for the molecular ions of heptane (C_7H_{16}) and cyclopropyl acetate ($C_5H_8O_2$), respectively. As similar as these values are, it is possible to distinguish between them using a *high-resolution mass spectrometer.* This means that the exact mass of a molecular ion can usually be translated into a unique molecular formula.

You can't duplicate these molecular weights for C_7H_{16} and $C_5H_8O_2$ by using the atomic weights given in the periodic table. Those values are for the natural-abundance mixture of isotopes. The exact values are 12.00000 for ^{12}C, 1.00783 for ^{1}H, and 15.9949 for ^{16}O.

Once we have the molecular formula, it can provide information that limits the amount of trial-and-error structure writing we have to do. Consider, for example, heptane and its molecular formula of C_7H_{16}. We know immediately that the molecular formula belongs to an alkane because it corresponds to C_nH_{2n+2}.

What about a substance with the molecular formula C_7H_{14}? This compound cannot be an alkane but may be either a cycloalkane or an alkene, because both these classes of hydrocarbons correspond to the general molecular formula C_nH_{2n}. *Any time a ring or a double bond is present in an organic molecule, its molecular formula has two fewer hydrogen atoms than that of an alkane with the same number of carbons.*

The relationship between molecular formulas, multiple bonds, and rings is referred to as the **index of hydrogen deficiency** and can be expressed by the equation:

Other terms that mean the same thing as the index of hydrogen deficiency include *elements of unsaturation, sites of unsaturation,* and *the sum of double bonds and rings.*

$$\text{Index of hydrogen deficiency} = \tfrac{1}{2}(C_nH_{2n+2} - C_nH_x)$$

where C_nH_x is the molecular formula of the compound.

A molecule that has a molecular formula of C_7H_{14} has an index of hydrogen deficiency of 1:

$$\text{Index of hydrogen deficiency} = \tfrac{1}{2}(C_7H_{16} - C_7H_{14})$$

$$\text{Index of hydrogen deficiency} = \tfrac{1}{2}(2) = 1$$

Thus, the compound has one ring or one double bond. It can't have a triple bond.

A molecule of molecular formula C_7H_{12} has four fewer hydrogens than the corresponding alkane. It has an index of hydrogen deficiency of 2 and can have two rings, two double bonds, one ring and one double bond, or one triple bond.

What about substances other than hydrocarbons, 1-heptanol [$CH_3(CH_2)_5CH_2OH$], for example? Its molecular formula ($C_7H_{16}O$) contains the same carbon-to-hydrogen ratio as heptane and, like heptane, it has no double bonds or rings. Cyclopropyl acetate ($C_5H_8O_2$), the structure of which was given at the beginning of this section, has one ring and one double bond and an index of hydrogen deficiency of 2. *Oxygen atoms have no effect on the index of hydrogen deficiency.*

A halogen substituent, like hydrogen, is monovalent and when present in a molecular formula is treated as if it were hydrogen for counting purposes. If a nitrogen is present, one hydrogen is taken away from the formula. For example, $C_5H_{11}N$ is treated as C_5H_{10} when calculating the index of hydrogen deficiency.

How does one distinguish between rings and double bonds? This additional piece of information comes from catalytic hydrogenation experiments in which the amount of hydrogen consumed is measured exactly. Each of a molecule's double bonds consumes one molar equivalent of hydrogen, but rings are unaffected. For example, a substance with a hydrogen deficiency of 5 that takes up 3 mol of hydrogen must have two rings.

Problem 13.30

How many rings are present in each of the following compounds? Each consumes 2 mol of hydrogen on catalytic hydrogenation.

(a) $C_{10}H_{18}$
(b) C_8H_8
(c) $C_8H_8Cl_2$
(d) C_8H_8O
(e) $C_8H_{10}O_2$
(f) C_8H_9ClO
(g) C_3H_5N
(h) C_4H_5N

Sample Solution (a) The molecular formula $C_{10}H_{18}$ contains four fewer hydrogens than the alkane having the same number of carbon atoms ($C_{10}H_{22}$). Therefore, the index of hydrogen deficiency of this compound is 2. Because it consumes two molar equivalents of hydrogen on catalytic hydrogenation, it must have either a triple bond or two double bonds and no rings.

13.26 SUMMARY

Section 13.1 Structure determination in modern organic chemistry relies heavily on instrumental methods. Several of the most widely used ones depend on the absorption of electromagnetic radiation.

Section 13.2 Absorption of electromagnetic radiation causes a molecule to be excited from its most stable state (the *ground* state) to a higher energy state (an *excited* state).

Spectroscopic method	*Transitions between*
Nuclear magnetic resonance	Spin states of an atom's nucleus
Infrared	Vibrational states
Ultraviolet-visible	Electronic states

Mass spectrometry is not based on absorption of electromagnetic radiation, but monitors what happens when a substance is ionized by collision with a high-energy electron.

¹H Nuclear Magnetic Resonance Spectroscopy

Section 13.3 In the presence of an external magnetic field, the $+\frac{1}{2}$ and $-\frac{1}{2}$ nuclear spin states of a proton have slightly different energies.

Section 13.4 The energy required to "flip" the spin of a proton from the lower energy spin state to the higher state depends on the extent to which a nucleus is shielded from the external magnetic field by the molecule's electrons.

Section 13.5 Protons in different environments within a molecule have different **chemical shifts;** that is, they experience different degrees of shielding. Chemical shifts (δ) are reported in parts per million (ppm) from tetramethylsilane (TMS). Table 13.1 lists characteristic chemical shifts for various types of protons.

Section 13.6 In addition to *chemical shift,* a ^{1}H NMR spectrum provides structural information based on:

Number of signals, which tells how many different kinds of protons there are

Integrated areas, which tells the ratios of the various kinds of protons

Splitting pattern, which gives information about the number of protons that are within two or three bonds of the one giving the signal

Section 13.7 **Spin–spin splitting** of NMR signals results from coupling of the nuclear spins that are separated by two bonds (*geminal coupling*) or three bonds (*vicinal coupling*).

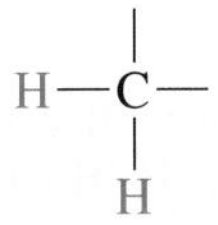

Geminal hydrogens are separated by two bonds

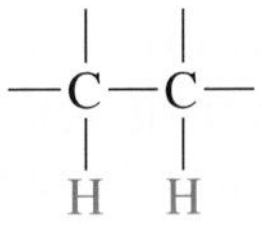

Vicinal hydrogens are separated by three bonds

In the simplest cases, the number of peaks into which a signal is split is equal to $n + 1$, where n is the number of protons to which the proton in question is coupled. *Protons that have the same chemical shift do not split each other's signal.*

Section 13.8 The methyl protons of an ethyl group appear as a *triplet* and the methylene protons as a *quartet* in compounds of the type CH_3CH_2X.

Section 13.9 The methyl protons of an isopropyl group appear as a *doublet* and the methine proton as a *septet* in compounds of the type $(CH_3)_2CHX$.

Section 13.10 A *pair of doublets* characterizes the signals for the protons of the type shown (where W, X, Y, and Z are not H or atoms that split H themselves).

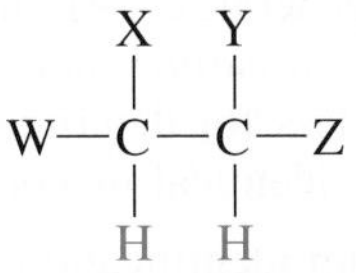

Section 13.11 Complicated splitting patterns can result when a proton is unequally coupled to two or more protons that are different from one another.

Section 13.12 Splitting resulting from coupling to the O—H proton of alcohols is not normally observed, because the hydroxyl proton undergoes rapid intermolecular exchange with other alcohol molecules, which "decouples" it from other protons in the molecule.

Section 13.13 Many processes such as conformational changes take place faster than they can be detected by NMR. Consequently, NMR provides information about the *average* environment of a proton. For example, cyclohexane gives a single peak for its 12 protons even though, at any instant, 6 are axial and 6 are equatorial.

^{13}C Nuclear Magnetic Resonance Spectroscopy

Section 13.14 ^{13}C has a nuclear spin of $\pm\frac{1}{2}$ but only about 1% of all the carbons in a sample are ^{13}C. Nevertheless, high-quality ^{13}C NMR spectra can be obtained by pulse FT techniques and are a useful complement to ^{1}H NMR spectra.

Section 13.15 ^{13}C signals are more widely separated from one another than proton signals, and ^{13}C NMR spectra are relatively easy to interpret. Table 13.3 gives chemical shift values for carbon in various environments.

Section 13.16 ^{13}C NMR spectra are rarely integrated because the pulse FT technique distorts the signal intensities.

Section 13.17 Carbon signals normally appear as singlets, but several techniques are available that allow one to distinguish among the various kinds of carbons shown.

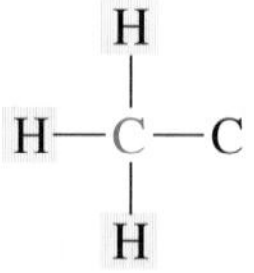

3 attached hydrogens (Primary carbon)

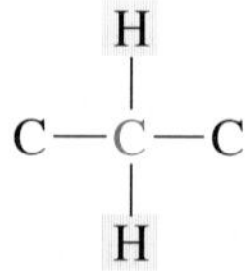

2 attached hydrogens (Secondary carbon)

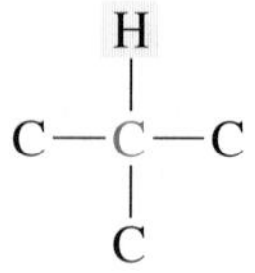

1 attached hydrogen (Tertiary carbon)

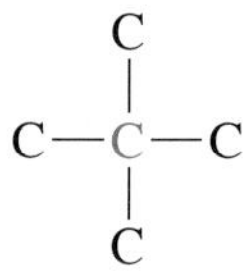

no attached hydrogens (Quaternary carbon)

Section 13.18 One of the special techniques for distinguishing carbons according to the number of their attached hydrogens is called **DEPT.** A series of NMR measurements using different pulse sequences gives normal, nulled, and inverted peaks that allow assignment of primary, secondary, tertiary, and quaternary carbons.

Section 13.19 2D NMR techniques are enhancements that are sometimes useful in gaining additional structural information. A ^1H-^1H COSY spectrum reveals which protons are spin-coupled to other protons, which helps in determining connectivity. A HETCOR spectrum shows the C—H connections by correlating ^{13}C and ^1H chemical shifts.

Infrared Spectroscopy

Section 13.20 IR spectroscopy probes molecular structure by examining transitions between quantized vibrational energy levels using electromagnetic radiation in the 625–4000-cm^{-1}range, where cm^{-1} are units of **wavenumbers,** defined as λ^{-1}. Wavenumbers are proportional to frequency. The simplest vibration is the stretching of the bond between two atoms, but more complex vibrations can involve movement of many of a molecule's atoms.

Section 13.21 IR spectra are commonly regarded as consisting of a functional-group region (1500–4000 cm^{-1}) and a fingerprint region (500–1500 cm^{-1}). Included in the functional-group region are absorptions due to C—H stretching. In general, C—H stretching frequencies lie below 3000 cm^{-1} for sp^3-hybridized carbon and above 3000 cm^{-1} for sp^2. The fingerprint region is used less for determining structure than for verifying whether two compounds are identical or not.

Section 13.22 Functional-group identification is the main contribution of IR spectroscopy to organic chemistry. Various classes of compounds exhibit peaks at particular frequencies characteristic of the functional groups they contain. (Table 13.4).

Ultraviolet-Visible Spectroscopy

Section 13.23 Transitions between electronic energy levels involving electromagnetic radiation in the 200–800-nm range form the basis of UV-VIS spectroscopy. The absorption peaks tend to be broad but are often useful in indicating the presence of particular π electron systems within a molecule.

Mass Spectrometry

Section 13.24 Mass spectrometry exploits the information obtained when a molecule is ionized by electron impact and then dissociates to smaller fragments. Positive ions are separated and detected according to their mass-to-charge (m/z) ratio. By examining the fragments and by knowing how classes of molecules dissociate on electron impact, one can deduce the structure of a compound. Mass spectrometry is quite sensitive; as little as 10^{-9} g of compound is sufficient for analysis.

Section 13.25 A compound's molecular formula gives information about the number of double bonds and rings it contains and is a useful complement to spectroscopic methods of structure determination.

PROBLEMS

13.31 Each of the following compounds is characterized by a ^{1}H NMR spectrum that consists of only a single peak having the chemical shift indicated. Identify each compound.

(a) C_8H_{18}; δ 0.9
(b) C_5H_{10}; δ 1.5
(c) C_8H_8; δ 5.8
(d) C_4H_9Br; δ 1.8
(e) $C_2H_4Cl_2$; δ 3.7
(f) $C_2H_3Cl_3$; δ 2.7
(g) $C_5H_8Cl_4$; δ 3.7
(h) $C_{12}H_{18}$; δ 2.2
(i) $C_3H_6Br_2$; δ 2.6

13.32 Deduce the structure of each of the following compounds on the basis of their ^{1}H NMR spectra and molecular formulas:

(a) C_8H_{10}; δ 1.2 (triplet, 3H)
δ 2.6 (quartet, 2H)
δ 7.1 (broad singlet, 5H)

(b) $C_{10}H_{14}$; δ 1.3 (singlet, 9H)
δ 7.0 to 7.5 (multiplet, 5H)

(c) C_6H_{14}; δ 0.8 (doublet, 12H)
δ 1.4 (septet, 2H)

(d) C_6H_{12}; δ 0.9 (triplet, 3H)
δ 1.6 (singlet, 3H)
δ 1.7 (singlet, 3H)
δ 2.0 (quintet, 2H)
δ 5.1 (triplet, 1H)

(e) $C_4H_6Cl_4$; δ 3.9 (doublet, 4H)
δ 4.6 (triplet, 2H)

(f) $C_4H_6Cl_2$; δ 2.2 (singlet, 3H)
δ 4.1 (doublet, 2H)
δ 5.7 (triplet, 1H)

(g) C_3H_7ClO; δ 2.0 (quintet, 2H)
δ 2.8 (singlet, 1H)
δ 3.7 (triplet, 2H)
δ 3.8 (triplet, 2H)

(h) $C_{14}H_{14}$; δ 2.9 (singlet, 4H)
δ 7.1 (broad singlet, 10H)

13.33 From among the isomeric compounds of molecular formula C_4H_9Cl, choose the one having a ^{1}H NMR spectrum that

(a) Contains only a single peak
(b) Has several peaks including a doublet at δ 3.4
(c) Has several peaks including a triplet at δ 3.5
(d) Has several peaks including two distinct three-proton signals, one of them a triplet at δ 1.0 and the other a doublet at δ 1.5

13.34 Identify the C_3H_5Br isomers on the basis of the following information:

(a) Isomer A has the ^{1}H NMR spectrum shown in Figure 13.46.
(b) Isomer B has three peaks in its ^{13}C NMR spectrum: δ 32.6 (CH_2); 118.8 (CH_2); and 134.2 (CH).
(c) Isomer C has two peaks in its ^{13}C NMR spectrum: δ 12.0 (CH_2) and 16.8 (CH). The peak at lower field is only half as intense as the one at higher field.

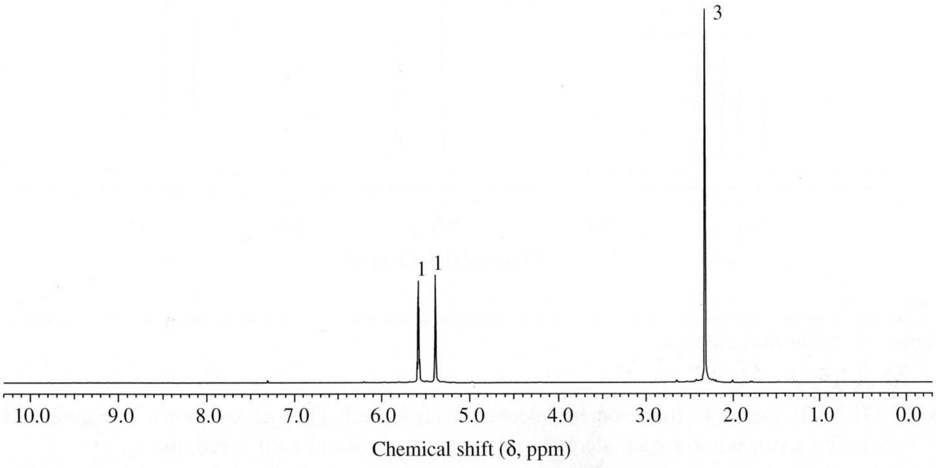

Figure 13.46
The 200-MHz ^{1}H NMR spectrum of isomer A (Problem 13.34a).

13.35 Identify each of the $C_4H_{10}O$ isomers on the basis of their ^{13}C NMR spectra:

(a) δ 18.9 (CH_3) (two carbons)
δ 30.8 (CH) (one carbon)
δ 69.4 (CH_2) (one carbon)

(b) δ 10.0 (CH_3)
δ 22.7 (CH_3)
δ 32.0 (CH_2)
δ 69.2 (CH)

(c) δ 31.2 (CH_3) (three carbons)
δ 68.9 (C) (one carbon)

13.36 A compound ($C_3H_7ClO_2$) exhibited three peaks in its ^{13}C NMR spectrum at δ 46.8 (CH_2), δ 63.5 (CH_2), and δ 72.0 (CH). Excluding compounds that have Cl and OH on the same carbon, which are unstable, what is the most reasonable structure for this compound?

13.37 Label nonequivalent carbons in the following compounds.

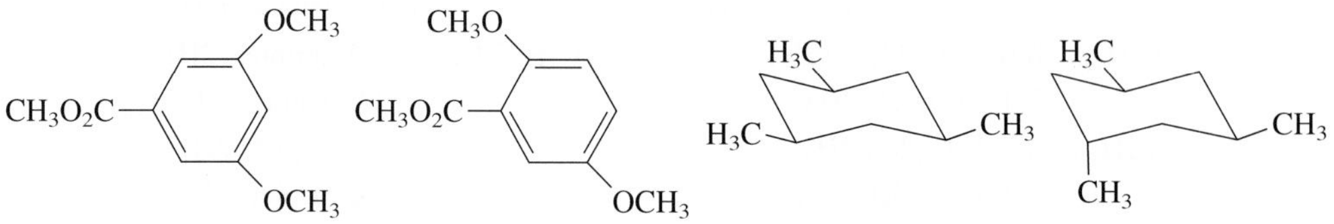
OCH3
CH3O2C
OCH3
CH3O
CH3O2C
OCH3
H3C
H3C
CH3
H3C
CH3
CH3

(*a*) (*b*) (*c*) (*d*)

13.38 The ^1H NMR spectrum of fluorene has signals at δ 3.8 and δ 7.2–7.7 in a 1:4 ratio. After heating with $NaOCH_3$ in CH_3OD at reflux for 15 minutes the signals at δ 7.2–7.7 remained, but the one at δ 3.8 had disappeared. Suggest an explanation and write a mechanism for this observation.

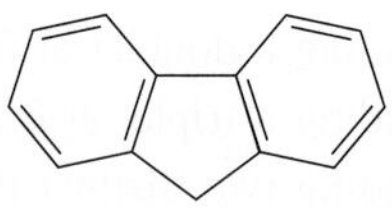

Fluorene

13.39 The vinyl proton region of the ^1H NMR spectrum of phenyl vinyl sulfoxide is shown in Figure 13.47. Construct a splitting diagram similar to the one in Figure 13.21 and label each of the coupling constants $J_{a,b}$, $J_{b,c}$, and $J_{a,c}$.

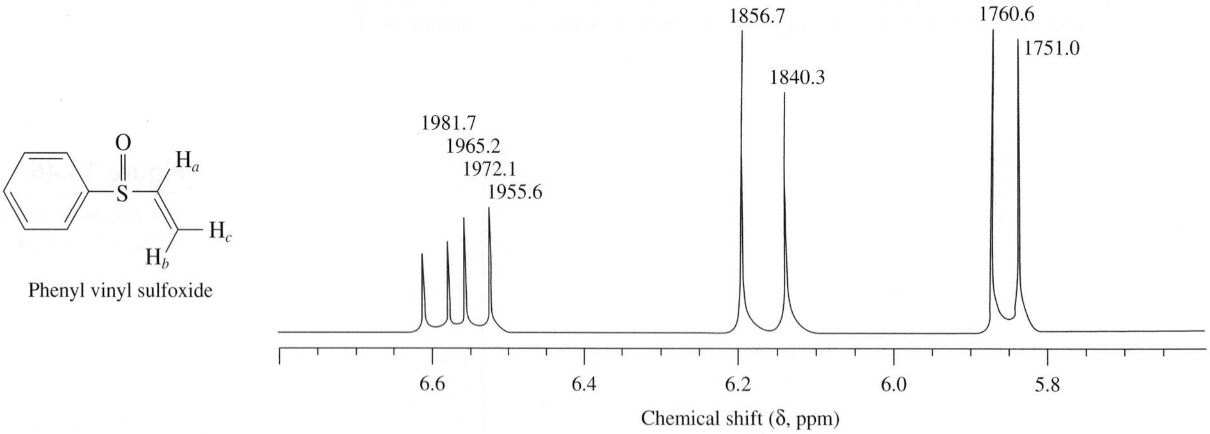
O
S
H_a
H_c
H_b
Phenyl vinyl sulfoxide
1981.7
1965.2
1972.1
1955.6
1856.7
1840.3
1760.6
1751.0
6.6
6.4
6.2
6.0
5.8
Chemical shift (δ, ppm)

Figure 13.47

Vinyl proton region of the 300-MHz ^1H NMR spectrum of phenyl vinyl sulfoxide.

13.40 ^1H NMR spectra of four isomeric alcohols with formula $C_9H_{12}O$ are shown in Figure 13.48. Assign a structure for each alcohol and assign the peaks in each spectrum.

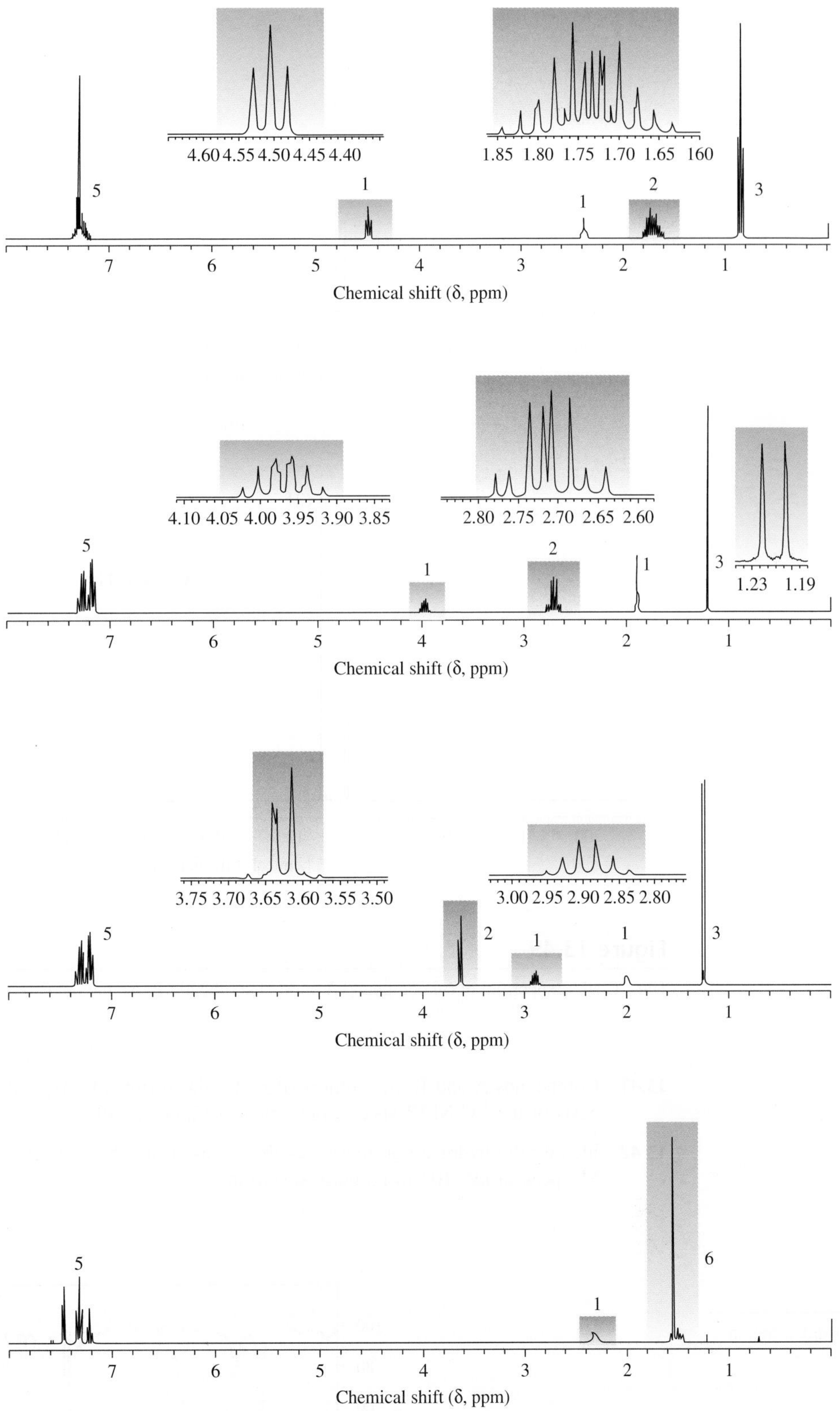

Figure 13.48

300-MHz ^{1}H NMR spectra of alcohols for Problem 13.40.

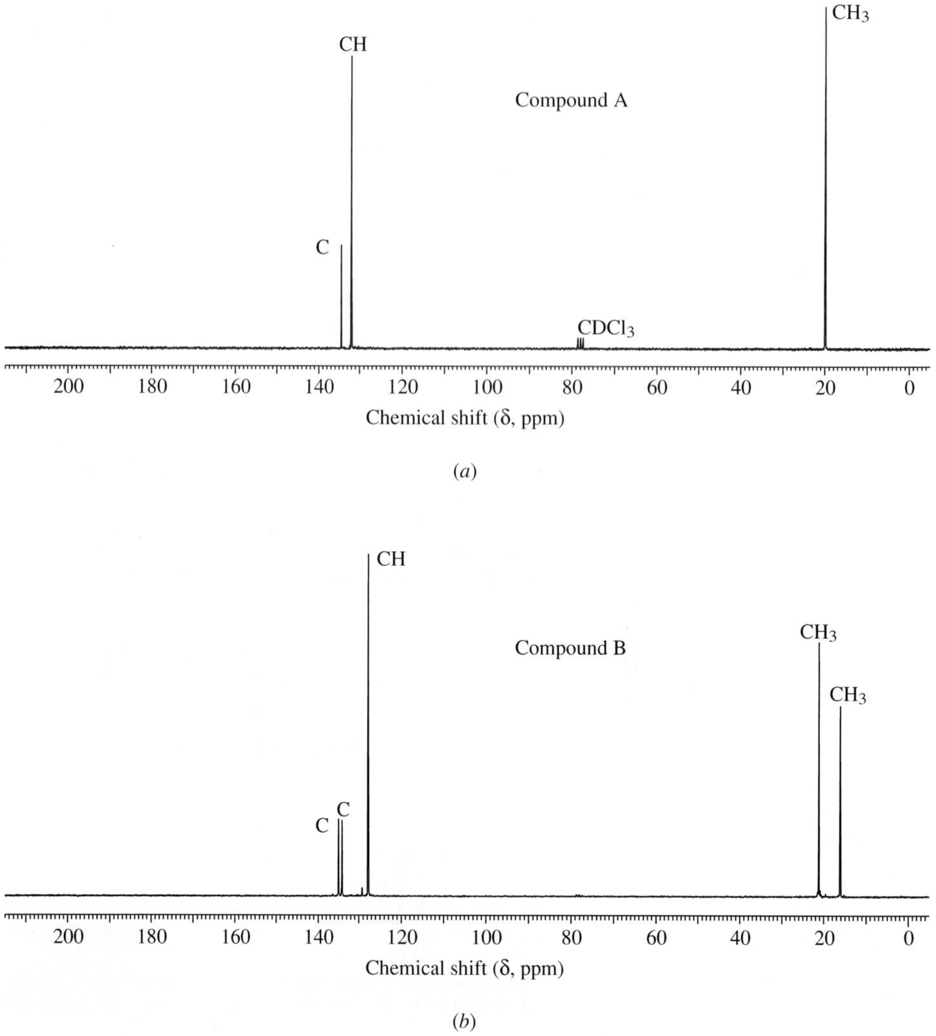

Figure 13.49

The ^{13}C NMR spectrum of (*a*) compound A and (*b*) compound B, isomers of $C_{10}H_{14}$ (Problem 13.41).

13.41 Compounds A and B are isomers of molecular formula $C_{10}H_{14}$. Identify each one on the basis of the ^{13}C NMR spectra presented in Figure 13.49.

13.42 Identify the hydrocarbon that gives the IR spectrum shown in Figure 13.50 and has an M^+ peak at *m/z* 102 in its mass spectrum.

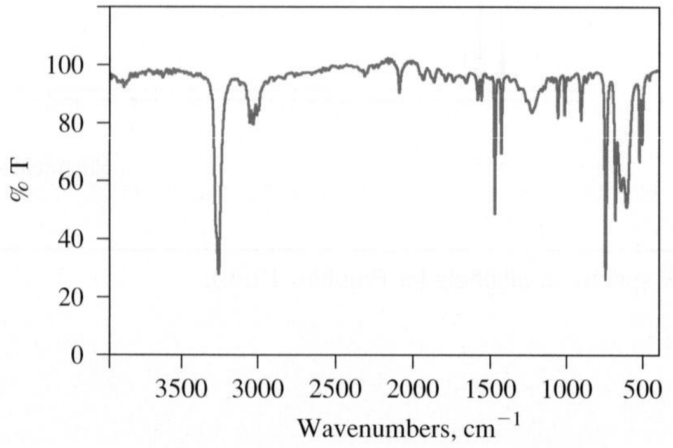

Figure 13.50

The IR spectrum of the hydrocarbon in Problem 13.42.

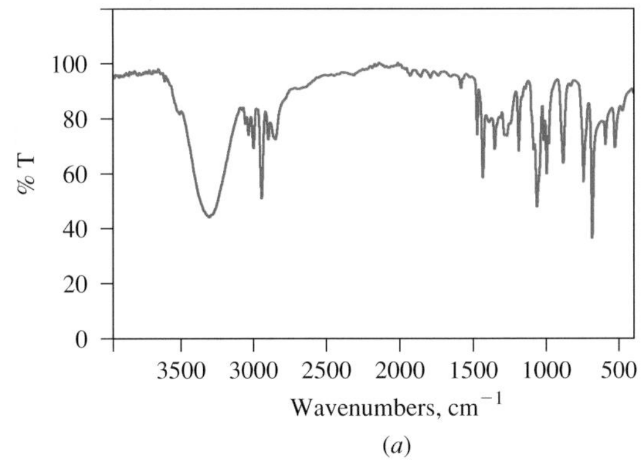

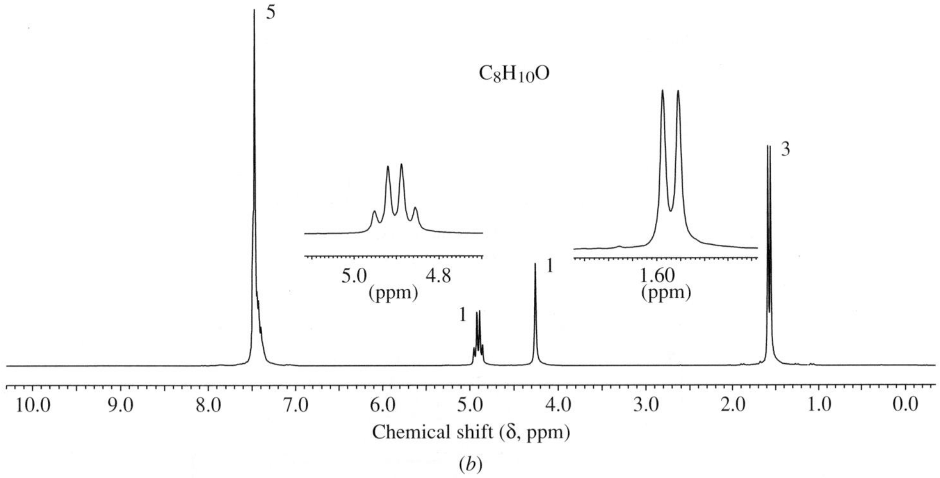

Figure 13.51

(a) IR and (b) 200-MHz ^{1}H NMR spectra of a compound $C_8H_{10}O$ (Problem 13.43).

13.43 A compound ($C_8H_{10}O$) has the IR and ^{1}H NMR spectra presented in Figure 13.51. What is its structure?

13.44 Deduce the structure of a compound having the mass, IR, and ^{1}H NMR spectra presented in Figure 13.52 (page 598).

13.45 ^{13}C NMR spectra for four isomeric alkyl bromides with the formula $C_5H_{11}Br$ are shown in Figure 13.53. Multiplicities obtained from DEPT analysis are shown above each peak. Assign structures to each of the alkyl bromides and assign the peaks in each spectrum.

13.46 Figure 13.54 (page 600) presents IR, ^{1}H NMR, ^{13}C NMR, and mass spectra for a particular compound. What is it?

13.47 Which would you predict to be more shielded, the inner or outer protons of [24]annulene?

13.48 ^{19}F is the only isotope of fluorine that occurs naturally, and it has a nuclear spin of $\pm\frac{1}{2}$.

(a) Into how many peaks will the proton signal in the ^{1}H NMR spectrum of methyl fluoride be split?

(b) Into how many peaks will the fluorine signal in the ^{19}F NMR spectrum of methyl fluoride be split?

(c) The chemical shift of the protons in methyl fluoride is δ 4.3. Given that the geminal ^{1}H—^{19}F coupling constant is 45 Hz, specify the δ values at which peaks are observed in the proton spectrum of this compound at 200 MHz.

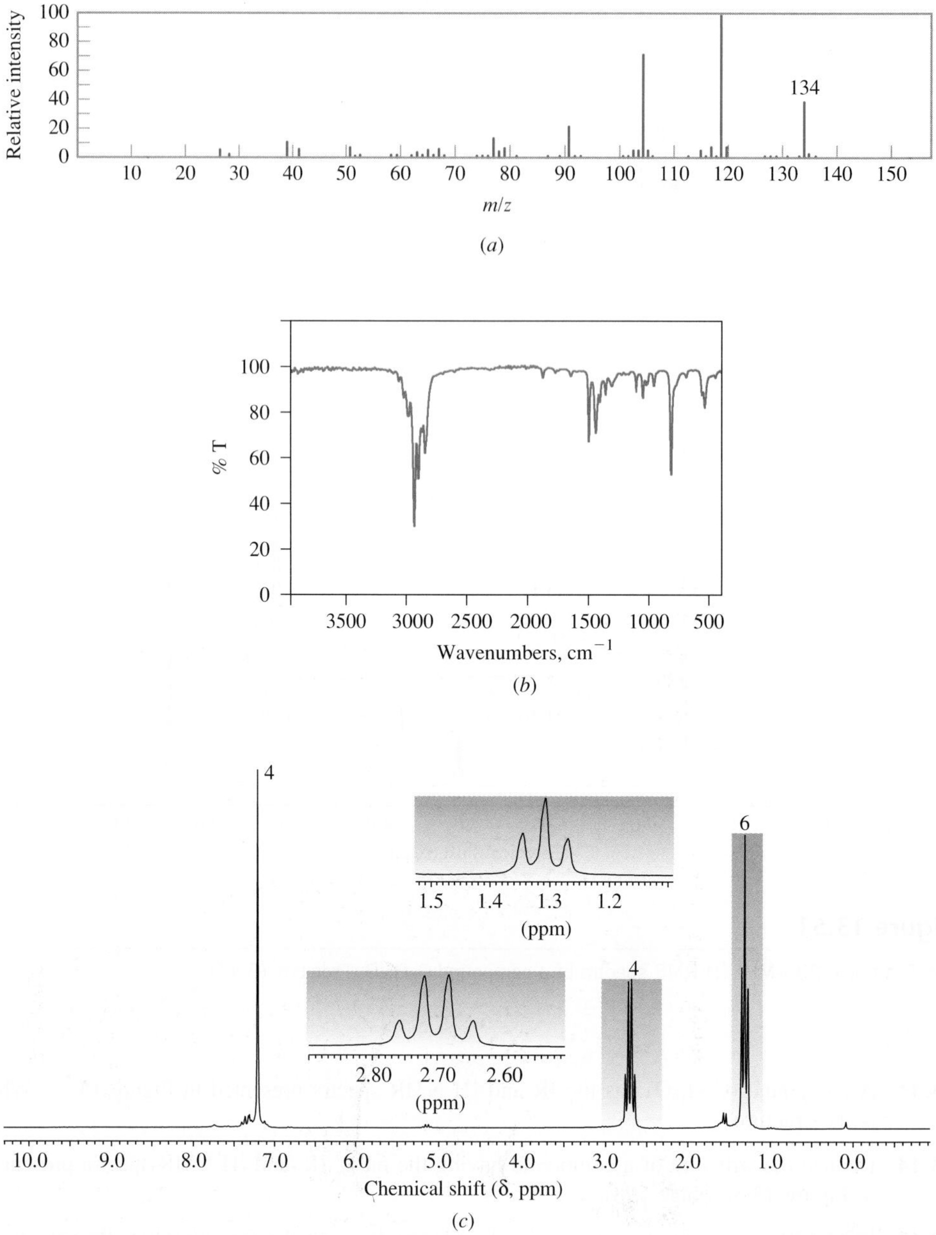

Figure 13.52

(*a*) Mass, (*b*) IR, and (*c*) 200-MHz ^{1}H NMR spectra of a compound (Problem 13.44).

13.49 ^{31}P is the only phosphorus isotope present at natural abundance and has a nuclear spin of $\pm\frac{1}{2}$. The ^{1}H NMR spectrum of trimethyl phosphite, $(CH_3O)_3P$, exhibits a doublet for the methyl protons with a splitting of 12 Hz.

(a) Into how many peaks is the ^{31}P signal split?

(b) What is the difference in chemical shift (in hertz) between the lowest and highest field peaks of the ^{31}P multiplet?

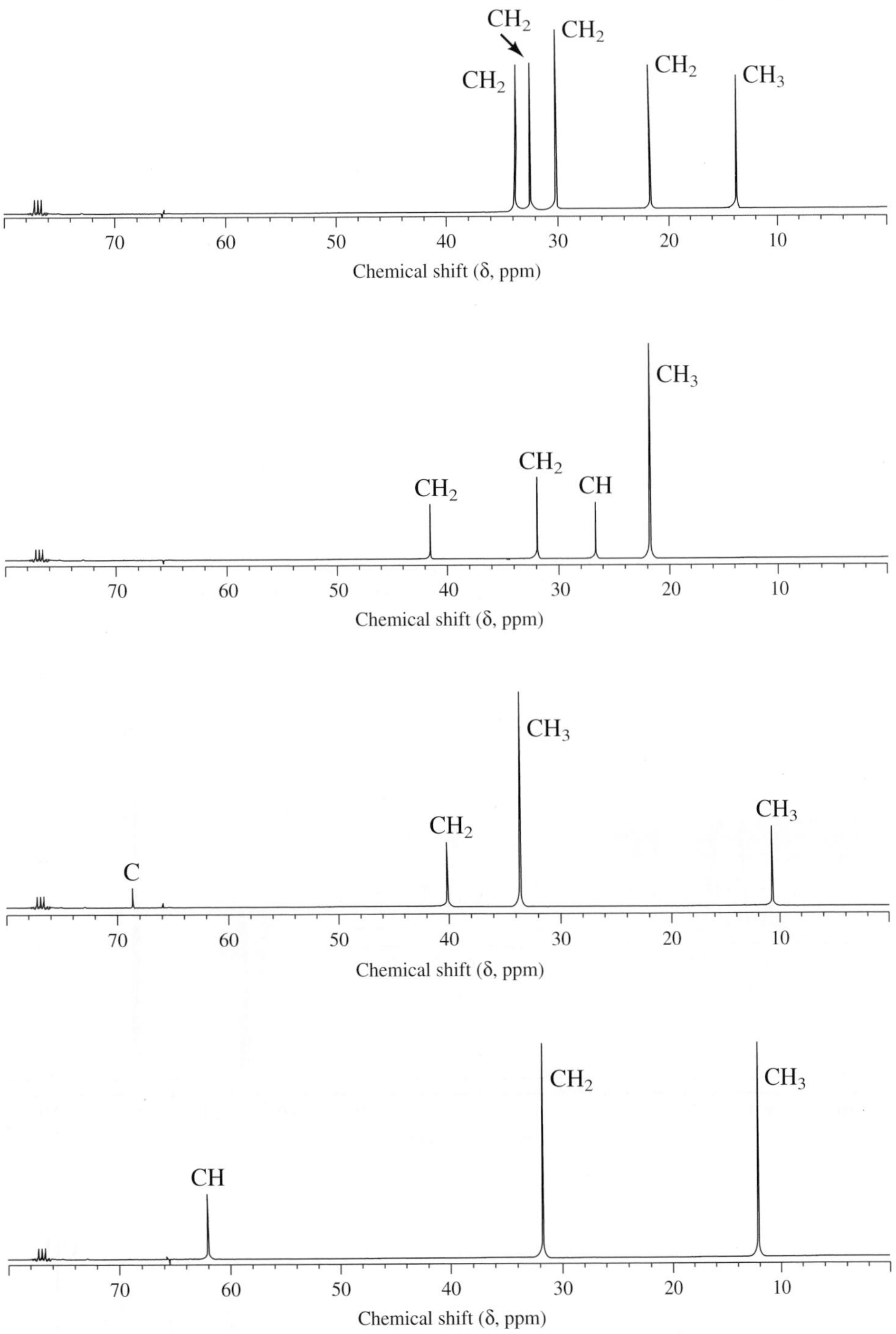

Figure 13.53

^{13}C NMR spectra for isomeric alkyl bromides in Problem 13.45.

13.50 We noted in Section 13.13 that an NMR spectrum is an average spectrum of the conformations populated by a molecule. From the following data, estimate the percentages of axial and equatorial bromine present in bromocyclohexane.

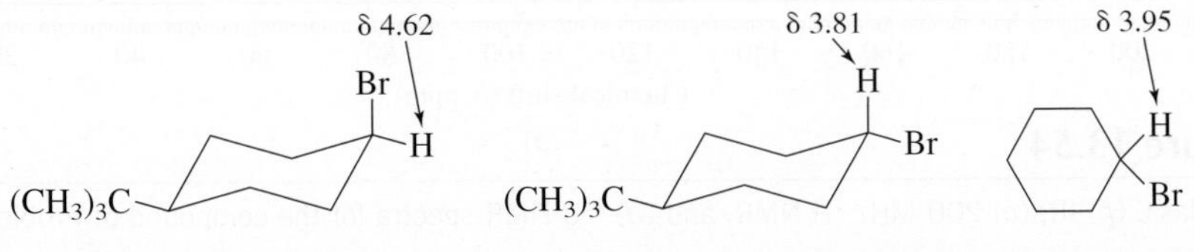

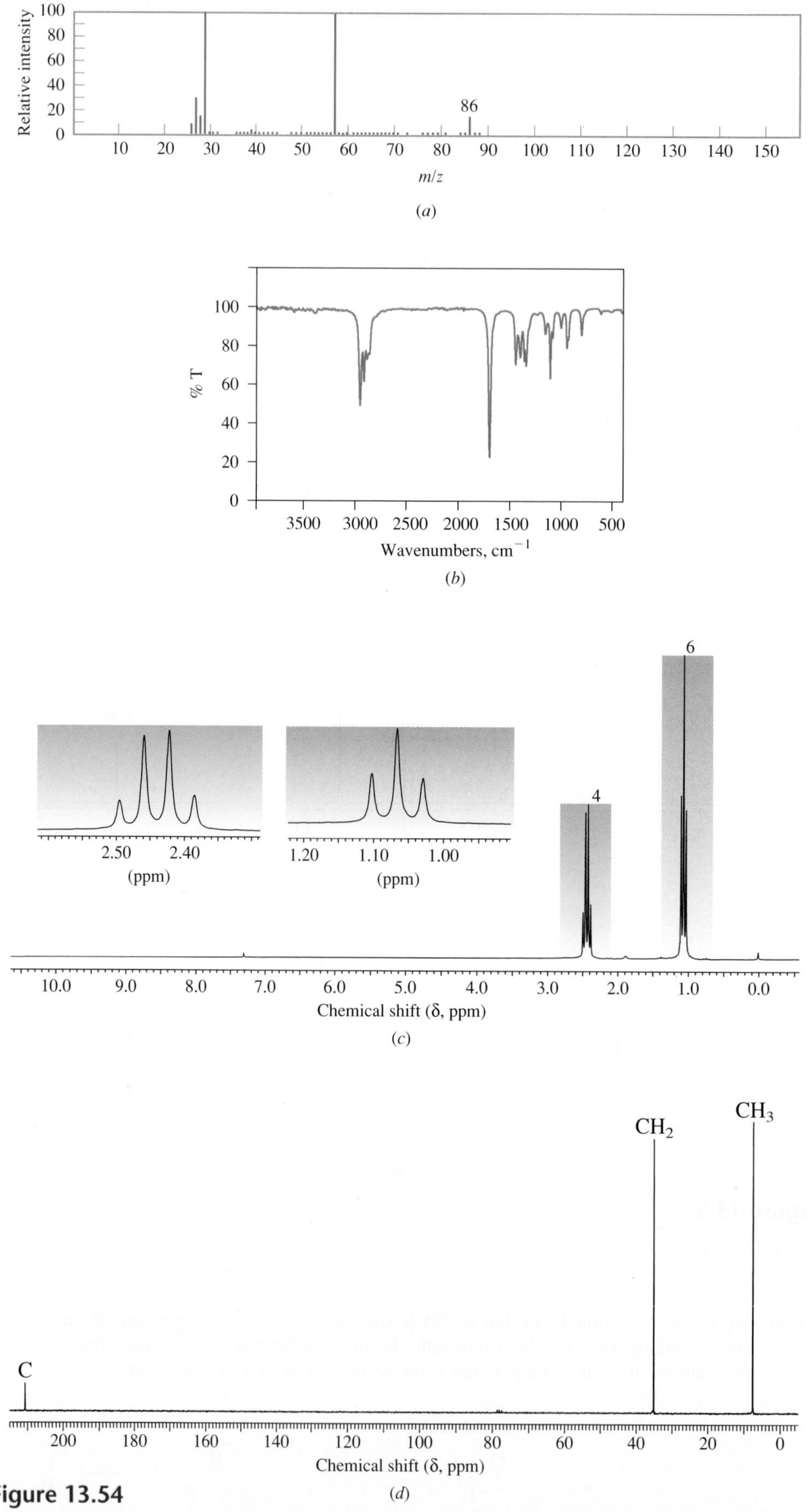

Figure 13.54

(*a*) Mass, (*b*) IR, (*c*) 200-MHz ^{1}H NMR, and (*d*) ^{13}C NMR spectra for the compound of Problem 13.46.

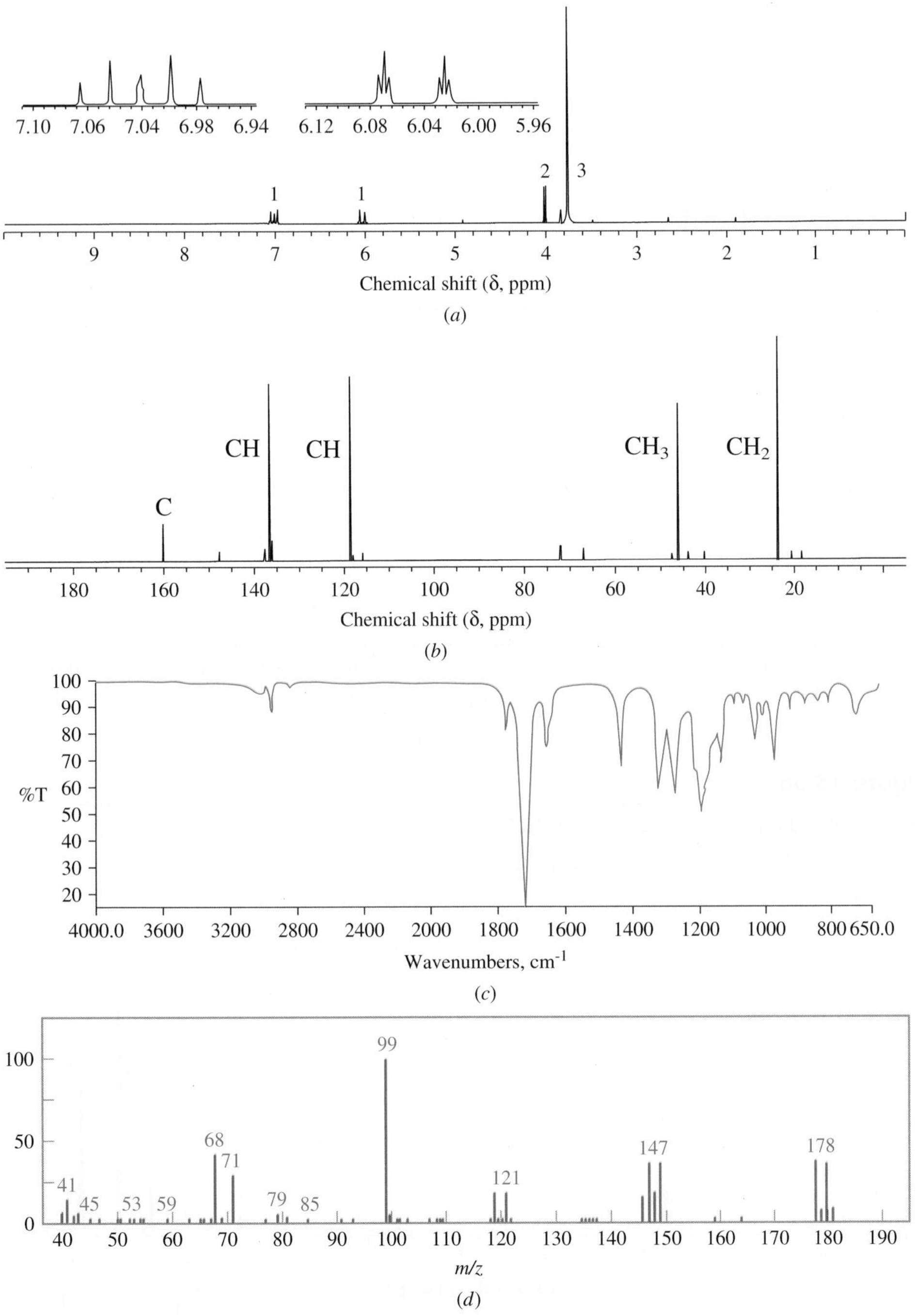

Figure 13.55

(*a*) ^{1}H NMR, (*b*) ^{13}C NMR, (*c*) IR, and (*d*) mass spectra Problem 13.51.

13.51 ^{1}H NMR, ^{13}C NMR, IR, and mass spectra are shown for a compound in Figure 13.55. Propose a structure and explain your answer based on spectral assignments.

13.52 ^{1}H NMR and IR spectra for a compound with the formula $C_7H_7NO_3$ are shown in Figure 13.56. Assign a structure and explain your reasoning.

13.53 Friedel–Crafts alkylation of benzene with 1-chlorobutane gave a product for which the ^{1}H and ^{13}C NMR spectra are shown in Figure 13.57. Peak multiplicities from DEPT analysis are indicated on the ^{13}C NMR spectrum. Assign a structure to the product.

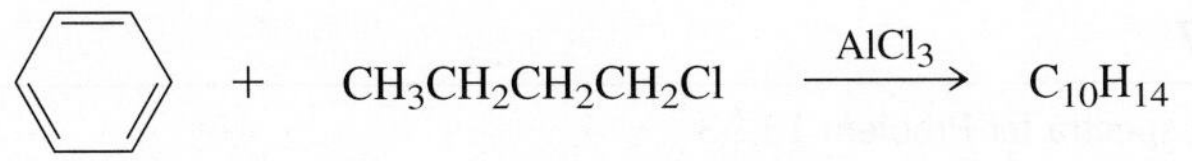

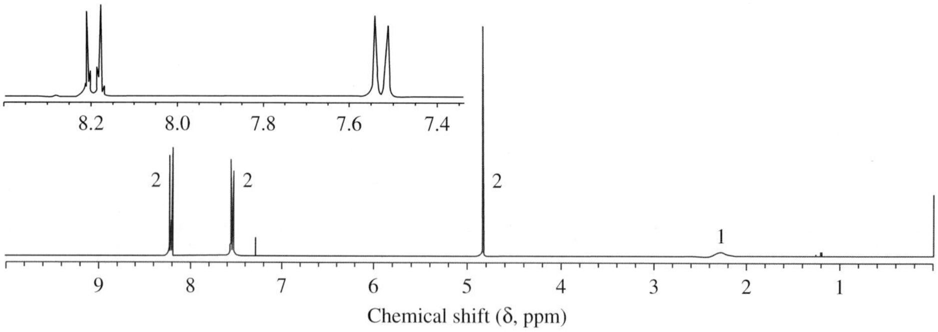

(*a*)

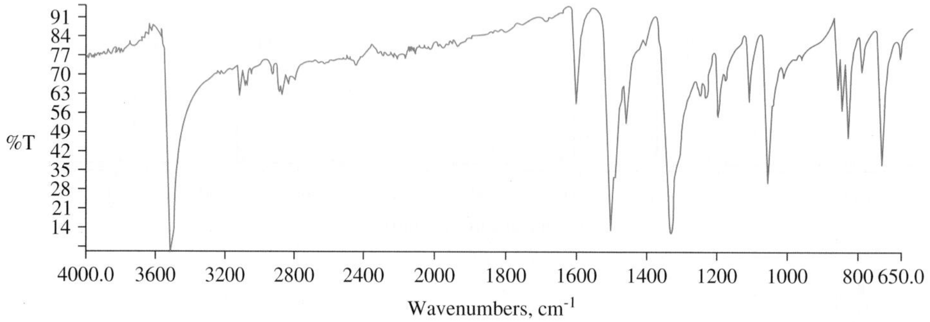

(*b*)

Figure 13.56

(*a*) ^{1}H NMR and (*b*) IR spectra for Problem 13.52.

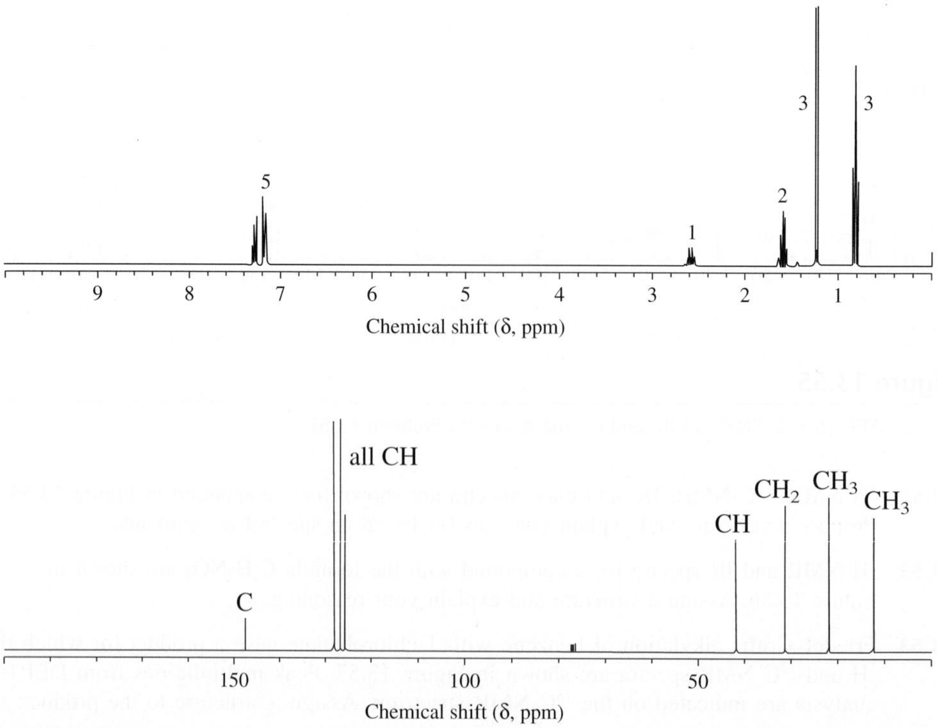

Figure 13.57

^{1}H and ^{13}C NMR spectra for Problem 13.53.

Descriptive Passage and Interpretive Problems 13

Calculating Aromatic ^{13}C Chemical Shifts

Although chemical-shift tables such as Table 13.3 (see Section 13.15) are useful guides to where we expect to find peaks for various structural units, they quote ranges rather than specific values. The value cited for aromatic ring carbons, for example, is 110–175 ppm. Information this general cannot discriminate between isomers such as *o*- and *m*-nitrotoluene solely on the basis of their ^{13}C NMR spectra. Both isomers have six nonequivalent ring carbons and similar, though not identical, ^{13}C NMR spectra.

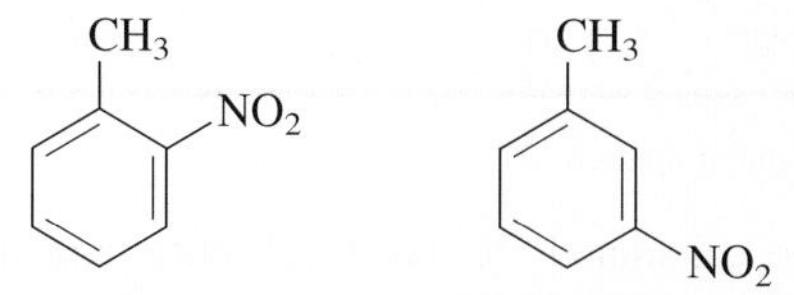

o-Nitrotoluene *m*-Nitrotoluene

This passage describes a simple method for predicting chemical shifts for carbons on a substituted benzene ring, based on the ^{13}C chemical shift of benzene (δ 128.5) as a standard. Substituents increase (+) or decrease (−) that value for the carbon to which they are attached and the carbons ortho, meta, and para to it. The direction (+ or −) and size (Δ) of substituent effects are determined experimentally; values for some common substituents are given in Table 13.6. When using this table, C-1 refers specifically to the carbon for which the chemical shift is being calculated without regard to its IUPAC number.

TABLE 13.6 Incremental ^{13}C Chemical Shift Effects of Substituents (δ), ppm

Substituent	$\Delta_{C\text{-}1}$	Δ_{ortho}	Δ_{meta}	Δ_{para}
H	0	0	0	0
CH_3	+9.1	+0.7	−0.1	−3.0
Cl	+5.3	+0.4	+1.4	−1.9
OH	+28.8	−12.8	+1.4	−7.4
OCH_3	+33.5	−14.4	+1.0	−7.7
NH_2	+18.2	−13.4	+0.8	−10.0
CO_2CH_3	+2.0	+1.2	−0.1	+4.3
C(O)Cl	+4.7	+2.7	+0.3	+6.5
$C(O)NH_2$	+5.0	−1.2	+0.1	+3.4
NO_2	+19.9	−4.9	+0.9	+6.1

Values calculated from www.stephanbird.org.uk/Chemistry/Carbon/AreneNMR-C13.html

^{13}C-substituent effects are additive for as many substituents as are present on the ring. Thus, for the carbon that bears the methyl group in *o*-nitrotoluene:

$$\delta = 128.5 + (\Delta \text{ for C-1} = CH_3) + (\Delta \text{ for ortho } NO_2)$$
$$\delta = 128.5 + (+9.1) + (-4.9) = 132.7$$

Likewise for the nitro-bearing carbon in *o*-nitrotoluene:

$$\delta = 128.5 + (\Delta \text{ for C-1} = NO_2) + (\Delta \text{ for ortho } CH_3)$$
$$\delta = 128.5 + (+19.9) + (+0.7) = 149.1$$

Analogous arithmetic gives the calculated chemical shifts for all the ring carbons (Table 13.7).

TABLE 13.7 Calculated and Observed ^{13}C Chemical Shifts for the Ring Carbons in *o*- and *m*-Nitrotoluene

		^{13}C Chemical Shift of Carbon, δ*					
		1	2	3	4	5	6
o-Nitrotoluene	Calculated	132.7	149.1	123.5	126.4	134.5	130.1
	Observed	133.5	149.4	124.6	126.9	133.0	132.8
m-Nitrotoluene	Calculated	139.6	124.3	148.3	120.6	129.3	135.3
	Observed	139.9	123.8	148.4	120.6	129.1	135.4

*Column numbers 1–6 are IUPAC locants of ring carbons.

The match between the calculated ^{13}C chemical shifts and those actually observed for the two isomers is quite good. The major difference between the two isomers is the chemical shift of the methyl-bearing carbon, which is both large enough and in the direction predicted by the calculation to serve to identify each isomer.

The URL referenced as a footnote in Table 13.6 provides an easy, automatic way to do these calculations. One simply selects substituents at the carbons of the benzene ring on the screen and is rewarded with the chemical shifts of all six ring carbons.

The problems that follow illustrate some of the ways that chemical shift calculations can assist in structure determination among compounds that contain benzene rings. The purpose of the first problem is to acquaint you with the online calculator at the URL referenced in Table 13.6. You will need the calculator for the substituent effect Δ of fluorine. With the exception of fluorine, substituent effects for the other atoms and groups are listed in the table.

13.54 Which carbons of 1-chloro-4-fluorobenzene are the most shielded?

A. C-1 C. C-3 and C-5
B. C-2 and C-6 D. C-4

13.55 A chlorinated derivative of benzene had only two peaks for aromatic carbons in its ^{13}C NMR spectrum. Of the following, which compound can be eliminated on the basis of this information?

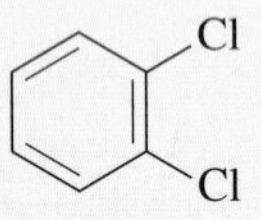

A.

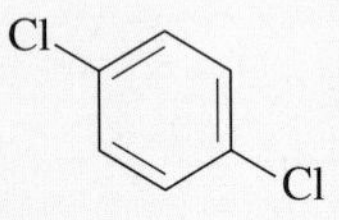

B.

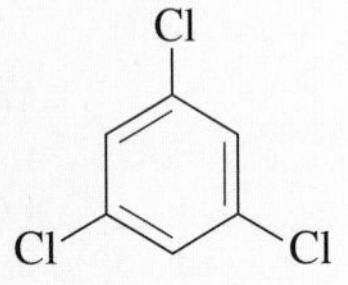

C.

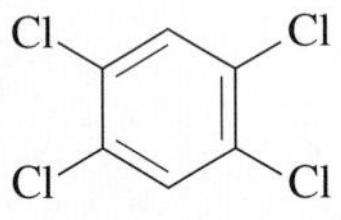

D.

13.56 Of the remaining possible compounds from Problem 13.55, which one is most consistent with the observed ^{13}C shifts of δ 127.2 and 135.6?

13.57 The ^{13}C NMR spectrum of a C_7H_7ClO isomer has peaks for a methyl carbon and six aromatic ring carbons. Which compound can you exclude based solely on the number of peaks?

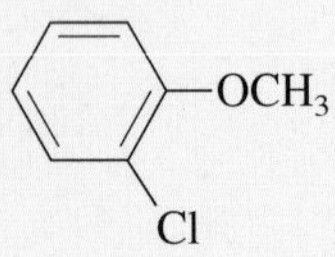

A.

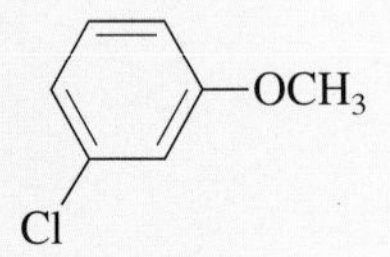

B.

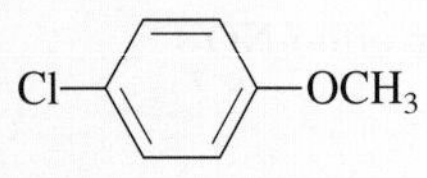

C.

13.58 The ^{13}C NMR spectrum of one of the chloroanisole isomers shown in the preceding problem has peaks at δ 112.6, 114.4, 120.9, 130.2, 135.0, and 160.5. Which isomer fits the data best?

13.59 A compound of molecular formula $C_8H_9NO_2$ had peaks for its benzene ring carbons at δ 113.8, 119.3, 131.6, and 151.4. No other peaks for benzene ring carbons were present. Which compound is most consistent with the ^{13}C NMR data?

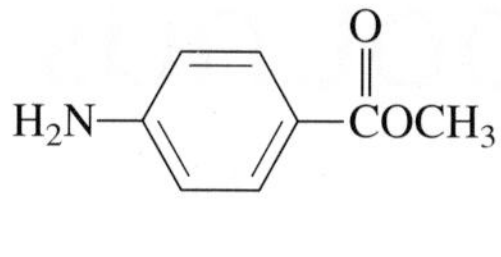

A.

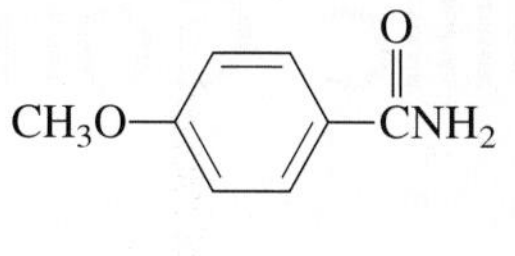

B.

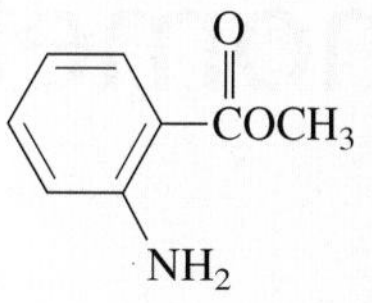

C.

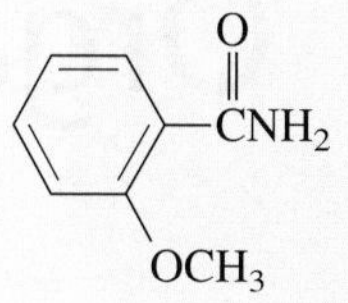

D.

13.60 A compound of molecular formula $C_8H_7ClO_2$ had peaks for its benzene ring carbons at δ 114.3, 125.3, 134.0, and 165.5. No other peaks for benzene ring carbons were present. Which compound is most consistent with the ^{13}C NMR data?

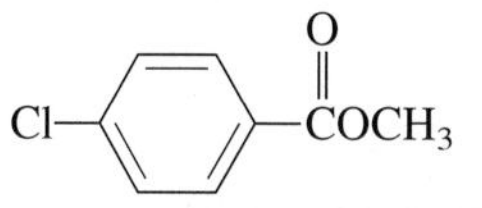

A.

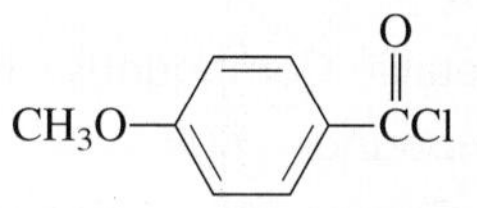

B.

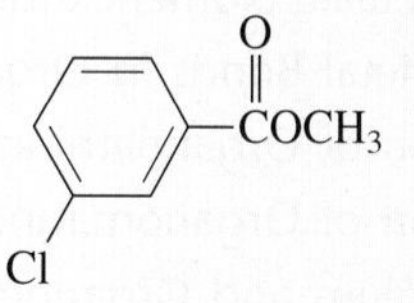

C.

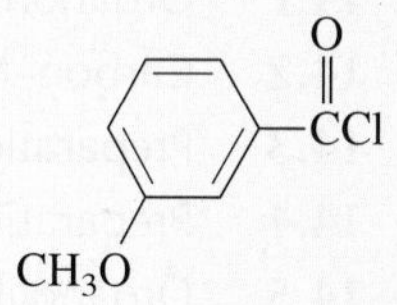

D.

13.61 There is one peak too few, and one of the peaks is too large in the ^{13}C NMR spectrum of 2-chloro-5-methylphenol. The reason for this is that two of the carbons, although nonequivalent, have the same chemical shift and appear as a single peak. For which two carbons are the calculated chemical shifts the closest?

A. C-1 and C-6
B. C-2 and C-6
C. C-1 and C-3
D. C-4 and C-5

14 Organometallic Compounds

Chapter Outline

Mechanisms

Steel towers support the steel cables that hold the bridge. Steel is iron to which some carbon has been added. An iron atom lies between two carbon-based rings in a ferrocene molecule.

ORGANOMETALLIC COMPOUNDS ARE *compounds that have a carbon–metal bond;* they lie at the place where organic and inorganic chemistry meet. You are already familiar with at least one organometallic compound, sodium acetylide ($NaC{\equiv}CH$), which has an ionic bond between carbon and sodium. But just because a compound contains both a metal and carbon isn't enough to classify it as organometallic. Like sodium acetylide, sodium methoxide ($NaOCH_3$) is an ionic compound. Unlike sodium acetylide, however, the negative charge in sodium methoxide resides on oxygen, not carbon.

$Na^+ \; :\bar{C}{\equiv}CH$	$Na^+ \; ^-:\ddot{\underset{..}{O}}CH_3$
Sodium acetylide	Sodium methoxide
(has a carbon-to-metal bond)	(does not have a carbon-to-metal bond)

The properties of organometallic compounds are much different from those of the other classes we have studied so far and differ among themselves according to the metal, its oxidation state, and the groups attached to the metal. Many organometallic compounds are sources of nucleophilic carbon, a quality that makes them especially valuable to the synthetic organic chemist who needs to make carbon–carbon bonds. For example, the preparation of alkynes by the reaction of sodium acetylide with alkyl halides (Section 9.6) depends on the presence of a negatively charged, nucleophilic carbon in acetylide ion. Conversely, certain other metals give compounds that behave as electrophiles.

A comprehensive treatment of organometallic chemistry would require a book of its own. In this chapter the preparation, properties, and usefulness of some of the most common organometallic reagents, those based on magnesium and lithium, are described in some detail. Other organometallic compounds, those derived from zinc, copper, and several less familiar metals, are introduced by highlighting some of their synthetic applications. We will also continue the story of Ziegler–Natta catalysis of alkene polymerization begun in Chapters 6 and 7 by exploring its mechanism.

14.1 Organometallic Nomenclature

Organometallic compounds are named as substituted derivatives of metals. The metal is the parent, and the attached alkyl groups are identified by the appropriate prefix.

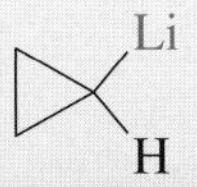

(structure)	$H_2C{=}CHNa$	$(CH_3CH_2)_2Mg$
Cyclopropyllithium	Vinylsodium	Diethylmagnesium

When the metal bears a substituent other than carbon, the substituent is treated as if it were an anion and named separately.

CH_3MgI	$(CH_3CH_2)_2AlCl$
Methylmagnesium iodide	Diethylaluminum chloride

Problem 14.1

Both of the following organometallic reagents will be encountered later in this chapter. Suggest a suitable name for each.

(a) $(CH_3)_3CLi$

(b) 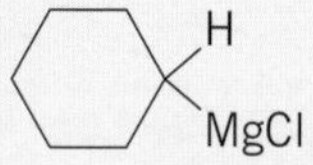

Sample Solution (a) The metal lithium provides the base name for $(CH_3)_3CLi$. The alkyl group to which lithium is bonded is *tert*-butyl, and so the name of this organometallic compound is *tert*-butyllithium. An alternative, equally correct name is 1,1-dimethylethyllithium.

An exception to this type of nomenclature is NaC≡CH, which is normally referred to as *sodium acetylide.* Both sodium acetylide and ethynylsodium are acceptable IUPAC names.

14.2 Carbon–Metal Bonds in Organometallic Compounds

With an electronegativity of 2.5 (Figure 14.1), carbon is neither strongly electropositive nor strongly electronegative. When carbon is bonded to an element more electronegative than itself, such as oxygen or chlorine, the electron distribution in the bond is polarized so that carbon is slightly positive and the more electronegative atom is slightly negative.

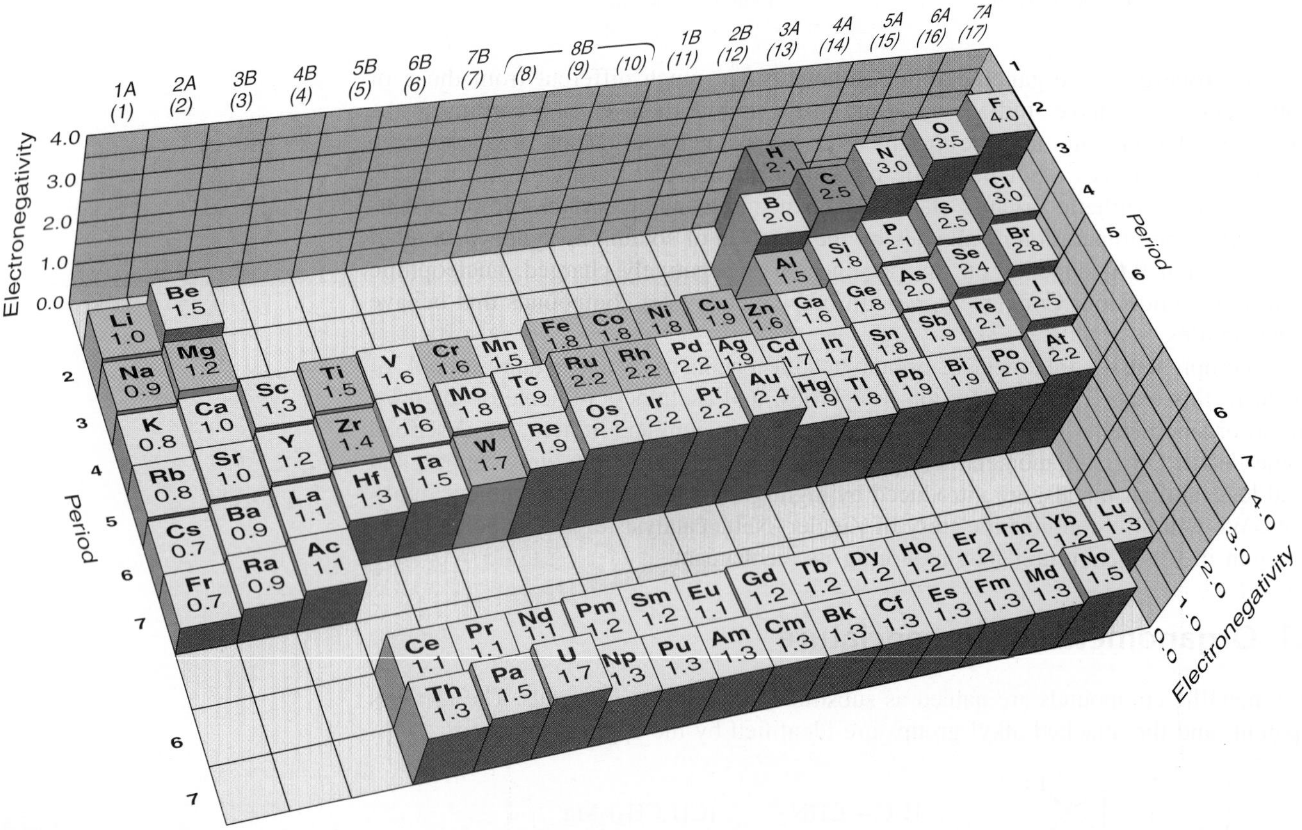

Figure 14.1

Electronegativities of the elements on the Pauling scale. The metals that appear in this chapter are shown in blue. Hydrogen and carbon are red. Adapted from Silberberg, *Chemistry,* 3/e, McGraw-Hill Higher Education, 2003, p. 344.

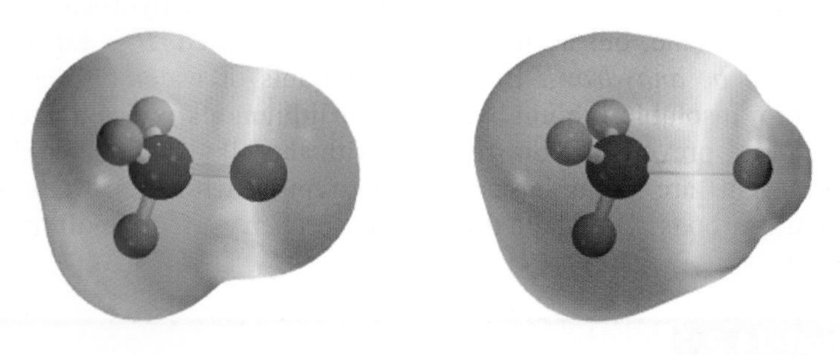

(*a*) Methyl fluoride (*b*) Methyllithium

Figure 14.2

Electrostatic potential maps of (*a*) methyl fluoride and (*b*) methyllithium. The electron distribution is reversed in the two compounds. Carbon is electron-poor (blue) in methyl fluoride, but electron-rich (red) in methyllithium.

Conversely, when carbon is bonded to a less electronegative element, such as a metal, the electrons in the bond are more strongly attracted toward carbon.

δ+ δ−
C—X
δ− δ+
C—M

X is more electronegative than carbon | M is less electronegative than carbon

Figure 14.2 uses electrostatic potential maps to show how different the electron distribution is between methyl fluoride (CH_3F) and methyllithium (CH_3Li).

An anion that contains a negatively charged carbon is referred to as a **carbanion.** Metals are less electronegative than carbon, and organometallic compounds have *carbanionic character.* As the metal becomes more electropositive, the ionic character of the carbon–metal bond becomes more pronounced. Organosodium and organopotassium compounds have ionic carbon–metal bonds; organolithium and organomagnesium compounds tend to have covalent, but rather polar, carbon–metal bonds with significant carbanionic character. *It is the carbanionic character of such compounds that is responsible for their usefulness as synthetic reagents.*

14.3 Preparation of Organolithium Compounds

Before we describe the applications of organometallic reagents to organic synthesis, let us examine their preparation. Organolithium compounds and other Group 1 organometallic compounds are prepared by the reaction of an alkyl halide with the appropriate metal.

The reaction of an alkyl halide with lithium is an *oxidation–reduction* reaction. Group 1 metals are powerful reducing agents.

$$RX + 2M \longrightarrow RM + M^+X^-$$

Alkyl halide | Group 1 metal | Group 1 organometallic compound | Metal halide

$$(CH_3)_3CCl + 2Li \xrightarrow[-30°C]{\text{diethyl ether}} (CH_3)_3CLi + LiCl$$

tert-Butyl chloride | Lithium | *tert*-Butyllithium (75%) | Lithium chloride

The alkyl halide can be primary, secondary, or tertiary. Alkyl iodides are the most reactive, followed by bromides, then chlorides. Fluorides are relatively unreactive.

Unlike elimination and nucleophilic substitution reactions, formation of organolithium compounds does not require that the halogen be bonded to sp^3-hybridized carbon. Compounds such as vinyl halides and aryl halides, in which the halogen is bonded to sp^2-hybridized carbon, react in the same way as alkyl halides, but at somewhat slower rates.

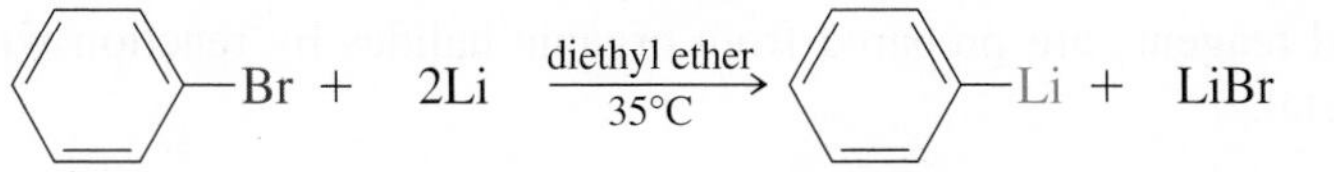
Br +
2Li
diethyl ether
35°C
Li + LiBr

Bromobenzene | Lithium | Phenyllithium (95–99%) | Lithium bromide

Organolithium compounds are sometimes prepared in hydrocarbon solvents such as pentane and hexane, but normally diethyl ether is used. *It is especially important that the solvent be anhydrous.* Even trace amounts of water or alcohols react with lithium to form insoluble lithium hydroxide or lithium alkoxides that coat the surface of the metal and prevent it from reacting with the alkyl halide. Furthermore, organolithium reagents are strong bases and react rapidly with even weak proton sources to form hydrocarbons. We shall discuss this property of organolithium reagents in Section 14.5.

Problem 14.2

Write an equation showing the formation of each of the following from the appropriate bromide:

(a) Isopropenyllithium (b) *sec*-Butyllithium

Sample Solution (a) In the preparation of organolithium compounds from organic halides, lithium becomes bonded to the carbon that bore the halogen. Therefore, isopropenyllithium must arise from isopropenyl bromide.

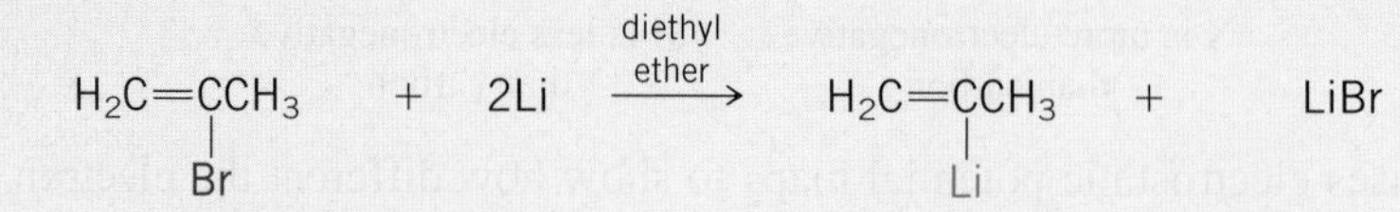

Isopropenyl bromide Lithium Isopropenyllithium Lithium bromide

Reaction with an alkyl halide takes place at the metal surface. In the first step, an electron is transferred from lithium to the alkyl halide.

$$R{:}\ddot{\underset{..}{X}}{:} \quad + \quad Li\cdot \quad \longrightarrow \quad [R{:}\ddot{\underset{..}{X}}{:}]^{\overset{\cdot}{-}} \quad + \quad Li^+$$

Alkyl halide Lithium Anion radical Lithium cation

Having gained one electron, the alkyl halide is now negatively charged and has an odd number of electrons. It is an *anion radical.* The extra electron occupies an antibonding orbital. This anion radical dissociates to an alkyl radical and a halide anion.

$$[R{:}\ddot{\underset{..}{X}}{:}]^{\overset{\cdot}{-}} \quad \longrightarrow \quad R\cdot \quad + \quad {:}\ddot{\underset{..}{X}}{:}^-$$

Anion radical Alkyl radical Halide anion

Following dissociation, the alkyl radical rapidly combines with a lithium atom to form the organometallic compound.

$$R\cdot \quad + \quad Li\cdot \quad \longrightarrow \quad R{:}Li$$

Alkyl radical Lithium Alkyllithium

14.4 Preparation of Organomagnesium Compounds: Grignard Reagents

The most important organometallic reagents in organic chemistry are organomagnesium compounds. They are called **Grignard reagents** after the French chemist Victor Grignard. Grignard developed efficient methods for the preparation of organic derivatives of magnesium and demonstrated their application in the synthesis of alcohols. For these achievements he was a corecipient of the 1912 Nobel Prize in Chemistry.

Grignard shared the prize with Paul Sabatier, who showed that finely divided nickel could be used to catalyze the hydrogenation of alkenes.

Grignard reagents are prepared from organic halides by reaction with magnesium, a Group 2 metal.

$$RX \quad + \quad Mg \quad \longrightarrow \quad RMgX$$

Organic halide Magnesium Organomagnesium halide

(R may be methyl or primary, secondary, or tertiary alkyl; it may also be a cycloalkyl, alkenyl, or aryl group.)

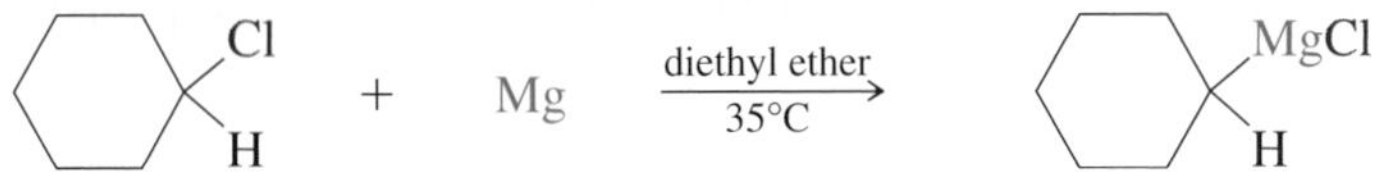

Cyclohexyl chloride Magnesium Cyclohexylmagnesium chloride (96%)

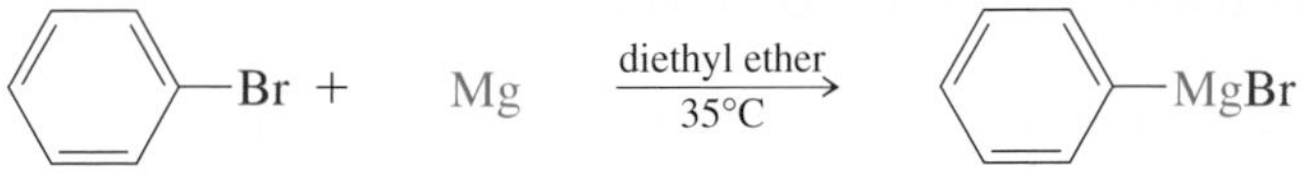

Bromobenzene Magnesium Phenylmagnesium bromide (95%)

Anhydrous diethyl ether is the customary solvent used when preparing organomagnesium compounds. Sometimes the reaction does not begin readily, but once started, it is exothermic and maintains the temperature of the reaction mixture at the boiling point of diethyl ether (35°C).

The order of halide reactivity is I > Br > Cl > F, and alkyl halides are more reactive than aryl and vinyl halides. Indeed, aryl and vinyl chlorides do not form Grignard reagents in diethyl ether. When more vigorous reaction conditions are required, tetrahydrofuran (THF) is used as the solvent.

Recall the structure of tetrahydrofuran from Section 3.15:

$$H_2C{=}CHCl \xrightarrow[\text{THF, 60°C}]{\text{Mg}} H_2C{=}CHMgCl$$

Vinyl chloride Vinylmagnesium chloride (92%)

THF forms a more stable complex with the Grignard reagent and, with a boiling point of 66°C, allows the reaction to be carried out at a higher temperature.

Problem 14.3

Write the structure of the Grignard reagent formed from each of the following compounds on reaction with magnesium in diethyl ether:

(a) *p*-Bromofluorobenzene
(b) Allyl chloride
(c) Iodocyclobutane
(d) 1-Bromocyclohexene

Sample Solution (a) Of the two halogen substituents on the aromatic ring, bromine reacts much faster than fluorine with magnesium. Therefore, fluorine is left intact on the ring, but the carbon–bromine bond is converted to a carbon–magnesium bond.

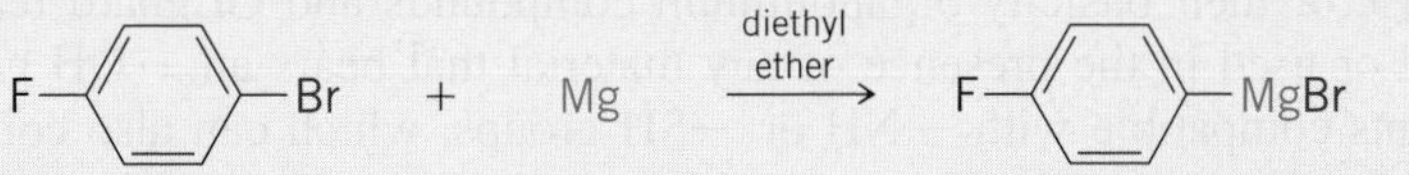

p-Bromofluorobenzene Magnesium *p*-Fluorophenylmagnesium bromide

The formation of a Grignard reagent is analogous to that of organolithium reagents except that each magnesium atom can participate in two separate one-electron transfer steps:

$$R{:}\ddot{\underset{..}{X}}{:} \quad + \quad Mg{:} \longrightarrow [R{:}\ddot{\underset{..}{X}}{:}]^{\overline{\cdot}} \quad + \overset{+}{Mg}\cdot$$

Alkyl halide Magnesium Anion radical

$$[R{:}\ddot{\underset{..}{X}}{:}]^{\overline{\cdot}} \longrightarrow R\cdot \quad + \quad {:}\ddot{\underset{..}{X}}{:}^{-} \xrightarrow{\overset{+}{Mg}\cdot} R{:}Mg^{+}{:}\ddot{\underset{..}{X}}{:}^{-}$$

Anion radical Alkyl radical Halide ion Alkylmagnesium halide

Organolithium and organomagnesium compounds find their chief use in the preparation of alcohols by reaction with aldehydes and ketones. Before discussing these reactions, let us first examine the reactions of these organometallic compounds with proton donors.

14.5 Organolithium and Organomagnesium Compounds as Brønsted Bases

Organolithium and organomagnesium compounds are stable species when prepared in suitable solvents such as diethyl ether. They are strongly basic, however, and react instantly with proton donors even as weakly acidic as water and alcohols. A proton is transferred from the hydroxyl group to the negatively polarized carbon of the organometallic compound to form a hydrocarbon.

$$\underset{\text{stronger base}}{R{-}M} + \underset{\text{stronger acid}}{R'{-}\ddot{O}{-}H} \rightleftharpoons \underset{\text{weaker acid}}{R{-}H} + \underset{\text{weaker base}}{R'{-}\ddot{O}{:}^-\ M^+}$$

via

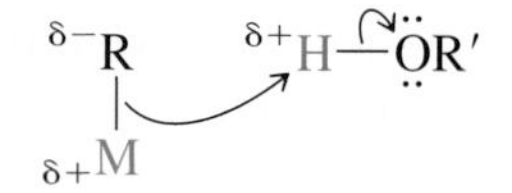

δ−R
δ+H—ÖR′
δ+M

$$\underset{\text{Butyllithium}}{CH_3CH_2CH_2CH_2Li} + \underset{\text{Water}}{H_2O} \longrightarrow \underset{\text{Butane (100\%)}}{CH_3CH_2CH_2CH_3} + \underset{\text{Lithium hydroxide}}{LiOH}$$

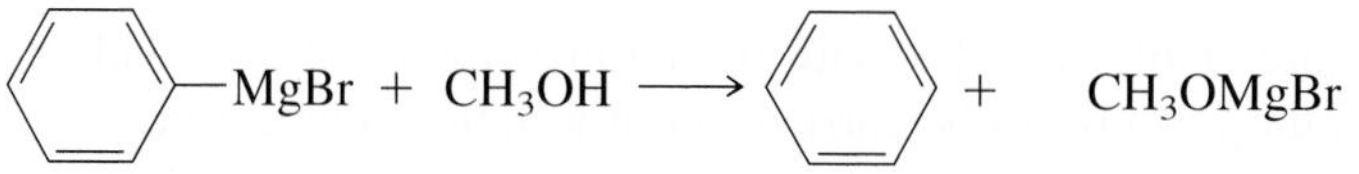

—MgBr + CH₃OH
+ CH₃OMgBr

Phenylmagnesium bromide | Methanol | Benzene (100%) | Methoxymagnesium bromide

Problem 14.4

Use curved arrows to show the flow of electrons in the reaction of butyllithium with water, and that of phenylmagnesium bromide with methanol.

Because of their basicity organolithium compounds and Grignard reagents cannot be prepared or used in the presence of any material that bears an —OH group. Nor are these reagents compatible with —NH or —SH groups, which can also convert an organolithium or organomagnesium compound to a hydrocarbon by proton transfer.

The carbon–metal bonds of organolithium and organomagnesium compounds have appreciable carbanionic character. Carbanions rank among the strongest bases that we'll see in this text. Their conjugate acids are hydrocarbons—very weak acids indeed, with pK_a's in the 25–70 range.

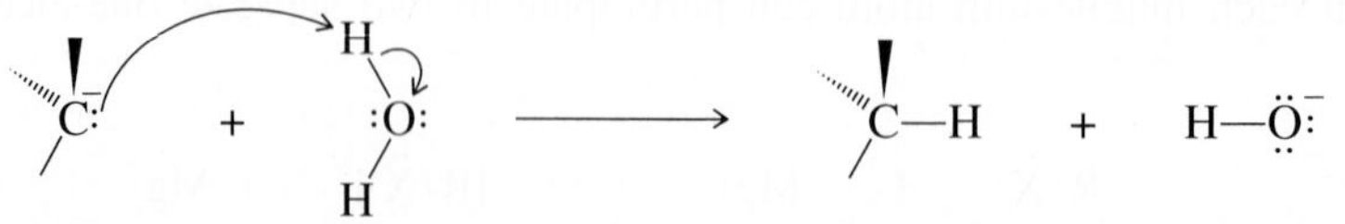

+
H
:O:
H
C—H
+
H—Ö:

Carbanion (very strong base) | Water (weak acid: $pK_a = 15.7$) | Hydrocarbon (very weak acid: pK_a ~25–70) | Hydroxide ion (strong base)

Table 14.1 repeats some approximate data presented earlier in Table 1.8 for the acid strengths of representative hydrocarbons and reference compounds.

TABLE 14.1 Approximate Acidities of Some Hydrocarbons and Reference Materials

Compound	pK_a	Formula*	Conjugate base
Methanol	15.2	$CH_3O{-}H$	$CH_3\ddot{O}:^-$
Water	15.7	$HO{-}H$	$H\ddot{O}:^-$
Ethanol	16	$CH_3CH_2O{-}H$	$CH_3CH_2\ddot{O}:^-$
Acetylene	26	$HC{\equiv}C{-}H$	$HC{\equiv}C:^-$
Ammonia	36	$H_2N{-}H$	$H_2\ddot{N}:^-$
Diisopropylamine	36	$[(CH_3)_2CH]_2N{-}H$	$[(CH_3)_2CH]_2\ddot{N}:^-$
Benzene	43	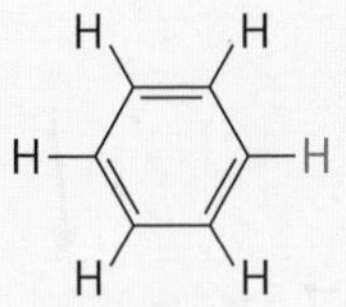	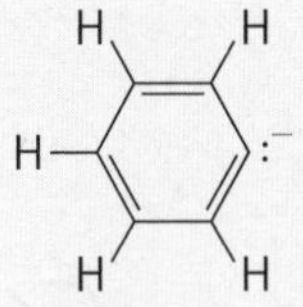
Ethylene	45	$H_2C{=}CH{-}H$	$H_2C{=}\ddot{C}H^-$
Methane	60	$H_3C{-}H$	$H_3C:^-$
Ethane	62	$CH_3CH_2{-}H$	$CH_3\ddot{C}H_2^-$
2-Methylpropane	71	$(CH_3)_3C{-}H$	$(CH_3)_3C:^-$

*The acidic proton in each compound is marked in red.

Acidity decreases from the top of Table 14.1 to the bottom. An acid will transfer a proton to the conjugate base of any acid below it in the table. Organolithium compounds and Grignard reagents act like carbanions and will abstract a proton from any substance more acidic than a hydrocarbon. Thus, N—H groups and terminal alkynes ($RC{\equiv}C{-}H$) are converted to their conjugate bases by proton transfer to organolithium and organomagnesium compounds.

CH_3Li	+	NH_3	$\longrightarrow$	CH_4	+	$LiNH_2$
Methyllithium (stronger base)		Ammonia (stronger acid: pK_a = 36)		Methane (weaker acid: pK_a = 60)		Lithium amide (weaker base)

CH_3CH_2MgBr	+	$HC{\equiv}CH$	$\longrightarrow$	CH_3CH_3	+	$HC{\equiv}CMgBr$
Ethylmagnesium bromide (stronger base)		Acetylene (stronger acid: pK_a = 26)		Ethane (weaker acid: pK_a = 62)		Ethynylmagnesium bromide (weaker base)

Problem 14.5

(a) Lithium diisopropylamide $[(CH_3)_2CH]_2NLi$, referred to as LDA, enjoys many uses as a strong base in synthetic organic chemistry (Section 20.1). It is customarily prepared by the reaction of diisopropylamine $[(CH_3)_2CH]_2NH$ with butyllithium. Write a chemical equation for the Brønsted acid–base reaction that occurs. Represent butyllithium as if it were the carbanion $:\bar{C}H_2CH_2CH_2CH_3$, and use

Continued

curved arrows to show the flow of electrons. Identify the acid, base, conjugate acid, and conjugate base.

Sample Solution

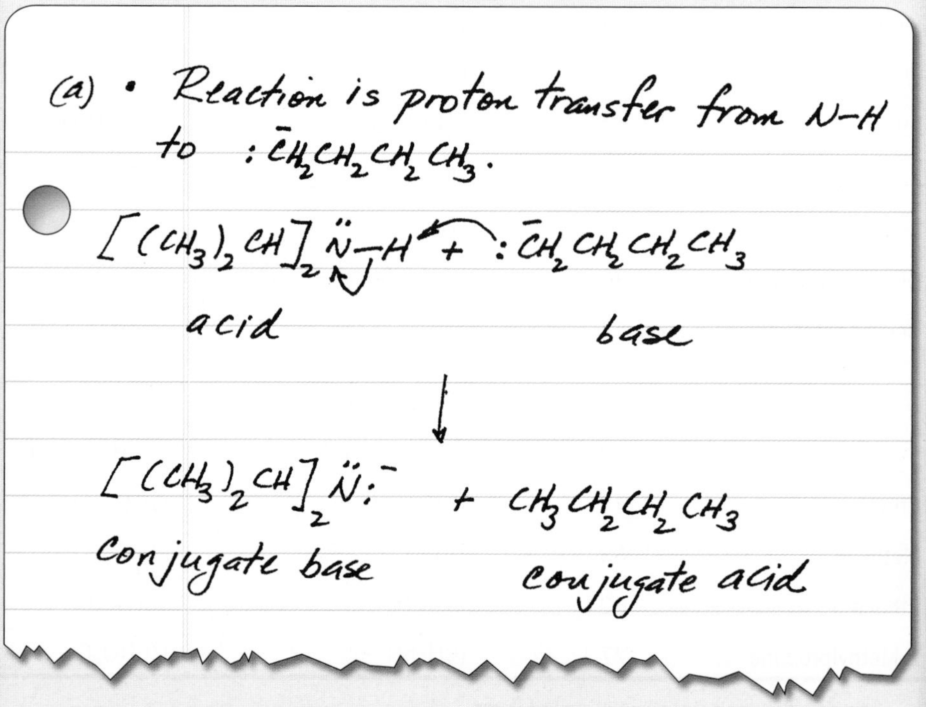

(b) Calculate an equilibrium constant K_{eq} for the equation in part (a).

Deuterium is the mass-2 isotope of hydrogen.

It is sometimes necessary in a synthesis to reduce an alkyl halide to a hydrocarbon. In such cases converting the halide to a Grignard reagent and then adding water or an alcohol as a proton source is a satisfactory procedure. Adding D_2O to a Grignard reagent is a commonly used method for introducing deuterium into a molecule at a specific location.

$$CH_3CH{=}CHBr \xrightarrow[THF]{Mg} CH_3CH{=}CHMgBr \xrightarrow{D_2O} CH_3CH{=}CHD$$

1-Bromopropene — Propenylmagnesium bromide — 1-Deuteriopropene (70%)

14.6 Synthesis of Alcohols Using Grignard Reagents

The main synthetic application of Grignard reagents is their reaction with carbonyl-containing compounds to produce alcohols. Carbon–carbon bond formation is rapid and exothermic when a Grignard reagent reacts with an aldehyde or ketone.

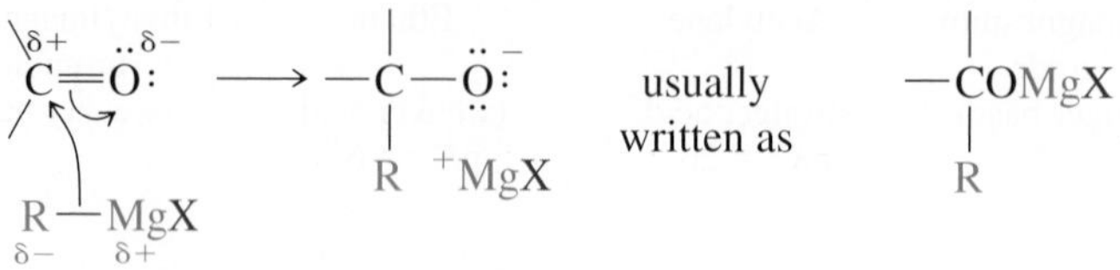

A carbonyl group is quite polar, and its carbon atom is electrophilic. Grignard reagents are nucleophilic and add to carbonyl groups, forming a new carbon–carbon bond. This addition step leads to an alkoxymagnesium halide, which in the second stage of the synthesis is converted to an alcohol by adding aqueous acid.

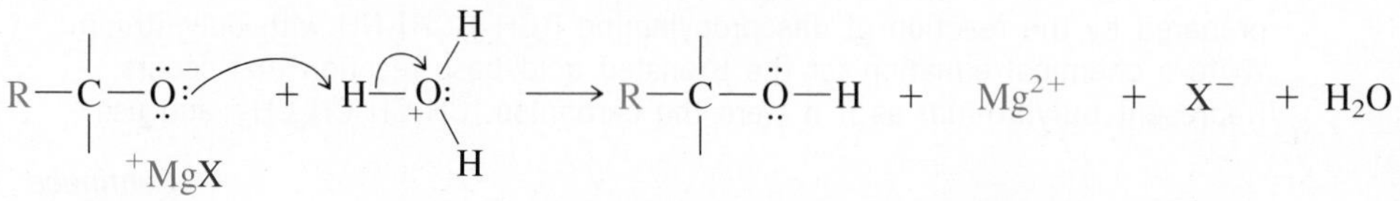

Alkoxymagnesium halide — Hydronium ion — Alcohol — Magnesium ion — Halide ion — Water

TABLE 14.2 Reactions of Grignard Reagents with Aldehydes and Ketones

Reaction	General equation and specific example
Reaction with formaldehyde Grignard reagents react with formaldehyde ($H_2C{=}O$) to give *primary alcohols* having one more carbon than the Grignard reagent.	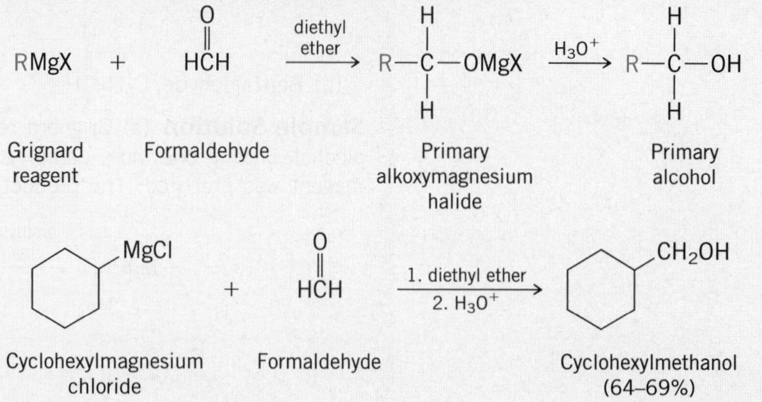
Reaction with aldehydes Grignard reagents react with aldehydes ($R'CH{=}O$) to give *secondary alcohols.*	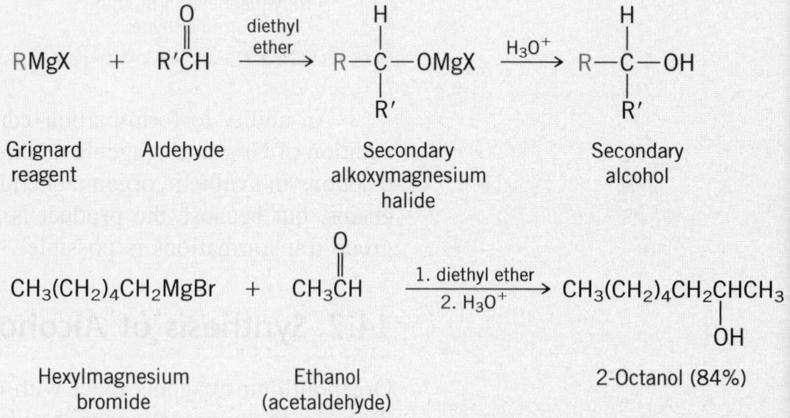
Reaction with ketones Grignard reagents react with ketones ($R'\overset{O}{\overset{\|}{C}}R''$) to give *tertiary alcohols.*	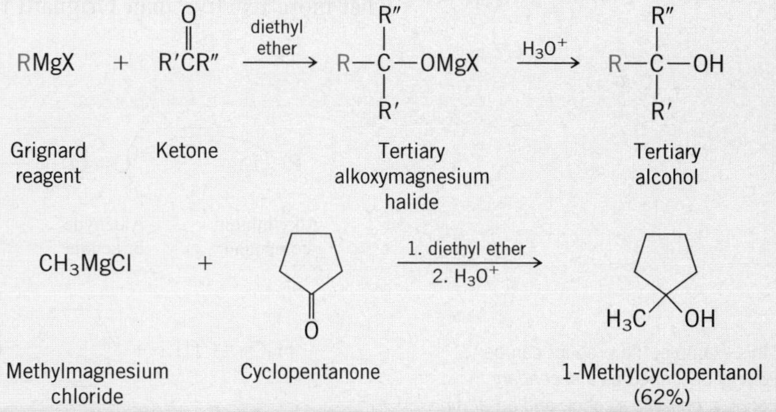

The type of alcohol produced depends on the carbonyl compound. Substituents present on the carbonyl group of an aldehyde or ketone stay there—they become substituents on the carbon that bears the hydroxyl group in the product. Thus as shown in Table 14.2, formaldehyde reacts with Grignard reagents to yield primary alcohols, aldehydes yield secondary alcohols, and ketones yield tertiary alcohols.

Problem 14.6

Write the structure of the product of the reaction of propylmagnesium bromide with each of the following. Assume that the reactions are worked up by the addition of dilute aqueous acid.

Continued

(a) Formaldehyde, HCH (with $C{=}O$)

(b) Benzaldehyde, C_6H_5CH (with $C{=}O$)

(c) Cyclohexanone, 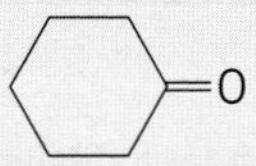

(d) 2-Butanone, $CH_3CCH_2CH_3$ (with $C{=}O$)

Sample Solution (a) Grignard reagents react with formaldehyde to give primary alcohols having one more carbon atom than the alkyl halide from which the Grignard reagent was prepared. The product is 1-butanol.

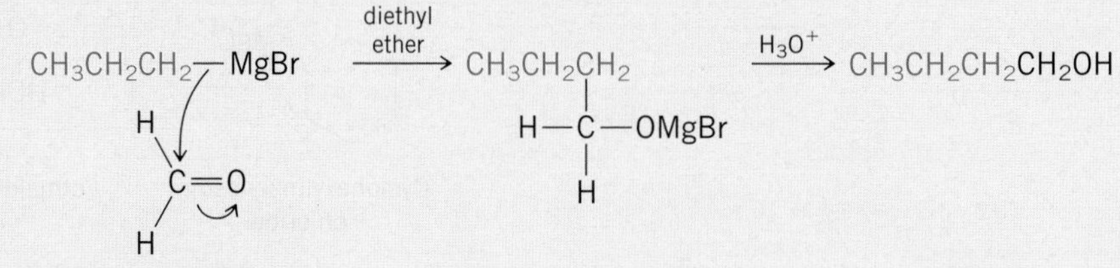

Propylmagnesium bromide + formaldehyde

1-Butanol

An ability to form carbon–carbon bonds is fundamental to organic synthesis. The addition of Grignard reagents to aldehydes and ketones is one of the most frequently used reactions in synthetic organic chemistry. Not only does it permit the extension of carbon chains, but because the product is an alcohol, a wide variety of subsequent functional group transformations is possible.

14.7 Synthesis of Alcohols Using Organolithium Reagents

Organolithium reagents react with carbonyl groups in the same way that Grignard reagents do. In their reactions with aldehydes and ketones, organolithium reagents are somewhat more reactive than Grignard reagents.

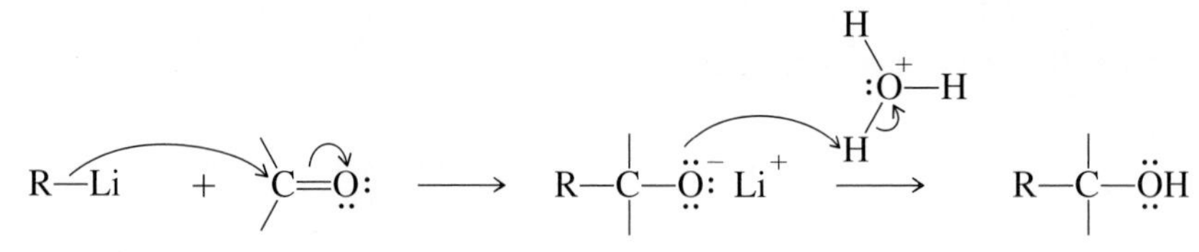

Alkyllithium compound — Aldehyde or ketone — Lithium alkoxide — Alcohol

In this example, the product can be variously described as a *secondary* alcohol, a *benzylic* alcohol, and an *allylic* alcohol. Can you identify the structural reason for each classification?

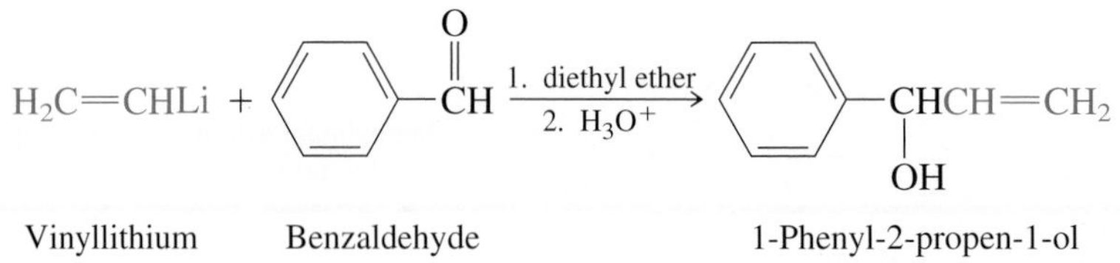

Vinyllithium — Benzaldehyde — 1-Phenyl-2-propen-1-ol (76%)

14.8 Synthesis of Acetylenic Alcohols

The first organometallic compounds we encountered were compounds of the type $RC{\equiv}CNa$ obtained by treatment of terminal alkynes with sodium amide in liquid ammonia (Section 9.5):

$$\underset{\text{Terminal alkyne}}{RC{\equiv}CH} + \underset{\text{Sodium amide}}{NaNH_2} \xrightarrow[-33°C]{NH_3} \underset{\text{Sodium alkynide}}{RC{\equiv}CNa} + \underset{\text{Ammonia}}{NH_3}$$

These compounds are sources of the nucleophilic anion RC≡C:$^-$, and their reaction with primary alkyl halides provides an effective synthesis of alkynes (see Section 9.6). The nucleophilicity of acetylide anions is also evident in their reactions with aldehydes and ketones, which are entirely analogous to those of Grignard and organolithium reagents.

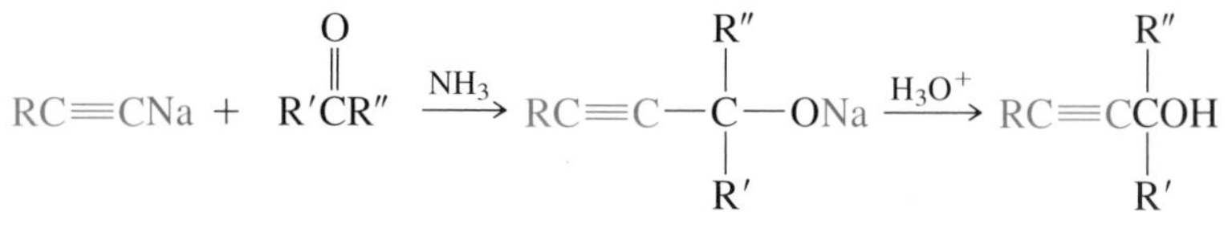

RC≡CNa + R′CR″
NH$_3$
RC≡C—C—ONa
H$_3$O$^+$
RC≡CCOH

Sodium alkynide | Aldehyde or ketone | Sodium salt of an alkynyl alcohol | Alkynyl alcohol

These reactions are normally carried out in liquid ammonia because that is the solvent in which the sodium salt of the alkyne is prepared.

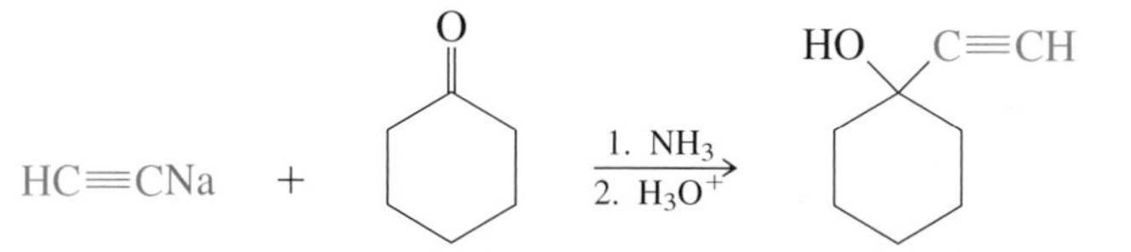

HC≡CNa +
1. NH$_3$
2. H$_3$O$^+$
HO C≡CH

Sodium acetylide | Cyclohexanone | 1-Ethynylcyclohexanol (65–75%)

Acetylenic Grignard reagents of the type RC≡CMgBr are prepared, not from an acetylenic halide, but by an acid–base reaction in which a Grignard reagent abstracts a proton from a terminal alkyne.

$$CH_3(CH_2)_3C{\equiv}CH + CH_3CH_2MgBr \xrightarrow{\text{diethyl ether}} CH_3(CH_2)_3C{\equiv}CMgBr + CH_3CH_3$$

1-Hexyne | Ethylmagnesium bromide | 1-Hexynylmagnesium bromide | Ethane

Which is the stronger acid in this reaction? The weaker acid?

$$CH_3(CH_2)_3C{\equiv}CMgBr + \overset{O}{\overset{\|}{HCH}} \xrightarrow[\text{2. } H_3O^+]{\text{1. diethyl ether}} CH_3(CH_2)_3C{\equiv}CCH_2OH$$

1-Hexynylmagnesium bromide | Formaldehyde | 2-Heptyn-1-ol (82%)

The corresponding acetylenic organolithium reagents are prepared by the reaction of terminal alkynes with methyllithium or butyllithium.

Problem 14.7

Write the equation for the reaction of 1-hexyne with ethylmagnesium bromide as if it involved ethyl anion ($CH_3\ddot{C}H_2^-$) instead of CH_3CH_2MgBr and use curved arrows to represent the flow of electrons.

14.9 Retrosynthetic Analysis

A critical feature of planning a synthesis is reasoning backward from the target molecule to suitable starting materials. This approach was used informally for many years until it was transformed, largely through the efforts of E. J. Corey of Harvard University, into a systematic strategy for synthetic planning that he called **retrosynthetic analysis.**

Corey was honored with the 1990 Nobel Prize in Chemistry for his achievements in synthetic organic chemistry.

A symbol used to indicate a retrosynthetic step is an open arrow written from product to suitable precursors or fragments of those precursors.

Target molecule ⇨ precursors

Often the precursor is not defined completely, but rather its chemical nature is emphasized by writing it as a species to which it is equivalent for synthetic purposes. Thus, a Grignard reagent or an organolithium reagent might be considered synthetically equivalent to a carbanion:

Problem 14.7 at the end of the preceding section introduced this idea with the suggestion that ethylmagnesium bromide be represented as ethyl anion.

RMgX or RLi is synthetically equivalent to R:$^-$

Figure 14.3

A retrosynthetic analysis of alcohol preparation by way of the addition of a Grignard reagent to an aldehyde or ketone.

Step 1: Locate the hydroxyl-bearing carbon.

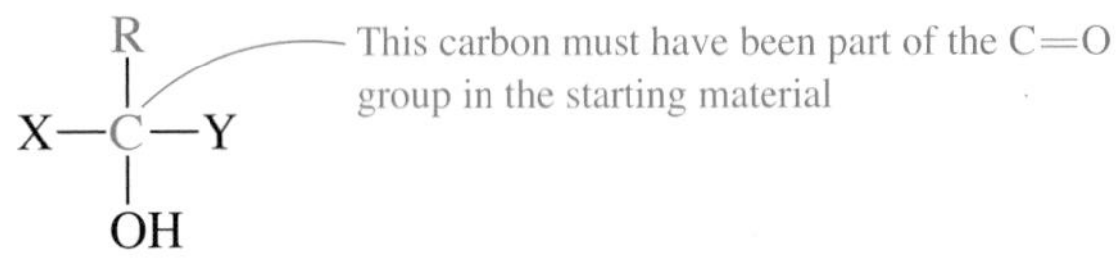

R
X—C—Y
OH
This carbon must have been part of the C=O group in the starting material

Step 2: Disconnect one of the organic substituents attached to the carbon that bears the hydroxyl group.

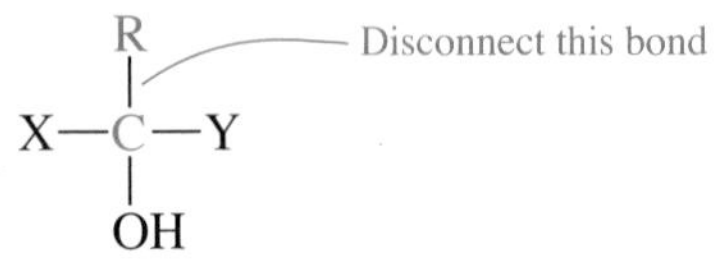

R
X—C—Y
OH
Disconnect this bond

Step 3: Steps 1 and 2 reveal the carbonyl-containing substrate and the carbanionic fragment.

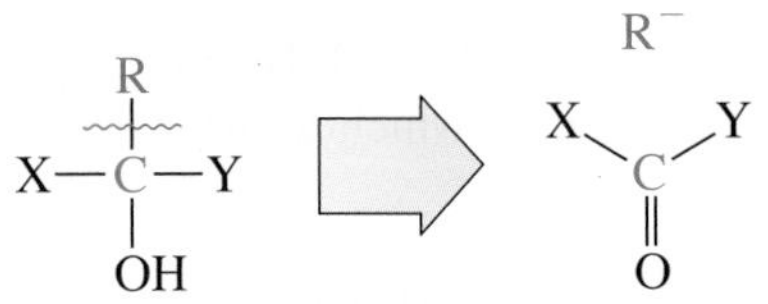

R
X—C—Y
OH
R⁻
X
Y
C
O

Step 4: Because a Grignard reagent may be considered as synthetically equivalent to a carbanion, this suggests the synthesis shown.

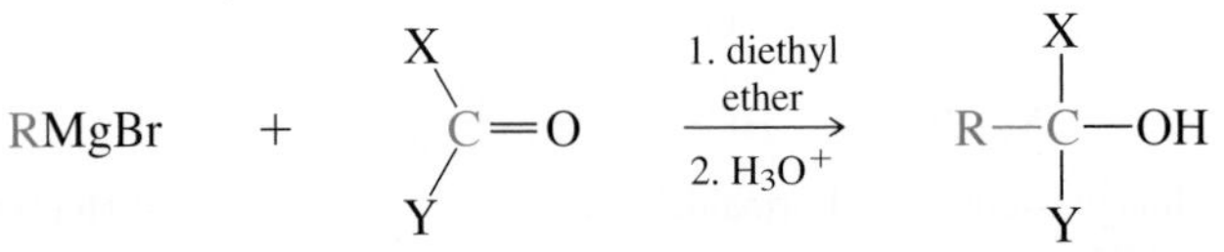

RMgBr
+
X
C=O
Y
1. diethyl ether
2. H3O+
X
R—C—OH
Y

Figure 14.3 illustrates how retrosynthetic analysis can guide you in planning the synthesis of alcohols by identifying suitable Grignard reagent and carbonyl-containing precursors. In the first step, locate the carbon of the target alcohol that bears the hydroxyl group, remembering that this carbon originated in the C=O group. Next, as shown in step 2, mentally disconnect a bond between that carbon and one of its attached groups (other than hydrogen). The attached group is the one that is to be transferred from the Grignard reagent. Once you recognize these two structural fragments, the carbonyl partner and the carbanion that attacks it (step 3), you can readily determine the synthetic mode wherein a Grignard reagent is used as the synthetic equivalent of a carbanion (step 4).

Primary alcohols, by this analysis, are seen to be the products of Grignard addition to formaldehyde:

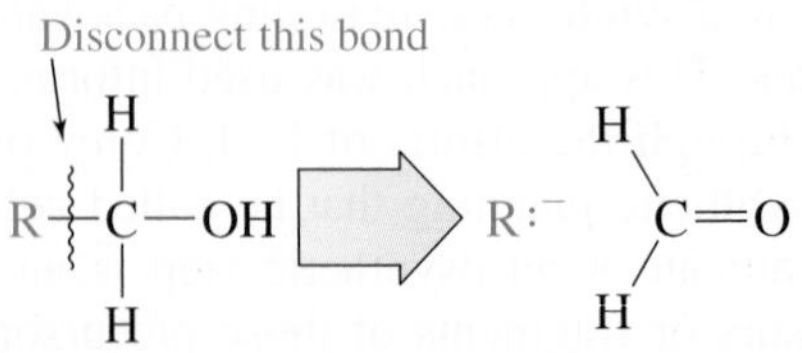

Disconnect this bond
H
R—C—OH
H
R:⁻
H
C=O
H

Secondary alcohols may be prepared by *two* different combinations of Grignard reagent and aldehyde:

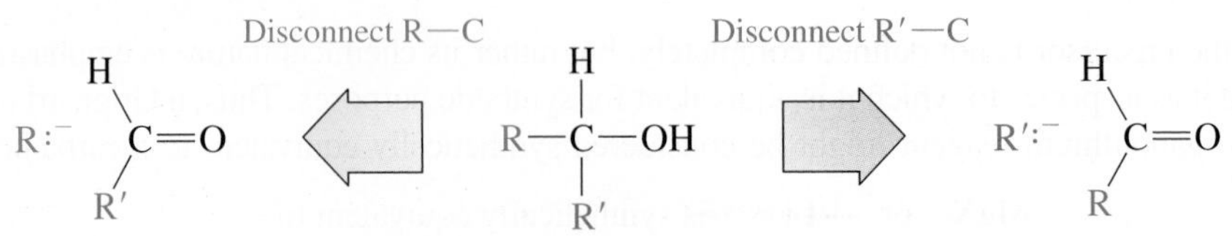

Disconnect R—C
Disconnect R′—C
H
R:⁻
C=O
R′
H
R—C—OH
R′
H
R′:⁻
C=O
R

Three combinations of Grignard reagent and ketone give rise to tertiary alcohols:

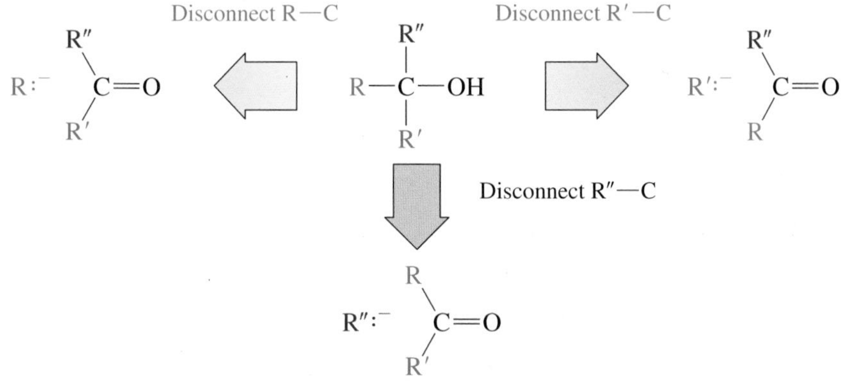

Disconnect R—C
Disconnect R′—C
Disconnect R″—C

Usually, there is little advantage in choosing one route over another when preparing a particular target alcohol. For example, all three of the following combinations have been used to prepare the tertiary alcohol 2-phenyl-2-butanol:

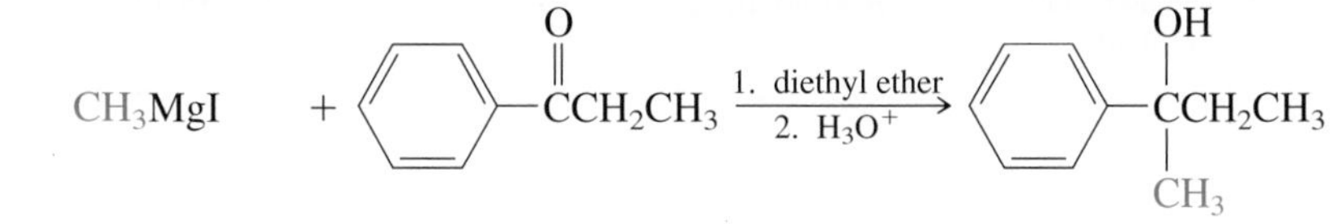

Methylmagnesium iodide | 1-Phenyl-1-propanone | 2-Phenyl-2-butanol

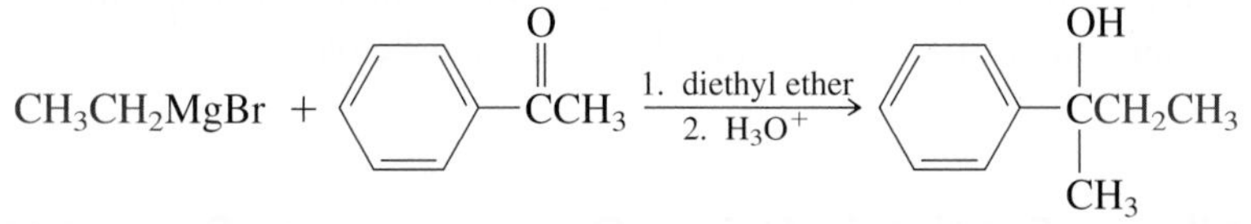

Ethylmagnesium bromide | Acetophenone | 2-Phenyl-2-butanol

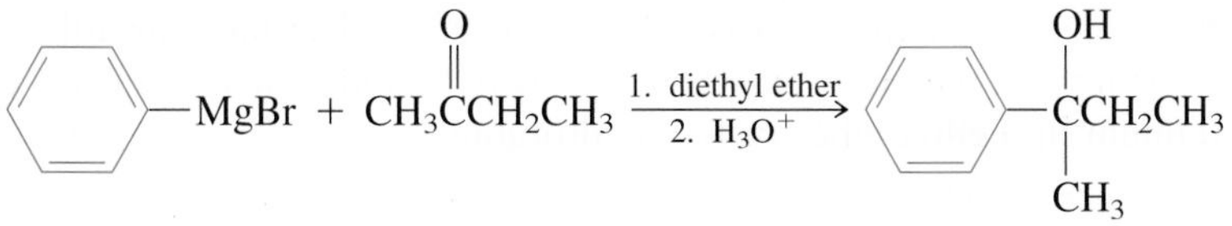

Phenylmagnesium bromide | 2-Butanone | 2-Phenyl-2-butanol

Problem 14.8

Suggest two ways in which each of the following alcohols might be prepared by using a Grignard reagent:

(a) 2-Hexanol, $CH_3\underset{OH}{CH}CH_2CH_2CH_2CH_3$ (b) 2-Phenyl-2-propanol, $C_6H_5\underset{OH}{C}(CH_3)_2$

Sample Solution (a) Because 2-hexanol is a secondary alcohol, we consider the reaction of a Grignard reagent with an aldehyde. Disconnection of bonds to the hydroxyl-bearing carbon generates two pairs of structural fragments:

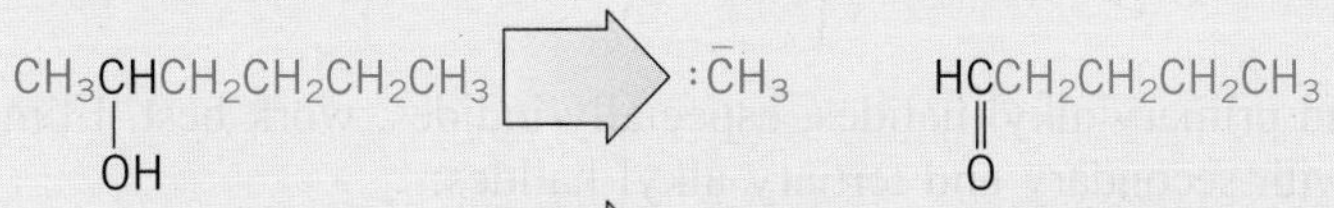

and

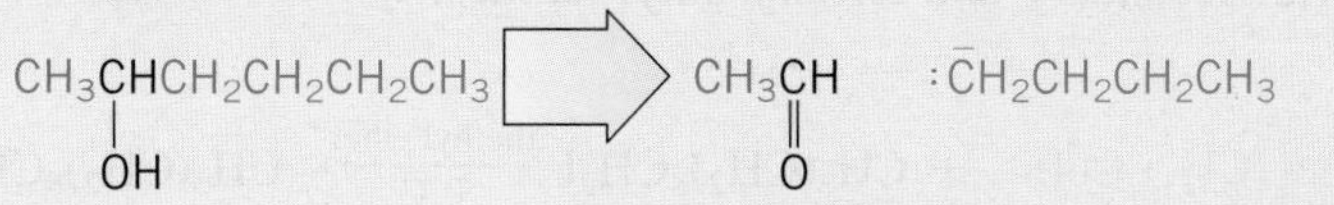

Continued

Therefore, one route involves the addition of a methyl Grignard reagent to a five-carbon aldehyde:

$$\underset{\text{Methylmagnesium iodide}}{CH_3MgI} + \underset{\text{Pentanal}}{CH_3CH_2CH_2CH_2\overset{\overset{O}{\|}}{C}H} \xrightarrow[\text{2. } H_3O^+]{\text{1. diethyl ether}} \underset{\text{2-Hexanol}}{CH_3CH_2CH_2CH_2\underset{\underset{OH}{|}}{C}HCH_3}$$

The other requires addition of a butylmagnesium halide to a two-carbon aldehyde:

$$\underset{\text{Butylmagnesium bromide}}{CH_3CH_2CH_2CH_2MgBr} + \underset{\text{Acetaldehyde}}{CH_3\overset{\overset{O}{\|}}{C}H} \xrightarrow[\text{2. } H_3O^+]{\text{1. diethyl ether}} \underset{\text{2-Hexanol}}{CH_3CH_2CH_2CH_2\underset{\underset{OH}{|}}{C}HCH_3}$$

All that has been said in this section applies with equal force to organolithium reagents. Grignard reagents are one source of nucleophilic carbon; organolithium reagents are another. Both have substantial carbanionic character in their carbon–metal bonds and undergo the same kind of reaction with aldehydes and ketones.

14.10 Alkane Synthesis Using Organocopper Reagents

Organocopper compounds used for carbon–carbon bond formation are called *Gilman reagents* in honor of Henry Gilman who first studied them. Gilman's career in teaching and research at Iowa State spanned more than half a century (1919–1975).

Organometallic compounds of copper were known for a long time before their versatility in synthetic organic chemistry was fully appreciated. The most useful ones are the lithium dialkylcuprates, which result when a copper(I) halide reacts with two equivalents of an alkyllithium in diethyl ether or tetrahydrofuran.

$$\underset{\text{Alkyllithium}}{2RLi} + \underset{\substack{\text{Cu(I) halide}\\(X = Cl, Br, I)}}{CuX} \xrightarrow[\text{or THF}]{\text{diethyl ether}} \underset{\substack{\text{Lithium}\\\text{dialkylcuprate}}}{R_2CuLi} + \underset{\substack{\text{Lithium}\\\text{halide}}}{LiX}$$

Mechanism 14.1 shows the formation of a lithium dialkylcuprate.

Lithium dialkylcuprates react with alkyl halides to produce alkanes by carbon–carbon bond formation between the alkyl group of the alkyl halide and the alkyl group of the dialkylcuprate:

$$\underset{\substack{\text{Lithium}\\\text{dialkylcuprate}}}{R_2CuLi} + \underset{\text{Alkyl halide}}{R'X} \longrightarrow \underset{\text{Alkane}}{R{-}R'} + \underset{\text{Alkylcopper}}{RCu} + \underset{\substack{\text{Lithium}\\\text{halide}}}{LiX}$$

Methyl and primary alkyl halides, especially iodides, work best. Elimination becomes a problem with secondary and tertiary alkyl halides:

$$\underset{\substack{\text{Lithium}\\\text{dimethylcuprate}}}{(CH_3)_2CuLi} + \underset{\text{1-Iodododecane}}{CH_3(CH_2)_8CH_2I} \xrightarrow[0°C]{\text{diethyl ether}} \underset{\text{Undecane (90\%)}}{CH_3(CH_2)_8CH_2CH_3}$$

Mechanism 14.1

Formation of a Lithium Dialkylcuprate (Gilman Reagent)

Step 1: One molar equivalent of an alkyllithium reagent displaces iodide from copper(I) iodide to give an alkylcopper(I) species.

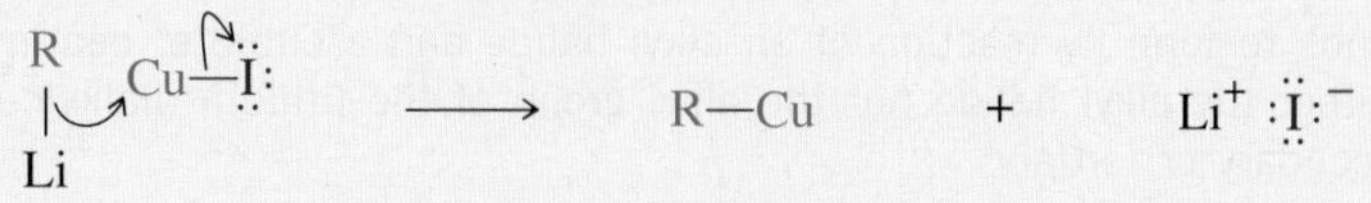

Alkyllithium Copper(I) iodide Alkylcopper Lithium iodide

Step 2: The second molar equivalent of the alkyllithium adds to the alkylcopper to give a negatively charged R_2Cu^- species called a *dialkylcuprate.* It is formed as its lithium salt, a lithium dialkylcuprate.

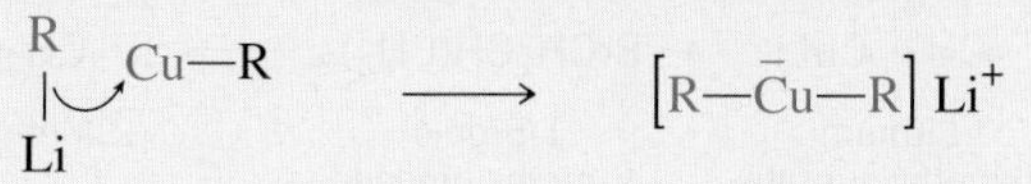

Alkyllithium Alkylcopper Lithium dialkylcuprate

The reaction is carried out in diethyl ether or tetrahydrofuran solution. The lithium dialkylcuprate is soluble under these conditions and used directly.

Lithium diarylcuprates are prepared in the same way as lithium dialkylcuprates and undergo comparable reactions with primary alkyl halides:

$$(C_6H_5)_2CuLi + ICH_2(CH_2)_6CH_3 \xrightarrow{\text{diethyl ether}} C_6H_5CH_2(CH_2)_6CH_3$$

Lithium diphenylcuprate 1-Iodooctane 1-Phenyloctane (99%)

The most frequently used organocuprates are those in which the alkyl group is primary. Steric hindrance makes secondary and tertiary dialkylcuprates less reactive, and they tend to decompose before they react with the alkyl halide. The reaction of cuprate reagents with alkyl halides follows the usual S_N2 order: CH_3 > primary > secondary > tertiary, and I > Br > Cl > F. *p*-Toluenesulfonates are somewhat more reactive than halides. Because the alkyl halide and dialkylcuprate reagent should both be primary in order to produce satisfactory yields of coupled products, the reaction is most often used for the formation of $RCH_2—CH_2R'$ and $RCH_2—CH_3$ bonds in alkanes.

A key step in the reaction mechanism appears to be nucleophilic attack on the alkyl halide by the negatively charged copper atom, but the details of the mechanism are not well understood. Indeed, there is probably more than one mechanism by which cuprates react with organic halogen compounds. Vinyl halides and aryl halides are known to be very unreactive in S_N2 reactions, yet react with lithium dialkylcuprates:

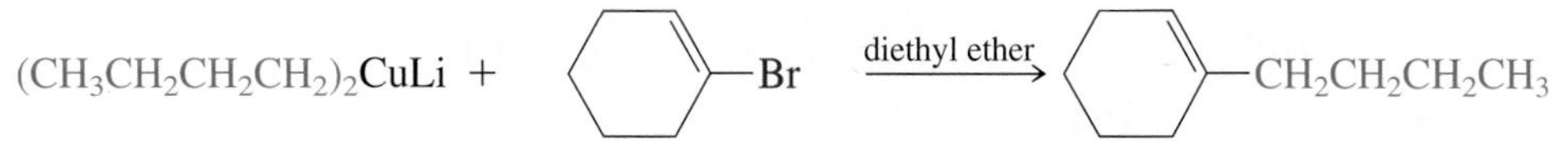

Lithium dibutylcuprate 1-Bromocyclohexene 1-Butylcyclohexene (80%)

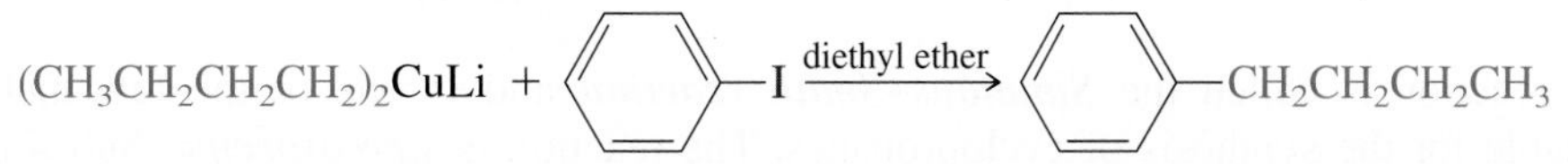

Lithium dibutylcuprate Iodobenzene Butylbenzene (75%)

Problem 14.9

Suggest a combination of organic halide and cuprate reagent appropriate for the preparation of each of the following compounds:

(a) 2-Methylbutane (b) 1,3,3-Trimethylcyclopentene

Sample Solution (a) First inspect the target molecule to see which are the best bonds to form by reaction of an alkyl halide and a cuprate, bearing in mind that neither the alkyl halide nor the alkyl group of the lithium dialkylcuprate should be secondary or tertiary.

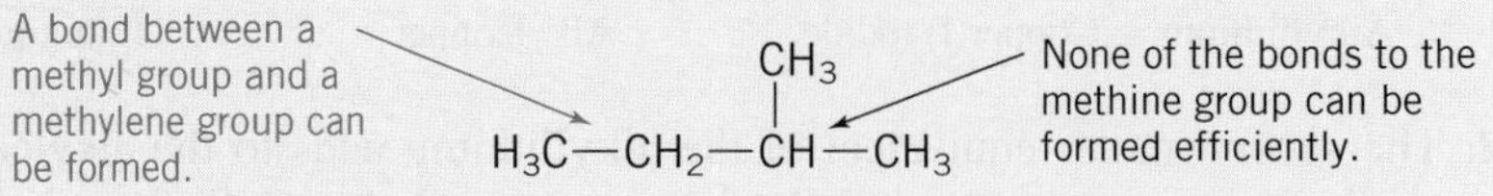
A bond between a methyl group and a methylene group can be formed.
CH3
H3C—CH2—CH—CH3
None of the bonds to the methine group can be formed efficiently.

There are two combinations, both acceptable, that give the $H_3C—CH_2$ bond:

$$(CH_3)_2CuLi + BrCH_2CH(CH_3)_2 \longrightarrow CH_3CH_2CH(CH_3)_2$$

Lithium dimethylcuprate; 1-Bromo-2-methylpropane; 2-Methylbutane

$$CH_3I + LiCu[CH_2CH(CH_3)_2]_2 \longrightarrow CH_3CH_2CH(CH_3)_2$$

Iodomethane; Lithium diisobutylcuprate; 2-Methylbutane

14.11 An Organozinc Reagent for Cyclopropane Synthesis

Zinc reacts with alkyl halides in a manner similar to that of magnesium.

$$RX + Zn \xrightarrow{\text{diethyl ether}} RZnX$$

Alkyl halide; Zinc; Alkylzinc halide

Victor Grignard was led to study organomagnesium compounds because of earlier work he performed with organic derivatives of zinc.

Zinc is less electropositive than lithium and magnesium, and the carbon–zinc bond is less polar. Organozinc reagents are not nearly as reactive toward aldehydes and ketones as Grignard reagents and organolithium compounds.

An organozinc compound that occupies a special niche in organic synthesis is *iodomethylzinc iodide* (ICH_2ZnI). It is prepared by the reaction of zinc–copper couple [Zn(Cu), zinc that has had its surface activated with a little copper] with diiodomethane in diethyl ether.

Iodomethylzinc iodide is known as the *Simmons–Smith reagent,* after Howard E. Simmons and Ronald D. Smith of DuPont, who first described its use in the preparation of cyclopropanes.

$$CH_2I_2 + Zn \xrightarrow[\text{Cu}]{\text{diethyl ether}} ICH_2ZnI$$

Diiodomethane; Zinc; Iodomethylzinc iodide

What makes iodomethylzinc iodide such a useful reagent is that it reacts with alkenes to give cyclopropanes.

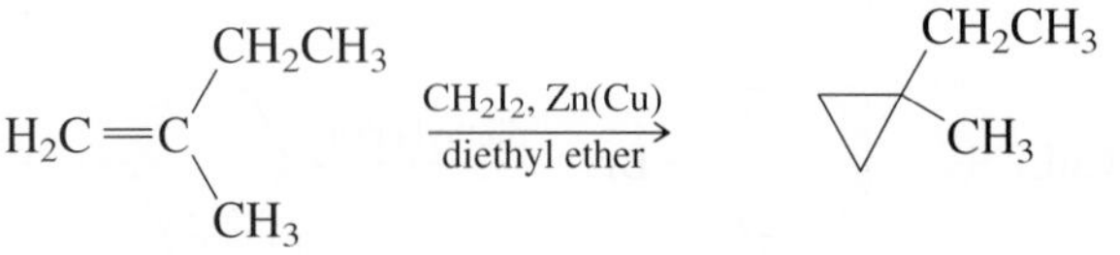
CH2CH3
H2C=C
CH3
CH2I2, Zn(Cu)
diethyl ether
CH2CH3
CH3

2-Methyl-1-butene → 1-Ethyl-1-methylcyclopropane (79%)

This reaction is called the ***Simmons–Smith reaction*** and is one of the few methods available for the synthesis of cyclopropanes. The reaction is *stereospecific.* Substituents that were cis in the alkene remain cis in the cyclopropane.

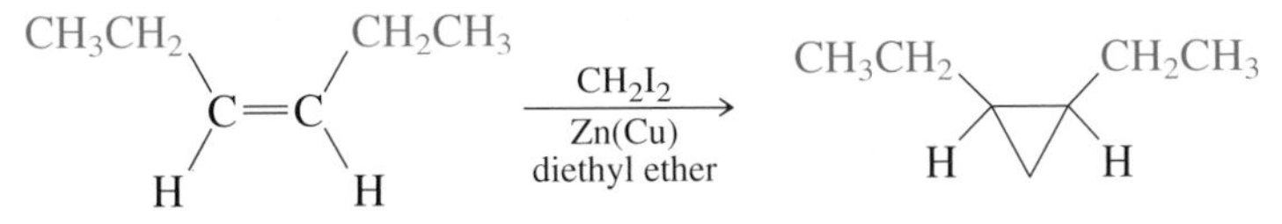

CH_3CH_2
CH_2CH_3
C=C
H
H
CH_2I_2
Zn(Cu)
diethyl ether
CH_3CH_2
CH_2CH_3
H
H

(*Z*)-3-Hexene *cis*-1,2-Diethylcyclopropane (34%)

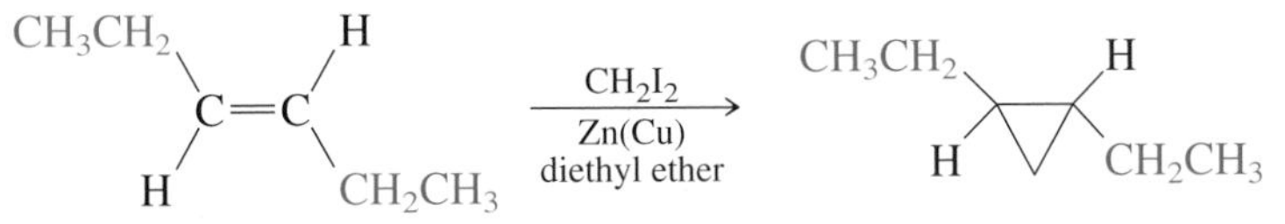

CH_3CH_2
H
C=C
H
CH_2CH_3
CH_2I_2
Zn(Cu)
diethyl ether
CH_3CH_2
H
H
CH_2CH_3

(*E*)-3-Hexene *trans*-1,2-Diethylcyclopropane (15%)

Problem 14.10

What alkenes would you choose as starting materials in order to prepare each of the following cyclopropane derivatives by reaction with iodomethylzinc iodide?

(a)

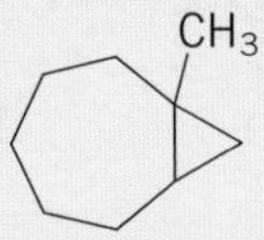

CH_3

(b)

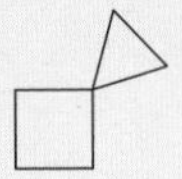

Sample Solution (a) In a cyclopropane synthesis using the Simmons–Smith reagent, you should remember that a CH_2 unit is transferred. Therefore, retrosynthetically disconnect the bonds to a CH_2 group of a three-membered ring to identify the starting alkene.

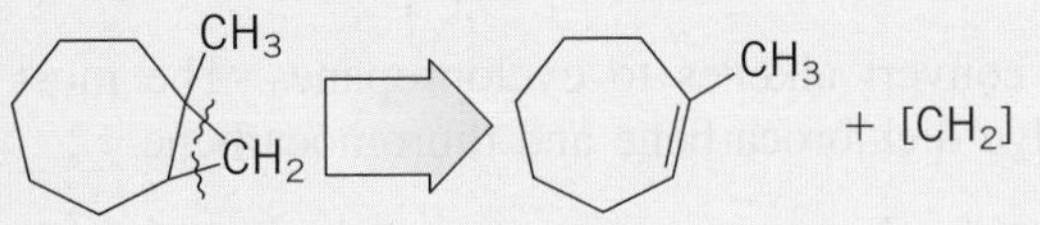

CH_3
CH_2
CH_3
+ $[CH_2]$

The complete synthesis is

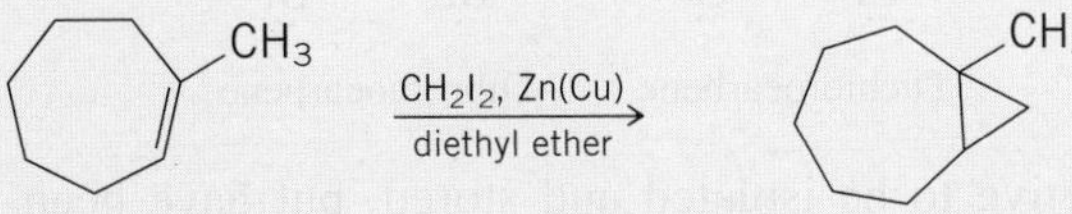

CH_3
CH_2I_2, Zn(Cu)
diethyl ether
CH_3

1-Methylcycloheptene 1-Methylbicyclo[5.1.0]octane (55%)

Cyclopropanation of alkenes with the Simmons–Smith reagent bears some similarity to epoxidation. Both reactions are stereospecific cycloadditions, and iodomethylzinc iodide behaves, like peroxy acids, as a weak electrophile. Both cycloadditions take place faster with more highly substituted double bonds than less substituted ones, but are sensitive to steric hindrance in the alkene. These similarities are reflected in the mechanisms proposed for the two reactions shown in Mechanism 14.2. Both are believed to be concerted.

Iodomethylzinc iodide breaks the pattern we have seen so far in organometallic reactivity. Unlike organolithium, Grignard, and organocopper reagents, all of which are nucleophilic, iodomethylzinc iodide is electrophilic.

14.12 Carbenes and Carbenoids

Iodomethylzinc iodide is often referred to as a **carbenoid,** meaning that it resembles a **carbene** in its chemical reactions. Carbenes are neutral molecules in which one of the carbon atoms has six valence electrons. Such carbons are *divalent;* they are directly bonded to only two other atoms and have no multiple bonds. Iodomethylzinc iodide reacts as if it were a source of the carbene H—$\ddot{C}$—H.

It is clear that free :CH_2 is not involved in the Simmons–Smith reaction, but there is substantial evidence to indicate that carbenes are formed as intermediates in certain

Divalent carbon species first received attention with the work of the Swiss-American chemist J. U. Nef in the late nineteenth century; they were then largely ignored until the 1950s.

Mechanism 14.2

Similarities Between the Mechanisms of Reaction of an Alkene with Iodomethylzinc Iodide and a Peroxy Acid

Cyclopropanation

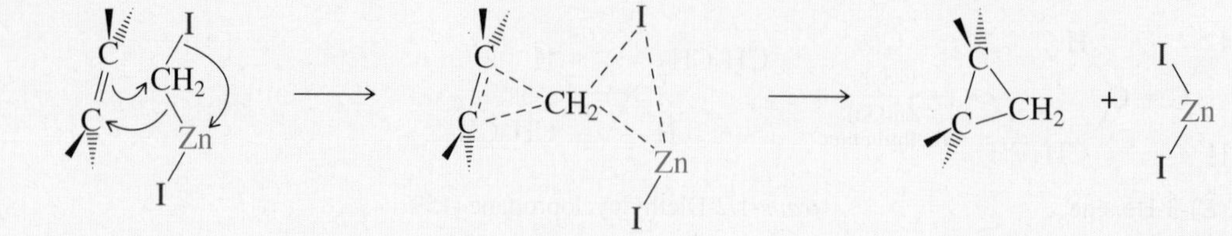

Alkene + Iodomethylzinc iodide — Transition state — A cyclopropane + Zinc iodide

Epoxidation

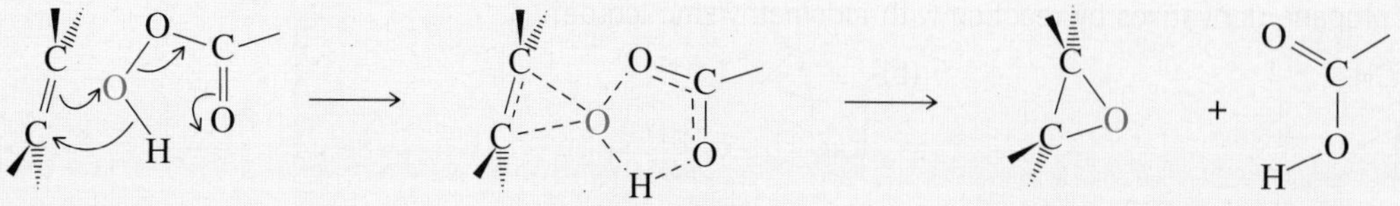

Alkene + Peroxy acid — Transition state — Epoxide + Carboxylic acid

other reactions that convert alkenes to cyclopropanes. The most studied examples of these reactions involve dichlorocarbene and dibromocarbene.

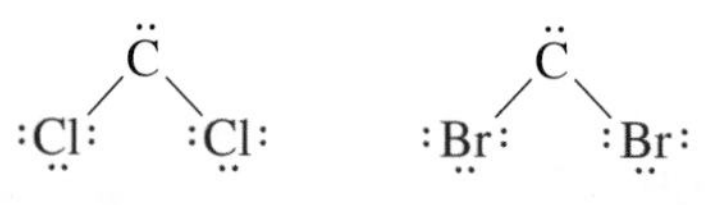

Dichlorocarbene   Dibromocarbene

Carbenes are too reactive to be isolated and stored, but have been trapped in frozen argon for spectroscopic study at very low temperatures. Bonding in dihalocarbenes is based on sp^2 hybridization of carbon, shown for CCl_2 in Figure 14.4*a*. Two of carbon's sp^2 hybrid orbitals are involved in σ bonds to the halogen. The third sp^2 orbital contains the unshared electron pair, and the unhybridized 2*p* orbital is vacant. The electrostatic potential map in Figure 14.4*b* illustrates this nicely with the highest negative character (red) concentrated in the region of the lone pair orbital, and the region of highest positive charge (blue) situated above and below the plane of the carbene.

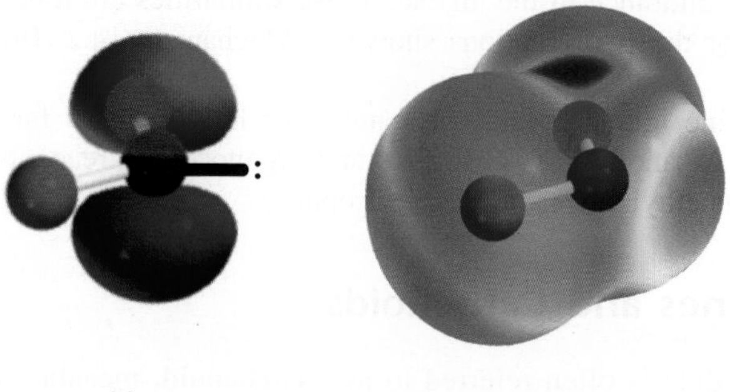

(*a*)   (*b*)

Figure 14.4

(*a*) The unshared electron pair occupies an sp^2-hybridized orbital in dichlorocarbene. There are no electrons in the unhybridized *p* orbital. (*b*) An electrostatic potential map of dichlorocarbene shows negative charge is concentrated in the region of the unshared pair, and positive charge above and below the carbon.

14.13 Transition-Metal Organometallic Compounds

A large number of organometallic compounds are based on transition metals, and many important industrial processes are catalyzed by transition metals or their complexes. Before we look at these processes, a few words about the structures of transition-metal complexes are in order.

We'll start with nickel carbonyl, an intermediate in the purification of nickel first prepared over a hundred years ago. It is a neutral molecule (boiling point: 43°C) that forms spontaneously when carbon monoxide is passed over elemental nickel.

$$\underset{\text{Nickel}}{Ni} \quad + \quad \underset{\text{Carbon monoxide}}{4CO} \quad \longrightarrow \quad \underset{\text{Nickel carbonyl}}{Ni(CO)_4}$$

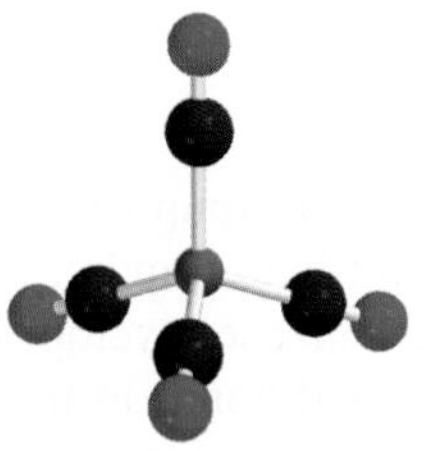

In a transition-metal complex, the groups attached to the metal are called **ligands** and can be an element (O_2, N_2), a compound (CO, $H_2C{=}CH_2$), or an ion (CN^-) and can be inorganic or organic. The bonds between nickel and its carbon monoxide ligands in $Ni(CO)_4$ are covalent. The electron pair in each Ni—C bond originates in a lone pair of carbon of CO, which is best thought of in terms of the resonance contributor $:\overset{-}{C}{\equiv}\overset{+}{O}:$. Other species that are common electron-pair donors in transition-metal complexes include:

$:H^-$	$:\ddot{Cl}:^-$	$:NH_3$	$:P(C_6H_5)_3$	$:R^-$
Hydride	Chloride	Ammonia	Triphenylphosphine	Alkyl

Many transition-metal complexes, including $Ni(CO)_4$, obey the **18-electron rule,** which is to transition-metal complexes as the octet rule is to main-group elements like carbon and oxygen. It states that

> For transition-metal complexes, *the number of ligands that can be attached to a metal will be such that the sum of the electrons brought by the ligands plus the valence electrons of the metal equals 18.*

The valence electrons of the metal are given by the group number of the metal in the periodic table. The relevant group numbers are those shown within parenthesis as (3), (4), (5), . . . (12) at the top of the table found on the inside back cover and are equal to the number of valence electrons of the neutral transition metal. Thus Ni, which is a group (10) transition metal, contributes ten electrons. The four CO ligands contribute 2 electrons each for a total of eight, which, when combined with the 10 valence electrons from Ni, satisfies the 18-electron rule.

Problem 14.11

Like nickel, iron reacts with carbon monoxide to form a compound having the formula $M(CO)_n$ that obeys the 18-electron rule. What is the value of n in the formula $Fe(CO)_n$?

A useful point to remember about the 18-electron rule when we discuss some reactions of transition-metal complexes is that if the number is less than 18, the metal is considered *coordinatively unsaturated* and can accept additional ligands.

Not all ligands use just two electrons to bond to transition metals. Benzene uses its six π electrons in organometallic compounds such as (benzene)tricarbonylchromium.

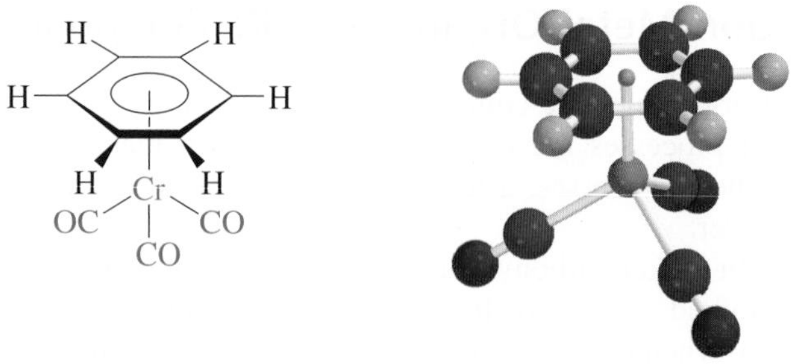

(Benzene)tricarbonylchromium

Chromium is a group (6) transition metal, so contributes 6 valence electrons, three CO ligands contribute 6 more, and the 6 π electrons of benzene bring the total to 18.

Like benzene, cyclopentadienide anion is an aromatic six π-electron system (Section 11.21) and bonds to transition metals in a similar way. Ferrocene is the best known example, with a structure aptly described as a *sandwich.*

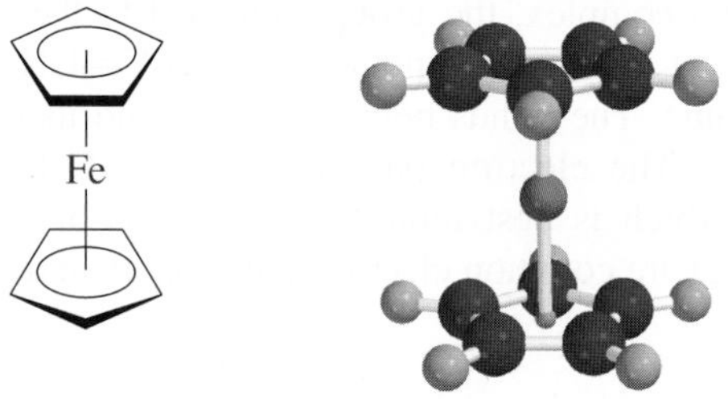

Ferrocene

Although iron is a group (8) transition metal, it is in the +2 oxidation state in ferrocene so contributes six, rather than eight, electrons. These 6, plus 12 from the two cyclopentadienide rings, bring the total to 18.

The preparation and structure determination of ferrocene marked the beginning of metallocene chemistry. **Metallocenes** are organometallic compounds that bear cyclopentadienide ligands. A large number are known, even some in which uranium is the metal. Metallocenes are not only stucturally interesting, but many of them have useful applications as catalysts for industrial processes. Zirconium-based metallocenes, for example, are the most widely used catalysts for Ziegler–Natta polymerization of alkenes. We'll have more to say about them in Section 14.16 and Chapter 27.

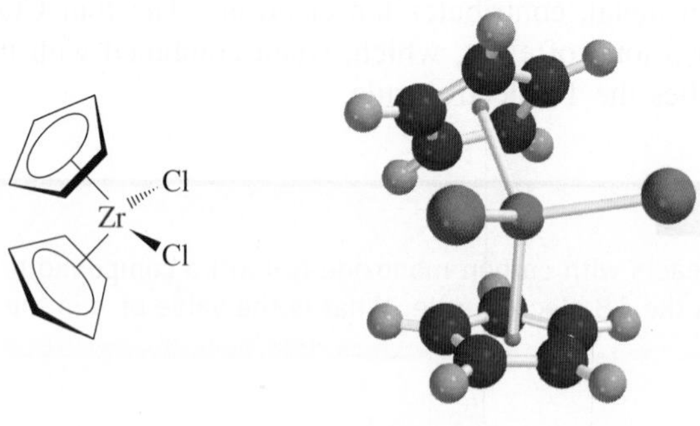

(Bis)-cyclopentadienylzirconium dichloride
(Zirconocene dichloride)

Naturally occurring compounds with carbon–metal bonds are very rare, coenzyme B_{12} being the best example (see the boxed essay *An Organometallic Compound That Occurs Naturally: Coenzyme B_{12}* accompanying this section). Among other biochemical

An Organometallic Compound That Occurs Naturally: Coenzyme B_{12}

Pernicious anemia is a disease characterized, as are all anemias, by a deficiency of red blood cells. Unlike ordinary anemia, pernicious anemia does not respond to treatment with sources of iron, and before effective treatments were developed, was often fatal. Injection of liver extracts was one such treatment, and in 1948 chemists succeeded in isolating the "antipernicious anemia factor" from beef liver as a red crystalline compound, which they called *vitamin B_{12}*. This compound had the formula $C_{63}H_{88}CoN_{14}O_{14}P$. Its complexity precluded structure determination by classical degradation techniques, and spectroscopic methods were too primitive to be of much help. The structure was solved by Dorothy Crowfoot Hodgkin of Oxford University in 1955 using X-ray diffraction techniques and is shown in Figure 14.5*a*. Structure determination by X-ray crystallography can be superficially considered as taking a photograph of a molecule with X-rays. It is a demanding task and earned Hodgkin the 1964 Nobel Prize in Chemistry. Modern structural studies by X-ray crystallography use computers to collect and analyze the diffraction data and take only a fraction of the time required years ago to solve the vitamin B_{12} structure.

The structure of vitamin B_{12} is interesting in that it contains a central cobalt atom that is surrounded by six atoms in an octahedral geometry. One substituent, the cyano (—CN) group, is what is known as an "artifact." It appears to be introduced into the molecule during the isolation process and leads to the synonym *cyanocobalamin* for vitamin B_{12}. This is the material used to treat pernicious anemia, but is not the form in which it exerts its activity. The biologically active substance is called *coenzyme B_{12}* and differs from vitamin B_{12} in the ligand attached to cobalt (Figure 14.5*b*). Coenzyme B_{12} is the only known naturally occurring substance that has a carbon–metal bond. Moreover, coenzyme B_{12} was discovered before any compound containing an alkyl group σ-bonded to cobalt had ever been isolated in the laboratory!

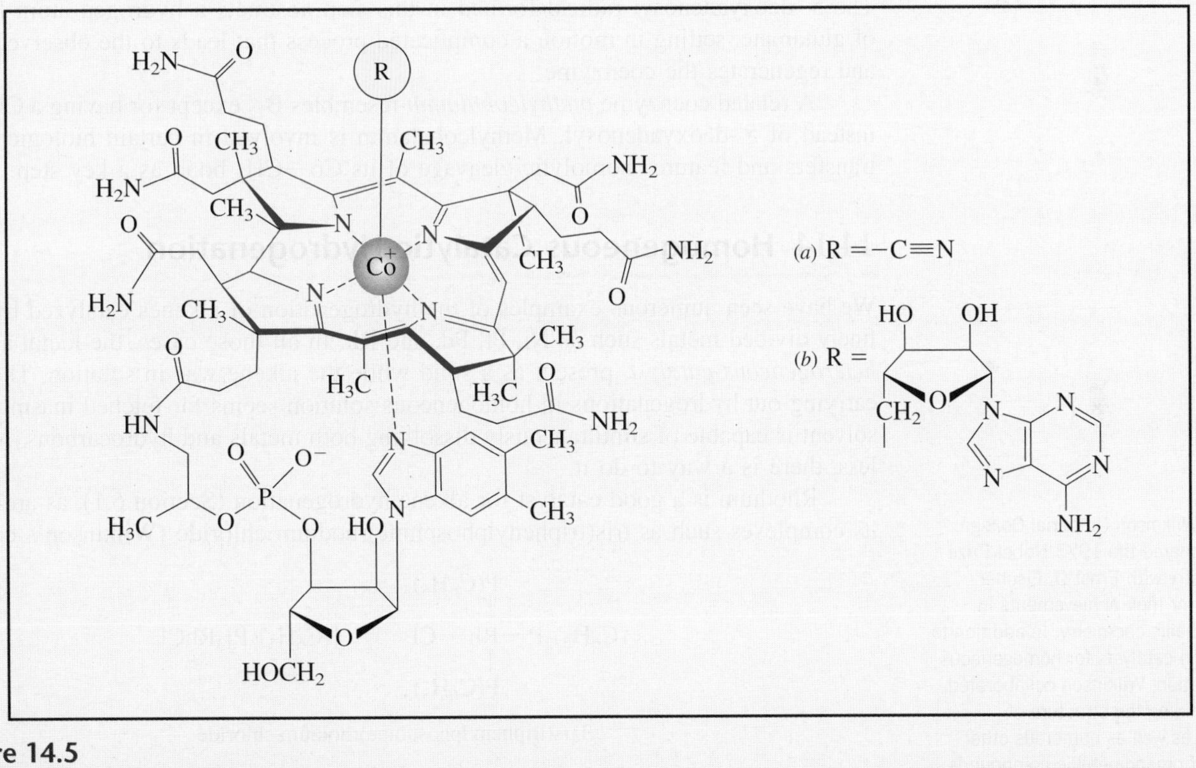

Figure 14.5

The structures of (*a*) vitamin B_{12} and (*b*) coenzyme B_{12}.

reactions, coenzyme B_{12} participates in enzyme-catalyzed rearrangements of a type in which an atom or group X switches places with H of an adjacent atom.

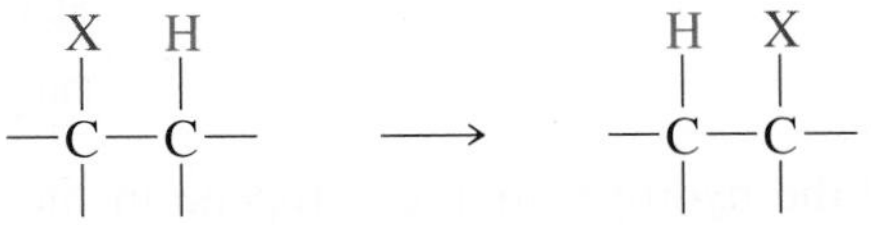

The enzyme *glutamate mutase* catalyzes such a rearrangement in which carboxylate (CO_2^-) and a neighboring hydrogen of glutamate exchange their points of attachment.

Glutamate and 3-methylaspartate are the conjugate bases of glutamic acid and 3-methylaspartic acid, respectively.

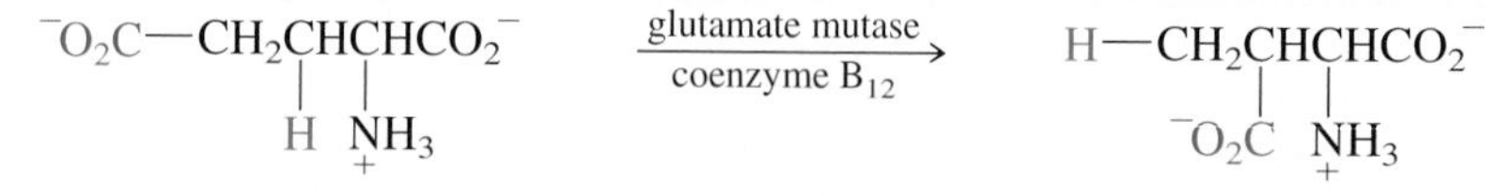

Glutamate

3-Methylaspartate

Reactions such as this that are promoted by coenzyme B_{12} follow free-radical pathways that begin with homolytic cleavage of the weak carbon–cobalt σ bond.

The shorthand representation of coenzyme B_{12} used in the equation helps us focus on the reactive portion of the molecule.

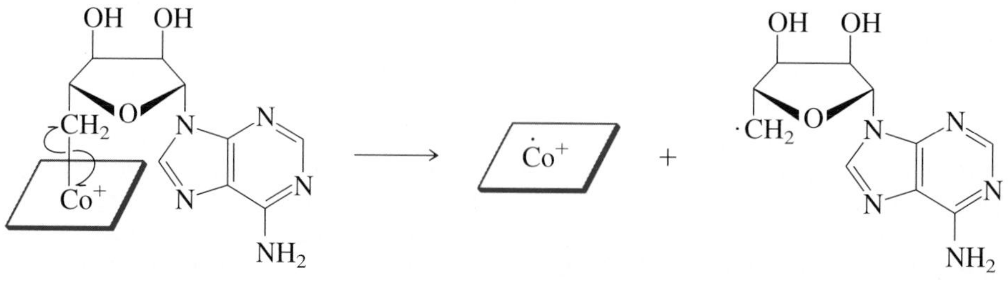

Coenzyme B_{12} (abbreviated from Figure 14.5)

Cobalamin

5′-Deoxyadenosyl radical

The 5′-deoxyadenosyl radical formed in this step abstracts a hydrogen atom from C-3 of glutamate, setting in motion a complicated process that leads to the observed product and regenerates the coenzyme.

A related coenzyme *methylcobalamin* resembles B_{12} except for having a CH_3 ligand instead of 5′-deoxyadenosyl. Methylcobalamin is involved in certain biological methyl transfers and features homolytic cleavage of its Co—CH_3 bond as a key step.

14.14 Homogeneous Catalytic Hydrogenation

We have seen numerous examples of the hydrogenation of alkenes catalyzed by various finely divided metals such as Ni, Pt, Pd, and Rh. In all those cases, the metal acted as a *heterogeneous catalyst,* present as a solid while the alkene was in solution. The idea of carrying out hydrogenations in homogeneous solution seems far-fetched inasmuch as no solvent is capable of simultaneously dissolving both metals and hydrocarbons. Nevertheless, there is a way to do it.

Rhodium is a good catalyst for alkene hydrogenation (Section 6.1), as are many of its complexes such as tris(triphenylphosphine)rhodium chloride (Wilkinson's catalyst).

Geoffrey Wilkinson (Imperial College, London) shared the 1973 Nobel Prize in Chemistry with Ernst O. Fischer (Munich) for their achievements in organometallic chemistry. In addition to his work on catalysts for homogeneous hydrogenation, Wilkinson collaborated on determining the structure of ferrocene as well as numerous other aspects of organometallic compounds.

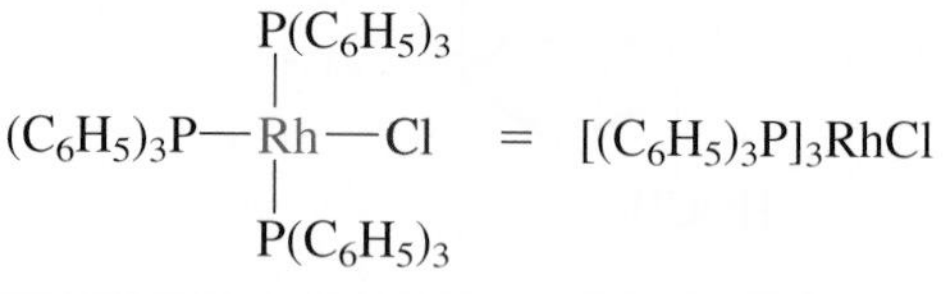

Tris(triphenylphosphine)rhodium chloride

Like rhodium itself, Wilkinson's catalyst is an effective catalyst for alkene hydrogenation. It is selective, reducing less-substituted double bonds faster than more-substituted ones and C═C in preference to C═O. Unlike rhodium metal, however, Wilkinson's catalyst is soluble in many organic solvents.

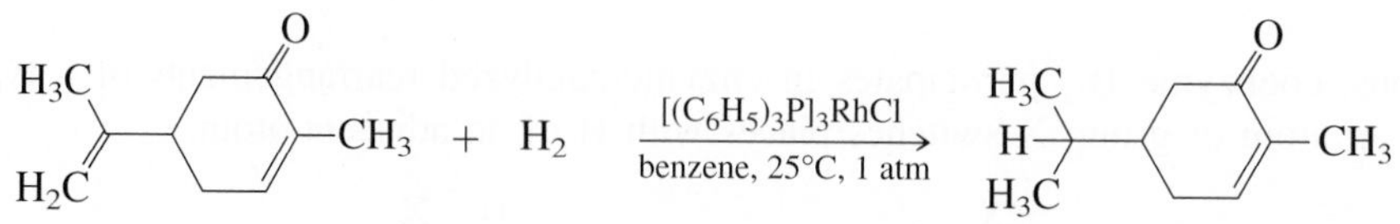

Carvone

Dihydrocarvone (90–94%)

The mechanism of the hydrogenation of propene in the presence of Wilkinson's catalyst is shown in Mechanism 14.3.

Mechanism 14.3

Homogeneous Hydrogenation of Propene in the Presence of Wilkinson's Catalyst

THE OVERALL REACTION:

$$H_2C{=}CHCH_3 \quad + \quad H_2 \xrightarrow{[(C_6H_5)_3P]_3RhCl} CH_3CH_2CH_3$$

Propene Hydrogen Propane

Steps 1 and 2: The catalyst is converted to its active form by reaction with H_2 and loss of triphenylphosphine in separate steps. These two steps are capable of occurring in either order.

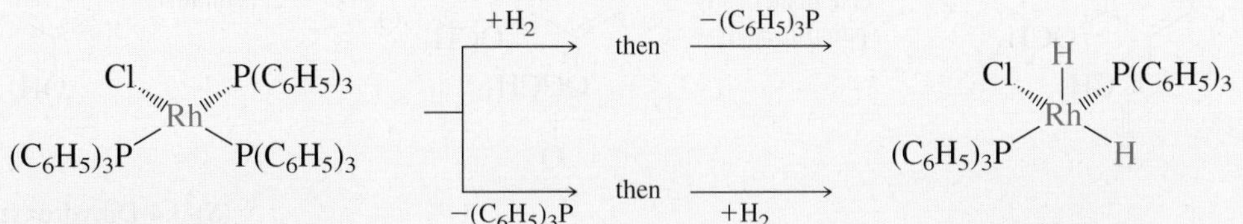

Tris(triphenylphosphine)rhodium chloride

Dihydride of bis(triphenylphosphine) complex

Step 3: Like Wilkinson's catalyst, the product of step 2 is a 16-electron complex of rhodium and is coordinatively unsaturated. It is capable of adding an additional ligand, such as an alkene, to satisfy the 18-electron rule. Propene uses its pair of π electrons to bond to rhodium.

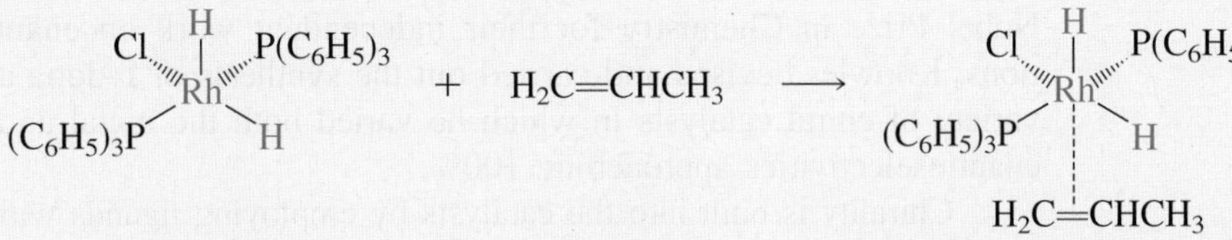

Dihydride of bis(triphenylphosphine) complex Propene Rhodium-propene complex

Step 4: The rhodium-propene complex formed in step 3 rearranges. The π electrons of the alkene ligand are used to form a σ bond between Rh and the less substituted carbon while hydride migrates from Rh to the more substituted carbon.

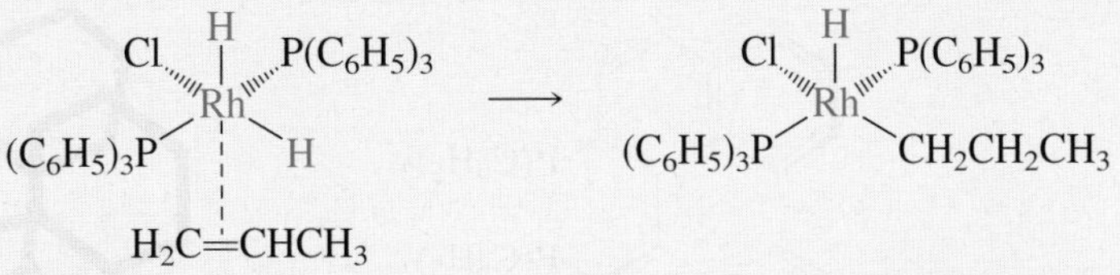

Rhodium–propene complex Propylrhodium complex

Step 5: Hydride migrates from Rh to carbon and the complex dissociates, releasing propane.

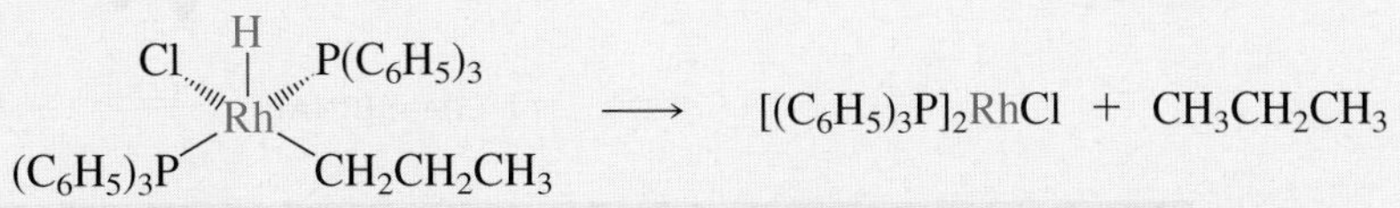

Propylrhodium complex "Bis-complex" Propane

Step 6: The "bis-complex" formed in step 5 reacts with H_2 to give the active form of the catalyst. This is the same substance shown as the product of steps 1 and 2. Steps 3–6 repeat over and over.

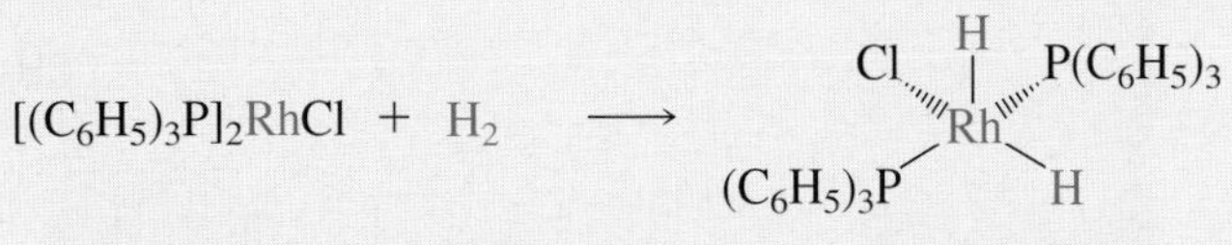

"Bis-complex" Hydrogen Dihydride of bis(triphenylphosphine) complex

The effect that homogeneous transition-metal catalysis has had on stereoselective synthesis is especially impressive. Using chiral ligands, it is possible to control hydrogenation of double bonds so that new chirality centers have a particular configuration. The drug L-dopa, used to treat Parkinsonism, is prepared in multiton quantities by enantioselective hydrogenation catalyzed by an enantiomerically pure chiral rhodium complex.

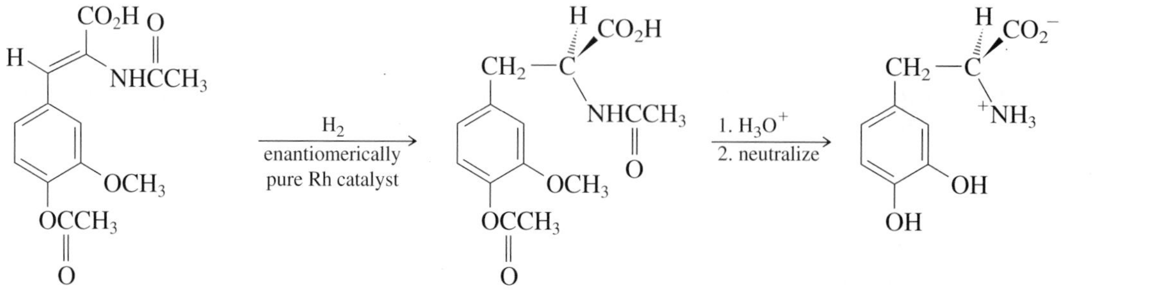

(*S*)-3,4-Dihydroxyphenylalanine (L-dopa)
(100% yield; 95% enantiomeric excess)

The synthesis of L-dopa was one of the earliest of what has become an important advance in the pharmaceutical industry—the preparation and marketing of chiral drugs as single enantiomers (see the boxed essay, *Chiral Drugs,* in Chapter 7). William S. Knowles (Monsanto) and Ryoji Noyori (Nagoya, Japan) shared one half of the 2001 Nobel Prize in Chemistry for their independent work on enantioselective hydrogenations. Knowles devised and carried out the synthesis of L-dopa and Noyori developed a variety of chiral catalysts in which he varied both the metal and the ligands to achieve enantioselectivities approaching 100%.

BINAP is an abbreviation for 2,2′-bis(diphenylphosphino)-1,1′-binaphthyl.

Chirality is built into the catalysts by employing ligands with either chirality centers or axes (Section 7.9). Noyori's widely used BINAP has a chirality axis, and crowding prevents interconversion of enantiomers by restricting rotation around the bond connecting the naphthalene rings. The metal, usually ruthenium, is held in place by the two phosphorus atoms (yellow) in a chiral environment. The steric demands in the cavity occupied by the metal in Ru-BINAP cause reaction to occur preferentially at one face of the double bond.

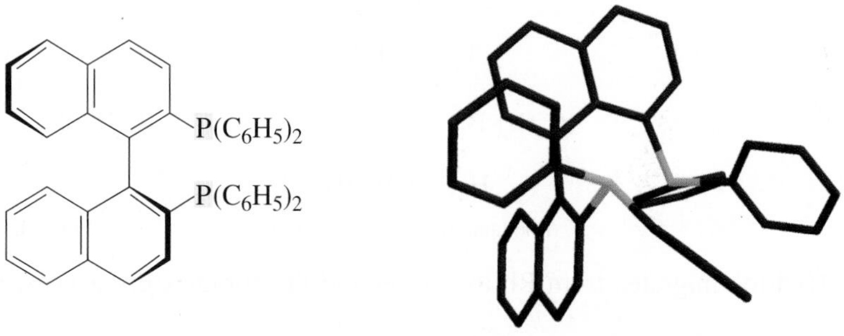

(*S*)-(–)-BINAP

Problem 14.12

The antiinflammatory drug *naproxen* is sold as its (*S*)-enantiomer. One large-scale synthesis uses a Ru-BINAP hydrogenation catalyst. What compound would you hydrogenate to prepare naproxen?

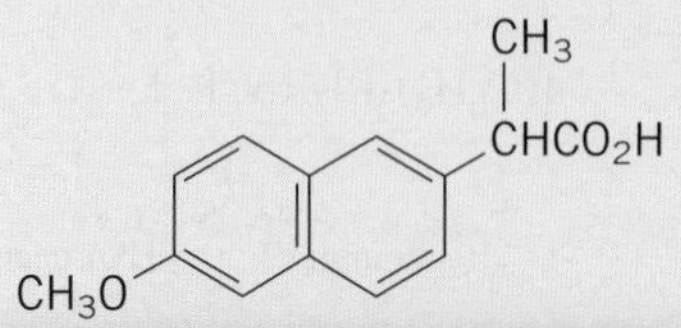

A large number of enantioselective transition-metal catalysts have been developed, not just for hydrogenation but for other reactions as well. The opportunities for fine-tuning their properties by varying the metal, its oxidation state, and the ligands are almost limitless.

14.15 Olefin Metathesis

The 2005 Nobel Prize in Chemistry was jointly awarded to Robert H. Grubbs (Caltech), Yves Chauvin (French Petroleum Institute), and Richard R. Schrock (MIT) for establishing **olefin metathesis** as a reaction of synthetic versatility and contributing to an understanding of the mechanism of this novel process. Olefin metathesis first surfaced in the late 1950s when industrial researchers found that alkenes underwent a novel reaction when passed over a heated bed of mixed metal oxides. Propene, for example, was converted to a mixture of ethylene and 2-butene (cis + trans).

The word "metathesis" refers to an interchange, or transposition, of objects.

$$\underset{\text{Propene}}{2CH_3CH{=}CH_2} \xrightleftharpoons{\text{catalyst}} \underset{\text{Ethylene}}{H_2C{=}CH_2} + \underset{\textit{cis-} + \textit{trans-}\text{2-butene}}{CH_3CH{=}CHCH_3}$$

This same transformation was subsequently duplicated at lower temperatures by homogeneous transition-metal catalysis. An equilibrium is established, and the same mixture is obtained regardless of whether propene or a 1:1 mixture of ethylene and 2-butene is subjected to the reaction conditions. This type of olefin metathesis is called a *cross-metathesis.*

When cross-metathesis was first discovered, propene enjoyed only limited use and the reaction was viewed as a potential source of ethylene. Once methods were developed for the preparation of stereoregular polypropylene, however, propene became more valuable and cross-metathesis of ethylene and 2-butene now serves as a source of propene.

The relationship between reactants and products in cross-metathesis can be analyzed retrosynthetically by joining the double bonds in two reactant molecules by dotted lines, then disconnecting in the other direction.

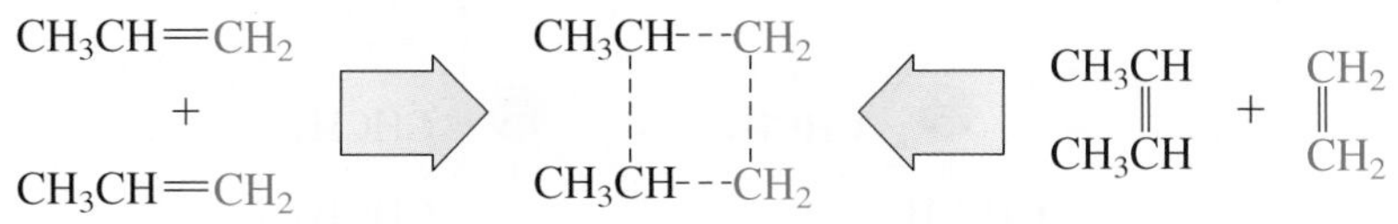

Two propene molecules

2-Butene + Ethylene

Although this representation helps us relate products and reactants, *it is not related to the mechanism. Nothing containing a ring of four carbons is an intermediate in olefin cross-metathesis.*

Problem 14.13

What alkenes are formed from 2-pentene by olefin cross-metathesis?

The generally accepted mechanism for olefin cross-metathesis is outlined for the case of propene in Mechanism 14.4. The catalyst belongs to a class of organometallics known as a *metallocarbene, carbene complex,* or *alkylidene complex.* Its structure is characterized by a carbon–metal double bond. In olefin metathesis the metal is typically ruthenium (Ru), tungsten (W), or molybdenum (Mo). Transition-metal carbene complexes were first prepared by Ernst O. Fischer (Munich) who shared the 1973 Nobel Prize in Chemistry with Geoffrey Wilkinson.

Mechanism 14.4

Olefin Cross-Metathesis

THE OVERALL REACTION:

$$2CH_3CH{=}CH_2 \xrightleftharpoons{\text{catalyst}} H_2C{=}CH_2 + CH_3CH{=}CHCH_3$$

Propene — Ethylene — 2-Butene (cis + trans)

THE MECHANISM:

To simplify the presentation of the mechanism, the symbol (M) stands for the transition metal and its ligands. Steps have been omitted in which ligands leave or become attached to the metal; therefore, the number of ligands is not necessarily the same throughout a stage.

Step 1: In this stage the sp^2-hybridized carbons of the alkene, with their attached groups, replace the benzylidene group of the catalyst. In the case of an unsymmetrical alkene such as propene, the two newly formed carbene complexes (A and B) are different.

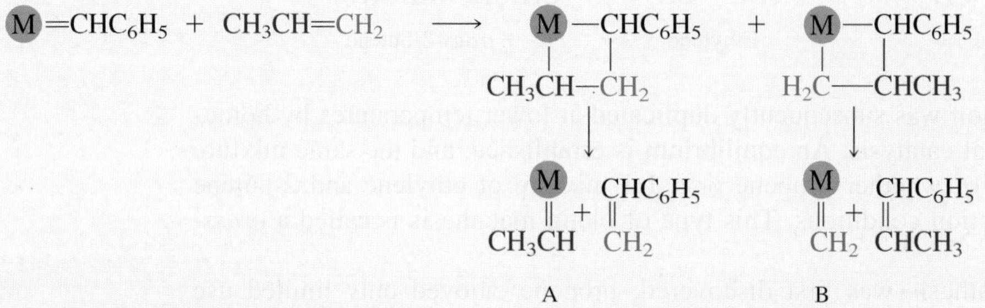
M=CHC6H5 + CH3CH=CH2
M—CHC6H5
CH3CH—CH2
M—CHC6H5
H2C—CHCH3
A
B

Step 2: *Complex A:* Propene adds to the double bond of the carbene complex to give metallocyclobutanes C and D. Dissociation of C gives propene + A. Dissociation of D gives 2-butene + B.

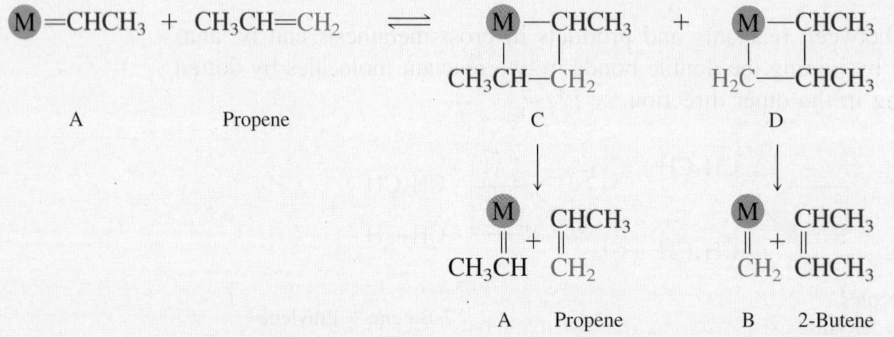
M=CHCH3 + CH3CH=CH2
A
Propene
C
D
A
Propene
B
2-Butene

Complex B: Propene adds to the double bond of B to give metallocyclobutanes E and F. Dissociation of E gives ethylene + A. Dissociation of F gives propene + B.

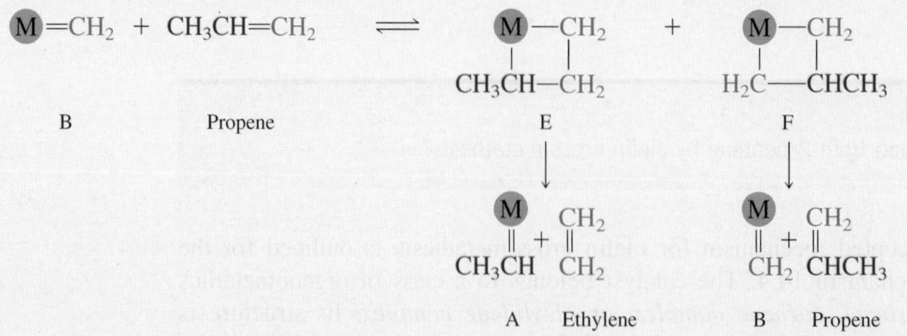
M=CH2 + CH3CH=CH2
B
Propene
E
F
A
Ethylene
B
Propene

Step 3: The two complexes A and B that react in stage 2 are also regenerated in the same stage. Thus, stage 3 is simply a repeat of stage 2 and the process continues.

One of the most widely used catalysts for olefin metathesis is the ruthenium complex shown. It is called *Grubb's catalyst* and abbreviated $Cl_2(PCy_3)_2Ru{=}CHC_6H_5$.

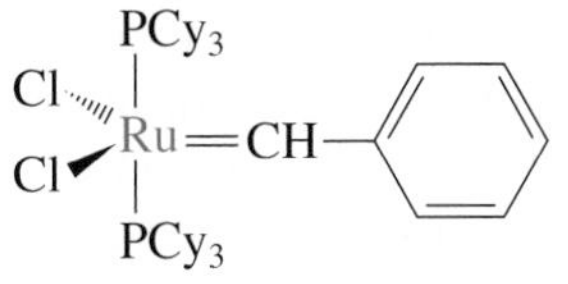

Cy = cyclohexyl

Grubb's catalyst

Olefin cross-metathesis is an intermolecular reaction between double bonds in separate molecules. Intramolecular metatheses in which two double bonds belong to the same molecule are also common and lead to ring formation. The process is called *ring-closing metathesis.*

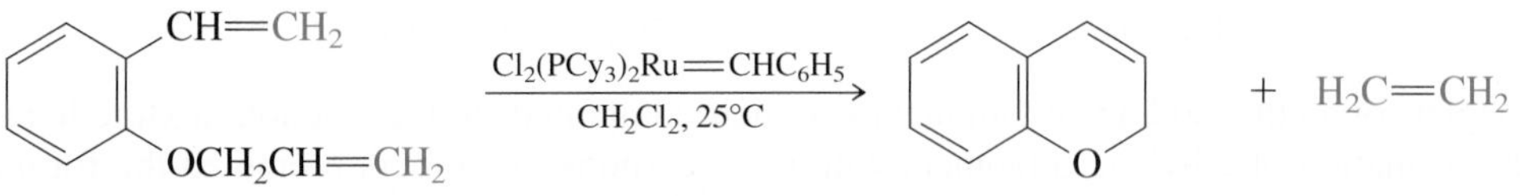

Allyl *o*-vinylphenyl ether 2H-1-Benzopyran (95%) Ethylene

Although olefin metathesis is an equilibrium process, it can give high yields of the desired product when ethylene is formed as the other alkene. Being a gas, ethylene escapes from the reaction mixture, and the equilibrium shifts to the right in accordance with Le Châtelier's principle. Ring-closing metathesis has been widely and imaginatively applied to the synthesis of natural products. It occurs under mild conditions and tolerates the presence of numerous functional groups.

Problem 14.14

The product of the following reaction was isolated in 99% yield. What is it?

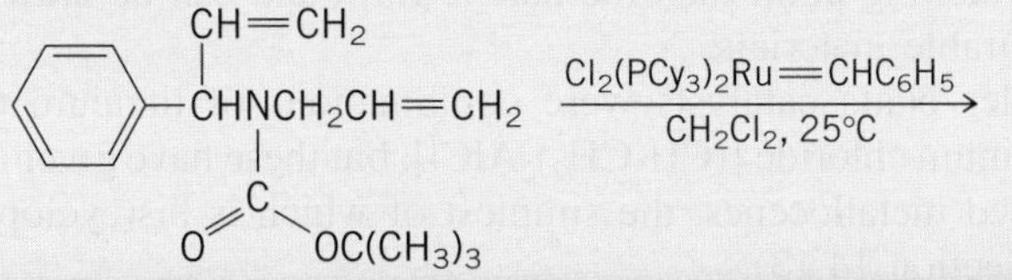

Ring-opening metathesis is the converse of ring-closing metathesis and holds promise as a polymerization method. It is applied most often when ring opening is accompanied by relief of strain as in, for example, bicyclic alkenes.

Norbornene is a common name for bicyclo[2.2.1]hept-2-ene

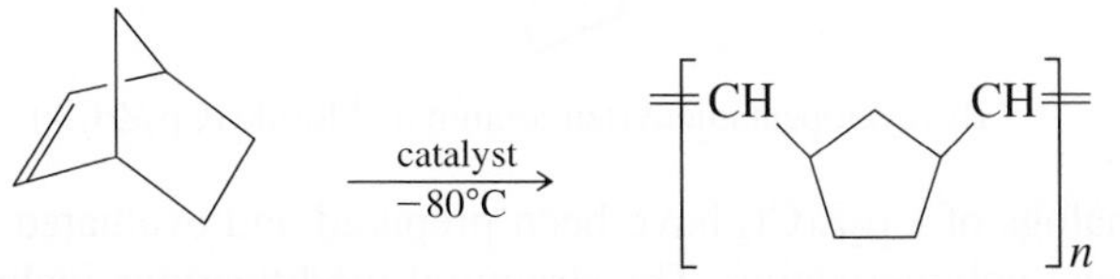

Bicyclo[2.2.1]hept-2-ene Polynorbornene

The catalyst for this and many other *ring-opening metathesis polymerizations* is a carbene complex of tungsten (W).

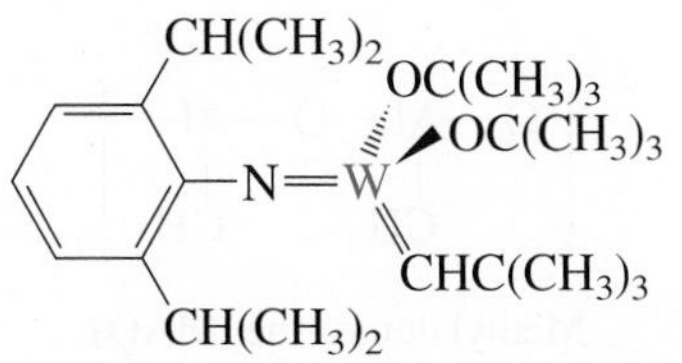

14.16 Ziegler–Natta Catalysis of Alkene Polymerization

In Section 6.21 we listed three main methods for polymerizing alkenes: cationic, free-radical, and coordination polymerization. In Section 7.16 we extended our knowledge of polymers to their stereochemical aspects by noting that although free-radical polymerization of propene gives atactic polypropylene, coordination polymerization produces a stereoregular polymer with superior physical properties. Because the catalysts responsible for coordination polymerization are organometallic compounds, we are now in a position to examine coordination polymerization in more detail, especially with respect to how the catalyst works.

In the early 1950s, Karl Ziegler, then at the Max Planck Institute for Coal Research in Germany, was studying the use of aluminum compounds as catalysts for the oligomerization of ethylene.

$$\underset{\text{Ethylene}}{nH_2C{=}CH_2} \xrightarrow{Al(CH_2CH_3)_3} \underset{\text{Ethylene oligomers}}{CH_3CH_2(CH_2CH_2)_{n-2}CH{=}CH_2}$$

Ziegler found that adding certain metals or their compounds to the reaction mixture led to the formation of ethylene oligomers with 6–18 carbons, but others promoted the formation of very long carbon chains giving polyethylene. Both were major discoveries. The 6–18 carbon ethylene oligomers constitute a class of industrial organic chemicals known as *linear α olefins* that are produced at a rate of 3×10^9 pounds/year in the United States. The Ziegler route to polyethylene is even more important because it occurs at modest temperatures and pressures and gives *high-density polyethylene,* which has properties superior to the low-density material formed by the free-radical polymerization described in Section 6.21.

Ziegler had a working relationship with the Italian chemical company Montecatini, for which Giulio Natta of the Milan Polytechnic Institute was a consultant. When Natta used Ziegler's catalyst to polymerize propene, he discovered that the catalyst was not only effective but that it gave mainly isotactic polypropylene. (Recall from Section 7.16 that free-radical polymerization of propene gives atactic polypropylene.) Isotactic polypropylene has a higher melting point than the atactic form and can be drawn into fibers or molded into hard, durable materials.

The earliest Ziegler–Natta catalysts were combinations of titanium tetrachloride ($TiCl_4$) and diethylaluminum chloride [$(CH_3CH_2)_2AlCl$], but these have given way to more effective zirconium-based metallocenes, the simplest of which is bis(cyclopentadienyl)-zirconium dichloride (Section 14.13).

Zirconium lies below titanium in the periodic table, so was an obvious choice in the search for other Ziegler–Natta catalysts.

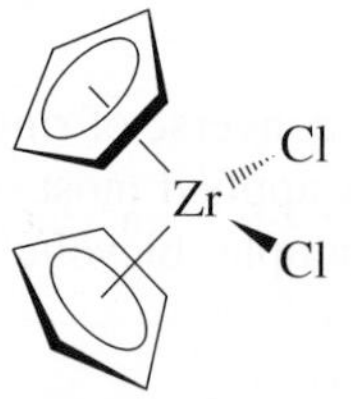

Bis(cyclopentadienyl)zirconium dichloride (Cp_2ZrCl_2)

Hundreds of analogs of Cp_2ZrCl_2 have been prepared and evaluated as catalysts for ethylene and propene polymerization. The structural modifications include replacing one or both of the cyclopentadienyl ligands by variously substituted cyclopentadienyl groups, linking the two rings with carbon chains, and so on. Some modifications give syndiotactic polypropylene, others give isotactic.

The metallocene catalyst is used in combination with a promoter, usually methylalumoxane (MAO).

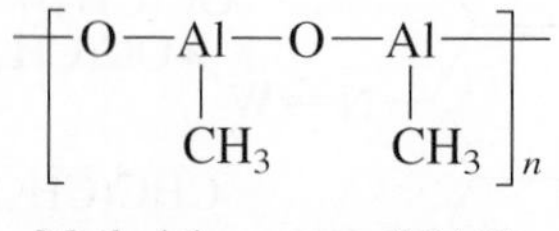

Methylalumoxane (MAO)

Mechanism 14.5

Polymerization of Ethylene in the Presence of Ziegler–Natta Catalyst

Step 1: Cp_2ZrCl_2 is converted to the active catalyst by reaction with the promoter methylalumoxane (MAO). A methyl group from MAO displaces one of the chlorine ligands of Cp_2ZrCl_2. The second chlorine is lost as chloride by ionization, giving a positively charged metallocene.

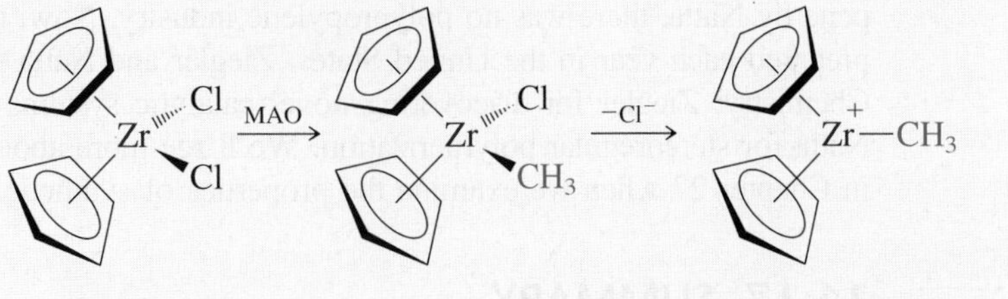

Cp_2ZrCl_2 Active form of catalyst

Step 2: Ethylene reacts with the active form of the catalyst. The two π electrons of ethylene are used to bind it as a ligand to zirconium.

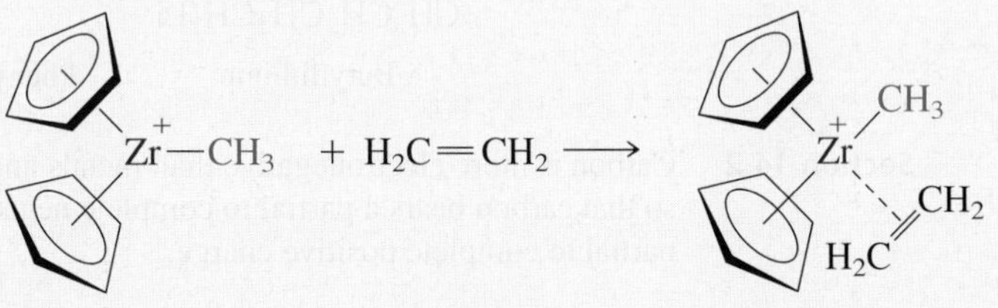

Active form of catalyst Ethylene Ethylene–catalyst complex

Step 3: The methyl group migrates from zirconium to one of the carbons of the ethylene ligand. At the same time, the π electrons of the ethylene ligand are used to form a σ bond between the other carbon and zirconium.

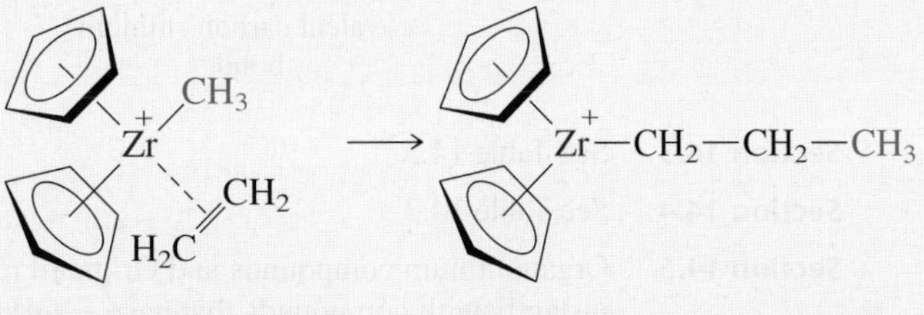

Ethylene–catalyst complex Chain-extended form of catalyst

Step 4: The catalyst now has a propyl group on zirconium instead of a methyl group. Repeating steps 2 and 3 converts the propyl group to a pentyl group, then a heptyl group, and so on. After thousands of repetitions, polyethylene results.

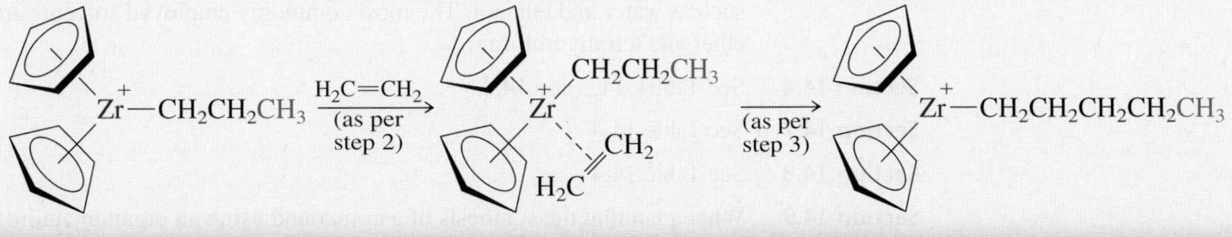

Mechanism 14.5 outlines ethylene polymerization in the presence of Cp_2ZrCl_2. Step 1 describes the purpose of the MAO promoter, which is to transfer a methyl group to the metallocene to convert it to its catalytically active form. This methyl group will be incorporated into the growing polymer chain—indeed, it will be the end from which the rest of the chain grows.

The active form of the catalyst, having one less ligand and being positively charged, acts as an electrophile toward ethylene in step 2.

With electrons flowing from ethylene to zirconium, the Zr—CH_3 bond weakens, the carbons of ethylene become positively polarized, and the methyl group migrates from zirconium to one of the carbons of ethylene. Cleavage of the Zr—CH_3 bond is accompanied by formation of a σ bond between zirconium and one of the carbons of ethylene in step 3. The product of this step is a chain-extended form of the active catalyst, ready to accept another ethylene ligand and repeat the chain-extending steps.

Before coordination polymerization was discovered by Ziegler and applied to propene by Natta, there was no polypropylene industry. Now, more than 10^{10} pounds of it are prepared each year in the United States. Ziegler and Natta shared the 1963 Nobel Prize in Chemistry: Ziegler for discovering novel catalytic systems for alkene polymerization and Natta for stereoregular polymerization. We'll see more about Ziegler–Natta polymerization in Chapter 27 when we examine the properties of synthetic polymers in more detail.

14.17 SUMMARY

Section 14.1 Organometallic compounds contain a carbon–metal bond. They are named as alkyl (or aryl) derivatives of metals.

$CH_3CH_2CH_2CH_2Li$ — Butyllithium

C_6H_5MgBr — Phenylmagnesium bromide

Section 14.2 Carbon is more electronegative than metals and carbon–metal bonds are polarized so that carbon bears a partial to complete negative charge and the metal bears a partial to complete positive charge.

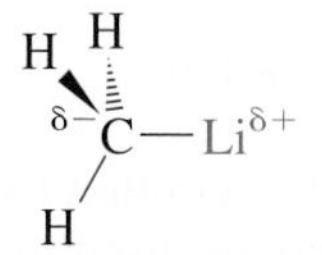

Methyllithium has a polar covalent carbon–lithium bond.

$HC{\equiv}\bar{C}{:}\ Na^+$

Sodium acetylide has an ionic bond between carbon and sodium.

Section 14.3 See Table 14.3

Section 14.4 See Table 14.3

Section 14.5 Organolithium compounds and Grignard reagents are strong bases and react instantly with compounds that have —OH groups.

$$R{-}M + H{-}O{-}R' \longrightarrow R{-}H + M^+ \ {}^-O{-}R'$$

These organometallic compounds cannot therefore be formed or used in solvents such as water and ethanol. The most commonly employed solvents are diethyl ether and tetrahydrofuran.

Section 14.6 See Tables 14.2 and 14.4

Section 14.7 See Table 14.4

Section 14.8 See Table 14.4

Section 14.9 When planning the synthesis of a compound using an organometallic reagent, or indeed any synthesis, the best approach is to reason backward from the product. This method is called **retrosynthetic analysis.** Retrosynthetic analysis of 1-methylcyclohexanol suggests it can be prepared by the reaction of methylmagnesium bromide and cyclohexanone.

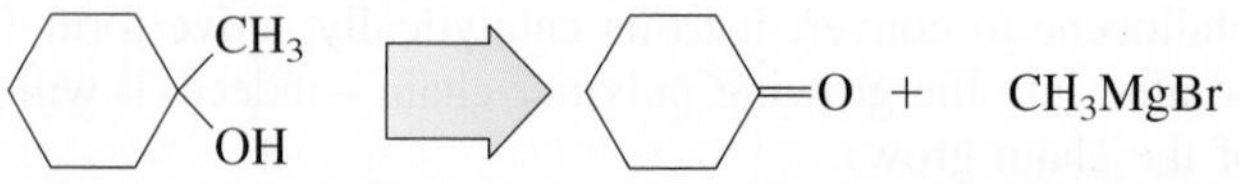

1-Methylcyclohexanol   Cyclohexanone   Methylmagnesium bromide

TABLE 14.3 Preparation of Organometallic Reagents Used in Synthesis

Type of organometallic reagent (section) and comments	General equation for preparation and specific example
Organolithium reagents (Section 14.3) Lithium metal reacts with organic halides to produce organolithium compounds. The organic halide may be alkyl, alkenyl, or aryl. Iodides react most and fluorides least readily; bromides are used most often. Suitable solvents include hexane, diethyl ether, and tetrahydrofuran.	$RX + 2Li \longrightarrow RLi + LiX$ Alkyl halide, Lithium, Alkyllithium, Lithium halide $CH_3CH_2CH_2Br \xrightarrow[\text{diethyl ether}]{Li} CH_3CH_2CH_2Li$ Propyl bromide → Propyllithium (78%)
Grignard reagents (Section 14.4) Grignard reagents are prepared in a manner similar to that used for organolithium compounds. Diethyl ether and tetrahydrofuran are appropriate solvents.	$RX + Mg \longrightarrow RMgX$ Alkyl halide, Magnesium, Alkylmagnesium halide (Grignard reagent) $C_6H_5CH_2Cl \xrightarrow[\text{diethyl ether}]{Mg} C_6H_5CH_2MgCl$ Benzyl chloride → Benzylmagnesium chloride (93%)
Lithium dialkylcuprates (Section 14.10) These reagents contain a negatively charged copper atom and are formed by the reaction of a copper(I) salt with two equivalents of an organolithium reagent.	$2RLi + CuX \longrightarrow R_2CuLi + LiX$ Alkyllithium, Copper(I) halide, Lithium dialkylcuprate, Lithium halide $2CH_3Li + CuI \xrightarrow{\text{diethyl ether}} (CH_3)_2CuLi + LiI$ Methyllithium, Copper(I) iodide, Lithium dimethylcuprate, Lithium iodide
Iodomethylzinc iodide (Section 14.11) This is the Simmons–Smith reagent. It is prepared by the reaction of zinc (usually in the presence of copper) with diiodomethane.	$CH_2I_2 + Zn \xrightarrow[Cu]{\text{diethyl ether}} ICH_2ZnI$ Diiodomethane, Zinc, Iodomethylzinc iodide

Section 14.10 See Tables 14.3 and 14.4

Section 14.11 See Tables 14.3 and 14.4

Section 14.12 Carbenes are species that contain a *divalent carbon;* that is, a carbon with only two bonds. One of the characteristic reactions of carbenes is with alkenes to give cyclopropane derivatives.

Certain organometallic compounds resemble carbenes in their reactions and are referred to as **carbenoids.** Iodomethylzinc iodide (Section 14.11) is an example.

TABLE 14.4 Carbon–Carbon Bond-Forming Reactions of Organometallic Reagents

Reaction (section) and comments	General equation and specific example
Alcohol synthesis via the reaction of Grignard reagents with carbonyl compounds (Section 14.6) This is one of the most useful reactions in synthetic organic chemistry. Grignard reagents react with formaldehyde to yield primary alcohols, with aldehydes to give secondary alcohols, and with ketones to form tertiary alcohols.	$RMgX + R'\overset{O}{\overset{\Vert}{C}}R'' \xrightarrow[2.\ H_3O^+]{1.\ \text{diethyl ether}} R\overset{R'}{\underset{R''}{C}}OH$ Grignard reagent, Aldehyde or ketone, Alcohol $CH_3MgI + CH_3CH_2CH_2\overset{O}{\overset{\Vert}{C}}H \xrightarrow[2.\ H_3O^+]{1.\ \text{diethyl ether}} CH_3CH_2CH_2\underset{OH}{CH}CH_3$ Methylmagnesium iodide, Butanal, 2-Pentanol (82%)

Continued

TABLE 14.4 Carbon–Carbon Bond-Forming Reactions of Organometallic Reagents (*Continued*)

Reaction (section) and comments	General equation and specific example
Synthesis of alcohols using organolithium reagents (Section 14.7) Organolithium reagents react with aldehydes and ketones in a manner similar to that of Grignard reagents to produce alcohols.	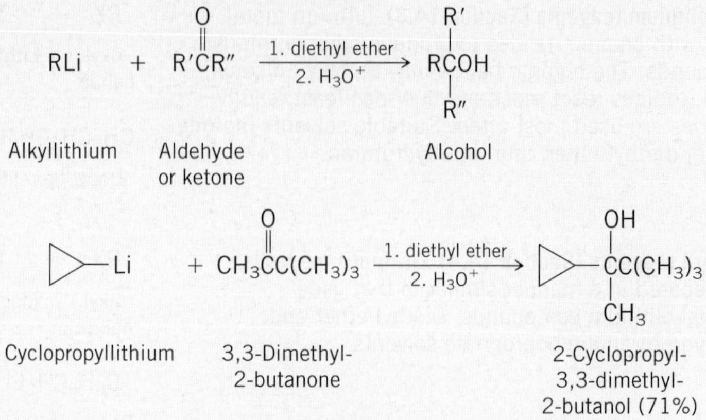
Synthesis of acetylenic alcohols (Section 14.8) Sodium acetylide and acetylenic Grignard reagents react with aldehydes and ketones to give alcohols of the type C≡C—C—OH.	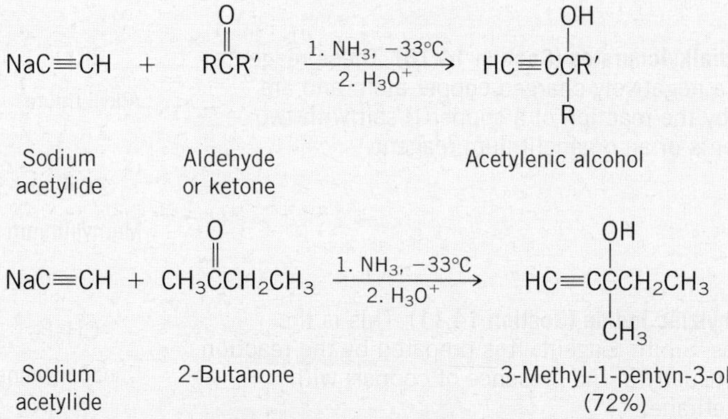
Preparation of alkanes using lithium dialkylcuprates (Section 14.10) Two alkyl groups may be coupled together to form an alkane by the reaction of an alkyl halide with a lithium dialkylcuprate. Both alkyl groups must be primary (or methyl). Aryl and vinyl halides may be used in place of alkyl halides.	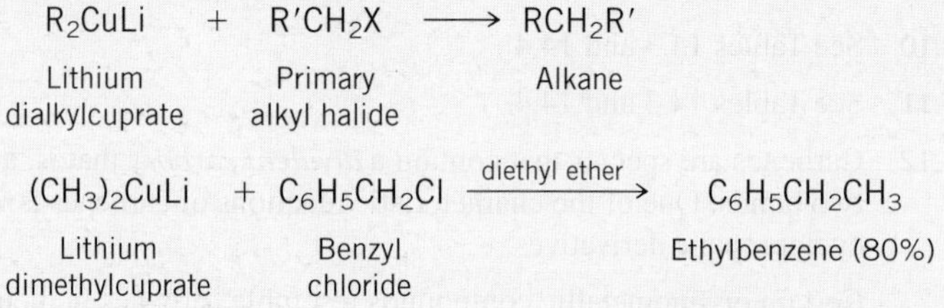
The Simmons–Smith reaction (Section 14.11) Methylene transfer from iodomethylzinc iodide converts alkenes to cyclopropanes. The reaction is a stereospecific syn addition of a CH_2 group to the double bond.	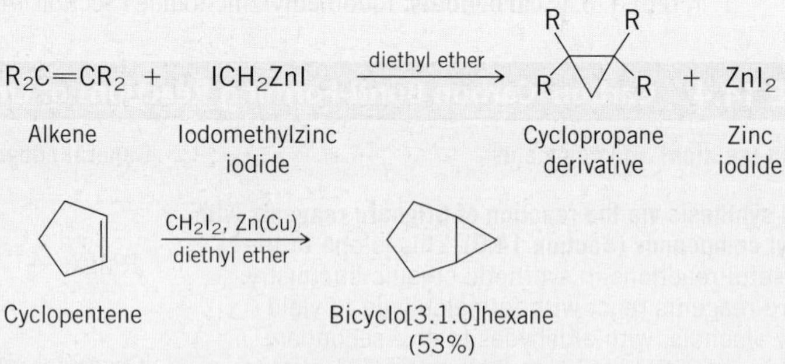

Section 14.13 Transition-metal complexes that contain one or more organic ligands offer a rich variety of structural types and reactivity. Organic ligands can be bonded to a metal by a σ bond or through its π system. **Metallocenes** are transition-metal complexes in which one or more of the ligands is a cyclopentadienyl ring. Ferrocene was the first metallocene synthesized; its electrostatic potential map opens this chapter.

Section 14.14 Organometallic compounds based on transition metals, especially rhodium and ruthenium, can catalyze the hydrogenation of alkenes under homogeneous conditions.

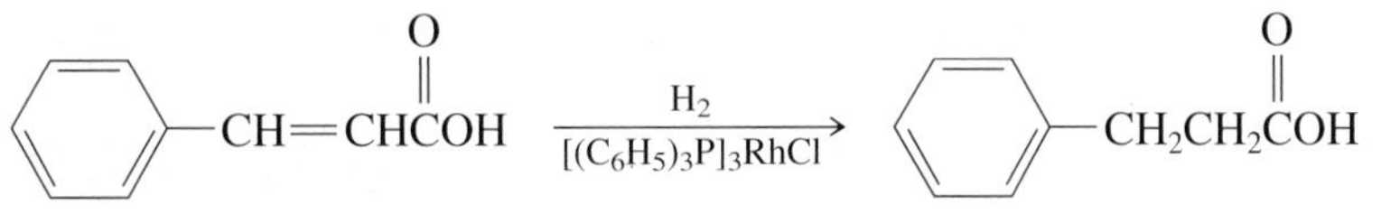

Cinnamic acid   3-Phenylpropanoic acid (90%)

When a single enantiomer of a chiral catalyst is used, hydrogenations can be carried out with high enantioselectivity.

Section 14.15 The doubly bonded carbons of two alkenes exchange partners on treatment with transition-metal carbene complexes, especially those derived from ruthenium and tungsten.

$$2R_2C{=}CR'_2 \xrightarrow[\text{catalyst}]{\text{Metallocarbene}} R_2C{=}CR_2 + R'_2C{=}CR'_2$$

Among other applications **olefin metathesis** is useful in the synthesis of cyclic alkenes, the industrial preparation of propene, and in polymerization.

Section 14.16 Coordination polymerization of ethylene and propene has the biggest economic impact of any organic chemical process. Ziegler–Natta polymerization is carried out using catalysts derived from transition metals such as titanium and zirconium. π-Bonded and σ-bonded organometallic compounds are intermediates in coordination polymerization.

PROBLEMS

14.15 Write structural formulas for each of the following compounds. Specify which compounds qualify as organometallic compounds.

(a) Cyclopentyllithium
(b) Ethoxymagnesium chloride
(c) 2-Phenylethylmagnesium iodide
(d) Lithium divinylcuprate
(e) Sodium carbonate
(f) Benzylpotassium
(g) Lithium diisopropylamide

14.16 Suggest appropriate methods for preparing each of the following compounds from the starting material of your choice.

(a) $CH_3CH_2CH_2CH_2CH_2MgI$
(b) $CH_3CH_2C{\equiv}CMgI$
(c) $CH_3CH_2CH_2CH_2CH_2Li$
(d) $(CH_3CH_2CH_2CH_2CH_2)_2CuLi$

14.17 Which compound in each of the following pairs would you expect to have the more polar carbon–metal bond?

(a) CH_3CH_2Li or $(CH_3CH_2)_3Al$
(b) $(CH_3)_2Zn$ or $(CH_3)_2Mg$
(c) CH_3CH_2MgBr or $HC{\equiv}CMgBr$

14.18 Write the structure of the principal organic product of each of the following reactions:

(a) 1-Bromopropane with lithium in diethyl ether
(b) 1-Bromopropane with magnesium in diethyl ether
(c) 2-Iodopropane with lithium in diethyl ether
(d) 2-Iodopropane with magnesium in diethyl ether
(e) Product of part (a) with copper(I) iodide
(f) Product of part (e) with 1-bromobutane
(g) Product of part (e) with iodobenzene
(h) Product of part (b) with D_2O and DCl
(i) Product of part (c) with D_2O and DCl
(j) Product of part (a) with formaldehyde in diethyl ether, followed by dilute acid
(k) Product of part (b) with benzaldehyde in diethyl ether, followed by dilute acid

(l) Product of part (c) with cycloheptanone in diethyl ether, followed by dilute acid

(m) Product of part (d) with $CH_3\overset{O}{\overset{\|}{C}}CH_2CH_3$ in diethyl ether, followed by dilute acid

(n) 1-Octene with diiodomethane and zinc–copper couple in diethyl ether

(o) (*E*)-2-Decene with diiodomethane and zinc–copper couple in diethyl ether

(p) (*Z*)-3-Decene with diiodomethane and zinc–copper couple in diethyl ether

14.19 Using 1-bromobutane and any necessary organic or inorganic reagents, suggest efficient syntheses of each of the following alcohols:

(a) 1-Pentanol
(b) 2-Hexanol
(c) 1-Phenyl-1-pentanol
(d) 1-Butylcyclobutanol

14.20 Using bromobenzene and any necessary organic or inorganic reagents, suggest efficient syntheses of each of the following:

(a) Benzyl alcohol
(b) 1-Phenyl-1-hexanol
(c) Bromodiphenylmethane
(d) 4-Phenyl-4-heptanol
(e) 1-Phenylcyclooctanol
(f) *trans*-2-Phenylcyclooctanol

14.21 Analyze the following structures so as to determine all the practical combinations of Grignard reagent and carbonyl compound that will give rise to each:

(a) $CH_3CH_2\underset{OH}{\underset{|}{C}}HCH_2CH(CH_3)_2$

(b) 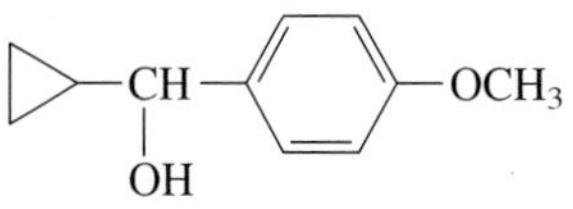
CH
OH
OCH3

(c) $(CH_3)_3CCH_2OH$

(d) 6-Methyl-5-hepten-2-ol

(e) 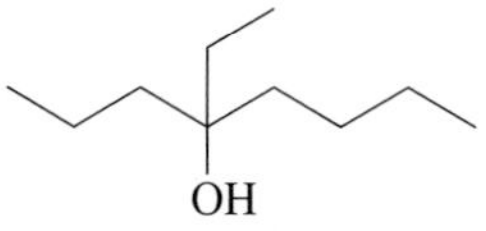
OH

14.22 A number of drugs are prepared by reactions in which carbon–carbon bond formation is the last step. Indicate what you believe would be a reasonable last step in the synthesis of each of the following:

(a) $CH_3CH_2\overset{OH}{\overset{|}{\underset{CH_3}{\underset{|}{C}}}}C{\equiv}CH$ *Meparfynol,* a mild hypnotic or sleep-inducing agent

(b) 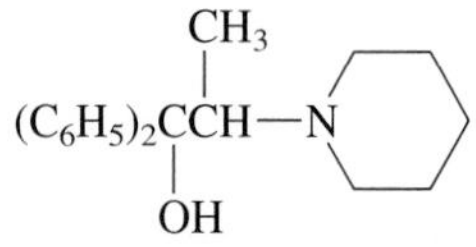
CH3
(C6H5)2CCH—N
OH
Diphepanol, an antitussive (cough suppressant)

(c) 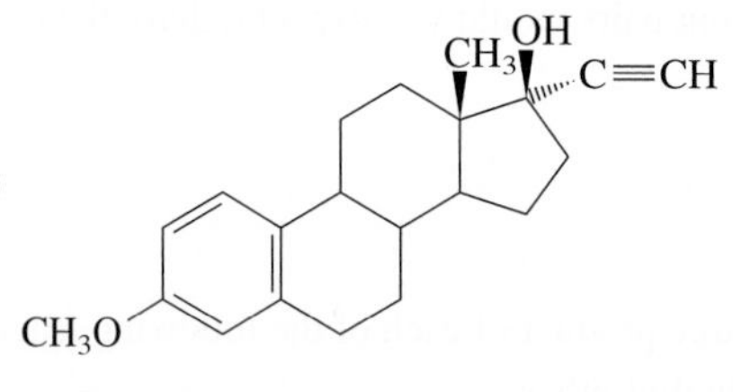
CH3
OH
C≡CH
CH3O
Mestranol, an estrogenic component of oral contraceptive drugs

14.23 Sometimes the strongly basic properties of Grignard reagents can be turned to synthetic advantage. A chemist needed samples of butane specifically labeled with deuterium, the mass-2 isotope of hydrogen, as shown:

(a) $CH_3CH_2CH_2CH_2D$
(b) $CH_3CHDCH_2CH_3$

Suggest methods for the preparation of each of these using D_2O as the source of deuterium, butanols of your choice, and any necessary organic or inorganic reagents.

14.24 Predict the principal organic product of each of the following reactions:

(a) 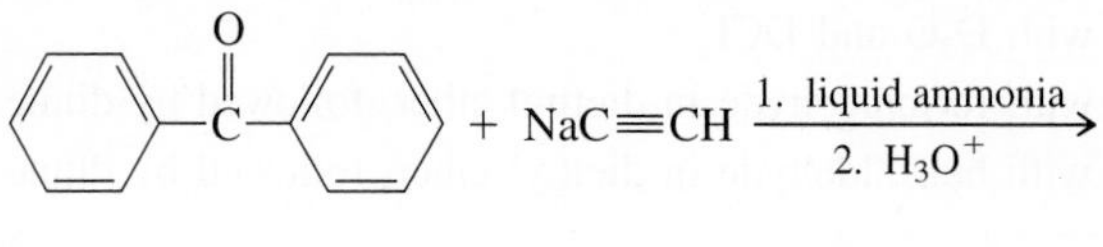
O
C
+ NaC≡CH
1. liquid ammonia
2. H3O+

(b) 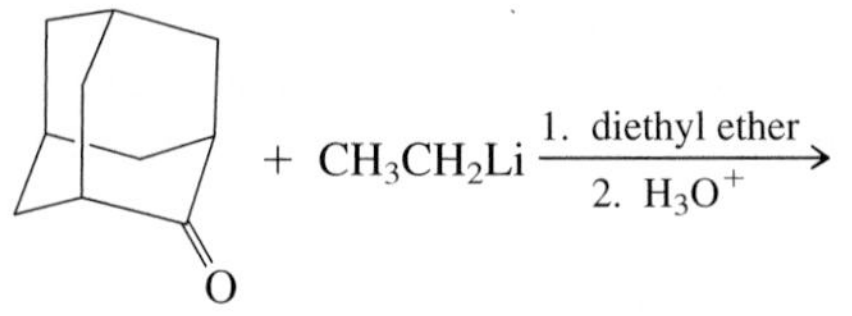
+ CH3CH2Li
1. diethyl ether
2. H3O+
O

(c) 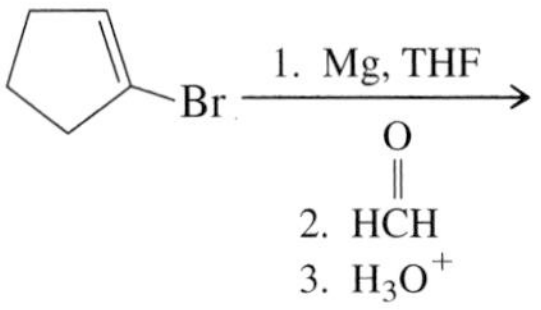
Br
1. Mg, THF
O
2. HCH
3. H3O+

(d) 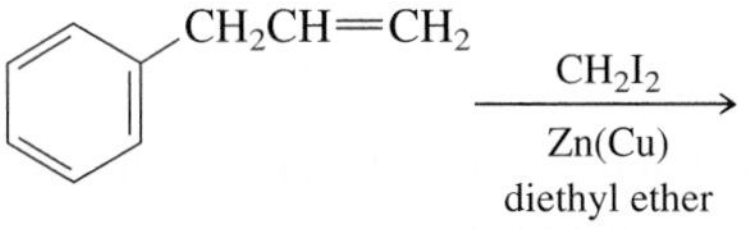
CH2CH=CH2
CH2I2
Zn(Cu)
diethyl ether

(e) 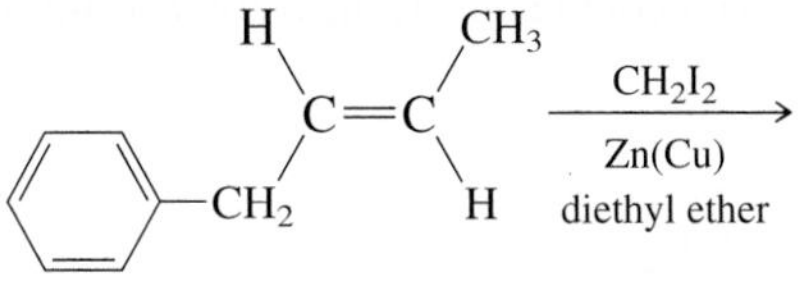
H
CH3
C=C
CH2
H
CH2I2
Zn(Cu)
diethyl ether

(f) 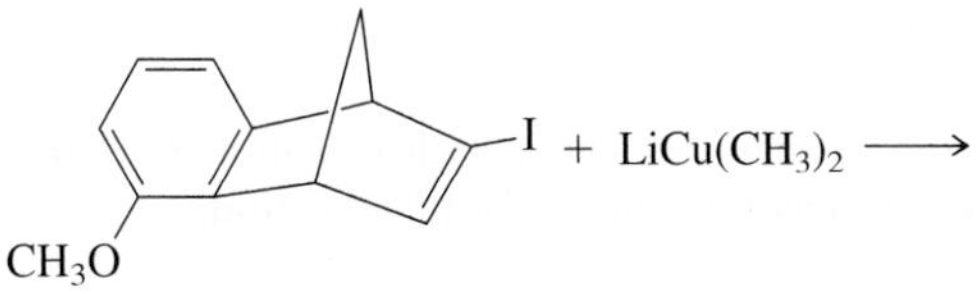
I
+ LiCu(CH3)2
CH3O

(g) 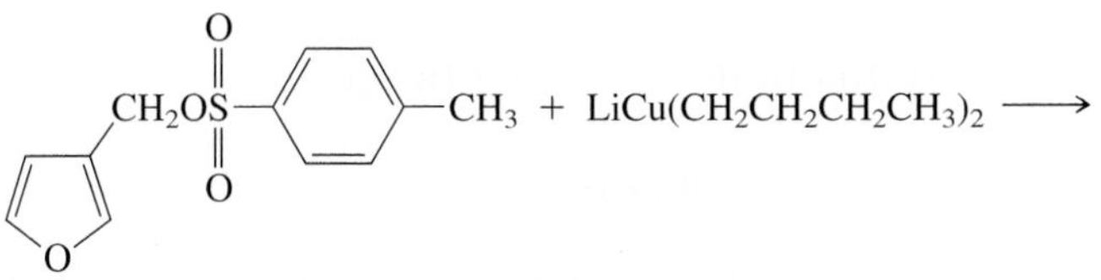
O
CH2OS
O
CH3 + LiCu(CH2CH2CH2CH3)2
O

14.25 Addition of phenylmagnesium bromide to 4-*tert*-butylcyclohexanone gives two isomeric tertiary alcohols as products. Both alcohols yield the same alkene when subjected to acid-catalyzed dehydration. Suggest reasonable structures for these two alcohols.

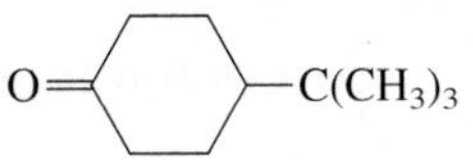
O
C(CH3)3

4-*tert*-Butylcyclohexanone

14.26 Reaction of lithium diphenylcuprate with optically active 2-bromobutane yields 2-phenylbutane, with high net inversion of configuration. When the 2-bromobutane used has the absolute configuration shown, will the 2-phenylbutane formed have the *R* or the *S* configuration?

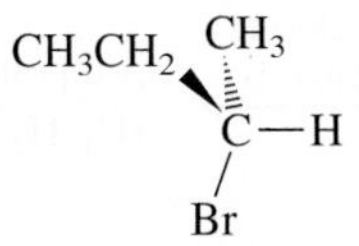
CH3CH2
CH3
C—H
Br

14.27 Suggest reasonable structures for compounds A, B, and C in the following reactions:

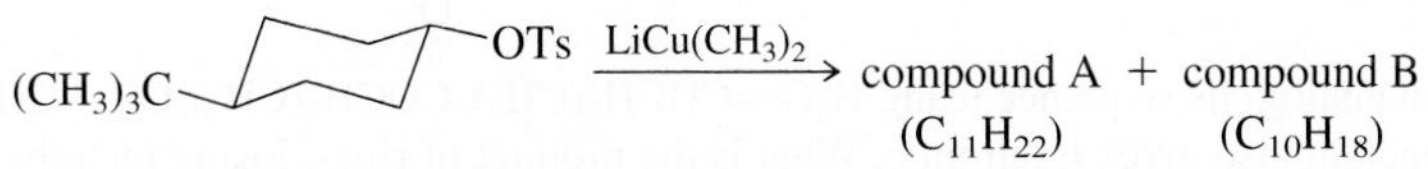
(CH3)3C
OTs
LiCu(CH3)2
compound A + compound B
(C11H22)
(C10H18)

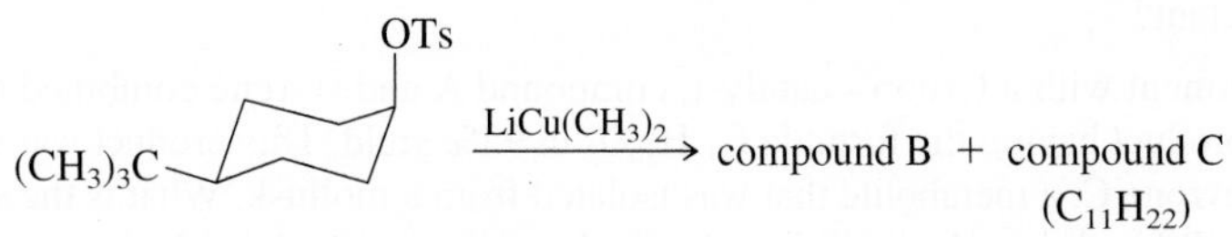
OTs
(CH3)3C
LiCu(CH3)2
compound B + compound C
(C11H22)

Compound C is more stable than compound A.

14.28 Isonitriles are stable, often naturally occurring, compounds that contain a divalent carbon. An example is axisonitrile-3, which can be isolated from a species of sponge and possesses antimalarial activity. Write a resonance form for axisonitrile-3 that satisfies the octet rule. Don't forget to include formal charges.

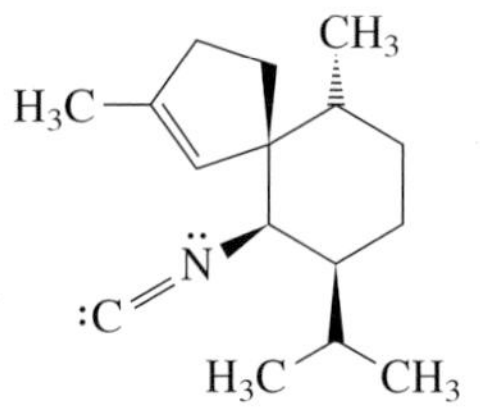

Axisonitrile-3

14.29 The following conversion has been reported in the chemical literature. It was carried out in two steps, the first of which involved formation of a *p*-toluenesulfonate. Indicate the reagents for this step, and show how you could convert the *p*-toluenesulfonate to the desired product.

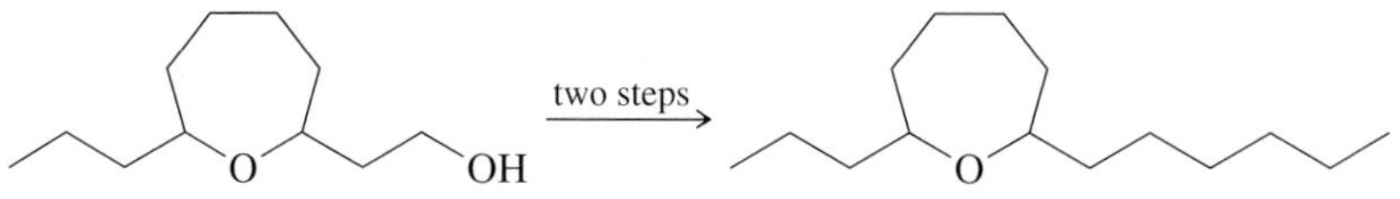

14.30 (*S*)-(+)-Ibuprofen can be prepared by enantioselective hydrogenation. Give the structure of the $C_{13}H_{16}O_2$ isomer you would select as a candidate for this reaction.

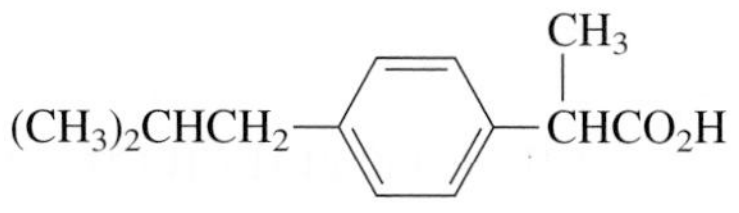

Ibuprofen

14.31 A compound having the molecular formula $C_{22}H_{32}O_2$ was isolated in 66% yield in the following reaction. Suggest a reasonable structure for this compound. What other organic compound is formed in this reaction?

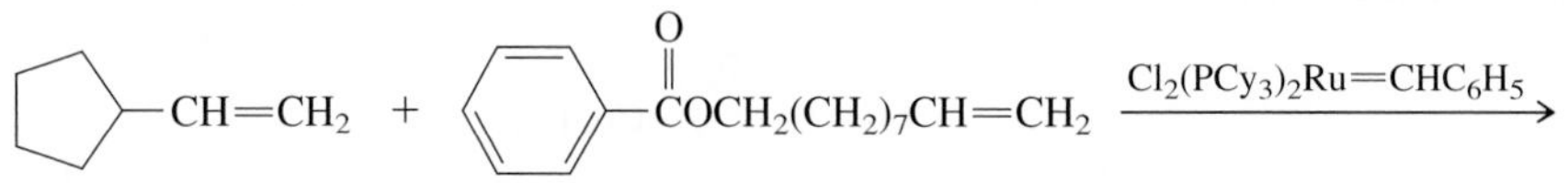

14.32 (a) *Exaltolide,* a musk substance, has been prepared by the reaction sequence shown. What is compound A?

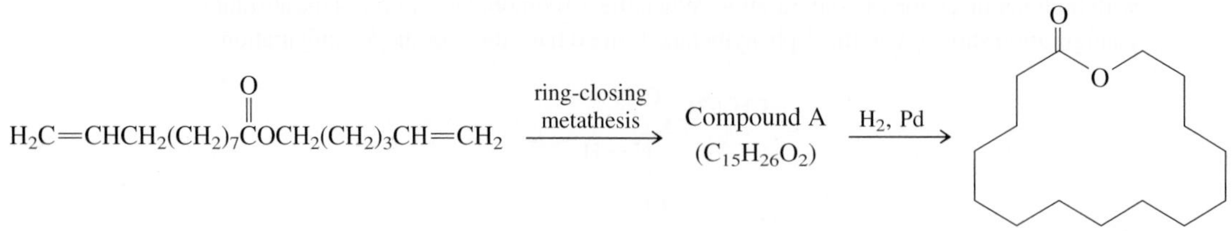

Exaltolide

(b) An analogous sequence using $H_2C{=}CHCH_2(CH_2)_2\overset{O}{\overset{\|}{C}}OCH_2(CH_2)_8CH{=}CH_2$ as the reactant also gives Exaltolide. What is the product of ring-closing metathesis of this reactant?

14.33 On treatment with a Grubb's catalyst, compound A and styrene combined to give a single product having the formula $C_{13}H_{20}O_2$ in 95% yield. This product was converted to baconipyrone C, a metabolite that was isolated from a mollusk. What is the substituent labeled "R" in the product? ("L" in the catalyst refers to the ligands that were used in this modified version of the original Grubb's catalyst.)

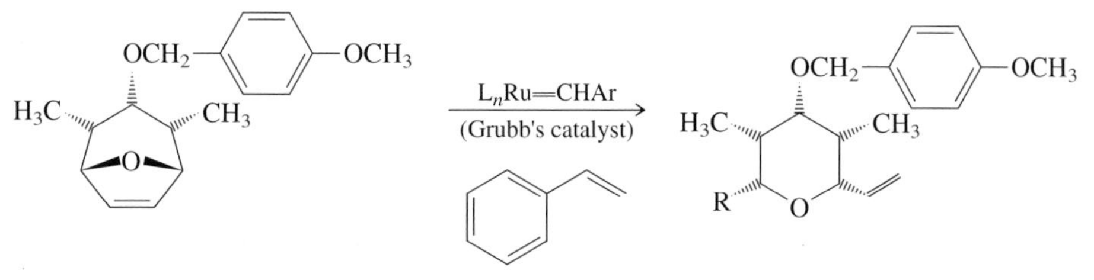

14.34 One synthetic advantage of olefin metathesis is that the catalyst tolerates the presence of a variety of functional groups in the reactant. In a synthesis of the antiinfluenza drug Tamiflu (Section 23.25), for example, ring-closing metathesis was used to prepare the highly functionalized cyclohexene derivative shown in the equation. What was the reactant?

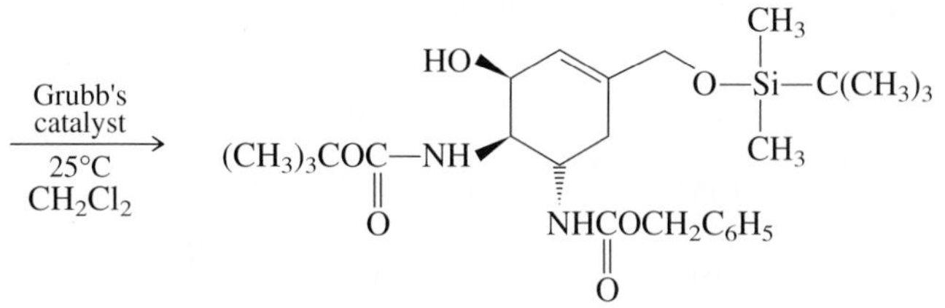

Descriptive Passage and Interpretive Problems 14

The Heck Reaction

Palladium-catalyzed substitutions of aryl halides have emerged as a powerful tool for making carbon–carbon bonds. The most widely used of these *arylations* leads to substitution of a vinylic hydrogen of an alkene by an aryl group and is called the *Heck reaction* after Richard F. Heck (University of Delaware) who pioneered its development.

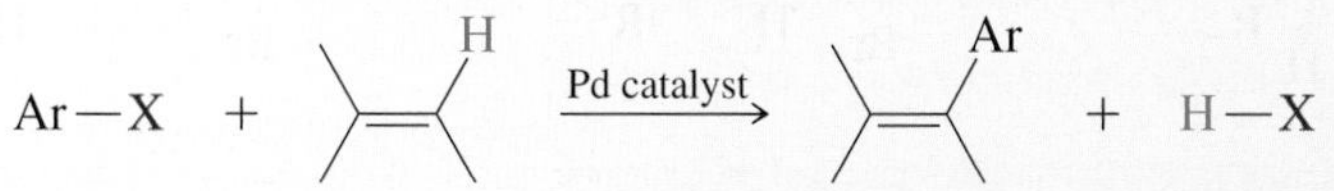

The source of palladium is often palladium(II) acetate [$Pd(OAc)_2$], but the active catalyst is Pd(0), which is formed under the reaction conditions. Only catalytic amounts of palladium are needed; a typical molar ratio of aryl halide to $Pd(OAc)_2$ is 100:1.

As is the case with most reactions of aryl halides, iodides are the most reactive, followed by aryl bromides. When bromides are used a triaryl- or trialkylphosphine (Ar_3P or R_3P) is included in the reaction mixture. Phosphines coordinate with Pd(0) to give a more active catalyst of the type $(Ar_3P)_4Pd$ or $(R_3P)_4Pd$. Typical Heck reaction conditions for aryl iodides and bromides are:

$$\text{Aryl iodide} + \text{Alkene} \xrightarrow[\text{triethylamine or tributylamine, } 80\text{–}120°\text{C}]{Pd(OAc)_2}$$

$$\text{Aryl bromide} + \text{Alkene} \xrightarrow[\text{triethylamine or tributylamine, } 80\text{–}120°\text{C}]{Pd(OAc)_2,\ (C_6H_5)_3P}$$

Heck reactions tolerate the presence of many of the most common functional groups, both on the aryl halide and the alkene. Substituents on the aromatic ring exercise no directing effect, substitution occurring at the carbon of the aryl halide that bears the halogen. The regioselectivity with respect to the alkene is governed mainly by steric considerations, with the aryl group becoming attached to the less substituted carbon of the alkene double bond.

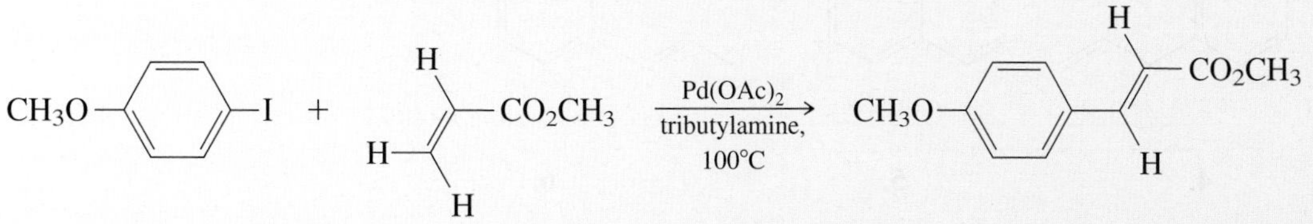

As with other transition metal-catalyzed reactions (Ziegler–Natta polymerization of alkenes, olefin metathesis), the mechanism of the Heck reaction is complicated. In brief, the species that reacts with the aryl halide is L_2Pd, where L is a ligand such as triphenylphosphine. By a process known as *oxidative addition,* palladium inserts into the carbon–halogen bond of the aryl halide.

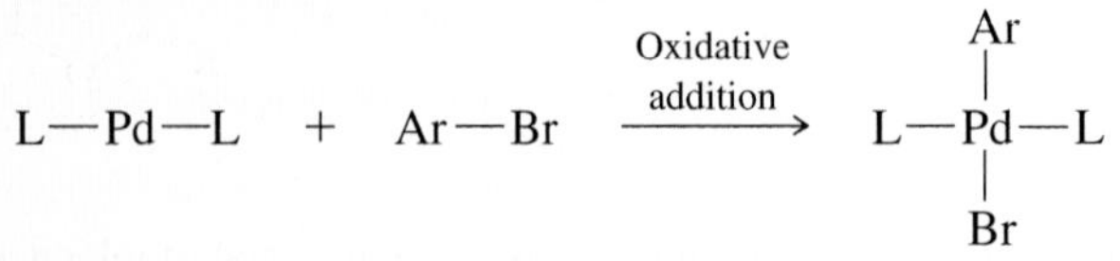

Using its π electrons, the alkene substitutes for one of the ligands attached to palladium, giving a π complex, which then rearranges. A palladium-to-carbon σ bond forms as the aryl group migrates to the less-crowded carbon of the alkene. The net result of these steps is syn addition of the aryl group and LBrPd to the doubly bonded carbons of the alkene.

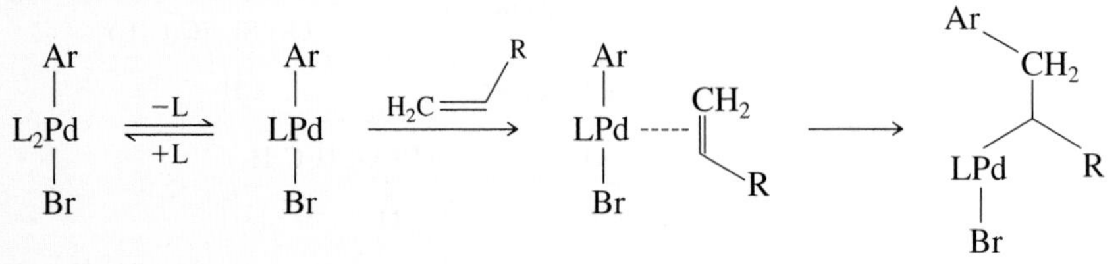

The arylated alkene product is formed by syn elimination of LBrPd—H. When elimination is from a CH_2 group, the lowest energy syn pathway leads to a trans double bond via a conformation in which Ar and R are remote from each other.

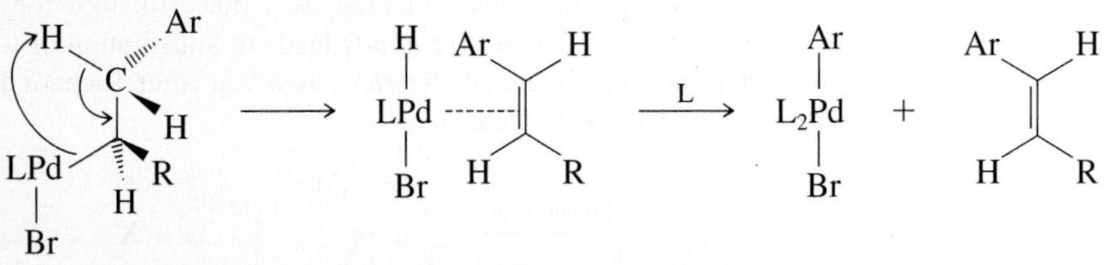

The process concludes with regeneration of the catalytically active L_2Pd by reaction with the base employed in the reaction mixture. For the case of triethylamine as the base:

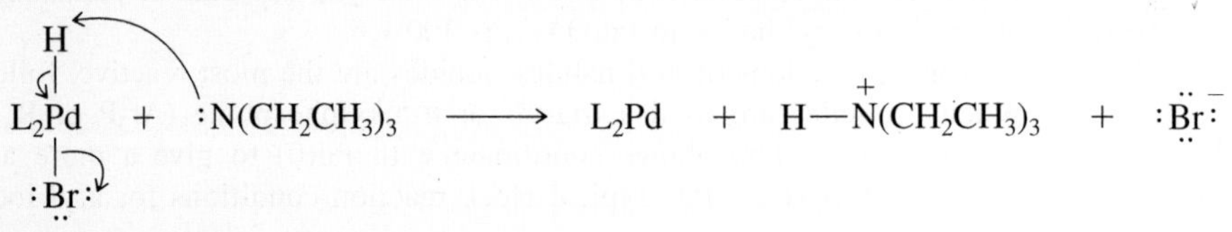

14.35 What combination of aryl halide (**1–3**) and alkene (**4–6**) is the best choice to prepare the compound shown by a Heck reaction?

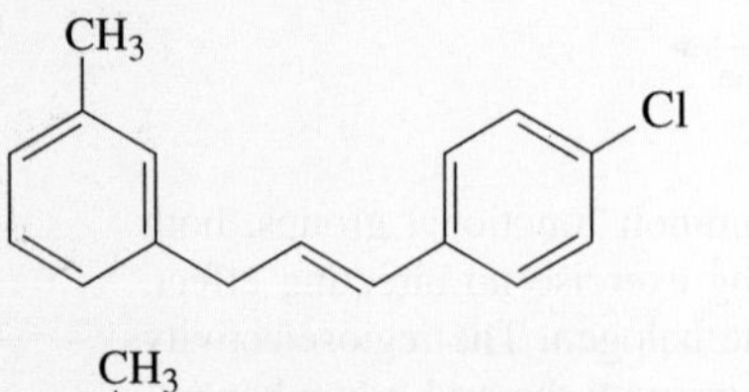

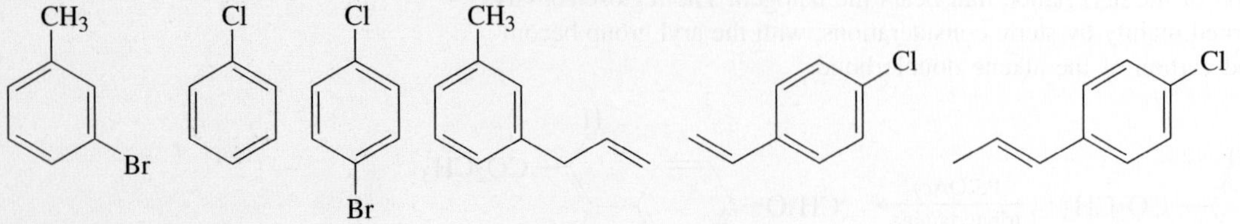

1. **2.** **3.** **4.** **5.** **6.**

A. **1 + 5** B. **1 + 6** C. **2 + 4** D. **3 + 4**

14.36 A single monosubstitution product was isolated in 82% yield in the following reaction. What was that product?

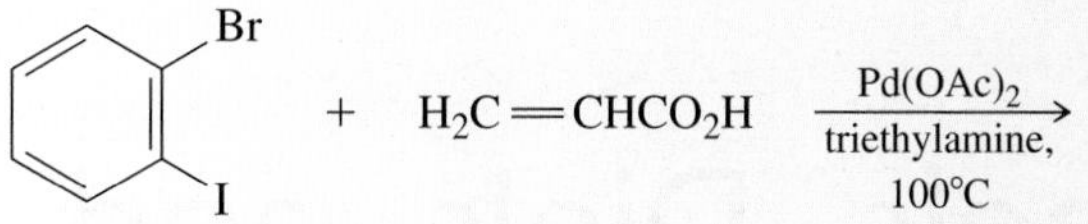

A. 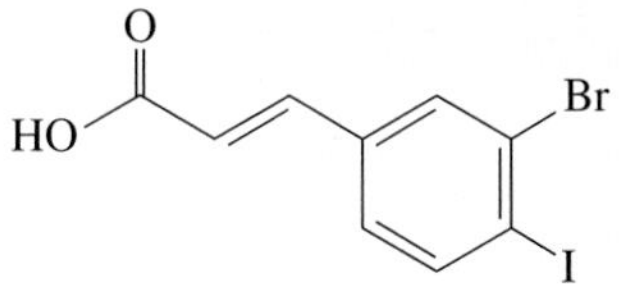 B. 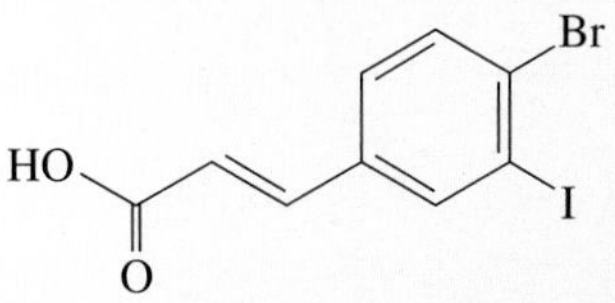 C. 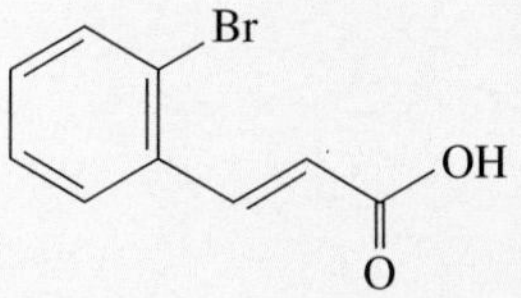 D. 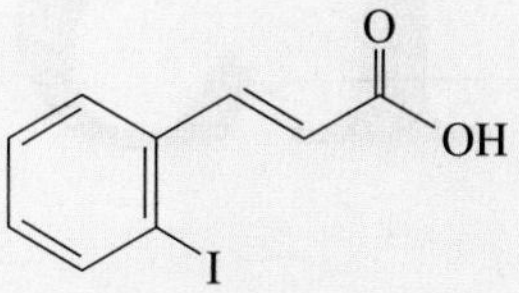

14.37 The Heck reaction between the aryl halide and enol ether shown is neither very regioselective nor very stereoselective and gives three isomers in comparable amounts.

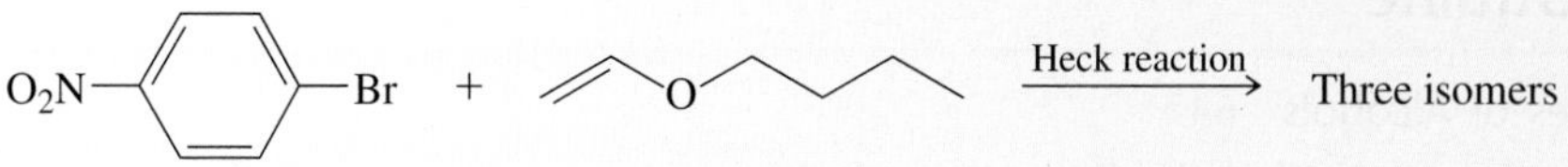

Which one of the following compounds is *the least likely* product?

A. 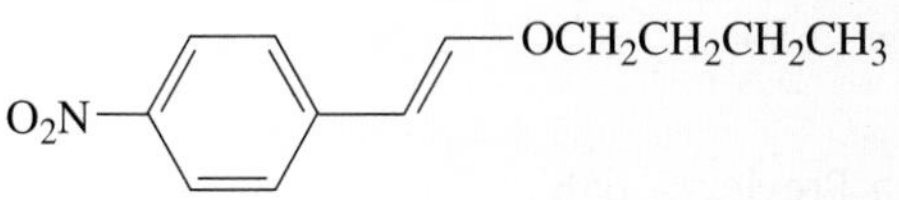

C. 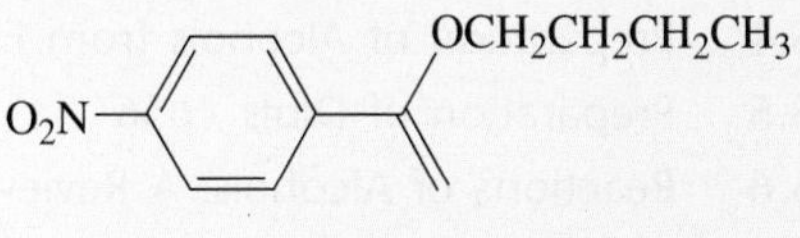

B. 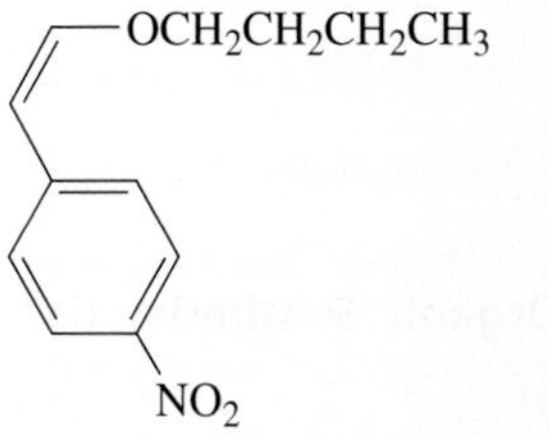

D. 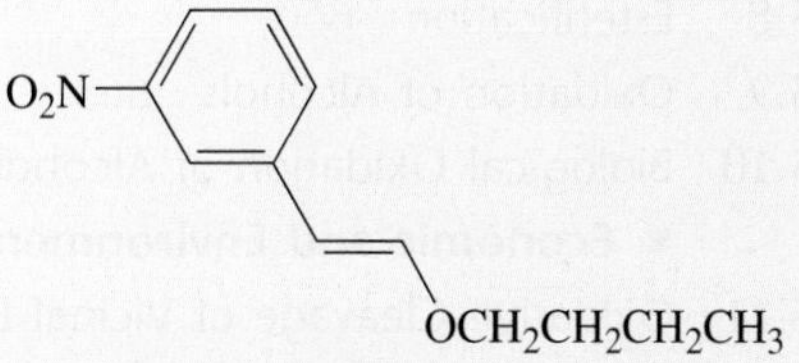

14.38 If the organopalladium complex from bromobenzene reacts with *trans*-2-butene by a stereospecific syn addition and the loss of LBrPd—H is a stereospecific syn elimination, what stereoisomer of the product should be formed?

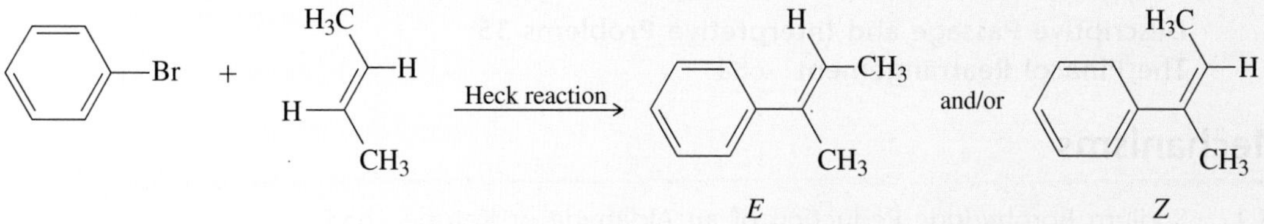

A. *E* B. *Z*

15 Alcohols, Diols, and Thiols

Chapter Outline

Mechanisms

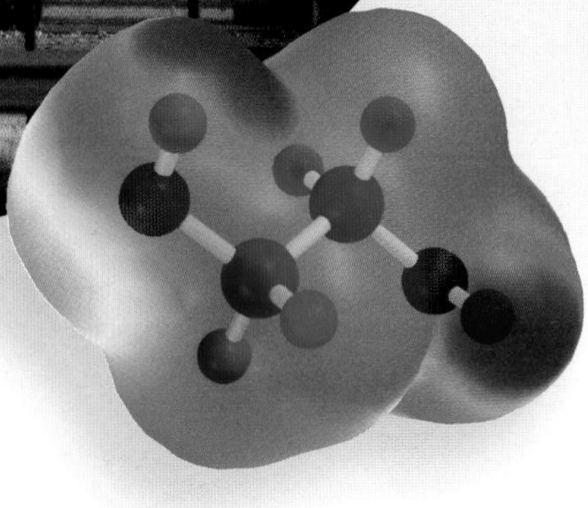

Ice on airplane wings is removed prior to takeoff by spraying with a mixture of ethylene glycol ($HOCH_2CH_2OH$) and propylene glycol ($HOCH_2CH(OH)CH_3$). The electrostatic potential map is ethylene glycol.

THE NEXT SEVERAL chapters deal with the chemistry of various oxygen-containing functional groups. The interplay of these important classes of compounds—alcohols, ethers, aldehydes, ketones, carboxylic acids, and derivatives of carboxylic acids—is fundamental to organic chemistry and biochemistry.

ROH	ROR′	$\overset{\text{O}}{\overset{\Vert}{\text{RCH}}}$	$\overset{\text{O}}{\overset{\Vert}{\text{RCR}'}}$	$\overset{\text{O}}{\overset{\Vert}{\text{RCOH}}}$
Alcohol	Ether	Aldehyde	Ketone	Carboxylic acid

We'll start by discussing in more detail a class of compounds already familiar to us, *alcohols.* Alcohols were introduced in Chapter 4 and have appeared regularly since then. With this chapter we extend our knowledge of alcohols, particularly with respect to their relationship to carbonyl-containing compounds. In the course of studying alcohols, we shall also look at some relatives. **Diols** are alcohols in which two hydroxyl groups (—OH) are present; **thiols** are compounds that contain an —SH group. **Phenols,** compounds of the type ArOH, share many properties in common with alcohols but are sufficiently different from them to warrant separate discussion in Chapter 22.

This chapter is a transitional one. It ties together much of the material encountered earlier and sets the stage for our study of other oxygen-containing functional groups in the chapters that follow.

15.1 Sources of Alcohols

At one time, the major source of *methanol* was as a byproduct in the production of charcoal from wood—hence, the name *wood alcohol.* Now, most of the more than 10 billion lb of methanol used annually in the United States is synthetic, prepared by reduction of carbon monoxide with hydrogen. Carbon monoxide is normally made from methane.

$$\underset{\text{Carbon monoxide}}{CO} + \underset{\text{Hydrogen}}{2H_2} \xrightarrow[400°C]{ZnO/Cr_2O_3} \underset{\text{Methanol}}{CH_3OH}$$

The major uses of methanol are in the preparation of formaldehyde and *tert*-butyl methyl ether (known commercially as MTBE). Formaldehyde is a starting material for various resins and plastics, including the first completely synthetic plastic Bakelite. MTBE is an effective gasoline additive, but problems with it leaking from underground tanks and contaminating groundwater make it unsuitable for continued use.

Methanol is a colorless liquid, boiling at 65°C, and is miscible with water in all proportions. It is poisonous; drinking as little as 30 mL has been fatal. Smaller amounts can produce blindness.

When vegetable matter ferments, its carbohydrates are converted to *ethanol* and carbon dioxide by enzymes present in yeast. Fermentation of barley produces beer; grapes give wine. The maximum ethanol content is on the order of 15%, because higher concentrations inactivate the enzymes, halting fermentation. Because ethanol boils at 78°C and water at 100°C, distillation of the fermentation broth gives "distilled spirits" of increased ethanol content. Whiskey is the aged distillate of fermented grain and contains slightly less than 50% ethanol. Brandy and cognac are made by aging the distilled spirits from fermented grapes and other fruits. The characteristic flavors, odors, and colors of the various alcoholic beverages depend on both their origin and the way they are aged.

Synthetic ethanol is derived from petroleum by hydration of ethylene. In the United States, some 700 million lb of synthetic ethanol is produced annually. It is relatively inexpensive and useful for industrial applications. To make it unfit for drinking, it is *denatured* by adding any of a number of noxious materials, exempting it from the high taxes most governments impose on ethanol used in beverages.

Some of the substances used to denature ethanol include methanol, benzene, pyridine, castor oil, and gasoline.

Our bodies are reasonably well equipped to metabolize ethanol, making it less dangerous than methanol. Alcohol abuse and alcoholism, however, have been and remain persistent problems.

Ethanol and methanol both have potential as fuels. Gasohol contains ethanol as an additive, and the use of ethanol alone as a fuel is increasing. Beginning in 2007, the Indy Racing League began using ethanol as a fuel. George Olah, winner of the 1994 Nobel Prize in Chemistry, and his colleagues have written about the use of methanol as an alternative to fossil fuels in their book *Beyond Oil and Gas: The Methanol Economy.*

Isopropyl alcohol is prepared from petroleum by hydration of propene. With a boiling point of 82°C, isopropyl alcohol evaporates quickly from the skin, producing a cooling effect. Often containing dissolved oils and fragrances, it is the major component of rubbing alcohol. Isopropyl alcohol possesses weak antibacterial properties and is used to maintain medical instruments in a sterile condition and to clean the skin before minor surgery.

Most alcohols of six carbons or fewer, as well as many higher alcohols, are commercially available at low cost. Some occur naturally; others are the products of efficient syntheses. Figure 15.1 presents the structures of a few naturally occurring alcohols. Table 15.1 (page 650) summarizes the reactions encountered in earlier chapters that give alcohols and illustrates a thread that runs through the fabric of organic chemistry: *a reaction that is characteristic of one functional group often serves as a synthetic method for preparing another.*

As Table 15.1 indicates, reactions leading to alcohols are not in short supply. Nevertheless, several more will be added to the list in the present chapter—testimony to the importance of alcohols in synthetic organic chemistry. Some of these methods involve reduction of carbonyl groups:

Recall from Section 2.19 that reduction corresponds to a decrease in the number of bonds between carbon and oxygen or an increase in the number of bonds between carbon and hydrogen (or both).

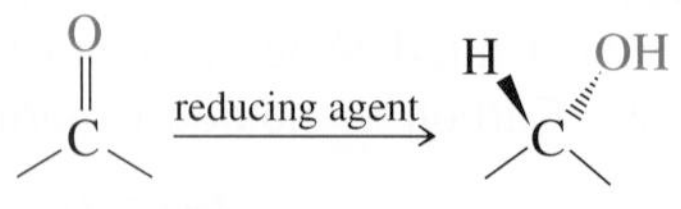

We will begin with the reduction of aldehydes and ketones.

15.2 Preparation of Alcohols by Reduction of Aldehydes and Ketones

The most obvious way to reduce an aldehyde or a ketone to an alcohol is by hydrogenation of the carbon–oxygen double bond. Like the hydrogenation of alkenes, the reaction is exothermic but exceedingly slow in the absence of a catalyst. Finely divided

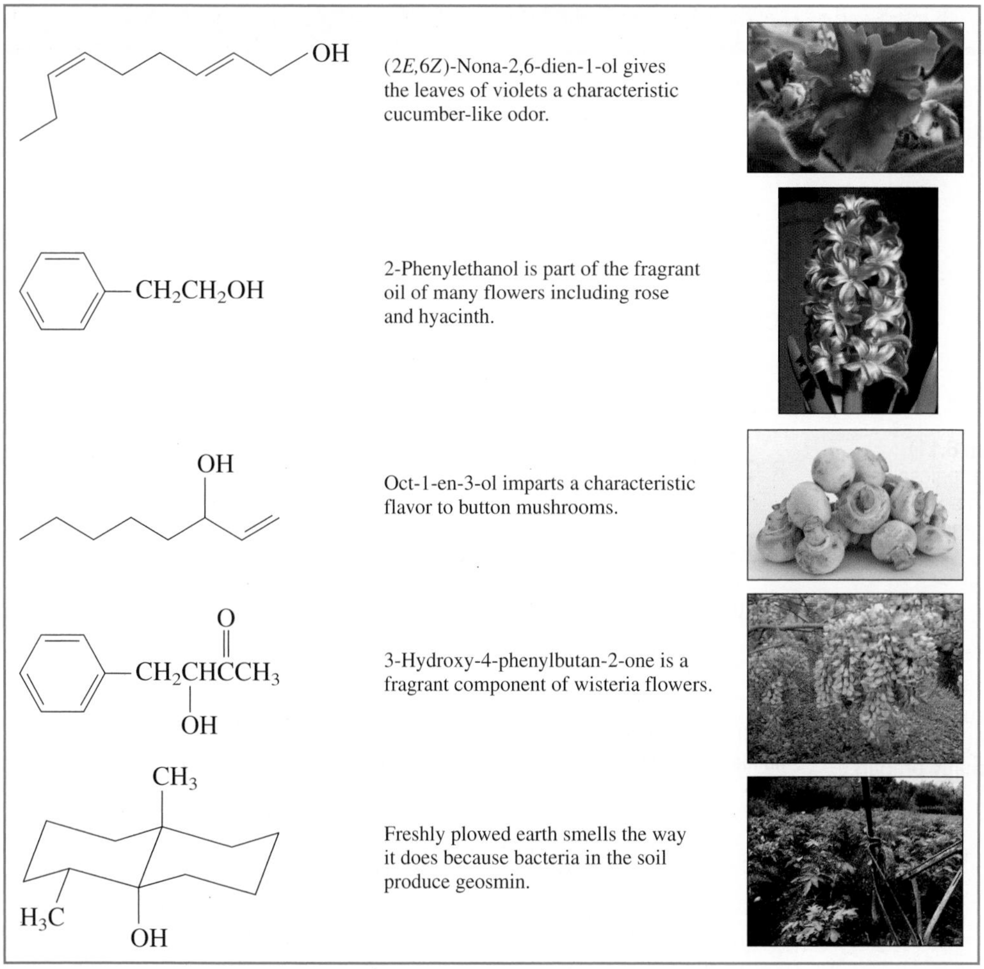
OH
(2E,6Z)-Nona-2,6-dien-1-ol gives the leaves of violets a characteristic cucumber-like odor.
CH2CH2OH
2-Phenylethanol is part of the fragrant oil of many flowers including rose and hyacinth.
OH
Oct-1-en-3-ol imparts a characteristic flavor to button mushrooms.
O
CH2CHCCH3
OH
3-Hydroxy-4-phenylbutan-2-one is a fragrant component of wisteria flowers.
CH3
H3C
OH
Freshly plowed earth smells the way it does because bacteria in the soil produce geosmin.

Figure 15.1

Several of the countless naturally occurring alcohols that stimulate our senses.

metals such as platinum, palladium, nickel, and ruthenium are effective catalysts for the hydrogenation of aldehydes and ketones. Aldehydes yield primary alcohols:

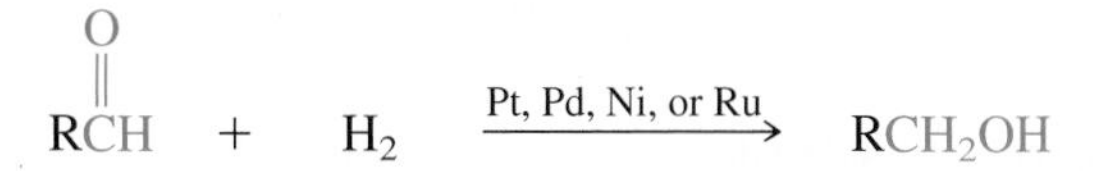

Aldehyde Hydrogen Primary alcohol

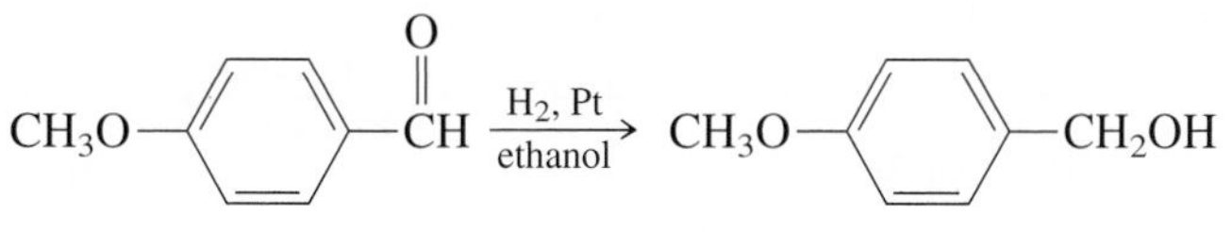

p-Methoxybenzaldehyde *p*-Methoxybenzyl alcohol (92%)

Ketones yield secondary alcohols:

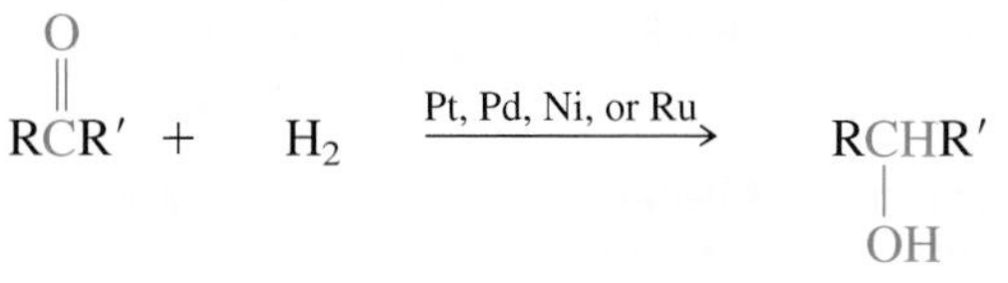

Ketone Hydrogen Secondary alcohol

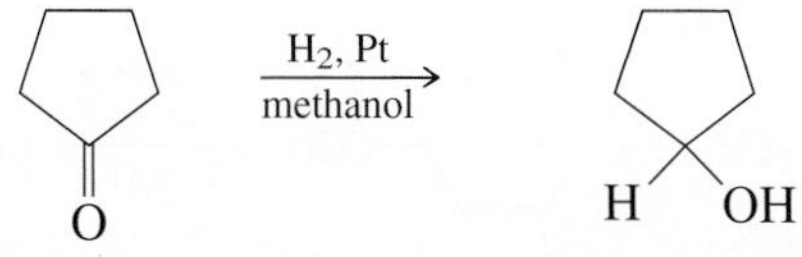

Cyclopentanone Cyclopentanol (93–95%)

TABLE 15.1 Summary of Reactions Discussed in Earlier Chapters That Yield Alcohols

Reaction (section) and comments	General equation and specific example
Acid-catalyzed hydration of alkenes (Section 6.9) Water adds to the double bond in accordance with Markovnikov's rule.	$R_2C{=}CR_2 + H_2O \xrightarrow{H^+} R_2CHCR_2(OH)$ Alkene, Water, Alcohol $(CH_3)_2C{=}CHCH_3 \xrightarrow[H_2SO_4]{H_2O} CH_3C(CH_3)(OH)CH_2CH_3$ 2-Methyl-2-butene → 2-Methyl-2-butanol (90%)
Hydroboration-oxidation of alkenes (Section 6.11) H and OH add to the double bond with a regioselectivity opposite to that of Markovnikov's rule. This is a very good synthetic method; addition is syn, and no rearrangements are observed.	$R_2C{=}CR_2 \xrightarrow[2.\ H_2O_2,\ HO^-]{1.\ B_2H_6} R_2CHCR_2(OH)$ Alkene, Alcohol $CH_3(CH_2)_7CH{=}CH_2 \xrightarrow[2.\ H_2O_2,\ HO^-]{1.\ B_2H_6,\ diglyme} CH_3(CH_2)_7CH_2CH_2OH$ 1-Decene → 1-Decanol (93%)
Hydrolysis of alkyl halides (Section 8.1) A reaction useful only with substrates that do not undergo E2 elimination readily. It is rarely used for the synthesis of alcohols, since alkyl halides are normally prepared from alcohols.	$RX + HO^- \longrightarrow ROH + X^-$ Alkyl halide, Hydroxide ion, Alcohol, Halide ion 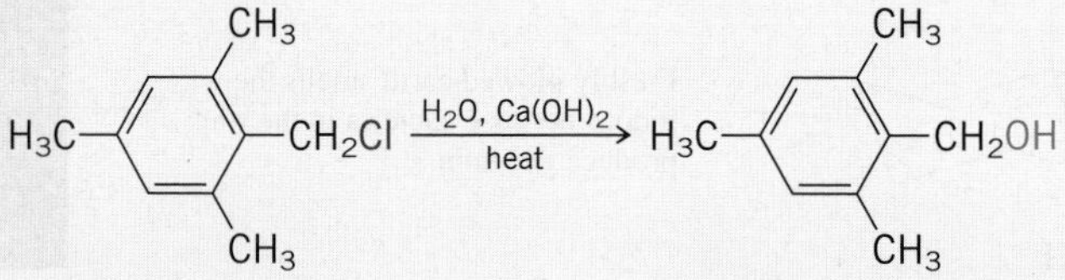 2,4,6-Trimethylbenzyl chloride → 2,4,6-Trimethylbenzyl alcohol (78%)
Reaction of Grignard reagents with aldehydes and ketones (Section 14.6) A method that allows for alcohol preparation with formation of new carbon–carbon bonds. Primary, secondary, and tertiary alcohols can all be prepared.	$RMgX + R'C(O)R'' \xrightarrow[2.\ H_3O^+]{1.\ diethyl\ ether} RC(R')(R'')OH$ Grignard reagent, Aldehyde or ketone, Alcohol 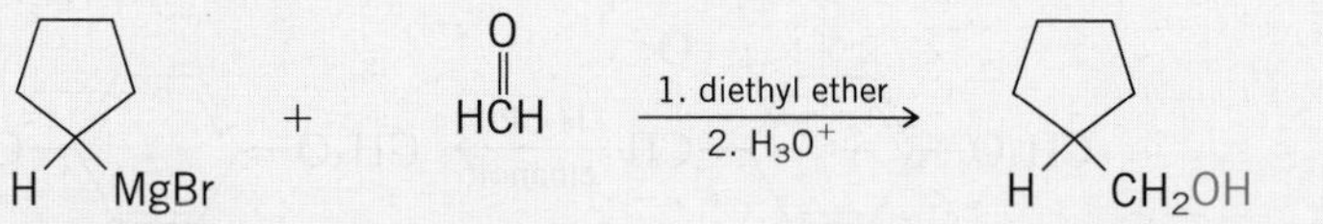 Cyclopentylmagnesium bromide + Formaldehyde → Cyclopentylmethanol (62–64%)
Reaction of organolithium reagents with aldehydes and ketones (Section 14.7) Organolithium reagents react with aldehydes and ketones in a manner similar to that of Grignard reagents to form alcohols.	$RLi + R'C(O)R'' \xrightarrow[2.\ H_3O^+]{1.\ diethyl\ ether} RC(R')(R'')OH$ Organolithium reagent, Aldehyde or ketone, Alcohol 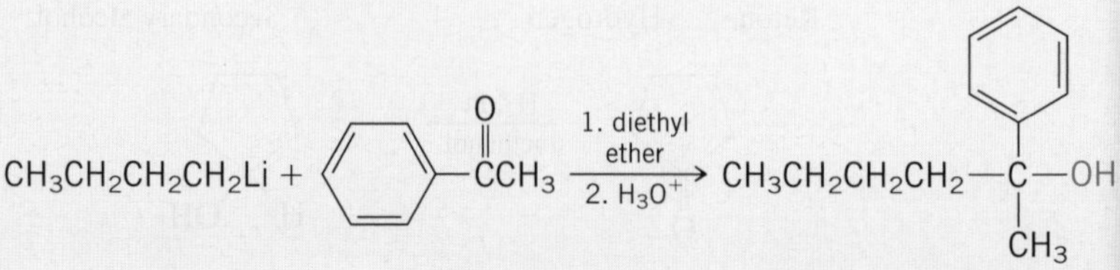 Butyllithium + Acetophenone → 2-Phenyl-2-hexanol (67%)

Problem 15.1

Which of the isomeric $C_4H_{10}O$ alcohols can be prepared by hydrogenation of aldehydes? Which can be prepared by hydrogenation of ketones? Which cannot be prepared by hydrogenation of a carbonyl compound?

For most laboratory-scale reductions of aldehydes and ketones, catalytic hydrogenation has been replaced by methods based on metal hydride reducing agents. The two most common reagents are sodium borohydride and lithium aluminum hydride.

Sodium borohydride ($NaBH_4$) Lithium aluminum hydride ($LiAlH_4$)

Sodium borohydride is especially easy to use, needing only to be added to an aqueous or alcoholic solution of an aldehyde or a ketone:

$$\underset{\text{Aldehyde}}{\overset{\overset{\displaystyle O}{\|}}{RCH}} \xrightarrow[\text{water, methanol, or ethanol}]{NaBH_4} \underset{\text{Primary alcohol}}{RCH_2OH}$$

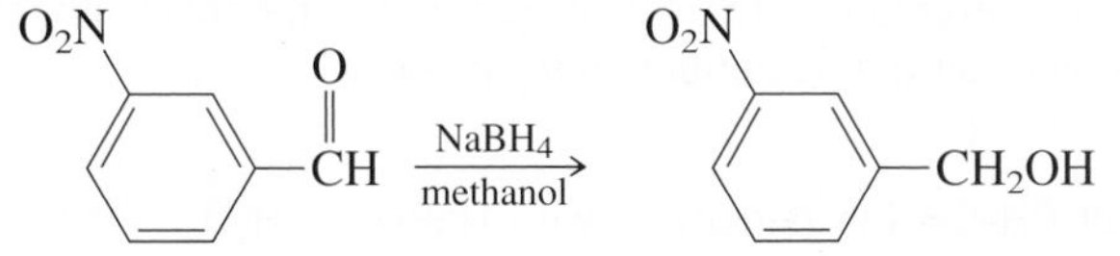

m-Nitrobenzaldehyde *m*-Nitrobenzyl alcohol (82%)

$$\underset{\text{Ketone}}{\overset{\overset{\displaystyle O}{\|}}{RCR'}} \xrightarrow[\text{water, methanol, or ethanol}]{NaBH_4} \underset{\text{Secondary alcohol}}{\overset{\overset{\displaystyle OH}{|}}{RCHR'}}$$

$$\underset{\text{4,4-Dimethyl-2-pentanone}}{CH_3\overset{\overset{\displaystyle O}{\|}}{C}CH_2C(CH_3)_3} \xrightarrow[\text{ethanol}]{NaBH_4} \underset{\text{4,4-Dimethyl-2-pentanol (85\%)}}{CH_3\overset{\overset{\displaystyle OH}{|}}{C}HCH_2C(CH_3)_3}$$

Lithium aluminum hydride reacts violently with water and alcohols, so it must be used in solvents such as anhydrous diethyl ether or tetrahydrofuran. Following reduction, a separate hydrolysis step is required to liberate the alcohol product:

The same kinds of aprotic solvents are used for $LiAlH_4$ as for Grignard reagents.

$$\underset{\text{Aldehyde}}{\overset{\overset{\displaystyle O}{\|}}{RCH}} \xrightarrow[\text{2. } H_2O]{\text{1. } LiAlH_4\text{, diethyl ether}} \underset{\text{Primary alcohol}}{RCH_2OH}$$

$$\underset{\text{Heptanal}}{CH_3(CH_2)_5\overset{\overset{\displaystyle O}{\|}}{C}H} \xrightarrow[\text{2. } H_2O]{\text{1. } LiAlH_4\text{, diethyl ether}} \underset{\text{1-Heptanol (86\%)}}{CH_3(CH_2)_5CH_2OH}$$

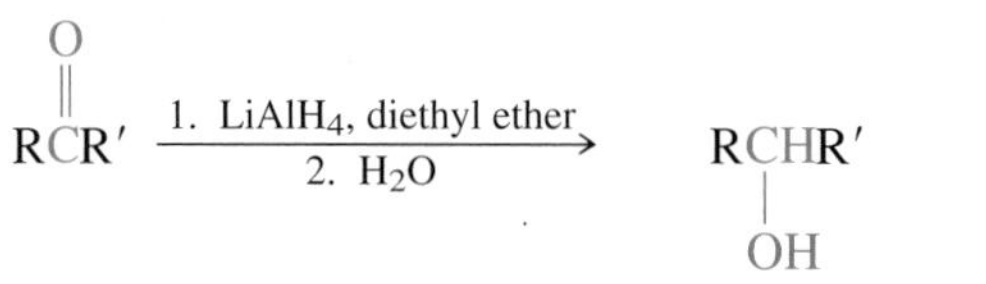

Ketone Secondary alcohol

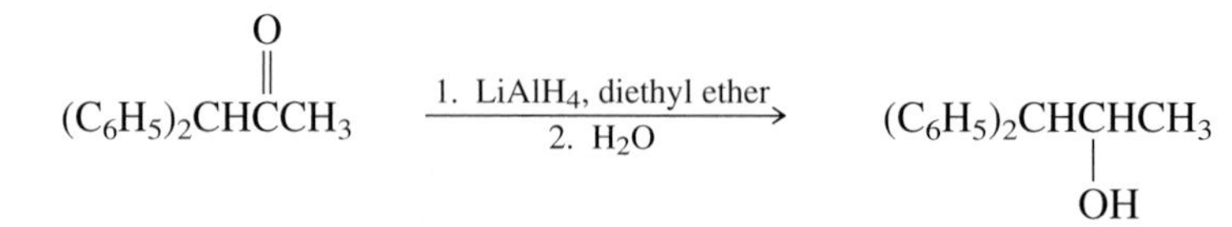

1,1-Diphenyl-2-propanone 1,1-Diphenyl-2-propanol (84%)

Sodium borohydride and lithium aluminum hydride react with carbonyl compounds in much the same way that Grignard reagents do, except that they function as *hydride donors* rather than as carbanion sources. Mechanism 15.1 outlines the general mechanism for the sodium borohydride reduction of an aldehyde or ketone ($R_2C{=}O$). Two points are especially important about this process.

1. At no point is H_2 involved. The reducing agent is borohydride ion (BH_4^-).
2. In the reduction $R_2C{=}O \rightarrow R_2CHOH$, the hydrogen bonded to carbon comes from BH_4^-; the hydrogen on oxygen comes from an OH group of the solvent (water, methanol, or ethanol).

Problem 15.2

Sodium borodeuteride ($NaBD_4$) and lithium aluminum deuteride ($LiAlD_4$) are convenient reagents for introducing deuterium, the mass-2 isotope of hydrogen, into organic compounds. Write the structure of the organic product of the following reactions, clearly showing the position of all the deuterium atoms in each:

(a) Reduction of $CH_3\overset{O}{\overset{\|}{C}}H$ (acetaldehyde) with $NaBD_4$ in H_2O

(b) Reduction of $CH_3\overset{O}{\overset{\|}{C}}CH_3$ (acetone) with $NaBD_4$ in CH_3OD

(c) Reduction of $C_6H_5\overset{O}{\overset{\|}{C}}H$ (benzaldehyde) with $NaBD_4$ in CD_3OH

(d) Reduction of $H\overset{O}{\overset{\|}{C}}H$ (formaldehyde) with $LiAlD_4$ in diethyl ether, followed by addition of D_2O

Sample Solution (a) Sodium borodeuteride transfers deuterium to the carbonyl group of acetaldehyde, forming a C—D bond.

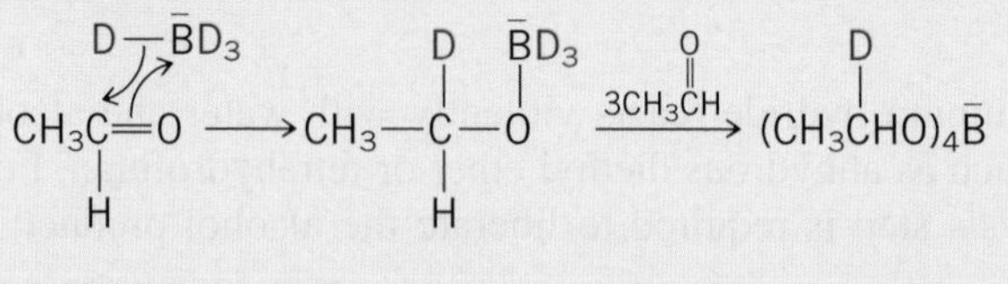

Hydrolysis of $(CH_3CHDO)_4B^-$ in H_2O leads to the formation of ethanol, retaining the C—D bond formed in the preceding step while forming an O—H bond.

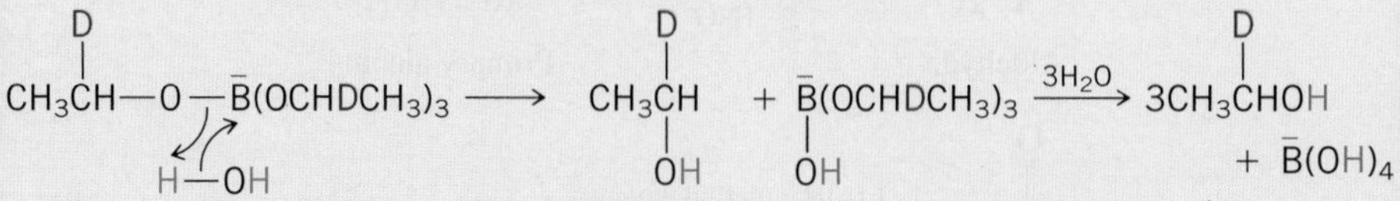

Ethanol-1-*d*

Mechanism 15.1

Sodium Borohydride Reduction of an Aldehyde or Ketone

THE OVERALL REACTION:

$$4R_2C{=}O \quad + \quad BH_4^- \quad + \quad 4H_2O \longrightarrow 4R_2CHOH \quad + \quad B(OH)_4^-$$

Aldehyde or ketone	Borohydride ion	Water	Alcohol	Borate ion

THE MECHANISM:

Step 1: Hydride (hydrogen + two electrons) is transferred from boron to the positively polarized carbon of the carbonyl group. The carbonyl oxygen bonds to boron.

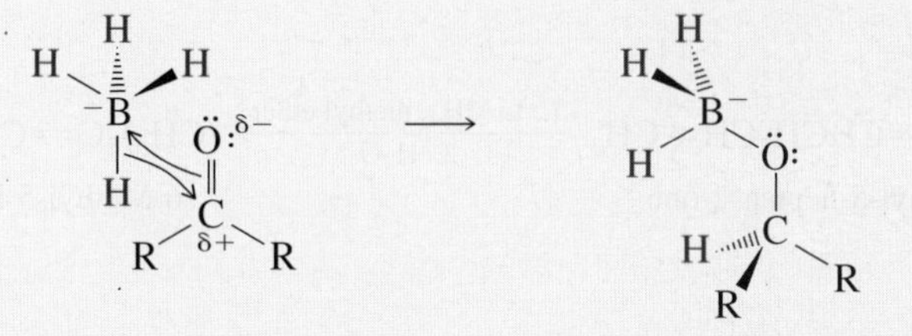

Borohydride ion + aldehyde or ketone — Alkoxyborohydride

Steps 2–4: The alkoxyborohydride formed in the first step contains three more hydrogens that can be donated to carbonyl groups. It reacts with three more molecules of the starting aldehyde or ketone.

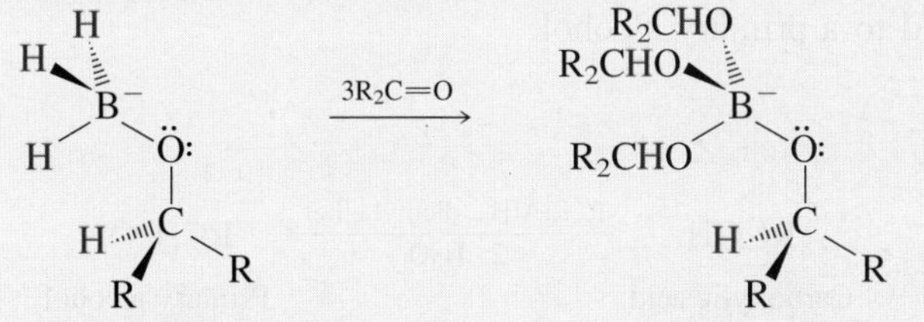

Alkoxyborohydride — Tetraalkoxyborate

Step 5: When the reaction is carried out in water as the solvent, the tetraalkoxyborate undergoes hydrolysis.

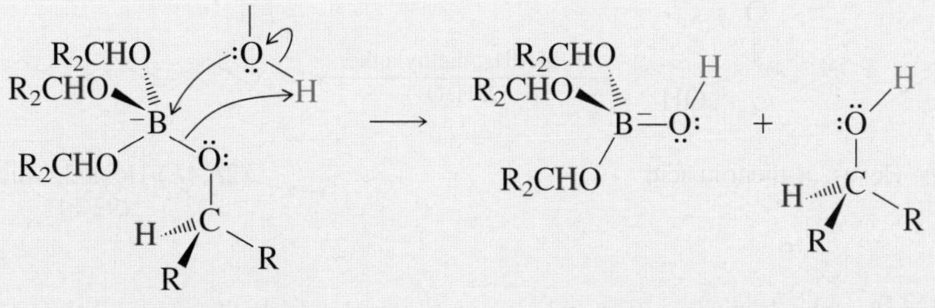

Tetraalkoxyborate + water — Trialkoxyborate — Alcohol

Steps 6–8: Three more hydrolysis steps convert the trialkoxyborate to three more molecules of R_2CHOH and $(HO)_4B^-$.

The mechanism of lithium aluminum hydride reduction of aldehydes and ketones is analogous to that of sodium borohydride except that the reduction and hydrolysis stages are independent operations. The reduction is carried out in diethyl ether, followed by a separate hydrolysis step when water is added to the reaction mixture.

$$\underset{\text{Aldehyde or ketone}}{4R_2C{=}O} \xrightarrow[\text{diethyl ether}]{LiAlH_4} \underset{\text{Tetraalkoxyaluminate}}{(R_2CHO)_4Al^-} \xrightarrow{4H_2O} \underset{\text{Alcohol}}{4R_2CHOH} + Al(OH)_4^-$$

Neither sodium borohydride nor lithium aluminum hydride reduces isolated carbon–carbon double bonds. This makes possible the selective reduction of a carbonyl group in a molecule that contains both carbon–carbon and carbon–oxygen double bonds.

Catalytic hydrogenation would not be suitable for this transformation, because H_2 adds to carbon–carbon double bonds faster than it reduces carbonyl groups.

$$\underset{\text{6-Methyl-5-hepten-2-one}}{(CH_3)_2C{=}CHCH_2CH_2\overset{\overset{\displaystyle O}{\|}}{C}CH_3} \xrightarrow[\text{2. } H_2O]{\text{1. } LiAlH_4\text{, diethyl ether}} \underset{\text{6-Methyl-5-hepten-2-ol (90\%)}}{(CH_3)_2C{=}CHCH_2CH_2\overset{\overset{\displaystyle OH}{|}}{C}HCH_3}$$

15.3 Preparation of Alcohols by Reduction of Carboxylic Acids

Carboxylic acids are exceedingly difficult to reduce. Acetic acid, for example, is often used as a solvent in catalytic hydrogenations because it is inert under the reaction conditions. Lithium aluminum hydride is one of the few reducing agents capable of reducing a carboxylic acid to a primary alcohol.

$$\underset{\text{Carboxylic acid}}{R\overset{\overset{\displaystyle O}{\|}}{C}OH} \xrightarrow[\text{2. } H_2O]{\text{1. } LiAlH_4\text{, diethyl ether}} \underset{\text{Primary alcohol}}{RCH_2OH}$$

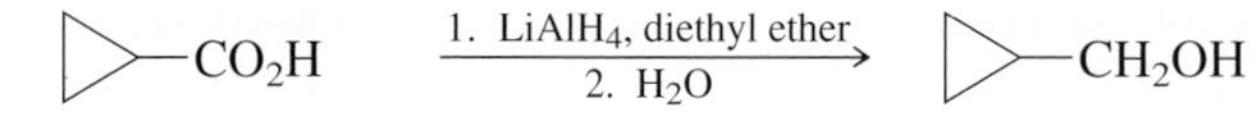

Cyclopropanecarboxylic acid → Cyclopropylmethanol (78%)

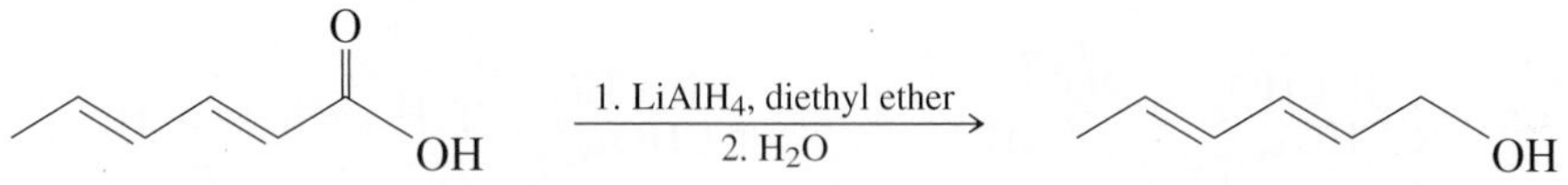

(2*E*,4*E*)-Hexa-2,4-dienoic acid → (2*E*,4*E*)-Hexa-2,4-dien-1-ol (92%)

Esters can also be reduced to alcohols with lithium aluminum hydride. We will examine this reaction in detail in Chapter 19.

Sodium borohydride is not nearly as potent a hydride donor as lithium aluminum hydride and does not reduce carboxylic acids.

15.4 Preparation of Alcohols from Epoxides

Although the chemical reactions of epoxides will not be covered in detail until the following chapter, we shall introduce their use in the synthesis of alcohols here.

Grignard reagents react with ethylene oxide to yield primary alcohols containing two more carbon atoms than the alkyl halide from which the organometallic compound was prepared.

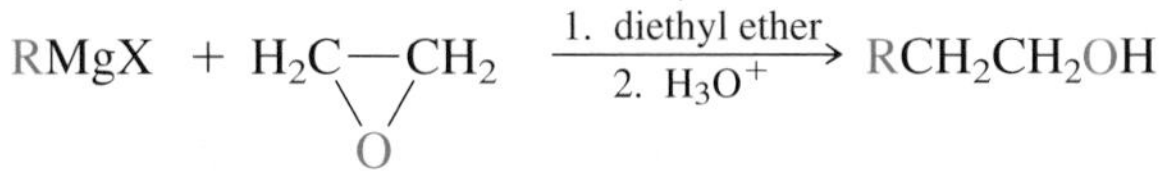

Grignard reagent Ethylene oxide Primary alcohol

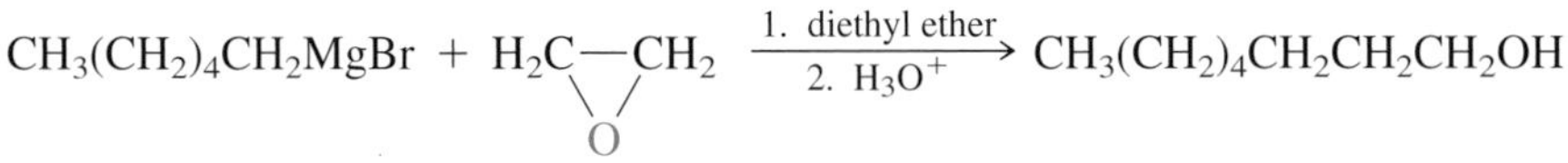

Hexylmagnesium bromide Ethylene oxide 1-Octanol (71%)

Organolithium reagents react with epoxides in a similar manner.

Problem 15.3

Each of the following alcohols has been prepared by reaction of a Grignard reagent with ethylene oxide. Select the appropriate Grignard reagent in each case.

Sample Solution (a) Reaction with ethylene oxide results in the addition of a —CH_2CH_2OH unit to the Grignard reagent. The Grignard reagent derived from *o*-bromotoluene (or *o*-chlorotoluene or *o*-iodotoluene) is appropriate here.

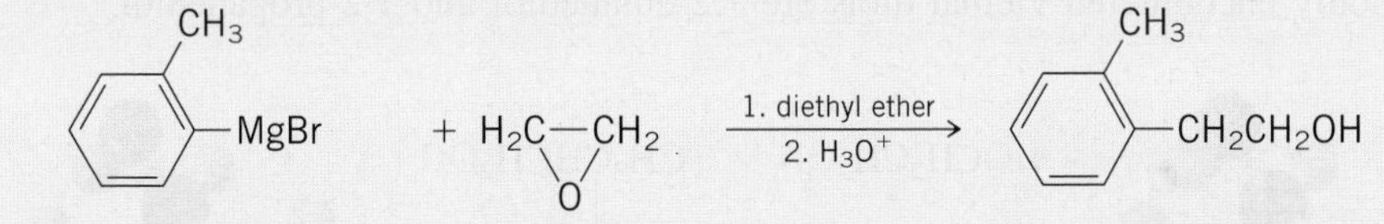

o-Methylphenylmagnesium bromide Ethylene oxide 2-(*o*-Methylphenyl)ethanol (66%)

Epoxide rings are readily opened with cleavage of the carbon–oxygen bond when attacked by nucleophiles. Grignard reagents and organolithium reagents react with ethylene oxide by serving as sources of nucleophilic carbon. The mechanism resembles an S_N2 reaction (Section 8.3). Cleavage of the epoxide C—O bond is analogous to the cleavage of the bond between the carbon and the leaving group.

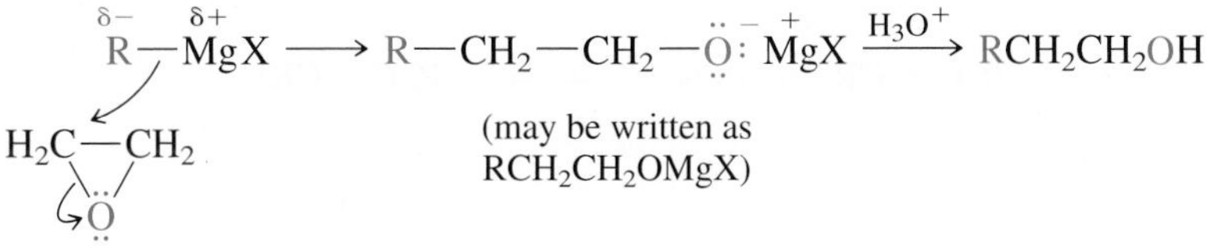

(may be written as RCH_2CH_2OMgX)

This kind of chemical reactivity of epoxides is rather general. Nucleophiles other than Grignard reagents react with epoxides, and epoxides more elaborate than ethylene oxide may be used. These features of epoxide chemistry will be discussed in Sections 16.11–16.13.

15.5 Preparation of Diols

Much of the chemistry of diols—compounds that bear two hydroxyl groups—is analogous to that of alcohols. Diols may be prepared, for example, from compounds that contain two carbonyl groups, using the same reducing agents employed in the preparation of alcohols. The following example shows the conversion of a dialdehyde to a diol by catalytic hydrogenation. Alternatively, the same transformation can be achieved by reduction with sodium borohydride or lithium aluminum hydride.

The 2004 IUPAC name for this diol is 3-methylpentane-1,5-diol.

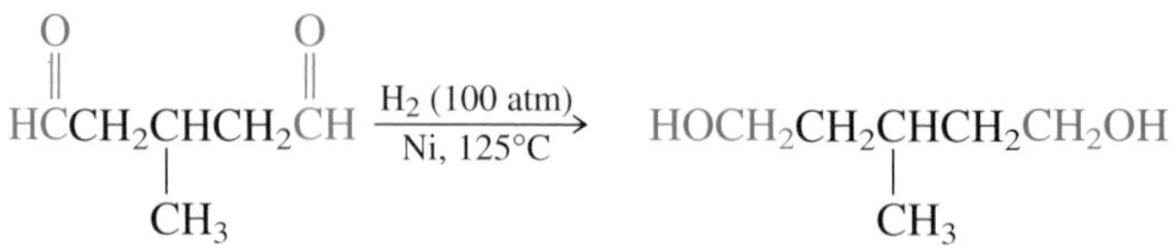
HCCH2CHCH2CH
O
O
CH3
H2 (100 atm)
Ni, 125°C
HOCH2CH2CHCH2CH2OH
CH3

3-Methylpentanedial 3-Methyl-1,5-pentanediol (81–83%)

As can be seen in the preceding equation, the nomenclature of diols is similar to that of alcohols. The suffix *-diol* replaces *-ol,* and two locants, one for each hydroxyl group, are required. Note that the final *-e* of the parent alkane name is retained when the suffix begins with a consonant *(-diol),* but dropped when the suffix begins with a vowel *(-ol).*

Problem 15.4

Write equations showing how 3-methyl-1,5-pentanediol could be prepared from a dicarboxylic acid.

Vicinal diols are diols that have their hydroxyl groups on adjacent carbons. Two commonly encountered vicinal diols are 1,2-ethanediol and 1,2-propanediol.

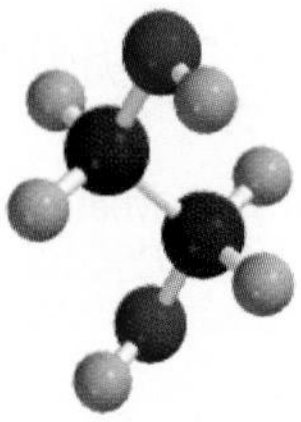

$HOCH_2CH_2OH$

1,2-Ethanediol (ethylene glycol)

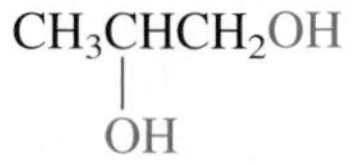
CH3CHCH2OH
OH

1,2-Propanediol (propylene glycol)

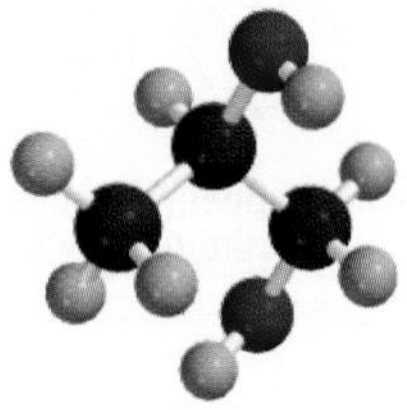

Ethylene glycol and propylene glycol are prepared industrially from the corresponding alkenes by way of their epoxides. Some applications were given in the box in Section 6.21.

Ethylene glycol and *propylene glycol* are common names for these two diols and are acceptable IUPAC names. Aside from these two compounds, the IUPAC system does not use the word *glycol* for naming diols.

In the laboratory, vicinal diols are normally prepared from alkenes using the reagent *osmium tetraoxide* (OsO_4). Osmium tetraoxide reacts rapidly with alkenes to give cyclic osmate esters.

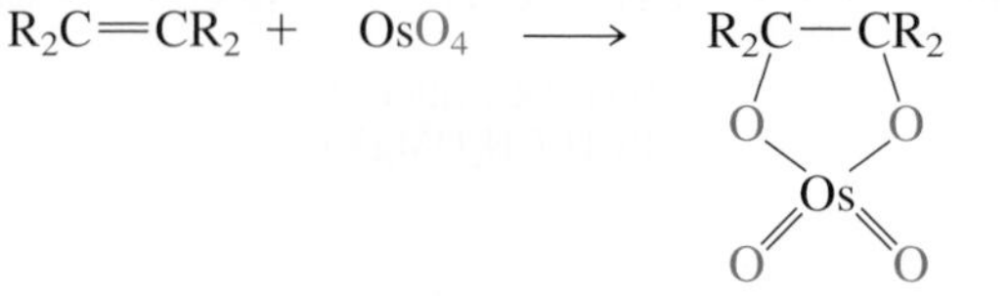
R2C=CR2 + OsO4
R2C—CR2
O
O
Os
O
O

Alkene Osmium tetraoxide Cyclic osmate ester

Osmate esters are fairly stable but are readily cleaved in the presence of an oxidizing agent such as *tert*-butyl hydroperoxide.

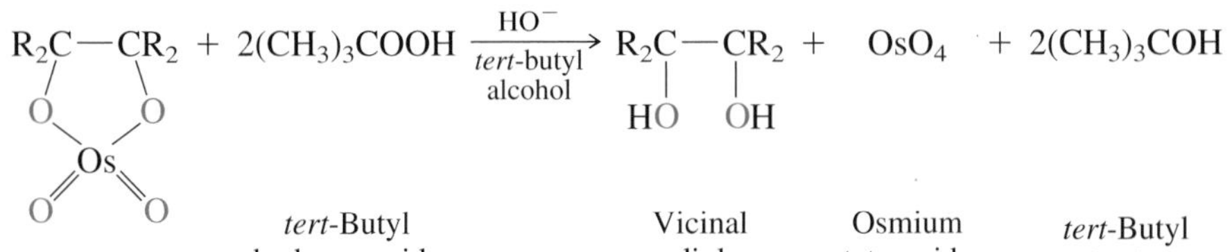

tert-Butyl hydroperoxide; Vicinal diol; Osmium tetraoxide; *tert*-Butyl alcohol

Because osmium tetraoxide is regenerated in this step, alkenes can be converted to vicinal diols using only catalytic amounts of osmium tetraoxide, which is both toxic and expensive. The entire process is performed in a single operation by simply allowing a solution of the alkene and *tert*-butyl hydroperoxide in *tert*-butyl alcohol containing a small amount of osmium tetraoxide and base to stand for several hours.

$$CH_3(CH_2)_7CH{=}CH_2 \xrightarrow[\textit{tert}\text{-butyl alcohol, } HO^-]{(CH_3)_3COOH,\ OsO_4(cat)} CH_3(CH_2)_7\underset{\displaystyle OH}{\underset{|}{C}}HCH_2OH$$

1-Decene → 1,2-Decanediol (73%)

Overall, the reaction leads to addition of two hydroxyl groups to the double bond and is referred to as **dihydroxylation.** Both oxygens of the diol come from osmium tetraoxide via the cyclic osmate ester. The reaction of OsO_4 with the alkene is a syn addition, and the conversion of the cyclic osmate to the diol involves cleavage of the bonds between oxygen and osmium. Thus, both hydroxyl groups of the diol become attached to the same face of the double bond; *syn dihydroxylation of the alkene is observed.*

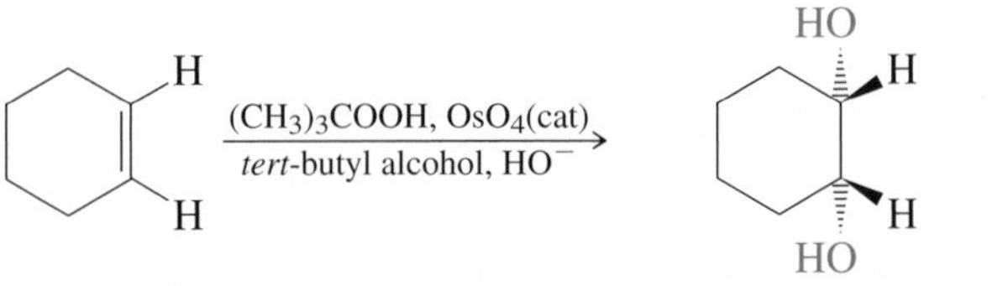

Cyclohexene → *cis*-1,2-Cyclohexanediol (62%)

Problem 15.5

Give the structures, including stereochemistry, for the diols obtained by dihydroxylation of *cis*-2-butene and *trans*-2-butene.

The osmium tetraoxide-catalyzed dihydroxylation of alkenes just described was developed by Professor K. Barry Sharpless (MIT, Stanford, Scripps Institute). He subsequently extended the method to the enantioselective synthesis of chiral diols using catalyst systems based on $K_2OsO_2(OH)_4$ as the osmium source and $K_3Fe(CN)_6$ as the oxidant. Including enantiomerically pure naturally occurring alkaloids such as dihydroquinidine as components of the reaction mixture makes these chiral reagents that are capable of dihydroxylating alkenes with high enantioselectivity.

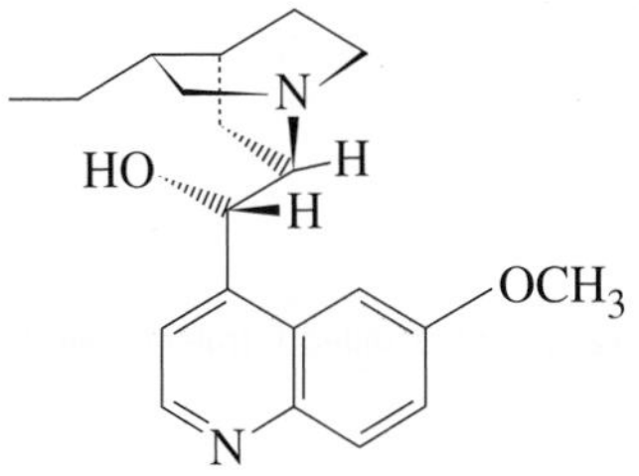

Dihydroquinidine

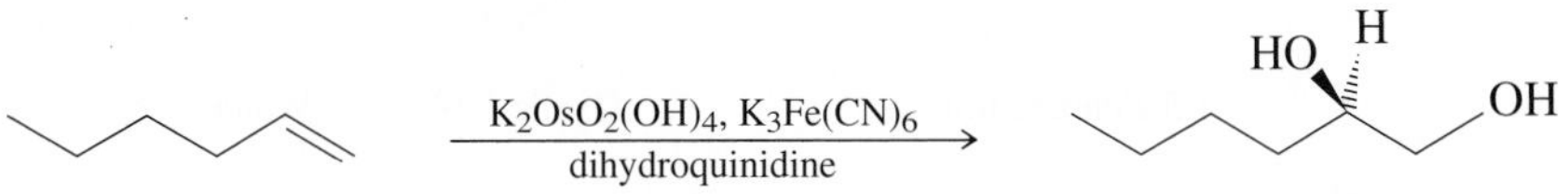

1-Hexene → (*R*)-1,2-Hexanediol (80% enantiomeric excess)

Problem 15.6

trans-2-Butene was subjected to enantioselective dihydroxylation by Sharpless's method. The 2,3-butanediol that was formed had the (*R*)-configuration at one carbon. What was the configuration at the other?

In addition to enantioselective dihydroxylation, Professor Sharpless devised a widely used method for enantioselective epoxidation (Section 16.9). The 2001 Nobel Prize in Chemistry was shared among Sharpless for enantioselective oxidations and Knowles and Noyori (Section 14.14) for enantioselective hydrogenations.

15.6 Reactions of Alcohols: A Review and a Preview

Alcohols are versatile starting materials for the preparation of a variety of organic functional groups. Several reactions of alcohols have already been seen in earlier chapters and are summarized in Table 15.2. The remaining sections of this chapter add to the list.

15.7 Conversion of Alcohols to Ethers

Primary alcohols are converted to ethers on heating in the presence of an acid catalyst, usually sulfuric acid.

$$\underset{\text{Primary alcohol}}{2RCH_2OH} \xrightarrow{H^+,\ heat} \underset{\text{Dialkyl ether}}{RCH_2OCH_2R} + \underset{\text{Water}}{H_2O}$$

This kind of reaction is called a *condensation.* A **condensation** is a reaction in which two molecules combine to form a larger one while liberating a small molecule. In this case two alcohol molecules combine to give an ether and water.

$$\underset{\text{1-Butanol}}{2CH_3CH_2CH_2CH_2OH} \xrightarrow[130°C]{H_2SO_4} \underset{\text{Dibutyl ether (60\%)}}{CH_3CH_2CH_2CH_2OCH_2CH_2CH_2CH_3} + \underset{\text{Water}}{H_2O}$$

When applied to the synthesis of ethers, the reaction is effective only with primary alcohols. Elimination to form alkenes predominates with secondary and tertiary alcohols.

Diethyl ether is prepared on an industrial scale by heating ethanol with sulfuric acid at 140°C. At higher temperatures elimination predominates, and ethylene becomes the major product. The formation of diethyl ether is outlined in Mechanism 15.2. The individual steps of this mechanism are analogous to those seen earlier. Nucleophilic attack on a protonated alcohol was encountered in the reaction of primary alcohols with hydrogen halides (Section 4.12), and the nucleophilic properties of alcohols were discussed in the context of solvolysis reactions (Section 8.5). Both the first and the last steps are proton-transfer reactions between oxygens.

Diols react intramolecularly to form cyclic ethers when a five-membered or six-membered ring can result.

$$\underset{\text{1,5-Pentanediol}}{HOCH_2CH_2CH_2CH_2CH_2OH} \xrightarrow[heat]{H_2SO_4}$$  $$+ \underset{\text{Water}}{H_2O}$$

Oxane (76%)

Oxane is also called tetrahydropyran.

In these intramolecular ether-forming reactions, the alcohol may be primary, secondary, or tertiary.

Problem 15.7

On the basis of the acid-catalyzed formation of diethyl ether from ethanol in Mechanism 15.2, write a stepwise mechanism for the formation of oxane from 1,5-pentanediol.

TABLE 15.2 Summary of Reactions of Alcohols Discussed in Earlier Chapters

Reaction (section) and comments	General equation and specific example
Reaction with hydrogen halides (Section 4.7) The order of alcohol reactivity parallels the order of carbocation stability: $R_3C^+ > R_2CH^+ > RCH_2^+$. Benzylic alcohols react readily.	$ROH + HX \longrightarrow RX + H_2O$ Alcohol, Hydrogen halide, Alkyl halide, Water 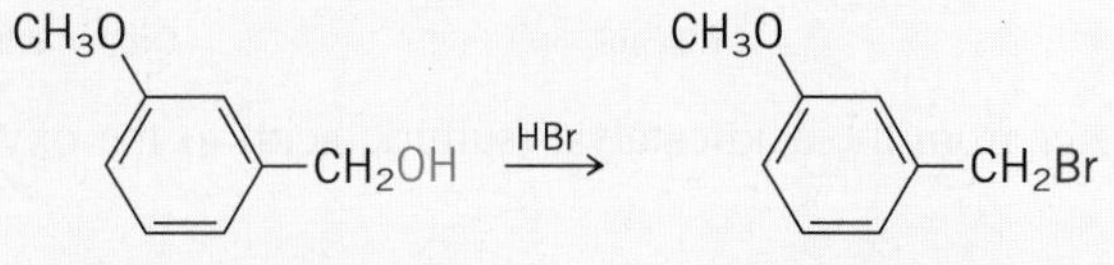 *m*-Methoxybenzyl alcohol → *m*-Methoxybenzyl bromide (98%)
Reaction with thionyl chloride (Section 4.14) Thionyl chloride converts alcohols to alkyl chlorides.	$ROH + SOCl_2 \longrightarrow RCl + SO_2 + HCl$ Alcohol, Thionyl chloride, Alkyl chloride, Sulfur dioxide, Hydrogen chloride $(CH_3)_2C{=}CHCH_2CH_2CH(OH)CH_3 \xrightarrow[\text{diethyl ether}]{SOCl_2,\ \text{pyridine}} (CH_3)_2C{=}CHCH_2CH_2CH(Cl)CH_3$ 6-Methyl-5-hepten-2-ol → 6-Chloro-2-methyl-2-heptene (67%)
Reaction with phosphorus tribromide (Section 4.14) Phosphorus tribromide converts alcohols to alkyl bromides.	$3ROH + PBr_3 \longrightarrow 3RBr + H_3PO_3$ Alcohol, Phosphorus tribromide, Alkyl bromide, Phosphorous acid 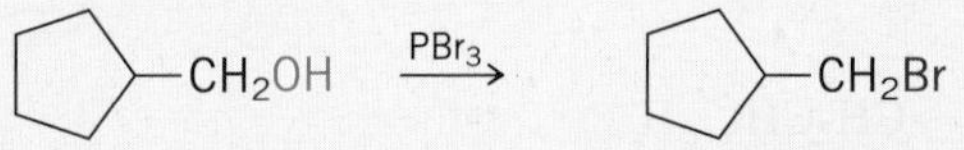 Cyclopentylmethanol → (Bromomethyl)cyclopentane (50%)
Acid-catalyzed dehydration (Section 5.9) This is a frequently used procedure for the preparation of alkenes. The order of alcohol reactivity parallels the order of carbocation stability: $R_3C^+ > R_2CH^+ > RCH_2^+$. Benzylic alcohols react readily. Rearrangements are sometimes observed.	$R_2C(OH)CHR_2 \xrightarrow[\text{heat}]{H^+} R_2C{=}CR_2 + H_2O$ Alcohol, Alkene, Water 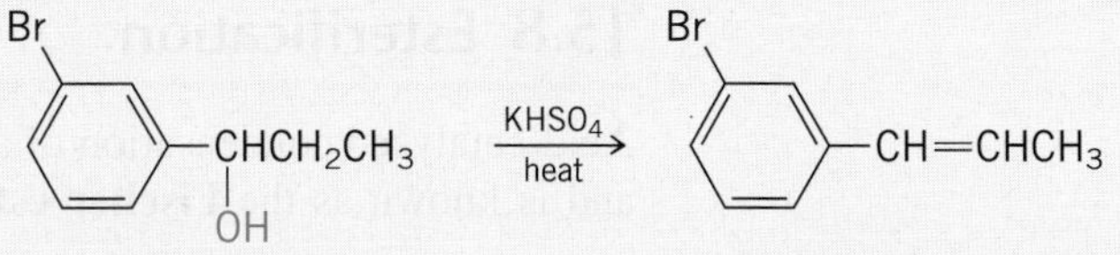 1-(*m*-Bromophenyl)-1-propanol → 1-(*m*-Bromophenyl)propene (71%)
Conversion to *p*-toluenesulfonates (Section 8.12) Alcohols react with *p*-toluenesulfonyl chloride to give *p*-toluenesulfonates. Sulfonates are reactive substrates for nucleophilic substitution and elimination.	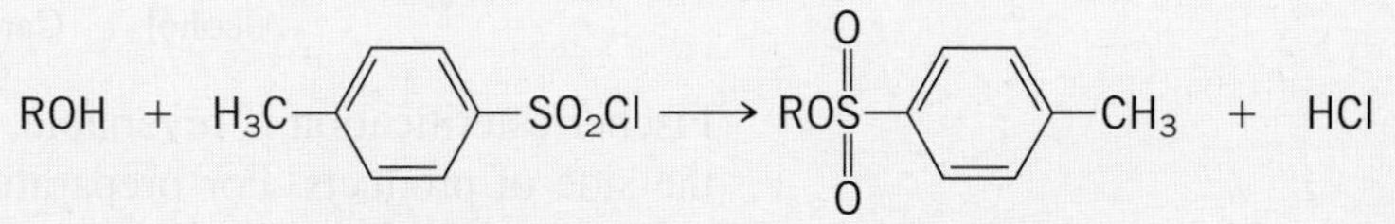 Alcohol, *p*-Toluenesulfonyl chloride, Alkyl *p*-toluenesulfonate, Hydrogen chloride 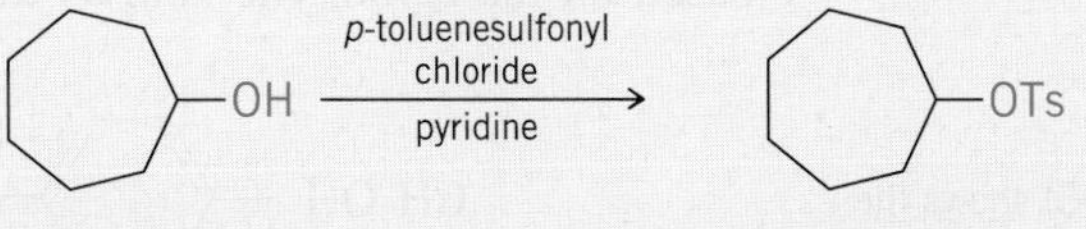 Cycloheptanol → Cycloheptyl *p*-toluenesulfonate (83%)

Mechanism 15.2

Acid-Catalyzed Formation of Diethyl Ether from Ethyl Alcohol

THE OVERALL REACTION:

$$2CH_3CH_2OH \xrightarrow[140°C]{H_2SO_4} CH_3CH_2OCH_2CH_3 + H_2O$$

Ethanol — Diethyl ether — Water

Step 1: Proton transfer from the acid catalyst (sulfuric acid) to the oxygen of the alcohol to produce an alkyloxonium ion.

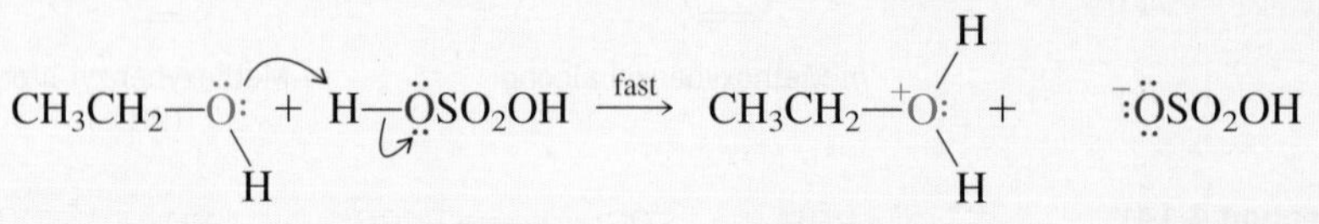

Ethanol — Sulfuric acid — Ethyloxonium ion — Hydrogen sulfate ion

Step 2: Nucleophilic attack by a molecule of alcohol on the alkyloxonium ion formed in step 1.

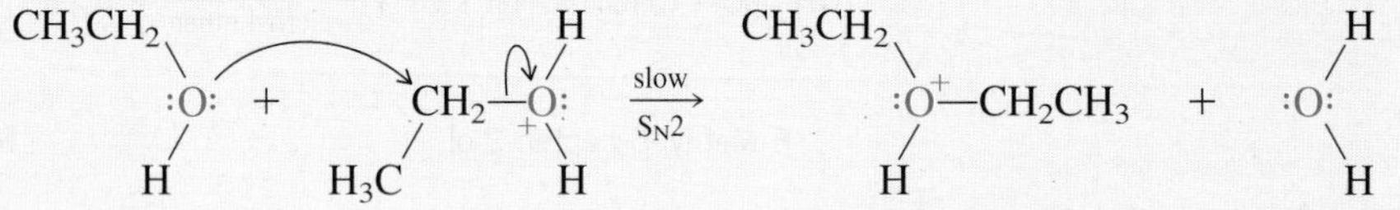

Ethanol — Ethyloxonium ion — Diethyloxonium ion — Water

Step 3: The product of step 2 is the conjugate acid of the dialkyl ether. It is deprotonated in the final step of the process to give the ether.

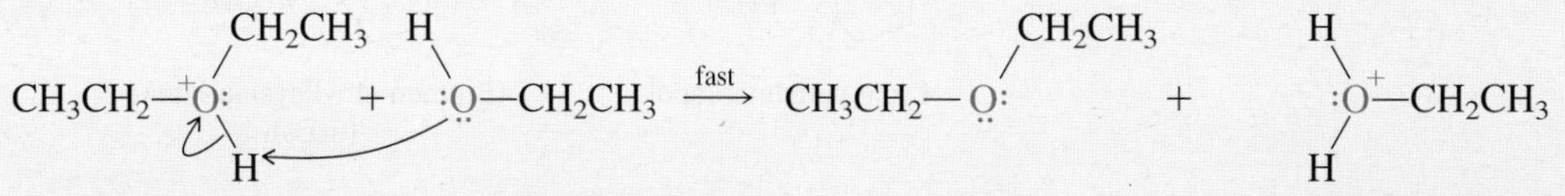

Diethyloxonium ion — Ethanol — Diethyl ether — Ethyloxonium ion

15.8 Esterification

Acid-catalyzed condensation of an alcohol and a carboxylic acid yields an ester and water and is known as the **Fischer esterification.**

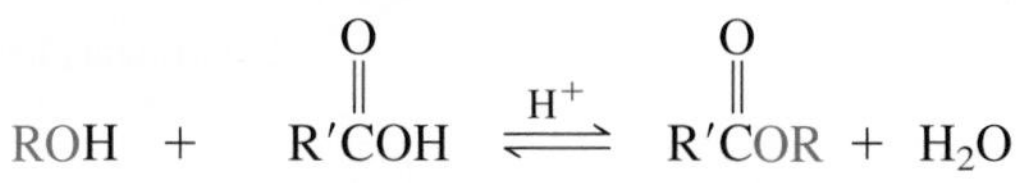

Alcohol — Carboxylic acid — Ester — Water

Fischer esterification is reversible, and the position of equilibrium usually lies slightly to the side of products. For preparative purposes, the position of equilibrium can be made more favorable by using either the alcohol or the carboxylic acid in excess. In the following example, in which an excess of the alcohol was employed, the yield indicated is based on the carboxylic acid as the limiting reactant.

Mechanism 18.1 (page 796) shows the mechanism of this reaction.

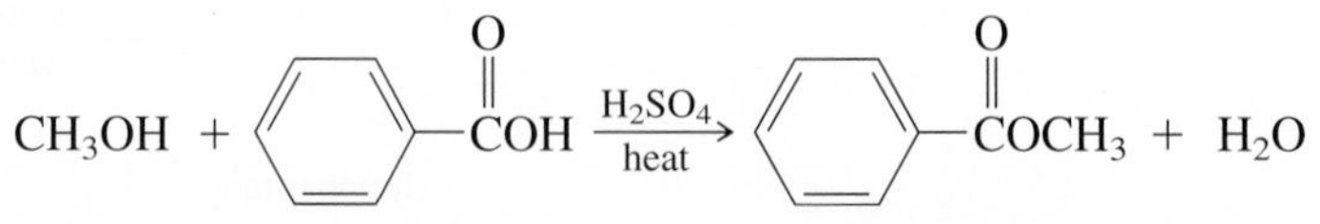

Methanol (0.6 mol) — Benzoic acid (0.1 mol) — Methyl benzoate (isolated in 70% yield based on benzoic acid) — Water

Another way to shift the position of equilibrium to favor the formation of ester is to remove water from the reaction mixture by using benzene as a cosolvent and distilling the azeotropic mixture of benzene and water. This can be accomplished in the laboratory with a Dean–Stark trap.

An *azeotropic mixture* contains two or more substances that distill together at a constant boiling point. The benzene–water azeotrope contains 9% water and boils at 69°C.

$$CH_3CH(OH)CH_2CH_3 + CH_3\overset{O}{\overset{\|}{C}}OH \xrightarrow[\text{benzene, heat}]{H^+} CH_3\overset{O}{\overset{\|}{C}}OCH(CH_3)CH_2CH_3 + H_2O$$

sec-Butyl alcohol (0.20 mol) | Acetic acid (0.25 mol) | *sec*-Butyl acetate (isolated in 71% yield based on *sec*-butyl alcohol) | Water (codistills with benzene)

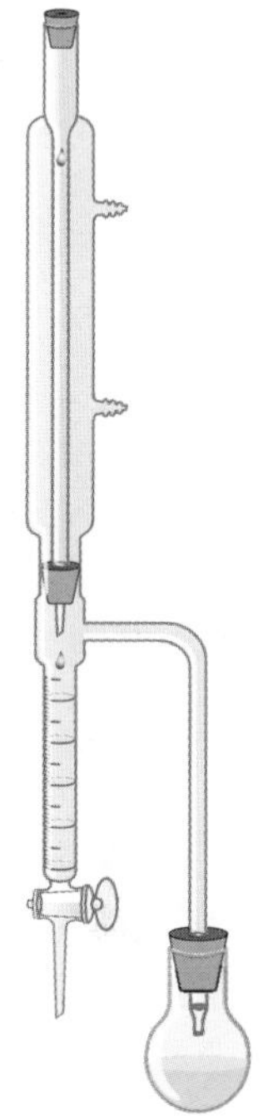

A reaction apparatus with a Dean–Stark trap. The water is denser than the ester–benzene mixture, and collects in the side arm of the trap.

Problem 15.8

Write the structure of the ester formed in each of the following reactions:

(a) $CH_3CH_2CH_2CH_2OH + CH_3CH_2\overset{O}{\overset{\|}{C}}OH \xrightarrow[\text{heat}]{H_2SO_4}$

(b) 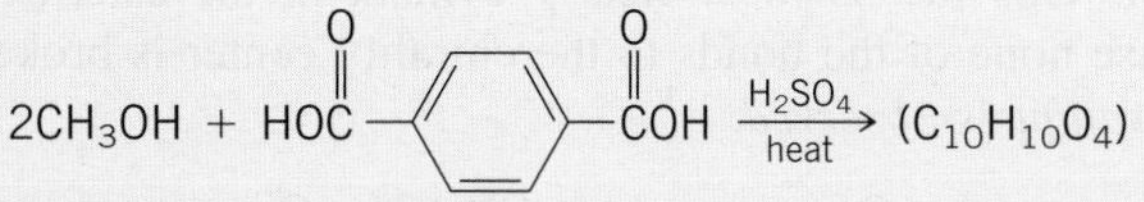

Sample Solution (a) By analogy to the general equation and to the examples cited in this section, we can write the equation

$$CH_3CH_2CH_2CH_2OH + CH_3CH_2\overset{O}{\overset{\|}{C}}OH \xrightarrow[\text{heat}]{H_2SO_4} CH_3CH_2\overset{O}{\overset{\|}{C}}OCH_2CH_2CH_2CH_3 + H_2O$$

1-Butanol | Propanoic acid | Butyl propanoate | Water

As actually carried out in the laboratory, 3 mol of propanoic acid was used per mole of 1-butanol, and the desired ester was obtained in 78% yield.

Esters are also formed by the reaction of alcohols with acyl chlorides:

$$ROH + R'\overset{O}{\overset{\|}{C}}Cl \longrightarrow R'\overset{O}{\overset{\|}{C}}OR + HCl$$

Alcohol | Acyl chloride | Ester | Hydrogen chloride

This reaction is normally carried out in the presence of a weak base such as pyridine, which reacts with the hydrogen chloride that is formed.

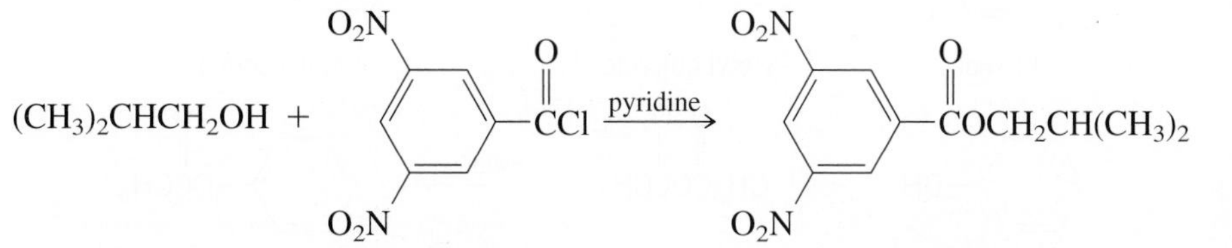
(CH3)2CHCH2OH +
O2N
O
CCl
pyridine
O2N
O2N
O
COCH2CH(CH3)2
O2N

Isobutyl alcohol | 3,5-Dinitrobenzoyl chloride | Isobutyl 3,5-dinitrobenzoate (86%)

Acetic anhydride is

$$\mathrm{CH_3\overset{O}{\overset{\|}{C}}O\overset{O}{\overset{\|}{C}}CH_3}$$

Acid anhydrides react similarly to acyl chlorides.

$$\underset{\text{Alcohol}}{\mathrm{ROH}} + \underset{\text{Acid anhydride}}{\mathrm{R'\overset{O}{\overset{\|}{C}}O\overset{O}{\overset{\|}{C}}R'}} \longrightarrow \underset{\text{Ester}}{\mathrm{R'\overset{O}{\overset{\|}{C}}OR}} + \underset{\text{Carboxylic acid}}{\mathrm{R'\overset{O}{\overset{\|}{C}}OH}}$$

$$\underset{\text{2-Phenylethanol}}{\mathrm{C_6H_5CH_2CH_2OH}} + \underset{\text{Trifluoroacetic anhydride}}{\mathrm{CF_3\overset{O}{\overset{\|}{C}}O\overset{O}{\overset{\|}{C}}CF_3}} \xrightarrow{\text{pyridine}} \underset{\text{2-Phenylethyl trifluoroacetate (83\%)}}{\mathrm{C_6H_5CH_2CH_2O\overset{O}{\overset{\|}{C}}CF_3}} + \underset{\text{Trifluoroacetic acid}}{\mathrm{CF_3\overset{O}{\overset{\|}{C}}OH}}$$

The mechanisms of the Fischer esterification and the reactions of alcohols with acyl chlorides and acid anhydrides will be discussed in detail in Chapters 18 and 19 after some fundamental principles of carbonyl group reactivity have been developed. For the present, it is sufficient to point out that most of the reactions that convert alcohols to esters leave the C—O bond of the alcohol intact.

$$\mathrm{H{-}O{-}R} \longrightarrow \mathrm{R'\overset{O}{\overset{\|}{C}}{-}O{-}R}$$

This is the same oxygen that was attached to the group R in the starting alcohol.

The acyl group of the carboxylic acid, acyl chloride, or acid anhydride is transferred to the oxygen of the alcohol. This fact is most clearly evident in the esterification of chiral alcohols, where, because none of the bonds to the chirality center is broken in the process, *retention of configuration is observed.*

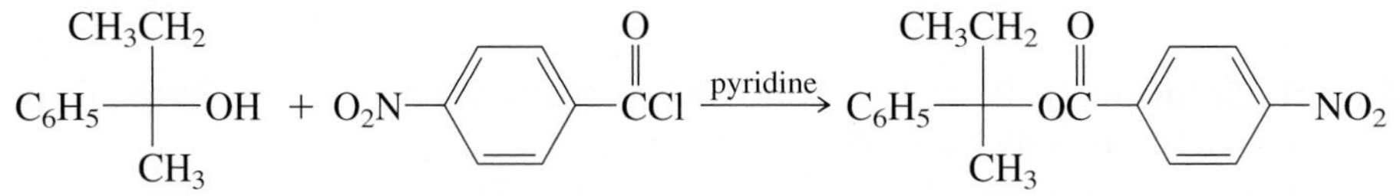
CH3CH2
C6H5
OH
CH3
O2N
O
CCl
pyridine
CH3CH2
O
C6H5
OC
NO2
CH3

(*R*)-(+)-2-Phenyl-2-butanol — *p*-Nitrobenzoyl chloride — (*R*)-(−)-1-Methyl-1-phenylpropyl *p*-nitrobenzoate (63% yield)

Problem 15.9

From what alcohol and acyl chloride can the following esters be synthesized? From what alcohol and acid anhydride?

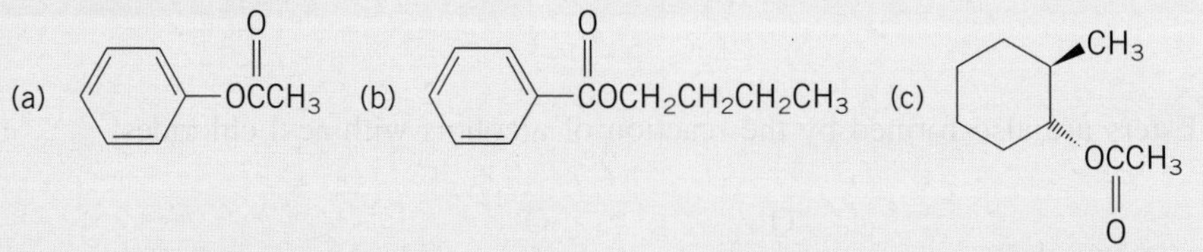
(a)
O
OCCH3
(b)
O
COCH2CH2CH2CH3
(c)
CH3
OCCH3
O

Sample Solution (a) The oxygen that has a single bond to the carbonyl carbon is the alcohol oxygen, and the carbonyl carbon is part of the acyl chloride or anhydride. The compound in part (a) is phenyl acetate, and it can be prepared from phenol and acetyl chloride, or acetic anhydride.

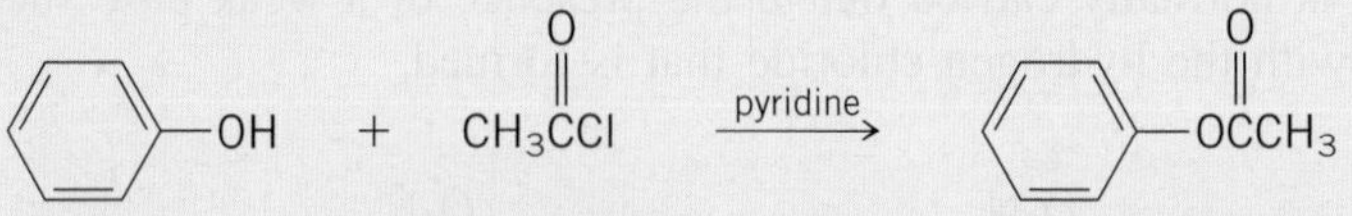
OH
+
O
CH3CCl
pyridine
O
OCCH3

Phenol — Acetyl chloride — Phenyl acetate

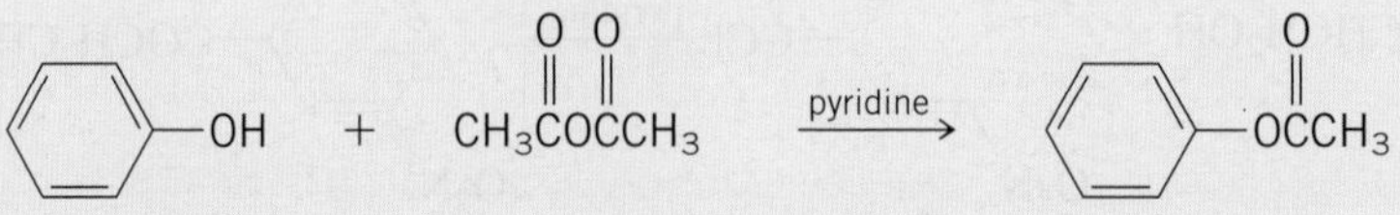
OH
+
O O
CH3COCCH3
pyridine
O
OCCH3

Phenol — Acetic anhydride — Phenyl acetate

15.9 Oxidation of Alcohols

Oxidation of an alcohol yields a carbonyl compound. Whether the resulting carbonyl compound is an aldehyde, a ketone, or a carboxylic acid depends on the alcohol and on the oxidizing agent.

Primary alcohols are oxidized either to an aldehyde or to a carboxylic acid:

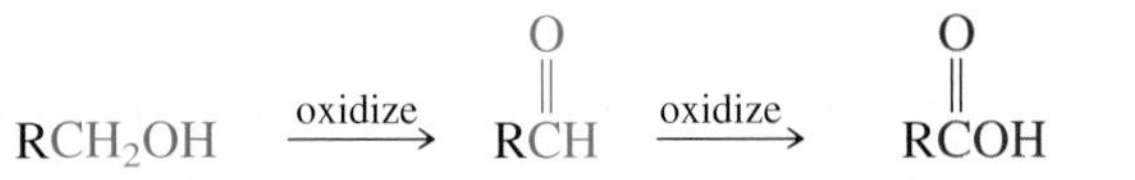

Primary alcohol Aldehyde Carboxylic acid

Vigorous oxidation leads to the formation of a carboxylic acid, but a number of methods permit us to stop the oxidation at the intermediate aldehyde stage. The reagents most commonly used for oxidizing alcohols are based on high-oxidation-state transition metals, particularly chromium(VI).

Chromic acid (H_2CrO_4) is a good oxidizing agent and is formed when solutions containing chromate (CrO_4^{2-}) or dichromate ($Cr_2O_7^{2-}$) are acidified. Sometimes it is possible to obtain aldehydes in satisfactory yield before they are further oxidized, but in most cases carboxylic acids are the major products isolated on treatment of primary alcohols with chromic acid.

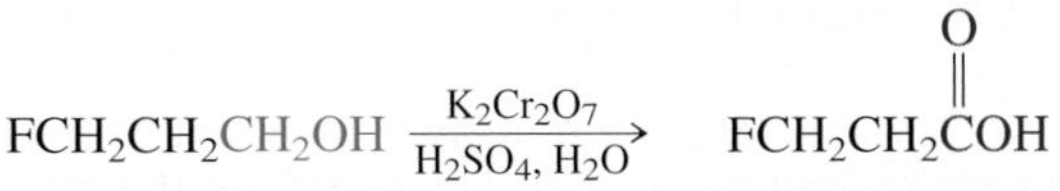

3-Fluoro-1-propanol 3-Fluoropropanoic acid (74%)

Conditions that do permit the easy isolation of aldehydes in good yield by oxidation of primary alcohols employ various Cr(VI) species as the oxidant in *anhydrous* media. Two such reagents are pyridinium chlorochromate (PCC), $C_5H_5NH^+\ ClCrO_3^-$, and pyridinium dichromate (PDC), $(C_5H_5NH)_2^{2+}\ Cr_2O_7^{2-}$; both are used in dichloromethane.

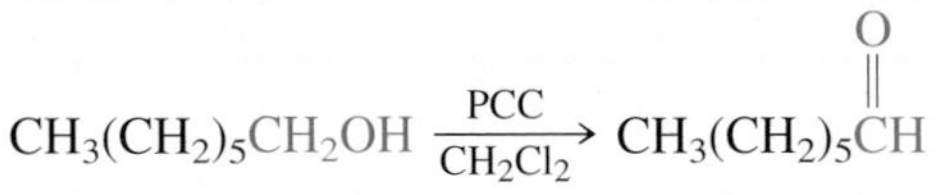

1-Heptanol Heptanal (78%)

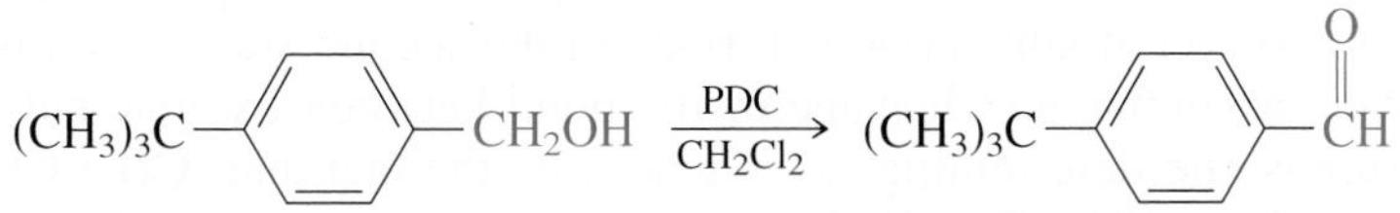

p-tert-Butylbenzyl alcohol *p-tert*-Butylbenzaldehyde (94%)

Secondary alcohols are oxidized to ketones by the same reagents that oxidize primary alcohols:

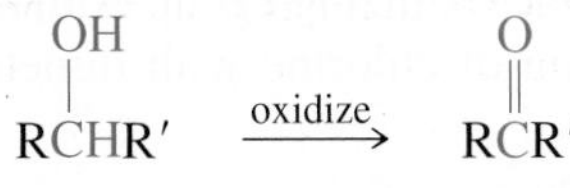

Secondary alcohol Ketone

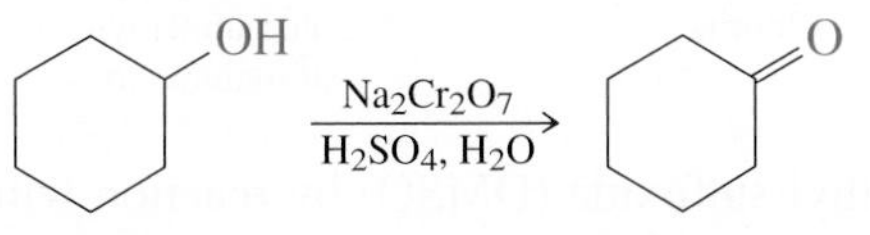

Cyclohexanol Cyclohexanone (85%)

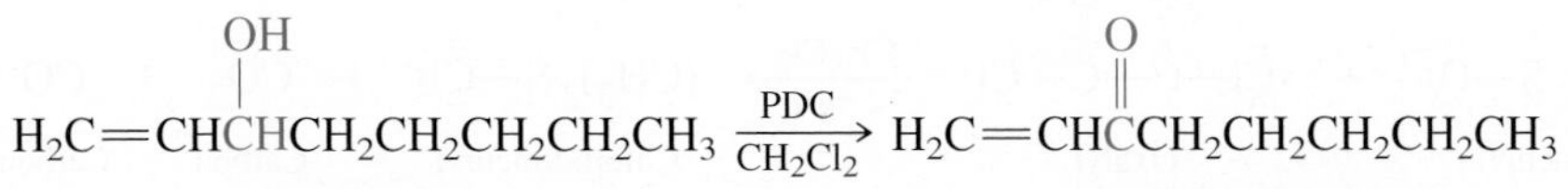

1-Octen-3-ol 1-Octen-3-one (80%)

Tertiary alcohols have no hydrogen on their hydroxyl-bearing carbon and do not undergo oxidation readily:

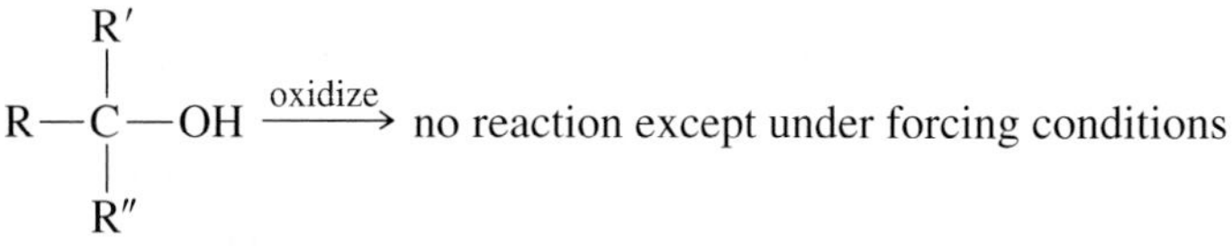
R′
R—C—OH
R″
oxidize
no reaction except under forcing conditions

In the presence of strong oxidizing agents at elevated temperatures, oxidation of tertiary alcohols leads to cleavage of the various carbon–carbon bonds at the hydroxyl-bearing carbon atom, and a complex mixture of products results.

Problem 15.10

Predict the principal organic product of each of the following reactions:

(a) $ClCH_2CH_2CH_2CH_2OH \xrightarrow[H_2SO_4,\ H_2O]{K_2Cr_2O_7}$

(b) $CH_3CHCH_2CH_2CH_2CH_2CH_2CH_3 \xrightarrow[H_2SO_4,\ H_2O]{Na_2Cr_2O_7}$ (OH on C-2)

(c) $CH_3CH_2CH_2CH_2CH_2CH_2CH_2OH \xrightarrow[CH_2Cl_2]{PCC}$

Sample Solution (a) The reactant is a primary alcohol and so can be oxidized either to an aldehyde or to a carboxylic acid. Aldehydes are the major products only when the oxidation is carried out in anhydrous media. Carboxylic acids are formed when water is present. The reaction shown produced 4-chlorobutanoic acid in 56% yield.

$$ClCH_2CH_2CH_2CH_2OH \xrightarrow[H_2SO_4,\ H_2O]{K_2Cr_2O_7} ClCH_2CH_2CH_2\overset{O}{\overset{\|}{C}}OH$$

4-Chloro-1-butanol → 4-Chlorobutanoic acid

Mechanism 15.3 outlines the mechanism of chromic acid oxidation of 2-propanol to acetone. The alcohol reacts with chromic acid in the first step to give a chromate ester. A carbon–oxygen double bond is formed in the second step when loss of a proton from carbon accompanies cleavage of the bond between oxygen and chromium. The second step is rate-determining as evidenced by the fact that $(CH_3)_2CHOH$ reacts 6.7 times faster than $(CH_3)_2CDOH$. If the second step were faster than the first, no deuterium isotope effect (Section 5.17) would have been observed.

Chromic acid and other Cr(VI)-containing compounds are toxic, so chemists have developed other reagents for oxidizing alcohols. Several are based on chlorodimethylsulfonium ion $[(CH_3)_2\overset{+}{S}Cl]$ as the oxidizing agent. Although chlorodimethylsulfonium ion can be prepared by the reaction of chlorine with dimethyl sulfide:

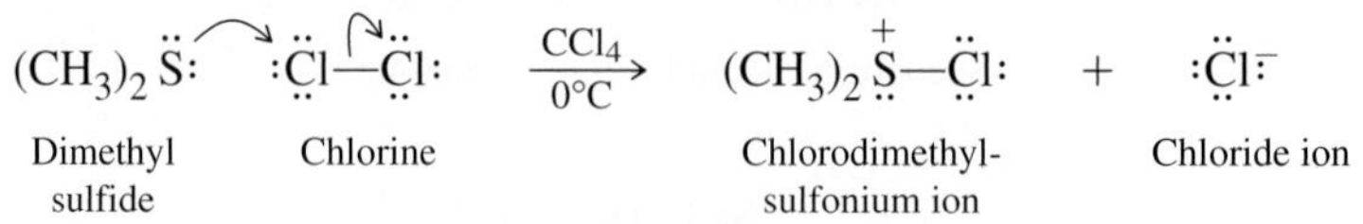
CCl4
0°C
Dimethyl sulfide
Chlorine
Chlorodimethyl-sulfonium ion
Chloride ion

its formation from dimethyl sulfoxide (DMSO) by reaction with oxalyl chloride is more convenient.

Oxidations of alcohols using the dimethyl sulfoxide–oxalyl chloride combination are known as *Swern oxidations* after Daniel Swern of Temple University who developed the method.

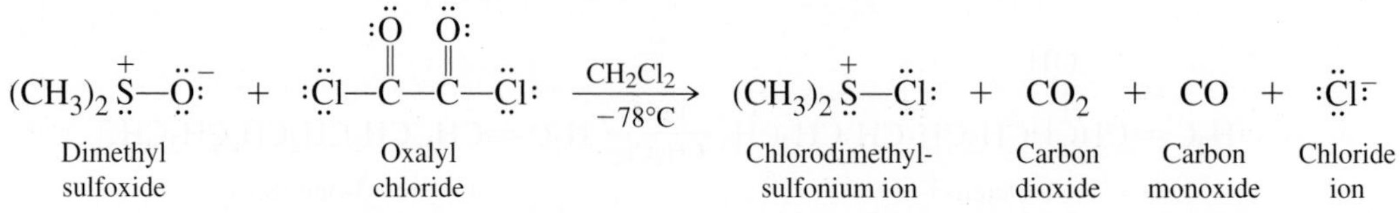
CH2Cl2
−78°C
Dimethyl sulfoxide
Oxalyl chloride
Chlorodimethyl-sulfonium ion
Carbon dioxide
Carbon monoxide
Chloride ion

Mechanism 15.3

Chromic Acid Oxidation of 2-Propanol

Step 1: Reaction of the alcohol with chromic acid gives an alkyl chromate.

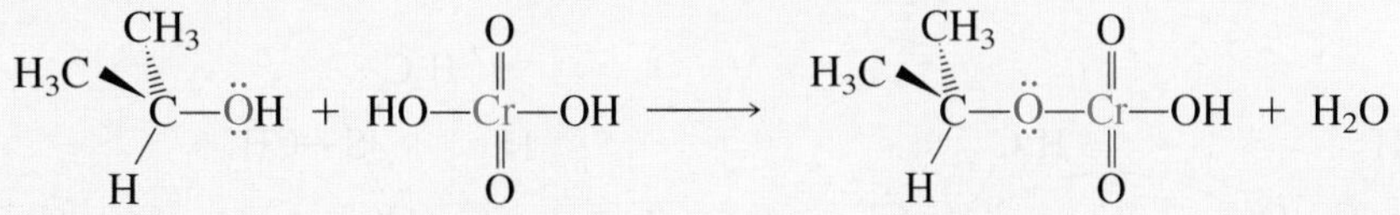
2-Propanol
Chromic acid
Isopropyl chromate
Water

Step 2: The oxidation step can be viewed as a β elimination. Water acts as a base to remove a proton from carbon while the O—Cr bond breaks.

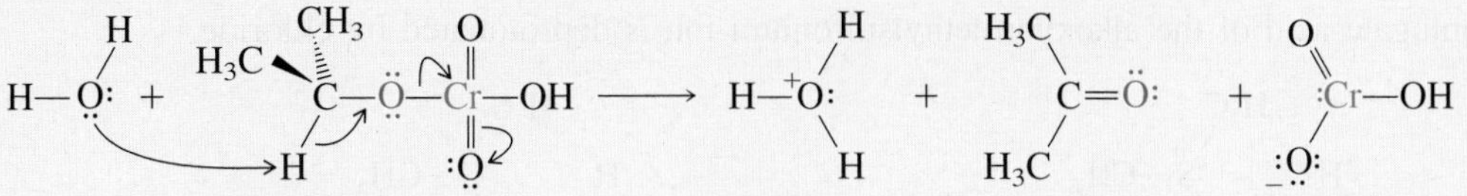
Water
Isopropyl chromate
Hydronium ion
Acetone
Chromite ion

Step 3: A series of redox reactions converts chromium from the 4+ oxidation state in $HCrO_3^-$ to the 3+ oxidation state.

Typical conditions involve formation of the chlorodimethylsulfonium ion in dichloromethane at low temperature, followed by addition of the alcohol to be oxidized, then treatment with a weak base such as triethylamine. Primary alcohols yield aldehydes; secondary alcohols yield ketones.

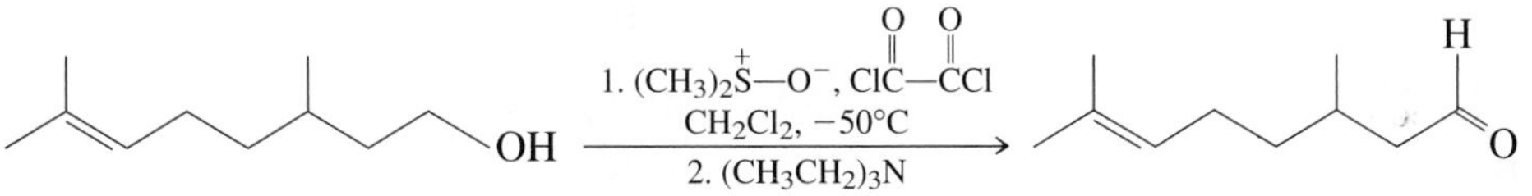
OH
1. $(CH_3)_2\overset{+}{S}—O^-$, ClC(=O)—C(=O)Cl
CH_2Cl_2, −50°C
2. $(CH_3CH_2)_3N$
H
O
Citronellol
Citronellal (83%)

Mechanism 15.4 outlines the mechanism, starting with reaction of the alcohol with chlorodimethylsulfonium ion (steps 1–2). Steps 3–4 describe what happens after the amine base is added.

Problem 15.11

Only one step in Mechanism 15.4 involves oxidation–reduction. Which one? Which atom is oxidized? Which atom is reduced?

Problem 15.12

Steps 3–4 in Mechanism 15.4 were suggested on the basis of deuterium-labeling studies carried out in a related procedure. If steps 3–4 are correct, what is the expected distribution of deuterium in the products of oxidation of $CH_3CH_2CH_2CD_2OH$?

Mechanism 15.4

Dimethyl Sulfoxide Oxidation of an Alcohol

Step 1: The alcohol reacts with the active oxidizing agent, chlorodimethylsulfonium ion, previously generated by the reaction between dimethyl sulfoxide and oxalyl chloride. The oxygen of the alcohol displaces chloride from sulfur.

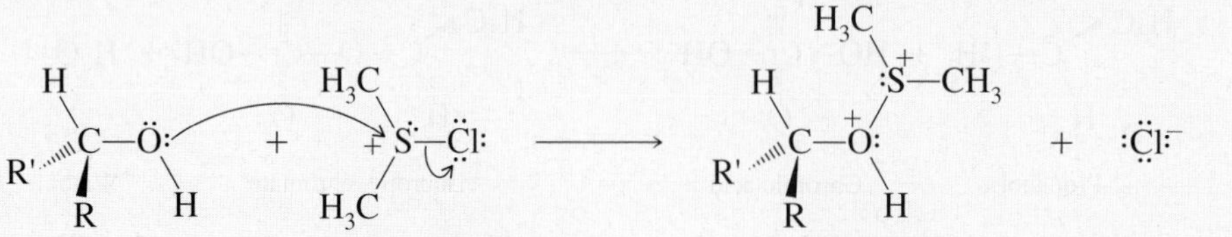

Alcohol — Chlorodimethylsulfonium ion — Conjugate acid of an alkoxydimethylsulfonium ion — Chloride ion

Step 2: The conjugate acid of the alkoxydimethylsulfonium ion is deprotonated by chloride.

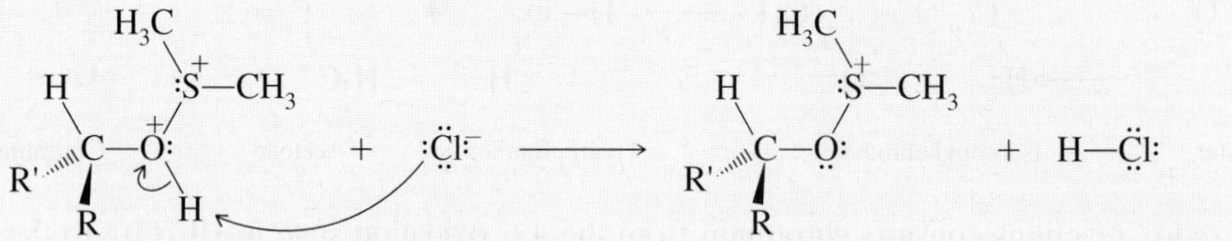

Conjugate acid of an alkoxydimethylsulfonium ion — Chloride ion — Alkoxydimethylsulfonium ion — Hydrogen chloride

Step 3: Triethylamine is added to the reaction mixture and acts as a base to remove a proton from one of the *S*-methyl groups of the alkoxydimethylsulfonium ion. The product of this reaction is an *ylide*—a neutral molecule in which two directly bonded atoms, each with an octet of electrons, are oppositely charged.

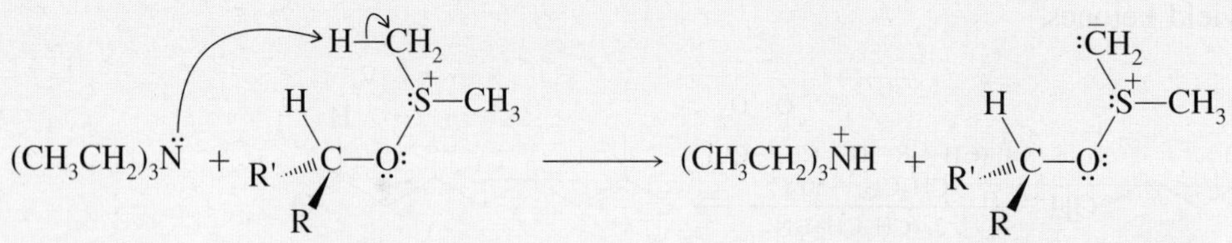

Triethylamine — Alkoxydimethylsulfonium ion — Triethylammonium ion — *S*-Alkoxy-*S*-methylsulfonium methylide

Step 4: Intramolecular proton abstraction from the H—C—O unit by the negatively charged carbon triggers dissociation of the ylide to an aldehyde or ketone and dimethyl sulfide.

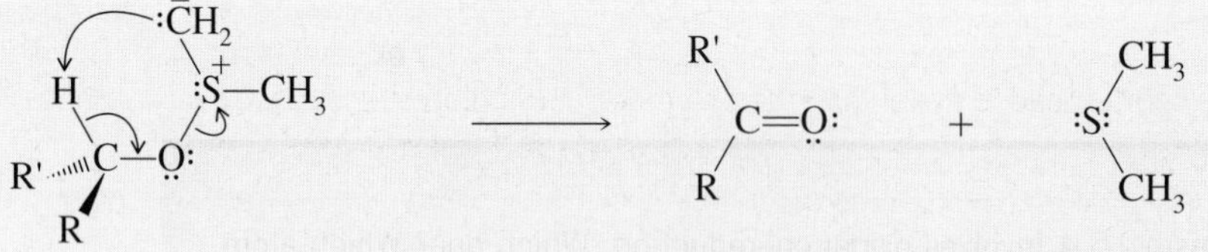

S-Alkoxy-*S*-methylsulfonium methylide — Aldehyde or ketone — Dimethyl sulfide

15.10 Biological Oxidation of Alcohols

Many biological processes involve oxidation of alcohols to carbonyl compounds or the reverse process, reduction of carbonyl compounds to alcohols. Ethanol, for example, is metabolized in the liver to acetaldehyde in a reaction catalyzed by the enzyme *alcohol dehydrogenase.*

Economic and Environmental Factors in Organic Synthesis

Beyond the obvious difference in scale that is evident when one compares preparing tons of a compound versus preparing just a few grams of it, there are sharp distinctions between "industrial" and "laboratory" syntheses. On a laboratory scale, a chemist is normally concerned only with obtaining a modest amount of a substance. Sometimes making the compound is an end in itself, but on other occasions the compound is needed for some further study of its physical, chemical, or biological properties. Considerations such as the cost of reagents and solvents tend to play only a minor role when planning most laboratory syntheses. Faced with a choice between two synthetic routes to a particular compound, one based on the cost of chemicals and the other on the efficient use of a chemist's time, the decision is almost always made in favor of the latter.

Not so for synthesis in the chemical industry, where a compound must be prepared not only on a large scale, but at low cost. There is a pronounced bias toward reactants and reagents that are both abundant and inexpensive. The oxidizing agent of choice in the chemical industry, for example, is O_2, and extensive research has been devoted to developing catalysts for preparing various compounds by air oxidation of readily available starting materials. To illustrate, air and ethylene are the reactants for the industrial preparation of both acetaldehyde and ethylene oxide. Which of the two products is obtained depends on the catalyst employed.

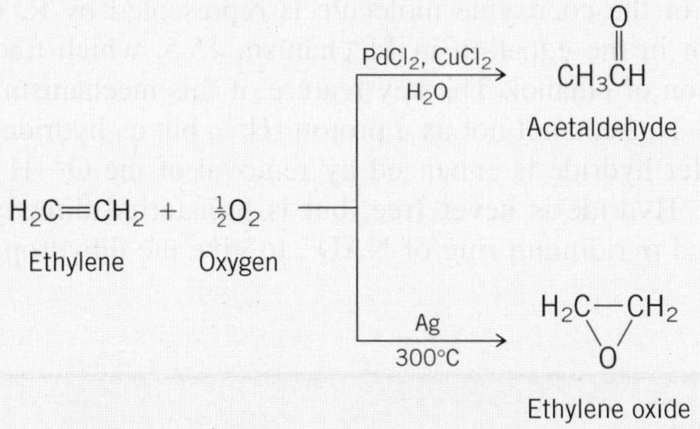

H2C=CH2 + ½O2
Ethylene
Oxygen
PdCl2, CuCl2
H2O
CH3CH
Acetaldehyde
Ag
300°C
H2C—CH2
O
Ethylene oxide

Dating approximately from the creation of the U.S. Environmental Protection Agency (EPA) in 1970, dealing with the byproducts of synthetic procedures has become an increasingly important consideration in designing a chemical synthesis. In terms of changing the strategy of synthetic planning, the chemical industry actually had a shorter road to travel than the pharmaceutical industry, academic laboratories, and research institutes. Simple business principles had long dictated that waste chemicals represented wasted opportunities. It made better sense for a chemical company to recover the solvent from a reaction and use it again than to throw it away and buy more. Similarly, it was far better to find a "value-added" use for a byproduct from a reaction than to throw it away. By raising the cost of generating chemical waste, environmental regulations increased the economic incentive to design processes that produced less of it.

The terms *green chemistry* and *environmentally benign synthesis* have been coined to refer to procedures explicitly designed to minimize the formation of byproducts that present disposal problems. Both the National Science Foundation and the Environmental Protection Agency have allocated a portion of their grant budgets to encourage efforts in this vein.

The application of environmentally benign principles to laboratory-scale synthesis can be illustrated by revisiting the oxidation of alcohols. As noted in Section 15.9, an alternative to chromium (VI)-based oxidants is the Swern oxidation. Another method is one that uses sodium hypochlorite. Aqueous solutions of sodium hypochlorite are available as "swimming-pool chlorine," and procedures for their use in oxidizing secondary alcohols to ketones have been developed.

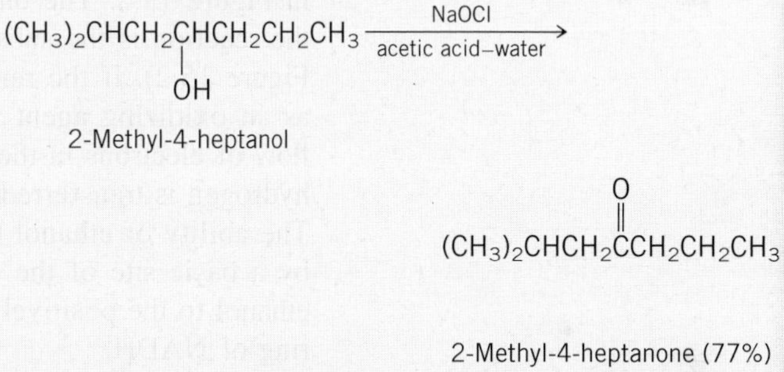

(CH3)2CHCH2CHCH2CH2CH3
OH
2-Methyl-4-heptanol
NaOCl
acetic acid–water
(CH3)2CHCH2CCH2CH2CH3
O
2-Methyl-4-heptanone (77%)

There is a curious irony in the nomination of hypochlorite as an environmentally benign oxidizing agent. It comes at a time of increasing pressure to eliminate chlorine and chlorine-containing compounds from the environment to as great a degree as possible. Any all-inclusive assault on chlorine needs to be carefully scrutinized, especially when one remembers that chlorination of the water supply has probably done more to extend human life than any other public health measure ever undertaken. (The role of chlorine in the formation of chlorinated hydrocarbons in water is discussed in Section 20.14.)

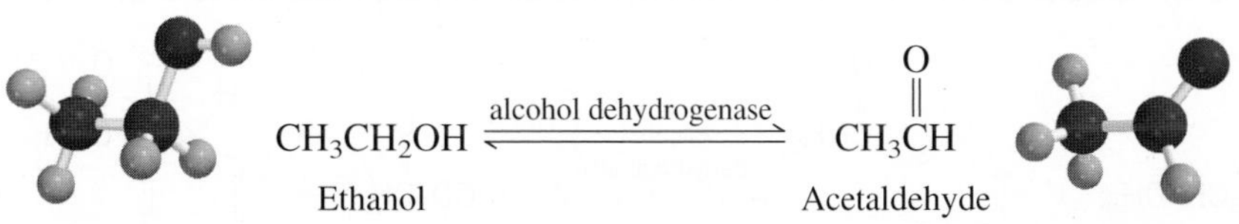

CH3CH2OH
Ethanol
alcohol dehydrogenase
CH3CH
O
Acetaldehyde

In addition to enzymes, biological oxidations require substances known as *coenzymes.* Coenzymes are organic molecules that, in concert with an enzyme, act on a substrate to bring about chemical change. Most vitamins are coenzymes. A coenzyme contains a functional group that is complementary to a functional group of the substrate;

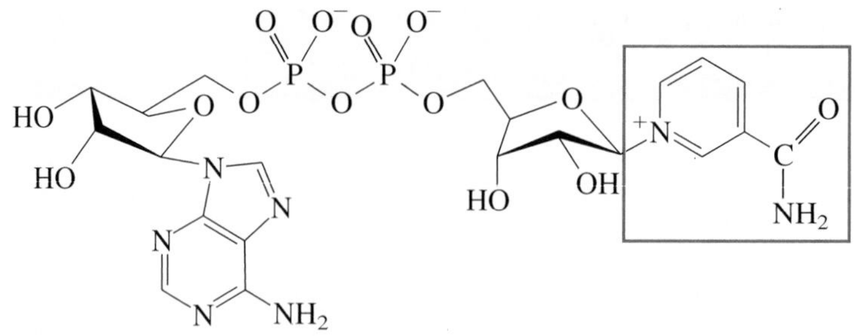

Figure 15.2

Structure of NAD^+, the oxidized form of the coenzyme nicotinamide adenine dinucleotide. The functional part of the coenzyme is framed in red.

the enzyme catalyzes the interaction of these mutually complementary functional groups. If ethanol is oxidized, some other substance must be reduced. This other substance is the oxidized form of the coenzyme *nicotinamide adenine dinucleotide* (NAD). By representing the oxidized form as NAD^+ and the reduced form as NADH, the chemical equation for the biological oxidation of ethanol may be written:

$$\underset{\text{Ethanol}}{CH_3CH_2OH} + \underset{\text{Oxidized form of NAD coenzyme}}{NAD^+} \xrightleftharpoons{\text{alcohol dehydrogenase}} \underset{\text{Acetaldehyde}}{CH_3\overset{O}{\overset{\|}{C}}H} + \underset{\text{Reduced form of NAD coenzyme}}{NADH} + H^+$$

The structure of the oxidized form of nicotinamide adenine dinucleotide is shown in Figure 15.2. The only portion of the coenzyme that undergoes chemical change in the reaction is the substituted pyridine ring of the nicotinamide unit (framed in red in Figure 15.2). If the remainder of the coenzyme molecule is represented by R, its role as an oxidizing agent is shown in the equation in Mechanism 15.5, which tracks the flow of electrons in the oxidation of ethanol. The key feature of this mechanism is that hydrogen is transferred from C-1 of ethanol not as a proton (H^+), but as hydride ($H:^-$). The ability of ethanol to transfer hydride is enhanced by removal of the O—H proton by a basic site of the enzyme. Hydride is never free, but is transferred directly from ethanol to the positively charged pyridinium ring of NAD^+ to give the dihydropyridine ring of NADH.

Problem 15.13

The mechanism of enzymatic oxidation has been studied by isotopic labeling with the aid of deuterated derivatives of ethanol. Specify the number of deuterium atoms that you would expect to find attached to the dihydropyridine ring of NADH following enzymatic oxidation of each of the alcohols given:

(a) CD_3CH_2OH (b) CH_3CD_2OH (c) CH_3CH_2OD

Sample Solution According to the proposed mechanism for biological oxidation of ethanol, the hydrogen that is transferred to the coenzyme comes from C-1 of ethanol. Therefore, the dihydropyridine ring will bear no deuterium atoms when CD_3CH_2OH is oxidized, because all the deuterium atoms of the alcohol are attached to C-2.

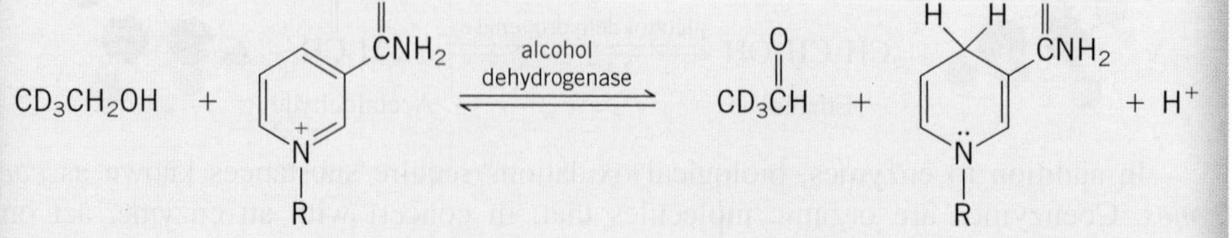

2,2,2-Tri-deuterioethanol | NAD^+ | 2,2,2-Tri-deuterioethanal | NADH

Mechanism 15.5

Oxidation of Ethanol by NAD⁺

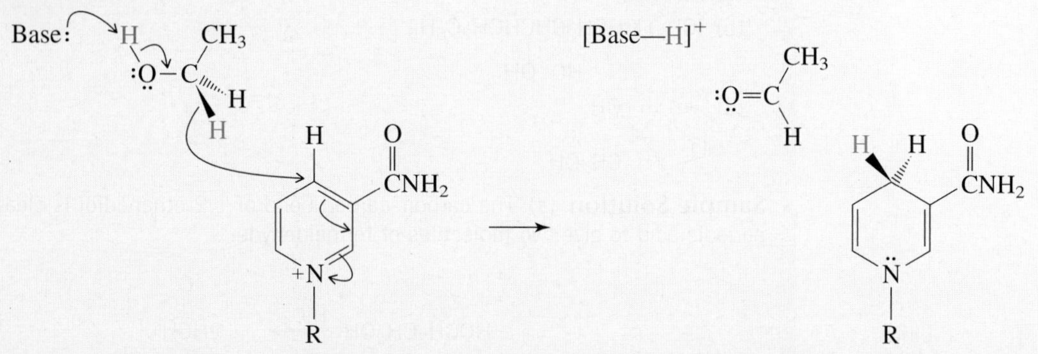

Ethanol NAD$^+$ Acetaldehyde NADH

The reverse reaction also occurs in living systems; NADH reduces acetaldehyde to ethanol in the presence of alcohol dehydrogenase. In this process, NADH serves as a hydride donor and is oxidized to NAD$^+$ while acetaldehyde is reduced.

The NAD$^+$–NADH coenzyme system is involved in a large number of biological oxidation–reductions. Another reaction similar to the ethanol–acetaldehyde conversion is the oxidation of lactic acid to pyruvic acid by NAD$^+$ and the enzyme *lactic acid dehydrogenase:*

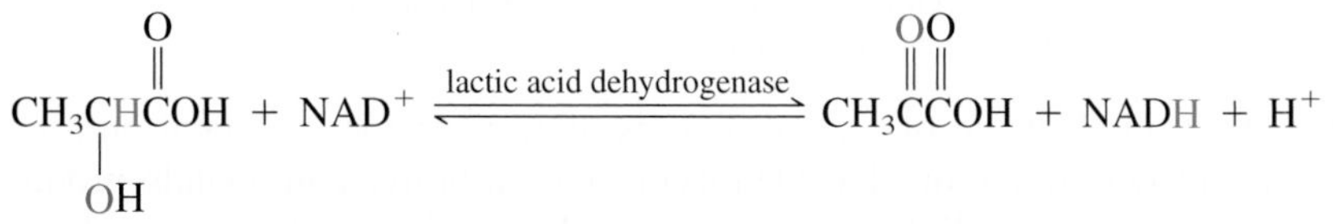

Lactic acid Pyruvic acid

We shall encounter other biological processes in which the NAD$^+$ $\rightleftharpoons$ NADH interconversion plays a prominent role in biological oxidation–reduction.

15.11 Oxidative Cleavage of Vicinal Diols

A reaction characteristic of vicinal diols is their oxidative cleavage on treatment with periodic acid (HIO_4). The carbon–carbon bond of the vicinal diol unit is broken and two carbonyl groups result. Periodic acid is reduced to iodic acid (HIO_3).

What is the oxidation state of iodine in HIO_4? In HIO_3?

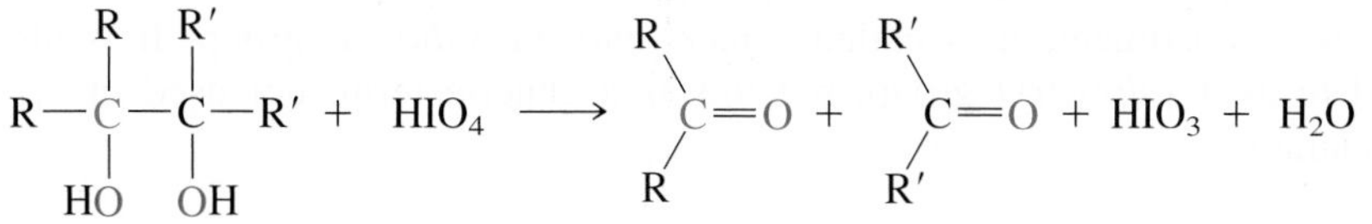

Vicinal diol | Periodic acid | Aldehyde or ketone | Aldehyde or ketone | Iodic acid | Water

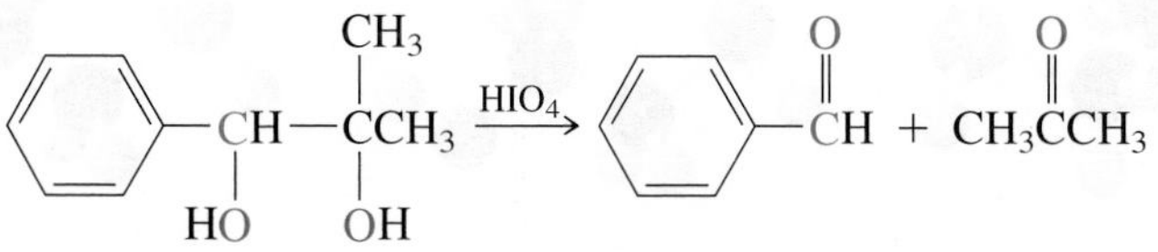

2-Methyl-1-phenyl-1,2-propanediol | Benzaldehyde (83%) | Acetone

Can you remember what reaction of an alkene would give the same products as the periodic acid cleavage shown here?

This reaction occurs only when the hydroxyl groups are on adjacent carbons.

Problem 15.14

Predict the products formed on oxidation of each of the following with periodic acid:

(a) $HOCH_2CH_2OH$

(b) $(CH_3)_2CHCH_2\underset{HO}{\underset{|}{C}}H\underset{OH}{\underset{|}{C}}HCH_2C_6H_5$

(c) 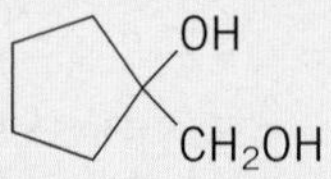

Sample Solution (a) The carbon–carbon bond of 1,2-ethanediol is cleaved by periodic acid to give two molecules of formaldehyde:

$$\underset{\text{1,2-Ethanediol}}{HOCH_2CH_2OH} \xrightarrow{HIO_4} \underset{\text{Formaldehyde}}{2H\overset{O}{\overset{\|}{C}}H}$$

Cyclic diols give dicarbonyl compounds. The reactions are faster when the hydroxyl groups are cis than when they are trans, but both stereoisomers are oxidized by periodic acid.

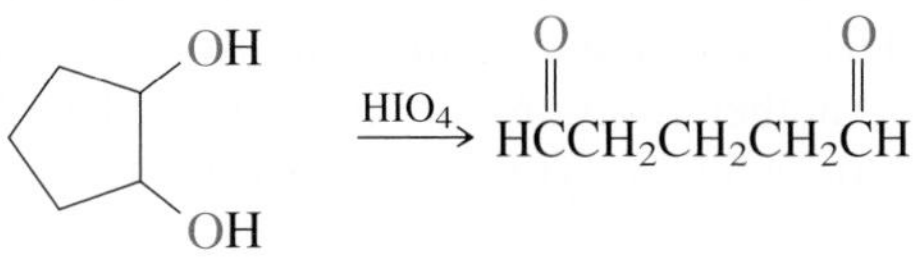

1,2-Cyclopentanediol (cis or trans stereoisomer) Pentanedial

Periodic acid cleavage of vicinal diols is often used for analytical purposes as an aid in structure determination. By identifying the carbonyl compounds produced, the constitution of the starting diol may be deduced. This technique finds its widest application with carbohydrates and will be discussed more fully in Chapter 23.

15.12 Thiols

Sulfur lies just below oxygen in the periodic table, and many oxygen-containing organic compounds have sulfur analogs. The sulfur analogs of alcohols (ROH) are **thiols (RSH).** Thiols are given substitutive IUPAC names by appending the suffix *-thiol* to the name of the corresponding alkane, numbering the chain in the direction that gives the lower locant to the carbon that bears the —SH group. As with diols (Section 15.5), the final *-e* of the alkane name is retained. When the —SH group is named as a substituent, it is called a *mercapto,* or *sulfanyl,* group. It is also often referred to as a *sulfhydryl* group, but this is a generic term, not used in systematic nomenclature.

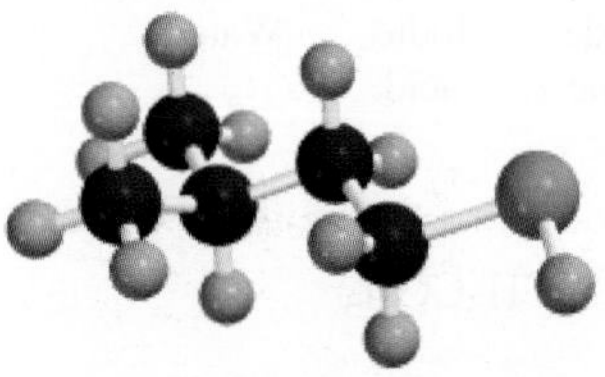

$(CH_3)_2CHCH_2CH_2SH$

3-Methyl-1-butanethiol
2004 names: 3-Methylbutane-1-thiol

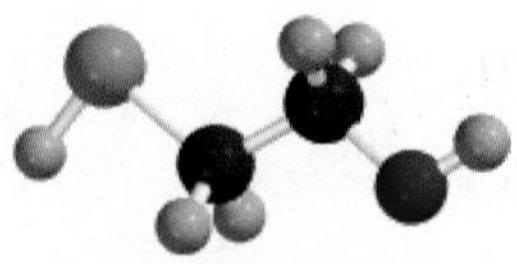

$HSCH_2CH_2OH$

2-Mercaptoethanol
2-Sulfanylethanol

$HSCH_2CH_2CH_2SH$

1,3-Propanedithiol
Propane-1,3-dithiol

At one time thiols were named **mercaptans.** Thus, CH_3CH_2SH was called "ethyl mercaptan" according to this system. This nomenclature was abandoned beginning with the 1965 revision of the IUPAC rules but is still sometimes encountered.

Thiols have a marked tendency to bond to mercury, and the word *mercaptan* comes from the Latin *mercurium captans,* which means "seizing mercury." The drug *dimercaprol* is used to treat mercury and lead poisoning; it is 2,3-dimercapto-1-propanol.

The most obvious property of a low-molecular-weight thiol is its foul odor. Ethanethiol is added to natural gas so that leaks can be detected without special equipment—your nose is so sensitive that it can detect less than one part of ethanethiol in 10,000,000,000 parts of air! The odor of thiols weakens with the number of carbons, because both the volatility and the sulfur content decrease. 1-Dodecanethiol, for example, has only a faint odor. On the positive side, of the hundreds of substances that contribute to the aroma of freshly brewed coffee, the one most responsible for its characteristic odor is the thiol 2-(mercaptomethyl)furan. Likewise, the contribution of *p*-1-menthene-8-thiol to the taste and odor of freshly squeezed grapefruit juice far exceeds that of most of the more than 260 other volatile components so far identified.

p-1-Menthene-8-thiol is a common name, not an IUPAC name.

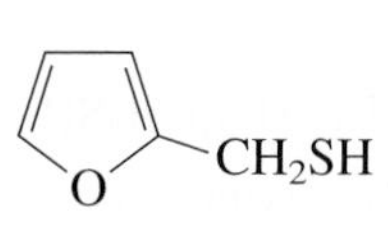

2-(Mercaptomethyl)furan

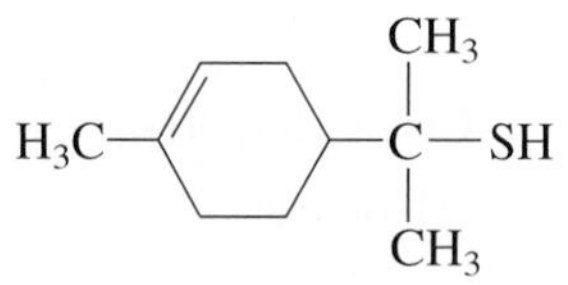

p-1-Menthene-8-thiol

Problem 15.15

Two major components of a skunk's scent fluid are 3-methyl-1-butanethiol and *trans*-2-butene-1-thiol. Write structural formulas for each of these compounds.

The S—H bond is less polar than the O—H bond, as is evident in the electrostatic potential maps of Figure 15.3. The decreased polarity of the S—H bond, especially the decreased positive character of the proton, causes hydrogen bonding to be absent in thiols. Thus, methanethiol (CH_3SH) is a gas at room temperature (bp 6°C), whereas methanol (CH_3OH) is a liquid (bp 65°C).

Compare the boiling points of H_2S (–60°C) and H_2O (100°C).

In spite of S—H bonds being less polar than O—H bonds, thiols are stronger acids than alcohols. This is largely because S—H bonds are weaker than O—H bonds. We have seen that most alcohols have pK_a's of 16–18. The corresponding value for a thiol is about 11. The significance of this difference is that a thiol can be quantitatively converted to its conjugate base (RS^-), called an **alkanethiolate,** by hydroxide. Consequently, thiols dissolve in aqueous base.

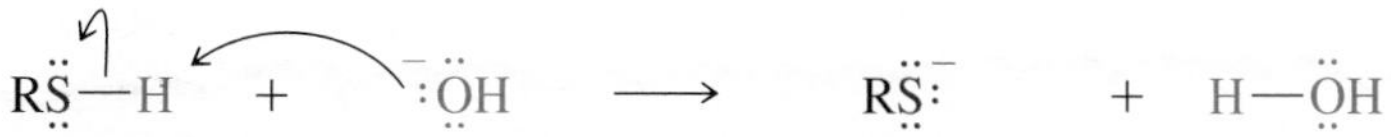

Alkanethiol	Hydroxide ion	Alkanethiolate ion	Water
(stronger acid)	(stronger base)	(weaker base)	(weaker acid)
(pK_a = 11)			(pK_a = 15.7)

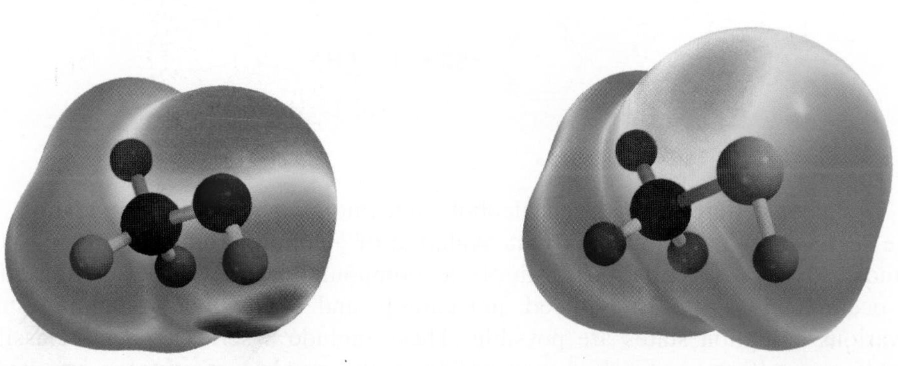
(*a*) Methanol (CH_3OH) (*b*) Methanethiol (CH_3SH)

Figure 15.3

Electrostatic potential maps of (*a*) methanol, and (*b*) methanethiol. The color scales were adjusted to be the same for both molecules to allow for direct comparison. The development of charge is more pronounced in the region surrounding the —OH group in methanol than it is for the —SH group in methanethiol.

Recall from Section 8.11 that the major pathway for reaction of *alkoxide* ions with secondary alkyl halides is E2, not S_N2.

Alkanethiolate ions (RS^-) are weaker bases than alkoxide ions (RO^-), but they are powerful nucleophiles and undergo synthetically useful S_N2 reactions even with secondary alkyl halides.

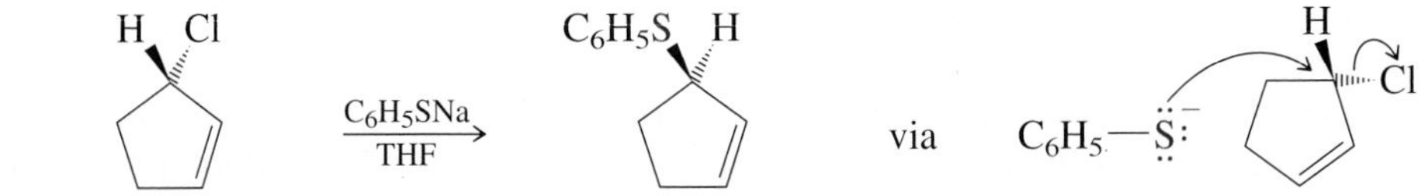

H Cl
C6H5SNa
THF
C6H5S H
via
C6H5—S:
H
Cl

3-Chlorocyclopentene → 2-Cyclopentenyl phenyl sulfide (75%)

Thiols themselves are sometimes prepared by nucleophilic substitution using the conjugate base of H_2S.

$$CH_3(CH_2)_4CH_2Br \xrightarrow[\text{ethanol}]{\text{KSH}} CH_3(CH_2)_4CH_2SH$$

1-Bromohexane → 1-Hexanethiol (67%)

Problem 15.16

(a) Outline a synthesis of 1-hexanethiol from 1-hexanol.

Sample Solution

- Hexanethiol is made by S_N2 reaction

$$H\text{-}\ddot{S}:^- + R\text{-}Br: \longrightarrow H\text{-}S\text{-}R + :Br:^-$$

- Therefore, we first need to convert 1-hexanol to 1-bromohexane, then do S_N2 with NaSH or KSH.

$$CH_3(CH_2)_4CH_2OH \xrightarrow[\text{heat}]{\text{HBr}} CH_3(CH_2)_4CH_2Br \xrightarrow{\text{NaSH}} CH_3(CH_2)_4CH_2SH$$

(b) Outline a synthesis of the following compound from allyl bromide and any other reagents.

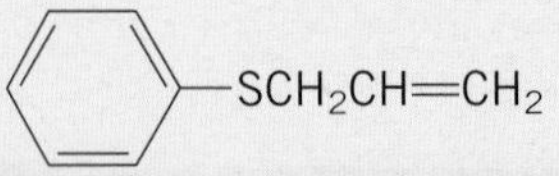

SCH2CH=CH2

A major difference between alcohols and thiols concerns their oxidation. We have seen earlier in this chapter that oxidation of alcohols produces carbonyl compounds. Analogous oxidation of thiols to compounds with C=S functions does *not* occur. Only sulfur is oxidized, not carbon, and compounds containing sulfur in various oxidation states are possible. These include a series of acids classified as *sulfenic, sulfinic,* and *sulfonic* according to the number of oxygens attached to sulfur.

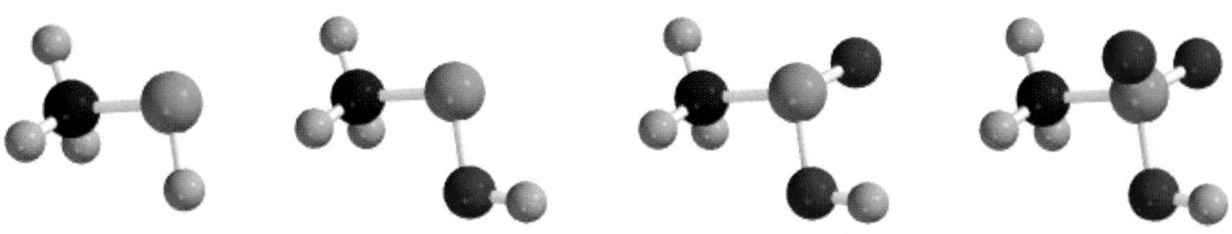

$CH_3SH \longrightarrow CH_3SOH \longrightarrow CH_3S(O)OH \longrightarrow CH_3SO_2OH$

Methanethiol — Methanesulfenic acid — Methanesulfinic acid — Methanesulfonic acid

Of these the most important are the sulfonic acids. In general though, sulfonic acids are not prepared by oxidation of thiols. Benzenesulfonic acid ($C_6H_5SO_2OH$), for example, is prepared by sulfonation of benzene (see Section 12.4).

From a biochemical perspective the most important oxidation is the conversion of thiols to **disulfides.**

$$\underset{\text{Thiol}}{2RSH} \underset{\text{reduction}}{\overset{\text{oxidation}}{\rightleftharpoons}} \underset{\text{Disulfide}}{RSSR}$$

Although a variety of oxidizing agents are available for this transformation, it occurs so readily that thiols are slowly converted to disulfides by the oxygen in air. Dithiols give cyclic disulfides by intramolecular sulfur–sulfur bond formation. An example of a cyclic disulfide is the coenzyme *α-lipoic acid.* The last step in the laboratory synthesis of α-lipoic acid is an iron(III)-catalyzed oxidation of the dithiol shown:

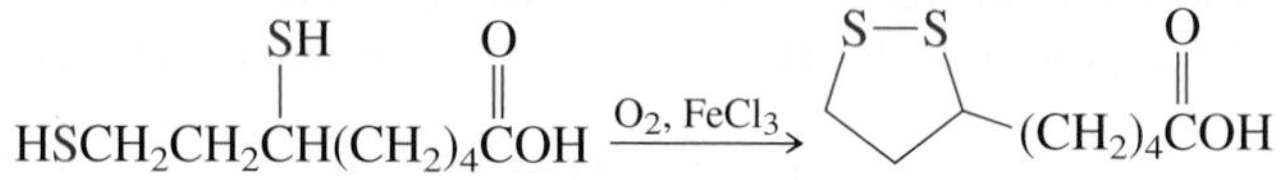

6,8-Dimercaptooctanoic acid — α-Lipoic acid (78%)

Rapid and reversible making and breaking of the sulfur–sulfur bond is essential to the biological function of α-lipoic acid.

The S—S bonds in disulfides are intermediate in strength between typical covalent bonds and weaker interactions such as hydrogen bonds. Covalent bonds involving C, H, N, and O have bond strengths on the order of 330–420 kJ/mol. The S—S bond energy is about 220 kJ/mol, and hydrogen bond strengths are usually less than 30 kJ/mol. Thus S—S bonds provide more structural stability than a hydrogen bond, but can be broken while leaving the covalent framework intact.

All mammalian cells contain a thiol called *glutathione.* Glutathione protects the cell by scavenging harmful oxidants. It reacts with these oxidants by forming a disulfide, which is eventually converted back to glutathione.

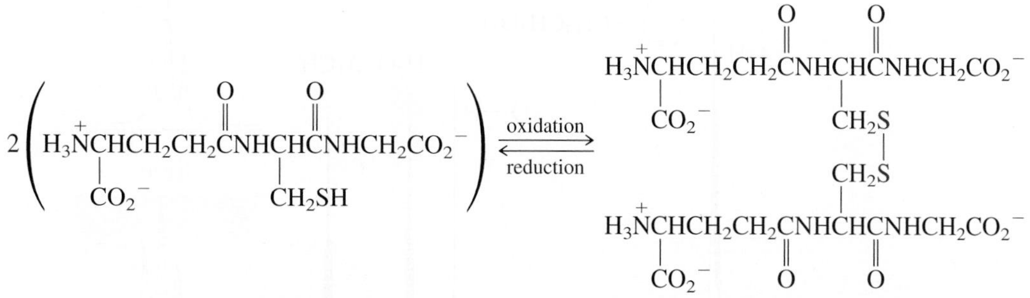

Glutathione (reduced form) — Glutathione (oxidized form)

The three-dimensional shapes of many proteins are governed and stabilized by S—S bonds connecting what would ordinarily be remote segments of the molecule. We'll have more to say about these *disulfide bridges* in Chapter 25.

Figure 15.4

The infrared spectrum of cyclohexanol.

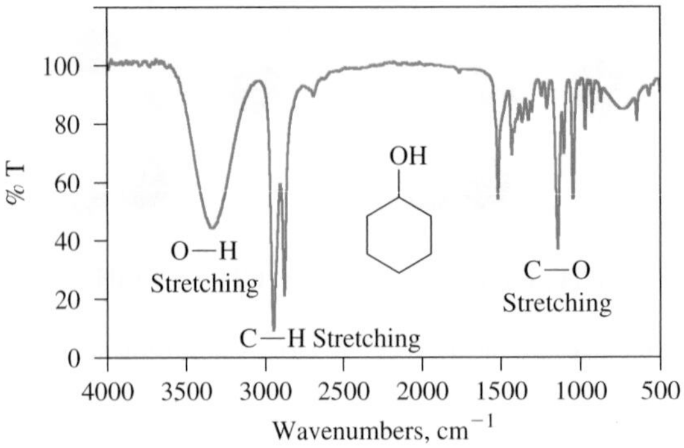

15.13 Spectroscopic Analysis of Alcohols and Thiols

Infrared: We discussed the most characteristic features of the infrared spectra of *alcohols* earlier (Section 13.22). The O—H stretching vibration is especially easy to identify, appearing in the 3200–3650 cm^{-1} region. As the infrared spectrum of cyclohexanol, presented in Figure 15.4 demonstrates, this peak is seen as a broad absorption of moderate intensity. The C—O bond stretching of alcohols gives rise to a moderate to strong absorbance between 1025 and 1200 cm^{-1}. It appears at 1065 cm^{-1} in cyclohexanol, a typical secondary alcohol, but is shifted to slightly higher energy in tertiary alcohols and slightly lower energy in primary alcohols.

The S—H stretching frequency of *thiols* gives rise to a weak band in the range 2550–2700 cm^{-1}.

^{1}H NMR: The most helpful signals in the ^{1}H NMR spectrum of *alcohols* result from the O—**H** proton and the proton in the **H**—C—O unit of primary and secondary alcohols.

H—C—O—H

δ 3.3–4.0 δ 0.5–5

The chemical shift of the hydroxyl proton signal is variable, depending on solvent, temperature, and concentration. Its precise position is not particularly significant in structure determination. Because the signals due to hydroxyl protons are not usually split by other protons in the molecule and are often rather broad, they are often fairly easy to identify. To illustrate, Figure 15.5 shows the ^{1}H NMR spectrum of 2-phenylethanol, in which the hydroxyl proton signal appears as a singlet at δ 4.5. Of the two triplets in this spectrum, the one at lower field (δ 4.0) corresponds to the protons of the CH_2O unit. The higher-field triplet at δ 3.1 arises from the benzylic CH_2 group. The assignment of a particular signal to the hydroxyl proton can be confirmed

Figure 15.5

The 200-MHz ^{1}H NMR spectrum of 2-phenylethanol ($C_6H_5CH_2CH_2OH$).

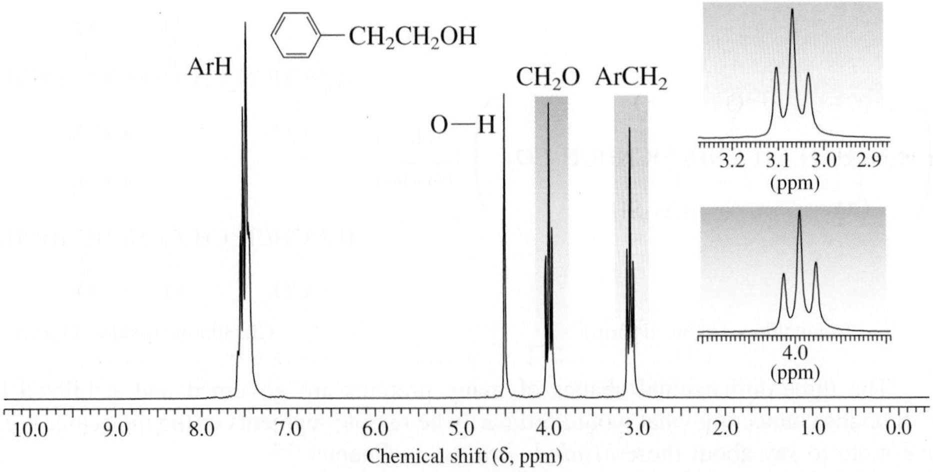

by adding D_2O. The hydroxyl proton is replaced by deuterium, and its ^1H NMR signal disappears.

Because of its lower electronegativity, sulfur deshields neighboring protons less than oxygen does. Thus, the protons of a CH_2S group appear at higher field than those of a CH_2OH group.

	$CH_3CH_2CH_2—CH_2—OH$	$CH_3CH_2CH_2—CH_2—SH$
^1H Chemical shift:	δ 3.6	δ 2.5

^{13}C NMR: The electronegative oxygen of an *alcohol* decreases the shielding of the carbon to which it is attached. The chemical shift for the carbon of the C—OH is 60–75 ppm for most alcohols. Carbon of a C—S group is more shielded than carbon of C—O.

	CH_3	—CH_2	—CH_2	—CH_2—OH	CH_3	—CH_2	—CH_2	—CH_2—SH
^{13}C Chemical shift:	δ 14	δ 19	δ 35	δ 62	δ 13	δ 21	δ 36	δ 24

UV-VIS: Unless the molecule has other chromophores, alcohols are transparent above about 200 nm; λ_{max} for methanol, for example, is 177 nm.

Mass Spectrometry: The molecular ion peak is usually quite small in the mass spectrum of an alcohol. A peak corresponding to loss of water is often evident. Alcohols also fragment readily by a pathway in which the molecular ion loses an alkyl group from the hydroxyl-bearing carbon to form a stable cation. Thus, the mass spectra of most primary alcohols exhibit a prominent peak at *m/z* 31.

$$RCH_2\ddot{O}H \longrightarrow R—CH_2—\dot{O}H^{+} \longrightarrow R\cdot + H_2C=\overset{+}{O}H$$

Primary alcohol; Molecular ion; Alkyl radical; Conjugate acid of formaldehyde, *m/z* 31

Problem 15.17

Three of the most intense peaks in the mass spectrum of 2-methyl-2-butanol appear at *m/z* 59, 70, and 73. Explain the origin of these peaks.

Interpreting the mass spectra of sulfur compounds is aided by the observation of an M+2 peak because of the presence of the mass-34 isotope of sulfur. The major cleavage pathway of *thiols* is analogous to that of alcohols.

15.14 SUMMARY

Section 15.1 Functional group interconversions involving alcohols either as reactants or as products are the focus of this chapter. Alcohols are commonplace natural products. Table 15.1 summarizes reactions discussed in earlier sections that can be used to prepare alcohols.

Section 15.2 Alcohols can be prepared from carbonyl compounds by reduction of aldehydes and ketones. See Table 15.3.

Section 15.3 Alcohols can be prepared from carbonyl compounds by reduction of carboxylic acids. See Table 15.3.

Section 15.4 Grignard and organolithium reagents react with ethylene oxide to give primary alcohols.

TABLE 15.3 Preparation of Alcohols by Reduction of Carbonyl Functional Groups

	Product of reduction of carbonyl compound by specified reducing agent		
Carbonyl compound	**Lithium aluminum hydride ($LiAlH_4$)**	**Sodium borohydride ($NaBH_4$)**	**Hydrogen (in the presence of a catalyst)**
Aldehyde RCH (with C=O) (Section 15.2)	Primary alcohol RCH_2OH	Primary alcohol RCH_2OH	Primary alcohol RCH_2OH
Ketone RCR' (with C=O) (Section 15.2)	Secondary alcohol $RCHR'$ (with OH)	Secondary alcohol $RCHR'$ (with OH)	Secondary alcohol $RCHR'$ (with OH)
Carboxylic acid $RCOH$ (with C=O) (Section 15.3)	Primary alcohol RCH_2OH	Not reduced	Not reduced

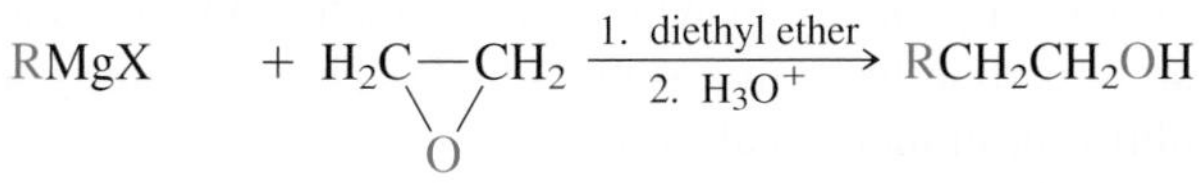

Grignard reagent Ethylene oxide Primary alcohol

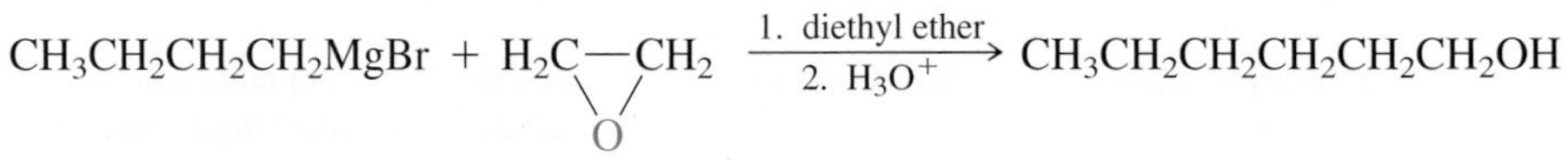

Butylmagnesium bromide Ethylene oxide 1-Hexanol (60–62%)

Section 15.5 Osmium tetraoxide is a key reagent in the conversion of alkenes to vicinal diols.

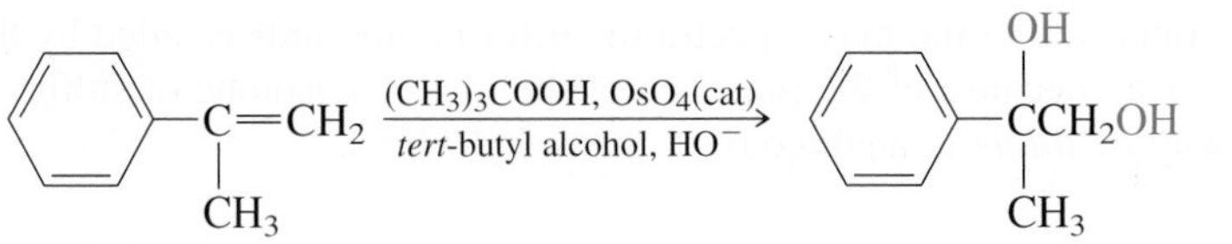

2-Phenylpropene 2-Phenyl-1,2-propanediol (71%)

The reaction is called **dihydroxylation** and proceeds by syn addition to the double bond. Osmium-based reagents that bear chiral ligands catalyze enantioselective dihydroxylation of alkenes.

Section 15.6 Table 15.2 summarizes reactions of alcohols that were introduced in earlier chapters.

Section 15.7 See Table 15.4

Section 15.8 See Table 15.4

Section 15.9 See Table 15.5

TABLE 15.4 Summary of Reactions of Alcohols Presented in This Chapter

Reaction (section) and comments	General equation and specific example
Conversion to dialkyl ethers (Section 15.7) On being heated in the presence of an acid catalyst, two molecules of a primary alcohol combine to form an ether and water. Diols can undergo an intramolecular condensation if a five-membered or six-membered cyclic ether results.	$2RCH_2OH \xrightarrow[\text{heat}]{H^+} RCH_2OCH_2R + H_2O$ Alcohol; Dialkyl ether; Water $2(CH_3)_2CHCH_2CH_2OH \xrightarrow[150°C]{H_2SO_4} (CH_3)_2CHCH_2CH_2OCH_2CH_2CH(CH_3)_2$ 3-Methyl-1-butanol; Di-(3-methylbutyl) ether (27%)
Fischer esterification (Section 15.8) Alcohols and carboxylic acids yield an ester and water in the presence of an acid catalyst. The reaction is an equilibrium process that can be driven to completion by using either the alcohol or the acid in excess or by removing the water as it is formed.	$ROH + R'\overset{O}{\overset{\parallel}{C}}OH \xrightarrow{H^+} R'\overset{O}{\overset{\parallel}{C}}OR + H_2O$ Alcohol; Carboxylic acid; Ester; Water $CH_3CH_2CH_2CH_2CH_2OH + CH_3\overset{O}{\overset{\parallel}{C}}OH \xrightarrow{H^+} CH_3\overset{O}{\overset{\parallel}{C}}OCH_2CH_2CH_2CH_2CH_3$ 1-Pentanol; Acetic acid; Pentyl acetate (71%)
Esterification with acyl chlorides (Section 15.8) Acyl chlorides react with alcohols to give esters. The reaction is usually carried out in the presence of pyridine.	$ROH + R'\overset{O}{\overset{\parallel}{C}}Cl \longrightarrow R'\overset{O}{\overset{\parallel}{C}}OR + HCl$ Alcohol; Acyl chloride; Ester; Hydrogen chloride $(CH_3)_3COH + CH_3\overset{O}{\overset{\parallel}{C}}Cl \xrightarrow{\text{pyridine}} CH_3\overset{O}{\overset{\parallel}{C}}OC(CH_3)_3$ *tert*-Butyl alcohol; Acetyl chloride; *tert*-Butyl acetate (62%)
Esterification with acid anhydrides (Section 15.8) Acid anhydrides react with alcohols to form esters in the same way that acyl chlorides do.	$ROH + R'\overset{O}{\overset{\parallel}{C}}O\overset{O}{\overset{\parallel}{C}}R' \longrightarrow R'\overset{O}{\overset{\parallel}{C}}OR + R'\overset{O}{\overset{\parallel}{C}}OH$ Alcohol; Acid anhydride; Ester; Carboxylic acid 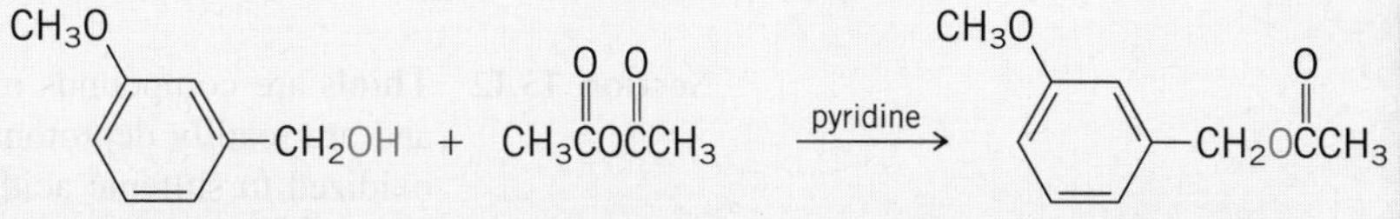 *m*-Methoxybenzyl alcohol; Acetic anhydride; *m*-Methoxybenzyl acetate (99%)

Section 15.10 Oxidation of alcohols to aldehydes and ketones is a common biological reaction. Most require a coenzyme such as the oxidized form of nicotinamide adenine dinucleotide (NAD^+).

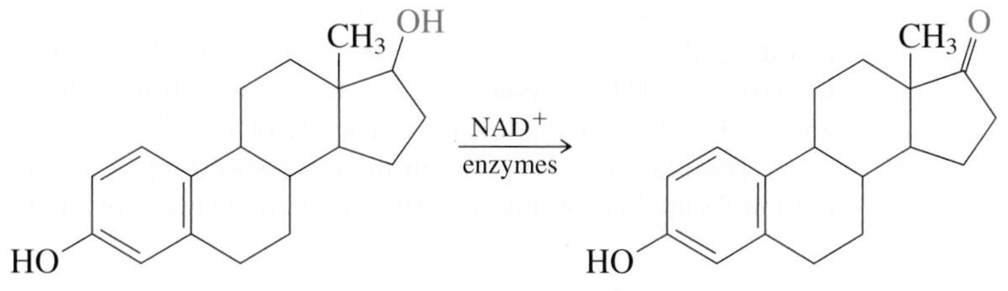

Estradiol Estrone

TABLE 15.5 Oxidation of Alcohols

Class of alcohol	Desired product	Suitable oxidizing agent(s)
Primary, RCH_2OH	Aldehyde $R\overset{\overset{O}{\|}}{C}H$	PCC* PDC* $DMSO/(COCl)_2$; $(CH_3CH_2)_3N$
Primary, RCH_2OH	Carboxylic acid $R\overset{\overset{O}{\|}}{C}OH$	$Na_2Cr_2O_7$, H_2SO_4, H_2O H_2CrO_4
Secondary, $R\underset{\underset{OH}{\|}}{C}HR'$	Ketone $R\overset{\overset{O}{\|}}{C}R'$	PCC* PDC* $Na_2Cr_2O_7$, H_2SO_4, H_2O H_2CrO_4 $DMSO/(COCl)_2$; $(CH_3CH_2)_3N$

*PCC is pyridinium chlorochromate; PDC is pyridinium dichromate. Both are used in dichloromethane.

Section 15.11 Periodic acid cleaves vicinal diols; two aldehydes, two ketones, or an aldehyde and a ketone are formed.

$$\underset{\text{Diol}}{R_2\underset{\underset{HO}{|}}{C}-\underset{\underset{OH}{|}}{C}R_2} \xrightarrow{HIO_4} \underset{\text{Two carbonyl-containing compounds}}{R_2C{=}O + O{=}CR_2}$$

$$\underset{\text{9,10-Dihydroxyoctadecanoic acid}}{CH_3(CH_2)_7\underset{\underset{HO}{|}}{C}H-\underset{\underset{OH}{|}}{C}H(CH_2)_7\overset{\overset{O}{\|}}{C}OH} \xrightarrow{HIO_4} \underset{\text{Nonanal (89\%)}}{CH_3(CH_2)_7\overset{\overset{O}{\|}}{C}H} + \underset{\text{9-Oxononanoic acid (76\%)}}{H\overset{\overset{O}{\|}}{C}(CH_2)_7\overset{\overset{O}{\|}}{C}OH}$$

Section 15.12 **Thiols** are compounds of the type RSH. They are more acidic than alcohols and are readily deprotonated by reaction with aqueous base. Thiols can be oxidized to sulfenic acids (RSOH), sulfinic acids (RSO_2H), and sulfonic acids (RSO_3H). The redox relationship between thiols and disulfides is important in certain biochemical processes.

$$\underset{\text{Thiol}}{2RSH} \underset{\text{reduction}}{\overset{\text{oxidation}}{\rightleftharpoons}} \underset{\text{Disulfide}}{RSSR}$$

Section 15.13 The hydroxyl group of an alcohol has its O—H and C—O stretching vibrations at 3200–3650 and 1025–1200 cm^{-1}, respectively.

The chemical shift of the proton of an O—H group is variable (δ 1–5) and depends on concentration, temperature, and solvent. Oxygen deshields both the proton and the carbon of an H—C—O unit. Typical NMR chemical shifts are δ 3.3–4.0 for ^1H and δ 60–75 for ^{13}C of H—C—O.

The most intense peaks in the mass spectrum of an alcohol correspond to the ion formed according to carbon–carbon cleavage of the type shown:

$$R-\overset{|}{\underset{|}{C}}-\overset{+\cdot}{\ddot{O}H} \longrightarrow R\cdot + \rangle C{=}\overset{+}{\ddot{O}H}$$

PROBLEMS

15.18 Write chemical equations, showing all necessary reagents, for the preparation of 1-butanol by each of the following methods:

(a) Hydroboration–oxidation of an alkene
(b) Use of a Grignard reagent
(c) Use of a Grignard reagent in a way different from part (b)
(d) Reduction of a carboxylic acid
(e) Hydrogenation of an aldehyde
(f) Reduction with sodium borohydride

15.19 Write chemical equations, showing all necessary reagents, for the preparation of 2-butanol by each of the following methods:

(a) Hydroboration–oxidation of an alkene
(b) Use of a Grignard reagent
(c) Use of a Grignard reagent different from that used in part (b)
(d–f) Three different methods for reducing a ketone

15.20 Which of the isomeric $C_5H_{12}O$ alcohols can be prepared by lithium aluminum hydride reduction of:

(a) An aldehyde
(b) A ketone
(c) A carboxylic acid
(d) An ester of the type $RC(=O)OCH_3$

15.21 Sorbitol is a sweetener often substituted for cane sugar, because it is better tolerated by diabetics. It is also an intermediate in the commercial synthesis of vitamin C. Sorbitol is prepared by high-pressure hydrogenation of glucose over a nickel catalyst. What is the structure (including stereochemistry) of sorbitol?

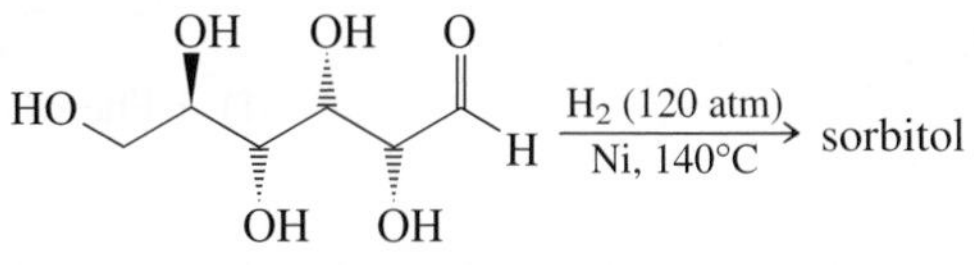

Glucose

15.22 Write equations showing how 1-phenylethanol ($C_6H_5CH(OH)CH_3$) could be prepared from each of the following starting materials:

(a) Bromobenzene
(b) Benzaldehyde
(c) Benzyl alcohol
(d) Acetophenone
(e) Benzene

15.23 Write equations showing how 2-phenylethanol ($C_6H_5CH_2CH_2OH$) could be prepared from each of the following starting materials:

(a) Bromobenzene
(b) Styrene
(c) 2-Phenylethanal ($C_6H_5CH_2CHO$)
(d) 2-Phenylethanoic acid ($C_6H_5CH_2CO_2H$)

15.24 Outline a brief synthesis of each of the compounds shown on the left from the indicated starting materials and any other necessary organic or inorganic reagents.

(a) 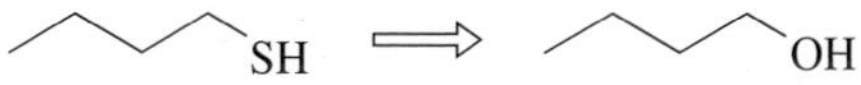

(b) 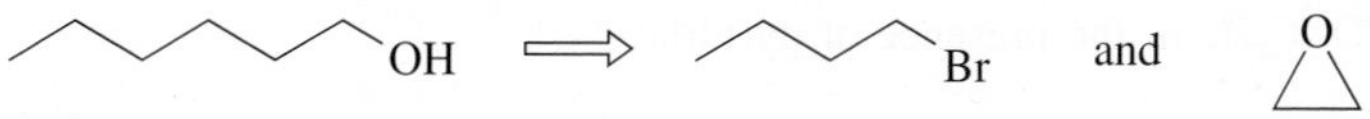

(c) 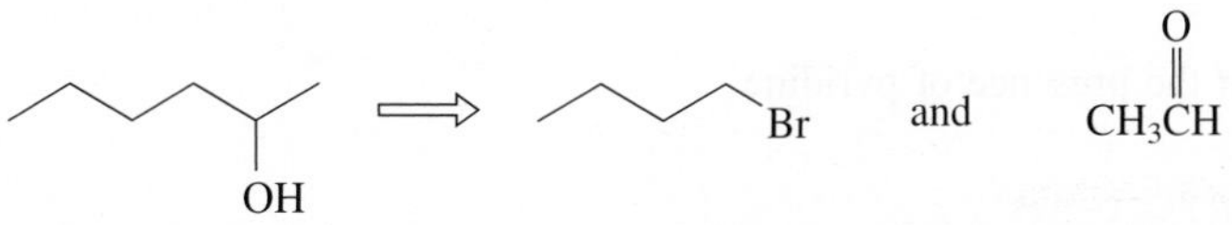

(d) 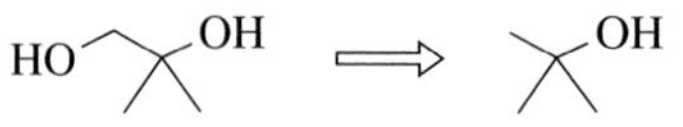

(e) 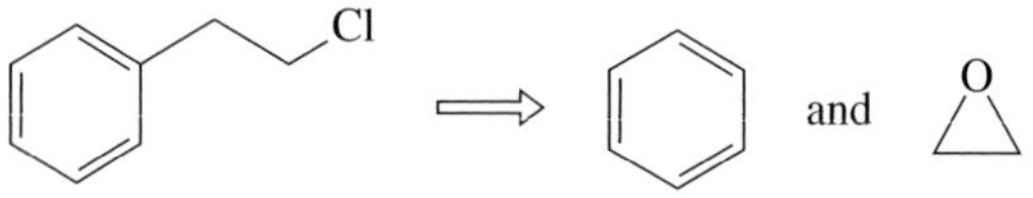

15.25 Several oxidizing reagents for alcohols were described in this chapter. Suggest one for each of the following oxidations.

(a) 

(b) 

(c) 

(d) 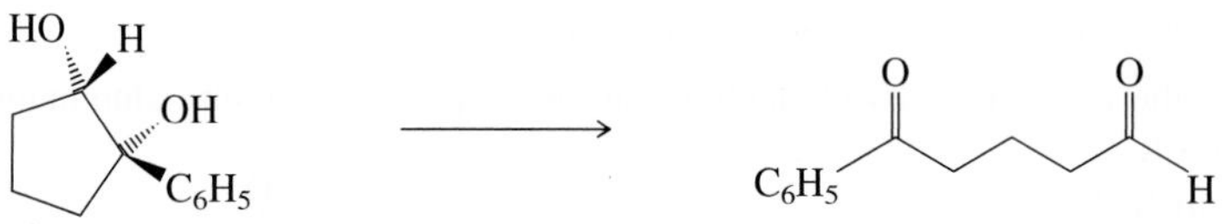

15.26 Show how each of the following compounds can be synthesized from cyclopentanol and any necessary organic or inorganic reagents. In many cases the desired compound can be made from one prepared in an earlier part of the problem.

(a) 1-Phenylcyclopentanol

(b) 1-Phenylcyclopentene

(c) *trans*-2-Phenylcyclopentanol

(d) 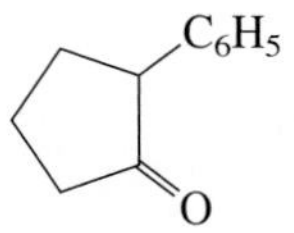

(e) 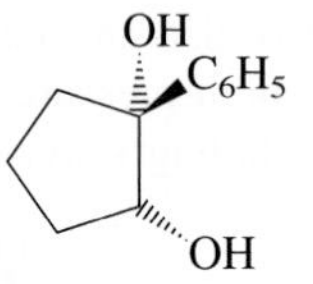

(f) 1-Phenyl-1,5-pentanediol

15.27 Write the structure of the principal organic product formed in the reaction of 1-propanol with each of the following reagents:

(a) Sulfuric acid (catalytic amount), heat at 140°C

(b) Sulfuric acid (catalytic amount), heat at 200°C

(c) Dimethyl sulfoxide (DMSO), oxalyl chloride [$(COCl)_2$], triethylamine [$N(CH_2CH_3)_3$]

(d) Pyridinium chlorochromate (PCC) in dichloromethane

(e) Potassium dichromate ($K_2Cr_2O_7$) in aqueous sulfuric acid, heat

(f) Sodium amide ($NaNH_2$)

(g) Acetic acid ($CH_3\overset{\displaystyle O}{\overset{\|}{C}}OH$) in the presence of dissolved hydrogen chloride

(h) 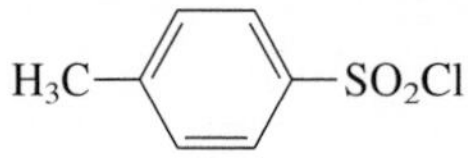 in the presence of pyridine

(i) 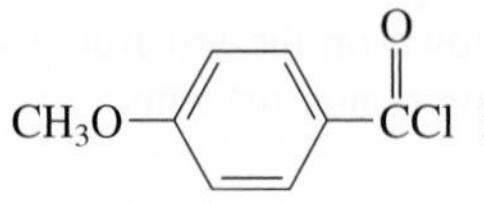 in the presence of pyridine

(j) $C_6H_5\overset{\displaystyle O}{\overset{\|}{C}}O\overset{\displaystyle O}{\overset{\|}{C}}C_6H_5$ in the presence of pyridine

(k) 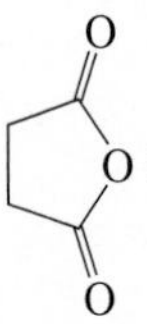 in the presence of pyridine

15.28 Each of the following reactions has been reported in the chemical literature. Predict the product in each case, showing stereochemistry where appropriate.

(a) 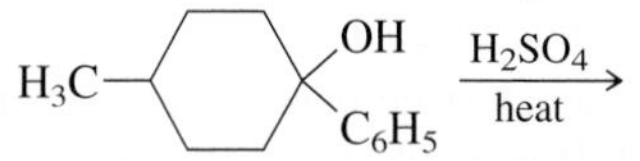

(b) $(CH_3)_2C{=}C(CH_3)_2 \xrightarrow[(CH_3)_3COH,\ HO^-]{(CH_3)_3COOH,\ OsO_4(\text{cat})}$

(c) 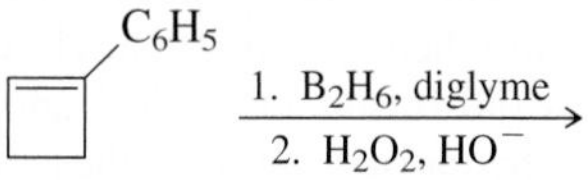

(d) 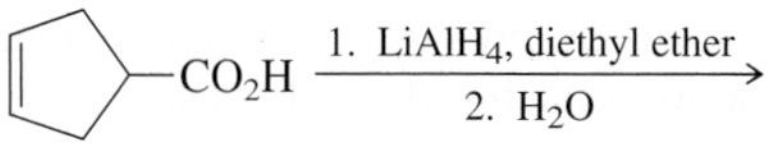

(e) $\underset{\displaystyle OH}{CH_3CHC}{\equiv}C(CH_2)_3CH_3 \xrightarrow[H_2SO_4,\ H_2O,\ \text{acetone}]{H_2CrO_4}$

(f) $CH_3\overset{O}{\overset{\|}{C}}CH_2CH{=}CHCH_2\overset{O}{\overset{\|}{C}}CH_3 \xrightarrow[2.\ H_2O]{1.\ LiAlH_4,\ \text{diethyl ether}}$

(g) 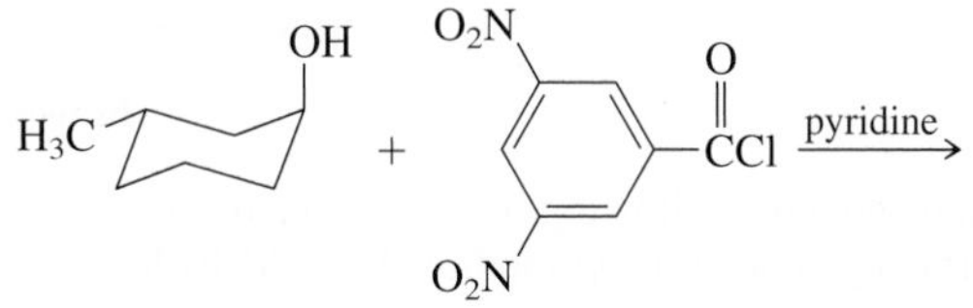

(h) 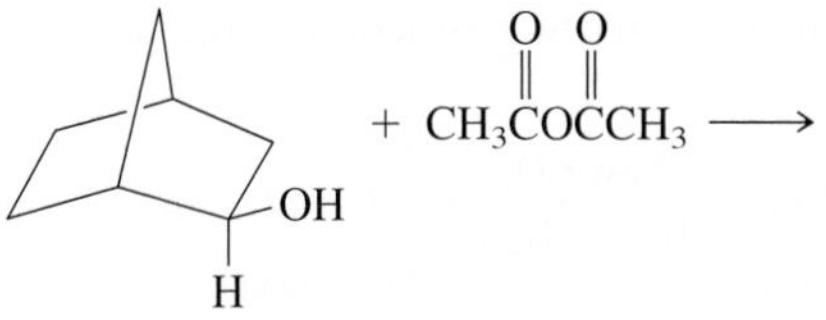

(i) 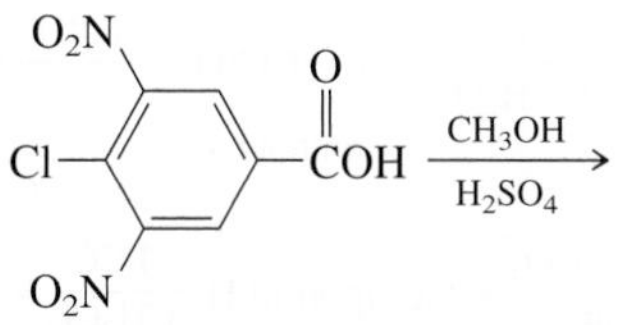

15.29 On heating 1,2,4-butanetriol in the presence of an acid catalyst, a cyclic ether of molecular formula $C_4H_8O_2$ was obtained in 81–88% yield. Suggest a reasonable structure for this product.

15.30 Suggest reaction sequences and reagents suitable for carrying out each of the following conversions. Two synthetic operations are required in each case.

(a) 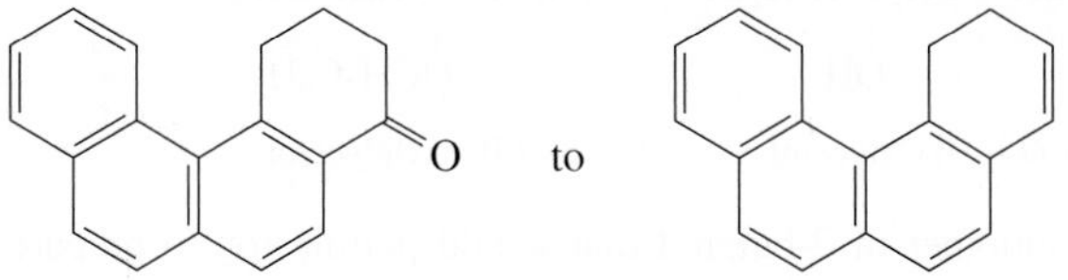

(b) 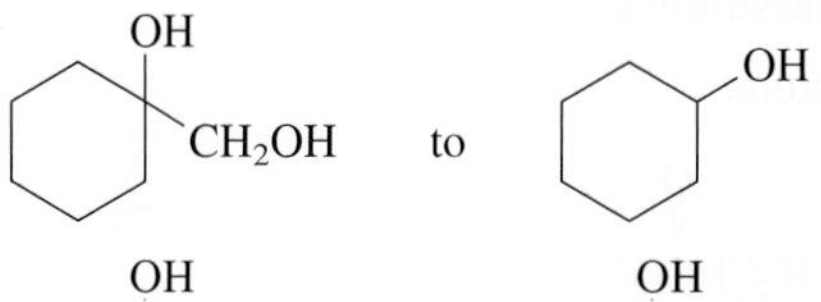

(c) 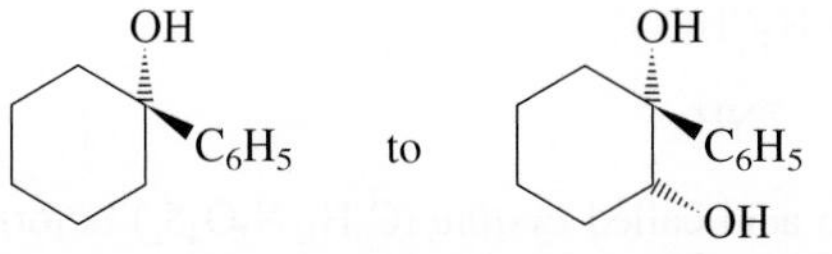 (racemic)

15.31 The fungus responsible for Dutch elm disease is spread by European bark beetles when they burrow into the tree. Other beetles congregate at the site, attracted by the scent of a mixture of chemicals, some emitted by other beetles and some coming from the tree. One of the compounds given off by female bark beetles is 4-methyl-3-heptanol. Suggest an efficient synthesis of this pheromone from alcohols of five carbon atoms or fewer.

15.32 (a) The cis isomer of 3-hexen-1-ol ($CH_3CH_2CH{=}CHCH_2CH_2OH$) has the characteristic odor of green leaves and grass. Suggest a synthesis for this compound from acetylene and any necessary organic or inorganic reagents.

(b) One of the compounds responsible for the characteristic odor of ripe tomatoes is the cis isomer of $CH_3CH_2CH{=}CHCH_2CH{=}O$. How could you prepare this compound?

15.33 R. B. Woodward was one of the leading organic chemists of the middle part of the twentieth century. Known primarily for his achievements in the synthesis of complex natural products, he was awarded the Nobel Prize in Chemistry in 1965. He entered Massachusetts Institute of Technology as a 16-year-old freshman in 1933 and four years later was awarded the Ph.D. While a student there he carried out a synthesis of *estrone,* a female sex hormone. The early stages of Woodward's estrone synthesis required the conversion of *m*-methoxybenzaldehyde to *m*-methoxybenzyl cyanide, which was accomplished in three steps:

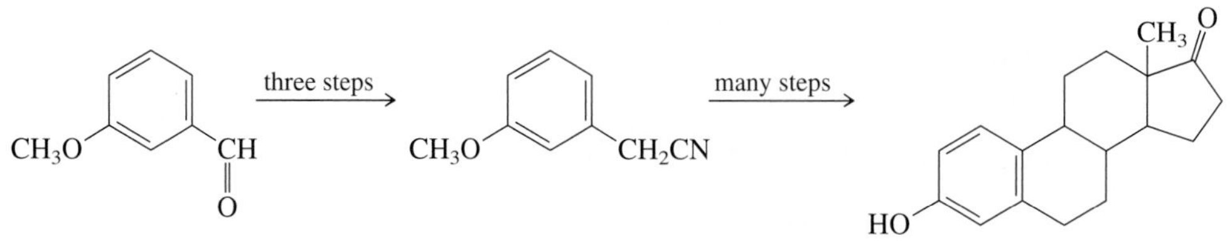

three steps
many steps
CH₃O
CH
O
CH₃O
CH₂CN
CH₃
O
HO

Estrone

Suggest a reasonable three-step sequence, showing all necessary reagents, for the preparation of *m*-methoxybenzyl cyanide from *m*-methoxybenzaldehyde.

15.34 Complete each of the following equations by writing structural formulas for compounds A through I:

(a)  $\xrightarrow{HCl}$ C_5H_7Cl (Compound A) $\xrightarrow[H_2O]{NaHCO_3}$ C_5H_8O (Compound B) $\xrightarrow[H_2SO_4,\ H_2O]{Na_2Cr_2O_7}$ C_5H_6O (Compound C)

(b) $H_2C{=}CHCH_2CH_2CHCH_3$ (with OH on the CH) $\xrightarrow[pyridine]{SOCl_2}$ $C_6H_{11}Cl$ (Compound D) $\xrightarrow[2.\ Zn,\ H_2O]{1.\ O_3}$ C_5H_9ClO (Compound E) $\xrightarrow{NaBH_4}$ $C_5H_{11}ClO$ (Compound F)

(c) 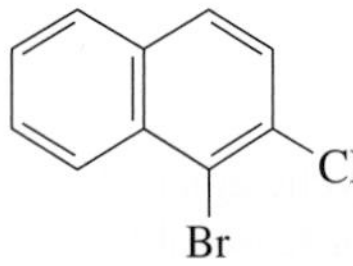 $\xrightarrow[\text{benzoyl peroxide, heat}]{NBS}$ Compound G $\xrightarrow[heat]{H_2O,\ CaCO_3}$ Compound H $\xrightarrow[CH_2Cl_2]{PCC}$ ($C_{11}H_7BrO$) Compound I

15.35 Suggest a chemical test that would permit you to distinguish between the two glycerol monobenzyl ethers shown.

$C_6H_5CH_2OCH_2CHCH_2OH$ (OH on the CH) — 1-*O*-Benzylglycerol

$HOCH_2CHCH_2OH$ ($OCH_2C_6H_5$ on the CH) — 2-*O*-Benzylglycerol

15.36 Choose the correct enantiomer of 2-butanol that would permit you to prepare (*R*)-2-butanethiol by way of a *p*-toluenesulfonate.

15.37 The amino acid *cysteine* has the structure.

$HSCH_2CHCO^-$ (C=O on the carbonyl carbon; $^+NH_3$ on the CH)

(a) A second sulfur-containing amino acid called *cystine* ($C_6H_{12}N_2O_4S_2$) is formed when cysteine undergoes biological oxidation. Suggest a reasonable structure for cystine.

(b) Another metabolic pathway converts cysteine to *cysteine sulfinic acid* ($C_3H_7NO_4S$), then to *cysteic acid* ($C_3H_7NO_5S$). What are the structures of these two compounds?

15.38 A diol ($C_8H_{18}O_2$) does not react with periodic acid. Its ^1H NMR spectrum is shown in Figure 15.6. What is the structure of this diol?

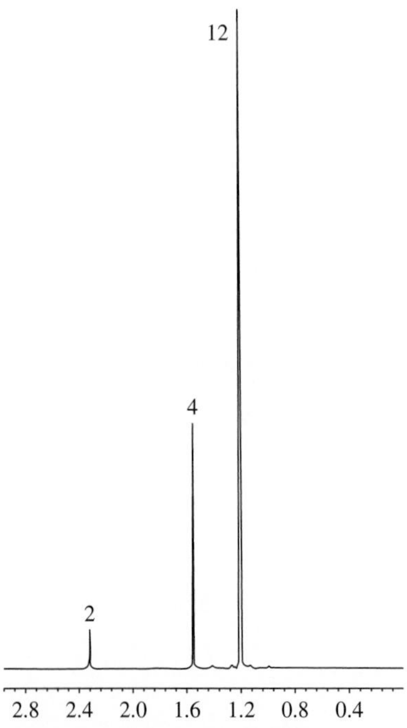

12
4
2
2.8 2.4 2.0 1.6 1.2 0.8 0.4
Chemical shift (δ, ppm)

Figure 15.6

^1H NMR spectrum of the diol in Problem 15.38

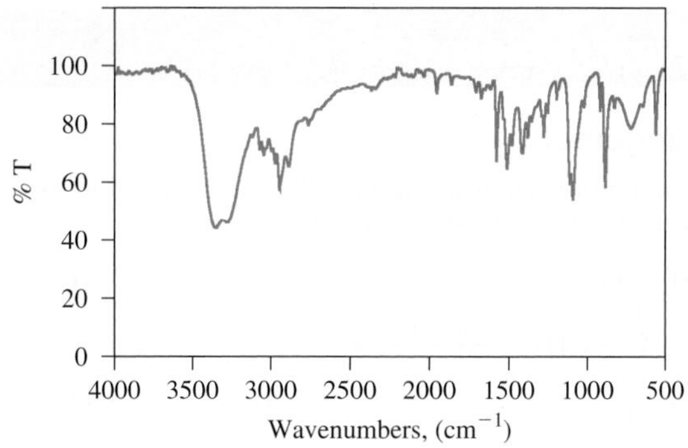

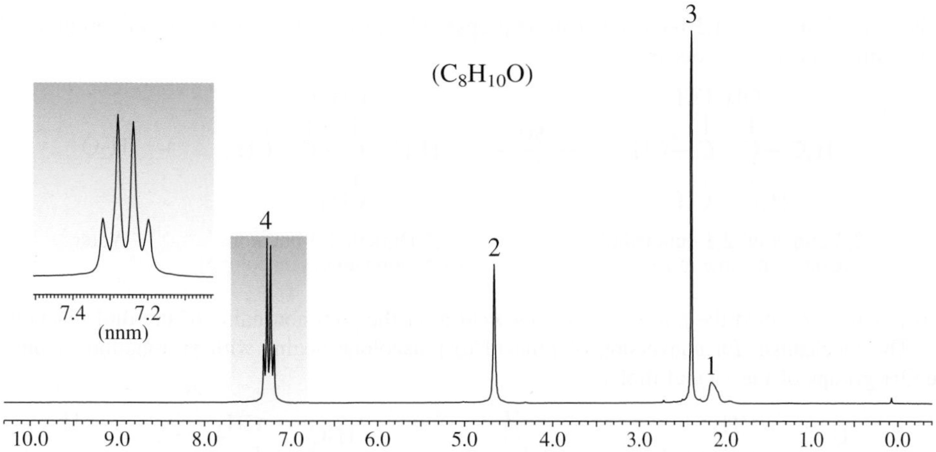

Figure 15.7

The IR (a) and 200-MHz ^1H NMR (b) spectra of a compound $C_8H_{10}O$ (Problem 15.39).

15.39 Identify the compound $C_8H_{10}O$ on the basis of its IR and ^1H NMR spectra (Figure 15.7). The broad peak at δ 2.1 in the NMR spectrum disappears when D_2O is added.

15.40 Identify each of the following $C_4H_{10}O$ isomers on the basis of their ^{13}C NMR spectra:

(a) δ 31.2: CH_3
δ 68.9: C

(b) δ 10.0: CH_3
δ 22.7: CH_3
δ 32.0: CH_2
δ 69.2: CH

(c) δ 18.9: CH_3, area 2
δ 30.8: CH, area 1
δ 69.4: CH_2, area 1

15.41 A compound $C_3H_7ClO_2$ exhibited three peaks in its ^{13}C NMR spectrum at δ 46.8 (CH_2), δ 63.5 (CH_2), and δ 72.0 (CH). What is the structure of this compound?

15.42 A compound $C_6H_{14}O$ has the ^{13}C NMR spectrum shown in Figure 15.8. Its mass spectrum has a prominent peak at *m/z* 31. Suggest a reasonable structure for this compound.

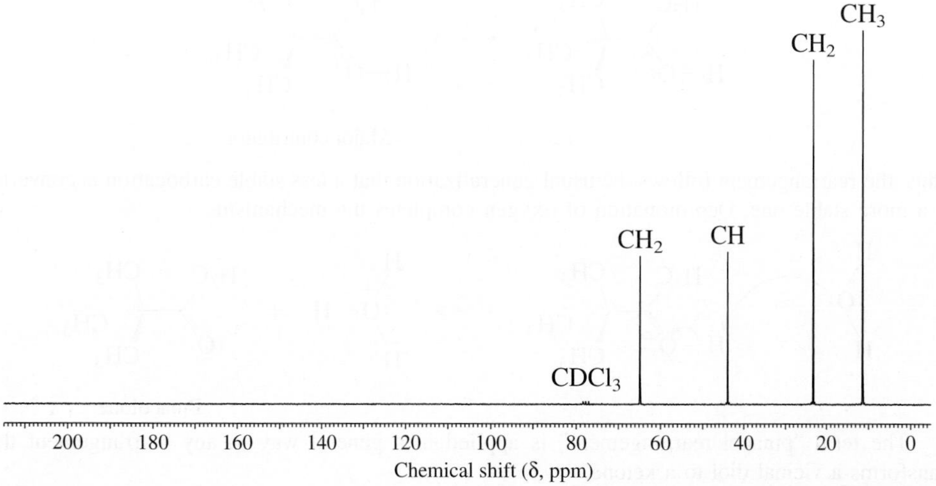

Figure 15.8

The ^{13}C NMR spectrum of the compound $C_6H_{14}O$ (Problem 15.42).

Descriptive Passage and Interpretive Problems 15

The Pinacol Rearrangement

We would expect a vicinal diol such as 2,3-dimethyl-2,3-butanediol to give a conjugated diene by double dehydration on treatment with an acid catalyst.

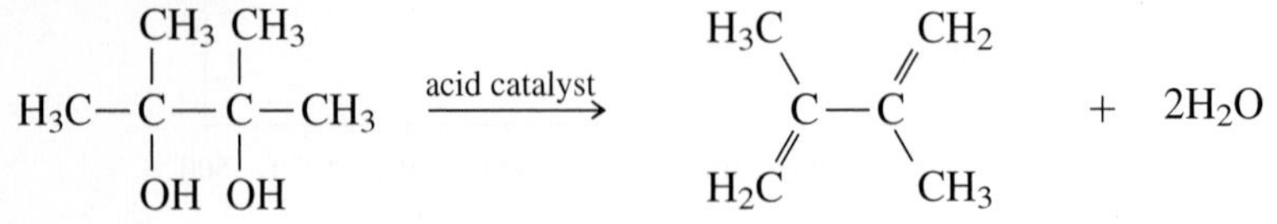

2,3-Dimethyl-2,3-butanediol — 2,3-Dimethyl-1,3-butadiene — Water

Although 2,3-dimethyl-1,3-butadiene can be prepared by just such a process, under certain conditions a different reaction occurs.

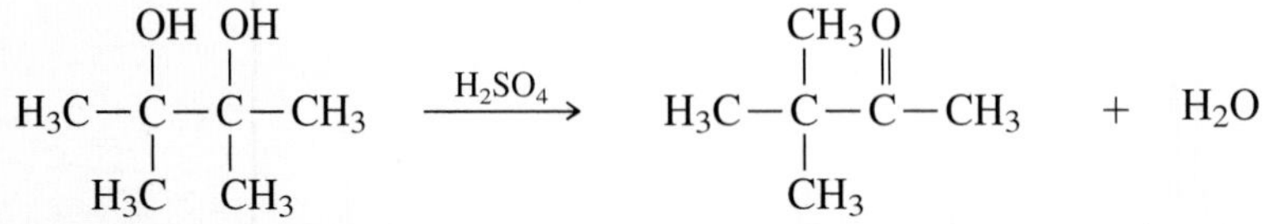

2,3-Dimethyl-2,3-butanediol (Common name: pinacol) — 3,3-Dimethyl-2-butanone (Common name: pinacolone) — Water

This reaction is called the *pinacol rearrangement* after the common name of the diol reactant.

The mechanism for conversion of pinacol to pinacolone begins with protonation of one of the OH groups of the vicinal diol.

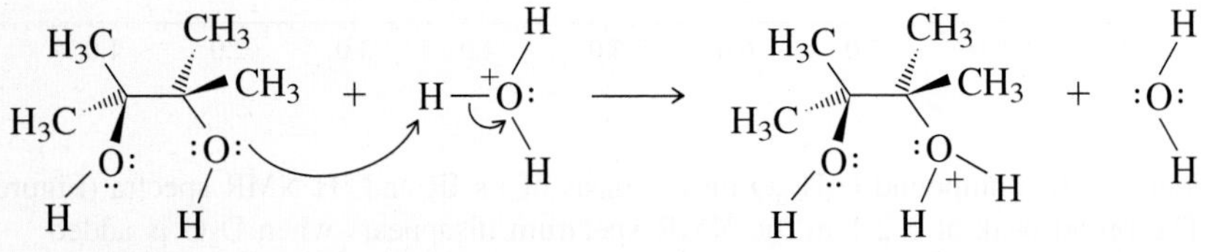

Pinacol

This is followed by loss of water and migration of a methyl group, usually in a single step in which the group anti to the departing water migrates.

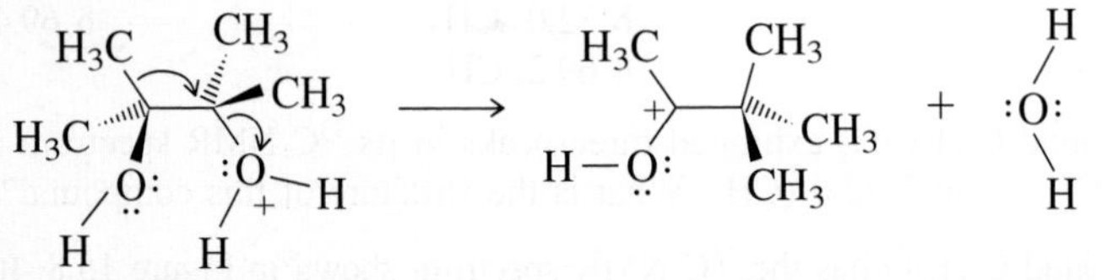

A key to understanding this migration is to recall that carbocations are stabilized by delocalization of a lone pair of an attached oxygen.

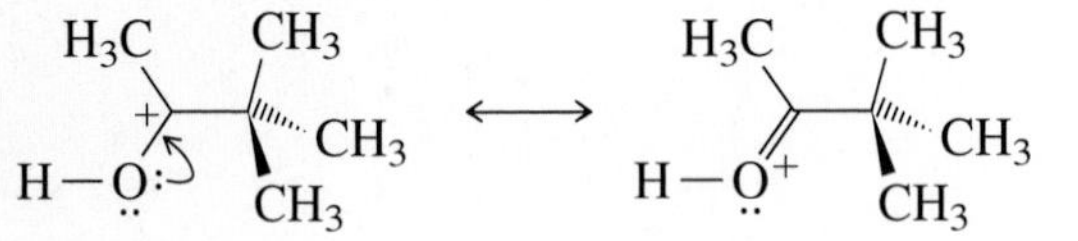

Major contributor

Thus, the rearrangement follows the usual generalization that a less stable carbocation is converted to a more stable one. Deprotonation of oxygen completes the mechanism.

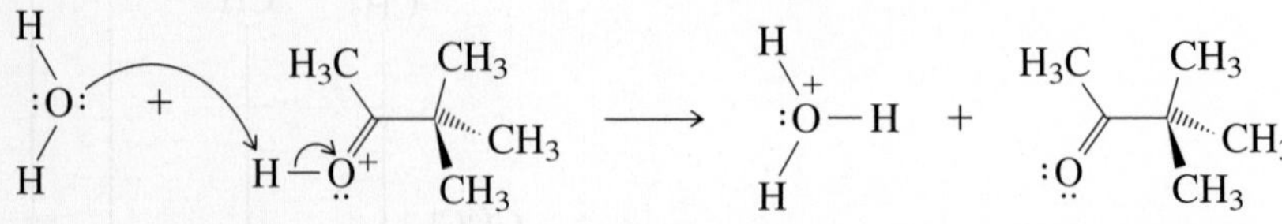

Pinacolone

The term "pinacol rearrangement" is applied in a general way to any rearrangement that transforms a vicinal diol to a ketone.

15.43 Which word or phrase best describes the stereochemistry of the product formed in the pinacol rearrangement of the diol shown?

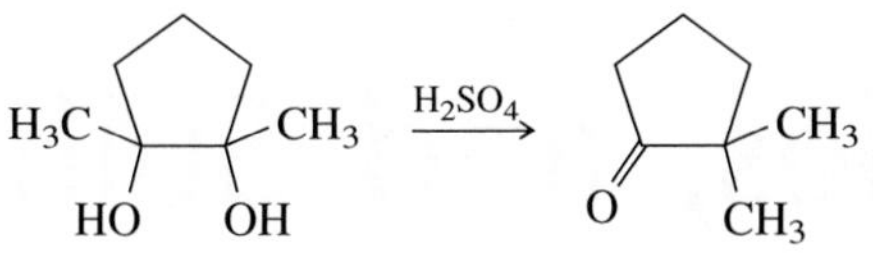

A. Achiral

B. A single enantiomer of a chiral molecule

C. Chiral, but racemic

D. Two diastereomers

15.44–15.45 Consider the two diols (**1** and **2**) and the two ketones (**3** and **4**).

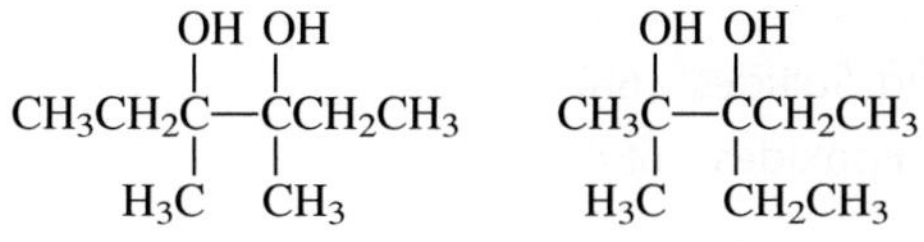

1 **2**

Vicinal diols

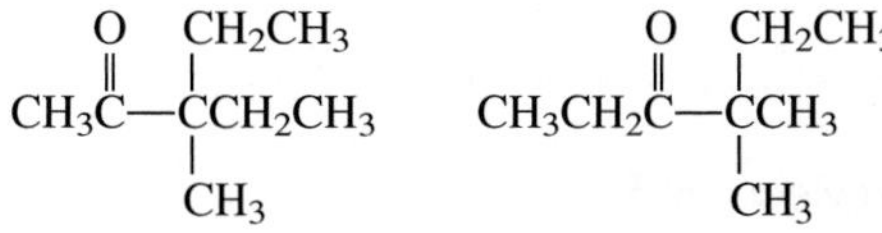

3 **4**

Ketones

A mixture of **3** and **4** is formed by pinacol rearrangement of either **1** or **2**. Given that an ethyl migrates in preference to methyl in pinacol rearrangements, predict the major ketone formed by rearrangement of each diol.

15.44 Diol **1** gives predominantly

A. Ketone **3**

B. Ketone **4**

15.45 Diol **2** gives predominantly

A. Ketone **3**

B. Ketone **4**

15.46 What is the product of the following reaction?

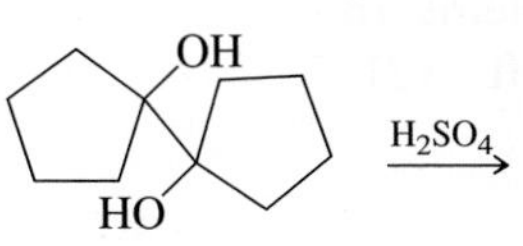

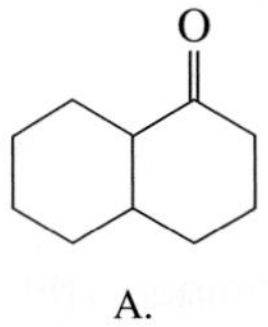

A.

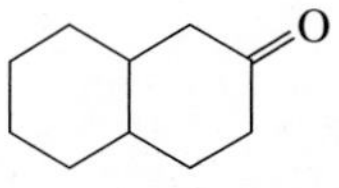

B.

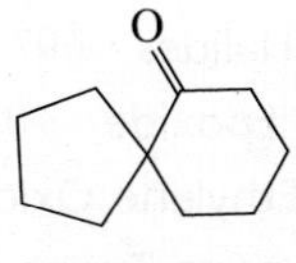

C.

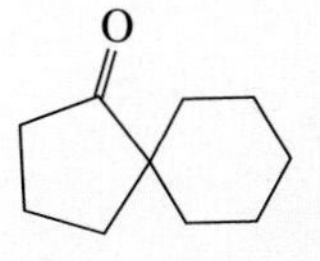

D.

15.47 The group that is anti to oxygen migrates in the pinacol rearrangement of the diol shown. What is the product?

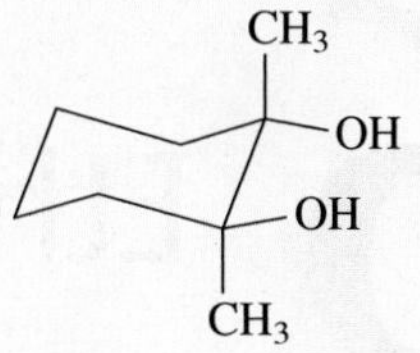

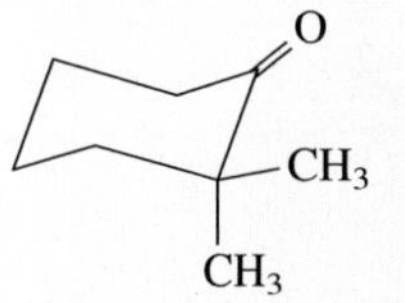

A.

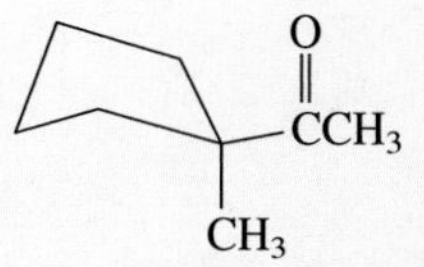

B.

15.48 Rather than following a mechanism in which a group migrates in the same step as water departs, the pinacol rearrangement of the vicinal diol shown proceeds by way of the more stable of two possible carbocations. A single ketone is formed in 73% yield. What is the structure of this ketone?

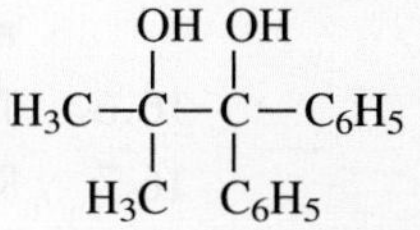

2-Methyl-1,1-diphenyl-1,2-propanediol

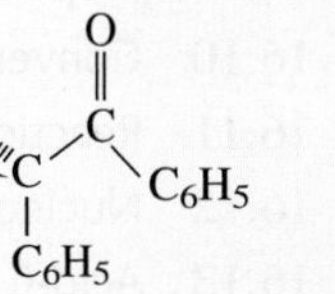

A.

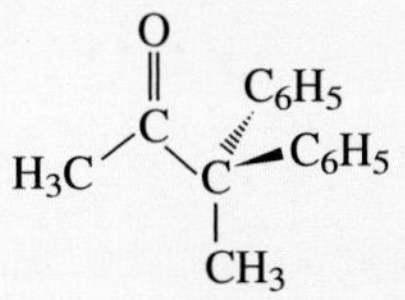

B.

15.49 Like the pinacol rearrangement in the preceding problem, this one also begins with the formation of the more stable of two possible carbocations from a vicinal diol. A 99% yield of a single ketone was isolated. What is this ketone?

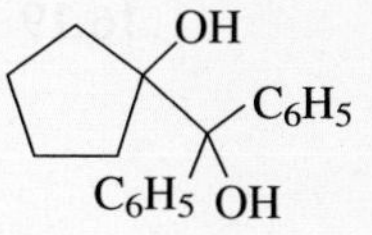

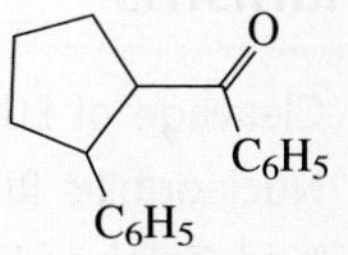

A.

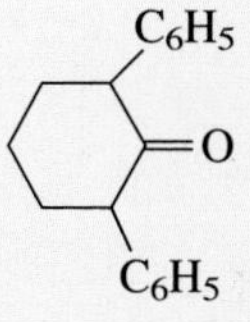

C.

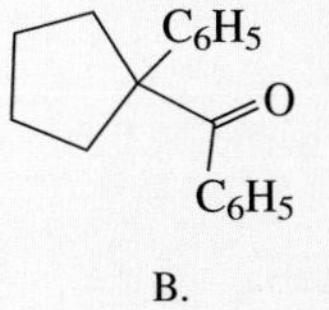

B.

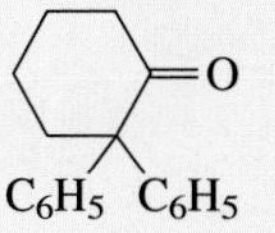

D.

16 Ethers, Epoxides, and Sulfides

Chapter Outline

Mechanisms

To organic chemists the term "crown" suggests a donut more than a symbol of royalty. The hole in the middle of this crown ether can accommodate K^+ quite comfortably and carry it from a polar solvent to a nonpolar one.

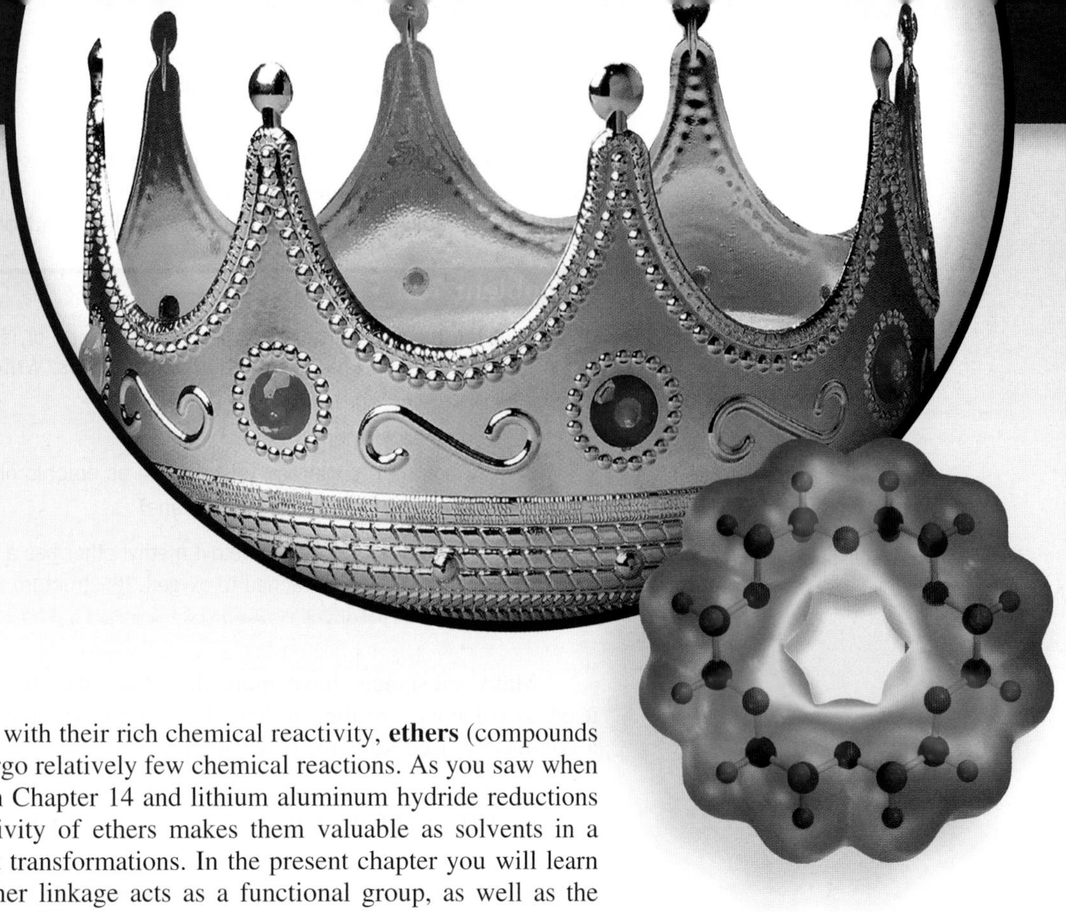

IN CONTRAST TO ALCOHOLS with their rich chemical reactivity, **ethers** (compounds containing a C—O—C unit) undergo relatively few chemical reactions. As you saw when we discussed Grignard reagents in Chapter 14 and lithium aluminum hydride reductions in Chapter 15, this lack of reactivity of ethers makes them valuable as solvents in a number of synthetically important transformations. In the present chapter you will learn of the conditions in which an ether linkage acts as a functional group, as well as the methods by which ethers are prepared.

Unlike most ethers, **epoxides** (compounds in which the C—O—C unit forms a three-membered ring) are very reactive substances. The principles of nucleophilic substitution are important in understanding the preparation and properties of epoxides.

Sulfides (RSR′) are the sulfur analogs of ethers. Just as in the preceding chapter, where we saw that the properties of thiols (RSH) are different from those of alcohols, we will explore differences between sulfides and ethers in this chapter.

16.1 Nomenclature of Ethers, Epoxides, and Sulfides

Ethers are named, in substitutive IUPAC nomenclature, as *alkoxy* derivatives of alkanes. Functional class IUPAC names of ethers are derived by listing the two alkyl groups in the general structure ROR′ in alphabetical order as separate words, and then adding the word *ether* at the end. When both alkyl groups are the same, the prefix *di-* precedes the name of the alkyl group.

	$CH_3CH_2OCH_2CH_3$	$CH_3CH_2OCH_3$	$CH_3CH_2OCH_2CH_2CH_2Cl$
Substitutive IUPAC name:	Ethoxyethane	Methoxyethane	3-Chloro-1-ethoxypropane
Functional class IUPAC name:	Diethyl ether	Ethyl methyl ether	3-Chloropropyl ethyl ether

Ethers are described as *symmetrical* or *unsymmetrical* depending on whether the two groups bonded to oxygen are the same or different. Diethyl ether is a symmetrical ether; ethyl methyl ether is an unsymmetrical ether.

Cyclic ethers have their oxygen as part of a ring—they are *heterocyclic compounds* (Section 3.15). Several have specific IUPAC names.

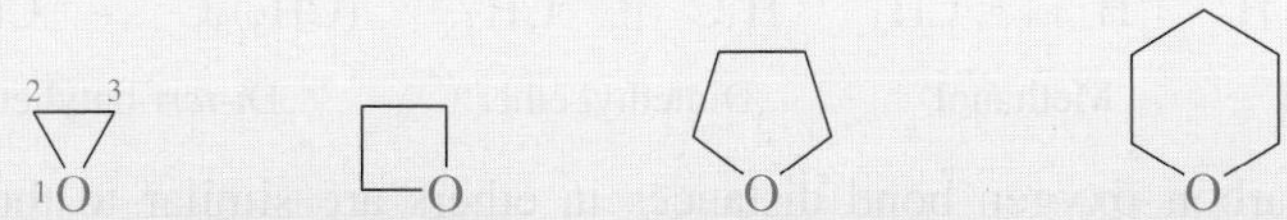

Oxirane (Ethylene oxide) | Oxetane | Oxolane (Tetrahydrofuran) | Oxane (Tetrahydropyran)

In each case the ring is numbered starting at the oxygen. The IUPAC rules also permit oxirane (without substituents) to be called *ethylene oxide*. *Tetrahydrofuran* and *tetrahydropyran* are acceptable synonyms for oxolane and oxane, respectively.

Problem 16.1

Each of the following ethers has been shown to be or is suspected to be a *mutagen*, which means it can induce mutations in test cells. Write the structure of each of these ethers.

(a) Chloromethyl methyl ether
(b) 2-(Chloromethyl)oxirane (also known as epichlorohydrin)
(c) 3,4-Epoxy-1-butene (2-vinyloxirane)

Sample Solution (a) Chloromethyl methyl ether has a chloromethyl group ($ClCH_2$—) and a methyl group (H_3C—) attached to oxygen. Its structure is $ClCH_2OCH_3$.

Recall from Section 6.19 that epoxides may be named as *-epoxy* derivatives of alkanes in substitutive IUPAC nomenclature.

Many substances have more than one ether linkage. Two such compounds, often used as solvents, are the *diethers* 1,2-dimethoxyethane and 1,4-dioxane. Diglyme, also a commonly used solvent, is a *triether*.

$CH_3OCH_2CH_2OCH_3$ — 1,2-Dimethoxyethane

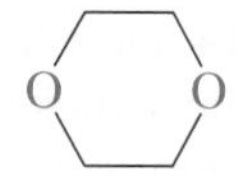
O
O

1,4-Dioxane

$CH_3OCH_2CH_2OCH_2CH_2OCH_3$ — Diethylene glycol dimethyl ether (diglyme)

Molecules that contain several ether functions are referred to as *polyethers*. Polyethers have some novel properties and will appear in Section 16.4.

The sulfur analogs (RS—) of alkoxy groups are called *alkylthio* groups. The first two of the following examples illustrate the use of alkylthio prefixes in substitutive nomenclature of sulfides. Functional class IUPAC names of sulfides are derived in exactly the same way as those of ethers but end in the word *sulfide*. Sulfur heterocycles have names analogous to their oxygen relatives, except that *ox-* is replaced by *thi-*. Thus the sulfur heterocycles containing three-, four-, five-, and six-membered rings are named *thiirane, thietane, thiolane,* and *thiane,* respectively.

Sulfides are sometimes informally referred to as *thioethers*, but this term is not part of systematic IUPAC nomenclature.

$CH_3CH_2SCH_2CH_3$ — Ethylthioethane (Diethyl sulfide)

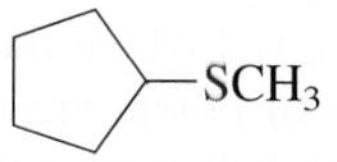
SCH3

(Methylthio)cyclopentane
Cyclopentyl methyl sulfide

S

Thiirane

16.2 Structure and Bonding in Ethers and Epoxides

Bonding in ethers is readily understood by comparing ethers with water and alcohols. Van der Waals strain involving alkyl groups causes the bond angle at oxygen to be larger in ethers than in alcohols, and larger in alcohols than in water. An extreme example is di-*tert*-butyl ether, where steric hindrance between the *tert*-butyl groups is responsible for a dramatic increase in the C—O—C bond angle.

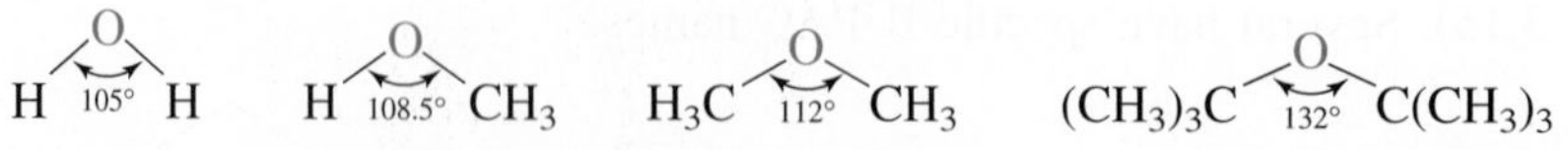
O
H 105° H
O
H 108.5° CH3
O
H3C 112° CH3
O
(CH3)3C 132° C(CH3)3

Water — Methanol — Dimethyl ether — Di-*tert*-butyl ether

Typical carbon–oxygen bond distances in ethers are similar to those of alcohols (≈142 pm) and are shorter than carbon–carbon bond distances in alkanes (≈153 pm).

An ether oxygen affects the conformation of a molecule in much the same way that a CH_2 unit does. The most stable conformation of diethyl ether is the all-staggered anti conformation. Tetrahydropyran is most stable in the chair conformation—a fact that has an important bearing on the structures of many carbohydrates.

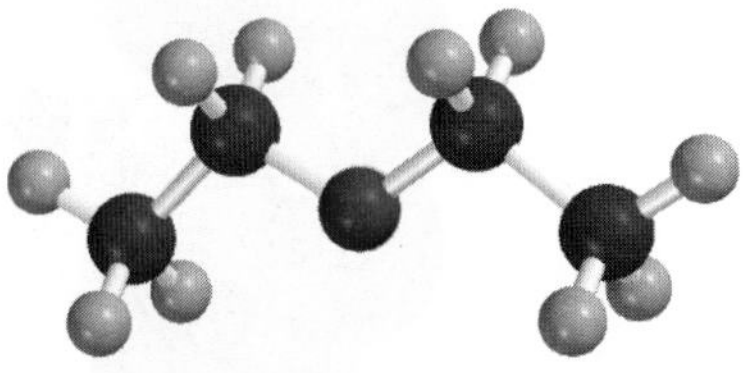

Anti conformation of diethyl ether

Chair conformation of tetrahydropyran

Incorporating an oxygen atom into a three-membered ring requires its bond angle to be seriously distorted from the normal tetrahedral value. In ethylene oxide, for example, the bond angle at oxygen is 61.5°.

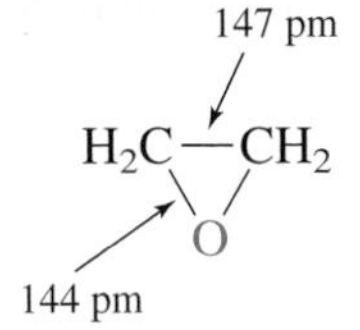
147 pm
H₂C—CH₂
O
144 pm

C—O—C angle 61.5°

C—C—O angle 59.2°

Thus epoxides, like cyclopropanes, have significant angle strain. They tend to undergo reactions that open the three-membered ring by cleaving one of the carbon–oxygen bonds.

Problem 16.2

The heats of combustion of 1,2-epoxybutane (2-ethyloxirane) and tetrahydrofuran have been measured: one is 2499 kJ/mol; the other is 2546 kJ/mol. Match the heats of combustion with the respective compounds.

16.3 Physical Properties of Ethers

Table 16.1 compares the physical properties of diethyl ether to those of an alkane (pentane) and an alcohol (1-butanol) of similar size and shape. With respect to boiling point, diethyl ether resembles pentane more than 1-butanol. With respect to dipole moment and solubility in water, the reverse is true.

TABLE 16.1 Physical Properties of Diethyl Ether, Pentane, and 1-Butanol

	Compound		Dipole moment, D	Boiling point, °C	Solubility in water, g/100 mL
Diethyl ether	$CH_3CH_2OCH_2CH_3$		1.2	35	7.5
Pentane	$CH_3CH_2CH_2CH_2CH_3$		0	36	≈0
1-Butanol	$CH_3CH_2CH_2CH_2OH$	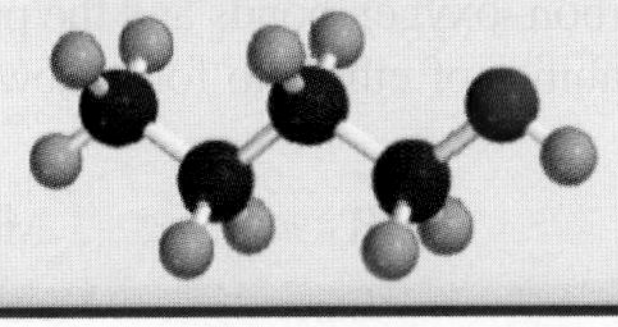	1.7	117	9

Figure 16.1

Hydrogen bonding between diethyl ether and water results from the attractive force between the negatively polarized oxygen of diethyl ether and the positively polarized hydrogen of water. The color ranges of the three electrostatic potential maps are the same.

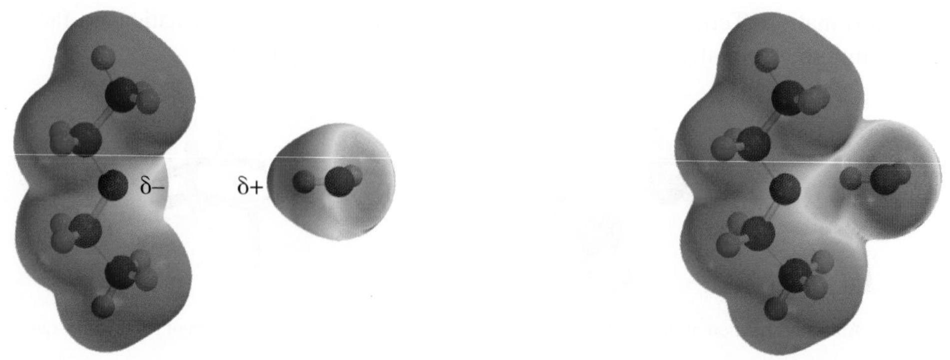

(*a*) Diethyl ether and water as separate molecules (*b*) Hydrogen-bonded complex

As we have seen before, alcohols have unusually high boiling points because of hydrogen bonding between —OH groups.

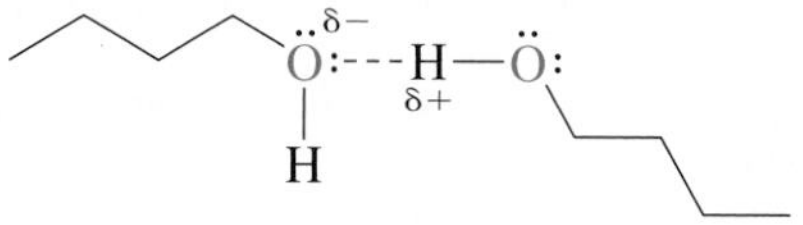

Intermolecular hydrogen bonding in 1-butanol

Lacking —OH groups, ethers resemble alkanes in that dispersion forces are the major contributors to intermolecular attractions. Although ethers have significant dipole moments, the fact that their boiling points are closer to alkanes than to alcohols tells us that dipole–dipole attractive forces are minor contributors.

On the other hand, ethers have a negatively polarized oxygen that can hydrogen bond to an —OH proton of water.

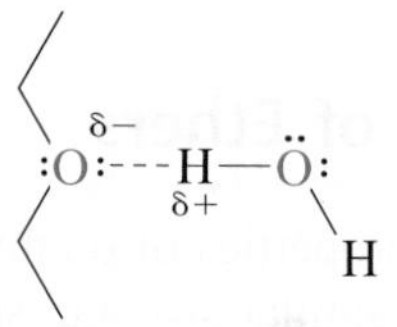

Hydrogen bonding between diethyl ether and water

Such hydrogen bonding causes ethers to dissolve in water to approximately the same extent as alcohols of similar size and shape. Alkanes cannot engage in hydrogen bonding to water. Figure 16.1 shows electrostatic potential maps of diethyl ether, water, and the hydrogen-bonded complex formed between them.

Problem 16.3

Of the two compounds cyclopentane and tetrahydrofuran, one has a boiling point of 49°C and is insoluble in water; the other has a boiling point of 65°C and is miscible with water in all proportions. Match the properties to the appropriate compound. In which property of which compound is hydrogen bonding important? Sketch the hydrogen-bonding interaction.

16.4 Crown Ethers

Their polar carbon–oxygen bonds and the presence of unshared electron pairs at oxygen contribute to the ability of ethers to form Lewis acid/Lewis base complexes with metal ions.

$$R_2\ddot{O}: \quad + \quad M^+ \quad \rightleftharpoons \quad R_2\overset{+}{\ddot{O}}—M$$

Ether (Lewis base) Metal ion (Lewis acid) Ether–metal ion complex

The strength of this bonding depends on the kind of ether. Simple ethers form relatively weak complexes with metal ions, but Charles J. Pedersen of DuPont discovered that certain *polyethers* form much more stable complexes with metal ions than do simple ethers.

Pedersen was a corecipient of the 1987 Nobel Prize in Chemistry.

Pedersen prepared a series of *macrocyclic polyethers,* cyclic compounds containing four or more oxygens in a ring of 12 or more atoms. He called these compounds **crown ethers,** because their molecular models resemble crowns. Systematic nomenclature of crown ethers is somewhat cumbersome, so Pedersen devised a shorthand description whereby the word *crown* is preceded by the total number of atoms in the ring and is followed by the number of oxygen atoms.

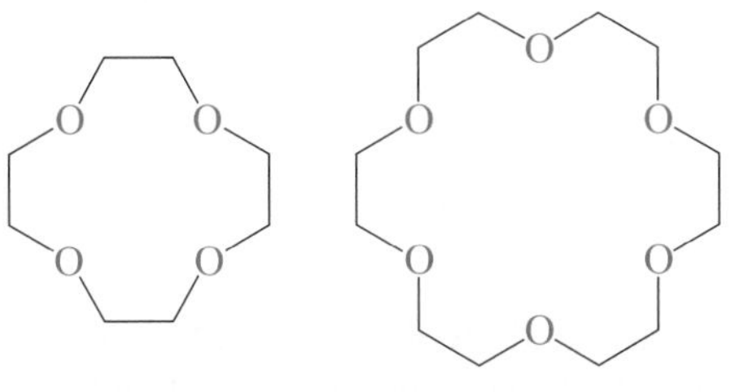

12-Crown-4 18-Crown-6

12-Crown-4 and 18-crown-6 are a cyclic tetramer and hexamer, respectively, of repeating —OCH_2CH_2— units; they are polyethers based on ethylene glycol ($HOCH_2CH_2OH$) as the parent alcohol.

Problem 16.4

What organic compound mentioned earlier in this chapter is a cyclic dimer of —OCH_2CH_2— units?

The metal–ion complexing properties of crown ethers are clearly evident in their effects on the solubility and reactivity of ionic compounds in nonpolar media. Potassium fluoride (KF) is ionic and practically insoluble in benzene alone, but dissolves in it when 18-crown-6 is present. This happens because of the electron distribution of 18-crown-6 as shown in Figure 16.2*a*. The electrostatic potential surface consists of essentially two regions: an electron-rich interior associated with the oxygens and a hydrocarbon-like exterior associated with the CH_2 groups. When KF is added to a solution of 18-crown-6 in benzene, potassium ion (K^+) interacts with the oxygens of the crown ether to form a Lewis acid/Lewis base complex. As can be seen in the space-filling model of this complex (Figure 16.2*b*), K^+, with a diameter of 266 pm, fits comfortably within the 260–320 pm internal cavity of 18-crown-6. Nonpolar CH_2 groups dominate the outer surface of the

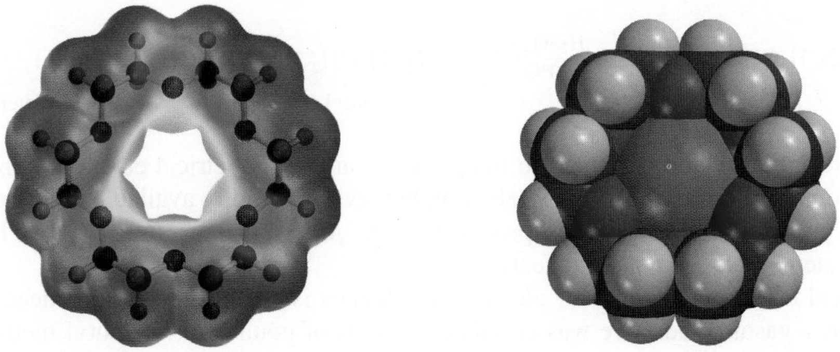

(*a*) (*b*)

Figure 16.2

(*a*) An electrostatic potential map of 18-crown-6. The region of highest electron density (red) is associated with the negatively polarized oxygens and their lone pairs. The outer periphery of the crown ether (blue) is relatively nonpolar (hydrocarbon-like) and causes the molecule to be soluble in nonpolar solvents such as benzene. (*b*) A space-filling model of the complex formed between 18-crown-6 and potassium ion (K^+). K^+ fits into the cavity of the crown ether where it is bound by a Lewis acid/Lewis base interaction with the oxygens.

complex, mask its polar interior, and permit the complex to dissolve in nonpolar solvents. Every K^+ that is carried into benzene brings a fluoride ion with it, resulting in a solution containing strongly complexed potassium ions and relatively unsolvated fluoride ions.

18-Crown-6 | Potassium fluoride (solid) | 18-Crown-6/potassium fluoride complex (in solution)

In solvents such as water and alcohols, fluoride ion is strongly solvated by ion–dipole forces and is neither very basic nor very nucleophilic. On the other hand, the poorly solvated, or "naked," fluoride ions that are present when potassium fluoride dissolves in benzene in the presence of a crown ether are better able to express their anionic reactivity. Thus, alkyl halides react with potassium fluoride in benzene containing 18-crown-6, thereby providing a method for the preparation of otherwise difficultly accessible alkyl fluorides. No reaction is observed in the absence of the crown ether.

The reaction proceeds in the direction indicated because a C—F bond is much stronger than a C—Br bond.

$$\underset{\text{1-Bromooctane}}{CH_3(CH_2)_6CH_2Br} \xrightarrow[\text{18-crown-6}]{\text{KF, benzene, 90°C}} \underset{\text{1-Fluorooctane (92\%)}}{CH_3(CH_2)_6CH_2F}$$

Catalysis by crown ethers has been used to advantage to increase the rate of many organic reactions that involve anions as reactants. Just as important, though, is the increased understanding that studies of crown ether catalysis have brought to our knowledge of biological processes in which metal ions, including Na^+ and K^+, are transported through the nonpolar interiors of cell membranes.

16.5 Preparation of Ethers

The mechanism for the formation of diethyl ether from ethanol under conditions of acid catalysis was shown in Mechanism 15.2 (p. 660).

Because they are widely used as solvents, many simple dialkyl ethers are commercially available. Diethyl ether and dibutyl ether, for example, are prepared by acid-catalyzed condensation of the corresponding alcohols, as described earlier in Section 15.7.

$$\underset{\text{1-Butanol}}{2CH_3CH_2CH_2CH_2OH} \xrightarrow[130°C]{H_2SO_4} \underset{\text{Dibutyl ether (60\%)}}{CH_3CH_2CH_2CH_2OCH_2CH_2CH_2CH_3} + \underset{\text{Water}}{H_2O}$$

In general, this method is limited to the preparation of symmetrical ethers in which both alkyl groups are primary. Isopropyl alcohol, however, is readily available at low cost and gives high enough yields of diisopropyl ether to justify making $(CH_3)_2CHOCH(CH_3)_2$ by this method on an industrial scale.

Acid-catalyzed addition of alcohols to alkenes is sometimes used. Indeed, before its use as a gasoline additive was curtailed, billions of pounds of *tert*-butyl methyl ether (MTBE) were prepared by the reaction:

tert-Butyl methyl ether is often referred to as MTBE, standing for the incorrect name "methyl *tert*-butyl ether." Remember, italicized prefixes are ignored when alphabetizing, and *tert*-butyl precedes methyl.

$$\underset{\text{2-Methylpropene}}{(CH_3)_2C{=}CH_2} + \underset{\text{Methanol}}{CH_3OH} \xrightarrow{H^+} \underset{\textit{tert}\text{-Butyl methyl ether}}{(CH_3)_3COCH_3}$$

Small amounts of *tert*-butyl methyl ether increase the octane rating of gasoline. Before environmental concerns placed limits on its use, the demand for MTBE exceeded the supply.

Polyether Antibiotics

One way in which pharmaceutical companies search for new drugs is by growing colonies of microorganisms in nutrient broths and assaying the substances produced for their biological activity. This method has yielded thousands of antibiotic substances, of which hundreds have been developed into effective drugs. Antibiotics are, by definition, toxic (*anti* = "against"; *bios* = "life"), and the goal is to find substances that are more toxic to infectious organisms than to their human hosts.

Since 1950, a number of polyether antibiotics have been discovered using fermentation technology. They are characterized by the presence of several cyclic ether structural units, as illustrated for the case of *monensin* in Figure 16.3*a*. Monensin and other naturally occurring polyethers are similar to crown ethers in their ability to form stable complexes with metal ions. The structure of the sodium salt of monensin is depicted in Figure 16.3*b*, where it can be seen that four ether oxygens and two hydroxyl groups surround a sodium ion. The alkyl groups are oriented toward the outside of the complex, and the polar oxygens and the metal ion are on the inside. The hydrocarbon-like surface of the complex permits it to carry its sodium ion through the hydrocarbon-like interior of a cell membrane. This disrupts the normal balance of sodium ions within the cell and interferes with important processes of cellular respiration. Small amounts of monensin are added to poultry feed to kill parasites that live in the intestines of chickens. Compounds such as monensin and the crown ethers that affect metal ion transport are referred to as *ionophores* ("ion carriers").

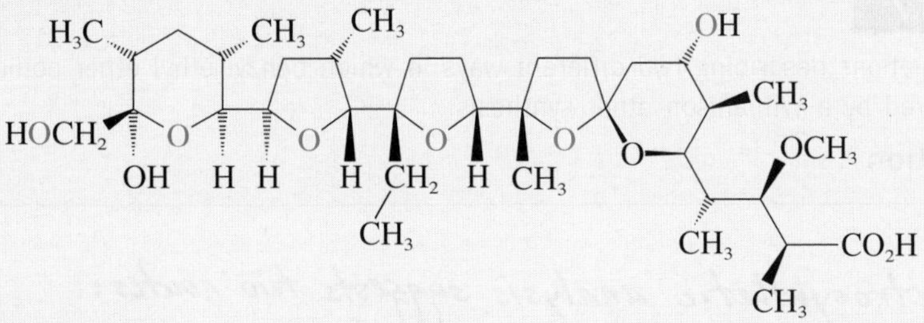

(*a*)

(*b*)

Figure 16.3

(*a*) The structure of monensin; (*b*) The structure of the sodium salt of monensin showing coordination of Na^+ (yellow) to the six oxygens shown in red in (*a*). Hydrogen atoms have been omitted from the model for clarity.

Problem 16.5

Outline a reasonable mechanism for the formation of *tert*-butyl methyl ether according to the preceding equation.

In a reaction resembling halohydrin formation (Section 6.17), vicinal haloethers are prepared from alkenes by reaction with an alcohol in the presence of halogens—usually bromine or iodine. This *haleotherification* proceeds through a cyclic halonium ion, which reacts with the alcohol. 1-Methylcyclohexene undergoes iodoetherification with ethanol in the presence of iodine to give *trans*-1-ethoxy-2-iodo-1-methylcyclohexane.

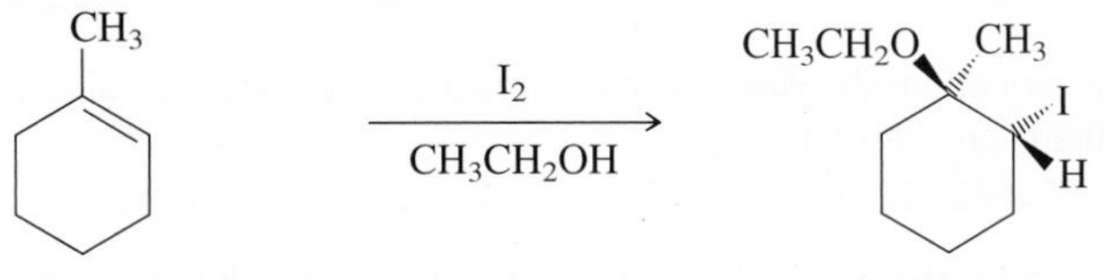

1-Methylcyclohexene → *trans*-1-Ethoxy-2-iodo-1-methylcyclohexane (67%)

Problem 16.6

Write a stepwise reaction mechanism for the formation of *trans*-1-ethoxy-2-iodo-1-methylcyclohexane by reaction of 1-methylcyclohexene with iodine and ethanol.

16.6 The Williamson Ether Synthesis

The reaction is named for Alexander Williamson, a British chemist who used it to prepare diethyl ether in 1850.

A long-standing method for the preparation of ethers is the **Williamson ether synthesis.** Nucleophilic substitution of an alkyl halide by an alkoxide gives the carbon–oxygen bond of an ether:

$$RO^{-} + R'—X \xrightarrow{S_N2} ROR' + X^{-}$$

Alkoxide ion | Alkyl halide | Ether | Halide ion

Preparation of ethers by the Williamson ether synthesis is most successful with methyl and primary alkyl halides.

$$CH_3CH_2CH_2CH_2ONa + CH_3CH_2I \longrightarrow CH_3CH_2CH_2CH_2OCH_2CH_3 + NaI$$

Sodium butoxide | Iodoethane | Butyl ethyl ether (71%) | Sodium iodide

Problem 16.7

(a) Write equations describing two different ways in which benzyl ethyl ether could be prepared by a Williamson ether synthesis.

Sample Solution

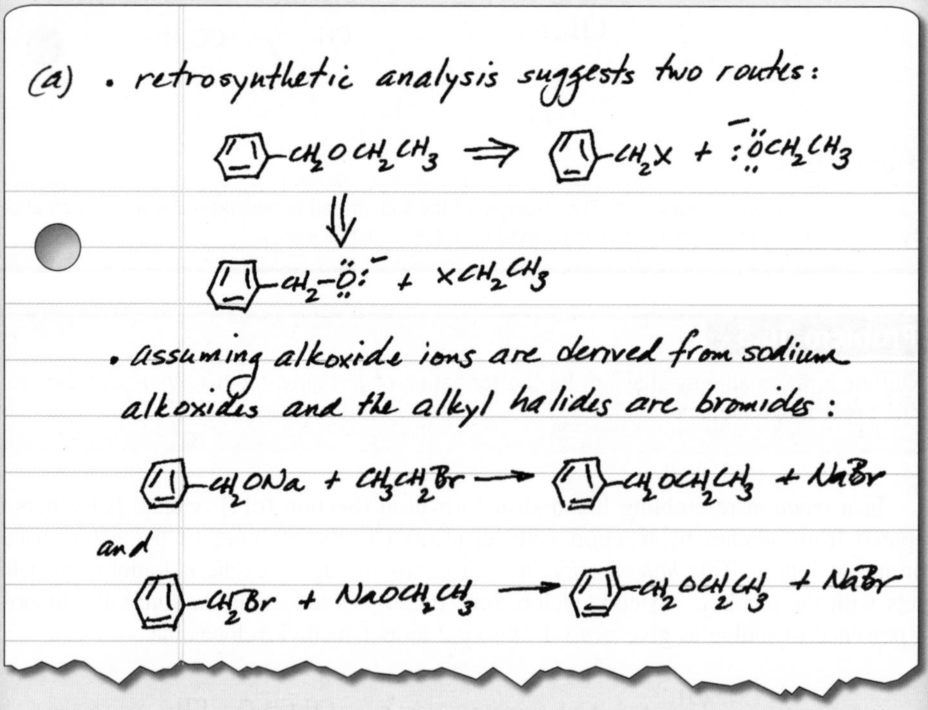

(b) Write an equation showing the most practical synthesis of allyl phenyl ether by the Williamson method.

Secondary and tertiary *alkyl halides* are not suitable, because they react with alkoxide bases by E2 elimination rather than by S_N2 substitution. Whether the *alkoxide base* is primary, secondary, or tertiary is much less important than the nature of the alkyl halide.

Thus benzyl isopropyl ether is prepared in high yield from benzyl chloride, a primary chloride that is incapable of undergoing elimination, and sodium isopropoxide:

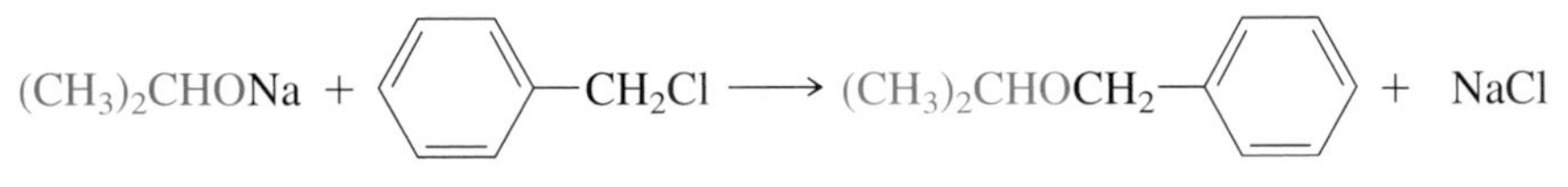

Sodium isopropoxide | Benzyl chloride | Benzyl isopropyl ether (84%) | Sodium chloride

The alternative synthetic route using the sodium salt of benzyl alcohol and an isopropyl halide would be much less effective, because of increased competition from elimination as the alkyl halide becomes more sterically hindered.

Problem 16.8

Only one combination of alkyl halide and alkoxide is appropriate for the preparation of each of the following ethers by the Williamson ether synthesis. What is the correct combination in each case?

(a) 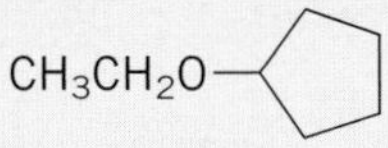

(b) $H_2C{=}CHCH_2OCH(CH_3)_2$

(c) $(CH_3)_3COCH_2C_6H_5$

Sample Solution (a) The ether linkage of cyclopentyl ethyl ether involves a primary carbon and a secondary one. Choose the alkyl halide corresponding to the primary alkyl group, leaving the secondary alkyl group to arise from the alkoxide nucleophile.

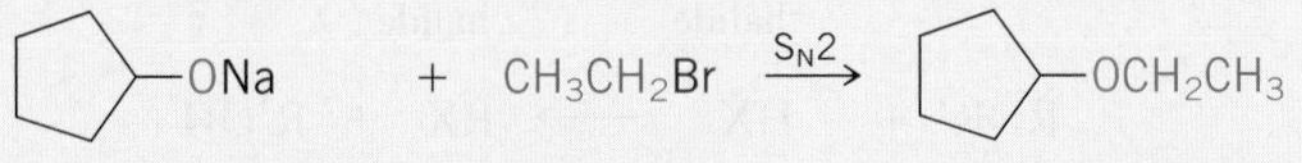

Sodium cyclopentanolate | Ethyl bromide | Cyclopentyl ethyl ether

The alternative combination, cyclopentyl bromide and sodium ethoxide, is not appropriate because elimination will be the major reaction:

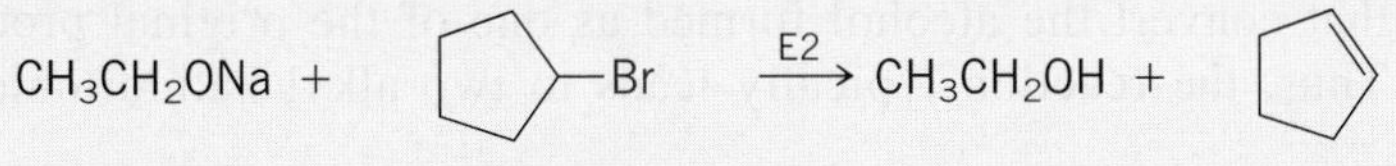

Sodium ethoxide | Cyclopentyl bromide | Ethanol | Cyclopentene (major products)

Both reactants in the Williamson ether synthesis usually come from alcohols. Sodium and potassium alkoxides are prepared by reaction of an alcohol with the appropriate metal, and alkyl halides are most commonly made from alcohols by reaction with a hydrogen halide, thionyl chloride, or phosphorus tribromide (Sections 4.7 and 4.14). Alternatively, alkyl *p*-toluenesulfonates may be used in place of alkyl halides; alkyl *p*-toluenesulfonates are also prepared from alcohols (Section 8.12).

16.7 Reactions of Ethers: A Review and a Preview

Up to this point, we haven't seen any reactions of dialkyl ethers. Indeed, ethers are one of the least reactive of the functional groups we shall study. It is this low level of reactivity, along with an ability to dissolve nonpolar substances, that makes ethers so often used as solvents when carrying out organic reactions. Nevertheless, most ethers are hazardous materials, and precautions must be taken when using them. Diethyl ether is extremely flammable and because of its high volatility can form explosive mixtures in air relatively quickly. Open flames must never be present in laboratories where diethyl ether is being used. Other low-molecular-weight ethers must also be treated as fire hazards.

Another dangerous property of ethers is the ease with which they undergo oxidation in air to form explosive peroxides. Air oxidation of diisopropyl ether proceeds according to the equation

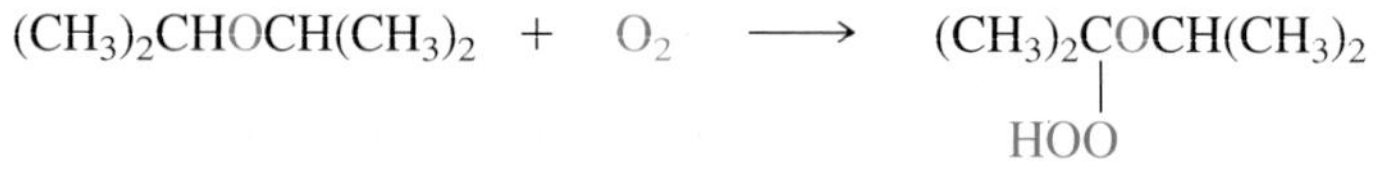
(CH3)2CHOCH(CH3)2 + O2 → (CH3)2COCH(CH3)2
HOO

Diisopropyl ether — Oxygen — Diisopropyl ether hydroperoxide

The reaction follows a free-radical mechanism and gives a hydroperoxide, a compound of the type ROOH. Hydroperoxides tend to be unstable and shock-sensitive. On standing, they form related peroxidic derivatives, which are also prone to violent decomposition. Air oxidation leads to peroxides within a few days if ethers are even briefly exposed to atmospheric oxygen. For this reason, one should never use old bottles of dialkyl ethers, and extreme care must be exercised in their disposal.

16.8 Acid-Catalyzed Cleavage of Ethers

Just as the carbon–oxygen bond of alcohols is cleaved on reaction with hydrogen halides, so too is an ether linkage broken:

$$\underset{\text{Alcohol}}{ROH} + \underset{\text{Hydrogen halide}}{HX} \longrightarrow \underset{\text{Alkyl halide}}{RX} + \underset{\text{Water}}{H_2O}$$

$$\underset{\text{Ether}}{ROR'} + \underset{\text{Hydrogen halide}}{HX} \longrightarrow \underset{\text{Alkyl halide}}{RX} + \underset{\text{Alcohol}}{R'OH}$$

The cleavage of ethers is normally carried out under conditions (excess hydrogen halide, heat) that convert the alcohol formed as one of the original products to an alkyl halide. Thus, the reaction typically leads to two alkyl halide molecules plus water.

$$\underset{\text{Ether}}{ROR'} + \underset{\text{Hydrogen halide}}{2HX} \xrightarrow{\text{heat}} \underset{\text{Two alkyl halides}}{RX + R'X} + \underset{\text{Water}}{H_2O}$$

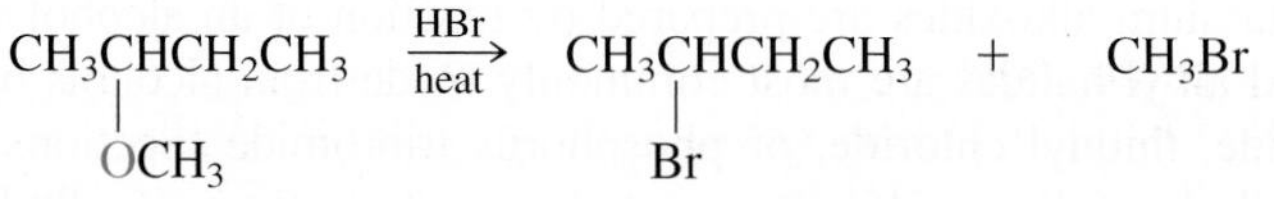
CH3CHCH2CH3
OCH3
HBr
heat
CH3CHCH2CH3
Br
+
CH3Br

sec-Butyl methyl ether — 2-Bromobutane (81%) — Bromomethane

The order of hydrogen halide reactivity is HI > HBr >> HCl. Hydrogen fluoride is not effective.

Problem 16.9

A series of dialkyl ethers was allowed to react with excess hydrogen bromide, with the following results. Identify the ether in each case.

(a) One ether gave a mixture of bromocyclopentane and 1-bromobutane.
(b) Another ether gave only benzyl bromide.
(c) A third ether gave one mole of 1,5-dibromopentane per mole of ether.

Sample Solution (a) In the reaction of dialkyl ethers with excess hydrogen bromide, each alkyl group of the ether function is cleaved and forms an alkyl bromide. Because bromocyclopentane and 1-bromobutane are the products, the starting ether must be butyl cyclopentyl ether.

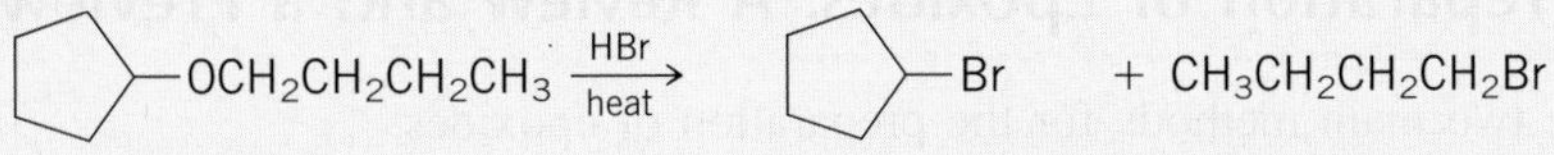

Butyl cyclopentyl ether   Bromocyclopentane   1-Bromobutane

The cleavage of diethyl ether by hydrogen bromide is outlined in Mechanism 16.1. The key step is an S_N2-like attack on a dialkyloxonium ion by bromide (step 2).

Problem 16.10

Adapt Mechanism 16.1 to the reaction:

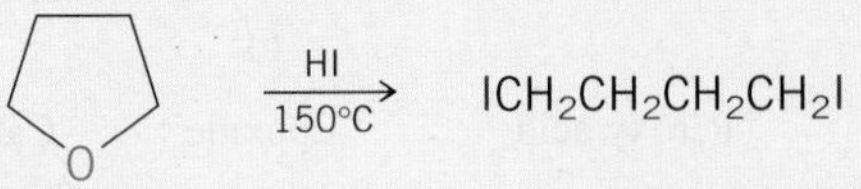

Tetrahydrofuran   1,4-Diiodobutane (65%)

Mechanism 16.1

Cleavage of Ethers by Hydrogen Halides

THE OVERALL REACTION:

$$CH_3CH_2OCH_2CH_3 + HBr \xrightarrow{\text{heat}} 2CH_3CH_2Br + H_2O$$

Diethyl ether   Hydrogen bromide   Ethyl bromide   Water

THE MECHANISM:

Step 1: Proton transfer to the oxygen of the ether to give a dialkyloxonium ion.

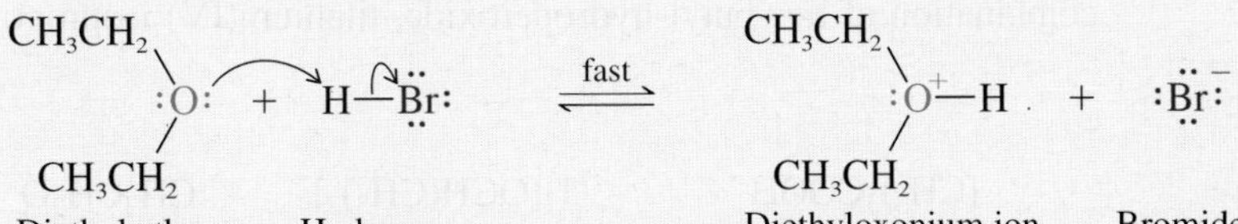

Diethyl ether   Hydrogen bromide   Diethyloxonium ion   Bromide ion

Step 2: Nucleophilic attack of the halide anion on carbon of the dialkyloxonium ion. This step gives one molecule of an alkyl halide and one molecule of an alcohol.

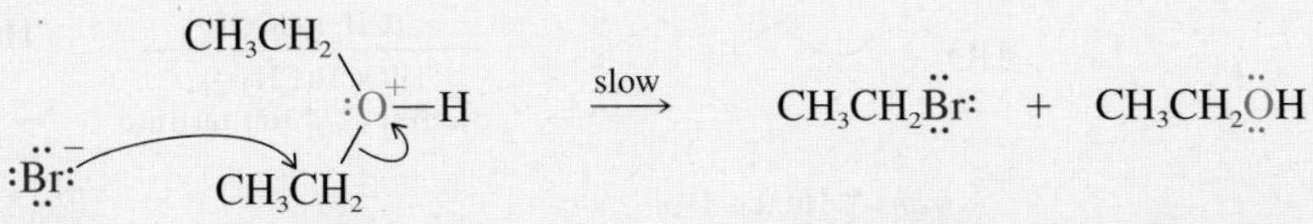

Bromide ion   Diethyloxonium ion   Ethyl bromide   Ethanol

Steps 3 and 4: These two steps do not involve an ether at all. They correspond to those in which an alcohol is converted to an alkyl halide (Sections 4.8–4.12).

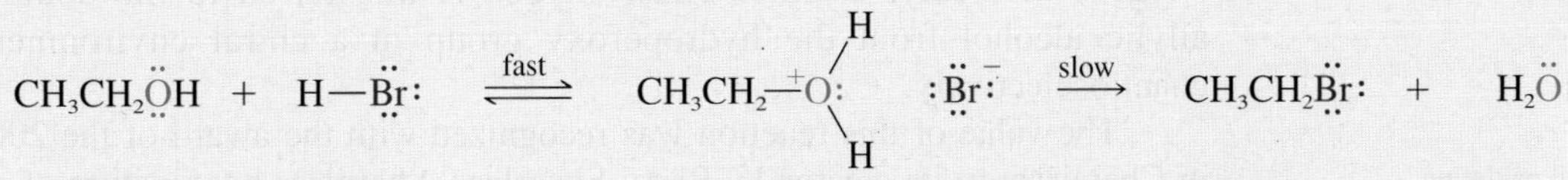

Ethanol   Hydrogen bromide   Ethyl bromide   Water

With ethers of the type ROR′ (R≠R′), the question of which carbon–oxygen bond is broken first is not one that we need examine at our level of study. Note also that ethers of tertiary alcohols would react with hydrogen halides by an S_N1 mechanism.

16.9 Preparation of Epoxides: A Review and a Preview

There are two main methods for the preparation of epoxides:

1. Epoxidation of alkenes
2. Base-promoted ring closure of vicinal halohydrins

Epoxidation of alkenes with peroxy acids was discussed in Section 6.19 and is represented by the general equation

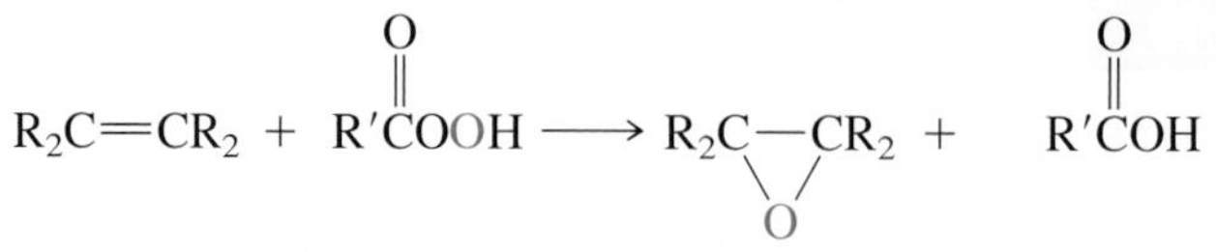

Alkene Peroxy acid Epoxide Carboxylic acid

The reaction is easy to carry out, and yields are usually high. Epoxidation is a stereospecific syn addition. (*E*)-1,2-Diphenylethene gives *trans*-2,3-diphenyloxirane. The *Z* alkene produces the cis epoxide.

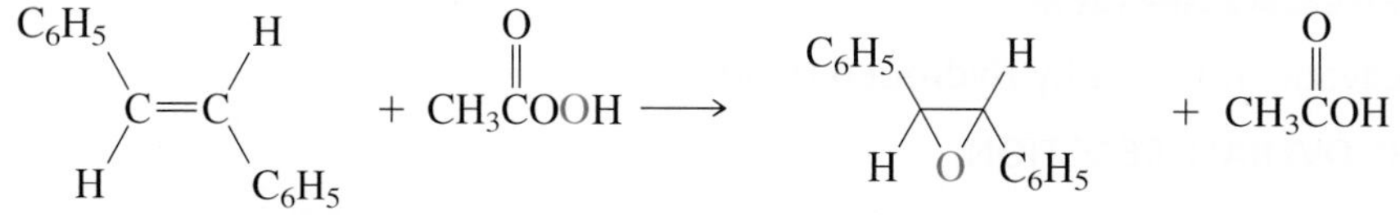

(*E*)-1,2-Diphenylethene Peroxyacetic acid *trans*-2,3-Diphenyloxirane (78–83%) Acetic acid

Allylic alcohols are converted to epoxides by oxidation with *tert*-butyl hydroperoxide in the presence of certain transition metals. The most significant aspect of this reaction—called the **Sharpless epoxidation**—is its high enantioselectivity when carried out using a combination of *tert*-butyl hydroperoxide, titanium(IV) isopropoxide, and diethyl tartrate.

$(CH_3)_3COOH$ — *tert*-Butyl hydroperoxide

$Ti[OCH(CH_3)_2]_4$ — Titanium(IV) isopropoxide

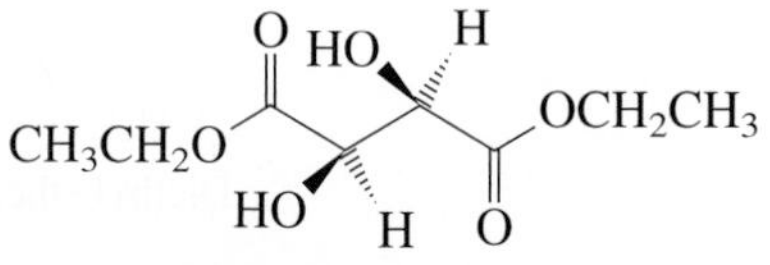

Diethyl (2*R*,3*R*)-tartrate

Diethyl (2*R*,3*R*)-tartrate is the diethyl ester of tartaric acid, a chiral molecule that was discussed in Section 7.15.

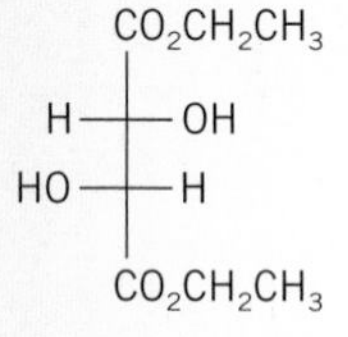

Diethyl (2*R*,3*R*)-tartrate

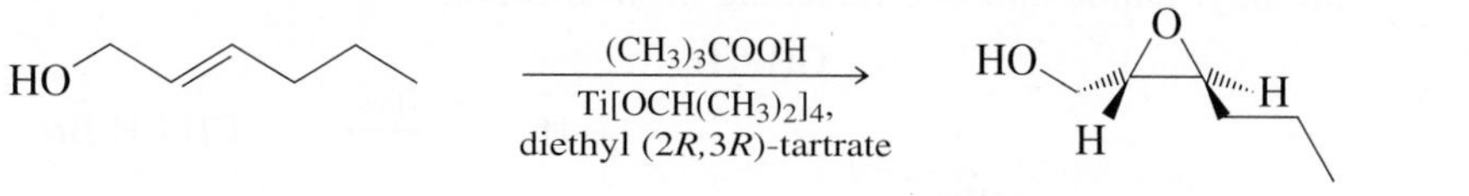

trans-2-Hexen-1-ol (2*S*,3*S*)-2,3-Epoxy-1-hexanol (78% yield; 98% enantiomeric excess)

An intermediate in the reaction contains Ti—O bonds to the oxygen of *tert*-butylperoxy [$(CH_3)_3COO$—], the oxygen of the allylic alcohol, and to both hydroxyl oxygens of diethyl tartrate. Thus, oxygen is transferred to the double bond of the allylic alcohol from the hydroperoxy group in a chiral environment and occurs enantioselectively.

The value of this reaction was recognized with the award of the 2001 Nobel Prize in Chemistry to its creator K. Barry Sharpless. Sharpless epoxidation of allylic alcohols can be carried out with catalytic amounts of titanium(IV) isopropoxide and, because both enantiomers of diethyl tartrate are readily available, can be applied to the synthesis of either enantiomer of a desired epoxy alcohol.

Sharpless's work in oxidation also included methods for the enantioselective dihydroxylation of alkenes (see Section 15.5).

Problem 16.11

What would be the absolute configuration of the 2,3-epoxy-1-hexanol produced in the preceding reaction if diethyl (2*S*,3*S*)-tartrate were used instead of (2*R*,3*R*)?

More than a laboratory synthesis, Sharpless epoxidation has been adapted to the large-scale preparation of (+)-disparlure, a sex pheromone used to control gypsy moth infestations, and of (*R*)-glycidol, an intermediate in the synthesis of cardiac drugs known as beta-blockers.

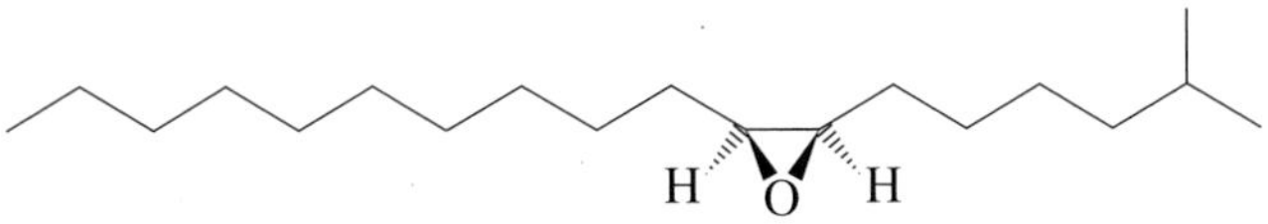

(+)-Disparlure
[(7*R*,8*S*)-(+)-7,8-Epoxy-2-methyloctadecane]

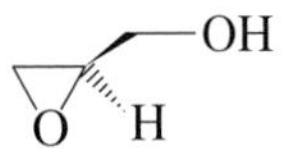

(*R*)-(+)-Glycidol
[(*R*)-(+)-2,3-Epoxy-1-propanol]

The following section describes the preparation of epoxides by the base-promoted ring closure of vicinal halohydrins. Because vicinal halohydrins are customarily prepared from alkenes (Section 6.17), both methods—epoxidation using peroxy acids and ring closure of halohydrins—are based on alkenes as the starting materials for preparing epoxides.

16.10 Conversion of Vicinal Halohydrins to Epoxides

The formation of vicinal halohydrins from alkenes was described in Section 6.17. Halohydrins are readily converted to epoxides on treatment with base:

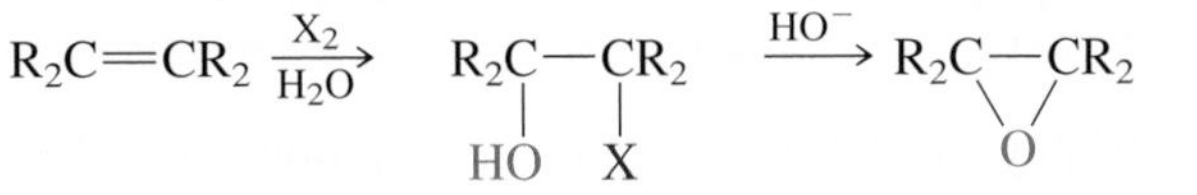

Alkene | Vicinal halohydrin | Epoxide

Reaction with base brings the alcohol function of the halohydrin into equilibrium with its corresponding alkoxide:

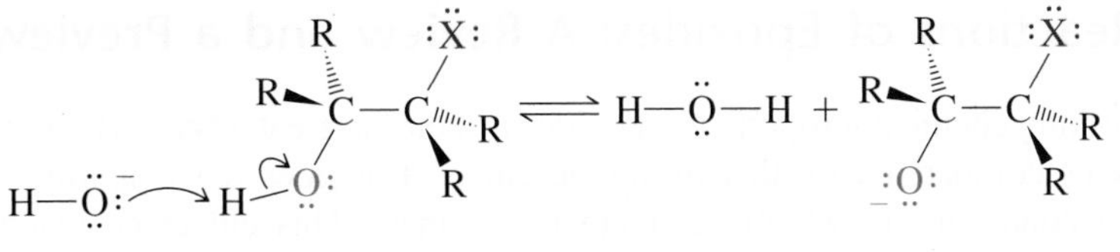

Vicinal halohydrin

Next, in what amounts to an *intramolecular* Williamson ether synthesis, the alkoxide oxygen attacks the carbon that bears the halide leaving group, giving an epoxide. As in other nucleophilic substitutions, the nucleophile approaches carbon from the side opposite the bond to the leaving group:

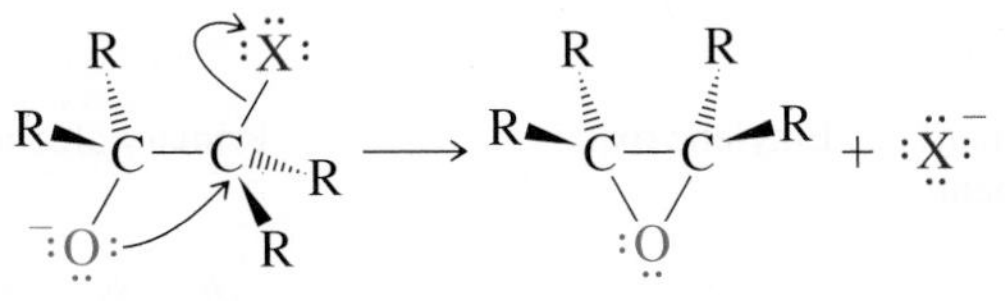

Epoxide

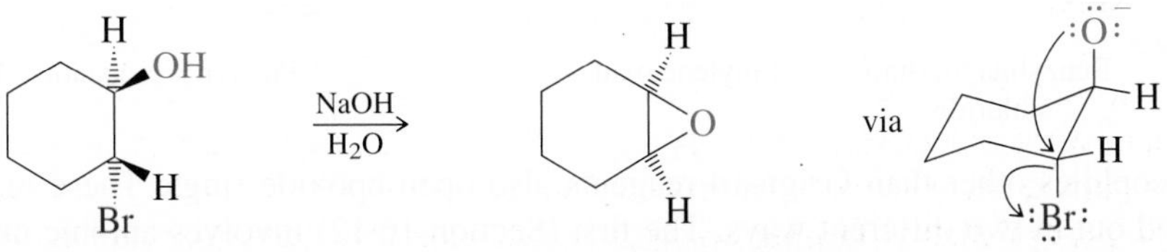

trans-2-Bromocyclohexanol | 1,2-Epoxycyclohexane (81%)

Overall, the stereospecificity of this method is the same as that observed in peroxy acid oxidation of alkenes. Substituents that are cis to each other in the alkene remain cis in the epoxide. This is because formation of the bromohydrin involves anti addition, and the ensuing intramolecular nucleophilic substitution reaction takes place with inversion of configuration at the carbon that bears the halide leaving group.

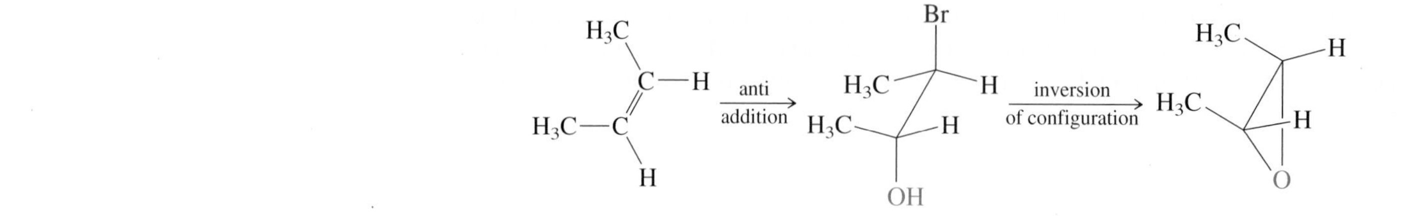

cis-2-Butene *cis*-2,3-Epoxybutane

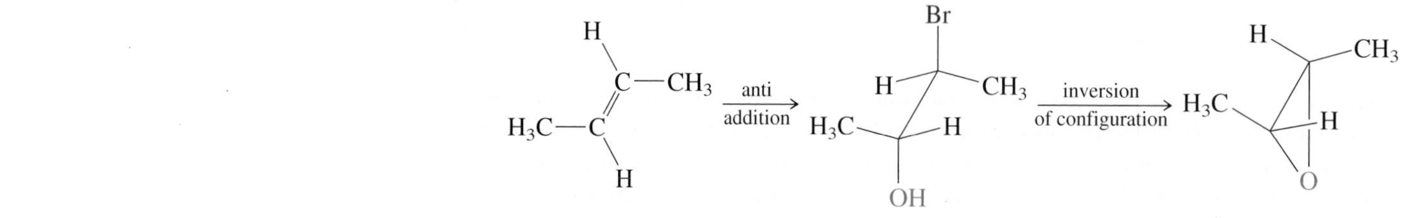

trans-2-Butene *trans*-2,3-Epoxybutane

Problem 16.12

Is either of the epoxides formed in the preceding reactions chiral? Is either epoxide optically active when prepared from the alkene by this method?

About 2 × 10^9 lb/year of 1,2-epoxypropane is produced in the United States as an intermediate in the preparation of various polymeric materials, including polyurethane plastics and foams and polyester resins. A large fraction of the 1,2-epoxypropane is made from propene by way of its chlorohydrin.

16.11 Reactions of Epoxides: A Review and a Preview

The most striking chemical property of epoxides is their far greater reactivity toward nucleophilic reagents compared with that of simple ethers. Epoxides react rapidly with nucleophiles under conditions in which other ethers are inert. This enhanced reactivity results from the angle strain of epoxides. Reactions that open the ring relieve this strain.

Angle strain is the main source of strain in epoxides, but torsional strain that results from the eclipsing of bonds on adjacent carbons is also present. Both kinds of strain are relieved when a ring-opening reaction occurs.

We saw an example of nucleophilic ring opening of epoxides in Section 15.4, where the reaction of Grignard reagents with ethylene oxide was described as a synthetic route to primary alcohols:

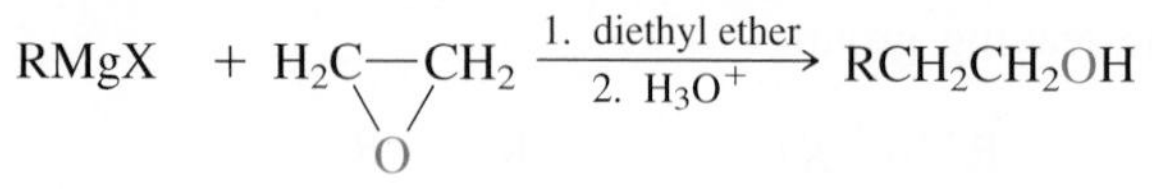

Grignard reagent Ethylene oxide Primary alcohol

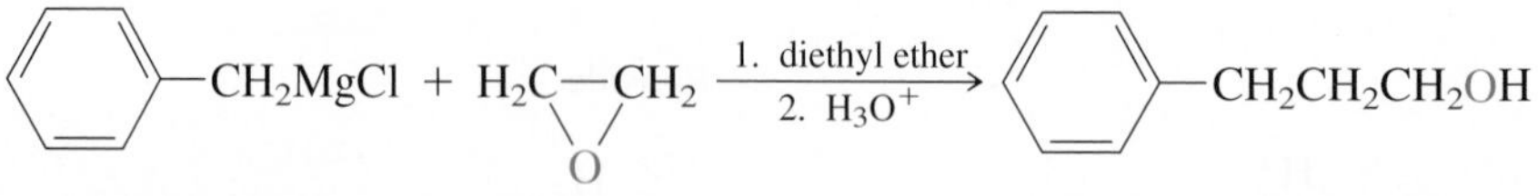

Benzylmagnesium chloride Ethylene oxide 3-Phenyl-1-propanol (71%)

Nucleophiles other than Grignard reagents also open epoxide rings. These reactions are carried out in two different ways. The first (Section 16.12) involves anionic nucleophiles in neutral or basic solution.

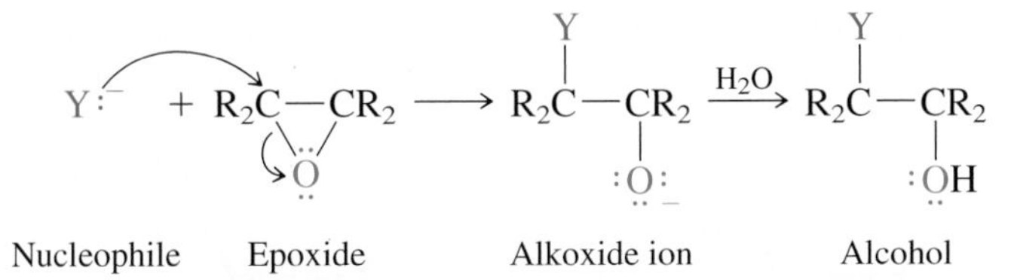
Nucleophile
Epoxide
Alkoxide ion
Alcohol

These reactions are usually performed in water or alcohols as solvents, and the alkoxide ion intermediate is rapidly transformed to an alcohol by proton transfer.

The other involves acid catalysis. Here the nucleophile is often a molecule of the solvent.

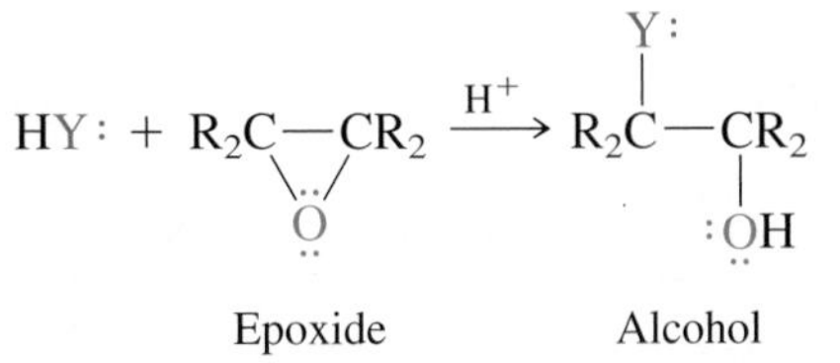
Epoxide
Alcohol

Acid-catalyzed ring opening of epoxides is discussed in Section 16.13.

There is an important difference in the regiochemistry of ring-opening reactions of epoxides depending on the reaction conditions. Unsymmetrically substituted epoxides tend to react with anionic nucleophiles at the less hindered carbon of the ring. Under conditions of acid catalysis, however, the more highly substituted carbon is attacked.

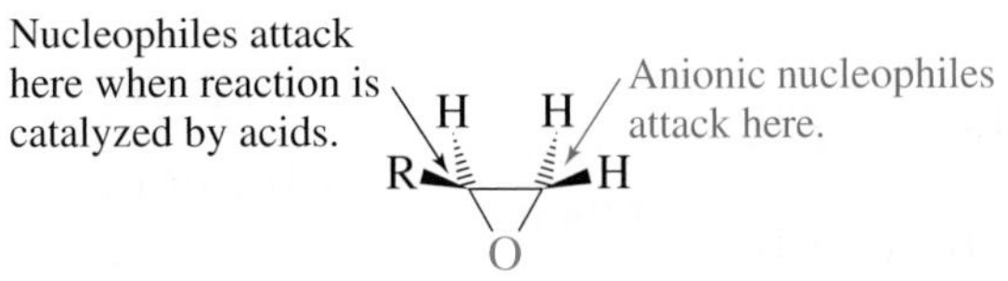
Nucleophiles attack here when reaction is catalyzed by acids.
Anionic nucleophiles attack here.

The underlying reasons for this difference in regioselectivity will be explained in Section 16.13.

16.12 Nucleophilic Ring Opening of Epoxides

Ethylene oxide is a very reactive substance. It reacts rapidly and exothermically with anionic nucleophiles to yield 2-substituted derivatives of ethanol by cleaving the carbon–oxygen bond of the ring:

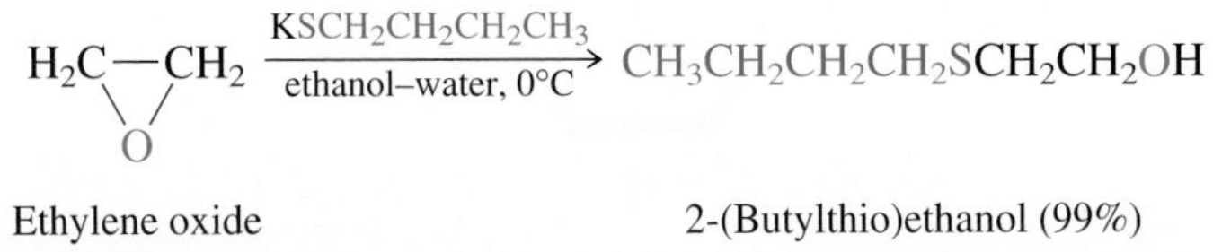
KSCH2CH2CH2CH3
ethanol–water, 0°C
CH3CH2CH2CH2SCH2CH2OH
Ethylene oxide
2-(Butylthio)ethanol (99%)

Problem 16.13

What is the principal organic product formed in the reaction of ethylene oxide with each of the following?

(a) Sodium cyanide (NaCN) in aqueous ethanol
(b) Sodium azide (NaN_3) in aqueous ethanol
(c) Sodium hydroxide (NaOH) in water
(d) Phenyllithium (C_6H_5Li) in diethyl ether, followed by addition of dilute sulfuric acid
(e) 1-Butynylsodium ($CH_3CH_2C{\equiv}CNa$) in liquid ammonia

Sample Solution (a) Sodium cyanide is a source of the nucleophilic cyanide anion. Cyanide ion attacks ethylene oxide, opening the ring and forming 2-cyanoethanol:

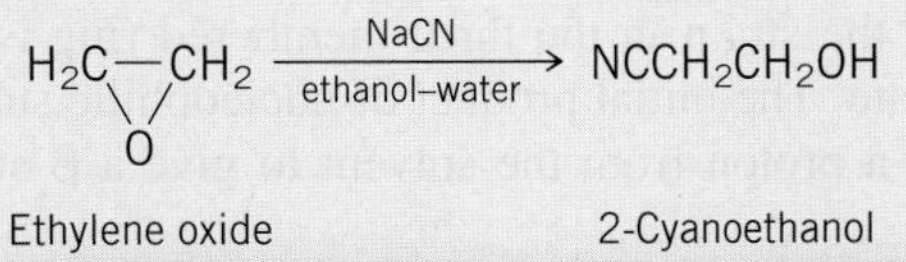
NaCN
ethanol–water
NCCH2CH2OH
Ethylene oxide
2-Cyanoethanol

Nucleophilic ring opening of epoxides has many of the features of an S_N2 reaction. Inversion of configuration is observed at the carbon at which substitution occurs.

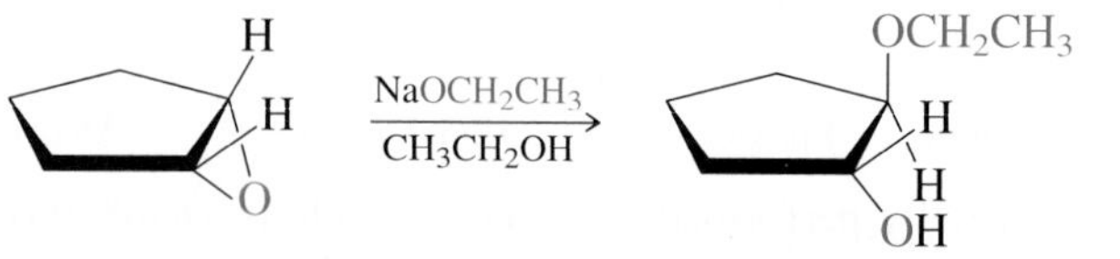
NaOCH2CH3
CH3CH2OH
OCH2CH3
OH

1,2-Epoxycyclopentane → *trans*-2-Ethoxycyclopentanol (67%)

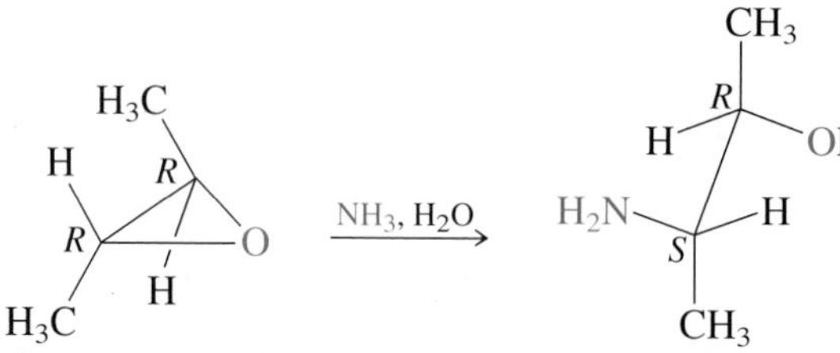
NH3, H2O
H2N
OH

(2*R*,3*R*)-2,3-Epoxybutane → (2*R*,3*S*)-3-Amino-2-butanol (70%)

Unsymmetrical epoxides are attacked at the less substituted, less sterically hindered carbon of the ring:

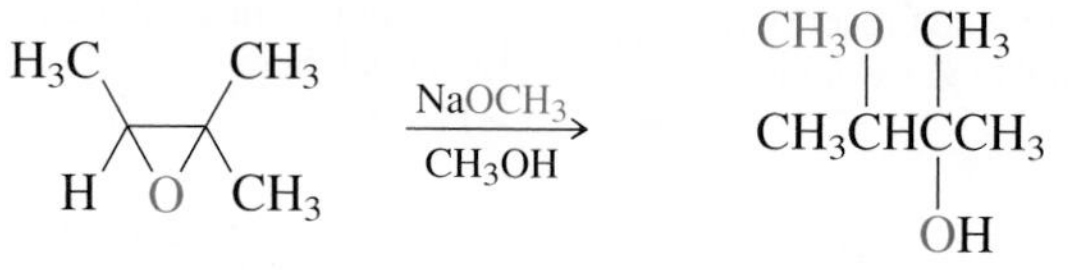
NaOCH3
CH3OH
CH3O CH3
CH3CHCCH3
OH

2,2,3-Trimethyloxirane → 3-Methoxy-2-methyl-2-butanol (53%)

Problem 16.14

Given the epoxide of absolute configuration as shown, decide which one of the compounds A through C correctly represents the product of its reaction with sodium methoxide in methanol.

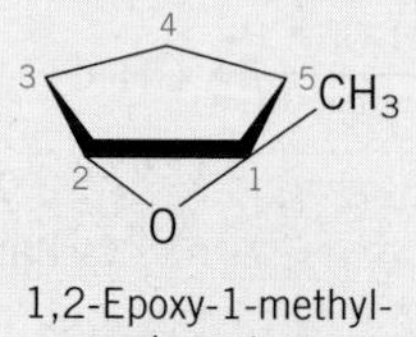
CH3

1,2-Epoxy-1-methyl-cyclopentane

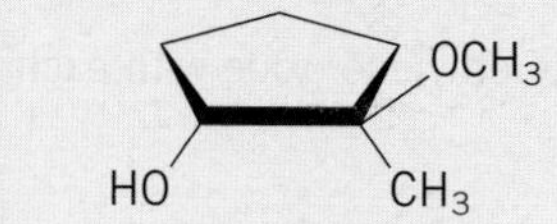
OCH3
HO
CH3

Compound A

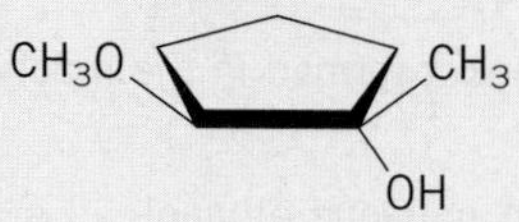
CH3O
CH3
OH

Compound B

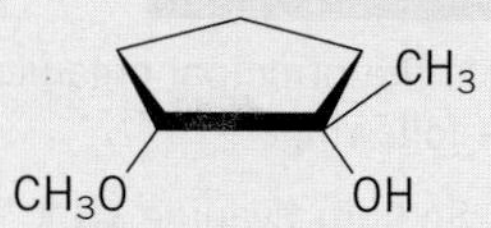
CH3
CH3O
OH

Compound C

The experimental observations combine with the principles of nucleophilic substitution to give the picture of epoxide ring opening shown in Mechanism 16.2. The nucleophile attacks the less crowded carbon from the side opposite the carbon–oxygen bond. Bond formation with the nucleophile accompanies carbon–oxygen bond breaking, and a substantial portion of the strain in the three-membered ring is relieved as it begins to open at the transition state. The initial product of nucleophilic substitution is an alkoxide, which rapidly abstracts a proton from the solvent to give a β-substituted alcohol as the isolated product.

Mechanism 16.2

Nucleophilic Ring Opening of an Epoxide

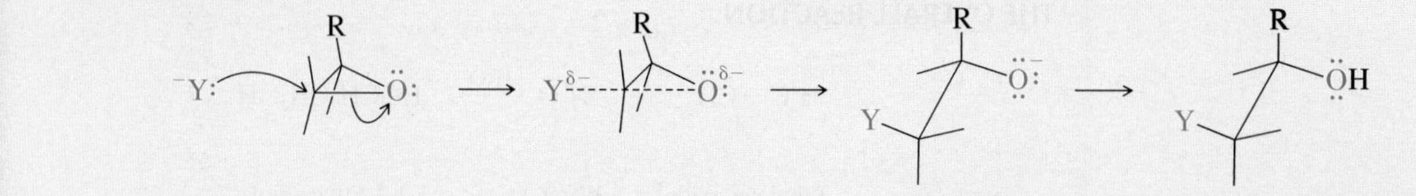
R
Y:
Ö:
δ−
Y
O
ÖH

Nucleophile Epoxide Transition state Alkoxide ion β-Substituted alcohol

The reaction of Grignard reagents with epoxides is regioselective in the same sense. Attack occurs at the less substituted carbon of the ring.

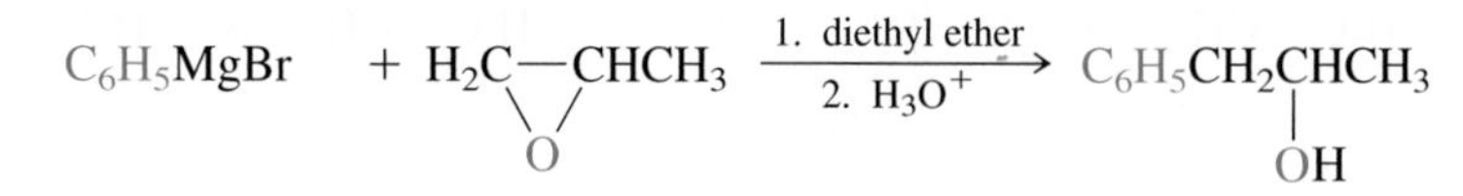
C6H5MgBr
+ H2C—CHCH3
O
1. diethyl ether
2. H3O+
C6H5CH2CHCH3
OH

Phenylmagnesium bromide 1,2-Epoxypropane 1-Phenyl-2-propanol (60%)

Epoxides are reduced to alcohols on treatment with lithium aluminum hydride. Hydride is transferred to the less substituted carbon.

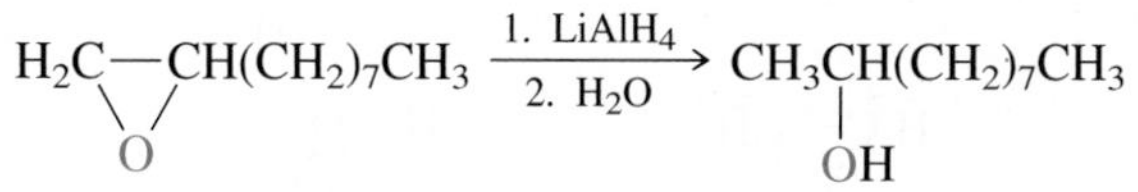
H2C—CH(CH2)7CH3
O
1. LiAlH4
2. H2O
CH3CH(CH2)7CH3
OH

1,2-Epoxydecane 2-Decanol (90%)

Epoxidation of an alkene, followed by lithium aluminum hydride reduction of the resulting epoxide, gives the same alcohol as would be obtained by hydration (Section 6.9) of an alkene in accordance with Markovnikov's rule.

16.13 Acid-Catalyzed Ring Opening of Epoxides

As we've just seen, nucleophilic ring opening of ethylene oxide yields 2-substituted derivatives of ethanol. Those reactions involved nucleophilic attack on the carbon of the ring under neutral or basic conditions. Other nucleophilic ring openings of epoxides likewise give 2-substituted derivatives of ethanol but involve an acid as either a reactant or a catalyst.

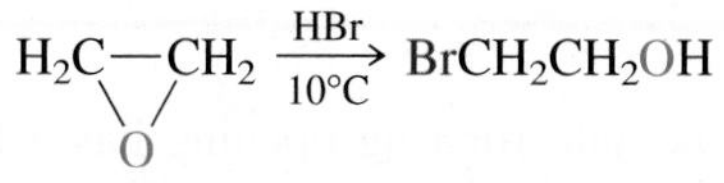
H2C—CH2
O
HBr
10°C
BrCH2CH2OH

Ethylene oxide 2-Bromoethanol (87–92%)

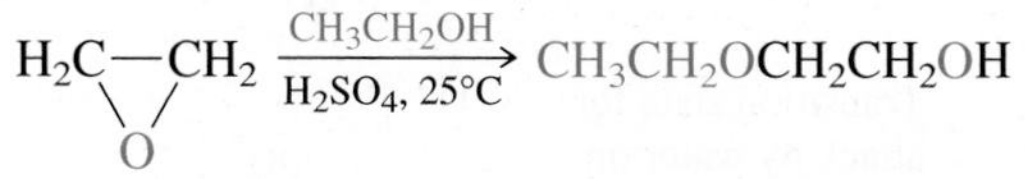
H2C—CH2
O
CH3CH2OH
H2SO4, 25°C
CH3CH2OCH2CH2OH

Ethylene oxide 2-Ethoxyethanol (85%)

A third example is the industrial preparation of ethylene glycol ($HOCH_2CH_2OH$) by hydrolysis of ethylene oxide in dilute sulfuric acid. This reaction and its mechanism (Mechanism 16.3) illustrate the difference between the ring openings of ethylene oxide discussed in the preceding section and the acid-catalyzed ones described here. In acid, the species that is attacked by the nucleophile is not the epoxide itself, but rather its

Mechanism 16.3

Acid-Catalyzed Ring Opening of Ethylene Oxide

THE OVERALL REACTION:

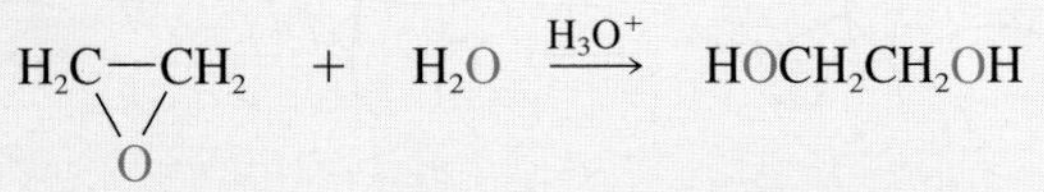

Ethylene oxide Water 1,2-Ethanediol (ethylene glycol)

THE MECHANISM:

Step 1: Proton transfer to the oxygen of the epoxide to give an oxonium ion.

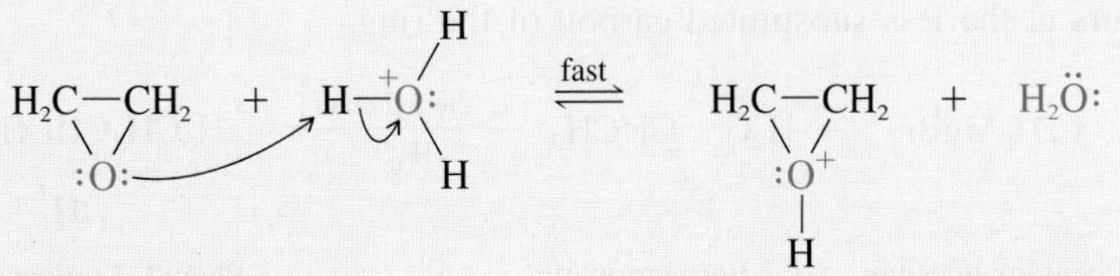

Ethylene oxide Hydronium ion Ethyleneoxonium ion Water

Step 2: Nucleophilic attack by water on carbon of the oxonium ion. The carbon–oxygen bond of the ring is broken in this step and the ring opens.

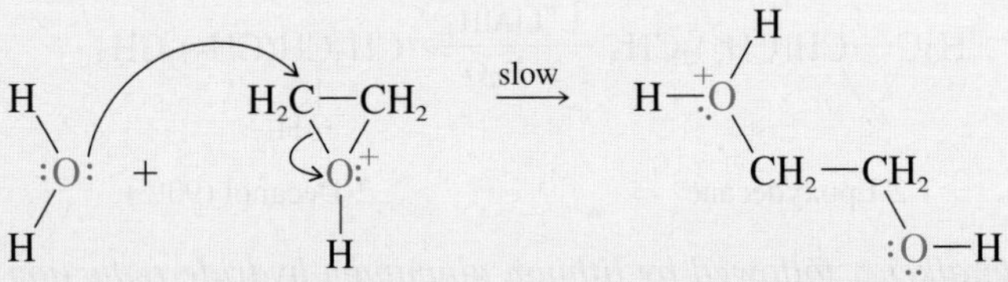

Water Ethyleneoxonium ion 2-Hydroxyethyloxonium ion

Step 3: Proton transfer to water completes the reaction and regenerates the acid catalyst.

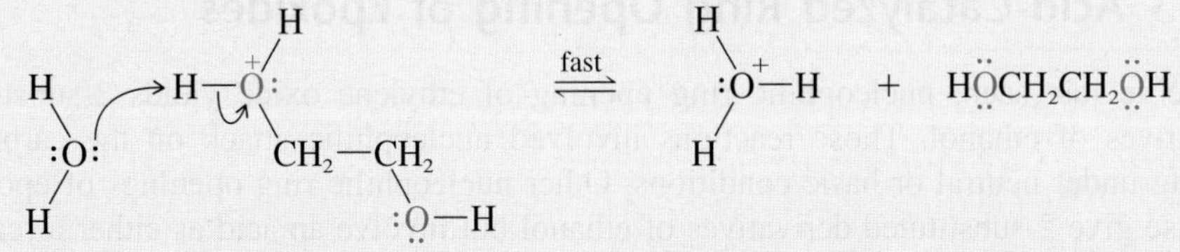

Water 2-Hydroxyethyloxonium ion Hydronium ion 1,2-Ethanediol

conjugate acid. The transition state for ring opening has a fair measure of carbocation character. Breaking of the ring carbon–oxygen bond is more advanced than formation of the bond to the nucleophile.

Transition state for attack by water on conjugate acid of ethylene oxide

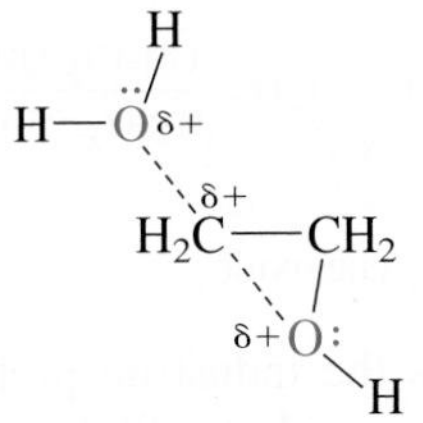

Because *carbocation* character develops at the transition state, substitution in unsymmetrical epoxides is favored at the carbon that can better support a developing positive charge. Thus, in contrast to the reaction of epoxides with relatively basic

nucleophiles, in which S_N2-like attack is faster at the less crowded carbon of the three-membered ring, acid catalysis promotes substitution at the position that bears the greater number of alkyl groups:

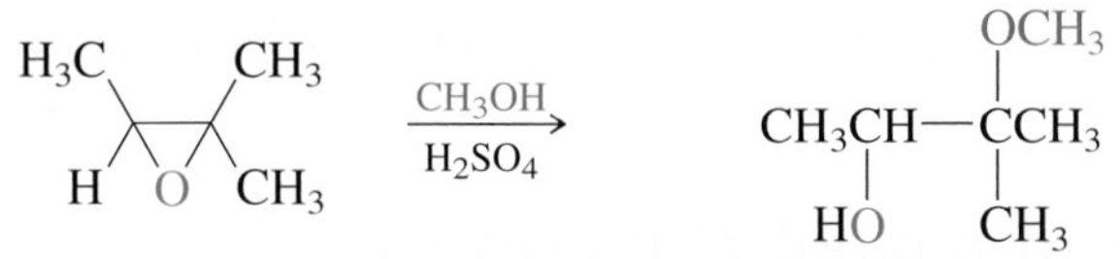

2,2,3-Trimethyloxirane 3-Methoxy-3-methyl-2-butanol (76%)

Although nucleophilic participation at the transition state is less than is usual under basic conditions, it is enough to ensure that substitution proceeds with inversion of configuration.

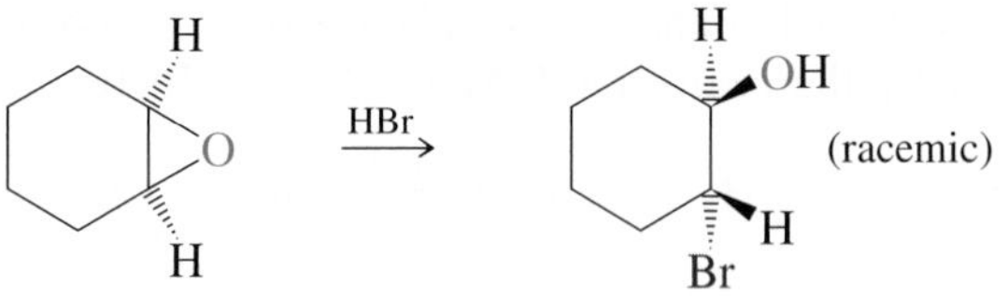

1,2-Epoxycyclohexane *trans*-2-Bromocyclohexanol (73%)

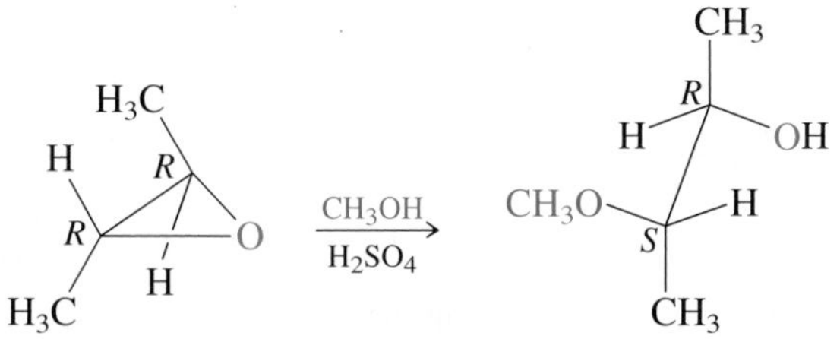

(2*R*,3*R*)-2,3-Epoxybutane (2*R*,3*S*)-3-Methoxy-2-butanol (57%)

Problem 16.15

Which product, compound A, B, or C, would you expect to be formed when the epoxide shown is allowed to stand in methanol containing a few drops of sulfuric acid? Compare your answer with that for Problem 16.14.

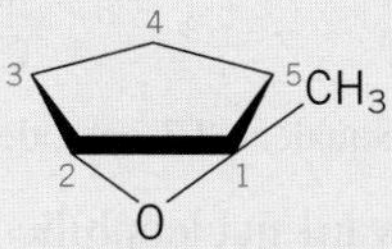

1,2-Epoxy-1-methylcyclopentane

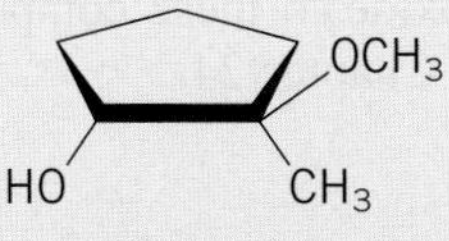

Compound A

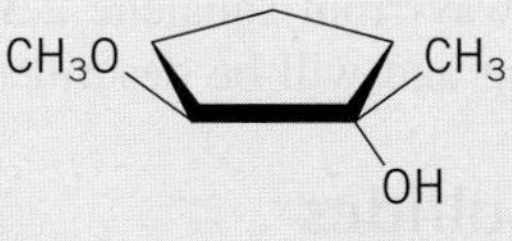

Compound B

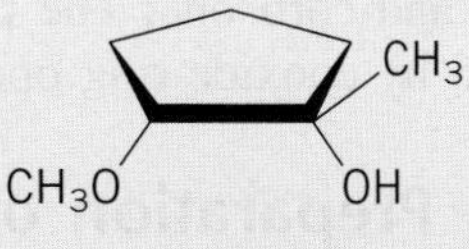

Compound C

A method for achieving net anti hydroxylation of alkenes combines two stereospecific processes: epoxidation of the double bond and hydrolysis of the derived epoxide.

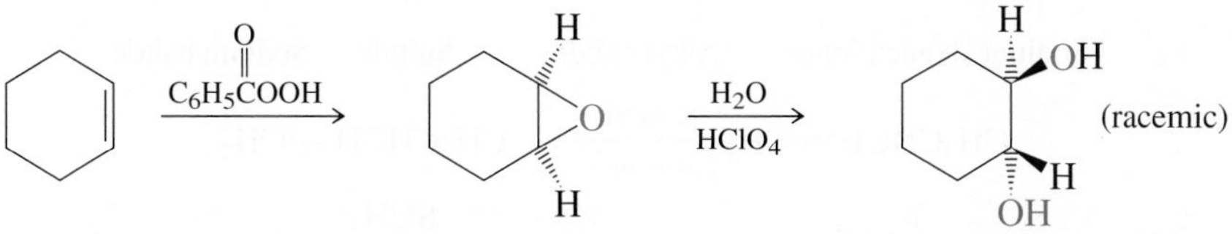

Cyclohexene 1,2-Epoxycyclohexane *trans*-1,2-Cyclohexanediol (80%)

Problem 16.16

Which alkene, *cis*-2-butene or *trans*-2-butene, would you choose in order to prepare *meso*-2,3-butanediol by epoxidation followed by acid-catalyzed hydrolysis? Which alkene would yield *meso*-2,3-butanediol by osmium tetraoxide dihydroxylation?

16.14 Epoxides in Biological Processes

Many naturally occurring substances are epoxides. In most cases, epoxides are biosynthesized by the enzyme-catalyzed transfer of one of the oxygen atoms of an O_2 molecule to an alkene. Because only one of the atoms of O_2 is transferred to the substrate, the enzymes that catalyze such transfers are classified as *monooxygenases.* A biological reducing agent, usually the coenzyme NADH (Section 15.10), is required as well.

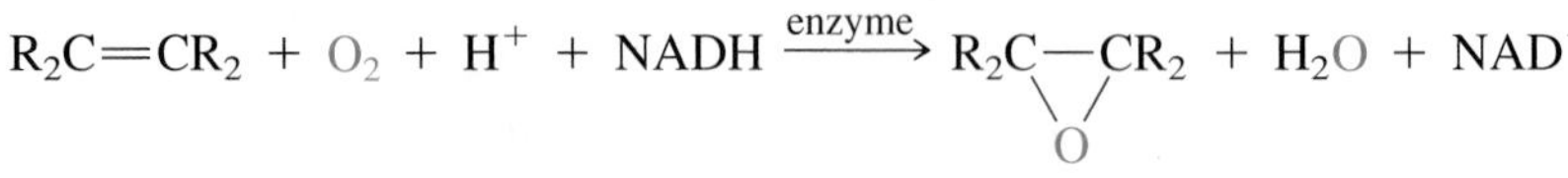
R2C=CR2 + O2 + H+ + NADH
enzyme
R2C—CR2 + H2O + NAD+
O

A prominent example of such a reaction is the biological epoxidation of the polyene squalene.

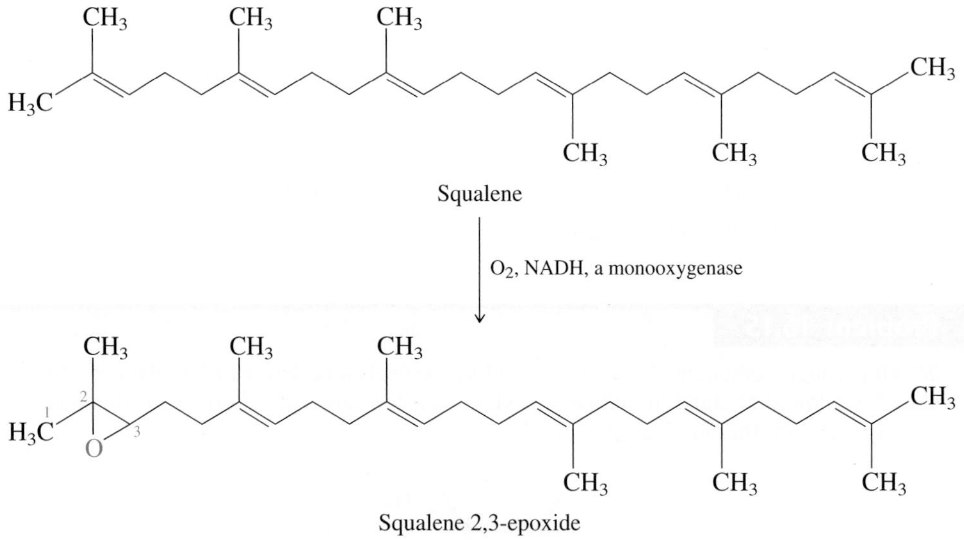
CH3
H3C
Squalene
O2, NADH, a monooxygenase
O
1
2
3
Squalene 2,3-epoxide

The reactivity of epoxides toward nucleophilic ring opening is responsible for one of the biological roles they play. Squalene 2,3-epoxide, for example, is the biological precursor to cholesterol and the steroid hormones, including testosterone, progesterone, estrone, and cortisone. The pathway from squalene 2,3-epoxide to these compounds is triggered by epoxide ring opening and will be described in Chapter 24.

16.15 Preparation of Sulfides

Sulfides, compounds of the type RSR′, are prepared by nucleophilic substitution. Treatment of a primary or secondary alkyl halide with an alkanethiolate ion (RS^-) gives a sulfide:

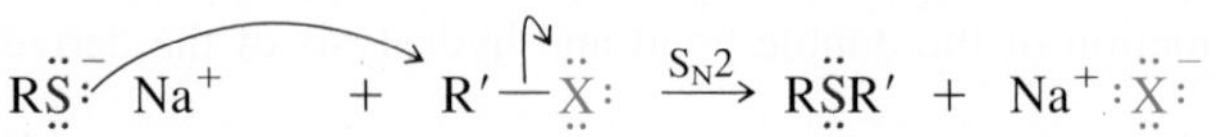
RS: Na+
R′—X:
SN2
RSR′ + Na+ :X:

Sodium alkanethiolate Alkyl halide Sulfide Sodium halide

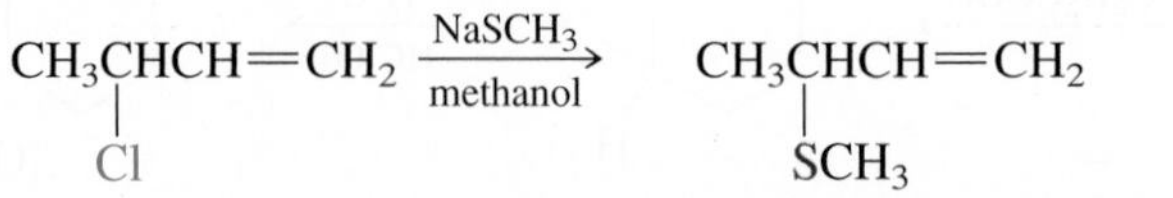
CH3CHCH=CH2
Cl
NaSCH3
methanol
CH3CHCH=CH2
SCH3

3-Chloro-1-butene Methyl 1-methylallyl sulfide (62%)

It is not necessary to prepare and isolate the sodium alkanethiolate in a separate operation. Because thiols are more acidic than water, they are quantitatively converted to their alkanethiolate anions by sodium hydroxide. Thus, all that is necessary is to add a thiol to sodium hydroxide in a suitable solvent (water or an alcohol) followed by the alkyl halide.

The pK_a for CH_3SH is 10.7.

Problem 16.17

The *p*-toluenesulfonate derived from (*R*)-2-octanol and *p*-toluenesulfonyl chloride was allowed to react with sodium benzenethiolate (C_6H_5SNa). Give the structure, including stereochemistry and the appropriate *R* or *S* descriptor, of the product.

16.16 Oxidation of Sulfides: Sulfoxides and Sulfones

We saw in Section 15.12 that thiols differ from alcohols with respect to their behavior toward oxidation. Similarly, sulfides differ from ethers in their behavior toward oxidizing agents. Whereas ethers tend to undergo oxidation at carbon to give hydroperoxides (Section 16.7), sulfides are oxidized at sulfur to give **sulfoxides.** If the oxidizing agent is strong enough and present in excess, oxidation can proceed further to give **sulfones.**

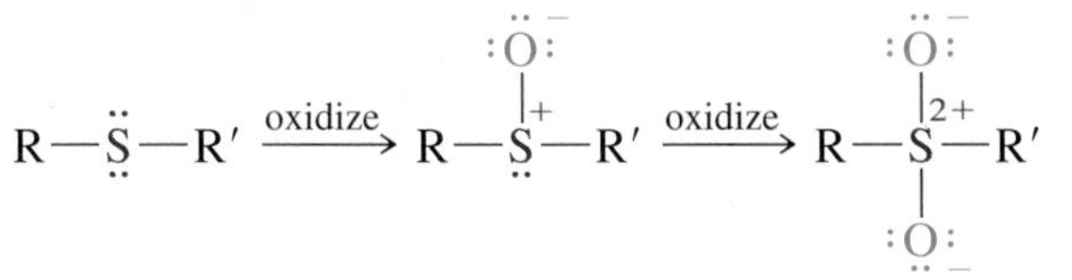

Sulfide Sulfoxide Sulfone

Third-row elements such as sulfur can expand their valence shell beyond eight electrons, and so sulfur–oxygen bonds in sulfoxides and sulfones are sometimes represented as double bonds.

When the desired product is a sulfoxide, sodium metaperiodate ($NaIO_4$) is an ideal reagent. It oxidizes sulfides to sulfoxides in high yield but shows no tendency to oxidize sulfoxides to sulfones.

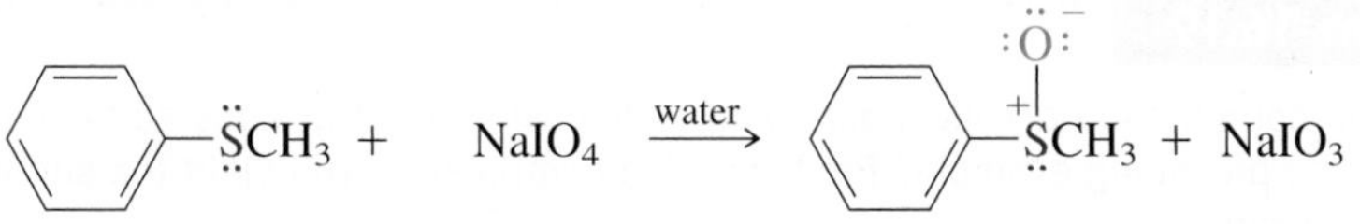

Methyl phenyl sulfide · Sodium metaperiodate · Methyl phenyl sulfoxide (91%) · Sodium iodate

Peroxy acids, usually in dichloromethane as the solvent, are also reliable reagents for converting sulfides to sulfoxides. One equivalent of a peroxy acid or of hydrogen peroxide converts sulfides to sulfoxides; two equivalents gives the corresponding sulfone.

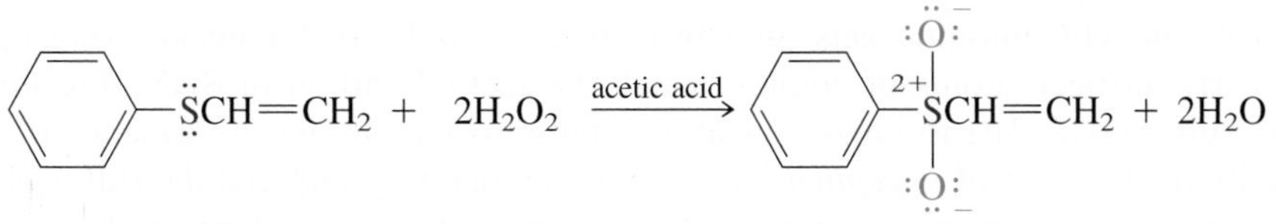

Phenyl vinyl sulfide · Hydrogen peroxide · Phenyl vinyl sulfone (74–78%) · Water

Problem 16.18

Prilosec and Nexium ("the little purple pill") are widely used to treat acid reflux and prevent the damage that stomach acid can do to the lining of the esophagus. Prilosec is the racemic form of omeprazole; Nexium is (*S*)-omeprazole. Write a structural formula for (*S*)-omeprazole clearly showing its stereochemistry. (For a hint, see Section 7.17.)

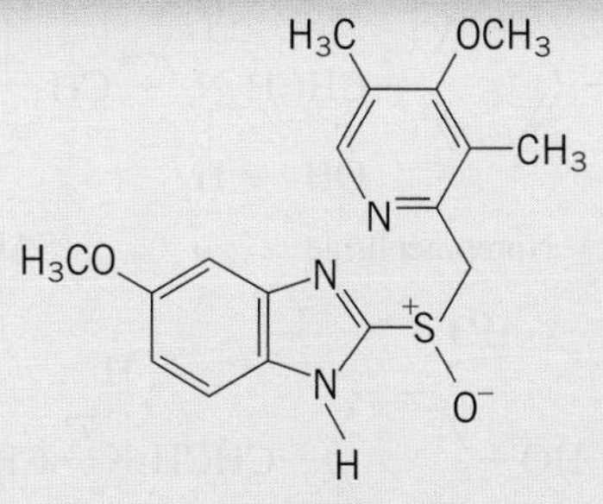

Omeprazole

Oxidation of sulfides occurs in living systems as well. Among naturally occurring sulfoxides, one that has received recent attention is *sulforaphane,* which is present in broccoli and other vegetables. Sulforaphane holds promise as a potential anticancer agent because, unlike most anticancer drugs, which act by killing rapidly dividing tumor cells faster than they kill normal cells, sulforaphane is nontoxic and may simply inhibit the formation of tumors.

The —N=C=S unit in sulforaphane is the *isothiocyanate* group. Isothiocyanates are among the key ingredients responsible for the flavor of wasabi.

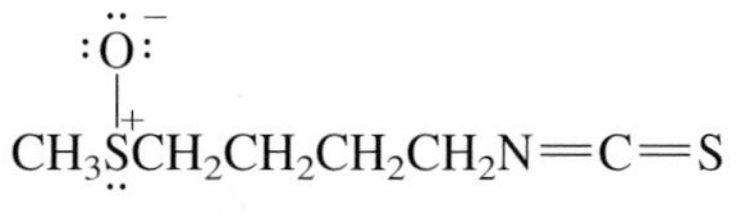
CH3SCH2CH2CH2CH2N=C=S

Sulforaphane

16.17 Alkylation of Sulfides: Sulfonium Salts

Sulfur is more nucleophilic than oxygen (Section 8.5), and sulfides react with alkyl halides much faster than do ethers. The products of these reactions, called *sulfonium salts,* are also more stable than the corresponding oxygen analogs.

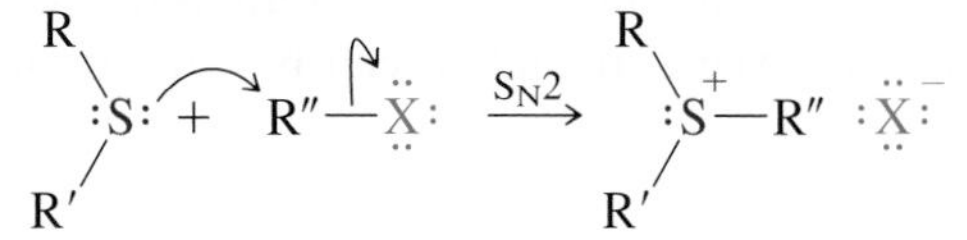
SN2

Sulfide Alkyl halide Sulfonium salt

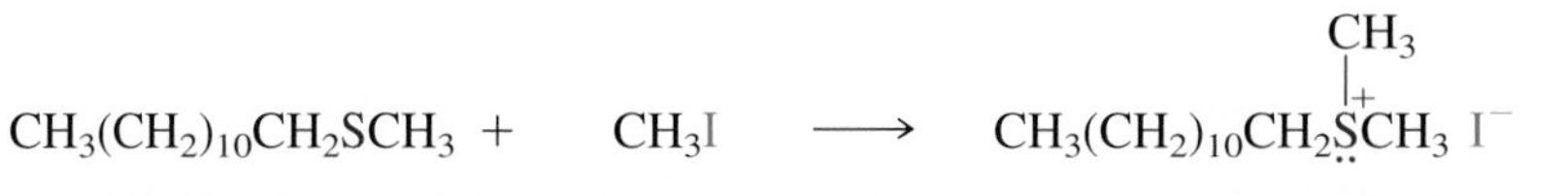

Dodecyl methyl sulfide Methyl iodide Dodecyldimethylsulfonium iodide

Problem 16.19

What other combination of alkyl halide and sulfide will yield the same sulfonium salt shown in the preceding example? Predict which combination will yield the sulfonium salt at the faster rate.

The *S* in *S*-adenosylmethionine indicates that the adenosyl group is bonded to sulfur. It does *not* stand for the Cahn–Ingold–Prelog stereochemical descriptor.

A naturally occurring sulfonium salt, *S-adenosylmethionine (SAM),* is a key substance in certain biological processes. It is formed by a nucleophilic substitution in which the sulfur atom of methionine attacks the primary carbon of adenosine triphosphate, displacing the triphosphate leaving group as shown in Mechanism 16.4.

S-Adenosylmethionine acts as a biological methyl-transfer agent. Nucleophiles, particularly nitrogen atoms of amines, attack the methyl carbon of SAM, breaking the carbon–sulfur bond. The following equation represents the biological formation of *epinephrine* by methylation of *norepinephrine.* Only the methyl group and the sulfur of SAM are shown explicitly in the equation to draw attention to the similarity of this reaction, which occurs in living systems, to the more familiar S_N2 reactions we have studied.

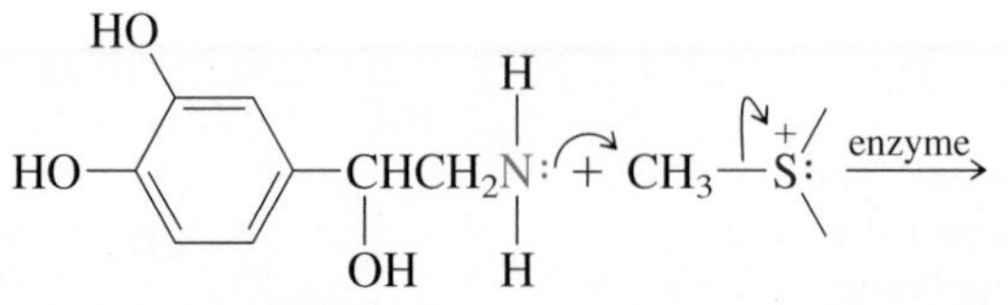
enzyme

Norepinephrine SAM

Epinephrine is also known as *adrenaline* and is a hormone with profound physiological effects designed to prepare the body for "fight or flight."

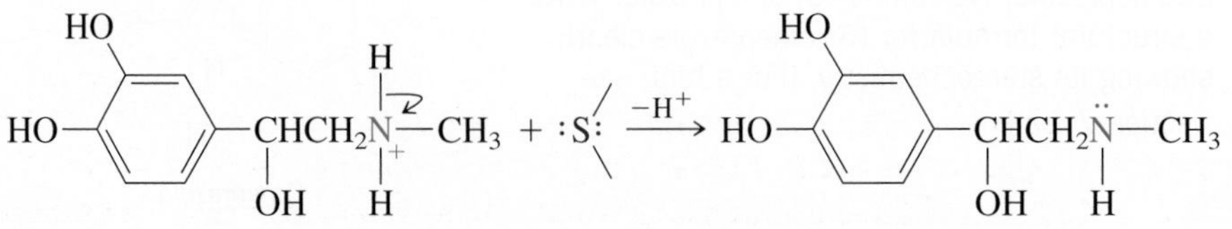
−H+

Epinephrine

Mechanism 16.4

Nucleophilic Substitution of Adenosine Triphosphate (ATP) by Methionine

Methionine:

ATP:

SAM:

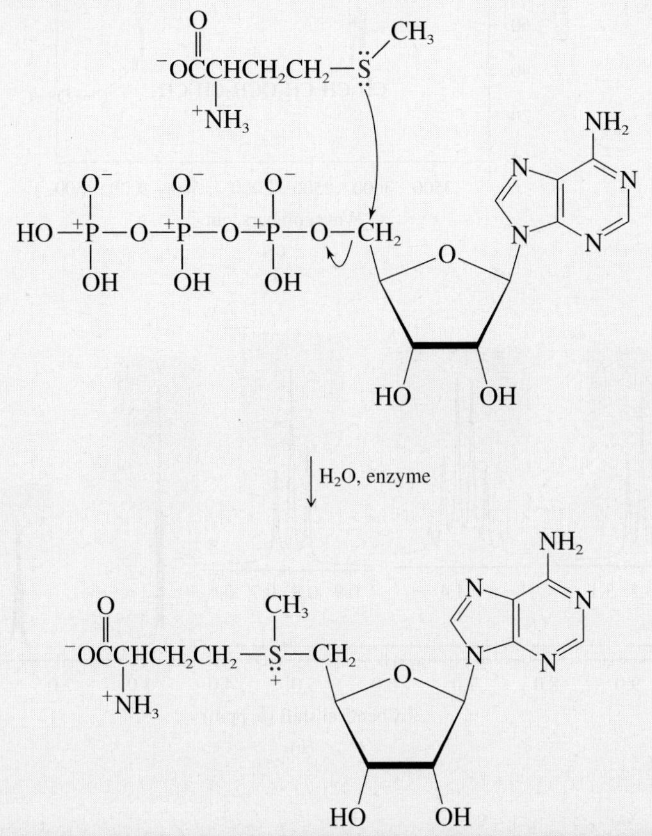

16.18 Spectroscopic Analysis of Ethers, Epoxides, and Sulfides

The IR, ^1H NMR, and ^{13}C NMR spectra of dipropyl ether, which appear in parts *a, b,* and *c,* respectively of Figure 16.4, illustrate some of the spectroscopic features of ethers.

Infrared: The infrared spectra of *ethers* are characterized by a strong, rather broad band due to antisymmetric C—O—C stretching between 1070 and 1150 cm^{1}. Dialkyl ethers exhibit this band consistently at 1120 cm^{1}, as shown in the IR spectrum of dipropyl ether.

$CH_3CH_2CH_2OCH_2CH_2CH_3$ — Dipropyl ether — C—O—C: $\nu = 1118\ cm^{-1}$

The analogous band in alkyl aryl ethers (ROAr) appears at 1200–1275 cm^{1} (Section 22.15).

Epoxides typically exhibit three bands. Two bands, one at 810–950 cm^{1} and the other near 1250 cm^{-1}, correspond to asymmetric and symmetric stretching of the ring, respectively. The third band appears in the range 750–840 cm^{1}.

1,2-Epoxydodecane

Epoxide vibrations: $\nu = 837, 917$, and $1265\ cm^{-1}$

The C—S—C stretching vibration of *sulfides* gives a weak peak in the 600–700 cm^{1} range. *Sulfoxides* show a strong peak due to S—O stretching at 1030–1070 cm^{1}. With two oxygens attached to sulfur, *sulfones* exhibit strong bands due to symmetric (1120–1160 cm^{1}) and asymmetric (1290–1350 cm^{-1}) S—O stretching.

Figure 16.4

The (*a*) infrared, (*b*) 200-MHz ^{1}H NMR, and (*c*) ^{13}C NMR spectra of dipropyl ether ($CH_3CH_2CH_2OCH_2CH_2CH_3$).

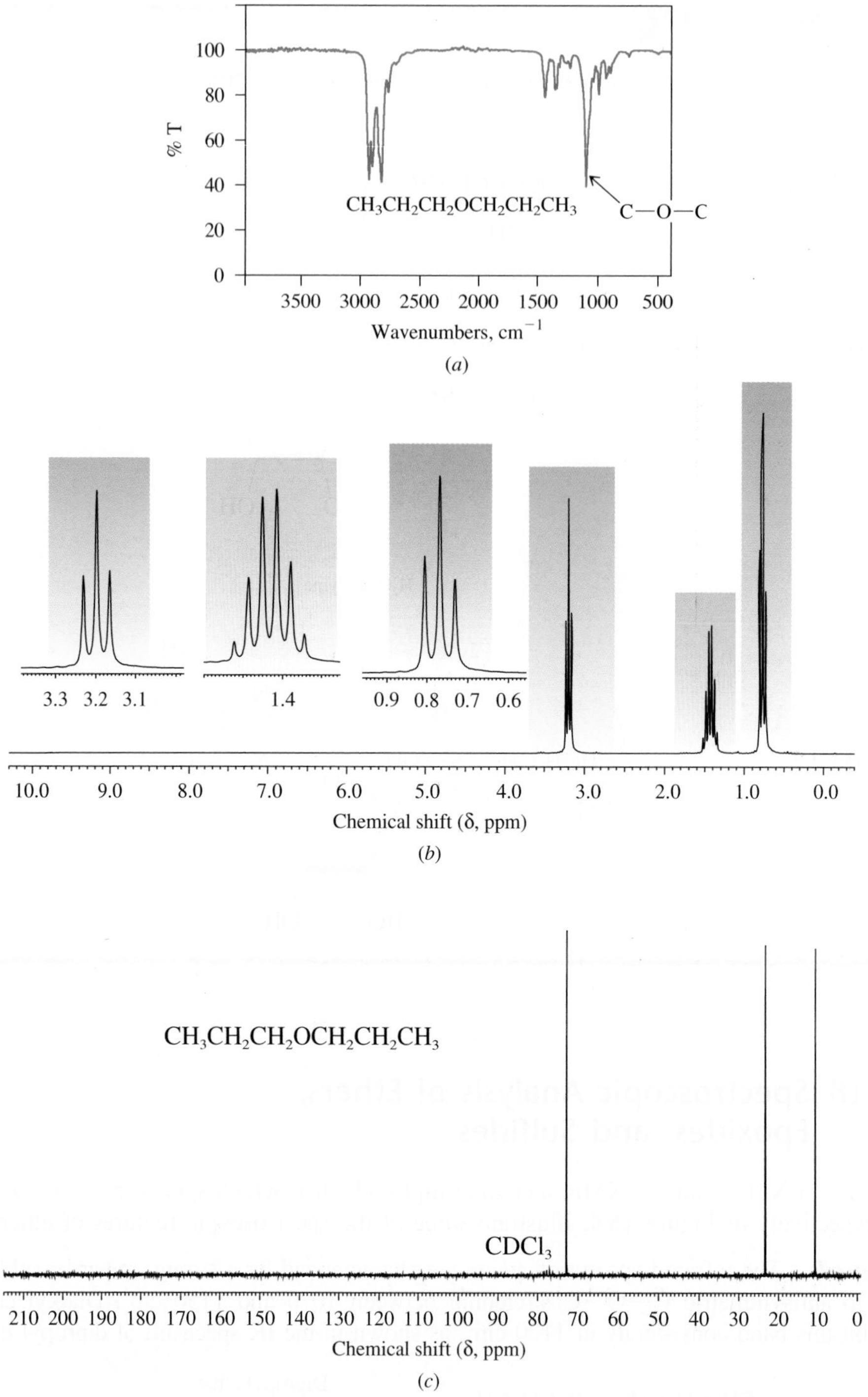

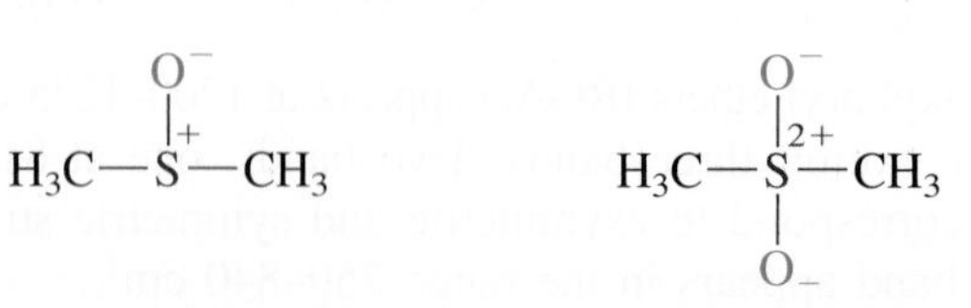

Dimethyl sulfoxide — S—O: ν = 1050 cm^{-1}

Dimethyl sulfone — S—O: ν = 1139 and 1298 cm^{-1}

^{1}H NMR: The chemical shift of the proton in the **H**—C—O—C unit of an *ether* is very similar to that of the proton in the **H**—C—OH unit of an alcohol. A range of δ 3.2–4.0 is typical. The proton in the **H**—C—S—C unit of a *sulfide* appears at higher field than the corresponding proton of an ether because sulfur is less electronegative than oxygen.

^{1}H Chemical shift (δ):

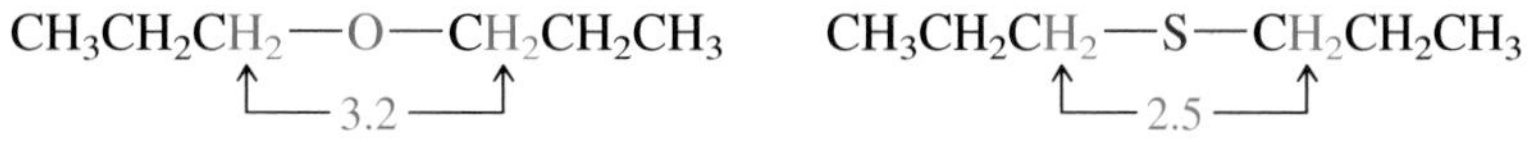
CH3CH2CH2—O—CH2CH2CH3
3.2
CH3CH2CH2—S—CH2CH2CH3
2.5

Oxidation of a sulfide to a *sulfoxide* or *sulfone* is accompanied by a decrease in shielding of the **H**—C—S—C proton by about 0.3–0.5 ppm for each oxidation.

Epoxides are unusual in that the protons on the ring are more shielded than expected. The protons in ethylene oxide, for example, appear at δ 2.5 instead of the δ 3.2–4.0 range just cited for dialkyl ethers.

^{13}C NMR: The carbons of the C—O—C group in an *ether* are about 10 ppm less shielded than those of an alcohol and appear in the range δ 57–87. The carbons of the C—S—C group in a *sulfide* are significantly more shielded than those of an ether.

^{13}C Chemical shift (δ):

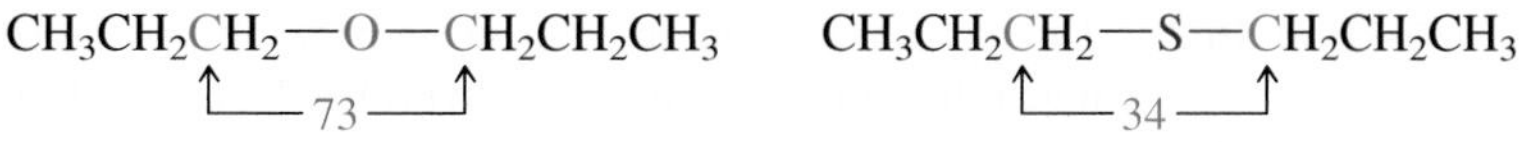
CH3CH2CH2—O—CH2CH2CH3
73
CH3CH2CH2—S—CH2CH2CH3
34

The ring carbons of an *epoxide* are somewhat more shielded than the carbons of a C—O—C unit of larger rings or dialkyl ethers.

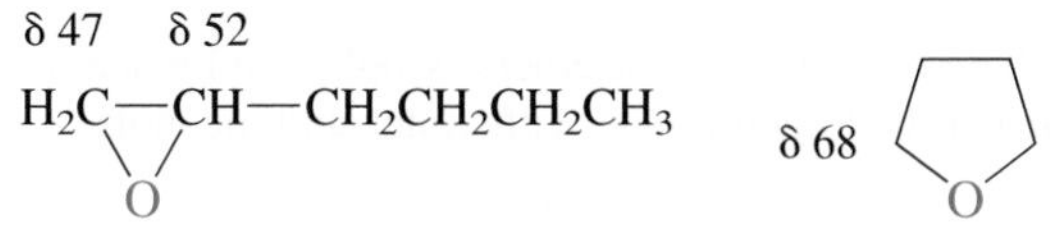
δ 47
δ 52
H2C—CH—CH2CH2CH2CH3
O
δ 68
O

UV-VIS: Simple ethers have their absorption maximum at about 185 nm and are transparent to ultraviolet radiation above about 220 nm.

Mass Spectrometry: Ethers, like alcohols, lose an alkyl radical from their molecular ion to give an oxygen-stabilized cation. Thus, *m/z* 73 and *m/z* 87 are both more abundant than the molecular ion in the mass spectrum of *sec*-butyl ethyl ether.

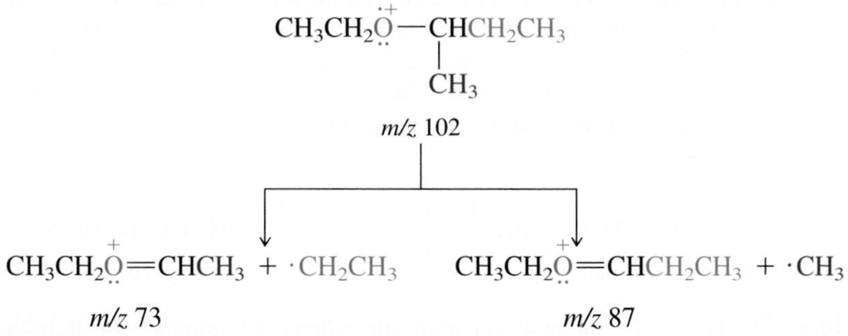
CH3CH2O—CHCH2CH3
CH3
m/z 102
CH3CH2O=CHCH3 + ·CH2CH3
m/z 73
CH3CH2O=CHCH2CH3 + ·CH3
m/z 87

Problem 16.20

There is another oxygen-stabilized cation of *m/z* 87 capable of being formed by fragmentation of the molecular ion in the mass spectrum of *sec*-butyl ethyl ether. Suggest a reasonable structure for this ion.

An analogous fragmentation process occurs in the mass spectra of sulfides. As with other sulfur-containing compounds, the presence of sulfur can be inferred by a peak at *m/z* of M+2.

16.19 SUMMARY

Section 16.1 **Ethers** are compounds that contain a C—O—C linkage. In substitutive IUPAC nomenclature, they are named as *alkoxy* derivatives of alkanes. In functional class IUPAC nomenclature, we name each alkyl group as a separate word (in alphabetical order) followed by the word *ether.*

$CH_3OCH_2CH_2CH_2CH_2CH_2CH_3$ **Substitutive IUPAC name:** 1-Methoxyhexane
Functional class name: Hexyl methyl ether

Epoxides are normally named as *epoxy* derivatives of alkanes or as substituted *oxiranes*.

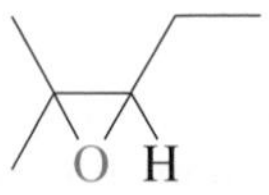

2,3-Epoxy-2-methylpentane
3-Ethyl-2,2-dimethyloxirane

Sulfides are sulfur analogs of ethers: they contain the C—S—C functional group. They are named as *alkylthio* derivatives of alkanes in substitutive IUPAC nomenclature. The functional class IUPAC names of sulfides are derived in the same manner as those of ethers, but the concluding word is *sulfide*.

$CH_3SCH_2CH_2CH_2CH_2CH_2CH_3$

Substitutive IUPAC name: 1-(Methylthio)hexane
Functional class name: Hexyl methyl sulfide

Section 16.2 The oxygen atom in an ether or epoxide affects the shape of the molecule in much the same way as an sp^3-hybridized carbon of an alkane or cycloalkane.

Pentane

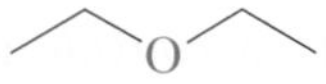

Diethyl ether

Section 16.3 The carbon–oxygen bond of ethers is polar, and ethers can act as proton *acceptors* in hydrogen bonds with water and alcohols.

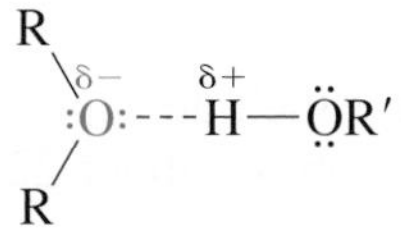

But ethers lack OH groups and cannot act as proton *donors* in forming hydrogen bonds.

Section 16.4 Ethers form Lewis acid/Lewis base complexes with metal ions. Certain cyclic polyethers, called **crown ethers,** are particularly effective in coordinating with Na^+ and K^+, and salts of these cations can be dissolved in nonpolar solvents when crown ethers are present. Under these conditions the rates of many reactions that involve anions are accelerated.

$$\underset{\text{1-Bromohexane}}{CH_3(CH_2)_4CH_2Br} \xrightarrow[\text{acetonitrile, heat}]{KOCCH_3 \text{ (C=O)}, \text{ 18-crown-6}} \underset{\text{Hexyl acetate (96\%)}}{CH_3(CH_2)_4CH_2OCCH_3 \text{ (C=O)}}$$

Sections 16.5 and 16.6 The two major methods for preparing ethers are summarized in Table 16.2.

Section 16.7 Dialkyl ethers are useful solvents for organic reactions, but must be used cautiously due to their tendency to form explosive hydroperoxides by air oxidation in opened bottles.

Section 16.8 The only important reaction of ethers is their cleavage by hydrogen halides.

$$\underset{\text{Ether}}{ROR'} + \underset{\text{Hydrogen halide}}{2HX} \longrightarrow \underset{\text{Alkyl halide}}{RX} + \underset{\text{Alkyl halide}}{R'X} + \underset{\text{Water}}{H_2O}$$

The order of hydrogen halide reactivity is HI > HBr > HCl.

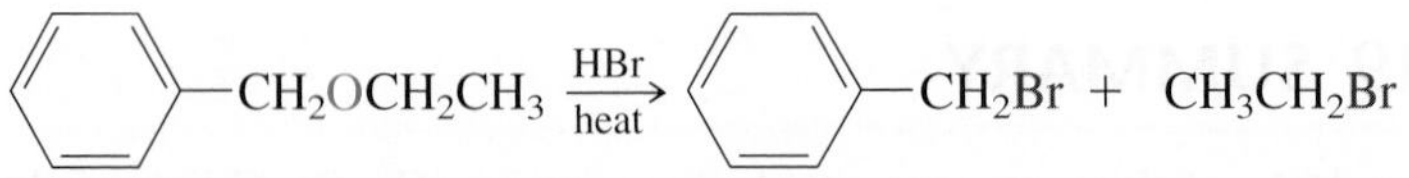

Benzyl ethyl ether Benzyl bromide Ethyl bromide

Sections 16.9 and 16.10 Epoxides are prepared by the methods listed in Table 16.3.

Section 16.11 Epoxides are much more reactive than ethers, especially in reactions that lead to cleavage of their three-membered ring.

TABLE 16.2 Preparation of Ethers

Reaction (section) and comments	General equation and specific example
Acid-catalyzed condensation of alcohols (Sections 15.7 and 16.5) Two molecules of an alcohol condense in the presence of an acid catalyst to yield a dialkyl ether and water. The reaction is limited to the synthesis of symmetrical ethers from primary alcohols.	$2RCH_2OH \xrightarrow{H^+} RCH_2OCH_2R + H_2O$ Alcohol; Ether; Water $CH_3CH_2CH_2OH \xrightarrow[heat]{H_2SO_4} CH_3CH_2CH_2OCH_2CH_2CH_3$ Propyl alcohol; Dipropyl ether
The Williamson ether synthesis (Section 16.6) An alkoxide ion displaces a halide or similar leaving group in an S_N2 reaction. The alkyl halide cannot be one that is prone to elimination, and so this reaction is limited to methyl and primary alkyl halides. There is no limitation on the alkoxide ion that can be used.	$RO^- + R'CH_2X \longrightarrow ROCH_2R' + X^-$ Alkoxide ion; Primary alkyl halide; Ether; Halide ion $(CH_3)_2CHCH_2ONa + CH_3CH_2Br \longrightarrow (CH_3)_2CHCH_2OCH_2CH_3 + NaBr$ Sodium isobutoxide; Ethyl bromide; Ethyl isobutyl ether (66%); Sodium bromide

TABLE 16.3 Preparation of Epoxides

Reaction (section) and comments	General equation and specific example
Peroxy acid oxidation of alkenes (Sections 6.19 and 16.9) Peroxy acids transfer oxygen to alkenes to yield epoxides. Stereospecific syn addition is observed.	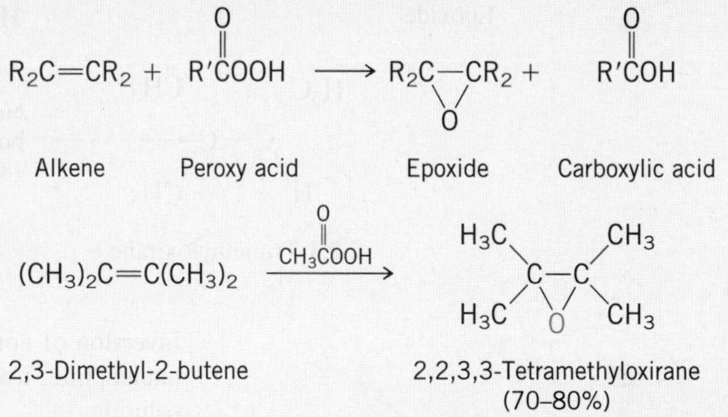 Alkene; Peroxy acid; Epoxide; Carboxylic acid; 2,3-Dimethyl-2-butene; 2,2,3,3-Tetramethyloxirane (70–80%)
Sharpless epoxidation (Section 16.9) Allylic alcohols are converted to epoxides by treatment with tert-butyl hydroperoxide and titanium(IV) alkoxides. The reaction is highly enantioselective in the presence of enantiomerically pure diethyl tartrate.	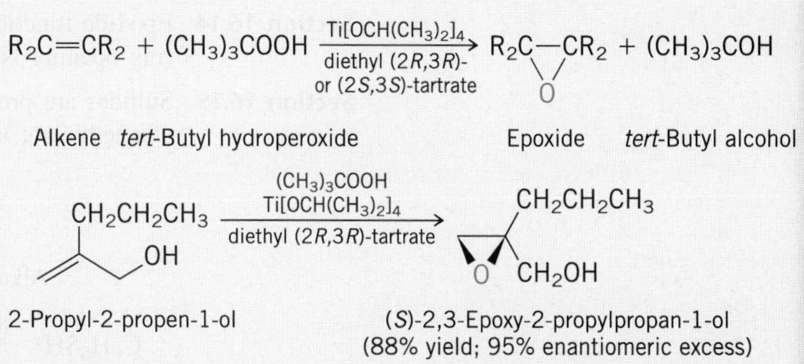 Alkene; *tert*-Butyl hydroperoxide; Epoxide; *tert*-Butyl alcohol; 2-Propyl-2-propen-1-ol; (*S*)-2,3-Epoxy-2-propylpropan-1-ol (88% yield; 95% enantiomeric excess)
Base-promoted cyclization of vicinal halohydrins (Section 16.10) This reaction is an intramolecular version of the Williamson ether synthesis. The alcohol function of a vicinal halohydrin is converted to its conjugate base, which then displaces halide from the adjacent carbon to give an epoxide.	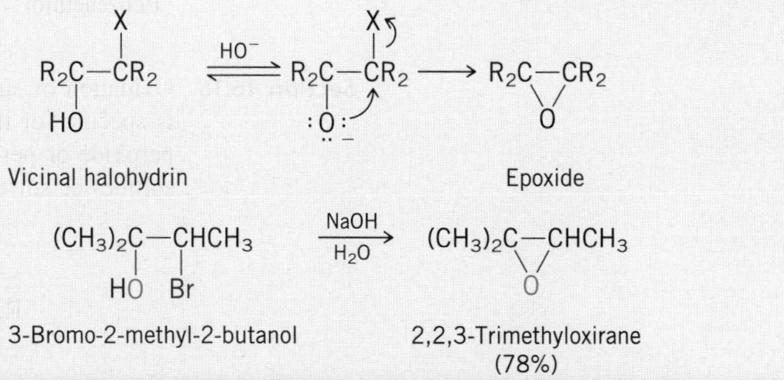 Vicinal halohydrin; Epoxide; 3-Bromo-2-methyl-2-butanol; 2,2,3-Trimethyloxirane (78%)

Section 16.12 Anionic nucleophiles usually attack the less substituted carbon of the epoxide in an S_N2-like fashion.

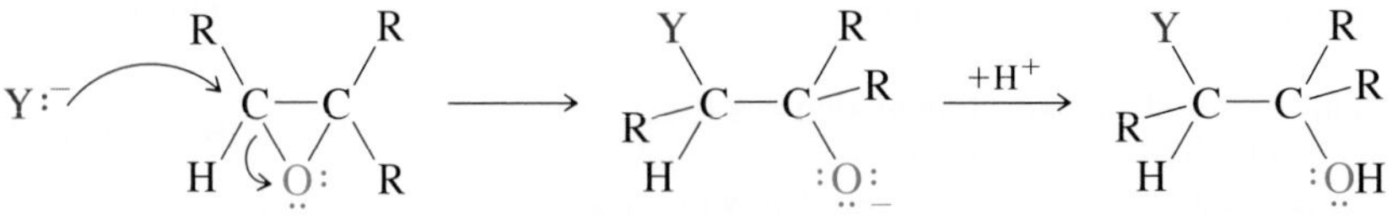

Nucleophile Epoxide β-Substituted alcohol

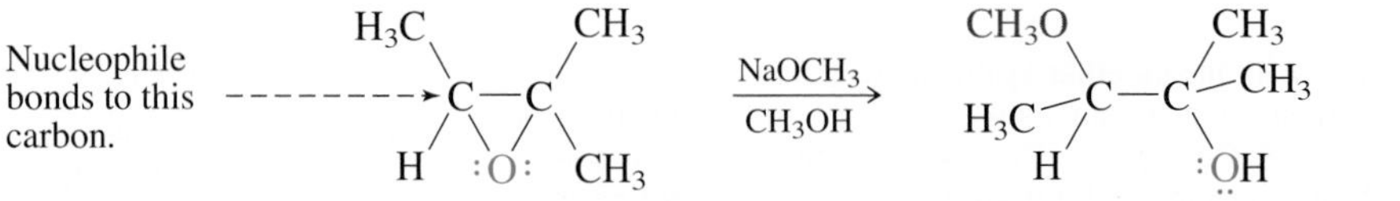

2,2,3-Trimethyloxirane 3-Methoxy-2-methyl-2-butanol (53%)

Section 16.13 Under conditions of acid catalysis, nucleophiles attack the carbon that can better support a positive charge. Carbocation character is developed in the transition state.

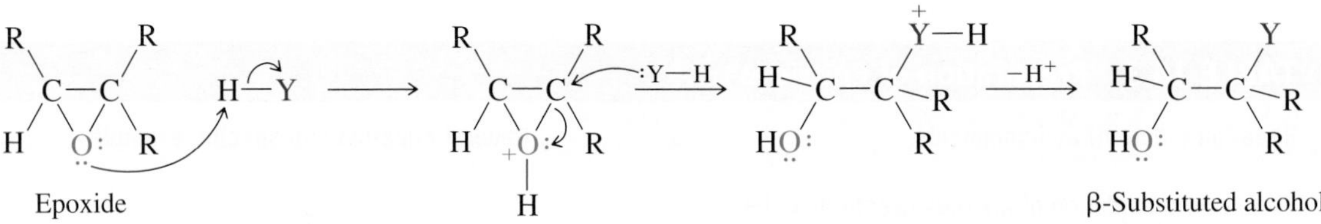

Epoxide β-Substituted alcohol

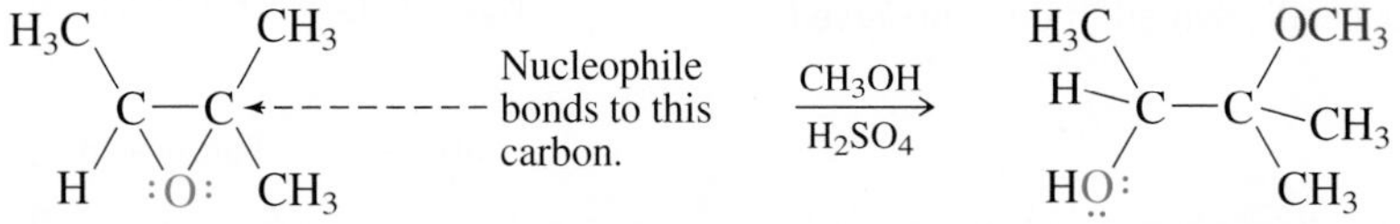

2,2,3-Trimethyloxirane 3-Methoxy-3-methyl-2-butanol (76%)

Inversion of configuration is observed at the carbon that is attacked by the nucleophile, irrespective of whether the reaction takes place in acidic or basic solution.

Section 16.14 Epoxide functions are present in a great many natural products, and epoxide ring opening is sometimes a key step in the biosynthesis of other substances.

Section 16.15 Sulfides are prepared by nucleophilic substitution (S_N2) in which an alkanethiolate ion reacts with an alkyl halide.

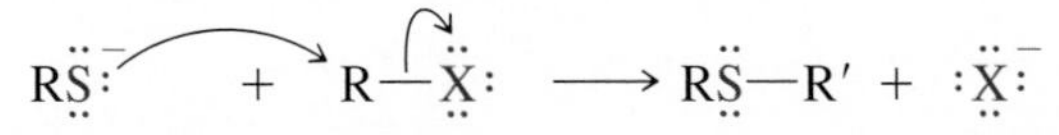

Alkanethiolate Alkyl halide Sulfide Halide

$$C_6H_5SH \xrightarrow{NaOCH_2CH_3} C_6H_5SNa \xrightarrow{C_6H_5CH_2Cl} C_6H_5SCH_2C_6H_5$$

Benzenethiol Sodium benzenethiolate Benzyl phenyl sulfide (60%)

Section 16.16 Oxidation of sulfides yields sulfoxides, then sulfones. Sodium metaperiodate is specific for the oxidation of sulfides to sulfoxides, and no further. Hydrogen peroxide or peroxy acids can yield sulfoxides (1 mol of oxidant per mole of sulfide) or sulfones (2 mol of oxidant per mole of sulfide).

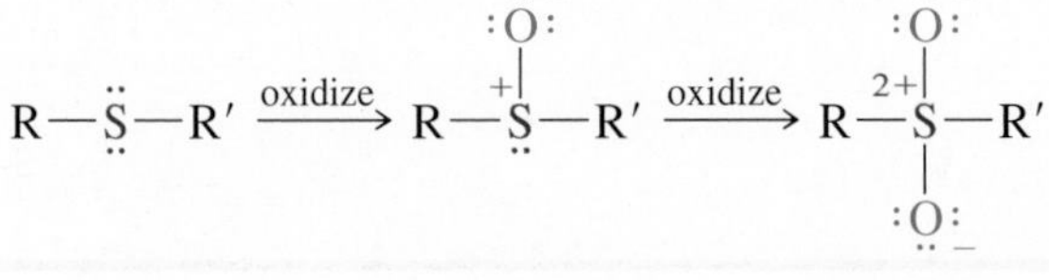

Sulfide Sulfoxide Sulfone

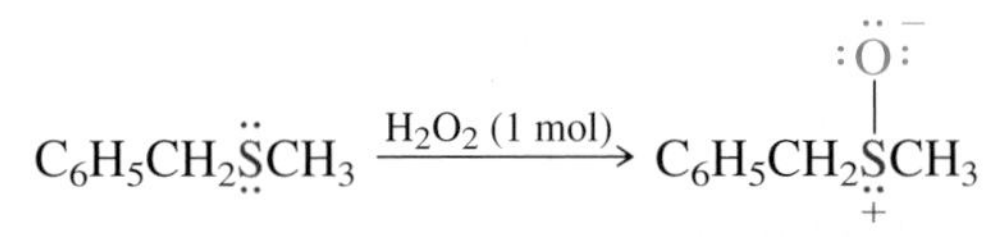

Benzyl methyl sulfide — Benzyl methyl sulfoxide (94%)

Section 16.17 Sulfides react with alkyl halides to give sulfonium salts.

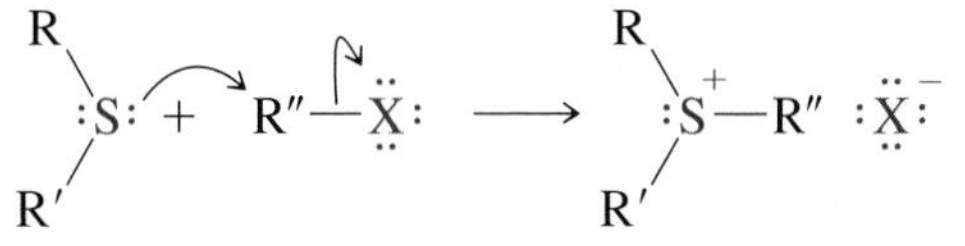

Sulfide — Alkyl halide — Sulfonium salt

Dimethyl sulfide — Methyl iodide — Trimethylsulfonium iodide (100%)

Section 16.18 An H—C—O—C structural unit in an ether resembles an H—C—O—H unit of an alcohol with respect to the C—O stretching frequency in its infrared spectrum and the H—C chemical shift in its ^{1}H NMR spectrum. Because sulfur is less electronegative than oxygen, the ^{1}H and ^{13}C chemical shifts of H—C—S—C units appear at higher field than those of H—C—O—C.

PROBLEMS

16.21 Write the structures of all the constitutionally isomeric ethers of molecular formula $C_5H_{12}O$, and give an acceptable name for each.

16.22 Many ethers, including diethyl ether, are effective as general anesthetics. Because simple ethers are quite flammable, their place in medical practice has been taken by highly halogenated nonflammable ethers. Two such general anesthetic agents are *isoflurane* and *enflurane*. These compounds are isomeric; isoflurane is 1-chloro-2,2,2-trifluoroethyl difluoromethyl ether; enflurane is 2-chloro-1,1,2-trifluoroethyl difluoromethyl ether. Write the structural formulas of isoflurane and enflurane.

16.23 Although epoxides are always considered to have their oxygen atom as part of a three-membered ring, the prefix *epoxy* in the IUPAC system of nomenclature can be used to denote a cyclic ether of various sizes. Thus

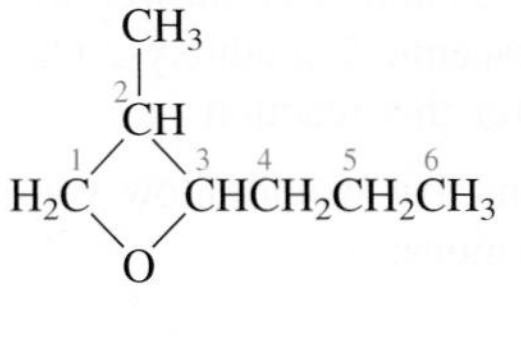

may be named 1,3-epoxy-2-methylhexane. Using the epoxy prefix in this way, name each of the following compounds:

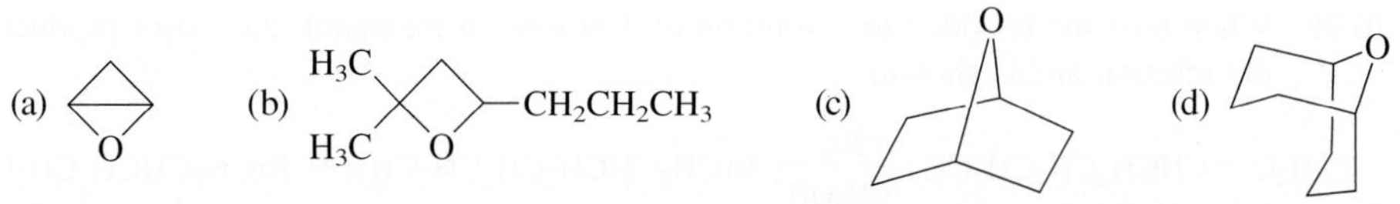

16.24 Outline the steps in the preparation of each of the constitutionally isomeric ethers of molecular formula $C_4H_{10}O$, starting with the appropriate alcohols. Use the Williamson ether synthesis as your key reaction.

16.25 Predict the principal organic product of each of the following reactions. Specify stereochemistry where appropriate.

(a) 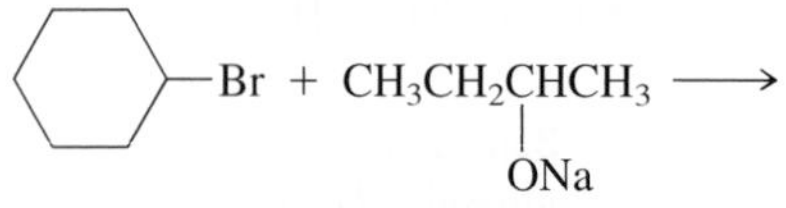

(b) 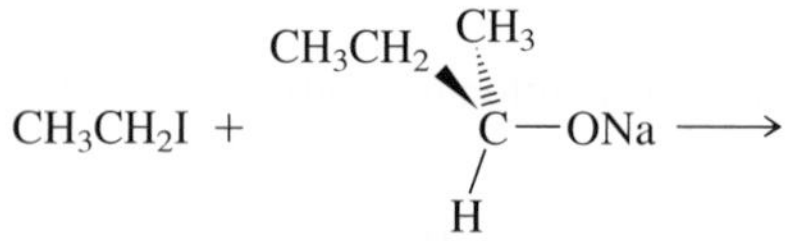

(c) $CH_3CH_2\underset{\displaystyle OH}{CH}CH_2Br \xrightarrow{NaOH}$

(d) 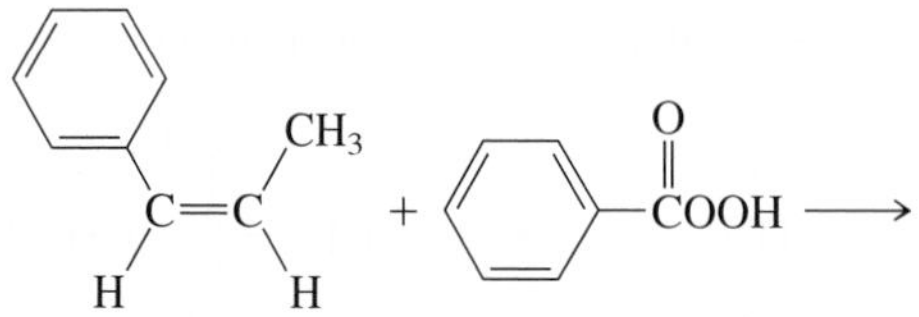

(e) 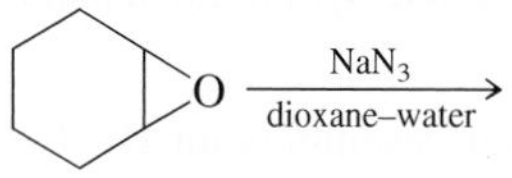

(f) 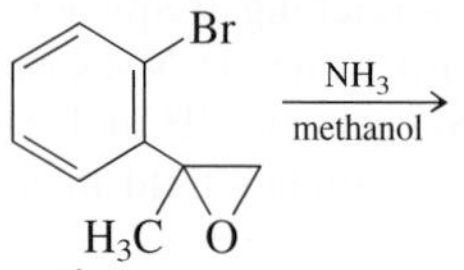

(g) 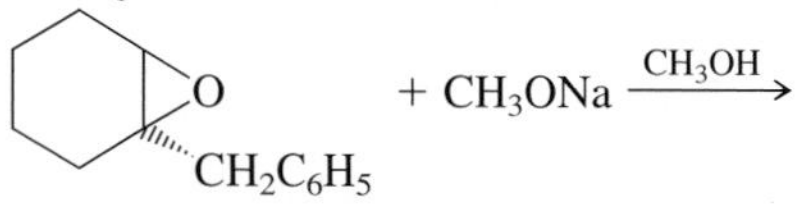

(h) 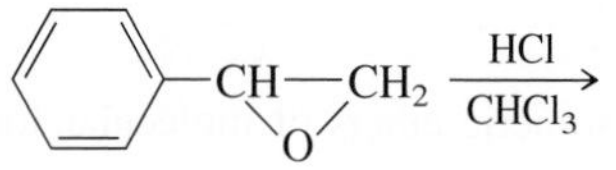

(i) 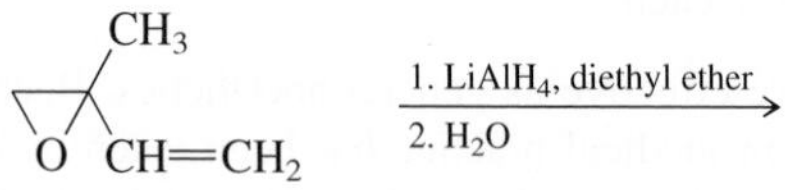

(j) $CH_3(CH_2)_{16}CH_2OTs + CH_3CH_2CH_2CH_2SNa \longrightarrow$

(k) 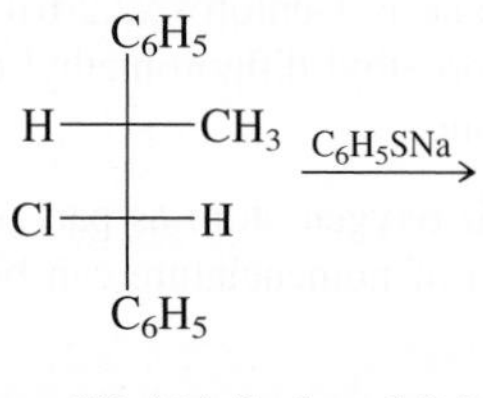

16.26 When (*R*)-(+)-2-phenyl-2-butanol is allowed to stand in methanol containing a few drops of sulfuric acid, racemic 2-methoxy-2-phenylbutane is formed. Suggest a reasonable mechanism for this reaction.

16.27 Select reaction conditions that would allow you to carry out each of the following stereospecific transformations:

(a) 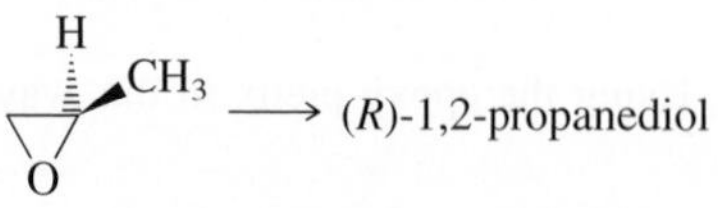 (b) 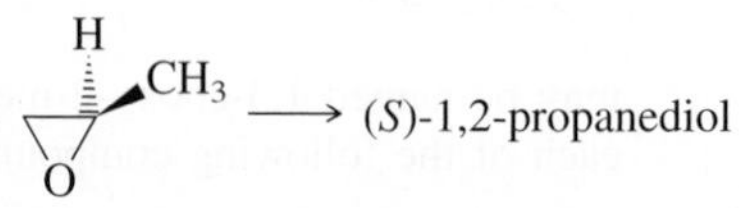

16.28 When bromine is added to a solution of 1-hexene in methanol, the major products of the reaction are as shown:

$$\underset{\text{1-Hexene}}{H_2C{=}CHCH_2CH_2CH_2CH_3} \xrightarrow[CH_3OH]{Br_2} \underset{\text{1,2-Dibromohexane}}{BrCH_2\underset{\displaystyle Br}{CH}CH_2CH_2CH_2CH_3} + \underset{\text{1-Bromo-2-methoxyhexane}}{BrCH_2\underset{\displaystyle OCH_3}{CH}CH_2CH_2CH_2CH_3}$$

1,2-Dibromohexane is not converted to 1-bromo-2-methoxyhexane under the reaction conditions. Suggest a reasonable explanation for the formation of 1-bromo-2-methoxyhexane.

16.29 Suggest short, efficient reaction sequences suitable for preparing each of the following compounds from the given starting materials and any necessary organic or inorganic reagents:

(a) 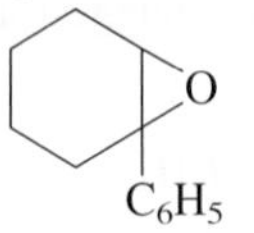 from bromobenzene and cyclohexanol

(b) $C_6H_5CH_2CHCH_3$ (with OH on the CH) from bromobenzene and isopropyl alcohol

(c) $C_6H_5CH_2CH_2CH_2OCH_2CH_3$ from benzyl alcohol and ethanol

(d) 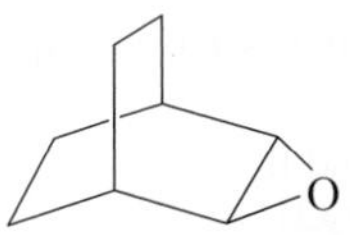 from 1,3-cyclohexadiene and ethanol

(e) $C_6H_5CHCH_2SCH_2CH_3$ (with OH on the CH) from styrene and ethanol

16.30 Propranolol is a drug prescribed to treat cardiac arrhythmia and angina pain and to lower blood pressure. It is chiral, and one enantiomer is responsible for its therapeutic effects. That enantiomer can be synthesized by a process, outlined in the following scheme, in which the first step is a Sharpless epoxidation to give (*S*)-glycidol. What is the configuration of the propranolol formed by this sequence? (No rearrangements occur.) A bond of the type ⌇ means unspecified stereochemistry.

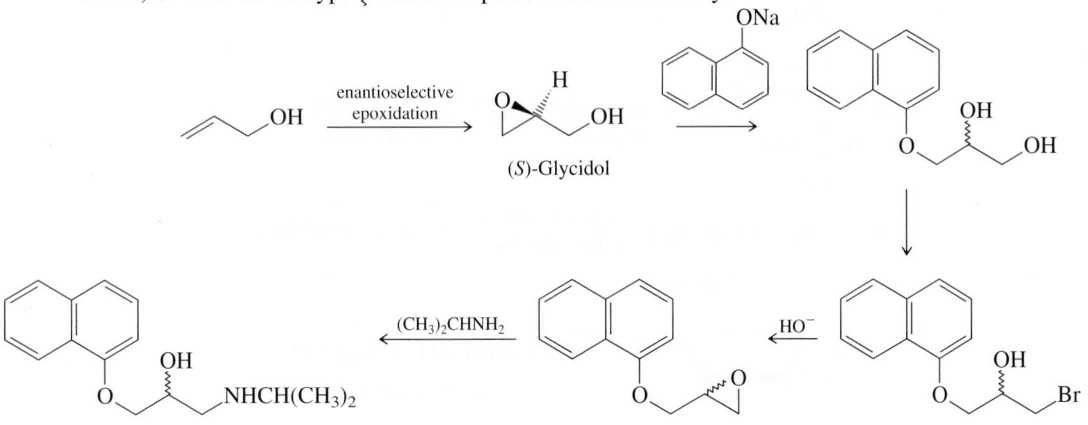

Propranolol

16.31 The growth of new blood vessels, angiogenesis, is crucial to wound healing and embryonic development. Abnormal angiogenesis is associated with tumor growth, suggesting that inhibition of angiogenesis may be an approach for the treatment of cancer. The diepoxide ovalicin is an angiogenesis inhibitor that was synthesized from compound C, which was in turn prepared from compound A by a two-step sequence. Can you suggest a structure for compound B?

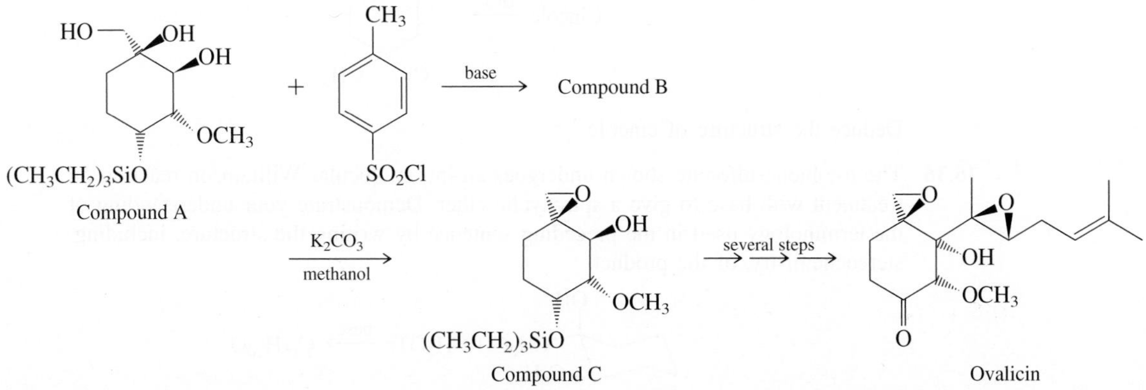

16.32 Write a mechanism for the following reaction.

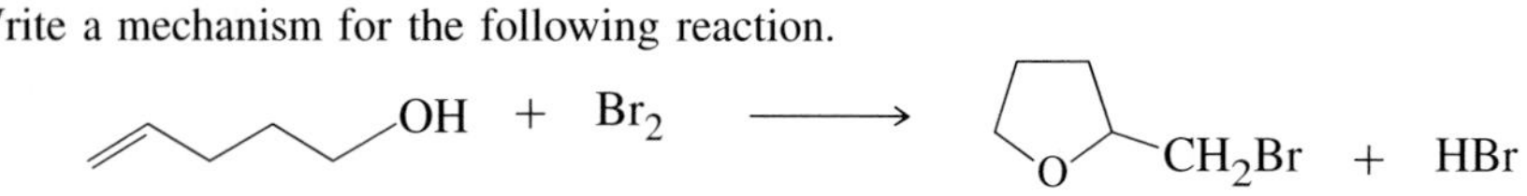

16.33 The following reaction has been reported in the chemical literature. Suggest a reasonable mechanism.

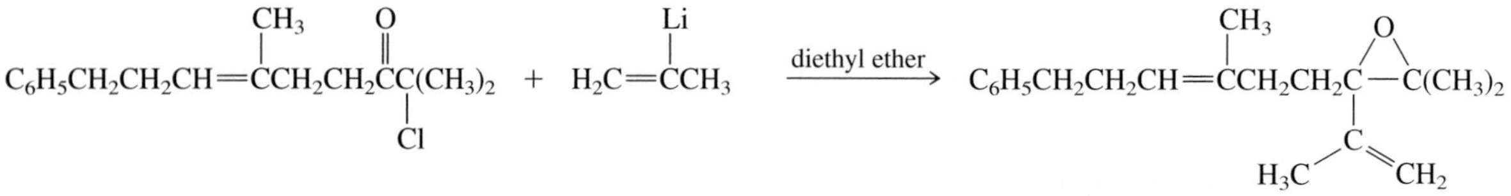

16.34 Deduce the identity of the missing compounds in the following reaction sequences. Show stereochemistry in parts (b) through (d).

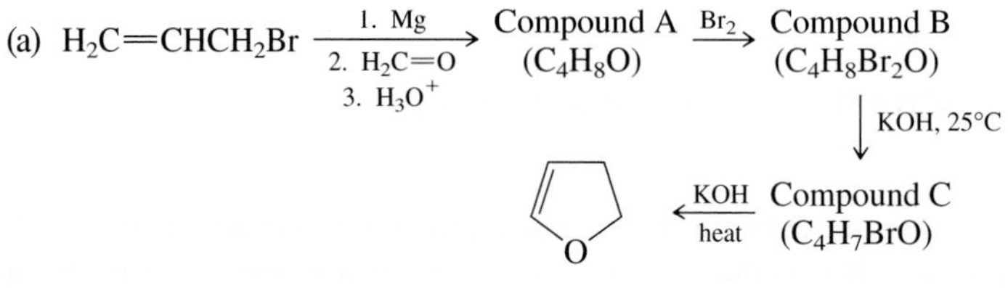

Compound D

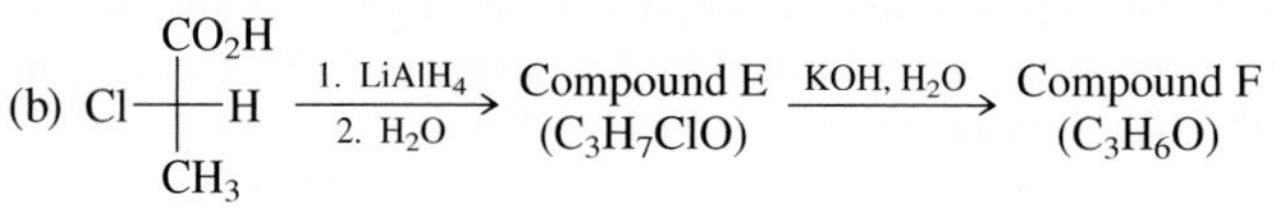

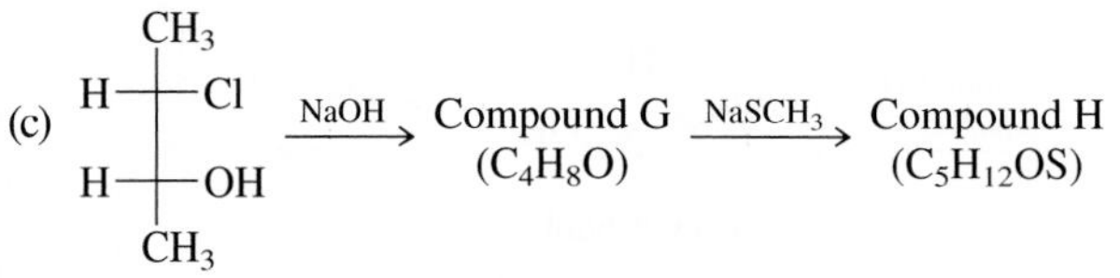

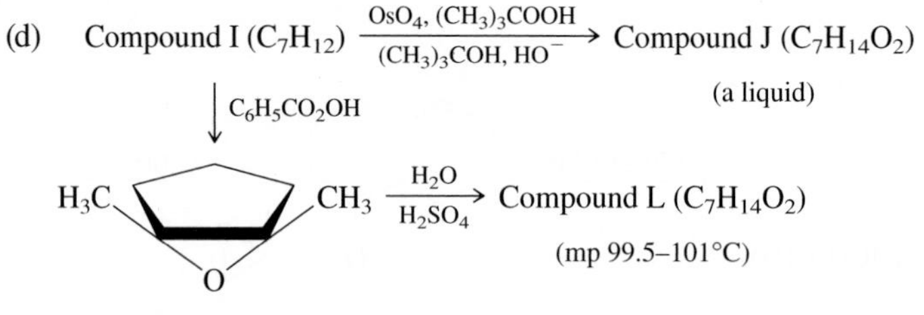

Compound K

16.35 Cineole is the chief component of eucalyptus oil; it has the molecular formula $C_{10}H_{18}O$ and contains no double or triple bonds. It reacts with hydrochloric acid to give the dichloride shown:

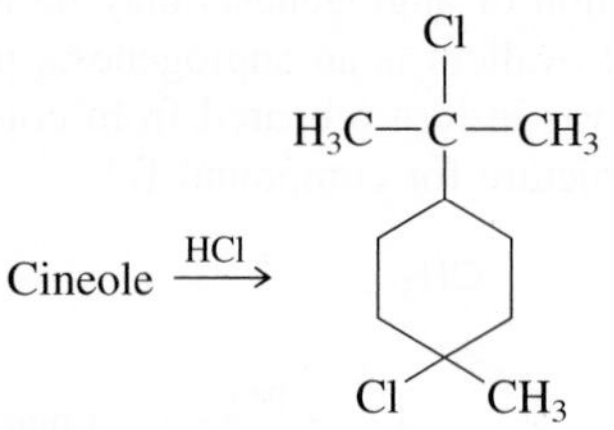

Deduce the structure of cineole.

16.36 The *p*-toluenesulfonate shown undergoes an intramolecular Williamson reaction on treatment with base to give a spirocyclic ether. Demonstrate your understanding of the terminology used in the preceding sentence by writing the structure, including stereochemistry, of the product.

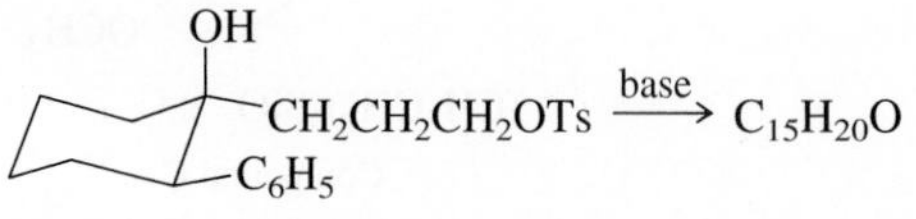

16.37 Given that:

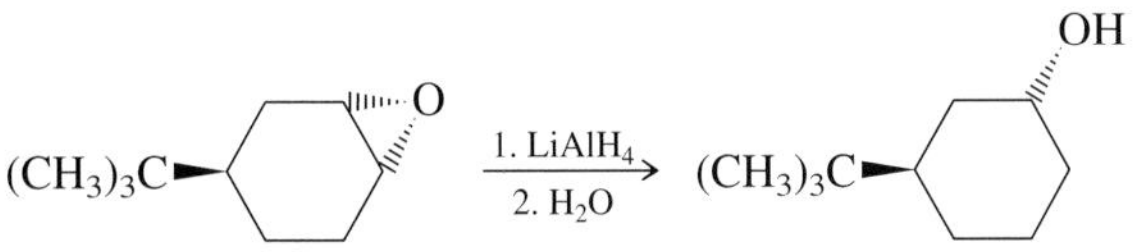

does the product of the analogous reaction using $LiAlD_4$ contain an axial or an equatorial deuterium?

16.38 The name of the parent six-membered sulfur-containing heterocycle is *thiane*. It is numbered beginning at sulfur. Multiple incorporation of sulfur in the ring is indicated by the prefixes *di-*, *tri-*, and so on.

(a) How many methyl-substituted thianes are there? Which ones are chiral?

(b) Write structural formulas for 1,4-dithiane and 1,3,5-trithiane.

(c) Which dithiane isomer (1,2-, 1,3-, or 1,4-) is a disulfide?

(d) Draw the two most stable conformations of the sulfoxide derived from thiane.

16.39 Oxidation of 4-*tert*-butylthiane (see Problem 16.38 for the structure of thiane) with sodium metaperiodate ($NaIO_4$) gives a mixture of two compounds of molecular formula $C_9H_{18}OS$. Both products give the same sulfone on further oxidation with hydrogen peroxide. What is the relationship between the two compounds?

16.40 This problem is adapted from an experiment designed for undergraduate organic chemistry laboratories.

(a) Reaction of (*E*)-1-(*p*-methoxyphenyl)propene with *m*-chloroperoxybenzoic acid converted the alkene to its corresponding epoxide. Give the structure, including stereochemistry, of this epoxide.

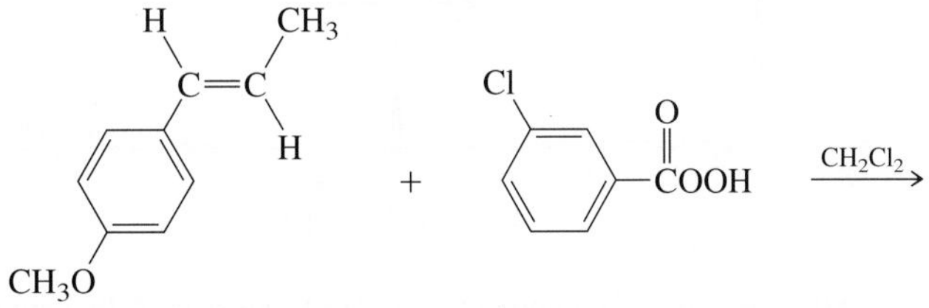

(*E*)-1-(*p*-Methoxyphenyl)propene   *m*-Chloroperoxybenzoic acid

(b) Assign the signals in the ^1H NMR spectrum of the epoxide to the appropriate hydrogens.

δ 1.4 (doublet, 3H)	δ 3.8 (singlet, 3H)
δ 3.0 (quartet of doublets, 1H)	δ 6.9 (doublet, 2H)
δ 3.5 (doublet, 1H)	δ 7.2 (doublet, 2H)

(c) Three signals appear in the range δ 55–60 in the ^{13}C NMR spectrum of the epoxide. To which carbons of the epoxide do these signals correspond?

(d) The epoxide is isolated only when the reaction is carried out under conditions (added Na_2CO_3) that ensure that the reaction mixture does not become acidic. Unless this precaution is taken, the isolated product has the molecular formula $C_{17}H_{17}O_4Cl$. Suggest a reasonable structure for this product and write a reasonable mechanism for its formation.

16.41 A different product is formed in each of the following reactions. Identify the product in each case from their ^1H NMR spectra in Figure 16.5 and suggest an explanation for the observed regioselectivity.

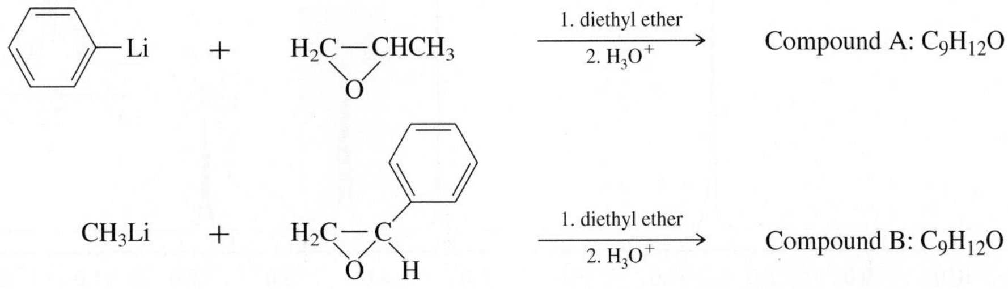

Figure 16.5

The 200-MHz ^{1}H NMR spectra of compounds formed by the reaction of (*a*) phenyllithium with 1,2-epoxypropane and (*b*) methyllithium with styrene oxide.

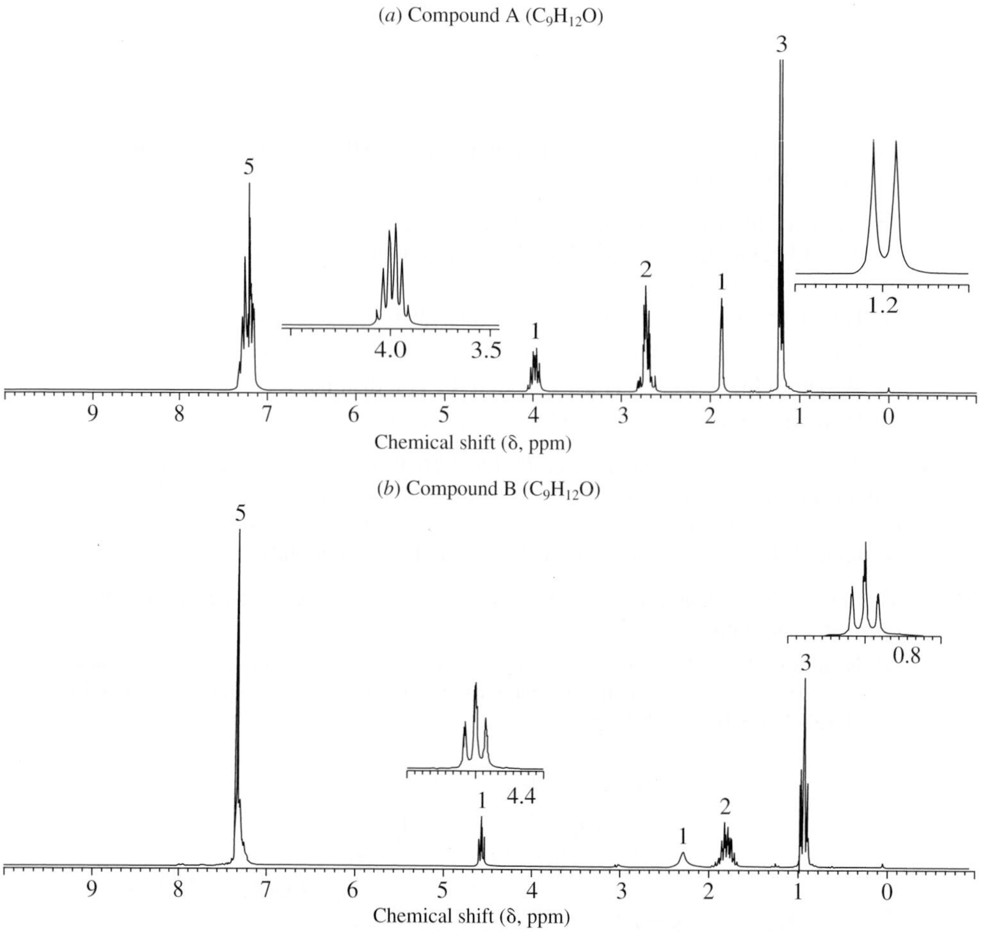

16.42 The ^{1}H NMR spectrum of compound A (C_8H_8O) consists of two singlets of equal area at δ 5.1 (sharp) and 7.2 ppm (broad). On treatment with excess hydrogen bromide, compound A is converted to a single dibromide ($C_8H_8Br_2$). The ^{1}H NMR spectrum of the dibromide is similar to that of A in that it exhibits two singlets of equal area at δ 4.7 (sharp) and 7.3 ppm (broad). Suggest reasonable structures for compound A and the dibromide derived from it.

16.43 The ^{1}H NMR spectrum of a compound ($C_{10}H_{13}BrO$) is shown in Figure 16.6. The compound gives benzyl bromide, along with a second compound $C_3H_6Br_2$, when heated with HBr. What is the first compound?

Figure 16.6

The 200-MHz ^{1}H NMR spectrum of a compound, $C_{10}H_{13}BrO$ (Problem 16.43). The integral ratios of the signals reading from left to right (low field to high field) are 5:2:2:2:2. The signals centered at δ 3.6 and δ 3.7 are two overlapping triplets.

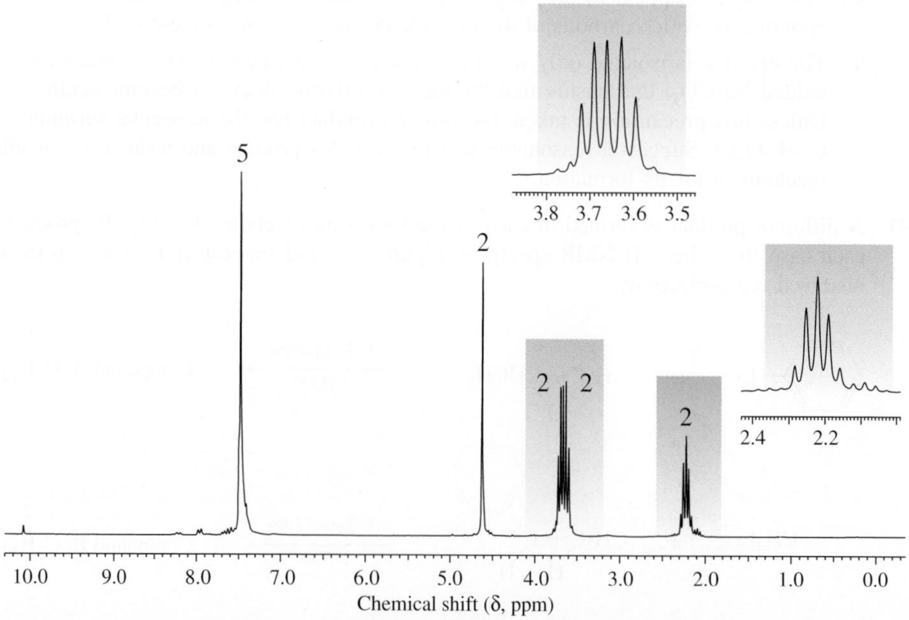

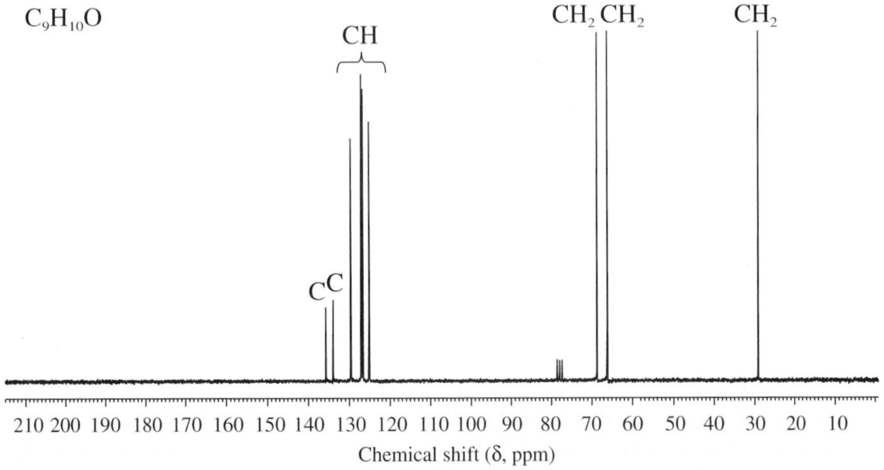

Figure 16.7

The ^{13}C NMR spectrum of a compound, $C_9H_{10}O$ (Problem 16.44).

16.44 A compound is a cyclic ether of molecular formula $C_9H_{10}O$. Its ^{13}C NMR spectrum is shown in Figure 16.7. Oxidation of the compound with sodium dichromate and sulfuric acid gave 1,2-benzenedicarboxylic acid. What is the compound?

Descriptive Passage and Interpretive Problems 16

Epoxide Rearrangements and the NIH Shift

This passage is about two seemingly unrelated aspects of epoxides:

1. epoxide rearrangements
2. arene oxides

These two topics merge in an important biological transformation in which neither the reactant nor the product is an epoxide—the conversion of the amino acid phenylalanine to tyrosine.

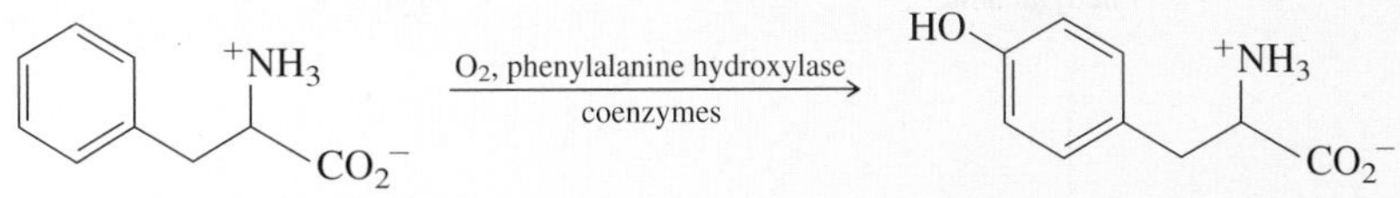

Phenylalanine Tyrosine

Epoxide rearrangements

In some epoxide ring-opening reactions C—O bond cleavage is accompanied by the development of enough carbocation character at carbon ($^{\delta+}$C---O) to allow rearrangement to occur. These reactions are typically promoted by protonation of the epoxide oxygen or by its coordination to Lewis acids such as boron trifluoride (BF_3) and aluminum chloride ($AlCl_3$).

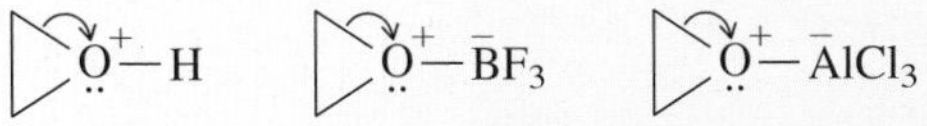

As positive charge develops on the ring carbon, one of the groups on the adjacent carbon migrates to it. This migration is assisted by electron-pair donation from oxygen. It is likely that all of this occurs in the same transition state. Subsequent deprotonation gives an aldehyde or ketone as the isolated product.

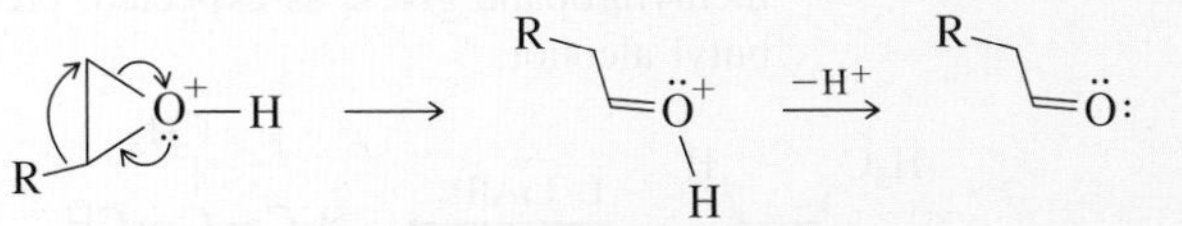

Overall, the reaction resembles the pinacol rearrangement of vicinal diols (see the Chapter 15 Descriptive Passage and Interpretive Problems) and takes place under similar conditions.

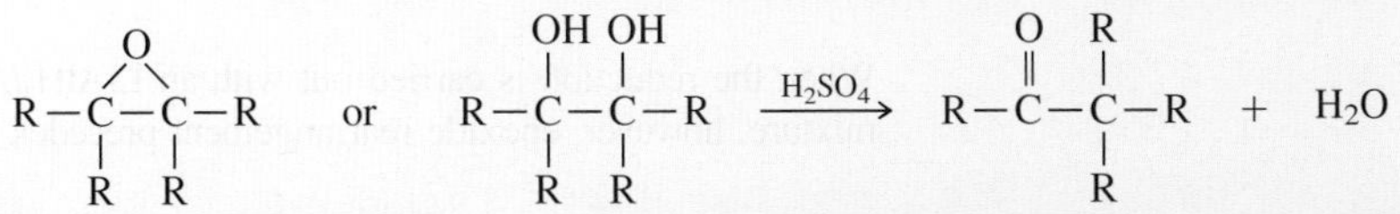

Arene Oxides

Aromatic rings are normally inert to the customary reagents that convert alkenes to epoxides, but arene oxides have been synthesized in the laboratory, often by indirect methods. Their chemical reactivity resembles that of other epoxides.

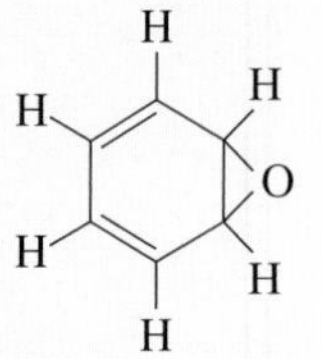

1,2-Epoxycyclohexa-3,5-diene is formally the epoxide of benzene and is the parent of the class of compounds known as arene oxides.

The most striking thing about arene oxides is their involvement in biological processes. Enzymes in the liver oxidize aromatic hydrocarbons to arene oxides, which then react with biological nucleophiles to give compounds used in subsequent reactions or to aid elimination of the arene oxide from the body. Some arene oxides, especially those from polycyclic aromatic hydrocarbons, are carcinogenic and react with nitrogen nucleophiles of DNA to induce mutations (Section 11.7).

The NIH shift

Although hydroxylation of phenylalanine to tyrosine looks like a typical electrophilic aromatic substitution, scientists at the U.S. National Institutes of Health discovered that the biochemical pathway combines epoxidation of the benzene ring followed by epoxide ring-opening with rearrangement. This rearrangement, which is the biochemical analog of the pinacol-type reactions described earlier, is known as the "NIH shift."

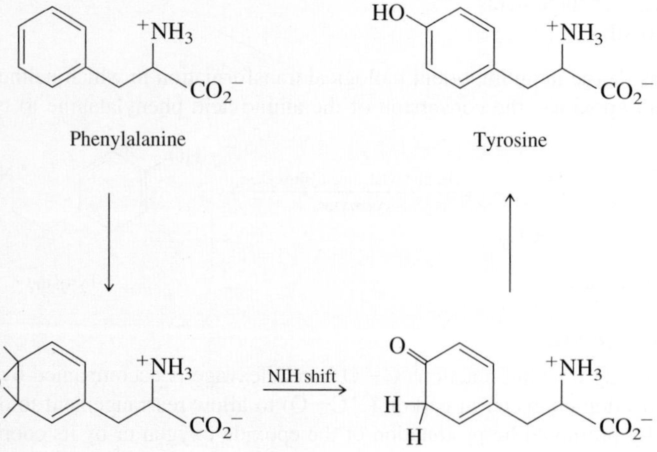

16.45 Epoxides X and Y give the same aldehyde ($C_{14}H_{12}O$) on BF_3-catalyzed rearrangement.

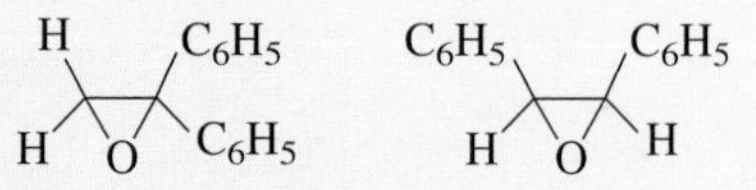

X Y

Which of the following best describes the rearrangement step?

A. H migrates in both X and Y.
B. C_6H_5 migrates in both X and Y.
C. H migrates in X; C_6H_5 migrates in Y.
D. C_6H_5 migrates in X; H migrates in Y.

16.46 Lithium aluminum hydride reduction of 1,2-epoxy-2-methylpropane gives, as expected, predominantly *tert*-butyl alcohol.

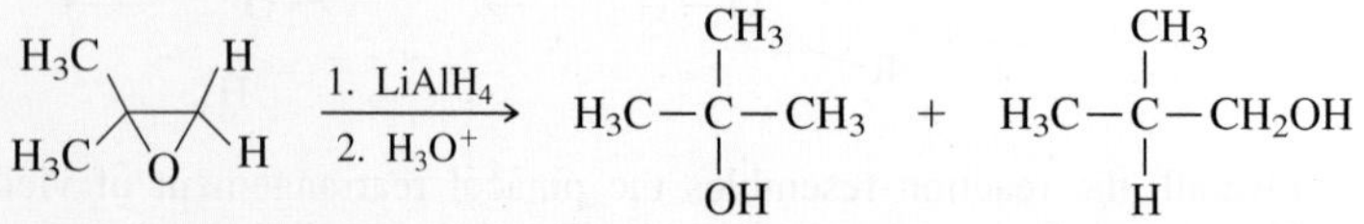

1,2-Epoxy-2-methylpropane — *tert*-Butyl alcohol (97%) — Isobutyl alcohol (3%)

When the reduction is carried out with an $LiAlH_4/AlCl_3$ mixture, however, epoxide rearrangement precedes reduction and isobutyl alcohol becomes the major product. This rearrangement was confirmed by a deuterium-labeling experiment in which an $LiAlD_4/AlCl_3$ mixture was used. Where was the deuterium located in the isobutyl alcohol product?

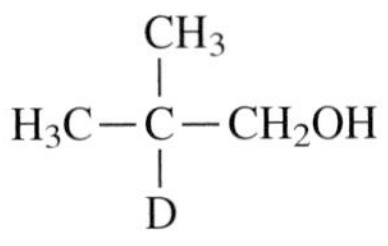
CH3
H3C—C—CH2OH
D
A.

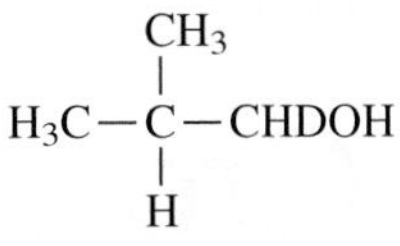
CH3
H3C—C—CHDOH
H
B.

16.47 The epoxide derived from benzene, 1,2-epoxycyclohexa-3,5-diene, exists in equilibrium with a monocyclic isomer oxepine.

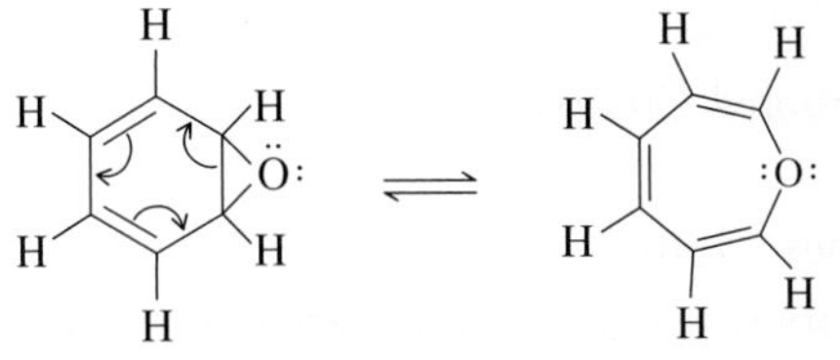
1,2-Epoxycyclohexa-3,5-diene — Oxepine

Which statement is correct concerning the aromaticity of these two isomers?

A. Both are aromatic.

B. Neither is aromatic.

C. 1,2-Epoxycyclohexa-3,5-diene is aromatic; oxepine is not aromatic.

D. Oxepine is aromatic; 1,2-epoxycyclohexa-3,5-diene is not aromatic.

16.48 Biological oxidation of naphthalene gives a trans vicinal diol by way of an epoxide intermediate. The diol formed is the most stable of the three isomers shown. Which diol is it?

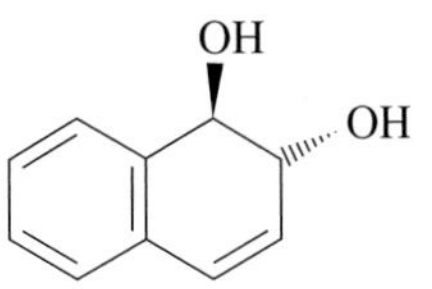
A.

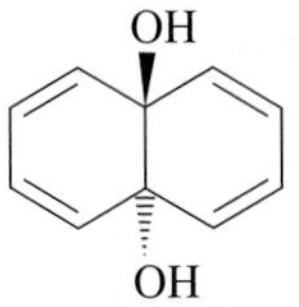
B.

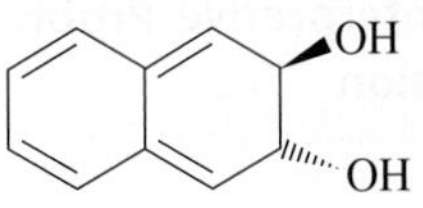
C.

16.49 Acetanilide, which has pain-relieving properties, undergoes a biochemical oxidation similar to that of the NIH shift that occurs with phenylalanine. The product formed from acetanilide is itself a pain-reliever. What is the structure of this substance (better known as Tylenol)?

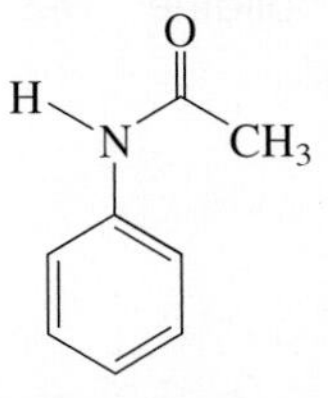
Acetanilide

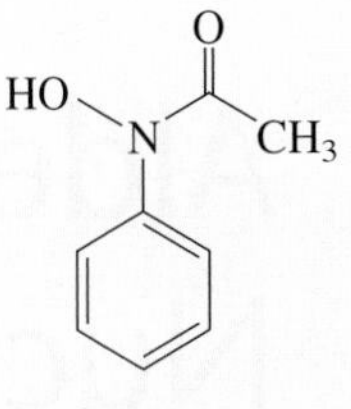
A.

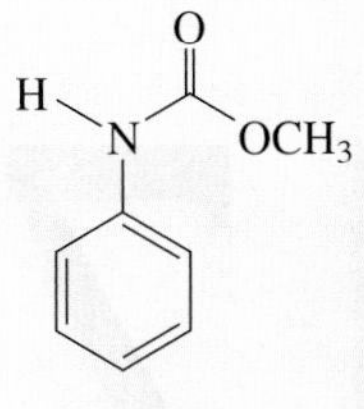
B.

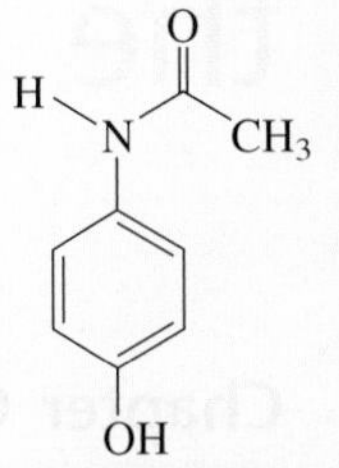
C.

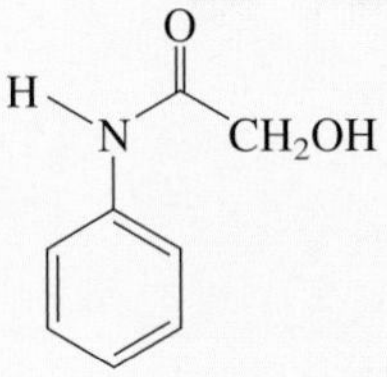
D.

16.50 The hormones serotonin and melatonin are biosynthesized from tryptophan by a series of reactions, including one that involves an NIH shift.

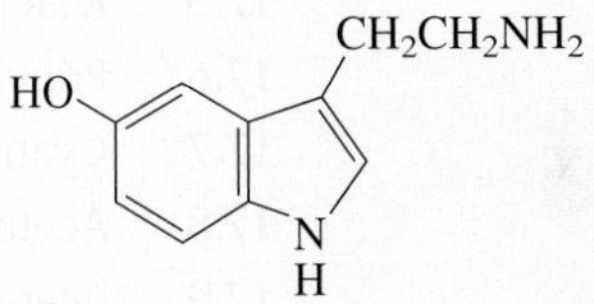
Serotonin

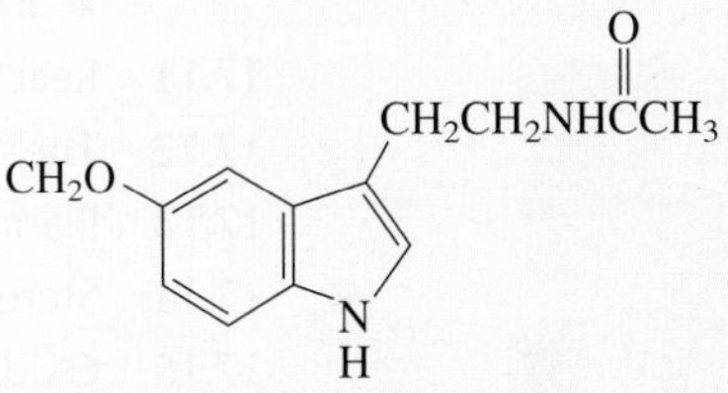
Melatonin

What is the most likely structure for tryptophan?

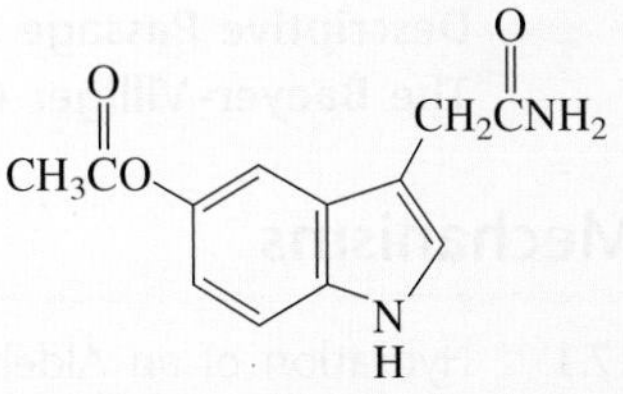
A.

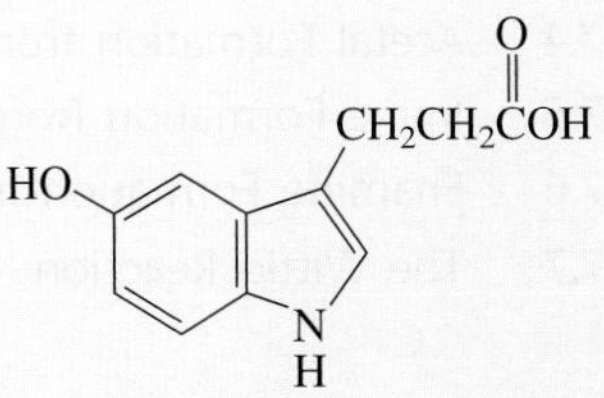
B.

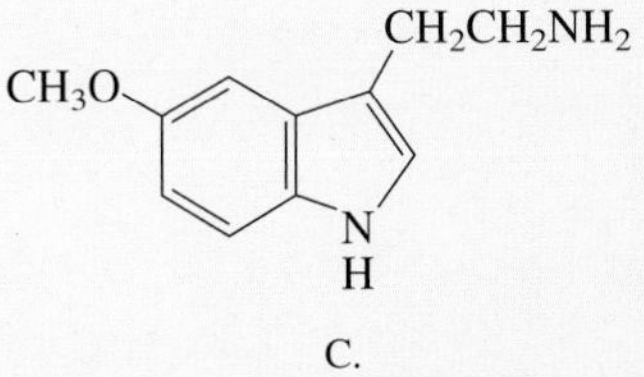
C.

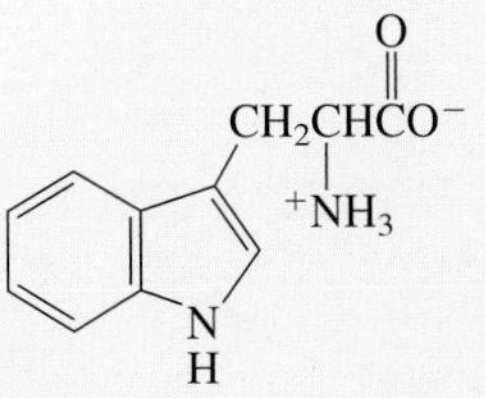
D.

17 Aldehydes and Ketones: Nucleophilic Addition to the Carbonyl Group

Chapter Outline

Mechanisms

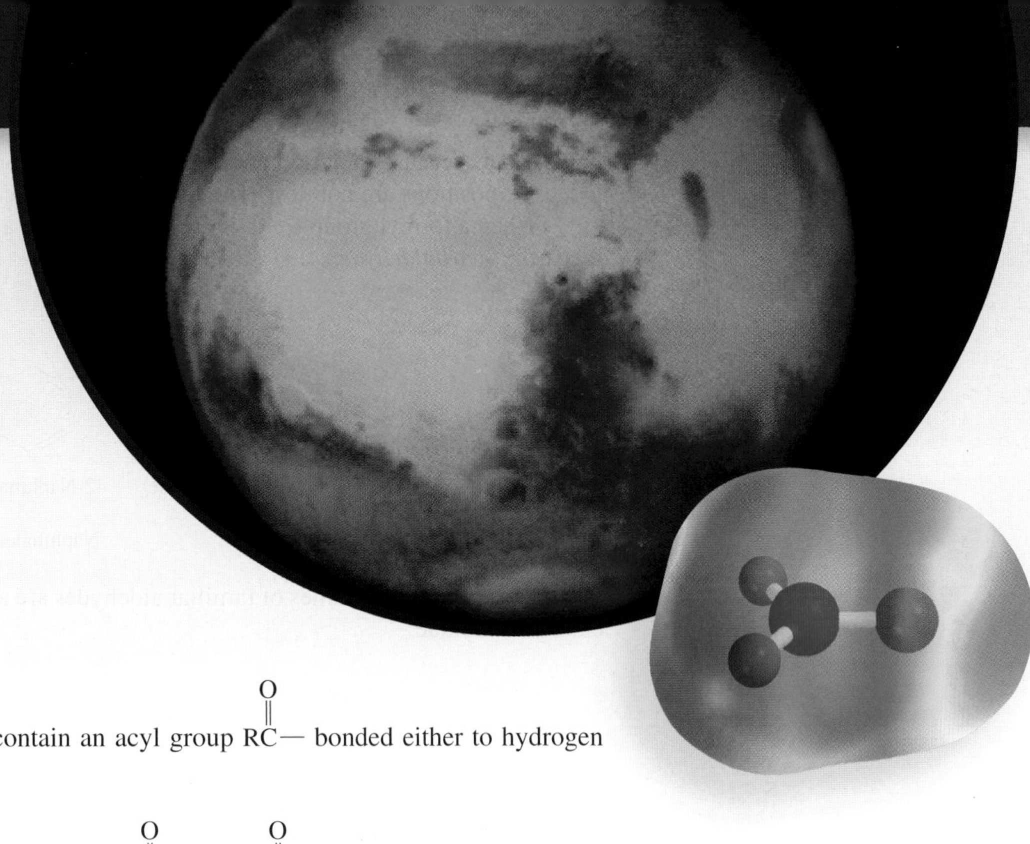

Is there formaldehyde in the Martian atmosphere? Maybe.

ALDEHYDES AND KETONES contain an acyl group RC(=O)— bonded either to hydrogen or to another carbon.

HC(=O)H	RC(=O)H	RC(=O)R′
Formaldehyde	Aldehyde	Ketone

Although the present chapter includes the usual collection of topics designed to acquaint us with a particular class of compounds, its central theme is a fundamental reaction type, *nucleophilic addition to carbonyl groups.* The principles of nucleophilic addition to aldehydes and ketones developed here will be seen to have broad applicability in later chapters when transformations of various derivatives of carboxylic acids are discussed.

17.1 Nomenclature

The longest continuous chain that contains the —C(=O)H group provides the base name for aldehydes. The *-e* ending of the corresponding alkane name is replaced by *-al,* and substituents are specified in the usual way. It is not necessary to specify the location of the —C(=O)H group in the name, because the chain must be numbered by starting with this group as C-1. The suffix *-dial* is added to the appropriate alkane name when the compound contains two aldehyde functions.*

$(CH_3)_3CCH_2CH_2CH{=}O$

4,4-Dimethylpentanal

$H_2C{=}CHCH_2CH_2CH_2CH{=}O$

5-Hexenal
or
Hex-5-enal

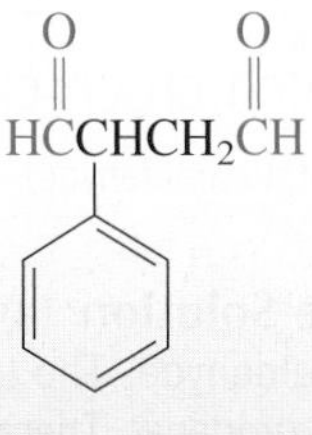

2-Phenylbutanedial

*The *-e* ending of an alkane name is dropped before a suffix beginning with a vowel (*-al*) and retained before one beginning with a consonant (*-dial*).

Notice that, because they define the ends of the carbon chain in 2-phenylbutanedial, the aldehyde positions are not designated by numerical locants in the name.

When a formyl group (—CH═O) is attached to a ring, the ring name is followed by the suffix *-carbaldehyde.*

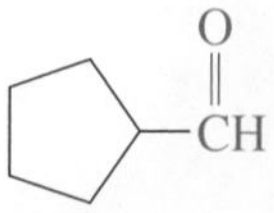
Cyclopentanecarbaldehyde

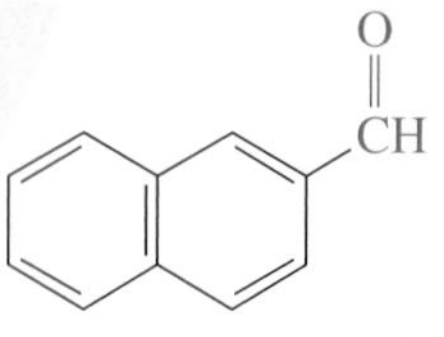
2-Naphthalenecarbaldehyde
or
Naphthalene-2-carbaldehyde

Certain common names of familiar aldehydes are acceptable as IUPAC names. A few examples include

Formaldehyde
(methanal)

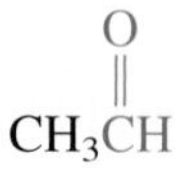
Acetaldehyde
(ethanal)

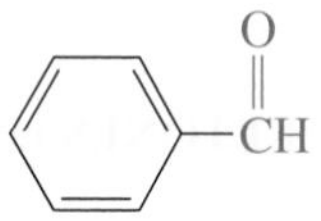
Benzaldehyde
(benzenecarbaldehyde)

Among oxygen-containing groups, a higher oxidation state takes precedence over a lower one in determining the suffix of the substitutive name. Thus, a compound that contains both an alcohol and an aldehyde function is named as an aldehyde.

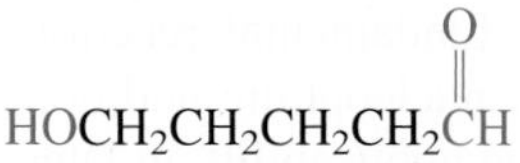
5-Hydroxypentanal

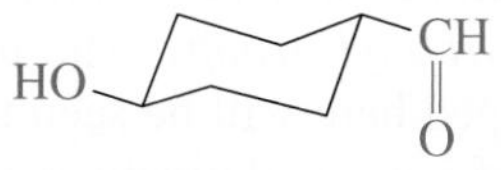
trans-4-Hydroxycyclohexanecarbaldehyde
or
trans-4-Hydroxycyclohexane-1-carbaldehyde

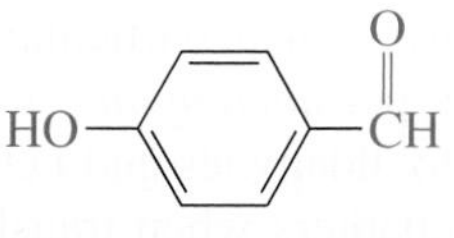
p-Hydroxybenzaldehyde

Problem 17.1

The common names and structural formulas of a few aldehydes follow. Provide an IUPAC name.

(a) 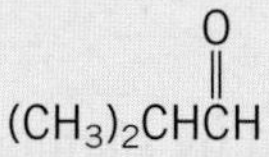
(isobutyraldehyde)

(b) 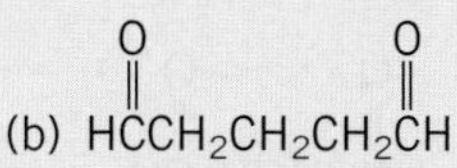
(glutaraldehyde)

(c) 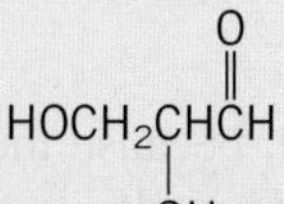
(glyceraldehyde)

(d) 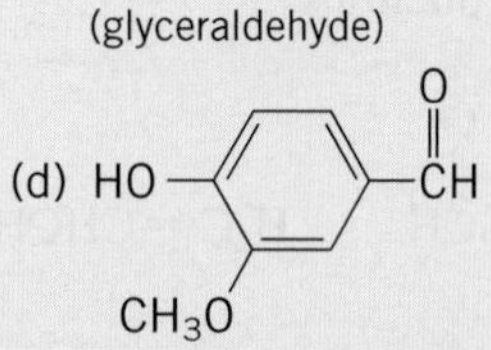
(vanillin)

Sample Solution (a) Don't be fooled by the fact that the common name is isobutyraldehyde. The longest continuous chain has three carbons, and so the base name is *propanal.* There is a methyl group at C-2; thus the compound is 2-methylpropanal.

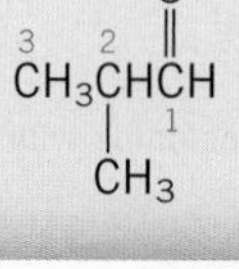
2-Methylpropanal
(isobutyraldehyde)

With ketones, the *-e* ending of an alkane is replaced by *-one* in the longest continuous chain containing the carbonyl group. The chain is numbered in the direction that provides the lower number for this group. The carbonyl carbon of a cyclic ketone is C-1 and the number does not appear in the name.

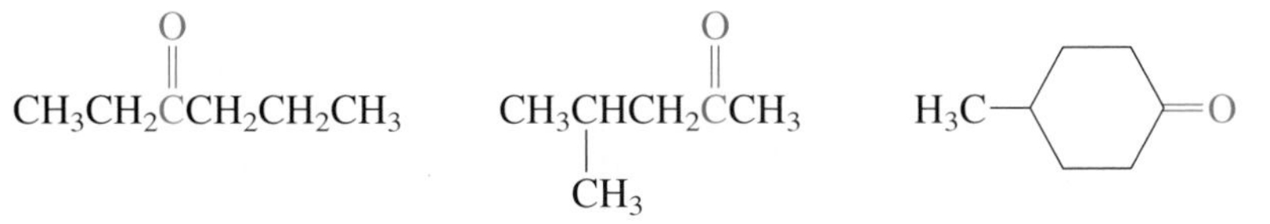

3-Hexanone or Hexan-3-one	4-Methyl-2-pentanone or 4-Methylpentan-2-one	4-Methylcyclohexanone

Like aldehydes, ketone functions take precedence over alcohol functions, double bonds, halogens, and alkyl groups in determining the parent name and direction of numbering. Aldehydes outrank ketones, however, and a compound that contains both an aldehyde and a ketone carbonyl group is named as an aldehyde. In such cases, the carbonyl oxygen of the ketone is considered an *oxo*-substituent on the main chain.

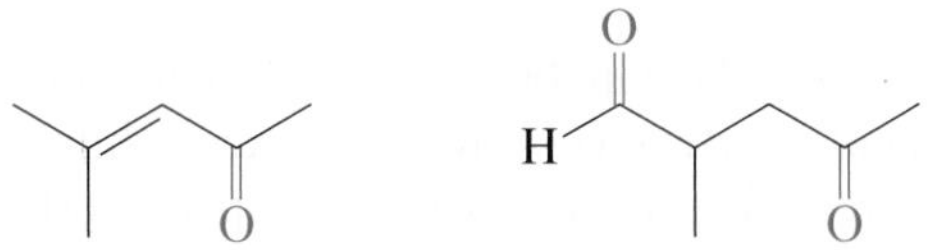

4-Methyl-3-penten-2-one or 4-Methylpent-3-en-2-one — 2-Methyl-4-oxopentanal

Although substitutive names of the type just described are preferred, the IUPAC rules also permit ketones to be named by functional class nomenclature. The groups attached to the carbonyl group are named as separate words followed by the word *ketone*. The groups are listed alphabetically.

There are no functional class names for aldehydes in the IUPAC system.

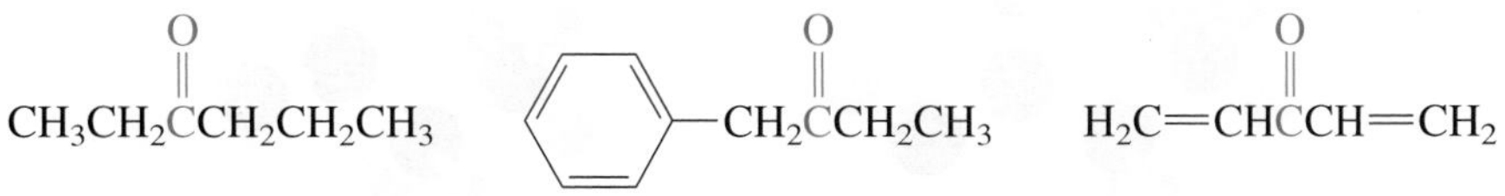

Ethyl propyl ketone — Benzyl ethyl ketone — Divinyl ketone

Problem 17.2

Convert each of the following functional class IUPAC names to a substitutive name.

(a) Dibenzyl ketone
(b) Ethyl isopropyl ketone
(c) Methyl 2,2-dimethylpropyl ketone
(d) Allyl methyl ketone

Sample Solution (a) First write the structure corresponding to the name. Dibenzyl ketone has two benzyl groups attached to a carbonyl.

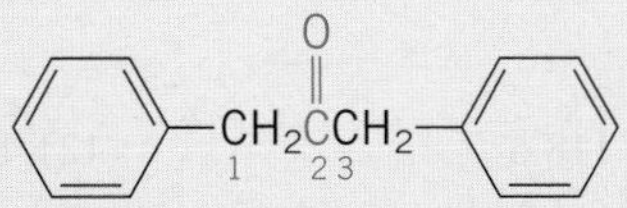

Dibenzyl ketone

The longest continuous chain contains three carbons, and C-2 is the carbon of the carbonyl group. The substitutive IUPAC name for this ketone is *1,3-diphenyl-2-propanone* or *1,3-diphenylpropan-2-one.*

A few of the common names acceptable for ketones in the IUPAC system are

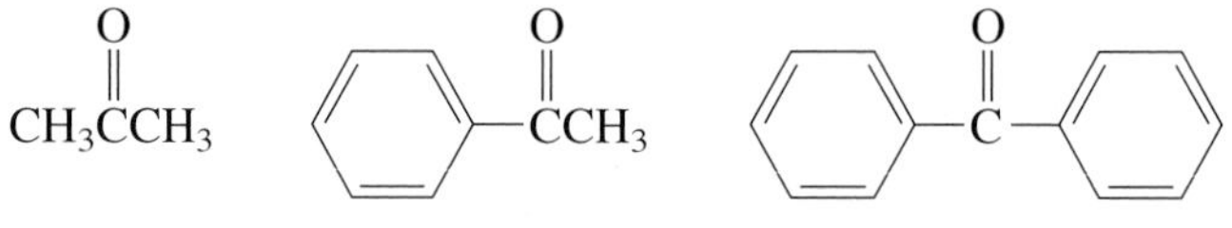

Acetone Acetophenone Benzophenone

(The suffix *-phenone* indicates that the acyl group is attached to a benzene ring.)

17.2 Structure and Bonding: The Carbonyl Group

Two notable aspects of the carbonyl group are its *geometry* and *polarity*. The coplanar geometry of the bonds to the carbonyl group is seen in the molecular models of formaldehyde, acetaldehyde, and acetone in Figure 17.1. The bond angles involving the carbonyl group are approximately 120°, but vary somewhat from compound to compound as shown by the examples in Figure 17.1. The C=O bond distance in aldehydes and ketones (122 pm) is significantly shorter than the typical C—O bond distance of 141 pm seen in alcohols and ethers.

Bonding in formaldehyde can be described according to an sp^2-hybridization model analogous to that of ethylene (Figure 17.2). According to this model, the carbon–oxygen double bond is viewed as one of the $\sigma + \pi$ type. Overlap of half-filled sp^2 hybrid orbitals of carbon and oxygen gives the σ component, whereas side-by-side overlap of half-filled $2p$ orbitals gives the π bond. The oxygen lone pairs occupy sp^2 hybrid orbitals, the axes of which lie in the plane of the molecule. The carbon–oxygen double bond of formaldehyde is both shorter and stronger than the carbon–carbon double bond of ethylene.

The carbonyl group makes aldehydes and ketones rather polar, with dipole moments that are substantially higher than alkenes.

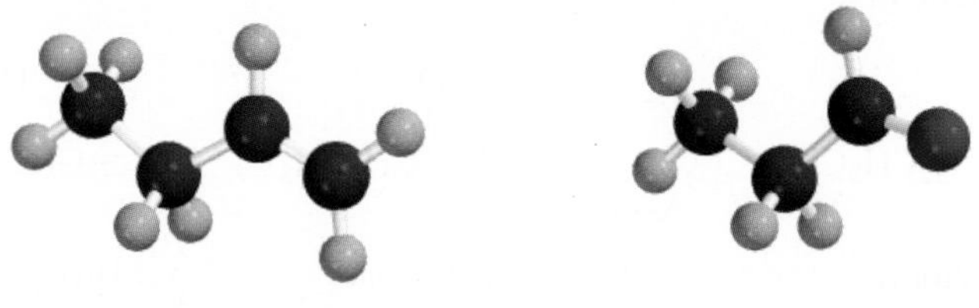

$CH_3CH_2CH{=}CH_2$ — 1-Butene — Dipole moment: 0.3 D

$CH_3CH_2CH{=}O$ — Propanal — Dipole moment: 2.5 D

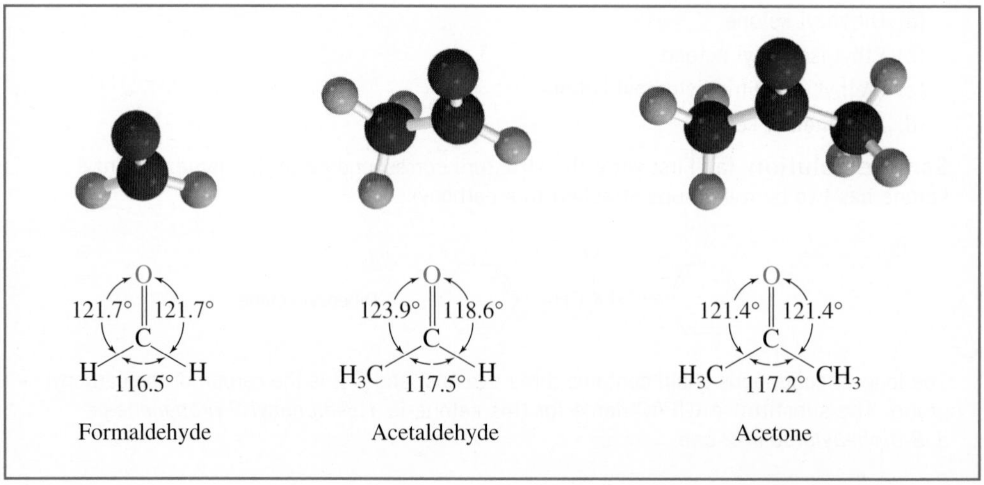

Figure 17.1

The bonds to the carbon of the carbonyl group lie in the same plane, and at angles of approximately 120° with respect to each other.

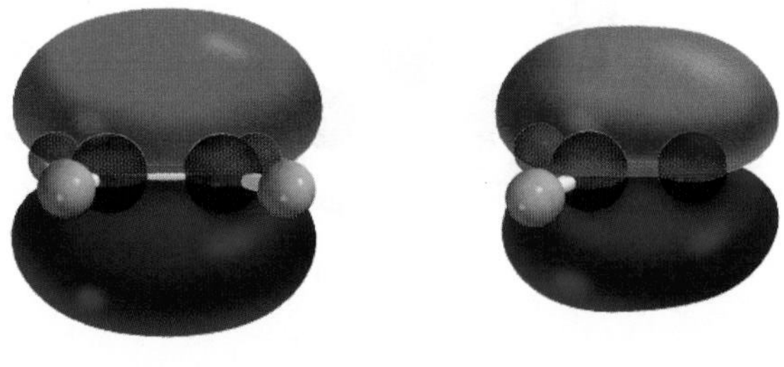

	(a) Ethylene	(b) Formaldehyde
Bond length (pm):	134	121
BDE (kJ/mol):	730	748
(kcal/mol):	172	179

Figure 17.2

Both (*a*) ethylene and (*b*) formaldehyde have the same number of electrons, and carbon is sp^2-hybridized in both. In formaldehyde, one of the carbons is replaced by an sp^2-hybridized oxygen. Like the carbon–carbon double bond of ethylene, the carbon–oxygen double bond of formaldehyde is composed of a σ component and a π component. The values given correspond to the C=C and the C=O units respectively.

How much a carbonyl group affects the charge distribution in a molecule is apparent in the electrostatic potential maps of 1-butene and propanal (Figure 17.3). When the color scale is adjusted to be the same for both molecules, the much greater separation of positive and negative charge in propanal relative to 1-butene is readily apparent. The carbonyl carbon of propanal is positively polarized and the oxygen is negatively polarized.

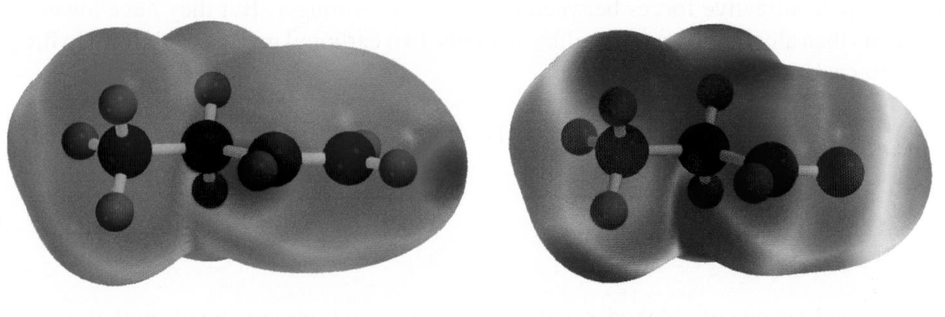

(*a*) 1-Butene ($CH_3CH_2CH{=}CH_2$) (*b*) Propanal ($CH_3CH_2CH{=}O$)

Figure 17.3

Electrostatic potential maps of (*a*) 1-butene and (*b*) propanal. The color ranges are adjusted to a common scale so that the charge distributions in the two compounds can be compared directly. The region of highest negative potential in 1-butene is associated with the π electrons of the double bond. The charge separation is greater in propanal. The carbon of the carbonyl group is a site of positive potential. The region of highest negative potential is near oxygen.

The various ways of representing this polarization include

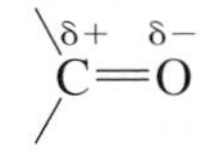 or 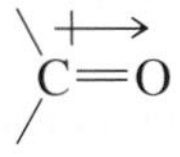 and

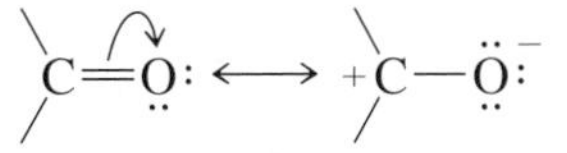

The structural features, especially the very polar nature of the carbonyl group, point clearly to the kind of chemistry we will see for aldehydes and ketones in this chapter. The partially positive carbon of C=O has carbocation character and is electrophilic. The planar arrangement of its bonds make this carbon relatively uncrowded and susceptible to attack by nucleophiles. Oxygen is partially negative and weakly basic.

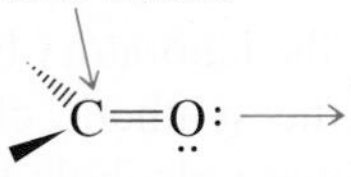
nucleophiles
bond to carbon
electrophiles, especially
protons, bond to oxygen

Alkyl substituents stabilize a carbonyl group in much the same way that they stabilize carbon–carbon double bonds and carbocations—by releasing electrons to sp^2-hybridized carbon. Thus, as their heats of combustion reveal, the ketone 2-butanone is more stable than its aldehyde isomer butanal.

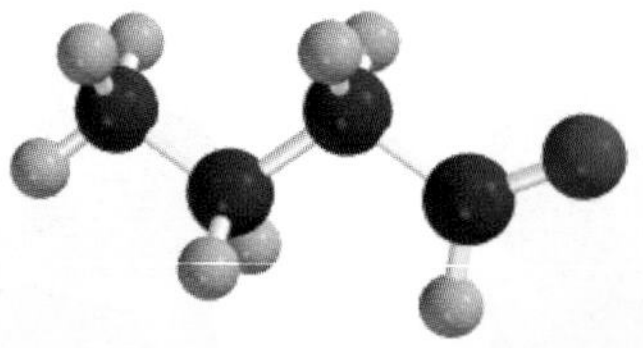

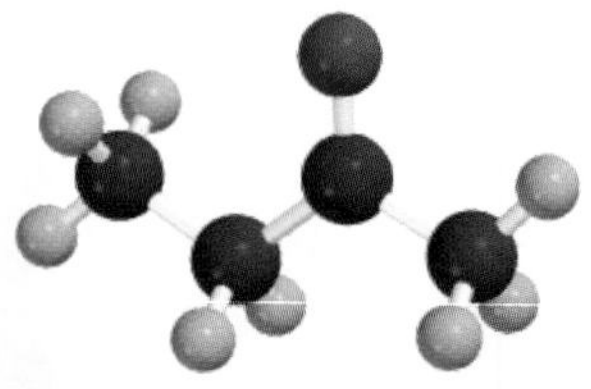

	Butanal	2-Butanone
	$CH_3CH_2CH_2CH{=}O$	$CH_3CH_2C({=}O)CH_3$
Heat of combustion:	2475 kJ/mol (592 kcal/mol)	2442 kJ/mol (584 kcal/mol)

The carbonyl carbon of a ketone bears two electron-releasing alkyl groups; an aldehyde carbonyl has only one. Just as a disubstituted double bond in an alkene is more stable than a monosubstituted double bond, a ketone carbonyl is more stable than an aldehyde carbonyl. We'll see later in this chapter that structural effects on the relative *stability* of carbonyl groups in aldehydes and ketones are an important factor in their relative *reactivity*.

17.3 Physical Properties

In general, aldehydes and ketones have higher boiling points than alkenes because the dipole–dipole attractive forces between molecules are stronger. But they have lower boiling points than alcohols because, unlike alcohols, two carbonyl groups can't form hydrogen bonds to each other.

	$CH_3CH_2CH{=}CH_2$ 1-Butene	$CH_3CH_2CH{=}O$ Propanal	$CH_3CH_2CH_2OH$ 1-Propanol
bp (1 atm)	−6°C	49°C	97°C
Solubility in water (g/100 mL)	Negligible	20	Miscible in all proportions

The carbonyl oxygen of aldehydes and ketones can form hydrogen bonds with the protons of OH groups. This makes them more soluble in water than alkenes, but less soluble than alcohols.

Problem 17.3

Sketch the hydrogen bonding between benzaldehyde and water.

17.4 Sources of Aldehydes and Ketones

As we'll see later in this chapter and the next, aldehydes and ketones are involved in many of the most used reactions in synthetic organic chemistry. Where do aldehydes and ketones themselves come from?

Many occur naturally. In terms of both variety and quantity, aldehydes and ketones rank among the most common and familiar natural products. Several are shown in Figure 17.4.

Many aldehydes and ketones are made in the laboratory by reactions that you already know about, summarized in Table 17.1. To the synthetic chemist, the most important of these are the last two: the oxidation of primary alcohols to aldehydes and secondary alcohols to ketones. *Indeed, when combined with reactions that yield alcohols, the oxidation methods are so versatile that it will not be necessary to introduce any new methods for preparing aldehydes and ketones in this chapter.* A few examples will illustrate this point.

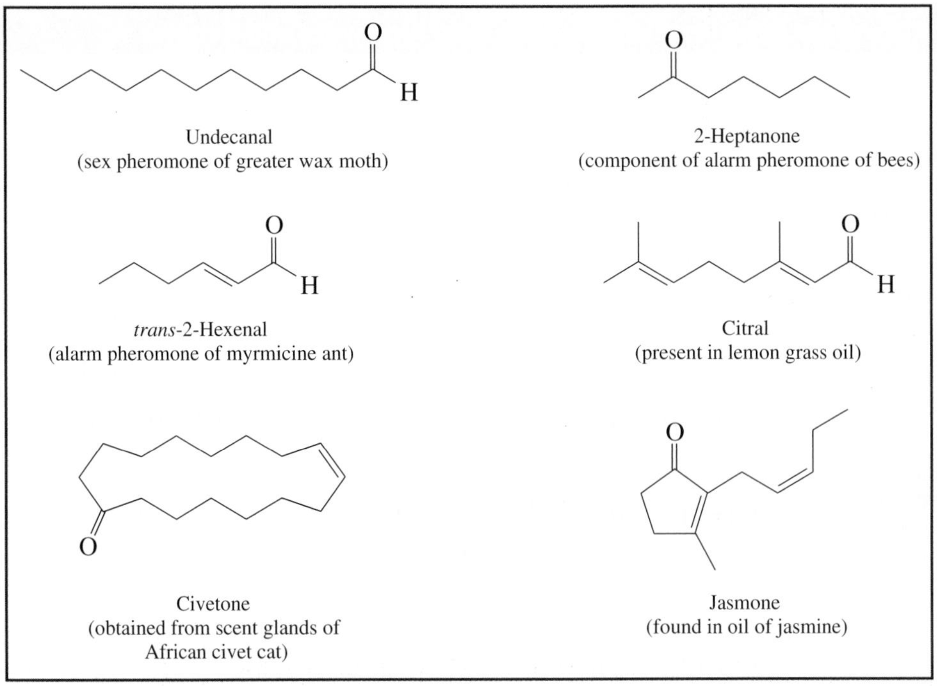

Undecanal
(sex pheromone of greater wax moth)
2-Heptanone
(component of alarm pheromone of bees)
trans-2-Hexenal
(alarm pheromone of myrmicine ant)
Citral
(present in lemon grass oil)
Civetone
(obtained from scent glands of African civet cat)
Jasmone
(found in oil of jasmine)

Figure 17.4

Some naturally occurring aldehydes and ketones.

Let's first consider how to prepare an aldehyde from a carboxylic acid. There are no good methods for going from RCO_2H to RCHO directly. Instead, we do it indirectly by first reducing the carboxylic acid to the corresponding primary alcohol, then oxidizing the primary alcohol to the aldehyde.

$$\underset{\text{Carboxylic acid}}{RCO_2H} \xrightarrow{\text{reduce}} \underset{\text{Primary alcohol}}{RCH_2OH} \xrightarrow{\text{oxidize}} \underset{\text{Aldehyde}}{R\overset{O}{\overset{\|}{C}}H}$$

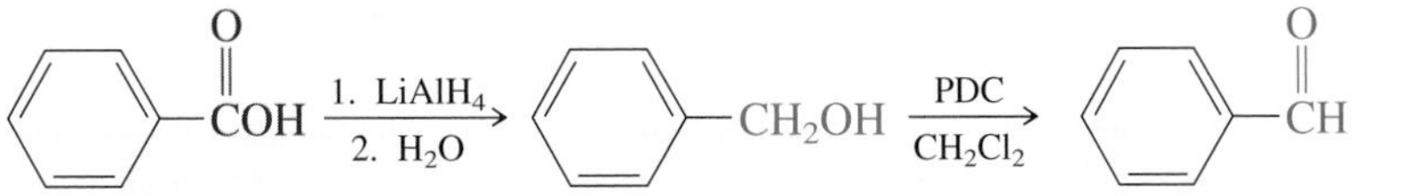

Benzoic acid　　Benzyl alcohol (81%)　　Benzaldehyde (83%)

Problem 17.4

Can catalytic hydrogenation be used to reduce a carboxylic acid to a primary alcohol in the first step of the sequence $RCO_2H \rightarrow RCH_2OH \rightarrow RCHO$?

It is often necessary to prepare ketones by processes involving carbon–carbon bond formation. In such cases the standard method combines addition of a Grignard reagent to an aldehyde with oxidation of the resulting secondary alcohol:

$$\underset{\text{Aldehyde}}{R\overset{O}{\overset{\|}{C}}H} \xrightarrow[\text{2. } H_3O^+]{\text{1. } R'MgX\text{, diethyl ether}} \underset{\text{Secondary alcohol}}{R\overset{OH}{\overset{|}{C}}HR'} \xrightarrow{\text{oxidize}} \underset{\text{Ketone}}{R\overset{O}{\overset{\|}{C}}R'}$$

$$\underset{\text{Propanal}}{CH_3CH_2\overset{O}{\overset{\|}{C}}H} \xrightarrow[\text{2. } H_3O^+]{\text{1. } CH_3(CH_2)_3MgBr \text{ diethyl ether}} \underset{\text{3-Heptanol}}{CH_3CH_2\overset{OH}{\overset{|}{C}}H(CH_2)_3CH_3} \xrightarrow{H_2CrO_4} \underset{\text{3-Heptanone (57\% from propanal)}}{CH_3CH_2\overset{O}{\overset{\|}{C}}(CH_2)_3CH_3}$$

TABLE 17.1 Summary of Reactions Discussed in Earlier Chapters That Yield Aldehydes and Ketones

Reaction (section) and comments	General equation and specific example
Ozonolysis of alkenes (Section 6.20) This cleavage reaction is more often seen in structural analysis than in synthesis. The substitution pattern around a double bond is revealed by identifying the carbonyl-containing compounds that make up the product. Hydrolysis of the ozonide intermediate in the presence of zinc (reductive workup) permits aldehyde products to be isolated without further oxidation.	$\mathrm{(R)(R')C{=}C(H)(R'')} \xrightarrow[2.\ \mathrm{H_2O,\ Zn}]{1.\ \mathrm{O_3}} \mathrm{R\overset{O}{\overset{\|}{C}}R' + R''\overset{O}{\overset{\|}{C}}H}$ Alkene → Two carbonyl compounds 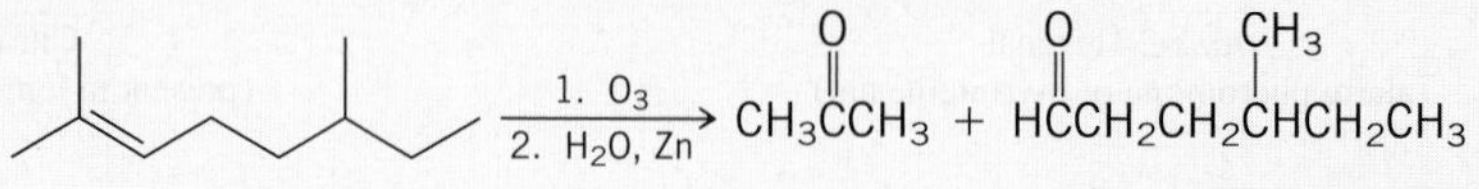 2,6-Dimethyl-2-octene → Acetone + 4-Methylhexanal (91%)
Hydration of alkynes (Section 9.12) Reaction occurs by way of an enol intermediate formed by Markovnikov addition of water to the triple bond.	$\mathrm{RC{\equiv}CR' + H_2O \xrightarrow[HgSO_4]{H_2SO_4} R\overset{O}{\overset{\|}{C}}CH_2R'}$ Alkyne → Ketone $\mathrm{HC{\equiv}C(CH_2)_5CH_3 + H_2O \xrightarrow[HgSO_4]{H_2SO_4} CH_3\overset{O}{\overset{\|}{C}}(CH_2)_5CH_3}$ 1-Octyne → 2-Octanone (91%)
Friedel–Crafts acylation of aromatic compounds (Section 12.7) Acyl chlorides and carboxylic acid anhydrides acylate aromatic rings in the presence of aluminum chloride. The reaction is electrophilic aromatic substitution in which acylium ions are generated and attack the ring.	$\mathrm{ArH + R\overset{O}{\overset{\|}{C}}Cl \xrightarrow{AlCl_3} Ar\overset{O}{\overset{\|}{C}}R + HCl}$ or $\mathrm{ArH + R\overset{O}{\overset{\|}{C}}O\overset{O}{\overset{\|}{C}}R \xrightarrow{AlCl_3} Ar\overset{O}{\overset{\|}{C}}R + RCO_2H}$ 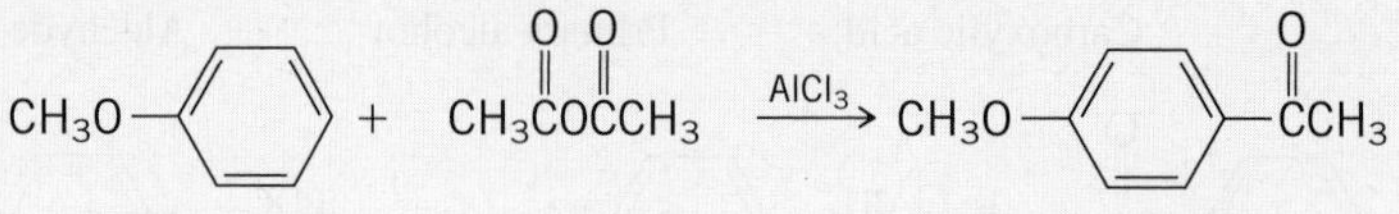 Anisole + Acetic anhydride → *p*-Methoxyacetophenone (90–94%)
Oxidation of primary alcohols to aldehydes (Section 15.9) Pyridinium dichromate (PDC) or pyridinium chlorochromate (PCC) in anhydrous media such as dichloromethane oxidizes primary alcohols to aldehydes while avoiding overoxidation to carboxylic acids.	$\mathrm{RCH_2OH \xrightarrow[CH_2Cl_2]{PDC\ or\ PCC} R\overset{O}{\overset{\|}{C}}H}$ Primary alcohol → Aldehyde $\mathrm{CH_3(CH_2)_8CH_2OH \xrightarrow[CH_2Cl_2]{PDC} CH_3(CH_2)_8\overset{O}{\overset{\|}{C}}H}$ 1-Decanol → Decanal (98%)
Oxidation of secondary alcohols to ketones (Section 15.9) Many oxidizing agents are available for converting secondary alcohols to ketones. PDC or PCC may be used, as well as other Cr(VI)-based agents such as chromic acid or potassium dichromate and sulfuric acid.	$\mathrm{R\underset{OH}{\underset{\|}{C}}HR' \xrightarrow{Cr(VI)} R\overset{O}{\overset{\|}{C}}R'}$ Secondary alcohol → Ketone $\mathrm{C_6H_5\underset{OH}{\underset{\|}{C}}HCH_2CH_2CH_2CH_3 \xrightarrow[acetic\ acid/\ water]{CrO_3} C_6H_5\overset{O}{\overset{\|}{C}}CH_2CH_2CH_2CH_3}$ 1-Phenyl-1-pentanol → 1-Phenyl-1-pentanone (93%)

Problem 17.5

(a) Show how 2-butanone could be prepared by a procedure in which all of the carbons originate in acetic acid (CH_3CO_2H).

Sample Solution

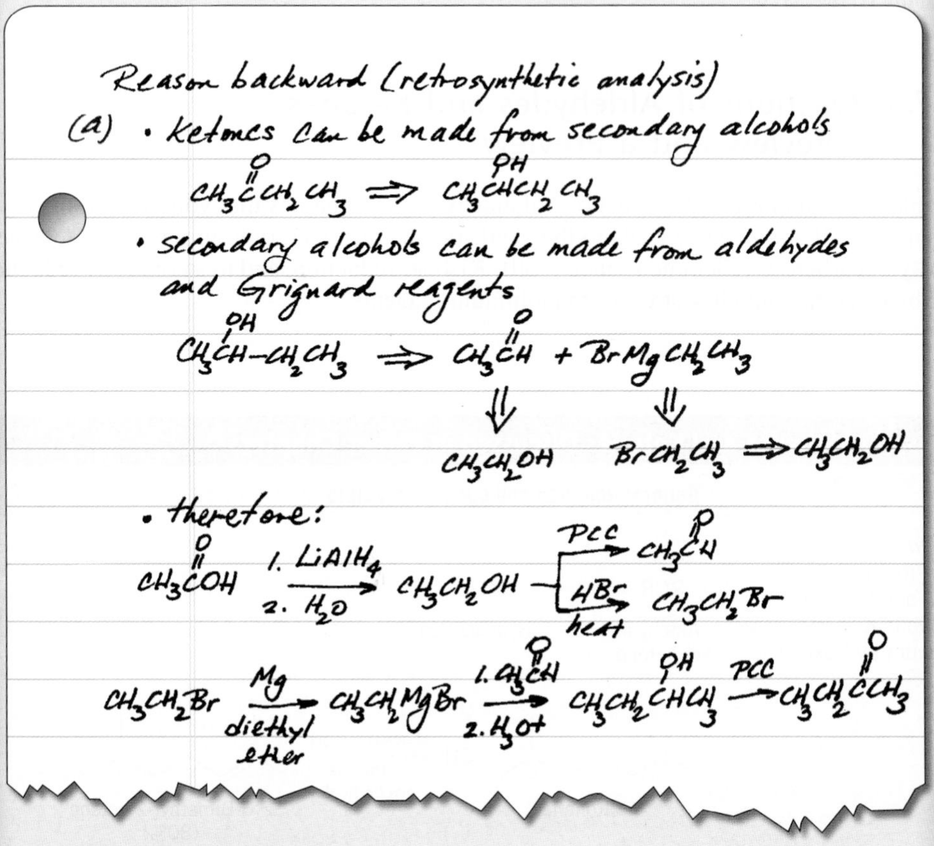

Reason backward (retrosynthetic analysis)
(A) • Ketones can be made from secondary alcohols
• secondary alcohols can be made from aldehydes and Grignard reagents
• therefore:
1. LiAlH4
2. H2O
PCC
HBr
heat
Mg
diethyl ether
1. CH3CH
2. H3O+

(b) Two species of ants found near the Mediterranean use 2-methyl-4-heptanone as an alarm pheromone. Suggest a synthesis of this compound from two 4-carbon alcohols.

Many low-molecular-weight aldehydes and ketones are important industrial chemicals. Formaldehyde, a starting material for a number of polymers, is prepared by oxidation of methanol over a silver or iron oxide/molybdenum oxide catalyst at elevated temperature.

The name *aldehyde* was invented to stand for *al*cohol *dehyd*rogenatum, indicating that aldehydes are related to alcohols by loss of hydrogen.

$$\underset{\text{Methanol}}{CH_3OH} + \underset{\text{Oxygen}}{\tfrac{1}{2}O_2} \xrightarrow[500°C]{\text{catalyst}} \underset{\text{Formaldehyde}}{\overset{\overset{\large O}{\|}}{HCH}} + \underset{\text{Water}}{H_2O}$$

Similar processes are used to convert ethanol to acetaldehyde and isopropyl alcohol to acetone.

The "linear α-olefins" described in Section 14.16 are starting materials for the preparation of a variety of aldehydes by reaction with carbon monoxide. The process is called **hydroformylation.**

$$\underset{\text{Alkene}}{RCH{=}CH_2} + \underset{\text{Carbon monoxide}}{CO} + \underset{\text{Hydrogen}}{H_2} \xrightarrow{Co_2(CO)_8} \underset{\text{Aldehyde}}{RCH_2CH_2\overset{\overset{\large O}{\|}}{C}H}$$

Excess hydrogen brings about the hydrogenation of the aldehyde and allows the process to be adapted to the preparation of primary alcohols. Over 2×10^9 lb/year of a variety of aldehydes and alcohols is prepared in the United States by hydroformylation.

A number of aldehydes and ketones are prepared both in industry and in the laboratory by a reaction known as the *aldol condensation,* which will be discussed in detail in Chapter 20.

17.5 Reactions of Aldehydes and Ketones: A Review and a Preview

Table 17.2 summarizes the reactions of aldehydes and ketones that you've seen in earlier chapters. All are valuable tools to the synthetic chemist. Carbonyl groups provide access to hydrocarbons by Clemmensen or Wolff–Kishner reduction, and to alcohols by reduction or by reaction with Grignard or organolithium reagents.

TABLE 17.2 Summary of Reactions of Aldehydes and Ketones Discussed in Earlier Chapters

Reaction (section) and comments	General equation and specific example
Reduction to hydrocarbons (Section 12.8) Two methods for converting carbonyl groups to methylene units are the Clemmensen reduction (zinc amalgam and concentrated hydrochloric acid) and the Wolff-Kishner reduction (heat with hydrazine and potassium hydroxide in a high-boiling alcohol).	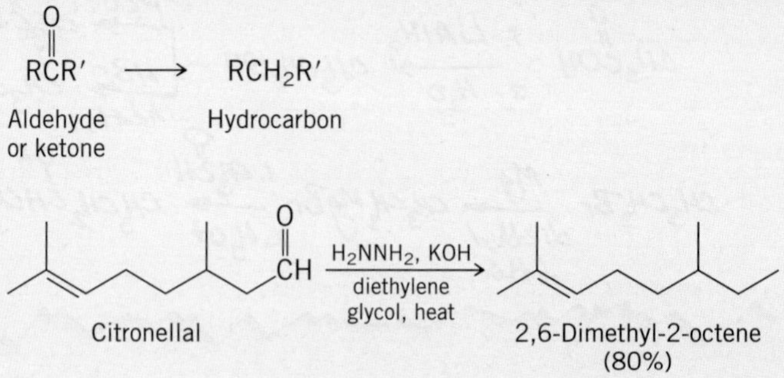
Reduction to alcohols (Section 15.2) Aldehydes are reduced to primary alcohols, and ketones are reduced to secondary alcohols by a variety of reducing agents. Catalytic hydrogenation over a metal catalyst and reduction with sodium borohydride or lithium aluminum hydride are general methods.	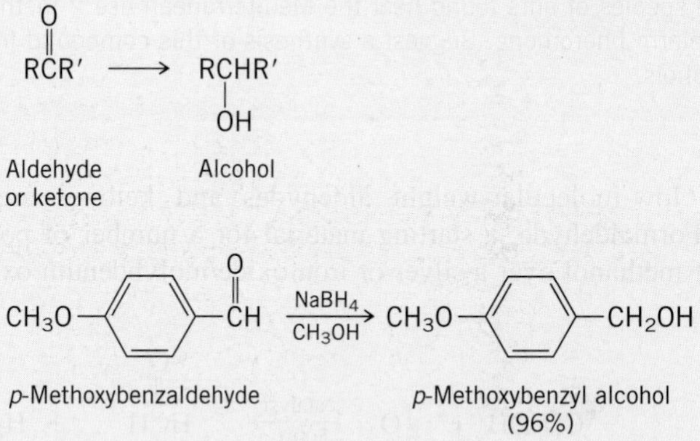
Addition of Grignard reagents and organolithium compounds (Sections 14.6–14.7) Aldehydes are converted to secondary alcohols and ketones to tertiary alcohols.	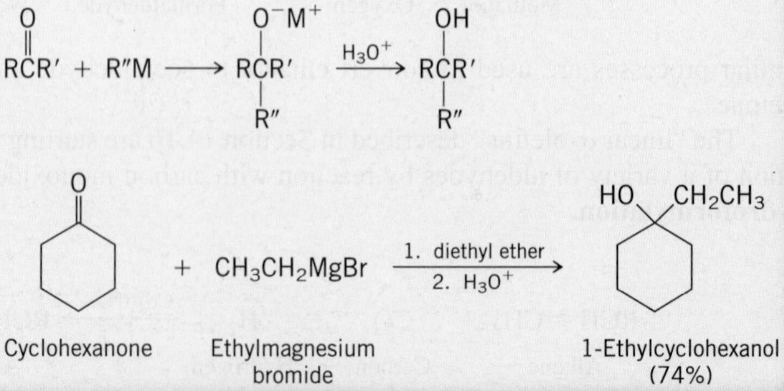

The most important chemical property of the carbonyl group is its tendency to undergo **nucleophilic addition** reactions of the type represented in the general equation:

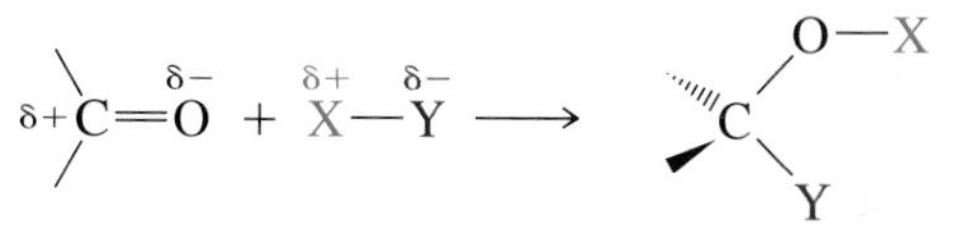

Aldehyde or ketone — Product of nucleophilic addition

A negatively polarized atom or group bonds to the positively polarized carbon of the carbonyl group in the rate-determining step of these reactions. Grignard reagents, organolithium reagents, lithium aluminum hydride, and sodium borohydride, for example, all react with carbonyl compounds by nucleophilic addition.

The next section explores the mechanism of nucleophilic addition to aldehydes and ketones. There we'll discuss their *hydration,* a reaction in which water adds to the C═O group. After we use this reaction to develop some general principles, we'll then survey a number of related reactions of synthetic, mechanistic, or biological interest.

17.6 Principles of Nucleophilic Addition: Hydration of Aldehydes and Ketones

Effects of Structure on Equilibrium: Aldehydes and ketones react with water in a rapid equilibrium. The product is a **geminal diol,** also called a hydrate.

The convention for writing equilibrium constant expressions without the solvent (water in this case) was discussed in Section 1.14.

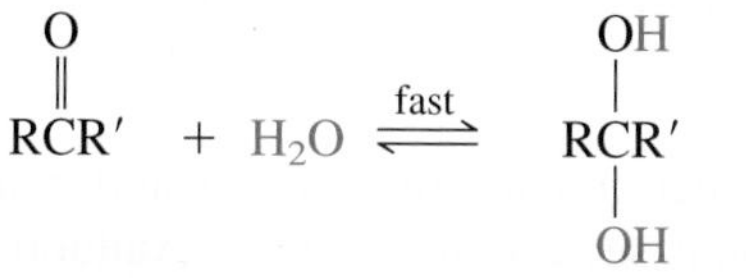

Aldehyde or ketone — Water — Geminal diol (hydrate)

$$K_{hydr} = \frac{[\text{hydrate}]}{[\text{carbonyl compound}]}$$

Overall, the reaction is classified as an *addition.* Water adds to the carbonyl group. Hydrogen becomes bonded to the negatively polarized carbonyl oxygen, hydroxyl to the positively polarized carbon.

Table 17.3 compares the equilibrium constants K_{hydr} for hydration of some simple aldehydes and ketones. The position of equilibrium depends on what groups are attached

TABLE 17.3 Equilibrium Constants (K_{hydr}) and Relative Rates of Hydration of Some Aldehydes and Ketones

Carbonyl compound	Hydrate	K_{hydr}*	Percent conversion to hydrate	Relative rate†
O ‖ HCH	$CH_2(OH)_2$	2300	>99.9	2200
O ‖ CH_3CH	$CH_3CH(OH)_2$	1.0	50	1.0
O ‖ $(CH_3)_3CCH$	$(CH_3)_3CCH(OH)_2$	0.2	17	0.09
O ‖ CH_3CCH_3	$(CH_3)_2C(OH)_2$	0.0014	0.14	0.0018

*$K_{hydr} = \frac{[\text{hydrate}]}{[\text{carbonyl compound}]}$

†Neutral solution, 25°C

to C═O and how they affect its *steric* and *electronic* environment. Both contribute, but the electronic effect controls K_{hydr} more than the steric effect.

Consider first the electronic effect of alkyl groups versus hydrogen atoms attached to C═O. Alkyl substituents stabilize C═O, making a ketone carbonyl more stable than an aldehyde carbonyl. As with all equilibria, factors that stabilize the reactants decrease the equilibrium constant. Thus, the extent of hydration decreases as the number of alkyl groups on the carbonyl increase.

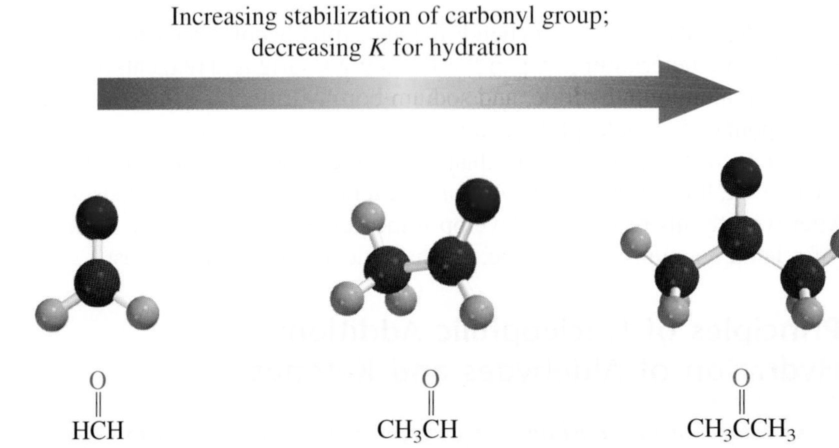
Increasing stabilization of carbonyl group;
decreasing K for hydration
O ‖ HCH
O ‖ CH_3CH
O ‖ CH_3CCH_3

Formaldehyde (almost completely hydrated in water) | Acetaldehyde (comparable amounts of aldehyde and hydrate present in water) | Acetone (hardly any hydrate present in water)

A striking example of an electronic effect on carbonyl group stability and its relation to the equilibrium constant for hydration is seen in the case of hexafluoroacetone. In contrast to the almost negligible hydration of acetone, hexafluoroacetone is completely hydrated.

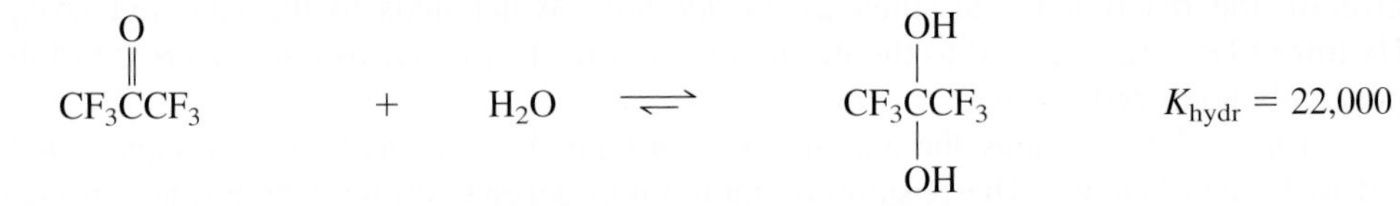
O ‖ CF_3CCF_3
+
H_2O
OH | CF_3CCF_3 | OH
$K_{hydr} = 22{,}000$

Hexafluoroacetone | Water | 1,1,1,3,3,3-Hexafluoro-2,2-propanediol

Instead of stabilizing the carbonyl group by electron donation as alkyl substituents do, trifluoromethyl groups destabilize it by withdrawing electrons. A less stabilized carbonyl group is associated with a greater equilibrium constant for addition.

Problem 17.6

Chloral is one of the common names for trichloroethanal. Its hydrate has featured prominently in countless detective stories as the notorious "Mickey Finn" knockout drops. Write a structural formula for chloral hydrate.

Now let's turn our attention to steric effects by looking at how the size of the groups that were attached to C═O affect K_{hydr}. The bond angles at carbon shrink from $\approx 120°$ to $\approx 109.5°$ as the hybridization changes from sp^2 in the reactant (aldehyde or ketone) to sp^3 in the product (hydrate). The increased crowding this produces in the hydrate is better tolerated, and K_{hydr} is greater when the groups are small (hydrogen) than when they are large (alkyl).

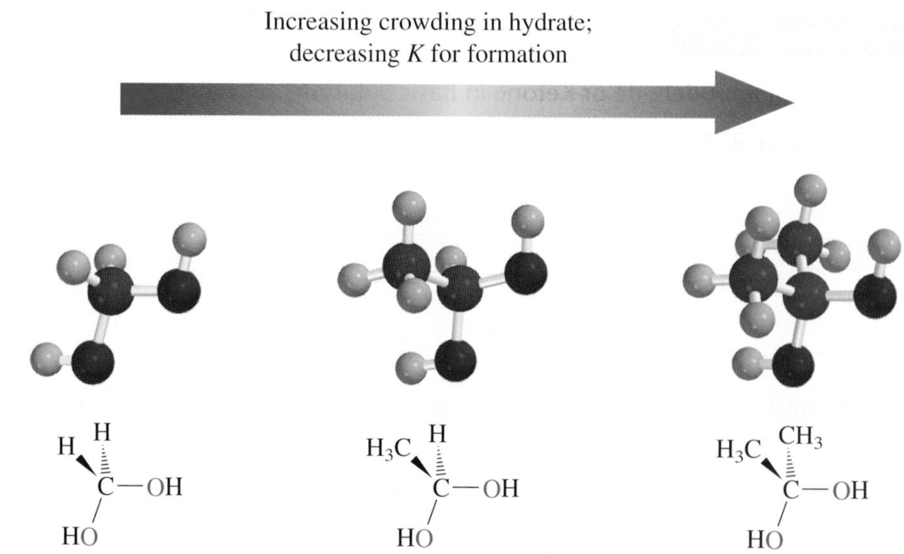
Increasing crowding in hydrate;
decreasing K for formation
H
H
C—OH
HO
H₃C
H
C—OH
HO
H₃C
CH₃
C—OH
HO

Hydrate of formaldehyde   Hydrate of acetaldehyde   Hydrate of acetone

Electronic and steric effects operate in the same direction. Both cause the equilibrium constants for hydration of aldehydes to be greater than those of ketones.

Effects of Structure on Rate: Electronic and steric effects influence the rate of hydration in the same way that they affect equilibrium. Indeed, the rate and equilibrium data of Table 17.3 parallel each other almost exactly.

Hydration of aldehydes and ketones is a rapid reaction, quickly reaching equilibrium, but faster in acid or base than in neutral solution. Thus, instead of a single mechanism for hydration, we'll look at two mechanisms, one for basic and the other for acidic solution.

Mechanism of Base-Catalyzed Hydration: The base-catalyzed mechanism (Mechanism 17.1) is a two-step process in which the first step is rate-determining. In step 1, the nucleophilic hydroxide ion bonds to the carbon of the carbonyl group. The alkoxide ion formed in step 1 abstracts a proton from water in step 2, yielding the geminal diol. The second step, like all other proton transfers between oxygen that we have seen, is fast.

The role of the basic catalyst (HO^-) is to increase the rate of the nucleophilic addition step. Hydroxide ion, the nucleophile in the base-catalyzed reaction, is much more reactive than a water molecule, the nucleophile in neutral solutions.

Aldehydes react faster than ketones for almost the same reasons that their equilibrium constants for hydration are more favorable. The $sp^2 \rightarrow sp^3$ hybridization change that the carbonyl carbon undergoes on hydration is partially developed in the transition state for the rate-determining nucleophilic addition step (Figure 17.5). Alkyl groups at the reaction site increase the activation energy by simultaneously lowering the energy of the starting state (ketones have a more stabilized carbonyl group than aldehydes) and raising the energy of the transition state (a steric crowding effect).

Mechanism of Acid-Catalyzed Hydration: Three steps are involved in acid-catalyzed hydration (Mechanism 17.2 on page 739). The first and last are rapid proton transfers between oxygens. The second is a nucleophilic addition. The acid catalyst activates the carbonyl group toward attack by a weakly nucleophilic water molecule. Protonation of oxygen makes the carbonyl carbon of an aldehyde or a ketone much more electrophilic. Expressed in resonance terms, the protonated carbonyl has a greater degree of carbocation character than an unprotonated carbonyl.

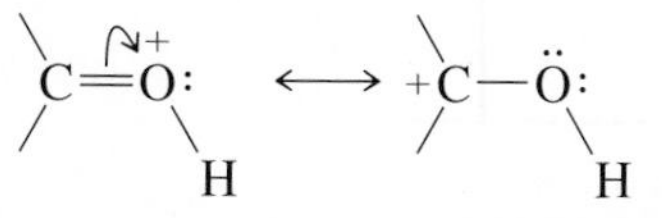

Mechanism 17.1

Hydration of an Aldehyde or Ketone in Basic Solution

THE OVERALL REACTION:

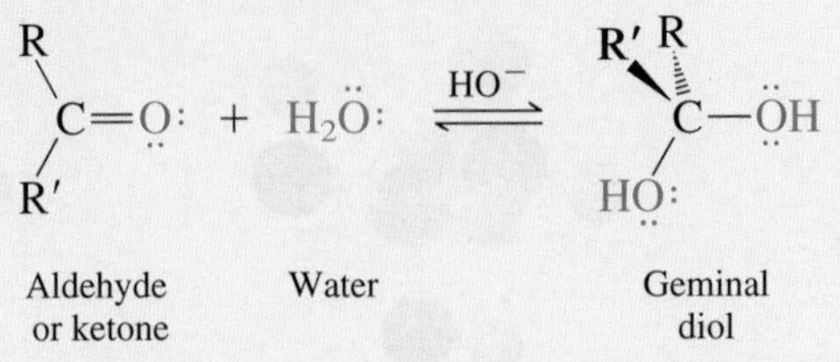

Step 1: Nucleophilic addition of hydroxide ion to the carbonyl group

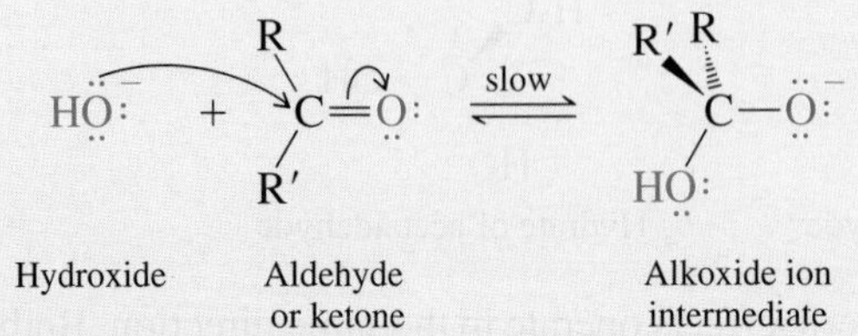

Step 2: Proton transfer from water to the intermediate formed in step 1

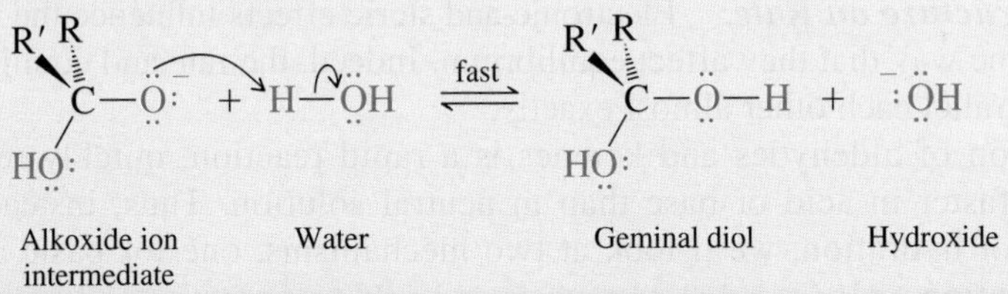

Steric and electronic effects influence the rate of nucleophilic addition to a protonated carbonyl group in much the same way as they do for the case of a neutral one, and protonated aldehydes react faster than protonated ketones.

With this as background, let us now examine how the principles of nucleophilic addition apply to the characteristic reactions of aldehydes and ketones. We'll begin with the addition of hydrogen cyanide.

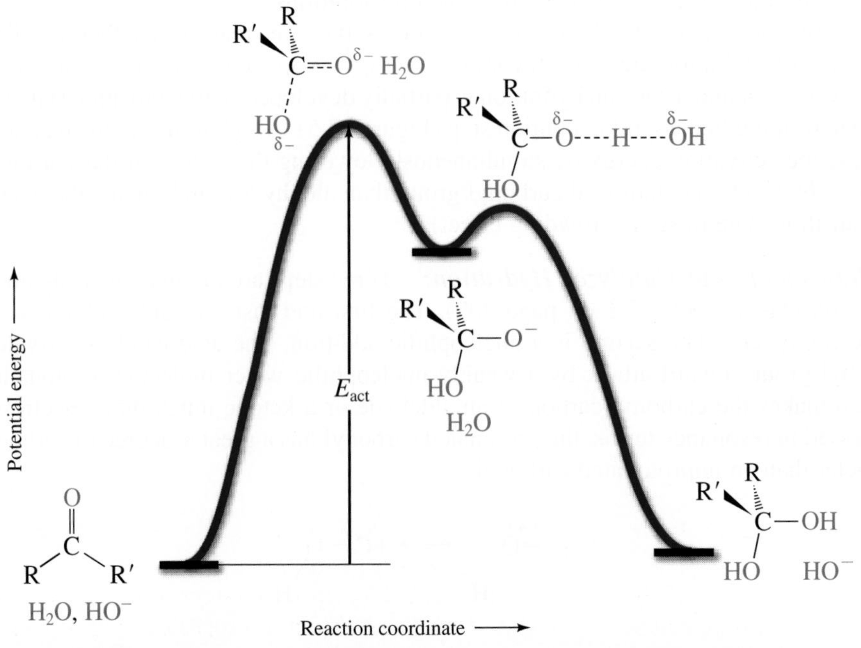

Figure 17.5

Potential energy diagram for base-catalyzed hydration of an aldehyde or ketone.

Mechanism 17.2

Hydration of an Aldehyde or Ketone in Acid Solution

THE OVERALL REACTION:

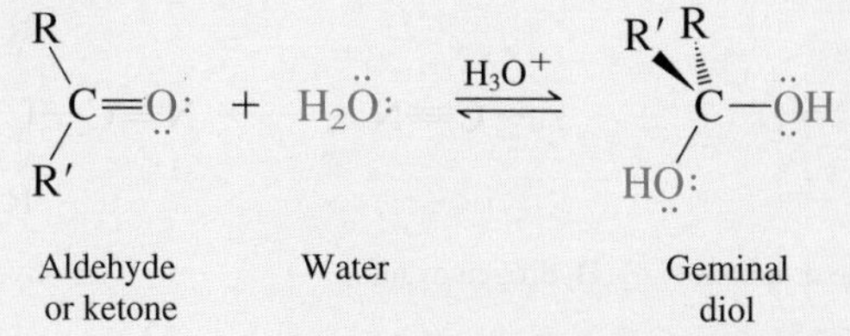

Aldehyde or ketone
Water
Geminal diol

Step 1: Protonation of the carbonyl oxygen

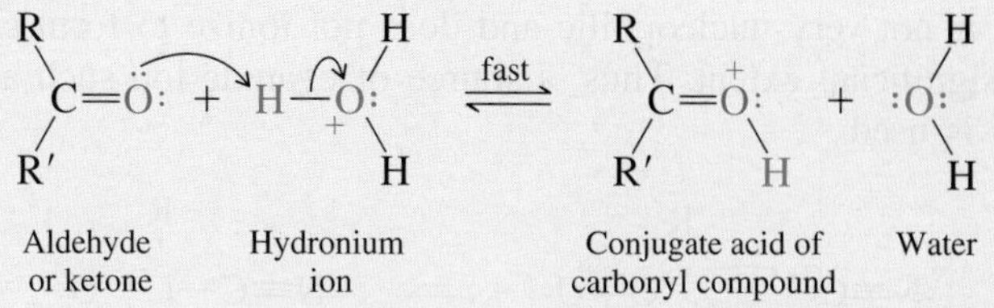

fast
Aldehyde or ketone
Hydronium ion
Conjugate acid of carbonyl compound
Water

Step 2: Nucleophilic addition to the protonated aldehyde or ketone

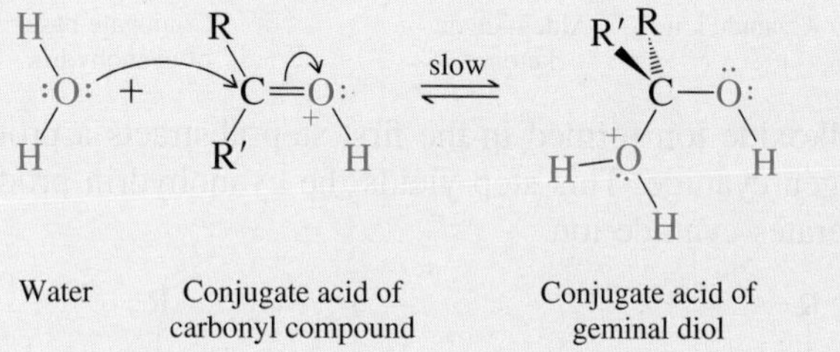

slow
Water
Conjugate acid of carbonyl compound
Conjugate acid of geminal diol

Step 3: Proton transfer from the conjugate acid of the geminal diol to a water molecule

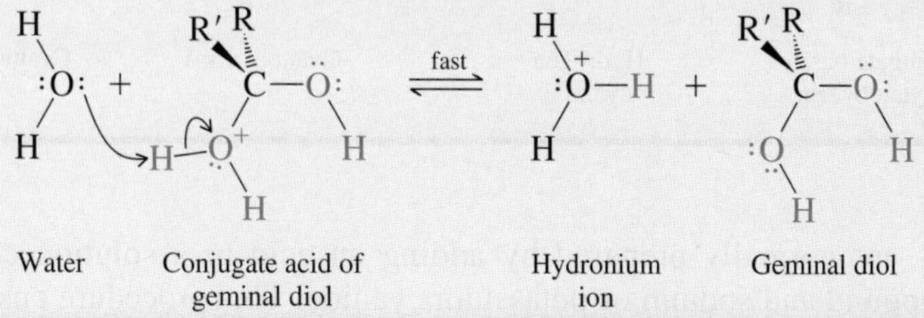

fast
Water
Conjugate acid of geminal diol
Hydronium ion
Geminal diol

17.7 Cyanohydrin Formation

The product of addition of hydrogen cyanide to an aldehyde or a ketone contains both a hydroxyl group and a cyano group bonded to the same carbon. Compounds of this type are called **cyanohydrins.**

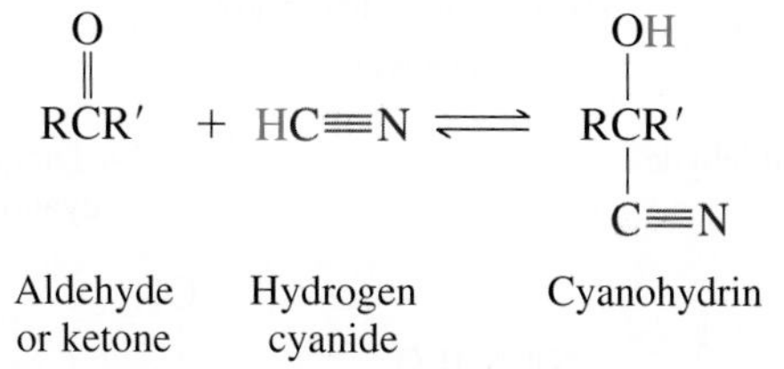

Aldehyde or ketone
Hydrogen cyanide
Cyanohydrin

Mechanism 17.3 describing cyanohydrin formation is analogous to the mechanism of base-catalyzed hydration. The nucleophile (cyanide ion) bonds to the carbonyl carbon in the first step of the reaction, followed by proton transfer to the carbonyl oxygen in the second step.

The addition of hydrogen cyanide is catalyzed by cyanide ion, but HCN is too weak an acid to provide enough $:\bar{C}{\equiv}N:$ for the reaction to proceed at a reasonable rate.

Mechanism 17.3

Cyanohydrin Formation

THE OVERALL REACTION:

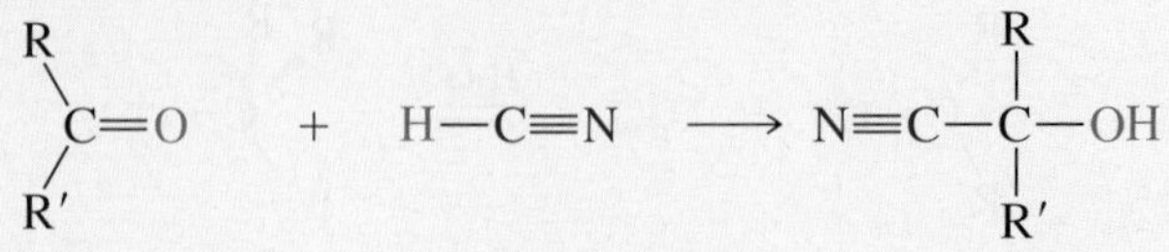

Aldehyde or ketone   Hydrogen cyanide   Cyanohydrin

Step 1: The negatively charged carbon of cyanide ion is nucleophilic and bonds to the carbonyl carbon of the aldehyde or ketone. Hydrogen cyanide itself is not very nucleophilic and does not ionize to form cyanide ion to a significant extent. Thus, a source of cyanide ion such as NaCN or KCN is used.

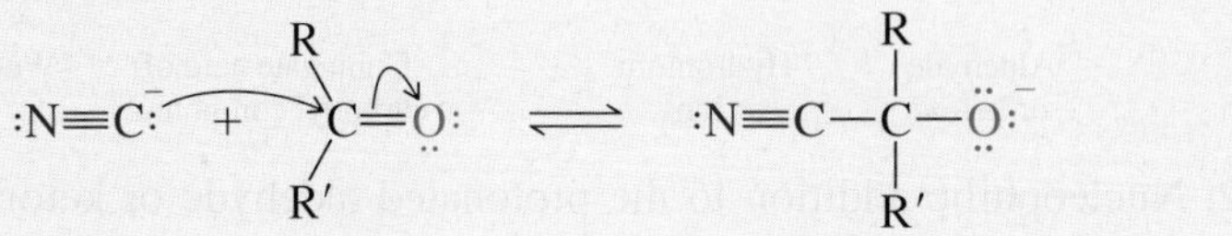

Cyanide ion   Aldehyde or ketone   Conjugate base of cyanohydrin

Step 2: The alkoxide ion formed in the first step abstracts a proton from hydrogen cyanide. This step yields the cyanohydrin product and regenerates cyanide ion.

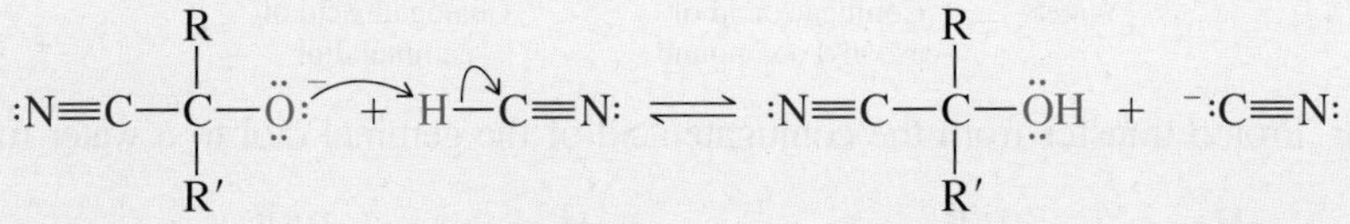

Conjugate base of cyanohydrin   Hydrogen cyanide   Cyanohydrin   Cyanide ion

Equilibrium Constants for Cyanohydrin Formation

$R\overset{O}{\overset{\|}{C}}R'$	K_{eq}
$CH_3\overset{O}{\overset{\|}{C}}H$	1.8×10^4
$CH_3\overset{O}{\overset{\|}{C}}CH_3$	14
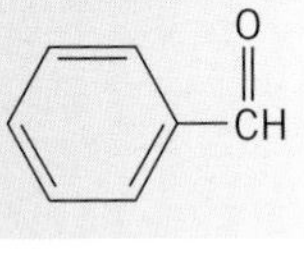	194

Cyanohydrins are normally prepared by adding an acid to a solution containing the carbonyl compound and sodium or potassium cyanide. This procedure ensures that free cyanide ion is always present in amounts sufficient to increase the rate of the reaction.

Cyanohydrin formation is reversible, and the position of equilibrium depends on the steric and electronic factors governing nucleophilic addition to carbonyl groups described in the preceding section. Aldehydes and unhindered ketones give good yields of cyanohydrins.

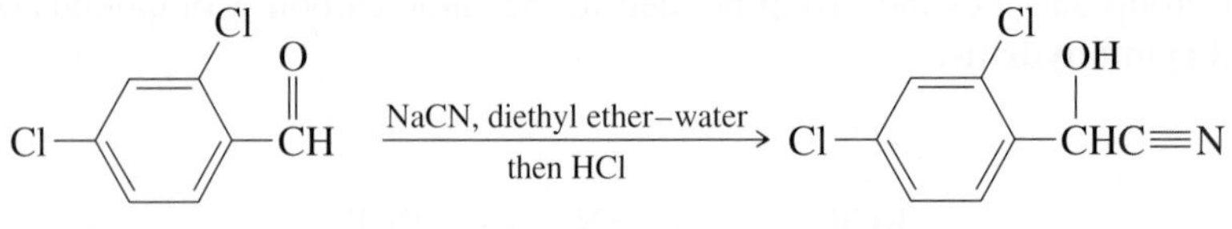

2,4-Dichlorobenzaldehyde   2,4-Dichlorobenzaldehyde cyanohydrin (100%)

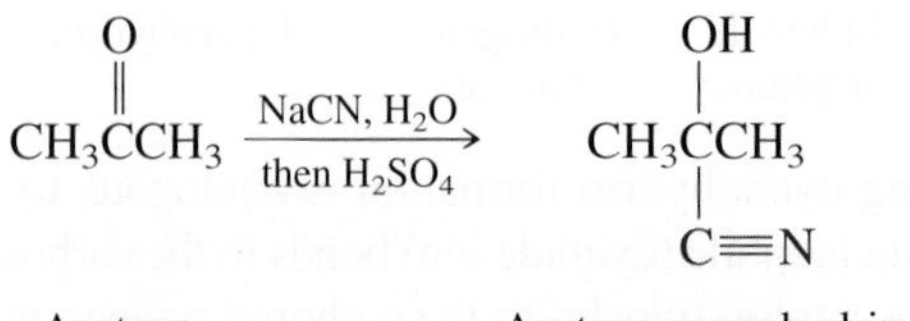

Acetone   Acetone cyanohydrin (77–78%)

In substitutive IUPAC nomenclature, cyanohydrins are named as hydroxy derivatives of nitriles. Because nitrile nomenclature will not be discussed until Section 19.1, we will refer to cyanohydrins as derivatives of the parent aldehyde or ketone as shown in the examples. This conforms to the practice of most chemists.

Problem 17.7

Cyanohydrin formation is reversible in base. Using sodium hydroxide as the base, use curved arrows to show the elimination of HCN from the cyanohydrin product in the presence of sodium hydroxide in step 2 in Mechanism 17.3.

Converting aldehydes and ketones to cyanohydrins is of synthetic value because:

1. A new carbon–carbon bond is formed.
2. The —C≡N group can be converted to $-\overset{O}{\overset{\|}{C}}OH$ (Section 18.12) and $-CH_2NH_2$ (Section 21.9).
3. The —OH group can undergo functional group transformations.

Problem 17.8

Methacrylonitrile is an industrial chemical used in the production of plastics and fibers. One method for its preparation is the acid-catalyzed dehydration of acetone cyanohydrin. Deduce the structure of *methacrylonitrile.*

Cyanohydrins occur naturally, often as derivatives in which the —OH group has been modified to —OR, where R is a carbohydrate unit. The compounds, called *cyanogenic glycosides,* are widespread in plants. Amygdalin, for example, is found in bitter almonds and in the kernels of peaches, plums, apricots, and related fruits. Its structure and behavior on hydrolysis are shown in Figure 17.6. Enzyme-catalyzed hydrolysis of amygdalin gives benzaldehyde cyanohydrin, which dissociates to benzaldehyde and hydrogen cyanide. Depending on the amount present and the manner in which food is prepared from plants containing cyanogenic glycosides, toxic levels of hydrogen cyanide can result.

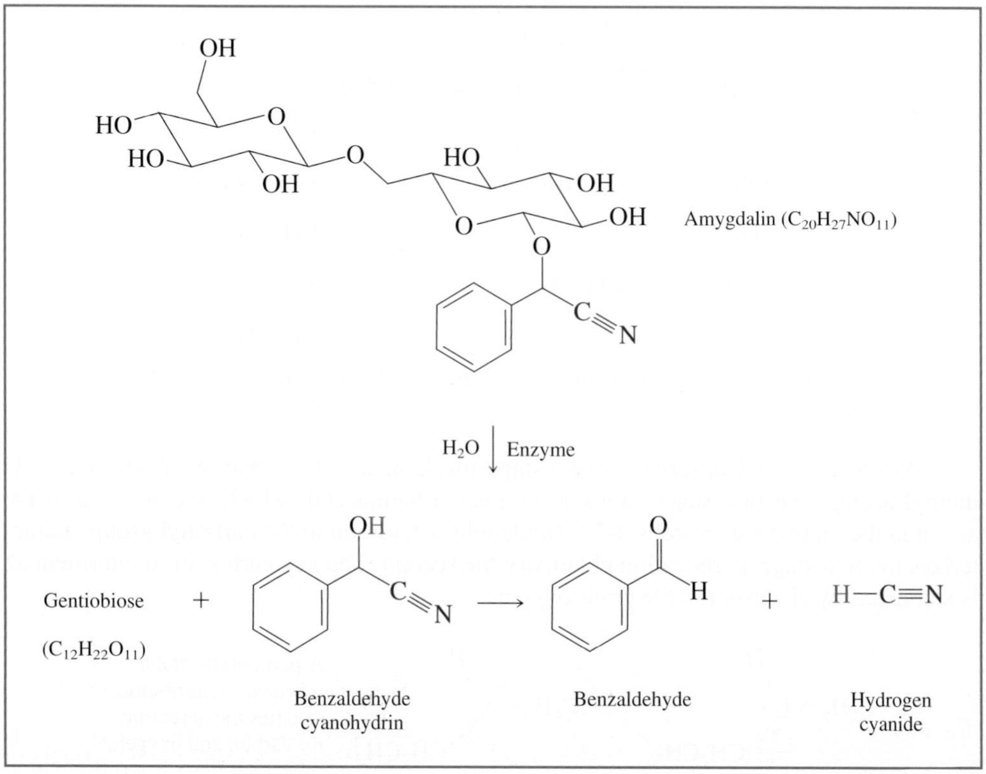
OH
HO
O
HO
O
OH
HO
OH
OH
O
O
C
N
Amygdalin ($C_{20}H_{27}NO_{11}$)
H_2O
Enzyme
Gentiobiose
($C_{12}H_{22}O_{11}$)
+
OH
C
N
O
H
+
H—C≡N
Benzaldehyde cyanohydrin
Benzaldehyde
Hydrogen cyanide

Figure 17.6

Hydrolysis of amygdalin gives benzaldehyde cyanohydrin, which then dissociates to give benzaldehyde and hydrogen cyanide.

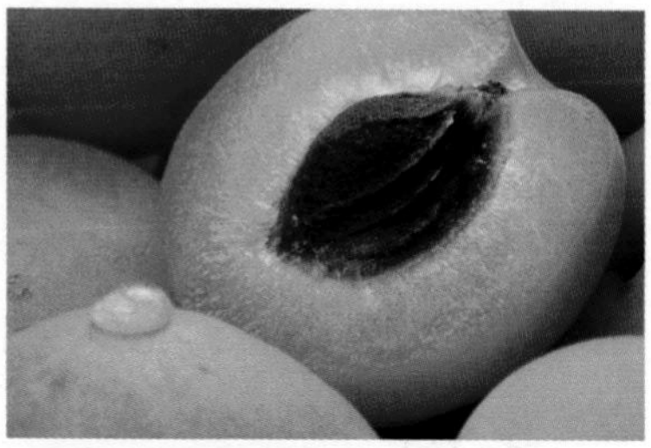

Apricot pits are the most common source of amygdalin.

Problem 17.9

Gynocardin is a naturally occurring cyanogenic glycoside having the structure shown. What cyanohydrin would you expect to be formed on hydrolysis of gynocardin, and to what ketone does this cyanohydrin correspond?

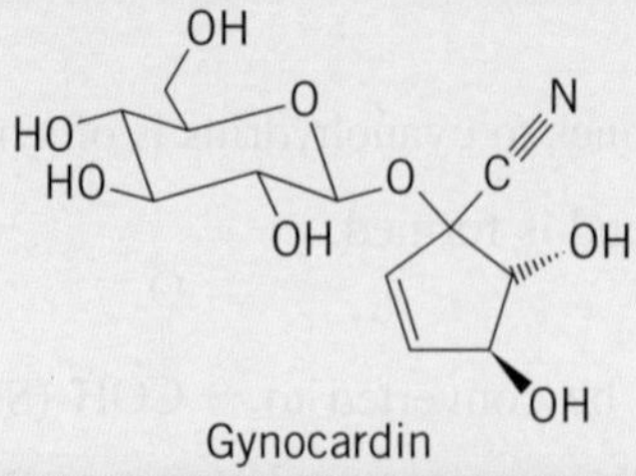
OH
HO
HO
O
O
OH
C
N
OH
OH

Gynocardin

Figure 17.7

When disturbed, many millipedes protect themselves by converting stored benzaldehyde cyanohydrin to hydrogen cyanide and benzaldehyde.

Cyanogenic compounds are not limited to plants. The defense secretion of many species of millipedes contains the products of cyanohydrin dissociation. These millipedes (Figure 17.7) store either benzaldehyde cyanohydrin or a derivative of it, and the enzyme that catalyzes its hydrolysis in separate chambers within their bodies. When the millipede is under stress, the contents of the two chambers are mixed and the hydrolysis products—including HCN—are released through the millipede's pores to deter predatory insects and birds.

17.8 Acetal Formation

Many of the most interesting and useful reactions of aldehydes and ketones involve transformation of the initial product of nucleophilic addition to some other substance under the reaction conditions. An example is the reaction of aldehydes with alcohols under conditions of acid catalysis. The expected product of nucleophilic addition of the alcohol to the carbonyl group is called a **hemiacetal.** The product actually isolated, however, corresponds to reaction of one mole of the aldehyde with *two* moles of alcohol to give *geminal diethers* known as **acetals:**

$$\mathrm{RCH{=}O} \xrightleftharpoons{\mathrm{R'OH,\ H^+}} \mathrm{RCH(OH)(OR')} \xrightleftharpoons{\mathrm{R'OH,\ H^+}} \mathrm{RCH(OR')_2} + \mathrm{H_2O}$$

Aldehyde Hemiacetal Acetal Water

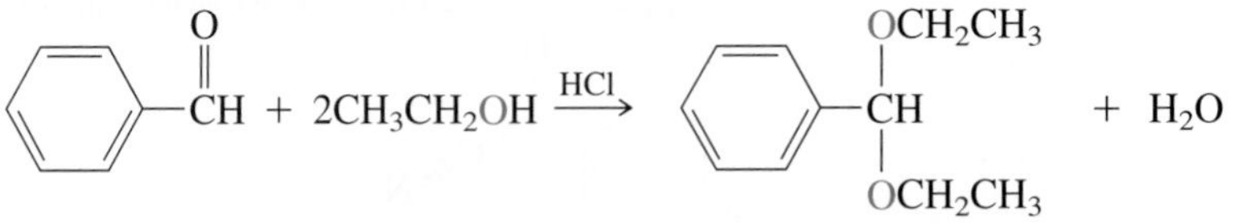
O
CH + 2CH3CH2OH
HCl
OCH2CH3
CH
OCH2CH3
+ H2O

Benzaldehyde Ethanol Benzaldehyde diethyl acetal (66%) Water

Mechanism 17.4 describes a two-stage mechanism for formation of benzaldehyde diethyl acetal. The first stage (steps 1–3) gives a hemiacetal, which is converted to the acetal in the second stage (steps 4–7). Nucleophilic addition to the carbonyl group characterizes the first stage, carbocation chemistry the second. The key carbocation intermediate is stabilized by electron release from oxygen.

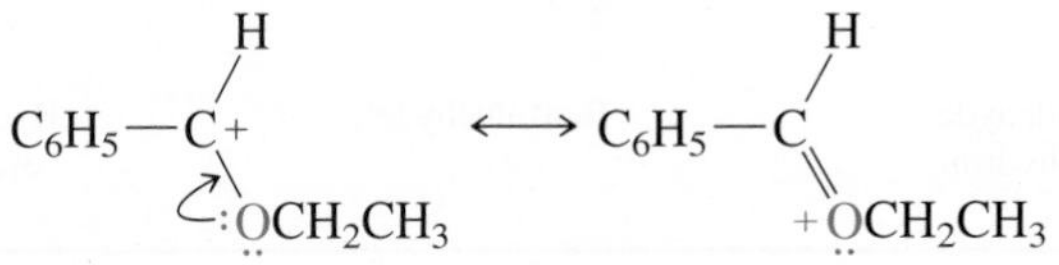
H
C6H5—C+
:OCH2CH3
H
C6H5—C
+OCH2CH3

A particularly stable resonance contributor; satisfies the octet rule for carbon and oxygen.

Mechanism 17.4

Acetal Formation from Benzaldehyde and Ethanol

THE OVERALL REACTION:

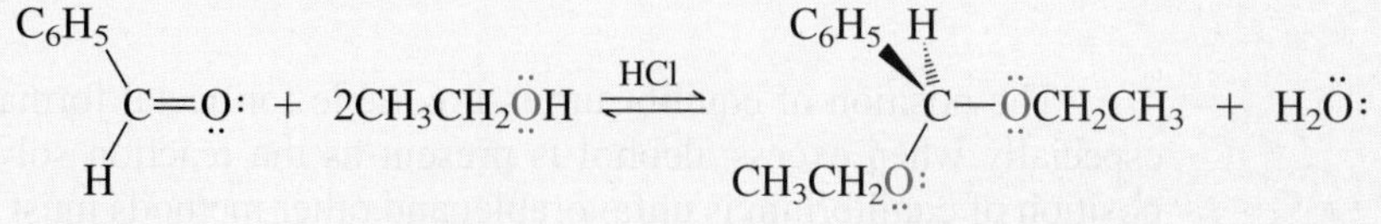

Benzaldehyde Ethanol Benzaldehyde diethyl acetal Water

Steps 1–3: Acid-catalyzed nucleophilic addition of 1 mole of ethanol to the carbonyl group. The details of these three steps are analogous to the three steps of acid-catalyzed hydration in Mechanism 17.2. The product of these three steps is a hemiacetal.

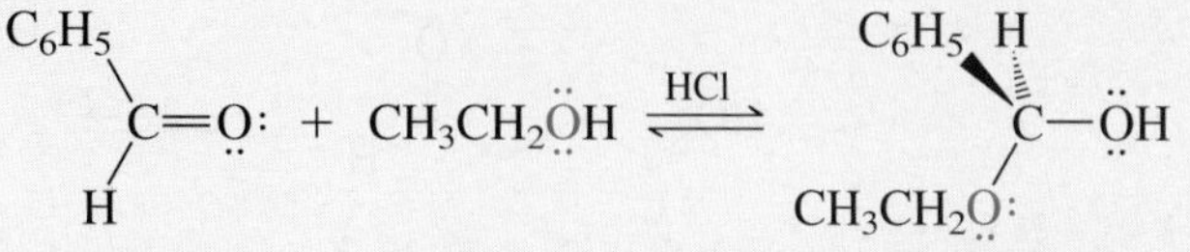

Benzaldehyde Ethanol Benzaldehyde ethyl hemiacetal

Step 4: Steps 4 and 5 are analogous to the two steps in the formation of carbocations in acid-catalyzed reactions of alcohols. Step 4 is proton-transfer from hydronium ion to the hydroxyl oxygen of the hemiacetal.

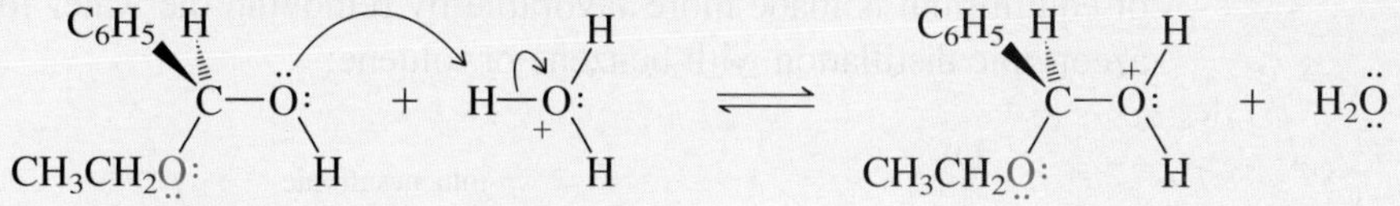

Benzaldehyde ethyl hemiacetal Hydronium ion Conjugate acid of benzaldehyde ethyl hemiacetal Water

Step 5: Loss of water from the protonated hemiacetal gives an oxygen-stabilized carbocation. Of the resonance structures shown, the more stable contributor satisfies the octet rule for both carbon and oxygen.

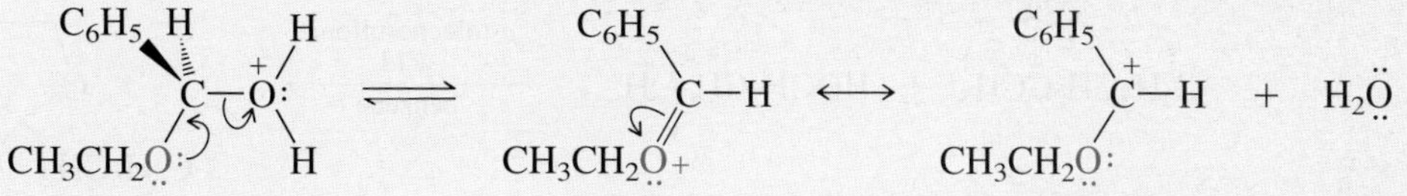

Conjugate acid of benzaldehyde ethyl hemiacetal More stable contributor Less stable contributor Water

Step 6: Nucleophilic attack of ethanol on the oxygen-stabilized carbocation

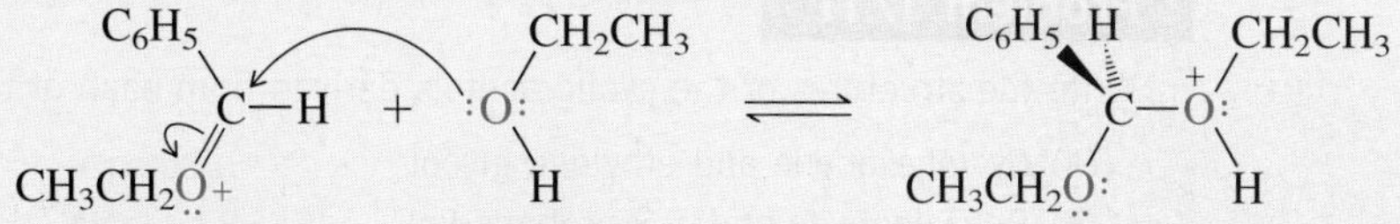

Oxygen-stabilized carbocation Ethanol Conjugate acid of acetal

Step 7: Proton transfer from the conjugate acid of the product to ethanol

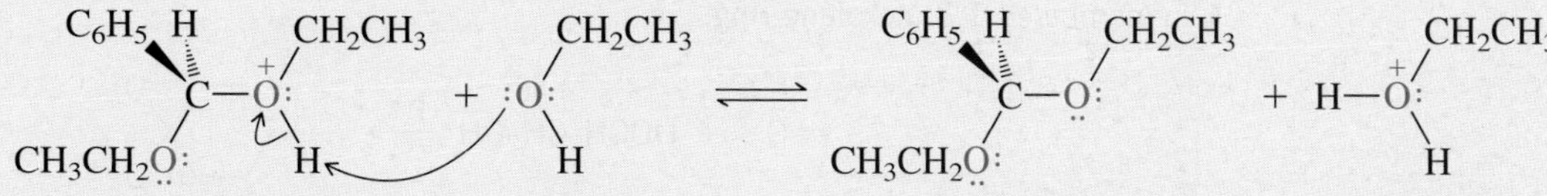

Conjugate acid of acetal Ethanol Benzaldehyde diethyl acetal Conjugate acid of ethanol

Problem 17.10

Be sure you fully understand Mechanism 17.4 by writing equations for steps 1–3. Use curved arrows to show electron flow.

The position of equilibrium is favorable for acetal formation from most aldehydes, especially when excess alcohol is present as the reaction solvent. For most ketones the position of equilibrium is unfavorable, and other methods must be used for the preparation of acetals from ketones.

Ketal is an acceptable term for acetals formed from ketones. It was dropped from IUPAC nomenclature, but continued to be so widely used that it was reinstated.

Compounds that contain both carbonyl and alcohol functional groups are often more stable as cyclic hemiacetals or acetals than as open-chain structures. An equilibrium mixture of 4-hydroxybutanal contains 11.4% of the open-chain hydroxy aldehyde and 88.6% of the cyclic hemiacetal.

HO
O
H
H
O
OH

4-Hydroxybutanal Tetrahydrofuran-2-ol

Diols that bear two hydroxyl groups in a 1,2 or 1,3 relationship to each other yield *cyclic acetals* with either aldehydes or ketones. The five-membered cyclic acetals derived from ethylene glycol are the most commonly encountered examples. Often the position of equilibrium is made more favorable by removing the water formed in the reaction by azeotropic distillation with benzene or toluene:

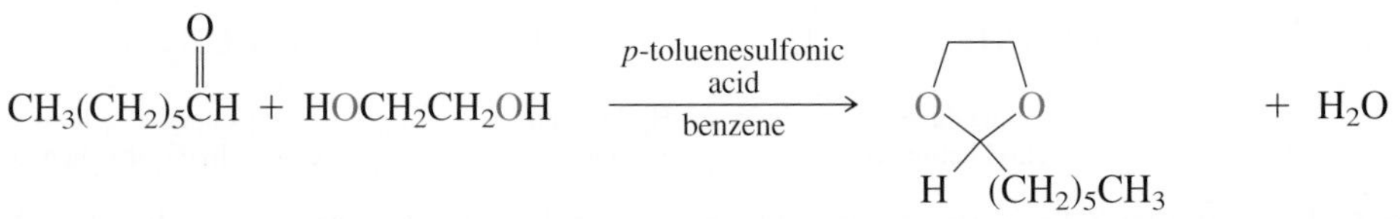
O
CH3(CH2)5CH + HOCH2CH2OH
p-toluenesulfonic acid
benzene
O O
H (CH2)5CH3
+ H2O

Heptanal Ethylene glycol 2-Hexyl-1,3-dioxolane (81%) Water

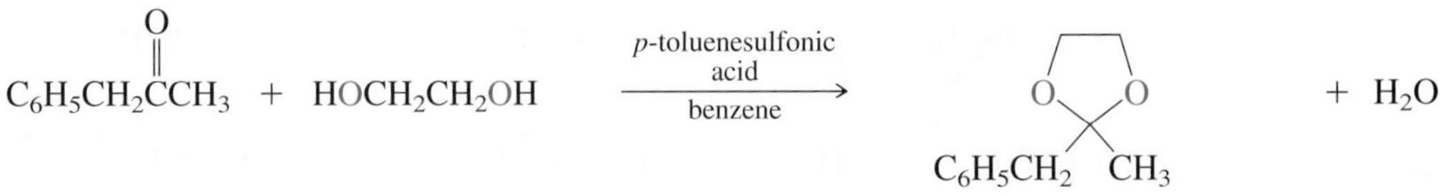
O
C6H5CH2CCH3 + HOCH2CH2OH
p-toluenesulfonic acid
benzene
O O
C6H5CH2 CH3
+ H2O

Benzyl methyl ketone Ethylene glycol 2-Benzyl-2-methyl-1,3-dioxolane (78%) Water

Problem 17.11

Write the structures of the cyclic acetals derived from each of the following.

(a) Cyclohexanone and ethylene glycol
(b) Benzaldehyde and 1,3-propanediol
(c) Isobutyl methyl ketone and ethylene glycol
(d) Isobutyl methyl ketone and 2,2-dimethyl-1,3-propanediol

Sample Solution (a) The cyclic acetals derived from ethylene glycol contain a five-membered 1,3-dioxolane ring.

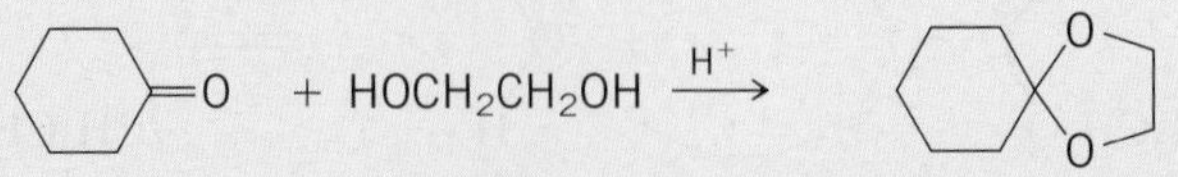
O
+ HOCH2CH2OH
H+
O
O

Cyclohexanone Ethylene glycol Acetal of cyclohexanone and ethylene glycol

Acetals are susceptible to hydrolysis in aqueous acid:

$$\underset{\text{Acetal}}{RC(OR'')_2R'} + \underset{\text{Water}}{H_2O} \overset{H^+}{\rightleftharpoons} \underset{\text{Aldehyde or ketone}}{RC(=O)R'} + \underset{\text{Alcohol}}{2R''OH}$$

This reaction is simply the reverse of the reaction by which acetals are formed—acetal formation is favored by excess alcohol, acetal hydrolysis by excess water. Acetal formation and acetal hydrolysis share the same mechanistic pathway but travel along that pathway in opposite directions. In the following section you'll see a clever way in which acetal formation and hydrolysis have been applied to synthetic organic chemistry.

Problem 17.12

Problem 17.10 asked you to write details of the mechanism describing formation of benzaldehyde diethyl acetal from benzaldehyde and ethanol. Write a stepwise mechanism for the acid hydrolysis of this acetal.

17.9 Acetals as Protecting Groups

In an organic synthesis, it sometimes happens that one of the reactants contains a functional group that is incompatible with the reaction conditions. Consider, for example, the conversion

$$\underset{\text{5-Hexyn-2-one}}{CH_3C(=O)CH_2CH_2C{\equiv}CH} \longrightarrow \underset{\text{5-Heptyn-2-one}}{CH_3C(=O)CH_2CH_2C{\equiv}CCH_3}$$

It looks as though all that is needed is to prepare the acetylenic anion $CH_3C(=O)CH_2CH_2C{\equiv}\bar{C}{:}$, then alkylate it with methyl iodide (Section 9.6). There is a complication, however. The carbonyl group in the starting alkyne will neither tolerate the strongly basic conditions required for anion formation nor survive in a solution containing carbanions. Acetylide ions add to carbonyl groups (Section 14.8). Thus, the necessary anion is inaccessible.

The strategy that is routinely followed is to *protect* the carbonyl group during the reactions with which it is incompatible and then to *remove* the protecting group in a subsequent step. Acetals, especially those derived from ethylene glycol, are among the most useful groups for carbonyl protection, because they can be introduced and removed readily. A key fact is that acetals resemble ethers in being inert to many of the reagents, such as hydride reducing agents and organometallic compounds, that react readily with carbonyl groups. The following sequence is the one that was actually used to bring about the desired transformation.

(a) Protection of carbonyl group

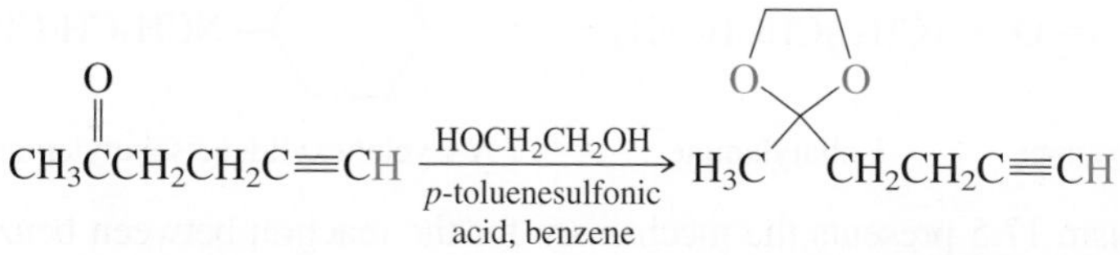

O
CH3CCH2CH2C≡CH
HOCH2CH2OH
p-toluenesulfonic acid, benzene
O O
H3C CH2CH2C≡CH

5-Hexyn-2-one — Acetal of reactant (80%)

(b) Alkylation of alkyne

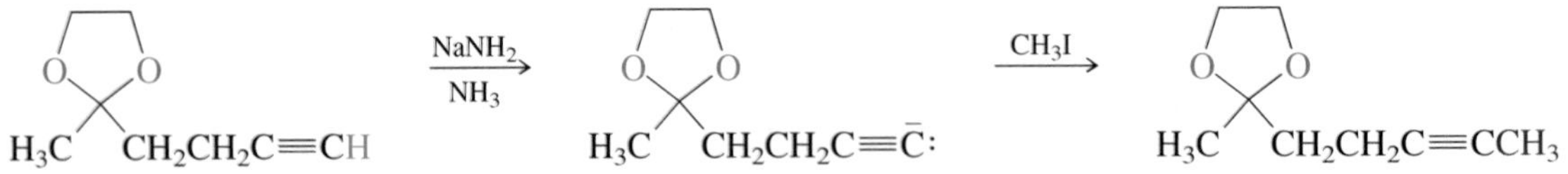

Acetal of product (78%)

(c) Removal of the protecting group by hydrolysis

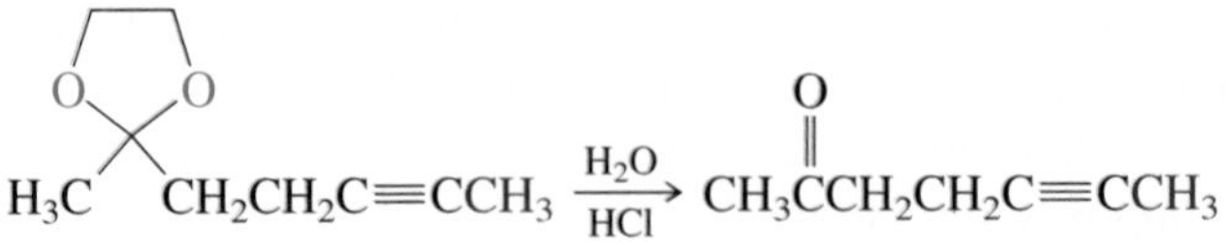

5-Heptyn-2-one (96%)

Although protecting and deprotecting the carbonyl group adds two steps to the synthetic procedure, both are essential to its success. The tactic of functional group protection is frequently encountered in preparative organic chemistry, and considerable attention has been paid to the design of effective protecting groups for a variety of functionalities.

Problem 17.13

Acetal formation is a characteristic reaction of aldehydes and ketones, but not of carboxylic acids. Show how you could advantageously use a cyclic acetal protecting group in the following synthesis:

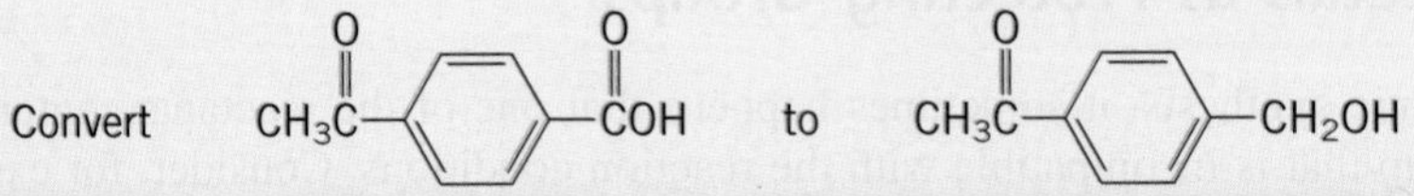

17.10 Reaction with Primary Amines: Imines

Like acetal formation, the reaction of aldehydes and ketones with primary amines—compounds of the type RNH_2 and $ArNH_2$—is a two-stage process. Its first stage is nucleophilic addition of the amine to the carbonyl group to give a **carbinolamine.** The second stage is the dehydration of the carbinolamine to yield the isolated product of the reaction, an *N*-alkyl or *N*-aryl-substituted **imine.**

N-substituted imines are sometimes called **Schiff's bases,** after Hugo Schiff, a German chemist who described their formation in 1864.

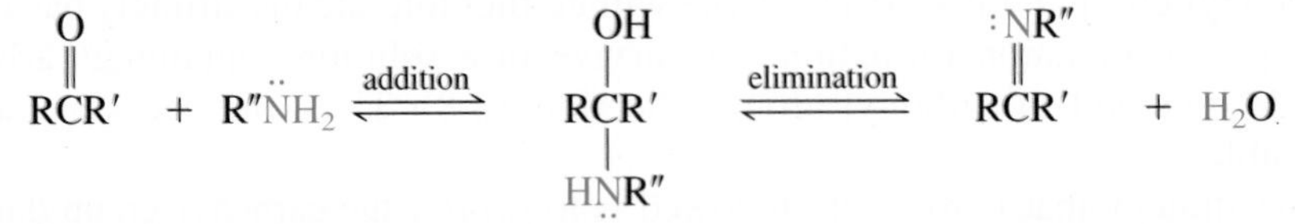

Aldehyde or ketone | Primary amine | Carbinolamine | *N*-substituted imine | Water

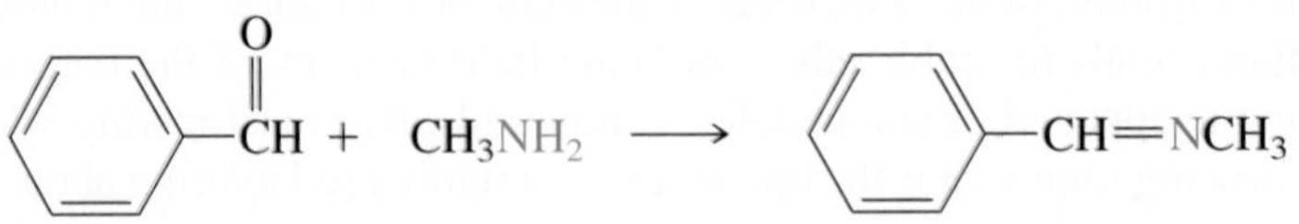

Benzaldehyde | Methylamine | *N*-Benzylidenemethylamine (70%)

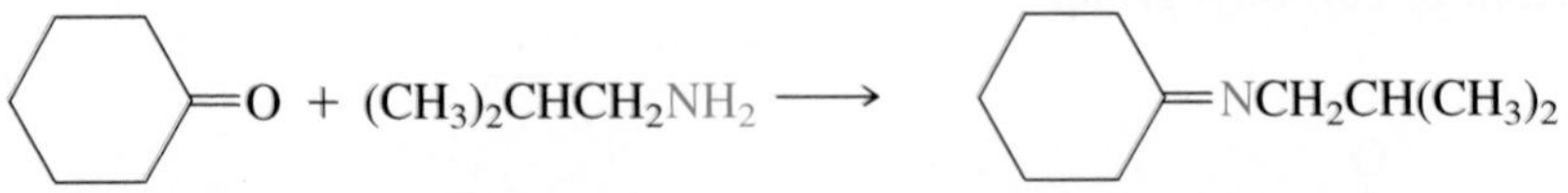

Cyclohexanone | Isobutylamine | *N*-Cyclohexylideneisobutylamine (79%)

Mechanism 17.5 presents the mechanism for the reaction between benzaldehyde and methylamine given in the first example. The first two steps lead to the carbinolamine; the

Mechanism 17.5

Imine Formation from Benzaldehyde and Methylamine

THE OVERALL REACTION:

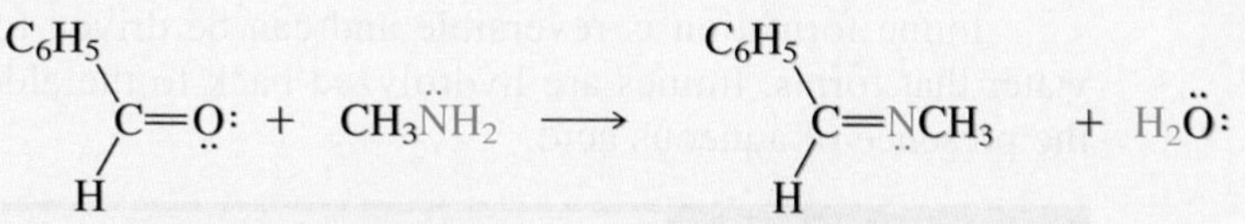

Benzaldehyde   Methylamine   *N*-Benzylidenemethylamine   Water

Step 1: The amine acts as a nucleophile, adding to the carbonyl group and forming a C—N bond.

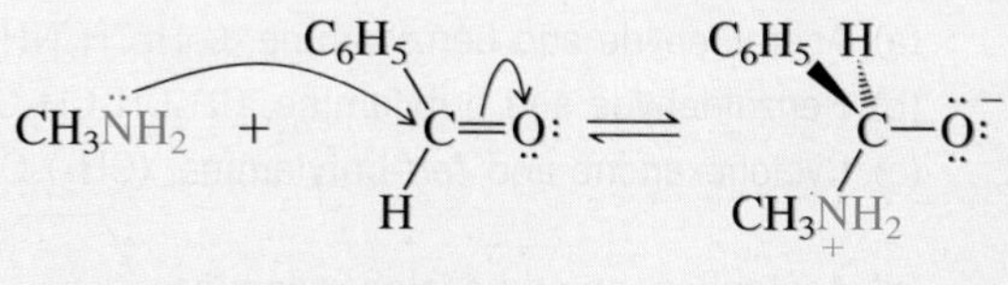

Methylamine   Benzaldehyde   First intermediate

Step 2: In a solvent such as water, proton transfers give the carbinolamine.

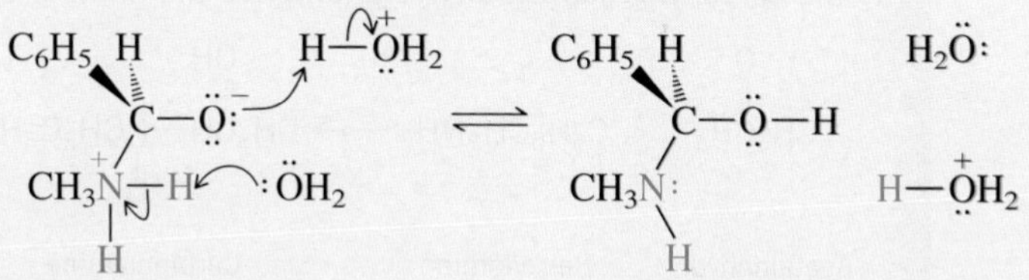

First intermediate   Carbinolamine

Step 3: The dehydration stage begins with protonation of the carbinolamine on oxygen.

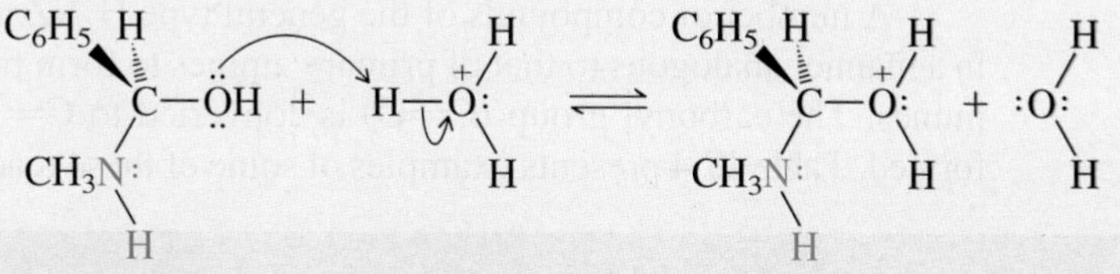

Carbinolamine intermediate   Hydronium ion   *O*-Protonated carbinolamine   Water

Step 4: The oxygen-protonated carbinolamine loses water to give a nitrogen-stabilized carbocation.

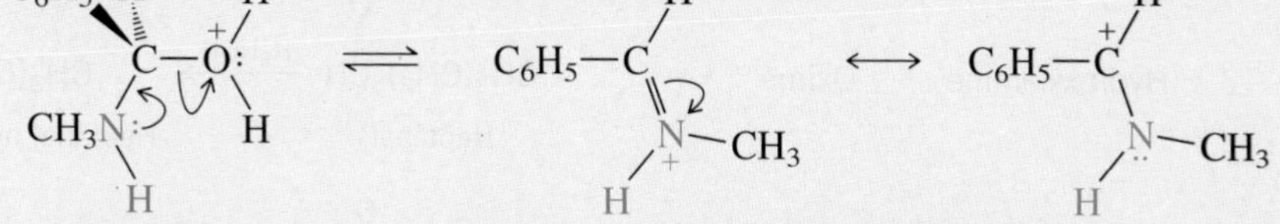

O-Protonated carbinolamine   Nitrogen-stabilized carbocation

Step 5: The nitrogen-stabilized carbocation is the conjugate acid of the imine. Proton transfer to water gives the imine.

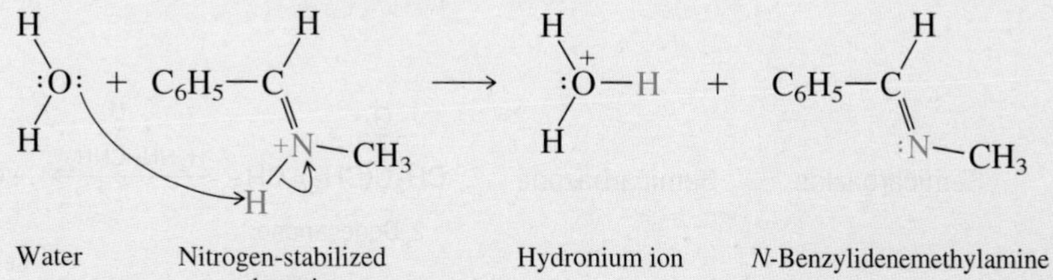

Water   Nitrogen-stabilized carbocation   Hydronium ion   *N*-Benzylidenemethylamine

last three show the dehydration of the carbinolamine to the imine. Step 4, the key step in the dehydration phase, is rate-determining when the reaction is carried out in acid solution. If the solution is too acidic, however, protonation of the amine blocks step 1. Therefore there is some optimum pH, usually about 5, at which the reaction rate is a maximum. Too basic a solution reduces the rate of step 4; too acidic a solution reduces the rate of step 1.

Imine formation is reversible and can be driven to completion by removing the water that forms. Imines are hydrolyzed back to the aldehyde or ketone and amine in the presence of aqueous acid.

Problem 17.14

Write the structure of the carbinolamine intermediate and the imine product formed in the reaction of each of the following:

(a) Acetaldehyde and benzylamine, $C_6H_5CH_2NH_2$
(b) Benzaldehyde and butylamine, $CH_3CH_2CH_2CH_2NH_2$
(c) Cyclohexanone and *tert*-butylamine, $(CH_3)_3CNH_2$
(d) Acetophenone and cyclohexylamine, 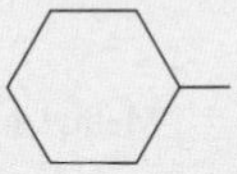 $-NH_2$

Sample Solution The carbinolamine is formed by nucleophilic addition of the amine to the carbonyl group. Its dehydration gives the imine product.

$$CH_3\overset{O}{\overset{\|}{C}}H + C_6H_5CH_2NH_2 \longrightarrow CH_3\overset{OH}{\overset{|}{\underset{H}{\underset{|}{C}}}}H{-}NCH_2C_6H_5 \xrightarrow{-H_2O} CH_3CH{=}NCH_2C_6H_5$$

Acetaldehyde; Benzylamine; Carbinolamine intermediate; Imine product (*N*-ethylidenebenzylamine)

A number of compounds of the general type H_2NZ react with aldehydes and ketones in a manner analogous to that of primary amines to form products that are more stable than imines. The carbonyl group (C=O) is converted to C=NZ, and a molecule of water is formed. Table 17.4 presents examples of some of these reactions. The mechanism by which

TABLE 17.4 Reaction of Aldehydes and Ketones with Derivatives of Ammonia: $R\overset{O}{\overset{\|}{C}}R' + H_2NZ \longrightarrow R\overset{NZ}{\overset{\|}{C}}R' + H_2O$

Reagent (H_2NZ)	Name of reagent	Type of product	Example
H_2NOH	Hydroxylamine	Oxime	$CH_3(CH_2)_5\overset{O}{\overset{\|}{C}}H \xrightarrow{H_2NOH} CH_3(CH_2)_5\overset{NOH}{\overset{\|}{C}}H$ Heptanal → Heptanal oxime (81–93%)
$H_2NNHC_6H_5$*	Phenylhydrazine	Phenylhydrazone	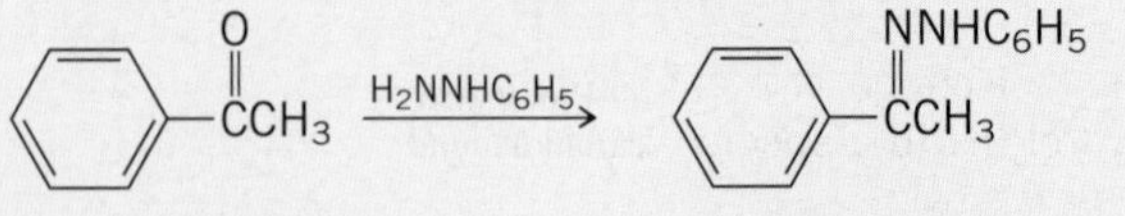 Acetophenone → Acetophenone phenylhydrazone (87–91%)
$H_2NNH\overset{O}{\overset{\|}{C}}NH_2$	Semicarbazide	Semicarbazone	$CH_3\overset{O}{\overset{\|}{C}}(CH_2)_9CH_3 \xrightarrow{H_2NNH\overset{O}{\overset{\|}{C}}NH_2} CH_3\overset{NNH\overset{O}{\overset{\|}{C}}NH_2}{\overset{\|}{C}}(CH_2)_9CH_3$ 2-Dodecanone → 2-Dodecanone semicarbazone (93%)

*Compounds related to phenylhydrazine react in an analogous way. *p*-Nitrophenylhydrazine yields *p*-nitrophenylhydrazones; 2,4-dinitrophenylhydrazine yields 2,4-dinitrophenylhydrazones.

each proceeds is similar to the nucleophilic addition–elimination mechanism described for the reaction of primary amines with aldehydes and ketones.

The reactions listed in Table 17.4 have been extensively studied from a mechanistic perspective because of their relevance to biological processes. Many biological reactions involve initial binding of a carbonyl compound to an enzyme or coenzyme via imine formation. The boxed essay *Imines in Biological Chemistry* gives some important examples.

Imines in Biological Chemistry

Many biological processes involve an "association" between two species in a step prior to some subsequent transformation. This association can take many forms. It can be a weak association of the attractive van der Waals type, or a stronger interaction such as a hydrogen bond. It can be an electrostatic attraction between a positively charged atom of one molecule and a negatively charged atom of another. Covalent bond formation between two species of complementary chemical reactivity represents an extreme kind of association. It often occurs in biological processes in which aldehydes or ketones react with amines via imine intermediates.

An example of a biologically important aldehyde is *pyridoxal phosphate,* which is the active form of *vitamin B_6* and a coenzyme for many of the reactions of α-amino acids. In these reactions the amino acid binds to the coenzyme by reacting with it to form an imine of the kind shown in the equation. Reactions then take place at the amino acid portion of the imine, modifying the amino acid. In the last step, enzyme-catalyzed hydrolysis cleaves the imine to pyridoxal and the modified amino acid.

A key step in the chemistry of vision is binding of an aldehyde to an enzyme via an imine. An outline of the steps involved is presented in Figure 17.8. It starts with *β-carotene,* a pigment that occurs naturally in several fruits and vegetables, including carrots. β-Carotene undergoes oxidative cleavage in the liver to give an alcohol known as *retinol,* or *vitamin A.* Oxidation of vitamin A, followed by isomerization of one of its double bonds, gives the aldehyde 11-*cis*-retinal. In the eye, the aldehyde function of 11-*cis*-retinal combines with an amino group of the protein *opsin* to form an imine called *rhodopsin.* When rhodopsin absorbs a photon of visible light, the cis double bond of the retinal unit undergoes a photochemical cis-to-trans isomerization, which is attended by a dramatic change in its shape and a change in the conformation of rhodopsin. This conformational change is translated into a nerve impulse perceived by the brain as a visual image. Enzyme-promoted hydrolysis of the photochemically isomerized rhodopsin regenerates opsin and a molecule of all-*trans*-retinal. Once all-*trans*-retinal has been enzymatically converted to its 11-cis isomer, it and opsin reenter the cycle.

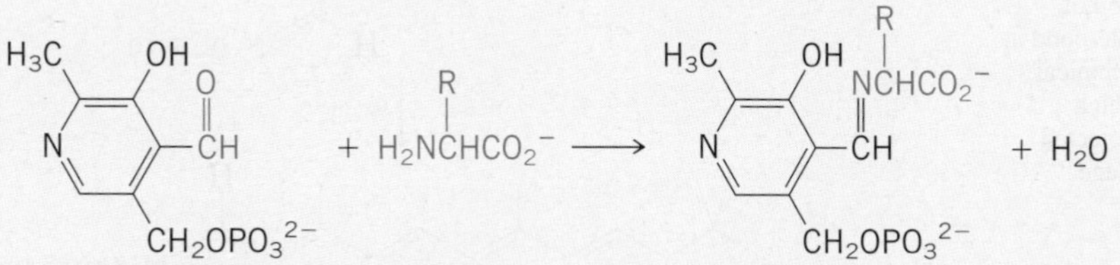

Problem 17.15

Not all biological reactions of amino acids involving imine intermediates require pyridoxal phosphate. The first step in the conversion of proline to glutamic acid is an oxidation giving the imine shown. Once formed, this imine undergoes hydrolysis to a species having the molecular formula $C_5H_9NO_3$, which then goes on to produce glutamic acid. Suggest a structure for the $C_5H_9NO_3$ species. (*Hint:* There are two reasonable possibilities; one is a carbinolamine, the other is not cyclic.)

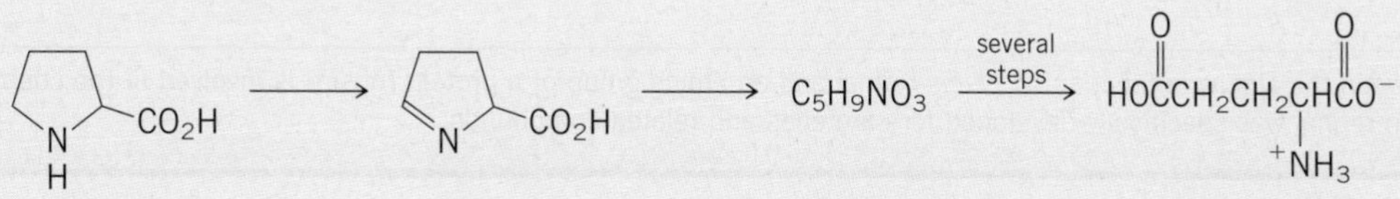

Continued

Continued

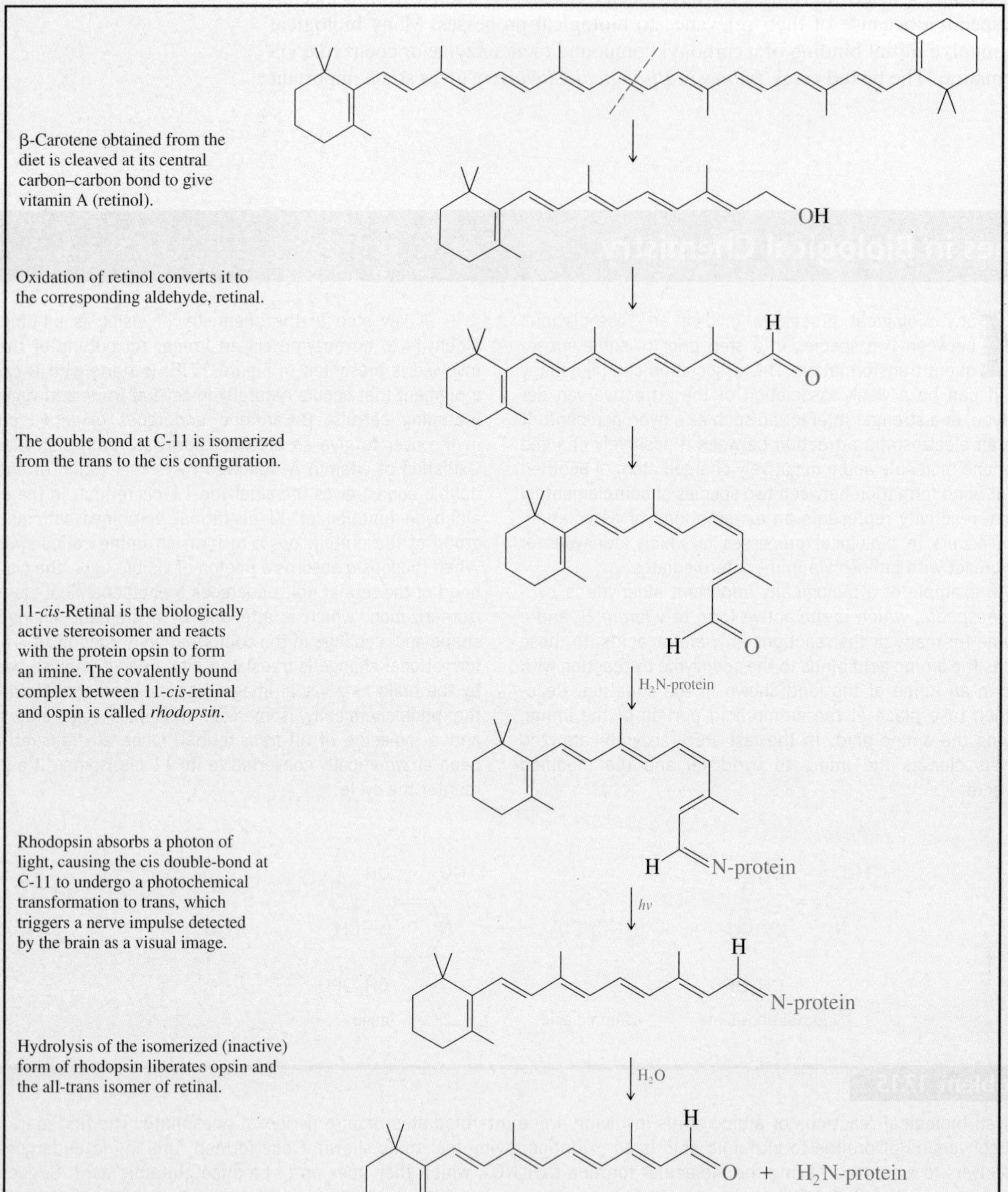
β-Carotene obtained from the diet is cleaved at its central carbon–carbon bond to give vitamin A (retinol).
OH
Oxidation of retinol converts it to the corresponding aldehyde, retinal.
H
O
The double bond at C-11 is isomerized from the trans to the cis configuration.
11-*cis*-Retinal is the biologically active stereoisomer and reacts with the protein opsin to form an imine. The covalently bound complex between 11-*cis*-retinal and ospin is called *rhodopsin*.
H
O
H₂N-protein
H
N-protein
Rhodopsin absorbs a photon of light, causing the cis double-bond at C-11 to undergo a photochemical transformation to trans, which triggers a nerve impulse detected by the brain as a visual image.
hν
H
N-protein
Hydrolysis of the isomerized (inactive) form of rhodopsin liberates opsin and the all-trans isomer of retinal.
H₂O
H
O + H₂N-protein

Figure 17.8

Imine formation between the aldehyde function of 11-*cis*-retinal and an amino group of a protein (opsin) is involved in the chemistry of vision. The numbering scheme in retinal was specifically developed for carotenes and related compounds.

17.11 Reaction with Secondary Amines: Enamines

Secondary amines are compounds of the type R_2NH. They add to aldehydes and ketones to form carbinolamines, but their carbinolamine intermediates can dehydrate to a stable product only in the direction that leads to a carbon–carbon double bond:

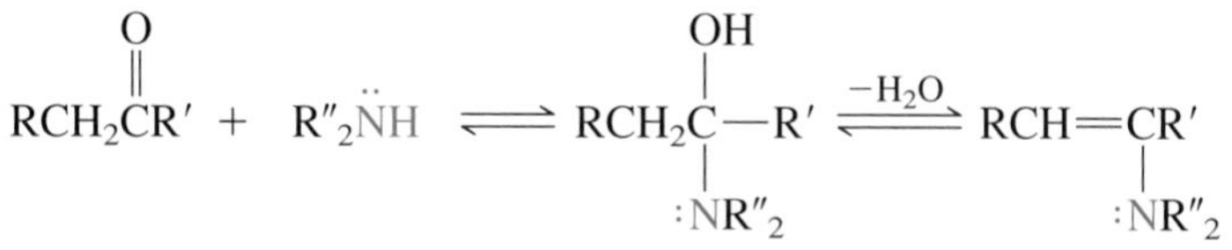

Aldehyde or ketone | Secondary amine | Carbinolamine | Enamine

The product is an alkenyl-substituted amine, or **enamine.**

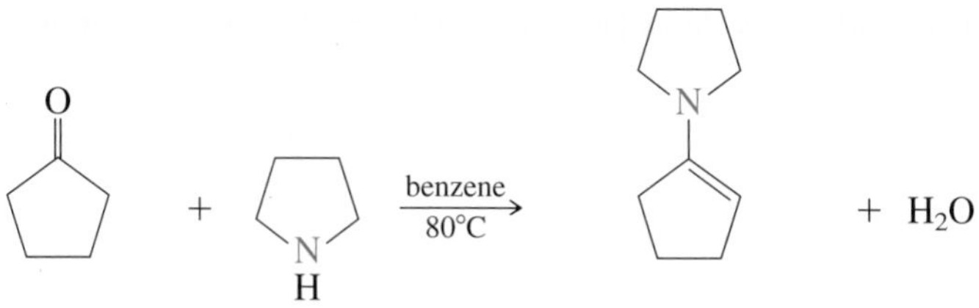

Cyclopentanone | Pyrrolidine | *N*-(1-Cyclopentenyl)-pyrrolidine (80–90%) | Water

The mechanism of enamine formation in this example is shown in Mechanism 17.6.

Problem 17.16

Write the structure of the carbinolamine intermediate and the enamine product formed in the reaction of each of the following:

(a) Propanal and dimethylamine, CH_3NHCH_3

(b) 3-Pentanone and pyrrolidine

(c) Acetophenone and HN 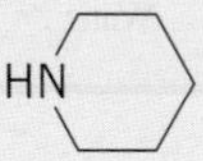

Sample Solution (a) Nucleophilic addition of dimethylamine to the carbonyl group of propanal produces a carbinolamine:

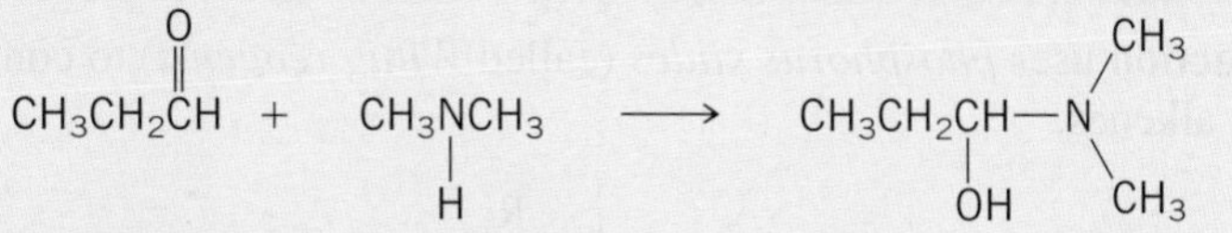

Propanal | Dimethylamine | Carbinolamine intermediate

Dehydration of this carbinolamine yields the enamine:

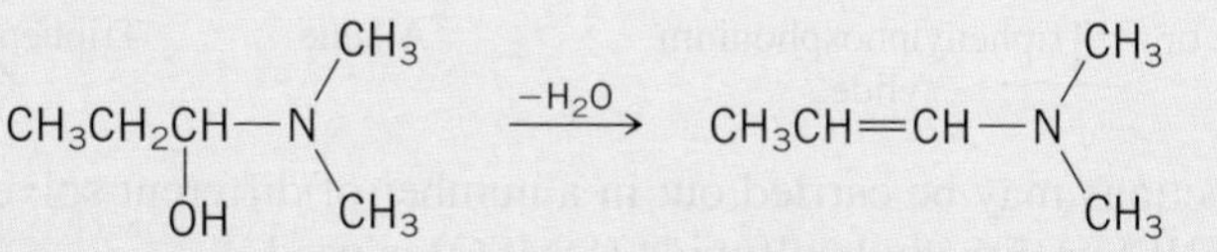

Carbinolamine intermediate | *N*-(1-Propenyl)dimethylamine

Enamines are used as reagents in synthetic organic chemistry and are involved in certain biochemical transformations.

Mechanism 17.6

Enamine Formation from Cyclopentanone and Pyrrolidine

Step 1: Pyrrolidine undergoes nucleophilic addition to cyclopentanone to give a carbinolamine. The mechanism is analogous to the addition of primary amines to aldehydes and ketones.

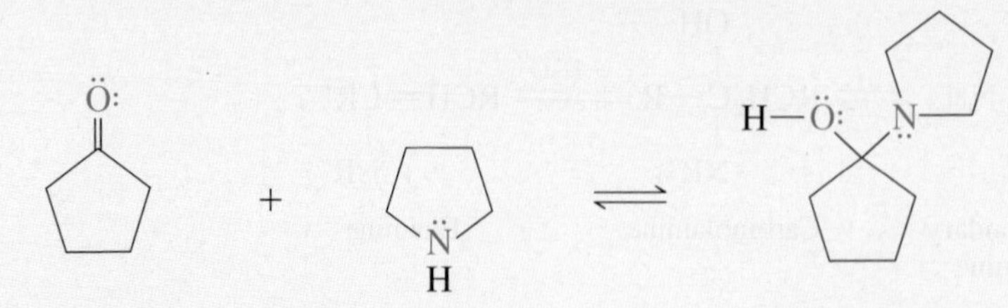

Cyclopentanone Pyrrolidine Carbinolamine intermediate

Step 2: The carbinolamine dissociates by loss of hydroxide. This dissociation is assisted by lone-pair donation from nitrogen.

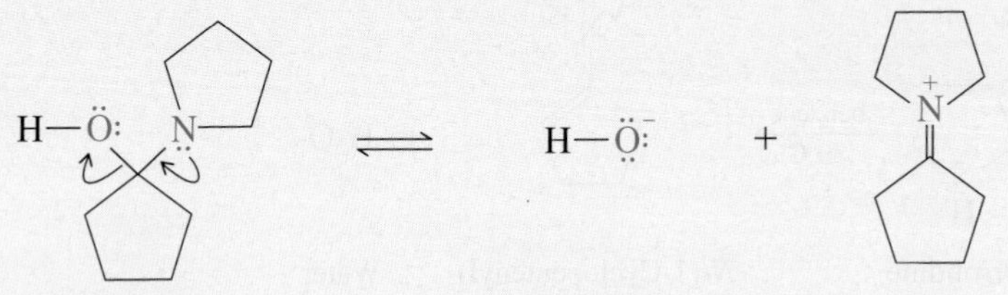

Carbinolamine intermediate Hydroxide ion Iminium ion

Step 3: The iminium ion is deprotonated in the direction that gives a carbon–carbon double bond.

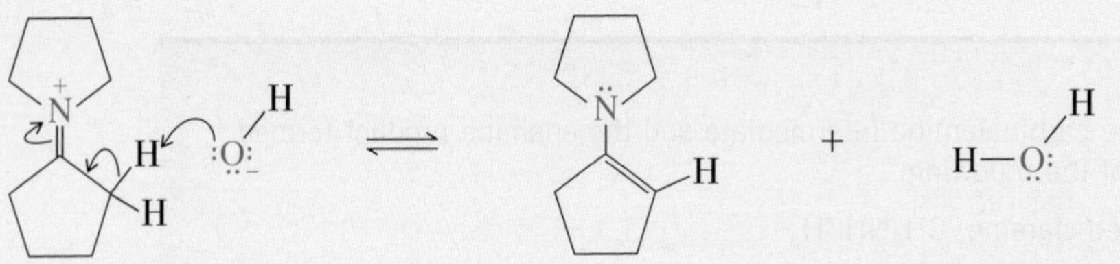

Iminium ion Hydroxide ion *N*-(1-Cyclopentenyl)pyrrolidine Water

17.12 The Wittig Reaction

The reaction is named after Georg Wittig, a German chemist who shared the 1979 Nobel Prize in Chemistry for demonstrating its synthetic potential.

The **Wittig reaction** uses *phosphorus ylides* (called *Wittig reagents*) to convert aldehydes and ketones to alkenes.

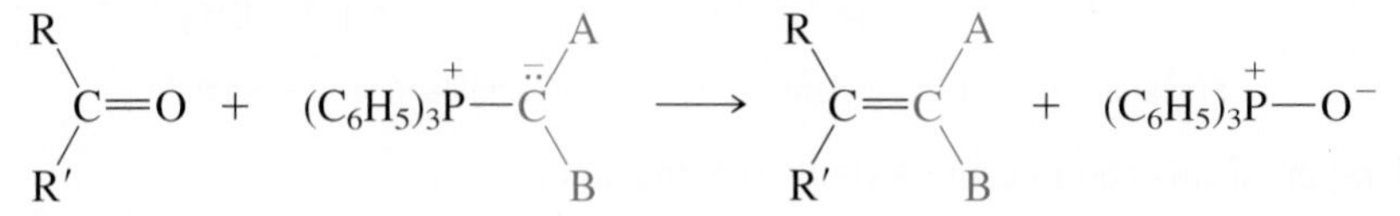

Aldehyde or ketone Triphenylphosphonium ylide Alkene Triphenylphosphine oxide

Wittig reactions may be carried out in a number of different solvents; normally tetrahydrofuran (THF) or dimethyl sulfoxide (DMSO) is used.

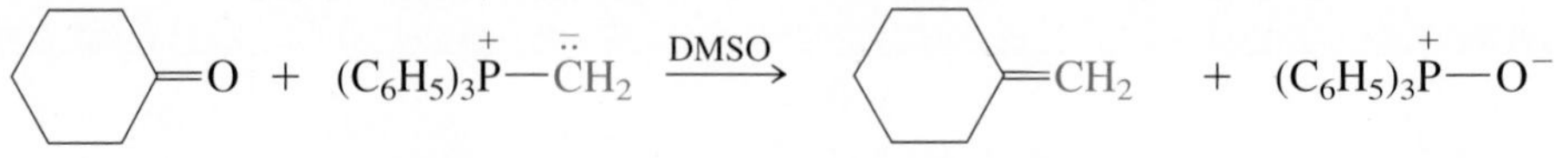

Cyclohexanone Methylenetriphenylphosphorane Methylenecyclohexane (86%) Triphenylphosphine oxide

The most attractive feature of the Wittig reaction is its regiospecificity. The location of the double bond is never in doubt. The double bond connects the carbon of the original C=O group of the aldehyde or ketone and the negatively charged carbon of the ylide.

Problem 17.17

Identify the alkene product in each of the following Wittig reactions:

(a) Benzaldehyde + 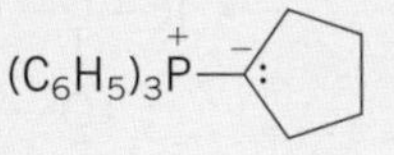

(b) Butanal + $(C_6H_5)_3\overset{+}{P}—\overset{-}{\ddot{C}}HCH{=}CH_2$

(c) Cyclohexyl methyl ketone + $(C_6H_5)_3\overset{+}{P}—\overset{-}{\ddot{C}}H_2$

Sample Solution (a) In a Wittig reaction the negatively charged substituent attached to phosphorus is transferred to the aldehyde or ketone, replacing the carbonyl oxygen. The reaction shown has been used to prepare the indicated alkene in 65% yield.

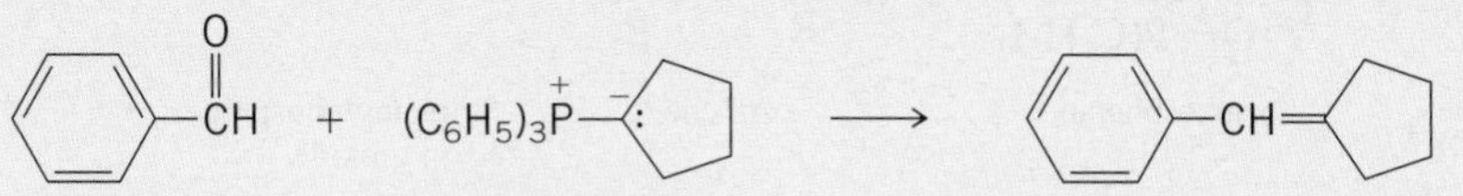

Benzaldehyde | Cyclopentylidenetriphenylphosphorane | Benzylidenecyclopentane (65%)

To understand the mechanism of the Wittig reaction, we need to examine the structure and properties of ylides. **Ylides** are neutral molecules that have two oppositely charged atoms, each with an octet of electrons, directly bonded to each other. In an ylide such as $(C_6H_5)_3\overset{+}{P}—\overset{-}{\ddot{C}}H_2$, phosphorus has eight electrons and is positively charged; its attached carbon has eight electrons and is negatively charged.

Problem 17.18

Can you write a resonance structure for $(C_6H_5)_3\overset{+}{P}—\overset{-}{\ddot{C}}H_2$ in which neither phosphorus nor carbon has a formal charge? (*Hint:* Remember phosphorus can have more than eight electrons in its valence shell.)

We can focus on the charge distribution in an ylide by examining the electrostatic potential map of $H_3\overset{+}{P}—\overset{-}{\ddot{C}}H_2$ in Figure 17.9, where it can be seen that the electron distribution is highly polarized in the direction that makes carbon electron-rich. The carbon has much of the character of a carbanion and can act as a nucleophile toward C=O.

Mechanism 17.7 outlines the Wittig reaction. The first stage is a cycloaddition in which the ylide reacts with the carbonyl group to give an intermediate containing a four-membered ring called an *oxaphosphetane.* This oxaphosphetane then dissociates to

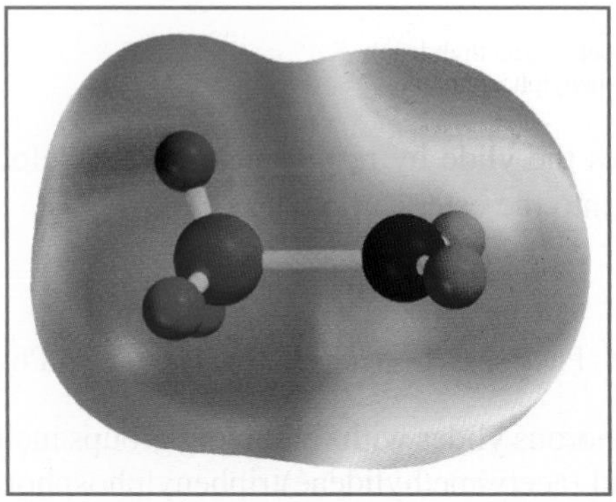

Figure 17.9

An electrostatic potential map of the ylide $H_3\overset{+}{P}—\overset{-}{\ddot{C}}H_2$. The region of greatest negative charge is concentrated at carbon.

Mechanism 17.7

The Wittig Reaction

Step 1: The ylide and the aldehyde or ketone combine to form an oxaphosphetane.

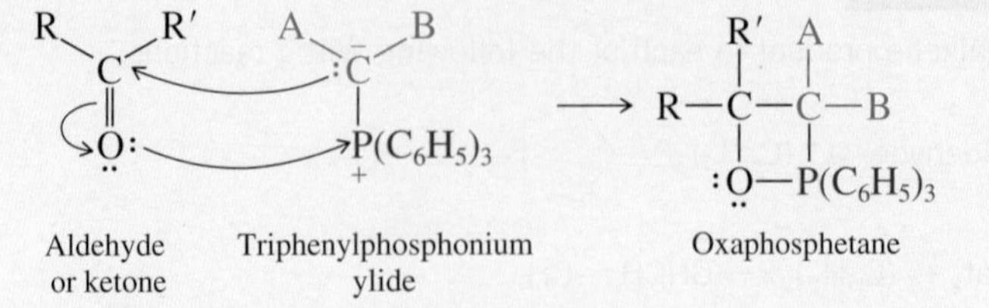

Step 2: The oxaphosphetane dissociates to an alkene and triphenylphosphine oxide.

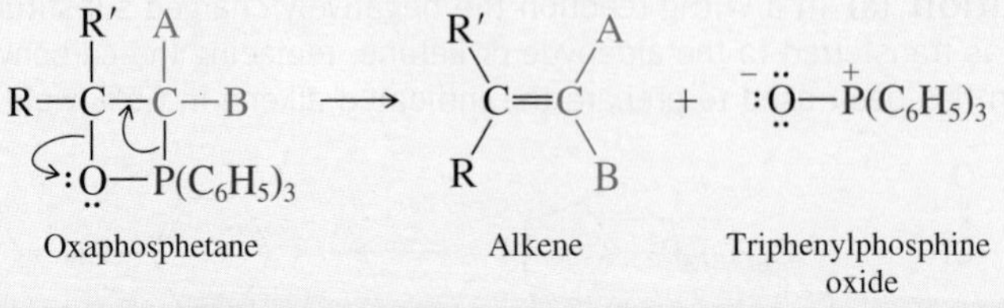

The Wittig reaction is one that is still undergoing mechanistic investigation. Another possibility is that the oxaphosphetane intermediate is formed by a two-step process, rather than the one-step process shown in Mechanism 17.7.

give an alkene and triphenylphosphine oxide. Presumably the direction of dissociation of the oxaphosphetane is dictated by the strong phosphorus–oxygen bond that results. The P—O bond strength in triphenylphosphine oxide has been estimated to be greater than 540 kJ/mol (130 kcal/mol).

The stereoselectivity of the Wittig reaction depends both on the structure of the phosphorous ylide and the reaction conditions. A useful generalization is that *unstabilized ylides give predominantly Z-alkenes, but stabilized ylides favor the E-alkene.* Unstabilized ylides are those that lack a carbanion-stabilizing group on the ylide carbon, for example ethylidenetriphenylphosphorane. Stabilized ylides contain an electron-withdrawing group on the ylide carbon, such as the ester group in (carboethoxymethylidene)triphenylphosphorane. Wittig reactions of these ylides with benzaldehyde give respectively, the *Z*- and *E*-alkenes as the major or exclusive products.

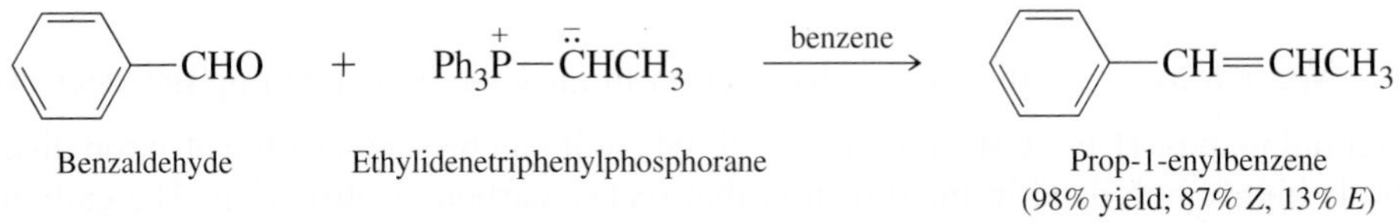

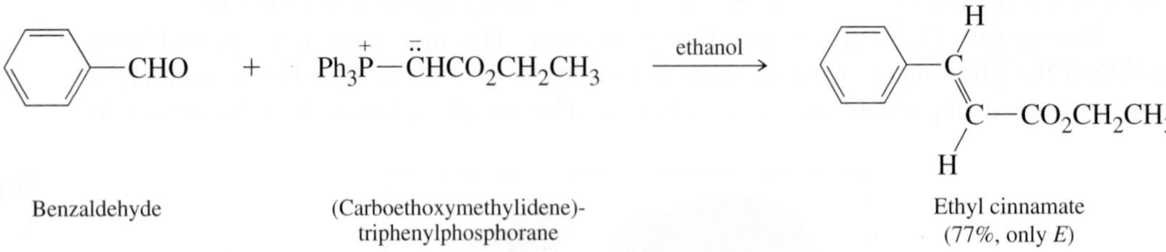

The ester group stabilizes the ylide by resonance, through delocalization of the electron pair on the ylide carbon into the ester carbonyl.

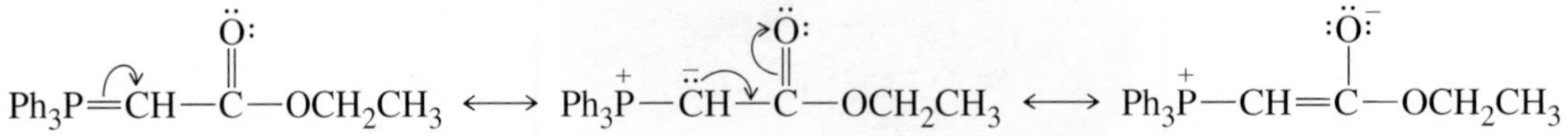

Examples of other phosphorous ylides with stabilizing groups include (cyanomethylidene)-triphenylphosphorane and (acetylmethylidene)triphenylphosphorane.

$Ph_3\overset{+}{P}-\overset{..}{\bar{C}}H-C\equiv N:$ (Cyanomethylidene)-triphenylphosphorane

$Ph_3\overset{+}{P}-\overset{..}{\bar{C}}H-\overset{\overset{..}{O}:}{\overset{\|}{C}}-CH_3$ (Acetylmethylidene)-triphenylphosphorane

Problem 17.19

Write two resonance structures for (a) (cyanomethylidene)triphenylphosphorane and (b) (acetylmethylidene)triphenylphosphorane.

Sample Solution Electron delocalization involving the cyano group generates an additional resonance structure for the ylide.

$$Ph_3P{=}CH-C\equiv N: \longleftrightarrow Ph_3\overset{+}{P}-\overset{..}{\bar{C}}H-C\equiv N: \longleftrightarrow Ph_3\overset{+}{P}-CH{=}C{=}\overset{-}{\underset{..}{N}}:$$

17.13 Planning an Alkene Synthesis via the Wittig Reaction

To identify the carbonyl compound and the ylide required to produce a given alkene, mentally disconnect the double bond so that one of its carbons is derived from a carbonyl group and the other is derived from an ylide. Taking styrene as a representative example, we see that two such disconnections are possible; either benzaldehyde or formaldehyde is an appropriate precursor.

$$\underset{\text{Styrene}}{C_6H_5CH{=}CH_2} \Longrightarrow \underset{\text{Benzaldehyde}}{C_6H_5\overset{O}{\overset{\|}{C}}H} + \underset{\text{Methylenetriphenylphosphorane}}{(C_6H_5)_3\overset{+}{P}-\overset{..}{\bar{C}}H_2}$$

$$\underset{\text{Styrene}}{C_6H_5CH{=}CH_2} \Longrightarrow \underset{\text{Benzylidenetriphenylphosphorane}}{(C_6H_5)_3\overset{+}{P}-\overset{..}{\bar{C}}HC_6H_5} + \underset{\text{Formaldehyde}}{H\overset{O}{\overset{\|}{C}}H}$$

Either route is feasible, and indeed styrene has been prepared from both combinations of reactants. Typically there will be two Wittig routes to an alkene, and any choice between them is made on the basis of availability of the particular starting materials.

Problem 17.20

What combinations of carbonyl compound and ylide could you use to prepare each of the following alkenes?

(a) $CH_3CH_2CH_2CH{=}\underset{\underset{CH_3}{|}}{C}CH_2CH_3$

(b) $CH_3CH_2CH_2CH{=}CH_2$

Sample Solution (a) Two Wittig reaction routes lead to the target molecule.

$$\underset{\text{3-Methyl-3-heptene}}{CH_3CH_2CH_2CH{=}\underset{\underset{CH_3}{|}}{C}CH_2CH_3} \Longrightarrow \underset{\text{Butanal}}{CH_3CH_2CH_2\overset{O}{\overset{\|}{C}}H} + \underset{\text{1-Methylpropylidenetriphenylphosphorane}}{(C_6H_5)_3\overset{+}{P}-\underset{\underset{CH_3}{|}}{\overset{..}{\bar{C}}}CH_2CH_3}$$

Continued

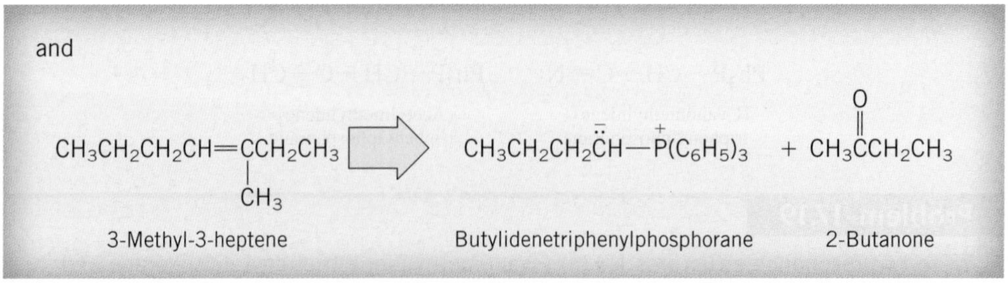

Phosphorus ylides are prepared from alkyl halides by a two-step sequence. The first step is a nucleophilic substitution of the S_N2 type by triphenylphosphine on an alkyl halide to give an alkyltriphenylphosphonium salt:

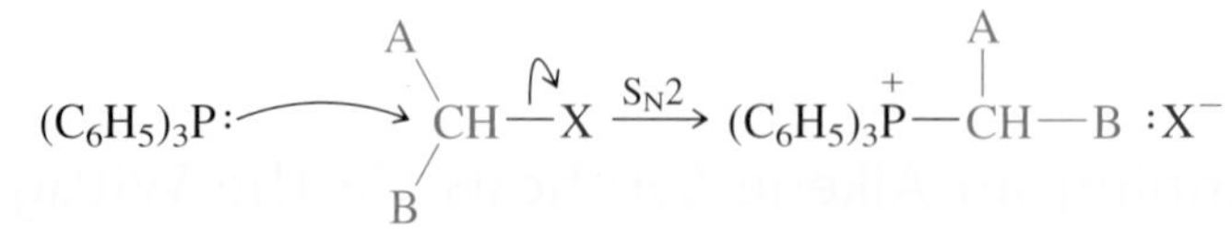

Triphenylphosphine   Alkyl halide   Alkyltriphenylphosphonium halide

Triphenylphosphine is a very powerful nucleophile, yet is not strongly basic. Methyl, primary, and secondary alkyl halides are all suitable substrates.

$$(C_6H_5)_3P: \quad + \quad CH_3Br \xrightarrow{\text{benzene}} (C_6H_5)_3\overset{+}{P}{-}CH_3 \; Br^-$$

Triphenylphosphine   Bromomethane   Methyltriphenylphosphonium bromide (99%)

The alkyltriphenylphosphonium salt products are ionic and crystallize in high yield from the nonpolar solvents in which they are prepared. After isolation, the alkyltriphenylphosphonium halide is converted to the desired ylide by deprotonation with a strong base:

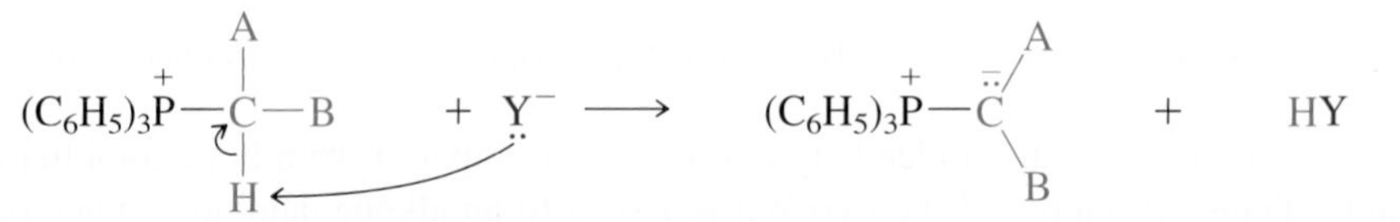

Alkyltriphenylphosphonium salt   Base   Triphenylphosphonium ylide   Conjugate acid of base used

Strong bases such as the sodium salt of dimethyl sulfoxide (in dimethyl sulfoxide as the solvent) and organolithium reagents (in diethyl ether or tetrahydrofuran) are used to generate unstabilized ylides.

$$(C_6H_5)_3\overset{+}{P}{-}CH_3 \; Br^- + NaCH_2\overset{O}{\overset{\|}{S}}CH_3 \xrightarrow{\text{DMSO}} (C_6H_5)_3\overset{+}{P}{-}\overset{..}{\overset{-}{C}}H_2 + CH_3\overset{O}{\overset{\|}{S}}CH_3 + NaBr$$

Methyltriphenylphosphonium bromide   Sodiomethyl methyl sulfoxide   Methylenetriphenylphosphorane   Dimethyl sulfoxide   Sodium bromide

Stabilized ylides can be prepared from the required phosphonium salts using weaker bases, including sodium hydroxide.

$$Ph_3\overset{+}{P}{-}CH_2CO_2CH_2CH_3 \; Br^- \xrightarrow[H_2O]{NaOH} Ph_3\overset{+}{P}{-}\overset{..}{\overset{-}{C}}HCO_2CH_2CH_3$$

(Carboethoxymethylidene)-triphenylphosphonium bromide   (Carboethoxymethylidene)-triphenylphosphorane

Problem 17.21

The sample solution to Problem 17.20(a) showed the preparation of 3-methyl-3-heptene by a Wittig reaction involving the ylide shown. Write equations showing the formation of this ylide beginning with 2-bromobutane.

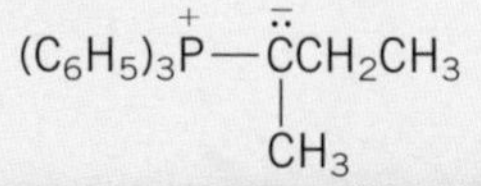
(C6H5)3P—CCH2CH3
CH3

17.14 Stereoselective Addition to Carbonyl Groups

Nucleophilic addition to carbonyl groups sometimes leads to a mixture of stereoisomeric products. The direction of addition is often controlled by steric factors, with the nucleophile approaching the carbonyl group at its less hindered face. Sodium borohydride reduction of 7,7-dimethylbicyclo[2.2.1]heptan-2-one illustrates this point:

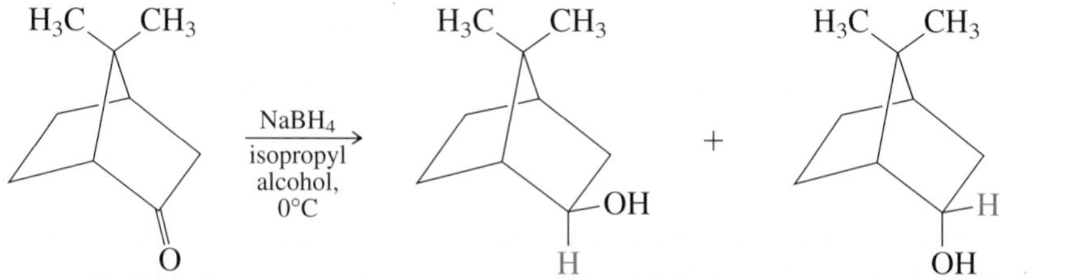
H3C
CH3
NaBH4
isopropyl alcohol, 0°C
OH
H
+
OH

7,7-Dimethylbicyclo-[2.2.1]heptan-2-one — *exo*-7,7-Dimethylbicyclo-[2.2.1]heptan-2-ol (80%) — *endo*-7,7-Dimethylbicyclo-[2.2.1]heptan-2-ol (20%)

Approach of borohydride to the top face of the carbonyl group is sterically hindered by one of the methyl groups. The bottom face of the carbonyl group is less congested, and the major product is formed by hydride transfer from this direction.

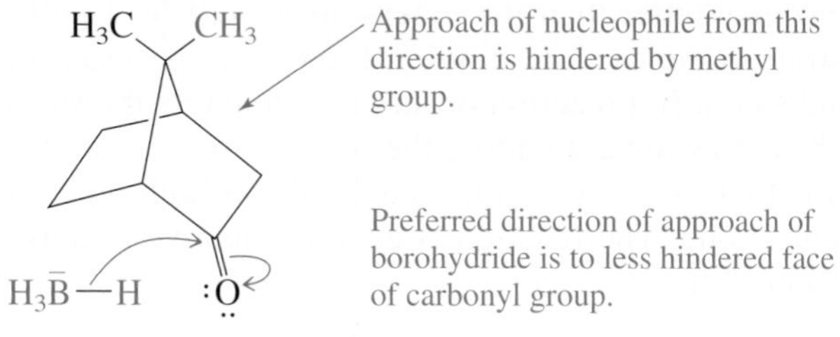
H3C
CH3
Approach of nucleophile from this direction is hindered by methyl group.
Preferred direction of approach of borohydride is to less hindered face of carbonyl group.
H3B—H
:O:

The reduction is *stereoselective.* A single starting material can form two stereoisomers of the product but yields one of them in greater amounts than the other or even to the exclusion of the other.

Problem 17.22

What is the relationship between the products of the reaction just described? Are they enantiomers or diastereomers? Is the reaction enantioselective or diastereoselective?

Enzyme-catalyzed reductions of carbonyl groups are, more often than not, completely stereoselective. Pyruvic acid, for example, is converted exclusively to (*S*)-(+)-lactic acid by the lactate dehydrogenase-NADH system (Section 15.10). The enantiomer (*R*)-(−)-lactic acid is not formed.

Figure 17.10

(*a*) Binding sites of enzyme discriminate between prochiral faces of substrate. One prochiral face can bind to the enzyme better than the other. (*b*) Reaction attaches fourth group to the top face of the substrate producing only one enantiomer of chiral product.

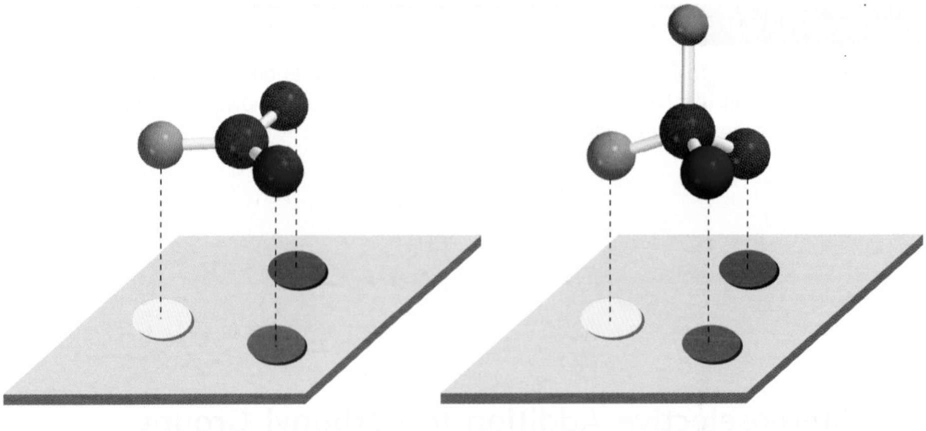

(*a*) (*b*)

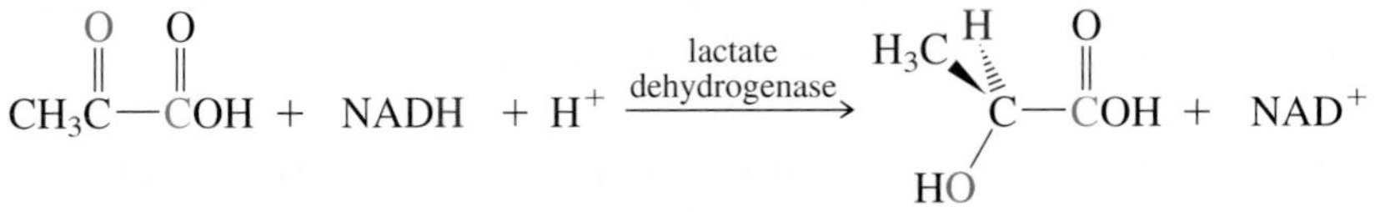

Pyruvic acid | Reduced form of coenzyme | (*S*)-(+)-lactic acid | Oxidized form of coenzyme

The enzyme is a single enantiomer of a chiral molecule and binds the coenzyme and substrate in such a way that hydride is transferred exclusively to the face of the carbonyl group that leads to (*S*)-(+)-lactic acid. Reduction of pyruvic acid in an achiral environment, say with sodium borohydride, also gives lactic acid but as a racemic mixture containing equal quantities of the *R* and *S* enantiomers.

Prochirality was introduced in Section 7.10 and is the topic of the Chapter 7 Descriptive Passage.

The enantioselectivity of enzyme-catalyzed reactions can be understood on the basis of a relatively simple model. Consider the case of an sp^2-hybridized carbon with prochiral faces as in Figure 17.10*a*. If structural features on the enzyme are complementary in some respect to the groups attached to this carbon, one prochiral face can bind to the enzyme better than the other—there will be a preferred geometry of the enzyme–substrate complex. The binding forces are the usual ones: electrostatic, van der Waals, and so on. If a reaction occurs that converts the sp^2-hybridized carbon to sp^3, there will be a bias toward adding the fourth group from a particular direction as shown in Figure 17.10*b*. As a result, an achiral molecule is converted to a single enantiomer of a chiral one. The reaction is enantioselective because it occurs preferentially at one prochiral face.

17.15 Oxidation of Aldehydes

Aldehydes are readily oxidized to carboxylic acids by a number of reagents, including those based on Cr(VI) in aqueous media.

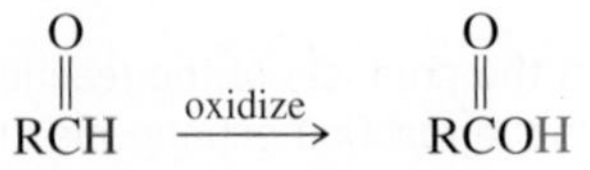

Aldehyde | Carboxylic acid

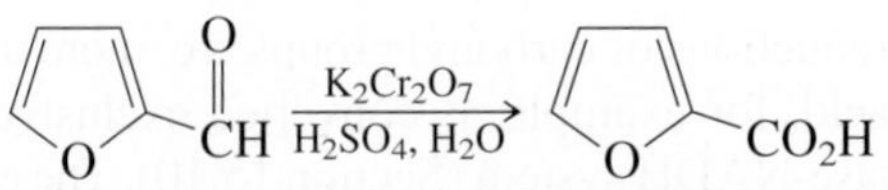

Furfural | Furoic acid (75%)

Mechanistically, these reactions probably proceed through the hydrate of the aldehyde and follow a course similar to that of alcohol oxidation.

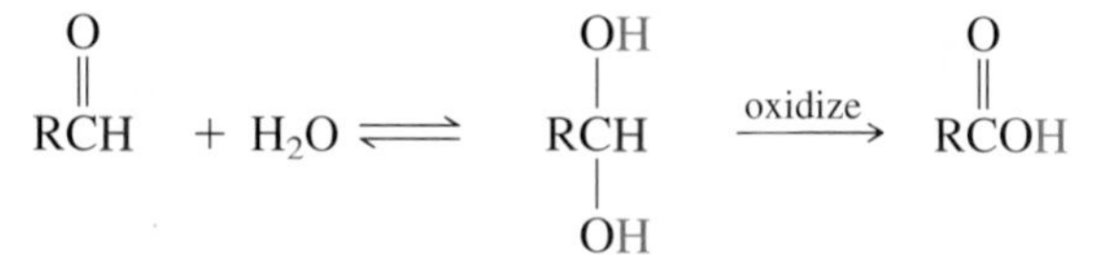

Aldehyde — Geminal diol (hydrate) — Carboxylic acid

Aldehydes are more easily oxidized than alcohols, which is why special reagents such as PCC and PDC (Section 15.9) have been developed for oxidizing primary alcohols to aldehydes and no further. PCC and PDC are effective not only because they are sources of Cr(VI), but also because they are used in nonaqueous media (dichloromethane). By keeping water out of the reaction mixture, the aldehyde is not converted to its hydrate, which is the necessary intermediate that leads to the carboxylic acid.

17.16 Spectroscopic Analysis of Aldehydes and Ketones

Infrared: Carbonyl groups are among the easiest functional groups to detect by IR spectroscopy. The C=O stretching vibration of aldehydes and ketones gives rise to strong absorption in the region 1710–1750 cm^{-1}, as illustrated for butanal in Figure 17.11. In addition to a peak for C=O stretching, the CH=O group of an aldehyde exhibits two weak bands for C—H stretching near 2720 and 2820 cm^{-1}.

^{1}H NMR: Aldehydes are readily identified by the presence of a signal for the hydrogen of CH=O at δ 9–10. This is a region where very few other protons ever appear. Figure 17.12 shows the ^{1}H NMR spectrum of 2-methylpropanal [$(CH_3)_2CHCH{=}O$)], where the large chemical shift difference between the aldehyde proton and the other protons in the molecule is clearly evident. As seen in the expanded-scale inset, the aldehyde proton is a doublet, split by the proton at C-2. Coupling between the protons in HC—CH=O is much smaller than typical vicinal couplings, making the multiplicity of the aldehyde peak difficult to see without expanding the scale.

Methyl ketones, such as 2-butanone in Figure 17.13, are characterized by sharp singlets near δ 2 for the protons of $CH_3C{=}O$. Similarly, the deshielding effect of the carbonyl causes the protons of $CH_2C{=}O$ to appear at lower field (δ 2.4) than in a CH_2 group of an alkane.

^{13}C NMR: The signal for the carbon of C=O in aldehydes and ketones appears at very low field, some 190–220 ppm downfield from tetramethylsilane. Figure 17.14 illustrates this for 3-heptanone, in which separate signals appear for each of the seven carbons. The six sp^3-hybridized carbons appear in the range δ 8–42, and the carbon of the C=O group is at δ 210. Note, too, that the intensity of the peak for the C=O carbon is much less

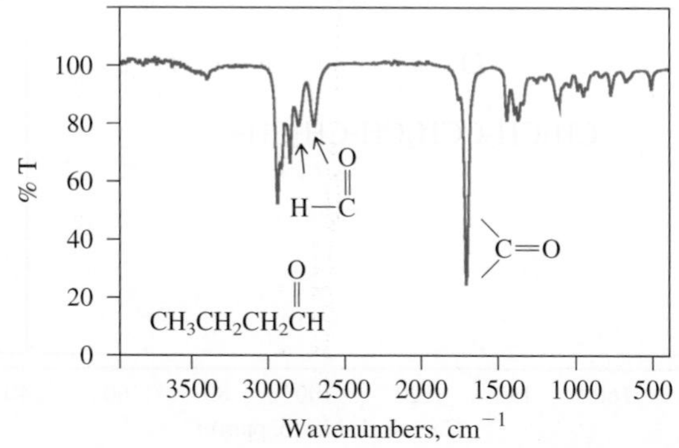

Figure 17.11

IR spectrum of butanal showing peaks characteristic of the CH=O unit at 2700 and 2800 cm^{-1} (C—H) and at 1720 cm^{-1} (C=O).

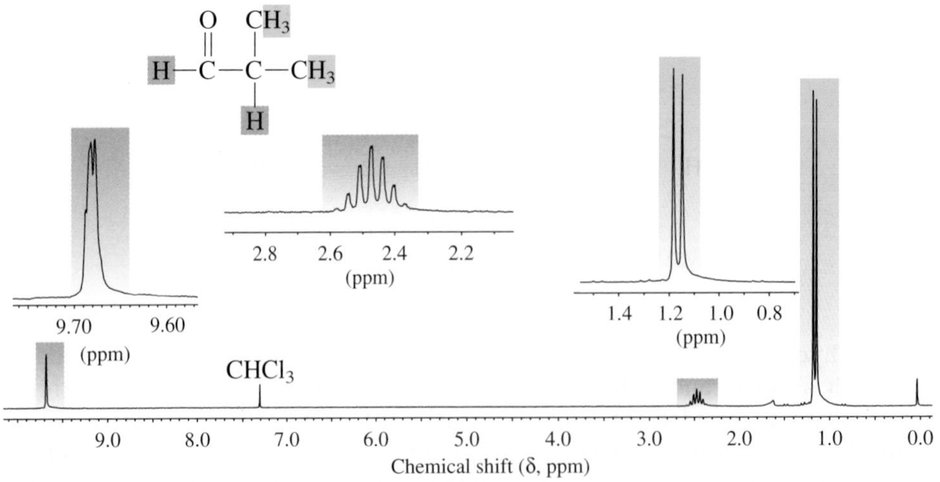

Figure 17.12

The 200-MHz ^{1}H NMR spectrum of 2-methylpropanal, showing the aldehyde proton as a doublet at low field (δ 9.7).

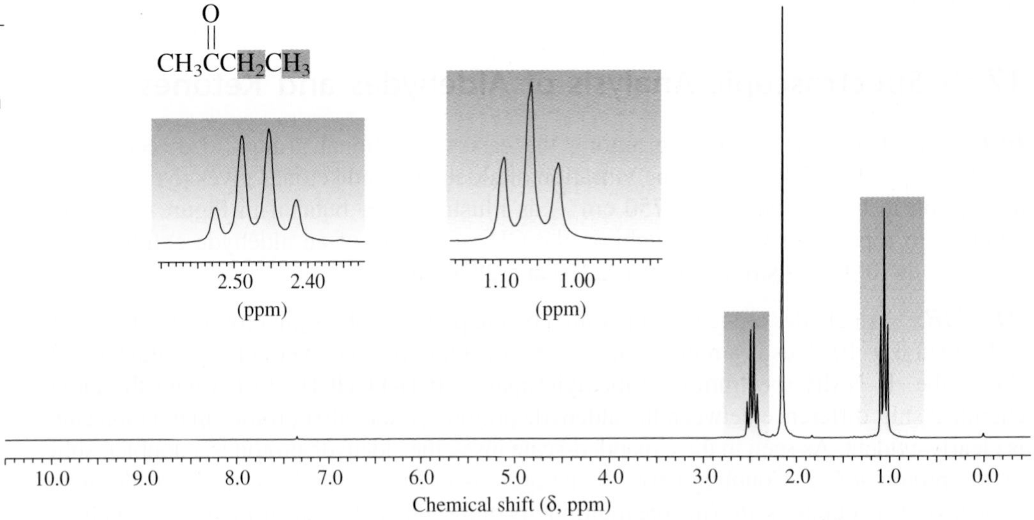

Figure 17.13

The 200-MHz ^{1}H NMR spectrum of 2-butanone. The triplet–quartet pattern of the ethyl group is seen more clearly in the expanded-scale insets.

than all the others, even though each peak corresponds to a single carbon. This decreased intensity is a characteristic of pulsed Fourier transform (FT) spectra for carbons that don't have attached hydrogens.

UV-VIS: Aldehydes and ketones have two absorption bands in the ultraviolet region. Both involve excitation of an electron to an antibonding π* orbital. In one, called a π→π* transition, the electron is one of the π electrons of the C=O group.

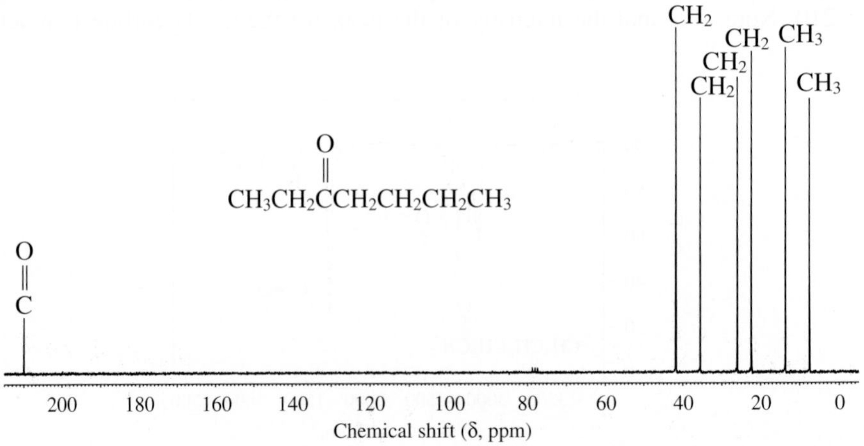

Figure 17.14

The ^{13}C NMR spectrum of 3-heptanone. Each signal corresponds to a single carbon. The carbonyl carbon is the least shielded and appears at δ 210.

In the other, called an $n \rightarrow \pi^*$ transition, it is one of the oxygen lone-pair electrons. Because the π electrons are more strongly held than the lone-pair electrons, the $\pi \rightarrow \pi^*$ transition is of higher energy and shorter wavelength than the $n \rightarrow \pi^*$ transition. For simple aldehydes and ketones, the $\pi \rightarrow \pi^*$ transition is below 200 nm and of little use in structure determination. The $n \rightarrow \pi^*$ transition, although weak, is of more diagnostic value.

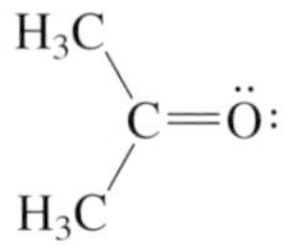

Acetone

$\pi \rightarrow \pi^*$ λ_{max} 187 nm
$n \rightarrow \pi^*$ λ_{max} 270 nm

Mass Spectrometry: Aldehydes and ketones typically give a prominent molecular ion peak in their mass spectra. Aldehydes also exhibit an M-1 peak. A major fragmentation pathway for both aldehydes and ketones leads to formation of acyl cations (acylium ions) by cleavage of an alkyl group from the carbonyl. The most intense peak in the mass spectrum of diethyl ketone, for example, is *m/z* 57, corresponding to loss of ethyl radical from the molecular ion.

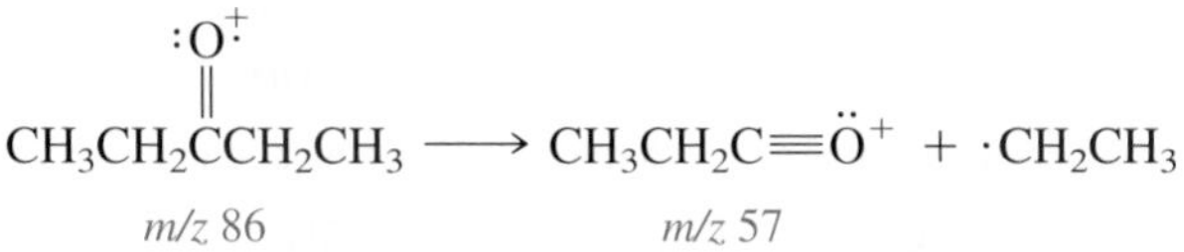

17.17 SUMMARY

The chemistry of the carbonyl group is probably the single most important aspect of organic chemical reactivity. Classes of compounds that contain the carbonyl group include many derived from carboxylic acids (acyl chlorides, acid anhydrides, esters, and amides) as well as the two related classes discussed in this chapter: *aldehydes* and *ketones.*

Section 17.1 The substitutive IUPAC names of aldehydes and ketones are developed by identifying the longest continuous chain that contains the carbonyl group and replacing the final *-e* of the corresponding alkane by *-al* for aldehydes and *-one* for ketones. The chain is numbered in the direction that gives the lowest locant to the carbon of the carbonyl group.

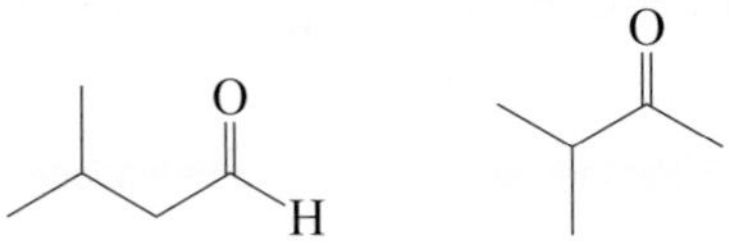

3-Methylbutanal   3-Methyl-2-butanone or 3-Methylbutan-2-one

Ketones may also be named using functional class IUPAC nomenclature by citing the two groups attached to the carbonyl in alphabetical order followed by the word *ketone.* Thus, 3-methyl-2-butanone (substitutive) becomes isopropyl methyl ketone (functional class).

Section 17.2 The carbonyl carbon is sp^2-hybridized, and it and the atoms attached to it are coplanar. Aldehydes and ketones are polar molecules. Nucleophiles attack C=O at carbon (positively polarized) and electrophiles, especially protons, attack oxygen (negatively polarized).

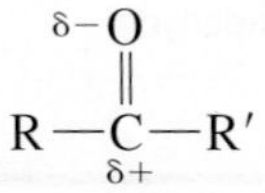

Section 17.3 Aldehydes and ketones have higher boiling points than hydrocarbons, but have lower boiling points than alcohols.

Section 17.4 The numerous reactions that yield aldehydes and ketones discussed in earlier chapters and reviewed in Table 17.1 are sufficient for most syntheses.

Sections 17.5–17.13 The characteristic reactions of aldehydes and ketones involve *nucleophilic addition* to the carbonyl group and are summarized in Table 17.5. Reagents of the type HY react according to the general equation

$$\overset{\delta+}{C}=\overset{\delta-}{O} + \overset{\delta+}{H}-\overset{\delta-}{Y} \rightleftharpoons Y-\overset{|}{\underset{|}{C}}-O-H$$

Aldehyde or ketone → Product of nucleophilic addition to carbonyl group

TABLE 17.5 Nucleophilic Addition to Aldehydes and Ketones

Reaction (section) and comments	General equation and typical example
Hydration (Section 17.6) Can be either acid- or base-catalyzed. Equilibrium constant is normally unfavorable for hydration of ketones unless R, R′, or both are strongly electron-withdrawing.	$RCR'(=O) + H_2O \rightleftharpoons RCR'(OH)_2$ Aldehyde or ketone; Water; Geminal diol $ClCH_2CCH_3(=O) \overset{H_2O}{\rightleftharpoons} ClCH_2C(OH)_2CH_3$ Chloroacetone (90% at equilibrium); Chloroacetone hydrate (10% at equilibrium)
Cyanohydrin formation (Section 17.7) Reaction is catalyzed by cyanide ion. Cyanohydrins are useful synthetic intermediates; cyano group can be hydrolyzed to $-CO_2H$ or reduced to $-CH_2NH_2$.	$RCR'(=O) + HCN \rightleftharpoons RCR'(OH)(CN)$ Aldehyde or ketone; Hydrogen cyanide; Cyanohydrin $CH_3CH_2CCH_2CH_3(=O) \xrightarrow[H^+]{KCN} CH_3CH_2C(OH)(CN)CH_2CH_3$ 3-Pentanone; 3-Pentanone cyanohydrin (75%)
Acetal formation (Sections 17.8–17.9) Reaction is acid-catalyzed. Equilibrium constant normally favorable for aldehydes, unfavorable for ketones. Cyclic acetals from vicinal diols form readily.	$RCR'(=O) + 2R''OH \overset{H^+}{\rightleftharpoons} RCR'(OR'')_2 + H_2O$ Aldehyde or ketone; Alcohol; Acetal; Water 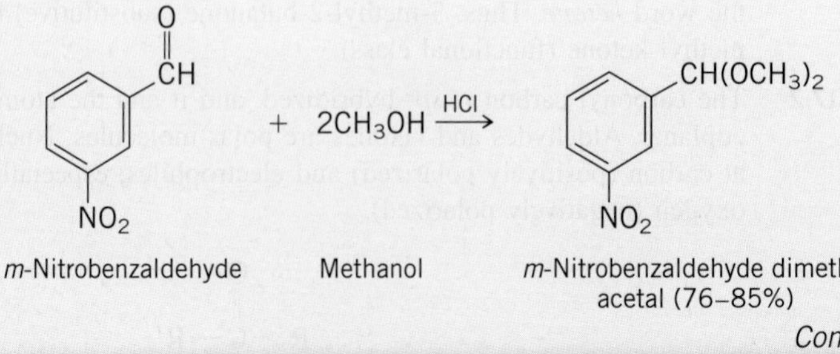 *m*-Nitrobenzaldehyde; Methanol; *m*-Nitrobenzaldehyde dimethyl acetal (76–85%)

Continued

TABLE 17.5 Nucleophilic Addition to Aldehydes and Ketones (*Continued*)

Reaction (section) and comments	General equation and typical example
Reaction with primary amines (Section 17.10) Isolated product is an imine (Schiff's base). A carbinolamine intermediate is formed, which undergoes dehydration to an imine.	$\overset{\text{O}}{\overset{\Vert}{\text{RCR}'}} + \text{R}''\text{NH}_2 \rightleftharpoons \overset{\text{NR}''}{\overset{\Vert}{\text{RCR}'}} + \text{H}_2\text{O}$ Aldehyde or ketone + Primary amine ⇌ Imine + Water $(\text{CH}_3)_2\text{CH}\overset{\text{O}}{\overset{\Vert}{\text{C}}}\text{H} + (\text{CH}_3)_3\text{CNH}_2 \longrightarrow (\text{CH}_3)_2\text{CHCH}{=}\text{NC}(\text{CH}_3)_3$ 2-Methylpropanal + *tert*-Butylamine → *N*-(2-Methyl-1-propylidene)-*tert*-butylamine (50%)
Reaction with secondary amines (Section 17.11) Isolated product is an enamine. Carbinolamine intermediate cannot dehydrate to a stable imine.	$\overset{\text{O}}{\overset{\Vert}{\text{R}}}\text{CCH}_2\text{R}' + \text{R}''_2\text{NH} \rightleftharpoons \text{R}\overset{\text{R}''\text{NR}''}{\overset{\vert}{\text{C}}}{=}\text{CHR}' + \text{H}_2\text{O}$ Aldehyde or ketone + Secondary amine ⇌ Enamine + Water 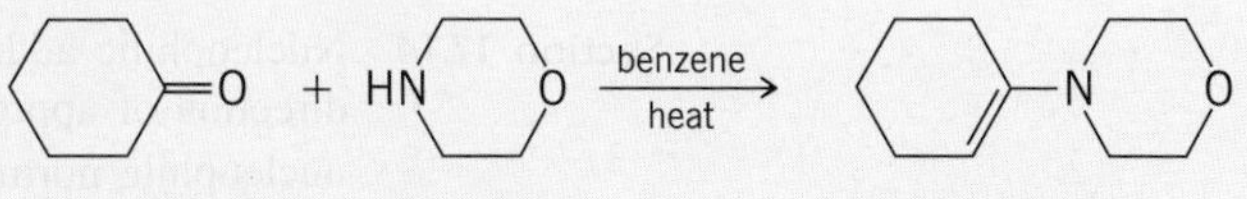 Cyclohexanone + Morpholine → 1-Morpholinocyclohexene (85%)
The Wittig reaction (Sections 17.12–17.13) Reaction of a phosphorus ylide with aldehydes and ketones leads to the formation of an alkene. A versatile method for the regiospecific preparation of alkenes.	$\overset{\text{O}}{\overset{\Vert}{\text{RCR}'}} + (\text{C}_6\text{H}_5)_3\overset{+}{\text{P}}{-}\overset{-}{\ddot{\text{C}}}\text{AB} \longrightarrow \text{RR}'\text{C}{=}\text{CAB} + (\text{C}_6\text{H}_5)_3\overset{+}{\text{P}}{-}\text{O}^-$ Aldehyde or ketone + Wittig reagent (an ylide) → Alkene + Triphenylphosphine oxide $\text{CH}_3\overset{\text{O}}{\overset{\Vert}{\text{C}}}\text{CH}_3 + (\text{C}_6\text{H}_5)_3\overset{+}{\text{P}}{-}\overset{-}{\ddot{\text{C}}}\text{HCH}_2\text{CH}_2\text{CH}_2\text{CH}_3 \xrightarrow{\text{DMSO}} (\text{CH}_3)_2\text{C}{=}\text{CHCH}_2\text{CH}_2\text{CH}_2\text{CH}_3 + (\text{C}_6\text{H}_5)_3\overset{+}{\text{P}}{-}\text{O}^-$ Acetone + 1-Pentylidenetriphenylphosphorane → 2-Methyl-2-heptene (56%) + Triphenylphosphine oxide

Aldehydes undergo nucleophilic addition more readily and have more favorable equilibrium constants for addition than do ketones.

The step in which the nucleophile attacks the carbonyl carbon is rate-determining in both base-catalyzed and acid-catalyzed nucleophilic addition. In the base-catalyzed mechanism this is the first step.

$$\text{Y}:^- + \text{C}{=}\ddot{\text{O}}: \underset{}{\overset{\text{slow}}{\rightleftharpoons}} \text{Y}{-}\text{C}{-}\ddot{\text{O}}:^-$$

Nucleophile, Aldehyde or ketone

$$\text{Y}{-}\text{C}{-}\ddot{\text{O}}:^- + \text{H}{-}\text{Y} \overset{\text{fast}}{\rightleftharpoons} \text{Y}{-}\text{C}{-}\ddot{\text{O}}\text{H} + \text{Y}:^-$$

Product of nucleophilic addition

Under conditions of acid catalysis, the nucleophilic addition step follows protonation of the carbonyl oxygen. Protonation increases the carbocation character of a carbonyl group and makes it more electrophilic.

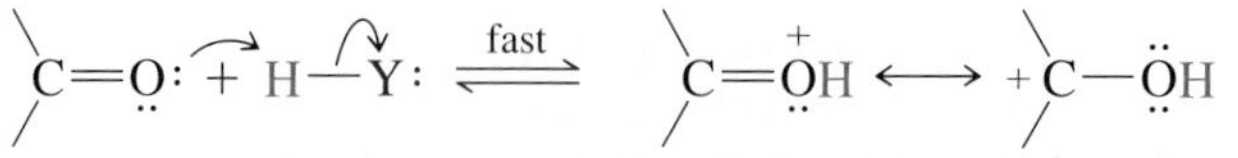

Aldehyde or ketone

Resonance contributors to protonated aldehyde or ketone

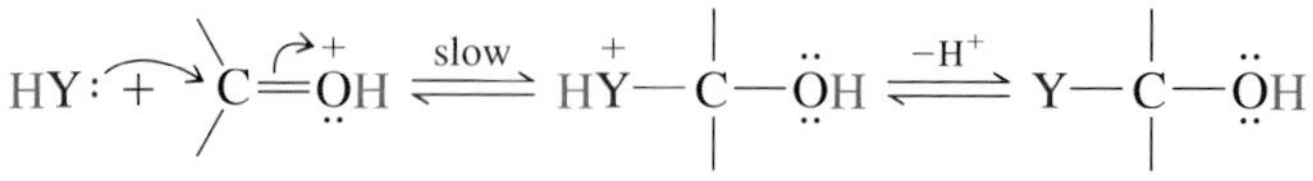

Product of nucleophilic addition

Often the product of nucleophilic addition is not isolated but is an intermediate leading to the ultimate product. Most of the reactions in Table 17.5 are of this type.

Section 17.14 Nucleophilic addition to the carbonyl group can be *stereoselective.* When one direction of approach to the carbonyl group is less hindered than the other, the nucleophile normally attacks at the less hindered face.

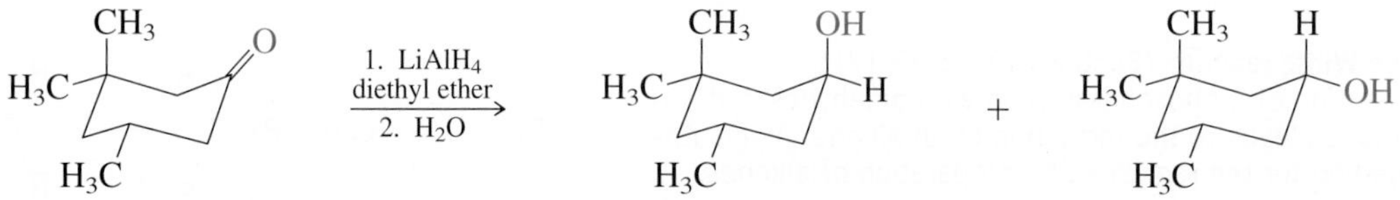

3,3,5-Trimethylcyclohexanone

trans-3,3,5-Trimethylcyclohexanol (83%)

cis-3,3,5-Trimethylcyclohexanol (17%)

Section 17.15 Aldehydes are easily oxidized to carboxylic acids.

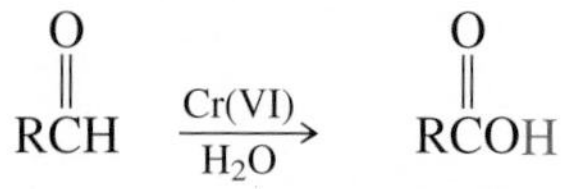

Aldehyde

Carboxylic acid

Section 17.16 A strong peak near 1700 cm^{-1} in the IR spectrum is characteristic of compounds that contain a C=O group. The ^1H and ^{13}C NMR spectra of aldehydes and ketones are affected by the deshielding of a C=O group. The proton of an H—C=O group appears in the δ 8–10 range. The carbon of a C=O group is at δ 190–210.

PROBLEMS

17.23 (a) Write structural formulas and provide IUPAC names for all the isomeric aldehydes and ketones that have the molecular formula $C_5H_{10}O$. Include stereoisomers.

(b) Which of the isomers in part (a) yield chiral alcohols on reaction with sodium borohydride?

(c) Which of the isomers in part (a) yield chiral alcohols on reaction with methylmagnesium iodide?

17.24 Each of the following aldehydes or ketones is known by a common name. Its substitutive IUPAC name is provided in parentheses. Write a structural formula for each one.

(a) Chloral (2,2,2-trichloroethanal)

(b) Pivaldehyde (2,2-dimethylpropanal)

(c) Acrolein (2-propenal)

(d) Crotonaldehyde [(*E*)-2-butenal]

(e) Citral [(*E*)-3,7-dimethyl-2,6-octadienal]

(f) Diacetone alcohol (4-hydroxy-4-methyl-2-pentanone)

(g) Carvone (5-isopropenyl-2-methyl-2-cyclohexenone)

(h) Biacetyl (2,3-butanedione)

17.25 The African dwarf crocodile secretes a volatile substance believed to be a sex pheromone. It is a mixture of two stereoisomers, one of which is shown:

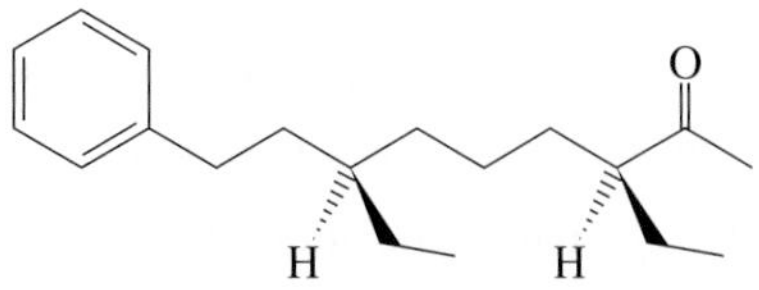

(a) Give the IUPAC name for this compound, including *R* and *S* descriptors for its chirality centers.

(b) One component of the scent substance has the *S* configuration at both chirality centers. How is this compound related to the one shown? Are the compounds enantiomers, or diastereomers?

17.26 Predict the product of the reaction of propanal with each of the following:

(a) Lithium aluminum hydride, followed by water

(b) Sodium borohydride, methanol

(c) Hydrogen (nickel catalyst)

(d) Methylmagnesium iodide, followed by dilute acid

(e) Sodium acetylide, followed by dilute acid

(f) Phenyllithium, followed by dilute acid

(g) Methanol containing dissolved hydrogen chloride

(h) Ethylene glycol, *p*-toluenesulfonic acid, benzene

(i) Aniline ($C_6H_5NH_2$)

(j) Dimethylamine, *p*-toluenesulfonic acid, benzene

(k) Hydroxylamine

(l) Hydrazine

(m) Product of part (l) heated in triethylene glycol with sodium hydroxide

(n) *p*-Nitrophenylhydrazine

(o) Semicarbazide

(p) Ethylidenetriphenylphosphorane [$(C_6H_5)_3\overset{+}{P}—\overset{..}{\overset{-}{C}}HCH_3$]

(q) Sodium cyanide with addition of sulfuric acid

(r) Chromic acid

17.27 Repeat the preceding problem for cyclopentanone instead of propanal.

17.28 Hydride reduction (with $LiAlH_4$ or $NaBH_4$) of each of the following ketones has been reported in the chemical literature and gives a mixture of two diastereomeric alcohols in each case. Give the structures of both alcohol products for each ketone.

(a) (*S*)-3-Phenyl-2-butanone

(b) 4-*tert*-Butylcyclohexanone

(c)

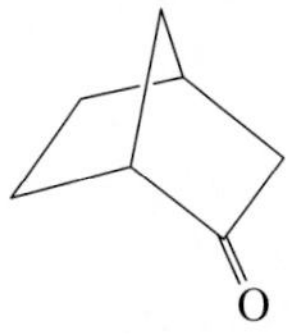

(d)

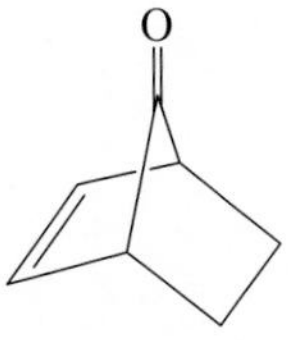

17.29 Choose which member in each of the following pairs reacts faster or has the more favorable equilibrium constant for reaction with the indicated reagent. Explain your reasoning.

(a) $C_6H_5\overset{O}{\overset{\|}{C}}H$ or $C_6H_5\overset{O}{\overset{\|}{C}}CH_3$ (rate of reduction with sodium borohydride)

(b) $Cl_3C\overset{O}{\overset{\|}{C}}H$ or $CH_3\overset{O}{\overset{\|}{C}}H$ (equilibrium constant for hydration)

(c) Acetone or 3,3-dimethyl-2-butanone (equilibrium constant for cyanohydrin formation)

(d) Acetone or 3,3-dimethyl-2-butanone (rate of reduction with sodium borohydride)

(e) $CH_2(OCH_2CH_3)_2$ or $(CH_3)_2C(OCH_2CH_3)_2$ (rate of acid-catalyzed hydrolysis)

17.30 Equilibrium constants for the dissociation (K_{diss}) of cyanohydrins according to the equation

$$R\underset{CN}{\underset{|}{\overset{OH}{\overset{|}{C}}}}R' \xrightleftharpoons{K_{diss}} R\overset{O}{\overset{\|}{C}}R' + HCN$$

Cyanohydrin — Aldehyde or ketone — Hydrogen cyanide

have been measured for a number of cyanohydrins. Which cyanohydrin in each of the following pairs has the greater dissociation constant?

(a) $CH_3CH_2\overset{OH}{\overset{|}{C}}HCN$ or $(CH_3)_2\overset{OH}{\overset{|}{C}}CN$

(b) $C_6H_5\overset{OH}{\overset{|}{C}}HCN$ or $C_6H_5\underset{CH_3}{\underset{|}{\overset{OH}{\overset{|}{C}}}}CN$

17.31 Each of the following reactions has been reported in the chemical literature and gives a single organic product in good yield. What is the principal product in each reaction?

(a) 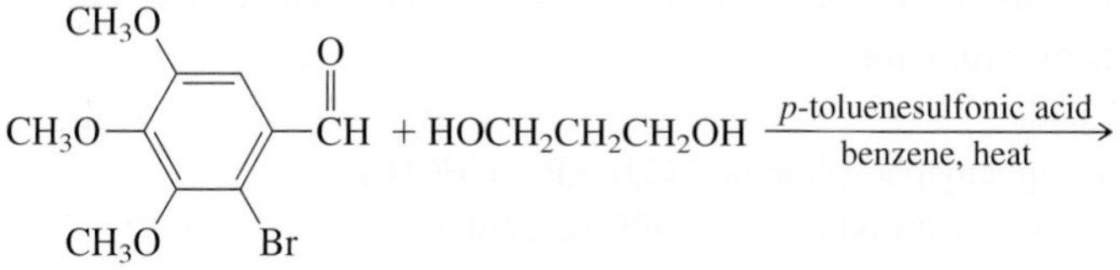

(b) 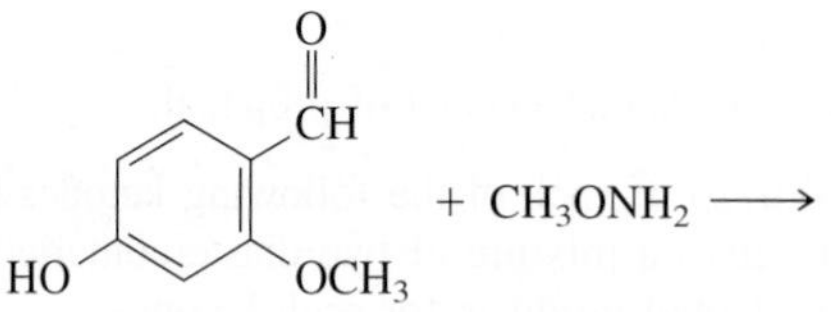

(c) $CH_3CH_2\overset{O}{\overset{\|}{C}}H + (CH_3)_2NNH_2 \longrightarrow$

(d) 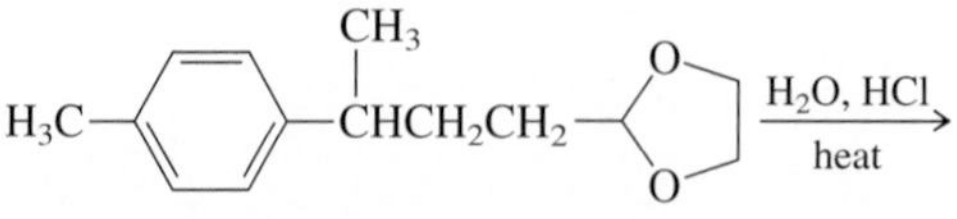

(e) $C_6H_5\overset{O}{\overset{\|}{C}}CH_3 \xrightarrow[HCl]{NaCN}$

(f) 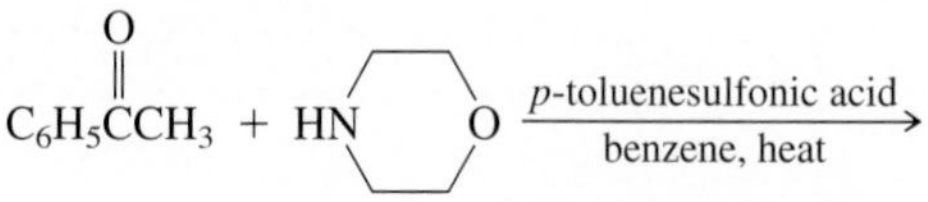

(g) $(CH_3)_2CH\overset{\overset{\displaystyle O}{\|}}{C}CH(CH_3)_2 + HOCH_2CH_2SH \xrightarrow[\text{benzene, heat}]{p\text{-toluenesulfonic acid}} C_9H_{18}OS$

17.32 Wolff–Kishner reduction (hydrazine, KOH, ethylene glycol, 130°C) of the compound shown gave compound A. Treatment of compound A with *m*-chloroperoxybenzoic acid (MCPBA) gave compound B, which on reduction with lithium aluminum hydride gave compound C. Oxidation of compound C with chromic acid gave compound D ($C_9H_{14}O$). Identify compounds A through D in this sequence.

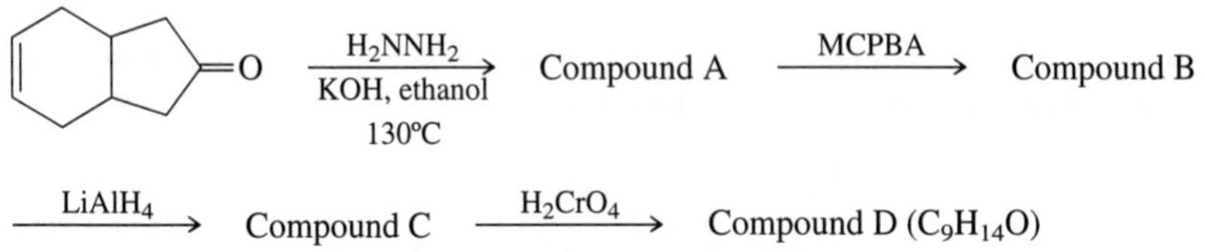

17.33 On standing in ^{17}O-labeled water, both formaldehyde and its hydrate are found to have incorporated the ^{17}O isotope of oxygen. Suggest a reasonable explanation for this observation.

17.34 Reaction of benzaldehyde with 1,2-octanediol in benzene containing a small amount of *p*-toluenesulfonic acid yields almost equal quantities of two products in a combined yield of 94%. Both products have the molecular formula $C_{15}H_{22}O_2$. Suggest reasonable structures for these products.

17.35 Compounds that contain both carbonyl and alcohol functional groups are often more stable as cyclic hemiacetals or cyclic acetals than as open-chain compounds. Examples of several of these are shown. Deduce the structure of the open-chain form of each.

(a) 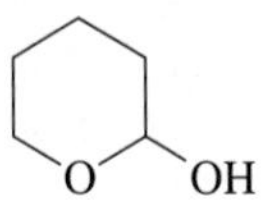

(b) 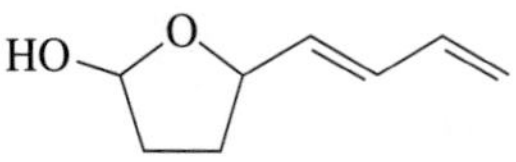

(c) 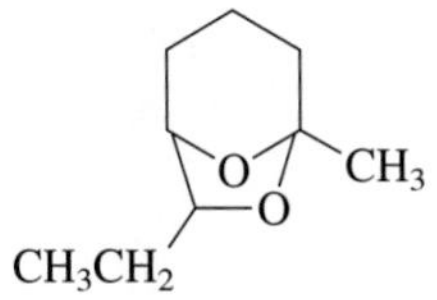

Brevicomin (sex attractant of Western pine beetle)

(d) 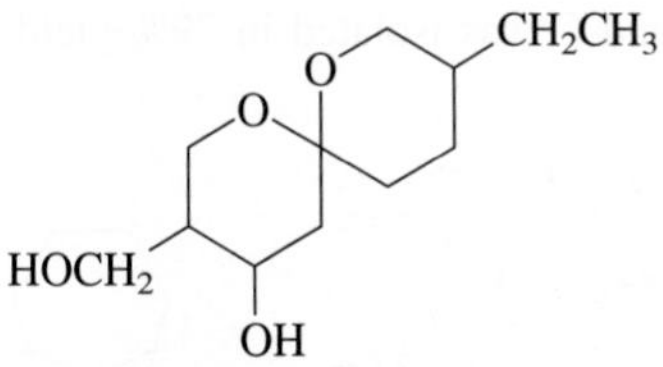

Talaromycin A (a toxic substance produced by a fungus that grows on poultry house litter)

17.36 The OH groups at C-4 and C-6 of methyl α-D-glucopyranoside can be protected by conversion to a benzylidene acetal. What reagents are needed for this conversion?

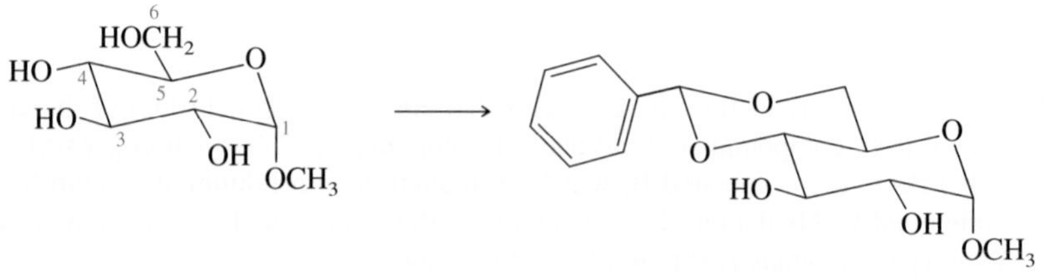
6
HOCH2
HO
HO
O
4
5
2
3
1
OH
OCH3
O
O
O
HO
OH
OCH3

Methyl α–D–glucopyranoside

Methyl 4,6-O-benzylidene-α-D-glucopyranoside

17.37 Compounds that contain a carbon–nitrogen double bond are capable of stereoisomerism much like that seen in alkenes. The structures

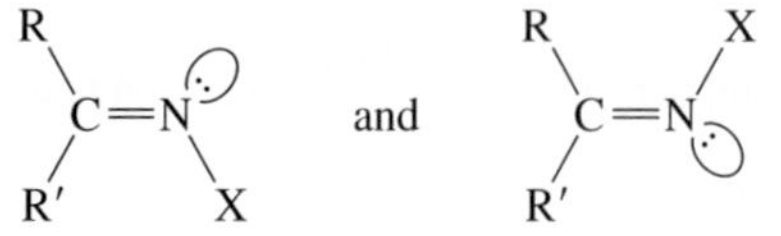
R
C=N
R′
X
and
R
X
C=N
R′

are stereoisomeric. Specifying stereochemistry in these systems is best done by using *E–Z* descriptors and considering the nitrogen lone pair to be the lowest priority group. Write the structures, clearly showing stereochemistry, of the following:

(a) (*Z*)-$CH_3CH{=}NCH_3$
(b) (*E*)-Acetaldehyde oxime
(c) (*Z*)-2-Butanone hydrazone
(d) (*E*)-Acetophenone semicarbazone

17.38 Suggest a reasonable mechanism for each of the following reactions:

(a)
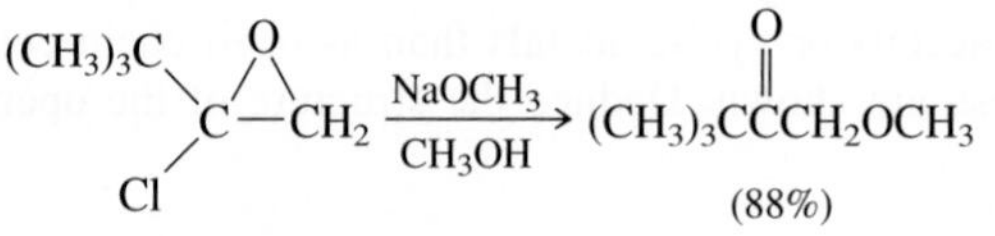
(CH3)3C
O
C—CH2
Cl
NaOCH3
CH3OH
O
(CH3)3CCCH2OCH3
(88%)

(b)
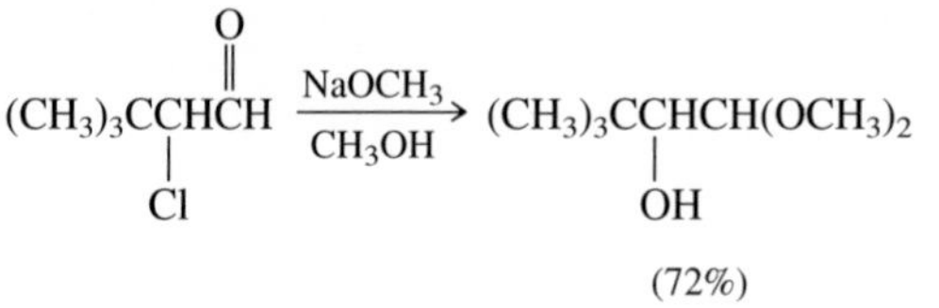
O
(CH3)3CCHCH
Cl
NaOCH3
CH3OH
(CH3)3CCHCH(OCH3)2
OH
(72%)

17.39 Describe reasonable syntheses of benzophenone, $C_6H_5\overset{\overset{O}{\|}}{C}C_6H_5$, from each of the following starting materials and any necessary inorganic reagents.

(a) Benzoyl chloride and benzene
(b) Benzyl alcohol and bromobenzene
(c) Bromodiphenylmethane, $(C_6H_5)_2CHBr$
(d) Dimethoxydiphenylmethane, $(C_6H_5)_2C(OCH_3)_2$
(e) 1,1,2,2-Tetraphenylethene, $(C_6H_5)_2C{=}C(C_6H_5)_2$

17.40 After heating compound A with a catalytic amount of *p*-toluenesulfonic acid and water in dichloromethane (45°C) for 24 hr, compound C was isolated in 79% yield.

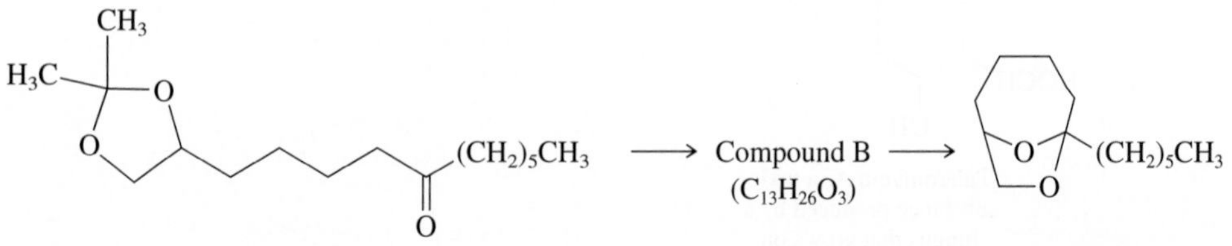
CH3
H3C
O
O
(CH2)5CH3
O
Compound B
(C13H26O3)
O
O
(CH2)5CH3

Compound A ($C_{16}H_{30}O_3$)

Compound C ($C_{13}H_{24}O_2$)

Demonstrate your understanding of the overall reaction by identifying the key intermediate (compound B). What other compound is formed in the reaction?

17.41 Studies of the sex pheromone of the Douglas fir tussock moth required the synthesis of (*E*)-1,6-henicosadien-11-one. Outline a synthesis of this ketone using (*E*)-5,10-undecadien-1-ol and 1-decanol as sources of all of the carbons.

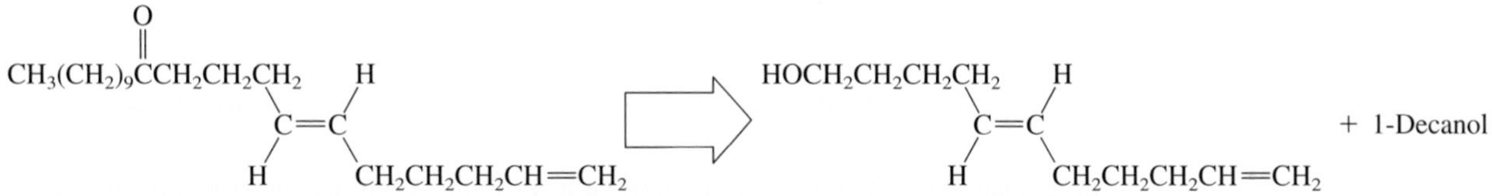

17.42 The sex attractant of the female winter moth has been identified as the tetraene $CH_3(CH_2)_8CH{=}CHCH_2CH{=}CHCH_2CH{=}CHCH{=}CH_2$. Devise a synthesis of this material from 3,6-hexadecadien-1-ol and allyl alcohol.

17.43 Leukotrienes are substances produced in the body that may be responsible for inflammatory effects (Section 24.6). In a synthesis of one of the members of this family of compounds, compound A reacted with (carboethoxymethylidene)triphenylphosphorane to give compound B as a mixture of 86:14 *E* and *Z* alkenes. Suggest a mechanism for this reaction.

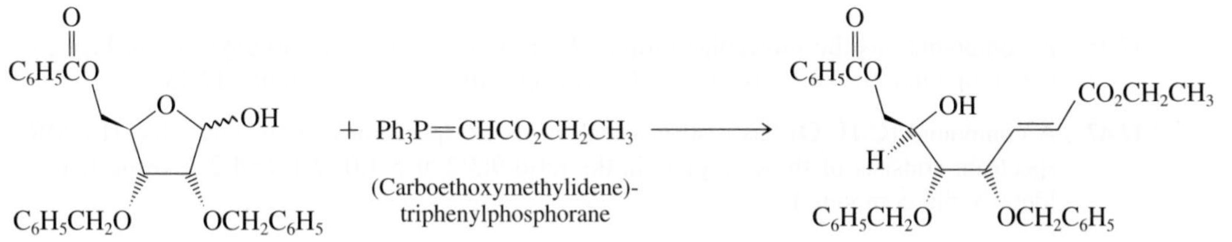

Compound A | Compound B

17.44 Syntheses of each of the following compounds have been reported in the chemical literature. Using the indicated starting material and any necessary organic or inorganic reagents, describe short sequences of reactions that would be appropriate for each transformation.

(a) 1,1,5-Trimethylcyclononane from 5,5-dimethylcyclononanone

(b) 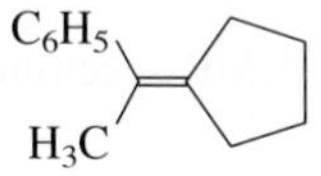 from 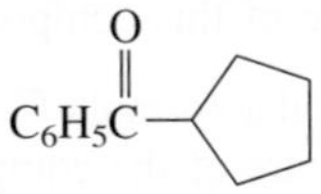

(c) 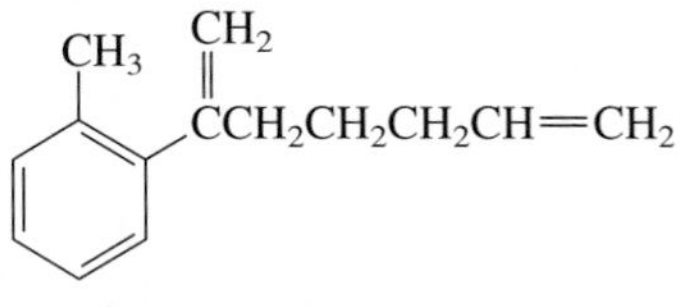 from *o*-bromotoluene and 5-hexenal

(d) $CH_3\overset{O}{\overset{\|}{C}}CH_2CH_2\overset{O}{\overset{\|}{C}}(CH_2)_5CH_3$ from $HC{\equiv}CCH_2CH_2CH_2OH$

(e) 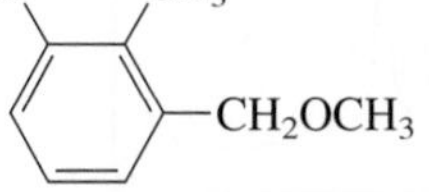 from 3-chloro-2-methylbenzaldehyde

(f) 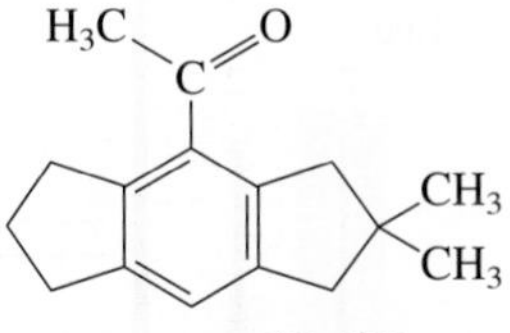 from 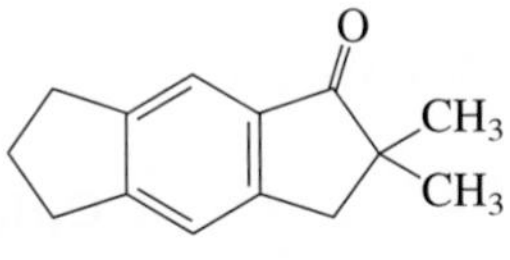

(g) 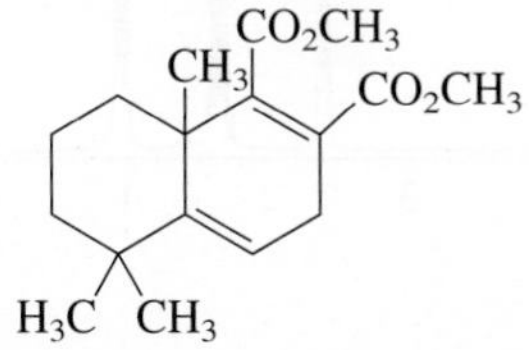 from 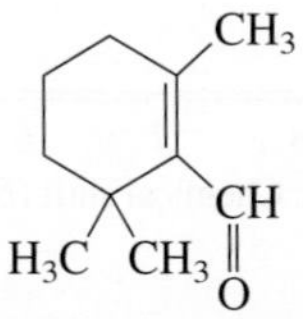

17.45 Alcohols react with dihydropyran in the presence of acid to give products of general structure I, rather than II. Note that I is an acetal.

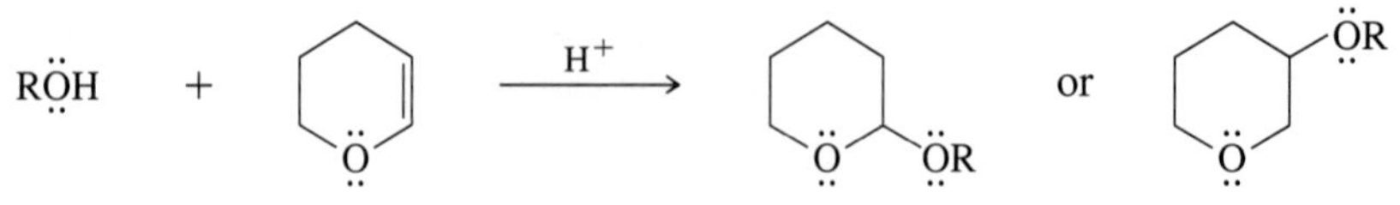

I II

Dihydropyran not formed

(a) Explain why I is formed in preference to II. (*Hint:* Consider the structures of the two carbocations that would form by protonation of dihydropyran at either alkene carbon.)

(b) From what alcohol and dihydropyran can the compound shown be synthesized?

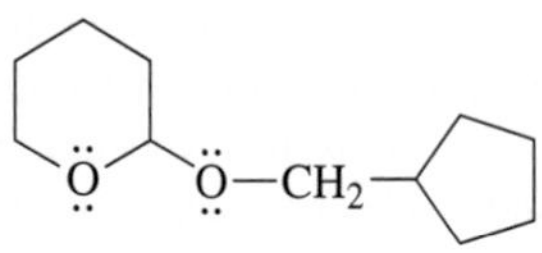

17.46 A compound has the molecular formula C_4H_8O and contains a carbonyl group. Identify the compound on the basis of its ^1H NMR spectrum shown in Figure 17.15.

17.47 A compound ($C_7H_{14}O$) has a strong peak in its IR spectrum at 1710 cm^{-1}. Its ^1H NMR spectrum consists of three singlets in the ratio 9:3:2 at δ 1.0, 2.1, and 2.3, respectively. Identify the compound.

17.48 Compounds A and B are isomeric diketones of molecular formula $C_6H_{10}O_2$. The ^1H NMR spectrum of compound A contains two signals, both singlets, at δ 2.2 (six protons) and 2.8 (four protons). The ^1H NMR spectrum of compound B contains two signals, one at δ 1.3 (triplet, six protons) and the other at δ 2.8 (quartet, four protons). What are the structures of compounds A and B?

17.49 A compound ($C_{11}H_{14}O$) has the (*a*) IR and (*b*) 200-MHz ^1H NMR spectra shown in Figure 17.16. What is the structure of this compound?

17.50 A compound is a ketone of molecular formula $C_7H_{14}O$. Its ^{13}C NMR spectrum is shown in Figure 17.17. What is the structure of the compound?

Figure 17.15

The 200-MHz ^1H NMR spectrum of a compound (C_4H_8O) (Problem 17.46).

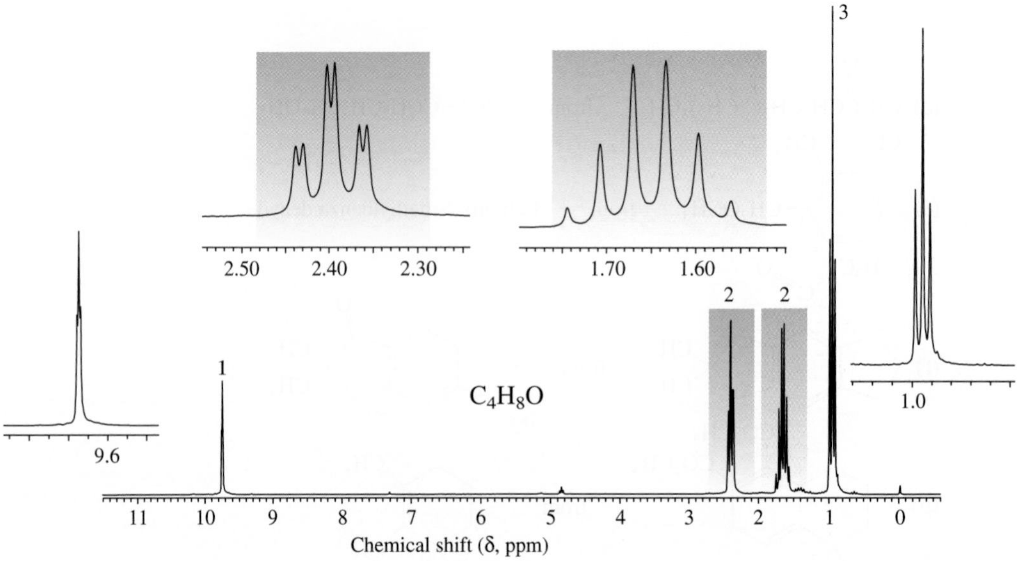

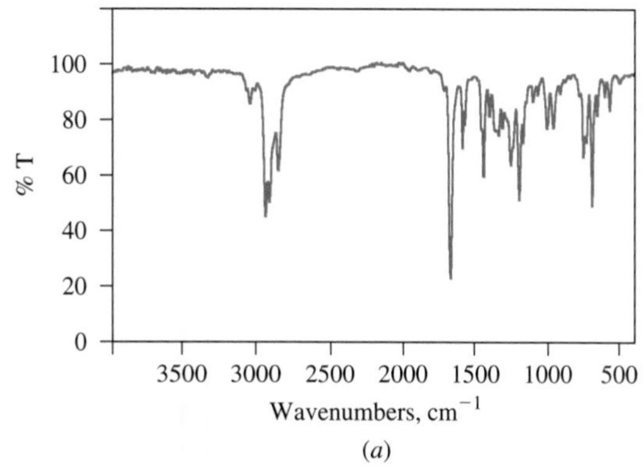

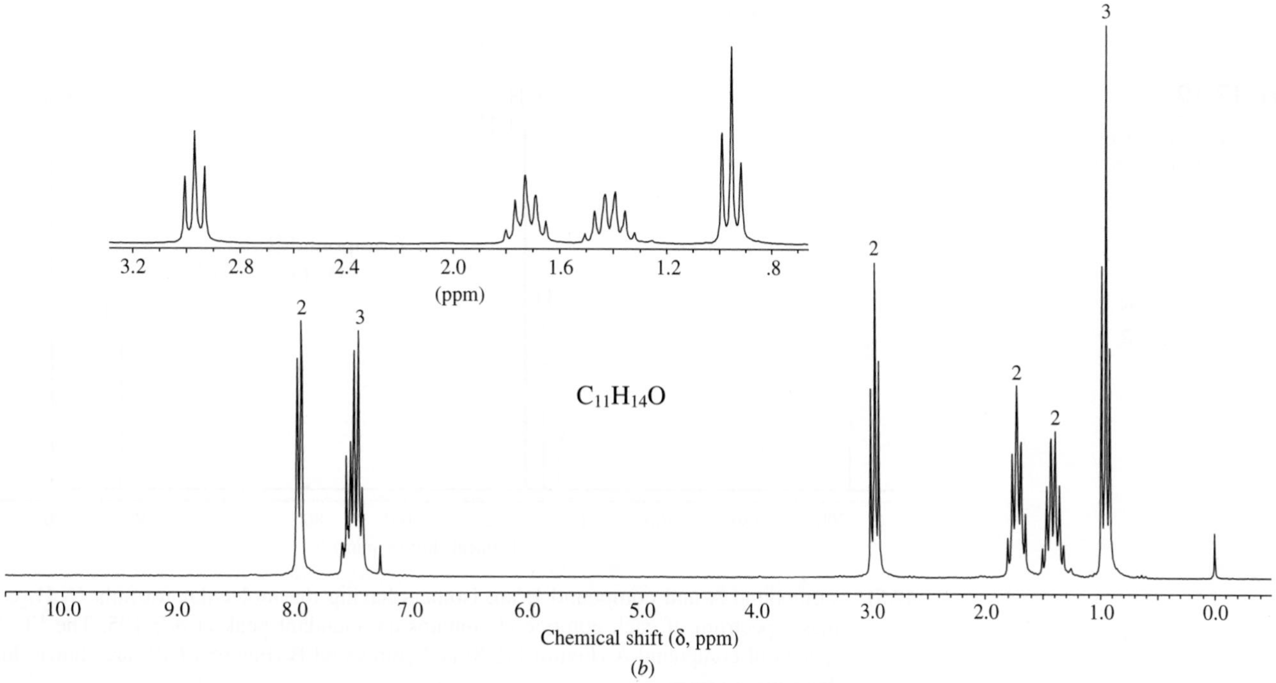

Figure 17.16

The (*a*) IR and (*b*) 200-MHz ^1H NMR spectra of a compound ($C_{11}H_{14}O$) (Problem 17.49).

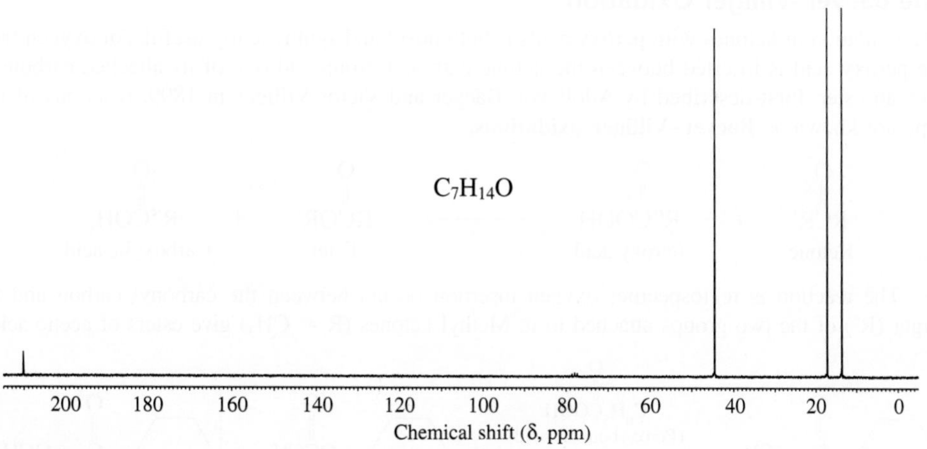

Figure 17.17

The ^{13}C NMR spectrum of a compound ($C_7H_{14}O$) (Problem 17.50).

Figure 17.18

The ^{13}C NMR spectrum of compound A ($C_{10}H_{12}O$) (Problem 17.51).

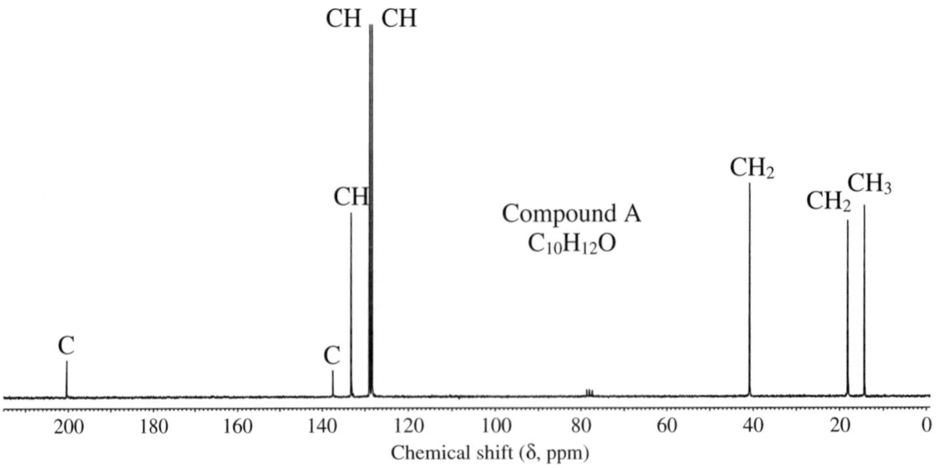

Figure 17.19

The ^{13}C NMR spectrum of compound B ($C_{10}H_{12}O$) (Problem 17.51).

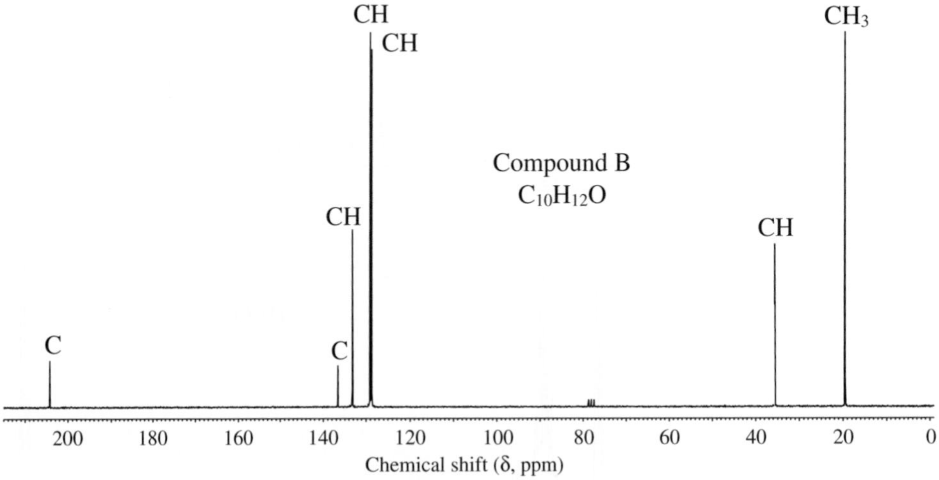

17.51 Compound A and compound B are isomers having the molecular formula $C_{10}H_{12}O$. The mass spectrum of each compound contains an abundant peak at *m/z* 105. The ^{13}C NMR spectra of compound A (Figure 17.18) and compound B (Figure 17.19) are shown. Identify these two isomers.

Descriptive Passage and Interpretive Problems 17

The Baeyer–Villiger Oxidation

The oxidation of ketones with peroxy acids is both novel and synthetically useful. An oxygen from the peroxy acid is inserted between the ketone carbonyl group and one of its attached carbons to give an ester. First described by Adolf von Baeyer and Victor Villiger in 1899, reactions of this type are known as **Baeyer–Villiger oxidations.**

$$\underset{\text{Ketone}}{R\overset{O}{\overset{\|}{C}}R'} + \underset{\text{Peroxy acid}}{R''\overset{O}{\overset{\|}{C}}OOH} \longrightarrow \underset{\text{Ester}}{R\overset{O}{\overset{\|}{C}}OR'} + \underset{\text{Carboxylic acid}}{R''\overset{O}{\overset{\|}{C}}OH}$$

The reaction is regiospecific; oxygen insertion occurs between the carbonyl carbon and the larger (R′) of the two groups attached to it. Methyl ketones (R = CH_3) give esters of acetic acid;

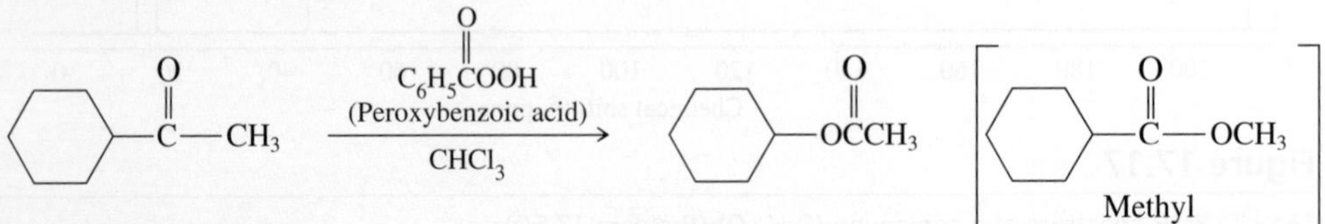

Cyclohexyl methyl ketone — Cyclohexyl acetate (67%) — Methyl cyclohexanecarboxylate (not observed)

The mechanism of the Baeyer–Villiger reaction begins with nucleophilic addition of the peroxy acid to the carbonyl group.

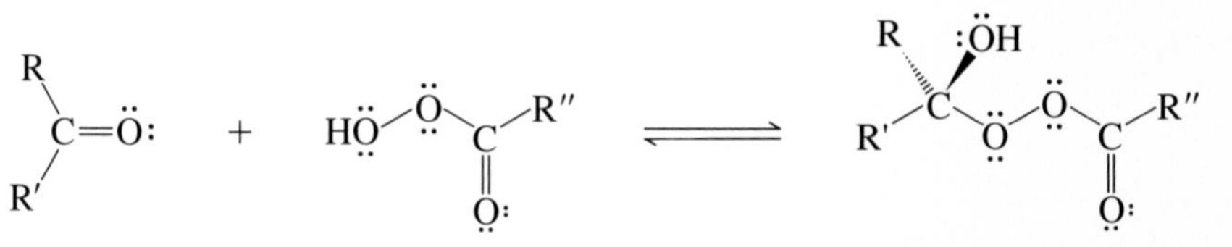

Ketone | Peroxy acid | Product of nucleophilic addition of peroxy acid to ketone

After protonation by an acid catalyst (either the peroxy acid or a carboxylic acid), the conjugate acid of the product of the first step rearranges by an alkyl group migration. Normally, it is the larger of the two groups originally bonded to the carbonyl group that migrates.

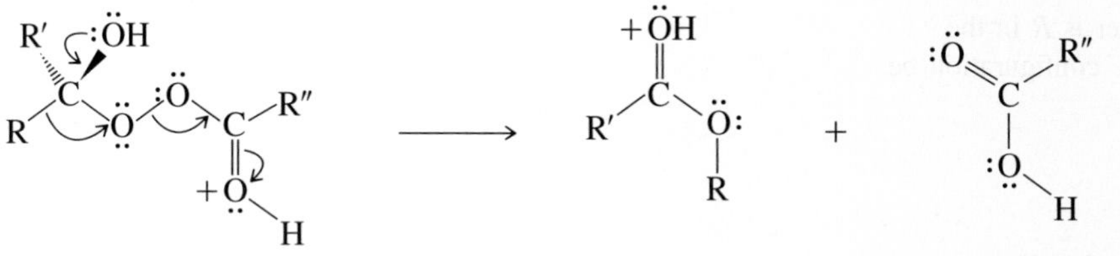

Conjugate acid of product of nucleophilic addition | Conjugate acid of ester | Carboxylic acid

The reaction is stereospecific; the alkyl group migrates with retention of configuration, as illustrated for the oxidation of *cis*-1-acetyl-2-methylcyclopentane; only the cis product is obtained.

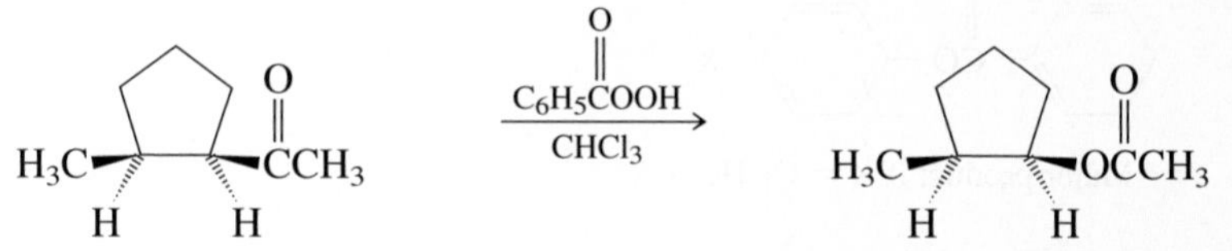

cis-1-Acetyl-2-methylcyclopentane | *cis*-2-Methylcyclopentyl acetate (66%)

When the ketone is cyclic, a cyclic ester, or *lactone,* is formed. Cyclobutanone is oxidized to a lactone by the Baeyer–Villiger reaction.

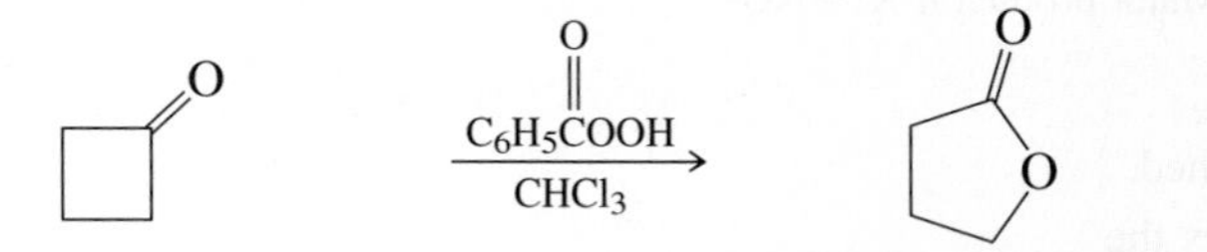

Cyclobutanone | γ-Butyrolactone (70%)

17.52 Which of the following are *not* intermediates in the Baeyer–Villiger oxidation of cyclohexyl methyl ketone with peroxybenzoic acid?

A. I and II
B. III and IV
C. I and III
D. II and IV

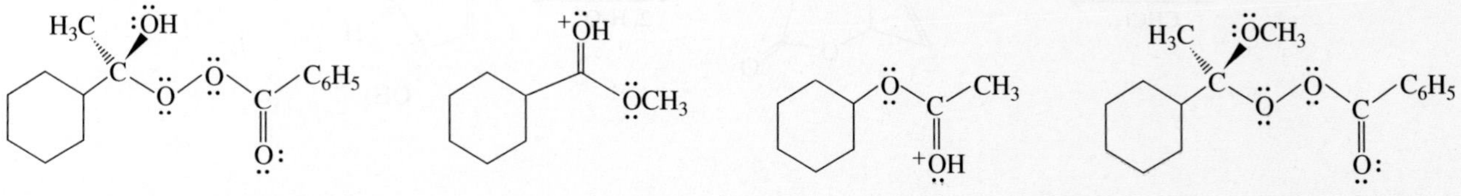

I. II. III. IV.

17.53 Which is the product of the following reaction?

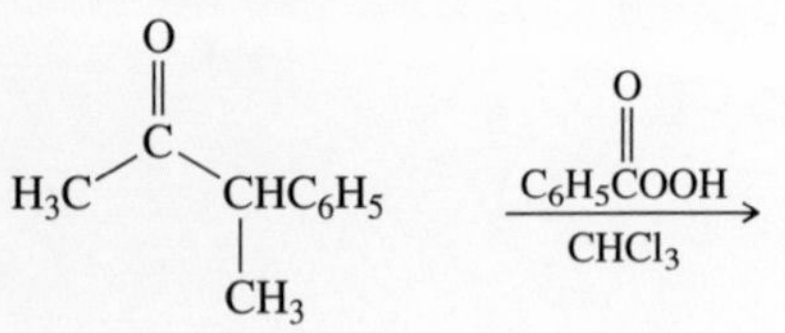

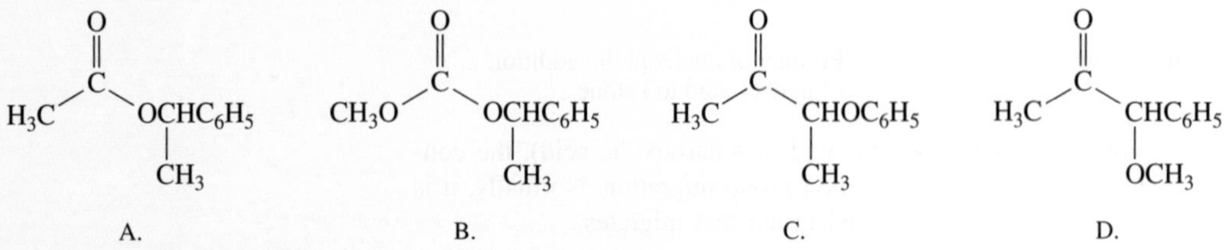

17.54 If the configuration of the chirality center is *R* in the reactant in Problem 17.53, what will the configuration be at this carbon in the product?

A. *R*
B. *S*
C. an equal mixture of *R* and *S*
D. an unequal mixture of *R* and *S*

17.55 The Baeyer–Villiger oxidations of the substituted diphenyl ketones proceeds as indicated because:

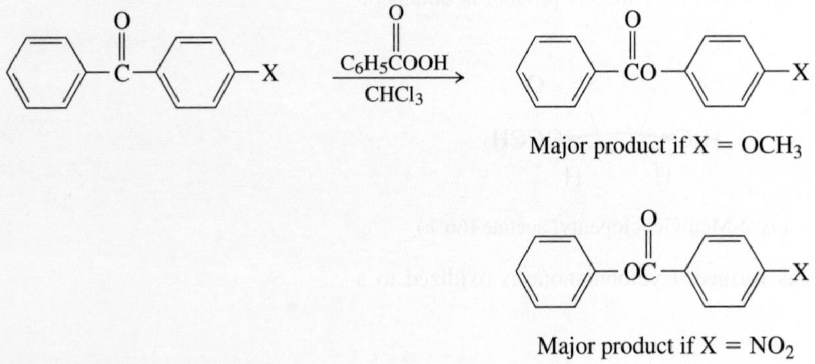

A. The electron-withdrawing nitro group retards the migration of the phenyl ring to which it is attached.
B. The electron-releasing methoxy group accelerates the migration of the phenyl ring to which it is attached.
C. Both A and B
D. Neither A nor B. The regiospecificity is due to steric effects in the migrating group.

17.56 A key step in the synthesis of an important class of lipids known as the prostaglandins involves the sequence shown here. What is the identity of compound X?

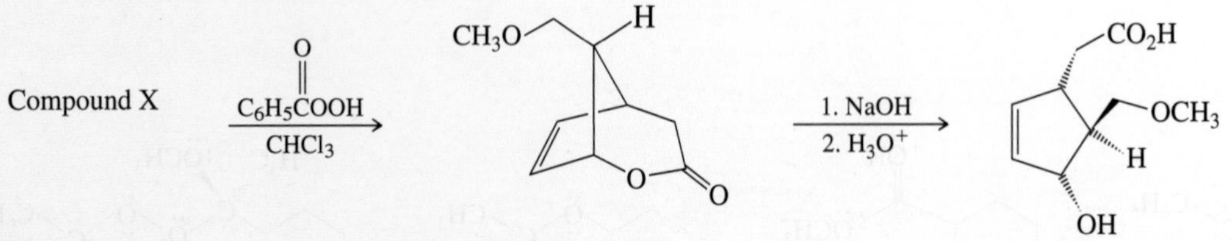

A. 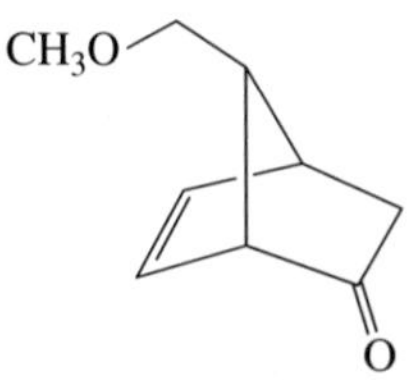

B. 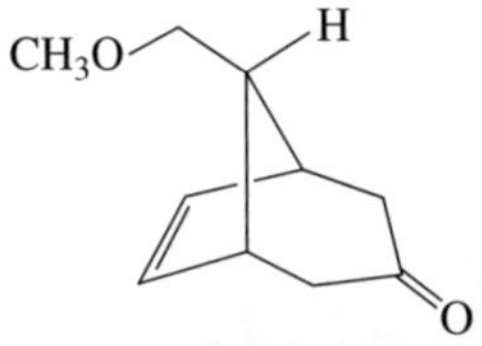

C. 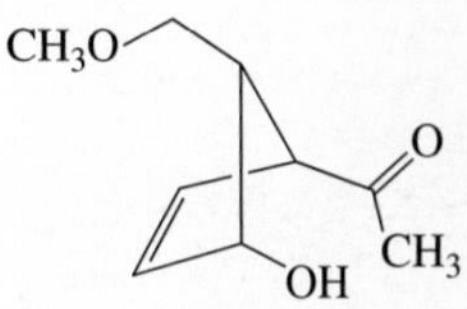

17.57 A reaction analogous to the Baeyer–Villiger reaction occurs in living systems through the action of enzymes in certain bacteria, for example, species of *Pseudomonas* and *Acinetobacter*. A preparation of the *S* enantiomer of compound Y has been described using a bacterial cyclohexanone monooxygenase enzyme system. What is compound X?

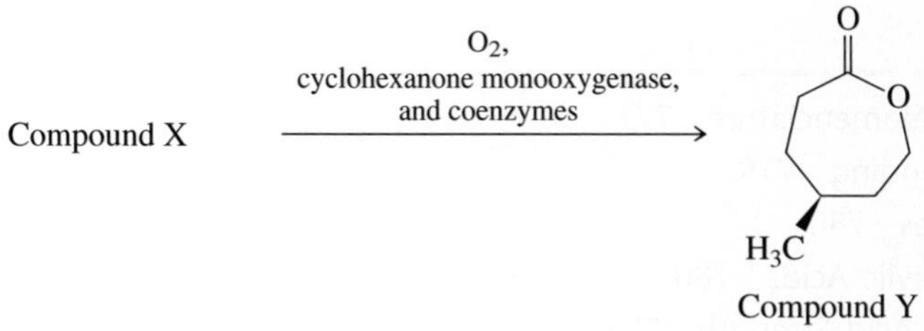

A. 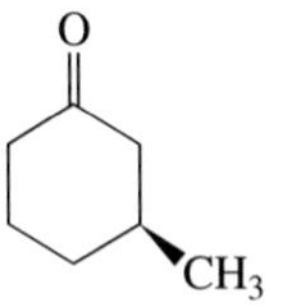

B. 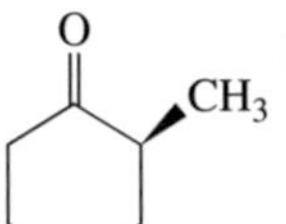

C. 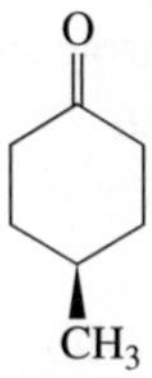

D. 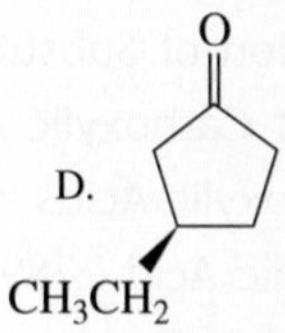

17.58 If compound Y is prepared by treatment of compound X with peroxybenzoic acid, compound Y would be obtained as:

A. Only the *S* enantiomer
B. Only the *R* enantiomer
C. A racemic mixture
D. An unequal mixture of *R* and *S* enantiomers

18 Carboxylic Acids

Chapter Outline

Mechanisms

This runner may experience discomfort from the lactic acid that formed in her muscles during her run. The discomfort will be gone in a day or so.

CARBOXYLIC ACIDS, compounds of the type $\mathrm{R\overset{O}{\overset{\|}{C}}OH}$, constitute one of the most frequently encountered classes of organic compounds. Countless natural products are carboxylic acids or are derived from them. Some carboxylic acids, such as acetic acid, have been known for centuries. Others, such as the prostaglandins, which are powerful regulators of numerous biological processes, remained unknown until relatively recently. Still others, aspirin for example, are the products of chemical synthesis. The therapeutic effects of aspirin, known for well over a century, are now understood to result from aspirin's ability to inhibit the biosynthesis of prostaglandins.

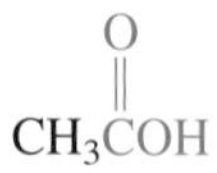

Acetic acid (present in vinegar)

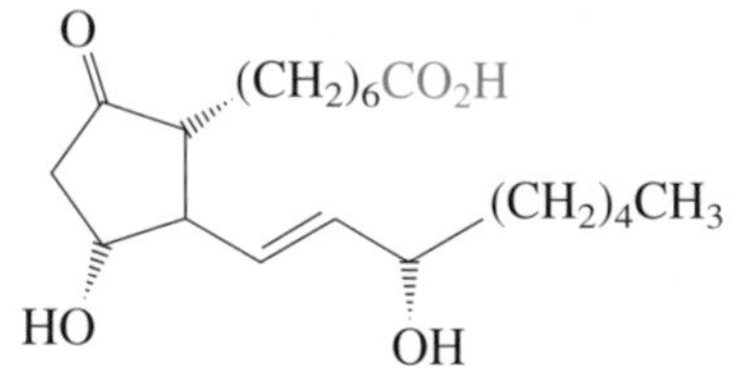

PGE_1 (a prostaglandin; a small amount of PGE_1 lowers blood pressure significantly)

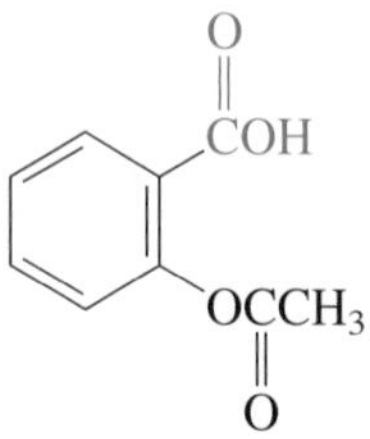

Aspirin

The importance of carboxylic acids is magnified when we realize that they are the parent compounds of a large group of derivatives that includes acyl chlorides, acid anhydrides, esters, and amides. Those classes of compounds will be discussed in Chapter 19. Together, this chapter and the next tell the story of some of the most fundamental structural types and functional group transformations in organic and biological chemistry.

18.1 Carboxylic Acid Nomenclature

It is hard to find a class of compounds in which the common names of its members have influenced organic nomenclature more than carboxylic acids. Not only are the common names of carboxylic acids themselves abundant and widely used, but the names of many other compounds are derived from them. Benzene took its name from benzoic acid and propane from *propionic* acid, not the other way around. The name butane comes from *butyric* acid, present in rancid butter. The common names of most aldehydes are derived from the common names of carboxylic acids—valeraldehyde from *valeric* acid, for example. Many carboxylic acids are better known by common names than by their systematic ones, and the framers of the IUPAC rules have taken

TABLE 18.1 Systematic and Common Names of Some Carboxylic Acids

	Structural formula	Systematic name	Common name*
1.	HCO_2H	Methanoic acid	Formic acid
2.	CH_3CO_2H	Ethanoic acid	Acetic acid
3.	$CH_3(CH_2)_3CO_2H$	Pentanoic acid	Valeric acid
4.	$CH_3(CH_2)_{16}CO_2H$	Octadecanoic acid	Stearic acid
5.	$CH_3\underset{OH}{\underset{\vert}{C}}HCO_2H$	2-Hydroxypropanoic acid	Lactic acid
6.	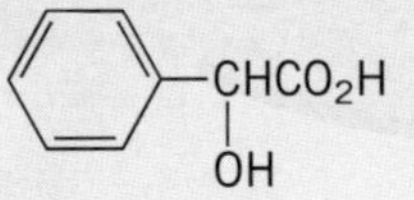	2-Hydroxy-2-phenylethanoic acid	Mandelic acid
7.	$H_2C{=}CHCO_2H$	Propenoic acid	Acrylic acid
8.	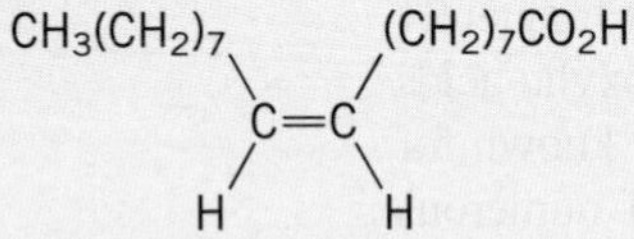	(*Z*)-9-Octadecenoic acid or (*Z*)-9-Octadec-9-enoic acid	Oleic acid
9.	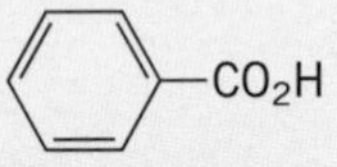	Benzenecarboxylic acid	Benzoic acid
10.	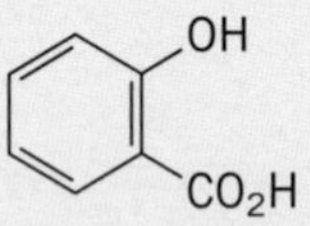	*o*-Hydroxybenzenecarboxylic acid	Salicylic acid
11.	$HO_2CCH_2CO_2H$	Propanedioic acid	Malonic acid
12.	$HO_2CCH_2CH_2CO_2H$	Butanedioic acid	Succinic acid
13.	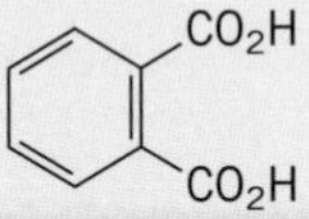	1,2-Benzenedicarboxylic acid	Phthalic acid

*Except for valeric, mandelic, and salicylic acid, all of the common names in this table are acceptable IUPAC names.

a liberal view toward accepting these common names as permissible alternatives to the systematic ones. Table 18.1 lists both common and systematic names for a number of important carboxylic acids.

Systematic names for carboxylic acids are derived by counting the number of carbons in the longest continuous chain that includes the carboxyl group and replacing the *-e* ending of the corresponding alkane by *-oic acid.* The first four acids in Table 18.1, methanoic (1 carbon), ethanoic (2 carbons), pentanoic (5 carbons) and octadecanoic acid (18 carbons), illustrate this point. When substituents are present, their locations are identified by number; numbering of the carbon chain always begins at the carboxyl group.

Notice that compounds 5 and 6 are named as hydroxy derivatives of carboxylic acids, rather than as carboxyl derivatives of alcohols. This parallels what we saw earlier in Section 17.1 where an aldehyde or ketone function took precedence over a hydroxyl group in defining the main chain. Carboxylic acids take precedence over all the common groups we have encountered to this point in respect to defining the main chain.

Double bonds in the main chain are signaled by the ending *-enoic acid,* and their position is designated by a numerical prefix. Entries 7 and 8 are representative carboxylic acids that contain double bonds. Double-bond stereochemistry is specified by using either the cis–trans or the *E–Z* notation.

When a carboxyl group is attached to a ring, the parent ring is named (retaining the final *-e*) and the suffix *-carboxylic acid* is added, as shown in entries 9 and 10.

Compounds with two carboxyl groups, as illustrated by entries 11 through 13, are distinguished by the suffix *-dioic acid* or *-dicarboxylic acid* as appropriate. The final *-e* in the name of the parent alkane is retained.

Problem 18.1

The list of carboxylic acids in Table 18.1 is by no means exhaustive insofar as common names are concerned. Many others are known by their common names, a few of which follow. Give a systematic IUPAC name for each.

(a) 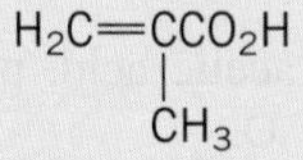

(Methacrylic acid)

(b) 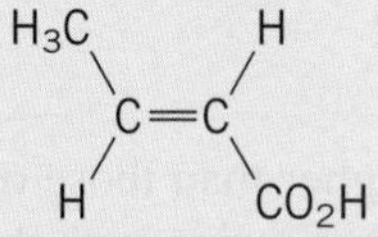

(Crotonic acid)

(c) HO_2CCO_2H

(Oxalic acid)

(d) 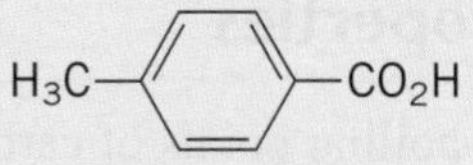

(*p*-Toluic acid)

Sample Solution (a) Methacrylic acid is an industrial chemical used in the preparation of transparent plastics such as *Lucite* and *Plexiglas.* The carbon chain that includes both the carboxylic acid and the double bond is three carbon atoms in length. The compound is named as a derivative of *propenoic acid.* The preferred 2004 IUPAC name, *2-methylprop-2-enoic acid,* contains locants for both the methyl group and the double bond. Older, but still permissible, IUPAC names omitted the double-bond locant because only one position for the double bond is structurally possible.

18.2 Structure and Bonding

The structural features of the carboxyl group are most apparent in formic acid, which is planar, with one of its carbon–oxygen bonds shorter than the other, and with bond angles at carbon close to 120°.

Bond Distances		***Bond Angles***	
C=O	120 pm	H—C=O	124°
C—O	134 pm	H—C—O	111°
		O—C=O	125°

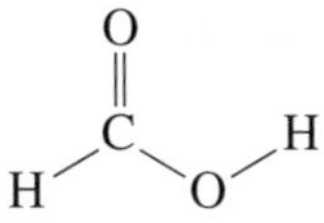

This suggests sp^2 hybridization at carbon, and a $\sigma + \pi$ carbon–oxygen double bond analogous to that of aldehydes and ketones.

Additionally, sp^2 hybridization of the hydroxyl oxygen allows one of its unshared electron pairs to be delocalized by orbital overlap with the π system of the carbonyl group (Figure 18.1*a*). In resonance terms, this electron delocalization is represented as:

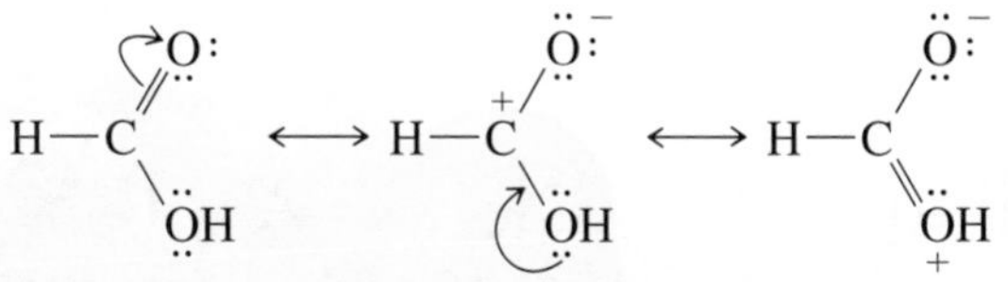

Lone-pair donation from the hydroxyl oxygen makes the carbonyl group less electrophilic than that of an aldehyde or ketone. The electrostatic potential map of formic acid (Figure 18.1*b*) shows the most electron-rich site to be the oxygen of the carbonyl group and the most electron-poor one to be, as expected, the OH hydrogen.

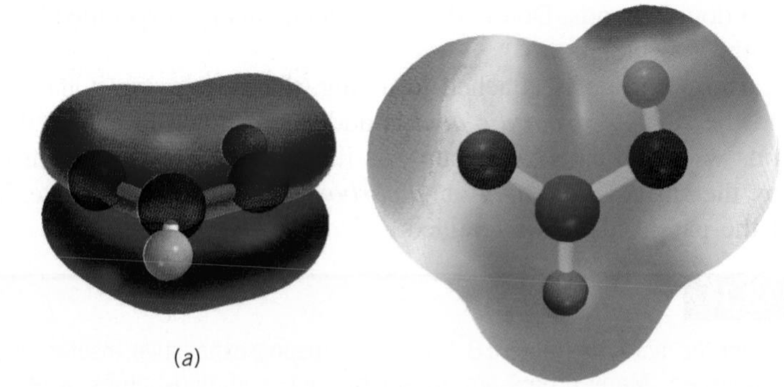

(a) (b)

Figure 18.1

(*a*) The *p* orbital of the OH group of formic acid overlaps with the π component of the double bond of the C=O group to form an extended π system that includes carbon and both oxygens. (*b*) The region of greatest negative charge (red) in formic acid is associated with the oxygen of C=O, and that of positive charge (blue) with the hydrogen of OH.

Carboxylic acids are fairly polar, and simple ones such as acetic acid, propanoic acid, and benzoic acid have dipole moments in the range 1.7–1.9 D.

18.3 Physical Properties

The melting points and boiling points of carboxylic acids are higher than those of hydrocarbons and oxygen-containing organic compounds of comparable size and shape and indicate strong intermolecular attractive forces.

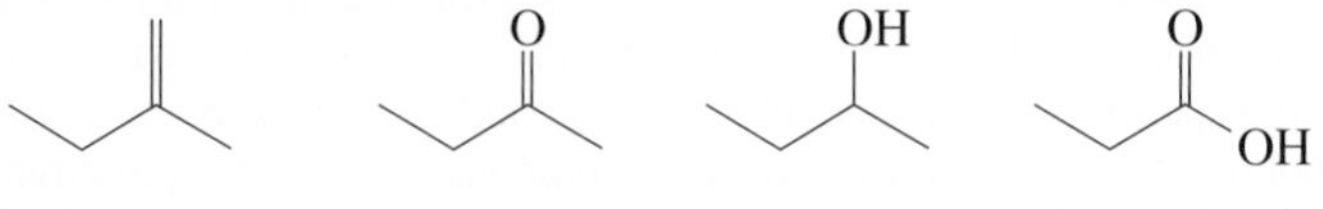

	2-Methyl-1-butene	2-Butanone	2-Butanol	Propanoic acid
bp (1 atm):	31°C	80°C	99°C	141°C

A unique hydrogen-bonding arrangement, shown in Figure 18.2, contributes to these attractive forces. The hydroxyl group of one carboxylic acid molecule acts as a proton donor toward the carbonyl oxygen of a second. In a reciprocal fashion, the hydroxyl proton of the second carboxyl function interacts with the carbonyl oxygen of the first. The result is that the two carboxylic acid molecules are held together by *two* hydrogen bonds. So efficient is this hydrogen bonding that some carboxylic acids exist as hydrogen-bonded dimers even in the gas phase. In the pure liquid a mixture of hydrogen-bonded dimers and higher aggregates is present.

In aqueous solution intermolecular association between carboxylic acid molecules is replaced by hydrogen bonding to water. The solubility properties of carboxylic acids are similar to those of alcohols. Carboxylic acids of four carbon atoms or fewer are miscible with water in all proportions.

18.4 Acidity of Carboxylic Acids

Carboxylic acids are the most acidic class of compounds that contain only carbon, hydrogen, and oxygen. With pK_a's of about 5, they are much stronger acids than water and alcohols. The case should not be overstated, however. Carboxylic acids are weak acids; a 0.1 M solution of acetic acid in water, for example, is only 1.3% ionized.

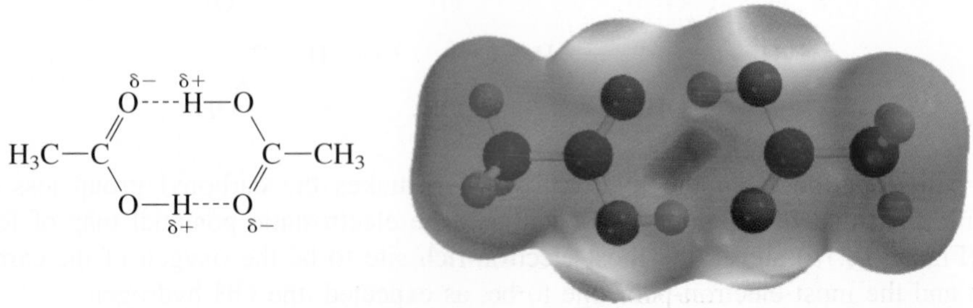

Figure 18.2

Hydrogen bonding between two acetic acid molecules.

To understand the greater acidity of carboxylic acids compared with water and alcohols, compare the structural changes that accompany the ionization of a representative alcohol (ethanol) and a representative carboxylic acid (acetic acid).

Ionization of ethanol

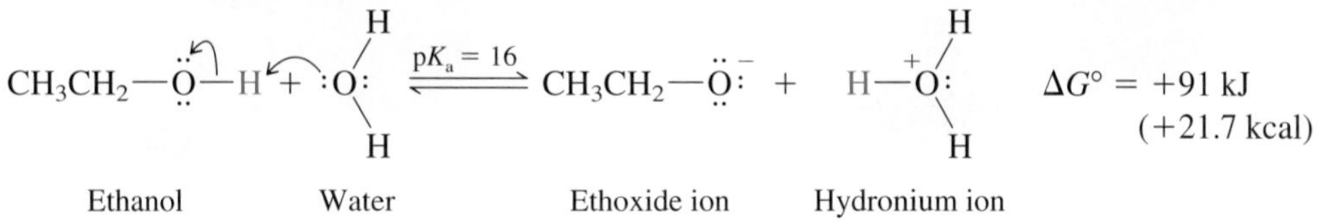

pKa = 16
ΔG° = +91 kJ
(+21.7 kcal)
Ethanol
Water
Ethoxide ion
Hydronium ion

Ionization of acetic acid

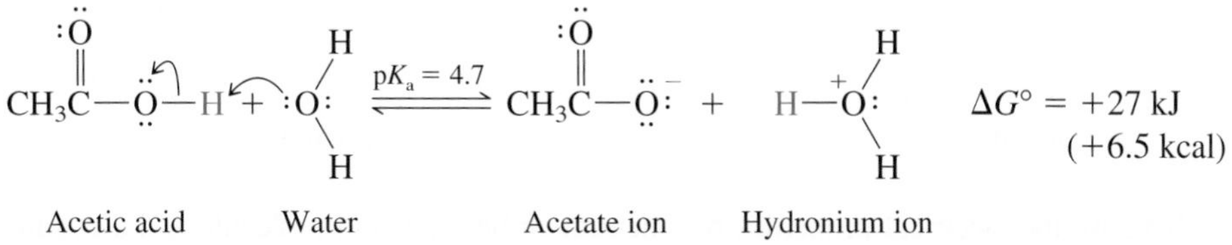

pKa = 4.7
ΔG° = +27 kJ
(+6.5 kcal)
Acetic acid
Water
Acetate ion
Hydronium ion

The large difference in the free energies of ionization of ethanol and acetic acid reflects a greater stabilization of acetate ion relative to ethoxide ion. Ionization of ethanol yields an alkoxide ion in which the negative charge is localized on oxygen. Solvation forces are the chief means by which ethoxide ion is stabilized. Acetate ion is also stabilized by solvation, but has two additional mechanisms for dispersing its negative charge that are not available to ethoxide ion:

1. *The inductive effect of the carbonyl group.* The carbonyl group of acetate ion is electron-withdrawing, and by attracting electrons away from the negatively charged oxygen, acetate anion is stabilized. This is an inductive effect, arising in the polarization of the electron distribution in the σ bond between the carbonyl carbon and the negatively charged oxygen.

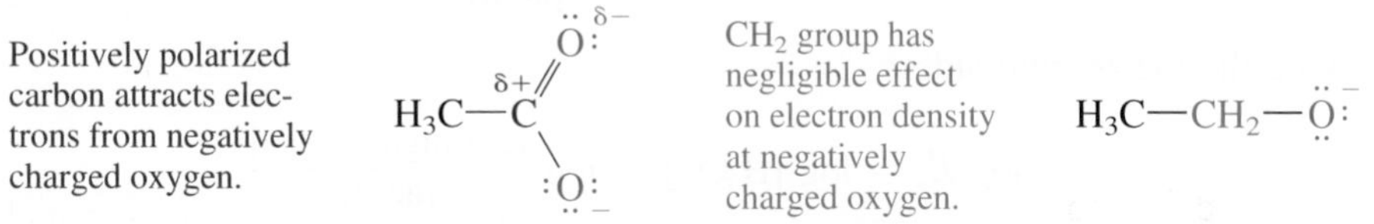

Positively polarized carbon attracts electrons from negatively charged oxygen.
CH2 group has negligible effect on electron density at negatively charged oxygen.

2. *The resonance effect of the carbonyl group.* Electron delocalization, expressed by resonance between the following Lewis structures, causes the negative charge in acetate to be shared equally by both oxygens. Electron delocalization of this type is not available to ethoxide ion.

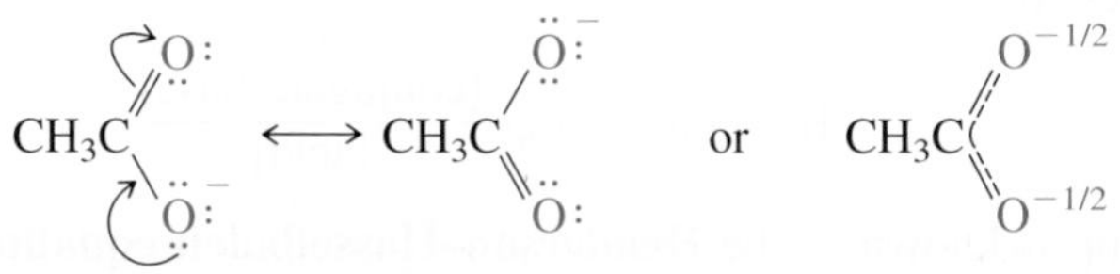

Figure 18.3 uses electrostatic potential maps to compare the localized negative charge in ethoxide ion with the delocalized charge in acetate.

Measured C—O bond distances also reflect the importance of electron delocalization in acetate ion. Acetic acid's bond distances are consistent with a short double bond (121 pm) and a long single bond (136 pm), whereas the two carbon–oxygen bond distances in acetate are the same (125 pm).

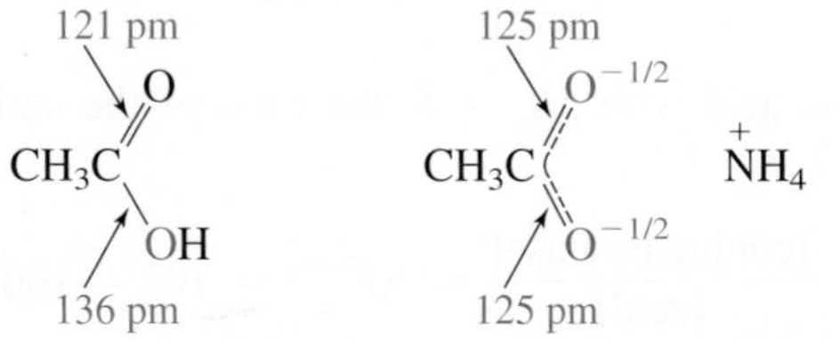

121 pm
136 pm
125 pm
125 pm

Figure 18.3

The negative charge in ethoxide (*a*) is localized on oxygen. Electron delocalization in acetate (*b*) causes the charge to be shared between two oxygens. The color scale is the same in both electrostatic potential maps.

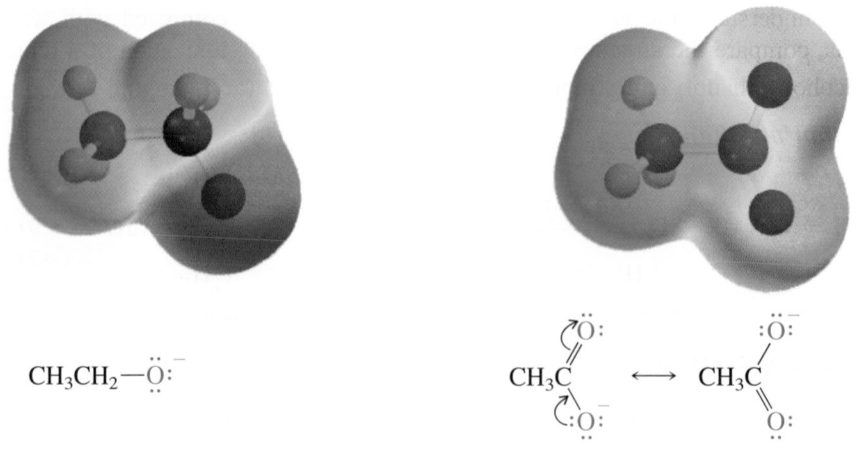

(*a*) Ethoxide (*b*) Acetate

Because the electrical properties of a neutral carboxylic acid molecule and a negatively charged carboxylate ion are so different, we need to be aware of which is the major form at the most commonly encountered pH values. For the ionization of a weak acid (HA) in water:

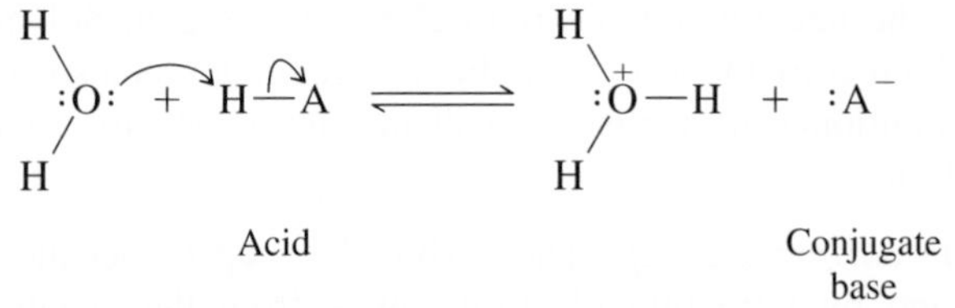

$$K_a = \frac{[H_3O^+][\text{conjugate base}]}{[\text{acid}]}$$

we can rewrite the expression for the equilibrium constant as:

$$K_a = [H_3O^+]\frac{[\text{conjugate base}]}{[\text{acid}]}$$

Taking the log of both sides:

$$\log K_a = \log [H_3O^+] + \log \frac{[\text{conjugate base}]}{[\text{acid}]}$$

and rearranging, we get:

$$-\log [H_3O^+] = -\log K_a + \log \frac{[\text{conjugate base}]}{[\text{acid}]}$$

which simplifies to:

$$\text{pH} = \text{p}K_a + \log \frac{[\text{conjugate base}]}{[\text{acid}]}$$

This relationship is known as the **Henderson–Hasselbalch equation.**

Beyond its usual application in calculating the pH of buffer solutions, the Henderson–Hasselbalch equation can be rearranged to tell us the ratio of concentrations of an acid and its conjugate base at a particular pH.

$$\log \frac{[\text{conjugate base}]}{[\text{acid}]} = \text{pH} - \text{p}K_a$$

$$\frac{[\text{conjugate base}]}{[\text{acid}]} = 10^{(\text{pH} - \text{p}K_a)}$$

For a typical carboxylic acid with $\text{p}K_a = 5$, the ratio of the carboxylate ion to the carboxylic acid at $\text{pH} = 7$ is:

$$\frac{[\text{conjugate base}]}{[\text{acid}]} = 10^{(7-5)} = 10^2 = 100$$

Thus, in a solution buffered at a pH of 7, the carboxylate concentration is 100 times greater than the concentration of the undissociated acid.

Notice that this ratio is for a solution at a specified pH, which is not the same as the pH that would result from dissolving a weak acid in pure (unbuffered) water. In the latter instance, ionization of the weak acid proceeds until equilibrium is established at some pH less than 7. Also note that the pH equals the pK_a when the carboxylic acid is 50% ionized (concentrations of the acid and its conjugate base are equal).

In most biochemical reactions the pH of the medium is close to 7. At this pH, carboxylic acids are nearly completely converted to their conjugate bases. Thus, it is common practice in biological chemistry to specify the derived carboxylate anion rather than the carboxylic acid itself. For example, we say that glycolysis leads to *lactate* by way of *pyruvate*.

Problem 18.2

(a) Lactic acid has a pK_a of 3.9. What is the [lactate]/[lactic acid] ratio at the pH of blood (7.4)?

(b) A 0.1 *M* solution of lactic acid in water has a pH of 2.5. What is the [lactate]/[lactic acid] ratio in this solution?

Sample Solution (a) Use the Henderson–Hasselbalch relationship to calculate the ratio of the concentration of the conjugate base (lactate) to the acid (lactic acid).

$$\frac{[\text{conjugate base}]}{[\text{acid}]} = 10^{(\text{pH}-\text{p}K_a)}$$

$$\frac{[\text{lactate}]}{[\text{lactic acid}]} = 10^{(7.4-3.9)} = 10^{3.5} = 3160$$

18.5 Substituents and Acid Strength

The effect of structure on acidity was introduced in Section 1.16 where we saw that electronegative substituents near an ionizable hydrogen increase its acidity. Substituent effects on the acidity of carboxylic acids have been extensively studied.

Alkyl groups have little effect. The ionization constants of all acids that have the general formula $C_nH_{2n+1}CO_2H$ are very similar to one another and equal approximately 10^{-5} (pK_a = 5). Table 18.2 gives a few examples.

An electronegative substituent, particularly if it is attached to the α carbon, increases the acidity of a carboxylic acid. All the monohaloacetic acids in Table 18.2 are about 100 times more acidic than acetic acid. Multiple halogen substitution increases the acidity even more; trichloroacetic acid is 7000 times more acidic than acetic acid!

The acid-strengthening effect of electronegative atoms or groups is easily seen as an inductive effect transmitted through the σ bonds of the molecule. According to this model, the σ electrons in the carbon–chlorine bond of chloroacetate ion are drawn toward chlorine, leaving the α-carbon atom with a slight positive charge. The α carbon, because of this positive character, attracts electrons from the negatively charged carboxylate, thus dispersing the charge and stabilizing the anion. The more stable the anion, the greater the equilibrium constant for its formation.

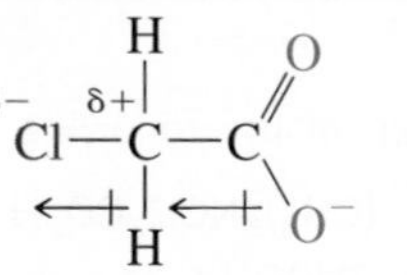

Chloroacetate anion is stabilized by electron-withdrawing effect of chlorine.

TABLE 18.2 Effect of Substituents on Acidity of Carboxylic Acids*

Name of acid	Structure	pK_a
Standard of comparison.		
Acetic acid	CH_3CO_2H	4.7
Alkyl substituents have a negligible effect on acidity.		
Propanoic acid	$CH_3CH_2CO_2H$	4.9
2-Methylpropanoic acid	$(CH_3)_2CHCO_2H$	4.8
2,2-Dimethylpropanoic acid	$(CH_3)_3CCO_2H$	5.1
Heptanoic acid	$CH_3(CH_2)_5CO_2H$	4.9
α-Halogen substituents increase acidity.		
Fluoroacetic acid	FCH_2CO_2H	2.6
Chloroacetic acid	$ClCH_2CO_2H$	2.9
Bromoacetic acid	$BrCH_2CO_2H$	2.9
Dichloroacetic acid	Cl_2CHCO_2H	1.3
Trichloroacetic acid	Cl_3CCO_2H	0.9
Electron-attracting groups increase acidity.		
Methoxyacetic acid	$CH_3OCH_2CO_2H$	3.6
Cyanoacetic acid	$N{\equiv}CCH_2CO_2H$	2.5
Nitroacetic acid	$O_2NCH_2CO_2H$	1.7

*In water at 25°C.

Inductive effects depend on the electronegativity of the substituent and the number of σ bonds between it and the affected site. As the number of bonds increases, the inductive effect decreases.

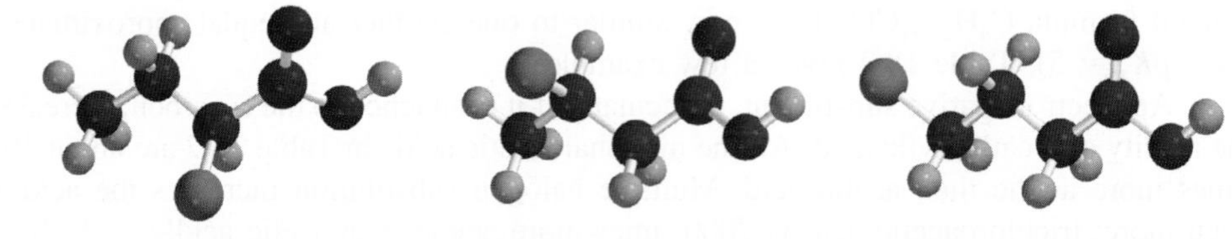

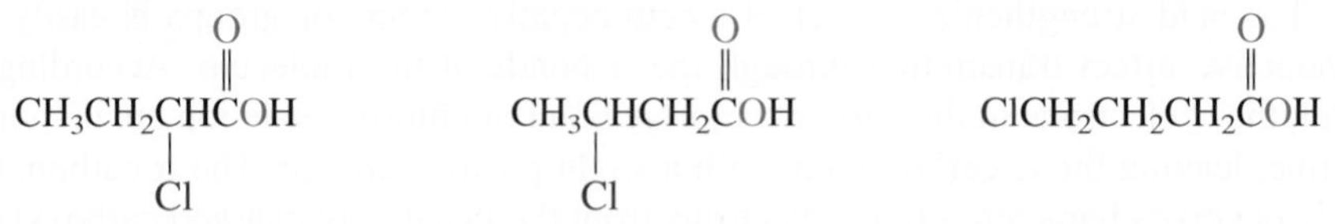

	2-Chlorobutanoic acid	3-Chlorobutanoic acid	4-Chlorobutanoic acid
pK_a	2.8	4.1	4.5

Problem 18.3

Which is the stronger acid in each of the following pairs?

(a) $(CH_3)_3CCH_2CO_2H$ or $(CH_3)_3\overset{+}{N}CH_2CO_2H$

(b) $CH_3CH_2CO_2H$ or $CH_3\underset{|}{C}HCO_2H$ (OH on the CH)

OH

(c) $CH_3\overset{O}{\overset{\|}{C}}CO_2H$ or $H_2C{=}CHCO_2H$

(d) $CH_3CH_2CH_2CO_2H$ or $CH_3\underset{\underset{O}{\|}}{\overset{\overset{O}{\|}}{S}}CH_2CO_2H$

Sample Solution (a) Think of the two compounds as substituted derivatives of acetic acid. A *tert*-butyl group is slightly electron-releasing and has only a modest effect on acidity. The compound $(CH_3)_3CCH_2CO_2H$ is expected to have an acid strength similar to that of acetic acid. A trimethylammonium substituent, on the other hand, is positively charged and is a powerful electron-withdrawing substituent. The compound $(CH_3)_3\overset{+}{N}CH_2CO_2H$ is expected to be a much stronger acid than $(CH_3)_3CCH_2CO_2H$. The measured ionization constants, shown as follows, confirm this prediction.

$(CH_3)_3CCH_2CO_2H$	$(CH_3)_3\overset{+}{N}CH_2CO_2H$
Weaker acid	Stronger acid
$pK_a = 5.3$	$pK_a = 1.8$

Closely related to the inductive effect, and operating in the same direction, is the **field effect.** In the field effect the electronegativity of a substituent is communicated, not by successive polarization of bonds but through the medium, usually the solvent. A substituent in a molecule polarizes surrounding solvent molecules and this polarization is transmitted through other solvent molecules to the remote site.

It is a curious fact that substituents affect the entropy of ionization more than they do the enthalpy term in the expression

$$\Delta G^\circ = \Delta H^\circ - T\Delta S^\circ$$

The enthalpy term ΔH° is close to zero for the ionization of most carboxylic acids, regardless of their strength. The free energy of ionization ΔG° is dominated by the $-T\Delta S^\circ$ term. Ionization is accompanied by an increase in solvation forces, leading to a decrease in the entropy of the system; ΔS° is negative, and $-T\Delta S^\circ$ is positive. Carboxylate ions with substituents capable of dispersing negative charge impose less order on the solvent (water), and less entropy is lost in their production.

18.6 Ionization of Substituted Benzoic Acids

A considerable body of data is available on the acidity of substituted benzoic acids. Benzoic acid itself is a somewhat stronger acid than acetic acid. Its carboxyl group is attached to an sp^2-hybridized carbon and ionizes to a greater extent than one that is attached to an sp^3-hybridized carbon. Remember, carbon becomes more electron-withdrawing as its *s* character increases.

CH_3CO_2H

Acetic acid
$pK_a = 4.7$

$H_2C{=}CHCO_2H$

Acrylic acid
$pK_a = 4.3$

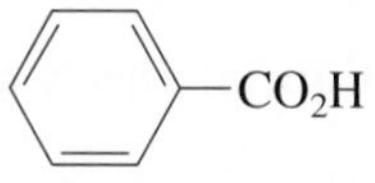

Benzoic acid
$pK_a = 4.2$

Problem 18.4

What is the most acidic neutral molecule characterized by the formula $C_3H_xO_2$?

Table 18.3 lists the ionization constants of some substituted benzoic acids. The largest effects are observed when strongly electron-withdrawing substituents are ortho to

TABLE 18.3 Acidity of Some Substituted Benzoic Acids*

Substituent in $XC_6H_4CO_2H$	pK_a for different positions of substituent X		
	Ortho	Meta	Para
H	4.2	4.2	4.2
CH_3	3.9	4.3	4.4
F	3.3	3.9	4.1
Cl	2.9	3.8	4.0
Br	2.8	3.8	4.0
I	2.9	3.9	4.0
CH_3O	4.1	4.1	4.5
O_2N	2.2	3.5	3.4

*In water at 25°C.

the carboxyl group. An *o*-nitro substituent, for example, increases the acidity of benzoic acid 100-fold. Substituent effects are small at positions meta and para to the carboxyl group. In those cases the pK_a values are clustered in the range 3.5–4.5.

18.7 Salts of Carboxylic Acids

In the presence of strong bases such as sodium hydroxide, carboxylic acids are neutralized rapidly and quantitatively:

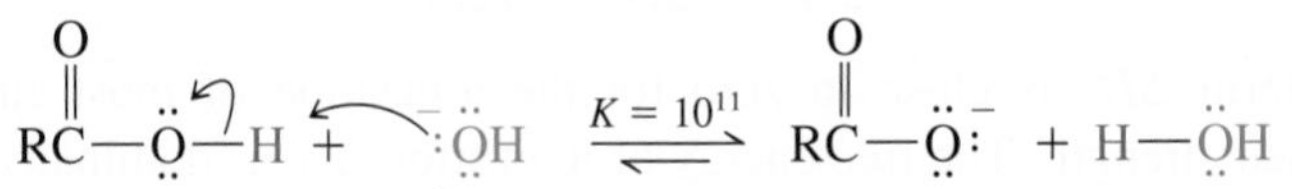

Carboxylic acid (stronger acid) + Hydroxide ion (stronger base) ⇌ Carboxylate ion (weaker base) + Water (weaker acid)

Problem 18.5

Write an ionic equation for the reaction of acetic acid with each of the following, and specify whether the equilibrium favors starting materials or products. What is the value of *K* for each?

(a) Sodium ethoxide
(b) Potassium *tert*-butoxide
(c) Sodium bromide
(d) Sodium acetylide
(e) Potassium nitrate
(f) Lithium amide

Sample Solution (a) This is an acid–base reaction; ethoxide ion is the base.

$$CH_3CO_2H + CH_3CH_2O^- \rightleftharpoons CH_3CO_2^- + CH_3CH_2OH$$

Acetic acid (stronger acid) + Ethoxide ion (stronger base) ⇌ Acetate ion (weaker base) + Ethanol (weaker acid)

The position of equilibrium lies well to the right. Ethanol, with $pK_a = 16$, is a much weaker acid than acetic acid ($pK_a = 4.7$). The equilibrium constant *K* is $10^{(16-4.7)}$ or $10^{11.3}$.

The salts formed on neutralization of carboxylic acids are named by first specifying the metal ion and then adding the name of the acid modified by replacing *-ic acid*

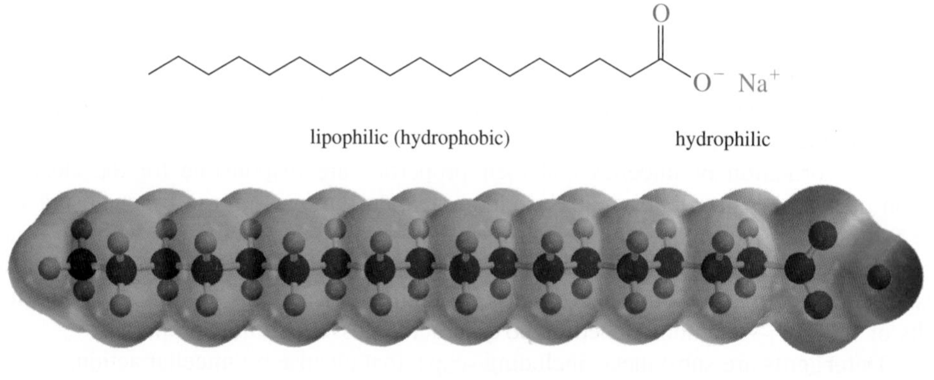
O
O⁻ Na⁺
lipophilic (hydrophobic)
hydrophilic

Sodium stearate [$CH_3(CH_2)_{16}CO_2Na$]

Figure 18.4

Structure and electrostatic potential map of sodium stearate.

by *-ate*. Monocarboxylate salts of diacids are designated by naming both the cation and the hydrogen of the CO_2H group.

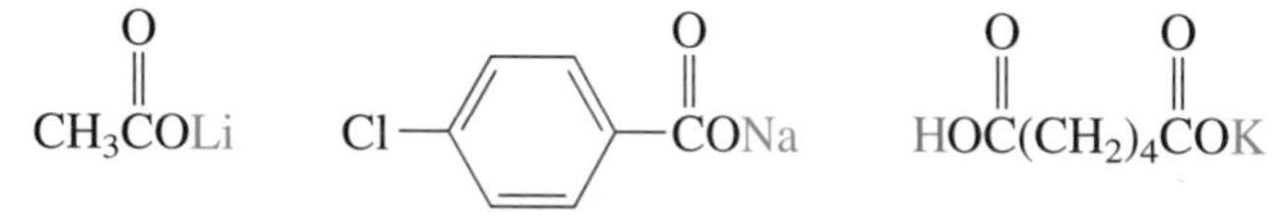

Lithium acetate   Sodium *p*-chlorobenzoate   Potassium hydrogen hexanedioate

Metal carboxylates are ionic, and when the molecular weight isn't too high, the sodium and potassium salts of carboxylic acids are soluble in water. Carboxylic acids therefore may be extracted from ether solutions into aqueous sodium or potassium hydroxide.

The solubility behavior of salts of carboxylic acids having 12–18 carbons is unusual and can be illustrated by considering sodium stearate (sodium octadecanoate). Stearate ion contains two very different structural units—a long nonpolar hydrocarbon chain and a polar carboxylate group. The electrostatic potential map of sodium stearate in Figure 18.4 illustrates how different most of the molecule is from its polar carboxylate end.

Carboxylate groups are **hydrophilic** ("water-loving") and tend to confer water solubility on species that contain them. Long hydrocarbon chains are **lipophilic** ("fat-loving") and tend to associate with other hydrocarbon chains. Sodium stearate is an example of an **amphiphilic** substance; both hydrophilic and lipophilic groups occur within the same molecule.

"Hydrophobic" is often used instead of "lipophilic."

When sodium stearate is placed in water, the hydrophilic carboxylate group encourages the formation of a solution; the lipophilic alkyl chain discourages it. The compromise achieved is to form a colloidal dispersion of aggregates called **micelles** (Figure 18.5). Micelles form spontaneously when the carboxylate concentration exceeds a certain minimum value called the **critical micelle concentration.** Each micelle is composed of 50–100 individual molecules, with the polar carboxylate groups directed toward its outside where they experience attractive forces with water and sodium ions. The nonpolar hydrocarbon chains are directed toward the interior of the micelle, where individually

Figure 18.5

Space-filling model of a micelle formed by association of carboxylate ions derived from a long-chain carboxylic acid. The hydrocarbon chains tend to be on the inside and the carboxylate ions on the surface where they are in contact with water molecules and metal cations.

weak but cumulatively significant induced-dipole/induced-dipole forces bind them together. Micelles are approximately spherical because a sphere exposes the minimum surface for a given volume of material and disrupts the water structure least. Because their surfaces are negatively charged, two micelles repel each other rather than clustering to form higher aggregates.

The formation of micelles and their properties are responsible for the cleansing action of soaps. Water that contains sodium stearate removes grease by enclosing it in the hydrocarbon-like interior of the micelles. The grease is washed away with the water, not because it dissolves in the water but because it dissolves in the micelles that are dispersed in the water. Sodium stearate is an example of a soap; sodium and potassium salts of other C_{12}–C_{18} unbranched carboxylic acids possess similar properties.

Detergents are substances, including soaps, that cleanse by micellar action. A large number of synthetic detergents are known. An example is sodium lauryl sulfate, which has a long hydrocarbon chain terminating in a polar sulfate ion and forms soap-like micelles in water.

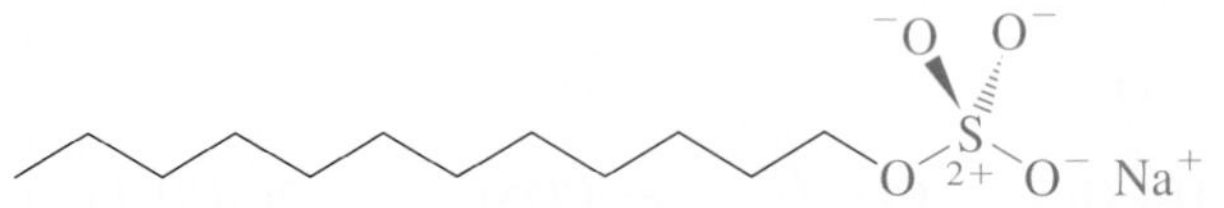

Sodium lauryl sulfate
(sodium dodecyl sulfate)

Detergents are designed to be effective in hard water, meaning water containing calcium salts that form insoluble calcium carboxylates with soaps. These precipitates rob the soap of its cleansing power and form an unpleasant scum. The calcium salts of synthetic detergents such as sodium lauryl sulfate, however, are soluble and retain their micelle-forming ability even in hard water.

18.8 Dicarboxylic Acids

Separate ionization constants, designated K_1 and K_2, respectively, characterize the two successive ionization steps of a dicarboxylic acid.

Oxalic acid is poisonous and occurs naturally in a number of plants including sorrel and begonia. It is a good idea to keep houseplants out of the reach of small children, who might be tempted to eat the leaves or berries.

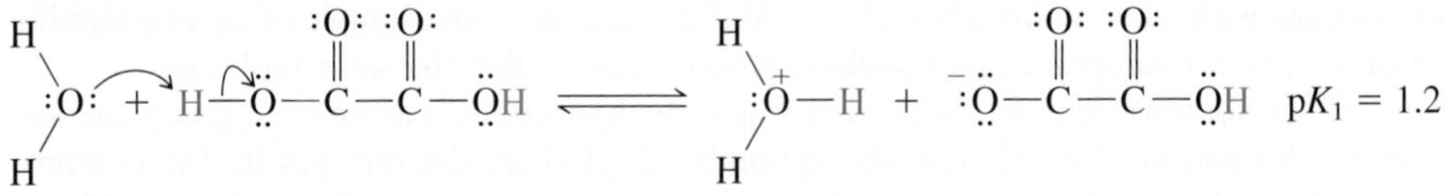

Water Oxalic acid Hydronium ion Hydrogen oxalate ion

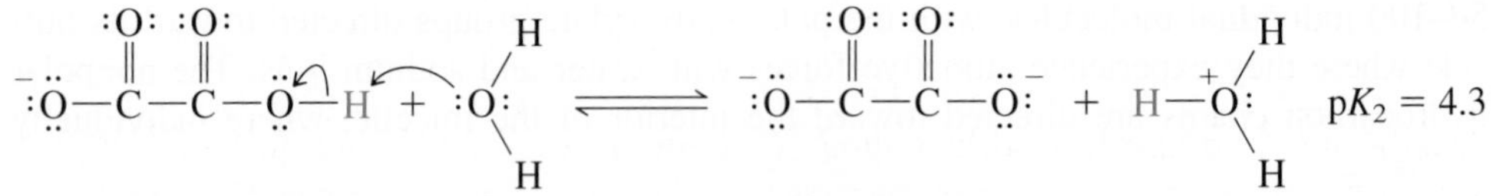

Hydrogen oxalate ion Water Oxalate ion Hydronium ion

The first ionization constant of dicarboxylic acids is larger than K_a for monocarboxylic analogs. One reason is statistical. There are two potential sites for ionization rather than one, making the effective concentration of carboxyl groups twice as large. Furthermore, one carboxyl group acts as an electron-withdrawing group to facilitate dissociation of the other. This is particularly noticeable when the two carboxyl groups are separated by only a few bonds. Oxalic and malonic acid, for example, are several orders of magnitude stronger than simple alkyl derivatives of acetic acid. Heptanedioic acid, in which the carboxyl groups are well separated from each other, is only slightly stronger than acetic acid.

HO_2CCO_2H	$HO_2CCH_2CO_2H$	$HO_2C(CH_2)_5CO_2H$
Oxalic acid	Malonic acid	Heptanedioic acid
$pK_1 = 1.2$	$pK_1 = 2.8$	$pK_1 = 4.3$

18.9 Carbonic Acid

Through an accident of history, the simplest dicarboxylic acid, carbonic acid, HOC(=O)OH, is not even classified as an organic compound. Because many minerals are carbonate salts, nineteenth-century chemists placed carbonates, bicarbonates, and carbon dioxide in the inorganic realm. Nevertheless, the essential features of carbonic acid and its salts are easily understood on the basis of our knowledge of carboxylic acids.

Carbonic acid is formed when carbon dioxide reacts with water. Hydration of carbon dioxide is far from complete, however. Almost all the carbon dioxide that is dissolved in water exists as carbon dioxide; only 0.3% of it is converted to carbonic acid. Carbonic acid is a weak acid and ionizes to a small extent to bicarbonate ion.

The systematic name for bicarbonate ion is *hydrogen carbonate.* Thus, the systematic name for sodium bicarbonate ($NaHCO_3$) is *sodium hydrogen carbonate.*

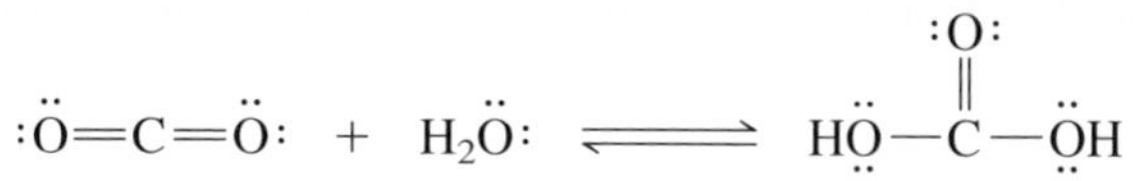

Carbon dioxide Water Carbonic acid

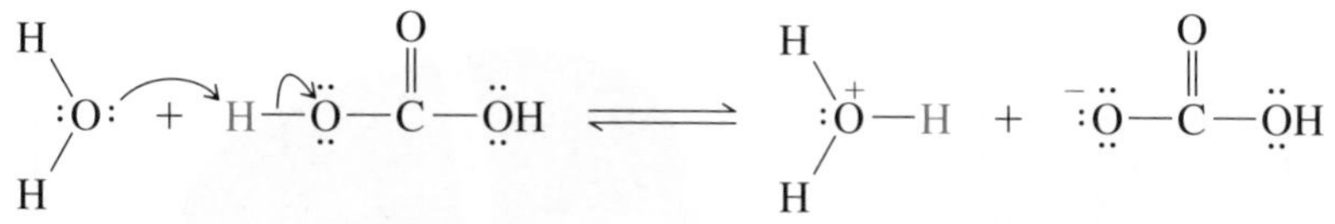

Water Carbonic acid Hydronium ion Bicarbonate ion

The equilibrium constant for the overall reaction is related to an apparent equilibrium constant K_1 for carbonic acid ionization by the expression

$$K_1 = \frac{[H_3O^+][HCO_3^-]}{[CO_2]} = 4.3 \times 10^{-7} \qquad pK_1 = 6.4$$

These equations tell us that the reverse process, proton transfer from acids to bicarbonate to form carbon dioxide, will be favorable when K_a of the acid exceeds 4.3×10^{-7} ($pK_a < 6.4$). Among compounds containing carbon, hydrogen, and oxygen, only carboxylic acids are acidic enough to meet this requirement. They dissolve in aqueous sodium bicarbonate with the evolution of carbon dioxide. This behavior is the basis of a qualitative test for carboxylic acids.

Problem 18.6

The value cited for the "apparent K_1" of carbonic acid, 4.3×10^{-7}, is the one normally given in reference books. It is determined by measuring the pH of water to which a known amount of carbon dioxide has been added. When we recall that only 0.3% of carbon dioxide is converted to carbonic acid in water, what is the "true K_1" of carbonic acid?

Carbonic anhydrase is an enzyme that catalyzes the hydration of carbon dioxide to bicarbonate. The uncatalyzed hydration of carbon dioxide is too slow to be effective in transporting carbon dioxide from the tissues to the lungs, and so animals have developed catalysts to speed this process. The activity of carbonic anhydrase is remarkable; it has been estimated that one molecule of this enzyme can catalyze the hydration of 3.6×10^7 molecules of carbon dioxide per minute.

As with other dicarboxylic acids, the second ionization constant of carbonic acid is far smaller than the first.

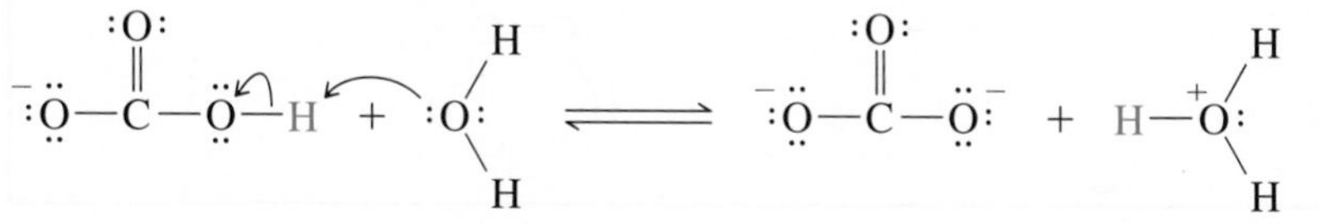

$pK_2 = 10.2$

Bicarbonate ion Water Carbonate ion Hydronium ion

Bicarbonate is a weaker acid than carboxylic acids but a stronger acid than water and alcohols.

18.10 Sources of Carboxylic Acids

Many carboxylic acids were first isolated from natural sources and were given common names based on their origin (Figure 18.6). Formic acid (Latin *formica,* meaning "ant") was obtained by distilling ants, but is found in some other insects as well. Since ancient times acetic acid (Latin *acetum,* for "vinegar") has been known to be present in wine that has turned sour. Butyric acid (Latin *butyrum,* meaning "butter") contributes to the odor of both rancid butter and ginkgo berries. Malic acid (Latin *malum,* meaning "apple") occurs in apples. Oleic acid (Latin *oleum,* "oil") takes its name from naturally occurring esters such as those that comprise the major portion of olive oil.

The large-scale preparation of carboxylic acids relies on chemical synthesis. Virtually none of the 3×10^9 lb of acetic acid produced in the United States each year is obtained from vinegar. Most of it comes from the reaction of methanol with carbon monoxide.

Where do we find carboxylic acids?

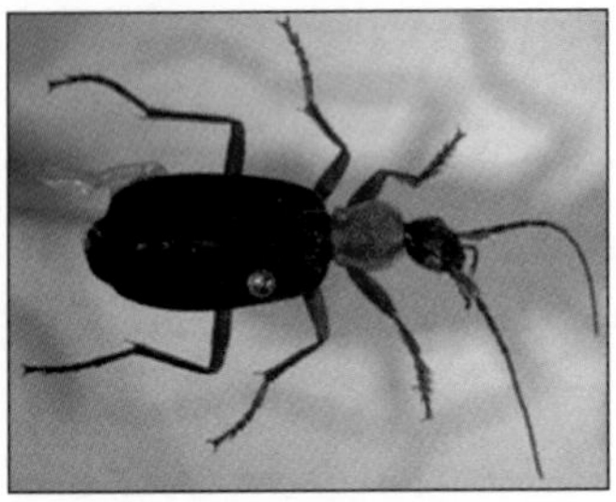

Ants aren't the only insects that use formic acid as a weapon. Some *Galerita* beetles spray attackers with an 80% solution of it.

HCO_2H

Ethanol is oxidized to acetic acid as wine becomes vinegar.

CH_3CO_2H

Malic acid and citric acid contribute to the tart taste of many fruits and vegetables.

$HO_2CCH_2CHCO_2H$ (with OH on the CH carbon)

$HO_2CCH_2CCH_2CO_2H$ (with OH and CO_2H on the central carbon)

Butanoic and hexanoic acid are responsible for the nasty odor of ginkgo seeds.

$CH_3CH_2CH_2CO_2H$

$CH_3CH_2CH_2CH_2CH_2CO_2H$

The oleic acid that forms during decomposition of dead ants is a chemical signal to other ants to carry them from the nest. In an experiment in which live ants had been coated with oleic acid, they were also removed.

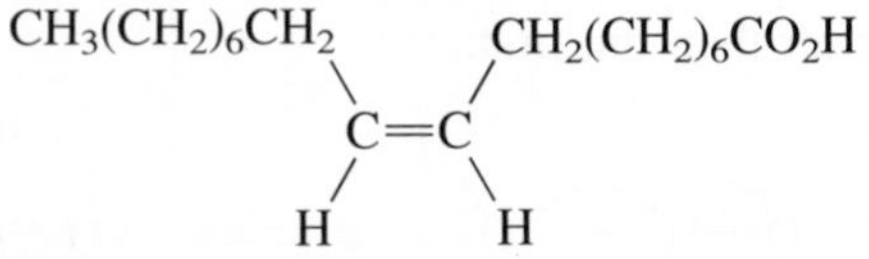
CH3(CH2)6CH2
CH2(CH2)6CO2H
C=C
H
H

Figure 18.6

Some naturally occurring carboxylic acids.

$$\underset{\text{Methanol}}{CH_3OH} + \underset{\text{Carbon monoxide}}{CO} \xrightarrow[\text{heat, pressure}]{\text{cobalt or rhodium catalyst}} \underset{\text{Acetic acid}}{CH_3CO_2H}$$

The principal end use of acetic acid is in the production of vinyl acetate for paints and adhesives.

The carboxylic acid produced in the greatest amounts is 1,4-benzenedicarboxylic acid (terephthalic acid). About 5×10^9 lb/year is produced in the United States as a starting material for the preparation of polyester fibers. One important process converts *p*-xylene to terephthalic acid by oxidation with nitric acid:

See Chapter 27 for more on polymers made from terephthalic acid.

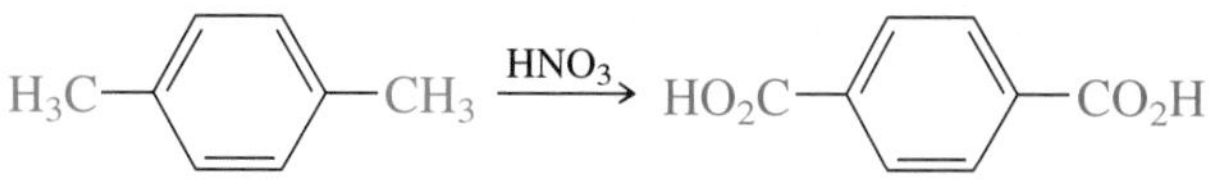

p-Xylene 1,4-Benzenedicarboxylic acid (terephthalic acid)

You will recognize the side-chain oxidation of *p*-xylene to terephthalic acid as a reaction type discussed previously (Section 11.12). Examples of other reactions encountered earlier that can be applied to the synthesis of carboxylic acids are collected in Table 18.4.

TABLE 18.4 Summary of Reactions Discussed in Earlier Chapters That Yield Carboxylic Acids

Reaction (section) and comments	General equation and specific example
Side-chain oxidation of alkylbenzenes (Section 11.12) A primary or secondary alkyl side chain on an aromatic ring is converted to a carboxyl group by reaction with a strong oxidizing agent such as potassium permanganate or chromic acid.	$\underset{\text{Alkylbenzene}}{ArCHR_2} \xrightarrow[K_2Cr_2O_7,\ H_2SO_4H_2O]{KMnO_4\ \text{or}} \underset{\text{Arenecarboxylic acid}}{ArCO_2H}$ 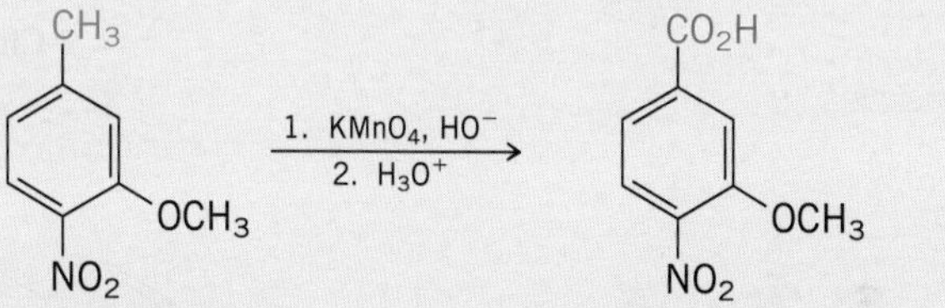 3-Methoxy-4-nitrotoluene → 3-Methoxy-4-nitrobenzoic acid (100%)
Oxidation of primary alcohols (Section 15.9) Potassium permanganate and chromic acid convert primary alcohols to carboxylic acids by way of the corresponding aldehyde.	$\underset{\text{Primary alcohol}}{RCH_2OH} \xrightarrow[K_2Cr_2O_7,\ H_2SO_4]{KMnO_4\ \text{or}} \underset{\text{Carboxylic acid}}{RCO_2H}$ $(CH_3)_3CC(CH_2OH)HC(CH_3)_3 \xrightarrow[H_2O,\ H_2SO_4]{H_2CrO_4} (CH_3)_3CC(CO_2H)HC(CH_3)_3$ 2-*tert*-Butyl-3,3-dimethyl-1-butanol → 2-*tert*-Butyl-3,3-dimethylbutanoic acid (82%)
Oxidation of aldehydes (Section 17.15) Aldehydes are particularly sensitive to oxidation and are converted to carboxylic acids by a number of oxidizing agents, including potassium permanganate and chromic acid.	$\underset{\text{Aldehyde}}{RCH{=}O} \xrightarrow{\text{oxidizing agent}} \underset{\text{Carboxylic acid}}{RCO_2H}$ 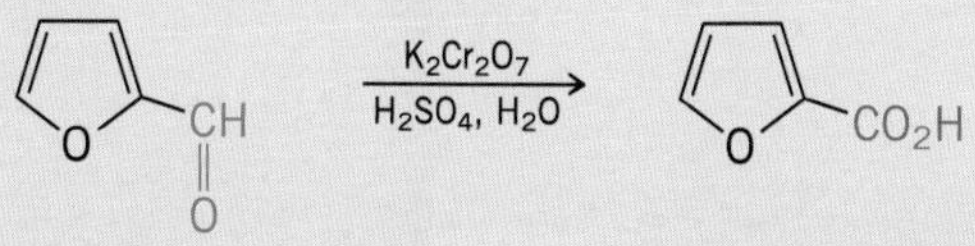 Furan-2-carbaldehyde (furfural) → Furan-2-carboxylic acid (furoic acid) (75%)

The examples in the table give carboxylic acids that have the same number of carbon atoms as the starting material. The reactions to be described in the next two sections permit carboxylic acids to be prepared by extending a chain by one carbon atom and are of great value in laboratory syntheses of carboxylic acids.

18.11 Synthesis of Carboxylic Acids by the Carboxylation of Grignard Reagents

We've seen how Grignard reagents add to the carbonyl group of aldehydes, ketones, and esters. Grignard reagents react in much the same way with *carbon dioxide* to yield magnesium salts of carboxylic acids. Acidification converts these magnesium salts to the desired carboxylic acids.

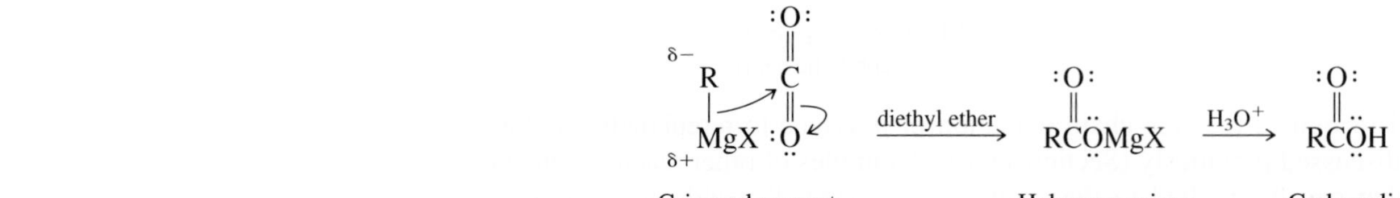
δ−
R
C
:O:
MgX :O
δ+
diethyl ether
RCOMgX
H_3O^+
RCOH

Grignard reagent acts as a nucleophile toward carbon dioxide | Halomagnesium carboxylate | Carboxylic acid

Overall, the carboxylation of Grignard reagents transforms an alkyl or aryl halide to a carboxylic acid in which the carbon skeleton has been extended by one carbon atom.

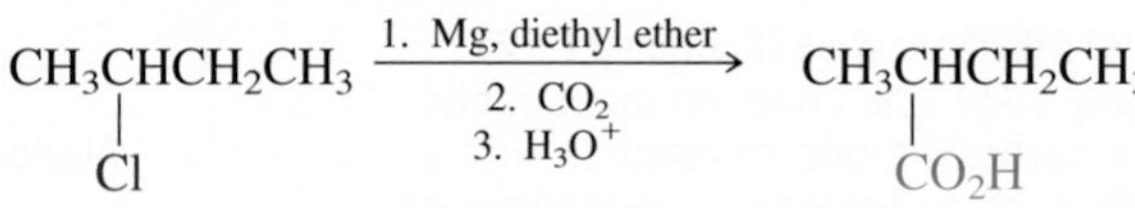
$CH_3CHCH_2CH_3$
Cl
1. Mg, diethyl ether
2. CO_2
3. H_3O^+
$CH_3CHCH_2CH_3$
CO_2H

2-Chlorobutane → 2-Methylbutanoic acid (76–86%)

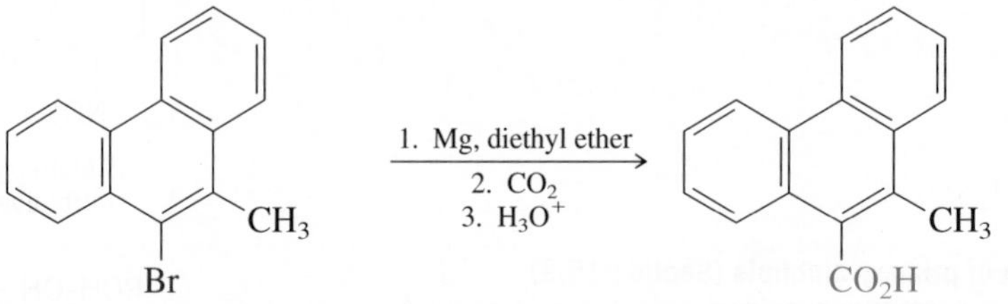
CH_3
Br
1. Mg, diethyl ether
2. CO_2
3. H_3O^+
CH_3
CO_2H

9-Bromo-10-methylphenanthrene → 10-Methylphenanthrene-9-carboxylic acid (82%)

The major limitation to this procedure is that the alkyl or aryl halide must not bear substituents that are incompatible with Grignard reagents, such as OH, NH, SH, or C=O.

Problem 18.7

2,6-Dimethoxybenzoic acid was needed for a synthesis of the β-lactam antibiotic methicillin. Show how this carboxylic acid could be synthesized from 2-bromo-1,3-benzenediol.

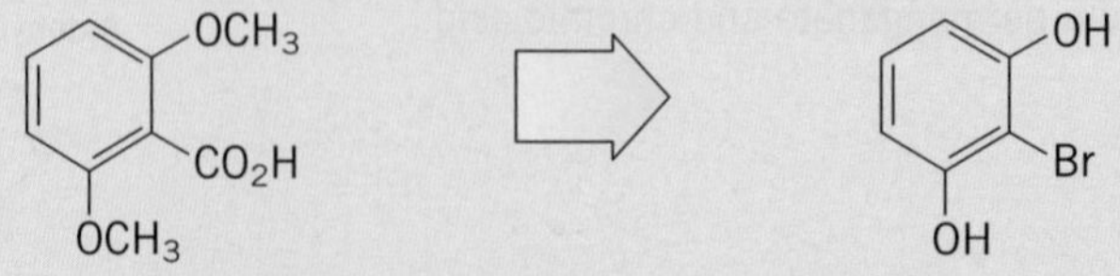
OCH_3
CO_2H
OCH_3
OH
Br
OH

2,6-Dimethoxybenzoic acid ⇐ 2-Bromo-1,3-benzenediol

18.12 Synthesis of Carboxylic Acids by the Preparation and Hydrolysis of Nitriles

Primary and secondary alkyl halides may be converted to the next higher carboxylic acid by a two-step synthetic sequence involving the preparation and hydrolysis of *nitriles.* Nitriles, also known as *alkyl cyanides,* are prepared by nucleophilic substitution.

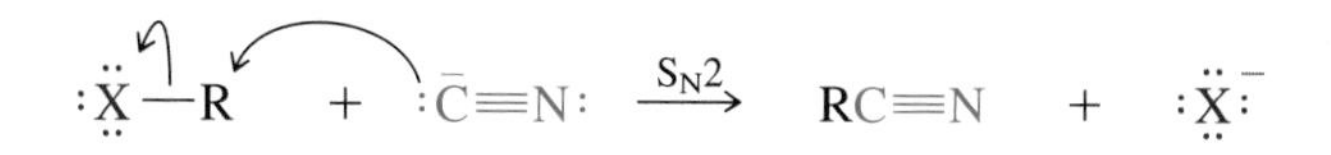

Primary or secondary alkyl halide — Cyanide ion — Nitrile (alkyl cyanide) — Halide ion

The reaction follows an S_N2 mechanism and works best with primary and secondary alkyl halides. Elimination is the only reaction observed with tertiary alkyl halides. Aryl and vinyl halides do not react. Dimethyl sulfoxide is the preferred solvent for this reaction, but alcohols and water–alcohol mixtures have also been used.

Once the cyano group has been introduced, the nitrile is subjected to hydrolysis. Usually this is carried out in aqueous acid at reflux.

$$\underset{\text{Nitrile}}{RC{\equiv}N} + \underset{\text{Water}}{2H_2O} + H^+ \xrightarrow{\text{heat}} \underset{\text{Carboxylic acid}}{RCOH} + \underset{\text{Ammonium ion}}{NH_4^+}$$

(RCOH: the carboxyl carbon bears =O.)

The mechanism of nitrile hydrolysis will be described in Section 19.18.

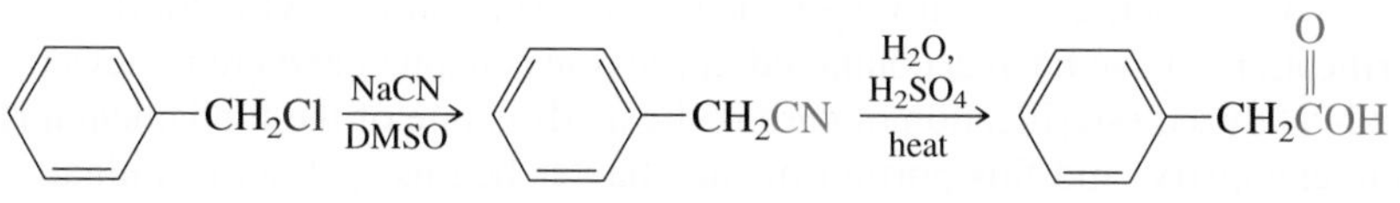

Benzyl chloride → Benzyl cyanide (92%) → Phenylacetic acid (77%)

Dicarboxylic acids have been prepared from dihalides by this method:

$$\underset{\text{1,3-Dibromopropane}}{BrCH_2CH_2CH_2Br} \xrightarrow[H_2O]{NaCN} \underset{\text{1,5-Pentanedinitrile (77–86\%)}}{NCCH_2CH_2CH_2CN} \xrightarrow[\text{heat}]{H_2O,\ HCl} \underset{\text{1,5-Pentanedioic acid (83–85\%)}}{HO\overset{O}{\overset{\|}{C}}CH_2CH_2CH_2\overset{O}{\overset{\|}{C}}OH}$$

Applications of dicarboxylic acids in the synthesis of nylon and other polymers are described in Sections 27.11 and 27.12.

Problem 18.8

Of the two procedures just described, preparation and carboxylation of a Grignard reagent or formation and hydrolysis of a nitrile, only one is appropriate to each of the following RX → RCO_2H conversions. Identify the correct procedure in each case, and specify why the other will fail.

(a) Bromobenzene → benzoic acid
(b) 2-Chloroethanol → 3-hydroxypropanoic acid
(c) *tert*-Butyl chloride → 2,2-dimethylpropanoic acid

Sample Solution (a) Bromobenzene is an aryl halide and is unreactive toward nucleophilic substitution by cyanide ion. The route $C_6H_5Br \rightarrow C_6H_5CN \rightarrow C_6H_5CO_2H$ fails because the first step fails. The route proceeding through the Grignard reagent

Continued

is perfectly satisfactory and appears as an experiment in a number of introductory organic chemistry laboratory texts.

Bromobenzene | Phenylmagnesium bromide | Benzoic acid

Recall the preparation of cyanohydrins in Section 17.7.

Nitrile groups in cyanohydrins are hydrolyzed under conditions similar to those of alkyl cyanides. Cyanohydrin formation followed by hydrolysis provides a route to the preparation of α-hydroxy carboxylic acids.

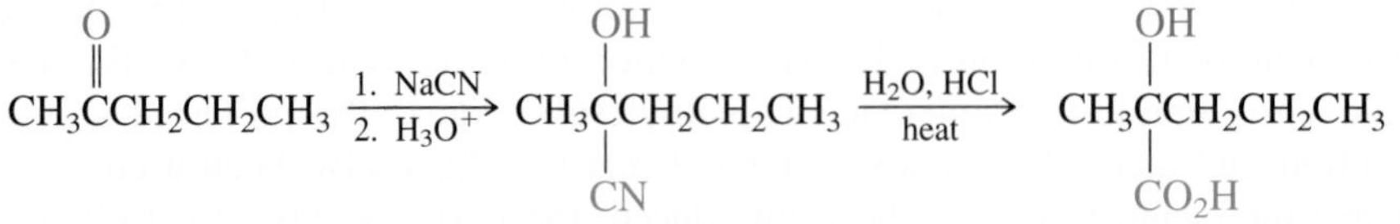

2-Pentanone | 2-Pentanone cyanohydrin | 2-Hydroxy-2-methylpentanoic acid (60% from 2-pentanone)

18.13 Reactions of Carboxylic Acids: A Review and a Preview

The most apparent chemical property of carboxylic acids, their acidity, has already been examined. Three reactions of carboxylic acids—conversion to acyl chlorides, reduction, and esterification—have been encountered in previous chapters and are reviewed in Table 18.5. Acid-catalyzed esterification of carboxylic acids is one of the fundamental reactions of organic chemistry, and this portion of the chapter begins with an examination of the mechanism by which it occurs.

18.14 Mechanism of Acid-Catalyzed Esterification

An important question about the mechanism of acid-catalyzed esterification concerns the origin of the alkoxy oxygen. For example, does the methoxy oxygen in methyl benzoate come from methanol, or is it derived from benzoic acid?

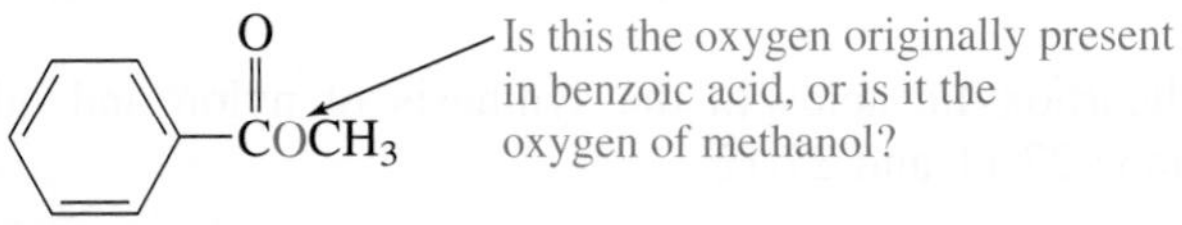

A clear-cut answer was provided by Irving Roberts and Harold C. Urey of Columbia University in 1938. They prepared methanol that had been enriched in the mass-18 isotope of oxygen. When this sample of methanol was esterified with benzoic acid, the methyl benzoate product contained all the ^{18}O label that was originally present in the methanol.

In this equation, the red O signifies oxygen enriched in its mass-18 isotope; analysis of isotopic enrichment was performed by mass spectrometry.

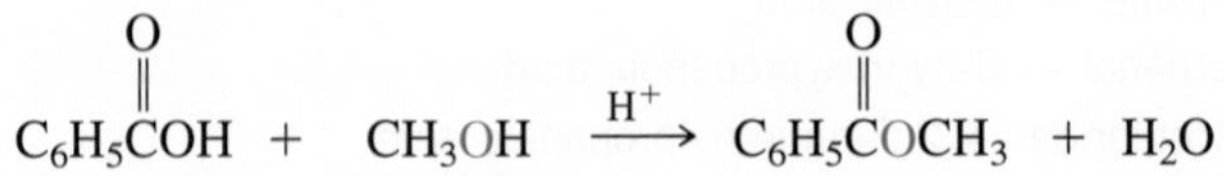

Benzoic acid | ^{18}O-enriched methanol | ^{18}O-enriched methyl benzoate | Water

The results of the Roberts–Urey experiment tell us that the C—O bond of the alcohol is preserved during esterification. The oxygen that is lost as a water molecule must come from the carboxylic acid.

TABLE 18.5 Summary of Reactions of Carboxylic Acids Discussed in Earlier Chapters

Reaction (section) and comments	General equation and specific example
Formation of acyl chlorides (Section 12.7) Thionyl chloride reacts with carboxylic acids to yield acyl chlorides.	$RCO_2H + SOCl_2 \longrightarrow RCCl(=O) + SO_2 + HCl$ Carboxylic acid, Thionyl chloride, Acyl chloride, Sulfur dioxide, Hydrogen chloride 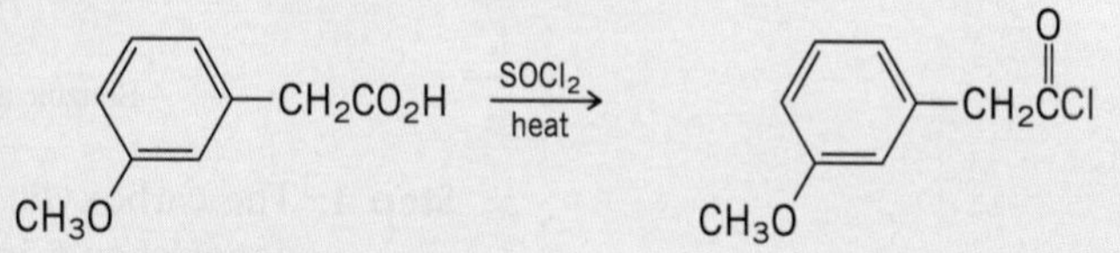 *m*-Methoxyphenylacetic acid → *m*-Methoxyphenylacetyl chloride (85%)
Lithium aluminum hydride reduction (Section 15.3) Carboxylic acids are reduced to primary alcohols by the powerful reducing agent lithium aluminum hydride.	$RCO_2H \xrightarrow[\text{2. } H_2O]{\text{1. } LiAlH_4\text{, diethyl ether}} RCH_2OH$ Carboxylic acid → Primary alcohol 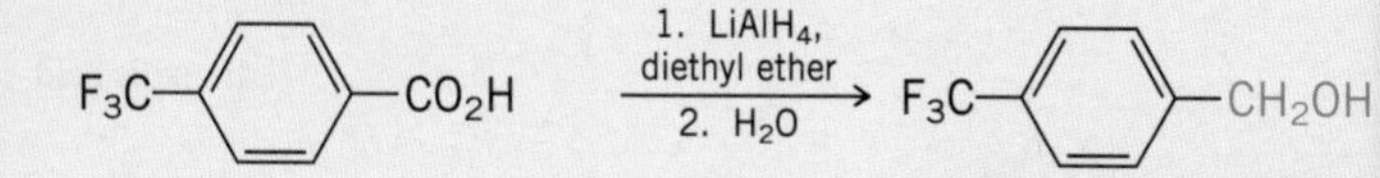 *p*-(Trifluoromethyl)benzoic acid → *p*-(Trifluoromethyl)benzyl alcohol (96%)
Esterification (Section 15.8) In the presence of an acid catalyst, carboxylic acids and alcohols react to form esters. The reaction is called the Fischer esterification. It is an equilibrium process but can be driven to favor the ester by removing the water that is formed.	$RCO_2H + R'OH \overset{H^+}{\rightleftharpoons} RCOR'(=O) + H_2O$ Carboxylic acid, Alcohol, Ester, Water 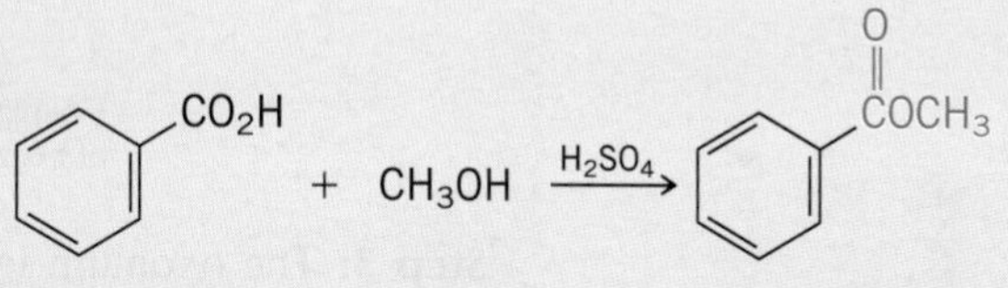 Benzoic acid, Methanol → Methyl benzoate (70%)

Mechanism 18.1 is consistent with these facts. The six steps are best viewed as a combination of two distinct stages. *Formation* of a **tetrahedral intermediate** characterizes the first stage (steps 1–3), and *dissociation* of this tetrahedral intermediate characterizes the second (steps 4–6).

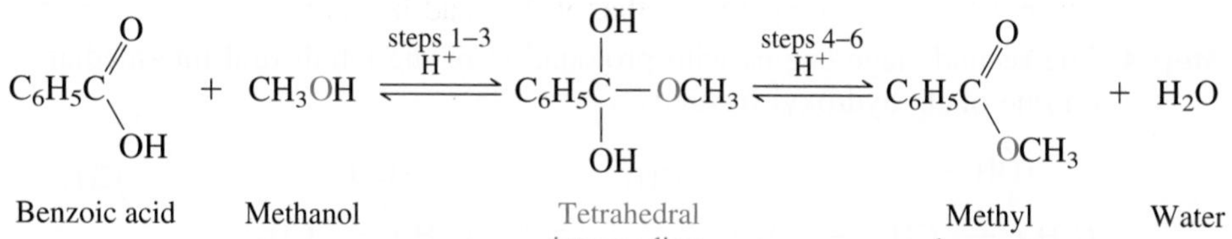

The species connecting the two stages is called a *tetrahedral intermediate* because the hybridization at carbon has changed from sp^2 in the carboxylic acid to sp^3 in the intermediate before returning to sp^2 in the ester product. *The tetrahedral intermediate is formed by nucleophilic addition of an alcohol to a carboxylic acid and is analogous to a hemiacetal formed by nucleophilic addition of an alcohol to an aldehyde or a ketone.* The three steps that lead to the tetrahedral intermediate in the first stage of esterification

Mechanism 18.1

Acid-Catalyzed Esterification of Benzoic Acid with Methanol

THE OVERALL REACTION:

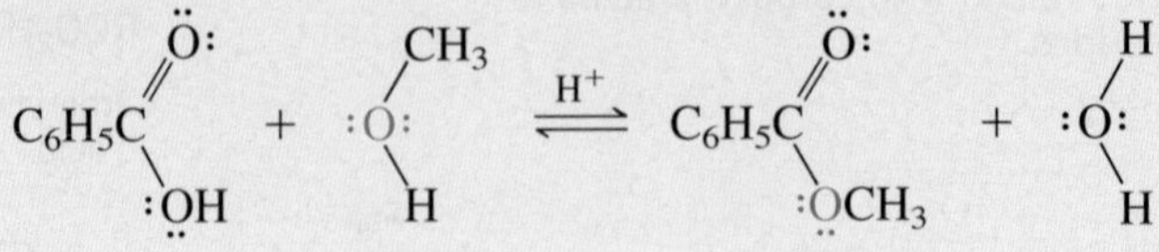

Benzoic acid — Methanol — Methyl benzoate — Water

Step 1: The carboxylic acid is protonated on its carbonyl oxygen. The proton donor shown in the equation for this step is an alkyloxonium ion formed by proton transfer from the acid catalyst to the alcohol.

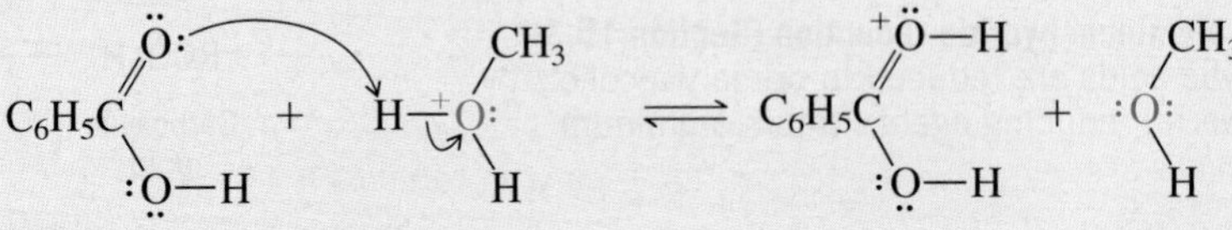

Benzoic acid — Methyloxonium ion — Conjugate acid of benzoic acid — Methanol

Step 2: Protonation of the carboxylic acid increases the positive character of its carbonyl group. A molecule of the alcohol acts as a nucleophile and bonds to the carbonyl carbon.

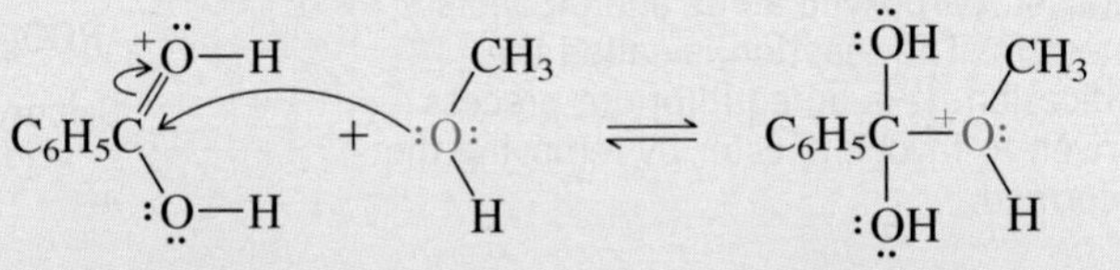

Conjugate acid of benzoic acid — Methanol — Protonated form of tetrahedral intermediate

Step 3: The oxonium ion formed in step 2 loses a proton to give the tetrahedral intermediate in its neutral form. The step concludes the first stage in the mechanism.

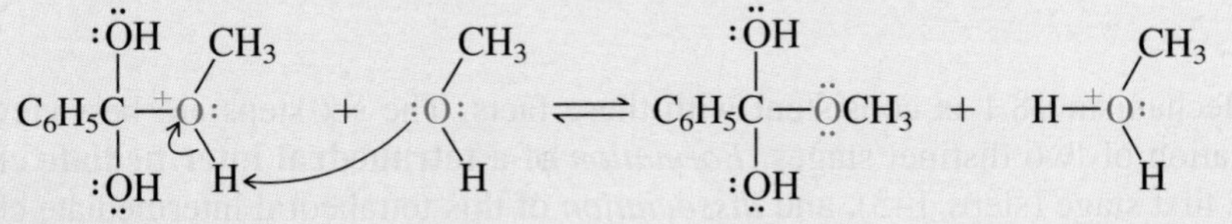

Protonated form of tetrahedral intermediate — Methanol — Tetrahedral intermediate — Methyloxonium ion

Step 4: The second stage begins with protonation of the tetrahedral intermediate on one of its hydroxyl oxygens.

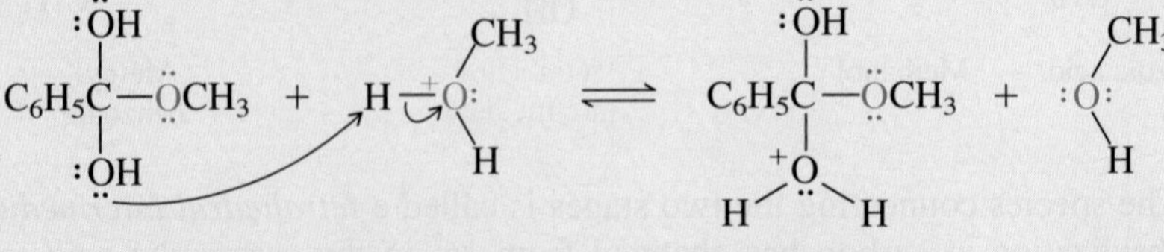

Tetrahedral intermediate — Methyloxonium ion — Hydroxyl-protonated tetrahedral intermediate — Methanol

Step 5: This intermediate loses a molecule of water to give the protonated form of the ester.

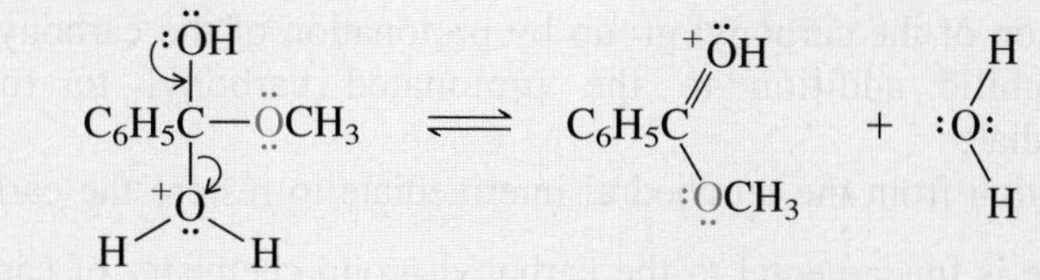

Hydroxyl-protonated tetrahedral intermediate — Conjugate acid of methyl benzoate — Water

Step 6: Deprotonation of the species formed in step 5 gives the neutral form of the ester product.

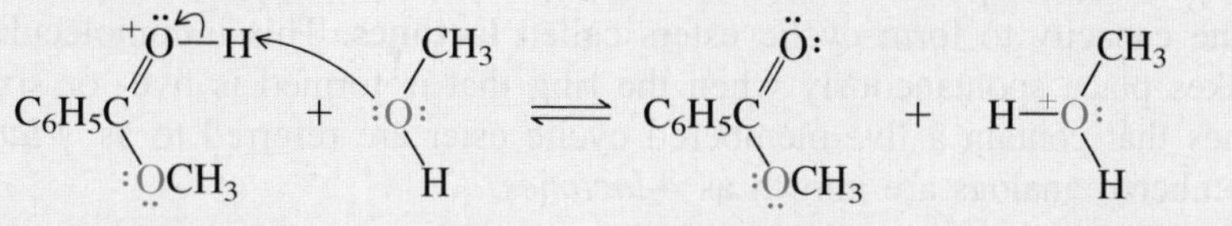

Conjugate acid of methyl benzoate — Methanol — Methyl benzoate — Methyloxonium ion

are analogous to those in the mechanism for acid-catalyzed nucleophilic addition of an alcohol to an aldehyde or a ketone (see Section 17.8). The tetrahedral intermediate cannot be isolated. It is unstable under the conditions of its formation and undergoes acid-catalyzed dehydration to form the ester.

Notice that the oxygen of methanol becomes incorporated into the methyl benzoate product according to Mechanism 18.1, as the results of the Roberts–Urey experiment require it to be.

Notice, too, that the carbonyl oxygen of the carboxylic acid is protonated in the first step and not the hydroxyl oxygen. The species formed by protonation of the carbonyl oxygen is more stable because it is stabilized by electron delocalization. The positive charge is shared equally by both oxygens.

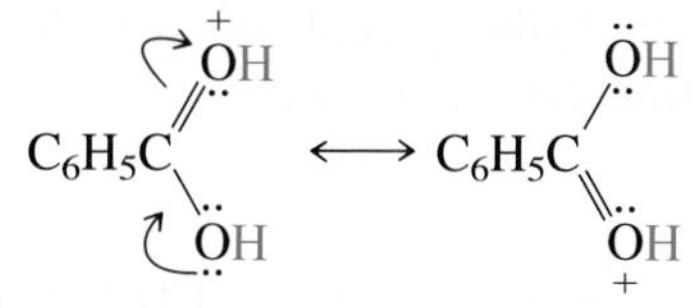

Electron delocalization in carbonyl-protonated benzoic acid

Protonation of the hydroxyl oxygen, on the other hand, yields a less stable cation:

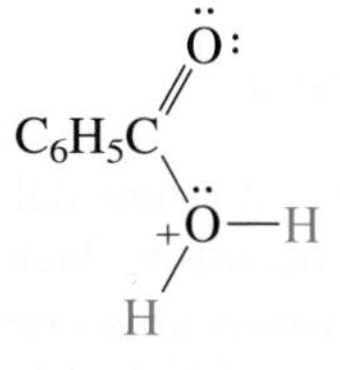

Localized positive charge in hydroxyl-protonated benzoic acid

The positive charge in this cation cannot be shared by the two oxygens; it is localized on one of them. Because protonation of the *carbonyl oxygen* gives a more stable cation, that cation is formed preferentially.

Problem 18.9

When benzoic acid is allowed to stand in water enriched in ^{18}O, the isotopic label becomes incorporated into the benzoic acid. The reaction is catalyzed by acids. Suggest an explanation for this observation.

In the next chapter the three elements of the mechanism just described will be seen again as part of the general theme that unites the chemistry of carboxylic acid derivatives. These elements are

1. Activation of the carbonyl group by protonation of the carbonyl oxygen
2. Nucleophilic addition to the protonated carbonyl to form a tetrahedral intermediate
3. Elimination from the tetrahedral intermediate to restore the carbonyl group

This sequence is fundamental to the carbonyl-group chemistry of carboxylic acids, acyl chlorides, anhydrides, esters, and amides.

18.15 Intramolecular Ester Formation: Lactones

Hydroxy acids, compounds that contain both a hydroxyl and a carboxylic acid function, have the capacity to form cyclic esters called **lactones.** This intramolecular esterification takes place spontaneously when the ring that is formed is five- or six-membered. Lactones that contain a five-membered cyclic ester are referred to as *γ-lactones;* their six-membered analogs are known as *δ-lactones.*

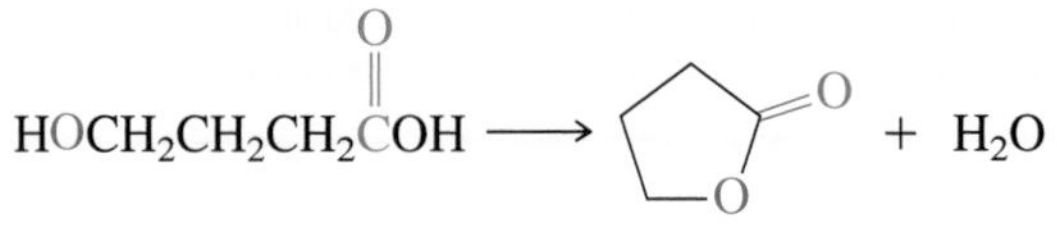

4-Hydroxybutanoic acid | 4-Butanolide (γ-Butyrolactone) | Water

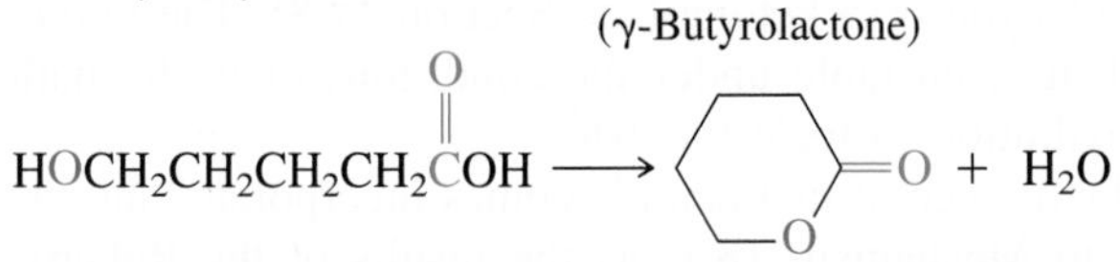

5-Hydroxypentanoic acid | 5-Pentanolide (δ-Valerolactone) | Water

Lactones are named by replacing the *-oic acid* ending of the parent carboxylic acid by *-olide* and identifying its oxygenated carbon by number as illustrated in the preceding equations.

Reactions that are expected to produce hydroxy acids often yield the derived lactones instead if a five- or six-membered ring can be formed.

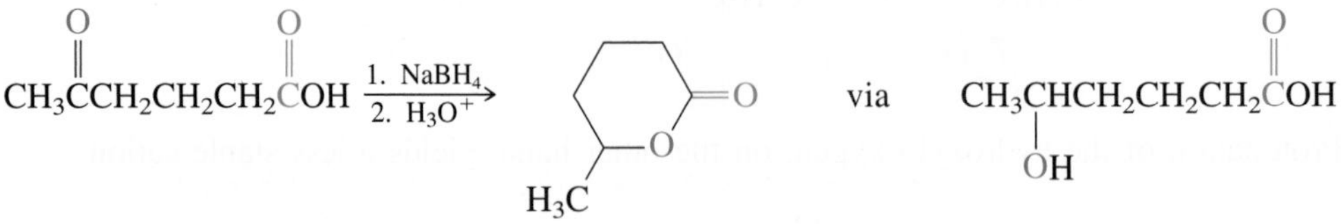

5-Oxohexanoic acid | 5-Hexanolide (78%) | 5-Hydroxyhexanoic acid

Many natural products are lactones, and it is not unusual to find examples in which the ring size is rather large. A few naturally occurring lactones are shown in Figure 18.7. The *macrolide antibiotics,* of which erythromycin is one example, are macrocyclic (large-ring) lactones. The lactone ring of erythromycin is 14-membered.

Problem 18.10

Write the structure of the hydroxy acid corresponding to each of the lactones shown in Figure 18.7.

(a) Mevalonolactone
(b) Pentadecanolide
(c) Vernolepin

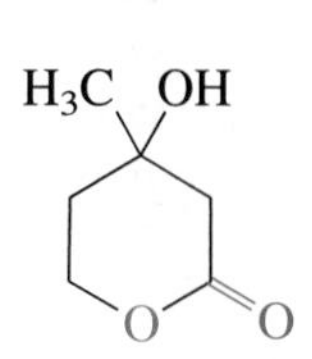

Mevalonolactone

(an intermediate in the biosynthesis of terpenes and steroids)

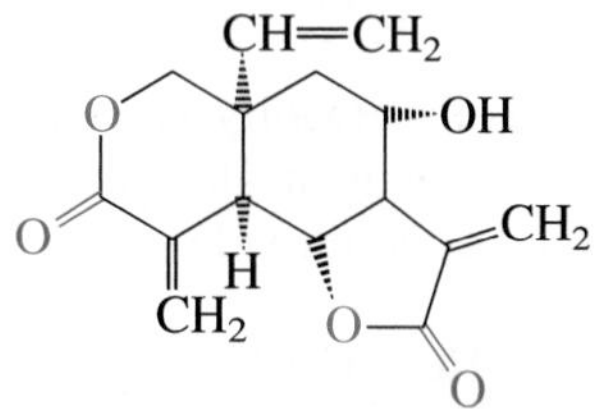

Vernolepin

(a tumor-inhibitory substance that incorporates both a γ-lactone and a δ-lactone into its tricyclic framework)

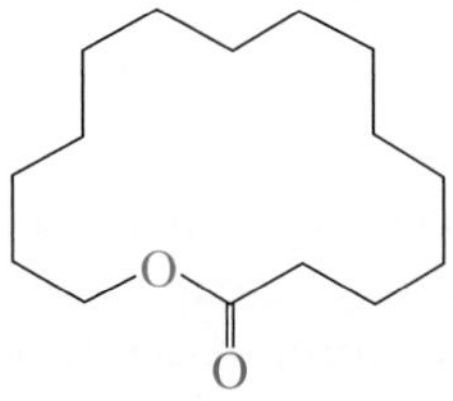

15-Pentadecanolide

(herbal musk found in Angelica and used in perfume)

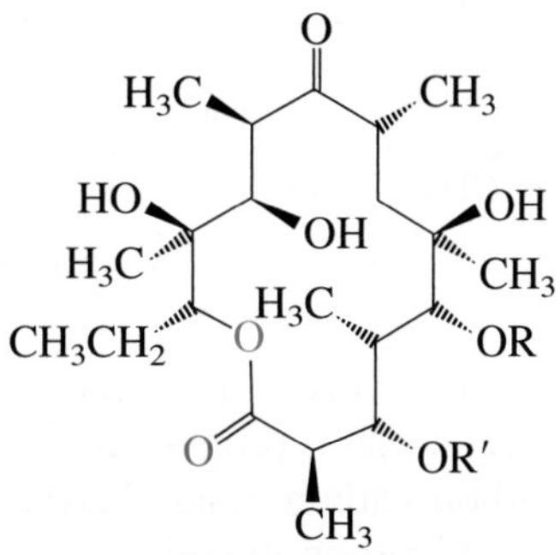

Erythromycin (R and R′ are carbohydrate units)

(a macrolide antibiotic; drug production is by fermentation processes, but the laboratory synthesis of this complex substance has been achieved)

Figure 18.7

Some naturally occurring lactones.

Sample Solution (a) The ring oxygen of the lactone is derived from the OH group of the hydroxy acid. To identify the hydroxy acid, disconnect the O—C(O) bond of the lactone.

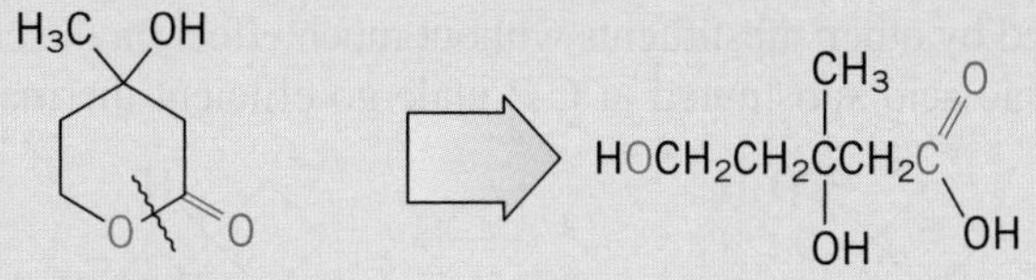

Mevalonolactone (disconnect bond indicated) | Mevalonic acid

Lactones with three- or four-membered rings (α-lactones and β-lactones) are very reactive, making their isolation difficult. Special methods are normally required for the laboratory synthesis of small-ring lactones as well as those that contain rings larger than six-membered.

18.16 Decarboxylation of Malonic Acid and Related Compounds

The loss of a molecule of carbon dioxide from a carboxylic acid is known as **decarboxylation.**

$$RCO_2H \longrightarrow RH + CO_2$$

Carboxylic acid — Alkane — Carbon dioxide

Decarboxylation of simple carboxylic acids takes place with great difficulty and is rarely encountered.

Compounds that readily undergo thermal decarboxylation include those related to malonic acid. On being heated above its melting point, malonic acid is converted to acetic acid and carbon dioxide.

$$HO_2CCH_2CO_2H \xrightarrow{150°C} CH_3CO_2H + CO_2$$

Malonic acid (propanedioic acid) — Acetic acid (ethanoic acid) — Carbon dioxide

It is important to recognize that only one carboxyl group is lost in this process. The second carboxyl group is retained. A mechanism recognizing the assistance that one carboxyl group gives to the departure of the other is represented by the equation

Keto-enol tautomerism was introduced in Chapter 9, Section 9.12.

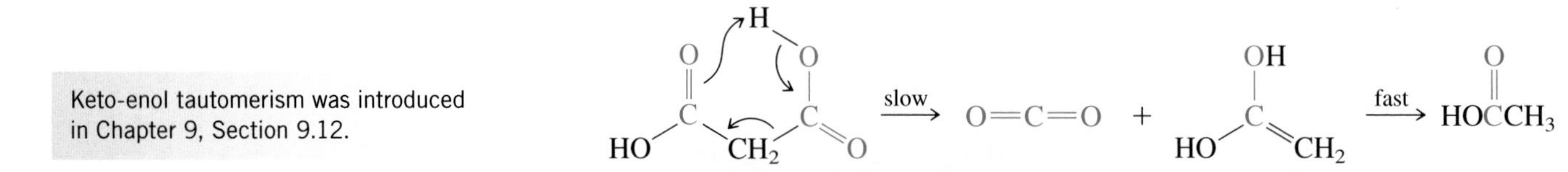

Carbon dioxide — Enol form of acetic acid — Acetic acid

The transition state involves the carbonyl oxygen of one carboxyl group—the one that stays behind—acting as a proton acceptor toward the hydroxyl group of the carboxyl that is lost. Carbon–carbon bond cleavage leads to the enol form of acetic acid, along with a molecule of carbon dioxide.

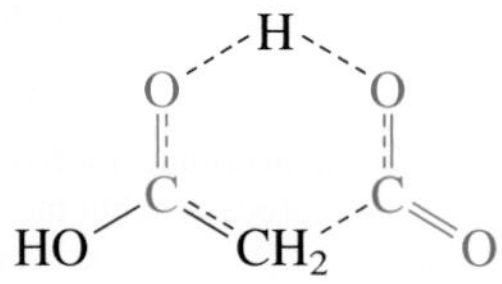

Transition state in thermal decarboxylation of malonic acid

The enol intermediate subsequently tautomerizes to acetic acid.

The protons attached to C-2 of malonic acid are not directly involved in the process and so may be replaced by other substituents without much effect on the ease of decarboxylation. Analogs of malonic acid substituted at C-2 undergo efficient thermal decarboxylation.

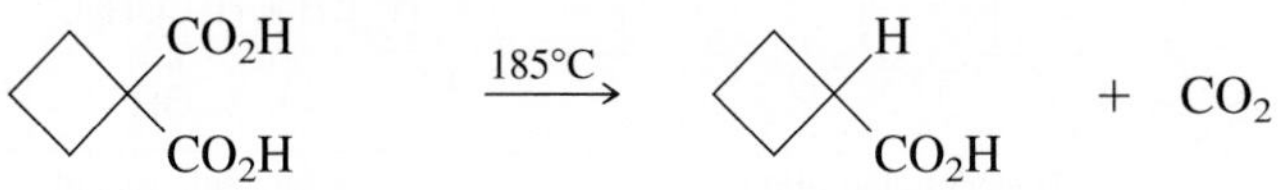

1,1-Cyclobutanedicarboxylic acid — Cyclobutanecarboxylic acid (74%) — Carbon dioxide

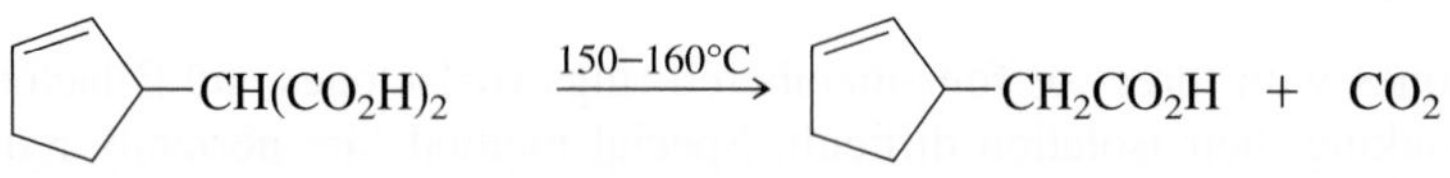

2-(2-Cyclopentenyl)malonic acid — (2-Cyclopentenyl)acetic acid (96–99%) — Carbon dioxide

Problem 18.11

What will be the product isolated after thermal decarboxylation of each of the following? Using curved arrows, represent the bond changes that take place at the transition state.

(a) $(CH_3)_2C(CO_2H)_2$

(b) $CH_3(CH_2)_6CHCO_2H$ with CO_2H on the CH carbon

(c) 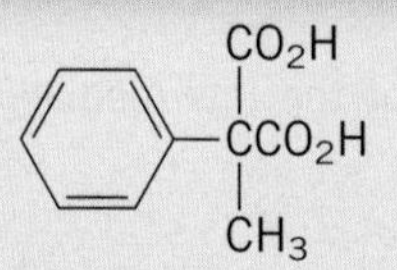

Sample Solution (a) Thermal decarboxylation of malonic acid derivatives leads to the replacement of one of the carboxyl groups by a hydrogen.

$$(CH_3)_2C(CO_2H)_2 \xrightarrow{\text{heat}} (CH_3)_2CHCO_2H + CO_2$$

2,2-Dimethylmalonic acid — 2-Methylpropanoic acid — Carbon dioxide

The transition state incorporates a cyclic array of six atoms:

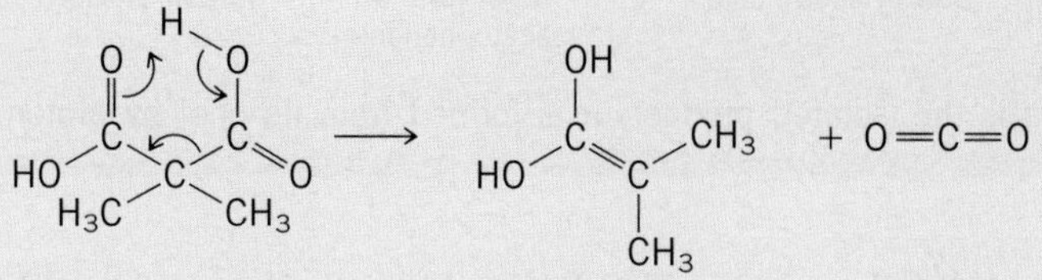

2,2-Dimethylmalonic acid — Enol form of 2-methylpropanoic acid — Carbon dioxide

Tautomerization of the enol form to 2-methylpropanoic acid completes the process.

The thermal decarboxylation of malonic acid derivatives is the last step in a multistep synthesis of carboxylic acids known as the *malonic ester synthesis.* This synthetic method will be described in Section 20.11.

Notice that the carboxyl group that stays behind during the decarboxylation of malonic acid has a hydroxyl function that is not directly involved in the process. Compounds that have substituents other than hydroxyl groups at this position undergo an analogous decarboxylation.

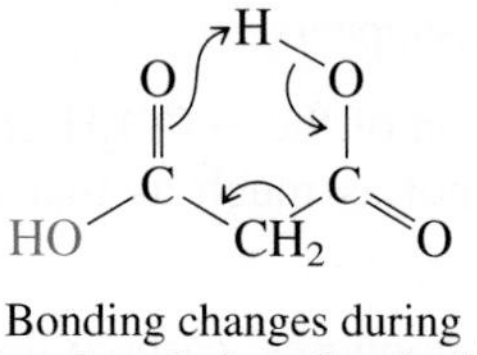

Bonding changes during decarboxylation of malonic acid

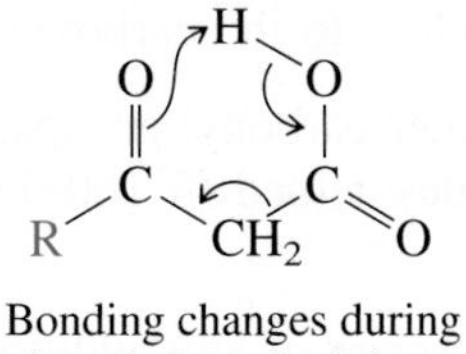

Bonding changes during decarboxylation of a β-keto acid

The compounds most frequently encountered in this reaction are β-keto acids, that is, carboxylic acids in which the β carbon is a carbonyl function. Decarboxylation of β-keto acids leads to ketones.

β-Keto acid — Carbon dioxide — Enol form of ketone — Ketone

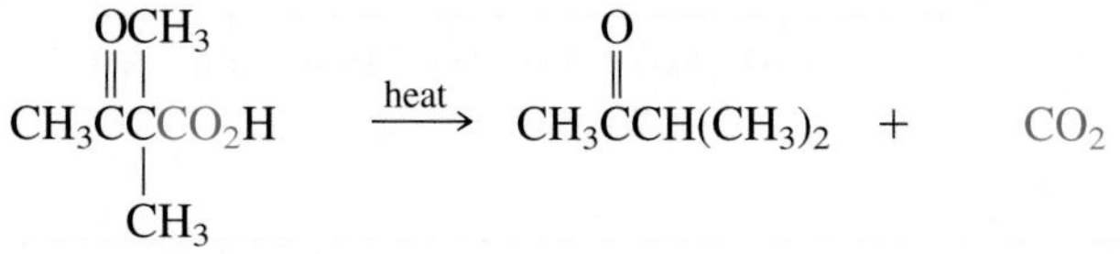

2,2-Dimethylacetoacetic acid — 3-Methyl-2-butanone — Carbon dioxide

Problem 18.12

Show the bonding changes that occur, and write the structure of the intermediate formed in the thermal decarboxylation of

(a) Benzoylacetic acid
(b) 2,2-Dimethylacetoacetic acid

Sample Solution (a) By analogy to the thermal decarboxylation of malonic acid, we represent the corresponding reaction of benzoylacetic acid as

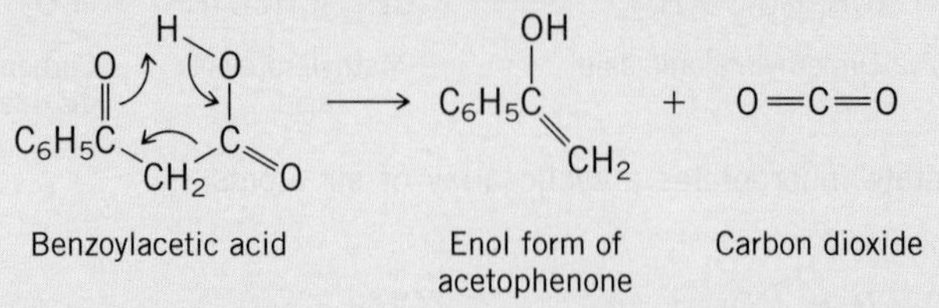

Acetophenone is the isolated product; it is formed from its enol by proton transfers.

The thermal decarboxylation of β-keto acids is the last step in a ketone synthesis known as the *acetoacetic ester synthesis,* which will be discussed in Section 20.10.

18.17 Spectroscopic Analysis of Carboxylic Acids

Infrared: The most characteristic peaks in the IR spectra of carboxylic acids are those of the hydroxyl and carbonyl groups. As shown in the IR spectrum of 4-phenylbutanoic acid (Figure 18.8) the O—H and C—H stretching frequencies overlap to produce a broad absorption in the 3500–2500 cm^{-1} region. The carbonyl group gives a strong band for C=O stretching at 1684 cm^{-1}.

^{1}H NMR: The hydroxyl proton of a CO_2H group is normally the least shielded of all the protons in an NMR spectrum, appearing 10–12 ppm downfield from tetramethylsilane, often as a broad peak. Figure 18.9 illustrates this for 4-phenylbutanoic acid. As with other hydroxyl protons, the proton of a carboxyl group can be identified by adding D_2O to the sample. Hydrogen–deuterium exchange converts $—CO_2H$ to $—CO_2D$, and the signal corresponding to the carboxyl group disappears.

^{13}C NMR: Like other carbonyl groups, the carbon of the $—CO_2H$ group of a carboxylic acid is strongly deshielded (δ 160–185), but not as much as that of an aldehyde or ketone (δ 190–215).

UV-VIS: In the absence of any additional chromophores, carboxylic acids absorb at a wavelength (210 nm) that is not very useful for diagnostic purposes.

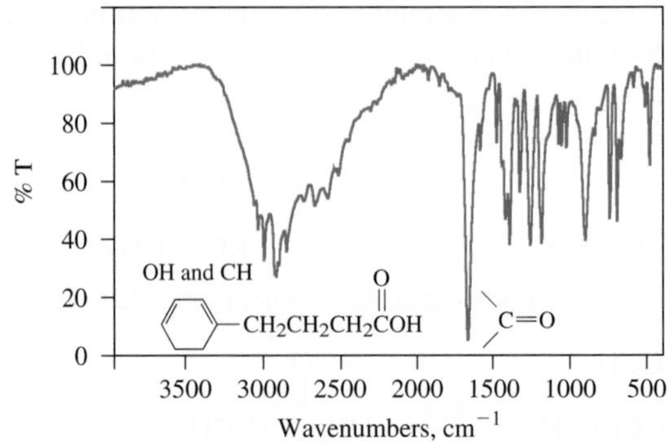

Figure 18.8

The IR spectrum of 4-phenylbutanoic acid.

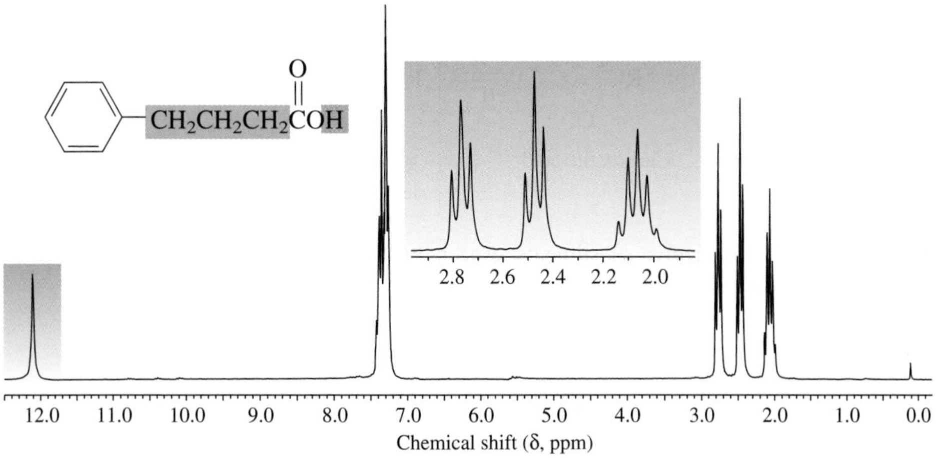

Figure 18.9

The 200-MHz ^{1}H NMR spectrum of 4-phenylbutanoic acid. The peak for the proton of the CO_2H group is at δ 12.

Mass Spectrometry: Aside from a peak for the molecular ion, which is normally easy to pick out, aliphatic carboxylic acids undergo a variety of fragmentation processes. The dominant fragmentation in aromatic acids corresponds to loss of OH, then loss of CO.

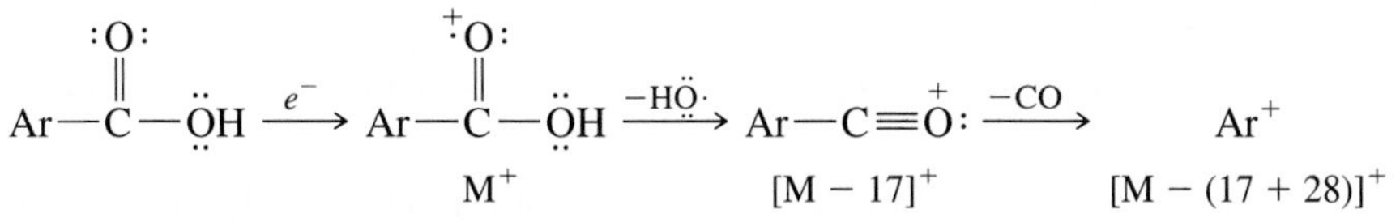

18.18 SUMMARY

Section 18.1 Carboxylic acids take their names from the alkane that contains the same number of carbons as the longest continuous chain that contains the —CO_2H group. The *-e* ending is replaced by *-oic acid.* Numbering begins at the carbon of the —CO_2H group.

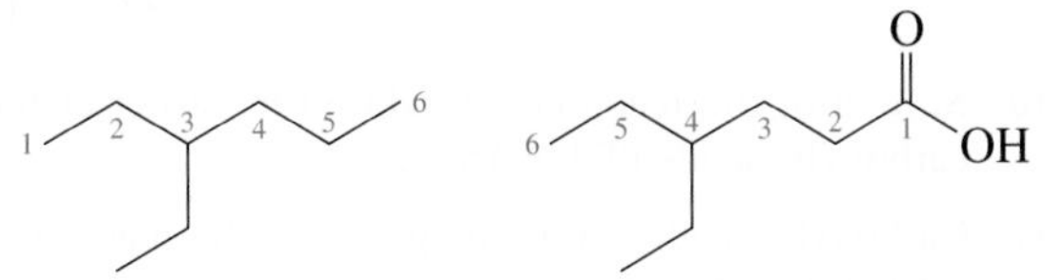

3-Ethylhexane 4-Ethylhexanoic acid

Section 18.2 Like the carbonyl group of aldehydes and ketones, the carbon of a C═O unit in a carboxylic acid is sp^2-hybridized. Compared with the carbonyl group of an aldehyde or ketone, the C═O unit of a carboxylic acid receives an extra degree of stabilization from its attached OH group.

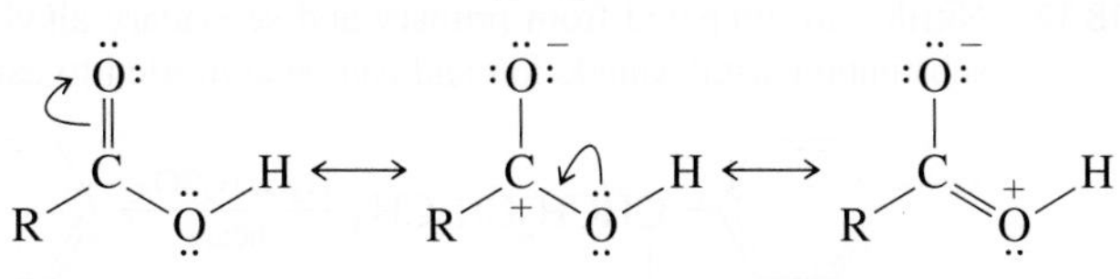

Section 18.3 Hydrogen bonding in carboxylic acids raises their melting points and boiling points above those of comparably constituted alkanes, alcohols, aldehydes, and ketones.

Section 18.4 Carboxylic acids are weak acids and, in the absence of electron-attracting substituents, have pK_a's of approximately 5. Carboxylic acids are much stronger acids than alcohols because of the electron-withdrawing power of the carbonyl group (inductive effect) and its ability to delocalize negative charge in the carboxylate anion (resonance effect).

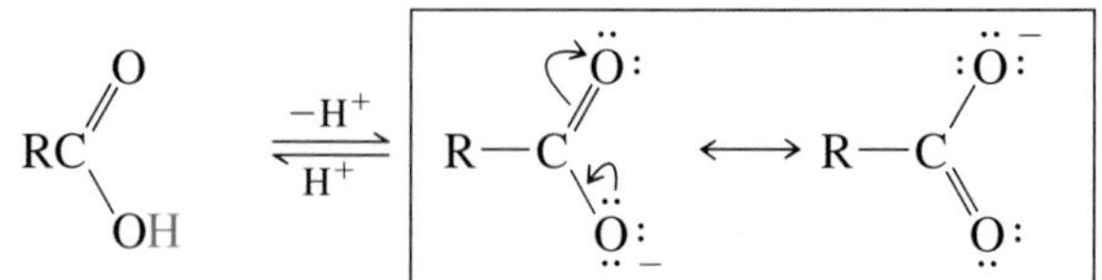

Carboxylic acid

Resonance description of electron delocalization in carboxylate anion

Sections 18.5–18.6 Electronegative substituents, especially those within a few bonds of the carboxyl group, increase the acidity of carboxylic acids.

CF_3CO_2H

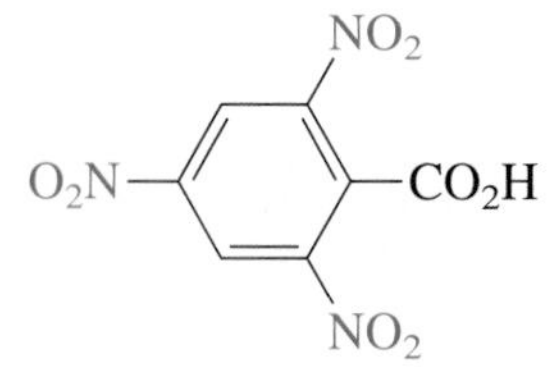

Trifluoroacetic acid $pK_a = 0.2$

2,4,6-Trinitrobenzoic acid $pK_a = 0.6$

Section 18.7 Although carboxylic acids dissociate to only a small extent in water, they are deprotonated almost completely in basic solution.

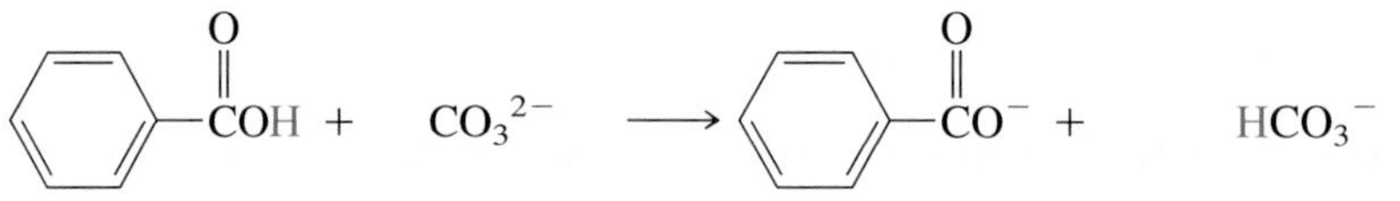

Benzoic acid $pK_a = 4.2$ (stronger acid) Carbonate ion Benzoate ion Hydrogen carbonate ion $pK_a = 10.2$ (weaker acid)

Section 18.8 Dicarboxylic acids have separate pK_a values for their first and second ionizations.

Section 18.9 Carbon dioxide and carbonic acid are in equilibrium in water. Carbon dioxide is the major component.

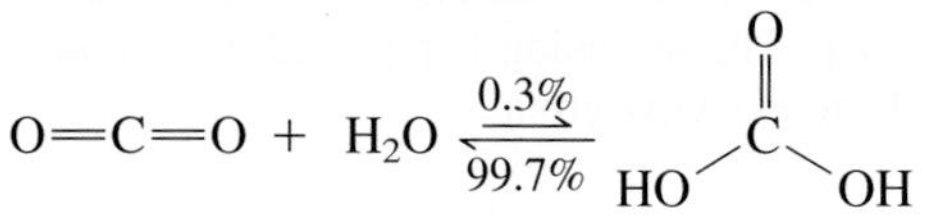

Section 18.10 Several of the reactions introduced in earlier chapters can be used to prepare carboxylic acids (Table 18.4).

Section 18.11 Carboxylic acids can be prepared by the reaction of Grignard reagents with carbon dioxide.

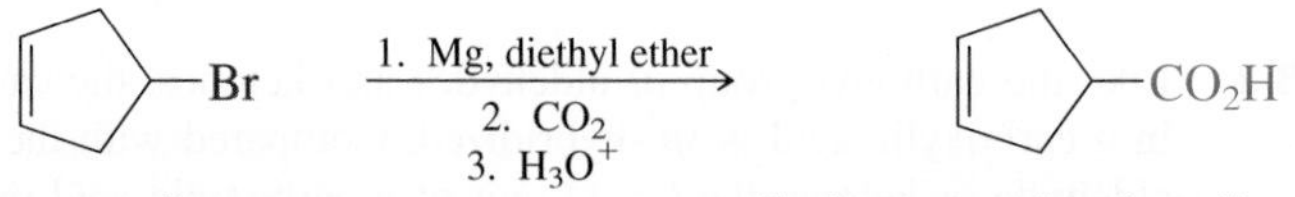

4-Bromocyclopentene Cyclopentene-4-carboxylic acid (66%)

Section 18.12 Nitriles are prepared from primary and secondary alkyl halides by nucleophilic substitution with cyanide ion and can be converted to carboxylic acids by hydrolysis.

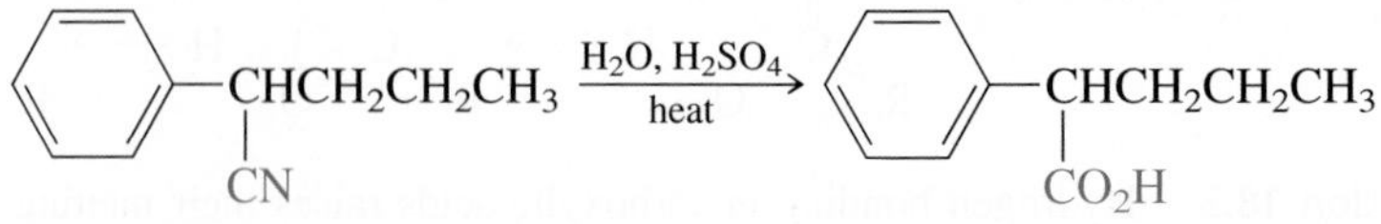

2-Phenylpentanenitrile 2-Phenylpentanoic acid (52%)

Likewise, the cyano group of a cyanohydrin can be hydrolyzed to —CO_2H.

Section 18.13 Among the reactions of carboxylic acids, their conversions to acyl chlorides, primary alcohols, and esters were introduced in earlier chapters and were reviewed in Table 18.5.

Section 18.14 The mechanism of acid-catalyzed esterification involves some key features that are fundamental to the chemistry of carboxylic acids and their derivatives.

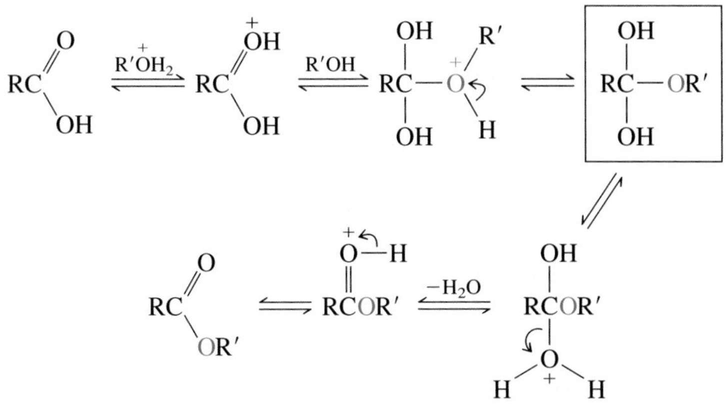

Protonation of the carbonyl oxygen activates the carbonyl group toward nucleophilic addition. Addition of an alcohol gives a tetrahedral intermediate (shown in the box in the preceding equation), which has the capacity to revert to starting materials or to undergo dehydration to yield an ester.

Section 18.15 An intramolecular esterification can occur when a molecule contains both a hydroxyl and a carboxyl group. Cyclic esters are called *lactones* and are most stable when the ring is five- or six-membered.

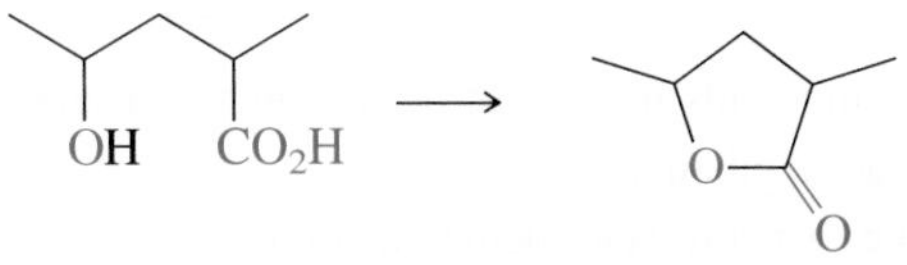

4-Hydroxy-2-methylpentanoic acid — 2-Methyl-4-pentanolide

Section 18.16 1,1-Dicarboxylic acids (malonic acids) and β-keto acids undergo thermal decarboxylation by a mechanism in which a β-carbonyl group assists the departure of carbon dioxide.

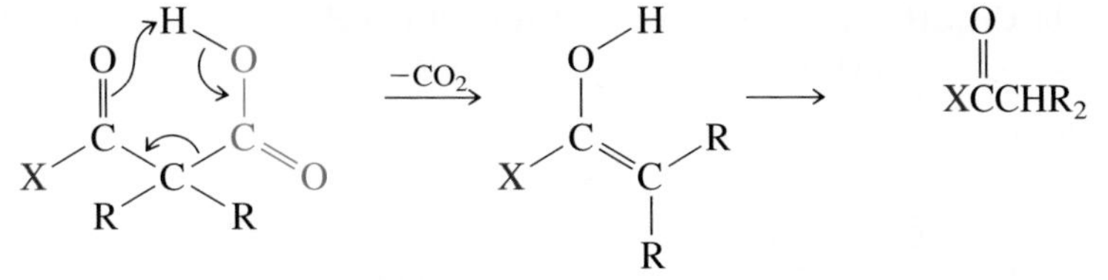

X = OH: malonic acid derivative
X = alkyl or aryl: β-keto acid

Enol form of product

X = OH: carboxylic acid
X = alkyl or aryl: ketone

Section 18.17 Carboxylic acids are readily identified by the presence of strong IR absorptions near 1700 cm^{-1} (C═O) and between 2500 and 3500 cm^{-1} (OH), a ^1H NMR signal for the hydroxyl proton at δ 10–12, and a ^{13}C signal for the carbonyl carbon near δ 180.

PROBLEMS

18.13 Many carboxylic acids are much better known by their common names than by their systematic names. Some of these follow. Provide a structural formula for each one on the basis of its systematic name.

(a) 2-Hydroxypropanoic acid (better known as *lactic acid,* it is found in sour milk and is formed in the muscles during exercise)

(b) 2-Hydroxy-2-phenylethanoic acid (also known as *mandelic acid,* it is obtained from plums, peaches, and other fruits)

(c) Tetradecanoic acid (also known as *myristic acid,* it can be obtained from a variety of fats)

(d) 10-Undecenoic acid (also called *undecylenic acid,* it is used, in combination with its zinc salt, to treat fungal infections such as athlete's foot)

(e) 3,5-Dihydroxy-3-methylpentanoic acid (also called *mevalonic acid,* it is an important intermediate in the biosynthesis of terpenes and steroids)

(f) (*E*)-2-Methyl-2-butenoic acid (also known as *tiglic acid,* it is a constituent of various natural oils)

(g) 2-Hydroxybutanedioic acid (also known as *malic acid,* it is found in apples and other fruits)

(h) 2-Hydroxy-1,2,3-propanetricarboxylic acid (better known as *citric acid,* it contributes to the tart taste of citrus fruits)

(i) 2-(*p*-Isobutylphenyl)propanoic acid (an antiinflammatory drug better known as *ibuprofen*)

(j) *o*-Hydroxybenzenecarboxylic acid (better known as *salicylic acid,* it is obtained from willow bark)

18.14 Give an acceptable IUPAC name for each of the following:

(a) $CH_3(CH_2)_6CO_2H$

(b) $CH_3(CH_2)_6CO_2K$

(c) $H_2C{=}CH(CH_2)_5CO_2H$

(d) 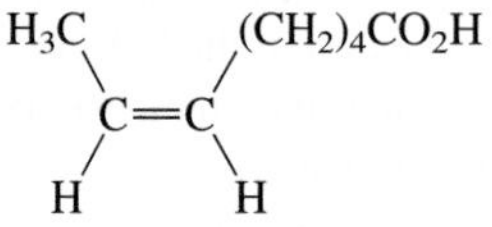

(e) $HO_2C(CH_2)_6CO_2H$

(f) $CH_3(CH_2)_4CH(CO_2H)_2$

(g) 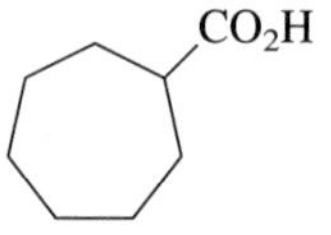

(h) 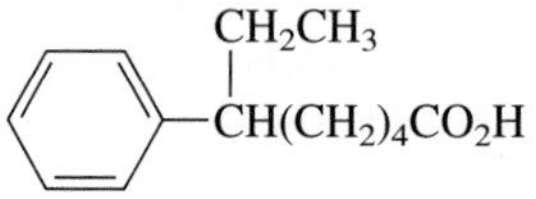

18.15 Rank the compounds in each of the following groups in order of decreasing acidity:

(a) Acetic acid, ethane, ethanol

(b) Benzene, benzoic acid, benzyl alcohol

(c) 1,3-Propanediol, propanedioic acid, propanoic acid

(d) Acetic acid, ethanol, trifluoroacetic acid, 2,2,2-trifluoroethanol, trifluoromethanesulfonic acid (CF_3SO_2OH)

18.16 Identify the more acidic compound in each of the following pairs:

(a) $CF_3CH_2CO_2H$ or $CF_3CH_2CH_2CO_2H$

(b) $CH_3CH_2CH_2CO_2H$ or $CH_3C{\equiv}CCO_2H$

(c) 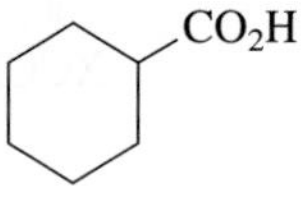 or 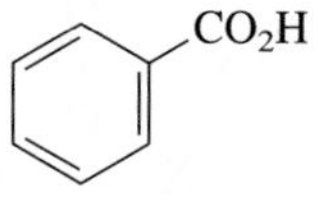

(d) 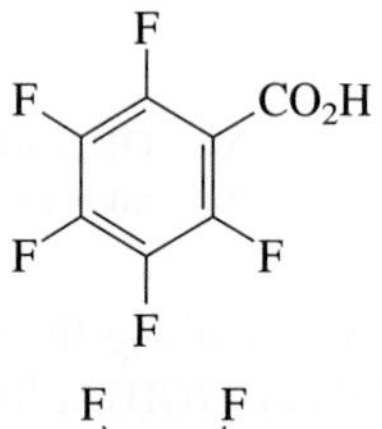 or 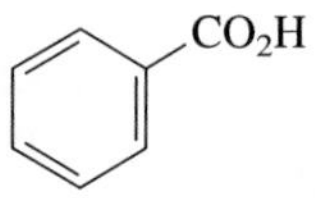

(e) 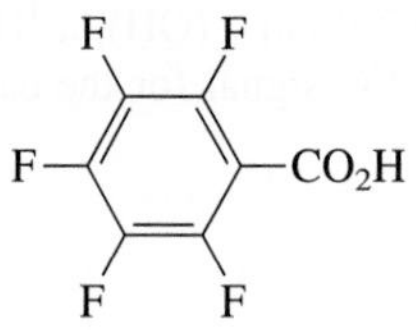 or 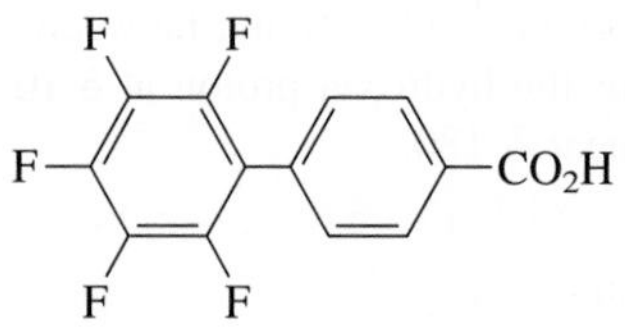

(f) 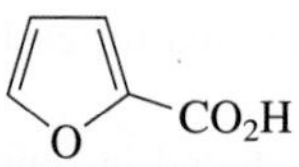 or 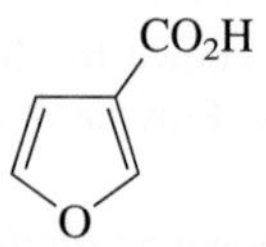

(g) 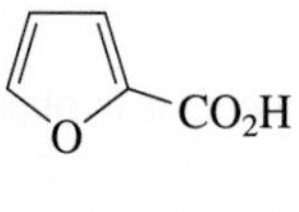 or 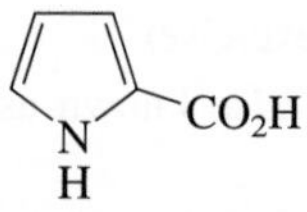

18.17 Propose methods for preparing butanoic acid from each of the following:

(a) 1-Butanol

(b) Butanal

(c) 1-Butene

(d) 1-Propanol

(e) 2-Propanol

(f) $CH_3CH_2CH(CO_2H)_2$

18.18 It is sometimes necessary to prepare isotopically labeled samples of organic substances for probing biological transformations and reaction mechanisms. Various sources of the radioactive mass-14 carbon isotope are available. Describe synthetic procedures by which benzoic acid, labeled with ^{14}C at its carbonyl carbon, could be prepared from benzene and the following ^{14}C-labeled precursors. You may use any necessary organic or inorganic reagents.

(a) CH_3Cl (b) $HC(=O)H$ (c) CO_2

18.19 Give the product of the reaction of pentanoic acid with each of the following reagents:

(a) Sodium hydroxide
(b) Sodium bicarbonate
(c) Thionyl chloride
(d) Phosphorus tribromide
(e) Benzyl alcohol, sulfuric acid (catalytic amount)
(f) Lithium aluminum hydride, then hydrolysis
(g) Phenylmagnesium bromide

18.20 Show how butanoic acid may be converted to each of the following compounds:

(a) 1-Butanol (c) 1-Chlorobutane (e) Phenyl propyl ketone
(b) Butanal (d) Butanoyl chloride (f) 4-Octanone

18.21 Each of the following reactions has been reported in the chemical literature and gives a single product in good yield. What is the product in each reaction?

(a) 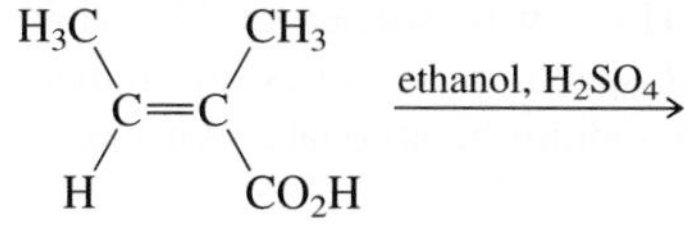

(d) 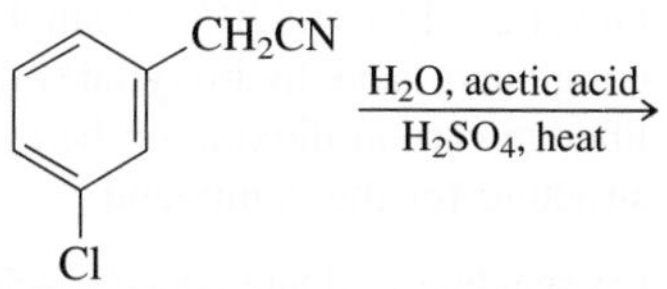

(b) 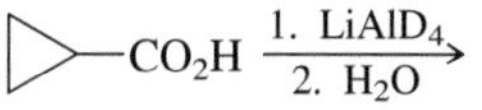

(e) $H_2C{=}CH(CH_2)_8CO_2H \xrightarrow[\text{benzoyl peroxide}]{\text{HBr}}$

(c) 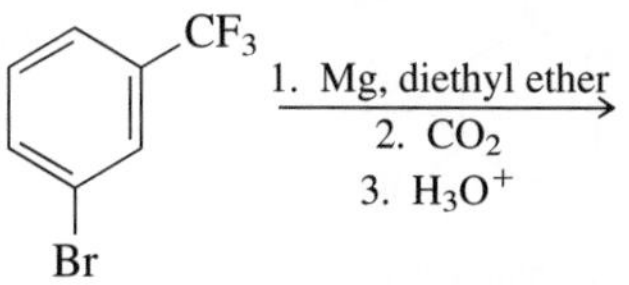

18.22 The compound shown was subjected to the following series of reactions to give a product having the molecular formula $C_9H_9ClO_3$. What is this product?

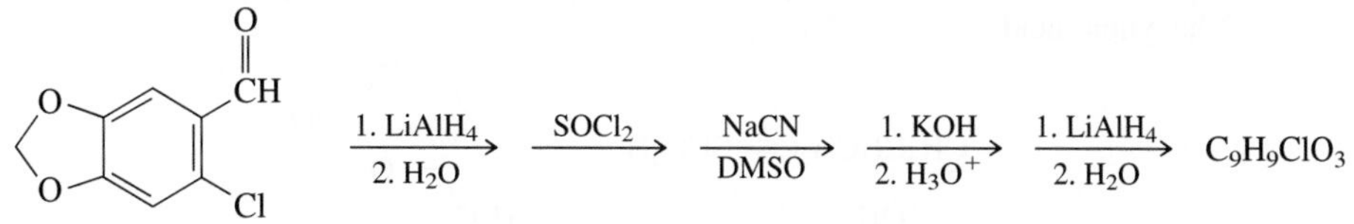

18.23 Show by a series of equations how you could synthesize each of the following compounds from the indicated starting material and any necessary organic or inorganic reagents:

(a) 2-Methylpropanoic acid from *tert*-butyl alcohol
(b) 3-Methylbutanoic acid from *tert*-butyl alcohol
(c) 3,3-Dimethylbutanoic acid from *tert*-butyl alcohol
(d) $HO_2C(CH_2)_5CO_2H$ from $HO_2C(CH_2)_3CO_2H$
(e) 3-Phenyl-1-butanol from $CH_3CH(C_6H_5)CH_2CN$
(f) 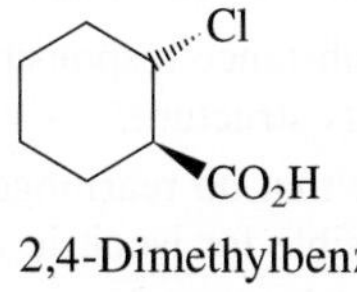 from (*E*)-$ClCH{=}CHCO_2H$
(g) 2,4-Dimethylbenzoic acid from *m*-xylene
(h) 4-Chloro-3-nitrobenzoic acid from *p*-chlorotoluene
(i) (*Z*)-$CH_3CH{=}CHCO_2H$ from propyne

18.24 (a) Which stereoisomer of 4-hydroxycyclohexanecarboxylic acid (cis or trans) can form a lactone? What is the conformation of the cyclohexane ring in the starting hydroxy acid? In the lactone?

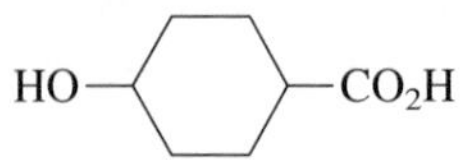
HO
CO_2H

(b) Repeat part (a) for the case of 3-hydroxycyclohexanecarboxylic acid.

18.25 When compound A is heated, two isomeric products are formed. What are these two products?

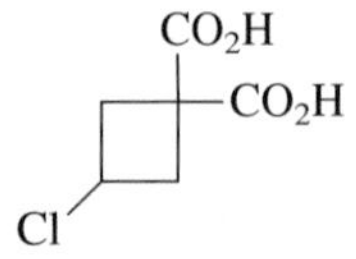
CO_2H
CO_2H
Cl

Compound A

18.26 A certain carboxylic acid ($C_{14}H_{26}O_2$), which can be isolated from whale blubber or sardine oil, yields nonanal and $O{=}CH(CH_2)_3CO_2H$ on ozonolysis. What is the structure of this acid?

18.27 When levulinic acid ($CH_3\overset{O}{\overset{\|}{C}}CH_2CH_2CO_2H$) was hydrogenated at high pressure over a nickel catalyst at 220°C, a single product, $C_5H_8O_2$, was isolated in 94% yield. This compound lacks hydroxyl absorption in its IR spectrum and does not immediately liberate carbon dioxide on being shaken with sodium bicarbonate. What is a reasonable structure for the compound?

18.28 On standing in dilute aqueous acid, compound A is smoothly converted to mevalonolactone.

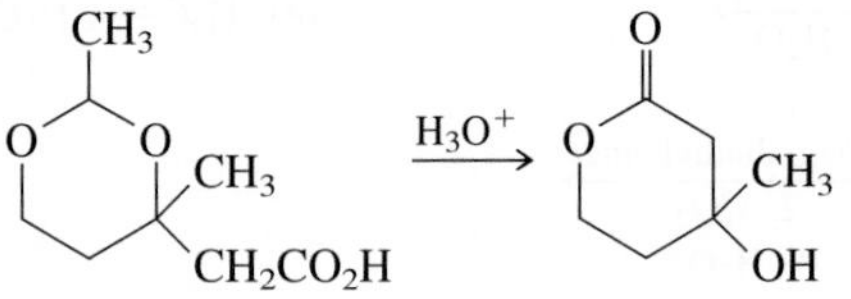
CH_3
O
O
CH_3
CH_2CO_2H
H_3O^+
O
O
CH_3
OH

Compound A — Mevalonolactone

Suggest a reasonable mechanism for this reaction. What other organic product is also formed?

18.29 Suggest reaction conditions suitable for the preparation of compound A from 5-hydroxy-2-hexynoic acid.

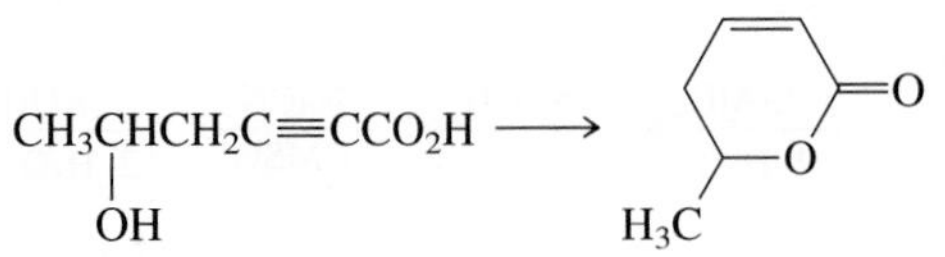
$CH_3CHCH_2C{\equiv}CCO_2H$
OH
O
O
H_3C

5-Hydroxy-2-hexynoic acid — Compound A

18.30 In the presence of the enzyme *aconitase,* the double bond of aconitic acid undergoes hydration. The reaction is reversible, and the following equilibrium is established:

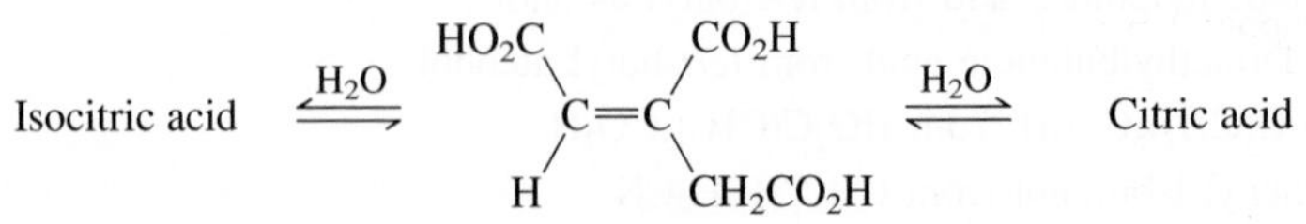
Isocitric acid
H_2O
HO_2C
CO_2H
C=C
H
CH_2CO_2H
H_2O
Citric acid

($C_6H_8O_7$) (6% at equilibrium) — Aconitic acid (4% at equilibrium) — ($C_6H_8O_7$) (90% at equilibrium)

(a) The major tricarboxylic acid present is *citric acid,* the substance responsible for the tart taste of citrus fruits. Citric acid is achiral. What is its structure?

(b) What must be the constitution of isocitric acid? (Assume that no rearrangements accompany hydration.) How many stereoisomers are possible for isocitric acid?

18.31 The ^1H NMR spectra of formic acid (HCO_2H), maleic acid (*cis*-$HO_2CCH{=}CHCO_2H$), and malonic acid ($HO_2CCH_2CO_2H$) are similar in that each is characterized by two

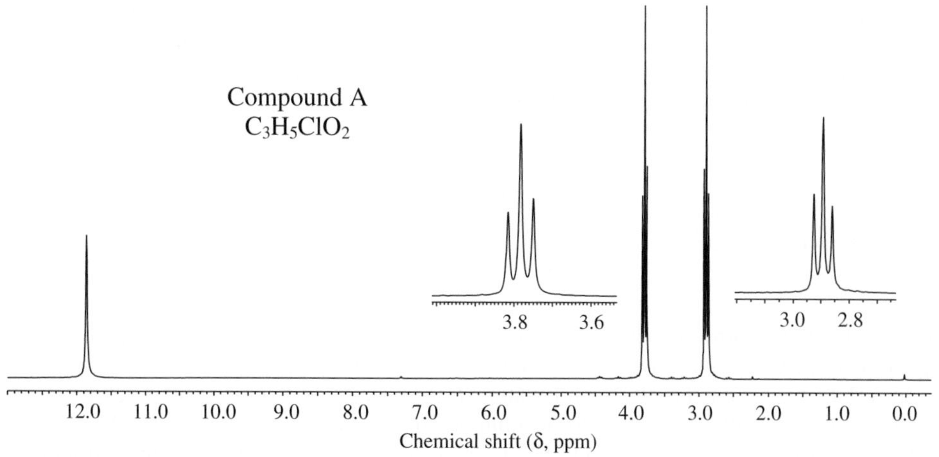

Figure 18.10

The 200-MHz ^{1}H NMR spectrum of compound A ($C_3H_5ClO_2$) (Problem 18.33a).

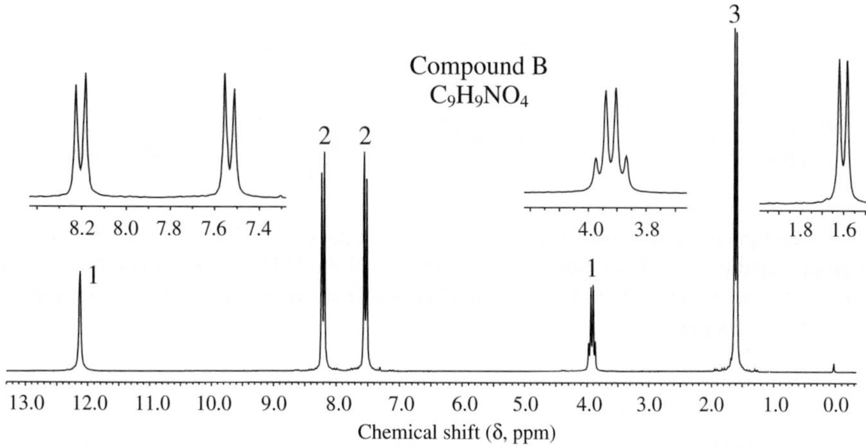

Figure 18.11

The 200-MHz ^{1}H NMR spectrum of compound B ($C_9H_9NO_4$) (Problem 18.33b).

singlets of equal intensity. Match these compounds with the designations A, B, and C on the basis of the appropriate ^{1}H NMR chemical shift data.

Compound A: signals at δ 3.2 and 12.1

Compound B: signals at δ 6.3 and 12.4

Compound C: signals at δ 8.0 and 11.4

18.32 Compounds A and B are isomers having the molecular formula $C_4H_8O_3$. Identify A and B on the basis of their ^{1}H NMR spectra.

Compound A: δ 1.3 (3H, triplet); 3.6 (2H, quartet); 4.1 (2H, singlet); 11.1 (1H, broad singlet)

Compound B: δ 2.6 (2H, triplet); 3.4 (3H, singlet); 3.7 (2H triplet); 11.3 (1H, broad singlet)

18.33 Compounds A and B are carboxylic acids. Identify each one on the basis of its ^{1}H NMR spectrum.

(a) Compound A ($C_3H_5ClO_2$) (Figure 18.10).

(b) Compound B ($C_9H_9NO_4$) has a nitro group attached to an aromatic ring (Figure 18.11).

Descriptive Passage and Interpretive Problems 18

Lactonization Methods

In Section 18.15 we saw that hydroxy-substituted carboxylic acids spontaneously cyclize to lactones if a five- or six-membered ring can be formed.

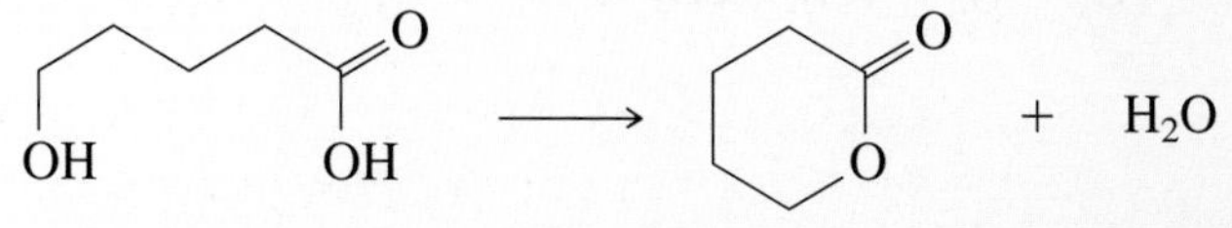

Many natural products are lactones, and chemists have directed substantial attention to developing alternative methods for their synthesis. The most successful of these efforts are based on electrophilic addition to the double bond of unsaturated carboxylic acids. For a generalized electrophilic reagent E—Y and 4-pentenoic acid, such reactions give a 5-substituted γ-lactone.

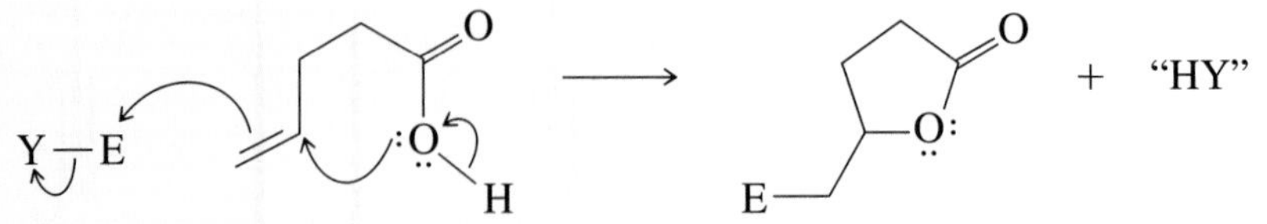

Although the curved arrows show the *overall* electron flow, the mechanism depends on the electrophilic reagent E—Y and normally involves more than one step.

In *iodolactonization* the electrophilic atom E = I, and E—Y represents a source of electrophilic iodine, usually I_2 or *N*-iodosuccinimide. In *phenylselenolactonization,* E = C_6H_5Se and E—Y is benzeneselenenyl chloride (C_6H_5SeCl). Anti addition is observed in both iodo- and phenylselenolactonization.

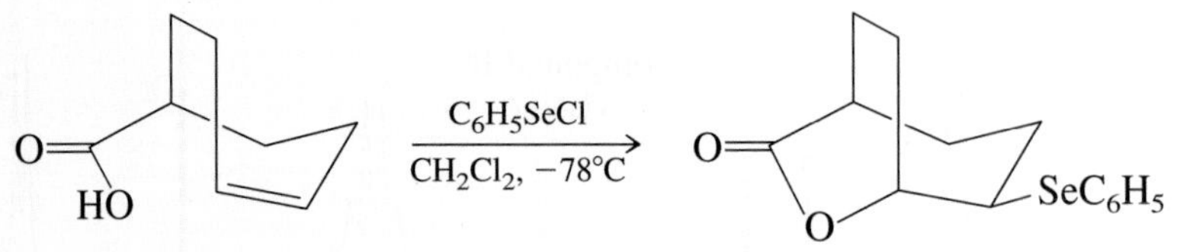

Both iodo- and phenylselenolactonization offer the advantage of giving a product containing a functional group capable of further modification. Oxidation of the C_6H_5Se substituent, for example, gives a selenoxide that undergoes elimination of C_6H_5SeOH at room temperature to introduce a double bond into the lactone.

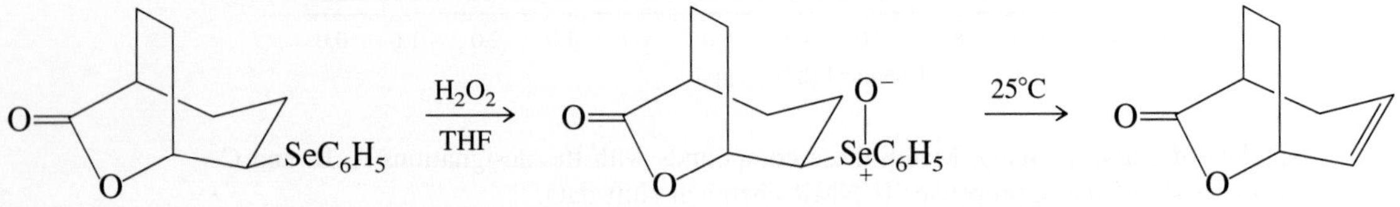

In eliminations of this type, H is always removed from the carbon β to selenium that is remote from the lactone oxygen. Elimination is syn.

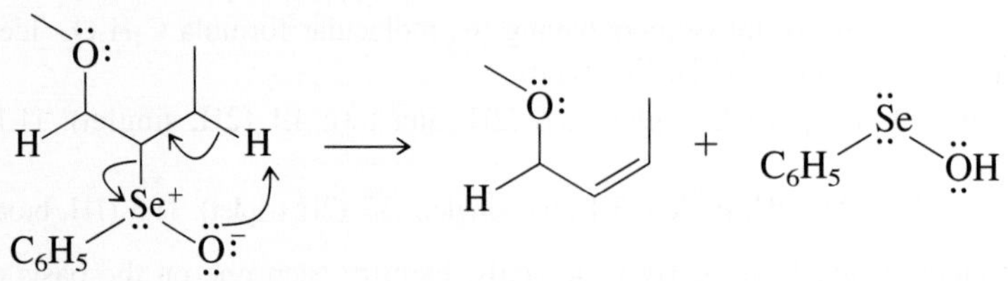

18.34 The dihydroxy acid shown was prepared as a single enantiomer and underwent spontaneous cyclization to give a δ-lactone, What are the *R–S* configurations of the chirality centers in this lactone? (No stereochemistry is implied in the structural drawing.)

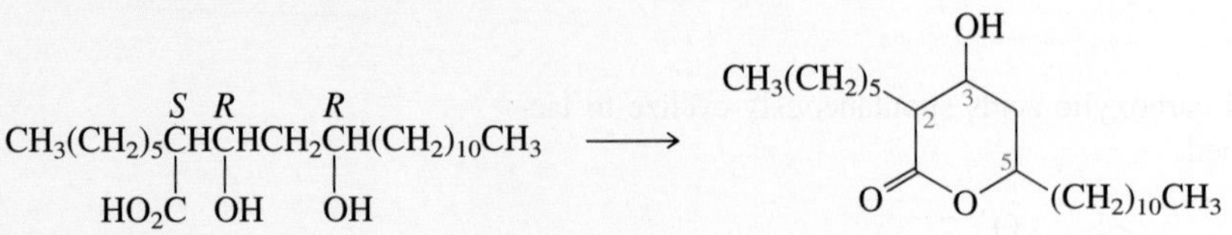

A. 2*R*, 3*R*, 5*R*
B. 2*R*, 3*R*, 5*S*
C. 2*S*, 3*R*, 5*R*
D. 2*S*, 3*R*, 5*S*

18.35 The product of the following reaction has the constitution shown. No stereochemistry is implied. Deduce the stereochemistry on the basis of the fact that iodolactonization is normally an anti addition and it was determined experimentally that the ring junction is cis.

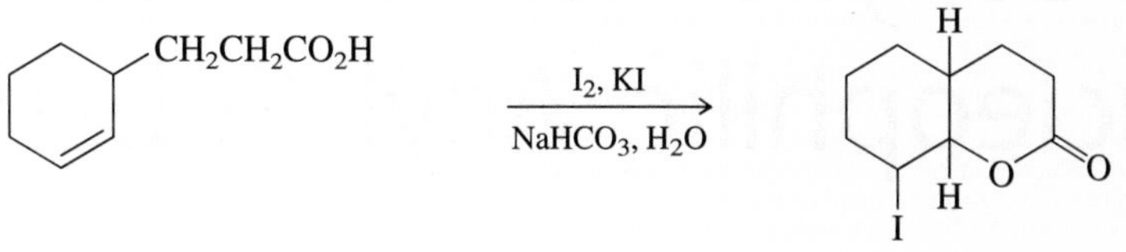

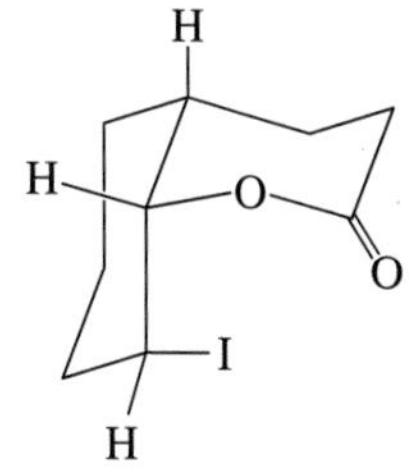

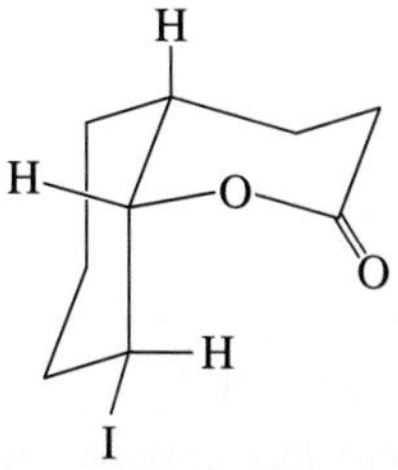

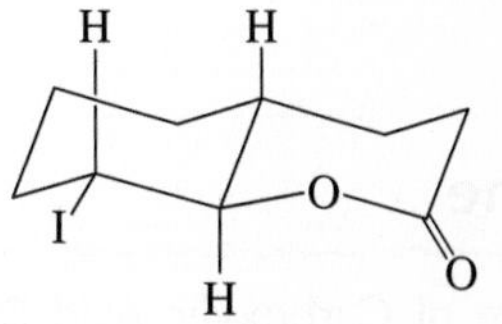

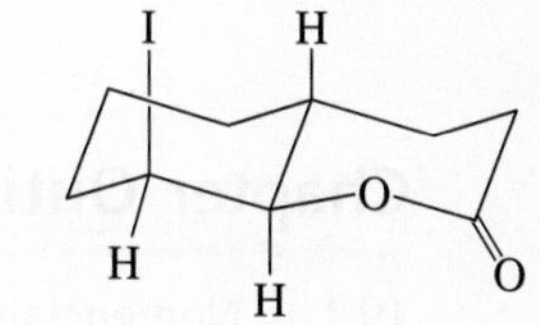

A. B. C. D.

18.36 What is the structure of the γ lactone formed by iodolactonization of 4-pentynoic acid ($HC{\equiv}CCH_2CH_2CO_2H$)? Anti addition to the triple bond occurs.

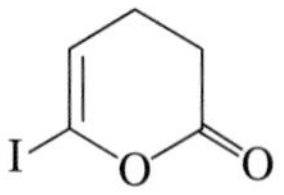

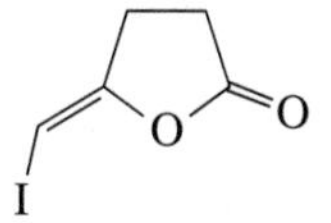

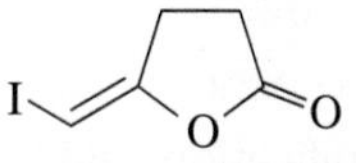

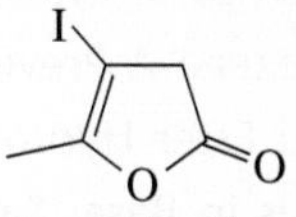

A. B. C. D.

18.37 Assume that the following reaction proceeds through a bridged selenonium ion.

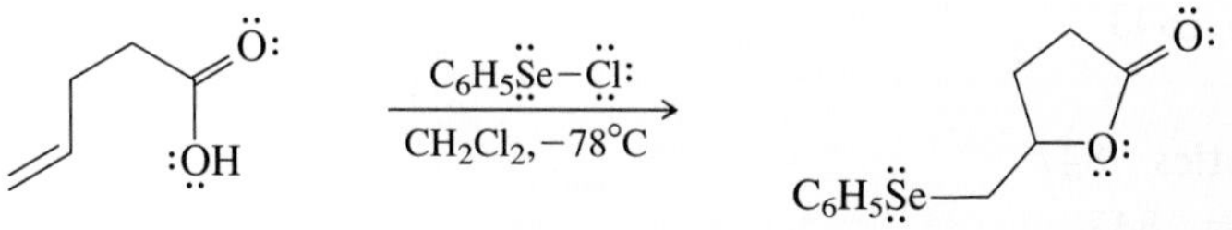

What Lewis structure best represents the bridged selenonium ion?

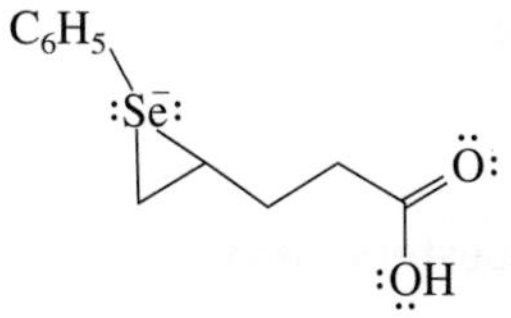

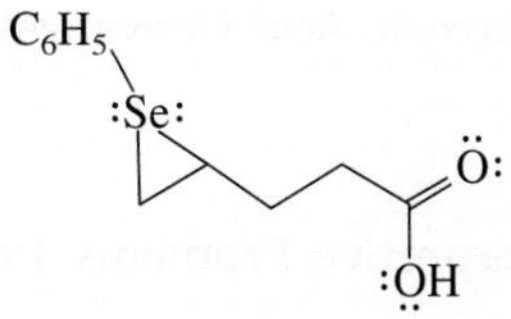

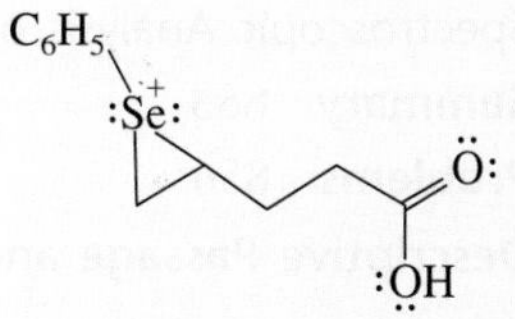

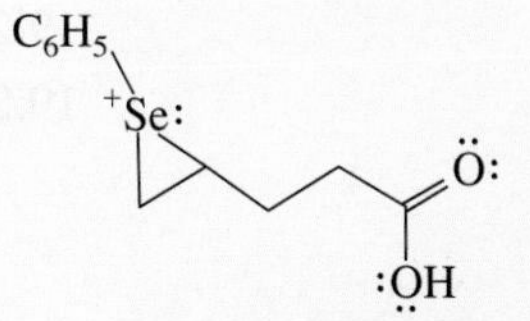

A. B. C. D.

18.38 What is compound X?

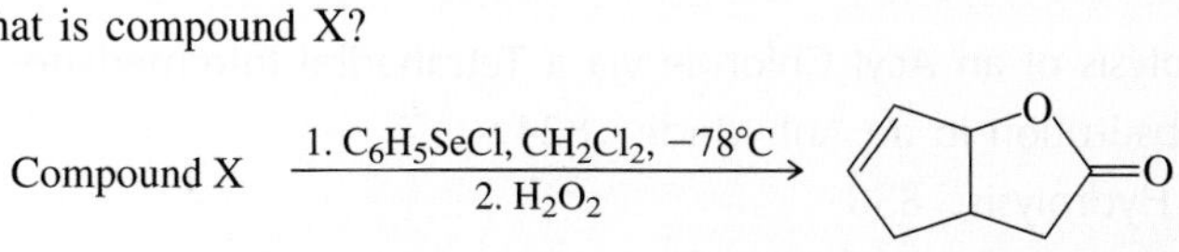

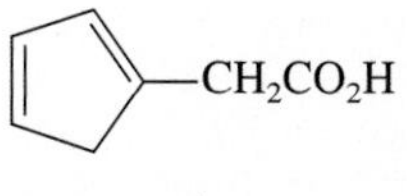

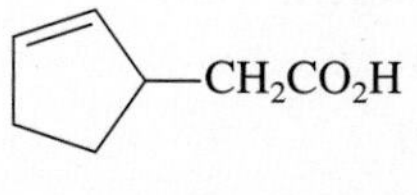

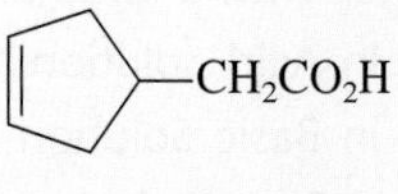

A. B. C.

19 Carboxylic Acid Derivatives: Nucleophilic Acyl Substitution

Chapter Outline

Mechanisms

3-Methylbutyl acetate (isoamyl acetate) is best known for the characteristic odor it gives to bananas. It is also one of the more than 40 compounds in the alarm pheromone a honeybee uses to alert other bees that an intruder has arrived.

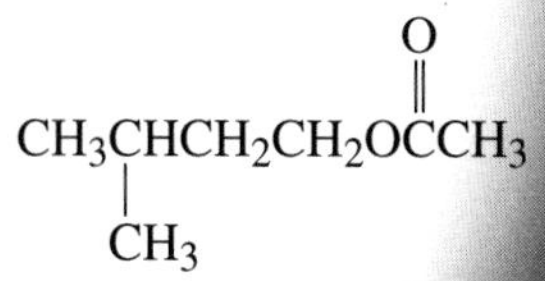

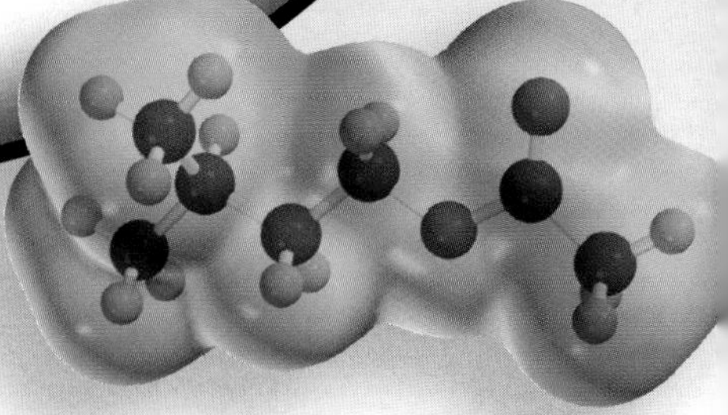

THIS CHAPTER DEALS with several related classes of compounds collectively referred to as **carboxylic acid derivatives.**

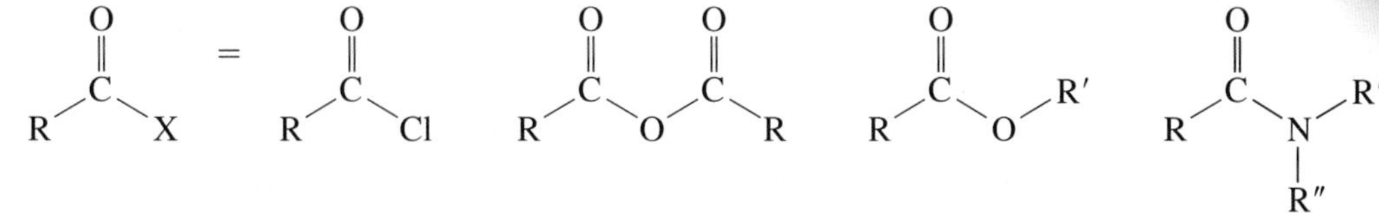

Carboxylic acid derivative = Acyl chloride, Acid anhydride, Ester, Amide

All have an acyl group RC(=O)— bonded to an electronegative element and have **nucleophilic acyl substitution** as their characteristic reaction type.

Nucleophilic Acyl Substitution:

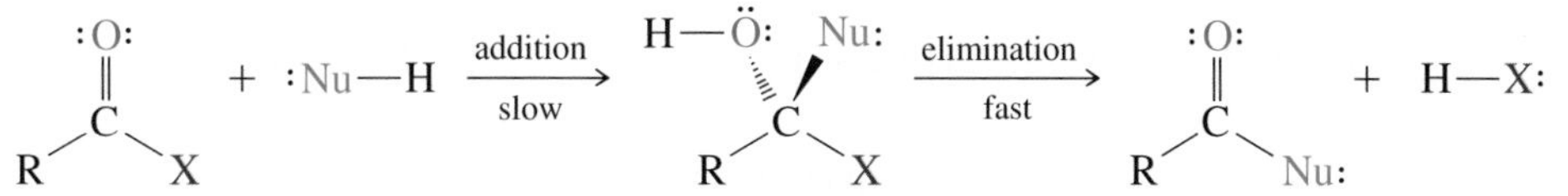

Reactants — Tetrahedral intermediate — Products

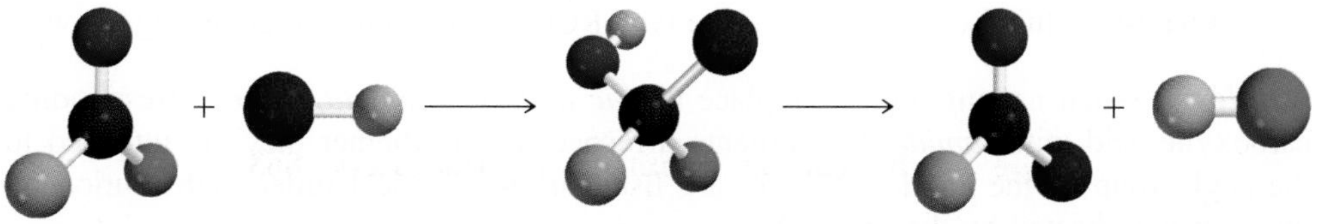

The mechanism of nucleophilic acyl substitution is a major emphasis of this chapter. As the preceding equation indicates, the reaction proceeds in two stages. In the first stage, nucleophilic *addition* to the carbonyl group occurs to give a **tetrahedral intermediate.** The second stage restores the carbonyl group by *elimination.* Experimental support exists for several different mechanisms of nucleophilic acyl substitution. Because most reactions of this type involve tetrahedral intermediates, only mechanisms based upon them will be presented in this chapter.

This chapter also explores how the mechanistic principles governing nucleophilic acyl substitution apply to many of the reactions of nitriles, compounds of the type R—C≡N:.

19.1 Nomenclature of Carboxylic Acid Derivatives

Formyl, acetyl, and benzoyl are preferred over methanoyl, ethanoyl, and benzenecarbonyl, respectively, according to the 2004 IUPAC recommendations.

Acyl Chlorides: Although acyl fluorides, bromides, and iodides are all known classes of organic compounds, they are not encountered nearly as often as acyl chlorides. Acyl chlorides, which will be the only acyl halides discussed in this chapter, are named by adding the word "chloride" after the name of the acyl group. To name an acyl group, replace the *-ic acid* ending of the IUPAC name of the corresponding carboxylic acid by *-yl.* The suffix *-carbonyl chloride* is used for attachments to rings other than benzene.

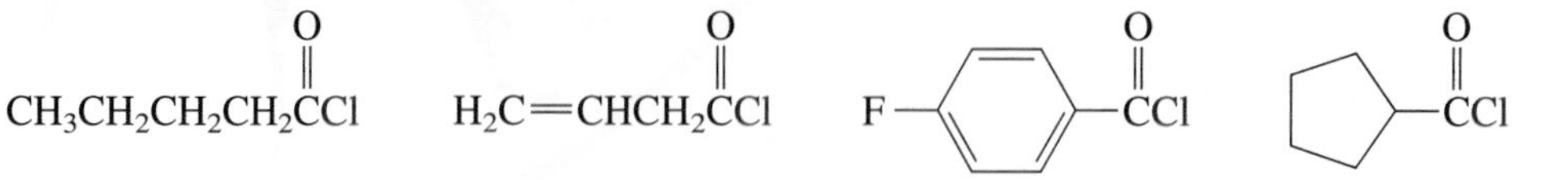

Pentanoyl chloride — 3-Butenoyl chloride or But-3-enoyl chloride — *p*-Fluorobenzoyl chloride or 4-Fluorobenzoyl chloride — Cyclopentanecarbonyl chloride

Acid Anhydrides: When both acyl groups are the same, the word "acid" in the corresponding carboxylic acid is replaced by "anhydride." When the two acyl groups are different, their corresponding carboxylic acids are cited in alphabetical order.

$CH_3\overset{O}{\overset{\|}{C}}O\overset{O}{\overset{\|}{C}}CH_3$	$C_6H_5\overset{O}{\overset{\|}{C}}O\overset{O}{\overset{\|}{C}}C_6H_5$	$C_6H_5\overset{O}{\overset{\|}{C}}O\overset{O}{\overset{\|}{C}}(CH_2)_5CH_3$
Acetic anhydride	Benzoic anhydride	Benzoic heptanoic anhydride

Esters: The alkyl group and the acyl group of an *ester* are specified independently. Esters are named as *alkyl alkanoates.* The alkyl group R′ of $R\overset{O}{\overset{\|}{C}}OR'$ is cited first, followed by the acyl portion $R\overset{O}{\overset{\|}{C}}—$. The acyl portion is named by substituting the suffix *-ate* for the *-ic acid* ending of the corresponding acid.

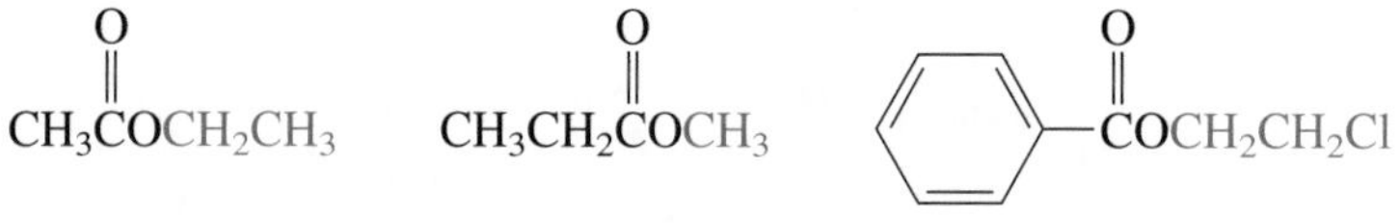

Ethyl acetate — Methyl propanoate — 2-Chloroethyl benzoate

Aryl esters, that is, compounds of the type $R\overset{O}{\overset{\|}{C}}OAr$, are named in an analogous way.

Amides: When naming amides, replace the *-ic acid* or *-oic acid* of the corresponding carboxylic acid with *-amide.* Substituents, irrespective of whether they are attached to the acyl group or the amide nitrogen, are listed in alphabetical order. Substitution on nitrogen is indicated by the locant *N-*.

$CH_3\underset{CH_3}{\underset{\|}{C}}HCH_2\overset{O}{\overset{\|}{C}}NH_2$	$CH_3\underset{CH_3}{\underset{\|}{C}}HCH_2\overset{O}{\overset{\|}{C}}NHCH_2CH_3$	$CH_3\underset{CH_3}{\underset{\|}{C}}HCH_2\overset{O}{\overset{\|}{C}}\underset{CH_3}{\underset{\|}{N}}CH_2CH_3$
3-Methylbutanamide	*N*-Ethyl-3-methylbutanamide	*N*-Ethyl-*N*,3-dimethylbutanamide

Similar to the *-carbonyl chloride* suffix for acyl chlorides, *-carboxamide* is used when an amide group is attached to a ring.

Nitriles: Substitutive IUPAC names for *nitriles* add the suffix *-nitrile* to the name of the parent hydrocarbon chain that includes the carbon of the cyano group. Nitriles may also be named by replacing the *-ic acid* or *-oic acid* ending of the corresponding carboxylic acid with *-onitrile.* Alternatively, they are sometimes given functional class IUPAC names as alkyl cyanides. The suffix *-carbonitrile* is used when a —CN group is attached to a ring.

$CH_3C{\equiv}N$

Ethanenitrile
or
Acetonitrile

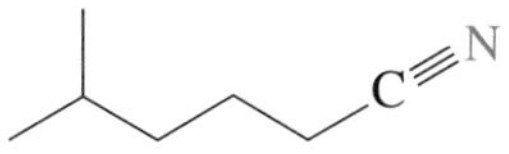

C
N

5-Methylhexanenitrile
or
4-Methylpentyl cyanide

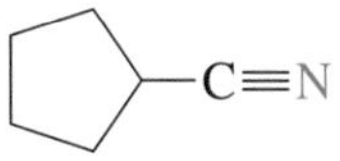

C≡N

Cyclopentanecarbonitrile
or
Cyclopentyl cyanide

Problem 19.1

Write a structural formula for each of the following compounds:

(a) 2-Phenylbutanoyl chloride
(b) 2-Phenylbutanoic anhydride
(c) Butyl 2-phenylbutanoate
(d) 2-Phenylbutyl butanoate
(e) 2-Phenylbutanamide
(f) *N*-Ethyl-2-phenylbutanamide
(g) 2-Phenylbutanenitrile

Sample Solution (a) A 2-phenylbutanoyl group is a four-carbon acyl unit that bears a phenyl substituent at C-2. When the name of an acyl group is followed by the name of a halide, it designates an *acyl halide.*

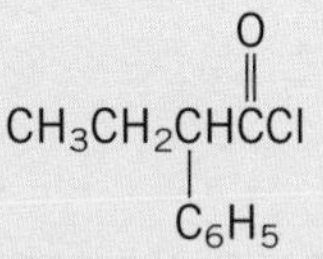

O
CH3CH2CHCCl
C6H5

2-Phenylbutanoyl chloride

19.2 Structure and Reactivity of Carboxylic Acid Derivatives

The number of reactions in this chapter is quite large and keeping track of them all can be difficult—or it can be manageable. The key to making it manageable is the same as always: *structure determines properties.*

Figure 19.1 shows the structures of various derivatives of acetic acid (acetyl chloride, acetic anhydride, ethyl acetate and acetamide) arranged in order of decreasing reactivity toward nucleophilic acyl substitution. Acyl chlorides are the most reactive, amides the least reactive. The reactivity order:

acyl chloride > anhydride > ester > amide

is general for nucleophilic acyl substitution and well worth remembering. The range of reactivities is quite large; a factor of about 10^{13} in relative rate separates acyl chlorides from amides.

This difference in reactivity, especially toward hydrolysis, has an important result. We'll see in Chapter 25 that the structure and function of proteins are critical to life itself. The bonds mainly responsible for the structure of proteins are amide bonds, which are about 100 times more stable to hydrolysis than ester bonds. These amide bonds are stable enough to maintain the structural integrity of proteins in an aqueous environment, but susceptible enough to hydrolysis to be broken when the occasion demands.

What structural features are responsible for the reactivity order of carboxylic acid derivatives? Like the other carbonyl-containing compounds that we've studied, they all have a planar arrangement of bonds to the carbonyl group. Thus, all are about the same

Figure 19.1

Structure, reactivity, and carbonyl-group stabilization in carboxylic acid derivatives. Acyl chlorides are the most reactive, amides the least reactive. Acyl chlorides have the least stabilized carbonyl group, amides the most. Conversion of one class of compounds to another is feasible only in the direction that leads to a more stabilized carbonyl group; that is, from more reactive to less reactive.

Reactivity	Compound	Structural formula	Molecular model	Stabilization of carbonyl group
Most reactive	Acetyl chloride	$CH_3C(=O)—Cl$	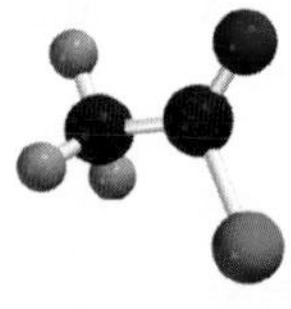	Least stabilized
↑	Acetic anhydride	$CH_3C(=O)—OC(=O)CH_3$	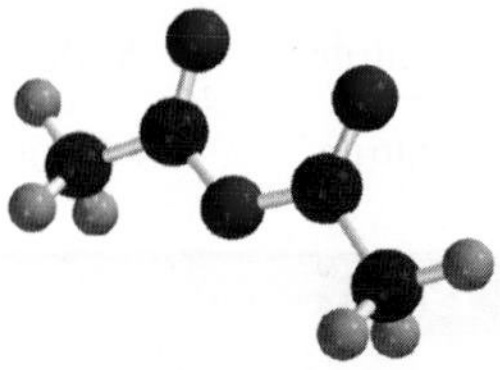	↓
↑	Ethyl acetate	$CH_3C(=O)—OCH_2CH_3$	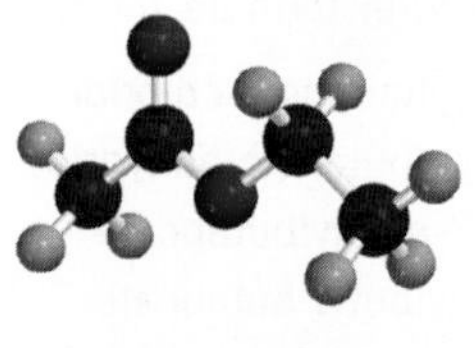	↓
Least reactive	Acetamide	$CH_3C(=O)—NH_2$	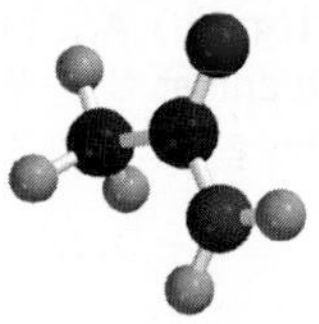	Most stabilized

in offering relatively unhindered access to the approach of a nucleophile. They differ in the degree to which the atom attached to the carbonyl group can stabilize the carbonyl group by electron donation.

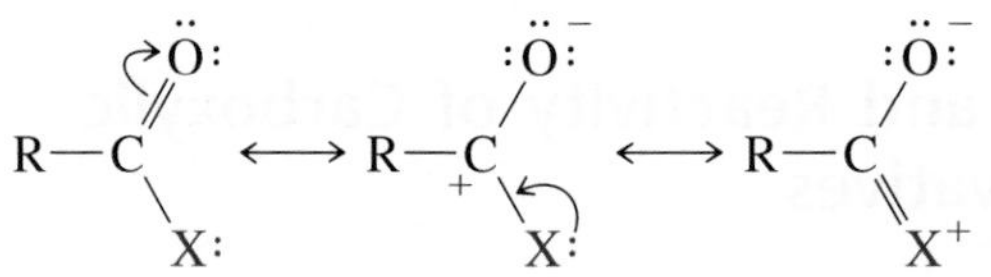

Electron release from the substituent X stabilizes the carbonyl group and makes it less electrophilic.

The order of reactivity of carboxylic acid derivatives toward nucleophilic acyl substitution can be explained on the basis of the electron-donating properties of substituent X. The greater the electron-donating powers of X, the slower the rate.

1. **Acyl chlorides:** Although chlorine has unshared electron pairs, it is a poor electron-pair donor in resonance of the type:

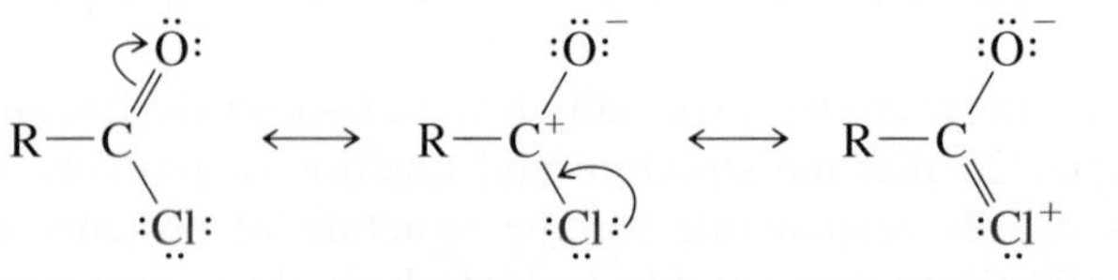

Lone-pair donation ineffective because of poor orbital overlap

Not a significant contributor

Because the C—Cl bond is so long, the lone-pair orbital ($3p$) of chlorine and the π orbital of the carbonyl group do not overlap sufficiently to permit delocalization of a chlorine unshared pair. Not only is the carbonyl group of an acyl chloride not stabilized by electron-pair donation, the electron-withdrawing inductive effect of chlorine makes it more electrophilic and more reactive toward nucleophiles.

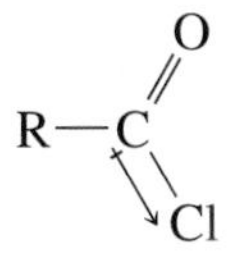

2. **Acid anhydrides:** The carbonyl group of an acid anhydride is better stabilized by electron donation than the carbonyl group of an acyl chloride. Even though oxygen is more electronegative than chlorine, it is a far better electron-pair donor toward sp^2-hybridized carbon.

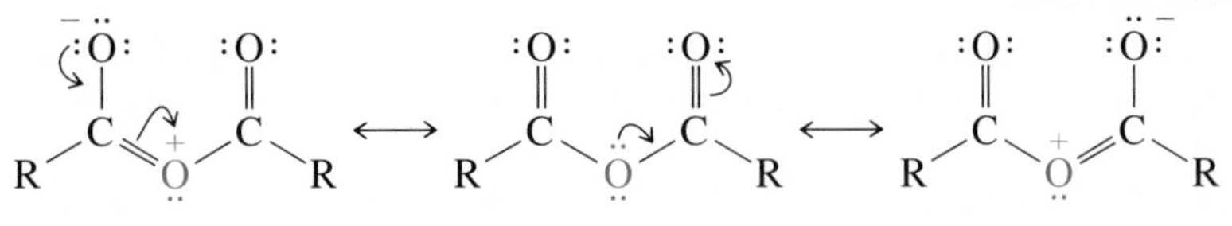

Working against this electron-delocalization is the fact that both carbonyl groups are competing for the same electron pair. Thus, the extent to which each one is stabilized is reduced.

3. **Esters:** Like acid anhydrides, the carbonyl group of an ester is stabilized by electron release from oxygen. Because there is only one carbonyl group, versus two in anhydrides, esters are stabilized more and are less reactive than anhydrides.

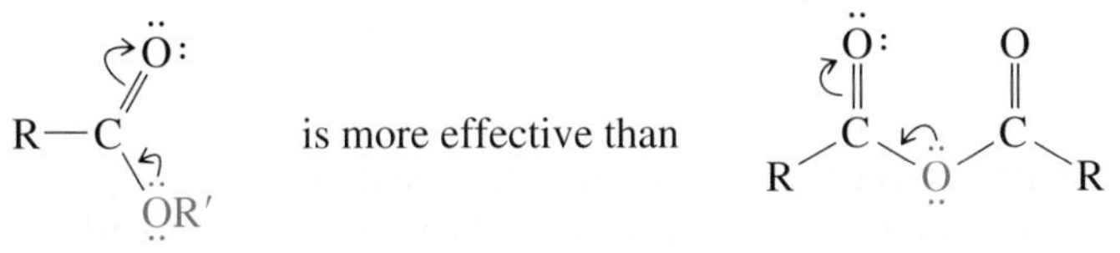

Ester Acid anhydride

4. **Amides:** Nitrogen is less electronegative than oxygen; therefore, the carbonyl group of an amide is stabilized more than that of an ester.

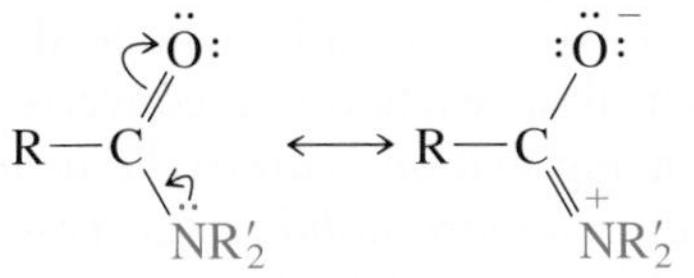

Very effective resonance stabilization

Amide resonance is a powerful stabilizing force and gives rise to a number of structural effects. Unlike the pyramidal arrangement of bonds in ammonia and amines, the bonds to nitrogen in amides lie in the same plane (Figure 19.2*a*). The carbon–nitrogen bond has considerable double-bond character and, at 135 pm, is substantially shorter than the normal 147 pm carbon–nitrogen single-bond distance observed in amines.

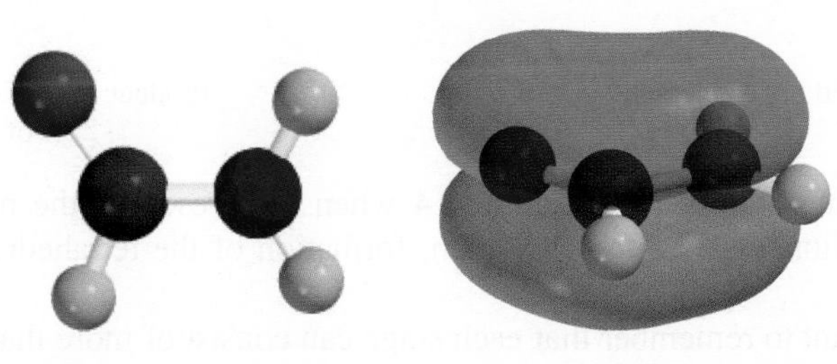

(*a*) (*b*)

Figure 19.2

(*a*) Formamide ($\overset{O}{\overset{\|}{HCNH_2}}$) is planar. Carbon and nitrogen are both sp^2-hybridized. (*b*) A π orbital generated by overlap of the $2p$ orbital of nitrogen and the π orbital of the carbonyl group allows delocalization of the nitrogen unshared pair.

The barrier to rotation about the carbon–nitrogen bond in amides is 75–85 kJ/mol (18–20 kcal/mol).

Recall (Section 3.1) that the rotational barrier in ethane is only 12 kJ/mol (3 kcal/mol).

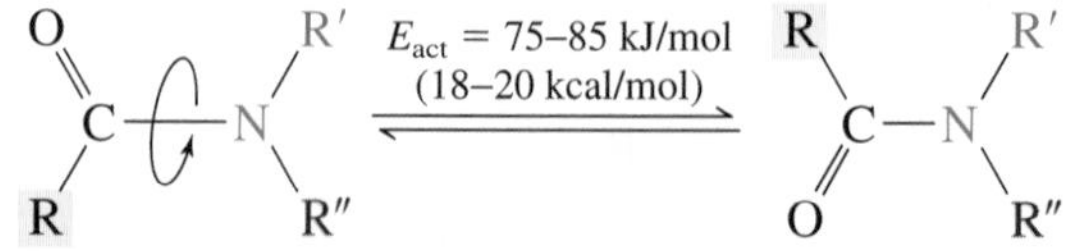

This is an unusually high rotational energy barrier for a single bond and indicates that the carbon–nitrogen bond has significant double-bond character, as the resonance and orbital overlap (Figure 19.2*b*) descriptions suggest.

Problem 19.2

Suggest an explanation for the fact that *N,N*-dimethylformamide [$(CH_3)_2NCH{=}O$] has signals for three nonequivalent carbons (δ 31.3, 36.4, and 162.6) in its ^{13}C NMR spectrum.

Electron release from nitrogen stabilizes the carbonyl group of amides and decreases the rate at which nucleophiles attack the carbonyl carbon.

An extreme example of carbonyl group stabilization is seen in carboxylate anions:

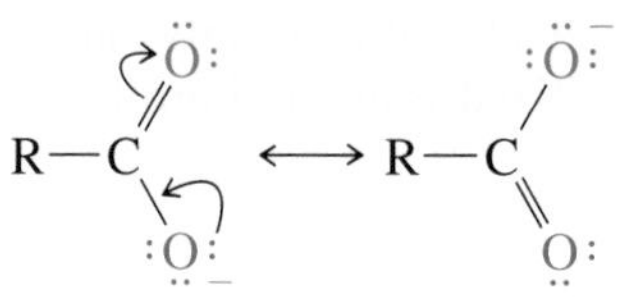

The negatively charged oxygen is a powerful electron donor to the carbonyl group. Resonance in carboxylate anions is more effective than resonance in carboxylic acids, acyl chlorides, anhydrides, esters, and amides. Carboxylate ions do not undergo nucleophilic acyl substitution.

Most methods for their preparation convert one class of carboxylic acid derivative to another by nucleophilic acyl substitution. The order of carbonyl group stabilization given in Figure 19.1 bears directly on the means by which these transformations may be achieved. A reaction that converts one carboxylic acid derivative to another that lies below it in the figure is practical; a reaction that converts it to one that lies above it is not. This is another way of saying that *one carboxylic acid derivative can be converted to another if the reaction leads to a more stabilized carbonyl group.* Numerous examples of reactions of this type will be presented in the sections that follow.

19.3 General Mechanism for Nucleophilic Acyl Substitution

Nucleophilic acyl substitutions follow a two-stage mechanism outlined in the introduction and proceed by way of a tetrahedral intermediate (TI).

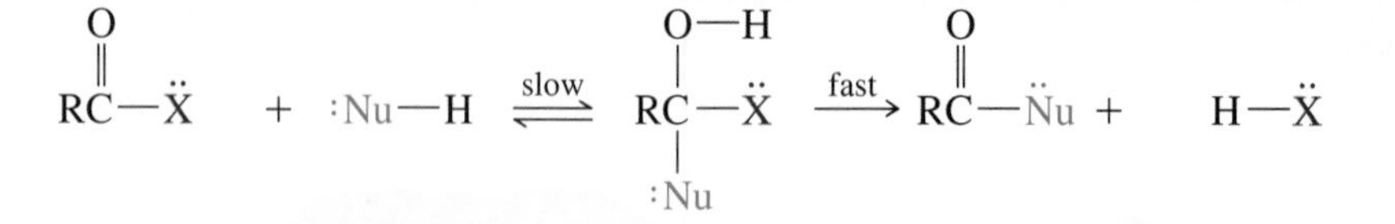

Carboxylic acid derivative | Nucleophile | Tetrahedral intermediate (TI) | Product | Conjugate acid of leaving group

We saw this theme before in Section 18.14 when we presented the mechanism of the Fischer esterification. As was the case then, formation of the tetrahedral intermediate is rate-determining.

It is important to remember that each stage can consist of more than one elementary step. Therefore, a complete mechanism can have many steps and look complicated if we try to absorb all of it at once. If we keep the two stages separate in our minds and build on what we already know, our job becomes easier. Two points are helpful:

1. The first stage of the mechanism for nucleophilic *acyl substitution* is exactly the same as for nucleophilic *addition* to the carbonyl group of an aldehyde or ketone. Many of the same nucleophiles that add to aldehydes and ketones—water (Section 17.6), alcohols (Section 17.8), amines (Sections 17.10–17.11)—add to the carbonyl groups of carboxylic acid derivatives.
2. The features that complicate the mechanism of nucleophilic acyl substitution are almost entirely related to acid–base chemistry. We try to keep track, as best we can, of the form in which the various species—reactants, intermediates, and products—exist under the reaction conditions.

With regard to the second point, we already know a good bit about the acid–base chemistry of the reactants and products; that of the tetrahedral intermediate is less familiar. We can, for example, imagine the following species in equilibrium with the tetrahedral intermediate (TI).

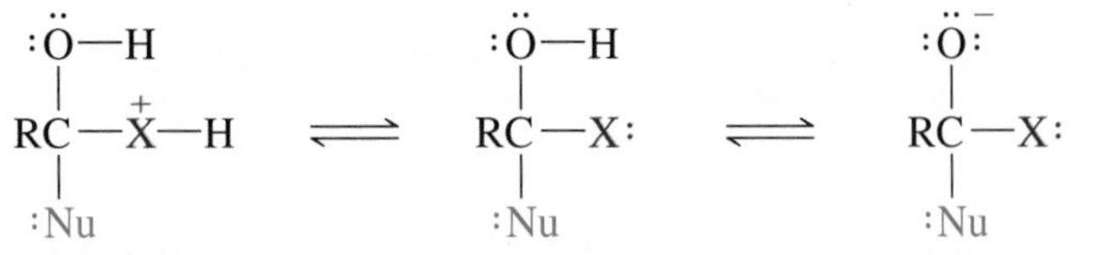

$TI—H^+$ Conjugate acid of tetrahedral intermediate — TI Tetrahedral intermediate — TI^- Conjugate base of tetrahedral intermediate

Each one of these can proceed to the product of nucleophilic acyl substitution.

Dissociation of TI—H⁺:

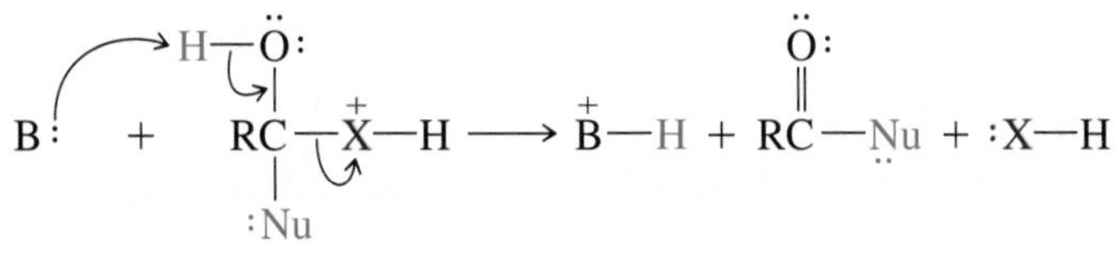

Brønsted base — $TI—H^+$ — Product

Dissociation of TI:

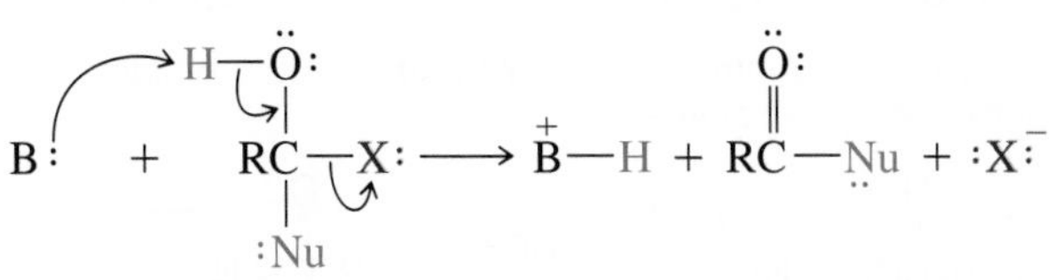

Brønsted base — TI — Product

Dissociation of TI⁻:

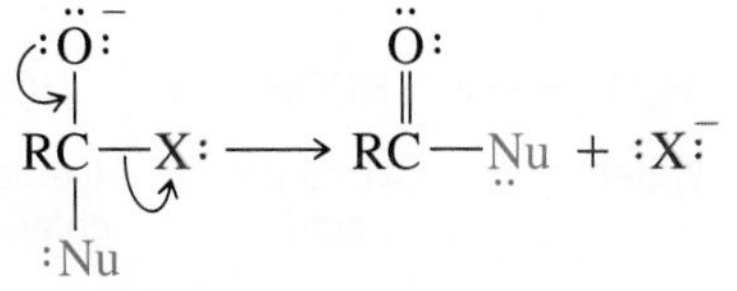

TI^- — Product

More than one form of the tetrahedral intermediate can be present at a particular pH, and the most abundant form need not be the one that gives most of the product. A less abundant form may react at a faster rate than a more abundant one.

Mechanisms for a number of nucleophilic acyl substitutions will appear in the sections that follow. It is better to look for the important ways in which they are similar than to search for details in which they differ.

One of the most useful reactions of acyl chlorides was presented in Section 12.7. Friedel-Crafts acylation of aromatic rings takes place when arenes are treated with acyl chlorides in the presence of aluminum chloride.

19.4 Nucleophilic Acyl Substitution in Acyl Chlorides

Among the various carboxylic acid derivatives, acyl chlorides are especially useful because they are readily converted to acid anhydrides, esters, and amides by nucleophilic acyl substitution (Table 19.1). Yields are high and the reaction rates are much

TABLE 19.1 Conversion of Acyl Chlorides to Other Carboxylic Acid Derivatives

Reaction (section) and comments	General equation and specific example
Reaction with carboxylic acids (Section 19.4) Acyl chlorides react with carboxylic acids to yield acid anhydrides. When this reaction is used for preparative purposes, a weak organic base such as pyridine is normally added. Pyridine is a catalyst for the reaction and also acts as a base to neutralize the hydrogen chloride that is formed.	$RCCl$ (O) + $R'COH$ (O) ⟶ $RCOCR'$ (O O) + HCl Acyl chloride; Carboxylic acid; Acid anhydride; Hydrogen chloride $CH_3(CH_2)_5CCl + CH_3(CH_2)_5COH \xrightarrow{\text{pyridine}} CH_3(CH_2)_5COC(CH_2)_5CH_3$ Heptanoyl chloride; Heptanoic acid; Heptanoic anhydride (78–83%)
Reaction with alcohols (Section 15.8) Acyl chlorides react with alcohols to form esters. The reaction is typically carried out in the presence of pyridine.	$RCCl$ (O) + $R'OH$ ⟶ $RCOR'$ (O) + HCl Acyl chloride; Alcohol; Ester; Hydrogen chloride $C_6H_5CCl + (CH_3)_3COH \xrightarrow{\text{pyridine}} C_6H_5COC(CH_3)_3$ Benzoyl chloride; *tert*-Butyl alcohol; *tert*-Butyl benzoate (80%)
Reaction with ammonia and amines (Section 19.14) Acyl chlorides react with ammonia and amines to form amides. A base such as sodium hydroxide is normally added to react with the hydrogen chloride produced.	$RCCl$ (O) + R'_2NH + HO^- ⟶ $RCNR'_2$ (O) + H_2O + Cl^- Acyl chloride; Ammonia or amine; Hydroxide; Amide; Water; Chloride ion 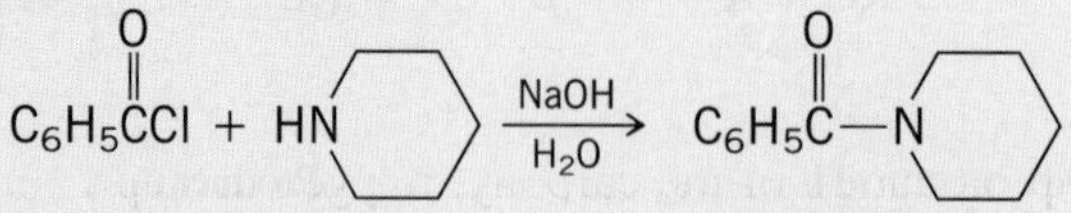 Benzoyl chloride; Piperidine; *N*-Benzoylpiperidine (87–91%)
Hydrolysis (Section 19.4) Acyl chlorides react with water to yield carboxylic acids. In base, the acid is converted to its carboxylate salt. The reaction has little preparative value because the acyl chloride is nearly always prepared from the carboxylic acid rather than vice versa.	$RCCl$ (O) + H_2O ⟶ $RCOH$ (O) + HCl Acyl chloride; Water; Carboxylic acid; Hydrogen chloride $C_6H_5CH_2CCl + H_2O \longrightarrow C_6H_5CH_2COH + HCl$ Phenylacetyl chloride; Water; Phenylacetic acid; Hydrogen chloride

greater than the corresponding rates of alkyl halides with the same nucleophiles. Benzoyl chloride, for example, is about 1,000 times more reactive than benzyl chloride toward hydrolysis at 25°C.

Problem 19.3

Use Table 19.1 to predict the major organic product obtained by reaction of benzoyl chloride with each of the following:

(a) Acetic acid
(b) Benzoic acid
(c) Ethanol
(d) Methylamine, CH_3NH_2
(e) Dimethylamine, $(CH_3)_2NH$
(f) Water

Sample Solution (a) As noted in Table 19.1, the reaction of an acyl chloride with a carboxylic acid yields an acid anhydride.

$$\underset{\text{Benzoyl chloride}}{C_6H_5\overset{O}{\overset{\|}{C}}Cl} + \underset{\text{Acetic acid}}{CH_3\overset{O}{\overset{\|}{C}}OH} \longrightarrow \underset{\text{Acetic benzoic anhydride}}{C_6H_5\overset{O}{\overset{\|}{C}}O\overset{O}{\overset{\|}{C}}CH_3}$$

The product is a mixed anhydride. Acetic acid acts as a nucleophile and substitutes for chloride on the benzoyl group.

On examining the specific examples in the table, we see that nucleophilic substitutions of acyl chlorides are often carried out in the presence of pyridine. Pyridine is both a catalyst and a weak base. As a catalyst it increases the rate of acylation. As a base it prevents the build-up of HCl, which is a strong acid.

How pyridine catalyzes acylation is the subject of end-of-chapter Problem 19.47.

Table 19.1 concludes with the hydrolysis of acyl chlorides. Because acyl chlorides are themselves prepared by the reaction of carboxylic acids with thionyl chloride (Section 12.7):

$$\underset{\text{Carboxylic acid}}{R\overset{O}{\overset{\|}{C}}OH} + \underset{\text{Thionyl chloride}}{SOCl_2} \longrightarrow \underset{\text{Acyl chloride}}{R\overset{O}{\overset{\|}{C}}Cl} + \underset{\text{Sulfur dioxide}}{SO_2} + \underset{\text{Hydrogen chloride}}{HCl}$$

their hydrolysis is of little synthetic value. It is instructive to examine its mechanism, however, in order to illustrate some of the principles described in the preceding section. The four steps of Mechanism 19.1 show the two stages of acid-catalyzed acyl halide hydrolysis—formation of a tetrahedral intermediate (TI) by nucleophilic addition to the carbonyl group, followed by restoration of the carbon-oxygen double bond by dissociation of the tetrahedral intermediate. Like the acid-catalyzed hydration of aldehydes and ketones (Section 17.6), protonation of the carbonyl oxygen in step 1 makes the carbonyl carbon more electrophilic, activating it for addition of a water molecule in step 2. The product of step 2 is the conjugate acid TI—H^+ of the tetrahedral intermediate (TI), to which it is converted by proton transfer to water in step 3. Departure of the leaving group (Cl^-) in step 4 is assisted by abstraction of the OH proton of TI by a water molecule. Chloride is lost from TI because it is the least basic leaving group and its departure completes the transformation of a less-stabilized C=O to a more-stabilized one.

Problem 19.4

Unlike acyl chlorides, aldehydes and ketones do not undergo nucleophilic acyl substitution. Explain.

Mechanism 19.1

Acid-Catalyzed Hydrolysis of an Acyl Chloride via a Tetrahedral Intermediate

THE OVERALL REACTION:

$$\underset{\text{Acyl chloride}}{\overset{\text{O}}{\overset{\|}{\text{RCCl}}}} + \underset{\text{Water}}{2H_2O} \longrightarrow \underset{\text{Carboxylic acid}}{\overset{\text{O}}{\overset{\|}{\text{RCOH}}}} + \underset{\text{Hydronium ion}}{H_3O^+} + \underset{\text{Chloride ion}}{Cl^-}$$

Steps 1–3: Acid-catalyzed addition of water to the carbonyl group: These three steps are the same as those for acid-catalyzed hydration of an aldehyde or ketone (Mechanism 17.2). Steps 1 and 3 are proton transfers between oxygens and are fast. Water acts as a nucleophile in step 2. Step 2 is rate-determining.

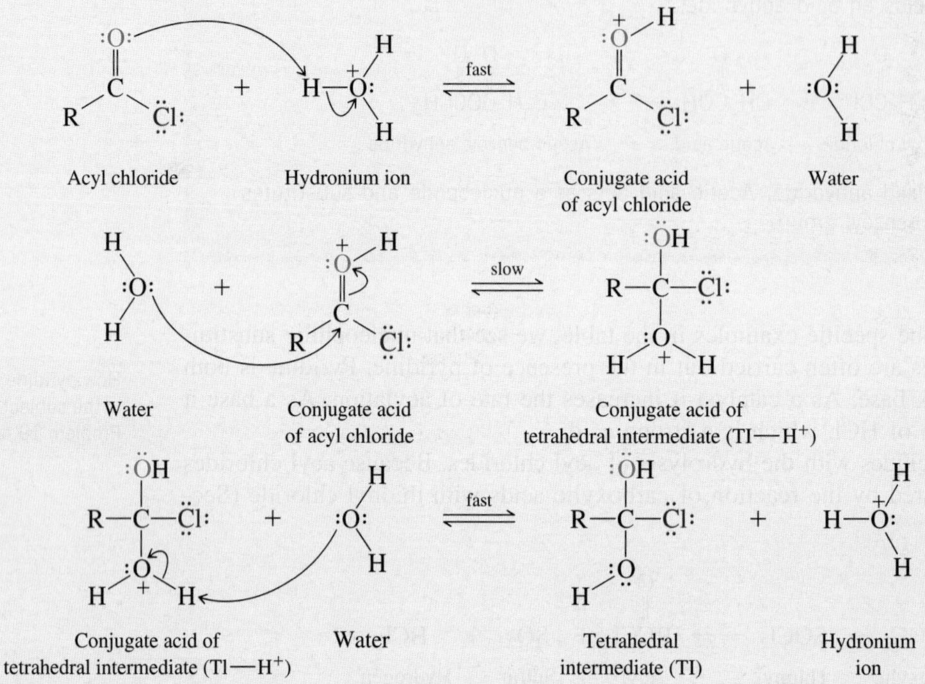

Step 4: Removal of a proton from oxygen and loss of chloride from carbon from the tetrahedral intermediate completes the mechanism.

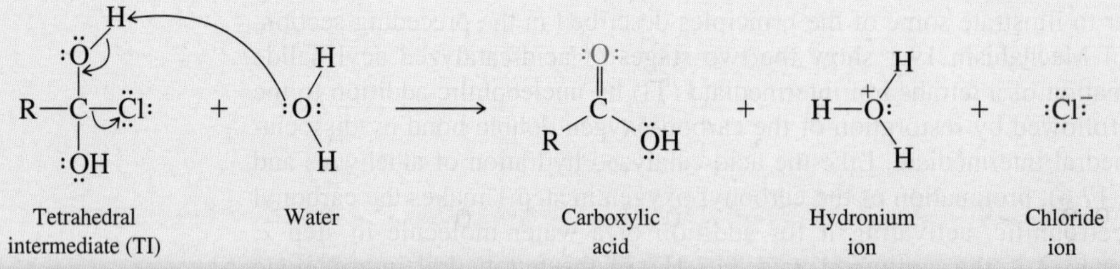

Problem 19.5

Write a structural formula for the tetrahedral intermediate formed in the rate-determining step of the hydrolysis of an acyl chloride in aqueous base according to the equation shown. Use curved arrows to show the flow of electrons in the dissociation of this intermediate.

$$\overset{\text{O}}{\overset{\|}{\text{RCCl}}} + 2HO^- \longrightarrow \overset{\text{O}}{\overset{\|}{\text{RCO}^-}} + H_2O + Cl^-$$

19.5 Nucleophilic Acyl Substitution in Acid Anhydrides

After acyl halides, acid anhydrides are the most reactive carboxylic acid derivatives. Although anhydrides can be prepared by reaction of carboxylic acids with acyl chlorides as was shown in Table 19.1, the three most commonly used anhydrides are industrial chemicals and are prepared by specialized methods. Phthalic anhydride and maleic anhydride, for example, are prepared from naphthalene and butane, respectively.

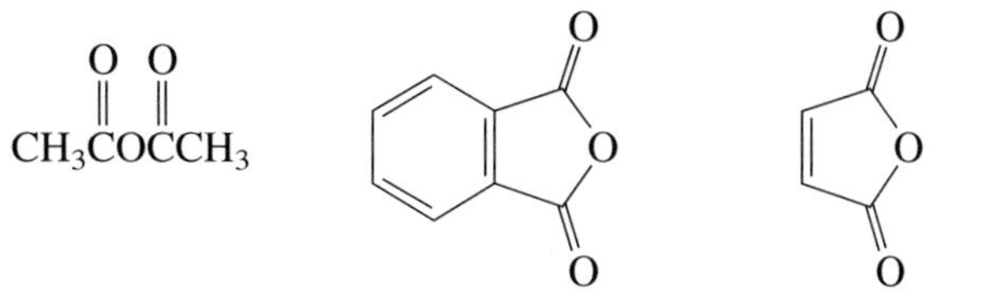

Acetic anhydride   Phthalic anhydride   Maleic anhydride

Acid anhydrides contain two acyl groups bonded to the same oxygen. In nucleophilic acyl substitution, one of these acyl groups becomes bonded to the nucleophilic atom. The other acyl group remains on oxygen to become part of a carboxylic acid.

$$\overset{O}{\overset{\|}{RC}}-\overset{O}{\overset{\|}{OCR}} + HY: \longrightarrow \overset{O}{\overset{\|}{RC}}-\ddot{Y} + HO\overset{O}{\overset{\|}{C}}R$$

Bond cleavage occurs here in an acid anhydride. | Nucleophile | Product of nucleophilic acyl substitution | Carboxylic acid

Acid anhydrides are more stable and less reactive than acyl chlorides. Acetyl chloride, for example, undergoes hydrolysis about 100,000 times more rapidly than acetic anhydride at 25°C.

Conversions of acid anhydrides to other carboxylic acid derivatives are illustrated in Table 19.2. Because a more highly stabilized carbonyl group must result in order for nucleophilic acyl substitution to be effective, acid anhydrides are readily converted to carboxylic acids, esters, and amides but not to acyl chlorides.

Problem 19.6

Use Table 19.2 to help you predict the major organic product of each of the following reactions:

(a) Benzoic anhydride + methanol $\xrightarrow{H_2SO_4}$
(b) Acetic anhydride + ammonia (2 mol) $\longrightarrow$
(c) Phthalic anhydride + $(CH_3)_2NH$ (2 mol) $\longrightarrow$
(d) Phthalic anhydride + sodium hydroxide (2 mol) $\longrightarrow$

Sample Solution (a) Nucleophilic acyl substitution by an alcohol on an acid anhydride yields an ester.

$$C_6H_5\overset{O}{\overset{\|}{C}}O\overset{O}{\overset{\|}{C}}C_6H_5 + CH_3OH \xrightarrow{H_2SO_4} C_6H_5\overset{O}{\overset{\|}{C}}OCH_3 + C_6H_5\overset{O}{\overset{\|}{C}}OH$$

Benzoic anhydride | Methanol | Methyl benzoate | Benzoic acid

Because phenols are stronger acids than alcohols, aryl acetates are conveniently prepared in aqueous media by first converting the phenol to its corresponding sodium salt with sodium hydroxide, followed by reaction with acetic anhydride.

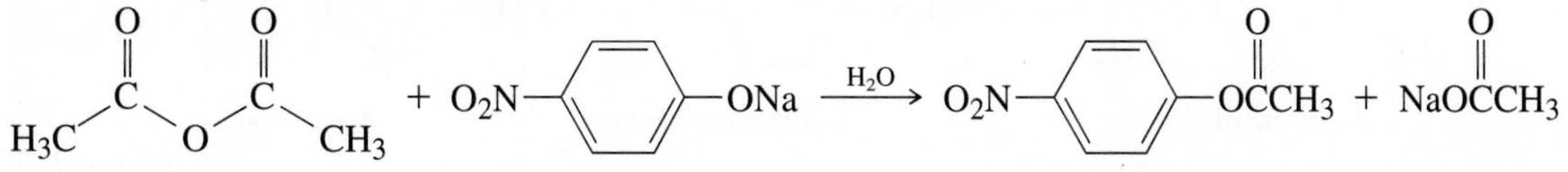

Acetic anhydride | Sodium *p*-nitrophenoxide | *p*-Nitrophenyl acetate (78%) | Sodium acetate

TABLE 19.2 Conversion of Acid Anhydrides to Other Carboxylic Acid Derivatives

Reaction (section) and comments	General equation and specific example
Reaction with alcohols (Section 15.8) Acid anhydrides react with alcohols to form esters. The reaction may be carried out in the presence of pyridine or it may be catalyzed by acids. Only one acyl group of acetic anhydride becomes incorporated into the ester; the other becomes the acyl group of an acetic acid molecule.	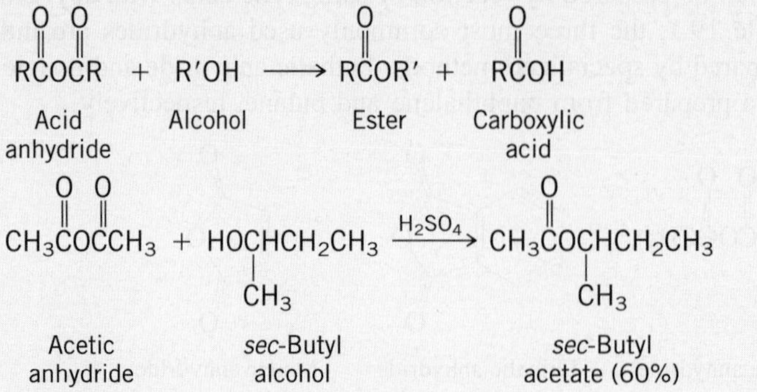 RCOCR + R′OH ⟶ RCOR′ + RCOH; Acid anhydride; Alcohol; Ester; Carboxylic acid; CH3COCCH3 + HOCHCH2CH3 (CH3) ⟶(H2SO4) CH3COCHCH2CH3 (CH3); Acetic anhydride; sec-Butyl alcohol; sec-Butyl acetate (60%)
Reaction with ammonia and amines (Section 19.14) Acid anhydrides react with ammonia and amines to form amides. Two molar equivalents of amine are required. Only one acyl group of acetic anhydride becomes incorporated into the amide; the other becomes the acyl group of the amine salt of acetic acid.	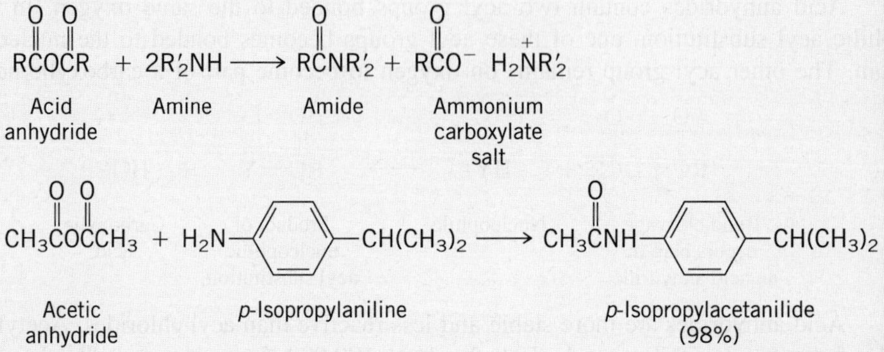 RCOCR + 2R′2NH ⟶ RCNR′2 + RCO⁻ H2NR′2 (+); Acid anhydride; Amine; Amide; Ammonium carboxylate salt; CH3COCCH3 + H2N–C6H4–CH(CH3)2 ⟶ CH3CNH–C6H4–CH(CH3)2; Acetic anhydride; p-Isopropylaniline; p-Isopropylacetanilide (98%)
Hydrolysis (Section 19.5) Acid anhydrides react with water to yield two carboxylic acids. Cyclic anhydrides yield dicarboxylic acids.	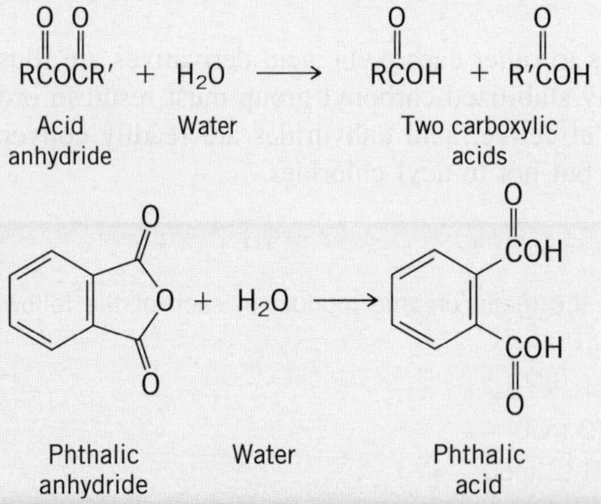 RCOCR′ + H2O ⟶ RCOH + R′COH; Acid anhydride; Water; Two carboxylic acids; + H2O ⟶; COH; COH; Phthalic anhydride; Water; Phthalic acid

The mechanism of this reaction (Mechanism 19.2) is straightforward and illustrates a nucleophilic acyl substitution that occurs in basic solution by way of a negatively charged tetrahedral intermediate (TI^-).

Mechanism 19.2

Nucleophilic Acyl Substitution in an Anhydride

THE OVERALL REACTION:

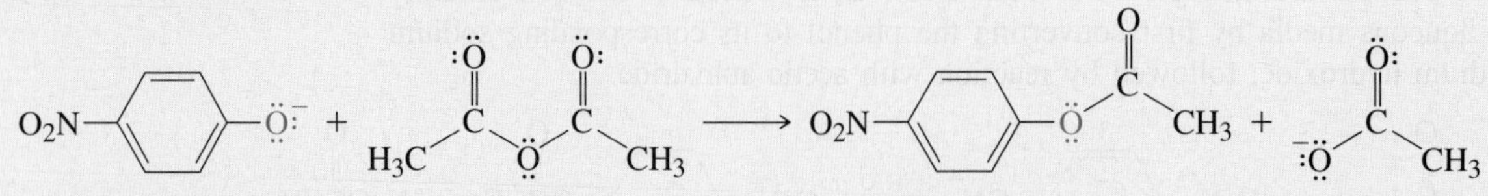
O2N
H3C
CH3
O2N
CH3
CH3

p-Nitrophenoxide ion | Acetic anhydride | *p*-Nitrophenyl acetate | Acetate ion

Continued

Step 1: Nucleophilic addition of *p*-nitrophenoxide to one of the carbonyl groups of the anhydride gives the conjugate base of the tetrahedral intermediate (TI^-).

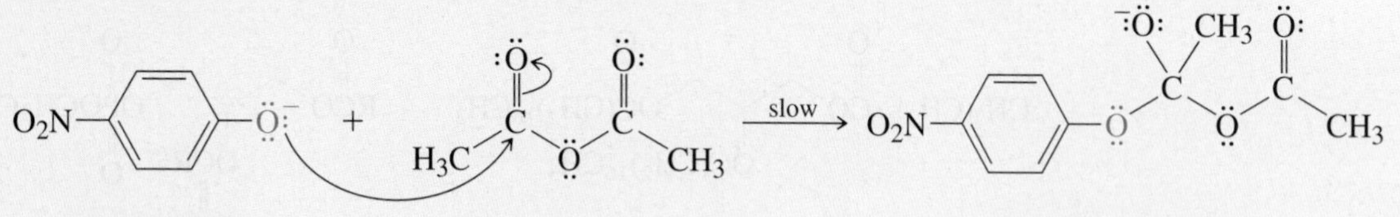

p-Nitrophenoxide ion — Acetic anhydride — Conjugate base of tetrahedral intermediate (TI^-)

Step 2: Expulsion of acetate from TI^- restores the carbonyl group.

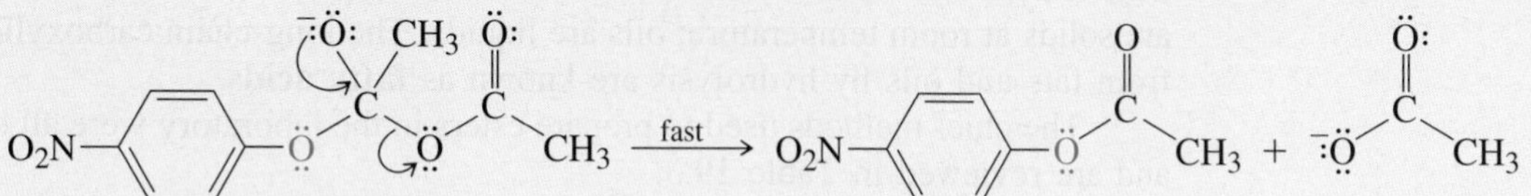

Conjugate base of tetrahedral intermediate (TI^-) — *p*-Nitrophenyl acetate — Acetate ion

19.6 Sources of Esters

Many esters occur naturally. Those of low molecular weight are fairly volatile, and many have pleasing odors. Esters often form a significant fraction of the fragrant oil of fruits and flowers. The aroma of oranges, for example, contains 30 different esters along with 10 carboxylic acids, 34 alcohols, 34 aldehydes and ketones, and 36 hydrocarbons.

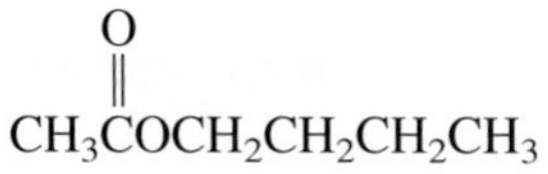

Butyl acetate
(contributes to characteristic pear odor)

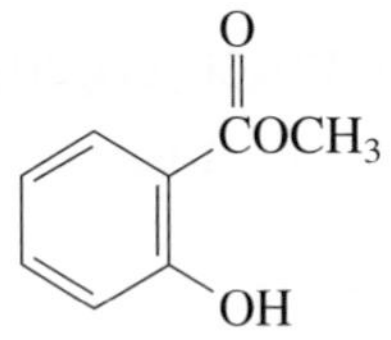

Methyl salicylate
(principal component of oil of wintergreen)

Among the chemicals used by insects to communicate with one another, esters occur frequently.

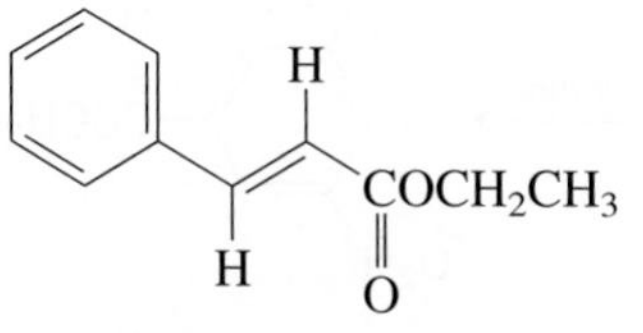

Ethyl cinnamate
(one of the constituents of the sex pheromone of the male oriental fruit moth)

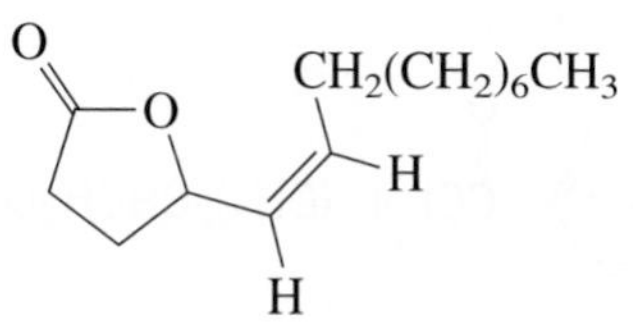

(*Z*)-5-Tetradecen-4-olide
(sex pheromone of female Japanese beetle)

Notice that (*Z*)-5-tetradecen-4-olide is a cyclic ester. Recall from Section 18.15 that cyclic esters are called *lactones* and that the suffix *-olide* is characteristic of IUPAC names for lactones.

Esters of glycerol, called *glycerol triesters, triacylglycerols,* or *triglycerides,* are abundant natural products. The most important group of glycerol triesters includes those in which each acyl group is unbranched and has 14 or more carbon atoms. Structurally related phosphatidylcholine is a component of cell membranes (Section 24.4).

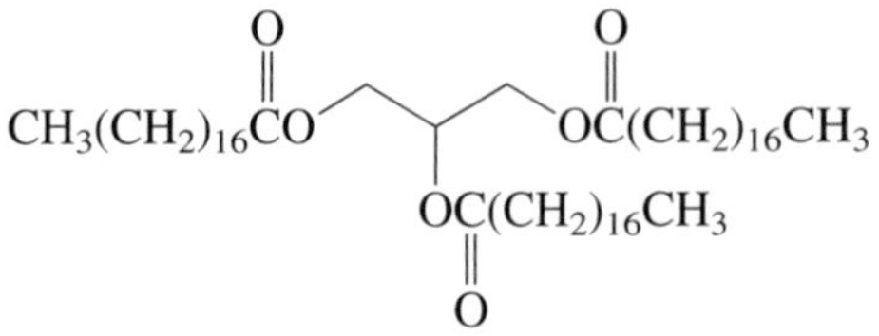

Tristearin, a trioctadecanoyl ester of glycerol found in many animal and vegetable fats

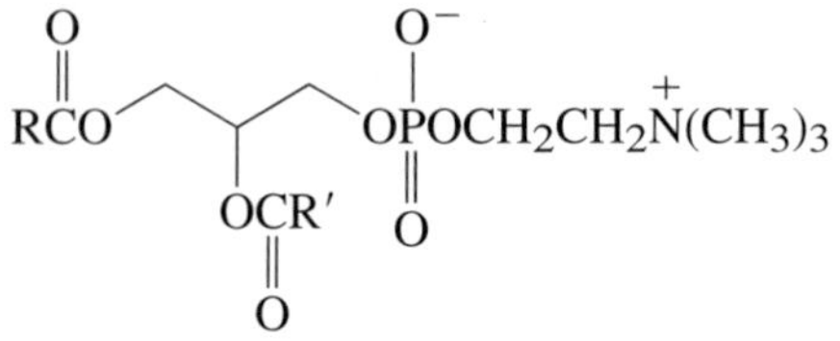

Phosphatidylcholine (R and R′ are carbon chains)

A molecular model of tristearin is shown in Figure 24.2.

Fats and **oils** are naturally occurring mixtures of glycerol triesters. Fats are mixtures that are solids at room temperature; oils are liquids. The long-chain carboxylic acids obtained from fats and oils by hydrolysis are known as **fatty acids.**

The chief methods used to prepare esters in the laboratory were all described earlier and are reviewed in Table 19.3.

TABLE 19.3 Preparation of Esters

Reaction (section) and comments	General equation and specific example
From carboxylic acids (Sections 15.8 and 18.14) In the presence of an acid catalyst, alcohols and carboxylic acids react to form an ester and water. This is the Fischer esterification.	$RCOH + R'OH \underset{}{\overset{H_2SO_4}{\rightleftharpoons}} RCOR' + H_2O$ (Carboxylic acid, Alcohol, Ester, Water) $CH_3CH_2COH + CH_3CH_2CH_2CH_2OH \xrightarrow{H_2SO_4} CH_3CH_2COCH_2CH_2CH_2CH_3 + H_2O$ (Propanoic acid, 1-Butanol, Butyl propanoate (85%), Water)
From acyl chlorides (Sections 15.8 and 19.4) Alcohols react with acyl chlorides by nucleophilic acyl substitution to yield esters. These reactions are typically performed in the presence of a weak base such as pyridine.	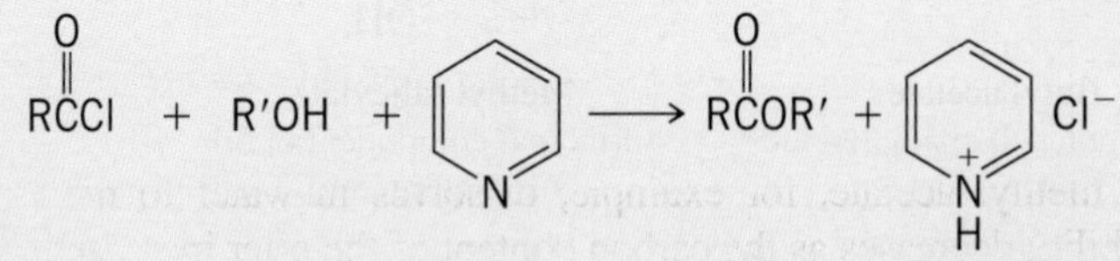 Acyl chloride, Alcohol, Pyridine, Ester, Pyridinium chloride 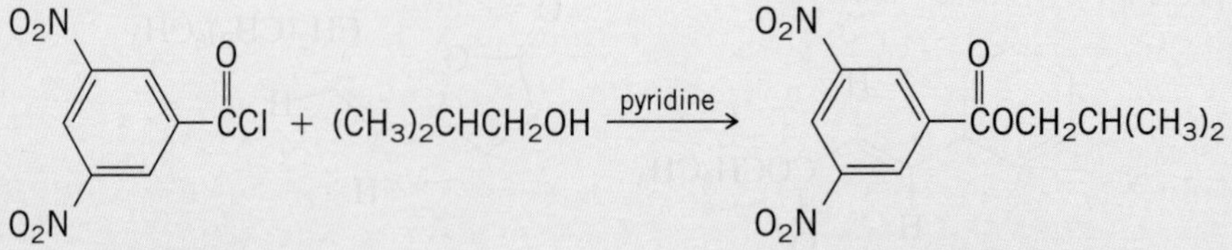 3,5-Dinitrobenzoyl chloride, Isobutyl alcohol, Isobutyl 3,5-dinitrobenzoate (85%)

Continued

TABLE 19.3 Preparation of Esters (*Continued*)

Reaction (section) and comments	General equation and specific example
From acid anhydrides (Sections 15.8 and 19.5) Acyl transfer from an acid anhydride to an alcohol is a standard method for the preparation of esters. The reaction is subject to catalysis by either acids (H_2SO_4) or bases (pyridine).	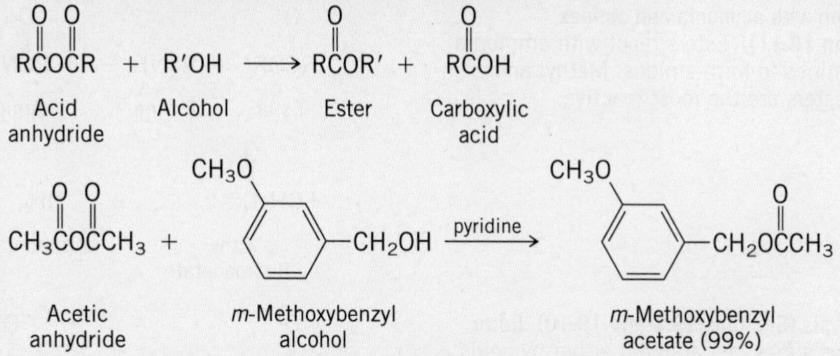
Baeyer–Villiger oxidation of ketones (Descriptive Passage 17) Ketones are converted to esters on treatment with peroxy acids. The reaction proceeds by migration of the group R′ from carbon to oxygen. It is the more highly substituted group that migrates. Methyl ketones give acetate esters.	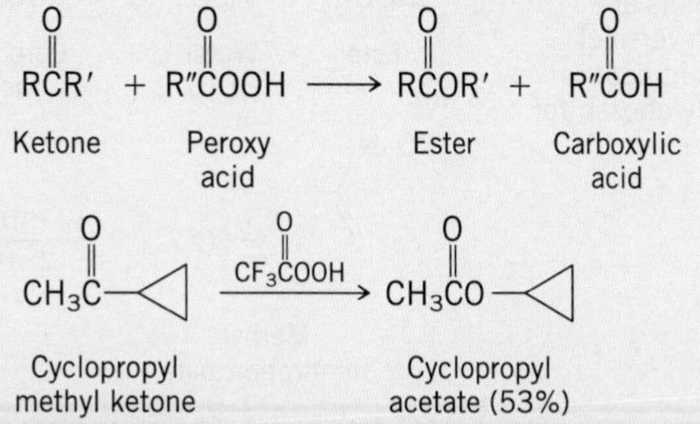

19.7 Physical Properties of Esters

Esters are moderately polar, with dipole moments in the 1.5 to 2.0-D range. Dipole–dipole attractive forces give esters higher boiling points than hydrocarbons of similar shape and molecular weight. Because they lack hydroxyl groups, however, ester molecules cannot form hydrogen bonds to each other; consequently, esters have lower boiling points than alcohols of comparable molecular weight.

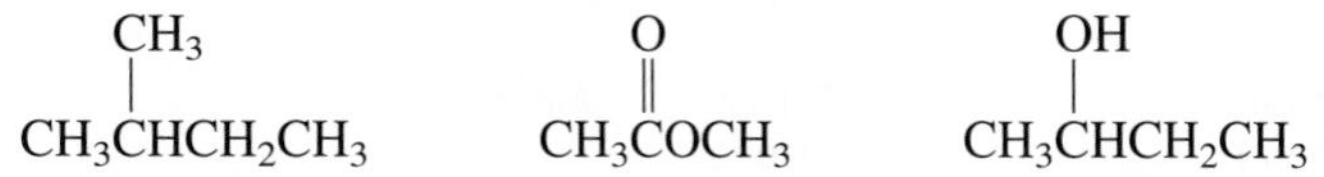

2-Methylbutane: mol wt 72, bp 28°C | Methyl acetate: mol wt 74, bp 57°C | 2-Butanol: mol wt 74, bp 99°C

Esters can participate in hydrogen bonds with substances that contain hydroxyl groups (water, alcohols, carboxylic acids). This confers some measure of water solubility on low-molecular-weight esters; methyl acetate, for example, dissolves in water to the extent of 33 g/100 mL. Water solubility decreases as the carbon content of the ester increases. Fats and oils, the glycerol esters of long-chain carboxylic acids, are practically insoluble in water.

19.8 Reactions of Esters: A Preview

Nucleophilic acyl substitutions at the ester carbonyl group are summarized in Table 19.4. Esters are less reactive than acyl chlorides and acid anhydrides. Nucleophilic acyl substitution in esters, especially ester hydrolysis, has been extensively investigated from a mechanistic perspective. Indeed, much of what we know concerning the general topic of nucleophilic acyl substitution comes from studies carried out on esters. The reactions of esters with Grignard reagents and with lithium aluminum hydride, both useful in the synthesis of alcohols, are summarized in Table 19.5. Although the products appear different from those shown in Table 19.4, in that they no longer contain a carbonyl group, you will see that nucleophilic acyl substitution plays an important role in both sets of reactions.

TABLE 19.4 Conversion of Esters to Other Carboxylic Acid Derivatives

Reaction (section) and comments	General equation and specific example
Reaction with ammonia and amines (Section 19.11)) Esters react with ammonia and amines to form amides. Methyl and ethyl esters are the most reactive.	$RCOR' + R''_2NH \longrightarrow RCNR''_2 + R'OH$ (with C=O in RCOR' and RCNR''₂ as drawn) Ester, Amine, Amide, Alcohol $FCH_2COCH_2CH_3 + NH_3 \xrightarrow{H_2O} FCH_2CNH_2 + CH_3CH_2OH$ Ethyl fluoroacetate, Ammonia, Fluoroacetamide (90%), Ethanol
Hydrolysis (Sections 19.9 and 19.10) Ester hydrolysis may be catalyzed either by acids or by bases. Acid-catalyzed hydrolysis is an equilibrium-controlled process, the reverse of the Fischer esterification. Hydrolysis in base is irreversible and is the method usually chosen for preparative purposes.	$RCOR' + H_2O \longrightarrow RCOH + R'OH$ Ester, Water, Carboxylic acid, Alcohol 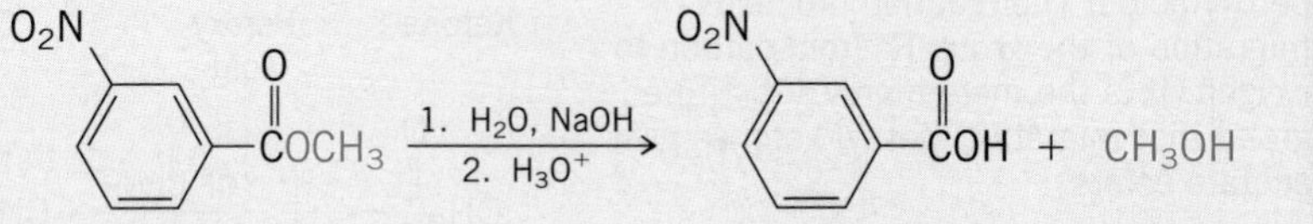 Methyl *m*-nitrobenzoate, *m*-Nitrobenzoic acid (90–96%), Methanol

TABLE 19.5 Reactions of Esters with Grignard Reagents and with Lithium Aluminum Hydride

Reaction (section) and comments	General equation and specific example
Reaction with Grignard reagents (Section 19.12) Esters react with two equivalents of a Grignard reagent to produce tertiary alcohols. Two of the groups bonded to the carbon that bears the hydroxyl group in the tertiary alcohol are derived from the Grignard reagent.	$RCOR' + 2R''MgX \xrightarrow[2.\ H_3O^+]{1.\ \text{diethyl ether}} RCR''(OH)R'' + R'OH$ Ester, Grignard reagent, Tertiary alcohol, Alcohol 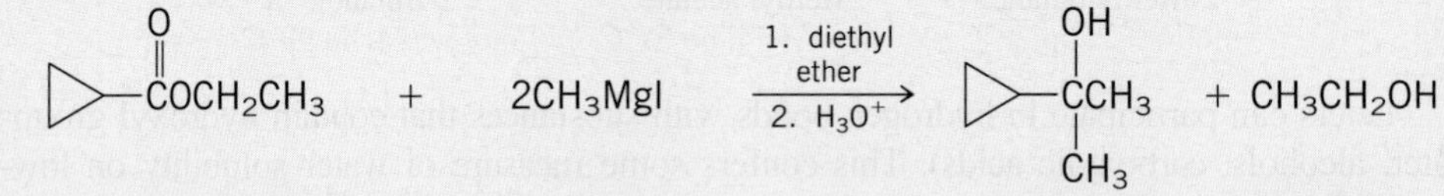 Ethyl cyclopropanecarboxylate, Methylmagnesium iodide, 2-Cyclopropyl-2-propanol (93%), Ethanol
Reduction with lithium aluminum hydride (Section 19.13) Lithium aluminum hydride cleaves esters to yield two alcohols.	$RCOR' \xrightarrow[2.\ H_2O]{1.\ LiAlH_4} RCH_2OH + R'OH$ Ester, Primary alcohol, Alcohol 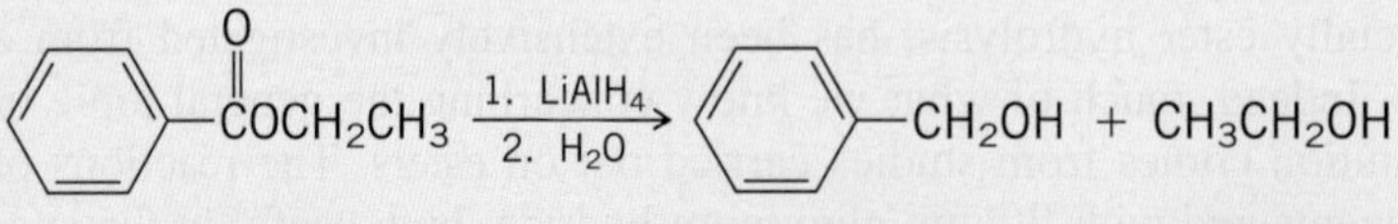 Ethyl benzoate, Benzyl alcohol (90%), Ethyl alcohol

19.9 Acid-Catalyzed Ester Hydrolysis

Ester hydrolysis is the most studied and best understood of all nucleophilic acyl substitutions. Esters are fairly stable in neutral aqueous media but are cleaved when heated with water in the presence of strong acids or bases. The hydrolysis of esters in dilute aqueous acid is the reverse of the Fischer esterification (Sections 15.8 and 18.14):

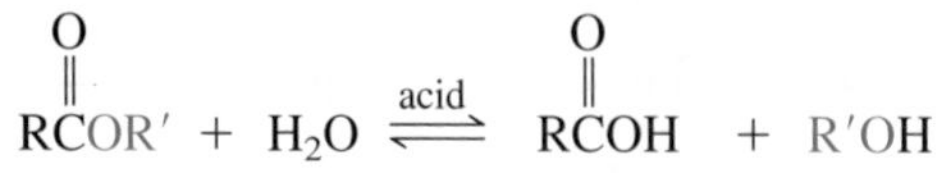

Ester Water Carboxylic acid Alcohol

When esterification is the objective, water is removed from the reaction mixture to encourage ester formation. When ester hydrolysis is the objective, the reaction is carried out in the presence of a generous excess of water. Both reactions illustrate the application of Le Châtelier's principle (Section 6.10) to organic synthesis.

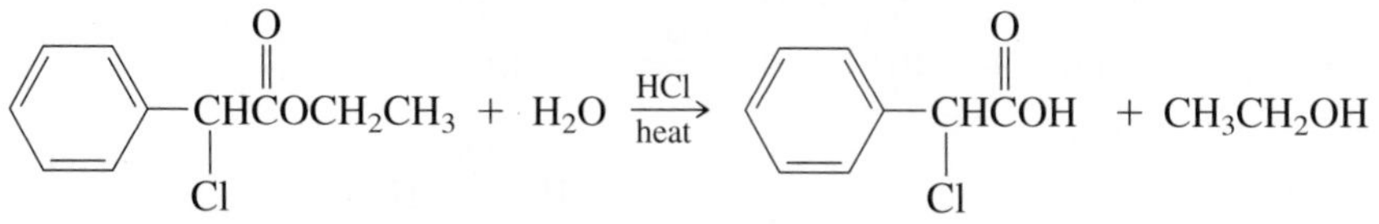

Ethyl 2-chloro-2-phenylacetate Water 2-Chloro-2-phenylacetic acid (80–82%) Ethyl alcohol

Problem 19.7

The compound having the structure shown was heated with dilute sulfuric acid to give a product having the molecular formula $C_5H_{12}O_3$ in 63–71% yield. Propose a reasonable structure for this product. What other organic compound is formed in this reaction?

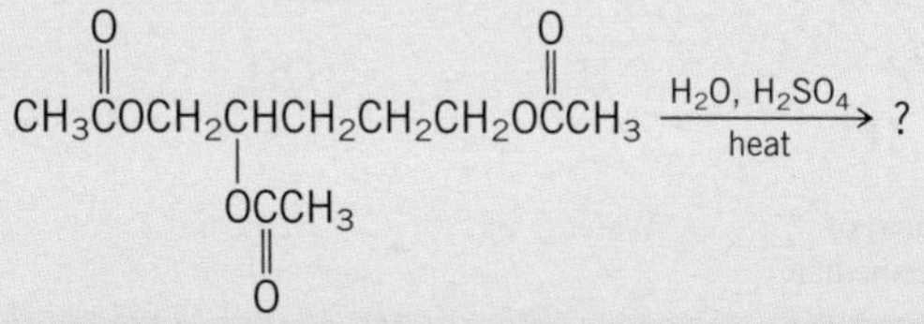

The pathway for acid-catalyzed ester hydrolysis is given in Mechanism 19.3. It is precisely the reverse of the mechanism given for acid-catalyzed ester formation in Section 18.14. Like other nucleophilic acyl substitutions, it proceeds in two stages. A tetrahedral intermediate is formed in the first stage, and this tetrahedral intermediate dissociates to products in the second stage.

A key feature of the first stage (steps 1–3) is the site at which the starting ester is protonated. Protonation of the carbonyl oxygen, as shown in step 1 of Mechanism 19.3, gives a cation that is stabilized by electron delocalization. The alternative site of protonation, the alkoxy oxygen, gives rise to a much less stable cation.

Protonation of carbonyl oxygen ***Protonation of alkoxy oxygen***

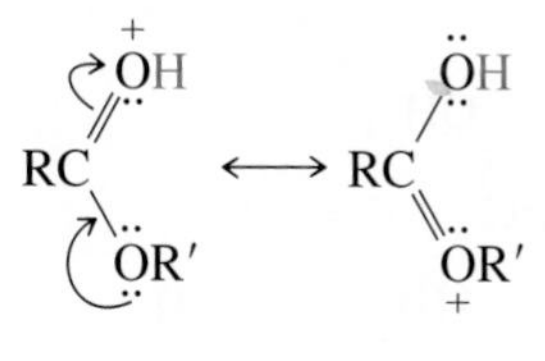

Positive charge is delocalized.

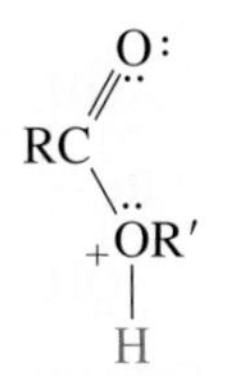

Positive charge is localized on a single oxygen.

Mechanism 19.3

Acid-Catalyzed Ester Hydrolysis

Step 1: Protonation of the carbonyl oxygen of the ester

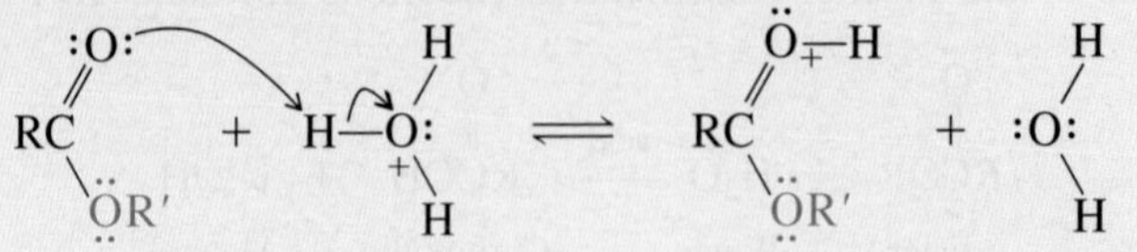

Ester | Hydronium ion | Protonated ester | Water

Step 2: Nucleophilic addition of water to protonated ester

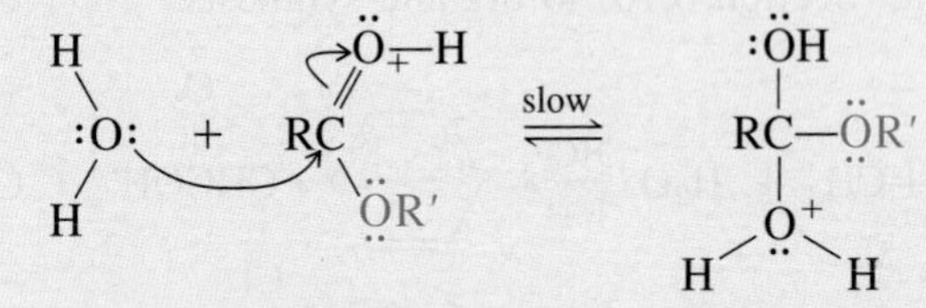

Water | Protonated ester | Conjugate acid of tetrahedral intermediate (TI-H$^+$)

Step 3: Deprotonation of the oxonium ion to give the neutral form of the tetrahedral intermediate

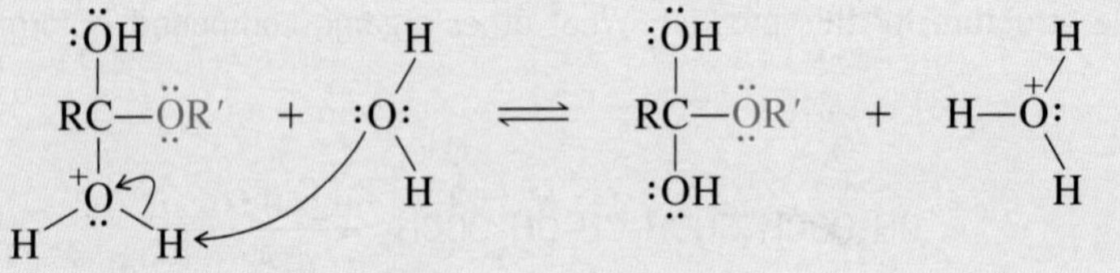

Conjugate acid of tetrahedral intermediate | Water | Tetrahedral intermediate (TI) | Hydronium ion

Step 4: Protonation of the tetrahedral intermediate at its alkoxy oxygen

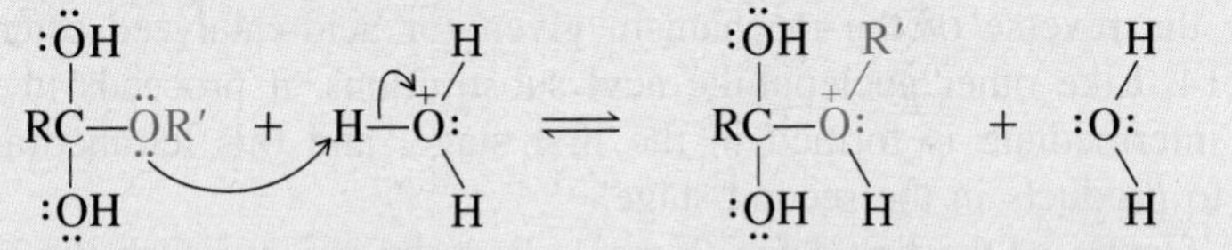

Tetrahedral intermediate | Hydronium ion | Conjugate acid of tetrahedral intermediate (TI-H$^+$) | Water

Step 5: Dissociation of the protonated form of the tetrahedral intermediate to an alcohol and the protonated carboxylic acid

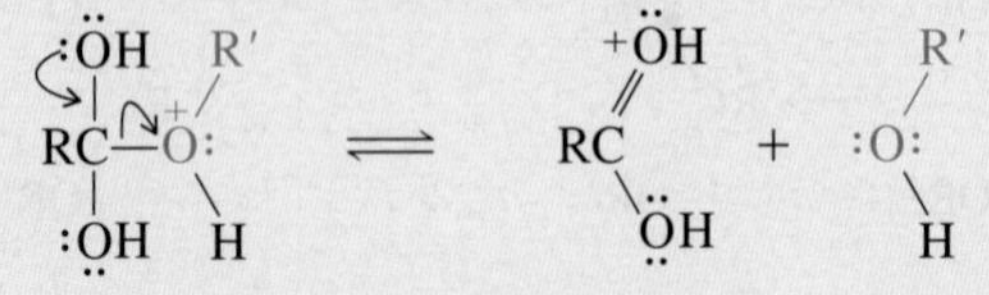

Conjugate acid of tetrahedral intermediate | Protonated carboxylic acid | Alcohol

Step 6: Deprotonation of the protonated carboxylic acid

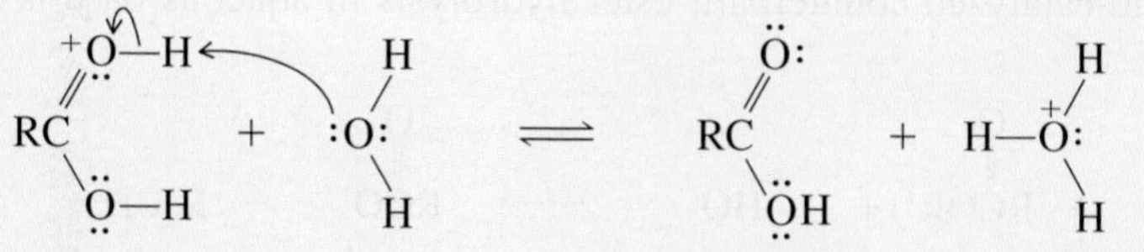

Protonated carboxylic acid | Water | Carboxylic acid | Hydronium ion

Protonation of the carbonyl oxygen, as emphasized earlier, makes the carbonyl group more electrophilic. A water molecule adds to the carbonyl group of the protonated ester in step 2. Loss of a proton from the resulting alkyloxonium ion gives the neutral form of the tetrahedral intermediate in step 3 and completes the first stage of the mechanism. In step 4 of Mechanism 19.3, protonation of the tetrahedral intermediate at its alkoxy oxygen gives a new oxonium ion, which loses a molecule of alcohol in step 5. Along with the alcohol, the protonated form of the carboxylic acid arises by dissociation of the tetrahedral intermediate. Its deprotonation in step 6 completes the process.

Problem 19.8

On the basis of the general mechanism for acid-catalyzed ester hydrolysis shown in Mechanism 19.3, write an analogous sequence of steps for the specific case of ethyl benzoate hydrolysis.

The most important species in the mechanism for ester hydrolysis is the tetrahedral intermediate. Evidence in support of the existence of the tetrahedral intermediate was developed by Professor Myron Bender on the basis of isotopic labeling experiments he carried out at the University of Chicago. Bender prepared ethyl benzoate, labeled with the mass-18 isotope of oxygen at the carbonyl oxygen, then subjected it to acid-catalyzed hydrolysis in ordinary (unlabeled) water. He found that ethyl benzoate, recovered from the reaction before hydrolysis was complete, had lost a portion of its isotopic label. This observation is consistent only with the reversible formation of a tetrahedral intermediate under the reaction conditions:

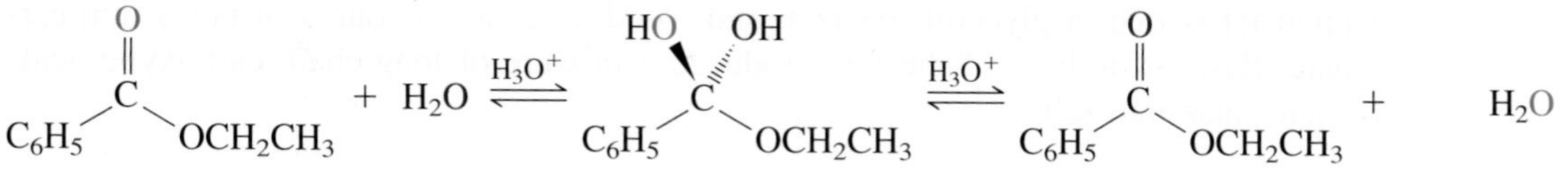

Ethyl benzoate (labeled with ^{18}O) | Water | Tetrahedral intermediate | Ethyl benzoate | Water (labeled with ^{18}O)

The two OH groups in the tetrahedral intermediate are equivalent, and so either the labeled or the unlabeled one can be lost when the tetrahedral intermediate reverts to ethyl benzoate. Both are retained when the tetrahedral intermediate goes on to form benzoic acid.

Problem 19.9

In a similar experiment, unlabeled 4-butanolide was allowed to stand in an acidic solution in which the water had been labeled with ^{18}O. When the lactone was extracted from the solution after four days, it was found to contain ^{18}O. Which oxygen of the lactone do you think became isotopically labeled?

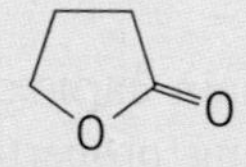

4-Butanolide

19.10 Ester Hydrolysis in Base: Saponification

Unlike its acid-catalyzed counterpart, ester hydrolysis in aqueous base is *irreversible.*

Because it is consumed, hydroxide ion is a reactant, not a catalyst.

$$\underset{\text{Ester}}{\overset{O}{\overset{\|}{RCOR'}}} + \underset{\text{Hydroxide ion}}{HO^-} \longrightarrow \underset{\text{Carboxylate ion}}{\overset{O}{\overset{\|}{RCO^-}}} + \underset{\text{Alcohol}}{R'OH}$$

This is because carboxylic acids are converted to their corresponding carboxylate anions, which are stable under the reaction conditions.

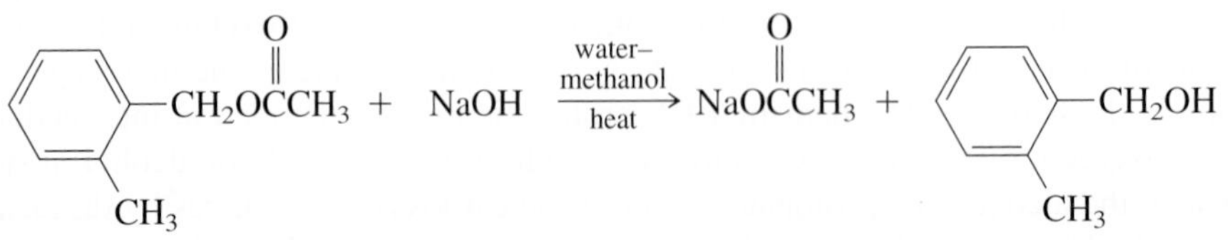

o-Methylbenzyl acetate — Sodium hydroxide — Sodium acetate — *o*-Methylbenzyl alcohol (95–97%)

To isolate the carboxylic acid, a separate acidification step following hydrolysis is necessary. Acidification converts the carboxylate salt to the free acid.

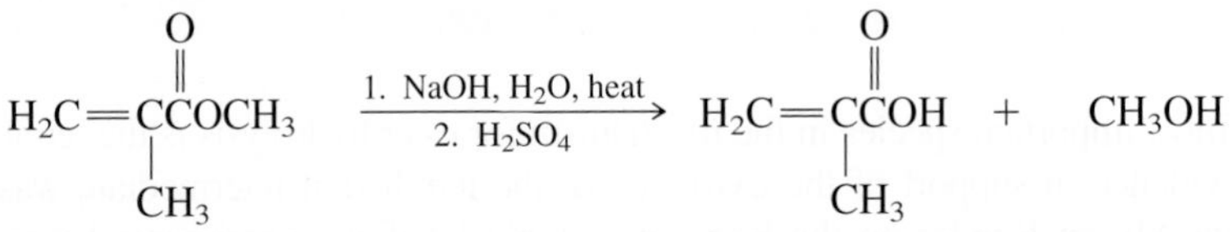

Methyl 2-methylpropenoate (methyl methacrylate) — 2-Methylpropenoic acid (87%) (methacrylic acid) — Methyl alcohol

Ester hydrolysis in base is called **saponification,** which means "soap making." Over 2000 years ago, the Phoenicians made soap by heating animal fat with wood ashes. Animal fat is rich in glycerol triesters, and wood ashes are a source of potassium carbonate. Basic hydrolysis of the fats produced a mixture of long-chain carboxylic acids as their potassium salts.

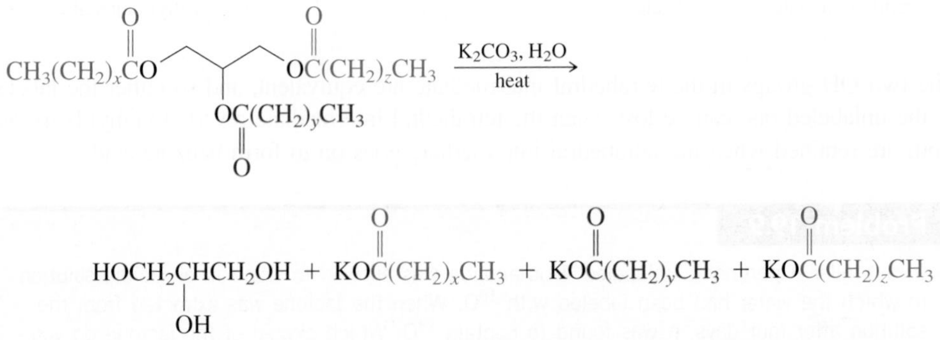

Glycerol — Potassium carboxylate salts

Potassium and sodium salts of long-chain carboxylic acids form micelles that dissolve grease (Section 18.7) and have cleansing properties. The carboxylic acids obtained by saponification of fats and oils are called *fatty acids.*

Problem 19.10

Trimyristin is obtained from coconut oil and has the molecular formula $C_{45}H_{86}O_6$. On being heated with aqueous sodium hydroxide followed by acidification, trimyristin was converted to glycerol and tetradecanoic acid as the only products. What is the structure of trimyristin?

In one of the earliest kinetic studies of an organic reaction, carried out in the nineteenth century, the rate of hydrolysis of ethyl acetate in aqueous sodium hydroxide was found to be first order in ester and first order in base.

$$\underset{\text{Ethyl acetate}}{CH_3\overset{O}{\overset{\|}{C}}OCH_2CH_3} + \underset{\text{Sodium hydroxide}}{NaOH} \longrightarrow \underset{\text{Sodium acetate}}{CH_3\overset{O}{\overset{\|}{C}}ONa} + \underset{\text{Ethanol}}{CH_3CH_2OH}$$

$$\text{Rate} = k[CH_3\overset{O}{\overset{\|}{C}}OCH_2CH_3][NaOH]$$

Overall, the reaction exhibits second-order kinetics. Both the ester and the base are involved in the rate-determining step or in a rapid step that precedes it.

Two processes consistent with second-order kinetics both involve hydroxide ion as a nucleophile but differ in the site of nucleophilic attack. One is an S_N2 reaction, the other is nucleophilic acyl substitution.

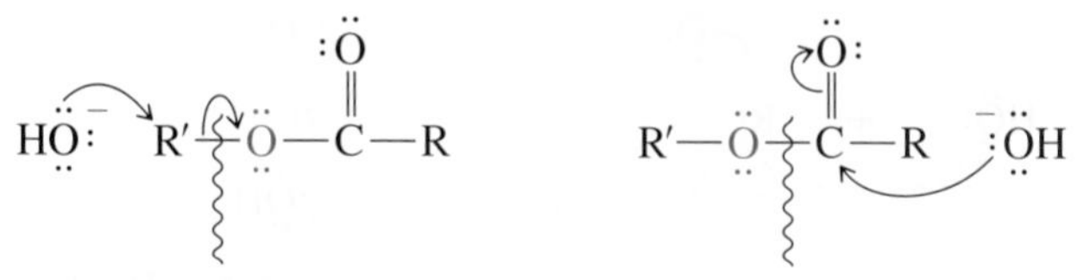

Bond between O and alkyl group R′ breaks

Bond between O and acyl group RC=O breaks

Convincing evidence that ester hydrolysis in base proceeds by a *nucleophilic acyl substitution* mechanism has been obtained from several sources. In one experiment, ethyl propanoate labeled with ^{18}O in the ethoxy group was hydrolyzed. On isolating the products, all the ^{18}O was found in the ethyl alcohol; none was in the sodium propanoate.

$$\underset{^{18}\text{O-labeled ethyl propanoate}}{CH_3CH_2\overset{O}{\overset{\|}{C}}OCH_2CH_3} + \underset{\text{Sodium hydroxide}}{NaOH} \longrightarrow \underset{\text{Sodium propanoate}}{CH_3CH_2\overset{O}{\overset{\|}{C}}ONa} + \underset{^{18}\text{O-labeled ethyl alcohol}}{CH_3CH_2OH}$$

The carbon–oxygen bond broken in the process is therefore the one between oxygen and the acyl group. The bond between oxygen and the ethyl group remains intact. An S_N2 reaction at the ethyl group would have broken this bond.

Problem 19.11

In a similar experiment, pentyl acetate was subjected to saponification with ^{18}O-labeled hydroxide in ^{18}O-labeled water. What product do you think became isotopically labeled here, acetate ion or 1-pentanol?

Identical conclusions come from stereochemical studies. Saponification of esters of optically active alcohols proceeds with *retention of configuration.*

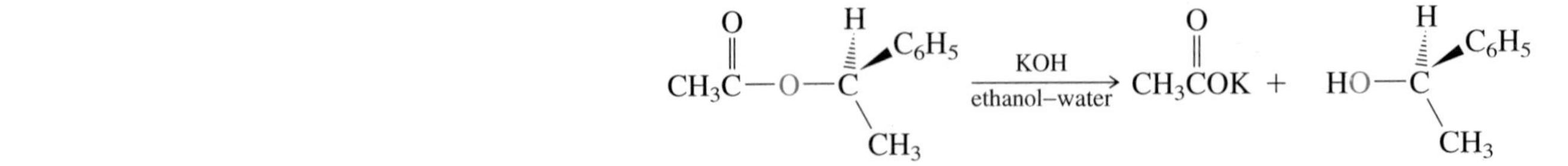

(*R*)-(+)-1-Phenylethyl acetate — Potassium acetate — (*R*)-(+)-1-Phenylethyl alcohol (80% yield; same optical purity as ester)

None of the bonds to the chirality center is broken when hydroxide attacks the carbonyl group. Had an S_N2 reaction occurred instead, inversion of configuration at the chirality center would have taken place to give (*S*)-(−)-1-phenylethyl alcohol.

In an extension of the work described in the preceding section, Bender showed that basic ester hydrolysis, like acid hydrolysis, takes place by way of a tetrahedral intermediate. The nature of the experiment was the same, and the results were similar to those observed in the acid-catalyzed reaction.

The observation of second-order kinetics, nucleophilic attack at the carbonyl group, and the involvement of a tetrahedral intermediate are accommodated by Mechanism 19.4.

Mechanism 19.4

Ester Hydrolysis in Basic Solution

Step 1: Nucleophilic addition of hydroxide ion to the carbonyl group

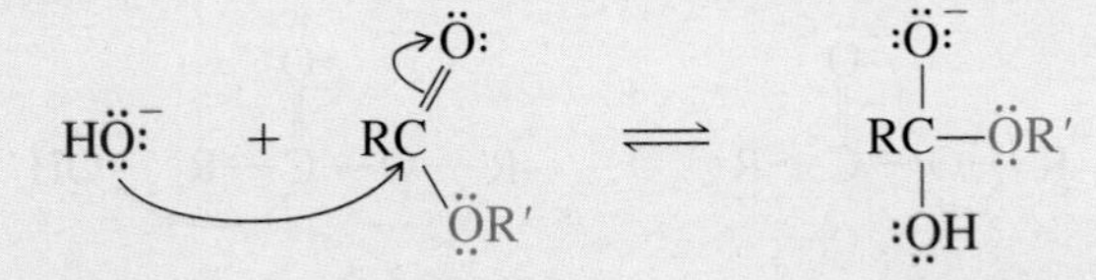

Hydroxide ion — Ester — Conjugate base TI^- of tetrahedral intermediate

Step 2: Dissociation of TI^-

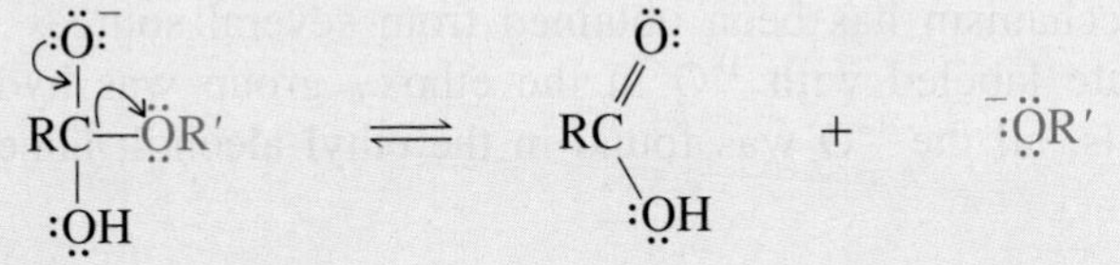

Conjugate base TI^- of tetrahedral intermediate — Carboxylic acid — Alkoxide ion

Step 3: Proton transfers yield an alcohol and a carboxylate ion

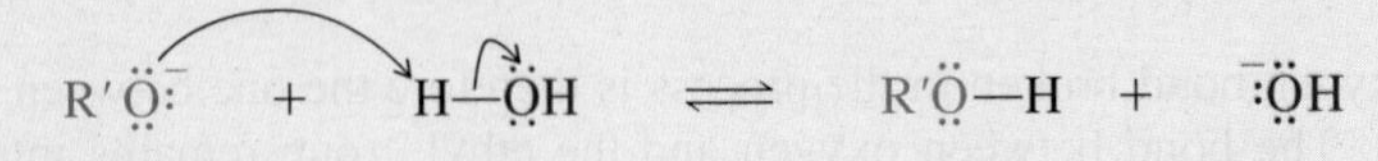

Alkoxide ion — Water — Alcohol — Hydroxide ion

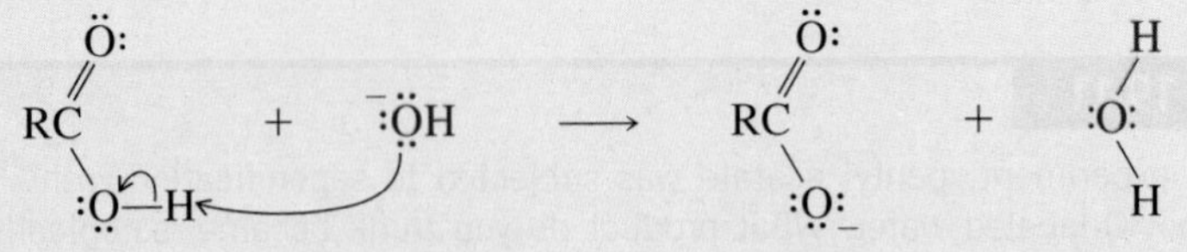

Carboxylic acid (stronger acid) — Hydroxide ion (stronger base) — Carboxylate ion (weaker base) — Water (weaker acid)

Like the acid-catalyzed mechanism, it has two distinct stages, namely, formation of the tetrahedral intermediate and its subsequent dissociation. Nucleophilic addition to the carbonyl group has a higher activation energy than dissociation of the tetrahedral intermediate; step 1 is rate-determining. All the steps are reversible except the last one. The equilibrium constant for proton abstraction from the carboxylic acid by hydroxide in step 3 is so large that it makes the overall reaction irreversible.

Problem 19.12

On the basis of the general mechanism for basic ester hydrolysis shown in Mechanism 19.4, write an analogous sequence of steps for the saponification of ethyl benzoate.

Problem 19.13

Which ester in each pair would be expected to undergo saponification at the faster rate?

(a) 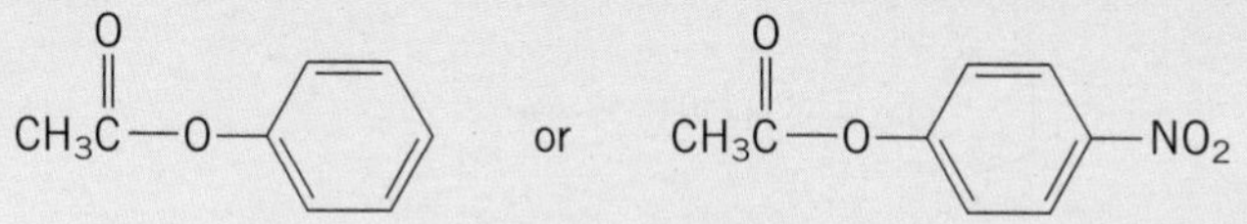

(b) 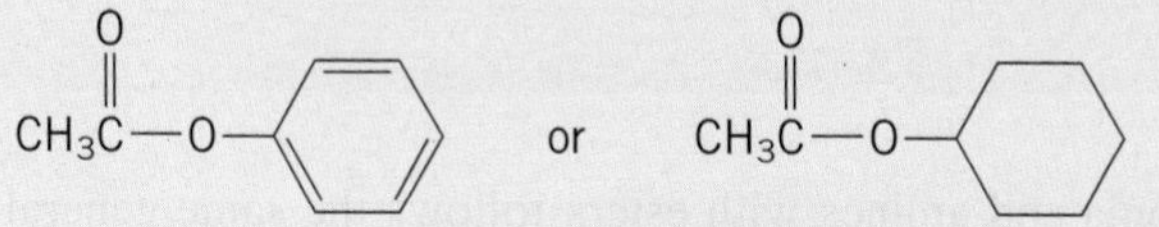

(c) $CH_3\overset{O}{\overset{\|}{C}}-OCH_2CH_3$ or $CF_3\overset{O}{\overset{\|}{C}}-OCH_2CH_3$

(d) $CH_3\overset{O}{\overset{\|}{C}}-OCH_2CH_3$ or $CH_3\overset{O}{\overset{\|}{C}}-OC(CH_3)_3$

Sample Solution (a) *p*-Nitrophenyl acetate reacts faster. A *p*-nitrophenyl group withdraws electrons from the ester oxygen which decreases its ability to stabilize the carbonyl group. A less-stabilized carbonyl is more reactive than a more-stabilized one.

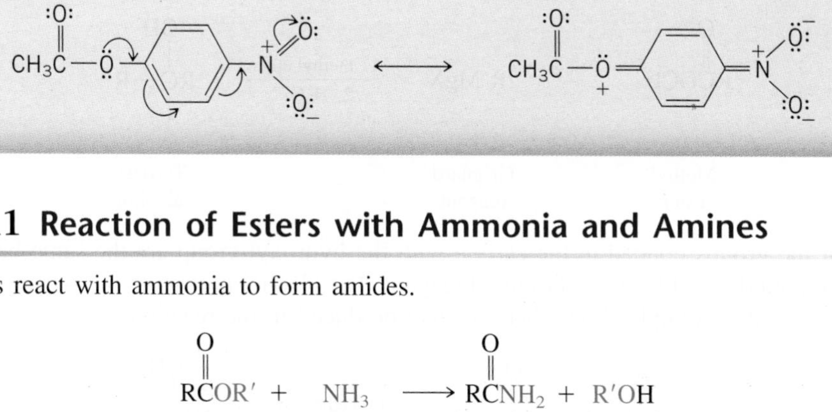

19.11 Reaction of Esters with Ammonia and Amines

Esters react with ammonia to form amides.

$$\underset{\text{Ester}}{R\overset{O}{\overset{\|}{C}}OR'} + \underset{\text{Ammonia}}{NH_3} \longrightarrow \underset{\text{Amide}}{R\overset{O}{\overset{\|}{C}}NH_2} + \underset{\text{Alcohol}}{R'OH}$$

Ammonia is more nucleophilic than water, making it possible to carry out this reaction using aqueous ammonia.

$$\underset{\text{Methyl 2-methylpropenoate}}{H_2C{=}\underset{CH_3}{\underset{|}{\overset{O}{\overset{\|}{C}}}}COCH_3} + \underset{\text{Ammonia}}{NH_3} \xrightarrow{H_2O} \underset{\text{2-Methylpropenamide (75\%)}}{H_2C{=}\underset{CH_3}{\underset{|}{\overset{O}{\overset{\|}{C}}}}CNH_2} + \underset{\text{Methyl alcohol}}{CH_3OH}$$

Amines react similarly:

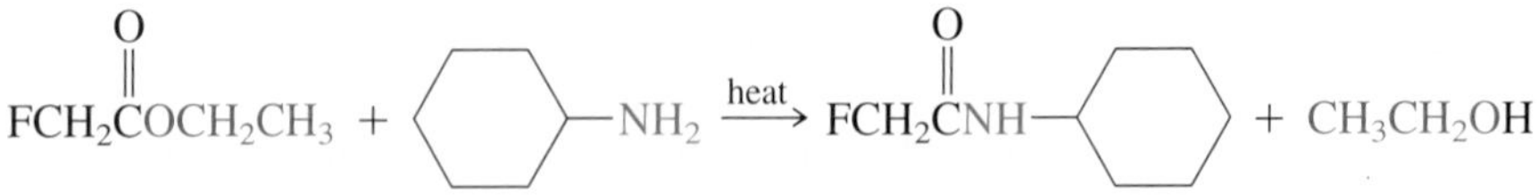

Ethyl fluoroacetate | Cyclohexylamine | *N*-Cyclohexylfluoroacetamide (61%) | Ethyl alcohol

The amine must be primary (RNH_2) or secondary (R_2NH). Tertiary amines (R_3N) cannot form amides because they have no proton on nitrogen that can be replaced by an acyl group.

Problem 19.14

Give the structure of the expected product of the following reaction:

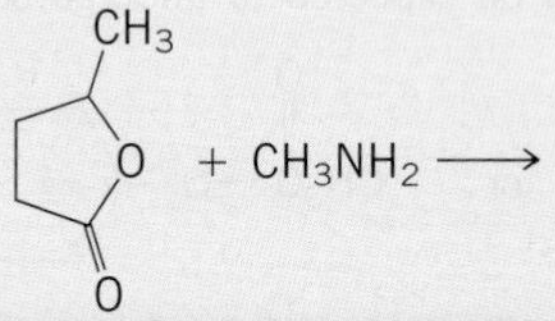

The reaction of ammonia and amines with esters follows the same general mechanistic course as other nucleophilic acyl substitution reactions. A tetrahedral intermediate is formed in the first stage of the process and dissociates in the second stage.

19.12 Reaction of Esters with Grignard Reagents: Synthesis of Tertiary Alcohols

Esters react with two equivalents of a Grignard reagent to form tertiary alcohols. Methyl and ethyl esters are readily available and are the types most often used.

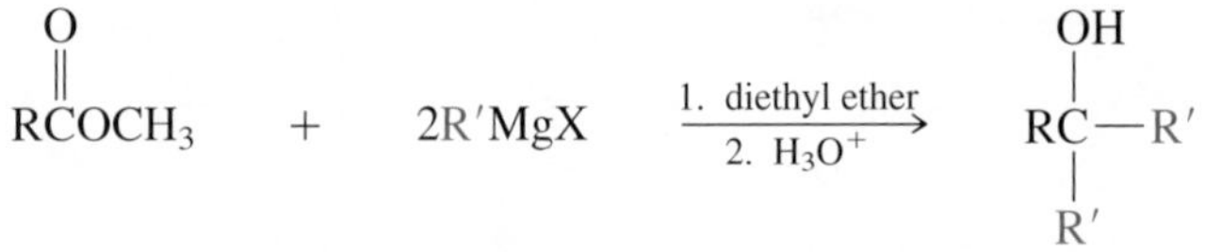

Methyl ester | Grignard reagent | Tertiary alcohol

Two of the groups bonded to the carbon with the hydroxyl group are the same because they are both derived from the Grignard reagent. The alcohol portion of the starting ester, methanol in the example shown here, is also produced in the reaction.

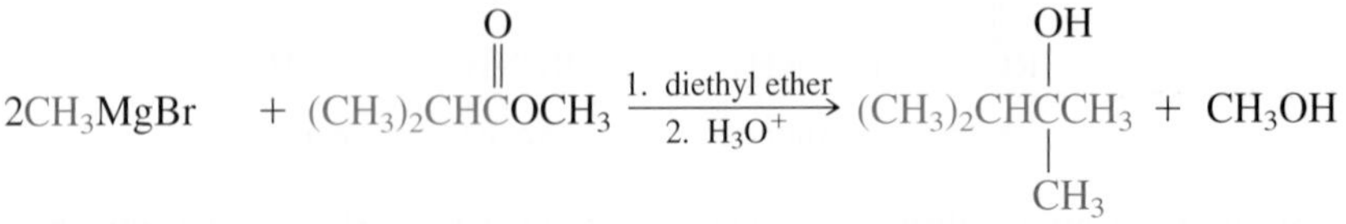

Methylmagnesium bromide | Methyl 2-methylpropanoate | 2,3-Dimethyl-2-butanol (73%) | Methanol

The mechanism of the reaction of a Grignard reagent with an ester first involves nucleophilic addition to the carbonyl carbon of the ester, forming a tetrahedral intermediate, illustrated in step 1 in Mechanism 19.5. The tetrahedral intermediate is unstable and breaks down to form a ketone (step 2), which rapidly undergoes addition of a second equivalent of the Grignard reagent (step 3). Subsequent hydrolysis gives the tertiary alcohol (step 4).

Mechanism 19.5

Reaction of an Ester with a Grignard Reagent

THE OVERALL REACTION:

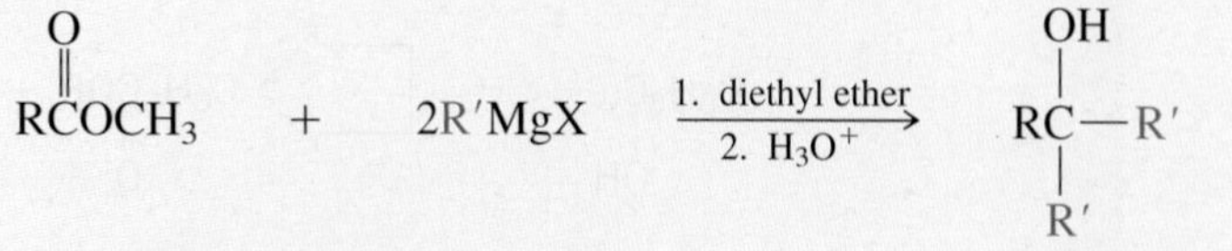

Methyl ester | Grignard reagent | Tertiary alcohol

Step 1: Nucleophilic addition of the Grignard reagent to the ester carbonyl group gives a tetrahedral intermediate.

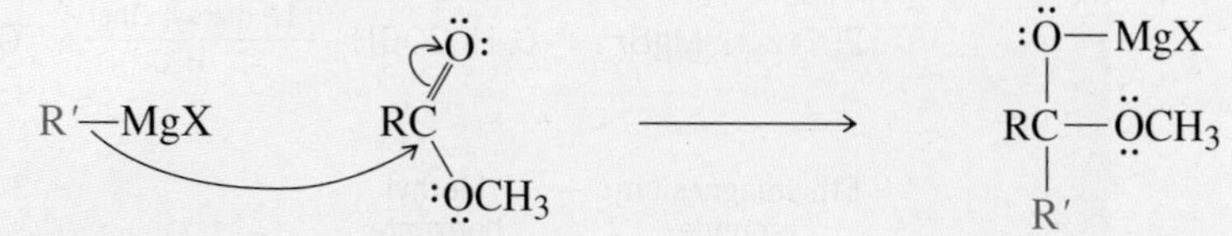

Grignard reagent | Methyl ester | Tetrahedral intermediate

Step 2: The tetrahedral intermediate undergoes elimination of methoxymagnesium halide and a ketone is formed.

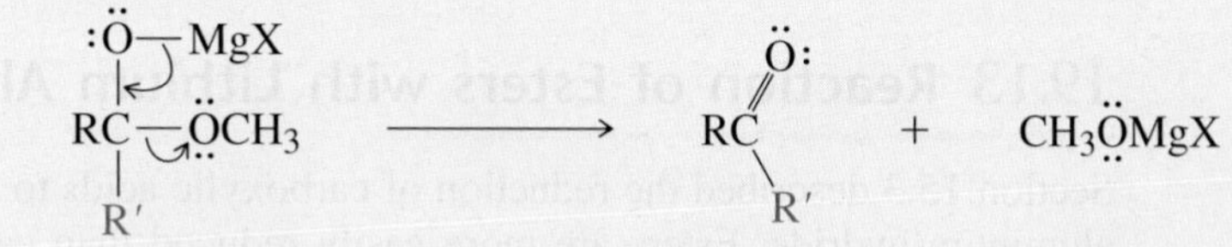

Tetrahedral intermediate | Ketone | Methoxymagnesium halide

Step 3: The ketone reacts with a second equivalent of the Grignard reagent.

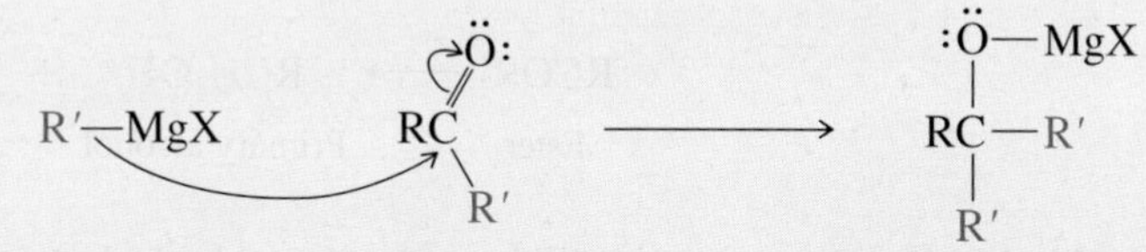

Grignard reagent | Ketone | Alkoxymagnesium halide

Step 4: In a separate operation, hydrolysis converts the alkoxymagnesium halide to a tertiary alcohol.

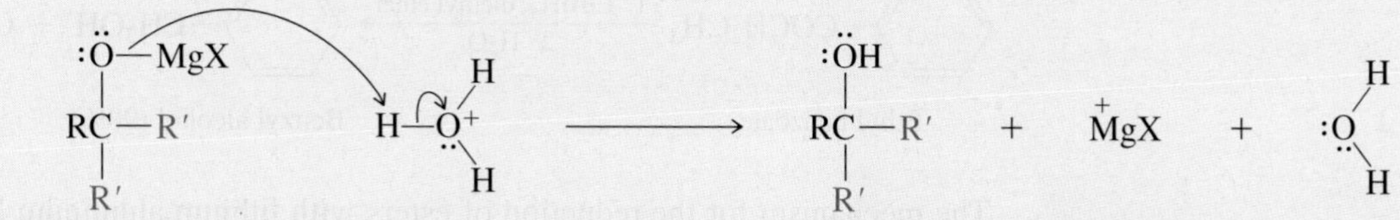

Alkoxymagnesium halide | Tertiary alcohol | Halomagnesium ion | Water

Problem 19.15

What combination of ester and Grignard reagent could you use to prepare each of the following tertiary alcohols?

(a) 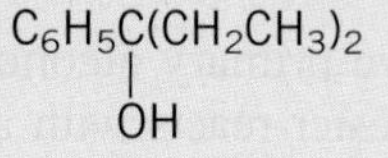 (b) 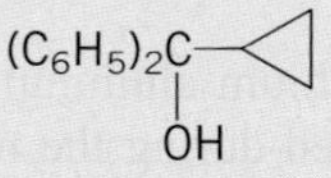

Continued

Sample Solution (a) To apply the principles of retrosynthetic analysis to this case, we disconnect both ethyl groups from the tertiary carbon and identify them as arising from the Grignard reagent. The phenyl group originates in an ester of the type $C_6H_5CO_2R$.

$$C_6H_5\underset{\displaystyle OH}{\underset{|}{C}}(CH_2CH_3)_2 \Longrightarrow C_6H_5\underset{\displaystyle O}{\underset{\|}{C}}OR + 2CH_3CH_2MgX$$

An appropriate synthesis would be

$$2CH_3CH_2MgBr + C_6H_5\overset{\displaystyle O}{\overset{\|}{C}}OCH_3 \xrightarrow[2.\ H_3O^+]{1.\ \text{diethyl ether}} C_6H_5\underset{\displaystyle OH}{\underset{|}{C}}(CH_2CH_3)_2$$

Ethylmagnesium bromide | Methyl benzoate | 3-Phenyl-3-pentanol

19.13 Reaction of Esters with Lithium Aluminum Hydride

Section 15.3 described the reduction of carboxylic acids to primary alcohols with lithium aluminum hydride. Esters are more easily reduced than carboxylic acids. Two alcohols are formed from each ester molecule, with cleavage of the acyl group of the ester giving a primary alcohol.

$$R\overset{\displaystyle O}{\overset{\|}{C}}OR' \longrightarrow RCH_2OH + R'OH$$

Ester | Primary alcohol | Alcohol

Ethyl benzoate reacts with lithium aluminum hydride to give benzyl alcohol and ethanol.

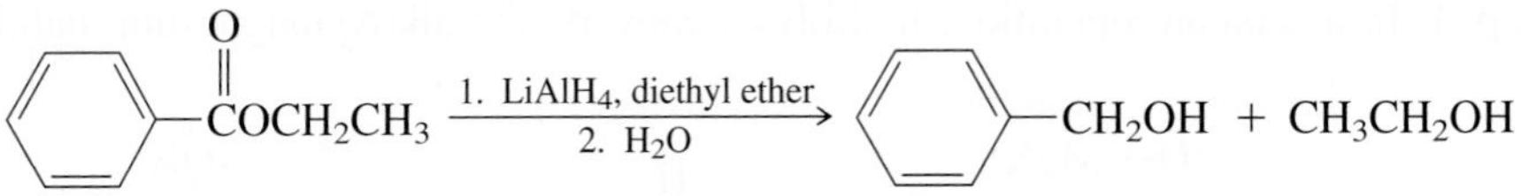

Ethyl benzoate | Benzyl alcohol (90%) | Ethanol

The mechanism for the reduction of esters with lithium aluminum hydride involves the addition of hydride to the ester carbonyl group to give a tetrahedral intermediate, which undergoes elimination to produce an aldehyde.

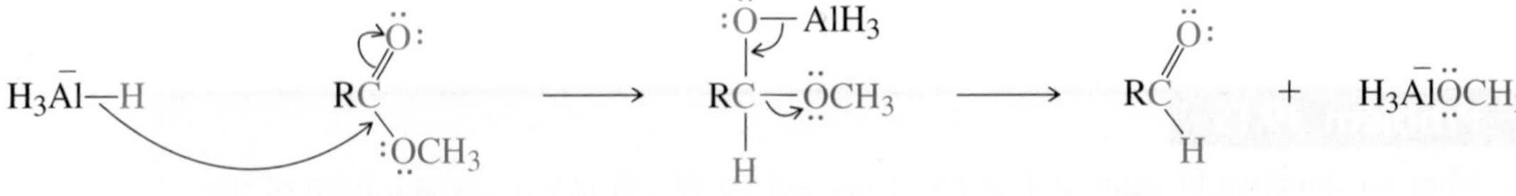

Ester | Aldehyde

Aldehydes react rapidly with lithium aluminum hydride to give primary alcohols (Section 15.2). The aldehyde produced during the reduction of the ester reacts with a second equivalent of hydride, to give the primary alcohol. One mole of lithium aluminum hydride reduces two moles of ester to alcohol.

Problem 19.16

Which aldehyde is an intermediate in the reduction of ethyl benzoate with lithium aluminum hydride?

Problem 19.17

Give the structure of an ester that will yield a mixture containing equimolar amounts of 1-propanol and 2-propanol on reduction with lithium aluminum hydride.

19.14 Amides

Physical Properties of Amides: Earlier in this chapter we connected the structure of amides to their reactivity in nucleophilic acyl substitution by emphasizing the interaction of the nitrogen lone pair with the carbonyl group.

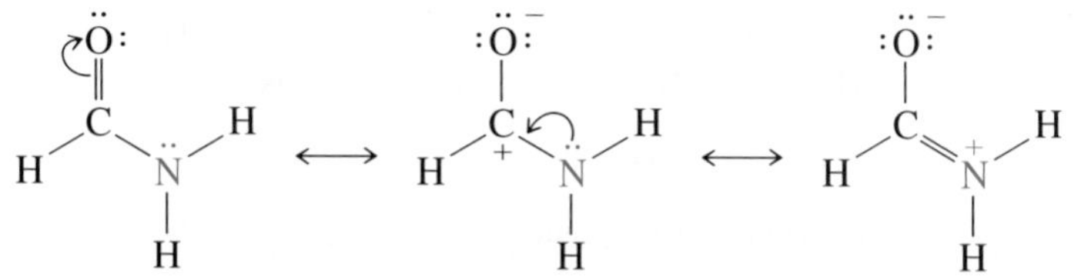

Formamide

Delocalization of the nitrogen lone pair stabilizes the carbonyl group and makes amides less reactive than other carboxylic acid derivatives toward nucleophilic acyl substitution.

In addition to influencing chemical properties, electron delocalization in amides affects their physical properties. Formamide, for example, has a dipole moment of 3.7 D, with its positive charge concentrated at the hydrogens of the NH_2 group and negative charge at the carbonyl oxygen. This combination is well suited for hydrogen bonding between the N—H bond of one molecule and the carbonyl oxygen of another. Figure 19.3 illustrates hydrogen bonding of one formamide molecule to three others.

Intermolecular forces, especially hydrogen bonding, are often evident when comparing the boiling points of compounds in which they are possible to those in which they are not. Comparing acetamide to 2-methylpropene and acetic acid we see evidence for significantly stronger intermolecular forces in acetamide and attribute it mainly to hydrogen bonding, stronger even than in acetic acid.

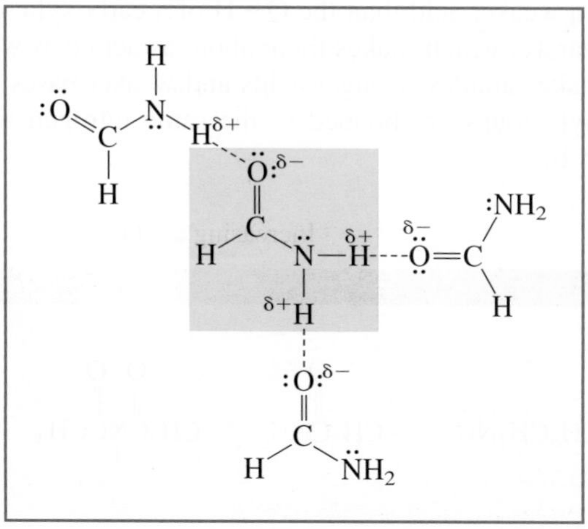

Figure 19.3

Hydrogen bonding in formamide. The formamide molecule framed in yellow can form hydrogen bonds to three other formamide molecules.

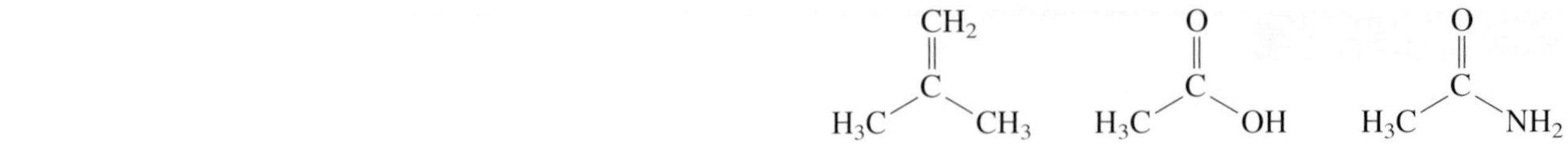

	2-Methylpropene	Acetic acid	Acetamide
Dipole moment, D	0.5	1.9	3.9
Boiling point, °C	−7	118	221
Intermolecular hydrogen bonding possible?	No	Yes	Yes

The number of substituents on nitrogen determines how many hydrogen bonds are possible. *N,N*-Dimethylacetamide, for example, lacks N—H bonds, and their absence causes its melting point and boiling point to be lower than *N*-methylacetamide. Continuing the progression, *N*-methylacetamide with one N—H bond has lower melting and boiling points than acetamide, which has two.

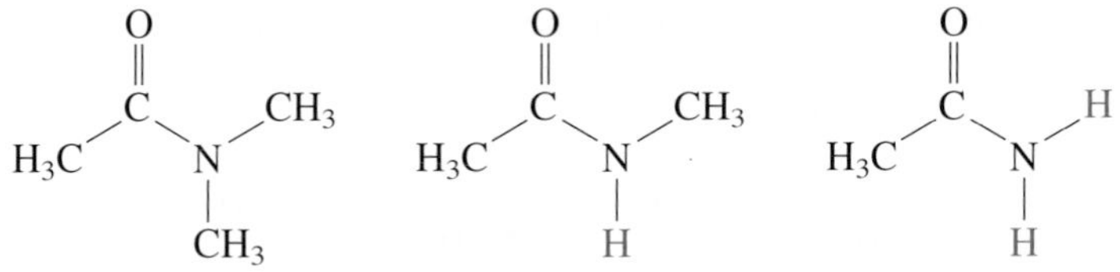

	N,N-Dimethylacetamide	*N*-Methylacetamide	Acetamide
Dipole moment, D	3.8	4.4	3.9
Melting point, °C	−20	31	80
Boiling point, °C	165	206	221
Number of N—H bonds	0	1	2

Problem 19.18

Compare *N*-methylacetamide (see the preceding paragraph) with its amide isomers propanamide and *N,N*-dimethylformamide. Which do you predict has the highest boiling point? The lowest?

Intermolecular hydrogen bonding in amides, along with the planar geometry of the amide functional group, are the two most important factors governing the conformation of protein chains. We'll learn more about this in Chapter 25.

Acidity of Amides: Because nitrogen is less electronegative than oxygen, the N—H group of an amide is a weaker acid than the O—H of a carboxylic acid. Typical primary amides have pK_a's near 16, which makes them about as acidic as water. The presence of the carbonyl group makes amides stronger acids and weaker bases than amines. Amides in which two carbonyl groups are bonded to the same nitrogen are called **imides** and have pK_a values near 10.

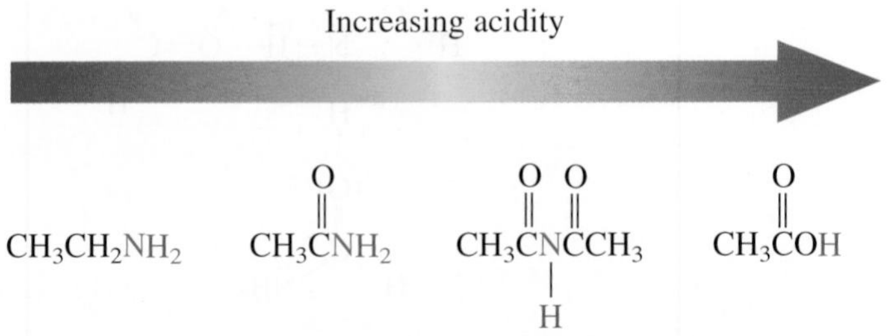

pK_a:	36	15	10	5

Problem 19.19

The pyrimidine thymine, present in DNA, was once thought to be A because A is analogous to benzene. In fact, thymine is B, which is also aromatic. Explain how B satisfies Hückel's rule, and write a contributing resonance structure for B that has a benzene-like ring.

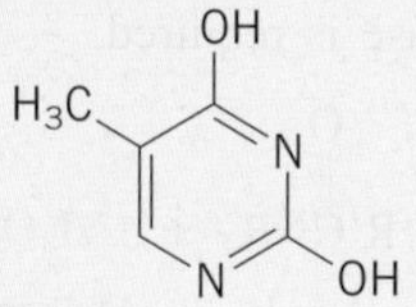

A.

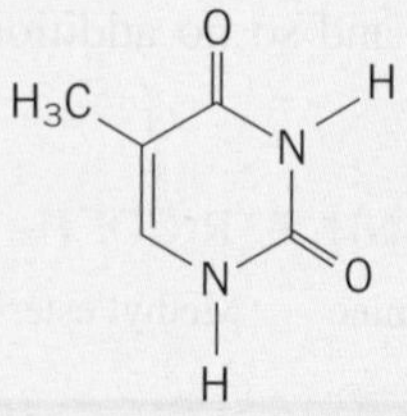

B.

- B is planar and monocyclic.
- Completely conjugated. All ring atoms are sp^2 hybridized.

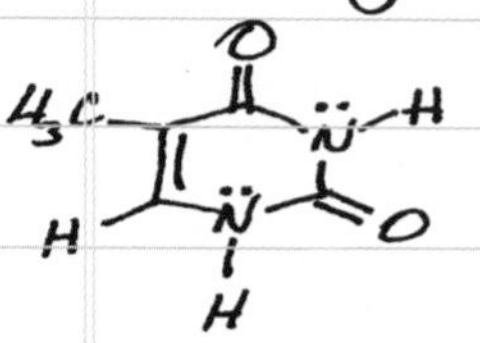

2π electrons from double bond

2π electrons from each N

- 6 π electrons satisfies Hückel's rule.
- Resonance

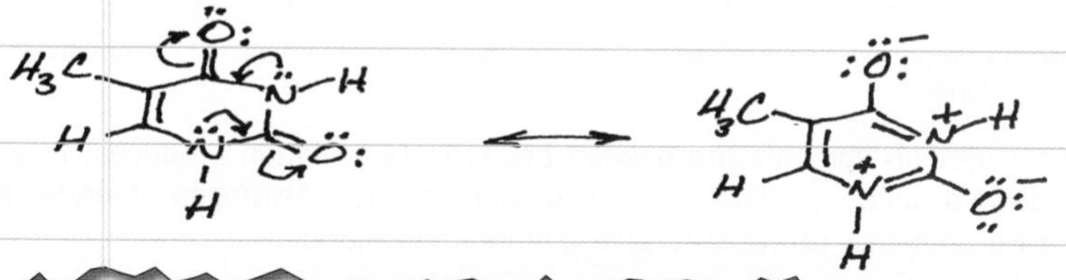

Synthesis of Amides: Tables 19.1, 19.2, and 19.4 included nucleophilic acyl substitutions that are useful for preparing amides by the reaction of amines with acyl chlorides, anhydrides, and esters, respectively. These are the most common methods for the laboratory synthesis of amides.

Because acylation of amines with acyl chlorides and anhydrides yields an acid as one of the products (HCl from acyl chlorides, a carboxylic acid from an anhydride), the efficient synthesis of amides requires some attention to stoichiometry.

Two molar equivalents of amine are frequently used in the reaction with acyl chlorides and acid anhydrides; one molecule of amine acts as a nucleophile, the second as a Brønsted base.

$$\underset{\text{Amine}}{2R_2NH} + \underset{\text{Acyl chloride}}{R'\overset{\overset{O}{\|}}{C}Cl} \longrightarrow \underset{\text{Amide}}{R'\overset{\overset{O}{\|}}{C}NR_2} + \underset{\text{Hydrochloride salt of amine}}{R_2\overset{+}{N}H_2\ Cl^-}$$

$$\underset{\text{Amine}}{2R_2NH} + \underset{\text{Acid anhydride}}{R'\overset{\overset{O}{\|}}{C}O\overset{\overset{O}{\|}}{C}R'} \longrightarrow \underset{\text{Amide}}{R'\overset{\overset{O}{\|}}{C}NR_2} + \underset{\text{Carboxylate salt of amine}}{R_2\overset{+}{N}H_2\ {}^-O\overset{\overset{O}{\|}}{C}R'}$$

It is possible to use only one molar equivalent of amine in these reactions if some other base, such as sodium hydroxide, is present in the reaction mixture to react with the hydrogen chloride or carboxylic acid that is formed. This is a useful procedure in those cases in which the amine is a valuable one or is available only in small quantities.

Esters and amines react in a 1:1 molar ratio to give amides. No acidic product is formed from the ester, and so no additional base is required.

$$\underset{\text{Amine}}{R_2NH} + \underset{\text{Methyl ester}}{R'\overset{\overset{O}{\|}}{C}OCH_3} \longrightarrow \underset{\text{Amide}}{R'\overset{\overset{O}{\|}}{C}NR_2} + \underset{\text{Methanol}}{CH_3OH}$$

Problem 19.20

Write an equation showing the preparation of the following amides from the indicated carboxylic acid derivative:

(a) $(CH_3)_2CH\overset{\overset{O}{\|}}{C}NH_2$ from an acyl chloride

(b) $CH_3\overset{\overset{O}{\|}}{C}NHCH_3$ from an acid anhydride

(c) $H\overset{\overset{O}{\|}}{C}N(CH_3)_2$ from a methyl ester

Sample Solution (a) Amides of the type $R\overset{\overset{O}{\|}}{C}NH_2$ are derived by acylation of ammonia.

$$\underset{\text{2-Methylpropanoyl chloride}}{(CH_3)_2CH\overset{\overset{O}{\|}}{C}Cl} + \underset{\text{Ammonia}}{2NH_3} \longrightarrow \underset{\text{2-Methylpropanamide}}{(CH_3)_2CH\overset{\overset{O}{\|}}{C}NH_2} + \underset{\text{Ammonium chloride}}{NH_4Cl}$$

Two molecules of ammonia are needed because its acylation produces, in addition to the desired amide, a molecule of hydrogen chloride. Hydrogen chloride (an acid) reacts with ammonia (a base) to give ammonium chloride.

All these reactions proceed by nucleophilic addition of the amine to the carbonyl group. Dissociation of the tetrahedral intermediate proceeds in the direction that leads to an amide.

$$\underset{\text{Acylating agent}}{R\overset{\overset{:O:}{\|}}{C}X} + \underset{\text{Amine}}{R'_2\ddot{N}H} \rightleftharpoons \underset{\text{Tetrahedral intermediate}}{R\overset{\overset{:\ddot{O}-H}{|}}{\underset{\underset{:NR'_2}{|}}{C}}-X} \rightleftharpoons \underset{\text{Amide}}{R\overset{\overset{:O:}{\|}}{C}\ddot{N}R'_2} + \underset{\text{Conjugate acid of leaving group}}{HX}$$

The carbonyl group of an amide is stabilized to a greater extent than that of an acyl chloride, acid anhydride, or ester; amides are formed rapidly and in high yield from each of these carboxylic acid derivatives.

Amides are sometimes prepared directly from carboxylic acids and amines by a two-step process. The first step is an acid–base reaction in which the acid and the amine combine to form an ammonium carboxylate salt. On heating, the ammonium carboxylate salt loses water to form an amide.

$$\underset{\text{Carboxylic acid}}{R\overset{\overset{O}{\|}}{C}OH} + \underset{\text{Amine}}{R'_2NH} \longrightarrow \underset{\text{Ammonium carboxylate salt}}{R\overset{\overset{O}{\|}}{C}O^- \ R'_2\overset{+}{N}H_2} \xrightarrow{\text{heat}} \underset{\text{Amide}}{R\overset{\overset{O}{\|}}{C}NR'_2} + \underset{\text{Water}}{H_2O}$$

In practice, both steps may be combined in a single operation by simply heating a carboxylic acid and an amine together:

$$C_6H_5\overset{O}{\overset{\|}{C}}OH + C_6H_5NH_2 \xrightarrow{225°C} C_6H_5\overset{O}{\overset{\|}{C}}NHC_6H_5 + H_2O$$

Benzoic acid | Aniline | *N*-Phenylbenzamide (80–84%) | Water

These thermal methods for preparing amides are limited in their generality. Most often amides are prepared in the laboratory from acyl chlorides, acid anhydrides, or esters, and these are the methods that you should apply to solving synthetic problems.

Section 25.17 describes specialized reagents used in peptide synthesis that bring about the formation of amides directly from amines and carboxylic acids.

19.15 Hydrolysis of Amides

Amides are the least reactive carboxylic acid derivative, and the only nucleophilic acyl substitution reaction they undergo is hydrolysis. Amides are fairly stable in water, but the amide bond is cleaved on heating in the presence of strong acids or bases. Nominally, this cleavage produces an amine and a carboxylic acid. In acid, however, the amine is protonated, giving an ammonium ion:

$$R\overset{O}{\overset{\|}{C}}\ddot{N}R'_2 + H_3O^+ \longrightarrow R\overset{O}{\overset{\|}{C}}OH + R'_2\overset{+}{N}H_2$$

Amide | Hydronium ion | Carboxylic acid | Ammonium ion

In base the carboxylic acid is deprotonated, giving a carboxylate ion:

$$R\overset{O}{\overset{\|}{C}}\ddot{N}R'_2 + HO^- \longrightarrow R\overset{O}{\overset{\|}{C}}O^- + R'_2\ddot{N}H$$

Amide | Hydroxide ion | Carboxylate ion | Amine

The acid–base reactions that occur after the amide bond is broken make the overall hydrolysis irreversible in both cases. The amine product is protonated in acid; the carboxylic acid is deprotonated in base.

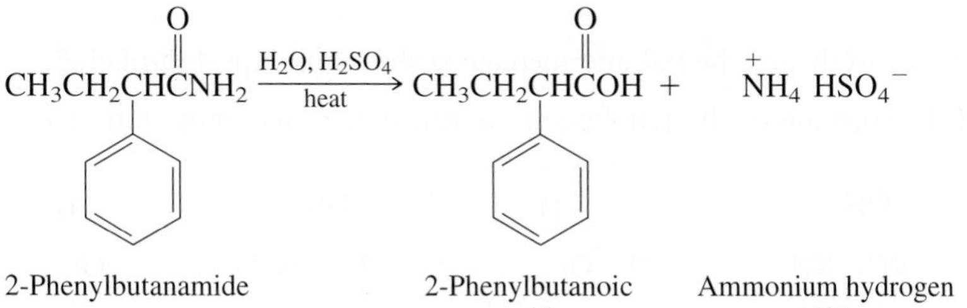

CH3CH2CHCNH2
H2O, H2SO4
heat
CH3CH2CHCOH +
NH4 HSO4

2-Phenylbutanamide | 2-Phenylbutanoic acid (88–90%) | Ammonium hydrogen sulfate

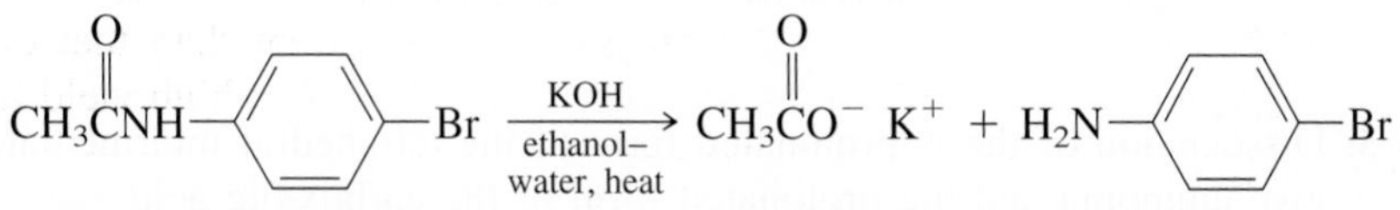

CH3CNH
Br
KOH
ethanol–
water, heat
CH3CO⁻ K⁺ + H2N
Br

N-(4-Bromophenyl)acetamide (*p*-bromoacetanilide) | Potassium acetate | *p*-Bromoaniline (95%)

Mechanistically, amide hydrolysis is similar to the hydrolysis of other carboxylic acid derivatives. The mechanism of hydrolysis in acid is presented in Mechanism 19.6. It proceeds in two stages; a tetrahedral intermediate is formed in the first stage and dissociates in the second.

The amide is activated toward nucleophilic attack by protonation of its carbonyl oxygen. The cation produced in this step is stabilized by resonance involving the nitrogen

Mechanism 19.6

Amide Hydrolysis in Acid Solution

Steps 1 through 3 show the formation of the tetrahedral intermediate.

Step 1: Protonation of the carbonyl oxygen of the amide

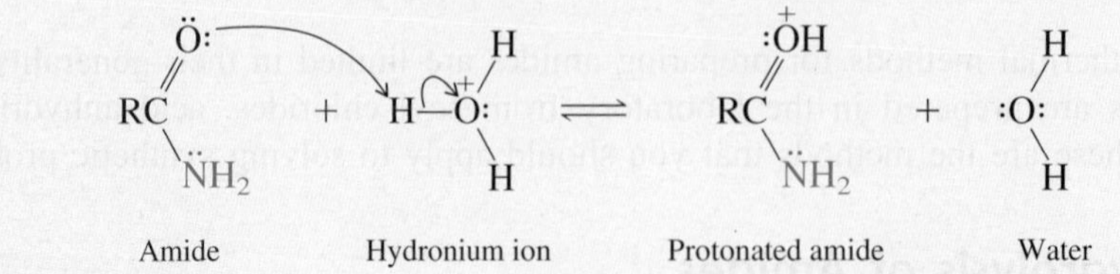
Amide
Hydronium ion
Protonated amide
Water

Step 2: Nucleophilic addition of water to the protonated amide

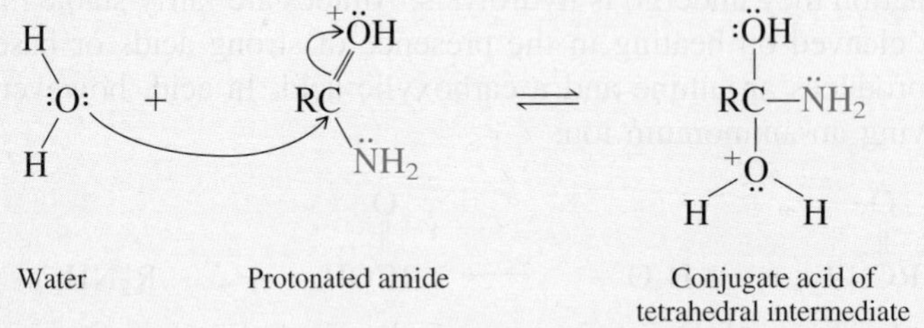
Water
Protonated amide
Conjugate acid of tetrahedral intermediate

Step 3: Deprotonation of the oxonium ion to give the neutral form of the tetrahedral intermediate

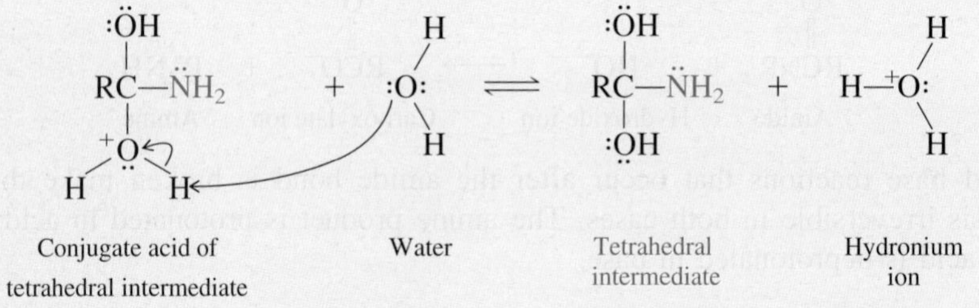
Conjugate acid of tetrahedral intermediate
Water
Tetrahedral intermediate
Hydronium ion

Dissociation of the tetrahedral intermediate is shown in steps 4 through 6.

Step 4: Protonation of the tetrahedral intermediate at its amino nitrogen

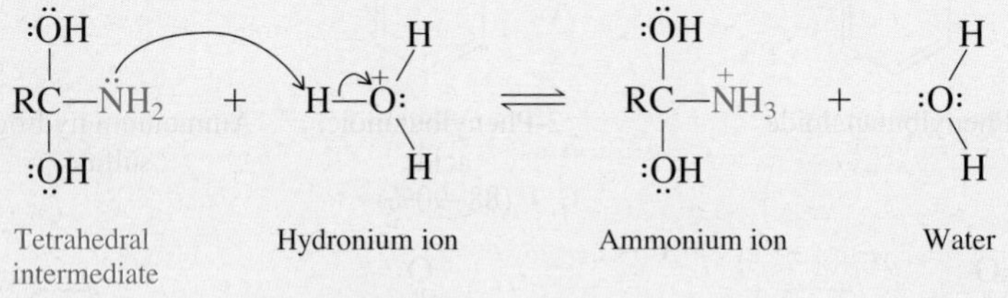
Tetrahedral intermediate
Hydronium ion
Ammonium ion
Water

Step 5: Dissociation of the *N*-protonated form of the tetrahedral intermediate to give ammonia and the protonated form of the carboxylic acid

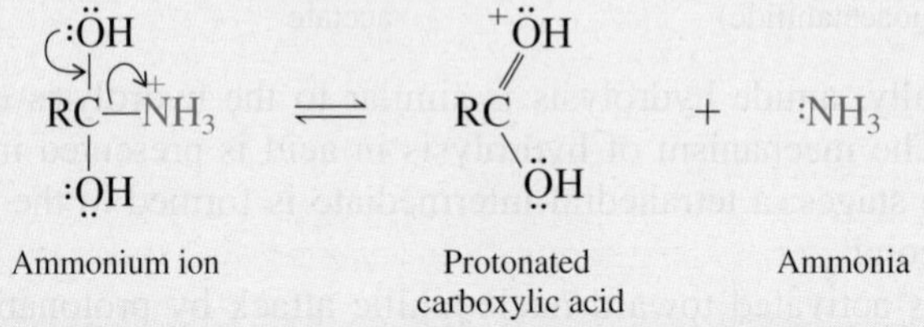
Ammonium ion
Protonated carboxylic acid
Ammonia

Step 6: Proton transfer processes yielding ammonium ion and the carboxylic acid

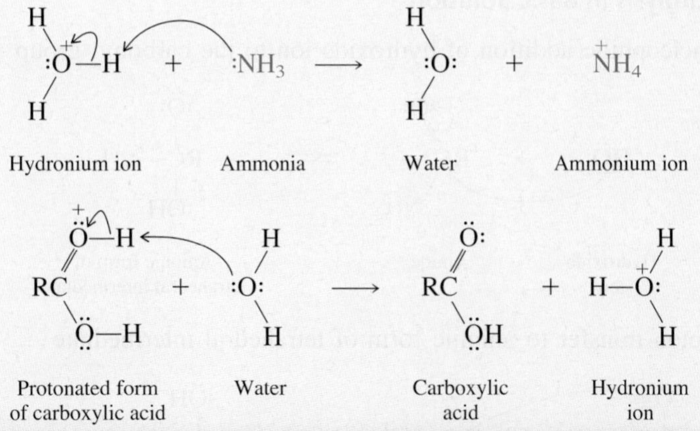
Hydronium ion
Ammonia
Water
Ammonium ion
Protonated form of carboxylic acid
Water
Carboxylic acid
Hydronium ion

lone pair and is more stable than the intermediate in which the amide nitrogen is protonated.

Protonation of carbonyl oxygen ***Protonation of amide nitrogen***

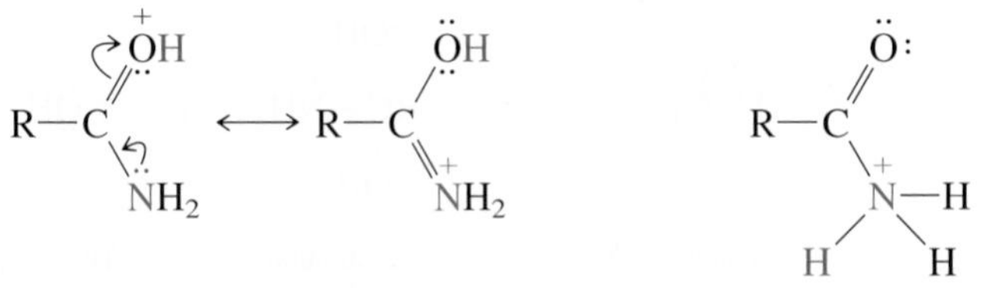

Most stable resonance contributors of an *O*-protonated amide

An acylammonium ion; the positive charge is localized on nitrogen

Once formed, the *O*-protonated intermediate is attacked by a water molecule in step 2. The intermediate formed in this step loses a proton in step 3 to give the neutral form of the tetrahedral intermediate. The tetrahedral intermediate has its amino group ($—NH_2$) attached to an sp^3-hybridized carbon, and this amino group is the site at which protonation occurs in step 4. Cleavage of the carbon–nitrogen bond in step 5 yields the protonated form of the carboxylic acid, along with a molecule of ammonia. In acid solution ammonia is immediately protonated to give ammonium ion, as shown in step 6.

The protonation of ammonia in step 6 has such a large equilibrium constant that it makes the overall reaction irreversible.

Problem 19.21

On the basis of the general mechanism for amide hydrolysis in acidic solution shown in Mechanism 19.6, write an analogous sequence of steps for the hydrolysis of acetanilide, $CH_3\overset{\overset{O}{\|}}{C}NHC_6H_5$.

In base the tetrahedral intermediate is formed in a manner analogous to that proposed for ester saponification. Steps 1 and 2 in Mechanism 19.7 show the formation of the tetrahedral intermediate in the basic hydrolysis of amides. In step 3 the basic amino group of the tetrahedral intermediate abstracts a proton from water, and in step 4 the derived ammonium ion dissociates. Conversion of the carboxylic acid to its corresponding carboxylate anion in step 5 completes the process and renders the overall reaction irreversible.

Mechanism 19.7

Amide Hydrolysis in Basic Solution

Step 1: Nucleophilic addition of hydroxide ion to the carbonyl group

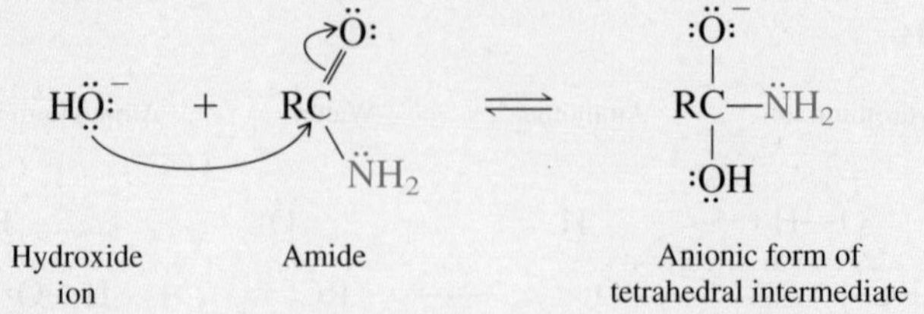
Hydroxide ion
Amide
Anionic form of tetrahedral intermediate

Step 2: Proton transfer to anionic form of tetrahedral intermediate

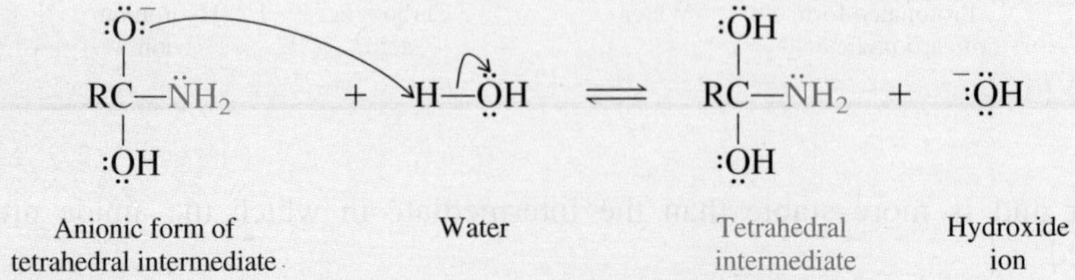
Anionic form of tetrahedral intermediate
Water
Tetrahedral intermediate
Hydroxide ion

Step 3: Protonation of amino nitrogen of tetrahedral intermediate

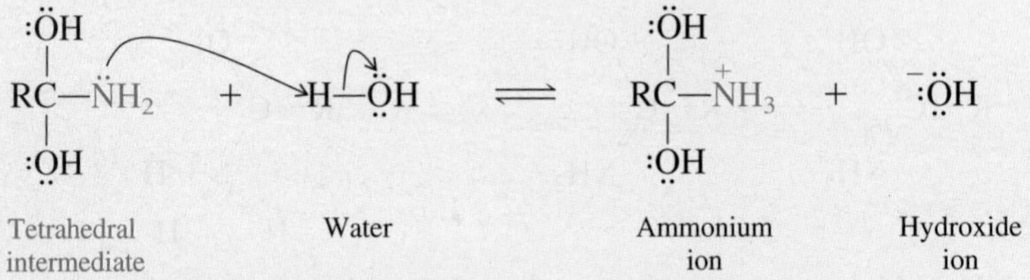
Tetrahedral intermediate
Water
Ammonium ion
Hydroxide ion

Step 4: Dissociation of *N*-protonated form of tetrahedral intermediate

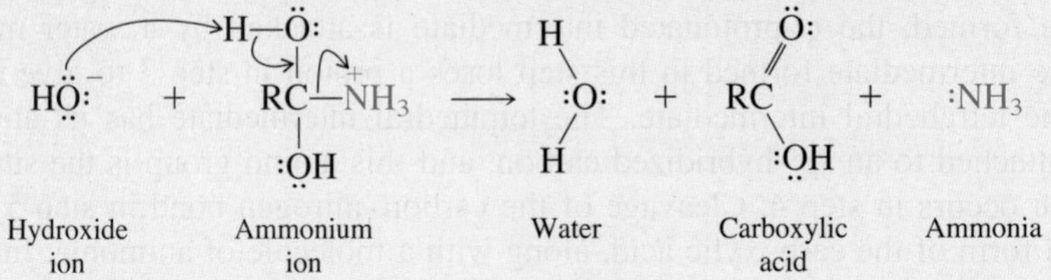
Hydroxide ion
Ammonium ion
Water
Carboxylic acid
Ammonia

Step 5: Irreversible formation of carboxylate anion

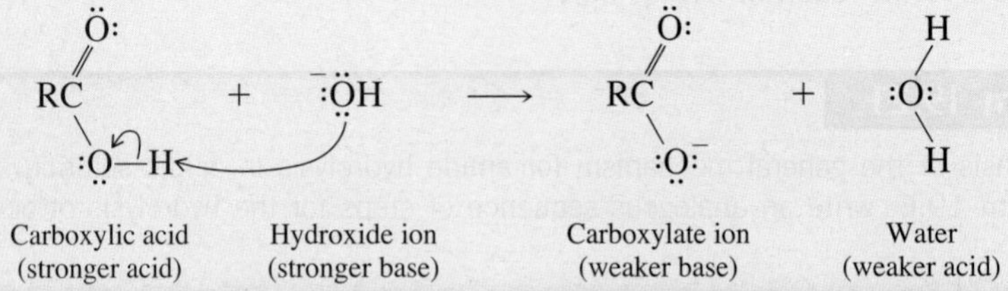
Carboxylic acid (stronger acid)
Hydroxide ion (stronger base)
Carboxylate ion (weaker base)
Water (weaker acid)

Problem 19.22

On the basis of the general mechanism for basic hydrolysis shown in Mechanism 19.7, write an analogous sequence for the hydrolysis of *N,N*-dimethylformamide, $HC(=O)N(CH_3)_2$.

19.16 Lactams

Lactams are cyclic amides and are analogous to lactones, which are cyclic esters. Most lactams are known by their common names, as the examples shown illustrate.

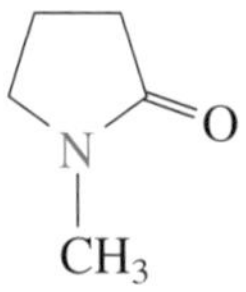

N-Methylpyrrolidone
(a polar aprotic solvent)

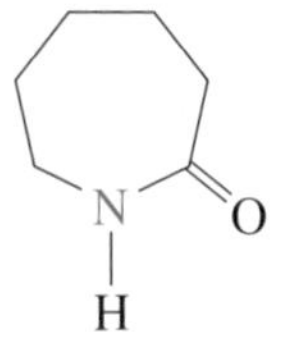

ε-Caprolactam
(industrial chemical used to prepare a type of nylon)

Just as amides are more stable than esters, lactams are more stable than lactones. Thus, although β-lactones are rare (Section 18.15), β-lactams are among the best known products of the pharmaceutical industry. The penicillin and cephalosporin antibiotics, which are so useful in treating bacterial infections, are β-lactams and are discussed in the boxed essay accompanying this section.

β-Lactam Antibiotics

It may never be known just how spores of *Penicillium notatum* found their way to a Petri dish containing *Staphylococcus* in Alexander Fleming's laboratory at St. Mary's Hospital in London during the summer of 1928. But they did, and the mold they produced made a substance that stopped the *Staphylococcus* colony from growing. His curiosity aroused, Fleming systematically challenged the substance he called "penicillin" with other bacteria and, in addition to *Staphylococcus,* found impressive activity against *Streptococcus* as well as the bacteria that cause diphtheria, meningitis, and pneumonia. Fleming published his findings in 1929, but his efforts to isolate the active substance responsible for penicillin's antibacterial properties were unsuccessful.

By 1938, Fleming had moved on to other research, and Howard Florey and Ernst Chain of the School of Pathology at Oxford were just beginning a program aimed at developing antibacterial agents from natural sources. A candidate that especially appealed to them was Fleming's penicillin.

Their most daunting initial problem was making enough penicillin. Enter Norman Heatley, of whom it has been said; ". . . without Heatley, no penicillin."* Heatley, an inventive and careful experimentalist, devised procedures to make and isolate penicillin on a scale sufficient to begin testing. By 1941, Florey, Chain, and Heatley had a drug that was both effective and safe.

England was at war, and the United States soon would be; the need for large amounts of penicillin was obvious. Working with U.S. Department of Agriculture, Heatley and Andrew J. Moyer of the USDA laboratories in Peoria, Illinois, found better penicillium sources and developed novel fermentation methods to produce ever-increasing amounts of penicillin. Treatment of wounded soldiers with penicillin became possible early in 1943 and was widely practiced before the war ended in August 1945. Four months later, Fleming, Florey, and Chain traveled to Stockholm to accept that year's Nobel Prize for Physiology or Medicine. Heatley was not included because custom dictates that a Nobel Prize can be divided among no more than three persons.

The structure of penicillin is unusual because it contains an amide function as part of a four-membered ring (a **β-lactam**). Various penicillins differ in respect to substituent groups and their effectiveness against different strains of bacteria. Penicillin G originated in a strain obtained from a rotting cantaloupe in Peoria and was the first penicillin made on a large scale (Figure 19.4). Fleming's original penicillin (now called penicillin F) bears a

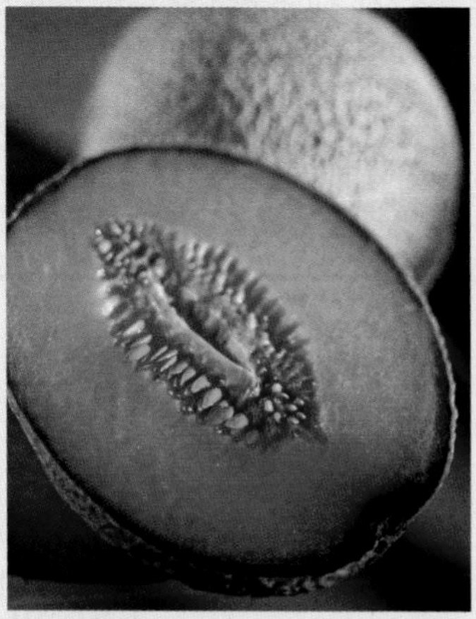

Figure 19.4

Penicillin G originated in a mold (*Penicillium chrysogenum*) found on a cantaloupe in Peoria, Illinois.

*H. Harris, "The Florey Centenary Lecture and the Development of Penicillin," as quoted in E. Lax, *The Mold in Dr. Florey's Coat,* Henry Holt and Company, New York, 2004, page 89.

Continued

$CH_3CH_2CH{=}CHCH_2$—group in place of the $C_6H_5CH_2$—of penicillin G. A different class of β-lactam antibiotics, the cephalosporins, are similar in structure to the penicillins but have a six-membered instead of a five-membered sulfur-containing ring.

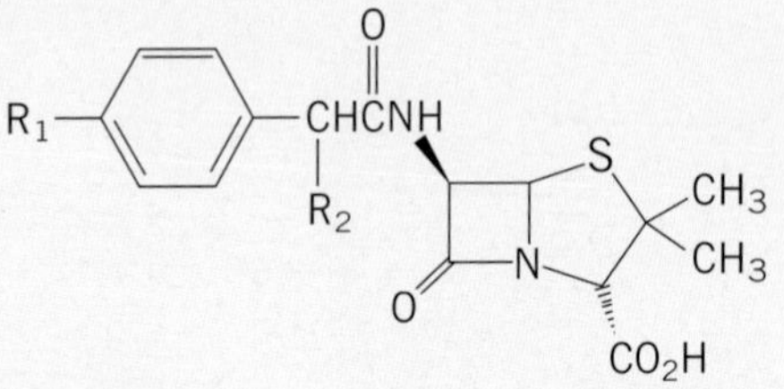

Penicillin G: $R_1 = H$; $R_2 = H$

Ampicillin: $R_1 = H$; $R_2 = NH_2$

Amoxicillin: $R_1 = OH$; $R_2 = NH_2$

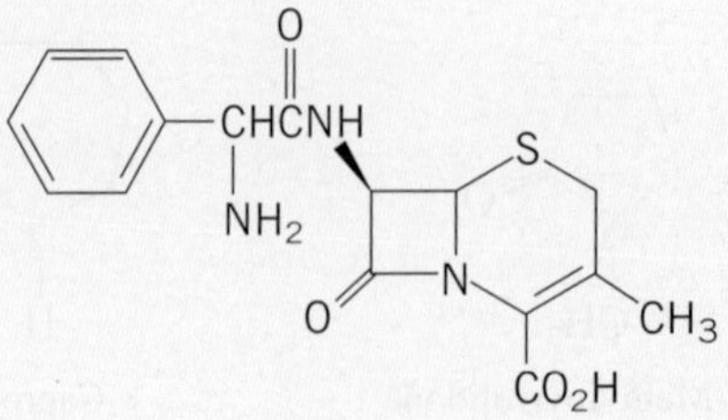

Cephalexin

Although their strained four-membered ring makes β-lactam antibiotics susceptible to hydrolysis, this same elevated reactivity toward nucleophilic acyl substitution is responsible for their antibacterial properties. β-Lactams act by deactivating an enzyme, *transpeptidase,* required for the biosynthesis of bacterial cell walls. The active site of transpeptidase contains a key hydroxyl group, which is converted to an ester by a nucleophilic acyl substitution that cleaves the β-lactam ring. With the acylated form of the enzyme unable to catalyze cell-wall biosynthesis, further bacterial growth is brought under control, and the body's immune system does the rest.

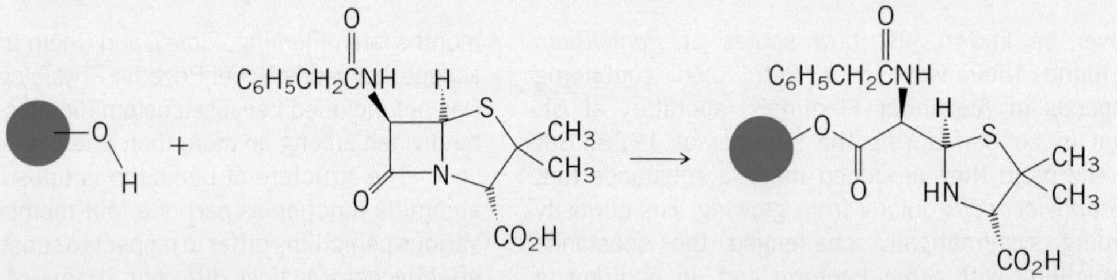

Active form of transpeptidase Penicillin G Inactive ester of transpeptidase

Problem 19.23

(a) Penicillin-resistant strains of bacteria contain *β-lactamases,* enzymes that catalyze the hydrolysis of a penicillin before the penicillin can acylate *transpeptidase*. Suggest a reasonable structure for the product $C_{16}H_{20}N_2O_5S$ formed by β-lactamase-catalyzed hydrolysis of penicillin G.

(b) 6-Aminopenicillanic acid ($C_8H_{12}N_2O_3S$), a key compound in the preparation of "semisynthetic" penicillins, is prepared from penicillin G by *penicillin acyl transferase*-catalyzed hydrolysis. Suggest a reasonable structure for 6-aminopenicillanic acid.

19.17 Preparation of Nitriles

We have already discussed two procedures by which nitriles are prepared, namely, nucleophilic substitution of alkyl halides by cyanide and conversion of aldehydes and ketones to cyanohydrins. Table 19.6 reviews aspects of these reactions. Neither of the reactions in Table 19.6 is suitable for aryl nitriles ($ArC{\equiv}N$); these compounds are normally prepared by a reaction to be discussed in Section 21.17.

Both alkyl and aryl nitriles are accessible by dehydration of amides.

$$\underset{\substack{\text{Amide}\\ \text{(R may be alkyl}\\ \text{or aryl)}}}{RC(=O)NH_2} \longrightarrow \underset{\substack{\text{Nitrile}\\ \text{(R may be alkyl}\\ \text{or aryl)}}}{RC{\equiv}N} + \underset{\text{Water}}{H_2O}$$

TABLE 19.6 Preparation of Nitriles

Reaction (section) and comments	General equation and specific example
Nitrophilic substitution by cyanide ion (Sections 8.1, 8.11) Cyanide ion is a good nucleophile and reacts with alkyl halides to give nitriles. The reaction is of the S_N2 type and is limited to primary and secondary alkyl halides. Tertiary alkyl halides undergo elimination; aryl and vinyl halides do not react.	$:N{\equiv}\bar{C}: + R{-}X \longrightarrow RC{\equiv}N + X^-$ Cyanide ion, Alkyl halide, Nitrile, Halide ion $CH_3(CH_2)_8CH_2Cl \xrightarrow[\text{ethanol–water}]{KCN} CH_3(CH_2)_8CH_2CN$ 1-Chlorodecane → Undecanenitrile (95%)
Cyanohydrin formation (Section 17.7) Hydrogen cyanide adds to the carbonyl group of aldehydes and ketones.	$RC(=O)R' + HCN \longrightarrow RC(OH)(C{\equiv}N)R'$ Aldehyde or ketone, Hydrogen cyanide, Cyanohydrin $CH_3CH_2C(=O)CH_2CH_3 \xrightarrow[H_3O^+]{KCN} CH_3CH_2C(OH)(CN)CH_2CH_3$ 3-Pentanone → 3-Pentanone cyanohydrin (75%)

Among the reagents used for this dehydration is P_4O_{10}, known by the common name *phosphorus pentoxide* because it was once thought to have the molecular formula P_2O_5. Phosphorus pentoxide is the anhydride of phosphoric acid and is used in a number of reactions requiring dehydrating agents.

$$(CH_3)_2CHC(=O)NH_2 \xrightarrow[200°C]{P_4O_{10}} (CH_3)_2CHC{\equiv}N$$

2-Methylpropanamide → 2-Methylpropanenitrile (69–86%)

Problem 19.24

Show how ethyl alcohol could be used to prepare (a) CH_3CN and (b) CH_3CH_2CN. Along with ethyl alcohol you may use any necessary inorganic reagents.

19.18 Hydrolysis of Nitriles

Nitriles are classified as carboxylic acid derivatives because they are converted to carboxylic acids on hydrolysis. The conditions required are similar to those for the hydrolysis of amides, namely, heating in aqueous acid or base for several hours. Like the hydrolysis of amides, nitrile hydrolysis is irreversible in the presence of acids or bases. Acid hydrolysis yields ammonium ion and a carboxylic acid.

$$RC{\equiv}N + H_2O + H_3O^+ \longrightarrow RC(=O)OH + \overset{+}{N}H_4$$

Nitrile, Water, Hydronium ion, Carboxylic acid, Ammonium ion

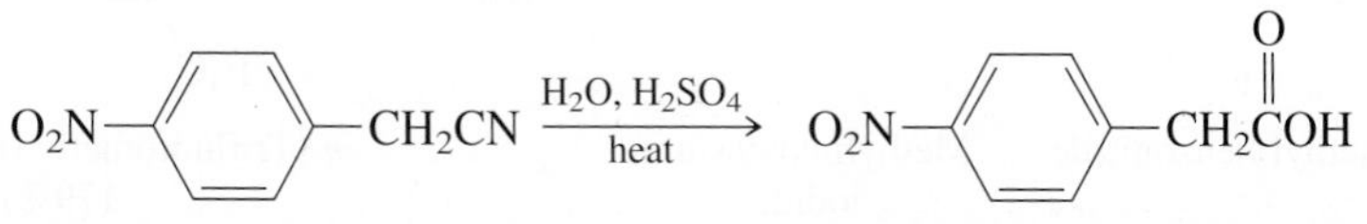

O₂N
CH₂CN
H₂O, H₂SO₄
heat
O₂N
O
CH₂COH

p-Nitrobenzyl cyanide → *p*-Nitrophenylacetic acid (92–95%)

In aqueous base, hydroxide ion abstracts a proton from the carboxylic acid. Isolating the acid requires a subsequent acidification step.

$$\underset{\text{Nitrile}}{RC{\equiv}N} + \underset{\text{Water}}{H_2O} + \underset{\text{Hydroxide ion}}{HO^-} \longrightarrow \underset{\text{Carboxylate ion}}{R\overset{O}{\overset{\|}{C}}O^-} + \underset{\text{Ammonia}}{NH_3}$$

$$\underset{\text{Undecanenitrile}}{CH_3(CH_2)_9CN} \xrightarrow[2.\ H_3O^+]{1.\ KOH,\ H_2O,\ heat} \underset{\text{Undecanoic acid (80\%)}}{CH_3(CH_2)_9\overset{O}{\overset{\|}{C}}OH}$$

The first four steps of the mechanism for hydrolysis of nitriles in basic solution are given in Mechanism 19.8. These steps convert the nitrile to an amide, which then proceeds to the hydrolysis products according to the mechanism of amide hydrolysis in Mechanism 19.7 (page 846).

The acid-catalyzed mechanism for nitrile hydrolysis also goes through the amide as an intermediate. Problem 19.25 encourages you to propose a mechanism for that process.

Problem 19.25

Suggest a reasonable mechanism for the conversion of a nitrile (RCN) to the corresponding amide in aqueous acid.

Nucleophiles other than water can also add to the carbon–nitrogen triple bond of nitriles. In the following section we will see a synthetic application of such a nucleophilic addition.

19.19 Addition of Grignard Reagents to Nitriles

The carbon–nitrogen triple bond of nitriles is much less reactive toward nucleophilic addition than is the carbon–oxygen double bond of aldehydes and ketones. Strongly basic nucleophiles such as Grignard reagents, however, do react with nitriles in a reaction that is of synthetic value:

$$\underset{\text{Nitrile}}{RC{\equiv}N} + \underset{\text{Grignard reagent}}{R'MgX} \xrightarrow[2.\ H_3O^+]{1.\ diethyl\ ether} \underset{\text{Imine}}{R\overset{NH}{\overset{\|}{C}}R'} \xrightarrow[heat]{H_3O^+} \underset{\text{Ketone}}{R\overset{O}{\overset{\|}{C}}R'}$$

The imine formed by nucleophilic addition of the Grignard reagent to the nitrile is normally not isolated but is hydrolyzed directly to a ketone. The overall sequence is used as a means of preparing ketones.

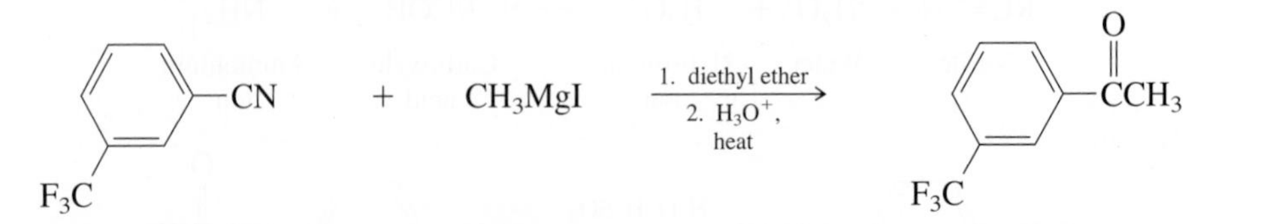

m-(Trifluoromethyl)benzonitrile — Methylmagnesium iodide — *m*-(Trifluoromethyl)acetophenone (79%)

Mechanism 19.8

Nitrile Hydrolysis in Basic Solution

THE OVERALL REACTION: Nitriles are hydrolyzed in base to give ammonia and a carboxylate ion. An amide is an intermediate.

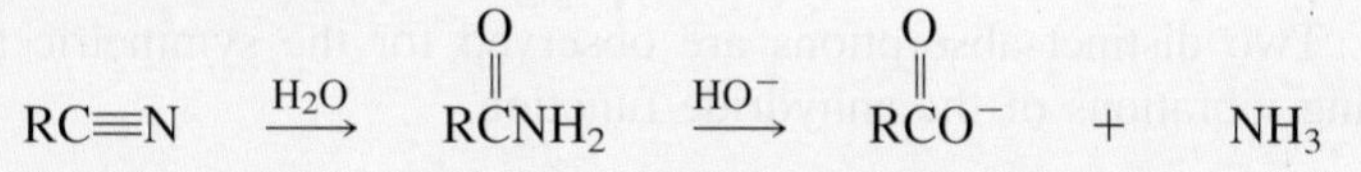

Nitrile Amide Carboxylate ion Ammonia

Step 1: Hydroxide adds to the carbon–nitrogen triple bond. This step is analogous to nucleophilic addition to a carbonyl group.

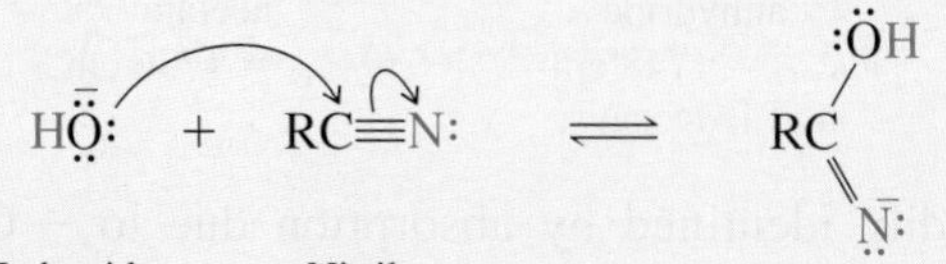

Hydroxide Nitrile

Step 2: The product of step 1 is the conjugate base of an imino acid to which it is converted by proton abstraction from water.

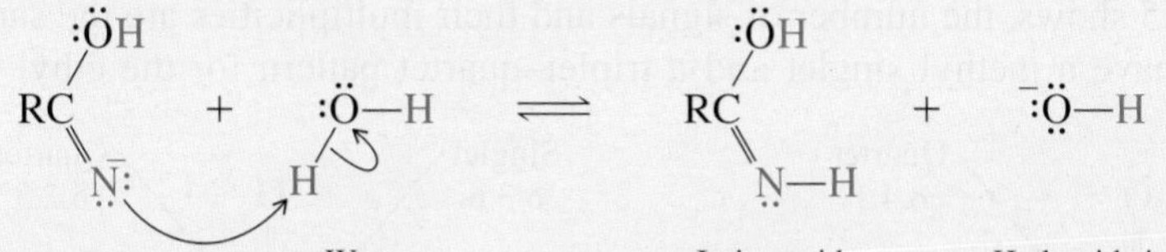

Water Imino acid Hydroxide ion

Step 3: Proton abstraction from oxygen of the imino acid gives the conjugate base of an amide.

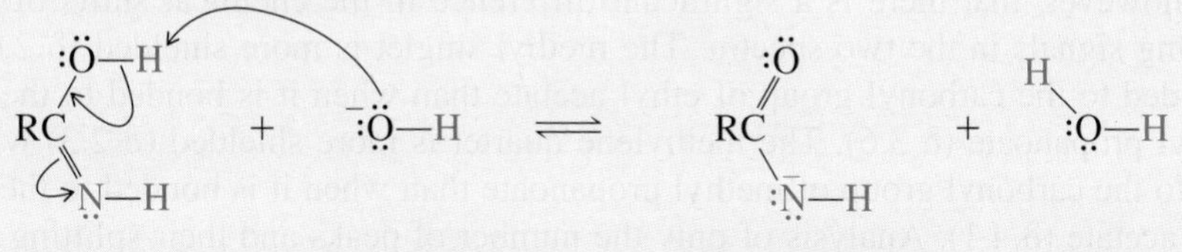

Imino acid Hydroxide ion Conjugate base of amide Water

Step 4: The conjugate base of the amide abstracts a proton from water.

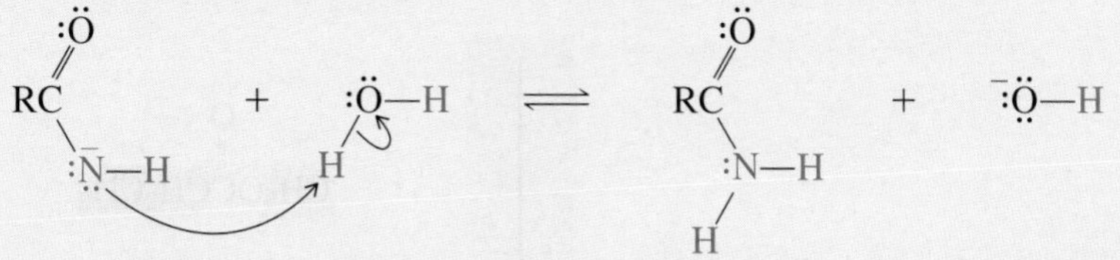

Conjugate base of amide Water Amide Hydroxide ion

The amide formed in this step then undergoes basic hydrolysis according to the process shown in Mechanism 19.7.

Problem 19.26

Write an equation showing how you could prepare ethyl phenyl ketone from propanenitrile and a Grignard reagent. What is the structure of the imine intermediate?

Organolithium reagents react in the same way and are often used instead of Grignard reagents.

19.20 Spectroscopic Analysis of Carboxylic Acid Derivatives

Infrared: IR spectroscopy is quite useful in identifying carboxylic acid derivatives. The carbonyl stretching vibration is very strong, and its position is sensitive to the nature of the carbonyl group. In general, electron donation from the substituent decreases the double-bond character of the bond between carbon and oxygen and decreases the stretching frequency. Two distinct absorptions are observed for the symmetric and antisymmetric stretching vibrations of the anhydride function.

$CH_3\overset{O}{\overset{\|}{C}}Cl$	$CH_3\overset{O}{\overset{\|}{C}}O\overset{O}{\overset{\|}{C}}CH_3$	$CH_3\overset{O}{\overset{\|}{C}}OCH_3$	$CH_3\overset{O}{\overset{\|}{C}}NH_2$
Acetyl chloride	Acetic anhydride	Methyl acetate	Acetamide
$\nu_{C=O} = 1822\ cm^{-1}$	$\nu_{C=O} = 1748\ cm^{-1}$ and $1815\ cm^{-1}$	$\nu_{C=O} = 1736\ cm^{-1}$	$\nu_{C=O} = 1694\ cm^{-1}$

Nitriles are readily identified by absorption due to —C≡N stretching in the 2210–2260 cm^{-1} region.

¹H NMR: Chemical-shift differences in their ^1H NMR spectra aid the structure determination of esters. Consider the two isomeric esters: ethyl acetate and methyl propanoate. As Figure 19.5 shows, the number of signals and their multiplicities are the same for both esters. Both have a methyl singlet and a triplet–quartet pattern for the ethyl group.

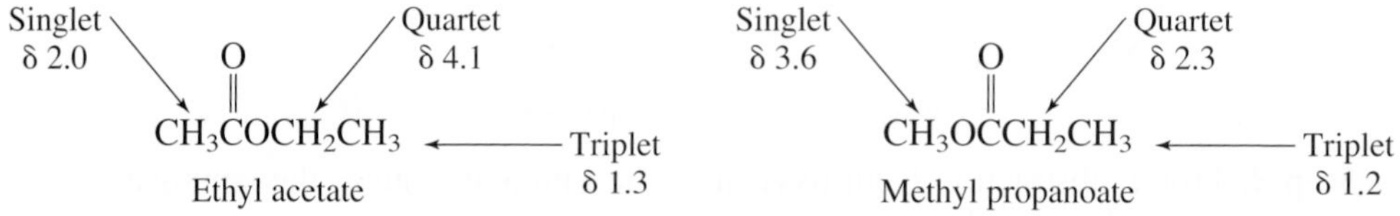
Singlet
δ 2.0
Quartet
δ 4.1
O
CH₃COCH₂CH₃
Triplet
δ 1.3
Ethyl acetate
Singlet
δ 3.6
Quartet
δ 2.3
O
CH₃OCCH₂CH₃
Triplet
δ 1.2
Methyl propanoate

Notice, however, that there is a significant difference in the chemical shifts of the corresponding signals in the two spectra. The methyl singlet is more shielded (δ 2.0) when it is bonded to the carbonyl group of ethyl acetate than when it is bonded to the oxygen of methyl propanoate (δ 3.6). The methylene quartet is more shielded (δ 2.3) when it is bonded to the carbonyl group of methyl propanoate than when it is bonded to the oxygen of ethyl acetate (δ 4.1). Analysis of only the number of peaks and their splitting patterns does not provide an unambiguous answer to structure assignment in esters; chemical-shift data such as that just described must also be considered.

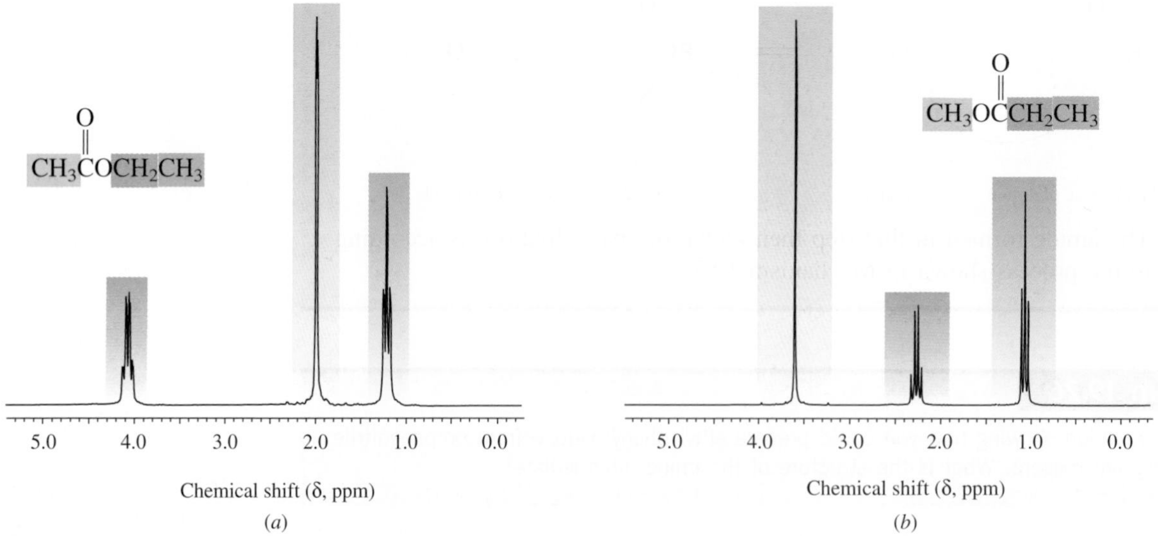
O
CH₃COCH₂CH₃
5.0
4.0
3.0
2.0
1.0
0.0
Chemical shift (δ, ppm)
(a)
O
CH₃OCCH₂CH₃
5.0
4.0
3.0
2.0
1.0
0.0
Chemical shift (δ, ppm)
(b)

Figure 19.5

The 200-MHz ^1H NMR spectra of (*a*) ethyl acetate and (*b*) methyl propanoate.

The chemical shift of the N—H proton of amides appears in the range δ 5–8. It is often a very broad peak; sometimes it is so broad that it does not rise much over the baseline and is easily overlooked.

^{13}C NMR: The ^{13}C NMR spectra of carboxylic acid derivatives, like the spectra of carboxylic acids themselves, are characterized by a low-field resonance for the carbonyl carbon in the range δ 160–180. The carbonyl carbons of carboxylic acid derivatives are more shielded than those of aldehydes and ketones, but less shielded than the sp^2-hybridized carbons of alkenes and arenes.

The carbon of a C≡N group appears near δ 120.

UV-VIS: The following values are typical for the $n \rightarrow \pi^*$ absorption associated with the C=O group of carboxylic acid derivatives.

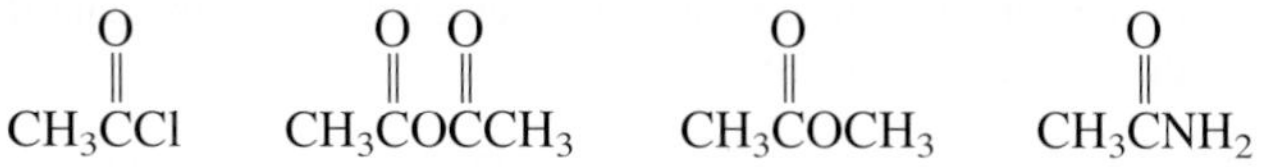

	Acetyl chloride	Acetic anhydride	Methyl acetate	Acetamide
λ_{max}	235 nm	225 nm	207 nm	214 nm

Mass Spectrometry: A prominent peak in the mass spectra of most carboxylic acid derivatives corresponds to an acylium ion derived by cleavage of the bond to the carbonyl group:

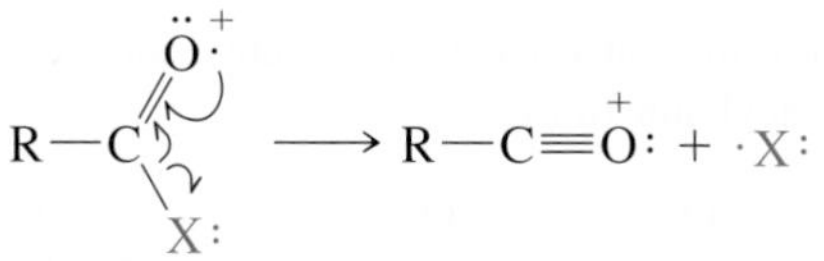

Amides, however, tend to cleave in the opposite direction to produce a nitrogen-stabilized acylium ion:

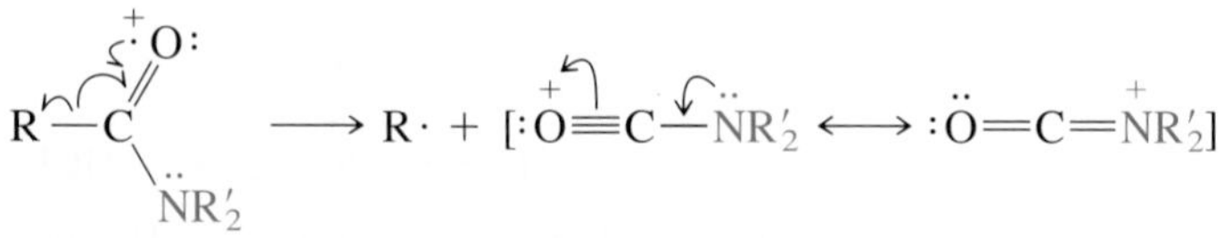

19.21 SUMMARY

Section 19.1 This chapter concerns the preparation and reactions of *acyl chlorides, acid anhydrides, esters, amides,* and *nitriles.* These compounds are generally classified as carboxylic acid derivatives, and their nomenclature is based on that of carboxylic acids.

Acyl chloride	Acid anhydride	Ester	Amide	Nitrile

Section 19.2 The structure and reactivity of carboxylic acid derivatives depend on how well the atom bonded to the carbonyl group donates electrons to it.

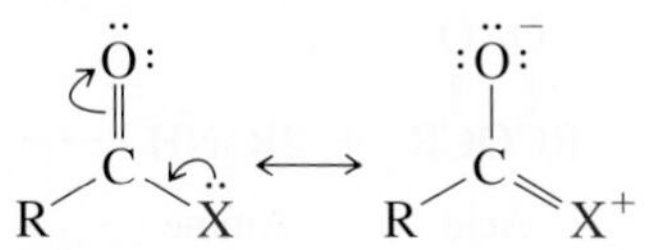

Electron-pair donation stabilizes the carbonyl group and makes it less reactive toward nucleophilic acyl substitution.

Most reactive — Least reactive

$$\mathrm{R\overset{O}{\overset{\|}{C}}Cl} > \mathrm{R\overset{O}{\overset{\|}{C}}O\overset{O}{\overset{\|}{C}}R} > \mathrm{R\overset{O}{\overset{\|}{C}}OR'} > \mathrm{R\overset{O}{\overset{\|}{C}}NR'_2}$$

Least stabilized carbonyl group — Most stabilized carbonyl group

Nitrogen is a better electron-pair donor than oxygen, and amides have a more stabilized carbonyl group than esters and anhydrides. Chlorine is the poorest electron-pair donor, and acyl chlorides have the least stabilized carbonyl group and are the most reactive.

Section 19.3 The characteristic reaction of acyl chlorides, acid anhydrides, esters, and amides is **nucleophilic acyl substitution.** Addition of a nucleophilic reagent :Nu—H to the carbonyl group leads to a tetrahedral intermediate that dissociates to give the product of substitution:

$$\mathrm{R\overset{O}{\overset{\|}{C}}{-}X} + \mathrm{:Nu{-}H} \rightleftharpoons \mathrm{R\overset{OH}{\underset{:Nu}{C}}{-}X} \rightleftharpoons \mathrm{R\overset{O}{\overset{\|}{C}}{-}\ddot{N}u} + \mathrm{HX:}$$

Carboxylic acid derivative	Nucleophile	Tetrahedral intermediate	Product of nucleophilic acyl substitution	Conjugate acid of leaving group

Section 19.4 Acyl chlorides are converted to acid anhydrides, esters, and amides by nucleophilic acyl substitution.

$$\mathrm{R\overset{O}{\overset{\|}{C}}Cl} + \mathrm{R'\overset{O}{\overset{\|}{C}}OH} \longrightarrow \mathrm{R\overset{O}{\overset{\|}{C}}O\overset{O}{\overset{\|}{C}}R'} + \mathrm{HCl}$$

Acyl chloride	Carboxylic acid	Acid anhydride	Hydrogen chloride

$$\mathrm{R\overset{O}{\overset{\|}{C}}Cl} + \mathrm{R'OH} \longrightarrow \mathrm{R\overset{O}{\overset{\|}{C}}OR'} + \mathrm{HCl}$$

Acyl chloride	Alcohol	Ester	Hydrogen chloride

$$\mathrm{R\overset{O}{\overset{\|}{C}}Cl} + \mathrm{2R'_2NH} \longrightarrow \mathrm{R\overset{O}{\overset{\|}{C}}NR'_2} + \mathrm{R'_2\overset{+}{N}H_2\ Cl^-}$$

Acyl chloride	Amine	Amide	Ammonium chloride salt

Examples of each of these reactions may be found in Table 19.1.

Section 19.5 Acid anhydrides are less reactive toward nucleophilic acyl substitution than acyl chlorides, but are useful reagents for preparing esters and amides.

$$\mathrm{R\overset{O}{\overset{\|}{C}}O\overset{O}{\overset{\|}{C}}R} + \mathrm{R'OH} \longrightarrow \mathrm{R\overset{O}{\overset{\|}{C}}OR'} + \mathrm{R\overset{O}{\overset{\|}{C}}OH}$$

Acid anhydride	Alcohol	Ester	Carboxylic acid

$$\mathrm{R\overset{O}{\overset{\|}{C}}O\overset{O}{\overset{\|}{C}}R} + \mathrm{2R'_2NH} \longrightarrow \mathrm{R\overset{O}{\overset{\|}{C}}NR'_2} + \mathrm{R'_2\overset{+}{N}H_2\ {}^-O\overset{O}{\overset{\|}{C}}R}$$

Acid anhydride	Amine	Amide	Ammonium carboxylate salt

Table 19.2 presents examples of these reactions.

Section 19.6 Esters occur naturally or are prepared from alcohols by Fischer esterification or by acylation with acyl chlorides or acid anhydrides (Table 19.3).

Section 19.7 Esters are polar and have higher boiling points than alkanes of comparable size and shape. Esters don't form hydrogen bonds to other ester molecules so have lower boiling points than analogous alcohols. They can form hydrogen bonds to water and so are comparable to alcohols in their solubility in water.

Section 19.8 Esters give amides on reaction with ammonia and amines and are cleaved to a carboxylic acid and an alcohol on hydrolysis (Table 19.4). Esters react with Grignard reagents and are reduced by lithium aluminum hydride (Table 19.5).

Section 19.9 Ester hydrolysis can be catalyzed by acids and its mechanism (Mechanism 19.3) is the reverse of the mechanism for Fischer esterification. The reaction proceeds via a tetrahedral intermediate.

$$R-\underset{\displaystyle OH}{\overset{\displaystyle OH}{\underset{|}{\overset{|}{C}}}}-OR'$$

Tetrahedral intermediate in ester hydrolysis

Section 19.10 Ester hydrolysis in basic solution is called *saponification* and proceeds through the same tetrahedral intermediate (Mechanism 19.4) as in acid-catalyzed hydrolysis. Unlike acid-catalyzed hydrolysis, saponification is irreversible because the carboxylic acid is deprotonated under the reaction conditions.

$$\underset{\text{Ester}}{R\overset{O}{\overset{\|}{C}}OR'} + \underset{\text{Hydroxide ion}}{HO^-} \longrightarrow \underset{\text{Carboxylate ion}}{R\overset{O}{\overset{\|}{C}}O^-} + \underset{\text{Alcohol}}{R'OH}$$

Section 19.11 Esters react with amines to give amides.

$$\underset{\text{Ester}}{R\overset{O}{\overset{\|}{C}}OR'} + \underset{\text{Amine}}{R''_2NH} \longrightarrow \underset{\text{Amide}}{R\overset{O}{\overset{\|}{C}}NR''_2} + \underset{\text{Alcohol}}{R'OH}$$

Section 19.12 Esters react with two equivalents of a Grignard reagent to form tertiary alcohols.

$$\underset{\text{Methyl ester}}{R\overset{O}{\overset{\|}{C}}OCH_3} + \underset{\text{Grignard reagent}}{2R'MgX} \xrightarrow[\text{2. } H_3O^+]{\text{1. diethyl ether}} \underset{\text{Tertiary alcohol}}{R\underset{\displaystyle R'}{\overset{\displaystyle OH}{\underset{|}{\overset{|}{C}}}}-R'}$$

Section 19.13 Lithium aluminum hydride reduces esters to alcohols. Two alcohols are formed; the acyl group is reduced to the primary alcohol.

$$\underset{\text{Ester}}{R\overset{O}{\overset{\|}{C}}OR'} \longrightarrow \underset{\text{Primary alcohol}}{RCH_2OH} + \underset{\text{Alcohol}}{R'OH}$$

Section 19.14 Amides having at least one N—H unit can form intermolecular hydrogen bonds with other amide molecules. Compounds of this type have higher melting and boiling points than comparable compounds in which N—H bonds are absent.

Amides are normally prepared by the reaction of amines with acyl chlorides, anhydrides, or esters.

Section 19.15 Like ester hydrolysis, amide hydrolysis can be achieved in either aqueous acid or aqueous base. The process is irreversible in both media. In base, the carboxylic acid is converted to the carboxylate anion; in acid, the amine is protonated to an ammonium ion:

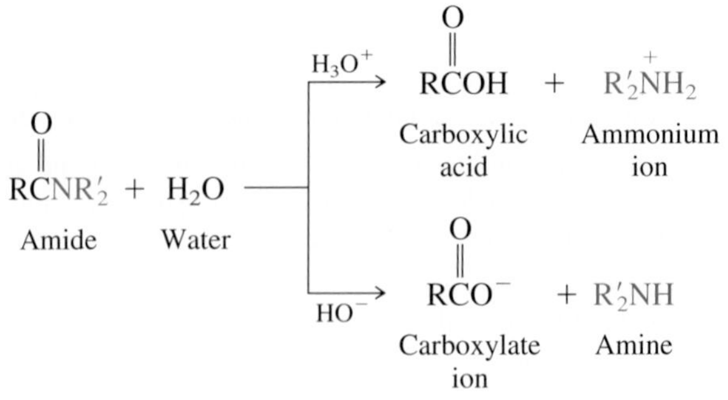
RCNR′₂ + H₂O
Amide
Water
H₃O⁺
RCOH + R′₂NH₂
Carboxylic acid
Ammonium ion
HO⁻
RCO⁻ + R′₂NH
Carboxylate ion
Amine

Section 19.16 Lactams are cyclic amides.

Section 19.17 Nitriles are prepared by nucleophilic substitution (S_N2) of alkyl halides with cyanide ion, by converting aldehydes or ketones to cyanohydrins (Table 19.6), or by dehydration of amides.

Section 19.18 The hydrolysis of nitriles to carboxylic acids is irreversible in both acidic and basic solution.

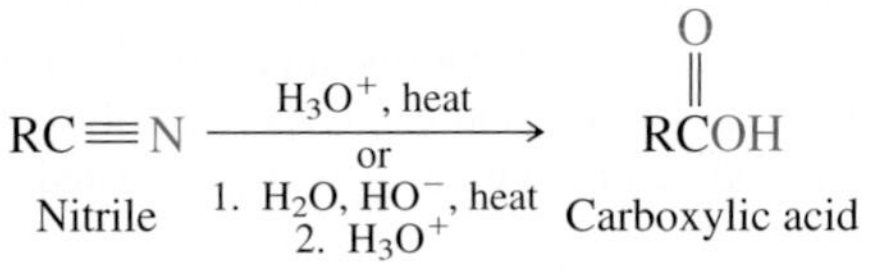
RC≡N
Nitrile
H₃O⁺, heat
or
1. H₂O, HO⁻, heat
2. H₃O⁺
RCOH
Carboxylic acid

Section 19.19 Nitriles are useful starting materials for the preparation of ketones by reaction with Grignard reagents.

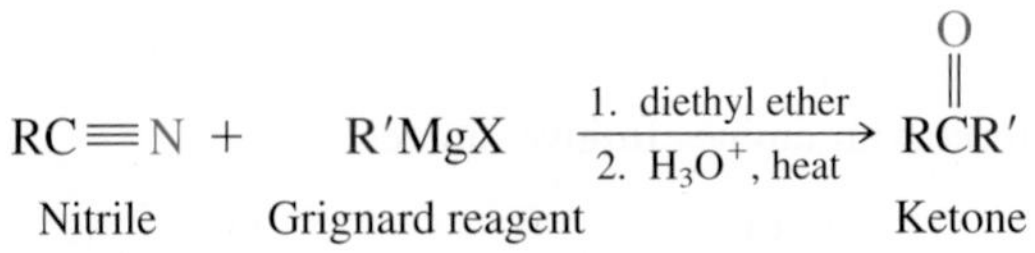
RC≡N +
Nitrile
R′MgX
Grignard reagent
1. diethyl ether
2. H₃O⁺, heat
RCR′
Ketone

Section 19.20 Acyl chlorides, anhydrides, esters, and amides all show a strong band for C=O stretching in the infrared. The range extends from about 1820 cm^{-1} (acyl chlorides) to 1690 cm^{-1} (amides). Their ^{13}C NMR spectra are characterized by a peak near δ 180 for the carbonyl carbon. ^1H NMR spectroscopy is useful for distinguishing between the groups R and R′ in esters (RCO_2R'). The protons on the carbon bonded to O in R′ appear at lower field (less shielded) than those on the carbon bonded to C=O.

PROBLEMS

19.27 Write a structural formula for each of the following compounds:

(a) *m*-Chlorobenzoyl chloride
(b) Trifluoroacetic anhydride
(c) *cis*-1,2-Cyclopropanedicarboxylic anhydride
(d) Ethyl cycloheptanecarboxylate
(e) 1-Phenylethyl acetate
(f) 2-Phenylethyl acetate
(g) *p*-Ethylbenzamide
(h) *N*-Ethylbenzamide
(i) 2-Methylhexanenitrile

19.28 Give an acceptable IUPAC name for each of the following compounds:

(a) 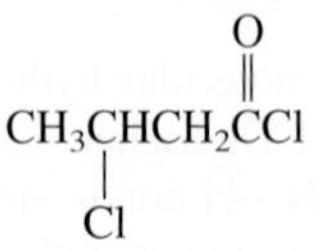
CH₃CHCH₂CCl
Cl
O

(b) 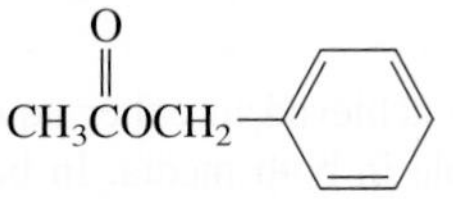
CH₃COCH₂
O

(c) 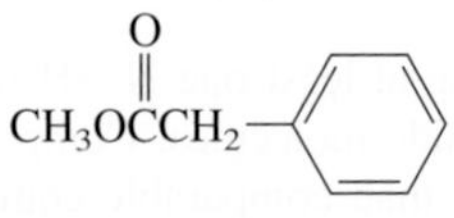
CH₃OCCH₂
O

(d) 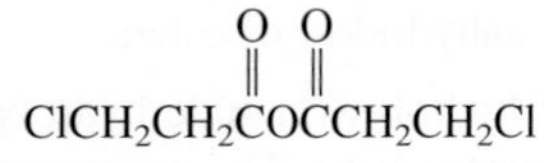
ClCH₂CH₂COCCH₂CH₂Cl
O O

(e) 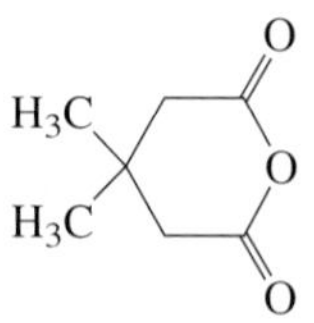

(f) $(CH_3)_2CHCH_2CH_2C{\equiv}N$

(g) $(CH_3)_2CHCH_2CH_2\overset{O}{\overset{\|}{C}}NH_2$

(h) $(CH_3)_2CHCH_2CH_2\overset{O}{\overset{\|}{C}}NHCH_3$

(i) $(CH_3)_2CHCH_2CH_2\overset{O}{\overset{\|}{C}}N(CH_3)_2$

19.29 Write a structural formula for the principal organic product or products of each of the following reactions:

(a) Propanoyl chloride and sodium propanoate

(b) Butanoyl chloride and benzyl alcohol

(c) *p*-Chlorobenzoyl chloride and ammonia

(d) 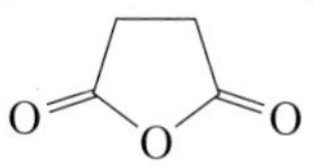 and water

(e) 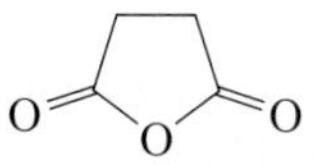 and aqueous sodium hydroxide to give $C_4H_4Na_2O_4$

(f) 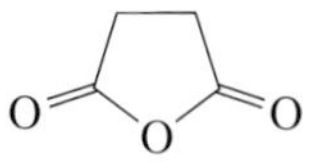 and aqueous ammonia to give $C_4H_{10}N_2O_3$

(g) Methyl benzoate and excess phenylmagnesium bromide, then H_3O^+

(h) Acetic anhydride and 3-pentanol

(i) Ethyl phenylacetate and lithium aluminum hydride, then H_3O^+

(j) 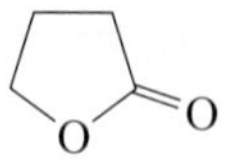 and aqueous sodium hydroxide to give $C_4H_7NaO_3$

(k) 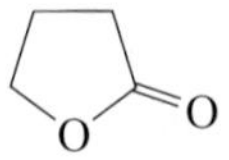 and aqueous ammonia

(l) 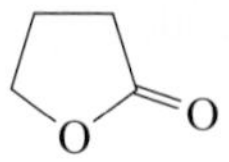 and lithium aluminum hydride, then H_2O

(m) 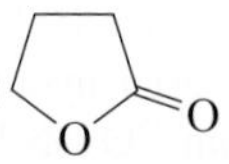 and excess methylmagnesium bromide, then H_3O^+

(n) Ethyl phenylacetate and methylamine (CH_3NH_2)

(o) 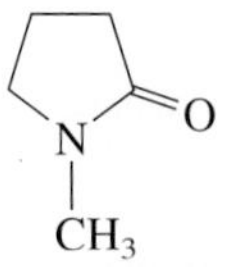 and aqueous sodium hydroxide to give $C_5H_{10}NNaO_2$

(p) 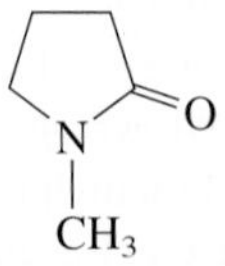 and aqueous hydrochloric acid, heat to give $[C_5H_{12}NO_2]^+$

(q) $C_6H_5NH\overset{O}{\overset{\|}{C}}CH_3$ and aqueous hydrochloric acid, heat to give $C_2H_4O_2 + C_6H_8ClN$

(r) $C_6H_5\overset{O}{\overset{\|}{C}}NHCH_3$ and aqueous sulfuric acid, heat to give $CH_7NO_4S + C_7H_6O_2$

(s) 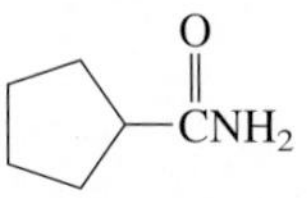 and P_4O_{10}

(t) $(CH_3)_2CHCH_2C{\equiv}N$ and aqueous hydrochloric acid, heat

(u) *p*-Methoxybenzonitrile and aqueous sodium hydroxide, heat to give $C_8H_7NaO_3$

(v) Propanenitrile and methylmagnesium bromide, then H_3O^+, heat

19.30 (a) Unlike other esters which react with Grignard reagents to give tertiary alcohols, ethyl formate ($\mathrm{H\overset{\overset{O}{\|}}{C}OCH_2CH_3}$) yields a different class of alcohols on treatment with Grignard reagents. What kind of alcohol is formed in this case and why?

(b) Diethyl carbonate ($\mathrm{CH_3CH_2O\overset{\overset{O}{\|}}{C}OCH_2CH_3}$) reacts with excess Grignard reagent to yield alcohols of a particular type. What is the structural feature that characterizes alcohols prepared in this way?

19.31 On being heated with a dilute solution of sulfuric acid in water, the compound shown was observed to liberate CO_2. What organic compound is formed during the reaction?

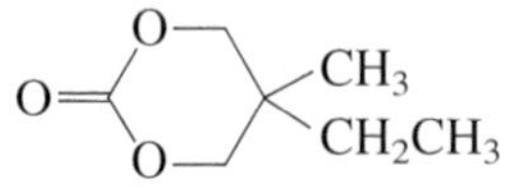

19.32 Using ethanol and sodium or potassium cyanide as the sources of the carbon atoms, along with any necessary inorganic reagents, show how you could prepare each of the following:

(a) Acetyl chloride
(b) Acetic anhydride
(c) Ethyl acetate
(d) Acetamide
(e) 2-Hydroxypropanoic acid

19.33 Using toluene, sodium cyanide, and carbon dioxide as the sources of the carbon atoms, along with any necessary inorganic reagents, show how you could prepare each of the following:

(a) Benzoyl chloride
(b) Benzoic anhydride
(c) Benzyl benzoate
(d) Benzamide
(e) Benzonitrile
(f) Benzyl cyanide
(g) Phenylacetic acid
(h) *p*-Nitrobenzoyl chloride
(i) *m*-Nitrobenzoyl chloride

19.34 The saponification of ^{18}O-labeled ethyl propanoate was described in Section 19.10 as one of the significant experiments that demonstrated acyl–oxygen cleavage in ester hydrolysis. The ^{18}O-labeled ethyl propanoate used in this experiment was prepared from ^{18}O-labeled ethyl alcohol, which in turn was obtained from acetaldehyde and ^{18}O-enriched water. Write a series of equations showing the preparation of $\mathrm{CH_3CH_2\overset{\overset{O}{\|}}{C}OCH_2CH_3}$ (where O$=^{18}$O) from these starting materials.

19.35 Suggest a reasonable explanation for each of the following observations:

(a) The second-order rate constant k for saponification (basic hydrolysis) of ethyl trifluoroacetate is over 1 million times greater than that for ethyl acetate (25°C).

(b) The second-order rate constant for saponification of ethyl 2,2-dimethylpropanoate, $(CH_3)_3CCO_2CH_2CH_3$, is almost 100 times smaller than that for ethyl acetate (30°C).

(c) The second-order rate constant k for saponification of methyl acetate is 100 times greater than that for *tert*-butyl acetate (25°C).

(d) The second-order rate constant k for saponification of methyl *m*-nitrobenzoate is 40 times greater than that for methyl benzoate (25°C).

(e) The second-order rate constant k for saponification of 5-pentanolide is over 20 times greater than that for 4-butanolide (25°C).

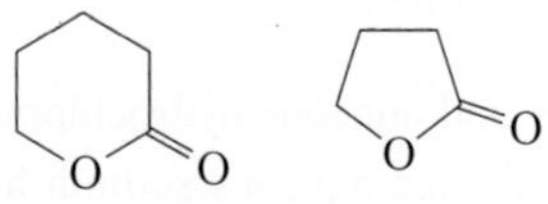

5-Pentanolide 4-Butanolide

(f) The second-order rate constant k for saponification of ethyl *trans*-4-*tert*-butylcyclohexanecarboxylate is 20 times greater than that for its cis diastereomer (25°C).

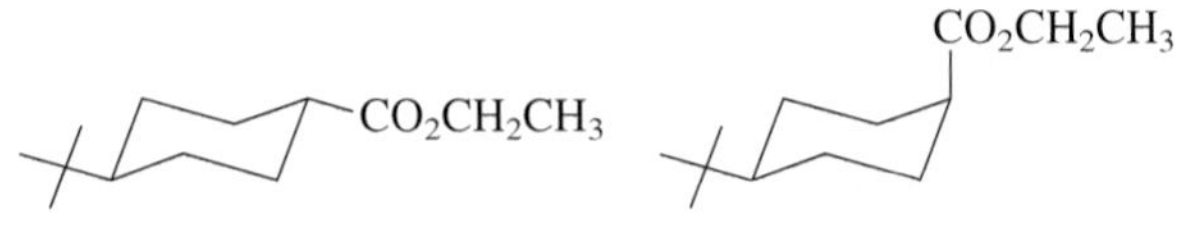

Ethyl *trans*-4-*tert*-butylcyclohexanecarboxylate — Ethyl *cis*-4-*tert*-butylcyclohexanecarboxylate

19.36 The hydrolysis of *tert*-butyl acetate occurs at a faster rate in aqueous acid than that of ethyl acetate. Explain. (*Hint:* The mechanism for this reaction is different from the one shown in Mechanism 19.3—it does not involve a tetrahedral intermediate. Also, review Section 4.8.)

$$CH_3\overset{O}{\overset{\|}{C}}{-}O{-}C(CH_3)_3 \xrightarrow{H_3O^+} CH_3\overset{O}{\overset{\|}{C}}{-}OH + HO{-}C(CH_3)_3$$

tert-Butyl acetate — Acetic acid — *tert*-Butyl alcohol

19.37 The preparation of *cis*-4-*tert*-butylcyclohexanol from its trans stereoisomer was carried out by the following sequence of steps. Write structural formulas, including stereochemistry, for compounds A and B.

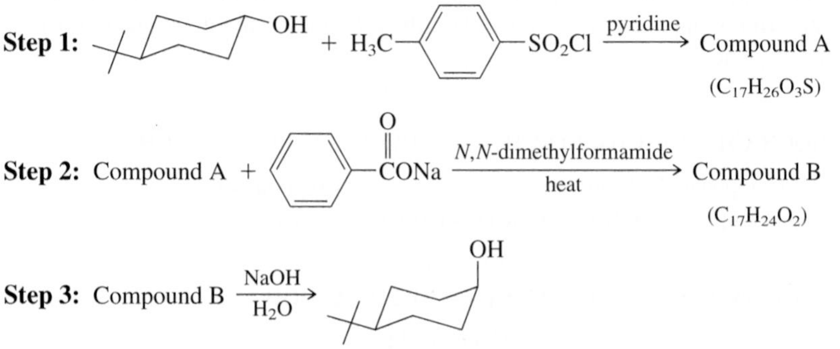

19.38 The ketone shown was prepared in a three-step sequence from ethyl trifluoroacetate. The first step in the sequence involved treating ethyl trifluoroacetate with ammonia to give compound A. Compound A was in turn converted to the desired ketone by way of compound B. Fill in the missing reagents in the sequence shown, and give the structures of compounds A and B.

$$CF_3\overset{O}{\overset{\|}{C}}OCH_2CH_3 \xrightarrow{NH_3} \text{Compound A} \longrightarrow \text{Compound B} \longrightarrow CF_3\overset{O}{\overset{\|}{C}}C(CH_3)_3$$

19.39 Compound A is a derivative of the carbohydrate perosamine, which is found in the antibiotic perimycin. When A is treated with excess acetic anhydride in methanol, a *mono*-acyl derivative B ($C_9H_{17}NO_5$) is obtained in 73% yield.

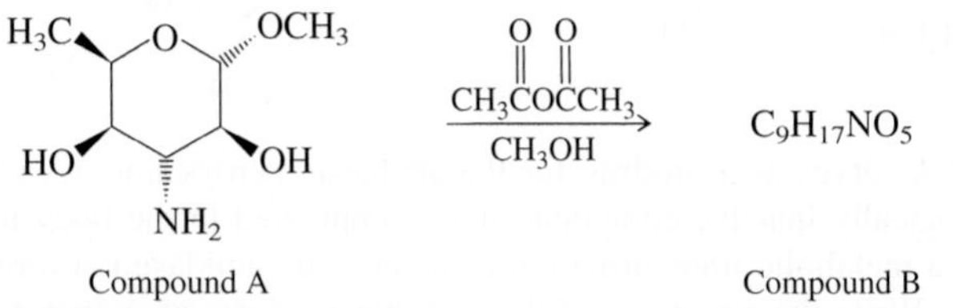

Compound A — Compound B

(a) What is its structure? (*Hint:* Consider that methanol reacts with acetic anhydride.)

(b) Explain the selectivity of this reaction.

(c) What is the expected product if methanol is omitted and the reaction is conducted with excess acetic anhydride in pyridine?

19.40 *Ambrettolide* is obtained from hibiscus and has a musk-like odor. Its preparation from compound A is outlined in the table that follows. Write structural formulas, ignoring stereochemistry, for compounds B through G in this synthesis. (*Hint:* Zinc, as used in step 4, converts vicinal dibromides to alkenes.)

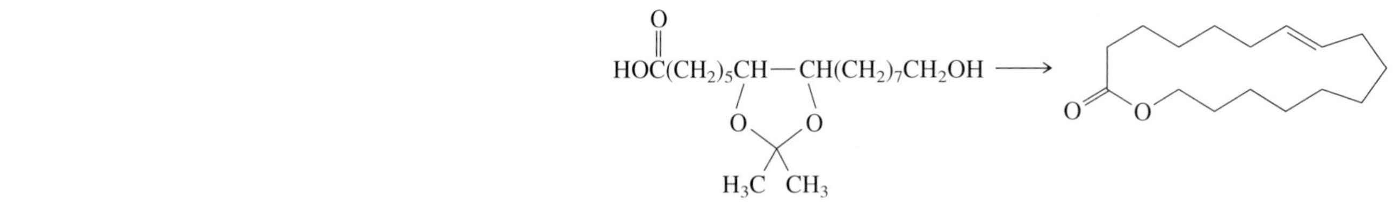

Compound A   Ambrettolide

Step	Reactant	Reagents	Product
1.	Compound A	H_2O, H^+ heat	Compound B ($C_{16}H_{32}O_5$)
2.	Compound B	HBr	Compound C ($C_{16}H_{29}Br_3O_2$)
3.	Compound C	Ethanol, H_2SO_4	Compound D ($C_{18}H_{33}Br_3O_2$)
4.	Compound D	Zinc, ethanol	Compound E ($C_{18}H_{33}BrO_2$)
5.	Compound E	Sodium acetate, acetic acid	Compound F ($C_{20}H_{36}O_4$)
6.	Compound F	KOH, ethanol, then H^+	Compound G ($C_{16}H_{30}O_3$)
7.	Compound G	Heat	Ambrettolide ($C_{16}H_{28}O_2$)

19.41 The preparation of the sex pheromone of the bollworm moth, (*E*)-9,11-dodecadien-1-yl acetate, from compound A has been described. Suggest suitable reagents for each step in this sequence.

(a) $HOCH_2CH{=}CH(CH_2)_7CO_2CH_3 \longrightarrow \overset{O}{\overset{\|}{HC}}CH{=}CH(CH_2)_7CO_2CH_3$

Compound A (*E* isomer)   Compound B

(b) Compound B $\longrightarrow H_2C{=}CHCH{=}CH(CH_2)_7CO_2CH_3$

Compound C

(c) Compound C $\longrightarrow H_2C{=}CHCH{=}CH(CH_2)_7CH_2OH$

Compound D

(d) Compound D $\longrightarrow H_2C{=}CHCH{=}CH(CH_2)_7CH_2O\overset{O}{\overset{\|}{C}}CH_3$

(*E*)-9,11-Dodecadien-1-yl acetate

19.42 Outline reasonable mechanisms for each of the following reactions:

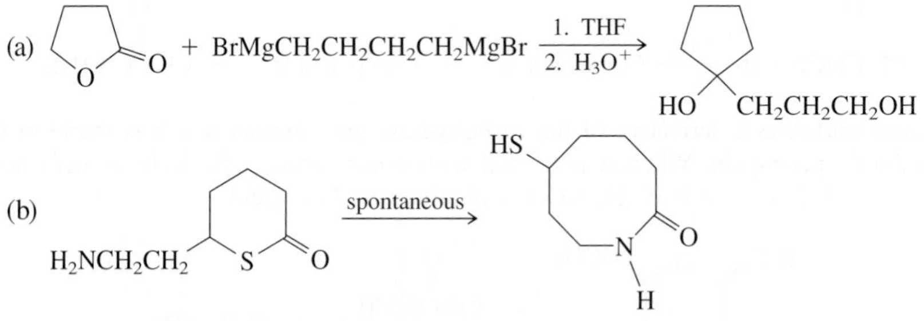

19.43 Compound A serves as a prodrug for the analgesic benzocaine. (A prodrug is a pharmacologically inactive compound that is converted in the body to an active drug, usually by a metabolic transformation.) The enzyme amidase converts compound A into benzocaine. Write the structures of the possible products of A that might be formed by hydrolysis in aqueous HCl.

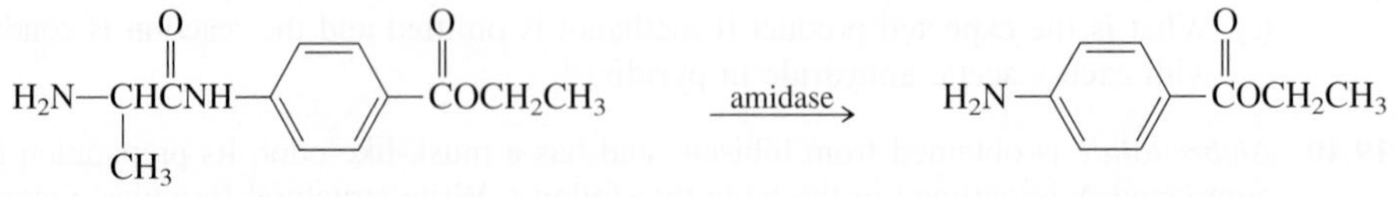

Compound A   Benzocaine

19.44 The serum cholesterol-lowering agent mevinolin (lovastatin) is shown here. Identify the ester and lactone functional groups of lovastatin, and for each, write the structures of the carboxylic acid and alcohol from which the ester and lactone are formed.

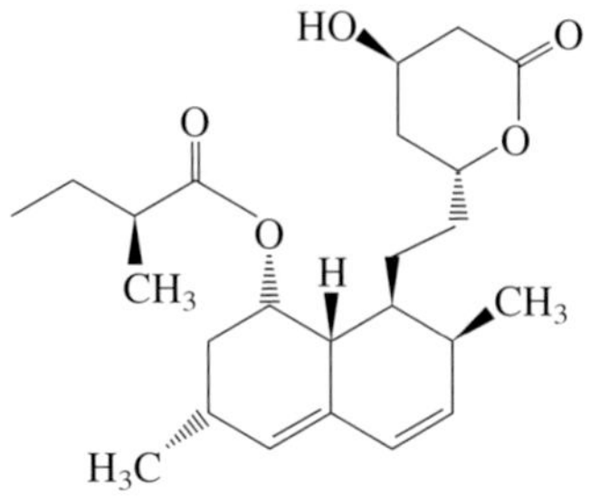
HO
O
O
O
CH3
H
CH3
H3C

Lovastatin

19.45 When compounds of the type represented by A are allowed to stand in pentane, they are converted to a constitutional isomer B.

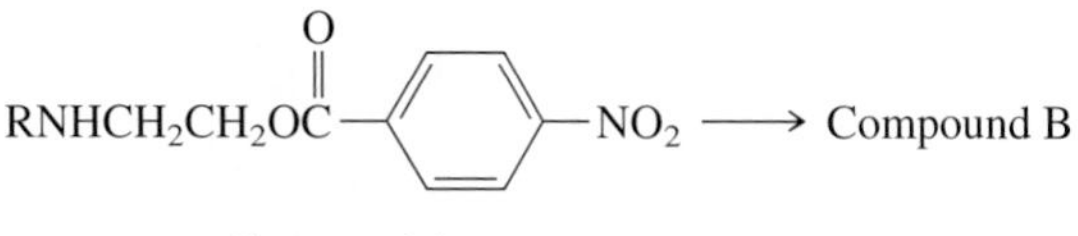
O
RNHCH2CH2OC
NO2
Compound B

Compound A

Hydrolysis of either A or B yields $RNHCH_2CH_2OH$ and *p*-nitrobenzoic acid. Suggest a reasonable structure for compound B, and demonstrate your understanding of the mechanism of this reaction by writing the structure of the key intermediate in the conversion of compound A to compound B.

19.46 (a) In the presence of dilute hydrochloric acid, compound A is converted to a constitutional isomer, compound B.

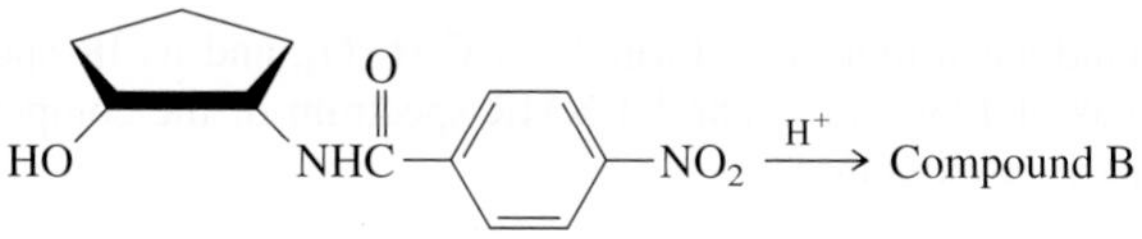
O
HO
NHC
NO2
H+
Compound B

Compound A

Suggest a reasonable structure for compound B.

(b) The trans stereoisomer of compound A is stable under the reaction conditions. Why does it not rearrange?

19.47 In the presence of pyridine, acyl chlorides react with alcohols to give esters according to the following overall equation:

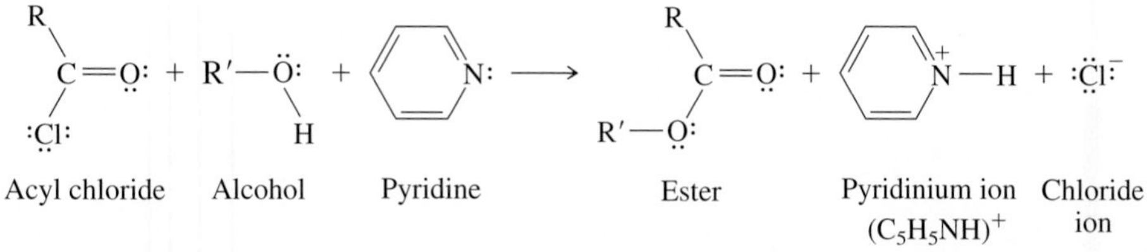
R
C=O:
:Cl:
R'—O:
H
N:
R
C=O:
R'—O:
N—H
:Cl:
Acyl chloride
Alcohol
Pyridine
Ester
Pyridinium ion
(C5H5NH)+
Chloride
ion

(a) The mechanism of this reaction involves a number of steps and begins with the reaction of pyridine with the acyl chloride to give an acylpyridinium ion.

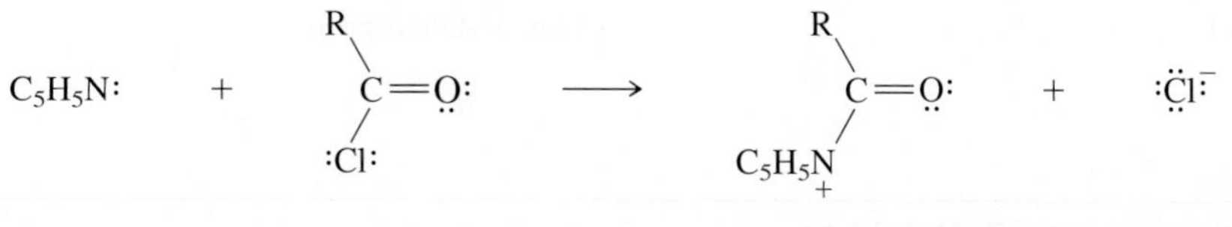
C5H5N:
R
C=O:
:Cl:
R
C=O:
C5H5N
:Cl:

Write a two-step mechanism for the formation of the acylpyridinium ion, using curved arrows to show the flow of electrons.

(b) Show the electron flow in the next step in which the alcohol reacts with the acylpyridinium ion.

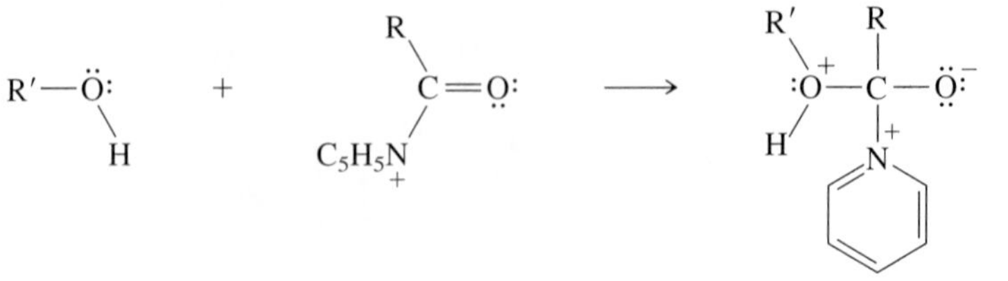

(c) Next, pyridine acts as a Brønsted base toward the product of part (b). What are the products of this reaction?

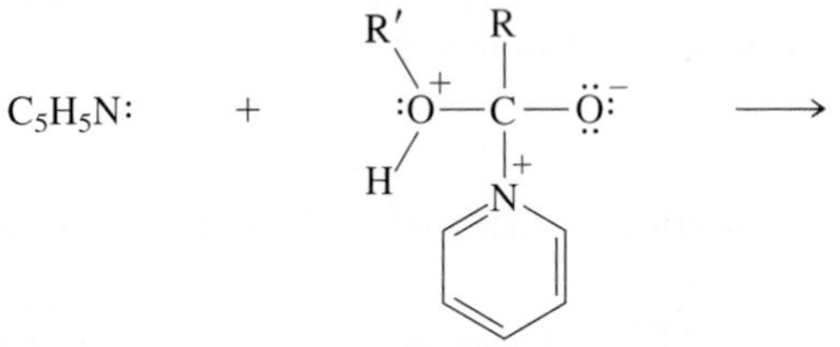

(d) A final step completes the mechanism. Write an equation for this step showing the products and the electron flow that leads to them.

(e) Does the sum of the individual steps of the mechanism correspond to the overall equation?

19.48 A certain compound has a molecular weight of 83 and contains nitrogen. Its infrared spectrum contains a moderately strong peak at 2270 cm^{-1}. Its ^{1}H and ^{13}C NMR spectra are shown in Figure 19.6. What is the structure of this compound?

19.49 A compound has a molecular formula of $C_8H_{14}O_4$, and its IR spectrum contains an intense peak at 1730 cm^{-1}. The ^{1}H NMR spectrum of the compound is shown in Figure 19.7. What is its structure?

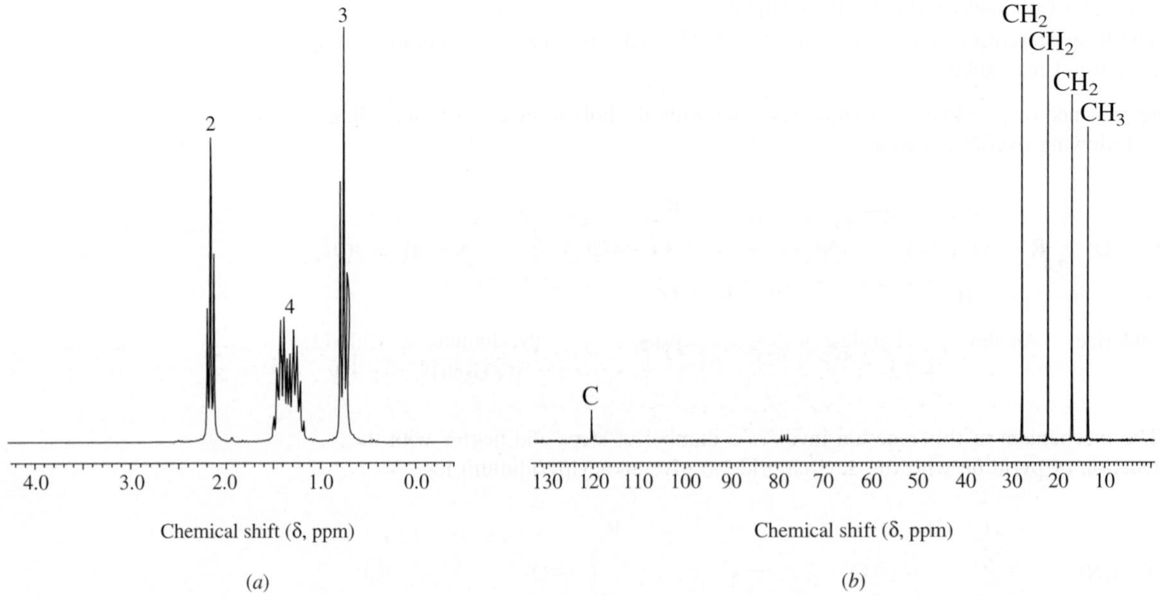

Figure 19.6

The (*a*) 200-MHz ^{1}H and (*b*) ^{13}C NMR spectra of the compound in Problem 19.48.

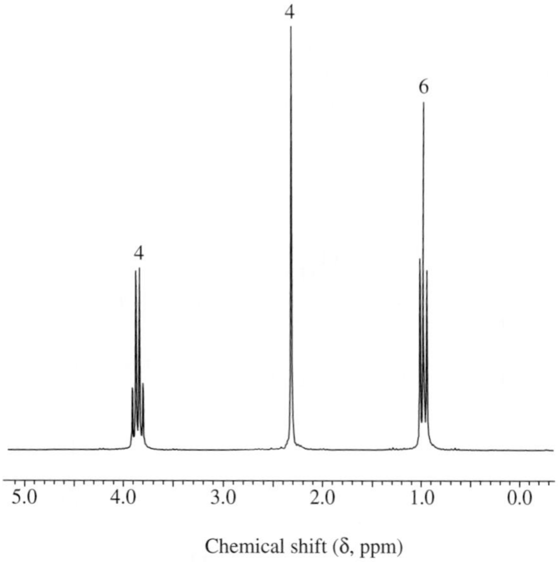
4
6
4
5.0
4.0
3.0
2.0
1.0
0.0
Chemical shift (δ, ppm)

Figure 19.7

The 200-MHz ^1H NMR spectrum of the compound $C_8H_{14}O_4$ in Problem 19.49.

19.50 A compound ($C_4H_6O_2$) has a strong band in the infrared at 1760 cm^{-1}. Its ^{13}C NMR spectrum exhibits signals at δ 20.2 (CH_3), 96.8 (CH_2), 141.8 (CH), and 167.6 (C). The ^1H NMR spectrum of the compound has a three-proton singlet at δ 2.1 along with three other signals, each of which is a doublet of doublets, at δ 4.7, 4.9, and 7.3. What is the structure of the compound?

Descriptive Passage and Interpretive Problems 19

Thioesters

Thioesters have the general formula $R\overset{O}{\overset{\|}{C}}SR'$. They resemble their oxygen counterparts $R\overset{O}{\overset{\|}{C}}OR'$ (oxoesters) in structure and reactivity more than other carboxylic acid derivatives such as acyl chlorides, acid anhydrides, and amides. Thioesters can be prepared from thiols by reaction with acyl chlorides or acid anhydrides in much the same way as oxoesters are prepared from alcohols.

$$\underset{\text{1-Butanethiol}}{CH_3CH_2CH_2CH_2SH} + \underset{\text{Acetyl chloride}}{CH_3\overset{O}{\overset{\|}{C}}Cl} \longrightarrow \underset{\text{Butyl thioacetate (91\%)}}{CH_3\overset{O}{\overset{\|}{C}}SCH_2CH_2CH_2CH_3}$$

The preparation of thioesters by Fischer esterification is not very effective, however, because the equilibrium is normally unfavorable. Under conditions in which ethanol is converted to ethyl benzoate to the extent of 68%, ethanethiol gives only 15% ethyl thiobenzoate.

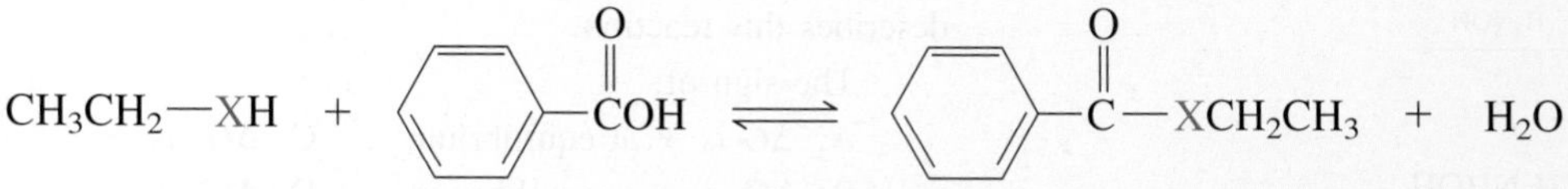
CH3CH2—XH +
COH
C—XCH2CH3 + H2O

Ethanol: X = O — At equilibrium: X = O; 68%

Ethanethiol: X = S — X = S; 15%

This, and numerous other observations, indicates that S—C═O is less stabilized than O—C═O. Like chlorine, sulfur is a third-row element and does not act as an electron-pair donor to the carbonyl group as well as oxygen.

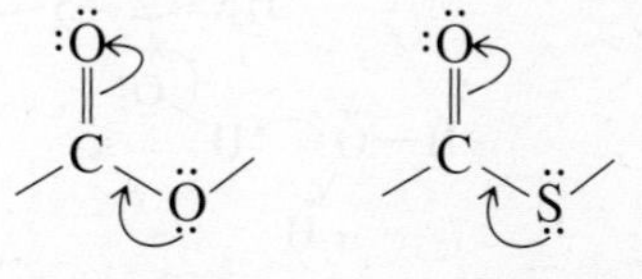

More effective — Less effective

Thioesters and oxoesters are similar in their rates of nucleophilic acyl substitution, except with amine nucleophiles for which thioesters are much more reactive. Many biological reactions involve nucleophilic acyl substitutions referred to as **acyl transfer** reactions. The thioester *acetyl coenzyme A* is an acetyl group donor to alcohols, amines, and assorted other biological nucleophiles.

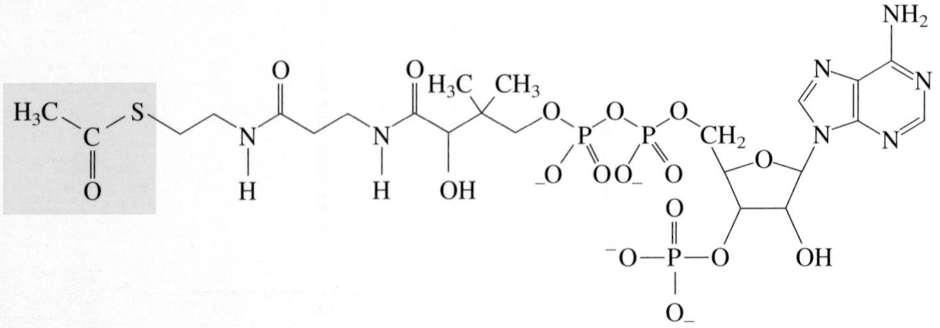

Acetyl coenzyme A ($CH_3\overset{O}{\overset{\|}{C}}SCoA$)

Melatonin, a hormone secreted by the pineal gland that regulates circadian rhythms, including wake–sleep cycles, is biosynthesized by a process in which the first step is an enzyme-catalyzed transfer of the acetyl group from sulfur of acetyl coenzyme A to the —NH_2 group of serotonin.

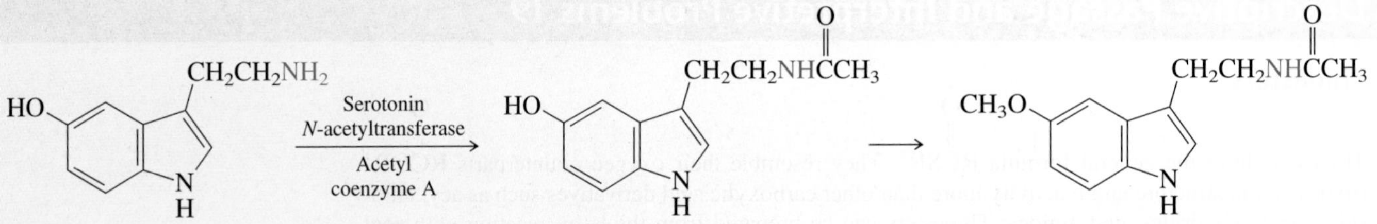
Serotonin
N-acetyltransferase
Acetyl
coenzyme A

Serotonin | *N*-Acetylserotonin | Melatonin

19.51 Thioesters react with hydroxylamine by nucleophilic acyl substitution to give hydroxamic acids. What is the structure of the hydroxamic acid formed in the following reaction?

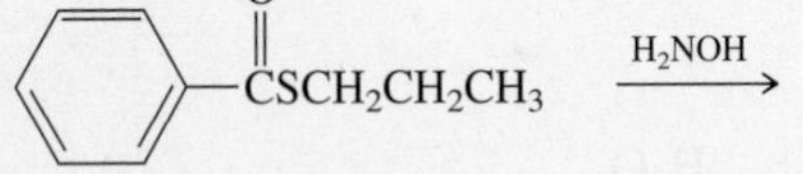
H_2NOH

A. $CH_3CH_2CH_2ONH_2$

B. $CH_3CH_2CH_2NHOH$

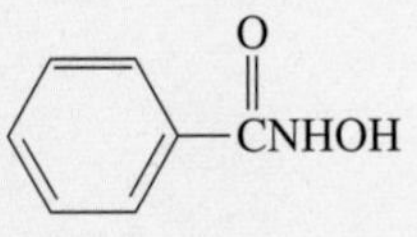

C.

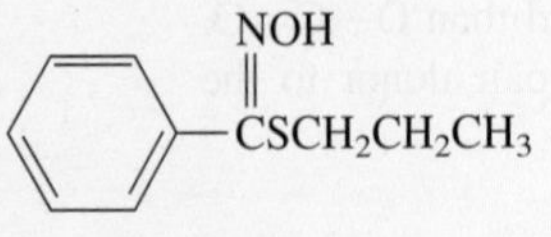

D.

19.52 The equilibrium constant K equals 56 for the reaction shown.

$$HOCH_2CH_2S\overset{O}{\overset{\|}{C}}CH_3 \overset{K}{\rightleftharpoons} CH_3\overset{O}{\overset{\|}{C}}OCH_2CH_2SH$$

Complete the following statement so that it correctly describes this reaction.

The sign of:

A. ΔG is + at equilibrium C. $\Delta G°$ is +

B. ΔG is − at equilibrium D. $\Delta G°$ is −

19.53 For the reaction shown in Problem 19.52, which of the following better represents the flow of electrons for the step in the mechanism leading to the isomer present in greatest amount at equilibrium?

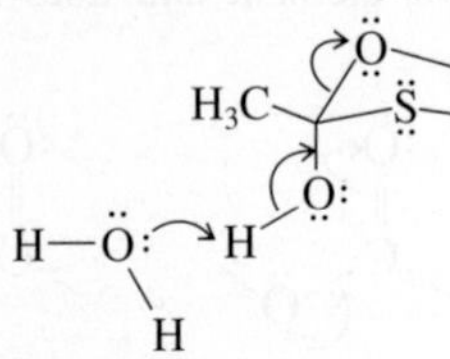

A.

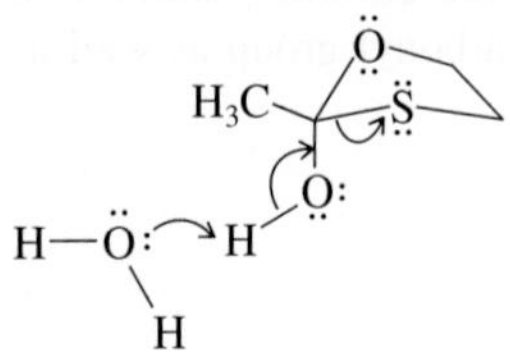

B.

19.54 Which reaction occurs at the fastest rate?

A. $\mathrm{CH_3\overset{O}{\overset{\|}{C}}SCH_2CH_3 + CH_3NH_2 \longrightarrow CH_3\overset{O}{\overset{\|}{C}}NHCH_3 + CH_3CH_2SH}$

B. $\mathrm{CH_3\overset{O}{\overset{\|}{C}}OCH_2CH_3 + CH_3NH_2 \longrightarrow CH_3\overset{O}{\overset{\|}{C}}NHCH_3 + CH_3CH_2OH}$

C. $\mathrm{CH_3\overset{O}{\overset{\|}{C}}SCH_2CH_3 + H_2O \longrightarrow CH_3\overset{O}{\overset{\|}{C}}OH + CH_3CH_2SH}$

D. $\mathrm{CH_3\overset{O}{\overset{\|}{C}}OCH_2CH_3 + H_2O \longrightarrow CH_3\overset{O}{\overset{\|}{C}}OH + CH_3CH_2OH}$

19.55 Which one of the reactions in the preceding problem has the most negative value of $\Delta G°$?

19.56 Acetylcholine is a neurotransmitter formed in nerve cells by the enzyme-catalyzed reaction of choline with acetyl coenzyme A.

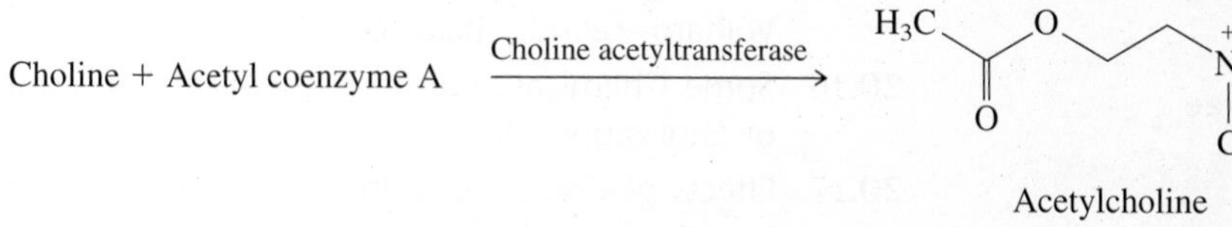
Choline + Acetyl coenzyme A
Choline acetyltransferase
H₃C
O
O
CH₃
N
CH₃
CH₃
Acetylcholine

What is the most reasonable structure for choline?

A. $\mathrm{HSCH_2CH_2\overset{+}{N}(CH_3)_3}$ B. $\mathrm{HOCH_2CH_2\overset{+}{N}(CH_3)_3}$ C. $\mathrm{CH_3\overset{O}{\overset{\|}{C}}OCH_2CH_2NH_2}$ D. $\mathrm{CH_3\overset{O}{\overset{\|}{C}}SCH_2CH_2\overset{+}{N}(CH_3)_3}$

19.57 Thiane was prepared in 76% yield from 5-chloro-1-pentene by the procedure shown. Deduce the structure of compound X in this synthesis.

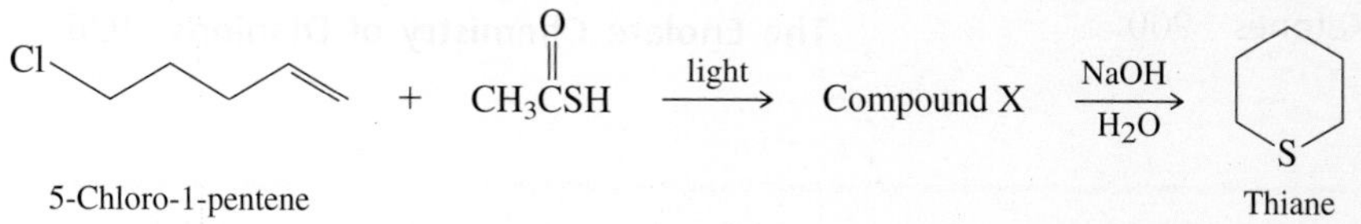
Cl
+
CH₃CSH
O
light
Compound X
NaOH
H₂O
S
5-Chloro-1-pentene
Thiane

A. 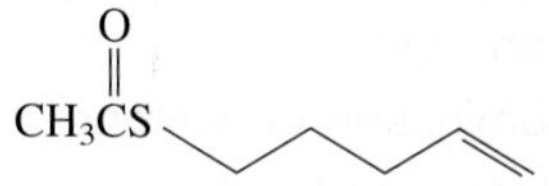
CH₃CS
O

C. 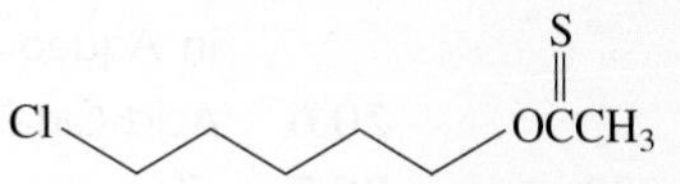
Cl
S
OCCH₃

B. 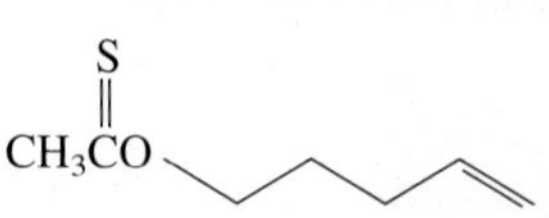
CH₃CO
S

D. 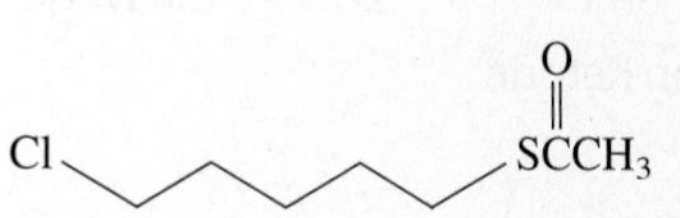
Cl
O
SCCH₃

20 Enols and Enolates

Chapter Outline

Mechanisms

This electrostatic potential map shows the enolate of diethyl malonate. This enolate has been used in a chemical synthesis of capsaicin, the substance most responsible for the fiery flavor of cayenne peppers. This synthesis is the basis of Problem 20.63 at the end of the chapter.

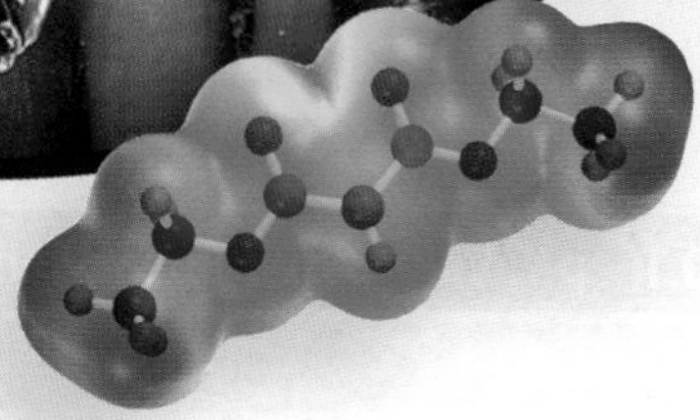

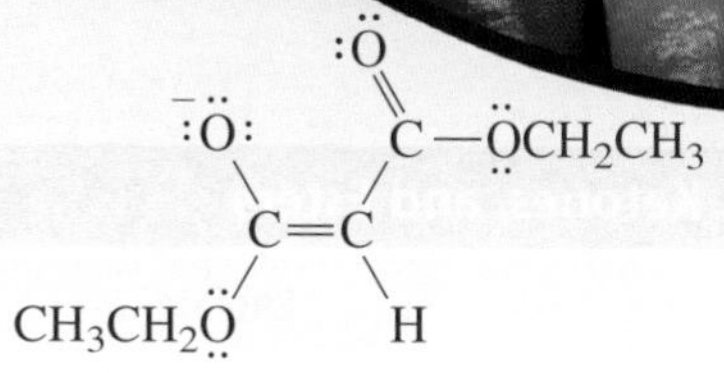

IN CHAPTER 17 you learned that nucleophilic addition to the carbonyl group of aldehydes and ketones is one of the fundamental types of reactions in organic chemistry. In Chapter 19 you saw how addition to the carbonyl group of esters is a key step in nucleophilic acyl substitution. In the present chapter, you'll encounter yet another pattern of reactivity of these compounds involving **enols** and the **enolates** derived from them.

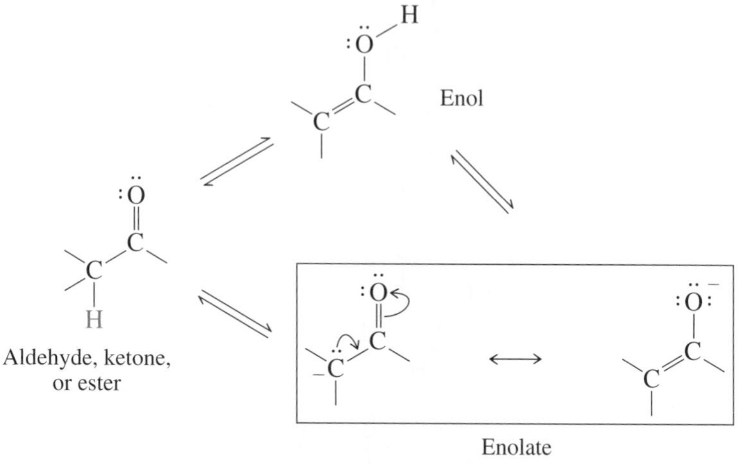

We encountered enols earlier as intermediates in the hydration of alkynes (see Mechanism 9.2). Enolates, represented as a hybrid of the resonance structures shown, are the conjugate bases of enols. The major enolate contributor is the structure with the negative charge on oxygen. It is, however, the carbanionic character of the α carbon that is responsible for the importance of enolates in organic synthesis, and we will sometimes write the enolate in the form that has the negative charge on carbon to emphasize this.

Enols and enolates are useful in synthesis as reagents for making carbon–carbon bonds and appear as intermediates in certain biological processes. We'll start with enolates, their formation and synthetic applications, then examine enols.

20.1 Aldehyde, Ketone, and Ester Enolates

It is convenient to use Greek letters to designate atoms in relation to the carbonyl group in an aldehyde, ketone, or ester. The carbon adjacent to the carbonyl carbon is the α carbon atom, the next one down the chain is the β carbon, and so on. Butanal, for example, has an α carbon, a β carbon, and a γ carbon.

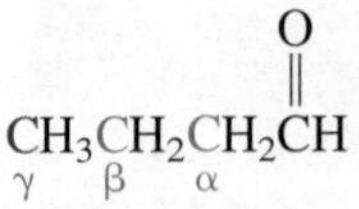

The carbonyl carbon is the reference atom; no Greek letter is assigned to it.

Substituents take the same Greek letter as the carbon atom to which they are attached. A hydrogen connected to the α-carbon atom is an α hydrogen. Butanal has two α hydrogens, two β hydrogens, and three γ hydrogens. No Greek letter is assigned to the hydrogen attached directly to the carbonyl group of an aldehyde.

Our experience to this point has been that C—H bonds are not very acidic. Alkanes, for example, have pK_a's of approximately 60. Compared with them, however, aldehydes, ketones, and esters have relatively acidic hydrogens on their α-carbon atoms (Table 20.1).

TABLE 20.1 pK_a Values of Some Aldehydes, Ketones, and Esters

Compound	Structure	Enolate	pK_a
Ethyl acetate	$CH_3\overset{O}{\overset{\|}{C}}OCH_2CH_3$	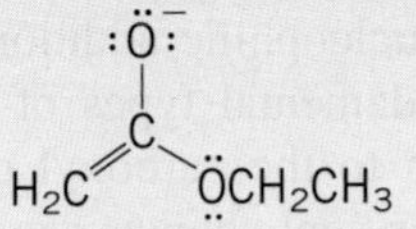	25.6
Acetone	$CH_3\overset{O}{\overset{\|}{C}}CH_3$	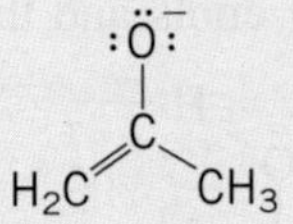	19.1
Acetaldehyde	$CH_3\overset{O}{\overset{\|}{C}}H$	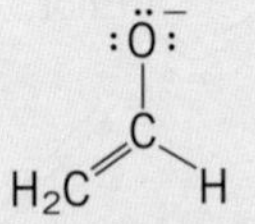	16.7
2-Methylpropanal	$(CH_3)_2CH\overset{O}{\overset{\|}{C}}H$	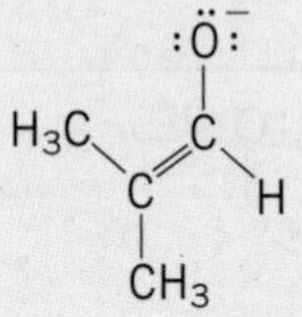	15.5
Diethyl malonate	$CH_3CH_2O\overset{O}{\overset{\|}{C}}CH_2\overset{O}{\overset{\|}{C}}OCH_2CH_3$	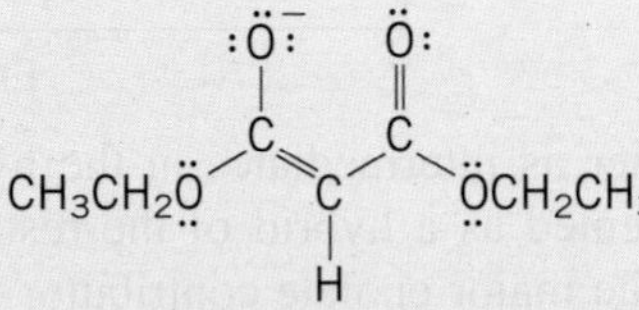	13
Ethyl acetoacetate	$CH_3\overset{O}{\overset{\|}{C}}CH_2\overset{O}{\overset{\|}{C}}OCH_2CH_3$	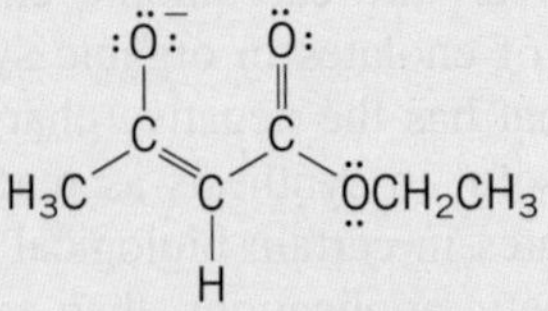	11
2,4-Pentanedione	$CH_3\overset{O}{\overset{\|}{C}}CH_2\overset{O}{\overset{\|}{C}}CH_3$	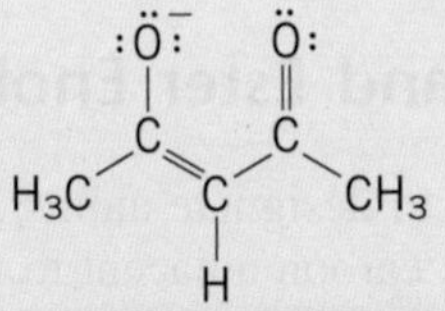	9

*The acidic hydrogens are bonded to the α carbon and are shown in red.

Problem 20.1

Find the most acidic hydrogen in each of the following and write a chemical equation for the proton-transfer process that occurs on reaction with hydroxide ion. Use curved arrows to show electron flow and label the acid, base, conjugate acid, and conjugate base.

(a) *tert*-Butyl methyl ketone (c) Methyl propanoate
(b) 3-Methylbutanal

Sample Solution (a) The only α hydrogens in *tert*-butyl methyl ketone are those of the methyl group attached to the carbonyl. Only α hydrogens are acidic enough to be removed by hydroxide. None of the hydrogens of the *tert*-butyl group are α hydrogens.

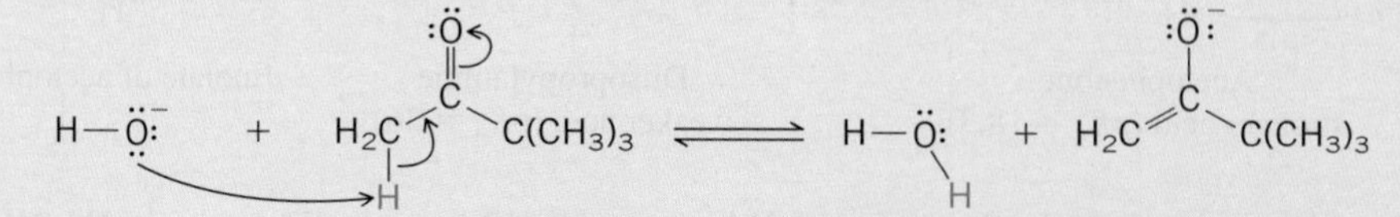

Hydroxide ion (base) | *tert*-Butyl methyl ketone (acid) | Water (conjugate acid) | Enolate ion (conjugate base)

The remainder of this section discusses and illustrates the following points to take from Table 20.1:

1. The acidic hydrogen is attached to an α carbon.
2. Simple aldehydes and ketones have pK_a's in the range 16–20, which makes an α C—H similar in acidity to the O—H group of water and most alcohols.
3. An α hydrogen of an ester is less acidic than an α hydrogen of an aldehyde or ketone.
4. A C—H group that is α to two carbonyl groups is more acidic than one that is α to only one carbonyl.

Two factors—one an electron-withdrawing inductive effect, the other electron delocalization—combine to make an H—C—C=O unit of an aldehyde, ketone, or ester relatively acidic compared with most other C—H bonds. The inductive effect of the carbonyl group increases the positive character of the α hydrogen, and resonance stabilizes the conjugate base.

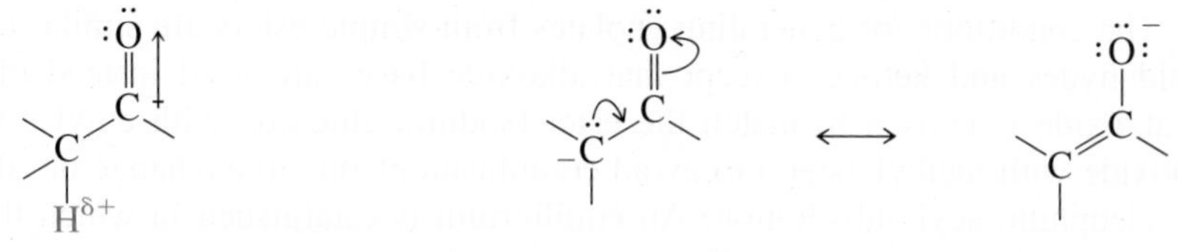

Inductive effect increases positive character of α hydrogen

Electron delocalization stabilizes enolate

The fact that the pK_a's of most simple aldehydes and ketones are about 16–20 means that both the carbonyl compound and its enolate are present when an acid–base equilibrium is established with hydroxide ion. The pK_a's of 2-methylpropanal and water, for example, are so similar that the equilibrium constant for enolate formation is approximately 1.

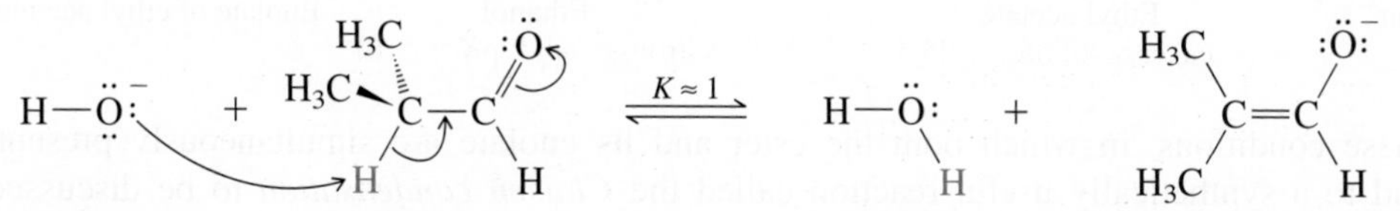

Hydroxide ion | 2-Methylpropanal $pK_a = 15.5$ | Water $pK_a = 15.7$ | Enolate of 2-methylpropanal

When an aldehyde or ketone and its enolate are both present in solution they can react with each other in what is called an *aldol condensation*. Such reactions are an important part of this chapter. The aldol condensation will be discussed in Sections 20.3 and 20.4.

Aldehydes and ketones can be converted *completely* to their enolates by using very strong bases such as lithium diisopropylamide $[(CH_3)_2CH]_2NLi$.

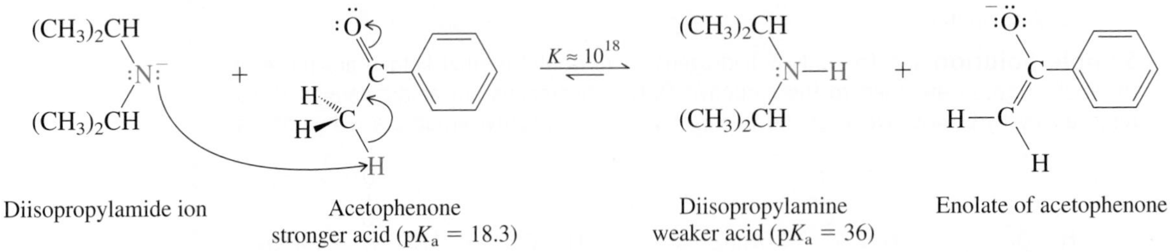
(CH3)2CH
:N:
(CH3)2CH
+
:O:
C
H
H
C
H
K ≈ 10^18
(CH3)2CH
:N—H
(CH3)2CH
+
:O:
C
H—C
H
Diisopropylamide ion
Acetophenone
stronger acid (pKa = 18.3)
Diisopropylamine
weaker acid (pKa = 36)
Enolate of acetophenone

Lithium diisopropylamide (known as **LDA**) is comparable to sodium amide ($NaNH_2$) in basicity, but, unlike $NaNH_2$, is too sterically hindered to undergo competing nucleophilic addition to the carbonyl group.

Problem 20.2

Methyllithium is a stronger base than lithium diisopropylamide but would not be a good choice for converting aldehydes and ketones to their enolates. Why?

The decreased acidity of ester α protons compared with those of aldehydes and ketones reflects the decreased electron-withdrawing ability of an ester carbonyl. Electron delocalization of the type:

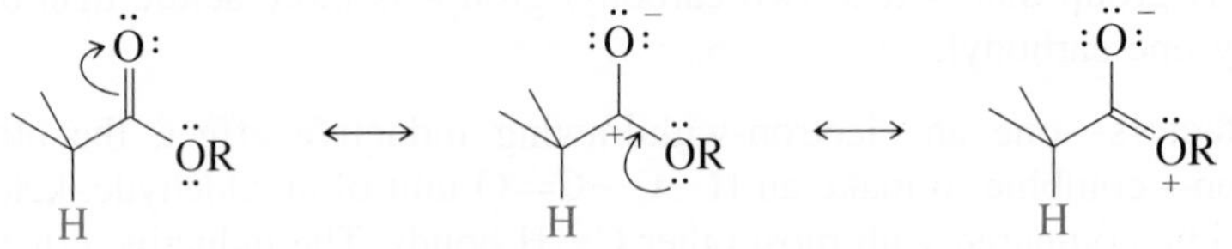
:O:
OR
H
:O:
OR
H
:O:
OR
H

decreases the positive character of an ester carbonyl group and reduces its ability to withdraw electrons away from the α hydrogen.

The conditions for generating enolates from simple esters are similar to those used for aldehydes and ketones except that alkoxide bases are used instead of hydroxide. The alkoxide is chosen to match the ester (sodium ethoxide with ethyl esters, sodium methoxide with methyl esters) to avoid complications due to exchange of alkoxy groups by nucleophilic acyl substitution. An equilibrium is established in which the ester predominates and only a very small amount of enolate is present.

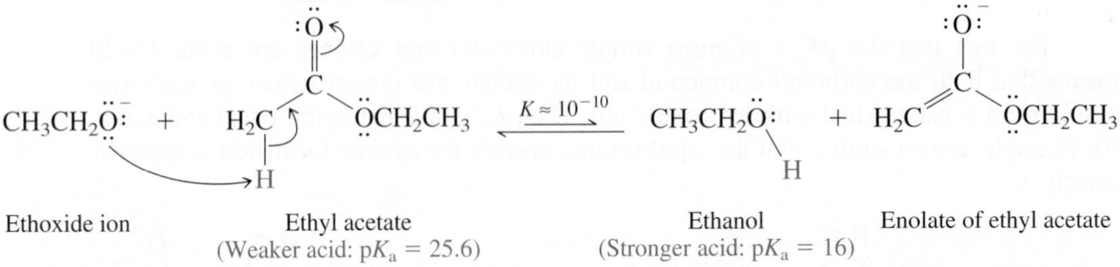
CH3CH2O:
+
:O:
C
H2C
OCH2CH3
H
K ≈ 10^−10
CH3CH2O:
H
+
H2C
C
:O:
OCH2CH3
Ethoxide ion
Ethyl acetate
(Weaker acid: pKa = 25.6)
Ethanol
(Stronger acid: pKa = 16)
Enolate of ethyl acetate

These conditions, in which both the ester and its enolate are simultaneously present, lead to a synthetically useful reaction called the *Claisen condensation* to be discussed in Section 20.5.

Esters can be converted entirely to their enolates by reaction with very strong bases such as lithium diisopropylamide.

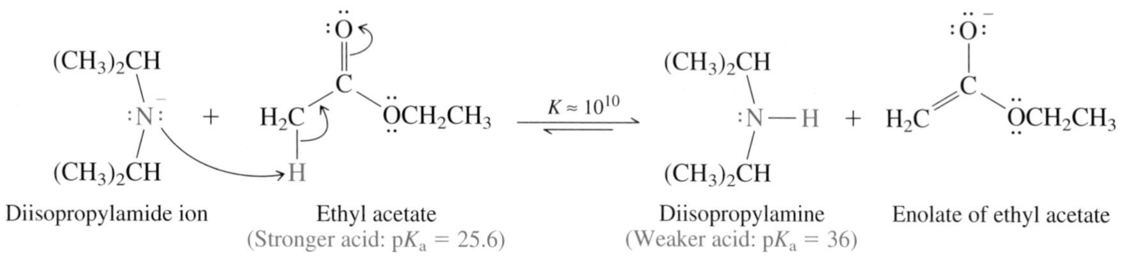

Dicarbonyl compounds such as β-diketones, β-keto esters, and diesters of malonic acid that have two carbonyl groups attached to the same carbon have pK_a's in the 9–13 range and are essentially completely converted to their enolates by hydroxide and alkoxides.

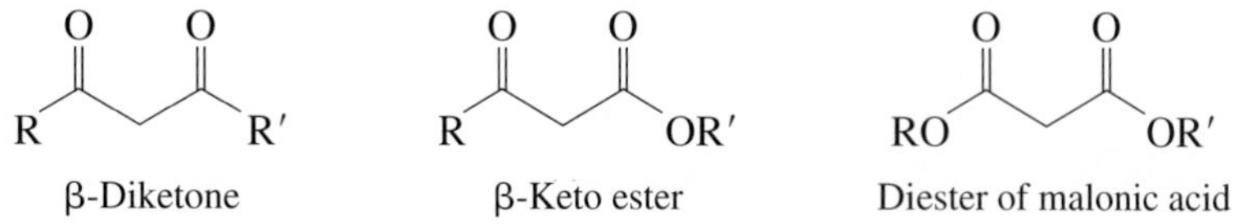

The equation shows enolate formation from 2,4-pentanedione, the β-diketone cited in Table 20.1.

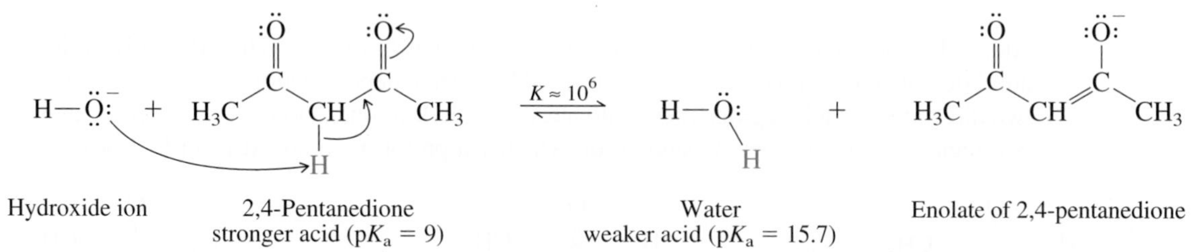

Both carbonyl groups participate in stabilizing the enolate by delocalizing its negative charge.

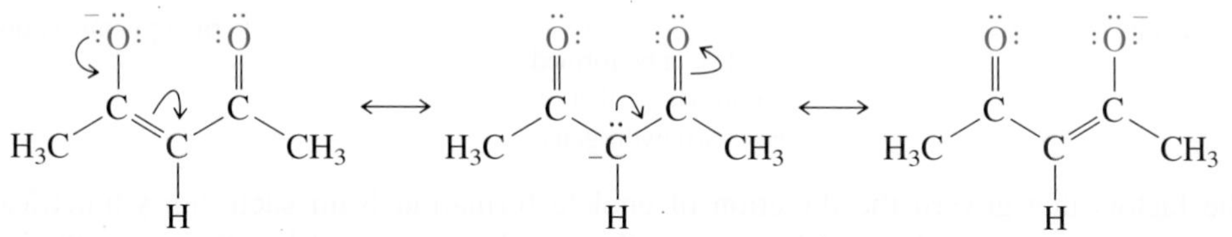

Analogous equations apply to ethyl acetoacetate and diethyl malonate—the β-keto ester and malonate diester, respectively, that are also cited in the table.

Problem 20.3

Write the structure of the enolate derived from each of the following. Give the three major resonance contributors of each enolate.

(a) 2-Methyl-1,3-cyclopentanedione

(b) 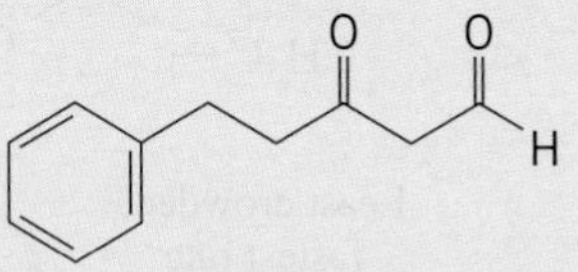

(c) 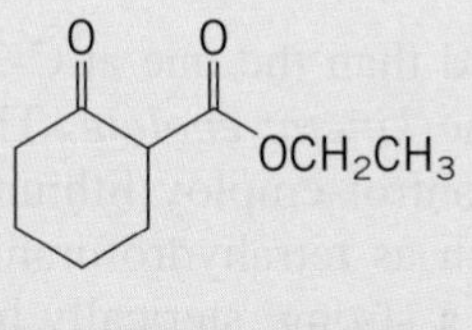

Continued

Sample Solution (a) First identify the proton that is removed by the base. It is on the carbon between the two carbonyl groups.

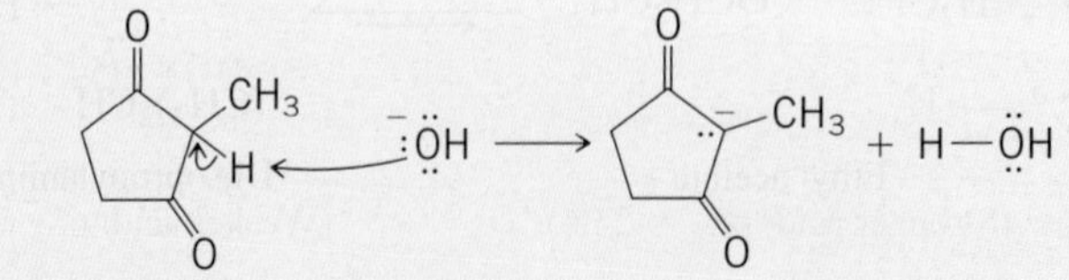

The three major resonance contributors of this anion are

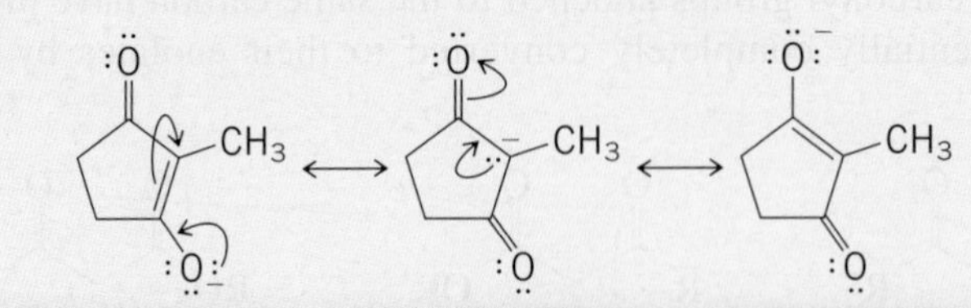

20.2 Enolate Regiochemistry

Aldehydes and esters have a single α carbon, but ketones have two. When R and R′ in RC(=O)R′ are different and each has at least one attached hydrogen, deprotonation can occur in either of two directions to yield regioisomeric enolates. 2-Methylcyclohexanone, for example, can give two regioisomeric enolates depending on whether a proton is removed from C-2 or C-6.

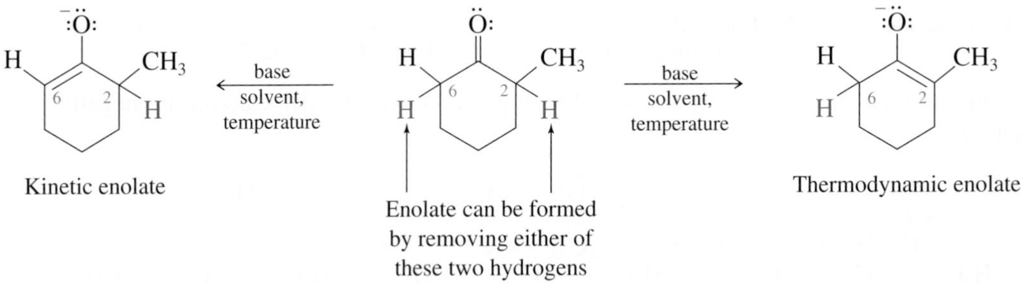

base
solvent,
temperature
Kinetic enolate
Enolate can be formed by removing either of these two hydrogens
Thermodynamic enolate

The factors that govern the direction of enolate formation from such "unsymmetrical" ketones are numerous, but can be organized using the concept of kinetic versus thermodynamic control.

Kinetic Control. A less crowded α proton is removed faster than a more crowded one. Thus, the rates of α-proton abstraction are:

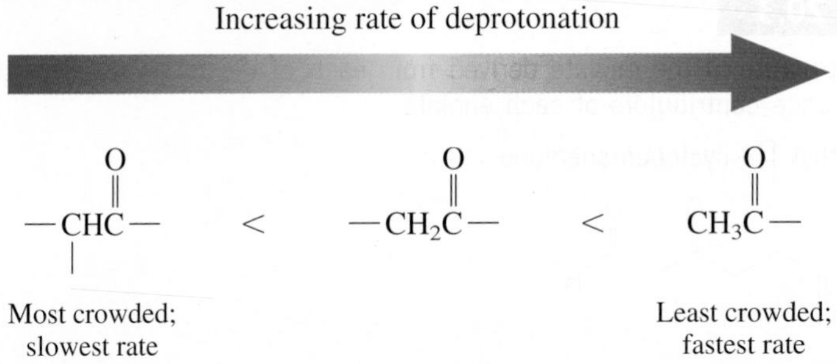

Increasing rate of deprotonation
Most crowded; slowest rate
Least crowded; fastest rate

A proton at C-6 in 2-methylcyclohexanone is less crowded than the one at C-2 and is removed faster. The resulting enolate is referred to as the *kinetic enolate*. The most common experimental conditions for achieving kinetic control employ lithium diisopropylamide (LDA) as the base in an aprotic solvent such as tetrahydrofuran (THF). The temperature is kept low, typically at −78°C. LDA is a strong, sterically hindered base, so proton abstraction at C-6 is favored.

Thermodynamic Control. Treatment of a ketone with a weaker base, such as sodium hydroxide or alkoxide, in the presence of a protic solvent such as an alcohol provides an equilibrium mixture containing the ketone and its enolates. Of the two regioisomeric enolates, the more stable one, termed the *thermodynamic enolate,* is present in greater amounts. For 2-methylcyclohexanone, the thermodynamic enolate is the one with the more substituted double bond, resulting from α-proton abstraction at C-2.

One of the important distinctions between kinetic and thermodynamic control of enolate formation is that conditions of kinetic control give rapid and complete enolate formation. Formation of the kinetic enolate is essentially irreversible under these conditions. The conditions of thermodynamic control establish an equilibrium containing both enolate and ketone. Some consequences of this distinction will be seen in later sections of this chapter.

Problem 20.4

Write the structures of all possible enolates of each of the following ketones. Which is kinetically favored? Thermodynamically favored?

(a) $(CH_3)_2CH\overset{O}{\overset{\|}{C}}CH_3$ (b) 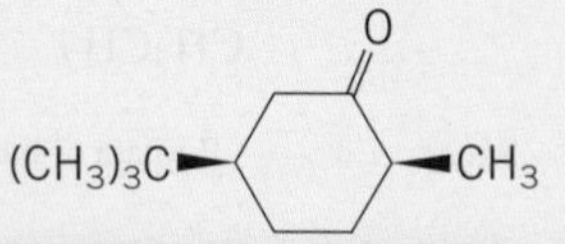 (c) 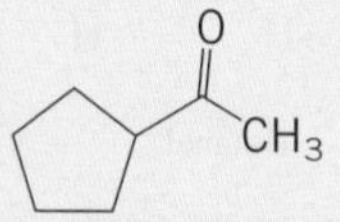

20.3 The Aldol Condensation

We have just seen that treatment of most aldehydes and ketones (those having α hydrogens with a pK_a of 16–20) with bases such as hydroxide and alkoxide gives a solution containing significant quantities of both the aldehyde or ketone and its enolate. Instead of simply maintaining an equilibrium between the carbonyl compound and its enolate, however, carbon–carbon bond formation can occur.

$$2RCH_2\overset{O}{\overset{\|}{C}}H \xrightleftharpoons{HO^-,\ H_2O} RCH_2\overset{OH}{\overset{|}{C}}H-\underset{R}{\underset{|}{C}}H\overset{O}{\overset{\|}{C}}H$$

Aldehyde β-Hydroxy aldehyde (aldol product)

The resulting equilibrium lies to the right for many aldehydes and to the left for most ketones. The β-hydroxy aldehyde product is called an *aldol* because it contains both an aldehyde and an alcohol function (*ald* + *ol* = *aldol*), and the carbon–carbon bond-forming reaction is referred to as **aldol addition.**

$$2CH_3CH_2CH_2\overset{O}{\overset{\|}{C}}H \xrightarrow[6\text{–}8°C]{KOH,\ H_2O} CH_3CH_2CH_2\underset{HO}{\underset{|}{C}}H\underset{CH_2CH_3}{\underset{|}{C}}H\overset{O}{\overset{\|}{C}}H$$

Butanal 2-Ethyl-3-hydroxyhexanal (75%)

The aldol addition of butanal is shown in Mechanism 20.1. The enolate is formed in the first step by deprotonation of the α carbon. At this point, the reaction mixture contains both the aldehyde and its enolate. The carbonyl group of the aldehyde is electrophilic; the enolate is nucleophilic. This complementary reactivity leads to nucleophilic addition of the enolate to the carbonyl group (step 2). This is the step in which the new carbon–carbon bond forms to give the alkoxide ion corresponding to the aldol. Proton transfer from the solvent (water) completes the process (step 3). The product of the aldol addition of butanal contains two chirality centers; however, it is racemic because the reactants are achiral.

Mechanism 20.1

Aldol Addition of Butanal

THE OVERALL REACTION:

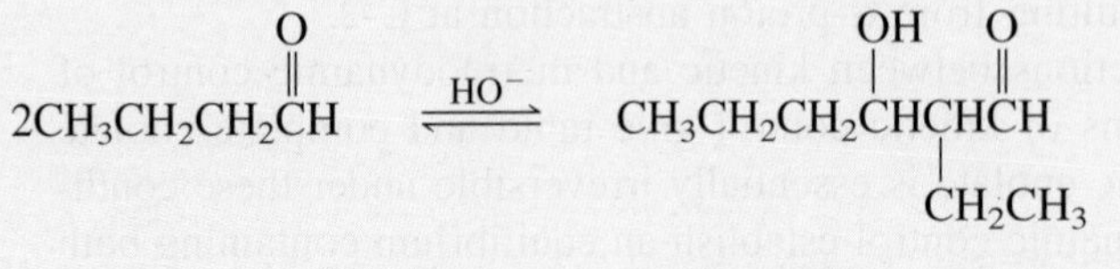

Butanal 2-Ethyl-3-hydroxyhexanal

Step 1: The base, in this case hydroxide ion, converts a portion of butanal to its enolate by abstracting a proton from the α carbon.

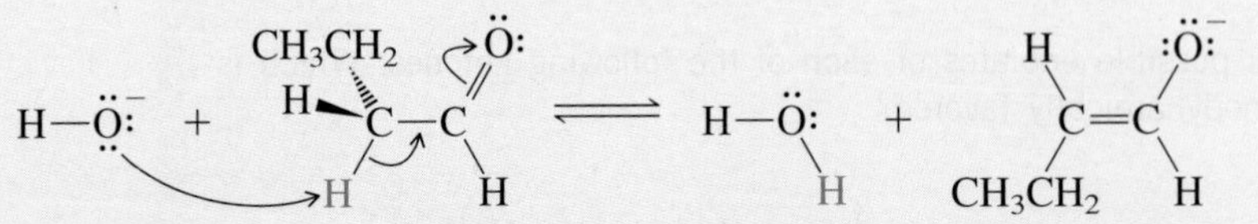

Hydroxide ion Butanal Water Enolate of butanal

Step 2: The enolate undergoes nucleophilic addition to the carbonyl group.

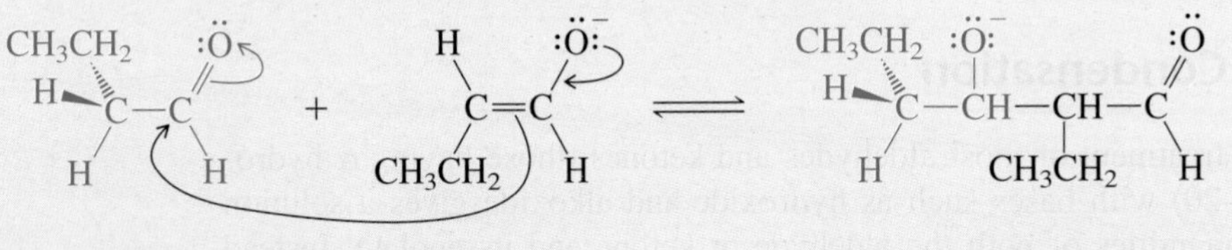

Butanal (electrophile) Enolate of butanal (nucleophile) Alkoxide from nucleophilic addition

Step 3: The alkoxide ion formed in step 2 abstracts a proton from water to give the product of aldol addition.

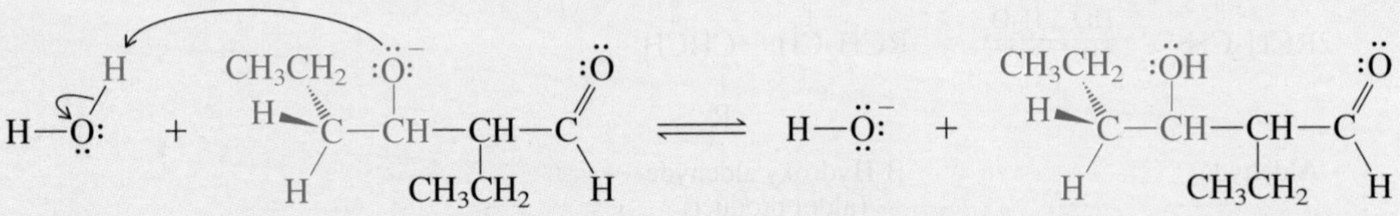

Water Alkoxide from nucleophilic addition Hydroxide ion 2-Ethyl-3-hydroxyhexanal

Problem 20.5

Write the structure of the aldol addition product of

(a) Pentanal, $CH_3CH_2CH_2CH_2\overset{\overset{\displaystyle O}{\|}}{C}H$

(b) 2-Methylbutanal, $CH_3CH_2\underset{\underset{\displaystyle CH_3}{|}}{C}H\overset{\overset{\displaystyle O}{\|}}{C}H$

(c) 3-Methylbutanal, $(CH_3)_2CHCH_2\overset{\overset{\displaystyle O}{\|}}{C}H$

Sample Solution (a) A good way to correctly identify the aldol addition product of any aldehyde is to work through the process mechanistically. Remember that the first step is enolate formation and that this *must* involve proton abstraction from the α carbon.

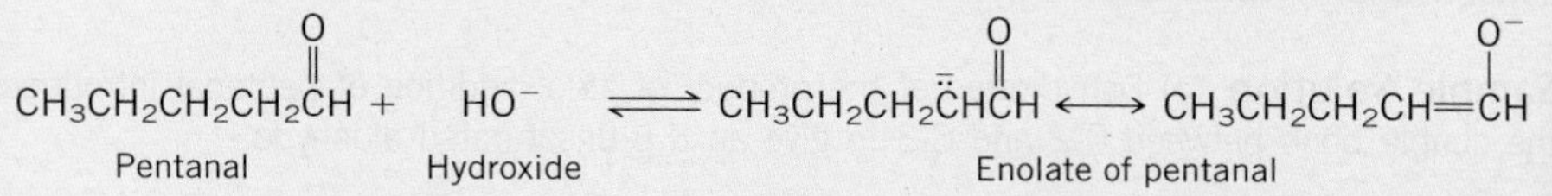

Pentanal
Hydroxide
Enolate of pentanal

Now use the negatively charged α carbon of the enolate to form a new carbon–carbon bond to the carbonyl group. Proton transfer from the solvent completes the process.

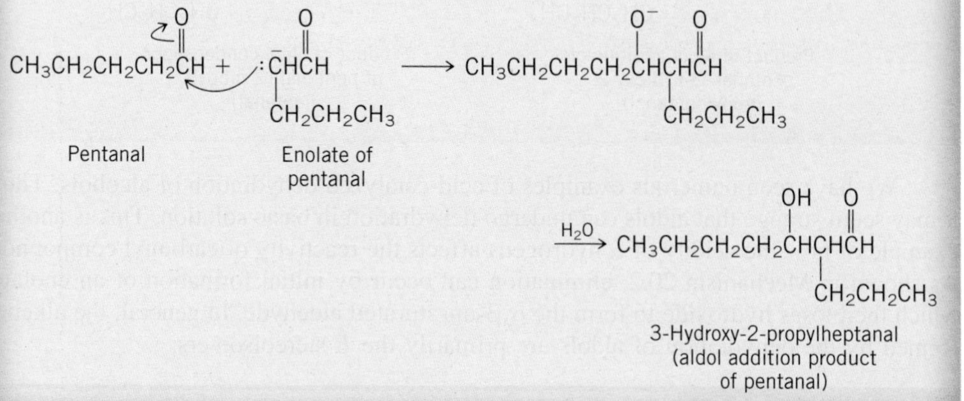

Pentanal
Enolate of pentanal
3-Hydroxy-2-propylheptanal (aldol addition product of pentanal)

The products of aldol addition undergo dehydration on heating, to yield *α,β-unsaturated aldehydes:*

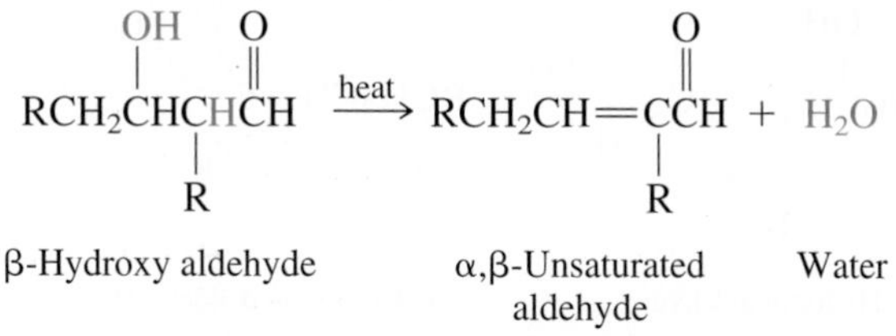

heat
β-Hydroxy aldehyde
α,β-Unsaturated aldehyde
Water

Conjugation of the newly formed double bond with the carbonyl group stabilizes the α,β-unsaturated aldehyde, provides the driving force for the dehydration, and controls its regioselectivity. Dehydration can be effected by heating the aldol with acid or base. Normally, if the α,β-unsaturated aldehyde is the desired product, all that is done is to carry out the base-catalyzed aldol addition reaction at elevated temperature. Under these conditions, once the aldol addition product is formed, it rapidly loses water to form the α,β-unsaturated aldehyde.

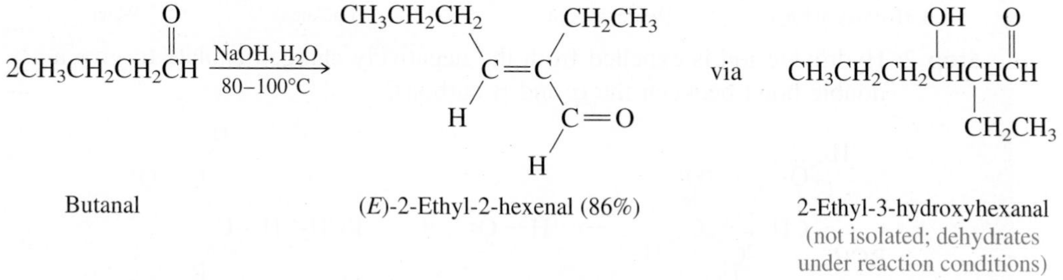

NaOH, H2O
80–100°C
via
Butanal
(E)-2-Ethyl-2-hexenal (86%)
2-Ethyl-3-hydroxyhexanal (not isolated; dehydrates under reaction conditions)

Reactions in which two molecules of an aldehyde combine to form an α,β-unsaturated aldehyde and a molecule of water are called **aldol condensations.**

Recall from Section 15.7 that a condensation is a reaction in which two molecules combine to give a product along with some small (usually inorganic) molecule such as water.

Problem 20.6

Write the structure of the aldol condensation product of each of the aldehydes in Problem 20.5. One of these aldehydes can undergo aldol addition, but not aldol condensation. Which one? Why?

Sample Solution (a) Dehydration of the product of aldol addition of pentanal introduces the double bond between C-2 and C-3 to give an α,β-unsaturated aldehyde.

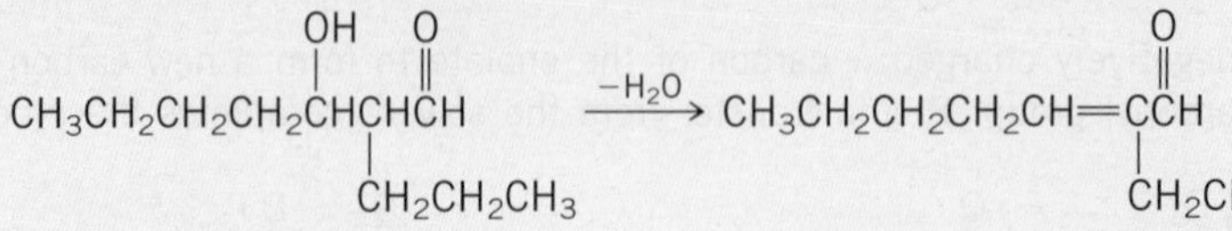

Product of aldol addition of pentanal (3-hydroxy-2-propylheptanal)

Product of aldol condensation of pentanal (2-propyl-2-heptenal)

Mechanism 20.2 is classified as E1cb (*elimination-unimolecular, conjugate base*) because the conjugate base of the reactant undergoes unimolecular dissociation.

We have seen numerous examples of acid-catalyzed dehydration of alcohols. Thus, it may seem strange that aldols can undergo dehydration in basic solution. This is another example of how the acidity of α hydrogens affects the reactivity of carbonyl compounds. As shown in Mechanism 20.2, elimination can occur by initial formation of an enolate, which then loses hydroxide to form the α,β-unsaturated aldehyde. In general, the alkenes formed by the dehydration of aldols are primarily the *E* stereoisomers.

Mechanism 20.2

Dehydration in a Base-Catalyzed Aldol Condensation

THE OVERALL REACTION:

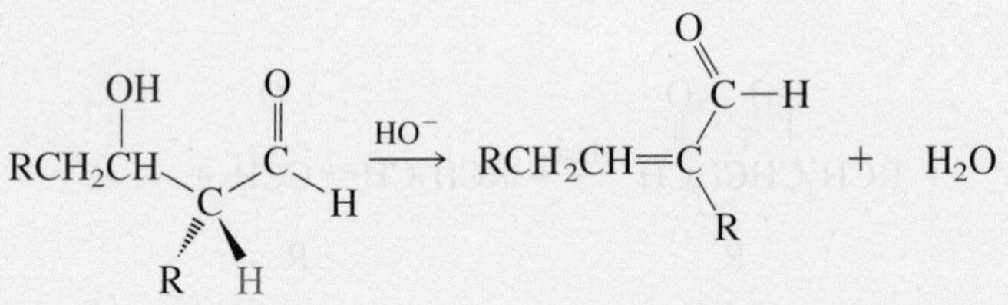

β-Hydroxy aldehyde α,β-Unsaturated aldehyde Water

Step 1: Hydroxide ion abstracts a proton from the α-carbon atom of the β-hydroxy aldehyde to form the corresponding enolate.

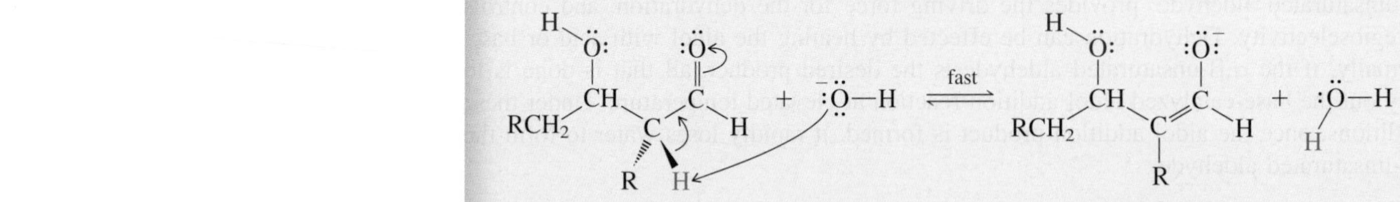

β-Hydroxy aldehyde Hydroxide ion Enolate Water

Step 2: Hydroxide ion is expelled from the negatively charged enolate to give a double bond between the α and β carbons.

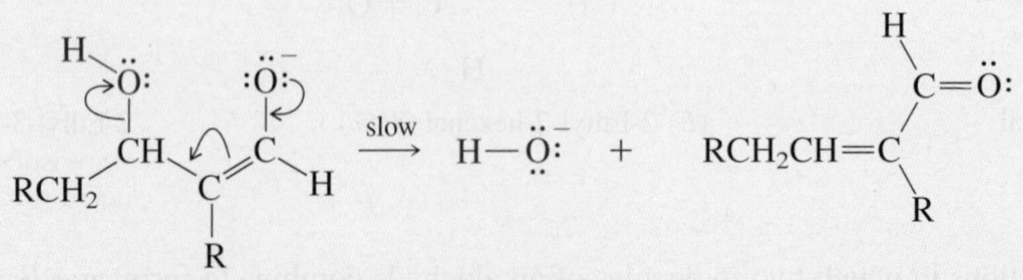

Enolate Hydroxide α,β-Unsaturated aldehyde

As with other reversible nucleophilic addition reactions, the equilibria for aldol additions are less favorable for ketones than for aldehydes. For example, only 2% of the aldol addition product of acetone is present at equilibrium.

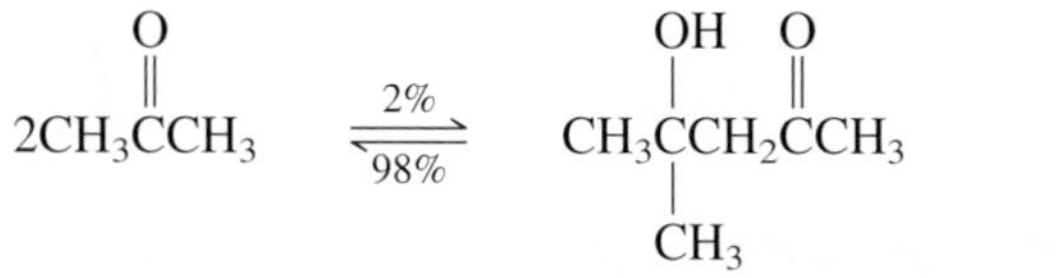
2CH3CCH3
2%
98%
CH3CCH2CCH3
CH3

Acetone   4-Hydroxy-4-methyl-2-pentanone

The situation is similar for other ketones. Special procedures for aldol addition and self-condensation of ketones have been developed, but are rarely used.

Aldol condensations of dicarbonyl compounds—even diketones—occur intramolecularly when five- or six-membered rings are possible.

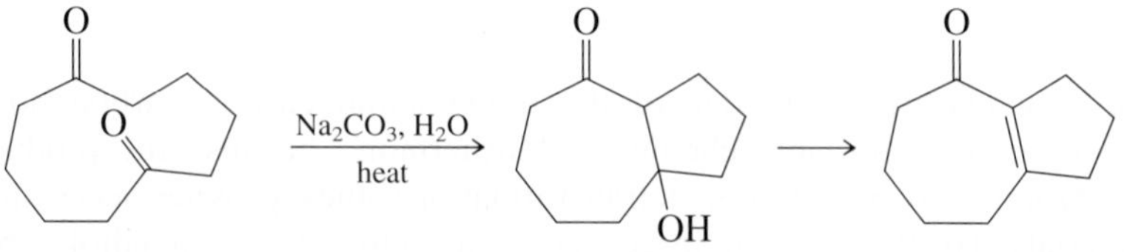
Na2CO3, H2O
heat
OH

1,6-Cyclodecanedione   Not isolated; dehydrates under reaction conditions   Bicyclo[5.3.0]dec-1(7)-en-2-one (96%)

Problem 20.7

Each of the following can be prepared by an intramolecular aldol condensation of a diketone. Deduce the structure of the diketone in each case. (*Hint:* Apply retrosynthetic analysis, starting with disconnection of C=C.)

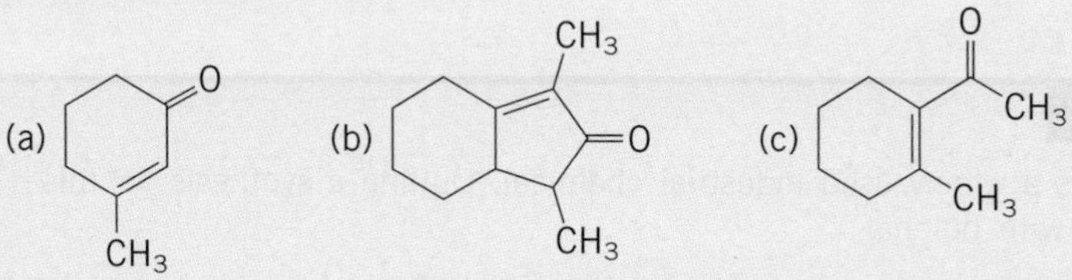
(a)
CH3
(b)
CH3
CH3
(c)
CH3
CH3

Sample Solution

(a) • Apply retrosynthetic analysis.
• To reveal an aldol condensation route to an α,β-unsaturated ketone, disconnect C=C so that the α carbon stays with C=O and the β carbon becomes the second C=O.

Continued

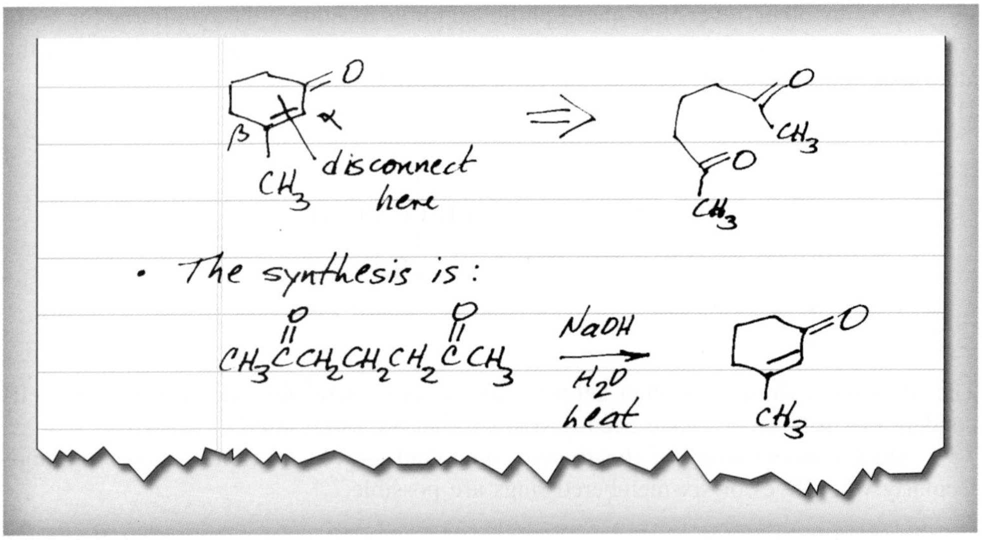
disconnect here
The synthesis is:
NaOH
H_2O
heat

Aldol condensations are one of the fundamental carbon–carbon bond-forming processes of synthetic organic chemistry. Furthermore, because the products contain functional groups capable of further transformations, they provide access to a host of useful materials. To illustrate, the insect repellent 2-ethyl-1,3-hexanediol is prepared by reduction of the aldol addition product of butanal.

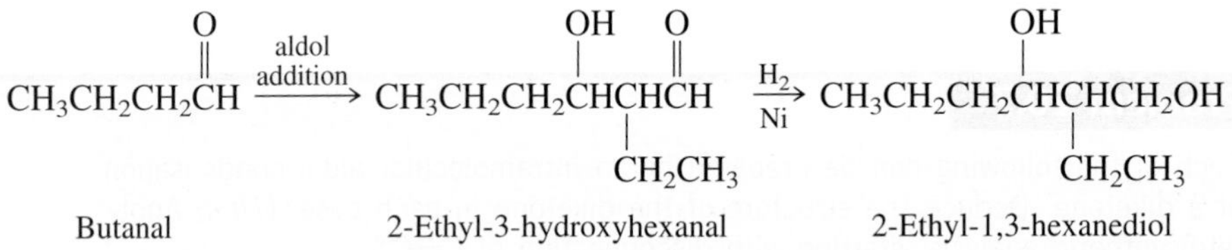
$CH_3CH_2CH_2CH$
aldol addition
$CH_3CH_2CH_2CHCHCH$
CH_2CH_3
H_2
Ni
$CH_3CH_2CH_2CHCHCH_2OH$
CH_2CH_3
Butanal
2-Ethyl-3-hydroxyhexanal
2-Ethyl-1,3-hexanediol

Problem 20.8

2-Ethyl-1-hexanol is a widely used industrial chemical. Outline a synthesis for this compound starting with butanal.

20.4 Mixed Aldol Condensations

Mixed aldol condensations can be effective only if we limit the number of reaction possibilities. It would not be useful, for example, to treat a solution of acetaldehyde and propanal with base. A mixture of four aldol addition products forms under these conditions. Two of the products are those of self-addition:

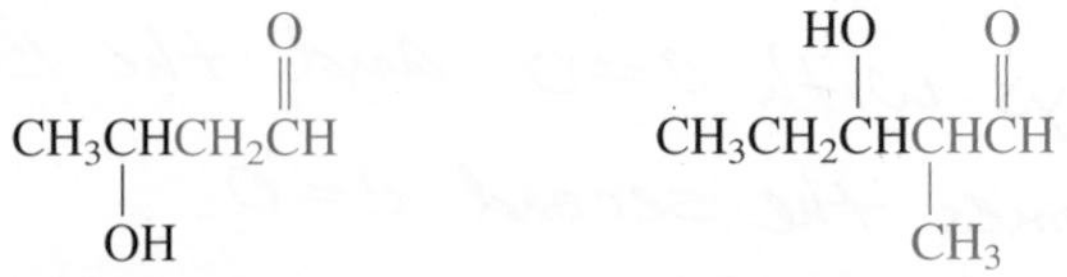
CH_3CHCH_2CH
OH
HO
$CH_3CH_2CHCHCH$
CH_3

3-Hydroxybutanal (from addition of enolate of acetaldehyde to acetaldehyde)

3-Hydroxy-2-methylpentanal (from addition of enolate of propanal to propanal)

Two are the products of mixed addition:

3-Hydroxy-2-methylbutanal (from addition of enolate of propanal to acetaldehyde)

3-Hydroxypentanal (from addition of enolate of acetaldehyde to propanal)

Problem 20.9

Use curved arrows to show the carbon–carbon bond-forming processes that lead to the four aldol addition products just shown.

The mixed aldol condensations that are the most synthetically useful are those in which:

1. Only one of the reactants can form an enolate; or
2. One of the reactants is more reactive toward nucleophilic addition to its carbonyl group than the other.

Formaldehyde, for example, cannot form an enolate but can react with the enolate of some other aldehyde or ketone.

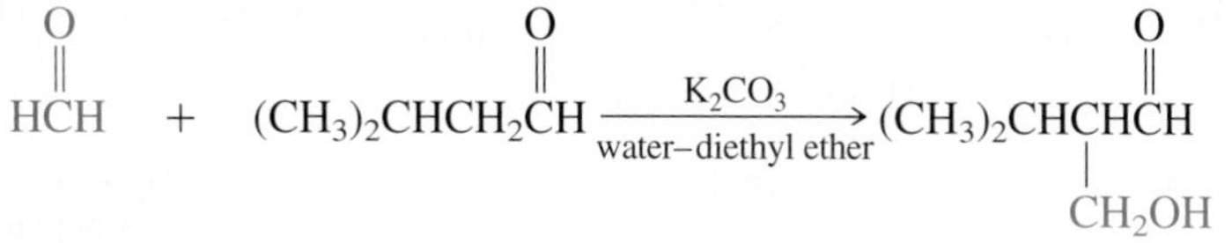

Formaldehyde

3-Methylbutanal

2-Hydroxymethyl-3-methylbutanal (52%)

Moreover, formaldehyde is so reactive toward nucleophilic addition that it suppresses the self-condensation of the other component by reacting rapidly with any enolate present.

Aromatic aldehydes cannot form enolates, and a large number of mixed aldol condensations have been carried out in which an aromatic aldehyde reacts with an enolate.

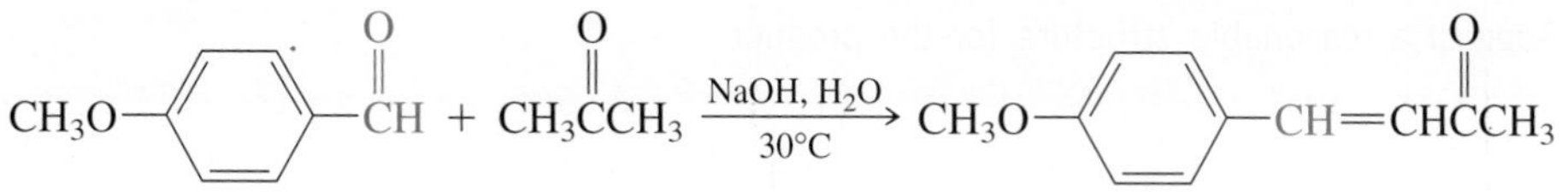

p-Methoxybenzaldehyde

Acetone

4-*p*-Methoxyphenyl-3-buten-2-one (83%)

Recall that ketones do not readily undergo self-condensation. Thus, in the preceding example, the enolate of acetone reacts preferentially with the aromatic aldehyde and gives the mixed aldol condensation product in good yield. Mixed aldol condensations using aromatic aldehydes normally involve spontaneous dehydration of the product of mixed addition and yield a product in which the double bond is conjugated to both the aromatic ring and the carbonyl group.

Mixed aldol condensations in which a ketone reacts with an aromatic aldehyde are known as **Claisen–Schmidt condensations.**

Problem 20.10

Give the structure of the mixed aldol condensation product of benzaldehyde with

(a) Acetophenone, $C_6H_5\overset{O}{\overset{\|}{C}}CH_3$

(b) *tert*-Butyl methyl ketone, $(CH_3)_3\overset{O}{\overset{\|}{C}}CH_3$

(c) Cyclohexanone

Continued

Sample Solution (a) The enolate of acetophenone reacts with benzaldehyde to yield the product of mixed addition. Dehydration of the intermediate occurs, giving the α,β-unsaturated ketone.

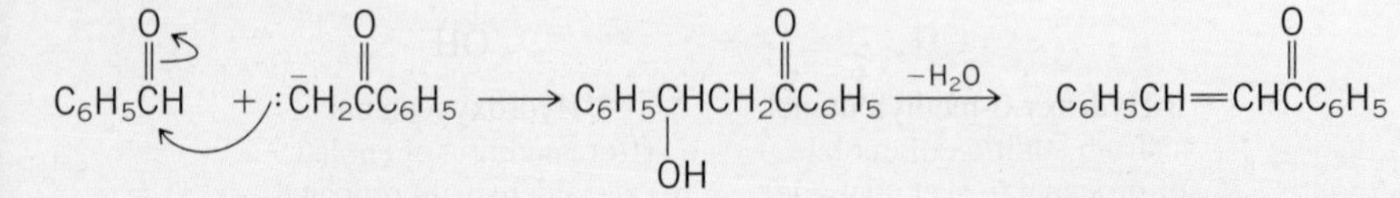
C_6H_5CH
$:\bar{C}H_2CC_6H_5$
$C_6H_5CHCH_2CC_6H_5$
OH
$-H_2O$
$C_6H_5CH{=}CHCC_6H_5$

Benzaldehyde Enolate of acetophenone 1,3-Diphenyl-2-propen-1-one

As actually carried out, the mixed aldol condensation product, 1,3-diphenyl-2-propen-1-one, has been isolated in 85% yield on treating benzaldehyde with acetophenone in an aqueous ethanol solution of sodium hydroxide at 15–30°C.

Another way to ensure that only one enolate is present is by using lithium diisopropylamide (LDA; see Section 20.1) as the base to remove the α proton. LDA is such a strong base that enolate formation is virtually instantaneous and quantitative. Experimentally, a ketone is added to a solution of LDA in a suitable solvent, followed by the compound with which the enolate is to react.

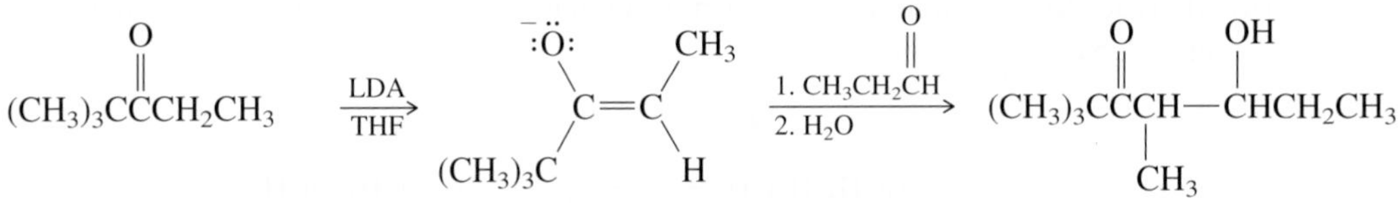
$(CH_3)_3CCCH_2CH_3$
LDA
THF
CH_3
$(CH_3)_3C$
H
1. CH_3CH_2CH
2. H_2O
$(CH_3)_3CCCH{-}CHCH_2CH_3$
OH
CH_3

2,2-Dimethyl-3-pentanone Enolate 5-Hydroxy-2,2,4-trimethyl-3-heptanone (81%)

Reactions of this type are referred to as *directed aldol additions*.

Problem 20.11

When a reaction analogous to the one just shown was carried out using 2-pentanone as the starting ketone, a single β-hydroxy alcohol was isolated in 65% yield. Is this product derived from the thermodynamic or the kinetic enolate of 2-pentanone? Suggest a reasonable structure for the product.

Chalcones: From the Mulberry Tree to Cancer Chemotherapy

Excluding certain types of skin cancer, breast cancer is the most common cancer among women. The chance of developing invasive breast cancer during a woman's lifetime is approximately 1 in 8, and postmenopausal women are at greater risk for developing it because of changes in the production of the hormone estrogen. One approach to the treatment of breast cancer is through the use of drugs such as *tamoxifen* that restrict tumor growth by binding to an estrogen receptor in tumor cells.

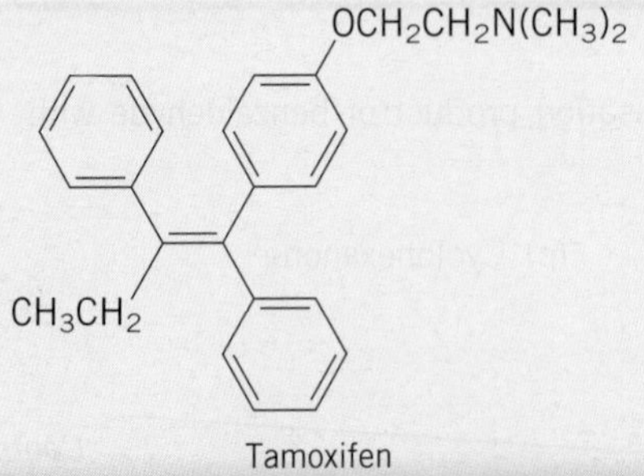
$OCH_2CH_2N(CH_3)_2$
CH_3CH_2

Tamoxifen

An alternative strategy is to reduce the body's production of estrogen by inhibiting the pathway for its biosynthesis. Aromatase inhibitors, which block the final step in the biosynthetic pathway to the estrogens estrone and estradiol, constitute one such class of anticancer drugs. For example, the enzyme aromatase catalyzes the conversion of androstenedione to estrone. By blocking the action of this enzyme, estrone production is inhibited, but the production of other necessary steroids is not affected.

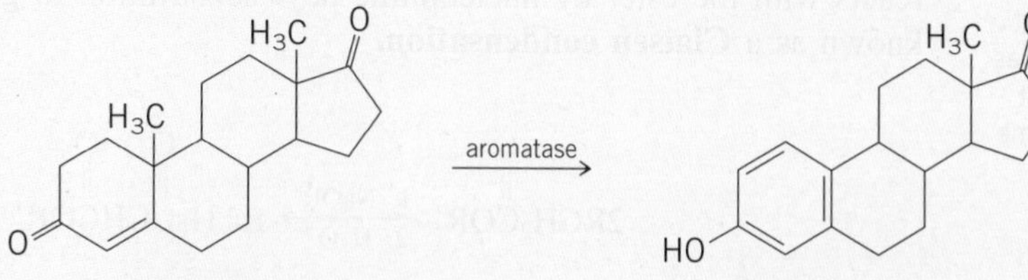

The potential of aromatase inhibitors in the treatment of breast cancer has spurred efforts to find more potent drugs. Bioassays of over 4000 plants led to the discovery of potent aromatase inhibitors from organic extracts of the paper mulberry tree, *Broussonetia papyrifera*. One of these compounds is a chalcone called morachalcone A. Chalcone and its derivatives are present in many plants, have a long history in folk medicine, and continue to be widely used as alternatives to conventional therapies.

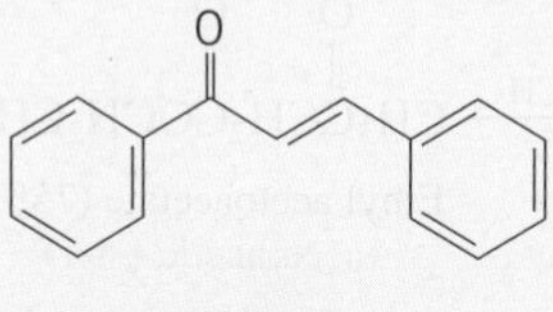

Chalcone

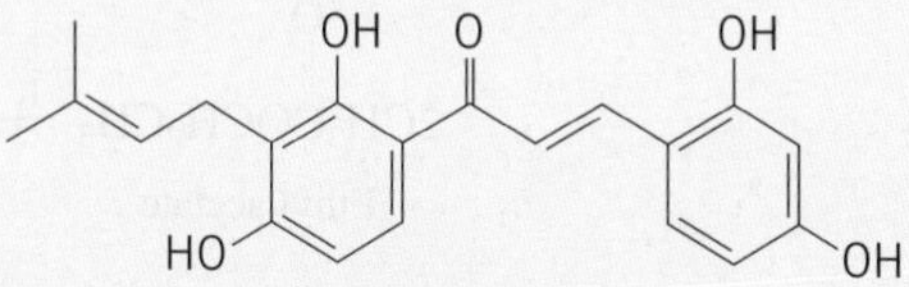

Morachalcone A

In one synthesis of morachalcone A, an aldol condensation between an aldehyde and a ketone was used to prepare the compound shown. The remaining steps involved adding the five-carbon side chain and removing the three groups shown in red.

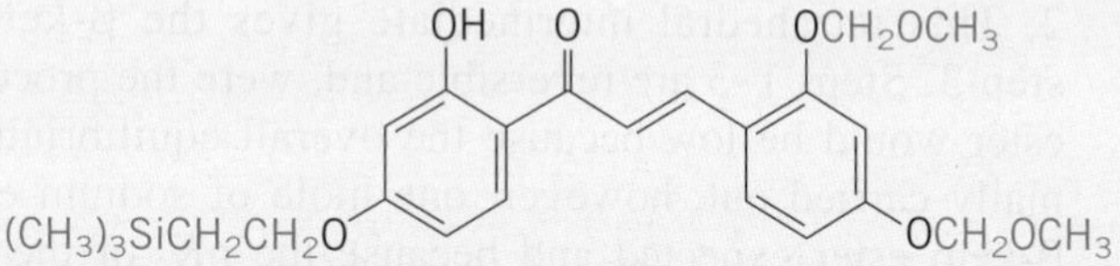

Problem 20.12

Two carbonyl compounds react in ethanol in the presence of KOH to yield the morachalcone precursor just shown. Write structural formulas for these two compounds.

The paper mulberry tree *Broussonetia papyrifera* is a source of a compound that holds promise as an anticancer drug.

20.5 The Claisen Condensation

Aldol additions and condensations begin with the reaction of an enolate derived from an aldehyde or ketone with the carbonyl group of a second molecule of aldehyde or ketone. A related reaction takes place between esters and their enolates when both are present in the same solution. In ethanol containing sodium ethoxide, for example, the enolate reacts with the ester by nucleophilic acyl substitution to give a β-keto ester in what is known as a **Claisen condensation.**

Ludwig Claisen was a German chemist who worked during the last two decades of the nineteenth century and the first two decades of the twentieth. His name is associated with three reactions. The *Claisen-Schmidt reaction* was presented in Section 20.4, the *Claisen condensation* is discussed here, and the *Claisen rearrangement* will be introduced in Section 22.13.

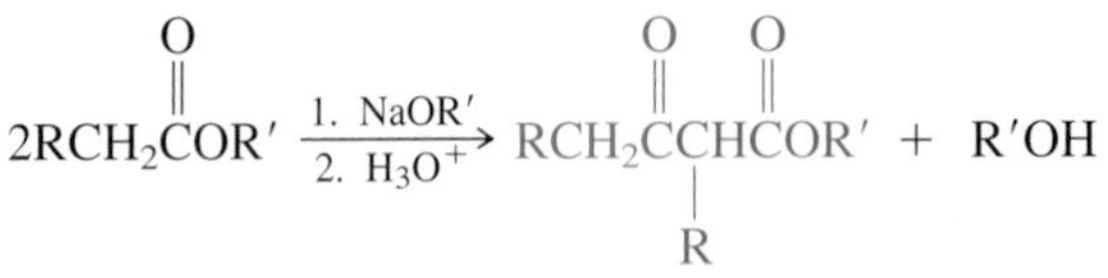

Ester β-Keto ester Alcohol

The Claisen condensation is the main method for preparing β-keto esters. The simplest example is the preparation of ethyl acetoacetate from ethyl acetate.

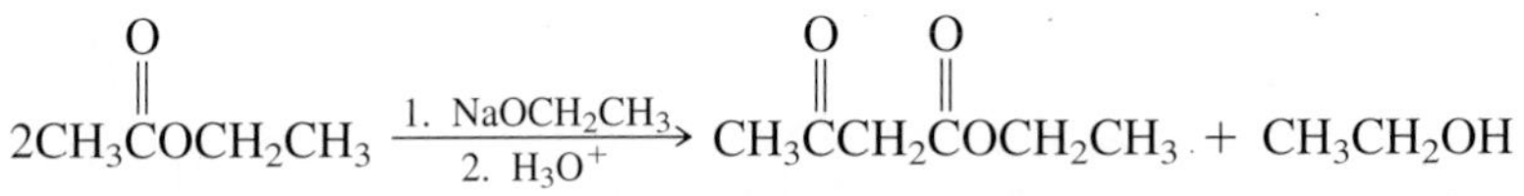

Ethyl acetate Ethyl acetoacetate (75%) (acetoacetic ester) Ethanol

Ethyl acetoacetate is also called *acetoacetic ester.* Its systematic IUPAC name is *ethyl 3-oxobutanoate,* where *"3-oxo"* signifies that C-3 is a carbonyl group.

Mechanism 20.3 outlines the steps in this reaction. The enolate formed in step 1 adds to the carbonyl group of the ester to give a tetrahedral intermediate in step 2. This tetrahedral intermediate gives the β-keto ester by expelling ethoxide in step 3. Steps 1–3 are reversible and, were the process to end here, the yield of β-keto ester would be low because the overall equilibrium constant is unfavorable. As normally carried out, however, one mole of sodium ethoxide is used for every mole of β-keto ester expected and because the pK_a of the β-keto ester is approximately 11, its deprotonation in step 4 drives the equilibrium to favor condensation. Subsequently (step 5), the reaction product is acidified to convert the enolate of the β-keto ester to its neutral form.

The ester enolate in the Claisen condensation undergoes nucleophilic acyl substitution, whereas the aldehyde or ketone enolate in the aldol condensation reacts by nucleophilic addition.

Like aldol condensations, Claisen condensations always involve bond formation between the α-carbon atom of one molecule and the carbonyl carbon of another:

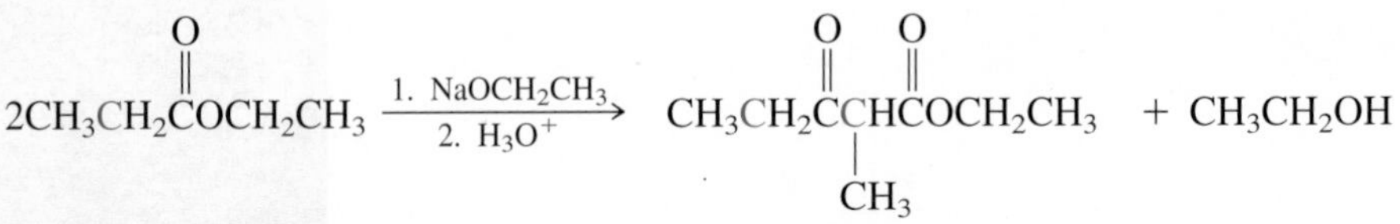

Ethyl propanoate Ethyl 2-methyl-3-oxopentanoate (81%) Ethanol

Unless the β-keto ester can form a stable anion by deprotonation as in step 4 of Mechanism 20.3, the Claisen condensation product is present in only trace amounts at equilibrium. Ethyl 2-methylpropanoate, for example, does not give any of its condensation product under the customary conditions of the Claisen condensation.

Mechanism 20.3

The Claisen Condensation of Ethyl Acetate

THE OVERALL REACTION:

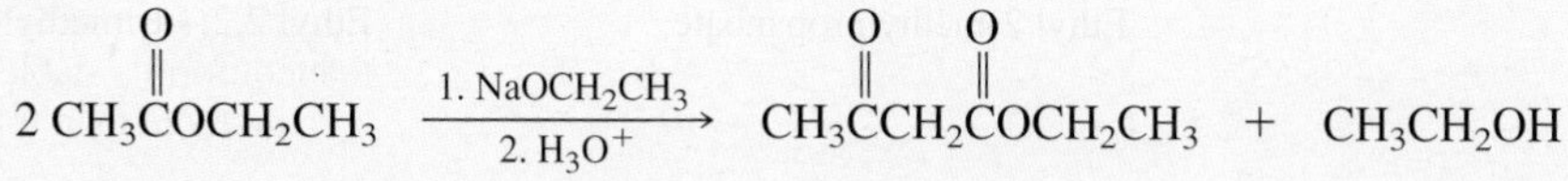

Ethyl acetate — Ethyl 3-oxobutanoate (ethyl acetoacetate) — Ethanol

Step 1: Proton abstraction from the α carbon atom of ethyl acetate to give the corresponding enolate.

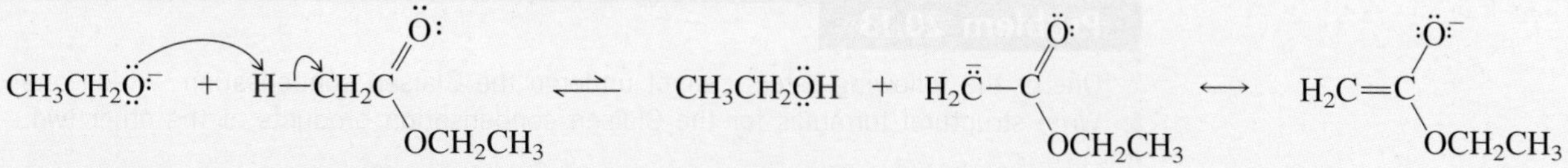

Ethoxide — Ethyl acetate (weaker acid: pK_a = 25.6) — Ethanol (stronger acid: pK_a = 16) — Enolate of ethyl acetate

Step 2: Nucleophilic addition of the ester enolate to the carbonyl group of the neutral ester. The product is the anionic form of the tetrahedral intermediate.

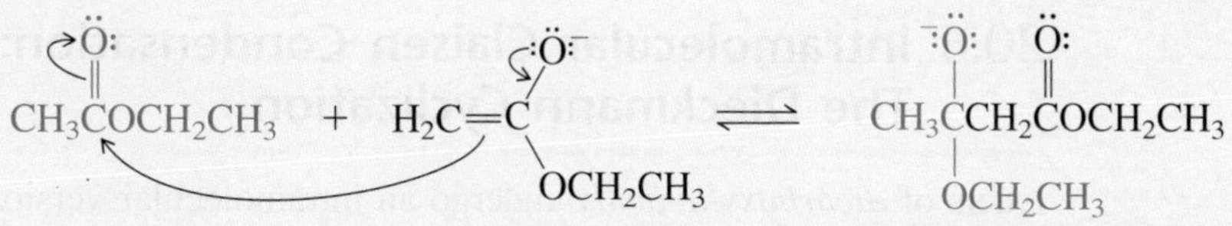

Ethyl acetate — Enolate of ethyl acetate — Anionic form of tetrahedral intermediate

Step 3: Dissociation of the tetrahedral intermediate.

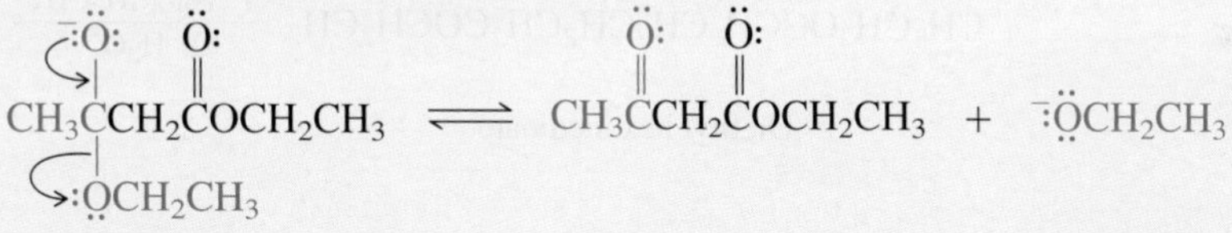

Anionic form of tetrahedral intermediate — Ethyl 3-oxobutanoate — Ethoxide ion

Step 4: Deprotonation of the β-keto ester product.

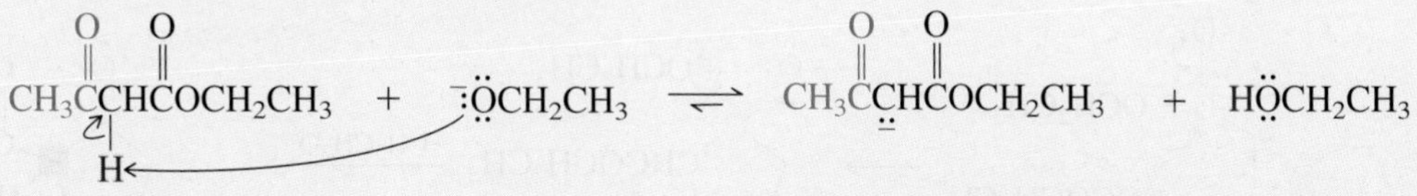

Ethyl 3-oxobutanoate (stronger acid: pK_a = 11) — Ethoxide ion — Conjugate base of ethyl 3-oxobutanoate — Ethanol (weaker acid: pK_a = 16)

Step 5: Acidification of the reaction mixture. This is performed in a separate synthetic operation to give the product in its neutral form for eventual isolation.

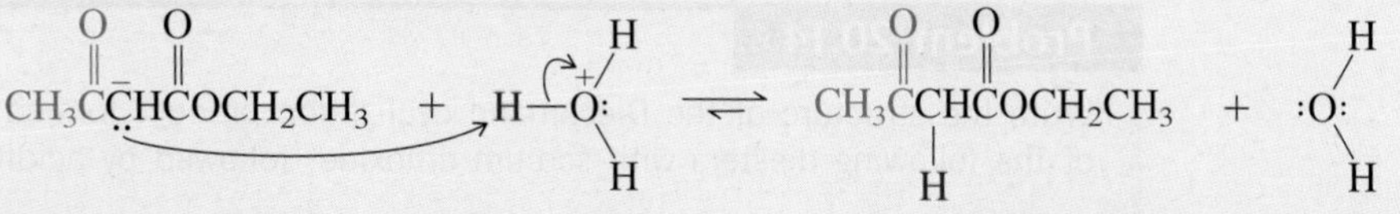

Conjugate base of ethyl 3-oxobutanoate — Hydronium ion (stronger acid: pK_a = −1.7) — Ethyl 3-oxobutanoate (weaker acid: pK_a = 11) — Water

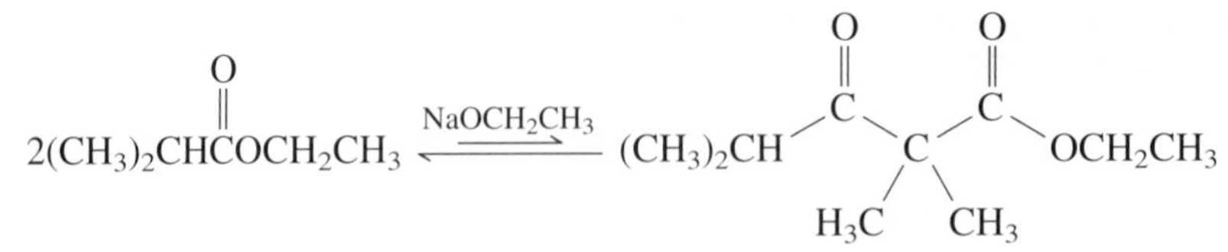

Ethyl 2-methylpropanoate

Ethyl 2,2,4-trimethyl-3-oxopentanoate (cannot form a stable anion; formed in no more than trace amounts)

At least two protons must be present at the α carbon of the starting ester for the equilibrium to favor product formation. Claisen condensation is possible for esters of the type $RC\mathbf{H}_2CO_2R'$, but not for $R_2C\mathbf{H}CO_2R'$.

Problem 20.13

One of the following esters cannot undergo the Claisen condensation. Which one? Write structural formulas for the Claisen condensation products of the other two.

$CH_3CH_2CH_2CH_2CO_2CH_2CH_3$	$C_6H_5CO_2CH_2CH_3$	$C_6H_5CH_2CO_2CH_2CH_3$
Ethyl pentanoate	Ethyl benzoate	Ethyl phenylacetate

20.6 Intramolecular Claisen Condensation: The Dieckmann Cyclization

Esters of *dicarboxylic acids* undergo an intramolecular version of the Claisen condensation when a five- or six-membered ring can be formed.

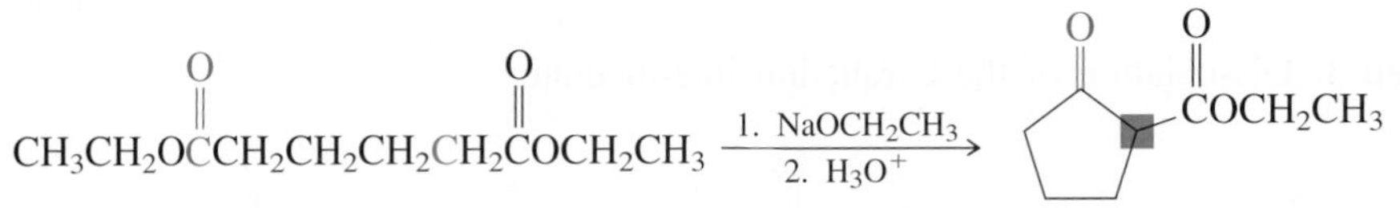

Diethyl hexanedioate

Ethyl (2-oxocyclopentane)-carboxylate (74–81%)

Walter Dieckmann was a German chemist and a contemporary of Claisen.

This reaction is an example of a **Dieckmann cyclization.** The anion formed by proton abstraction at the carbon α to one carbonyl group attacks the other carbonyl to form a five-membered ring.

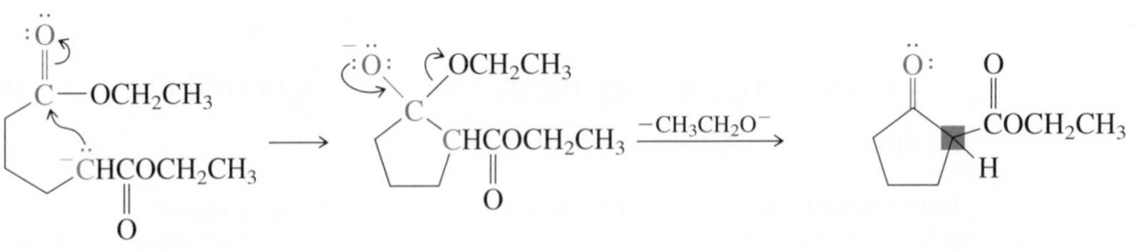

Enolate of diethyl hexanedioate

Ethyl (2-oxocyclopentane)carboxylate

Problem 20.14

Write the structure of the Dieckmann cyclization product formed on treatment of each of the following diesters with sodium ethoxide, followed by acidification.

(a) $CH_3CH_2O\overset{O}{\overset{\|}{C}}CH_2CH_2CH_2CH_2CH_2\overset{O}{\overset{\|}{C}}OCH_2CH_3$

(b) 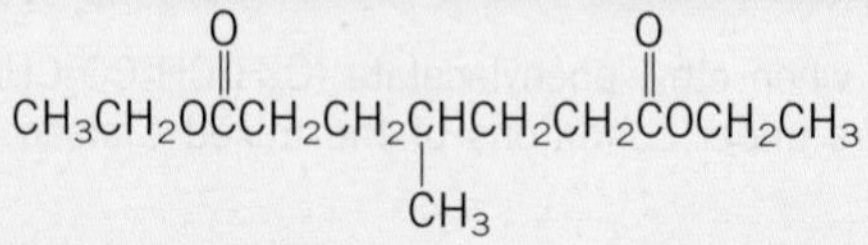

(c) 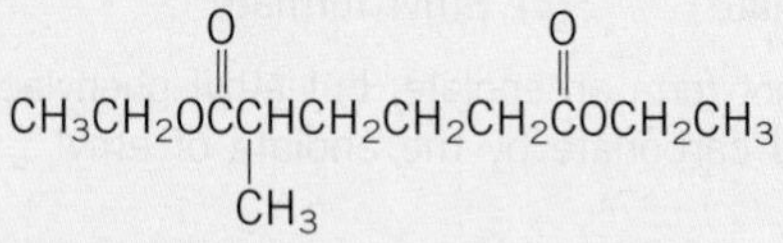

Sample Solution (a) Diethyl heptanedioate has one more methylene group in its chain than the diester cited in the example (diethyl hexanedioate). Its Dieckmann cyclization product contains a six-membered ring instead of the five-membered ring formed from diethyl hexanedioate.

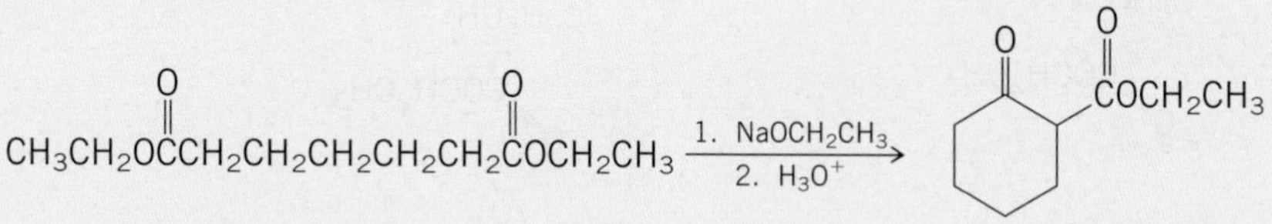

Diethyl heptanedioate → Ethyl (2-oxocyclohexane)-carboxylate

20.7 Mixed Claisen Condensations

Mixed Claisen condensations are analogous to mixed aldol condensations and involve carbon–carbon bond formation between the α-carbon atom of one ester and the carbonyl carbon of another.

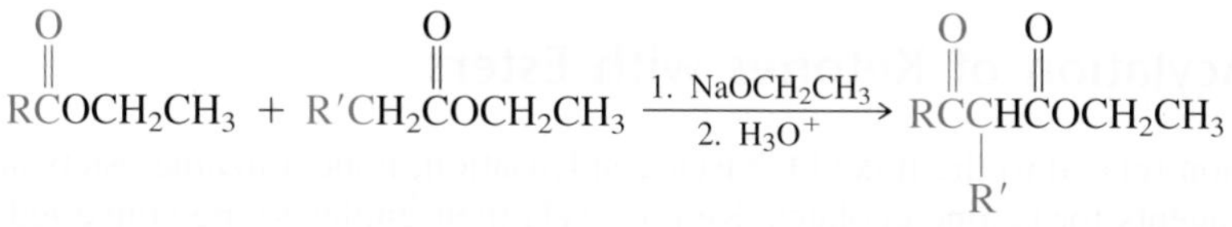

Ester + Another ester → β-Keto ester

The best results are obtained when one of the esters is incapable of forming an enolate. Examples of esters of this type include:

Ethyl formate

Diethyl carbonate

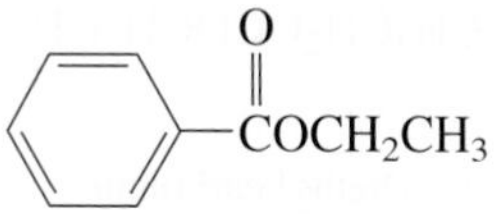

Ethyl benzoate

The following equation shows an example of a mixed Claisen condensation in which a benzoate ester is used as the nonenolizable component:

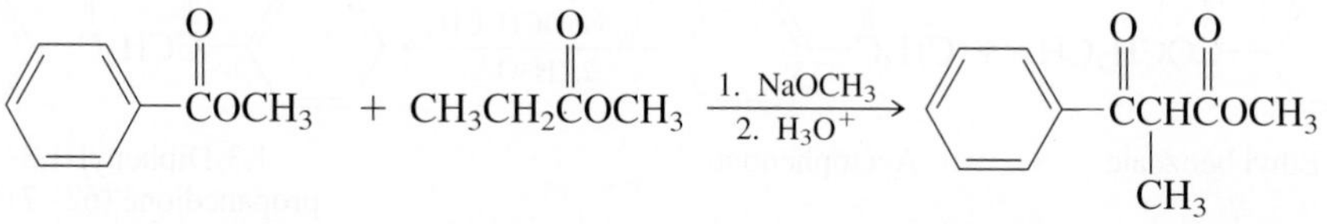

Methyl benzoate (cannot form an enolate) + Methyl propanoate → Methyl 2-methyl-3-oxo-3-phenylpropanoate (60%)

Problem 20.15

Give the structure of the product obtained when ethyl phenylacetate ($C_6H_5CH_2CO_2CH_2CH_3$) is treated with each of the following esters under conditions of the mixed Claisen condensation:

(a) Diethyl carbonate (b) Diethyl oxalate (c) Ethyl formate

Sample Solution (a) Diethyl carbonate cannot form an enolate, but ethyl phenylacetate can. Nucleophilic acyl substitution on diethyl carbonate by the enolate of ethyl phenylacetate yields a *diester.*

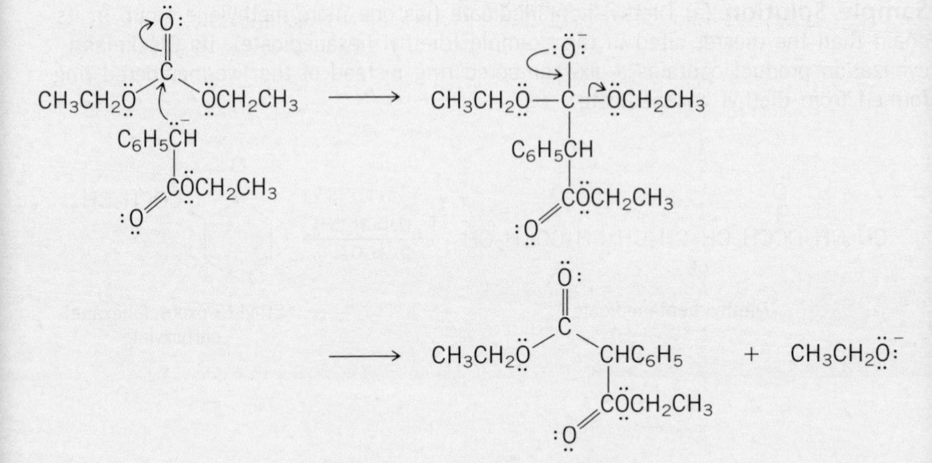

Diethyl phenylmalonate

The reaction proceeds in good yield (86%), and the product, known by its common name diethyl phenylmalonate, is useful in further synthetic transformations of the type to be described in Section 20.11.

20.8 Acylation of Ketones with Esters

In a reaction related to the mixed Claisen condensation, nonenolizable esters are used as acylating agents for ketone enolates. Ketones (via their enolates) are converted to β-keto esters by reaction with diethyl carbonate.

Sodium hydride was used as the base in this example. It is often used instead of sodium ethoxide in these reactions.

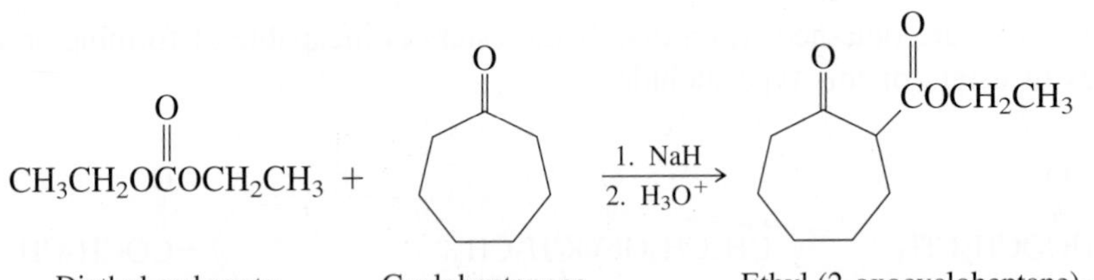

Diethyl carbonate Cycloheptanone Ethyl (2-oxocycloheptane)-carboxylate (91–94%)

Esters of nonenolizable monocarboxylic acids such as ethyl benzoate give β-diketones on reaction with ketone enolates:

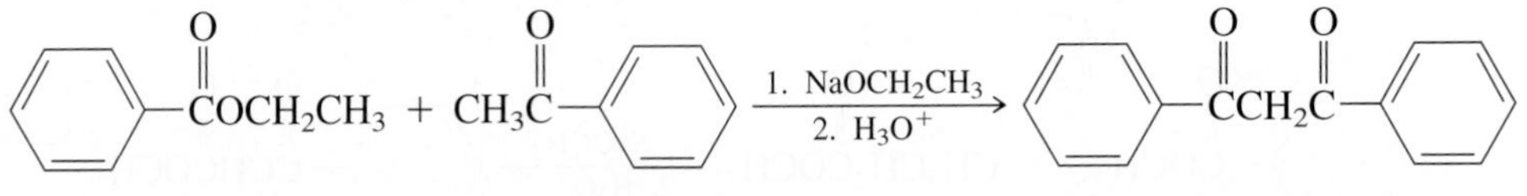

Ethyl benzoate Acetophenone 1,3-Diphenyl-1,3-propanedione (62–71%)

Intramolecular acylation of ketones yields cyclic β-diketones when the ring that is formed is five- or six-membered.

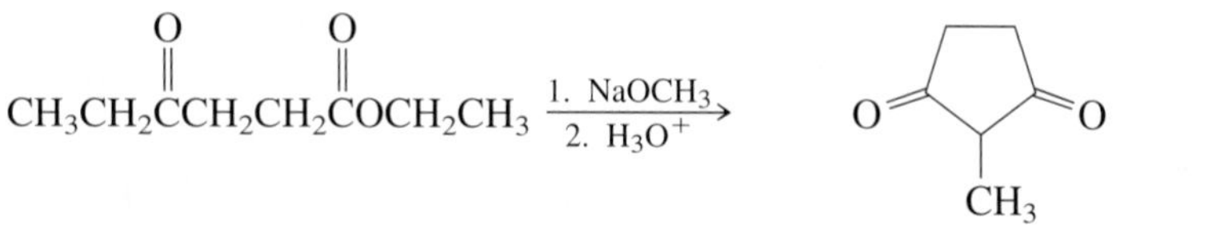

Ethyl 4-oxohexanoate

2-Methyl-1,3-cyclopentanedione (70–71%)

Problem 20.16

Write an equation for the carbon–carbon bond-forming step in the cyclization just cited. Show clearly the structure of the enolate ion, and use curved arrows to represent its nucleophilic addition to the appropriate carbonyl group. Write a second equation showing dissociation of the tetrahedral intermediate formed in the carbon–carbon bond-forming step.

Even though ketones have the potential to react with themselves by aldol addition, recall that the position of equilibrium for such reactions lies to the side of the starting materials (Section 20.3). On the other hand, acylation of ketone enolates gives products (β-keto esters or β-diketones) that are converted to stabilized anions under the reaction conditions. Consequently, ketone *acylation* is observed to the exclusion of aldol addition when ketones are treated with base in the presence of esters.

20.9 Alkylation of Enolates

As sources of nucleophilic carbon, enolates can be alkylated at the α carbon by reaction with alkyl halides.

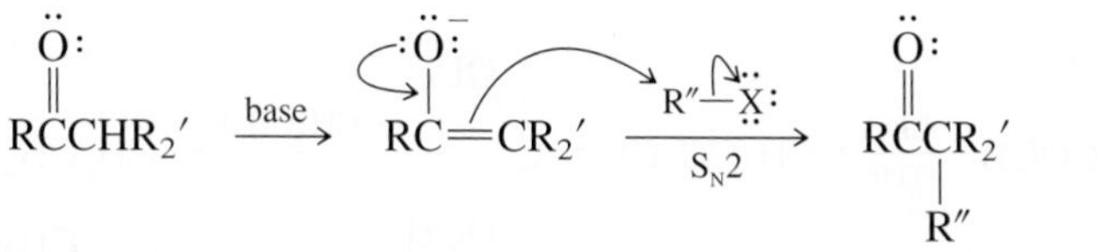

Aldehyde or ketone — Enolate — α-Alkylated aldehyde or ketone

Alkylation occurs by an S_N2 mechanism. The alkyl halide should be methyl or primary, and preferably allylic or benzylic. Secondary and tertiary alkyl halides undergo elimination under these conditions.

Enolate alkylation can be difficult to carry out with simple aldehydes and ketones. It is not always possible to limit the reaction to monoalkylation, and aldol condensation competes with alkylation, especially with aldehydes. The formation of regioisomeric alkylation products is an issue with unsymmetrical ketones but can be minimized by selecting reaction conditions that favor either kinetic or thermodynamic control of enolate formation. The *kinetic enolate* of 2-methylcyclohexanone, for example, was prepared by deprotonation with lithium diisopropylamide then treated with benzyl bromide to give predominantly 2-benzyl-6-methylcyclohexanone,

Some of the stereochemical aspects of the alkylation of 2-methylcyclohexanone are treated in Problem 20.74.

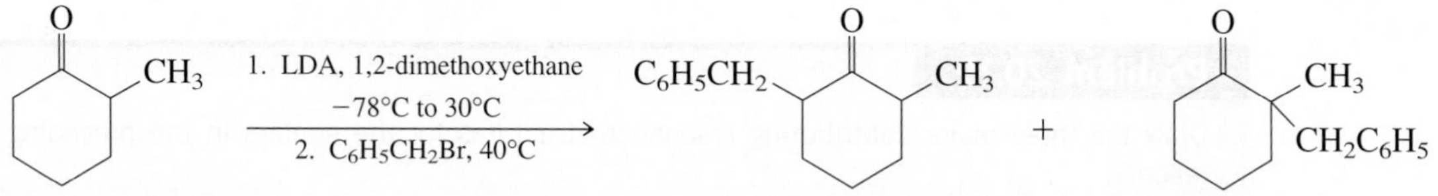

2-Methylcyclohexanone

2-Benzyl-6-methylcyclohexanone (≈52% yield)

2-Benzyl-2-methylcyclohexanone (≈7% yield)

Under similar conditions, but with the ketone present in slightly greater amounts than the base, enolate equilibration causes the *thermodynamic enolate* to predominate. This procedure was used in the first step of a laboratory synthesis of gymnomitrol, a natural product isolated from liverwort.

Note that the alkylating agent in this reaction contains two bromines: one is vinylic, the other allylic. Allylic halides react readily in S_N2 reactions; vinylic halides do not (Section 8.1).

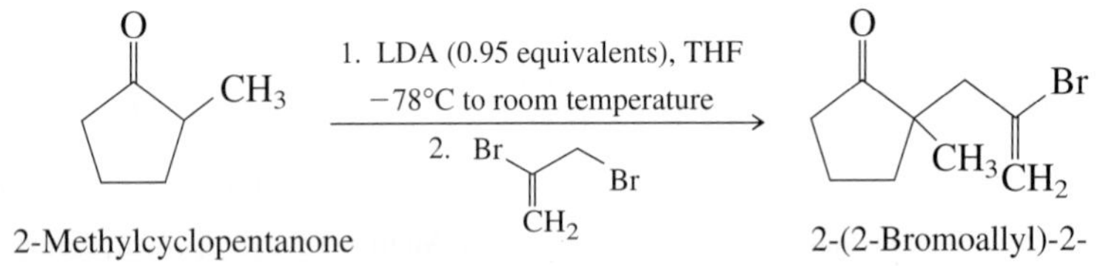

2-Methylcyclopentanone → 2-(2-Bromoallyl)-2-methylcyclopentanone (79%)

Problem 20.17

Give the structure of the major enolate generated in each of the two preceding examples.

Problem 20.18

What combination of ketone, base, and alkyl halide would you choose to prepare each of the following compounds by enolate alkylation?

(a) 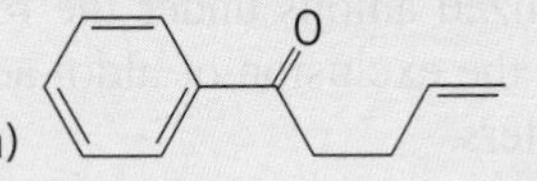

(b) 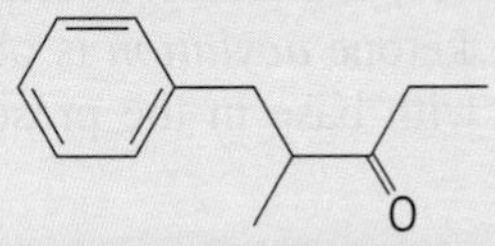

α Alkylation of esters is carried out by procedures analogous to those used for the α alkylation of ketones.

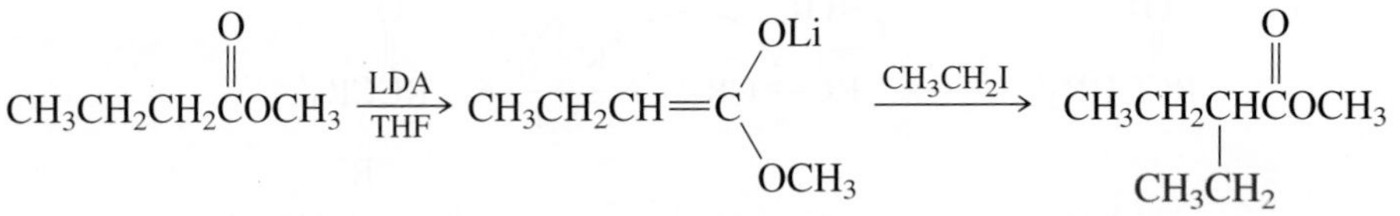

Methyl butanoate → Lithium enolate of methyl butanoate → Methyl 2-ethylbutanoate (92%)

Alkylation of β-dicarbonyl compounds is highly regioselective for the α carbon flanked by two carbonyl groups. A proton at this carbon in a β-ketone is relatively acidic ($pK_a \approx 9$) and is removed by bases even as weak as carbonate.

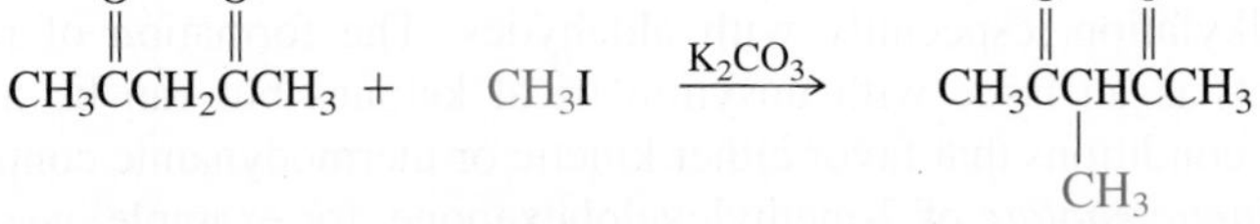

2,4-Pentanedione + Iodomethane → 3-Methyl-2,4-pentanedione (75–77%)

Problem 20.19

Draw the three major contributing resonance structures for the enolate in the preceding reaction.

Other β-dicarbonyl compounds react similarly to β-diketones. The enolates of β-keto esters and 1,3-diesters especially have long occupied an important place in synthetic organic chemistry. Some of their applications are described in the next two sections.

Problem 20.20

You can anticipate some of the material in the next section by completing the following equation.

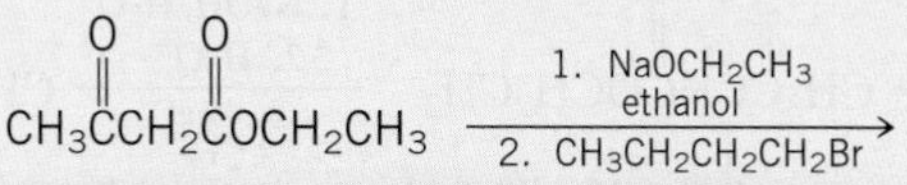

20.10 The Acetoacetic Ester Synthesis

Ethyl acetoacetate (acetoacetic ester), available by the Claisen condensation of ethyl acetate, has properties that make it a useful starting material for the preparation of ketones. These properties are

1. The acidity of the α hydrogen
2. The ease with which acetoacetic acid undergoes thermal decarboxylation

Ethyl acetoacetate is quantitatively converted to its enolate on treatment with sodium ethoxide in ethanol.

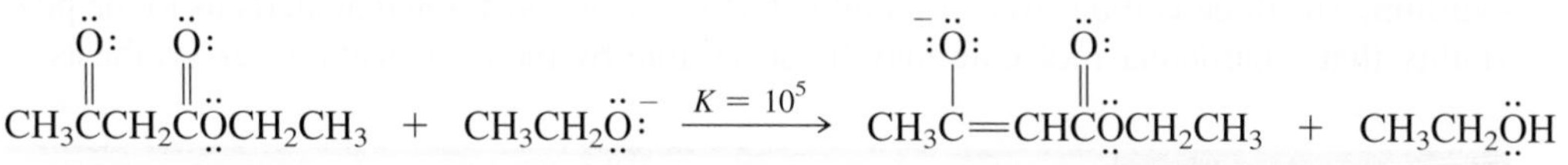

Ethyl acetoacetate (stronger acid: pK_a = 11) — Ethoxide ion — Enolate of ethyl acetoacetate — Ethanol (weaker acid: pK_a = 16)

Adding an alkyl halide to the enolate leads to alkylation at the α carbon by an S_N2 reaction.

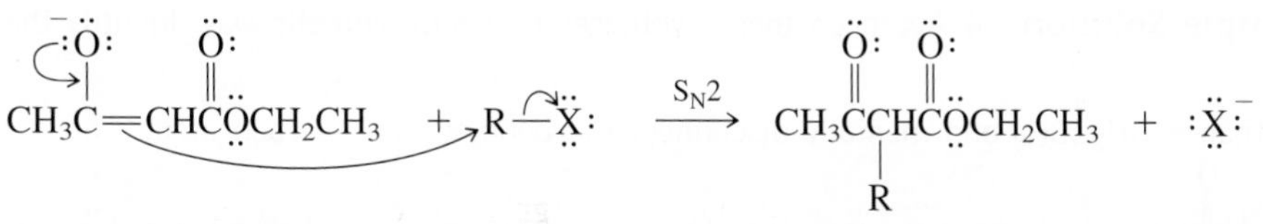

Enolate of ethyl acetoacetate — Alkyl halide — α-Alkyl derivative of ethyl acetoacetate — Halide ion

Methyl and primary alkyl halides work best; secondary alkyl halides give lower yields. Tertiary alkyl halides fail, reacting only by elimination, not substitution.

Saponification and decarboxylation of the α-alkyl derivative of ethyl acetoacetate yields a ketone.

Saponification of esters is covered in Section 19.10. Decarboxylation and its mechanism are discussed in Section 18.16.

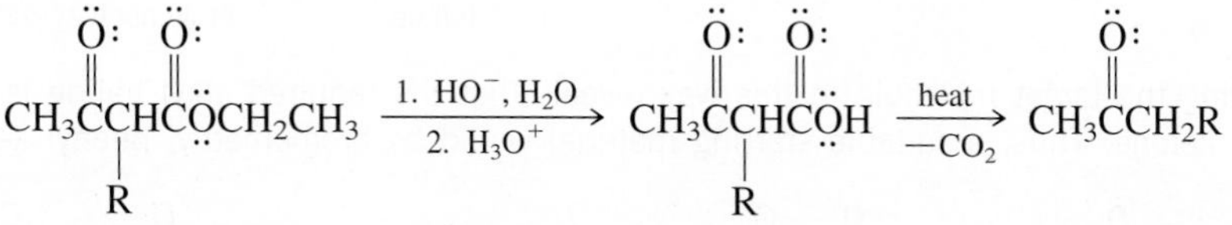

α-Alkyl derivative of ethyl acetoacetate — α-Alkyl derivative of acetoacetic acid — Ketone

Problem 20.21

Draw a curved-arrow representation of the bonding changes in the decarboxylation step of the reaction just shown. What is the relationship of the species formed in this step to the product ketone? (*Hint:* See Section 18.16.)

This reaction sequence is called the **acetoacetic ester synthesis.** It is a standard procedure for the preparation of ketones from alkyl halides, as the conversion of 1-bromobutane to 2-heptanone illustrates.

$$\underset{\text{Ethyl acetoacetate}}{CH_3\overset{O}{\overset{\|}{C}}CH_2\overset{O}{\overset{\|}{C}}OCH_2CH_3} \xrightarrow[\text{2. } CH_3CH_2CH_2CH_2Br]{\text{1. } NaOCH_2CH_3, \text{ ethanol}} \underset{\text{Ethyl 2-butyl-3-oxobutanoate (70\%)}}{CH_3\overset{O}{\overset{\|}{C}}\underset{CH_2CH_2CH_2CH_3}{\underset{|}{C}H}\overset{O}{\overset{\|}{C}}OCH_2CH_3} \xrightarrow[\text{3. heat } -CO_2]{\text{1. NaOH, } H_2O \text{ 2. } H_3O^+} \underset{\text{2-Heptanone (60\%)}}{CH_3\overset{O}{\overset{\|}{C}}CH_2CH_2CH_2CH_2CH_3}$$

The acetoacetic ester synthesis brings about the overall transformation of an alkyl halide to an alkyl derivative of acetone.

$$\underset{\text{Primary or secondary alkyl halide}}{R{-}X} \longrightarrow \underset{\alpha\text{-Alkylated derivative of acetone}}{R{-}CH_2\overset{O}{\overset{\|}{C}}CH_3}$$

We call a structural unit in a molecule that is related to a synthetic operation a **synthon.** The three-carbon structural unit $-CH_2\overset{O}{\overset{\|}{C}}CH_3$ is a synthon that alerts us to the possibility that a particular molecule may be accessible by the acetoacetic ester synthesis.

E. J. Corey (Section 14.9) invented the word *synthon* in connection with his efforts to formalize synthetic planning.

Problem 20.22

Show how you could prepare each of the following ketones from ethyl acetoacetate and any necessary organic or inorganic reagents:

(a) 1-Phenyl-1,4-pentanedione
(b) 4-Phenyl-2-butanone
(c) 5-Hexen-2-one

Sample Solution (a) Approach these syntheses in a retrosynthetic way. Identify the synthon $-CH_2\overset{O}{\overset{\|}{C}}CH_3$ and mentally disconnect the bond to the α-carbon atom. The $-CH_2\overset{O}{\overset{\|}{C}}CH_3$ synthon is derived from ethyl acetoacetate; the remainder of the molecule originates in the alkyl halide.

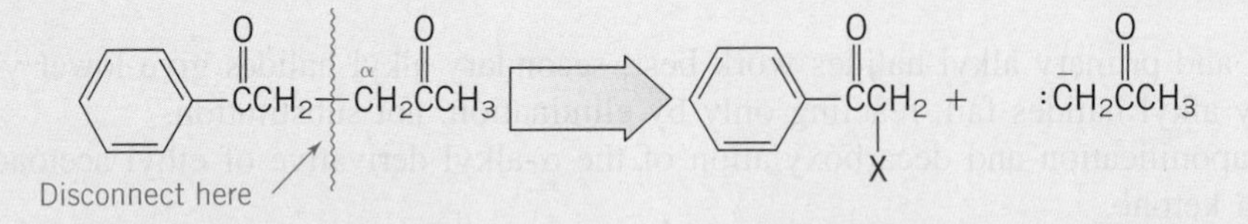

1-Phenyl-1,4-pentanedione   Required alkyl halide   Derived from ethyl acetoacetate

Analyzing the target molecule in this way reveals that the required alkyl halide is an α-halo ketone. Thus, a suitable starting material would be bromomethyl phenyl ketone.

α-Halo ketones are discussed in Section 20.14.

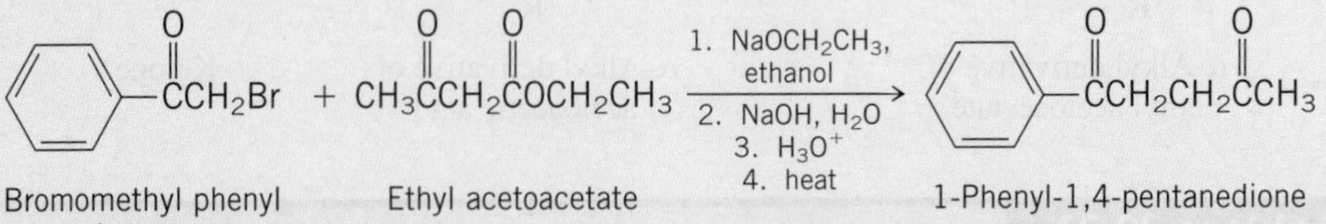

Bromomethyl phenyl ketone   Ethyl acetoacetate   1-Phenyl-1,4-pentanedione

Dialkylation of ethyl acetoacetate can also be accomplished, opening the way to ketones with two alkyl substituents at the α carbon:

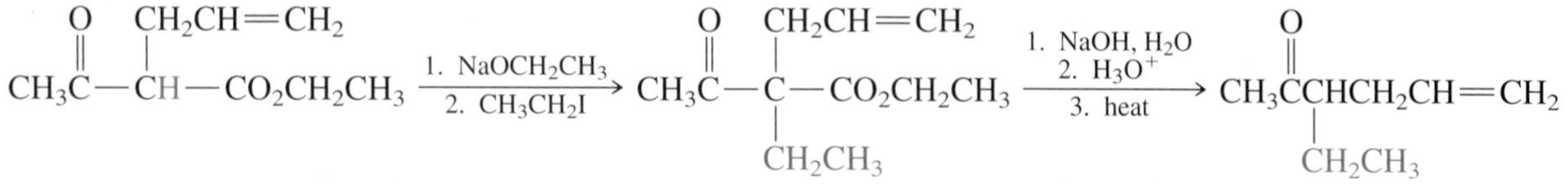

Ethyl 2-allylacetoacetate — Ethyl 2-allyl-2-ethylacetoacetate (75%) — 3-Ethyl-5-hexen-2-one (48%)

Problem 20.23

Show how you could prepare 3-methyl-2-butanone by double alkylation of ethyl acetoacetate.

The starting material in the preceding example is obtained by alkylation of ethyl acetoacetate with allyl bromide.

Recognize, too, that the reaction sequence is one that is characteristic of β-keto esters in general and not limited to just ethyl acetoacetate and its derivatives. Thus,

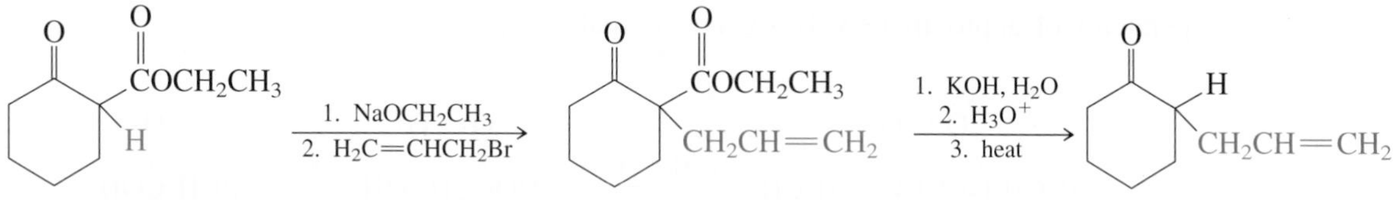

Ethyl 2-oxocyclohexanecarboxylate — Ethyl 1-allyl-2-oxocyclohexanecarboxylate (89%) — 2-Allylcyclohexanone (66%)

The starting material in this reaction is the β-keto ester product formed by the Dieckmann cyclization seen in part (a) of Problem 20.14. The β-keto esters formed by Claisen condensations (Sections 20.5 and 20.7) and the acylation of ketones with esters (Section 20.8) can also be employed as reactants in analogous alkylation–saponification–decarboxylation sequences.

20.11 The Malonic Ester Synthesis

The **malonic ester synthesis** is a method for preparing carboxylic acids. It is similar in concept to the acetoacetic ester synthesis and is represented by the general equation:

$$\underset{\substack{\text{Alkyl}\\\text{halide}}}{RX} + \underset{\substack{\text{Diethyl malonate}\\\text{(malonic ester)}}}{CH_2(COOCH_2CH_3)_2} \xrightarrow[\text{ethanol}]{NaOCH_2CH_3} \underset{\substack{\alpha\text{-Alkylated}\\\text{derivative of}\\\text{diethyl malonate}}}{RCH(COOCH_2CH_3)_2} \xrightarrow[\text{3. heat}]{\text{1. } HO^-,\ H_2O \quad \text{2. } H_3O^+} \underset{\text{Carboxylic acid}}{RCH_2COOH}$$

The overall transformation is:

$$\underset{\substack{\text{Primary or secondary}\\\text{alkyl halide}}}{R{-}X} \longrightarrow \underset{\substack{\alpha\text{-Alkylated derivative}\\\text{of acetic acid}}}{R{-}CH_2\overset{\overset{O}{\|}}{C}OH}$$

Diethyl malonate (also known as malonic ester) serves as a source of the synthon $-CH_2\overset{\overset{O}{\|}}{C}OH$ in the same way that the ethyl acetoacetate serves as a source of the synthon $-CH_2\overset{\overset{O}{\|}}{C}CH_3$.

The properties of diethyl malonate that make the malonic ester synthesis a useful procedure are the same as those responsible for the synthetic value of ethyl acetoacetate. The hydrogens at C-2 of diethyl malonate are relatively acidic, and one is readily removed on treatment with sodium ethoxide.

$$\underset{\substack{\text{Diethyl malonate}\\\text{(stronger acid: } pK_a = 13)}}{CH_3CH_2O\overset{\overset{O}{\|}}{C}CH_2\overset{\overset{O}{\|}}{C}OCH_2CH_3} + \underset{\text{Ethoxide ion}}{CH_3CH_2\ddot{\underset{..}{O}}{:}^-} \xrightarrow{K = 10^3} \underset{\substack{\text{Enolate of diethyl}\\\text{malonate}}}{CH_3CH_2O\overset{\overset{:\ddot{O}:^-}{|}}{C}{=}CH\overset{\overset{O}{\|}}{C}OCH_2CH_3} + \underset{\substack{\text{Ethanol}\\\text{(weaker acid: } pK_a = 16)}}{CH_3CH_2\ddot{O}H}$$

Reaction of the enolate of diethyl malonate with alkyl halides leads to alkylation at C-2.

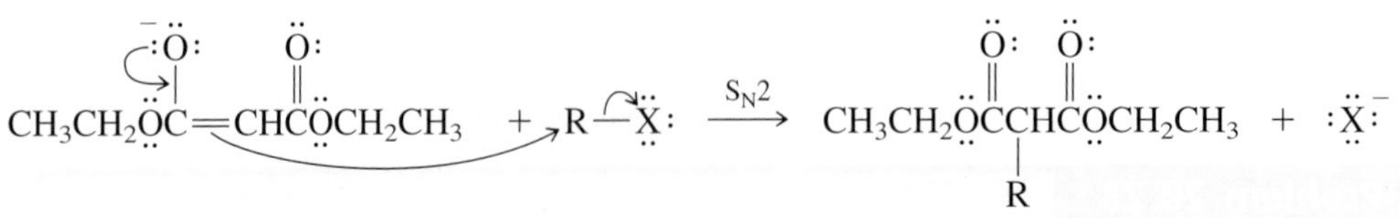

Enolate of diethyl malonate — Alkyl halide — 2-Alkyl derivative of diethyl malonate — Halide ion

Converting the C-2 alkylated derivative to the corresponding malonic acid derivative by ester hydrolysis gives a compound susceptible to thermal decarboxylation. Temperatures of approximately 180°C are normally required.

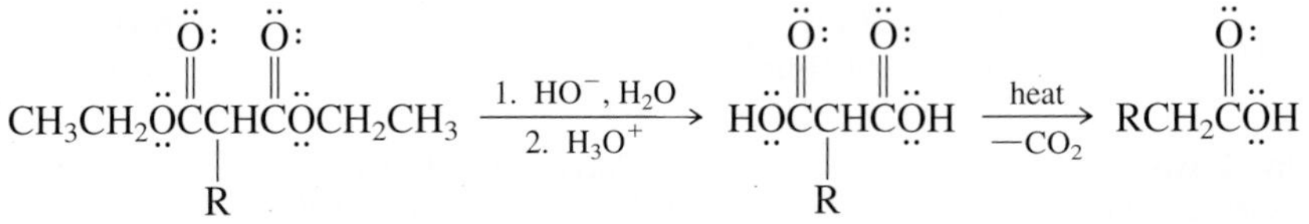

2-Alkyl derivative of diethyl malonate — 2-Alkyl derivative of malonic acid — Carboxylic acid

In a typical example of the malonic ester synthesis, 6-heptenoic acid has been prepared from 5-bromo-1-pentene.

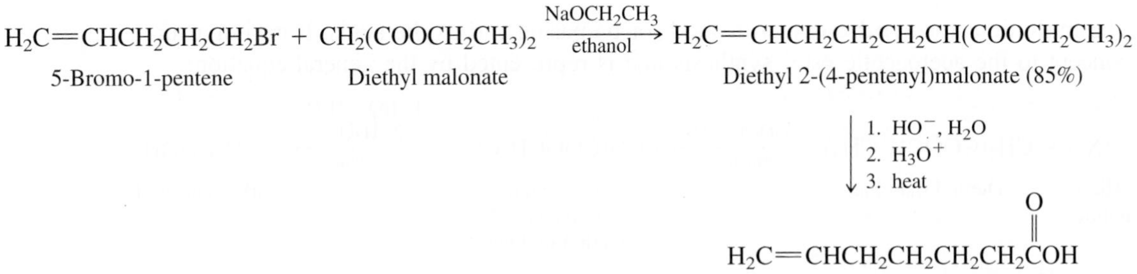

5-Bromo-1-pentene — Diethyl malonate — Diethyl 2-(4-pentenyl)malonate (85%)

6-Heptenoic acid (75%)

Problem 20.24

Show how you could prepare each of the following carboxylic acids from diethyl malonate and any necessary organic or inorganic reagents:

(a) 3-Methylpentanoic acid
(b) Nonanoic acid
(c) 4-Methylhexanoic acid
(d) 3-Phenylpropanoic acid

Sample Solution (a) Analyze the target molecule retrosynthetically by mentally disconnecting a bond to the α-carbon atom.

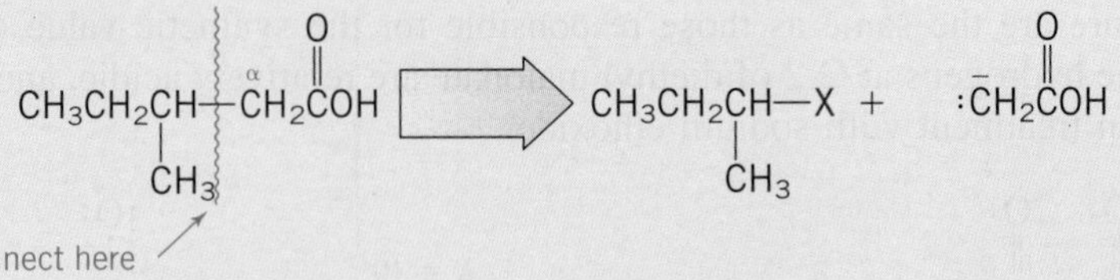

3-Methylpentanoic acid — Required alkyl halide — Derived from diethyl malonate

We see that a secondary alkyl halide is needed as the alkylating agent. The anion of diethyl malonate is a weaker base than ethoxide ion and reacts with secondary alkyl halides by substitution rather than elimination. Thus, the synthesis of 3-methylpentanoic acid begins with the alkylation of the anion of diethyl malonate by 2-bromobutane.

$$\underset{\text{2-Bromobutane}}{CH_3CH_2CH(CH_3)Br} + \underset{\text{Diethyl malonate}}{CH_2(COOCH_2CH_3)_2} \xrightarrow[\text{3. } H_3O^+ \text{ 4. heat}]{\text{1. } NaOCH_2CH_3\text{, ethanol 2. NaOH, } H_2O} \underset{\text{3-Methylpentanoic acid}}{CH_3CH_2CH(CH_3)CH_2COOH}$$

As actually carried out and reported in the chemical literature, diethyl malonate has been alkylated with 2-bromobutane in 83–84% yield and the product of that reaction converted to 3-methylpentanoic acid by saponification, acidification, and decarboxylation in 62–65% yield.

By performing two successive alkylation steps, the malonic ester synthesis can be applied to the synthesis of α,α-disubstituted derivatives of acetic acid:

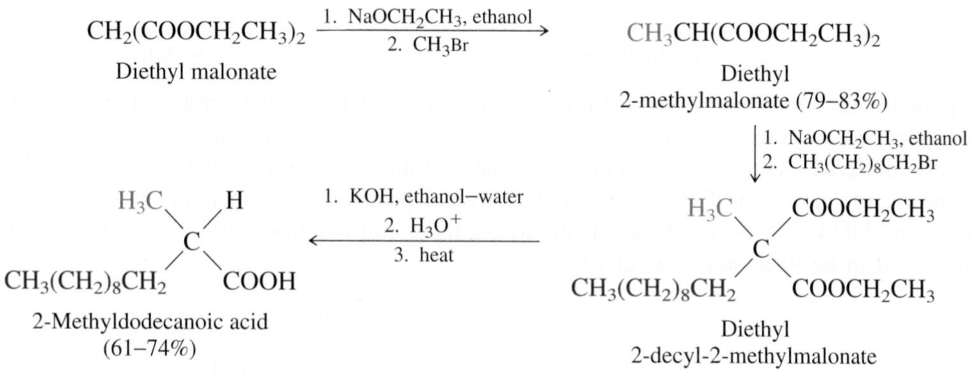
CH2(COOCH2CH3)2
Diethyl malonate
1. NaOCH2CH3, ethanol
2. CH3Br
CH3CH(COOCH2CH3)2
Diethyl
2-methylmalonate (79–83%)
1. NaOCH2CH3, ethanol
2. CH3(CH2)8CH2Br
H3C
COOCH2CH3
C
CH3(CH2)8CH2
COOCH2CH3
Diethyl
2-decyl-2-methylmalonate
1. KOH, ethanol–water
2. H3O+
3. heat
H3C
H
C
CH3(CH2)8CH2
COOH
2-Methyldodecanoic acid
(61–74%)

20.12 Alkylation of Chiral Enolates

Enolate alkylations can be carried out stereoselectively to give enantiomerically enriched α-alkyl carbonyl compounds. Particular success has been achieved with the aid of *chiral auxiliaries,* enantiomerically pure chiral compounds that can be attached to the carbonyl group to control the direction of substitution at the α carbon, then removed to leave the desired α-alkyl carbonyl compound with high enantioselectivity.

One class of chiral auxiliary is based on the heterocyclic compound 2-oxazolidinone. Oxazolidinone itself is achiral, but derivatives such as (4*R*,5*S*)-4-methyl-5-phenyl-2-oxazolidinone are readily prepared as single enantiomers from naturally occurring chiral compounds.

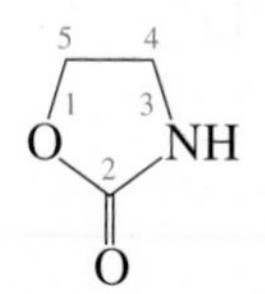
5
4
1
3
O
2
NH
O

2-Oxazolidinone

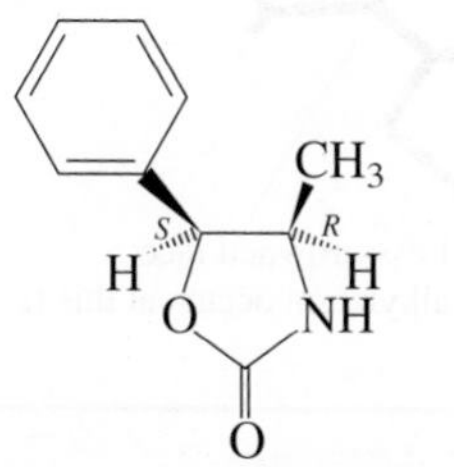
CH3
S
R
H
H
O
NH
O

(4*R*,5*S*)-4-Methyl-5-phenyl-2-oxazolidinone
(a chiral auxiliary)

This chiral auxiliary has been applied to the enantioselective synthesis of α-alkyl carboxylic acids such as (*S*)-2-methyl-4-pentenoic acid. A retrosynthetic analysis of the carbon skeleton of 2-methyl-4-pentenoic acid suggests preparation by alkylation of the enolate of a propanoyl group with an allyl halide.

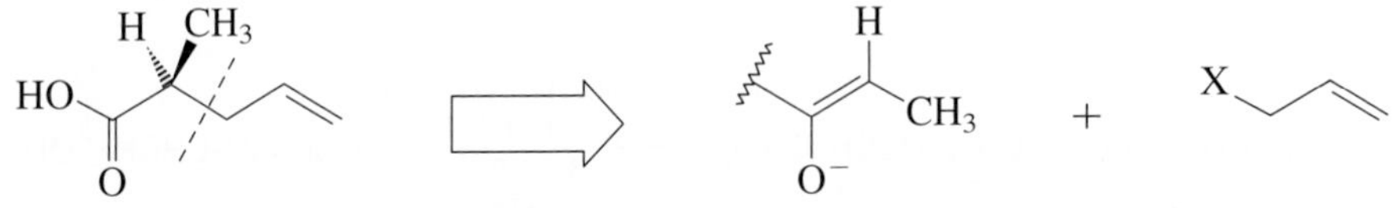

(*S*)-2-Methyl-4-pentenoic acid — Enolate of a propanoyl group — Allyl halide

As actually carried out, the chiral auxiliary was acylated at nitrogen with propanoyl chloride and the propanoyl group then deprotonated at its α carbon with lithium diisopropylamide forming an enolate.

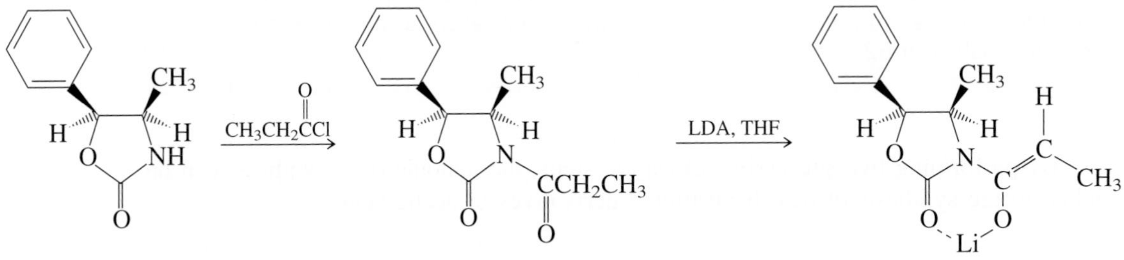

Chiral auxiliary — *N*-Acyl derivative of chiral auxiliary — Lithium enolate of *N*-acyl derivative

Spontaneous coordination of lithium to both the enolate oxygen and to the carbonyl oxygen of the oxazolidinone gives a rigid, cyclic structure, one face of which is sterically hindered. As shown in Figure 20.1, the bottom face of the enolate is less crowded than the top face. The CH_3 group of the original propanoyl group is cis to the enolate oxygen. Alkylation occurs faster at the lower face, and reaction with allyl bromide was observed to be 98% stereoselective.

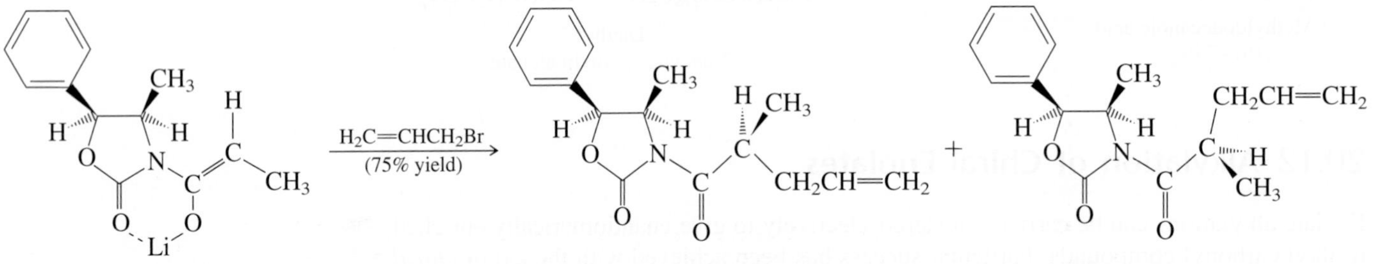

Lithium enolate of *N*-acyl derivative — Major product (98%) — Minor product (2%)

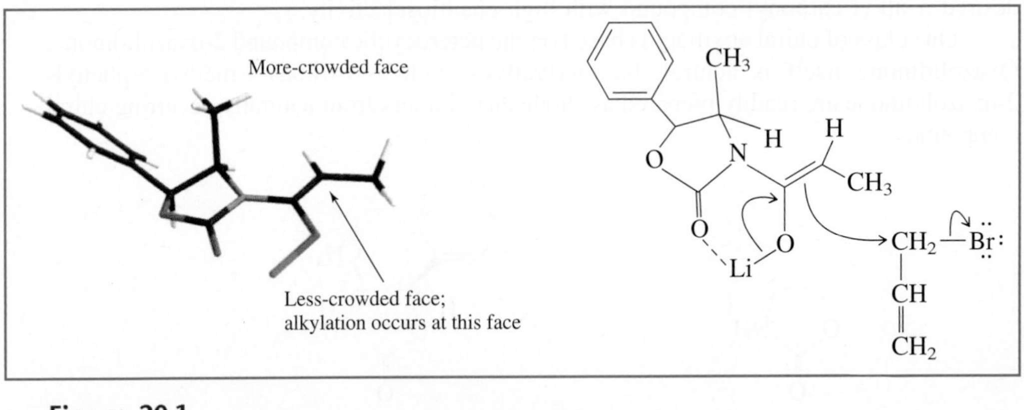

Figure 20.1

The top face of the enolate is more crowded than the lower face. Alkylation occurs preferentially at the lower face.

Problem 20.25

Is the product a mixture of enantiomers or diastereomers? Is the reaction enantioselective or diastereoselective?

To complete the synthesis, the major and minor products were separated, and the major stereoisomer hydrolyzed (nucleophilic acyl substitution) to give the desired enantiomerically pure (*S*)-2-methyl-4-pentenoic acid.

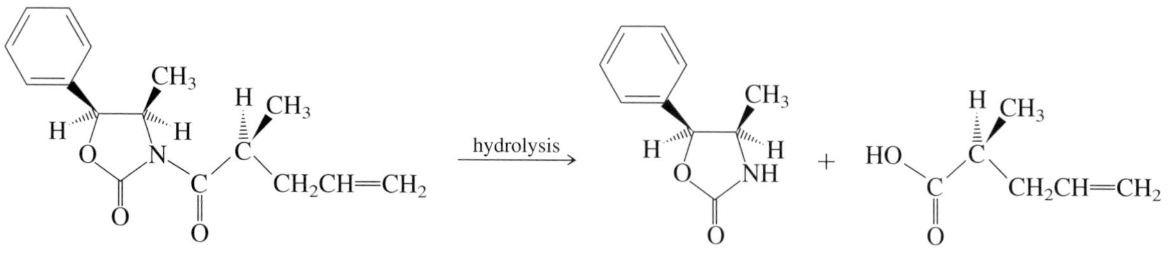

Major product

Recovered chiral auxiliary

(*S*)-2-Methyl-4-pentenoic acid

The other product of hydrolysis is the original chiral auxiliary; it is recovered to be used another day.

The kind of sequence described here is often termed an **asymmetric synthesis,** the stereoselective introduction of a chirality center in a reactant in which the stereoisomeric products are formed in unequal amounts.

20.13 Enolization and Enol Content

As Sections 20.3 through 20.12 demonstrate, the importance of enolates in organic chemistry lies in their synthetic applications, especially in procedures involving the formation of carbon–carbon bonds. What about *enols*—the conjugate acids of enolates?

Although some synthetic applications of enols are known, their main role is as reactive intermediates. Enols are intermediates in the hydration of alkynes (Section 9.12) and the decarboxylation of β-keto acids and malonic acid derivatives (Section 18.16), for example. They are intermediates in a number of biochemical processes including glucose metabolism and malonyl coenzyme A biosynthesis.

This section explores the equilibrium established between a carbonyl compound, the enolic forms available to it, and the mechanisms of enolization. The two sections following this one describe the characteristic reactions of enols.

The amount of enol present at equilibrium, the *enol content,* is quite small for simple carbonyl compounds. Table 20.2 gives equilibrium constants *K* for some representative examples. The table is organized according to the name of the compound and shows its "keto" and "enol" forms, which is how chemists differentiate the two. They are often called **tautomers**—isomers that differ by the placement of an atom or group.

Problem 20.26

Compare Tables 20.1 and 20.2. Can you make a generalization between enol content and α-proton acidity?

TABLE 20.2 Enolization Equilibria of Some Carbonyl Compounds

Compound	Keto form	Enol form	Equilibrium constant, *K**
2,4-Pentanedione	$CH_3\overset{\overset{O}{\|}}{C}CH_2\overset{\overset{O}{\|}}{C}CH_3$	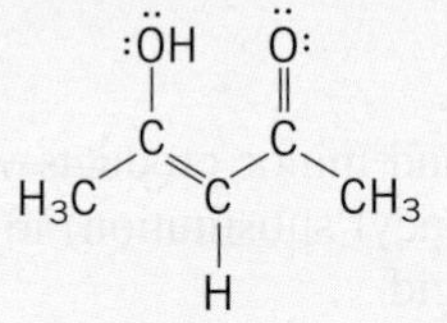	2×10^{-1}
Ethyl acetoacetate	$CH_3\overset{\overset{O}{\|}}{C}CH_2\overset{\overset{O}{\|}}{C}OCH_2CH_3$	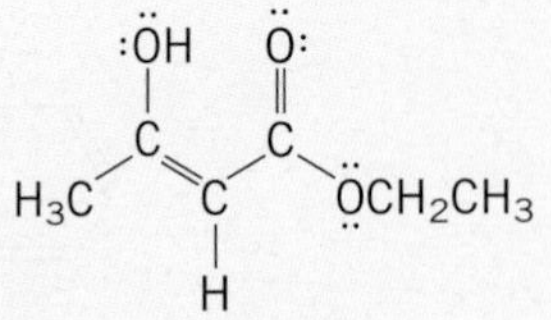	7×10^{-2}
2-Methylpropanal	$(CH_3)_2CH\overset{\overset{O}{\|}}{C}H$	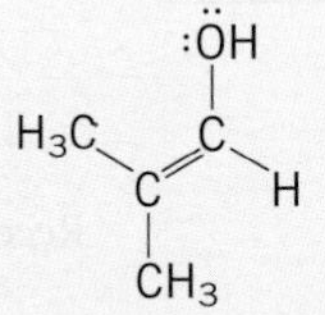	1.4×10^{-4}
Acetaldehyde	$CH_3\overset{\overset{O}{\|}}{C}H$	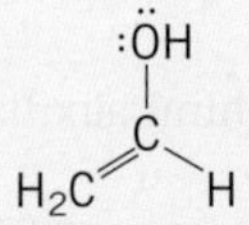	6×10^{-7}
Acetone	$CH_3\overset{\overset{O}{\|}}{C}CH_3$	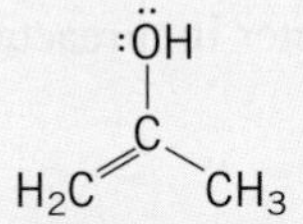	6×10^{-9}
Methyl acetate	$CH_3\overset{\overset{O}{\|}}{C}OCH_3$	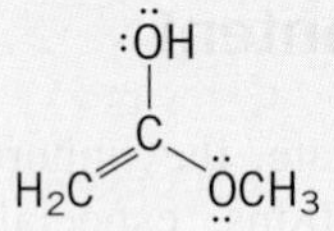	*est.* $10^{-19} - 10^{-24}$
Acetic acid	$CH_3\overset{\overset{O}{\|}}{C}OH$	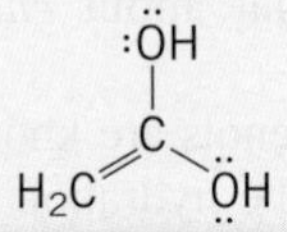	$\approx 10^{-20}$

*In water, 25°C.

As Table 20.2 illustrates, the enol content of simple carbonyl compounds is quite small. Most exist almost entirely in their keto forms. Acetaldehyde contains less than 1 ppm of its enol, and acetone 100 times less than that. Each of the three "simple" aldehydes and ketones in the table (acetaldehyde, acetone, and 2-methylpropanal) has a single enol isomer. In many other cases stereoisomeric and constitutionally isomeric enols are possible.

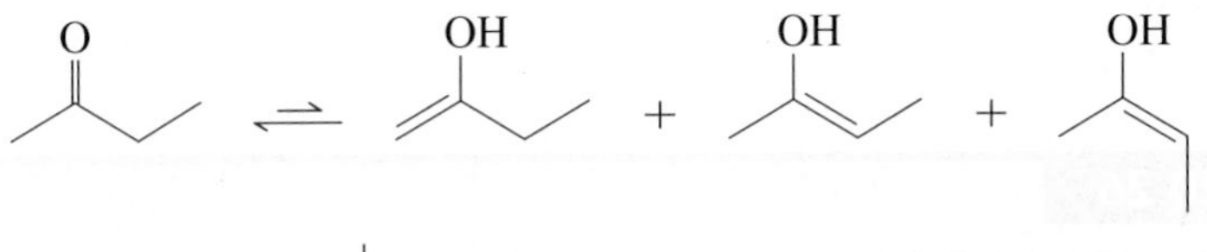

2-Butanone (keto form) | 1-Buten-2-ol | (*Z*)-2-Buten-2-ol | (*E*)-2-Buten-2-ol (enol forms)

Problem 20.27

Write structural formulas for the enols formed from:

(a) 2,2-Dimethyl-3-pentanone
(b) Acetophenone
(c) 2-Methylcyclohexanone
(d) Methyl vinyl ketone

Sample Solution (a) Only one of the α carbons of 2,2-dimethyl-3-pentanone has an attached hydrogen, so only one constitutional isomer is possible for the enol. *E* and *Z* stereoisomers are possible.

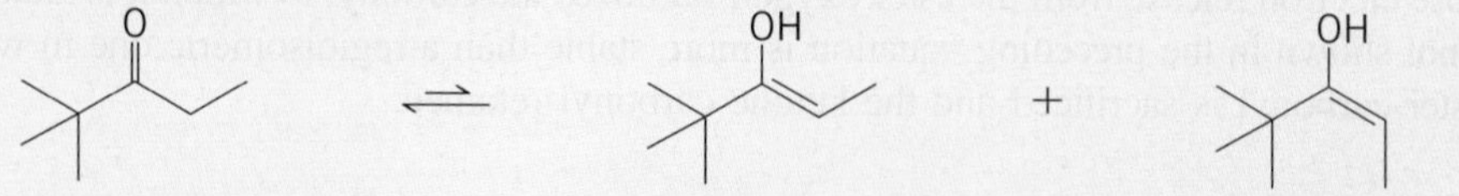

2,2-Dimethyl-3-pentanone (keto form) ⇌ (*Z*)-4,4-Dimethyl-2-penten-3-ol (enol form) + (*E*)-4,4-Dimethyl-2-penten-3-ol (enol form)

Carboxylic acids and esters contain far less enol than aldehydes and ketones. So little enol is present that it is difficult to measure, and the $\approx 10^{-20}$ values for the enolization equilibrium constants of acetic acid and methyl acetate given in Table 20.2 are only approximate. The main reason for the decreased tendency of carboxylic acids and esters to enolize appears to be the stabilization of the carbonyl group of the keto form by electron release from the alkoxy oxygen.

On the other hand, the enol content is much higher in β-dicarbonyl compounds than in simple aldehydes and ketones. As shown in Table 20.2, the enols of 2,4-pentanedione and ethyl acetoacetate are present in the amount of almost 20% and 7%, respectively. In these cases, structural features that stabilize the enol are important. The two most important ones are:

1. Conjugation of the carbon–carbon double bond of the enol with the carbonyl group, and
2. Intramolecular hydrogen bonding of the enolic —OH with the carbonyl oxygen.

Both features are apparent in the structure of the enol of 2,4-pentanedione shown in Figure 20.2.

Problem 20.28

Give the structure of a second enol of 2,4-pentanedione that is a constitutional isomer of the one shown in Figure 20.2. Is it more stable or less stable? Why?

The same features—conjugation of C=C and C=O and intramolecular hydrogen bonding—stabilize the enol forms of ethyl acetoacetate and other β-keto esters.

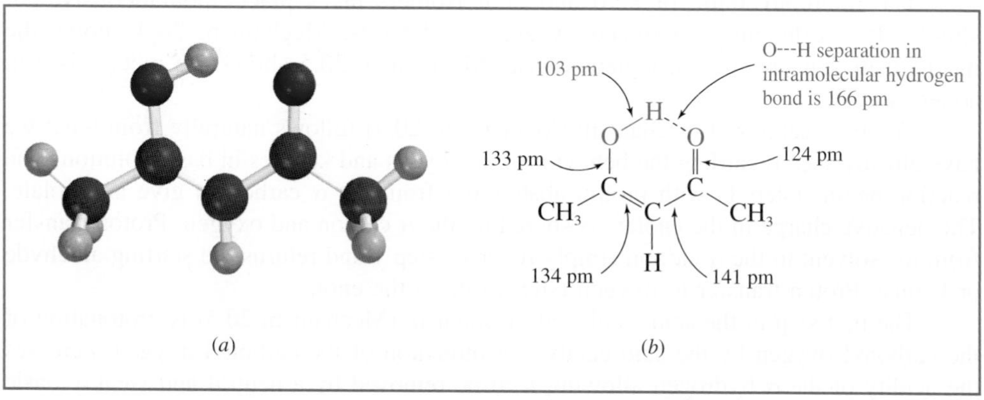

Figure 20.2

(*a*) A molecular model and (*b*) bond distances in the enol of 2,4-pentanedione.

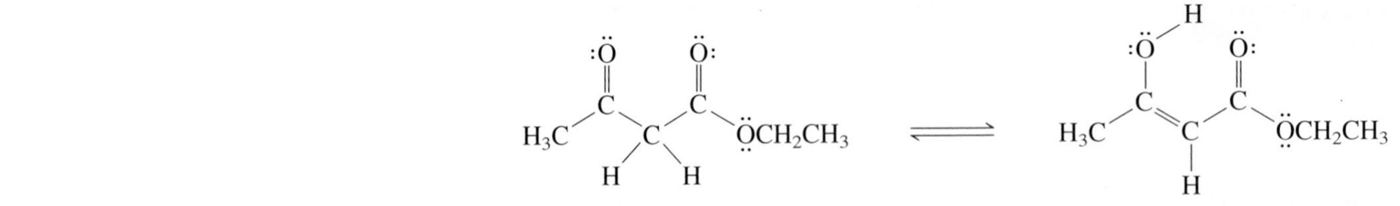

Ethyl acetoacetate — Enol of ethyl acetoacetate

Because electron release from the ester oxygen stabilizes the carbonyl to which it is attached, this enol shown in the preceding equation is more stable than a regioisomeric one in which the ester carbonyl is sacrificed and the ketone carbonyl retained.

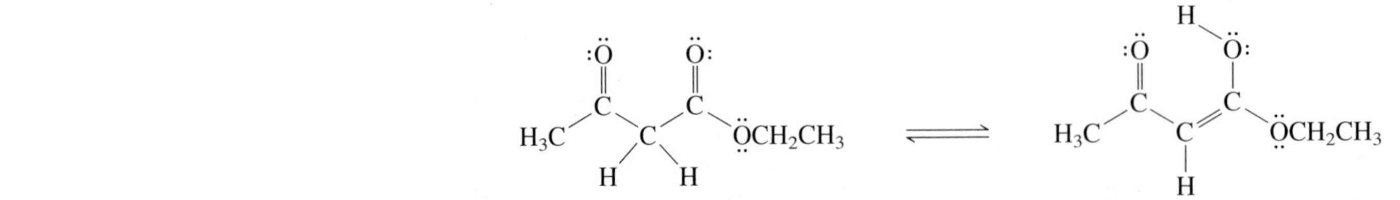

Ethyl acetoacetate — Less stable enol of ethyl acetoacetate

Problem 20.29

Give the structure, including stereochemistry, of the two most stable enol isomers of

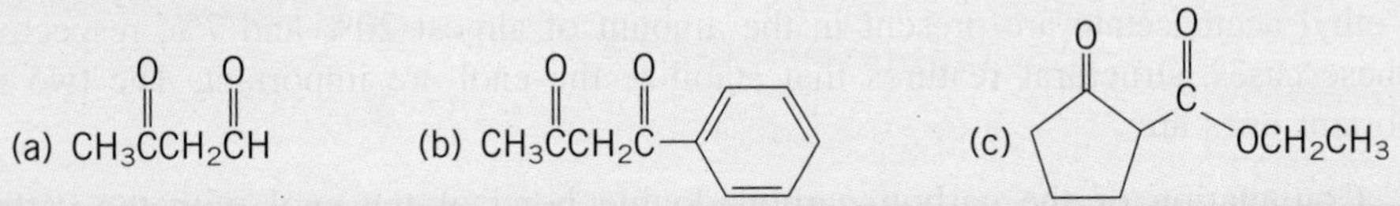
(a) CH3CCH2CH
(b) CH3CCH2C
(c)
OCH2CH3

Problem 20.30

Suggest an explanation for the fact that 2,4-cyclohexadienone exists almost entirely in its enol form.

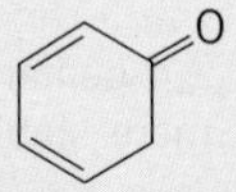

2,4-Cyclohexadienone

The interconversion of keto and enol isomers takes place spontaneously, but slowly. It is efficiently catalyzed by acids and bases. Mechanism 20.4 shows the mechanism that operates in aqueous base. Mechanism 20.5 shows the mechanism in aqueous acid.

The base-catalyzed mechanism (Mechanism 20.4) follows naturally from what we have already seen regarding the behavior of aldehydes and ketones in basic solution. The reaction begins (step 1) with proton abstraction from the α carbon to give an enolate. The negative charge in the enolate is shared by the α carbon and oxygen. Proton transfer from the solvent to the α carbon simply reverses step 1 and returns the starting aldehyde or ketone. Proton transfer to oxygen (step 2) gives the enol.

The first step in the acid-catalyzed mechanism (Mechanism 20.5) is protonation of the carbonyl oxygen by the acid catalyst. Protonation of the carbonyl oxygen increases the acidity of the α hydrogen allowing it to be removed by a neutral and weakly basic solvent molecule in the second step. The product is the enol.

Mechanism 20.4

Base-Catalyzed Enolization of an Aldehyde or Ketone in Aqueous Solution

THE OVERALL REACTION:

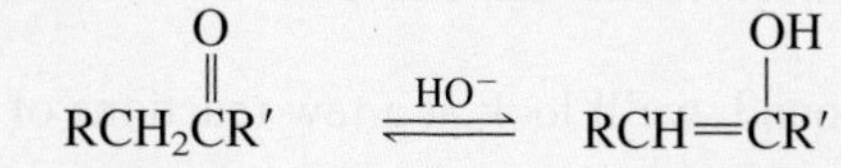

Aldehyde or ketone — Enol

Step 1: A proton is abstracted by hydroxide ion from the α-carbon atom of the carbonyl compound.

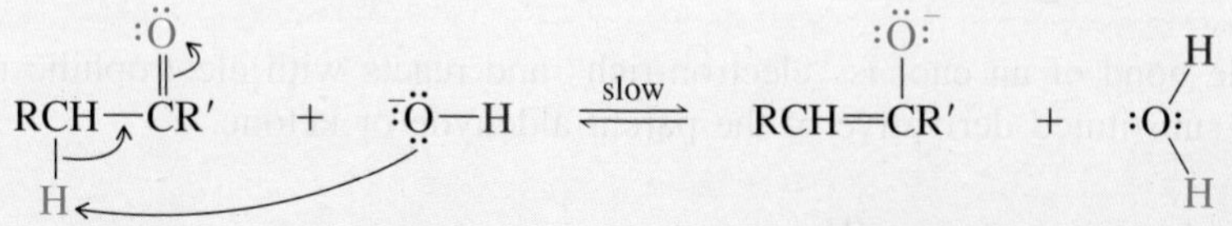

Aldehyde or ketone — Hydroxide ion — Conjugate base of carbonyl compound — Water

Step 2: A water molecule acts as a Brønsted acid to transfer a proton to the oxygen of the enolate ion.

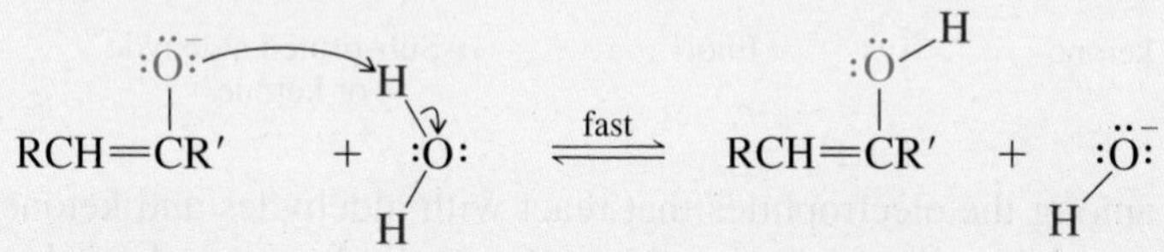

Conjugate base of carbonyl compound — Water — Enol — Hydroxide ion

Mechanism 20.5

Acid-Catalyzed Enolization of an Aldehyde or Ketone in Aqueous Solution

THE OVERALL REACTION:

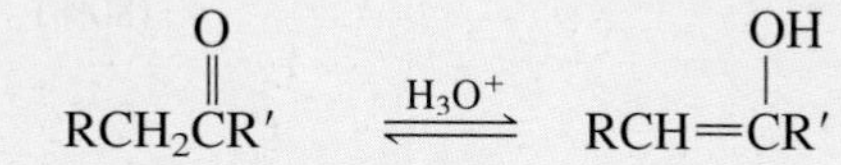

Aldehyde or ketone — Enol

Step 1: A proton is transferred from the acid catalyst to the carbonyl oxygen.

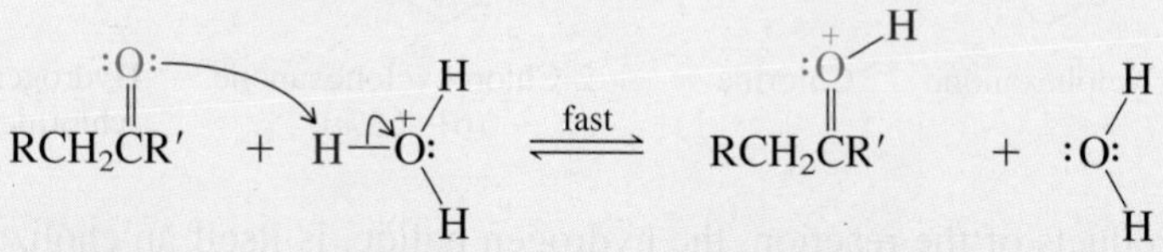

Aldehyde or ketone — Hydronium ion — Conjugate acid of carbonyl compound — Water

Step 2: A water molecule acts as a Brønsted base to remove a proton from the α-carbon atom of the protonated aldehyde or ketone.

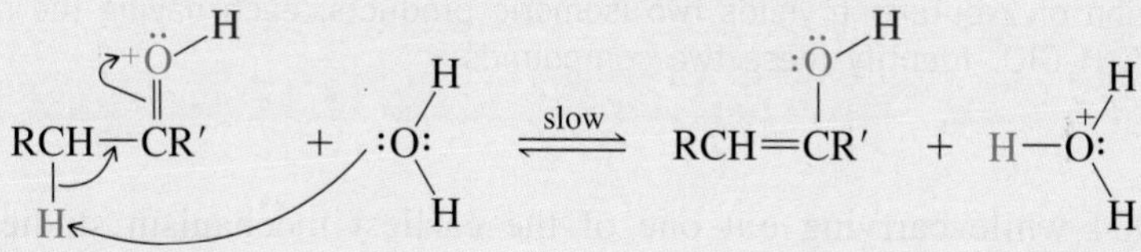

Conjugate acid of carbonyl compound — Water — Enol — Hydronium ion

Both mechanisms, acid-catalyzed and base-catalyzed, consist of two proton-transfer steps: proton abstraction from carbon and proton transfer to oxygen. The difference between the two is that the sequence of steps is reversed. Proton abstraction from the α carbon is the first step in the base-catalyzed mechanism; it is the second step in the acid-catalyzed one. In each mechanism, proton abstraction from the α carbon is rate-determining.

With this as background, we'll look at a few reactions of carbonyl compounds that proceed by way of enol intermediates.

20.14 α Halogenation of Aldehydes and Ketones

The double bond of an enol is "electron-rich" and reacts with electrophilic reagents to give an α-substituted derivative of the parent aldehyde or ketone.

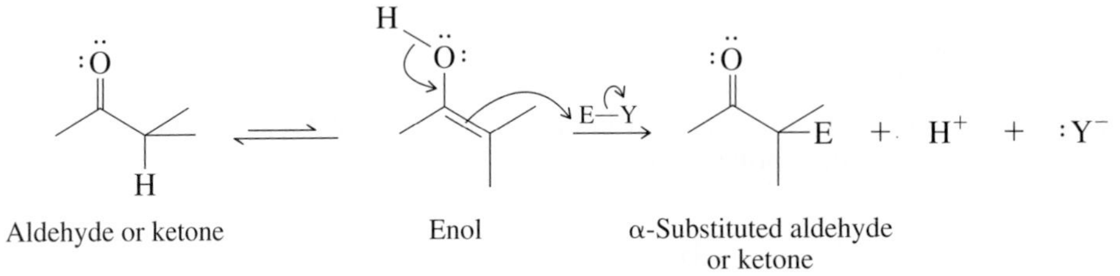

E—Y
E
+ H⁺ + :Y⁻
Aldehyde or ketone
Enol
α-Substituted aldehyde or ketone

Halogens are among the electrophiles that react with aldehydes and ketones in this way. The reaction is regiospecific for replacement of an α hydrogen and can be carried out in a variety of solvents such as water, chloroform, acetic acid, and diethyl ether.

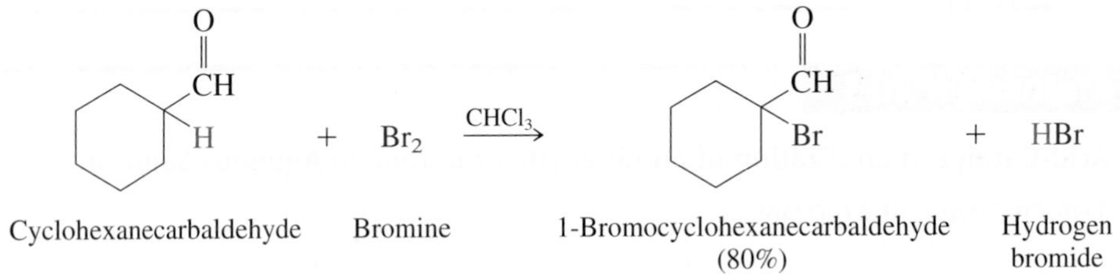

+ Br₂
CHCl₃
+ HBr
Cyclohexanecarbaldehyde
Bromine
1-Bromocyclohexanecarbaldehyde (80%)
Hydrogen bromide

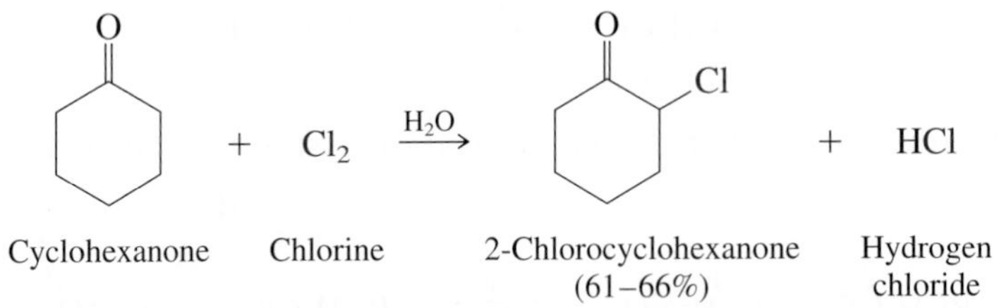

+ Cl₂
H₂O
+ HCl
Cyclohexanone
Chlorine
2-Chlorocyclohexanone (61–66%)
Hydrogen chloride

One of the products of the reaction, the hydrogen halide, is itself an enolization catalyst; therefore, the reaction requires no additional catalyst—it is *autocatalytic.*

Problem 20.31

Chlorination of 2-butanone yields two isomeric products, each having the molecular formula C_4H_7ClO. Identify these two compounds.

In 1904 while carrying out one of the earliest mechanism studies in organic chemistry, Arthur Lapworth found that the rates of α chlorination and bromination of acetone were the same. He later showed that iodination proceeded at the same rate, and that all the rates were independent of the halogen concentration. Lapworth

Mechanism 20.6

Acid-Catalyzed Bromination of Acetone

THE OVERALL REACTION:

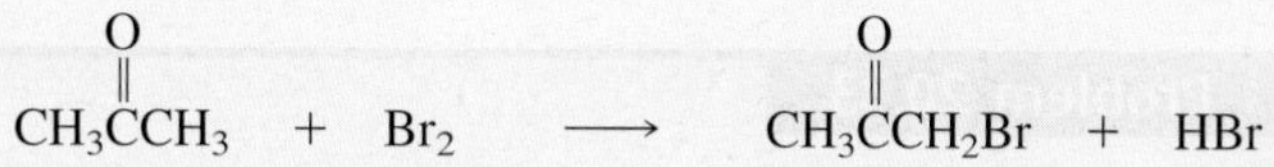

Acetone Bromine Bromoacetone Hydrogen bromide

Steps 1 and 2: The first two steps establish the equilibrium between acetone and its enol. They correspond to the two steps in the acid-catalyzed mechanism for enolization in Mechanism 20.5.

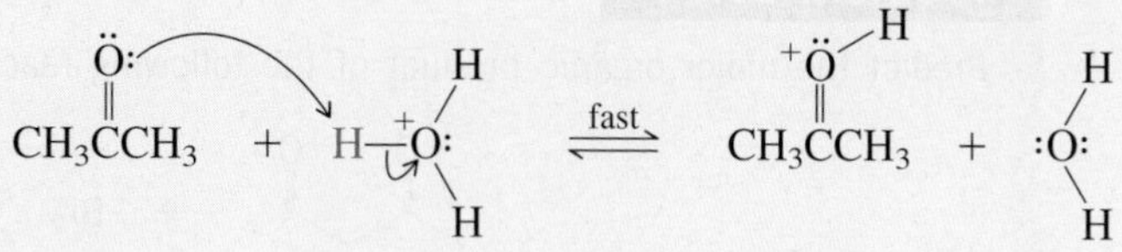

Acetone Hydronium ion Conjugate acid of acetone Water

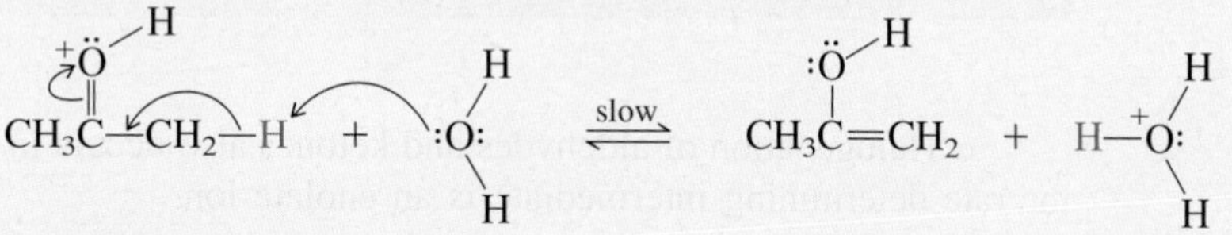

Conjugate acid of acetone Water Enol of acetone Hydronium ion

Step 3: An enolic double bond is "electron-rich" and acts as a nucleophile toward bromine, displacing bromide ion and forming a C—Br bond.

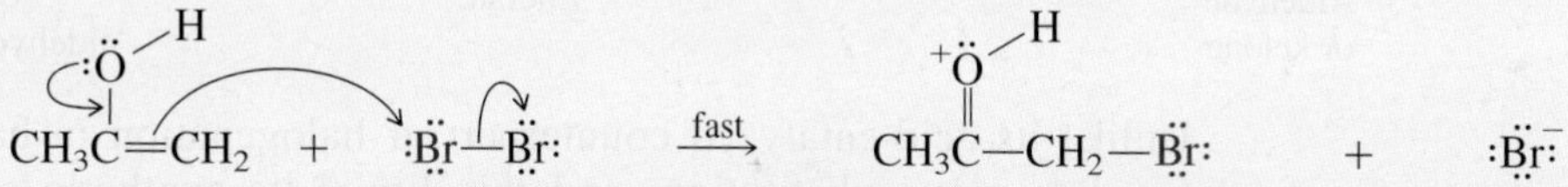

Enol of acetone Bromine Conjugate acid of bromoacetone Bromide ion

Step 4: The product of step 3 is converted to the α-bromo ketone by transferring a proton to water (or bromide ion).

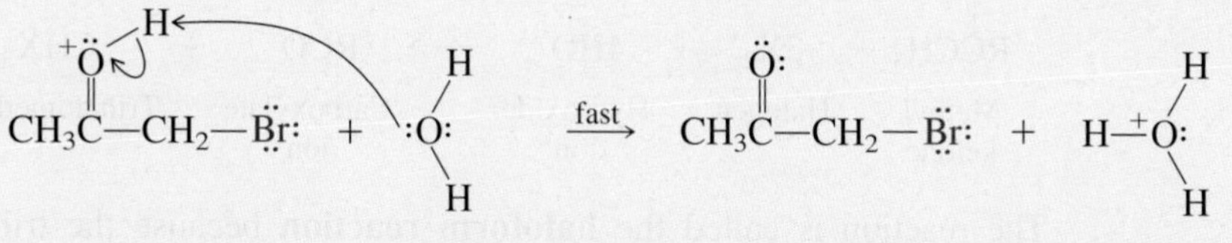

Conjugate acid of bromoacetone Water Bromoacetone Hydronium ion

concluded that *the halogen does not participate in the reaction until after the rate-determining step.* To account for these kinetic observations as well as the α regioselectivity, he proposed the sequence outlined in Mechanism 20.6 for the acid-catalyzed bromination of acetone.

The first two steps of the Lapworth mechanism are the same as the general mechanism of acid-catalyzed enolization (Mechanism 20.5, p. 899). Formation of the enol is rate-determining. Once formed, the enol reacts rapidly with bromine in step 3 to give, after transferring a proton to water in step 4, the α-bromo ketone.

Problem 20.32

Write structural formulas for the enols of 2-butanone that react with chlorine to give 1-chloro-2-butanone and 3-chloro-2-butanone.

Problem 20.33

Using curved arrows, show the flow of electrons in the reaction of each of the enols of 2-butanone with Cl_2.

Problem 20.34

Predict the major organic product of the following reaction.

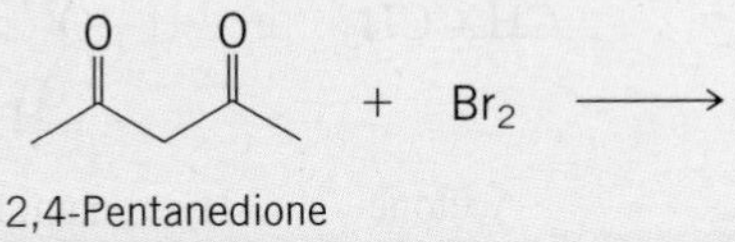

2,4-Pentanedione

α Halogenation of aldehydes and ketones also occurs in basic solution. In this case, the rate-determining intermediate is an enolate ion.

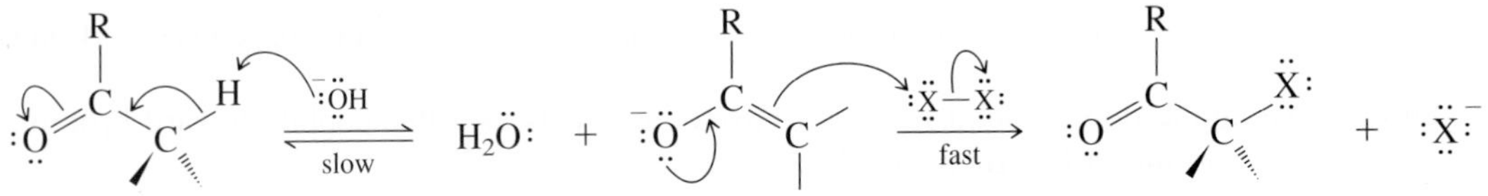

Aldehyde or ketone — Enolate — α-Halo aldehyde or ketone

Unlike its acid-catalyzed counterpart, α halogenation in base cannot normally be limited to monosubstitution, and this limits its synthetic applications. Methyl ketones, however, undergo a novel C—C cleavage on treatment with halogens in aqueous base.

$$\underset{\text{Methyl ketone}}{\overset{O}{\overset{\|}{RCCH_3}}} + \underset{\text{Halogen}}{3X_2} + \underset{\text{Hydroxide ion}}{4HO^-} \longrightarrow \underset{\text{Carboxylate ion}}{\overset{O}{\overset{\|}{RCO^-}}} + \underset{\text{Trihalomethane}}{CHX_3} + \underset{\text{Halide ion}}{3X^-} + \underset{\text{Water}}{3H_2O}$$

The reaction is called the **haloform reaction** because the trihalomethane produced is chloroform ($CHCl_3$), bromoform ($CHBr_3$), or iodoform (CHI_3), depending on the halogen used.

Mechanism 20.7 outlines the steps involved in the haloform cleavage of a methyl ketone. The boxed essay *The Haloform Reaction and the Biosynthesis of Trihalomethanes* describes its involvement as an environmental source of chloroform and bromoform.

The haloform reaction is sometimes used for the preparation of carboxylic acids from methyl ketones.

$$\underset{\text{3,3-Dimethyl-2-butanone}}{(CH_3)_3\overset{O}{\overset{\|}{C}}CH_3} \xrightarrow[\text{2. } H_3O^+]{\text{1. } Br_2,\ NaOH,\ H_2O} \underset{\text{2,2-Dimethylpropanoic acid (71–74\%)}}{(CH_3)_3\overset{O}{\overset{\|}{C}}OH} + \underset{\text{Tribromomethane (bromoform)}}{CHBr_3}$$

Mechanism 20.7

Cleavage of a Tribromomethyl Ketone

Step 1: Hydroxide ion acts as a nucleophile and adds to the carbonyl group of the 1,1,1-tribromomethyl ketone.

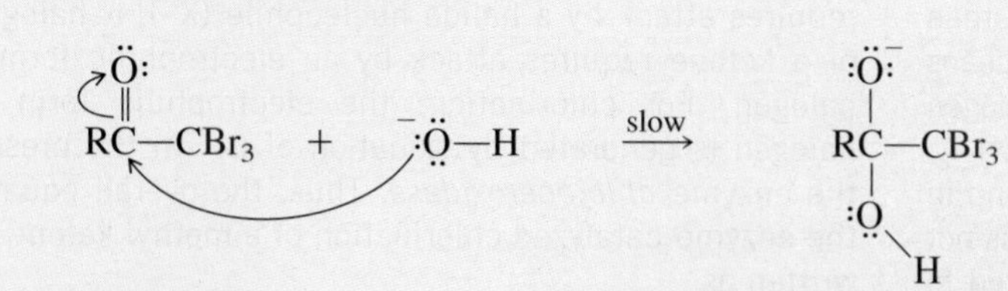

1,1,1-Tribromomethyl Ketone | Hydroxide ion | Product of nucleophilic addition

Step 2: The carbon–oxygen double bond is restored by expelling the tribromomethide ion.

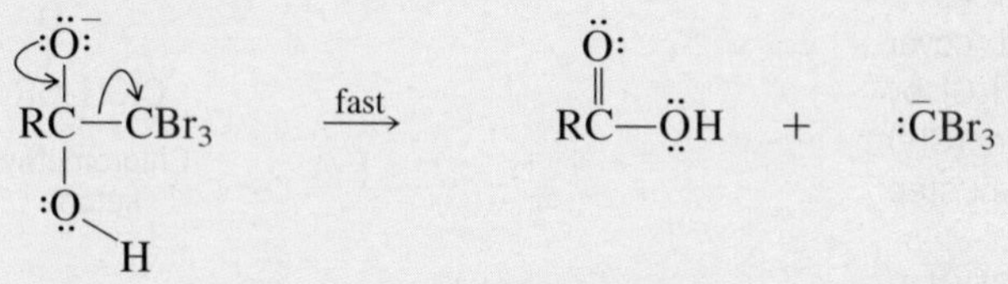

Product of nucleophilic addition | Carboxylic acid | Tribromomethide ion

Step 3: Proton transfer reactions convert the acid to carboxylate and tribromomethide ion to tribromomethane (bromoform).

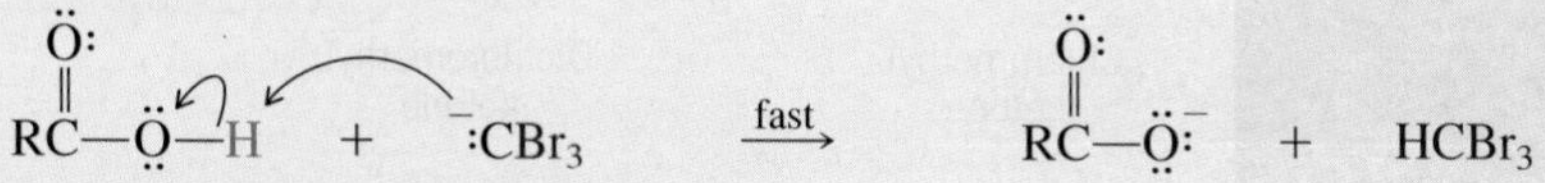

Acid | Tribromomethide ion | Carboxylate ion | Bromoform

The methyl ketone shown in the example can enolize in only one direction and typifies the kind of reactant that can be converted to a carboxylic acid in synthetically acceptable yield by the haloform reaction. Methyl ketones of the type $RCH_2C(O)CH_3$ and $R_2CHC(O)CH_3$ undergo nonregioselective α halogenation to yield a mixture of products.

Problem 20.35

Which of the following would be the most suitable starting material for the preparation of a carboxylic acid by the haloform reaction? Give the structure of the carboxylic acid.

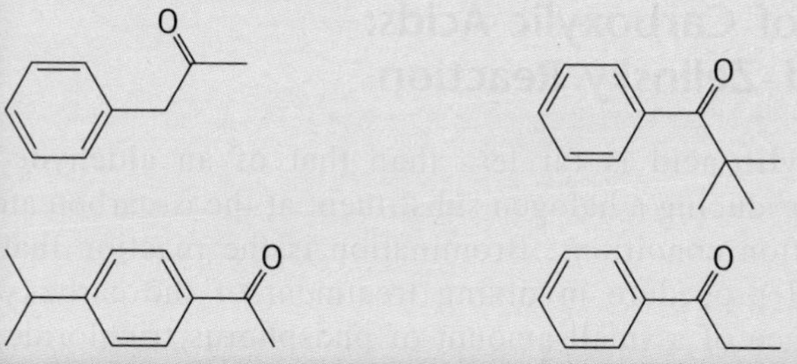

The Haloform Reaction and the Biosynthesis of Trihalomethanes

Until scientists started looking specifically for them, it was widely believed that naturally occurring organohalogen compounds were rare. We now know that more than 4000 such compounds occur naturally, with the oceans being a particularly rich source. Over 50 organohalogen compounds, including $CHBr_3$, $CHBrClI$, $BrCH_2CH_2I$, CH_2I_2, $Br_2CHCH{=}O$, I_2CHCO_2H, and $(Cl_3C)_2C{=}O$, have been found in a single species of Hawaiian red seaweed, for example. It is not surprising that organisms living in the oceans have adapted to their halide-rich environment by incorporating chlorine, bromine, and iodine into their metabolic processes. Chloromethane (CH_3Cl), bromomethane (CH_3Br), and iodomethane (CH_3I) are all produced by marine algae and kelp, but land-based plants and fungi also contribute their share to the more than 5 million tons of the methyl halides formed each year by living systems. The ice plant, which grows in arid regions throughout the world and is cultivated as a ground cover along coastal highways in California, biosynthesizes CH_3Cl by a process in which nucleophilic attack by chloride ion (Cl^-) on the methyl group of *S*-adenosylmethionine is the key step (Section 16.17).

Interestingly, the trihalomethanes chloroform ($CHCl_3$), bromoform ($CHBr_3$), and iodoform (CHI_3) are biosynthesized by an entirely different process, one that is equivalent to the haloform reaction and begins with the formation of an α-halo ketone. Unlike the biosynthesis of methyl halides, which requires attack by a halide nucleophile (X^-), α halogenation of a ketone requires attack by an electrophilic form of the halogen. For chlorination, the electrophilic form of the halogen is generated by oxidation of Cl^- in the presence of the enzyme *chloroperoxidase*. Thus, the overall equation for the enzyme-catalyzed chlorination of a methyl ketone may be written as

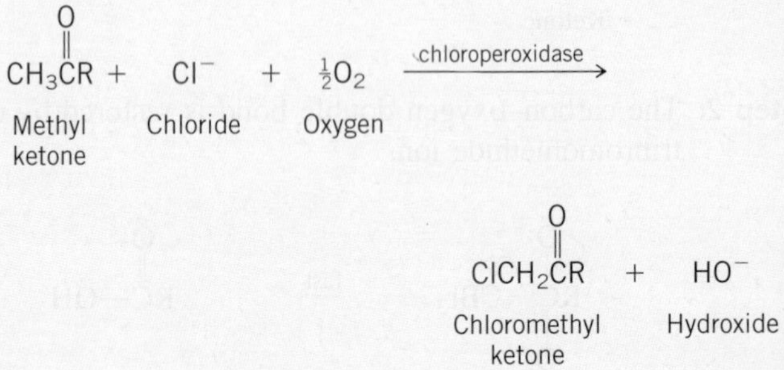
CH₃CR + Cl⁻ + ½O₂
chloroperoxidase
Methyl ketone
Chloride
Oxygen
ClCH₂CR + HO⁻
Chloromethyl ketone
Hydroxide

Further chlorination of the chloromethyl ketone gives the corresponding trichloromethyl ketone, which then undergoes hydrolysis to form chloroform.

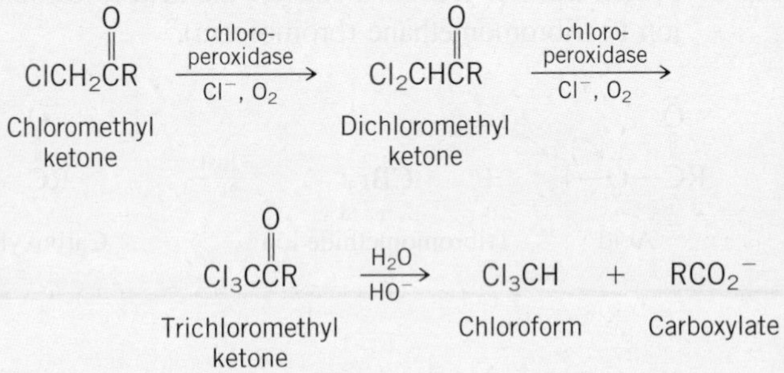
ClCH₂CR
chloro-peroxidase
Cl⁻, O₂
Cl₂CHCR
chloro-peroxidase
Cl⁻, O₂
Chloromethyl ketone
Dichloromethyl ketone
Cl₃CCR
H₂O
HO⁻
Cl₃CH + RCO₂⁻
Trichloromethyl ketone
Chloroform
Carboxylate

Purification of drinking water, by adding Cl_2 to kill bacteria, is a source of electrophilic chlorine and contributes a nonenzymatic pathway for α chlorination and subsequent chloroform formation. Although some of the odor associated with tap water may be due to chloroform, more of it probably results from chlorination of algae-produced organic compounds.

The ice plant generates chloromethane naturally.

20.15 α Halogenation of Carboxylic Acids: The Hell–Volhard–Zelinsky Reaction

The enol content of a carboxylic acid is far less than that of an aldehyde or ketone (see Table 20.2), and introducing a halogen substituent at the α-carbon atom requires a different set of reaction conditions. Bromination is the reaction that is normally carried out, the usual procedure involving treatment of the carboxylic acid with bromine in the presence of a small amount of phosphorus trichloride as a catalyst.

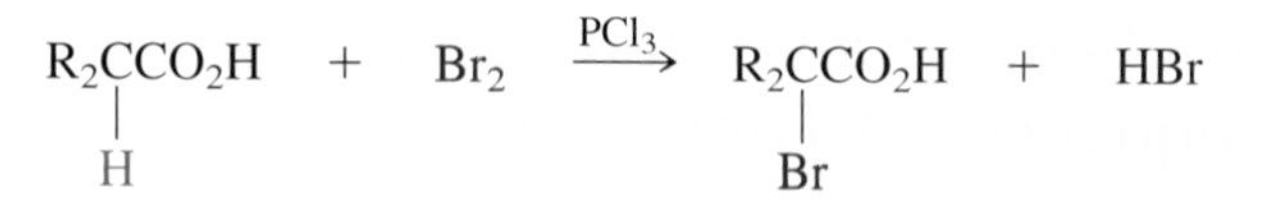

Carboxylic acid | Bromine | α-Bromo carboxylic acid | Hydrogen bromide

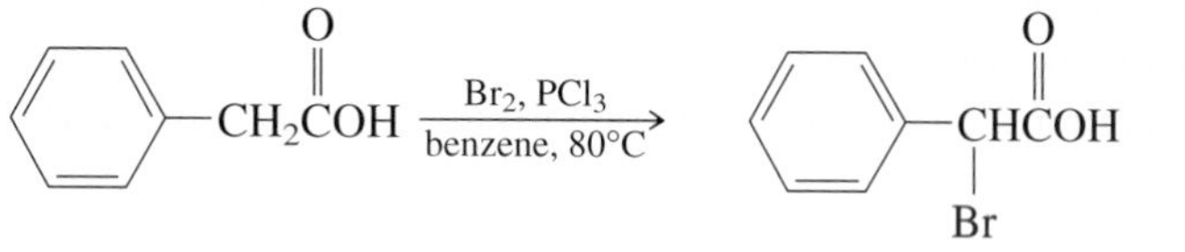

Phenylacetic acid | 2-Bromo-2-phenylacetic acid (60–62%)

This method of α bromination of carboxylic acids is called the **Hell–Volhard–Zelinsky reaction.** This reaction is sometimes carried out by using a small amount of phosphorus instead of phosphorus trichloride. Phosphorus reacts with bromine to yield phosphorus tribromide as the active catalyst under these conditions. The catalyst, PCl_3 or PBr_3, acts by converting some of the carboxylic acid to the corresponding acyl halide, which has a higher enol content than the parent carboxylic acid.

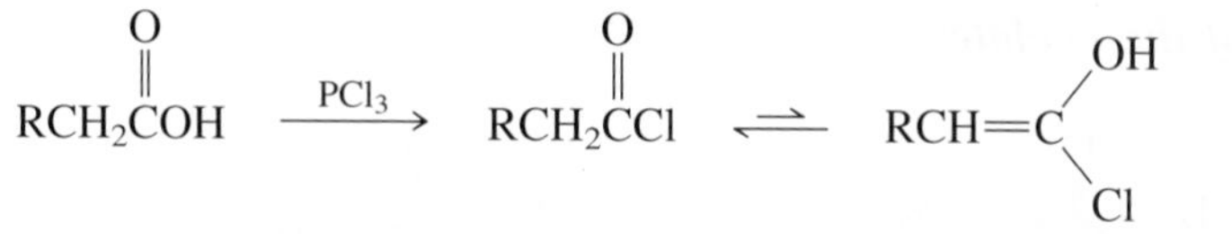

Carboxylic acid | Acyl chloride | Enol form of acyl chloride

The enol then reacts with Br_2.

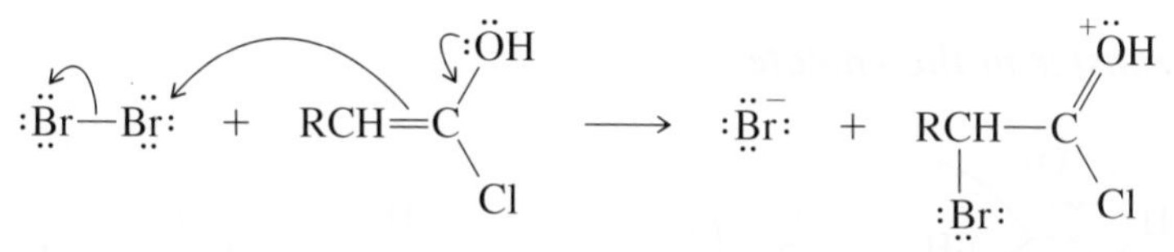

Bromine | Enol form of acyl chloride | Bromide ion | Conjugate acid of α-bromoacyl chloride

After deprotonation of its conjugate acid, the α-bromoacyl chloride reacts with a molecule of the starting carboxylic acid to give the product plus another molecule of acyl chloride, which then undergoes enolization, bromination, and so on.

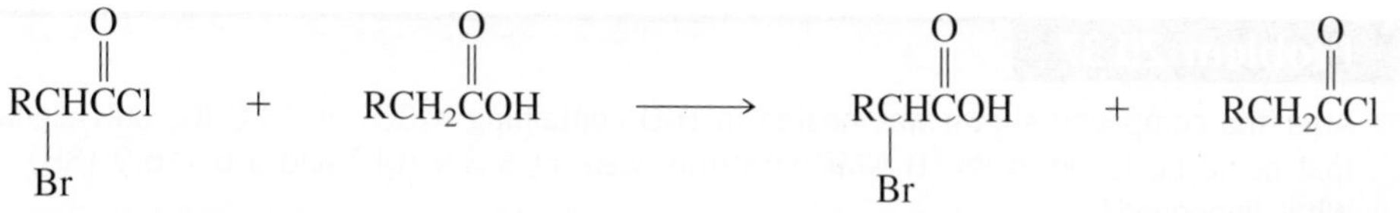

α-Bromoacyl chloride | Carboxylic acid | α-Bromocarboxylic acid | Acyl chloride

Problem 20.36

The Hell–Volhard–Zelinsky reaction has been used to prepare the amino acid valine from 3-methylbutanoic acid by the following procedure.

$$(CH_3)_2CHCH_2CO_2H \xrightarrow[\text{2. } NH_3,\ H_2O]{\text{1. } Br_2,\ PCl_3} C_5H_{11}NO_2$$

3-Methylbutanoic acid | Valine

Deduce the structure of valine. Natural valine that is isolated from proteins has a specific rotation [α] of +26°. What is the rotation of the valine prepared by the Hell–Volhard–Zelinsky method?

20.16 Some Chemical and Stereochemical Consequences of Enolization

A number of novel reactions involving the α-carbon atom of aldehydes and ketones involve enol and enolate anion intermediates.

Substitution of deuterium for hydrogen at the α-carbon atom of an aldehyde or a ketone is a convenient way to introduce an isotopic label into a molecule and is readily carried out by treating the carbonyl compound with deuterium oxide (D_2O) and base.

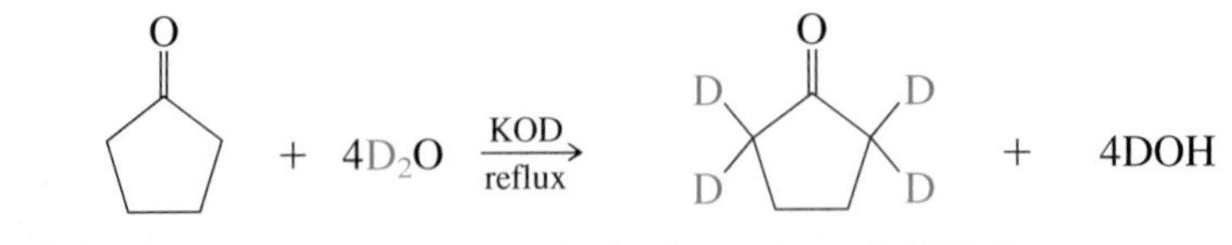

Cyclopentanone Cyclopentanone-2,2,5,5-d_4

Only the α hydrogens are replaced by deuterium in this reaction. The key intermediate is the enolate ion formed by proton abstraction from the α-carbon atom of cyclopentanone. Transfer of deuterium from the solvent D_2O to the enolate gives cyclopentanone containing a deuterium atom in place of one of the hydrogens at the α carbon.

Formation of the enolate

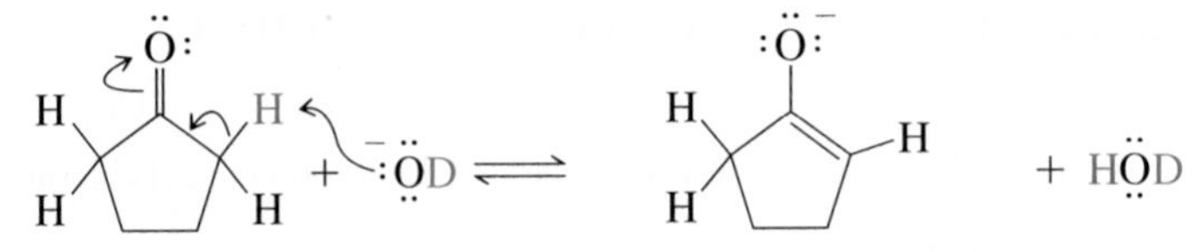

Cyclopentanone Enolate of cyclopentanone

Deuterium transfer to the enolate

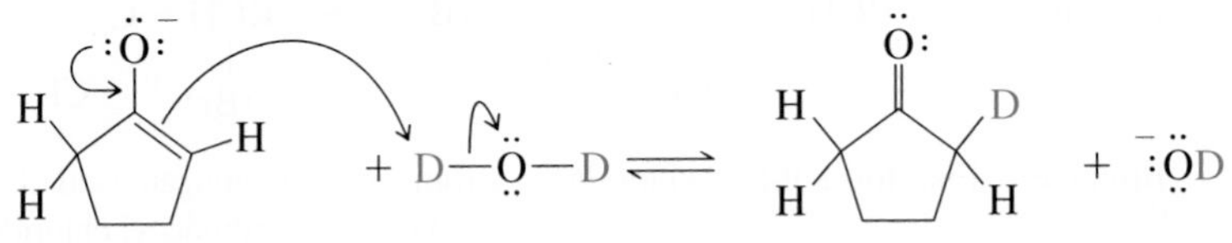

Enolate of cyclopentanone Cyclopentanone-2-d_1

In excess D_2O the process continues until all four α protons are eventually replaced by deuterium.

Problem 20.37

After the compound shown was heated in D_2O containing K_2CO_3 at 70°C the only signals that could be found in its ^1H NMR spectrum were at δ 3.9 (6H) and δ 6.7–6.9 (3H). What happened?

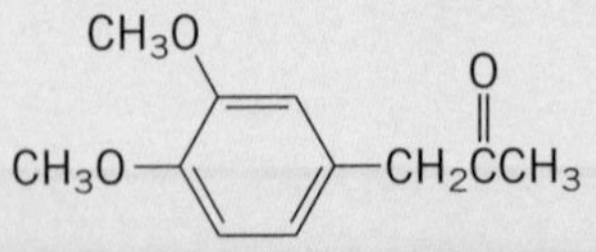

If the α-carbon atom of an aldehyde or a ketone is a chirality center, its stereochemical integrity is lost on enolization. Enolization of optically active *sec*-butyl phenyl ketone leads to its racemization by way of the achiral enol form.

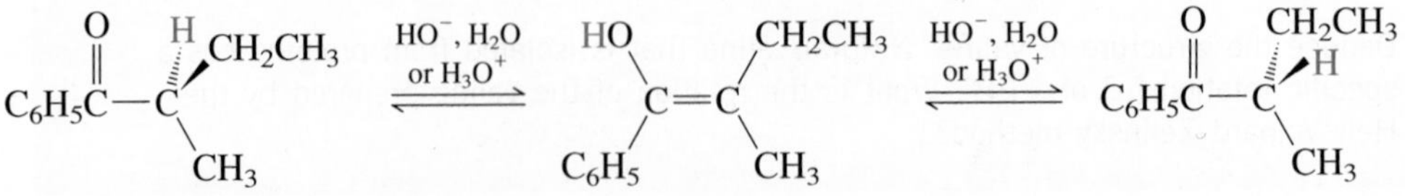

(*R*)-*sec*-Butyl phenyl ketone Achiral enol (*S*)-*sec*-Butyl phenyl ketone

Each proton abstraction from the α carbon converts a chiral molecule to an achiral enol or enolate ion. The sp^3-hybridized carbon that is the chirality center in the starting ketone becomes sp^2-hybridized in the enol or enolate. Careful kinetic studies have established that the rate of loss of optical activity of *sec*-butyl phenyl ketone is equal to its rate of hydrogen–deuterium exchange, its rate of bromination, and its rate of iodination. In each case, the rate-determining step is conversion of the starting ketone to the enol or enolate anion.

Problem 20.38

Is the product from the α chlorination of (*R*)-*sec*-butyl phenyl ketone with Cl_2 in acetic acid chiral? Is it optically active?

Problem 20.39

Explain why optically active piperitone is converted to racemic piperitone on standing in a solution of sodium ethoxide in ethanol.

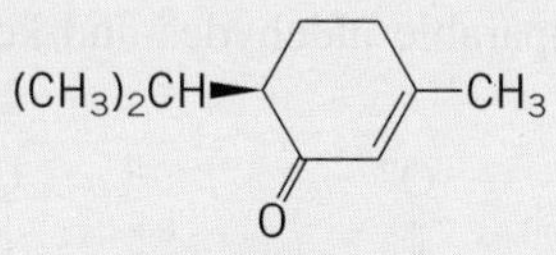

Piperitone

20.17 Effects of Conjugation in α,β-Unsaturated Aldehydes and Ketones

Aldol condensation offers an effective route to α,β-unsaturated aldehydes and ketones. These compounds have some interesting properties that result from conjugation of the carbon–carbon double bond with the carbonyl group. As shown in Figure 20.3, the π systems of the carbon–carbon and carbon–oxygen double bonds overlap to form an extended π system that permits increased electron delocalization.

This electron delocalization stabilizes a conjugated system. Under conditions chosen to bring about their interconversion, the equilibrium between a β,γ-unsaturated ketone and an α,β-unsaturated analog favors the conjugated isomer.

$$CH_3CH{=}CHCH_2\overset{\overset{\displaystyle O}{\|}}{C}CH_3 \underset{25°C}{\overset{K = 4.8}{\rightleftharpoons}} CH_3CH_2CH{=}CH\overset{\overset{\displaystyle O}{\|}}{C}CH_3$$

4-Hexen-2-one (17%) (β,γ-unsaturated ketone) — 3-Hexen-2-one (83%) (α,β-unsaturated ketone)

Figure 3.19 (page 118) shows how the composition of an equilibrium mixture of two components varies according to the free-energy difference between them. For the equilibrium shown in the accompanying equation, $\Delta G° = -4$ kJ/mol (−1 kcal/mol).

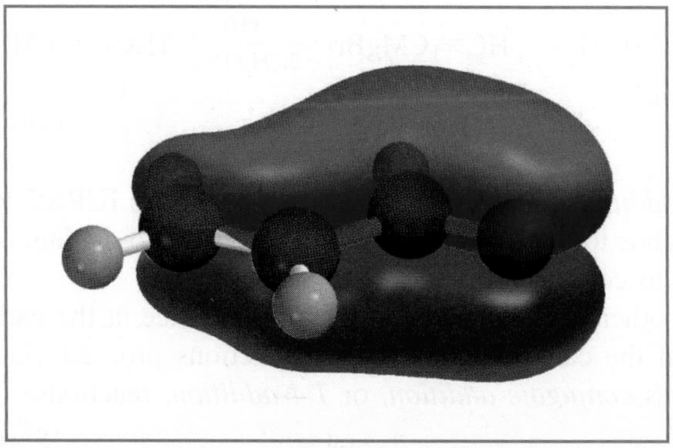

Figure 20.3

Acrolein ($H_2C{=}CHCH{=}O$) is a planar molecule. Oxygen and all three carbons are sp^2-hybridized, and each contributes one electron to a conjugated π electron system analogous to that of 1,3-butadiene.

Problem 20.40

Commercial mesityl oxide, $(CH_3)_2C{=}CH\overset{O}{\overset{\|}{C}}CH_3$, is often contaminated with about 10% of an isomer having the same carbon skeleton. What is a likely structure for this compound?

Electron delocalization in α,β-unsaturated carbonyl compounds is represented by three principal resonance contributors:

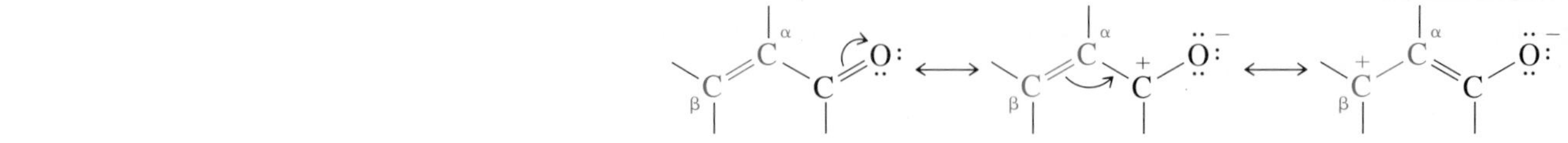

Most stable contributor

The carbonyl group withdraws π electron density from the double bond, and both the carbonyl carbon and the β carbon are positively polarized. Their greater degree of charge separation makes the dipole moments of α,β-unsaturated carbonyl compounds significantly larger than those of comparable aldehydes and ketones.

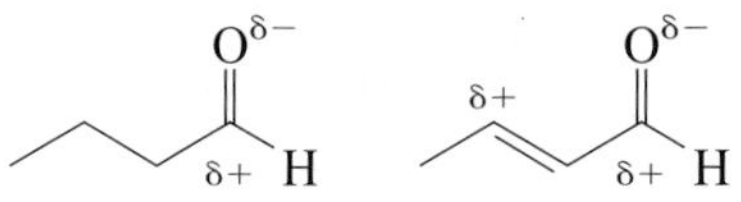

Butanal
μ = 2.7 D

trans-2-Butenal
μ = 3.7 D

The diminished π electron density in the double bond makes α,β-unsaturated aldehydes and ketones less reactive than alkenes toward electrophilic addition. Electrophilic reagents—bromine and peroxy acids, for example—react more slowly with the carbon–carbon double bond of α,β-unsaturated carbonyl compounds than with simple alkenes.

On the other hand, the polarization of electron density in α,β-unsaturated carbonyl compounds makes their β-carbon atoms rather electrophilic. Some chemical consequences of this enhanced electrophilicity are described in the following section.

20.18 Conjugate Addition to α,β-Unsaturated Carbonyl Compounds

α,β-Unsaturated carbonyl compounds contain two electrophilic sites: the carbonyl carbon and the carbon atom that is β to it. Nucleophiles such as organolithium and Grignard reagents and lithium aluminum hydride tend to react by nucleophilic addition to the carbonyl group, as shown in the following example:

$$\underset{\text{2-Butenal}}{CH_3CH{=}CH\overset{O}{\overset{\|}{C}}H} + \underset{\text{Ethynylmagnesium bromide}}{HC{\equiv}CMgBr} \xrightarrow[\text{2. } H_3O^+]{\text{1. THF}} \underset{\text{4-Hexen-1-yn-3-ol (84\%)}}{CH_3CH{=}CH\overset{OH}{\overset{|}{C}}HC{\equiv}CH}$$

This is called *1,2-addition.* (The "1" and "2" do not refer to IUPAC locants but are used in a manner analogous to that employed in Section 10.13 to distinguish between the two modes of addition to conjugated dienes.)

With certain other nucleophiles, addition takes place at the carbon–carbon double bond rather than at the carbonyl group. Such reactions proceed via enol intermediates and are described as *conjugate addition,* or *1,4-addition,* reactions.

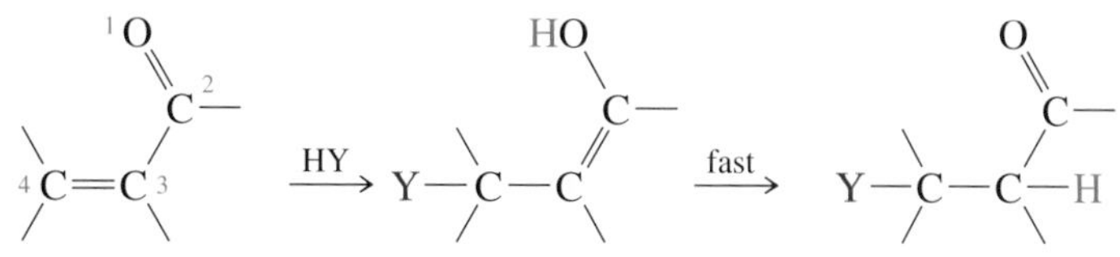

α,β-Unsaturated aldehyde or ketone — Enol formed by 1,4-addition — Isolated product of 1,4-addition

The nucleophilic portion of the reagent (Y in HY) becomes bonded to the β carbon. For reactions carried out under conditions in which the nucleophile is an anion, an enolate ion precedes the enol.

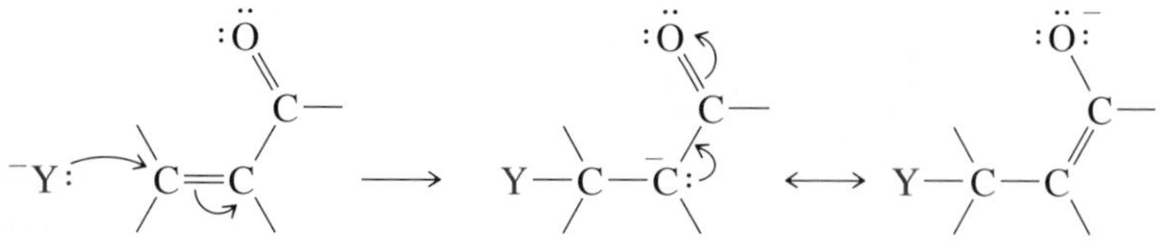

Enolate ion formed by nucleophilic addition of :Y$^-$ to β carbon

Ordinarily, nucleophilic addition to the carbon–carbon double bond of an alkene is very rare. It occurs with α,β-unsaturated carbonyl compounds because the carbanion that results is an enolate, which is more stable than a simple alkyl anion.

Conjugate addition is most often observed when the nucleophile (Y:$^-$) is weakly basic. The nucleophiles in the two examples that follow are $^-$:C≡N: and $C_6H_5CH_2\ddot{S}:^-$, respectively. Both are much weaker bases than acetylide ion, which was the nucleophile used in the example illustrating 1,2-addition.

Hydrogen cyanide and alkanethiols have pK_a values in the 9 to 10 range, and the pK_a for acetylene is 26.

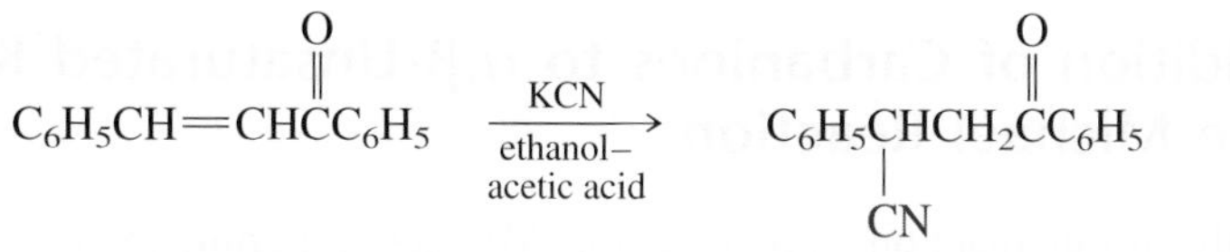

1,3-Diphenyl-2-propen-1-one → 4-Oxo-2,4-diphenylbutanenitrile (93–96%)

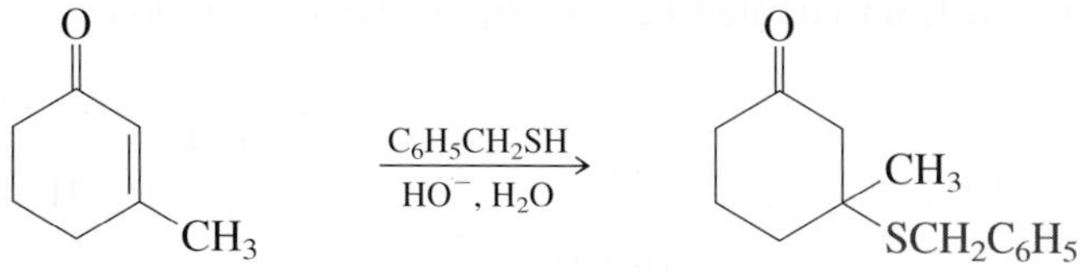

3-Methyl-2-cyclohexenone → 3-Benzylthio-3-methylcyclohexanone (58%)

As presented in Figure 20.4, nucleophilic addition to α,β-unsaturated aldehydes and ketones may be governed either by *kinetic control* or by *thermodynamic control* (Section 10.13). Under conditions in which the 1,2- and 1,4-addition products do not equilibrate, 1,2-addition predominates because it is faster than 1,4-addition. Kinetic control operates with strongly basic nucleophiles to give the 1,2-addition product. A weakly basic nucleophile, however, goes on and off the carbonyl carbon readily and permits the 1,2-addition product to equilibrate with the more slowly formed, but more stable, 1,4-addition product. Thermodynamic control is observed with weakly basic nucleophiles. The product of 1,4-addition, which retains the carbon–oxygen double bond, is more stable than the product of 1,2-addition, which retains the carbon–carbon double bond. In general, carbon–oxygen double bonds are stronger than carbon–carbon double bonds because the greater electronegativity of oxygen permits the π electrons to be bound more strongly.

Figure 20.4

Nucleophilic addition to α,β-unsaturated aldehydes and ketones may take place either in a 1,2- or 1,4-mode. 1,2-Addition occurs faster than 1,4-addition, but the product of 1,4-addition is more stable. The product of 1,4-addition retains the carbon–oxygen double bond, which is more stable than the carbon–carbon double bond.

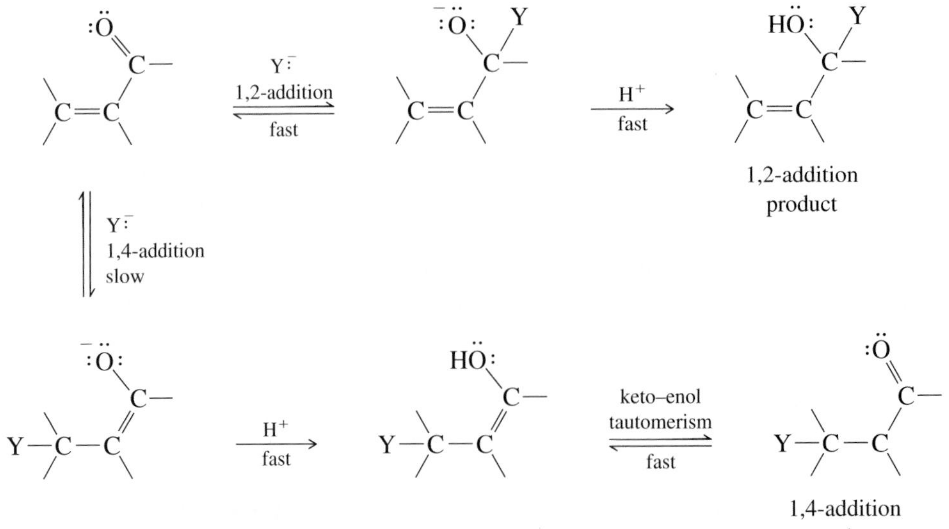

Problem 20.41

Acrolein ($H_2C{=}CHCH{=}O$) reacts with sodium azide (NaN_3) in aqueous acetic acid to form a compound, $C_3H_5N_3O$ in 71% yield. Propanal ($CH_3CH_2CH{=}O$), when subjected to the same reaction conditions, is recovered unchanged. Suggest a structure for the product formed from acrolein, and offer an explanation for the difference in reactivity between acrolein and propanal.

20.19 Addition of Carbanions to α,β-Unsaturated Ketones: The Michael Reaction

Arthur Michael, for whom the reaction is named, was an American chemist whose career spanned the period between the 1870s and the 1930s. He was independently wealthy and did much of his research in his own private laboratory.

A synthetically useful reaction known as the **Michael reaction** involves nucleophilic addition of carbanions to α,β-unsaturated ketones. The most common types of carbanions used are enolate ions derived from β-diketones. These enolates are weak bases (Section 20.1) and react with α,β-unsaturated ketones by *conjugate addition.*

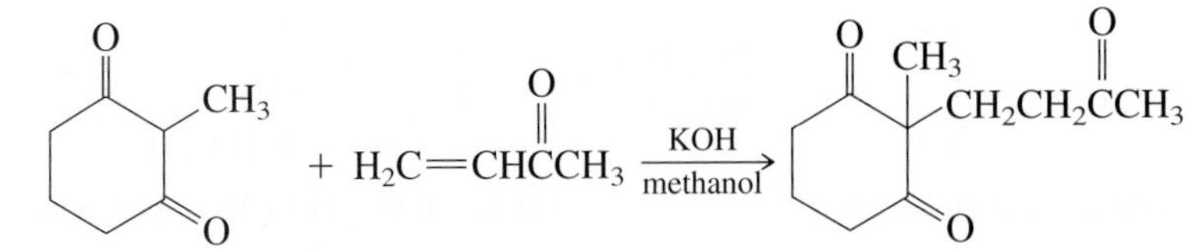

2-Methyl-1,3-cyclohexanedione + Methyl vinyl ketone → 2-Methyl-2-(3′-oxobutyl)-1,3-cyclohexanedione (85%)

The product of this Michael addition has the necessary functionality to undergo an intramolecular aldol condensation:

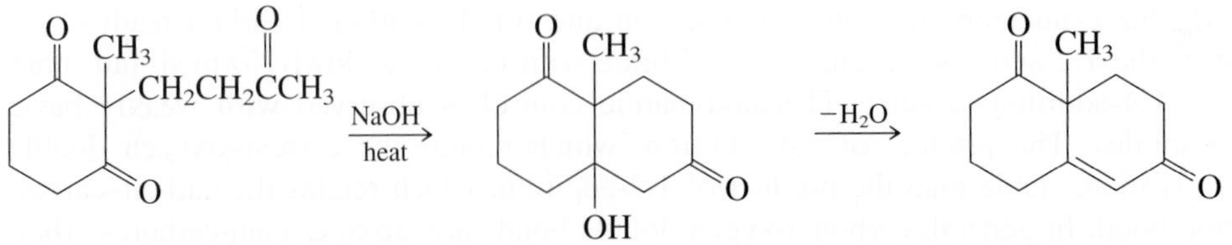

2-Methyl-2-(3′-oxobutyl)-1,3-cyclohexanedione → Intramolecular aldol addition product; not isolated → Δ^4-9-Methyloctalin-3,8-dione

The synthesis of cyclohexenone derivatives by Michael addition followed by intramolecular aldol condensation is called the **Robinson annulation,** after Sir Robert Robinson, who popularized its use. By *annulation* we mean the building of a ring onto some starting molecule.

Problem 20.42

Both the conjugate addition step and the intramolecular aldol condensation step can be carried out in one synthetic operation without isolating any of the intermediates along the way. For example, consider the reaction

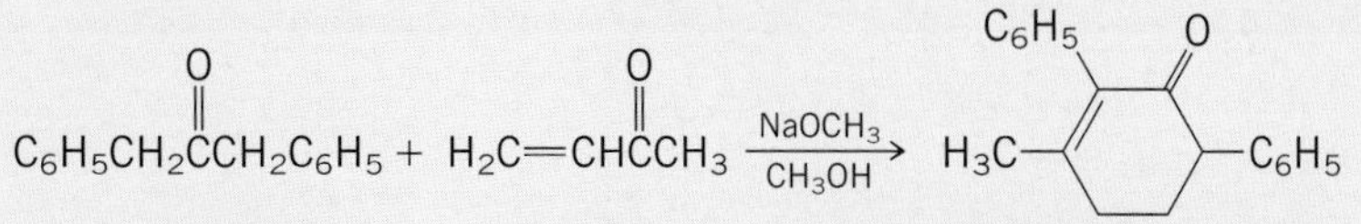

O
O
C6H5CH2CCH2C6H5 + H2C=CHCCH3
NaOCH3
CH3OH
C6H5
O
H3C
C6H5

Dibenzyl ketone — Methyl vinyl ketone — 3-Methyl-2,6-diphenyl-2-cyclohexenone (55%)

Write structural formulas corresponding to the intermediates formed in the conjugate addition step and in the aldol addition step.

The enolates of ethyl acetoacetate and diethyl malonate also undergo Michael addition to the β-carbon atom of α,β-unsaturated aldehydes, ketones, and esters.

$$CH_3\overset{O}{\overset{\|}{C}}CH{=}CH_2 + CH_2(COOCH_2CH_3)_2 \xrightarrow[\text{ethanol}]{\text{KOH}} CH_3\overset{O}{\overset{\|}{C}}CH_2CH_2CH(COOCH_2CH_3)_2$$

Methyl vinyl ketone — Diethyl malonate — Diethyl 2-(3-oxobutyl)malonate (83%)

In this reaction the enolate of diethyl malonate adds to the β carbon of methyl vinyl ketone.

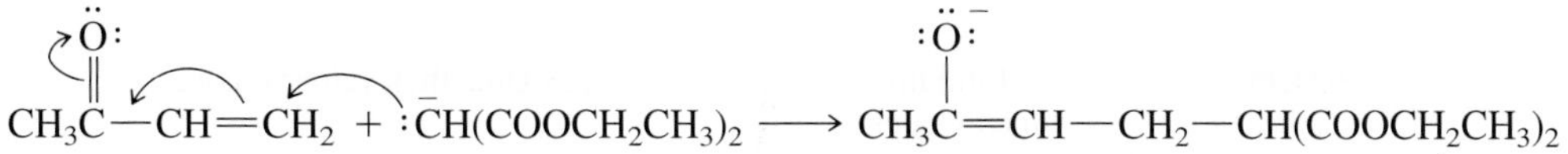

The intermediate formed in the nucleophilic addition step abstracts a proton from the solvent to give the observed product.

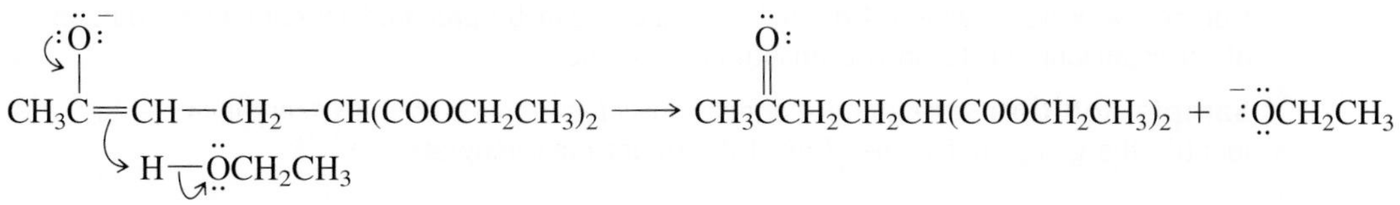

After isolation, the Michael adduct may be subjected to ester hydrolysis and decarboxylation. When α,β-unsaturated ketones are carried through this sequence, the final products are 5-keto acids (δ-keto acids).

$$CH_3\overset{O}{\overset{\|}{C}}CH_2CH_2CH(COOCH_2CH_3)_2 \xrightarrow[\text{2. } H_3O^+ \text{ 3. heat}]{\text{1. KOH, ethanol–water}} CH_3\overset{O}{\overset{\|}{C}}CH_2CH_2CH_2\overset{O}{\overset{\|}{C}}OH$$

Diethyl 2-(3-oxobutyl)malonate (from diethyl malonate and methyl vinyl ketone) — 5-Oxohexanoic acid (42%)

Problem 20.43

Ethyl acetoacetate behaves similarly to diethyl malonate in its reactivity toward α,β-unsaturated carbonyl compounds. Give the structure of the product of the following reaction sequence:

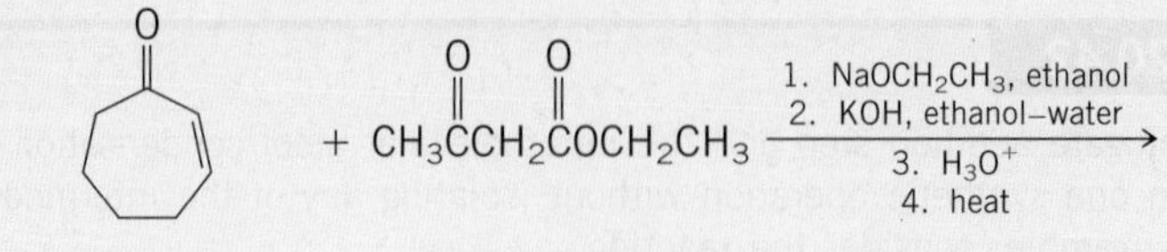
O
O O
+ CH3CCH2COCH2CH3
1. NaOCH2CH3, ethanol
2. KOH, ethanol–water
3. H3O+
4. heat

2-Cycloheptenone Ethyl acetoacetate

20.20 Conjugate Addition of Organocopper Reagents to α,β-Unsaturated Carbonyl Compounds

The preparation and some synthetic applications of lithium dialkylcuprates were described earlier (Section 14.10). The most prominent feature of these reagents is their capacity to undergo conjugate addition to α,β-unsaturated aldehydes and ketones.

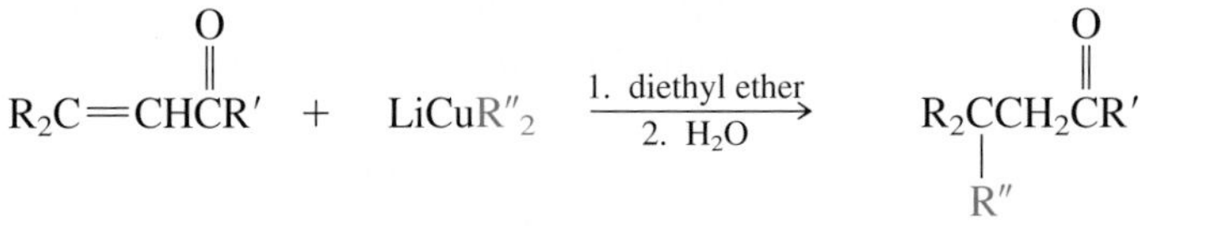
R2C=CHCR′ + LiCuR″2
1. diethyl ether
2. H2O
R2CCH2CR′
R″

α,β-Unsaturated aldehyde or ketone | Lithium dialkylcuprate | Aldehyde or ketone alkylated at the β position

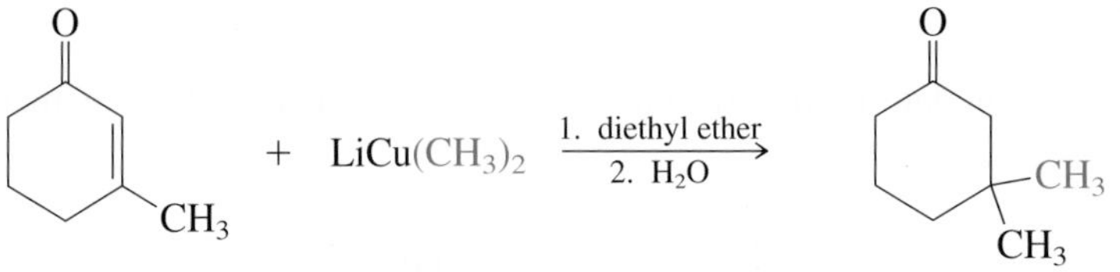
CH3
+ LiCu(CH3)2
1. diethyl ether
2. H2O
CH3
CH3

3-Methyl-2-cyclohexenone | Lithium dimethylcuprate | 3,3-Dimethylcyclohexanone (98%)

Problem 20.44

Outline two ways in which 4-methyl-2-octanone can be prepared by conjugate addition of an organocuprate to an α,β-unsaturated ketone.

Sample Solution Mentally disconnect one of the bonds to the β carbon so as to identify the group that comes from the lithium dialkylcuprate.

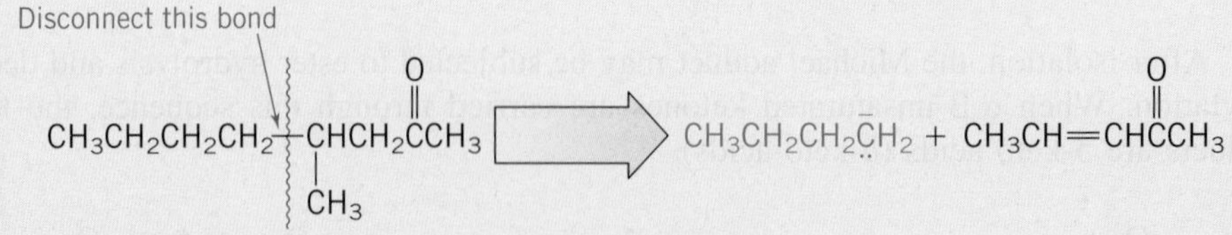
Disconnect this bond
CH3CH2CH2CH2—CHCH2CCH3
CH3
CH3CH2CH2CH2 + CH3CH=CHCCH3

4-Methyl-2-octanone

According to this disconnection, the butyl group is derived from lithium dibutylcuprate. A suitable preparation is

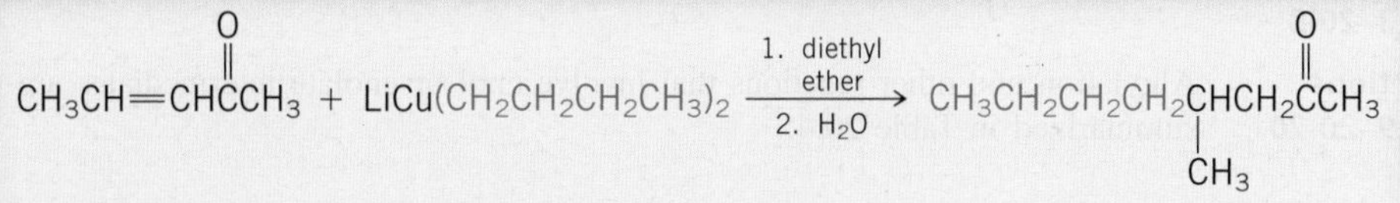

3-Penten-2-one Lithium dibutylcuprate 4-Methyl-2-octanone

Now see if you can identify the second possibility.

Like other carbon–carbon bond-forming reactions, organocuprate addition to enones is a powerful tool in organic synthesis.

20.21 SUMMARY

Section 20.1 An α hydrogen of an aldehyde or ketone is more acidic than most other protons bound to carbon and has a pK_a in the range of 16–20. Their enhanced acidity is due to the electron-withdrawing effect of the carbonyl group and the resonance stabilization of the enolate.

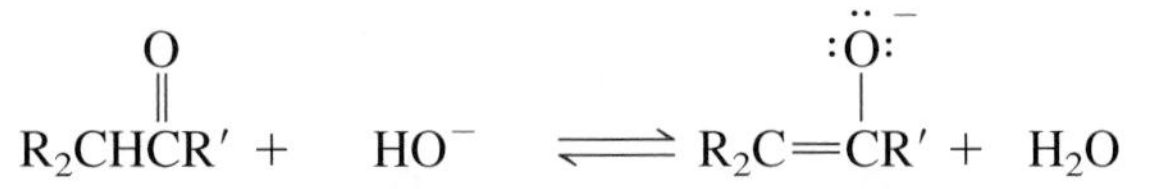

Aldehyde or ketone Hydroxide ion Enolate anion Water

The α hydrogen of an ester is also acidic, but less than that of an aldehyde or ketone. The pK_a of esters is around 24. When treated with alkoxide bases an ester enolate is formed that is in equilibrium with the starting ester. The ester is the major component of the equilibrium.

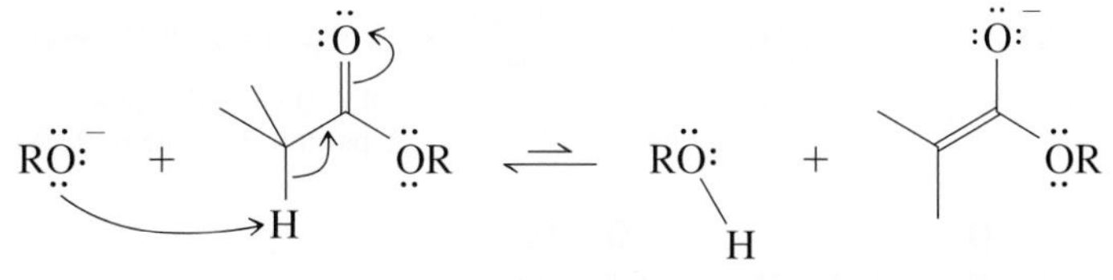

Ester Ester enolate

A C—H group that is α to two carbonyl groups is more acidic than one that is α to one carbonyl group. β-Keto esters have pK_a values of approximately 11 and are converted completely to their enolates by alkoxide bases.

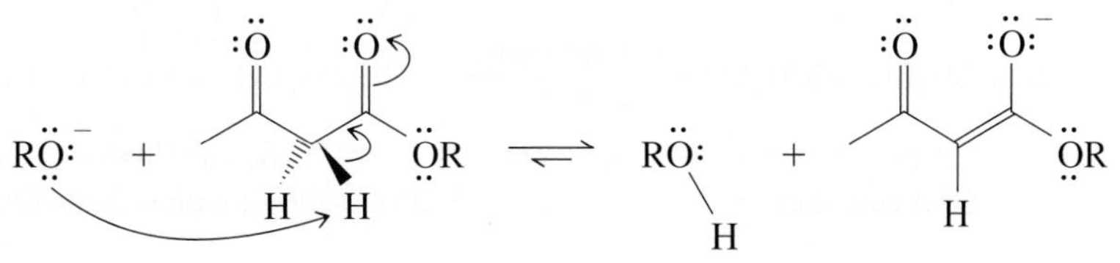

β-Keto ester β-Keto ester enolate

Section 20.2 Unsymmetric ketones give rise to regioisomeric enolates.

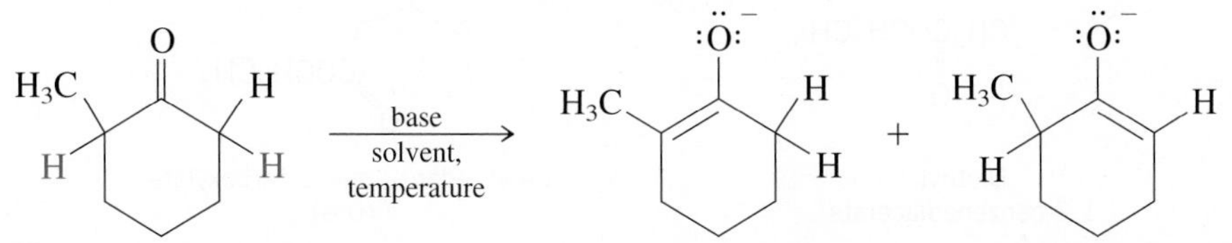

Thermodynamic enolate Kinetic enolate

Sections 20.3–20.8 Carbonyl condensations of enolates are summarized in Table 20.3.

Sections 20.9–20.20 Alkylation and other reactions that involve enol or enolate intermediates are summarized in Table 20.4.

TABLE 20.3 Carbonyl Condensations

Reaction (section) and comments	General equation and typical example
Aldol condensation (Section 20.3) A reaction of great synthetic value for carbon–carbon bond formation. Nucleophilic addition of an enolate ion to a carbonyl group, followed by dehydration of the β-hydroxy aldehyde, yields an α,β-unsaturated aldehyde.	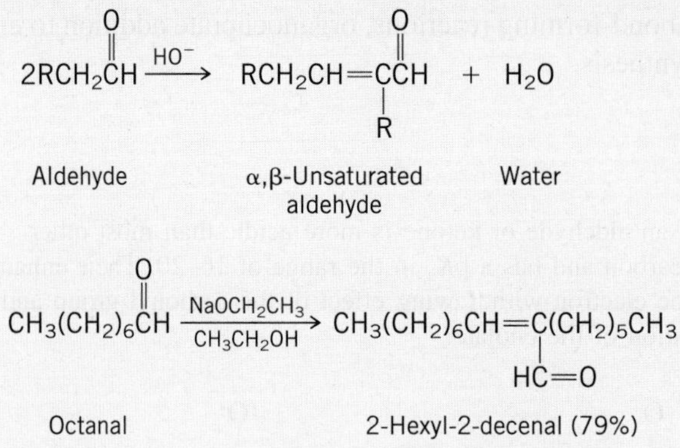2RCH2CH HO− RCH2CH=CCH + H2O R Aldehyde α,β-Unsaturated aldehyde Water CH3(CH2)6CH NaOCH2CH3 CH3CH2OH CH3(CH2)6CH=C(CH2)5CH3 HC=O Octanal 2-Hexyl-2-decenal (79%)
Claisen–Schmidt reaction (Section 20.4) A mixed aldol condensation in which an aromatic aldehyde reacts with an enolizable aldehyde or ketone.	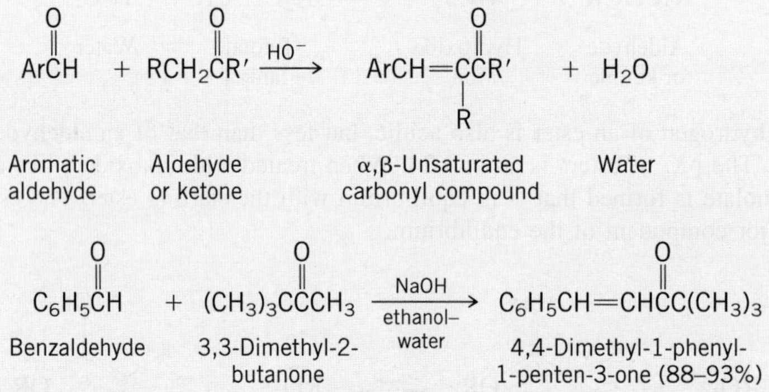ArCH + RCH2CR′ HO− ArCH=CCR′ + H2O R Aromatic aldehyde Aldehyde or ketone α,β-Unsaturated carbonyl compound Water C6H5CH + (CH3)3CCCH3 NaOH ethanol–water C6H5CH=CHCC(CH3)3 Benzaldehyde 3,3-Dimethyl-2-butanone 4,4-Dimethyl-1-phenyl-1-penten-3-one (88–93%)
Claisen condensation (Section 20.5) Esters of the type $RCH_2\overset{O}{\overset{\Vert}{C}}OR'$ are converted to β-keto esters on treatment with alkoxide bases. One molecule of an ester is converted to its enolate; a second molecule of ester acts as an acylating agent toward the enolate.	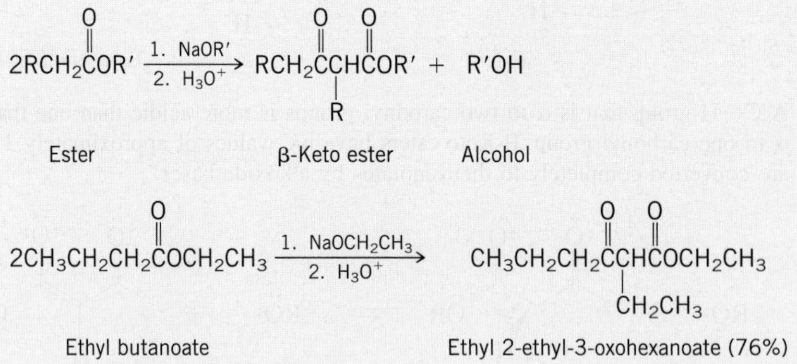2RCH2COR′ 1. NaOR′ 2. H3O+ RCH2CCHCOR′ + R′OH R Ester β-Keto ester Alcohol 2CH3CH2CH2COCH2CH3 1. NaOCH2CH3 2. H3O+ CH3CH2CH2CCHCOCH2CH3 CH2CH3 Ethyl butanoate Ethyl 2-ethyl-3-oxohexanoate (76%)
Dieckmann cyclization (Section 20.6) An intramolecular analog of the Claisen condensation. Cyclic β-keto esters in which the ring is five- to seven-membered may be formed by using this reaction.	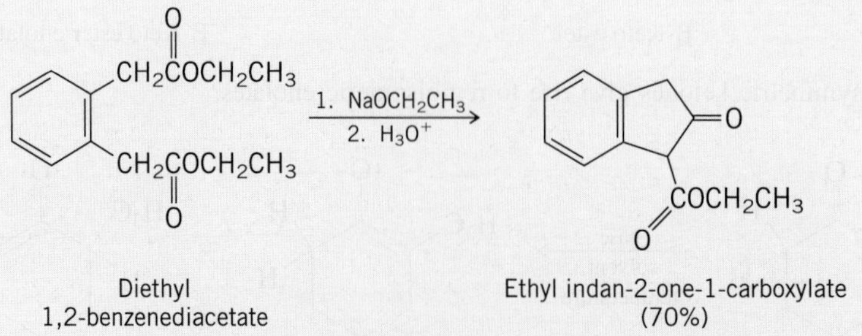CH2COCH2CH3 CH2COCH2CH3 1. NaOCH2CH3 2. H3O+ =O COCH2CH3 Diethyl 1,2-benzenediacetate Ethyl indan-2-one-1-carboxylate (70%)

Continued

TABLE 20.3 Carbonyl Condensations (*Continued*)	
Reaction (section) and comments	**General equation and typical example**
Mixed Claisen condensations (Section 20.7) Diethyl carbonate, diethyl oxalate, ethyl formate, and benzoate esters cannot form ester enolates but can act as acylating agents toward other ester enolates.	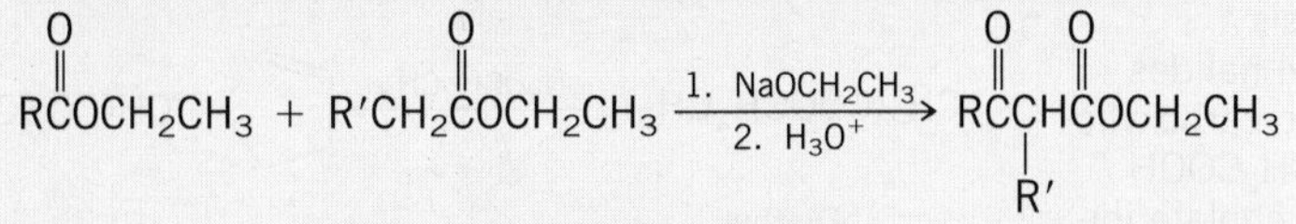 Ester · Another ester · β-Keto ester 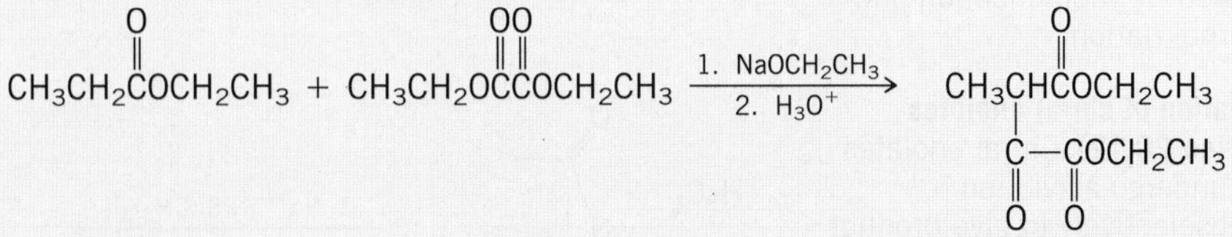 Ethyl propanoate · Diethyl oxalate · Diethyl 3-methyl-2-oxobutanedioate (60–70%)
Acylation of ketones (Section 20.8) Diethyl carbonate and diethyl oxalate can be used to acylate ketone enolates to give β-keto esters.	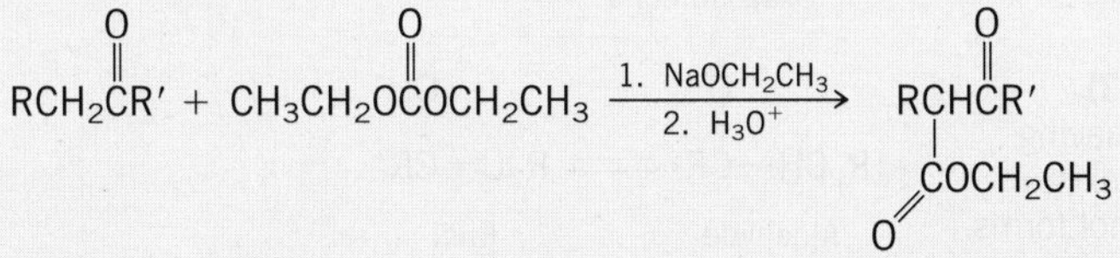 Ketone · Diethyl carbonate · β-Keto ester 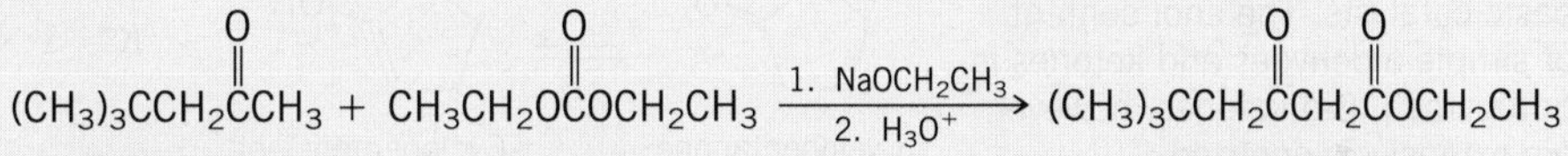 4,4-Dimethyl-2-pentanone · Diethyl carbonate · Ethyl 5,5-dimethyl-3-oxohexanoate (66%)

TABLE 20.4 Alkylation and Other Reactions That Involve Enol or Enolate Intermediates	
Reaction (section) and comments	**General equation and typical example**
α Alkylation of aldehydes and ketones (Section 20.9) Alkylation of simple aldehydes and ketones via their enolates is difficult. β-Diketones can be converted quantitatively to their enolate anions, which react efficiently with primary alkyl halides.	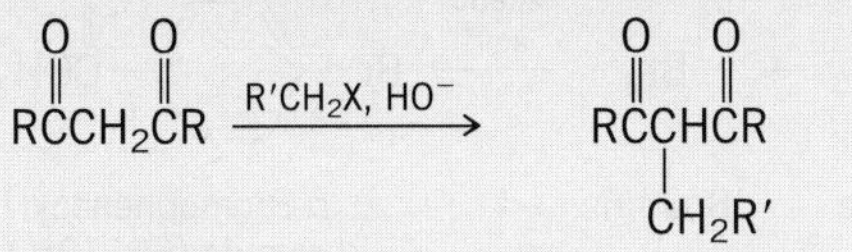 β-Diketone · α-Alkyl-β-diketone 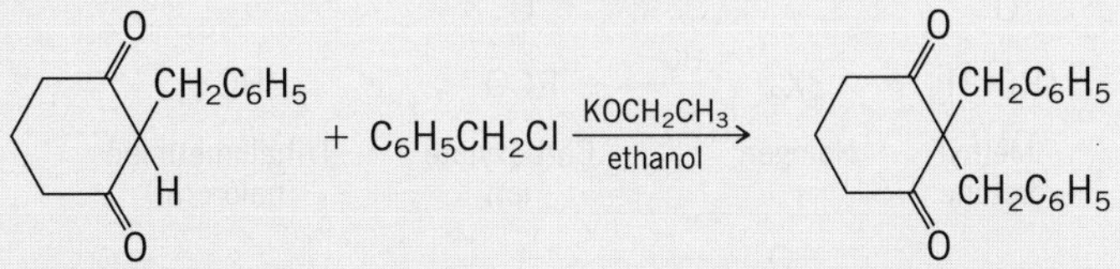 2-Benzyl-1,3-cyclohexanedione · Benzyl chloride · 2,2-Dibenzyl-1,3-cyclohexanedione (69%)
Acetoacetic ester synthesis (Section 20.10) Ethyl acetoacetate is alkylated with an alkyl halide as the first step in the preparation of ketones of the type $CH_3\overset{O}{\overset{\Vert}{C}}CH_2R$.	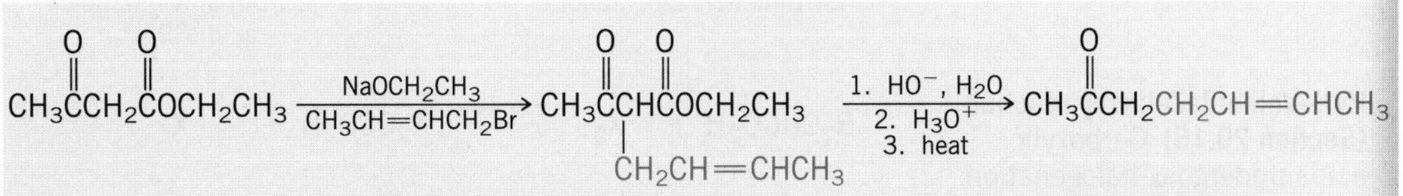 Ethyl acetoacetate · 5-Hepten-2-one (81%)

Continued

TABLE 20.4 Alkylation and Other Reactions That Involve Enol or Enolate Intermediates (*Continued*)

Reaction (section) and comments	General equation and typical example
Malonic ester synthesis (Section 20.11) Alkyl halides (RX) are converted to carboxylic acids of the type RCH_2COOH by reaction with the enolate ion derived from diethyl malonate, followed by saponification and decarboxylation.	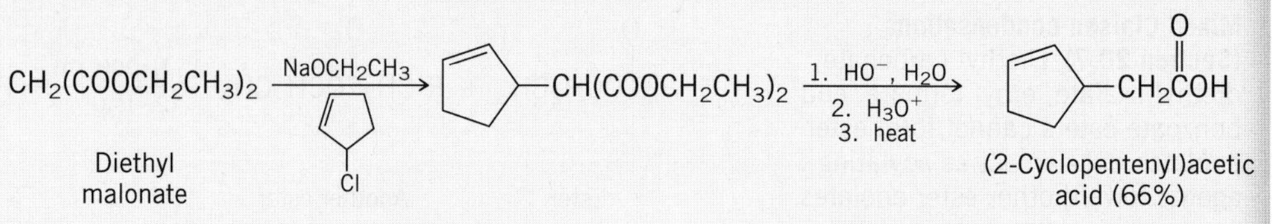 $CH_2(COOCH_2CH_3)_2$ $NaOCH_2CH_3$ Cl $CH(COOCH_2CH_3)_2$ 1. HO^-, H_2O 2. H_3O^+ 3. heat O $CH_2\overset{\|}{C}OH$ Diethyl malonate (2-Cyclopentenyl)acetic acid (66%)
Alkylation of chiral enolates (Section 20.12) Chiral enolates may undergo alkylation stereoselectively to give product enriched in one enantiomer.	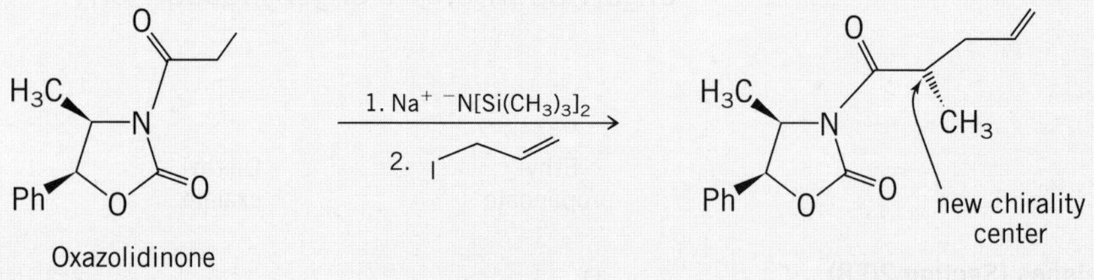 O H_3C N Ph O O 1. Na^+ $^-N[Si(CH_3)_3]_2$ 2. I CH_3 new chirality center Oxazolidinone
Enolization (Section 20.13) Aldehydes and ketones having at least one α hydrogen exist in equilibrium with their enol forms. The rate at which equilibrium is achieved is increased by acidic or basic catalysts. The enol content of simple aldehydes and ketones is quite small; β-diketones, however, are extensively enolized.	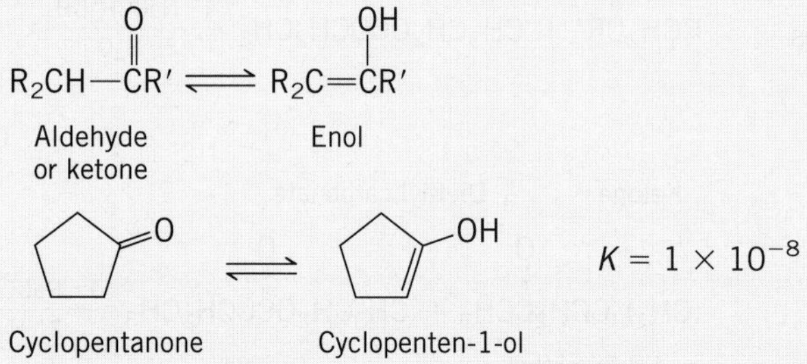 $R_2CH—\overset{O}{\overset{\|}{C}}R'$ $R_2C═\overset{OH}{\overset{\|}{C}}R'$ Aldehyde or ketone Enol O OH $K = 1 \times 10^{-8}$ Cyclopentanone Cyclopenten-1-ol
α Halogenation (Section 20.14) Halogens react with aldehydes and ketones by substitution; an α hydrogen is replaced by a halogen. Reaction occurs by electrophilic attack of the halogen on the carbon–carbon double bond of the enol form of the aldehyde or ketone. An acid catalyst increases the rate of enolization, which is the rate-determining step.	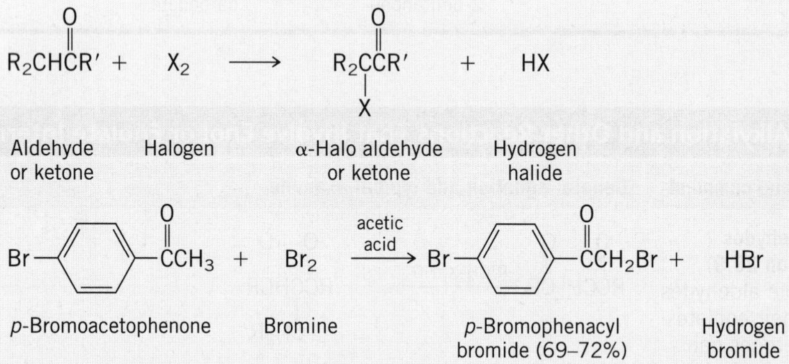 $R_2CH\overset{O}{\overset{\|}{C}}R'$ + X_2 $R_2\underset{X}{C}\overset{O}{\overset{\|}{C}}R'$ + HX Aldehyde or ketone Halogen α-Halo aldehyde or ketone Hydrogen halide Br $\overset{O}{\overset{\|}{C}}CH_3$ + Br_2 acetic acid Br $\overset{O}{\overset{\|}{C}}CH_2Br$ + HBr *p*-Bromoacetophenone Bromine *p*-Bromophenacyl bromide (69–72%) Hydrogen bromide
Haloform reaction (Section 20.14) Methyl ketones are cleaved on reaction with excess halogen in the presence of base. The products are a trihalomethane (haloform) and a carboxylate salt.	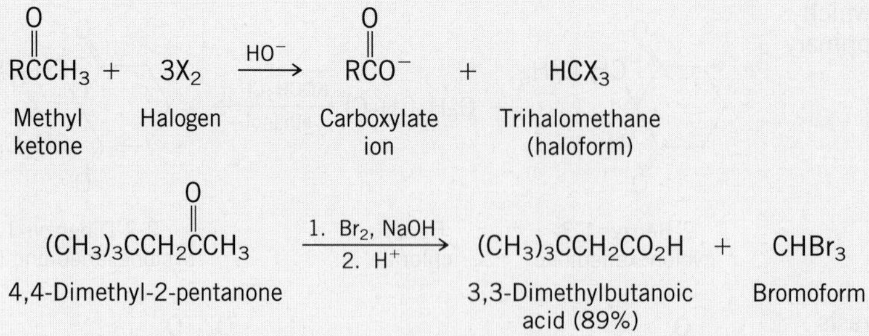 $R\overset{O}{\overset{\|}{C}}CH_3$ + $3X_2$ HO^- $R\overset{O}{\overset{\|}{C}}O^-$ + HCX_3 Methyl ketone Halogen Carboxylate ion Trihalomethane (haloform) $(CH_3)_3CCH_2\overset{O}{\overset{\|}{C}}CH_3$ 1. Br_2, NaOH 2. H^+ $(CH_3)_3CCH_2CO_2H$ + $CHBr_3$ 4,4-Dimethyl-2-pentanone 3,3-Dimethylbutanoic acid (89%) Bromoform
Hell–Volhard–Zelinsky reaction (Section 20.15) Carboxylic acids undergo α-halogenation on reaction with Cl_2 or Br_2 in the presence of phosphorus or a phosphorus trihalide as a catalyst.	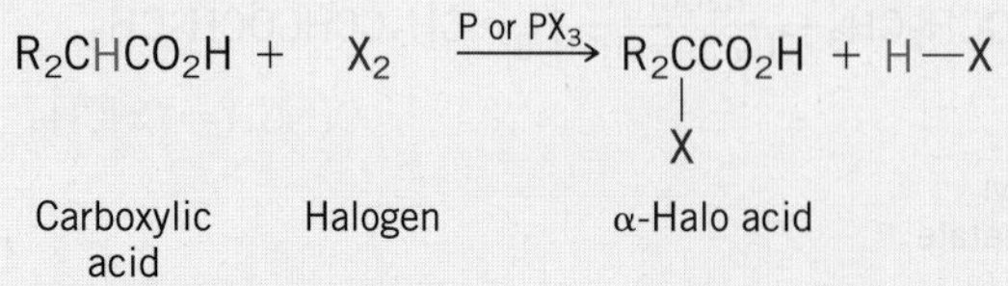 R_2CHCO_2H + X_2 P or PX_3 $R_2\underset{X}{C}CO_2H$ + H—X Carboxylic acid Halogen α-Halo acid

Continued

TABLE 20.4 Alkylation and Other Reactions That Involve Enol or Enolate Intermediates (*Continued*)

Reaction (section) and comments	General equation and typical example
Conjugate addition to α,β-unsaturated carbonyl compounds (Sections 20.17–20.20) The β-carbon atom of an α,β-unsaturated carbonyl compound is electrophilic; nucleophiles, especially weakly basic ones, yield the products of conjugate addition to α,β-unsaturated aldehydes and ketones.	$R_2C{=}CH\overset{O}{C}R' + HY{:} \longrightarrow R_2\underset{Y}{C}CH_2\overset{O}{C}R'$ α,β-Unsaturated aldehyde or ketone; Nucleophile; Product of conjugate addition $(CH_3)_2C{=}CH\overset{O}{C}CH_3 \xrightarrow[H_2O]{NH_3} (CH_3)_2\underset{NH_2}{C}CH_2\overset{O}{C}CH_3$ 4-Methyl-3-penten-2-one; 4-Amino-4-methyl-2-pentanone (63–70%)
Michael addition (Section 20.19) α,β-Unsaturated carbonyl compounds undergo conjugate addition on reaction with weakly basic nucleophiles, including enolates.	$CH_2(COOCH_2CH_3)_2 + CH_3CH{=}CH\overset{O}{C}OCH_2CH_3 \xrightarrow[CH_3CH_2OH]{NaOCH_2CH_3} CH_3\underset{CH(COOCH_2CH_3)_2}{C}HCH_2\overset{O}{C}OCH_2CH_3$ Diethyl malonate; Ethyl 2-butenoate; Triethyl 2-methylpropane-1,1,3-tricarboxylate (95%)
Robinson annulation (Section 20.19) A combination of conjugate addition of an enolate anion to an α,β-unsaturated ketone with subsequent intramolecular aldol condensation.	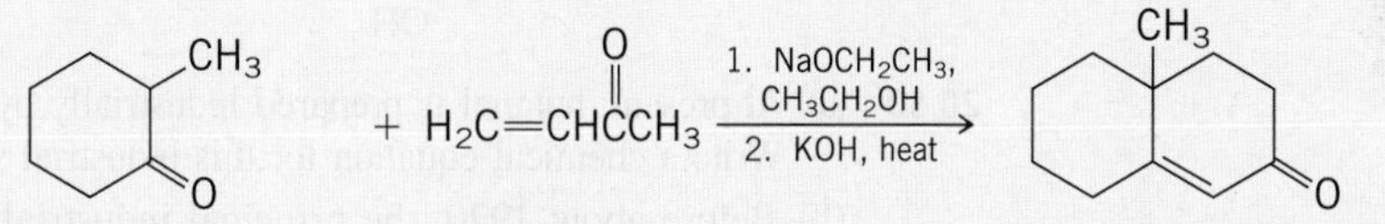 2-Methylcyclohexanone; Methyl vinyl ketone; 6-Methylbicyclo[4.4.0]-1-decen-3-one (46%)
Conjugate addition of organocopper compounds (Section 20.20) The principal synthetic application of lithium dialkylcuprate reagents is their reaction with α,β-unsaturated carbonyl compounds. Alkylation of the β carbon occurs.	$R_2C{=}CH\overset{O}{C}R' + R''_2CuLi \xrightarrow[2.\ H_2O]{1.\ diethyl\ ether} R_2\underset{R''}{C}{-}CH_2\overset{O}{C}R'$ α,β-Unsaturated aldehyde or ketone; Lithium dialkylcuprate; β-Alkyl aldehyde or ketone 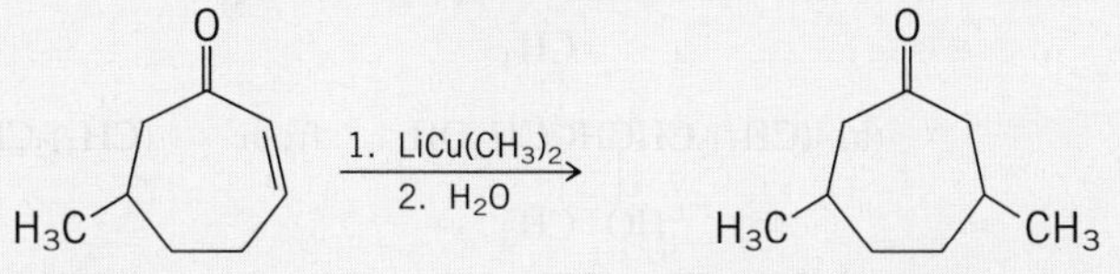 6-Methylcyclohept-2-enone; 3,6-Dimethylcycloheptanone (85%)

PROBLEMS

20.45 (a) Write structural formulas for all the noncyclic aldehydes and ketones of molecular formula C_4H_6O.

(b) Are any of these compounds stereoisomers?

(c) Are any of these compounds chiral?

(d) Which of these are α,β-unsaturated aldehydes or α,β-unsaturated ketones?

(e) Which of these can be prepared by a simple (i.e., not mixed) aldol condensation?

20.46 The main flavor component of the hazelnut is (2*E*,5*S*)-5-methyl-2-hepten-4-one. Write a structural formula showing its stereochemistry.

20.47 (a) Arrange the following in order of decreasing acidity.

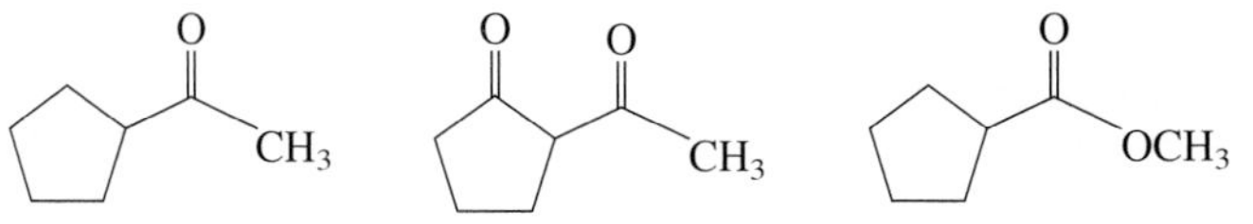

I. II. III.

(b) Write the structures of the kinetic and thermodynamic enolates of compound I.

20.48 Certain functional groups other than carbonyl groups can enhance the acidity of α hydrogens. Three such groups are nitro, cyano, and sulfonyl. Write resonance structures for the anions derived from each of the following compounds by loss of an α hydrogen.

(a) $CH_3C{\equiv}N:$ (b) 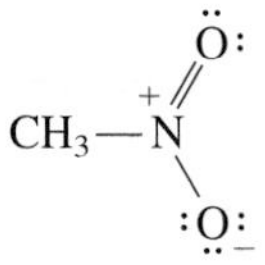 (c) 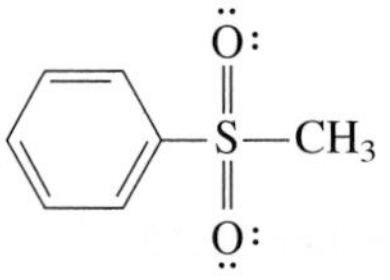

20.49 The simplest α,β-unsaturated aldehyde *acrolein* is prepared by heating glycerol with an acid catalyst. Suggest a mechanism for this reaction.

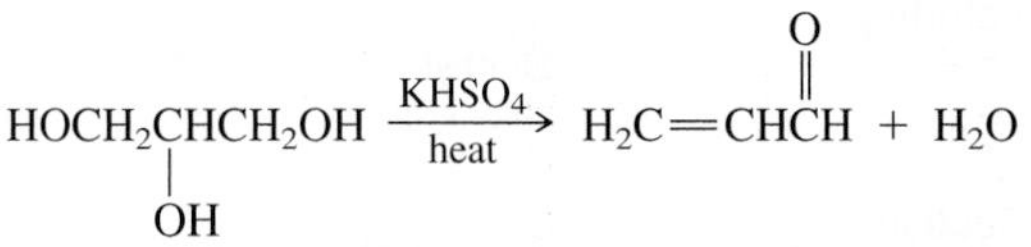

20.50 (a) At present, butanal is prepared industrially by hydroformylation of propene (Section 17.4). Write a chemical equation for this industrial synthesis.

(b) Before about 1970, the principal industrial preparation of butanal was from acetaldehyde. Outline a practical synthesis of butanal from acetaldehyde.

20.51 Show how each of the following compounds could be prepared from 3-pentanone. In most cases more than one synthetic transformation will be necessary.

(a) 2-Bromo-3-pentanone
(b) 1-Penten-3-one
(c) 1-Penten-3-ol
(d) 3-Hexanone
(e) 2-Methyl-1-phenyl-1-penten-3-one

20.52 Prepare each of the following compounds from the starting materials given and any necessary organic or inorganic reagents:

(a) 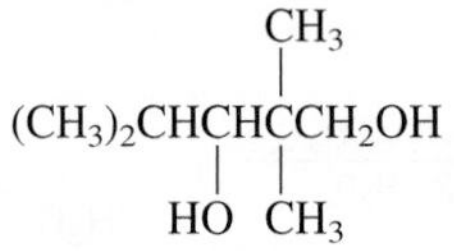 from $(CH_3)_2CHCH_2OH$

(b) 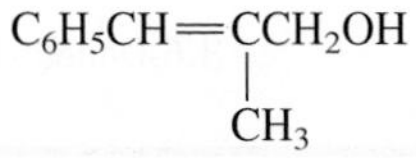 from benzyl alcohol and 1-propanol

(c) 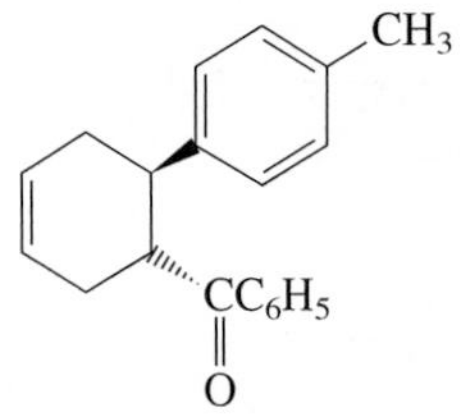 from acetophenone, 4-methylbenzyl alcohol, and 1,3-butadiene

20.53 The fragrance *cis*-jasmone can be prepared by an intramolecular aldol condensation of a diketone.

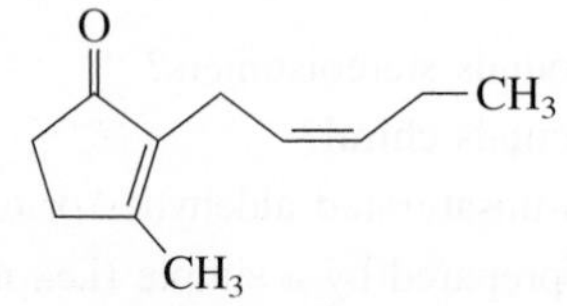

(a) Write the structure of the diketone.

(b) What other cyclic aldol condensation product might you expect to be formed in this reaction?

20.54 The following questions pertain to the esters shown and their behavior under conditions of the Claisen condensation.

$CH_3CH_2CH_2CH_2\overset{O}{\overset{\|}{C}}OCH_2CH_3$	$CH_3CH_2\underset{CH_3}{\underset{\|}{C}}H\overset{O}{\overset{\|}{C}}OCH_2CH_3$	$CH_3\underset{CH_3}{\underset{\|}{C}}HCH_2\overset{O}{\overset{\|}{C}}OCH_2CH_3$	$(CH_3)_3C\overset{O}{\overset{\|}{C}}OCH_2CH_3$
Ethyl pentanoate	Ethyl 2-methylbutanoate	Ethyl 3-methylbutanoate	Ethyl 2,2-dimethylpropanoate

(a) Two of these esters are converted to β-keto esters in good yield on treatment with sodium ethoxide and subsequent acidification of the reaction mixture. Which two are these? Write the structure of the Claisen condensation product of each one.

(b) One ester is capable of being converted to a β-keto ester on treatment with sodium ethoxide, but the amount of β-keto ester than can be isolated after acidification of the reaction mixture is quite small. Which ester is this?

(c) One ester is incapable of reaction under conditions of the Claisen condensation. Which one? Why?

20.55 (a) Give the structure of the Claisen condensation product of ethyl phenylacetate ($C_6H_5CH_2COOCH_2CH_3$).

(b) What ketone would you isolate after saponification and decarboxylation of this Claisen condensation product?

(c) What ketone would you isolate after treatment of the Claisen condensation product of ethyl phenylacetate with sodium ethoxide and allyl bromide, followed by saponification and decarboxylation?

(d) Give the structure of the mixed Claisen condensation product of ethyl phenylacetate and ethyl benzoate.

(e) What ketone would you isolate after saponification and decarboxylation of the product in part (d)?

(f) What ketone would you isolate after treatment of the product in part (d) with sodium ethoxide and allyl bromide, followed by saponification and decarboxylation?

20.56 All the following questions concern ethyl (2-oxocyclohexane)carboxylate.

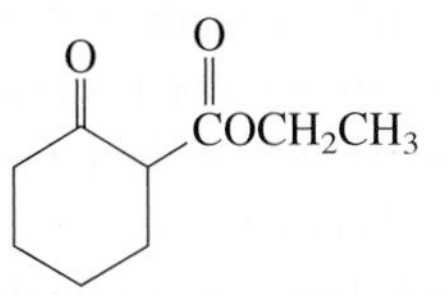

Ethyl (2-oxocyclohexane)carboxylate

(a) Write a chemical equation showing how you could prepare ethyl (2-oxocyclohexane)carboxylate by a Dieckmann cyclization.

(b) Write a chemical equation showing how you could prepare ethyl (2-oxocyclohexane)carboxylate by acylation of a ketone.

(c) Write structural formulas for the two most stable enol forms of ethyl (2-oxocyclohexane)carboxylate.

(d) Write the three most stable resonance contributors to the most stable enolate derived from ethyl (2-oxocyclohexane)carboxylate.

(e) Show how you could use ethyl (2-oxocyclohexane)carboxylate to prepare 2-methylcyclohexanone.

(f) Give the structure of the product formed on treatment of ethyl (2-oxocyclohexane)-carboxylate with acrolein ($H_2C{=}CH\overset{O}{\overset{\|}{C}}H$) in ethanol in the presence of sodium ethoxide.

20.57 The α-methylene ketone sarkomycin has an inhibitory effect on certain types of tumors. A key step in the synthesis of sarkomycin is the reaction of lactone ester A with potassium *tert*-butoxide in tetrahydrofuran to give the bicyclic compound B. Write a mechanism for this reaction.

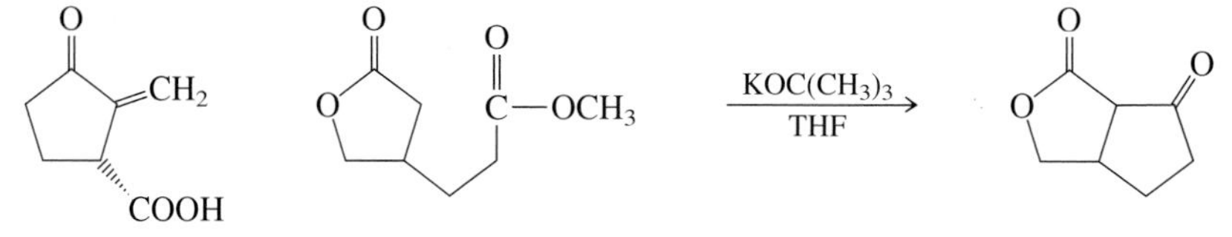

Sarkomycin A B

20.58 By applying the type of carbon–carbon bond-forming process used in the Claisen and Dieckmann reactions, coupled with ester hydrolysis and decarboxylation, show how 5-nonanone could be synthesized from ethyl pentanoate.

$$CH_3CH_2CH_2CH_2\overset{\overset{\large O}{\|}}{C}OCH_2CH_3 \longrightarrow \longrightarrow \longrightarrow CH_3CH_2CH_2CH_2\overset{\overset{\large O}{\|}}{C}CH_2CH_2CH_2CH_3$$

Ethyl pentanoate 5-Nonanone

20.59 β-Lactones can be prepared in good yield from thioester enolates. Suggest a mechanism for the reaction shown.

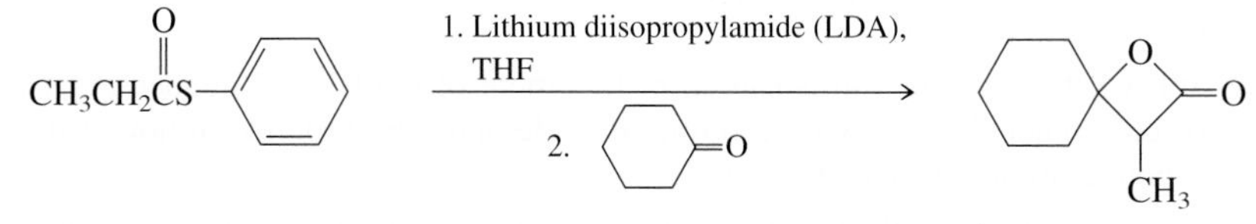

(92%)

20.60 Give the structure of the product formed on reaction of ethyl acetoacetate with each of the following:

(a) 1-Bromopentane and sodium ethoxide
(b) Saponification (basic hydrolysis) and decarboxylation of the product in part (a)
(c) Methyl iodide and the product in part (a) treated with sodium ethoxide
(d) Saponification and decarboxylation of the product in part (c)
(e) 1-Bromo-3-chloropropane and one equivalent of sodium ethoxide
(f) Product in part (e) treated with a second equivalent of sodium ethoxide
(g) Saponification and decarboxylation of the product in part (f)
(h) Phenyl vinyl ketone and sodium ethoxide
(i) Saponification and decarboxylation of the product in part (h)

20.61 Repeat the preceding problem for diethyl malonate.

20.62 Give the structure of the principal organic product of each of the following reactions:

(a)

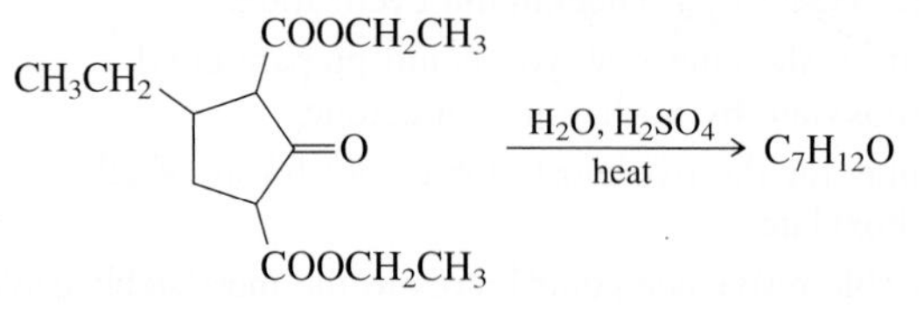

(b)

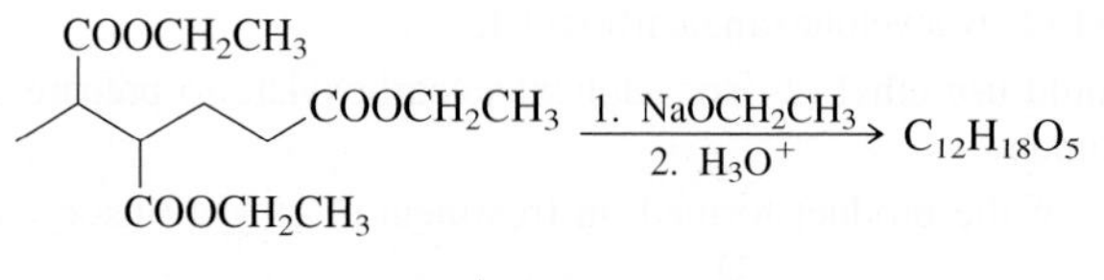

(c) Product of part (b) $\xrightarrow[\text{heat}]{H_3O^+}$ $C_7H_{10}O_3$

(d) 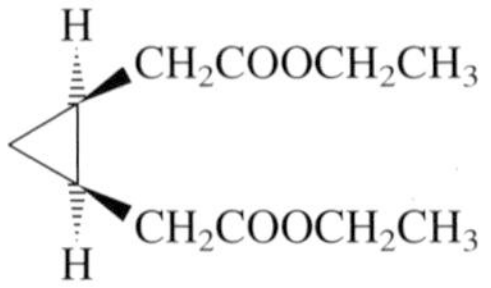 $\xrightarrow[\text{2. } H_3O^+]{\text{1. } NaOCH_2CH_3}$ $C_9H_{12}O_3$

(e) Product of part (d) $\xrightarrow[\text{2. } H_3O^+ \text{ 3. heat}]{\text{1. } HO^-, H_2O}$ C_6H_8O

20.63 The spicy flavor of cayenne pepper is due mainly to a substance called *capsaicin.* See if you can deduce the structure of capsaicin on the basis of its laboratory synthesis:

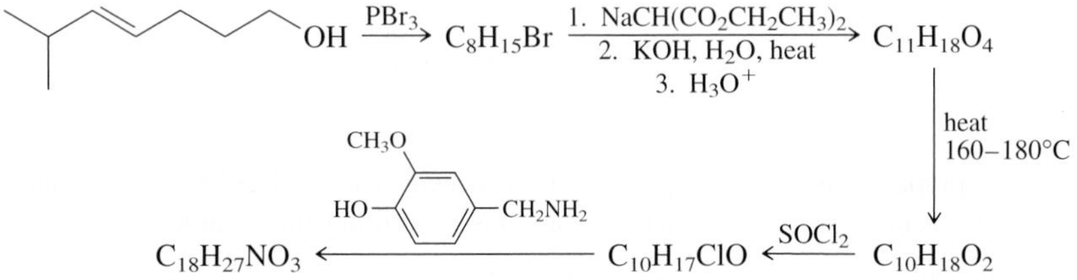
OH
PBr3
C8H15Br
1. NaCH(CO2CH2CH3)2
2. KOH, H2O, heat
3. H3O+
C11H18O4
heat
160–180°C
C10H18O2
SOCl2
C10H17ClO
CH3O
HO
CH2NH2
C18H27NO3
Capsaicin

20.64 Show how you could prepare each of the following compounds. Use the starting material indicated along with ethyl acetoacetate or diethyl malonate and any necessary inorganic reagents. Assume also that the customary organic solvents are freely available.

(a) 4-Phenyl-2-butanone from benzyl alcohol
(b) 3-Phenylpropanoic acid from benzyl alcohol
(c) 2-Allyl-1,3-propanediol from propene
(d) 4-Penten-1-ol from propene
(e) 5-Hexen-2-ol from propene
(f) Cyclopropanecarboxylic acid from 1,2-dibromoethane
(g) 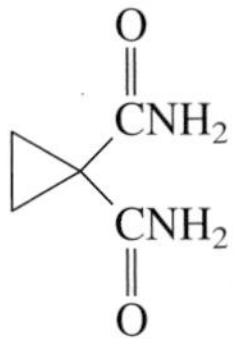 from 1,2-dibromoethane
(h) $HO_2C(CH_2)_{10}CO_2H$ from $HO_2C(CH_2)_6CO_2H$

20.65 *Diphenadione* inhibits the clotting of blood; that is, it is an *anticoagulant.* It is used to control vampire bat populations in South America by a "Trojan horse" strategy. A few bats are trapped, smeared with diphenadione, and then released back into their normal environment. Other bats, in the course of grooming these diphenadione-coated bats, ingest the anticoagulant and bleed to death, either internally or through accidental bites and scratches.

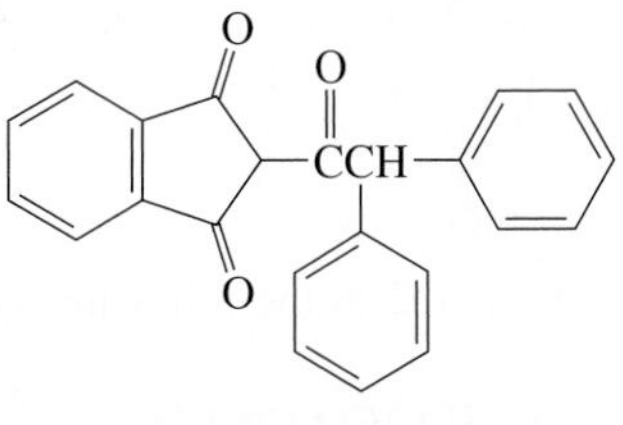

Diphenadione

Suggest a synthesis of diphenadione from 1,1-diphenylacetone and dimethyl 1,2-benzenedicarboxylate.

20.66 The use of epoxides as alkylating agents for diethyl malonate provides a useful route to γ-lactones. Write equations illustrating such a sequence for styrene oxide as the starting epoxide. Is the lactone formed by this reaction 3-phenylbutanolide, or is it 4-phenylbutanolide?

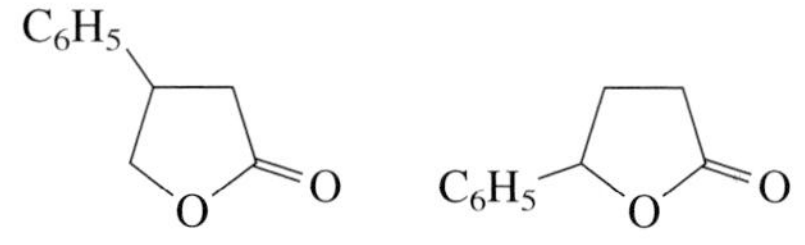
C_6H_5
O
O
C_6H_5
O
O

3-Phenylbutanolide 4-Phenylbutanolide

20.67 Diethyl malonate is prepared commercially by hydrolysis and esterification of ethyl cyanoacetate.

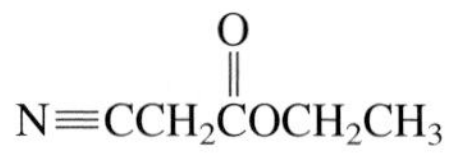
O
$N{\equiv}CCH_2COCH_2CH_3$

Ethyl cyanoacetate

The preparation of ethyl cyanoacetate proceeds via ethyl chloroacetate and begins with acetic acid. Write a sequence of reactions describing this synthesis.

20.68 When the compound shown was heated in refluxing aqueous hydrochloric acid for 60 hours, a product with the molecular formula $C_5H_6O_3$ was isolated in 97% yield. Identify this product. Along with this product, three other carbon-containing substances are formed. What are they?

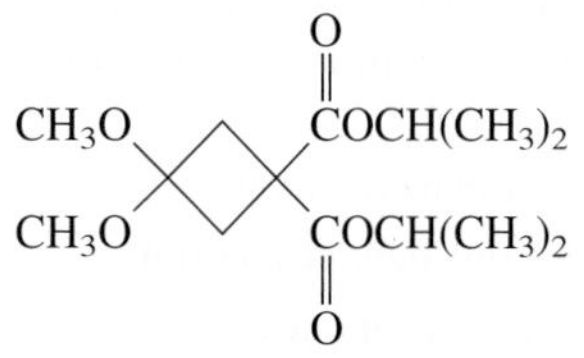
CH_3O
CH_3O
O
$COCH(CH_3)_2$
$COCH(CH_3)_2$
O

20.69 In each of the following pairs of compounds, choose the one that has the greater enol content, and write the structure of its enol form:

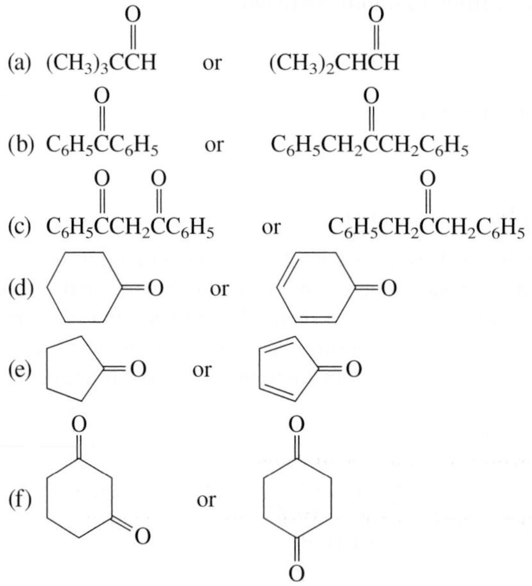
(a) $(CH_3)_3CCH$ or $(CH_3)_2CHCH$
(b) $C_6H_5CC_6H_5$ or $C_6H_5CH_2CCH_2C_6H_5$
(c) $C_6H_5CCH_2CC_6H_5$ or $C_6H_5CH_2CCH_2C_6H_5$
(d) or
(e) or
(f) or

20.70 Identify compounds A, B, and C in the following synthetic sequence.

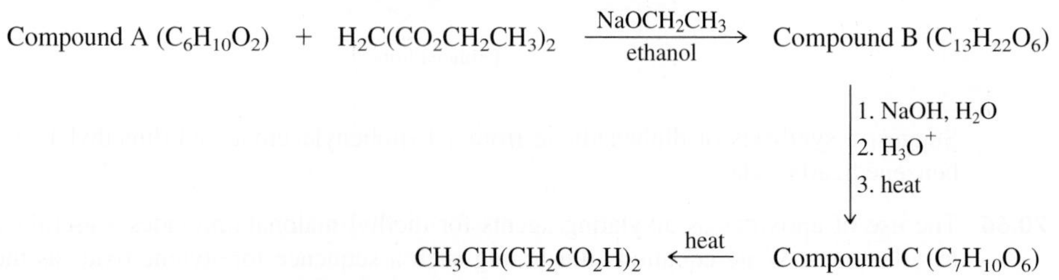
Compound A ($C_6H_{10}O_2$) + $H_2C(CO_2CH_2CH_3)_2$
$NaOCH_2CH_3$
ethanol
Compound B ($C_{13}H_{22}O_6$)
1. NaOH, H_2O
2. H_3O^+
3. heat
$CH_3CH(CH_2CO_2H)_2$
heat
Compound C ($C_7H_{10}O_6$)

20.71 *Terreic acid* is a naturally occurring antibiotic substance. Its actual structure is an enol isomer of the structure shown. Write the two most stable enol forms of terreic acid, and choose which of those two is more stable.

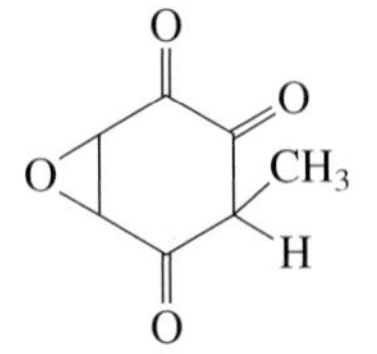

20.72 In each of the following, the indicated observations were made before any of the starting material was transformed to aldol addition or condensation products:

(a) In aqueous acid, only 17% of $(C_6H_5)_2CHCH{=}O$ is present as the aldehyde; 2% of the enol is present. Some other species accounts for 81% of the material. What is it?

(b) In aqueous base, 97% of $(C_6H_5)_2CHCH{=}O$ is present as a species different from any of those in part (a). What is this species?

20.73 For a long time attempts to prepare compound A were thwarted by its ready isomerization to compound B. The isomerization is efficiently catalyzed by traces of base. Write a reasonable mechanism for this isomerization.

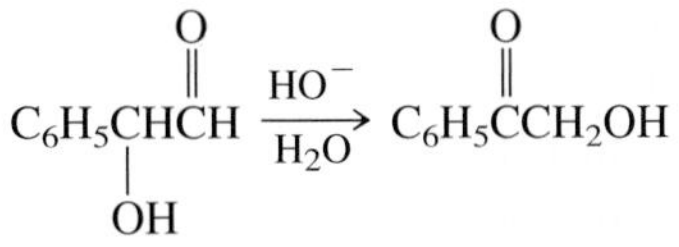

Compound A Compound B

20.74 The following questions address stereochemical aspects of the alkylation of 2-methylcyclohexanone, which was described in Section 20.9.

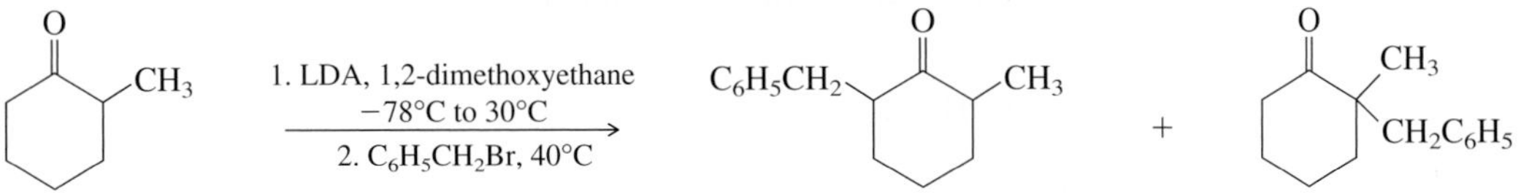

2-Methylcyclohexanone

2-Benzyl-6-methylcyclohexanone (≈52% yield)

2-Benzyl-2-methylcyclohexanone (≈7% yield)

(a) Which of the products is (are) chiral?

(b) How many stereoisomers, including enantiomers and diastereomers, are possible for each of the two products?

(c) If the reactant is a racemic mixture, will either of the products be optically active? If so, which one(s)?

(d) If the reactant is a nonracemic mixture, will either of the products be optically active? If so, which one(s)?

20.75 (a) Only a small amount (less than 0.01%) of the enol form of diethyl malonate is present at equilibrium. Write a structural formula for this enol.

(b) Enol forms are present to the extent of about 8% in ethyl acetoacetate. There are three constitutionally isomeric enols possible. Write structural formulas for these three enols. Which one do you think is the most stable? The least stable? Why?

(c) Bromine reacts rapidly with both diethyl malonate and ethyl acetoacetate. The reaction is acid-catalyzed and liberates hydrogen bromide. What is the product formed in each reaction?

20.76 (a) On addition of one equivalent of methylmagnesium iodide to ethyl acetoacetate, the Grignard reagent is consumed, but the only organic product obtained after working up the reaction mixture is ethyl acetoacetate. Why? What happens to the Grignard reagent?

(b) On repeating the reaction but using D_2O and DCl to work up the reaction mixture, it is found that the recovered ethyl acetoacetate contains deuterium. Where is this deuterium located?

20.77 Give the structure of the principal organic product of each of the following reactions:

(a) Ethyl acetoacetate + 1-bromobutane $\xrightarrow{\text{NaOCH}_2\text{CH}_3\text{, ethanol}}$

(b) Product of part (a) $\xrightarrow[\text{2. H}_3\text{O}^+ \quad \text{3. heat}]{\text{1. NaOH, H}_2\text{O}}$

(c) Acetophenone + diethyl carbonate $\xrightarrow[\text{2. H}_3\text{O}^+]{\text{1. NaOCH}_2\text{CH}_3}$

(d) Acetone + diethyl oxalate $\xrightarrow[\text{2. H}_3\text{O}^+]{\text{1. NaOCH}_2\text{CH}_3}$

(e) Diethyl malonate + 1-bromo-2-methylbutane $\xrightarrow{\text{NaOCH}_2\text{CH}_3\text{, ethanol}}$

(f) Product of part (e) $\xrightarrow[\text{2. H}_3\text{O}^+ \quad \text{3. heat}]{\text{1. NaOH, H}_2\text{O}}$

(g) Diethyl malonate + 6-methyl-2-cyclohexenone $\xrightarrow{\text{NaOCH}_2\text{CH}_3\text{, ethanol}}$

(h) Product of part (g) $\xrightarrow{\text{H}_2\text{O, HCl, heat}}$

(i) *tert*-Butyl acetate $\xrightarrow[\text{2. benzaldehyde} \quad \text{3. H}_3\text{O}^+]{\text{1. [(CH}_3)_2\text{CH]}_2\text{NLi, THF}}$

20.78 Give the structure of the expected organic product in the reaction of 3-phenylpropanal with each of the following:

(a) Chlorine in acetic acid

(b) Sodium hydroxide in ethanol, 10°C

(c) Sodium hydroxide in ethanol, 70°C

(d) Product of part (c) with lithium aluminum hydride; then H_2O

(e) Product of part (c) with sodium cyanide in acidic ethanol

20.79 Each of the following reactions has been reported in the chemical literature. Write the structure of the product(s) formed in each case.

(a) 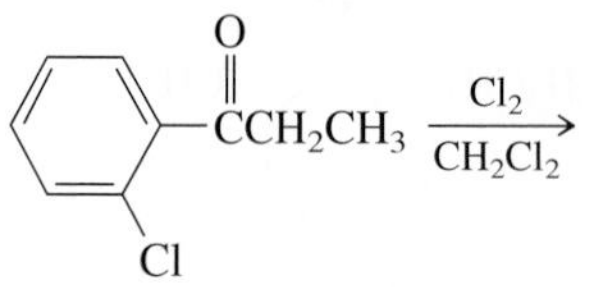

(b) 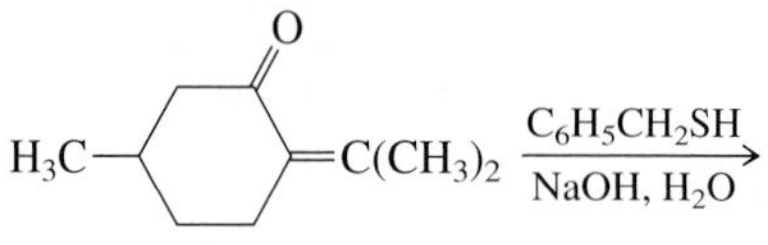

(c) 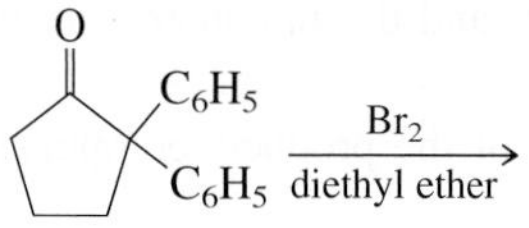

(d) 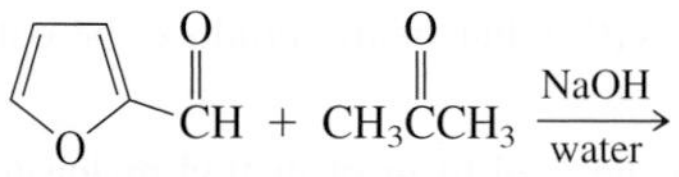

(e) 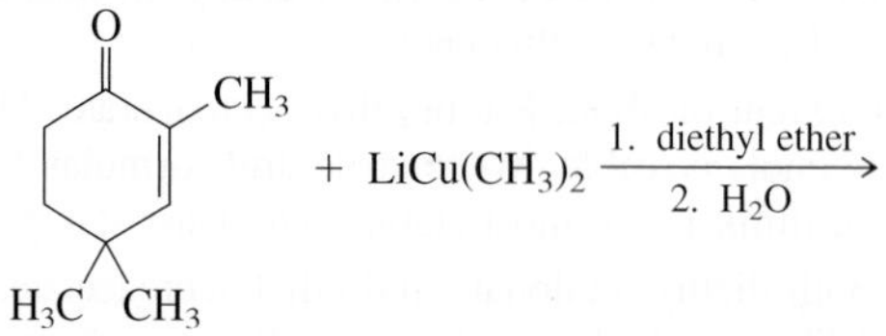

(f) 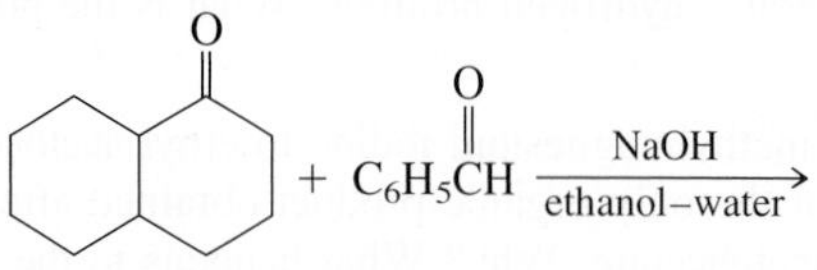

(g) 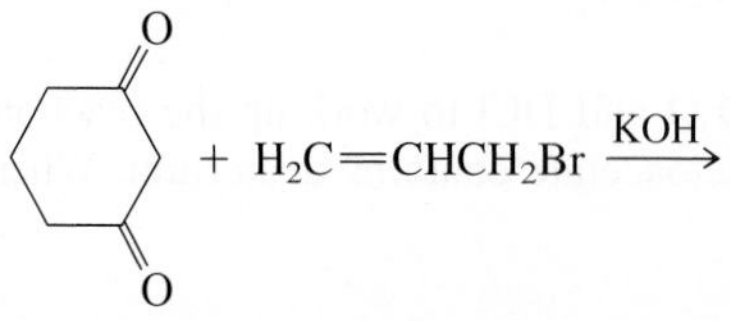

20.80 (a) A synthesis that begins with 3,3-dimethyl-2-butanone gives the epoxide shown. Suggest reagents appropriate for each step in the synthesis.

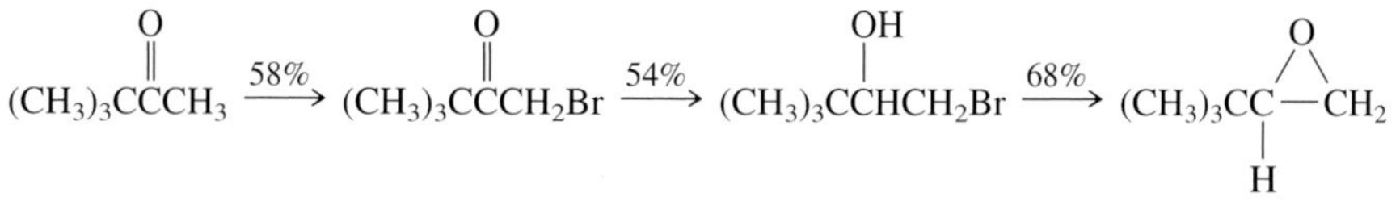

(b) The yield for each step as actually carried out in the laboratory is given above each arrow. What is the overall yield for the three-step sequence?

20.81 Using benzene, acetic anhydride, and 1-propanethiol as the source of all the carbon atoms, along with any necessary inorganic reagents, outline a synthesis of the compound shown.

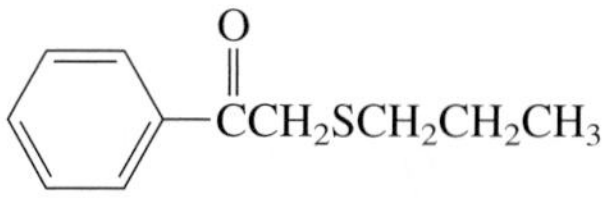

20.82 Identify the reagents appropriate for each step in the following syntheses:

(a)

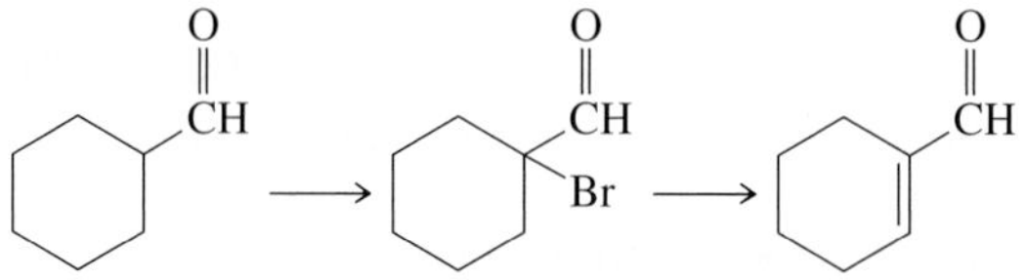

(b)

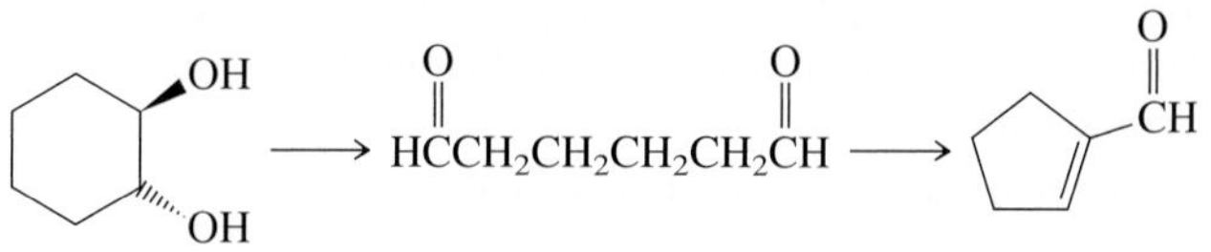

(c)

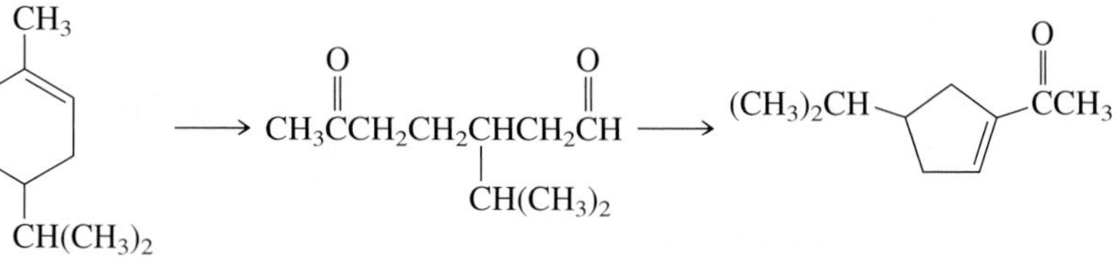

20.83 Consider the ketones menthone and isomenthone.

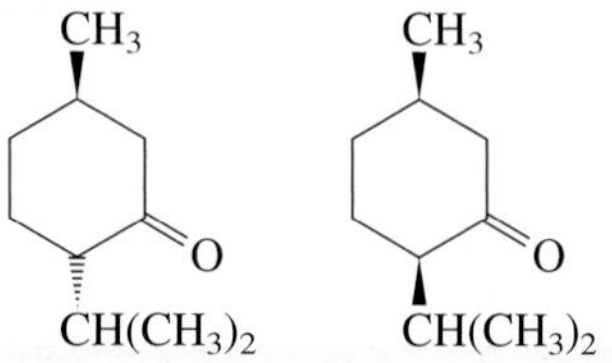

Menthone Isomenthone

Suggest an explanation for the observation that menthone is converted to a mixture of menthone and isomenthone on treatment with 90% sulfuric acid.

20.84 Outline reasonable mechanisms for each of the following reactions:

(a)

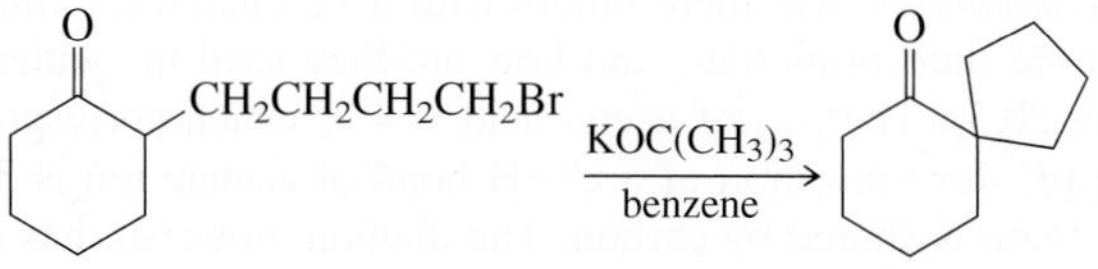

(76%)

(b)

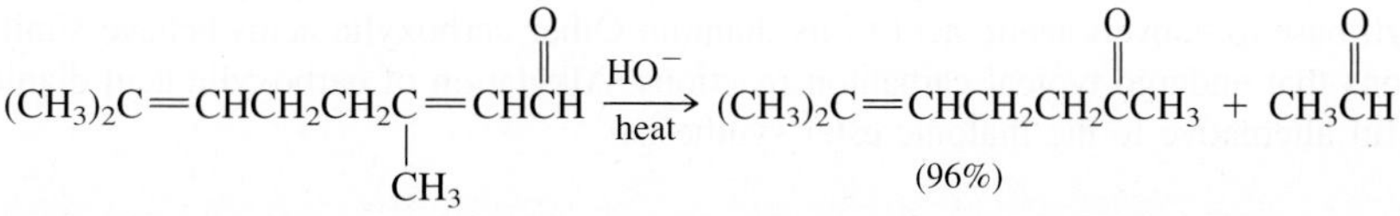

(96%)

(c) 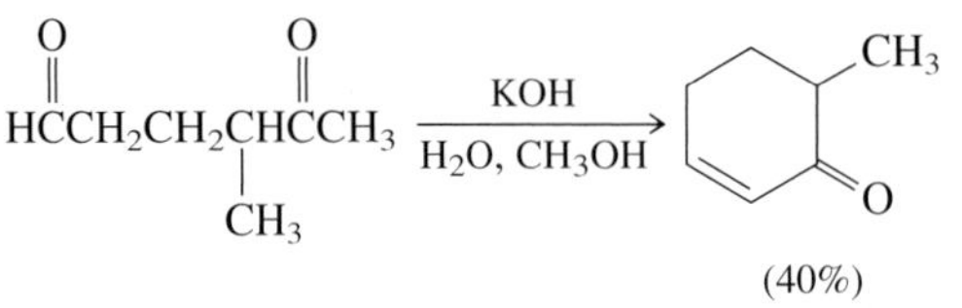

(d) 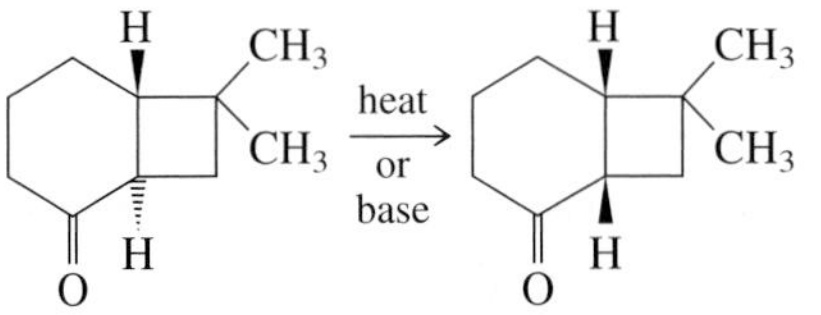

20.85 The addition of diethyl malonate to benzaldehyde in the presence of the basic amine piperidine gives the unsaturated diester, diethyl 2-benzylidenemalonate. This is an example of the *Knoevenagel* reaction, a synthetically useful carbonyl condensation reaction, which is related to the aldol condensation.

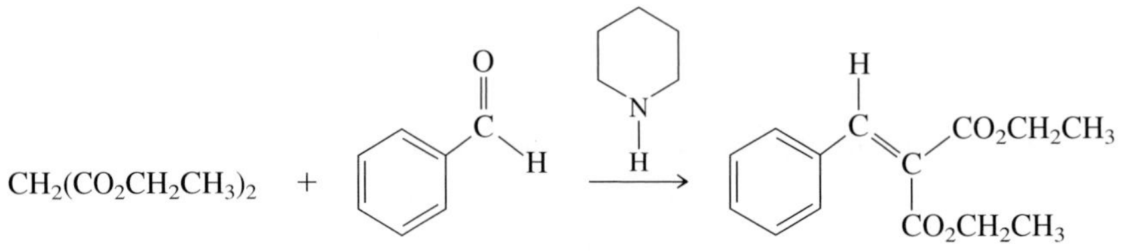

Diethyl malonate Benzaldehyde Diethyl 2-benzylidenemalonate

(a) What is the nucleophile in the Knoevenagel reaction?

(b) The unsaturated diester A in part (c) was prepared by a Knoevenagel reaction and used to synthesize the anticonvulsant drug gabapentin. From which two reactants is the diester synthesized?

(c) Fill in the missing reagents in the reaction scheme.

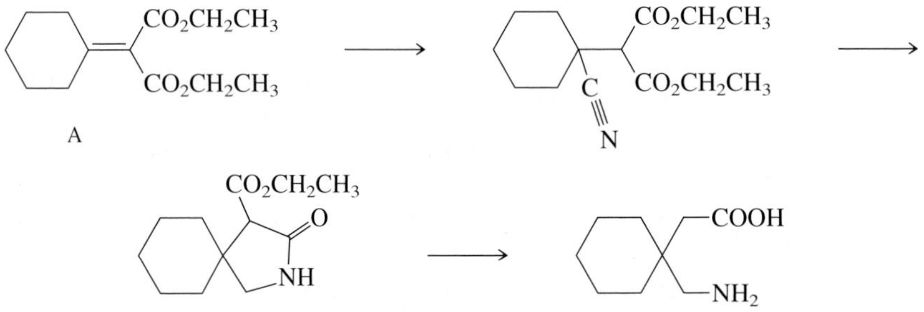

Gabapentin

Descriptive Passage and Interpretive Problems 20

The Enolate Chemistry of Dianions

The synthetic applications of carbanions as reagents for carbon–carbon bond formation have been highlighted numerous times throughout this text. All of the reagents covered so far have a net charge of −1; that is, they are *monoanions.* Are there others with a −2 charge (*dianions*), and, if so, how are they prepared, what are their properties, and how are they used in synthesis?

Consider acetic acid (Figure 20.5). The pK_a of acetic acid is 4.7, which corresponds to ionization of the O—H group. The pK_a for ionization of a C—H bond of acetate ion is 33. None of the negative charge of the monoanion is shared by carbon. The dianion, however, has carbanionic character and the potential to act as a nucleophile in carbon–carbon bond-forming reactions.

Diisopropylamine has a pK_a of 36, which makes lithium diisopropylamide (LDA) a strong enough base to convert acetic acid to its dianion. Other carboxylic acids behave similarly to give dianions that undergo typical carbanion reactions. Alkylation of carboxylic acid dianions provide a useful alternative to the malonic ester synthesis.

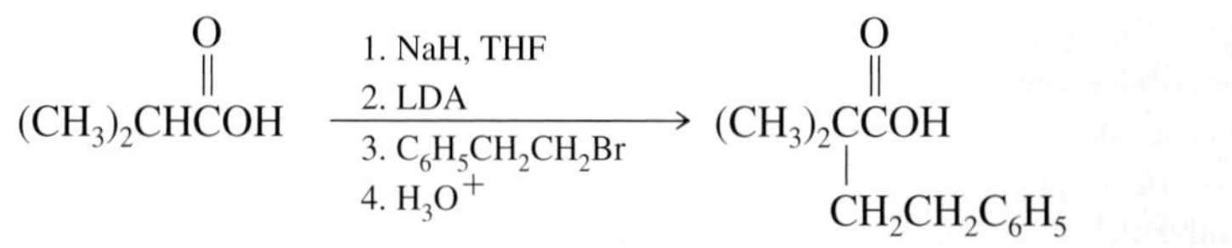

2-Methylpropanoic acid → 2,2-Dimethyl-4-phenylbutanoic acid (70–76%)

Experimentally, as in this example, it is sometimes useful to convert the carboxylic acid to its carboxylate (monoanion) with sodium hydride (NaH, step 1) before treating with LDA (step 2). Because the dianion is a strong base, the alkyl halide used in step 3 must be methyl or primary. A pH adjustment (step 4) converts the resulting carboxylate salt to the desired carboxylic acid.

The dianions of α-halocarboxylic acids give epoxy acids (called *glycidic acids*) on reaction with aldehydes and ketones.

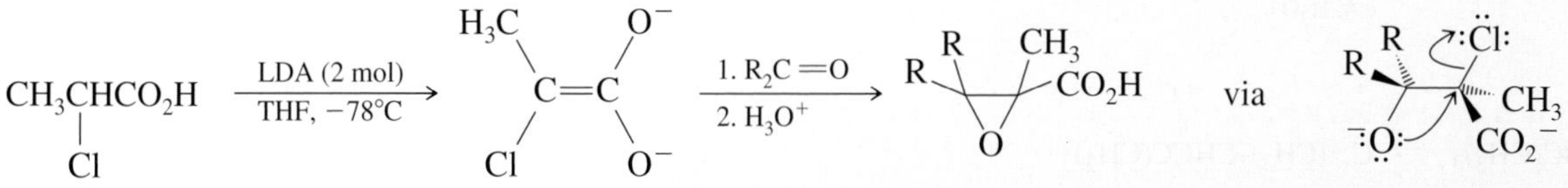

Dianions have been prepared from β-keto esters by double deprotonation using a number of strong bases including LDA.

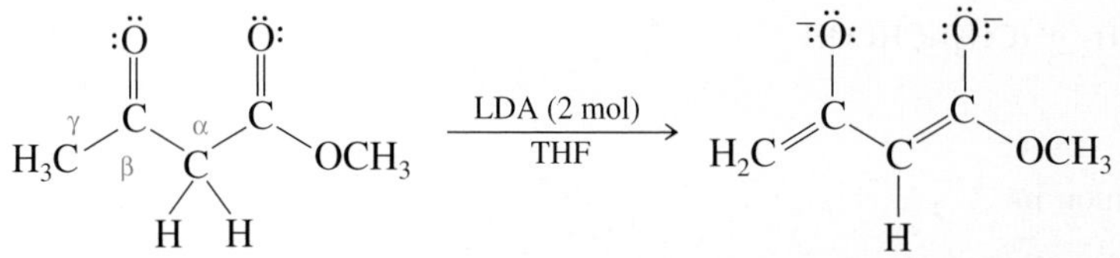

Protons on the α carbon of β-keto esters are flanked by two carbonyl groups and are far more acidic than those on the γ carbon. In the dianion, therefore, the γ carbon is more basic and more nucleophilic. Alkylation of the monoanion of β-keto acids occurs at the α carbon. Alkylation of the dianion occurs at the γ carbon.

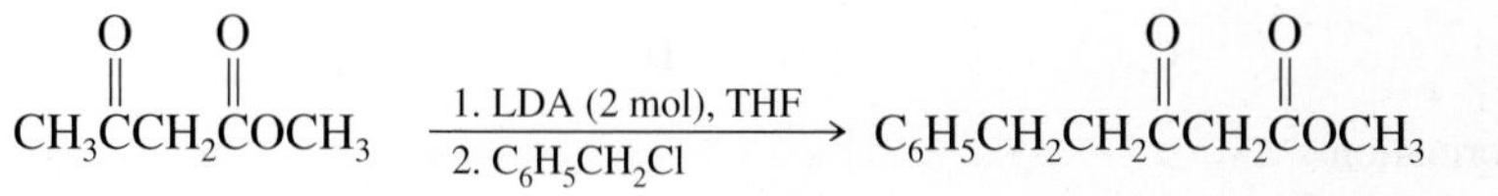

Methyl acetoacetate → Methyl 3-oxo-5-phenylpentanoate (83%)

β-Diketones behave similarly to β-keto esters. Sodium or potassium amide in liquid ammonia is a suitable base/solvent system in this case.

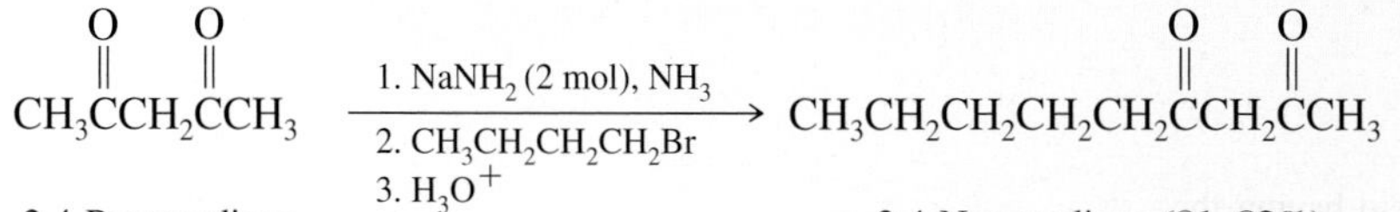

2,4-Pentanedione → 2,4-Nonanedione (81–82%)

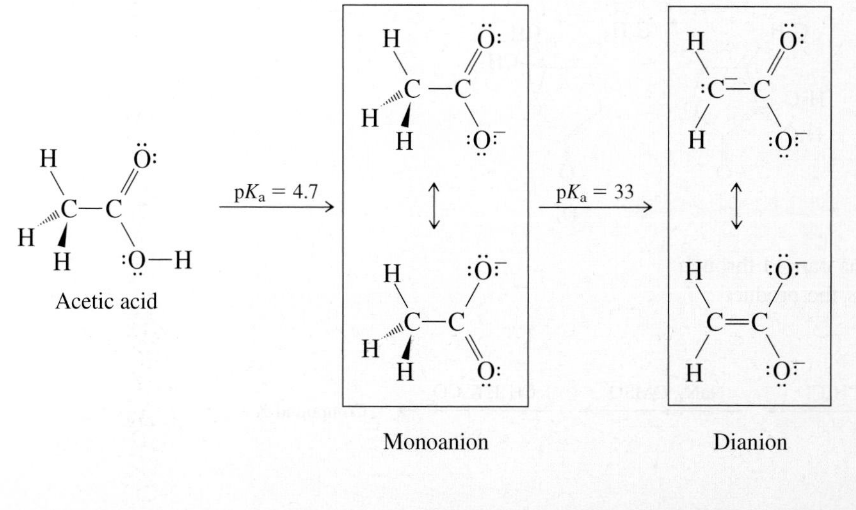

Acetic acid, Monoanion, Dianion

Figure 20.5

Acetic acid, its monoanion and dianion.

20.86 The first reaction shown in the passage described the synthesis of a carboxylic acid (2,2-dimethyl-4-phenylbutanoic acid) in which the α carbon is quaternary. What is the maximum degree of substitution at the α carbon of a carboxylic acid prepared by the malonic ester synthesis?

A. Primary
B. Secondary
C. Tertiary
D. Quaternary

20.87 Predict the major organic product(s) of the following reaction.

$$(CH_3)_2CH\overset{O}{\overset{\|}{C}}OH \xrightarrow[\substack{3.\ (CH_3)_3CCl \\ 4.\ H_3O^+}]{\substack{1.\ NaH,\ THF \\ 2.\ LDA}}$$

A. $(CH_3)_2CH\overset{O}{\overset{\|}{C}}OC(CH_3)_3$

B. $(CH_3)_2\underset{C(CH_3)_3}{\underset{|}{C}}\overset{O}{\overset{\|}{C}}OH$

C. $(CH_3)_2CH\overset{O}{\overset{\|}{C}}C(CH_3)_3$

D. $(CH_3)_2C{=}CH_2 + (CH_3)_2CH\overset{O}{\overset{\|}{C}}OH$

20.88 What is the product of the reaction of the dianion in Problem 20.87 with cyclohexanone?

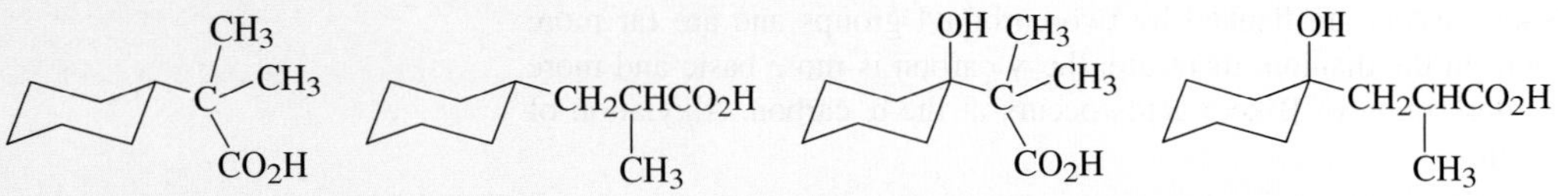
CH3
CH3
C
CO2H
CH2CHCO2H
CH3
OH CH3
CH3
C
CO2H
OH
CH2CHCO2H
CH3

A. B. C. D.

20.89 The regiochemistry of carbon–carbon bond formation between the dianion shown and styrene oxide is as indicated by the curved arrows:

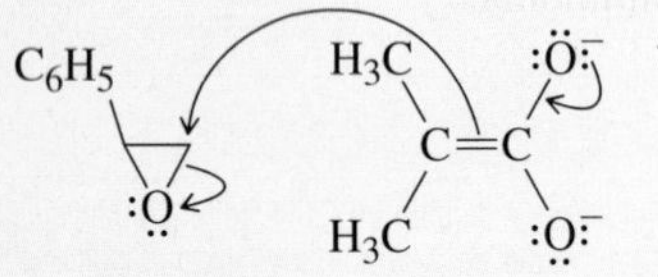
C6H5
H3C
:Ö:
C=C
:O:
H3C
:O:

The product of this reaction is a hydroxy acid having the molecular formula $C_{12}H_{16}O_3$, which cyclizes readily to give a lactone ($C_{12}H_{14}O_2$). What is the structure of this lactone?

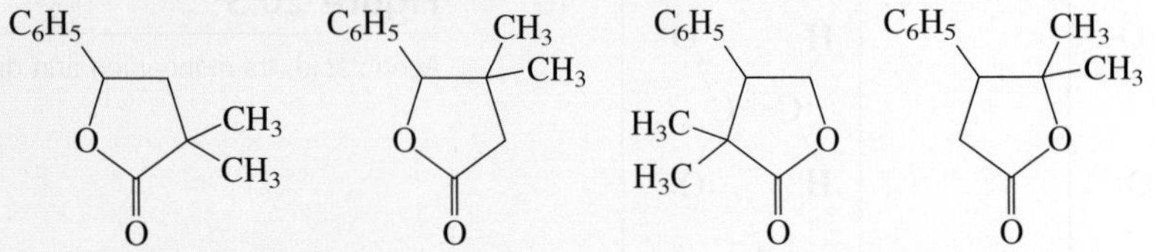
C6H5
O
CH3
CH3
O
C6H5
CH3
CH3
O
O
C6H5
H3C
H3C
O
O
C6H5
CH3
CH3
O
O

A. B. C. D.

20.90 The dianion from ethyl acetoacetate was carried through the reaction sequence shown. What was the product (compound X)?

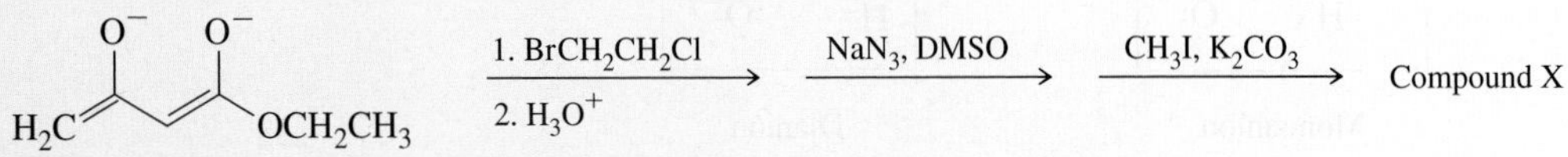
O−
O−
H2C
OCH2CH3
1. BrCH2CH2Cl
2. H3O+
NaN3, DMSO
CH3I, K2CO3
Compound X

A. 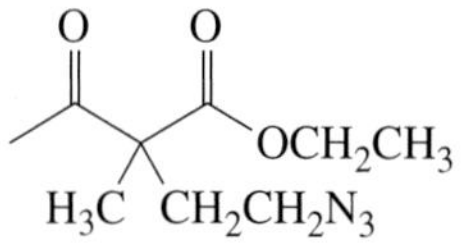

C. 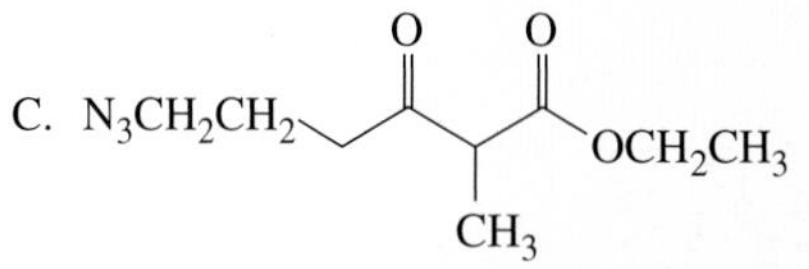

B. 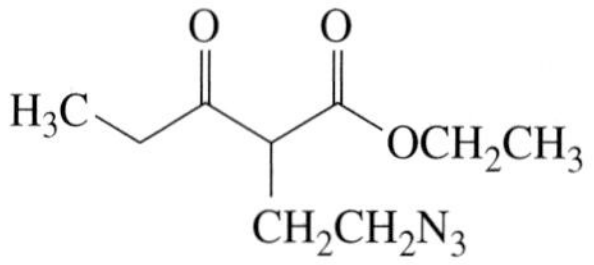

D. 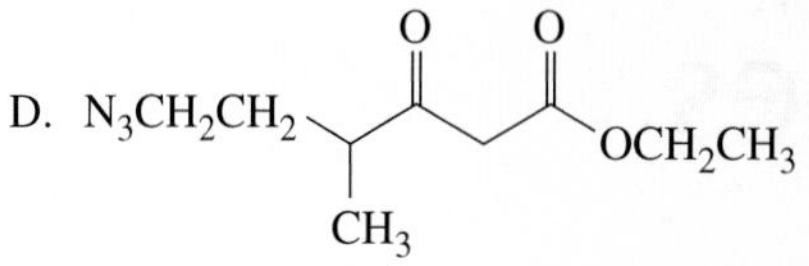

20.91 What pair of compounds is the most reasonable source of all of the carbon atoms if you wished to prepare the epoxy acid shown via a dianion?

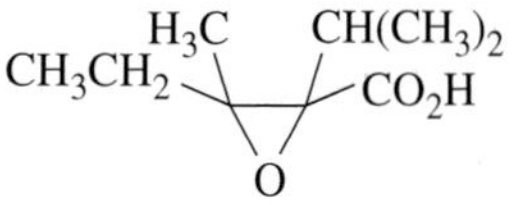

(mixture of stereoisomers)

A. $(CH_3)_2CHCHCO_2H$ (Cl) and $CH_3CCH_2CH_3$ (C=O)

B. $CH_3CH_2CHCO_2H$ (Cl) and $CH_3CCH(CH_3)_2$ (C=O)

C. $(CH_3)_2CCH_2CO_2H$ (Cl) and $CH_3CCH_2CH_3$ (C=O)

D. CH_3CHCO_2H (Cl) and $CH_3CH_2CCH(CH_3)_2$ (C=O)

20.92 Two dianions A and B are capable of being formed in the following reaction, but only a single alkylation product, isolated in 62% yield, was obtained. This product is not capable of cis–trans isomerism. Was this product formed from dianion A or dianion B?

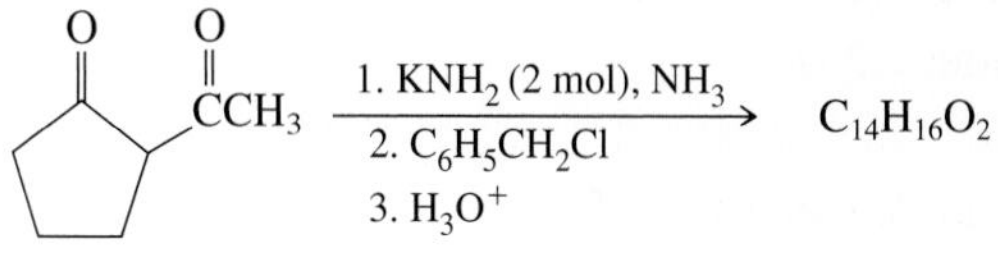

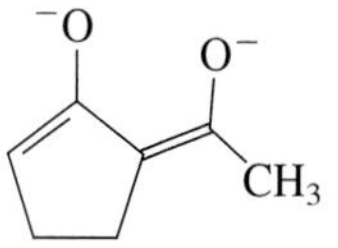

A.

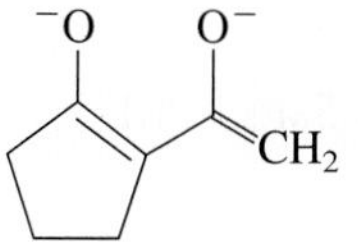

B.

21 Amines

Chapter Outline

Mechanisms

In 1917, Robert Robinson verified his ideas about the biosynthesis of alkaloids by combining methylamine with the compounds shown. The reaction worked as Robinson planned and is recognized as the first chemical synthesis inspired by biochemical thinking.

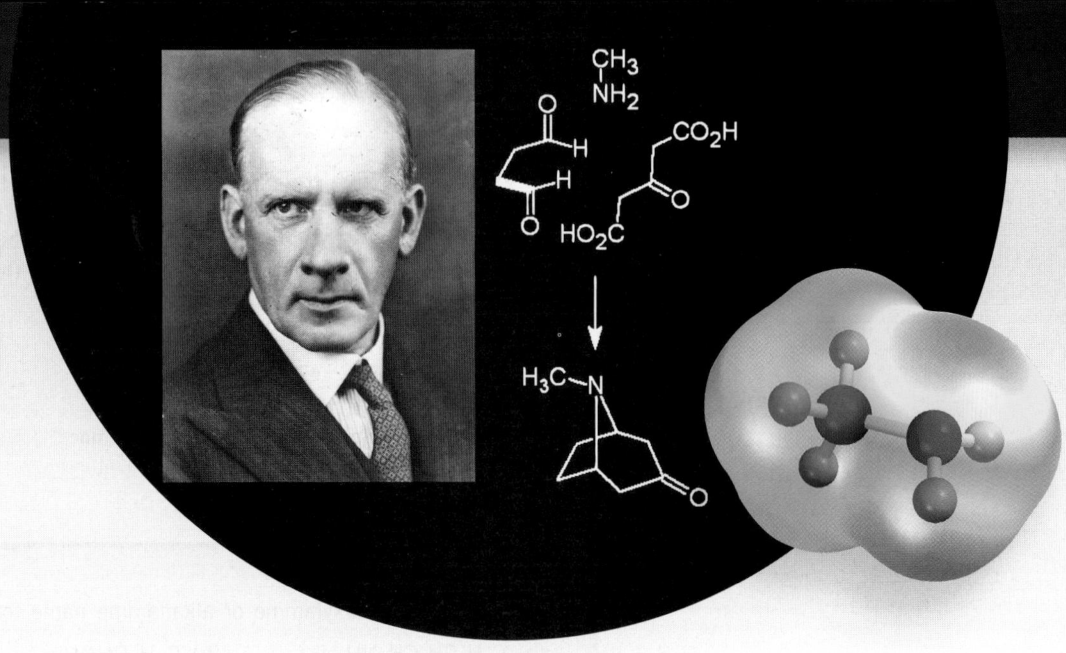

NITROGEN-CONTAINING compounds are essential to life. Their ultimate source is atmospheric nitrogen that, by a process known as *nitrogen fixation,* is reduced to ammonia, then converted to organic nitrogen compounds. This chapter describes the chemistry of **amines,** organic derivatives of ammonia. **Alkylamines** have their nitrogen attached to sp^3-hybridized carbon; **arylamines** have their nitrogen attached to an sp^2-hybridized carbon of a benzene or benzene-like ring.

R = alkyl group: alkylamine — Ar = aryl group: arylamine

Amines, like ammonia, are weak bases. They are, however, the strongest uncharged bases found in significant quantities under physiological conditions. Amines are usually the bases involved in biological acid–base reactions; they are often the nucleophiles in biological nucleophilic substitutions.

Our word *vitamin* was coined in 1912 in the belief that the substances present in the diet that prevented scurvy, pellagra, beriberi, rickets, and other diseases were "vital amines." In many cases, that belief was confirmed; certain vitamins did prove to be amines. In many other cases, however, vitamins were not amines. Nevertheless, the name *vitamin* entered our language and stands as a reminder that early chemists recognized the crucial place occupied by amines in biological processes.

21.1 Amine Nomenclature

Unlike alcohols and alkyl halides, which are classified as primary, secondary, or tertiary according to the degree of substitution at the carbon that bears the functional group, amines are classified according to their *degree of substitution at nitrogen.* An amine with one carbon attached to nitrogen is a **primary amine,** an amine with two is a **secondary amine,** and an amine with three is a **tertiary amine.**

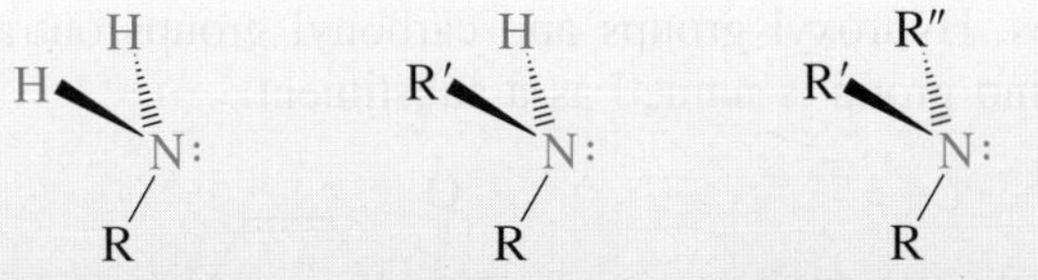

Primary amine — Secondary amine — Tertiary amine

The groups attached to nitrogen may be any combination of alkyl or aryl groups.

Amines are named in two main ways in the IUPAC system, either as *alkylamines* or as *alkanamines*. When primary amines are named as alkylamines, the ending *-amine* is added to the name of the alkyl group that bears the nitrogen. When named as alkanamines, the alkyl group is named as an alkane and the *-e* ending replaced by *-amine*.

$CH_3CH_2NH_2$

Ethylamine
(ethanamine)

Cyclohexylamine
(cyclohexanamine)

$CH_3CHCH_2CH_2CH_3$ (with NH_2 on C-2)

1-Methylbutylamine
(2-pentanamine
or pentan-2-amine)

Problem 21.1

Give an acceptable alkylamine or alkanamine name for each of the following amines:

(a) $C_6H_5CH_2CH_2NH_2$ (b) $C_6H_5CHNH_2$ with CH_3 on the CH (c) $H_2C{=}CHCH_2NH_2$

Sample Solution (a) The amino substituent is bonded to an ethyl group that bears a phenyl substituent at C-2. The compound $C_6H_5CH_2CH_2NH_2$ may be named as either 2-phenylethylamine or 2-phenylethanamine.

Aniline was first isolated in 1826 as a degradation product of indigo, a dark blue dye obtained from the West Indian plant *Indigofera anil,* from which the name *aniline* is derived.

Aniline is the parent IUPAC name for amino-substituted derivatives of benzene. Substituted derivatives of aniline are numbered beginning at the carbon that bears the amino group. Substituents are listed in alphabetical order, and the direction of numbering is governed by the usual "first point of difference" rule.

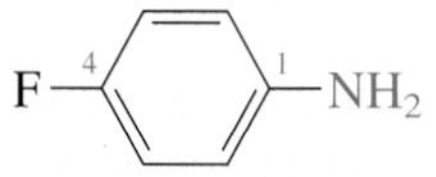

p-Fluoroaniline
or
4-Fluoroaniline

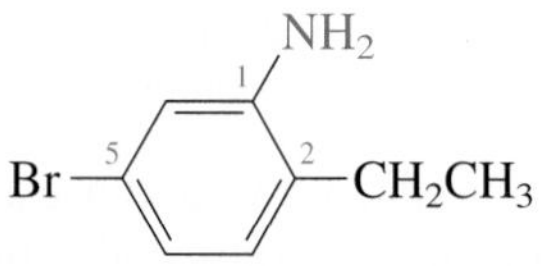

5-Bromo-2-ethylaniline

Arylamines may also be named as *arenamines*. Thus, *benzenamine* is an alternative, but rarely used, name for aniline.

Compounds with two amino groups are named by adding the suffix *-diamine* to the name of the corresponding alkane or arene. The final *-e* of the parent hydrocarbon is retained.

$H_2NCH_2CHCH_3$ (with NH_2 on C-2)

1,2-Propanediamine
or
Propane-1,2-diamine

$H_2NCH_2CH_2CH_2CH_2CH_2CH_2NH_2$

1,6-Hexanediamine
or
Hexane-1,6-diamine

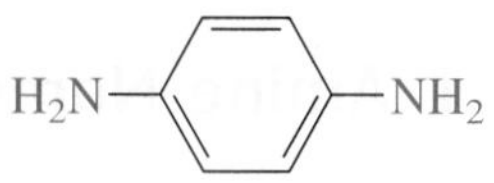

1,4-Benzenediamine
or
Benzene-1,4-diamine

Amino groups rank rather low in seniority when the parent compound is identified for naming purposes. Hydroxyl groups and carbonyl groups outrank amino groups. In these cases, the amino group is named as a substituent.

$HOCH_2CH_2NH_2$

2-Aminoethanol

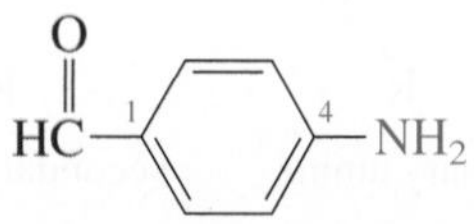

p-Aminobenzaldehyde
(4-Aminobenzenecarbaldehyde)

Secondary and tertiary amines are named as *N*-substituted derivatives of primary amines. The parent primary amine is taken to be the one with the longest carbon chain. Rings, however, take precedence over chains. The prefix *N*- is added as a locant to identify substituents on the amine nitrogen.

$CH_3NHCH_2CH_3$

N-Methylethylamine
(a secondary amine)

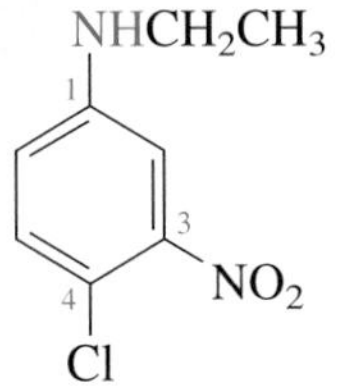

4-Chloro-*N*-ethyl-3-nitroaniline
(a secondary amine)

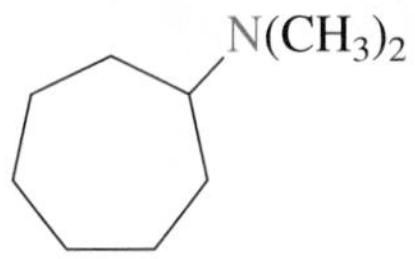

N,*N*-Dimethylcycloheptylamine
(a tertiary amine)

Problem 21.2

Assign alkanamine names to *N*-methylethylamine and to *N*,*N*-dimethylcycloheptylamine.

Sample Solution *N*-Methylethylamine (given as $CH_3NHCH_2CH_3$ in the preceding example) is an *N*-substituted derivative of ethanamine; it is *N*-methylethanamine.

Problem 21.3

Classify the following amine as primary, secondary, or tertiary, and give it an acceptable IUPAC name.

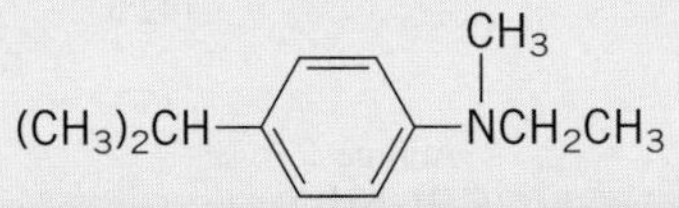

A nitrogen that bears four substituents is positively charged and is named as an *ammonium* ion. The anion that is associated with it is also identified in the name.

$CH_3\overset{+}{N}H_3\ Cl^-$

Methylammonium chloride

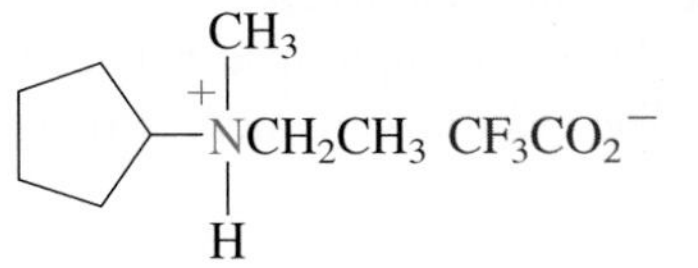

N-Ethyl-*N*-methylcyclopentylammonium trifluoroacetate

$C_6H_5CH_2\overset{+}{N}(CH_3)_3\ I^-$

Benzyltrimethylammonium iodide
(a quaternary ammonium salt)

Ammonium salts that have four alkyl groups bonded to nitrogen are called **quaternary ammonium salts.**

21.2 Structure and Bonding

Alkylamines: As shown in Figure 21.1, methylamine, like ammonia, has a pyramidal arrangement of bonds to nitrogen. Its H—N—H angles (106°) are slightly smaller than the tetrahedral value of 109.5°, whereas the C—N—H angle (112°) is slightly larger. The C—N bond distance of 147 pm lies between typical C—C bond distances in alkanes (153 pm) and C—O bond distances in alcohols (143 pm).

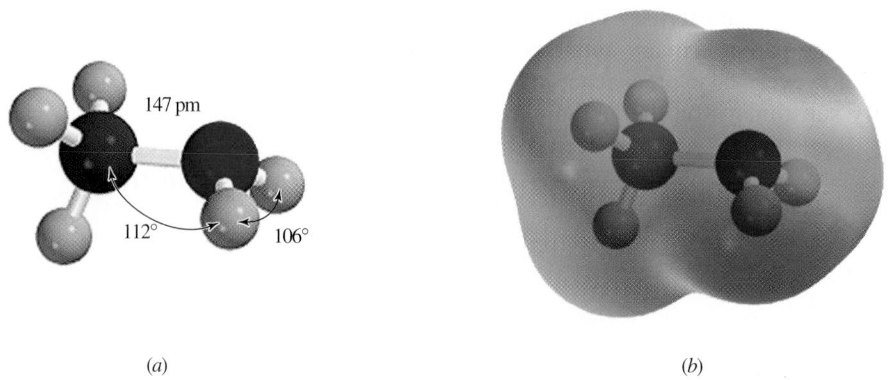

Figure 21.1

Methylamine. (*a*) Bond angles at nitrogen and C—N bond distance. (*b*) The unshared electron pair of nitrogen is a major contributor to the concentration of negative charge indicated by the red region in the electrostatic potential map.

Amines that are substituted with three different groups on the nitrogen are chiral but cannot be resolved into enantiomers because of the low energy barrier for racemization by inversion (Section 7.17).

Nitrogen and carbon are both sp^3-hybridized and are joined by a σ bond in methylamine. The unshared electron pair on nitrogen occupies an sp^3-hybridized orbital. This lone pair is involved in reactions in which amines act as bases or nucleophiles. The electrostatic potential map clearly shows the concentration of electron density at nitrogen in methylamine.

Arylamines: Aniline, like alkylamines, has a pyramidal arrangement of bonds around nitrogen, but its pyramid is somewhat shallower. One measure of the extent of this flattening is given by the angle between the carbon–nitrogen bond and the bisector of the H—N—H angle.

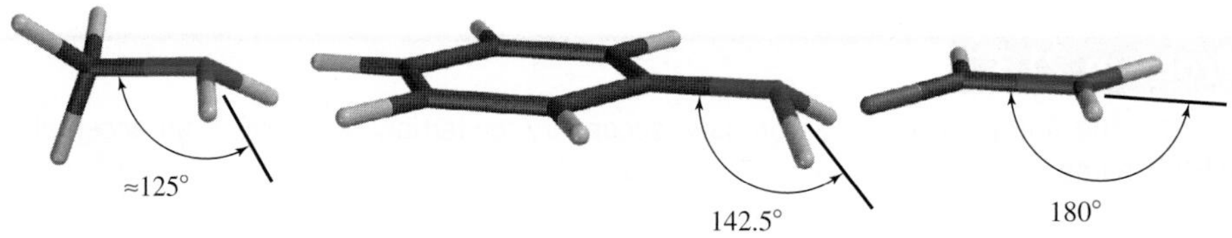

Methylamine (CH_3NH_2) Aniline ($C_6H_5NH_2$) Formamide ($O{=}CHNH_2$)

For sp^3-hybridized nitrogen, this angle (not the same as the C—N—H bond angle) is 125°, and the measured angles in simple alkylamines are close to that. The corresponding angle for sp^2 hybridization at nitrogen with a planar arrangement of bonds, as in amides, for example, is 180°. The measured value for this angle in aniline is 142.5°, suggesting a hybridization somewhat closer to sp^3 than to sp^2.

The structure of aniline reflects a compromise between two modes of binding the nitrogen lone pair (Figure 21.2). The electrons are more strongly attracted to nitrogen when

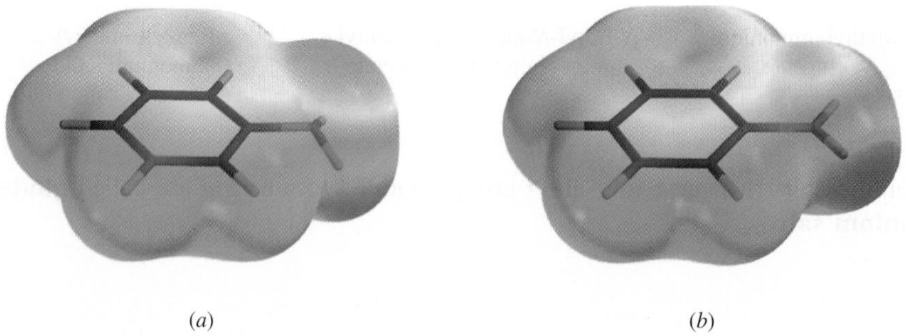

Figure 21.2

Electrostatic potential maps of aniline in which the geometry at nitrogen is (*a*) nonplanar and (*b*) planar. In the nonplanar geometry, the unshared pair occupies an sp^3 hybrid orbital of nitrogen. The region of highest electron density in (*a*) is associated with nitrogen. In the planar geometry, nitrogen is sp^2-hybridized, and the electron pair is delocalized between a *p* orbital of nitrogen and the π system of the ring. The region of highest electron density in (*b*) encompasses both the ring and nitrogen. The actual structure combines features of both; nitrogen adopts a hybridization state between sp^3 and sp^2. The color scale is the same for both models.

they are in an orbital with some *s* character—an sp^3-hybridized orbital, for example—than when they are in a *p* orbital. On the other hand, delocalization of these electrons into the aromatic π system is better achieved if they occupy a *p* orbital. A *p* orbital of nitrogen is better aligned for overlap with the *p* orbitals of the benzene ring to form an extended π system than is an sp^3-hybridized orbital. As a result of these two opposing forces, nitrogen adopts an orbital hybridization that is between sp^3 and sp^2.

The corresponding resonance description shows the delocalization of the nitrogen lone-pair electrons in terms of contributions from dipolar structures.

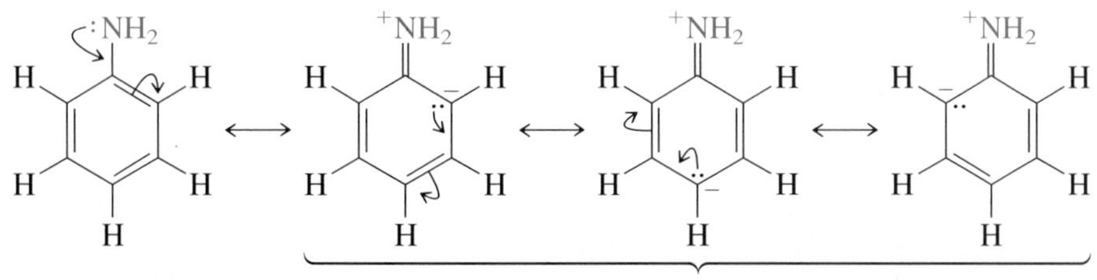

Most stable Lewis structure for aniline

Dipolar resonance contributors of aniline

Delocalization of the nitrogen lone pair decreases the electron density at nitrogen while increasing it in the π system of the aromatic ring. We've already seen one chemical consequence of this in the high level of reactivity of aniline in electrophilic aromatic substitution reactions (Section 12.12). Other ways in which electron delocalization affects the properties of arylamines are described in later sections of this chapter.

Problem 21.4

As the extent of electron delocalization into the ring increases, the geometry at nitrogen flattens. *p*-Nitroaniline, for example, is planar. Write a resonance contributor for *p*-nitroaniline that shows how the nitro group increases electron delocalization.

21.3 Physical Properties

We have often seen that the polar nature of a substance can affect physical properties such as boiling point. This is true for amines, which are more polar than alkanes but less polar than alcohols. For similarly constituted compounds, alkylamines have boiling points higher than those of alkanes but lower than those of alcohols.

Most commonly encountered alkylamines are liquids with unpleasant, "fishy" odors.

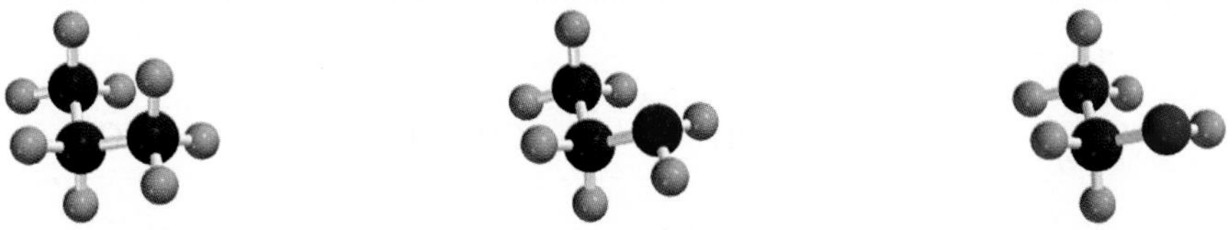

$CH_3CH_2CH_3$	$CH_3CH_2NH_2$	CH_3CH_2OH
Propane	Ethylamine	Ethanol
$\mu = 0$ D	$\mu = 1.2$ D	$\mu = 1.7$ D
bp −42°C	bp 17°C	bp 78°C

Dipole–dipole interactions, especially hydrogen bonding, are present in amines but absent in alkanes. But because nitrogen is less electronegative than oxygen, an N—H bond is less polar than an O—H bond and hydrogen bonding is weaker in amines than in alcohols.

Among isomeric amines, primary amines have the highest boiling points, and tertiary amines the lowest.

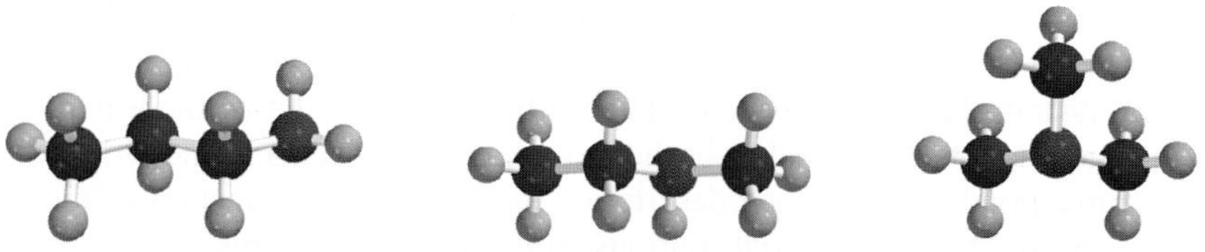

$CH_3CH_2CH_2NH_2$	$CH_3CH_2NHCH_3$	$(CH_3)_3N$
Propylamine (primary amine)	*N*-Methylethylamine (secondary amine)	Trimethylamine (tertiary amine)
bp 50°C	bp 34°C	bp 3°C

Primary and secondary amines can participate in intermolecular hydrogen bonding, but tertiary amines lack N—H bonds and so cannot.

Amines that have fewer than six or seven carbon atoms are soluble in water. All amines, even tertiary amines, can act as proton acceptors in hydrogen bonding to water molecules.

21.4 Basicity of Amines

As we discussed in Section 1.15, it is more useful to describe the basicity of amines in terms of the pK_a's of their conjugate acids than as basicity constants K_b. Always bear in mind that:

The more basic the amine, the weaker its conjugate acid.

The more basic the amine, the larger the pK_a *of its conjugate acid.*

Citing amine basicity according to the pK_a of the conjugate acid makes it possible to analyze acid–base reactions of amines according to the usual Brønsted relationships. For example, we see that amines are converted to ammonium ions by acids even as weak as acetic acid:

Recall that acid–base reactions are favorable when the stronger acid is on the left and the weaker acid on the right.

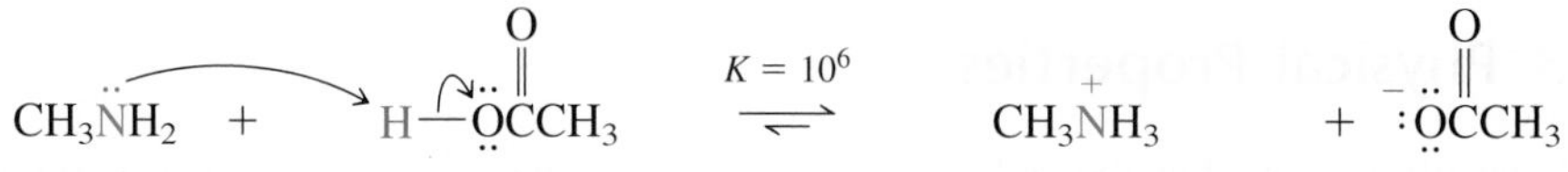

Methylamine; Acetic acid (stronger acid; pK_a = 4.7); Methylammonium ion (weaker acid; pK_a = 10.7); Acetate ion

Conversely, adding sodium hydroxide to an ammonium salt converts it to the free amine:

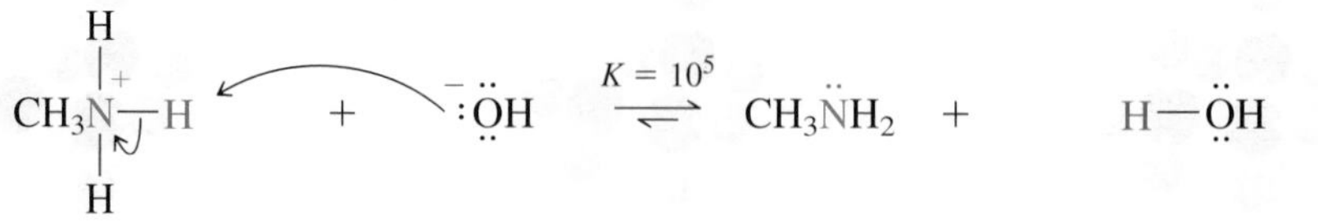

Methylammonium ion (stronger acid; pK_a = 10.7); Hydroxide ion; Methylamine; Water (weaker acid; pK_a = 15.7)

Problem 21.5

Apply the Henderson–Hasselbalch equation (Section 18.4) to calculate the $CH_3NH_3^+$/CH_3NH_2 ratio in water buffered at pH 7.

Their basicity provides a means by which amines may be separated from neutral organic compounds. A mixture containing an amine is dissolved in diethyl ether and shaken with dilute hydrochloric acid to convert the amine to an ammonium salt. The ammonium salt, being ionic, dissolves in the aqueous phase, which is separated from the ether layer. Adding sodium hydroxide to the aqueous layer converts the ammonium salt back to the free amine, which is then removed from the aqueous phase by extraction with a fresh portion of ether.

Amines are weak bases, but as a class, *amines are the strongest bases of all neutral molecules.* Table 21.1 lists basicity data for a number of amines. The most important relationships to be drawn from the data are:

1. Alkylamines are slightly stronger bases than ammonia.
2. Alkylamines differ very little among themselves in basicity. Their basicities cover a range of less than 10 in equilibrium constant (1 pK unit).
3. Arylamines are about 1 million times (6 pK units) weaker bases than ammonia and alkylamines.

The small differences in basicity between ammonia and alkylamines, and among the various classes of alkylamines (primary, secondary, tertiary), come from a mix of effects. Replacing hydrogens of ammonia by alkyl groups affects both sides of the acid–base equilibrium in ways that largely cancel.

Replacing hydrogens by aryl groups is a different story, however. An aryl group affects the base much more than the conjugate acid, and the overall effect is large. One way to compare alkylamines and arylamines is by examining the Brønsted equilibrium for proton transfer *to* an alkylamine *from* the conjugate acid of an arylamine.

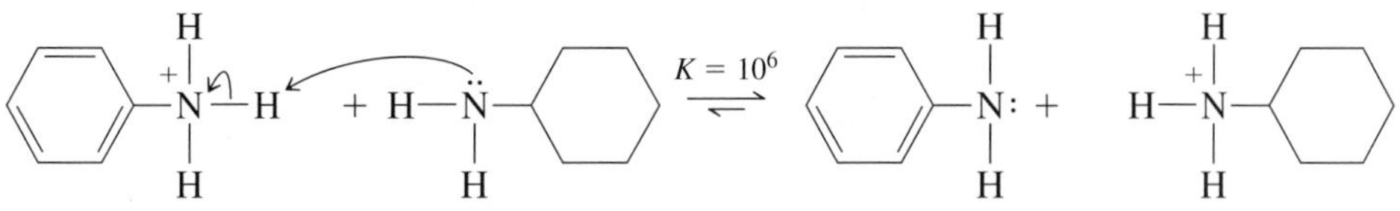

Anilinium ion (stronger acid; pK_a = 4.6) + Cyclohexylamine ⇌ Aniline + Cyclohexylammonium ion (weaker acid; pK_a = 10.6)

TABLE 21.1 Basicity of Amines As Measured by the pK_a of Their Conjugate Acids*

Compound	Structure	pK_a of conjugate acid
Ammonia	NH_3	9.3
Primary amines		
Methylamine	CH_3NH_2	10.6
Ethylamine	$CH_3CH_2NH_2$	10.8
Isopropylamine	$(CH_3)_2CHNH_2$	10.6
tert-Butylamine	$(CH_3)_3CNH_2$	10.4
Aniline	$C_6H_5NH_2$	4.6
Secondary amines		
Dimethylamine	$(CH_3)_2NH$	10.7
Diethylamine	$(CH_3CH_2)_2NH$	11.1
N-Methylaniline	$C_6H_5NHCH_3$	4.8
Tertiary amines		
Trimethylamine	$(CH_3)_3N$	9.7
Triethylamine	$(CH_3CH_2)_3N$	10.8
N,N-Dimethylaniline	$C_6H_5N(CH_3)_2$	5.1

*In water, 25°C.

The equilibrium shown in the equation lies to the right. $K_{eq} = 10^6$ for proton transfer from the conjugate acid of aniline to cyclohexylamine, making cyclohexylamine 1,000,000 times more basic than aniline.

Reading the equation from left to right, we can say that anilinium ion is a stronger acid than cyclohexylammonium ion because loss of a proton from anilinium ion creates an unshared electron pair of aniline. Conjugation of this unshared pair with the aromatic ring stabilizes aniline and biases the equilibrium toward the right.

Reading the equation from right to left, we can say that aniline is a weaker base than cyclohexylamine because the electron pair on nitrogen of aniline is strongly held by virtue of being delocalized into the π system of the aromatic ring. The unshared pair in cyclohexylamine is localized on nitrogen, less strongly held, and therefore "more available" in an acid–base reaction.

Even though they are weaker bases, arylamines, like alkylamines, can be completely protonated by strong acids. Aniline is extracted from an ether solution into 1 M hydrochloric acid by being completely converted to a water-soluble anilinium salt under these conditions.

Problem 21.6

The pK_a's of the conjugate acids of the two amines shown differ by a factor of 40,000. Which amine is the stronger base? Why?

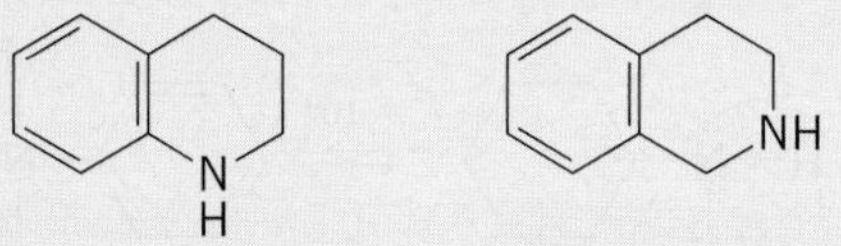

Tetrahydroquinoline Tetrahydroisoquinoline

Conjugation of the amino group of an arylamine with a second aromatic ring, then a third, reduces its basicity even further. Diphenylamine is 6300 times less basic than aniline, whereas triphenylamine is scarcely a base at all, being estimated as 10^{10} times less basic than aniline and 10^{14} times less basic than ammonia.

	$C_6H_5NH_2$	$(C_6H_5)_2NH$	$(C_6H_5)_3N$
	Aniline	Diphenylamine	Triphenylamine
pK_a of conjugate acid:	4.6	0.8	≈ −5

In general, electron-donating substituents on the aromatic ring increase the basicity of arylamines only slightly. Thus, as shown in Table 21.2, an electron-donating methyl group in the para position *increases* the basicity of aniline by less than 1 pK unit. Electron-withdrawing groups are base-weakening and can exert large effects. A *p*-trifluoromethyl group *decreases* the basicity of aniline by a factor of 200 and a *p*-nitro group by a factor of 3800. In the case of *p*-nitroaniline a resonance interaction of the type shown provides for extensive delocalization of the unshared electron pair of the amine group.

TABLE 21.2 Effect of para Substituents on the Basicity of Aniline

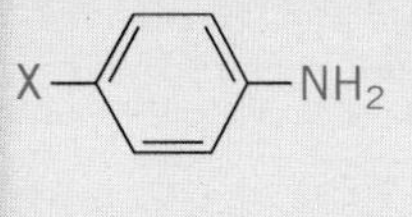

X	pK_a of conjugate acid
H	4.6
CH_3	5.3
CF_3	2.5
O_2N	1.0

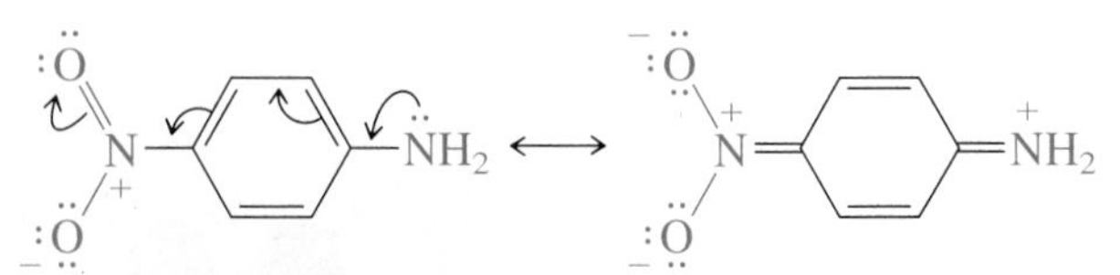

Electron delocalization in *p*-nitroaniline

Just as aniline is much less basic than alkylamines because the unshared electron pair of nitrogen is delocalized into the π system of the ring, *p*-nitroaniline is even less basic because the extent of this delocalization is greater and involves the oxygens of the nitro group.

Problem 21.7

Each of the following is a much weaker base than aniline. Present a resonance argument to explain the effect of the substituent in each case.

(a) *o*-Cyanoaniline

(b) $C_6H_5NH\overset{O}{\overset{\|}{C}}CH_3$

(c) *p*-Aminoacetophenone

Sample Solution (a) A cyano substituent is strongly electron-withdrawing. When present at a position ortho to an amino group on an aromatic ring, a cyano substituent increases the delocalization of the amine lone-pair electrons by a direct resonance interaction.

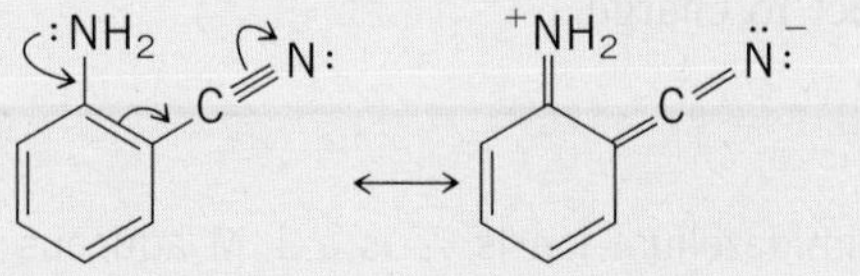

This resonance stabilization is lost when the amine group becomes protonated, and *o*-cyanoaniline is therefore a weaker base than aniline.

Multiple substitution by strongly electron-withdrawing groups diminishes the basicity of arylamines still more. Aniline is 3800 times as strong a base as *p*-nitroaniline and 10^9 times more basic than 2,4-dinitroaniline. A practical consequence of this is that arylamines that bear two or more strongly electron-withdrawing groups are often not capable of being extracted from a diethyl ether solution into dilute aqueous acid.

Nonaromatic heterocyclic compounds, piperidine, for example, are similar in basicity to alkylamines. When nitrogen is part of an aromatic ring, however, its basicity decreases markedly. Pyridine, for example, resembles arylamines in being almost 1 million times less basic than piperidine.

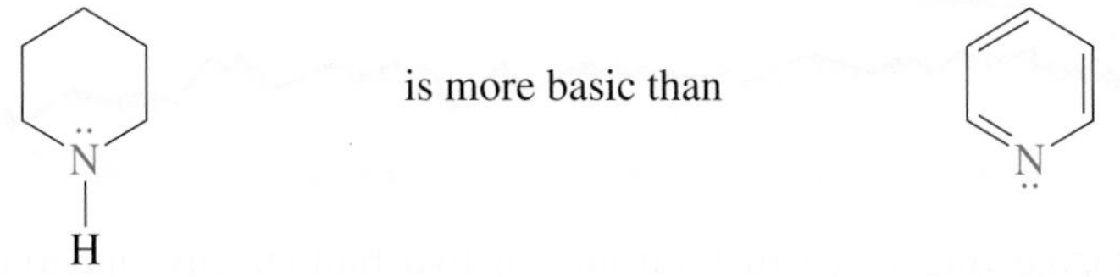

Piperidine is more basic than Pyridine

Piperidine
pK_a of conjugate acid = 11.2

Pyridine
pK_a of conjugate acid = 5.2

The difference between the two lies in the fact that the nitrogen lone pair occupies an sp^3-hybridized orbital in piperidine versus an sp^2-hybridized one in pyridine. As we have noted on several occasions, electrons in orbitals with more *s* character are more strongly held than those with less *s* character. For this reason, nitrogen holds on to its unshared pair more strongly in pyridine than in piperidine and is less basic.

Pyridine and imidazole were two of the heterocyclic aromatic compounds described in Section 11.22.

Imidazole and its derivatives form an interesting and important class of heterocyclic aromatic amines. Imidazole is approximately 100 times more basic than pyridine.

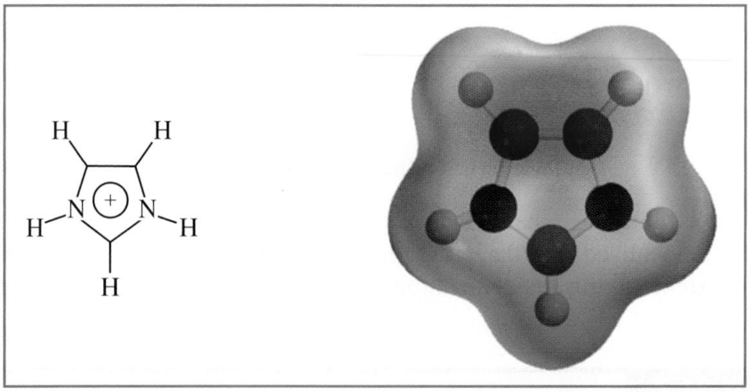

Figure 21.3

Electrostatic potential map of imidazolium ion showing equal distribution of charge between both nitrogens.

Protonation of imidazole yields an ion that is stabilized by the electron delocalization represented in the resonance structures shown:

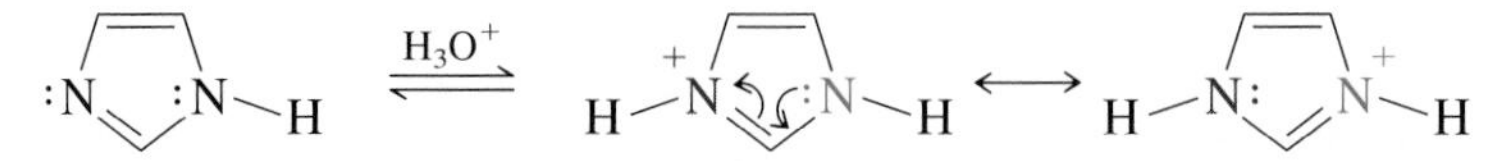

Imidazole
pK_a of conjugate acid = 7

Imidazolium ion

As seen in Figure 21.3, the electrostatic potential map of the conjugate acid of imidazole (imidazolium ion) is consistent with the resonance description in showing both nitrogens as equivalent with respect to charge.

Problem 21.8

Given that the pK_a of imidazolium ion is 7, is a 1 M aqueous solution of imidazolium chloride acidic, basic, or neutral? (b) What about a 1 M solution of imidazole? (c) A solution containing equal molar quantities of imidazole and imidazolium chloride?

Sample Solution

(a)

is a stronger acid than H_2O

$pK_a = 7$ $pKa = 15.7$

Therefore, the H_3O^+ concentration is greater in a solution of imidazolium ion in water than it is in pure H_2O – solution of imidazolium chloride is <u>acidic</u>.

An imidazole ring is a structural unit in two biologically important compounds, *histidine* and *histamine*. Histidine is one of the amino acid building blocks of proteins and is directly involved in key proton-transfer processes. The drop in blood pressure associated with shock is a result of the formation of histamine, which stimulates the dilation of blood vessels.

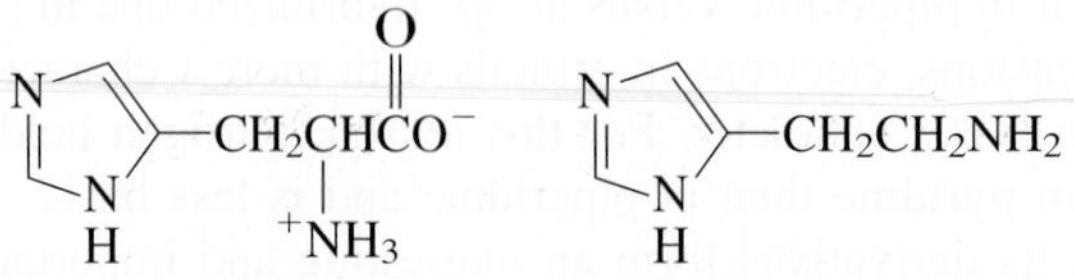

Histidine Histamine

Amines as Natural Products

The ease with which amines are extracted into aqueous acid, combined with their regeneration on treatment with base, makes it a simple matter to separate amines from other plant materials, and nitrogen-containing natural products were among the earliest organic compounds to be studied. Their basic (alkaline) properties led amines obtained from plants to be called **alkaloids.** The number of known alkaloids exceeds 5000. They are of special interest because most are characterized by a high level of biological activity. Some examples include *cocaine, coniine,* and *morphine.*

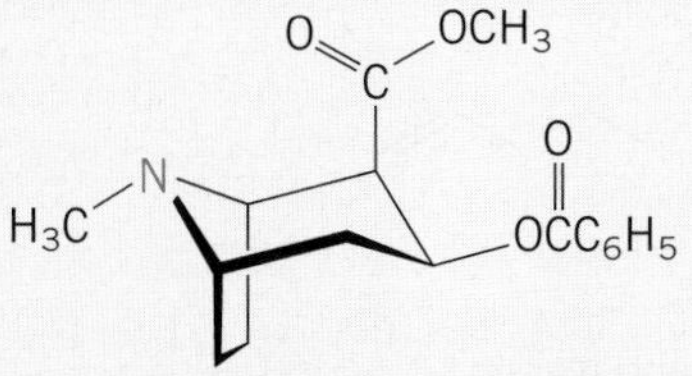

Cocaine

(A central nervous system stimulant obtained from the leaves of the coca plant.)

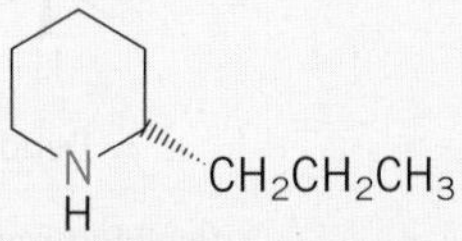

Coniine

(Present along with other alkaloids in the hemlock extract used to poison Socrates.)

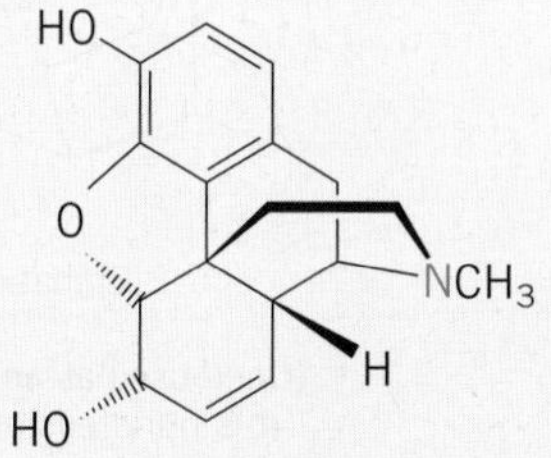

Morphine

(An opium alkaloid. Although it is an excellent analgesic, its use is restricted because of the potential for addiction. Heroin is the diacetate ester of morphine.)

Many alkaloids, such as *nicotine* and *quinine,* contain two (or more) nitrogen atoms. The nitrogens shown in red in quinine and nicotine are part of a substituted quinoline and pyridine ring, respectively.

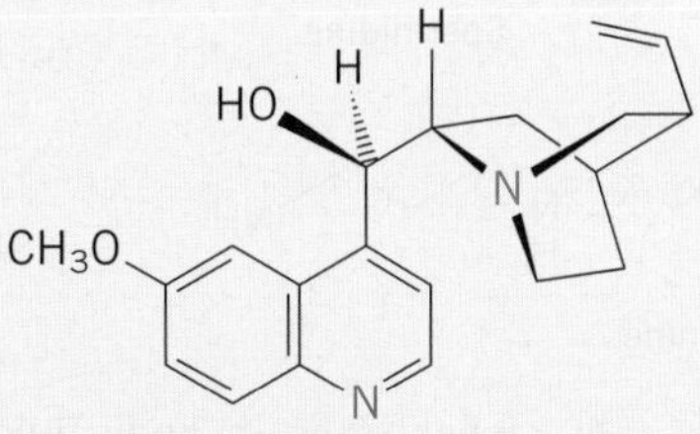

Quinine

(Alkaloid of cinchona bark used to treat malaria)

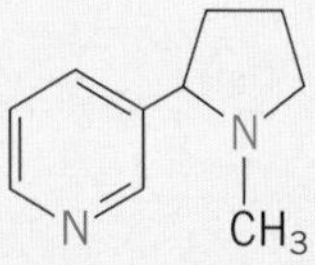

Nicotine

(An alkaloid present in tobacco; a very toxic compound sometimes used as an insecticide)

Problem 21.9

Estimate the pK_a of the conjugate acid of nicotine.

Several naturally occurring amines mediate the transmission of nerve impulses and are referred to as **neurotransmitters.** Two examples are *epinephrine* and *serotonin.* (Strictly speaking these compounds are not classified as alkaloids because they are not isolated from plants.)

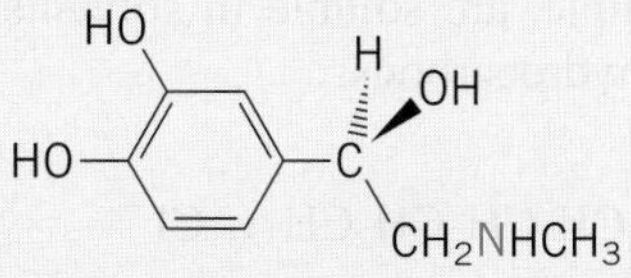

Epinephrine

(Also called *adrenaline;* a hormone secreted by the adrenal gland that prepares the organism for "flight or fight.")

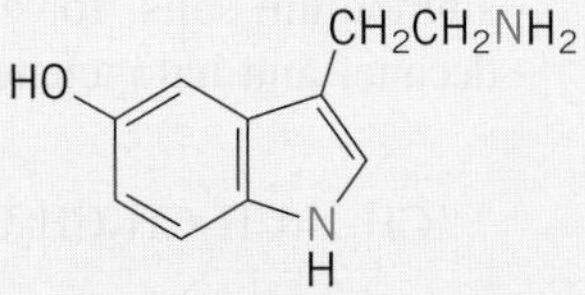

Serotonin

(A hormone synthesized in the pineal gland. Certain mental disorders are believed to be related to serotonin levels in the brain.)

Continued

(Continued)

Bioactive amines are also widespread in animals. A variety of structures and properties have been found in substances isolated from frogs, for example. One, called *epibatidine,* is a naturally occurring painkiller isolated from the skin of an Ecuadoran frog (*Epipedobates tricolor*). Another family of frogs produces a toxic mixture of several stereoisomeric amines, called *dendrobines,* on their skin that protects them from attack.

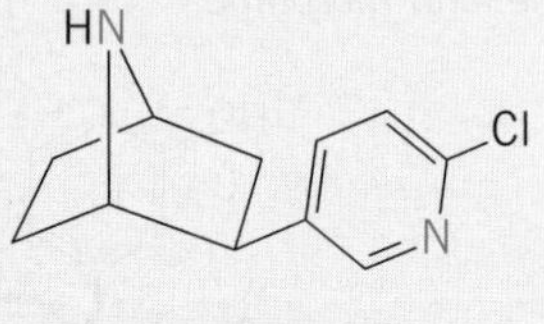

Epibatidine

(Once used as an arrow poison, it is hundreds of times more powerful than morphine in relieving pain. It is too toxic to be used as a drug, however.)

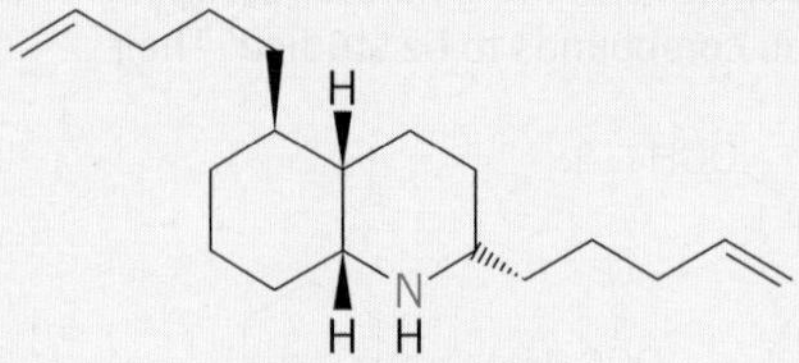

Dendrobine

(Isolated from frogs of the Dendrobatidae family. Related compounds have also been isolated from certain ants.)

Among the more important amine derivatives found in the body are a group of compounds known as **polyamines,** which contain two to four nitrogen atoms separated by several methylene units:

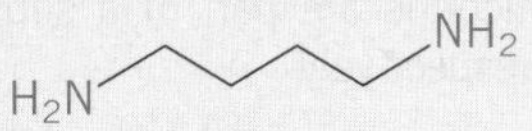

Putrescine

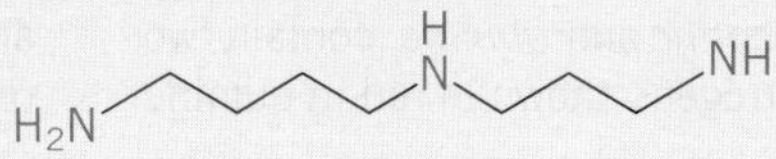

Spermidine

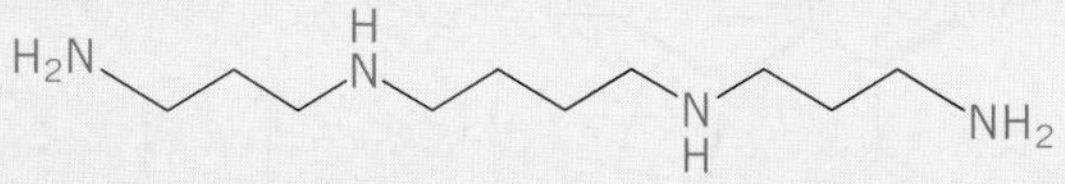

Spermine

These compounds are present in almost all mammalian cells, where they are believed to be involved in cell differentiation and proliferation. Because each nitrogen of a polyamine is protonated at physiological pH (7.4), putrescine, spermidine, and spermine exist as cations with a charge of +2, +3, and +4, respectively, in body fluids. Structural studies suggest that these polyammonium ions affect the conformation of biological macromolecules by electrostatic binding to specific anionic sites—the negatively charged phosphate groups of DNA, for example.

21.5 Tetraalkylammonium Salts as Phase-Transfer Catalysts

In spite of being ionic, many quaternary ammonium salts dissolve in nonpolar media. The four alkyl groups attached to nitrogen shield its positive charge and impart *lipophilic* (hydrophobic) character to the tetraalkylammonium ion. The following two quaternary ammonium salts, for example, are soluble in solvents of low polarity such as benzene, decane, and halogenated hydrocarbons:

$$CH_3\overset{+}{N}(CH_2CH_2CH_2CH_2CH_2CH_2CH_2CH_3)_3\ Cl^-$$

Methyltrioctylammonium chloride

Benzyltriethylammonium chloride

This property of quaternary ammonium salts is used to advantage in an experimental technique known as **phase-transfer catalysis.** Imagine that you wish to carry out the reaction

$$\underset{\text{Butyl bromide}}{CH_3CH_2CH_2CH_2Br} + \underset{\text{Sodium cyanide}}{NaCN} \longrightarrow \underset{\text{Pentanenitrile}}{CH_3CH_2CH_2CH_2CN} + \underset{\text{Sodium bromide}}{NaBr}$$

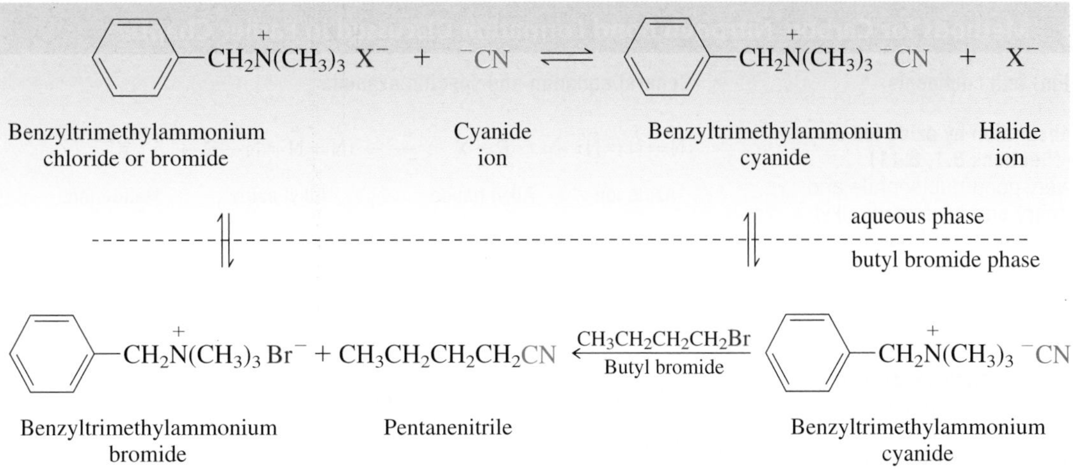

Figure 21.4

Phase-transfer catalysis. Nucleophilic cyanide ion is transferred from the aqueous to the organic phase as benzyltrimethylammonium cyanide.

Sodium cyanide does not dissolve in butyl bromide. The two reactants contact each other only at the surface of the solid sodium cyanide, and the rate of reaction under these conditions is too slow to be of synthetic value. Dissolving the sodium cyanide in water is of little help because butyl bromide is not soluble in water and reaction can occur only at the interface between the two phases. Adding a small amount of benzyltrimethylammonium chloride, however, causes pentanenitrile to form rapidly even at room temperature. The quaternary ammonium salt is acting as a *catalyst;* it increases the reaction rate. How?

Quaternary ammonium salts catalyze the reaction between an anion and an organic substrate by transferring the anion from the aqueous phase, where it cannot contact the substrate, to the organic phase. In the cycle shown in Figure 21.4, the first step occurs in the aqueous phase and is an exchange of the anionic partner of the quaternary ammonium salt for cyanide ion.

The benzyltrimethylammonium ion migrates to the butyl bromide phase, carrying a cyanide ion along with it. Once in the organic phase, cyanide ion is only weakly solvated and is far more reactive than it is in water or ethanol, where it is strongly solvated by hydrogen bonding. Nucleophilic substitution takes place rapidly. The benzyltrimethylammonium bromide formed in the substitution step returns to the aqueous phase, where it can repeat the cycle.

21.6 Reactions That Lead to Amines: A Review and a Preview

Methods for preparing amines address either or both of the following questions:

1. How is the required carbon–nitrogen bond to be formed?
2. Given a nitrogen-containing organic compound such as an amide, a nitrile, or a nitro compound, how is the correct oxidation state of the desired amine to be achieved?

A number of reactions that lead to carbon–nitrogen bond formation were presented in earlier chapters and are summarized in Table 21.3. Among the reactions in the table, the nucleophilic ring opening of epoxides and the reaction of α-halo acids with ammonia give amines directly. The other reactions in Table 21.3 yield products that are converted to amines by some subsequent procedure. As these procedures are described in the following sections, you will see that they are largely applications of principles that you've already learned. You will encounter some new reagents and some new uses for familiar reagents, but very little in the way of new reaction types is involved.

TABLE 21.3 Methods for Carbon–Nitrogen Bond Formation Discussed in Earlier Chapters

Reaction (section) and comments	General equation and specific example
Nucleophilic substitution by azide ion on an alkyl halide (Sections 8.1, 8.11) Azide ion is a very good nucleophile and reacts with primary and secondary alkyl halides to give alkyl azides. Phase-transfer catalysts accelerate the rate of reaction.	$:\ddot{N}^{-}{=}\overset{+}{N}{=}\ddot{N}:^{-} + R{-}X \longrightarrow :\ddot{N}^{-}{=}\overset{+}{N}{=}\ddot{N}{-}R + X^{-}$ Azide ion, Alkyl halide, Alkyl azide, Halide ion $CH_3CH_2CH_2CH_2CH_2Br \xrightarrow[\text{phase-transfer catalyst}]{NaN_3} CH_3CH_2CH_2CH_2CH_2N_3$ Pentyl bromide (1-bromopentane) → Pentyl azide (89%) (1-azidopentane)
Nitration of arenes (Section 12.3) The standard method for introducing a nitrogen atom as a substituent on an aromatic ring is nitration with a mixture of nitric acid and sulfuric acid. The reaction proceeds by electrophilic aromatic substitution.	$ArH + HNO_3 \xrightarrow{H_2SO_4} ArNO_2 + H_2O$ Arene, Nitric acid, Nitroarene, Water 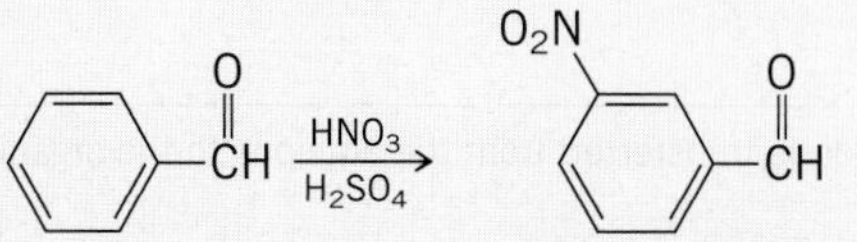 Benzaldehyde → *m*-Nitrobenzaldehyde (75–84%)
Nucleophilic ring opening of epoxides by ammonia (Section 16.12) The strained ring of an epoxide is opened on nucleophilic attack by ammonia and amines to give β-amino alcohols. Azide ion also reacts with epoxides; the products are β-azido alcohols.	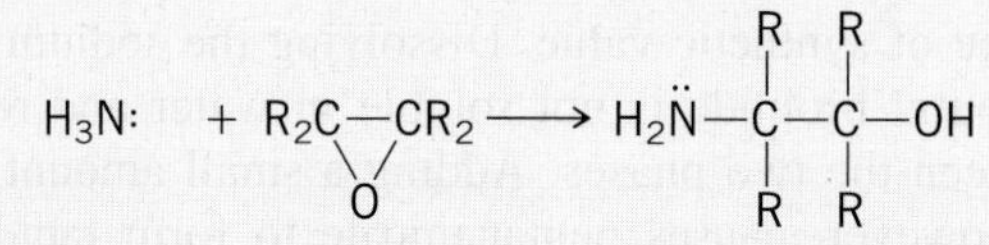 Ammonia, Epoxide, β-Amino alcohol 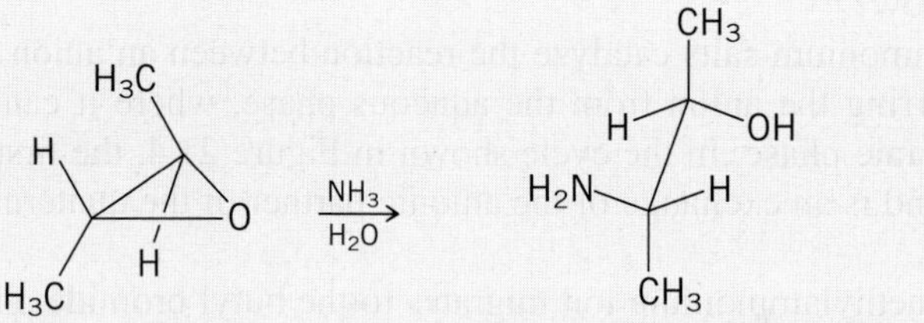 (2*R*,3*R*)-2,3-Epoxybutane → (2*R*,3*S*)-3-Amino-2-butanol (70%)
Nucleophilic addition of amines to aldehydes and ketones (Sections 17.10, 17.11) Primary amines undergo nucleophilic addition to the carbonyl group of aldehydes and ketones to form carbinolamines. These carbinolamines dehydrate under the conditions of their formation to give *N*-substituted imines. Secondary amines yield enamines.	$R\ddot{N}H_2 + R'\overset{O}{\overset{\Vert}{C}}R'' \longrightarrow R'\overset{:NR}{\overset{\Vert}{C}}R'' + H_2O$ Primary amine, Aldehyde or ketone, Imine, Water $CH_3NH_2 + C_6H_5\overset{O}{\overset{\Vert}{C}}H \longrightarrow C_6H_5CH{=}NCH_3$ Methylamine, Benzaldehyde, *N*-Benzylidenemethylamine (70%)
Nucleophilic substitution by ammonia on α-halo acids (Section 20.15) The α-halo acids obtained by halogenation of carboxylic acids under conditions of the Hell–Volhard–Zelinsky reaction are reactive substrates in nucleophilic substitution processes. A standard method for the preparation of α-amino acids is displacement of halide from α-halo acids by nucleophilic substitution using excess aqueous ammonia.	$H_3N: + R\underset{X}{CH}CO_2H \longrightarrow R\underset{^{+}NH_3}{CH}CO_2^{-} + NH_4X$ Ammonia (excess), α-Halo carboxylic acid, α-Amino acid, Ammonium halide $(CH_3)_2CH\underset{Br}{CH}CO_2H \xrightarrow[H_2O]{NH_3} (CH_3)_2CH\underset{^{+}NH_3}{CH}CO_2^{-}$ 2-Bromo-3-methylbutanoic acid → 2-Amino-3-methylbutanoic acid (47–48%)

Continued

TABLE 21.3 Methods for Carbon–Nitrogen Bond Formation Discussed in Earlier Chapters (*Continued*)

Reaction (section) and comments	General equation and specific example
Nucleophilic acyl substitution (Sections 19.4, 19.5, and 19.11) Acylation of ammonia and amines by an acyl chloride, acid anhydride, or ester is an exceptionally effective method for the formation of carbon–nitrogen bonds.	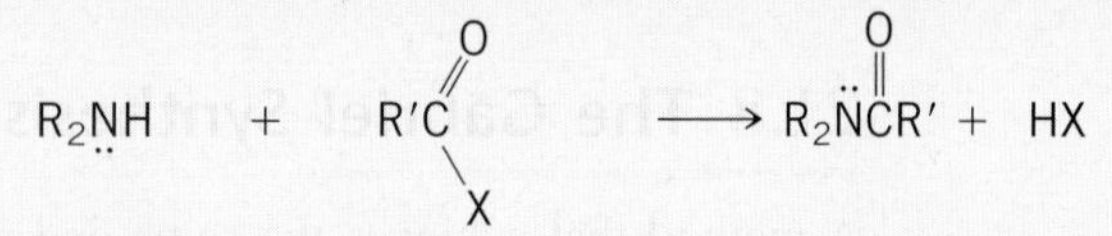 Primary or secondary amine, or ammonia; Acyl chloride, acid anhydride, or ester; Amide 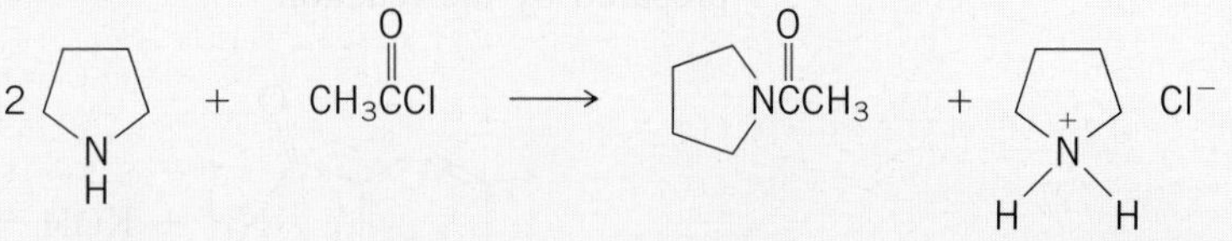 Pyrrolidine; Acetyl chloride; *N*-Acetylpyrrolidine (79%); Pyrrolidine hydrochloride

21.7 Preparation of Amines by Alkylation of Ammonia

Alkylamines are, in principle, capable of being prepared by nucleophilic substitution reactions of alkyl halides with ammonia.

$$\underset{\text{Alkyl halide}}{RX} + \underset{\text{Ammonia}}{2NH_3} \longrightarrow \underset{\text{Primary amine}}{RNH_2} + \underset{\text{Ammonium halide salt}}{\overset{+}{N}H_4\, X^-}$$

Although this reaction is useful for preparing α-amino acids (Table 21.3, fifth entry), it is *not* a general method for the synthesis of amines. Its major limitation is that the expected primary amine product is itself a nucleophile and competes with ammonia for the alkyl halide.

$$\underset{\text{Alkyl halide}}{RX} + \underset{\text{Primary amine}}{RNH_2} + \underset{\text{Ammonia}}{NH_3} \longrightarrow \underset{\text{Secondary amine}}{RNHR} + \underset{\text{Ammonium halide salt}}{\overset{+}{N}H_4\, X^-}$$

When 1-bromooctane, for example, is allowed to react with ammonia, both the primary amine and the secondary amine are isolated in comparable amounts.

$$\underset{\substack{\text{1-Bromooctane}\\ \text{(1 mol)}}}{CH_3(CH_2)_6CH_2Br} \xrightarrow{NH_3\text{ (2 mol)}} \underset{\substack{\text{Octylamine}\\ (45\%)}}{CH_3(CH_2)_6CH_2NH_2} + \underset{\substack{N,N\text{-Dioctylamine}\\ (43\%)}}{[CH_3(CH_2)_6CH_2]_2NH}$$

Competitive alkylation may continue, resulting in the formation of tertiary amines and quaternary ammonium salts.

Alkylation of ammonia is used to prepare primary amines only when the starting alkyl halide is not particularly expensive and the desired amine can be easily separated from the other components of the reaction mixture.

Problem 21.10

Alkylation of ammonia is sometimes employed in industrial processes; the resulting mixture of amines is separated by distillation. The ultimate starting materials for the industrial preparation of allylamine are propene, chlorine, and ammonia. Write a series of equations showing the industrial preparation of allylamine from these starting materials. (Allylamine has a number of uses, including the preparation of the diuretic drugs *meralluride* and *mercaptomerin.*)

Aryl halides that are substituted with electron-withdrawing groups react with amines by nucleophilic aromatic substitution (Section 12.20).

21.8 The Gabriel Synthesis of Primary Alkylamines

A method that achieves the same end result as that of alkylation of ammonia but which avoids the formation of secondary and tertiary amines as byproducts is the **Gabriel synthesis.** Alkyl halides are converted to primary alkylamines without contamination by secondary or tertiary amines. The key reagent is the potassium salt of phthalimide, prepared by the reaction

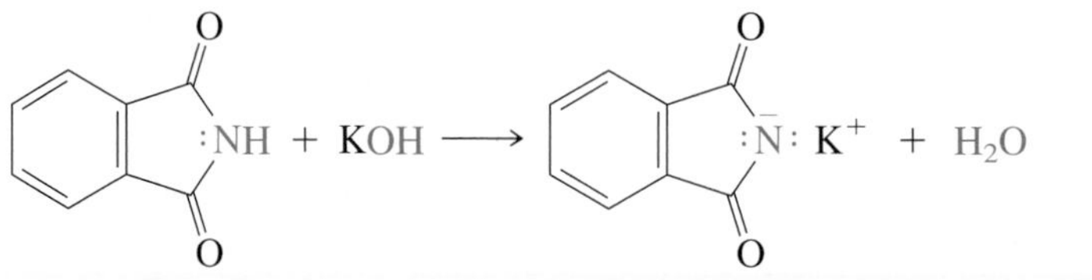

Phthalimide ($pK_a = 8.3$)   *N*-Potassiophthalimide   Water ($pK_a = 15.7$)

Phthalimide, with a pK_a of 8.3, can be quantitatively converted to its potassium salt with potassium hydroxide. The potassium salt of phthalimide has a negatively charged nitrogen atom, which acts as a nucleophile toward primary alkyl halides in a bimolecular nucleophilic substitution (S_N2) process.

DMF is an abbreviation for *N,N*-dimethylformamide, $HC(=O)N(CH_3)_2$. DMF is a polar aprotic solvent (Section 8.10) and an excellent medium for S_N2 reactions.

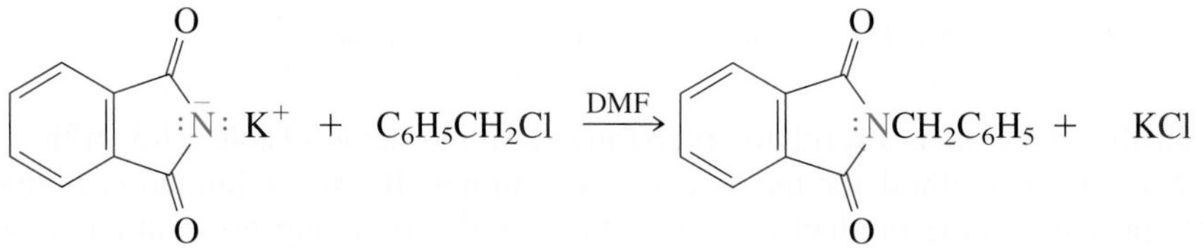

N-Potassiophthalimide   Benzyl chloride   *N*-Benzylphthalimide (74%)   Potassium chloride

The product of this reaction is an imide, a diacyl derivative of an amine. Either aqueous acid or aqueous base can be used to hydrolyze its two amide bonds and liberate the desired primary amine. A more effective method of cleaving the two amide bonds is by acyl transfer to hydrazine:

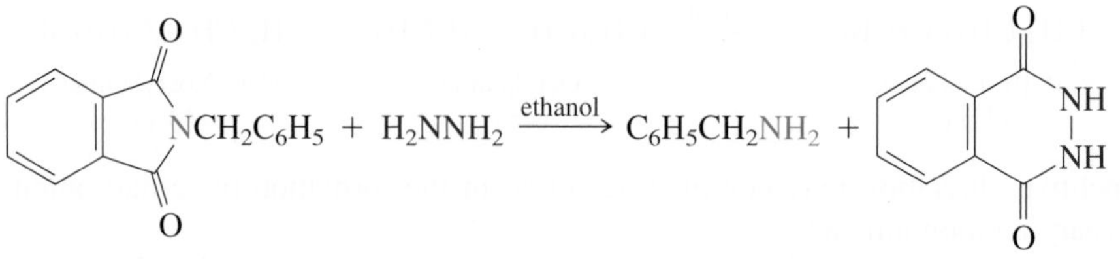

N-Benzylphthalimide   Hydrazine   Benzylamine (97%)   Phthalhydrazide

Aryl halides cannot be converted to arylamines by the Gabriel synthesis because they do not undergo nucleophilic substitution with *N*-potassiophthalimide in the first step of the procedure.

Among compounds other than simple alkyl halides, α-halo ketones, α-halo esters, and alkyl *p*-toluenesulfonates have also been used. Because phthalimide can undergo only a single alkylation, the formation of secondary and tertiary amines does not occur, and the Gabriel synthesis is a valuable procedure for the laboratory preparation of primary amines.

Problem 21.11

Three of the following amines can be prepared by the Gabriel synthesis; three cannot. Write equations showing the successful applications of this method.

(a) Butylamine
(b) Isobutylamine
(c) *tert*-Butylamine
(d) 2-Phenylethylamine
(e) *N*-Methylbenzylamine
(f) Aniline

Sample Solution

(a) • Gabriel synthesis makes primary amines (RNH_2).

• C—N bond is made by S_N2 reaction. Therefore, works best with primary alkyl halides.

• Use $CH_3CH_2CH_2CH_2Br$ to make $CH_3CH_2CH_2CH_2NH_2$.

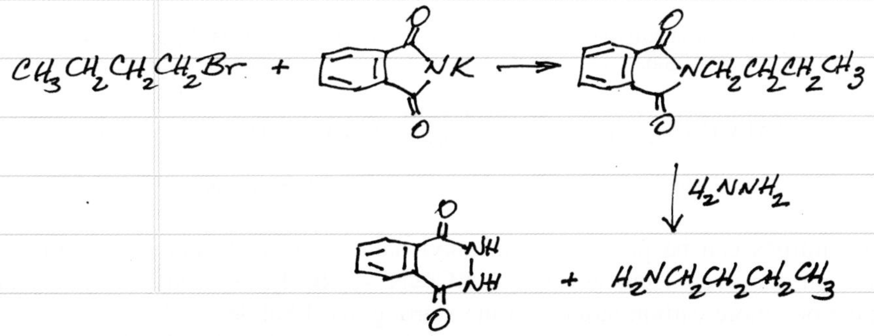

21.9 Preparation of Amines by Reduction

Almost any nitrogen-containing organic compound can be reduced to an amine. The synthesis of amines then becomes a question of the availability of suitable precursors and the choice of an appropriate reducing agent.

Alkyl *azides,* prepared by nucleophilic substitution of alkyl halides by sodium azide, as shown in the first entry of Table 21.3, are reduced to alkylamines by a variety of reagents, including lithium aluminum hydride.

$$R{-}\ddot{N}{=}\overset{+}{N}{=}\ddot{N}\!:^{-} \xrightarrow{\text{reduce}} R\ddot{N}H_2$$

Alkyl azide — Primary amine

$$C_6H_5CH_2CH_2N_3 \xrightarrow[\text{2. } H_2O]{\text{1. } LiAlH_4,\ \text{diethyl ether}} C_6H_5CH_2CH_2NH_2$$

2-Phenylethyl azide — 2-Phenylethylamine (89%)

Catalytic hydrogenation is also effective:

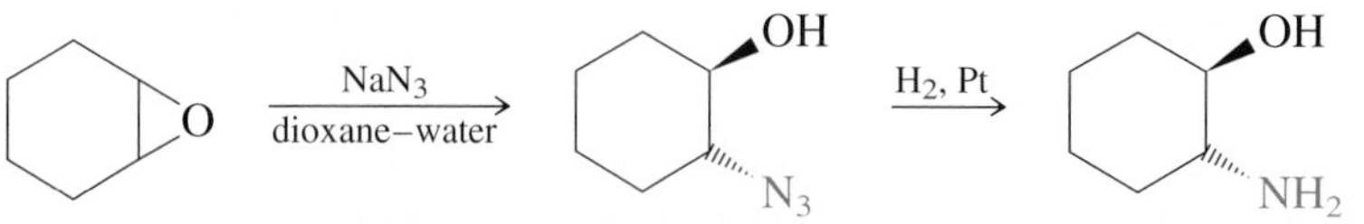

1,2-Epoxycyclohexane — *trans*-2-Azidocyclohexanol (61%) — *trans*-2-Aminocyclohexanol (81%)

In its overall design, this procedure is similar to the Gabriel synthesis; a nitrogen nucleophile is used in a carbon–nitrogen bond-forming operation and then converted to an amino group in a subsequent transformation.

The same reduction methods may be applied to the conversion of *nitriles* to primary amines.

$$RC{\equiv}N \xrightarrow[H_2,\ \text{catalyst}]{LiAlH_4\ \text{or}} RCH_2NH_2$$

Nitrile — Primary amine

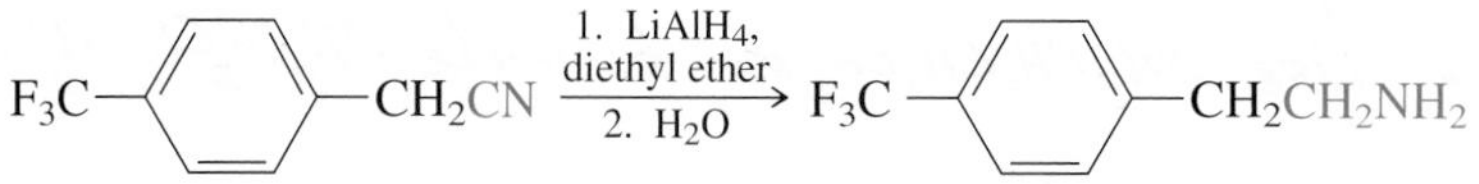

p-(Trifluoromethyl)benzyl cyanide — 2-(*p*-Trifluoromethyl)phenylethylamine (53%)

The preparation of pentanenitrile under phase-transfer conditions was described in Section 21.5.

$$CH_3CH_2CH_2CH_2CN \xrightarrow[\text{diethyl ether}]{H_2\ (100\ \text{atm}),\ Ni} CH_3CH_2CH_2CH_2CH_2NH_2$$

Pentanenitrile — 1-Pentanamine (56%)

Because nitriles can be prepared from alkyl halides by nucleophilic substitution with cyanide ion, the overall process $RX \rightarrow RC{\equiv}N \rightarrow RCH_2NH_2$ leads to primary amines that have one more carbon atom than the starting alkyl halide.

Cyano groups in *cyanohydrins* (Section 17.7) are reduced under the same reaction conditions.

Nitro groups are readily reduced to primary amines by a variety of methods. Catalytic hydrogenation over platinum, palladium, or nickel is often used, as is reduction by iron or tin in hydrochloric acid. The ease with which nitro groups are reduced is especially useful in the preparation of arylamines, where the sequence $ArH \rightarrow ArNO_2 \rightarrow ArNH_2$ is the standard route to these compounds.

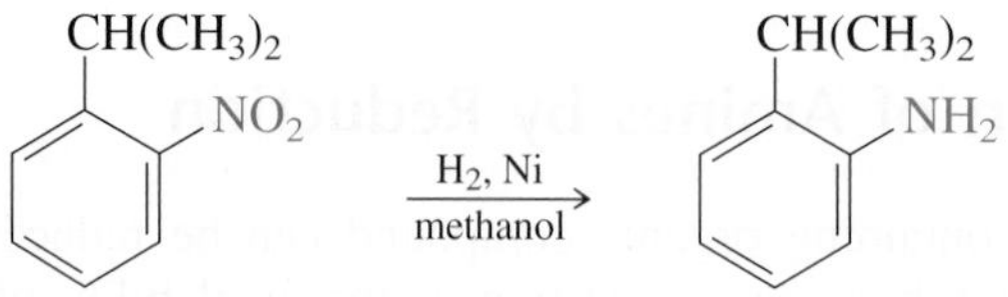

o-Isopropylnitrobenzene — *o*-Isopropylaniline (92%)

For reductions carried out in acidic media, a pH adjustment with sodium hydroxide is required in the last step in order to convert $ArNH_3^+$ to $ArNH_2$.

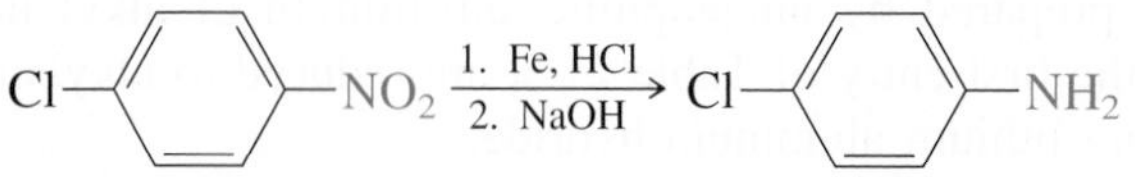

p-Chloronitrobenzene — *p*-Chloroaniline (95%)

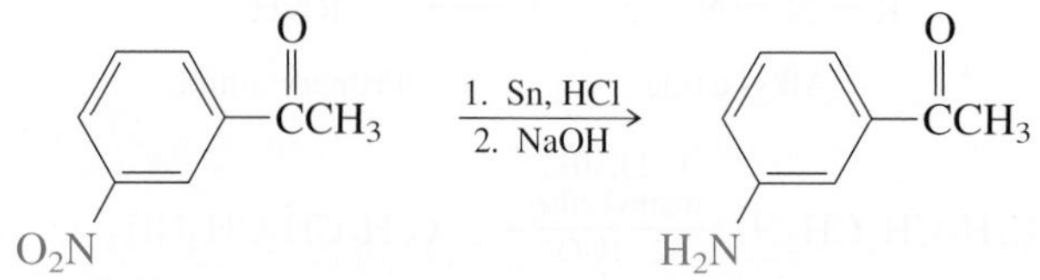

m-Nitroacetophenone — *m*-Aminoacetophenone (82%)

Problem 21.12

Outline the synthesis of each of the following arylamines from benzene:

(a) *o*-Isopropylaniline
(b) *p*-Isopropylaniline
(c) 4-Isopropyl-1,3-benzenediamine
(d) *p*-Chloroaniline
(e) *m*-Aminoacetophenone

Sample Solution (a) The last step in the synthesis of *o*-isopropylaniline, the reduction of the corresponding nitro compound by catalytic hydrogenation, is given as one of the three preceding examples. The necessary nitroarene is obtained by fractional distillation of the ortho–para mixture formed during nitration of isopropylbenzene.

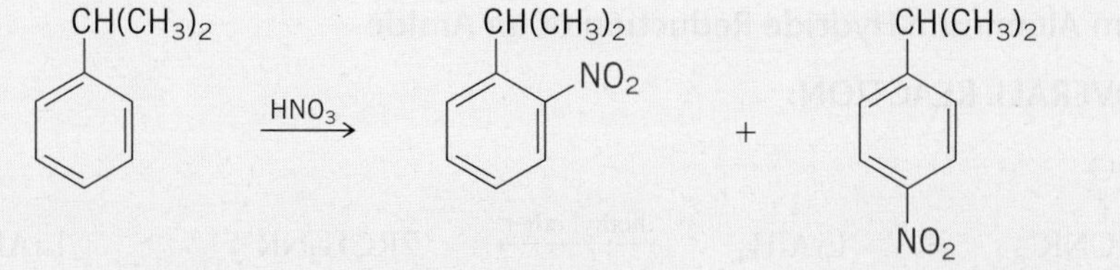
CH(CH3)2
HNO3
CH(CH3)2
NO2
+
CH(CH3)2
NO2

Isopropylbenzene — *o*-Isopropylnitrobenzene (bp 110°C) — *p*-Isopropylnitrobenzene (bp 131°C)

As actually performed, a 62% yield of a mixture of ortho and para nitration products has been obtained with an ortho–para ratio of about 1:3.

Isopropylbenzene is prepared by the Friedel–Crafts alkylation of benzene using isopropyl chloride and aluminum chloride (Section 12.6).

Reduction of an azide, a nitrile, or a nitro compound furnishes a primary amine. A method that provides access to primary, secondary, or tertiary amines is reduction of the carbonyl group of an amide by lithium aluminum hydride.

$$\underset{\text{Amide}}{\mathrm{RC(=O)NR'_2}} \xrightarrow[\text{2. } H_2O]{\text{1. } LiAlH_4} \underset{\text{Amine}}{\mathrm{RCH_2NR'_2}}$$

In this general equation, R and R′ may be either alkyl or aryl groups. When R′ = H, the product is a primary amine:

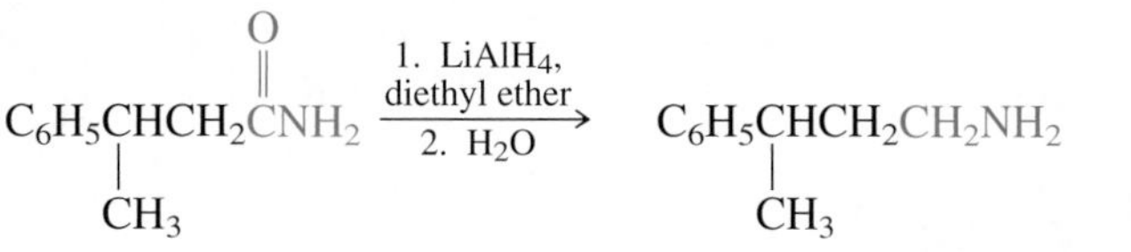
C6H5CHCH2CNH2
O
CH3
1. LiAlH4,
diethyl ether
2. H2O
C6H5CHCH2CH2NH2
CH3

3-Phenylbutanamide — 3-Phenyl-1-butanamine (59%)

N-Substituted amides yield secondary amines:

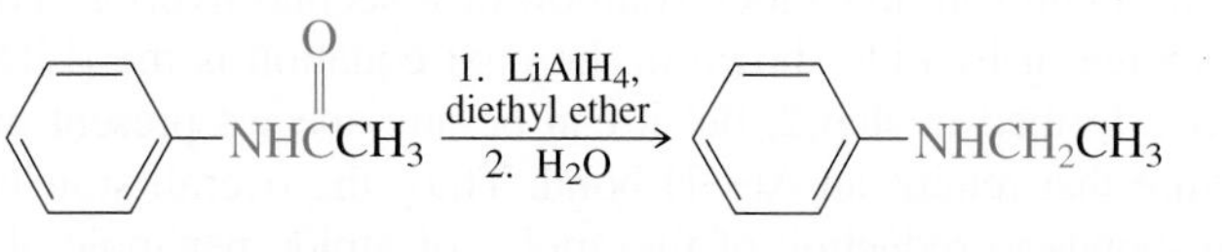
NHCCH3
O
1. LiAlH4,
diethyl ether
2. H2O
NHCH2CH3

Acetanilide — *N*-Ethylaniline (92%)

Acetanilide is an acceptable IUPAC synonym for *N*-phenylethanamide.

N,N-Disubstituted amides yield tertiary amines:

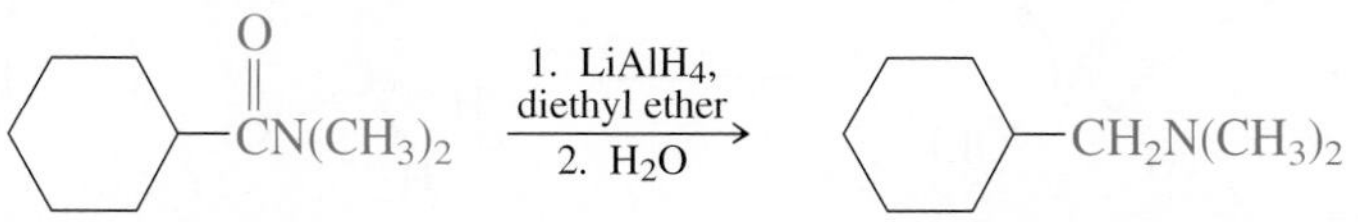
CN(CH3)2
O
1. LiAlH4,
diethyl ether
2. H2O
CH2N(CH3)2

N,N-Dimethylcyclohexanecarboxamide — *N,N*-Dimethyl(cyclohexylmethyl)amine (88%)

Mechanism 21.1 shows the reduction of amides with lithium aluminum hydride. The reduction of amides follows a similar course to the reduction of esters (Section 19.13). A tetrahedral intermediate is formed by the addition of hydride (step 1), and undergoes elimination (step 2). In the case of an ester, the alkoxy group is lost to give an intermediate aldehyde. Amides, on the other hand, retain the nitrogen and lose the oxygen from the tetrahedral intermediate. The iminium ion formed undergoes addition of a second hydride and is reduced to the amine in step 3.

Mechanism 21.1

Lithium Aluminum Hydride Reduction of an Amide

THE OVERALL REACTION:

$$\mathrm{2R\overset{\overset{O}{\|}}{C}NR'_2 \;+\; LiAlH_4 \xrightarrow{\text{diethyl ether}} 2RCH_2NR'_2 \;+\; LiAlO_2}$$

Amide   Amine

THE MECHANISM:

Step 1: Hydride is transferred from aluminum to the carbonyl carbon of the amide to give a tetrahedral intermediate. The carbonyl oxygen becomes bound to aluminum.

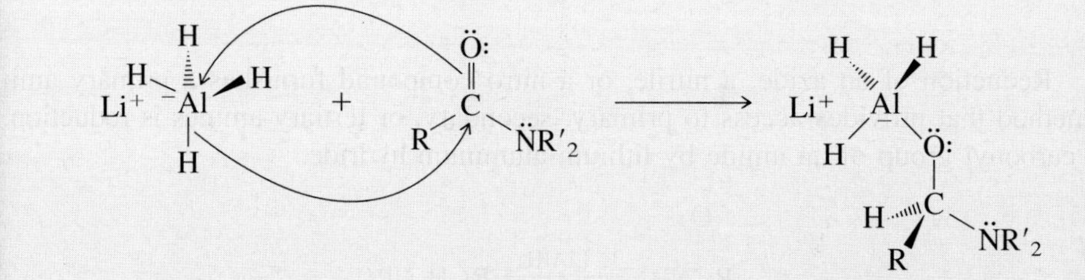

Lithium aluminum hydride   Amide   Tetrahedral intermediate

Step 2: The tetrahedral intermediate undergoes elimination to form an iminium ion.

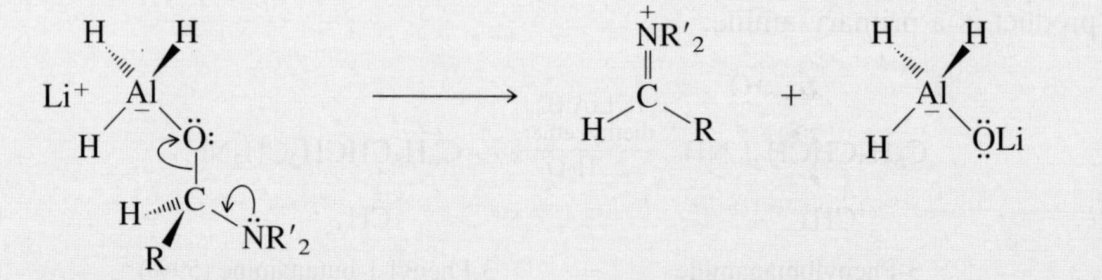

Tetrahedral intermediate   Iminium ion

Step 3: The iminium ion undergoes addition of a second hydride. For simplicity, the source of hydride shown in the next equation is the $[LiOAlH_3]^-$ species formed in step 2, but it can be any species present in the reaction mixture that retains an Al—H bond. Thus, the overall stoichiometry corresponds to reduction of two moles of amide per mole of $LiAlH_4$ and the final inorganic product is $LiAlO_2$.

Iminium ion   Amine

Because amides are so easy to prepare, this is a versatile method for the preparation of amines.

The preparation of amines by the methods described in this section involves the prior synthesis and isolation of some reducible material that has a carbon–nitrogen bond: an azide, a nitrile, a nitro-substituted arene, or an amide. The following section describes a method that combines the two steps of carbon–nitrogen bond formation and reduction into a single operation. Like the reduction of amides, it offers the possibility of preparing primary, secondary, or tertiary amines by proper choice of starting materials.

21.10 Reductive Amination

A class of nitrogen-containing compounds that was omitted from the section just discussed includes *imines* and their derivatives. Imines are formed by the reaction of aldehydes and ketones with ammonia (Section 17.10). Imines can be reduced to primary amines by catalytic hydrogenation.

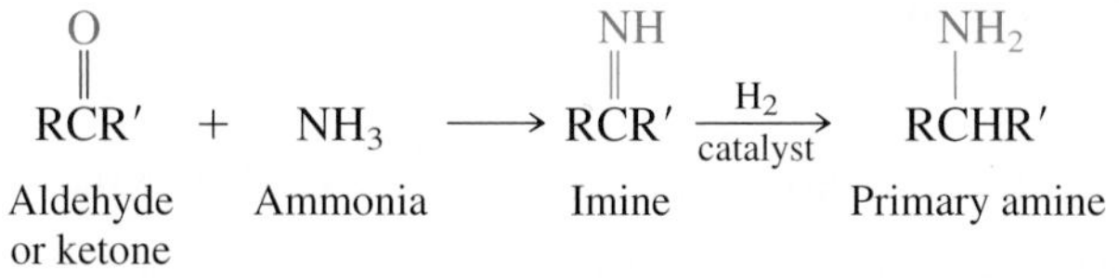
RCR′
NH3
RCR′
H2
catalyst
RCHR′
Aldehyde or ketone
Ammonia
Imine
Primary amine

The overall reaction converts a carbonyl compound to an amine by carbon–nitrogen bond formation and reduction; it is commonly known as **reductive amination.** What makes it a particularly valuable synthetic procedure is that it can be carried out in a single operation by hydrogenation of a solution containing both ammonia and the carbonyl compound along with a hydrogenation catalyst. The intermediate imine is not isolated but undergoes reduction under the conditions of its formation. Also, the reaction is broader in scope than implied by the preceding equation. All classes of amines—primary, secondary, and tertiary—may be prepared by reductive amination.

When primary amines are desired, the reaction is carried out as just described:

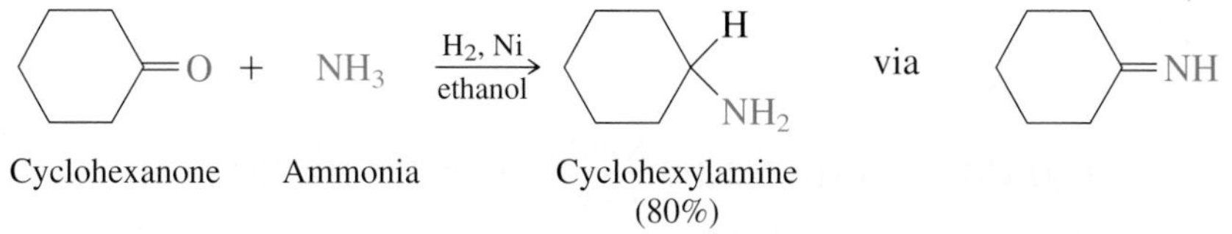
NH3
H2, Ni
ethanol
via
NH
Cyclohexanone
Ammonia
Cyclohexylamine (80%)

Secondary amines are prepared by hydrogenation of a carbonyl compound in the presence of a primary amine. An *N*-substituted imine, or *Schiff's base,* is an intermediate:

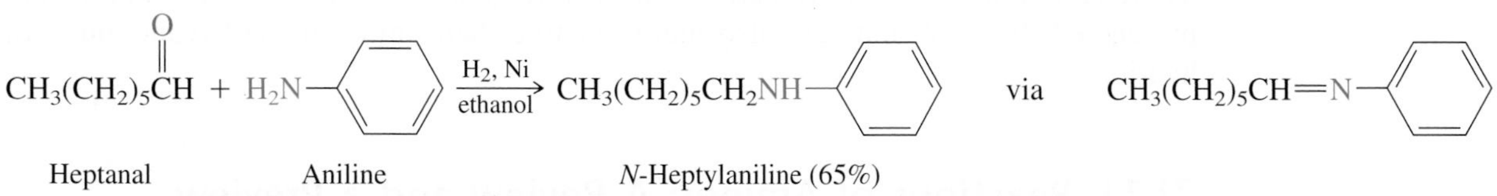
CH3(CH2)5CH
H2N
H2, Ni
ethanol
CH3(CH2)5CH2NH
via
CH3(CH2)5CH=N
Heptanal
Aniline
N-Heptylaniline (65%)

Reductive amination has been successfully applied to the preparation of tertiary amines from carbonyl compounds and secondary amines even though a neutral imine is not possible in this case.

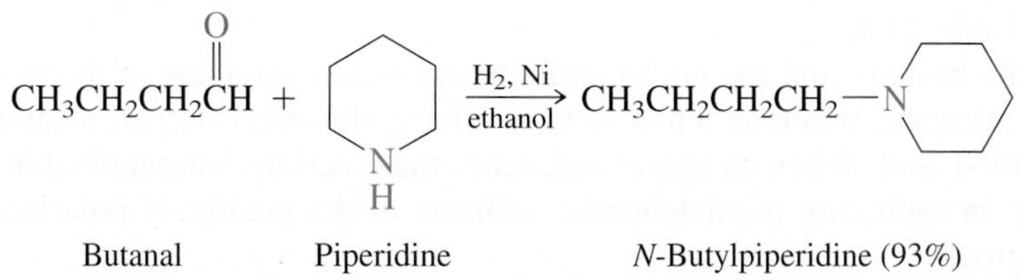
CH3CH2CH2CH
H2, Ni
ethanol
CH3CH2CH2CH2—N
Butanal
Piperidine
N-Butylpiperidine (93%)

Presumably, the species that undergoes reduction here is a carbinolamine, an iminium ion derived from it, or an enamine.

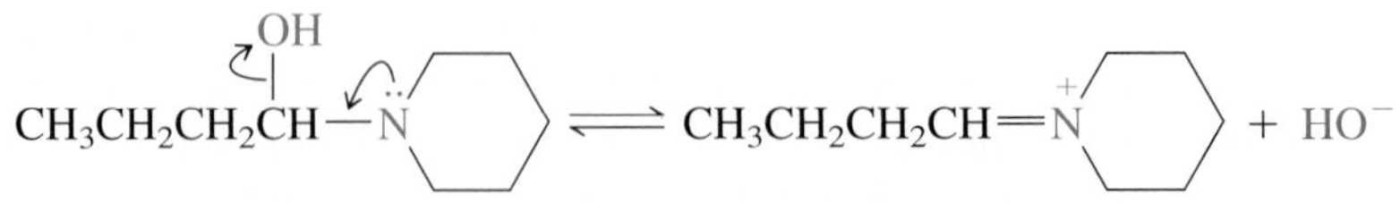

Carbinolamine Iminium ion

Problem 21.13

Show how you could prepare each of the following amines from benzaldehyde by reductive amination:

(a) Benzylamine (c) *N,N*-Dimethylbenzylamine
(b) Dibenzylamine (d) *N*-Benzylpiperidine

Sample Solution (a) Because benzylamine is a primary amine, it is derived from ammonia and benzaldehyde.

$$\underset{\text{Benzaldehyde}}{C_6H_5\overset{O}{\overset{\|}{C}}H} + \underset{\text{Ammonia}}{NH_3} + \underset{\text{Hydrogen}}{H_2} \xrightarrow{Ni} \underset{\substack{\text{Benzylamine}\\(89\%)}}{C_6H_5CH_2NH_2} + \underset{\text{Water}}{H_2O}$$

The reaction proceeds by initial formation of the imine $C_6H_5CH{=}NH$, followed by its hydrogenation.

A variation of the classical reductive amination procedure uses sodium cyanoborohydride ($NaBH_3CN$) instead of hydrogen as the reducing agent and is better suited to amine syntheses in which only a few grams of material are needed. All that is required is to add sodium cyanoborohydride to an alcohol solution of the carbonyl compound and an amine.

$$\underset{\text{Benzaldehyde}}{C_6H_5\overset{O}{\overset{\|}{C}}H} + \underset{\text{Ethylamine}}{CH_3CH_2NH_2} \xrightarrow[\text{methanol}]{NaBH_3CN} \underset{N\text{-Ethylbenzylamine (91\%)}}{C_6H_5CH_2NHCH_2CH_3}$$

Sodium cyanoborohydride reduces aldehydes and ketones less rapidly than sodium borohydride, but it reduces iminium ions rapidly. To take advantage of this selectivity, reductive aminations are carried out at mildly acidic pH, where the imines are protonated. Iminium ions are also more reactive than imines toward reduction with hydride.

21.11 Reactions of Amines: A Review and a Preview

The noteworthy properties of amines are their *basicity* and their *nucleophilicity*. The basicity of amines has been discussed in Section 21.4. Several reactions in which amines act as nucleophiles have already been encountered in earlier chapters. These are summarized in Table 21.4.

Both the basicity and the nucleophilicity of amines originate in the unshared electron pair of nitrogen. When an amine acts as a base, this electron pair abstracts a proton from a Brønsted acid. When an amine undergoes the reactions summarized in Table 21.4, the first step in each case is nucleophilic addition to the positively polarized carbon of a carbonyl group.

TABLE 21.4 Reactions of Amines Discussed in Previous Chapters*

Reaction (section) and comments	General equation and specific example
Reaction of primary amines with aldehydes and ketones (Section 17.10) Imines are formed by nucleophilic addition of a primary amine to the carbonyl group of an aldehyde or a ketone. The key step is formation of a carbinolamine intermediate, which then dehydrates to the imine.	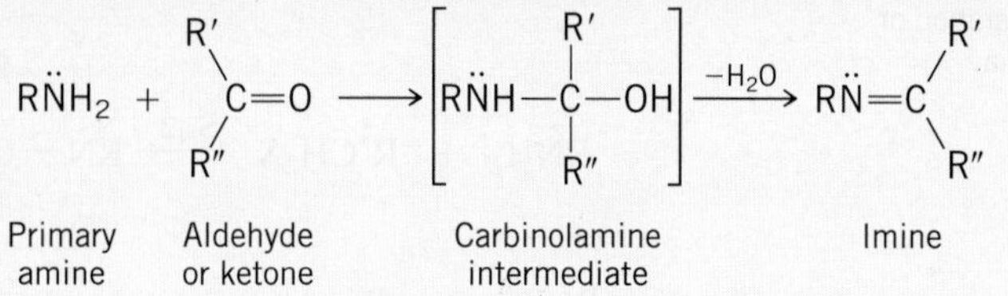 Primary amine Aldehyde or ketone Carbinolamine intermediate Imine 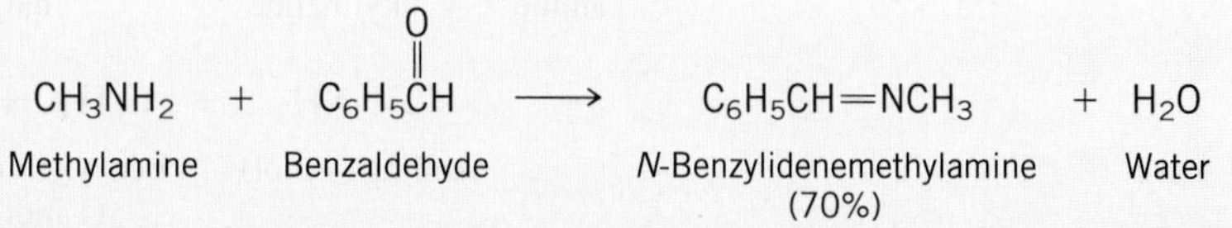 Methylamine Benzaldehyde N-Benzylidenemethylamine (70%) Water
Reaction of secondary amines with aldehydes and ketones (Section 17.11) Enamines are formed in the corresponding reaction of secondary amines with aldehydes and ketones.	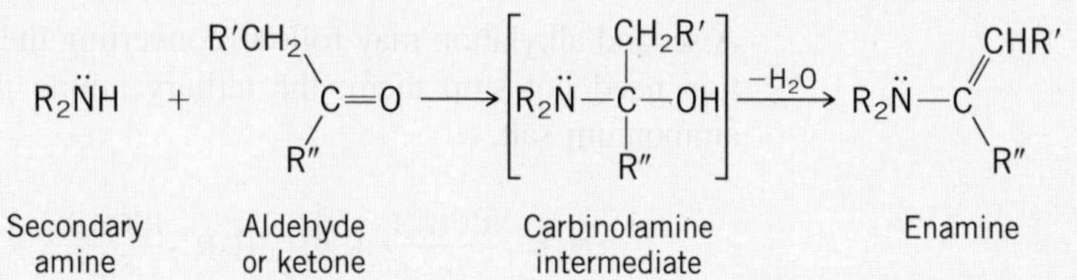 Secondary amine Aldehyde or ketone Carbinolamine intermediate Enamine 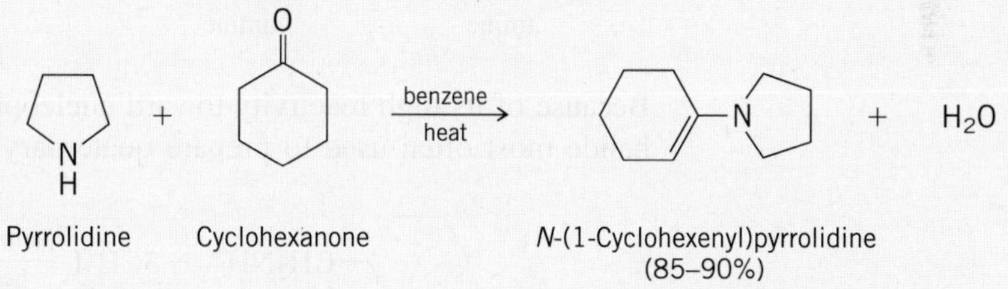 benzene heat Pyrrolidine Cyclohexanone N-(1-Cyclohexenyl)pyrrolidine (85–90%)
Reaction of amines with acyl chlorides (Section 19.4) Amines are converted to amides on reaction with acyl chlorides. Other acylating agents, such as acid anhydrides and esters, may also be used but are less reactive.	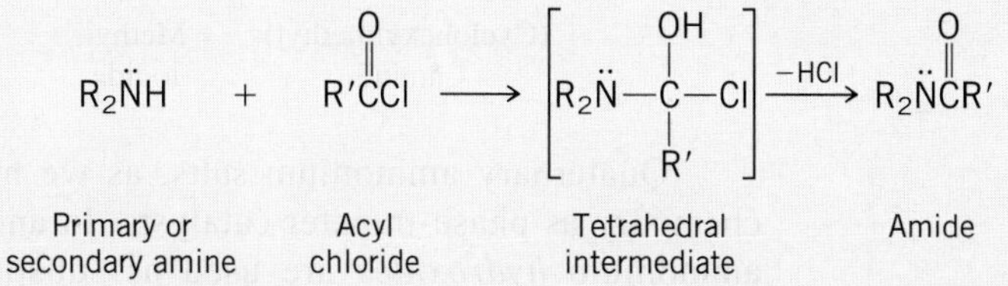 Primary or secondary amine Acyl chloride Tetrahedral intermediate Amide 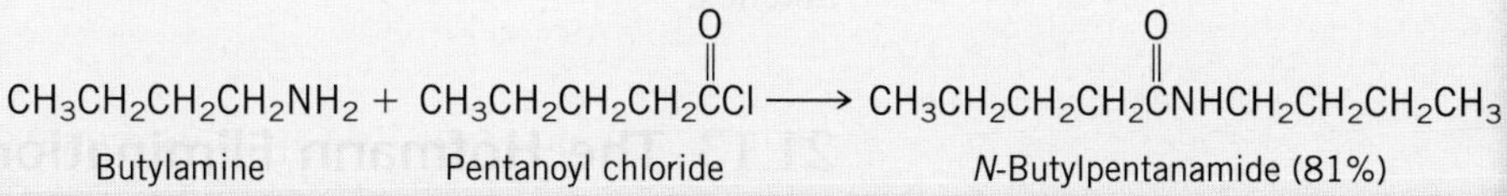 Butylamine Pentanoyl chloride N-Butylpentanamide (81%)

*Both alkylamines and arylamines undergo these reactions.

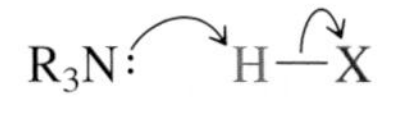

Amine acting as a base

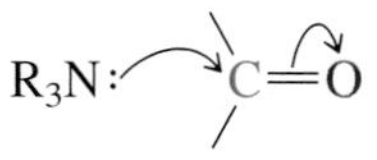

Amine acting as a nucleophile

In addition to being more basic than arylamines, alkylamines are also more nucleophilic. All the reactions in Table 21.4 take place faster with alkylamines than with arylamines.

The sections that follow introduce some additional reactions of amines. In all cases our understanding of how these reactions take place starts with a consideration of the role of the unshared electron pair of nitrogen.

We will begin with an examination of the reactivity of amines as nucleophiles in S_N2 reactions.

21.12 Reaction of Amines with Alkyl Halides

The reaction of amines with alkyl halides was seen earlier (Section 21.7) as a complicating factor in the preparation of amines by alkylation of ammonia.

Nucleophilic substitution results when primary alkyl halides are treated with amines.

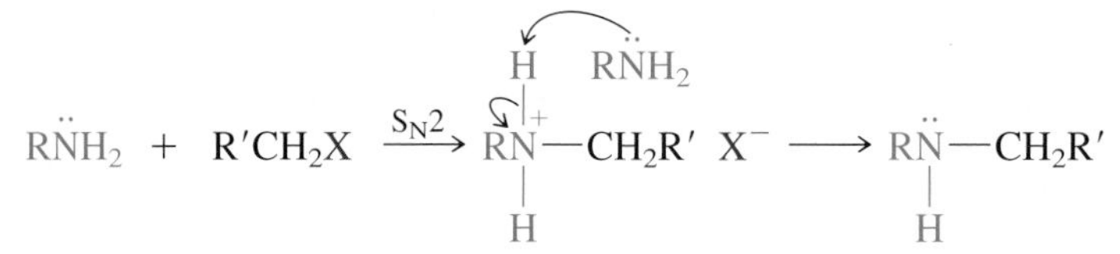

Primary amine | Primary alkyl halide | Ammonium halide salt | Secondary amine

$$C_6H_5NH_2 + C_6H_5CH_2Cl \xrightarrow[90°C]{NaHCO_3} C_6H_5NHCH_2C_6H_5$$

Aniline (4 mol) | Benzyl chloride (1 mol) | *N*-Benzylaniline (85–87%)

A second alkylation may follow, converting the secondary amine to a tertiary amine. Alkylation need not stop there; the tertiary amine may itself be alkylated, giving a quaternary ammonium salt.

$$R\ddot{N}H_2 \xrightarrow[S_N2]{R'CH_2X} R\ddot{N}HCH_2R' \xrightarrow[S_N2]{R'CH_2X} R\ddot{N}(CH_2R')_2 \xrightarrow[S_N2]{R'CH_2X} R\overset{+}{N}(CH_2R')_3\ X^-$$

Primary amine | Secondary amine | Tertiary amine | Quaternary ammonium salt

Because of its high reactivity toward nucleophilic substitution, methyl iodide is the alkyl halide most often used to prepare quaternary ammonium salts.

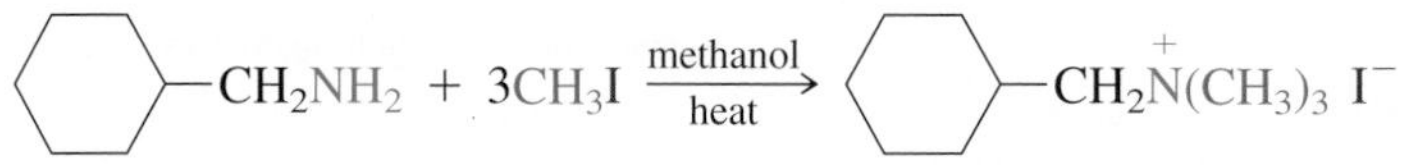

(Cyclohexylmethyl)amine | Methyl iodide | (Cyclohexylmethyl)trimethylammonium iodide (99%)

Quaternary ammonium salts, as we have seen, are useful in synthetic organic chemistry as phase-transfer catalysts. In another, more direct application, quaternary ammonium *hydroxides* are used as substrates in an elimination reaction to form alkenes.

21.13 The Hofmann Elimination

The halide anion of quaternary ammonium iodides may be replaced by hydroxide by treatment with an aqueous slurry of silver oxide. Silver iodide precipitates, and a solution of the quaternary ammonium hydroxide is formed.

$$2(R_4\overset{+}{N}\ I^-) + Ag_2O + H_2O \longrightarrow 2(R_4\overset{+}{N}\ \overset{-}{O}H) + 2AgI$$

Quaternary ammonium iodide | Silver oxide | Water | Quaternary ammonium hydroxide | Silver iodide

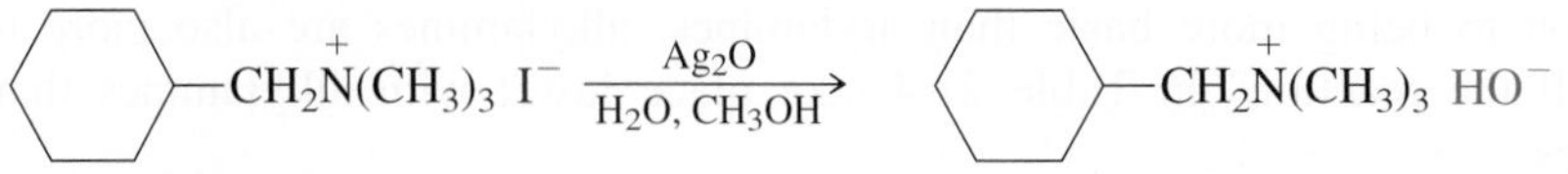

(Cyclohexylmethyl)trimethylammonium iodide | (Cyclohexylmethyl)trimethylammonium hydroxide

When quaternary ammonium hydroxides are heated, they undergo β elimination to form an alkene and an amine.

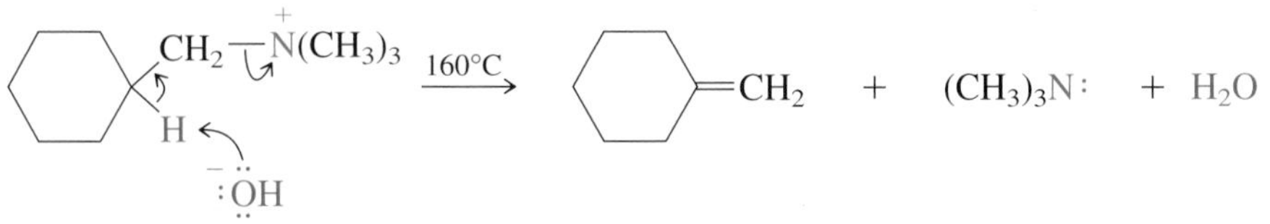

(Cyclohexylmethyl)trimethylammonium hydroxide — Methylenecyclohexane (69%) — Trimethylamine — Water

This reaction is known as the **Hofmann elimination;** it was developed by August W. Hofmann in the middle of the nineteenth century and is both a synthetic method to prepare alkenes and an analytical tool for structure determination.

A novel aspect of the Hofmann elimination is its regioselectivity. Elimination in alkyltrimethylammonium hydroxides proceeds in the direction that gives the *less* substituted alkene.

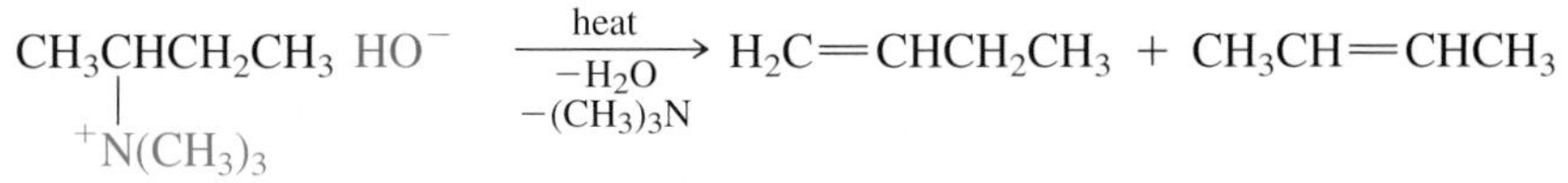

sec-Butyltrimethylammonium hydroxide — 1-Butene (95%) — 2-Butene (5%) (cis and trans)

The least sterically hindered β hydrogen is removed by the base in Hofmann elimination reactions. Methyl groups are deprotonated in preference to methylene groups, and methylene groups are deprotonated in preference to methines. The regioselectivity of Hofmann elimination is opposite to that predicted by the Zaitsev rule (Section 5.10). Elimination reactions of alkyltrimethylammonium hydroxides are said to obey the **Hofmann rule;** they yield the less substituted alkene.

Problem 21.14

Give the structure of the major alkene formed when the hydroxide of each of the following quaternary ammonium ions is heated.

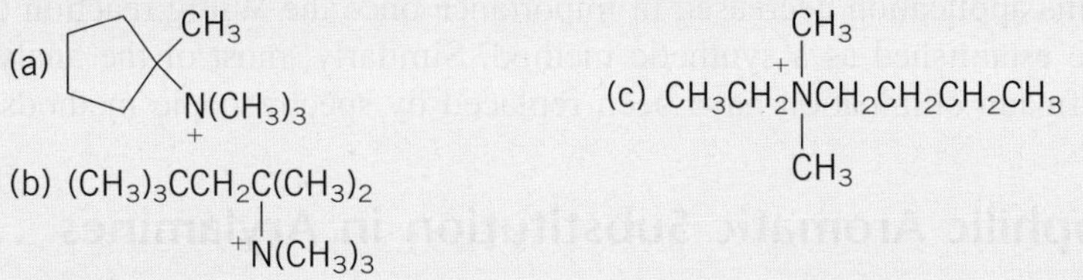

Sample Solution (a) Two alkenes are capable of being formed by β elimination: methylenecyclopentane and 1-methylcyclopentane.

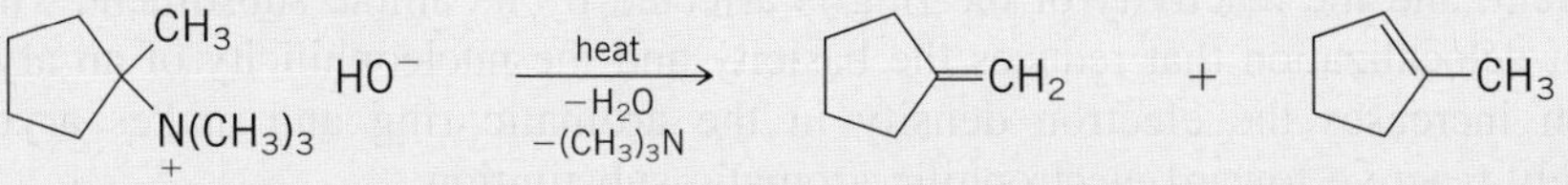

(1-Methylcyclopentyl)trimethylammonium hydroxide — Methylenecyclopentane — 1-Methylcyclopentene

Methylenecyclopentane has the less substituted double bond and is the major product. The reported isomer distribution is 91% methylenecyclopentane and 9% 1-methylcyclopentene.

We can understand the regioselectivity of the Hofmann elimination by comparing steric effects in the E2 transition states for formation of 1-butene and *trans*-2-butene from *sec*-butyltrimethylammonium hydroxide. In terms of its size, $(CH_3)_3\overset{+}{N}—$ (trimethylammonio) is comparable to $(CH_3)_3C—$ (*tert*-butyl). As Figure 21.5 illustrates, the E2 transition state requires an anti relationship between the proton that is removed and the trimethylammonio group. No serious van der Waals repulsions are evident in the transition state geometry for formation of 1-butene. The conformation leading

Figure 21.5

Newman projections showing the conformations leading to (*a*) 1-butene, and (*b*) *trans*-2-butene by Hofmann elimination of *sec*-butyltrimethylammonium hydroxide. The major product is 1-butene.

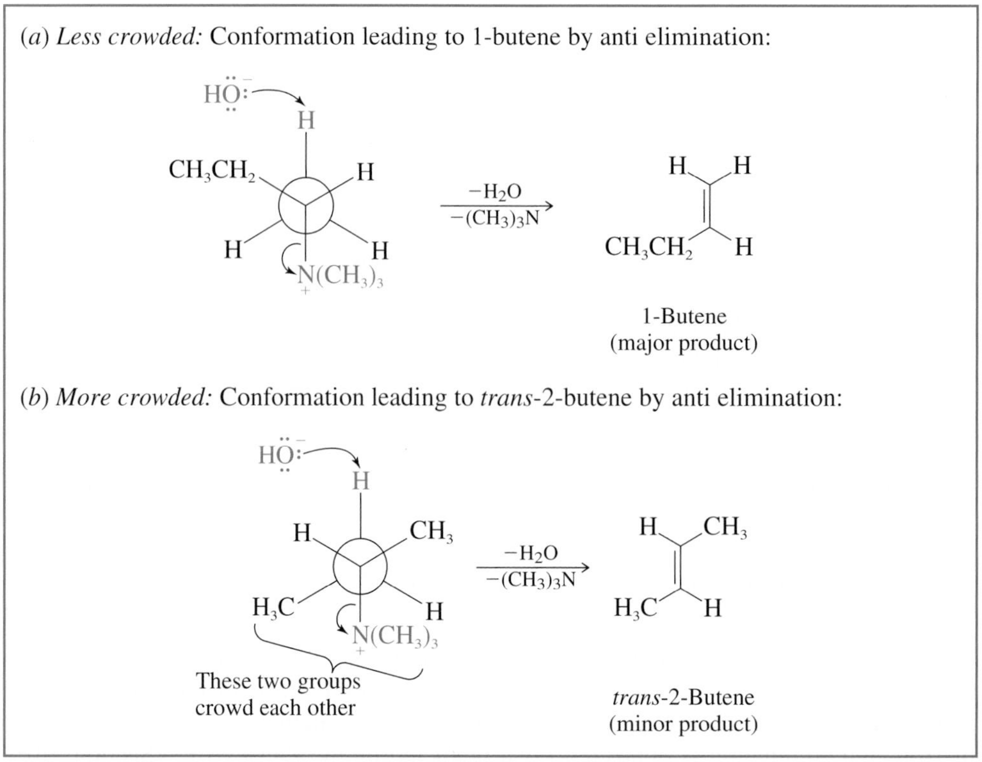
(a) Less crowded: Conformation leading to 1-butene by anti elimination:
HO:
H
CH3CH2
H
H
H
N(CH3)3
−H2O
−(CH3)3N
H
H
CH3CH2
H
1-Butene
(major product)
(b) More crowded: Conformation leading to trans-2-butene by anti elimination:
HO:
H
H
CH3
H3C
H
N(CH3)3
These two groups
crowd each other
−H2O
−(CH3)3N
H
CH3
H3C
H
trans-2-Butene
(minor product)

to *trans*-2-butene, however, is destabilized by van der Waals strain between the trimethylammonio group and a methyl group gauche to it. Thus, the activation energy for formation of *trans*-2-butene exceeds that of 1-butene, which becomes the major product because it is formed faster.

With a regioselectivity opposite to that of the Zaitsev rule, the Hofmann elimination is sometimes used in synthesis to prepare alkenes not accessible by dehydrohalogenation of alkyl halides. This application decreased in importance once the Wittig reaction (Section 17.12) became established as a synthetic method. Similarly, most of the analytical applications of Hofmann elimination have been replaced by spectroscopic methods.

21.14 Electrophilic Aromatic Substitution in Arylamines

Arylamines contain two functional groups, the amine group and the aromatic ring; they are *difunctional compounds.* The reactivity of the amine group is affected by its aryl substituent, and the reactivity of the ring is affected by its amine substituent. The same electron delocalization that reduces the basicity and the nucleophilicity of an arylamine nitrogen increases the electron density in the aromatic ring and makes arylamines extremely reactive toward electrophilic aromatic substitution.

The reactivity of arylamines was noted in Section 12.12, where it was pointed out that $—\ddot{N}H_2$, $—\ddot{N}HR$, and $—\ddot{N}R_2$ are ortho, para-directing and exceedingly powerful activating groups. These substituents are such powerful activators that electrophilic aromatic substitution is only rarely performed directly on arylamines.

Direct nitration of aniline and other arylamines fails because oxidation leads to the formation of dark-colored “tars.” As a solution to this problem it is standard practice to first protect the amino group by acylation with either acetyl chloride or acetic anhydride.

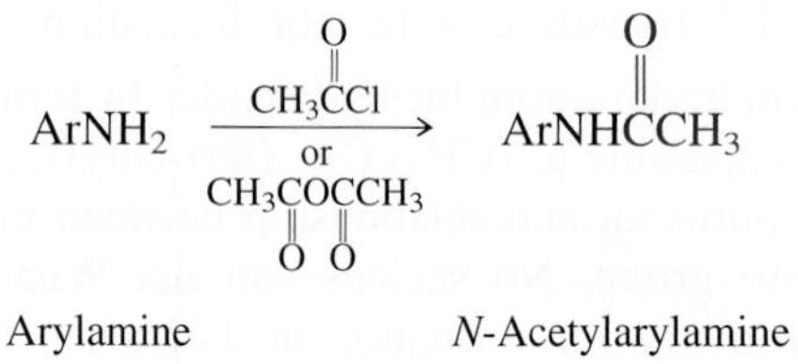
ArNH2
O
CH3CCl
or
CH3COCCH3
O O
O
ArNHCCH3
Arylamine
N-Acetylarylamine

Amide resonance within the *N*-acetyl group competes with delocalization of the nitrogen lone pair into the ring. Protecting the amino group of an arylamine in this way moderates its reactivity and permits nitration of the ring. The acetamido group is activating toward electrophilic aromatic substitution and is ortho, para-directing. After the *N*-acetyl-protecting group has served its purpose, it may be removed by hydrolysis, restoring the amino group:

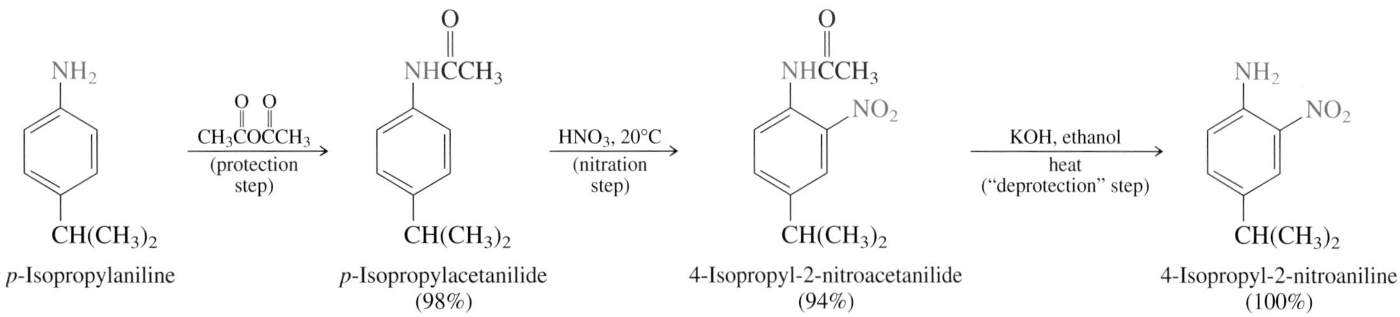

NH2
CH3COCCH3
(protection step)
NHCCH3
HNO3, 20°C
(nitration step)
NO2
KOH, ethanol
heat
("deprotection" step)
CH(CH3)2
p-Isopropylaniline
p-Isopropylacetanilide (98%)
4-Isopropyl-2-nitroacetanilide (94%)
4-Isopropyl-2-nitroaniline (100%)

The net effect of the sequence *protect–nitrate–deprotect* is the same as if the substrate had been nitrated directly. Because direct nitration is impossible, however, the indirect route is the only practical method.

Problem 21.15

Outline syntheses of each of the following from aniline and any necessary organic or inorganic reagents.

(a) *p*-Nitroaniline
(b) 2,4-Dinitroaniline
(c) *p*-Aminoacetanilide

Sample Solution (a) Because direct nitration of aniline is not a practical reaction, the amino group must first be protected as its *N*-acetyl derivative.

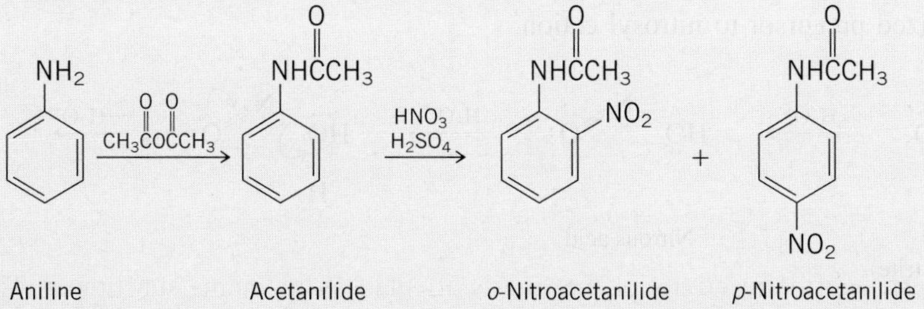

NH2
CH3COCCH3
NHCCH3
HNO3
H2SO4
NO2
+
Aniline
Acetanilide
o-Nitroacetanilide
p-Nitroacetanilide

Nitration of acetanilide yields a mixture of ortho and para substitution products. The para isomer is separated, then subjected to hydrolysis to give *p*-nitroaniline.

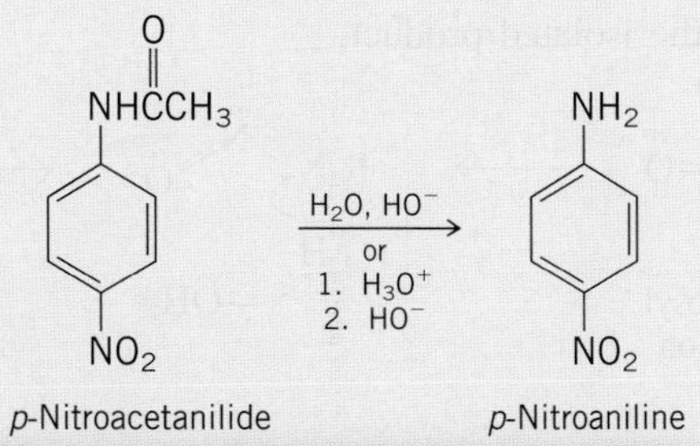

NHCCH3
H2O, HO−
or
1. H3O+
2. HO−
NH2
NO2
p-Nitroacetanilide
p-Nitroaniline

Unprotected arylamines are so reactive that it is difficult to limit halogenation to monosubstitution. Generally, halogenation proceeds rapidly to replace all the available hydrogens that are ortho or para to the amino group.

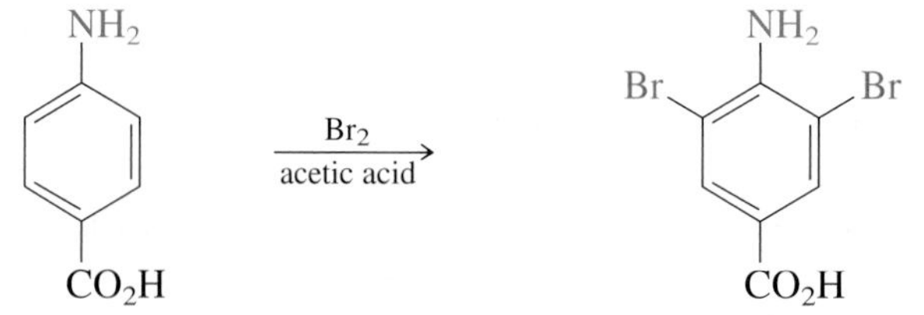

p-Aminobenzoic acid   4-Amino-3,5-dibromobenzoic acid (82%)

Decreasing the electron-donating ability of an amino group by acylation makes monohalogenation possible.

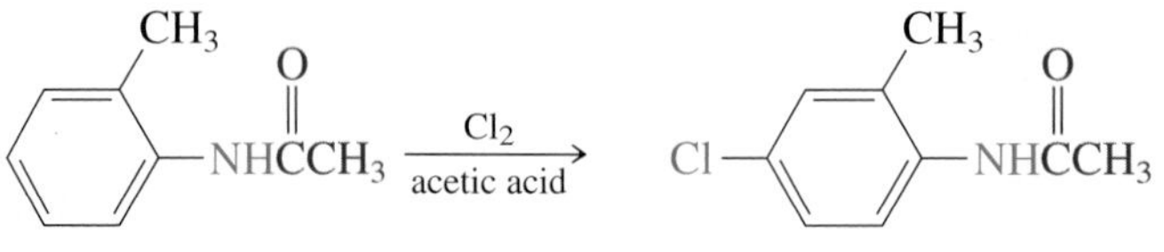

2-Methylacetanilide   4-Chloro-2-methylacetanilide (74%)

Friedel–Crafts reactions are normally not successful when attempted on an arylamine, but can be carried out readily once the amino group is protected.

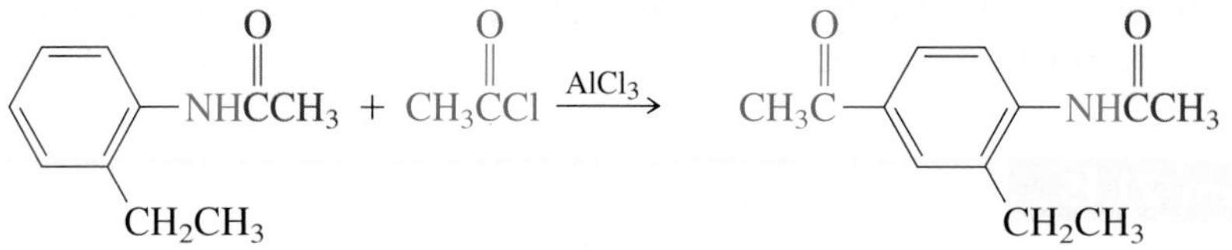

2-Ethylacetanilide   4-Acetamido-3-ethylacetophenone (57%)

21.15 Nitrosation of Alkylamines

When solutions of sodium nitrite ($NaNO_2$) are acidified, a number of species are formed that act as sources of nitrosyl cation, $:\overset{+}{N}{=}\ddot{O}:$. For simplicity, organic chemists group all these species together and speak of the chemistry of one of them, *nitrous acid,* as a generalized precursor to nitrosyl cation.

Nitrosyl cation is also called *nitrosonium* ion. It can be represented by the two resonance structures

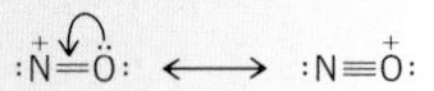

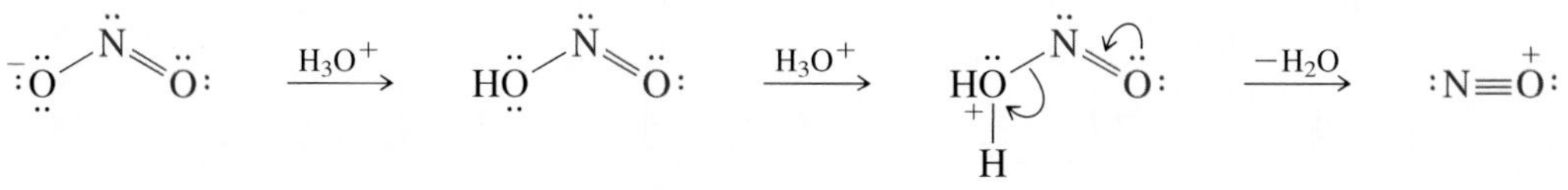

Nitrite ion (from sodium nitrite)   Nitrous acid   Nitrosyl cation

Nitrosation of amines is best illustrated by examining what happens when a secondary amine "reacts with nitrous acid." The amine acts as a nucleophile toward the nitrogen of nitrosyl cation. The intermediate that is formed in the first step loses a proton to give an **N-nitroso amine** as the isolated product.

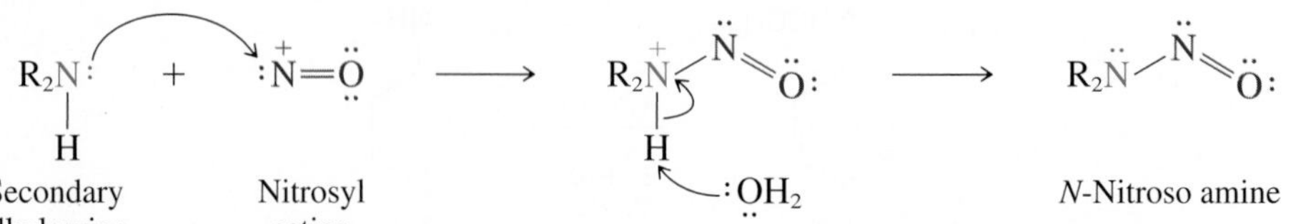

Secondary alkylamine   Nitrosyl cation   N-Nitroso amine

For example,

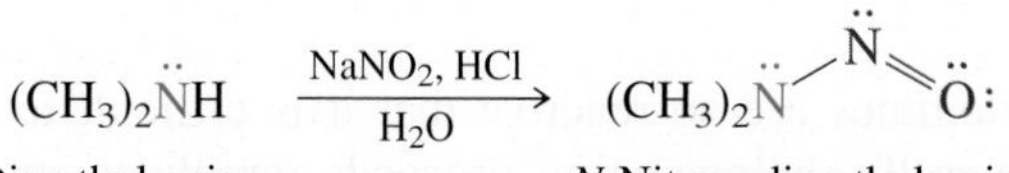

Dimethylamine   *N*-Nitrosodimethylamine (88–90%)

Problem 21.16

N-Nitroso amines are stabilized by electron delocalization. Write the two most stable resonance contributors of *N*-nitrosodimethylamine, $(CH_3)_2NNO$.

N-Nitroso amines are more often called **nitrosamines,** and because many of them are potent carcinogens, they have been the object of much investigation. We encounter nitrosamines in the environment on a daily basis. A few of these, all of which are known carcinogens, are:

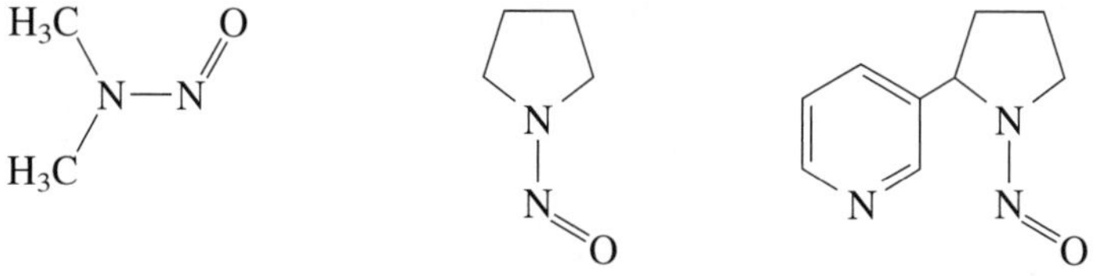

N-Nitrosodimethylamine (formed during tanning of leather; also found in beer and herbicides) | *N*-Nitrosopyrrolidine (formed when bacon that has been cured with sodium nitrite is fried) | *N*-Nitrosonornicotine (present in tobacco smoke)

Nitrosamines are formed whenever nitrosating agents come in contact with secondary amines, and more are probably synthesized within our body than enter it by environmental contamination. Enzyme-catalyzed reduction of nitrate (NO_3^-) produces nitrite (NO_2^-), which combines with amines present in the body to form *N*-nitroso amines.

When primary amines are nitrosated, their *N*-nitroso compounds can't be isolated because they react further.

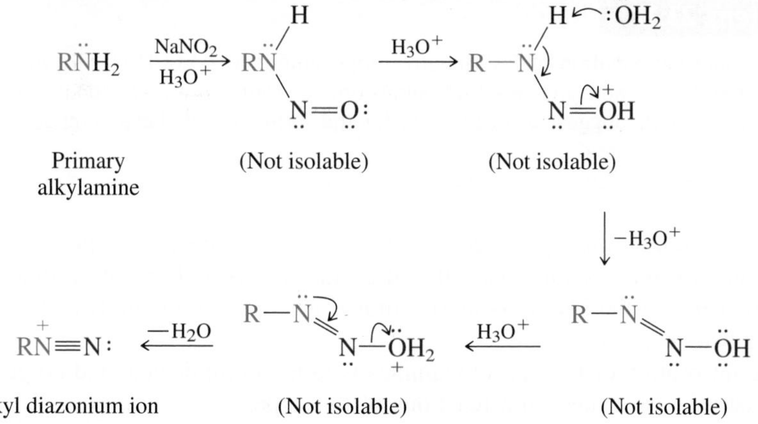
Primary alkylamine
(Not isolable)
(Not isolable)
Alkyl diazonium ion
(Not isolable)
(Not isolable)

The product of this series of steps is an alkyl **diazonium ion,** and the amine is said to have been **diazotized.** Alkyl diazonium ions are not very stable, decomposing rapidly under the conditions of their formation. Molecular nitrogen is a leaving group par excellence, and the reaction products arise by solvolysis of the diazonium ion. Usually, a carbocation intermediate is involved.

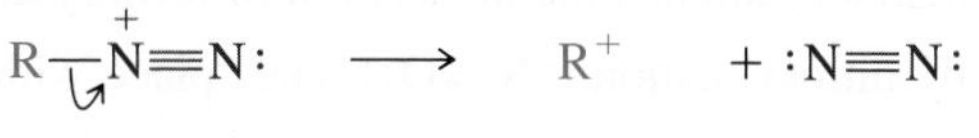

Alkyl diazonium ion | Carbocation | Nitrogen

Recall from Section 8.12 that decreasing basicity is associated with increasing leaving-group ability. Molecular nitrogen is an exceedingly weak base and an excellent leaving group.

Mechanism 21.2 shows what happens when a typical primary alkylamine reacts with nitrous acid.

Because nitrogen-free products result from the formation and decomposition of diazonium ions, these reactions are often referred to as *deamination reactions.* Alkyl diazonium ions are rarely used in synthetic work but have been studied extensively to probe the behavior of carbocations generated under conditions in which the leaving group is lost rapidly and irreversibly.

Mechanism 21.2

Reactions of an Alkyl Diazonium Ion

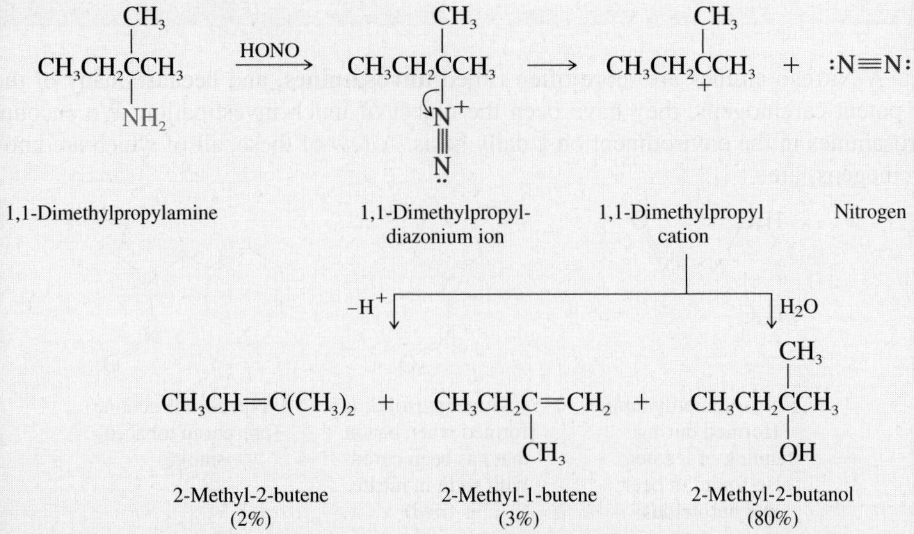
HONO
1,1-Dimethylpropylamine
1,1-Dimethylpropyl-diazonium ion
1,1-Dimethylpropyl cation
Nitrogen
$-H^+$
H_2O
2-Methyl-2-butene (2%)
2-Methyl-1-butene (3%)
2-Methyl-2-butanol (80%)

The diazonium ion generated by treatment of a primary alkylamine with nitrous acid loses nitrogen to give a carbocation. The isolated products are derived from the carbocation and include, in this example, alkenes (by loss of a proton) and an alcohol (nucleophilic capture by water).

Problem 21.17

Nitrous acid deamination of 2,2-dimethylpropylamine, $(CH_3)_3CCH_2NH_2$, gives the same products as were indicated as being formed from 1,1-dimethylpropylamine in Mechanism 21.2. Suggest a mechanism for the formation of these compounds from 2,2-dimethylpropylamine.

Aryl diazonium ions, prepared by nitrous acid diazotization of primary arylamines, are substantially more stable than alkyl diazonium ions and are of enormous synthetic value. Their use in the synthesis of substituted aromatic compounds is described in the following two sections.

The nitrosation of tertiary alkylamines is rather complicated, and no generally useful chemistry is associated with reactions of this type.

21.16 Nitrosation of Arylamines

We learned in the preceding section that different reactions are observed when the various classes of alkylamines—primary, secondary, and tertiary—react with nitrosating agents. Although no useful chemistry attends the nitrosation of tertiary alkylamines, electrophilic aromatic substitution by nitrosyl cation ($:N{\equiv}\overset{+}{O}:$) takes place with *N,N*-dialkylarylamines.

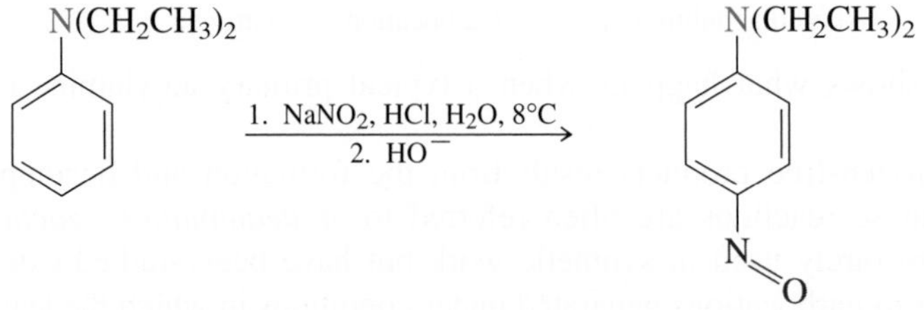
$N(CH_2CH_3)_2$
1. $NaNO_2$, HCl, H_2O, 8°C
2. HO^-
$N(CH_2CH_3)_2$
N
O

N,N-Diethylaniline *N,N*-Diethyl-*p*-nitrosoaniline (95%)

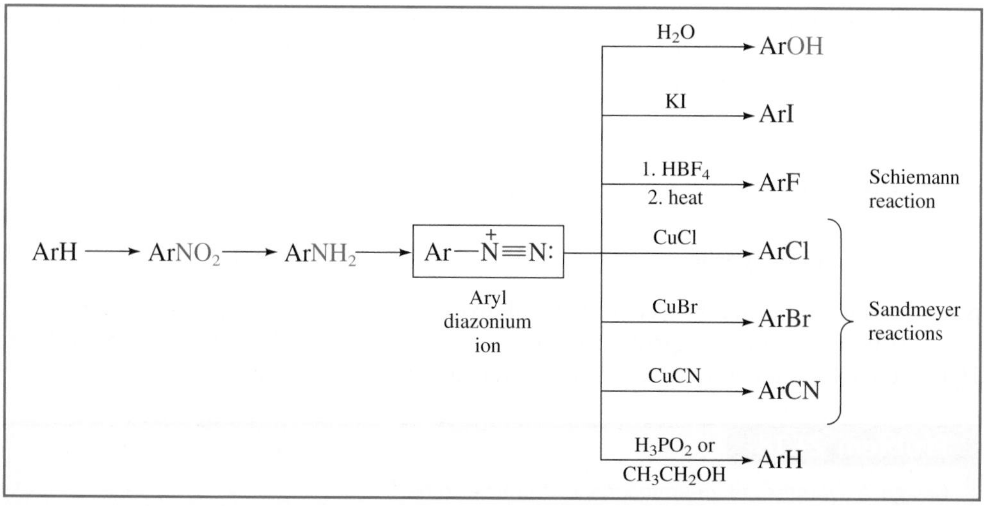

Figure 21.6

The synthetic origin of aryl diazonium ions and their most useful transformations.

Nitrosyl cation is a relatively weak electrophile and attacks only very strongly activated aromatic rings.

N-Alkylarylamines resemble secondary alkylamines in that they form *N*-nitroso compounds on reaction with nitrous acid.

$$C_6H_5NHCH_3 \xrightarrow[H_2O,\ 10°C]{NaNO_2,\ HCl} C_6H_5N(CH_3)—N{=}O$$

N-Methylaniline → *N*-Methyl-*N*-nitrosoaniline (87–93%)

Primary arylamines, like primary alkylamines, form diazonium ion salts on nitrosation. Whereas alkyl diazonium ions decompose under the conditions of their formation, aryl diazonium salts are considerably more stable and can be stored in aqueous solution at 0–5°C for a reasonable time. Loss of nitrogen from an aryl diazonium ion generates an unstable aryl cation and is much slower than loss of nitrogen from an alkyl diazonium ion.

$$C_6H_5NH_2 \xrightarrow[H_2O,\ 0\text{–}5°C]{NaNO_2,\ HCl} C_6H_5\overset{+}{N}{\equiv}N\colon\ Cl^-$$

Aniline → Benzenediazonium chloride

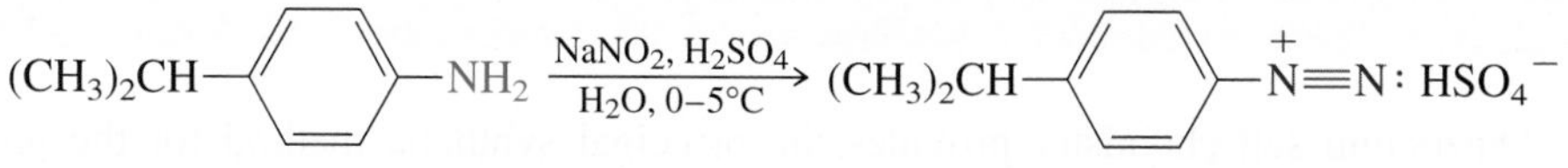

p-Isopropylaniline → *p*-Isopropylbenzenediazonium hydrogen sulfate

Aryl diazonium ions undergo a variety of reactions that make them versatile intermediates for preparing a host of ring-substituted aromatic compounds. In these reactions, summarized in Figure 21.6 and discussed individually in the following section, molecular nitrogen acts as a leaving group and is replaced by another atom or group. All the reactions are regiospecific; the entering group becomes bonded to the same carbon from which nitrogen departs.

21.17 Synthetic Transformations of Aryl Diazonium Salts

An important reaction of aryl diazonium ions is their conversion to *phenols* by hydrolysis:

$$Ar\overset{+}{N}{\equiv}N\colon + H_2O \longrightarrow ArOH + H^+ + \colon N{\equiv}N\colon$$

Aryl diazonium ion + Water → A phenol + Nitrogen

This is the most general method for preparing phenols. It is easily performed; the aqueous acidic solution in which the diazonium salt is prepared is heated and gives the phenol directly. An aryl cation is probably generated, which is then captured by water acting as a nucleophile.

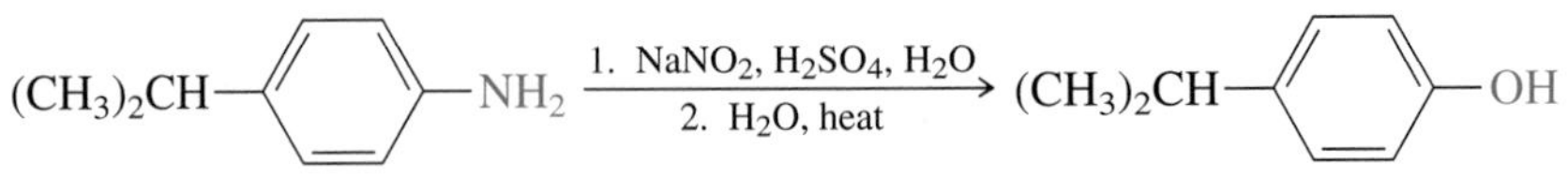
(CH3)2CH
NH2
1. NaNO2, H2SO4, H2O
2. H2O, heat
(CH3)2CH
OH

p-Isopropylaniline *p*-Isopropylphenol (73%)

Sulfuric acid is normally used instead of hydrochloric acid in the diazotization step so as to minimize the competition with water for capture of the cationic intermediate. Hydrogen sulfate anion (HSO_4^-) is less nucleophilic than chloride.

Problem 21.18

Design a synthesis of *m*-bromophenol from benzene.

The reaction of an aryl diazonium salt with potassium iodide is the standard method for the preparation of *aryl iodides.* The diazonium salt is prepared from a primary aromatic amine in the usual way, a solution of potassium iodide is then added, and the reaction mixture is brought to room temperature or heated to accelerate the reaction.

$$Ar{-}\overset{+}{N}{\equiv}N: \; + \; I^- \longrightarrow ArI \; + :N{\equiv}N:$$

Aryl diazonium ion | Iodide ion | Aryl iodide | Nitrogen

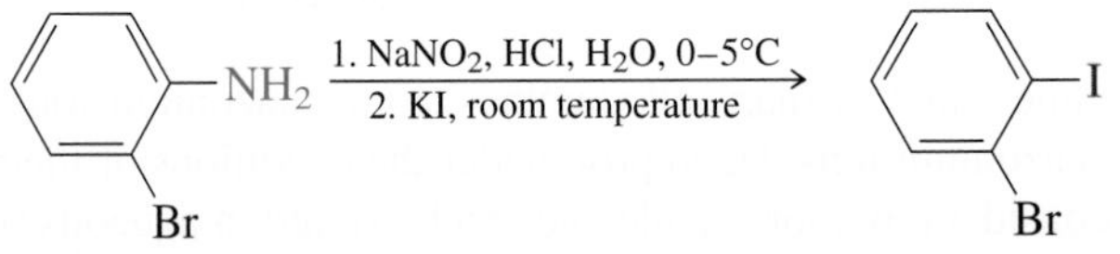
NH2
1. NaNO2, HCl, H2O, 0–5°C
2. KI, room temperature
I
Br
Br

o-Bromoaniline *o*-Bromoiodobenzene (72–83%)

Problem 21.19

Show how you could prepare *m*-bromoiodobenzene from benzene.

Diazonium salt chemistry provides the principal synthetic method for the preparation of *aryl fluorides* through a process known as the **Schiemann reaction.** In this procedure the aryl diazonium ion is isolated as its fluoroborate salt, which then yields the desired aryl fluoride on being heated.

$$Ar{-}\overset{+}{N}{\equiv}N: \overset{-}{B}F_4 \xrightarrow{\text{heat}} ArF \; + \; BF_3 \; + :N{\equiv}N:$$

Aryl diazonium fluoroborate | Aryl fluoride | Boron trifluoride | Nitrogen

A standard way to form the aryl diazonium fluoroborate salt is to add fluoroboric acid (HBF_4) or a fluoroborate salt to the diazotization medium.

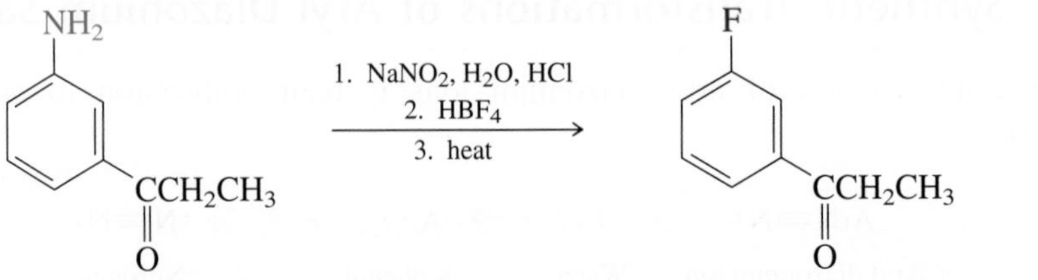
NH2
1. NaNO2, H2O, HCl
2. HBF4
3. heat
F
CCH2CH3
O
CCH2CH3
O

m-Aminophenyl ethyl ketone Ethyl *m*-fluorophenyl ketone (68%)

Problem 21.20

Show the proper sequence of synthetic transformations in the conversion of benzene to ethyl *m*-fluorophenyl ketone.

Although it is possible to prepare *aryl chlorides* and *aryl bromides* by electrophilic aromatic substitution, it is often necessary to prepare these compounds from an aromatic amine. The amine is converted to the corresponding diazonium salt and then treated with copper(I) chloride or copper(I) bromide as appropriate.

$$Ar{-}\overset{+}{N}{\equiv}N{:} \xrightarrow{CuX} ArX + {:}N{\equiv}N{:}$$

Aryl diazonium ion — Aryl chloride or bromide — Nitrogen

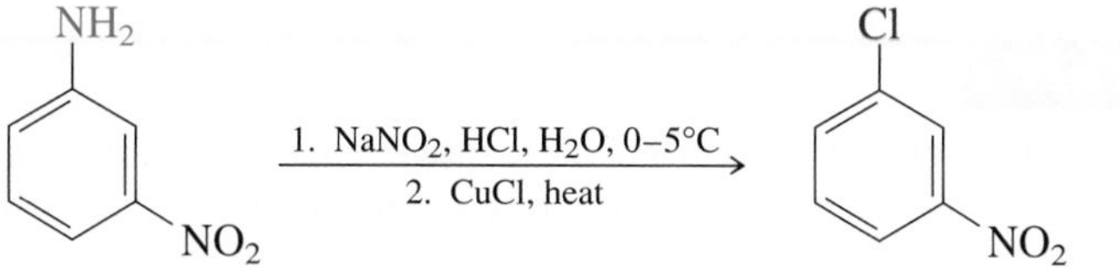

m-Nitroaniline → *m*-Chloronitrobenzene (68–71%)

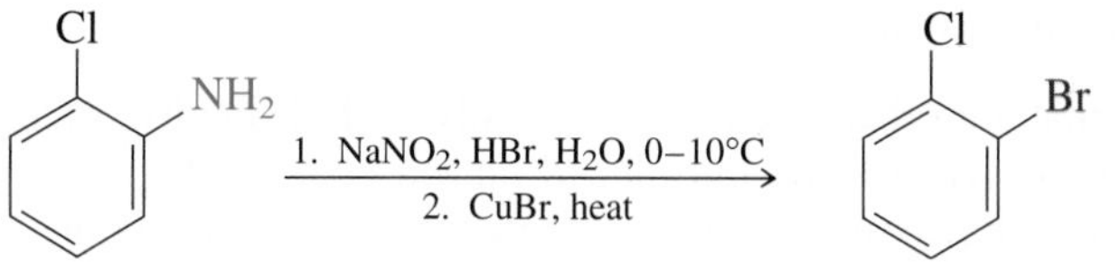

o-Chloroaniline → *o*-Bromochlorobenzene (89–95%)

Reactions that use copper(I) salts to replace nitrogen in diazonium salts are called **Sandmeyer reactions.** The Sandmeyer reaction using copper(I) cyanide is a good method for the preparation of aromatic *nitriles:*

$$Ar{-}\overset{+}{N}{\equiv}N{:} \xrightarrow{CuCN} ArCN + {:}N{\equiv}N{:}$$

Aryl diazonium ion — Aryl nitrile — Nitrogen

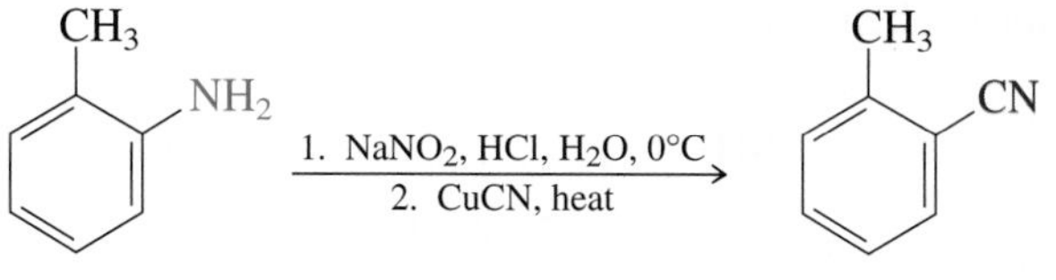

o-Toluidine → *o*-Methylbenzonitrile (64–70%)

Because cyano groups may be hydrolyzed to carboxylic acids (Section 19.18), the Sandmeyer preparation of aryl nitriles is a key step in the conversion of arylamines to substituted benzoic acids. In the example just cited, the *o*-methylbenzonitrile that was formed was subsequently subjected to acid-catalyzed hydrolysis to give *o*-methylbenzoic acid in 80–89% yield.

The preparation of aryl chlorides, bromides, and nitriles by the Sandmeyer reaction is mechanistically complicated and may involve arylcopper intermediates.

It is possible to replace amino groups on an aromatic ring by hydrogen by reducing a diazonium salt with hypophosphorous acid (H_3PO_2) or with ethanol. These reductions are free-radical reactions in which ethanol or hypophosphorous acid acts as a hydrogen atom donor:

$$Ar{-}\overset{+}{N}{\equiv}N{:} \xrightarrow[CH_3CH_2OH]{H_3PO_2\text{ or}} ArH + {:}N{\equiv}N{:}$$

Aryl diazonium ion — Arene — Nitrogen

Reactions of this type are called *reductive deaminations.*

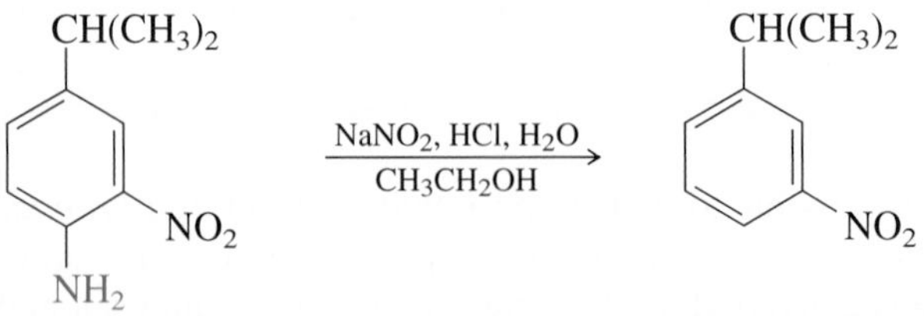
CH(CH3)2
NO2
NH2
NaNO2, HCl, H2O
CH3CH2OH
CH(CH3)2
NO2

4-Isopropyl-2-nitroaniline

m-Isopropylnitrobenzene (59%)

Sodium borohydride has also been used to reduce aryl diazonium salts in reductive deamination reactions.

Problem 21.21

Cumene (isopropylbenzene) is a relatively inexpensive commercially available starting material. Show how you could prepare *m*-isopropylnitrobenzene from cumene.

The value of diazonium salts in synthetic organic chemistry rests on two main points. Through the use of diazonium salt chemistry:

1. Substituents that are otherwise accessible only with difficulty, such as fluoro, iodo, cyano, and hydroxyl, may be introduced onto a benzene ring.
2. Compounds that have substitution patterns not directly available by electrophilic aromatic substitution can be prepared.

The first of these two features is readily apparent and is illustrated by Problems 21.18 to 21.20. If you have not done these problems yet, you are strongly encouraged to attempt them now.

The second point is somewhat less obvious but is illustrated by the synthesis of 1,3,5-tribromobenzene. This particular substitution pattern cannot be obtained by direct bromination of benzene because bromine is an ortho, para director. Instead, advantage is taken of the powerful activating and ortho, para-directing effects of the amino group in aniline. Bromination of aniline yields 2,4,6-tribromoaniline in quantitative yield. Diazotization of the resulting 2,4,6-tribromoaniline and reduction of the diazonium salt gives the desired 1,3,5-tribromobenzene.

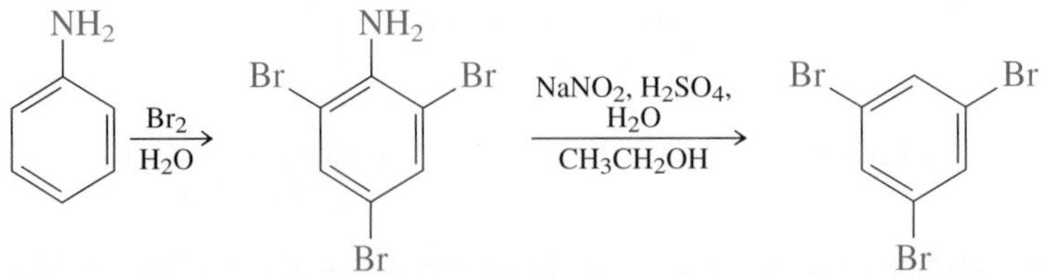
NH2
Br2
H2O
NH2
Br
Br
Br
NaNO2, H2SO4,
H2O
CH3CH2OH
Br
Br
Br

Aniline 2,4,6-Tribromoaniline (100%) 1,3,5-Tribromobenzene (74–77%)

To exploit the synthetic versatility of aryl diazonium salts, be prepared to reason backward. When you see a fluorine attached to a benzene ring, for example, realize that it probably will have to be introduced by a Schiemann reaction of an arylamine; realize that the required arylamine is derived from a nitroarene, and that the nitro group is introduced by nitration. Be aware that an unsubstituted position of a benzene ring need not have always been that way. It might once have borne an amino group that was used to control the orientation of electrophilic aromatic substitution reactions before being removed by reductive deamination. The strategy of synthesis is intellectually demanding, and a considerable sharpening of your reasoning power can be gained by attacking the synthesis problems at the end of each chapter. Remember, plan your sequence of accessible intermediates by reasoning backward from the target; then fill in the details on how each transformation is to be carried out.

21.18 Azo Coupling

A reaction of aryl diazonium salts that does not involve loss of nitrogen takes place when they react with phenols and arylamines. Aryl diazonium ions are relatively weak electrophiles but have sufficient reactivity to attack strongly activated aromatic rings. The reaction is known as *azo coupling;* two aryl groups are joined together by an azo (—N=N—) function.

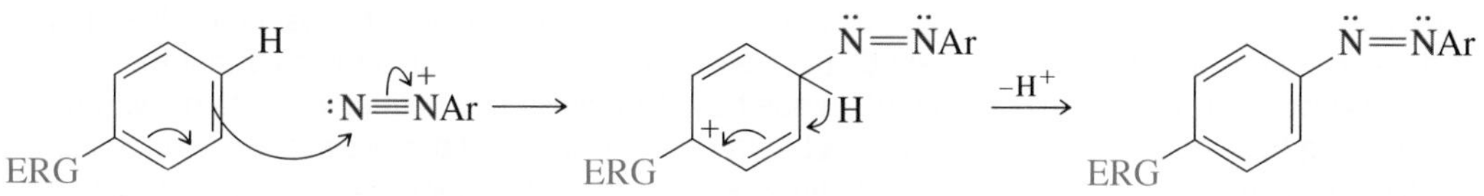

(ERG is a powerful electron-releasing group such as —OH or —NR_2) | Aryl diazonium ion | Intermediate in electrophilic aromatic substitution | Azo compound

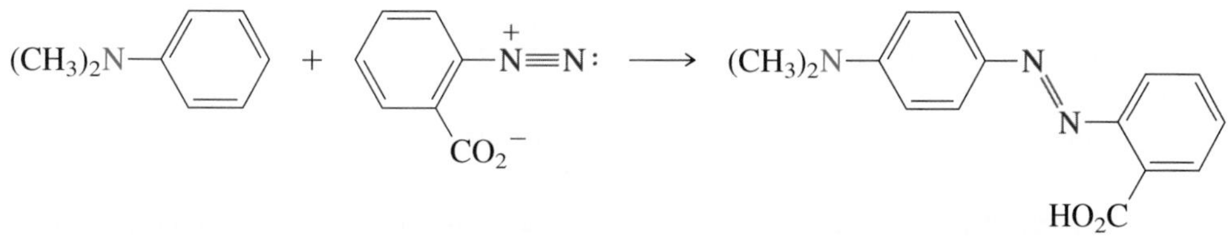

N,N-Dimethylaniline | Diazonium ion from *o*-aminobenzoic acid | Methyl red (62–66%)

The product of this reaction, as with many azo couplings, is highly colored. It is called *methyl red* and was a familiar acid–base indicator before the days of pH meters. It is red in solutions of pH 4 and below, yellow above pH 6.

Soon after azo coupling was discovered in the mid-nineteenth century, the reaction received major attention as a method for preparing dyes. Azo dyes first became commercially available in the 1870s and remain widely used, with more than 50% of the synthetic dye market. Chrysoidine, an azo dye for silk, cotton, and wool, first came on the market in 1876 and remains in use today.

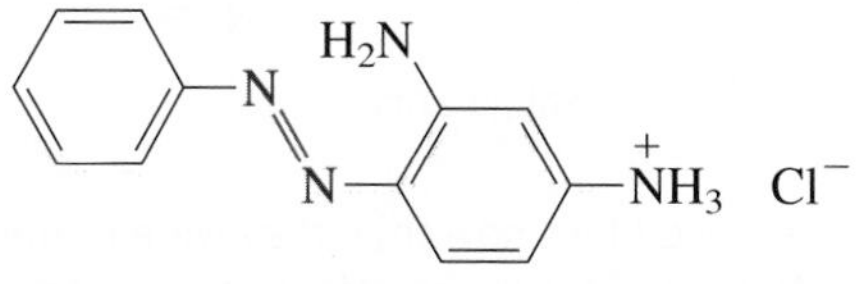

Chrysoidine

Problem 21.22

What amine and what diazonium salt would you use to prepare chrysoidine?

Dyes are regulated in the United States by the Food and Drug Administration (FDA). Over the years FDA has removed a number of dyes formerly approved for use in food and cosmetics because of concerns about toxicity, cancer-causing potential, or because they are skin irritants. Naturally occurring pigments, too

From Dyes to Sulfa Drugs

The medicine cabinet was virtually bare of antibacterial agents until *sulfa drugs* burst on the scene in the 1930s. Before sulfa drugs became available, bacterial infection might transform a small cut or puncture wound to a life-threatening event. The story of how sulfa drugs were developed is an interesting example of being right for the wrong reasons. It was known that many bacteria absorbed dyes, and staining was a standard method for making bacteria more visible under the microscope. Might there not be some dye that is both absorbed by bacteria and toxic to them? Acting on this hypothesis, scientists at the German dyestuff manufacturer I. G. Farbenindustrie undertook a program to test the thousands of compounds in their collection for their antibacterial properties.

In general, in vitro testing of drugs precedes in vivo testing. The two terms mean, respectively, "in glass" and "in life." In vitro testing of antibiotics is carried out using bacterial cultures in test tubes or Petri dishes. Drugs that are found to be active in vitro progress to the stage of in vivo testing. In vivo testing is carried out in living organisms: laboratory animals or human volunteers. The I. G. Farben scientists found that some dyes did possess antibacterial properties, both in vitro and in vivo. Others were active in vitro but were converted to inactive substances in vivo and therefore of no use as drugs. Unexpectedly, an azo dye called *Prontosil* was inactive in vitro but active in vivo.

In 1932, a member of the I. G. Farben research group, Gerhard Domagk used Prontosil to treat a young child suffering from a serious, potentially fatal staphylococcal infection. According to many accounts, the child was Domagk's own daughter; her infection was cured and her recovery was rapid and complete. Systematic testing followed and Domagk was awarded the 1939 Nobel Prize in Medicine or Physiology.

In spite of the rationale on which the testing of dyestuffs as antibiotics rested, subsequent research revealed that the antibacterial properties of Prontosil had nothing at all to do with its being a dye! In the body, Prontosil undergoes a reductive cleavage of its azo linkage to form *sulfanilamide,* which is the substance actually responsible for the observed biological activity. This is why Prontosil is active in vivo, but not in vitro.

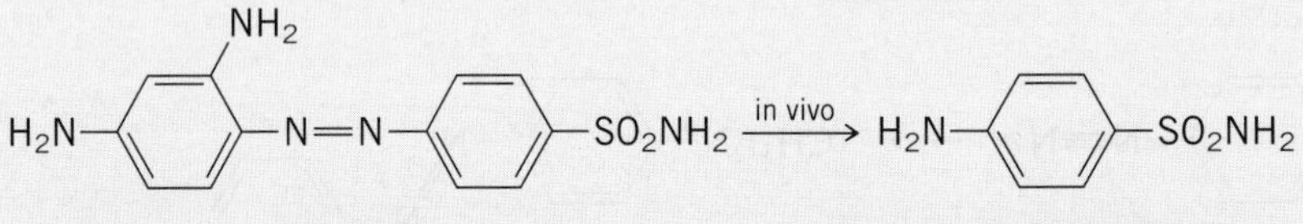

Prontosil Sulfanilamide

Bacteria require *p*-aminobenzoic acid to biosynthesize *folic acid,* a growth factor. Structurally, sulfanilamide resembles *p*-aminobenzoic acid and is mistaken for it by the bacteria. Folic acid biosynthesis is inhibited and bacterial growth is slowed sufficiently to allow the body's natural defenses to effect a cure. Because animals do not biosynthesize folic acid but obtain it in their food, sulfanilamide halts the growth of bacteria without harm to the host.

Identification of the mechanism by which Prontosil combats bacterial infections was an early triumph of *pharmacology,* a branch of science at the interface of physiology and biochemistry that studies the mechanism of drug action. By recognizing that sulfanilamide was the active agent, the task of preparing structurally modified analogs with potentially superior properties was considerably simplified. Instead of preparing Prontosil analogs, chemists synthesized sulfanilamide analogs. They did this with a vengeance; over 5000 compounds related to sulfanilamide were prepared during the period 1935–1946. Two of the most widely used sulfa drugs are *sulfathiazole* and *sulfadiazine.*

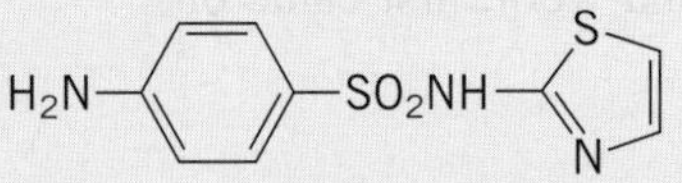

Sulfathiazole

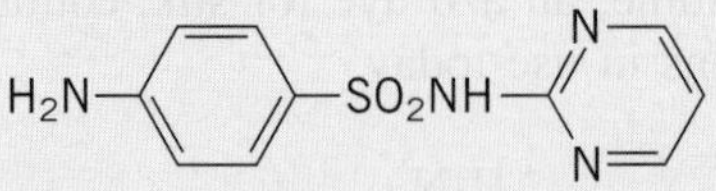

Sulfadiazine

We tend to take the efficacy of modern drugs for granted. One comparison with the not-too-distant past might put this view into better perspective. Once sulfa drugs were introduced in the United States, the number of pneumonia deaths alone decreased by an estimated 25,000 per year. The sulfa drugs are used less now than they were in the mid-twentieth century. Not only are more-effective, less-toxic antibiotics available, such as the penicillins and tetracyclines, but many bacteria that were once susceptible to sulfa drugs have become resistant.

numerous to count (saffron, turmeric, fruit colors, for example), are exempt from the approval process.

Of the seven synthetic dyes presently approved for food use, the three shown in Figure 21.7 are azo dyes. Red dye #40, which provides the red color to cherry-flavored foods, is the most popular. Not only is red dye #40 used to color foods, but you may

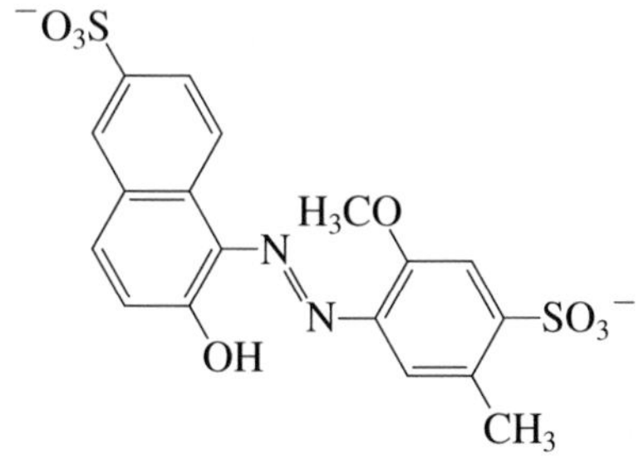

λ_{max} 507 nm

Red Dye #40

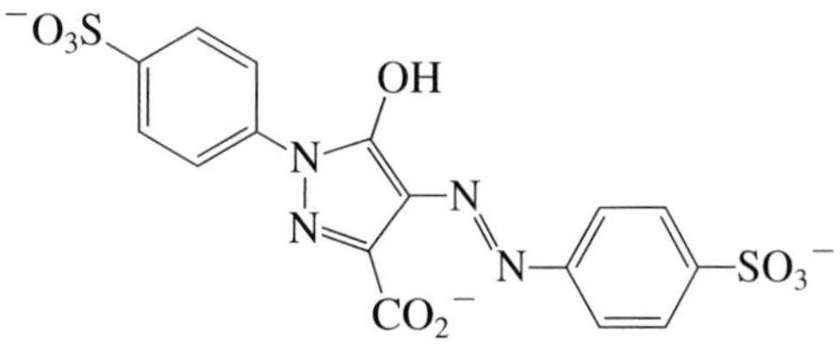

λ_{max} 426 nm

Yellow #5

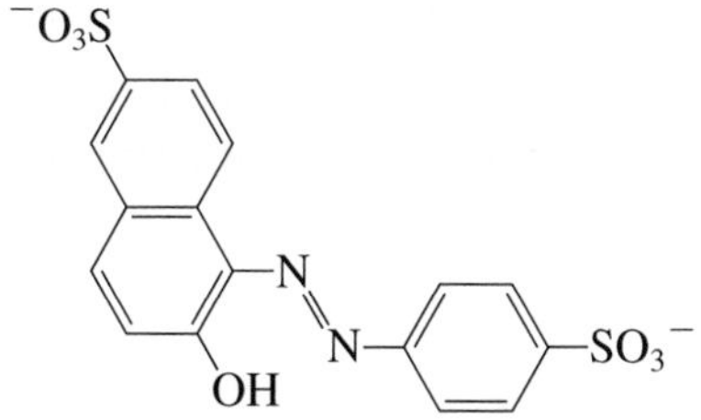

λ_{max} 480 nm

Yellow #6

Figure 21.7

Of the seven dyes approved for coloring foods, these three are azo dyes. All are sold as their sodium salts.

have noticed that almost every over-the-counter cold medicine is a red liquid or comes in a red capsule. The color is red dye #40 and is there by custom more than necessity. Yellow #5 is a lemon color; yellow #6 is orange. The highly conjugated azo linkage and combination of electron-donating and electron-attracting groups are responsible for the intense absorption of visible light by these molecules. Substituents affect the wavelengths absorbed and ultimately the color. Red #40, yellow #5, and yellow #6 all are sodium salts of sulfonic acids, which confers on them the water solubility they need to be effective food colors.

The official names of these dyes are FD&C Red No. 40, FD&C Yellow No. 5, and FD&C Yellow No. 6, where FD&C stands for "Food, Drug, and Cosmetic," which is both the name of the law under which these dyes are regulated and the purposes for which they are approved.

21.19 Spectroscopic Analysis of Amines

Infrared: The absorptions of interest in the IR spectra of amines are those associated with N—H vibrations. Primary alkyl- and arylamines exhibit two peaks in the range 3000–3500 cm^{-1}, which are due to symmetric and antisymmetric N—H stretching modes.

Symmetric N—H stretching of a primary amine

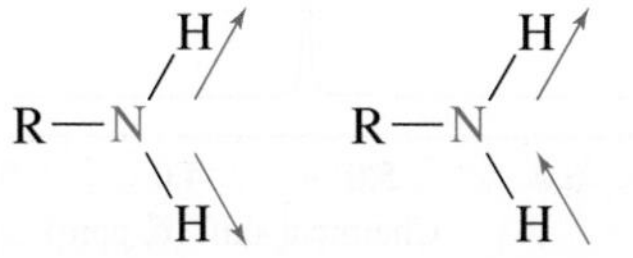

Antisymmetric N—H stretching of a primary amine

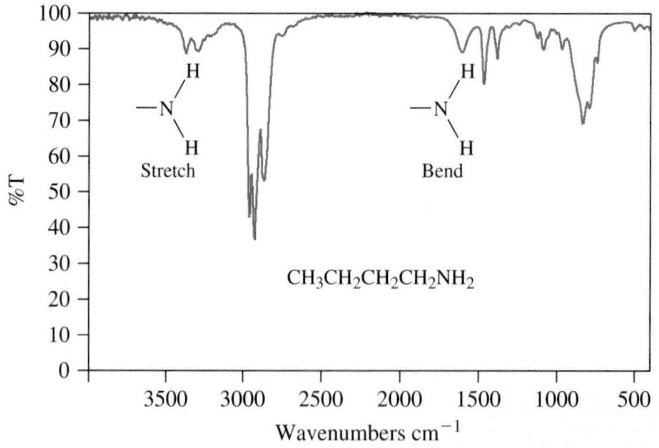

Figure 21.8

The infrared spectrum of butylamine has peaks for N—H stretching at 3290 and 3370 cm^{-1}. One corresponds to a symmetrical stretch of the two N—H bonds, the other to an antisymmetrical stretch. The peak at 1600 cm^{-1} is for NH_2 bending (scissoring).

These two vibrations are clearly visible at 3290 and 3370 cm^{-1} in the IR spectrum of butylamine, shown in Figure 21.8. Secondary amines such as diethylamine, exhibit only one peak due to N—H stretching. Tertiary amines, of course, are transparent in this region because they have no N—H bonds.

^{1}H NMR: Characteristics of the nuclear magnetic resonance spectra of amines may be illustrated by comparing 4-methylbenzylamine (Figure 21.9*a*) with 4-methylbenzyl

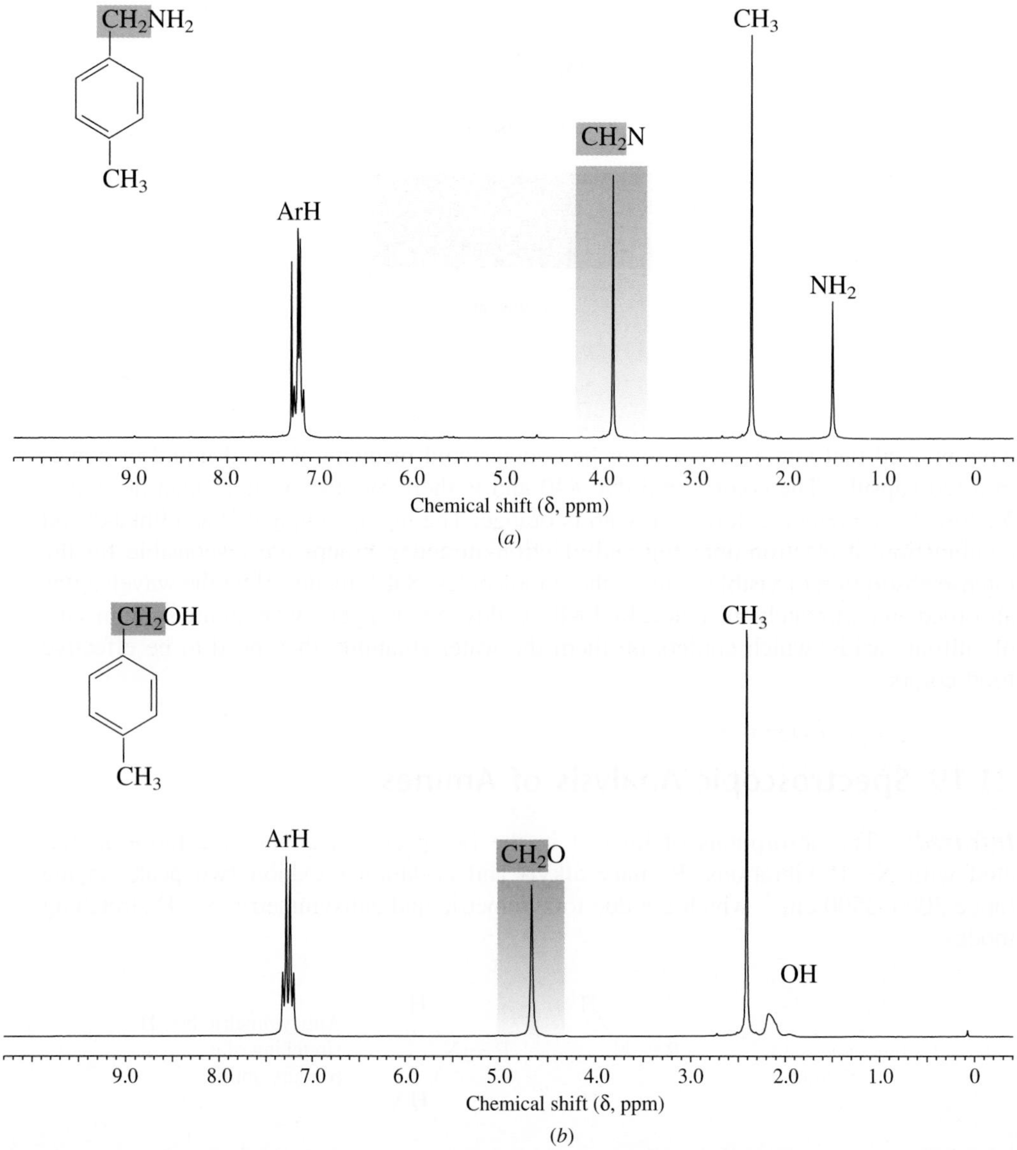

Figure 21.9

The 200-MHz ^{1}H NMR spectra of (*a*) 4-methylbenzylamine and (*b*) 4-methylbenzyl alcohol. The singlet corresponding to CH_2N in (*a*) is more shielded than that of CH_2O in (*b*).

alcohol (Figure 21.9*b*). Nitrogen is less electronegative than oxygen and so shields neighboring nuclei to a greater extent. The benzylic methylene group attached to nitrogen in 4-methylbenzylamine appears at higher field (δ 3.8) than the benzylic methylene of 4-methylbenzyl alcohol (δ 4.6). The N—H protons are somewhat more shielded than the O—H protons of an alcohol. In 4-methylbenzylamine the protons of the amino group correspond to the signal at δ 1.5, whereas the hydroxyl proton signal of 4-methylbenzyl alcohol is found at δ 2.1. The chemical shifts of amino group protons, like those of hydroxyl protons, are variable and are sensitive to solvent, concentration, and temperature.

^{13}C NMR: Similarly, carbons that are bonded to nitrogen are more shielded than those bonded to oxygen, as revealed by comparing the ^{13}C chemical shifts of methylamine and methanol.

δ 26.9 CH_3NH_2 — Methylamine δ 48.0 CH_3OH — Methanol

UV-VIS: In the absence of any other chromophore, the UV-VIS spectrum of an alkylamine is not very informative. The longest wavelength absorption involves promoting one of the unshared electrons of nitrogen to an antibonding σ* orbital (n → σ*) with a λ_{max} in the relatively inaccessible region near 200 nm. In arylamines the interaction of the nitrogen lone pair with the π-electron system of the ring shifts the ring's absorptions to longer wavelength. Tying up the lone pair by protonation causes the UV-VIS spectrum of anilinium ion to resemble benzene.

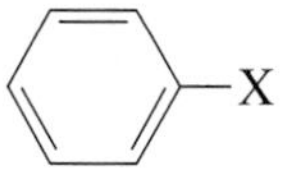
X

	X	λ_{max} (*nm*)
Benzene	H	204, 256
Aniline	NH_2	230, 280
Anilinium ion	NH_3^+	203, 254

Mass Spectrometry: A number of features make amines easily identifiable by mass spectrometry.

First, the peak for the molecular ion M^+ for all compounds that contain only carbon, hydrogen, and oxygen has an *m/z* value that is an even number. The presence of a nitrogen atom in the molecule requires that the *m/z* value for the molecular ion be odd. An odd number of nitrogens corresponds to an odd value of the molecular weight; an even number of nitrogens corresponds to an even molecular weight.

Recall the "nitrogen rule" from Section 13.25.

Second, nitrogen is exceptionally good at stabilizing adjacent carbocation sites. The fragmentation pattern seen in the mass spectra of amines is dominated by cleavage of groups from the carbon atom attached to the nitrogen, as the data for the following pair of constitutionally isomeric amines illustrate:

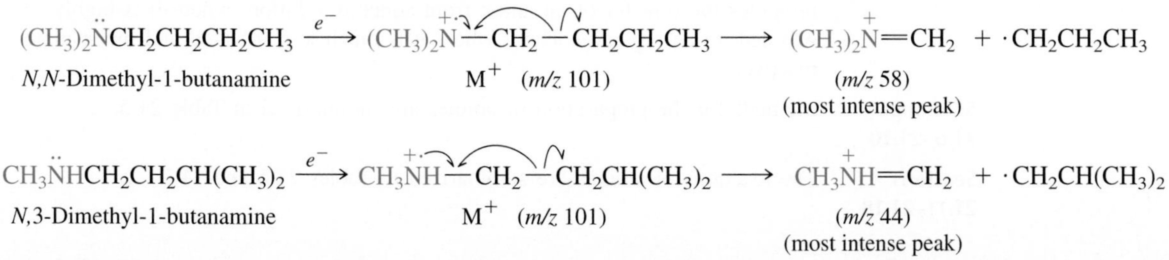
(CH3)2NCH2CH2CH2CH3
e−
(CH3)2N—CH2—CH2CH2CH3
(CH3)2N=CH2 + ·CH2CH2CH3
N,N-Dimethyl-1-butanamine
M+ (m/z 101)
(m/z 58)
(most intense peak)
CH3NHCH2CH2CH(CH3)2
e−
CH3NH—CH2—CH2CH(CH3)2
CH3NH=CH2 + ·CH2CH(CH3)2
N,3-Dimethyl-1-butanamine
M+ (m/z 101)
(m/z 44)
(most intense peak)

21.20 SUMMARY

Section 21.1 Alkylamines are compounds of the type shown, where R, R′, and R″ are alkyl groups. One or more of these groups is an aryl group in arylamines.

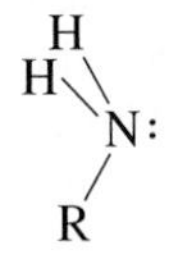

Primary amine

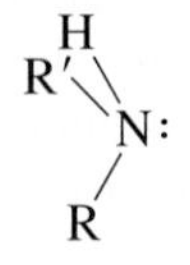

Secondary amine

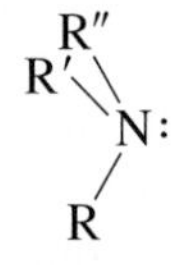

Tertiary amine

Alkylamines are named in two ways. One method adds the ending *-amine* to the name of the alkyl group. The other applies the principles of substitutive nomenclature by replacing the *-e* ending of an alkane name by *-amine* and uses appropriate locants to identify the position of the amino group. Arylamines are named as derivatives of aniline.

Section 21.2 Nitrogen's unshared electron pair is of major importance in understanding the structure and properties of amines. Alkylamines have a pyramidal arrangement of bonds to nitrogen, with an unshared electron pair in an sp^3-hybridized orbital. The geometry at nitrogen in arylamines is somewhat flatter, and the unshared electron pair is delocalized into the π system of the ring. Delocalization binds the electron pair more strongly in arylamines than in alkylamines. Arylamines are less basic and less nucleophilic than alkylamines.

Section 21.3 Amines are less polar than alcohols. Hydrogen bonding in amines is weaker than in alcohols because nitrogen is less electronegative than oxygen. Amines have lower boiling points than alcohols, but higher boiling points than alkanes. Primary amines have higher boiling points than isomeric secondary amines; tertiary amines, which cannot form intermolecular hydrogen bonds, have the lowest boiling points. Amines resemble alcohols in their solubility in water.

Section 21.4 The basicity of amines is conveniently expressed in terms of the pK_a of their conjugate acids.

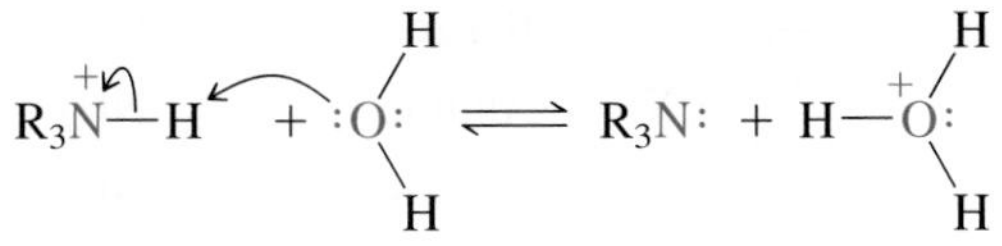

Conjugate acid of amine — Amine

The stronger base is associated with the weaker conjugate acid. The greater the pK_a of the conjugate acid, the stronger the base. The pK_a's of the conjugate acids of alkylamines lie in the 9–11 range. Arylamines are much weaker bases than alkylamines. The pK_a's of the conjugate acids of arylamines are usually 3–5. Strong electron-withdrawing groups can weaken the basicity of arylamines even more.

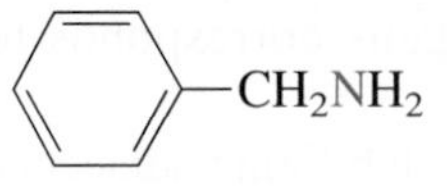

Benzylamine
(alkylamine: pK_a of conjugate acid = 9.3)

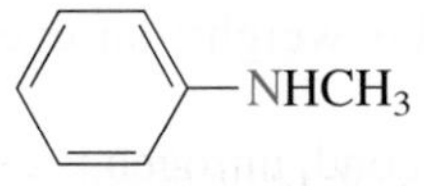

N-Methylaniline
(arylamine: pK_a of conjugate acid = 4.8)

Section 21.5 Quaternary ammonium salts, compounds of the type $R_4N^+ X^-$, find application as **phase-transfer catalysts.** A small amount of a quaternary ammonium salt promotes the transfer of an anion from aqueous solution, where it is highly solvated, to an organic solvent, where it is much less solvated and much more reactive.

Sections 21.6–21.10 Methods for the preparation of amines are summarized in Table 21.5.

Sections 21.11–21.18 The reactions of amines are summarized in Tables 21.6 and 21.7.

TABLE 21.5 Preparation of Amines

Reaction (section) and comments	General equation and specific example
Alkylation methods	
Alkylation of ammonia (Section 21.7) Ammonia can act as a nucleophile toward primary and some secondary alkyl halides to give primary alkylamines. Yields tend to be modest because the primary amine is itself a nucleophile and undergoes alkylation. Alkylation of ammonia can lead to a mixture containing a primary amine, a secondary amine, a tertiary amine, and a quaternary ammonium salt.	$RX + 2NH_3 \longrightarrow RNH_2 + NH_4X$ Alkyl halide, Ammonia, Alkylamine, Ammonium halide $C_6H_5CH_2Cl \xrightarrow{NH_3\ (8\ mol)} C_6H_5CH_2NH_2 + (C_6H_5CH_2)_2NH$ Benzyl chloride (1 mol), Benzylamine (53%), Dibenzylamine (39%)
Alkylation of phthalimide. The Gabriel synthesis (Section 21.8) The potassium salt of phthalimide reacts with alkyl halides to give *N*-alkylphthalimide derivatives. Hydrolysis or hydrazinolysis of this derivative yields a primary alkylamine.	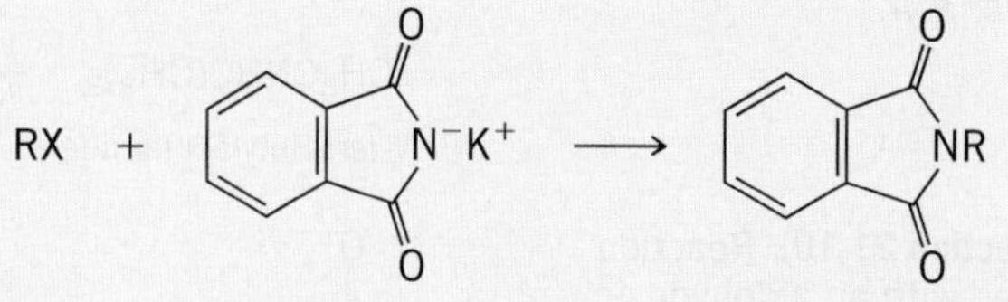 Alkyl halide, *N*-Potassiophthalimide, *N*-Alkylphthalimide 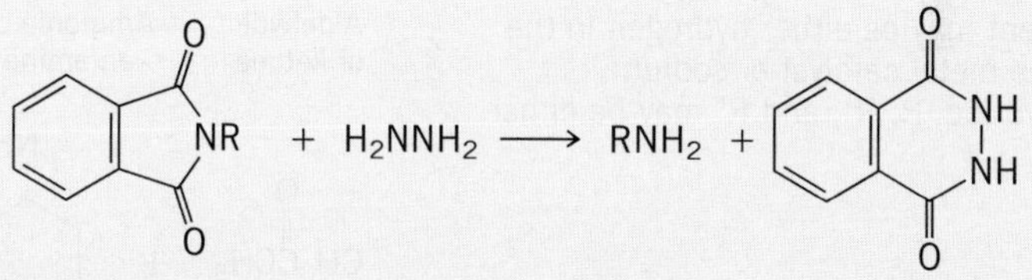 *N*-Alkylphthalimide, Hydrazine, Primary amine, Phthalhydrazide $CH_3CH{=}CHCH_2Cl \xrightarrow[2.\ H_2NNH_2,\ ethanol]{1.\ N\text{-potassiophthalimide, DMF}} CH_3CH{=}CHCH_2NH_2$ 1-Chloro-2-butene, 2-Buten-1-amine (95%)
Reduction methods	
Reduction of alkyl azides (Section 21.9) Alkyl azides, prepared by nucleophilic substitution by azide ion in primary or secondary alkyl halides, are reduced to primary alkylamines by lithium aluminum hydride or by catalytic hydrogenation.	$R\ddot{N}{=}\overset{+}{N}{=}\ddot{N}:^{-} \xrightarrow{reduce} R\ddot{N}H_2$ Alkyl azide, Primary amine $CF_3CH_2CH(N_3)CO_2CH_2CH_3 \xrightarrow{H_2,\ Pd} CF_3CH_2CH(NH_2)CO_2CH_2CH_3$ Ethyl 2-azido-4,4,4-trifluorobutanoate, Ethyl 2-amino-4,4,4-trifluorobutanoate (96%)
Reduction of nitriles (Section 21.9) Nitriles are reduced to primary amines by lithium aluminum hydride or by catalytic hydrogenation.	$RC{\equiv}N \xrightarrow{reduce} RCH_2NH_2$ Nitrile, Primary amine 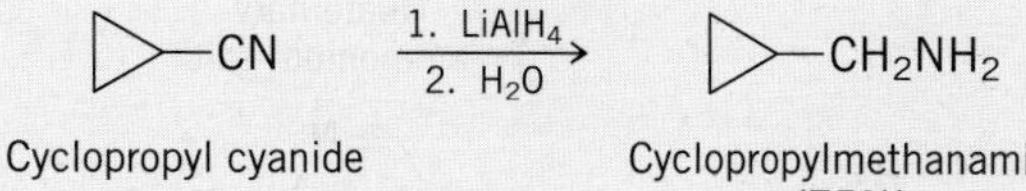 Cyclopropyl cyanide, Cyclopropylmethanamine (75%)

Continued

TABLE 21.5 Preparation of Amines (*Continued*)

Reaction (section) and comments	General equation and specific example
Reduction of aryl nitro compounds (Section 21.9) The standard method for the preparation of an arylamine is by nitration of an aromatic ring, followed by reduction of the nitro group. Typical reducing agents include iron or tin in hydrochloric acid or catalytic hydrogenation.	$ArNO_2 \xrightarrow{\text{reduce}} ArNH_2$ Nitroarene Arylamine $C_6H_5NO_2 \xrightarrow[2.\ HO^-]{1.\ Fe,\ HCl} C_6H_5NH_2$ Nitrobenzene Aniline (97%)
Reduction of amides (Section 21.9) Lithium aluminum hydride reduces the carbonyl group of an amide to a methylene group. Primary, secondary, or tertiary amines may be prepared by proper choice of the starting amide. R and R′ may be either alkyl or aryl.	$R\overset{O}{\overset{\Vert}{C}}NR'_2 \xrightarrow{\text{reduce}} RCH_2NR'_2$ Amide Amine $CH_3\overset{O}{\overset{\Vert}{C}}NHC(CH_3)_3 \xrightarrow[2.\ H_2O]{1.\ LiAlH_4} CH_3CH_2NHC(CH_3)_3$ *N-tert*-Butylacetamide *N*-Ethyl-*tert*-butylamine (60%)
Reductive amination (Section 21.10) Reaction of ammonia or an amine with an aldehyde or a ketone in the presence of a reducing agent is an effective method for the preparation of primary, secondary, or tertiary amines. The reducing agent may be either hydrogen in the presence of a metal catalyst or sodium cyanoborohydride. R, R′, and R″ may be either alkyl or aryl.	$R\overset{O}{\overset{\Vert}{C}}R' + R''_2NH \xrightarrow{\text{reducing agent}} R\underset{H}{\underset{\vert}{\overset{NR''_2}{\overset{\vert}{C}}}}R'$ Aldehyde or ketone Ammonia or an amine Amine 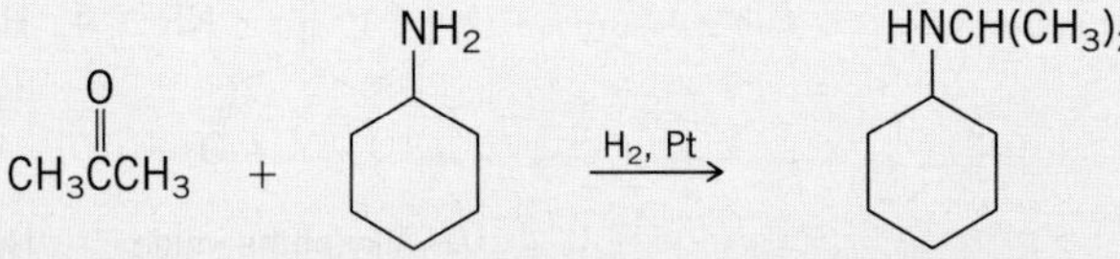 Acetone Cyclohexylamine *N*-Isopropylcyclohexylamine (79%)

TABLE 21.6 Reactions of Amines Discussed in This Chapter

Reaction (section) and comments	General equation and specific example
Alkylation (Section 21.12) Amines act as nucleophiles toward alkyl halides. Primary amines yield secondary amines, secondary amines yield tertiary amines, and tertiary amines yield quaternary ammonium salts.	$R\ddot{N}H_2 \xrightarrow{R'CH_2X} R\ddot{N}HCH_2R'$ Primary amine Secondary amine $\downarrow R'CH_2X$ $R\overset{+}{N}(CH_2R')_3\ X^- \xleftarrow{R'CH_2X} R\ddot{N}(CH_2R')_2$ Quaternary ammonium salt Tertiary amine 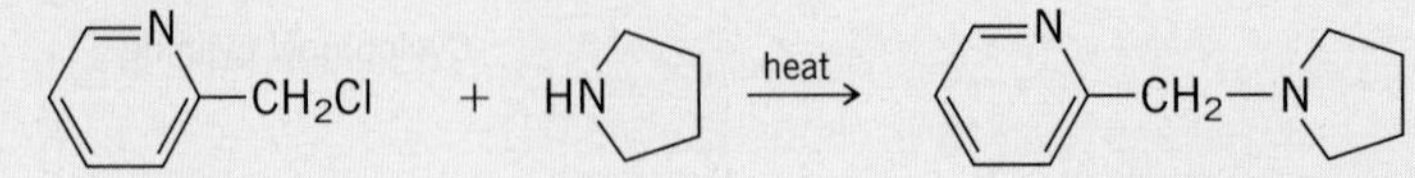 2-Chloromethylpyridine Pyrrolidine 2-(Pyrrolidinylmethyl)pyridine (93%)

Continued

TABLE 21.6 Reactions of Amines Discussed in This Chapter (*Continued*)

Reaction (section) and comments	General equation and specific example
Hofmann elimination (Section 21.13) Quaternary ammonium hydroxides undergo elimination on being heated. It is an anti elimination of the E2 type. The regioselectivity of the Hofmann elimination is opposite to that of the Zaitsev rule and leads to the less highly substituted alkene.	$RCH_2CHR'(\overset{+}{N}(CH_3)_3)\ HO^- \xrightarrow{\text{heat}} RCH{=}CHR' + :N(CH_3)_3 + H_2O$ Alkyltrimethylammonium hydroxide; Alkene; Trimethylamine; Water 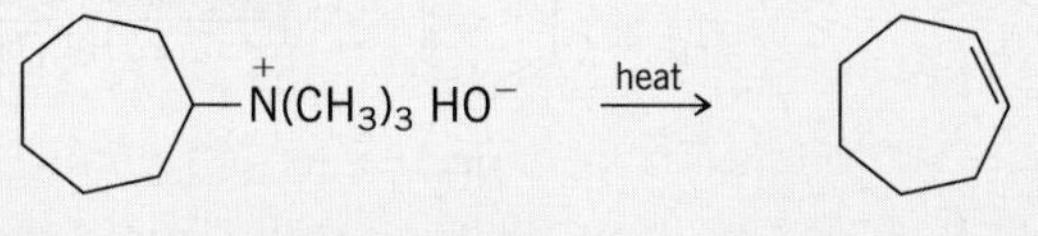 Cycloheptyltrimethylammonium hydroxide → Cycloheptene (87%)
Electrophilic aromatic substitution (Section 21.14) Arylamines are very reactive toward electrophilic aromatic substitution. It is customary to protect arylamines as their *N*-acyl derivatives before carrying out ring nitration, chlorination, bromination, sulfonation, or Friedel Crafts reactions.	$ArH + E^+ \longrightarrow ArE + H^+$ Arylamine; Electrophile; Product of electrophilic aromatic substitution; Proton 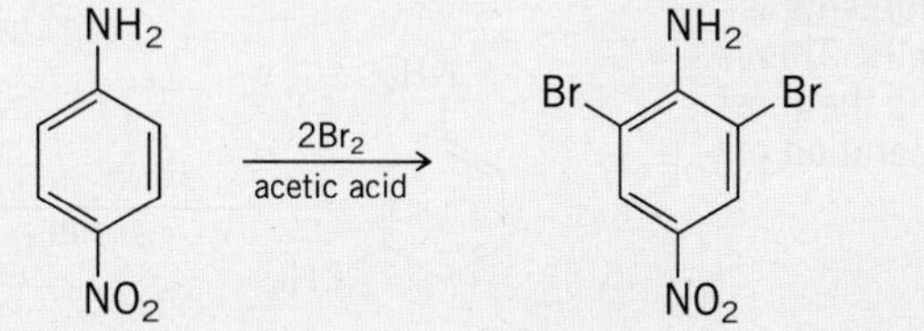 *p*-Nitroaniline → 2,6-Dibromo-4-nitroaniline (95%)
Nitrosation (Sections 21.15–16) Nitrosation of amines occurs when sodium nitrite is added to a solution containing an amine and an acid. *Primary amines* yield alkyl diazonium salts. Alkyl diazonium salts are very unstable and yield carbocation-derived products. Aryl diazonium salts are exceedingly useful synthetic intermediates. Their reactions are described in Table 21.7.	$RNH_2 \xrightarrow[H_3O^+]{NaNO_2} R\overset{+}{N}{\equiv}N:$ Primary amine; Diazonium ion 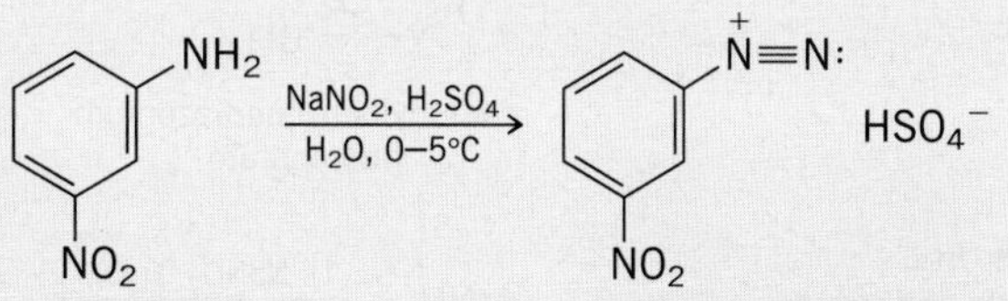 *m*-Nitroaniline → *m*-Nitrobenzenediazonium hydrogen sulfate
Secondary alkylamines and *secondary arylamines* yield *N*-nitroso amines.	$R_2NH \xrightarrow[H_3O^+]{NaNO_2} R_2N{-}N{=}O$ Secondary amine; *N*-Nitroso amine 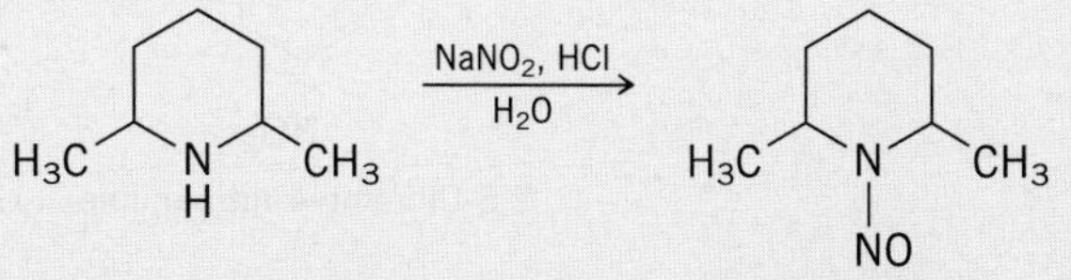 2,6-Dimethylpiperidine → 2,6-Dimethyl-*N*-nitrosopiperidine (72%)
Tertiary alkylamines illustrate no useful chemistry on nitrosation. *Tertiary arylamines* undergo nitrosation of the ring by electrophilic aromatic substitution.	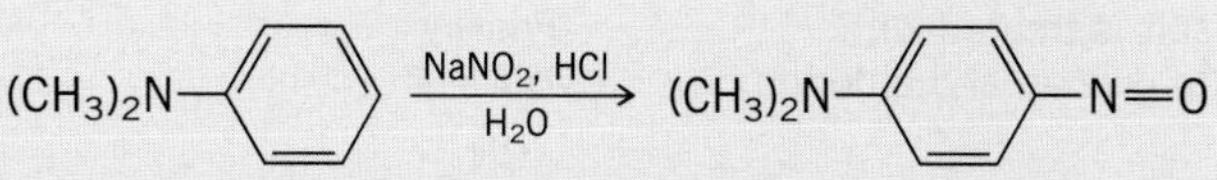 *N*,*N*-Dimethylaniline → *N*,*N*-Dimethyl-4-nitrosoaniline (80–89%)

TABLE 21.7 Synthetically Useful Transformations Involving Aryl Diazonium Ions (Section 21.17)

<table>
<tr><th>Reaction and comments</th><th>General equation and specific example</th></tr>
<tr><td>Preparation of phenols Heating its aqueous acidic solution converts a diazonium salt to a phenol. This is the most general method for the synthesis of phenols.</td><td>

$$\underset{\text{Primary arylamine}}{ArNH_2} \xrightarrow[\text{2. } H_2O\text{, heat}]{\text{1. } NaNO_2\text{, } H_2SO_4\text{, } H_2O} \underset{\text{Phenol}}{ArOH}$$

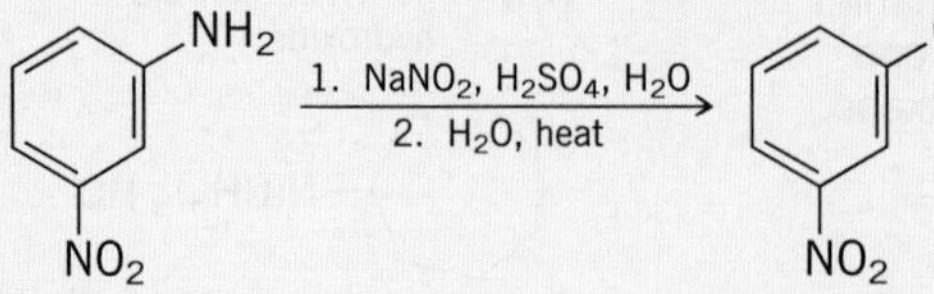

m-Nitroaniline — *m*-Nitrophenol (81–86%)
</td></tr>
<tr><td>Preparation of aryl fluorides Addition of fluoroboric acid to a solution of a diazonium salt causes the precipitation of an aryl diazonium fluoroborate. When the dry aryl diazonium fluoroborate is heated, an aryl fluoride results. This is the Schiemann reaction; it is the most general method for the preparation of aryl fluorides.</td><td>

$$\underset{\text{Primary arylamine}}{ArNH_2} \xrightarrow[\text{2. } HBF_4]{\text{1. } NaNO_2\text{, } H_3O^+} \underset{\text{Aryl diazonium fluoroborate}}{Ar\overset{+}{N}{\equiv}N:\ \overset{-}{B}F_4} \xrightarrow{\text{heat}} \underset{\text{Aryl fluoride}}{ArF}$$

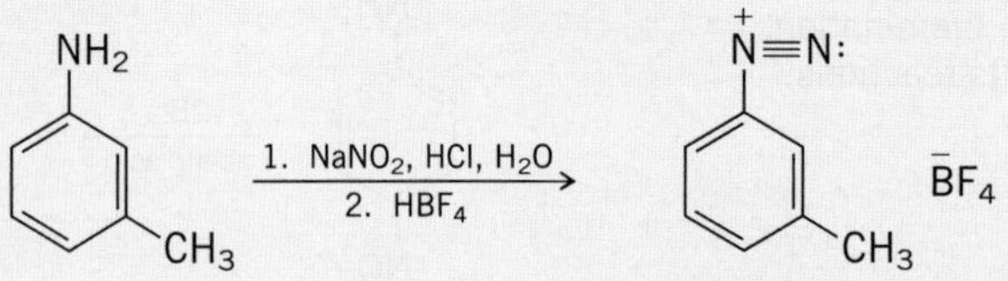

m-Toluidine — *m*-Methylbenzenediazonium fluoroborate (76–84%)

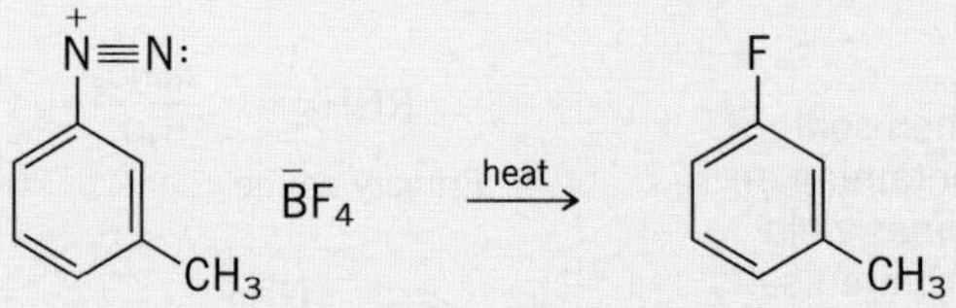

m-Methylbenzenediazonium fluoroborate — *m*-Fluorotoluene (89%)
</td></tr>
<tr><td>Preparation of aryl iodides Aryl diazonium salts react with sodium or potassium iodide to form aryl iodides. This is the most general method for the synthesis of aryl iodides.</td><td>

$$\underset{\text{Primary arylamine}}{ArNH_2} \xrightarrow[\text{2. NaI or KI}]{\text{1. } NaNO_2\text{, } H_3O^+} \underset{\text{Aryl iodide}}{ArI}$$

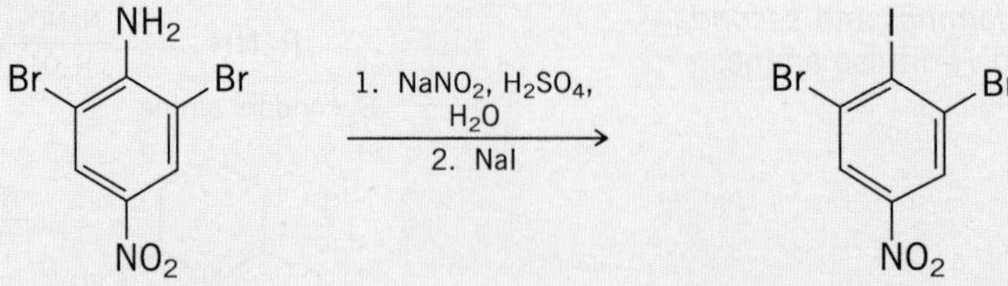

2,6-Dibromo-4-nitroaniline — 1,3-Dibromo-2-iodo-5-nitrobenzene (84–88%)
</td></tr>
<tr><td>Preparation of aryl chlorides In the Sandmeyer reaction, a solution containing an aryl diazonium salt is treated with copper(I) chloride to give an aryl chloride.</td><td>

$$\underset{\text{Primary arylamine}}{ArNH_2} \xrightarrow[\text{2. CuCl}]{\text{1. } NaNO_2\text{, HCl, } H_2O} \underset{\text{Aryl chloride}}{ArCl}$$

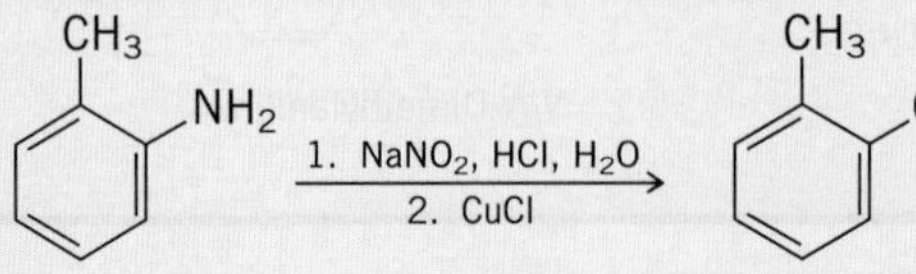

o-Toluidine — *o*-Chlorotoluene (74–79%)
</td></tr>
</table>

Continued

TABLE 21.7 Synthetically Useful Transformations Involving Aryl Diazonium Ions (Section 21.17) (*Continued*)

Reaction and comments	General equation and specific example
Preparation of aryl bromides The Sandmeyer reaction using copper(I) bromide converts primary arylamines to aryl bromides.	$ArNH_2 \xrightarrow[\text{2. CuBr}]{\text{1. NaNO}_2\text{, HBr, H}_2\text{O}} ArBr$ Primary arylamine — Aryl bromide 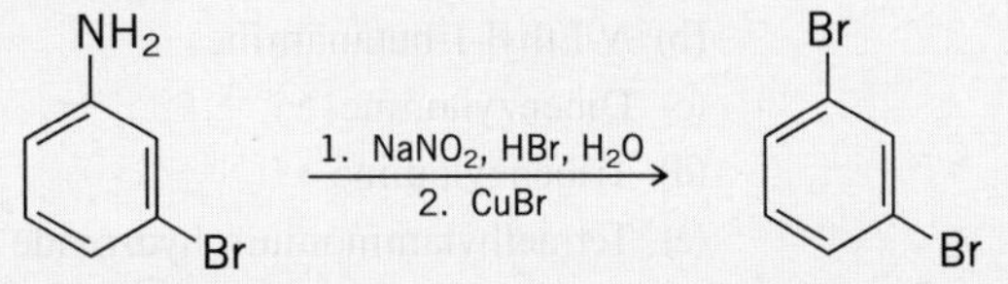 *m*-Bromoaniline — *m*-Dibromobenzene (80–87%)
Preparation of aryl nitriles Copper(I) cyanide converts aryl diazonium salts to aryl nitriles.	$ArNH_2 \xrightarrow[\text{2. CuCN}]{\text{1. NaNO}_2\text{, H}_2\text{O}} ArCN$ Primary arylamine — Aryl nitrile 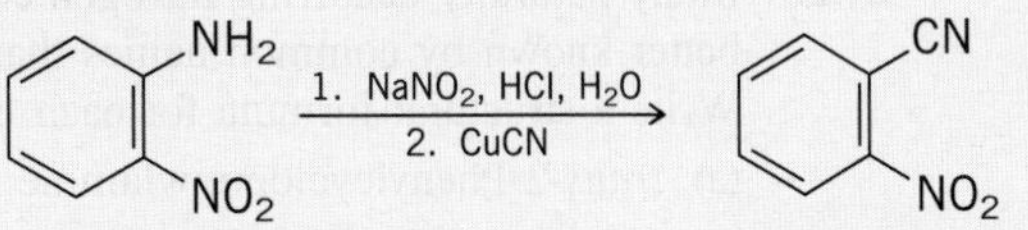 *o*-Nitroaniline — *o*-Nitrobenzonitrile (87%)
Reductive deamination of primary arylamines The amino group of an arylamine can be replaced by hydrogen by treatment of its diazonium salt with ethanol or with hypophosphorous acid.	$ArNH_2 \xrightarrow[\text{2. CH}_3\text{CH}_2\text{OH or H}_3\text{PO}_2]{\text{1. NaNO}_2\text{, H}_3\text{O}^+} ArH$ Primary arylamine — Arene 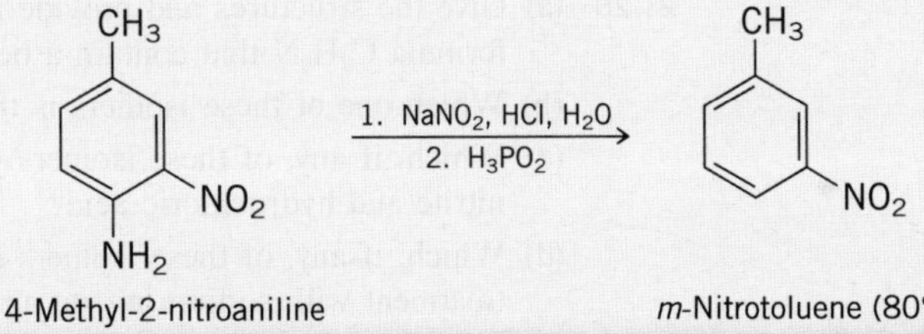 4-Methyl-2-nitroaniline — *m*-Nitrotoluene (80%)

Section 21.19 The N—H stretching frequency of primary and secondary amines appears in the infrared in the 3000–3500 cm^{-1} region. In the NMR spectra of amines, protons and carbons of the type H—C—N are more shielded than H—C—O.

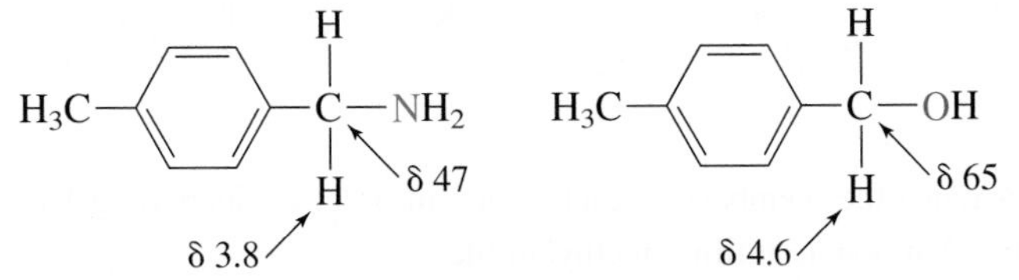

Amines have odd-numbered molecular weights, which helps identify them by mass spectrometry. Fragmentation tends to be controlled by the formation of a nitrogen-stabilized cation.

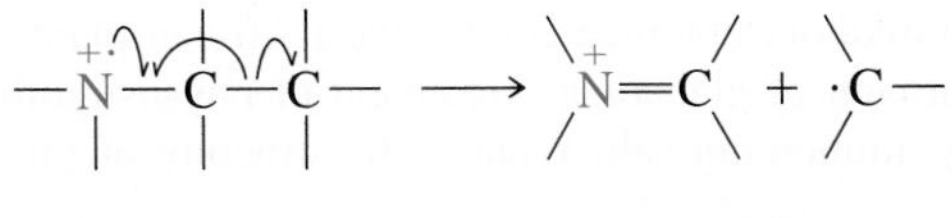

PROBLEMS

21.23 Write structural formulas for all the amines of molecular formula $C_4H_{11}N$. Give an acceptable name for each one, and classify it as a primary, secondary, or tertiary amine.

21.24 Provide a structural formula for each of the following compounds:

(a) 2-Ethyl-1-butanamine
(b) *N*-Ethyl-1-butanamine
(c) Dibenzylamine
(d) Tribenzylamine
(e) Tetraethylammonium hydroxide
(f) *N*-Allylcyclohexylamine
(g) *N*-Allylpiperidine
(h) Benzyl 2-aminopropanoate
(i) 4-(*N,N*-Dimethylamino)cyclohexanone
(j) 2,2-Dimethyl-1,3-propanediamine

21.25 Many naturally occurring nitrogen compounds and many nitrogen-containing drugs are better known by common names than by their systematic names. A few of these follow. Write a structural formula for each one.

(a) *trans*-2-Phenylcyclopropylamine, better known as *tranylcypromine:* an antidepressant drug
(b) *N*-Benzyl-*N*-methyl-2-propynylamine, better known as *pargyline:* a drug used to treat high blood pressure
(c) 1-Phenyl-2-propanamine, better known as *amphetamine:* a stimulant
(d) 1-(*m*-Hydroxyphenyl)-2-(methylamino)ethanol: better known as *phenylephrine:* a nasal decongestant

21.26 (a) Give the structures and provide an acceptable name for all the isomers of molecular formula C_7H_9N that contain a benzene ring.
(b) Which one of these isomers is the strongest base?
(c) Which, if any, of these isomers yield an *N*-nitroso amine on treatment with sodium nitrite and hydrochloric acid?
(d) Which, if any, of these isomers undergo nitrosation of their benzene ring on treatment with sodium nitrite and hydrochloric acid?

21.27 Arrange the following compounds or anions in each group in order of decreasing basicity:

(a) H_3C^-, H_2N^-, HO^-, F^-
(b) H_2O, NH_3, HO^-, H_2N^-
(c) HO^-, H_2N^-, $:\bar{C}{\equiv}N:$, NO_3^-
(d)

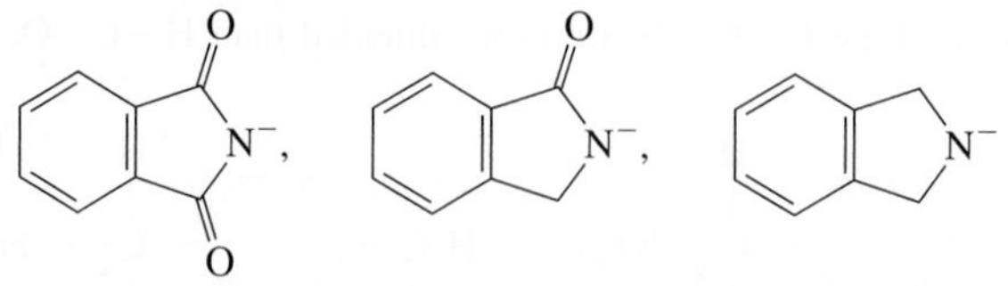

21.28 Arrange the members of each group in order of decreasing basicity:

(a) Ammonia, aniline, methylamine
(b) Acetanilide, aniline, *N*-methylaniline
(c) 2,4-Dichloroaniline, 2,4-dimethylaniline, 2,4-dinitroaniline
(d) 3,4-Dichloroaniline, 4-chloro-2-nitroaniline, 4-chloro-3-nitroaniline
(e) Dimethylamine, diphenylamine, *N*-methylaniline

21.29 *Physostigmine,* an alkaloid obtained from a West African plant (*Physotigma venenosum*), is used in the treatment of glaucoma. Treatment of physostigmine with methyl iodide gives a quaternary ammonium salt. What is the structure of this salt?

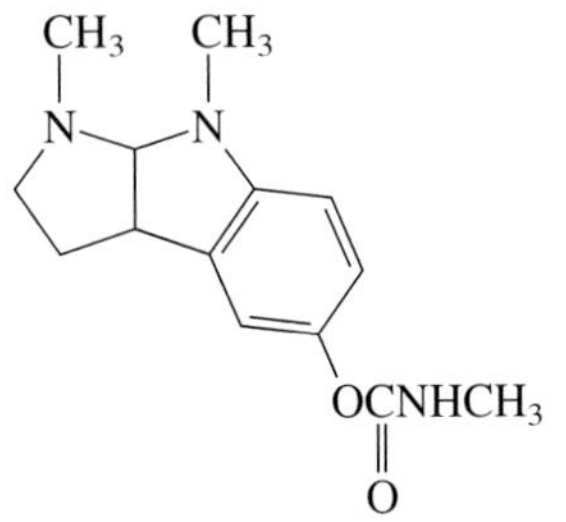

Physostigmine

21.30 Carnosine, found in muscle and brain tissue, acts as a buffer to neutralize small amounts of acid. The pK_a of the conjugate acid of carnosine is close to 7.0. What is its structure?

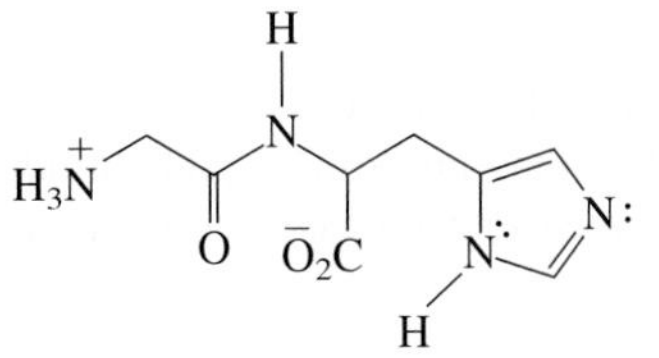

Carnosine

21.31 9-Aminofluorene has applications in the structural analysis of proteins and carbohydrates. Write a stepwise procedure with equations to show how to separate a mixture of 9-aminofluorene and fluorene in diethyl ether solution.

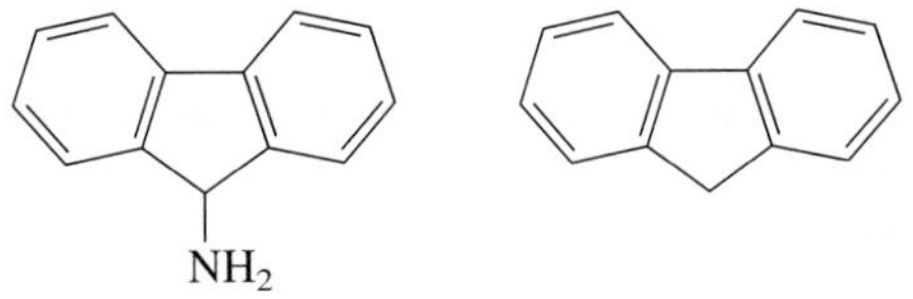

9-Aminofluorene Fluorene

21.32 Both alkyl- and arylamines have a low barrier for inversion at nitrogen, which prevents the separation of chiral amines into their enantiomers. The barrier for inversion at nitrogen in alkylamines is approximately 25 kJ/mol (6 kcal/mol), whereas for arylamines it is much lower, on the order of 6.3 kJ/mol (1.5 kcal/mol). Can you suggest a reason for the difference?

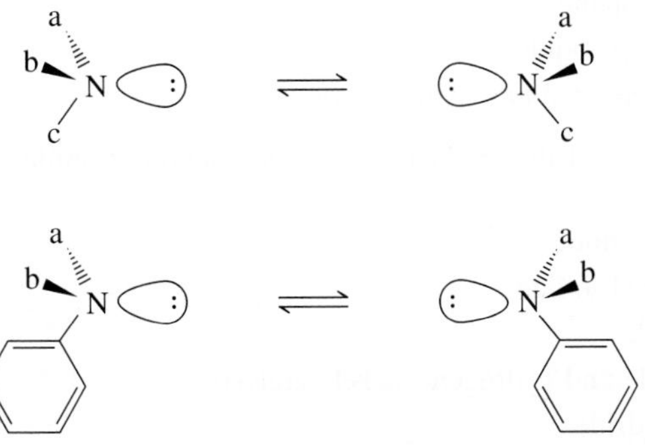

21.33 Describe procedures for preparing each of the following compounds, using ethanol as the source of all their carbon atoms. Once you prepare a compound, you need not repeat its synthesis in a subsequent part of this problem.

(a) Ethylamine
(b) *N*-Ethylacetamide
(c) Diethylamine
(d) *N,N*-Diethylacetamide
(e) Triethylamine
(f) Tetraethylammonium bromide

21.34 Show by writing the appropriate sequence of equations how you could carry out each of the following transformations:

(a) 1-Butanol to 1-pentanamine
(b) *tert*-Butyl chloride to 2,2-dimethyl-1-propanamine
(c) Cyclohexanol to *N*-methylcyclohexylamine
(d) Isopropyl alcohol to 1-amino-2-methyl-2-propanol
(e) Isopropyl alcohol to 1-amino-2-propanol
(f) Isopropyl alcohol to 1-(*N,N*-dimethylamino)-2-propanol
(g) 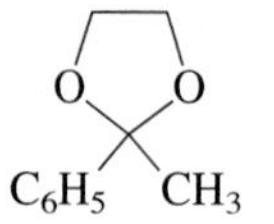

to 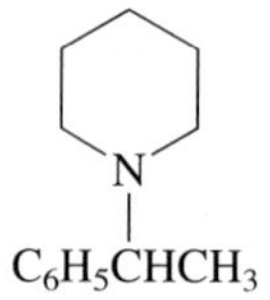

21.35 Each of the following dihaloalkanes gives an *N*-(haloalkyl)phthalimide on reaction with one equivalent of the potassium salt of phthalimide. Write the structure of the phthalimide derivative formed in each case and explain the basis for your answer.

(a) FCH_2CH_2Br
(b) 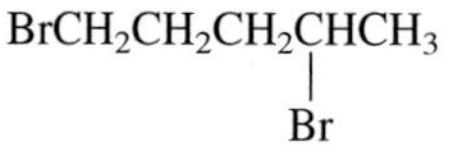

(c) 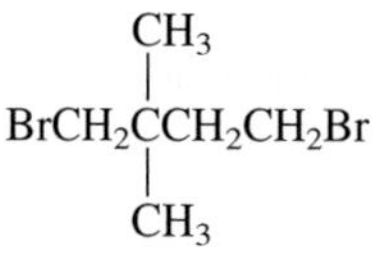

21.36 Give the structure of the expected product formed when benzylamine reacts with each of the following reagents:

(a) Hydrogen bromide
(b) Sulfuric acid
(c) Acetic acid
(d) Acetyl chloride
(e) Acetic anhydride
(f) Acetone
(g) Acetone and hydrogen (nickel catalyst)
(h) Ethylene oxide
(i) 1,2-Epoxypropane
(j) Excess methyl iodide
(k) Sodium nitrite in dilute hydrochloric acid

21.37 Write the structure of the product formed on reaction of aniline with each of the following:

(a) Hydrogen bromide
(b) Excess methyl iodide
(c) Acetaldehyde
(d) Acetaldehyde and hydrogen (nickel catalyst)
(e) Acetic anhydride
(f) Benzoyl chloride
(g) Sodium nitrite, aqueous sulfuric acid, 0–5°C

21.38 Write the structure of the product formed on reaction of acetanilide with each of the following:

(a) Lithium aluminum hydride, followed by water
(b) Nitric acid and sulfuric acid
(c) Sulfur trioxide and sulfuric acid
(d) Bromine in acetic acid
(e) *tert*-Butyl chloride, aluminum chloride

(f) Acetyl chloride, aluminum chloride
(g) 6 M hydrochloric acid, reflux
(h) Aqueous sodium hydroxide, reflux

21.39 Identify the principal organic products of each of the following reactions:

(a) Cyclohexanone + cyclohexylamine $\xrightarrow{H_2,\ Ni}$

(b) 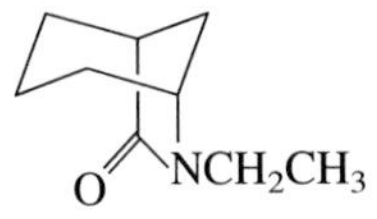 $\xrightarrow[\text{2. } H_2O]{\text{1. } LiAlH_4}$

(c) $C_6H_5CH_2CH_2CH_2OH \xrightarrow[\text{2. } (CH_3)_2NH \text{ (excess)}]{\text{1. } p\text{-toluenesulfonyl chloride, pyridine}}$

(d) $(CH_3)_2CHNH_2$ + 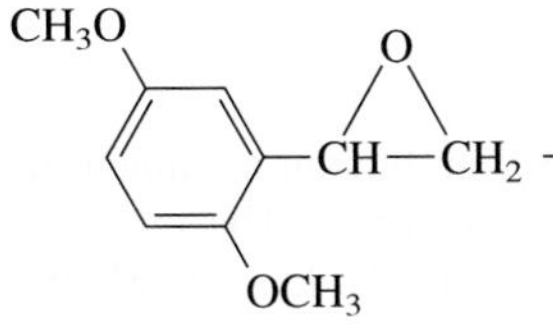 $\longrightarrow$

(e) $(C_6H_5CH_2)_2NH + CH_3\overset{O}{\overset{\|}{C}}CH_2Cl \xrightarrow[\text{THF}]{\text{triethylamine}}$

(f) 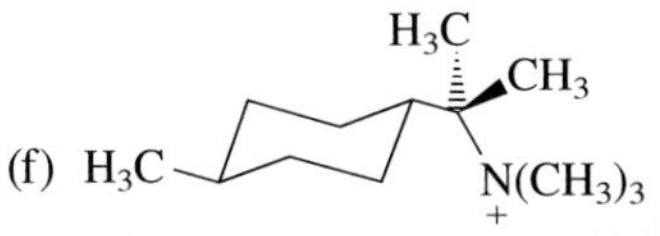 $HO^- \xrightarrow{\text{heat}}$

(g) $(CH_3)_2CHNHCH(CH_3)_2 \xrightarrow[HCl,\ H_2O]{NaNO_2}$

21.40 Each of the following reactions has been reported in the chemical literature and proceeds in good yield. Identify the principal organic product of each reaction.

(a) 1,2-Diethyl-4-nitrobenzene $\xrightarrow[\text{ethanol}]{H_2,\ Pt}$

(b) 1,3-Dimethyl-2-nitrobenzene $\xrightarrow[\text{2. } HO^-]{\text{1. } SnCl_2,\ HCl}$

(c) Product of part (b) + $ClCH_2\overset{O}{\overset{\|}{C}}Cl \longrightarrow$

(d) Product of part (c) + $(CH_3CH_2)_2NH \longrightarrow$

(e) Product of part (d) + $HCl \longrightarrow$

(f) $C_6H_5NH\overset{O}{\overset{\|}{C}}CH_2CH_2CH_3 \xrightarrow[\text{2. } H_2O]{\text{1. } LiAlH_4}$

(g) Aniline + heptanal $\xrightarrow{H_2,\ Ni}$

(h) Acetanilide + $ClCH_2\overset{O}{\overset{\|}{C}}Cl \xrightarrow{AlCl_3}$

(i) 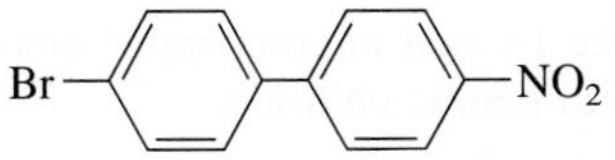 $\xrightarrow[\text{2. } HO^-]{\text{1. } Fe,\ HCl}$

(j) Product of part (i) $\xrightarrow[\text{2. } H_2O,\ \text{heat}]{\text{1. } NaNO_2,\ H_2SO_4,\ H_2O}$

(k) 2,6-Dinitroaniline $\xrightarrow[\text{2. } CuCl]{\text{1. } NaNO_2,\ H_2SO_4,\ H_2O}$

(l) *m*-Bromoaniline $\xrightarrow[\text{2. } CuBr]{\text{1. } NaNO_2,\ HBr,\ H_2O}$

(m) *o*-Nitroaniline $\xrightarrow[\text{2. } CuCN]{\text{1. } NaNO_2,\ HCl,\ H_2O}$

(n) 2,6-Diiodo-4-nitroaniline $\xrightarrow[\text{2. KI}]{\text{1. NaNO}_2\text{, H}_2\text{SO}_4\text{, H}_2\text{O}}$

(o) 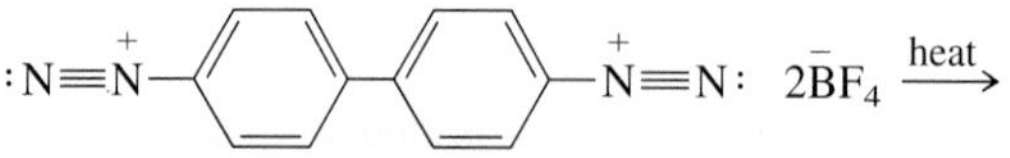

(p) 2,4,6-Trinitroaniline $\xrightarrow[\text{H}_2\text{O, H}_3\text{PO}_2]{\text{NaNO}_2\text{, H}_2\text{SO}_4}$

(q) 2-Amino-5-iodobenzoic acid $\xrightarrow[\text{2. CH}_3\text{CH}_2\text{OH}]{\text{1. NaNO}_2\text{, HCl, H}_2\text{O}}$

(r) Aniline $\xrightarrow[\text{2. 2,3,6-trimethylphenol}]{\text{1. NaNO}_2\text{, H}_2\text{SO}_4\text{, H}_2\text{O}}$

(s) 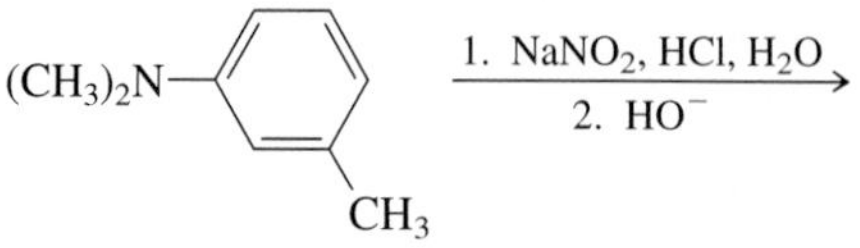

21.41 Most amides are reduced to amines with lithium aluminum hydride (Section 21.9); however, *N*-methoxy-*N*-methylamides are an exception. The initial reduction product is a cyclic intermediate that is hydrolyzed to an aldehyde on workup.

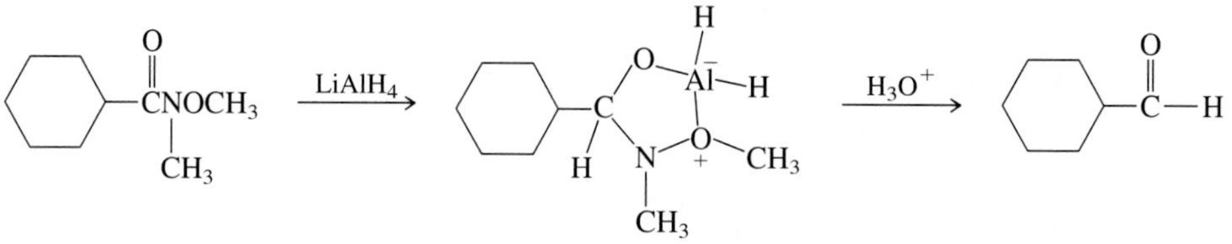

N-Methoxy-*N*-methylcyclohexanecarboxamide — Cyclic intermediate — Cyclohexanecarbaldehyde

N-methoxy-*N*-methylamides are readily synthesized from carboxylic acids, by reaction of the acyl chloride with *N,O*-dimethylhydroxylamine (CH_3ONHCH_3).

(a) How many equivalents of $LiAlH_4$ are required for the reduction of the amide?

(b) Write a series of equations to show the preparation of cyclohexanecarbaldehyde from the appropriate carboxylic acid and any other necessary reagents.

(c) Would you expect the amide shown here to undergo reduction with lithium aluminum hydride to give the same product? Why or why not?

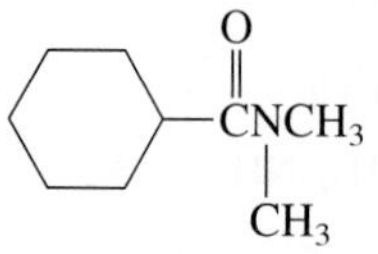

21.42 Provide a reasonable explanation for each of the following observations:

(a) 4-Methylpiperidine has a higher boiling point than *N*-methylpiperidine.

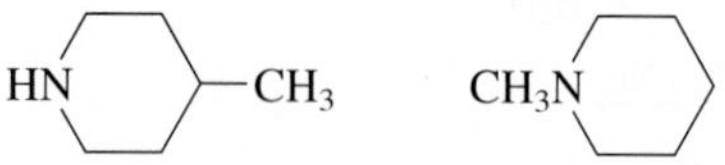

4-Methylpiperidine (bp 129°C) — *N*-Methylpiperidine (bp 106°C)

(b) Two isomeric quaternary ammonium salts are formed in comparable amounts when 4-*tert*-butyl-*N*-methylpiperidine is treated with benzyl chloride.

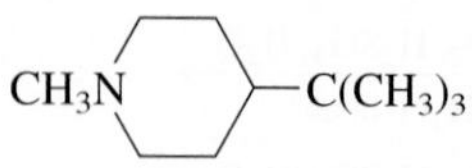

4-*tert*-Butyl-*N*-methylpiperidine

(c) When tetramethylammonium hydroxide is heated at 130°C, trimethylamine and methanol are formed.

(d) The major product formed on treatment of 1-propanamine with sodium nitrite in dilute hydrochloric acid is 2-propanol.

21.43 Give the structures, including stereochemistry, of compounds A through C.

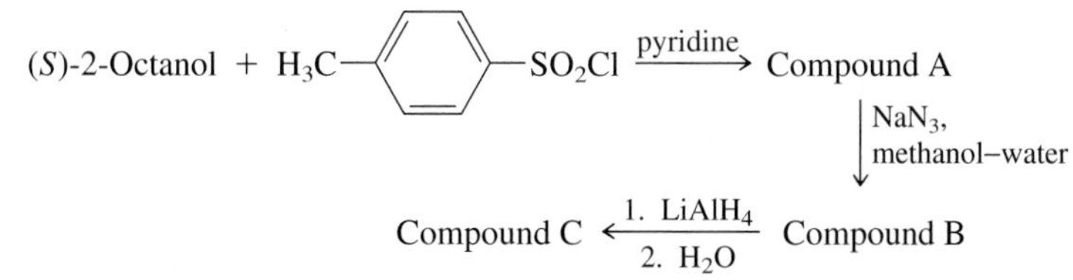
(S)-2-Octanol + H₃C—⟨⟩—SO₂Cl
pyridine
Compound A
NaN₃,
methanol–water
Compound C
1. LiAlH₄
2. H₂O
Compound B

21.44 Devise efficient syntheses of each of the following compounds from the designated starting materials. You may also use any necessary organic or inorganic reagents.

(a) 3,3-Dimethyl-1-butanamine from 1-bromo-2,2-dimethylpropane

(b) 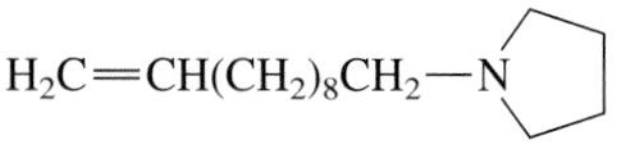 from 10-undecenoic acid and pyrrolidine

(c) 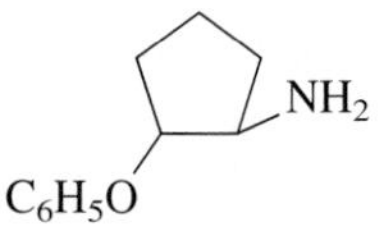 from 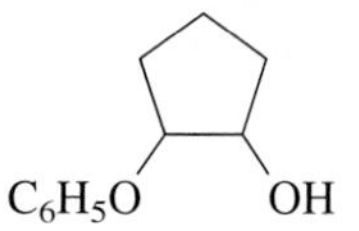

(d) $C_6H_5CH_2\underset{CH_3}{N}CH_2CH_2CH_2CH_2NH_2$ from $C_6H_5CH_2NHCH_3$ and $BrCH_2CH_2CH_2CN$

(e) 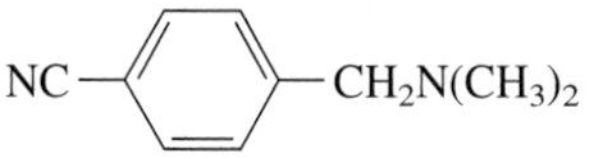 from 

21.45 Each of the following compounds has been prepared from *p*-nitroaniline. Outline a reasonable series of steps leading to each one.

(a) *p*-Nitrobenzonitrile
(b) 3,4,5-Trichloroaniline
(c) 1,3-Dibromo-5-nitrobenzene
(d) 3,5-Dibromoaniline
(e) *p*-Acetamidophenol (*acetaminophen*)

21.46 Each of the following compounds has been prepared from *o*-anisidine (*o*-methoxyaniline). Outline a series of steps leading to each one.

(a) *o*-Bromoanisole
(b) *o*-Fluoroanisole
(c) 3-Fluoro-4-methoxyacetophenone
(d) 3-Fluoro-4-methoxybenzonitrile
(e) 3-Fluoro-4-methoxyphenol

21.47 (a) Outline a synthesis of the following compound from nitrobenzene, *p*-nitrobenzyl alcohol, and any necessary organic or inorganic reagents.

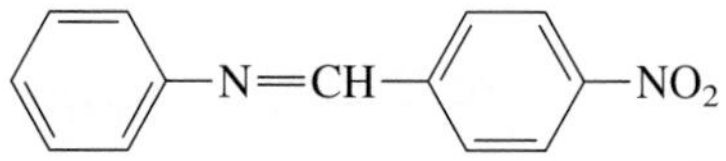

(b) How would you modify the synthesis if you had to start with *p*-nitrotoluene instead of *p*-nitrobenzyl alcohol?

21.48 Design syntheses of each of the following compounds from the indicated starting material and any necessary organic or inorganic reagents:

(a) *p*-Aminobenzoic acid from *p*-methylaniline

(b) $p\text{-}FC_6H_4\overset{\overset{O}{\|}}{C}CH_2CH_3$ from benzene

(c) 1-Bromo-2-fluoro-3,5-dimethylbenzene from *m*-xylene

(d) 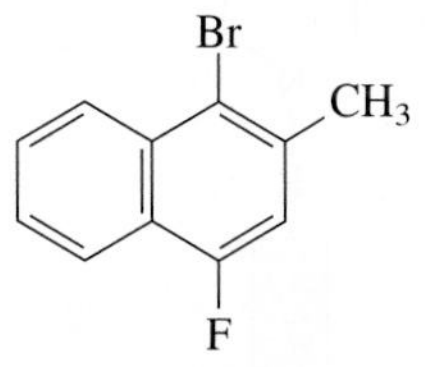 from 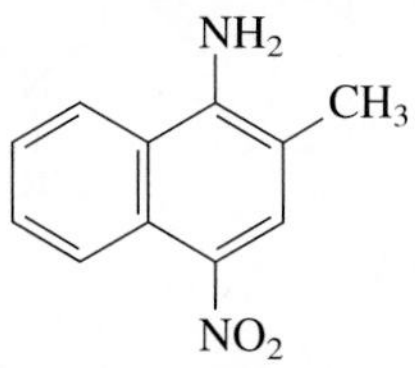

(e) o-$BrC_6H_4C(CH_3)_3$ from p-$O_2NC_6H_4C(CH_3)_3$

(f) m-$ClC_6H_4C(CH_3)_3$ from p-$O_2NC_6H_4C(CH_3)_3$

(g) 1-Bromo-3,5-diethylbenzene from *m*-diethylbenzene

(h) 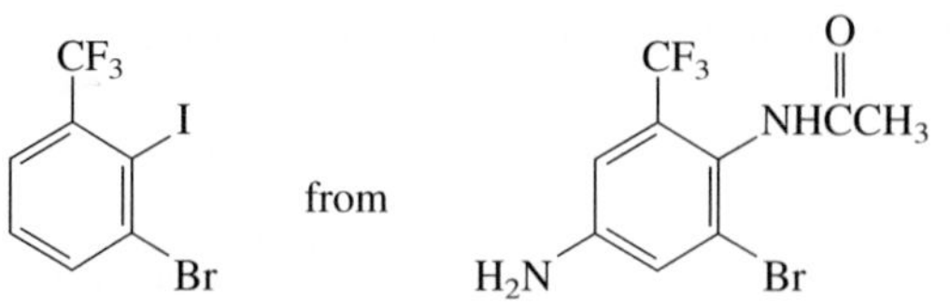

(i) 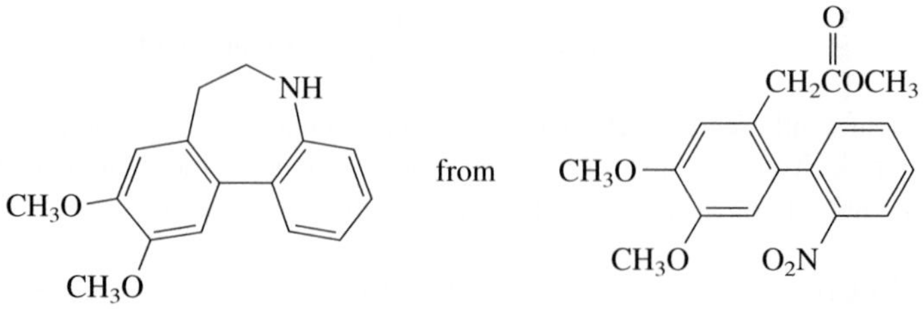

21.49 Show how 2-(2-bromophenyl)ethanamine could be prepared by the Gabriel amine synthesis, from *N*-potassiophthalimide and compound A, an alkyl halide.

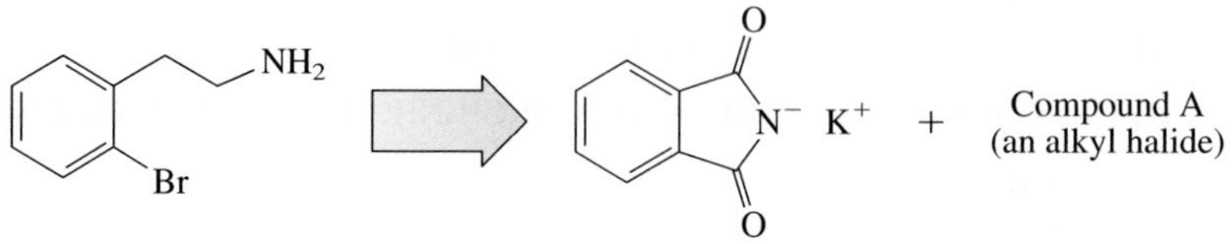
Compound A
(an alkyl halide)

2-(2-Bromophenyl)-ethanamine

N-Potassiophthalimide

21.50 Ammonia and amines undergo conjugate addition to α,β-unsaturated carbonyl compounds (Section 20.18). On the basis of this information, predict the principal organic product of each of the following reactions:

(a) $(CH_3)_2C{=}CH\overset{\displaystyle O}{\overset{\|}{C}}CH_3 + NH_3 \longrightarrow$

(b) 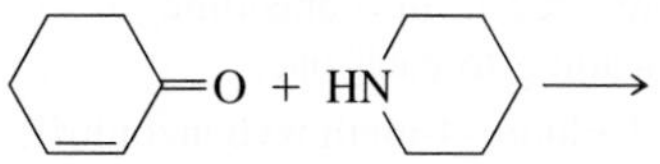

(c) 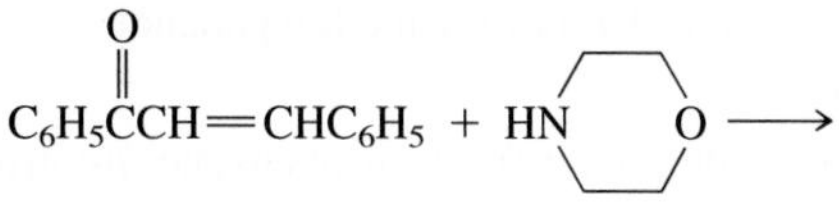

(d) 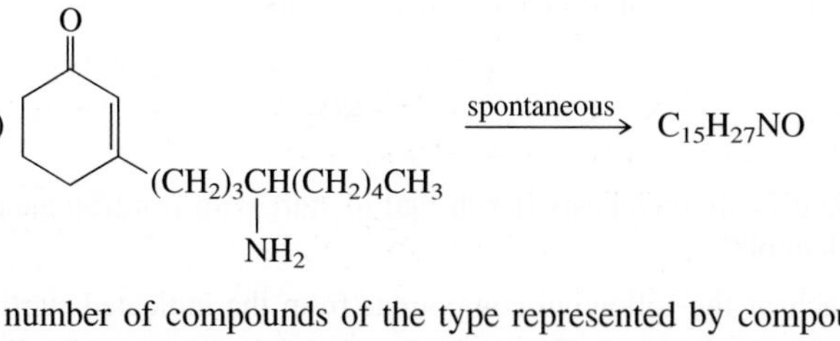
spontaneous
$C_{15}H_{27}NO$

21.51 A number of compounds of the type represented by compound A were prepared for evaluation as potential analgesic drugs. Their preparation is described in a retrosynthetic format as shown.

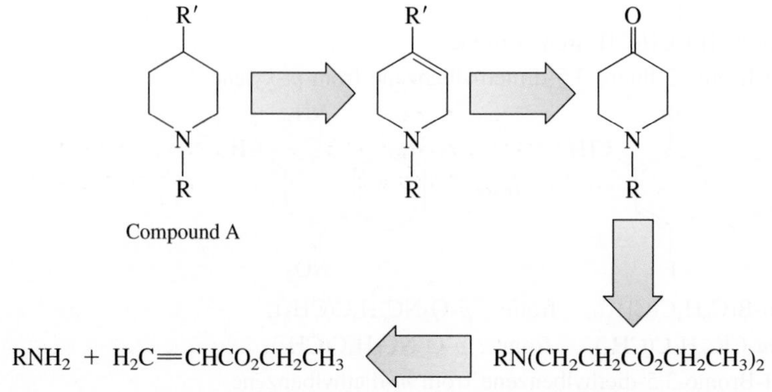
Compound A
$RN(CH_2CH_2CO_2CH_2CH_3)_2$
$RNH_2 + H_2C{=}CHCO_2CH_2CH_3$

On the basis of this retrosynthetic analysis, design a synthesis of *N*-methyl-4-phenylpiperidine (compound A, where R=CH_3, R′=C_6H_5). Present your answer as a series of equations, showing all necessary reagents and isolated intermediates.

21.52 A key step in the synthesis of the analgesic brifentanil is the reductive amination of 1-benzyl-3-methylpiperidin-4-one with 2-fluoroaniline. Write structural formulas for compound A and bifentanil.

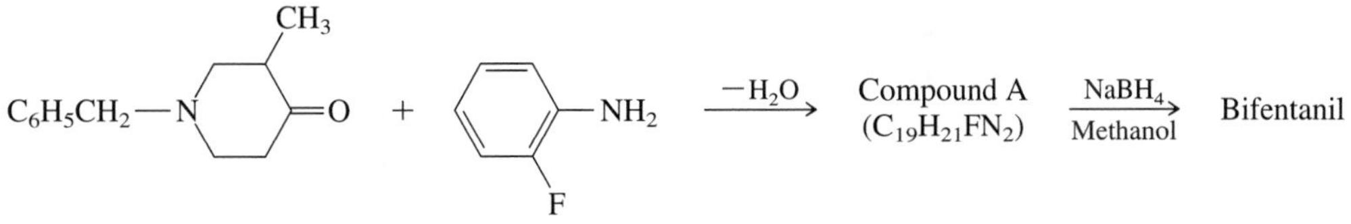
CH3
C6H5CH2—N
=O
+
NH2
F
−H2O
Compound A
(C19H21FN2)
NaBH4
Methanol
Bifentanil

1-Benzyl-3-methylpiperidin-4-one   2-Fluoroaniline

21.53 *N,N*-Dimethylaniline and pyridine are similar in basicity, whereas 4-(*N,N*-dimethylamino)pyridine is considerably more basic than either.

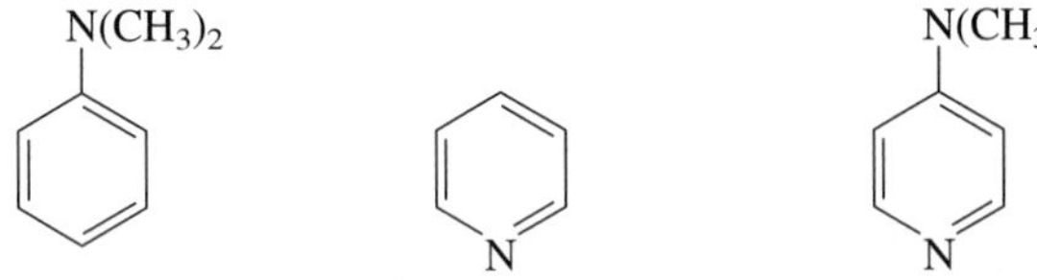
N(CH3)2
N
N(CH3)2
N

N,N-Dimethylaniline
pK_a of conjugate acid = 5.1

Pyridine
pK_a of conjugate acid = 5.3

4-(*N,N*-Dimethylamino)pyridine
pK_a of conjugate acid = 9.7

Apply resonance principles to identify the more basic of the two nitrogens of 4-(*N,N*-dimethylamino)pyridine, and suggest an explanation for its enhanced basicity.

21.54 The compound shown is a somewhat stronger base than ammonia. Which nitrogen do you think is protonated when it is treated with an acid? Write a structural formula for the species that results.

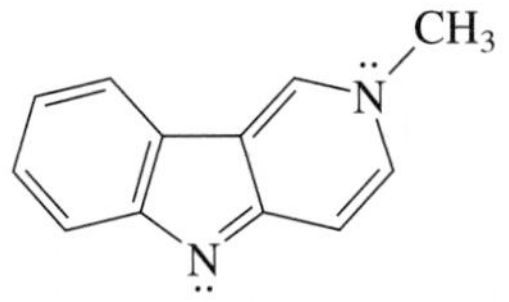
CH3
N
N

5-Methyl-γ-carboline
pK_a of conjugate acid = 10.5

21.55 Compounds A and B are isomeric amines of molecular formula $C_8H_{11}N$. Identify each isomer on the basis of the ^1H NMR spectra given in Figure 21.10.

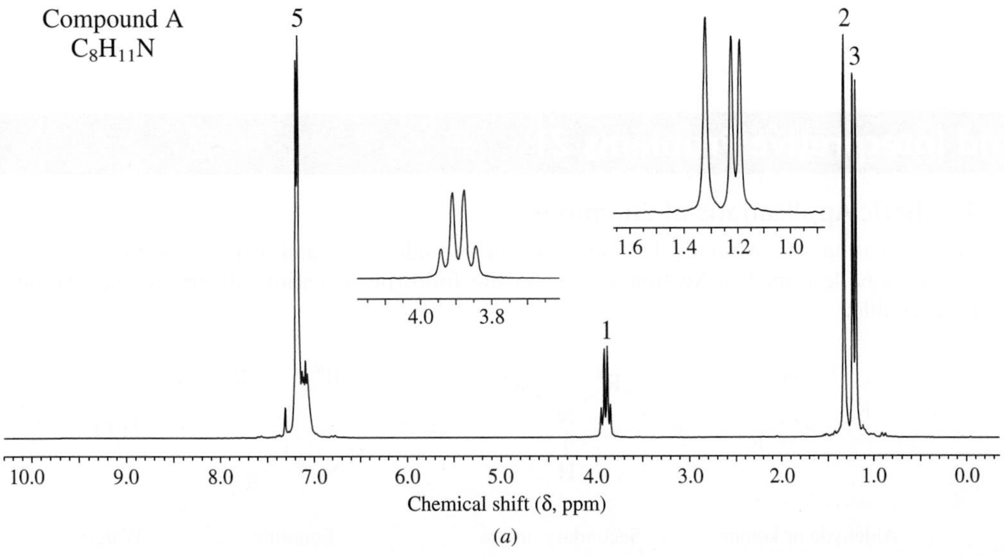
Compound A
C8H11N
5
1
2
3
4.0 3.8
1.6 1.4 1.2 1.0
10.0 9.0 8.0 7.0 6.0 5.0 4.0 3.0 2.0 1.0 0.0
Chemical shift (δ, ppm)

(*a*)

Figure 21.10

The 200-MHz ^1H NMR spectra of (*a*) compound A and (*b*) compound B (Problem 21.55).

Figure 21.10

Continued

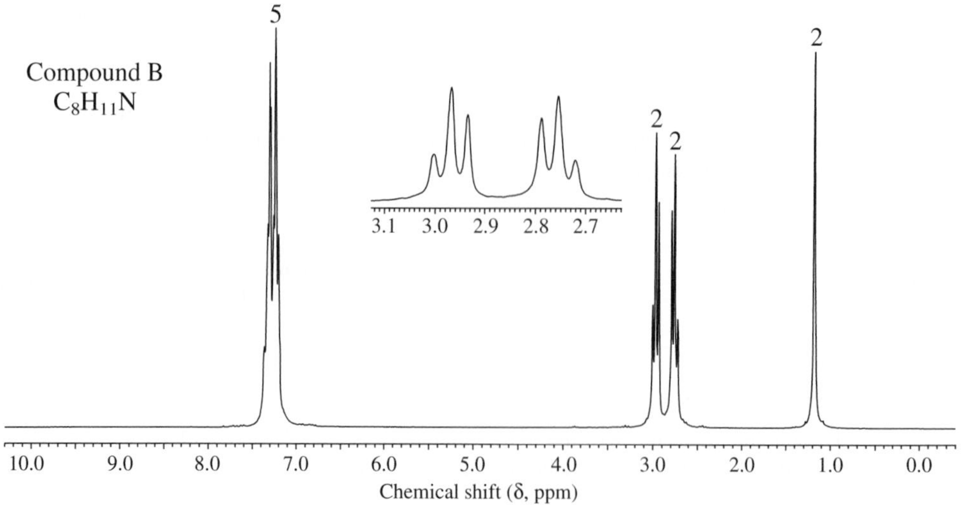
5
2
Compound B
$C_8H_{11}N$
3.1 3.0 2.9 2.8 2.7
2
2
10.0 9.0 8.0 7.0 6.0 5.0 4.0 3.0 2.0 1.0 0.0
Chemical shift (δ, ppm)

(*b*)

21.56 Does the ^{13}C NMR spectrum shown in Figure 21.11 correspond to that of 1-amino-2-methyl-2-propanol or to 2-amino-2-methyl-1-propanol? Could this compound be prepared by reaction of an epoxide with ammonia?

Figure 21.11

The ^{13}C NMR spectrum of the compound described in Problem 21.56.

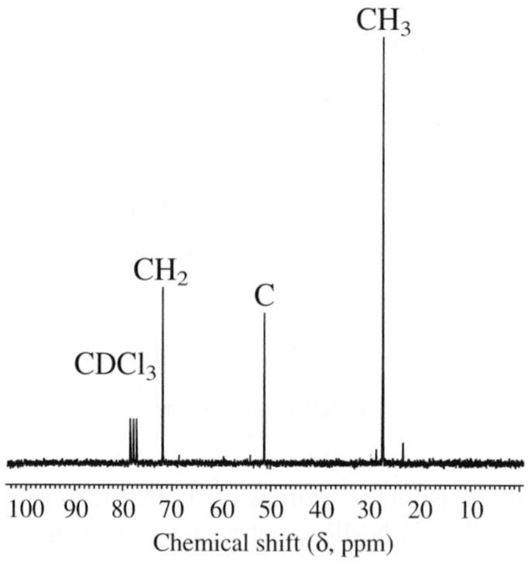
CH_3
CH_2
C
$CDCl_3$
100 90 80 70 60 50 40 30 20 10
Chemical shift (δ, ppm)

Descriptive Passage and Interpretive Problems 21

Synthetic Applications of Enamines

The formation of enamines by the reaction of aldehydes and ketones with secondary amines was described in Section 17.11. As the following equation illustrates, the reaction is reversible.

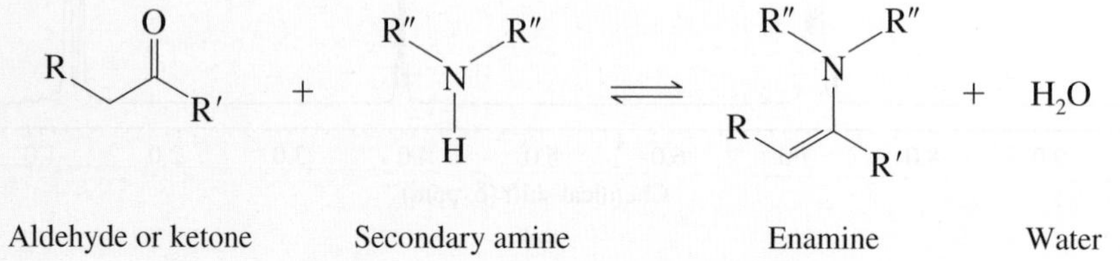
Aldehyde or ketone
Secondary amine
Enamine
Water

When preparing enamines, the reaction is normally carried out by heating in benzene as the solvent. No catalyst is necessary, but *p*-toluenesulfonic acid is sometimes added. The water formed is removed by distillation of its azeotropic mixture with benzene, which shifts the position of equilibrium to the right to give the enamine in high yield. Conversely, enamines can be hydrolyzed in aqueous acid to aldehydes and ketones.

Enamines resemble enols in that electron-pair donation makes their double bond electron-rich and nucleophilic.

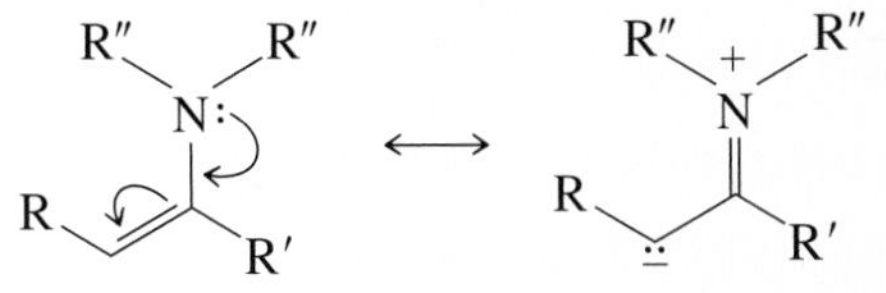

Because nitrogen is a better electron-pair donor than oxygen, an enamine is more nucleophilic than an enol. Enamines, being neutral molecules are, however, less nucleophilic than enolates, which are anions.

Reactions of enamines with electrophiles (E^+) lead to carbon–carbon bond formation. Subsequent hydrolysis gives an α-substituted derivative of the original aldehyde or ketone.

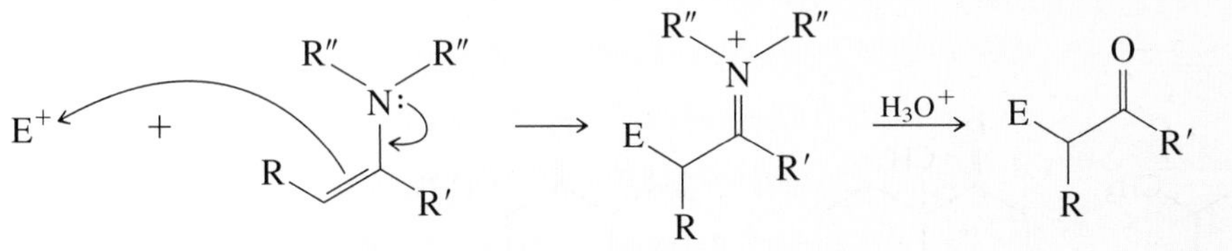

Electrophile + enamine α-Substituted aldehyde or ketone

Pyrrolidine is the secondary amine used most often for making enamines from aldehydes and ketones.

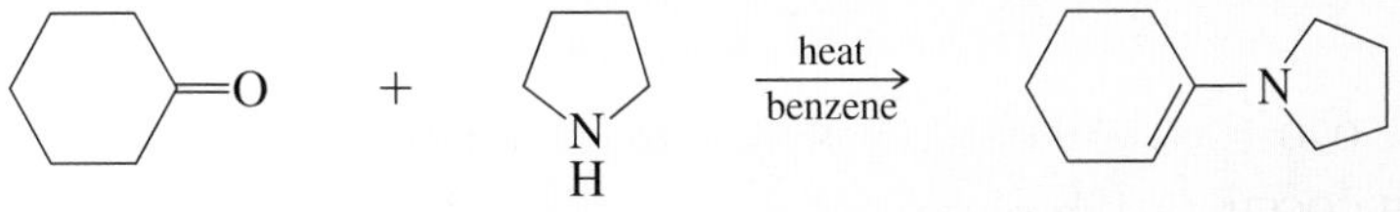

Cyclohexanone Pyrrolidine 1-Pyrrolidinocyclohexene (85–90%)

For synthetic purposes, the electrophilic reagents that give the best yields of α-substituted aldehydes and ketones on reactions with enamines are the following:

1. Alkyl halides that are very reactive in S_N2 reactions such as primary allylic and benzylic halides, α-halo ethers, α-halo esters, and α-halo nitriles.
2. Acyl chlorides and acid anhydrides.
3. Michael acceptors: α,β-unsaturated nitriles, esters, and ketones.

21.57 One of the following is often used to prepare enamines from aldehydes and ketones. The others do not yield enamines. Identify the enamine-forming compound.

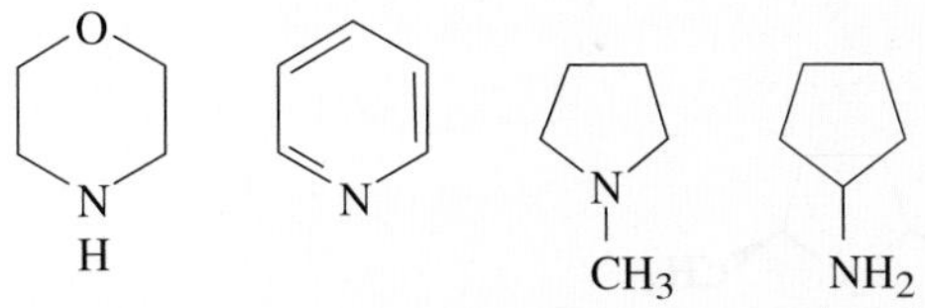

A. B. C. D.

21.58 What is the product of the following reaction?

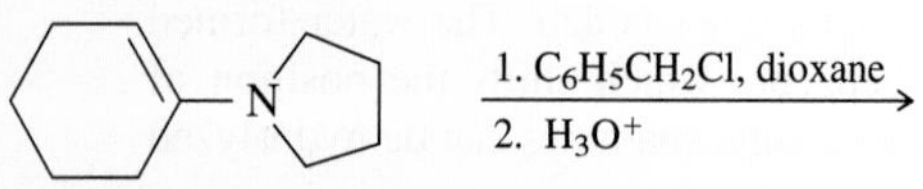

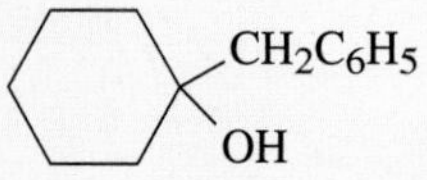

A.

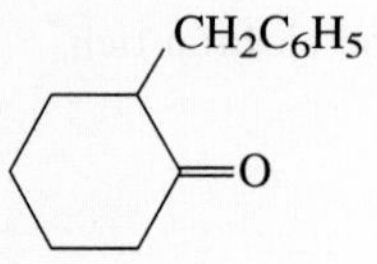

B.

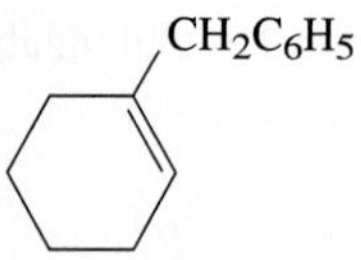

C.

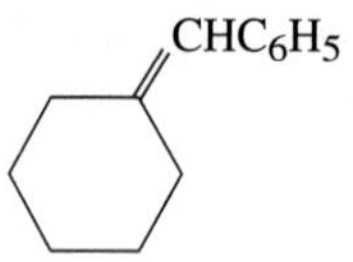

D.

21.59 Unsymmetrical ketones give a mixture of two pyrrolidine enamines in which the enamine with the less-substituted double bond predominates. What is the major product of the following reaction sequence?

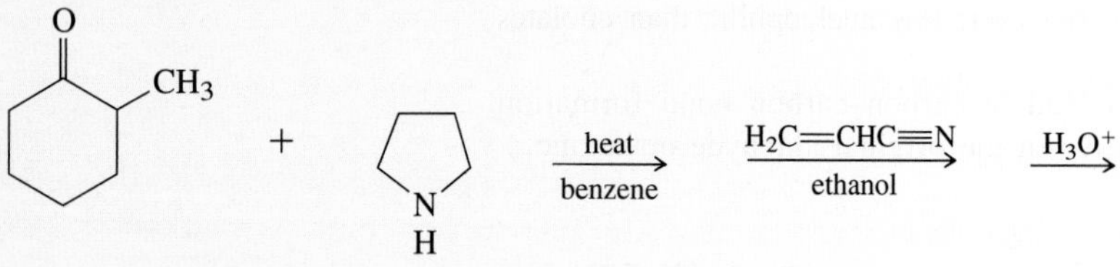

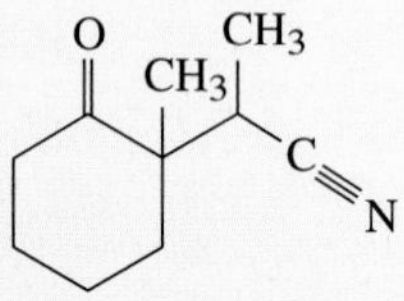

A.

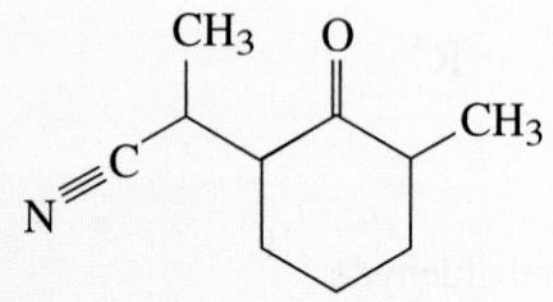

(*cis* + *trans*)

B.

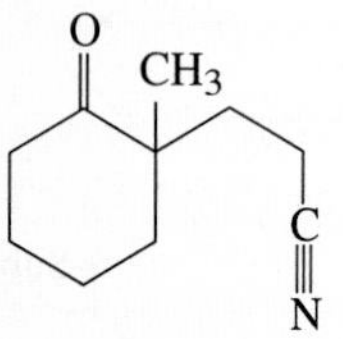

C.

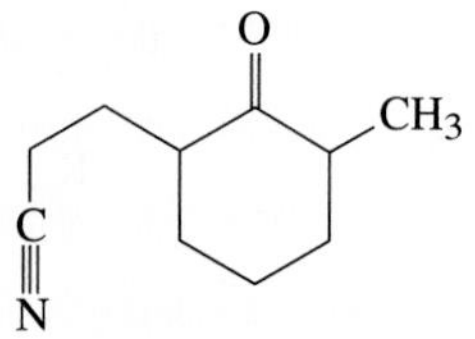

(*cis* + *trans*)

D.

21.60 What is the product of the following reaction sequence?

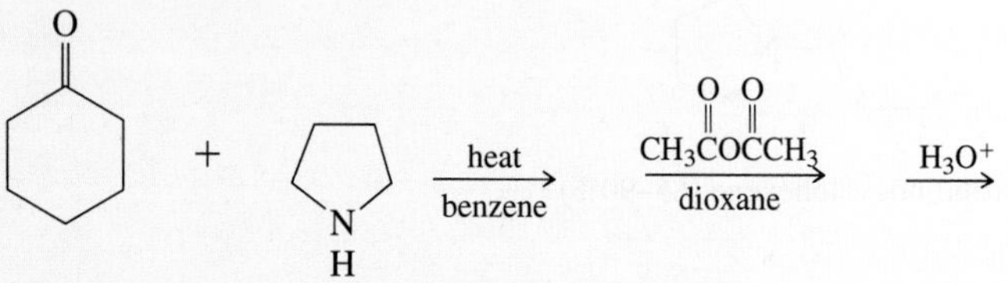

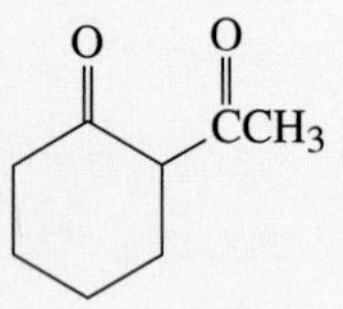

A.

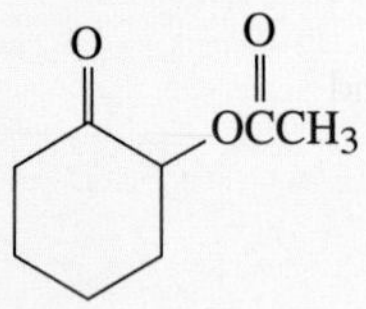

B.

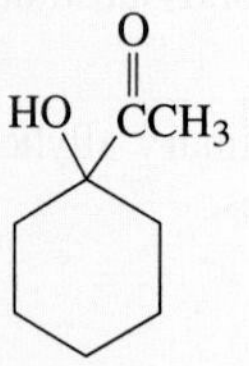

C.

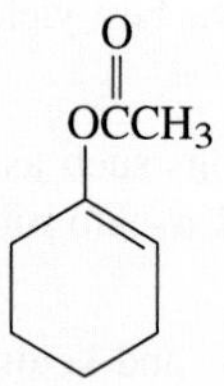

D.

21.61 (+)-2-Allylcyclohexanone has been prepared in 82% enantiomeric excess by alkylation of the optically active enamine prepared from cyclohexanone and an enantiomerically pure pyrrolidine derivative. Of the following, which one is the best pyrrolidine derivative to use in this enantioselective synthesis?

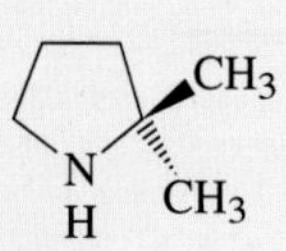

A.

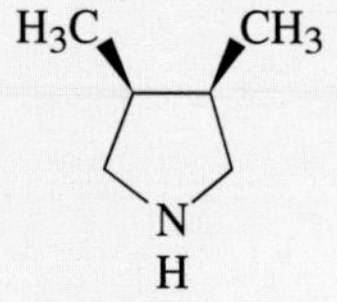

B.

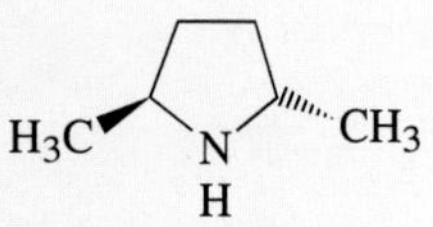

C.

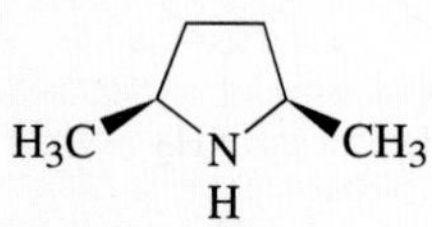

D.

21.62 Cyclooctanecarboxaldehyde was converted to a ketone having the molecular formula $C_{13}H_{20}O$ via its piperidine enamine as shown. What is the structure of the ketone?

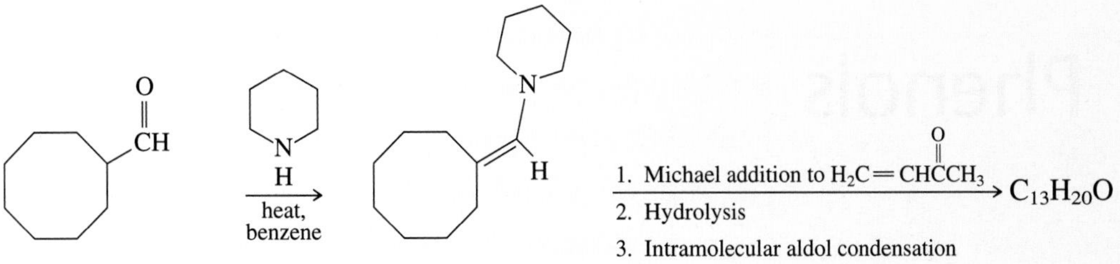
O
CH
N
H
heat,
benzene
N
H
1. Michael addition to $H_2C{=}CHCCH_3$
2. Hydrolysis
3. Intramolecular aldol condensation
$C_{13}H_{20}O$

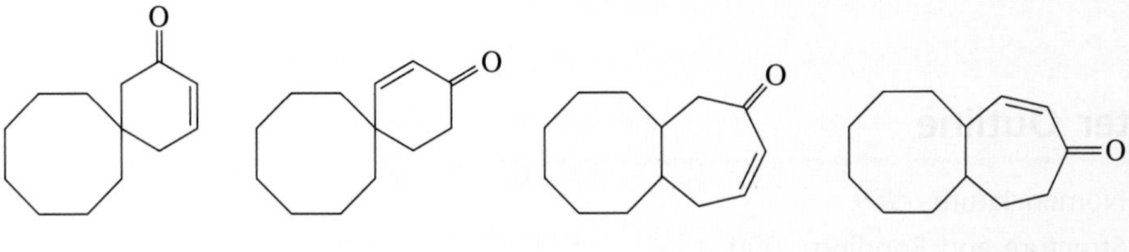

A. B. C. D.

22 Phenols

Chapter Outline

The skins of red grapes are a source of *resveratrol,* a polyphenol having antioxidant properties that is widely used as a nutritional supplement. For more on antioxidants, see the essay "James Bond, Oxidative Stress, and Antioxidant Phenols" on page 1004.

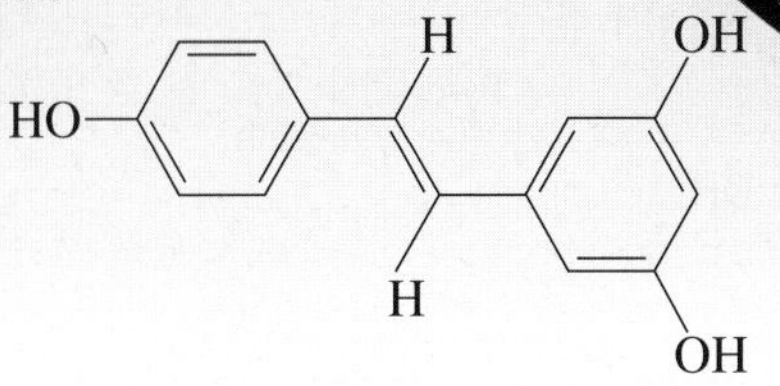

PHENOLS ARE compounds that have a hydroxyl group bonded directly to a benzene or benzenoid ring. The parent compound of this group, C_6H_5OH, called simply *phenol,* is an important industrial chemical. Many of the properties of phenols are analogous to those of alcohols, but this similarity is something of an oversimplification. Like arylamines, phenols are difunctional compounds; the hydroxyl group and the aromatic ring interact strongly, affecting each other's reactivity. This interaction leads to some novel and useful properties of phenols. A key step in the synthesis of aspirin, for example, is without parallel in the reactions of either alcohols or arenes. With periodic reminders of the ways in which phenols resemble alcohols and arenes, this chapter emphasizes the ways in which phenols are unique.

22.1 Nomenclature

An old name for benzene was *phene,* and its hydroxyl derivative came to be called *phenol.* Phenol is not only an acceptable IUPAC name, it is the preferred 2004 name. More highly substituted compounds are named as derivatives of phenol. Numbering of the ring begins at the hydroxyl-substituted carbon and proceeds in the direction that gives the lower number to the next substituted carbon. Substituents are cited in alphabetical order.

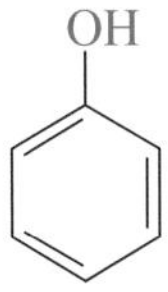

Phenol

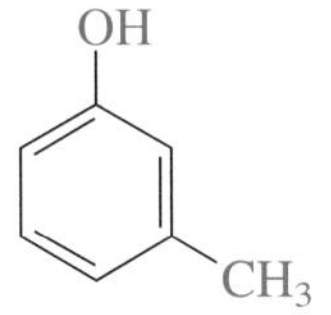

3-Methylphenol

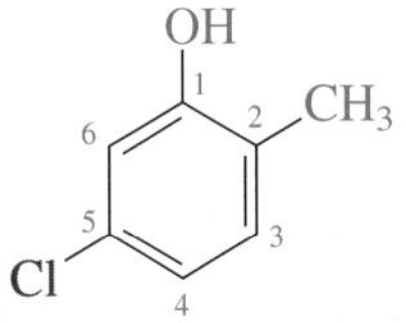

5-Chloro-2-methylphenol

The common names of 2-,3-, and 4-methylphenol are *o*-, *m*-, and *p*-cresol, respectively.

The three dihydroxy derivatives of benzene may be named as 1,2-, 1,3-, and 1,4-benzenediol, respectively, but each is more familiarly known by the common name indicated in parentheses below the structures shown here. These common names were permissible IUPAC names prior to the 2004 recommendations.

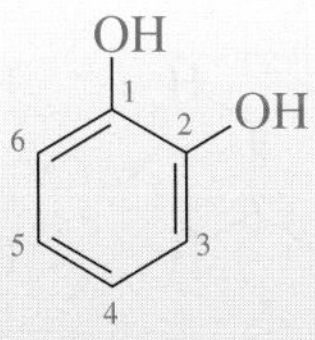

1,2-Benzenediol
(pyrocatechol)

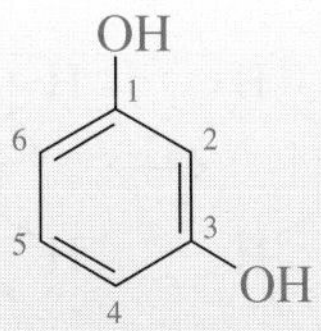

1,3-Benzenediol
(resorcinol)

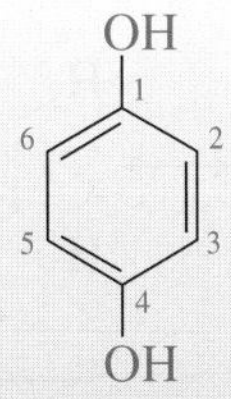

1,4-Benzenediol
(hydroquinone)

Pyrocatechol is often called *catechol.*

The common names for the two hydroxy derivatives of naphthalene are 1-naphthol and 2-naphthol. The 2004 recommended names are naphthalen-1-ol and naphthalen-2-ol.

Problem 22.1

Write structural formulas for each of the following compounds:

(a) Pyrogallol (1,2,3-benzenetriol) (b) *o*-Benzylphenol (c) 3-Nitro-1-naphthol

Sample Solution (a) Like the dihydroxybenzenes, the isomeric trihydroxybenzenes have unique names. Pyrogallol, used as a developer of photographic film, is 1,2,3-benzenetriol. The three hydroxyl groups occupy adjacent positions on a benzene ring.

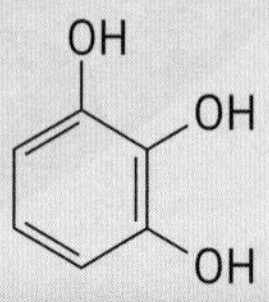

Pyrogallol (1,2,3-benzenetriol)

Carboxyl and acyl groups take precedence over the phenolic hydroxyl in determining the base name. The hydroxyl is treated as a substituent in these cases.

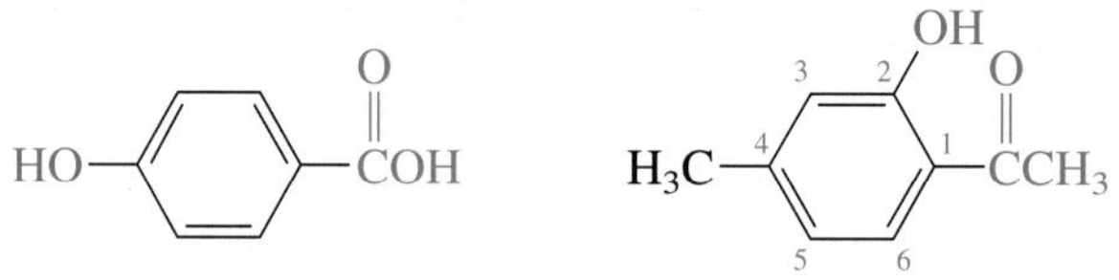

p-Hydroxybenzoic acid 2-Hydroxy-4-methylacetophenone

22.2 Structure and Bonding

Phenol is planar, with a C—O—H angle of 109°, almost the same as the tetrahedral angle and not much different from the 108.5° C—O—H angle of methanol:

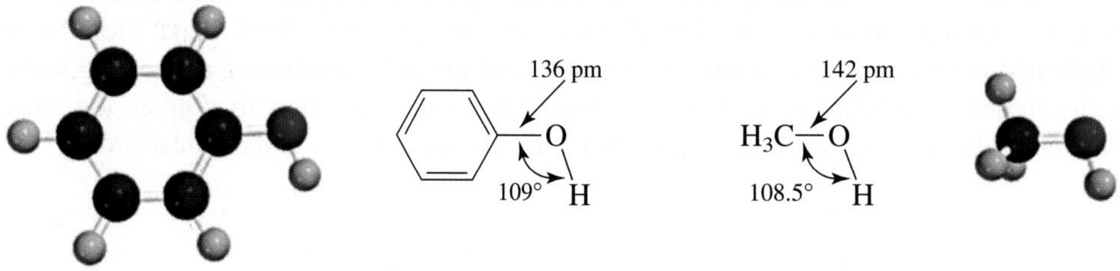

Phenol Methanol

As we've seen on a number of occasions, bonds to sp^2-hybridized carbon are shorter than those to sp^3-hybridized carbon, and the case of phenols is no exception. The carbon–oxygen bond distance in phenol is slightly less than that in methanol.

In resonance terms, the shorter carbon–oxygen bond distance in phenol is attributed to the partial double-bond character that results from conjugation of the unshared electron pair of oxygen with the aromatic ring.

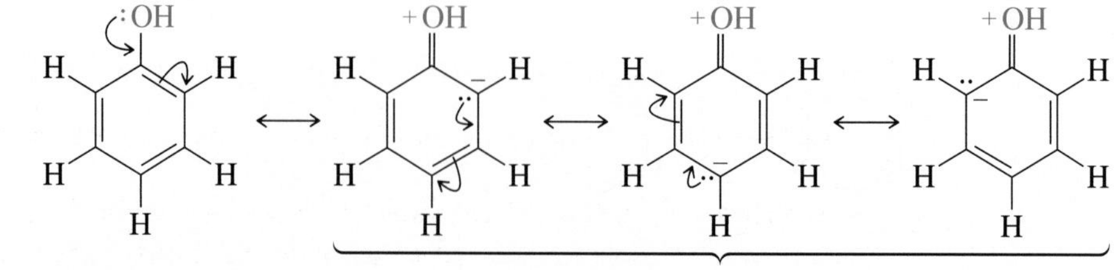

Most stable Lewis structure for phenol

Dipolar resonance contributors of phenol

Many of the properties of phenols reflect the polarization implied by the contributing structures. The hydroxyl oxygen is less basic, and the hydroxyl proton more acidic, in phenols than in alcohols. Electrophilic aromatic substitution in phenols is much faster than in benzene, indicating that the ring, especially at the positions ortho and para to the hydroxyl group, is relatively "electron-rich."

22.3 Physical Properties

The physical properties of phenols are strongly influenced by the hydroxyl group, which permits phenols to form hydrogen bonds with other phenol molecules (Figure 22.1*a*) and with water (Figure 22.1*b*). Thus, phenols have higher melting points and boiling points and are more soluble in water than arenes and aryl halides of comparable molecular weight. Table 22.1 compares phenol, toluene, and fluorobenzene with regard to these physical properties.

Some ortho-substituted phenols, such as *o*-nitrophenol, have significantly lower boiling points than those of the meta and para isomers. This is because the *intramolecular* hydrogen bond that forms between the hydroxyl group and the substituent partially compensates for the energy required to go from the liquid state to the vapor.

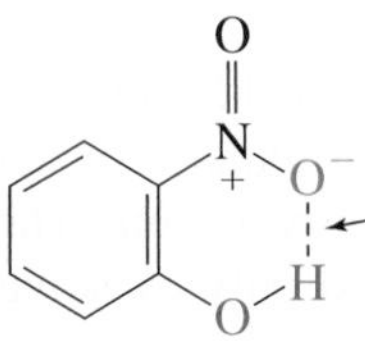

Problem 22.2

One of the hydroxybenzoic acids is known by the common name *salicylic acid.* Its methyl ester, methyl salicylate, occurs in oil of wintergreen. Methyl salicylate boils over 50°C lower than either of the other two methyl hydroxybenzoates. What is the structure of methyl salicylate? Why is its boiling point so much lower than that of either of its regioisomers?

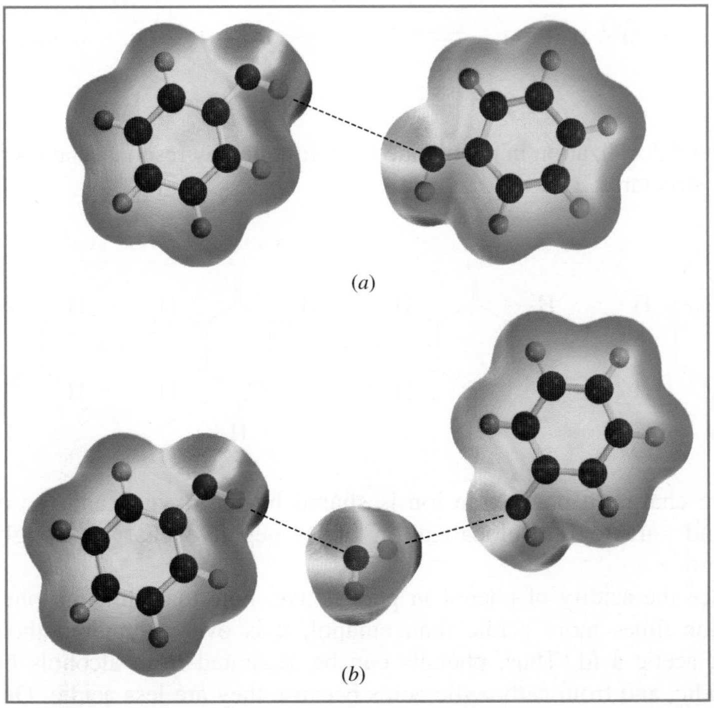

Figure 22.1

(*a*) A hydrogen bond between two phenol molecules; (*b*) hydrogen bonds between water and phenol molecules.

TABLE 22.1 Comparison of Physical Properties of an Arene, a Phenol, and an Aryl Halide

	Compound		
Physical property	**Toluene, $C_6H_5CH_3$**	**Phenol, C_6H_5OH**	**Fluorobenzene, C_6H_5F**
Molecular weight	92	94	96
Melting point	−95°C	43°C	−41°C
Boiling point (1 atm)	111°C	132°C	85°C
Solubility in water (25°C)	0.05 g/100 mL	8.2 g/100 mL	0.2 g/100 mL

22.4 Acidity of Phenols

Because of its acidity, phenol was known as *carbolic acid* when Joseph Lister introduced it as an antiseptic in 1865 to prevent postoperative bacterial infections that were then a life-threatening hazard in even minor surgical procedures.

The most characteristic property of phenols is their acidity. Phenols are more acidic than alcohols but less acidic than carboxylic acids. Recall that carboxylic acids have pK_a's of approximately 5, whereas the pK_a's of alcohols are in the 16–20 range. The pK_a for most phenols is about 10.

To help us understand why phenols are more acidic than alcohols, compare the ionization equilibria for phenol and ethanol. In particular, consider the differences in charge delocalization in ethoxide ion and in phenoxide ion. The negative charge in ethoxide ion is localized on oxygen and is stabilized only by solvation forces.

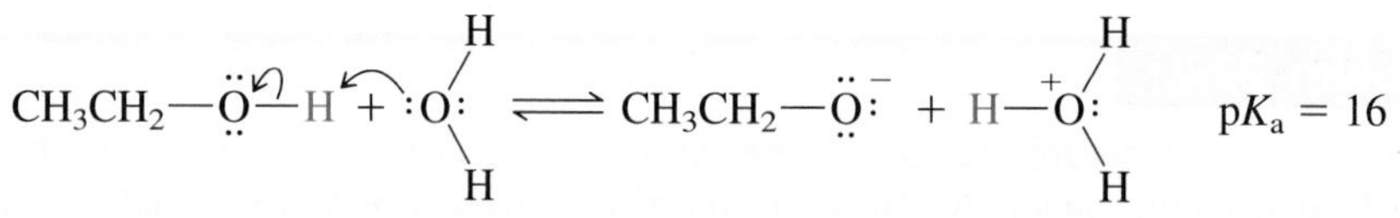

CH₃CH₂—O—H + :O:
H
H
CH₃CH₂—O: + H—O:
H
H
pKa = 16

The negative charge in phenoxide ion is stabilized both by solvation and by electron delocalization into the ring.

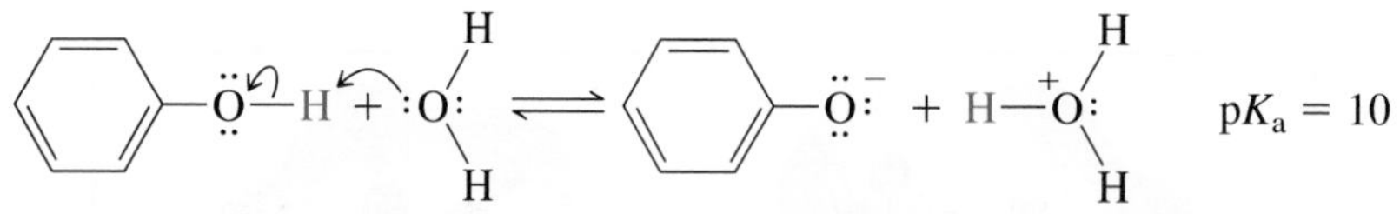

H
H
H
H
pKa = 10

Electron delocalization in phenoxide is represented by resonance among the various contributing structures:

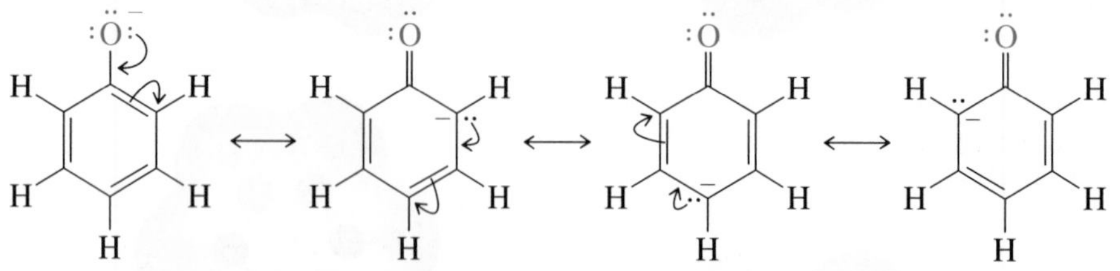

The negative charge in phenoxide ion is shared by the oxygen and the carbons that are ortho and para to it. Delocalization of its negative charge strongly stabilizes phenoxide ion.

To place the acidity of phenol in perspective, note that although phenol is more than a million times more acidic than ethanol, it is over a hundred thousand times weaker than acetic acid. Thus, phenols can be separated from alcohols because they are more acidic, and from carboxylic acids because they are less acidic. On shaking an ether solution containing both an alcohol and a phenol with dilute sodium hydroxide, the phenol is converted quantitatively to its sodium salt, which is extracted into the aqueous phase. The alcohol remains in the ether phase.

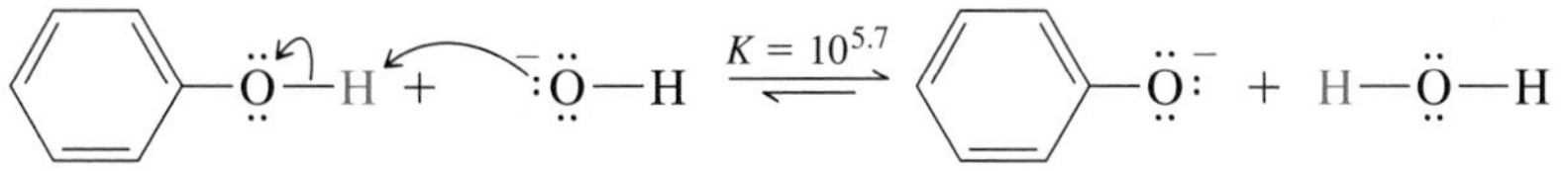
$K = 10^{5.7}$

Phenol (stronger acid; $pK_a = 10$) | Hydroxide ion (stronger base) | Phenoxide ion (weaker base) | Water (weaker acid; $pK_a = 15.7$)

On shaking an ether solution of a phenol and a carboxylic acid with dilute sodium bicarbonate, the carboxylic acid is converted quantitatively to its sodium salt and extracted into the aqueous phase. The phenol remains in the ether phase.

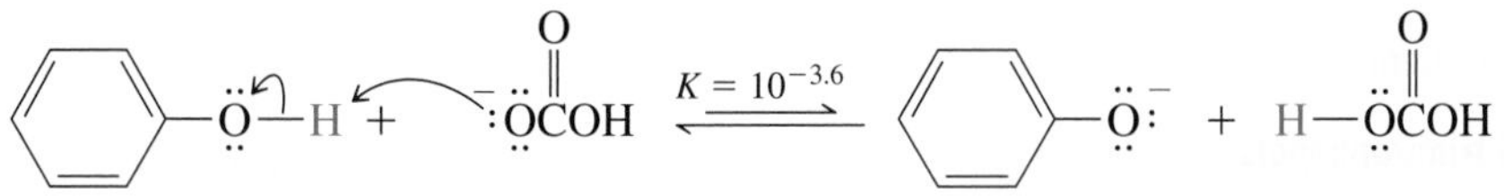
$K = 10^{-3.6}$

Phenol (weaker acid; $pK_a = 10$) | Bicarbonate ion (weaker base) | Phenoxide ion (stronger base) | Carbonic acid (stronger acid; $pK_a = 6.4$)

It is necessary to keep the acidity of phenols in mind when we discuss preparation and reactions. Reactions that produce phenols, when carried out in basic solution, require an acidification step to convert the phenoxide ion to the neutral form of the phenol.

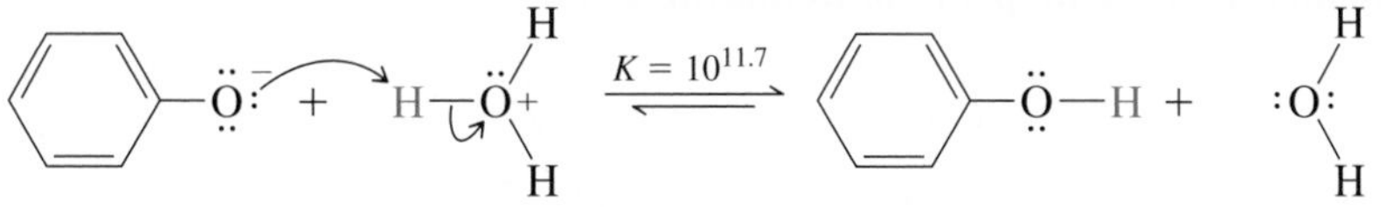
$K = 10^{11.7}$

Phenoxide ion (stronger base) | Hydronium ion (stronger acid; $pK_a = -1.7$) | Phenol (weaker acid; $pK_a = 10$) | Water (weaker base)

Many synthetic reactions involving phenols as nucleophiles are carried out in the presence of sodium or potassium hydroxide. Under these conditions the phenol is converted to the corresponding phenoxide ion, which is a far better nucleophile.

22.5 Substituent Effects on the Acidity of Phenols

As Table 22.2 shows, most phenols have ionization constants similar to that of phenol itself. Substituent effects, in general, are small.

Alkyl substitution produces negligible changes in acidities, as do weakly electronegative groups attached to the ring.

Only when the substituent is strongly electron-withdrawing, as is a nitro group, is a substantial change in acidity noted. The ionization constants of *o*- and *p*-nitrophenol are several hundred times greater than that of phenol. An ortho- or para-nitro group greatly stabilizes the phenoxide ion by permitting a portion of the negative charge to be carried by its own oxygens.

Electron delocalization in o-nitrophenoxide ion

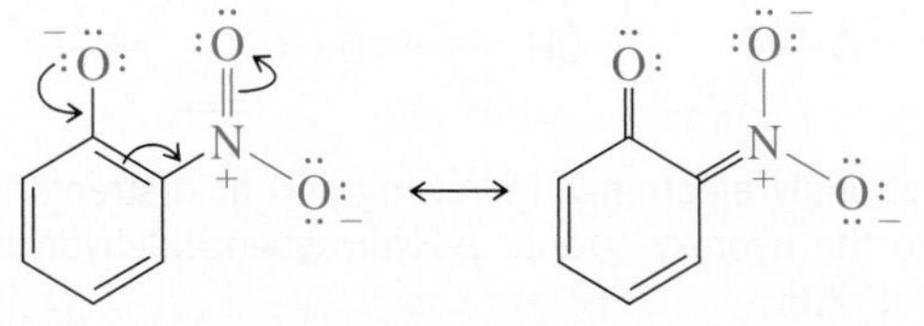

Recall from Section 22.1 that cresols are methyl-substituted derivatives of phenol.

TABLE 22.2 Acidities of Some Phenols

Compound name	pK_a	Compound name	pK_a
Monosubstituted phenols			
Phenol	10.0	*o*-Methoxyphenol	10.0
o-Cresol	10.3	*m*-Methoxyphenol	9.6
m-Cresol	10.1	*p*-Methoxyphenol	10.2
p-Cresol	10.3	*o*-Nitrophenol	7.2
o-Chlorophenol	8.6	*m*-Nitrophenol	8.4
m-Chlorophenol	9.1	*p*-Nitrophenol	7.2
p-Chlorophenol	9.4		
Di- and trinitrophenols			
2,4-Dinitrophenol	4.0	2,4,6-Trinitrophenol	0.4
3,5-Dinitrophenol	6.7		
Naphthols			
1-Naphthol	9.2	2-Naphthol	9.5

Electron delocalization in* p*-nitrophenoxide ion

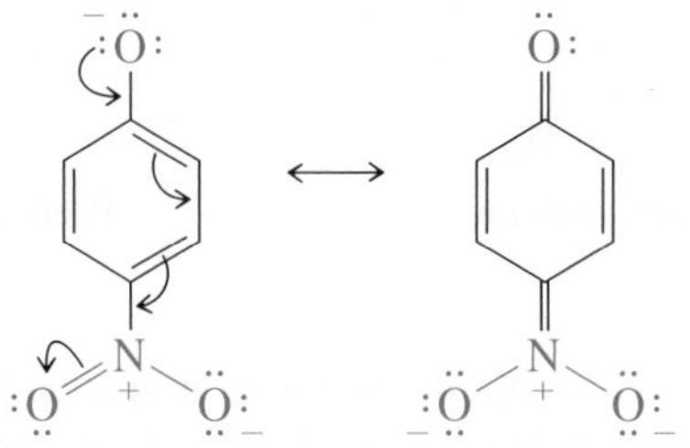

A meta-nitro group is not directly conjugated to the phenoxide oxygen and thus stabilizes a phenoxide ion to a smaller extent. *m*-Nitrophenol is more acidic than phenol but less acidic than either *o*- or *p*-nitrophenol.

Problem 22.3

Which is the stronger acid in each of the following pairs? Explain your reasoning.

(a) Phenol or *p*-hydroxybenzaldehyde
(b) *m*-Cyanophenol or *p*-cyanophenol
(c) *o*-Fluorophenol or *p*-fluorophenol

Sample Solution (a) The best approach when comparing the acidities of different phenols is to assess opportunities for stabilization of negative charge in their anions. Electron delocalization in the anion of *p*-hydroxybenzaldehyde is very effective because of conjugation.

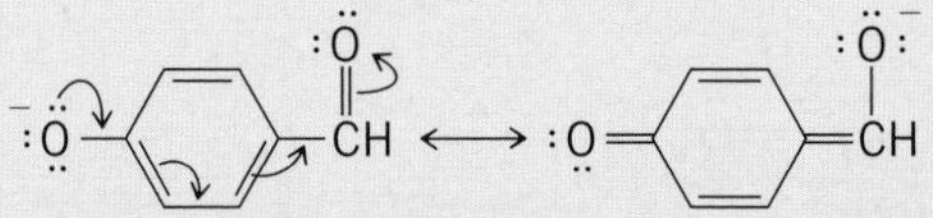

A carbonyl group is strongly electron-withdrawing and acid-strengthening, especially when ortho or para to the hydroxyl group. *p*-Hydroxybenzaldehyde is a stronger acid than phenol. Its pK_a is 7.6.

Multiple substitution by strongly electron-withdrawing groups greatly increases the acidity of phenols, as the pK_a values for 2,4-dinitrophenol (4.0) and 2,4,6-trinitrophenol (0.4) in Table 22.2 attest.

22.6 Sources of Phenols

Phenol was first isolated in the early nineteenth century from coal tar, and a small portion of the more than 4 billion pounds of phenol produced in the United States each year comes from this source. Although significant quantities of phenol are used to prepare aspirin and dyes, most of it is converted to phenolic resins used in adhesives and plastics. Almost all the phenol produced commercially is synthetic, with several different processes in current use. These are summarized in Table 22.3.

The reaction of benzenesulfonic acid with sodium hydroxide (first entry in Table 22.3) proceeds by the addition–elimination mechanism of nucleophilic aromatic substitution (see Section 12.21). Hydroxide replaces sulfite ion (SO_3^{2-}) at the carbon atom that bears the leaving group. Thus, *p*-toluenesulfonic acid is converted exclusively to *p*-cresol:

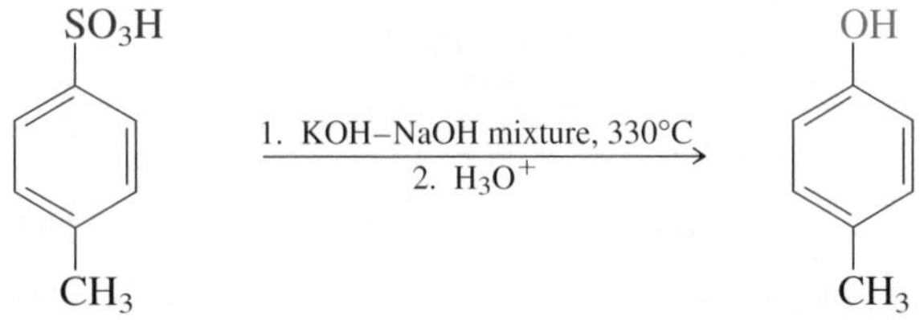
SO3H
1. KOH–NaOH mixture, 330°C
2. H3O+
OH
CH3
CH3

p-Toluenesulfonic acid *p*-Cresol (63–72%)

Can you recall how to prepare *p*-toluenesulfonic acid?

Problem 22.4

Write a stepwise mechanism for the conversion of *p*-toluenesulfonic acid to *p*-cresol under the conditions shown in the preceding equation.

TABLE 22.3 Industrial Syntheses of Phenol

Reaction and comments	Chemical equation
Reaction of benzenesulfonic acid with sodium hydroxide This is the oldest method for the preparation of phenol. Benzene is sulfonated and the benzenesulfonic acid heated with molten sodium hydroxide. Acidification of the reaction mixture gives phenol.	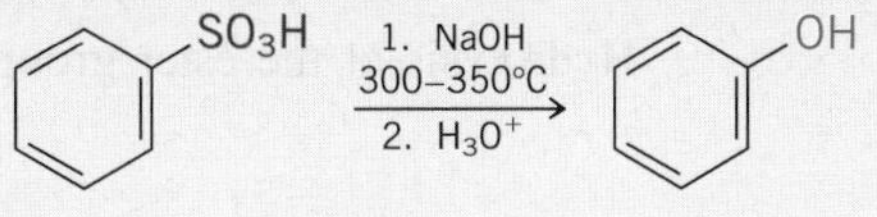 Benzenesulfonic acid → Phenol
Hydrolysis of chlorobenzene Heating chlorobenzene with aqueous sodium hydroxide at high pressure gives phenol after acidification.	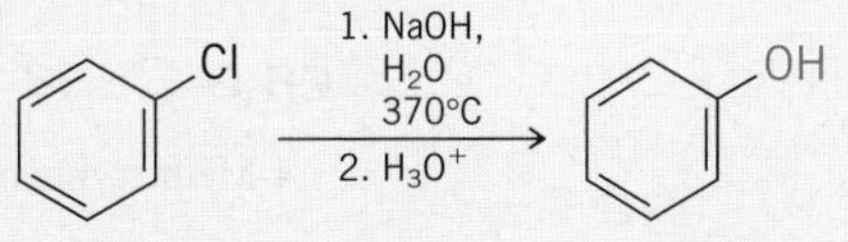 Chlorobenzene → Phenol
From cumene Almost all the phenol produced in the United States is prepared by this method. Oxidation of cumene takes place at the benzylic position to give a hydroperoxide. On treatment with dilute sulfuric acid, this hydroperoxide is converted to phenol and acetone.	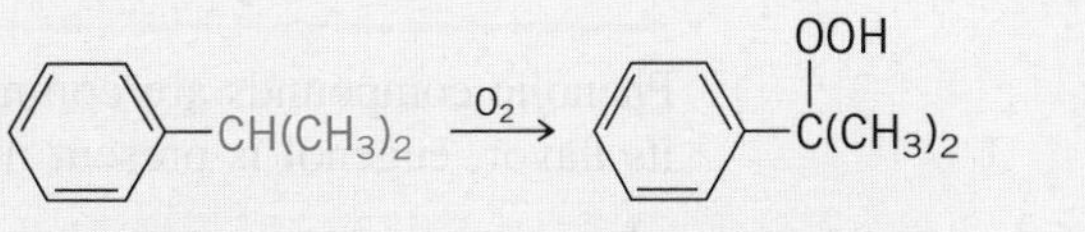 Isopropylbenzene (cumene) → 1-Methyl-1-phenylethyl hydroperoxide 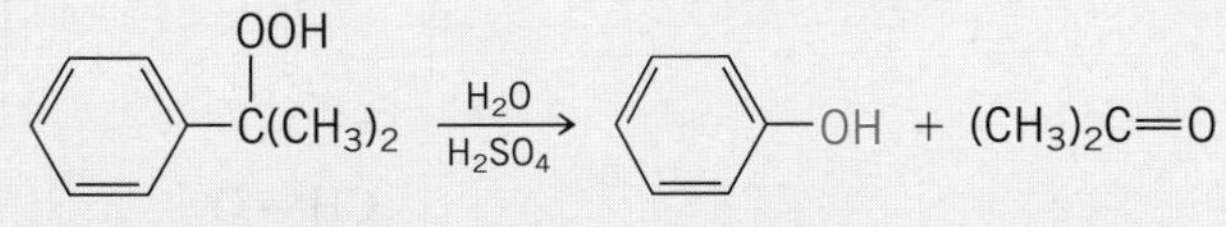 1-Methyl-1-phenylethyl hydroperoxide → Phenol + Acetone

Can you recall how to prepare chlorobenzene?

On the other hand, ^{14}C-labeling studies have shown that the base-promoted hydrolysis of chlorobenzene (second entry in Table 22.3) proceeds by the elimination–addition mechanism and involves a novel intermediate called *benzyne*.

Problem 22.5

Refer to Descriptive Passage and Interpretive Problems 12: Benzyne and suggest a mechanism for the hydrolysis of chlorobenzene under the conditions shown in Table 22.3.

Can you recall how to prepare isopropylbenzene?

The most widely used industrial synthesis of phenol is based on isopropylbenzene (cumene) as the starting material and is shown in the third entry of Table 22.3. The economically attractive features of this process are its use of cheap reagents (oxygen and sulfuric acid) and the fact that it yields two high-volume industrial chemicals: phenol and acetone. The mechanism of this novel synthesis forms the basis of Problem 22.34 at the end of this chapter.

The most important synthesis of phenols in the laboratory is from amines by hydrolysis of their corresponding diazonium salts, as described in Section 21.17.

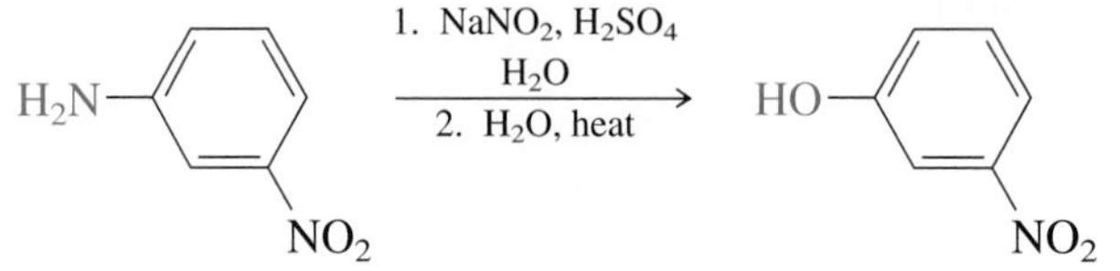

m-Nitroaniline   *m*-Nitrophenol (81–86%)

The Baeyer-Villiger oxidation is described in Descriptive Passage and Interpretive Problems 17.

Phenols can also be prepared by combining two reactions that you have seen previously: Baeyer–Villiger oxidation and ester hydrolysis. Oxidation of *p*-methoxyacetophenone with trifluoroperoxyacetic acid gives 4-methoxyphenyl acetate as the major product and methyl 4-methoxybenzoate as the minor one.

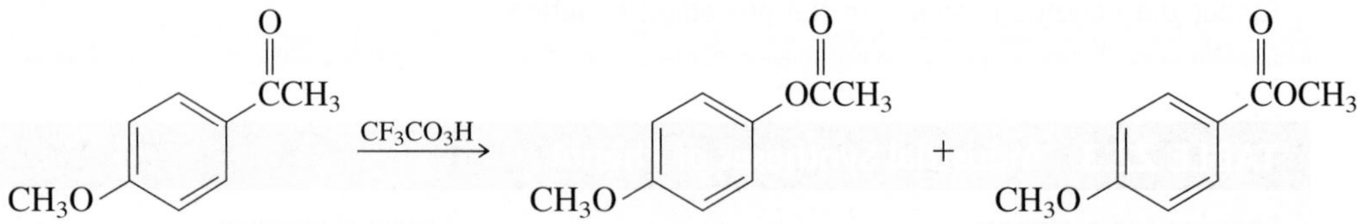

p-Methoxyacetophenone   4-Methoxyphenyl acetate (66%)   Methyl 4-methoxybenzoate (9%)

Hydrolysis of the ester group of 4-methoxyphenyl acetate gives 4-methoxyphenol.

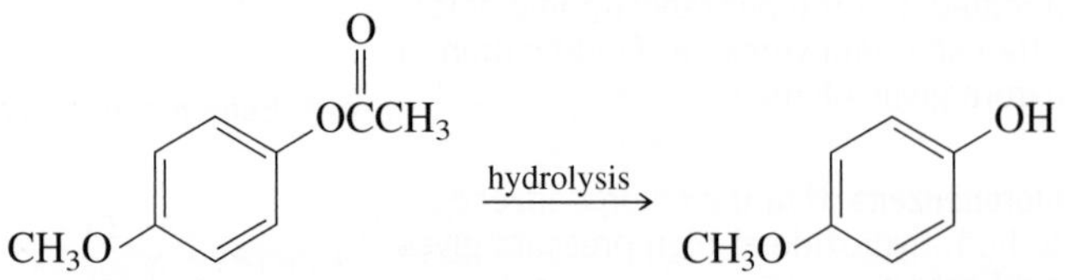

4-Methoxyphenyl acetate   4-Methoxyphenol

22.7 Naturally Occurring Phenols

Phenolic compounds are commonplace natural products. Vanillin gives the vanilla bean its flavor, eugenol is present in the oil of cloves, and thymol in thyme.

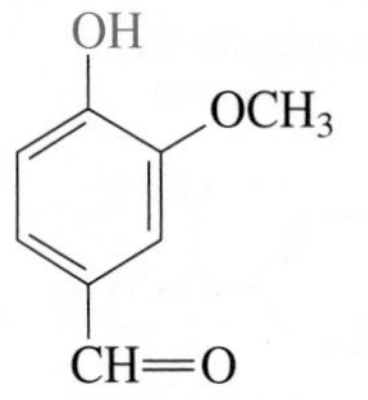

Vanillin

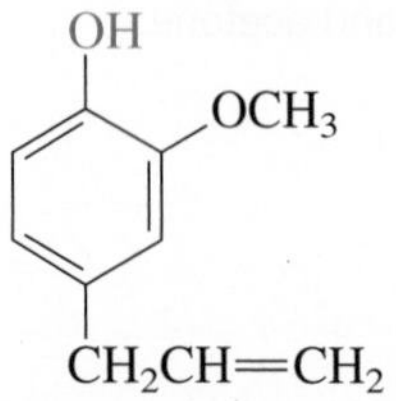

Eugenol

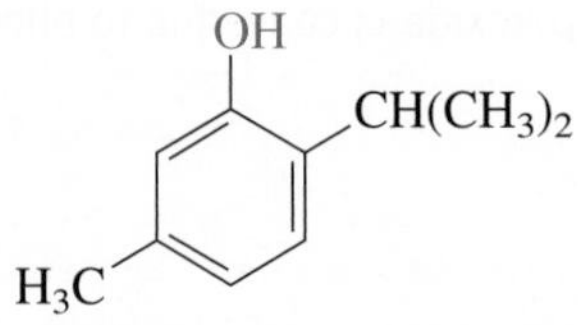

Thymol

2,5-Dichlorophenol has been isolated from the defensive substance of a species of grasshopper, Δ^9-tetrahydrocannabinol is the psychoactive material in marijuana, and tyrosine is the only phenol represented among the 20 amino acid components of proteins.

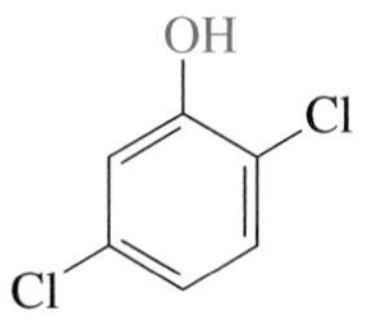

2,5-Dichlorophenol

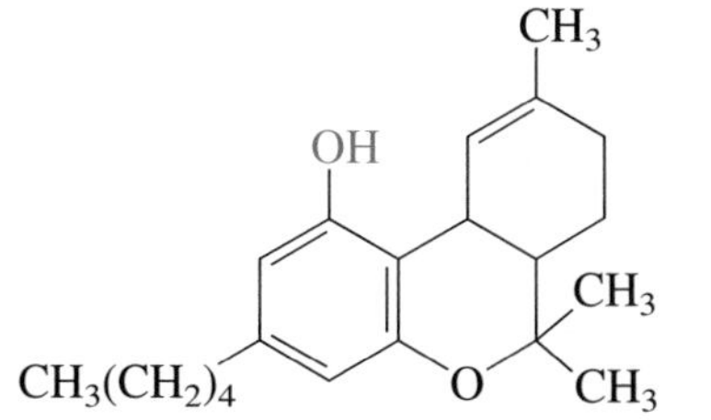

Δ^9-Tetrahydrocannabinol

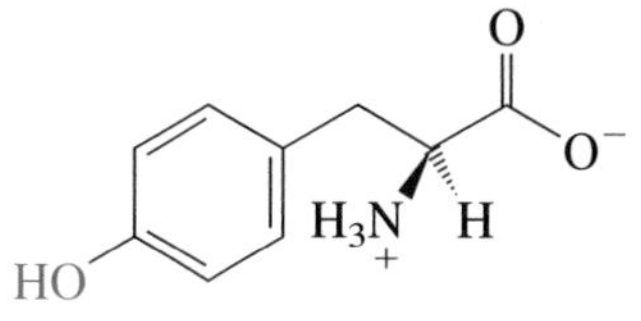

(*S*)-Tyrosine

Figure 22.2

Spotted knapweed produces a phenolic compound that kills other plants.

Many plant pigments are tricyclic phenols called flavanoids, which among their other properties are antioxidants. A flavanoid in green tea and red wine, (+)-catechin may play a role in the low incidence of atherosclerosis in Japan and France.

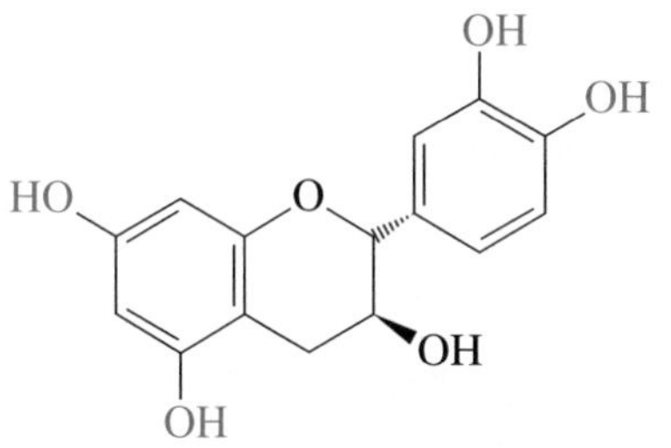

(+)-Catechin

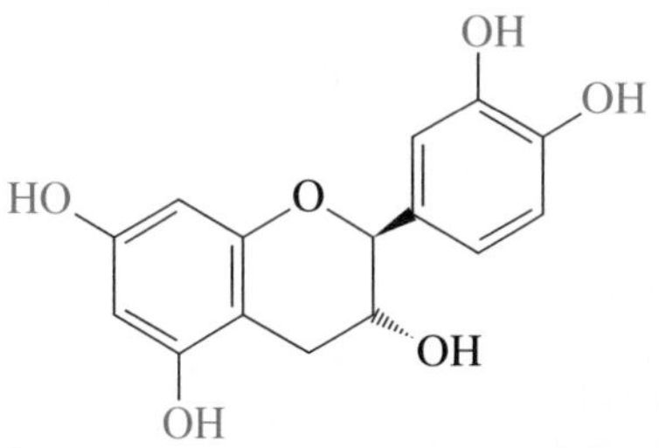

(−)-Catechin

Both catechin enantiomers are secreted as a racemic mixture through the roots of spotted knapweed (Figure 22.2). (+)-Catechin has antibacterial properties, but (−)-catechin kills other plants with which it comes in contact. In the approximately 100 years since it was accidentally introduced to the United States, spotted knapweed has spread rapidly, replacing native plants over millions of acres and forcing grazing animals to search elsewhere for food.

22.8 Reactions of Phenols: Electrophilic Aromatic Substitution

In most of their reactions phenols behave as nucleophiles, and the reagents that act on them are electrophiles. Either the hydroxyl oxygen or the aromatic ring may be the site of nucleophilic reactivity in a phenol. Reactions that take place on the ring lead to electrophilic aromatic substitution; Table 22.4 summarizes the behavior of phenols in reactions of this type.

A hydroxyl group is a very powerful activating substituent, and electrophilic aromatic substitution in phenols occurs far faster, and under milder conditions, than in benzene. The first entry in Table 22.4, for example, shows the monobromination of phenol in high yield at low temperature and in the absence of any catalyst. In this case, the reaction was carried out in the nonpolar solvent 1,2-dichloroethane. In polar solvents such as water it is difficult to limit the bromination of phenols to monosubstitution. In the following example, all three positions that are ortho or para to the hydroxyl undergo rapid substitution:

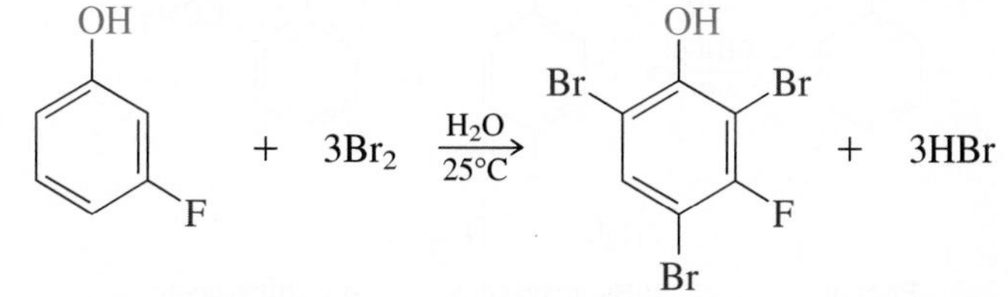

m-Fluorophenol — Bromine — 2,4,6-Tribromo-3-fluorophenol (95%) — Hydrogen bromide

TABLE 22.4 Electrophilic Aromatic Substitution Reactions of Phenols

Reaction and comments	Specific example
Halogenation Bromination and chlorination of phenols occur readily even in the absence of a catalyst. Substitution occurs primarily at the position para to the hydroxyl group. When the para position is blocked, ortho substitution is observed.	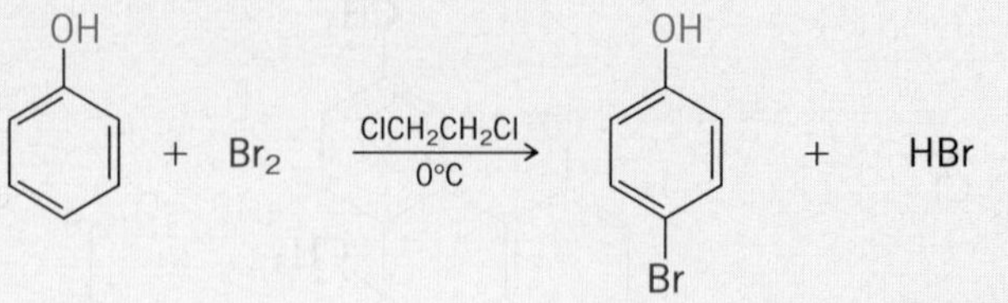 Phenol; Bromine; *p*-Bromophenol (93%); Hydrogen bromide
Nitration Phenols are nitrated on treatment with a dilute solution of nitric acid in either water or acetic acid. It is not necessary to use mixtures of nitric and sulfuric acids, because of the high reactivity of phenols.	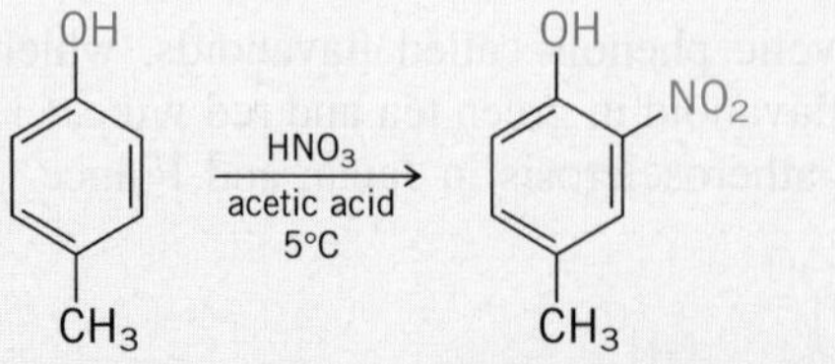 *p*-Cresol; 4-Methyl-2-nitrophenol (73–77%)
Nitrosation On acidification of aqueous solutions of sodium nitrite, the nitrosyl cation ($:\overset{+}{N}\equiv O:$) is formed, which is a weak electrophile and attacks the strongly activated ring of a phenol. The product is a nitroso phenol.	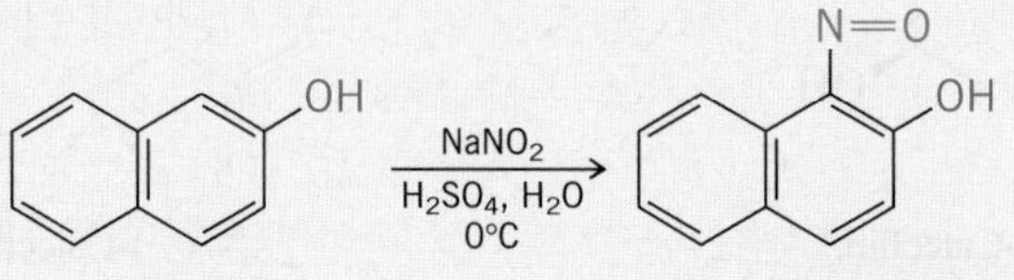 2-Naphthol; 1-Nitroso-2-naphthol (99%)
Sulfonation Heating a phenol with concentrated sulfuric acid causes sulfonation of the ring.	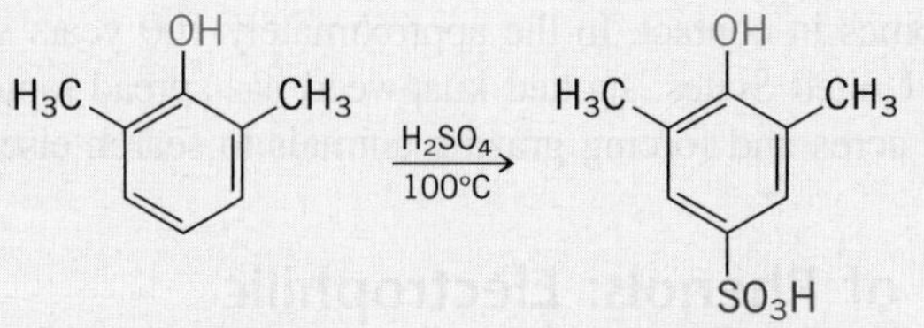 2,6-Dimethylphenol; 4-Hydroxy-3,5-dimethylbenzenesulfonic acid (69%)
Friedel–Crafts alkylation Alcohols in combination with acids serve as sources of carbocations. Attack of a carbocation on the electron-rich ring of a phenol brings about its alkylation.	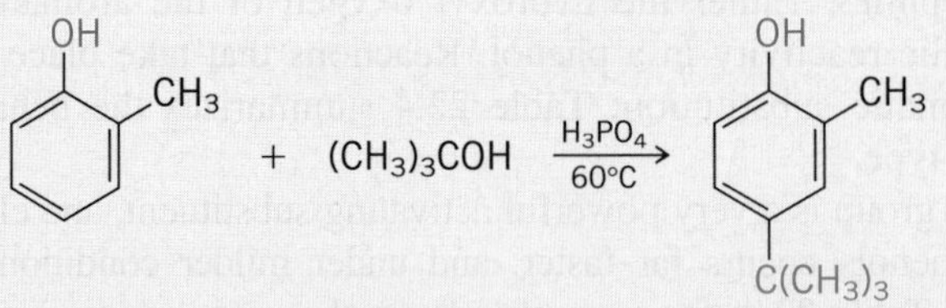 *o*-Cresol; *tert*-Butyl alcohol; 4-*tert*-Butyl-2-methylphenol (63%)
Friedel–Crafts acylation In the presence of aluminum chloride, acyl chlorides and acid anhydrides acylate the aromatic ring of phenols.	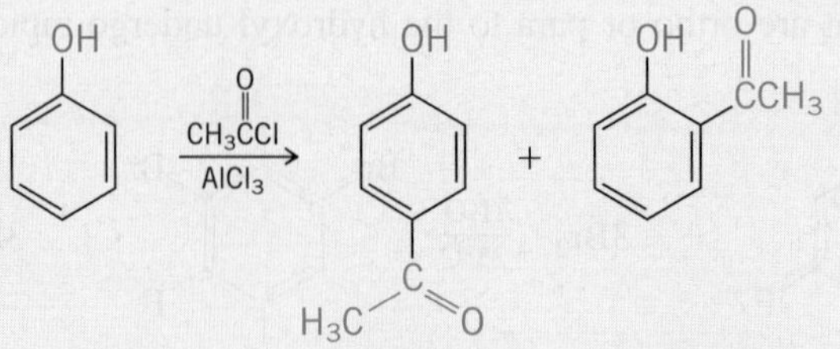 Phenol; *p*-Hydroxyacetophenone (74%); *o*-Hydroxyacetophenone (16%)

TABLE 22.4 Electrophilic Aromatic Substitution Reactions of Phenols (*Continued*)

Reaction and comments	Specific example
Reaction with arenediazonium salts Adding a phenol to a solution of a diazonium salt formed from a primary aromatic amine leads to formation of an azo compound. The reaction is carried out at a pH such that a significant portion of the phenol is present as its phenoxide ion. The diazonium ion acts as an electrophile toward the strongly activated ring of the phenoxide ion.	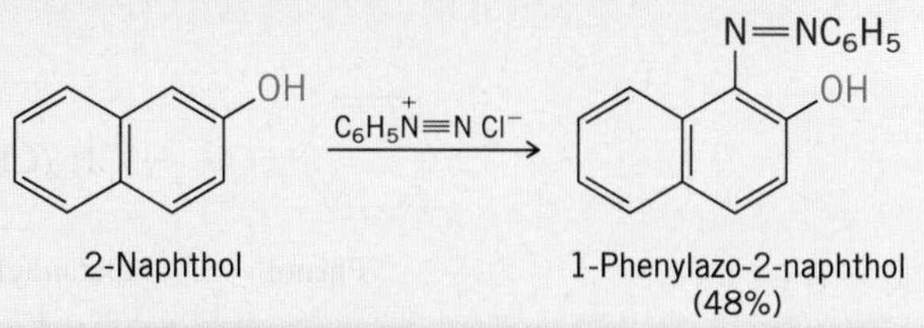

Other typical electrophilic aromatic substitution reactions—nitration (second entry), sulfonation (fourth entry), and Friedel–Crafts alkylation and acylation (fifth and sixth entries)—take place readily and are synthetically useful. Phenols also undergo electrophilic substitution reactions that are limited to only the most active aromatic compounds; these include nitrosation (third entry) and coupling with diazonium salts (seventh entry).

Problem 22.6

Each of the following reactions has been reported in the chemical literature and gives a single organic product in high yield. Identify the product in each case.

(a) 3-Benzyl-2,6-dimethylphenol treated with bromine in chloroform
(b) 4-Bromo-2-methylphenol treated with 2-methylpropene and sulfuric acid
(c) 2-Isopropyl-5-methylphenol (thymol) treated with sodium nitrite and dilute hydrochloric acid
(d) *p*-Cresol treated with propanoyl chloride and aluminum chloride

Sample Solution (a) The ring that bears the hydroxyl group is much more reactive than the other ring. In electrophilic aromatic substitution reactions of rings that bear several substituents, it is the most activating substituent that controls the orientation. Bromination occurs para to the hydroxyl group.

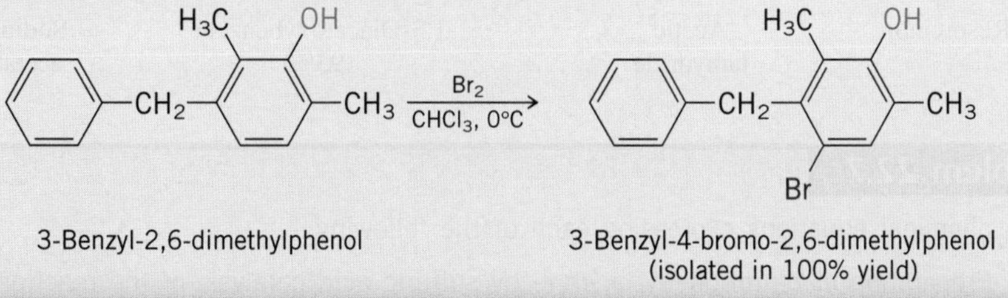
H3C
OH
CH2
CH3
Br2
CHCl3, 0°C
Br
3-Benzyl-2,6-dimethylphenol
3-Benzyl-4-bromo-2,6-dimethylphenol
(isolated in 100% yield)

The aromatic ring of a phenol, like that of an arylamine, is seen as an electron-rich functional unit and is capable of a variety of reactions. In some cases, however, it is the hydroxyl oxygen that reacts instead. An example of this kind of chemical reactivity is described in the following section.

22.9 Acylation of Phenols

Acylating agents, such as acyl chlorides and acid anhydrides, can react with phenols either at the aromatic ring (C-acylation) or at the hydroxyl oxygen (O-acylation):

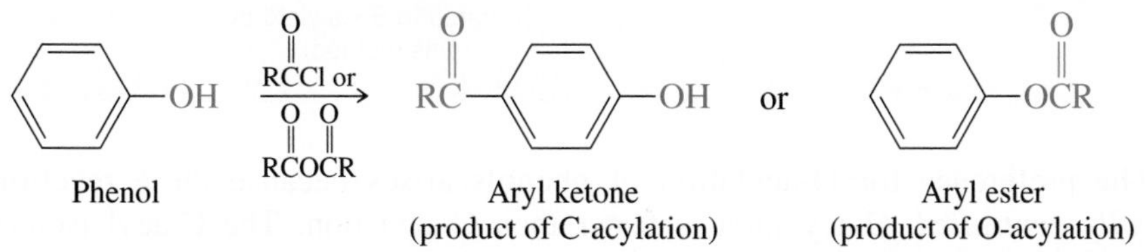
OH
RCCl or
RCOCR
RC
OH
or
OCR
Phenol
Aryl ketone
(product of C-acylation)
Aryl ester
(product of O-acylation)

As shown in the sixth entry of Table 22.4, C-acylation of phenols is observed under the customary conditions of the Friedel–Crafts reaction (treatment with an acyl chloride or acid anhydride in the presence of aluminum chloride). In the absence of aluminum chloride, however, O-acylation occurs instead.

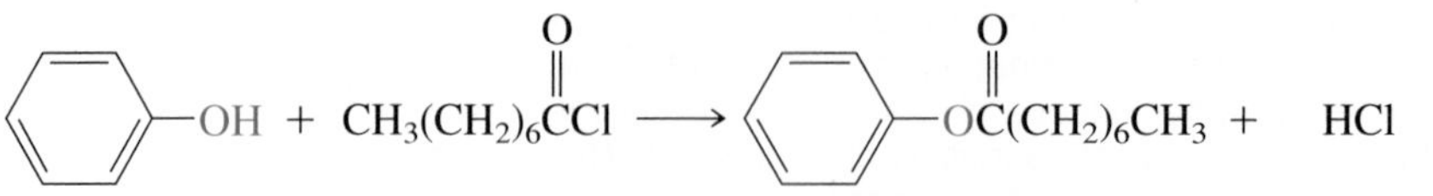

Phenol | Octanoyl chloride | Phenyl octanoate (95%) | Hydrogen chloride

The O-acylation of phenols with acid anhydrides can be conveniently catalyzed in either of two ways. One method involves converting the acid anhydride to a more powerful acylating agent by protonation of one of its carbonyl oxygens. Addition of a few drops of sulfuric acid is usually sufficient.

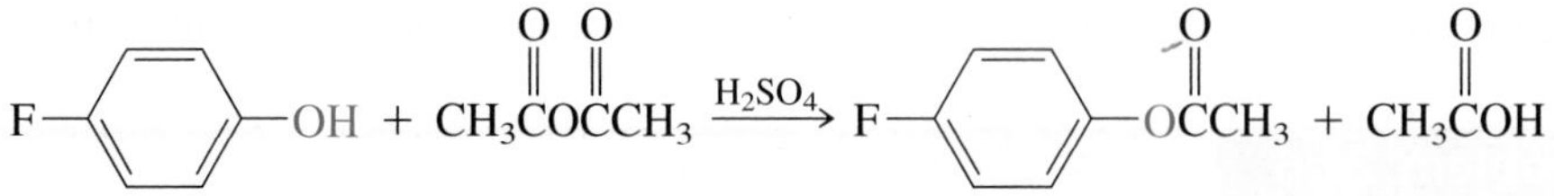

p-Fluorophenol | Acetic anhydride | *p*-Fluorophenyl acetate (81%) | Acetic acid

An alternative approach is to increase the nucleophilicity of the phenol by converting it to its phenoxide anion in basic solution:

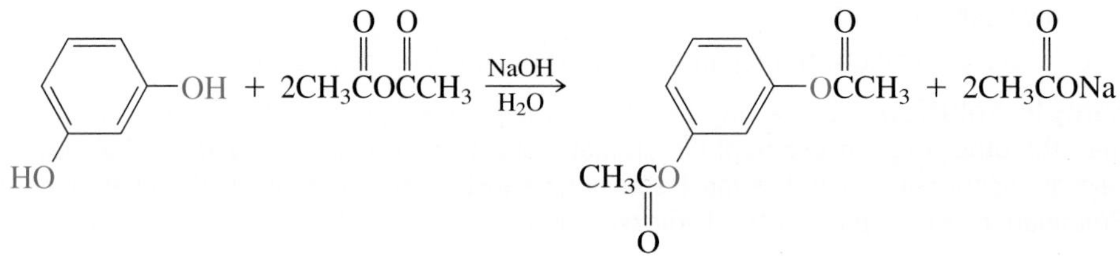

Resorcinol | Acetic anhydride | 1,3-Diacetoxybenzene (93%) | Sodium acetate

Problem 22.7

Write chemical equations expressing each of the following:

(a) Preparation of *o*-nitrophenyl acetate by sulfuric acid catalysis of the reaction between a phenol and an acid anhydride.

(b) Esterification of 2-naphthol with acetic anhydride in aqueous sodium hydroxide

(c) Reaction of phenol with benzoyl chloride

Sample Solution (a) The problem specifies that an acid anhydride be used; therefore, use acetic anhydride to prepare the acetate ester of *o*-nitrophenol:

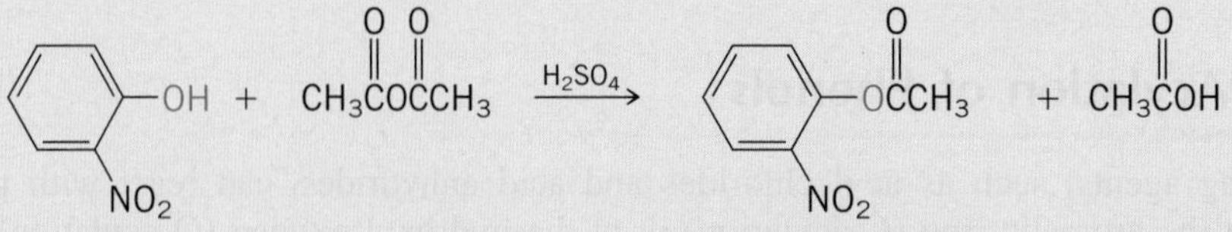

o-Nitrophenol | Acetic anhydride | *o*-Nitrophenyl acetate (isolated in 93% yield by this method) | Acetic acid

The preference for O-acylation of phenols arises because these reactions are *kinetically controlled.* O-acylation is faster than C-acylation. The C-acyl isomers are more stable, however, and it is known that aluminum chloride is a very effective catalyst for the conversion of aryl esters to aryl ketones. This isomerization is called the **Fries rearrangement.**

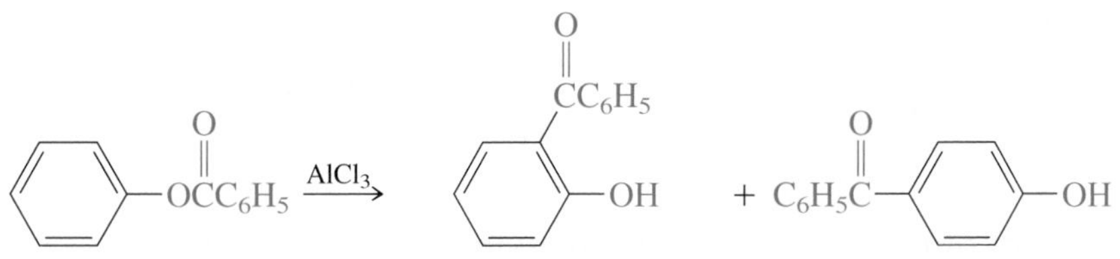

Phenyl benzoate — *o*-Hydroxybenzophenone (9%) — *p*-Hydroxybenzophenone (64%)

Thus, ring acylation of phenols is observed under Friedel–Crafts conditions because the presence of aluminum chloride causes that reaction to be subject to *thermodynamic (equilibrium) control.*

Fischer esterification, in which a phenol and a carboxylic acid condense in the presence of an acid catalyst, is not used to prepare aryl esters.

22.10 Carboxylation of Phenols: Aspirin and the Kolbe–Schmitt Reaction

The best known aryl ester is *O*-acetylsalicylic acid, better known as *aspirin.* It is prepared by acetylation of the phenolic hydroxyl group of salicylic acid:

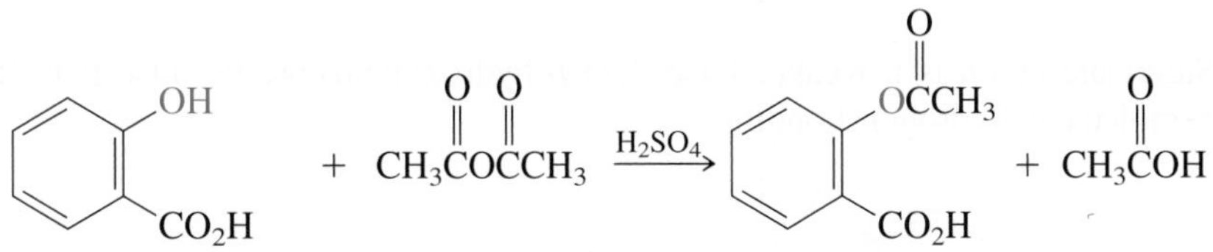

Salicylic acid (*o*-hydroxybenzoic acid) — Acetic anhydride — *O*-Acetylsalicylic acid (aspirin) — Acetic acid

Aspirin possesses a number of properties that make it an often-recommended drug. It is an analgesic, effective in relieving headache pain. It is also an antiinflammatory agent, providing some relief from the swelling associated with arthritis and minor injuries. Aspirin is an antipyretic compound; that is, it reduces fever. How aspirin does all this was once a mystery but is now better understood and will be discussed in Section 24.6. Until recently, more than 40 million lb of aspirin were produced in the United States, a rate equal to 300 tablets per year for every man, woman, and child.

The key compound in the synthesis of aspirin, salicylic acid, is prepared from phenol by a process discovered in the nineteenth century by the German chemist Hermann Kolbe. In the Kolbe synthesis, also known as the **Kolbe–Schmitt reaction,** sodium phenoxide is heated with carbon dioxide under pressure, and the reaction mixture is subsequently acidified to yield salicylic acid:

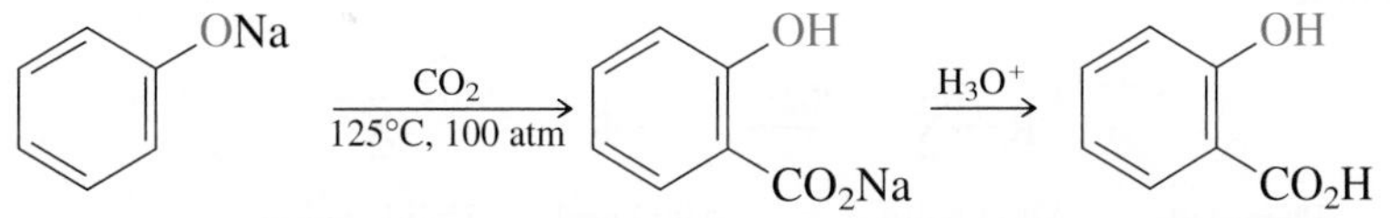

Sodium phenoxide — Sodium salicylate — Salicylic acid (79%)

Although a hydroxyl group strongly activates an aromatic ring toward electrophilic attack, an oxyanion substituent is an even more powerful activator. Electron delocalization in phenoxide anion leads to increased electron density at the positions ortho and para to oxygen. The increased nucleophilicity of the ring permits it to react with carbon dioxide. An intermediate is formed that is simply the keto form of salicylate anion:

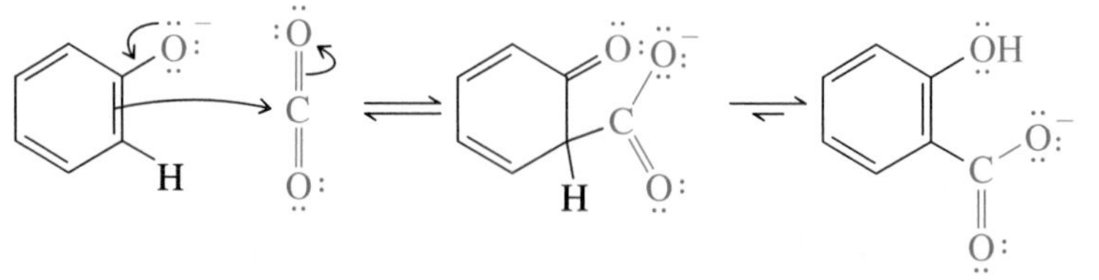

Phenoxide anion (stronger base) | Carbon dioxide | Cyclohexadienone intermediate | Salicylate anion (weaker base)

The Kolbe–Schmitt reaction is an equilibrium process governed by thermodynamic control. The position of equilibrium favors formation of the weaker base (salicylate ion) at the expense of the stronger one (phenoxide ion). Thermodynamic control is also responsible for the pronounced bias toward ortho over para substitution. Salicylate anion is a weaker base than *p*-hydroxybenzoate and predominates at equilibrium.

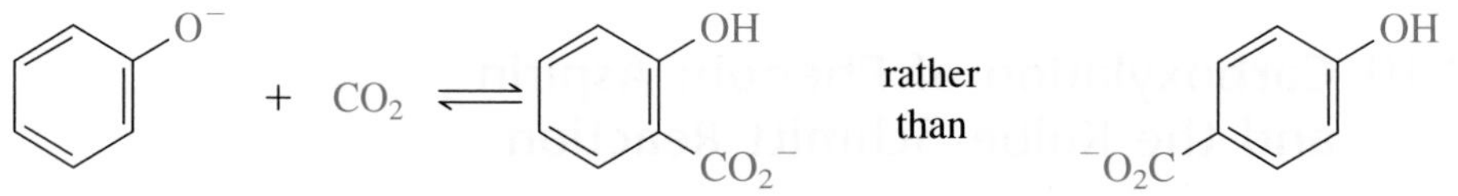

Phenoxide ion (strongest base; pK_a of conjugate acid, 10) | Carbon dioxide | Salicylate anion (weakest base; pK_a of conjugate acid, 3) | *p*-Hydroxybenzoate anion (pK_a of conjugate acid, 4.5)

Salicylate anion is a weaker base than *p*-hydroxybenzoate because it is stabilized by intramolecular hydrogen bonding.

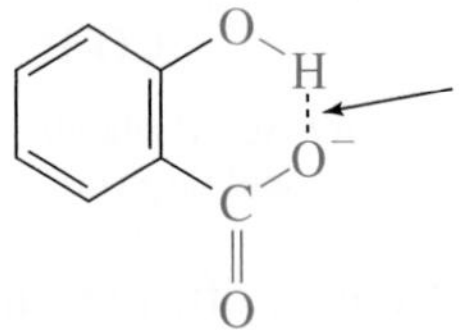

Intramolecular hydrogen bonding in salicylate anion

The Kolbe–Schmitt reaction has been applied to the preparation of other *o*-hydroxybenzoic acids. Alkyl derivatives of phenol behave very much like phenol itself. Phenols that bear strongly electron-withdrawing substituents usually give low yields of carboxylated products; their derived phenoxide anions are less basic, and the equilibrium constants for their carboxylation are smaller.

22.11 Preparation of Aryl Ethers

Aryl ethers are best prepared by the Williamson method (Section 16.6). Alkylation of the hydroxyl oxygen of a phenol takes place readily when a phenoxide anion reacts with an alkyl halide.

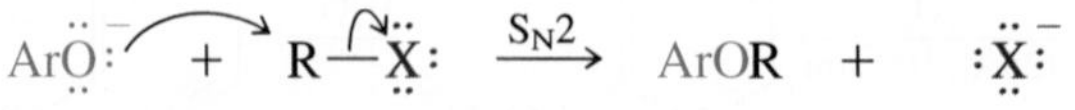

Phenoxide anion | Alkyl halide | Alkyl aryl ether | Halide anion

This is an example of an S_N2 reaction in a polar aprotic solvent.

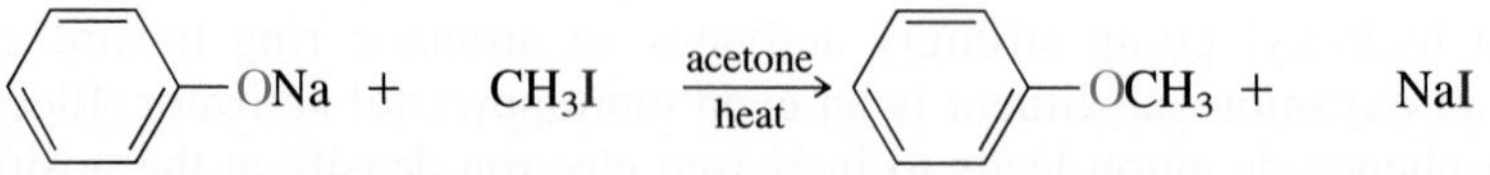

Sodium phenoxide | Iodomethane | Anisole (95%) | Sodium iodide

As the synthesis is normally performed, a solution of the phenol and alkyl halide is simply heated in the presence of a suitable base such as potassium carbonate:

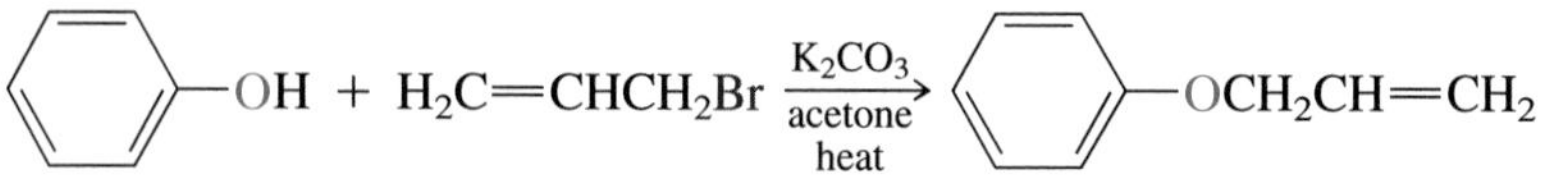

Phenol Allyl bromide Allyl phenyl ether (86%)

The alkyl halide must be one that reacts readily by an S_N2 mechanism. Thus, methyl and primary alkyl halides are the most effective alkylating agents. Elimination competes with substitution when secondary alkyl halides are used and is the only reaction observed with tertiary alkyl halides.

Problem 22.8

The aryl ethers shown here are intermediates that have been used to make pharmaceuticals. In each case, provide the structures of the phenol and the alkyl halide from which these compounds can be synthesized.

(a) 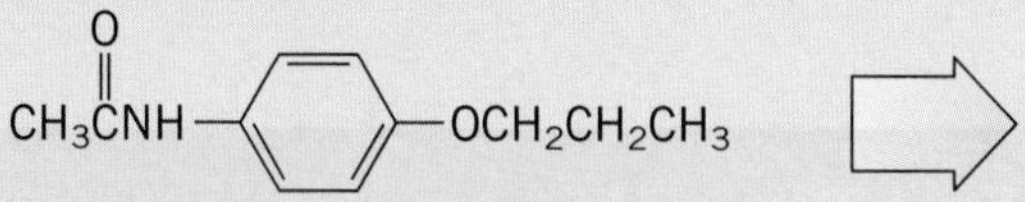

(b) 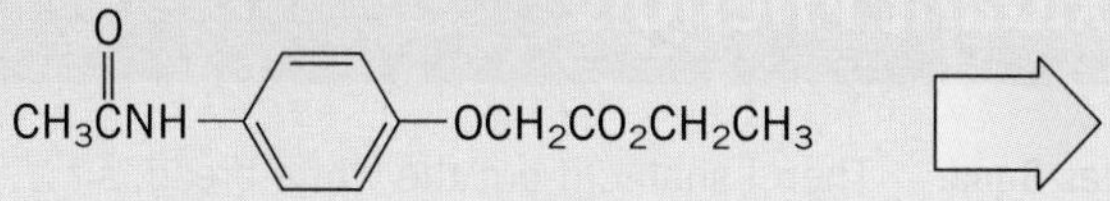

(c) 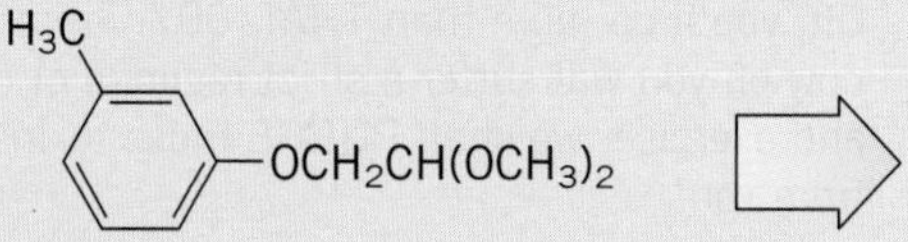

Sample Solution

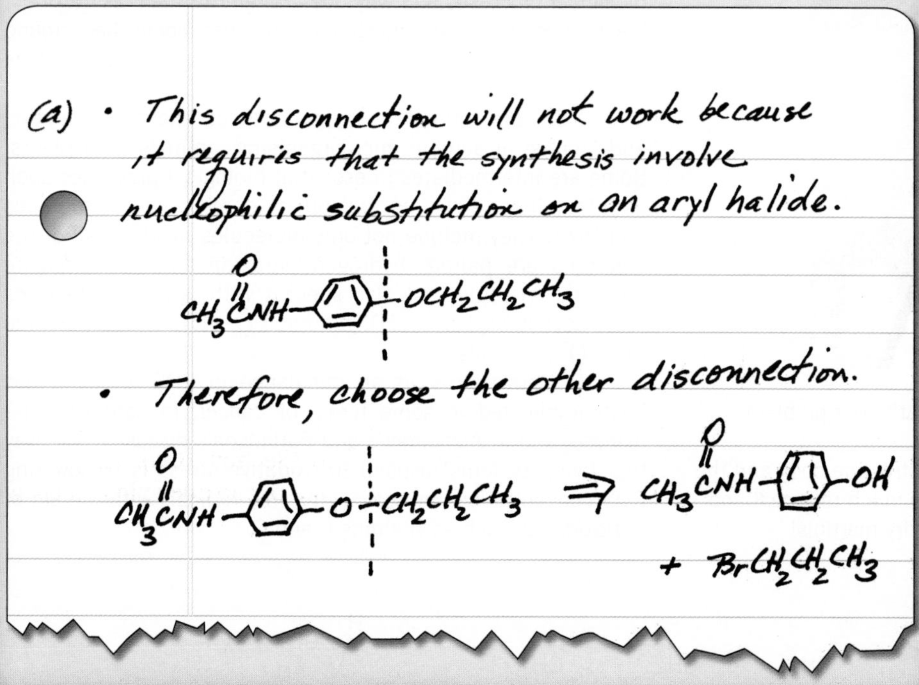

The reaction between an alkoxide ion and an aryl halide can be used to prepare alkyl aryl ethers only when the aryl halide is one that reacts rapidly by the addition–elimination mechanism of nucleophilic aromatic substitution (Section 12.21).

p-Fluoronitrobenzene *p*-Nitroanisole (93%)

Problem 22.9

Which of the following two combinations of reactants is more appropriate for the preparation of *p*-nitrophenyl phenyl ether?

Fluorobenzene and sodium *p*-nitrophenoxide, or *p*-fluoronitrobenzene and sodium phenoxide

James Bond, Oxidative Stress, and Antioxidant Phenols

In the film *Never Say Never Again* James Bond has the following conversation with M:

M: Too many free radicals. That's your problem.
James Bond: "Free radicals," sir?
M: Yes. They're toxins that destroy the body and the brain, caused by eating too much red meat and white bread and too many dry martinis!
James Bond: Then I shall cut out the white bread, sir.
M: Oh, you'll do more than THAT, 007. From now on you will suffer a strict regimen of diet and exercise; we shall PURGE those toxins from you!

We're familiar with free radicals as reactive, short-lived intermediates in chemical reactions. How are they involved in biological processes and why does M refer to them as "toxins"? We'll consider these questions, but first some background on *reactive oxygen species* (ROS) and *oxidative stress* will be helpful.

ROS are formed as byproducts of energy production and storage in aerobic microorganisms, animals, and plants. Some are intermediates in essential biological processes such as intracellular signaling, but others damage cell membranes and DNA. They include not only molecules in which all of the electrons are paired, hydrogen peroxide and its conjugate base, for example, but also species with unpaired electrons such as hydroxyl (·Ö—H), alkoxyl (·Ö—R), and superoxide (·Ö—Ö:$^-$) radicals.

Oxidative stress is an imbalance in ROS levels. It has been implicated in some forms of cancer, in cardiovascular disease, and in Alzheimer's and Parkinson's diseases. One way that living systems respond to oxidative stress is by lowering ROS concentrations by scavenging free radicals with the aid of *antioxidants,* such as vitamins E and C.

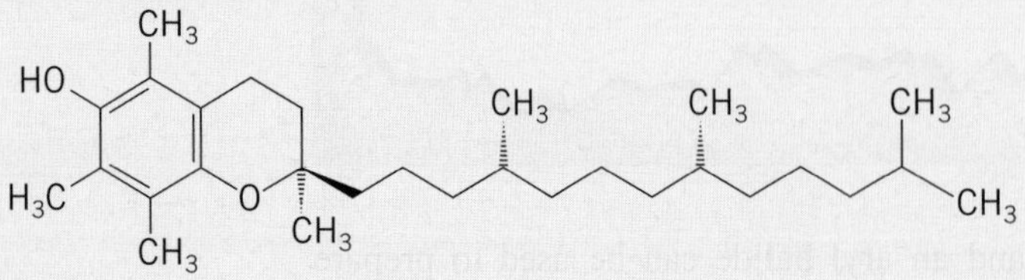

Vitamin E (α-tocopherol)

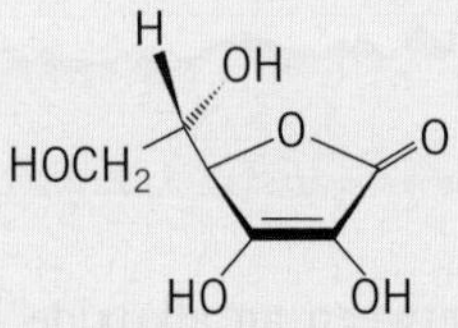

Vitamin C (L-ascorbic acid)

Vitamin E is present in the nonpolar interior of cell membranes, and vitamin C in the water-rich cytosol. Vitamin E protects against oxidation of cell membrane lipids by a process that you have seen before; namely, termination of a radical chain reaction.

Consider a cell membrane lipid derived from linoleic acid, which is an unsaturated fatty acid. Hydrogen atom abstraction of an allylic hydrogen by a hydroxyl radical leads to a resonance-stabilized free allylic radical in which the unpaired electron is delocalized over C-9, C-11, and C-13.

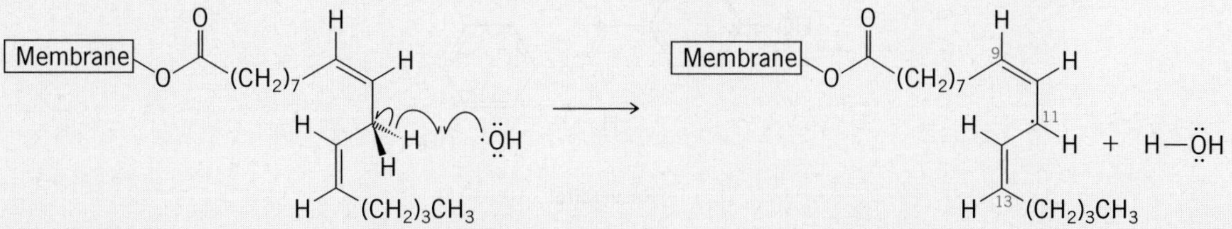

Linoleic portion of a cell membrane + hydroxyl radical — An allylic radical — Water

Reaction of O_2 with this allylic radical is possible at any of the three carbons that share the unpaired electron to give a peroxyl radical according to the general equation:

Allylic radical from preceding equation — Oxygen — Peroxyl radical

The peroxyl radical can then go on to abstract an allylic hydrogen from another molecule of lipid as part of a chain reaction that can damage the cell membrane.

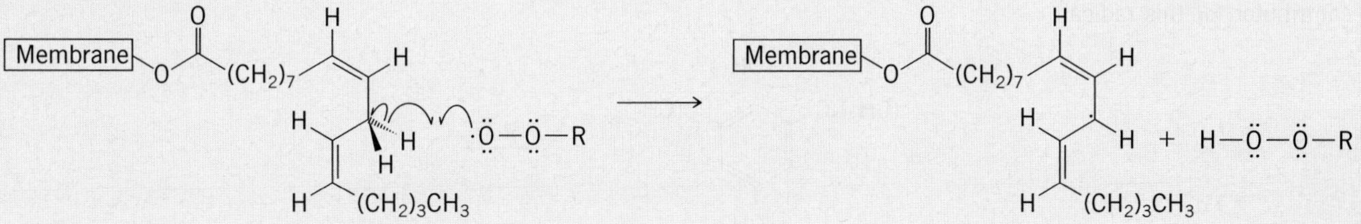

Linoleic portion of a cell membrane + peroxyl radical — An allylic radical — A hydroperoxide

Vitamin E interferes with the chain-propagating step described in the preceding equation by transferring a hydrogen atom to the peroxyl radical, converting it to a less-reactive hydroperoxide and preventing further damage to the cell membrane.

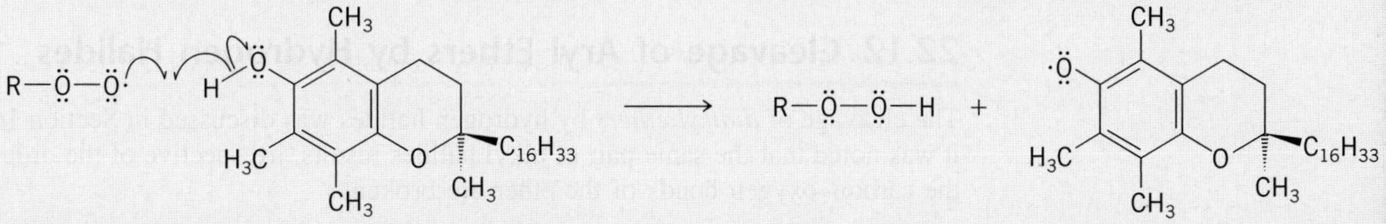

Peroxyl radical — Vitamin E — A hydroperoxide — Vitamin E radical

The vitamin E radical is stabilized by delocalization of the unpaired electron into the benzene ring and does not abstract hydrogen atoms from the lipid.

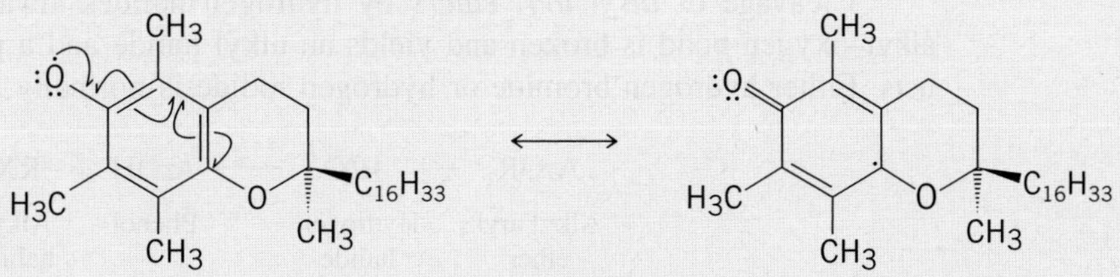

Vitamin E radical

Continued

Vitamin E is found in nuts, seeds, and vegetable and fish oils. Citrus fruits and juices are rich in vitamin C. In addition to vitamins C and E, it is estimated that over 4000 polyphenolic natural products with antioxidant properties are found in fruits and vegetables as well as in coffee, tea, red wine, and chocolate. The beneficial effects of resveratrol, a phenolic compound present in the skins of red grapes and other foods, are thought to originate in resveratrol's ability to act as an antioxidant.

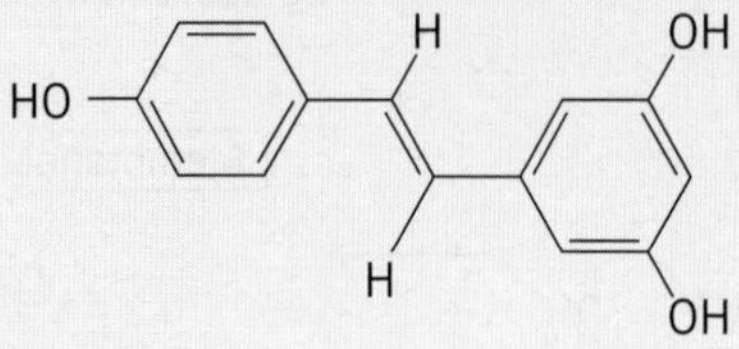

Resveratrol

Excessive alcohol consumption as well as smoking are thought to increase oxidative stress, whereas regular exercise may actually enhance the body's antioxidant defense systems. So, it looks like it's more exercise, a better diet, and fewer martinis for 007!

Problem 22.10

The allyl radical of linoleic acid is stabilized by resonance involving the C-9–10 and C-12–13 double bonds. Show structures for each of these resonance contributors.

Problem 22.11

Phenolic compounds such as BHT (butylated hydroxytoluene) are added to food products to retard spoilage. What is the structure of the radical that would be formed from BHT when it reacts with an alkyl radical R·? Write a resonance contributor for this radical.

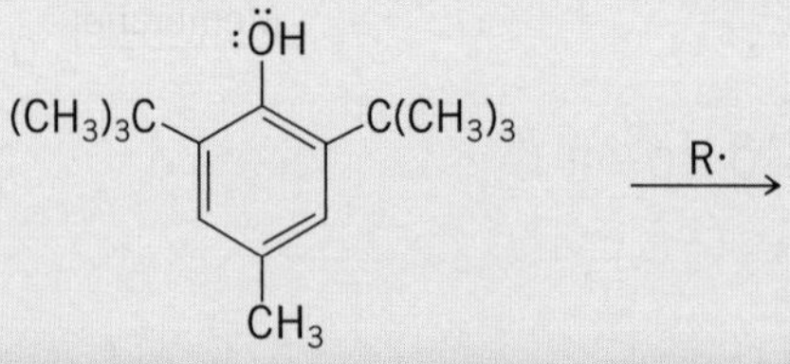

22.12 Cleavage of Aryl Ethers by Hydrogen Halides

The cleavage of *dialkyl ethers* by hydrogen halides was discussed in Section 16.8, where it was noted that the same pair of alkyl halides results, irrespective of the order in which the carbon–oxygen bonds of the ether are broken.

$$\underset{\text{Dialkyl ether}}{ROR'} + \underset{\text{Hydrogen halide}}{2HX} \longrightarrow \underset{\text{Two alkyl halides}}{RX + R'X} + \underset{\text{Water}}{H_2O}$$

Cleavage of *alkyl aryl ethers* by hydrogen halides always proceeds so that the alkyl–oxygen bond is broken and yields an alkyl halide and a phenol as the *final* products. Either hydrogen bromide or hydrogen iodide is normally used.

$$\underset{\text{Alkyl aryl ether}}{ArOR} + \underset{\text{Hydrogen halide}}{HX} \longrightarrow \underset{\text{Phenol}}{ArOH} + \underset{\text{Alkyl halide}}{RX}$$

Because phenols are not converted to aryl halides by reaction with hydrogen halides, the reaction proceeds no further than shown in the preceding general equation. For example,

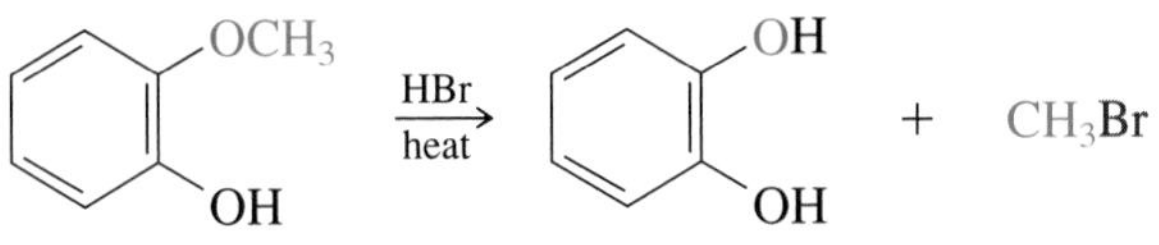

Guaiacol — Pyrocatechol (85–87%) — Methyl bromide (57–72%)

Guaiacol is obtained by chemical treatment of *lignum vitae,* the wood from a species of tree that grows in warm climates. It is sometimes used as an expectorant to help relieve bronchial congestion.

The first step in the reaction of an alkyl aryl ether with a hydrogen halide is protonation of oxygen to form an alkylaryloxonium ion:

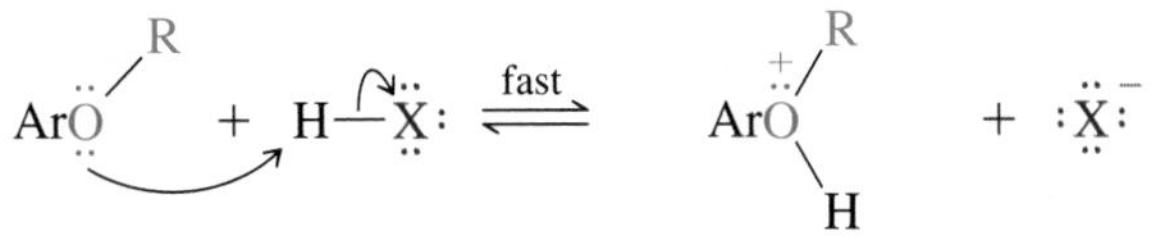

Alkyl aryl ether — Hydrogen halide — Alkylaryloxonium ion — Halide ion

This is followed by a nucleophilic substitution step, which is S_N2-like if the alkyl group is primary or secondary.

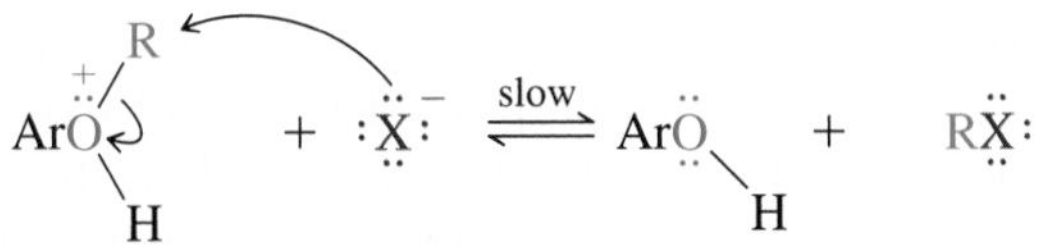

Alkylaryloxonium ion — Halide ion — Phenol — Alkyl halide

Attack by the halide nucleophile always occurs at the sp^3-hybridized carbon of the alkyl group and is analogous to what takes place in the cleavage of dialkyl ethers. Nucleophilic *aromatic* substitution does not occur under these conditions.

22.13 Claisen Rearrangement of Allyl Aryl Ethers

Allyl aryl ethers undergo an interesting reaction, called the **Claisen rearrangement,** on being heated. The allyl group migrates from oxygen to the ring carbon ortho to it.

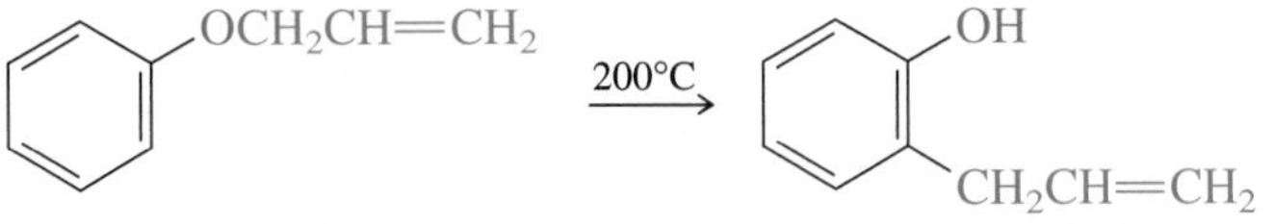

Allyl phenyl ether — *o*-Allylphenol (73%)

Allyl phenyl ether is prepared by the reaction of phenol with allyl bromide, as described in Section 22.11.

Carbon-14 labeling of the allyl group reveals that the terminal carbon of the allyl group is the one that becomes bonded to the ring and suggests a mechanism involving a concerted electron reorganization in the first step. This step is followed by enolization of the resulting cyclohexadienone to regenerate the aromatic ring.

$* = {}^{14}C$

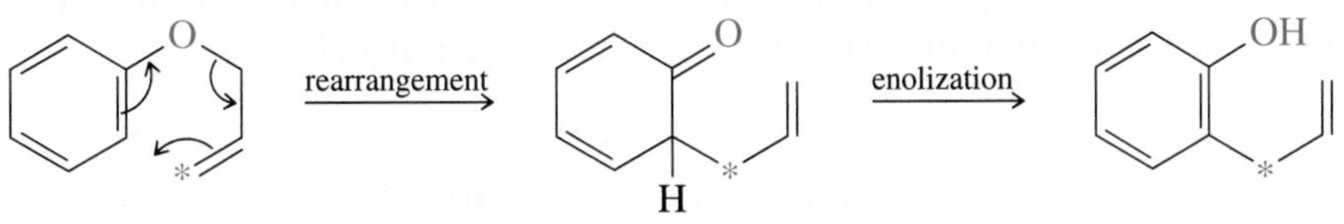

Allyl phenyl ether — 6-Allyl-2,4-cyclohexadienone — *o*-Allylphenol

Problem 22.12

The mechanism of the Claisen rearrangement of other allylic ethers of phenol is analogous to that of allyl phenyl ether. What is the product of the Claisen rearrangement of $C_6H_5OCH_2CH{=}CHCH_3$?

The transition state for the first step of the Claisen rearrangement bears much in common with the transition state for the Diels–Alder cycloaddition. Both involve a concerted six-electron reorganization.

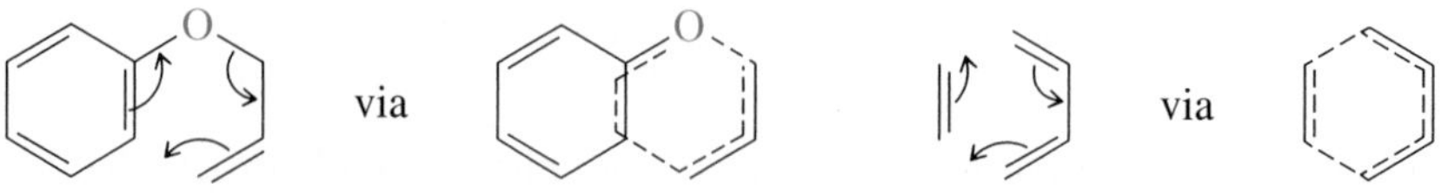
via
via

Claisen rearrangement Diels–Alder cycloaddition

The Claisen rearrangement is an example of a **sigmatropic rearrangement.** A sigmatropic rearrangement is characterized by a transition state in which a σ bond migrates from one end of a conjugated π electron system to the other. In this case the σ bond to oxygen at one end of an allyl unit is broken and replaced by a σ bond to the ring carbon at the other end.

22.14 Oxidation of Phenols: Quinones

Phenols are more easily oxidized than alcohols, and a large number of inorganic oxidizing agents have been used for this purpose. The phenol oxidations that are of the most use to the organic chemist are those involving derivatives of 1,2-benzenediol (pyrocatechol) and 1,4-benzenediol (hydroquinone). Oxidation of compounds of this type with silver oxide or with chromic acid yields conjugated dicarbonyl compounds called **quinones.**

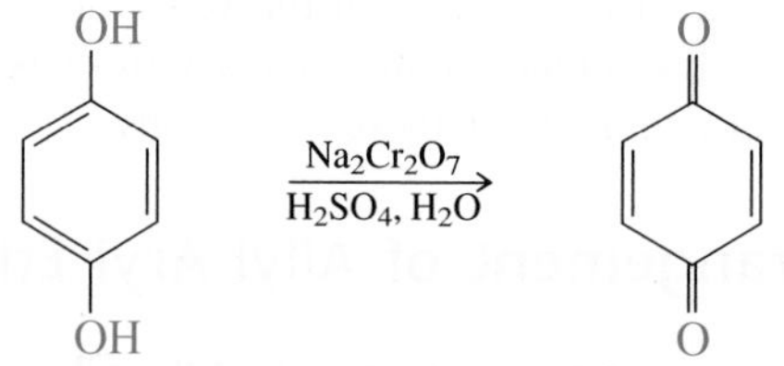
OH
OH
$Na_2Cr_2O_7$
H_2SO_4, H_2O
O
O

Hydroquinone *p*-Benzoquinone (76–81%)

Silver oxide is a weak oxidizing agent.

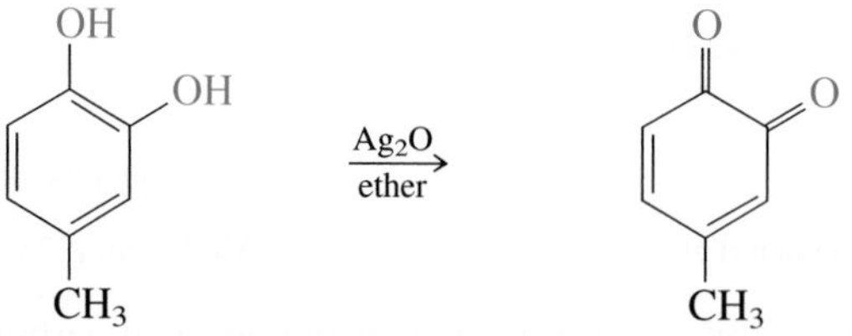
OH
OH
CH_3
Ag_2O
ether
O
O
CH_3

4-Methylpyrocatechol (4-methyl-1,2-benzenediol) 4-Methyl-1,2-benzoquinone (68%)

Quinones are colored; *p*-benzoquinone, for example, is yellow. Many occur naturally and have been used as dyes. *Alizarin* is a red pigment extracted from the roots of the madder plant. Its preparation from anthracene, a coal tar derivative, in 1868 was a significant step in the development of the synthetic dyestuff industry.

Quinones that are based on the anthracene ring system are called *anthraquinones.* Alizarin is one example of an *anthraquinone dye.*

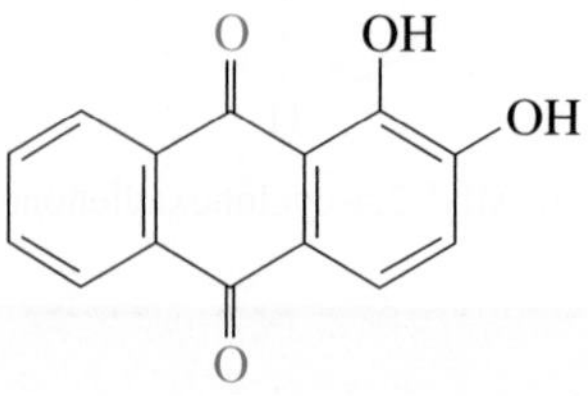
O
OH
OH
O

Alizarin

The oxidation–reduction process that connects hydroquinone and benzoquinone involves two 1-electron transfers:

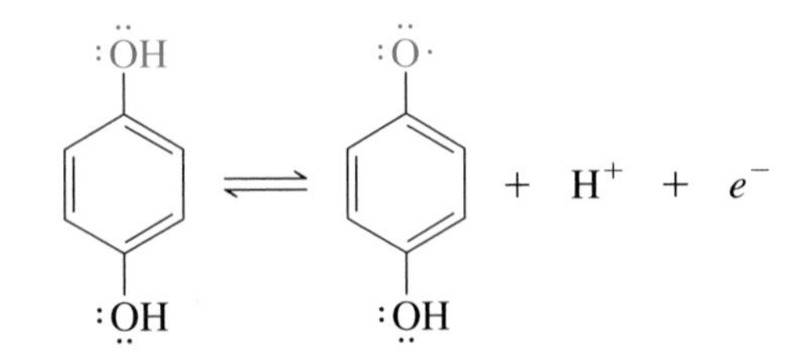

Hydroquinone

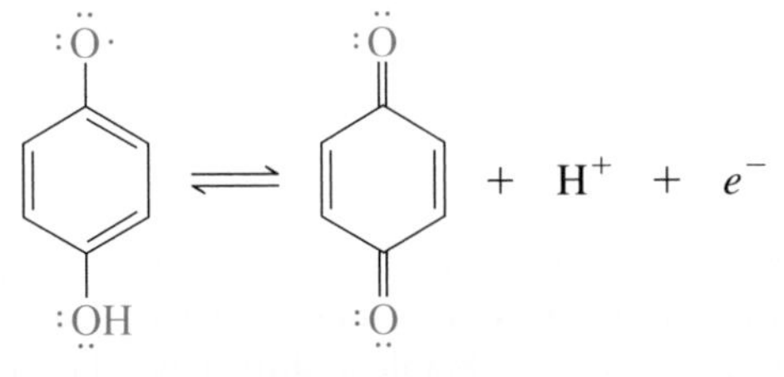

Benzoquinone

The ready reversibility of this reaction is essential to the role that quinones play in cellular respiration, the process by which an organism uses molecular oxygen to convert its food to carbon dioxide, water, and energy. Electrons are not transferred directly from the substrate molecule to oxygen but instead are transferred by way of an *electron transport chain* involving a succession of oxidation–reduction reactions. A key component of this electron transport chain is the substance known as *ubiquinone,* or **coenzyme Q:**

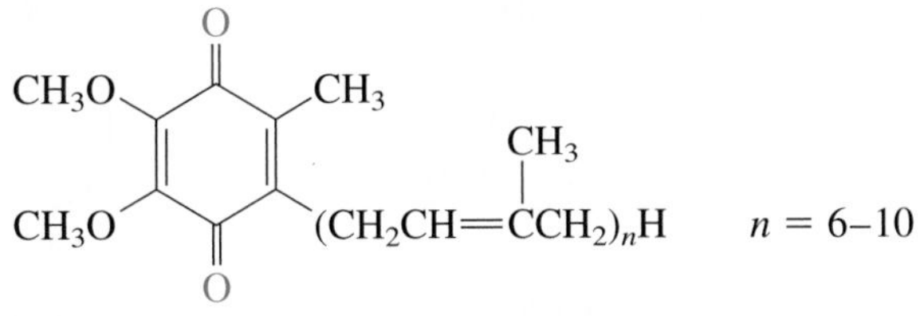

Ubiquinone (coenzyme Q)

The name *ubiquinone* is a shortened form of *ubiquitous quinone,* a term coined to describe the observation that this substance can be found in all cells. The length of its side chain varies among different organisms; the most common form in vertebrates has $n = 10$, and ubiquinones in which $n = 6$ to 9 are found in yeasts and plants.

Another physiologically important quinone is vitamin K. Here "K" stands for *koagulation* (Danish) because this substance was first identified as essential for the normal clotting of blood.

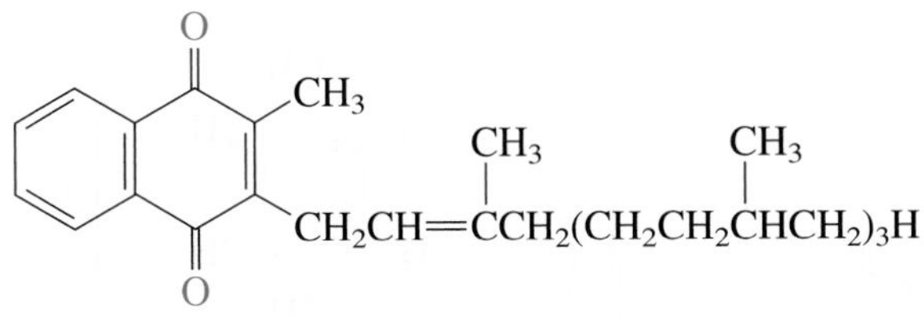

Vitamin K

Some vitamin K is provided in the normal diet, but a large proportion of that required by humans is produced by their intestinal flora.

Intestinal flora is a general term for the bacteria, yeast, and fungi that live in the large intestine.

22.15 Spectroscopic Analysis of Phenols

Infrared: The IR spectra of phenols combine features of those of alcohols and aromatic compounds. Hydroxyl absorbances resulting from O—H stretching are found in the 3600-cm^{-1} region, and the peak due to C—O stretching appears around 1200–1250 cm^{-1}. These features can be seen in the IR spectrum of *p*-cresol, shown in Figure 22.3.

Figure 22.3

The infrared spectrum of *p*-cresol.

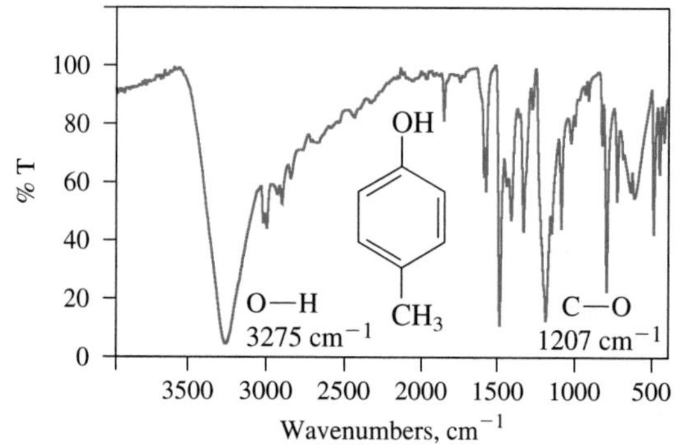

^{1}H NMR: The ^{1}H NMR signals for the hydroxyl protons of phenols are often broad, and their chemical shift, like their acidity, lies between alcohols and carboxylic acids. The range is δ 4–12, with the exact chemical shift depending on the concentration, the solvent, and the temperature. The phenolic proton in the ^{1}H NMR spectrum shown for *p*-cresol, for example, appears at δ 5.1 (Figure 22.4).

^{13}C NMR: The —OH group of a phenol has its largest effects on the carbon to which it is attached, and those ortho to it. The —OH group *deshields* the carbon to which it is attached by about 25 ppm, while *shielding* the ortho carbon by about 14 ppm. Aryl ethers behave similarly.

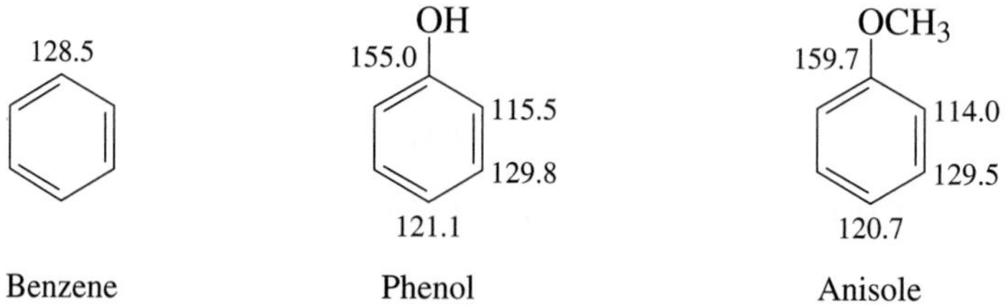

^{13}C Chemical shifts δ (ppm)

Notice, too, that the most shielded carbons of the aromatic ring are the ones that are ortho and para to the hydroxyl group in keeping with our experience that the OH group donates electrons preferentially to these positions.

UV-VIS: Just as with arylamines (Section 21.19), it is informative to look at the UV-VIS behavior of phenols in terms of how the OH group affects the benzene chromophore.

Figure 22.4

The 200-MHz ^{1}H NMR spectrum of *p*-cresol.

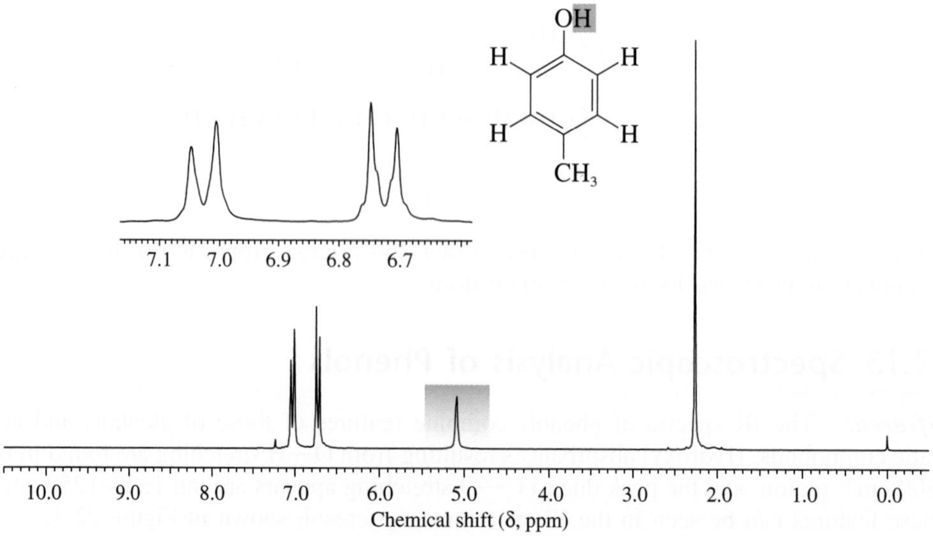

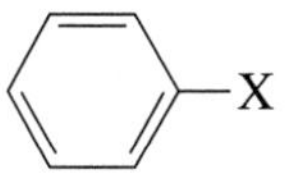

	X	λ_{max} (*nm*)
Benzene	H	204, 256
Aniline	NH_2	230, 280
Phenol	OH	210, 270
Phenoxide ion	O^-	235, 287

An OH group affects the UV-VIS spectrum of benzene in a way similar to that of an NH_2 group, but to a smaller extent. In basic solution, in which OH is converted to O^-, however, the shift to longer wavelengths exceeds that of an NH_2 group.

Mass Spectrometry: A peak for the molecular ion is usually quite prominent in the mass spectra of phenols. It is, for example, the most intense peak in phenol.

22.16 SUMMARY

Section 22.1 Phenol is both an important industrial chemical and the parent of a large class of compounds widely distributed as natural products. Although *benzenol* is the systematic name for C_6H_5OH, the IUPAC rules permit *phenol* to be used instead. Substituted derivatives are named on the basis of phenol as the parent compound.

Section 22.2 Phenols are polar compounds, but less polar than alcohols. They resemble arylamines in having an electron-rich aromatic ring.

Section 22.3 The —OH group of phenols makes it possible for them to participate in hydrogen bonding. This contributes to the higher boiling points and greater water-solubility of phenolic compounds compared with arenes and aryl halides.

Section 22.4 With pK_a's of approximately 10, phenols are stronger acids than alcohols, but weaker than carboxylic acids. They are converted quantitatively to phenoxide anions on treatment with aqueous sodium hydroxide.

$$ArOH + NaOH \rightarrow ArONa + H_2O$$

Section 22.5 Electron-releasing substituents attached to the ring have a negligible effect on the acidity of phenols. Strongly electron-withdrawing groups increase the acidity. The compound 4-nitro-3-(trifluoromethyl)phenol, for example, is 10,000 times more acidic than phenol.

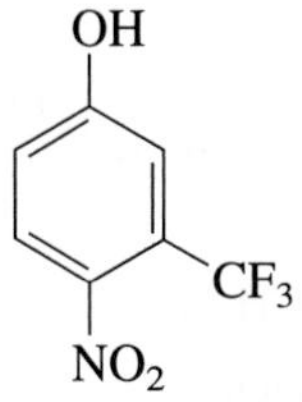

4-Nitro-3-(trifluoromethyl)phenol:
pK_a = 6.0

Section 22.6 Table 22.3 listed the main industrial methods for the preparation of phenol. Laboratory syntheses of phenols are usually carried out by hydrolysis of aryl diazonium salts.

$$\underset{\text{Arylamine}}{ArNH_2} \xrightarrow{NaNO_2,\ H_3O^+} \underset{\text{Aryl diazonium ion}}{Ar\overset{+}{N}{\equiv}N:} \xrightarrow[\text{heat}]{H_2O} \underset{\text{A phenol}}{ArOH}$$

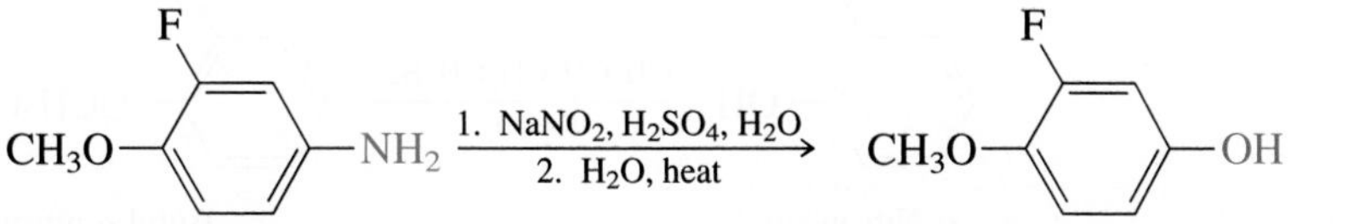

3-Fluoro-4-methoxyaniline → 3-Fluoro-4-methoxyphenol (70%)

Section 22.7 Many phenols occur naturally.

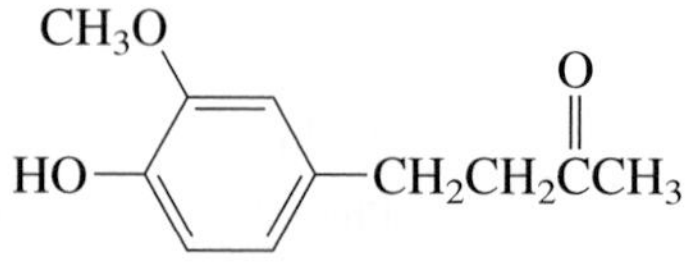

Zingerone
(responsible for spicy taste of ginger)

Section 22.8 The hydroxyl group of a phenol is a strongly activating substituent, and electrophilic aromatic substitution occurs readily in phenol and its derivatives. Typical examples were presented in Table 22.4.

Section 22.9 On reaction with acyl chlorides and acid anhydrides, phenols may undergo either acylation of the hydroxyl group (O-acylation) or acylation of the ring (C-acylation). The product of C-acylation is more stable and predominates under conditions of thermodynamic control when aluminum chloride is present (see entry 6 in Table 22.4, Section 22.8). O-acylation is faster than C-acylation, and aryl esters are formed under conditions of kinetic control.

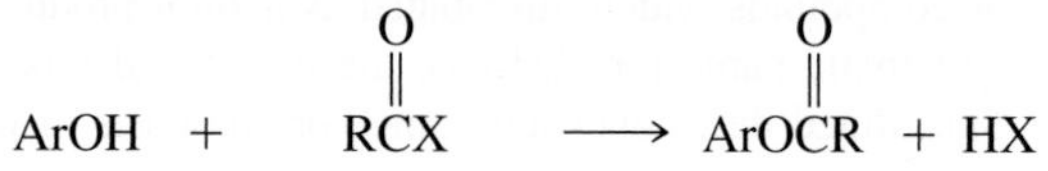

A phenol — Acylating agent — Aryl ester

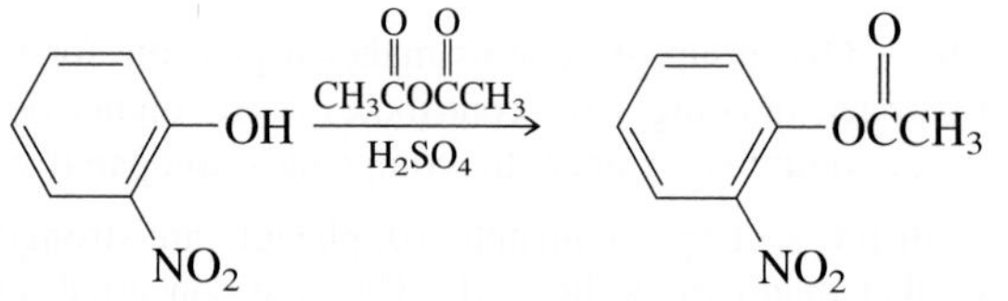

o-Nitrophenol — *o*-Nitrophenyl acetate (93%)

Section 22.10 The Kolbe–Schmitt synthesis of salicylic acid is a vital step in the preparation of aspirin. Phenols, as their sodium salts, undergo highly regioselective ortho carboxylation on treatment with carbon dioxide at elevated temperature and pressure.

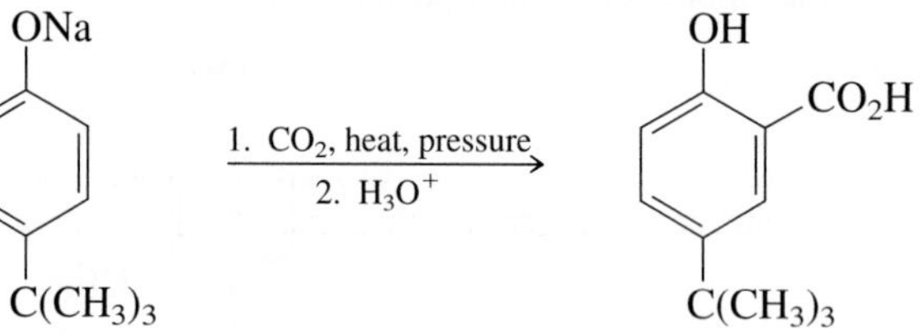

Sodium *p-tert*-butylphenoxide — 5-*tert*-Butyl-2-hydroxybenzoic acid (74%)

Section 22.11 Phenoxide anions are nucleophilic toward alkyl halides, and the preparation of alkyl aryl ethers is easily achieved under S_N2 conditions.

$$\mathrm{ArO^-} + \mathrm{RX} \longrightarrow \mathrm{ArOR} + \mathrm{X^-}$$

Phenoxide anion — Alkyl halide — Alkyl aryl ether — Halide anion

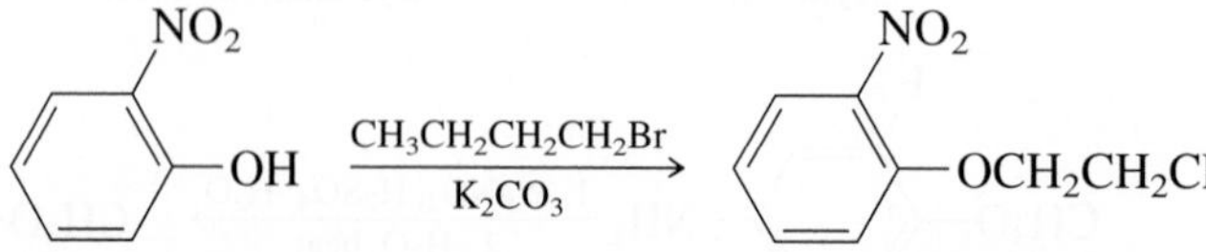

o-Nitrophenol — Butyl *o*-nitrophenyl ether (75-80%)

Section 22.12 The cleavage of alkyl aryl ethers by hydrogen halides yields a phenol and an alkyl halide.

$$\underset{\text{Alkyl aryl ether}}{\text{ArOR}} + \underset{\text{Hydrogen halide}}{\text{HX}} \xrightarrow{\text{heat}} \underset{\text{A phenol}}{\text{ArOH}} + \underset{\text{Alkyl halide}}{\text{RX}}$$

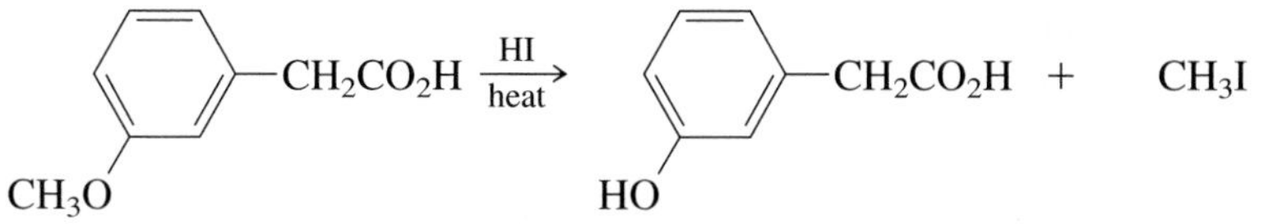

m-Methoxyphenylacetic acid — *m*-Hydroxyphenylacetic acid (72%) — Methyl iodide

Section 22.13 On being heated, allyl aryl ethers undergo a **Claisen rearrangement** to form *o*-allylphenols. A cyclohexadienone, formed by a concerted six-π-electron reorganization, is an intermediate.

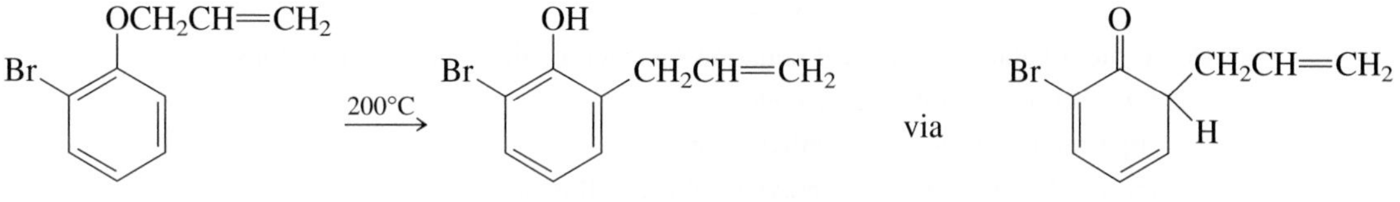

Allyl *o*-bromophenyl ether — 2-Allyl-6-bromophenol (82%) — 6-Allyl-2-bromo-2,4-cyclohexadienone

Section 22.14 Oxidation of 1,2- and 1,4-benzenediols gives colored compounds known as **quinones.**

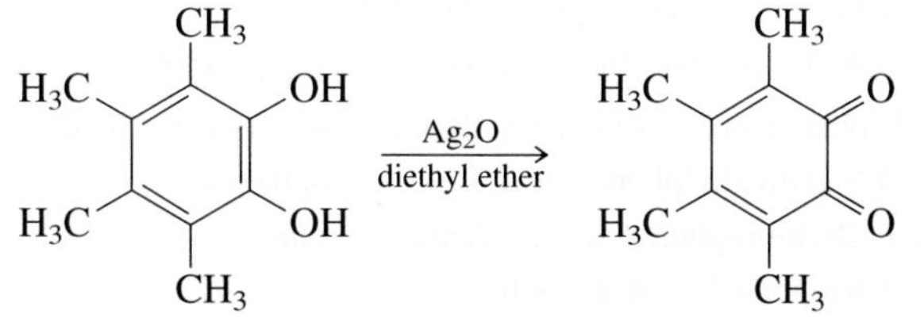

3,4,5,6-Tetramethyl-1,2-benzenediol — 3,4,5,6-Tetramethyl-1,2-benzoquinone (81%)

Section 22.15 The IR and ^{1}H NMR spectra of phenols are similar to those for alcohols, except that the OH proton is somewhat less shielded in a phenol than in an alcohol. In ^{13}C NMR, an OH group deshields the carbon of an aromatic ring to which it is attached. An OH group causes a shift in the UV-VIS spectrum of benzene to longer wavelengths. The effect is quite large in basic solution because of conversion of OH to O^-.

PROBLEMS

22.13 The IUPAC rules permit the use of common names for a number of familiar phenols and aryl ethers. These common names are listed here along with their systematic names. Write the structure of each compound.

(a) *Vanillin* (4-hydroxy-3-methoxybenzaldehyde): a component of vanilla bean oil, which contributes to its characteristic flavor

(b) *Thymol* (2-isopropyl-5-methylphenol): obtained from oil of thyme

(c) *Carvacrol* (5-isopropyl-2-methylphenol): present in oil of thyme and marjoram

(d) *Eugenol* (4-allyl-2-methoxyphenol): obtained from oil of cloves

(e) *Gallic acid* (3,4,5-trihydroxybenzoic acid): prepared by hydrolysis of tannins derived from plants

(f) *Salicyl alcohol* (*o*-hydroxybenzyl alcohol): obtained from bark of poplar and willow trees

22.14 Name each of the following compounds:

(a) 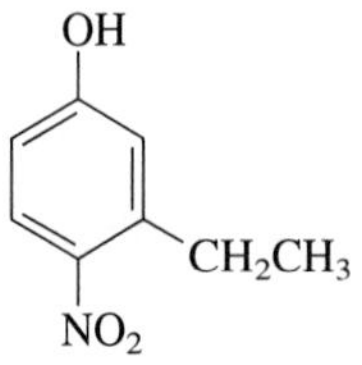

(b) 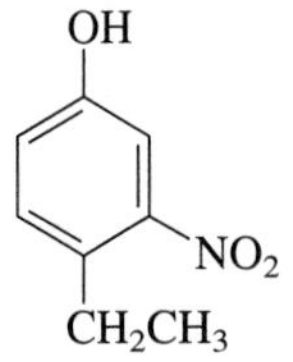

(c) 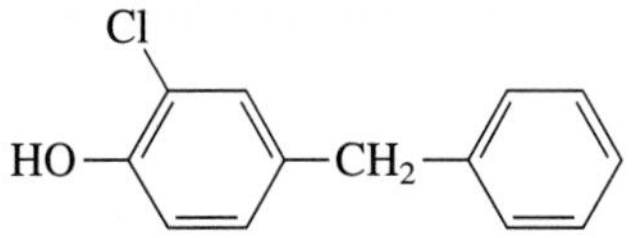

(d) 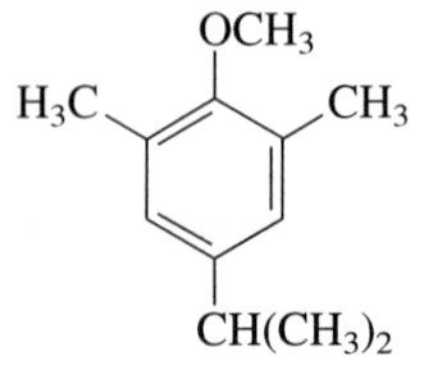

(e) 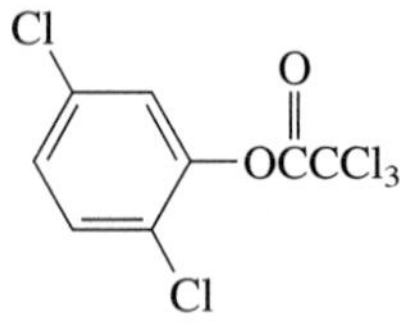

22.15 Write a balanced chemical equation for each of the following reactions:

(a) Phenol + sodium hydroxide
(b) Product of part (a) + ethyl bromide
(c) Product of part (a) + butyl *p*-toluenesulfonate
(d) Product of part (a) + acetic anhydride
(e) *o*-Cresol + benzoyl chloride
(f) *m*-Cresol + ethylene oxide
(g) 2,6-Dichlorophenol + bromine
(h) *p*-Cresol + excess aqueous bromine
(i) Isopropyl phenyl ether + excess hydrogen bromide + heat

22.16 Which phenol in each of the following pairs is more acidic? Justify your choice.

(a) 2,4,6-Trimethylphenol or 2,4,6-trinitrophenol
(b) 2,6-Dichlorophenol or 3,5-dichlorophenol
(c) 3-Nitrophenol or 4-nitrophenol
(d) Phenol or 4-cyanophenol
(e) 2,5-Dinitrophenol or 2,6-dinitrophenol

22.17 Choose the reaction in each of the following pairs that proceeds at the faster rate. Explain your reasoning.

(a) Basic hydrolysis of phenyl acetate or *m*-nitrophenyl acetate
(b) Basic hydrolysis of *m*-nitrophenyl acetate or *p*-nitrophenyl acetate
(c) Reaction of ethyl bromide with phenol or with the sodium salt of phenol
(d) Reaction of ethylene oxide with the sodium salt of phenol or with the sodium salt of *p*-nitrophenol
(e) Bromination of phenol or phenyl acetate

22.18 Pentafluorophenol is readily prepared by heating hexafluorobenzene with potassium hydroxide in *tert*-butyl alcohol:

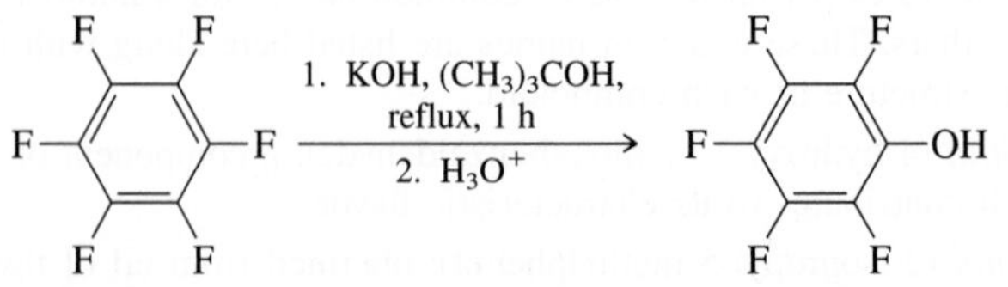

Hexafluorobenzene → Pentafluorophenol (71%)

What is the most reasonable mechanism for this reaction? Comment on the comparative ease with which this conversion occurs.

22.19 Each of the following reactions has been reported in the chemical literature and proceeds cleanly in good yield. Identify the principal organic product in each case.

(a) 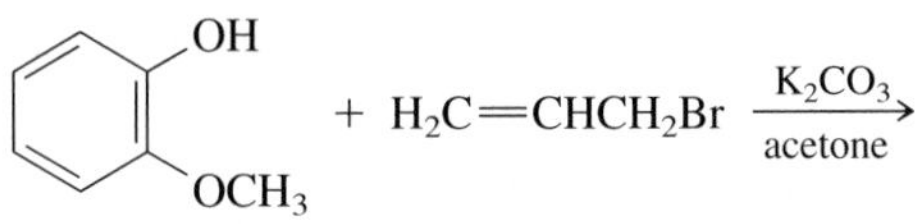
OH
OCH3
+ H2C=CHCH2Br
K2CO3
acetone

(b) 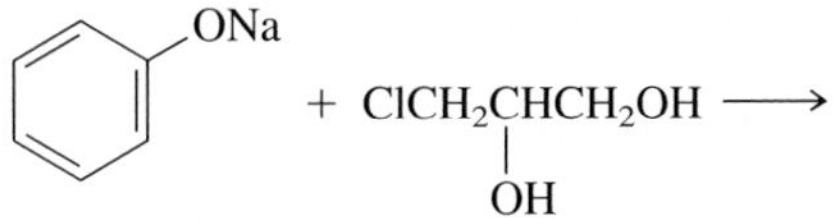
ONa
+ ClCH2CHCH2OH
OH

(c) 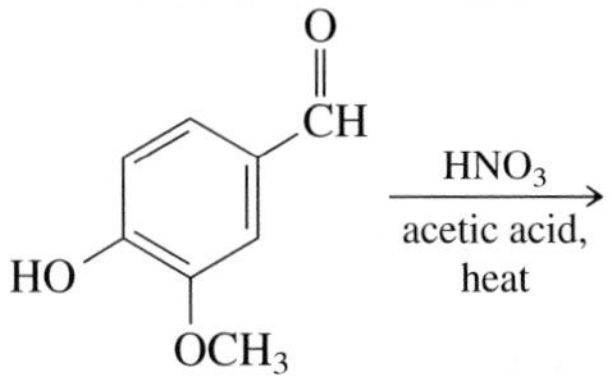
O
CH
HO
OCH3
HNO3
acetic acid,
heat

(d) 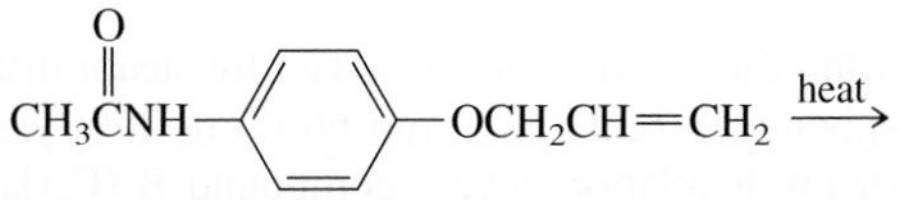
O
CH3CNH
OCH2CH=CH2
heat

(e) 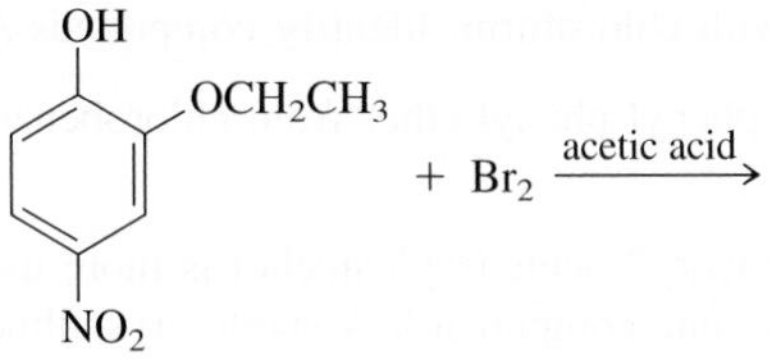
OH
OCH2CH3
NO2
+ Br2
acetic acid

(f) 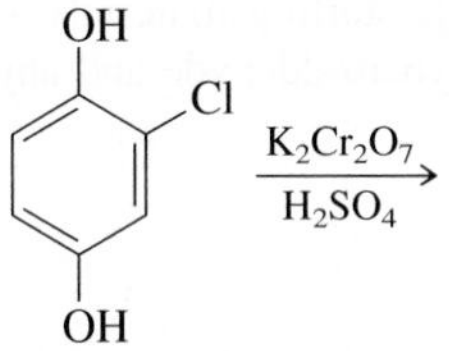
OH
Cl
OH
K2Cr2O7
H2SO4

(g) 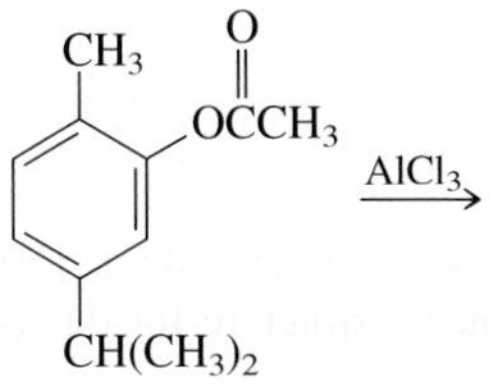
CH3
O
OCCH3
CH(CH3)2
AlCl3

(h) 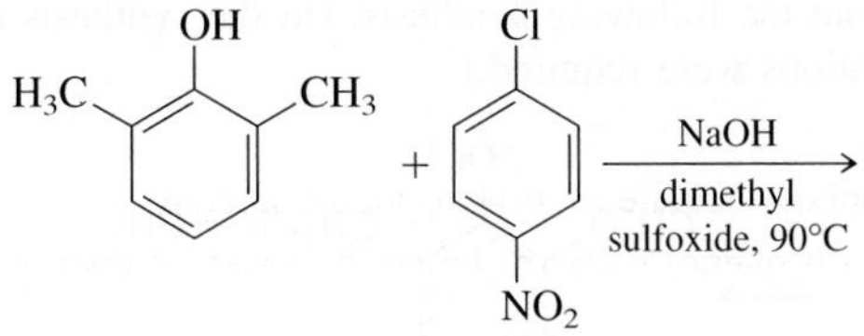
OH
H3C
CH3
+
Cl
NO2
NaOH
dimethyl
sulfoxide, 90°C

(i) 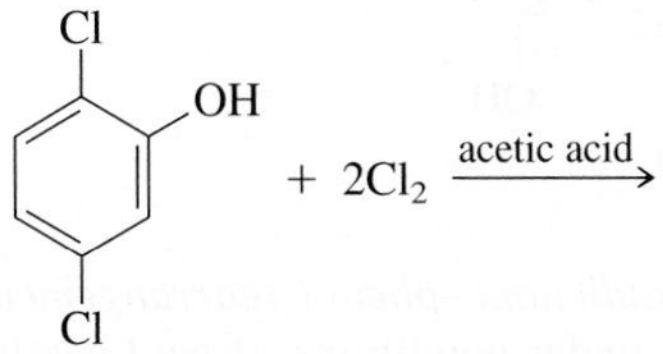
Cl
OH
Cl
+ 2Cl2
acetic acid

(j) 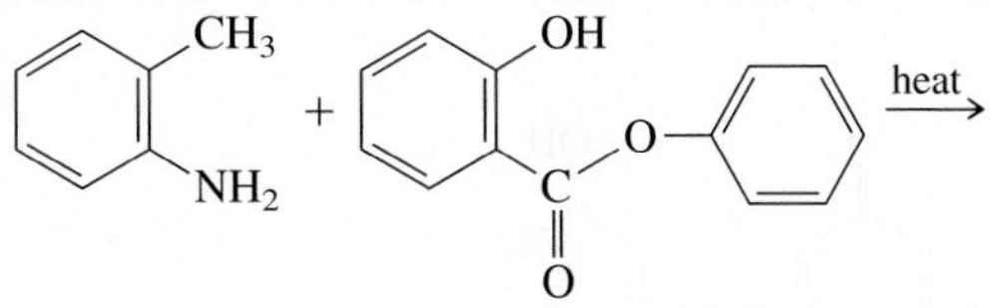
CH3
NH2
+
OH
C
O
O
heat

(k) 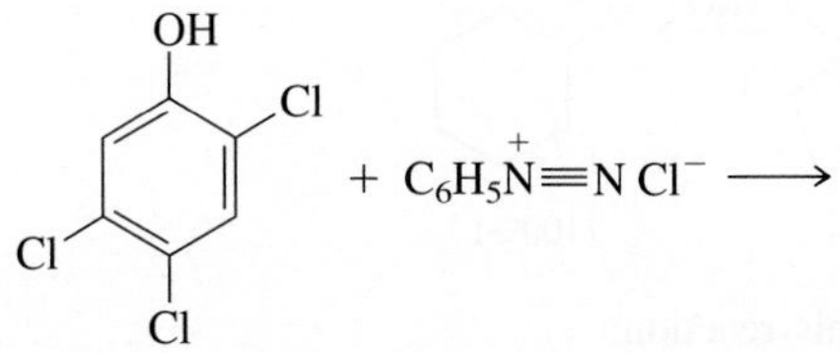
OH
Cl
Cl
Cl
+ C6H5N≡N Cl−

22.20 A synthesis of the pain reliever *phenacetin* is outlined in the following equation. What is the structure of phenacetin?

$$p\text{-Nitrophenol} \xrightarrow[\text{2. Fe, HCl; then HO}^- \quad \text{3. } CH_3\overset{O}{\overset{\|}{C}}O\overset{O}{\overset{\|}{C}}CH_3]{\text{1. } CH_3CH_2Br,\ NaOH} \text{Phenacetin}$$

22.21 Identify compounds A through C in the synthetic sequence represented by equations (a) through (c).

(a) $\text{Phenol} + H_2SO_4 \xrightarrow{\text{heat}} \text{Compound A } (C_6H_6O_7S_2)$

(b) $\text{Compound A} + Br_2 \xrightarrow[\text{2. H}^+]{\text{1. HO}^-} \text{Compound B } (C_6H_5BrO_7S_2)$

(c) $\text{Compound B} + H_2O \xrightarrow[\text{heat}]{\text{H}^+} \text{Compound C } (C_6H_5BrO)$

22.22 Treatment of 3,5-dimethylphenol with dilute nitric acid, followed by steam distillation of the reaction mixture, gave a compound A ($C_8H_9NO_3$, mp 66°C) in 36% yield. The nonvolatile residue from the steam distillation gave a compound B ($C_8H_9NO_3$, mp 108°C) in 25% yield on extraction with chloroform. Identify compounds A and B.

22.23 Outline a reasonable synthesis of 4-nitrophenyl phenyl ether from chlorobenzene and phenol.

22.24 As an allergen for testing purposes, synthetic 3-pentadecylcatechol is more useful than natural poison ivy extracts (of which it is one component). A stable crystalline solid, it is efficiently prepared in pure form from readily available starting materials. Outline a reasonable synthesis of this compound from 2,3-dimethoxybenzaldehyde and any necessary organic or inorganic reagents.

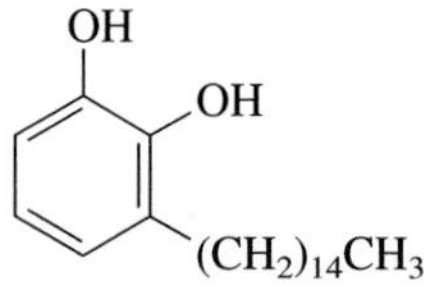
OH
OH
(CH2)14CH3

3-Pentadecylcatechol

22.25 Reaction of phenol with 1,2-epoxypropane in aqueous sodium hydroxide at 150°C gives a single product, $C_9H_{12}O_2$, in 90% yield. Suggest a reasonable structure for this compound.

22.26 Describe a scheme for carrying out the following synthesis. (In the synthesis reported in the literature, four separate operations were required.)

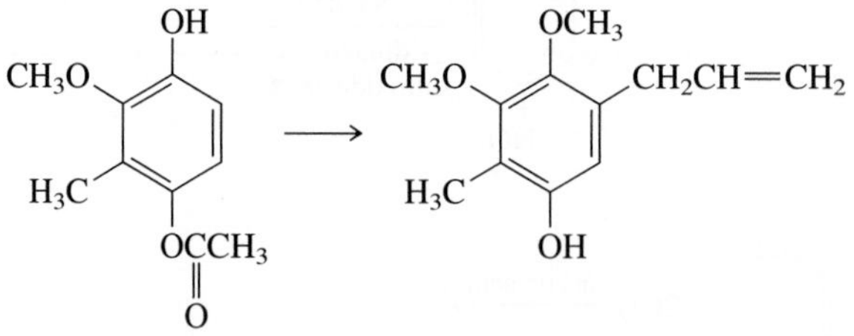
OH
CH3O
H3C
OCCH3
O
OCH3
CH3O
CH2CH=CH2
H3C
OH

22.27 In a general reaction known as the *cyclohexadienone–phenol rearrangement,* cyclohexadienones are converted to phenols under conditions of acid catalysis. An example is

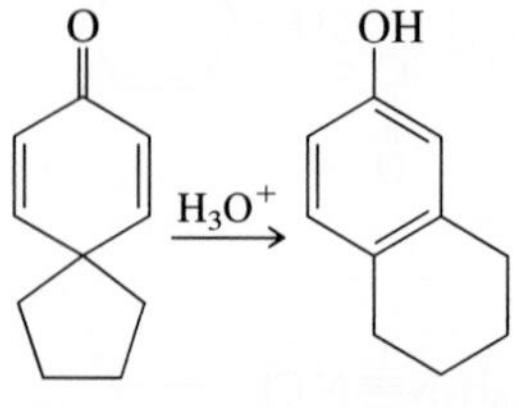
O
H3O+
OH

(100%)

Write a reasonable mechanism for this reaction.

22.28 Treatment of *p*-hydroxybenzoic acid with aqueous bromine leads to the evolution of carbon dioxide and the formation of 2,4,6-tribromophenol. Explain.

22.29 Treatment of phenol with excess aqueous bromine is actually more complicated than expected. A white precipitate forms rapidly, which on closer examination is not 2,4,6-tribromophenol but is instead 2,4,4,6-tetrabromocyclohexadienone. Explain the formation of this product.

22.30 Treatment of 2,4,6-tri-*tert*-butylphenol with bromine in cold acetic acid gives the compound $C_{18}H_{29}BrO$ in quantitative yield. The infrared spectrum of this compound contains absorptions at 1630 and 1655 cm^{-1}. Its ^1H NMR spectrum shows only three peaks (all singlets), at δ 1.2, 1.3, and 6.9, in the ratio 9:18:2. What is a reasonable structure for the compound?

22.31 Compound A undergoes hydrolysis of its acetal function in dilute sulfuric acid to yield 1,2-ethanediol and compound B ($C_6H_6O_2$), mp 54°C. Compound B exhibits a carbonyl stretching band in the infrared at 1690 cm^{-1} and has two singlets in its ^1H NMR spectrum, at δ 2.9 and 6.7, in the ratio 2:1. On standing in water or ethanol, compound B is converted cleanly to an isomeric substance, compound C, mp 172–173°C. Compound C has no peaks attributable to carbonyl groups in its infrared spectrum. Identify compounds B and C.

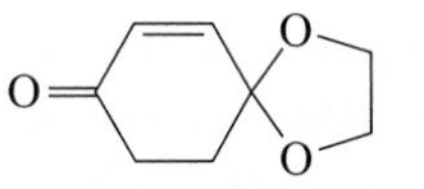

Compound A

22.32 In a study aimed at the synthesis of analogs of anthracycline antibiotics, 4,8-dimethoxynaphthalen-1-ol was converted to 4,8-dimethoxy-2-propanoylnaphthalen-1-ol by the following sequence of reactions. What is compound A?

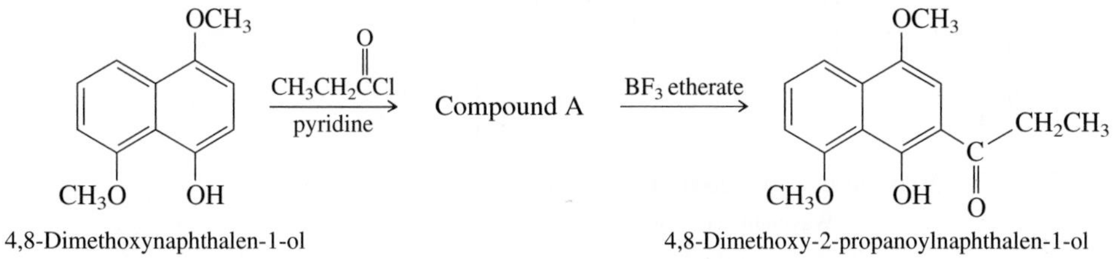

22.33 Tamoxifen is an estrogen receptor modulator that is used in the treatment of breast cancer. Provide the missing reagents and the structure of compound A in the synthesis of tamoxifen.

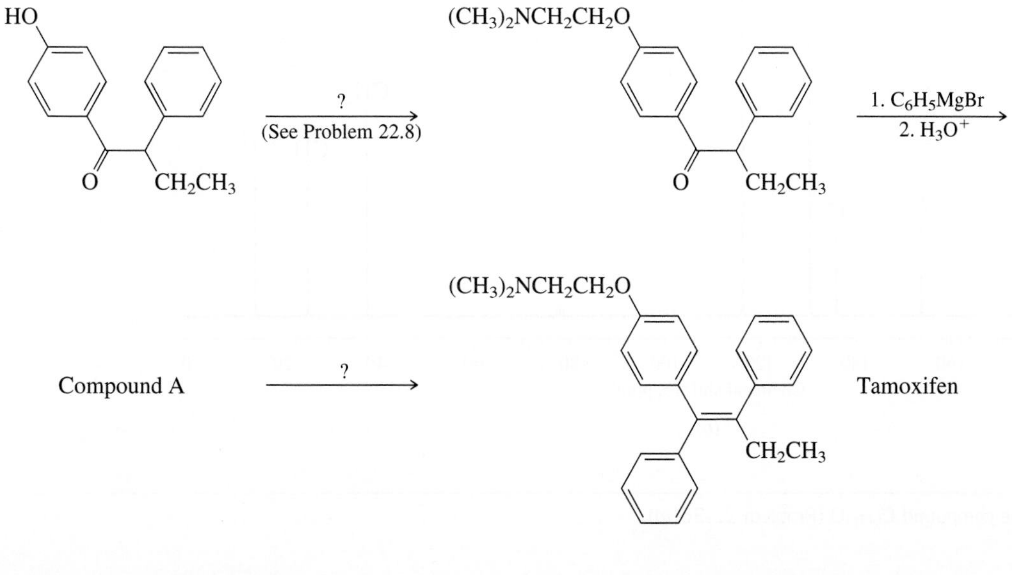

22.34 One of the industrial processes for the preparation of phenol, discussed in Section 22.6, includes an acid-catalyzed rearrangement of cumene hydroperoxide as a key step. This reaction proceeds by way of an intermediate hemiacetal:

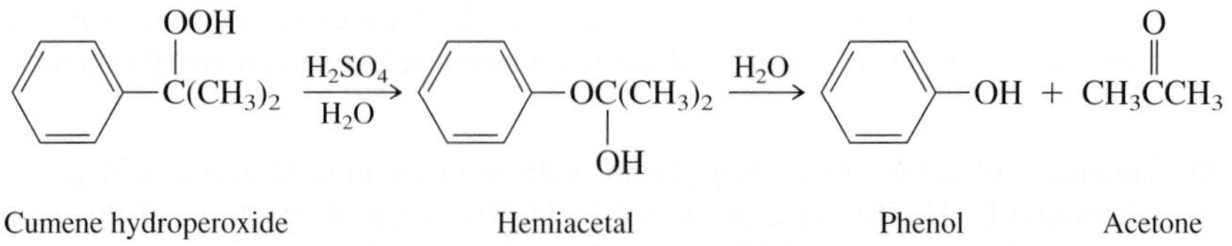
OOH
C(CH3)2
H2SO4
H2O
OC(CH3)2
OH
H2O
OH + CH3CCH3
O
Cumene hydroperoxide
Hemiacetal
Phenol
Acetone

You learned in Section 17.8 of the relationship among hemiacetals, ketones, and alcohols; the formation of phenol and acetone is an example of hemiacetal hydrolysis. The formation of the hemiacetal intermediate is a key step in the synthetic procedure; it is the step in which the aryl–oxygen bond is generated. Can you suggest a reasonable mechanism for this step?

22.35 Devise a synthesis of the compound shown on the left from the indicated starting materials and any other necessary reagents.

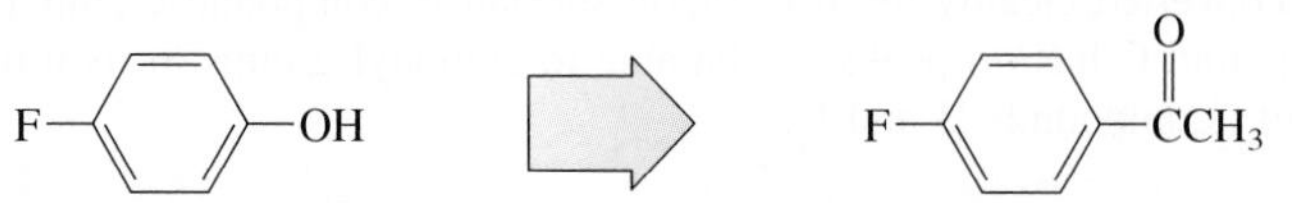
F
OH
F
O
CCH3

22.36 Identify the following compounds on the basis of the information provided:

(a) $C_9H_{12}O$: Its IR and ^{13}C NMR spectra are shown in Figure 22.5.

(b) $C_9H_{11}BrO$: Its IR and ^{13}C NMR spectra are shown in Figure 22.6.

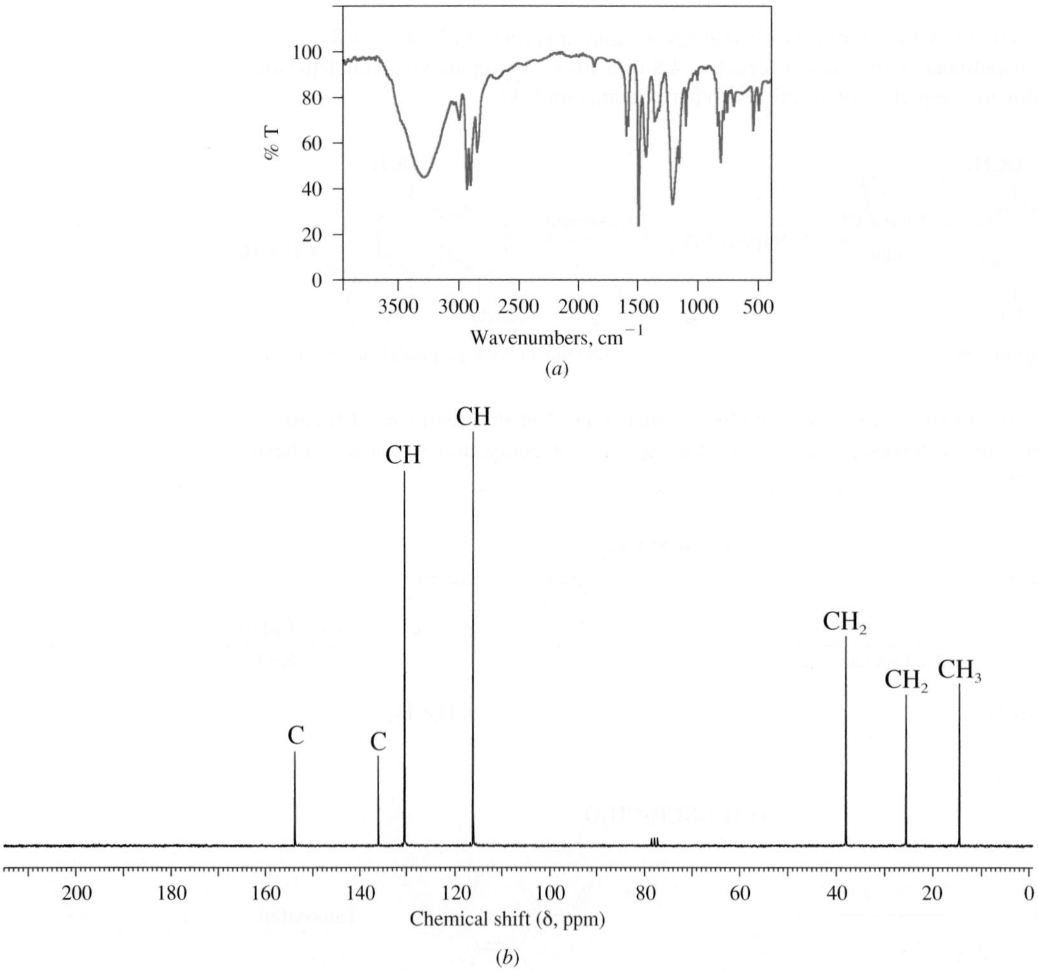
100
80
60
40
20
0
% T
3500 3000 2500 2000 1500 1000 500
Wavenumbers, cm⁻¹
(a)
CH
CH
C
C
CH2
CH2
CH3
200 180 160 140 120 100 80 60 40 20 0
Chemical shift (δ, ppm)
(b)

Figure 22.5

(*a*) Infrared and (*b*) ^{13}C NMR spectra of the compound $C_9H_{12}O$ (Problem 22.36(*a*)).

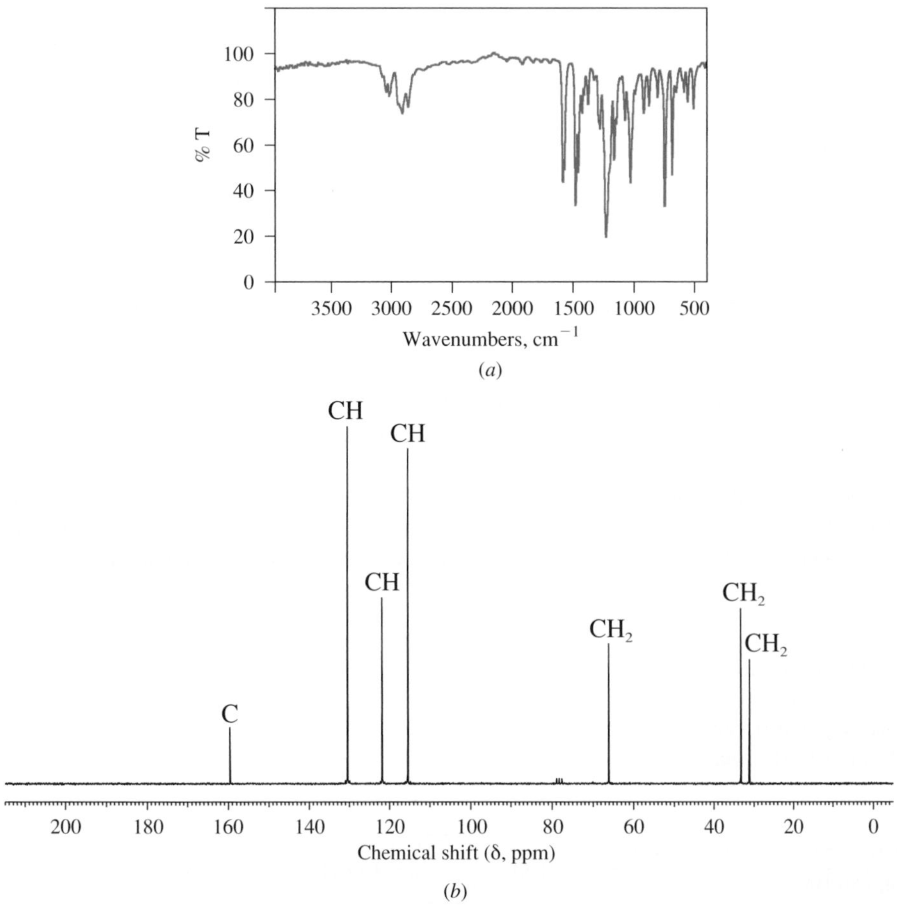
100
80
60
40
20
0
% T
3500 3000 2500 2000 1500 1000 500
Wavenumbers, cm^{-1}
(a)
CH
CH
CH
C
CH_2
CH_2
CH_2
200 180 160 140 120 100 80 60 40 20 0
Chemical shift (δ, ppm)
(b)

Figure 22.6

(*a*) Infrared and (*b*) ^{13}C NMR spectra of the compound $C_9H_{11}BrO$ (Problem 22.36(*b*)).

Descriptive Passage and Interpretive Problems 22

Directed Metalation of Aryl Ethers

Aryllithium reagents are familiar to us as the products of the reaction of aryl halides with lithium. A second route to aryllithiums is by deprotonation of an aromatic ring with an alkyllithium reagent. This process is called *metalation*.

$$\underset{\substack{\text{Arene}\\(\mathrm{p}K_a \approx 43)}}{\text{Ar—H}} + \underset{\text{Alkyllithium}}{\text{R—Li}} \longrightarrow \underset{\text{Aryllithium}}{\text{Ar—Li}} + \underset{\substack{\text{Alkane}\\(\mathrm{p}K_a \approx 60\text{-}70)}}{\text{R—H}}$$

Although reasonable from an acid–base perspective (stronger acid on the left, weaker acid on the right), reactions such as this are inconveniently slow and not practical as a general method for making aryllithiums. Certain substitutents on an aromatic ring, however, both promote metalation and control its regioselectivity.

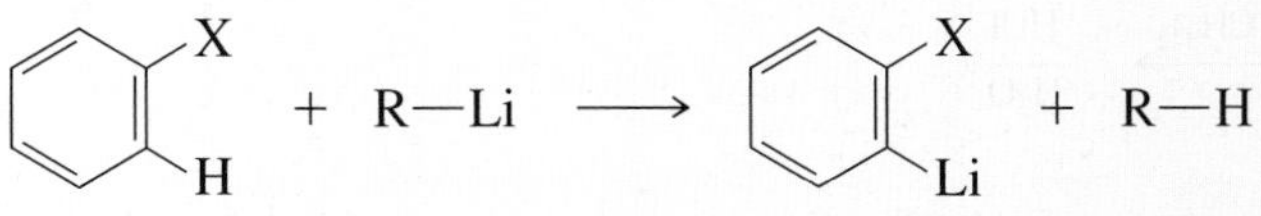
X
H
+ R—Li
X
Li
+ R—H

$X = OCH_3, OCH_2OCH_3, CH_2OH, CO_2^-, N(CH_3)_2, CH_2N(CH_3)_2, \overset{O}{\overset{\|}{C}}NHR$, and others

These *directed metalations* are regioselective for lithiation at carbons ortho to the substituent. The substituents contain atoms, especially oxygen and nitrogen, capable of coordinating to lithium. This coordination is reflected in the transition state for proton abstraction and causes lithiation to be fastest at the carbons ortho to the substituent.

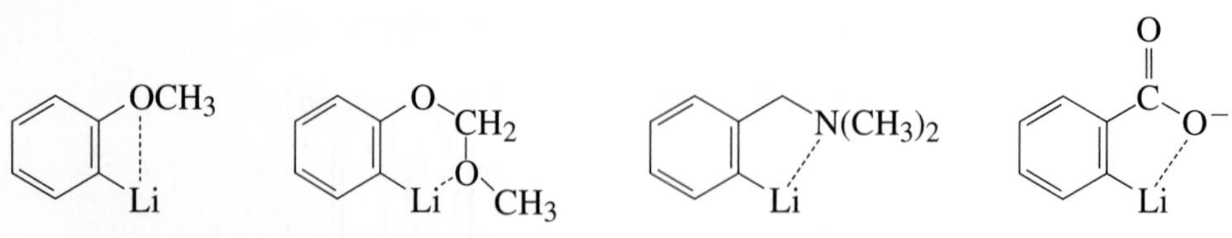

X = OCH_3 X = OCH_2OCH_3 X = $CH_2N(CH_3)_2$ X = CO_2^-

Typical conditions for metalation involve treating a diethyl ether or tetrahydrofuran solution of the aryl derivative with an alkyllithium in hexane. *N,N,N′,N′*-Tetramethylethylenediamine [$(CH_3)_2NCH_2CH_2N(CH_3)_2$, TMEDA] is sometimes added to increase the rate of metalation.

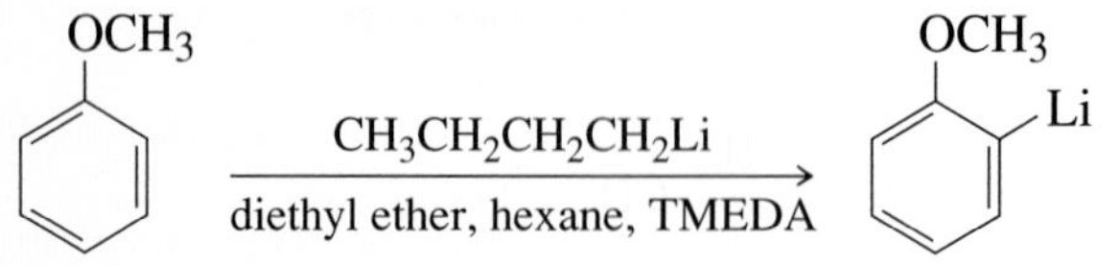

Once formed, the ortho-metalated derivative undergoes the usual reactions of organolithium reagents.

The directing effects associated with electrophilic aromatic substitution play no role in directed metalation. Likewise, steric effects are not very important. It is common, for example, for metalation to occur at C-2 of a 1,3-disubstituted benzene.

22.37 4-Methoxybenzoic acid was metalated and allowed to react with methyl iodide as shown in the equation. (Two molar equivalents of *sec*-butyllithium are required because one equivalent is needed to deprotonate the CO_2H group.)

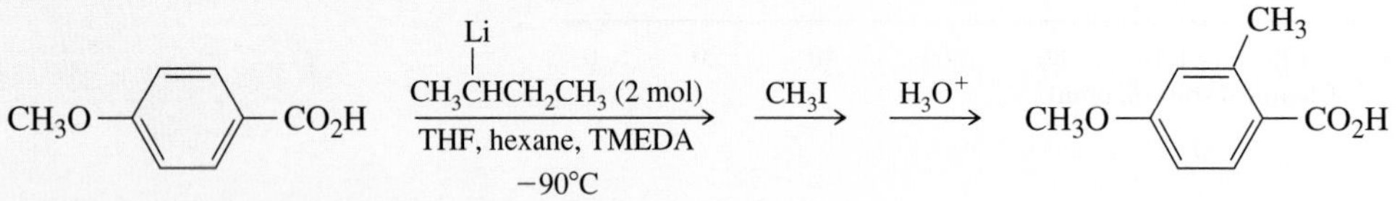

(major product)

Predict the major product of the analogous methylation of 2-methoxybenzoic acid.

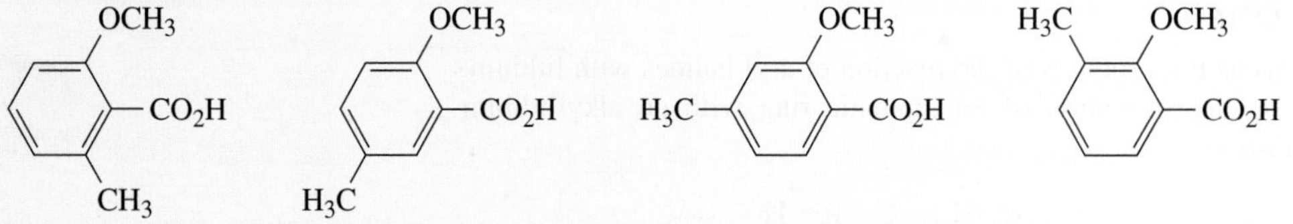

A. B. C. D.

22.38 Even though the following sequence, reported in the chemical literature, includes an unfamiliar reaction, you should be able to deduce the structure of the product from among the possible choices.

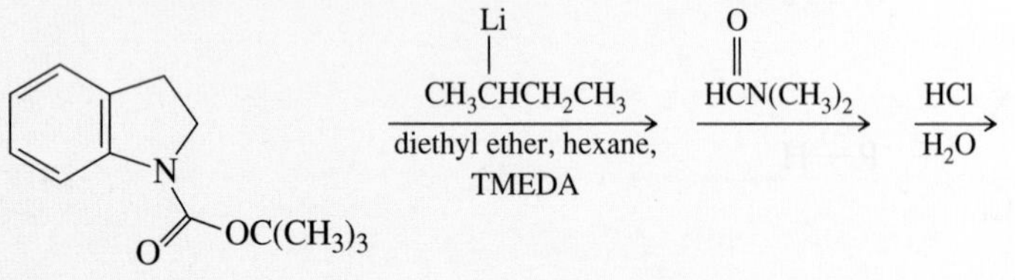

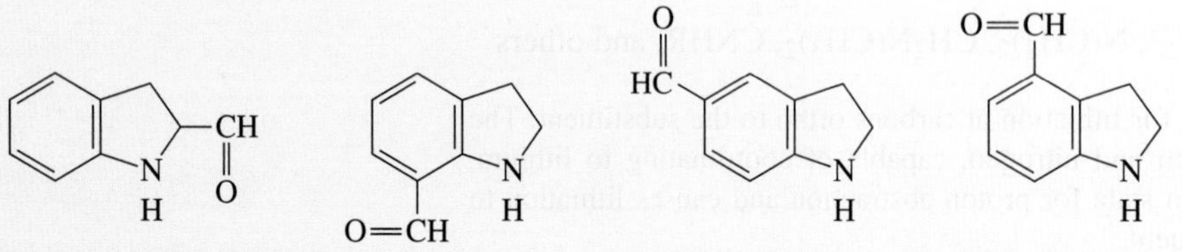

A. B. C. D.

22.39 The methoxymethyl group is an acetal, easily removed by acid hydrolysis. What are the products of the following hydrolysis?

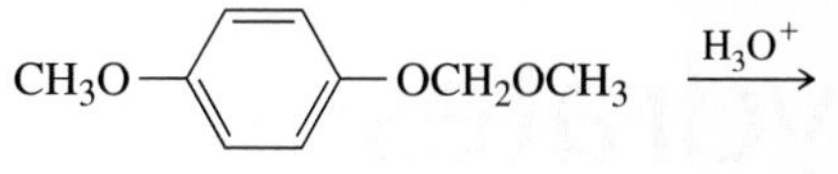

A. 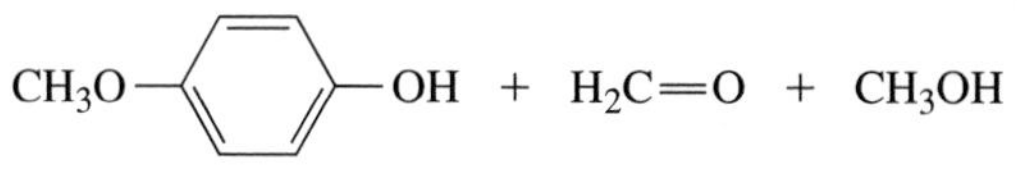

B. 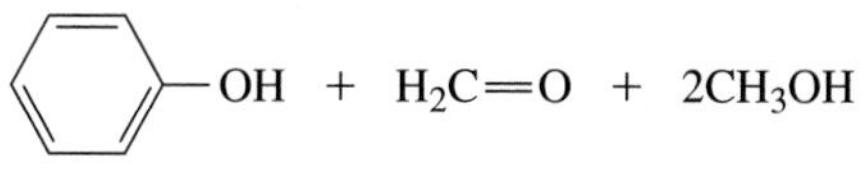

C. 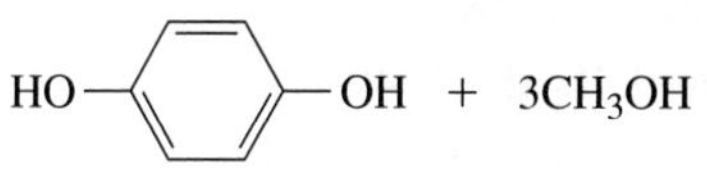

D. 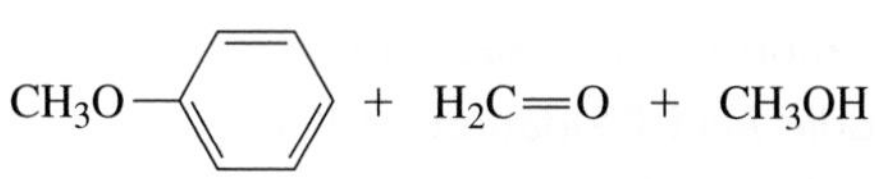

22.40 A synthesis of a natural product (broussonin A) began with metalation and trimethylsilylation of compound **1** to give **2.** Compound **2** was metalated and the resulting organolithium reagent treated with 3-(*p*-benzyloxy)propanal to give compound **3.** Subsequent transformations of **3** gave broussonin A ($C_{16}H_{18}O_3$). Which of the choices is most reasonable for compound **3**?

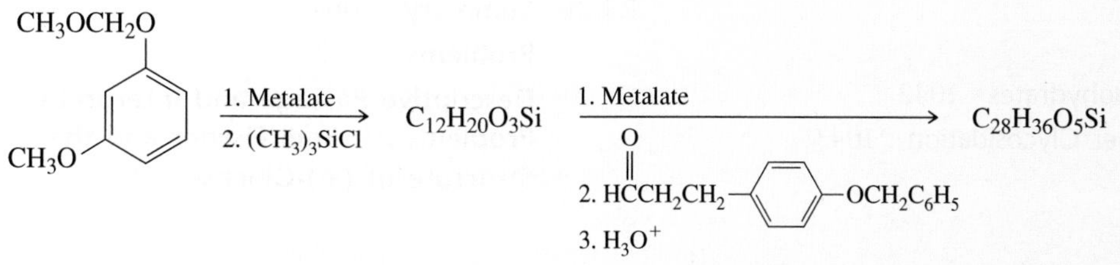

1 **2** **3**

A. 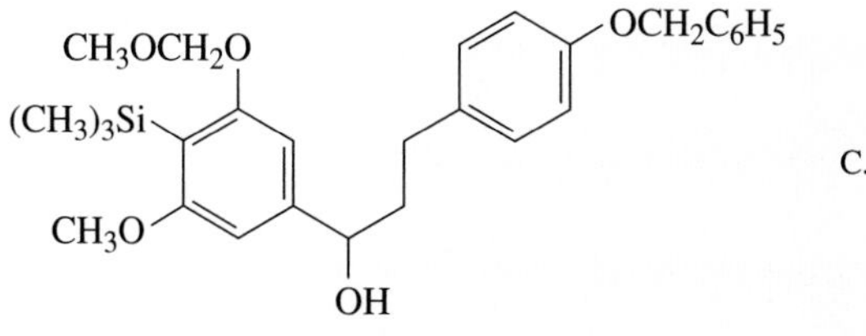

C. 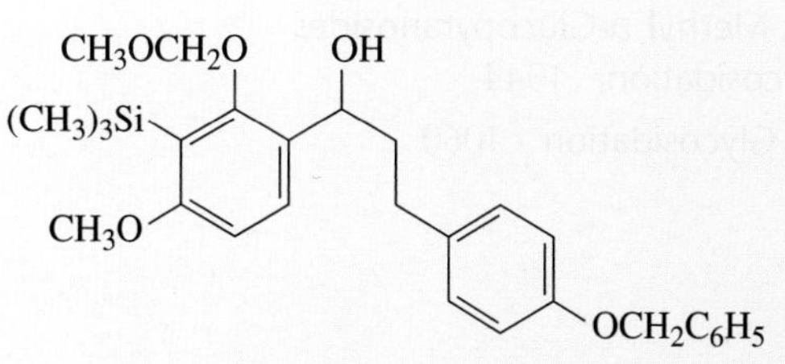

B. 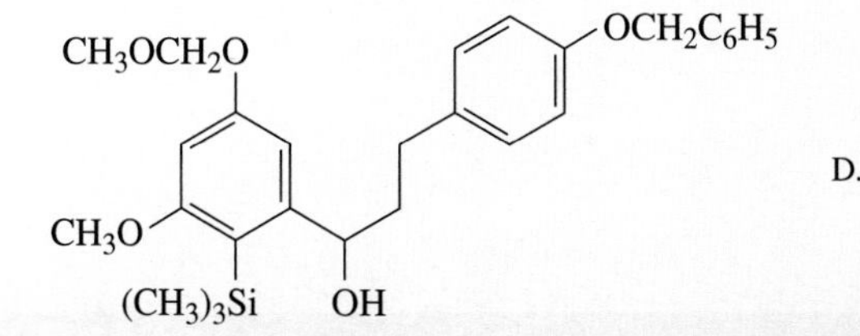

D. 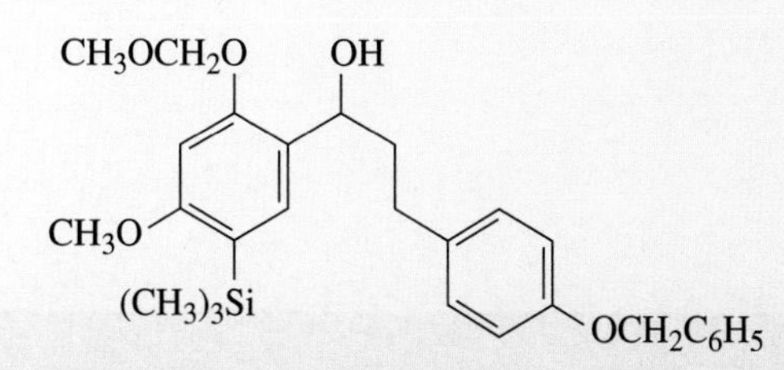

23 Carbohydrates

Chapter Outline

Mechanisms

Hummingbirds receive nourishment from flower nectar. Nectar contains glucose (shown), fructose, and sucrose.

THE MAJOR CLASSES of organic compounds common to living systems are *lipids, proteins, nucleic acids,* and *carbohydrates.* Carbohydrates are very familiar to us—we call many of them "sugars." They make up a substantial portion of the food we eat and provide most of the energy that keeps the human engine running. Carbohydrates are structural components of the walls of plant cells and the wood of trees; they are also major components of the exoskeletons of insects, crabs, and lobsters. Carbohydrates are found on every cell surface, where they provide the molecular basis for cell-to-cell communication. Genetic information is stored and transferred by way of nucleic acids, specialized derivatives of carbohydrates, which we'll examine in more detail in Chapter 26.

Historically, carbohydrates were once considered to be "hydrates of carbon" because their molecular formulas in many (but not all) cases correspond to $C_n(H_2O)_m$. It is more realistic to define a carbohydrate as a *polyhydroxy aldehyde* or *polyhydroxy ketone,* a point of view closer to structural reality and more suggestive of chemical reactivity.

This chapter is divided into two parts. The first, and major, portion is devoted to carbohydrate *structure.* You will see how the principles of stereochemistry and conformational analysis combine to aid our understanding of this complex subject. The remainder of the chapter describes chemical *reactions* of carbohydrates. Most of these reactions are simply extensions of what you have already learned concerning alcohols, aldehydes, ketones, and acetals. Finally, we will consider the role of carbohydrates in the emerging field of glycobiology.

23.1 Classification of Carbohydrates

The Latin word for *sugar* is *saccharum,* and the derived term *saccharide* is the basis of a system of carbohydrate classification. A **monosaccharide** is a simple carbohydrate, one that on attempted hydrolysis is not cleaved to smaller carbohydrates. *Glucose* ($C_6H_{12}O_6$), for example, is a monosaccharide. A **disaccharide** on hydrolysis is cleaved to two monosaccharides, which may be the same or different. *Sucrose*—common table sugar—is a disaccharide that yields one molecule of glucose and one of fructose on hydrolysis.

Sugar is a combination of the Sanskrit words *su* (sweet) and *gar* (sand). Thus, its literal meaning is "sweet sand."

$$\text{Sucrose } (C_{12}H_{22}O_{11}) + H_2O \longrightarrow \text{glucose } (C_6H_{12}O_6) + \text{fructose } (C_6H_{12}O_6)$$

An **oligosaccharide** (*oligos* is a Greek word that in its plural form means "few") yields two or more monosaccharides on hydrolysis. Thus, the IUPAC classifies disaccharides, trisaccharides, and so on as subcategories of oligosaccharides.

TABLE 23.1 Some Classes of Monosaccharides

Number of carbon atoms	Aldose	Ketose
Four	Aldotetrose	Ketotetrose
Five	Aldopentose	Ketopentose
Six	Aldohexose	Ketohexose
Seven	Aldoheptose	Ketoheptose
Eight	Aldooctose	Ketooctose

Polysaccharides are hydrolyzed to "many" monosaccharides. The IUPAC has chosen not to specify the number of monosaccharide components that separates oligosaccharides from polysaccharides. The standard is a more practical one; it notes that an oligosaccharide is homogeneous. Each molecule of a particular oligosaccharide has the same number of monosaccharide units joined together in the same order as every other molecule of the same oligosaccharide. Polysaccharides are almost always mixtures of molecules having similar, but not necessarily the same, chain length. *Cellulose,* for example, is a polysaccharide that gives thousands of glucose molecules on hydrolysis but only a small fraction of the cellulose chains contain exactly the same number of glucose units.

Over 200 different monosaccharides are known. They can be grouped according to the number of carbon atoms they contain and whether they are polyhydroxy aldehydes or polyhydroxy ketones. Monosaccharides that are polyhydroxy aldehydes are called **aldoses;** those that are polyhydroxy ketones are **ketoses.** Aldoses and ketoses are further classified according to the number of carbon atoms in the main chain. Table 23.1 lists the terms applied to monosaccharides having four to eight carbon atoms.

23.2 Fischer Projections and D,L Notation

Fischer determined the structure of glucose in 1900 and won the Nobel Prize in Chemistry in 1902.

Stereochemistry is the key to understanding carbohydrate structure, a fact that was clearly appreciated by the German chemist Emil Fischer. The projection formulas used by Fischer to represent stereochemistry in chiral molecules (Section 7.7) are particularly well-suited to studying carbohydrates. Figure 23.1 illustrates their application to the enantiomers of *glyceraldehyde* (2,3-dihydroxypropanal), a fundamental molecule in carbohydrate stereochemistry. When the Fischer projection is oriented as shown in the figure, with the carbon chain vertical and the aldehyde carbon at the top, the C-2 hydroxyl group points to the right in (+)-glyceraldehyde and to the left in (−)-glyceraldehyde.

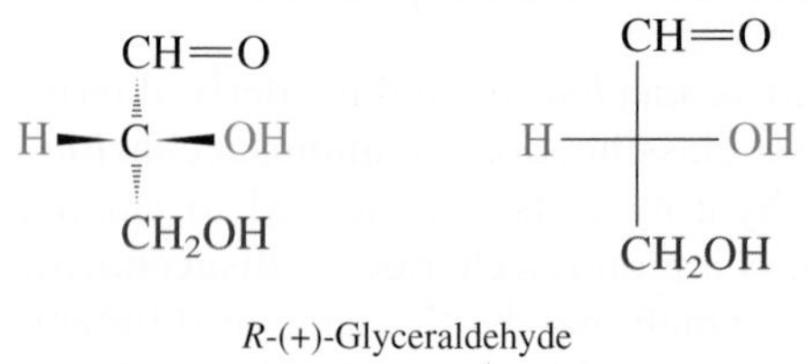
CH=O
H
C
OH
CH₂OH
CH=O
H
OH
CH₂OH

R-(+)-Glyceraldehyde

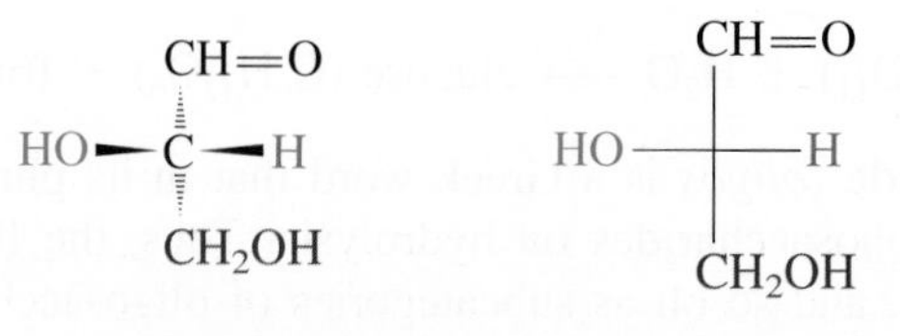
CH=O
HO
C
H
CH₂OH
CH=O
HO
H
CH₂OH

S-(−)-Glyceraldehyde

Figure 23.1

Three-dimensional representations and Fischer projections of the enantiomers of glyceraldehyde.

Techniques for determining the absolute configuration of chiral molecules were not developed until the 1950s, and so it was not possible for Fischer and his contemporaries to relate the sign of rotation of any substance to its absolute configuration. A system evolved based on the arbitrary assumption, later shown to be correct, that the enantiomers of glyceraldehyde have the signs of rotation and absolute configurations shown in Figure 23.1. Two stereochemical descriptors were defined: D and L: D from the Latin (dexter) for right and L from the Latin (laevus) for left. The absolute configuration of (+)-glyceraldehyde was said to be D and that of its enantiomer, (−)-glyceraldehyde, L, as depicted in Figure 23.1 with the hydroxyl groups on the right and left, respectively. Compounds that had a spatial arrangement of substituents analogous to (+)-D- or (−)-L-glyceraldehyde were said to have the D or L configurations.

Adopting the enantiomers of glyceraldehyde as stereochemical reference compounds originated with proposals made in 1906 by M. A. Rosanoff, a chemist at New York University.

Problem 23.1

Identify each of the following as either D- or L-glyceraldehyde:

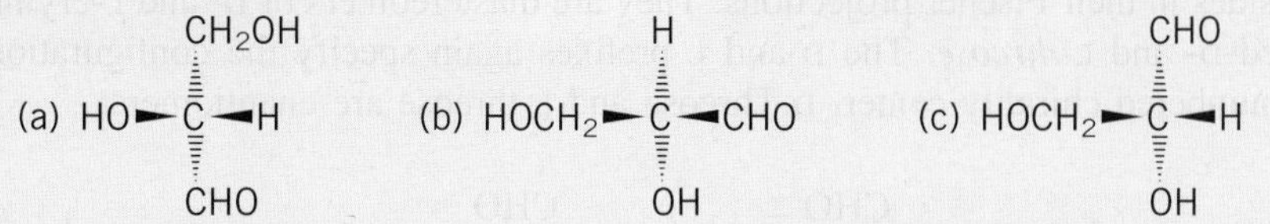

Sample Solution (a) To compare the structure given to glyceraldehyde most easily, turn it 180° in the plane of the page so that CHO is at the top and CH_2OH is at the bottom. Rotation in this sense keeps the horizontal bonds pointing forward, and the vertical bonds pointing back making it an easy matter to convert the structural drawing to a Fischer projection.

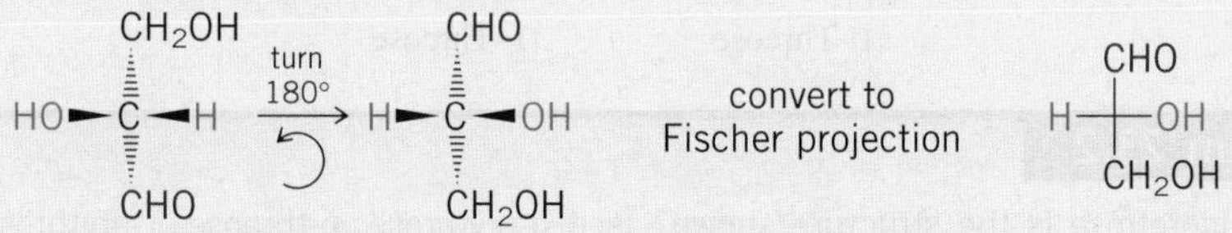

The structure is the same as that of (+)-glyceraldehyde in Figure 23.1. It is D-glyceraldehyde.

Fischer projections and D,L notation have proved to be so helpful in representing carbohydrate stereochemistry that the chemical and biochemical literature is replete with their use. To read that literature you need to be acquainted with these devices, as well as the more modern Cahn–Ingold–Prelog *R,S* system.

23.3 The Aldotetroses

Glyceraldehyde can be thought of as the simplest chiral carbohydrate. It is an *aldotriose* and, because it contains one chirality center, exists in two stereoisomeric forms: the D and L enantiomers. Moving up the ladder in complexity, next come the *aldotetroses.* Examining their structures illustrates the application of the Fischer system to compounds that contain more than one chirality center.

The aldotetroses are the four stereoisomers of 2,3,4-trihydroxybutanal. Fischer projections are constructed by orienting the molecule in an eclipsed conformation with the aldehyde group at the top. The four carbon atoms define the main chain of the Fischer projection and are arranged vertically. Horizontal bonds point outward, vertical bonds back.

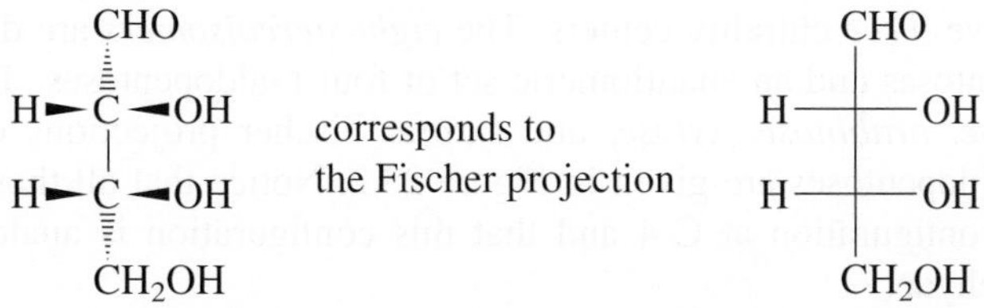

The particular aldotetrose just shown is called D-*erythrose.* The prefix D tells us that the configuration at the *highest numbered chirality center* is analogous to that of (+)-D-glyceraldehyde. Its mirror image is L-erythrose.

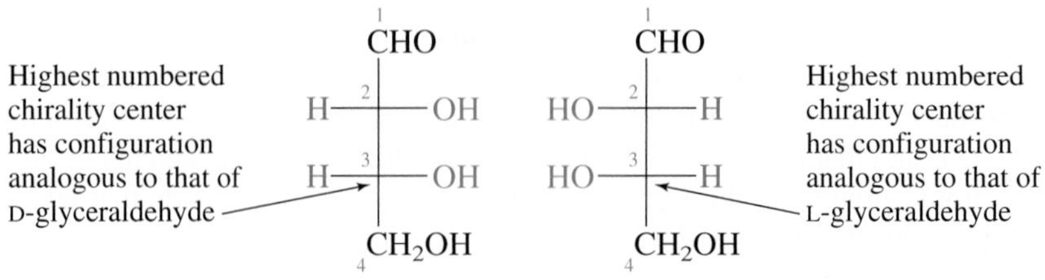

D-Erythrose L-Erythrose

Relative to each other, both hydroxyl groups are on the same side in Fischer projections of the erythrose enantiomers. The remaining two stereoisomers have hydroxyl groups on opposite sides in their Fischer projections. They are diastereomers of D- and L-erythrose and are called D- and L-*threose.* The D and L prefixes again specify the configuration of the highest numbered chirality center. D-Threose and L-threose are enantiomers:

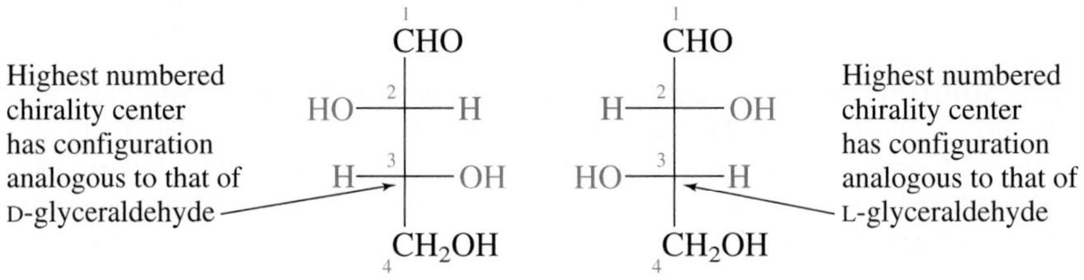

D-Threose L-Threose

Problem 23.2

Which aldotetrose is the structure shown? Is it D-erythrose, D-threose, L-erythrose, or L-threose? (Be careful! The conformation given is not the same as that used to generate a Fischer projection.)

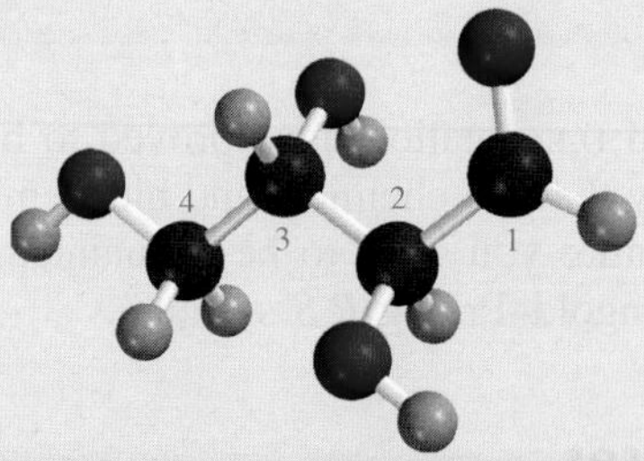

As shown for the aldotetroses, an aldose belongs to the D or the L series according to the configuration of the chirality center farthest removed from the aldehyde function. Individual names, such as erythrose and threose, specify the particular arrangement of chirality centers within the molecule relative to each other. Optical activities cannot be determined directly from the D and L prefixes. As it turns out, both D-erythrose and D-threose are levorotatory, but D-glyceraldehyde is dextrorotatory.

Dextrorotatory and levorotatory are older terms for (+) and (−) optical rotation, respectively.

23.4 Aldopentoses and Aldohexoses

$2^3 = 8$

Aldopentoses have *three* chirality centers. The *eight stereoisomers* are divided into a set of four D-aldopentoses and an enantiomeric set of four L-aldopentoses. The aldopentoses are named *ribose, arabinose, xylose,* and *lyxose.* Fischer projections of the D stereoisomers of the aldopentoses are given in Figure 23.2. Notice that all these diastereomers have the same configuration at C-4 and that this configuration is analogous to that of (+)-D-glyceraldehyde.

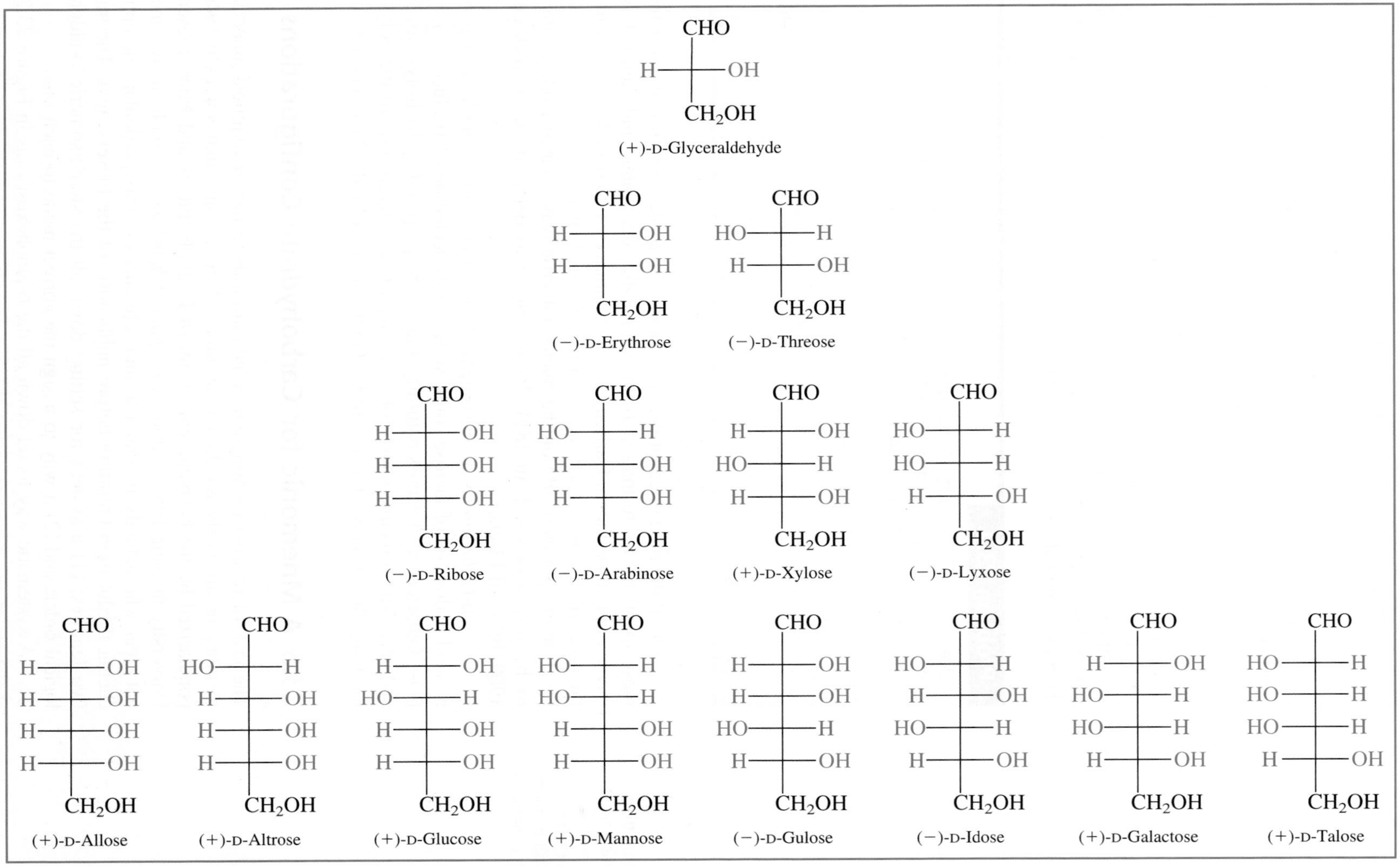

Figure 23.2

Configurations of the D series of aldoses containing three through six carbon atoms.

Problem 23.3

(+)-L-Arabinose is a naturally occurring L sugar. It is obtained by acid hydrolysis of the polysaccharide present in mesquite gum. Write a Fischer projection for (+)-L-arabinose.

Among the aldopentoses, D-ribose is a component of many biologically important substances, most notably the ribonucleic acids. D-Xylose is very abundant and is isolated by hydrolysis of the polysaccharides present in corncobs and the wood of trees.

$2^4 = 16$

The aldohexoses include some of the most familiar of the monosaccharides, as well as one of the most abundant organic compounds on Earth, (+)-D-glucose. With *four* chirality centers, *16* stereoisomeric aldohexoses are possible; 8 belong to the D series and 8 to the L series. All are known, either as naturally occurring substances or as the products of synthesis. The eight D-aldohexoses are given in Figure 23.2; the spatial arrangement at C-5, hydrogen to the left in a Fischer projection and hydroxyl to the right, identifies them as carbohydrates of the D series.

Problem 23.4

Use Figure 23.2 as a guide to help you name the aldose shown. What is the D,L configuration at the highest numbered chirality center? The *R,S* configuration? What is its sign of rotation?

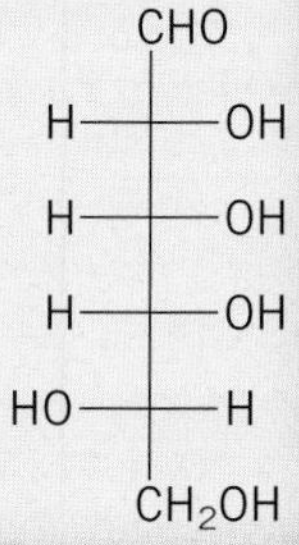
CHO
H
OH
H
OH
H
OH
HO
H
CH₂OH

Cellulose is more abundant than glucose, but each cellulose molecule is a polysaccharide composed of thousands of glucose units (Section 23.16). Methane may also be more abundant, but most of the methane comes from glucose.

Of all the monosaccharides, (+)-D-*glucose* is the best known, most important, and most abundant. Its formation from carbon dioxide, water, and sunlight is the central theme of photosynthesis. Carbohydrate formation by photosynthesis is estimated to be on the order of 10^{11} tons per year, a source of stored energy utilized, directly or indirectly, by all higher forms of life on the planet. Glucose was isolated from raisins in 1747 and by hydrolysis of starch in 1811. Its structure was determined, in work culminating in 1900, by Emil Fischer.

(+)-D-*Galactose* is a constituent of numerous polysaccharides. It is best obtained by acid hydrolysis of lactose (milk sugar), a disaccharide of D-glucose and D-galactose. (−)-L-Galactose also occurs naturally and can be prepared by hydrolysis of flaxseed gum and agar. The principal source of (+)-D-*mannose* is hydrolysis of the polysaccharide of the ivory nut, a large, nut-like seed obtained from a South American palm.

23.5 A Mnemonic for Carbohydrate Configurations

The task of relating carbohydrate configurations to names requires either a world-class memory or an easily recalled mnemonic. A mnemonic that serves us well here was popularized by the husband–wife team of Louis F. Fieser and Mary Fieser of Harvard University in their 1956 textbook, *Organic Chemistry*. As with many mnemonics, it's not clear who actually invented it, and references to this particular one appeared in the chemical education literature before publication of the Fiesers' text. The mnemonic has two features: (1) a system for setting down all the stereoisomeric D-aldohexoses in a logical order; and (2) a way to assign the correct name to each one.

A systematic way to set down all the D-aldohexoses (as in Figure 23.2) is to draw skeletons of the necessary eight Fischer projections, placing the C-5 hydroxyl group to the right in each so as to guarantee that they all belong to the D series. Working up the carbon chain, place the C-4 hydroxyl group to the right in the first four structures, and to the left in the next four. In each of these two sets of four, place the C-3 hydroxyl group to the right in the first two and to the left in the next two; in each of the resulting four sets of two, place the C-2 hydroxyl group to the right in the first one and to the left in the second.

Once the eight Fischer projections have been written, they are named in order with the aid of the sentence: "All altruists gladly make gum in gallon tanks." The words of the sentence stand for *allose, altrose, glucose, mannose, gulose, idose, galactose, talose.*

An analogous pattern of configurations can be seen in the aldopentoses when they are arranged in the order *ribose, arabinose, xylose, lyxose.* (RAXL is an easily remembered nonsense word that gives the correct sequence.) This pattern is discernible even in the aldotetroses erythrose and threose.

23.6 Cyclic Forms of Carbohydrates: Furanose Forms

Aldoses incorporate two functional groups, C=O and OH, which are capable of reacting with each other. We saw in Section 17.8 that nucleophilic addition of an alcohol function to a carbonyl group gives a hemiacetal. When the hydroxyl and carbonyl groups are part of the same molecule, a *cyclic hemiacetal* results, as illustrated in Figure 23.3.

Cyclic hemiacetal formation is most common when the ring that results is five- or six-membered. Five-membered cyclic hemiacetals of carbohydrates are called **furanose** forms; six-membered ones are called **pyranose** forms. The ring carbon that is derived from the carbonyl group, the one that bears two oxygen substituents, is called the **anomeric carbon.**

Aldoses exist almost exclusively as their cyclic hemiacetals; very little of the open-chain form is present at equilibrium. To understand their structures and chemical reactions, we need to be able to translate Fischer projections of carbohydrates into their cyclic hemiacetal forms. Consider first cyclic hemiacetal formation in D-erythrose. To visualize furanose ring formation more clearly, redraw the Fischer projection in a

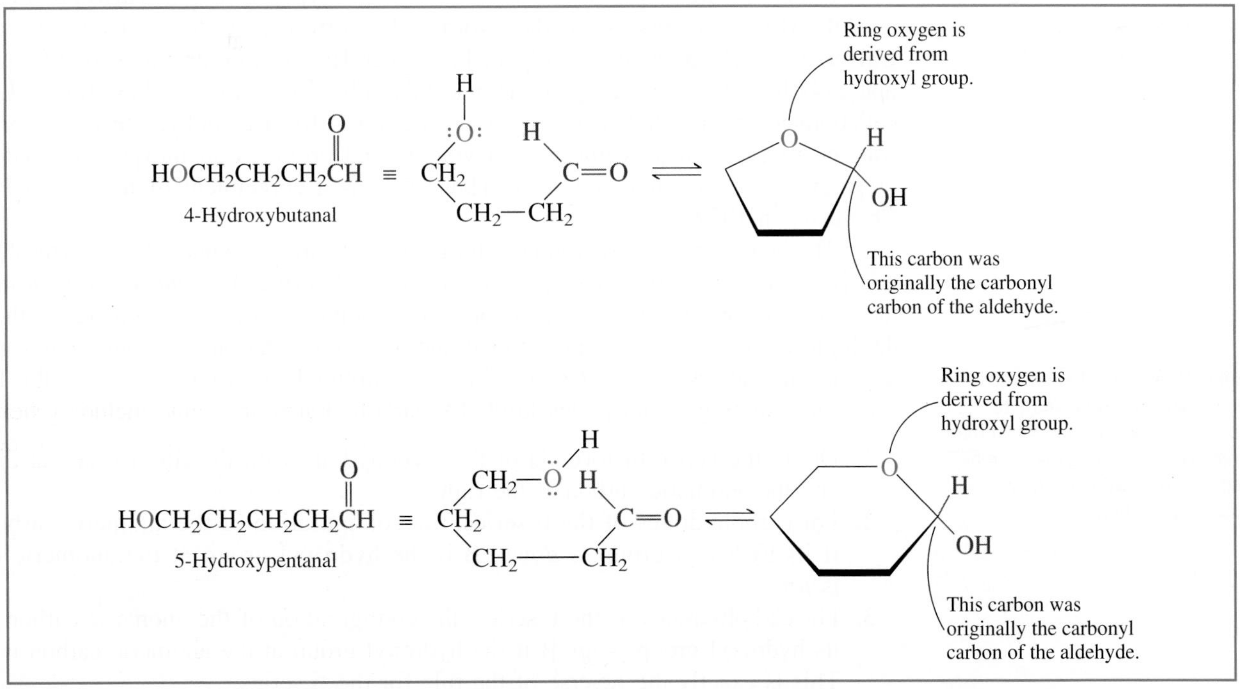

Figure 23.3

Cyclic hemiacetal formation in 4-hydroxybutanal and 5-hydroxypentanal.

form more suited to cyclization, being careful to maintain the stereochemistry at each chirality center.

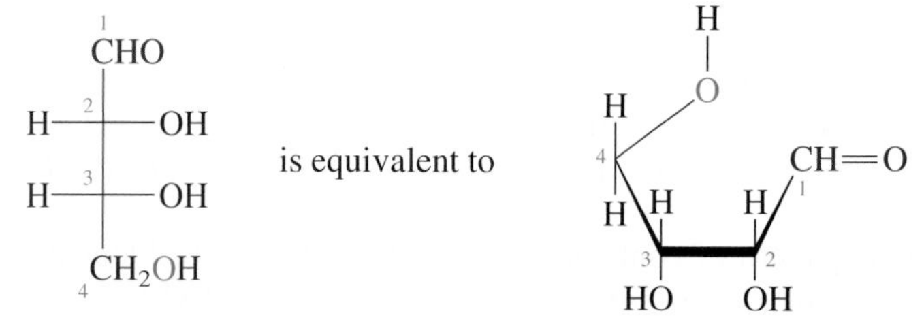
is equivalent to

D-Erythrose

Eclipsed conformation of D-erythrose showing C-4 hydroxyl group in position to add to carbonyl group

Hemiacetal formation between the carbonyl group and the C-4 hydroxyl yields the five-membered furanose ring form. The anomeric carbon is a new chirality center; its hydroxyl group can be either cis or trans to the other hydroxyl groups. The two cyclic forms are diastereomers and are referred to as anomers because they have different configurations at the anomeric carbon.

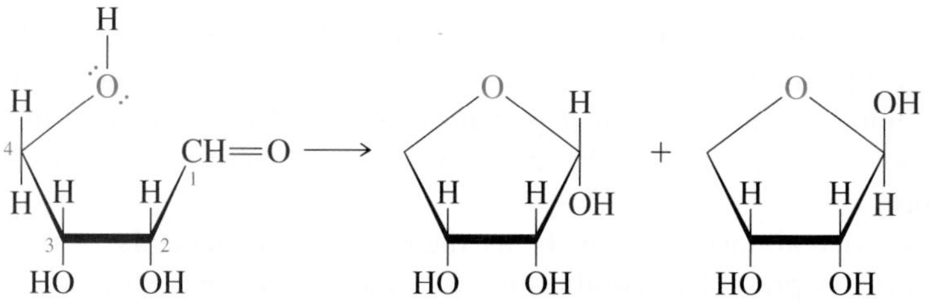

D-Erythrose

α-D-Erythrofuranose (hydroxyl group at anomeric carbon is down)

β-D-Erythrofuranose (hydroxyl group at anomeric carbon is up)

Structural drawings of carbohydrates of this type are called **Haworth formulas,** after the British chemist Sir Walter Norman Haworth (St. Andrew's University and the University of Birmingham). Early in his career Haworth contributed to the discovery that carbohydrates exist as cyclic hemiacetals rather than in open-chain forms. Later he collaborated on an efficient synthesis of vitamin C from carbohydrate precursors. This was the first chemical synthesis of a vitamin and provided an inexpensive route to its preparation on a commercial scale. Haworth was a corecipient of the Nobel Prize in Chemistry in 1937.

The two stereoisomeric furanose forms of D-erythrose are named α-D-erythrofuranose and β-D-erythrofuranose. The prefixes α and β describe the *relative configuration* of the anomeric carbon. The configuration of the anomeric carbon is compared with that of the highest numbered chirality center in the molecule—the one that determines whether the carbohydrate is D or L. Chemists use a simplified, informal version of the IUPAC rules for assigning α and β that holds for carbohydrates up to and including hexoses.

The formal IUPAC rules for α and β notation in carbohydrates are more detailed and less easily understood than our purposes require. These rules can be accessed at http://www.chem.qmw.ac.uk/iupac/2carb/06n07.html.

1. Orient the Haworth formula of the carbohydrate with the ring oxygen at the back and the anomeric carbon at the right.
2. For carbohydrates of the D series, the configuration of the anomeric carbon is α if its hydroxyl group is *down,* β if the hydroxyl group at the anomeric carbon is *up*.
3. For carbohydrates of the L series, the configuration of the anomeric carbon is α if its hydroxyl group is *up,* β if the hydroxyl group at the anomeric carbon is *down.* This is exactly the reverse of the rule for the D series.

Substituents that are to the right in a Fischer projection are "down" in the corresponding Haworth formula; those to the left are "up."

Problem 23.5

The structures shown are the four stereoisomeric threofuranoses. Assign the proper D, L and α, β stereochemical descriptors to each.

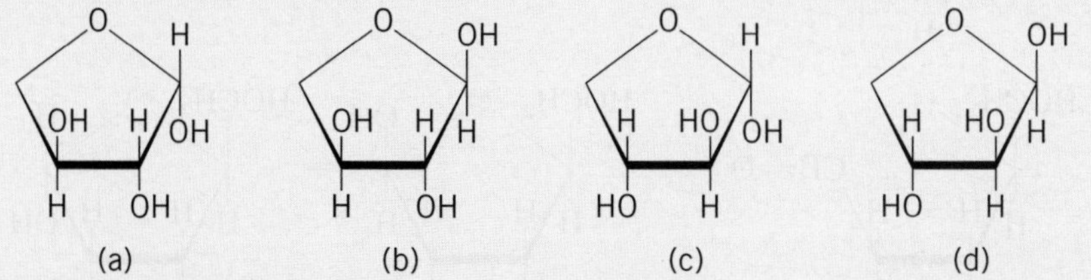

Sample Solution (a) The —OH group at the highest-numbered chirality center (C-3) is up, which places it to the left in the Fischer projection of the open-chain form. The stereoisomer belongs to the L series. The —OH group at the anomeric carbon (C-1) is down, making this the β-furanose form.

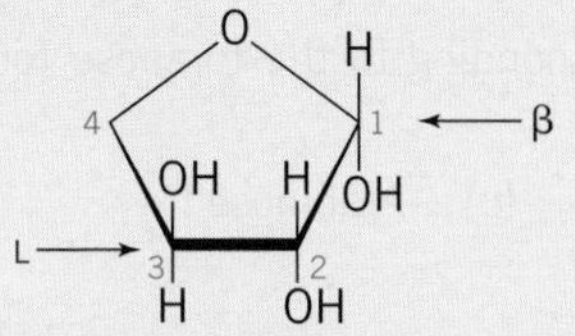

β-L-Threofuranose

Generating Haworth formulas to show stereochemistry in furanose forms of higher aldoses is slightly more complicated and requires an additional operation. Furanose forms of D-ribose are frequently encountered building blocks in biologically important organic molecules. They result from hemiacetal formation between the aldehyde group and the C-4 hydroxyl:

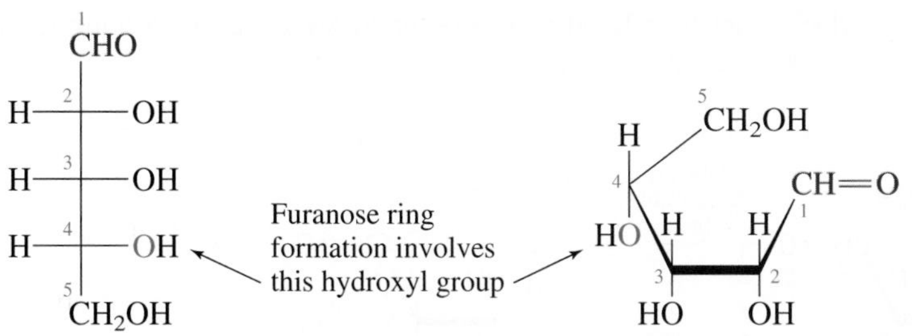
Furanose ring formation involves this hydroxyl group

D-Ribose

Eclipsed conformation of D-ribose

Notice that the eclipsed conformation of D-ribose derived directly from the Fischer projection does not have its C-4 hydroxyl group properly oriented for furanose ring formation. We must redraw it in a conformation that permits the five-membered cyclic hemiacetal to form. This is accomplished by rotation about the C(3)—C(4) bond, taking care that the configuration at C-4 is not changed.

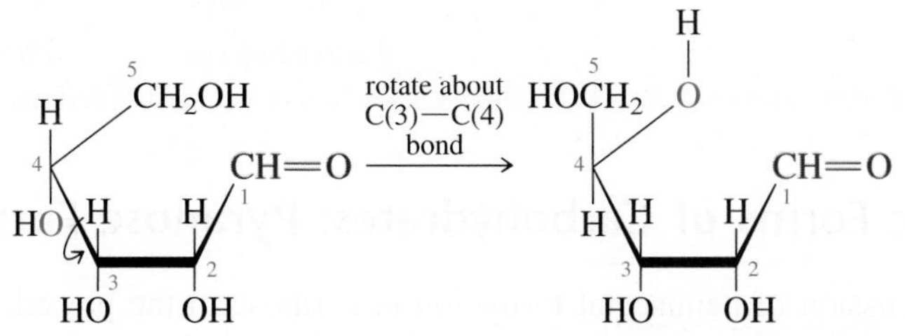
rotate about C(3)—C(4) bond

Conformation of D-ribose suitable for furanose ring formation

As viewed in the drawing, a 120° counterclockwise rotation of C-4 places its hydroxyl group in the proper position. At the same time, this rotation moves the CH_2OH group to a position such that it will become a substituent that is "up" on the five-membered ring. The hydrogen at C-4 then will be "down" in the furanose form.

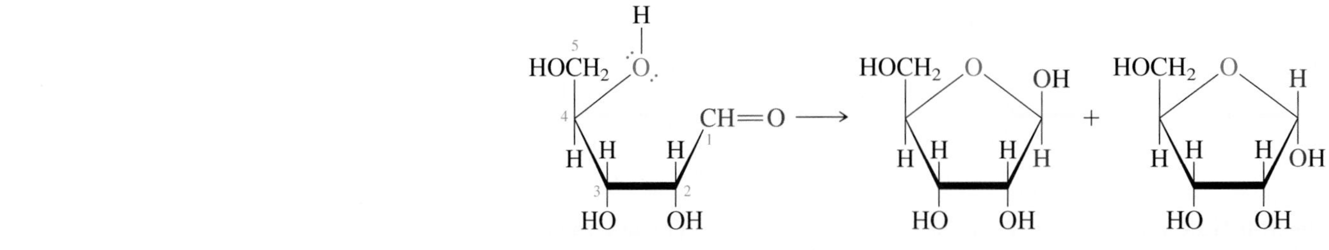

β-D-Ribofuranose α-D-Ribofuranose

Problem 23.6

Write Haworth formulas corresponding to the furanose forms of each of the following carbohydrates:

(a) D-Xylose
(b) D-Arabinose
(c) L-Arabinose

Sample Solution (a) The Fischer projection of D-xylose is given in Figure 23.2.

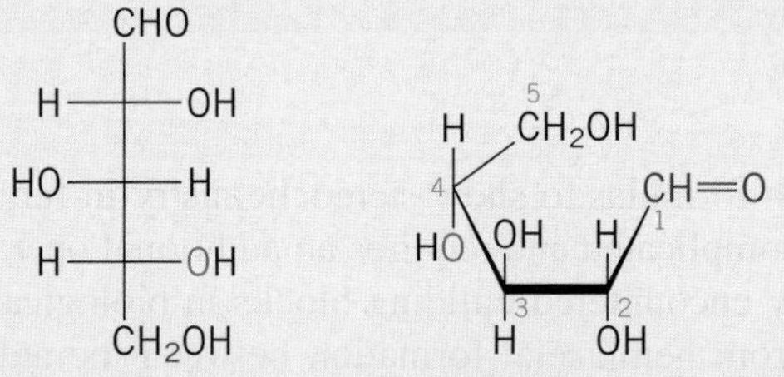

D-Xylose Eclipsed conformation of D-xylose

Carbon-4 of D-xylose must be rotated in a counterclockwise sense to bring its hydroxyl group into the proper orientation for furanose ring formation.

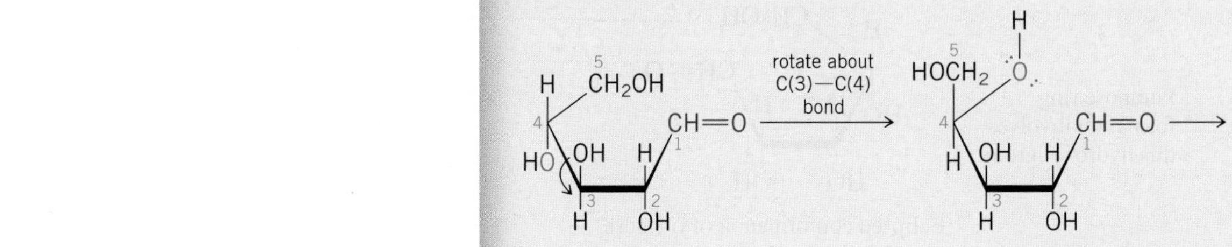

D-Xylose

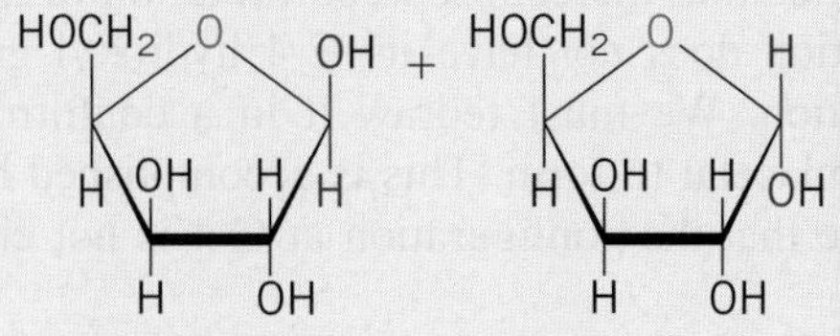

β-D-Xylofuranose α-D-Xylofuranose

23.7 Cyclic Forms of Carbohydrates: Pyranose Forms

During the discussion of hemiacetal formation in D-ribose in the preceding section, you may have noticed that aldopentoses can potentially form a six-membered cyclic hemiacetal via addition of the C-5 hydroxyl to the carbonyl group. This mode of ring closure leads to α- and β-*pyranose* forms:

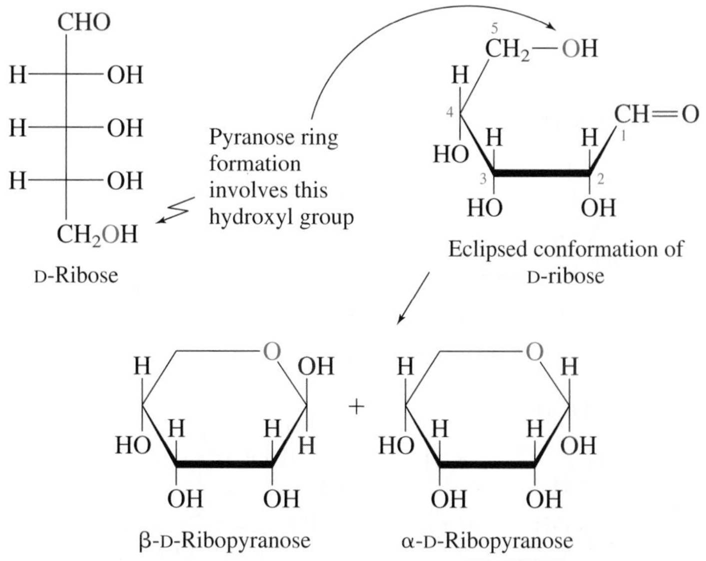

Like aldopentoses, aldohexoses such as D-glucose are capable of forming two furanose forms (α and β) and two pyranose forms (α and β). The Haworth representations of the pyranose forms of D-glucose are constructed as shown in Figure 23.4; each has a CH_2OH group as a substituent on the six-membered ring.

Haworth formulas are satisfactory for representing *configurational* relationships in pyranose forms but are uninformative as to carbohydrate *conformations*. X-ray

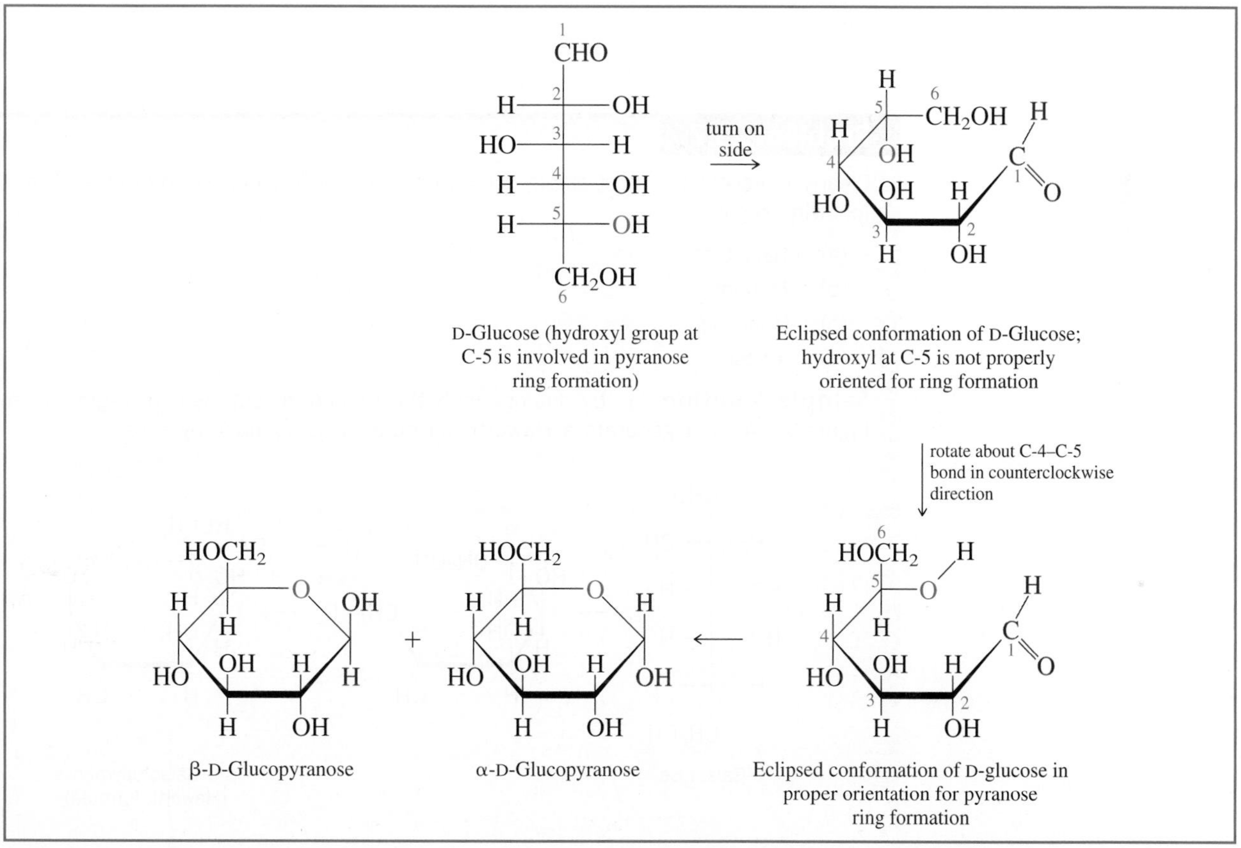

Figure 23.4

Haworth formulas for α- and β-pyranose forms of D-glucose.

crystallographic studies of a large number of carbohydrates reveal that the six-membered pyranose ring of D-glucose adopts a chair conformation:

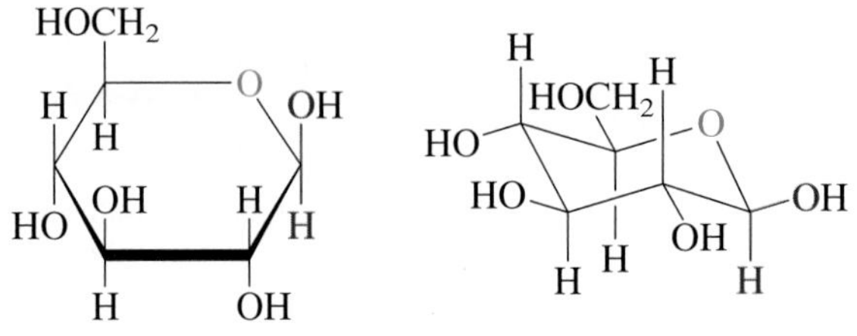

β-D-Glucopyranose

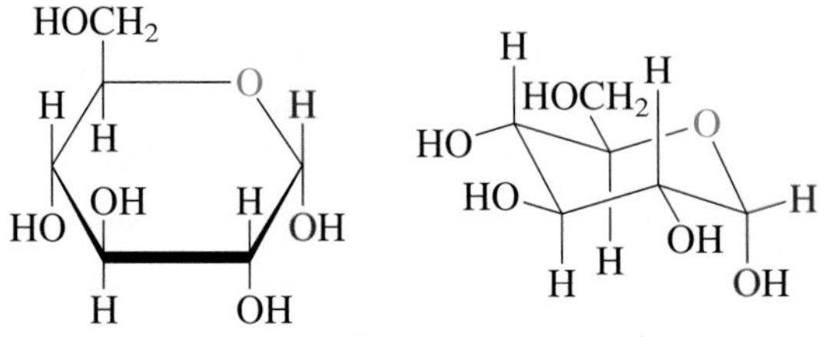

α-D-Glucopyranose

All the ring substituents in β-D-glucopyranose are equatorial in the most stable chair conformation. Only the anomeric hydroxyl group is axial in the α isomer; all the other substituents are equatorial.

Other aldohexoses behave similarly in adopting chair conformations that permit the CH_2OH substituent to occupy an equatorial orientation. Normally the CH_2OH group is the bulkiest, most conformationally demanding substituent in the pyranose form of a hexose.

Problem 23.7

Clearly represent the most stable conformation of the β-pyranose form of each of the following sugars:

(a) D-Galactose
(b) D-Mannose
(c) L-Mannose
(d) L-Ribose

Sample Solution (a) By analogy with the procedure outlined for D-glucose in Figure 23.4, first generate a Haworth formula for β-D-galactopyranose:

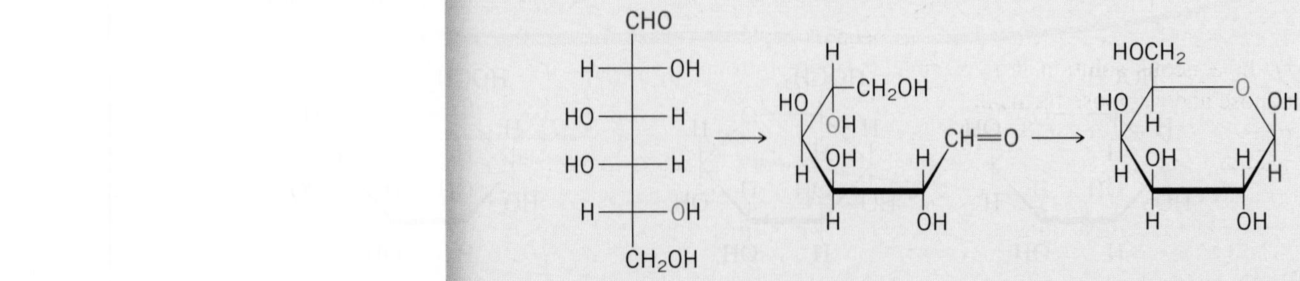

D-Galactose

β-D-Galactopyranose
(Haworth formula)

Next, convert the Haworth formula to the chair conformation that has the CH_2OH group equatorial.

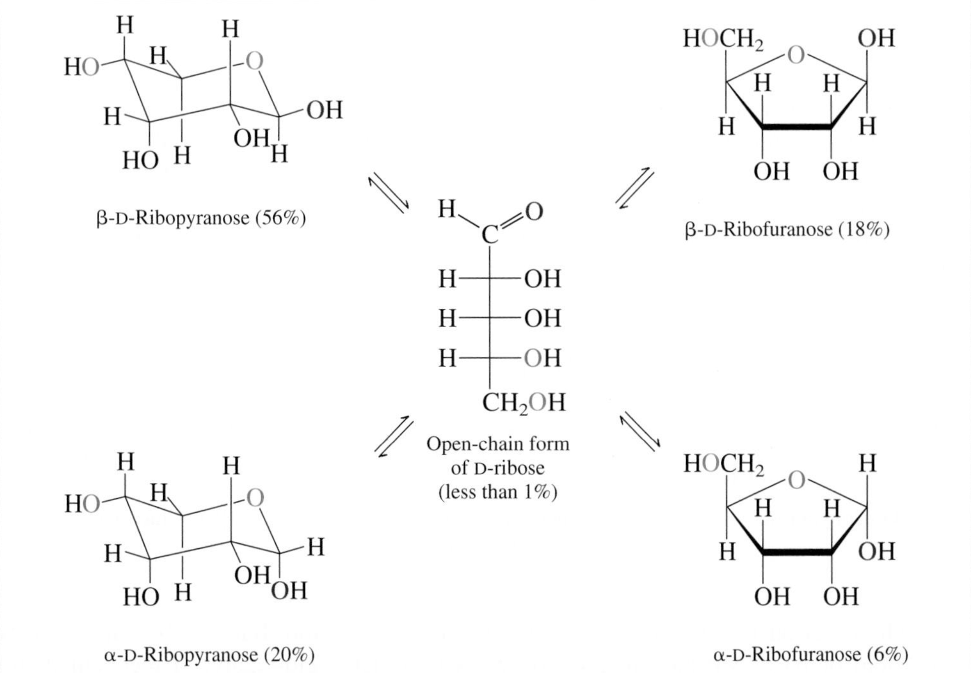

β-D-Ribopyranose (56%)
β-D-Ribofuranose (18%)
Open-chain form of D-ribose (less than 1%)
α-D-Ribopyranose (20%)
α-D-Ribofuranose (6%)

Figure 23.5

Distribution of furanose, pyranose, and open-chain forms of D-ribose in aqueous solution as measured by ^1H and ^{13}C NMR spectroscopy.

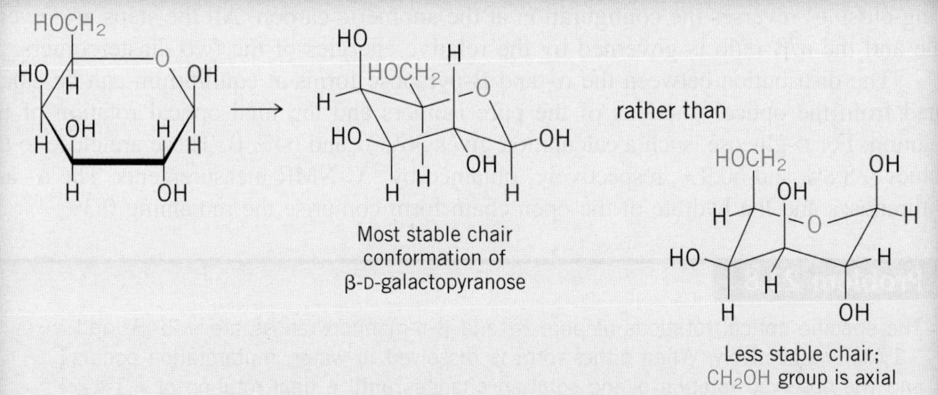

rather than
Most stable chair conformation of β-D-galactopyranose
Less stable chair; CH_2OH group is axial

Galactose differs from glucose in configuration at C-4. The C-4 hydroxyl is axial in β-D-galactopyranose, but it is equatorial in β-D-glucopyranose.

Because six-membered rings are normally less strained than five-membered ones, pyranose forms are usually present in greater amounts than furanose forms at equilibrium, and the concentration of the open-chain form is quite small. The distribution of carbohydrates among their various hemiacetal forms has been examined by using ^1H and ^{13}C NMR spectroscopy. In aqueous solution, for example, D-ribose is found to contain the various α- and β-furanose and pyranose forms in the amounts shown in Figure 23.5. The concentration of the open-chain form at equilibrium is too small to measure directly. Nevertheless, it occupies a central position, in that interconversions of α and β anomers and furanose and pyranose forms take place by way of the open-chain form as an intermediate. As will be seen later, certain chemical reactions also proceed by way of the open-chain form.

23.8 Mutarotation

The α and β stereoisomeric forms of carbohydrates are capable of independent existence, and many have been isolated in pure form as stable, crystalline solids. When crystallized from ethanol, for example, D-glucose yields α-D-glucopyranose, mp 146°C, $[\alpha]_D$ +112.2°.

Crystallization from a water–ethanol mixture produces β-D-glucopyranose, mp 148–155°C, $[\alpha]_D$ +18.7°. In the solid state the two forms do not interconvert and are stable indefinitely. Their structures have been unambiguously confirmed by X-ray crystallography.

The optical rotations just cited for each isomer are those measured immediately after each one is dissolved in water. On standing, the rotation of the solution containing the α isomer decreases from +112.2° to +52.5°; the rotation of the solution of the β isomer increases from +18.7° to the same value of +52.5°. This phenomenon is called **mutarotation.** What is happening is that each solution, initially containing only one anomeric form, undergoes equilibration to the same mixture of α- and β-pyranose forms. The open-chain form is an intermediate in the process.

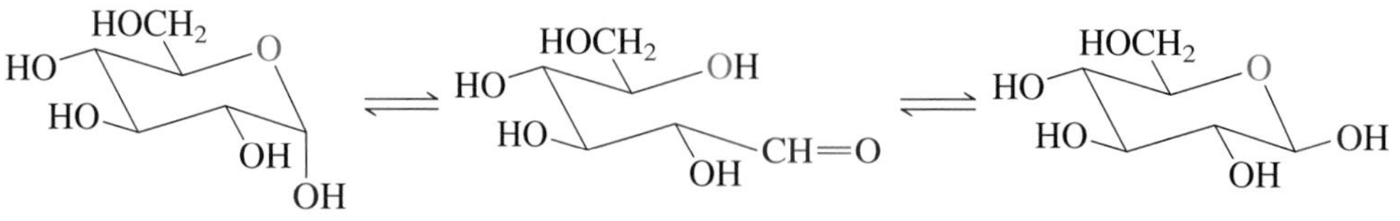

α-D-Glucopyranose (mp 146°C; $[\alpha]_D$ +112.2°) | Open-chain form of D-glucose | β-D-Glucopyranose (mp 148–155°C; $[\alpha]_D$ +18.7°)

Mutarotation occurs slowly in neutral aqueous solution, but can be catalyzed by either acid or base. Mechanism 23.1 shows a four-step, acid-catalyzed mechanism for mutarotation starting with α-D-glucopyranose. Steps 1 and 4 are proton transfers and describe the role of the acid catalyst. The combination of step 2 (ring-opening) and step 3 (ring-closing) reverses the configuration at the anomeric carbon. All the steps are reversible and the α/β ratio is governed by the relative energies of the two diastereomers.

The distribution between the α- and β-pyranose forms at equilibrium can be calculated from the optical rotations of the pure isomers and the final optical rotation of the solution. For D-glucose, such a calculation gives 36% α and 64% β. These are close to the values (38.8% and 60.9%, respectively) obtained by ^{13}C NMR measurements. The α- and β-furanoses and the hydrate of the open-chain form comprise the remaining 0.3%.

Problem 23.8

The specific optical rotations of pure α- and β-D-mannopyranose are +29.3° and −17.0°, respectively. When either form is dissolved in water, mutarotation occurs, and the observed rotation of the solution changes until a final rotation of +14.2° is observed. Assuming that only α- and β-pyranose forms are present, calculate the percent of each isomer at equilibrium.

It is not possible to tell by inspection which pyranose form of a particular carbohydrate—α or β—predominates at equilibrium. As just described, the β-pyranose is the major species present in an aqueous solution of D-glucose, whereas the α-pyranose predominates in a solution of D-mannose (Problem 23.8). In certain other carbohydrates, D-ribose for example, furanose and pyranose forms are both well represented at equilibrium (Figure 23.5).

Problem 23.9

Write a four-step mechanism for the mutarotation of D-glucopyranose in aqueous base. Use curved arrows to track electron flow. The first step is:

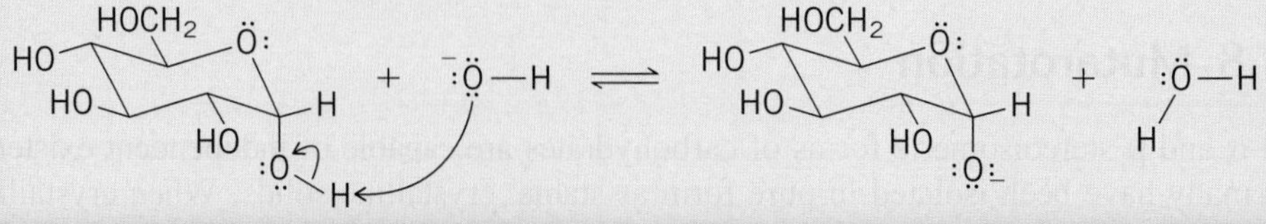

Mechanism 23.1

Acid-Catalyzed Mutarotation of D-Glucopyranose

THE OVERALL REACTION:

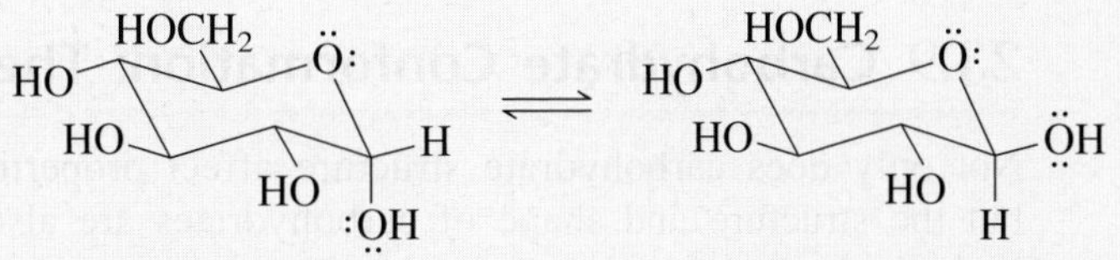

α-D-Glucopyranose ⇌ β-D-Glucopyranose

Step 1: Protonation of the oxygen of the pyranose ring by the acid catalyst. In aqueous solution, the acid catalyst is the hydronium ion.

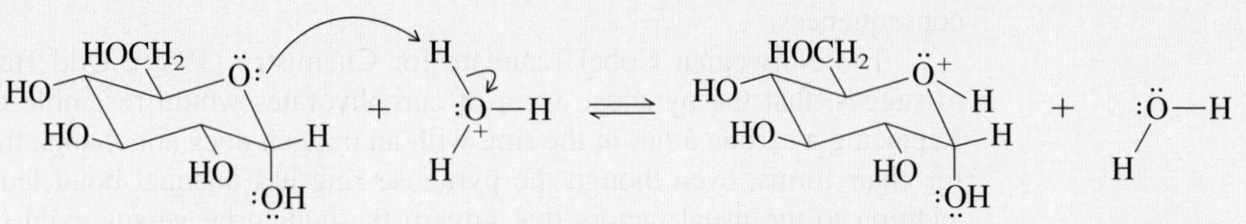

α-D-Glucopyranose + Hydronium ion ⇌ Conjugate acid of α-D-glucopyranose + Water

Step 2: The pyranose ring opens by cleaving the bond between the anomeric carbon and the positively charged oxygen. This ring opening is facilitated by electron release from the OH group at the anomeric carbon and gives the conjugate acid of the open-chain form of D-glucose.

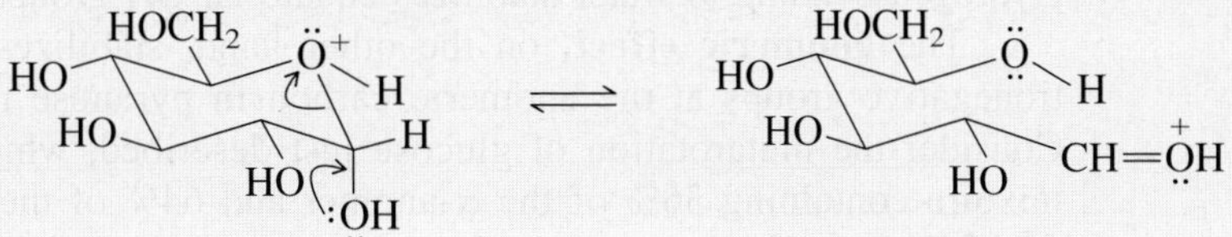

Conjugate acid of α-D-glucopyranose ⇌ Conjugate acid of open-chain form of D-glucose

Step 3: The species formed in the preceding step cyclizes to give the conjugate acid of β-D-glucopyranose. This cyclization is analogous to the acid-catalyzed nucleophilic additions to aldehydes and ketones in Chapter 17.

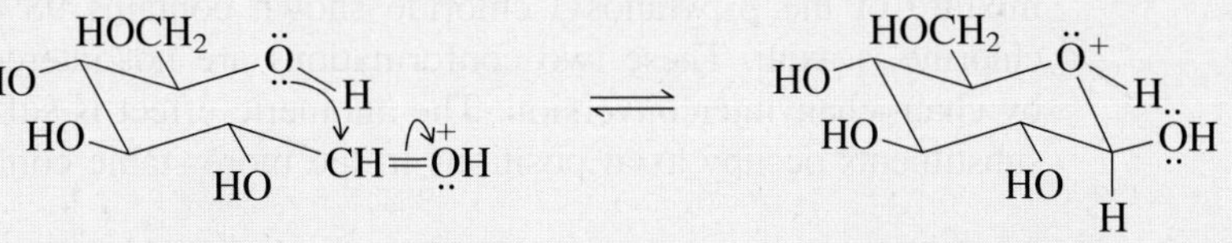

Conjugate acid of open-chain form of D-glucose ⇌ Conjugate acid of β-D-glucopyranose

Step 4: The product of step 3 transfers a proton to water to regenerate the acid catalyst and yield the neutral form of the product.

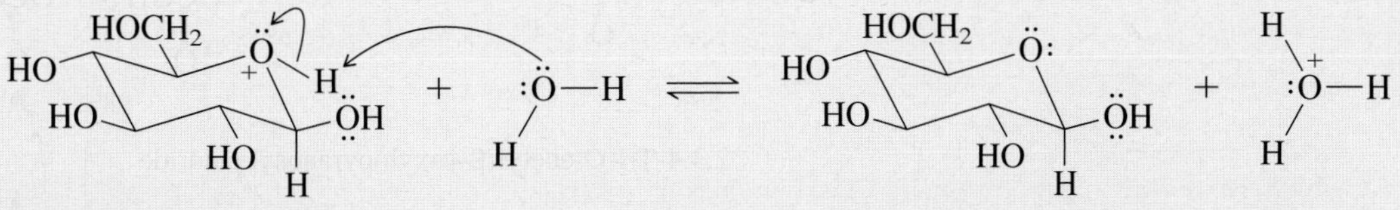

Conjugate acid of β-D-glucopyranose + Water ⇌ β-D-Glucopyranose + Hydronium ion

The factors that control the equilibrium composition of sugars in solution are complex. Although the well-established preference for substituents in six-membered rings to be equatorial rather than axial is important, it is not always the overriding factor. The next section introduces a new structural feature that plays a significant part in determining carbohydrate conformations and α/β anomeric ratios.

23.9 Carbohydrate Conformation: The Anomeric Effect

Not only does carbohydrate structure affect properties such as chemical reactivity, but the structure and shape of carbohydrates are also major factors in a number of biological processes that depend on interactions between molecules—a phenomenon known as *molecular recognition.* In this section, we will consider mainly the conformations of carbohydrates in their pyranose forms. We will return to some familiar concepts of chair conformations and axial versus equatorial groups, but you will see that the presence of an oxygen atom in a six-membered ring leads to some surprising consequences.

The Norwegian Nobel Laureate for Chemistry (1969), Odd Hassel, was the first to suggest that the pyranose form of carbohydrates would resemble chair cyclohexane. Replacing a carbon atom in the ring with an oxygen does not change the basic preference for chair forms, even though the pyranose ring has unequal bond lengths. However, in addition to the usual factors that govern the equatorial versus axial orientation of substituents on a six-membered ring, two other factors are important:

1. an equatorial OH is less crowded and better solvated by water than an axial one
2. the anomeric effect

The first of these is straightforward and alerts us to the fact that the relative energies of two species may be different in solution than in the solid state or the gas phase. Hydrogen bonding to water stabilizes equatorial OH groups better than axial ones.

The **anomeric effect,** on the other hand, stabilizes *axial* OH and other electronegative groups at the anomeric carbon in pyranose rings better than equatorial. Consider the mutarotation of glucose just described, which produces an equilibrium mixture containing 36% of the α-anomer and 64% of the β. If we consider only the destabilizing effect of a solvated axial hydroxyl group in the axial position, we would expect only 11% α and 89% β. The presence of more axial hydroxyl than expected results from the contribution of the anomeric effect to the free energy difference between these two *stereoisomers*.

The anomeric effect also influences the *conformational* equilibria in pyranoses with an electronegative atom, usually oxygen or halogen, at C-l. For example, the equilibrium mixture of the β-pyranosyl chloride shown contains 98% of the conformer in which chlorine is axial. These two conformations are not interconverted by mutarotation but by chair–chair interconversion. The anomeric effect is sufficiently large so that all four substituents occupy axial positions in the more stable conformer.

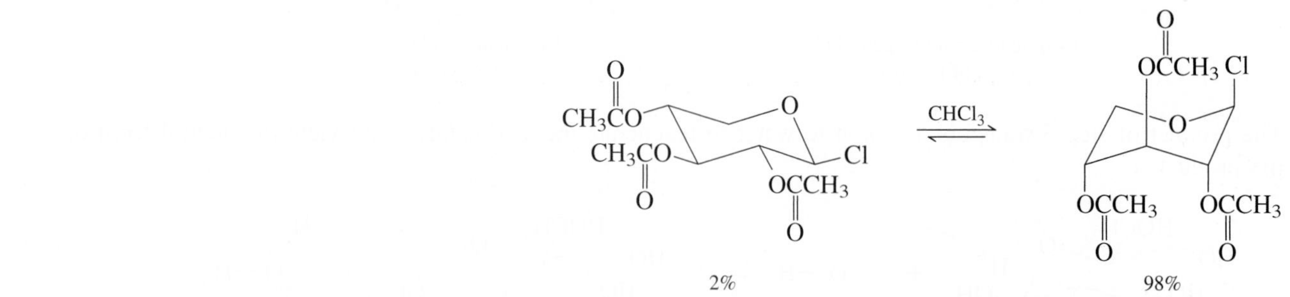

2,3,4-Tri-*O*-acetyl-β-D-xylopyranosyl chloride

What is responsible for the anomeric effect? Chemists continue to debate this question, and a number of explanations have emerged, one of which is detailed in Figure 23.6. The structure depicted in (a), in which substituent X is axial, is stabilized

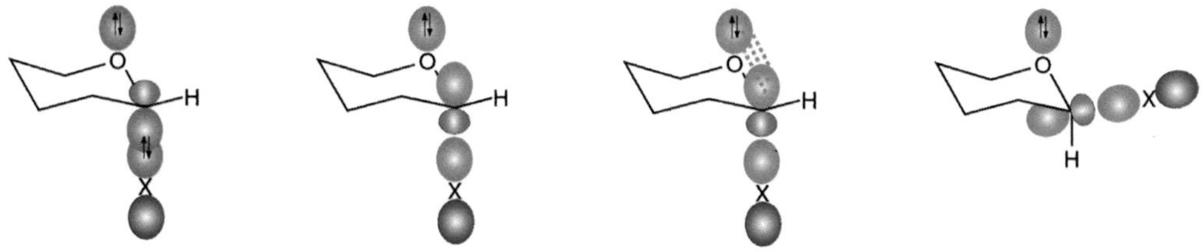

(a) (b) (c) (d)

Figure 23.6

Delocalization of an oxygen lone pair stabilizes the conformation in which an electronegative substituent X at the anomeric carbon is axial. (a) Atom X is connected to the ring by a σ bond formed by overlap of an sp^3 hybrid orbital of carbon and a *p* orbital of X. It is doubly occupied and is therefore incapable of sharing oxygen's unshared pair. (b) The antibonding C—X orbital σ^* is unoccupied and can accommodate two electrons. (c) Overlap of the oxygen 2*p* orbital with the C—X σ^* orbital allows oxygen's unshared electron pair to be delocalized. Delocalization is maximized when the oxygen 2*p* orbital and σ^* are anti coplanar; that is when X is axial. (d) When X is equatorial, the axis of its C—X σ^* orbital is gauche to that of the oxygen 2*p* orbital and does not allow oxygen's unshared electron pair to be delocalized as well as in (c).

by the interaction of an unshared electron pair on the ring oxygen with the antibonding σ^* orbital of the C—X bond [(b) and (c)]. This interaction is greater when X is axial and anti coplanar to the nonbonding electron pair than it is when X is equatorial (d). The transfer of electron density toward the anomeric carbon is facilitated when X is an electronegative substituent, so the effect is often seen in pyranoses with oxygen or halogen at the anomeric carbon. This model also accounts for the shortening of the O—C-1 bond seen in pyranoses that bear an axial electronegative substituent at the anomeric carbon.

Because five-membered rings are more flexible than six-membered, the anomeric effect is less important in furanose than in pyranose forms.

The anomeric effect is a general property of structural units of the type X—C—Y—R where X and Y are electronegative and Y has at least one unshared pair. Of the conformations about the Y—C bond, gauche is normally more stable than anti.

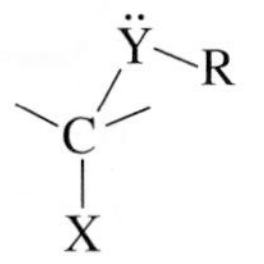

is more stable than

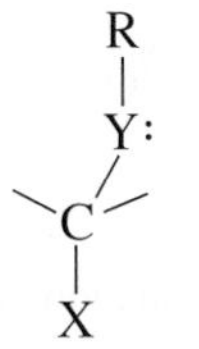

C—X and Y—R bonds are gauche

C—X and Y—R bonds are anti

For a simple structure such as chloromethyl methyl ether ($ClCH_2OCH_3$), the gauche conformation has been estimated to be about 8 kJ/mol (2 kcal/mol) more stable than the anti.

Problem 23.10

Sketch the most stable conformation of chloromethyl methyl ether.

23.10 Ketoses

Up to this point all our attention has been directed toward aldoses, carbohydrates having an aldehyde function in their open-chain form. Aldoses are more common than ketoses, and their role in biological processes has been more thoroughly studied. Nevertheless, a large number of ketoses are known, and several of them are pivotal intermediates in carbohydrate biosynthesis and metabolism. Examples of some ketoses include D-*ribulose,* L-*xylulose,* and D-*fructose:*

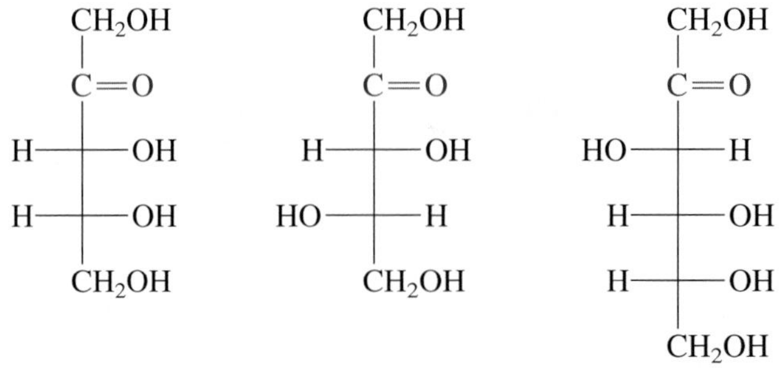
CH2OH
C=O
H
OH
CH2OH
HO
D-Ribulose
L-Xylulose
D-Fructose

In all three the carbonyl group is at C-2, which is the most common case for naturally occurring ketoses. D-Ribulose is a key intermediate in photosynthesis, the process by which energy from sunlight drives the formation of D-glucose from carbon dioxide and water. L-Xylulose is a product of the abnormal metabolism of xylitol in persons who lack a particular enzyme. D-Fructose is the most familiar ketose; it is present in fruits and honey and is sweeter than sucrose.

Problem 23.11

How many ketotetroses are possible? Write Fischer projections for each.

Ketoses, like aldoses, exist mainly as cyclic hemiacetals. In the case of D-ribulose, furanose forms result from addition of the C-5 hydroxyl to the carbonyl group.

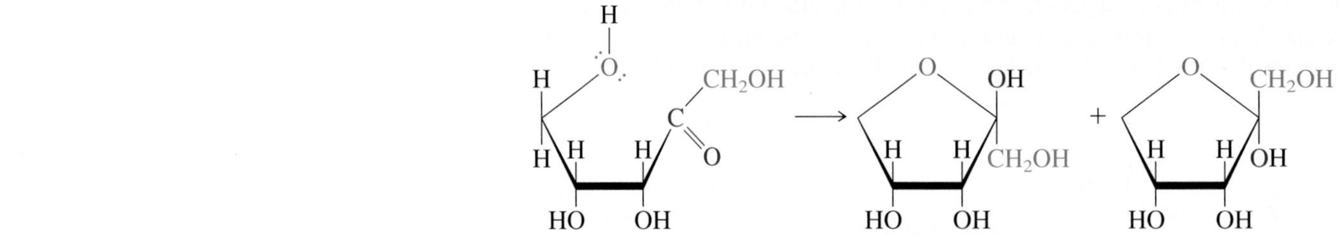
H
O
CH2OH
C
O
H
HO
OH
+

Eclipsed conformation of D-ribulose

β-D-Ribulofuranose

α-D-Ribulofuranose

The anomeric carbon of a furanose or pyranose form of a ketose bears both a hydroxyl group and a carbon substituent. In the case of 2-ketoses, this substituent is a CH_2OH group. As with aldoses, the anomeric carbon of a cyclic hemiacetal is readily identifiable because it is bonded to two oxygens.

Problem 23.12

Use the method outlined in Figure 23.4 to write a Haworth formula for the β-furanose form of D-fructose.

23.11 Deoxy Sugars

A common variation on the general pattern seen in carbohydrate structure is the replacement of one or more of the hydroxyl substituents by some other atom or group. In **deoxy sugars** the hydroxyl group is replaced by hydrogen. Two examples of deoxy sugars are 2-deoxy-D-ribose and L-fucose:

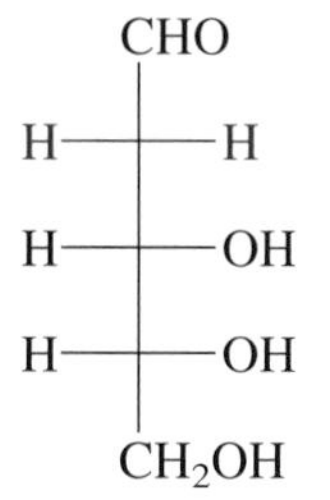

2-Deoxy-D-ribose

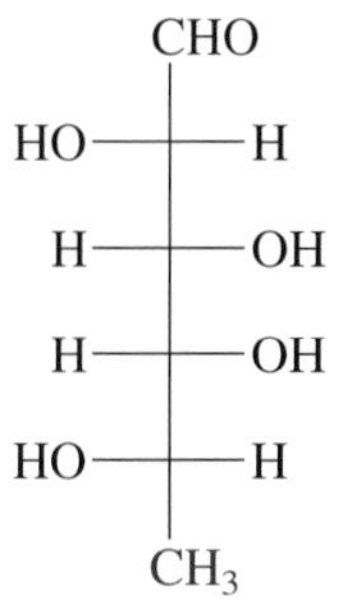

L-Fucose
(6-deoxy-L-galactose)

The hydroxyl at C-2 in D-ribose is absent in 2-deoxy-D-ribose. In Chapter 26 we shall see how derivatives of 2-deoxy-D-ribose, called *deoxyribonucleotides,* are the fundamental building blocks of deoxyribonucleic acid (DNA), the material responsible for storing genetic information. L-Fucose, the carbon chain of which terminates in a methyl rather than a CH_2OH group, is often found as one of the carbohydrates in glycoproteins, such as those on the surface of red blood cells that determine blood type (Section 23.25).

Problem 23.13

Write Fischer projections of

(a) *Cordycepose* (3-deoxy-D-ribose): a deoxy sugar isolated by hydrolysis of the antibiotic substance *cordycepin*
(b) L-*Rhamnose* (6-deoxy-L-mannose): found in plants

Sample Solution (a) The hydroxyl group at C-3 in D-ribose is replaced by hydrogen in 3-deoxy-D-ribose.

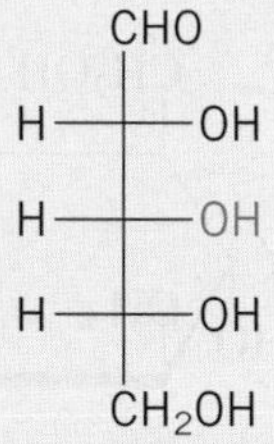

D-Ribose
(from Figure 23.2)

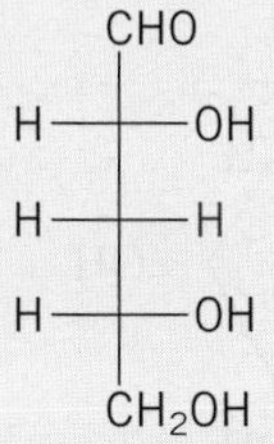

3-Deoxy-D-ribose
(cordycepose)

23.12 Amino Sugars

Another structural variation is the replacement of a hydroxyl group in a carbohydrate by an amino group to give an **amino sugar.** The most abundant amino sugar is one of the oldest and most abundant organic compounds on Earth. *N*-Acetyl-D-glucosamine is the main component of the polysaccharide in *chitin,* the substance that makes up the tough outer skeleton of arthropods and insects. Chitin has been isolated from a 25-million-year-old beetle fossil, and more than 10^9 tons of chitin is produced in the biosphere each year. Lobster shells, for example, are mainly chitin (Figure 23.7). More than 60 amino sugars are known, many of them having been isolated and identified only recently as components of antibiotics. The anticancer drug doxorubicin hydrochloride (Adriamycin), for example, contains the amino sugar L-daunosamine as one of its structural units.

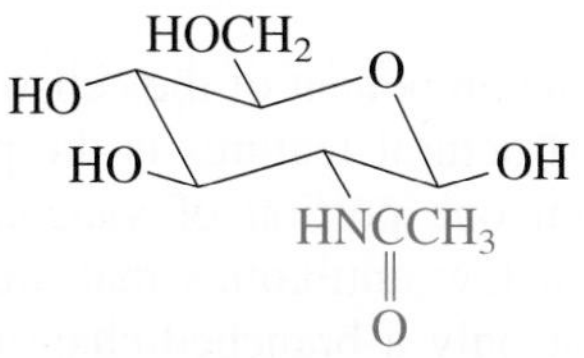

N-Acetyl-D-glucosamine

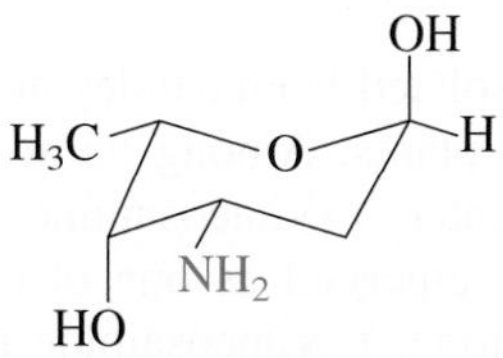

L-Daunosamine

Figure 23.7

The shells of lobsters are mainly chitin, a polymer of *N*-acetyl-D-glucosamine.

Sialic acids are a group of carbohydrates that have the interesting structural feature of being amino-substituted derivatives of a nine-carbon ketose. *N*-Acetylneuraminic acid can be considered the parent.

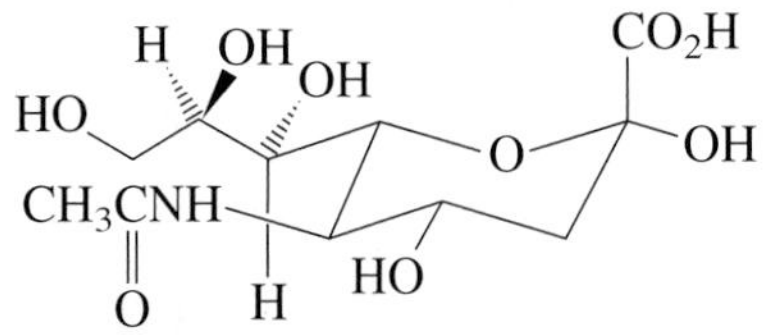

N-Acetylneuraminic acid

More than 40 structurally related sialic acids occur naturally where they play a number of roles. As covalently bound components of glycolipids and glycoproteins, they are intimately involved in cell recognition processes.

Problem 23.14

We've included *N*-acetylneuraminic acid in the section on amino sugars and described it as a ketose. We could also call it a deoxy sugar. Locate reasons for these classifications in the structural formula. Number the carbon atoms in the nine-carbon chain. What is the configuration (D or L) at the highest numbered chirality center?

Nitrogen-containing sugars in which nitrogen replaces the ring oxygen are known as imino sugars. An example of an imino sugar is nojirimicin, which occurs naturally. A synthetic derivative, *N*-butyl-1-deoxynojirimicin, is used in the treatment of a lipid metabolism disorder known as Gaucher's disease.

The symbol ~~~ is used to represent a bond of undefined stereochemistry.

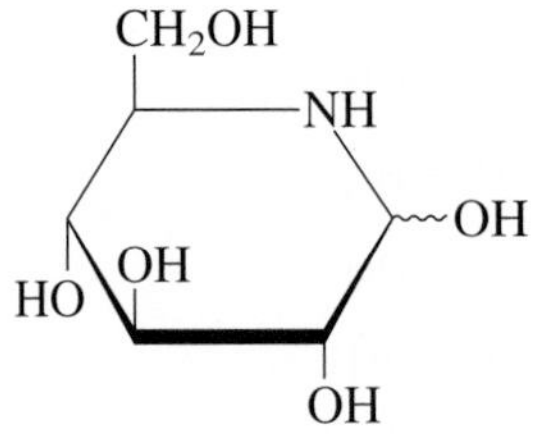

Nojirimicin

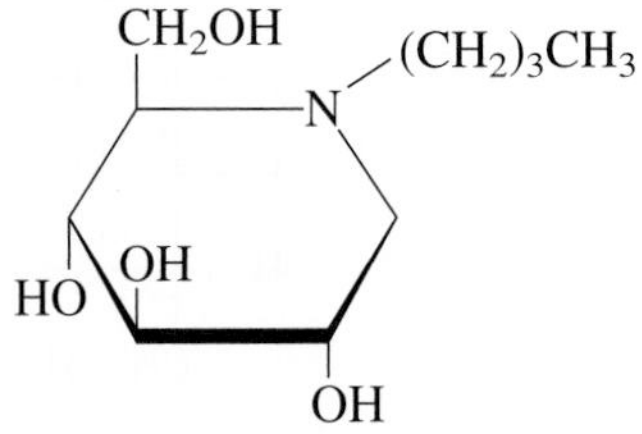

N-Butyl-1-deoxynojirimicin

23.13 Branched-Chain Carbohydrates

Carbohydrates that have a carbon substituent attached to the main chain are said to have a **branched chain.** D-Apiose and L-vancosamine are representative branched-chain carbohydrates:

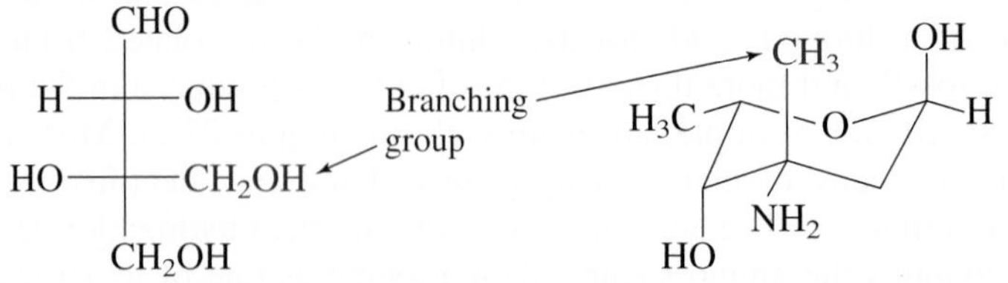

D-Apiose L-Vancosamine

D-Apiose can be isolated from parsley and is a component of the cell-wall polysaccharide of various marine plants. Among its novel structural features is the presence of only a single chirality center. L-Vancosamine is but one portion of vancomycin, a powerful antibiotic that has emerged as one of only a few antibiotics that are effective against drug-resistant bacteria. L-Vancosamine is not only a branched-chain carbohydrate, it is a deoxy sugar and an amino sugar as well.

23.14 Glycosides: The Fischer Glycosidation

Glycosides are a large and important class of carbohydrate derivatives characterized by the replacement of the anomeric hydroxyl group by some other substituent. Glycosides are termed *O*-glycosides when the atom attached to the anomeric carbon is oxygen. If the atom is sulfur, the names *S*-glycoside and thioglycoside are both used. Glycosides in which the atom attached to the anomeric carbon is nitrogen are named as glycosylamines.

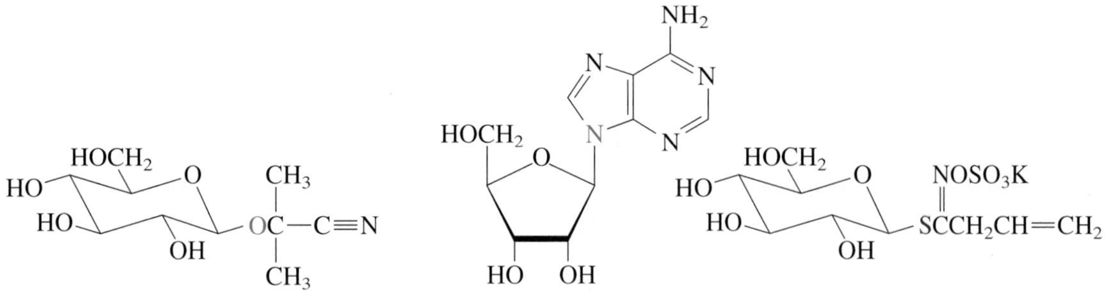

Linamarin Adenosine Sinigrin

Linamarin is an *O*-glycoside of D-glucose and acetone cyanohydrin. It is present in manioc (cassava), a tuberous food plant grown in tropical climates and is just one of many cyanogenic glycosides. Nucleosides such as adenosine are glycosylamines of heterocyclic aromatic compounds. The most important ones are those derived from D-ribose and 2-deoxy-D-ribose. Sinigrin is an *S*-glycoside that contributes to the characteristic flavor of mustard and horseradish. All three of the glycosides shown have the β configuration at their anomeric carbon. Many antibiotics occur as glycosides. The most common are *O*-glycosides, such as erythromycin (Figure 18.15).

The term *glycoside* without a prefix is taken to mean an *O*-glycoside. *O*-Glycosides bear an alkoxy group —OR instead of —OH at the anomeric carbon. Structurally, they are mixed acetals. Recall the sequence of intermediates in acetal formation (Section 17.8):

Aldehyde or ketone Hemiacetal Acetal

If the aldehyde or ketone bears a γ or δ OH group, the first step takes place intramolecularly to yield a *cyclic* hemiacetal. The second step is intermolecular and requires an alcohol ROH as a reactant.

4-Hydroxybutanal Cyclic hemiacetal Acetal

In this illustration only a five-membered cyclic hemiacetal, analogous to the furanose form of a carbohydrate, is possible from the γ-hydroxy aldehyde. The final acetal is analogous to a glycoside; in this case, a furanoside. The corresponding products from an aldehyde with a δ —OH group would be a pyranose (cyclic hemiacetal) and a pyranoside (acetal).

In a reaction known as the **Fischer glycosidation,** glycosides are prepared by simply allowing a carbohydrate to react with an alcohol in the presence of an acid catalyst.

The reaction is *thermodynamically controlled,* and the major product is the most stable glycoside; for the reaction of D-glucose with methanol this is methyl α-D-glucopyranoside. Six-membered rings are more stable than five-membered ones, and the anomeric effect stabilizes an axial —OCH_3 group.

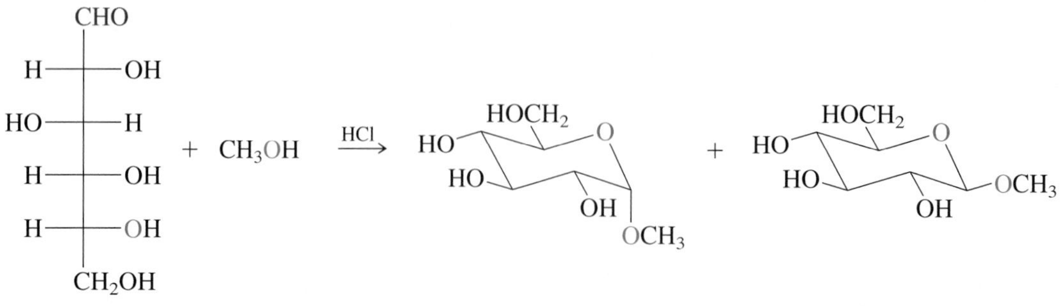

D-Glucose | Methanol | Methyl α-D-glucopyranoside (major product; isolated in 49% yield) | Methyl β-D-glucopyranoside (minor product)

Problem 23.15

Write structural formulas for the α- and β-methyl pyranosides formed by reaction of D-galactose with methanol in the presence of hydrogen chloride.

Experimental observations suggest that the methyl glycosides are formed by more than one mechanism and can involve formation of a hemiacetal, acetal, or oxonium ion as an intermediate, followed by its cyclization. For the reaction of D-glucose with methanol these key intermediates are:

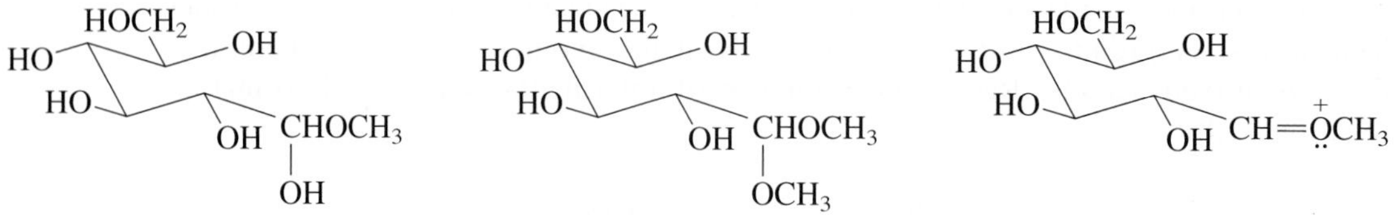

Hemiacetal | Dimethyl acetal | Oxonium ion

Cyclization can lead to the α- or β-furanoside or α- or β-pyranoside. The furanosides are the kinetic products of the Fischer glycosidation and they can be isolated if the reaction is stopped prior to equilibrium. Mechanism 23.2 describes the initial formation of the methyl hemiacetal of D-glucose and its cyclization to a mixture of methyl α-D-glucopyranoside and its β anomer.

Mechanism 23.2

Preparation of Methyl D-Glucopyranosides by Fischer Glycosidation

Steps 1–3: Acid-catalyzed nucleophilic addition of methanol to the carbonyl group of D-glucose. (See Mechanisms 17.2 and 17.4 for details of acid-catalyzed addition to aldehydes and ketones.)

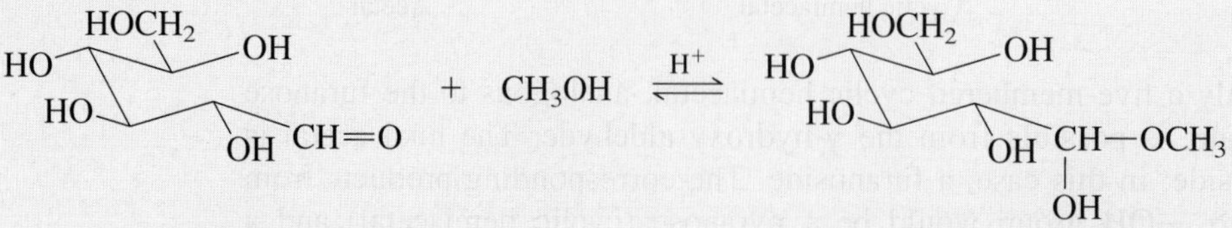

D-Glucose | Methanol | D-Glucose methyl hemiacetal

Step 4: Protonation of the —OH group of the hemiacetal unit. The proton donor is shown as the conjugate acid of methanol. It was formed by proton transfer from the acid catalyst to methanol.

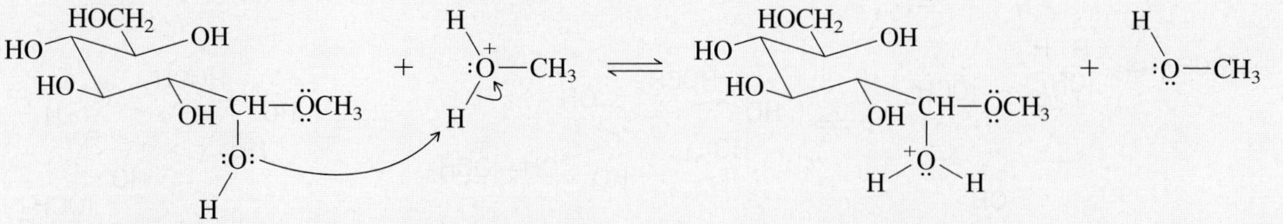

D-Glucose methyl hemiacetal — Methyloxonium ion — Conjugate acid of D-glucose methyl hemiacetal — Methanol

Step 5: Loss of water from the protonated hemiacetal to give an oxonium ion.

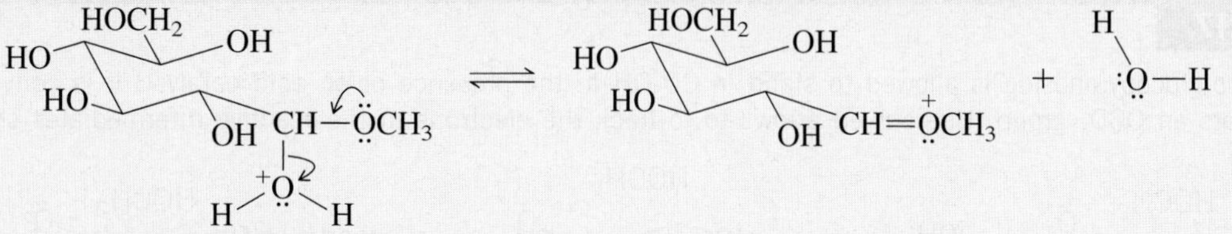

Conjugate acid of D-glucose methyl hemiacetal — Oxonium ion — Water

Step 6: Cyclization of the oxonium ion. An unshared electron pair of the C-5 oxygen is used to form a bond to C-1, forming the six-membered ring of the glycopyranoside. Both the α and β stereoisomers are formed in this reaction with the α stereoisomer (axial OCH_3) predominating.

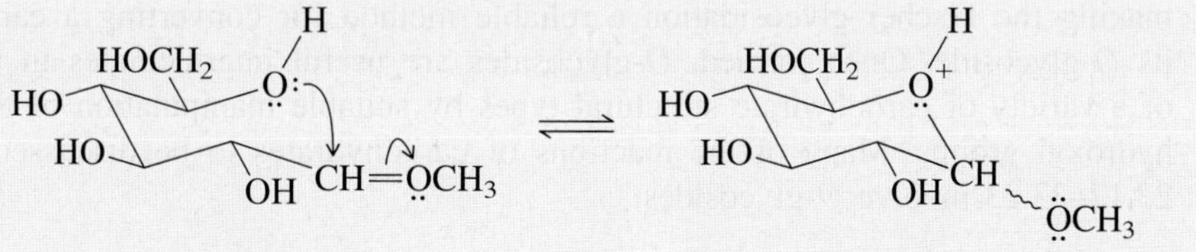

Oxonium ion — Conjugate acid of methyl D-glucopyranoside (α + β)

The symbol ∿∿∿ is used to represent a bond of undefined stereochemistry.

Step 7: Proton transfer from the positively charged ring oxygen to the oxygen of methanol giving a mixture of methyl α- and β-D-glucopyranoside. The acid catalyst is regenerated in this step.

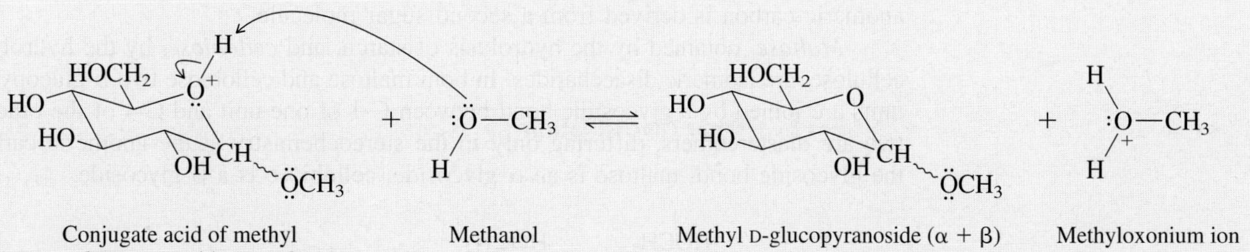

Conjugate acid of methyl D-glucopyranoside (α + β) — Methanol — Methyl D-glucopyranoside (α + β) — Methyloxonium ion

A process similar to that of Mechanism 23.2 involving the —OH group at C-4 gives the methyl α- and β-furanosides. These then undergo subsequent conversion to the more stable pyranosides by a mechanism not requiring reversion to D-glucose itself.

Problem 23.16

Add curved arrows to the following sequence to show how the conjugate acid of methyl β-D-glucofuranoside is converted to the corresponding pyranoside.

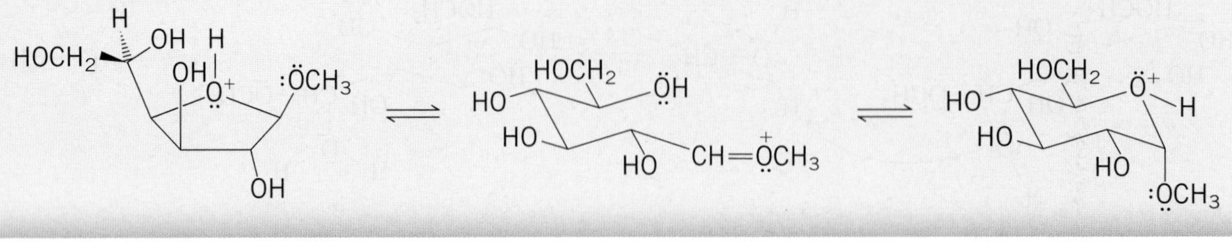

Still another mechanism, one involving a cyclic oxonium ion, is believed to interconvert the α and β anomers of methyl D-glucopyranoside.

Problem 23.17

When methyl β-D-glucopyranoside is allowed to stand in CD_3OH in the presence of an acid catalyst, it is converted to an α anomer that bears an OCD_3 group. Use curved arrows to to track the electrons in the reactive intermediates shown.

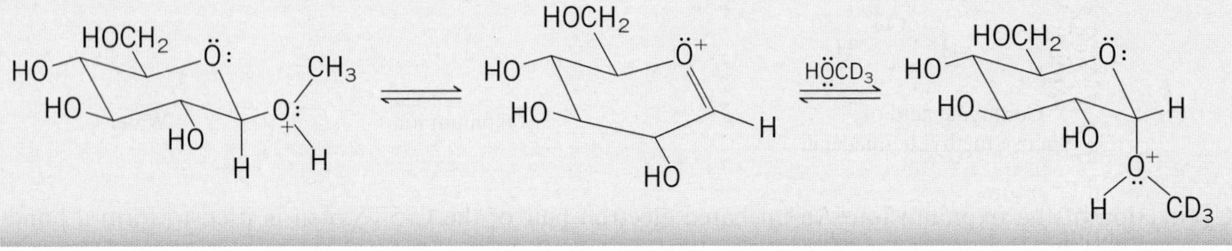

In spite of its mechanistic complexity, equilibrium is established rapidly thereby making the Fischer glycosidation a reliable method for converting a carbohydrate to its *O*-glycoside. Once formed, *O*-glycosides are useful intermediates in the synthesis of a variety of carbohydrate structural types by suitable manipulation of the remaining hydroxyl groups. Many of the reactions of carbohydrates to be discussed in Sections 23.17–23.23 involve *O*-glycosides.

23.15 Disaccharides

Disaccharides are carbohydrates that yield two monosaccharide molecules on hydrolysis. Structurally, disaccharides are *glycosides* in which the alkoxy group attached to the anomeric carbon is derived from a second sugar molecule.

Maltose, obtained by the hydrolysis of starch, and *cellobiose,* by the hydrolysis of cellulose, are isomeric disaccharides. In both maltose and cellobiose two D-glucopyranose units are joined by a glycosidic bond between C-1 of one unit and C-4 of the other. The two are diastereomers, differing only in the stereochemistry at the anomeric carbon of the glycoside bond; maltose is an α-glycoside, cellobiose is a β-glycoside.

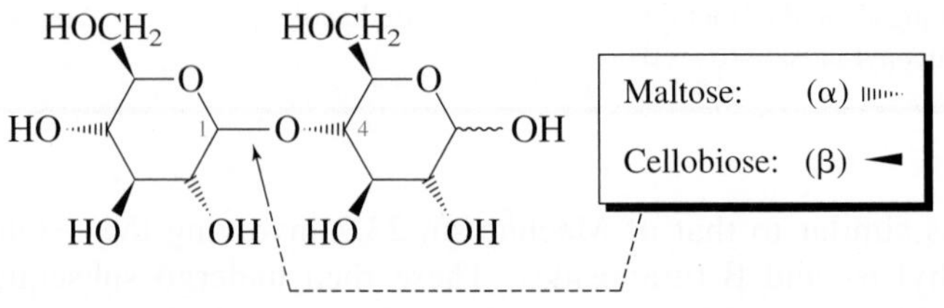

The stereochemistry and points of connection of glycosidic bonds are commonly designated by symbols such as α-(1→4) for maltose and β-(1→4) for cellobiose; α and

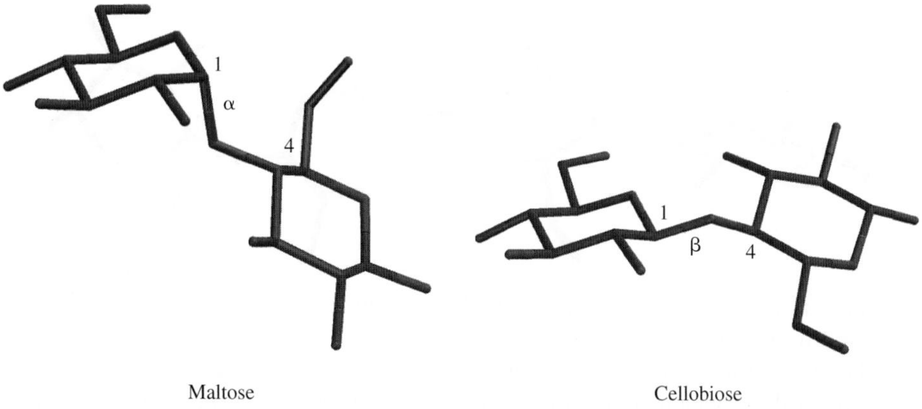

Figure 23.8

Molecular models of the disaccharides maltose and cellobiose. Two D-glucopyranose units are connected by a glycoside linkage between C-1 and C-4. The glycosidic bond has the α orientation in maltose and is β in cellobiose. Maltose and cellobiose are diastereomers.

β designate the stereochemistry at the anomeric position; the numerals specify the ring carbons involved.

Both maltose and cellobiose have a free anomeric hydroxyl group that is not involved in a glycoside bond. The configuration at the free anomeric center is variable and may be either α or β. Indeed, two stereoisomeric forms of maltose have been isolated: one has its anomeric hydroxyl group in an equatorial orientation; the other has an axial anomeric hydroxyl.

Problem 23.18

The two stereoisomeric forms of maltose just mentioned undergo mutarotation when dissolved in water. What is the structure of the key intermediate in this process?

The single difference in their structures, the stereochemistry of the glycosidic bond, causes maltose and cellobiose to differ significantly in their three-dimensional shape, as the molecular models of Figure 23.8 illustrate. This difference in shape affects the way in which maltose and cellobiose interact with other chiral molecules such as proteins, causing them to behave much differently toward enzyme-catalyzed hydrolysis. The enzyme *maltase* catalyzes the hydrolytic cleavage of the α-glycosidic bond of maltose but not the β-glycosidic bond of cellobiose. A different enzyme, *emulsin,* produces the opposite result: emulsin catalyzes the hydrolysis of cellobiose but not maltose. The behavior of each enzyme is general for glucosides (glycosides of glucose). Maltase catalyzes the hydrolysis of α-glucosides and is also known as *α-glucosidase,* whereas emulsin catalyzes the hydrolysis of β-glucosides and is known as *β-glucosidase.* The specificity of these enzymes offers a useful tool for structure determination because it allows the stereochemistry of glycosidic linkages to be assigned.

Lactose is a disaccharide constituting 2–6% of milk and is known as *milk sugar.* It differs from maltose and cellobiose in that only one of its monosaccharide units is D-glucose. The other monosaccharide unit, the one that contributes its anomeric carbon to the glycoside bond, is D-galactose. Like cellobiose, lactose is a β-glycoside.

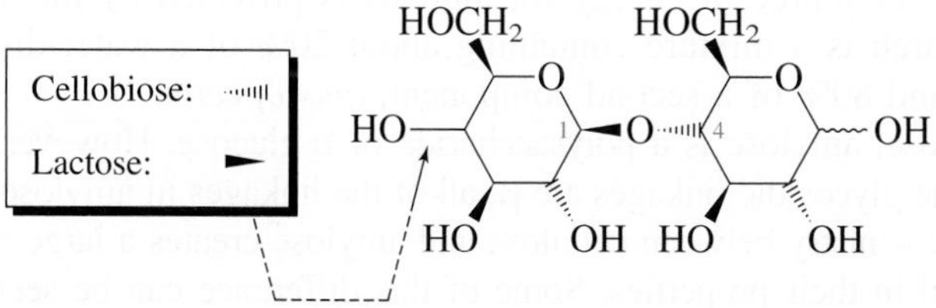

Digestion of lactose is facilitated by the β-glycosidase *lactase.* A deficiency of this enzyme makes it difficult to digest lactose and causes abdominal discomfort. Lactose

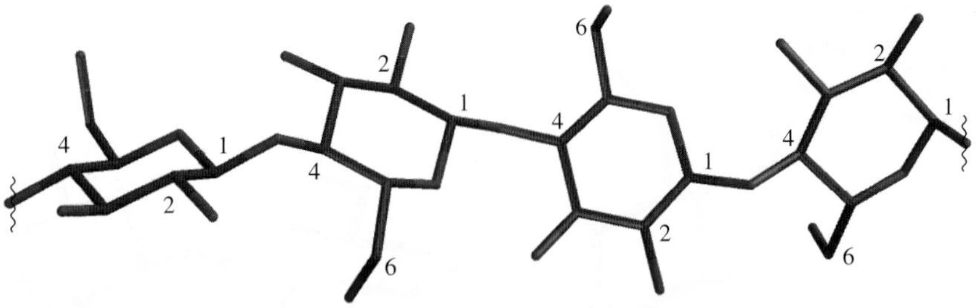

Figure 23.9

Cellulose is a polysaccharide in which D-glucose units are connected by β-(1→4)-glycoside linkages analogous to cellobiose. Hydrogen bonding, especially between the C-2 and C-6 hydroxyl groups, causes adjacent glucose units to tilt at an angle of 180° with each other.

intolerance is a genetic trait; it is treatable through over-the-counter formulations of lactase and by limiting the amount of milk in the diet.

The most familiar of all the carbohydrates is *sucrose*—common table sugar. Sucrose is a disaccharide in which D-glucose and D-fructose are joined at their anomeric carbons by a glycosidic bond.

D-Glucose portion of molecule

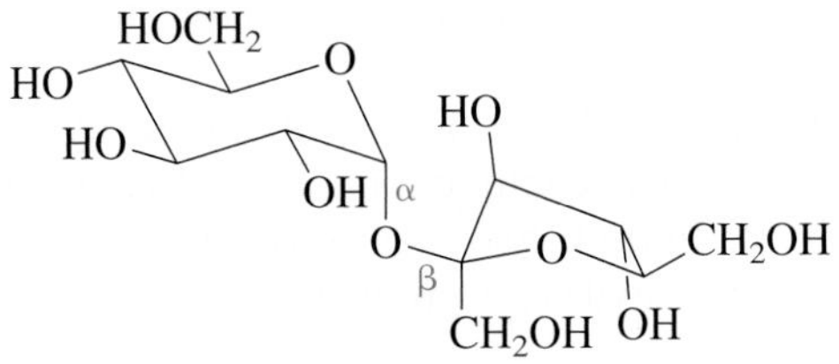

D-Fructose portion of molecule

Sucrose

Its chemical composition is the same irrespective of its source; sucrose from cane and sucrose from sugar beets are identical. Because sucrose does not have a free anomeric hydroxyl group, it does not undergo mutarotation. Hydrolysis of sucrose, catalyzed either by acid or by the enzyme *invertase,* gives a 1:1 mixture of D-glucose and D-fructose, which is sweeter than sucrose. The mixture prepared this way is called "invert sugar" because the sign of rotation of the aqueous solution in which it is carried out "inverts" from + to – as sucrose is converted to the glucose–fructose mixture.

23.16 Polysaccharides

Cellulose is the principal structural component of vegetable matter. Wood is 30–40% cellulose, cotton over 90%. Photosynthesis in plants is responsible for the formation of 10^9 tons per year of cellulose. Structurally, cellulose is a polysaccharide composed of D-glucose units joined by β-(1→4)-glycosidic linkages (Figure 23.9). The average is 7000 glucose units, but can be as many as 12,000. Complete hydrolysis of all the glycosidic bonds of cellulose yields D-glucose. The disaccharide fraction that results from partial hydrolysis is cellobiose.

As Figure 23.9 shows, the glucose units of cellulose are turned with respect to each other. The overall shape of the chain, however, is close to linear. Consequently, neighboring chains can pack together in bundles where networks of hydrogen bonds stabilize the structure and impart strength to cellulose fibers.

Animals lack the enzymes necessary to catalyze the hydrolysis of cellulose and so can't digest it. Cattle and other ruminants use cellulose as a food source indirectly. Colonies of bacteria that live in their digestive tract consume cellulose and in the process convert it to other substances that the animal can digest.

A more direct source of energy for animals is provided by the starches found in many plants. Starch is a mixture containing about 20% of a water-dispersible fraction called *amylose* and 80% of a second component, *amylopectin.*

Like cellulose, amylose is a polysaccharide of D-glucose. However, unlike cellulose in which all of the glycosidic linkages are β, all of the linkages in amylose are α. The small change in stereochemistry between cellulose and amylose creates a large difference in their overall shape and in their properties. Some of this difference can be seen in the structure of a short portion of amylose in Figure 23.10. The presence of the α-glycosidic linkages imparts a twist to the amylose chain. Where the main chain is roughly linear in cellulose, it is helical in amylose. Attractive forces *between* chains are weaker in amylose, and amylose does not form the same kind of strong fibers that cellulose does.

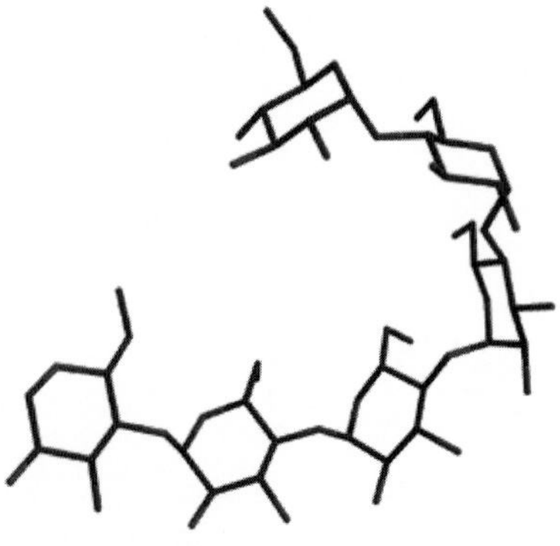

Figure 23.10

Amylose is a polysaccharide in which D-glucose units are connected by α-(1→4)-glycoside linkages analogous to maltose. The geometry of the glycoside linkage is responsible for the left-hand helical twist of the chain.

How Sweet It Is!

How sweet is it?

There is no shortage of compounds, natural or synthetic, that taste sweet. The most familiar, sucrose, glucose, and fructose, all occur naturally with worldwide production of sucrose from cane and sugar beets exceeding 100 million tons per year. Glucose is prepared by the enzymatic hydrolysis of starch, and fructose is made by the isomerization of glucose.

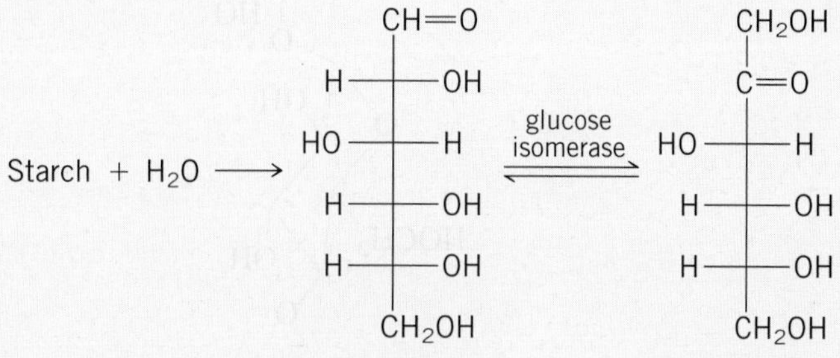

(+)-D-Glucose (−)-D-Fructose

Among sucrose, glucose, and fructose, fructose is the sweetest. Honey is sweeter than table sugar because it contains fructose formed by the isomerization of glucose as shown in the equation.

You may have noticed that most soft drinks contain "high-fructose corn syrup." Corn starch is hydrolyzed to glucose, which is then treated with glucose isomerase to produce a fructose-rich mixture. The enhanced sweetness permits less to be used, reducing the cost of production. Using less carbohydrate-based sweetener also reduces the number of calories.

Artificial sweeteners are a billion-dollar-per-year industry. The primary goal is, of course, to maximize sweetness and minimize calories. We'll look at the following sweeteners to give us an overview of the field.

Saccharin

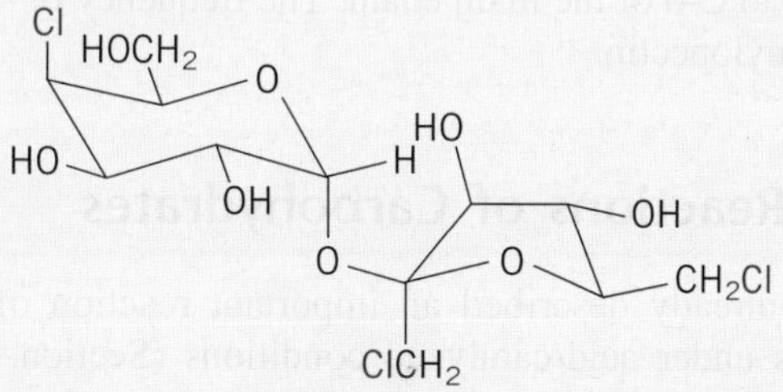

Sucralose

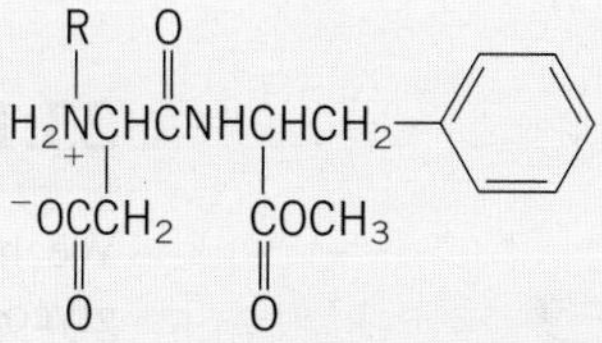

Aspartame, R = H
Neotame, R = $CH_2CH_2C(CH_3)_3$

All of these are hundreds of times sweeter than sucrose and variously described as "low-calorie" or "nonnutritive" sweeteners.

Saccharin was discovered at Johns Hopkins University in 1879 in the course of research on coal-tar derivatives and is the oldest artificial sweetener. In spite of its name, which comes from the Latin word for sugar, saccharin bears no structural relationship to any sugar. Nor is saccharin itself very soluble in water. The proton bonded to nitrogen, however, is fairly acidic and saccharin is normally marketed as its water-soluble sodium or calcium salt. Its earliest applications were not in weight control, but as a replacement for sugar in the diet of diabetics before insulin became widely available.

Sucralose has the structure most similar to sucrose. Galactose replaces the glucose unit of sucrose, and chlorines replace three of the hydroxyl groups. The three chlorine substituents do not diminish sweetness, but do interfere with the ability of the body to metabolize sucralose. It, therefore, has no food value and is "noncaloric."

Aspartame is the market leader among artificial sweeteners. It is a methyl ester of a dipeptide, unrelated to any carbohydrate. A recently approved relative, Neotame, is even sweeter than aspartame.

Saccharin, sucralose, and aspartame illustrate the diversity of structural types that taste sweet, and the vitality and continuing development of the industry of which they are a part.

Amylopectin resembles amylose in being a polysaccharide built on a framework of α-(1→4)-linked D-glucose units. In addition to this main framework, however, amylose incorporates polysaccharide branches of 24–30 glucose units joined by α-(1→4)-glycosidic bonds. These branches sprout from C-6 of glucose units at various points along the main framework, connected to it by α-(1→6)-glycosidic bonds (Figure 23.11).

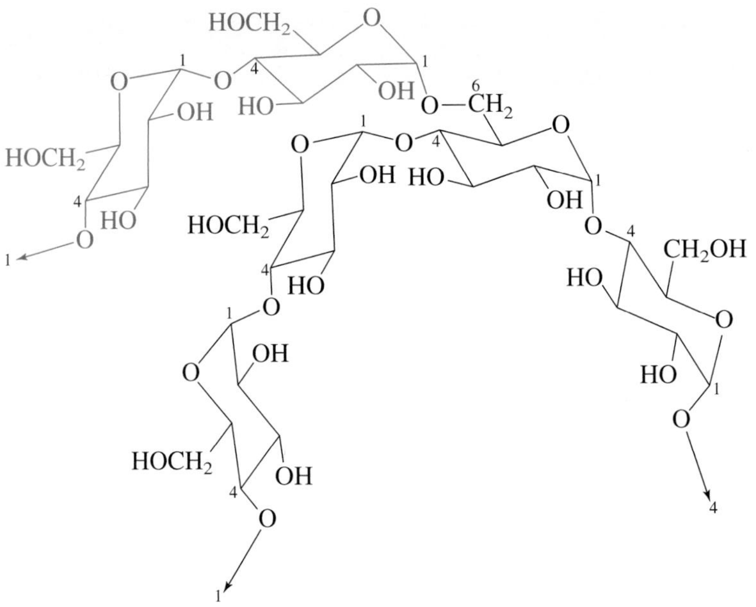
HOCH$_2$
O
1
4
OH
O
HO
OH
6
O—CH$_2$
O
1
O
O
4
HOCH$_2$
OH
HO
1
OH
4
HO
O
4
CH$_2$OH
HOCH$_2$
HO
1
O
HO
O
OH
O
HO
1
O
HOCH$_2$
OH
4
O
1
4

Figure 23.11

Amylopectin. The main chain (black) is the same as in amylose. Amylopectin differs from amylose in having branches (red) linked to the main chain by α-(1→6) glycosidic bonds. Except for the glycoside bonds connecting the branches to the main chain, all other glycoside bonds are α-(1→4).

One of the most important differences between cellulose and starch is that animals can digest starch. Because the glycosidic linkages in starch are α, an animal's α-glycosidase enzymes can catalyze their hydrolysis to glucose. When more glucose is available than is needed as fuel, animals store some of it as glycogen. Glycogen resembles amylopectin in that it is a branched polysaccharide of α-(1→4)-linked D-glucose units with branches connected to C-6 of the main chain. The frequency of such branches is greater in glycogen than in amylopectin.

23.17 Reactions of Carbohydrates

We have already described an important reaction of carbohydrates—the formation of glycosides under acid-catalyzed conditions (Section 23.14). Glycoside formation drew our attention to the fact that an OH group at the anomeric carbon of a furanose or pyranose form differs in reactivity from the other OH groups of a carbohydrate. It also demonstrated that what looks like a new reaction is one we saw before in a different guise. Mechanistically, glycoside formation is just a structural variation on the aldehyde → hemiacetal → acetal theme we saw when discussing the reactions of aldehydes and ketones.

In the remaining sections of this chapter we'll explore several reactions of carbohydrates. In some cases our emphasis will be on the similarity of these reactions to those of simple classes of compounds. In others, we will point out where a particular reaction is especially useful in carbohydrate chemistry.

23.18 Reduction of Monosaccharides

Although they exist almost entirely as cyclic hemiacetals, carbohydrates are in rapid equilibrium with their open-chain forms, and most of the reducing agents that convert aldehydes to alcohols do the same with aldoses.

Typical reducing agents include catalytic hydrogenation and sodium borohydride. Lithium aluminum hydride is not suitable, because it is not compatible with the solvents (water, alcohols) that are required to dissolve carbohydrates. The products of carbohydrate reduction are called **alditols.** Because these alditols lack a carbonyl group, they are, of course, incapable of forming cyclic hemiacetals and exist exclusively in noncyclic forms.

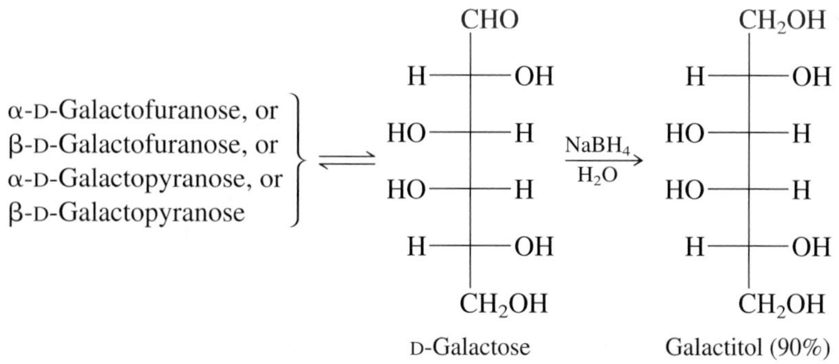
α-D-Galactofuranose, or
β-D-Galactofuranose, or
α-D-Galactopyranose, or
β-D-Galactopyranose
CHO
H OH
HO H
HO H
H OH
CH2OH
NaBH4
H2O
CH2OH
H OH
HO H
HO H
H OH
CH2OH
D-Galactose
Galactitol (90%)

Problem 23.19

Is the galactitol formed by the sodium borohydride reduction of D-galactose in the preceding equation optically active? Explain.

Another name for glucitol, obtained by reduction of D-glucose, is *sorbitol;* it is used as a sweetener, especially in special diets required to be low in sugar. Reduction of D-fructose yields a mixture of glucitol and mannitol, corresponding to the two possible configurations at the newly generated chirality center at C-2.

23.19 Oxidation of Monosaccharides

The most easily oxidized groups in an aldose are the aldehyde at one end and the primary alcohol at the other. Oxidation of CH=O gives an **aldonic acid;** oxidation of CH_2OH gives a **uronic acid,** and oxidation of both gives an **aldaric acid.**

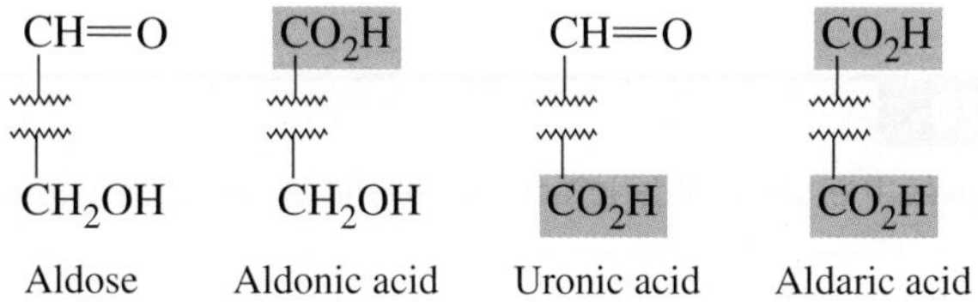
CH=O
CO2H
CH=O
CO2H
CH2OH
CH2OH
CO2H
CO2H
Aldose
Aldonic acid
Uronic acid
Aldaric acid

Aldonic acids are named by replacing the *–ose* ending of the aldose by *–onic acid.* Similarly, the endings *-uronic acid* and *-aric acid* are used in uronic and aldaric acids, respectively.

The most commonly used method for preparing aldonic acids is by oxidation with bromine in aqueous solution. The species that is oxidized is a furanose or pyranose form of the carbohydrate.

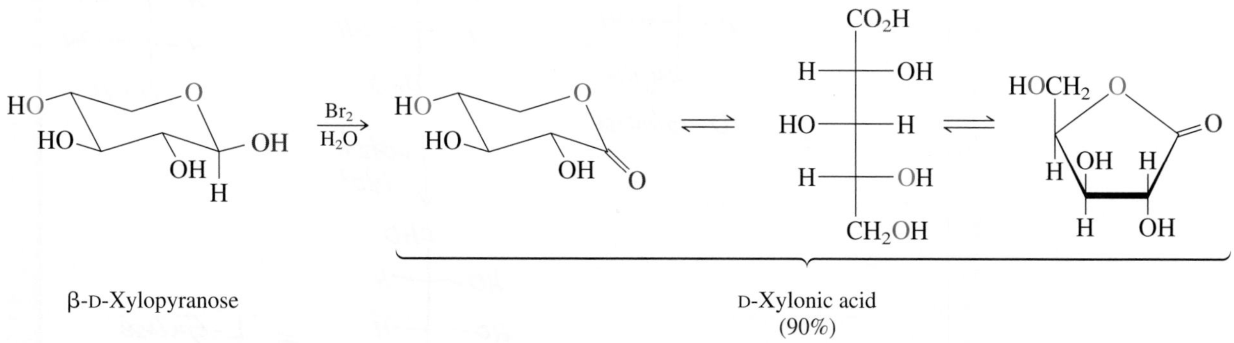
HO
HO
O
OH
OH
H
Br2
H2O
HO
HO
O
OH
O
CO2H
H OH
HO H
H OH
CH2OH
HOCH2 O
O
H OH H
H OH
β-D-Xylopyranose
D-Xylonic acid
(90%)

Aldonic acids spontaneously form five-membered (γ) or six-membered (δ) lactones. γ-Lactones are normally more stable than δ-lactones.

Direct oxidation of CH_2OH in the presence of CH=O is not practical, so laboratory preparations of uronic acids are limited to processes that include appropriate protection–deprotection strategies. D-Glucuronic acid, an intermediate in the biosynthesis of vitamin C and in various metabolic pathways, is biosynthesized by an enzyme-catalyzed NAD^+ oxidation of a derivative of D-glucose.

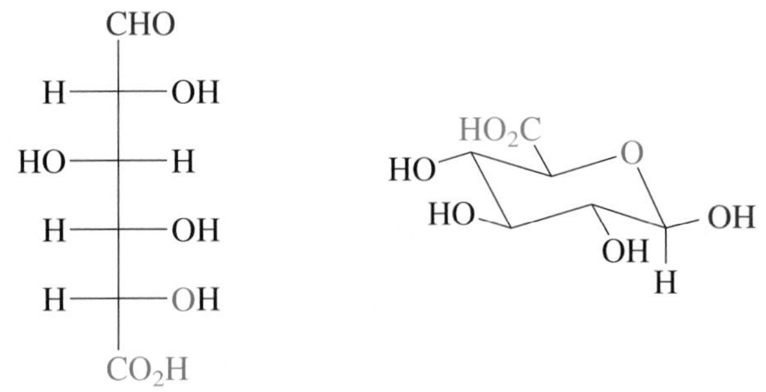

Fischer projection of D-glucuronic acid

β-Pyranose form of D-glucuronic acid

The tartaric acids are aldaric acids.

Nitric acid oxidizes both the aldehyde and the terminal primary alcohol functions of an aldose to carboxylic acid groups giving an aldaric acid.

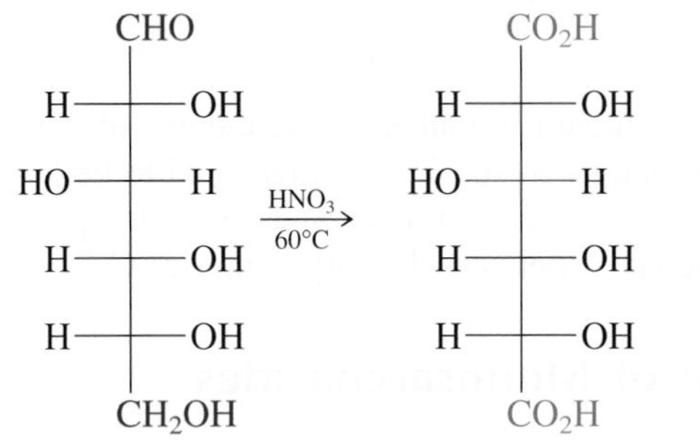

D-Glucose D-Glucaric acid (41%)

Like aldonic acids, aldaric acids exist mainly as lactones when the rings are five-membered or six-membered.

Problem 23.20

Another hexose gives the same aldaric acid on oxidation as does D-glucose. Which one?

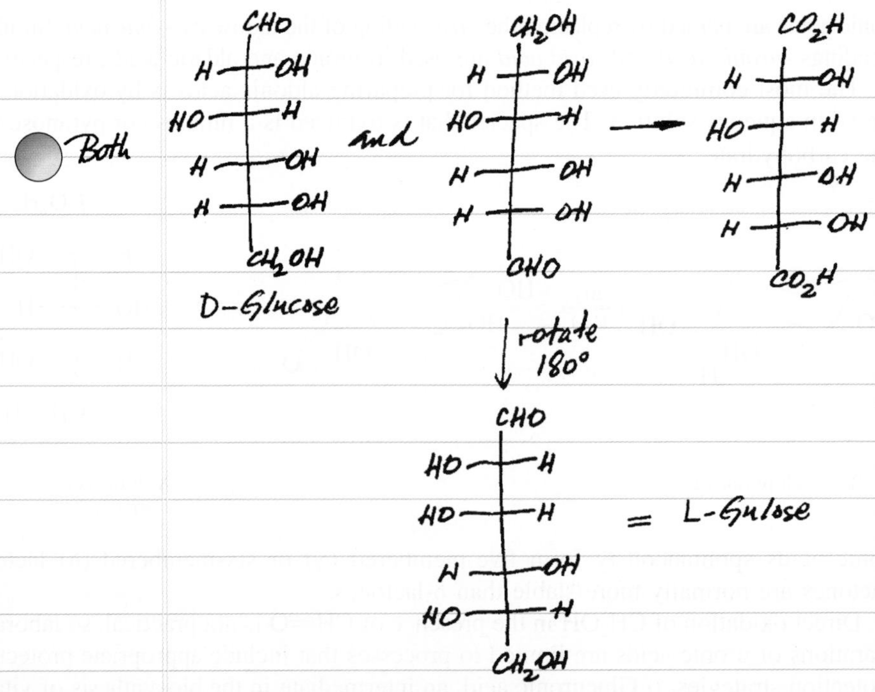

23.20 Periodic Acid Oxidation

Periodic acid oxidation (Section 15.11) finds use as an analytical method in carbohydrate chemistry. Structural information is obtained by measuring the number of equivalents of periodic acid that react with a given compound and by identifying the reaction products. A vicinal diol consumes one equivalent of periodate and is cleaved to two carbonyl compounds:

$$\underset{\text{Vicinal diol}}{R_2C(OH){-}C(OH)R'_2} + \underset{\text{Periodic acid}}{HIO_4} \longrightarrow \underset{\text{Two carbonyl compounds}}{R_2C{=}O + R'_2C{=}O} + \underset{\text{Iodic acid}}{HIO_3} + \underset{\text{Water}}{H_2O}$$

α-Hydroxy carbonyl compounds are cleaved to a carboxylic acid and a carbonyl compound:

$$\underset{\alpha\text{-Hydroxy carbonyl compound}}{RC({=}O)C(OH)R'_2} + \underset{\text{Periodic acid}}{HIO_4} \longrightarrow \underset{\text{Carboxylic acid}}{RC({=}O)OH} + \underset{\text{Carbonyl compound}}{R'_2C{=}O} + \underset{\text{Iodic acid}}{HIO_3}$$

When three contiguous carbons bear hydroxyl groups, two moles of periodate are consumed per mole of carbohydrate, and the central carbon is oxidized to a molecule of formic acid:

$$R_2C(OH){\,\wr\,}CH(OH){\,\wr\,}C(OH)R'_2 + \underset{\text{Periodic acid}}{2HIO_4} \longrightarrow \underset{\text{Carbonyl compound}}{R_2C{=}O} + \underset{\text{Formic acid}}{HC({=}O)OH} + \underset{\text{Carbonyl compound}}{R'_2C{=}O} + \underset{\text{Iodic acid}}{2HIO_3}$$

(Points at which cleavage occurs: the C–C bonds marked ≀.)

Ether and acetal functions are not affected by the reagent.

The use of periodic acid oxidation in structure determination can be illustrated by a case in which a previously unknown methyl glycoside was obtained by the reaction of D-arabinose with methanol and hydrogen chloride. The size of the ring was identified as five-membered because only one mole of periodic acid was consumed per mole of glycoside and no formic acid was produced. Were the ring six-membered, two moles of periodic acid would be required per mole of glycoside and one mole of formic acid would be produced.

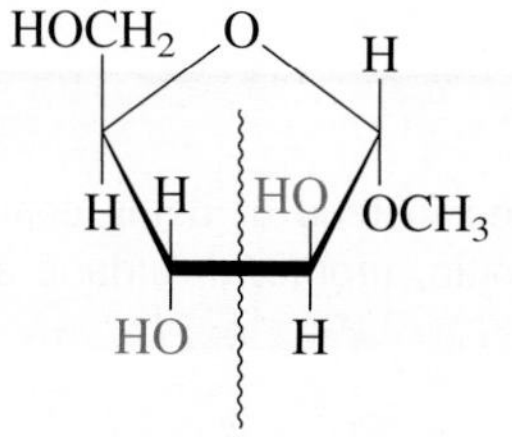

Only one site for periodic acid cleavage in methyl α-D-arabinofuranoside

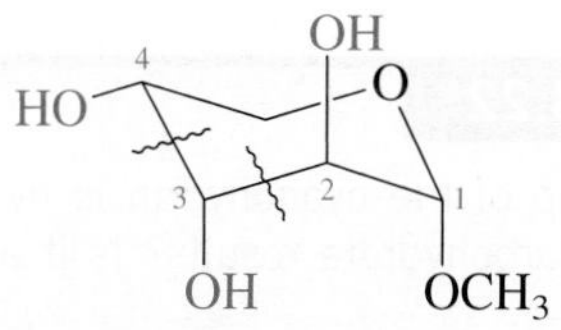

Two sites of periodic acid cleavage in methyl α-D-arabinopyranoside, C-3 lost as formic acid

Problem 23.21

The α-methyl glycosides of the pyranose forms of all four D-aldopentoses give the same product on periodic acid oxidation. What is this product?

23.21 Cyanohydrin Formation and Chain Extension

The presence of an aldehyde function in their open-chain forms makes aldoses reactive toward nucleophilic addition of hydrogen cyanide. Addition yields a mixture of diastereomeric cyanohydrins.

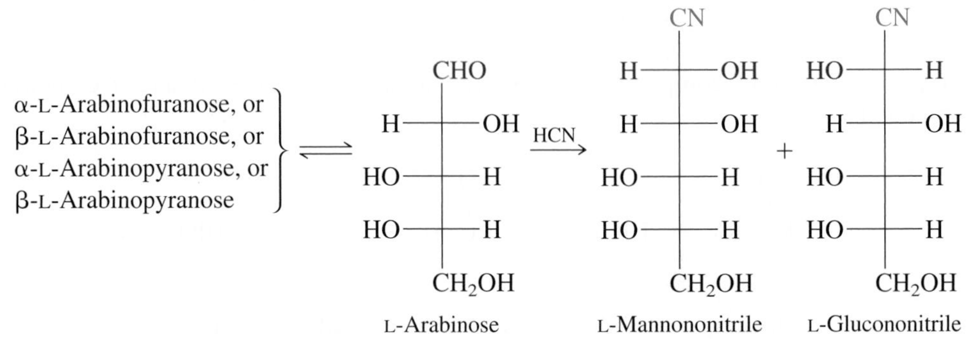

The reaction is used for the chain extension of aldoses in the synthesis of new or unusual sugars. In this case, the starting material, L-arabinose, is an abundant natural product and possesses the correct configurations at its three chirality centers for elaboration to the relatively rare L-enantiomers of glucose and mannose. After cyanohydrin formation, the cyano groups are converted to aldehyde functions by hydrogenation in aqueous solution. Under these conditions, —C≡N is reduced to —CH═NH and hydrolyzes rapidly to —CH═O. Use of a poisoned palladium-on-barium sulfate catalyst prevents further reduction to the alditols.

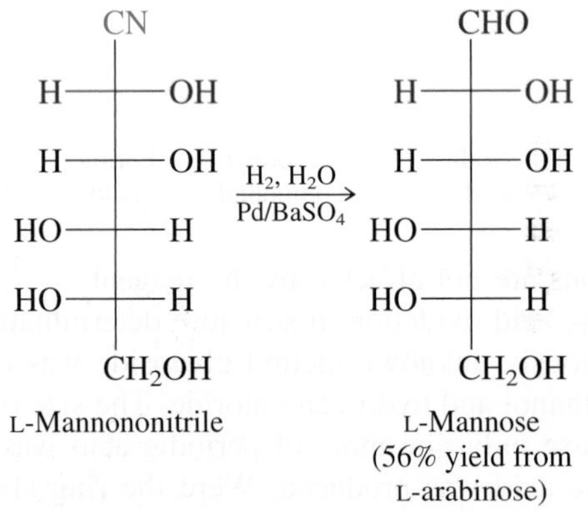

Similarly, L-glucononitrile has been reduced to L-glucose; its yield was 26% from L-arabinose.

Problem 23.22

If the CN group of the cyanohydrin is hydrolyzed instead of being converted to CH═O, what kind of carbohydrate results? Is it an aldonic, uronic, or aldaric acid?

An older version of the chain-extension sequence is called the **Kiliani–Fischer synthesis.** It, too, proceeds through a cyanohydrin, but it uses a less efficient method for converting the cyano group to the required aldehyde.

23.22 Epimerization, Isomerization, and Retro-Aldol Cleavage

Carbohydrates undergo a number of isomerization and degradation reactions under both laboratory and physiological conditions. For example, a mixture of glucose, fructose, and mannose results when any one of them is treated with aqueous base. This reaction can be understood by examining the consequences of enolization of glucose:

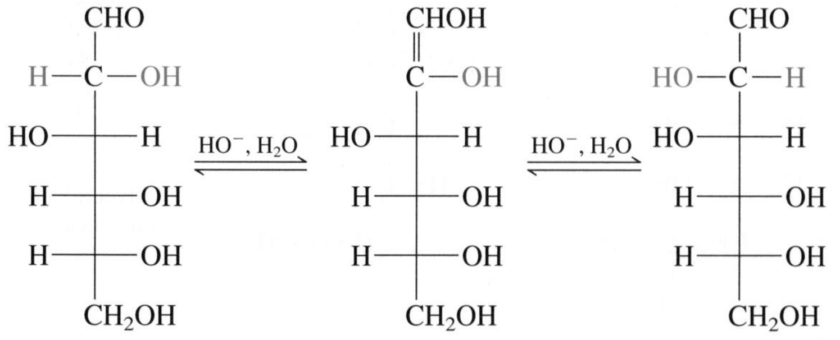

D-Glucose (*R* configuration at C-2) | Enediol (C-2 not a chirality center) | D-Mannose (*S* configuration at C-2)

Because the configuration at C-2 is lost on enolization, the enediol intermediate can revert either to D-glucose or to D-mannose. Two stereoisomers that have multiple chirality centers but differ in configuration at only one of them are referred to as **epimers.** Glucose and mannose are epimeric at C-2. Under these conditions epimerization occurs only at C-2 because it alone is α to the carbonyl group.

Epimers are a type of diastereomer.

There is another reaction available to the enediol intermediate. Proton transfer from water to C-1 converts the enediol not to an aldose but to the ketose D-fructose:

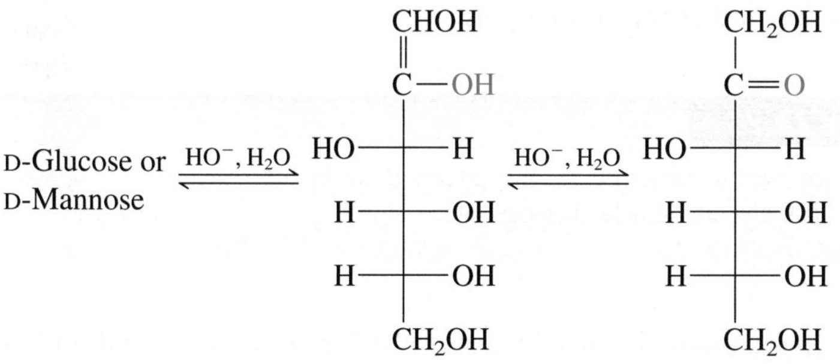

Enediol | D-Fructose

See the boxed essay *How Sweet It Is!* (page 1049) for more on this process.

The isomerization of D-glucose to D-fructose by way of an enediol intermediate is an important step in **glycolysis,** a complex process (11 steps) by which an organism converts glucose to chemical energy. The substrate is not glucose itself but its 6-phosphate ester. The enzyme that catalyzes the isomerization is called *phosphoglucose isomerase.*

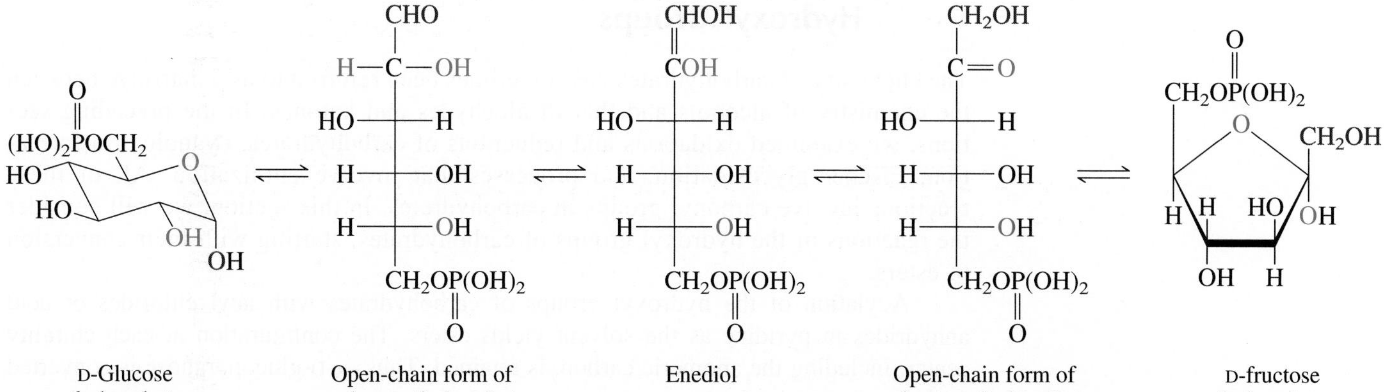

D-Glucose 6-phosphate | Open-chain form of D-glucose 6-phosphate | Enediol | Open-chain form of D-fructose 6-phosphate | D-fructose 6-phosphate

Following its formation, D-fructose 6-phosphate is converted to its corresponding 1,6-phosphate diester, which is then cleaved to two 3-carbon fragments under the influence of the enzyme *aldolase:*

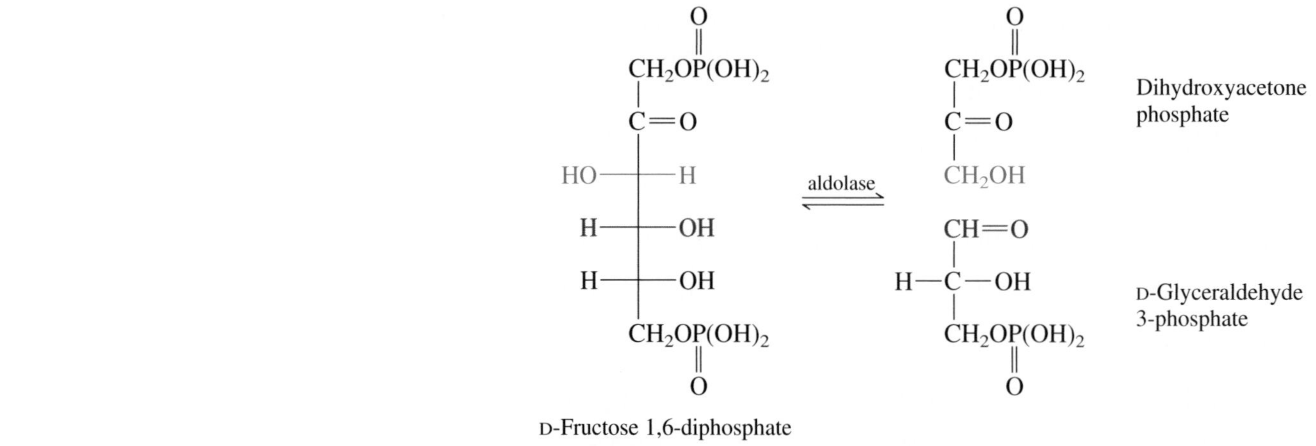

This cleavage is a *retro-aldol* reaction. It is the reverse of the process by which D-fructose 1,6-diphosphate would be formed by aldol addition of the enolate of dihydroxyacetone phosphate to D-glyceraldehyde 3-phosphate. The enzyme aldolase catalyzes both the aldol addition of the two components and, in glycolysis, the retro-aldol cleavage of D-fructose 1,6-diphosphate.

Further steps in glycolysis use the D-glyceraldehyde 3-phosphate formed in the aldolase-catalyzed cleavage reaction as a substrate. Its coproduct, dihydroxyacetone phosphate, is not wasted, however. The enzyme *triose phosphate isomerase* converts dihydroxyacetone phosphate to D-glyceraldehyde 3-phosphate, which enters the glycolysis pathway for further transformations.

Problem 23.23

Suggest a reasonable structure for the intermediate in the conversion of dihydroxyacetone phosphate to D-glyceraldehyde 3-phosphate.

Cleavage reactions of carbohydrates also occur on treatment with aqueous base for prolonged periods as a consequence of base-catalyzed retro-aldol reactions. As pointed out in Section 20.3, aldol addition is a reversible process, and β-hydroxy carbonyl compounds can be cleaved to an enolate and either an aldehyde or a ketone.

23.23 Acylation and Alkylation of Carbohydrate Hydroxyl Groups

The chemistry of carbohydrates has sometimes been referred to as a marriage between the chemistry of alcohols and that of aldehydes and ketones. In the preceding sections, we examined oxidations and reductions of carbohydrates, cyanohydrin formation, Fischer glycosidation, and processes that involve enolization. All of these reactions involve carbonyl groups in carbohydrates. In this section, we will consider the reactions of the hydroxyl groups of carbohydrates, starting with their conversion to esters.

Acylation of the hydroxyl groups of carbohydrates with acyl chlorides or acid anhydrides in pyridine as the solvent yields esters. The configuration at each chirality center, including the anomeric carbon, is retained. Thus, α-D-glucopyranose is converted to the pentaacetate shown.

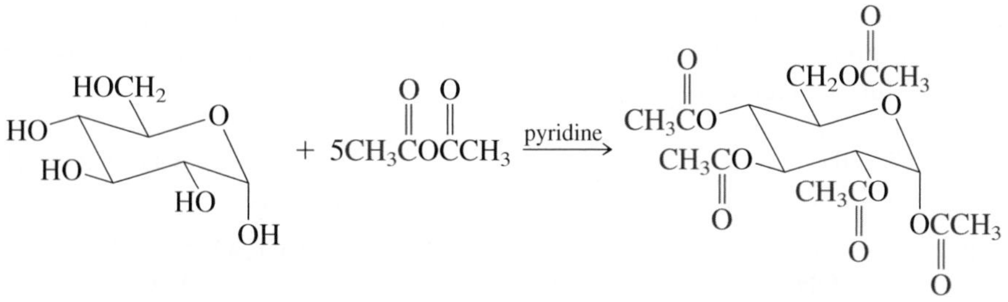

α-D-Glucopyranose — Acetic anhydride — 1,2,3,4,6-Penta-*O*-acetyl-α-D-glucopyranose (88%)

Ethers are formed under conditions of the Williamson ether synthesis. Methyl ethers of carbohydrates are efficiently prepared by alkylation with methyl iodide in the presence of silver oxide.

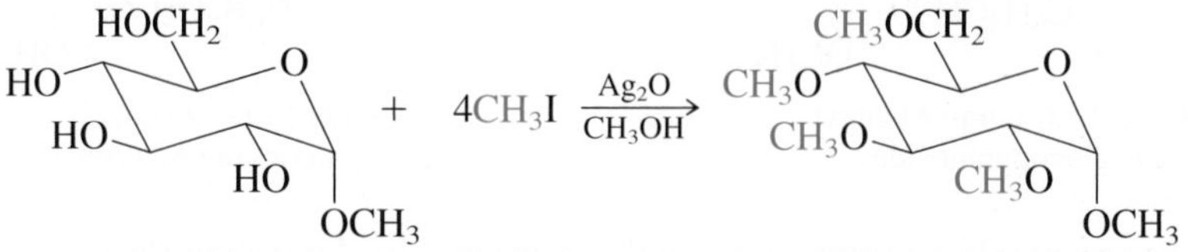

Methyl α-D-glucopyranoside — Methyl iodide — Methyl 2,3,4,6-tetra-*O*-methyl-α-D-glucopyranoside (97%)

This reaction has been used in an imaginative way to determine the ring size of glycosides. Once all the free hydroxyl groups of a glycoside have been methylated, the glycoside is subjected to acid-catalyzed hydrolysis. Only the anomeric methoxy group is hydrolyzed under these conditions—another example of the ease of carbocation formation at the anomeric position.

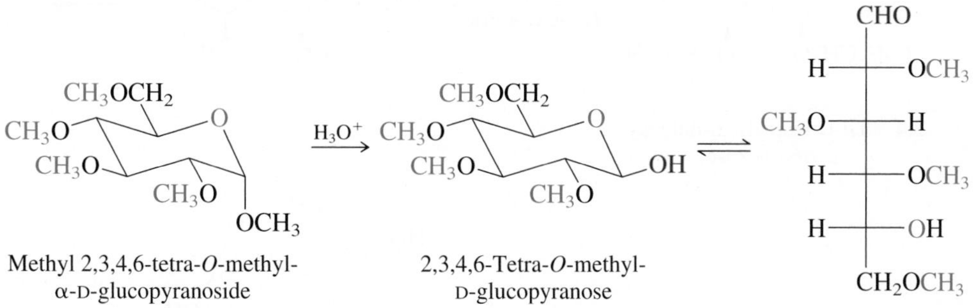

Methyl 2,3,4,6-tetra-*O*-methyl-α-D-glucopyranoside — 2,3,4,6-Tetra-*O*-methyl-D-glucopyranose

Notice that all the hydroxyl groups in the free sugar except C-5 are methylated. Carbon-5 is not methylated, because it was originally the site of the ring oxygen in the methyl glycoside. Once the position of the hydroxyl group in the free sugar has been determined, either by spectroscopy or by converting the sugar to a known compound, the ring size stands revealed.

The resistance of methyl ethers to acid hydrolysis makes them useful for the determination of ring size, but not as protecting groups, because they are difficult to remove. If it is necessary to carry out a reaction at only one of the hydroxyl groups in a carbohydrate, then the other hydroxyl groups must be protected, so that they do not interfere. You were introduced to protecting groups in Section 17.9, and the same principles apply here; namely, that the protecting group must be stable to the conditions of the reaction of interest, and we must be able to remove the protecting group at the end of the synthesis. An alternative type of ether that is more versatile as a protecting group than the methyl ether is the benzyl ether. Benzyl ethers are prepared by the reaction of the carbohydrate with a base and a benzyl halide, usually benzyl chloride or benzyl bromide. Benzyl ethers are stable to acid and base hydrolysis, organometallic reagents, and other reaction conditions. Hydrolysis of the methyl glycoside in the benzylation product shown in the following reactions gives the unmasked hemiacetal, with the anomeric hydroxyl group now free for further reactions.

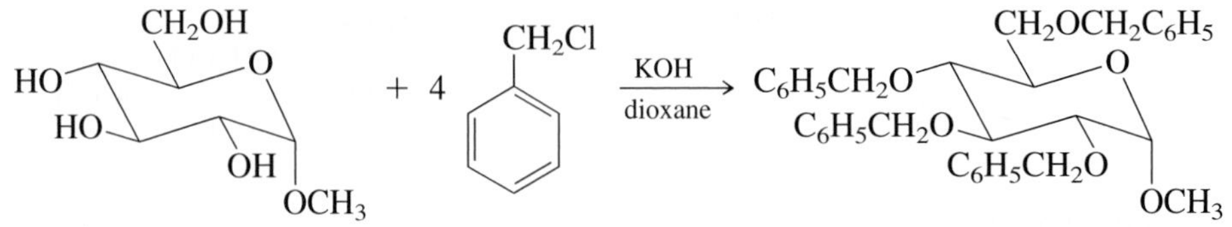

Methyl α-D-glucopyranoside

Methyl 2,3,4,6-tetra-*O*-benzyl-α-D-glucopyranoside (95%)

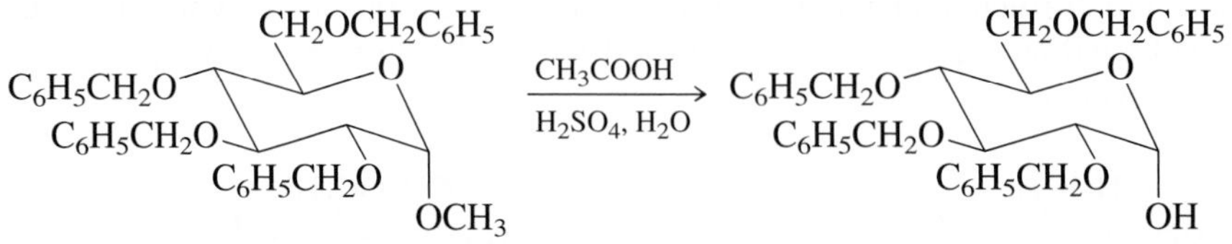

Methyl 2,3,4,6-tetra-*O*-benzyl-α-D-glucopyranoside

2,3,4,6-Tetra-*O*-benzyl-α-D-glucopyranose (70%)

Benzyl ether protecting groups are usually removed by *hydrogenolysis,* a reaction in which hydrogen in the presence of a metal catalyst cleaves the $C_6H_5CH_2$—O bond to give an OH group and toluene. Cleavage of the three benzyl ether groups from the tribenzyl ether shown gives the deprotected aldehyde, which exists mainly in the cyclic hemiacetal form. This product is the carbohydrate known as digitoxose, and is a component of the drug digoxin that is used to treat heart failure and irregular heartbeat.

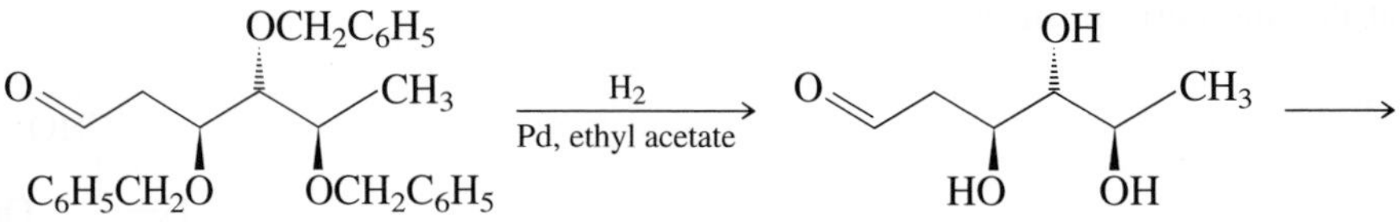

3,4,5-Tri-*O*-benzyl-2,6-dideoxy-D-*ribo*-hexose

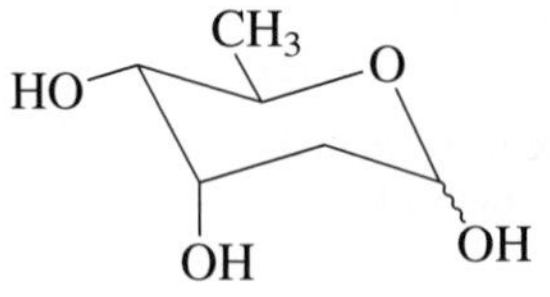

D-Digitoxose (40%)

23.24 Glycosides: Synthesis of Oligosaccharides

As we saw in Section 23.14, the preparation of glycosides by acid-catalyzed condensation of a carbohydrate with a simple alcohol—the **Fischer glycosidation**—is thermodynamically controlled and favors the formation of pyranose over furanose rings. The anomeric effect causes the α stereoisomer to predominate over the β.

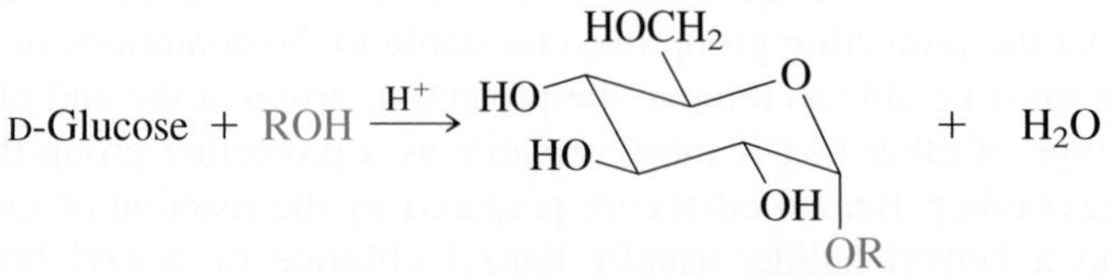

An alcohol An α-D-glucopyranoside Water

When the desired glycoside is a disaccharide, however, the "alcohol" is no longer simple; it is a carbohydrate with more than one OH group capable of bonding to the anomeric carbon of the other carbohydrate. Thus, constitutionally isomeric as well as stereoisomeric pyranosides are possible.

Consider, for example, a disaccharide such as gentiobiose in which both carbohydrate units are pyranosyl forms of D-glucose. Using the notation introduced in Section 23.15, gentiobiose is a β-(1→6) glycoside. An oxygen atom is bonded to the anomeric carbon of one D-glucopyranose and C-6 of a second D-glucopyranose.

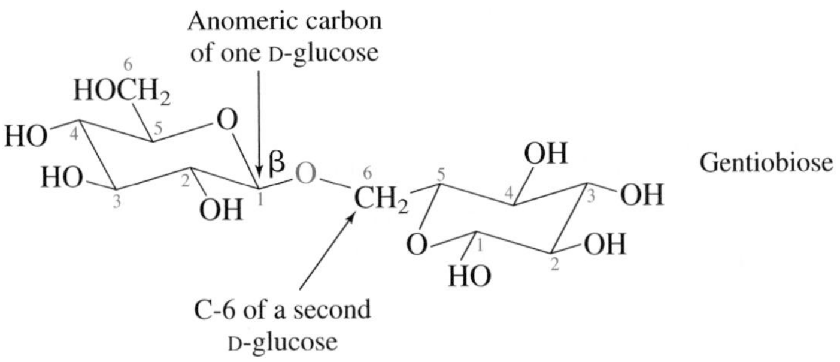

Gentiobiose occurs naturally and has been isolated from gentian root, saffron, and numerous other plant materials. For more, see the boxed essay *Crocuses Make Saffron from Carotenes* (page 1105) in Chapter 24.

Although the stereochemistry of the glycosidic linkage is a concern, the first problem to be addressed in a chemical synthesis of a disaccharide such as gentiobiose is achieving the desired 1→6 connectivity. The general strategy involves three stages:

1. Preparation of a suitably protected *glycosyl donor* and *glycosyl acceptor.* A glycosyl donor contains a leaving group at the anomeric carbon. The glycosyl acceptor contains a nucleophilic group, hydroxyl in this case, at the desired carbon.
2. Formation of the glycosidic C—O bond by a nucleophilic substitution in which an OH group of the glycosyl acceptor acts as the nucleophile toward the anomeric carbon of the glycosyl donor.
3. Removal of the protecting groups from the protected disaccharide formed in stage 2.

The glycosyl donor in the synthesis of gentiobiose is 2,3,4,6-tetra-*O*-benzoyl-α-D-glucopyranosyl bromide. It is prepared by treating D-glucose with benzoyl chloride in pyridine, then hydrogen bromide. This reaction places a bromide leaving group on the anomeric carbon and protects the four remaining OH groups as benzoate esters. Like the Fischer glycosylation, the α stereochemistry at the anomeric carbon results from thermodynamic control. The glycosyl acceptor is methyl 2,3,4-tri-*O*-acetyl-β-D-glucopyranoside, in which all of the hydroxyl groups are protected except the one at C-6.

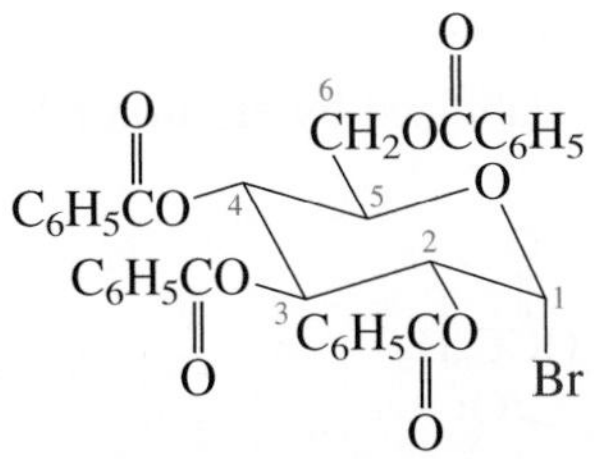

2,3,4,6-Tetra-*O*-benzoyl-α-D-glucopyranosyl bromide

Glycosyl donor

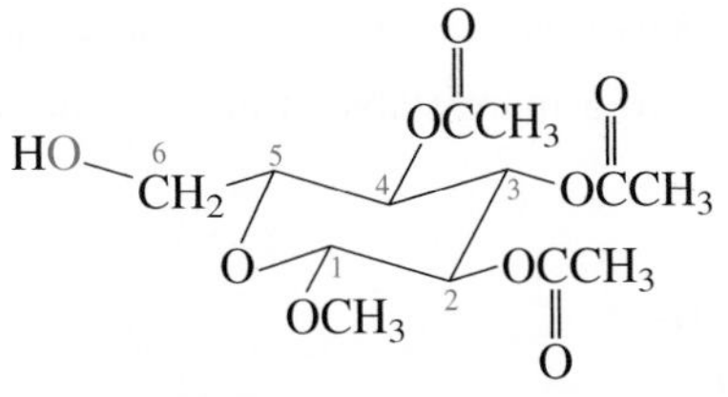

Methyl 2,3,4,-tri-*O*-acetyl-β-D-glucopyranoside

Glycosyl acceptor

Coupling of the glycosyl donor and acceptor takes place in the presence of silver trifluoromethanesulfonate ($AgOSO_2CF_3$) as a source of Ag^+, which activates the pyranosyl bromide toward nucleophilic substitution. A weak base such as 2,4,6-trimethylpyridine is included in order to react with the trifluoromethanesulfonic acid that is produced in the reaction.

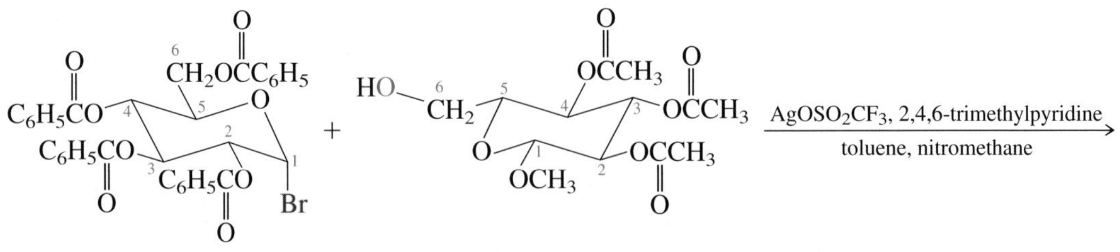

2,3,4,6-Tetra-*O*-benzoyl-α-D-glucopyranosyl bromide

Methyl 2,3,4,-tri-*O*-acetyl-β-D-glucopyranoside

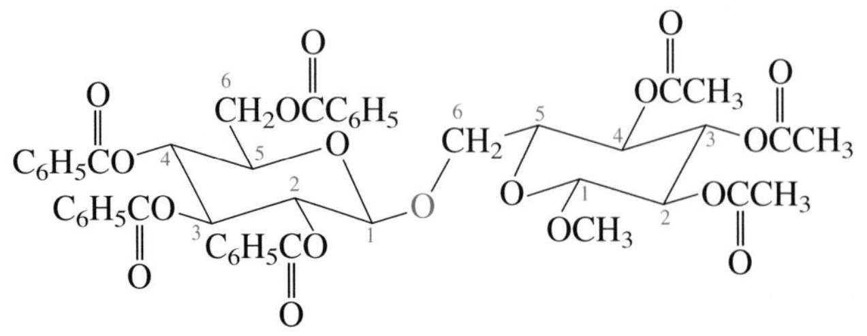

Methyl 2,3,4-tri-*O*-acetyl-2',3',4',6'-tetra-*O*-benzoyl-β-gentiobioside (91%)

This method of silver-assisted glycoside synthesis is a variation of the *Koenigs–Knorr reaction,* which uses silver carbonate. William Koenigs and Edward Knorr first reported their method in 1901.

The coupling reaction is stereoselective for formation of the β-disaccharide. Removal of the ester protecting groups and hydrolysis of the methyl glycoside are required to complete the synthesis of gentiobiose.

The mechanism of the coupling reaction resembles an S_N1 reaction and begins with a silver-ion-assisted ionization of the carbon-bromine bond of the glycosyl donor to give a carbocation (Mechanism 23.3).

Mechanism 23.3

Silver-Assisted Glycosidation

Step 1: Silver ion acts as a Lewis acid to promote loss of bromide from the anomeric carbon giving a carbocation.

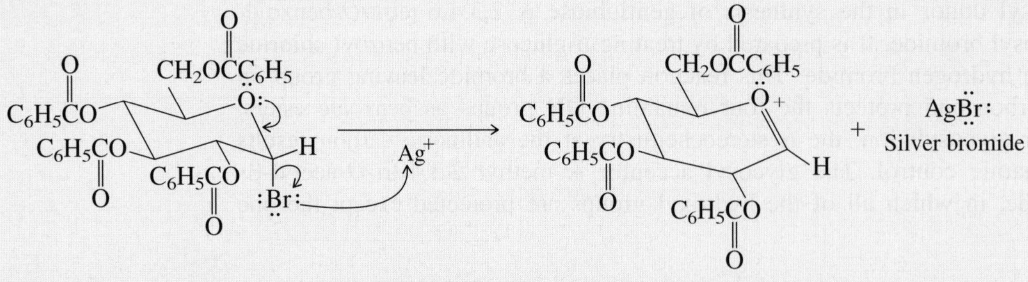

Glycosyl bromide Silver ion Oxygen-stabilized carbocation

Step 2: Once formed, this carbocation rearranges to a more stable structure, one that is stabilized by electron release from two oxygens.

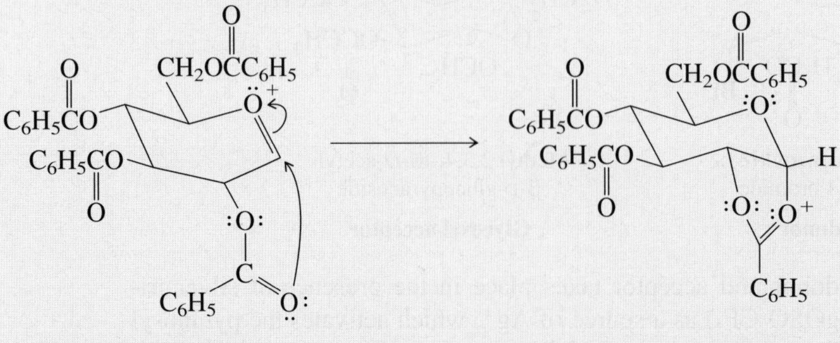

Carbocation (less stable) A dioxolenium ion (more stable)

Continued

Step 3: The dioxolenium ion reacts with the free hydroxyl group of the glycosyl acceptor to give the conjugate acid of the disaccharide. Attack of the hydroxyl group occurs stereoselectively from the direction opposite the dioxolane ring.

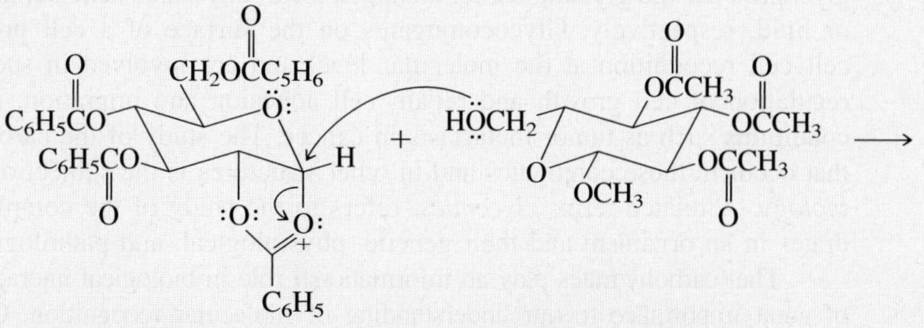

Dioxolenium ion

Methyl 2,3,4-tri-*O*-acetyl-β-D-glucopyranoside

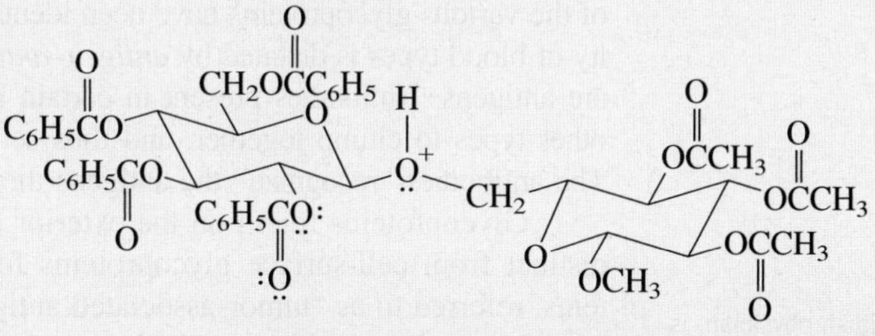

Conjugate acid of the β-disaccharide

Step 4: Subsequent to its formation, the conjugate acid is converted to the disaccharide by proton transfer to 2,4,6-trimethylpyridine, which forms a salt with the trifluoromethanesulfonate that is produced in step 1.

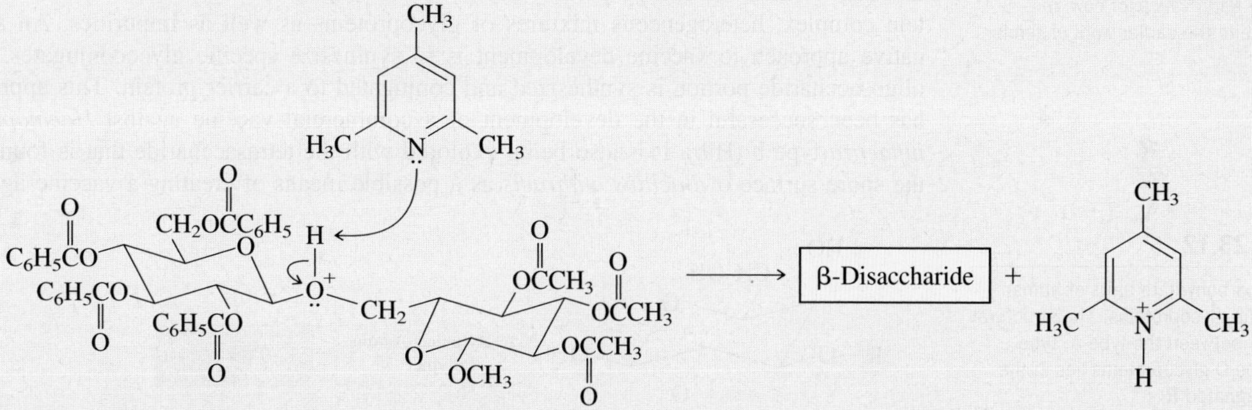

Conjugate acid of the β-disaccharide

2,4,6-Trimethylpyridinium ion

Problem 23.24

Write resonance structures that show how the two oxygens of the five-membered ring stabilize the dioxolenium ion.

In the 1970s, the synthesis of di- and trisaccharides was considered a major achievement, but the advent of new synthetic methods has made it possible to construct more complex oligosaccharides, such as those found on cell surfaces. In 2006, an oligosaccharide containing 28 sugar residues was synthesized. Advances in oligosaccharide synthesis are critical to the development of emerging areas of research in carbohydrate chemistry and biology. An exciting new area is **glycobiology,** described in the following section.

Other methods of oligosaccharide synthesis are considered in Problems 23.43 and 23.44 in this chapter and in Problem 26.28.

23.25 Glycobiology

Carbohydrates are often linked, that is conjugated, to other types of biomolecules. In glycoproteins and glycolipids, for example, a carbohydrate is covalently bound to a protein or lipid, respectively. Glycoconjugates on the surface of a cell provide receptors for cell–cell recognition at the molecular level and are involved in such processes as the regulation of cell growth and repair, cell adhesion and migration, and in pathological conditions such as tumor metastasis in cancer. The study of the carbohydrates in nature that occur in these conjugates and in other structures is the subject of the field of *glycobiology*. A related term, *glycomics,* refers to the study of the complete set of carbohydrates in an organism and their genetic, physiological, and pathological roles.

That carbohydrates play an informational role in biological interactions is a revelation of great importance to our understanding of molecular recognition. One example of the informational role of cell-surface carbohydrates occurs in the distinctions among human blood groups. The structure of the glycoproteins attached to the surface of blood cells determines whether the blood is type A, B, AB, or O. Differences among the carbohydrate components of the various glycoproteins have been identified and are shown in Figure 23.12. Compatibility of blood types is dictated by *antigen–antibody* interactions. The cell-surface glycoproteins are antigens. Antibodies present in certain blood types can cause the blood cells of certain other types to clump together, and thus set practical limitations on transfusion procedures. The antibodies "recognize" the antigens they act on by their terminal saccharide units.

Glycoproteins found on the exterior of bacteria, parasites, or viruses are sometimes distinct from cell-surface glycoproteins found in the human body, and some glycoproteins, referred to as "tumor-associated antigens" are more highly expressed on the surface of cancer cells. Antigen–antibody interactions are the fundamental basis by which the immune system functions and are also the basis for vaccines, which rely on the recognition of cell-specific markers. Vaccines are usually prepared from either killed microorganisms or from live, but attenuated microorganisms that have been cultivated under conditions that disable their virulent properties. Such vaccine preparations from natural sources contain complex, heterogeneous mixtures of glycoproteins as well as impurities. An alternative approach to vaccine development is to synthesize specific glycoconjugates. The oligosaccharide portion is synthesized and conjugated to a carrier protein. This approach has been successful in the development of a commercial vaccine against *Haemophilus influenza* type b (Hib). It is also being explored with the tetrasaccharide that is found on the spore surface of *Bacillus anthracis* as a possible means of creating a vaccine against

Edward Jenner, an English physician, is credited with the development of a vaccine for smallpox in the 1790s. Louis Pasteur developed a vaccine against anthrax for use in cattle in the 1870s and coined the term *vaccine* from the Latin *vacca* for cow, in reference to the earlier work of Jenner.

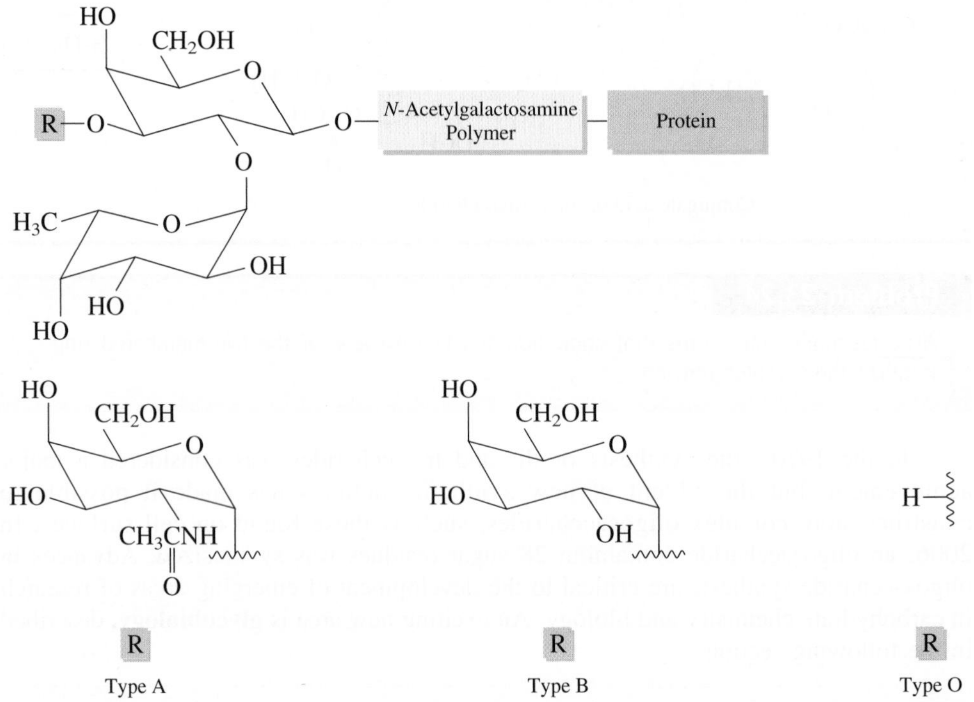

Figure 23.12

Terminal carbohydrate units of human blood-group glycoproteins. The structural difference between the type A, type B, and type O glycoproteins lies in the group designated R.

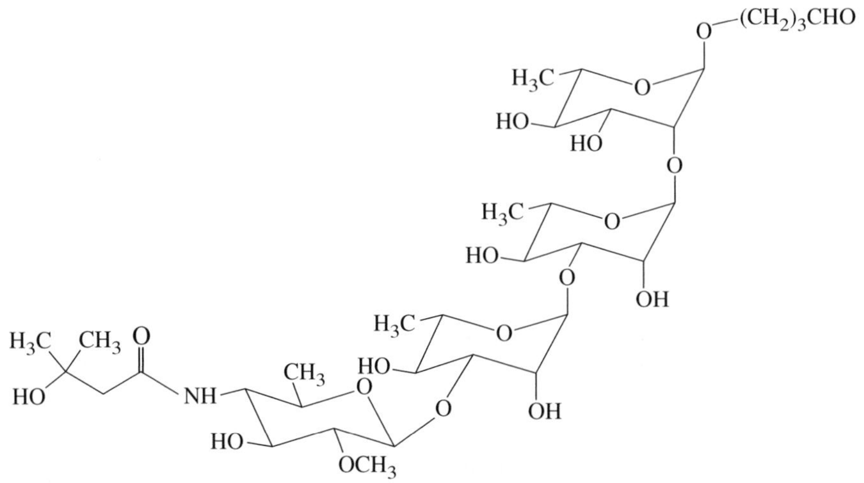

Figure 23.13

The tetrasaccharide found on the surface of anthrax spores. The aldehyde group is used to attach the tetrasaccharide covalently to a carrier protein.

anthrax (Figure 23.13). The tetrasaccharide–protein conjugate may be useful in not only developing a vaccine but also as a method for detecting anthrax. The $(CH_2)_3CHO$ chain attached to the first carbohydrate provides an aldehyde function that is used to attach the tetrasaccharide to the protein carrier by reductive amination (Section 21.10).

Problem 23.25

Label the glycosidic linkages between the carbohydrates in the anthrax spore tetrasaccharide, using the system introduced in Section 23.15.

New drugs to treat influenza were developed based on an understanding of the oligosaccharide structure on the surface of the virus. The key player is *N*-acetylneuraminic acid which is the *N*-acetyl derivative of a nine-carbon monosaccharide. *N*-Acetylneuraminic acid is part of the cell-surface glycoprotein that is recognized by an invading influenza virus (Figure 23.14).

N-Acetylneuraminic acid also forms a coating on newly emerging virus particles that must be removed from the exterior before the virus can adhere to and infect a new cell. This neuraminic acid is removed by an enzyme, neuraminidase, that the virus carries on its surface. Chemists have found that by inhibiting this enzyme, the virus cannot shed its coating of neuraminic acid and is unable to infect new cells. Two drugs that inhibit the activity of neuraminidase are *oseltamivir* (Tamiflu) and *zanamivir* (Relenza).

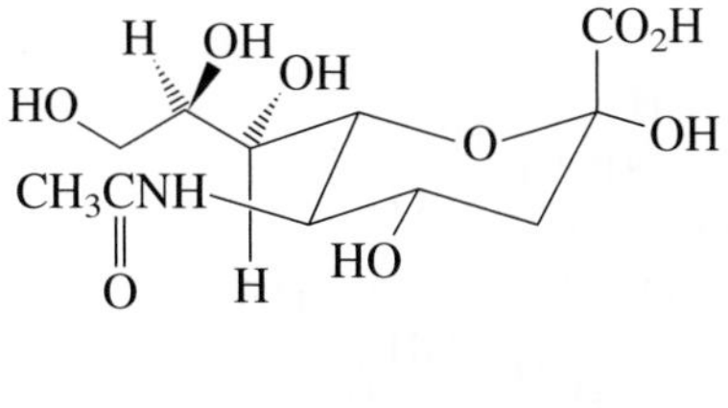

N-Acetylneuraminic acid

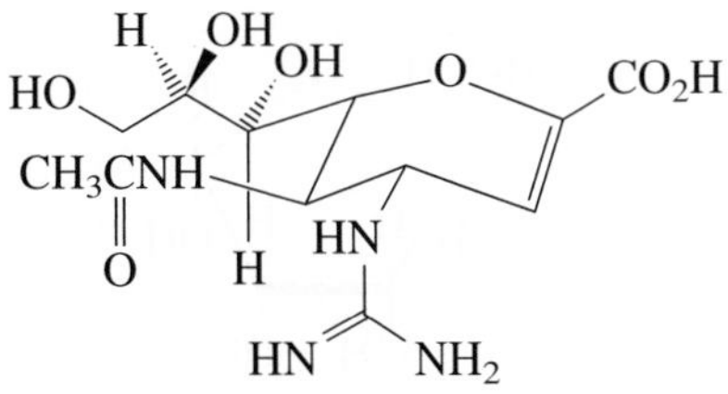

Zanamivir

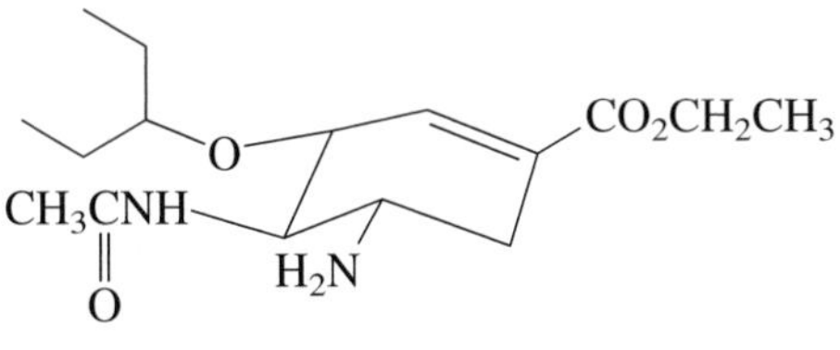

Oseltamivir

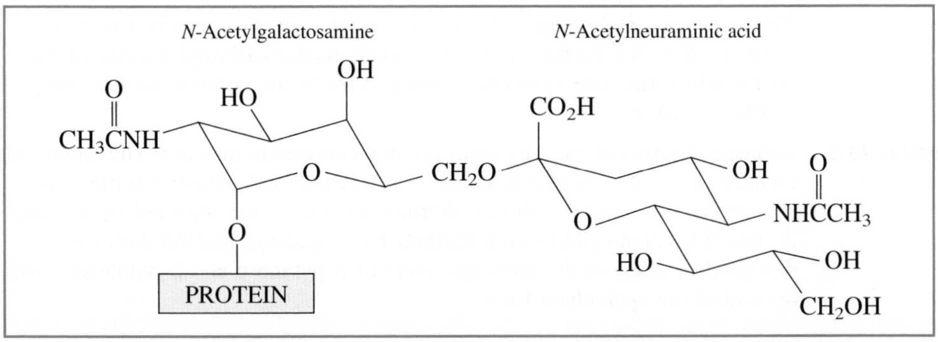

Figure 23.14

Diagram of a cell-surface glycoprotein, showing the disaccharide unit that is recognized by an invading influenza virus.

23.26 SUMMARY

Section 23.1 Carbohydrates are marvelous molecules! In most of them, every carbon bears a functional group, and the nature of the functional groups changes as the molecule interconverts between open-chain and cyclic hemiacetal forms. Any approach to understanding carbohydrates must begin with structure.

Carbohydrates are polyhydroxy aldehydes and ketones. Those derived from aldehydes are classified as **aldoses;** those derived from ketones are **ketoses.**

Section 23.2 Fischer projections and D,L notation are commonly used to describe carbohydrate stereochemistry. The standards are the enantiomers of glyceraldehyde.

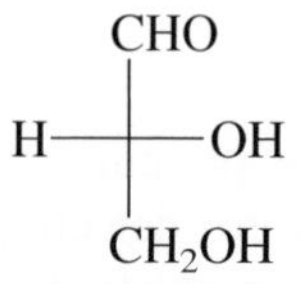
CHO
H
OH
CH2OH

(+)-D-Glyceraldehyde

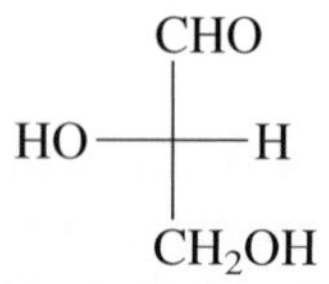
CHO
HO
H
CH2OH

(−)-L-Glyceraldehyde

Section 23.3 Aldotetroses have two chirality centers, so four stereoisomers are possible. They are assigned to the D or the L series according to whether the configuration at their highest numbered chirality center is analogous to D- or L-glyceraldehyde, respectively. Both hydroxyl groups are on the same side of the Fischer projection in erythrose, but on opposite sides in threose. The Fischer projections of D-erythrose and D-threose are shown in Figure 23.2.

Section 23.4 Of the eight stereoisomeric aldopentoses, Figure 23.2 shows the Fischer projections of the D-enantiomers (D-ribose, D-arabinose, D-xylose, and D-lyxose). Likewise, Figure 23.2 gives the Fischer projections of the eight D-aldohexoses.

Section 23.5 The aldohexoses are allose, altrose, glucose, mannose, gulose, idose, galactose, and talose. The mnemonic "All altruists gladly make gum in gallon tanks" is helpful in writing the correct Fischer projection for each one.

Sections 23.6–23.7 Most carbohydrates exist as cyclic hemiacetals. Those with five-membered rings are called **furanose** forms; those with six-membered rings are called **pyranose** forms.

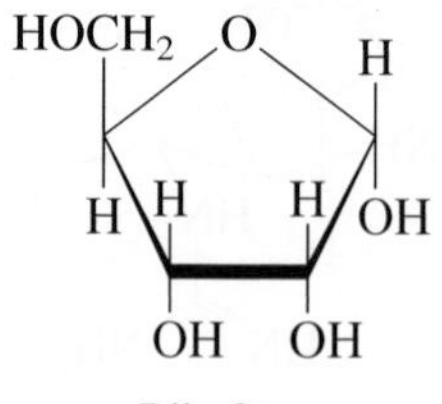
HOCH2
O
H
H
H
H
OH
OH
OH

α-D-Ribofuranose

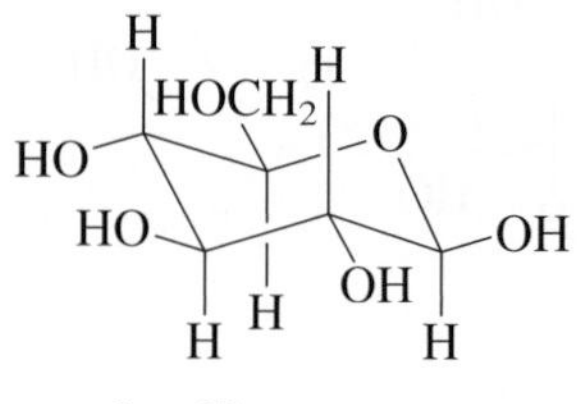
H
H
HOCH2
O
HO
HO
OH
OH
H
H
H

β-D-Glucopyranose

The **anomeric carbon** in a cyclic hemiacetal is the one attached to *two* oxygens. It is the carbon that corresponds to the carbonyl carbon in the open-chain form. The symbols α and β refer to the configuration at the anomeric carbon.

Section 23.8 Hemiacetal forms of carbohydrates are interconvertible in water. The equilibrium mixture can contain α and β anomers of furanose and pyranose forms. The change from one form to the equilibrium mixture is accompanied by a change in optical rotation called **mutarotation.** For D-glucose, mutarotation can be described in terms of the interconversion of α-pyranose and β-pyranose forms by way of the open-chain form.

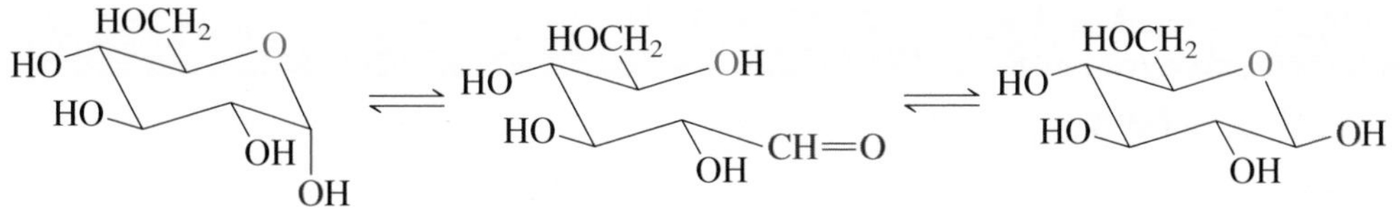

α-D-Glucopyranose　　Open-chain form　　β-D-Glucopyranose

Section 23.9 Pyranose forms of carbohydrates resemble cyclohexane in their conformational preference for chair forms. The **anomeric effect** causes an electronegative substituent at the anomeric (C-1) carbon to be more stable when it is axial than when it is equatorial. The effect is believed to result from the delocalization of an electron pair of the ring oxygen into an antibonding orbital of the anomeric substituent.

Section 23.10 Most naturally occurring ketoses have their carbonyl group located at C-2. Like aldoses, ketoses cyclize to hemiacetals and exist as furanose or pyranose forms.

Sections 23.11–23.13 Structurally modified carbohydrates include **deoxy sugars, amino sugars,** and **branched-chain carbohydrates.**

Section 23.14 Glycosides are acetals, compounds in which the anomeric hydroxyl group has been replaced by an alkoxy group. Glycosides are easily prepared by allowing a carbohydrate and an alcohol to stand in the presence of an acid catalyst.

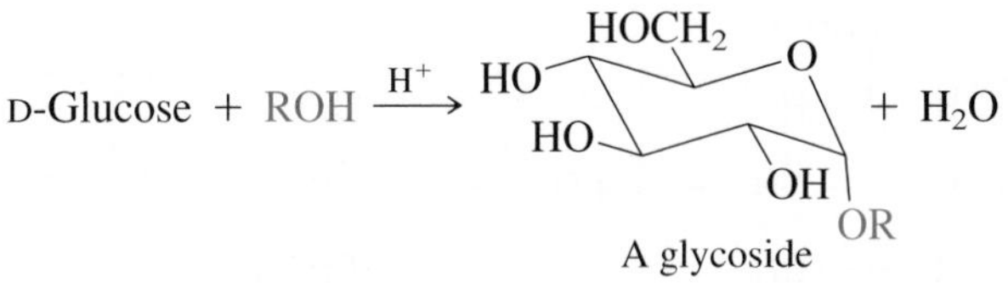

A glycoside

Sections 23.15–23.16 **Disaccharides** are carbohydrates in which two monosaccharides are joined by a glycosidic bond. **Polysaccharides** have many monosaccharide units connected through glycosidic linkages. Complete hydrolysis of disaccharides and polysaccharides cleaves the glycoside bonds, yielding the free monosaccharide components.

Sections 23.17–23.23 Carbohydrates undergo chemical reactions characteristic of aldehydes and ketones, alcohols, diols, and other classes of compounds, depending on their structure. A review of the reactions described in this chapter is presented in Table 23.2.

Section 23.24 Carbohydrates linked by *O*-glycosidic linkages such as those found in disaccharides are synthesized by the reaction of a glycosyl donor and a glycosyl acceptor.

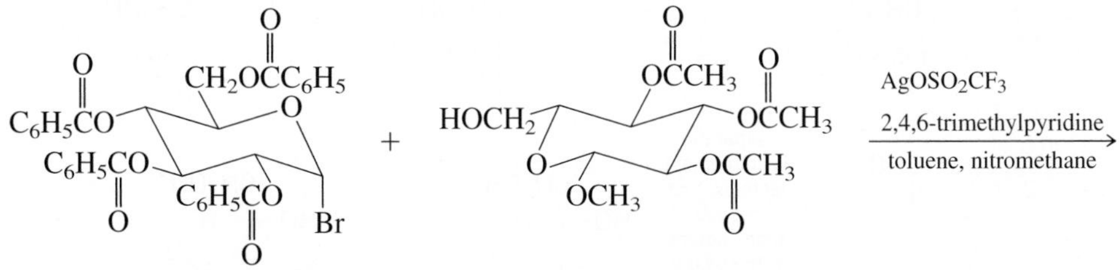

2,3,4,6-Tetra-*O*-benzoyl-α-D-glucopyranosyl bromide (Glycosyl donor)

Methyl 2,3,4-tri-*O*-acetyl-β-D-glucopyranoside (Glycosyl acceptor)

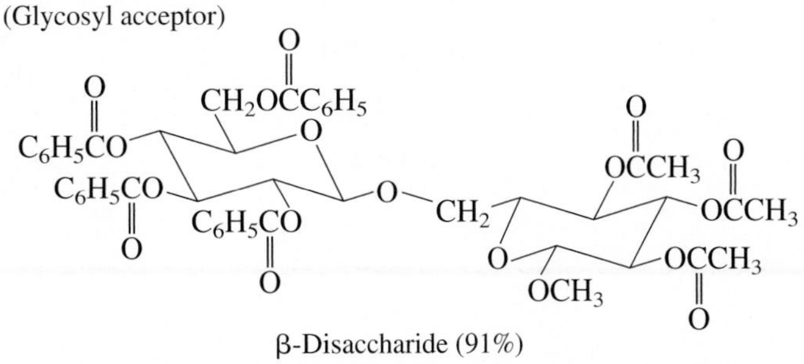

β-Disaccharide (91%)

TABLE 23.2 Summary of Reactions of Carbohydrates

Reaction (section) and comments	Example
Reduction (Section 23.18) The carbonyl group of aldoses and ketoses is reduced by sodium borohydride or by catalytic hydrogenation. The products are called *alditols.*	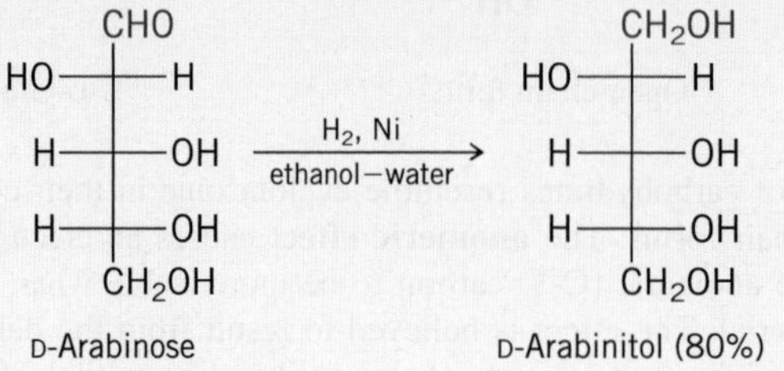
Oxidation (Section 23.19) Bromine oxidizes the aldehyde function of an aldose to a carboxylic acid. The product is an aldonic acid that normally exists as a lactone. The main path to uronic acids, carbohydrates that bear a CO_2H group instead of CH_2OH, is biosynthetic. Nitric acid oxidizes both the CHO and CH_2OH ends of an aldose to CO_2H. These compounds are called *aldaric acids.*	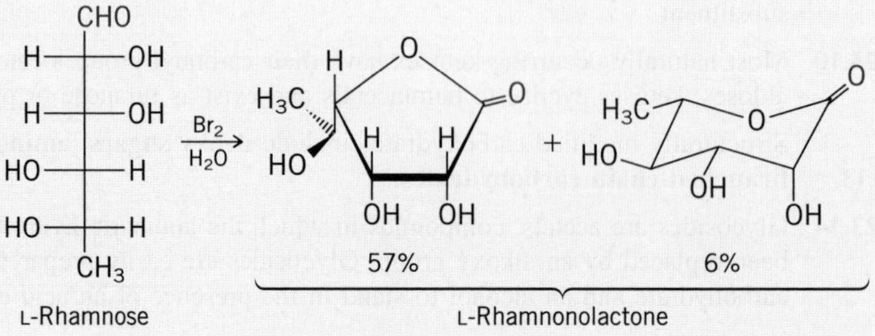
Periodic acid oxidation (Section 23.20) Vicinal diol and α-hydroxy carbonyl functions in carbohydrates are cleaved by periodic acid.	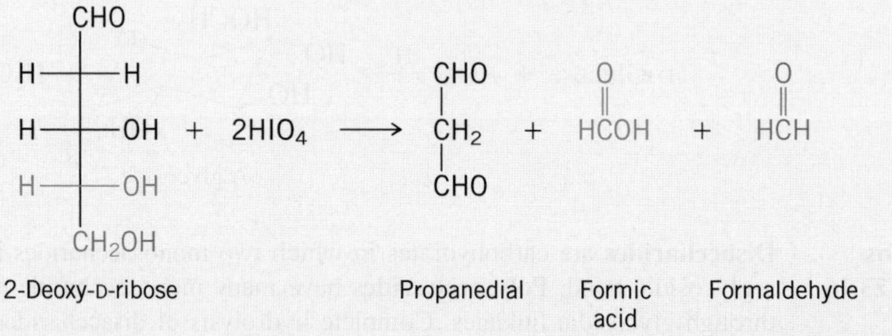
Chain extension (Section 23.21) The Kiliani–Fischer synthesis proceeds by nucleophilic addition of HCN to an aldose, followed by conversion of the cyano group to an aldehyde. A mixture of stereoisomers results; the two aldoses are epimeric at C-2. Section 23.21 describes the modern version of the Kiliani–Fischer synthesis. The example at the right illustrates the classical version.	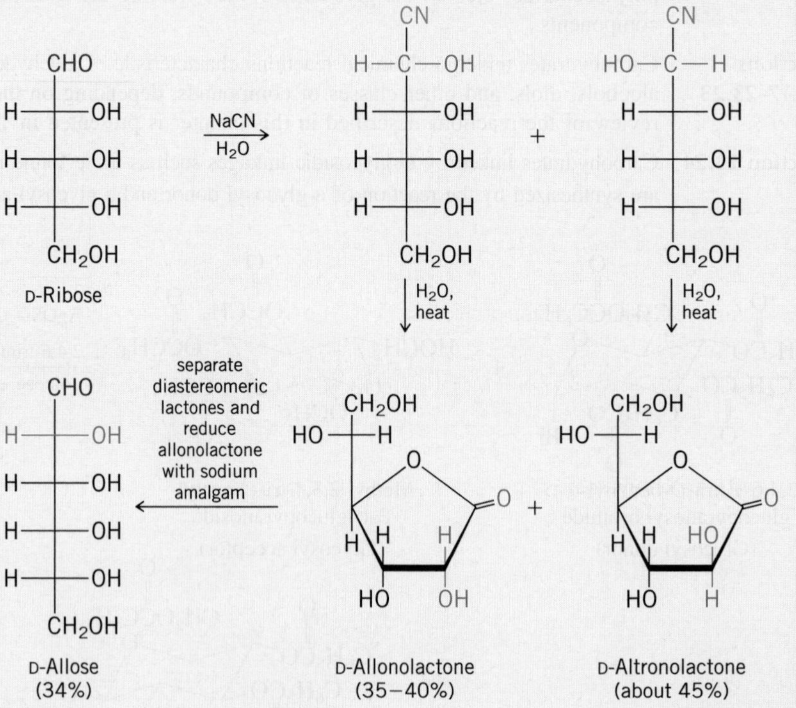

(*Continued*)

TABLE 23.2 Summary of Reactions of Carbohydrates (*Continued*)

Reaction (section) and comments	Example
Enediol formation (Section 23.22) Enolization of an aldose or a ketose gives an enediol. Enediols can revert to aldoses with loss of stereochemical integrity at the α-carbon atom or isomerize to ketoses.	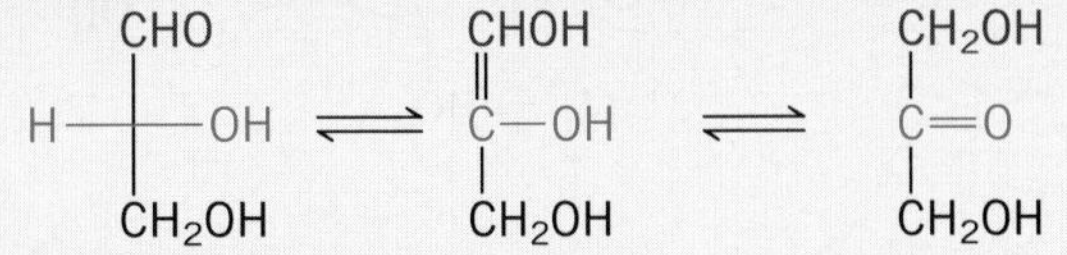 D-Glyceraldehyde ⇌ Enediol ⇌ 1,3-Dihydroxyacetone
Acylation (Section 23.23) Esterification of the available hydroxyl groups occurs when carbohydrates are treated with acylating agents.	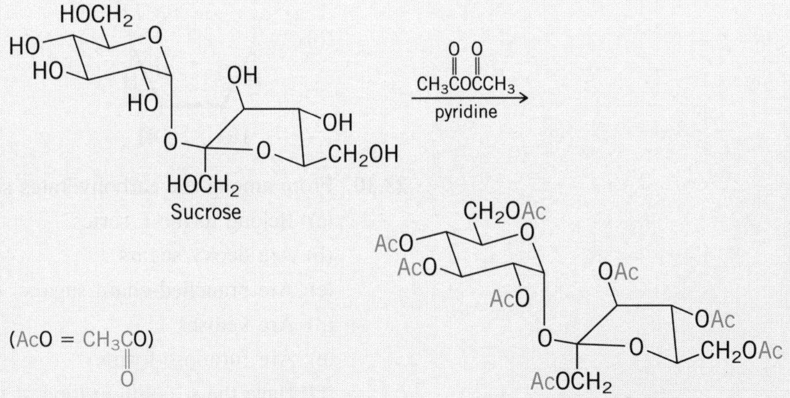 Sucrose → Sucrose octaacetate (66%) (AcO = $CH_3\overset{\displaystyle O}{\overset{\|}{C}}O$)
Alkylation (Section 23.23) Alkyl halides react with carbohydrates to form ethers at the available hydroxyl groups, an application of the Williamson ether synthesis to carbohydrates.	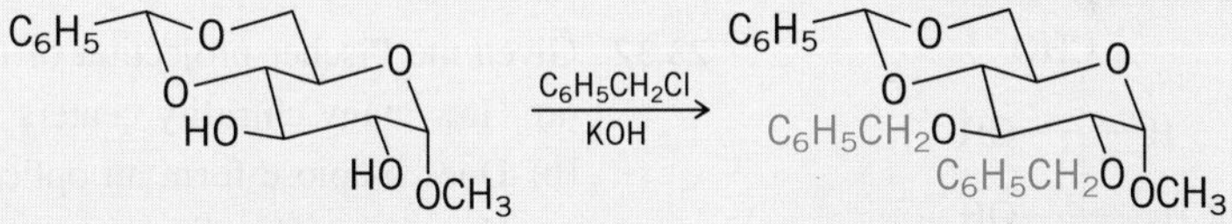 Methyl 4,6-*O*-benzylidene-α-D-glucopyranoside → Methyl 2,3-di-*O*-benzyl-4,6-*O*-benzylidene-α-D-glucopyranoside (92%)

Section 23.25 Carbohydrates linked to proteins, termed glycoproteins, are one example of glycoconjugates in which a carbohydrate is attached to some other biomolecule. Cell-surface glycoproteins are involved in molecular recognition and are important in the immune system as well as in certain diseases.

PROBLEMS

23.26 Refer to the Fischer projection of (+)-D-xylose and give structural formulas for
(a) (–)-Xylose (Fischer projection)
(b) Xylitol
(c) β-D-Xylopyranose
(d) α-L-Xylofuranose
(e) Methyl α-L-xylofuranoside
(f) D-Xylonic acid (open-chain Fischer projection)
(g) δ-Lactone of D-xylonic acid
(h) γ-Lactone of D-xylonic acid
(i) Xylaric acid (open-chain Fischer projection)

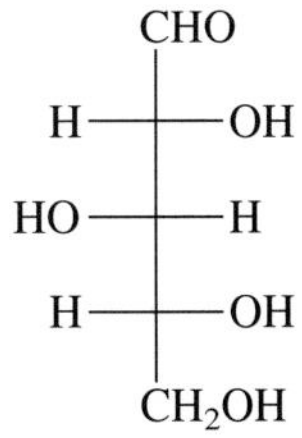
(+)-D-Xylose

23.27 What are the *R,S* configurations of the three chirality centers in D-xylose?

23.28 From among the carbohydrates shown in Figure 23.2, choose the D-aldohexoses that yield
(a) An optically inactive product on reduction with sodium borohydride
(b) An optically inactive product on oxidation with bromine
(c) An optically inactive product on oxidation with nitric acid
(d) The same enediol

23.29 Write the Fischer projection of the open-chain form of each of the following:

(a) 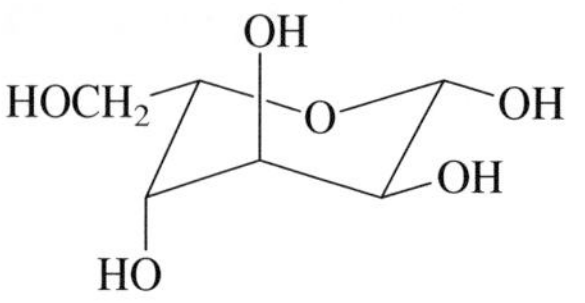

(c) 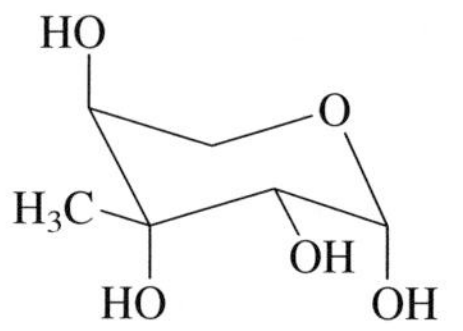

(b) 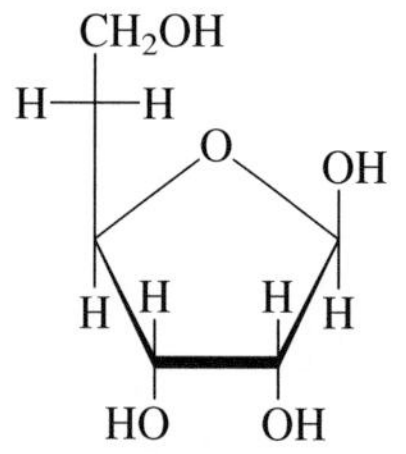

(d) 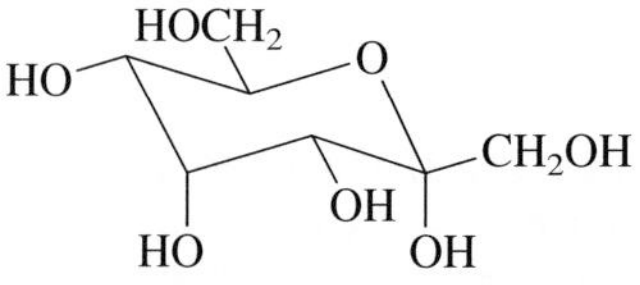

23.30 From among the carbohydrates shown in Problem 23.29, choose the one(s) that

(a) Belong to the L series
(b) Are deoxy sugars
(c) Are branched-chain sugars
(d) Are ketoses
(e) Are furanose forms
(f) Have the α configuration at their anomeric carbon

23.31 How many ketopentoses are possible? Write their Fischer projections.

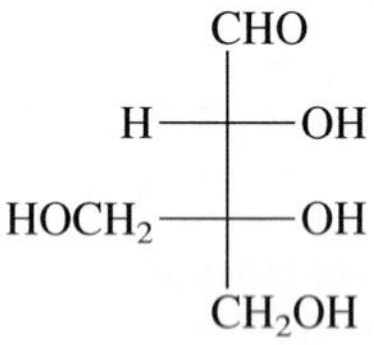

(+)-D-Apiose

23.32 Given the Fischer projection of the branched-chain carbohydrate (+)-D-apiose:

(a) How many chirality centers are in the open-chain form of D-apiose?
(b) Does D-apiose form an optically active alditol on reduction?
(c) How many chirality centers are in the furanose forms of D-apiose?
(d) How many stereoisomeric furanose forms of D-apiose are possible? Write their Haworth formulas.

23.33 Treatment of D-mannose with methanol in the presence of an acid catalyst yields four isomeric products having the molecular formula $C_7H_{14}O_6$. What are these four products?

23.34 Give the products of periodic acid oxidation of each of the following. How many moles of reagent will be consumed per mole of substrate in each case?

(a) D-Arabinose
(b) D-Ribose
(c) Methyl β-D-glucopyranoside
(d) 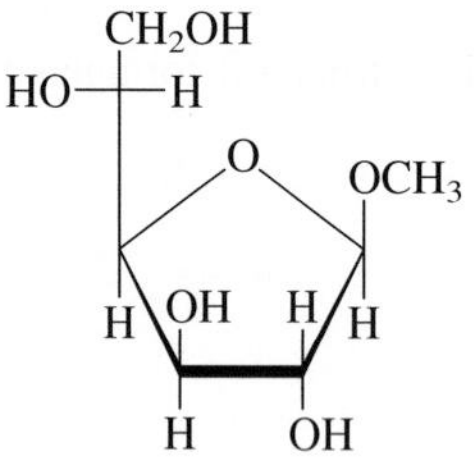

23.35 Protecting groups are often used in carbohydrate synthetic work. Deduce the structure of the carbohydrate derivative from the description given.

(a) Triphenylmethyl chloride reacts with primary alcohols faster than secondary. What is the product of the reaction of triphenylmethyl chloride with methyl α-D-glucopyranoside?

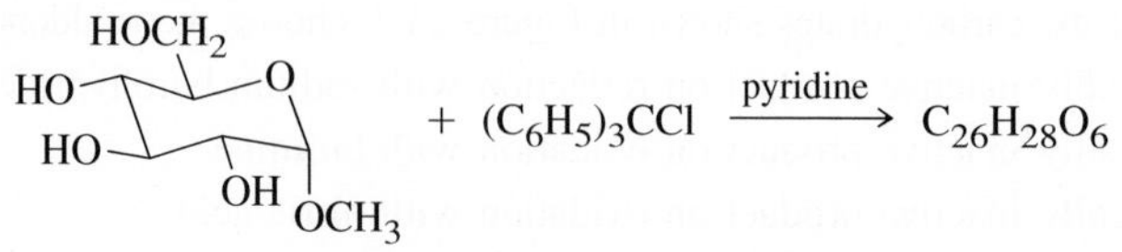
+ $(C_6H_5)_3CCl$
pyridine
$C_{26}H_{28}O_6$

(b) The product of the reaction between methyl α-D-glucopyranoside and benzaldehyde is an acetal with a framework analogous to *trans*-decalin and an equatorial phenyl group. It consumes one mole of periodic acid. Of the four OH groups of methyl α-D-glucopyranoside, this reaction protects two while leaving two others available for further transformations. What is the product?

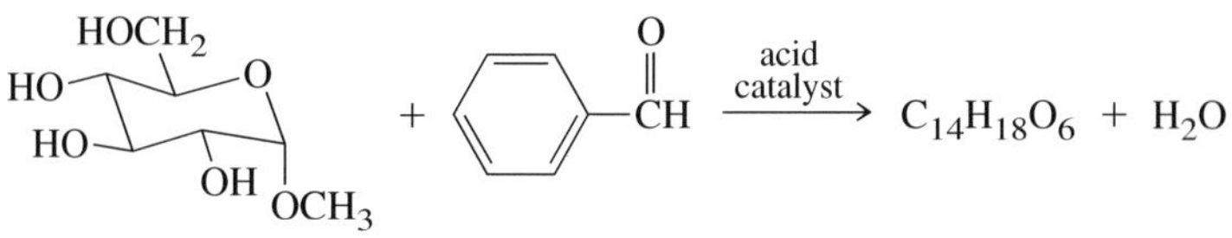

23.36 Compound A was oxidized with periodic acid to give B, which after acid hydrolysis gave C. Bromine oxidation of C gave D. Suggest structural formulas, including stereochemistry, for compounds B, C, and D.

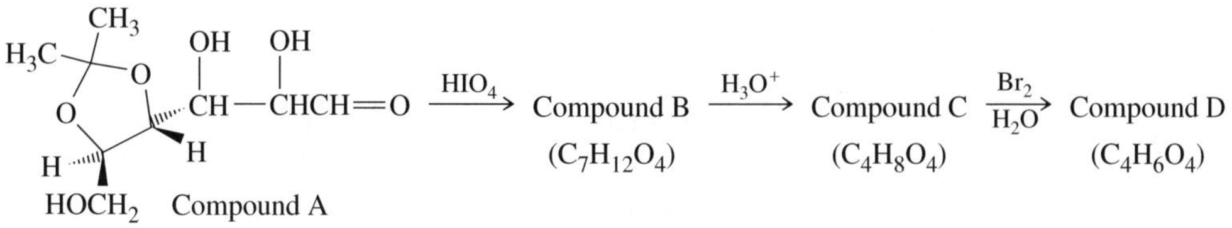

23.37 The γ-lactone of D-gulonic acid was prepared by way of a cyanohydrin derived from an aldopentose.

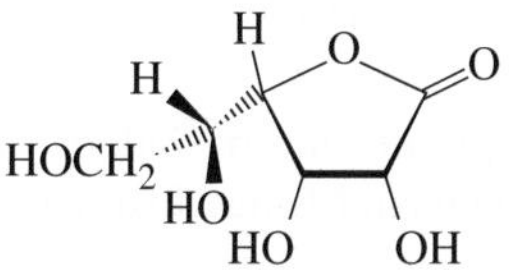

Identify the aldopentose subjected to this chain extension.

23.38 Methyl glycosides of 2-deoxy sugars have been prepared by the acid-catalyzed addition of methanol to unsaturated carbohydrates known as *glycals*.

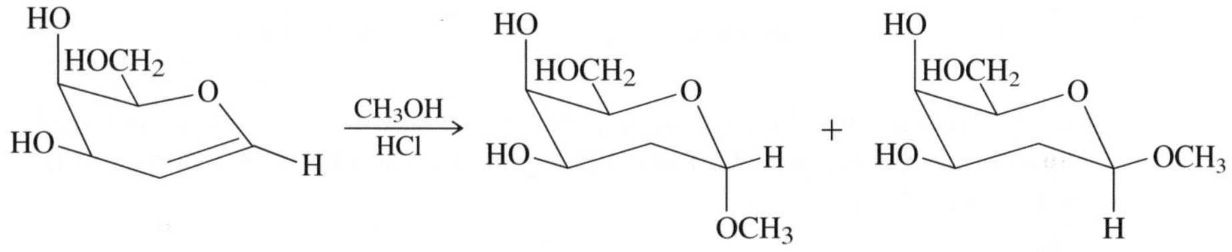

D-Galactal

Methyl 2-deoxy-α-D-lyxohexopyranoside (38%)

Methyl 2-deoxy-β-D-lyxohexopyranoside (36%)

Suggest a reasonable mechanism for this reaction.

23.39 The following are the more stable anomers of the pyranose forms of D-glucose, D-mannose, and D-galactose:

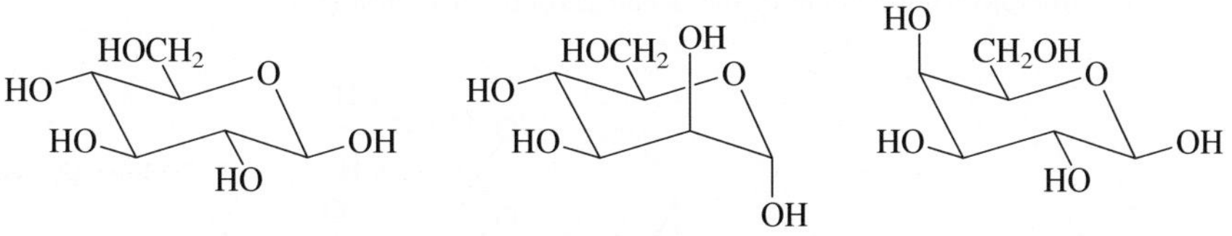

β-D-Glucopyranose (64% at equilibrium)

α-D-Mannopyranose (68% at equilibrium)

β-D-Galactopyranose (64% at equilibrium)

On the basis of these empirical observations and your own knowledge of steric effects in six-membered rings, predict the preferred form (α- or β-pyranose) at equilibrium in aqueous solution for each of the following:

(a) D-Gulose
(c) D-Xylose
(b) D-Talose
(d) D-Lyxose

23.40 Basing your answers on the general mechanism for the first stage of acid-catalyzed acetal hydrolysis

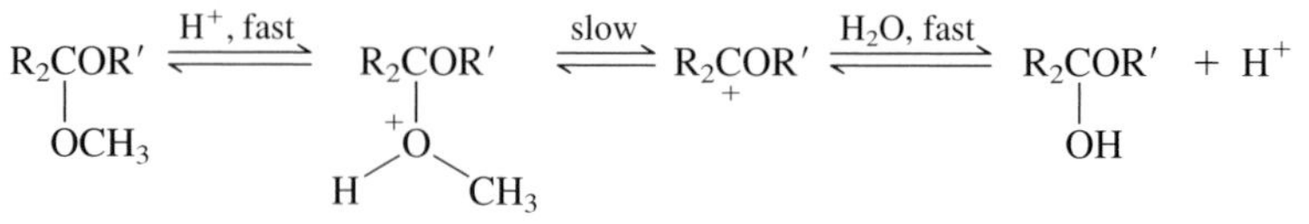

Acetal Hemiacetal

suggest reasonable explanations for the following observations:

(a) Methyl α-D-fructofuranoside (compound A) undergoes acid-catalyzed hydrolysis some 10^5 times faster than methyl α-D-glucofuranoside (compound B).

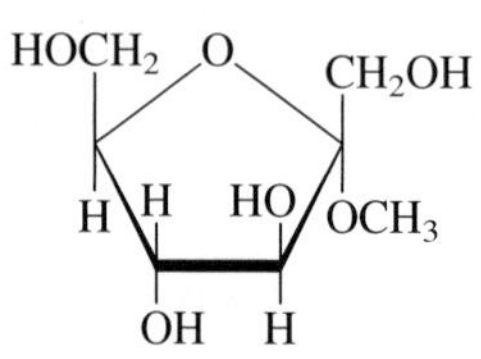

Compound A

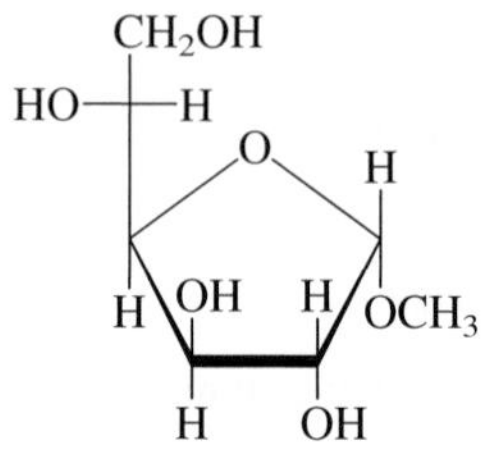

Compound B

(b) The β-methyl glucopyranoside of 2-deoxy-D-glucose (compound C) undergoes hydrolysis several thousand times faster than that of D-glucose (compound D).

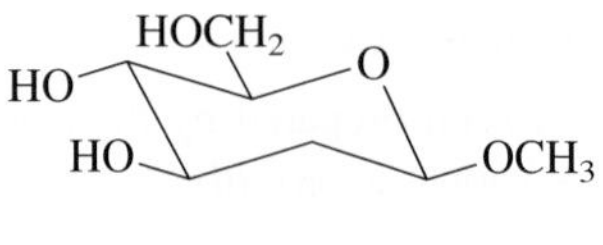

Compound C

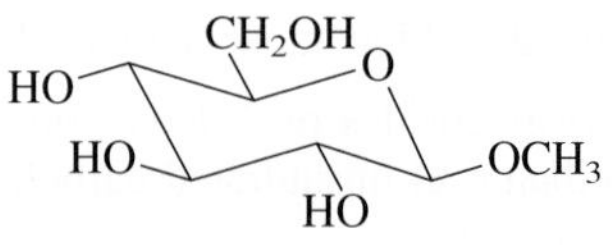

Compound D

(c) Using Mechanism 23.2 as a guide, write a mechanism for the acid-catalyzed hydrolysis of compound D (methyl β-D-glucopyranoside, shown in part b) to D-glucose.

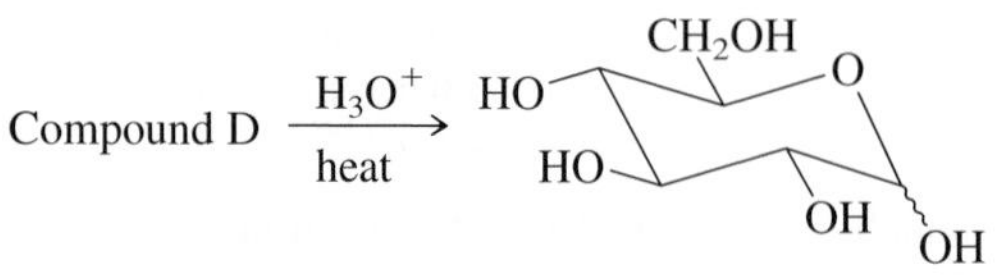

23.41 The compound shown here is the anticonvulsant drug known as topiramate. It is a derivative of D-fructopyranose. Identify the acetal carbons in topiramate.

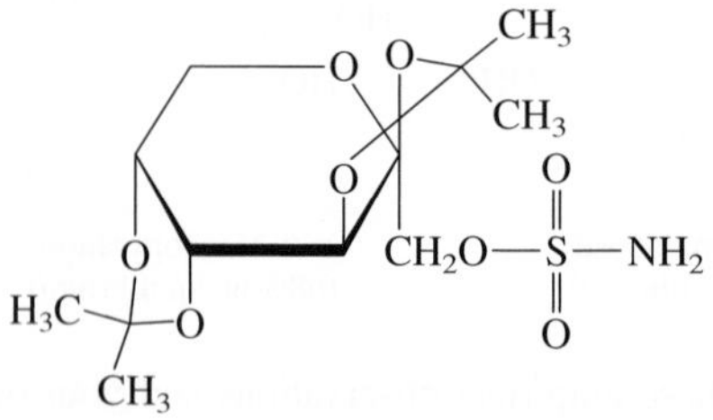

Topiramate

23.42 Acetone reacts with carbohydrates in the presence of an acid catalyst to form products that are commonly referred to as "isopropylidene" or "acetonide" derivatives. The carbohydrate D-ribono-(1,4)-lactone reacts with acetone in the presence of hydrochloric acid to give the acetonide shown here. Write a mechanism for this reaction. (*Hint:* Review Section 17.9 and Problem 17.10.)

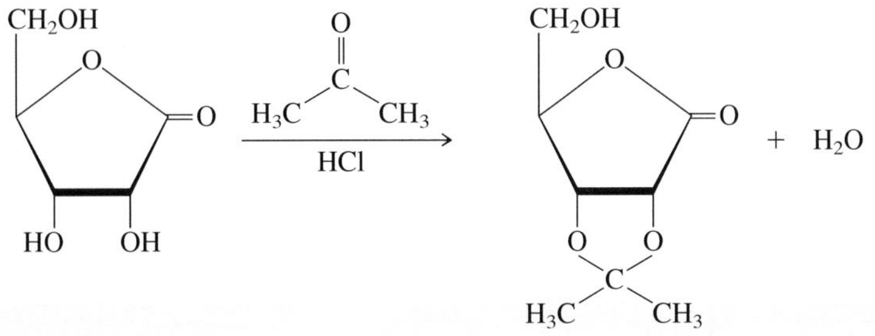

23.43 In each of the following reactions, the glycosyl donor is activated by reaction with an electrophilic reagent by the general pathway:

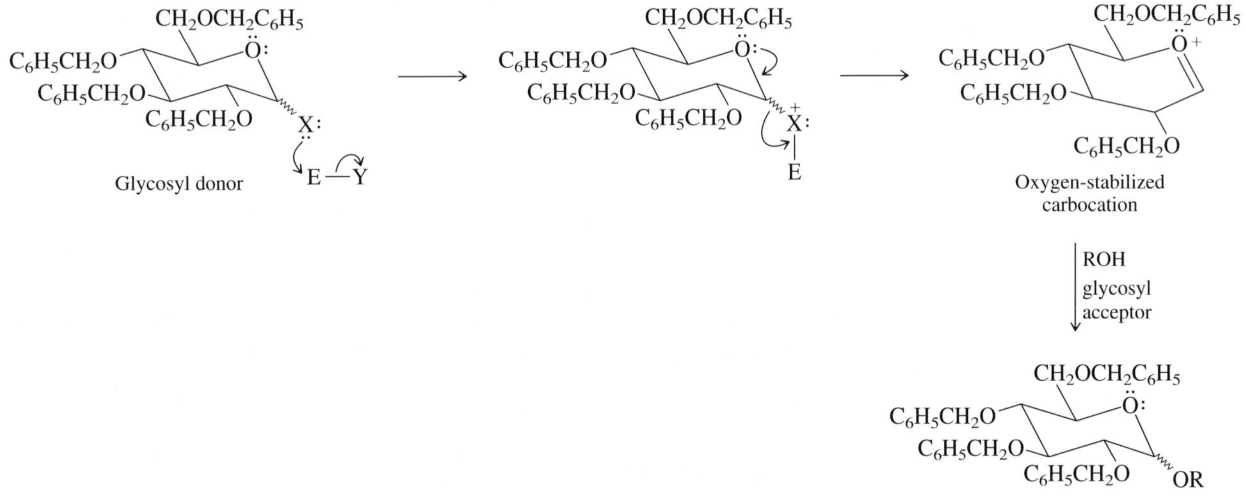
Glycosyl donor
Oxygen-stabilized carbocation
ROH glycosyl acceptor

The oxygen-stabilized carbocation that is formed reacts with the glycosyl acceptor ROH to form an anomeric mixture of *O*-glycosides.

Using curved arrows, write a mechanism to show how each of the glycosyl donors produces the oxygen-stabilized carbocation. For part (a) consult Section 16.17 on sulfides and 8.12 on triflates; for part (b) examine step 2 in Mechanism 19.8, and assume that silylation occurs at nitrogen in the glycosyl imidate ester.

(a)

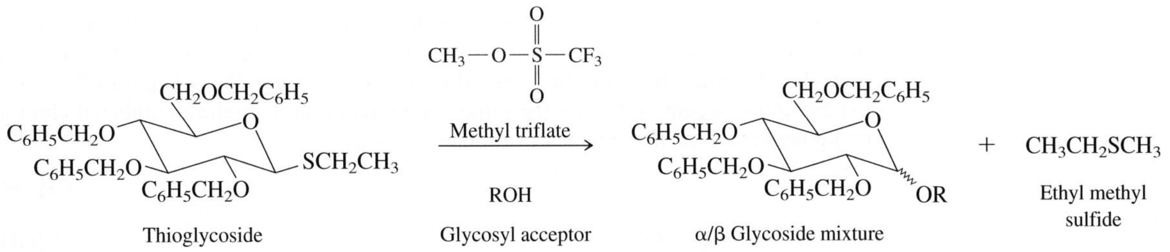
Methyl triflate
ROH
Thioglycoside
Glycosyl acceptor
α/β Glycoside mixture
Ethyl methyl sulfide

(b)

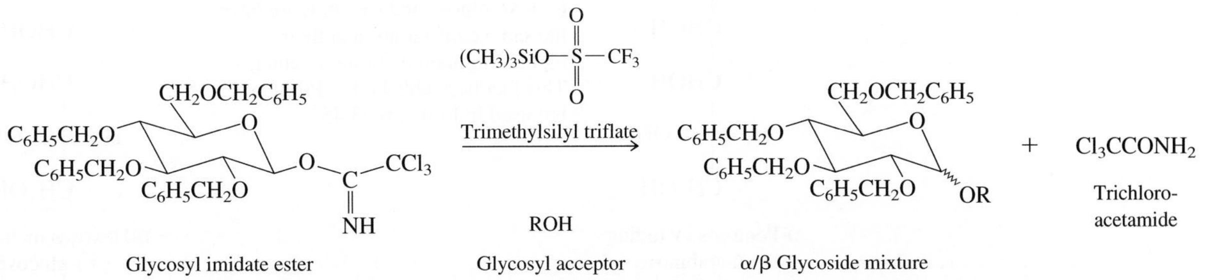
Trimethylsilyl triflate
ROH
Glycosyl imidate ester
Glycosyl acceptor
α/β Glycoside mixture
Trichloro-acetamide

23.44 Pentenyl glycosides are glycosyl donors as illustrated by their conversion to glycosyl bromides with bromine. Write a mechanism for this reaction. Why is the α-anomer formed selectively?

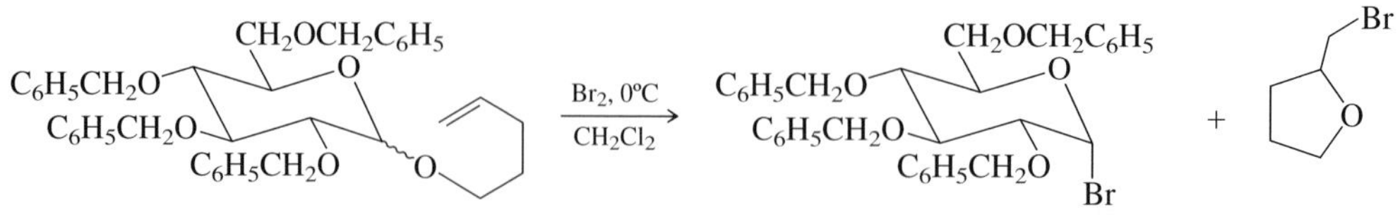

Pentenyl glycoside α Glycosyl bromide (90%) 2-(Bromomethyl)-tetrahydrofuran

Descriptive Passage and Interpretive Problems 23

Emil Fischer and the Structure of (+)-Glucose

Emil Fischer's determination of the structure of glucose was carried out as the nineteenth century ended and the twentieth began. The structure of no other sugar was known at that time, and the spectroscopic techniques that now aid organic analysis were not yet available. All Fischer had was information from chemical transformations, polarimetry, and his own intellect.

Fischer knew that (+)-glucose was one of 16 possible stereoisomers having the constitution:

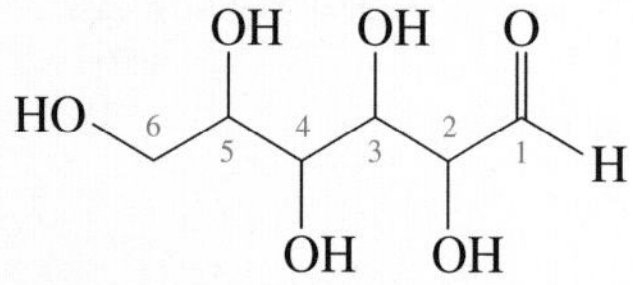

C-2, 3, 4, and 5 are chirality centers of unknown configuration.

By arbitrarily assigning a particular configuration to the chirality center at C-5, Fischer realized that he could determine the configurations of C-2, C-3, and C-4 *relative* to C-5. This reduces the number of structural possibilities to the eight that we now call D-hexoses.

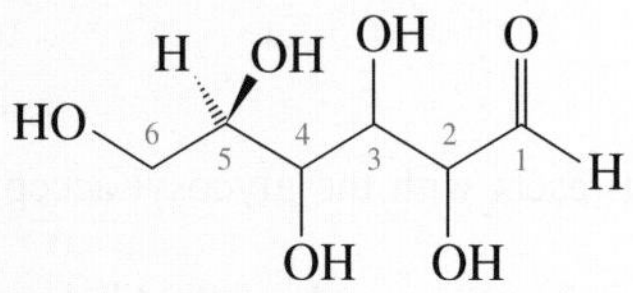

or, as in a Fischer projection

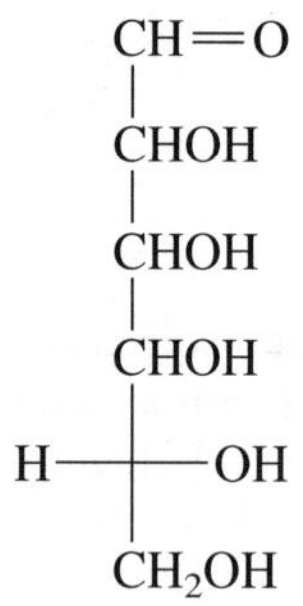

Eventually, Fischer's arbitrary assignment proved correct, which made his stereochemical assignments for all of the chirality centers of (+)-glucose correct in an absolute as well as a relative sense.

The following problems lead you through Fischer's interpretation of the information available to him in determining the structure of (+)-glucose. The order in which the facts are presented is modified slightly from Fischer's, but the logic is the same. We'll begin in Problem 23.45 with (−)-arabinose, a pentose having the same configuration at its highest numbered chirality center as (+)-glucose, a fact that emerges in Problem 23.46.

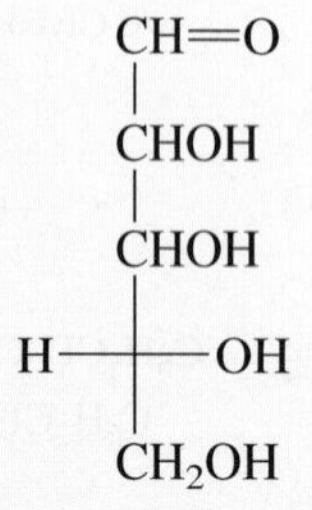

D-Pentoses including (−)-arabinose

(−)-Arabinose and (+)-glucose have the same configuration at their highest numbered chirality center. This fact is established in Problem 23.46 but used in Problem 23.45.

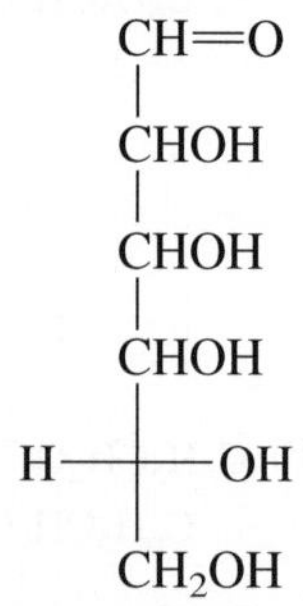

D-Hexoses including (+)-glucose

23.45 Oxidation of (−)-arabinose with warm nitric acid gave an optically active aldaric acid. In this reaction, both C-1 and C-5 are oxidized to CO_2H. Assuming the C-4 OH is to the right, which two of the structures shown are possible for (−)-arabinose?

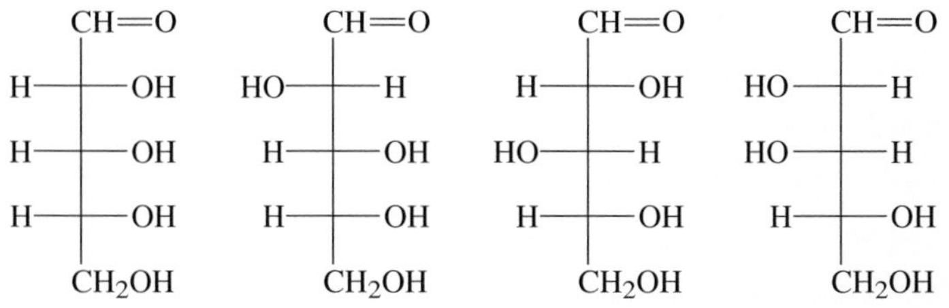

1 2 3 4

A. **1** and **2**
B. **1** and **3**
C. **1** and **4**
D. **2** and **3**
E. **2** and **4**
F. **3** and **4**

23.46 Chain extension of (−)-arabinose by way of its derived cyanohydrin gave a mixture of (+)-glucose and (+)-mannose. Based on this observation *and* your answer to the preceding problem, which pairs are possible for (+)-mannose and (+)-glucose?

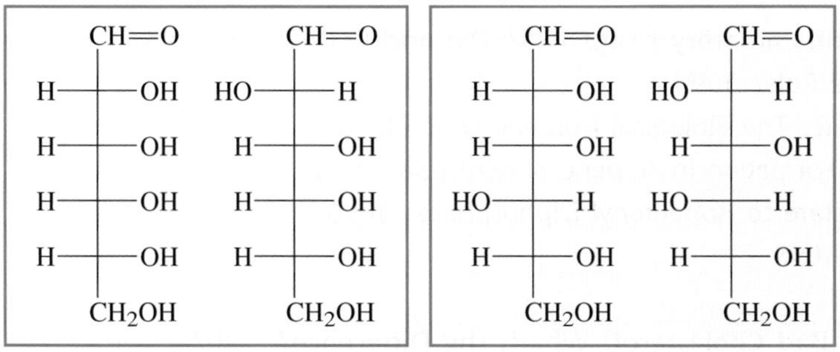

Pair 1 Pair 2

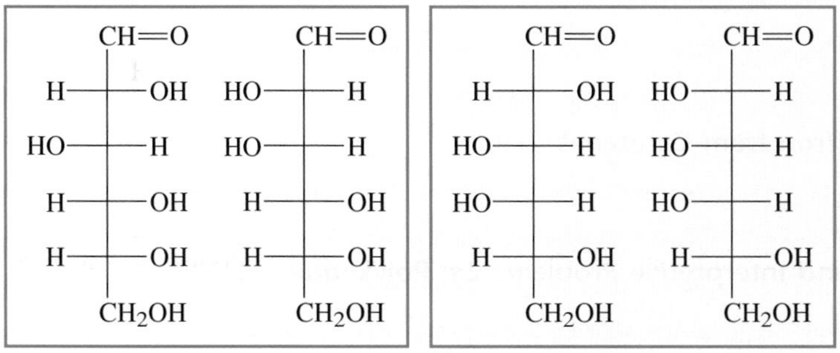

Pair 3 Pair 4

A. Pair 1 and pair 2
B. Pair 1 and pair 3
C. Pair 1 and pair 4
D. Pair 2 and pair 3
E. Pair 2 and pair 4
F. Pair 3 and pair 4

23.47 Both (+)-glucose and (+)-mannose were oxidized to optically active dicarboxylic acids (aldaric acids) ($C_6H_{14}O_4$) with nitric acid. Of the pairs remaining after solving the preceding problem, which one is the (+)-glucose/(+)-mannose pair?

A. Pair 1
B. Pair 2
C. Pair 3
D. Pair 4

In order to do the next problem, you need to know that pair 3 is the correct answer to Problem 23.46. If you are not certain about how this answer is arrived at, it would be a good idea to review the previous questions.

23.48 Because both C-1 and C-6 are oxidized to —CO_2H groups by nitric acid, Fischer recognized that two diastereomeric hexoses could give the same aldaric acid.

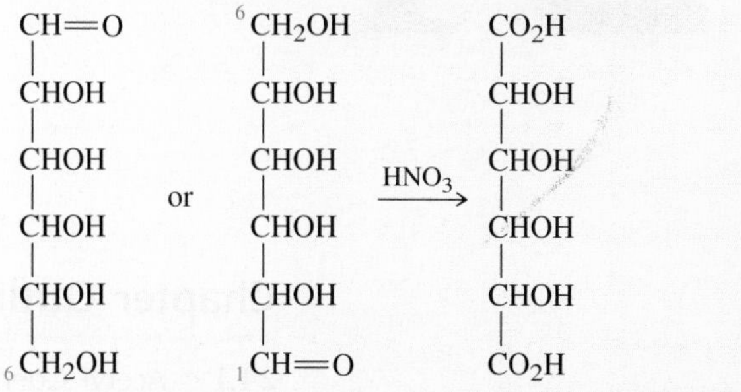

Of the (+)-glucose/(+)-mannose pair (pair 3 in Problem 23.46), only (+)-glucose has a diastereomeric hexose that gives the same aldaric acid. Fischer synthesized that specific diastereomer and found it gave the same aldaric acid as (+)-glucose. Thus, he was able to determine that (+)-glucose and (+)-mannose are:

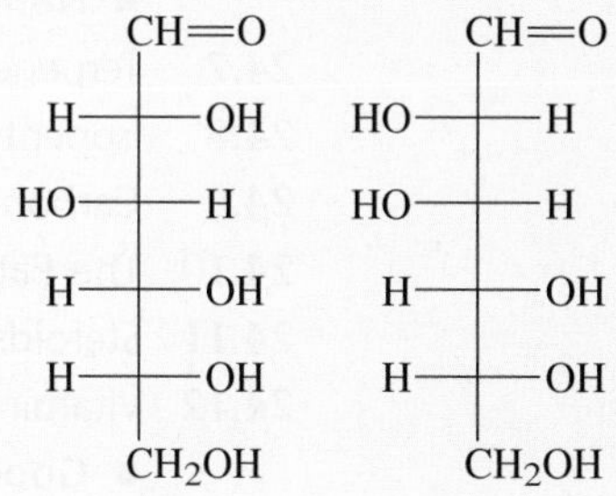

(+)-Glucose (+)-Mannose

Which hexose did Fischer synthesize that gave the same aldaric acid as (+)-glucose?

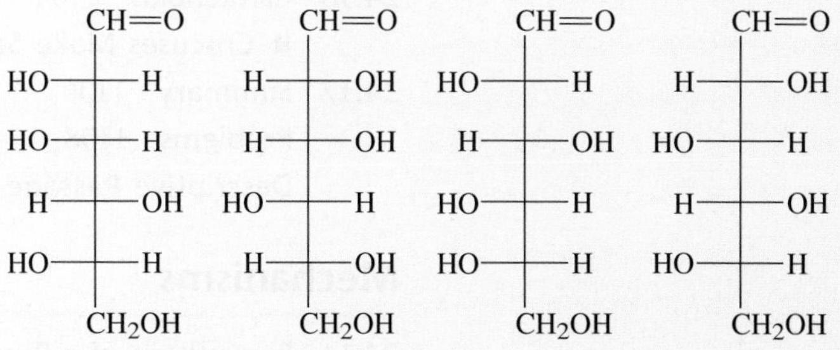

A. B. C. D.

23.49 Refer to Table 23.2 to identify the hexose that is the answer to Problem 23.48.

A. (+)-D-Altrose
B. (+)-D-Galactose
C. (−)-L-Glucose
D. (+)-L-Gulose

24 Lipids

Chapter Outline

Mechanisms

Digitoxigenin, a steroid, is found in digoxin, a cardiac glycoside found in the foxglove plant. Digoxin is used to treat heart conditions. Like other steroids, digitoxigenin is a lipid.

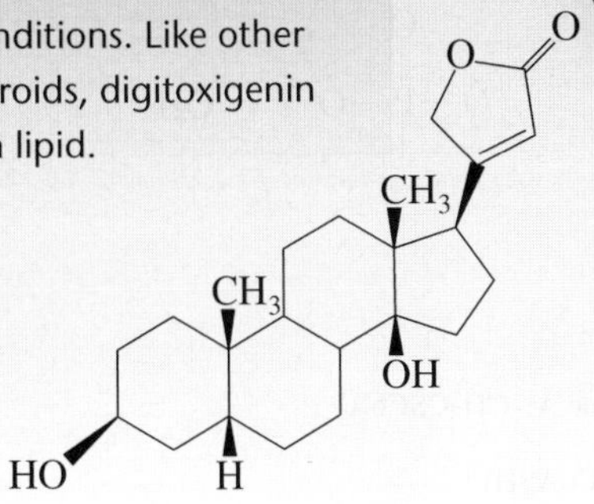

Digitoxigenin

LIPIDS DIFFER from the other classes of naturally occurring biomolecules (carbohydrates, proteins, and nucleic acids) in that they are more soluble in nonpolar to weakly polar solvents (diethyl ether, hexane, dichloromethane) than they are in water. They include a variety of structural types, a collection of which is introduced in this chapter.

A common theme in this chapter is biosynthesis. Although it may seem that molecules such as fatty acids, terpenes, and steroids are very different from one another, lipids share a common biosynthetic origin in that they are ultimately derived from glucose. During one stage of carbohydrate metabolism, called *glycolysis,* glucose is converted to lactic acid. Pyruvic acid is an intermediate.

$$\underset{\text{Glucose}}{C_6H_{12}O_6} \longrightarrow \underset{\text{Pyruvic acid}}{CH_3\overset{\overset{\displaystyle O}{\|}}{C}CO_2H} \longrightarrow \underset{\text{Lactic acid}}{CH_3\overset{\overset{\displaystyle OH}{|}}{C}HCO_2H}$$

Pyruvate is used by living systems in a number of different ways. One pathway, the one leading to lactate and beyond, is concerned with energy storage and production. Lactate is not the only destination of pyruvate, however. A significant fraction of pyruvate is converted to acetate for use as a starting material in the biosynthesis of more complex substances, especially lipids. This chapter is organized around that theme. We'll begin by looking at the reaction in which acetate (two carbons) is formed from pyruvate (three carbons).

24.1 Acetyl Coenzyme A

The form in which acetate is used in most of its important biochemical reactions is the thioester **acetyl coenzyme A** (Figure 24.1*a*). Its formation from pyruvate involves several steps, all of which are catalyzed by enzymes. The overall process is summarized as:

$$\underset{\substack{\text{Pyruvic}\\ \text{acid}}}{CH_3\overset{\overset{\displaystyle O}{\|}}{C}\overset{\overset{\displaystyle O}{\|}}{C}OH} + \underset{\text{Coenzyme A}}{CoASH} + \underset{\substack{\text{Oxidized}\\ \text{form of}\\ \text{nicotinamide}\\ \text{adenine}\\ \text{dinucleotide}}}{NAD^+} \longrightarrow \underset{\substack{\text{Acetyl}\\ \text{coenzyme A}}}{CH_3\overset{\overset{\displaystyle O}{\|}}{C}SCoA} + \underset{\substack{\text{Reduced}\\ \text{form of}\\ \text{nicotinamide}\\ \text{adenine}\\ \text{dinucleotide}}}{NADH} + \underset{\substack{\text{Carbon}\\ \text{dioxide}}}{CO_2} + \underset{\text{Proton}}{H^+}$$

Figure 24.1

Structures of (*a*) acetyl coenzyme A and (*b*) coenzyme A.

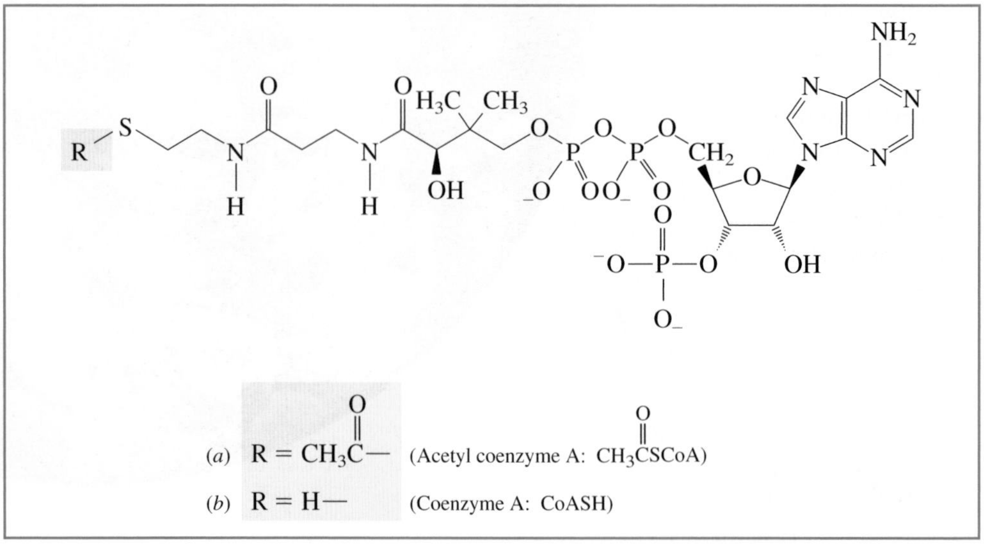

Coenzyme A was isolated and identified by Fritz Lipmann, an American biochemist. Lipmann shared the 1953 Nobel Prize in Physiology or Medicine for this work.

NAD^+ (Section 15.10) is required as an oxidizing agent, and coenzyme A (Figure 24.1*b*) is the acetyl group acceptor. Coenzyme A is a *thiol;* its chain terminates in a *sulfhydryl* (—SH) group. Acetylation of the sulfhydryl group of coenzyme A gives acetyl coenzyme A.

Thioesters are more reactive than ordinary esters toward nucleophilic acyl substitution. They also contain a greater proportion of enol at equilibrium. Both properties are apparent in the properties of acetyl coenzyme A. In some reactions it is the carbonyl group of acetyl coenzyme A that reacts; in others it is the α-carbon atom.

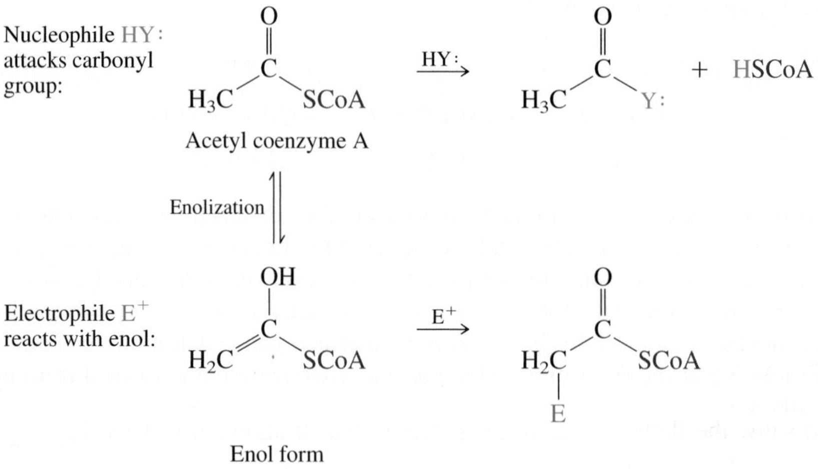

Biochemical examples of these two modes of reactivity are

Nucleophilic Acyl Substitution

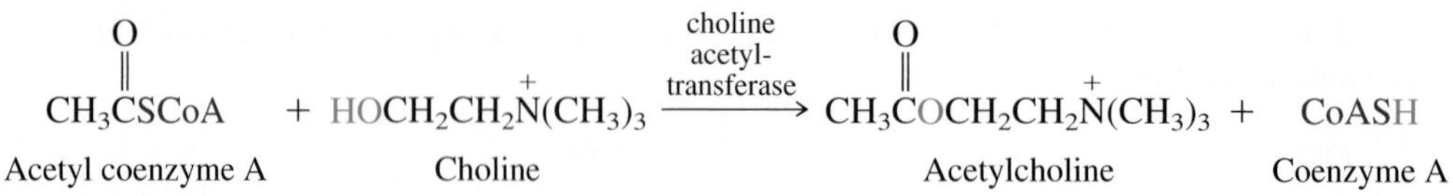

Reaction at α Carbon

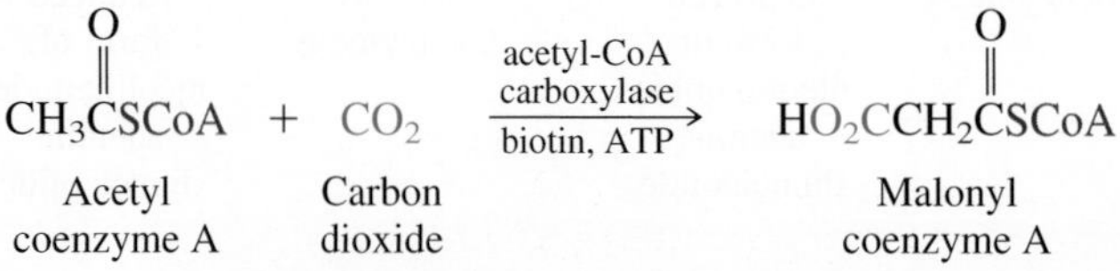

We'll see numerous examples of both reaction types in the following sections. Even though these reactions are enzyme-catalyzed and occur at far greater rates than those for the same transformations carried out in their absence, the types of reactions are essentially the same as the fundamental processes of organic chemistry described throughout this text.

Fats are one type of lipid. They have a number of functions in living systems, including that of energy storage. Although carbohydrates also serve as a source of readily available energy, an equal mass of fat delivers over twice the amount of energy. It is more efficient for an organism to store energy in the form of fat because it requires less mass than storing the same amount of energy in carbohydrates or proteins.

How living systems convert acetate to fats is an exceedingly complex story, one that is well understood in broad outline and becoming increasingly clear in detail. We will examine several aspects of this topic in the next few sections, focusing mostly on its structural and chemical features.

24.2 Fats, Oils, and Fatty Acids

Fats and oils are naturally occurring mixtures of **triacylglycerols,** also called *triglycerides.* They differ in that fats are solids at room temperature and oils are liquids. We generally ignore this distinction and refer to both groups as fats.

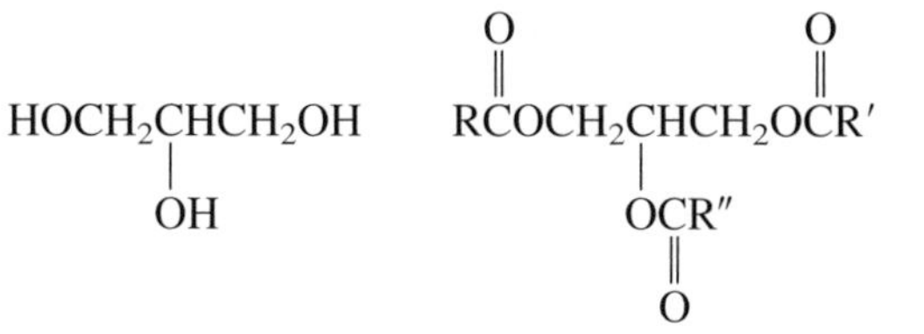

Glycerol A triacylglycerol

All three acyl groups in a triacylglycerol may be the same, all three may be different, or one may be different from the other two.

Figure 24.2 shows the structures of two typical triacylglycerols, 2-oleyl-1,3-distearylglycerol (Figure 24.2*a*) and tristearin (Figure 24.2*b*). Both occur naturally—in cocoa butter, for example. All three acyl groups in tristearin are stearyl (octadecanoyl) groups. In 2-oleyl-1,3-distearylglycerol, two of the acyl groups are stearyl, but the one in the middle is oleyl (*cis*-9-octadecenoyl). As the figure shows, tristearin can be prepared by catalytic hydrogenation of the carbon–carbon double bond of 2-oleyl-1,3-distearylglycerol. Hydrogenation raises the melting point from 43°C in 2-oleyl-1,3-distearylglycerol to 72°C in tristearin and is a standard technique in the food industry for converting liquid vegetable oils to solid "shortenings." The space-filling models of the two show the flatter structure of tristearin, which allows it to pack better in a crystal lattice than the more irregular shape of 2-oleyl-1,3-distearyl-glycerol permits. This irregular shape is a direct result of the cis double bond in the side chain.

Hydrolysis of fats yields glycerol and long-chain **fatty acids.** Thus, tristearin gives glycerol and three molecules of stearic acid on hydrolysis. Table 24.1 lists a few representative fatty acids. As these examples indicate, most naturally occurring fatty acids possess an even number of carbon atoms and an unbranched carbon chain. The carbon chain may be saturated or it can contain one or more double bonds. When double bonds are present, they are almost always cis. Acyl groups containing 14–20 carbon atoms are the most abundant in triacylglycerols.

The term *fatty acid* originally referred to those carboxylic acids that occur naturally in triacylglycerols. Its use has expanded to include all unbranched carboxylic acids, irrespective of their origin and chain length.

Problem 24.1

What fatty acids are produced on hydrolysis of 2-oleyl-1,3-distearylglycerol? What other triacylglycerol gives the same fatty acids and in the same proportions as 2-oleyl-1,3-distearylglycerol?

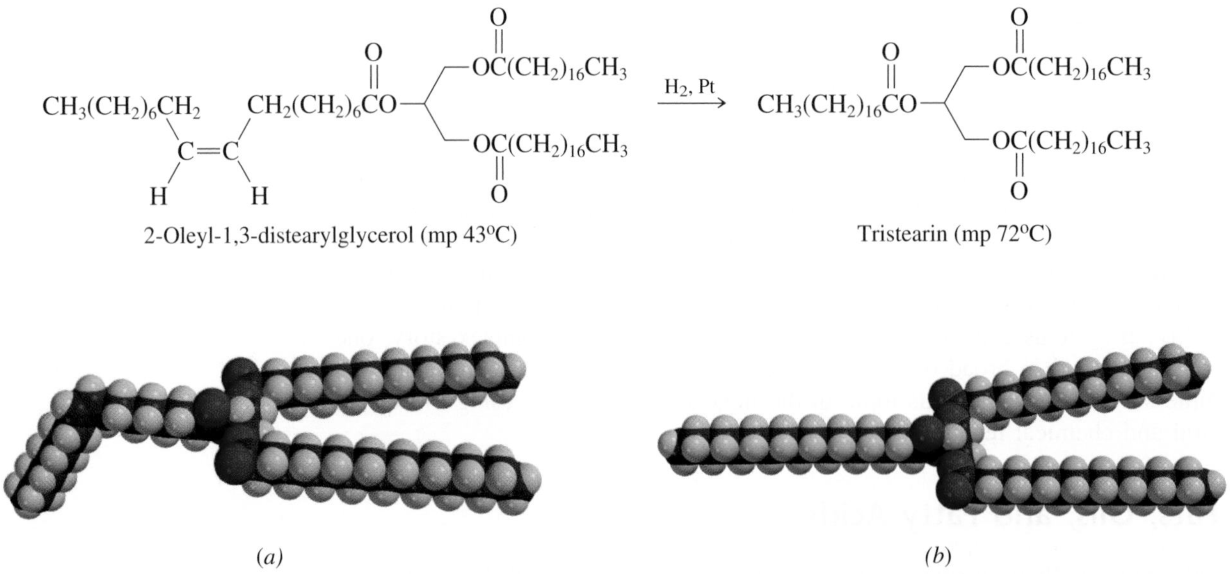
H2, Pt
2-Oleyl-1,3-distearylglycerol (mp 43°C)
Tristearin (mp 72°C)
(a)
(b)

Figure 24.2

The structures of two typical triacylglycerols. (*a*) 2-Oleyl-1,3-distearylglycerol is a naturally occurring triacylglycerol found in cocoa butter. The cis double bond of its oleyl group gives the molecule a shape that interferes with efficient crystal packing. (*b*) Catalytic hydrogenation converts 2-oleyl-1,3-distearylglycerol to tristearin. Tristearin has a higher melting point than 2-oleyl-1,3-distearylglycerol.

A few fatty acids with trans double bonds (trans fatty acids) occur naturally, but the major source of trans fats comes from partial hydrogenation of vegetable oils in, for example, the preparation of margarine. The same catalysts that catalyze the hydrogenation of the double bonds in a triacylglycerol also catalyze their stereoisomerization. The mechanism for conversion of a cis to a trans double bond follows directly from the mechanism of catalytic hydrogenation (see Section 6.1) once one realizes that all of the steps in the mechanism are reversible.

Palmitic acid is the most abundant naturally occurring fatty acid. It is present in many fats and oils and is best known as the major fatty acid component of palm oil.

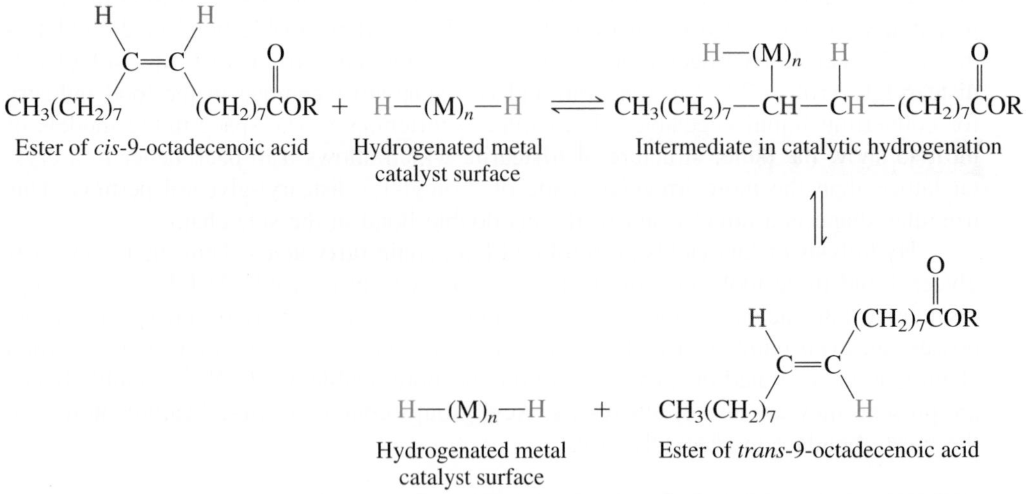
Ester of cis-9-octadecenoic acid
Hydrogenated metal catalyst surface
Intermediate in catalytic hydrogenation
Hydrogenated metal catalyst surface
Ester of trans-9-octadecenoic acid

The intermediate in hydrogenation, formed by reaction of the unsaturated ester with the hydrogenated surface of the metal catalyst, not only can proceed to the saturated fatty acid ester, but also can dissociate to the original ester having a cis double bond or to its trans stereoisomer. Unlike polyunsaturated vegetable oils, which tend to

TABLE 24.1 Some Representative Fatty Acids

Number of carbons	Common name	Systematic name	Structural formula	Melting point, °C
Saturated fatty acids				
12	Lauric acid	Dodecanoic acid	$CH_3(CH_2)_{10}CO_2H$	44
14	Myristic acid	Tetradecanoic acid	$CH_3(CH_2)_{12}CO_2H$	58.5
16	Palmitic acid	Hexadecanoic acid	$CH_3(CH_2)_{14}CO_2H$	63
18	Stearic acid	Octadecanoic acid	$CH_3(CH_2)_{16}CO_2H$	69
20	Arachidic acid	Icosanoic acid	$CH_3(CH_2)_{18}CO_2H$	75
Unsaturated fatty acids				
18	Oleic acid	*cis*-9-Octadecenoic acid	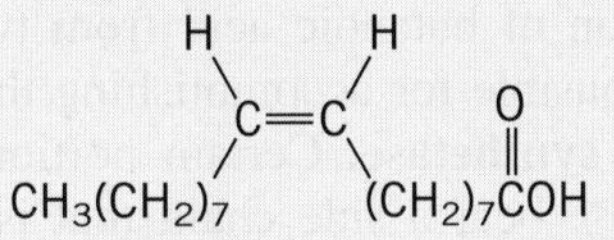	4
18	Linoleic acid	*cis,cis*-9,12-Octadecadienoic acid	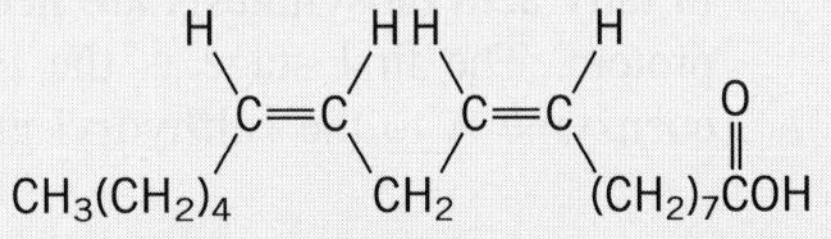	−12
18	Linolenic acid	*cis,cis,cis*-9,12,15-Octadecatrienoic acid	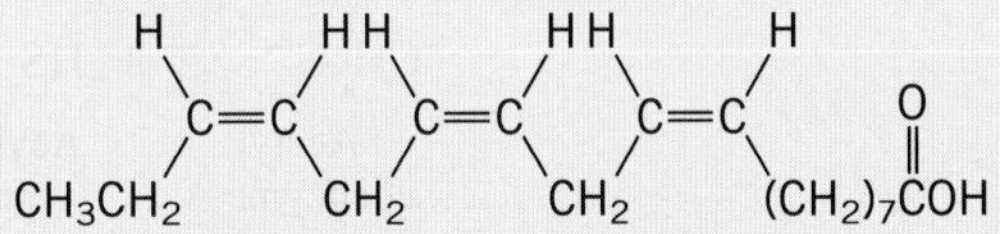	−11
20	Arachidonic acid	*cis,cis,cis,cis*-5,8,11,14-Icosatetraenoic acid	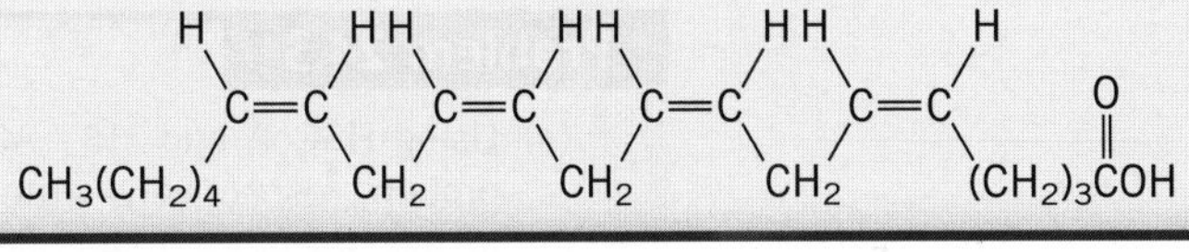	−49

reduce serum cholesterol levels, the trans fats produced by stereoisomerization during partial hydrogenation have cholesterol-raising effects similar to those of saturated fats. Increased consumption of trans fats has been linked to higher levels of coronary artery disease.

Problem 24.2

In addition to the stereoisomerism during catalytic hydrogenation, double bond migration is also observed. Show how esters of *cis*-9-octadecenoic acid are converted to *cis*- and *trans*-8-octadecenoic esters under these conditions.

Fatty acids occur naturally in forms other than as triacylglycerols, and we'll see numerous examples as we go through the chapter. *Anandamide,* for example, is an amide of arachidonic acid (Table 24.1).

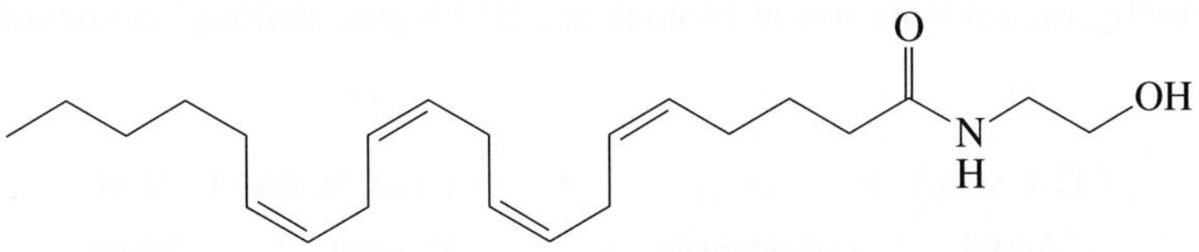

Anandamide

The cannabinoid receptor belongs to a large family of receptor proteins that span the cell membrane, known as G-coupled protein receptors. Membrane receptor proteins are illustrated in Figure 24.4.

Anandamide was isolated from pig's brain and identified as the substance that normally binds to the "cannabinoid receptor." The active component of marijuana, Δ^9-tetrahydrocannabinol (THC, see Section 22.7), must exert its effect by binding to a receptor, and scientists had long wondered what compound in the body was the natural substrate for this binding site. Anandamide is that compound, and it is now probably more appropriate to speak of cannabinoids binding to the anandamide receptor instead of vice versa. Anandamide seems to be involved in moderating pain. Once the identity of the "endogenous cannabinoid" was known, scientists looked specifically for it and found it in some surprising places—chocolate, for example.

Fatty acids are biosynthesized by way of acetyl coenzyme A. The following section outlines the mechanism of fatty acid biosynthesis.

24.3 Fatty Acid Biosynthesis

We can describe the major elements of fatty acid biosynthesis by considering the formation of butanoic acid from two molecules of acetyl coenzyme A. The "machinery" responsible for accomplishing this conversion is a complex of enzymes known as **fatty acid synthetase.** Certain portions of this complex, referred to as *acyl carrier protein (ACP),* bear a side chain that is structurally similar to coenzyme A. Two early stages in fatty acid biosynthesis are needed to assemble the building blocks on the acyl carrier protein. The first stage is the transfer of the acetyl group from a molecule of acetyl coenzyme A to the sulfhydryl group of acyl carrier protein.

$$\underset{\text{Acetyl coenzyme A}}{CH_3\overset{O}{\overset{\|}{C}}SCoA} + \underset{\text{Acyl carrier protein}}{HS{-}ACP} \longrightarrow \underset{S\text{-Acetyl acyl carrier protein}}{CH_3\overset{O}{\overset{\|}{C}}S{-}ACP} + \underset{\text{Coenzyme A}}{HSCoA}$$

Problem 24.3

(a) Using HSCoA and HS—ACP as abbreviations for coenzyme A and acyl carrier protein, respectively, write a structural formula for the tetrahedral intermediate in the preceding reaction.

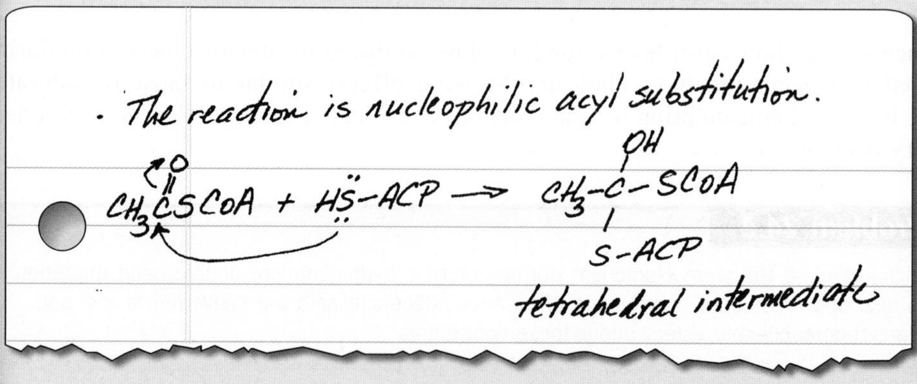

(b) Use curved arrows to show the conversion of the tetrahedral intermediate to products.

In the second stage, another molecule of acetyl coenzyme A reacts with carbon dioxide (actually bicarbonate ion at biological pH) to give malonyl coenzyme A:

$$\underset{\text{Acetyl coenzyme A}}{CH_3\overset{O}{\overset{\|}{C}}SCoA} + \underset{\text{Bicarbonate}}{HCO_3^-} \longrightarrow \underset{\text{Malonyl coenzyme A}}{{}^-O\overset{O}{\overset{\|}{C}}CH_2\overset{O}{\overset{\|}{C}}SCoA} + \underset{\text{Water}}{H_2O}$$

Formation of malonyl coenzyme A is followed by a nucleophilic acyl substitution, which transfers the malonyl group to the acyl carrier protein as a thioester.

$$\underset{\text{Malonyl coenzyme A}}{{}^{-}\mathrm{O\overset{O}{\overset{\|}{C}}CH_2\overset{O}{\overset{\|}{C}}SCoA}} + \underset{\text{Acyl carrier protein}}{\mathrm{HS{-}ACP}} \longrightarrow \underset{\text{S-Malonyl acyl carrier protein}}{{}^{-}\mathrm{O\overset{O}{\overset{\|}{C}}CH_2\overset{O}{\overset{\|}{C}}S{-}ACP}} + \underset{\text{Coenzyme A}}{\mathrm{HSCoA}}$$

When both building blocks are in place on the acyl carrier protein, carbon–carbon bond formation occurs between the α-carbon atom of the malonyl group and the carbonyl carbon of the acetyl group. This is shown in step 1 of Mechanism 24.1. Carbon–carbon bond formation is accompanied by decarboxylation and produces a four-carbon acetoacetyl (3-oxobutanoyl) group bound to acyl carrier protein.

The acetoacetyl group is then transformed to a butanoyl group by the reaction sequence illustrated in steps 2 to 4. Steps 2 and 4 are reductions; step 3 is an alcohol dehydration. The combination of these three steps is the reduction of a ketone C=O to CH_2.

The four carbon atoms of the butanoyl group originate in two molecules of acetyl coenzyme A. Carbon dioxide assists the reaction but is not incorporated into the product. The same carbon dioxide that is used to convert one molecule of acetyl coenzyme A to malonyl coenzyme A is regenerated in the decarboxylation step that accompanies carbon–carbon bond formation.

Successive repetitions of the steps shown in Mechanism 24.1 give unbranched acyl groups having 6, 8, 10, 12, 14, and 16 carbon atoms. In each case, chain extension occurs by reaction with a malonyl group bound to the acyl carrier protein. Thus, the biosynthesis of the 16-carbon acyl group of hexadecanoic (palmitic) acid can be represented by the overall equation:

$$\underset{\text{S-Acetyl acyl carrier protein}}{\mathrm{CH_3\overset{O}{\overset{\|}{C}}S{-}ACP}} + \underset{\text{S-Malonyl acyl carrier protein}}{\mathrm{7HO\overset{O}{\overset{\|}{C}}CH_2\overset{O}{\overset{\|}{C}}S{-}ACP}} + \underset{\text{Reduced form of coenzyme}}{\mathrm{14NADPH}} + \underset{\text{Hydronium ion}}{\mathrm{14H_3O^+}} \longrightarrow$$

$$\underset{\text{S-Hexadecanoyl acyl carrier protein}}{\mathrm{CH_3(CH_2)_{14}\overset{O}{\overset{\|}{C}}S{-}ACP}} + \underset{\text{Carbon dioxide}}{\mathrm{7CO_2}} + \underset{\text{Acyl carrier protein}}{\mathrm{7HS{-}ACP}} + \underset{\text{Oxidized form of coenzyme}}{\mathrm{14NADP^+}} + \underset{\text{Water}}{\mathrm{21H_2O}}$$

Problem 24.4

By analogy to the intermediates given in steps 1–4 of Mechanism 24.1, write the sequence of acyl groups that are attached to the acyl carrier protein in the conversion of

$$\mathrm{CH_3(CH_2)_{12}\overset{O}{\overset{\|}{C}}S{-}ACP} \quad \text{to} \quad \mathrm{CH_3(CH_2)_{14}\overset{O}{\overset{\|}{C}}S{-}ACP}$$

This phase of fatty acid biosynthesis concludes with the transfer of the acyl group from acyl carrier protein to coenzyme A. The resulting acyl coenzyme A molecules can then undergo a number of subsequent biological transformations. One such transformation is chain extension, leading to acyl groups with more than 16 carbons. Another is the introduction of one or more carbon–carbon double bonds. A third is acyl transfer from sulfur to oxygen to form esters such as triacylglycerols. The process by which acyl coenzyme A molecules are converted to triacylglycerols involves a type of intermediate called a *phospholipid* and is discussed in the following section.

Mechanism 24.1

Biosynthesis of a Butanoyl Group from Acetyl and Malonyl Building Blocks

Step 1: An acetyl group is transferred to the α-carbon atom of the malonyl group with evolution of carbon dioxide. Presumably decarboxylation gives an enol, which attacks the acetyl group.

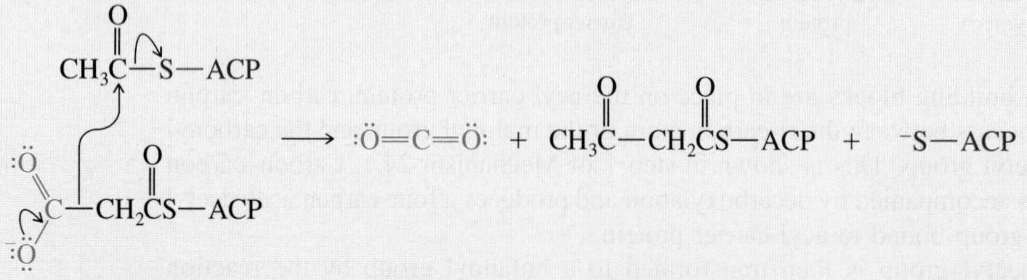

Acetyl and malonyl groups bound to acyl carrier protein — Carbon dioxide — *S*-Acetoacetyl acyl carrier protein — Acyl carrier protein (anionic form)

Step 2: The ketone carbonyl of the acetoacetyl group is reduced to an alcohol function. This reduction requires NADPH as a coenzyme. (NADPH is the phosphate ester of NADH and reacts similarly to it.)

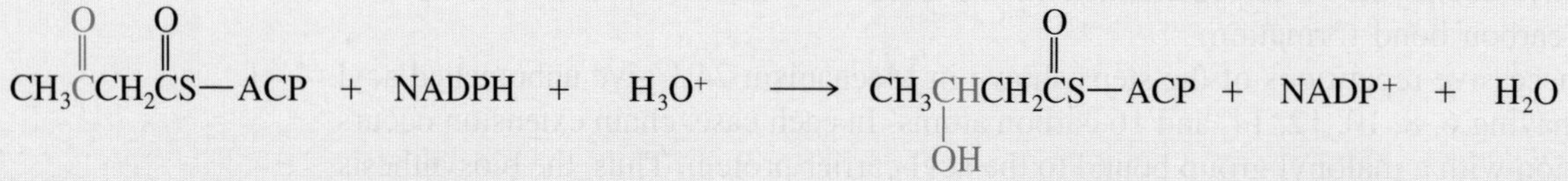

S-Acetoacetyl acyl carrier protein — Reduced form of coenzyme — Hydronium ion — *S*-3-Hydroxybutanoyl acyl carrier protein — Oxidized form of coenzyme — Water

Step 3: Dehydration of the β-hydroxy acyl group.

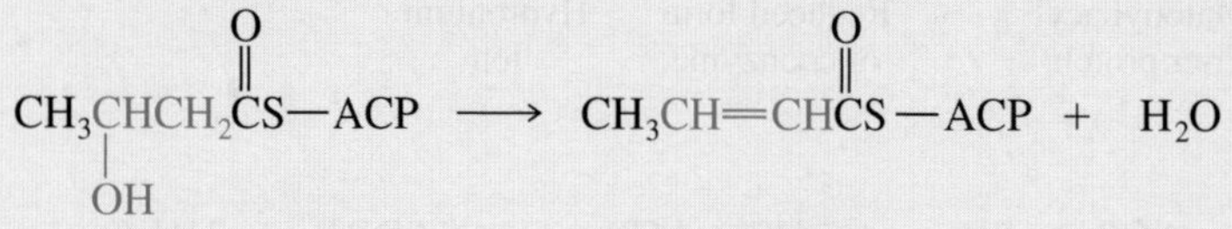

S-3-Hydroxybutanoyl acyl carrier protein — *S*-2-Butenoyl acyl carrier protein — Water

Step 4: Reduction of the double bond of the α,β-unsaturated acyl group. This step requires NADPH as a coenzyme.

S-2-Butenoyl acyl carrier protein — Reduced form of coenzyme — Hydronium ion — *S*-Butanoyl acyl carrier protein — Oxidized form of coenzyme — Water

24.4 Phospholipids

Triacylglycerols arise, not by acylation of glycerol itself, but by a sequence of steps in which the first stage is acyl transfer to L-glycerol 3-phosphate (from reduction of dihydroxyacetone 3-phosphate, formed as described in Section 23.22). The product of this stage is called a **phosphatidic acid.**

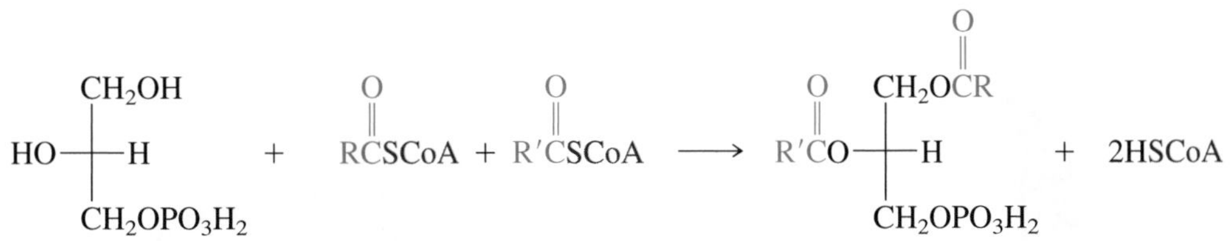

L-Glycerol 3-phosphate — Two acyl coenzyme A molecules (R and R′ may be the same or they may be different) — Phosphatidic acid — Coenzyme A

Problem 24.5

What is the absolute configuration (*R* or *S*) of L-glycerol 3-phosphate? What must be the absolute configuration of the naturally occurring phosphatidic acids biosynthesized from it?

Hydrolysis of the phosphate ester function of the phosphatidic acid gives a diacylglycerol, which then reacts with a third acyl coenzyme A molecule to produce a triacylglycerol.

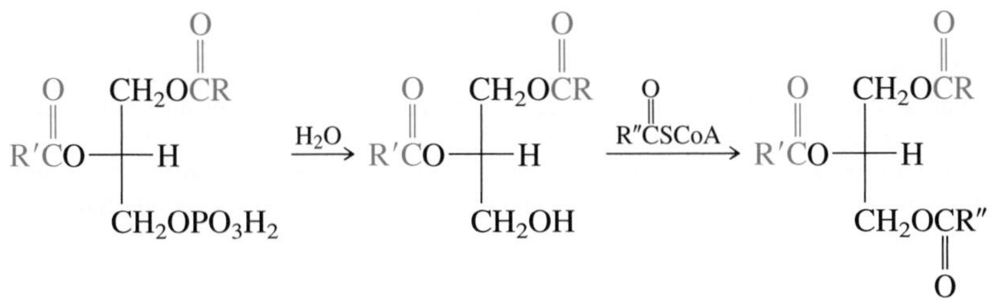

Phosphatidic acid — Diacylglycerol — Triacylglycerol

Phosphatidic acids not only are intermediates in the biosynthesis of triacylglycerols but also are biosynthetic precursors of other members of a group of compounds called *phosphoglycerides* or *glycerol phosphatides.* Phosphorus-containing derivatives of lipids are known as **phospholipids,** and phosphoglycerides are one type of phospholipid.

One important phospholipid is **phosphatidylcholine,** also called *lecithin.* Phosphatidylcholine is a mixture of diesters of phosphoric acid. One ester function is derived from a diacylglycerol, whereas the other is a choline [$—OCH_2CH_2\overset{+}{N}(CH_3)_3$] unit.

Lecithin is added to foods such as mayonnaise as an emulsifying agent to prevent the fat and water from separating into two layers.

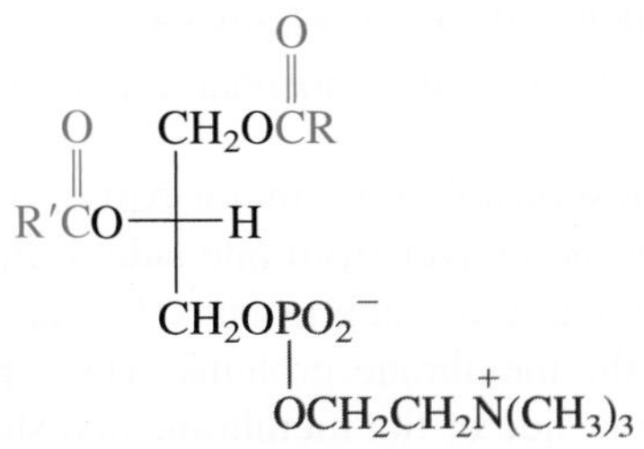

Phosphatidylcholine (R and R′ are usually different)

Phosphatidylcholine possesses a hydrophilic polar "head group" (the positively charged choline and negatively charged phosphate units) and two lipophilic (hydrophobic) nonpolar "tails" (the acyl groups). Under certain conditions, such as at the interface of two aqueous phases, phosphatidylcholine forms what is called a *lipid bilayer,* as

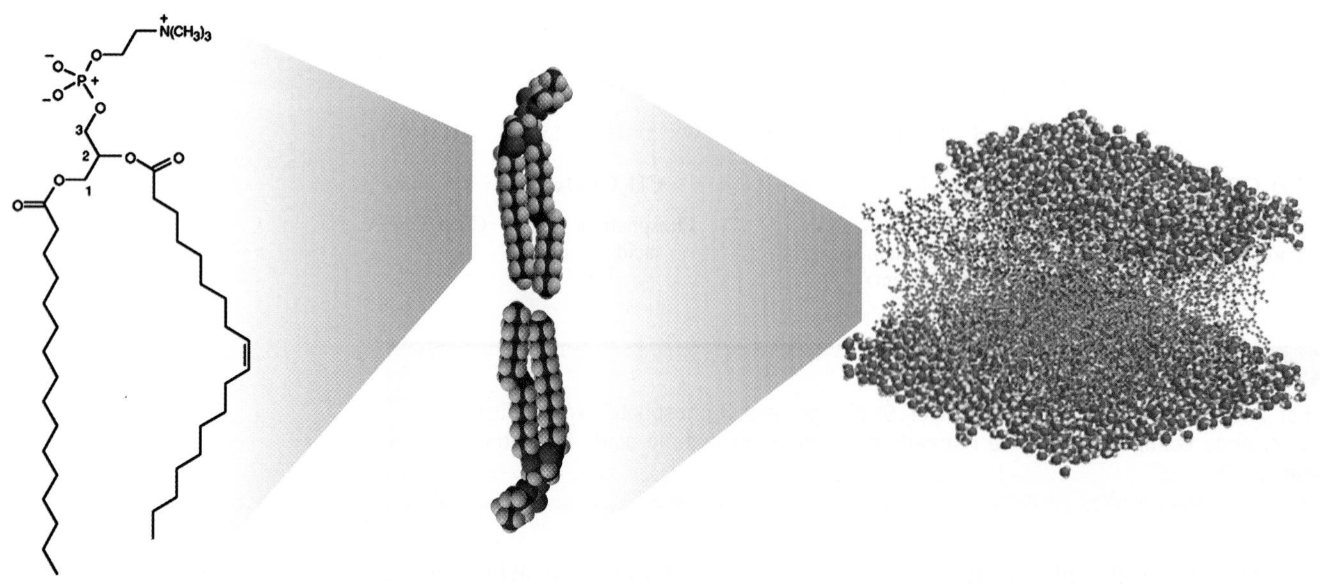

Figure 24.3

(*a*) Phosphatidylcholine. The C-1 and C-2 oxygens of glycerol bear hexadecanoyl (palmityl) and *cis*-9-octadecenoyl (oleyl) groups, respectively; C-3 bears the phosphate ester of choline. (*b*) Two space-filling models of (*a*) oriented so that the polar head group of one points up and the other down. (*c*) A simulation of a phospholipid bilayer. The space-filling models at the top and bottom are water molecules. The polar head groups are in contact with water molecules. The hydrocarbon chains are grey and shown as ball-and-spoke models with the hydrogens omitted. Water molecules are omitted at the upper left corner to make the head groups visible. The simulation is based on the coordinates of H. Heller, M. Schaefer, and K. Schulten, "Molecular Dynamics Simulation of a Bilayer of 200 Lipids in the Gel and in the Liquid-Crystal Phases," *Journal of Physical Chemistry,* 97, 8343–8360 (1993) and taken from an interactive animated tutorial by E. Martz and A. Herráez, "Lipid Bilayers and the Gramicidin Channel" [http://molvis.sdsc.edu/bilayers/index.htm (2001)] by courtesy of Professor Martz.

shown in Figure 24.3. Because there are two long-chain acyl groups in each molecule, the most stable assembly has the polar groups solvated by water molecules at the top and bottom surfaces and the lipophilic acyl groups directed toward the interior of the bilayer.

Phosphatidylcholine is an important component of cell membranes, but cell membranes are more than simply lipid bilayers. Although their composition varies with their source, a typical membrane contains about equal amounts of lipid and protein, and the amount of cholesterol in the lipid fraction can approximate that of phosphatidylcholine. A schematic representation of a cell membrane is shown in Figure 24.4. This diagram is known as the **fluid mosaic model.**

The lipid fraction is responsible for the structure of the membrane. Phosphatidylcholine provides the bilayer that is the barrier between what is inside the cell and what is outside. Cholesterol intermingles with the phosphatidylcholine to confer an extra measure of rigidity to the membrane.

The protein fraction is responsible for a major part of membrane function. Nonpolar materials can diffuse through the bilayer from one side to the other relatively easily, but polar materials, particularly metal ions such as Na^+, K^+, and Ca^{2+} cannot. The transport of metal ions is assisted by the membrane proteins. These proteins pick up a metal ion from the aqueous phase on one side of the membrane and shield it from the hydrophobic environment within the membrane while transporting it to the aqueous phase on the other side of the membrane. Ionophore antibiotics such as monensin (see Section 16.5, Figure 16.3) disrupt the normal functioning of cells by facilitating metal ion transport across cell membranes.

Like cells, only much smaller, spherical objects called **liposomes** are enclosed by a phospholipid bilayer that separates a watery interior from an external (also watery) environment. Liposomes occur naturally but can also be prepared from lecithin as a phosphatidylcholine source. Following their chance discovery in 1961, liposomes

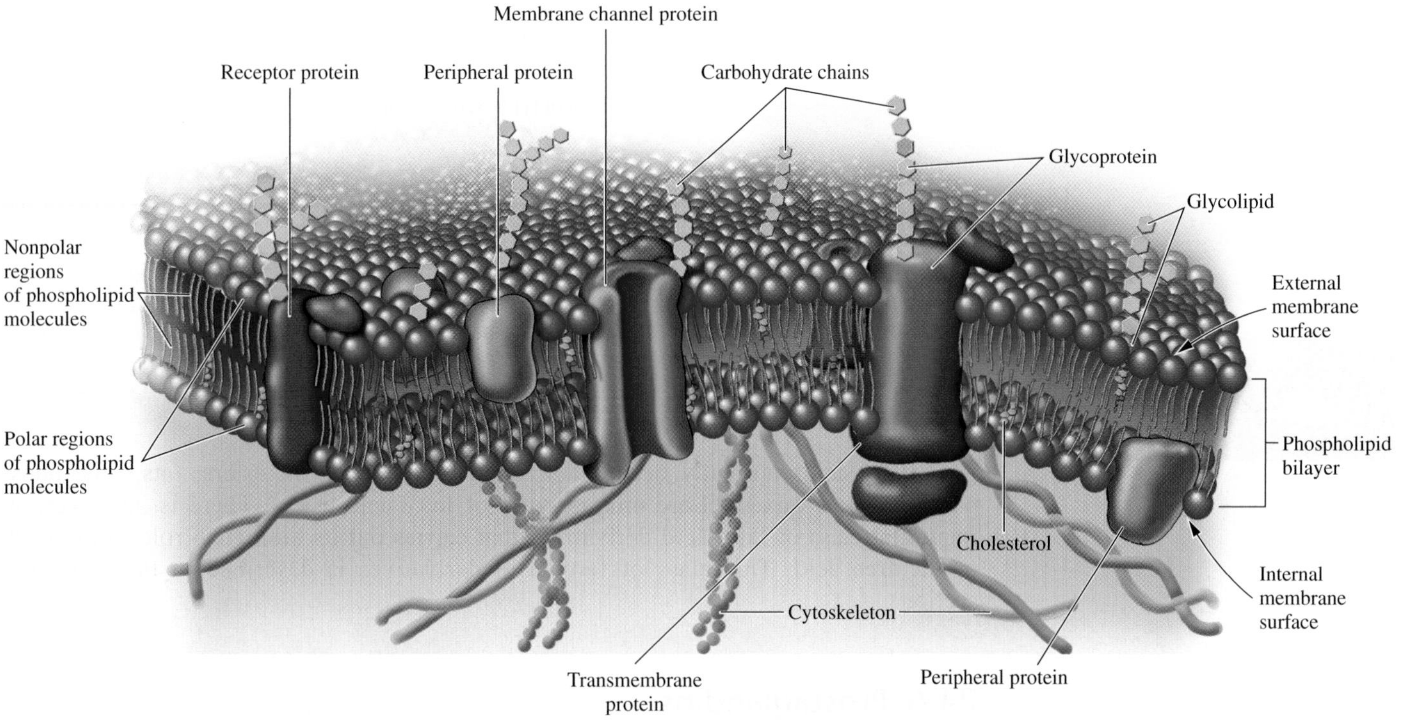

Figure 24.4

Fluid mosaic model of a cell membrane.
Receptor protein: A protein that acts as a receptor toward a hormone, neurotransmitter, or other molecule that can serve as a ligand.
Peripheral protein: A protein that adheres temporarily to the membrane.
Membrane channel protein: Proteins that can form a pore through the membrane, through which ions or other solutes may flow.

originally received attention as models for membrane structure. Subsequently, their use as novel vehicles for drug delivery was demonstrated and has led to important applications in medicine.

Generating a liposome in an aqueous solution containing a water-soluble drug yields the species illustrated in Figure 24.5 in which the drug is encapsulated within the interior of the liposome. When given to a patient, the drug-carrying liposome binds to one of the patient's cells and transfers the drug into the cell. Antitumor agents for cancer chemotherapy and zidovudine (AZT) (Section 26.13) for treatment of HIV-AIDS are among the drugs that are often administered using this technique.

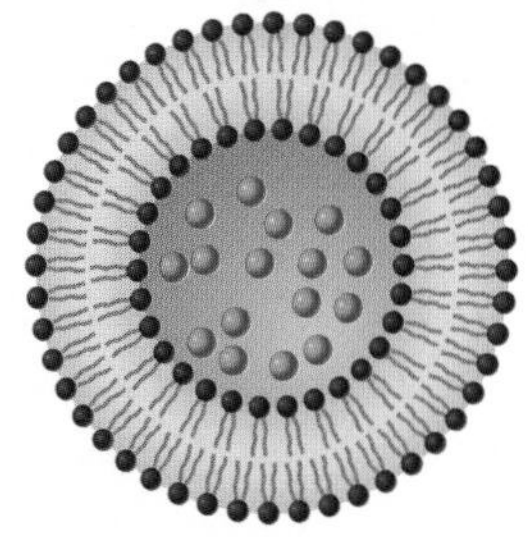

Figure 24.5

A spherical liposome shown in cross section. The membrane is a phospholipid bilayer. The interior is water. The blue spheres represent a water-soluble drug.

Problem 24.6

In addition to bilayers in cell membranes and liposomes, phosphatidylcholine also forms micelles. Compare a typical fatty acid micelle (see Section 18.7) and a phosphatidylcholine micelle, both at pH 7. What is the most significant difference in their surface properties?

24.5 Waxes

Waxes are water-repelling solids that are part of the protective coatings of a number of living things, including the leaves of plants, the fur of animals, and the feathers of birds. They are usually mixtures of esters in which both the alkyl and acyl group are unbranched and contain a dozen or more carbon atoms. Beeswax, for example, contains the ester triacontyl hexadecanoate as one component of a complex mixture of hydrocarbons, alcohols, and esters.

$$CH_3(CH_2)_{14}\overset{\overset{\displaystyle O}{\|}}{C}OCH_2(CH_2)_{28}CH_3$$

Triacontyl hexadecanoate

Problem 24.7

Spermaceti is a wax obtained from the sperm whale. It contains, among other materials, an ester known as *cetyl palmitate,* which is used as an emollient in a number of soaps and cosmetics. The systematic name for cetyl palmitate is *hexadecyl hexadecanoate.* Write a structural formula for this substance.

Fatty acids normally occur naturally as components of esters; fats, oils, phospholipids, and waxes all are unique types of fatty acid esters. There is, however, an important class of fatty acid derivatives that carries out its biological role in the form of the free acid. This class of fatty acid derivatives is described in the following section.

24.6 Prostaglandins

Research in physiology carried out in the 1930s established that the lipid fraction of semen contains small amounts of substances that exert powerful effects on smooth muscle. Sheep prostate glands proved to be a convenient source of this material and yielded a mixture of structurally related substances referred to collectively as **prostaglandins.** We now know that prostaglandins are present in almost all animal tissues, where they carry out a variety of regulatory functions. Among these functions are relaxation or constriction of bronchial muscles, platelet aggregation or disaggregation, induction of labor, and the regulation of inflammation.

Much of the fundamental work on prostaglandins was carried out by Sune Bergström and Bengt Samuelsson of the Karolinska Institute (Sweden) and by Sir John Vane of the Wellcome Foundation (Great Britain). These three shared the Nobel Prize for Physiology or Medicine in 1982.

Prostaglandins are extremely potent substances and exert their physiological effects at very small concentrations. Because of this, their isolation was difficult, and it was not until 1960 that the first members of this class, designated PGE_1 and $PGF_{1\alpha}$ (see Figure 24.6), were obtained as pure compounds. More than a dozen structurally related prostaglandins have since been isolated and identified. All the prostaglandins are 20-carbon carboxylic acids and contain a cyclopentane ring. All have hydroxyl groups at C-11 and C-15 (for the numbering of the positions in prostaglandins, Figure 24.6). Prostaglandins belonging to the F series have an additional hydroxyl group at C-9, and a carbonyl function is present at this position in the various PGEs. The subscript numerals in their abbreviated names indicate the number of double bonds.

Arachidonic acid gets its name from *arachidic acid,* the saturated C_{20} fatty acid isolated from peanut (*Arachis hypogaea*) oil.

Prostaglandins arise from unsaturated C_{20}-carboxylic acids such as arachidonic acid (see Table 24.1). Mammals cannot biosynthesize arachidonic acid directly. They obtain linoleic acid (see Table 24.1) from vegetable oils in their diet and extend the carbon chain of linoleic acid from 18 to 20 carbons while introducing two more

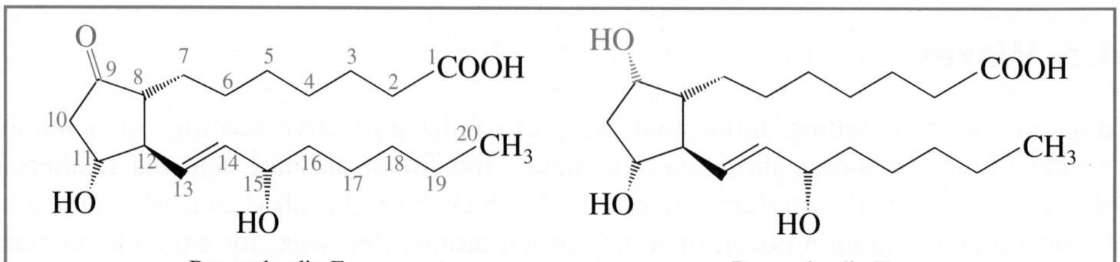

Prostaglandin E_1 (PGE_1)

Prostaglandin $F_{1\alpha}$ ($PGF_{1\alpha}$)

Figure 24.6

Structures of two representative prostaglandins. The numbering scheme is illustrated in the structure of PGE_1.

double bonds. Linoleic acid is said to be an **essential fatty acid,** forming part of the dietary requirement of mammals. Animals fed on diets that are deficient in linoleic acid grow poorly and suffer a number of other disorders, some of which are reversed on feeding them vegetable oils rich in linoleic acid and other *polyunsaturated fatty acids.* One function of these substances is to provide the raw materials for prostaglandin biosynthesis.

Studies of the biosynthesis of PGE_2 from arachidonic acid have shown that the three new oxygens come from O_2. The enzyme involved, *prostaglandin endoperoxide synthase,* has *cyclooxygenase* (COX) activity and catalyzes the reaction of arachidonic acids with O_2 to give an *endoperoxide* (PGG_2).

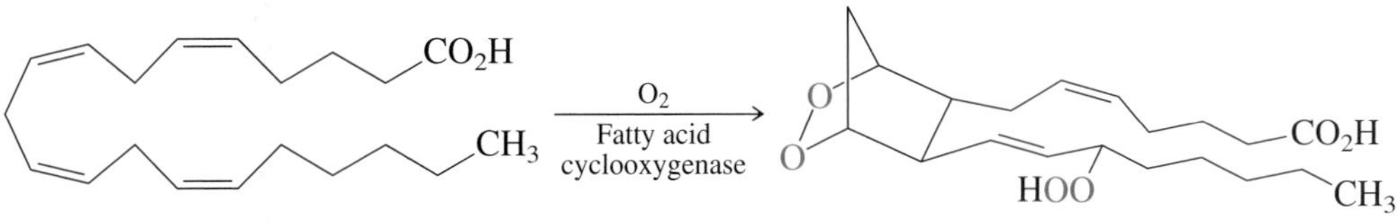

Arachidonic acid $\quad$ PGG_2

In the next step, the —OOH group of PGG_2 is reduced to an alcohol function. Again, prostaglandin endoperoxide synthase is the enzyme responsible. The product of this step is called PGH_2.

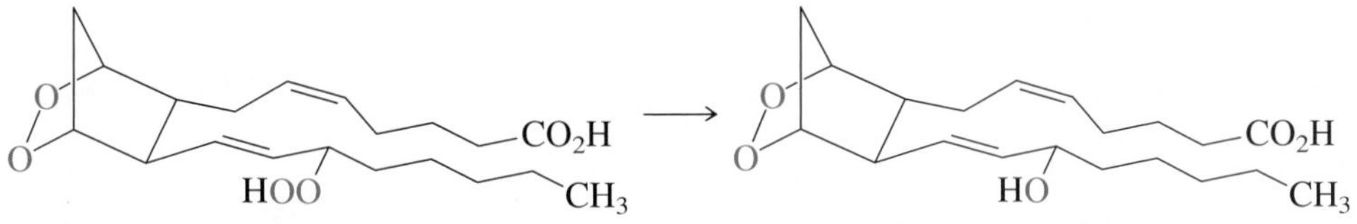

PGG_2 $\quad$ PGH_2

PGH_2 is the precursor to a number of prostaglandins and related compounds, depending on the enzyme that acts on it. One of these cleaves the O—O bond of the endoperoxide and gives PGE_2.

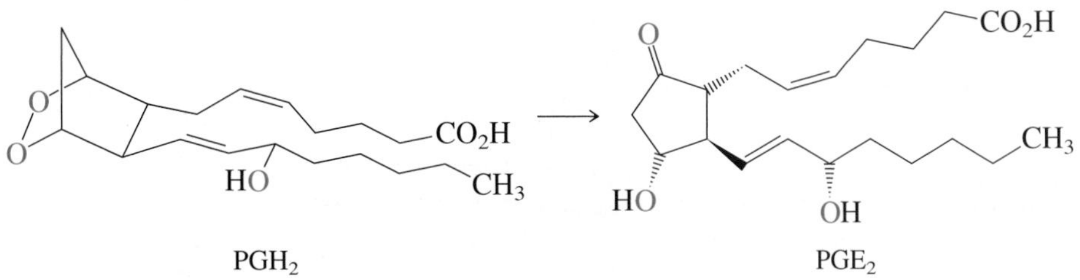

PGH_2 $\quad$ PGE_2

Before leaving this biosynthetic scheme, notice that PGE_2 has four chirality centers. Even though arachidonic acid is achiral, only the stereoisomer shown as PGE_2 in the equation is formed. Moreover, it is formed as a single enantiomer. Like most enzyme-catalyzed reactions, the transformations in the biosynthesis of PGE_2 are enantioselective.

Problem 24.8

Write the structural formula and give the IUPAC name for the fatty acid from which PGE_1 is biosynthesized. The structure of PGE_1 is shown in Figure 24.6

Prostaglandins belong to a group of compounds that, because they are related to icosanoic acid [$CH_3(CH_2)_{18}CO_2H$)], are collectively known as **icosanoids.** The other icosanoids are *thromboxanes, prostacyclins,* and *leukotrienes.*

Older versions of the IUPAC rules called the unbranched carboxylic acid with 20 carbon atoms *eicosanoic acid.* Consequently, icosanoids are often referred to as *eicosanoids.*

Nonsteroidal Antiinflammatory Drugs (NSAIDs) and COX-2 Inhibitors

An injection of the steroid cortisone (Section 24.14) is often effective for reducing the pain and inflammation that comes from an injury. But chronic pain and inflammation, such as occurs with arthritis, is better managed with an orally administered remedy. Enter nonsteroidal antiinflammatory drugs (NSAIDs).

Aspirin (Section 22.10) is the oldest and best known NSAID. Over the years it has been joined by many others, a few of which are:

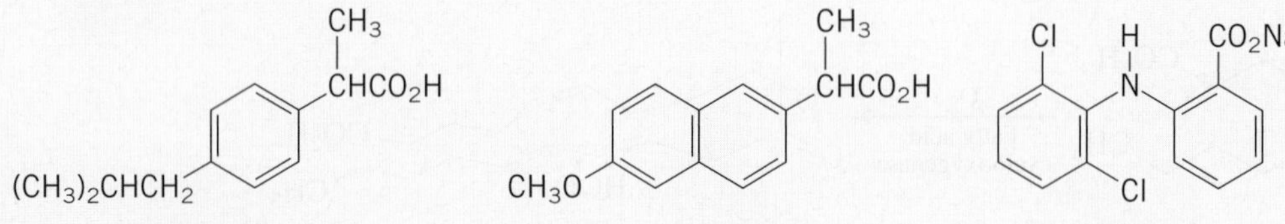

Ibuprofen Naproxen Diclofenac

The long-standing question of how aspirin works has been answered in terms of its effect on prostaglandin biosynthesis. Prostaglandins are made continuously in all mammalian cells and serve a variety of functions. They are biosynthesized in greater amounts at sites of tissue damage, and it is they that cause the pain and inflammation we feel. Cells contain two forms of the cyclooxygenase enzyme, COX-1 and COX-2; both catalyze prostaglandin biosynthesis. Some of the prostaglandins produced with the aid of COX-1 are involved in protecting the stomach and kidneys. COX-2 is concentrated in injured tissue where it works to catalyze the biosynthesis of the prostaglandins responsible for inflammation. Aspirin inhibits prostaglandin biosynthesis by inactivating both COX-1 and COX-2. Although inhibition of COX-2 has the desired effect of relieving pain and inflammation, inhibition of COX-1 causes irritation of the stomach lining.

A good antiinflammatory drug, therefore, will selectively inactivate COX-2 while leaving COX-1 untouched. Aspirin fails this test. In fact, aspirin is about ten times more effective toward inactivating the "wrong" COX. Beginning in the late 1990s, new antiinflammatory drugs became available that selectively inactivate the right one, COX-2. Two of these COX-2 inhibitors are *rofecoxib* and *celecoxib.*

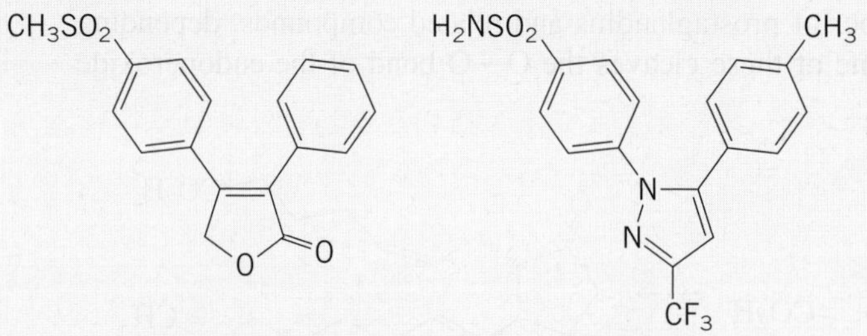

Rofecoxib (Vioxx) Celecoxib (Celebrex)

None of the NSAIDs mentioned so far has a structure that bears any resemblance to a typical lipid, yet all interact with enzymes that have lipids as their substrates. The classical period of drug development emphasized testing a large number of unrelated compounds for biological activity, identifying the structural features believed to be associated with the desired activity, then synthesizing and testing numerous analogs. The most recently developed NSAIDs, the COX-2 inhibitors rofecoxib and celecoxib, were developed with the intent of targeting the COX-2 enzyme for inactivation. They emerged by a combination of the classical "prepare and test" strategy and molecular modeling. Models of the three-dimensional structures of COX-1 and COX-2 were examined to guide thinking about the kinds of structural units that a drug should have to selectively inactivate COX-2. Although the goal of COX-2 inhibition was achieved and led to widely prescribed drugs, the situation changed abruptly in the fall of 2004 when Merck withdrew Vioxx from the market because of evidence indicating it increased the risk of heart attack and stroke. An ideal drug in this class will need to have greater specificity for target tissues, such as the joints in arthritis.

Thromboxane A_2 (TXA_2) promotes platelet aggregation and blood clotting. The biosynthetic pathway to TXA_2 is the same as that of PGE_2 up to PGH_2. At that point separate pathways lead to PGE_2 and TXA_2.

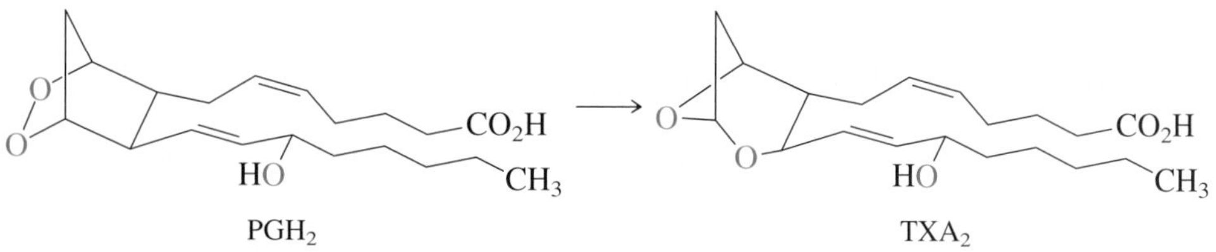

Prostacyclin I_2 (PGI_2) inhibits platelet aggregation and relaxes coronary arteries. Like PGE_2 and TXA_2, it is formed from arachidonic acid via PGH_2.

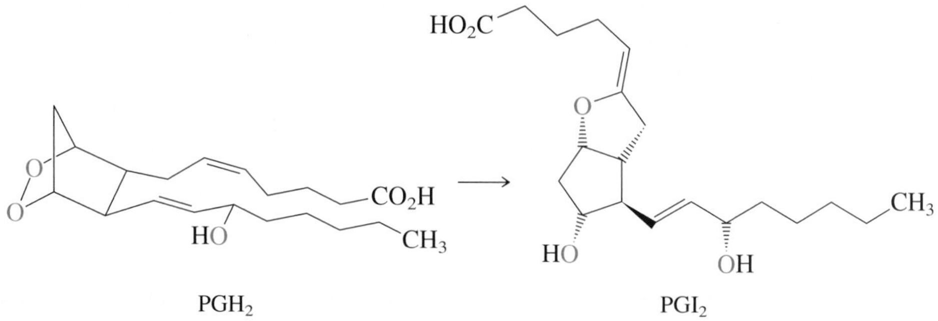

Leukotrienes are the substances most responsible for the constriction of bronchial passages during asthma attacks. They arise from arachidonic acid by a pathway different from the one that leads to prostaglandins and related compounds. The pathway to leukotrienes does not involve cyclooxygenation. Instead, oxidation simply introduces —OH groups at specific carbons along the chain. Allylic radicals are involved and some of the double bonds in the product are in different locations than those in arachidonic acid. The enzymes involved are known as *lipoxygenases* and are differentiated according to the carbon of the chain that is oxidized. The biosynthesis of the leukotriene shown begins with a 5-lipoxygenase-catalyzed oxidation of arachidonic acid.

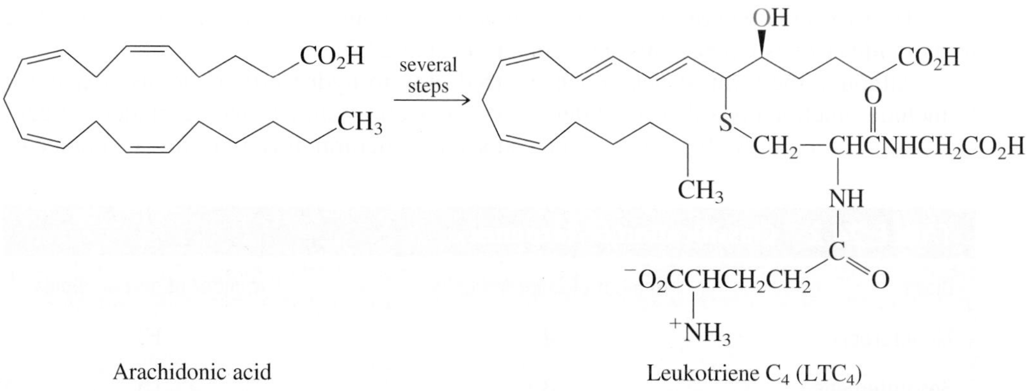

Problem 24.9

The carbon-sulfur bond in LTC_4 is formed by the reaction of glutathione (see Section 15.12) with leukotriene A_4 (LTA_4). LTA_4 is an epoxide. Suggest a reasonable structure for LTA_4.

Most of the drugs such as epinephrine and albuterol used to treat asthma attacks are *bronchodilators*—substances that expand the bronchial passages. Newer drugs are designed to either inhibit the 5-lipoxygenase enzyme, which acts on arachidonic acid in the first stage of leukotriene biosynthesis, or to block leukotriene receptors.

24.7 Terpenes: The Isoprene Rule

A bayberry (wax myrtle) plant.

The word *essential* as applied to naturally occurring organic substances can have two different meanings. With respect to fatty acids, *essential* means "necessary." Linoleic acid is an "essential" fatty acid; it must be included in the diet for animals to grow properly because they lack the ability to biosynthesize it directly.

Essential is also used as the adjective form of the noun *essence*. The mixtures of substances that make up the fragrant material of plants are called **essential oils** because they contain the essence, that is, the odor, of the plant. The study of the composition of essential oils ranks as one of the oldest areas of organic chemical research. Very often, the principal volatile component of an essential oil belongs to a class of chemical substances called the **terpenes.**

Myrcene, a hydrocarbon isolated from bayberry oil, is a typical terpene:

$$(CH_3)_2C{=}CHCH_2CH_2\overset{\overset{\displaystyle CH_2}{\|}}{C}CH{=}CH_2 \equiv$$

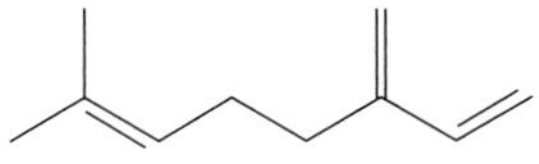

Myrcene

The structural feature that distinguishes terpenes from other natural products is the **isoprene unit.** The carbon skeleton of myrcene (exclusive of its double bonds) corresponds to the head-to-tail union of two isoprene units.

$$H_2C{=}\overset{\overset{\displaystyle CH_3}{|}}{C}{-}CH{=}CH_2 \equiv$$

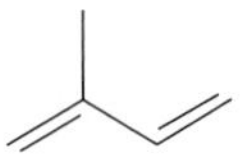

Isoprene
(2-methyl-1,3-butadiene)

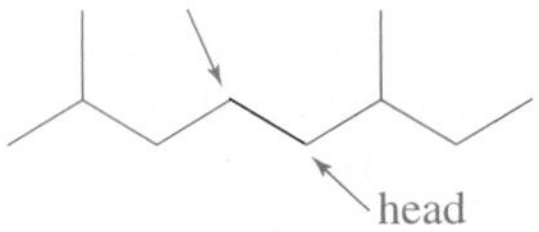

Two isoprene units
linked head-to-tail

Wallach was awarded the 1910 Nobel Prize in Chemistry.

The German chemist Otto Wallach determined the structures of many terpenes and is credited with setting forth the **isoprene rule:** terpenes are repeating assemblies of isoprene units, normally joined head-to-tail.

Terpenes are often referred to as *isoprenoid* compounds and are classified according to the number of isoprene units they contain (Table 24.2).

There are more than 23,000 known isoprenoid compounds.

Although the term *terpene* once referred only to hydrocarbons, its use expanded to include functionally substituted derivatives as well, grouped together under the general term *isoprenoids.* Figure 24.7 presents the structural formulas for a number of

TABLE 24.2 Classification of Terpenes

Class	Number of isoprene units	Number of carbon atoms
Monoterpene	2	10
Sesquiterpene	3	15
Diterpene	4	20
Sesterpene	5	25
Triterpene	6	30
Tetraterpene	8	40

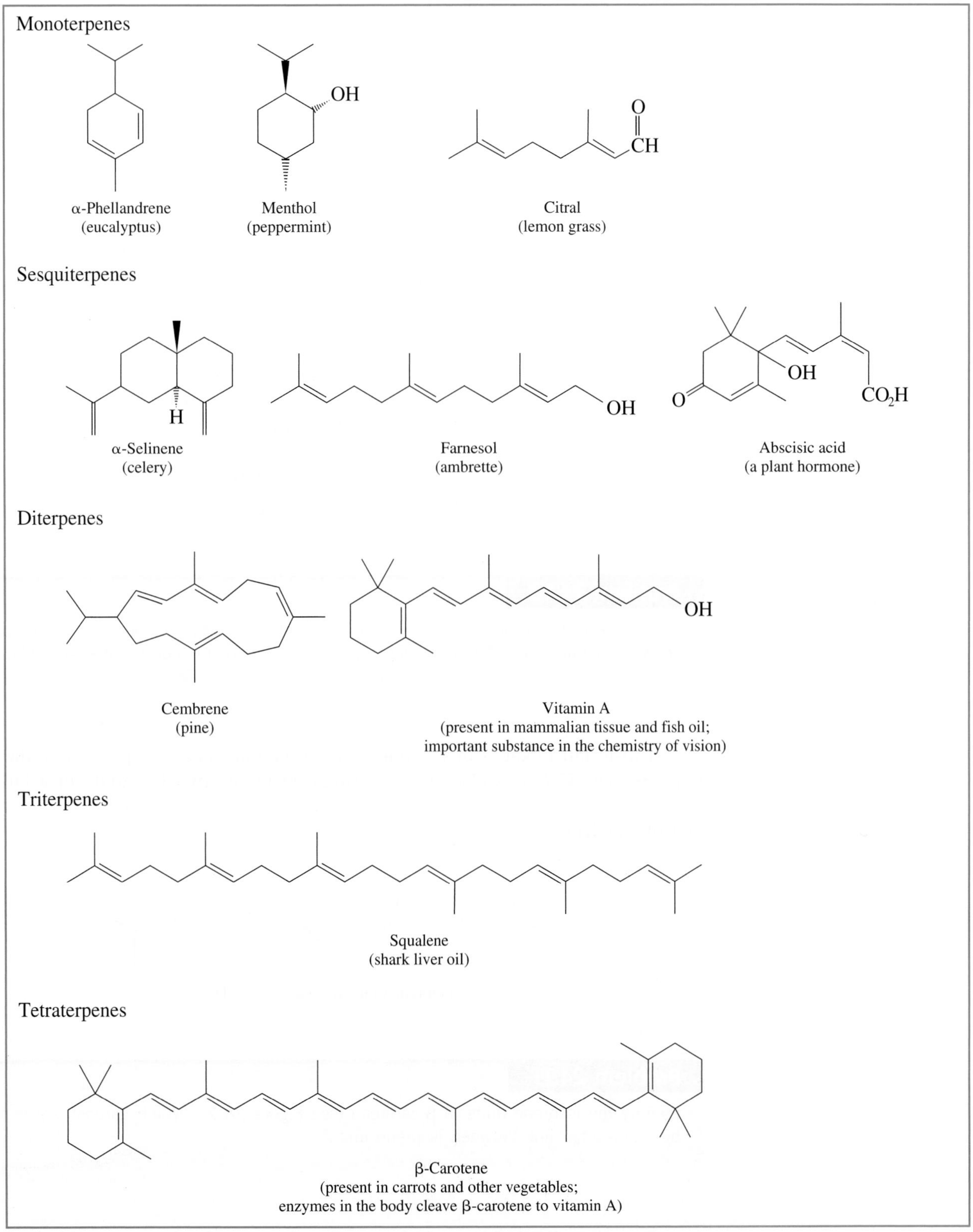

Figure 24.7

Some representative terpenes and related natural products. Structures are customarily depicted as carbon skeleton formulas when describing compounds of isoprenoid origin.

representative examples. The isoprene units in some of these are relatively easy to identify. The three isoprene units in the sesquiterpene *farnesol,* for example, are indicated as follows in color. They are joined in a head-to-tail fashion.

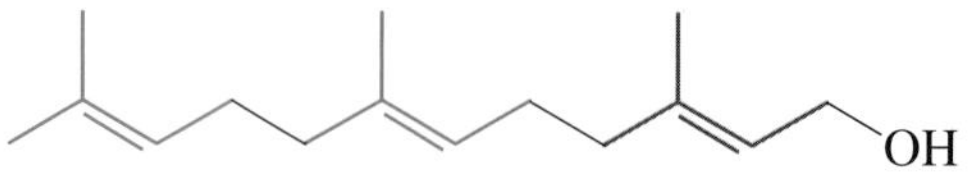

Isoprene units in farnesol ($C_{15}H_{26}O$)

Many terpenes contain one or more rings, but these also can be viewed as collections of isoprene units. An example is α-selinene. Like farnesol, it is made up of three isoprene units linked head-to-tail.

In locating isoprene units within a given carbon skeleton, keep in mind that the double bonds may no longer be present.

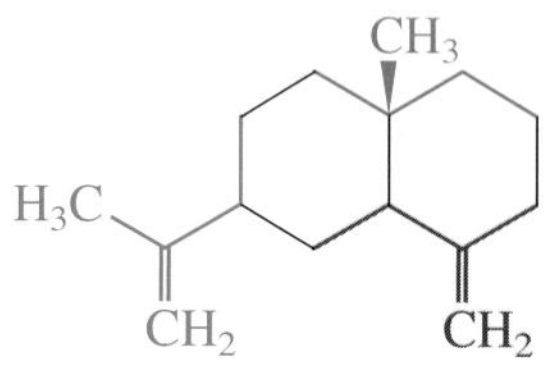

Isoprene units in α-selinene ($C_{15}H_{24}$)

Problem 24.10

Locate the isoprene units in each of the monoterpenes, sesquiterpenes, and diterpenes shown in Figure 24.7. (In some cases there are two equally correct arrangements.)

Tail-to-tail linkages of isoprene units sometimes occur, especially in the higher terpenes. The C(12)—C(13) bond of squalene unites two C_{15} units in a tail-to-tail manner. Notice, however, that isoprene units are joined head-to-tail within each C_{15} unit of squalene.

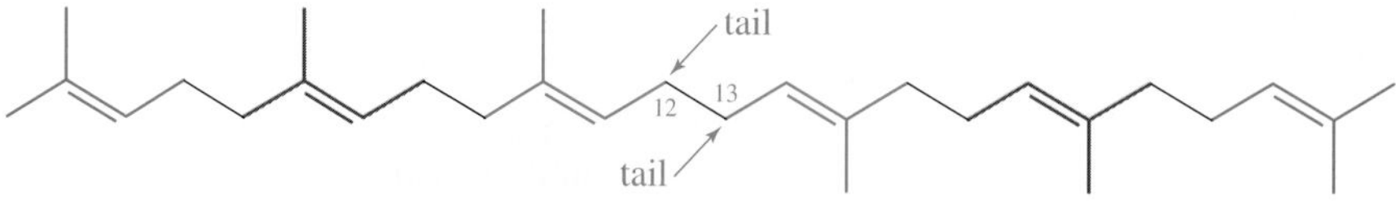

Isoprene units in squalene ($C_{30}H_{50}$)

Problem 24.11

Identify the isoprene units in β-carotene (see Figure 24.7). Which carbons are joined by a tail-to-tail link between isoprene units?

Ruzicka was a corecipient of the 1939 Nobel Prize in Chemistry.

Over time, Wallach's original isoprene rule was refined, most notably by Leopold Ruzicka of the Swiss Federal Institute of Technology (Zürich), who put forward a *biological isoprene rule* in which he connected the various classes of terpenes according to their biological precursors. Thus arose the idea of the *biological isoprene unit.* Isoprene is the fundamental structural unit of terpenes and related compounds, but isoprene does not occur naturally—at least in places where biosynthesis is going on. What then is the biological isoprene unit, how is this unit itself biosynthesized, and how do individual isoprene units combine to give terpenes?

24.8 Isopentenyl Diphosphate: The Biological Isoprene Unit

Isoprenoid compounds are biosynthesized from acetate by a process that involves several stages. The first stage is the formation of mevalonic acid from three molecules of acetic acid:

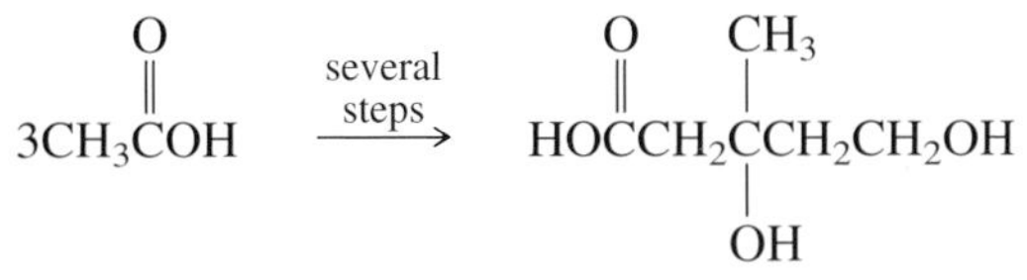

Acetic acid Mevalonic acid

In the second stage, mevalonic acid is converted to isopentenyl diphosphate:

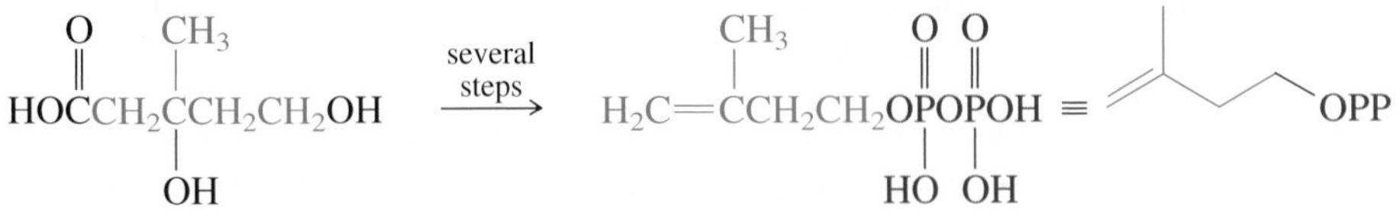

Mevalonic acid Isopentenyl diphosphate

It is convenient to use the symbol —OPP to represent the diphosphate group. Diphosphate is also known as pyrophosphate.

Isopentenyl diphosphate is the biological isoprene unit; it contains five carbon atoms connected in the same order as in isoprene.

In the presence of the enzyme *isopentenyl diphosphate isomerase,* isopentenyl diphosphate is converted to dimethylallyl diphosphate. The isomerization involves two successive proton transfers: one from an acidic site of the enzyme (Enz—H) to the double bond to give a tertiary carbocation; the other is deprotonation of the carbocation by a basic site of the enzyme to generate the double bond of dimethylallyl diphosphate.

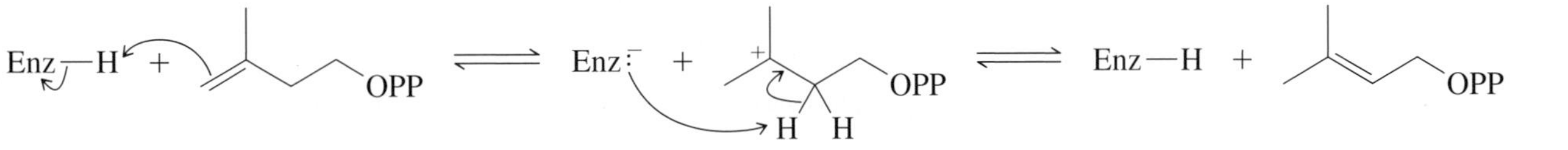

Isopentenyl diphosphate Carbocation intermediate Dimethylallyl diphosphate

Isopentenyl diphosphate and dimethylallyl diphosphate are structurally similar—both contain a double bond and a diphosphate ester unit—but the chemical reactivity expressed by each is different. The principal site of reaction in dimethylallyl diphosphate is the carbon that bears the diphosphate group. Diphosphate is a reasonably good leaving group in nucleophilic substitution reactions, especially when, as in dimethylallyl diphosphate, it is located at an allylic carbon. Isopentenyl diphosphate, on the other hand, does not have its leaving group attached to an allylic carbon and is far less reactive than dimethylallyl diphosphate toward nucleophilic reagents. The principal site of reaction in isopentenyl diphosphate is the carbon–carbon double bond, which, like the double bonds of simple alkenes, is reactive toward electrophiles.

24.9 Carbon–Carbon Bond Formation in Terpene Biosynthesis

The chemical properties of isopentenyl diphosphate and dimethylallyl diphosphate are complementary in a way that permits them to react with each other to form a carbon–carbon bond that unites two isoprene units. In broad outline, the enzyme-catalyzed process involves bond formation between the allylic CH_2 of dimethylallyl diphosphate and the vinylic CH_2 of isopentenyl diphosphate. Diphosphate is the leaving group and a tertiary carbocation results.

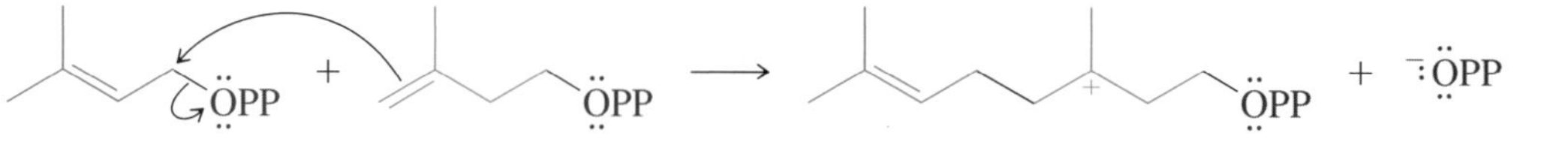

Dimethylallyl diphosphate · Isopentenyl diphosphate · Ten-carbon carbocation · Diphosphate ion

Alternatively, ionization of dimethylallyl diphosphate could precede carbon–carbon bond formation.

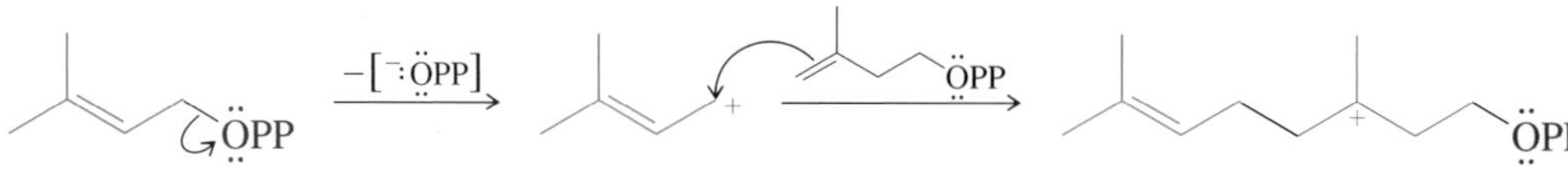

Dimethylallyl diphosphate · Dimethylallyl cation · Ten-carbon carbocation

The ten-carbon carbocation that results is the same regardless of whether it is formed in one step or two. Once formed it can react in several different ways, all of which are familiar to us as typical carbocation processes. One is deprotonation to give the carbon–carbon double bond of *geranyl diphosphate.*

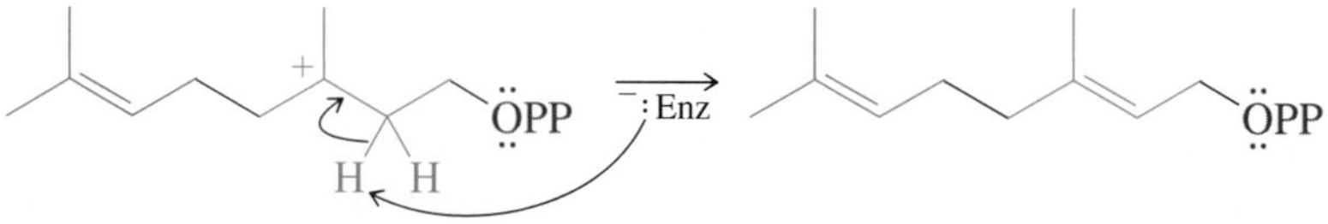

Ten-carbon carbocation · Geranyl diphosphate

Hydrolysis of geranyl diphosphate gives *geraniol,* a pleasant-smelling monoterpene found in rose oil.

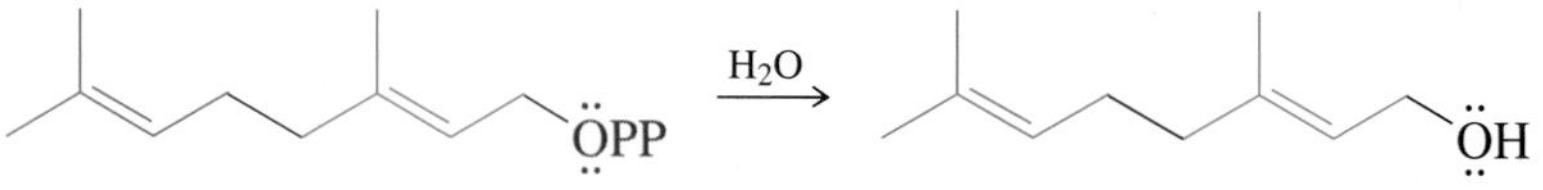

Geranyl diphosphate · Geraniol

Geranyl diphosphate is an allylic diphosphate and, like dimethylallyl diphosphate, can react with isopentenyl diphosphate. A 15-carbon carbocation is formed, which on deprotonation gives *farnesyl diphosphate.* Hydrolysis of farnesyl diphosphate gives the sesquiterpene *farnesol.*

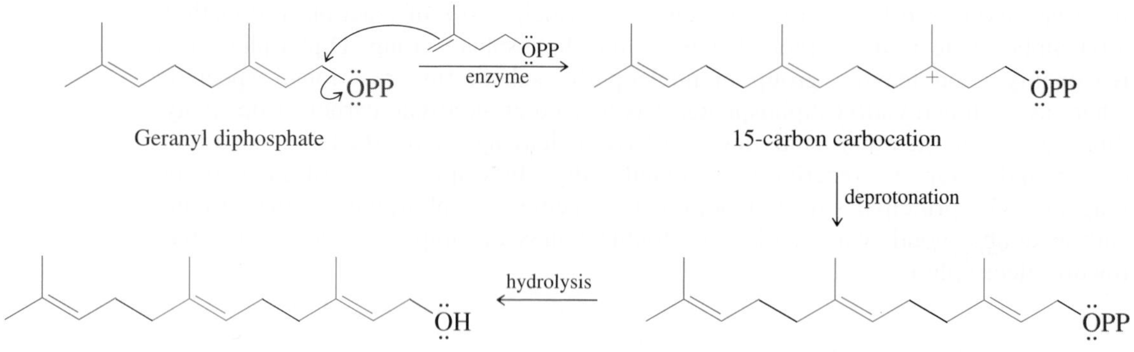

Geranyl diphosphate · 15-carbon carbocation · Farnesyl diphosphate · Farnesol

Repeating the process produces the diterpene geranylgeraniol from farnesyl diphosphate.

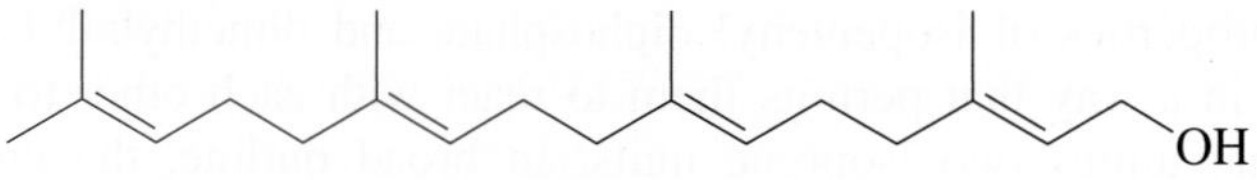

Geranylgeraniol

Problem 24.12

Write a sequence of reactions that describes the formation of geranylgeraniol from farnesyl diphosphate.

Geraniol, farnesol, and geranylgeraniol are classified as **prenols,** compounds of the type:

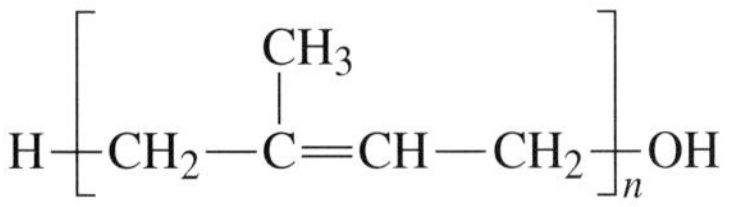

The group to which the OH (or other substituent) is attached is called a **prenyl** group.

The higher terpenes are formed not by successive addition of C_5 units but by the coupling of simpler terpenes. Thus, the triterpenes (C_{30}) are derived from two molecules of farnesyl diphosphate, and the tetraterpenes (C_{40}) from two molecules of geranylgeranyl diphosphate. These carbon–carbon bond-forming processes involve tail-to-tail couplings and proceed by a more complicated mechanism than that just described.

The enzyme-catalyzed reactions that lead to geraniol and farnesol (as their diphosphate esters) are mechanistically related to the acid-catalyzed dimerization of alkenes discussed in Section 6.21. The reaction of an allylic diphosphate or a carbocation with a source of π electrons is a recurring theme in terpene biosynthesis and is invoked to explain the origin of more complicated structural types. Consider, for example, the formation of cyclic monoterpenes. *Neryl diphosphate,* formed by an enzyme-catalyzed isomerization of the *E* double bond in geranyl diphosphate, has the proper geometry to form a six-membered ring via intramolecular attack of the double bond on the allylic diphosphate unit.

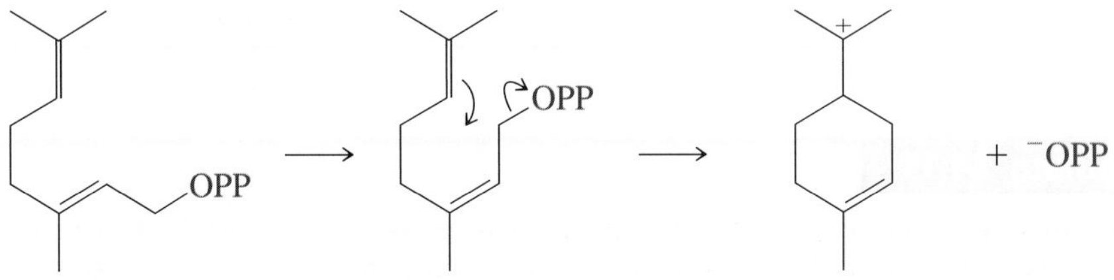

Geranyl diphosphate Neryl diphosphate Tertiary carbocation

Loss of a proton from the tertiary carbocation formed in this step gives *limonene,* an abundant natural product found in many citrus fruits. Capture of the carbocation by water gives *α-terpineol,* also a known natural product.

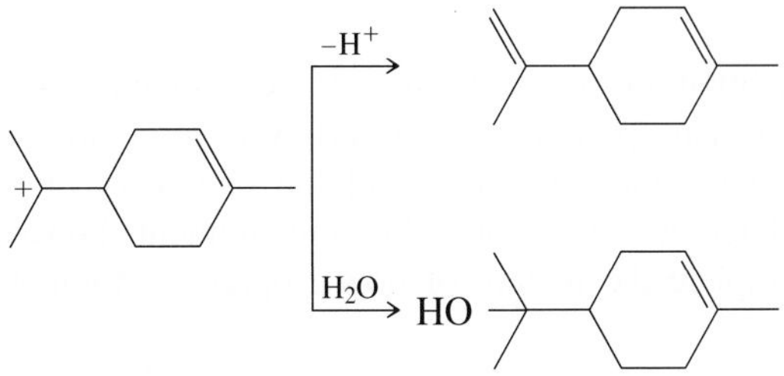

Limonene

α-Terpineol

The same tertiary carbocation serves as the precursor to numerous bicyclic monoterpenes. A carbocation having a bicyclic skeleton is formed by intramolecular attack of the π electrons of the double bond on the positively charged carbon.

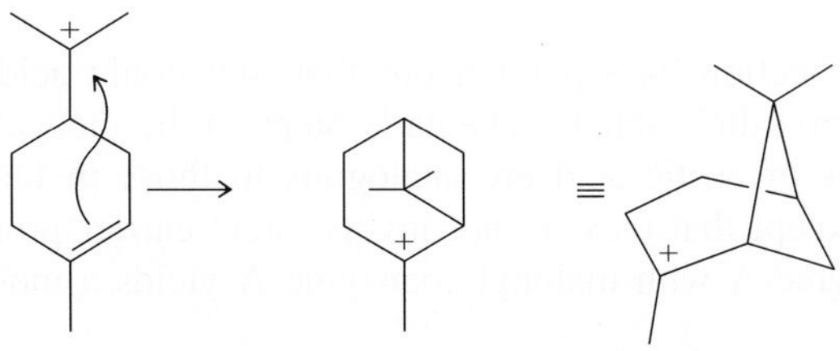

Bicyclic carbocation

This bicyclic carbocation then undergoes many reactions typical of carbocation intermediates to provide a variety of bicyclic monoterpenes, as outlined in Figure 24.8.

Figure 24.8

Two of the reaction pathways available to the C_{10} bicyclic carbocation formed from neryl diphosphate. The same carbocation can lead to monoterpenes based on either the bicyclo[3.1.1] or the bicyclo[2.2.1] carbon skeleton.

A. Loss of a proton from the bicyclic carbocation yields α-pinene and β-pinene. The pinenes are the most abundant of the monoterpenes. They are the main constituents of turpentine.

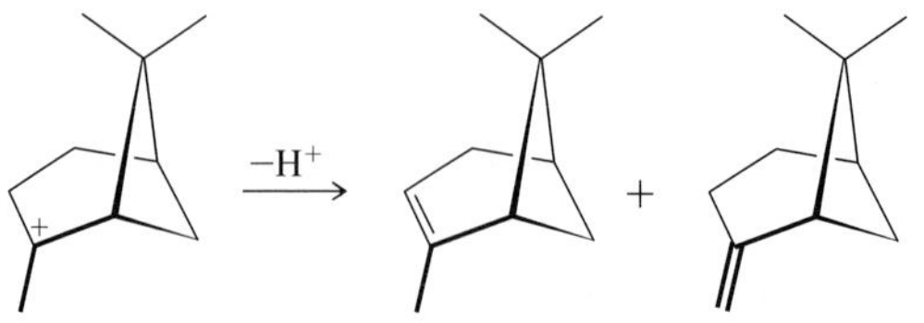

B. Capture of the carbocation by water, accompanied by rearrangement of the bicyclo-[3.1.1] carbon skeleton to a bicyclo[2.2.1] unit, yields borneol. Borneol is found in the essential oil of certain trees that grow in Indonesia.

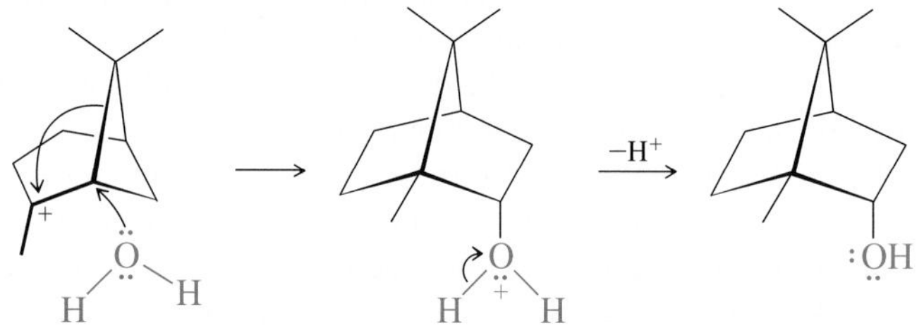

Problem 24.13

The structure of the bicyclic monoterpene borneol is shown in Figure 24.8. Isoborneol, a stereoisomer of borneol, can be prepared in the laboratory by a two-step sequence. In the first step, borneol is oxidized to camphor by treatment with chromic acid. In the second step, camphor is reduced with sodium borohydride to a mixture of 85% isoborneol and 15% borneol. On the basis of these transformations, deduce structural formulas for isoborneol and camphor.

Analogous processes involving cyclizations and rearrangements of carbocations derived from farnesyl diphosphate produce a rich variety of structural types in the sesquiterpene series. We will have more to say about the chemistry of higher terpenes, especially the triterpenes, later in this chapter. For the moment, however, let's return to smaller molecules to complete the picture of how isoprenoid compounds arise from acetate.

24.10 The Pathway from Acetate to Isopentenyl Diphosphate

The introduction to Section 24.8 pointed out that mevalonic acid is the biosynthetic precursor of isopentenyl diphosphate. The early steps in the biosynthesis of mevalonate from three molecules of acetic acid are analogous to those in fatty acid biosynthesis (see Section 24.3) except that they do not involve acyl carrier protein. Thus, the reaction of acetyl coenzyme A with malonyl coenzyme A yields a molecule of acetoacetyl coenzyme A.

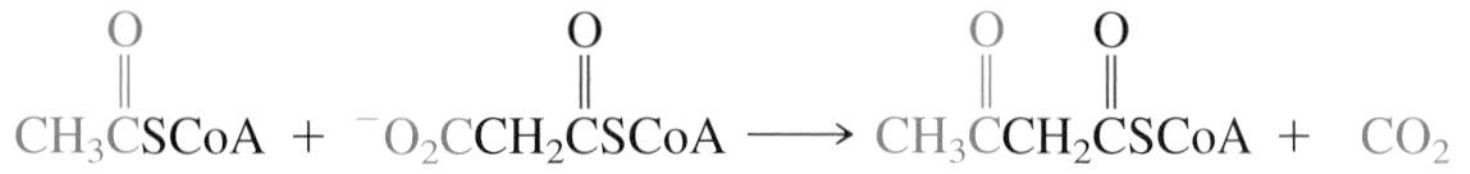

Acetyl coenzyme A | Malonyl coenzyme A | Acetoacetyl coenzyme A | Carbon dioxide

Carbon–carbon bond formation then occurs between the ketone carbonyl of acetoacetyl coenzyme A and the α carbon of a molecule of acetyl coenzyme A.

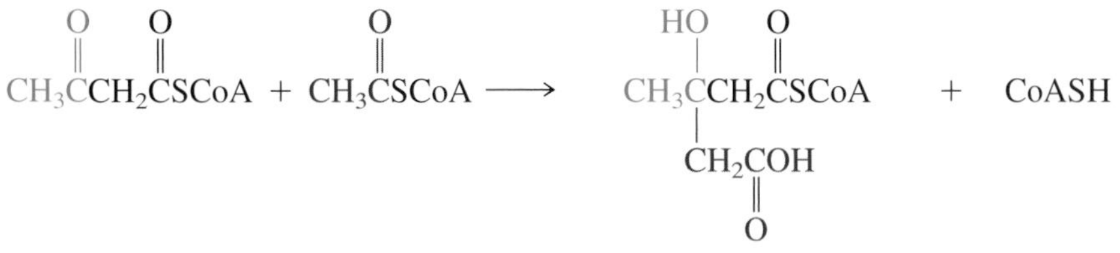

Acetoacetyl coenzyme A | Acetyl coenzyme A | 3-Hydroxy-3-methylglutaryl coenzyme A (HMG CoA) | Coenzyme A

The product of this reaction, 3-hydroxy-3-methylglutaryl coenzyme A (HMG CoA), has the carbon skeleton of mevalonic acid and is converted to it by enzyme-catalyzed reduction.

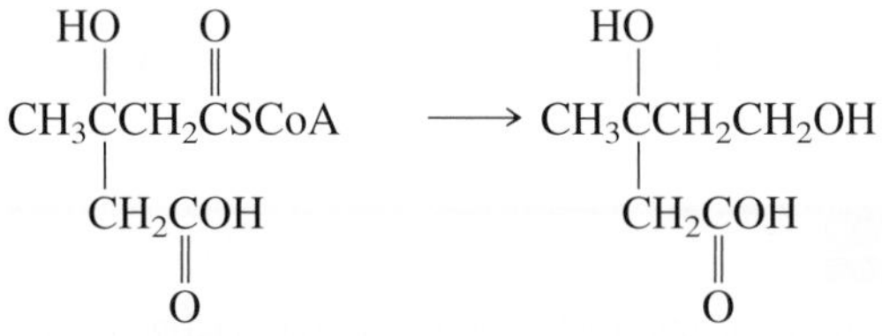

3-Hydroxy-3-methylglutaryl coenzyme A (HMG CoA) | Mevalonic acid

Some of the most effective cholesterol-lowering drugs act by inhibiting the enzyme that catalyzes this reaction.

In keeping with its biogenetic origin in three molecules of acetic acid, mevalonic acid has six carbon atoms. The conversion of mevalonate to isopentenyl diphosphate involves loss of the "extra" carbon as carbon dioxide. First, the alcohol hydroxyl groups of mevalonate are converted to phosphate ester functions—they are enzymatically *phosphorylated,* with introduction of a simple phosphate at the tertiary site and a diphosphate at the primary site. Decarboxylation, in concert with loss of the tertiary phosphate, introduces a carbon–carbon double bond and gives isopentenyl diphosphate, the fundamental building block for formation of isoprenoid natural products.

Some bacteria, algae, and plants make isopentenyl diphosphate by a different route.

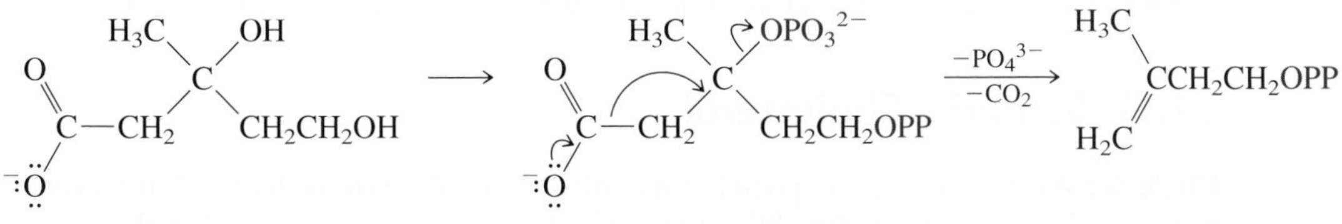

Mevalonate | (Unstable; undergoes rapid decarboxylation with loss of phosphate) | Isopentenyl diphosphate

Much of what we know concerning the pathway from acetate to mevalonate to isopentenyl diphosphate to terpenes comes from "feeding" experiments, in which plants are grown in the presence of radioactively labeled organic substances and the distribution of the radioactive label is determined in the products of biosynthesis. To illustrate, eucalyptus plants were allowed to grow in a medium containing acetic acid enriched with ^{14}C in its methyl group. *Citronellal* was isolated from the mixture of monoterpenes produced by the plants and shown, by a series of chemical degradations, to contain the radioactive ^{14}C label at carbons 2, 4, 6, and 8, as well as at the carbons of both branching methyl groups.

Citronellal occurs naturally as the principal component of citronella oil and is used as an insect repellent.

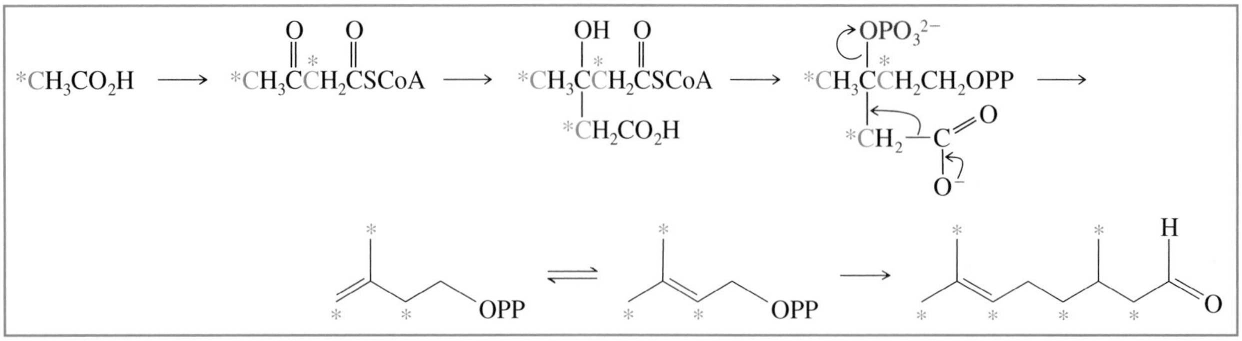

Figure 24.9

The distribution of the ^{14}C label in citronellal biosynthesized from acetate in which the methyl carbon was isotopically enriched with ^{14}C.

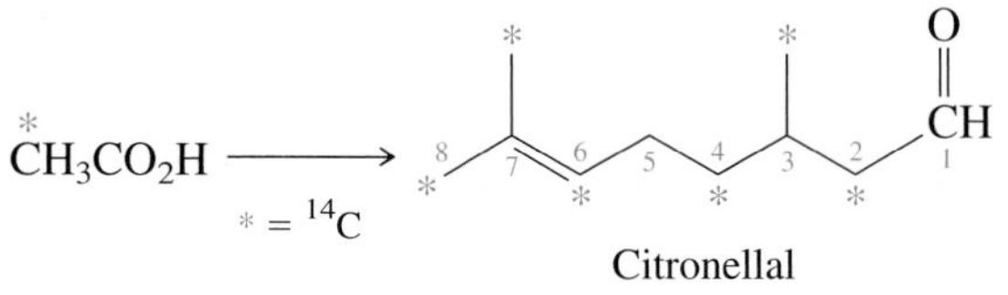

Figure 24.9 traces the ^{14}C label from its origin in acetic acid to its experimentally determined distribution in citronellal.

Problem 24.14

How many carbon atoms of citronellal would be radioactively labeled if the acetic acid used in the experiment were enriched with ^{14}C at C-1 instead of at C-2? Identify these carbon atoms.

A more recent experimental technique employs ^{13}C as the isotopic label. Instead of locating the position of a ^{14}C label by a laborious degradation procedure, the ^{13}C NMR spectrum of the natural product is recorded. The signals for the carbons that are enriched in ^{13}C are far more intense than those corresponding to carbons in which ^{13}C is present only at the natural abundance level.

Isotope incorporation experiments have demonstrated the essential correctness of the scheme presented in this and preceding sections for terpene biosynthesis. Considerable effort has been expended toward its detailed elaboration because of the common biosynthetic origin of terpenes and another class of acetate-derived natural products, the steroids.

24.11 Steroids: Cholesterol

Cholesterol is the central compound in any discussion of **steroids.** Its name is a combination of the Greek words for "bile" (*chole*) and "solid" (*stereos*) preceding the characteristic alcohol suffix *-ol.* It is the most abundant steroid present in humans and the most important one as well because all other steroids arise from it. An average adult has over 200 g of cholesterol; it is found in almost all body tissues, with relatively large amounts present in the brain and spinal cord and in gallstones. Cholesterol is the chief constituent of the plaque that builds up on the walls of arteries in atherosclerosis.

Cholesterol was isolated in the eighteenth century, but its structure is so complex that its correct constitution was not determined until 1932 and its stereochemistry not verified until 1955. Steroids are characterized by the tetracyclic ring system shown in Figure 24.10*a*. As shown in Figure 24.10*b*, cholesterol contains this tetracyclic skeleton modified to include an alcohol function at C-3, a double bond at C-5, methyl groups at C-10 and C-13, and a C_8H_{17} side chain at C-17. Isoprene units may be discerned in various portions of the cholesterol molecule, but the overall correspondence with the

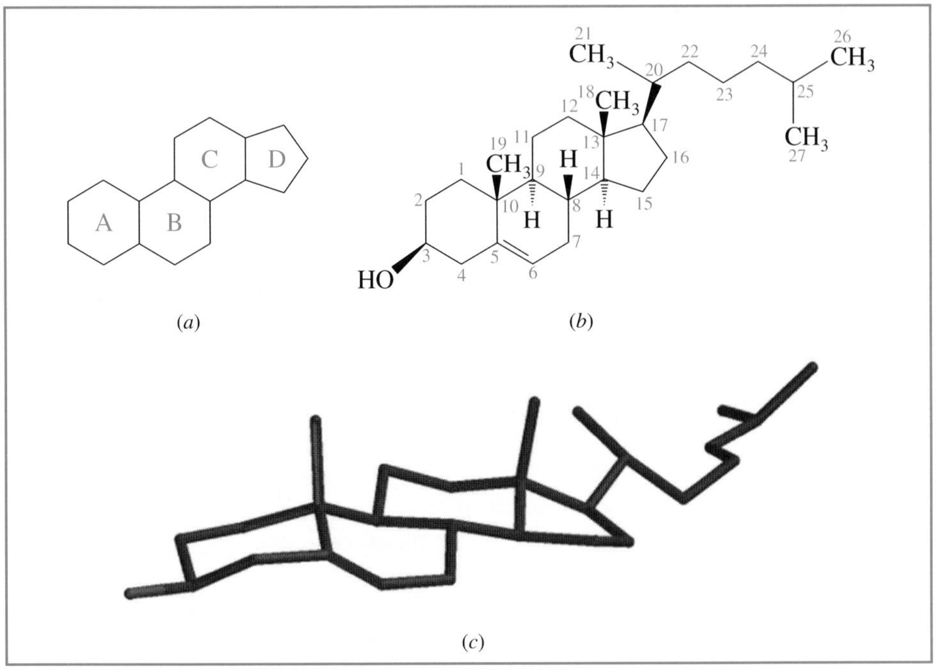

Figure 24.10

(*a*) The tetracyclic ring system characteristic of steroids. The rings are designated A, B, C, and D as shown. (*b*) and (*c*) The structure of cholesterol. A unique numbering system is used for steroids and is indicated in the structural formula.

isoprene rule is far from perfect. Indeed, cholesterol has only 27 carbon atoms, three too few for it to be classed as a triterpene.

Animals accumulate cholesterol from their diet, but are also able to biosynthesize it from acetate. The pioneering work that identified the key intermediates in the complicated pathway of cholesterol biosynthesis was carried out by Konrad Bloch (Harvard) and Feodor Lynen (Munich). An important discovery was that the triterpene **squalene** (see Figure 24.7) is an intermediate in the formation of cholesterol from acetate. Thus, *the early stages of cholesterol biosynthesis are the same as those of terpene biosynthesis* described in Sections 24.8–24.10. In fact, a significant fraction of our knowledge of terpene biosynthesis is a direct result of experiments carried out in the area of steroid biosynthesis.

Bloch and Lynen shared the 1964 Nobel Prize for Physiology or Medicine.

How does the tetracyclic steroid cholesterol arise from the acyclic triterpene squalene? It begins with the epoxidation of squalene described earlier in Section 16.14 and continues from that point in Mechanism 24.2. Step 1 is an enzyme-catalyzed electrophilic ring opening of squalene 2,3-epoxide. Epoxide ring opening triggers a series of carbocation reactions. These carbocation processes involve cyclization via carbon–carbon bond formation (step 1), ring expansion via a carbocation rearrangement (step 2), another cyclization (step 3), followed by a cascade of methyl group migrations and hydride shifts (step 4). The result of all these steps is the tetracyclic triterpene *lanosterol.* Step 5 of Mechanism 24.2 summarizes the numerous remaining transformations by which lanosterol is converted to cholesterol.

Lanosterol is one component of lanolin, a mixture of many substances that coats the wool of sheep.

Problem 24.15

The biosynthesis of cholesterol as outlined in Mechanism 24.2 is admittedly quite complicated. It will aid your understanding of the process if you consider the following questions:

(a) Which carbon atoms of squalene 2,3-epoxide correspond to the doubly bonded carbons of cholesterol?

(b) Which two hydrogen atoms of squalene 2,3-epoxide are the ones that migrate in step 4?

(c) Which methyl group of squalene 2,3-epoxide becomes the methyl group at the C,D ring junction of cholesterol?

(d) What three methyl groups of squalene 2,3-epoxide are lost during the conversion of lanosterol to cholesterol?

Continued

Mechanism 24.2

Biosynthesis of Cholesterol from Squalene

The biosynthetic conversion of squalene to cholesterol proceeds through lanosterol. Lanosterol is formed by enzyme-catalyzed cyclization of the 2,3-epoxide of squalene.

Step 1: An electrophilic species, shown here as $^{+}$Enz—H, catalyzes ring opening of squalene 2,3-epoxide. Ring opening is accompanied by cyclization to give a tricyclic tertiary carbocation. It is not known whether formation of the three new carbon–carbon bonds occurs in a single step or a series of steps.

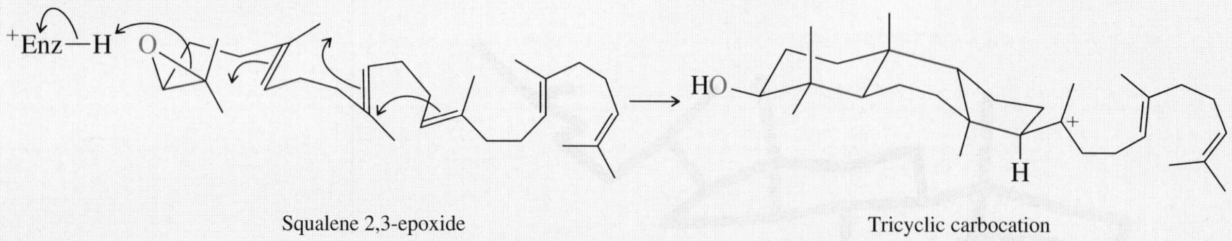

Step 2: Ring expansion converts the five-membered ring of the carbocation formed in step 1 to a six-membered ring.

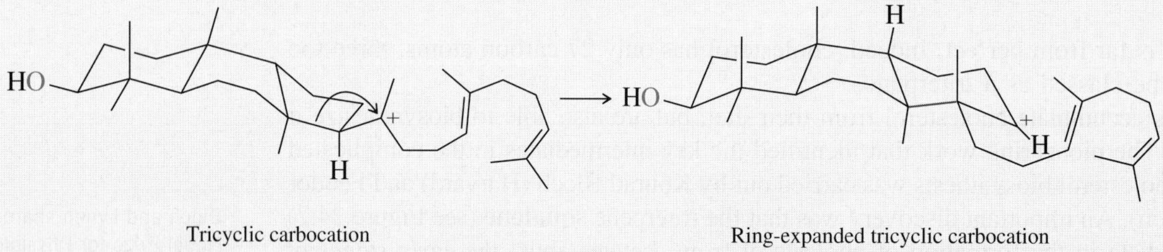

Step 3: Cyclization of the carbocation formed in step 2 gives a tetracyclic carbocation (*protosteryl cation*).

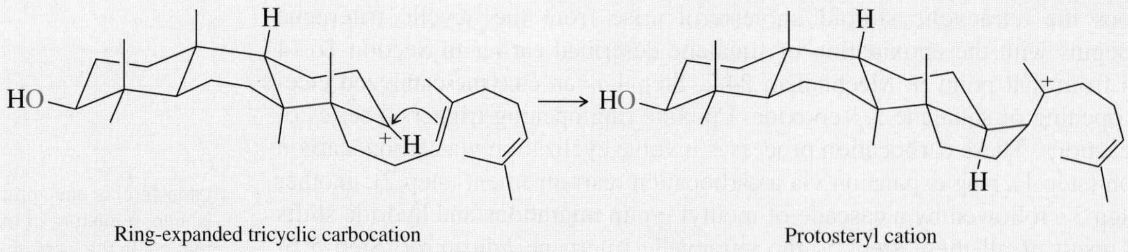

Step 4: Rearrangement and deprotonation of protosteryl cation gives the tetracyclic triterpene lanosterol.

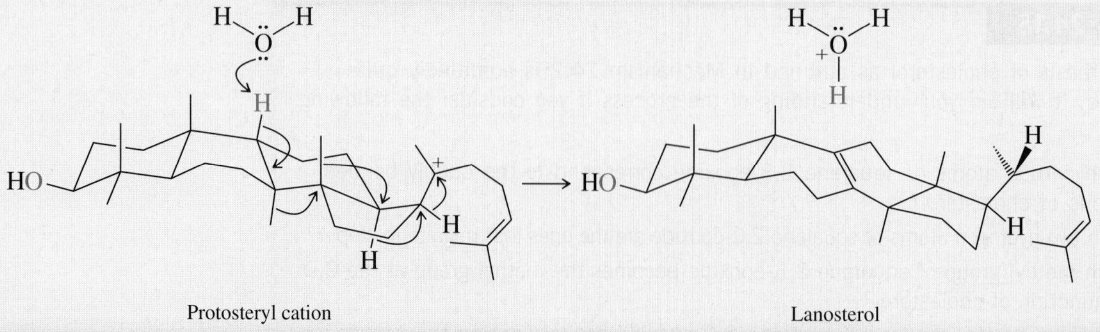

Step 5: A series of enzyme-catalyzed reactions converts lanosterol to cholesterol. The methyl groups at C-4 and C-14 are lost, the C-8 and C-24 double bonds are reduced, and a new double bond is introduced at C-5.

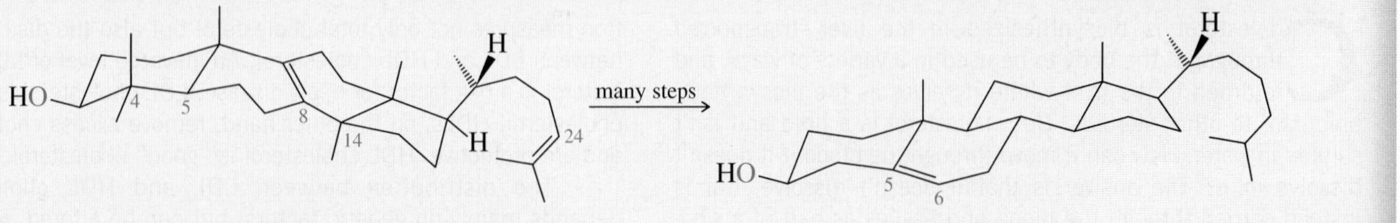

Lanosterol Cholesterol

Sample Solution (a) As the structural formula in step 5 of Mechanism 24.2 indicates, the double bond of cholesterol unites C-5 and C-6 (steroid numbering). The corresponding carbons in the cyclization reaction of step 1 in the figure may be identified as C-7 and C-8 of squalene 2,3-epoxide (systematic IUPAC numbering).

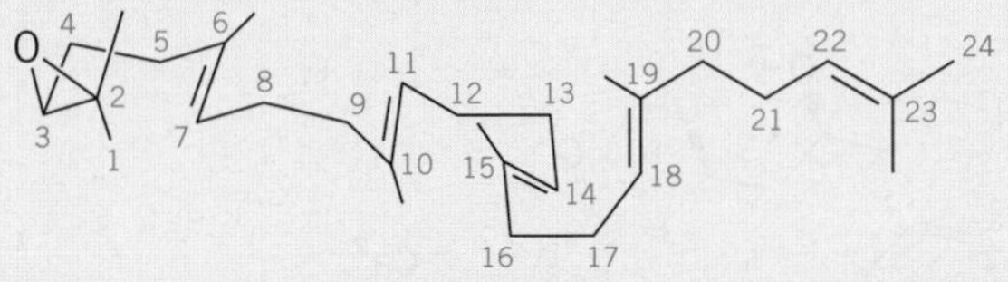

Squalene 2,3-epoxide

Problem 24.16

The biosynthetic pathway shown in Mechanism 24.2 was developed with the aid of isotopic labeling experiments. Which carbon atoms of cholesterol would you expect to be labeled when acetate enriched with ^{14}C in its methyl group ($^{14}CH_3COOH$) is used as the carbon source?

Once formed, cholesterol undergoes a number of biochemical transformations. A very common one is acylation of its C-3 hydroxyl group by reaction with coenzyme A derivatives of fatty acids. Other processes convert cholesterol to the biologically important steroids described in the following sections.

24.12 Vitamin D

A steroid very closely related structurally to cholesterol is its 7-dehydro derivative. 7-Dehydrocholesterol is formed by enzymatic oxidation of cholesterol and has a conjugated diene unit in its B ring. 7-Dehydrocholesterol is present in the tissues of the skin, where it is transformed to vitamin D_3 by a sunlight-induced photochemical reaction.

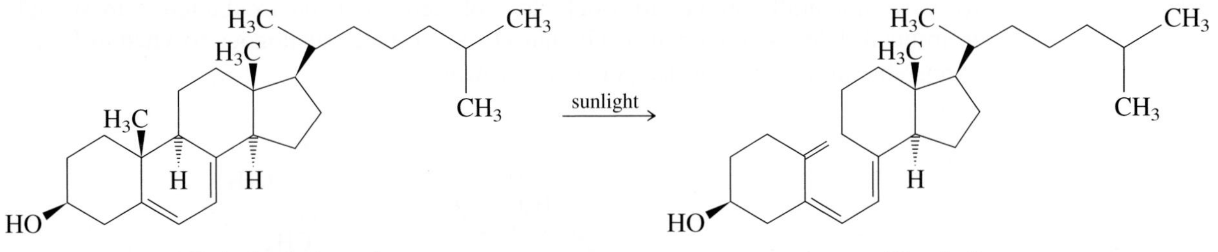

7-Dehydrocholesterol Vitamin D_3

Vitamin D_3 is a key compound in the process by which Ca^{2+} is absorbed from the intestine. Low levels of vitamin D_3 lead to Ca^{2+} concentrations in the body that are insufficient to support proper bone growth, resulting in the bone disease called rickets.

Good Cholesterol? Bad Cholesterol? What's the Difference?

Cholesterol is biosynthesized in the liver, transported throughout the body to be used in a variety of ways, and returned to the liver where it serves as the biosynthetic precursor to other steroids. But cholesterol is a lipid and isn't soluble in water. How can it move through the blood if it doesn't dissolve in it? The answer is that it doesn't dissolve, but is instead carried through the blood and tissues as part of a *lipoprotein* (lipid + protein = lipoprotein).

The proteins that carry cholesterol from the liver are called *low-density lipoproteins,* or LDLs; those that return it to the liver are the *high-density lipoproteins,* or HDLs. If too much cholesterol is being transported by LDL, or too little by HDL, the extra cholesterol builds up on the walls of the arteries, causing atherosclerosis. Blood work done as part of a thorough physical examination measures not only total cholesterol but also the distribution between LDL and HDL cholesterol. An elevated level of LDL cholesterol is a risk factor for heart disease. LDL cholesterol is "bad" cholesterol. HDLs, on the other hand, remove excess cholesterol and are protective. HDL cholesterol is "good" cholesterol.

The distribution between LDL and HDL cholesterol depends mainly on genetic factors, but can be altered. Regular exercise increases HDL and reduces LDL cholesterol, as does limiting the amount of saturated fat in the diet. Much progress has been made in developing new drugs to lower cholesterol. The *statin* class, beginning with lovastatin in 1988, has proven especially effective. The most prescribed cholesterol-lowering drug is atorvastatin (as its calcium salt). A chiral drug, atorvastatin was introduced in 1997 and is sold as a single enantiomer.

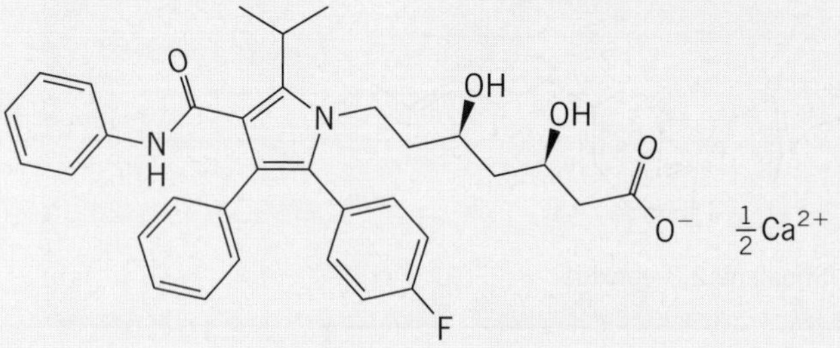

Atorvastatin calcium (Lipitor)

The statins lower cholesterol by inhibiting the enzyme 3-hydroxy-3-methylglutaryl coenzyme A reductase, which is required for the biosynthesis of mevalonic acid (see Section 24.10). Mevalonic acid is an obligatory precursor to cholesterol, so less mevalonic acid translates into less cholesterol.

Rickets was once more widespread than it is now. It was thought to be a dietary deficiency disease because it could be prevented in children by feeding them fish liver oil. Actually, rickets is an environmental disease brought about by a deficiency of sunlight. Where the winter sun is weak, children may not be exposed to enough of its light to convert the 7-dehydrocholesterol in their skin to vitamin D_3 at levels sufficient to promote the growth of strong bones. Fish have adapted to an environment that screens them from sunlight, and so they are not directly dependent on photochemistry for their vitamin D_3 and accumulate it by a different process. Although fish liver oil is a good source of vitamin D_3, it is not very palatable. Synthetic vitamin D_3, prepared from cholesterol, is often added to milk and other foods to ensure that children receive enough of the vitamin for their bones to develop properly. *Irradiated ergosterol* is another dietary supplement added to milk and other foods for the same purpose. Ergosterol, a steroid obtained from yeast, is structurally similar to 7-dehydrocholesterol and, on irradiation with sunlight or artificial light, is converted to vitamin D_2, a substance analogous to vitamin D_3 and comparable in its ability to support bone growth.

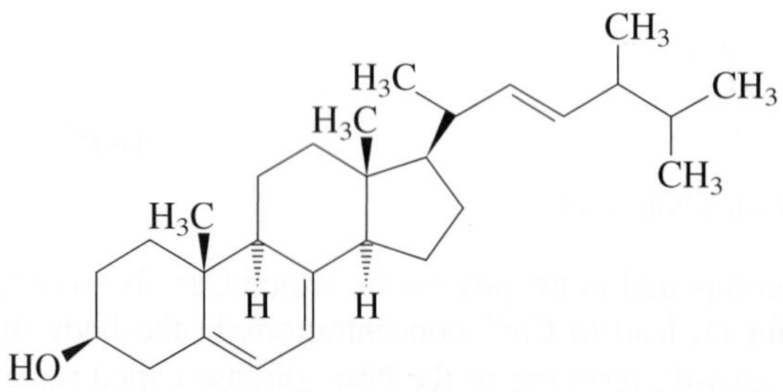

Ergosterol

Problem 24.17

Suggest a reasonable structure for vitamin D_2.

24.13 Bile Acids

A significant fraction of the body's cholesterol is used to form **bile acids.** Oxidation in the liver removes a portion of the C_8H_{17} side chain, and additional hydroxyl groups are introduced at various positions on the steroid nucleus. *Cholic acid* is the most abundant of the bile acids. In the form of certain amide derivatives such as *sodium taurocholate,* bile acids act as emulsifying agents to aid the digestion of fats.

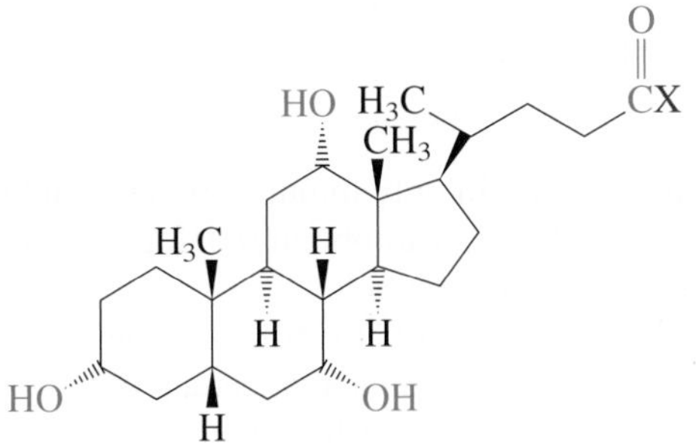

X = OH; cholic acid
X = $NHCH_2CH_2SO_3Na$; sodium taurocholate

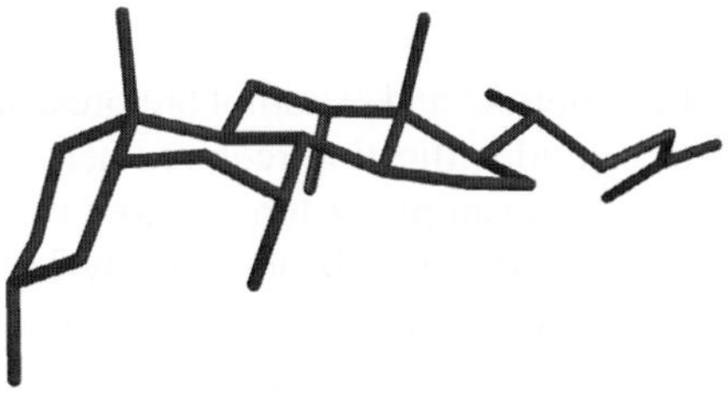

Cholic acid

The structure of cholic acid helps us understand how bile salts such as sodium taurocholate promote the transport of lipids through a water-rich environment. The bottom face of the molecule bears all of the polar groups, and the top face is exclusively hydrocarbon-like. Bile salts emulsify fats by forming micelles in which the fats are on the inside and the bile salts are on the outside. The hydrophobic face of the bile salt associates with the fat that is inside the micelle; the hydrophilic face is in contact with water on the outside.

24.14 Corticosteroids

The outer layer, or *cortex,* of the adrenal gland is the source of a large group of substances known as **corticosteroids.** Like the bile acids, they are derived from cholesterol by oxidation, with cleavage of a portion of the alkyl substituent on the D ring. *Cortisol* is the most abundant of the corticosteroids, but *cortisone* is probably the best known. Cortisone is commonly prescribed as an antiinflammatory drug, especially in the treatment of rheumatoid arthritis.

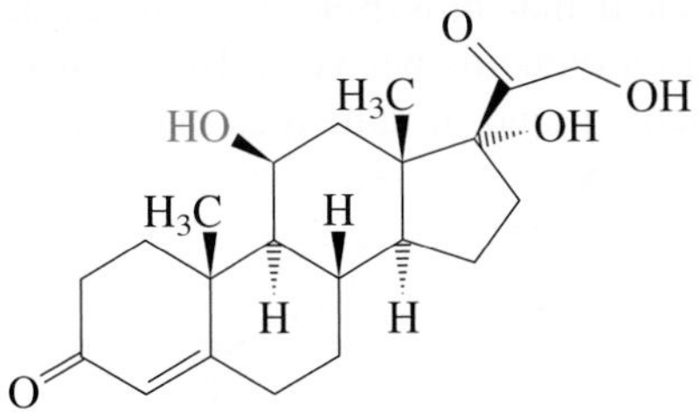

Cortisol

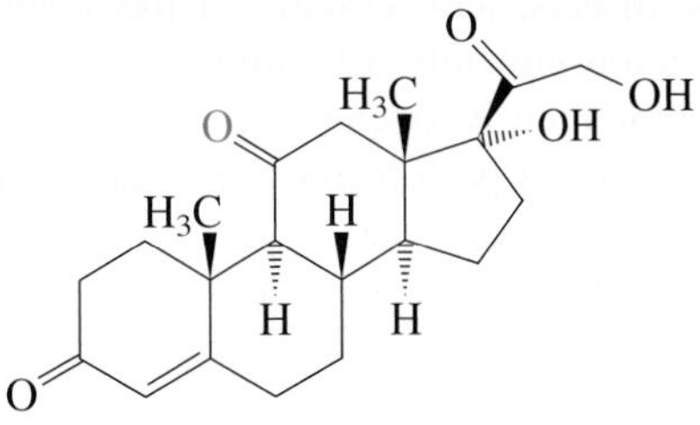

Cortisone

Corticosteroids exhibit a wide range of physiological effects. One important function is to assist in maintaining the proper electrolyte balance in body fluids. They also play a vital regulatory role in the metabolism of carbohydrates and in mediating the allergic response.

Many antiitch remedies contain dihydrocortisone.

24.15 Sex Hormones

Hormones are the chemical messengers of the body; they are secreted by the endocrine glands and regulate biological processes. Corticosteroids, described in the preceding section, are hormones produced by the adrenal glands. The sex glands—testes in males, ovaries in females—secrete a number of hormones that are involved in sexual development and reproduction. *Testosterone* is the principal male sex hormone; it is an **androgen.** Testosterone promotes muscle growth, deepening of the voice, the growth of body hair, and other male secondary sex characteristics. Testosterone is formed from cholesterol and is the biosynthetic precursor of estradiol, the principal female sex hormone, or **estrogen.** *Estradiol* is a key substance in the regulation of the menstrual cycle and the reproductive process. It is the hormone most responsible for the development of female secondary sex characteristics.

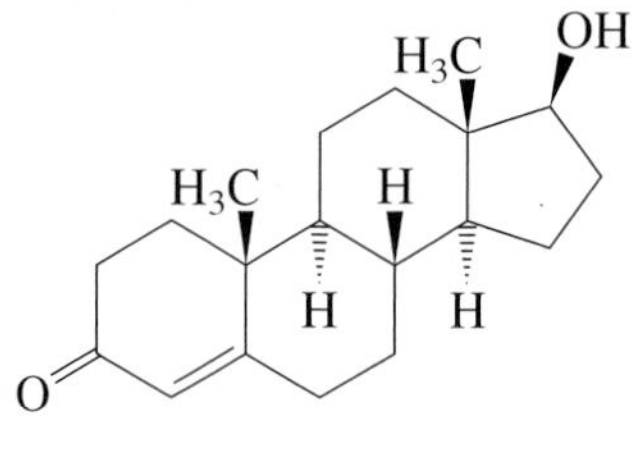

Testosterone

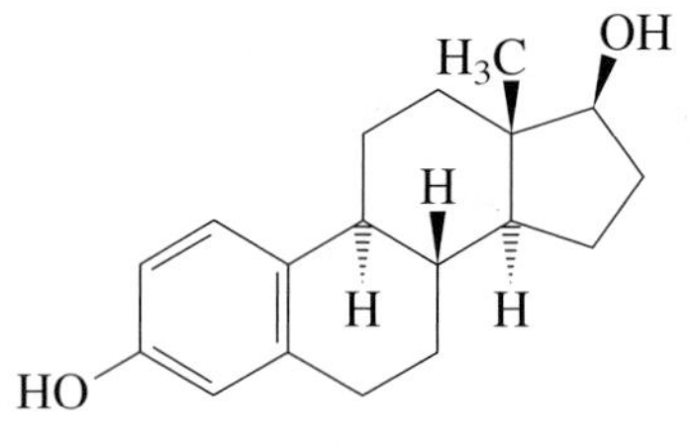

Estradiol

Testosterone and estradiol are present in the body in only minute amounts, and their isolation and identification required heroic efforts. In order to obtain 0.012 g of estradiol for study, for example, 4 tons of sow ovaries had to be extracted!

A separate biosynthetic pathway leads from cholesterol to *progesterone,* a female sex hormone. One function of progesterone is to suppress ovulation at certain stages of the menstrual cycle and during pregnancy. Synthetic substances, such as *norethindrone,* have been developed that are superior to progesterone when taken orally to "turn off" ovulation. By inducing temporary infertility, they form the basis of most oral contraceptive agents.

The color of a flamingo's feathers comes from the carotenes in the brine shrimp they eat.

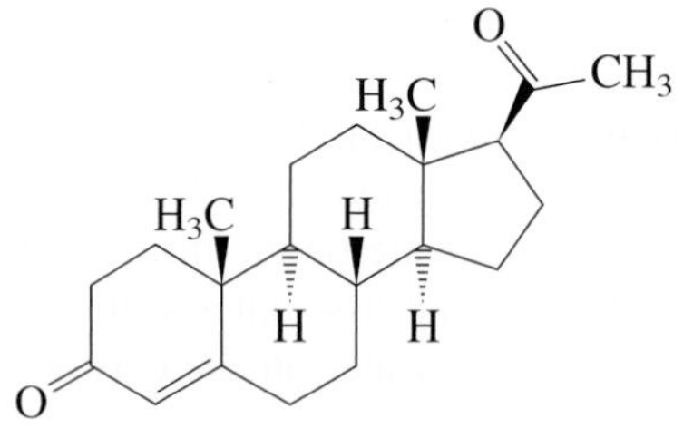

Progesterone

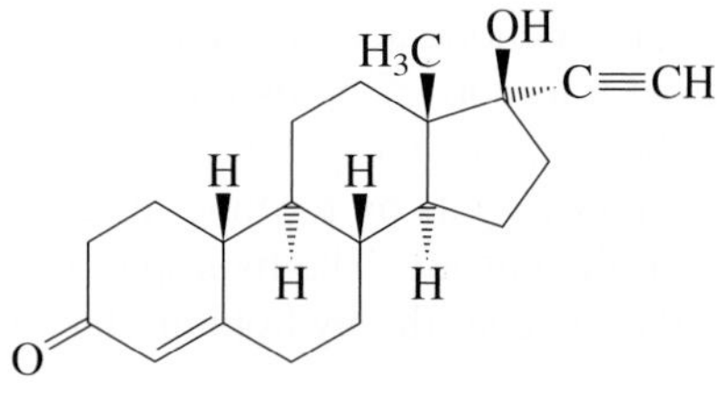

Norethindrone

24.16 Carotenoids

Carotenoids are natural pigments characterized by a tail-to-tail linkage between two C_{20} units and an extended conjugated system of double bonds. They are the most widely distributed of the substances that give color to our world and occur in flowers, fruits, plants, insects, and animals. It has been estimated that biosynthesis from acetate produces approximately a hundred million tons of carotenoids per year. The most familiar carotenoids are lycopene and β-carotene, pigments found in numerous plants and easily isolable from ripe tomatoes and carrots, respectively.

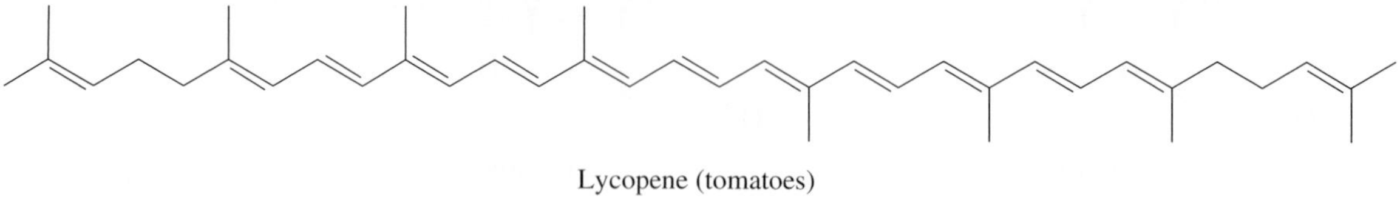

Lycopene (tomatoes)

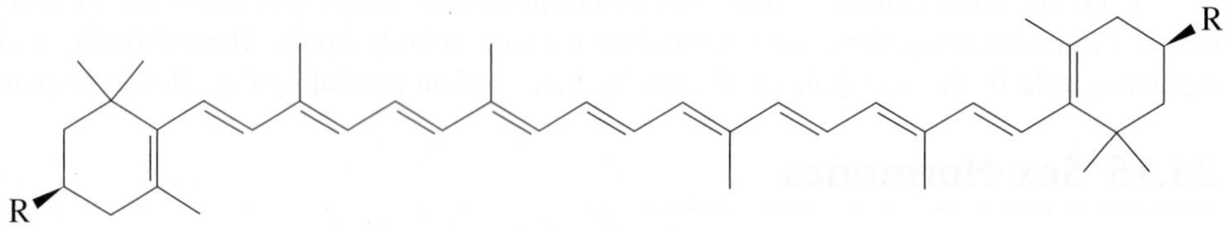

R = H; β-Carotene (carrots)
R = OH; Zeaxanthyn (yellow corn)

Crocuses Make Saffron from Carotenes

The flowers of *Crocus sativus* are not only pretty, they are valuable. The saffron crocus is cultivated on a large scale because the three gold-colored filaments in each bloom are the source of *saffron,* a dye and a spice that has been used for thousands of years. The amount is small; 75,000 flowers are needed to provide 1 pound of saffron, yet 300 tons of it reach the worldwide market each year.

Saffron is a mixture of substances. Those that make it desirable as a spice and dye are among the ones the plant uses to attract insects. Two of them, *crocetin* and *crocin,* are mainly responsible for its color, another (*safranal*) its odor, and another (*picrocrocin*) its taste. The same 20-carbon conjugated polyene unit is the chromophore that gives crocetin and crocin their yellow color. The difference between the two is that crocin is a glycoside in which both carboxylic acid functions of crocetin are attached to a disaccharide (*gentiobiose*) by ester linkages.

Crocetin

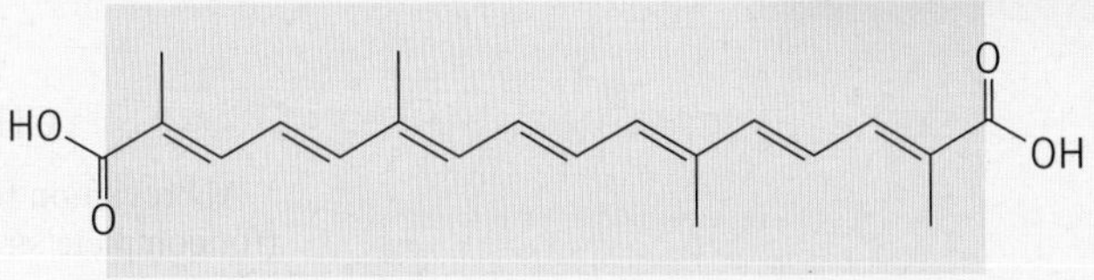

Crocin

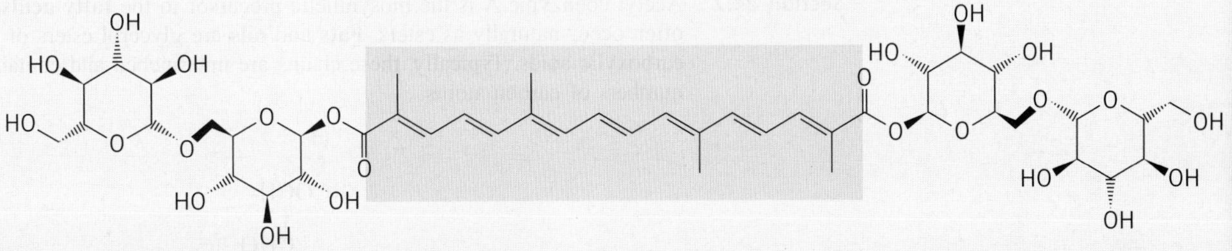

The 20-carbon chromophore originates in biochemical degradation of β-carotene and related carotenoids. Enzyme-catalyzed oxidation cleaves the double bonds at the points indicated to give crocetin.

β-Carotene

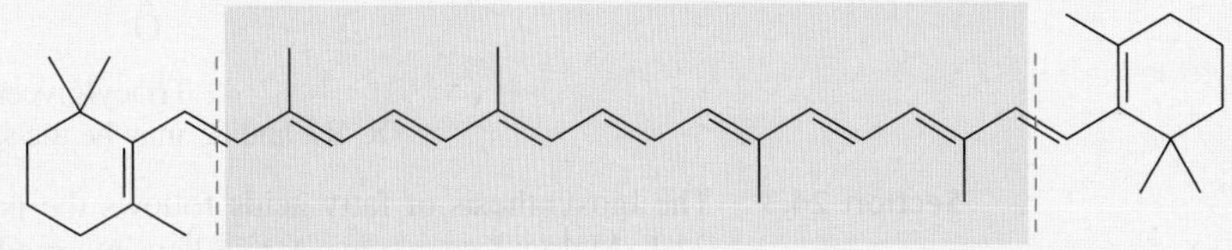

Safranal and picrocrocin are both aldehydes. Their structures suggest that they too come from carotenoid precursors. Because it is volatile, safranal contributes to the odor that attracts insects to the flowers. Picrocrocin is a glycoside. Its ability to participate in hydrogen bonding makes it nonvolatile and allows it to remain in place within the flowers where it provides the characteristic saffron flavor.

Safranal

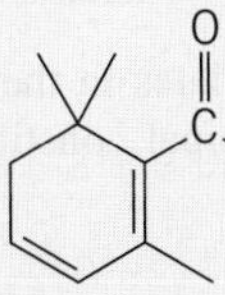

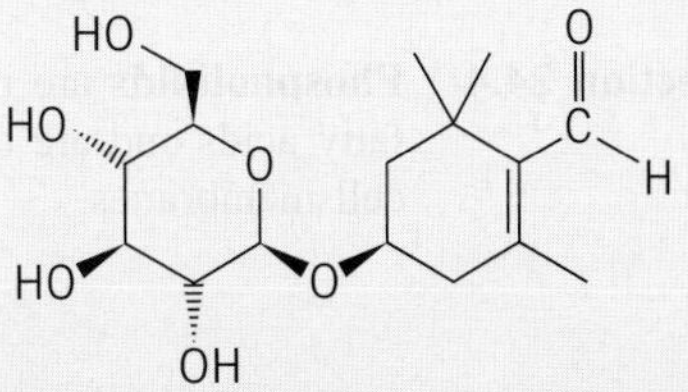

Picrocrocin

Problem 24.18

Can you find the isoprene units in crocetin, crocin, safranal, and picrocrocin?

The structural chemistry of the visual process, beginning with β-carotene, was described in the boxed essay entitled *Imines in Biological Chemistry* in Chapter 17.

Not all carotenoids are hydrocarbons. Oxygen-containing carotenes called *xanthophylls,* which are often the pigments responsible for the yellow color of flowers, are especially abundant.

Carotenoids absorb visible light (see Section 13.23) and dissipate its energy as heat, providing protection from the potentially harmful effects of sunlight-induced photochemistry. They are also indirectly involved in the chemistry of vision, owing to the fact that β-carotene is the biosynthetic precursor of vitamin A, also known as retinol, a key substance in the visual process.

24.17 SUMMARY

Section 24.1 Chemists and biochemists find it convenient to divide the principal organic substances present in cells into four main groups: *carbohydrates, proteins, nucleic acids,* and **lipids.** Structural differences separate carbohydrates from proteins, and both of these are structurally distinct from nucleic acids. Lipids, on the other hand, are characterized by a *physical property,* their solubility in nonpolar solvents, rather than by their structure. In this chapter we have examined lipid molecules that share a common biosynthetic origin in that all their carbons are derived from acetic acid (acetate). The form in which acetate occurs in many of these processes is a thioester called acetyl coenzyme A.

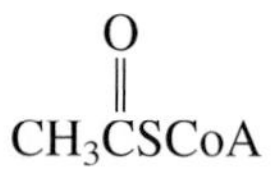

Abbreviation for acetyl coenzyme A
(for complete structure, see Figure 24.1)

Section 24.2 Acetyl coenzyme A is the biosynthetic precursor to the **fatty acids,** which most often occur naturally as esters. **Fats** and **oils** are glycerol esters of long-chain carboxylic acids. Typically, these chains are unbranched and contain even numbers of carbon atoms.

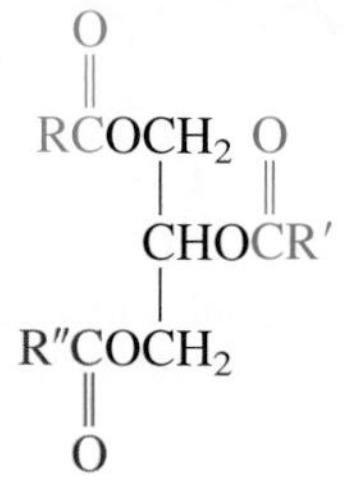

Triacylglycerol
(R, R′, and R″ may be the same or different)

Section 24.3 The biosynthesis of fatty acids follows the pathway outlined in Mechanism 24.1. Malonyl coenzyme A is a key intermediate.

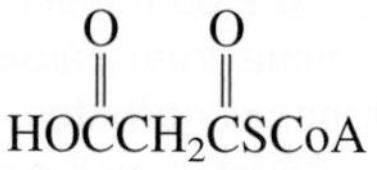

Malonyl coenzyme A

Section 24.4 **Phospholipids** are intermediates in the biosynthesis of triacylglycerols from fatty acids and are the principal constituents of the lipid bilayer component of cell membranes.

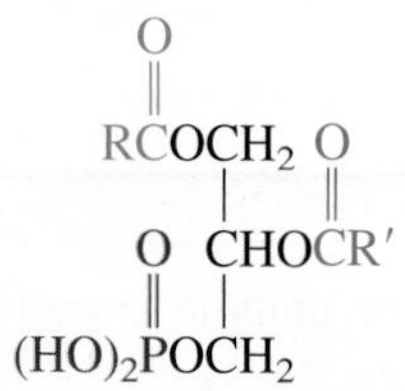

A phospholipid

Section 24.5 **Waxes** are mixtures of substances that usually contain esters of fatty acids and long-chain alcohols.

Section 24.6 **Icosanoids** are a group of naturally occurring compounds derived from unsaturated C_{20} carboxylic acids. Icosanoids include *prostaglandins, prostacyclins, thromboxanes,* and *leukotrienes.* Although present in very small amounts, icosanoids play regulatory roles in a very large number of biological processes.

Section 24.7 **Terpenes** have structures that follow the isoprene rule in that they can be viewed as collections of isoprene units.

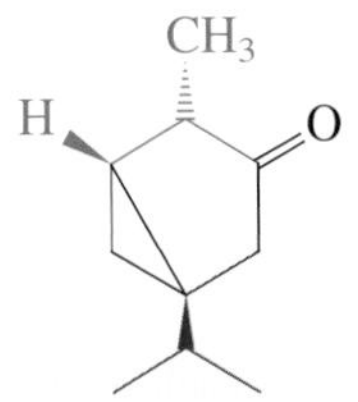
CH3
H
O

β-Thujone: a toxic monoterpene present in absinthe

Section 24.8 Terpenes and related *isoprenoid* compounds are biosynthesized from *isopentenyl diphosphate.*

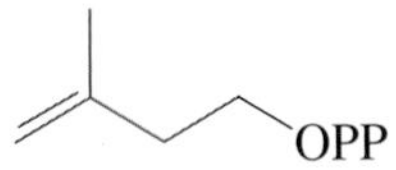
OPP

Isopentenyl diphosphate is the "biological isoprene unit."

Section 24.9 Carbon–carbon bond formation between isoprene units can be understood on the basis of nucleophilic attack of the π electrons of a double bond on a carbocation or an allylic carbon that bears a diphosphate leaving group.

OPP
+
OPP
OPP

Section 24.10 The biosynthesis of isopentenyl diphosphate begins with acetate and proceeds by way of *mevalonic acid.*

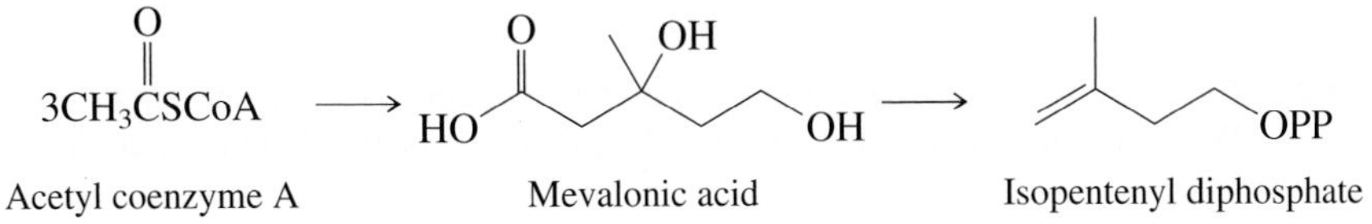
3CH3CSCoA
O
HO
O
OH
OH
OPP
Acetyl coenzyme A
Mevalonic acid
Isopentenyl diphosphate

Section 24.11 The triterpene *squalene* is the biosynthetic precursor to cholesterol by the pathway shown in Mechanism 24.2.

Sections 24.12–24.15 Most of the steroids in animals are formed by biological transformations of cholesterol.

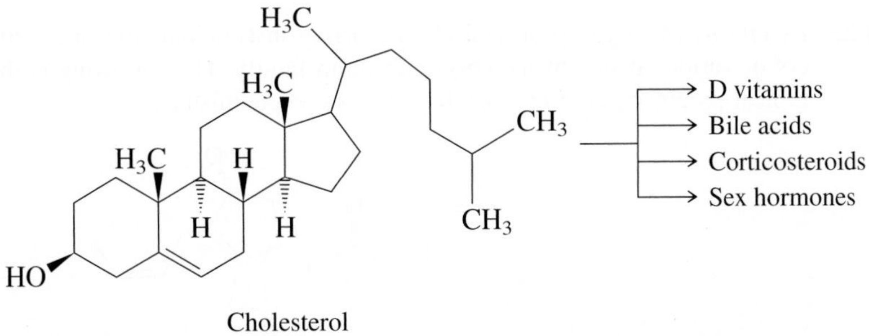
H3C
H3C
H3C
H
H
H
CH3
CH3
HO
D vitamins
Bile acids
Corticosteroids
Sex hormones

Cholesterol

Section 24.16 **Carotenoids** are tetraterpenes. They have 40 carbons and numerous double bonds. Many of the double bonds are conjugated, causing carotenes to absorb visible light and be brightly colored. They are often plant pigments.

PROBLEMS

24.19 The structures of each of the following are given within the chapter. Identify the carbon atoms expected to be labeled with ^{14}C when each is biosynthesized from acetate enriched with ^{14}C in its methyl group.

(a) Palmitic acid

(b) PGE_2

(c) PGI_2

(d) Limonene

(e) β-Carotene

24.20 Identify the isoprene units in each of the following naturally occurring substances:

(a) *Ascaridole,* a naturally occurring peroxide present in chenopodium oil:

(b) *Dendrolasin,* a constituent of the defense secretion of a species of ant:

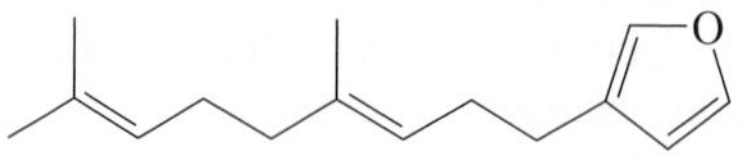

(c) γ-*Bisabolene,* a sesquiterpene found in the essential oils of a large number of plants:

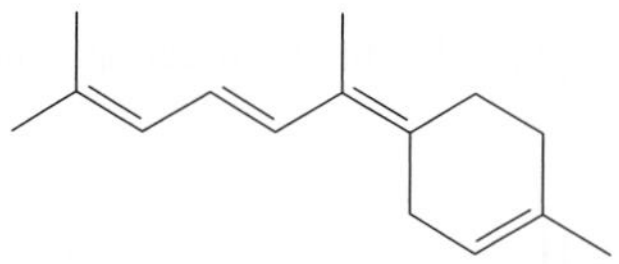

(d) α-*Santonin,* a lactone found in artemisia flowers:

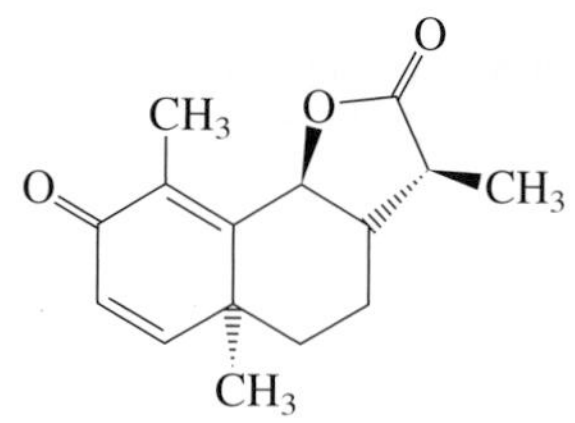

(e) *Tetrahymanol,* a pentacyclic triterpene isolated from a species of protozoans:

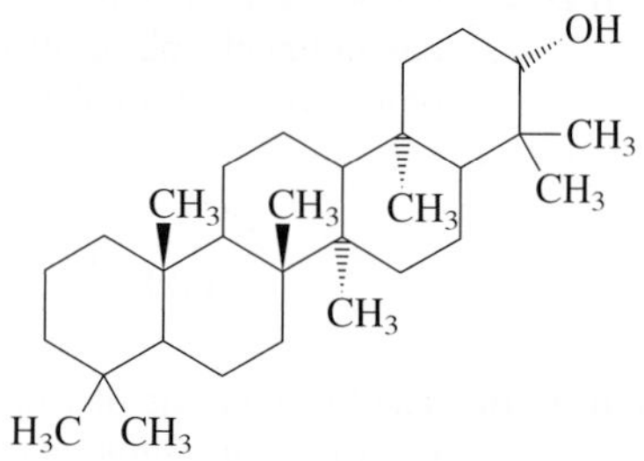

24.21 *Cubitene* is a diterpene present in the defense secretion of a species of African termite. What unusual feature characterizes the joining of isoprene units in cubitene?

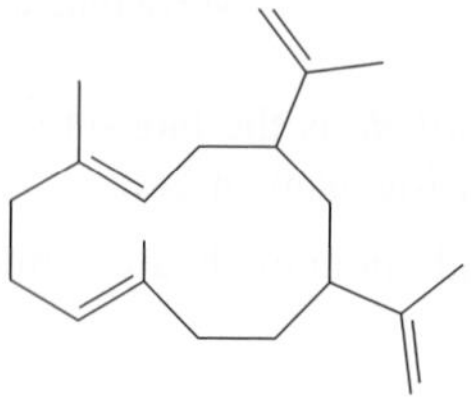

24.22 *Pyrethrins* are a group of naturally occurring insecticidal substances found in the flowers of various plants of the chrysanthemum family. The following is the structure of a typical pyrethrin, *cinerin I* (exclusive of stereochemistry):

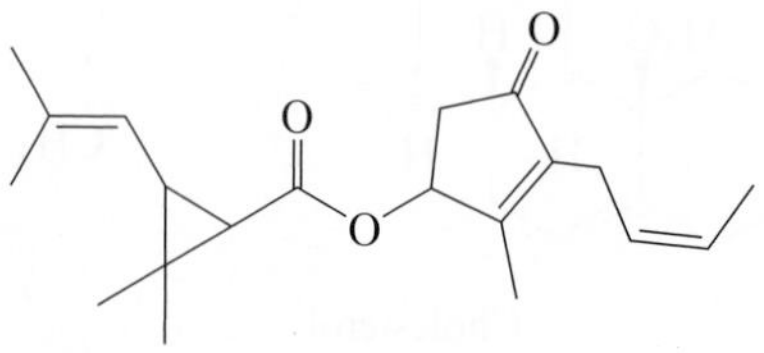

(a) Locate any isoprene units present in cinerin I.

(b) Hydrolysis of cinerin I gives an optically active carboxylic acid, (+)-chrysanthemic acid. Ozonolysis of (+)-chrysanthemic acid, followed by oxidation, gives acetone and an optically active dicarboxylic acid, (–)-caronic acid ($C_7H_{10}O_4$). What is the structure of (–)-caronic acid? Are the two carboxyl groups cis or trans to each other? What does this information tell you about the structure of (+)-chrysanthemic acid?

24.23 *Cerebrosides* are found in the brain and in the myelin sheath of nerve tissue. The structure of the cerebroside *phrenosine* is

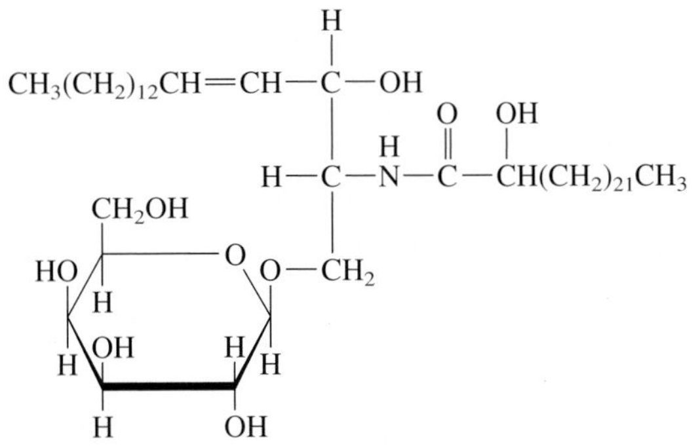

(a) What hexose is formed on hydrolysis of the glycoside bond of phrenosine? Is phrenosine an α- or a β-glycoside?

(b) Hydrolysis of phrenosine gives, in addition to the hexose in part (a), a fatty acid called *cerebronic acid,* along with a third substance called *sphingosine.* Write structural formulas for both cerebronic acid and sphingosine.

24.24 Each of the following reactions has been reported in the chemical literature and proceeds in good yield. What are the principal organic products of each reaction? In some of the exercises more than one diastereomer may be theoretically possible, but in such instances one diastereomer is either the major product or the only product. For those reactions in which one diastereomer is formed preferentially, indicate its expected stereochemistry.

(a) $CH_3(CH_2)_7C{\equiv}C(CH_2)_7COOH + H_2 \xrightarrow{\text{Lindlar Pd}}$

(b) $CH_3(CH_2)_7C{\equiv}C(CH_2)_7COOH \xrightarrow[\text{2. } H^+]{\text{1. Li, } NH_3}$

(c) (Z)-$CH_3(CH_2)_7CH{=}CH(CH_2)_7\overset{O}{\overset{\|}{C}}OCH_2CH_3 + H_2 \xrightarrow{\text{Pt}}$

(d) (Z)-$CH_3(CH_2)_5\underset{OH}{\underset{|}{C}}HCH_2CH{=}CH(CH_2)_7\overset{O}{\overset{\|}{C}}OCH_3 \xrightarrow[\text{2. } H_2O]{\text{1. } LiAlH_4}$

(e) (Z)-$CH_3(CH_2)_7CH{=}CH(CH_2)_7COOH + C_6H_5\overset{O}{\overset{\|}{C}}OOH \longrightarrow$

(f) Product of part (e) + $H_3O^+ \longrightarrow$

(g) (Z)-$CH_3(CH_2)_7CH{=}CH(CH_2)_7COOH \xrightarrow[\text{2. } H^+]{\text{1. } OsO_4, (CH_3)_3COOH, HO^-}$

(h) 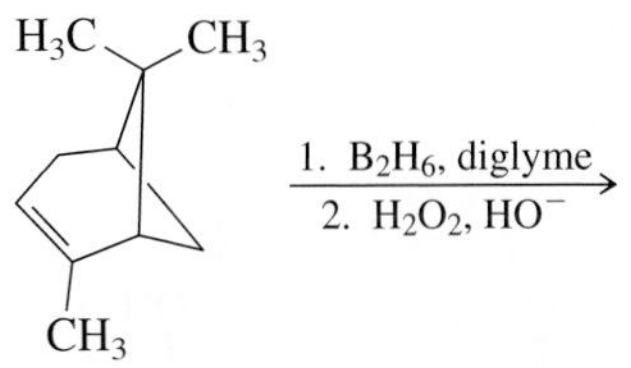

(i) 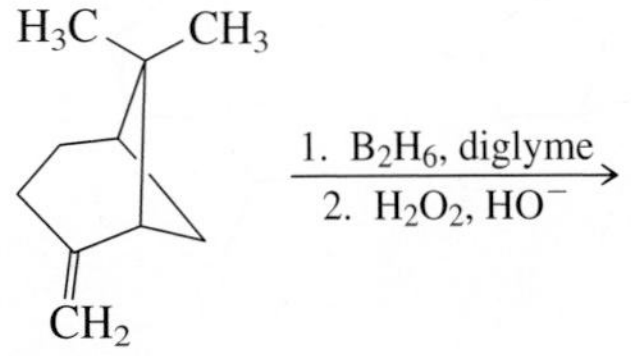

(j) 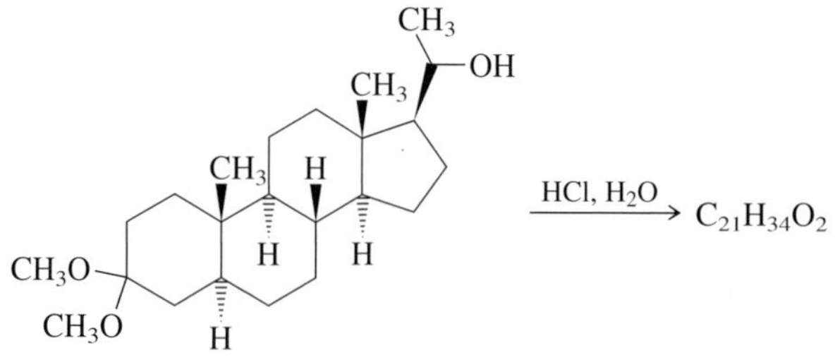

24.25 Describe an efficient synthesis of each of the following compounds from octadecanoic (stearic) acid using any necessary organic or inorganic reagents:

(a) Octadecane
(b) 1-Phenyloctadecane
(c) 3-Ethylicosane
(d) Icosanoic acid
(e) 1-Octadecanamine
(f) 1-Nonadecanamine

24.26 A synthesis of triacylglycerols has been described that begins with the substance shown.

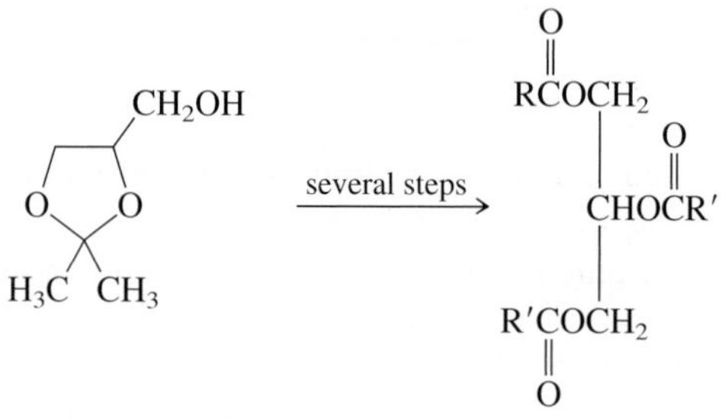

4-(Hydroxymethyl)-
2,2-dimethyl-1,3-dioxolane

Triacylglycerol

Outline a series of reactions suitable for the preparation of a triacylglycerol of the type illustrated in the equation, where R and R′ are different.

24.27 The isoprenoid compound shown is a scent marker present in the urine of the red fox. Suggest a reasonable synthesis for this substance from 3-methyl-3-buten-1-ol and any necessary organic or inorganic reagents.

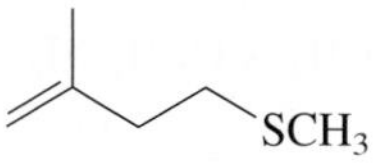

24.28 *Sabinene* is a monoterpene found in the oil of citrus fruits and plants. It has been synthesized from 6-methyl-2,5-heptanedione by the sequence that follows. Suggest reagents suitable for carrying out each of the indicated transformations.

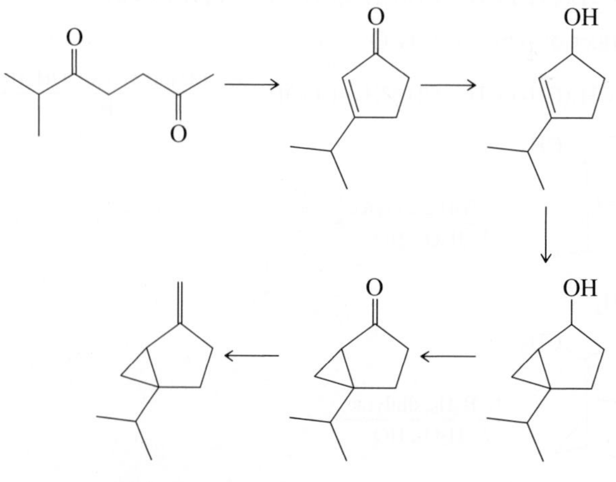

Sabinene

24.29 Isoprene has sometimes been used as a starting material in the laboratory synthesis of terpenes. In one such synthesis, the first step is the electrophilic addition of 2 mol of hydrogen bromide to isoprene to give 1,3-dibromo-3-methylbutane.

$$H_2C{=}\overset{CH_3}{\overset{|}{C}}CH{=}CH_2 + 2HBr \longrightarrow (CH_3)_2\overset{Br}{\overset{|}{C}}CH_2CH_2Br$$

2-Methyl-1,3-butadiene (isoprene) — Hydrogen bromide — 1,3-Dibromo-3-methylbutane

Write a series of equations describing the mechanism of this reaction.

24.30 The ionones are fragrant substances present in the scent of iris and are used in perfume. A mixture of α- and β-ionone can be prepared by treatment of pseudoionone with sulfuric acid.

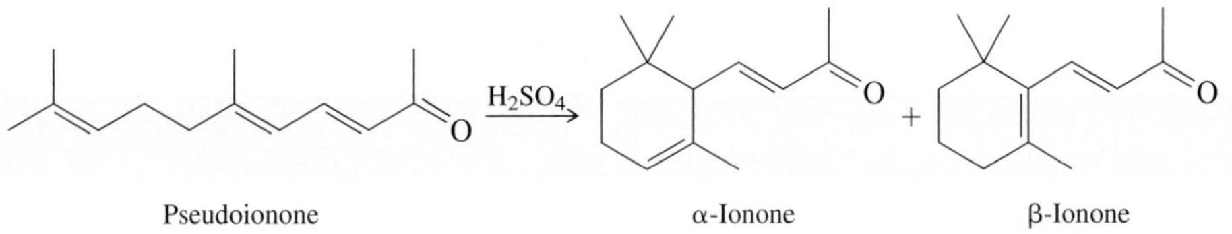

Pseudoionone — α-Ionone — β-Ionone

Write a stepwise mechanism for this reaction.

24.31 β,γ-Unsaturated steroidal ketones represented by the partial structure shown here are readily converted in acid to their α,β-unsaturated isomers. Write a stepwise mechanism for this reaction.

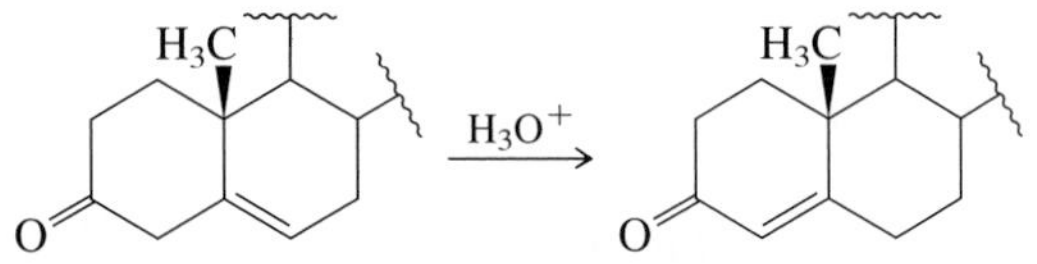

24.32 (a) Suggest a mechanism for the following reaction.

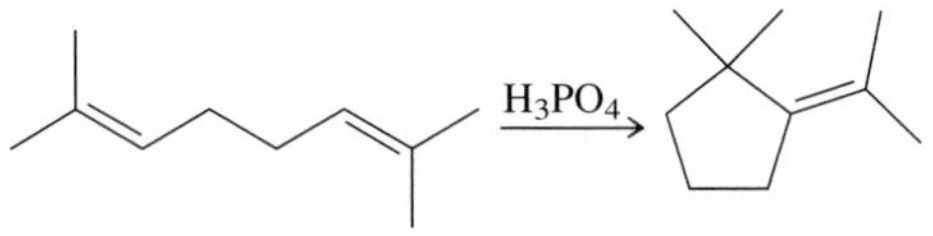

(b) The following two compounds are also formed in the reaction given in part (a). How are these two products formed?

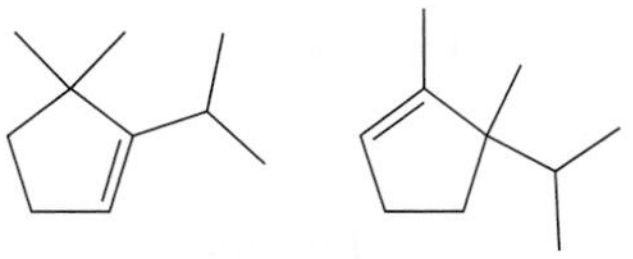

24.33 The following transformation was carried out as part of a multistep synthesis of digitoxigenin. Propose a mechanism.

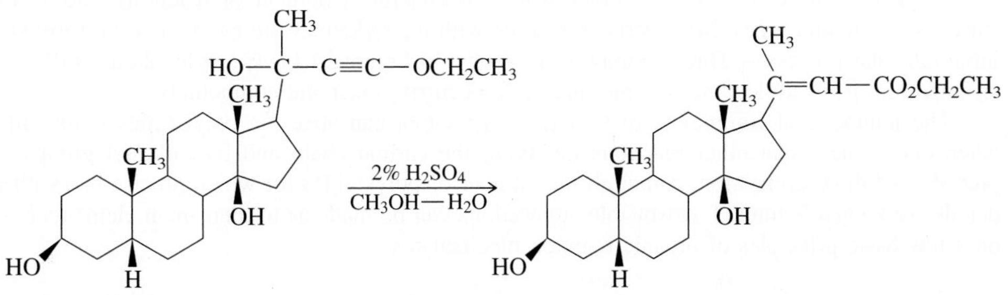

24.34 The glaucoma drug *bimatoprost* is synthesized in two steps from another prostaglandin. Can you suggest a method for this conversion?

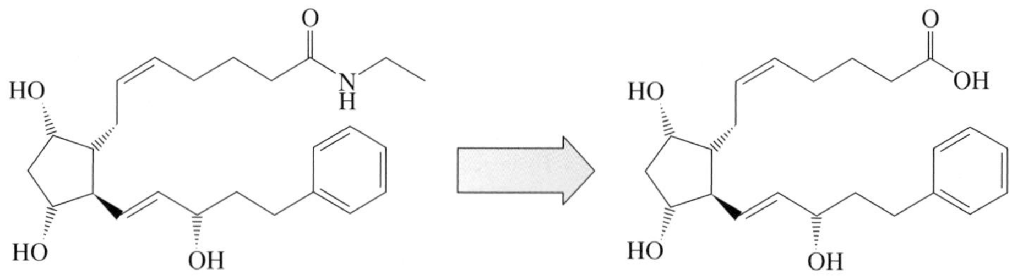

Bimatoprost

Descriptive Passage and Interpretive Problems 24

Polyketides

We have seen in this chapter that acetoacetate is an intermediate in both fatty acid and terpene biosynthesis. It is also an intermediate in the biosynthesis of the **polyketides,** a class of compounds of which more than 7000 are known to occur naturally. Polyketides are composed of alternating C═O and CH_2 groups as well as compounds derived from them. Their biosynthesis resembles fatty acid biosynthesis except that many of the carbonyl groups destined for reduction during fatty acid biosynthesis are retained in polyketide biosynthesis.

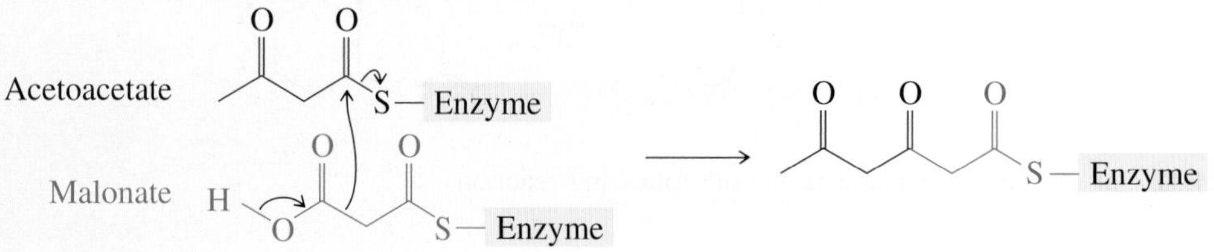

Many polyketides have one or more methyl substituents on their carbon chain. In some cases *S*-adenosylmethionine is the source of a methyl group; in others methylmalonyl CoA (from propanoic acid) substitutes for acetate during chain assembly.

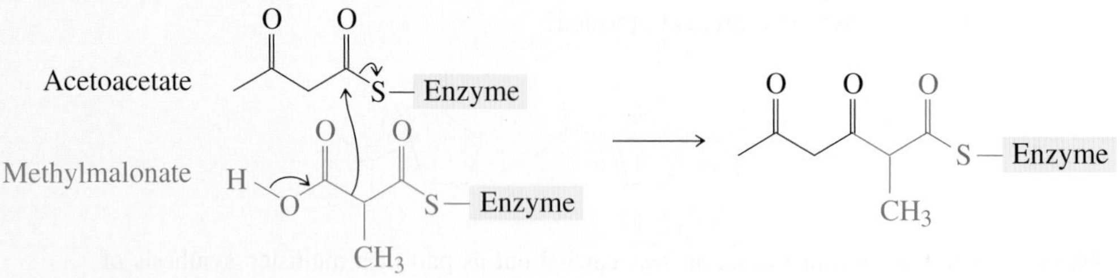

Polyketides have the requisite functionality to undergo a number of reactions leading to a variety of structural types. Some structural units within polyketides are cyclic and are formed by intramolecular processes. Three examples are shown in Figure 24.11; examples **2** and **3** illustrate an important pathway leading to many naturally occurring phenolic compounds.

The number and complexity of structural types that can arise via polyketides is magnified when one realizes that other reactions involving the carbon chain and its carbonyl groups can precede or follow cyclization. Although the number of polyketides for which precise biosynthetic details are known is limited, reasonable suggestions can be made as to their main elements based on a few basic principles of organic reaction mechanisms.

1. An enolic OH derived from the β-diketone structural unit can act as the nucleophile in a nucleophilic acyl substitution to give a six-membered oxygen heterocycle known as a pyrone.

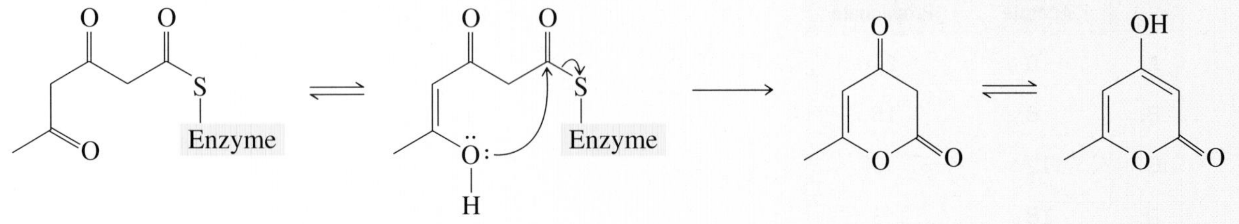

2. Intramolecular Claisen condensation gives 1,3,5-trihydroxybenzene.

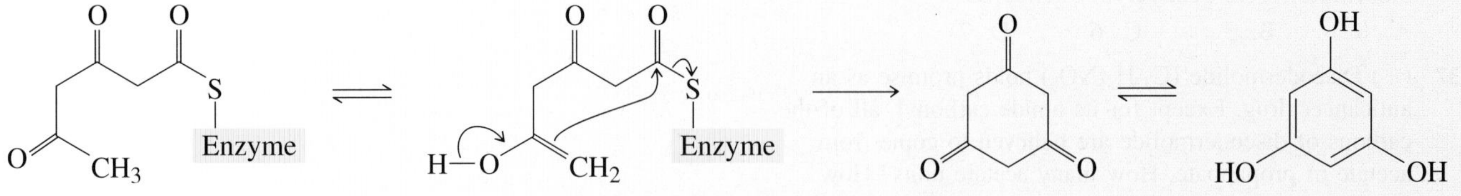

3. Intramolecular aldol condensation of a slightly longer polyketide chain gives orsellinic acid.

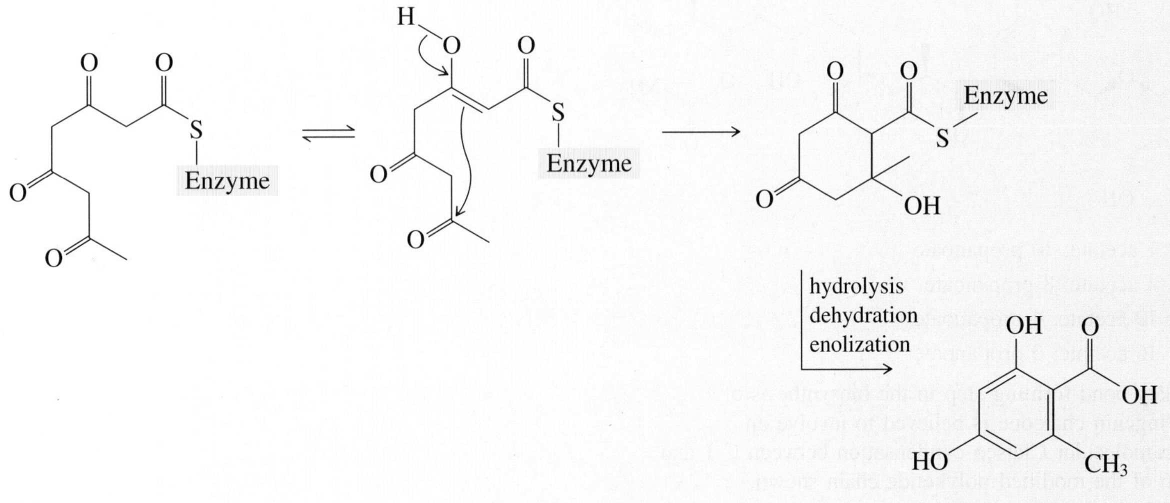

Figure 24.11

Some cyclization reactions of polyketide chains.

The first enzyme-free intermediate in the biosynthesis of erythromycin is the polyketide 6-deoxyerythronolide B. All of the carbons come from either acetate or propanoate.

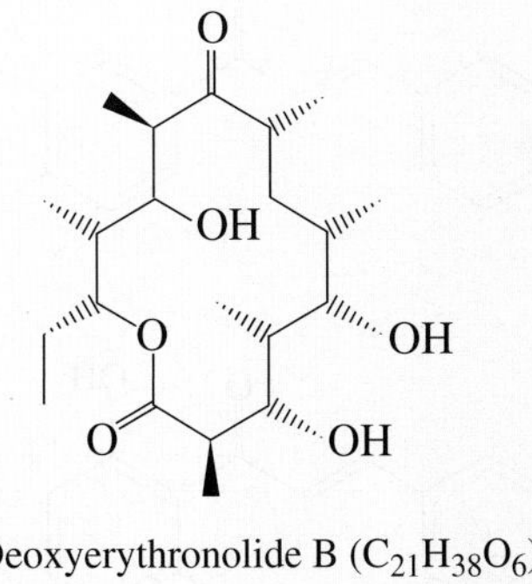

6-Deoxyerythronolide B ($C_{21}H_{38}O_6$)

24.35 How many of the carbons come from acetate? From propanoate?

	Acetate	Propanoate
A.	0	21
B.	6	15
C.	12	9
D.	18	3

24.36 How many methylmalonates are involved in the biosynthesis of 6-deoxyerythronolide B?

A. 0 B. 3 C. 6 D. 7

24.37 (+)-Discodermolide ($C_{33}H_{55}NO_8$) holds promise as an anticancer drug. Except for its amide carbonyl, all of the carbons of discodermolide are believed to come from acetate or propanoate. How many acetate units? How many propanoate units?

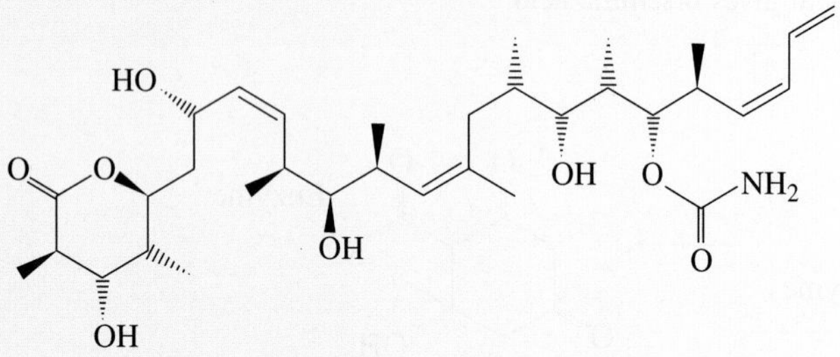

A. 1 acetate; 10 propanoate
B. 4 acetate; 8 propanoate
C. 10 acetate; 4 propanoate
D. 16 acetate; 0 propanoate

24.38 A key bond-forming step in the biosynthesis of naringenin chalcone is believed to involve an intramolecular Claisen condensation between C-1 and C-6 of the modified polyketide chain shown.

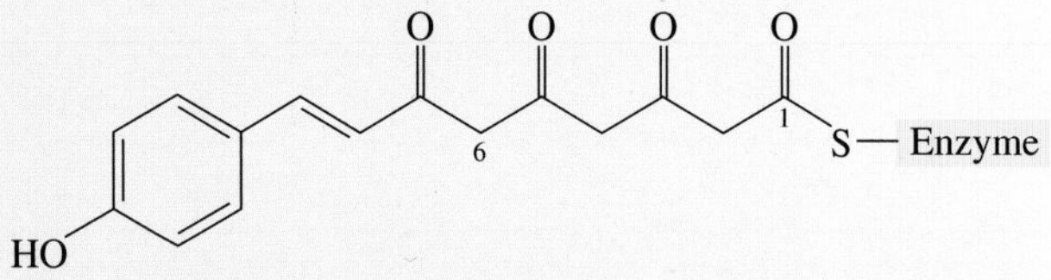

Which of the following is the most reasonable structure of naringenin chalcone based on this hypothesis?

A. 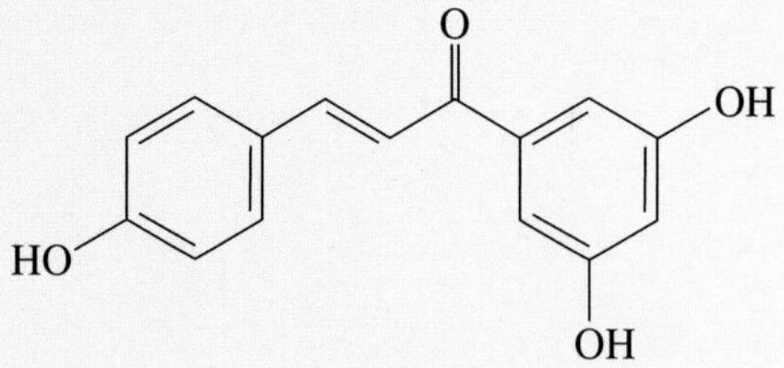

C. 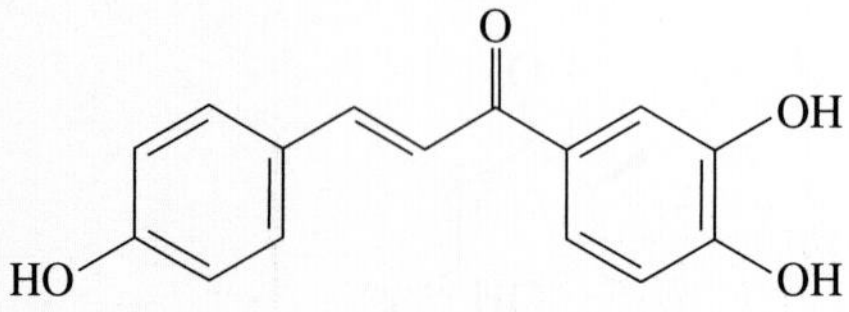

B. 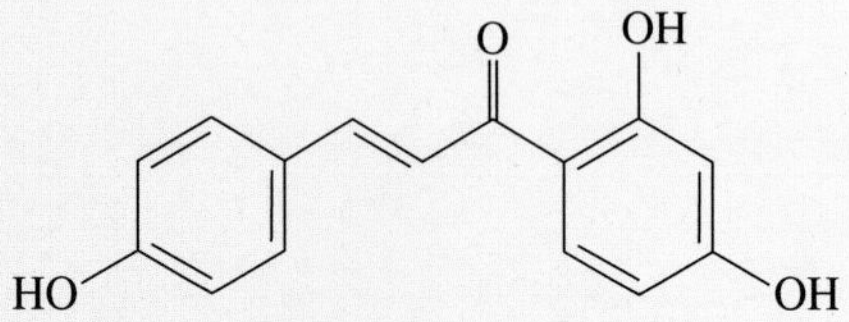

D. 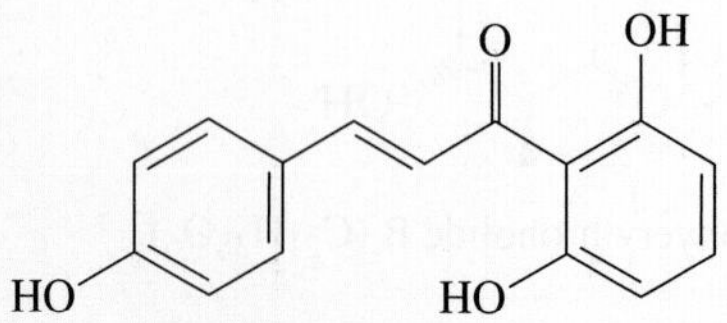

24.39 Carbon–carbon bond formation in the 14-carbon polyketo chain is suggested to be a key biosynthetic step leading to the compound shown.

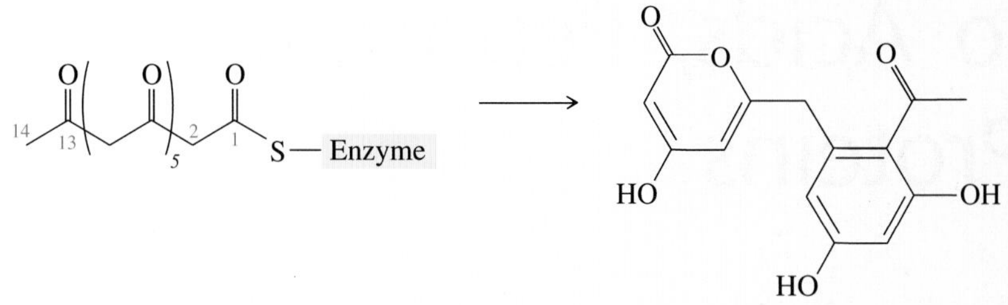

What two carbons are involved in this carbon–carbon bond-forming step?

A. C-1 and C-5

B. C-2 and C-14

C. C-7 and C-12

D. C-8 and C-13

24.40 Alternariol is a toxin produced by a mold that grows on agricultural products. It is a polyketide derived from seven acetate units. Which of the following is the most reasonable structure for alternariol?

A.

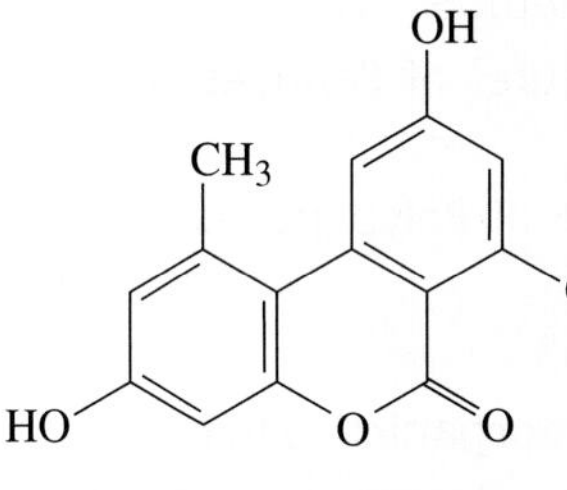

C.

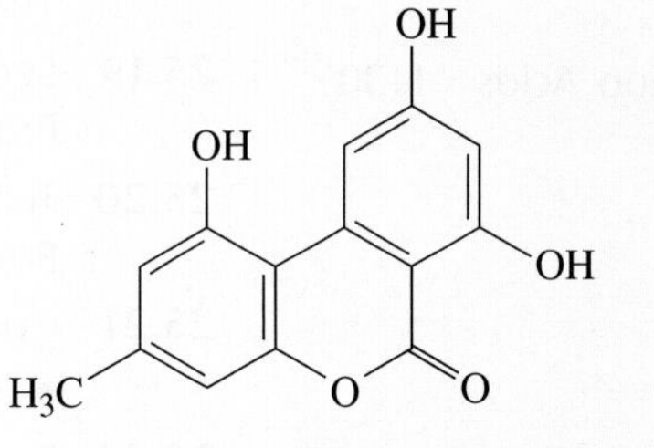

B.

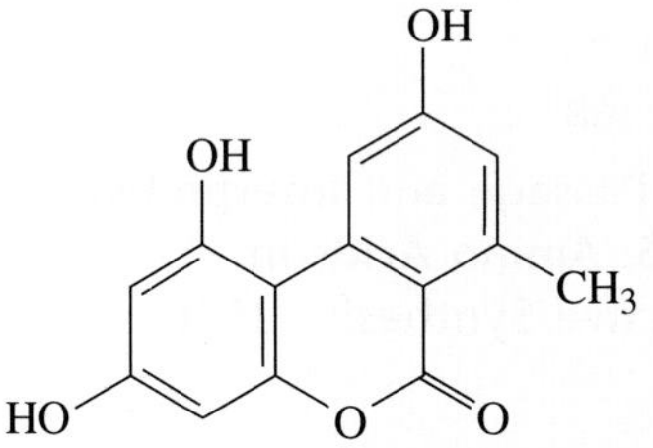

D.

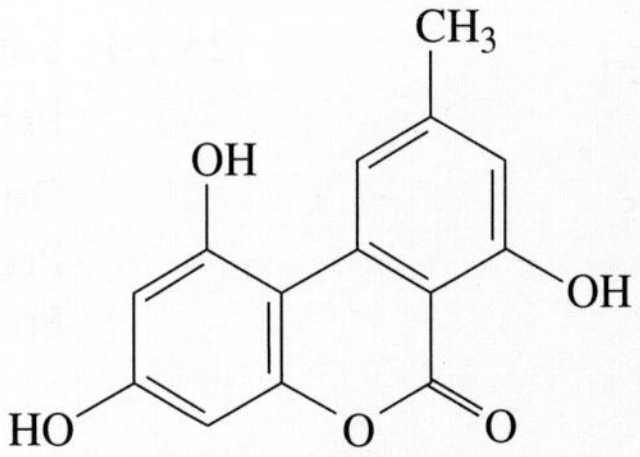

25 Amino Acids, Peptides, and Proteins

Chapter Outline

Mechanisms

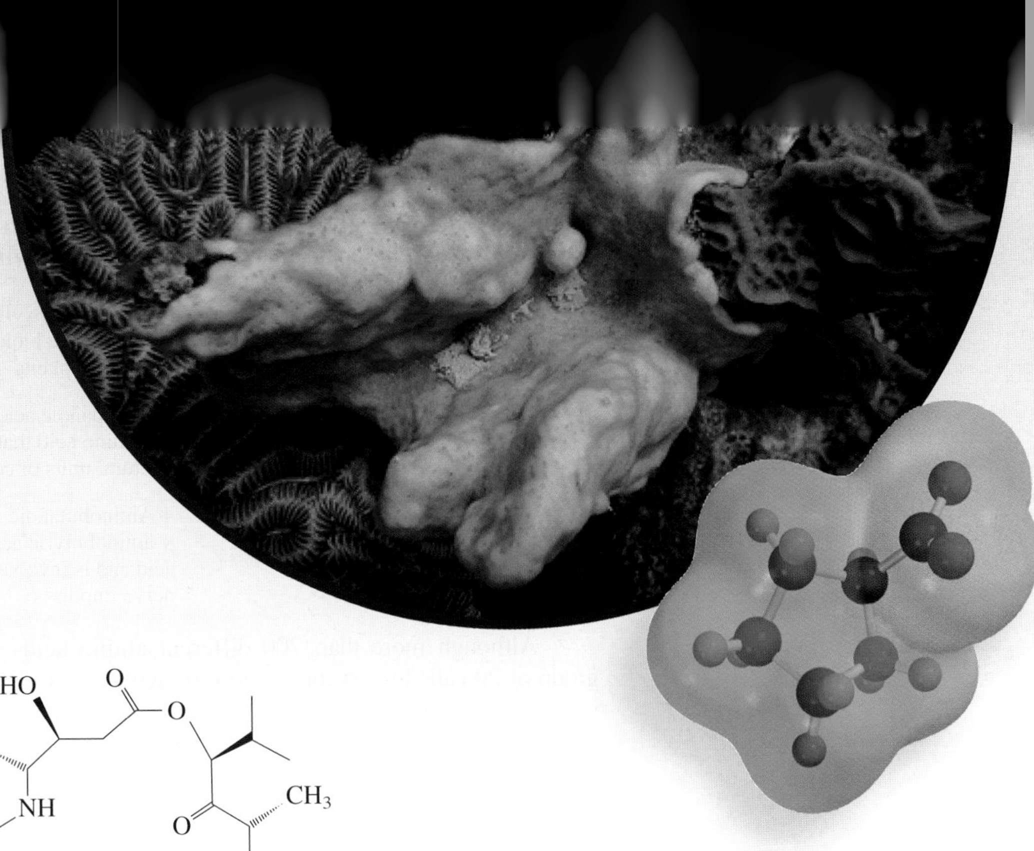

The didemnins are cyclic molecules composed of amino acids that were isolated from the marine invertebrate *Trididemnum solidum*. One of the amino acids is proline. The didemnins exhibit antitumor, antiviral, and immunosuppressive activity. The group labeled "R" in the structure is a side chain that contains additional amino acids.

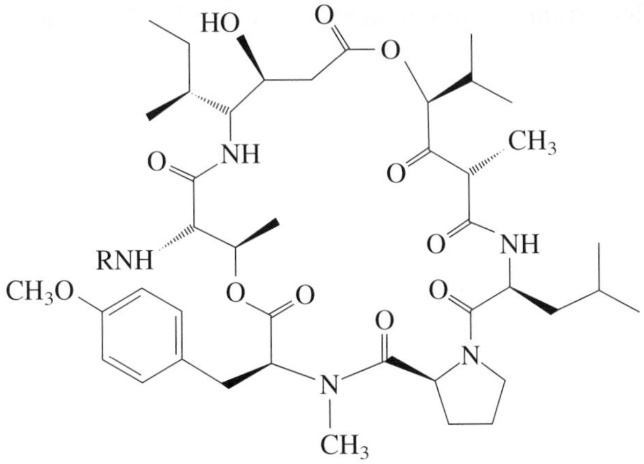

THE RELATIONSHIP between structure and function reaches its ultimate expression in the chemistry of amino acids, peptides, and proteins.

Amino acids are carboxylic acids that contain an amine function. An amide bond between the carboxylic acid function of one amino acid and the amino nitrogen of another is called a **peptide bond.**

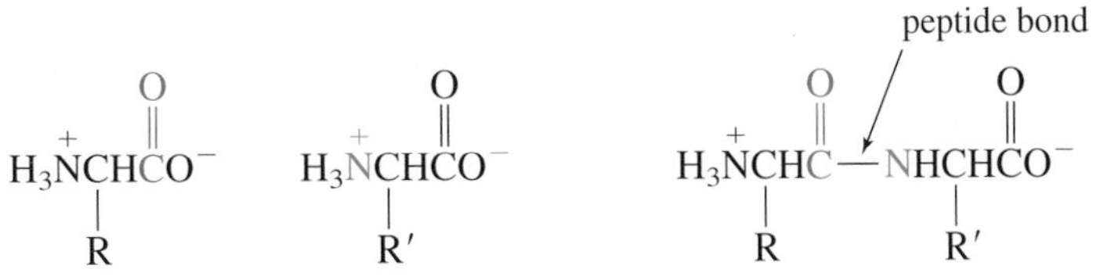

Two α-amino acids

Dipeptide

A **dipeptide** is a molecule consisting of two amino acids joined by a peptide bond. A **tripeptide** has three amino acids joined by two peptide bonds, a **tetrapeptide** has four amino acids, and so on. Peptides with more than 30–50 amino acids are **polypeptides. Proteins** are polypeptides that have some biological function.

The most striking thing about proteins is the diversity of their roles in living systems: silk is a protein, skin and hair are mostly proteins, many hormones are proteins, a protein carries oxygen from the lungs to the tissues where it is stored by another protein, and all enzymes are proteins.

As in most aspects of chemistry and biochemistry, structure is the key to function. We'll explore the structure of proteins by first concentrating on their fundamental building block units, the α-amino acids. Then, after developing the principles of peptide structure, we'll see how the insights gained from these smaller molecules aid our understanding of proteins.

25.1 Classification of Amino Acids

Amino acids are classified as α, β, γ, and so on, according to the location of the amine group on the carbon chain that contains the carboxylic acid function.

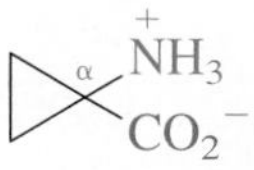

1-Aminocyclopropanecarboxylic acid: an α-amino acid that is the biological precursor to ethylene in plants

$H_3\overset{+}{N}\underset{\beta}{CH_2}\underset{\alpha}{CH_2}CO_2^-$

3-Aminopropanoic acid: known as β-alanine, it is a β-amino acid that makes up one of the structural units of coenzyme A

$H_3\overset{+}{N}\underset{\gamma}{CH_2}\underset{\beta}{CH_2}\underset{\alpha}{CH_2}CO_2^-$

4-Aminobutanoic acid: known as γ-aminobutyric acid (GABA), it is a γ-amino acid and is involved in the transmission of nerve impulses

Although more than 700 different amino acids are known to occur naturally, the group of 20 called the **standard amino acids** listed in Table 25.1 (pages 1120–1121) commands special attention. These 20 are the amino acids coded for in DNA-directed protein synthesis. All are α-amino acids, and all but one contain a primary amino function.

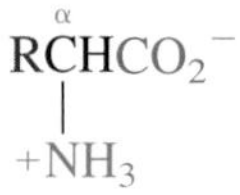

The one exception is proline, a secondary amine in which the amino nitrogen is incorporated into a five-membered ring.

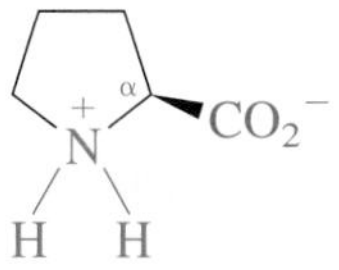

Proline

Table 25.1 includes three-letter and one-letter abbreviations for the amino acids. Both enjoy wide use.

Our bodies make some of the amino acids shown in the table. The others, which are called **essential amino acids,** we have to obtain from our diet.

When a protein contains an amino acid different from those in the table, it normally is formed by modification of one of the 20 rather than being coded for in DNA. Two exceptions have recently been discovered, selenocysteine (1986) and pyrrolysine (2002). They have been referred to as the "twenty-first and twenty-second amino acids," respectively.

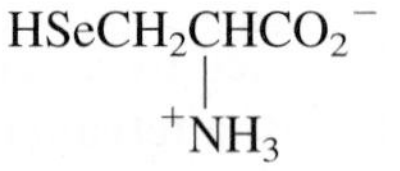

Selenocysteine

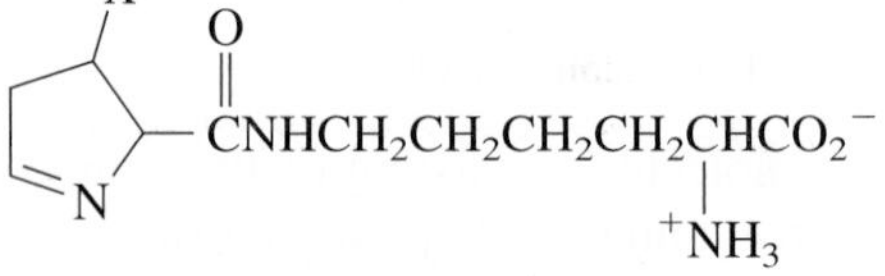

Pyrrolysine
(X is uncertain; CH_3, NH_2, or OH)

The most important aspect of Table 25.1 is that while the 20 amino acids share the common feature of being α-amino acids, their side chains differ in respect to their:

1. Size and shape
2. Electronic characteristics, acid–base properties, and ability to engage in ionic bonding, covalent bonding, hydrogen bonding, and van der Waals forces.

Table 25.1 shows the amino acids in the form in which they exist at a pH of 7: amine groups as positively charged ammonium ions, and carboxylic acid groups as negatively charged carboxylates. The electrostatic potential is mapped on the van der Waals surface of the molecule and so displays the charge distribution, size, and shape at the same time.

Nonpolar Side Chains: *Glycine* is the smallest amino acid because it has no side chain. The main service it offers is to the polypeptide chain itself. It can add length and flexibility to a polypeptide without sacrificing strength or making spatial demands of its own.

$$H_3\overset{+}{N}CH_2CO_2^-$$

Glycine

After glycine, the next four amino acids in the figure all have alkyl groups (R) as side chains: *alanine* (R = methyl), *valine* (R = isopropyl), *leucine* (R = isobutyl), and *isoleucine* (R = *sec*-butyl). All are hydrophobic side chains and although electronically similar, they differ in size. Alanine is slightly larger than glycine, valine slightly larger than alanine, leucine slightly larger than valine, and isoleucine somewhat more spherical than leucine.

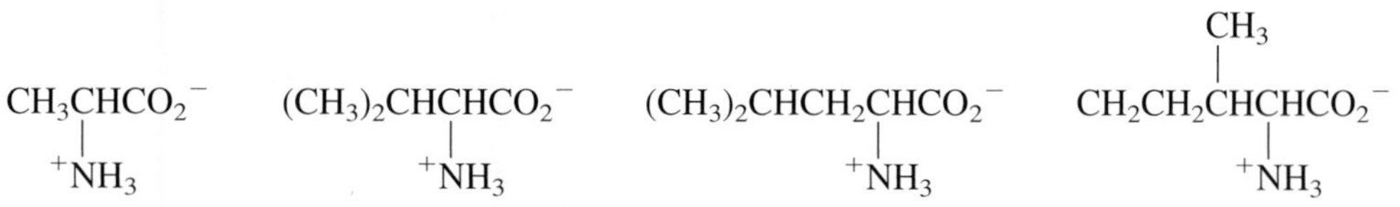

Alanine Valine Leucine Isoleucine

Compared with these, the presence of sulfur in its side chain makes *methionine* somewhat more polarizable and increases its ability to participate in dispersion forces.

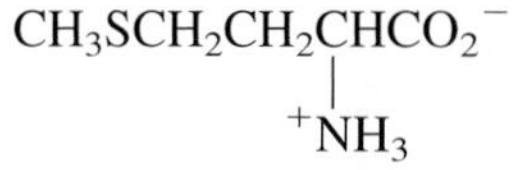

Methionine

Proline is relatively compact and has limited conformational flexibility because its side chain is cyclic. Moreover, amides of proline lack N—H bonds so cannot form hydrogen bonds. Consequently, the presence of proline affects the shape of a peptide chain more than most other amino acids.

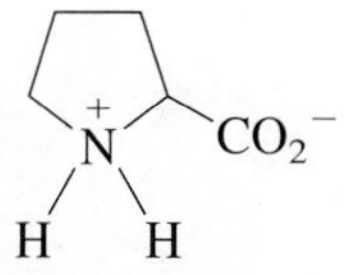

Proline

Phenylalanine and *tryptophan* have side chains that incorporate aromatic rings, which are large and hydrophobic. In addition to being larger than phenylalanine, tryptophan has a more electron-rich aromatic ring and is more polarizable. Its role is more specialized, and it is less abundant in proteins than most of the other amino acids.

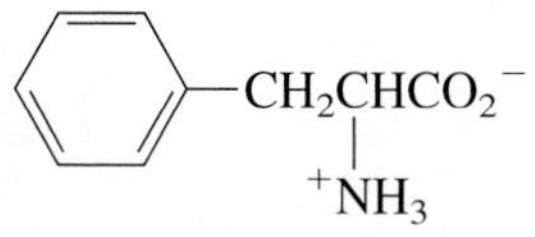

Phenylalanine

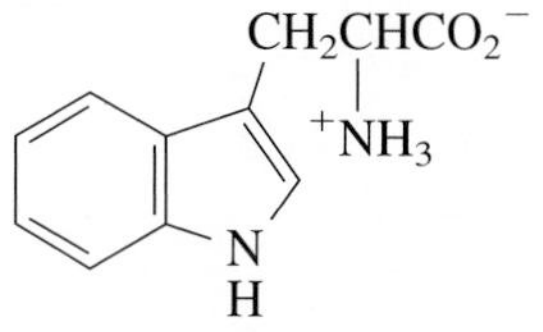

Tryptophan

Amino Acids with Polar but Nonionized Side Chains: Among amino acids with polar side chains, *serine* is the smallest; it is not much larger than alanine. With a $—CH_2OH$ side chain, serine participates well in hydrogen bonding and often occurs in regions of a peptide that are exposed to water. *Threonine* has a methyl group in place of one of the hydrogens of the $—CH_2OH$ group of serine, sterically hindering the OH group and making it less effective in hydrogen bonding.

TABLE 25.1 The Standard Amino Acids

Name and abbreviation	Structural formula*	Electrostatic potential map	Name and abbreviation	Structural formula*	Electrostatic potential map
Amino acids with nonpolar side chains					
Glycine Gly (G)	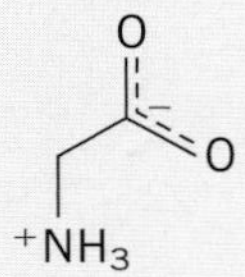	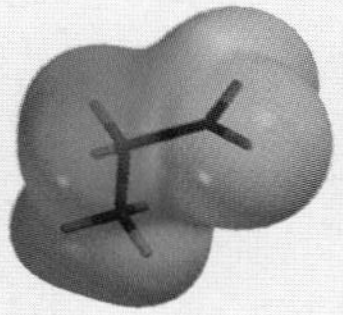	Methionine[†] Met (M)	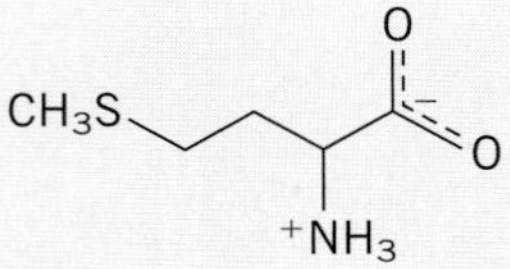	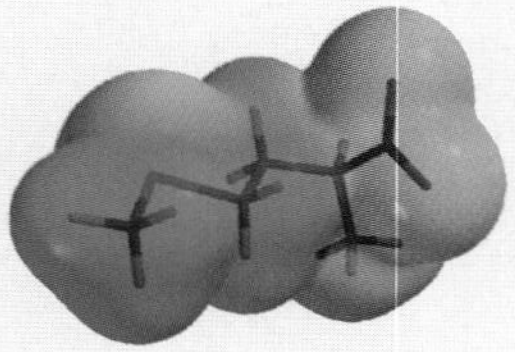
Alanine Ala (A)	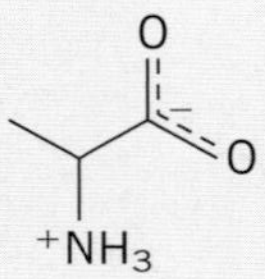	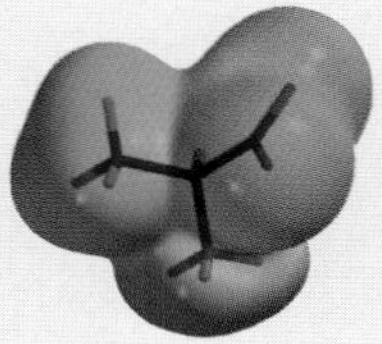	Proline Pro (P)	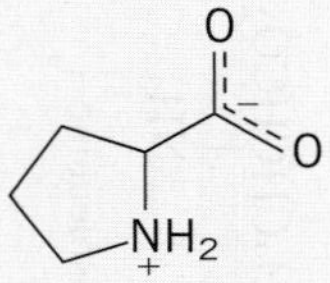	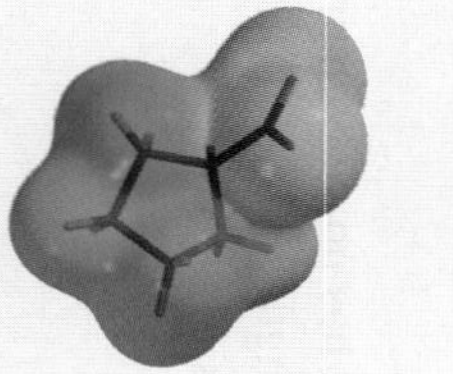
Valine[†] Val (V)	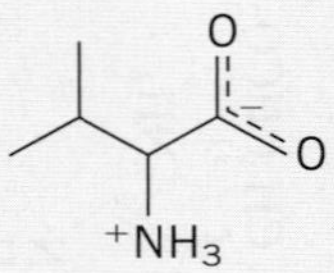	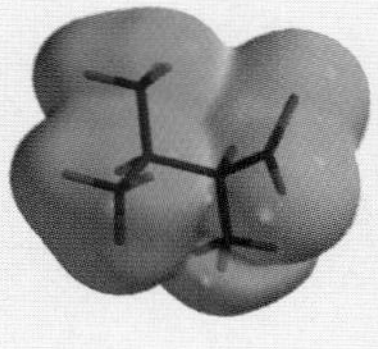	Phenylalanine[†] Phe (F)	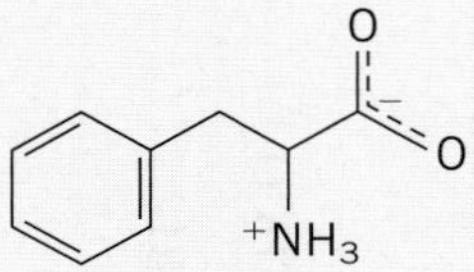	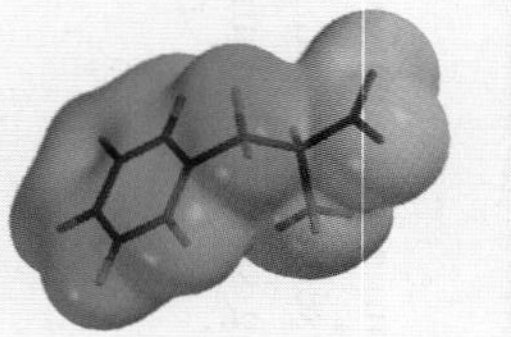
Leucine[†] Leu (L)	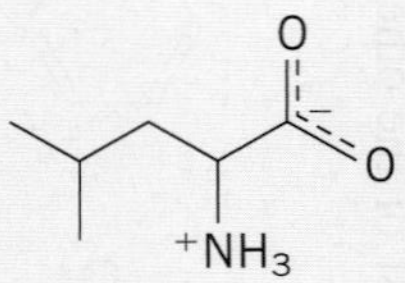	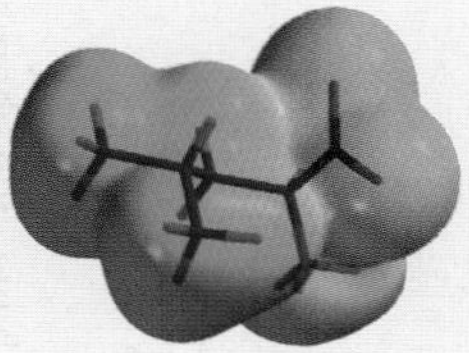	Tryptophan[†] Trp (W)	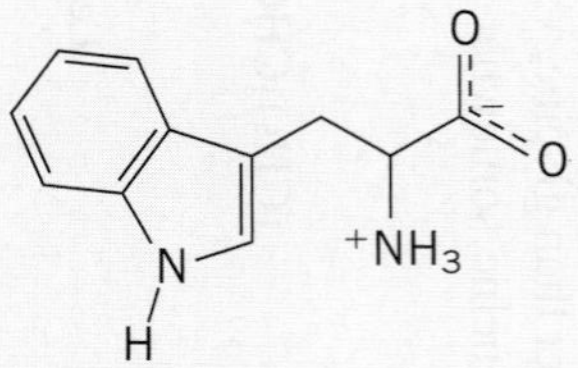	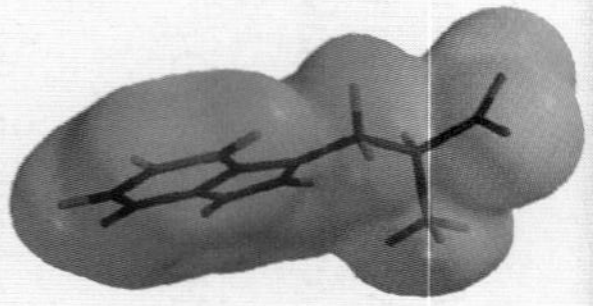
			Amino acids with polar but nonionized side chains		
Isoleucine[†] Ile (I)	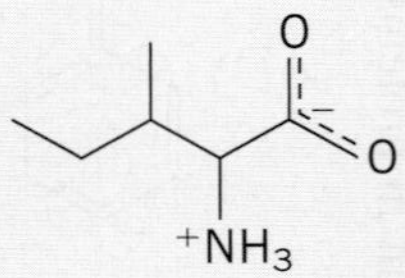	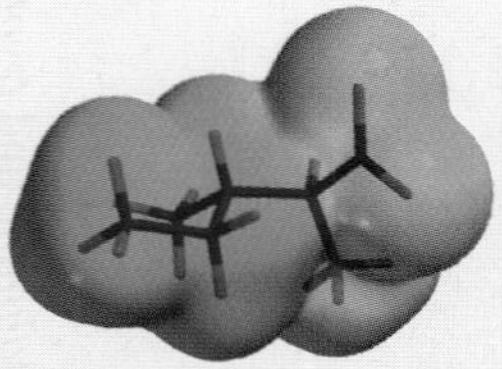	Asparagine Asn (N)	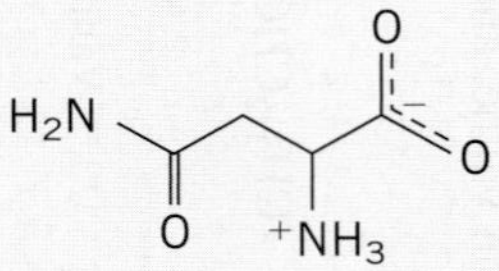	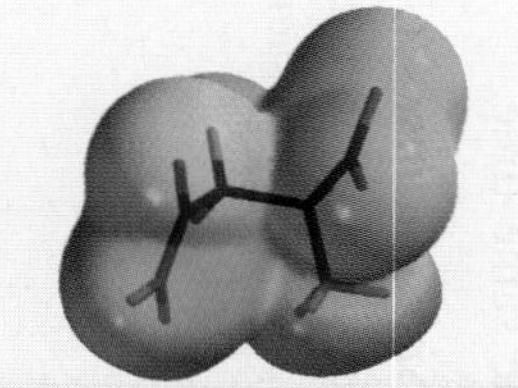

*All amino acids are shown in the form present in greatest concentration at pH 7.
[†]An essential amino acid, which must be present in the diet of animals to ensure normal growth.

TABLE 25.1 The Standard Amino Acids (*Continued*)

Name and abbreviation	Structural formula*	Electrostatic potential map	Name and abbreviation	Structural formula*	Electrostatic potential map
Amino acids with polar but nonionized side chains (*continued*)			**Amino acids with acidic side chains**		
Glutamine Gln (N)	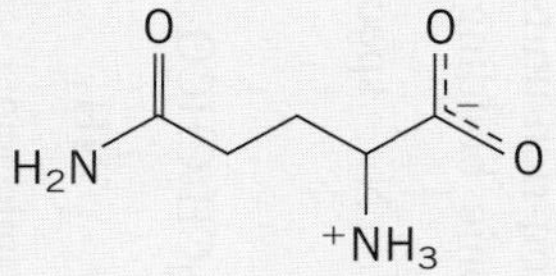	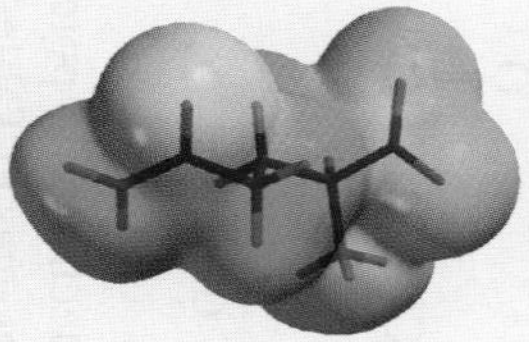	Aspartic acid Asp (D)	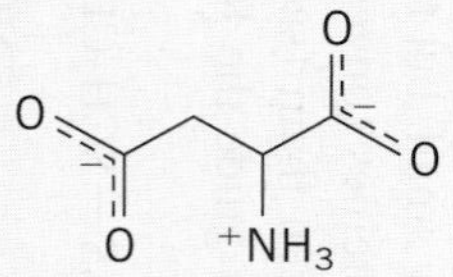	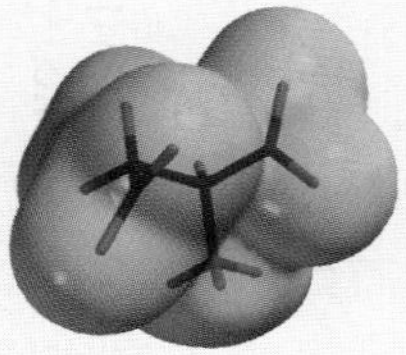
Serine Ser (S)	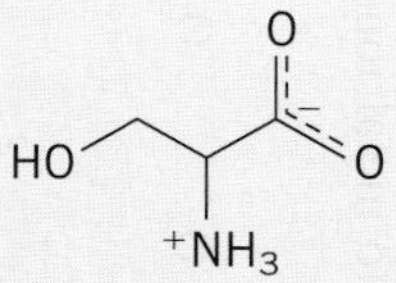	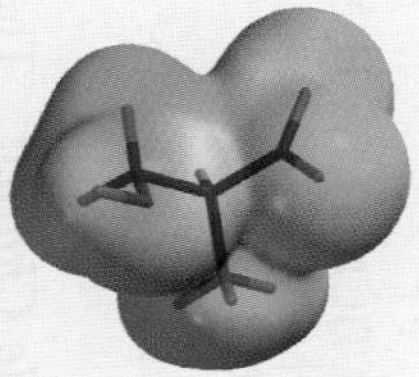	Glutamic acid Glu (E)	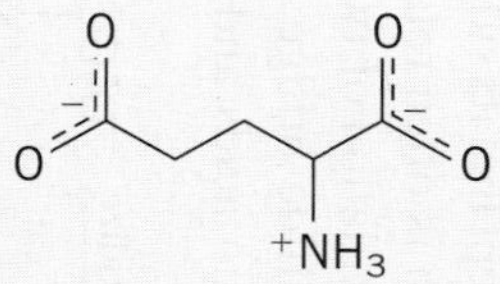	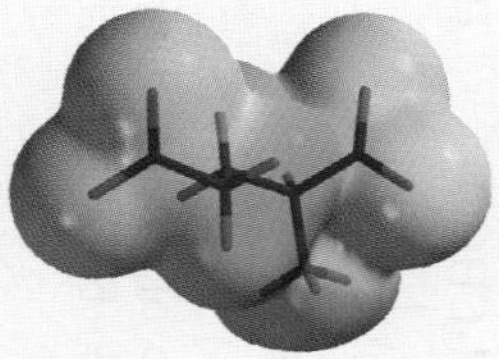
			Amino acids with basic side chains		
Threonine Thr (T)	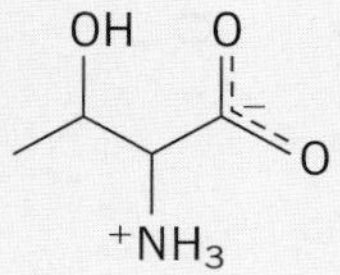	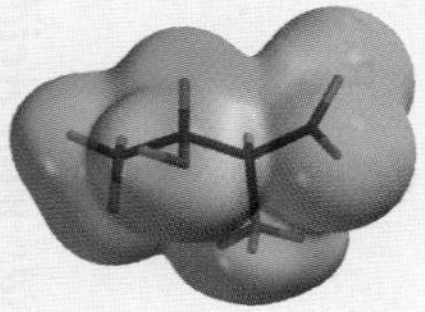	Lysine Lys (K)	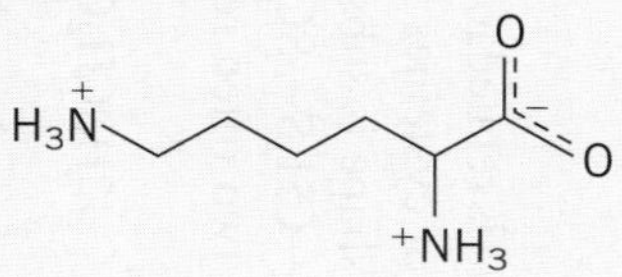	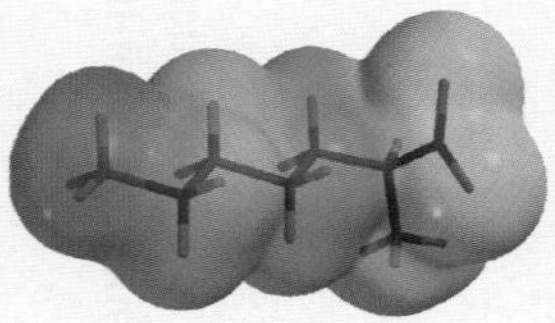
Tyrosine Tyr (Y)	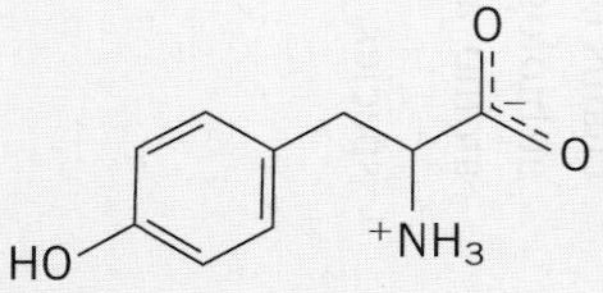	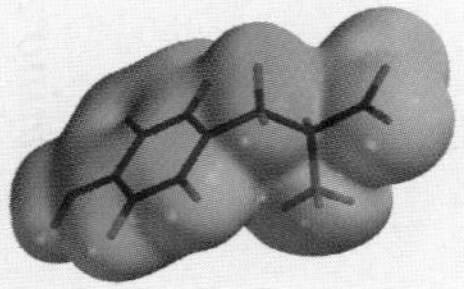	Arginine[†] Arg (R)	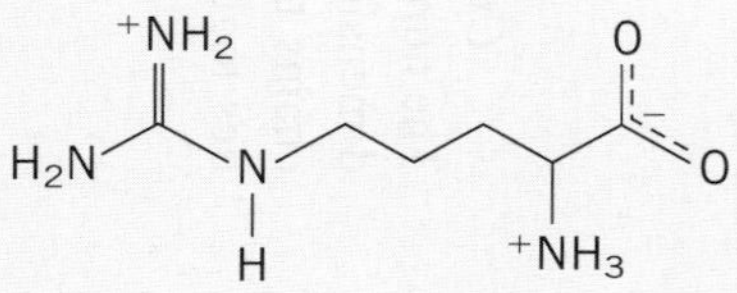	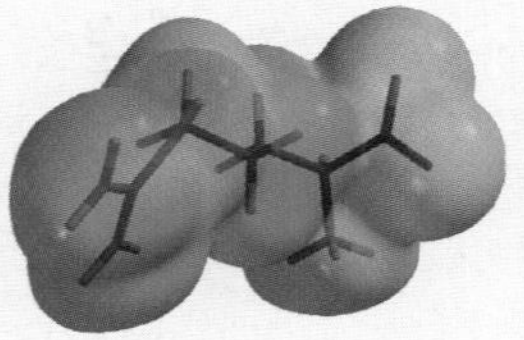
Cysteine Cys (C)	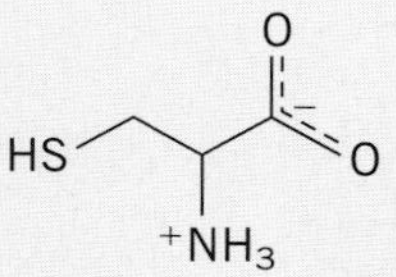	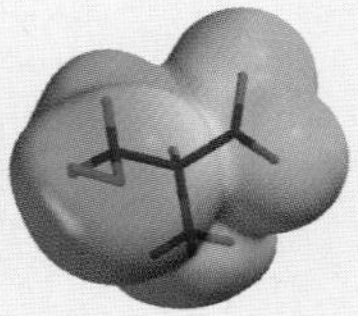	Histidine[†] His (H)	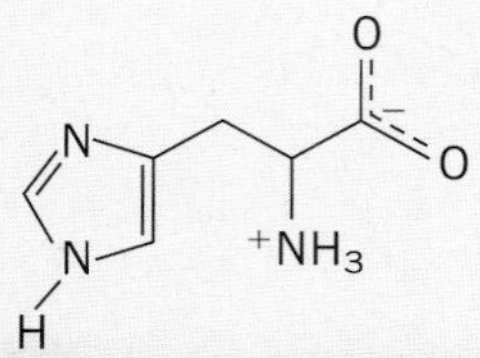	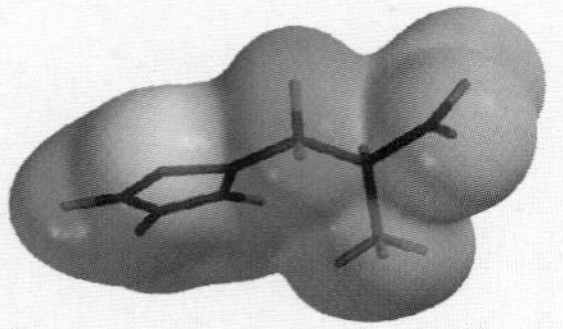

*All amino acids are shown in the form present in greatest concentration at pH 7.

[†]An essential amino acid, which must be present in the diet of animals to ensure normal growth.

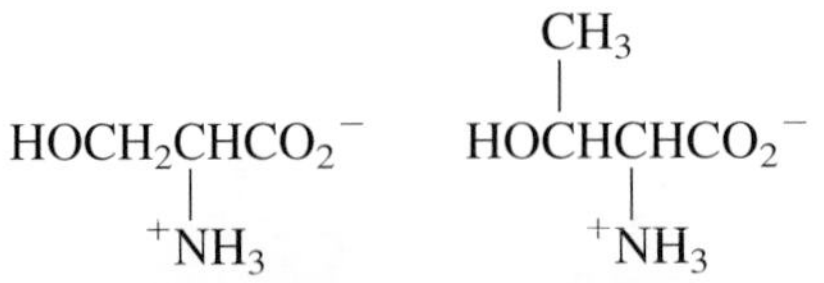

Serine Threonine

As a *p*-hydroxy derivative of phenylalanine, *tyrosine* has properties similar to it plus the ability to form hydrogen bonds involving its phenolic —OH group.

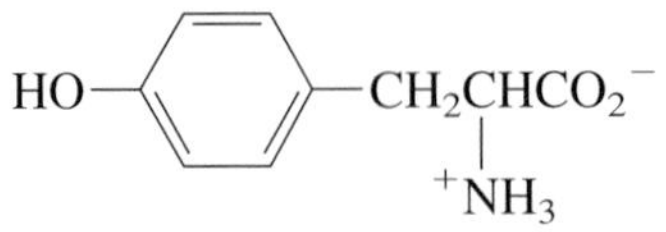

Tyrosine

Cysteine is related to serine except that its side chain is $—CH_2SH$ rather than $—CH_2OH$. The number of cysteines in a protein is often relatively small, but their effect on its three-dimensional shape is substantial. Oxidation of two cysteines converts their $—CH_2SH$ side chains to a $—CH_2S—SCH_2—$ disulfide bridge between them (see Section 15.12). This ties together two often remote amino acids and helps guide the folding of the protein.

Cysteine Cystine

Asparagine and *glutamine* are amides. The side chains of both terminate in $—\overset{O}{\overset{\|}{C}}NH_2$ and differ by only a single CH_2 group. Amide functions are quite polar and interact strongly with water molecules by hydrogen bonding. Like serine, asparagine and glutamine are often found in regions of a peptide that are in contact with water.

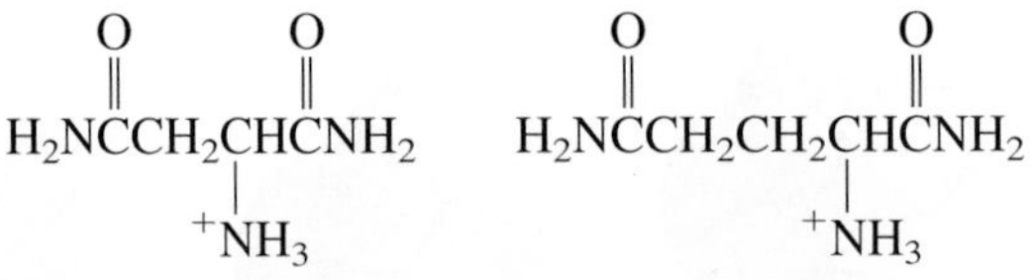

Asparagine Glutamine

Amino Acids with Acidic Side Chains: The electrostatic potential maps of *aspartic acid* and *glutamic acid* are two of the most prominent ones in Table 25.1. The $—CO_2H$ side chains of both aspartic and glutamic acid are almost completely deprotonated to $—CO_2^-$ at biological pH, giving these species the most electron-rich units of all of the common amino acids. Their most important function is in ionic bonding to positively charged species. These ionic bonds can be to metal ions and $—\underset{|}{\overset{|}{N}}{}^{+}—$ species among others.

Aspartic acid Glutamic acid

Amino Acids with Basic Side Chains: Basic amino acids are the opposite of acidic amino acids. Their most important role is to form ionic bonds to negative ions—phosphate and the like. Lysine is a simple example. The side chain contains four CH_2 groups and terminates in $—NH_3^+$. Arginine has an even more basic, if somewhat more complicated and larger, side chain. Conversely, the side chain of histidine is not as basic as that of lysine and the concentrations of the unprotonated and protonated forms of histidine are almost equal at biological pH. Strong Lewis acid/Lewis base complexes between the unprotonated form of histidine and metal ions is very common in proteins. Histidine side chains are also involved in moving protons from one atom to another.

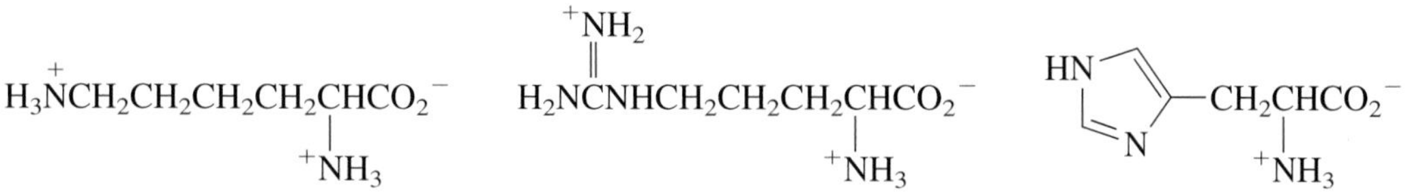

Lysine Arginine Histidine

What is remarkable is not only the range of properties that are covered with just 20 amino acids, but also the fine tuning with respect to a particular property.

25.2 Stereochemistry of Amino Acids

Glycine is the only amino acid in Table 25.1 that is achiral. The α-carbon atom is a chirality center in all the others. Configurations in amino acids are normally specified by the D, L notational system. All the chiral amino acids obtained from proteins have the L configuration at their α-carbon atom, meaning that the amine group is at the left when a Fischer projection is arranged so the carboxyl group is at the top.

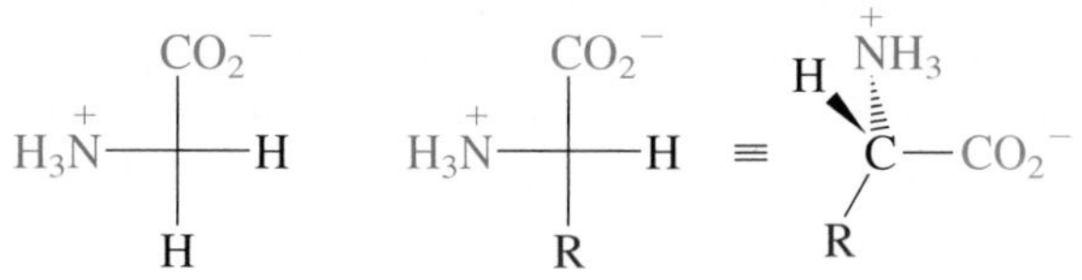

Glycine (achiral)

Fischer projection of an L-amino acid

Problem 25.1

What is the absolute configuration (*R* or *S*) at the α-carbon atom in each of the following L-amino acids?

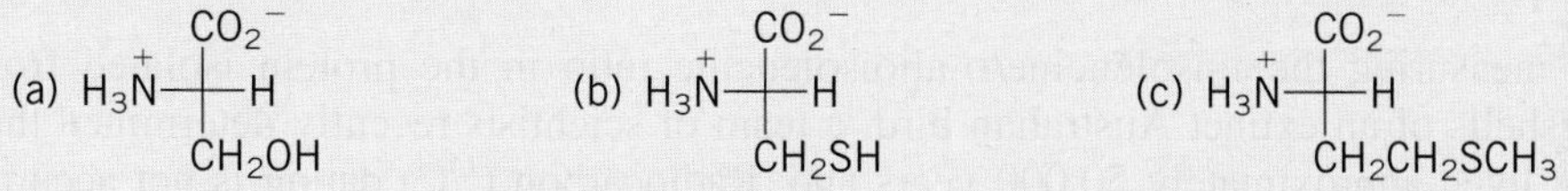

L-Serine L-Cysteine L-Methionine

Sample Solution (a) First identify the four groups attached directly to the chirality center, and rank them in order of decreasing sequence rule precedence. For L-serine these groups are

$$H_3\overset{+}{N}\text{—} > \text{—}CO_2^- > \text{—}CH_2OH > H$$

Highest ranked Lowest ranked

Next, translate the Fischer projection of L-serine to a three-dimensional representation, and orient it so that the lowest ranked substituent at the chirality center is directed away from you.

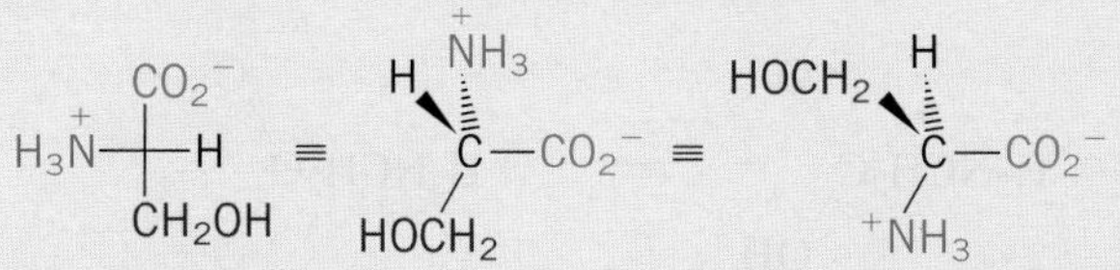

Continued

In order of decreasing precedence the three highest ranked groups trace a counterclockwise path.

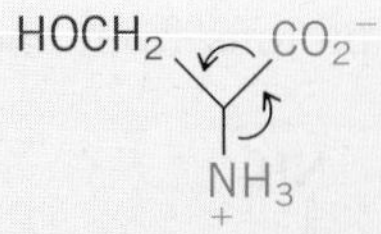

The absolute configuration of L-serine is *S*.

Problem 25.2

The amino acid L-threonine is (2*S*,3*R*)-2-amino-3-hydroxybutanoic acid. Draw a Fischer projection for L-threonine.

Although all the chiral amino acids obtained from proteins have the L configuration at their α carbon, that should not be taken to mean that D-amino acids are unknown. In fact, quite a number of D-amino acids occur naturally. D-Alanine, for example, is a constituent of bacterial cell walls and D-serine occurs in brain tissue. The point is that D-amino acids are not coded for by DNA.

A novel technique for dating archaeological samples called **amino acid racemization (AAR)** is based on the stereochemistry of amino acids. Over time, the configuration at the α-carbon atom of a protein's amino acids is lost in a reaction that follows first-order kinetics. When the α carbon is the only chirality center, this process corresponds to racemization. For an amino acid with two chirality centers, changing the configuration of the α carbon from L to D gives a diastereomer. In the case of isoleucine, for example, the diastereomer is an amino acid not normally present in proteins, called *alloisoleucine.*

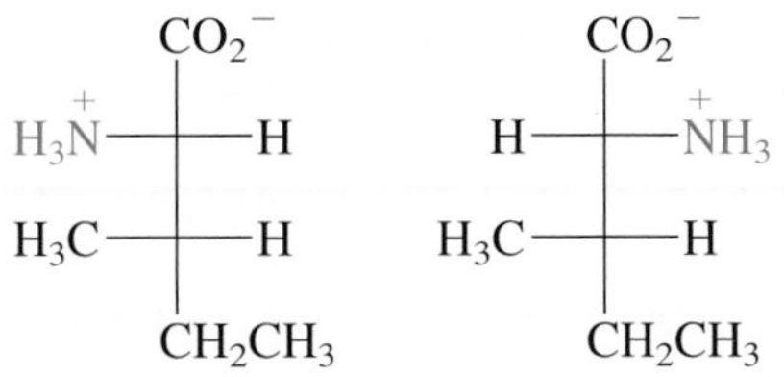

L-Isoleucine D-Alloisoleucine

By measuring the L-isoleucine/D-alloisoleucine ratio in the protein isolated from the eggshells of an extinct Australian bird, a team of scientists recently determined that this bird lived approximately 50,000 years ago. Radiocarbon (^{14}C) dating is not accurate for samples older than about 35,000 years, so AAR is a useful addition to the tools available to paleontologists.

25.3 Acid–Base Behavior of Amino Acids

The physical properties of a typical amino acid such as glycine suggest that it is a very polar substance, much more polar than would be expected on the basis of its formulation as $H_2NCH_2CO_2H$. Glycine is a crystalline solid; it does not melt, but on being heated it eventually decomposes at 233°C. It is very soluble in water but practically insoluble in nonpolar organic solvents. These properties are attributed to the fact that the stable form of glycine is a **zwitterion.**

The zwitterion is also often referred to as a *dipolar ion*. Note, however, that it is not an ion, but a neutral molecule.

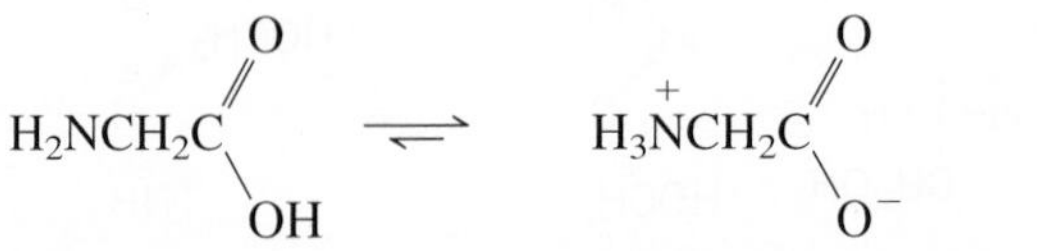

Zwitterionic form of glycine

Stronger acid $pK_a \approx 5$ Weaker acid $pK_a \approx 9$

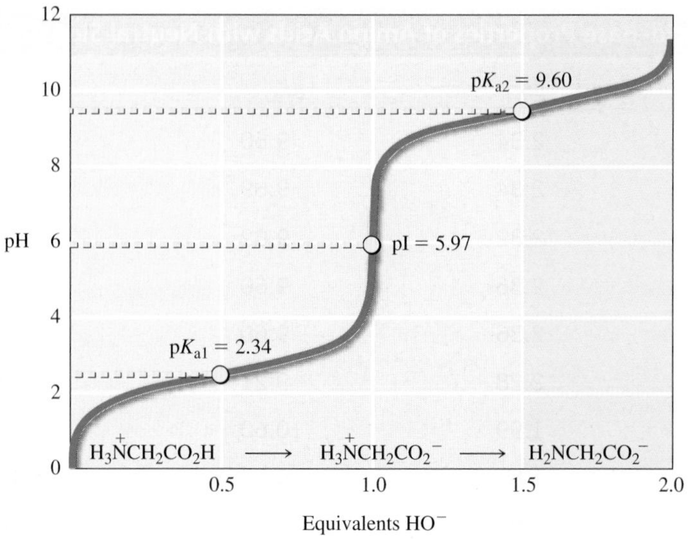

Figure 25.1

Titration curve of glycine. At pH values lower than 2.34, $H_3\overset{+}{N}CH_2CO_2H$ is the major species. At pH = 2.34 $[H_3\overset{+}{N}CH_2CO_2H] = [H_3\overset{+}{N}CH_2CO_2^-]$. Between pH = 2.34 and 9.60, $H_3\overset{+}{N}CH_2CO_2^-$ is the major species. Its concentration is a maximum at the isoelectric point (pI = 5.97). At pH = 9.60, $[H_3\overset{+}{N}CH_2CO_2^-] = [H_2NCH_2CO_2^-]$. Above pH = 9.60, $H_2NCH_2CO_2^-$ is the predominant species.

The equilibrium expressed by the preceding equation lies overwhelmingly to the side of the zwitterion.

Glycine, as well as other amino acids, is *amphoteric,* meaning it contains an acidic functional group and a basic functional group. The acidic functional group is the ammonium ion $H_3\overset{+}{N}$—; the basic functional group is the carboxylate ion $—CO_2^-$. How do we know this? Aside from its physical properties, the acid–base properties of glycine, as illustrated by the titration curve in Figure 25.1, require it. In a strongly acidic medium the species present is the cation $H_3\overset{+}{N}CH_2CO_2H$. As the pH is raised, its most acidic proton is removed. Is this proton removed from the positively charged nitrogen or from the carboxyl group? We know what to expect for the relative acid strengths of $R\overset{+}{N}H_3$ and RCO_2H. A typical ammonium ion has $pK_a \approx 9$, and a typical carboxylic acid has $pK_a \approx 5$. The measured pK_a for the conjugate acid of glycine is 2.34, a value closer to that expected for deprotonation of the carboxyl group. As the pH is raised, a second deprotonation step, corresponding to removal of a proton from nitrogen of the zwitterion, is observed. The pK_a associated with this step is 9.60, much like that of typical alkylammonium ions.

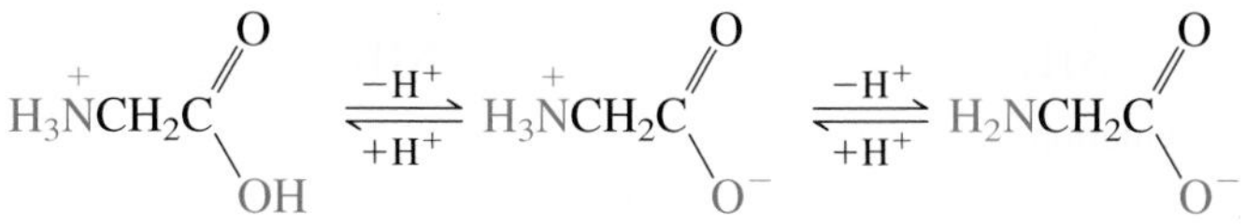

Species present in strong acid

Zwitterion

Species present in strong base

Thus, glycine is characterized by two pK_a values: the one corresponding to the more acidic site is designated pK_{a1}, the one corresponding to the less acidic site is designated pK_{a2}. Table 25.2 lists pK_{a1} and pK_{a2} values for the α-amino acids that have neutral side chains, which are the first two groups of amino acids given in Table 25.1. In all cases their pK_a values are similar to those of glycine.

Table 25.2 includes a column labeled pI, which is the *isoelectric point* of the amino acid. The **isoelectric point,** also called the **isoionic point,** is the pH at which the amino acid has no net charge. It is the pH at which the concentration of the zwitterion is a maximum. At a pH lower than pI, the amino acid is positively charged; at a pH higher than pI, the amino acid is negatively charged. For the amino acids in Table 25.2, pI is the average of pK_{a1} and pK_{a2} and lies slightly to the acid side of neutrality.

Some amino acids have side chains that bear acidic or basic groups. As Table 25.3 indicates, these amino acids are characterized by three pK_a values. The third pK_a reflects the nature of the side chain. Acidic amino acids (aspartic and glutamic acid) have acidic side chains; basic amino acids (lysine, arginine, and histidine) have basic side chains.

The isoelectric points of the amino acids in Table 25.3 are midway between the pK_a values of the zwitterion and its conjugate acid. Take two examples: aspartic acid and lysine.

TABLE 25.2 Acid–Base Properties of Amino Acids with Neutral Side Chains

Amino acid	pK_{a1}*	pK_{a2}*	pI
Glycine	2.34	9.60	5.97
Alanine	2.34	9.69	6.00
Valine	2.32	9.62	5.96
Leucine	2.36	9.60	5.98
Isoleucine	2.36	9.60	6.02
Methionine	2.28	9.21	5.74
Proline	1.99	10.60	6.30
Phenylalanine	1.83	9.13	5.48
Tryptophan	2.83	9.39	5.89
Asparagine	2.02	8.80	5.41
Glutamine	2.17	9.13	5.65
Serine	2.21	9.15	5.68
Threonine	2.09	9.10	5.60
Tyrosine	2.20	9.11	5.66

*In all cases, pK_{a1} corresponds to ionization of the carboxyl group; pK_{a2} corresponds to deprotonation of the ammonium ion.

Aspartic acid has an acidic side chain and a pI of 2.77. Lysine has a basic side chain and a pI of 9.74.

Aspartic Acid:

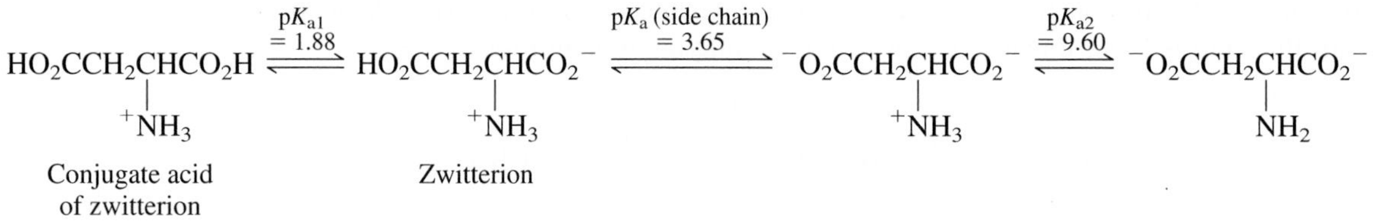
pK_{a1} = 1.88
pK_a (side chain) = 3.65
pK_{a2} = 9.60
$HO_2CCH_2CHCO_2H$
$^+NH_3$
$HO_2CCH_2CHCO_2^-$
$^+NH_3$
$^-O_2CCH_2CHCO_2^-$
$^+NH_3$
$^-O_2CCH_2CHCO_2^-$
NH_2
Conjugate acid of zwitterion
Zwitterion

The pI of aspartic acid is the average of pK_{a1} (1.88) and the pK_a of the side chain (3.65) or 2.77.

TABLE 25.3 Acid–Base Properties of Amino Acids with Ionizable Side Chains

Amino acid	pK_{a1}*	pK_{a2}	pK_a of side chain	pI
Aspartic acid	1.88	9.60	3.65	2.77
Glutamic acid	2.19	9.67	4.25	3.22
Lysine	2.18	8.95	10.53	9.74
Arginine	2.17	9.04	12.48	10.76
Histidine	1.82	9.17	6.00	7.59

*In all cases, pK_{a1} corresponds to ionization of the carboxyl group of $R\underset{\substack{|\\ \overset{+}{N}H_3}}{C}HCO_2H$ and pK_{a2} to ionization of the ammonium ion.

Lysine:

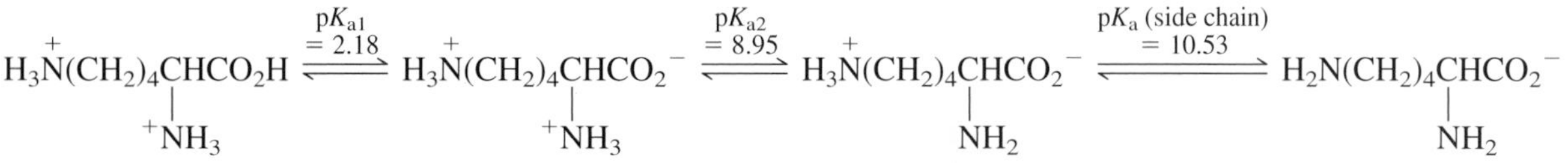

Conjugate acid of zwitterion

Zwitterion

The pI of lysine is the average of pK_{a2} (8.95) and the pK_a of the side chain (10.53) or 9.74.

Problem 25.3

Cysteine has pK_{a1} = 1.96 and pK_{a2} = 10.28. The pK_a for ionization of the —SH group of the side chain is 8.18. What is the isoelectric point of cysteine?

Problem 25.4

Above a pH of about 10, the major species present in a solution of tyrosine has a net charge of −2. Suggest a reasonable structure for this species.

Individual amino acids differ in their acid–base properties. This is important in peptides and proteins, where the properties of the substance depend on its amino acid constituents, especially on the nature of the side chains. It is also important in analyses in which a complex mixture of amino acids is separated into its components by taking advantage of the differences in their proton-donating and accepting power.

Electrophoresis

Electrophoresis is a method for separation and purification that depends on the movement of charged particles in an electric field. Its principles can be introduced by considering some representative amino acids. The medium is a cellulose acetate strip that is moistened with an aqueous solution buffered at a particular pH. The opposite ends of the strip are placed in separate compartments containing the buffer, and each compartment is connected to a source of direct electric current (Figure 25.2*a*). If the buffer solution is more acidic than the isoelectric point (pI) of the amino acid, the amino acid has a net positive charge and migrates toward the negatively charged electrode. Conversely, when the buffer is more basic than the pI of the amino acid, the amino acid has a net negative charge and migrates toward the positively charged electrode. When the pH of the buffer corresponds to the pI, the amino acid has no net charge and does not migrate from the origin.

Thus if a mixture containing alanine, aspartic acid, and lysine is subjected to electrophoresis in a buffer that matches the isoelectric point of alanine (pH 6.0), aspartic acid (pI = 2.8) migrates toward the positive electrode, alanine remains at the origin, and lysine (pI = 9.7) migrates toward the negative electrode (Figure 25.2*b*).

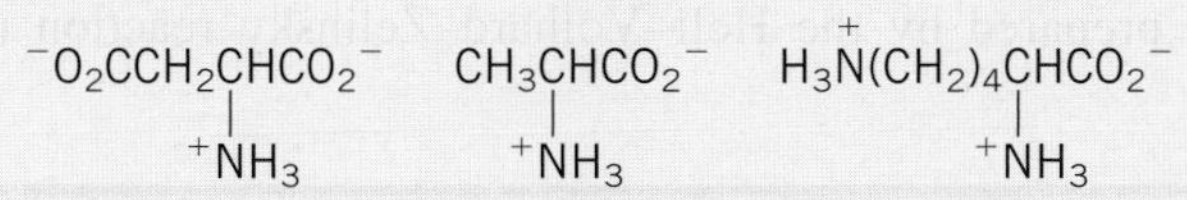

Aspartic acid (−1 ion) | Alanine (neutral) | Lysine (+1 ion)

Electrophoresis is used primarily to analyze mixtures of peptides and proteins, rather than individual amino acids, but analogous principles apply. Because they incorporate different numbers of amino acids and because their side chains are different, two peptides will have slightly different acid–base properties and slightly different net charges at a particular pH. Thus, their mobilities in an electric field will be different, and electrophoresis can be used to separate them. The medium used to separate peptides and proteins is typically a polyacrylamide gel, leading to the term *gel electrophoresis* for this technique.

A second factor that governs the rate of migration during electrophoresis is the size (length and shape) of the peptide or protein. Larger molecules move through the polyacrylamide gel more slowly than smaller ones. In current practice, the experiment is modified to exploit differences in size more than differences in net charge, especially in the *SDS gel electrophoresis* of proteins. Approximately 1.5 g of the detergent *sodium dodecyl sulfate* (SDS, page 788) per gram of protein is added to the aqueous buffer. SDS binds to the protein, causing the protein to unfold so that it is roughly rod-shaped with the $CH_3(CH_2)_{10}CH_2$— groups of SDS associated with the lipophilic (hydrophobic) portions of the protein. The negatively charged sulfate groups are exposed to the water. The SDS molecules that they carry ensure that all the protein molecules are negatively charged and migrate toward the positive electrode. Further more, all the proteins in the mixture now have similar shapes and tend to travel at rates proportional

Continued

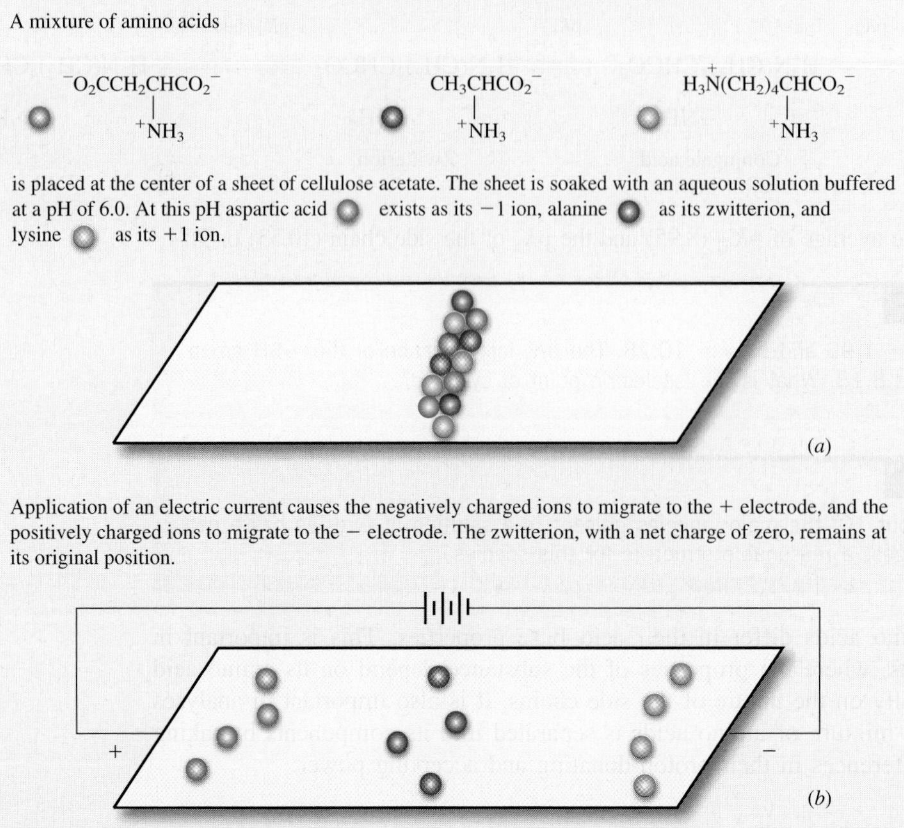
A mixture of amino acids
$^{-}O_2CCH_2CHCO_2^{-}$ ($^{+}NH_3$)
$CH_3CHCO_2^{-}$ ($^{+}NH_3$)
$H_3\overset{+}{N}(CH_2)_4CHCO_2^{-}$ ($^{+}NH_3$)
is placed at the center of a sheet of cellulose acetate. The sheet is soaked with an aqueous solution buffered at a pH of 6.0. At this pH aspartic acid exists as its −1 ion, alanine as its zwitterion, and lysine as its +1 ion.
(a)
Application of an electric current causes the negatively charged ions to migrate to the + electrode, and the positively charged ions to migrate to the − electrode. The zwitterion, with a net charge of zero, remains at its original position.
+
−
(b)

Figure 25.2

Application of electrophoresis to the separation of aspartic acid, alanine, and lysine according to their charge type at a pH corresponding to the isoelectric point (pI) of alanine.

to their chain length. Thus, when carried out on a preparative scale, SDS gel electrophoresis permits proteins in a mixture to be separated according to their molecular weight. On an analytical scale, it is used to estimate the molecular weight of a protein by comparing its electrophoretic mobility with that of proteins of known molecular weight.

Later, in Chapter 26, we will see how gel electrophoresis is used in nucleic acid chemistry.

25.4 Synthesis of Amino Acids

One of the earliest methods for the synthesis of amino acids dates back to the nineteenth century and is simply a nucleophilic substitution in which ammonia reacts with an α-halo carboxylic acid.

$$\underset{\text{2-Bromopropanoic acid}}{CH_3\underset{\displaystyle Br}{CH}CO_2H} + \underset{\text{Ammonia}}{2NH_3} \xrightarrow{H_2O} \underset{\text{Alanine (65–70\%)}}{CH_3\underset{\displaystyle \overset{+}{N}H_3}{CH}CO_2^{-}} + \underset{\text{Ammonium bromide}}{NH_4Br}$$

The α-halo acid is normally prepared by the Hell–Volhard–Zelinsky reaction (Section 20.15).

Problem 25.5

Outline the steps in a synthesis of valine from 3-methylbutanoic acid.

In the **Strecker synthesis** an aldehyde is converted to an α-amino acid with one more carbon atom by a two-stage procedure in which an α-amino nitrile is an intermediate. The α-amino nitrile is formed by reaction of the aldehyde with ammonia or an ammonium salt and a source of cyanide ion. Hydrolysis of the nitrile group to a carboxylic acid function completes the synthesis.

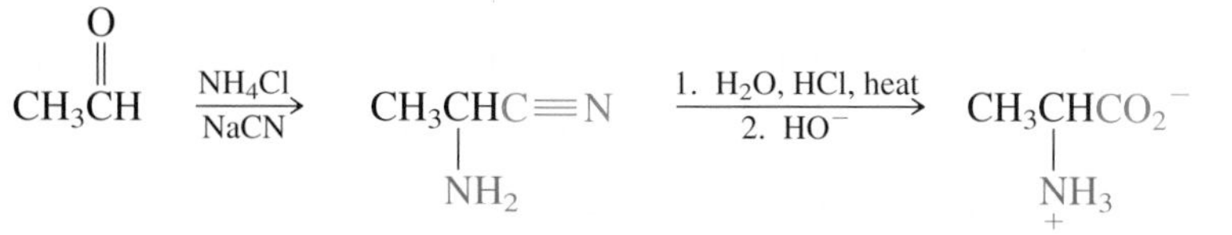

Acetaldehyde 2-Aminopropanenitrile Alanine (52–60%)

The synthesis of alanine was described by Adolf Strecker of the University of Würzburg (Germany) in a paper published in 1850.

Problem 25.6

Outline the steps in the preparation of valine by the Strecker synthesis.

A widely used method for the laboratory synthesis of α-amino acids is a modification of the malonic ester synthesis (see Section 20.11). The key reagent is *diethyl acetamidomalonate,* a derivative of malonic ester that already has the critical nitrogen substituent in place at the α-carbon atom. The side chain is introduced by alkylating diethyl acetamidomalonate in the same way as diethyl malonate itself is alkylated.

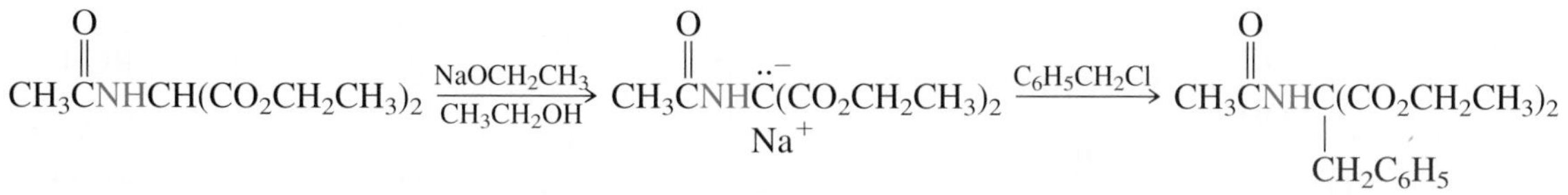

Diethyl acetamidomalonate — Sodium salt of diethyl acetamidomalonate — Diethyl acetamidobenzylmalonate (90%)

Hydrolysis removes the acetyl group from nitrogen and converts the two ester functions to carboxyl groups. Decarboxylation gives the desired product.

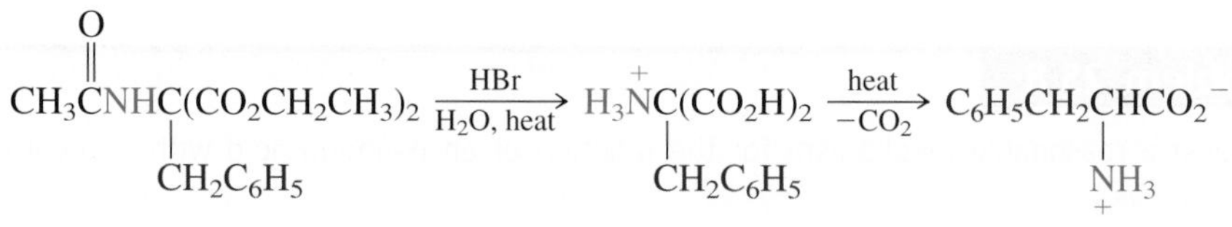

Diethyl acetamidobenzylmalonate — (not isolated) — Phenylalanine (65%)

Problem 25.7

Outline the steps in the synthesis of valine from diethyl acetamidomalonate. The overall yield of valine by this method is reported to be rather low (31%). Can you think of a reason why this synthesis is not very efficient?

The α-amino acids prepared by the synthetic methods just described are racemic unless a resolution step is included, enantiomerically enriched reactants are used, or the reaction is modified so as to become enantioselective. Considerable progress has been made in the last of these methods, allowing chemists to prepare not only L-amino acids, but also their much rarer D-enantiomers. We have already seen one example of this approach in the synthesis of the anti-parkinsonism drug L-dopa by enantioselective hydrogenation (see Section 14.14). A variation of the Strecker synthesis using a chiral catalyst has recently been developed that gives α-amino acids with greater than 99% enantioselectivity.

25.5 Reactions of Amino Acids

Amino acids undergo reactions characteristic of both their amine and carboxylic acid functional groups. Acylation is a typical reaction of the amino group.

$$\overset{+}{H_3N}CH_2CO_2^- + CH_3\overset{O}{\overset{\|}{C}}O\overset{O}{\overset{\|}{C}}CH_3 \longrightarrow CH_3\overset{O}{\overset{\|}{C}}NHCH_2CO_2H + CH_3CO_2H$$

Glycine | Acetic anhydride | *N*-Acetylglycine (89–92%) | Acetic acid

Ester formation is a typical reaction of the carboxyl group.

$$CH_3\underset{\underset{+}{NH_3}}{\underset{|}{C}}HCO_2^- + CH_3CH_2OH \xrightarrow{HCl} CH_3\underset{\underset{+}{NH_3}}{\underset{|}{C}}H\overset{O}{\overset{\|}{C}}OCH_2CH_3\ Cl^-$$

Alanine | Ethanol | Hydrochloride salt of alanine ethyl ester (90–95%)

Ninhydrin is used to detect fingerprints.

The presence of amino acids can be detected by the formation of a purple color on treatment with *ninhydrin.* The same compound responsible for the purple color is formed from all amino acids in which the α-amino group is primary.

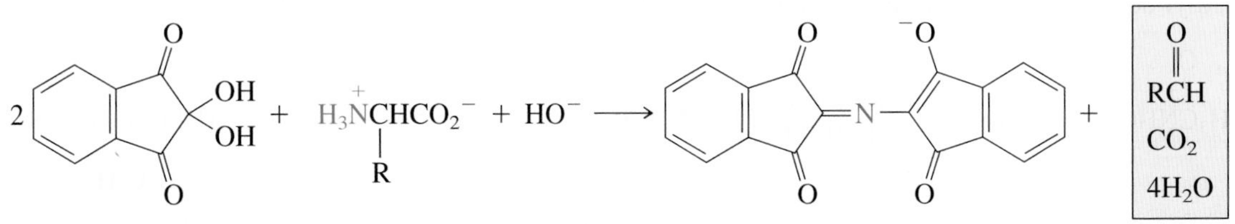

Ninhydrin | Violet dye ("Ruhemann's purple") | (Formed, but not normally isolated)

Proline, in which the α-amino group is secondary, gives an orange compound on reaction with ninhydrin.

Problem 25.8

Suggest a reasonable mechanism for the reaction of an α-amino acid with ninhydrin.

25.6 Some Biochemical Reactions of Amino Acids

The fact that we focus on the 20 (or 21 or 22) amino acids that are coded for by DNA indicates the importance we place on protein biosynthesis as a reaction of amino acids. In addition to serving as building blocks for proteins, amino acids are involved in numerous other biochemical processes. They store energy, although less efficiently than carbohydrates and lipids, and are starting materials for the biosynthesis of other amino acids, amines, alkaloids, and neurotransmitters.

Many of the biochemical reactions of amino acids require *pyridoxal 5′-phosphate* (PLP), the active form of vitamin B_6, as a coenzyme. Before acting on an amino acid, PLP uses its aldehyde function to form an imine with the amino group of a lysine side chain of a protein.

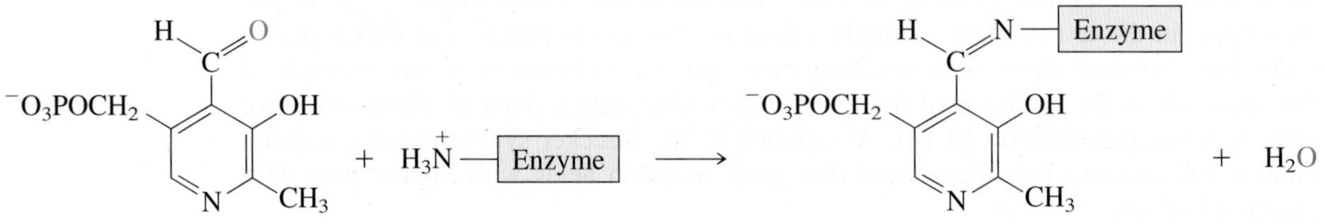

Pyridoxal phosphate (PLP) | Enzyme | Enzyme-bound PLP | Water

Reaction of the enzyme-bound PLP with an amino acid connects the amino acid and PLP by a new imine linkage.

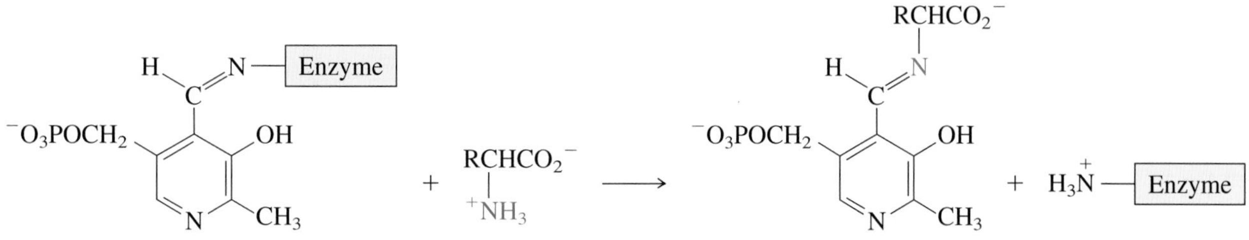

Enzyme-bound PLP + Amino acid → Imine + Enzyme

The pyridine ring of PLP, especially when protonated, facilitates several kinds of reactions at the amino acid's α carbon by acting as an electron-withdrawing group. One is decarboxylation.

$$\underset{\text{Amino acid}}{RCH(\overset{+}{N}H_3)CO_2^-} \xrightarrow[\text{amino acid decarboxylase}]{\text{pyridoxal phosphate}} \underset{\text{Amine}}{RCH_2NH_2} + \underset{\text{Carbon dioxide}}{CO_2}$$

Mechanism 25.1 outlines the mechanism of decarboxylation, showing the role played by the coenzyme.

Mechanism 25.1

Pyridoxal 5′-Phosphate-Mediated Decarboxylation of an α-Amino Acid

THE OVERALL REACTION:

$$\underset{\text{Amino acid}}{RCH(\overset{+}{N}H_3)CO_2^-} \xrightarrow[\text{amino acid decarboxylase}]{\text{pyridoxal 5'-phosphate}} \underset{\text{Amine}}{RCH_2NH_2} + \underset{\text{Carbon dioxide}}{CO_2}$$

THE MECHANISM: Each stage is enzyme-catalyzed and can involve more than one elementary step.

Stage 1: The amino acid reacts with enzyme-bound pyridoxal 5′-phosphate (PLP). An imine linkage (C=N) between the amino acid and PLP forms, and the enzyme is displaced.

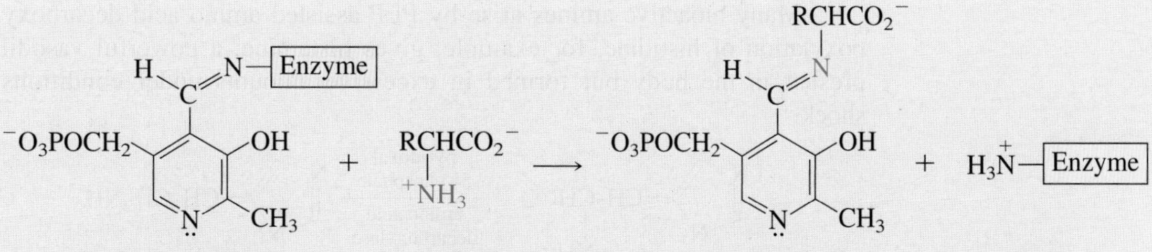

Enzyme-bound PLP + Amino acid → Imine + Enzyme

Continued

Stage 2: When the pyridine ring is protonated on nitrogen, it becomes a stronger electron-withdrawing group, and decarboxylation is facilitated by charge neutralization.

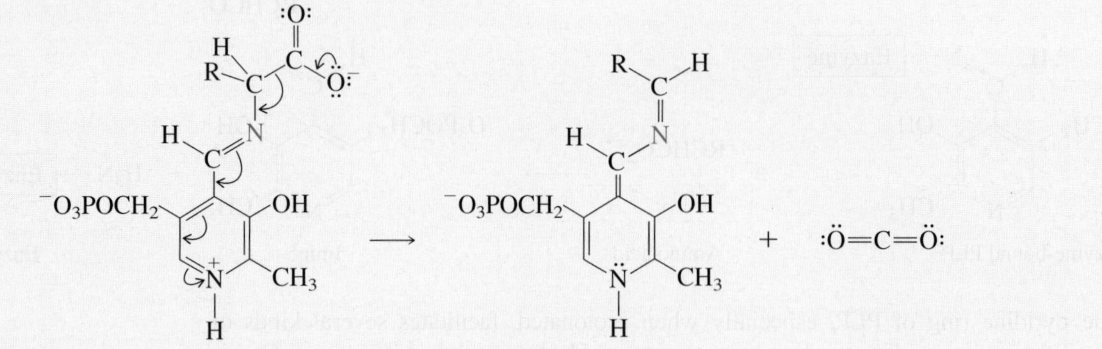

Imine — Decarboxylated imine — Carbon dioxide

Stage 3: Proton transfer to the α carbon and abstraction of a proton from the pyridine nitrogen brings about rearomatization of the pyridine ring.

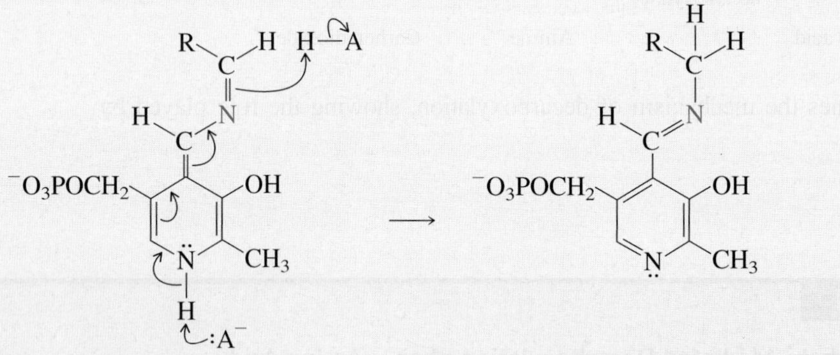

Decarboxylated imine — PLP-bound imine

Stage 4: Reaction of the PLP-bound imine with the enzyme liberates the amine and restores the enzyme-bound coenzyme.

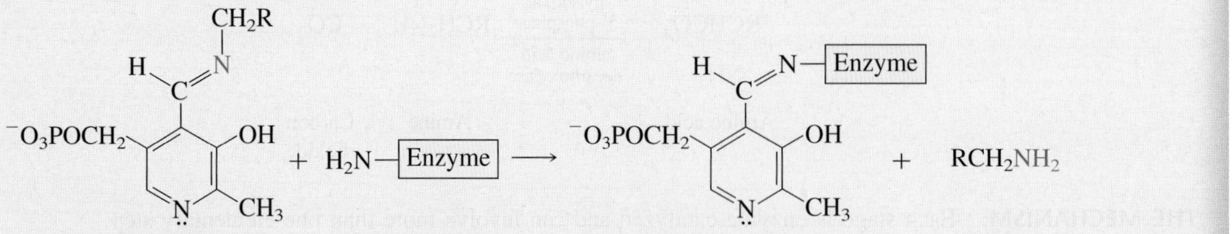

PLP-bound imine — Enzyme — Enzyme-bound PLP — Amine

Many bioactive amines arise by PLP-assisted amino acid decarboxylation. Decarboxylation of histidine, for example, gives histamine, a powerful vasodilator normally present in the body but formed in excessive amounts under conditions of traumatic shock.

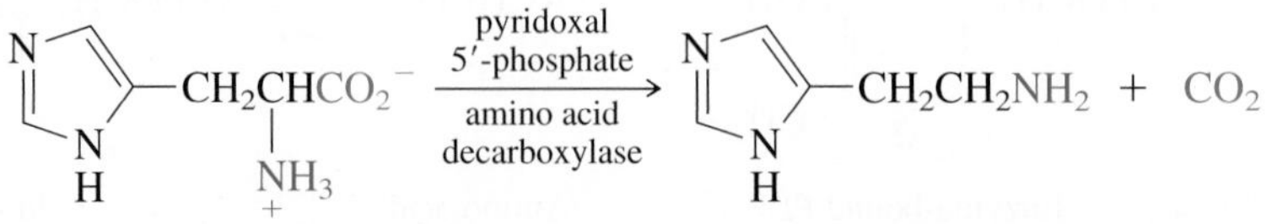

Histidine — Histamine — Carbon dioxide

Histamine is present in various tissues and produces different effects depending on the kind of receptor it binds to. Binding of histamine to H_1 receptors in mast cells triggers, for example, the sneezing and watery eyes of hay fever and the itching of mosquito bites. The H_2 receptors in the cells that line the stomach regulate the secretion of gastric acid. The present generation of antiallergy and antiulcer drugs bind to H_1 and H_2, respectively, and act by denying histamine access to these receptors.

Problem 25.9

One of the amino acids in Table 25.1 is the biological precursor to γ-aminobutyric acid (4-aminobutanoic acid), which it forms by a decarboxylation reaction. Which amino acid is this?

The chemistry of the brain and central nervous system is affected by a group of substances called **neurotransmitters,** substances that carry messages across a synapse from one neuron to another. Several of these neurotransmitters arise from L-tyrosine by structural modification and decarboxylation, as outlined in Figure 25.3.

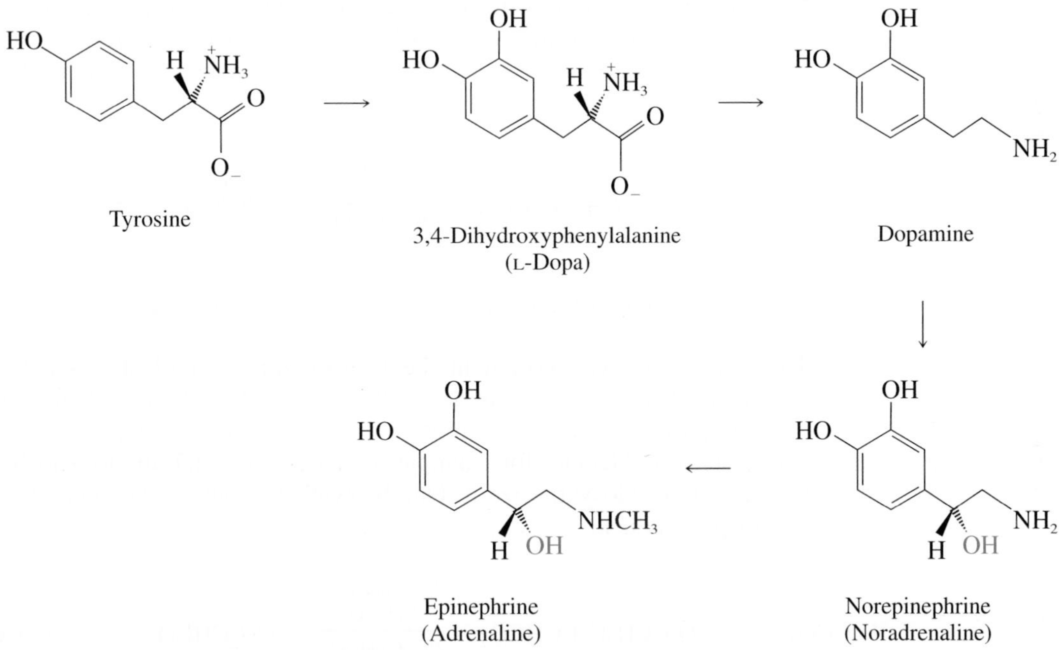

Figure 25.3

Tyrosine is the biosynthetic precursor to a number of neurotransmitters. Each transformation is enzyme-catalyzed.

Problem 25.10

Which of the transformations in Figure 25.3 is catalyzed by an amino acid decarboxylase?

Many of the drugs prescribed to treat anxiety, depression, or attention deficit disorder are "reuptake" inhibitors. They increase the concentration in the brain of a necessary neurotransmitter such as dopamine or epinephrine by slowing the rate at which it is reabsorbed.

Pyridoxal 5′-phosphate is also a coenzyme for the enzyme-catalyzed racemization of amino acids. The key reaction is proton abstraction from the α carbon of the amino acid imine of PLP. This step converts the α carbon, which is a chirality center, from sp^3 to sp^2.

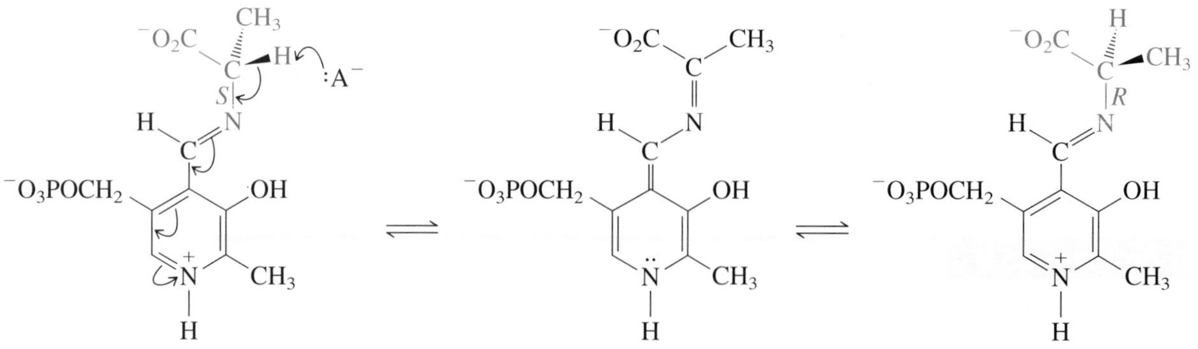

PLP imine of L-alanine — Achiral intermediate — PLP imine of D-alanine

Proton transfer to the imine carbon of the achiral intermediate gives equal amounts of both enantiomers of the PLP imine. The equation illustrates the racemization of L-alanine, which is catalyzed by the PLP-dependent enzyme *alanine racemase.* Because D-alanine is an essential component of bacterial cell walls, there is considerable interest in designing inhibitors of alanine racemase as potential antibacterial drugs.

In addition to amino acid decarboxylation and racemization, PLP is a coenzyme for **transamination**—the transfer of an amino group from one compound to another. The enzymes that catalyze transaminations are called *aminotransferases* or *transaminases.* Many transaminations involve two compounds: α-ketoglutaric acid and L-glutamic acid.

$$RCH(\overset{+}{N}H_3)CO_2^- + {}^-O_2CCH_2CH_2C(=O)CO_2^- \xrightarrow[\text{aminotransferase}]{\text{pyridoxal 5'-phosphate}} RC(=O)CO_2^- + {}^-O_2CCH_2CH_2CH(\overset{+}{N}H_3)CO_2^-$$

Amino acid — α-Ketoglutaric acid — α-Keto acid — L-Glutamic acid

The reaction shown, written in the forward direction illustrates a feature of amino acid metabolism, the breaking down of amino acids and using their structural units for other purposes. Written in the reverse direction, it illustrates a biosynthetic pathway to amino acids. L-Alanine, for example, is not an essential amino acid because we have the capacity to biosynthesize it. One biosynthetic route to L-alanine is the transamination of pyruvic acid.

$$CH_3C(=O)CO_2^- + {}^-O_2CCH_2CH_2CH(\overset{+}{N}H_3)CO_2^- \xrightarrow[\text{aminotransferase}]{\text{pyridoxal 5'-phosphate}} CH_3CH(\overset{+}{N}H_3)CO_2^- + {}^-O_2CCH_2CH_2C(=O)CO_2^-$$

Pyruvic acid — L-Glutamic acid — L-Alanine — α-Ketoglutaric acid

Although the equation shows a single transamination, the mechanism actually involves two, each of which is understandable in light of what we have already seen concerning the way PLP functions as a coenzyme. In the first, as outlined in the first four stages of Mechanism 25.2, the amino group of L-glutamic acid is transferred to the coenzyme PLP to give pyridoxamine 5′-phosphate (PMP). The second transamination shown in abbreviated form as stage 5, is analogous to the first but followed in reverse order.

Problem 25.11

α-Ketoglutaric acid undergoes a transamination reaction with L-aspartic acid (see Table 25.1), converting it to a compound known as oxaloacetic acid. What is the structure of oxaloacetic acid?

Mechanism 25.2

Transamination: Biosynthesis of L-Alanine from L-Glutamic Acid and Pyruvic Acid

THE OVERALL REACTION:

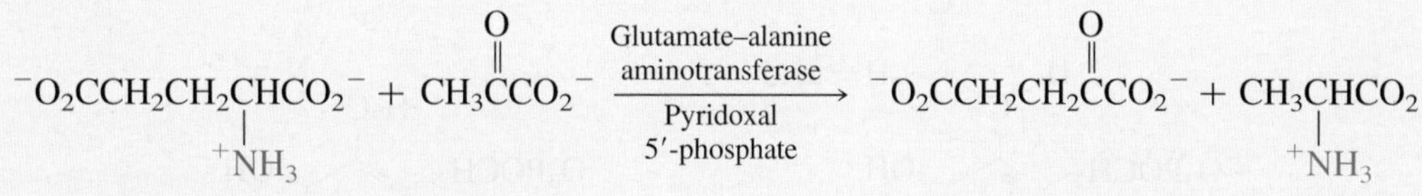

L-Glutamic acid Pyruvic acid α-Ketoglutaric acid L-Alanine

THE MECHANISM: Each stage can involve more than one elementary step. Each reaction is enzyme-catalyzed. Stages 1–4 show the transfer of the amino group of L-glutamic acid to pyridoxal 5′-phosphate to give α-ketoglutaric acid and pyridoxamine 5′-phosphate (PMP). PMP reacts with pyruvic acid to give an imine, which then follows stages analogous to 1–4, but in reverse order, to give L-alanine and PLP. These stages are summarized as stage 5.

Stage 1: L-Glutamic acid forms an imine bond to the coenzyme PLP by reaction with imine formed between PLP and the enzyme.

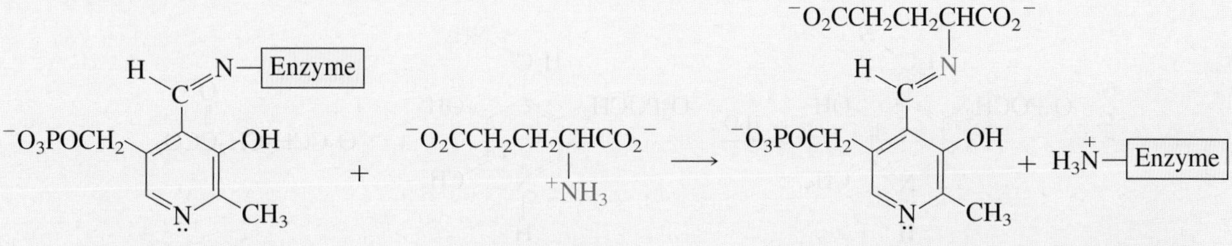

Enzyme-bound PLP L-Glutamic acid PLP-glutamic acid imine Enzyme

Stage 2: The electron-withdrawing effect of the pyridinium ring stabilizes the conjugate base formed by proton abstraction from the α carbon of the imine.

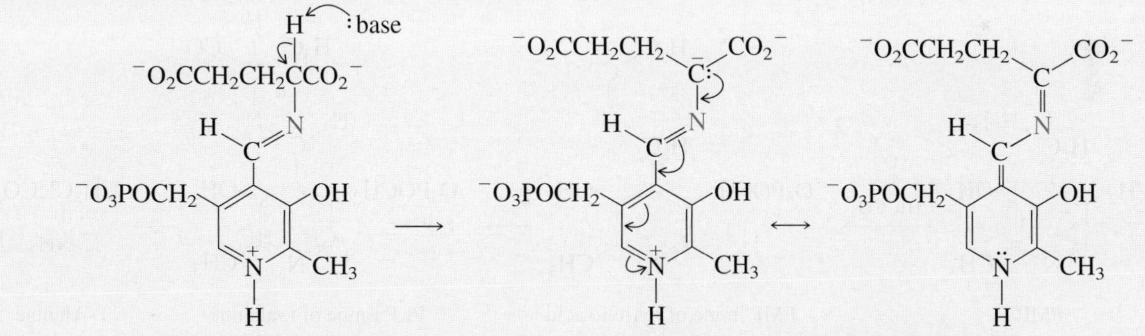

PLP-glutamic acid imine Conjugate base of PLP-glutamic acid imine

Continued

Peptide-bond formation and transamination are the most general reactions of the standard amino acids, but individual amino acids often undergo reactions of more limited scope. One of the biosynthetic pathways to L-tyrosine is oxidation of L-phenylalanine. An *arene oxide* is an intermediate.

For more on this reaction, see Descriptive Passage and Interpretive Problems 16: Epoxide Rearrangements and the NIH Shift.

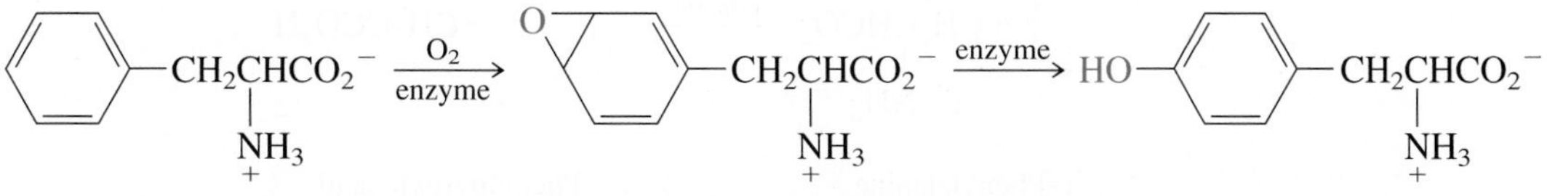

L-Phenylalanine Arene oxide intermediate L-Tyrosine

Stage 3: Electron reorganization and protonation of carbon restores the aromaticity of the pyridine ring while converting a PLP imine to a PMP imine.

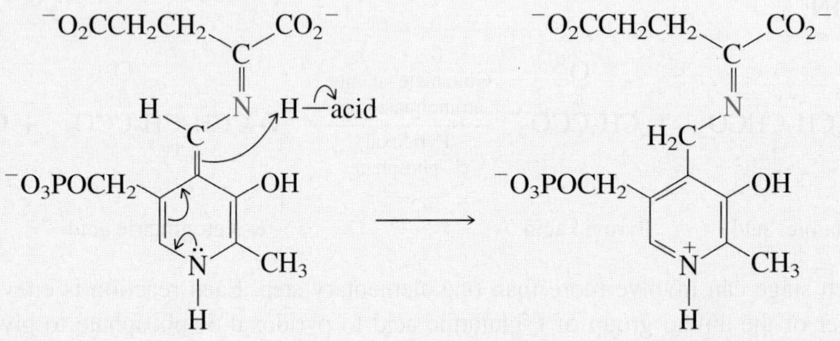

Conjugate base of PLP-glutamic acid imine — PMP imine of α-ketoglutaric acid

Stage 4: Cleavage of the PMP imine, shown here as a hydrolysis, gives pyridoxamine and α-ketoglutaric acid.

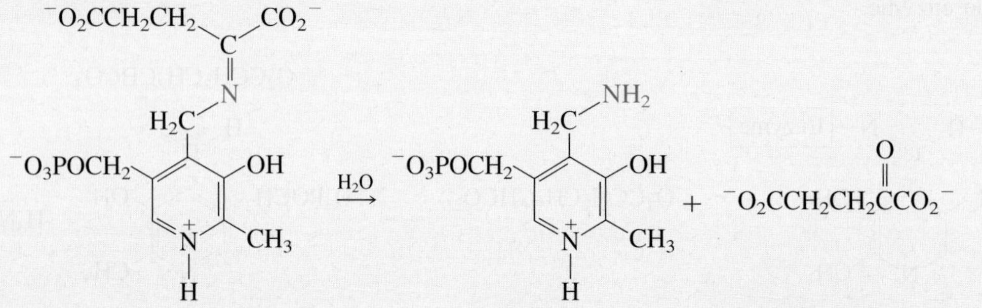

PMP imine of α-ketoglutaric acid — PMP — α-Ketoglutaric acid

Stage 5: Formation of the imine from PMP and pyruvic acid sets the stage for the conversion of pyruvic acid to L-alanine.

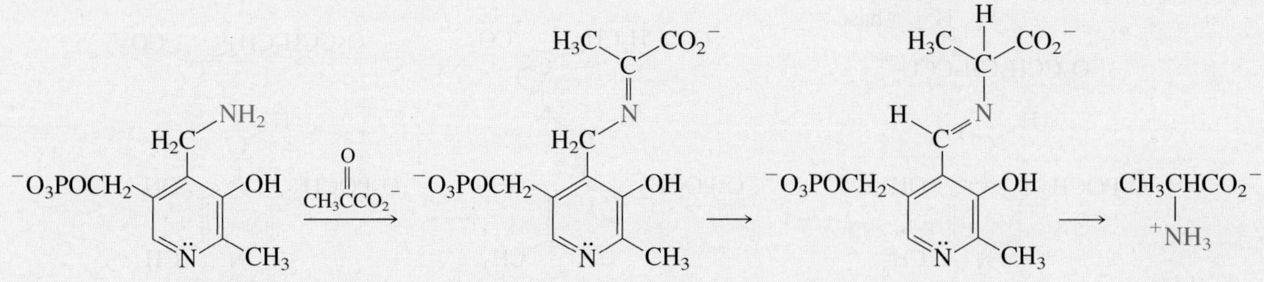

PMP — PMP imine of pyruvic acid — PLP imine of L-alanine — L-Alanine

Some individuals lack the enzyme *phenylalanine hydroxylase* required for this conversion, and any L-phenylalanine that would ordinarily be converted to L-tyrosine is converted to phenylpyruvic acid by transamination.

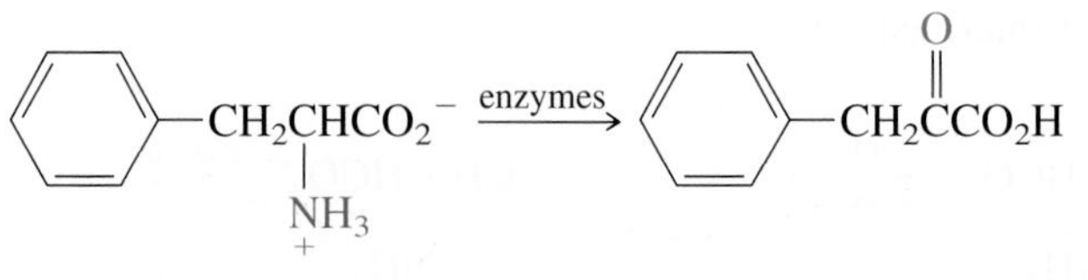

L-Phenylalanine — Phenylpyruvic acid

Too much phenylpyruvic acid causes *phenylketonuria* (PKU disease), which can lead to mental retardation in growing children. Infants are routinely screened for PKU disease within a few days of birth. PKU disease cannot be cured, but is controlled by restricting the dietary intake of foods, such as meat, that are rich in L-phenylalanine.

Foods sweetened with aspartame (page 1048) contain a PKU warning. Can you see why?

25.7 Peptides

A key biochemical reaction of amino acids is their conversion to peptides, polypeptides, and proteins. In all these substances amino acids are linked together by amide bonds. The amide bond between the amino group of one amino acid and the carboxyl of another is called a **peptide bond.** Alanylglycine is a representative dipeptide.

N-terminal amino acid

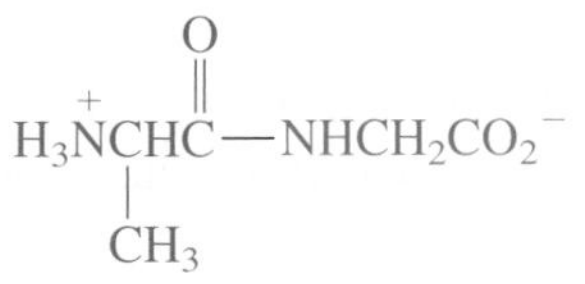

C-terminal amino acid

Alanylglycine
(Ala-Gly or AG)

It is understood that α-amino acids occur as their L stereoisomers unless otherwise indicated. The D notation is explicitly shown when a D amino acid is present, and a racemic amino acid is identified by the prefix DL.

By agreement, peptide structures are written so that the amino group (as $H_3\overset{+}{N}$— or H_2N—) is at the left and the carboxyl group (as CO_2^- or CO_2H) is at the right. The left and right ends of the peptide are referred to as the **N terminus** (or amino terminus) and the **C terminus** (or carboxyl terminus), respectively. Alanine is the N-terminal amino acid in alanylglycine; glycine is the C-terminal amino acid. A dipeptide is named as an acyl derivative of the C-terminal amino acid. The precise order of bonding in a peptide (its amino acid *sequence*) is conveniently specified by using the three-letter amino acid abbreviations for the respective amino acids and connecting them by hyphens. One-letter abbreviations are used without punctuation. Individual amino acid components of peptides are often referred to as **amino acid residues.**

Problem 25.12

Write structural formulas showing the constitution of each of the following dipeptides. Rewrite each sequence using one-letter abbreviations for the amino acids.

(a) Gly-Ala
(b) Ala-Phe
(c) Phe-Ala
(d) Gly-Glu
(e) Lys-Gly
(f) D-Ala-D-Ala

Sample Solution (a) Gly-Ala is a constitutional isomer of Ala-Gly. Glycine is the N-terminal amino acid in Gly-Ala; alanine is the C-terminal amino acid.

N-terminal amino acid

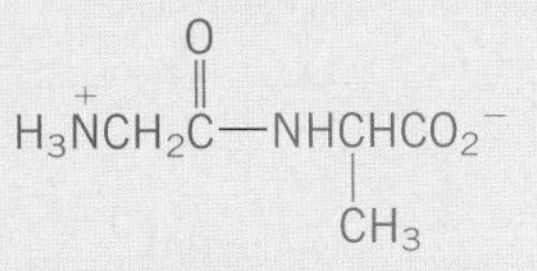

C-terminal amino acid

Glycylalanine (GA)

Figure 25.4 shows the structure of Ala-Gly as determined by X-ray crystallography. An important feature is the planar geometry of the peptide bond, and the most stable conformation with respect to this bond has the two α-carbon atoms anti to each other. Rotation about the amide bond is slow because delocalization of the unshared electron pair of nitrogen into the carbonyl group gives partial double-bond character to the carbon–nitrogen bond.

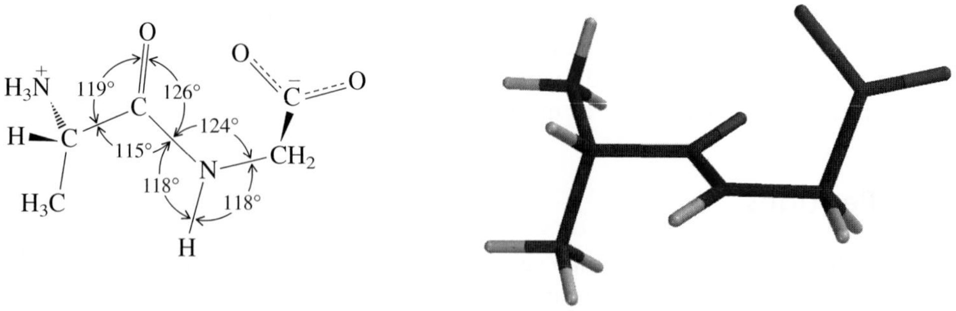

(*a*) (*b*)

Figure 25.4

Structural features of the dipeptide L-alanylglycine as determined by X-ray crystallography. All of the bonds of the peptide linkage lie in the same plane and both α carbons are anti to each other.

In addition to its planar geometry, the amide bond affects the structure of peptides in another important way. The N—H and the C=O units are candidates for hydrogen bonding with other peptide linkages both within the same and with adjacent polypeptide chains.

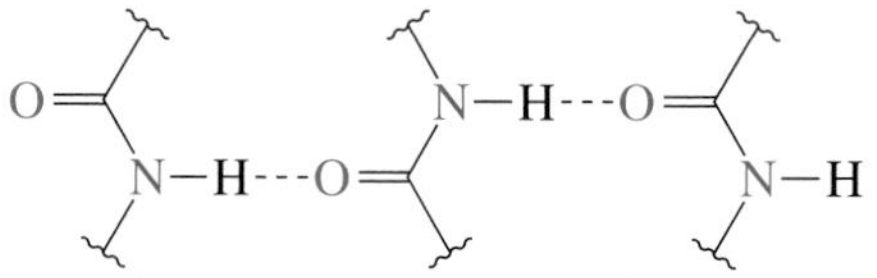

As the only secondary amine among the standard amino acids, L-proline is an exception in that its amides lack an N—H bond.

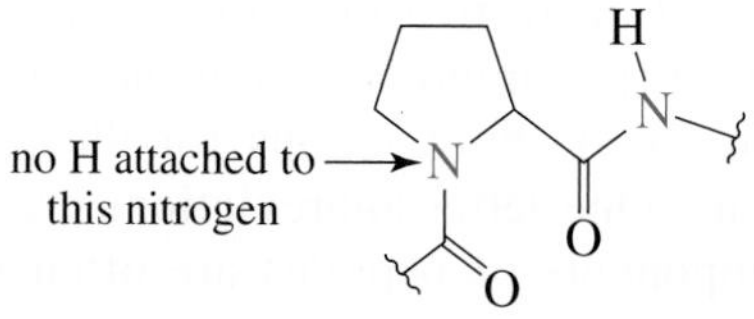

This structural feature of L-proline affects the three-dimensional shape of peptides that contain it by limiting the number of hydrogen-bonding opportunities.

Problem 25.13

Expand your answer to Problem 25.12 by showing the structural formula for each dipeptide in a manner that reveals the stereochemistry at the α-carbon atom.

Sample Solution (a) Glycine is achiral, and so Gly-Ala has only one chirality center, the α-carbon atom of the L-alanine residue. When the carbon chain is drawn in an extended zigzag fashion and L-alanine is the C terminus, its structure is as shown:

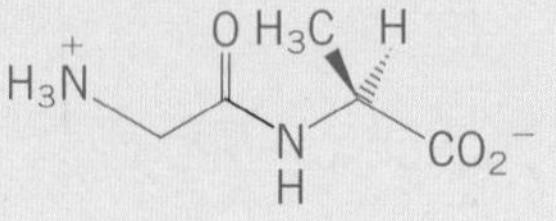

Gly-Ala (GA)

The structures of higher peptides are extensions of the structural features of dipeptides. Figure 25.5 gives the structural formula and amino acid sequence of the naturally occurring pentapeptide *leucine enkephalin.* Enkephalins are pentapeptide components of *endorphins,* polypeptides present in the brain that act as the body's own painkillers. A second substance, *methionine enkephalin,* is also present in endorphins. Methionine enkephalin is about 20 times more potent than leucine enkephalin. It differs from leucine enkephalin only in having methionine instead of leucine as its C-terminal amino acid.

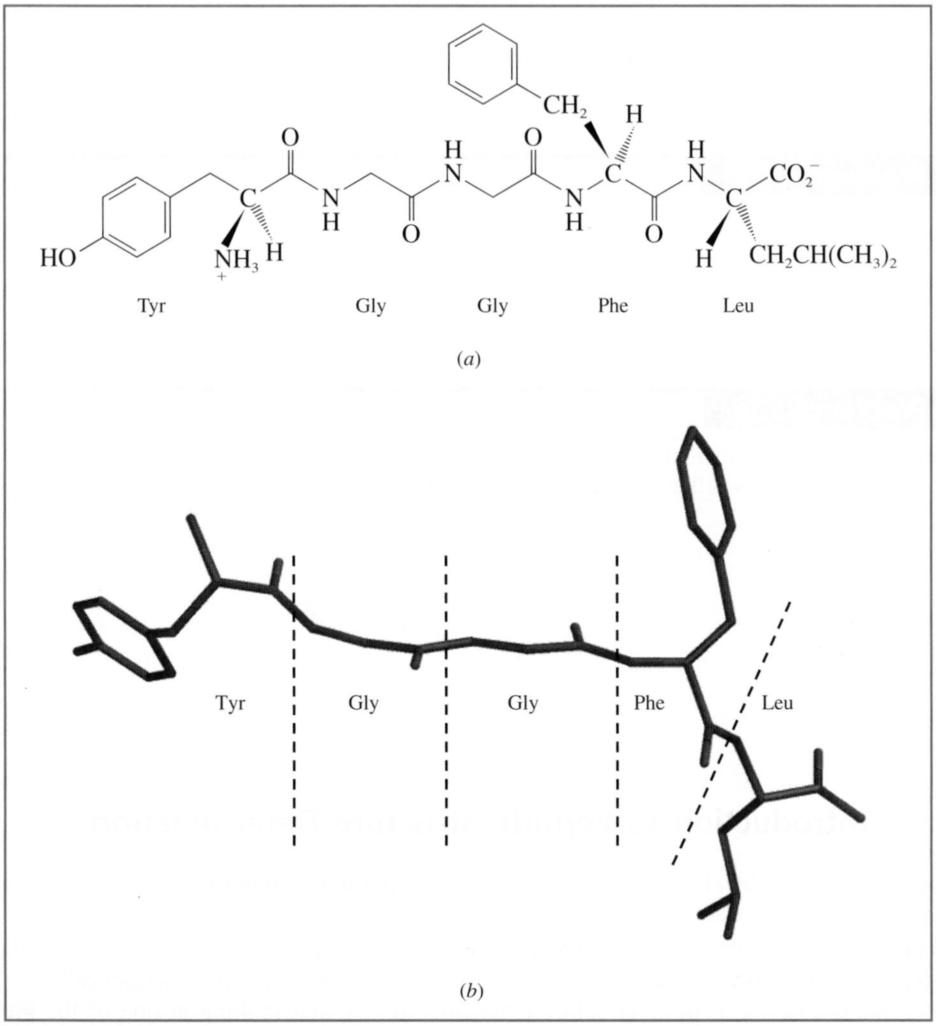

Figure 25.5

The structure of the pentapeptide leucine enkephalin shown as (*a*) a structural drawing and (*b*) as a molecular model. The shape of the molecular model was determined by X-ray crystallography. Hydrogens have been omitted for clarity.

Problem 25.14

What is the amino acid sequence (using three-letter abbreviations) of methionine enkephalin? Also show it using one-letter abbreviations.

Peptides having structures slightly different from those described to this point are known. One such variation is seen in the nonapeptide *oxytocin,* shown in Figure 25.6. Oxytocin is a hormone secreted by the pituitary gland that stimulates uterine contractions during childbirth and promotes lactation. Rather than terminating in a carboxyl group, the C-terminal glycine residue in oxytocin has been modified to become its corresponding amide. Two

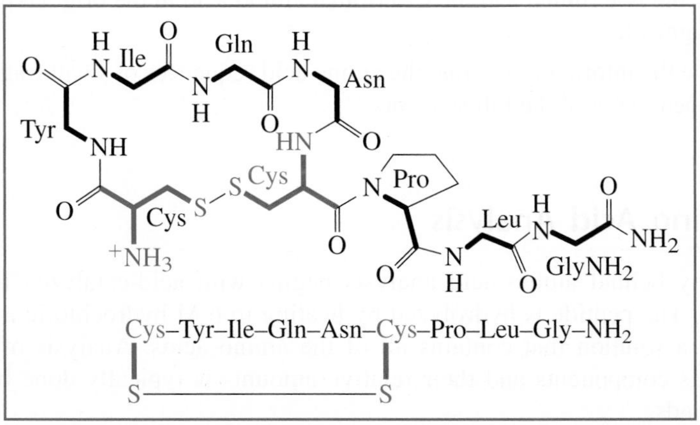

Figure 25.6

The connectivity of oxytocin with most of the side chains omitted for clarity. A disulfide bond connects the two cysteines. The cysteine shown in blue is the N-terminal amino acid; the one in red is the fourth amino acid beginning at the C-terminus. The C-terminus is the amide of glycine.

Recall from Section 15.12 that compounds of the type RSH are readily oxidized to RSSR.

cysteine units, one of them the N-terminal amino acid, are joined by the sulfur–sulfur bond of a large-ring cyclic disulfide unit. This is a common structural modification in polypeptides and proteins that contain cysteine residues. It provides a covalent bond between regions of peptide chains that may be many amino acid residues removed from each other.

Problem 25.15

What is the net charge of oxytocin at pH = 7?

Cyclic peptides in which the backbone is made entirely of amide bonds are also well known. Examples include the didemnins, which were featured in our chapter opener.

Problem 25.16

The side chain in didemnin B has two amino acids that resemble those in Table 25.1. Which two? How are they different from those in the table?

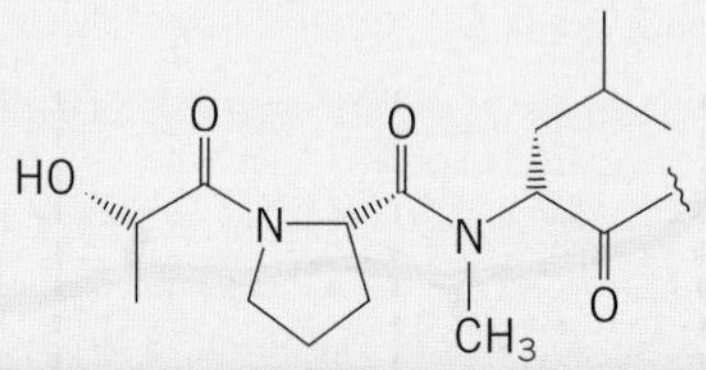

25.8 Introduction to Peptide Structure Determination

There are several levels of peptide structure. The **primary structure** is the amino acid sequence plus any disulfide links. With the 20 amino acids of Table 25.1 as building blocks, 20^2 dipeptides, 20^3 tripeptides, 20^4 tetrapeptides, and so on, are possible. Given a peptide of unknown structure, how do we determine its amino acid sequence?

We'll describe peptide structure determination by first looking at one of the great achievements of biochemistry, the determination of the amino acid sequence of insulin by Frederick Sanger of Cambridge University (England). Sanger was awarded the 1958 Nobel Prize in Chemistry for this work, which he began in 1944 and completed 10 years later. The methods used by Sanger and his coworkers are, of course, dated by now, but the overall logic hasn't changed very much. We'll use Sanger's insulin strategy to orient us, then show how current methods of protein sequencing have evolved from it.

Sanger was a corecipient of a second Nobel Prize in 1980 for devising methods for sequencing nucleic acids. Sanger's strategy for nucleic acid sequencing will be described in Section 26.14.

Sanger's strategy can be outlined as follows:

1. Determine what amino acids are present and their molar ratios.
2. Cleave the peptide into smaller fragments, separate these fragments, and determine the amino acid composition of the fragments.
3. Identify the N-terminal and the C-terminal amino acid in the original peptide and in each fragment.
4. Organize the information so that the amino acid sequences of small fragments can be overlapped to reveal the full sequence.

25.9 Amino Acid Analysis

The chemistry behind amino acid analysis begins with acid-catalyzed hydrolysis of amide bonds. The peptide is hydrolyzed by heating in 6 M hydrochloric acid for about 24 h to give a solution that contains all of the amino acids. Analysis of the mixture in terms of its components and their relative amounts is typically done by chromatographic methods.

These methods flow from the work of Stanford Moore and William H. Stein of Rockefeller University who developed automated techniques for separating and identifying amino acids. In their original work, Moore and Stein used ion-exchange chromatography. Modern methods based on high-performance liquid chromatography (HPLC) are both faster and more selective for separating the individual amino acids in a mixture. Either before or after their separation, the amino acids are allowed to react ("tagged") with a substance that bears a group—a naphthalene ring, for example that fluoresces. The fluorescence is strong enough so that modern analyzers can detect the amino acids obtained from 10^{-5} to 10^{-7}g of peptide.

Moore and Stein shared one half of the 1972 Nobel Prize in Chemistry.

Fluorescence is the emission of radiation by a substance after it has absorbed radiation of a higher frequency.

Problem 25.17

Amino acid analysis of a certain tetrapeptide gave alanine, glycine, phenylalanine, and valine in equimolar amounts. What amino acid sequences are possible for this tetrapeptide?

25.10 Partial Hydrolysis of Peptides

Whereas acid-catalyzed hydrolysis of peptides cleaves amide bonds indiscriminately and eventually breaks all of them, enzymatic hydrolysis is much more selective and is the method used to convert a peptide into smaller fragments.

The enzymes that catalyze the hydrolysis of peptide bonds are called **peptidases.** *Trypsin,* a digestive enzyme present in the intestine, catalyzes only the hydrolysis of peptide bonds involving the carboxyl group of a lysine or arginine residue. *Chymotrypsin,* another digestive enzyme, is selective for peptide bonds involving the carboxyl group of amino acids with aromatic side chains (phenylalanine, tyrosine, tryptophan). One group of pancreatic enzymes, known as *carboxypeptidases,* catalyzes only the hydrolysis of the peptide bond to the C-terminal amino acid. In addition to these, many other digestive enzymes are known and their selectivity exploited in the selective hydrolysis of peptides.

Papain, the active component of most meat tenderizers, is a peptidase.

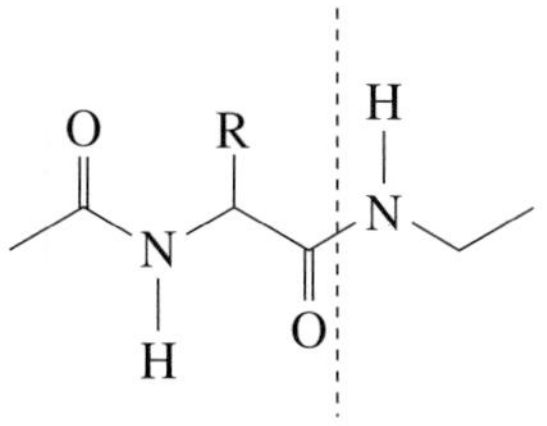

Trypsin cleaves here when
R = side chain of lysine or arginine

Chymotrypsin cleaves here when
R = side chain of phenylalanine, tyrosine, or tryptophan

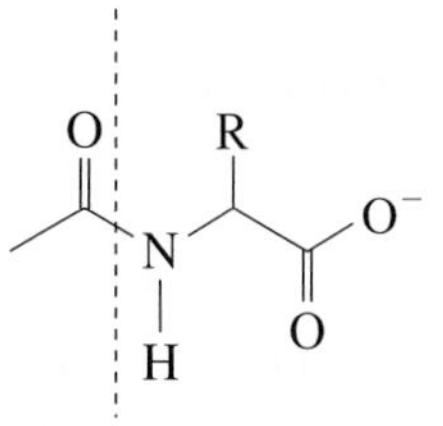

Carboxypeptidase cleaves the peptide bond of the C-terminal amino acid

Problem 25.18

Digestion of the tetrapeptide of Problem 25.17 with chymotrypsin gave a dipeptide that on amino acid analysis gave phenylalanine and valine in equimolar amounts. What amino acid sequences are possible for the tetrapeptide?

25.11 End Group Analysis

An amino acid sequence is ambiguous unless we know the direction in which to read it—left to right, or right to left. We need to know which end is the N terminus and which is the C terminus. As we saw in the preceding section, carboxypeptidase-catalyzed hydrolysis cleaves the C-terminal amino acid and so can be used to identify it. What about the N terminus?

Several chemical methods have been devised for identifying the N-terminal amino acid. They all take advantage of the fact that the N-terminal amino group is free and can act as a nucleophile. The α-amino groups of all the other amino acids are part of amide linkages, are not free, and are much less nucleophilic. Sanger's method for N-terminal residue analysis involves treating a peptide with 1-fluoro-2,4-dinitrobenzene, which is very reactive toward nucleophilic aromatic substitution (see Chapter 12).

1-Fluoro-2,4-dinitrobenzene is commonly referred to as *Sanger's reagent.*

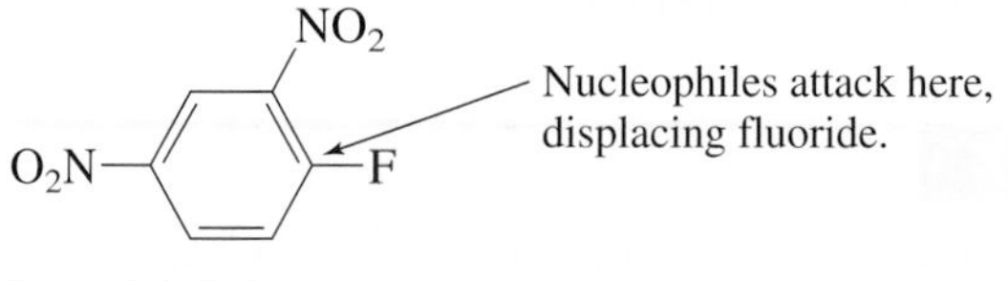

1-Fluoro-2,4-dinitrobenzene

The amino group of the N-terminal amino acid displaces fluoride from 1-fluoro-2,4-dinitrobenzene and gives a peptide in which the N-terminal nitrogen is labeled with a 2,4-dinitrophenyl (DNP) group. This is shown for the case of Val-Phe-Gly-Ala in Figure 25.7. The 2,4-dinitrophenyl-labeled peptide DNP-Val-Phe-Gly-Ala is isolated and subjected to hydrolysis, after which the 2,4-dinitrophenyl derivative of the N-terminal amino acid is isolated and identified as DNP-Val by comparing its chromatographic behavior with that of standard samples of 2,4-dinitrophenyl-labeled amino acids. None

The reaction is carried out by mixing the peptide and 1-fluoro-2,4-dinitrobenzene in the presence of a weak base such as sodium carbonate. In the first step the base abstracts a proton from the terminal $H_3\overset{+}{N}$ group to give a free amino function. The nucleophilic amino group attacks 1-fluoro-2,4-dinitrobenzene, displacing fluoride.

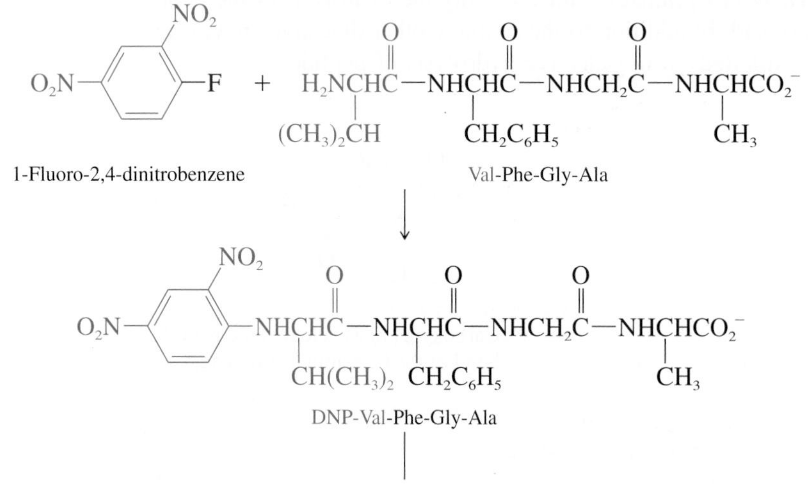

Acid hydrolysis cleaves the amide bonds of the 2,4-dinitrophenyl-labeled peptide, giving the 2,4-dinitrophenyl-labeled N-terminal amino acid and a mixture of unlabeled amino acids.

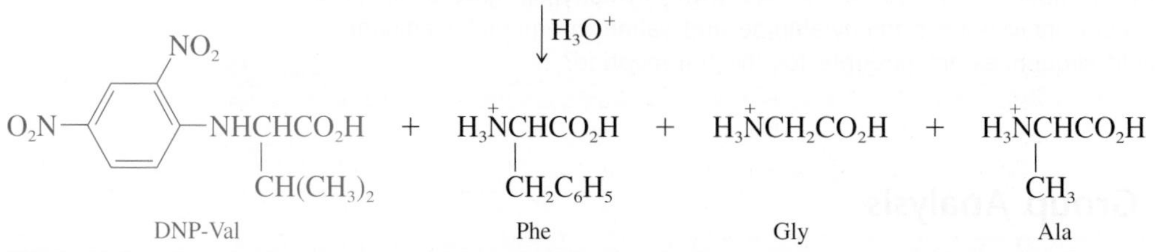

Figure 25.7

Use of 1-fluoro-2,4-dinitrobenzene to identify the N-terminal amino acid of a peptide.

of the other amino acid residues bear a 2,4-dinitrophenyl group; they appear in the hydrolysis product as the free amino acids.

Labeling the N-terminal amino acid as its DNP derivative is mainly of historical interest and has been replaced by other methods. We'll discuss one of these—the Edman degradation—in Section 25.13. First, though, we'll complete our review of the general strategy for peptide sequencing by seeing how Sanger tied all of the information together into a structure for insulin.

25.12 Insulin

Sanger worked with insulin from cows, which has 51 amino acids, divided between two chains. One of these, the A chain, has 21 amino acids; the other, the B chain, has 30. The A and B chains are joined by disulfide bonds between cysteine residues (Cys-Cys). Figure 25.8 shows some of the information that defines the amino acid sequence of the B chain.

- Reaction of the B chain peptide with 1-fluoro-2,4-dinitrobenzene established that phenylalanine is the N terminus.
- Pepsin-catalyzed hydrolysis gave the four peptides shown in blue in Figure 25.8. (Their sequences were determined in separate experiments.) These four peptides contain 27 of the 30 amino acids in the B chain, but there are no points of overlap between them.
- The sequences of the four tetrapeptides shown in red in Figure 25.8 bridge the gaps between three of the four "blue" peptides to give an unbroken sequence from 1 through 24.
- The peptide shown in green was isolated by trypsin-catalyzed hydrolysis and has an amino acid sequence that completes the remaining overlaps.

The collection of sequenced fragments constitutes the **peptide map** for insulin.

1 2 3 4 5 6 7 8 9 10 11
Phe-Val-Asn-Gln-His-Leu-Cys-Gly-Ser-His-Leu

Ser-His-Leu-Val

Leu-Val-Glu-Ala

12 13 14 15
Val-Glu-Ala-Leu

Ala-Leu-Tyr

16 17
Tyr-Leu-Val-Cys

18 19 20 21 22 23 24
Val-Cys-Gly-Glu-Arg-Gly-Phe

25
Gly-Phe-Phe-Tyr-Thr-Pro-Lys

26 27 28 29 30
Tyr-Thr-Pro-Lys-Ala

1 5 10 15 20 25 30
Phe-Val-Asn-Gln-His-Leu-Cys-Gly-Ser-His-Leu-Val-Glu-Ala-Leu-Tyr-Leu-Val-Cys-Gly-Glu-Arg-Gly-Phe-Phe-Tyr-Thr-Pro-Lys-Ala

Figure 25.8

Diagram showing how the amino acid sequence of the B chain of bovine insulin can be determined by overlap of peptide fragments. Pepsin-catalyzed hydrolysis produced the fragments shown in blue, trypsin produced the one shown in green, and acid-catalyzed hydrolysis gave many fragments, including the four shown in red.

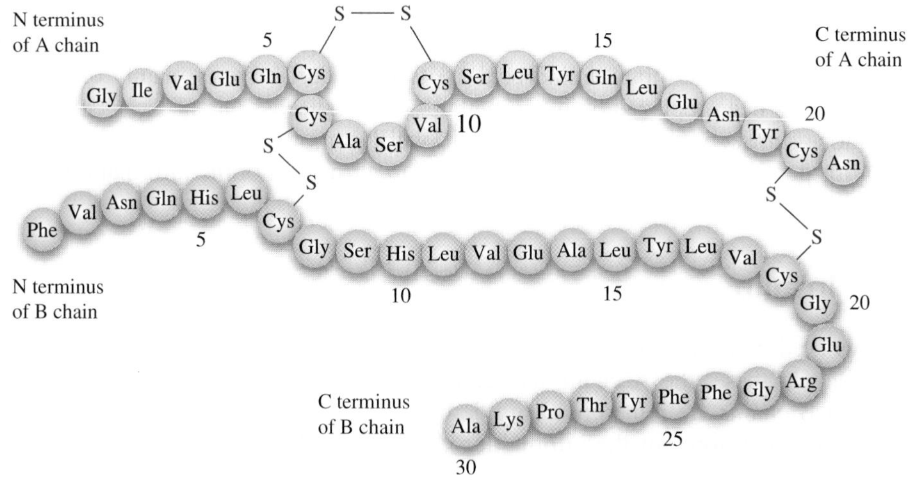

Figure 25.9

The amino acid sequence in bovine insulin. The A chain is joined to the B chain by two disulfide units (shown in green). There is also a disulfide bond linking cysteines 6 and 11 in the A chain. Human insulin has threonine and isoleucine at residues 8 and 10, respectively, in the A′ chain and threonine as the C-terminal amino acid in the B chain.

Sanger also determined the sequence of the A chain and identified the cysteine residues involved in disulfide bonds between the A and B chains as well as in the disulfide linkage within the A chain. The complete insulin structure is shown in Figure 25.9. The structure shown is that of bovine insulin. The A chains of human insulin and bovine insulin differ in only two amino acid residues; their B chains are identical except for the amino acid at the C terminus.

25.13 The Edman Degradation and Automated Sequencing of Peptides

When Sanger's method for N-terminal residue analysis was discussed, you may have wondered why it was not done sequentially. Simply start at the N terminus and work steadily back to the C terminus identifying one amino acid after another. The idea is fine, but it just doesn't work well in practice, at least with 1-fluoro-2,4-dinitrobenzene.

A major advance was devised by Pehr Edman (University of Lund, Sweden) that has become the standard method for N-terminal residue analysis. The **Edman degradation** is based on the chemistry shown in Mechanism 25.3. A peptide reacts with phenyl isothiocyanate to give a *phenylthiocarbamoyl* (PTC) derivative, as shown in the first step. This PTC derivative is then treated with an acid in an *anhydrous* medium (Edman used nitromethane saturated with hydrogen chloride) to cleave the amide bond between the N-terminal amino acid and the remainder of the peptide. No other peptide bonds are cleaved in this step as amide bond hydrolysis requires water. When the PTC derivative is treated with acid in an anhydrous medium, the sulfur atom of the C═S unit acts as an internal nucleophile, and the only amide bond cleaved under these conditions is the one to the N-terminal amino acid. The product of this cleavage, called a *thiazolone,* is unstable under the conditions of its formation and rearranges to a *phenylthiohydantoin* (PTH), which is isolated and identified by comparing it with standard samples of PTH derivatives of known amino acids. This is normally done by chromatographic methods, but mass spectrometry has also been used.

Only the N-terminal amide bond is broken in the Edman degradation; the rest of the peptide chain remains intact. It can be isolated and subjected to a second Edman procedure to determine its new N terminus. We can proceed along a peptide chain by beginning with the N terminus and determining each amino acid in order. The sequence is given directly by the structure of the PTH derivative formed in each successive degradation.

Problem 25.19

Give the structure of the PTH derivative isolated in the second Edman cycle of the tetrapeptide Val-Phe-Gly-Ala.

Mechanism 25.3

The Edman Degradation

Step 1: A peptide is treated with phenyl isothiocyanate to give a phenylthiocarbamoyl (PTC) derivative.

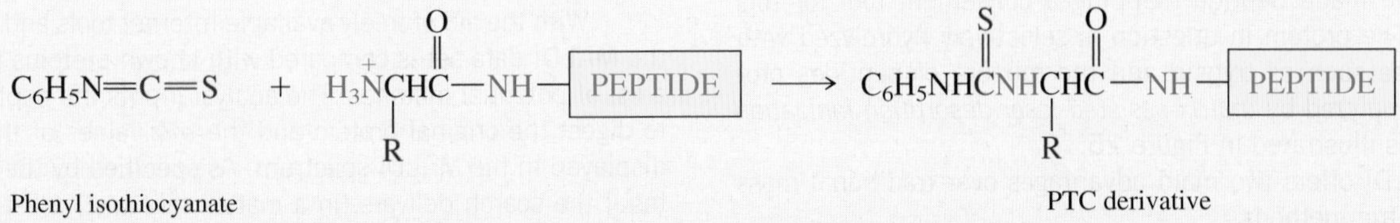

Step 2: On reaction with hydrogen chloride in an anhydrous solvent, the thiocarbonyl sulfur of the PTC derivative attacks the carbonyl carbon of the N-terminal amino acid. The N-terminal amino acid is cleaved as a thiazolone derivative from the remainder of the peptide.

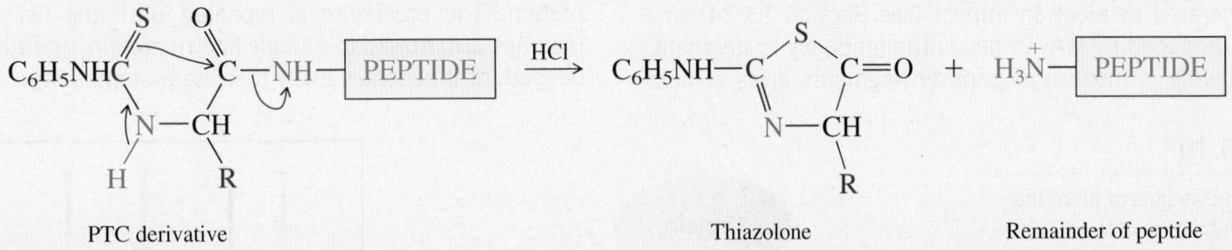

Step 3: Once formed, the thiazolone derivative isomerizes to a more stable phenylthiohydantoin (PTH) derivative, which is isolated and characterized, thereby providing identification of the N-terminal amino acid. The remainder of the peptide (formed in step 2) can be isolated and subjected to a second Edman degradation.

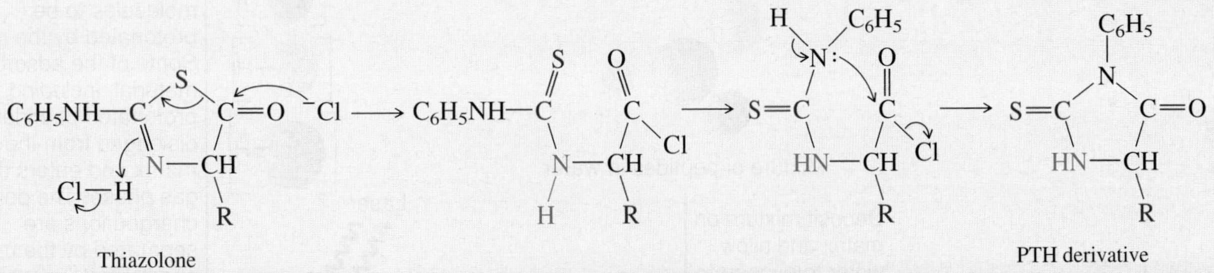

Ideally, one could determine the primary structure of even the largest protein by repeating the Edman procedure. Because anything less than 100% conversion in any single Edman degradation gives a mixture containing some of the original peptide along with the degraded one, two different PTH derivatives are formed in the next Edman cycle, and the ideal is not realized in practice. However, it is a fairly routine matter to sequence the first 20 amino acids from the N terminus by repetitive Edman cycles, and even 60 residues have been determined on a single sample of the protein myoglobin. The entire procedure has been automated and incorporated into a device called an *Edman sequenator*. The amount of sample required is quite small; as little as 10^{-10} mol is typical.

So many peptides and proteins have been sequenced now that it is impossible to give an accurate count. What was Nobel Prize-winning work in 1958 is routine today. Nor has the story ended. Sequencing of *nucleic acids* has advanced so dramatically that it is possible to clone the gene that codes for a particular protein, sequence its DNA, and deduce the structure of the protein from the nucleotide sequence of the DNA. We'll have more to say about DNA sequencing in the next chapter.

Peptide Mapping and MALDI Mass Spectrometry

Biological materials often contain proteins that must be identified. Recent advances in mass spectrometry have made peptide mapping a convenient tool for this purpose. The protein in question is selectively hydrolyzed with a peptidase such as trypsin and the mixture of peptides produced is analyzed by *matrix-assisted laser desorption ionization* (MALDI) as illustrated in Figure 25.10.

MALDI offers two main advantages over traditional mass spectrometric methods.

1. Substances, such as peptides, that lack sufficient volatility to be vaporized for analysis by conventional mass spectrometry can be vaporized by MALDI.
2. The species analyzed by the mass spectrometer is the conjugate acid of a peptide (peptide + H^+). Unlike the highly energetic ions generated by electron impact (see Section 13.24), the cations produced by MALDI have little tendency to fragment. Consequently, a mixture of peptide fragments gives a mass spectrum dominated by peaks with *m/z* values corresponding to those of the individual protonated peptides.

With the aid of freely available Internet tools and databases, the MALDI data set is compared with known proteins to generate a list of potential matches. The analyst inputs the peptidase used to digest the original protein and the *m/z* values of the peptides displayed in the MALDI spectrum. As specified by the search criteria, the search delivers (in a matter seconds!) a list of peptide sequences and the proteins these sequences contain. Next, a different peptidase is used to hydrolyze the protein, to provide a second set of peptides that are also analyzed by MALDI and matched against the database. MALDI mass spectrometry compares the amino acid *composition* of unsequenced peptides with the amino acid *sequence* of known proteins in order to identify an unknown protein. The procedure is repeated until the list of potential matches is narrowed to a single known protein, additional data are needed, or it becomes likely that the protein is new.

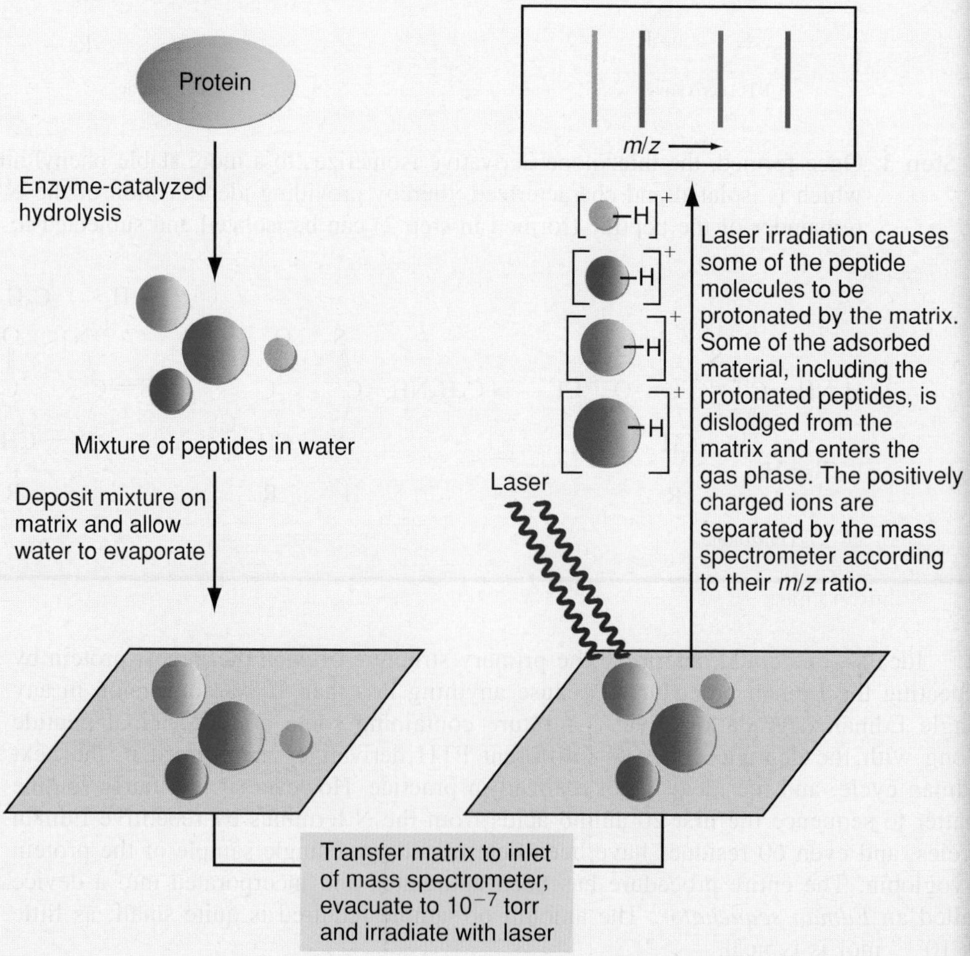
Protein
Enzyme-catalyzed hydrolysis
Mixture of peptides in water
Deposit mixture on matrix and allow water to evaporate
Transfer matrix to inlet of mass spectrometer, evacuate to 10^{-7} torr and irradiate with laser
Laser
m/z
Laser irradiation causes some of the peptide molecules to be protonated by the matrix. Some of the adsorbed material, including the protonated peptides, is dislodged from the matrix and enters the gas phase. The positively charged ions are separated by the mass spectrometer according to their m/z ratio.

Figure 25.10

The molecular weights of all of the peptides in a mixture obtained by the enzyme-catalyzed hydrolysis of a protein are determined simultaneously by mass spectrometry using matrix-assisted laser desorption ionization (MALDI).

Problem 25.20

Which two of the standard amino acids cannot be differentiated on the basis of their *m/z* ratio?

25.14 The Strategy of Peptide Synthesis

One way to confirm the structure proposed for a peptide is to synthesize a peptide having a specific sequence of amino acids and compare the two. This was done, for example, in the case of *bradykinin,* a peptide present in blood that acts to lower blood pressure. Excess bradykinin, formed as a response to the sting of wasps and other insects containing substances in their venom that stimulate bradykinin release, causes severe local pain. Bradykinin was originally believed to be an octapeptide containing two proline residues; however, a nonapeptide containing three prolines in the following sequence was synthesized and determined to be identical with natural bradykinin in every respect, including biological activity:

Arg-Pro-Pro-Gly-Phe-Ser-Pro-Phe-Arg

Bradykinin

A reevaluation of the original sequence data established that natural bradykinin was indeed the nonapeptide shown. Here the synthesis of a peptide did more than confirm structure; synthesis was instrumental in determining structure.

The synthesis of peptides is an important area of drug development. For many years, the high cost of peptide synthesis and the rapid degradation of peptides when administered in the body hampered the progress of peptide drug discovery. Recent advances in peptide manufacturing and drug delivery have increased the use of peptide-based pharmaceuticals. Two of the most prominent peptide drugs are insulin and calcitonin.

A promising method of drug delivery uses liposomes (see Section 24.4).

The insulin required for the treatment of diabetes used to be obtained by extraction from the pancreas glands of cows and pigs. Since the early 1980s, this "natural" insulin has been replaced by "synthetic" human insulin prepared by recombinant DNA technology (Section 26.17). Synthetic insulin is not only identical to human insulin, it is both safer and less expensive than insulin obtained from animals. A somewhat smaller polypeptide, *calcitonin* with 32 amino acids, is prepared by more traditional methods of synthetic organic chemistry. Synthetic calcitonin is identical to that obtained from salmon and is widely used for the treatment of osteoporosis. How calcitonin acts remains uncertain, but one possibility is that it maintains bone mass, not by increasing the rate of bone growth, but by decreasing the rate of bone loss. The biological receptor for calcitonin is located on a type of bone cell known as an osteoclast.

Other than the biochemical methods typified by the synthesis of insulin, there are two major approaches to peptide synthesis:

1. Solution phase
2. Solid phase

Although the two approaches differ in respect to the phase in which the synthesis is carried out, the overall strategy is the same in both.

The objective in peptide synthesis may be simply stated: to connect amino acids in a prescribed sequence by amide bond formation between them. A number of very effective methods and reagents have been designed for peptide bond formation, so that the joining together of amino acids by amide linkages is not difficult. The real difficulty lies in ensuring that the correct sequence is obtained. This can be illustrated by considering the synthesis of a representative dipeptide, Phe-Gly. Random peptide bond formation in a mixture containing phenylalanine and glycine would be expected to lead to four dipeptides:

$$H_3\overset{+}{N}CHCO_2^- \text{ (with } CH_2C_6H_5 \text{ on CH)} + H_3\overset{+}{N}CH_2CO_2^- \longrightarrow \text{Phe-Gly} + \text{Phe-Phe} + \text{Gly-Phe} + \text{Gly-Gly}$$

Phenylalanine Glycine

To direct the synthesis so that only Phe-Gly is formed, the amino group of phenylalanine and the carboxyl group of glycine must be protected so that they cannot react under the conditions of peptide bond formation. We can represent the peptide bond formation step by the following equation, where X and Y are amine- and carboxyl-protecting groups, respectively:

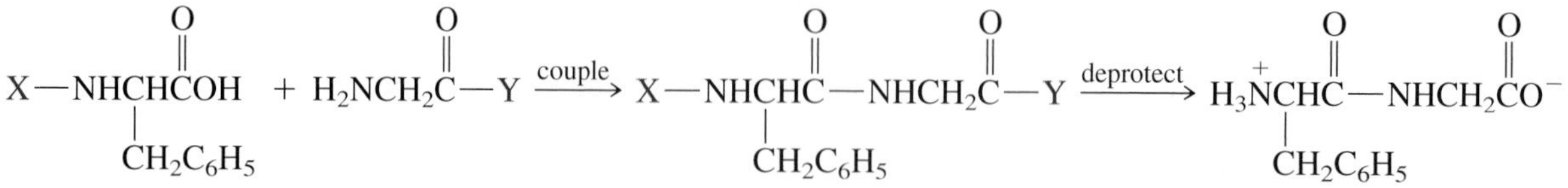

N-Protected phenylalanine — C-Protected glycine — Protected Phe-Gly — Phe-Gly

Thus, the synthesis of a dipeptide of prescribed sequence requires at least three operations:

1. *Protect* the amino group of the N-terminal amino acid and the carboxyl group of the C-terminal amino acid.
2. *Couple* the two protected amino acids by amide bond formation between them.
3. *Deprotect* the amino group at the N terminus and the carboxyl group at the C terminus.

Higher peptides are prepared in an analogous way by a direct extension of the logic just outlined for the synthesis of dipeptides.

Sections 25.15 through 25.17 describe the chemistry associated with the protection and deprotection of amino and carboxyl functions, along with methods for peptide bond formation. The focus in those sections is on solution-phase peptide synthesis. Section 25.18 shows how these methods are adapted to solid-phase synthesis.

25.15 Amino Group Protection

The reactivity of an amino group is suppressed by converting it to an amide, and amino groups are most often protected by acylation. The benzyloxycarbonyl group ($C_6H_5CH_2O\overset{\overset{O}{\|}}{C}—$) is one of the most often used amino-protecting groups. It is attached by acylation of an amino acid with benzyloxycarbonyl chloride.

Another name for the benzyloxycarbonyl group is *carbobenzoxy*. This name, and its abbreviation Cbz, are often found in the older literature, but are no longer a part of IUPAC nomenclature.

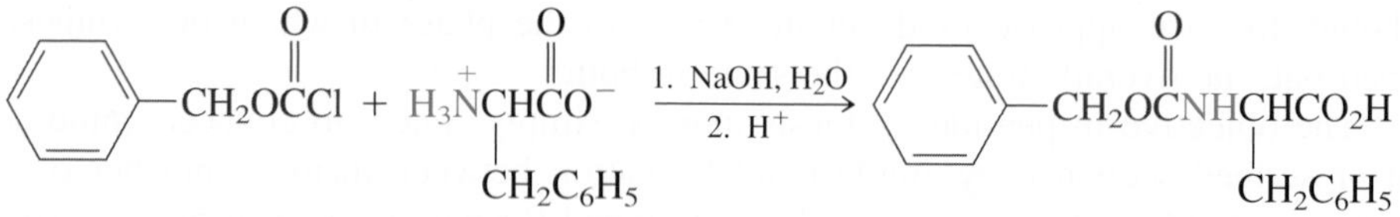

Benzyloxycarbonyl chloride — Phenylalanine — *N*-Benzyloxycarbonylphenylalanine (82–87%)

Problem 25.21

Lysine reacts with two equivalents of benzyloxycarbonyl chloride to give a derivative containing two benzyloxycarbonyl groups. What is the structure of this compound?

Just as it is customary to identify individual amino acids by abbreviations, so too with protected amino acids. The approved abbreviation for a benzyloxycarbonyl group is the letter *Z*. Thus, *N*-benzyloxycarbonylphenylalanine is represented as

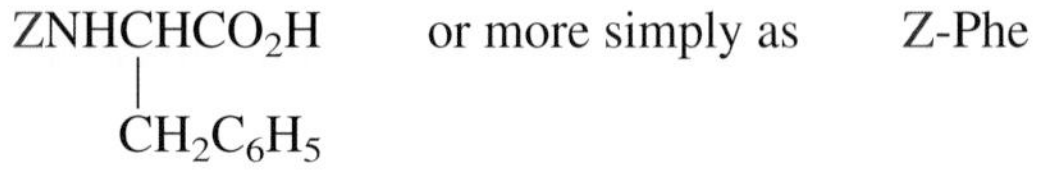

The value of the benzyloxycarbonyl protecting group is that it is easily removed by reactions other than hydrolysis. In peptide synthesis, amide bonds are formed. We protect the N terminus as an amide but need to remove the protecting group without cleaving the very amide bonds we labored so hard to construct. Removing the protecting group by hydrolysis would surely bring about cleavage of peptide bonds as well. One advantage that the benzyloxycarbonyl protecting group enjoys over more familiar acyl groups such as acetyl is that it can be removed by *hydrogenolysis* in the presence of palladium. The following equation illustrates this for the removal of the benzyloxycarbonyl protecting group from the ethyl ester of Z-Phe-Gly:

Hydrogenolysis refers to the cleavage of a molecule under conditions of catalytic hydrogenation.

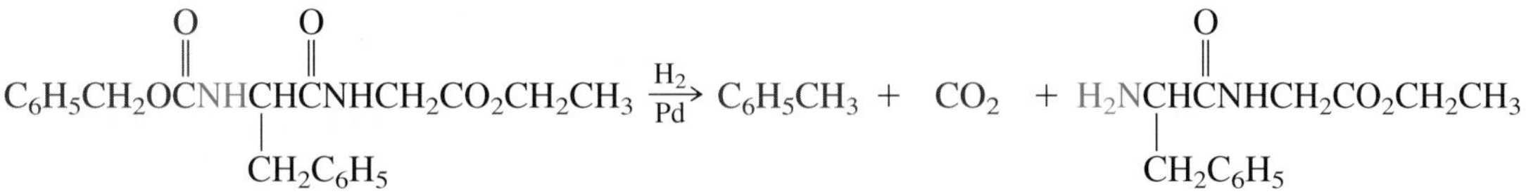

N-Benzyloxycarbonylphenylalanylglycine ethyl ester — Toluene — Carbon dioxide — Phenylalanylglycine ethyl ester (100%)

Alternatively, the benzyloxycarbonyl protecting group may be removed by treatment with hydrogen bromide in acetic acid:

N-Benzyloxycarbonylphenylalanylglycine ethyl ester — Benzyl bromide — Carbon dioxide — Phenylalanylglycine ethyl ester hydrobromide (82%)

Deprotection by this method rests on the ease with which benzyl esters are cleaved by nucleophilic attack at the benzylic carbon in the presence of strong acids. Bromide ion is the nucleophile.

A related N-terminal-protecting group is *tert*-butoxycarbonyl, abbreviated *Boc:*

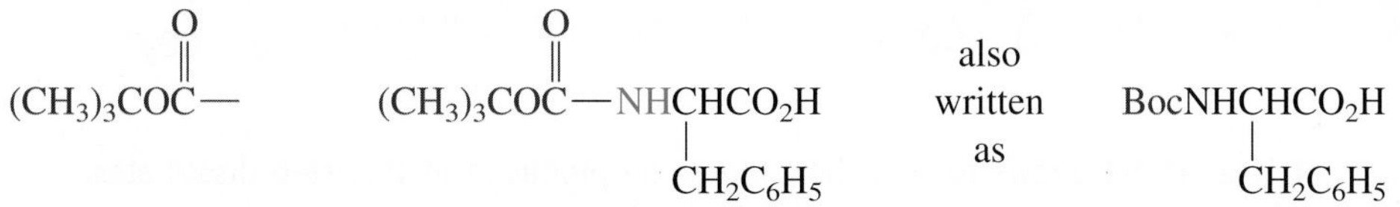

tert-Butoxycarbonyl (Boc-) — *N-tert*-Butoxycarbonylphenylalanine — Boc-Phe

Like the benzyloxycarbonyl protecting group, the Boc group may be removed by treatment with hydrogen bromide (it is stable to hydrogenolysis, however):

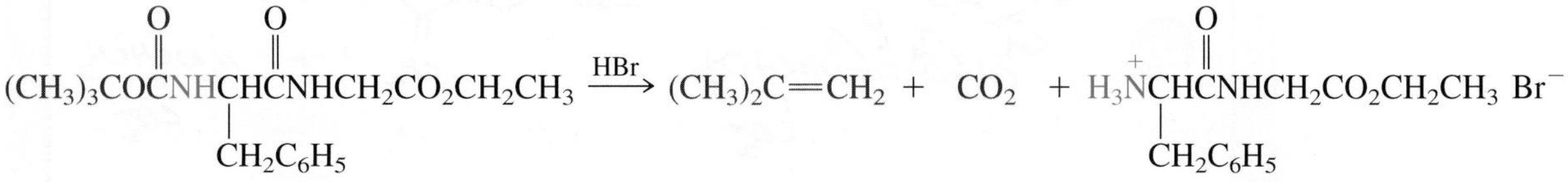

N-tert-Butoxycarbonylphenylalanylglycine ethyl ester — 2-Methylpropene — Carbon dioxide — Phenylalanylglycine ethyl ester hydrobromide (86%)

The *tert*-butyl group is cleaved as the corresponding carbocation. Loss of a proton from *tert*-butyl cation converts it to 2-methylpropene. Because of the ease with which a *tert*-butyl group is cleaved as a carbocation, other acidic reagents, such as trifluoroacetic acid, may also be used.

A third N-protecting group option is 9-fluorenylmethoxycarbonyl (FMOC).

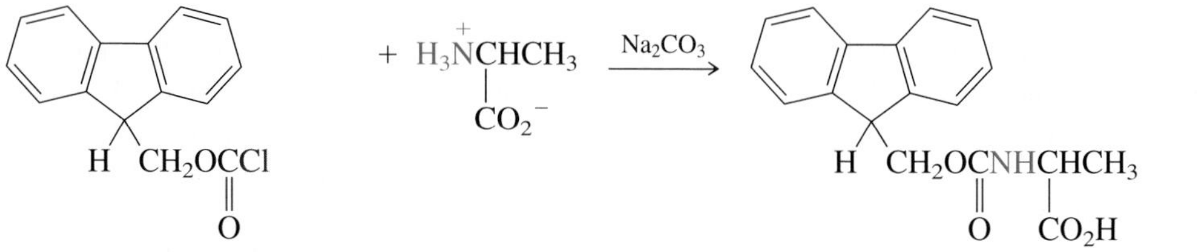
H CH2OCCl
O
+ H3NCHCH3
CO2−
Na2CO3
H CH2OCNHCHCH3
O CO2H

9-Fluorenylmethoxycarbonyl chloride (FMOC-Cl) | Alanine | 9-Fluorenylmethoxycarbonylalanine (88%)

FMOC differs from *Z* and Boc in that it is removed under basic conditions.

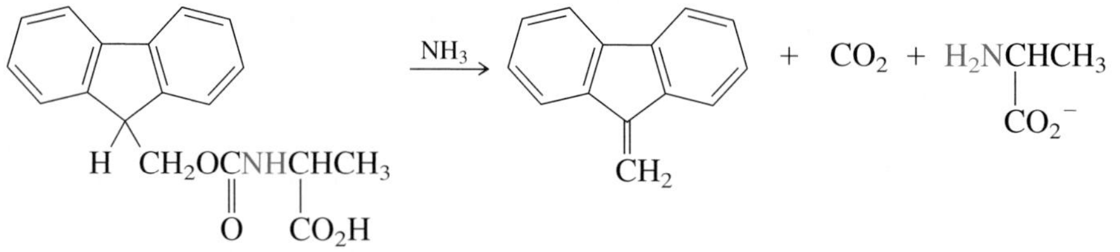
NH3
H CH2OCNHCHCH3
O CO2H
CH2
+ CO2 + H2NCHCH3
CO2−

9-Fluorenylmethoxycarbonylalanine | Dibenzofulvene | Carbon dioxide | Alanine (100%)

Problem 25.22

The mechanism of removing the FMOC protecting group rests on the fact its five-membered ring is structurally analogous to 1,3-cyclopentadiene, which makes its hydrogen relatively acidic. Ammonia removes this hydrogen to give an anion:

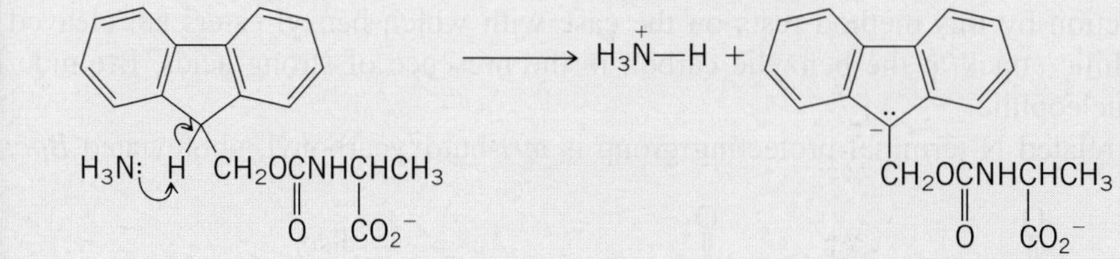
H3N: H CH2OCNHCHCH3
O CO2−
H3N—H +
CH2OCNHCHCH3
O CO2−

(a) Use curved arrows to show how the anion produced in this step dissociates.

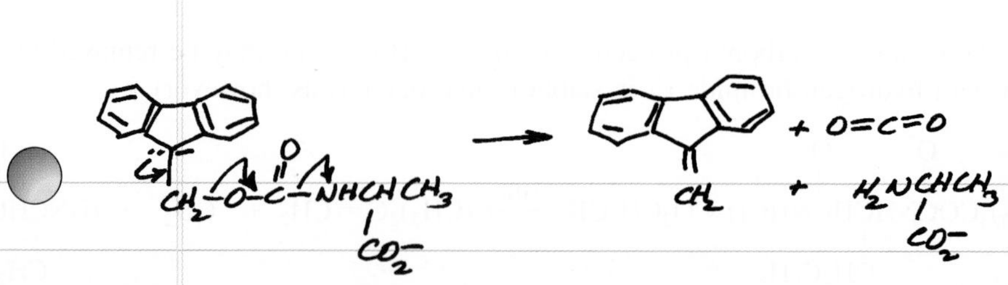
CH2—O—C—NHCHCH3
CO2−
CH2
+ O=C=O
+ H2NCHCH3
CO2−

(b) The dibromo analog of the FMOC protecting group can be removed with pyridine, which is a weaker base than is used to remove the FMOC group. Explain why the dibromo derivative is more easily removed by base than FMOC.

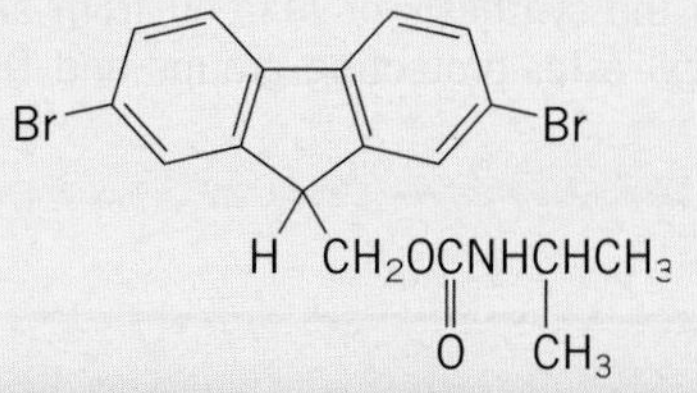

25.16 Carboxyl Group Protection

Carboxyl groups of amino acids and peptides are normally protected as esters. Methyl and ethyl esters are prepared by Fischer esterification. Deprotection of methyl and ethyl esters is accomplished by hydrolysis in base. Benzyl esters are a popular choice because they can also be removed by hydrogenolysis. Thus a synthetic peptide, protected at its N terminus with a Z group and at its C terminus as a benzyl ester, can be completely deprotected in a single operation.

$$C_6H_5CH_2O\overset{O}{\overset{\|}{C}}NH\underset{CH_2C_6H_5}{\underset{|}{C}H}\overset{O}{\overset{\|}{C}}NHCH_2CO_2CH_2C_6H_5 \xrightarrow[Pd]{H_2} H_3\overset{+}{N}\underset{CH_2C_6H_5}{\underset{|}{C}H}\overset{O}{\overset{\|}{C}}NHCH_2CO_2^- + 2C_6H_5CH_3 + CO_2$$

N-Benzyloxycarbonylphenylalanylglycine benzyl ester → Phenylalanylglycine (87%) + Toluene + Carbon dioxide

Several of the amino acids listed in Table 25.1 bear side-chain functional groups, which must also be protected during peptide synthesis. In most cases, protecting groups are available that can be removed by hydrogenolysis.

25.17 Peptide Bond Formation

To form a peptide bond between two suitably protected amino acids, the free carboxyl group of one of them must be *activated* so that it is a reactive acylating agent. The most familiar acylating agents are acyl chlorides, and they were once extensively used to couple amino acids. Certain drawbacks to this approach, however, led chemists to seek alternative methods.

In one method, treatment of a solution containing the N-protected and the C-protected amino acids with *N,N'*-dicyclohexylcarbodiimide (DCCI) leads directly to peptide bond formation:

$$ZNH\underset{CH_2C_6H_5}{\underset{|}{C}H}\overset{O}{\overset{\|}{C}}OH + H_2NCH_2\overset{O}{\overset{\|}{C}}OCH_2CH_3 \xrightarrow[chloroform]{DCCI} ZNH\underset{CH_2C_6H_5}{\underset{|}{C}H}\overset{O}{\overset{\|}{C}}—NHCH_2\overset{O}{\overset{\|}{C}}OCH_2CH_3$$

Z-Protected phenylalanine + Glycine ethyl ester → Z-Protected Phe-Gly ethyl ester (83%)

N,N'-Dicyclohexylcarbodiimide has the structure:

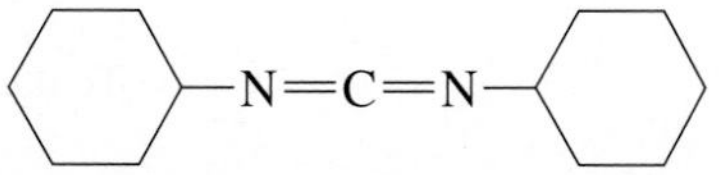

N,N'-Dicyclohexylcarbodiimide (DCCI)

Mechanism 25.4 shows how DCCI promotes the formation of the peptide bond shown in the preceding equation.

Problem 25.23

Show the steps involved in the synthesis of Ala-Leu from alanine and leucine using benzyloxycarbonyl and benzyl ester protecting groups and DCCI-promoted peptide bond formation.

Mechanism 25.4

Amide Bond Formation Between a Carboxylic Acid and an Amine Using *N,N'*-Dicyclohexylcarbodiimide

THE OVERALL REACTION:

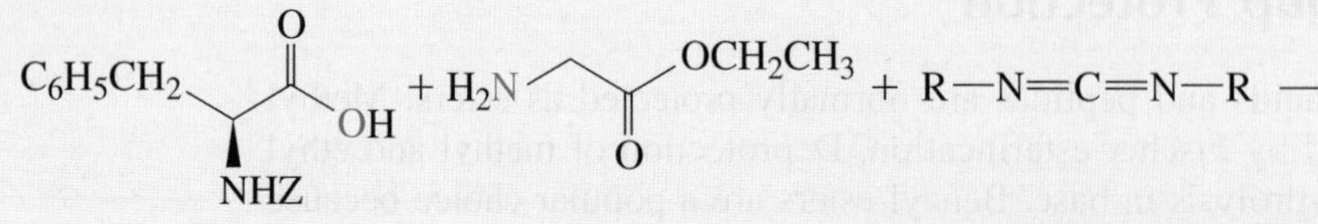

Z-Protected phenylalanine (Z = Benzyloxycarbonyl)   Glycine ethyl ester   *N,N'*-Dicyclohexylcarbodiimide (DCCI) (R = cyclohexyl)

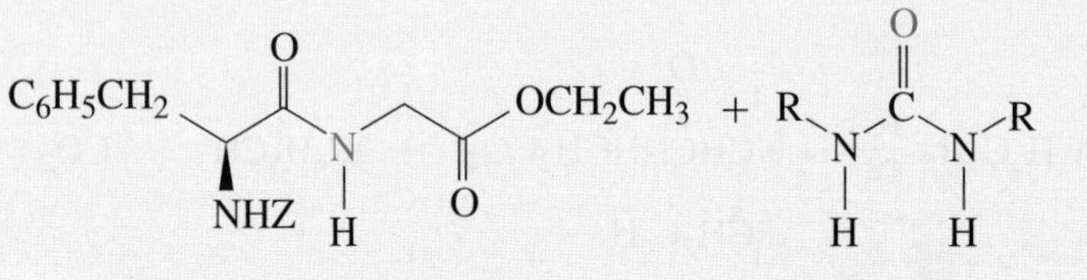

Z-Protected Phe-Gly ethyl ester   *N,N'*-Dicyclohexylurea

THE MECHANISM:

Step 1: In the first stage of the reaction, the carboxylic acid adds to one of the double bonds of DCCI to give an *O*-acylisourea.

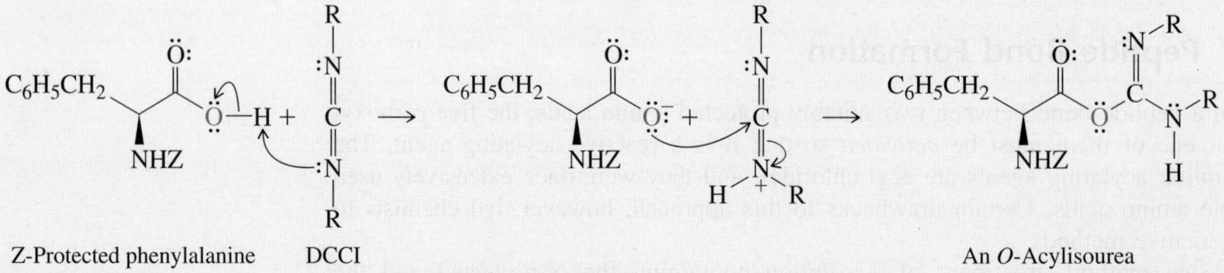

Z-Protected phenylalanine   DCCI   An *O*-Acylisourea

Step 2: Structurally, *O*-acylisoureas resemble acid anhydrides and are powerful acylating agents. In the reaction's second stage the amine adds to the carbonyl group of the *O*-acylisourea to give a tetrahedral intermediate.

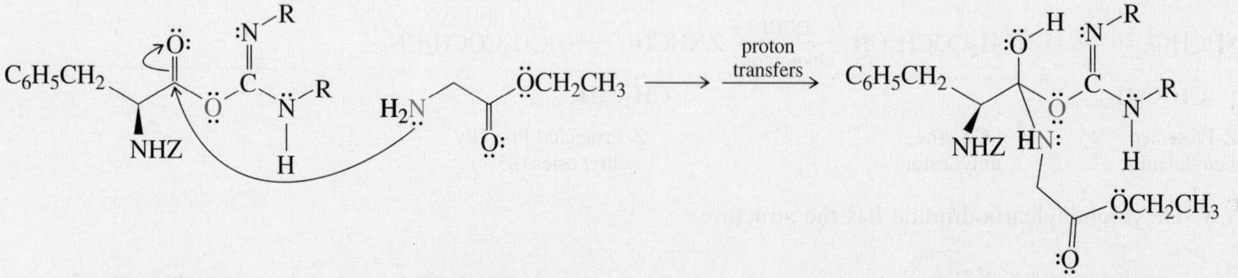

An *O*-acylisourea   Glycine ethyl ester   Tetrahedral intermediate

Continued

Step 3: The tetrahedral intermediate dissociates to an amide and *N,N′*-dicyclohexylurea.

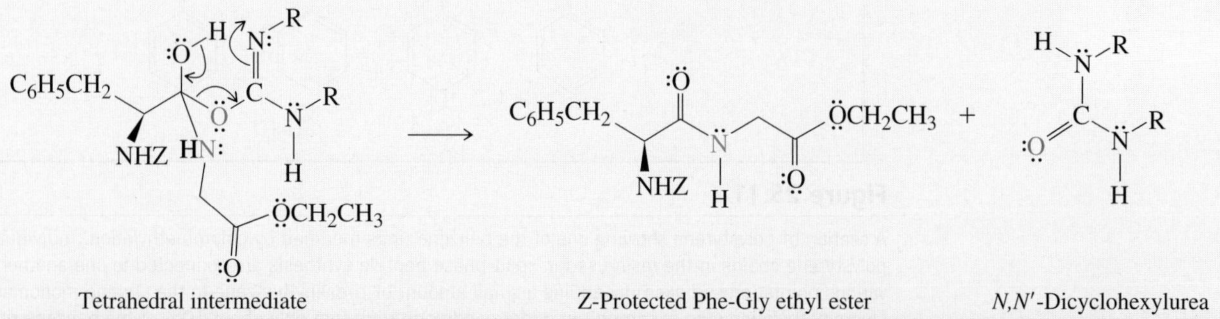

The *N,N′*-dicyclohexylurea that is produced in peptide couplings with DCCI is soluble in the same solvents as the product, making it more difficult to separate the by-products during purification. An alternative carbodiimide that is used in peptide synthesis is *N*-(3-dimethylaminopropyl)-*N′*-ethylcarbodiimide, known as EDCI. Both EDCI and the urea that is produced from it are water-soluble and can be more easily separated from the peptide product. EDCI can also be used in organic solvents.

$$CH_3CH_2N{=}C{=}NCH_2CH_2CH_2N(CH_3)_2$$

N-(3-dimethylaminopropyl)-*N′*-ethylcarbodiimide (EDCI)

Problem 25.24

What is the structure of the urea that is produced in the coupling reaction of a protected amino acid and a carboxylic acid with EDCI?

Higher peptides are prepared either by stepwise extension of peptide chains, one amino acid at a time, or by coupling of fragments containing several residues (the *fragment condensation* approach). Human pituitary adrenocorticotropic hormone (ACTH), for example, has 39 amino acids and was synthesized by coupling of smaller peptides containing residues 1–10, 11–16, 17–24, and 25–39. An attractive feature of this approach is that the various protected peptide fragments may be individually purified, which simplifies the purification of the final product. Among the substances that have been synthesized by fragment condensation are insulin (51 amino acids) and the protein ribonuclease A (124 amino acids). In the stepwise extension approach, the starting peptide in a particular step differs from the coupling product by only one amino acid residue and the properties of the two peptides may be so similar as to make purification by conventional techniques all but impossible. The solid-phase method described in the following section overcomes many of the difficulties involved in the purification of intermediates.

25.18 Solid-Phase Peptide Synthesis: The Merrifield Method

In 1962, R. Bruce Merrifield of Rockefeller University reported the synthesis of the nonapeptide bradykinin by a novel method. In Merrifield's method, peptide coupling and deprotection are carried out not in homogeneous solution but at the surface of an insoluble polymer, or *solid support*. Beads of a copolymer prepared from styrene containing about 2% divinylbenzene are treated with chloromethyl methyl ether and tin(IV) chloride to give a resin in which about 10% of the aromatic rings bear —CH_2Cl groups (Figure 25.11). The growing peptide is anchored to this polymer, and excess reagents, impurities, and by-products are removed by thorough washing after each operation. This greatly simplifies the purification of intermediates.

Merrifield was awarded the 1984 Nobel Prize in Chemistry for developing the solid-phase method of peptide synthesis.

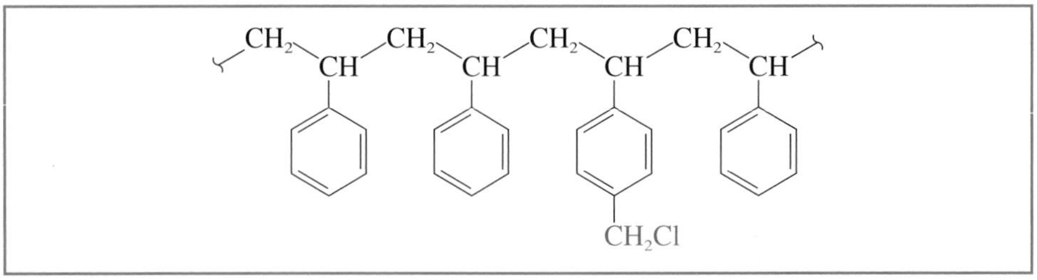

Figure 25.11

A section of polystyrene showing one of the benzene rings modified by chloromethylation. Individual polystyrene chains in the resin used in solid-phase peptide synthesis are connected to one another at various points (cross-linked) by adding a small amount of *p*-divinylbenzene to the styrene monomer. The chloromethylation step is carried out under conditions such that only about 10% of the benzene rings bear —CH_2Cl groups.

The Merrifield procedure has been adapted to accommodate FMOC as well as Boc protecting groups.

The actual process of **solid-phase peptide synthesis,** outlined in Figure 25.12, begins with the attachment of the C-terminal amino acid to the chloromethylated polymer in step 1. Nucleophilic substitution by the carboxylate anion of an *N*-Boc-protected C-terminal amino acid displaces chloride from the chloromethyl group of the polymer to form an ester, protecting the C terminus while anchoring it to a solid support. Next, the Boc group is removed by treatment with acid (step 2), and the polymer containing the unmasked N terminus is washed with a series of organic solvents. By-products are removed, and only the polymer and its attached C-terminal amino acid residue remain. Next (step 3), a peptide bond to an *N*-Boc-protected amino acid is formed by condensation in the presence of *N,N'*-dicyclohexylcarbodiimide. Again, the polymer is washed thoroughly. The Boc protecting group is then removed by acid treatment (step 4), and after washing, the polymer is now ready for the addition of another amino acid residue by a repetition of the cycle. When all the amino acids have been added, the synthetic peptide is removed from the polymeric support by treatment with hydrogen bromide in trifluoroacetic acid.

By successively adding amino acid residues to the C-terminal amino acid, it took Merrifield only eight days to synthesize the nonapeptide bradykinin in 68% yield. The biological activity of synthetic bradykinin was identical with that of natural material.

Problem 25.25

Starting with phenylalanine and glycine, outline the steps in the preparation of Phe-Gly by the Merrifield method.

Merrifield successfully automated all the steps in solid-phase peptide synthesis, and computer-controlled equipment is commercially available to perform this synthesis. Using an early version of his "peptide synthesizer," in collaboration with coworker Bernd Gutte, Merrifield reported the synthesis of the enzyme ribonuclease in 1969. It took them only six weeks to perform the 369 reactions and 11,391 steps necessary to assemble the sequence of 124 amino acids of ribonuclease.

Solid-phase peptide synthesis does not solve all purification problems, however. Even if every coupling step in the ribonuclease synthesis proceeded in 99% yield, the product would be contaminated with many different peptides containing 123 amino acids, 122 amino acids, and so on. Thus, Merrifield and Gutte's six weeks of synthesis was followed by four months spent in purifying the final product. The technique has since been refined to the point that yields at the 99% level and greater are achieved with current instrumentation, and thousands of peptides and peptide analogs have been prepared by the solid-phase method.

Merrifield's concept of a solid-phase method for peptide synthesis and his development of methods for carrying it out set the stage for an entirely new way to do chemical reactions. Solid-phase synthesis has been extended to include numerous other classes of

Step 1: The Boc-protected amino acid is anchored to the resin. Nucleophilic substitution of the benzylic chloride by the carboxylate anion gives an ester.

Step 2: The Boc protecting group is removed by treatment with hydrochloric acid in dilute acetic acid. After the resin has been washed, the C-terminal amino acid is ready for coupling.

Step 3: The resin-bound C-terminal amino acid is coupled to an N-protected amino acid by using *N,N′*-dicyclohexylcarbodiimide. Excess reagent and *N,N′*-dicyclohexylurea are washed away from the resin after coupling is complete.

Step 4: The Boc protecting group is removed as in step 2. If desired, steps 3 and 4 may be repeated to introduce as many amino acid residues as desired.

Step *n*: When the peptide is completely assembled, it is removed from the resin by treatment with hydrogen bromide in trifluoroacetic acid.

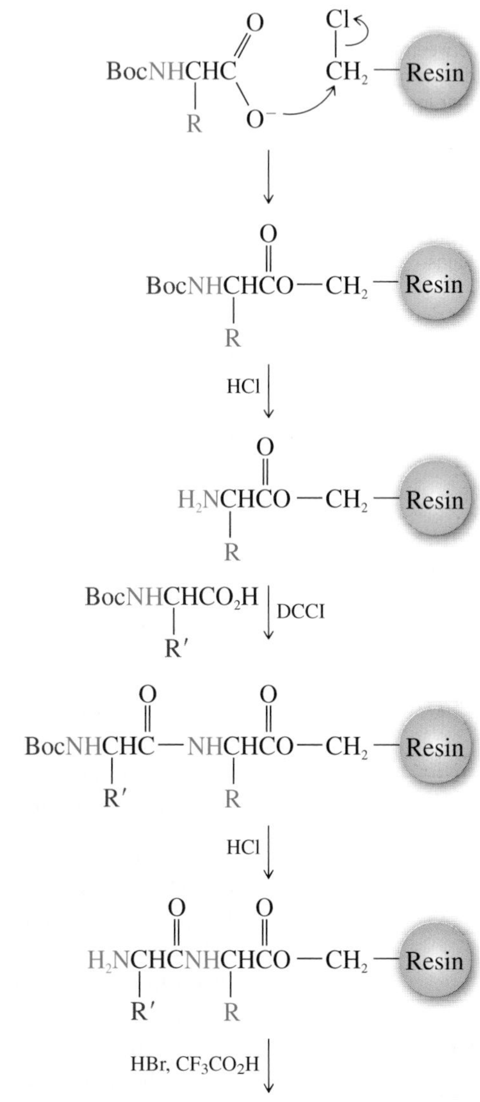

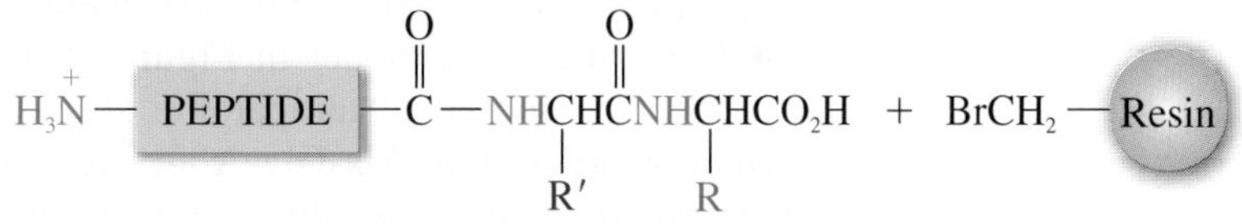

Figure 25.12

Peptide synthesis by the solid-phase method. Amino acid residues are attached sequentially beginning at the C terminus.

compounds and has helped spawn a whole new field called **combinatorial chemistry.** Combinatorial synthesis allows a chemist, using solid-phase techniques, to prepare hundreds of related compounds (called *libraries*) at a time.

25.19 Secondary Structures of Peptides and Proteins

The primary structure of a peptide is its amino acid sequence. The **secondary structure** is the conformational relationship of nearest neighbor amino acids with respect to each other. On the basis of X-ray crystallographic studies and careful examination of molecular models, Linus Pauling and Robert B. Corey of the California Institute of Technology showed that certain peptide conformations were more stable than others. Two arrangements, the

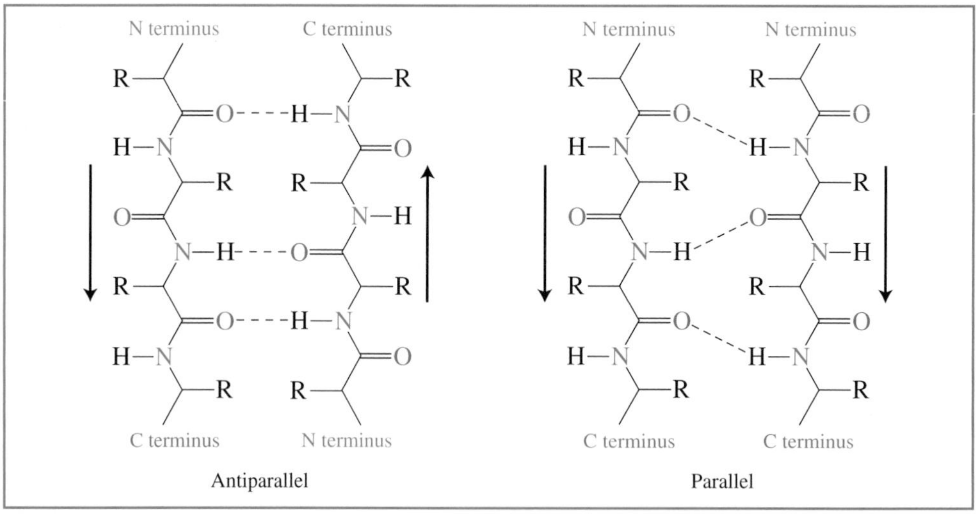

Figure 25.13

Hydrogen bonding between the carbonyl oxygen of one peptide chain and the amide N—H of another in a β-pleated sheet. In the antiparallel arrangement, the N-terminus → C-terminus direction of one chain is opposite to that of the other. In the parallel arrangement, the N-terminus → C-terminus direction is the same for both chains.

α helix and the **β sheet,** stand out as secondary structural units that are both particularly stable and commonly encountered. Both of these incorporate two important features:

1. The geometry of the peptide bond is planar and the main chain is arranged in an anti conformation (see Section 25.7).
2. Hydrogen bonding can occur when the N—H group of one amino acid unit and the C═O group of another are close in space; conformations that maximize the number of these hydrogen bonds are stabilized by them.

Chains in a β sheet exist in an extended conformation with hydrogen bonds between a carbonyl oxygen of one chain and an amide N—H of another (Figure 25.13). Both the parallel and antiparallel arrangements of chains occur in proteins. Some of the space between peptide chains is occupied by the amino acid side chains, represented by R in Figure 25.13. Van der Waals repulsive forces involving these substituents cause the chains to rotate with respect to one another, giving a rippled effect known as a *β-pleated sheet* (Figure 25.14).

The β-pleated sheet is an important secondary structure in proteins that are rich in amino acids with small side chains such as H (glycine), CH_3 (alanine), and CH_2OH (serine). The model in Figure 25.14 is a portion of the calculated structure for a sheet composed of antiparallel strands containing only glycine and alanine in alternating order (Gly-Ala-Gly-Ala-, etc.). It was designed to resemble *fibroin,* the major protein of silk. Fibroin is almost entirely pleated sheet, and over 80% of it is a repeating sequence of the six-residue unit -Gly-Ser-Gly-Ala-Gly-Ala-. Because the polypeptide backbone adopts an extended zigzag conformation, silk, unlike wool for example, resists stretching.

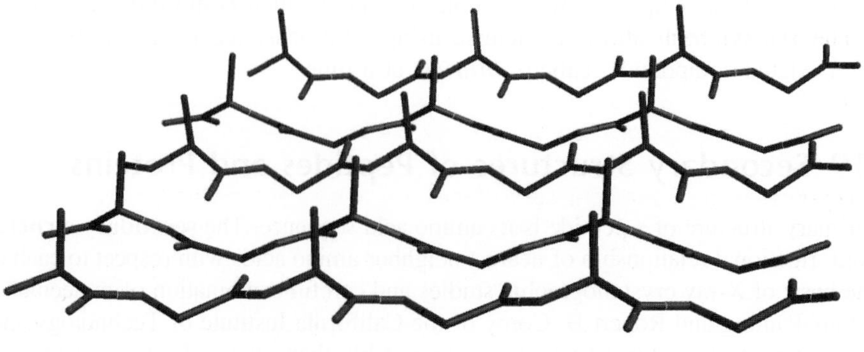

Figure 25.14

The β-pleated sheet secondary structure of a protein, composed of alternating glycine and alanine residues.

Problem 25.26

The methyl groups of the alanine residues of the β sheet in Figure 25.14 all point upward. If this pleated sheet were composed of only alanine residues instead of being Gly-Ala-Gly-Ala, etc. what would be the pattern of methyl groups? Would they all point up, alternate up and down, or be random?

The α helix is another commonly encountered secondary structure. Figure 25.15 gives three views of an α-helix model constructed from eight L-alanine residues. Part *a* of the figure is a ball-and-spoke model; part *b* is a view through the center of the helix along its axis. The helix is right-handed with about 3.6 amino acids per turn and is stabilized by hydrogen bonds between the carbonyl oxygens and N—H protons. View *b* shows how the methyl groups of L-alanine project outward from the main chain. This outward orientation of amino acid side chains makes them the points of contact with other amino acids of the same chain, with different protein chains, and with other biomolecules. Part *c* of the figure uses a ribbon to trace the peptide backbone. The ribbon helps distinguish front from back, makes the right-handedness of the helix more apparent, and is especially useful when looking at how proteins are folded.

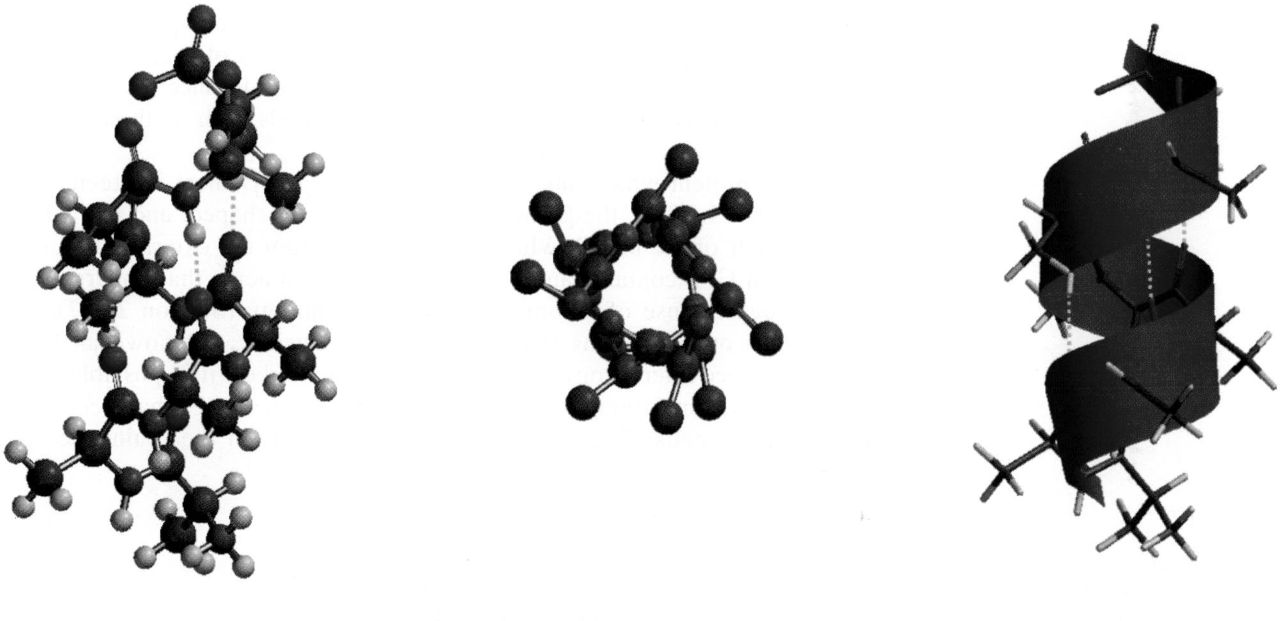

(*a*) (*b*) (*c*)

Figure 25.15

Molecular models of an α helix composed of eight alanine residues. The N terminus is at the bottom. (*a*) A ball-and-spoke model. Hydrogen bonds are shown as dashed lines. (*b*) The same model looking up the helical axis from the bottom. Hydrogens have been omitted for clarity. The helix is right-handed, and all of the methyl groups point outward. (*c*) A tube model framed in a ribbon that traces the path of the helix.

The protein components of muscle (*myosin*) and wool (*α-keratin*) contain high percentages of α helix. When wool is stretched, hydrogen bonds break and the peptide chain is elongated. Covalent S—S bonds between L-cysteine residues limit the extent to which the chain can be stretched, however, and once the stretching force is removed the hydrogen bonds re-form spontaneously.

Most proteins cannot be described in terms of a single secondary structure. Rather, most are mixtures of α helix and β sheet, interspersed with regions of **random coils** that have no regular pattern. Figure 25.16 shows a model of ribonuclease, an enzyme that catalyzes the hydrolysis of RNA. The helical regions are shown in red, the β sheets in yellow. Of the 124 amino acids in this protein, 24 are represented in three sections of

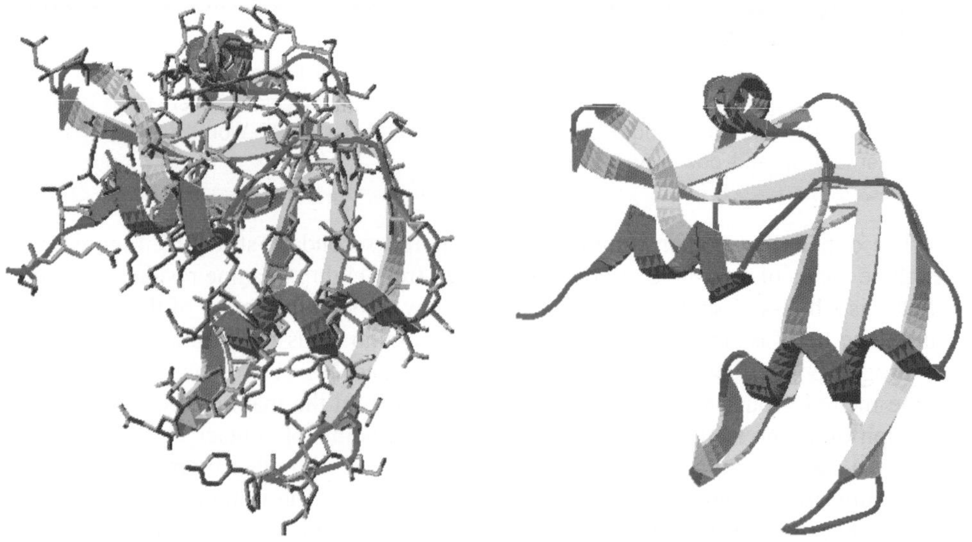

(a) (b)

Figure 25.16

Molecular models of ribonuclease. Red ribbons identify sequences where the secondary structure is a helix; yellow ribbons identify strands of β sheet. Arrowheads point in the direction from the N terminus to the C terminus. (*a*) Shows both a molecular model and ribbons. (*b*) Shows only the ribbons.

α helix. There are two β sheets, one with three strands accounting for 21 amino acids, the other with four strands and 20 amino acids. The strands of each β sheet belong to the same chain and are brought within hydrogen bonding distance because of how the chain is folded. Indeed, the formation of hydrogen bonds such as these is one of the factors that contributes to chain folding.

Another protein which has regions of α helix and β sheet is green fluorescent protein, or GFP. In GFP, the β sheet structure is barrel-shaped, and the α helix runs along the center of the barrel, where it is shielded from the exterior (Figure 25.17). The α helix in GFP contains an arrangement of amino acids that undergoes fluorescence. Through the use of recombinant DNA technology (Section 26.17), GFP can be attached to other proteins that are normally invisible, but now can be imaged using fluorescence microscopy. Using GFP as a biomarker, it is possible to monitor the roles of different proteins in the body, in part because GFP has very low toxicity toward living cells. The development of nerve cells in the brain, the spread of

Figure 25.17

Barrel-shaped green fluorescent protein (GFP) has an outer β-sheet structure and an α helix in the inner region.

cancer cells, and the damage to neurons that occurs during Alzheimer's disease can be followed by techniques that use GFP. The 2008 Nobel Prize in Chemistry was awarded to Osamu Shimomura, Martin Chalfie, and Roger Tsien for their work in the development of GFP.

25.20 Tertiary Structure of Polypeptides and Proteins

The way a protein chain is folded, its **tertiary structure,** affects both its physical properties and its biological function. The two main categories of protein tertiary structure are **fibrous** and **globular.**

1. Fibrous proteins are bundles of elongated filaments of protein chains and are insoluble in water.
2. Globular proteins are approximately spherical and are either soluble or form colloidal dispersions in water.

The primary structure of a protein, its amino acid sequence, is the main determinant of tertiary structure. Secondary structure also contributes by limiting the number of conformations available to a polypeptide chain.

Fibrous proteins, being insoluble in water often have a structural or protective function. The most familiar fibrous proteins are the keratins and collagen. α-Keratin (Figure 25.18) is based on the α-helix secondary structure and is the protein structural component of hair, wool, nails, claws, quills, horns, and the outer layer of skin. β-Keratin is based on the β-sheet secondary structure and occurs in silk as fibroin. L-Cysteine is especially abundant in keratins, where it can account for more than 20% of the amino acids present. Collagen occurs mainly in connective tissue (cartilage and tendons) and has a triple helix structure.

Globular proteins include most enzymes and function in aqueous environments. About 65% of the mass of most cells, for example, is water. When placed in water, nonpolar materials, including nonpolar amino acid side chains, cause nearby water molecules to adopt a more ordered arrangement, reducing the entropy of water. This is called the **hydrophobic effect.** The unfavorable negative ΔS is moderated if the protein adopts a spherical shape which places nonpolar side chains inside and polar ones on the surface. Of the various globular arrangements, the one that best offsets the entropy loss with attractive forces among the side chains is the tertiary structure adopted by the protein in its normal, or *native state.*

(a) *(b)* *(c)* *(d)*

Figure 25.18

α-Keratin. Two α helices (*a*) combine to give a coiled coil (*b*). A pair of coiled coils is a protofilament (*c*). Four protofilaments give a filament (*d*), which is the structural material from which the fibrous protein is assembled.

Table 25.4 lists the attractive forces that most influence protein tertiary structure. The strongest of these is the covalent S—S bond that unites two cysteine residues. This *disulfide bridge* can form between the $—CH_2SH$ groups of two cysteines, which, although they may be remote from each other in respect to the amino acid sequence, become neighbors when the chain is folded. Formation of the disulfide bond connecting the two stabilizes the local folded arrangement. A typical globular protein normally has only a small number of disulfide bridges. Of the 124 amino acids in ribonuclease (see Figure 25.16 in Section 25.19), 6 are cysteines and each participates in a disulfide bridge; one bridge unites Cys-26 and Cys-84, another Cys-58 and Cys-110, and a third Cys-65 and Cys-72.

The noncovalent interactions are all much weaker than the S—S covalent bond. Among them, the electrostatic attraction between positively and negatively charged side chains, called a salt bridge, is the strongest, followed by hydrogen bonding, then van der Waals forces. Keep in mind though, that the total contribution of the various forces depends not only on the magnitude of an interaction but also on their number.

TABLE 25.4 Covalent and Noncovalent Interactions Between Amino Acid Side Chains in Proteins

Description	Type of interaction	Example
Covalent		
Disulfide bridge	S—S bond between two cysteines	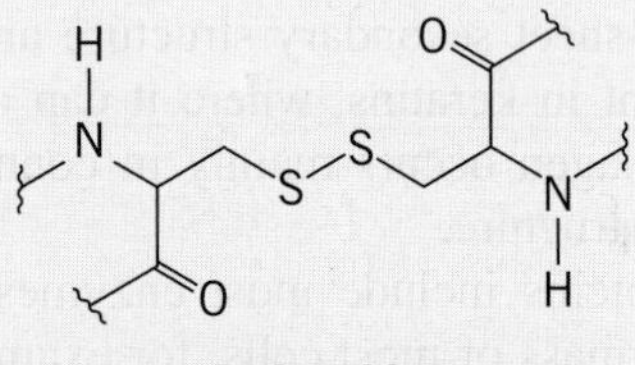 CysS — SCys
Noncovalent		
Salt bridge	Electrostatic attraction between oppositely charged ions	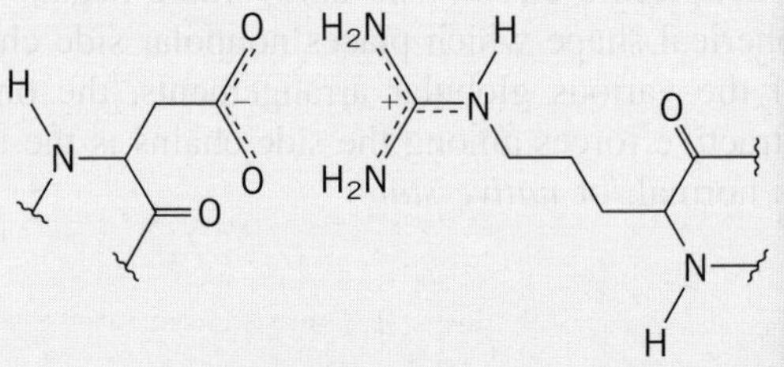 Asp---Arg
Hydrogen bond	Positively polarized H of O—H or N—H group interacts with an electronegative atom (O or N)	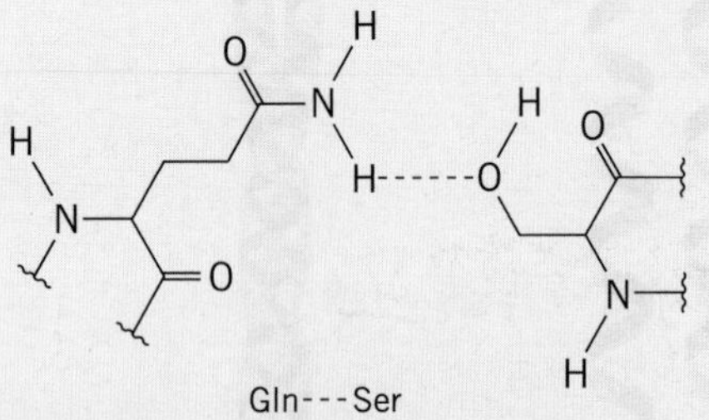 Gln---Ser
Van der Waals	Induced-dipole/induced-dipole attraction (dispersion force) between nonpolar side chains	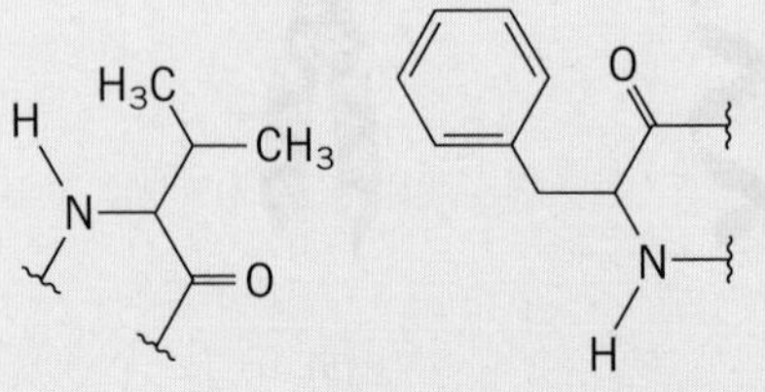 Val---Phe

Disulfide bridges may be strong, but there are usually only a few of them. Van der Waals forces are weak, but they outnumber all the other intermolecular attractive forces.

Problem 25.27

Table 25.4 shows a salt bridge between aspartic acid and arginine. Sketch the analogous electrostatic attraction between lysine and an amino acid other than aspartic acid from Table 25.1.

Knowing how the protein chain is folded is a key element in understanding how an **enzyme** catalyzes a reaction. Biochemical processes are usually related to the core reaction types of organic chemistry and involve similar key intermediates. The reactions, however, are much faster and more selective. In proposing an enzyme-catalyzed mechanism for a reaction such as amide or ester hydrolysis, it is customary to assume it proceeds by way of a tetrahedral intermediate, then modify the usual nucleophilic acyl substitution mechanism by assigning various catalytic functions to selected amino acid side chains of the enzyme.

We saw in Section 25.10 that carboxypeptidase A catalyzes the hydrolysis of the peptide bond to the C-terminal amino acid of polypeptides. Carboxypeptidase A is a *metalloprotein;* it contains a Zn^{2+} ion, which is essential for catalytic activity. The X-ray crystal structure of carboxypeptidase A (Figure 25.19) locates this Zn^{2+} ion in a hydrophobic cavity near the center of the enzyme, where it is held by coordination to a glutamic acid residue (Glu-72) and two histidines (His-69 and His-196). This is the same region, called the **active site,** where the substrate binds. The substrate in the case of carboxypeptidase is a peptide, especially a peptide with a hydrophobic C-terminal amino acid such as phenylalanine or tyrosine. In addition to being hydrophobic, as is the active site, the substrate is bound by an electrostatic attraction between its negatively charged carboxylate and the positively charged side chain of Arg-145. Mechanism 25.5 shows the interactions of the side chains of carboxypeptidase A with Zn^{2+} and a peptide, then describes the mechanism for cleaving the peptide bond to the terminal amino acid. Side chains other than those shown in Mechanism 25.5 have been implicated but have been omitted. The main feature of the mechanism is its relationship to the mechanism of nucleophilic acyl substitution (on which it was patterned). Not only does the enzyme bring the substrate and catalytically active functions together at the active site, but by stabilizing the tetrahedral intermediate it lowers the activation energy for its formation and increases the reaction rate.

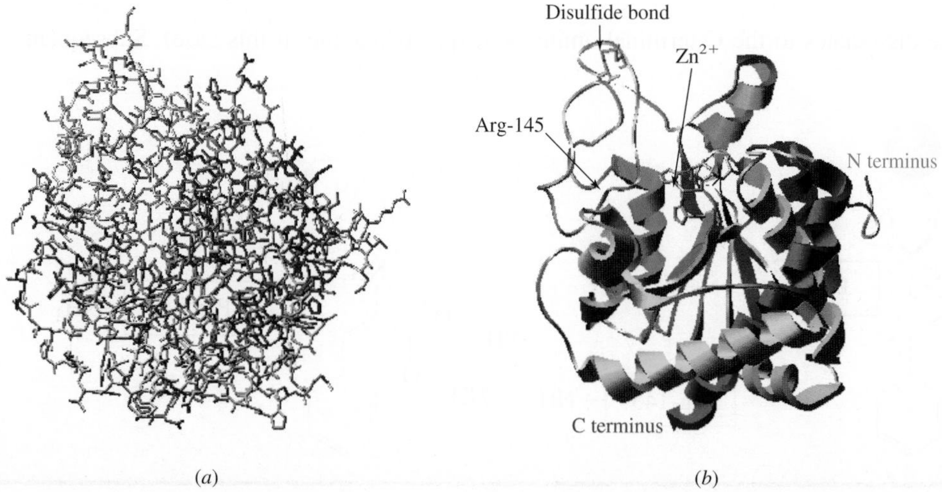

Figure 25.19

The structure of carboxypeptidase A displayed as (*a*) a tube model and (*b*) a ribbon diagram. The most evident feature illustrated by (*a*) is the globular shape of the enzyme. The ribbon diagram emphasizes the folding of the chain.

Mechanism 25.5

Carboxypeptidase-Catalyzed Hydrolysis

THE MECHANISM: The mechanism shown outlines the major stages in carboxypeptidase-catalyzed hydrolysis of a peptide in which the C-terminal amino acid is phenylalanine. Proton transfers accompany stages 2 and 3 but are not shown. Only the major interactions of the substrate with the carboxypeptidase side chains are shown although others may also be involved.

Stage 1: The peptide is positioned in the active site by an electrostatic bond between its negatively charged C-terminal carboxylate and a positively charged arginine side chain of the enzyme. Also at the active site, Zn^{2+} engages in Lewis acid/Lewis base interactions with His-69 and His-196 and an electrostatic attraction with the negatively charged carboxylate of Glu-72. These ligands are shown here but will be omitted in subsequent steps for simplicity.

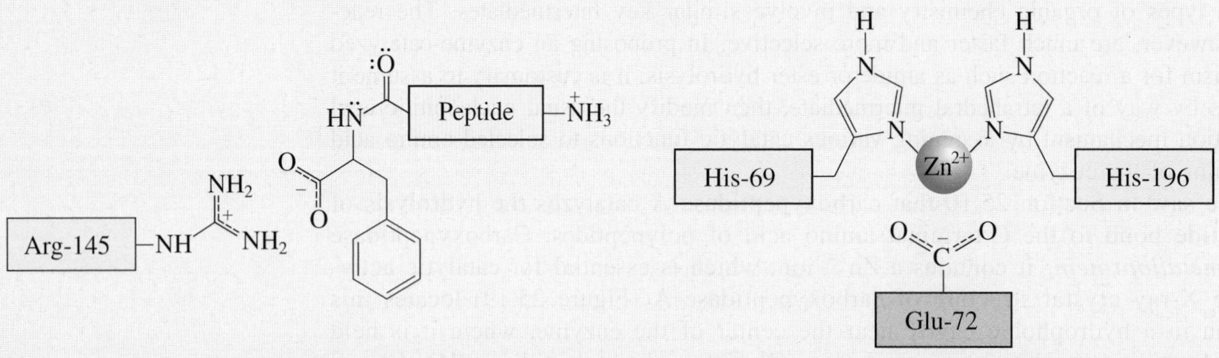

Stage 2: Water adds to the carbonyl group of the peptide bond. The rate of this nucleophilic addition is accelerated by coordination of the carbonyl oxygen to Zn^{2+} and/or to one of the N—H protons of Arg-127 (not shown). The product is a tetrahedral intermediate stabilized by coordination to zinc. Stabilization of the tetrahedral intermediate may be the major factor for the rapid rate of the carboxypeptidase-catalyzed hydrolysis.

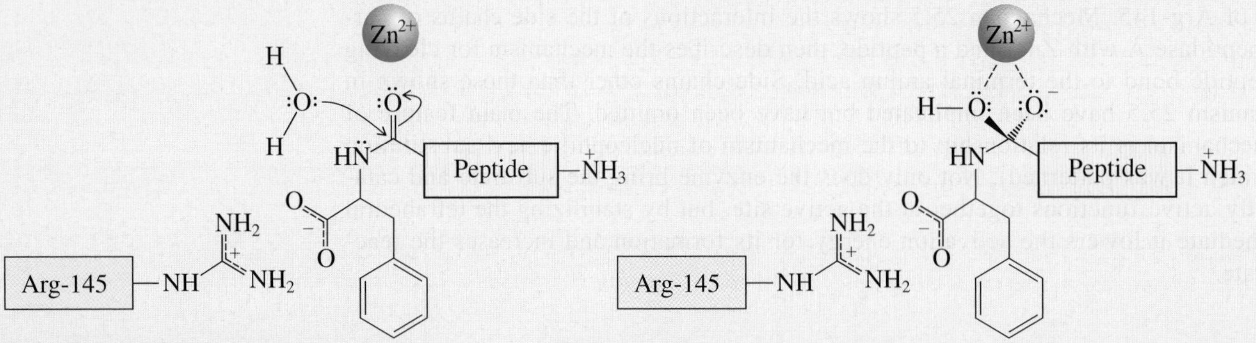

Stage 3: The tetrahedral intermediate dissociates to the C-terminal amino acid (phenylalanine in this case). Subsequent steps restore the active site.

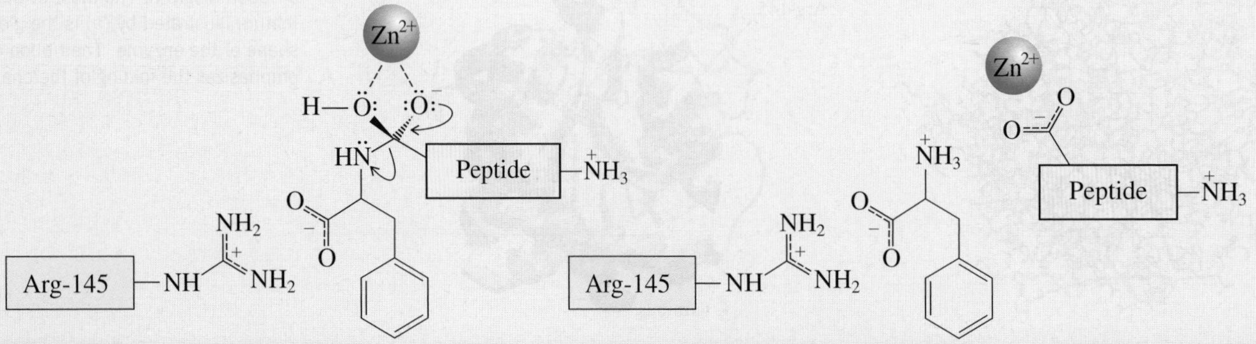

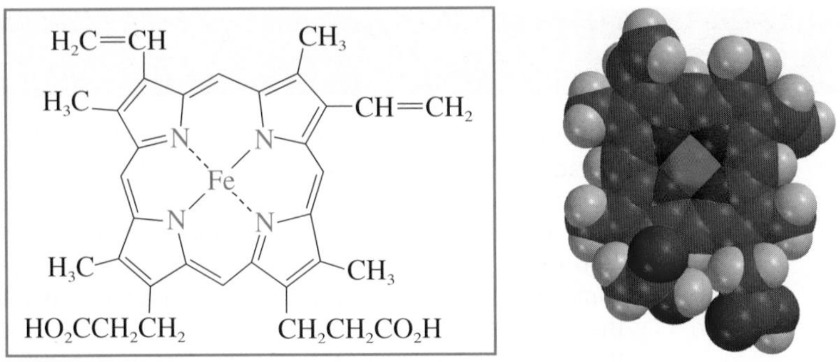
H2C=CH
CH3
H3C
CH=CH2
N
N
Fe
N
N
H3C
CH3
HO2CCH2CH2
CH2CH2CO2H

(*a*) (*b*)

Figure 25.20

Heme shown as (*a*) a structural drawing and as (*b*) a space-filling model. The space-filling model shows the coplanar arrangement of the groups surrounding iron.

25.21 Coenzymes

The number of chemical processes that protein side chains can engage in is rather limited. Most prominent among them are proton donation, proton abstraction, and nucleophilic addition to carbonyl groups. In many biological processes a richer variety of reactivity is required, and proteins often act in combination with substances other than proteins to carry out the necessary chemistry. These substances are called **cofactors,** and they can be organic or inorganic and strongly or weakly bound to the enzyme. Among cofactors that are organic molecules, the term **coenzyme** is applied to those that are not covalently bound to the enzyme, and **prosthetic group** to those that are. Acting alone, for example, proteins lack the necessary functionality to be effective oxidizing or reducing agents. They can catalyze biological oxidations and reductions, however, in the presence of a suitable coenzyme. In earlier sections we saw numerous examples of these reactions in which the coenzyme NAD^+ acted as an oxidizing agent, and others in which NADH acted as a reducing agent.

Pyridoxal 5′-phosphate (see Section 25.6) is a coenzyme.

Heme (Figure 25.20) is an important prosthetic group in which iron(II) is coordinated with the four nitrogen atoms of a type of tetracyclic aromatic substance known as a *porphyrin.* The oxygen-storing protein of muscle, myoglobin, represented schematically in Figure 25.21, consists of a heme group surrounded by a protein of 153 amino acids. Four of the six available coordination sites of Fe^{2+} are taken up by the nitrogens of the porphyrin, one by a histidine residue of the protein, and the last by a water molecule. Myoglobin stores oxygen obtained from the blood by formation of an $Fe{-}O_2$ complex. The oxygen displaces water as the sixth ligand on iron and is held there until needed. The protein serves as a container for the heme and prevents oxidation of Fe^{2+} to Fe^{3+}, an oxidation state in which iron lacks the ability to bind oxygen. Separately, neither heme nor the protein binds oxygen in aqueous solution; together, they do it very well.

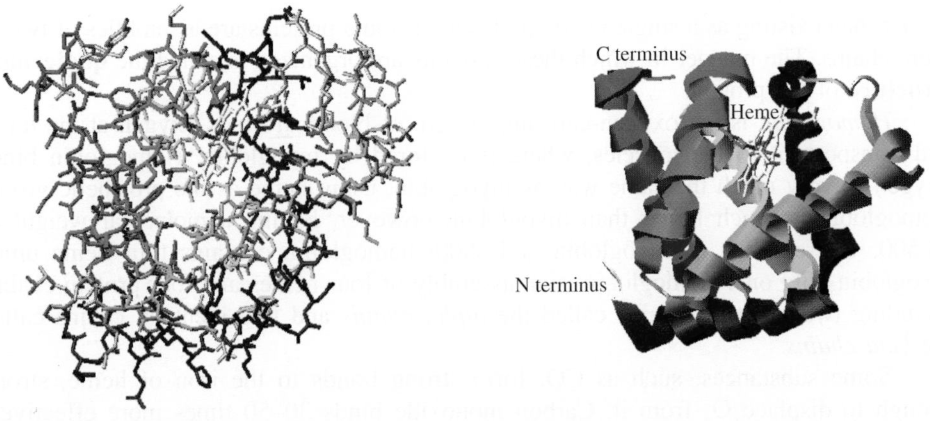
C terminus
Heme
N terminus

(*a*) (*b*)

Figure 25.21

The structure of sperm-whale myoglobin displayed as (*a*) a tube model and (*b*) a ribbon diagram. There are five separate regions of α helix in myoglobin, which are shown in different colors to distinguish them more clearly. The heme portion is included in both drawings, but is easier to locate in the ribbon diagram, as is the histidine side chain that is attached to the iron of heme.

Oh NO! It's Inorganic!

The amino acid L-arginine undergoes an interesting biochemical conversion.

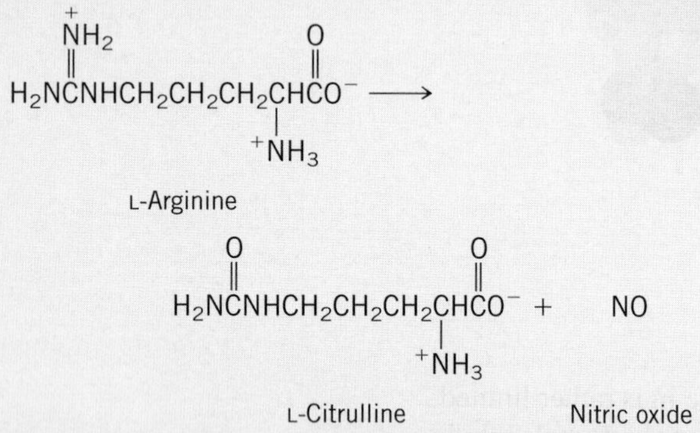

L-Arginine → L-Citrulline + Nitric oxide

Our experience conditions us to focus on the organic components of the reaction—L-arginine and L-citrulline—and to give less attention to the inorganic one—nitric oxide (nitrogen monoxide, NO). To do so, however, would lead us to overlook one of the most important discoveries in biology in the last quarter of the twentieth century.

Our story starts with the long-standing use of nitroglycerin to treat the chest pain that characterizes angina, a condition in diseases such as atherosclerosis in which restricted blood flow to the heart muscle itself causes it to receive an insufficient amount of oxygen. Placing a nitroglycerin tablet under the tongue provides rapid relief by expanding the blood vessels feeding the heart. A number of other nitrogen-containing compounds such as amyl nitrite and sodium nitroprusside exert a similar effect.

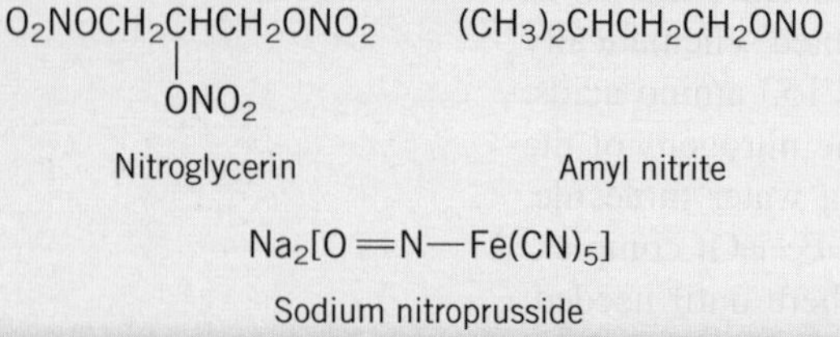

Nitroglycerin; Amyl nitrite; Sodium nitroprusside

A chemical basis for their action was proposed in 1977 by Ferid Murad who showed that all were sources of NO, thereby implicating it as the active agent.

Three years later, Robert F. Furchgott discovered that the relaxing of smooth muscles, such as blood vessel walls, was stimulated by an unknown substance produced in the lining of the blood vessels (the *endothelium*). He called this substance the *endothelium-dependent relaxing factor,* or EDRF and, in 1986, showed that EDRF was NO. Louis J. Ignarro reached the same conclusion at about the same time. Further support was provided by Salvador Moncada who showed that endothelial cells did indeed produce NO and that the L-arginine-to-L-citrulline conversion was responsible.

The initial skepticism that greeted the idea that NO, which is (a) a gas, (b) toxic, (c) inorganic, and (d) a free radical, could be a biochemical messenger was quickly overcome. An avalanche of results confirmed not only NO's role in smooth-muscle relaxation, but added more and more examples to an ever-expanding list of NO-stimulated biochemical processes. Digestion is facilitated by the action of NO on intestinal muscles. The drug Viagra (*sildenafil citrate*), prescribed to treat erectile dysfunction, works by increasing the concentration of a hormone, the release of which is signaled by NO. A theory that NO is involved in long-term memory receives support from the fact the brain is a rich source of the enzyme *nitric oxide synthase* (NOS), which catalyzes the formation of NO from L-arginine. NO even mediates the glow of fireflies. They glow nonstop when placed in a jar containing NO, but not at all when measures are taken to absorb NO.

Identifying NO as a signaling molecule in biological processes clearly justified a Nobel Prize. The only mystery was who would get it. Nobel Prizes are often shared, but never among more than three persons. Although four scientists—Murad, Furchgott, Ignarro, and Moncada—made important contributions, the Nobel committee followed tradition and recognized only the first three of them with the 1998 Nobel Prize in Physiology or Medicine.

25.22 Protein Quaternary Structure: Hemoglobin

Rather than existing as a single polypeptide chain, some proteins are assemblies of two or more chains. The manner in which these subunits are organized is called the **quaternary structure** of the protein.

Hemoglobin is the oxygen-carrying protein of blood. It binds oxygen at the lungs and transports it to the muscles, where it is stored by myoglobin. Hemoglobin binds oxygen in very much the same way as myoglobin, using heme as the prosthetic group. Hemoglobin is much larger than myoglobin, however, having a molecular weight of 64,500, whereas that of myoglobin is 17,500; hemoglobin contains four heme units, myoglobin only one. Hemoglobin is an assembly of four hemes and four protein chains, including two identical chains called the *alpha chains* and two identical chains called the *beta chains.*

Some substances, such as CO, form strong bonds to the iron of heme, strong enough to displace O_2 from it. Carbon monoxide binds 30–50 times more effectively than oxygen to myoglobin and hundreds of times better than oxygen to hemoglobin.

Strong binding of CO at the active site interferes with the ability of heme to perform its biological task of transporting and storing oxygen, with potentially lethal results.

How function depends on structure can be seen in the case of the genetic disorder *sickle cell anemia.* This is a debilitating, sometimes fatal, disease in which red blood cells become distorted ("sickle-shaped") and interfere with the flow of blood through the capillaries. This condition results from the presence of an abnormal hemoglobin in affected people. The primary structures of the beta chain of normal and sickle cell hemoglobin differ by a single amino acid out of 146; sickle cell hemoglobin has valine in place of glutamic acid as the sixth residue from the N terminus of the β chain. A tiny change in amino acid sequence can produce a life-threatening result! This modification is genetically controlled and probably became established in the gene pool because bearers of the trait have an increased resistance to malaria.

25.23 G-Coupled Protein Receptors

Biological receptors have been mentioned a few times in this book, in the context of organic molecules that bind to them. In Section 24.2 of the previous chapter, we described anandamide, a lipid that binds to the same receptor in the brain as the cannabinoids. In Section 25.14 of this chapter, we mentioned calcitonin, a peptide that regulates blood calcium levels and is used in the treatment of osteoporosis. The biological receptors for anandamide and for calcitonin both belong to a very large class of protein receptors known as **G-coupled protein receptors,** or GCPRs. The "G" stands for guanine in "guanine nucleotide-binding proteins." GCPRs occur throughout the body and function as "molecular switches" that regulate many physiological processes.

GCPRs span the cell membrane (see Figure 24.4). When GCPRs bind their specific ligand, such as a lipid, peptide, or ion, they undergo a conformational change, which results in the transduction of a signal across the membrane. The details of how signaling occurs are not completely understood. The conformational change may result in an interaction with a nearby G protein, which in turn can activate enzymes or ion channels (Figure 25.22).

In the biochemistry of vision (see Chapter 17), the interaction of rhodopsin, a GCPR, with light results in the photo-induced isomerization of cis retinal imine to trans retinal imine. This causes conformational changes in rhodopsin, which ultimately result in the *closing* of an ion channel, polarization of the cell membrame, and a nerve impulse that is transmitted to the brain in vision.

G-coupled protein receptors are involved in many diseases. It is estimated that nearly one-half of prescription drugs target GCPRs, in the treatment of cancer, cardiac malfunction, inflammation, pain, and disorders of the central nervous system.

Signal transduction pathways in biochemistry involve a relay of molecular interactions that ultimately produce an intracellular response. These pathways can serve as a connection between events at the cell surface and gene expression in the nucleus.

GDP and GTP are abbreviations for the nucleotides guanosine 5′-diphosphate and guanosine 5′-triphosphate. Nucleotides are discussed in Section 26.2.

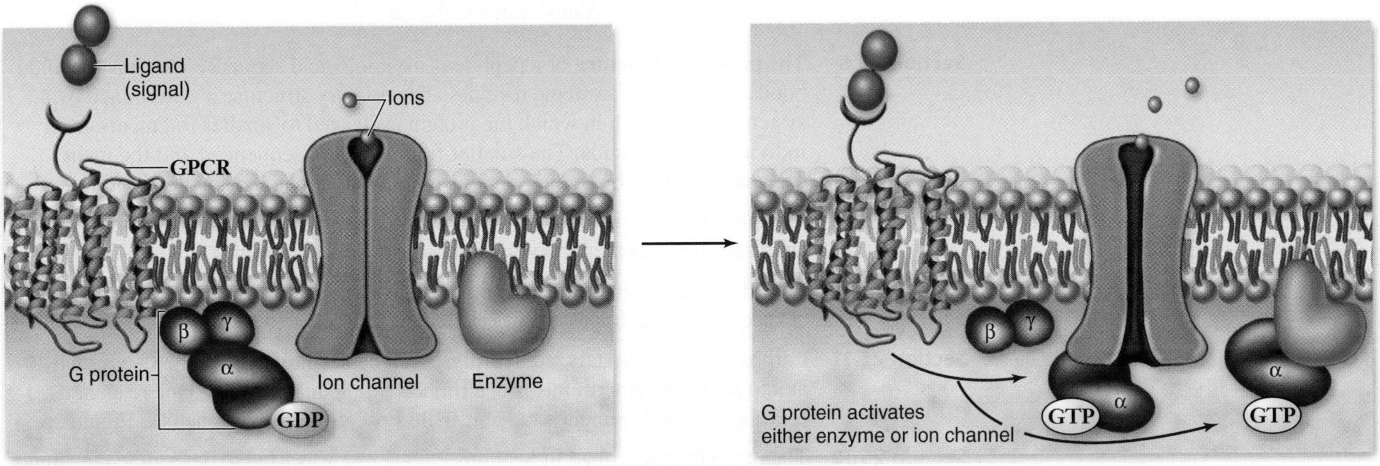

Figure 25.22

Signal transduction is initiated by the binding of a G-coupled protein receptor (GCPR) to the ligand on the exterior of the cell. Interactions with a nearby G protein inside the cell results in the exchange of bound GDP to GTP and in the release of one of its subunits in a GTP-bound form that activates an ion channel or enzyme.

25.24 SUMMARY

This chapter revolves around **proteins.** The first half describes the building blocks of proteins, progressing through **amino acids** and **peptides.** The second half deals with proteins themselves.

Section 25.1 The 20 amino acids listed in Table 25.1 are the building blocks of proteins. All are α-amino acids.

Section 25.2 Except for glycine, which is achiral, all of the α-amino acids in Table 25.1 are chiral and have the L configuration at the α carbon.

Section 25.3 The most stable structure of a neutral amino acid is a **zwitterion.** The pH of an aqueous solution at which the concentration of the zwitterion is a maximum is called the isoelectric point (pI).

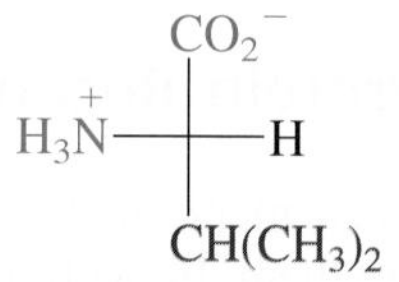

Fischer projection of L-valine in its zwitterionic form

Section 25.4 Amino acids are synthesized in the laboratory from

1. α-Halo acids by reaction with ammonia
2. Aldehydes by reaction with ammonia and cyanide ion (the Strecker synthesis)
3. Alkyl halides by reaction with the enolate anion derived from diethyl acetamidomalonate

Section 25.5 Amino acids undergo reactions characteristic of the amino group (e.g., amide formation) and the carboxyl group (e.g., esterification). Amino acid side chains undergo reactions characteristic of the functional groups they contain.

Section 25.6 Among the biochemical reactions of α-amino acids, several use pyridoxal 5′-phosphate as a coenzyme. These reactions involve bonds to the α carbon and include transamination, decarboxylation, and racemization.

Section 25.7 An amide linkage between two α-amino acids is called a **peptide bond.** By convention, peptides are named and written beginning at the N terminus.

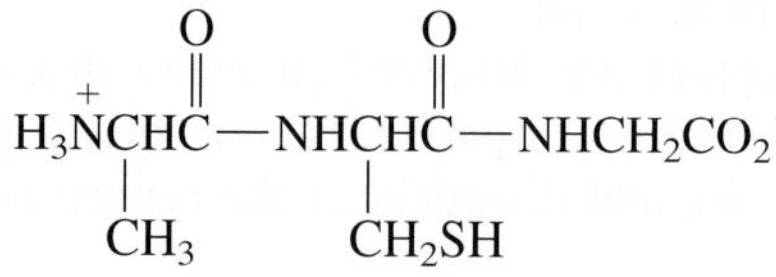

Ala-Cys-Gly (ACG)

Alanylcysteinylglycine

Section 25.8 The **primary structure** of a peptide is its amino acid sequence plus any disulfide bonds between two cysteine residues. The primary structure is determined by a systematic approach in which the protein is cleaved to smaller fragments, even individual amino acids. The smaller fragments are sequenced and the main sequence deduced by finding regions of overlap among the smaller peptides.

Section 25.9 Complete hydrolysis of a peptide gives a mixture of amino acids. An amino acid analyzer identifies the individual amino acids and determines their molar ratios.

Section 25.10 Selective hydrolysis can be accomplished by using enzymes to catalyze cleavage at specific peptide bonds.

Section 25.11 Carboxypeptidase-catalyzed hydrolysis can be used to identify the C-terminal amino acid. The N terminus is determined by chemical means. One reagent used for this purpose is Sanger's reagent, 1-fluoro-2,4-dinitrobenzene (see Figure 25.7).

Section 25.12 The procedure described in Sections 25.8–25.11 was used to determine the amino acid sequence of insulin.

Section 25.13 Modern methods of peptide sequencing follow a strategy similar to that used to sequence insulin, but are automated and can be carried out on a small scale. A key feature is repetitive N-terminal amino acid identification using the **Edman degradation.**

Section 25.14 Synthesis of a peptide of prescribed sequence requires the use of protecting groups to minimize the number of possible reactions.

Section 25.15 Amino-protecting groups include *benzyloxycarbonyl* (Z), tert-*butoxycarbonyl* (Boc), and *9-fluorenylmethoxycarbonyl* (FMOC).

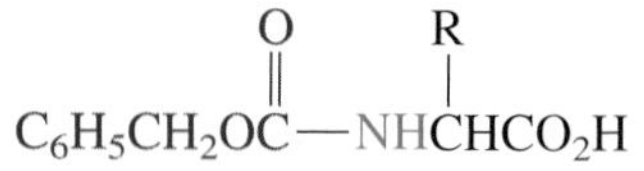

Benzyloxycarbonyl-protected amino acid

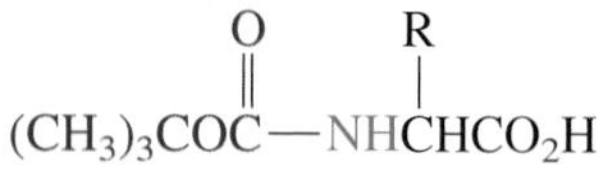

tert-Butoxycarbonyl-protected amino acid

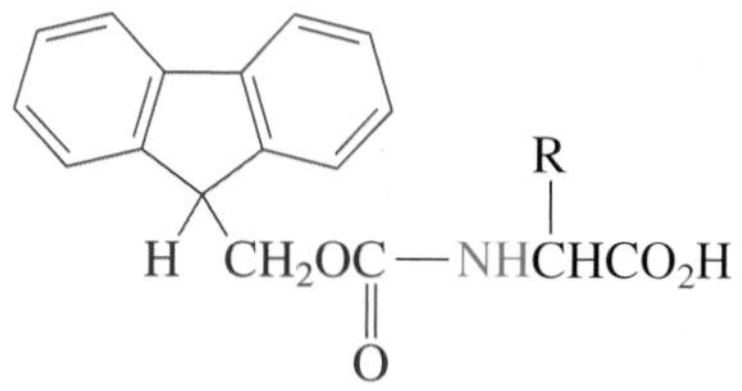

9-Fluorenylmethoxycarbonyl-protected amino acid

Hydrogen bromide may be used to remove either the benzyloxycarbonyl or *tert*-butoxycarbonyl protecting group. The benzyloxycarbonyl protecting group may also be removed by catalytic hydrogenolysis. FMOC is removed in base.

Section 25.16 Carboxyl groups are normally protected as benzyl, methyl, or ethyl esters. Hydrolysis in dilute base is normally used to deprotect methyl and ethyl esters. Benzyl protecting groups are removed by hydrogenolysis.

Section 25.17 Peptide bond formation between a protected amino acid having a free carboxyl group and a protected amino acid having a free amino group can be accomplished with the aid of *N,N′*-dicyclohexylcarbodiimide (DCCI).

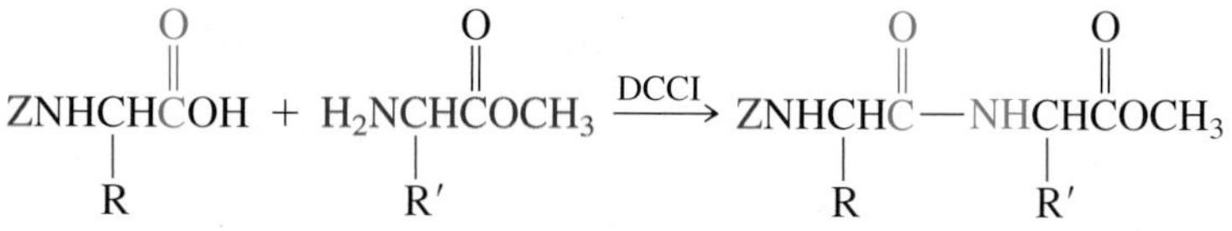

Section 25.18 In the Merrifield method the carboxyl group of an amino acid is anchored to a solid support and the chain extended one amino acid at a time. When all the amino acid residues have been added, the polypeptide is removed from the solid support.

Section 25.19 Two **secondary structures** of proteins are particularly prominent. The *pleated β sheet* is stabilized by hydrogen bonds between N—H and C═O groups of adjacent chains. The *α helix* is stabilized by hydrogen bonds within a single polypeptide chain.

Section 25.20 The folding of a peptide chain is its **tertiary structure.** The tertiary structure has a tremendous influence on the properties of the peptide and the biological role it plays. The tertiary structure is normally determined by X-ray crystallography.

Many globular proteins are enzymes. They accelerate the rates of chemical reactions in biological systems, but the kinds of reactions that take place are the fundamental reactions of organic chemistry. One way in which enzymes accelerate these reactions is by bringing reactive functions together in the presence of catalytically active functions of the protein.

Section 25.21 Often the catalytically active functions of an enzyme are nothing more than proton donors and proton acceptors. In many cases a protein acts in cooperation with a **coenzyme,** a small molecule having the proper functionality to carry out a chemical change not otherwise available to the protein itself.

Section 25.22 Many proteins consist of two or more chains, and the way in which the various units are assembled in the native state of the protein is called its **quaternary structure.**

Section 25.23 G-coupled protein receptors are transmembrane proteins that function as molecular switches in many physiological processes.

PROBLEMS

25.28 The imidazole ring of the histidine side chain acts as a proton acceptor in certain enzyme-catalyzed reactions. Which is the more stable protonated form of the histidine residue, A or B? Why?

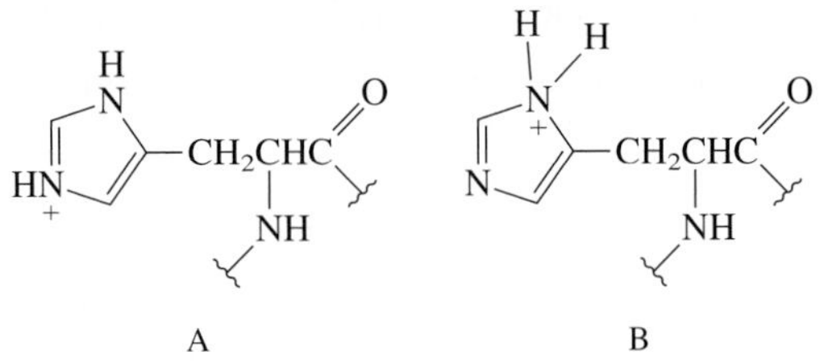

25.29 Which two α-amino acids are the biosynthetic precursors of the penicillins?

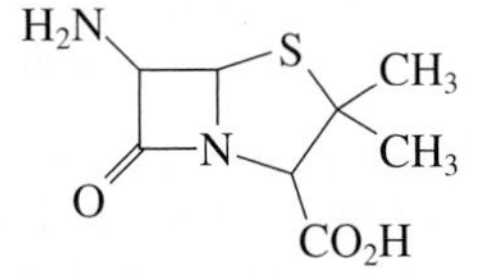

25.30 (a) Use the data in Table 25.2 and the Henderson–Hasselbalch equation to calculate the ratio $\frac{[A]}{[B]}$ at pH = 7.

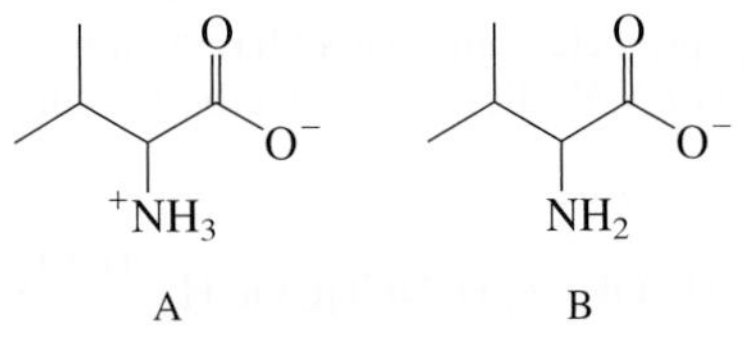

(b) At what pH is [A] the largest?

25.31 α-Amino acids are not the only compounds that exist as zwitterions. *p*-Aminobenzenesulfonic acid (sulfanilic acid) is normally written in the form shown but its zwitterionic form is more stable. Write a structural formula for the zwitterion.

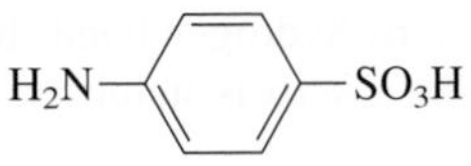

25.32 Putrescine, citrulline, and ornithine are products of arginine metabolism. The molecular formulas are given for the form in which each exists at pH = 7. The net charge corresponding to each formula is zero, +1, and +2. Suggest a reasonable structure for each species.

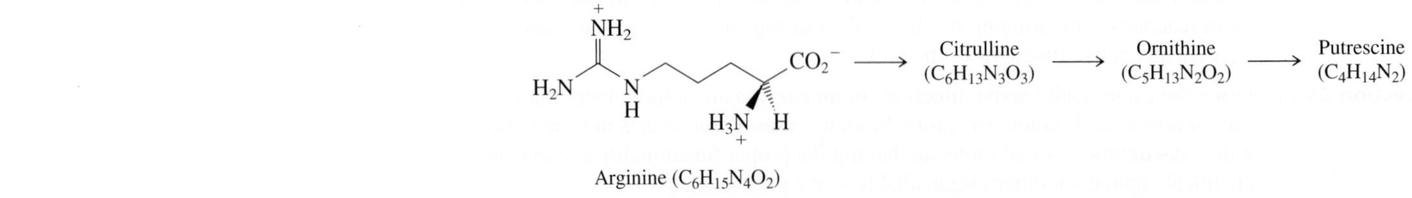

25.33 Acrylonitrile ($H_2C{=}CHC{\equiv}N$) readily undergoes conjugate addition when treated with nucleophilic reagents. Describe a synthesis of β-alanine ($H_3\overset{+}{N}CH_2CH_2CO_2^-$) that takes advantage of this fact.

25.34 (a) Isoleucine has been prepared by the following sequence of reactions. Give the structure of compounds A through D isolated as intermediates in this synthesis.

$$\underset{\displaystyle\ \ \ \ \ |\\ \ \ \ \ \ Br}{CH_3CH_2CHCH_3} \xrightarrow[\text{sodium ethoxide}]{\text{diethyl malonate}} A \xrightarrow[\text{2. HCl}]{\text{1. KOH}} B\ (C_7H_{12}O_4)$$

$$B \xrightarrow{Br_2} C\ (C_7H_{11}BrO_4) \xrightarrow{\text{heat}} D \xrightarrow[H_2O]{NH_3} \text{isoleucine (racemic)}$$

(b) An analogous procedure has been used to prepare phenylalanine. What alkyl halide would you choose as the starting material for this synthesis?

25.35 Hydrolysis of the following compound in concentrated hydrochloric acid for several hours at 100°C gives one of the amino acids in Table 25.1. Which one? Is it optically active?

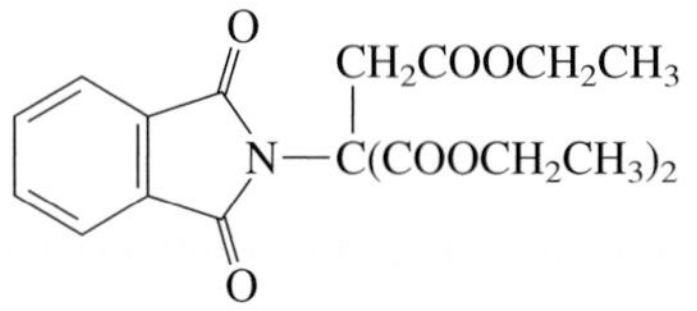
O
$CH_2COOCH_2CH_3$
N—C(COOCH$_2$CH$_3$)$_2$
O

25.36 Identify the major product in each of the following reactions involving amino acids.

(a)

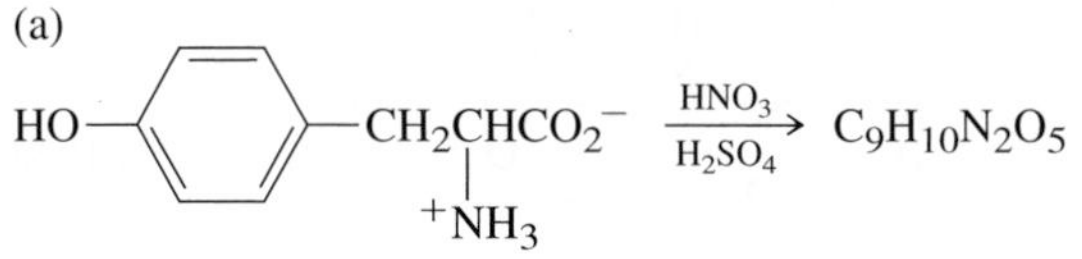
HO
$CH_2CHCO_2^-$
$^+NH_3$
HNO_3
H_2SO_4
$C_9H_{10}N_2O_5$

(b)

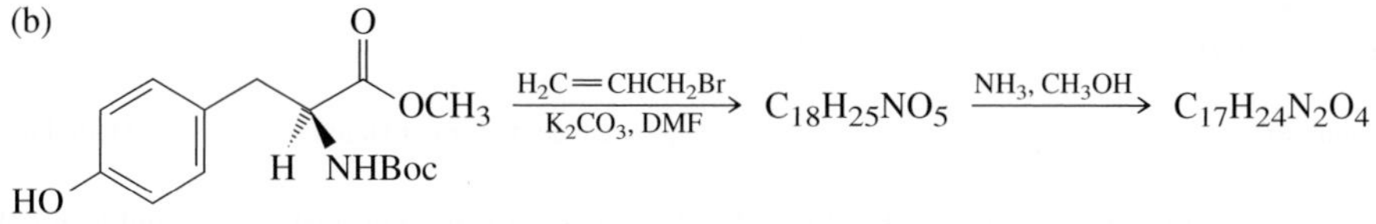
O
OCH_3
H NHBoc
HO
$H_2C{=}CHCH_2Br$
K_2CO_3, DMF
$C_{18}H_{25}NO_5$
NH_3, CH_3OH
$C_{17}H_{24}N_2O_4$

(c)

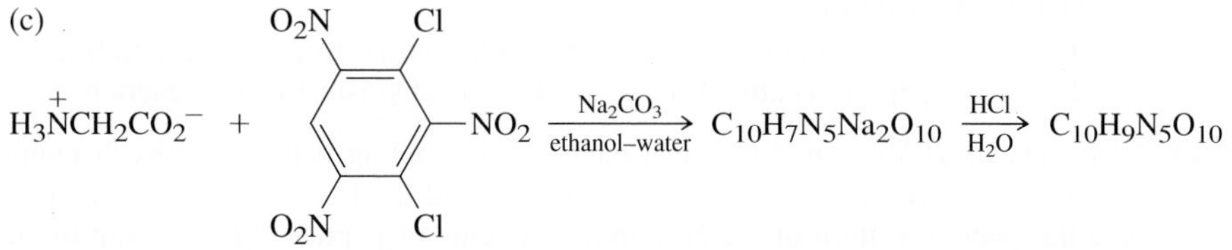
$H_3\overset{+}{N}CH_2CO_2^-$ +
O_2N
Cl
NO_2
O_2N
Cl
Na_2CO_3
ethanol–water
$C_{10}H_7N_5Na_2O_{10}$
HCl
H_2O
$C_{10}H_9N_5O_{10}$

25.37 The synthetic peptide shown here is an inhibitor of the enzyme β-secretase, which plays a role in the development of Alzheimer's disease. It contains five amino acids from Table 25.1. Which ones?

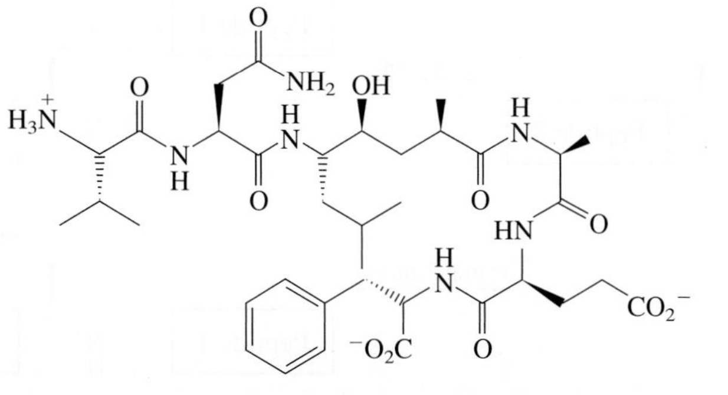
$H_3\overset{+}{N}$
NH_2 OH
HN
CO_2^-
^-O_2C

25.38 Cyanogen bromide (BrC≡N) cleaves peptides specifically between the carbonyl group of a methionine residue and the next amino acid.

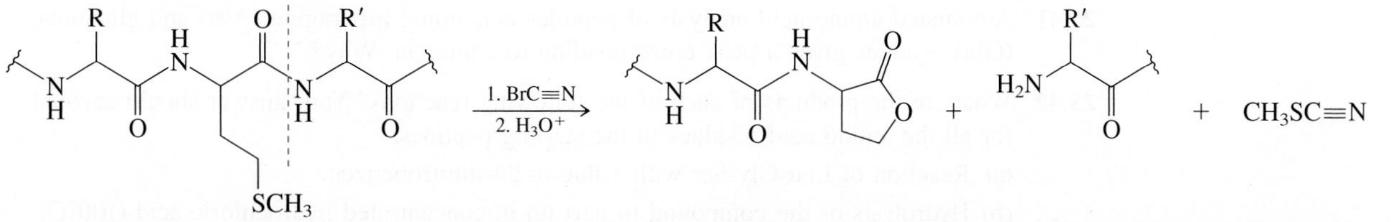
R
R′
SCH_3
1. BrC≡N
2. H_3O^+
R
R′
H_2N
+ $CH_3SC{\equiv}N$

Methionine-containing peptide — Lactone — C-Terminal fragment — Methyl thiocyanate

(a) The mechanism of this process begins with the reaction of cyanogen bromide with the peptide to give a sulfonium ion:

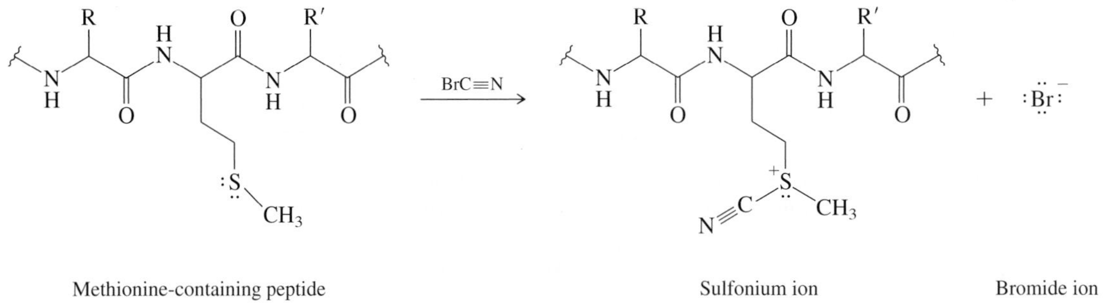

The sulfonium ion then cyclizes to an iminolactone with loss of methyl thiocyanate.

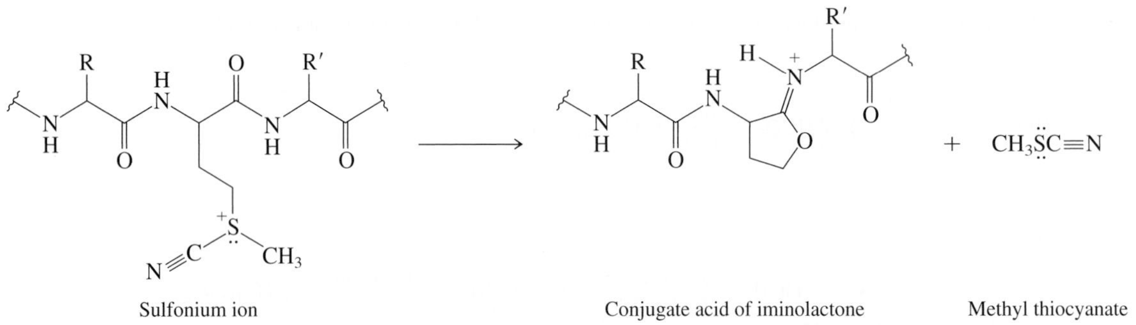

Using curved arrows to show the flow of electrons, suggest a reasonable mechanism for this cyclization.

(b) Hydrolysis of the conjugate acid of the iminolactone to the corresponding lactone cleaves the peptide chain. Write a stepwise mechanism for this reaction.

25.39 *Native chemical ligation* (NCL) is a method for coupling peptide chains. It requires that one of the peptides has an N-terminal cysteine, and the C terminus of the other peptide be a thioester. The thiol of the N-terminal cysteine of peptide 2 reacts with the thioester of peptide 1 to give an initial product that contains a "nonnative" thioester linkage. The initial product rearranges to a more stable, amide-linked (native) peptide. Write mechanisms for the coupling and rearrangement steps in NCL.

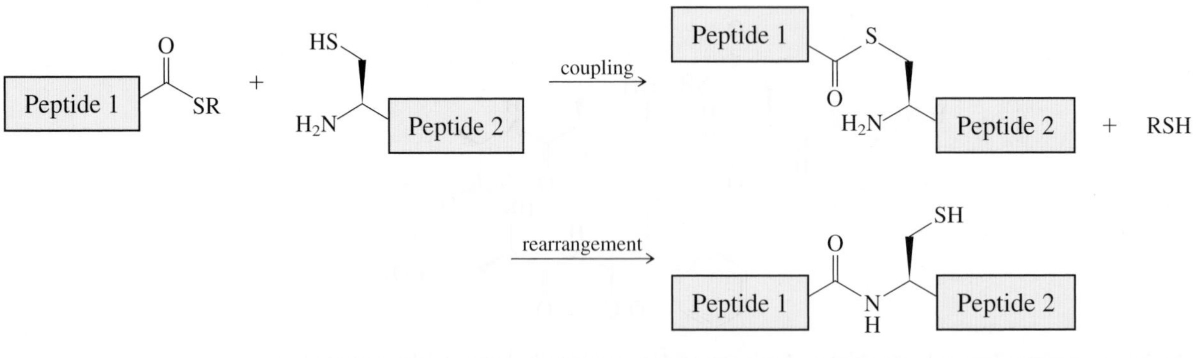

25.40 If you synthesized the tripeptide Leu-Phe-Ser from amino acids prepared by the Strecker synthesis, how many stereoisomers would you expect to be formed?

25.41 Automated amino acid analysis of peptides containing asparagine (Asn) and glutamine (Gln) residues gives a peak corresponding to ammonia. Why?

25.42 What are the products of each of the following reactions? Your answer should account for all the amino acid residues in the starting peptides.

(a) Reaction of Leu-Gly-Ser with 1-fluoro-2,4-dinitrobenzene

(b) Hydrolysis of the compound in part (a) in concentrated hydrochloric acid (100°C)

(c) Treatment of Ile-Glu-Phe with $C_6H_5N{=}C{=}S$, followed by hydrogen bromide in nitromethane

(d) Reaction of Asn-Ser-Ala with benzyloxycarbonyl chloride

(e) Reaction of the product of part (d) with *p*-nitrophenol and *N,N′*-dicyclohexylcarbodiimide

(f) Reaction of the product of part (e) with the ethyl ester of valine

(g) Hydrogenolysis of the product of part (f) by reaction with H_2 over palladium

25.43 The first 32 amino acids from the N terminus of the protein *bovine angiogenin* were determined by Edman degradation and have the sequence:

AQDDYRYIHFLTQHYDAKPKGRNDEYCFNMMK

(a) Identify the sites of cleavage during trypsin-catalyzed hydrolysis of this protein.

(b) What are the cleavage sites using chymotrypsin?

25.44 *Somatostatin* is a tetradecapeptide of the hypothalamus that inhibits the release of pituitary growth hormone. Its amino acid sequence has been determined by a combination of Edman degradations and enzymic hydrolysis experiments. On the basis of the following data, deduce the primary structure of somatostatin:

1. Edman degradation gave PTH-Ala.
2. Selective hydrolysis gave peptides having the following indicated sequences:

   Phe-Trp

   Thr-Ser-Cys

   Lys-Thr-Phe

   Thr-Phe-Thr-Ser-Cys

   Asn-Phe-Phe-Trp-Lys

   Ala-Gly-Cys-Lys-Asn-Phe
3. Somatostatin has a disulfide bridge.

25.45 What protected amino acid would you anchor to the solid support in the first step of a synthesis of oxytocin (see Figure 25.6) by the Merrifield method?

Descriptive Passage and Interpretive Problems 25

Amino Acids in Enantioselective Synthesis

Organic chemists speak of a "chiral pool," which comprises those naturally occurring compounds that are readily available as a single enantiomer and capable of being used as starting materials for the enantioselective synthesis of other chiral molecules. Amino acids are well represented in the chiral pool. All except glycine have at least one chirality center and, although L-amino acids are more abundant and less expensive than their D-enantiomers, both are available.

Most of the standard amino acids have served as starting materials for enantioselective syntheses. One of the most widely used is L-glutamic acid and its lactam (*S*)-pyroglutamic acid, which is easily prepared by heating an aqueous solution of L-glutamic acid in a sealed container.

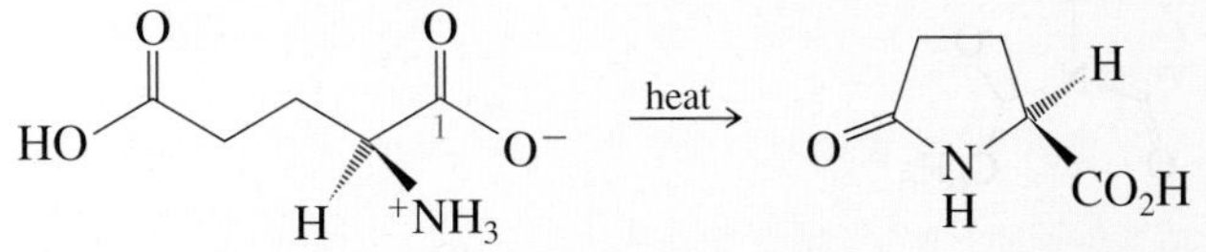

L-Glutamic acid (*S*)-Pyroglutamic acid

With three functional groups in a compound with only five carbons, L-glutamic acid and (*S*)-pyroglutamic acid provide access to more complex molecules via functional-group manipulation.

Many of the syntheses are lengthy, and most contain some specialized reactions. The following problems emphasize the planning aspect of amino acid-based enantioselective syntheses starting with either L-glutamic acid or (*S*)-pyroglutamic acid. The few reactions that are included are either familiar or similar to those covered in earlier chapters.

25.46 (*S*)-Pyroglutamic acid was used as the starting material in a synthesis of (+)-ipalbidine, an analgesic alkaloid obtained from the seeds of the white moonflower, *Ipomoea alba*. No bonds are made or broken to the chirality center of (*S*)-pyroglutamic acid in this synthesis. Which of the following is the structure of (+)-ipalbidine?

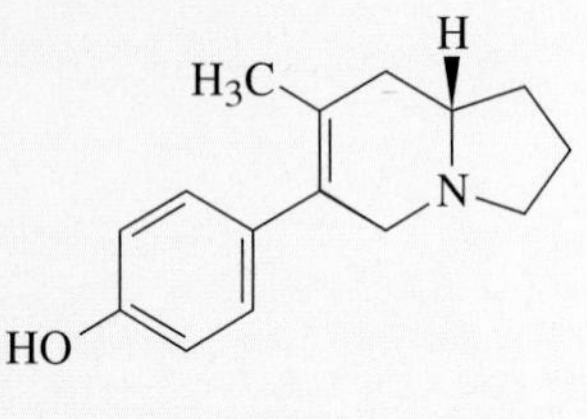

A.

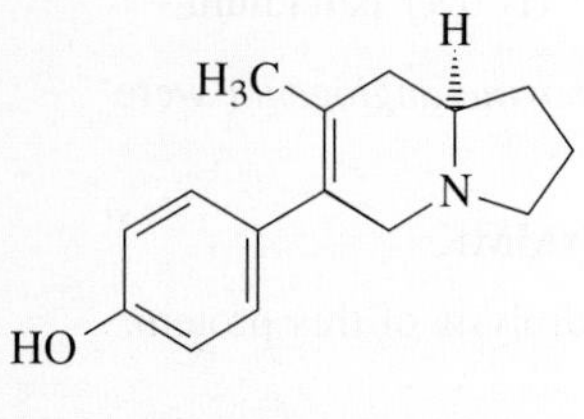

B.

25.47 A synthesis of poison-dart frog toxin has been described that begins with L-glutamic acid.

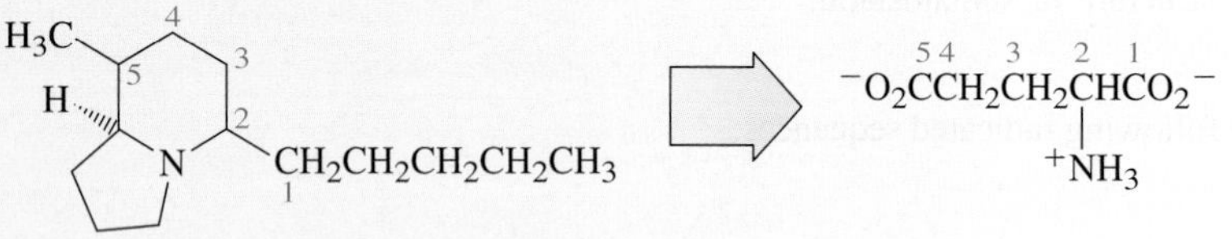

The numbered carbons in the product correspond to the same numbered carbons in L-glutamic acid. No bonds to carbon-2 are made or broken in the synthesis. If the configuration at carbon-5 in the product is *R*, which of the following best represents the stereochemistry of the frog toxin?

A. 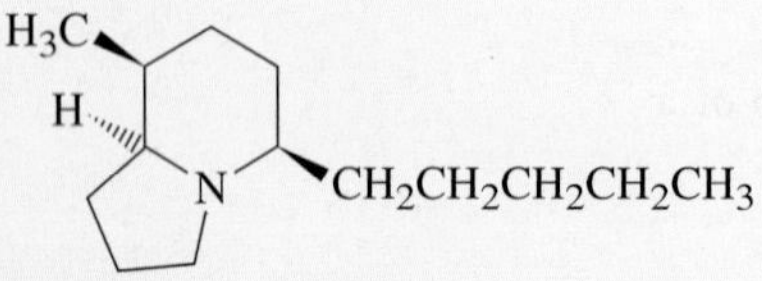

C. 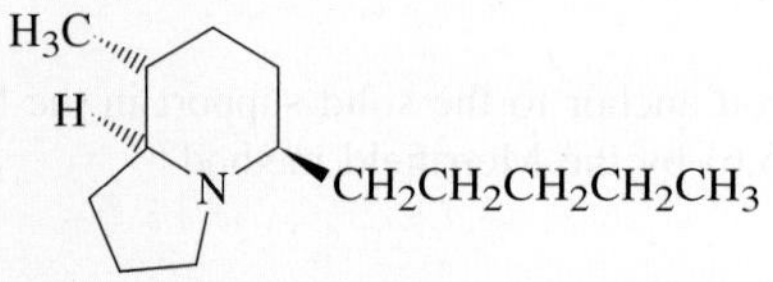

B. 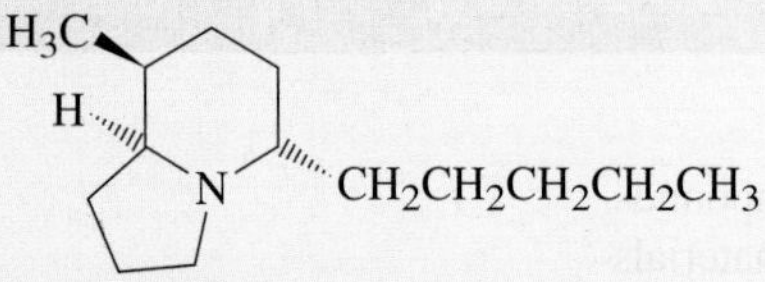

D. 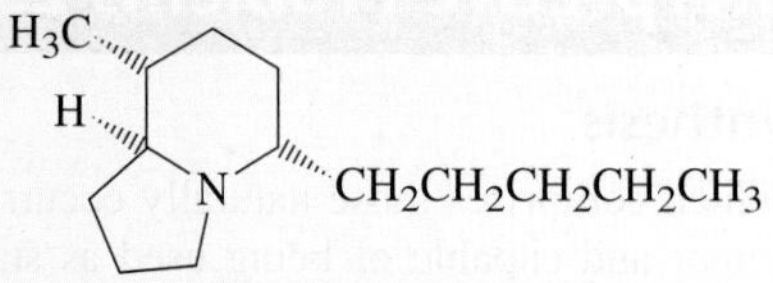

25.48 One synthesis of fosinopril, a drug used to combat high blood pressure, starts with the reduction of (*S*)-pyroglutamic acid to the primary alcohol, followed by protection of the OH and NH groups.

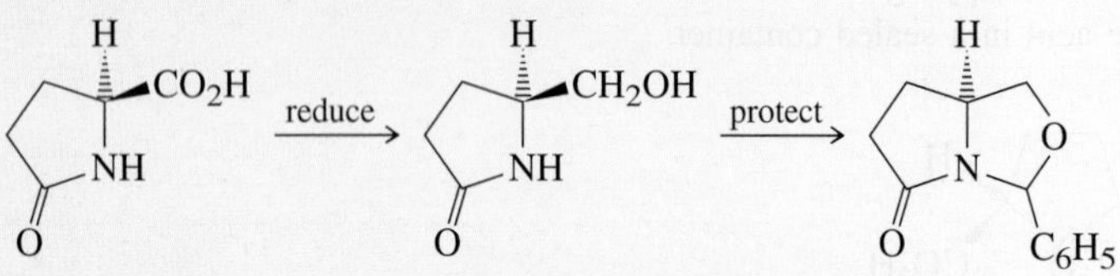

What reaction conditions are appropriate for the protection step?

A. $C_6H_5\overset{O}{\overset{\|}{C}}Cl$, pyridine
B. $C_6H_5CH{=}O$, *p*-toluenesulfonic acid, toluene, heat
C. $C_6H_5CO_2CH_3$
D. $C_6H_5CH_2Br$, Na_2CO_3, acetone

25.49 The *N,O*-protected compound formed in the preceding problem was alkylated with 3-bromocyclohexene.

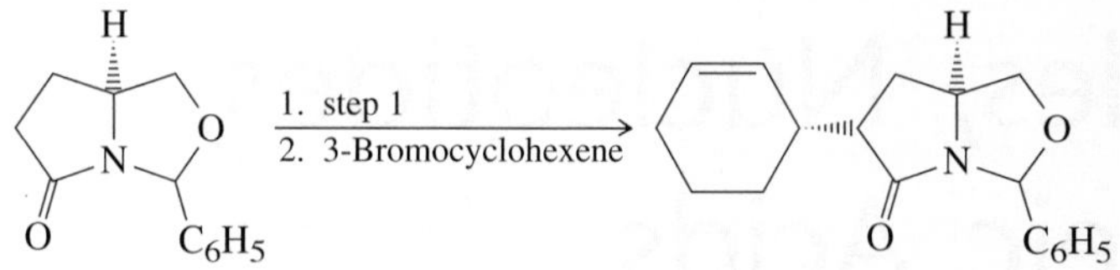

What reaction conditions are appropriate for step 1?

A. $LiAlH_4$, diethyl ether

B. $NaOCH_2CH_3$, ethanol

C. Mg, diethyl ether

D. $[(CH_3)_2CH]_2NLi$, tetrahydrofuran

25.50 α-Kainic acid is a neurotoxin produced by certain algae. Several enantioselective syntheses of it have been described, one of which is based on 1-bromo-3-methyl-2-butene and L-glutamic acid.

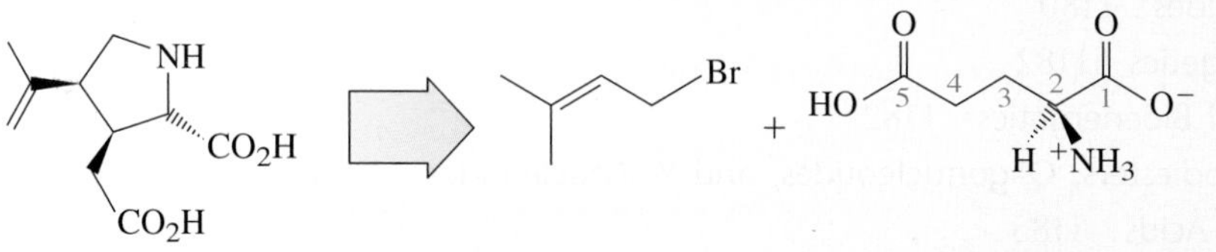

α-Kainic acid L-Glutamic acid

Which carbon of L-glutamic acid is involved in the only carbon–carbon bond forming step of the synthesis?

A. C-1 D. C-4

B. C-2 E. C-5

C. C-3

25.51 (*S*)-Tylophorine is an alkaloid isolated from a plant that grows in India and Southeast Asia, which is of interest as a potential antitumor drug. It has been synthesized by a multistep procedure based on L-glutamic acid and the phenanthrene derivative shown.

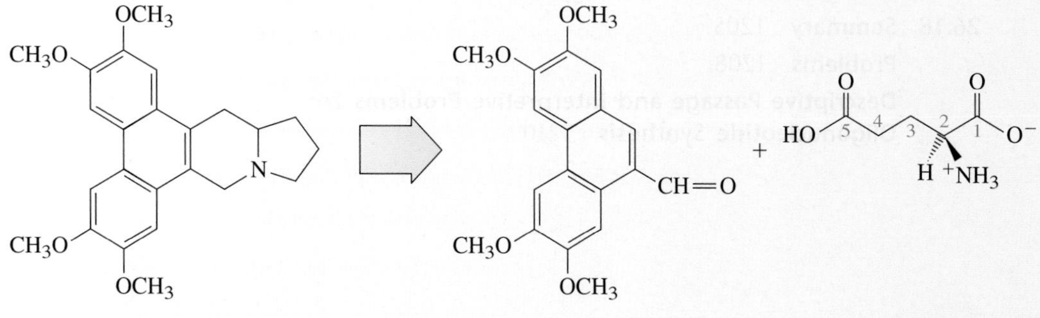

Tylophorine L-Glutamic acid

Which carbon of L-glutamic acid is involved in the only carbon–carbon bond forming step of the synthesis?

A. C-1 D. C-4

B. C-2 E. C-5

C. C-3

26 Nucleosides, Nucleotides, and Nucleic Acids

Chapter Outline

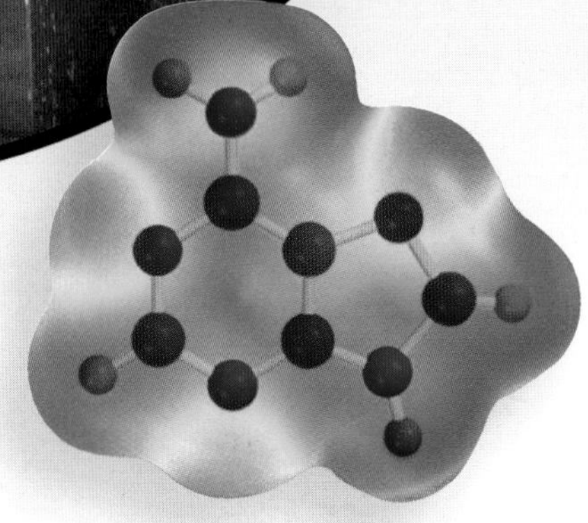

Gregor Mendel's studies of the inherited traits of garden peas received little attention during his lifetime. What he found is now recognized as the beginning of our understanding of genetics. Adenine, shown in the electrostatic potential map, is a "purine base" present in DNA and RNA.

IN CHAPTER 1 we saw that a major achievement of the first half of the twentieth century was the picture of atomic and molecular structure revealed by quantum mechanics. In this chapter we examine the major achievement of the second half of that century—a molecular view of genetics based on the structure and biochemistry of nucleic acids.

Nucleic acids are acidic substances present in the nuclei of cells and were known long before anyone suspected they were the primary substances involved in the storage, transmission, and processing of genetic information. There are two kinds of nucleic acids: ribonucleic acid (RNA) and deoxyribonucleic acid (DNA). Both are complicated biopolymers, based on three structural units: a carbohydrate, a phosphate ester linkage between carbohydrates, and a heterocyclic aromatic compound. The heterocyclic aromatic compounds are referred to as purine and pyrimidine bases. We'll begin with them and follow the structural thread:

Purine and pyrimidine bases $\rightarrow$ Nucleosides $\rightarrow$ Nucleotides $\rightarrow$ Nucleic acids

There will be a few pauses along the way as we stop to examine some biochemical roles played by these compounds independent of their genetic one.

26.1 Pyrimidines and Purines

Two nitrogen-containing heterocyclic aromatic compounds—**pyrimidine** and **purine**—are the parents of the "bases" that constitute a key structural unit of nucleic acids.

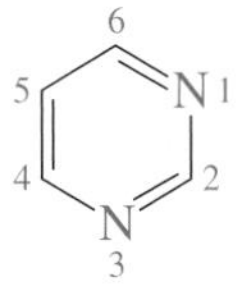

Pyrimidine

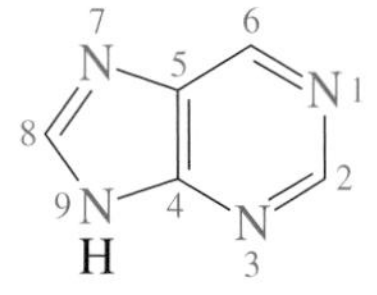

Purine

Both pyrimidine and purine are planar. You will see how important this flat shape is when we consider the structure of nucleic acids. In terms of their chemistry, pyrimidine and purine resemble pyridine. They are weak bases and relatively unreactive toward electrophilic aromatic substitution.

Pyrimidine and purine themselves do not occur naturally, but many of their derivatives do. Before going too far, we need to point out an important structural difference between derivatives that bear —OH groups and those with $—NH_2$ groups. The structure of a pyrimidine or purine that bears an $—NH_2$ group follows directly from the structure of the parent ring system.

6-Aminopurine is *adenine* and will appear numerous times in this chapter.

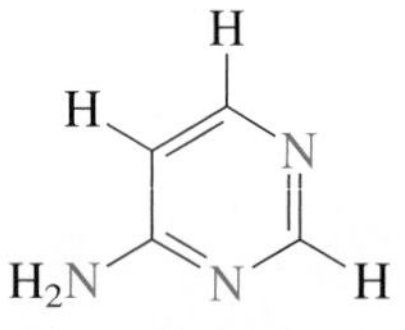

4-Aminopyrimidine

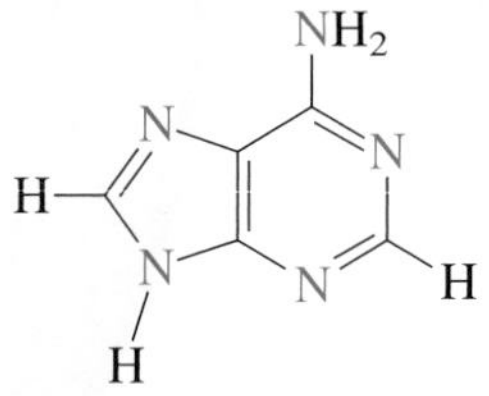

6-Aminopurine

However, the corresponding compounds that have an —OH group resemble enols:

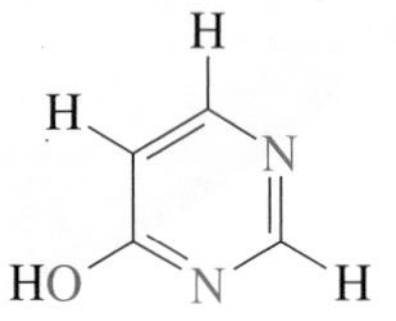

4-Hydroxypyrimidine

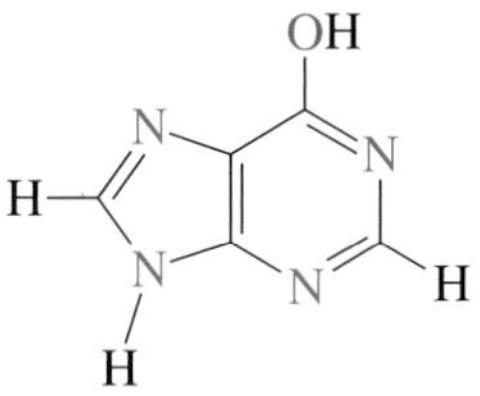

6-Hydroxypurine

but exist instead in their keto forms.

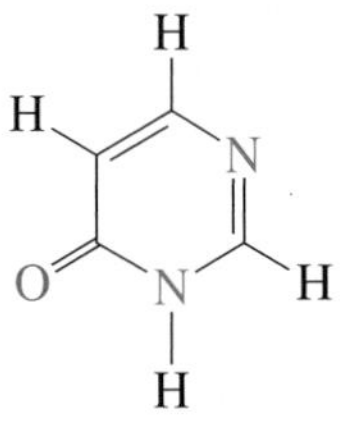

Keto form of 4-hydroxypyrimidine

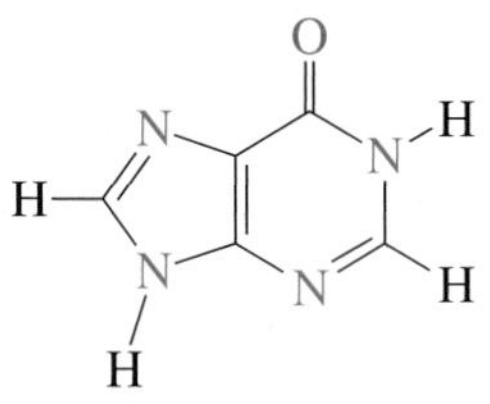

Keto form of 6-hydroxypurine

By analogy to phenols, we would expect the isomers with —OH groups on benzene-like rings to be more stable. This turns out not to be true because the keto forms are also aromatic owing to amide resonance.

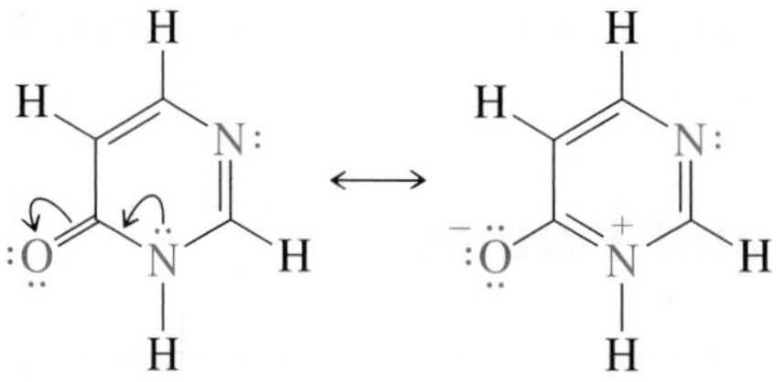

Resonance in keto form of 4-hydroxypyrimidine

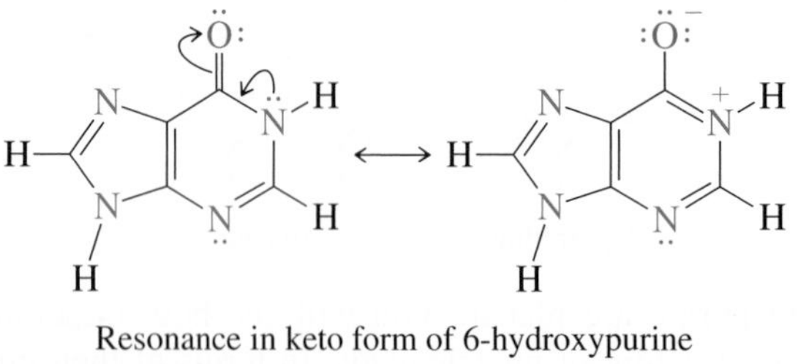

Resonance in keto form of 6-hydroxypurine

These relationships are general. Hydroxyl-substituted purines and pyrimidines exist in their keto forms; amino-substituted ones retain structures with an amino group on the ring. The pyrimidine and purine bases in DNA and RNA listed in Table 26.1 follow this general rule. Beginning in Section 26.7 we'll see how critical it is that we know the correct tautomeric forms of the nucleic acid bases.

Problem 26.1

Write a structural formula for the enol tautomer of cytosine (see Table 26.1).

TABLE 26.1 Pyrimidines and Purines That Occur in DNA and/or RNA

	Name	Structure	Occurrence
Pyrimidines			
	Cytosine	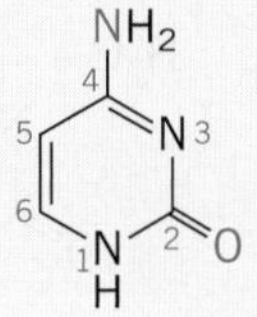	DNA and RNA
	Thymine	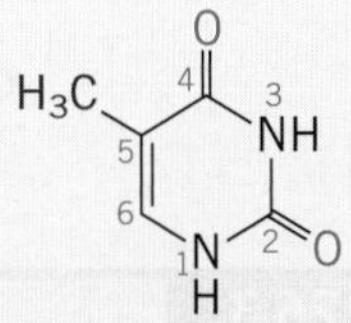	DNA
	Uracil	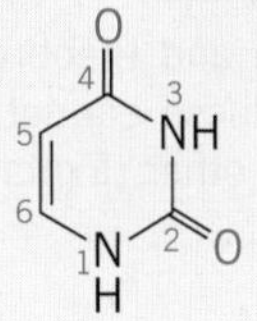	RNA
Purines			
	Adenine	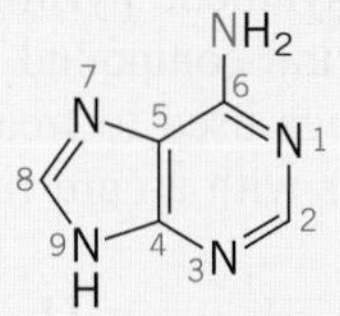	DNA and RNA
	Guanine	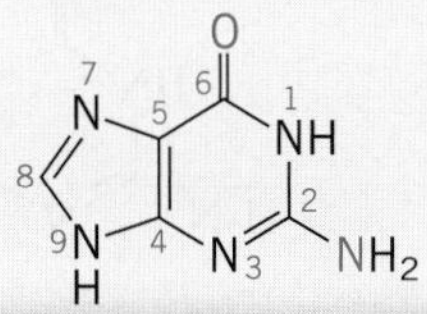	DNA and RNA

Problem 26.2

(a) Write a resonance form for guanine in which the six-membered ring has an electronic structure analogous to benzene. Show all unshared pairs and don't forget to include formal charges.

• Start with the structure of guanine from Table 26.1, add electron pairs, and move electrons to generate resonance form.

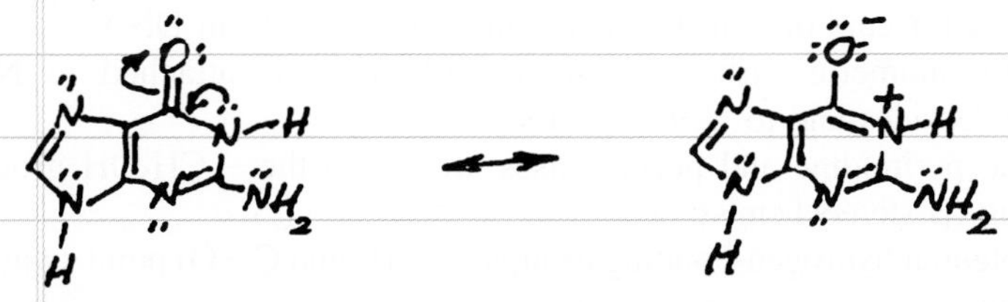

(b) Write three contributors for uracil that involve amide resonance.

Pyrimidines and purines occur naturally in substances other than nucleic acids. Coffee, for example, is a familiar source of caffeine. Tea contains both caffeine and theobromine.

The caffeine added to soft drinks is the caffeine that was removed when decaffeinating coffee and tea.

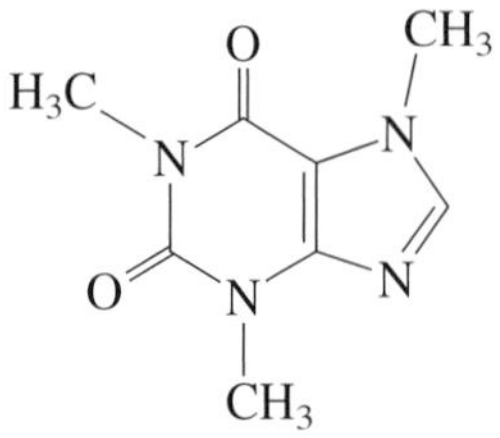
Caffeine

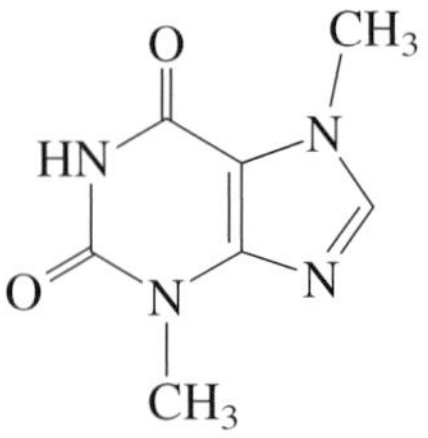
Theobromine

Problem 26.3

Classify caffeine and theobromine according to whether each is a pyrimidine or a purine. One of these cannot isomerize to an enolic form; two different enols are possible for the other. Explain and write structural formulas for the possible enols.

Several synthetic pyrimidines and purines are useful drugs. *Acyclovir* was the first effective antiviral compound and is used to treat herpes infections. 6-*Mercaptopurine* is one of the drugs used to treat childhood leukemia, which has become a very treatable form of cancer with a cure rate approaching 80%.

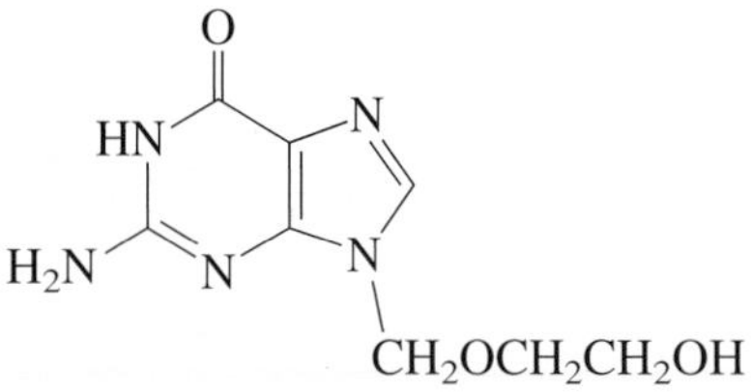
Acyclovir

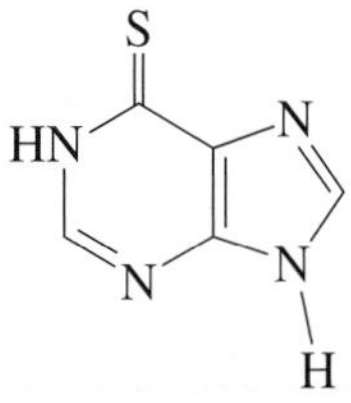
6-Mercaptopurine

26.2 Nucleosides

The most important derivatives of pyrimidines and purines are nucleosides. **Nucleosides** are glycosylamines in which a pyrimidine or purine nitrogen is bonded to the anomeric carbon of a carbohydrate. The nucleosides listed in Table 26.2 are the main building blocks of nucleic acids. In RNA the carbohydrate component is D-ribofuranose; in DNA it is 2-deoxy-D-ribofuranose.

Among the points to be made concerning Table 26.2 are the following:

1. Three of the bases (cytosine, adenine, and guanine) occur in both RNA and DNA.
2. Uracil occurs only in RNA; thymine occurs only in DNA.
3. The anomeric carbon of the carbohydrate is attached to N-1 in pyrimidine nucleosides and to N-9 in purines.
4. The pyrimidine and purine bases are cis to the —CH_2OH group of the furanose ring (β stereochemistry).
5. Potential hydrogen-bonding groups (—NH_2 and C=O) point away from the furanose ring.

The numbering scheme used for nucleosides maintains the independence of the two structural units. The pyrimidine or purine is numbered in the usual way. So is the

TABLE 26.2 The Major Pyrimidine and Purine Nucleosides in RNA and DNA

	Pyrimidines			Purines	
Name	**Cytidine**	**Thymidine**	**Uridine**	**Adenosine**	**Guanosine**
Abbreviation*	C	T	U	A	G
Systematic name	1-β-D-Ribo-furanosylcytosine	2′-Deoxy-1-β-D-ribo-furanosylthymine	1-β-D-Ribo-furanosyluracil	9-β-D-Ribo-furanosyladenine	9-β-D-Ribo-furanosylguanine
Structural formula	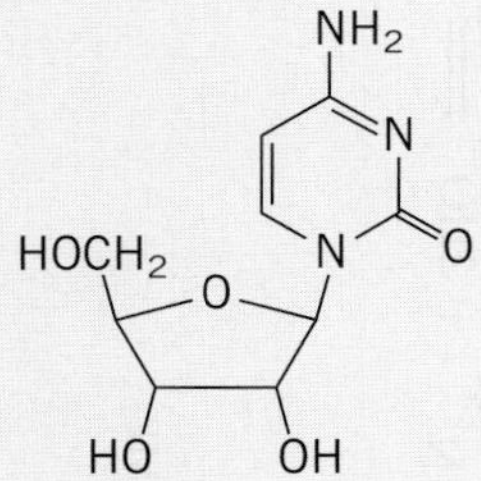	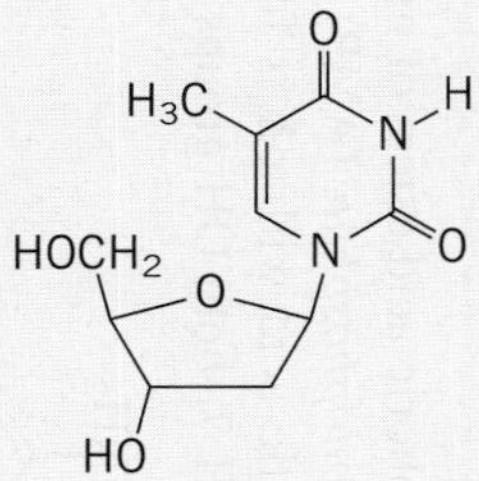	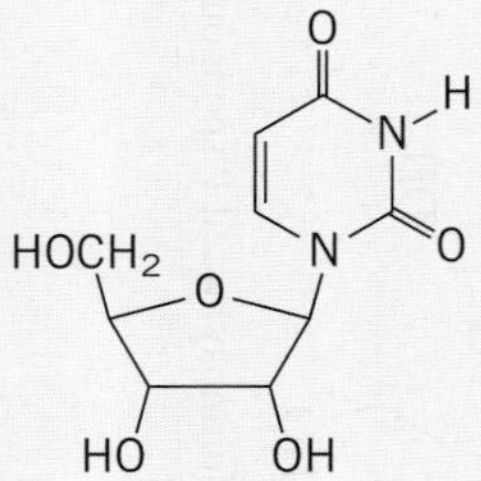	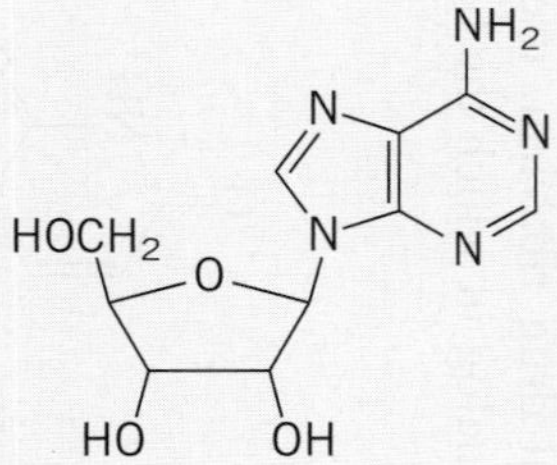	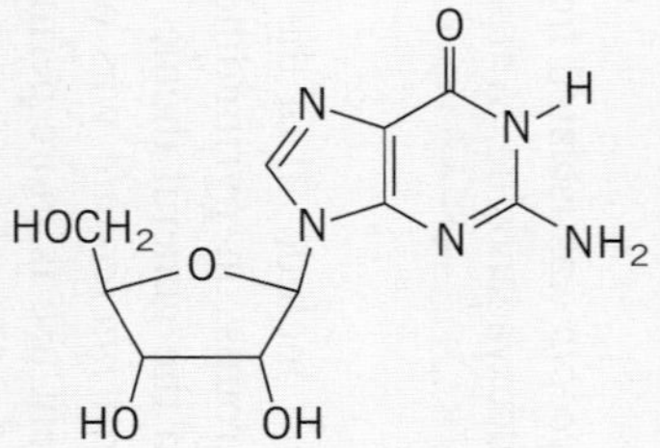
Molecular model	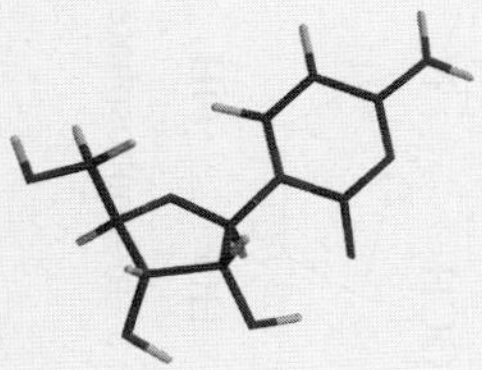	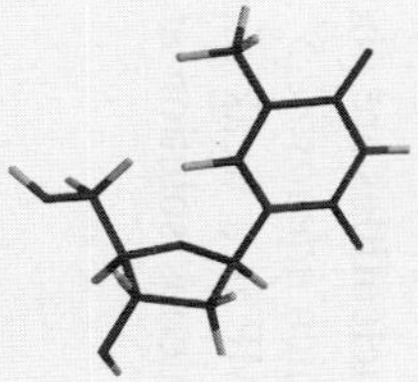	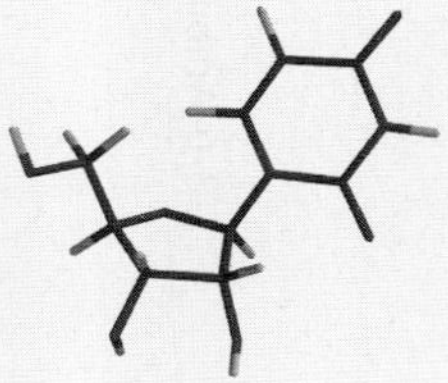	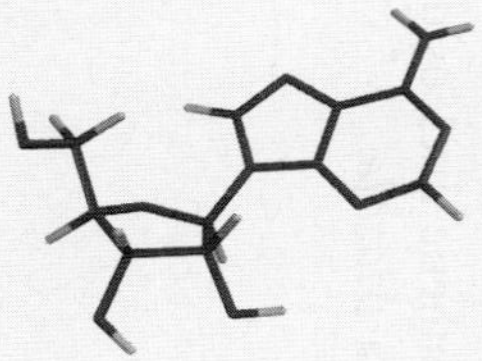	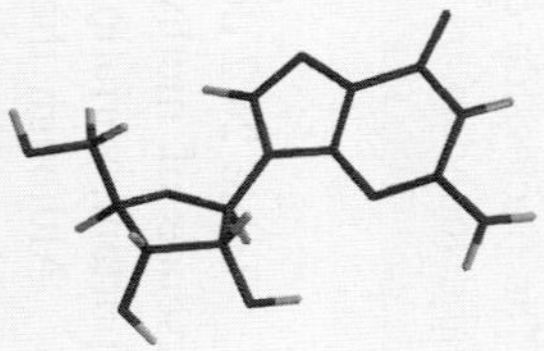
Found in	RNA 2′-Deoxy analog in DNA	DNA	RNA	RNA 2′-Deoxy analog in DNA	RNA 2′-Deoxy analog in DNA

*Sometimes the abbreviation applies to the pyrimidine or purine base, sometimes to the nucleoside. Though this may seem confusing, it is normally clear from the context what is intended and causes no confusion in practice.

carbohydrate, except that a prime symbol (′) follows each locant. Thus adenosine is a nucleoside of D-ribose, and 2′-deoxyadenosine is a nucleoside of 2-deoxy-D-ribose.

Problem 26.4

The nucleoside *cordycepin* was isolated from cultures of the fungus *Cordyceps militaris* and found to be 3′-deoxyadenosine. Write its structural formula.

Table 26.2 doesn't include all of the nucleoside components of nucleic acids. The presence of methyl groups on pyrimidine and purine rings is a common, and often important, variation on the general theme.

Although the term *nucleoside* was once limited to the compounds in Table 26.2 and a few others, current use is more permissive. Pyrimidine derivatives of D-arabinose, for example, occur in the free state in certain sponges and are called *spongonucleosides.* The powerful antiviral drug ribavirin, used to treat hepatitis C and Lassa fever, is a synthetic nucleoside analog in which the base, rather than being a pyrimidine or purine, is a *triazole.*

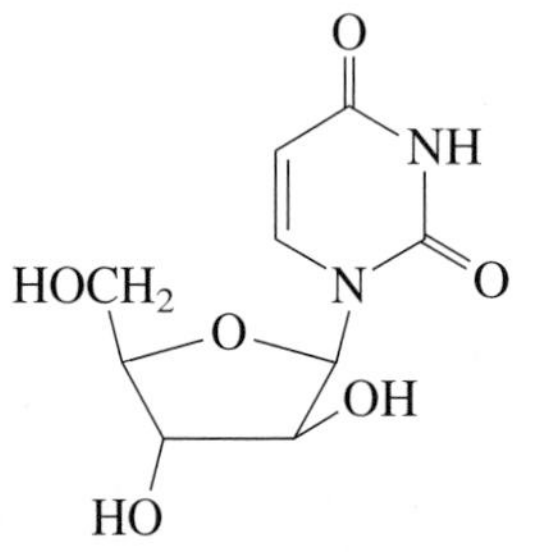

1-β-D-Arabinofuranosyluracil ("spongouridine")

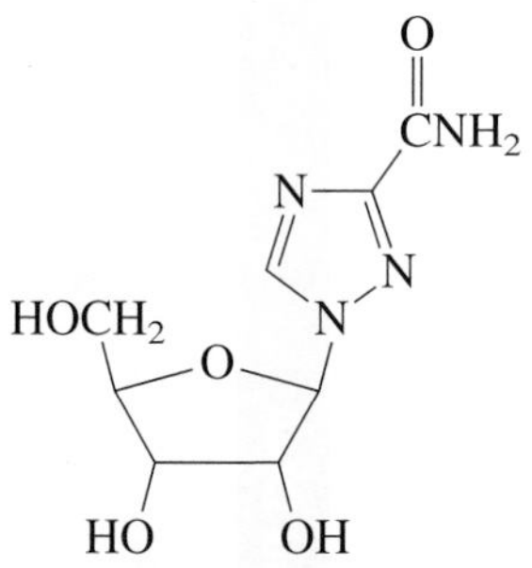

Ribavirin

26.3 Nucleotides

Nucleotides are phosphoric acid esters of nucleosides. Those derived from adenosine, of which *adenosine 5′-monophosphate* (AMP) is but one example, are especially prominent. AMP is a weak diprotic acid with pK_a's for ionization of 3.8 and 6.2, respectively. In aqueous solution at pH 7, both OH groups of the $P(O)(OH)_2$ unit are ionized.

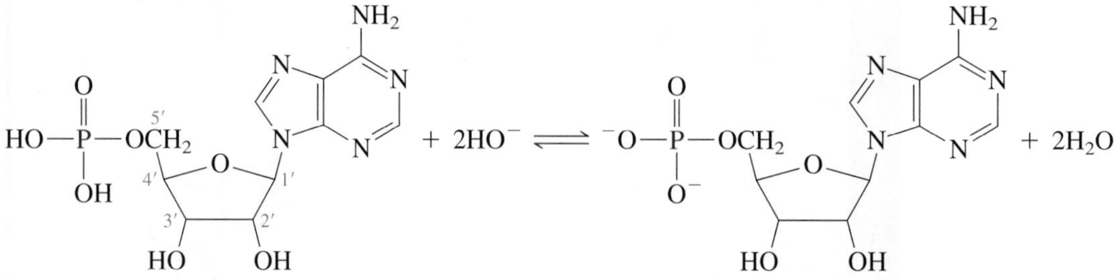

Adenosine 5′-monophosphate (AMP) Major species at pH 7

Problem 26.5

Write a structural formula for 2′-deoxycytidine 3′-monophosphate. You may wish to refer to Table 26.2 for the structure of cytidine.

Other important 5′-nucleotides of adenosine include **adenosine 5′-diphosphate** (ADP) and **adenosine 5′-triphosphate** (ATP):

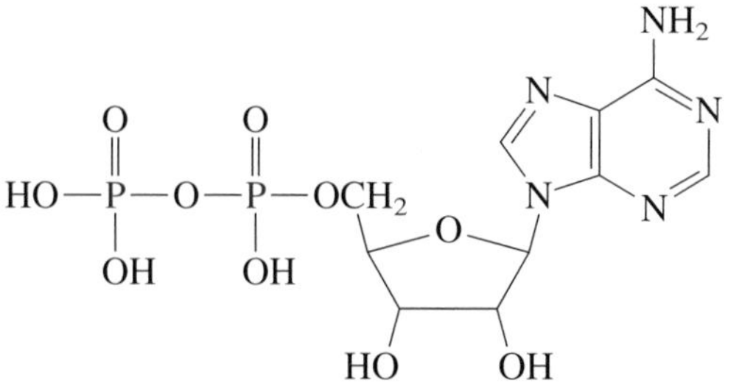

Adenosine 5′-diphosphate (ADP)

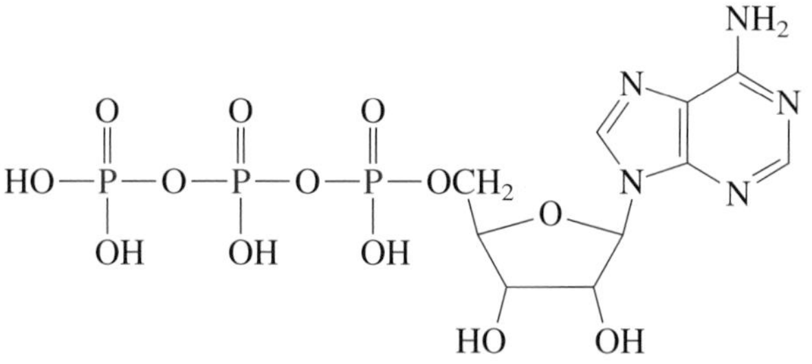

Adenosine 5′-triphosphate (ATP)

ATP is the main energy-storing molecule for practically every form of life on Earth. We often speak of ATP as a "high-energy compound" and its P—O bonds as "high-energy bonds." This topic is discussed in more detail in Sections 26.4 and 26.5.

The biological transformations that involve ATP are both numerous and fundamental. They include, for example, many *phosphorylation* reactions in which ATP transfers one of its phosphate units to the —OH of another molecule. These phosphorylations are catalyzed by enzymes called *kinases.* An example is the first step in the metabolism of glucose:

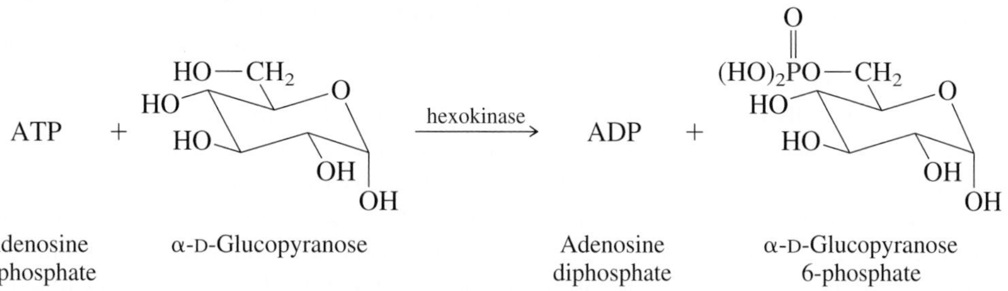
ATP
hexokinase
ADP
Adenosine triphosphate
α-D-Glucopyranose
Adenosine diphosphate
α-D-Glucopyranose 6-phosphate

Both adenosine and guanosine form cyclic monophosphates (*cyclic-AMP* or *cAMP* and *cyclic-GMP* or *cGMP,* respectively) that are involved in a large number of biological processes as "second messengers." Many hormones (the "first messengers") act by first stimulating the formation of cAMP or cGMP on a cell surface, which triggers a series of events characteristic of the organism's response to the hormone. Signalling by cAMP is also involved in the activation of G-coupled protein receptors (Section 25.23).

Earl Sutherland of Vanderbilt University won the 1971 Nobel Prize in Physiology or Medicine for uncovering the role of cAMP as a second messenger in connection with his studies of the "fight or flight" hormone epinephrine (Section 25.6).

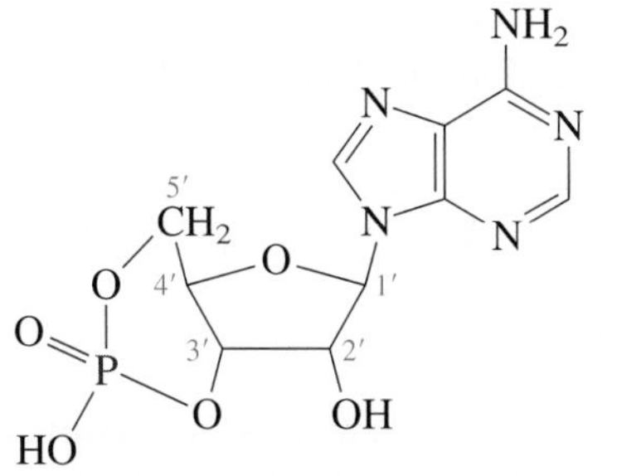

Adenosine 3′,5′-cyclic monophosphate (cAMP)

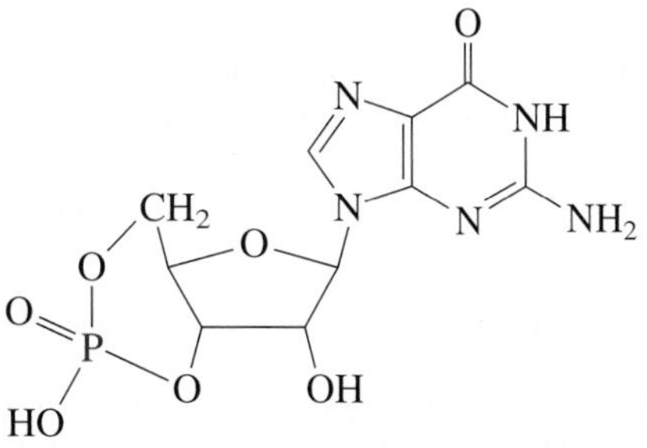

Guanosine 3′,5′-cyclic monophosphate (cGMP)

As we saw in the boxed essay *Oh NO! It's Inorganic!* in Chapter 25, nitric oxide (NO) expands blood vessels and increases blood flow. This process begins when NO stimulates the synthesis of cGMP as a second messenger. Erectile dysfunction drugs such as *sildenafil* (Viagra) increase the concentration of cGMP by inhibiting the enzyme that catalyzes hydrolysis of its cyclic phosphate unit.

Problem 26.6

Cyclic-AMP is formed from ATP in a reaction catalyzed by the enzyme *adenylate cyclase.* Assume that adenylate cyclase acts as a base to remove a proton from the 3′-hydroxyl group of ATP and write a mechanism for the formation of cAMP.

26.4 Bioenergetics

Recall that free energy is the energy available to do work. By focusing on free energy, we concern ourselves more directly with what is important to a living organism.

Bioenergetics is the study of the thermodynamics of biological processes, especially those that are important in energy storage and transfer. Some of its conventions are slightly different from those we are accustomed to. First, it is customary to focus on changes in free energy (ΔG) rather than changes in enthalpy (ΔH). Consider the reaction

$$m\text{A}(aq) \rightleftharpoons n\text{B}(aq)$$

where (aq) indicates that both A and B are in aqueous solution. The reaction is spontaneous in the direction written when ΔG is negative, nonspontaneous when ΔG is positive.

But spontaneity depends on the concentrations of reactants and products. If the ratio $[\text{B}]^n/[\text{A}]^m$ is less than a certain value, the reaction is spontaneous in the forward direction; if $[\text{B}]^n/[\text{A}]^m$ exceeds this value, the reaction is spontaneous in the reverse direction. Therefore, it is useful to define a **standard free-energy change ($\Delta G°$)** that applies to a standard state where [A] = [B] = 1 M.

$$m\text{A}(aq) \rightleftharpoons n\text{B}(aq)$$

standard state: 1 M 1 M

Reactions are classified as **exergonic** or **endergonic** according to the sign of $\Delta G°$. An exergonic reaction is one in which $\Delta G°$ is negative, an endergonic reaction has a positive value of $\Delta G°$.

Thus, ΔG tells us about the *reaction* with respect to the substances present and their concentrations. $\Delta G°$ focuses more clearly on the differences in free energy between the reactants and products by removing their concentrations from consideration.

The next point takes the standard-state idea and makes it more suitable for biological processes by defining a new $\Delta G°$, called $\Delta G°'$. This new standard state is one with a pH of 7. This is the standard state used most of the time for biochemical reactions and is the one we will use. Not only does it make a big difference in reactions in which H^+ is consumed or produced, it also requires us to be aware of the form in which various species exist at a pH of 7. A reaction that is endergonic at $[H^+]$ = 1 M can easily become exergonic at $[H^+] = 10^{-7}$ M (pH = 7) and vice versa.

26.5 ATP and Bioenergetics

The key reaction in bioenergetics is the interconversion of ATP and ADP, usually expressed in terms of the hydrolysis of ATP.

HPO_4^{2-} is often referred to as "inorganic phosphate" and abbreviated P_i.

$$\text{ATP} + H_2O \longrightarrow \text{ADP} + HPO_4^{2-} \qquad \Delta G°' = -31 \text{ kJ } (-7.4 \text{ kcal})$$

Adenosine triphosphate; Water; Adenosine diphosphate; Hydrogen phosphate

As written, the reaction is exergonic at pH = 7. The reverse process—conversion of ADP to ATP—is endergonic. Relative to ADP + HPO_4^{2-}, ATP is a "high-energy compound."

When coupled to some other process, the conversion of ATP to ADP can provide the free energy to transform an otherwise endergonic process to an exergonic one. Take, for example, the conversion of glutamic acid to glutamine at pH = 7.

Equation 1:

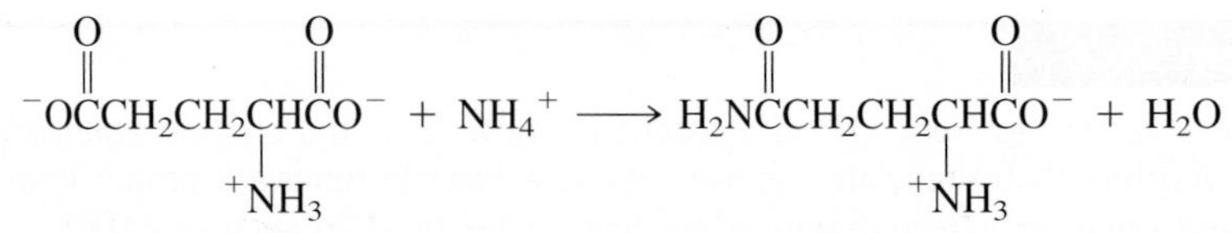

Glutamic acid Glutamine

Equation 1 has $\Delta G^{\circ\prime} = +14$ kJ and is endergonic. The main reason for this is that one of the very stable carboxylate groups of glutamic acid is converted to a less-stable amide function.

Nevertheless, the biosynthesis of glutamine proceeds from glutamic acid. The difference is that the endergonic process in Equation 1 is coupled with the strongly exergonic hydrolysis of ATP.

Equation 2:

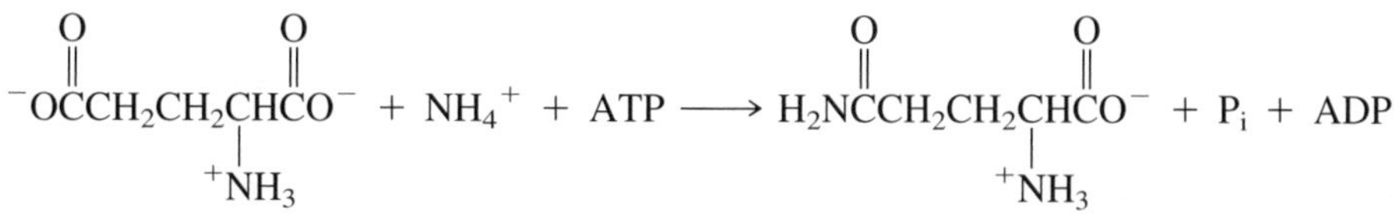

Glutamic acid Glutamine

Adding the value of $\Delta G^{\circ\prime}$ for the hydrolysis of ATP (-31 kJ) to that of Equation 1 ($+14$ kJ) gives $\Delta G^{\circ\prime} = -17$ kJ for Equation 2. The biosynthesis of glutamine from glutamic acid is exergonic because it is coupled to the hydrolysis of ATP.

Problem 26.7

Verify that Equation 2 is obtained by adding Equation 1 to the equation for the hydrolysis of ATP.

There is an important qualification to the idea that ATP can serve as a free-energy source for otherwise endergonic processes. There must be some mechanism by which ATP reacts with one or more species along the reaction pathway. Simply being present and undergoing independent hydrolysis isn't enough. More often than not, the mechanism involves transfer of a phosphate unit from ATP to some nucleophilic site. In the case of glutamine synthesis, this step is phosphate transfer to glutamic acid to give γ-glutamyl phosphate as a reactive intermediate.

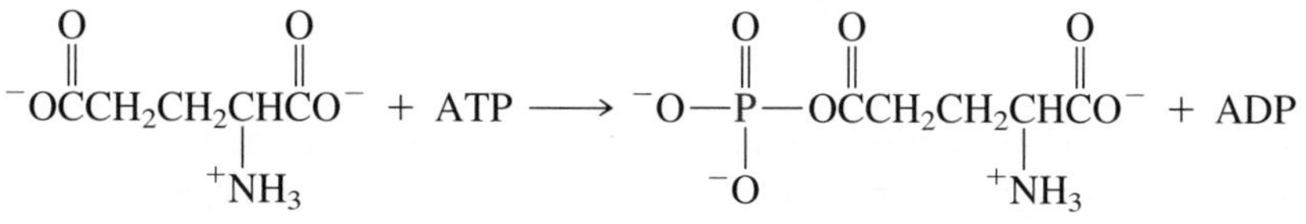

Glutamic acid γ-Glutamyl phosphate

The γ-glutamyl phosphate formed in this step is a mixed anhydride of glutamic acid and phosphoric acid. It is activated toward nucleophilic acyl substitution and gives glutamine when attacked by ammonia.

Problem 26.8

Write a stepwise mechanism for the formation of glutamine by attack of NH_3 on γ-glutamyl phosphate.

If free energy is stored and transferred by way of ATP, where does the ATP come from? It comes from ADP by the endergonic reaction

$$\underset{\text{Adenosine diphosphate}}{\text{ADP}} + \underset{\text{Hydrogen phosphate}}{HPO_4^{2-}} \longrightarrow \underset{\text{Adenosine triphosphate}}{\text{ATP}} + \underset{\text{Water}}{H_2O} \qquad \Delta G^{\circ\prime} = +31 \text{ kJ } (+7.4 \text{ kcal})$$

which you recognize as the reverse of the exergonic hydrolysis of ATP. The free energy to drive this endergonic reaction comes from the metabolism of energy sources such as fats and carbohydrates. In the metabolism of glucose during glycolysis (Section 23.22), for example, about one third of the free energy produced is used to convert ADP to ATP. Glycolysis produces phosphoenolpyruvate, which provides sufficient energy for the conversion of ADP to ATP. Energy-rich compounds are compared in terms of $\Delta G°'$ for their hydrolysis in Table 26.3.

TABLE 26.3 $\Delta G°'$ for the Hydrolysis of Bioenergetically Important Phosphates

Phosphate	$\Delta G°'$ kJ/mol; kcal/mol
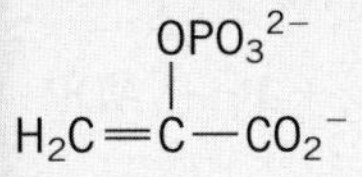 Phosphoenolpyruvate	−62.2 kJ/mol; −14.9 kcal/mol
ADP	−35 kJ/mol; −8.4 kcal/mol
ATP (page 1181)	−31 kJ/mol; −7.4 kcal/mol
Glucose-1-phosphate	−21.0 kJ/mol; −5.0 kcal/mol
Glucose-6-phosphate	−13.9 kJ/mol; −3.3 kcal/mol
AMP	−9.2 kJ/mol; −2.2 kcal/mol

Problem 26.9

Is $K > 1$ or $K < 1$ for the transfer of a phosphate group from ATP to glucose to give glucose 6-phosphate?

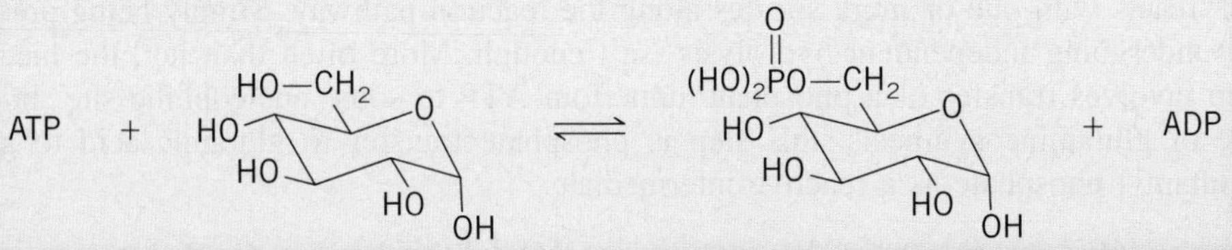

As important as nucleotides of adenosine are to bioenergetics, that is not the only indispensable part they play in biology. The remainder of this chapter describes how these and related nucleotides are the key compounds in storing and expressing genetic information.

26.6 Phosphodiesters, Oligonucleotides, and Polynucleotides

Just as amino acids can join together to give dipeptides, tripeptides, and so on up to polypeptides and proteins, so too can nucleotides join to form larger molecules. Analogous to the "peptide bond" that connects two amino acids, a **phosphodiester** joins two nucleosides. Figure 26.1 shows the structure and highlights the two phosphodiester units of a trinucleotide of 2′-deoxy-D-ribose in which the bases are adenine (A), thymine (T), and guanine (G). Phosphodiester units connect the 3′-oxygen of one nucleoside to the 5′-oxygen of the next. Nucleotide sequences are written with the free 5′ end at the left and the free 3′ end at the right. Thus, the trinucleotide sequence shown in Figure 26.1 is written as ATG.

The same kind of 5′→3′ phosphodiester units that join the 2′-deoxy-D-ribose units in Figure 26.1 are also responsible for connecting nucleosides of D-ribose.

Problem 26.10

How would the structures of the trinucleotides AUG and GUA in which all of the pentoses are D-ribose differ from the trinucleotide in Figure 26.1?

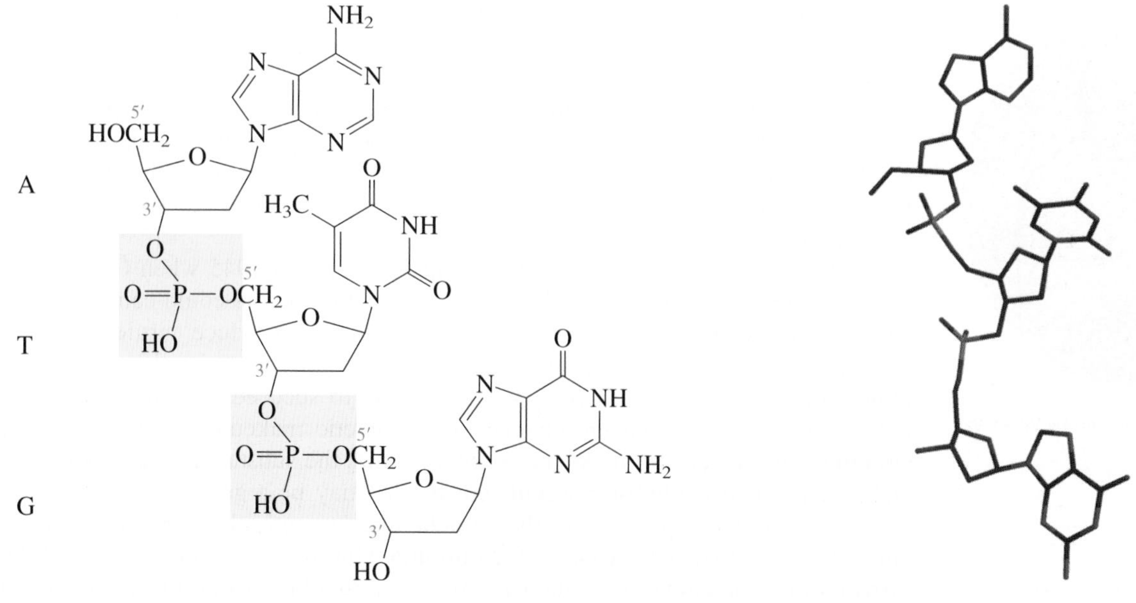

(*a*) (*b*)

Figure 26.1

(*a*) Structural formula and (*b*) molecular model of the trinucleotide ATG. The phosphodiester units highlighted in yellow in (*a*) join the oxygens at 3′ of one nucleoside to 5′ of the next. By convention, the sequence is read in the direction that starts at the free CH_2OH group (5′) and proceeds toward the free 3′ OH group at other end.

Adding nucleotides to the 3′-oxygen of an existing structure is called *elongation* and leads ultimately to a **polynucleotide.** The most important polynucleotides are ribonucleic acid (RNA) and deoxyribonucleic acid (DNA). As we shall see in later sections, the polynucleotide chains of DNA and some RNAs are quite long and contain hundreds of thousands of bases.

Polynucleotides of modest chain length, say 50 or fewer, are called **oligonucleotides.** With the growth of the biotechnology industry, the chemical synthesis of oligonucleotides has become a thriving business with hundreds of companies offering custom syntheses of oligonucleotides of prescribed sequences. Such oligonucleotides are required as "primers" in the polymerase chain reaction (Section 26.16) and as "probes" in DNA cloning and genetic engineering. Their synthesis is modeled after the Merrifield solid-phase method and, like it, is automated. The synthesis of a typical oligonucleotide containing 20–50 bases can be accomplished in a few hours.

Oligonucleotide synthesis is the subject of the Descriptive Passage at the end of this chapter.

26.7 Nucleic Acids

The nineteenth century saw three things happen that, taken together, prepared the way for our present understanding of genetics. In 1854, an Augustinian monk named Gregor Mendel began growing peas and soon discovered some fundamental relationships about their inherited characteristics. Mendel discovered two laws of heredity: segregation and independent assortment. His work demonstrated the existence of paired elementary units of heredity, and revealed statistical relationships that govern their expression. He described these at a scientific meeting in 1865 and sent copies of a paper describing his work to a number of prominent scientists. At about the same time (1859), Charles Darwin published his book *On the Origin of Species by Means of Natural Selection.* Mendel's work was ignored until it was rediscovered in 1900; Darwin's was widely known and vigorously debated. The third event occurred in 1869 when Johann Miescher isolated a material he called *nuclein* from the nuclei of white blood cells harvested from the pus of surgical bandages. Miescher's nuclein contained both a protein and an acidic, phosphorus-rich substance that, when eventually separated from the protein, was given the name *nucleic acid.*

Gregor Mendel systematically studied and statistically analyzed inherited traits in garden peas.

The Mendel Medal, awarded by Villanova University.

After 1900, genetic research—but not research on nucleic acids—blossomed. Nucleic acids were difficult to work with, hard to purify, and, even though they were present in all cells, did not seem to be very interesting. Early analyses, later shown to be incorrect, were interpreted to mean that nucleic acids were polymers consisting of repeats of some sequence of adenine (A), thymine (T), guanine (G), and cytosine (C) in a 1:1:1:1 ratio. Nucleic acids didn't seem to offer a rich enough alphabet from which to build a genetic dictionary. Most workers in the field believed proteins to be better candidates.

More attention began to be paid to nucleic acids in 1945 when Oswald Avery of the Rockefeller Institute for Medical Research found that he could cause a nonvirulent strain of a bacterium (*Streptococcus pneumoniae*) to produce virulent offspring by incubating them with a substance isolated from a virulent strain. What was especially important was that this virulence was passed on to succeeding generations and could only result from a permanent change in the genetic makeup—what we now call the **genome**—of the bacterium. Avery established that the substance responsible was DNA and in a letter to his brother speculated that it "may be a gene."

Avery's paper prompted other biochemists to rethink their ideas about DNA. One of them, Erwin Chargaff of Columbia University, soon discovered that the distribution of adenine, thymine, cytosine, and guanine differed from species to species, but was the same within a species and within all the cells of a species. Perhaps DNA did have the capacity to carry genetic information after all. Chargaff also found that regardless of the source of the DNA, half the bases were purines and the other half were pyrimidines. Significantly, the ratio of the purine adenine (A) to the pyrimidine thymine (T) was always close to 1:1. Likewise, the ratio of the purine guanine (G) to the pyrimidine cytosine (C) was also close to 1:1. For human DNA the values are:

Purine	Pyrimidine	Base ratio
Adenine (A) 30.3%	Thymine (T) 30.3%	A/T = 1.00
Guanine (G) 19.5%	Cytosine (C) 19.9%	G/C = 0.98
Total purines 49.8%	Total pyrimidines 50.2%	

Problem 26.11

Estimate the guanine content in turtle DNA if adenine = 28.7% and cytosine = 21.3%.

Avery's studies shed light on the *function* of DNA. Chargaff's touched on *structure* in that knowing the distribution of A, T, G, and C in DNA is analogous to knowing the amino acid composition of a protein, but not its sequence or three-dimensional shape.

The breakthrough came in 1953 when James D. Watson and Francis H. C. Crick proposed a structure for DNA. The Watson–Crick proposal ranks as one of the most important in all of science and has spurred a revolution in our understanding of genetics. The structure of DNA is detailed in the next section. The boxed essay *It Has Not Escaped Our Notice...* describes how it came about.

26.8 Secondary Structure of DNA: The Double Helix

Watson and Crick shared the 1962 Nobel Prize in Physiology or Medicine with Maurice Wilkins who, with Rosalind Franklin, was responsible for the X-ray crystallographic work.

Watson and Crick relied on molecular modeling to guide their thinking about the structure of DNA. Because X-ray crystallographic evidence suggested that DNA was composed of two polynucleotide chains running in opposite directions, they focused on the forces holding the two chains together. Hydrogen bonding between bases seemed the

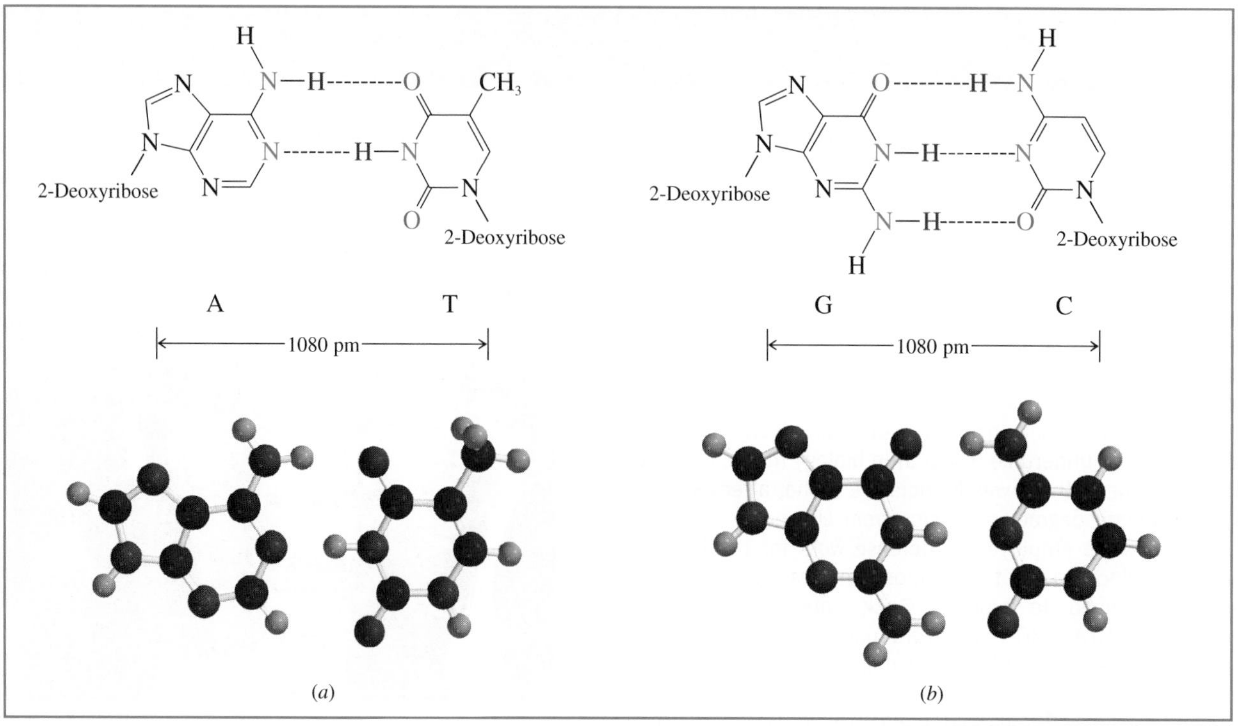

Figure 26.2

Hydrogen bonding between DNA bases shown as structural drawings of nucleosides (top) and as molecular models (bottom) of (*a*) adenine and thymine and (*b*) guanine and cytosine.

most likely candidate. After exploring a number of possibilities, Watson and Crick hit on the arrangement shown in Figure 26.2 in which adenine and thymine comprise one complementary *base pair* and guanine and cytosine another. This base-pairing scheme has several desirable features.

1. Pairing A with T and G with C gives the proper Chargaff ratios (A = T and G = C).
2. Each pair contains one purine and one pyrimidine base. This makes the A---T and G---C pairs approximately the same size and ensures a consistent distance between the two DNA strands.
3. Complementarity between A and T, and G and C suggests a mechanism for copying DNA. This is called replication and is discussed in Section 26.10.

Figure 26.4 supplements Figure 26.2 by showing portions of two DNA strands arranged side by side with the base pairs in the middle.

Hydrogen bonding between complementary bases is responsible for association between the strands, whereas conformational features of its carbohydrate–phosphate backbone and the orientation of the bases with respect to the furanose rings govern the overall shape of each strand. Using the X-ray crystallographic data available to them, Watson and Crick built a molecular model in which each strand took the shape of a right-handed helix. Joining two antiparallel strands by appropriate hydrogen bonds produced the **double helix** shown in the photograph (Figure 26.3). Figure 26.5 shows two modern renderings of DNA models.

In addition to hydrogen bonding between the two polynucleotide chains, the double-helical arrangement is stabilized by having its negatively charged phosphate groups on the outside where they are in contact with water and various cations, Na^+, Mg^{2+}, and ammonium ions, for example. Attractive van der Waals forces between the aromatic pyrimidine and purine rings, called π-*stacking,* stabilize the layered arrangement of the bases on the inside. Even though the bases are on the inside, they are accessible to

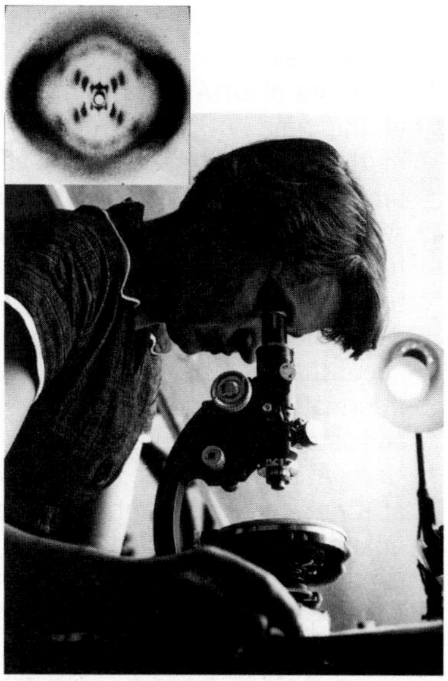

Rosalind Franklin (1920–1958). Her X-ray crystallographic data was used to solve the structure of DNA.

A helical structure for DNA strands had been suggested in 1949 by Sven Furberg in his Ph.D. dissertation at the University of London.

It Has Not Escaped Our Notice...

Our text began with an application of physics to chemistry when we described the electronic structure of atoms. We saw then that Erwin Schrödinger's introduction of wave mechanics figured prominently in developing the theories that form the basis for our present understanding. As we near the end of our text, we see applications of chemistry to areas of biology that are fundamental to life itself. Remarkably, Schrödinger appears again, albeit less directly. His 1944 book *What Is Life?* made the case for studying genes, their structure, and function.

Schrödinger's book inspired a number of physicists to change fields and undertake research in biology from a physics perspective. One of these was Francis Crick who, after earning an undergraduate degree in physics from University College, London, and while employed in defense work for the British government, decided that the most interesting scientific questions belonged to biology. Crick entered Cambridge University in 1949 as a 30-year-old graduate student, eventually settling on a research problem involving X-ray crystallography of proteins.

One year later, 22-year-old James Watson completed his Ph.D. studies on bacterial viruses at Indiana University and began postdoctoral research in biochemistry in Copenhagen. After a year at Copenhagen, Watson decided Cambridge was the place to be.

Thus it was that the paths of James Watson and Francis Crick crossed in the fall of 1951. One was a physicist, the other a biologist. Both were ambitious in the sense of wanting to do great things and shared a belief that the chemical structure of DNA was the most important scientific question of the time. At first, Watson and Crick talked about DNA in their spare time because each was working on another project. Soon, however, it became their major effort. Their sense of urgency grew when they learned that Linus Pauling, fresh from his proposal of helical protein structures, had turned his attention to DNA. Indeed, Watson and Crick were using the Pauling approach to structure—take what is known about the structure of small molecules, couple it to structural information about larger ones, and build molecular models consistent with the data.

At the same time, Maurice Wilkins and Rosalind Franklin at King's College, Cambridge, were beginning to obtain high-quality X-ray crystallographic data of DNA. Some of their results were presented in a seminar at King's attended by Watson, and even more were disclosed in a progress report to the Medical Research Council of the U.K. Armed with Chargaff's A = T and G = C relationships and Franklin's X-ray data, Watson and Crick began their model building. A key moment came when Jerry Donohue, a postdoctoral colleague from the United States, noticed that they were using the wrong structures for the pyrimidine and purine bases. Watson and Crick were using models of the enol forms of thymine, cytosine, and guanine, rather than the correct keto forms (recall Section 26.1). Once they fixed this error, the now-familiar model shown in Figure 26.3 emerged fairly quickly and they had the structure of DNA.

Figure 26.3

Molecular modeling—1953 style. James Watson (*left*) and Francis Crick (*right*) with their DNA model. © *A. Barrington Brown/Science Source Photo Researchers, Inc.*

Watson and Crick published their work in a paper entitled "A Structure for Deoxyribose Nucleic Acid" in the British journal *Nature* on April 25, 1953. In addition to being one of the most important papers of the twentieth century, it is also remembered for one brief sentence appearing near the end.

> "It has not escaped our notice that the specific pairing we have postulated immediately suggests a possible copying mechanism for the genetic material."

True to their word, Watson and Crick followed up their April 25 paper with another on May 30. This second paper, "Genetical Implications of the Structure of Deoxyribonucleic Acid," outlines a mechanism for DNA replication that is still accepted as essentially correct.

other substances through two grooves that run along the axis of the double helix. They are more accessible via the *major groove,* which is almost twice as wide as the *minor groove.* The grooves differ in size because of the way the bases are tilted with respect to the furanose ring.

The structure proposed by Watson and Crick was modeled to fit crystallographic data obtained on a sample of the most common form of DNA called B-DNA. Other

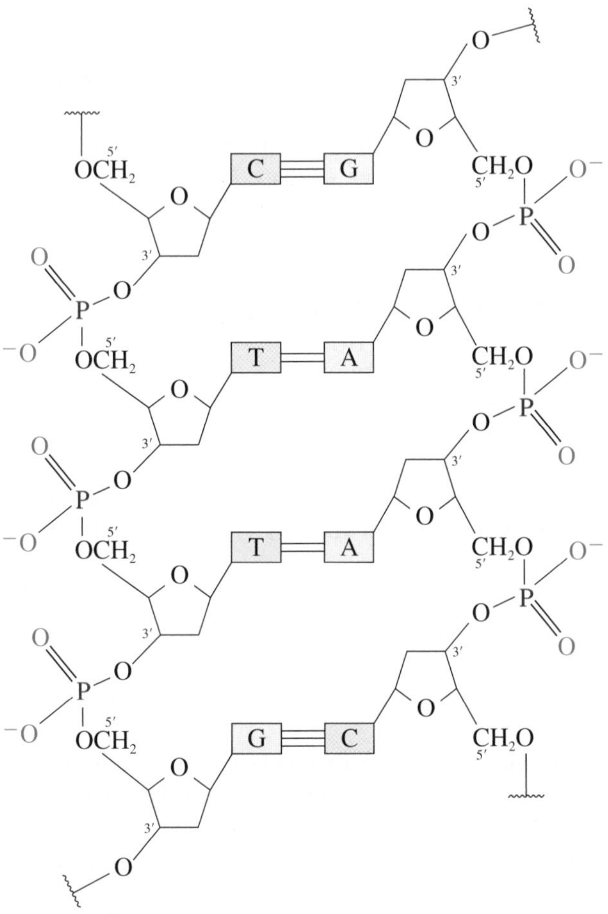

Figure 26.4

Hydrogen bonding between complementary bases (A and T, and G and C) permit pairing of two DNA strands. The strands are antiparallel; the 5′ end of the left strand is at the top, and the 5′ end of the right strand is at the bottom.

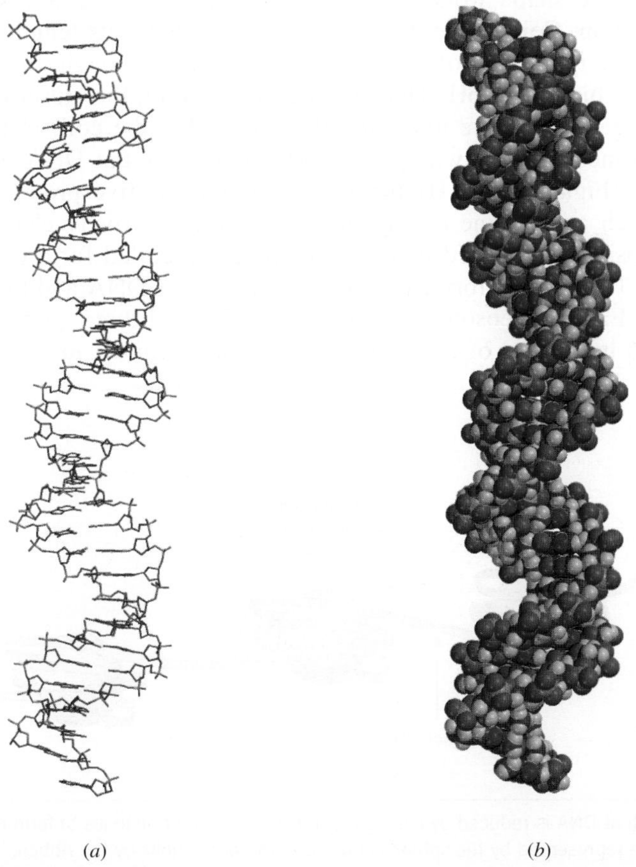

(*a*) (*b*)

Figure 26.5

(*a*) Tube and (*b*) space-filling models of a DNA double helix. The carbohydrate–phosphate "backbone" is on the outside and can be roughly traced in (*b*) by the red oxygen atoms. The blue atoms belong to the purine and pyrimidine bases and lie on the inside. The base-pairing is more clearly seen in (*a*).

forms include A-DNA, which is similar to, but more compact than B-DNA, and Z-DNA, which is a left-handed double helix.

By analogy to the levels of structure of proteins, the **primary structure** of DNA is the sequence of bases along the polynucleotide chain, and the A-DNA, B-DNA, and Z-DNA helices are varieties of **secondary structures.**

Not all DNAs are double helices (*duplex DNA*). Some types of viral DNA are single-stranded, and even a few triple and quadruple DNA helices are known.

26.9 Tertiary Structure of DNA: Supercoils

We have, so far, described the structure of DNA as an extended double helix. The crystallographic evidence that gave rise to this picture was obtained on a sample of DNA removed from the cell that contained it. Within a cell—its native state—DNA almost always adopts some shape other than an extended chain. We can understand why by doing a little arithmetic. Each helix of B-DNA makes a complete turn every 3.4×10^{-9} m and there are about 10 base pairs per turn. A typical human DNA contains 10^8 base pairs. Therefore,

$$\text{Length of DNA chain} = \frac{3.4 \times 10^{-9} \text{ m/turn}}{10 \text{ base pairs/turn}} \times 10^8 \text{ base pairs}$$

$$\text{Length of DNA chain} = 3.4 \times 10^{-2} \text{ m} = 3.4 \text{ cm}$$

For a 3-cm-long molecule of DNA to fit inside a cell so tiny that we can only see it with a microscope, the polynucleotide chain must be folded into a more compact form. Not only must the DNA be compacted, it must be folded in a way that allows it to carry out its main functions. The way the chain is folded defines the **tertiary structure** of a nucleic acid.

The compacting mechanism is a marvel of cellular engineering. A twisted tangle of indefinite shape would present serious problems as a vessel for storing genetic information. Coiling the duplex, however, reduces its length without blocking access to important parts of its structure. Remember though, that DNA is negatively charged at biological pH. Thus, the tighter the coil, the closer together are the negatively charged phosphate units and the less stable the coil. Nature solves this puzzle for chromosomes by wrapping short sections of the DNA around proteins called histones (Figure 26.6). **Histones** are a family of five proteins rich in basic amino acids such as arginine and lysine, which are positively charged at biological pH. The positively charged histones stabilize the coiled form of the negatively charged DNA. The species formed between a section of DNA and histones is called a **nucleosome.** Each nucleosome contains about one and three quarters turns of coil comprising 146 base pairs of DNA and is separated from the next nucleosome by a

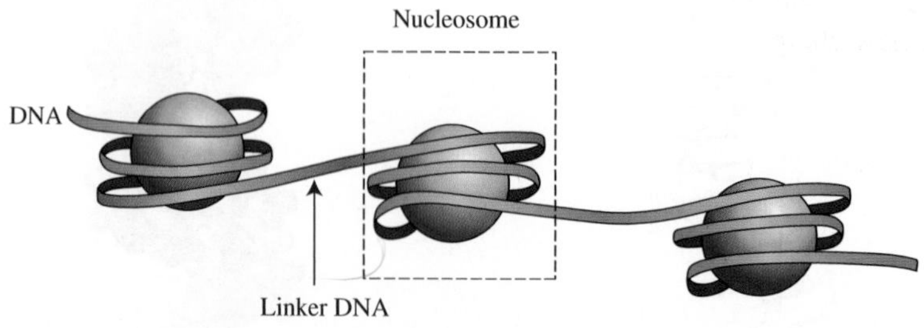

Figure 26.6

The effective length of DNA is reduced by coiling around the surface of histones to form nucleosomes. The histone proteins are represented by the spheres and the DNA double helix by the ribbon.

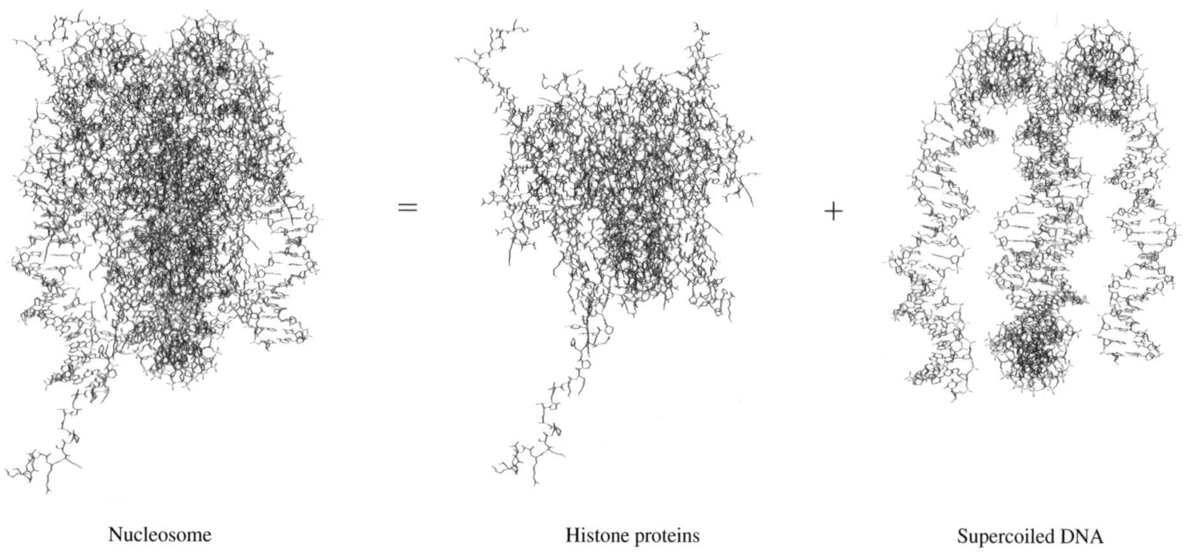
=
+
Nucleosome
Histone proteins
Supercoiled DNA

Figure 26.7

Molecular models of a nucleosome and its components. The nucleosome has a protein core around which is wound a supercoil of duplex DNA.

"linker" of about 50 base pairs of DNA. Figure 26.7 shows a molecular model of a single nucleosome.

Problem 26.12

Approximately how many nucleosomes are in a gene with 10,000 base pairs?

A single helix is a coil; a double helix is two nested coils. The tertiary structure of DNA in a nucleosome is a coiled coil. Coiled coils are referred to as **supercoils** and are quite common.

A coiled α-helix in a protein is another example of a supercoil.

26.10 Replication of DNA

Every time a cell divides, its DNA is duplicated so that the DNA in the new cell is identical to that in the original one. As Figure 26.8 shows, Watson–Crick base-pairing provides the key to understanding this process of DNA **replication.** During cell division the DNA double helix begins to unwind, generating a **replication fork** separating the two strands. Each strand serves as a template on which a new DNA strand is constructed. The A—T, G—C base-pairing requirement ensures that each new strand is the precise complement of its template strand. Each of the two new duplex DNA molecules contains one original and one new strand.

Both new chains grow in their 5′→3′ direction. Because of this, one grows toward the replication fork (the **leading strand**) and the other away from it (the **lagging strand**), making the details of chain extension somewhat different for the two. The fundamental chemistry, however, is straightforward (Figure 26.9). The hydroxyl group at the 3′ end of the growing polynucleotide chain acts as a nucleophile, attacking the 5′-triphosphate of 2′-deoxyadenosine, 2′-deoxyguanosine, 2′-deoxycytidine, or thymidine to form the new phosphodiester linkage. The enzyme that catalyzes phosphodiester bond formation is called *DNA polymerase;* different DNA polymerases operate on the leading strand and the lagging strand.

All of the steps, from the unwinding of the original DNA double helix to the supercoiling of the new DNAs, are catalyzed by enzymes.

1. The DNA to be copied is a double helix, shown here as flat for clarity.
2. The two strands begin to unwind. Each strand will become a template for construction of its complement.
3. As the strands unwind, the pyrimidine and purine bases become exposed. Notice that the bases are exposed in the 3′ → 5′ direction in one strand, and in the 5′ → 3′ direction in the other.
4. Two new strands form as nucleotides that are complementary to those of the original strands are joined by phosphodiester linkages. The sources of the new bases are dATP, dGTP, dCTP, and dTTP already present in the cell.
5. Because nucleotides are added in the 5′ → 3′ direction, the processes by which the two new chains grow are different. Chain growth can be continuous in the leading strand, but not in the lagging strand.
6. Two duplex DNA molecules result, each of which is identical to the original DNA.

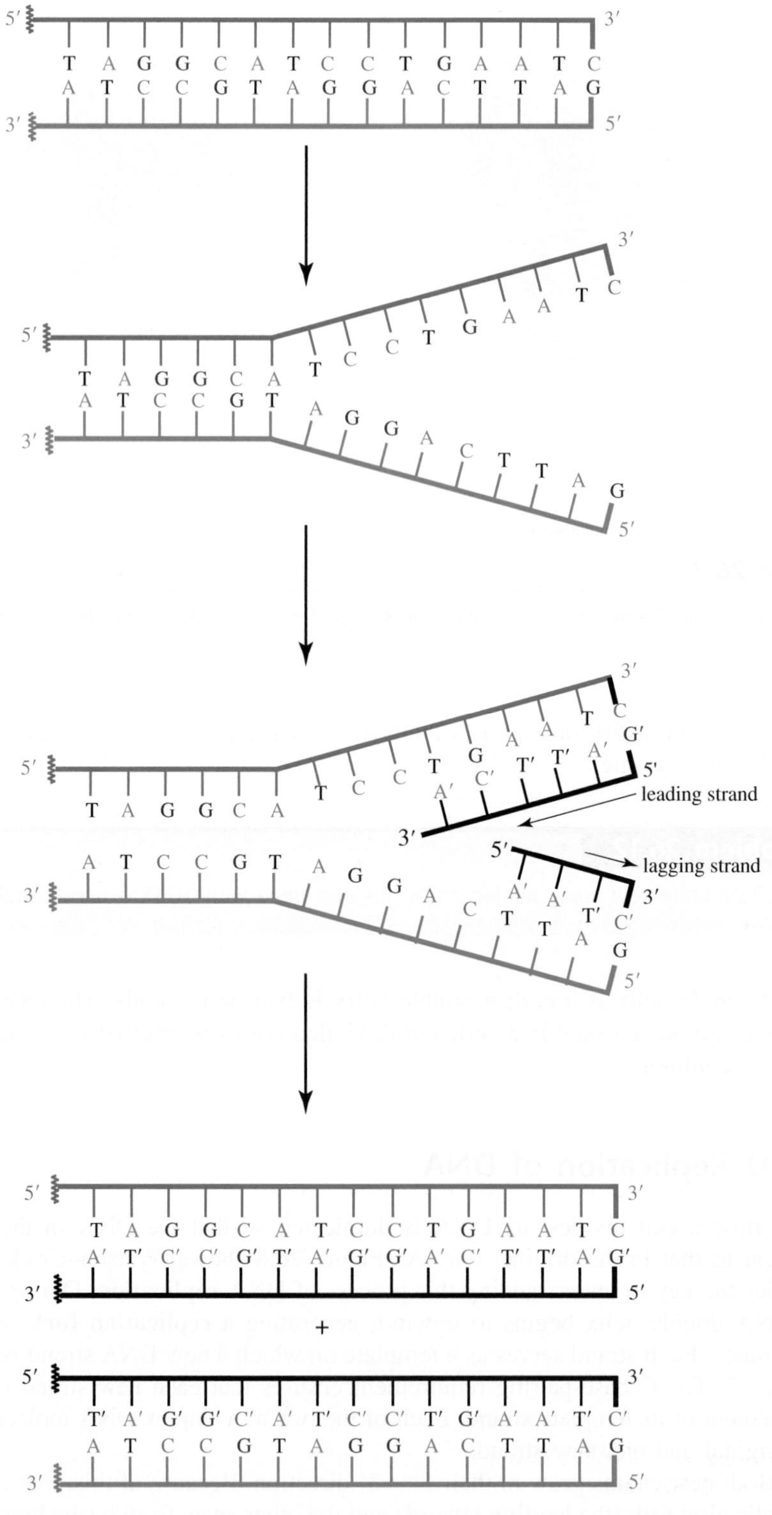

Figure 26.8

Outline of DNA replication. The original strands are shown in red and blue and are the templates from which the new strands, shown in black, are copied.

Genes are DNA and carry the inheritable characteristics of an organism and these characteristics are normally *expressed* at the molecular level via protein synthesis. Gene expression consists of two stages, **transcription** and **translation,** both of which involve RNAs. Sections 26.11 and 26.12 describe these RNAs and their roles in transcription and translation.

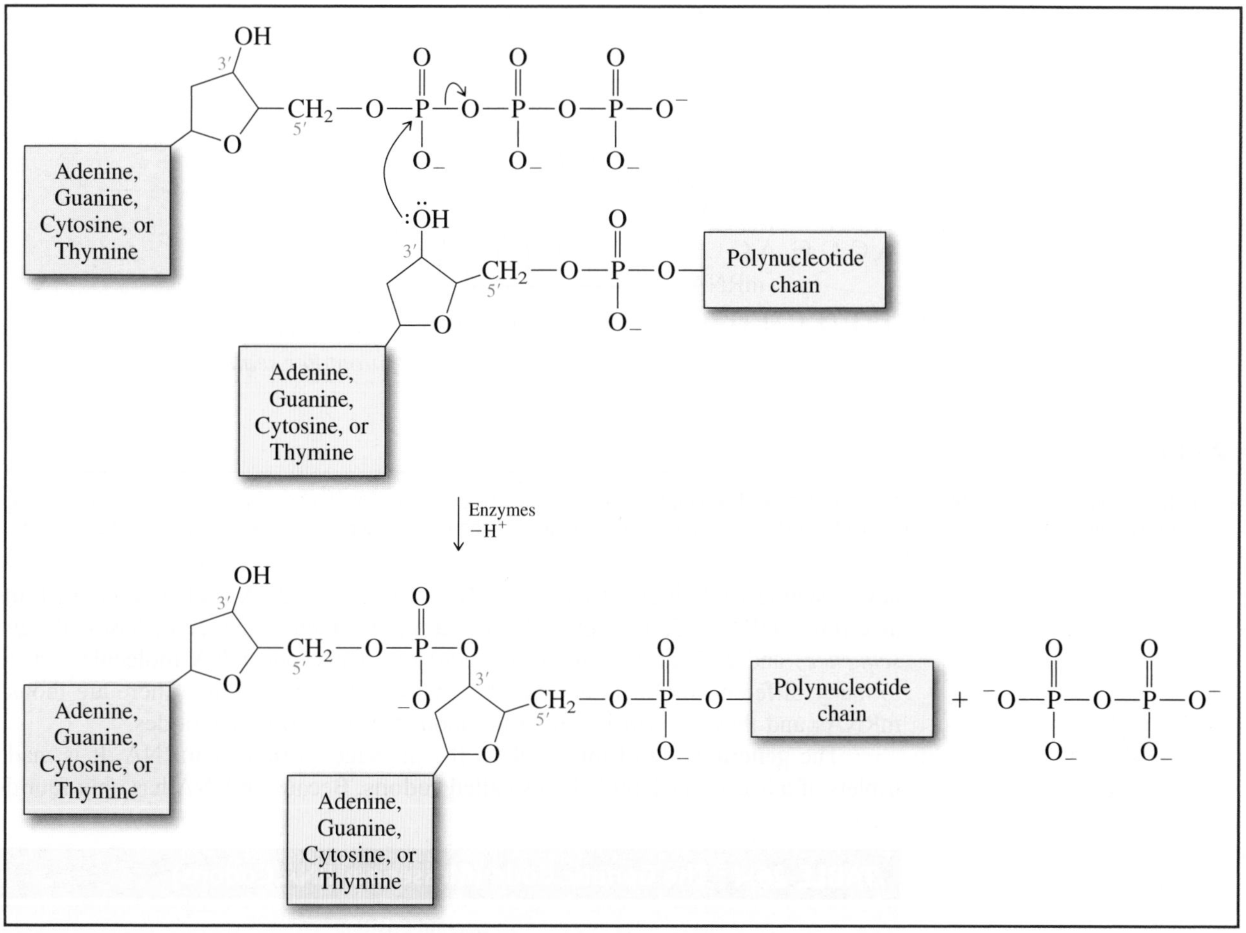

Figure 26.9

The new polynucleotide chain grows by reaction of its free 3′-OH group with the 5′-triphosphate of an appropriate 2′-deoxyribonucleoside.

26.11 Ribonucleic Acids

Unlike DNA, most of which is in the nucleus, RNA is found mostly in the cell's main compartment, the cytoplasm. There are three different kinds of RNA, which differ substantially from one another in both structure and function:

1. Messenger RNA (*mRNA*)
2. Transfer RNA (*tRNA*)
3. Ribosomal RNA (*rRNA*)

All are important in the biosynthesis of proteins.

Messenger RNA (mRNA): According to Crick, the so-called central dogma of molecular biology is "DNA makes RNA makes protein." The first part can be restated more exactly as "DNA makes mRNA." This is what transcription is—transcribing the message of DNA to a complementary RNA, in this case messenger RNA. mRNA is the least abundant of the RNAs and is the only one that is synthesized in the cell's nucleus. This transcription process is illustrated in Figure 26.10. Transcription resembles DNA replication in that a DNA strand serves as the template for construction of, in this case, a ribonucleic acid. mRNA synthesis begins at its 5′ end, and ribonucleotides complementary to the DNA strand being copied are added. The phosphodiester linkages are formed by reaction of the free 3′-OH group of the growing mRNA with ATP, GTP, CTP, or UTP (recall that uracil, not thymine, is the complement of adenine in RNA). The enzyme that catalyzes this reaction is *RNA polymerase.* Only a small section of about 10 base pairs of the DNA template is exposed at a time. As the synthesis zone moves down the DNA chain, restoration of hydrogen bonds between the two original DNA strands displaces the

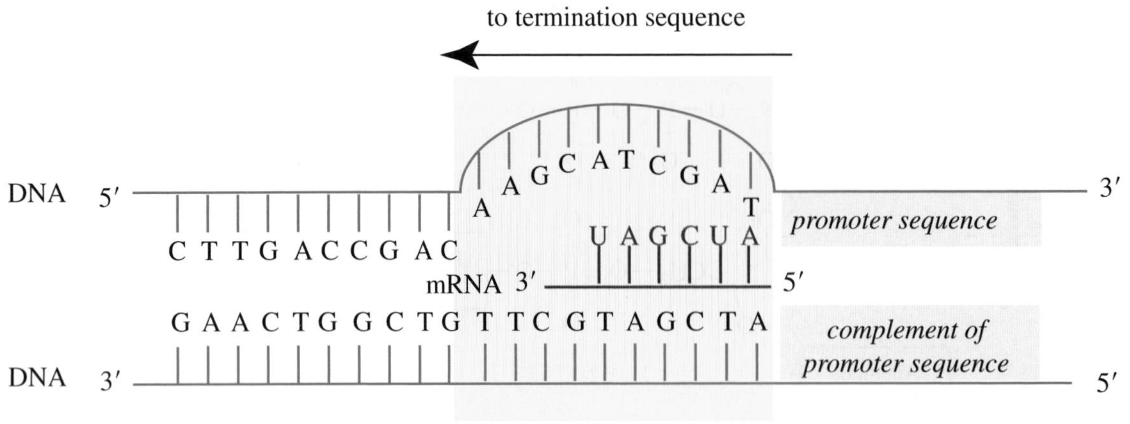

Figure 26.10

During transcription a molecule of mRNA is assembled from a DNA template. Transcription begins at a promoter sequence and proceeds in the 5′→3′ direction of the mRNA until a termination sequence of the DNA is reached. Only a region of about 10 base pairs is unwound at any time.

newly synthesized single-stranded mRNA. The entire DNA molecule is not transcribed as a single mRNA. Transcription begins at a prescribed sequence of bases (the *promoter sequence*) and ends at a *termination sequence.* Thus, one DNA molecule can give rise to many different mRNAs and code for many different proteins. There are thousands of mRNAs and they vary in length from about 500 to 6000 nucleotides.

The **genetic code** (Table 26.4) is the message carried *by* mRNA. It is made up of triplets of adjacent nucleotide bases called **codons.** Because mRNA has only four different

TABLE 26.4 The Genetic Code (Messenger RNA Codons)

First Position (5′ end)	Second Position: U	C	A	G	Third Position (3′ end)
U	UUU Phe	UCU Ser	UAU Tyr	UGU Cys	U
U	UUC Phe	UCC Ser	UAC Tyr	UGC Cys	C
U	UUA Leu	UCA Ser	UAA Stop	UGA Stop*	A
U	UUG Leu	UCG Ser	UAG Stop†	UGG Trp	G
C	CUU Leu	CCU Pro	CAU His	CGU Arg	U
C	CUC Leu	CCC Pro	CAC His	CGC Arg	C
C	CUA Leu	CCA Pro	CAA Gln	CGA Arg	A
C	CUG Leu	CCG Pro	CAG Gln	CGG Arg	G
A	AUU Ile	ACU Thr	AAU Asn	AGU Ser	U
A	AUC Ile	ACC Thr	AAC Asn	AGC Ser	C
A	AUA Ile	ACA Thr	AAA Lys	AGA Arg	A
A	AUG Met	ACG Thr	AAG Lys	AGG Arg	G
G	GUU Val	GCU Ala	GAU Asp	GGU Gly	U
G	GUC Val	GCC Ala	GAC Asp	GGC Gly	C
G	GUA Val	GCA Ala	GAA Glu	GGA Gly	A
G	GUG Val	GCG Ala	GAG Glu	GGG Gly	G

*UGA also codes for selenocysteine.
†UAG also codes for pyrrolysine.

bases and 20 amino acids must be coded for, codes using either one or two nucleotides per amino acid are inadequate. If nucleotides are read in sets of three, however, the four mRNA bases generate 64 possible "words," more than sufficient to code for 20 amino acids.

In addition to codons for amino acids, there are *start* and *stop* codons. Protein biosynthesis begins at a start codon and ends at a stop codon of mRNA. The start codon is the nucleotide triplet AUG, which is also the codon for methionine. The stop codons are UAA, UAG, and UGA. UAG and UGA can also code for pyrrolysine and selenocysteine, respectively. How these two "ambiguous" codons are read depends on the presence of specific genes.

Transfer RNA (tRNA): Transfer-RNAs are relatively small nucleic acids, containing only about 70 nucleotides. They get their name because they transfer amino acids to the ribosome for incorporation into a polypeptide. Although 20 amino acids need to be transferred, there are 50–60 tRNAs, some of which transfer the same amino acids. Figure 26.11 shows the structure of phenylalanine tRNA ($tRNA^{Phe}$). Like all tRNAs it is composed of a single strand, with a characteristic shape that results from the presence of paired bases in some regions and their absence in others.

The 1968 Nobel Prize in Physiology or Medicine was shared by Robert W. Holley of Cornell University for determining the nucleotide sequence of phenylalanine transfer RNA.

Among the 76 nucleotides of $tRNA^{Phe}$ are two sets of three that are especially important. The first is a group of three bases called the **anticodon,** which is complementary to the mRNA codon for the amino acid being transferred. Table 26.4 lists two mRNA codons for phenylalanine, UUU and UUC (reading in the 5′→3′ direction). Because base-pairing requires the mRNA and tRNA to be antiparallel, the two anticodons are read in the 3′→5′ direction as AAA and AAG.

3′ —A—A—G— 5′ tRNA anticodon

5′ —U—U—C— 3′ mRNA codon

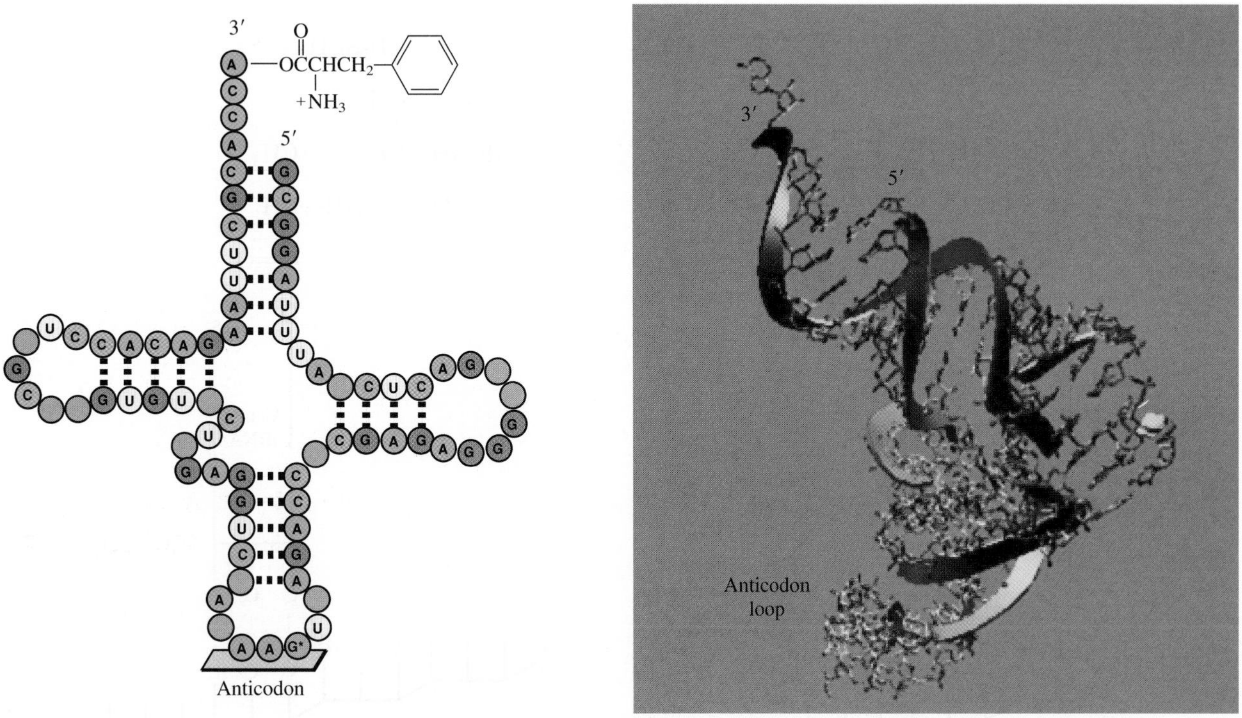

(*a*) (*b*)

Figure 26.11

Phenylalanine tRNA from yeast. (*a*) A schematic drawing showing the sequence of bases. Transfer RNAs usually contain a number of modified bases (gray circles). One of these is a modified guanosine (G*) in the anticodon. Hydrogen bonds, where present, are shown as dashed lines. (*b*) The structure of yeast $tRNA^{Phe}$ as determined by X-ray crystallography.

The other important sequence is the CCA triplet at the 3′ end. The amino acid that is to be transferred is attached through an ester linkage to the terminal 3′-oxygen of this sequence. *All tRNAs have a CCA sequence at their 3′ end.*

Transfer RNAs normally contain some bases other than A, U, G, and C. Of the 76 bases in $tRNA^{Phe}$, for example, 13 are of the modified variety. One of these, marked G* in Figure 26.11, is a modified guanosine in the anticodon. Many of the modified bases, including G*, are methylated derivatives of the customary RNA bases.

For their studies on the structure and mode of action of ribosomes, Venkatraman Ramakrishnan (Medical Research Council, UK), Thomas Steitz (Yale University, U.S.) and Ada Yonath (Weizmann Institute, Israel) were awarded the 2009 Nobel Prize in Chemistry.

Ribosomal RNA (rRNA): Ribosomes, which are about two thirds nucleic acid and about one third protein, constitute about 90% of a cell's RNA. A ribosome is made up of two subunits. The larger one contains two rRNAs, one with 122 nucleotides and the other with 2923; the smaller subunit contains one rRNA with 1500 nucleotides.

The ribosome is where the message carried by the mRNA is **translated** into the amino acid sequence of a protein. How it occurs is described in the next section. One of its most noteworthy aspects was discovered only recently. It was formerly believed that the RNA part of the ribosome was a structural component and the protein part was the catalyst for protein biosynthesis. Present thinking tilts toward reversing these two functions by ascribing the structural role to the protein and the catalytic one to rRNA. RNAs that catalyze biological processes are called **ribozymes.**

Sidney Altman (Yale University) and Thomas Cech (University of Colorado) shared the 1989 Nobel Prize in Chemistry for showing that RNAs could function as biological catalysts.

26.12 Protein Biosynthesis

As described in the preceding sections, protein synthesis involves transcription of the DNA to mRNA, followed by translation of the mRNA as an amino acid sequence. In addition to outlining the mechanics of transcription, we have described the relationship among mRNA codons, tRNA anticodons, and amino acids.

During translation the protein is synthesized beginning at its N-terminus (Figure 26.12). The mRNA is read in its 5′→3′ direction beginning at the start codon AUG and ending

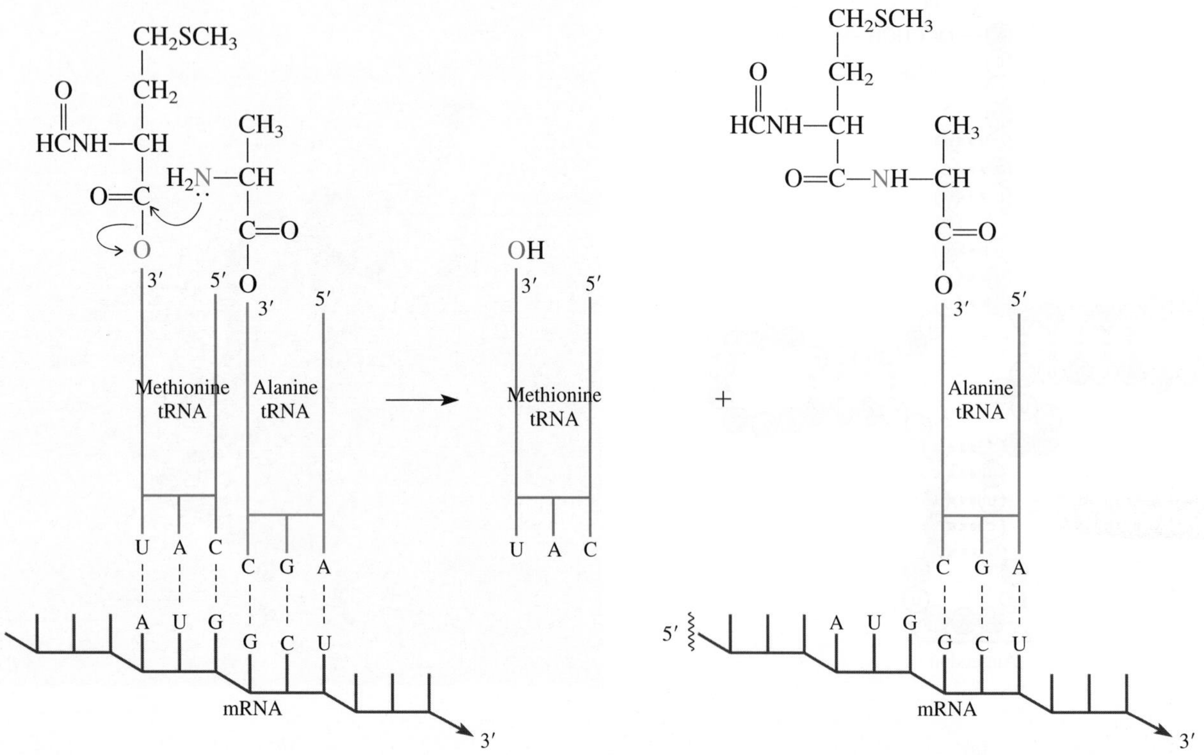

Figure 26.12

Translation of mRNA to an amino acid sequence of a protein starts at an mRNA codon for methionine. Nucleophilic acyl substitution transfers the *N*-formylmethionine residue from its tRNA to the amino group of the next amino acid (shown here as alanine). The process converts an ester to an amide.

at a stop codon (UAA, UAG, or UGA). Because the start codon is always AUG, the N-terminal amino acid is always methionine (as its *N*-formyl derivative). However, this *N*-formylmethionine residue is normally lost in a subsequent process and the N-terminus of the expressed protein is therefore determined by the second mRNA codon. The portion of the mRNA between the start and stop codons is called the coding sequence and is flanked on either side by noncoding regions.

In addition to illustrating the mechanics of translation, Figure 26.12 is important in that it shows the mechanism of peptide bond formation as a nucleophilic acyl substitution. Both methionine and alanine are attached to their respective tRNAs as esters. In a reaction apparently catalyzed by a ribozyme, the amino group of alanine attacks the methionine carbonyl, displacing methionine from its tRNA and converting the carbonyl group of methionine from an ester to an amide function.

Problem 26.13

Modify Figure 26.12 so that it corresponds to translation of an mRNA in which the sequence of the first six bases of the coding sequence are AUGUCU.

26.13 AIDS

The explosive growth of our knowledge of nucleic acid chemistry and its role in molecular biology in the 1980s coincided with the emergence of AIDS (acquired immune deficiency syndrome) as a major public health threat. In AIDS, a virus devastates the body's defenses to the extent that its victims can die from infections that are normally held in check by a healthy immune system. In the short time since its discovery in the early 1980s, AIDS has claimed the lives of over 25 million people, and current estimates place the number of those infected at more than 39 million. According to the World Health Organization (WHO), AIDS is now the fourth leading cause of death worldwide and the leading cause of death in Africa.

The viruses responsible for AIDS are human immunodeficiency virus 1 and 2 (HIV-1 and HIV-2). Both are **retroviruses,** meaning that their genetic material is RNA rather than DNA. HIVs require a host cell to reproduce, and the hosts in humans are the T4 lymphocytes, which are the cells primarily responsible for inducing the immune system to respond when provoked. The HIV penetrates the cell wall of a T4 lymphocyte and deposits both its RNA and an enzyme called *reverse transcriptase* inside. There, the reverse transcriptase catalyzes the formation of a DNA strand that is complementary to the viral RNA. The transcribed DNA then serves as the template from which the host lymphocyte produces copies of the virus, which then leave the host to infect other T4 cells. In the course of HIV reproduction, the ability of the T4 lymphocyte to reproduce itself is compromised. As the number of T4 cells decrease, so does the body's ability to combat infections.

Problem 26.14

When the RNA of a retrovirus is transcribed, what DNA base is the complement of the uracil in the viral RNA?

Although there is no known cure for AIDS, progress is being made in delaying the onset of symptoms and prolonging the lives of those infected with HIV. The first advance in treatment came with drugs such as the nucleoside *zidovudine,* also known as azidothymine, or AZT. During reverse transcription, AZT replaces thymidine in the DNA being copied from the viral RNA. AZT has a 5′-OH group, so can be incorporated into a growing polynucleotide chain. But because it lacks a 3′-OH group, the chain cannot be extended beyond it and synthesis of the viral DNA stops before the chain is complete.

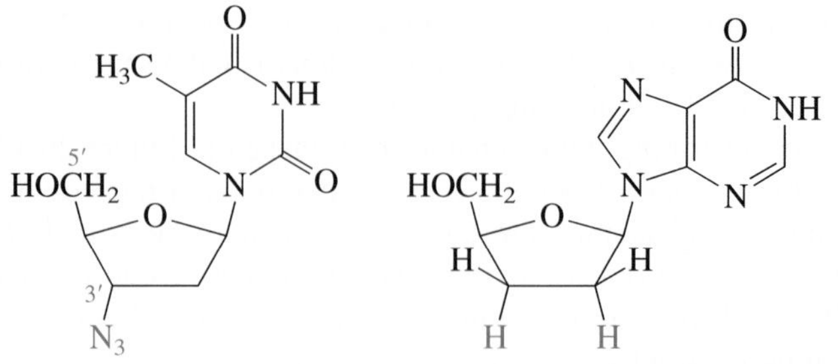

Zidovudine (AZT) 2′,3′-Dideoxyinosine (ddI)

Reverse transcriptase inhibitors are also used against certain viruses which, although they are not retroviruses, do require reverse transcriptase to reproduce. The virus that causes hepatitis B is an example.

Other nucleosides such as 2′,3′-dideoxyinosine (ddI) also block the action of reverse transcriptase and are often combined with AZT in "drug cocktails." Using a mixture of drugs makes it more difficult for a virus to develop resistance than using a single drug.

An advance in treating HIV infections has been to simultaneously attack the virus on a second front using a *protease inhibitor.* Recall from Section 25.10 that proteases are enzymes that catalyze the hydrolysis of proteins at specific points. When HIV uses a cell's DNA to synthesize its own proteins, the initial product is a long polypeptide that contains several different proteins joined together. To be useful, the individual proteins must be separated from the aggregate by protease-catalyzed hydrolysis of peptide bonds. Protease inhibitors prevent this hydrolysis and, in combination with reverse transcriptase inhibitors, slow the reproduction of HIV. Dramatic reductions in the "viral load" in HIV-infected patients have been achieved with this approach.

26.14 DNA Sequencing

Once the Watson–Crick structure was proposed, determining the nucleotide sequence of DNA emerged as an important area of research. Some difficulties were apparent from the beginning, especially if one draws comparisons to protein sequencing. First, most DNAs are much larger biopolymers than proteins. Not only does it take three nucleotides to code for a single amino acid, but vast regions of DNA don't seem to code for anything at all. A less obvious problem is that the DNA alphabet contains only four letters (A, G, C, and T) compared with the 20 amino acids from which proteins are built. Recall too that protein sequencing benefits from having proteases available that cleave the chain at specific amino acids. Not only are there no enzymes that cleave nucleic acids at specific bases but, with only four bases to work with, the resulting fragments would be too small to give useful information. In spite of this, DNA sequencing not only developed very quickly, but also has turned out to be much easier to do than protein sequencing.

Gel electrophoresis of proteins was described in the boxed essay accompanying Section 25.3.

To explain how DNA sequencing works, we must first mention **restriction enzymes.** Like all organisms, bacteria are subject to infection by external invaders (e.g., viruses and other bacteria) and possess defenses in the form of restriction enzymes that destroy the invader by cleaving its DNA. About 200 different restriction enzymes are known. Unlike proteases, which recognize a single amino acid, restriction enzymes recognize specific nucleotide *sequences*. Cleavage of the DNA at prescribed sequences gives fragments small enough to be sequenced conveniently. These smaller DNA fragments are separated and purified by gel electrophoresis. At a pH of 7.4, each phosphate link between adjacent nucleotides is ionized, giving the DNA fragments a negative charge and causing them to migrate to the positively charged electrode. Separation is size-dependent. Larger polynucleotides move more slowly through the polyacrylamide gel than smaller ones. The technique is so sensitive that two polynucleotides differing in length by only a single nucleotide can be separated from each other on polyacrylamide gels.

Once the DNA is separated into smaller fragments, each fragment is sequenced independently. Again, gel electrophoresis is used, this time as an analytical tool. In the technique devised by Frederick Sanger, the two strands of a sample of a small fragment of DNA, 100–200 base pairs in length, are separated and one strand is used as a template to create complements of itself. The single-stranded sample is divided among four test tubes, each of which contains the materials necessary for DNA synthesis. These materials include the four nucleosides present in DNA, 2′-deoxyadenosine (dA), 2′-deoxythymidine (dT), 2′-deoxyguanosine (dG), and 2′-deoxycytidine (dC) as their triphosphates dATP, dTTP, dGTP, and dCTP.

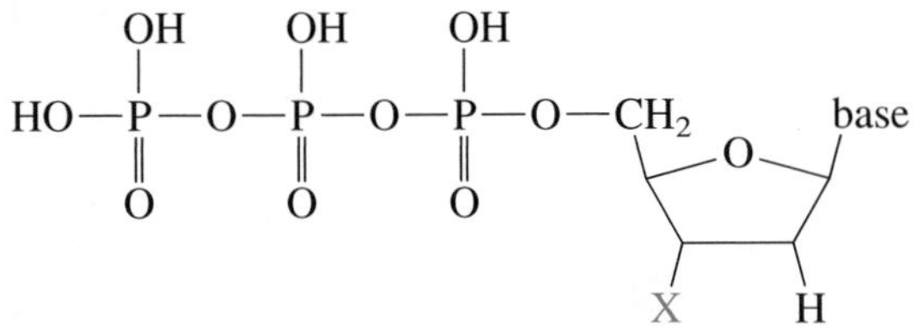

X = OH	X = H
dATP	ddATP
dTTP	ddTTP
dGTP	ddGTP
dCTP	ddCTP

Also present in the first test tube is a synthetic analog of ATP in which both the 2′- and 3′-hydroxyl groups have been replaced by hydrogens. This compound is called 2′,3′-dideoxyadenosine triphosphate (ddATP). Similarly, ddTTP is added to the second tube, ddGTP to the third, and ddCTP to the fourth. Each tube also contains a "primer." The primer is a short section of the complementary DNA strand, which has been labeled with a radioactive isotope of phosphorus (^{32}P). When the electrophoresis gel is examined at the end of the experiment, the positions of the DNAs formed by chain extension of the primer are located by a technique called *autoradiography,* which detects the particles emitted by the ^{32}P isotope.

As DNA synthesis proceeds, nucleotides from the solution are added to the growing polynucleotide chain. Chain extension takes place without complication as long as the incorporated nucleotides are derived from dATP, dTTP, dGTP, and dCTP. If, however, the incorporated species is derived from a dideoxy analog, chain extension stops. Because the dideoxy species ddA, ddT, ddG, and ddC lack hydroxyl groups at 3′, they cannot engage in the 3′→5′ phosphodiester linkage necessary for chain extension. Thus, the first tube—the one containing ddATP—contains a mixture of DNA fragments of different length, *all of which terminate in ddA.* Similarly, all the polynucleotides in the second tube terminate in ddT, those in the third tube terminate in ddG, and those in the fourth terminate in ddC.

The contents of each tube are then subjected to electrophoresis in separate lanes on the same sheet of polyacrylamide gel and the DNAs located by autoradiography. A typical electrophoresis gel of a DNA fragment containing 50 nucleotides will exhibit a pattern of 50 bands distributed among the four lanes with no overlaps. Each band corresponds to a polynucleotide that is one nucleotide longer than the one that precedes it (which may be in a different lane). One then simply "reads" the nucleotide sequence according to the lane in which each succeeding band appears.

The Sanger method for DNA sequencing is summarized in Figure 26.13. This work produced a second Nobel Prize for Sanger. (His first was for protein sequencing in 1958.) Sanger shared the 1980 chemistry prize with Walter Gilbert of Harvard University, who developed a chemical method for DNA sequencing (the Maxam–Gilbert method), and with Paul Berg of Stanford University, who was responsible for many of the most important techniques in nucleic acid chemistry and biology.

A modification of Sanger's method has resulted in the commercial availability of automated *DNA sequenators* based on Sanger's use of dideoxy analogs of nucleotides. Instead, however, of tagging a primer with ^{32}P, the purine and pyrimidine base portions of the dideoxynucleotides are each modified to contain a side chain that bears a different fluorescent dye, and all the dideoxy analogs are present in the same reaction. After electrophoretic separation of the products in a single lane, the gel is

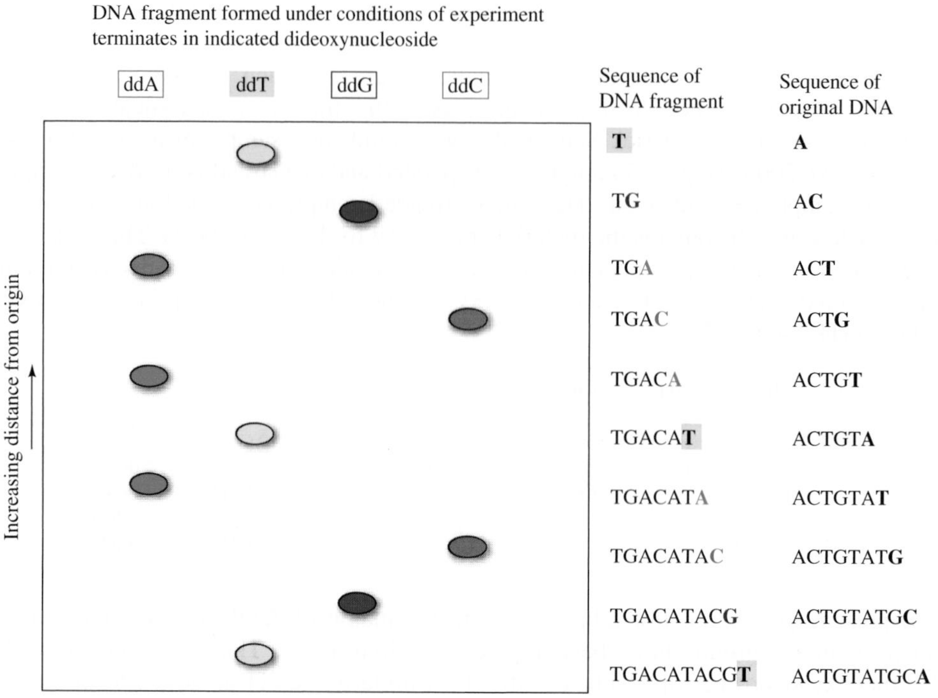
DNA fragment formed under conditions of experiment terminates in indicated dideoxynucleoside
ddA
ddT
ddG
ddC
Increasing distance from origin
Sequence of DNA fragment
Sequence of original DNA
T
A
TG
AC
TGA
ACT
TGAC
ACTG
TGACA
ACTGT
TGACAT
ACTGTA
TGACATA
ACTGTAT
TGACATAC
ACTGTATG
TGACATACG
ACTGTATGC
TGACATACGT
ACTGTATGCA

Figure 26.13

Sequencing of a short strand of DNA (10 bases) by Sanger's method using dideoxynucleotides to halt polynucleotide chain extension. Double-stranded DNA is separated and one of the strands used to produce complements of itself in four different tubes. All of the tubes contain a primer tagged with ^{32}P, dATP, dTTP, dGTP, and dCTP (see text for abbreviations). The first tube also contains ddATP, the second ddTTP, the third ddGTP, and the fourth ddCTP. All of the DNA fragments in the first tube terminate in A, those in the second terminate in T, those in the third terminate in G, and those in the fourth terminate in C. Location of the zones by autoradiographic detection of ^{32}P identifies the terminal nucleoside. The original DNA strand is its complement.

read by argon-laser irradiation at four different wavelengths. One wavelength causes the modified ddA-containing polynucleotides to fluoresce, another causes modified-ddT fluorescence, and so on. The data are stored and analyzed in a computer and printed out as the DNA sequence. A single instrument can sequence about 10,000 bases per day.

In addition to sequencing bits of DNA or individual genes, DNA sequencing has become so powerful a technique that the entire genomes of more than a thousand organisms have been sequenced. The first and largest number of these organisms were viruses—organisms with relatively small genomes. Then came a bacterium with 1.8 million base pairs, then baker's yeast with 12 million base pairs, followed by a roundworm with 97 million. The year 2000 brought announcements of the sequences of the 100 million base-pair genome of the wild mustard plant and the 180 million base-pair genome of the fruit fly. On the horizon was the 3-billion-base-pair human genome.

26.15 The Human Genome Project

In 1988, the National Research Council (NRC) recommended that the United States mount a program to map and then sequence the human genome. Shortly thereafter, the U.S. Congress authorized the first allocation of funds for what became a 15-year $3-billion-dollar project. Most of the NRC's recommendations for carrying out the project were adopted, including a strategy emphasizing technology development in the early stages followed by the sequencing of model organisms before attacking the human genome. The NRC's recommendation that the United States collaborate with other countries was also realized with the participation of teams from the United Kingdom, Japan, France, Germany, and China.

The International Human Genome Sequencing Consortium was headed by Francis S. Collins of the U.S. National Institutes of Health. J. Craig Venter led the Celera effort.

What was not anticipated was that in 1998 Celera Genomics of Rockville, Maryland, would undertake its own privately funded program toward the same goal. By 2000, the two groups agreed to some coordination of their efforts and published draft sequences in 2001 and final versions in 2003.

Because a fruit fly, for example, has about 13,000 genes, scientists expected humans to have on the order of 100,000 genes. The first surprise to emerge from the human genome sequence is that we have far fewer genes than we thought—only about 20,000–25,000. Because human DNA has more proteins to code for than fruit-fly DNA, gene expression must be more complicated than the phrase "one gene–one protein" suggests. Puzzles such as this belong to the new research field of **genomics**—the study of genome sequences and their function.

The human genome sequence has been called "the book of life" and, more modestly, a "tool box" and an "instruction manual." Regardless of what we call it, it promises a bright future for advances in medical science.

26.16 DNA Profiling and the Polymerase Chain Reaction

DNA sequencing and DNA profiling are different. The former, as we have seen, applies to procedures for determining the sequence of nucleotides in DNA. The latter is also a familiar term, usually encountered in connection with evidence in legal proceedings. In DNA profiling, the genes themselves are of little interest because their role in coding for proteins demands that they differ little, if at all, between individuals. But less than 2% of the human genome codes for proteins. Most of it lies in noncoding regions and this DNA does vary between individuals. Enzymatic cleavage of DNA produces a mixture of fragments that can be separated by electrophoresis to give a pattern of bands more likely to belong to one individual than others. Repeating the process with other cleaving enzymes gives a different pattern of bonds and increases the probability that the identification is correct. Until the 1980s, the limiting factor in both DNA profiling and sequencing was often the small amount of sample that was available. A major advance, called the **polymerase chain reaction (PCR),** effectively overcomes this obstacle and was recognized with the award of the 1993 Nobel Prize in Chemistry to its inventor Kary B. Mullis.

The main use of PCR is to "amplify," or make hundreds of thousands—even millions—of copies of a portion of the polynucleotide sequence in a sample of DNA. Suppose, for example, we wish to copy a 500-base-pair region of a sample of DNA that contains a total of 1 million base pairs. We would begin as described in Section 26.14 by cleaving the DNA into smaller fragments using restriction enzymes, then use PCR to make copies of the desired fragment.

Figure 26.14 illustrates how PCR works. In general, it involves multiple cycles of a three-step sequence. In working through Figure 26.14, be alert to the fact that the material we want does not arise until after the third cycle. After that, its contribution to the mixture of DNA fragments increases disproportionately. Repetitive PCR cycling increases both the amount of material and its homogeneity (Table 26.5). If every step proceeds in 100% yield, a greater than 1-billionfold amplification is possible after 30 cycles.

Each cycle incorporates three steps:

1. Denaturation
2. Annealing (also called priming)
3. Synthesis (also called extension or elongation)

All of the substances necessary for PCR are present throughout, and proceeding from one cycle to the next requires only changing the temperature after suitable time intervals. The entire process is carried out automatically, and 30 cycles can be completed within a few hours.

The double-stranded DNA shown in Figure 26.14(*a*) contains the polynucleotide sequence (the target region) we wish to amplify. The DNA is denatured by heating to ≈95°C, which causes the strands to separate by breaking the hydrogen bonds between them [Figure 26.14(*b*)].

Figure 26.14

The polymerase chain reaction (PCR). Three cycles are shown; the target region appears after the third cycle. Additional cycles lead to amplification of the target region.

(*a*) Consider double-stranded DNA containing a polynucleotide sequence (the **target region**) that you wish to amplify (make millions of copies of).

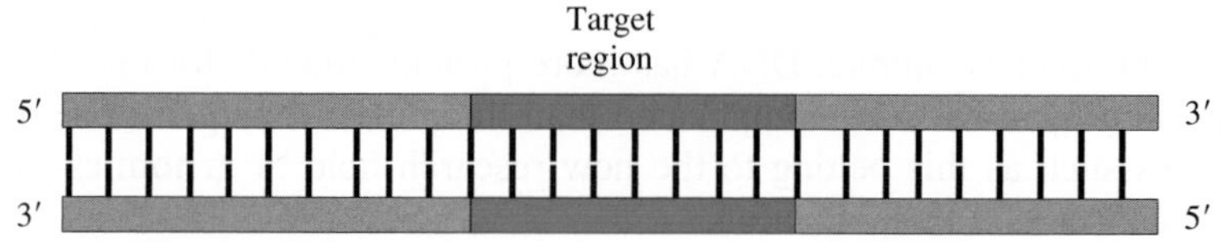

(*b*) Heating the DNA to ≈95°C causes the strands to separate. This is the denaturation step.

(*c*) Cooling the sample to ≈60°C causes one primer oligonucleotide to bind to one strand and the other primer to the other strand. This is the annealing step.

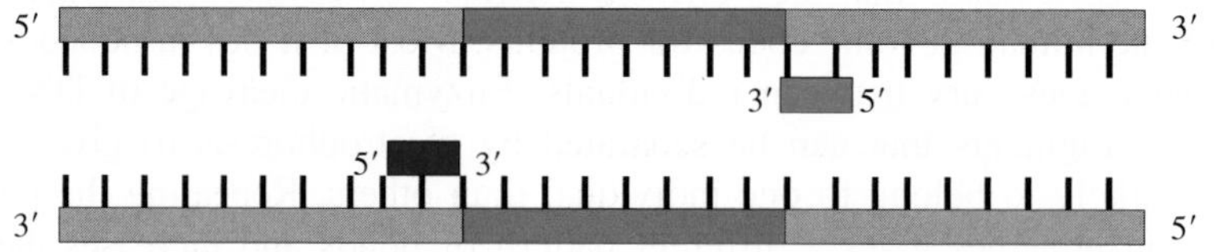

(*d*) In the presence of the four DNA nucleotides and the enzyme DNA polymerase, the primer is extended in its 3′ direction as it adds nucleotides that are complementary to the original DNA strand. This is the synthesis step and is carried out at ≈72°C.

(*e*) Steps (*a*)–(*d*) constitute one cycle of the polymerase chain reaction and produce two double-stranded DNA molecules from one. Denaturing the two DNAs and priming the four strands gives:

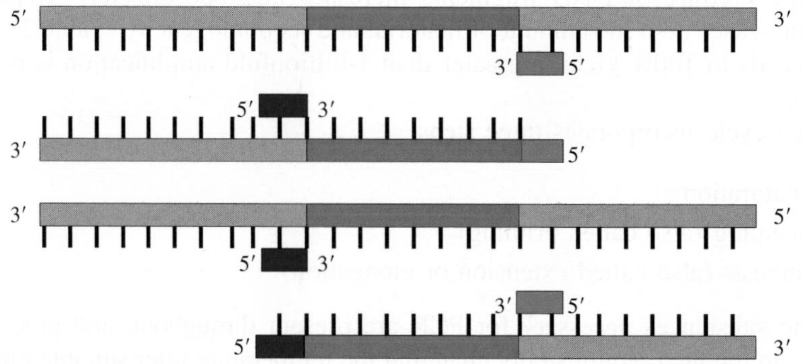

Continued

Figure 26.14

Continued

(*f*) Elongation of the primed polynucleotide fragments completes the second cycle and gives four DNAs.

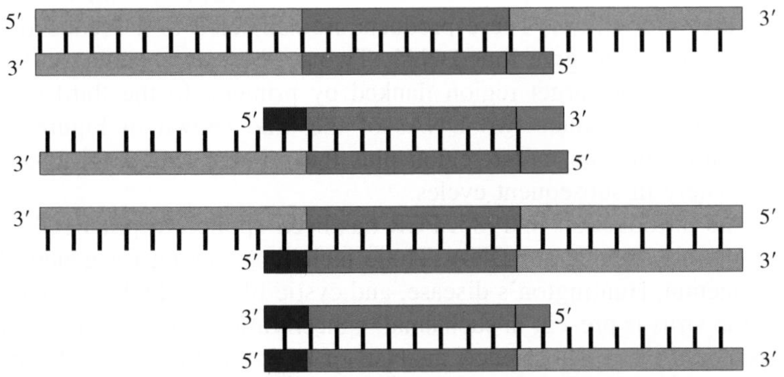

(*g*) Among the eight DNAs formed in the third cycle are two having the structure shown. This is the structure that increases disproportionately in the succeeding cycles.

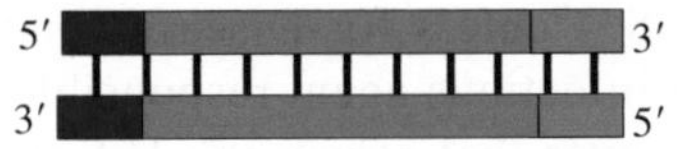

The solution is then cooled to ≈60°C, allowing new hydrogen bonds to form [Figure 26.14(*c*)]. However, the reaction mixture contains much larger concentrations of two primer molecules than DNA, and the new hydrogen bonds are between the separated DNA strands and the primers rather than between the two strands.

Each primer is a synthetic oligonucleotide of about 20 bases, prepared so that their sequences are complementary to the (previously determined) sequences that flank the target regions on opposite strands. Thus, one primer is annealed to one strand, the other to the other strand. The 3′-hydroxyl end of each primer points toward the target region.

The stage is now set for DNA synthesis to proceed from the 3′ end of each primer [Figure 26.14(*d*)]. The solution contains a DNA polymerase and Mg^{2+} in addition to the deoxynucleoside triphosphates dATP, dTTP, dGTP, and dCTP. The particular DNA

TABLE 26.5 Distribution of DNAs with Increasing Number of PCR Cycles

Cycle number	Total number of DNAs*	Number of DNAs containing only the target region
0 (start)	1	0
1	2	0
2	4	0
3	8	2
4	16	8
5	32	22
10	1,024	1,004
20	1,048,566	1,048,526
30	1,073,741,824	1,073,741,764

*Total number of DNAs is 2^n, where n = number of cycles

Taq polymerase was first found in a bacterium (*Thermus aquaticus*) that lives in hot springs in Yellowstone National Park. Bacteria of this type are called thermophiles because they thrive in warm environments.

polymerase used is one called *Taq polymerase* that is stable and active at the temperature at which the third step of the cycle is carried out (72°C).

The products of the first cycle are two DNAs, each of which is composed of a longer and a shorter strand. These products are subjected to a second three-step cycle [Figure 26.14(*e*)–(*f*)] to give four DNAs. Two of these four contain a "strand" that is nothing more than the target region flanked by primers. In the third cycle, these two ultrashort "strands" produce two DNAs of the kind shown in Figure 26.14(*g*). This product contains only the target region plus the primers and is the one that increases disproportionately in subsequent cycles.

Since its introduction in 1985, PCR has been applied to practically every type of study that requires samples of DNA. These include screening for genetic traits such as sickle cell anemia, Huntington's disease, and cystic fibrosis. PCR can detect HIV infection when the virus is present in such small concentrations that no AIDS symptoms have as yet appeared. In forensic science, analysis of PCR-amplified DNA from tiny amounts of blood or semen have helped convict the guilty and free the innocent. Anthropologists increasingly use information from DNA analysis to trace the origins of racial and ethnic groups but sometimes find it difficult, for cultural reasons, to convince individuals to volunteer blood samples. Thanks to PCR, a strand of hair is now sufficient.

Scientists at the U.S. Centers for Disease Control and Prevention (CDC) used PCR to help identify the infectious agent responsible for an outbreak of an especially dangerous hemorrhagic fever that struck the U.S. southwest in 1993. By annealing with synthetic oligonucleotide primers having sequences complementary to known hantaviruses, portions of the viral DNA obtained from those infected with the disease could be sucessfully amplified. Not only did this provide material for analysis, it also suggested that the new viral DNA had stretches where its sequence was the same as already known hantaviruses. Thus, the "Four Corners virus" was found to be a new strain of hantavirus and diagnostic procedures were developed specific for it.

"Four Corners" describes where the virus was first discovered. It is the region where Arizona, New Mexico, Colorado, and Utah meet.

More recently, PCR proved to be a valuable detection and analytical tool during the terrorist-inspired anthrax outbreak in the fall of 2001.

26.17 Recombinant DNA Technology

The use of restriction enzymes to cleave DNA at specific sequences was mentioned earlier in this chapter in the context of DNA sequence analysis. These enzymes are also important in the field of **recombinant DNA** technology. We will illustrate this application by describing a method for the production of human insulin.

The term *chimera* comes from Greek mythology and refers to a beast composed of the parts of different animals. Homer's *Iliad* describes such a creature: "...the Khimaira, of ghastly and inhuman origin, her forepart lionish, her tail a snake's, a she-goat in between. This thing exhaled in jets a rolling fire." (Translation by R. Fitzgerald, Book Six, line 210.) Farrar, Straus, and Giroux, 2004, New York

A plasmid, which is a circular DNA molecule separate from the chromosomal DNA, is obtained from bacterial cells such as *Escherischia coli* and treated with a restriction enzyme to snip the DNA at a specific site (Figure 26.15). The human DNA sequence that codes for the synthesis of insulin is then inserted into the plasmid to give a *recombinant DNA molecule*. The new DNA is the result of the recombination of DNA from the plasmid plus the sequence that codes for human insulin. The new plasmid is termed a *chimeric plasmid* because it contains DNA from two sources, bacterial and human. The plasmid is taken up by growing bacterial cells through a process called *transformation*. The chimeric plasma serves as a *cloning vector* because it serves as a vehicle to carry the recombinant DNA into *E. coli*. Transcription and translation of the insulin DNA then occur to produce human insulin. When the cells divide, the plasmids are divided between the daughter cells and they continue to produce clones. Insulin produced by recombinant DNA technology is commercially sold as Humulin.

The amplification of many other DNA sequences has been carried out by transfection of a cloning vector into a bacterial cell, making it possible to produce quantities of natural proteins that were not previously available, as well as unknown proteins. Green fluorescent protein (see Section 25.19) and its derivatives are produced by recombinant DNA technology. A recombinant human-platelet-derived growth factor (rh-PDGF) *becaplermin* (Regranex) is used clinically in the treatment of diabetic skin ulcers.

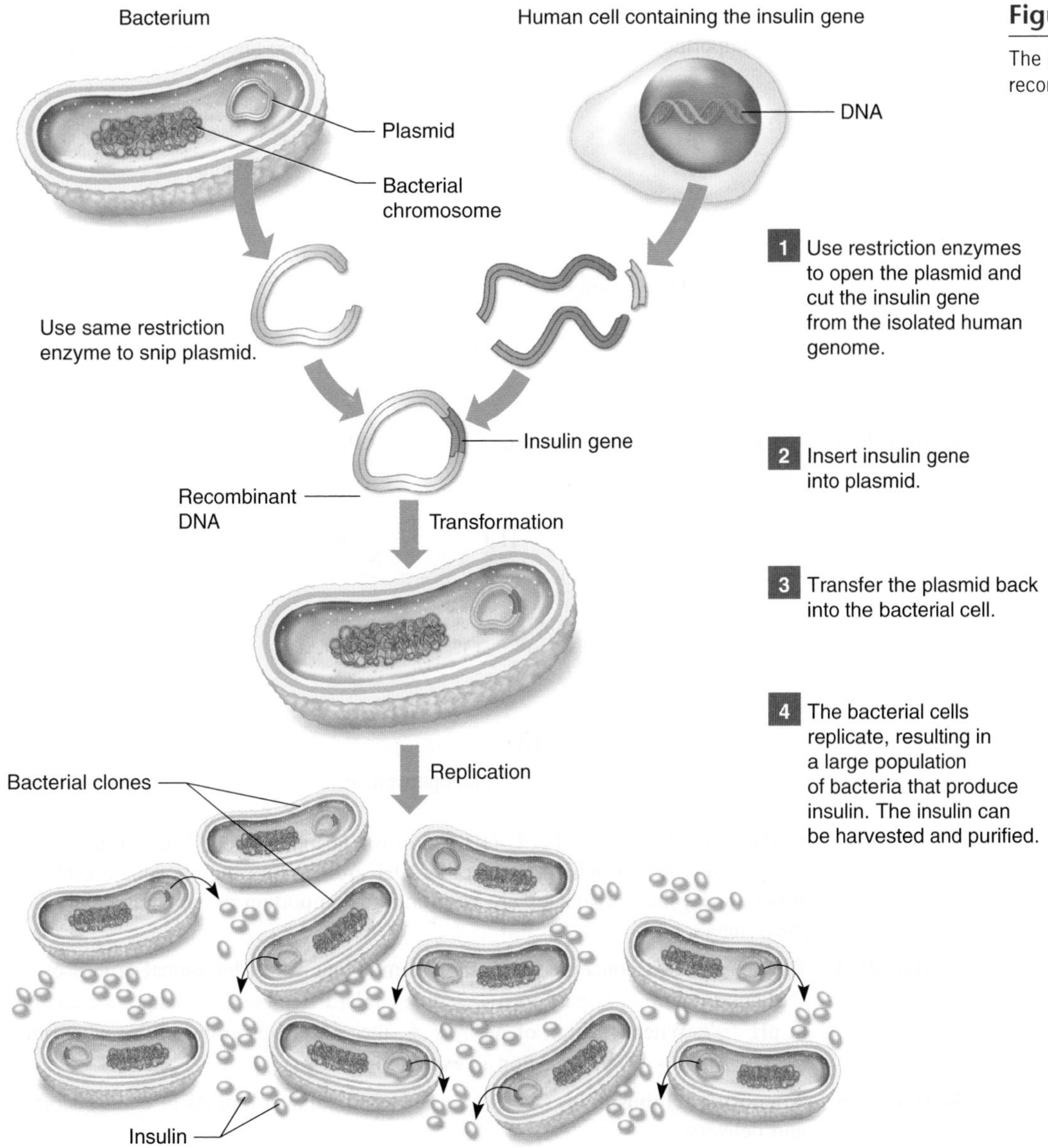
Bacterium
Human cell containing the insulin gene
Plasmid
Bacterial chromosome
DNA
1 Use restriction enzymes to open the plasmid and cut the insulin gene from the isolated human genome.
Use same restriction enzyme to snip plasmid.
Insulin gene
2 Insert insulin gene into plasmid.
Recombinant DNA
Transformation
3 Transfer the plasmid back into the bacterial cell.
4 The bacterial cells replicate, resulting in a large population of bacteria that produce insulin. The insulin can be harvested and purified.
Replication
Bacterial clones
Insulin

Figure 26.15

The production of human insulin by recombinant DNA technology.

26.18 SUMMARY

Section 26.1 Many biologically important compounds are related to the heterocyclic aromatic compounds pyrimidine and purine.

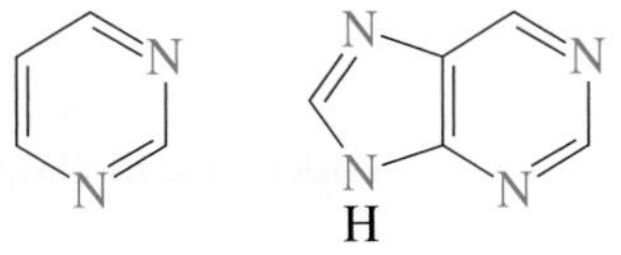

Pyrimidine Purine

The structure of guanine illustrates an important feature of substituted pyrimidines and purines. Oxygen substitution on the ring favors the keto form rather than the enol. Amino substitution does not.

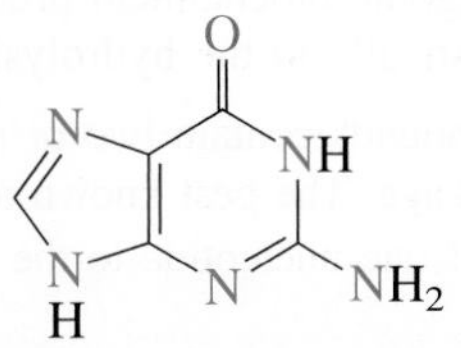

Guanine

Section 26.2 **Nucleosides** are carbohydrate derivatives of pyrimidine and purine bases. The most important nucleosides are derived from D-ribose and 2-deoxy-D-ribose.

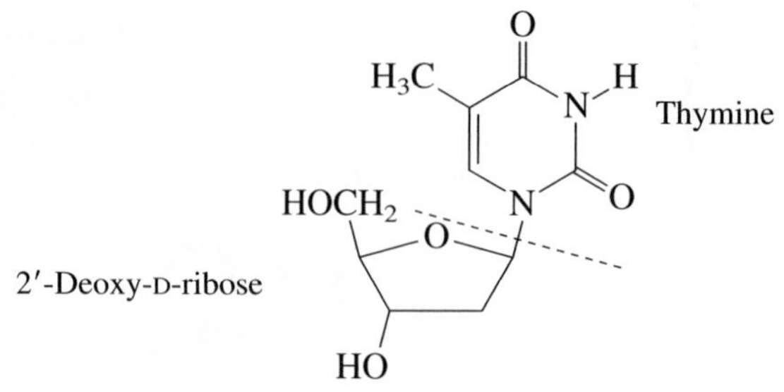

Thymidine

Section 26.3 **Nucleotides** are phosphate esters of nucleosides.

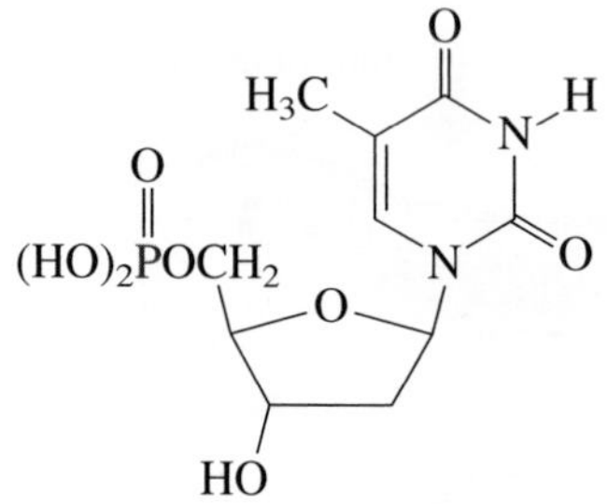

Thymidine 5′-monophosphate

In the example shown, the 5′-OH group is phosphorylated. Nucleotides are also possible in which some other OH group bears the phosphate ester function. Cyclic phosphates are common and important as biochemical messengers.

Section 26.4 **Bioenergetics** is concerned with the thermodynamics of biological processes. Particular attention is paid to $\Delta G^{\circ\prime}$, the standard free-energy change of reactions at pH = 7. When the sign of $\Delta G^{\circ\prime}$ is +, the reaction is **endergonic;** when the sign of $\Delta G^{\circ\prime}$ is −, the reaction is **exergonic.**

Section 26.5 **Adenosine triphosphate (ATP)** is a key compound in biological energy storage and delivery.

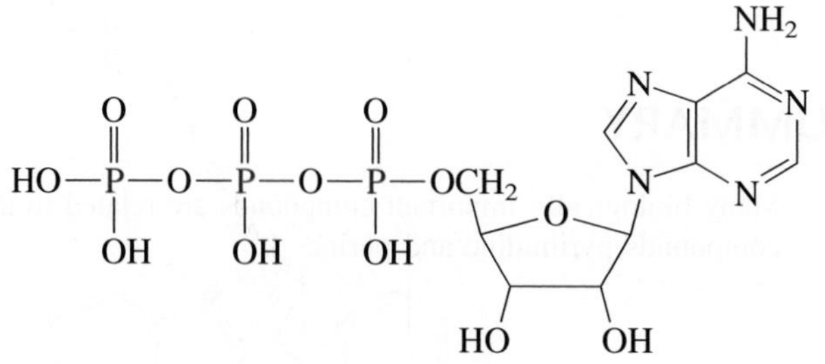

Adenosine triphosphate (ATP)

The hydrolysis of ATP to ADP and HPO_4^{2-} is exergonic.

$$\text{ATP} + H_2O \longrightarrow \text{ADP} + HPO_4^{2-} \qquad \Delta G^{\circ\prime} = -31 \text{ kJ } (-7.4 \text{ kcal})$$

Many formally endergonic biochemical processes become exergonic when they are coupled mechanistically to the hydrolysis of ATP.

Section 26.6 Many important compounds contain two or more nucleotides joined together by a **phosphodiester** linkage. The best known are those in which the phosphodiester joins the 5′-oxygen of one nucleotide to the 3′-oxygen of the other.

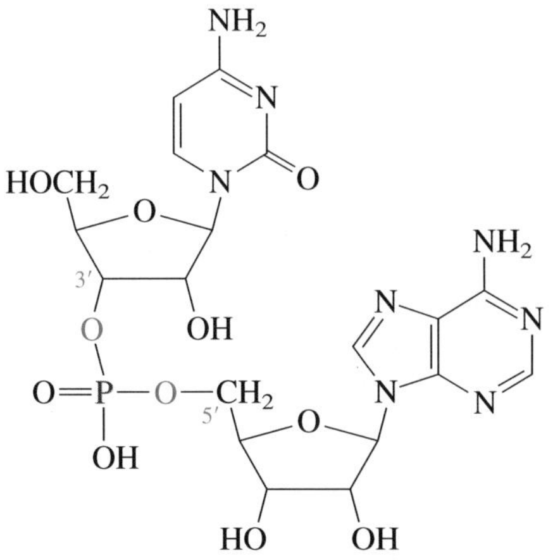

Oligonucleotides contain about 50 or fewer nucleotides held together by phosphodiester links; **polynucleotides** can contain thousands of nucleotides.

Section 26.7 **Nucleic acids** are polynucleotides present in cells. The carbohydrate component is D-ribose in ribonucleic acid (RNA) and 2-deoxy-D-ribose in deoxyribonucleic acid (DNA).

Section 26.8 The most common form of DNA is B-DNA, which exists as a right-handed double helix. The carbohydrate–phosphate backbone lies on the outside, the purine and pyrimidine bases on the inside. The double helix is stabilized by complementary hydrogen bonding (base pairing) between adenine (A) and thymine (T), and guanine (G) and cytosine (C).

Section 26.9 Within the cell nucleus, double-helical DNA adopts a **supercoiled** tertiary structure in which short sections are wound around proteins called **histones.** This reduces the effective length of the DNA and maintains it in an ordered arrangement.

Section 26.10 During DNA replication the two strands of the double helix begin to unwind, exposing the pyrimidine and purine bases in the interior. Nucleotides with complementary bases hydrogen bond to the original strands and are joined together by phosphodiester linkages with the aid of DNA polymerase. Each new strand grows in its $5' \rightarrow 3'$ direction.

Section 26.11 Three RNAs are involved in gene expression. In the **transcription** phase, a strand of **messenger RNA (mRNA)** is synthesized from a DNA template. The four bases A, G, C, and U, taken three at a time, generate 64 possible combinations called **codons.** These 64 codons comprise the **genetic code** and code for the 20 amino acids found in proteins plus start and stop signals. The mRNA sequence is **translated** into a prescribed protein sequence at the ribosomes. There, small polynucleotides called **transfer RNA (tRNA),** each of which contains an **anticodon** complementary to an mRNA codon, carries the correct amino acid for incorporation into the growing protein. **Ribosomal RNA (rRNA)** is the main constituent of ribosomes and appears to catalyze protein biosynthesis.

Section 26.12 The start codon for protein biosynthesis is AUG, which is the same as the codon for methionine. Thus, all proteins initially have methionine as their N-terminal amino acid, but lose it subsequent to their formation. The reaction responsible for extending the protein chain is nucleophilic acyl substitution.

Section 26.13 HIV, which causes AIDS, is a retrovirus. Its genetic material is RNA instead of DNA. HIV contains an enzyme called reverse transcriptase that allows its RNA to serve as a template for DNA synthesis in the host cell.

Section 26.14 The nucleotide sequence of DNA can be determined by a technique in which a short section of single-stranded DNA is allowed to produce its complement in the presence of dideoxy analogs of ATP, TTP, GTP, and CTP. DNA formation terminates when a dideoxy analog is incorporated into the growing polynucleotide chain. A mixture of polynucleotides differing from one another by an incremental nucleoside is produced and analyzed by electrophoresis. From the observed sequence of the complementary chain, the sequence of the original DNA is deduced.

Section 26.15 The sequence of nucleotides that make up the human genome has been completed. There is every reason to believe that the increased knowledge of human biology it offers will dramatically affect the practice of medicine.

Section 26.16 In DNA profiling the noncoding regions are cut into smaller fragments using enzymes that recognize specific sequences, and these smaller bits of DNA are then separated by electrophoresis. The observed pattern of DNA fragments is believed to be highly specific for the source of the DNA. Using the **polymerase chain reaction (PCR),** millions of copies of minute amounts of DNA can be produced in a relatively short time.

Section 26.17 DNA sequences that code for the synthesis of a specific protein can be inserted into a bacterial DNA plasmid. The growing bacteria then incorporate the **recombinant DNA** and produce the protein.

PROBLEMS

26.15 5-*Fluorouracil* is one component of a mixture of three drugs used in breast-cancer chemotherapy. What is its structure?

26.16 (a) Which isomer, the keto or enol form of cytosine, is the stronger acid?

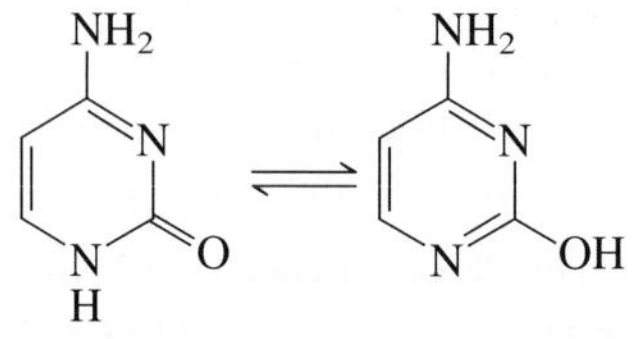

keto enol

(b) What is the relationship between the conjugate base of the keto form and the conjugate base of the enol form?

26.17 Birds excrete nitrogen as *uric acid.* Uric acid is a purine having the molecular formula $C_5H_4N_4O_3$; it has no C—H bonds. Write a structural formula for uric acid.

26.18 *Nebularine* is a toxic nucleoside isolated from a species of mushroom. Its systematic name is 9-β-D-ribofuranosylpurine. Write a structural formula for nebularine.

26.19 The D-arabinose analog of adenosine is an anitiviral agent (vidarabine) used to treat conjunctivitis and shingles. Write a structural formula for this compound.

26.20 Adenine is a weak base. Which one of the three nitrogens designated by arrows in the structural formula shown is protonated in acidic solution? Evaluation of the resonance contributors of the three protonated forms will tell you which one is the most stable.

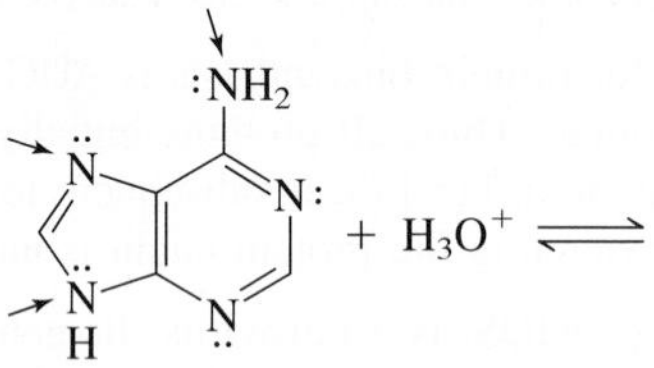

26.21 When 6-chloropurine is heated with aqueous sodium hydroxide, it is quantitatively converted to *hypoxanthine*. Suggest a reasonable mechanism for this reaction.

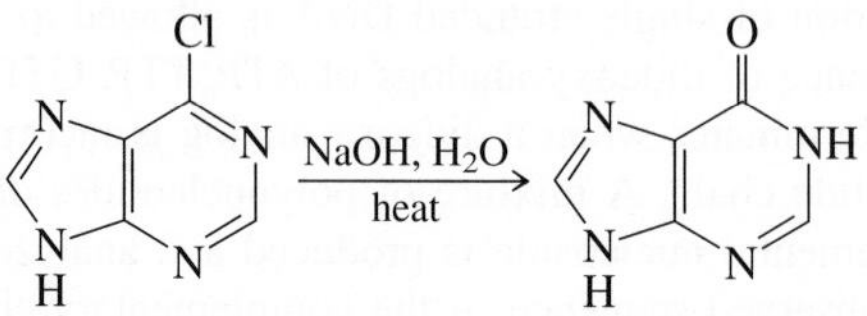

6-Chloropurine Hypoxanthine

26.22 Treatment of adenosine with nitrous acid gives a nucleoside known as *inosine*. Suggest a reasonable mechanism for this reaction.

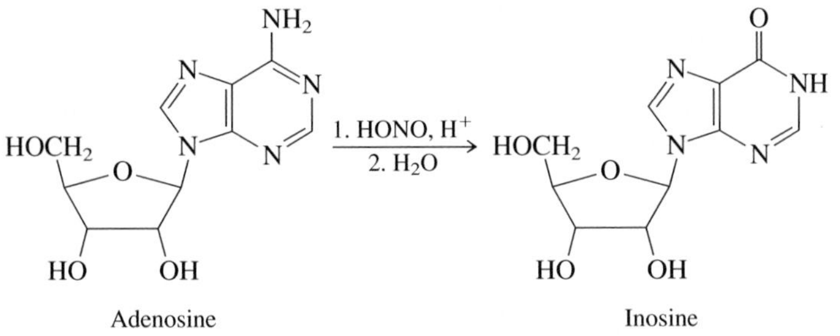

26.23 The 5′-nucleotide of inosine, *inosinic acid* ($C_{10}H_{13}N_4O_8P$) is added to foods as a flavor enhancer. What is the structure of inosinic acid? (The structure of inosine is given in Problem 26.22.)

26.24 The phosphorylation of α-D-glucopyranose by ATP (Section 26.3) has $\Delta G°' = -23$ kJ at 298 K.

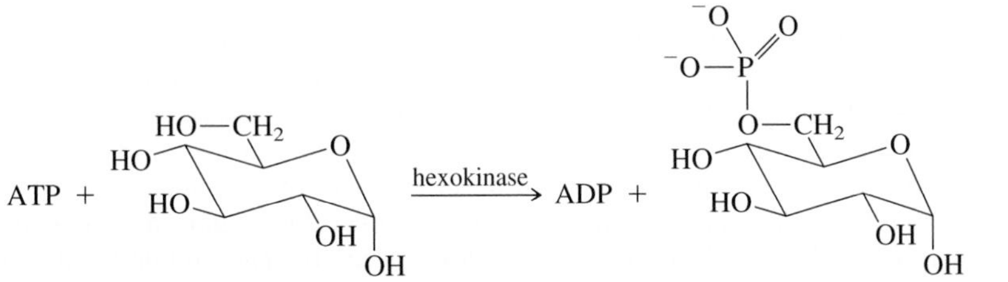

(a) Is this reaction exergonic or endergonic?

(b) How would the value of $\Delta G°'$ change in the absence of the enzyme hexokinase? Would it become more positive, more negative, or would it stay the same? Why?

(c) Use the value for the hydrolysis of ATP to ADP (Section 26.5) to calculate $\Delta G°'$ for the reaction of α-D-glucopyranose with inorganic phosphate. Is this reaction exergonic or endergonic?

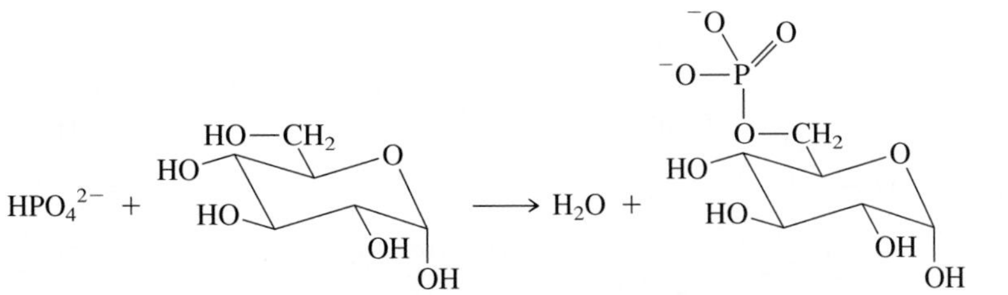

26.25 In one of the early experiments designed to elucidate the genetic code, Marshall Nirenberg of the U.S. National Institutes of Health (Nobel Prize in Physiology or Medicine, 1968) prepared a synthetic mRNA in which all the bases were uracil. He added this poly(U) to a cell-free system containing all the necessary materials for protein biosynthesis. A polymer of a single amino acid was obtained. What amino acid was polymerized?

26.26 (a) The two most acidic hydrogens of uracil have pK_a's of 9.5 and 14.2 respectively. Match these pK_a's with the hydrogens in the structural formula and provide structures for the most stable resonance contributors of the monoanion and the dianion.

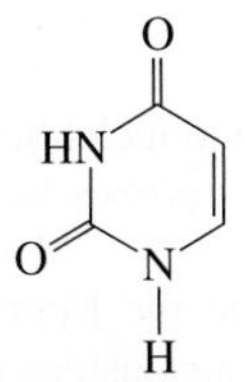

Uracil

(b) The pK_a of the conjugate acid of triethylamine is 10.4. Is triethylamine a strong enough base to convert uracil to its monoanion? To its dianion?

26.27 The coupling reaction of 2,6-dichloropurine with 1,2,3,4-tetra-*O*-acetyl-α,β-arabinofuranose takes place when the two are heated in the presence of an acid catalyst to give the nucleoside in 75% yield. The reaction is stereoselective for the formation of the α-anomer, even though the starting sugar is a mixture of anomers. Can you think of a reason for the stereoselectivity? (*Hint:* See Mechanism 23.3.)

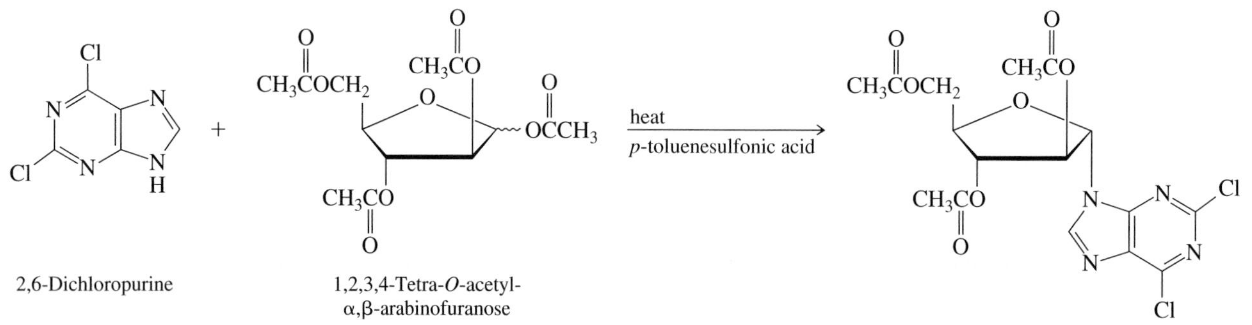

26.28 The descriptive passage in this chapter describes the solid-phase synthesis of oligonucleotides. The solid phase technique has also been applied to the automated synthesis of oligosaccharides in what is termed the glycal assembly method. An unsaturated carbohydrate known as a glycal is attached to the solid polystyrene support. The glycal is then converted to an epoxide by treatment with dimethyldioxirane (DMDO). The epoxide serves as the glycosyl donor and undergoes nucleophilic attack by the hydroxyl group of the glycosyl acceptor. The double bond of the new disaccharide is activated and coupled in the same way to allow extension of the oligosaccharide chain.

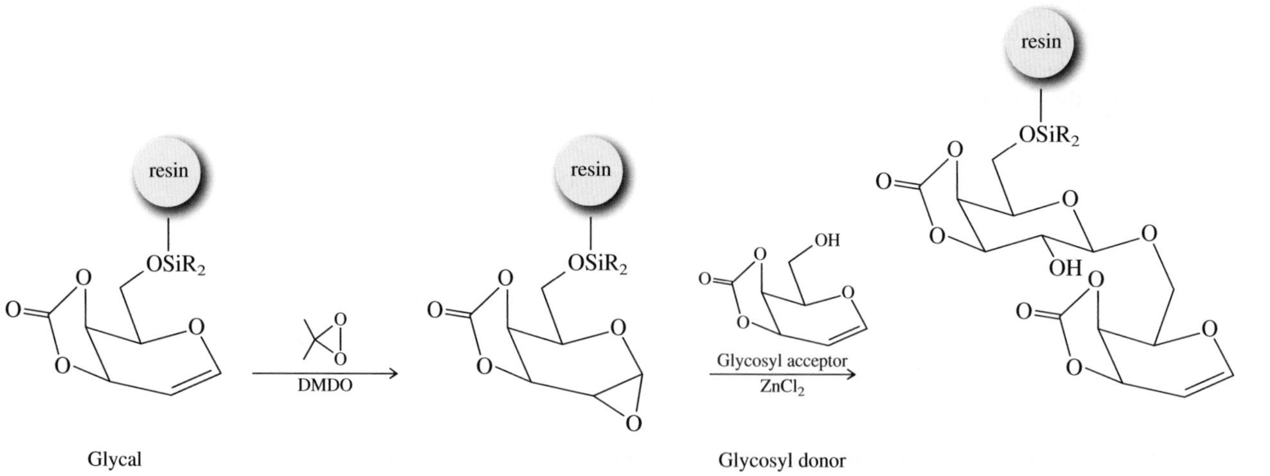

(a) Show a mechanism for the reaction of the glycosyl donor with the glycosyl acceptor. (Hint: Zn^{2+} acts as a catalyst.)

(b) Explain the regioselectivity of the reaction in (a).

Descriptive Passage and Interpretive Problems 26

Oligonucleotide Synthesis

In Section 26.6 we noted that synthetic oligonucleotides of defined sequence were commercially available for use as primers for PCR and as probes for cloning DNA. Here we will examine how these oligonucleotides are prepared.

The method bears many similarities to the Merrifield solid-phase synthesis of peptides. A starter unit is attached to a solid support, nucleosides are attached one-by-one until the sequence is complete, whereupon the target oligonucleotide is removed from the support and purified.

Like solid-phase peptide synthesis, the preparation of oligonucleotides relies heavily on protecting groups and bond-forming methods.

The starter units are nucleosides in which amine groups on the DNA bases have been protected by acylation.

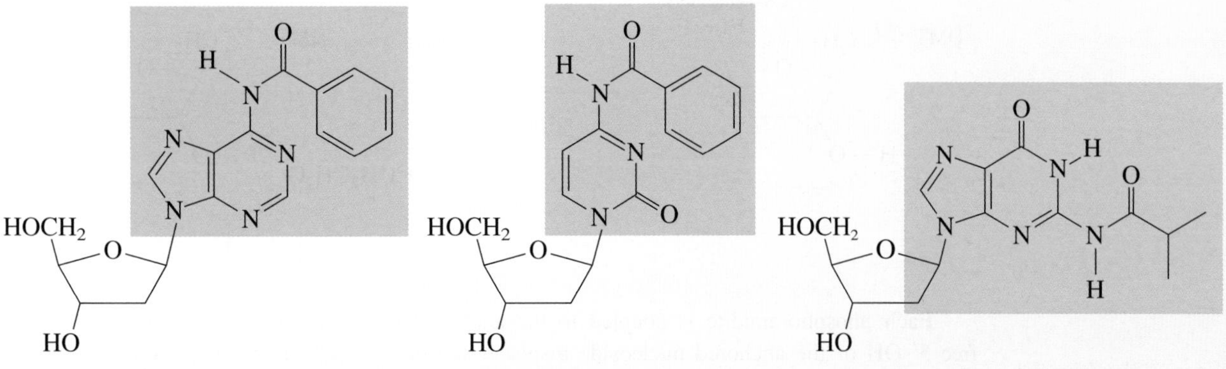
HOCH2
HO
N-Benzoyl-protected 2′-deoxyadenosine
N-Benzoyl-protected 2′-deoxycytidine
N-2-Methylpropanoyl-protected 2′-deoxyguanosine

Thymidine lacks an —NH_2 group, so needs no protecting group on its pyrimidine base.

These N-protecting groups remain in place throughout the synthesis. They are the first ones added and the last ones removed. None of the further "chemistry" that takes place involves the purine or pyrimidine rings.

The 5′-OH group of the 2′-deoxyribose portion of the nucleosides is primary and more reactive toward ether formation than the 3′-OH group, which is secondary. This difference allows selective protection of the 5′-OH as its 4,4′-dimethoxytriphenylmethyl (DMT) ether.

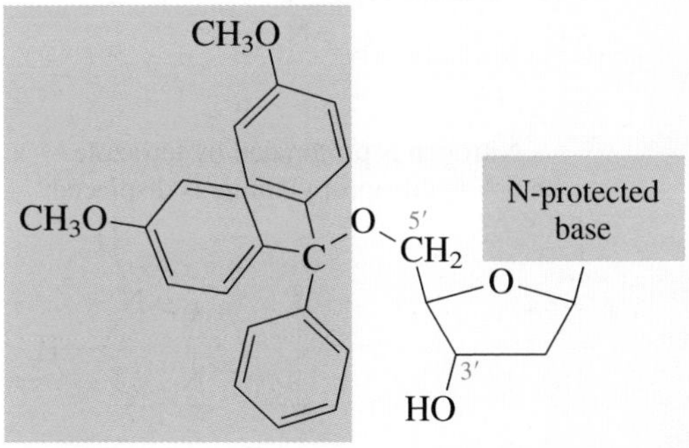
CH3O
CH3O
N-protected base
HO

The nucleoside that is to serve as the 3′ end of the final oligonucleotide is attached to a controlled-pore glass (CPG) bead by ester formation between its unprotected 3′-OH and a linker unit already attached to the CPG. In order for chain elongation to proceed in the 3′→5′ direction, the DMT group that protects the 5′-OH of the starter unit is removed by treatment with dichloroacetic acid.

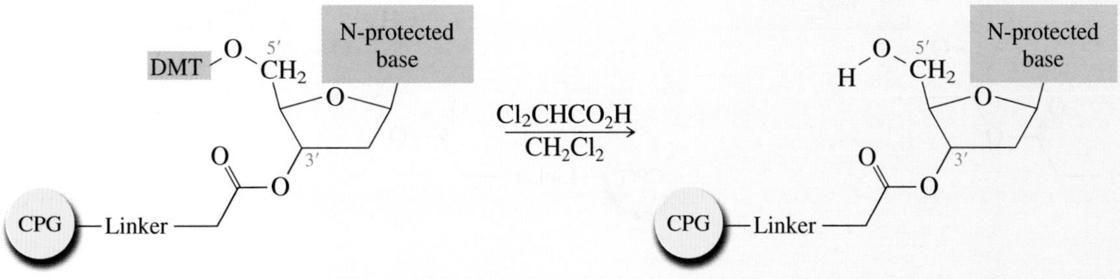
DMT
N-protected base
Cl2CHCO2H
CH2Cl2
CPG
Linker
N-protected base
CPG
Linker

The stage is now set for adding the second nucleoside. The four blocked nucleosides prepared earlier are converted to their corresponding 3′-phosphoramidite derivatives. An appropriate A, C, T, or G phosphoramidite is used in each successive stage of the elongation cycle.

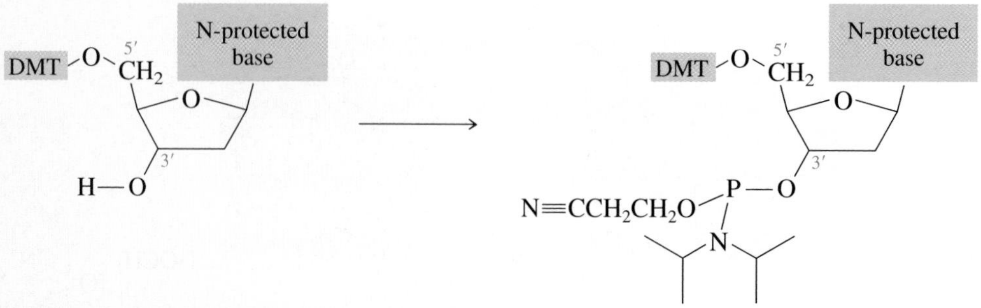

Each phosphoramidite is coupled to the anchored nucleoside by a reaction in which the free 5′-OH of the anchored nucleoside displaces the diisopropylamino group from phosphorus (Figure 26.16). The coupling is catalyzed by tetrazole, which acts as a weak acid to protonate the diisopropylamino group.

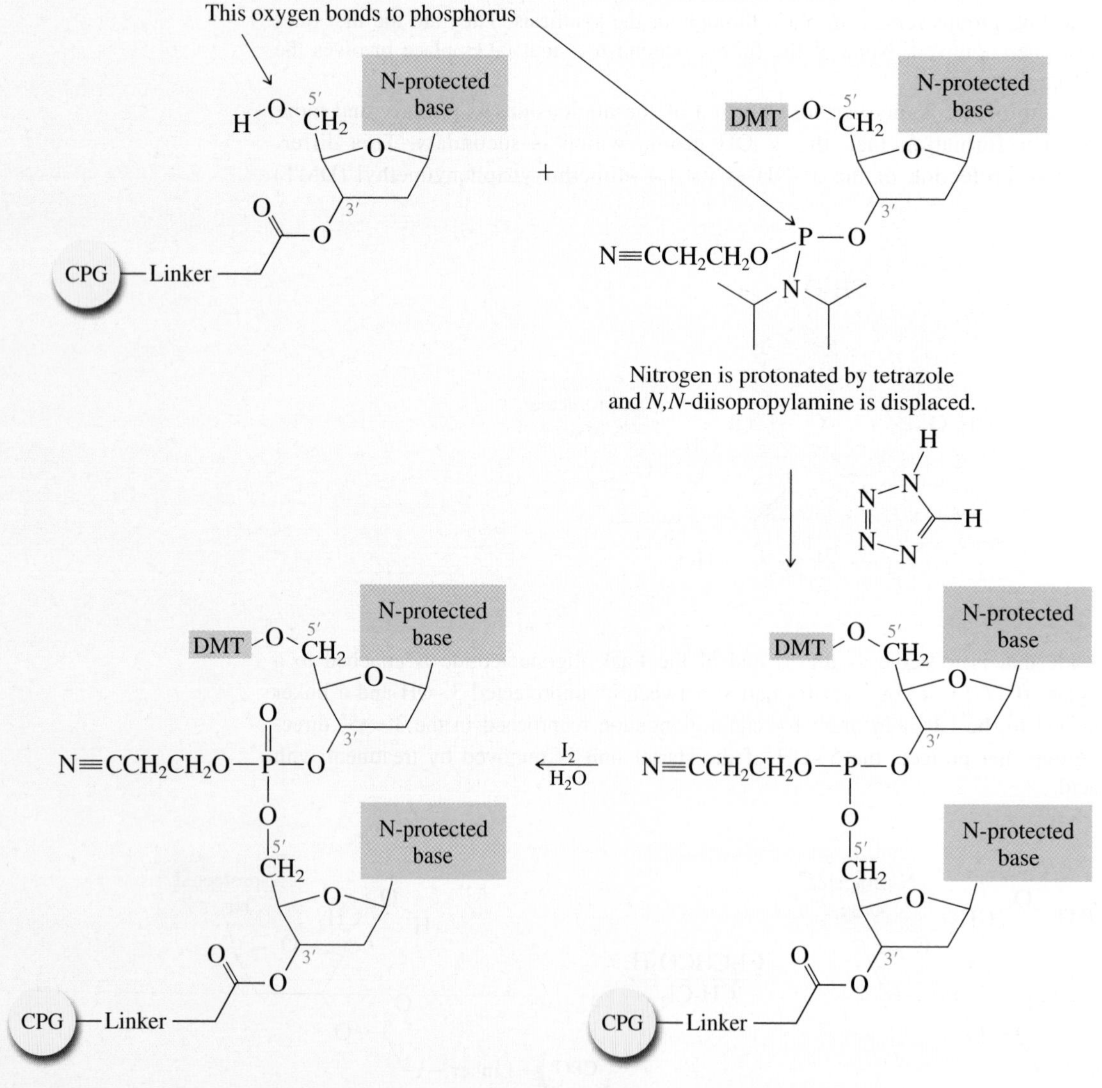

Figure 26.16

Coupling of a 3′-phosphoramidite derivative of a nucleoside to the starter unit nucleoside in a solid-phase oligonucleotide synthesis. Following the coupling, the resulting phosphite is oxidized to a phosphate.

The product of the coupling is a phosphite; it has the general formula $P(OR)_3$. It is oxidized to phosphate $[P(O)(OR)_3]$ in the last step of Figure 26.16.

The 5′-OH of the newly added nucleoside is then deprotected to prepare the bound dinucleotide for the next elongation cycle.

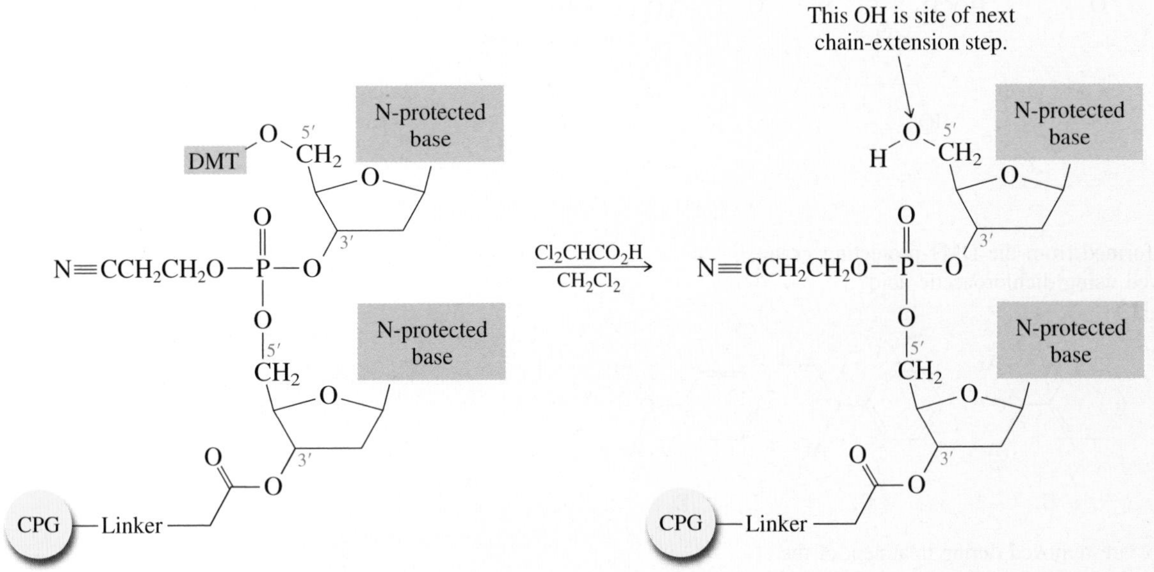

Once all the nucleosides are in place and the last DMT is removed, treatment with aqueous ammonia removes the acyl and cyanoethyl groups and cleaves the oligonucleotide from the CPG support.

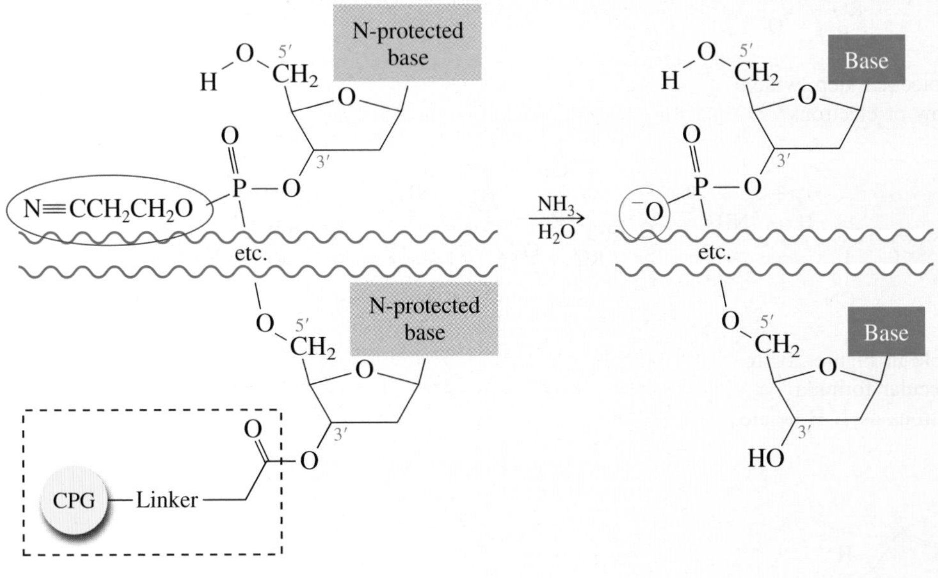

26.29 What is the product of the following reaction?

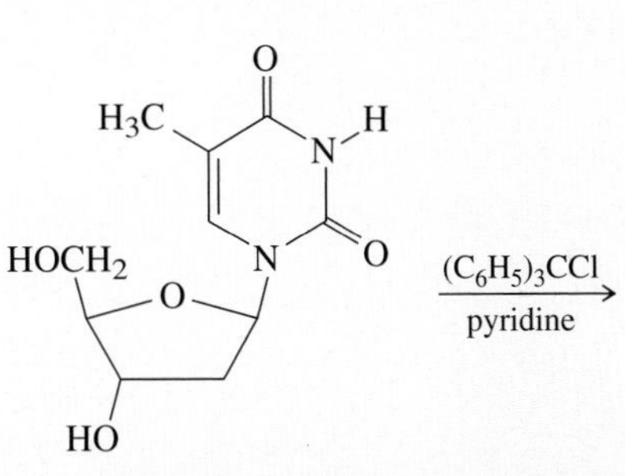

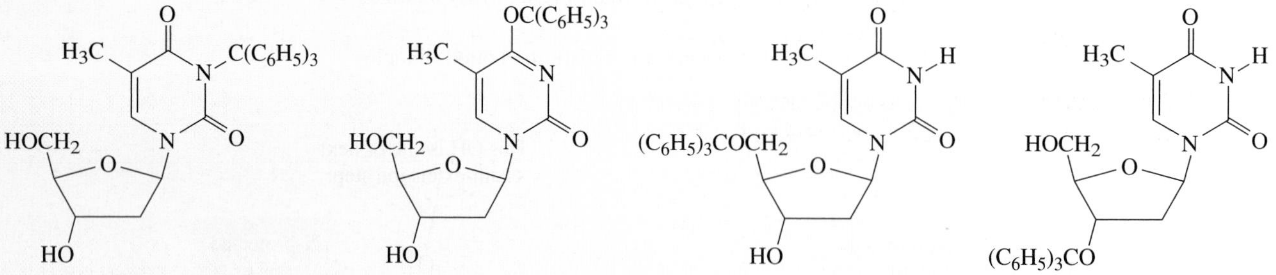

A. B. C. D.

26.30 What species is formed from the DMT-protecting group when it is removed using dichloroacetic acid? (Ar = *p*-$CH_3OC_6H_4$)

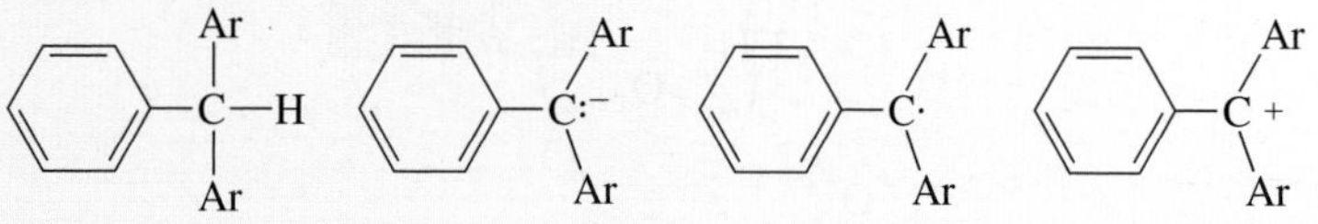

A. B. C. D.

26.31 Cyanoethyl groups are removed during treatment of the product with aqueous ammonia in the last stage of the synthesis.

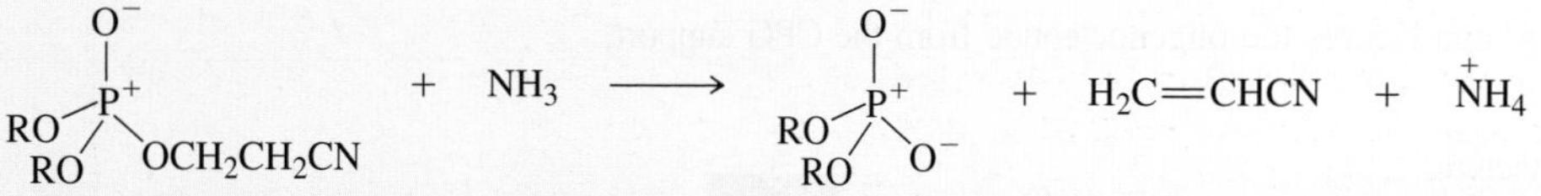

If this reaction occurs in a single bimolecular step, which of the following best represents the flow of electrons?

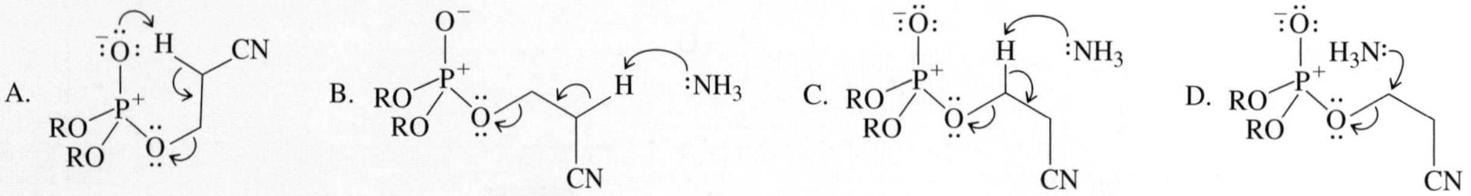

26.32 Structure **1** is the one given for tetrazole in Figure 26.16. Structures **2** and **3** have the same molecular formula (CH_2N_4) and the same number of electrons as **1**. How are these structures related?

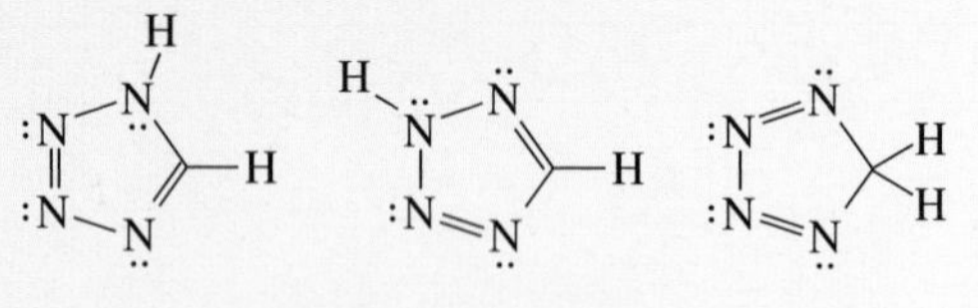

1 **2** **3**

A. **1**, **2**, and **3** are constitutional isomers.

B. **1**, **2**, and **3** are resonance contributors of the same compound.

C. **1** and **2** are resonance contributors of the same compound; **3** is an isomer of **1** and **2**.

D. **1** and **3** are resonance contributors of the same compound; **2** is an isomer of **1** and **3**.

26.33 Consider the conjugate bases of structures **1, 2,** and **3** in the preceding problem and choose the correct response.

A. **1, 2,** and **3** give different conjugate bases on deprotonation.

B. **1, 2,** and **3** give the same conjugate base on deprotonation.

C. **1** and **2** give the same conjugate base on deprotonation; the conjugate base of **3** is different.

D. **1** and **3** give the same conjugate base on deprotonation; the conjugate base of **2** is different.

26.34 Antisense oligonucleotides are a new class of synthetic drugs, one of which has been approved for use, with numerous others being developed and tested. An antisense drug is designed to have a sequence that is complementary to a portion of a messenger RNA of an organism connected with a disease. The rationale is that the oligonucleotide will bind to the mRNA and interfere with the biosynthesis of a particular protein. An antisense oligonucleotide proposed for treatment of ulcerative colitis has the sequence 5′-GCC CAA GCT GGC ATC GCT CA-3′. In the solid-phase synthesis of this drug, what nucleoside is attached to the controlled-pore glass bead?

A. A

B. T

C. C

D. G

27 Synthetic Polymers

Chapter Outline

Mechanisms

Like all soccer balls made since the 1980s, the official ball for the 2006 World Cup was constructed of polymeric materials. Four layers of polyurethane, with a total thickness of 1.1 mm, make up the outer covering. Polyurethanes are normally prepared from a diol and a diisocyanate, such as toluene diisocyanate.

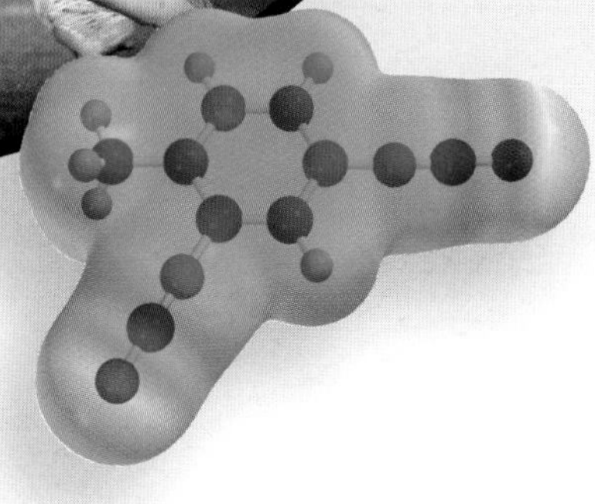

A **POLYMER** IS A substance composed of **macromolecules,** molecules that contain a very large number of atoms and have a high molecular weight. Starch, cellulose, silk, and DNA are examples of naturally occurring polymers. Synthetic polymers include nylon, polyethylene, and Bakelite, among countless others. Polymers need not be homogeneous, and most are not. Even one as simple as polyethylene is a mixture of macromolecules with different chain lengths and different degrees of branching.

This chapter is about synthetic polymers, many of which have been introduced in earlier chapters where we emphasized the connection between the reactions used to prepare polymers and the core reactions of organic chemistry. In this chapter, we will add new polymers and methods to those already introduced and expand our understanding of their synthesis, structure, and properties. As we do so, keep in mind that *the reactions used to prepare polymers are the same fundamental reactions that occur with simple organic compounds.*

27.1 Some Background

The earliest applications of polymer chemistry involved chemical modification designed to improve the physical properties of naturally occurring polymers. In 1839, Charles Goodyear transformed natural rubber, which is brittle when cold and tacky when warm, to a substance that maintains its elasticity over a wider temperature range by heating it with sulfur (vulcanization). The first synthetic fibers—called *rayons*—were made by chemical modification of cellulose near the end of the nineteenth century.

Vulcanization was summarized in the essay *Diene Polymers* in Chapter 10, p. 406.

Leo Baekeland patented the first totally synthetic polymer, which he called *Bakelite,* in 1910 (Figure 27.1). Bakelite is a versatile, durable material prepared from low-cost materials (phenol and formaldehyde) and was the most successful synthetic material of its kind for many years.

Figure 27.1

At one time, it almost always went without saying that anything plastic was made of Bakelite. Many Bakelite items are now sought after as collectibles.

These early successes notwithstanding, knowledge about polymer *structure* was meager. Most chemists believed that rubber, proteins, and the like were colloidal dispersions of small molecules. During the 1920s Hermann Staudinger, beginning at the Swiss Federal Institute of Technology and continuing at the University of Freiburg, argued that polymers were high-molecular-weight compounds held together by normal covalent bonds. Staudinger's views received convincing support in a 1929 paper by Wallace H. Carothers of Du Pont who reached similar conclusions.

Staudinger's studies of polymer structure and Carothers' achievements in polymer synthesis accelerated the development of polymer chemistry, especially its shift from

Staudinger received the 1953 Nobel Prize in Chemistry for his studies of polymers. Many believe that were it not for his untimely death in 1937, Carothers would likely have shared in the award.

chemical modification of natural polymers to the design and synthesis of new materials. Thousands of synthetic polymers are now known; some mimic the properties of natural materials, others have superior properties and have replaced natural materials.

27.2 Polymer Nomenclature

A monomer is any compound from which a polymer can be prepared.

Although the IUPAC has set forth rules for naming polymers according to structure, an alternative IUPAC *source-based* system that names polymers according to the **monomers** from which they are prepared is more widely used.

Source-based names are, for example, the ones we are accustomed to seeing for polymers such as polyethylene (see Section 6.21) and polystyrene (see Section 11.17). When the name of the monomer is a single word, the polymer derived from it is generated by simply adding the prefix *poly-*. When the name of the monomer consists of two words, both words are enclosed in parentheses immediately following *poly*. Thus, polyacrylonitrile and poly(vinyl chloride) are the polymers of acrylonitrile and vinyl chloride, respectively.

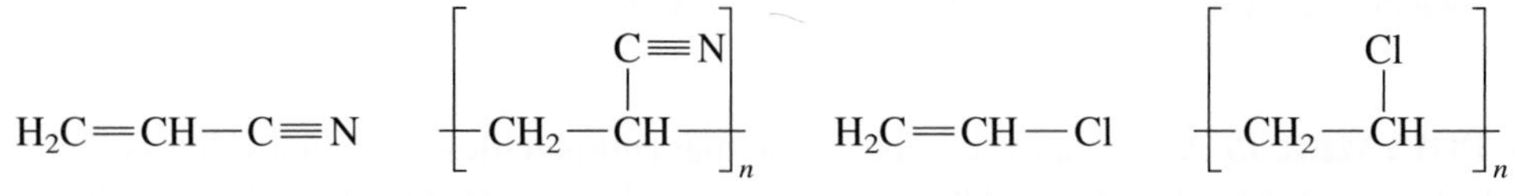

Acrylonitrile Polyacrylonitrile Vinyl chloride Poly(vinyl chloride)

The convention for writing polymer formulas is to enclose the **repeating unit** within brackets, followed by the letter n to indicate that the number of repeating units is not specified. It is, however, assumed to be large.

Problem 27.1

Structural formulas for acrylic and methacrylic acids are shown at the right. Give the names of the polymers requested in (a) and (b) and represent their structures in the bracketed repeating unit format.

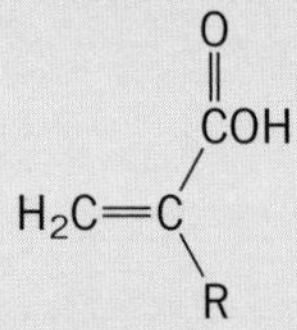

R = H; Acrylic acid

R = CH_3; Methacrylic acid

(a) The amide of acrylic acid (acrylamide)

(b) The methyl ester of methacrylic acid (methyl methacrylate)

Sample Solution (a) *Acrylamide* is one word; therefore, its polymer is *polyacrylamide*. The repeating unit follows the pattern illustrated for polyacrylonitrile and poly(vinyl chloride).

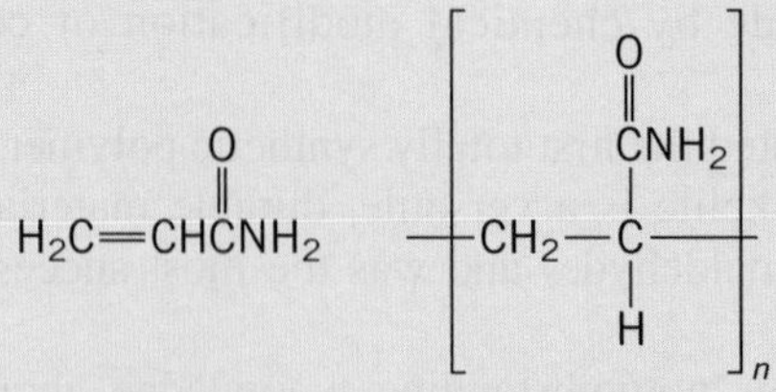

Acrylamide Polyacrylamide

Source-based nomenclature does not require that a particular polymer actually be made from the "source" monomer. Both poly(ethylene glycol) and poly(ethylene oxide), for example, are made from ethylene oxide and have the same repeating unit.

$$\left[CH_2CH_2O \right]_n$$

The structural difference between the two is that the value of n is larger for poly(ethylene oxide) than for poly(ethylene glycol). Therefore, their physical properties are different and they are known by different source-based names.

Many polymers are routinely referred to by their common names or trade names. The polymer $\left[CF_2CF_2 \right]_n$ is almost always called Teflon rather than polytetrafluoroethylene.

27.3 Classification of Polymers: Reaction Type

Structure, synthesis, production, and applications of polymers span so many disciplines that it is difficult to classify them in a way that serves every interest. Figure 27.2 compares some of the different ways. This section describes how polymers are classified according to the type of reaction—addition or condensation—that occurs.

Addition polymers are formed by reactions of the type:

$$A + B \longrightarrow A{-}B$$

where the product (A—B) retains all of the atoms of the reactants (A + B). In the general equation, A and B are monomers that react to give the polymer. When A = B, the resulting polymer is a **homopolymer.** Polystyrene is an example of a homopolymer.

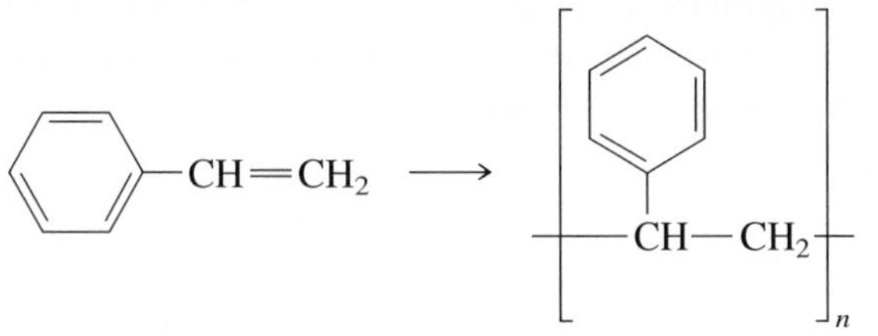

Styrene Polystyrene

When the two monomers are different, the polymer is a **copolymer.** Saran, used as a protective wrap for food, is a copolymer of vinylidene chloride and vinyl chloride.

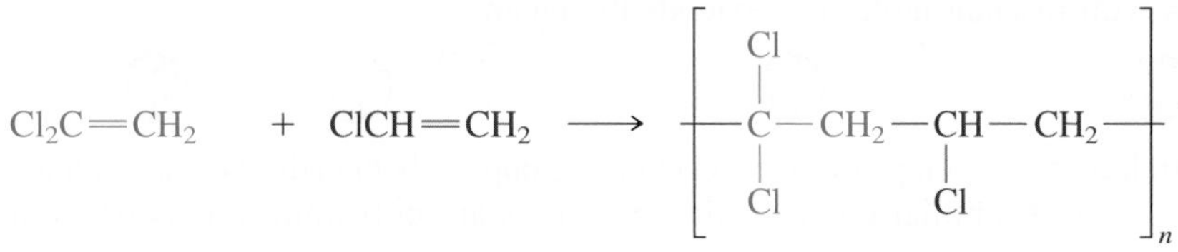

Vinylidene chloride Vinyl chloride Saran

The two components in a copolymer need not be present in equal-molar amounts. In a typical Saran formulation vinylidene chloride is the major monomer (about 85%), and vinyl chloride the minor one.

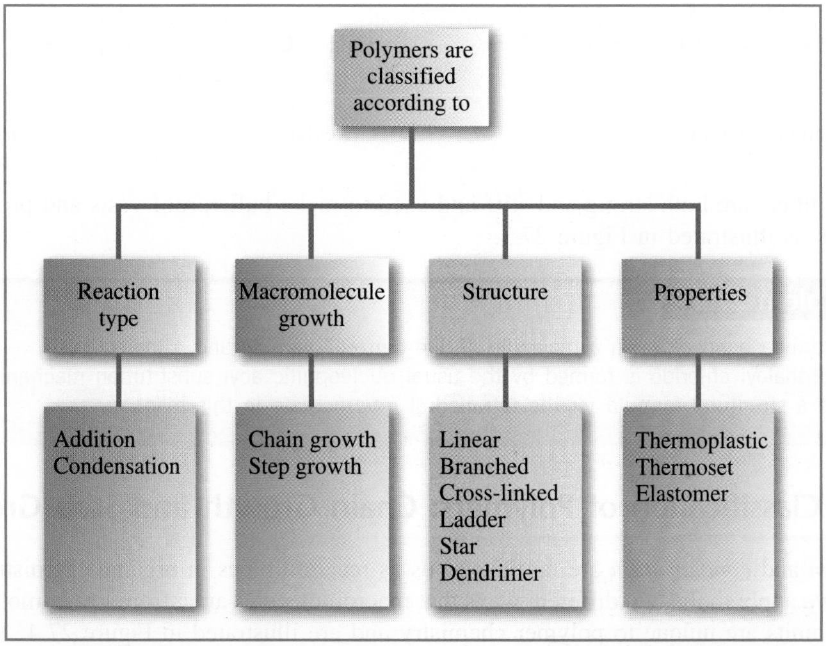

Figure 27.2

Classification of polymers.

Polymers prepared from alkenes (olefins), regardless of whether they are homopolymers or copolymers, are known as **polyolefins** and are the most familiar addition polymers.

Not all addition polymers are polyolefins. Formaldehyde, for example, polymerizes to give an addition polymer that retains all of the atoms of the monomer.

$$H_2C{=}O \rightleftharpoons \left[CH_2{-}O \right]_n$$

Formaldehyde — Polyformaldehyde

When monomeric formaldehyde is needed, to react with a Grignard reagent, for example, it is prepared as needed by heating the polymer in order to "depolymerize" it.

Problem 27.2

Under certain conditions formaldehyde forms a cyclic trimer ($C_3H_6O_3$) called *trioxane*. Suggest a structure for this compound.

Condensation polymers are prepared by covalent bond formation between monomers, accompanied by the loss of some small molecule such as water, an alcohol, or a hydrogen halide. The condensation reaction:

gives a condensation polymer when applied to difunctional reactants. The first condensation step:

gives a product that has reactive functional groups. Condensation of these functional groups with reactant molecules extends the chain.

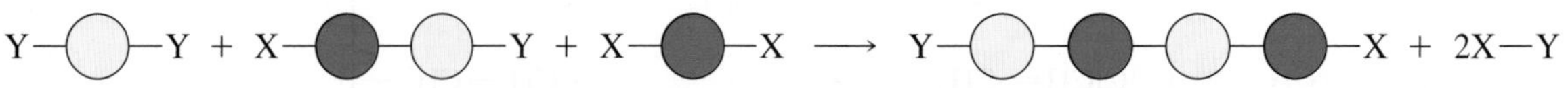

The product retains complementary functional groups at both ends and can continue to grow.

The most familiar condensation polymers are polyamides, polyesters, and polycarbonates.

The **aramids,** polyamides in which the amide bonds join aromatic rings, are one class of condensation polymer. Heating 1,4-benzenediamine and the acyl chloride of benzene-1,4-dicarboxylic acid (terephthalic acid) gives the aramid *Kevlar* with loss of hydrogen chloride.

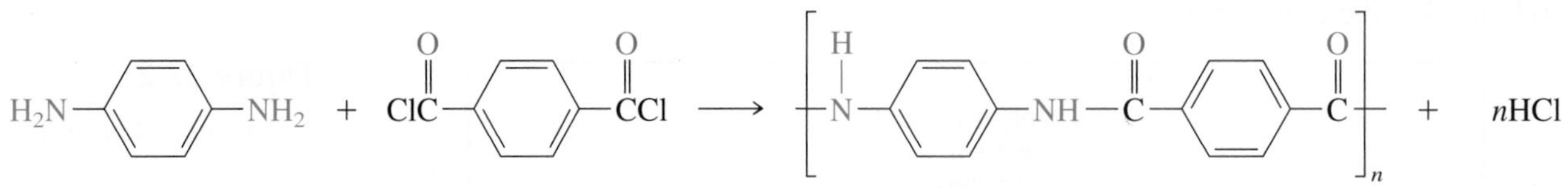

1,4-Benzenediamine — Terephthaloyl chloride — Kevlar — Hydrogen chloride

Kevlar fibers are both strong and stiff and used to make bulletproof vests and protective helmets as illustrated in Figure 27.3.

Figure 27.3

Police and the military depend on body armor and helmets made of Kevlar fibers. Kevlar protective equipment is more effective than steel, yet far lighter in weight.

Problem 27.3

The amide bond between a molecule of 1,4-benzenediamine and a molecule of terephthaloyl chloride is formed by the usual nucleophilic acyl substitution mechanism. Write a structural formula for the tetrahedral intermediate in this reaction.

27.4 Classification of Polymers: Chain Growth and Step Growth

Addition and *condensation* are familiar to us as reaction types in organic chemistry. The terms we apply to the two different ways that macromolecules arise from lower-molecular-weight units are unique to polymer chemistry and are illustrated in Figure 27.4.

(*a*) **Chain growth:** Monomers add one-by-one to the same end of a growing chain.

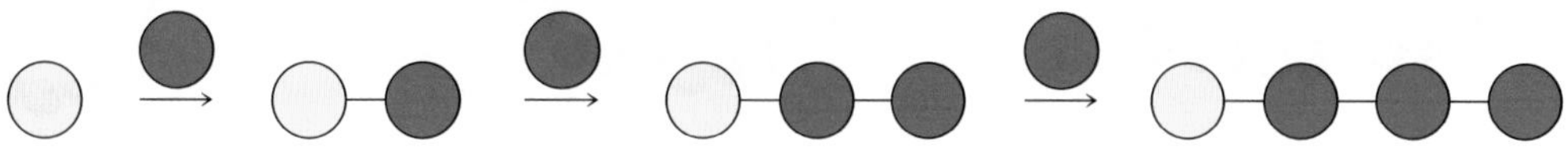

(*b*) **Step growth:** A mixture of polymers of intermediate length (oligomers) form. These oligomers react together to give longer chains.

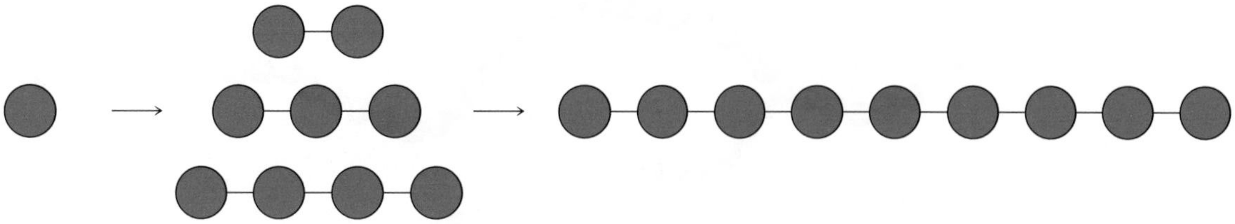

Figure 27.4

Chain-growth (*a*) and step-growth (*b*) polymerization. During chain growth, the amount of monomer remaining decreases gradually. In step growth, most of the monomer is consumed early and the molecular weight of the polymer increases as oligomers combine to form longer chains.

In a **chain-growth** process monomers add one-by-one to the same end of a growing chain (Figure 27.4*a*). Each chain has only one growth point. The concentration of monomer decreases gradually until it is depleted.

In a **step-growth** process (Figure 27.4*b*), chains have at least two growth points. Most of the monomer molecules are consumed early in the process to give a mixture of compounds of intermediate molecular weight called **oligomers.** These oligomers react with one another to form the polymer. The molecular weight continues to increase even after all the monomer molecules have reacted.

In general, chain growth is associated with addition polymerization and step growth with condensation polymerization. It is not always so, however. We'll see an example later in this chapter of an addition polymer in which step growth, not chain growth, characterizes macromolecule formation.

The terms *chain growth* and *step growth* are attributed to Paul Flory who was awarded the 1974 Nobel Prize in Chemistry for his studies on the physical chemistry of polymers.

Problem 27.4

We can anticipate this "later in the chapter" example by examining the reaction:

$$ROH + R'N{=}C{=}O \longrightarrow RO\overset{\overset{O}{\|}}{C}NHR'$$

Is this an addition reaction or a condensation?

27.5 Classification of Polymers: Structure

Polymers made from the same compounds can have different properties depending on how they are made. These differences in physical properties result from differences in the overall *structure* of the polymer chain. The three major structural types—linear, branched, and cross-linked—are illustrated in Figure 27.5. Other, more specialized, structural types—ladders, stars, and dendrimers—have unique properties and are under active investigation.

Linear polymers (Figure 27.5*a*) have a continuous chain of repeating units. The repeating units within the chain are subject to the usual conformational requirements of organic chemistry. The collection of chains can range from *random,* much like a bowl

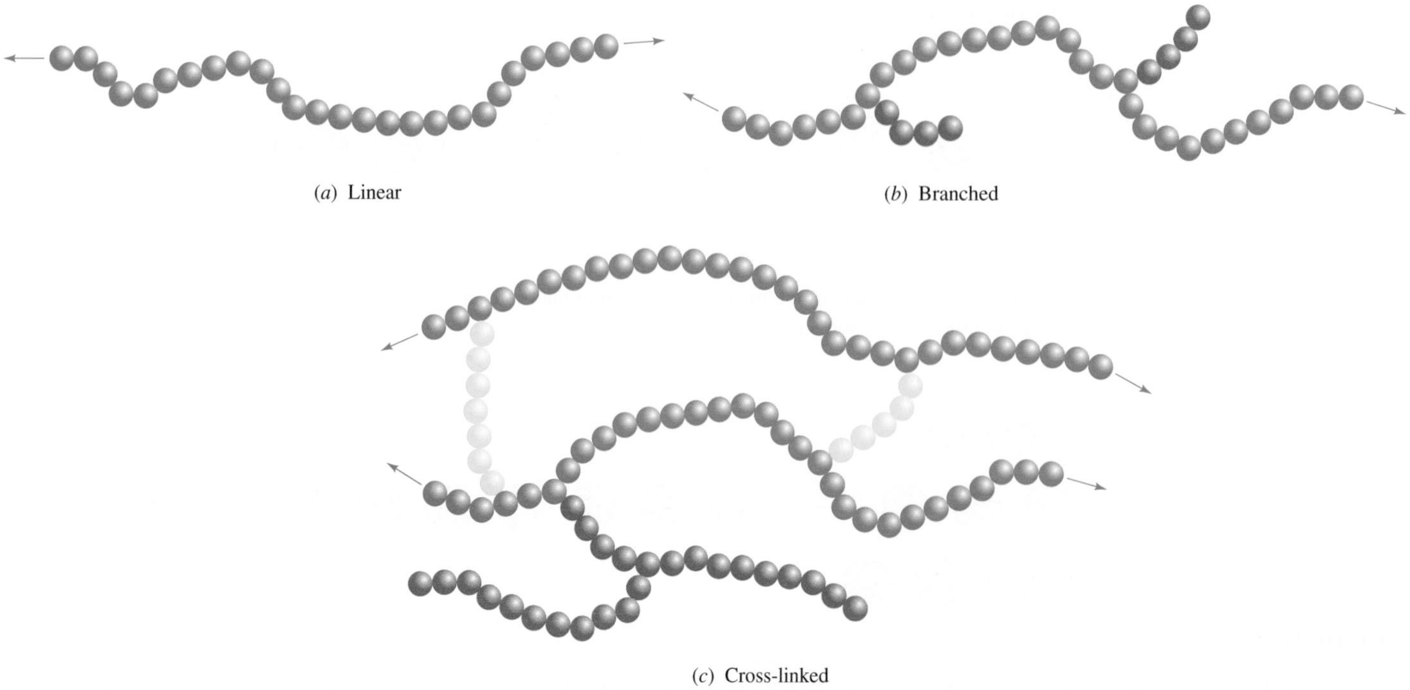
(a) Linear
(b) Branched
(c) Cross-linked

Figure 27.5

(*a*) A linear polymer has a continuous chain. (*b*) A branched polymer has relatively short branches connected to the main chain. (*c*) A cross-linked polymer has covalently bonded linking units between chains. The main chains are shown in blue, the branches in red, and the cross links in yellow.

of spaghetti, to *ordered*. We describe polymers at the random extreme as *amorphous* and those at the ordered extreme as *crystalline*.

Most polymers are a mixture of random tangles interspersed with crystalline domains called **crystallites** (Figure 27.6). The degree of crystallinity of a polymer, that is, the percentage of crystallites, depends on the strength of intermolecular forces between chains.

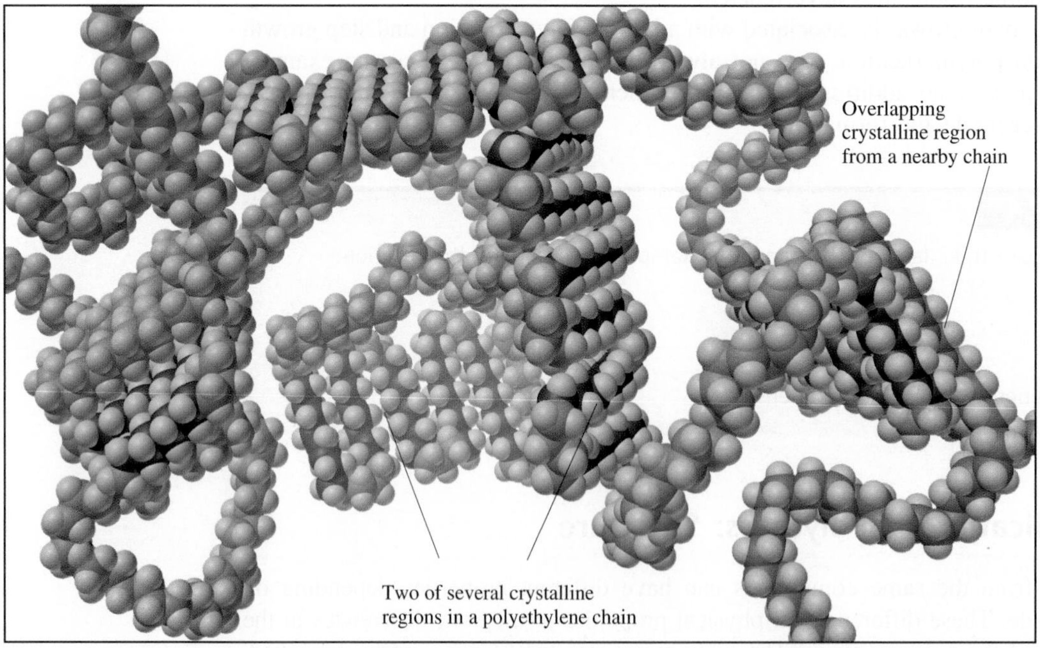
Overlapping crystalline region from a nearby chain
Two of several crystalline regions in a polyethylene chain

Figure 27.6

Polyethylene contains both randomly coiled (amorphous) and ordered (crystalline) regions. The ordered regions (crystallites) of one chain are shown in a darker color than the random main chain. Crystallites involving the main chain with neighboring ones are in red and yellow. Reprinted, with permission, from M. Silberberg, *Chemistry,* 5th ed., McGraw-Hill Higher Education, 2009. p. 486.

For a particular polymer, density increases with crystallinity because randomly coiled chains consume volume, while closer packing puts the same mass into a smaller volume. The efficiency with which the chains can pack together is strongly affected by the extent to which the chain is branched.

Branched polymers (Figure 27.5*b*) have branches extending from the main chain. In general, increased branching reduces the crystallinity of a polymer and alters properties such as density.

Contrast the properties of low-density polyethylene (LDPE) and high-density (HDPE), two of the six polymers familiar enough to have their own identifying codes for recycling (Table 27.1). Both are homopolymers of ethylene, but are prepared by different methods and have different properties and uses. As their names imply, LDPE has a lower density than HDPE (0.92 g/cm^3 versus 0.96 g/cm^3). LDPE is softer, HDPE more rigid. LDPE has a lower melting point than HDPE. LDPE is the plastic used for grocery store bags; HDPE is stronger and used for water bottles, milk jugs, and gasoline tanks.

The structural difference between the two is that LDPE is more branched, averaging about 20 branches for every thousand carbon atoms compared with about 5 per thousand for HDPE. The greater density of HDPE results from packing more mass into the same volume. Unbranched chains pack more efficiently than branched ones, which translates into stronger intermolecular forces, greater crystallinity, and a tougher, more durable material.

Like HDPE, isotactic polypropylene is highly crystalline with numerous uses, including fibers for rope and carpets. Atactic polypropylene, on the other hand, is much less crystalline and has few applications.

Stereoregular polymers including isotactic polypropylene were described in Section 7.16.

Chains in a **cross-linked** or **network polymer** (Figure 27.5*c*) are connected to one another by linking units, which may be long or short and composed of the same repeating units as the main chain or different ones. Vulcanization, for example, uses sulfur to cross-link the hydrocarbon chains of natural rubber. In general, cross-linking increases rigidity by restricting the movement of the polymer chains. Vulcanized rubber is a lightly cross-linked elastomer; Bakelite can be so highly cross-linked as to be considered a single molecule.

27.6 Classification of Polymers: Properties

How a polymer responds to changes in temperature is important not only with respect to the conditions under which it can be used, but also in the methods by which it is transformed into a commercial product.

Thermoplastic polymers are the most common and are those that soften when heated. At their *glass transition temperature* (T_g), thermoplastic polymers change from a glass to a flexible, rubbery state. Past this point amorphous polymers are gradually transformed to a liquid as the temperature is raised. Crystalline polymers undergo a second transition, liquefying only when the *melting temperature* (T_m) is reached. Compare the behaviors of atactic, isotactic, and syndiotactic poly(methyl methacrylate) on being heated.

Poly(methyl methacrylate)	T_g(°C)	T_m(°C)
atactic	114	—
isotactic	48	160
syndiotactic	126	200

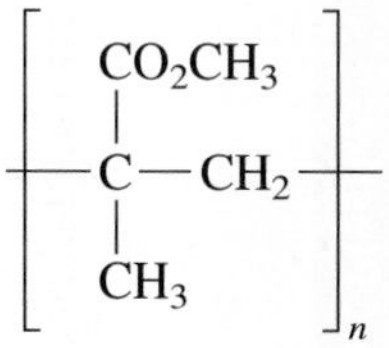

The atactic form of poly(methyl methacrylate) is amorphous and exhibits only one transition temperature (T_g). The stereoregular isotactic and syndiotactic forms are partially crystalline and undergo both a glass transition and melting.

The process that takes place at T_g is an increase in the conformational mobility of the polymer chains. At T_m, attractive forces in crystallites are broken and individual chains separate.

Melting temperature is an important factor in respect to how polymers are used. The relatively low T_m for low-density polyethylene (115°C) makes it an easy polymer to cast into the desired shape when melted, but at the same time limits its applications.

TABLE 27.1 Recycling of Plastics

Symbol	Polymer	Some uses* New	Some uses* Recycled
1 PETE	Poly(ethylene terephthalate)	Polyester textile fibers, tire cords, photographic film, soft drink and water bottles, food jars	Carpet fibers, detergent bottles, bathtubs, car parts, audio- and videotapes
2 HDPE	High-density polyethylene	Bottles, automobile fuel tanks, milk jugs, bags, cereal box liners	Plastic lumber for exterior uses (picnic tables, mailboxes, decks, trash bins, planters)
3 V	Poly(vinyl chloride)	Floor tiles, vinyl siding, plumbing pipe, gutters and downspouts, garden hoses, shower curtains, window frames, blister packs	Many of the uses of recycled poly(vinyl chloride) are the same as those of new material
4 LDPE	Low-density polyethylene	Trash bags, packaging, squeezable bottles, grocery bags	Packaging film and bags
5 PP	Polypropylene	Indoor–outdoor carpet, rope, medicine bottles, packaging	Indoor–outdoor carpet, rope, fishing nets, tarpaulins, auto parts
6 PS	Polystyrene	Television cabinets, luggage, egg cartons, toys, Styrofoam cups, appliances	Styrofoam insulation and packaging, coat hangers, containers
7 OTHER	Other (acrylics, nylon, polycarbonates, etc.)	5-Gallon reusable water bottles, automobile bumpers and other parts, tires, telephones, safety helmets	

*The uses of new and recycled plastics are often the same, and many products are a mixture of new and recycled material.

When, for example, a container is required that must be sterilized by heating, the higher T_m of HDPE (137°C) makes it a better choice than LDPE.

Unlike thermoplastic polymers that soften on heating, **thermosetting polymers** (also called *thermosetting resins*) pass through a liquid state then solidify ("cure") on continued heating. The solidified material is a **thermoset.** It is formed by irreversible chemical reactions that create cross links as the thermosetting polymer is heated. *Bakelite,* a highly cross-linked thermoset made from phenol and formaldehyde, is prepared in two stages. In the first stage, condensation between phenol and formaldehyde gives a polymer, which, in its fluid state, is cast in molds and heated, whereupon it solidifies to a hard, rigid mass. The chemical reactions that form the fluid polymer and the solid thermoset are the same kind of condensations; the difference is that there are more cross links in the thermoset. *Melamine* (used in plastic dinnerware) is another example of a thermoset.

Elastomers are flexible polymers that can be stretched but return to their original state when the stretching force is released. Most amorphous polymers become rubbery beyond their glass transition temperature, but not all rubbery polymers are elastic. Cross links in elastomers limit the extent to which elastomers can be deformed then encourage them to return to their original shape when they are relaxed.

27.7 Addition Polymers: A Review and a Preview

Addition polymers are most familiar to us in connection with the polymerization of alkenes.

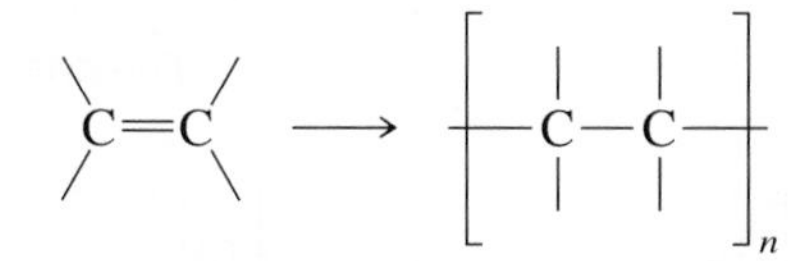

Table 27.2 reviews alkene polymerizations that proceed by free radicals and by coordination complexes of the Ziegler–Natta type. Both are chain-growth processes; their propagation steps were outlined in Mechanism 6.11 (page 265) and Mechanism 14.5 (page 635), respectively. The present section examines two other significant factors in alkene polymerization: initiation and termination.

Initiators of Alkene Polymerization: Whether free-radical or coordination polymerization occurs depends primarily on the substance used to initiate the reaction. Free-radical polymerization occurs when a compound is present that undergoes homolytic bond cleavage when heated. Two examples include

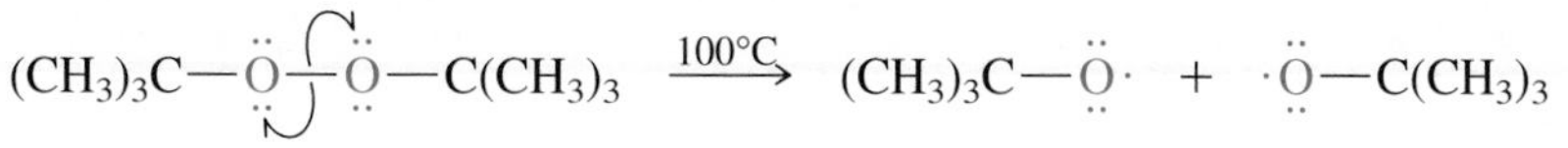

Di-*tert*-butyl peroxide   Two *tert*-butoxy radicals

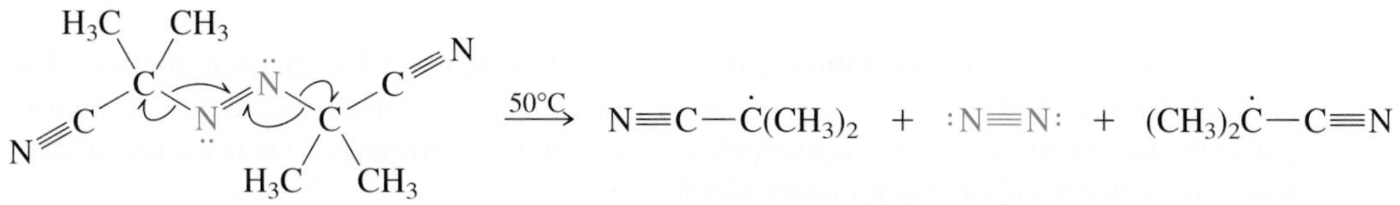

Azobisisobutyronitrile (AIBN)   Two 1-cyano-1-methylethyl radicals + nitrogen

Problem 27.5

(a) Write a chemical equation for the reaction in which *tert*-butoxy radical adds to vinyl chloride to initiate polymerization. Show the flow of electrons with curved arrows.

(b) Repeat part (a) for the polymerization of styrene using AIBN as an initiator.

Sample Solution (a) *tert*-Butoxy radical adds to the CH_2 group of vinyl chloride. The free radical formed in this process has its unpaired electron on the carbon bonded to chlorine.

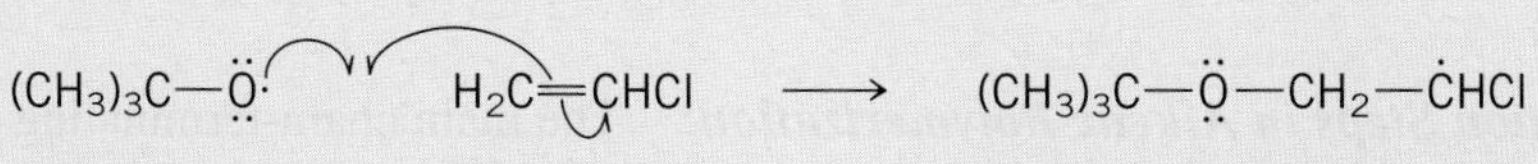

tert-Butoxy radical   Vinyl chloride   2-*tert*-Butoxy-1-chloroethyl radical

TABLE 27.2 Summary of Alkene Polymerizations Discussed in Earlier Chapters

Reaction (section) and comments	Example
Free-radical polymerization of alkenes (Section 6.21) Many alkenes polymerize when treated with free-radical initiators. A free-radical chain mechanism is followed and was illustrated for the case of ethylene in Mechanism 6.11 (page 264).	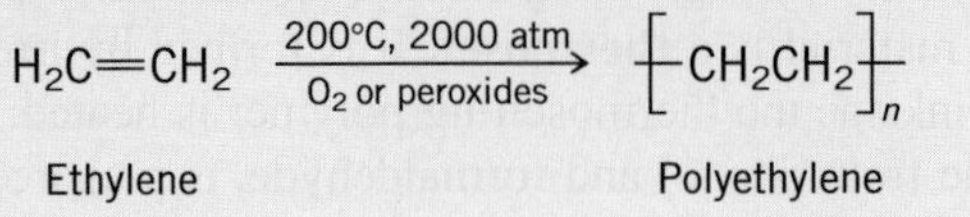200°C, 2000 atm O_2 or peroxides Ethylene Polyethylene
Free-radical polymerization of dienes (Section 10.14) Conjugated dienes undergo free-radical polymerization under conditions similar to those of alkenes. The major product corresponds to 1,4-addition.	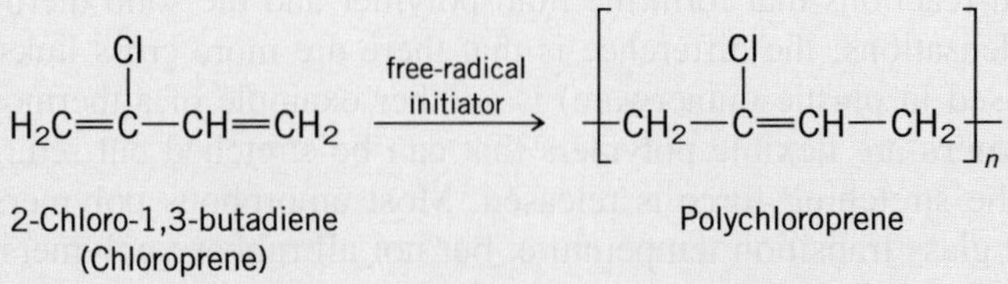free-radical initiator 2-Chloro-1,3-butadiene (Chloroprene) Polychloroprene
Free-radical polymerization of styrene (Section 11.17) Styrene can be polymerized under free-radical, cationic, anionic, and Ziegler–Natta conditions. The mechanism of the free-radical polymerization was shown in Mechanism 11.2 (page 453).	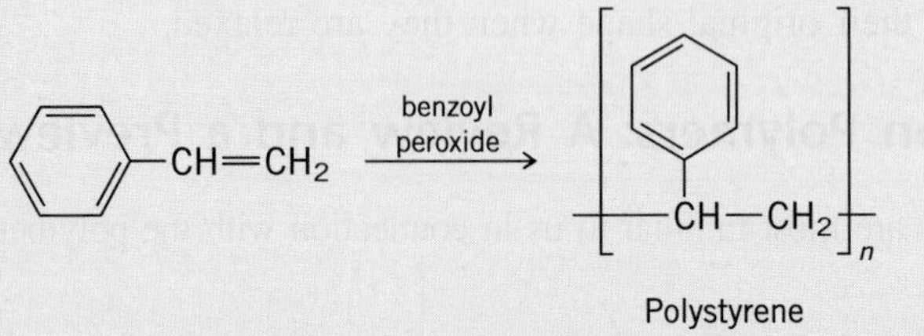benzoyl peroxide Polystyrene
Ring-opening metathesis polymerization (Section 14.15) The double bonds of strained cyclic alkenes are cleaved by certain carbene complexes of tungsten and, in the process, undergo polymerization.	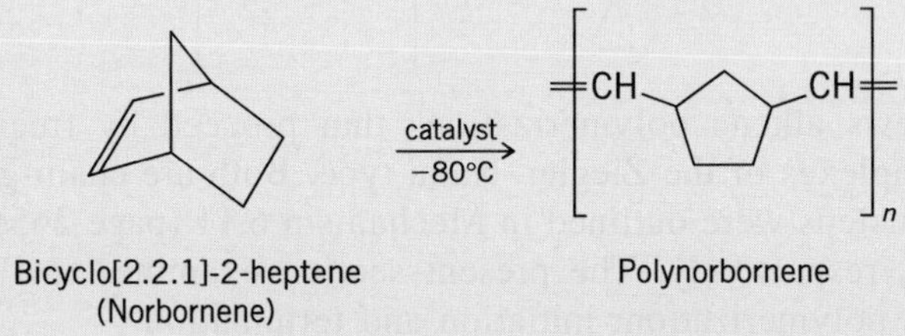catalyst −80°C Bicyclo[2.2.1]-2-heptene (Norbornene) Polynorbornene
Coordination polymerization (Section 14.16) Organometallic compounds such as bis(cyclopentadienyl)zirconium dichloride (Cp_2ZrCl_2) catalyze the polymerization of ethylene by the sequence of steps shown in Mechanism 14.5 (page 635).	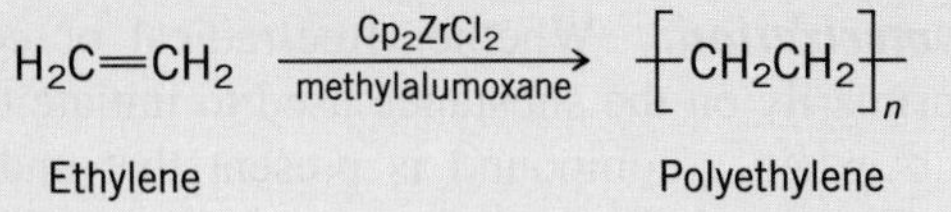Cp_2ZrCl_2 methylalumoxane Ethylene Polyethylene

Coordination polymerization catalysts are complexes of transition metals. The original Ziegler–Natta catalyst, a mixture of titanium tetrachloride and diethylaluminum chloride, has been joined by numerous organometallic complexes such as the widely used bis(cyclopentadienyl)zirconium dichloride.

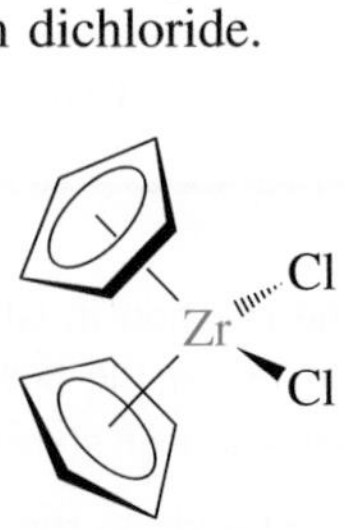
Zr
Cl
Cl

Bis(cyclopentadienyl)zirconium dichloride

Termination Steps in Alkene Polymerization: The main chain-terminating processes in free-radical polymerization are *combination* and *disproportionation*. In a combination, the pairing of the odd electron of one growing radical chain with that of another gives a stable macromolecule.

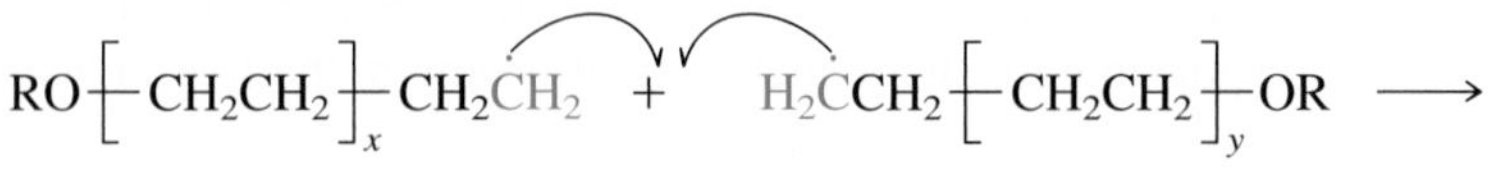

Two growing polyethylene chains

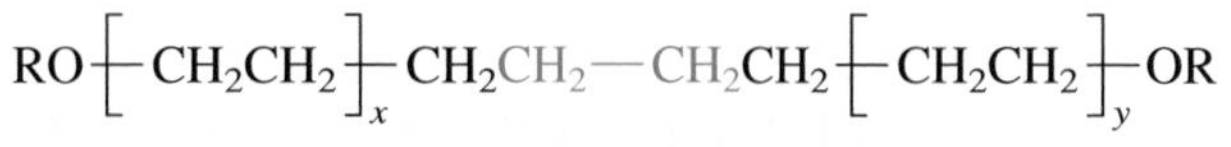

Terminated polyethylene

In disproportionation, two alkyl radicals react by hydrogen-atom transfer. Two stable molecules result; one terminates in a methyl group, the other in a double bond.

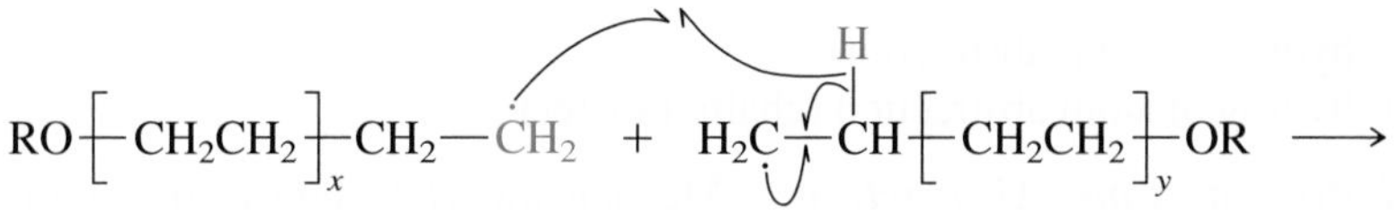

Two growing polyethylene chains

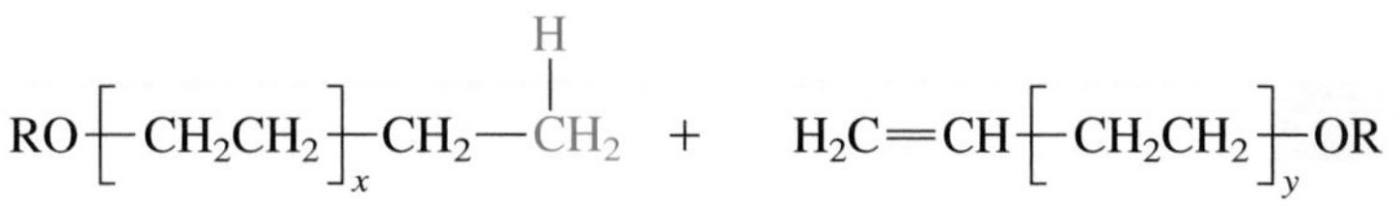

Methyl-terminated polyethylene Double-bond-terminated polyethylene

Both combination and disproportionation consume free radicals and decrease the number of growing chains. Because they require a reaction between two free radicals, each of which is present in low concentration, they have a low probability compared with chain growth, in which a radical reacts with a monomer. Combination involves only bond making and has a low activation energy; disproportionation has a higher activation energy because bond breaking accompanies bond making. Disproportionation has a more adverse effect on chain length and molecular weight than combination.

Problem 27.6

Other than combination, a macromolecule of the type $RO{+}CH_2CH_2{+}_x CH_2{-}CH_2{-}OR$ can arise by a different process, one which also terminates chain growth. Show a reasonable reaction and represent the flow of electrons by curved arrows.

Among several chain terminating reactions that can occur in coordination polymerization, a common one is an elimination in which a β-hydrogen is transferred to the metal.

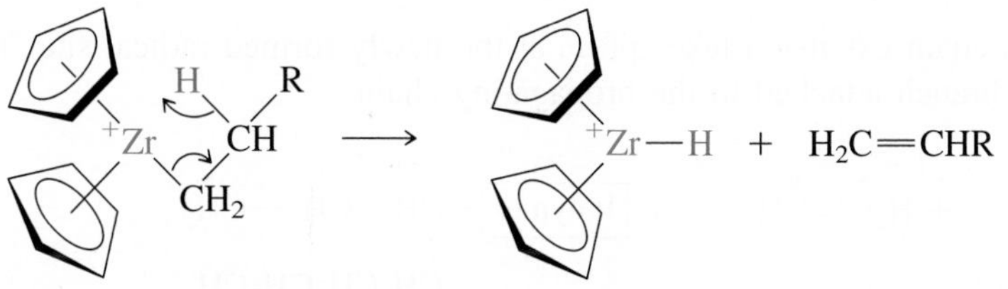

27.8 Chain Branching in Free-Radical Polymerization

Even with the same monomer, the properties of a polymer can vary significantly depending on how it is prepared. Free-radical polymerization of ethylene gives low-density polyethylene; coordination polymerization gives high-density polyethylene. The properties are different because the structures are different, and the difference in the structures comes from the mechanisms by which the polymerizations take place. Free-radical polymerization of ethylene gives a branched polymer, coordination polymerization gives a linear one.

What is the mechanism responsible for the branching that occurs in the free-radical polymerization of ethylene?

By itself, the propagation step in the free-radical polymerization of ethylene cannot produce branches.

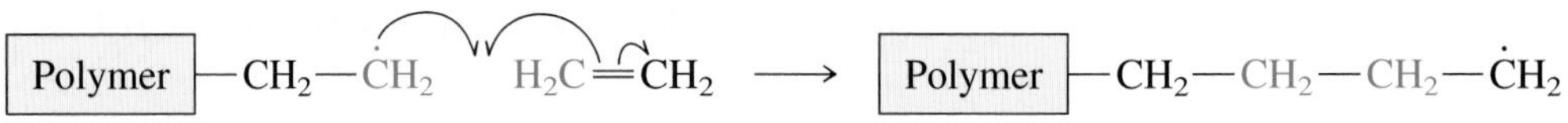

In order for the polymer to be branched, an additional process must occur involving a radical site somewhere other than at the end of the chain. The two main ways this can happen both involve hydrogen abstraction from within the polymer chain.

1. Intramolecular hydrogen atom abstraction
2. Intermolecular hydrogen atom abstraction (chain transfer)

Intramolecular Hydrogen Atom Abstraction: Mechanism 27.1 shows how intramolecular hydrogen atom abstraction can lead to the formation of a four-carbon branch.

Mechanism 27.1

Branching in Polyethylene Caused by Intramolecular Hydrogen Transfer

THE OVERALL REACTION:

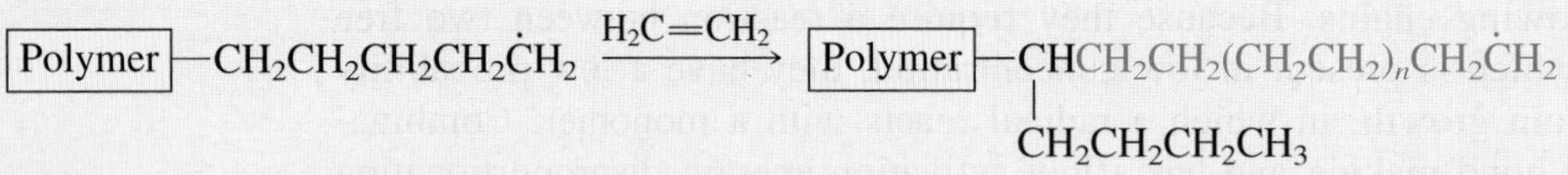

THE MECHANISM:

Step 1: The carbon at the end of the chain—the one with the unpaired electron—abstracts a hydrogen atom from the fifth carbon. The transition state is a cyclic arrangement of six atoms.

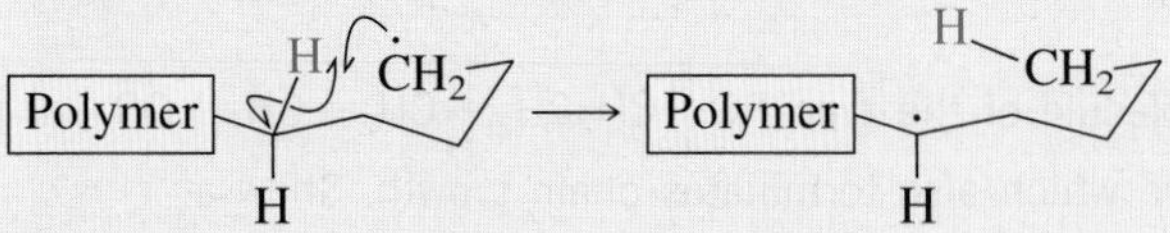

The resulting radical is secondary and more stable than the original primary radical. Therefore, the hydrogen atom abstraction is exothermic.

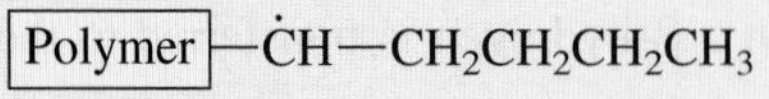

Step 2: When the radical reacts with ethylene, chain extension takes place at the newly formed radical site. The product of this step has a four-carbon branch attached to the propagating chain.

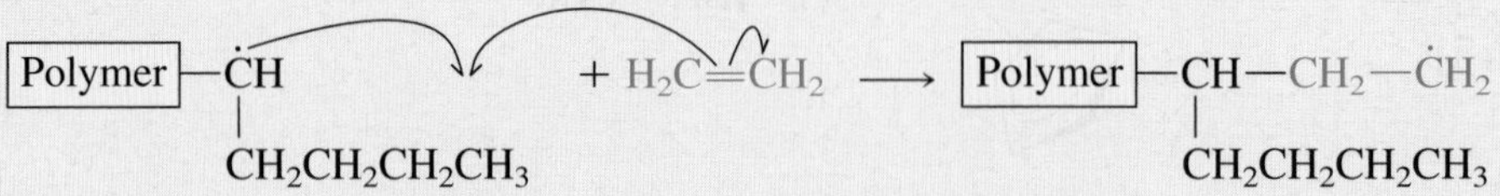

Step 3: Reaction with additional ethylene molecules extends the growing chain.

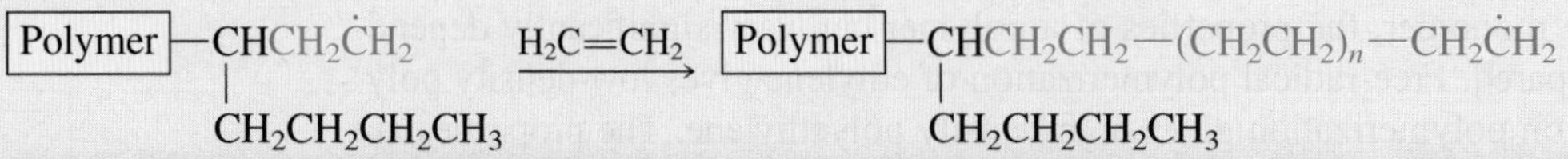

Recall that an intramolecular process takes place *within* a molecule, not *between* molecules. As the mechanism shows, the radical at the end of the growing polymer abstracts a hydrogen atom from the fifth carbon. Five carbons and one hydrogen comprise six atoms of a cyclic transition state. When a hydrogen atom is removed from the fifth carbon, a secondary radical is generated at that site. This, then, is the carbon that becomes the origin for further chain growth. Analogous mechanisms apply to branches shorter or longer than four carbons.

Problem 27.7

Suggest an explanation for the observation that branches shorter or longer than four carbons are found infrequently in polyethylene. Frame your explanation in terms of how ΔH and ΔS affect the activation energy for intramolecular hydrogen atom abstraction.

A comparable process cannot occur when Ziegler–Natta catalysts are used because free radicals are not intermediates in coordination polymerization.

Intermolecular Hydrogen Atom Abstraction (Chain Transfer): Mechanism 27.2 shows how a growing polymer chain abstracts a hydrogen atom from a terminated chain. The original growing chain is now terminated, and the original terminated chain is activated toward further growth. Chain growth, however, occurs at the branch point, not at the end of the chain. An already long chain adds a branch while terminating a (presumably shorter) growing chain. Chain transfer not only leads to branching, but also encourages disparity in chain lengths—more short chains and more long branched chains. Both decrease the crystallinity of the polymer and reduce its strength.

As in the case of intramolecular hydrogen abstraction, branching by chain transfer is not a problem when alkenes are polymerized under Ziegler–Natta conditions because free radicals are not intermediates in coordination polymerization.

Mechanism 27.2

Branching in Polyethylene Caused by Intermolecular Hydrogen Transfer

Step 1: A growing polymer chain abstracts a hydrogen atom from a terminated chain. This step terminates the growing chain and activates the terminated one.

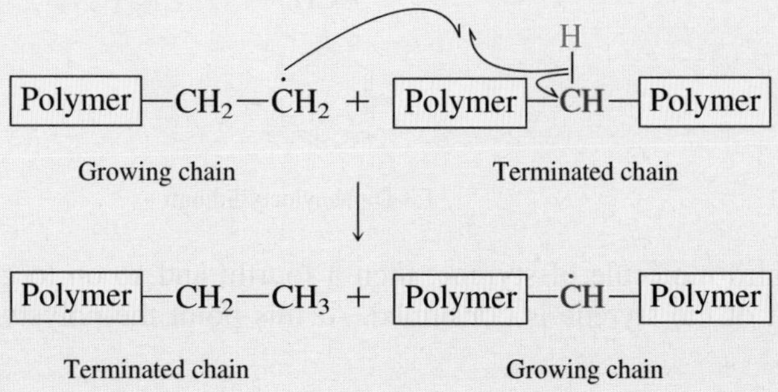

Step 2: Reaction of the new chain with monomer molecules produces a branch at which future growth occurs.

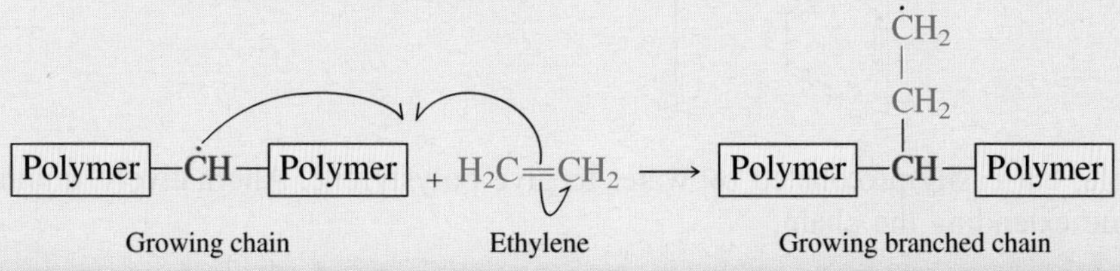

27.9 Anionic Polymerization: Living Polymers

Anionic polymerization is a useful alternative to free-radical and Ziegler–Natta procedures for certain polymers. Adding butyllithium to a solution of styrene in tetrahydrofuran (THF), for example, gives polystyrene.

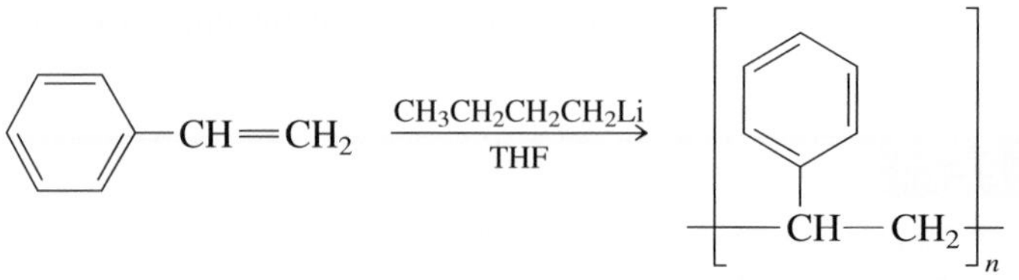

Styrene   Polystyrene

Mechanism 27.3 shows how addition of butyllithium to the double bond of styrene initiates polymerization. The product of this step is a benzylic carbanion that then adds to a second molecule of styrene to give another benzylic carbanion, and so on by a chain-growth process.

Polystyrene formed under these conditions has a narrower range of molecular weights than provided by other methods. Initiation of polymerization by addition of

Mechanism 27.3

Anionic Polymerization of Styrene

Step 1: Anionic polymerization of styrene is initiated by addition of butyllithium to the double bond. The regioselectivity of addition is governed by formation of the more stable carbanion, which in this case is benzylic.

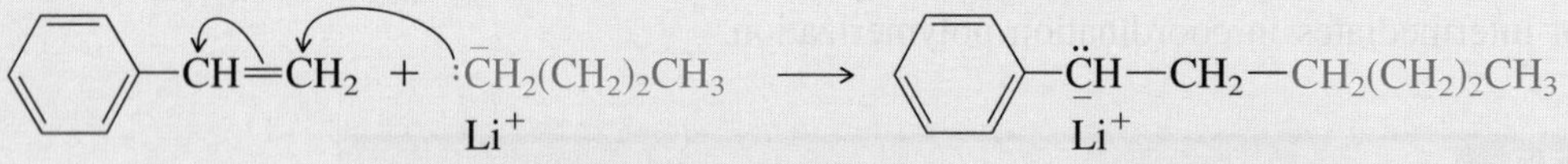

Styrene   Butyllithium   1-Phenylhexyllithium

Step 2: The product of the first step adds to a second molecule of styrene.

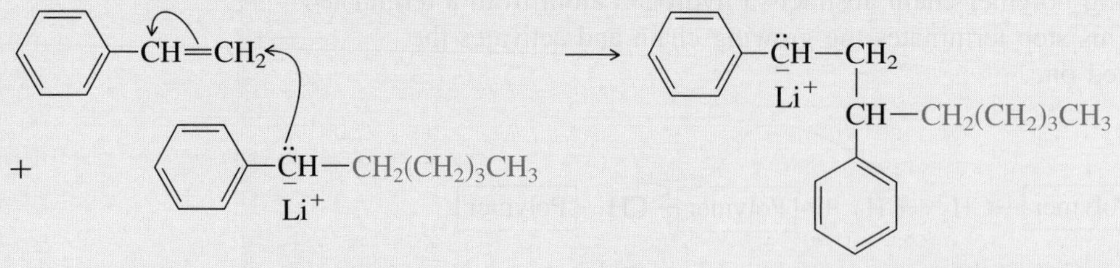

Styrene + 1-Phenylhexyllithium   1,3-Diphenyloctyllithium

Step 3: The product of the second step adds to a third molecule of styrene, then a fourth, and so on to give a macromolecule. Reaction continues until all of the styrene is consumed. At this point the polystyrene exists as an organolithium reagent.

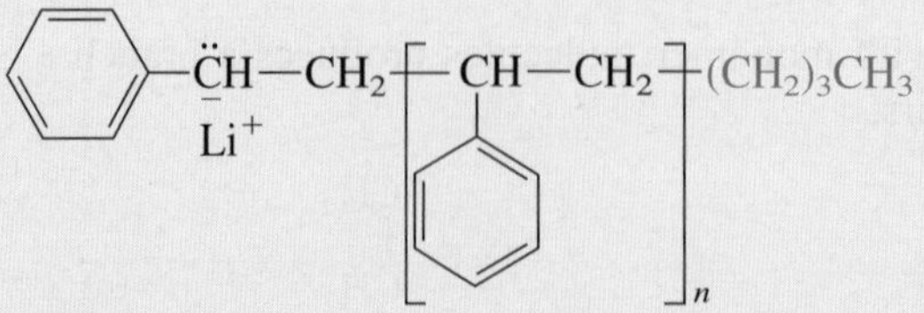

The organolithium reagent is stable, but easily protonated by water to give polystyrene. Alternatively, another monomer can be added to continue extending the chain.

butyllithium to styrene is much faster than subsequent chain growth. Thus, all the butyllithium is consumed and the number of chains is equal to the number of molecules of butyllithium used. These starter chains then grow at similar rates to produce similar chain lengths.

Problem 27.8

How will the average chain length of polystyrene vary with the amount of butyllithium used to initiate polymerization?

As shown in step 3 of Mechanism 27.3 once all of the monomer is consumed the polymer is present as its organolithium derivative. This material is referred to as a **living polymer** because more monomer can be added and anionic polymerization will continue until the added monomer is also consumed. Adding 1,3-butadiene, for example, to a living polymer of styrene gives a new living polymer containing sections ("blocks") of polystyrene and poly(1,3-butadiene).

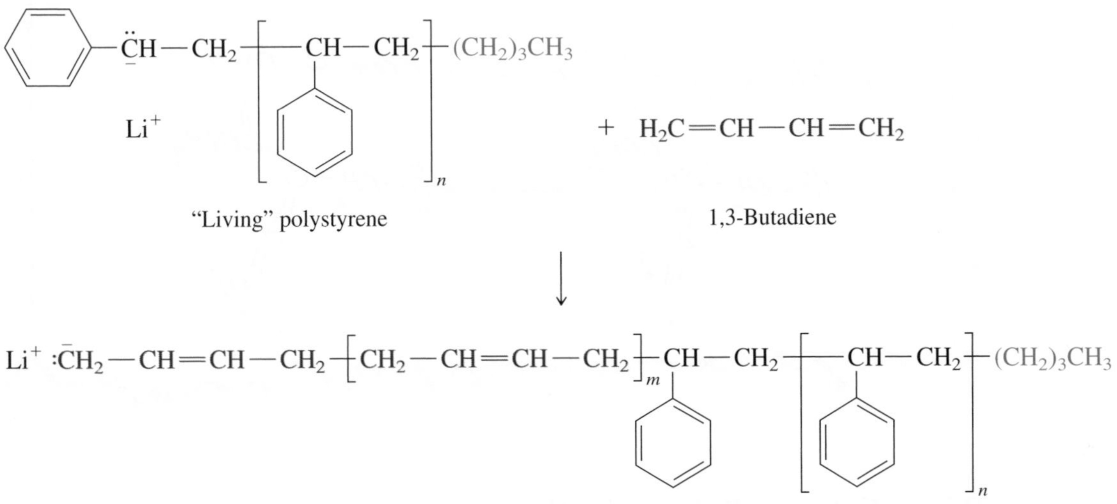

"Living" styrene-butadiene copolymer

Living polymerizations are characterized by the absence of efficient termination processes. They are normally terminated by intentionally adding a substance that reacts with carbanions such as an alcohol or carbon dioxide.

The kinds of vinyl monomers that are susceptible to anionic polymerization are those that bear electron-withdrawing groups such as $-C{\equiv}N$ and $-\overset{O}{\overset{\|}{C}}-$ on the double bond.

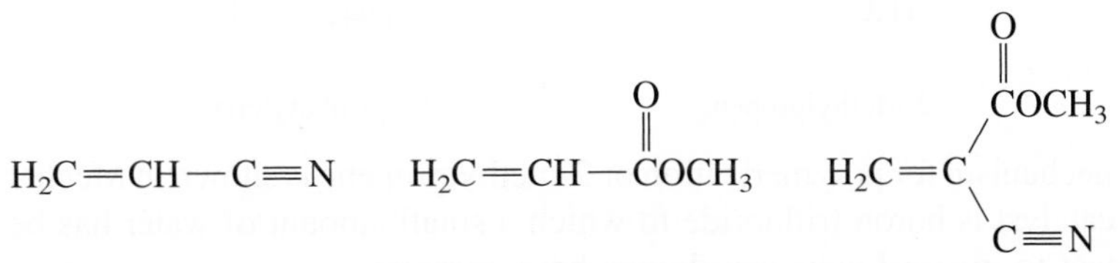

Acrylonitrile   Methyl acrylate   Methyl 2-cyanoacrylate

When a carbonyl and a cyano group are attached to the same carbon as in methyl 2-cyanoacrylate, the monomer that constitutes *Super Glue,* anionic polymerization can be initiated by even weak bases such as atmospheric moisture or normal skin dampness.

Problem 27.9

Write a structural formula for the carbanion formed by addition of hydroxide ion to methyl 2-cyanoacrylate. Accompany this structural formula by a contributing resonance structure that shows delocalization of the negative charge to oxygen, and another to nitrogen.

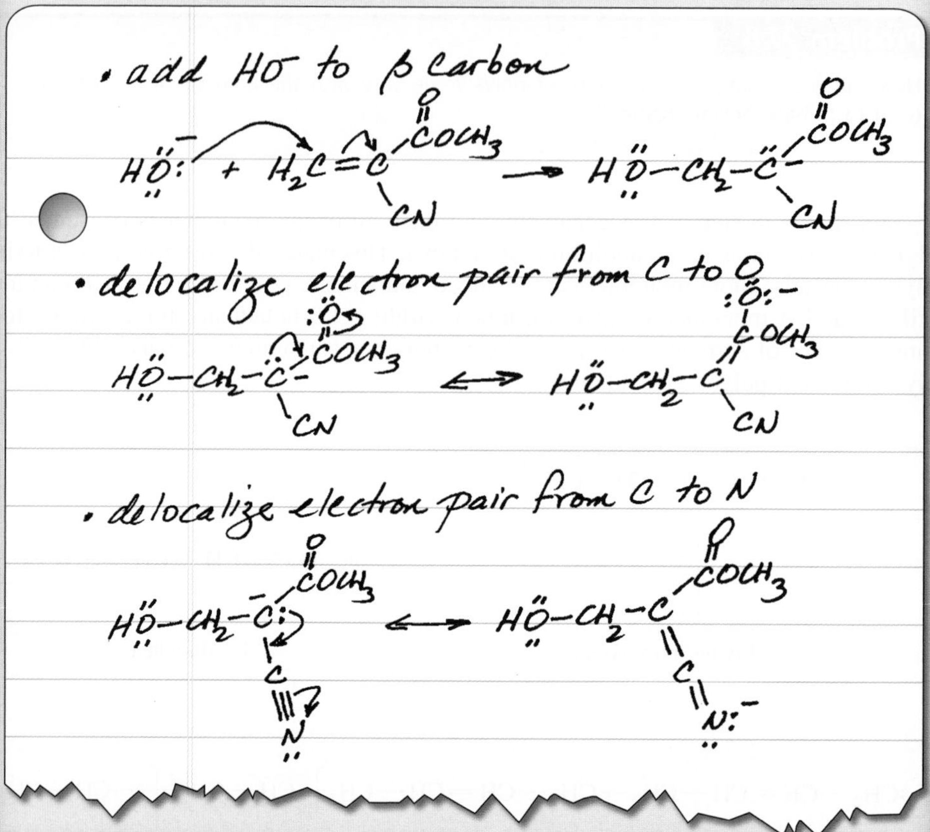

27.10 Cationic Polymerization

Analogous to the initiation of anionic polymerization by addition of nucleophiles to alkenes, cationic polymerization can be initiated by the addition of electrophiles. The alkenes that respond well to cationic polymerization are those that form relatively stable carbocations when protonated. Of these, the one used most often is 2-methylpropene, better known in polymer chemistry by its common name *isobutylene*.

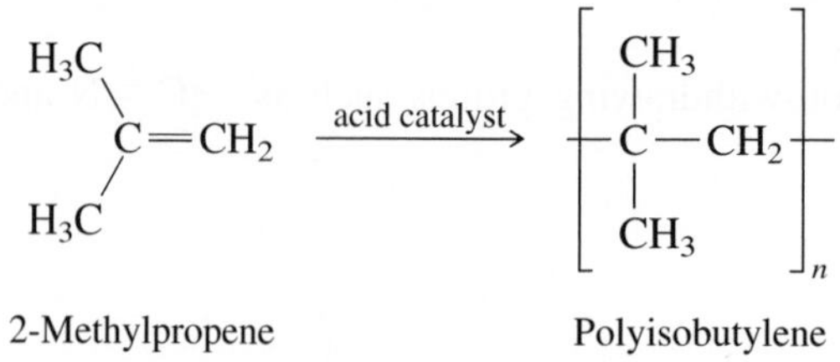

The mechanism for polymerization of 2-methylpropene is shown in Mechanism 27.4. The usual catalyst is boron trifluoride to which a small amount of water has been added. The two react to give a Lewis acid/Lewis base complex.

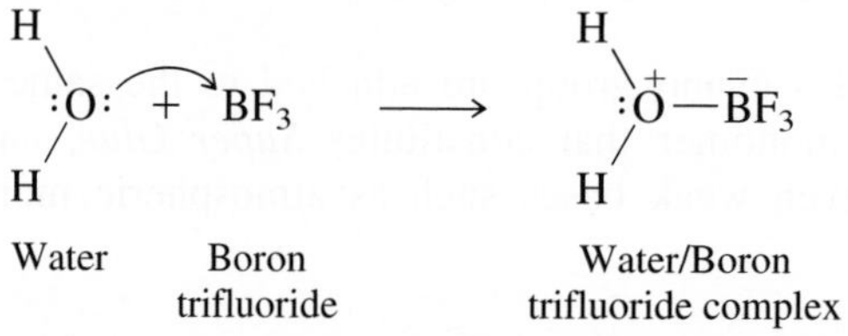

Mechanism 27.4

Cationic Polymerization of 2-Methylpropene

Step 1: The alkene is protonated, forming a carbocation.

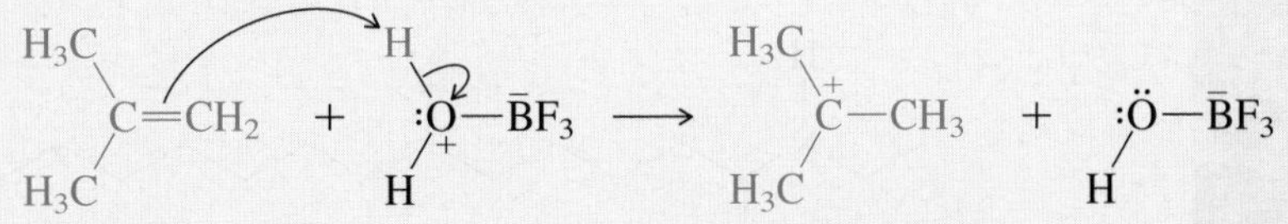

2-Methylpropene *tert*-Butyl cation

Step 2: The carbocation formed in the preceding step reacts with a molecule of the alkene, forming a new carbocation.

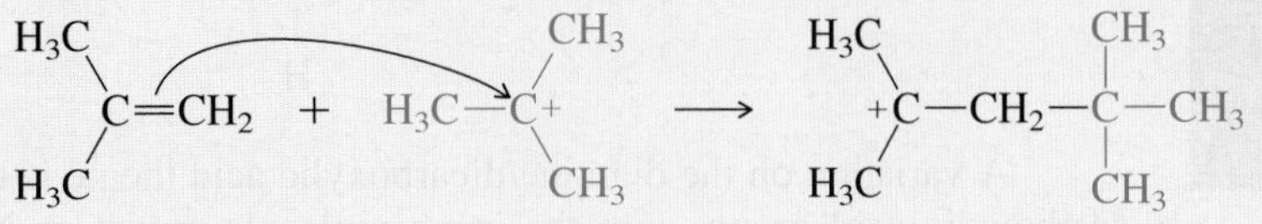

2-Methylpropene *tert*-Butyl cation 1,1,3,3-Tetramethylbutyl cation

Step 3: The process shown in step 2 continues, forming a chain-extended carbocation.

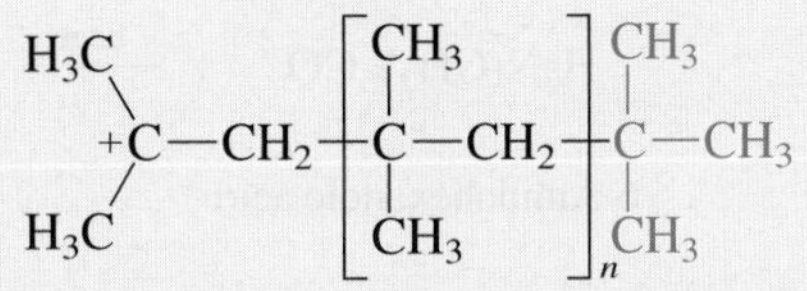

Step 4: One mechanism for chain termination is loss of a proton.

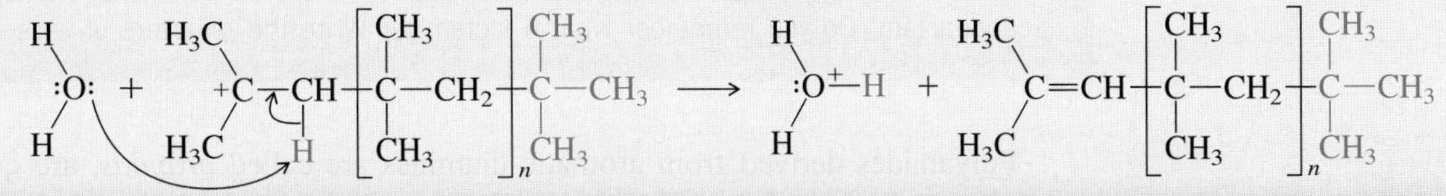

This complex is a strong Brønsted acid and protonates the double bond of 2-methylpropene in step 1 of the mechanism.

Polyisobutylene is the "butyl" in butyl rubber, one of the first synthetic rubber substitutes. Most inner tubes are a copolymer of 2-methylpropene (isobutylene) and 2-methyl-1,3-butadiene (isoprene).

27.11 Polyamides

The polyamide nylon 66 takes its name from the fact that it is prepared from a six-carbon dicarboxylic acid and a six-carbon diamine. The acid–base reaction between adipic acid and hexamethylenediamine gives a salt, which on heating undergoes condensation polymerization in which the two monomers are joined by amide bonds.

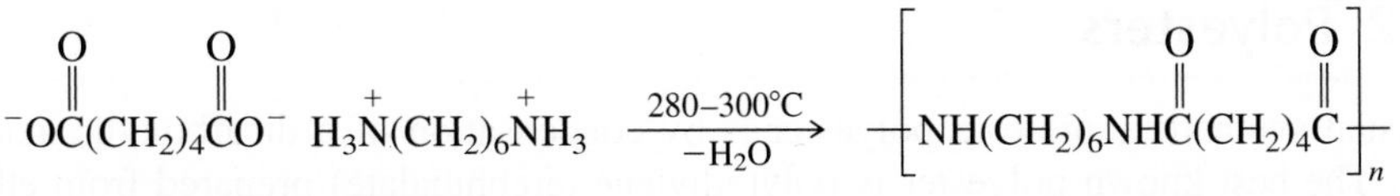

Salt of adipic acid and hexamethylenediamine Nylon 66

The systematic names of adipic acid and hexamethylenediamine are hexanedioic acid and 1,6-hexanediamine, respectively.

Figure 27.7

Skydivers' parachutes are made of nylon 66.

Nylon 66 was the first and remains the most commercially successful synthetic polyamide (Figure 27.7). Others have been developed by varying the number of carbons in the chains of the diamine and the dicarboxylic acid.

Nylon 66 resembles silk in both structure and properties. Both are polyamides in which hydrogen bonds provide an ordered arrangement of adjacent chains.

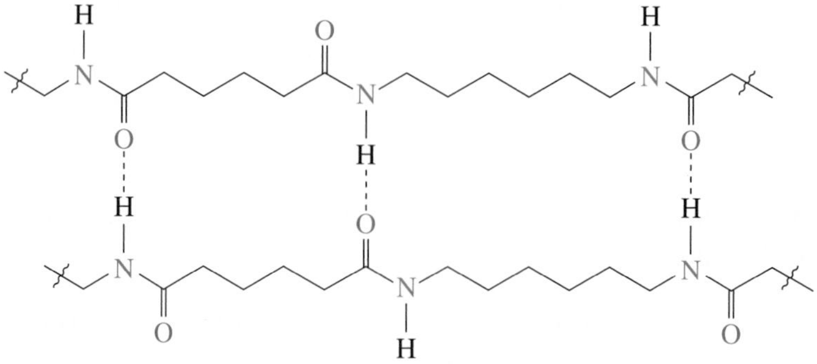

A variation on the diamine/dicarboxylic acid theme is to incorporate the amino and carboxylic acid groups into the same molecule, much as Nature does in amino acids. Nylon 6 is a polyamide derived by heating 6-aminohexanoic acid.

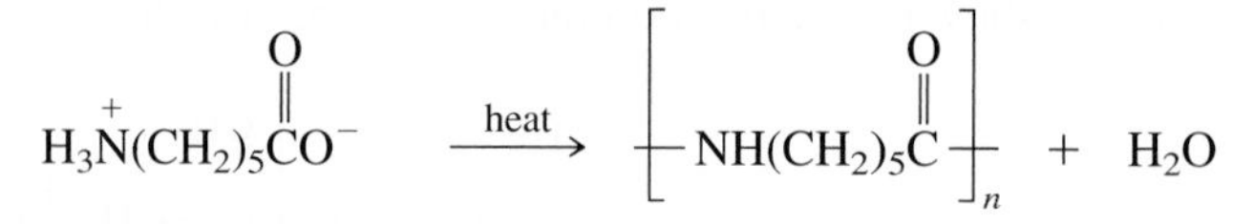

6-Aminohexanoic acid Nylon 6 Water

Problem 27.10

Nylon 6 is normally prepared from the lactam derived from 6-aminohexanoic acid, called ε-caprolactam. Do you remember what a lactam is? Write the structure of ε-caprolactam.

Polyamides derived from aromatic diamines are called *aramids,* are quite strong, and enjoy a number of uses. Protective clothing, including bullet-resistant vests, made from the aramid fiber *Kevlar,* for example, are effective yet light in weight.

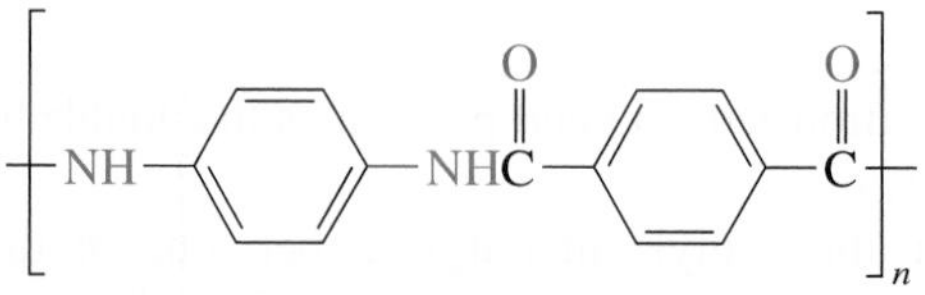

Kevlar

Problem 27.11

Nomex is an aramid fiber used for fire-resistant protective clothing. It is a polyamide prepared by condensation of 1,3-benzenediamine (*m*-phenylenediamine) and 1,3-benzenedicarboxylic acid (isophthalic acid). What is the repeating unit of Nomex?

27.12 Polyesters

The dimethyl ester of terephthalic acid is used in an analogous method.

The usual synthetic route to a polyester is by condensation of a dicarboxylic acid with a diol. The best known polyester is poly(ethylene terephthalate) prepared from ethylene glycol and terephthalic acid.

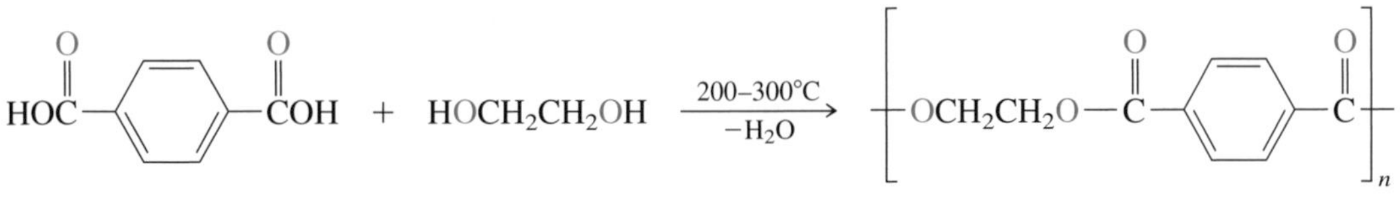

Terephthalic acid (Benzene-1,4-dicarboxylic acid) | Ethylene glycol | Poly(ethylene terephthalate)

The popularity of clothing made of polyester-cotton blends testifies to the economic impact of this polymer. Poly(ethylene terephthalate) is the PETE referred to in the recycling codes listed in Table 27.1. Plastic bottles for juice, ketchup, and soft drinks are usually made of PETE, as is Mylar film (Figure 27.8).

Alkyd resins number in the hundreds and are used in glossy paints and enamels—house, car, and artist's—as illustrated in Figure 27.9. Most are derived from benzene-1,2-dicarboxylic acid (*o*-phthalic acid) and 1,2,3-propanetriol (glycerol). Two of the hydroxyl groups of glycerol are converted to esters of *o*-phthalic acid; the third is esterified with an unsaturated fatty acid that forms cross links to other chains.

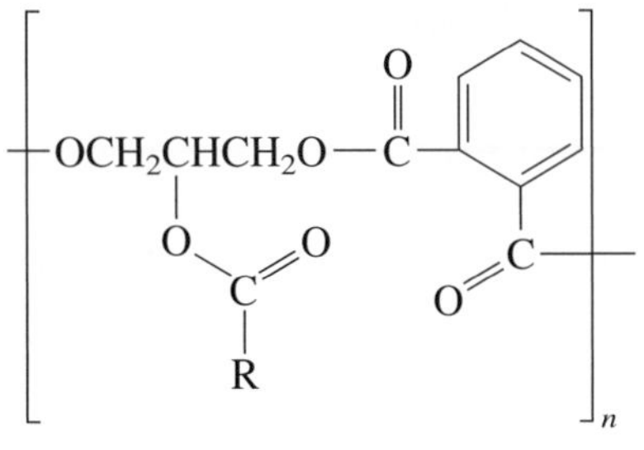

An alkyd resin

Figure 27.8

Mylar balloons add zest to a party and because they are PETE can be recycled.

With both a hydroxyl group and a carboxylic acid function in the same molecule, glycolic acid and lactic acid have the potential to form polyesters. Heating the α-hydroxy acid gives a cyclic diester, which, on treatment with a Lewis acid catalyst ($SnCl_2$ or SbF_3) yields the polymer.

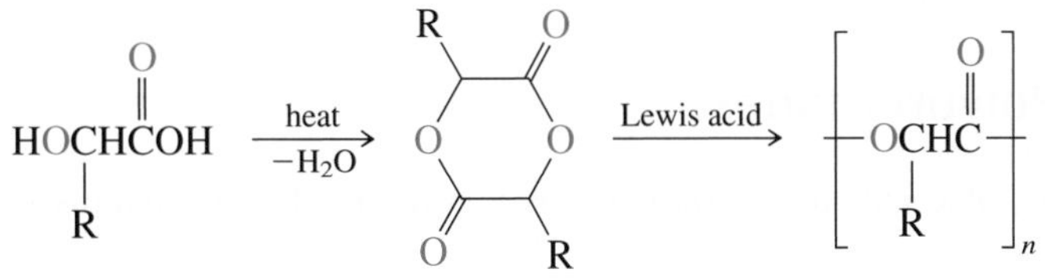

R = H	Glycolic acid	Glycolide	Poly(glycolic acid)
R = CH_3	Lactic acid	Lactide	Poly(lactic acid)

Surgical sutures made from poly(glycolic acid) and poly(lactic acid), while durable enough to substitute for ordinary stitches, are slowly degraded by ester hydrolysis and don't require a return visit for their removal. Poly(glycolic acid) fibers also hold promise as a scaffold upon which to grow skin cells. This "artificial skin" is then applied to a wound to promote healing.

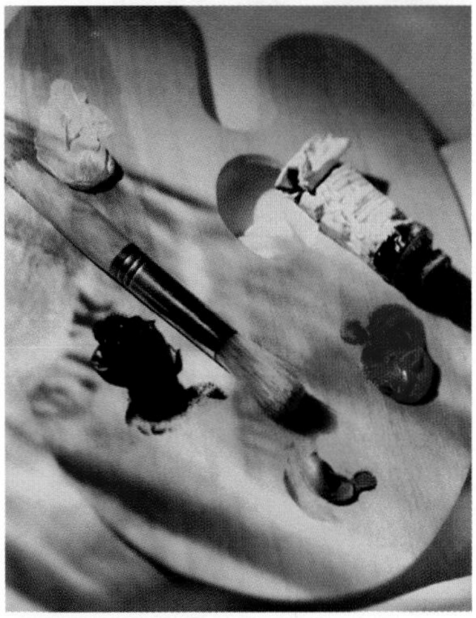

Figure 27.9

Alkyds are used for more than painting rooms. Artists use them too.

Problem 27.12

Another monomer from which surgical sutures are made is ε-caprolactone. What is the repeating unit of poly(ε-caprolactone)?

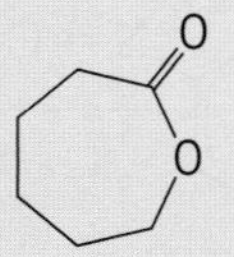

ε-Caprolactone

Polyesters are also used in controlled-release forms of drugs and agricultural products such as fertilizers and herbicides. By coating the active material with a polyester selected so as to degrade over time, the material is released gradually rather than all at once.

27.13 Polycarbonates

Polycarbonates are polyesters of carbonic acid. *Lexan* is the most important of the polycarbonates and is prepared from the diphenolic compound bisphenol A.

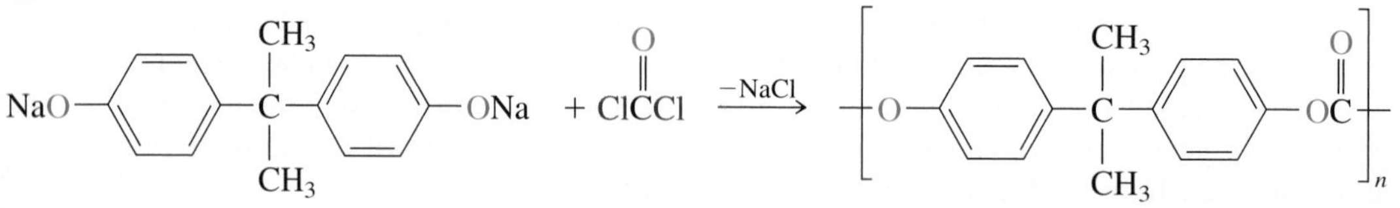

Disodium salt of bisphenol A — Phosgene — Bisphenol A polycarbonate (Lexan)

Bisphenol A is made from phenol and acetone. Industrial processes are usually very efficient. One process, described in Chapter 22, gives both phenol and acetone as products of the same reaction. Can you find it?

Problem 27.13

Write a mechanism for the reaction of one molecule of the disodium salt of bisphenol A with one molecule of phosgene.

Figure 27.10

The polycarbonate lenses in these protective glasses are lightweight, yet shatterproof.

Lexan is a clear, transparent, strong, and impact-resistant plastic with literally countless applications. It is used in both protective and everyday eyeglasses as illustrated in Figure 27.10. The Apollo 11 astronauts wore Lexan helmets with Lexan visors on their 1969 trip to the moon. CDs and DVDs are Lexan polycarbonate, as are many cell phones, automobile dashpanels, and headlight and taillight lenses.

27.14 Polyurethanes

A *urethane,* also called a *carbamate,* is a compound that contains the functional group $-\overset{\displaystyle O}{\overset{\|}{OC}}NH-$. Urethanes are normally prepared by the reaction of an alcohol and an isocyanate.

$$\underset{\text{Alcohol}}{ROH} + \underset{\text{Isocyanate}}{R'N{=}C{=}O} \longrightarrow \underset{\text{Urethane}}{RO\overset{\displaystyle O}{\overset{\|}{C}}NHR'}$$

Polyurethanes are the macromolecules formed from a diol and a diisocyanate. In most cases the diol is polymeric and the diisocyanate is a mixture of the "toluene diisocyanate" isomers.

$HOCH_2$—[Polymer]—CH_2OH

Polymeric diol

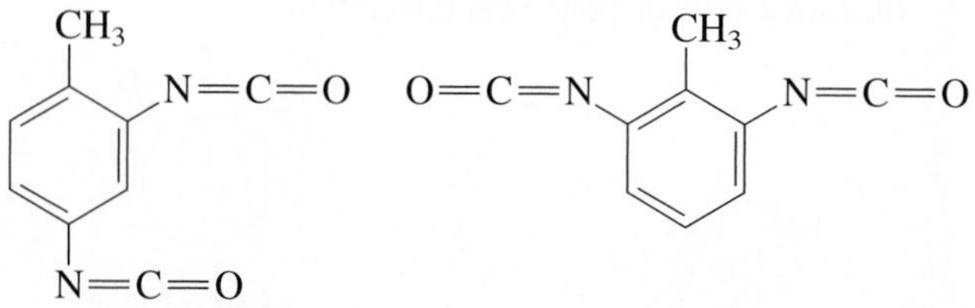

Mixture of "toluene diisocyanate" isomers

If, for example, only the 2,6-diisocyanate were present, the repeating unit of the resulting polyurethane would be

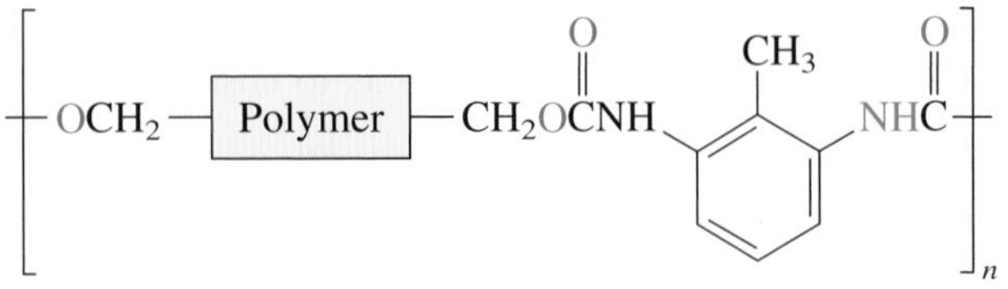

Because a mixture of diisocyanate isomers is actually used, a random mixture of 2,4- and 2,6-substitution patterns results.

Figure 27.11

Spandex skinsuits make speedskaters more aerodynamic.

Problem 27.14

Write the repeating unit of the "polymeric diol" if it is derived from 1,2-epoxypropane.

The reaction of an alcohol with an isocyanate is addition, not condensation. Therefore, polyurethanes are classified as addition polymers. But because the monomers are difunctional, the molecular weight increases by step growth rather than chain growth.

A major use of polyurethanes is in spandex fibers. Spandex, even when stretched several times its length, has the ability to return to its original state and is a superior substitute for rubber in elastic garments. Its most recognizable application is in athletic wear (swimming, cycling, running) where it is the fabric of choice for high-performance athletes (Figure 27.11).

Polyurethanes have many other applications, especially in paints, adhesives, and foams. Polyurethane foams, which can be rigid (insulation panels) or flexible (pillows, cushions, and mattresses) depending on their degree of cross linking, are prepared by adding foaming agents to the polymerization mixture. One method takes advantage of the reaction between isocyanates and water.

$$\underset{\text{Isocyanate}}{RN{=}C{=}O} + \underset{\text{Water}}{H_2O} \longrightarrow \underset{\text{Carbamic acid}}{RNH{-}\overset{\overset{\displaystyle O}{\|}}{C}{-}OH} \longrightarrow \underset{\text{Amine}}{RNH_2} + \underset{\text{Carbon dioxide}}{CO_2}$$

Although esters of carbamic acid (urethanes) are stable compounds, carbamic acid itself rapidly dissociates to an amine and carbon dioxide. Adding some water to the reactants during polymerization generates carbon dioxide bubbles which are trapped within the polymer.

27.15 Copolymers

Copolymers, polymers made from more than one monomer, are as common as homopolymers. The presence of more than one monomer in a chain makes some control of properties possible. Some structural units stiffen the chain, others make it more flexible. Often a second monomer is added to allow cross linking.

Copolymers are classified according to the distribution of monomers in the macromolecule.

1. Random
2. Block
3. Graft

Random Copolymers: As the name implies, there is no pattern to the distribution of monomer units in a random copolymer.

—A—A—B—A—B—B—A—A—B—B—A—B—A—B—

Styrene–butadiene rubber (SBR) for automobile tires is a random copolymer. It is prepared by two methods, free-radical and anionic polymerization, both of which are carried out on a mixture of styrene and 1,3-butadiene. Free-radical initiation is essentially nonselective and gives the random copolymer. Anionic initiation is carried out under conditions designed to equalize the reactivity of the two monomers so as to ensure randomness.

Block Copolymers: The main chain contains sections (blocks) of repeating units derived from different monomers. The sequence:

—A—A—A—A—A—B—B—B—B—B—B—B—B—B—

shows only two blocks, one derived from A and the other from B. A macromolecule derived from A and B can contain many blocks.

The living polymers generated by anionic polymerization are well suited to the preparation of block polymers. Adding 1,3-butadiene to a living polystyrene block sets the stage for attaching a poly(1,3-butadiene) block.

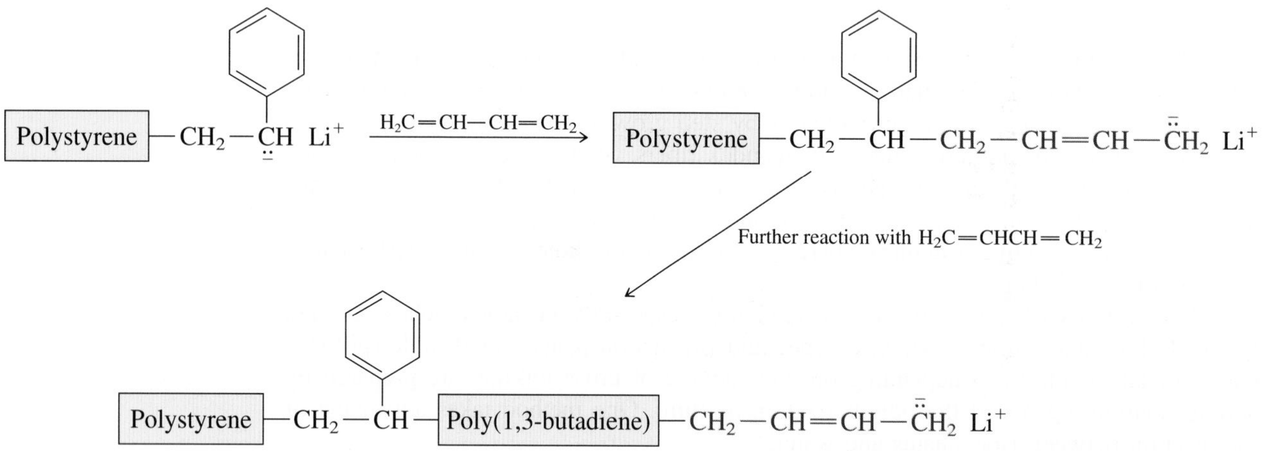

The properties of the block copolymer prepared by anionic living polymerization are different from the random styrene–butadiene copolymer.

Graft Copolymer: The main chain bears branches (grafts) that are derived from a different monomer.

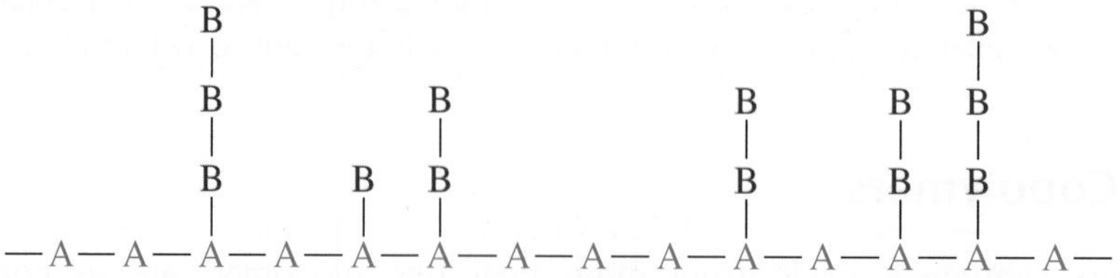

A graft copolymer of styrene and 1,3-butadiene is called "high-impact polystyrene" and is used, for example, in laptop computer cases. It is prepared by free-radical polymerization of styrene in the presence of poly(1,3-butadiene). Instead of reacting with styrene, the free-radical initiator abstracts an allylic hydrogen from poly(1,3-butadiene).

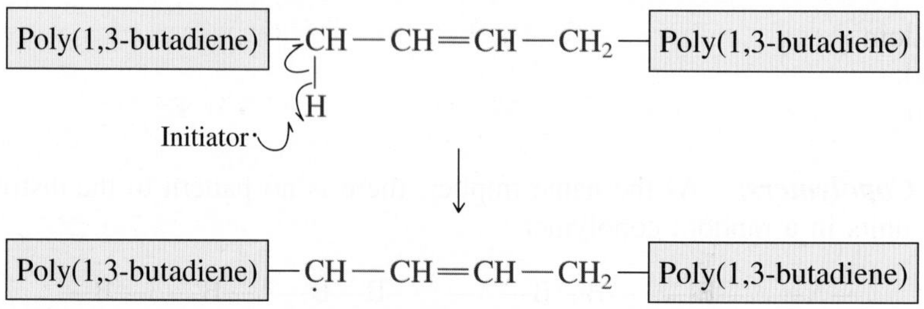

Polystyrene chain growth begins at the allylic radical site and proceeds in the usual way at this and random other allylic carbons of poly(1,3-butadiene).

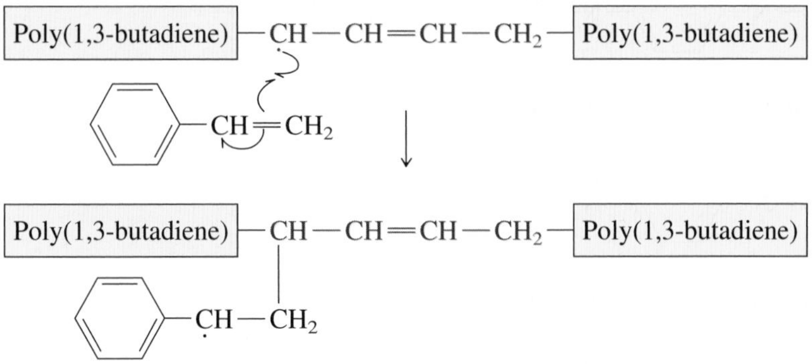

Polystyrene grafts on a poly(1,3-butadiene) chain are the result.

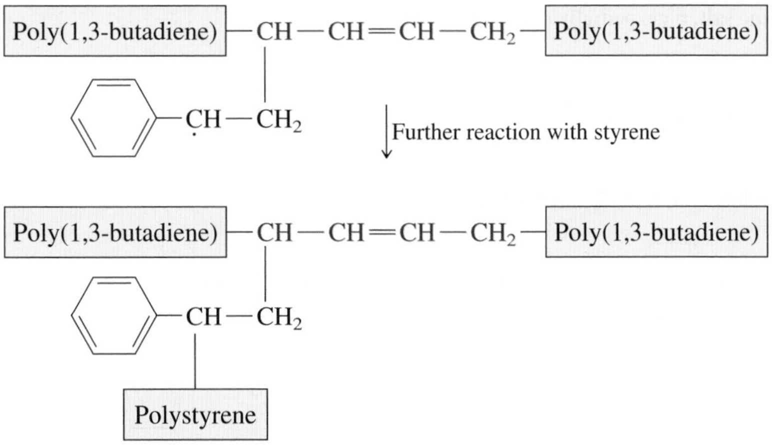

Polystyrene alone is brittle; poly(1,3-butadiene) alone is rubbery. The graft copolymer is strong, but absorbs shock without cracking because of the elasticity provided by its poly(1,3-butadiene) structural units.

Conducting Polymers

The notion that polymers can conduct electricity seems strange to most of us. After all, the plastic wrapped around the wires in our homes and automobiles serves as insulation. Do polymers exist that can conduct electricity? Even if such materials could be made, why would we be interested in them?

Henry Letheby, a lecturer in chemistry and toxicology at the College of London Hospital, obtained a partially conducting material in 1862 by the anodic oxidation of aniline in sulfuric acid. The material Letheby synthesized was a form of polyaniline. In the 1980s, the New Zealander Alan MacDiarmid of the University of Pennsylvania reinvestigated polyaniline, which is now a widely used conducting polymer. Polyaniline exists in a variety of oxidation states (Figure 27.12), each with different properties. The emaraldine salt is a conductor without the use of additives that enhance conductivity, but its conductivity is enhanced by adding a Brønsted acid that protonates the nitrogen atoms.

The synthesis of polyaniline can be carried out in aqueous HCl solution, by electrochemical oxidation, or in the presence of a chemical oxidant such as ammonium persulfate. The different forms of polyaniline can then be obtained by altering the current or the pH of the solution. The ability to tailor the process increases the potential for commercial application where the unique properties of a certain polyaniline are desired. Polyanilines are used as corrosion inhibitors and in the electromagnetic shielding of circuits, where they can protect against electrostatic discharge.

Continued

Figure 27.12

Polyaniline exists in different forms with varying states of oxidation. One of the forms is a conductor.

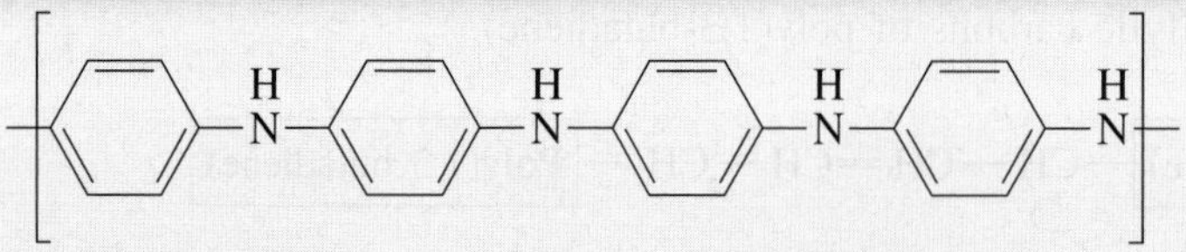

Leucoemaraldine, colorless, fully reduced, insulating

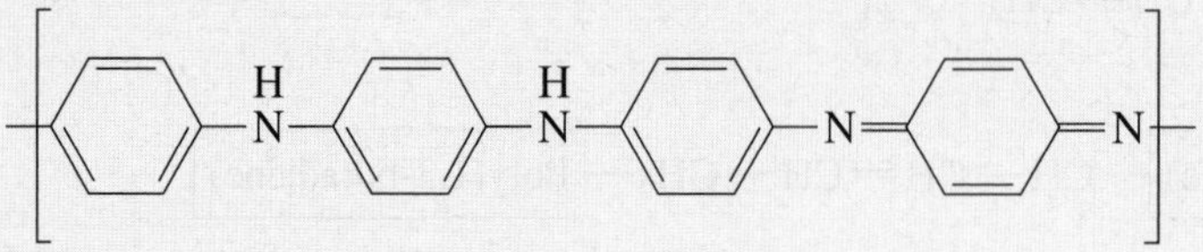

Emaraldine base, green, partially oxidized, insulating

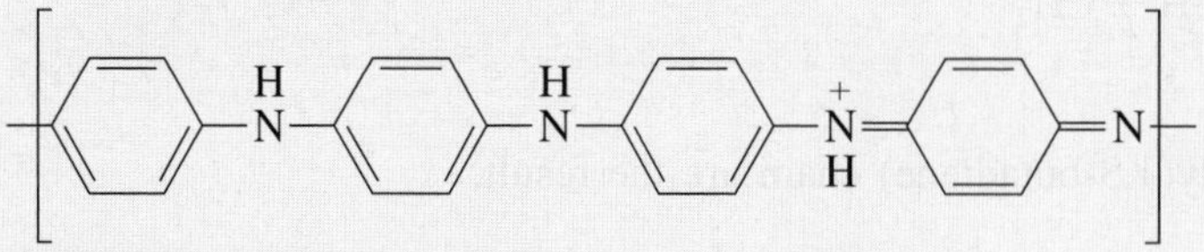

Emaraldine salt, blue, partially oxidized, conducting

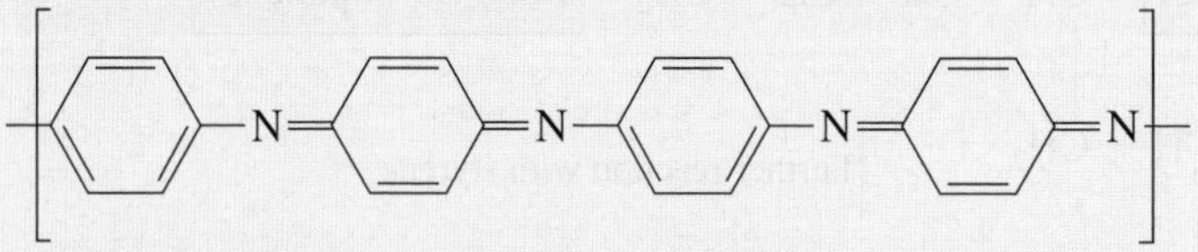

Pernigraniline, purple, fully oxidized, insulating

Figure 27.13

A mixture of poly(3,4-ethylenedioxythiophene) and poly(styrene sulfonate) is used in the manufacture of organic light-emitting diodes (OLEDs).

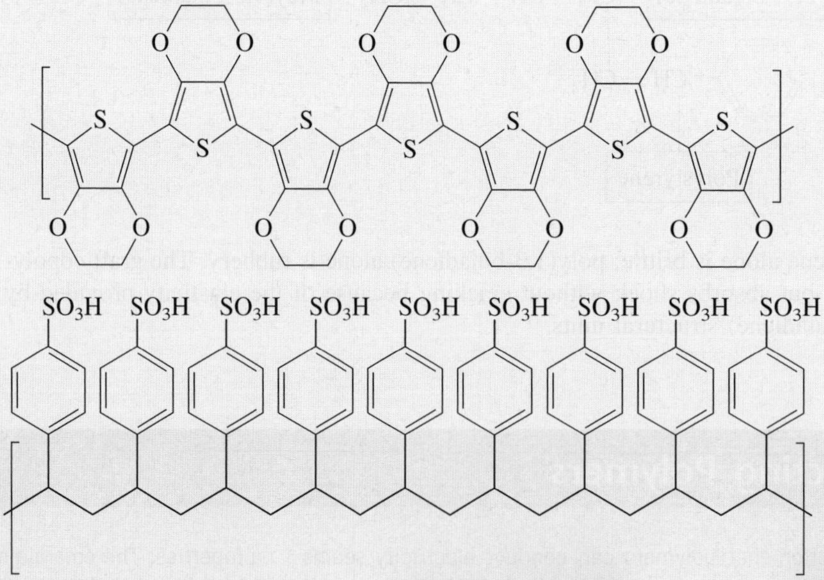

Another conducting polymer that has found commercial application is poly(3,4-ethylenedioxythiophene), PEDOT, which is marketed as a dispersion that contains poly(styrene sulfonate). This polymer dispersion is used in the manufacture of organic light-emitting diodes (OLEDs), which are materials that emit light when an electric current is applied to them (Figure 27.13). OLEDs are used for flat panel displays in televisions and cellular telephone displays.

The 2000 Nobel Prize in Chemistry was awarded to Alan Heeger (University of California Santa Barbara, U.S.A., Alan MacDiarmid (University of Pennsylvania, U.S.A., and Hideki Shirakawa (University of Tsukuba, Japan) for their "discovery and development of electrically conductive polymers."

SONY now markets a flat panel television with an OLED screen that is 3 mm thick.

This cellular telephone made by Nokia uses an OLED display.

27.16 SUMMARY

Section 27.1 Polymer chemistry dates to the nineteenth century with the chemical modification of polymeric natural products. Once the structural features of polymers were determined, polymer synthesis was placed on a rational basis.

Section 27.2 Polymers are usually named according to the monomers from which they are prepared (*source-based nomenclature*). When the name of the monomer is one word, the polymer is named by simply adding the prefix *poly-*. When the name of the monomer is two words, they are enclosed in parentheses and preceded by *poly*.

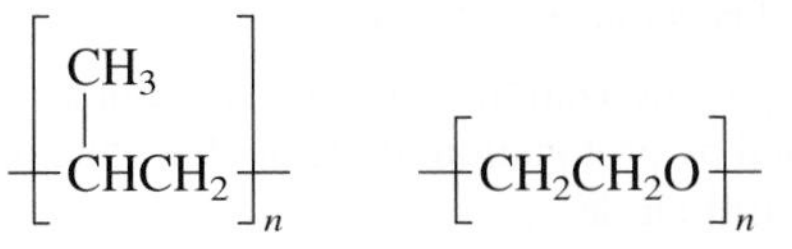

Polypropylene Poly(ethylene oxide)

Sections 27.3–27.6 Polymers may be classified in several different ways:

- Reaction type (addition and condensation)
- Chain-growth or step-growth
- Structure (linear, branched, cross-linked)
- Properties (thermoplastic, thermoset, or elastomer)

Section 27.7 This section emphasizes initiation and termination steps in alkene polymerization. The main terminating reactions in free-radical polymerization are the coupling of two radicals and disproportionation. *Coupling* of two radicals pairs the odd electrons and stops chain growth.

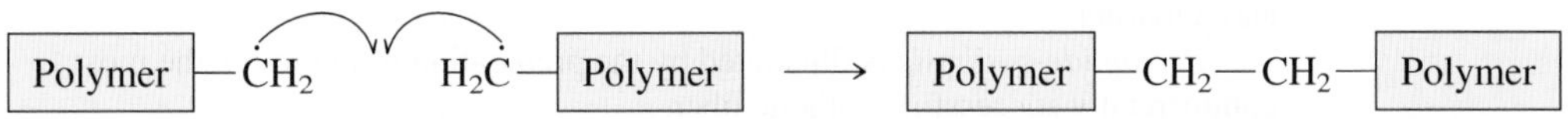

In *disproportionation,* a hydrogen atom is exchanged between two growing chains, terminating one in a double bond and the other in a new C—H bond.

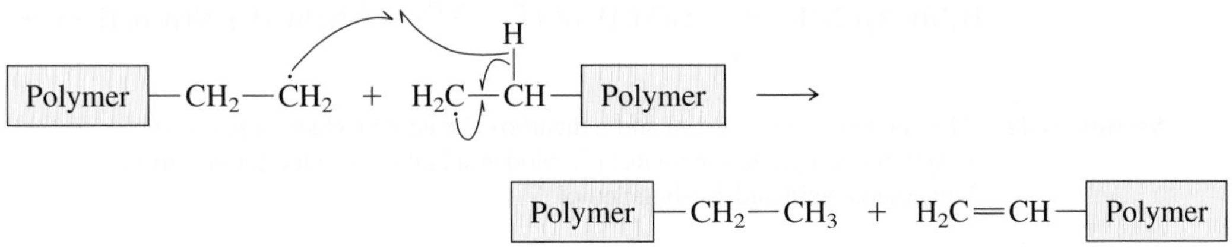

Section 27.8 Free-radical polymerization of alkenes usually gives branched polymers of low crystallinity. The two main mechanisms by which branches form both involve hydrogen atom abstraction by the radical site. In one, a growing chain abstracts a hydrogen atom from a terminated polymer.

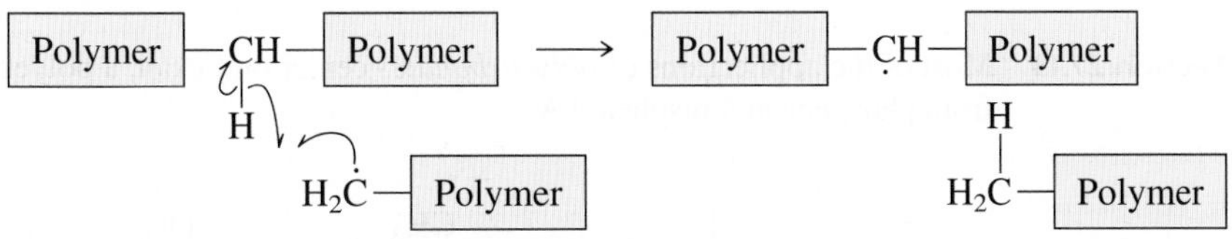

The other is an intramolecular hydrogen-atom abstraction. In most cases this reaction proceeds by a six-center transition state and moves the reactive site from the end of the growing chain to inside it.

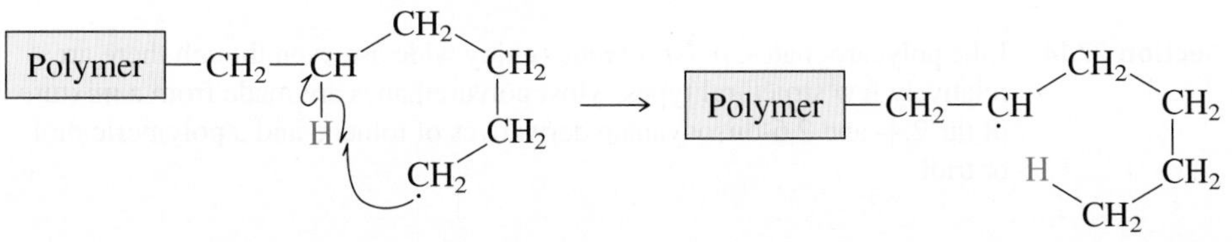

Section 27.9 Anionic polymerization of alkenes that bear a carbanion-stabilizing substituent (X) can be initiated by strong bases such as alkyllithium reagents.

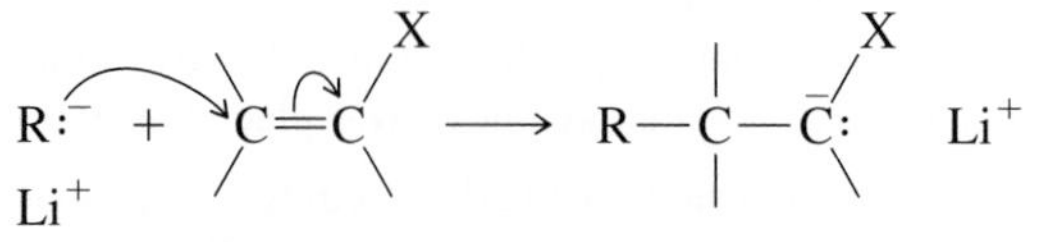

The product of this step is a new organolithium reagent that can react with a second monomer molecule, then a third, and so on. The growing organolithium chain is stable and is called a living polymer.

Section 27.10 Cationic polymerization of alkenes that can form relatively stable carbocations can be initiated by protonation of the double bond or coordination to Lewis acids such as boron trifluoride.

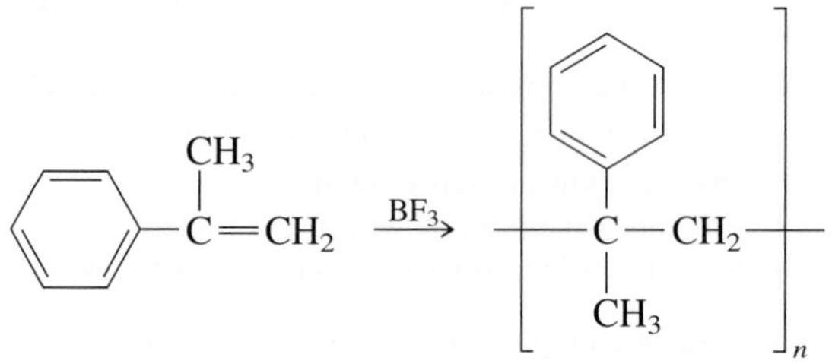

Section 27.11 The key bond-forming process in many polymerizations is a *condensation* reaction. The most common condensations are those that produce polyamides and polyesters.

Polyamide synthesis is illustrated by the preparation of nylon 66, the most commercially successful synthetic fiber.

$$H_3\overset{+}{N}(CH_2)_6\overset{+}{N}H_3 + {}^{-}O\overset{O}{\overset{\|}{C}}(CH_2)_4\overset{O}{\overset{\|}{C}}O^{-} \xrightarrow{\text{heat}} \left[NH(CH_2)_6NH\overset{O}{\overset{\|}{C}}(CH_2)_4\overset{O}{\overset{\|}{C}} \right]_n$$

Section 27.12 The condensation of a diol and a dicarboxylic acid produces a *polyester.* Poly(tetramethylene succinate) is a biodegradable polyester derived from butanedioic acid and 1,4-butanediol.

$$\left[OCH_2CH_2CH_2CH_2O\overset{O}{\overset{\|}{C}}CH_2CH_2\overset{O}{\overset{\|}{C}} \right]_n$$

Section 27.13 Most of the applications of *polycarbonates* center on Lexan, a polyester derived from phosgene and bisphenol A.

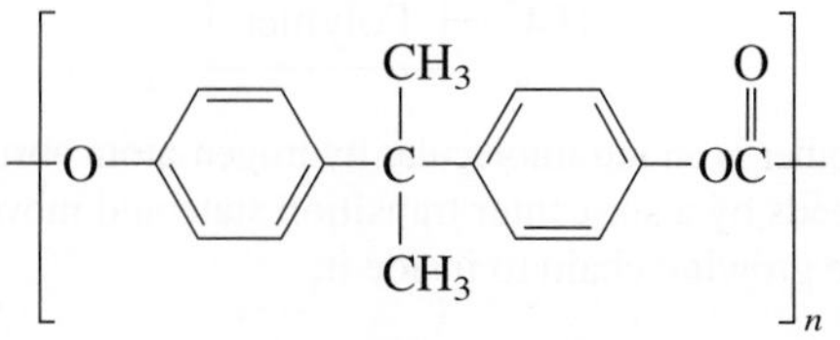

Section 27.14 Like polycarbonates, *polyurethanes* enjoy wide use even though there are relatively few structural types. Most polyurethanes are made from a mixture of the 2,4- and 2,6-diisocyanate derivatives of toluene and a polymeric diol or triol.

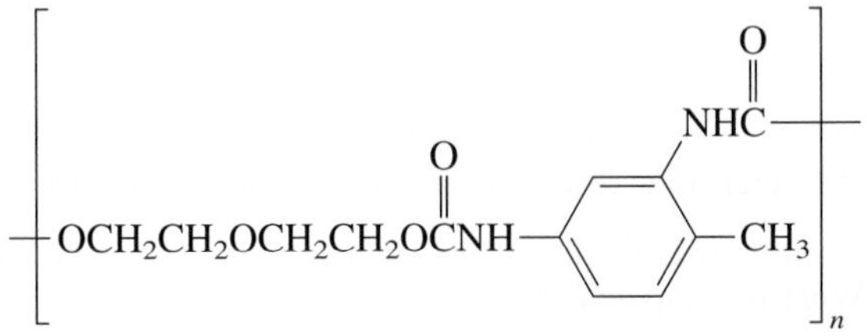

Section 27.15 *Copolymers* are the polymers formed when two or more monomers are present in the mixture to be polymerized. They are classified as random, block, or graft. A *random copolymer* lacks a regular sequence in respect to the appearance of the structural units of the components. A *block copolymer* of monomers A and B is composed of blocks of poly(A) and poly(B). A *graft copolymer* has a main chain of poly(A) to which are grafted branches of poly(B).

PROBLEMS

27.15 Nylon 11 is a polyamide used as fishing line and is prepared by heating 11-aminoundecanoic acid [$H_2N(CH_2)_{10}CO_2H$]. What is the repeating unit of nylon 11? Is it a condensation or an addition polymer? Chain-growth or step-growth?

27.16 Is protein biosynthesis as shown in Figure 26.12 (page 1196) step growth or chain growth? Is the protein that results an addition or a condensation polymer? Why?

27.17 *Pseudomonas oleovorans* oxidizes nonanoic acid, then stores the 3-hydroxynonanoic acid produced as a homopolymer. Write the formula for the repeating unit of this polyester.

27.18 From what monomer is the polymer with the repeating unit 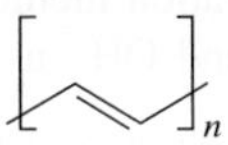 prepared? Suggest a source-based name.

27.19 Give the structure of the lactone from which $\left[OCH_2CH_2\overset{O}{\overset{\|}{C}} \right]_n$ is prepared.

27.20 Kodel fibers are made from the polymer shown. Suggest suitable monomers for its preparation.

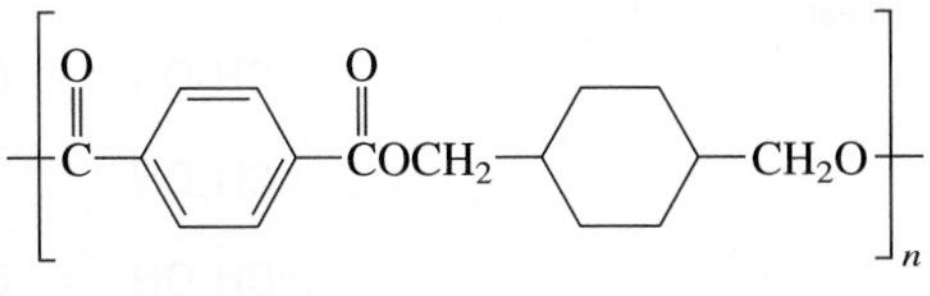

27.21 Of the following monomers, which one would undergo cationic polymerization most readily?

$H_2C{=}CHCH_3$ $\quad H_2C{=}CHCH{=}CH_2$ $\quad H_2C{=}CHC{\equiv}N$ $\quad H_2C{=}CHCl$

27.22 Of the following monomers, which one would undergo anionic polymerization most readily?

$H_2C{=}CHCH_3$ $\quad H_2C{=}CHO\overset{O}{\overset{\|}{C}}CH_3$ $\quad H_2C{=}CHC{\equiv}N$ $\quad H_2C{=}CHCl$

27.23 Polymerization of styrene can occur by a free-radical, cationic, anionic, or coordination mechanism. What mechanism will be followed when each of the compounds shown is used to initiate polymerization?

(a) $TiCl_4$, $(CH_3CH_2)_3Al$ (b) 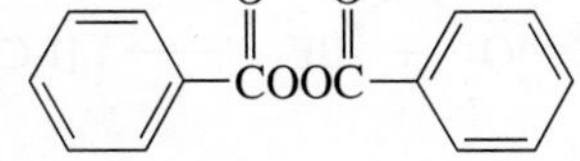 (c) BF_3

27.24 Styrene undergoes anionic polymerization at a faster rate than *p*-methoxystyrene. Suggest an explanation for this observation.

27.25 Given that —C≡N stabilizes carbanions better than phenyl, which monomer would you start with to prepare a copolymer of styrene and acrylonitrile?

27.26 *Poly(vinyl butyral)* is the inner liner in safety glass. It is prepared by the reaction shown. What is compound A?

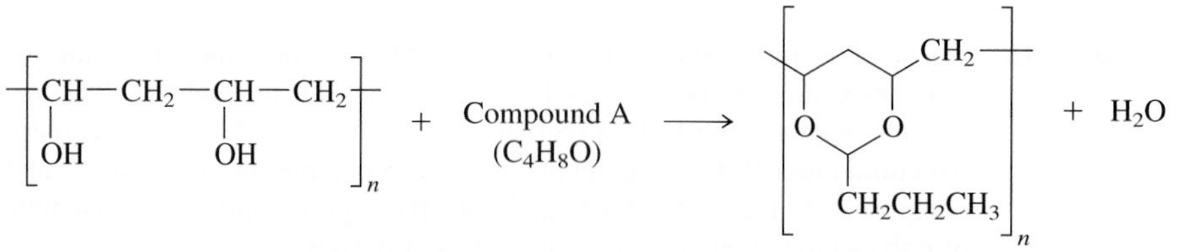

27.27 *Linear low-density polyethylene* is a copolymer in which ethylene is polymerized under Ziegler–Natta conditions in the presence of a smaller quantity of a second alkene such as 1-hexene. What structural feature characterizes the resulting polymer?

27.28 (a) Bisphenol A (shown) is made by the reaction of phenol and acetone. Suggest a mechanism for this reaction. Assume acid (H_3O^+) catalysis.

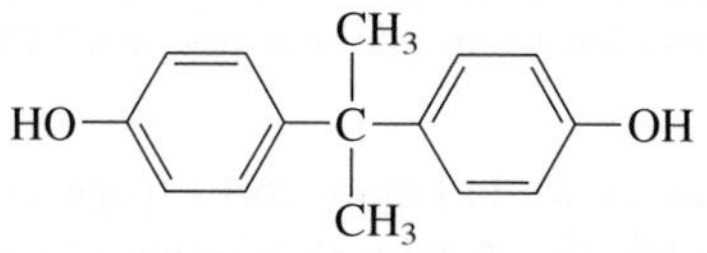

(b) Bisphenol B is made from phenol and 2-butanone. What is its structure?

27.29 Poly(ethylene oxide) can be prepared from ethylene oxide by either anionic or cationic polymerization methods. Write reaction mechanisms for both processes. Use H_3O^+ as the acid and OH^- as the base.

27.30 (a) The first step in the formation of Bakelite from phenol and formaldehyde introduces —CH_2OH groups onto the ring.

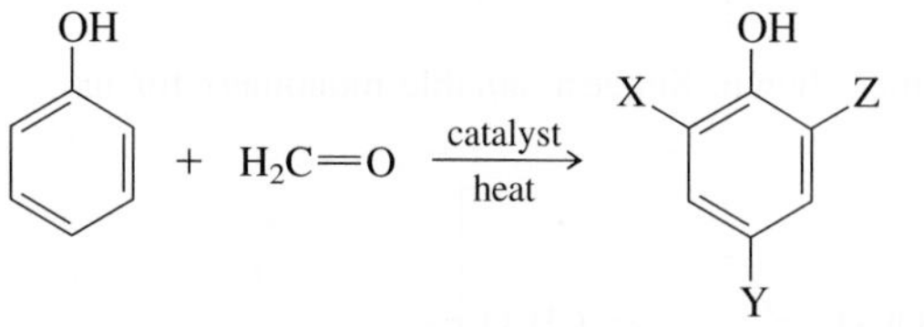

Phenol Formaldehyde

X	Y	Z
H	CH_2OH	H
CH_2OH	H	H
CH_2OH	CH_2OH	H
CH_2OH	H	CH_2OH
CH_2OH	CH_2OH	CH_2OH

Write a mechanism for the formation of *o*-hydroxybenzyl alcohol (X = CH_2OH, Y = Z = H) in this reaction. Assume the catalyst is H_3O^+.

(b) The second step links two of the aromatic rings by a CH_2 group. Write a mechanism for the example shown.

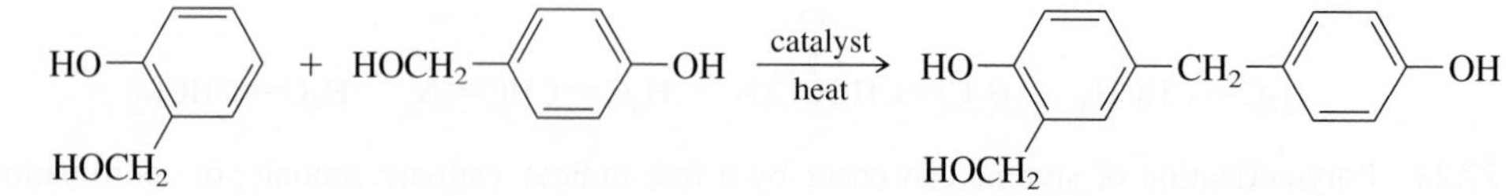

27.31 The first step in the mechanism of cationic polymerization of formaldehyde is:

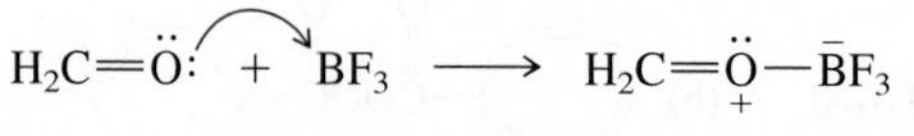

Write an equation for the second step using curved arrows to track electron movement.

Descriptive Passage and Interpretive Problems 27

Chemical Modification of Polymers

Many useful polymers are not themselves the initial products of polymerization but are prepared by chemically modifying the original polymer. Partially fluorinated polyethylene used for protective gloves and to coat automobile gasoline tanks is made by exposing polyethylene to F_2 diluted with nitrogen.

Partial fluorination gives a polymer that, like polyethylene, is easy to cast into films but with a greater resistance to oxidation and water penetration.

The solid support in Merrifield's synthesis of ribonuclease (see Section 25.18) was prepared by incorporating —CH_2Cl groups into a styrene/*p*-divinylbenzene copolymer by electrophilic aromatic substitution.

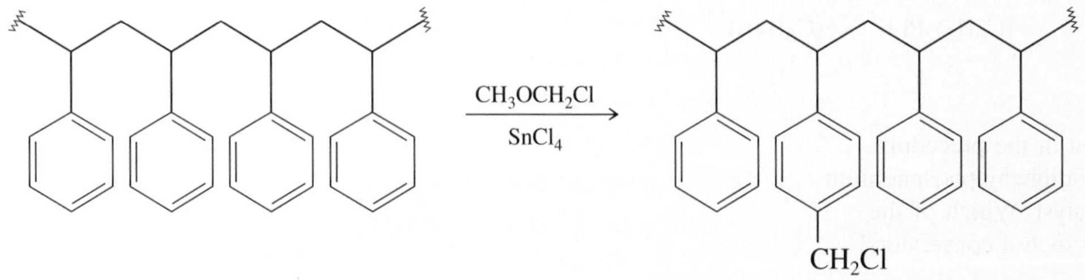

At the same time that Merrifield was developing his method for the solid-phase synthesis of peptides, Robert Letsinger (Northwestern University) was independently applying the same concept to polynucleotide synthesis. Modern methods for making oligonucleotides are direct descendants of Letsinger's method.

Today's chemists can buy Merrifield-type resins with varying degrees of chloromethyl substitution and cross linking tailored for specific purposes. Because the chlorine atom is primary and benzylic, these resins can be further modified by nucleophilic substitution.

$$\text{PS}-CH_2Cl + Nu{:}^- \longrightarrow \text{PS}-CH_2Nu + {:}\ddot{\underset{..}{Cl}}{:}^-$$

(In this and succeeding equations, the blue sphere represents a polymer bead and PS stands for polystyrene or a copolymer of polystyrene and *p*-divinylbenzene.)

The products of these reactions form the basis for an entire methodology—*polymer-supported chemical reactions*—wherein the modified polystyrene serves as a reactant, reagent, or catalyst. The reactions are the usual ones of organic chemistry. In the following equation, for example, the modified polystyrene serves as a phase-transfer catalyst (see Section 21.5). The main advantage of using a polymer-supported reagent, or in this case a polymer-supported catalyst, is that it makes isolation of the reaction product easier.

$$\underset{\text{(in toluene)}}{CH_3(CH_2)_6CH_2Br} + \underset{\text{(in water)}}{KCN} \xrightarrow{\text{PS}-CH_2\overset{+}{P}(Bu)_3\ Cl^-} \underset{\text{(in toluene)}}{CH_3(CH_2)_6CH_2CN} + \underset{\text{(in water)}}{KBr}$$

Cyanide ion from aqueous KCN exchanges with Cl^- of the polymer-supported phosphonium chloride and reacts with 1-bromooctane on the surface and within channels of the polymer support. When the reaction is judged to be complete, the polymer (insoluble in both toluene and water) is recovered by filtration and the aqueous layer removed. Distillation of the toluene solution of the product furnishes nonanenitrile, the product of nucleophilic substitution of cyanide for bromide.

The number of applications of chemically modified polymers as materials, reagents, and catalysts is extremely large. The following problems give a few examples.

27.32 Chemical modification of polymers is not always beneficial. Which of the following polymers will be adversely affected by air oxidation the most?

A. $\left[\!-CF_2-\right]_n$

B. $\left[\!-CH_2-\right]_n$

C. 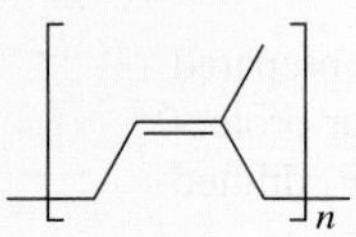

D. $\left[\!-CH_2CH(Cl)-\right]_n$

27.33 The living polymer formed by reaction of ethylene with butyllithium can be converted to a long-chain alkyldiphenylphosphine by reaction with compound X. The alkyldiphenylphosphine is used in the preparation of phase-transfer catalysts and as a ligand in polymer-supported organometallic compounds. What is compound X?

$$H_2C{=}CH_2 \xrightarrow{BuLi} Bu\left[\!-CH_2CH_2-\right]_n Li \xrightarrow{\text{compound X}} Bu\left[\!-CH_2CH_2-\right]_n P(C_6H_5)_2$$

A. $(C_6H_5)_2PH$ B. $(C_6H_5)_2PCl$ C. $(C_6H_5)_2PLi$ D. $(C_6H_5)_3P$

27.34 The alkyldiphenylphosphine formed in the preceding equation was converted to a dialkyldiphenylphosphonium salt for use as a phase-transfer catalyst. Which of the following is a suitable reactant for such a conversion?

A. $CH_3CH_2CH_2CH_2Br$ B. $CH_3CH_2CH_2CH_2Li$ C. $CH_3CH_2CH_2CH_2OH$ D. $CH_3CH_2CH_2CH_2ONa$

27.35 A copolymer of styrene and *p*-bromostyrene can be transformed into a living polymer as shown. The aryllithium sites then serve to start chain growth when a suitable monomer is added.

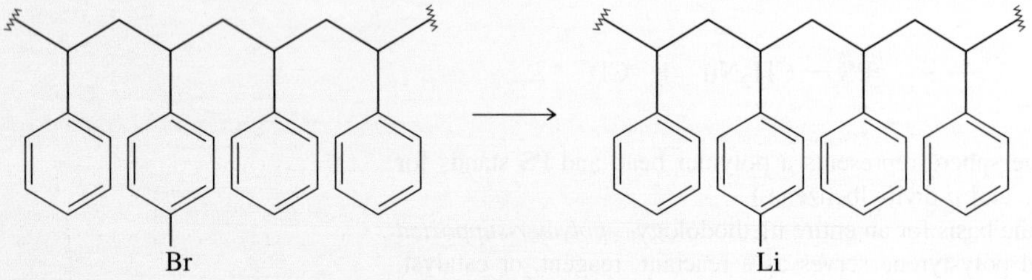
Br
Li

Which of the following is the most suitable for the transformation in the equation?

A. LiOH B. LiCl C. $LiCu(CH_3)_2$ D. Li

27.36 What is the polymer-containing product of the following reaction?

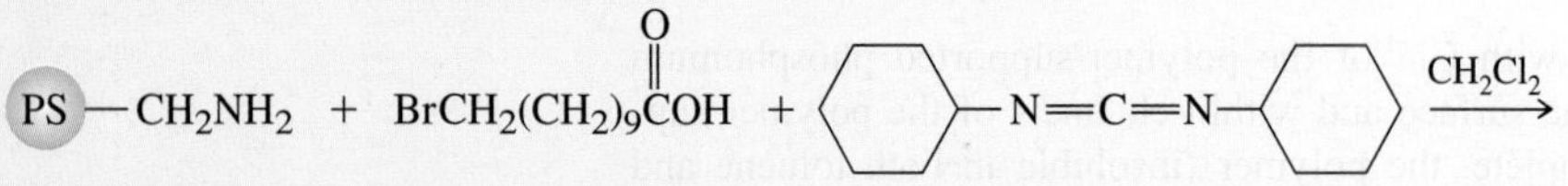
PS —CH₂NH₂ + BrCH₂(CH₂)₉COH (C=O) + N=C=N
CH₂Cl₂

A. $PS-CH_2NHCH_2(CH_2)_9\overset{O}{\overset{\|}{C}}OH$

B. $PS-CH_2NH\overset{O}{\overset{\|}{C}}(CH_2)_9CH_2Br$

27.37 The ethyl ester function in the *R*-BINAP derivative shown was used as the reactive "handle" to bind the chiral unit to polystyrene giving a ligand suitable for ruthenium-catalyzed enantioselective hydrogenation.

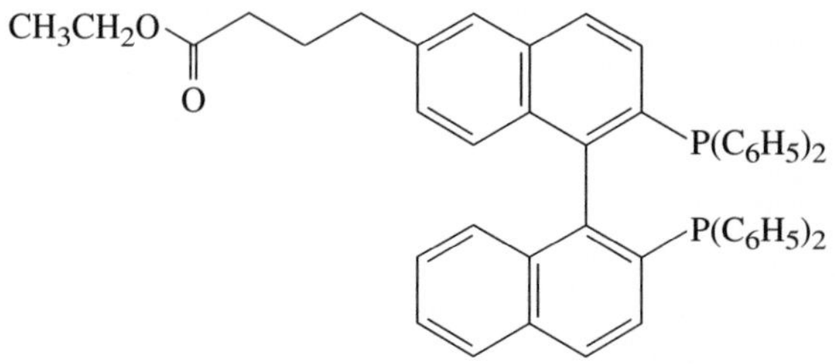

Which of the following has the proper functionality to react with this ester by nucleophilic acyl substitution to give a polystyrene-supported ligand?

A. PS—CH_2Cl

C. PS—$CH_2N(CH_3)_2$

B. PS—CH_2NH_2

D. PS—$CH_2\overset{+}{N}(CH_3)_3$ Cl^-

27.38 The polystyrene-supported quaternary ammonium chloride shown was treated with aqueous sodium hydroxide, then shaken with a solution of compound X and phenol in toluene at 90°C to give butyl phenyl ether in 97% yield. What is compound X?

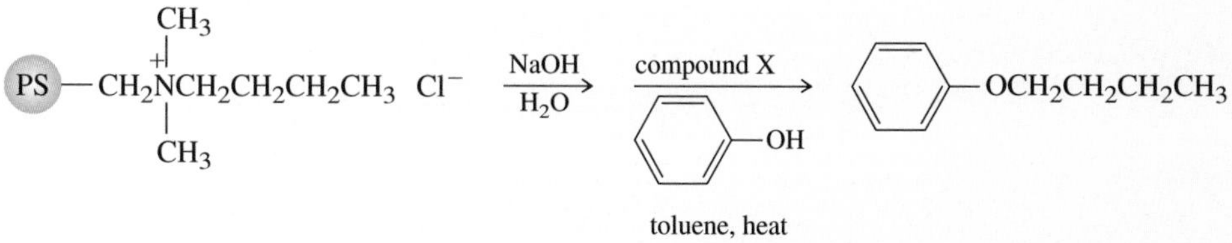

A. $CH_3CH_2CH_2CH_2OH$

C. $CH_3CH_2CH_2CH_2Br$

B. $CH_3CH_2CH{=}CH_2$

D. $CH_3CH_2CH_2CH_2NH_2$

Glossary

A

Absolute configuration: The three-dimensional arrangement of atoms or groups at a chirality center.

Absorbance: In UV-VIS spectroscopy, the value of $\log_{10}(I_0/I)$, where I_0 is the intensity of the incident radiation and I is the intensity of the beam after it has passed through the sample.

Acetal: Product of the reaction of an aldehyde or a ketone with two moles of an alcohol according to the equation

$$\mathrm{R\overset{O}{\overset{\|}{C}}R'} + 2\mathrm{R''OH} \xrightarrow{\mathrm{H^+}} \mathrm{R\overset{OR''}{\underset{OR''}{C}}R'} + \mathrm{H_2O}$$

Acetoacetic ester synthesis: A synthetic method for the preparation of ketones in which alkylation of the enolate of ethyl acetoacetate

$$\mathrm{CH_3\overset{O}{\overset{\|}{C}}CH_2\overset{O}{\overset{\|}{C}}OCH_2CH_3}$$

is the key carbon–carbon bond-forming step.

Acetyl coenzyme A: A thioester abbreviated as

$$\mathrm{CH_3\overset{O}{\overset{\|}{C}}SCoA}$$

that acts as the source of acetyl groups in biosynthetic processes involving acetate.

Acetylene: The simplest alkyne, $\mathrm{HC{\equiv}CH}$.

Achiral: Opposite of *chiral.* An achiral object is superimposable on its mirror image.

Acid: According to the Arrhenius definition, a substance that ionizes in water to produce protons. According to the Brønsted–Lowry definition, a substance that donates a proton to some other substance. According to the Lewis definition, an electron-pair acceptor.

Acid anhydride: Compound of the type

$$\mathrm{R\overset{O}{\overset{\|}{C}}O\overset{O}{\overset{\|}{C}}R}$$

Both R groups are usually the same, although they need not always be.

Acidity constant K_a: Equilibrium constant for dissociation of an acid:

$$K_a = \frac{[\mathrm{H^+}][\mathrm{A^-}]}{[\mathrm{HA}]}$$

Activating substituent: A group that when present in place of a hydrogen causes a particular reaction to occur faster. Term is most often applied to substituents that increase the rate of electrophilic aromatic substitution.

Activation energy: The minimum energy that a reacting system must possess above its most stable state in order to undergo a chemical or structural change.

Active site: The region of an enzyme at which the substrate is bound.

Acylation: Reaction in which an acyl group becomes attached to some structural unit in a molecule. Examples include the Friedel–Crafts acylation and the conversion of amines to amides.

Acyl cation: Synonymous with *acylium ion.*

Acyl chloride: Compound of the type

$$\mathrm{R\overset{O}{\overset{\|}{C}}Cl}$$

R may be alkyl or aryl.

Acyl group: The group

$$\mathrm{R\overset{O}{\overset{\|}{C}}{-}}$$

R may be alkyl or aryl.

Acylium ion: The cation $\mathrm{R{-}C{\equiv}\overset{+}{O}}$:

Acyl transfer: A nucleophilic acyl substitution. A reaction in which one type of carboxylic acid derivative is converted to another.

Addition: Reaction in which a reagent X—Y adds to a multiple bond so that X becomes attached to one of the carbons of the multiple bond and Y to the other.

1,2 Addition: Addition of reagents of the type X—Y to conjugated dienes in which X and Y add to adjacent doubly bonded carbons:

$$\mathrm{R_2C{=}CH{-}CH{=}CR_2} \xrightarrow{\mathrm{X{-}Y}} \mathrm{R_2\underset{X}{C}{-}\underset{Y}{C}H{-}CH{=}CR_2}$$

1,4 Addition: Addition of reagents of the type X—Y to conjugated dienes in which X and Y add to the termini of the diene system (see *conjugate addition*).

$$\mathrm{R_2C{=}CH{-}CH{=}CR_2} \xrightarrow{\mathrm{X{-}Y}} \mathrm{R_2\underset{X}{C}{-}CH{=}CH{-}\underset{Y}{C}R_2}$$

Addition–elimination mechanism: Two-stage mechanism for nucleophilic aromatic substitution. In the addition stage, the nucleophile adds to the carbon that bears the leaving group. In the elimination stage, the leaving group is expelled.

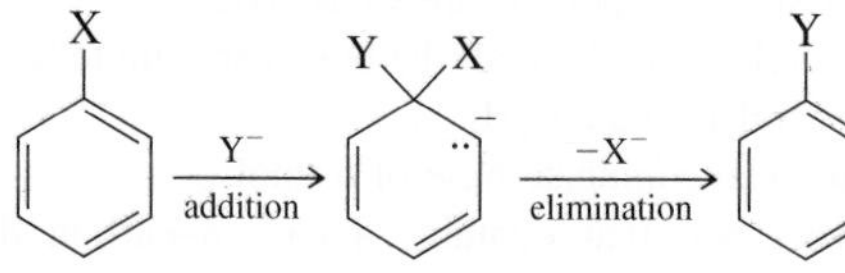

Addition polymer: A polymer formed by addition reactions of monomers.

Adenosine 5′-triphosphate (ATP): The main energy-storing compound in all living organisms.

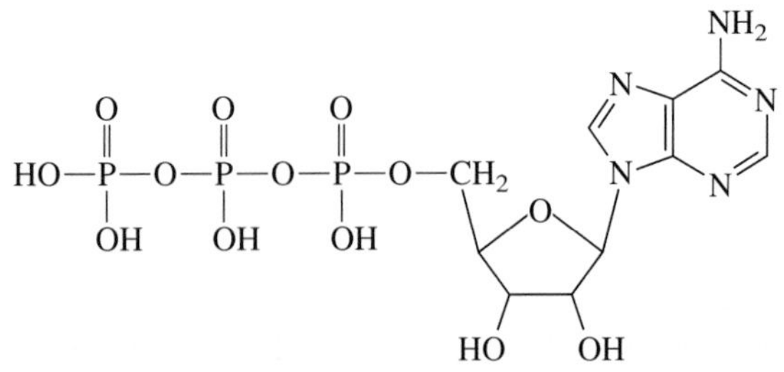

Alcohol: Compound of the type ROH.

Aldaric acid: Carbohydrate in which carboxylic acid functions are present at both ends of the chain. Aldaric acids are typically prepared by oxidation of aldoses with nitric acid.

Aldehyde: Compound of the type

$$\overset{O}{\overset{\|}{RCH}} \quad \text{or} \quad \overset{O}{\overset{\|}{ArCH}}$$

Alditol: The polyol obtained on reduction of the carbonyl group of a carbohydrate.

Aldol addition: Nucleophilic addition of an aldehyde or ketone enolate to the carbonyl group of an aldehyde or a ketone. The most typical case involves two molecules of an aldehyde, and is usually catalyzed by bases.

$$2RCH_2\overset{O}{\overset{\|}{C}}H \xrightarrow{HO^-} RCH_2\overset{OH}{\overset{|}{C}}H\underset{\underset{CH=O}{|}}{C}HR$$

Aldol condensation: When an aldol addition is carried out so that the β-hydroxy aldehyde or ketone dehydrates under the conditions of its formation, the product is described as arising by an aldol condensation.

$$2RCH_2\overset{O}{\overset{\|}{C}}H \xrightarrow[\text{heat}]{HO^-} RCH_2CH=\underset{\underset{CH=O}{|}}{C}R \quad + H_2O$$

Aldonic acid: Carboxylic acid obtained by oxidation of the aldehyde function of an aldose.

Aldose: Carbohydrate that contains an aldehyde carbonyl group in its open-chain form.

Aliphatic: Term applied to compounds that do not contain benzene or benzene-like rings as structural units. (Historically, *aliphatic* was used to describe compounds derived from fats and oils.)

Alkadiene: Hydrocarbon that contains two carbon–carbon double bonds; commonly referred to as a *diene.*

Alkaloid: Amine that occurs naturally in plants. The name derives from the fact that such compounds are weak bases.

Alkane: Hydrocarbon in which all the bonds are single bonds. Alkanes have the general formula C_nH_{2n+2}.

Alkanethiolate: The conjugate base of a thiol.

Alkene: Hydrocarbon that contains a carbon–carbon double bond (C=C); also known by the older name *olefin.*

Alkoxide ion: Conjugate base of an alcohol; a species of the type $R—\ddot{\underset{..}{O}}:^-$.

Alkylamine: Amine in which the organic groups attached to nitrogen are alkyl groups.

Alkylation: Reaction in which an alkyl group is attached to some structural unit in a molecule.

Alkyl group: Structural unit related to an alkane by replacing one of the hydrogens by a potential point of attachment to some other atom or group. The general symbol for an alkyl group is R—.

Alkyl halide: Compound of the type RX, in which X is a halogen substituent (F, Cl, Br, I).

Alkyloxonium ion: Positive ion of the type ROH_2^+.

Alkyne: Hydrocarbon that contains a carbon–carbon triple bond.

Allene: The compound $H_2C=C=CH_2$.

Allyl group: The group

$$H_2C=CHCH_2—$$

Allylic anion: A carbanion in which the negatively charged carbon is allylic.

Allylic carbocation: A carbocation in which the positively charged carbon is allylic.

Allylic carbon: The sp^3-hybridized carbon of a C=C—C unit. Atoms or groups attached to an allylic carbon are termed *allylic substituents.*

Allylic free radical: A free radical in which the unpaired electron is on an allylic carbon.

Allylic rearrangement: Functional group transformation in which double-bond migration has converted one allylic structural unit to another, as in:

$$R_2C=CHCH_2X \longrightarrow R_2\underset{\underset{Y}{|}}{C}CH=CH_2$$

Amide: Compound of the type $R\overset{O}{\overset{\|}{C}}NR'_2$.

Amine: Molecule in which a nitrogen-containing group of the type $—NH_2$, —NHR, or $—NR_2$ is attached to an alkyl or aryl group.

α-Amino acid: A carboxylic acid that contains an amino group at the α-carbon atom. α-Amino acids are the building blocks of peptides and proteins. An α-amino acid normally exists as a *zwitterion.*

$$R\underset{\underset{^+NH_3}{|}}{C}HCO_2^-$$

L-Amino acid: The Fischer projection of an L-amino acid has the amino group on the left when the carbon chain is vertical with the carboxyl group at the top.

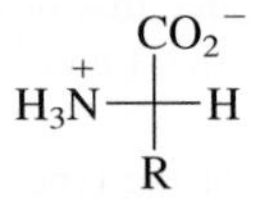

Amino acid racemization: A method for dating archeological samples based on the rate at which the stereochemistry at the α carbon of amino acid components is randomized. It is useful for samples too old to be reliably dated by ^{14}C decay.

Amino acid residues: Individual amino acid components of a peptide or protein.

Amino sugar: Carbohydrate in which one of the hydroxyl groups has been replaced by an amino group.

Amphiphilic: Possessing both hydrophilic and lipophilic properties within the same species.

Amylopectin: A polysaccharide present in starch. Amylopectin is a polymer of α-(1→4)-linked glucose units, as is amylose (see *amylose*). Unlike amylose, amylopectin contains branches of 24–30 glucose units connected to the main chain by an α-(1→6) linkage.

Amylose: The water-dispersible component of starch. It is a polymer of α-(1→4)-linked glucose units.

Androgen: A male sex hormone.

Angle strain: The strain a molecule possesses because its bond angles are distorted from their normal values.

Anion: Negatively charged ion.

Anionic polymerization: A polymerization in which the reactive intermediates are negatively charged.

Annulene: Monocyclic hydrocarbon characterized by a completely conjugated system of double bonds. Annulenes may or may not be aromatic.

[*x*]Annulene: An annulene in which the ring contains *x* carbons.

Anomeric carbon: The carbon atom in a furanose or pyranose form that is derived from the carbonyl carbon of the open-chain form. It is the ring carbon that is bonded to two oxygens.

Anomeric effect: The preference for an electronegative substituent, especially a hydroxyl group, to occupy an axial orientation when bonded to the anomeric carbon in the pyranose form of a carbohydrate.

Anti: Term describing relative position of two substituents on adjacent atoms when the angle between their bonds is on the order of 180°. Atoms X and Y in the structure shown are anti to each other.

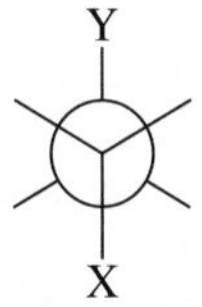

Anti addition: Addition reaction in which the two portions of the attacking reagent X—Y add to opposite faces of the double bond.

Antiaromatic: The quality of being destabilized by electron delocalization.

Antibonding orbital: An orbital in a molecule in which an electron is less stable than when localized on an isolated atom.

Anticodon: Sequence of three bases in a molecule of tRNA that is complementary to the codon of mRNA for a particular amino acid.

Aprotic solvent: A solvent that does not have easily exchangeable protons such as those bonded to oxygen of hydroxyl groups.

Aramid: A polyamide of a benzenedicarboxylic acid and a benzenediamine.

Arene: Aromatic hydrocarbon. Often abbreviated ArH.

Arenium ion: The carbocation intermediate formed by attack of an electrophile on an aromatic substrate in electrophilic aromatic substitution. See *cyclohexadienyl cation.*

Aromatic hydrocarbon: An electron-delocalized species that is much more stable than any structure written for it in which all the electrons are localized either in covalent bonds or as unshared electron pairs.

Aromaticity: Special stability associated with aromatic compounds.

Arylamine: An amine that has an aryl group attached to the amine nitrogen.

Aryne: A species that contains a triple bond within an aromatic ring (see *benzyne*).

Asymmetric: Lacking all significant symmetry elements; an asymmetric object does not have a plane, axis, or center of symmetry.

Asymmetric synthesis: The stereoselective introduction of a chirality center in a reactant in which the stereoisomeric products are formed in unequal amounts.

Atactic polymer: Polymer characterized by random stereochemistry at its chirality centers. An atactic polymer, unlike an isotactic or a syndiotactic polymer, is not a stereoregular polymer.

Atomic number: The number of protons in the nucleus of a particular atom. The symbol for atomic number is *Z*, and each element has a unique atomic number.

Atropisomers: Stereoisomers that result from restricted rotation about single bonds where the barrier for rotation is sufficient to allow isolation of the isomers.

Axial bond: A bond to a carbon in the chair conformation of cyclohexane oriented like the six "up-and-down" bonds in the following:

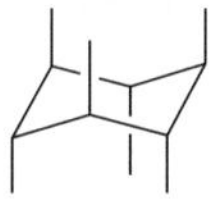

Azo coupling: Formation of a compound of the type ArN=NAr′ by reaction of an aryl diazonium salt with an arene. The arene must be strongly activated toward electrophilic aromatic substitution; that is, it must bear a powerful electron-releasing substituent such as —OH or $—NR_2$.

B

Baeyer–Villiger oxidation: Oxidation of an aldehyde or, more commonly, a ketone with a peroxy acid. The product of Baeyer–Villiger oxidation of a ketone is an ester.

$$\mathrm{R\overset{O}{\overset{\|}{C}}R'} \xrightarrow{\mathrm{R''\overset{O}{\overset{\|}{C}}OOH}} \mathrm{R\overset{O}{\overset{\|}{C}}OR'}$$

Base: According to the Arrhenius definition, a substance that ionizes in water to produce hydroxide ions. According to the Brønsted–Lowry definition, a substance that accepts a proton from some suitable donor. According to the Lewis definition, an electron-pair donor.

Base pair: Term given to the purine of a nucleotide and its complementary pyrimidine. Adenine (A) is complementary to thymine (T), and guanine (G) is complementary to cytosine (C).

Base peak: The most intense peak in a mass spectrum. The base peak is assigned a relative intensity of 100, and the intensities of all other peaks are cited as a percentage of the base peak.

Basicity constant K_b: A measure of base strength, especially of amines.

$$K_b = \frac{[R_3NH^+][HO^-]}{[R_3N]}$$

Bending vibration: The regular, repetitive motion of an atom or a group along an arc the radius of which is the bond connecting the atom or group to the rest of the molecule. Bending vibrations are one type of molecular motion that gives rise to a peak in the infrared spectrum.

Benzyl group: The group $C_6H_5CH_2—$.

Benzylic carbon: A carbon directly attached to a benzene ring. A hydrogen attached to a benzylic carbon is a benzylic hydrogen. A carbocation in which the benzylic carbon is positively charged is a benzylic carbocation. A free radical in which the benzylic carbon bears the unpaired electron is a benzylic radical.

Benzyne: Benzene that lacks two hydrogens.

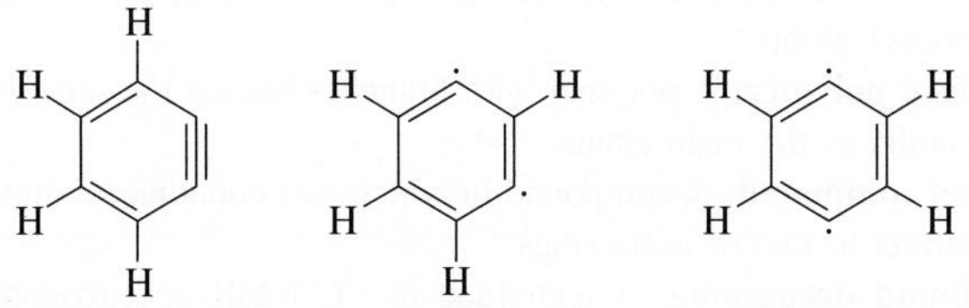

o-Benzyne *m*-Benzyne *p*-Benzyne

Biaryl: A Compound in which two aromatic rings are joined by a single bond.

Bile acids: Steroid derivatives biosynthesized in the liver that aid digestion by emulsifying fats.

Bimolecular: A process in which two particles react in the same elementary step.

Bioenergetics: The study of energy transfer in biological processes.

Biological isoprene unit: Isopentenyl diphosphate, the biological precursor to terpenes and steroids:

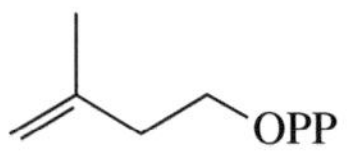

Birch reduction: Reduction of an aromatic ring to a 1,4-cyclohexadiene on treatment with a group 1 metal (Li, Na, K) and an alcohol in liquid ammonia.

Block copolymer: A copolymer of monomers A and B in which sections of poly-A and poly-B of variable length alternate.

Boat conformation: An unstable conformation of cyclohexane, depicted as

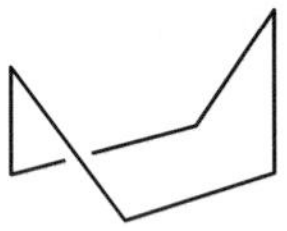

π bond: In alkenes, a bond formed by overlap of *p* orbitals in a side-by-side manner. A π bond is weaker than a σ bond. The carbon–carbon double bond in alkenes consists of two sp^2-hybridized carbons joined by a σ bond and a π bond.

σ bond: A connection between two atoms in which the orbitals involved overlap along the internuclear axis. A cross section perpendicular to the internuclear axis is a circle.

Bond dipole moment: The dipole moment of a bond between two atoms.

Bond dissociation enthalpy: For a substance A:B, the energy required to break the bond between A and B so that each retains one of the electrons in the bond.

Bonding orbital: An orbital in a molecule in which an electron is more stable than when localized on an isolated atom. All the bonding orbitals are normally doubly occupied in stable neutral molecules.

Bond-length distortion: The deviation of the length of a bond between two atoms from its normal value.

Bond-line formula: Formula in which connections between carbons are shown but individual carbons and hydrogens are not. The bond-line formula

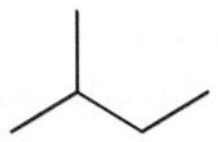

represents the compound $(CH_3)_2CHCH_2CH_3$.

Boundary surface: The surface that encloses the region where the probability of finding an electron is high (90–95%).

Branched-chain carbohydrate: Carbohydrate in which the main carbon chain bears a carbon substituent in place of a hydrogen or hydroxyl group.

Branched polymer: A polymer with branches having the same repeating units as the main chain.

Bridged compound: A compound in which two nonadjacent atoms are common to two or more rings.

Broadband decoupling: A technique in ^{13}C NMR spectroscopy that removes the splitting of ^{13}C signals caused by coupling of ^{13}C and ^{1}H nuclei. Thus, all of the ^{13}C signals appear as singlets.

Bromohydrin: A halohydrin in which the halogen is bromine (see *halohydrin*).

Bromonium ion: A halonium ion in which the halogen is bromine (see *halonium ion*).

Buckminsterfullerene: (Chapter 11, essay, "Carbon Clusters, Fullerenes, and Nanotubes"): Name given to the C_{60} cluster with structure resembling the geodesic domes of R. Buckminster Fuller.

***n*-Butane:** Common name for butane $CH_3CH_2CH_2CH_3$.

***n*-Butyl group:** The group $CH_3CH_2CH_2CH_2$—.

***sec*-Butyl group:** The group

$$CH_3CH_2\underset{|}{C}HCH_3$$

***tert*-Butyl group:** The group $(CH_3)_3C$—.

C

Cahn–Ingold–Prelog system: System for specifying absolute configuration as *R* or *S* on the basis of the order in which atoms or groups are attached to a chirality center. Groups are ranked in order of precedence according to rules based on atomic number.

Carbanion: Anion in which the negative charge is borne by carbon. An example is acetylide ion.

Carbene: A neutral species in which one of the carbon atoms is associated with six valence electrons.

Carbenoid: A compound, usually organometallic, that resembles a carbene in its chemical reactions.

Carbinolamine: Compound of the type

$$HO—\overset{|}{\underset{|}{C}}—NR_2$$

Carbinolamines are formed by nucleophilic addition of an amine to a carbonyl group and are intermediates in the formation of imines and enamines.

Carbocation: Positive ion in which the charge resides on carbon. An example is *tert*-butyl cation, $(CH_3)_3C^+$. Carbocations are unstable species that, though they cannot normally be isolated, are believed to be intermediates in certain reactions.

Carbon skeleton diagram: Synonymous with bond-line formula.

Carboxylic acid: Compound of the type $R\overset{\overset{O}{\|}}{C}OH$, also written as RCO_2H.

Carboxylic acid derivative: Compound that yields a carboxylic acid on hydrolysis. Carboxylic acid derivatives include acyl chlorides, acid anhydrides, esters, and amides.

Carcinogen: A cancer-causing substance.

Carotenoids: Naturally occurring tetraterpenoid compounds found in plants and animals.

Catalyst: (Introduction): A substance that increases the rate of a chemical reaction, but is not consumed by it.

Cation: Positively charged ion.

Cationic polymerization: A polymerization in which the reactive intermediates are carbocations.

Cation radical: A positively charged species that has an odd number of electrons.

Cellobiose: A disaccharide in which two glucose units are joined by a β-(1→4) linkage. Cellobiose is obtained by the hydrolysis of cellulose.

Cellulose: A polysaccharide in which thousands of glucose units are joined by β-(1→4) linkages.

Center of symmetry: A point in the center of a structure located so that a line drawn from it to any element of the structure, when extended an equal distance in the opposite direction, encounters an identical element. Benzene, for example, has a center of symmetry.

Chain-growth polymerization: Macromolecule formation by a process in which monomers add sequentially to one end of a chain.

Chain reaction: Reaction mechanism in which a sequence of individual steps repeats itself many times, usually because a reactive intermediate consumed in one step is regenerated in a subsequent step. The halogenation of alkanes is a chain reaction proceeding via free-radical intermediates.

Chain terminating step: A chemical reaction that stops further growth of a polymer chain.

Chain transfer: A reaction between a growing chain and a terminated chain that terminates the growing chain and activates the previously terminated chain to further growth.

Chair–chair interconversion: Synonymous with ring inversion of cyclohexane and related compounds.

Chair conformation: The most stable conformation of cyclohexane:

Characteristic absorption frequencies: The regions of the infrared (IR) spectrum where peaks characteristic of particular structural units are normally found.

Chemical bond: A connection between atoms.

Chemically nonequivalent: In NMR, synonymous with *chemical-shift-nonequivalent.*

Chemical shift: A measure of how shielded the nucleus of a particular atom is. Nuclei of different atoms have different chemical shifts, and nuclei of the same atom have chemical shifts that are sensitive to their molecular environment. In proton and carbon-13 NMR, chemical shifts are cited as δ, or parts per million (ppm), from the hydrogens or carbons, respectively, of tetramethylsilane.

Chemical-shift-nonequivalent: Nuclei with different chemical shifts in nuclear magnetic resonance (NMR).

Chiral: Term describing an object that is not superimposable on its mirror image.

Chirality axis: Line drawn through a molecule that is analogous to the long axis of a right-handed or left-handed screw or helix.

Chirality center: An atom that has four nonequivalent atoms or groups attached to it. At various times chirality centers have been called *asymmetric centers* or *stereogenic centers.*

Chlorohydrin: A halohydrin in which the halogen is chlorine (see *halohydrin*).

Chloronium ion: A halonium ion in which the halogen is chlorine (see *halonium ion*).

Cholesterol: The most abundant steroid in animals and the biological precursor to other naturally occurring steroids, including the bile acids, sex hormones, and corticosteroids.

Chromatography: A method for separation and analysis of mixtures based on the different rates at which different compounds are removed from a stationary phase by a moving phase.

Chromophore: The structural unit of a molecule responsible for absorption of radiation of a particular frequency; a term usually applied to ultraviolet-visible spectroscopy.

***cis*-:** Stereochemical prefix indicating that two substituents are on the same side of a ring or double bond. (Contrast with the prefix *trans*-.)

Claisen condensation: Reaction in which a β-keto ester is formed by condensation of two moles of an ester in base:

$$RCH_2\overset{O}{\overset{\|}{C}}OR' \xrightarrow[2.\ H_3O^+]{1.\ NaOR'} RCH_2\overset{O}{\overset{\|}{C}}\underset{R}{\underset{|}{C}}H\overset{O}{\overset{\|}{C}}OR' + R'OH$$

Claisen rearrangement: Thermal conversion of an allyl phenyl ether to an *o*-allyl phenol. The rearrangement proceeds via a cyclohexadienone intermediate.

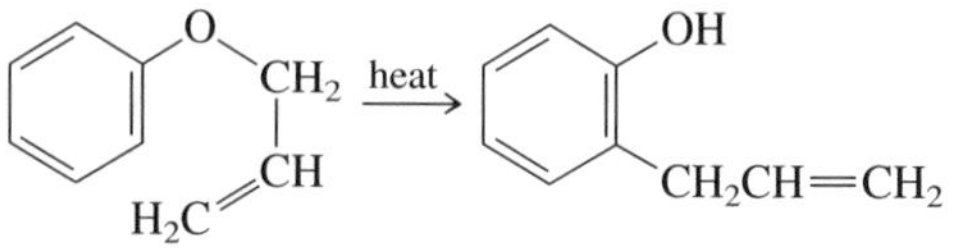

Claisen–Schmidt condensation: A mixed aldol condensation involving a ketone enolate and an aromatic aldehyde or ketone.

Clemmensen reduction: Method for reducing the carbonyl group of aldehydes and ketones to a methylene group ($C{=}O \rightarrow CH_2$) by treatment with zinc amalgam [Zn(Hg)] in concentrated hydrochloric acid.

Closed-shell electron configuration: Stable electron configuration in which all the lowest energy orbitals of an atom (in the case of the noble gases), an ion (e.g., Na^+), or a molecule (e.g., benzene) are filled.

^{13}C NMR: Nuclear magnetic resonance spectroscopy in which the environments of individual carbon atoms are examined via their mass-13 isotope.

Codon: Set of three successive nucleotides in mRNA that is unique for a particular amino acid. The 64 codons possible from combinations of A, T, G, and C code for the 20 amino acids from which proteins are constructed.

Coenzyme: A cofactor that is not chemically bonded to an enzyme.

Coenzyme Q: Naturally occurring group of related quinones involved in the chemistry of cellular respiration. Also known as ubiquinone.

Cofactor: A molecule that acts in combination with an enzyme to bring about a reaction. A cofactor may be either a coenzyme or a prosthetic group.

Combinatorial chemistry: A method for carrying out a large number of reactions on a small scale in the solid phase so as to generate a "library" of related compounds for further study, such as biological testing.

Combustion: Burning of a substance in the presence of oxygen. All hydrocarbons yield carbon dioxide and water when they undergo combustion.

Common name: Name given to a compound on some basis other than a systematic set of rules.

σ-Complex: Synonymous with *arenium ion.*

Compound: An assembly of two or more atoms with properties different from the individual atoms.

Concerted reaction: Reaction that occurs in a single elementary step.

Condensation polymer: Polymer in which the bonds that connect the monomers are formed by condensation reactions. Typical condensation polymers include polyesters and polyamides.

Condensation reaction: Reaction in which two molecules combine to give a product accompanied by the expulsion of some small stable molecule (such as water). An example is acid-catalyzed ether formation:

$$2ROH \xrightarrow{H_2SO_4} ROR + H_2O$$

Condensed formula: Structural formula in which subscripts are used to indicate replicated atoms or groups, as in $(CH_3)_2CHCH_2CH_3$.

Conformational analysis: Study of the conformations available to a molecule, their relative stability, and the role they play in defining the properties of the molecule.

Conformational enantiomers: Nonsuperimposable mirror-image conformations of a molecule.

Conformations: Nonidentical representations of a molecule generated by rotation about single bonds.

Conformers: Different conformations of a single molecule.

Conjugate acid: The species formed from a Brønsted base after it has accepted a proton.

Conjugate addition: Addition reaction in which the reagent adds to the termini of the conjugated system with migration of the double bond; synonymous with 1,4 addition. The most common examples include conjugate addition to 1,3-dienes and to α,β-unsaturated carbonyl compounds.

Conjugate base: The species formed from a Brønsted acid after it has donated a proton.

Conjugated diene: System of the type C=C—C=C, in which two pairs of doubly bonded carbons are joined by a single bond. The π electrons are delocalized over the unit of four consecutive sp^2-hybridized carbons.

Conjugated system: A structural arrangement in which electron delocalization permits two groups to interact so that the properties of the conjugated system are different from those of the separate groups.

Conjugation energy: Synonymous with *resonance energy.*

Connectivity: Order in which a molecule's atoms are connected. Synonymous with *constitution.*

Constitutional isomers: Isomers that differ in respect to the order in which the atoms are connected. Butane ($CH_3CH_2CH_2CH_3$) and isobutane [$(CH_3)_3CH$] are constitutional isomers.

Contributing structures: The various resonance structures that can be written for a molecule.

Coordination polymerization: A method of addition polymerization in which monomers are added to the growing chain on an active organometallic catalyst.

Copolymer: Polymer formed from two or more different monomers.

Corticosteroid: A steroid present in the outer layer, or *cortex,* of the adrenal gland.

COSY: A 2D NMR technique that correlates the chemical shifts of spin-coupled nuclei. COSY stands for correlated spectroscopy.

Coulombic attraction: The electrical attraction between opposite charges.

Coupling constant *J*: A measure of the extent to which two nuclear spins are coupled. In the simplest cases, it is equal to the distance between adjacent peaks in a split NMR signal.

Covalent bond: Chemical bond between two atoms that results from their sharing of two electrons.

COX-2: Cyclooxygenase-2, an enzyme that catalyzes the biosynthesis of prostaglandins. COX-2 inhibitors reduce pain and inflammation by blocking the activity of this enzyme.

Cracking: A key step in petroleum refining in which high-molecular-weight hydrocarbons are converted to lower molecular-weight ones by thermal or catalytic carbon–carbon bond cleavage.

Critical micelle concentration: Concentration above which substances such as salts of fatty acids aggregate to form micelles in aqueous solution.

Cross-linked polymer: A polymer in which two or more chains are covalently bonded.

Crown ether: A cyclic polyether that, via ion–dipole attractive forces, forms stable complexes with metal ions. Such complexes, along with their accompanying anion, are soluble in nonpolar solvents.

Crystallite: An ordered crystalline region within a polymer.

C terminus: The amino acid at the end of a peptide or protein chain that has its carboxyl group intact—that is, in which the carboxyl group is not part of a peptide bond.

Cumulated diene: Diene of the type C=C=C, in which one carbon has double bonds to two others.

Curved arrows: Arrows that show the direction of electron flow in chemical reactions; also used to show differences in electron placement between resonance forms.

Cyanohydrin: Compound of the type

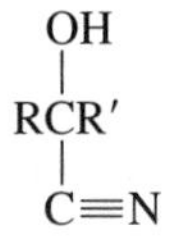

Cyanohydrins are formed by nucleophilic addition of HCN to the carbonyl group of an aldehyde or a ketone.

Cycloaddition: Addition, such as the Diels–Alder reaction, in which a ring is formed via a cyclic transition state.

Cycloalkane: An alkane in which a ring of carbon atoms is present.

Cycloalkene: A cyclic hydrocarbon characterized by a double bond between two of the ring carbons.

Cycloalkyne: A cyclic hydrocarbon characterized by a triple bond between two of the ring carbons.

Cyclohexadienyl anion: The key intermediate in nucleophilic aromatic substitution by the addition–elimination mechanism. It is represented by the general structure shown, where Y is the nucleophile and X is the leaving group.

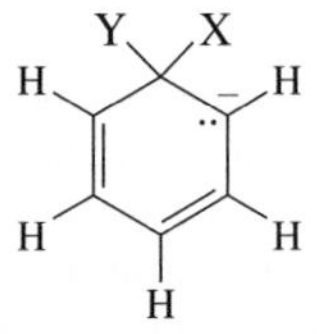

Cyclohexadienyl cation: The key intermediate in electrophilic aromatic substitution reactions. It is represented by the general structure

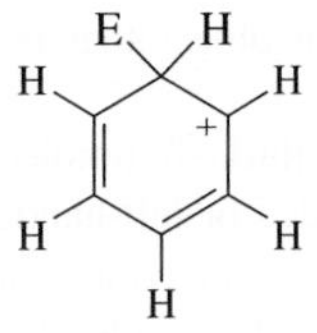

where E is derived from the electrophile that reacts with the ring.

D

Deactivating substituent: A group that when present in place of hydrogen causes a particular reaction to occur more slowly. The term is most often applied to the effect of substituents on the rate of electrophilic aromatic substitution.

Debye unit (D): Unit customarily used for measuring dipole moments:

$$1\text{D} = 1 \times 10^{-18} \text{ esu·cm.}$$

Decarboxylation: Reaction of the type $RCO_2H \rightarrow RH + CO_2$, in which carbon dioxide is lost from a carboxylic acid. Decarboxylation normally occurs readily only when the carboxylic acid is a 1,3-dicarboxylic acid or a β-keto acid.

Decoupling: In NMR spectroscopy, any process that destroys the coupling of nuclear spins between two nuclei. Two types of decoupling are employed in ^{13}C NMR spectroscopy. Broadband decoupling removes all the ^1H–^{13}C couplings; off-resonance decoupling removes all ^1H–^{13}C couplings except those between directly bonded atoms.

Dehydration: Removal of H and OH from adjacent atoms. The term is most commonly employed in the preparation of alkenes by heating alcohols in the presence of an acid catalyst.

1,2-, 1,3- and 1,4-Dehydrobenzene: See *benzyne*.

Dehydrogenation: Elimination in which H_2 is lost from adjacent atoms. The term is most commonly encountered in the industrial preparation of ethylene from ethane, propene from propane, 1,3-butadiene from butane, and styrene from ethylbenzene.

Dehydrohalogenation: Reaction in which an alkyl halide, on being treated with a base such as sodium ethoxide, is converted to an alkene by loss of a proton from one carbon and the halogen from the adjacent carbon.

Delocalization: Association of an electron with more than one atom. The simplest example is the shared electron pair (covalent) bond. Delocalization is important in conjugated π electron systems, where an electron may be associated with several carbon atoms.

Delocalization energy: Synonymous with *resonance energy.*

Deoxy sugar: A carbohydrate in which one of the hydroxyl groups has been replaced by a hydrogen.

DEPT: Abbreviation for *d*istortionless *e*nhancement of *p*olarization *t*ransfer. DEPT is an NMR technique that reveals the number of hydrogens directly attached to a carbon responsible for a particular signal.

Detergents: Substances that clean by micellar action. Although the term usually refers to a synthetic detergent, soaps are also detergents.

Deuterium isotope effect: The difference in a property, usually reaction rate, that results when one or more atoms of ^1H in a compound are replaced by ^2H.

Diastereomers: Stereoisomers that are not enantiomers—stereoisomers that are not mirror images of one another.

Diastereotopic: Describing two atoms or groups in a molecule that are attached to the same atom but are in stereochemically different environments that are not mirror images of each other. The two protons shown in bold in $\mathbf{H_2}C{=}CHCl$, for example, are diastereotopic. One is cis to chlorine, the other is trans.

1,3-Diaxial repulsion: Repulsive forces between axial substituents on the same side of a cyclohexane ring.

Diazonium ion: Ion of the type $R{-}\overset{+}{N}{\equiv}N\colon$. Aryl diazonium ions are formed by treatment of primary aromatic amines with nitrous acid. They are extremely useful in the preparation of aryl halides, phenols, and aryl cyanides.

Diazotization: The reaction by which a primary amine is converted to the corresponding diazonium ion by nitrosation.

Dieckmann cyclization: An intramolecular version of the Claisen condensation.

Dielectric constant: A measure of the ability of a material to disperse the force of attraction between oppositely charged particles. The symbol for dielectric constant is ϵ.

Diels–Alder reaction: Conjugate addition of an alkene to a conjugated diene to give a cyclohexene derivative. Diels–Alder reactions are extremely useful in synthesis.

Dienophile: The alkene that adds to the diene in a Diels–Alder reaction.

Dihydroxylation: Reaction or sequence of reactions in which an alkene is converted to a vicinal diol.

β-Diketone: Compound of the type

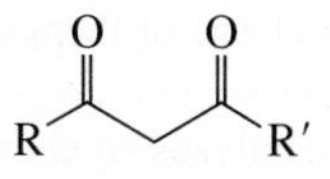

also referred to as a 1,3-diketone.

Dimer: Molecule formed by the combination of two identical molecules.

Diol: A compound with two alcohol functional groups.

Dipeptide: A compound in which two α-amino acids are linked by an amide bond between the amino group of one and the carboxyl group of the other:

$$H_3\overset{+}{N}\underset{R}{\underset{|}{C}}H\overset{O}{\overset{\|}{C}}{-}NH\underset{R'}{\underset{|}{C}}HCO_2^-$$

Dipole–dipole attractive force: A force of attraction between oppositely polarized atoms.

Dipole/induced-dipole force: A force of attraction that results when a species with a permanent dipole induces a complementary dipole in a second species.

Dipole moment: Product of the attractive force between two opposite charges and the distance between them. Dipole moment has the symbol μ and is measured in Debye units (D).

Direct addition: Synonymous with *1,2-addition.*

Disaccharide: A carbohydrate that yields two monosaccharide units (which may be the same or different) on hydrolysis.

Disproportionation: A reaction in which transfer of an atom from one growing polymer chain to another terminates both.

Disubstituted alkene: Alkene of the type $R_2C{=}CH_2$ or $RCH{=}CHR$. The groups R may be the same or different, they may be any length, and they may be branched or unbranched. The significant point is that there are two carbons *directly* bonded to the carbons of the double bond.

Disulfide: A compound of the type RSSR′.

Disulfide bridge: An S—S bond between the sulfur atoms of two cysteine residues in a peptide or protein.

DNA (deoxyribonucleic acid): A polynucleotide of 2′-deoxyribose present in the nuclei of cells that serves to store and replicate genetic information. Genes are DNA.

Double bond: Bond formed by the sharing of four electrons between two atoms.

Double dehydrohalogenation: Reaction in which a geminal dihalide or vicinal dihalide, on being treated with a very strong base such as sodium amide, is converted to an alkyne by loss of two protons and the two halogen substituents.

Double helix: The form in which DNA normally occurs in living systems. Two complementary strands of DNA are associated with each other by hydrogen bonds between their base pairs, and each DNA strand adopts a helical shape.

Downfield: The low-field region of an NMR spectrum. A signal that is downfield with respect to another lies to its left in the spectrum.

E

***E*-:** Stereochemical descriptor used when higher ranked substituents are on opposite sides of a double bond.

E1: See *Elimination unimolecular (E1) mechanism.*

E2: See *Elimination bimolecular (E2) mechanism.*

Eclipsed conformation: Conformation in which bonds on adjacent atoms are aligned with one another. For example, the C—H bonds indicated in the structure shown are eclipsed.

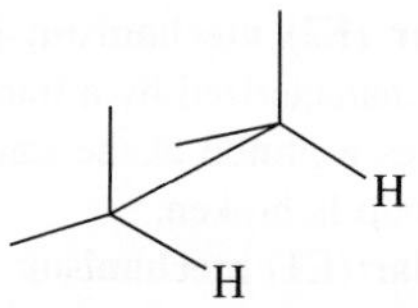

Edman degradation: Method for determining the N-terminal amino acid of a peptide or protein. It involves treating the material with phenyl isothiocyanate ($C_6H_5N{=}C{=}S$), cleaving with acid, and then identifying the phenylthiohydantoin (PTH derivative) produced.

Elastomer: A synthetic polymer that possesses elasticity.

Electromagnetic radiation: Various forms of radiation propagated at the speed of light. Electromagnetic radiation includes (among others) visible light; infrared, ultraviolet, and microwave radiation; and radio waves, cosmic rays, and X-rays.

Electron affinity: Energy change associated with the capture of an electron by an atom.

Electronegativity: A measure of the ability of an atom to attract the electrons in a covalent bond toward itself. Fluorine is the most electronegative element.

Electronic effect: An effect on structure or reactivity that is attributed to the change in electron distribution that a substituent causes in a molecule.

Electron impact: Method for producing positive ions in mass spectrometry whereby a molecule is bombarded by high-energy electrons.

Electron-releasing group: An atom or group that increases the electron density around another atom by an inductive or resonance effect.

18-Electron rule: The number of ligands that can be attached to a transition metal are such that the sum of the electrons brought by the ligands plus the valence electrons of the metal equals 18.

Electron-withdrawing group: An atom or group that decreases the electron density around another atom by an inductive or resonance effect.

Electrophile: A species (ion or compound) that can act as a Lewis acid, or electron pair acceptor; an "electron seeker." Carbocations are one type of electrophile.

Electrophilic addition: Mechanism of addition in which the species that first reacts with the multiple bond is an electrophile ("electron seeker").

Electrophilic aromatic substitution: Fundamental reaction type exhibited by aromatic compounds. An electrophilic species (E^+) replaces one of the hydrogens of an aromatic ring.

$$\text{Ar—H} + \text{E—Y} \longrightarrow \text{Ar—E} + \text{H—Y}$$

Electrostatic attraction: Force of attraction between oppositely charged particles.

Electrostatic potential map: The charge distribution in a molecule represented by mapping the interaction energy of a point positive charge with the molecule's electric field on the van der Waals surface.

Elementary step: A step in a reaction mechanism in which each species shown in the equation for this step participates in the same transition state. An elementary step is characterized by a single transition state.

Elements of unsaturation: See *index of hydrogen deficiency.*

β Elimination: Reaction in which a double or triple bond is formed by loss of atoms or groups from adjacent atoms. (See *dehydration, dehydrogenation, dehydrohalogenation,* and *double dehydrohalogenation.*)

Elimination–addition mechanism: Two-stage mechanism for nucleophilic aromatic substitution. In the first stage, an aryl halide undergoes elimination to form an aryne intermediate. In the second stage, nucleophilic addition to the aryne yields the product of the reaction.

Elimination bimolecular (E2) mechanism: Mechanism for elimination of alkyl halides characterized by a transition state in which the attacking base removes a proton at the same time that the bond to the halide leaving group is broken.

Elimination unimolecular (E1) mechanism: Mechanism for elimination characterized by the slow formation of a carbocation intermediate followed by rapid loss of a proton from the carbocation to form the alkene.

Enamine: Product of the reaction of a secondary amine and an aldehyde or a ketone. Enamines are characterized by the general structure

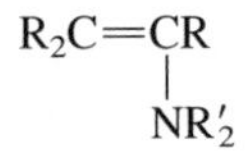

Enantiomeric excess: Difference between the percentage of the major enantiomer present in a mixture and the percentage of its mirror image. An optically pure material has an enantiomeric excess of 100%. A racemic mixture has an enantiomeric excess of zero.

Enantiomers: Stereoisomers that are related as an object and its nonsuperimposable mirror image.

Enantioselective synthesis: Reaction that converts an achiral or racemic starting material to a chiral product in which one enantiomer is present in excess of the other.

Enantiotopic: Describing two atoms or groups in a molecule whose environments are nonsuperimposable mirror images of each other. The two protons shown in bold in $CH_3\mathbf{CH_2}Cl$, for example, are enantiotopic. Replacement of first one, then the other, by some arbitrary test group yields compounds that are enantiomers of each other.

Endergonic: A process in which $\Delta G°$ is positive.

Endothermic: Term describing a process or reaction that absorbs heat.

Enediyne antibiotics: A family of tumor-inhibiting substances that is characterized by the presence of a C≡C—C=C—C≡C unit as part of a nine- or ten-membered ring.

Enol: Compound of the type

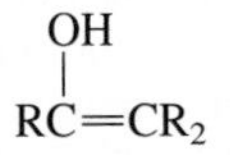

Enols are in equilibrium with an isomeric aldehyde or ketone, but are normally much less stable than aldehydes and ketones.

Enolate ion: The conjugate base of an enol. Enolate ions are stabilized by electron delocalization.

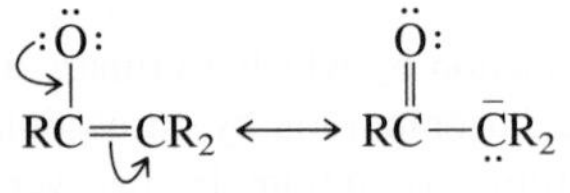

Enolization: A reaction of the type:

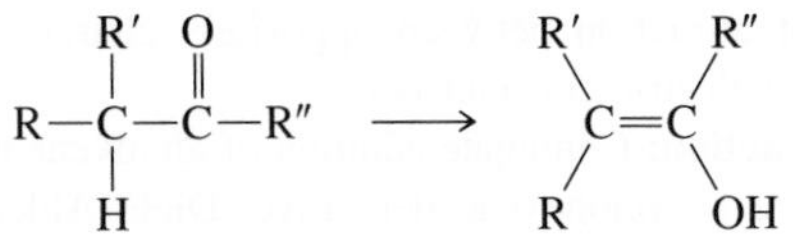

Entgegen: See *E-*.

Enthalpy: The heat content of a substance; symbol, *H*.

Envelope: One of the two most stable conformations of cyclopentane. Four of the carbons in the envelope conformation are coplanar; the fifth carbon lies above or below this plane.

Enzymatic resolution: Resolution of a mixture of enantiomers based on the selective reaction of one of them under conditions of enzyme catalysis.

Enzyme: A protein that catalyzes a chemical reaction in a living system.

Epimers: Diastereomers that differ in configuration at only one of their chirality centers.

Epoxidation: Conversion of an alkene to an epoxide, usually by treatment with a peroxy acid.

Epoxide: Compound of the type

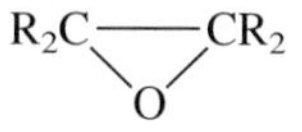
R2C
CR2
O

Equatorial bond: A bond to a carbon in the chair conformation of cyclohexane oriented approximately along the equator of the molecule.

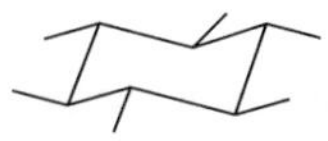

Erythro: Term applied to the relative configuration of two chirality centers within a molecule. The erythro stereoisomer has like substituents on the same side of a Fischer projection.

Essential amino acids: Amino acids that must be present in the diet for normal growth and good health.

Essential fatty acids: Fatty acids that must be present in the diet for normal growth and good health.

Essential oils: Pleasant-smelling oils of plants consisting of mixtures of terpenes, esters, alcohols, and other volatile organic substances.

Ester: Compound of the type

$$\overset{\displaystyle O}{\overset{\|}{RCOR'}}$$

Estrogen: A female sex hormone.

Ethene: IUPAC name for $H_2C{=}CH_2$. The common name ethylene, however, is used far more often, and the IUPAC rules permit its use.

Ether: Molecule that contains a C—O—C unit such as ROR′, ROAr, or ArOAr.

Ethylene: $H_2C{=}CH_2$, the simplest alkene and the most important industrial organic chemical.

Ethyl group: The group CH_3CH_2—.

Exergonic: A process in which $\Delta G°$ is negative.

Exothermic: Term describing a reaction or process that gives off heat.

Extinction coefficient: See *molar absorptivity.*

***E–Z* notation for alkenes:** System for specifying double-bond configuration that is an alternative to cis–trans notation. When higher ranked substituents are on the same side of the double bond, the configuration is *Z.* When higher ranked substituents are on opposite sides, the configuration is *E.* Rank is determined by the Cahn–Ingold–Prelog system.

F

Fats and oils: Triesters of glycerol. Fats are solids at room temperature, oils are liquids.

Fatty acid: Carboxylic acids obtained by hydrolysis of fats and oils. Fatty acids typically have unbranched chains and contain an even number of carbon atoms in the range of 12–20 carbons. They may include one or more double bonds.

Fatty acid synthetase: Complex of enzymes that catalyzes the biosynthesis of fatty acids from acetate.

Fibrous protein: A protein consisting of bundled chains of elongated filaments.

Field effect: An electronic effect in a molecule that is transmitted from a substituent to a reaction site via the medium (e.g., solvent).

Fingerprint region: The region 1500–500 cm^{-1} of an infrared spectrum. This region is less characteristic of functional groups than others, but varies so much from one molecule to another that it can be used to determine whether two substances are identical or not.

Fischer esterification: Acid-catalyzed ester formation between an alcohol and a carboxylic acid:

$$\overset{\displaystyle O}{\overset{\|}{RCOH}} + R'OH \xrightarrow{H^+} \overset{\displaystyle O}{\overset{\|}{RCOR'}} + H_2O$$

Fischer glycosidation: A reaction in which glycosides are formed by treating a carbohydrate with an alcohol in the presence of an acid catalyst.

Fischer projection: Method for representing stereochemical relationships. The four bonds to a tetrahedral carbon are represented by a cross. The horizontal bonds are understood to project toward the viewer and the vertical bonds away from the viewer.

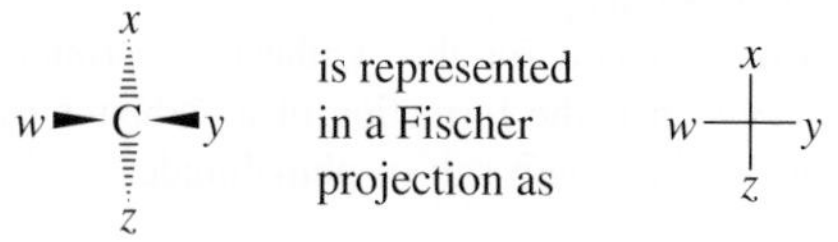
x
w
C
y
z
is represented
in a Fischer
projection as
x
w
y
z

Fluid mosaic model: A schematic representation of a cell membrane.

Formal charge: The charge, either positive or negative, on an atom calculated by subtracting from the number of valence electrons in the neutral atom a number equal to the sum of its unshared electrons plus half the electrons in its covalent bonds.

Fragmentation pattern: In mass spectrometry, the ions produced by dissociation of the molecular ion.

Free energy: The available energy of a system; symbol, *G.* See also *Gibbs energy.*

Free radical: Neutral species in which one of the electrons in the valence shell of carbon is unpaired. An example is methyl radical, $\cdot CH_3$.

Free-radical polymerization: An alkene polymerization proceeding via free-radical intermediates.

Frequency: Number of waves per unit time. Although often expressed in hertz (Hz), or cycles per second, the SI unit for frequency is s^{-1}.

Friedel–Crafts acylation: An electrophilic aromatic substitution in which an aromatic compound reacts with an acyl chloride or a carboxylic acid anhydride in the presence of aluminum chloride. An acyl group becomes bonded to the ring.

$$Ar{-}H + \overset{\displaystyle O}{\overset{\|}{RC}}{-}Cl \xrightarrow{AlCl_3} Ar{-}\overset{\displaystyle O}{\overset{\|}{C}}R$$

Friedel–Crafts alkylation: An electrophilic aromatic substitution in which an aromatic compound reacts with an alkyl halide in the presence of aluminum chloride. An alkyl group becomes bonded to the ring.

$$Ar{-}H + R{-}X \xrightarrow{AlCl_3} Ar{-}R$$

Fries rearrangement: Aluminum chloride-promoted rearrangement of an aryl ester to a ring-acylated derivative of phenol.

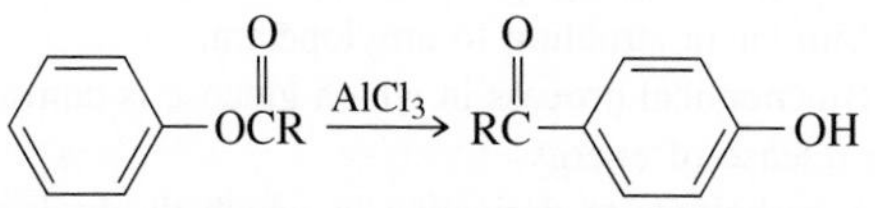
O
OCR
AlCl3
O
RC
OH

Frontier orbitals: Orbitals involved in a chemical reaction, usually the highest occupied molecular orbital of one reactant and the lowest unoccupied molecular orbital of the other.

Frost's circle: A mnemonic that gives the Hückel π MOs for cyclic conjugated molecules and ions.

Functional class nomenclature: Type of IUPAC nomenclature in which compounds are named according to functional group families. The last word in the name identifies the functional group; the first word designates the alkyl or aryl group that bears the functional group. Methyl bromide, ethyl alcohol, and diethyl ether are examples of functional class names.

Functional group: An atom or a group of atoms in a molecule responsible for its reactivity under a given set of conditions.

Furanose form: Five-membered ring arising via cyclic hemiacetal formation between the carbonyl group and a hydroxyl group of a carbohydrate.

G

***G*:** Symbol for Gibbs energy.

Gabriel synthesis: Method for the synthesis of primary alkylamines in which a key step is the formation of a carbon–nitrogen bond by alkylation of the potassium salt of phthalimide.

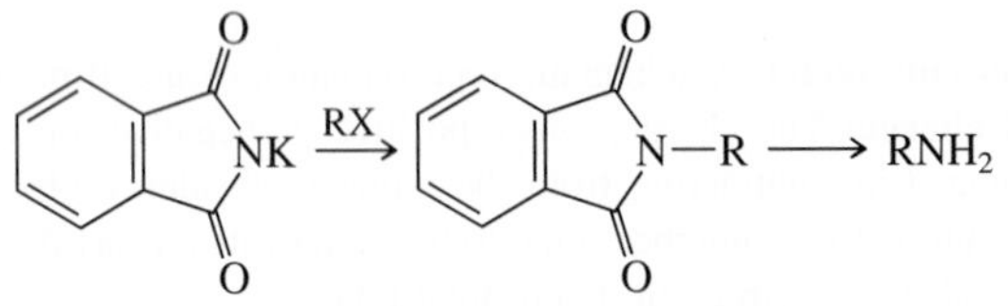

Gauche: Term describing the position relative to each other of two substituents on adjacent atoms when the angle between their bonds is on the order of 60°. Atoms X and Y in the structure shown are gauche to each other.

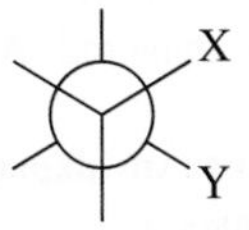

G-coupled protein receptors: A large family of protein receptors that function as transmembrane molecular switches to regulate many physiological processes.

Geminal dihalide: A dihalide of the form R_2CX_2, in which the two halogen substituents are located on the same carbon.

Geminal diol: The hydrate $R_2C(OH)_2$ of an aldehyde or a ketone.

Generic name: The name of a drug as designated by the U.S. Adopted Names Council.

Genetic code: The relationship between triplets of nucleotide bases in messenger RNA and the amino acids incorporated into a protein in DNA-directed protein biosynthesis.

Genome: The aggregate of all the genes that determine what an organism becomes.

Genomics: The study of genome sequences and their function.

Gibbs energy: The free energy (energy available to do work) of a system.

Globular protein: An approximately spherically shaped protein that forms a colloidal dispersion in water. Most enzymes are globular proteins.

Glycogen: A polysaccharide present in animals that is derived from glucose. Similar in structure to amylopectin.

Glycolysis: Biochemical process in which glucose is converted to pyruvate with release of energy.

Glycoside: A carbohydrate derivative in which the hydroxyl group at the anomeric position has been replaced by some other group. An *O*-glycoside is an ether of a carbohydrate in which the anomeric position bears an alkoxy group.

Graft copolymer: A copolymer of monomers A and B in which branches of poly-A are attached to a poly-B main chain.

Grain alcohol: A common name for ethanol (CH_3CH_2OH).

Grignard reagent: An organomagnesium compound of the type RMgX formed by the reaction of magnesium with an alkyl or aryl halide.

H

Half-chair: One of the two most stable conformations of cyclopentane. Three consecutive carbons in the half-chair conformation are coplanar. The fourth and fifth carbons lie, respectively, above and below the plane.

Haloform reaction: The formation of CHX_3 (X = Br, Cl, or I) brought about by cleavage of a methyl ketone on treatment with Br_2, Cl_2, or I_2 in aqueous base.

$$\mathrm{R\overset{O}{\overset{\|}{C}}CH_3} \xrightarrow[\mathrm{HO^-}]{\mathrm{X_2}} \mathrm{R\overset{O}{\overset{\|}{C}}O^-} + \mathrm{CHX_3}$$

Halogenation: Replacement of a hydrogen by a halogen. The most frequently encountered examples are the free-radical halogenation of alkanes and the halogenation of arenes by electrophilic aromatic substitution.

Halohydrin: A compound that contains both a halogen atom and a hydroxyl group. The term is most often used for compounds in which the halogen and the hydroxyl group are on adjacent atoms (vicinal halohydrins). The most commonly encountered halohydrins are chlorohydrins and bromohydrins.

Halonium ion: A species that incorporates a positively charged halogen. Bridged halonium ions are intermediates in the addition of halogens to the double bond of an alkene.

Hammond's postulate: Principle used to deduce the approximate structure of a transition state. If two states, such as a transition state and an unstable intermediate derived from it, are similar in energy, they are believed to be similar in structure.

Haworth formulas: Planar representations of furanose and pyranose forms of carbohydrates.

Heat of combustion: Heat evolved on combustion of a substance. It is the value of $-\Delta H°$ for the combustion reaction.

Standard Heat of formation: The value of $\Delta H°$ for formation of a substance from its elements.

Heat of hydrogenation: Heat evolved on hydrogenation of a substance. It is the value of $-\Delta H°$ for the addition of H_2 to a multiple bond.

α Helix: One type of protein secondary structure. It is a right-handed helix characterized by hydrogen bonds between NH and C=O groups. It contains approximately 3.6 amino acids per turn.

Hell–Volhard–Zelinsky reaction: The phosphorus trihalide-catalyzed α halogenation of a carboxylic acid:

$$\mathrm{R_2CHCO_2H} + \mathrm{X_2} \xrightarrow[\text{or } \mathrm{PX_3}]{\mathrm{P}} \mathrm{R_2\underset{\underset{X}{|}}{C}CO_2H} + \mathrm{HX}$$

Hemiacetal: Product of nucleophilic addition of one molecule of an alcohol to an aldehyde or a ketone. Hemiacetals are compounds of the type

$$\mathrm{R_2\overset{OH}{\overset{|}{C}}—OR'}$$

Hemiketal: A hemiacetal derived from a ketone.

Henderson–Hasselbalch equation: An equation that relates degree of dissociation of an acid at a particular pH to its pK_a.

$$\mathrm{pH} = pK_a + \log\frac{[\text{conjugate base}]}{[\text{acid}]}$$

HETCOR: A 2D NMR technique that correlates the ^{1}H chemical shift of a proton to the ^{13}C chemical shift of the carbon to which it is attached. HETCOR stands for *heteronuclear chemical shift correlation.*

Heteroatom: An atom in an organic molecule that is neither carbon nor hydrogen.

Heterocyclic aromatic compound: A heterocyclic compound in which the ring that contains the heteroatom is aromatic.

Heterocyclic compound: Cyclic compound in which one or more of the atoms in the ring are elements other than carbon. Heterocyclic compounds may or may not be aromatic.

Heterogeneous reaction: A reaction involving two or more substances present in different phases. Hydrogenation of alkenes is a heterogeneous reaction that takes place on the surface of an insoluble metal catalyst.

Heterolytic cleavage: Dissociation of a two-electron covalent bond in such a way that both electrons are retained by one of the initially bonded atoms.

Hexose: A carbohydrate with six carbon atoms.

Histones: Proteins that are associated with DNA in nucleosomes.

Hofmann elimination: Conversion of a quaternary ammonium hydroxide, especially an alkyltrimethylammonium hydroxide, to an alkene on heating. Elimination occurs in the direction that gives the less substituted double bond.

$$R_2CH-\underset{{}^{+}N(CH_3)_3}{\underset{|}{C}}R'_2\ \ HO^- \xrightarrow{\text{heat}} R_2C{=}CR'_2 + N(CH_3)_3 + H_2O$$

Hofmann rule: β-Elimination of quaternary ammonium hydroxides gives predominantly the alkene with the least substituted double bond.

HOMO: Highest occupied molecular orbital (the orbital of highest energy that contains at least one of a molecule's electrons).

Homogeneous hydrogenation: Hydrogenation of a double bond catalyzed by an organometallic compound that is soluble in the solvent in which the reaction is carried out.

Homologous series: Group of structurally related substances in which successive members differ by a CH_2 group.

Homolytic cleavage: Dissociation of a two-electron covalent bond in such a way that one electron is retained by each of the initially bonded atoms.

Homopolymer: A polymer formed from a single monomer.

Hückel's rule: Completely conjugated planar monocyclic hydrocarbons possess special stability when the number of their π electrons $= 4n + 2$, where n is an integer.

Hund's rule: When two orbitals are of equal energy, they are populated by electrons so that each is half-filled before either one is doubly occupied.

Hybrid orbital: An atomic orbital represented as a mixture of various contributions of that atom's *s, p, d,* etc., orbitals.

Hydration: Addition of the elements of water (H, OH) to a multiple bond.

Hydride shift: Migration of a hydrogen with a pair of electrons (H:) from one atom to another. Hydride shifts are most commonly seen in carbocation rearrangements.

Hydroboration–oxidation: Reaction sequence involving a separate hydroboration stage and oxidation stage. In the hydroboration stage, diborane adds to an alkene to give an alkylborane. In the oxidation stage, the alkylborane is oxidized with hydrogen peroxide to give an alcohol. The reaction product is an alcohol corresponding to the anti-Markovnikov, syn hydration of an alkene.

Hydrocarbon: A compound that contains only carbon and hydrogen.

Hydroformylation: An industrial process for preparing aldehydes ($RCH_2CH_2CH{=}O$) by the reaction of terminal alkenes ($RCH{=}CH_2$) with carbon monoxide.

Hydrogenation: Addition of H_2 to a multiple bond.

Hydrogen bonding: Type of dipole–dipole attractive force in which a positively polarized hydrogen of one molecule is weakly bonded to a negatively polarized atom of an adjacent molecule. Hydrogen bonds typically involve the hydrogen of one —OH or —NH group and the oxygen or nitrogen of another.

Hydrolysis: Water-induced cleavage of a bond.

Hydronium ion: The species H_3O^+.

Hydrophilic: Literally, "water-loving"; a term applied to substances that are soluble in water, usually because of their ability to form hydrogen bonds with water.

Hydrophobic: Literally, "water-hating"; a term applied to substances that are not soluble in water, but are soluble in nonpolar, hydrocarbon-like media.

Hydrophobic effect: The excluding of nonpolar molecules from water.

Hyperconjugation: Delocalization of σ electrons.

I

Icosanoids: A group of naturally occurring compounds derived from unsaturated C_{20} carboxylic acids.

Imide: A compound containing the group 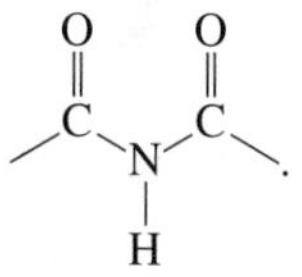.

Imine: Compound of the type $R_2C{=}NR'$ formed by the reaction of an aldehyde or a ketone with a primary amine ($R'NH_2$). Imines are sometimes called *Schiff's bases.*

Index of hydrogen deficiency: A measure of the total double bonds and rings a molecule contains. It is determined by comparing the molecular formula C_nH_x of the compound to that of an alkane that has the same number of carbons according to the equation:

$$\text{Index of hydrogen deficiency} = \frac{1}{2}(C_nH_{2n+2} - C_nH_x)$$

Induced-dipole/induced-dipole attractive force: Force of attraction resulting from a mutual and complementary polarization of one molecule by another. Also referred to as *London forces* or *dispersion forces.*

Inductive effect: An electronic effect transmitted by successive polarization of the σ bonds within a molecule or an ion.

Infrared (IR) spectroscopy: Analytical technique based on energy absorbed by a molecule as it vibrates by stretching and bending bonds. Infrared spectroscopy is useful for analyzing the functional groups in a molecule.

Initiation step: A process which causes a reaction, usually a free-radical reaction, to begin but which by itself is not the principal source of products. The initiation step in the halogenation of an alkane is the dissociation of a halogen molecule to two halogen atoms.

Integrated area: The relative area of a signal in an NMR spectrum. Areas are proportional to the number of equivalent protons responsible for the peak.

Intermediate: Transient species formed during a chemical reaction. Typically, an intermediate is not stable under the conditions of its formation and proceeds further to form the product. Unlike a transition state, which corresponds to a maximum along a potential energy surface, an intermediate lies at a potential energy minimum.

Intermolecular attractive forces: Forces, either attractive or repulsive, between two atoms or groups in *separate* molecules.

Intramolecular forces: Forces, either attractive or repulsive, between two atoms or groups *within* the same molecule.

Inversion of configuration: Reversal of the three-dimensional arrangement of the four bonds to sp^3-hybridized carbon. The representation shown illustrates inversion of configuration in a nucleophilic substitution where LG is the leaving group and Nu is the nucleophile.

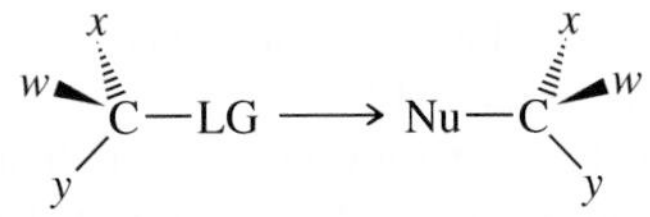

Ion: A charged particle.

Ionic bond: Chemical bond between oppositely charged particles that results from the electrostatic attraction between them.

Ionization energy: Amount of energy required to remove an electron from some species.

Isobutane: The common name for 2-methylpropane, $(CH_3)_3CH$.

Isobutyl group: The group $(CH_3)_2CHCH_2$—.

Isoelectric point: pH at which the concentration of the zwitterionic form of an amino acid is a maximum. At a pH below the isoelectric point the dominant species is a cation. At higher pH, an anion predominates. At the isoelectric point the amino acid has no net charge.

Isoionic point: Synonymous with *isoelectric point.*

Isolated diene: Diene of the type

$$C{=}C{-}(C)_x{-}C{=}C$$

in which the two double bonds are separated by one or more sp^3-hybridized carbons. Isolated dienes are slightly less stable than isomeric conjugated dienes.

Isomers: Different compounds that have the same molecular formula. Isomers may be either constitutional isomers or stereoisomers.

Isoprene rule: Terpenes are composed of repeating head-to-tail-linked isoprene units.

Isoprene unit: The characteristic five-carbon structural unit found in terpenes:

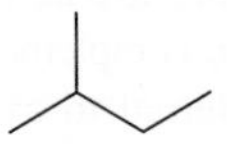

Isopropenyl group: The group $H_2C{=}\underset{\displaystyle CH_3}{\underset{|}{C}}{-}$.

Isopropyl group: The group $(CH_3)_2CH$—.

Isotactic polymer: A stereoregular polymer in which the substituent at each successive chirality center is on the same side of the zigzag carbon chain.

Isotope effect: The difference in a property, usually reaction rate, that is evident when isotopes of the same atom are compared.

Isotopic cluster: In mass spectrometry, a group of peaks that differ in *m/z* because they incorporate different isotopes of their component elements.

IUPAC rules: The most widely used method of naming organic compounds. It uses a set of rules proposed and periodically revised by the International Union of Pure and Applied Chemistry.

K

Kekulé structure: Structural formula for an aromatic compound that satisfies the customary rules of bonding and is usually characterized by a pattern of alternating single and double bonds. There are two Kekulé formulations for benzene:

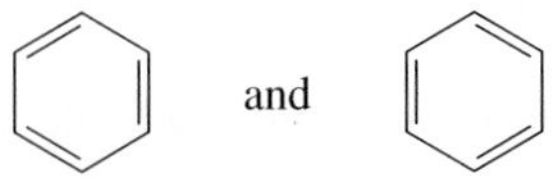

A single Kekulé structure does not completely describe the actual bonding in the molecule.

Ketal: An acetal derived from a ketone.

Keto-: A tautomeric form that contains a carbonyl group.

Keto–enol tautomerism: Process by which an aldehyde or a ketone and its enol equilibrate:

$$R\overset{\displaystyle O}{\overset{\|}{C}}{-}CHR_2 \rightleftharpoons R\overset{\displaystyle OH}{\overset{|}{C}}{=}CR_2$$

β-Keto ester: A compound of the type

$$R\overset{\displaystyle O}{\overset{\|}{C}}CH_2\overset{\displaystyle O}{\overset{\|}{C}}OR'$$

Ketone: A member of the family of compounds in which both atoms attached to a carbonyl group (C=O) are carbon, as in

$$R\overset{\displaystyle O}{\overset{\|}{C}}R \qquad R\overset{\displaystyle O}{\overset{\|}{C}}Ar \qquad Ar\overset{\displaystyle O}{\overset{\|}{C}}Ar$$

Ketose: A carbohydrate that contains a ketone carbonyl group in its open-chain form.

Kiliani–Fischer synthesis: A synthetic method for carbohydrate chain extension. The new carbon–carbon bond is formed by converting an aldose to its cyanohydrin. Reduction of the cyano group to an aldehyde function completes the synthesis.

Kinases: Enzymes that catalyze the transfer of phosphate from ATP to some other molecule.

Kinetically controlled reaction: Reaction in which the major product is the one that is formed at the fastest rate.

Kinetic isotope effect: An effect on reaction rate that depends on isotopic composition.

Kinetic resolution: Separation of enantiomers based on their unequal rates of reaction with a chiral reactant.

Kinetics: The study of reaction rates and the factors that influence them.

Kolbe–Schmitt reaction: The high-pressure reaction of the sodium salt of a phenol with carbon dioxide to give an *o*-hydroxybenzoic acid. The Kolbe–Schmitt reaction is used to prepare salicylic acid in the synthesis of aspirin.

L

Lactam: A cyclic amide.

β-Lactam: A cyclic amide in which the amide function is part of a four-membered ring. The antibiotic penicillin contains a β-lactam.

Lactone: A cyclic ester.

Lactose: Milk sugar; a disaccharide formed by a β-glycosidic linkage between C-4 of glucose and C-1 of galactose.

Lagging strand: In DNA replication, the strand that grows away from the replication fork.

LDA: Abbreviation for lithium diisopropylamide $LiN[CH(CH_3)_2]_2$. LDA is a strong, sterically hindered base.

Leading strand: In DNA replication, the strand that grows toward the replication fork.

Leaving group: The group, normally a halide ion, that is lost from carbon in a nucleophilic substitution or elimination.

Le Châtelier's principle: A reaction at equilibrium responds to any stress imposed on it by shifting the equilibrium in the direction that minimizes the stress.

Lewis acid/Lewis base complex: The species that results by covalent bond formation between a Lewis acid and a Lewis base.

Lewis structure: A chemical formula in which electrons are represented by dots. Two dots (or a line) between two atoms represent a covalent bond in a Lewis structure. Unshared electrons are explicitly shown, and stable Lewis structures are those in which the octet rule is satisfied.

Ligand: An atom or group attached to another atom, especially when the other atom is a metal.

Lindlar catalyst: A catalyst for the hydrogenation of alkynes to *cis*-alkenes. It is composed of palladium, which has been "poisoned" with lead(II) acetate and quinoline, supported on calcium carbonate.

Linear polymer: A polymer in which the chain of repeating units is not branched.

Lipid bilayer: Arrangement of two layers of phospholipids that constitutes cell membranes. The polar termini are located at the inner and outer membrane–water interfaces, and the lipophilic hydrocarbon tails cluster on the inside.

Lipids: Biologically important natural products characterized by high solubility in nonpolar organic solvents.

Lipophilic: Literally, "fat-loving"; synonymous in practice with *hydrophobic.*

Liposome: Spherical objects comprised of a phospholipid bilayer.

Living polymer: A polymer that retains active sites capable of further reaction on addition of more monomer.

Locant: In IUPAC nomenclature, a prefix that designates the atom that is associated with a particular structural unit. The locant is most often a number, and the structural unit is usually an attached substituent as in 2-chlorobutane.

LUMO: The orbital of lowest energy that contains none of a molecule's electrons; the lowest unoccupied molecular orbital.

M

Macromolecule: A substance containing a large number of atoms and having a high molecular weight.

Magnetic resonance imaging (MRI): A diagnostic method in medicine in which tissues are examined by NMR.

Malonic ester synthesis: Synthetic method for the preparation of carboxylic acids involving alkylation of the enolate of diethyl malonate

$$CH_3CH_2O\overset{\overset{\large O}{\|}}{C}CH_2\overset{\overset{\large O}{\|}}{C}OCH_2CH_3$$

as the key carbon–carbon bond-forming step.

Maltose: A disaccharide obtained from starch in which two glucose units are joined by an α-(1$\rightarrow$4)-glycosidic link.

Markovnikov's rule: An unsymmetrical reagent adds to an unsymmetrical double bond in the direction that places the positive part of the reagent on the carbon of the double bond that has the greater number of hydrogens.

Mass spectrometry: Analytical method in which a molecule is ionized and the various ions are examined on the basis of their mass-to-charge ratio.

Mechanism: The sequence of steps that describes how a chemical reaction occurs; a description of the intermediates and transition states that are involved during the transformation of reactants to products.

Mercaptan: An old name for the class of compounds now known as *thiols.*

Merrifield method: See *solid-phase peptide synthesis.*

Meso stereoisomer: An achiral molecule that has chirality centers. The most common kind of meso compound is a molecule with two chirality centers and a plane of symmetry.

Messenger RNA (mRNA): A polynucleotide of ribose that "reads" the sequence of bases in DNA and interacts with tRNAs in the ribosomes to promote protein biosynthesis.

Meta: Term describing a 1,3 relationship between substituents on a benzene ring.

Meta director: A group that when present on a benzene ring directs an incoming electrophile to a position meta to itself.

Metallocene: A transition metal complex that bears a cyclopentadienyl ligand.

Methine group: The group CH.

Methylene group: The group $—CH_2—$.

Methyl group: The group $—CH_3$.

Micelle: A spherical aggregate of species such as carboxylate salts of fatty acids that contain a lipophilic end and a hydrophilic end. Micelles containing 50–100 carboxylate salts of fatty acids are soaps.

Michael reaction: The conjugate addition of a carbanion (usually an enolate) to an α,β-unsaturated carbonyl compound.

Microscopic reversibility: The principle that the intermediates and transition states in the forward and backward stages of a reversible reaction are identical, but are encountered in the reverse order.

Molar absorptivity: A measure of the intensity of a peak, usually in UV-VIS spectroscopy.

Molecular dipole moment: The overall measured dipole moment of a molecule. It can be calculated as the resultant (or vector sum) of all the individual bond dipole moments.

Molecular formula: Chemical formula in which subscripts are used to indicate the number of atoms of each element present in one molecule. In organic compounds, carbon is cited first, hydrogen second, and the remaining elements in alphabetical order.

Molecular ion: In mass spectrometry, the species formed by loss of an electron from a molecule.

Molecularity: The number of species that react together in the same elementary step of a reaction mechanism.

Molecular orbital theory: Theory of chemical bonding in which electrons are assumed to occupy orbitals in molecules much as they occupy orbitals in atoms. The molecular orbitals are described as combinations of the orbitals of all of the atoms that make up the molecule.

Monomer: The simplest stable molecule from which a particular polymer may be prepared.

Monosaccharide: A carbohydrate that cannot be hydrolyzed further to yield a simpler carbohydrate.

Monosubstituted alkene: An alkene of the type $RCH{=}CH_2$, in which there is only one carbon directly bonded to the carbons of the double bond.

Multiplicity: The number of peaks into which a signal is split in nuclear magnetic resonance spectroscopy. Signals are described as singlets, doublets, triplets, and so on, according to the number of peaks into which they are split.

Mutarotation: The change in optical rotation that occurs when a single form of a carbohydrate is allowed to equilibrate to a mixture of isomeric hemiacetals.

N

Nanotube: A form of elemental carbon composed of a cylindrical cluster of carbon atoms.

Network polymer: Synonymous with *cross-linked polymer.*

Neurotransmitter: Substance, usually a naturally occurring amine, that mediates the transmission of nerve impulses.

Newman projection: Method for depicting conformations in which one sights down a carbon–carbon bond and represents the front carbon by a point and the back carbon by a circle.

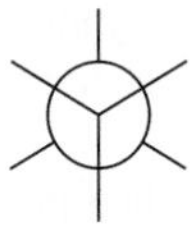

Nitration: Replacement of a hydrogen by an $—NO_2$ group. The term is usually used in connection with electrophilic aromatic substitution.

$$Ar—H \xrightarrow[H_2SO_4]{HNO_3} Ar—NO_2$$

Nitrile: A compound of the type $RC\equiv N$. R may be alkyl or aryl. Also known as alkyl or aryl cyanides.

Nitrogen rule: The molecular weight of a substance that contains C, H, O, and N is odd if the number of nitrogens is odd. The molecular weight is even if the number of nitrogens is even.

Nitrosamine: See N-*nitroso amine.*

Nitrosation: The reaction of a substance, usually an amine, with nitrous acid. Primary amines yield diazonium ions; secondary amines yield *N*-nitroso amines. Tertiary aromatic amines undergo nitrosation of their aromatic ring.

***N*-Nitroso amine:** A compound of the type $R_2N—N{=}O$. R may be alkyl or aryl groups, which may be the same or different. *N*-Nitroso amines are formed by nitrosation of secondary amines.

Noble gases: The elements in group 8A of the periodic table (helium, neon, argon, krypton, xenon, radon). Also known as the *rare gases,* they are, with few exceptions, chemically inert.

Nodal surface: A plane drawn through an orbital where the algebraic sign of a wave function changes. The probability of finding an electron at a node is zero.

Nonpolar solvent: A solvent with a low dielectric constant.

N terminus: The amino acid at the end of a peptide or protein chain that has its α-amino group intact; that is, the α-amino group is not part of a peptide bond.

Nuclear magnetic resonance (NMR) spectroscopy: A method for structure determination based on the effect of molecular environment on the energy required to promote a given nucleus from a lower energy spin state to a higher energy state.

Nucleic acid: A polynucleotide present in the nuclei of cells.

Nucleophile: An atom or ion that has an unshared electron pair which can be used to form a bond to carbon. Nucleophiles are Lewis bases.

Nucleophilic acyl substitution: Nucleophilic substitution at the carbon atom of an acyl group.

Nucleophilic addition: The characteristic reaction of an aldehyde or a ketone. An atom possessing an unshared electron pair bonds to the carbon of the C═O group, and some other species (normally hydrogen) bonds to the oxygen.

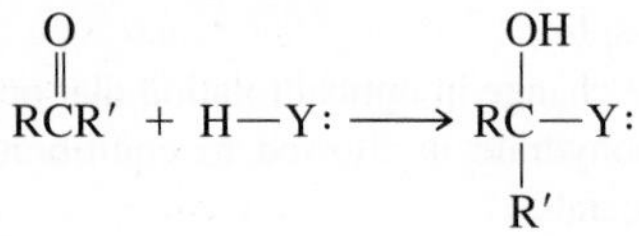

Nucleophilic aliphatic substitution: Reaction in which a nucleophile replaces a leaving group, usually a halide ion, from sp^3-hybridized carbon. Nucleophilic aliphatic substitution may proceed by either an S_N1 or an S_N2 mechanism.

Nucleophilic aromatic substitution: A reaction in which a nucleophile replaces a leaving group as a substituent on an aromatic ring. Substitution may proceed by an addition–elimination mechanism or an elimination–addition mechanism.

Nucleophilicity: A measure of the reactivity of a Lewis base in a nucleophilic substitution reaction.

Nucleoside: The combination of a purine or pyrimidine base and a carbohydrate, usually ribose or 2-deoxyribose.

Nucleosome: A DNA–protein complex by which DNA is stored in cells.

Nucleotide: The phosphate ester of a nucleoside.

O

Octane rating: The capacity of a sample of gasoline to resist "knocking," expressed as a number equal to the percentage of 2,2,4-trimethylpentane ("isooctane") in an isooctane–heptane mixture that has the same knocking characteristics.

Octet: A filled shell of eight electrons in an atom.

Octet rule: When forming compounds, atoms gain, lose, or share electrons so that the number of their valence electrons is the same as that of the nearest noble gas. For the elements carbon, nitrogen, oxygen, and the halogens, this number is 8.

Olefin metathesis: Exchange of substituents on the double bonds of two alkenes.

$$2R_2C{=}CR'_2 \longrightarrow R_2C{=}CR_2 + R'_2C{=}CR'_2$$

Oligomer: A molecule composed of too few monomer units for it to be classified as a polymer, but more than in a dimer, trimer, tetramer, etc.

Oligonucleotide: A polynucleotide containing a relatively small number of bases.

Oligosaccharide: A carbohydrate that gives three to ten monosaccharides on hydrolysis.

Optical activity: Ability of a substance to rotate the plane of polarized light. To be optically active, a substance must be chiral, and one enantiomer must be present in excess of the other.

Optically pure: Describing a chiral substance in which only a single enantiomer is present.

Optical rotation: The extent to which a chiral substance rotates the plane of plane-polarized light.

Orbital: Strictly speaking, a wave function ψ. It is convenient, however, to think of an orbital in terms of the probability ψ^2 of finding an electron at some point relative to the nucleus, as the volume inside the boundary surface of an atom, or the region in space where the probability of finding an electron is high.

σ Orbital: A bonding orbital characterized by rotational symmetry.

σ*Orbital: An antibonding orbital characterized by rotational symmetry.

Organometallic compound: A compound that contains a carbon-to-metal bond.

Ortho: Term describing a 1,2 relationship between substituents on a benzene ring.

Ortho, para director: A group that when present on a benzene ring directs an incoming electrophile to the positions ortho and para to itself.

Oxidation: A decrease in the number of electrons associated with an atom. In organic chemistry, oxidation of carbon occurs when a bond between carbon and an atom that is less electronegative than carbon is replaced by a bond to an atom that is more electronegative than carbon.

Oxidation number: The formal charge an atom has when the atoms in its covalent bonds are assigned to the more electronegative partner.

Oxidation–reduction: A reaction in which an electron is transferred from one atom to another so that each atom undergoes a change in oxidation number.

Oxidation state: See *oxidation number.*

Oxonium ion: The species H_3O^+ (also called *hydronium ion*).

Oxymercuration–demercuration: A two-stage procedure for alkene hydration.

Ozonide: A compound formed by the reaction of ozone with an alkene.

Ozonolysis: Ozone-induced cleavage of a carbon–carbon double or triple bond.

P

Para: Term describing a 1,4 relationship between substituents on a benzene ring.

Paraffin hydrocarbons: An old name for alkanes and cycloalkanes.

Partial rate factor: In electrophilic aromatic substitution, a number that compares the rate of attack at a particular ring carbon with the rate of attack at a single position of benzene.

Pauli exclusion principle: No two electrons can have the same set of four quantum numbers. An equivalent expression is that only two electrons can occupy the same orbital, and then only when they have opposite spins.

PCC: Abbreviation for pyridinium chlorochromate $C_5H_5NH^+\ ClCrO_3^-$. When used in an anhydrous medium, PCC oxidizes primary alcohols to aldehydes and secondary alcohols to ketones.

PDC: Abbreviation for pyridinium dichromate $(C_5H_5NH)_2^{2+}\ Cr_2O_7^{2-}$. Used in same manner and for same purposes as PCC (see preceding entry).

***n*-Pentane:** The common name for pentane, $CH_3CH_2CH_2CH_2CH_3$.

Pentose: A carbohydrate with five carbon atoms.

Peptide: Structurally, a molecule composed of two or more α-amino acids joined by peptide bonds.

Peptide bond: An amide bond between the carboxyl group of one α-amino acid and the amino group of another.

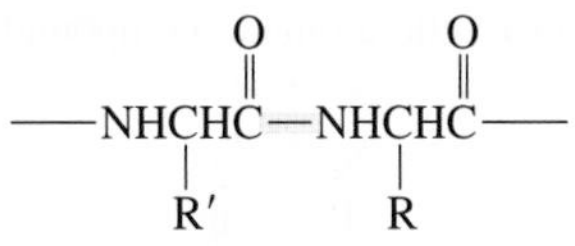

(The bond highlighted in yellow is the peptide bond.)

Peptide map: The collection of sequenced fragments of a protein from which its amino acid sequence is determined.

Pericyclic reaction: A reaction that proceeds through a cyclic transition state.

Period: A horizontal row of the periodic table.

Peroxide: A compound of the type ROOR.

Peroxide effect: Reversal of regioselectivity observed in the addition of hydrogen bromide to alkenes brought about by the presence of peroxides in the reaction mixture.

Phase-transfer catalysis: Method for increasing the rate of a chemical reaction by transporting an ionic reactant from an aqueous phase where it is solvated and less reactive to an organic phase where it is not solvated and is more reactive. Typically, the reactant is an anion that is carried to the organic phase as its quaternary ammonium salt.

Phenols: Family of compounds characterized by a hydroxyl substituent on an aromatic ring as in ArOH. *Phenol* is also the name of the parent compound, C_6H_5OH.

Phenyl group: The group

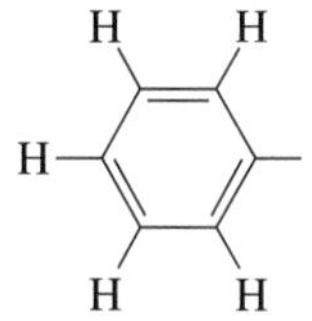

It is often abbreviated $C_6H_5—$.

Phosphatidic acid: A compound of the type shown, which is an intermediate in the biosynthesis of triacylglycerols.

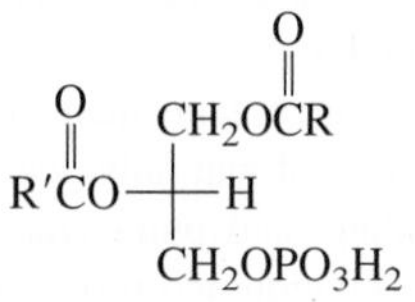

Phosphatidylcholine: One of a number of compounds of the type:

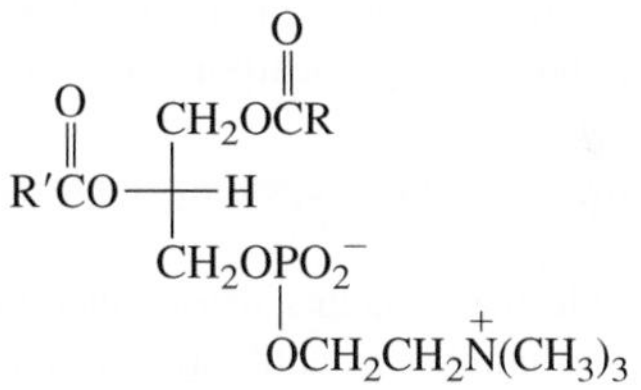

Phosphodiester: Compound of the type shown, especially when R and R′ are D-ribose or 2-deoxy-D-ribose.

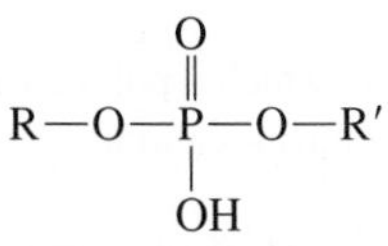

Phospholipid: A diacylglycerol bearing a choline-phosphate "head group." Also known as *phosphatidylcholine*.

Photochemical reaction: A chemical reaction that occurs when light is absorbed by a substance.

Photon: Term for an individual "bundle" of energy, or particle, of electromagnetic radiation.

Pi (π) bond: A bond in which the electron distribution is concentrated above and below the internuclear axis, rather than along it as in a σ bond. In organic chemistry π bonds are most often associated with a side-by-side overlap of *p* orbitals on adjacent atoms that are already connected by a σ bond.

Pi (π) electron: Electrons in a π bond or a π orbital.

PIN: Acronym for the name of a compound designated as its "preferred IUPAC name."

p*K*a: A measure of acid strength defined as $-\log K_a$. The stronger the acid, the smaller the value of pK_a.

Planck's constant: Constant of proportionality (h) in the equation $E = h\nu$, which relates the energy (E) to the frequency (ν) of electromagnetic radiation.

Plane of symmetry: A plane that bisects an object, such as a molecule, into two mirror-image halves; also called a mirror plane. When a line is drawn from any element in the object perpendicular to such a plane and extended an equal distance in the opposite direction, a duplicate of the element is encountered.

Pleated β sheet: Type of protein secondary structure characterized by hydrogen bonds between NH and C═O groups of adjacent parallel peptide chains. The individual chains are in an extended zigzag conformation.

Polar covalent bond: A shared electron pair bond in which the electrons are drawn more closely to one of the bonded atoms than the other.

Polarimeter: An instrument used to measure optical activity.

Polarizability: A measure of the ease of distortion of the electric field associated with an atom or a group. A fluorine atom in a molecule, for example, holds its electrons tightly and is very nonpolarizable. Iodine is very polarizable.

Polar solvent: A solvent with a high dielectric constant.

Polyamide: A polymer in which individual structural units are joined by amide bonds. Nylon is a synthetic polyamide; proteins are naturally occurring polyamides.

Polyamine: A compound that contains many amino groups. The term is usually applied to a group of naturally occurring substances, including spermine, spermidine, and putrescine, that are believed to be involved in cell differentiation and proliferation.

Polycarbonate: A polyester of carbonic acid.

Polycyclic aromatic hydrocarbon: An aromatic hydrocarbon characterized by the presence of two or more fused benzene rings.

Polycyclic hydrocarbon: A hydrocarbon in which two carbons are common to two or more rings.

Polyester: A polymer in which repeating units are joined by ester bonds.

Polyether: A molecule that contains many ether linkages. Polyethers occur naturally in a number of antibiotic substances.

Polyethylene: A polymer of ethylene.

Polymer: Large molecule formed by the repetitive combination of many smaller molecules (monomers).

Polymerase chain reaction: A laboratory method for making multiple copies of DNA.

Polymerization: Process by which a polymer is prepared. The principal processes include free-radical, cationic, coordination, and condensation polymerization.

Polynucleotide: A polymer in which phosphate ester units join an oxygen of the carbohydrate unit of one nucleoside to that of another.

Polyolefin: An addition polymer prepared from alkene monomers.

Polypeptide: A polymer made up of "many" (more than eight to ten) amino acid residues.

Polypropylene: A polymer of propene.

Polysaccharide: A carbohydrate that yields "many" monosaccharide units on hydrolysis.

Polyurethane: A polymer in which structural units are corrected by a linkage of the type $-NH\overset{\overset{\textstyle O}{\|}}{C}O-$.

Potential energy: The energy a system has exclusive of its kinetic energy.

Potential energy diagram: Plot of potential energy versus some arbitrary measure of the degree to which a reaction has proceeded (the reaction coordinate). The point of maximum potential energy is the transition state.

Primary alkyl group: Structural unit of the type RCH_2-, in which the point of attachment is to a primary carbon.

Primary amine: An amine with a single alkyl or aryl substituent and two hydrogens: an amine of the type RNH_2 (primary alkylamine) or $ArNH_2$ (primary arylamine).

Primary carbon: A carbon that is directly attached to only one other carbon.

Primary structure: The sequence of amino acids in a peptide or protein.

Principal quantum number: The quantum number (n) of an electron that describes its energy level. An electron with $n = 1$ must be an s electron; one with $n = 2$ has s and p states available.

Prochiral: The capacity of an achiral molecule to become chiral by replacement of an existing atom or group by a different one.

Prochirality center: An atom of a molecule that becomes a chirality center when one of its attached atoms or groups is replaced by a different atom or group.

Propagation steps: Elementary steps that repeat over and over again in a chain reaction. Almost all of the products in a chain reaction arise from the propagation steps.

Prostaglandin: One of a class of lipid hormones containing 20 carbons, 5 of which belong to an oxygenated cyclopentanoid ring; the remaining 15 carbons are incorporated into two unbranched side chains, adjacent to each other on the ring.

Prosthetic group: A cofactor that is covalently bonded to an enzyme.

Protease inhibitor: A substance that interferes with enzyme-catalyzed hydrolysis of peptide bonds.

Protecting group: A temporary alteration in the nature of a functional group so that it is rendered inert under the conditions in which reaction occurs somewhere else in the molecule. To be synthetically useful, a protecting group must be stable under a prescribed set of reaction conditions, yet be easily introduced and removed.

Protein: A naturally occurring polypeptide that has a biological function.

Protic solvent: A solvent that has easily exchangeable protons, especially protons bonded to oxygen as in hydroxyl groups.

Proton acceptor: A Brønsted base.

Proton donor: A Brønsted acid.

Purine: The heterocyclic aromatic compound

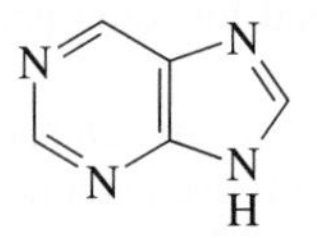

Pyranose form: Six-membered ring arising via cyclic hemiacetal formation between the carbonyl group and a hydroxyl group of a carbohydrate.

Pyrimidine: The heterocyclic aromatic compound

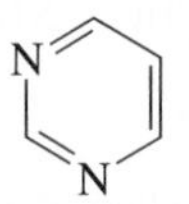

Quantized: Referring to states for which only certain energies are allowed. These states are governed by the relationship $E = nh\nu$, where n is an integer, h is Planck's constant, and ν is the frequency of electromagnetic radiation.

Quantum: The energy associated with a photon.

Quaternary ammonium salt: Salt of the type $R_4N^+ X^-$. The positively charged ion contains a nitrogen with a total of four organic substituents (any combination of alkyl and aryl groups).

Quaternary carbon: A carbon that is directly attached to four other carbons.

Quaternary structure: Description of the way in which two or more protein chains, not connected by chemical bonds, are organized in a larger protein.

Quinone: The product of oxidation of an ortho or para dihydroxybenzene derivative. Examples of quinones include

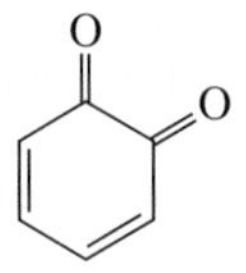

and

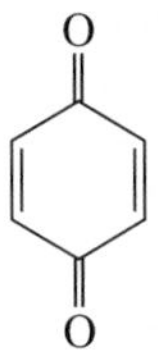

R

R: Symbol for an alkyl group.

Racemic mixture: Mixture containing equal quantities of enantiomers.

Random coil: A portion of a protein that lacks an ordered secondary structure.

Rare gases: Synonymous with noble gases (helium, neon, argon, krypton, and xenon).

Rate-determining step: Slowest step of a multistep reaction mechanism. The overall rate of a reaction can be no faster than its slowest step.

Rearrangement: Intramolecular migration of an atom, a group, or a bond from one atom to another.

Recombinant DNA: DNA molecules that are made by combining nucleotide sequences obtained from different sources. For example, the nucleotide sequence that codes for the synthesis of human insulin can be combined with a bacterial nucleotide sequence to produce recombinant DNA used in the production of insulin.

Reduction: Gain in the number of electrons associated with an atom. In organic chemistry, reduction of carbon occurs when a bond between carbon and an atom which is more electronegative than carbon is replaced by a bond to an atom which is less electronegative than carbon.

Reductive amination: Method for the preparation of amines in which an aldehyde or a ketone is treated with ammonia or an amine under conditions of catalytic hydrogenation.

Refining: Conversion of crude oil to useful materials, especially gasoline.

Reforming: Step in oil refining in which the proportion of aromatic and branched-chain hydrocarbons in petroleum is increased so as to improve the octane rating of gasoline.

Regioselective: Term describing a reaction that can produce two (or more) constitutional isomers but gives one of them in greater amounts than the other. A reaction that is 100% regioselective is termed regiospecific.

Relative configuration: Stereochemical configuration on a comparative, rather than an absolute, basis. Terms such as D, L, erythro, threo, α, and β describe relative configuration.

Repeating unit: The structural units that make up a polymer; usually written enclosed in brackets.

Replication: Biosynthetic copying of DNA.

Replication fork: Point at which strands of double-helical DNA separate.

Resolution: Separation of a racemic mixture into its enantiomers.

Resonance: Method by which electron delocalization may be shown using Lewis structures. The true electron distribution in a molecule is regarded as a hybrid of the various Lewis structures that can be written for it.

Resonance energy: Extent to which a substance is stabilized by electron delocalization. It is the difference in energy between the substance and a hypothetical model in which the electrons are localized.

Resonance hybrid: The collection of Lewis structures that, taken together, represent the electron distribution in a molecule.

Restriction enzymes: Enzymes that catalyze the cleavage of DNA at specific sites.

Retention of configuration: Stereochemical pathway observed when a new bond is made that has the same spatial orientation as the bond that was broken.

Retrosynthetic analysis: Technique for synthetic planning based on reasoning backward from the target molecule to appropriate starting materials. An arrow of the type ⇨ designates a retrosynthetic step.

Retrovirus: A virus for which the genetic material is RNA rather than DNA.

Ribosomal RNA (rRNA): The RNA in a cell's ribosomes.

Ribozyme: A polynucleotide that has catalytic activity.

Ring current: Electric field associated with circulating system of π electrons.

Ring flipping: Synonymous with *ring inversion* of cyclohexane and related compounds.

Ring inversion: Process by which a chair conformation of cyclohexane is converted to a mirror-image chair. All of the equatorial substituents become axial, and vice versa. Also called ring flipping, or chair–chair interconversion.

RNA (ribonucleic acid): A polynucleotide of ribose.

Robinson annulation: The combination of a Michael addition and an intramolecular aldol condensation used as a synthetic method for ring formation.

Row: Synonymous with *period* in the periodic table.

S

Sandmeyer reaction: Reaction of an aryl diazonium ion with CuCl, CuBr, or CuCN to give, respectively, an aryl chloride, aryl bromide, or aryl cyanide (nitrile).

Saponification: Hydrolysis of esters in basic solution. The products are an alcohol and a carboxylate salt. The term means "soap making" and derives from the process whereby animal fats were converted to soap by heating with wood ashes.

Saturated hydrocarbon: A hydrocarbon in which there are no multiple bonds.

Sawhorse formula: A representation of the three-dimensional arrangement of bonds in a molecule by a drawing of the type shown.

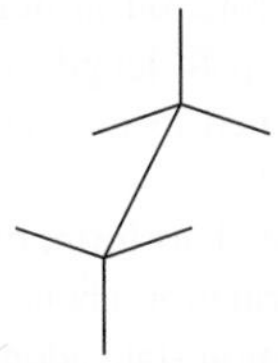

Schiemann reaction: Preparation of an aryl fluoride by heating the diazonium fluoroborate formed by addition of tetrafluoroboric acid (HBF_4) to a diazonium ion.

Schiff's base: Another name for an imine; a compound of the type $R_2C{=}NR'$.

Scientific method: A systematic approach to establishing new knowledge in which observations lead to laws, laws to theories, theories to testable hypotheses, and hypotheses to experiments.

Secondary alkyl group: Structural unit of the type R_2CH—, in which the point of attachment is to a secondary carbon.

Secondary amine: An amine with any combination of two alkyl or aryl substituents and one hydrogen on nitrogen; an amine of the type

RNHR′ or RNHAr or ArNHAr′

Secondary carbon: A carbon that is directly attached to two other carbons.

Secondary structure: The conformation with respect to nearest neighbor amino acids in a peptide or protein. The α helix and the pleated β sheet are examples of protein secondary structures.

Sequence rule: Foundation of the Cahn–Ingold–Prelog system. It is a procedure for ranking substituents on the basis of atomic number.

Shared-electron pair: Two electrons shared between two atoms.

Sharpless epoxidation: Epoxidation, especially enantioselective epoxidation, of an allylic alcohol by *tert*-butyl hydroperoxide in the presence of a Ti(IV) catalyst and diethyl tartrate.

β-Sheet: A type of protein secondary structure in which the C=O and N—H groups of adjacent chains, or regions of one chain, are hydrogen-bonded in a way that produces a sheet-like structure which may be flat or pleated.

Shell: The group of orbitals that have the same principal quantum number *n*.

Shielding: Effect of a molecule's electrons that decreases the strength of an external magnetic field felt by a proton or another nucleus.

Sigma (σ) bond: In valence-bond theory, a bond characterized by overlap of a half-filled orbital of one atom with a half-filled orbital of a second atom along a line connecting the two nuclei.

Sigmatropic rearrangement: Migration of a σ bond from one end of a conjugated π electron system to the other. The Claisen rearrangement is an example.

Simmons–Smith reaction: Reaction of an alkene with iodomethylzinc iodide to form a cyclopropane derivative.

Skew boat: A conformation of cyclohexane that is less stable than the chair, but slightly more stable than the boat.

Solid-phase peptide synthesis: Method for peptide synthesis in which the C-terminal amino acid is covalently attached to an inert solid support and successive amino acids are attached via peptide bond formation. At the completion of the synthesis the polypeptide is removed from the support.

Solvolysis reaction: Nucleophilic substitution in a medium in which the only nucleophiles present are the solvent and its conjugate base.

Specific rotation: Optical activity of a substance per unit concentration per unit path length:

$$[\alpha] = \frac{100\alpha}{cl}$$

where α is the observed rotation in degrees, *c* is the concentration in g/100 mL, and *l* is the path length in decimeters.

Spectrometer: Device designed to measure absorption of electromagnetic radiation by a sample.

Spectrum: Output, usually in chart form, of a spectrometer. Analysis of a spectrum provides information about molecular structure.

***sp* Hybridization:** Hybridization state adopted by carbon when it bonds to two other atoms as, for example, in alkynes. The *s* orbital and one of the 2*p* orbitals mix to form two equivalent *sp*-hybridized orbitals. A linear geometry is characteristic of *sp* hybridization.

sp^2 Hybridization: A model to describe the bonding of a carbon attached to three other atoms or groups. The carbon 2*s* orbital and the two 2*p* orbitals are combined to give a set of three equivalent sp^2 orbitals having 33.3% *s* character and 66.7% *p* character. One *p* orbital remains unhybridized. A trigonal planar geometry is characteristic of sp^2 hybridization.

sp^3 Hybridization: A model to describe the bonding of a carbon attached to four other atoms or groups. The carbon 2*s* orbital and the three 2*p* orbitals are combined to give a set of four equivalent orbitals having 25% *s* character and 75% *p* character. These orbitals are directed toward the corners of a tetrahedron.

Spin: Synonymous with *spin quantum number*.

Spin density: A measure of the unpaired electron distribution at the various atoms in a molecule.

Spin quantum number: One of the four quantum numbers that describe an electron. An electron may have either of two different spin quantum numbers, $+\frac{1}{2}$ or $-\frac{1}{2}$.

Spin–spin coupling: The communication of nuclear spin information between two nuclei.

Spin–spin splitting: The splitting of NMR signals caused by the coupling of nuclear spins. Only nonequivalent nuclei (such as protons with different chemical shifts) can split one another's signals.

Spiro compound: A compound in which a single carbon is common to two rings.

Spontaneous reaction: Among several definitions, the one most relevant to the material in this text defines a spontaneous reaction as one that proceeds with a decrease in free energy ($\Delta G < 0$). The "official" definition is that a spontaneous process is one in which the entropy of the universe increases.

Squalene: A naturally occurring triterpene from which steroids are biosynthesized.

Staggered conformation: Conformation of the type shown, in which the bonds on adjacent carbons are as far away from one another as possible.

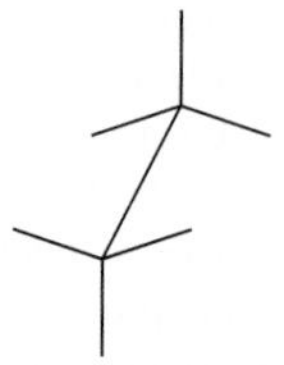

Standard amino acids: The 20 α-amino acids normally present in proteins.

Standard free-energy change ($\Delta G°$): The free-energy change ΔG for a reaction occurring under standard state conditions. The standard state is the state (solid, liquid, or gas) of a substance at a pressure of 1 atm. The standard state for a solution is 1 M.

Standard free-energy change ($\Delta G°'$): The value of $\Delta G°$ at pH=7.

Step-growth polymerization: Polymerization by a process in which monomers are first consumed in oligomer formation followed by subsequent reaction between oligomers to form macromolecules.

Stereochemistry: Chemistry in three dimensions; the relationship of physical and chemical properties to the spatial arrangement of the atoms in a molecule.

Stereoelectronic effect: An electronic effect that depends on the spatial arrangement between the orbitals of the electron donor and acceptor.

Stereoisomers: Isomers with the same constitution but that differ in respect to the arrangement of their atoms in space. Stereoisomers may be either *enantiomers* or *diastereomers.*

Stereoregular polymer: Polymer containing chirality centers according to a regular repeating pattern. Syndiotactic and isotactic polymers are stereoregular.

Stereoselective reaction: Reaction in which a single starting material has the capacity to form two or more stereoisomeric products but forms one of them in greater amounts than any of its stereoisomers. Terms such as "addition to the less hindered side" describe stereoselectivity.

Stereospecific reaction: Reaction in which stereoisomeric starting materials give stereoisomeric products. Terms such as *syn addition, anti elimination,* and *inversion of configuration* describe stereospecific reactions.

Steric effect: See *steric hindrance.*

Steric hindrance: An effect on structure or reactivity that depends on van der Waals repulsive forces.

Steric strain: Destabilization of a molecule as a result of van der Waals repulsion, distorted bond distances, bond angles, or torsion angles.

Steroid: Type of lipid present in both plants and animals characterized by a nucleus of four fused rings (three are six-membered, one is five-membered). Cholesterol is the most abundant steroid in animals.

Strain energy: Excess energy possessed by a species because of van der Waals repulsion, distorted bond lengths, bond angles, or torsion angles.

Strecker synthesis: Method for preparing amino acids in which the first step is reaction of an aldehyde with ammonia and hydrogen cyanide to give an amino nitrile, which is then hydrolyzed.

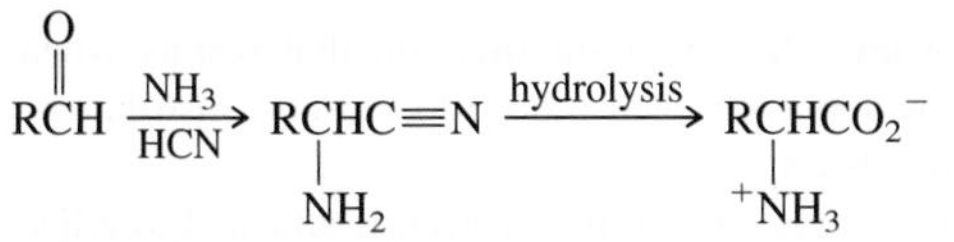

Stretching vibration: A regular, repetitive motion of two atoms or groups along the bond that connects them.

Structural isomer: Synonymous with *constitutional isomer.*

Structure: The sequence of connections that defines a molecule, including the spatial orientation of these connections.

Substitution: The replacement of an atom or group in a molecule by a different atom or group.

Substitution nucleophilic bimolecular (S_N2) mechanism: Concerted mechanism for nucleophilic substitution in which the nucleophile attacks carbon from the side opposite the bond to the leaving group and assists the departure of the leaving group.

Substitution nucleophilic unimolecular (S_N1) mechanism: Mechanism for nucleophilic substitution characterized by a two-step process. The first step is rate-determining and is the ionization of an alkyl halide to a carbocation and a halide ion.

Substitutive nomenclature: Type of IUPAC nomenclature in which a substance is identified by a name ending in a suffix characteristic of the type of compound. 2-Methylbutanol, 3-pentanone, and 2-phenylpropanoic acid are examples of substitutive names.

Sucrose: A disaccharide of glucose and fructose in which the two monosaccharides are joined at their anomeric positions.

Sulfide: A compound of the type RSR′. Sulfides are the sulfur analogs of ethers.

Sulfonation: Replacement of a hydrogen by an —SO_3H group. The term is usually used in connection with electrophilic aromatic substitution.

$$\text{Ar—H} \xrightarrow[H_2SO_4]{SO_3} \text{Ar—}SO_3H$$

Sulfone: Compound of the type

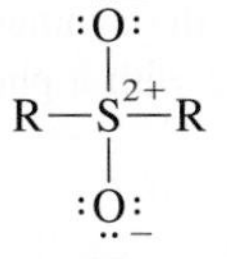

Sulfoxide: Compound of the type

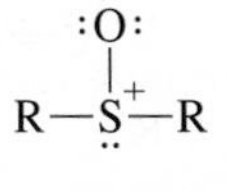

Supercoil: Coiled DNA helices.

Symmetry-allowed reaction: Concerted reaction in which the orbitals involved overlap in phase at all stages of the process.

Symmetry-forbidden reaction: Concerted reaction in which the orbitals involved do not overlap in phase at all stages of the process.

Syn addition: Addition reaction in which the two portions of the reagent that add to a multiple bond add from the same side.

Syndiotactic polymer: Stereoregular polymer in which the configuration of successive chirality centers alternates along the chain.

Synthon: A structural unit in a molecule that is related to a synthetic operation.

Systematic names: Names for chemical compounds that are developed on the basis of a prescribed set of rules. Usually the IUPAC system is meant when the term *systematic nomenclature* is used.

T

Tautomerism: Process by which two isomers are interconverted by the movement of an atom or a group. Enolization is a form of tautomerism.

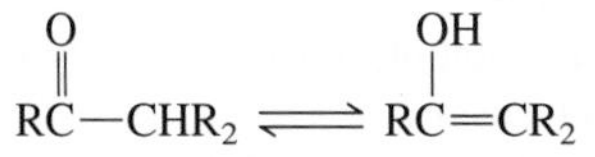

Tautomers: Constitutional isomers that interconvert by migration of an atom or group.

Terminal alkyne: Alkyne of the type RC≡CH, in which the triple bond appears at the end of the chain.

Termination steps: Reactions that halt a chain reaction. In a free-radical chain reaction, termination steps consume free radicals without generating new radicals to continue the chain.

Terpenes: Compounds that can be analyzed as clusters of isoprene units. Terpenes with 10 carbons are classified as monoterpenes, those with 15 are sesquiterpenes, those with 20 are diterpenes, and those with 30 are triterpenes.

Tertiary alkyl group: Structural unit of the type R_3C—, in which the point of attachment is to a tertiary carbon.

Tertiary amine: Amine of the type R_3N with any combination of three alkyl or aryl substituents on nitrogen.

Tertiary carbon: A carbon that is directly attached to three other carbons.

Tertiary structure: A description of how a protein chain is folded.

Tesla: SI unit for magnetic field strength.

Tetrahedral angle: The angle between one line directed from the center of a tetrahedron to a vertex and a second line from the center to a different vertex. This angle is 109° 28′.

Tetrahedral intermediate: The key intermediate in nucleophilic acyl substitution. Formed by nucleophilic addition to the carbonyl group of a carboxylic acid derivative.

Tetramethylsilane (TMS): The molecule $(CH_3)_4Si$, used as a standard to calibrate proton and carbon-13 NMR spectra.

Tetrapeptide: A compound composed of four α-amino acids connected by peptide bonds.

Tetrasubstituted alkene: Alkene of the type $R_2C{=}CR_2$, in which there are four carbons *directly* bonded to the carbons of the double bond. (The R groups may be the same or different.)

Tetrose: A carbohydrate with four carbon atoms.

Thermodynamically controlled reaction: Reaction in which the reaction conditions permit two or more products to equilibrate, giving a predominance of the most stable product.

Thermoplastic polymer: A polymer that softens or melts when heated.

Thermoset: The cross-linked product formed by heating a thermoplastic polymer.

Thermosetting polymer: A polymer that solidifies ("cures") when heated.

Thiol: Compound of the type RSH or ArSH.

Three-bond coupling: Synonymous with *vicinal coupling*.

Threo: Term applied to the relative configuration of two chirality centers within a molecule. The threo stereoisomer has like substituents on opposite sides of a Fischer projection.

Torsional strain: Decreased stability of a molecule associated with eclipsed bonds.

***trans*-:** Stereochemical prefix indicating that two substituents are on opposite sides of a ring or a double bond. (Contrast with the prefix *cis*-.)

Transamination: The transfer (usually biochemical) of an amino group from one compound to another.

Transcription: Construction of a strand of mRNA complementary to a DNA template.

Transfer RNA (tRNA): A polynucleotide of ribose that is bound at one end to a unique amino acid. This amino acid is incorporated into a growing peptide chain.

Transition state: The point of maximum energy in an elementary step of a reaction mechanism.

Translation: The "reading" of mRNA by various tRNAs, each one of which is unique for a particular amino acid.

Triacylglycerol: A derivative of glycerol (1,2,3-propanetriol) in which the three oxygens bear acyl groups derived from fatty acids.

Tripeptide: A compound in which three α-amino acids are linked by peptide bonds.

Triple bond: Bond formed by the sharing of six electrons between two atoms.

Trisubstituted alkene: Alkene of the type $R_2C{=}CHR$, in which there are three carbons *directly* bonded to the carbons of the double bond. (The R groups may be the same or different.)

Trivial nomenclature: Term synonymous with *common nomenclature*.

Twist boat: Synonymous with *skew boat*.

U

Ultraviolet-visible (UV-VIS) spectroscopy: Analytical method based on transitions between electronic energy states in molecules. Useful in studying conjugated systems such as polyenes.

Unimolecular: Describing a step in a reaction mechanism in which only one particle undergoes a chemical change at the transition state.

α, β-Unsaturated aldehyde or ketone: Aldehyde or ketone that bears a double bond between its α and β carbons as in

$$R_2C{=}CH\overset{\overset{\displaystyle O}{\|}}{C}R'$$

Unsaturated hydrocarbon: A hydrocarbon that can undergo addition reactions; that is, one that contains multiple bonds.

Unshared pair: In a Lewis structure, two valence electrons of an atom that are in the same orbital and not shared with any other atom.

Upfield: The high-field region of an NMR spectrum. A signal that is upfield with respect to another lies to its right on the spectrum.

Uronic acids: Carbohydrates that have an aldehyde function at one end of their carbon chain and a carboxylic acid group at the other.

V

Valence bond theory: Theory of chemical bonding based on overlap of half-filled atomic orbitals between two atoms. Orbital hybridization is an important element of valence bond theory.

Valence electrons: The outermost electrons of an atom. For second-row elements these are the $2s$ and $2p$ electrons.

Valence shell: The group of orbitals, filled and unfilled, responsible for the characteristic chemical properties of an atom.

Valence shell electron-pair repulsion (VSEPR) model: Method for predicting the shape of a molecule based on the notion that electron pairs surrounding a central atom repel one another. Four electron pairs will arrange themselves in a tetrahedral geometry, three will assume a trigonal planar geometry, and two electron pairs will adopt a linear arrangement.

Van der Waals forces: Intermolecular forces that do not involve ions (dipole–dipole, dipole/induced-dipole, and induced-dipole/induced-dipole forces).

Van der Waals radius: A measure of the effective size of an atom or a group. The repulsive force between two atoms increases rapidly when they approach each other at distances less than the sum of their van der Waals radii.

Van der Waals strain: Destabilization that results when two atoms or groups approach each other too closely. Also known as van der Waals repulsion.

Vicinal: Describing two atoms or groups attached to adjacent atoms.

Vicinal coupling: Coupling of the nuclear spins of atoms X and Y on adjacent atoms as in X—A—B—Y. Vicinal coupling is the most common cause of spin–spin splitting in ^1H NMR spectroscopy.

Vicinal dihalide: A compound containing two halogens on adjacent carbons.

Vicinal diol: Compound that has two hydroxyl (—OH) groups on adjacent sp^3-hybridized carbons.

Vicinal halohydrin: A compound containing a halogen and a hydroxyl group on adjacent carbons.

Vinyl group: The group $H_2C{=}CH{-}$.

Vinylic carbon: A carbon that is doubly bonded to another carbon. Atoms or groups attached to a vinylic carbon are termed *vinylic substituents*.

W

Wave functions: The solutions to arithmetic expressions that express the energy of an electron in an atom.

Wavelength: Distance between two successive maxima (peaks) or two successive minima (troughs) of a wave.

Wavenumbers: Conventional units in infrared spectroscopy that are proportional to frequency. Wavenumbers are cited in reciprocal centimeters (cm^{-1}).

Wax: A mixture of water-repellent substances that form a protective coating on the leaves of plants, the fur of animals, and the feathers of birds, among other things. A principal component of a wax is often an ester in which both the acyl portion and the alkyl portion are characterized by long carbon chains.

Williamson ether synthesis: Method for the preparation of ethers involving an S_N2 reaction between an alkoxide ion and a primary alkyl halide:

$$RONa + R'CH_2Br \longrightarrow R'CH_2OR + NaBr$$

Wittig reaction: Method for the synthesis of alkenes by the reaction of an aldehyde or a ketone with a phosphorus ylide.

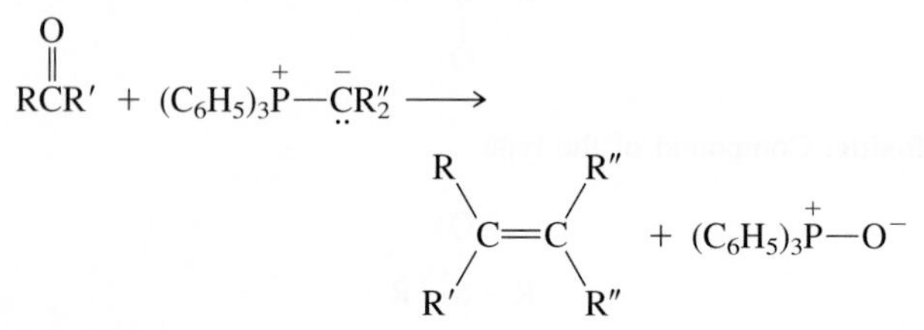

Wolff–Kishner reduction: Method for reducing the carbonyl group of aldehydes and ketones to a methylene group ($C{=}O \rightarrow CH_2$) by treatment with hydrazine (H_2NNH_2) and base (KOH) in a high-boiling alcohol solvent.

Wood alcohol: A common name for methanol, CH_3OH.

Y

Ylide: A neutral molecule in which two oppositely charged atoms, each with an octet of electrons, are directly bonded to each other. The compound

$$(C_6H_5)_3\overset{+}{P}—\overset{-}{\underset{\cdot\cdot}{C}}H_2$$

is an example of an ylide.

Z

Z-: Stereochemical descriptor used when higher ranked substituents are on the same side of a double bond.

Zaitsev's rule: When two or more alkenes are capable of being formed by an elimination reaction, the one with the more highly substituted double bond (the more stable alkene) is the major product.

Zusammen: See *Z*-.

Zwitterion: The form in which neutral amino acids actually exist. The amino group is in its protonated form and the carboxyl group is present as a carboxylate

$$\begin{array}{l}\mathrm{RCHCO_2^-}\\ \quad\,| \\ \mathrm{^+NH_3}\end{array}$$

Credits

Photographs

Chapter 1
Opener: © Joseph Sohm/Visions of America/Corbis.

Chapter 2
Opener: NASA; Fig. 2.8: USGS; Fig 2.12: © Mark Jones; Fig 2.14: © Werner Blessing/StockFood Creative/Getty.

Chapter 3
Opener: © William Whitehurst/Corbis; Fig. 3.13: © evirgen/iStock.com; Fig. 3.21: © Paul Truong/The Vetiver Network.

Chapter 4
Opener: © Thom Lang/Corbis.

Chapter 5
Opener: © BananaStock/PunchStock; p. 190 (top): © Corbis; p. 190 (bottom): © Photodisc Green/Getty.

Chapter 6
Opener: © Lee Snider/Photo Images/Corbis; p. 258: © Hans Reinhard/zefa/Corbis.

Chapter 7
Opener: © Tom Grill/Corbis; p. 292 (left): © DAJ/Getty; p. 292 (right): © The McGraw-Hill Companies/Elite Images; p. 308: © Russell Illig/ Photodisc/Getty.

Chapter 8
Opener: © iStock.com.

Chapter 9
Opener: © David Waldorf/Iconica/Getty; p. 361: © John Somerville.

Chapter 10
Opener: © CN Public Affairs; p. 389 (botom): © Stockbyte/Getty.

Chapter 11
Opener (chandelier): © Joel Schumacher Productions/The Kobal Collection.

Chapter 11
Opener (mask): Image courtesy of The Advertising Archives; p. 454: © Photodisc Blue/Getty.

Chapter 13
Opener: © iStock.com; 13.25a–c: J.M. Burns, R.H. Swerdlow: Right orbitofrontal tumor with pedophilia symptom and constructional aprazia sign. *Archives of Neurology,* vol. 60; 437–440; © 2003 American Medial Association. All rights reserved.

Chapter 14
Opener: © iStock.com.

Chapter 15
Opener: © Etienne Jong.

Chapter 15
Opener: © John Tobin/www.tobinphoto.com; 15.1 (all): © iStock.com.

Chapter 16
Opener: © iStock.com.

Chapter 17
Opener: NASA Goddard Space Flight Center; 17.6: © iStock.com; 17.7: © William Weber/Visuals Unlimited.

Chapter 18
Opener: Courtesy of Robert C. Atkins; 18.6 (beetle): © Kirk J. Larsen, Luther College Biology; 18.6 (vinegar): © Corbis; 18.6 (fruits and vegetables): © Photodisc/Getty; 18.6 (ginko seeds): © Barry Runkstan/Grant Heilman Photography; 18.6 (ants): © Digital Vision/Getty.

Chapter 19
Opener: © Tanja Schwander; 19.4: © Corbis.

Chapter 20
Opener: © iStock.com; p. 881: © Forest & Kim Starr; p. 904: © Steven Pinker.

Chapter 21
Opener: © Keystone/Getty.

Chapter 21
Opener: USDA; 22.2: © Steven Dewey/Utah State University; p. 1004: © Bob Penn/Sygma/Corbis.

Chapter 23
Opener: © Terry Sohl; 23.7: © Carmela Leszczynski/Animals Animals/Earth Scenes.

Chapter 24
Opener: © Royalty-Free/Corbis; p. 1078: © Philippe Colombi/Getty; p. 1083: © Steve Horrell/Photo Researchers; p. 1086: © DAJ/Getty Images; p. 1090: © USDA photo by J.S. Peterson; p. 1104: © Medioimages/PunchStock; p. 1105: © David Liebman.

Chapter 25
Opener: © Miriam Goldstein.

Chapter 26
Opener: Brian Mehler, Villanova University 2005; p. 1178: © Corbis; p. 1185, p. 1186: Courtesy of Villanova University; p. 1187 (top): © CSHL Archives/ Peter Arnold Inc.; p. 1187 (bottom): © NLM/Science Source/Photo Researchers; p. 1188: © A. Barrington Brown/Science Source/Photo Researchers.

Chapter 27
Opener: © DIPAG; 27.1: © Phil Nelson; 27.3: U.S. Air Force phote by Tech. Sgt. Jim Varthegyi; 27.7: © Jump Run Productions/Riser/Getty; 27.8, 27.9: © Corbis/RF; 27.10, 27.11: © Getty/RF; p. 1240 (left) Courtesy of Nokia; p. 1240 (right) AP Photo/Jae C. Hong;.

Line Art

Cover
Cover image courtesy of Robert R. Johnson, Johnson Group, University of Pennsylvania.

Chapter 8
Figure 8.4 adapted from crystallographic coordinates deposited with the Protein Data Bank. PDB ID: 1EDB. Verschueren, K.H.G., Kingma, J., Rozeboom, H.J., Kalk, K.H., Janssen, D.B., Dijkstra, B.W.D., Crystallographic and Fluorescence Studies of the Interaction of Haloalkane Dehalogenase with Halide Ions: Studies with Halide Cmpounds Reveal a Halide Binding Site in the Active Site.

Chapter 11
Figure 11.5 Courtesy of Dmitry Kazachkin, Ph.D.
Figure 11.6 Courtesy of Dmitry Kazachkin, Ph.D.
Figure 11.7 Courtesy of Dmitry Kazachkin, Ph.D.
Figure 11.8 Courtesy of Dr. Dirk Guldi, University of Erlangen, Germany and Dr. Maurizio Prato, University of Trieste Italy.

Chapter 13
Figure 13.1 From M. Silberberg, *Chemistry,* 3rd ed., p. 257, McGraw-Hill, New York, 2003.
Figure 13.33 Courtesy of Professor Craig Merlic, UCLA and Professor Bradley D. Smith, Department of Chemistry & Biochemistry, University of Notre Dame.
Figure 13.36 From M. Silberberg, *Chemistry,* 3rd ed., p. 53, McGraw-Hill, New York, 2003.

Chapter 14
Figure 14.1 From M. Silberberg, *Chemistry,* 3rd ed., p. 344, McGraw-Hill, New York, 2003.

Chapter 16
Figure 16.3 Adapted from crystallographic coordinates deposited with The Cambridge Crystallographic Data Centre, CCDC ID: NAMNSB. Duax, W.L., Smith, G.D., Strong, P.D. *Journal of the American Chemical Society,* 1980, 102, 6725.

Chapter 23
Figure 23.9 Adapted from crystallographic coordinates deposited with the Protein Data Bank. PDB ID: 4TF4. Sakon, J., Irwin, D. Wilson, D.B., Karplus, P.., Structure and Mechanism of Endo/ Exocellulase E4 from Thermomonospora Fusca.
Figure 23.10 is adapted from crystallographic coordinates deposited with The Protein Data Bank, PDB ID 1C58, Gessler, K., Uson, I., Takahan, T., Krauss, N., Smith, S.M., Okada, G.M., Sheldrick, G.M. Saenger, W., V-Amylose at Atomic Resolution: X-ray Structure of a Cycloamylose with 26 Glucose Residues (Cycloamaltohexaicosaose). *Proc. Nat. Acad. Sci. USA,* 1999, 96, 4246.

Chapter 24
Figure 24.3 The simulation is based on the coordinates of H. Heller, M. Schaefer, and K. Schulten, Molecular Dynamics Simulation of a Bilayer of 200 Lipids inm the Gel and in the Liquid-Crystal Phases, Journal of Physical Chemistry, 97, 8343–8360 (1993) and taken from an interactive animated tutorial by E. Martz and A. Herraez, "Lipid Bilayers and the Gramicidin Channel" (http://molvis.sds.edu/bilayers/index.htm (2001) by courtesy of Professor Martz.

Figure 24.10 Adapted from crystallographic coordinates deposited with the Protein Data Bank. PDB 1D: 1CLE, Ghosh, D. Wawrzak, Z., Pletnev, V.A., Li, N., Kaiser, R., Pangborn, W., Jornvall, H., Erman, M. Duax, W.L., Structure of Uncomplexed and Linoleate-Bound Candida Cholesterol Esterase.

Chapter 25

Figure 25.14 Adapted from crystallographic coordinates deposited with the Proten Data Bank PDB ID: 2SK, Fossey, S.A., Nemethy, G. Gibson, K.D., Scheraga, H.A. Conformational Energy Studies of Beta-Sheets of Model Silk Fibroin Peptides. I. Sheets of Poly (Ala-Gly) Chains. *Biopolymers* 31, 1529 (1991).
Figure 25.16 Adapted from crystallographic coordinates deposited with the Protein Data Bank, PDB ID: 1A5P, Pearson, M.A., Karplus, P.A., Dodge, R.W., Laity, J.H., Scherage, H.A., Crystal Structures of Two Mutants That Have Implications for the Folding of Bovine Pancreatic Ribonuclease A.
Figure 25.17 Adapted from crystal structure of the green fluorescent protein (GFP) variant YFP-H148Q with two bound iodides. PDB ID: 1F09. Wachter, R.M., Yarbrough, D., Kallio, K., Remington, S.J. (2000) Crystallographic and energetic analysis of binding of selected anions to the yellow variants of green fluorescent protein. *J.Mol.Biol. 301: 157–171*
Figure 25.18 Adapted from T. McKee and J. McKee, *Biochemistry: The Molecular Basis of Life,* 3rd ed., p. 141. McGraw-Hill, New York, 2003.
Figure 25.19 Adapted from crystallographic coordinates deposited with the Protein Data Bank. PDB ID: 2CTB. Teplyakov, A., Wilson, K. S., Orioli, P., Mangani S., The High Resolution Structure of the Complex between Carboxypeptidase A and L-Phenyl Lactate.
Figure 25.21 Adapted from crystallographic coordinates deposited with the Protein Data Bank. PDB ID: 1VXH. Yang, F., Phillips Jr., G.N., Structures of Co-, Deoxy- and met-Myoglobins at Various pH values.

Chapter 26

Figure 26.5 Adapted from crystallographic coordinates deposited with the Protein Data Bank. PDZB ID: 1DDN. White, A., Ding, X., Vanderspek, J.C., Murphy J.R., Ringe, D., Structure fo the Metal-Ion-Activated Diptheria Toxin Repressor/Tox Operator Complex. *Nature,* 394, 502 (1998).
Figure 26.7 Adapted from crystallographic coordinates deposited with The Protein Data Bank. PDB ID: 1A01. Luger, A., Mader, W. Richmond, R.K., Sargent, D.F., Richmond, T.J., Crystal Structure of the Nucleosome Core Particle at 2.8 A Resolution. *Nature,* 1997, V. 389, 251.
Figure 26.11 Adapted from crystallographic coordinates deposted with the Protein Data Bank. PDB ID: 6TNA. Sussman, J.L., Holbrook, S.R., Warrant, R.W., Church, G.M., Kim, S.H., Crystal Structure of Yeast Phenylalanine tRNA. I. Crystallographic Refinement, *J. Mol. Biol.,* 126, 607. (1978).

Chapter 27

Figure 27.6 From M. Silberberg, *Chemistry,* 3rd ed., p. 470, McGraw-Hill, New York, 2003.

Index

B

C

D

N

O

P

Q

R

S

T

U

V

SOME COMMONLY ENCOUNTERED GROUPS

Group	*Name**	*Group*	*Name**
$CH_3CH_2CH_2$—	Propyl or *n*-propyl		
$(CH_3)_2CH$—	1-Methylethyl or isopropyl	$CH_3C(=O)$—	Ethanoyl or acetyl
$CH_3CH_2CH_2CH_2$—	Butyl or *n*-butyl		
$CH_3CH(—)CH_2CH_3$	1-Methylpropyl or *sec*-butyl	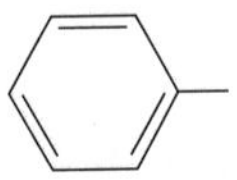	Phenyl
$(CH_3)_3C$—	1,1-Dimethylethyl or *tert*-butyl		
$(CH_3)_2CHCH_2$—	2-Methylpropyl or isobutyl	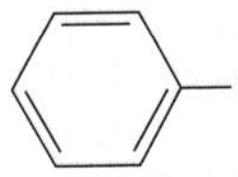 —CH_2—	Phenylmethyl or benzyl
$(CH_3)_3CCH_2$—	2,2-Dimethylpropyl or neopentyl		
$H_2C{=}CH$—	Ethenyl or vinyl		
$H_2C{=}CHCH_2$—	2-Propenyl or allyl	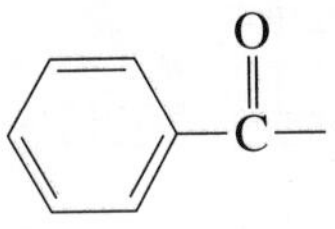	Benzenecarbonyl or benzoyl
$H_2C{=}C(—)CH_3$	1-Methylvinyl or isopropenyl		

*When two names are cited, either one is acceptable in IUPAC nomenclature.

COMMONLY ENCOUNTERED GROUPS LISTED IN ORDER OF INCREASING RANK IN THE CAHN–INGOLD–PRELOG SYSTEM

1. H—
2. CH_3—
3. CH_3CH_2—
4. $CH_3CH(CH_3)CH_2$—
5. $(CH_3)_3CCH_2$—
6. $(CH_3)_2CH$—
7. $H_2C{=}CH$—
8. $(CH_3)_3C$—
9. $HOCH_2$—
10. $HC(=O)$—
11. $CH_3C(=O)$—
12. $HOC(=O)$—
13. $CH_3OC(=O)$—
14. $HSCH_2$—
15. H_2N—
16. HO—
17. CH_3O—
18. CH_3CH_2O—
19. $HC(=O)O$—
20. $CH_3C(=O)O$—
21. F—
22. HS—
23. Cl—
24. Br—
25. I—

Periodic Table of the Elements

- Metals (main-group)
- Metals (transition)
- Metals (inner transition)
- Metalloids
- Nonmetals

MAIN–GROUP ELEMENTS (groups 1A–2A; 3A–8A) — TRANSITION ELEMENTS (groups 3B–2B)

Period	1A (1)	2A (2)	3B (3)	4B (4)	5B (5)	6B (6)	7B (7)	8B (8)	8B (9)	8B (10)	1B (11)	2B (12)	3A (13)	4A (14)	5A (15)	6A (16)	7A (17)	8A (18)
1	1 **H** 1.008																	2 **He** 4.003
2	3 **Li** 6.941	4 **Be** 9.012											5 **B** 10.81	6 **C** 12.01	7 **N** 14.01	8 **O** 16.00	9 **F** 19.00	10 **Ne** 20.18
3	11 **Na** 22.99	12 **Mg** 24.31											13 **Al** 26.98	14 **Si** 28.09	15 **P** 30.97	16 **S** 32.07	17 **Cl** 35.45	18 **Ar** 39.95
4	19 **K** 39.10	20 **Ca** 40.08	21 **Sc** 44.96	22 **Ti** 47.88	23 **V** 50.94	24 **Cr** 52.00	25 **Mn** 54.94	26 **Fe** 55.85	27 **Co** 58.93	28 **Ni** 58.69	29 **Cu** 63.55	30 **Zn** 65.41	31 **Ga** 69.72	32 **Ge** 72.61	33 **As** 74.92	34 **Se** 78.96	35 **Br** 79.90	36 **Kr** 83.80
5	37 **Rb** 85.47	38 **Sr** 87.62	39 **Y** 88.91	40 **Zr** 91.22	41 **Nb** 92.91	42 **Mo** 95.94	43 **Tc** (98)	44 **Ru** 101.1	45 **Rh** 102.9	46 **Pd** 106.4	47 **Ag** 107.9	48 **Cd** 112.4	49 **In** 114.8	50 **Sn** 118.7	51 **Sb** 121.8	52 **Te** 127.6	53 **I** 126.9	54 **Xe** 131.3
6	55 **Cs** 132.9	56 **Ba** 137.3	57 **La** 138.9	72 **Hf** 178.5	73 **Ta** 180.9	74 **W** 183.9	75 **Re** 186.2	76 **Os** 190.2	77 **Ir** 192.2	78 **Pt** 195.1	79 **Au** 197.0	80 **Hg** 200.6	81 **Tl** 204.4	82 **Pb** 207.2	83 **Bi** 209.0	84 **Po** (209)	85 **At** (210)	86 **Rn** (222)
7	87 **Fr** (223)	88 **Ra** (226)	89 **Ac** (227)	104 **Rf** (263)	105 **Db** (262)	106 **Sg** (266)	107 **Bh** (267)	108 **Hs** (277)	109 **Mt** (268)	110 **Ds** (281)	111 **Rg** (272)	112 (285)		114 (289)		116 (292)		

As of late 2006, elements 112, 114, and 116 have not been named.

INNER TRANSITION ELEMENTS

Period	Series														
6	Lanthanides	58 **Ce** 140.1	59 **Pr** 140.9	60 **Nd** 144.2	61 **Pm** (145)	62 **Sm** 150.4	63 **Eu** 152.0	64 **Gd** 157.3	65 **Tb** 158.9	66 **Dy** 162.5	67 **Ho** 164.9	68 **Er** 167.3	69 **Tm** 168.9	70 **Yb** 173.0	71 **Lu** 175.0
7	Actinides	90 **Th** 232.0	91 **Pa** (231)	92 **U** 238.0	93 **Np** (237)	94 **Pu** (242)	95 **Am** (243)	96 **Cm** (247)	97 **Bk** (247)	98 **Cf** (251)	99 **Es** (252)	100 **Fm** (257)	101 **Md** (258)	102 **No** (259)	103 **Lr** (260)